D1642327

TECHNIQUES IN THE BEHAVIORAL AND NEURAL SCIENCES

VOLUME 13

HANDBOOK OF MOLECULAR-GENETIC TECHNIQUES FOR BRAIN AND BEHAVIOR RESEARCH

Previously published in
TECHNIQUES IN THE BEHAVIORAL AND NEURAL SCIENCES

Volume 1: Feeding and Drinking, by F. Toates and N.E. Rowland (Eds), 1987, ISBN 0-444-80895-7

Volume 2: Distribution-Free Statistics: An Application-Oriented Approach, by J. Krauth, 1988, ISBN 0-444-80934-1, Paperback ISBN 0-444-80988-0

Volume 3: Molecular Neuroanatomy, by F.W. Van Leeuwen, R.M. Buijs, C.W. Pool and O. Pach (Eds), 1989, ISBN 0-444-81014-5, Paperback ISBN 0-444-81016-1

Volume 4: Manual of Microsurgery on the Laboratory Rat, Part 1, by J.J. van Dongen, R. Remie, J.W. Rensema and G.H.J. van Wunnik (Eds), 1990, ISBN 0-444-81138-9, Paperback ISBN 0-444-81139-7

Volume 5: Digital Biosignal Processing, by R. Weitkunat (Ed), 1991, ISBN 0-444-81140-0, Paperback ISBN 0-444-98144-7

Volume 6: Experimental Analysis of Behavior, by I.H. Iversen and K.A. Lattal (Eds), 1991, Part 1, ISBN 0-444-81251-2, Paperback ISBN 0-444-89160-9, Part 2, ISBN 0-444-89194-3, Paperback ISBN 0-444-89195-1

Volume 7: Microdialysis in the Neurosciences, by T.E. Robinson and J.B. Justice Jr. (Eds), 1991, ISBN 0-444-81194-X, Paperback ISBN 0-444-89375-X

Volume 8 Techniques for the Genetic Analysis of Brain and Behavior, by D. Goldowitz, D. Wahlsten and R.E. Wimer (Eds), 1992, ISBN 0-444-81249-0, Paperback ISBN 0-444-89682-1

Volume 9: Research Designs and Methods in Psychiatry, by M. Fava and J.F. Rosenbaum (Eds), 1992, ISBN 0-444-89595-7, Paperback ISBN 0-444-89594-9

Volume 10: Methods in Behavioral Pharmacology, by F. van Haaren (Ed), 1993, ISBN 0-444-81444-2, Paperback ISBN 0-444-81445-0

Volume 11: Methods in Neurotransmitter and Neuropeptide Research, by S.H. Parvez (Eds), 1993, Part 1, ISBN 0-444-81369-1, Paperback ISBN 0-444-81674-7, Part 2, ISBN 0-444-81368-3, Paperback ISBN 0-444-81675-5

Volume 12: Neglected Factors in Pharmacology and Neuroscience Research, by V. Claassen, 1994, ISBN 0-444-81871-5, Paperback ISBN 0-444-81907-X

Techniques in the Behavioral and Neural Sciences

Series Editor

J.P. HUSTON
Düsseldorf

Volume 13

HANDBOOK OF MOLECULAR-GENETIC TECHNIQUES FOR BRAIN AND BEHAVIOR RESEARCH

Editors

W.E. CRUSIO *(Orléans, France)*

and

R.T. GERLAI *(South San Francisco, CA, USA)*

1999

ELSEVIER

AMSTERDAM – LAUSANNE – NEW YORK – OXFORD – SHANNON – SINGAPORE – TOKYO

ELSEVIER SCIENCE B.V.
Sara Burgerhartstraat 25
P.O. Box 211, 1000 AE Amsterdam, The Netherlands

First edition 1999

Library of Congress Cataloging-in-Publication Data
Handbook of molecular-genetic techniques for brain and behavior
research / editors, W.E. Crusio and R.T. Gerlai. -- 1st ed.
p. cm. -- (Techniques in the behavioral and neural sciences ;
v. 13)
ISBN 0-444-50239-4 (alk. paper)
1. Neurogenetics Handbooks, manuals, etc. 2. Neuropsychology
Handbooks, manuals, etc. 3. Genetic psychology Handbooks, manuals,
etc. I. Crusio, W. E. II. Gerlai, R. T. III. Series.
QP356.22.H36 1999 99-38172
573.8'648–dc21 CIP

ISBN: 0-444-50239-4

∞ The paper used in this publication meets the requirements of ANSI/NISO Z39.48-1992 (Permanence of Paper). Printed in the Netherlands.

Dedication

We dedicate this book to our Fathers: to Kees Crusio (1929–1996), for showing Wim how to live and how to die, and to Gerlai Tamás (1933–1992), who taught Robert how to enjoy the challenges of science and of logical thinking.

Preface

Behavioral neurogenetics: the new synthesis

Wim E. Crusio[1,*] and Robert T. Gerlai[2]

[1]*Génétique, Neurogénétique et Comportement, CNRS UPR 9074, 3B Rue de la Férollerie, 45071 Orléans Cedex 02, France*
[2]*Neuroscience Department, Mailstop #72, GENENTECH Inc., 1 DNA Way, South San Francisco, CA 94080-4990, USA*

Ever since the dawn of scientific thinking, the biological bases of behavior have fascinated scientists, philosophers, and the public alike (De Wit 1982). Unraveling brain–behavior relationships is perhaps the most intriguing, and also the most challenging, problem in modern biology. In the second half of this century, researchers from different disciplines such as psychology, ethology, neuroanatomy, and physiology, started addressing these problems employing ever more sophisticated methods and theoretical concepts. During the seventies, this interdisciplinary approach gave rise to a new field of scientific endeavor, neuroscience, which has been growing at an enormous pace ever since.

At the same time, psychologists and ethologists began addressing the equally intriguing question of the genetic underpinnings of behavior and, some four decades ago, the field of behavior genetics was born (Fuller and Thompson 1960). At that time, studying the genetic bases of behavior was mainly restricted to classical and quantitative genetic analysis (Fuller and Thompson, 1960). By then it was already clear that behavior is influenced by genetic factors, as shown by a wealth of studies of inbred strains and their crosses, by the generation of artificially selected lines, and by the results of human twin studies. Using these quantitative genetic tools researchers could obtain information about the genetic architecture of the trait in question, including the number of segregating units (presumably genes), or the proportion of phenotypical variance attributable to additive genetic effects or to genes exhibiting dominance or epistasis, for example. However, with the exception of some neurological mutations with rather severe phenotypical effects, the genes themselves remained elusive. Questions concerning where the genes were located on the chromosomes, what proteins they encoded, and where in the nervous system they were expressed, remained a mystery.

* Corresponding author: Tel: +33 2 38 25 79 74; Fax: +33 2 38 25 79 79; e-mail: crusio@cnrs-orleans.fr

Asking mechanistic questions was difficult, and testing such hypotheses was even more difficult, if not impossible.

Furthermore, there was little emphasis on the neurobiological mechanisms underpinning the genetics of behavior. For example, in the index of the 1978 updated edition of Fuller and Thompson's classical 1960 book (Fuller and Thompson 1978) hardly any neuroscience-related keywords can be found. Nevertheless, in the 1970s a few scientists began to integrate behavioral, neurobiological, and genetic aspects of their research (see, for example, Van Abeelen 1974) and this integration has continued in both animal and human studies. Simultaneously, a radically different approach, recombinant DNA technology, was being born. This technology revolutionized research across a wide range of biology disciplines, one of them being behavior and brain research. In fact, some researchers would argue that the most important scientific advancement of this century comes not from quantum physics, or computer technology, but from our ability to harness the power of the gene.

Recombinant DNA technology allowed scientists to start studying the molecular mechanisms underlying neural function. With the ability to generate animals overexpressing or underexpressing certain gene products at will came the necessity of evaluating the effects of such genetic changes on the animals' phenotypes, not only on a neuronal level, but also on the behavioral level. So while behavior geneticists were moving closer to neuroscience, research in neuroscience and molecular biology was also moving towards behavior genetics. This marriage of disciplines with different histories and traditions has recently led to the emergence of the new and budding field of behavioral neurogenetics, which is rapidly coming of age with its own specialist society (the International Behavioural and Neural Genetics Society).

A theoretical and conceptual framework of this new field incorporates both the proximate (phenogenetic) and the ultimate (phylogenetic) aspects of causation. The former relates to the physiological mechanisms underlying a trait, whereas the latter concerns questions about its evolution and adaptive value. Most contemporary neurobehavioral genetic research, and all of the work described in this Handbook, are concerned mainly with phenogenetic (proximate) problems. It appears that current methods of analysis would add another, orthogonal dimension to this dichotomy: that of naturally polymorphic vs. monomorphic genes (cf. Table I in Crusio 1999). When studying genes that in nature are monomorphic, we will generally deal with underlying mechanisms common to most or even all members of a species. In contrast, we are investigating mechanisms underlying spontaneous individual differences when studying natural genetic variation, that is, polymorphic genes. Analysis of this natural genetic variation may thus enable us to identify genes that modify behavioral and neural function to a degree that is not grossly disadvantageous to the individual that carries such alleles. In short, whereas one type of question addresses, for example, how animals store information, the other type of question asks why some individuals perform better in a given task than others. Mutational analysis can be used to study both types of genes, but it should be realized that the results of KO or transgenesis studies do not contribute to the explanation of naturally occurring interindividual variation in case the genes investigated are monomorphic in nature. Indeed, most null mutations are not found to occur spontaneously in natural populations.

The new technologies have put expectations for progress in the near future very high, but, although probably a caveat, some warning against putting them too high should perhaps be made at this point. It is doubtful whether single-gene analysis, be it by means of mutational analysis or by investigating the genes underlying QTL effects, will ever help us to elucidate all aspects of, say, learning behavior. We and others (Rose 1998) would argue that it will not. Defining the function of the hippocampus, say, or to explain the existence of multiple memory systems would be a daunting task indeed if it were to be done by single-gene analysis only and would take reductionism too far. This appears akin to trying to deduce the orbit of the earth around the sun using only knowledge about subatomic particles. However, we are convinced of the manyfold advantages offered by modern behavioral-neurogenetic analysis. Sooner or later, genetic analysis will certainly help us clarify important basic cellular mechanisms and brain–behavior relationships. In addition, there is very clearly a great potential for the development of new therapeutic tools.

In 1990, US President George Bush proclaimed the last ten years of this century to be the "Decade of the Brain", an initiative that found positive acclaim all over the world. And we only just have started our quest. What lies ahead is unknown, but what is clear is that now we have the tools with which we can start dissecting the complex web of interactions from molecules and cells to neuronal networks and higher-order interactions at multiple levels of the biological organization of the brain. The blueprint of information that tells our cells what to do lies within our DNA. Genes may represent the foundation upon which all levels of the complex hierarchical biological organization is built. Recombinant DNA technology now allows us to discover and manipulate genes and study their effects on brain and behavior. It is important for scientists to understand these new techniques so that they can capitalize on the technological advances and the opportunities these offer to unravel the genetic bases of behavior and brain in a way that has never been possible before.

The time now appears ripe for a Handbook which provides up-to-date technical information on these new methods, one which focuses on the methodological aspects of molecular-genetic research in the analysis of brain and behavior, ranging from discovering new genes to making targeted mutations. Each chapter is written by internationally recognized experts in their respective fields, and the emphasis is on integrating molecular-genetic and behavioral neuroscience methodology.

The Handbook is divided into six sections, most of which start with basic descriptions of the technique and its strengths, how it can be used, and what the problems and pitfalls are. Next, some case studies are presented to illustrate these points. The aim is to enable the reader to obtain information about both the practical details of the methods and their applications in behavioral brain research. A variety of species are used, including worms (*Caenorhabditis elegans*), flies (*Drosophila melanogaster*), fish (*Danio rerio*), birds (*Aphelocoma coerulescens*), rodents (*Mus musculus domesticus*), and humans (*Homo sapiens*). The information presented covers most, if not all, aspects of modern behavioral neurogenetic research, from information gathering techniques and statistical considerations for experimental design, through molecular-genetic techniques for discovering novel genes and manipulating gene expression, to the analysis of behavioral phenotypes and the ethical implications of genetic research of behavior and the brain.

References

Crusio, W.E. (1999) Using spontaneous and induced mutations to dissect brain and behavior genetically. *Trends Neurosci.*, 22: 100–102.

De Wit, H.C.D. (1982) *Ontwikkelingsgeschiedenis van de Biologie*. Deel 1. Wageningen: Pudoc.

Fuller, J.L. and Thompson, W.R. (1960) *Behavior Genetics*. New York: John Wiley and Sons.

Fuller, J.L. and Thompson, W.R. (1978) *Foundations of Behavior Genetics*. Saint Louis: C.V. Mosby Company.

Rose, S. (1998) *Lifelines: Biology Beyond Determinism*. Oxford: Oxford University Press.

Van Abeelen, J.H.F. (1974) Genotype and the cholinergic control of exploratory behaviour in mice. In: J.H.F. Van Abeelen (Ed.), *The Genetics of Behaviour*, North-Holland, Amsterdam, pp. 347–374.

Acknowledgements

In the three years during which this book took form, we were fortunate to benefit from the advice and encouragement of many colleagues, to whom we extend our warmest thanks: Laura Almasy (San Antonio, TX, USA), Michael A. Arbib (Los Angeles, CA, USA), Jon Beckwith (Boston, MA, USA), Catherine Belzung (Tours, France), Seymour Benzer (Pasadena, CA, USA), Alexis Brice (Paris, France), Christine van Broeckhoven (Antwerp, Belgium), Erich Buchner (Würzburg, Germany), Michèle Carlier (Orléans, France), Sue Carter (College Park, MD, USA), Nicola S. Clayton (Davis, CA, USA), J. Steven de Belle (Las Vegas, NV, USA), Lynn DeLisi (Stony Brook, NY, USA), Catherine Derom (Leuven, Belgium), Didier Devys (Illkirch, France), Karen Duff (Orangeburg, NY, USA), Jean-Michel Elsen (Auzeville, France), Sharon Erickson (San Francisco, CA, USA), Karl-Friedrich Fischbach (Freiburg, Germany), Karl-Peter Giese (London, UK), Philip Gorwood (Colombes, France), Irving I. Gottesman (Charlottesville, VA, USA), Seth G.N. Grant (Edinburgh, UK), Ralph J. Greenspan (New York, NY, USA), Su Guo (San Francisco, CA, USA), Martin Hahn (Wayne, NJ, USA), Kristin Hamre (Memphis, TN, USA), John Hardy (Jacksonville, FL, USA), Byron C. Jones (University Park, PA, USA), Laura Karkowski (Richmond, VA, USA), Sandra Kelly (Columbia, SC, USA), Anu K. Kinnunnen (Helsinki, Finland), Susanne Kramer (Würzburg, Germany), Charalambos P. Kyriacou (Leicester, UK), Robert Lalonde (Mont-Saint-Aignan, France), Serge Laroche (Orsay, France), Marc Le Bert (Orléans, France), Marion Leboyer (Créteil, France), Pedro R. Lowenstein (Manchester, UK), Derek Choi-Lunndberg (San Francisco, CA, USA), Jean-Louis Mandel (Illkirch, France), Ron Mandel (Lund, Sweden), Isabelle Mansuy (Zurich, Switzerland), Jean Mariani (Paris, France), Alexander Marks (Toronto, Ontario, Canada), Stephen Maxson (Storrs, CT, USA), Mark Mayford (La Jolla, CA, USA), James McGaugh (Irvine, CA, USA), Wendi S. Neckameyer (St. Louis, MO, USA), Pat Nolan (Harwell, UK), Michael J. Owen (Cardiff, UK), Gert Pflugfelder (Würzburg, Germany), M. Ian Philips (Gainesville, FL, USA), Klaus Rajewski (Cologne, Germany), Arnon Rosenthal (San Francisco, CA, USA), Pierre L. Roubertoux (Orléans, France), Hans-Jörg Sauer (Baltimore, MD, USA), Günther Schütz (Heidelberg, Germany), Morris Soller (Jerusalem, Israel), Ortrud K. Steinlein (Bonn, Germany), Peter St. Georges-Hyslop (Toronto, Ont., Canada), Yvon Trottier (Illkirch, France), Joe Tsien (Princeton, NJ, USA), Derek Van Der Kooy (Toronto, Ontario, Canada), Claes Wahlestedt (Stockholm, Sweden), and Stephen Wilson (London, UK). RTG is grateful to Evelyn Berry (South San Francisco, CA, USA) for her secretarial help. We would like to express our gratitude to the many authors who devoted their efforts and expertise to the completion of this project and who, by-and-large, kept to the sometimes very stringent deadlines we had to impose on them to ensure timely publication of this Handbook. Special thanks are due to Joe Huston (Düsseldorf, Germany), who first suggested to us to undertake this project. Finally, we are greatly indebted to Hans (J.H.F.) van Abeelen (1936–1998), for his teaching and friendship and for showing us the way.

List of Contributors

N. Agopyan Samuel Lunenfeld Research Institute, Mount Sinai Hospital, 600 University Avenue, Toronto, Ont. M5G 1X5, Canada

L. Almasy Department of Genetics, Southwest Foundation for Biomedical Research, San Antonio, TX 78245-0549, USA

J.M. Aran Down Syndrome Research Group, Medical and Molecular Genetics Center-IRO, Hospital Duran i Reynals, Autov. Castelldefels km 2.7, 08907 L'Hospitalet de Llobregat, Barcelona, Spain

M. Arbonés Down Syndrome Research Group, Medical and Molecular Genetics Center-IRO, Hospital Duran i Reynals, Autov. Castelldefels km 2.7, 08907 L'Hospitalet de Llobregat, Barcelona, Spain

M. Armanini Department of Neuroscience, Genentech Inc., 1 DNA Way, South San Francisco, CA 94080-4990, USA

J.D. Armstrong Division of Molecular Genetics, University of Glasgow, 56 Dumbarton Road, Glasgow G11 6NU, UK

M. Barkats Laboratoire de Génétique Moléculaire de la Neurotransmission et des Processus Neurodégénératifs, CNRS, Hôpital de la Pitié Salpêtrière, 75013 Paris, France

C. Barlow Laboratory of Genetics, The Salk Institute for Biological Studies, 10010 North Torrey Pines Road, La Jolla, CA 92037, USA

L. Bate University College London, Department of Paediatrics, The Rayne Institute, University Street, London WC1E 6JJ, UK

J. Beckwith Department of Microbiology and Molecular Genetics, Harvard Medical School, 200 Longwood Avenue, Boston, MA 02115, USA

F. Bellivier Henri Mondor Hospital, Service de Psychiatrie Adulte, Avenue du Maréchal de Lattre de Tassigny, 94010 Créteil Cedex, France

C. Belzung Laboratoire d'Ethologie et de Pharmacologie du Comportement (LEPCO), Faculté des Sciences, Parc Grandmont, F-37200 Tours, France

J. Blangero Department of Genetics, Southwest Foundation for Biomedical Research, San Antonio, TX 78245-0549, USA

D.A. Blizard Center for Developmental and Health Genetics and Intercollege Graduate Program in Genetics, The Pennsylvania State University, University Park, PA 16802, USA

F.E. Bloom Department of Neuropharmacology, The Scripps Research Institute, La Jolla, CA 92037, USA

J. Bouchard Center for Advanced Biotechnology and Departments of Biomedical Engineering, Biology and Pharmacology, Boston University, 36 Cummington Street, Boston, MA 02215, USA

B. Cairns Department of Pathology, Genentech Inc., 1 DNA Way, South San Francisco, CA 94080-4990, USA

A. Chen Genetic Disease Research Branch, National Human Genome Research Institute, National Institute of Health, Bethesda, MD 20892, USA

N.S. Clayton Section of Neurobiology, Physiology and Behavior, Briggs Hall, University of California Davis, 1 Shields Avenue, Davis, CA 95616, USA

O. Corti Laboratoire de Génétique Moléculaire de la Neurotransmission et des Processus Neurodégénératifs, CNRS, Hôpital de la Pitié Salpêtrière, 75013 Paris, France

J.C. Crabbe Portland Alcohol Research Center, Portland, OR 97201, USA

J.N. Crawley Section on Behavioral Neuropharmacology, Experimental Therapeutics Branch, National Institute of Mental Health, National Institutes of Health, Bethesda, MD 20892b-1735, USA

W.E. Crusio Génétique, Neurogénétique et Comportement, CNRS UPR 9074, 3B Rue de la Férollerie, 45071 Orléans Cedex 02, France

C.L. Cunningham Portland Alcohol Research Center, Portland, OR 97201, USA

A. Darvasi Smith, Kline Beecham Pharmaceuticals, Harlow, Essex CM19 5AW, UK

B.J. Davies Centre for Genome Research and Centre for Neuroscience, University of Edinburgh, King's Buildings, West Mains Road, Edinburgh EH9 3JQ, UK

M. Dierssen Down Syndrome Research Group, Medical and Molecular Genetics Center-IRO, Hospital Duran i Reynals, Autov. Castelldefels km 2.7, 08907 L'Hospitalet de Llobregat, Barcelona, Spain

C. De Jonghe Department of Molecular Genetics, Flanders Interuniversity Institute for Biotechnology (VIB) and Neurogenetics Laboratory, Born-Bunge Foundation (BBS), University of Antwerp (UIA), Department of Biochemistry, Universiteitsplein 1, B-2610 Antwerp, Belgium

J.E. Dowling Department of Molecular and Cellular Biology, The Biological Laboratories, Harvard University, Cambridge, MA 02138, USA

W. Driever Biology 1, University of Freiburg, Hauptstrasse 1, D-79110 Freiburg, Germany

K. Duff Nathan Kline Institute, 140 Old Orangeburg Road, Orangeburg, NY 10962, USA

X. Estivill Down Syndrome Research Group, Medical and Molecular Genetics Center-IRO, Hospital Duran i Reynals, Autov. Castelldefels km 2.7, 08907 L'Hospitalet de Llobregat, Barcelona, Spain

A.E. Farmer Institute of Psychiatry, Social, Genetic and Developmental Psychiatry Research Centre, De Crespigny Park, Denmark Hill, London SE5 8AF, UK

I. Fenton Neuropsychiatric Genetics Unit, Divisions of Psychological Medicine and Medical Genetics, Tenovus Building, University of Wales College of Medicine, Heath Park, Cardiff CF4 4XN, UK

C. Fillat Down Syndrome Research Group, Medical and Molecular Genetics Center-IRO, Hospital Duran i Reynals, Autov. Castelldefels km 2.7, 08907 L'Hospitalet de Llobregat, Barcelona, Spain

A.S. Fleming Department of Psychology, University of Toronto at Mississauga, Mississauga, Ont. L5L 1C6, Canada

J. Flórez Department of Physiology and Pharmacology, Faculty of Medicine, University of Cantabria, 39011 Santander, Spain

C. Foulon Institute des Sciences et Techniques de L'Ingénieur d'Angers, 62 Rue Notre Dame du Lac, 49100 Angers, France

W.N. Frankel The Jackson Laboratory, 600 Main St, Bar Harbor, ME 04609, USA

F.H. Gage Laboratory of Genetics, The Salk Institute for Biological Studies, 10010 North Torrey Pines Road, La Jolla, CA 92037, USA

J.A. Galloway Department of Psychology, University of British Columbia, Vancouver, BC V6T 1Z4, Canada

W.-Q. Gao Department of Neuroscience, Genentech Inc., 1 DNA Way, South San Francisco, CA 94080-4990, USA

M. Gardiner University College London, Department of Paediatrics, The Rayne Institute, University Street, London WC1E 6JJ, UK

L. Garrett Genetic Disease Research Branch, National Human Genome Research Institute, National Institute of Health, Bethesda, MD 20892, USA

B. Geddes Department of Medicine, University of Bristol, Marlborough Street, Bristol BS2 8HW, UK

R.T. Gerlai Department of Neuroscience, Genentech Inc., 1 DNA Way, South San Francisco, CA 94080-4990, USA

K.P. Giese Department of Anatomy and Developmental Biology, University College London, London WC1E 6BT, UK

P. Gorwood INSERM 155 Unity, Genetics Epidemiology, Paris, France

S.G.N. Grant Centre for Genome Research and Centre for Neuroscience, University of Edinburgh, King's Building, West Mains Road, Edinburgh EH9 3JQ, UK

S. Guo Department of Neuroscience, 1 DNA Way, Genentech Inc., South San Francisco, CA 94080, USA

T. Harding Department of Medicine, University of Bristol, Marlborough Street, Bristol BS2 8HW, UK

M.M. Harpold National Center for Genome Resources, 1800-A Old Pecos Trail, Santa Fe, NM 87505, USA

A.C. Hart Massachusetts General Hospital Cancer Center, Harvard Medical School, Department of Pathology, CNY 149-7202 13th Street, Charlestown, MA 02129, USA

L.M. Houdebine Unité de Différenciation Cellulaire, Institut National de la Recherche Agronomique, 78352 Jouy-en-Josas Cedex, France

N.J.O. Jacobsen Neuropsychiatric Genetics Unit, Divisions of Psychological Medicine and Medical Genetics, Tenovus Building, University of Wales College of Medicine, Heath Park, Cardiff CF4 4XN, UK

Z. Jia Samuel Lunenfeld Research Institute, Mount Sinai Hospital, 600 University Avenue, Toronto, Ont. M5G 1X5, Canada

E.D. Jarvis Department of Neurobiology, Duke University Medical Center, Durham, NC 27710, USA

K. Kaiser Division of Molecular Genetics, University of Glasgow, 56 Dumbarton Road, Glasgow G11 6NU, UK

E.R. Kandel Howard Hughes Medical Institute, Center for Neurobiology and Behavior, College of Physicians and Surgeons of Columbia University and New York State Psychiatric Institute, 722 West 168th Street, New York, NY 10032, USA

L.J. Kriegsfeld Departments of Psychology, Neuroscience and Population Dynamics, The Johns Hopkins University, Baltimore, MD 21218-2686, USA

R. Lalonde Université de Rouen, Faculté des Sciences, UPRES PSY.CO-EA 1780, Laboratoire de Neurobiologie de l'Apprentissage, 76821 Mont-Saint-Aignan Cedex, France

R. Lathe Centre for Genome Research and Centre for Neuroscience, University of Edinburgh, King's Buildings, West Mains Road, Edinburgh EH9 3JQ, UK

M. Leboyer Albert Chenevier Hospital, Service de Psychiatrie Adulte, 40 Rue de Mesly, 94010 Créteil Cedex, France

I. Le Roy UPR CNRS 9074, Génétique, Neurogénétique, Comportement, Institut de Transgénose, 3B Rue de la Férollerie, 45071 Orléans Cedex 03, France

D.F. Levinson Department of Psychiatry, MCP Hahnemann University, MCPH-EPPI, Room 206, 3200 Henry Avenue, Philadelphia, PA 19129, USA

C. Li Department of Biology, Boston University, 5 Cummington Street, Boston, MA 02215, USA

N. Lijam Laboratory of Genetic Disease Research, National Human Genome Research Institute, National Institutes of Health, Bethesda, MD 20892, USA

K. Lindblad Department of Molecular Medicine, Neurogenetic Unit, CMM L8:00, Karolinska Hospital, 171 76 Stockholm, Sweden

Y.M. Lu Samuel Lunenfeld Research Institute, Mount Sinai Hospital, 600 University Avenue, Toronto, Ont. M5G 1X5, Canada

A. Maggi Institute of Pharmacological Sciences, University of Milan, Via Balzaretti 9, 20133 Milan, Italy

S. Malkani Program in Neuroscience and Department of Psychology, University of Delaware, Newark, DE 19716, USA

J. Mallet Laboratoire de Génétique Moléculaire de la Neurotransmission et des Processus Neurodégénératifs, CNRS, Hôpital de la Pitié Salpêtrière, 75013 Paris, France

I.M. Mansuy Institute of Cell Biology, Department of Biology, ETH Hönggerberg HPM D24, CH-8093 Zurich, Switzerland

J. Mariani Laboratoire de Neurobiologie du Développement, Institut des Neurosciences, Université Pierre et Marie Curie et CNRS UMR 7624, 75005 Paris, France

G.J.M. Martens Department of Molecular Animal Physiology, University of Nijmegen, Toernooiveld 1, 6525 ED Nijmegen, The Netherlands

M. Mayford Department of Neurosciences 0691, University of California San Diego, School of Medicine, 9500 Gilman Drive, La Jolla, CA 92093, USA

M.P. McDonald Section on Behavioral Neuropharmacology, Experimental Therapeutics Branch, National Institute of Mental Health, National Institutes of Health, Bethesda, MD 20892, USA

P. McGuffin Institute of Psychiatry, Social, Genetic and Developmental Psychiatry Research Centre, De Crespigny Park, Denmark Hill, London SE5 8AF, UK

C.V. Mello Laboratory of Animal Behavior, The Rockefeller University, 1230 York Avenue, New York, NY 10021, USA

P. Moran Department of Neuroscience, Genentech Inc., 1 DNA Way, South San Francisco, CA 94080-4990, USA

S. Mortaud UPR CNRS 9074, Génétique, Neurogénétique, Comportement, Institut de Transgénose, 3B Rue de la Férollerie, 45071 Orléans Cedex 02, France

R.J. Nelson Departments of Psychology, Neuroscience and Population Dynamics, The Johns Hopkins University, Baltimore, MD 21218-2686, USA

G.H. Nguyen Center for Advanced Biotechnology, Boston University, 36 Cummington Street, Boston, MA 02215, USA

M. Nosten-Bertrand Chargée de Recherche au CNRS, INSERM U 513, IM3, Faculté de Médecine, 8 Rue du Général Sarrail, 94000 Créteil Cedex, France

D.H. O'Day Department of Zoology, University of Toronto at Mississauga, Mississauga, Ont. L5L 1C6, Canada

B.A. Oostra Department of Clinical Genetics, Erasmus University Rotterdam, PO Box 1738, Dr Molewaterplein 50, 3000 DR Rotterdam, The Netherlands

M.J. Owen Neuropsychiatric Genetics Unit, Divisions of Psychological Medicine and Medical Genetics, Tenovus Building, University of Wales College of Medicine, Heath Park, Cardiff CF4 4XN, UK

R. Paylor Section on Behavioral Neuropharmacology, Experimental Therapeutics Branch, National Institute of Mental Health, National Institutes of Health, Bethesda, MD 20892, USA

F. Perez-Diaz UPR CNRS 9074, Génétique, Neurogénétique, Comportement, Institut de Transgénose, 3B Rue de la Férollerie, 45071 Orléans Cedex 02, France

K.R. Peters Department of Psychology, University of British Columbia, Vancouver, BC V6T 1Z4, Canada

G.O. Pflugfelder Lehrstuhl für Genetik, Theodor-Boveri-Institut, Biozentrum, Unversität Würzburg, Am Hubland, D-97074 Würzburg, Germany

H.S. Phillips Department of Neuroscience, Genentech Inc., 1 DNA Way, South San Francisco, CA 94080-4990, USA

B.S. Pickard Centre for Genome Research and Centre for Neuroscience, University of Edinburgh, King's Buildings, West Mains Road, Edinburgh EH9 3JQ, UK

L.H. Pinto Department of Neurobiology and Physiology, Northwestern University, 2153 North Campus Drive, Evanston, IL 60208-3520, USA

G. Pollio Institute of Pharmacological Sciences, University of Milan, Via Balzaretti 9, 20133 Milan, Italy

T. Préat Institut Alfred Fessard, CNRS, Bât. 33, 1 Avenue de la Terrasse, 91190 Gif-sur-Yvette, France

M. Pritchard IRD, Monash Medical Center, Clayton, Victoria, Australia

C.H. Rankin Department of Psychology, University of British Columbia, Vancouver, BC V6T 1Z4, Canada

C.A.L. Riedl Department of Biology, University of Toronto, Erindale Campus, Missassauga, Ont. L5L 1C6, Canada

J. Roder Samuel Lunenfeld Research Institute, Mount Sinai Hospital, 600 University Avenue, Toronto, Ont. M5G 1X5, Canada

P. Rosay Division of Molecular Genetics, University of Glasgow, 56 Dumbarton Road, Glasgow G11 6NU, UK

K.A. Rose Centre for Genome Research and Centre for Neuroscience, University of Edinburgh, King's Buildings, West Mains Road, Edinburgh EH9 3JQ, UK

J.B. Rosen Program in Neuroscience and Department of Psychology, University of Delaware, Newark, DE 19716, USA

A. Rosenthal Department of Neuroscience, Genentech Inc., 1 DNA Way, South San Francisco, CA 94080, USA

P.L. Roubertoux UPR CNRS 9074, Génétique, Neurogénétique, Comportement, Institut de Transgénose, 3B Rue de la Férollerie, 45071 Orléans Cedex 02, France

M.W. Salter Programmes in Brain and Behaviour and Cell Biology, Hospital for Sick Chidren, Department of Physiology, University of Toronto, Toronto, Ont. M5G 1X8, Canada

G. Sandberg Department of Molecular Medicine, Neurogenetic Unit, CMM L8:00, Karolinska Hospital, 171 76 Stockholm, Sweden

M. Schalling Department of Molecular Medicine, Neurogenetic Unit, CMM L8:00, Karolinska Hospital, 171 76 Stockholm, Sweden

M.-C. Senut Laboratory of Genetics, The Salk Institute for Biological Studies, 10010 North Torrey Pines Road, La Jolla, CA 92037, USA

D.L. Shelton Department of Neuroscience, Genentech Inc., 1 DNA Way, South San Francisco, CA 94080-4990, USA

A. Shih Department of Neuroscience, Genentech Inc., 1 DNA Way, South San Francisco, CA 94080-4990, USA

N. Shinsky Department of Neuroscience, Genentech Inc., 1 DNA Way, South San Francisco, CA 94080-4990, USA

C.L. Smith Center for Advanced Biotechnology and Departments of Biomedical Engineering, Biology and Pharmacology, Boston University, 36 Cummington Street, Boston, MA 02215, USA

B.W.S. Sobral National Center for Genome Resources, 1800-A Old Pecos Trail, Santa Fe, NM 87505, USA

M.B. Sokolowski Department of Biology, University of Toronto, Erindale Campus, Missassauga, Ont. L5L 1C6, Canada

C. Sotelo Neuromorphologie: Développement-Evolution, INSERM U. 106, Hôpital de la Salpêtrière, 75651 Paris Cedex 13, France

G. Stapleton Centre for Genome Research and Centre for Neuroscience, University of Edinburgh, King's Buildings, West Mains Road, Edinburgh EH9 3JQ, UK

M. Steele Centre for Genome Research and Centre for Neuroscience, University of Edinburgh, King's Buildings, West Mains Road, Edinburgh EH9 3JQ, UK

N. Storm Center for Advanced Biotechnology and Departments of Biomedical Engineering, Biology and Pharmacology, Boston University, 36 Cummington Street, Boston, MA 02215, USA

C. Strazielle Université de Rouen, Faculté des Sciences, UPRES PSY.CO-EA 1780, Laboratoire de Neurobiologie de l'Apprentissage, 76821 Mont-Saint-Aignan Cedex, France

S.T. Suhr Laboratory of Genetics, The Salk Institute for Biological Studies, 10010 North Torrey Pines Road, La Jolla, CA 92037, USA

D.J. Sussman Division of Human Genetics, University of Maryland School of Medicine, Baltimore, MD 21201, USA

J.S. Takahashi Howard Hughes Medical Institute, Northwestern University, 2153 North Campus Drive, Evanston, IL 60208-3520, USA

B.A. Taylor The Jackson Laboratory, 600 Main St, Bar Harbor, ME 04609, USA

B. Thompson Program in Neuroscience and Department of Psychology, University of Delaware, Newark, DE 19716, USA

S. Tordjman UPR CNRS 9074, Génétique, Neurogénétique, Comportement, Institut de Transgénose, 3B Rue de la Férollerie, 45071 Orléans Cedex 02, France

J.Z. Tsien Department of Molecular Biology and Neuroscience Program, Princeton University, Princeton, NJ 08544, USA

J.B. Uney Department of Medicine, University of Bristol, Marlborough Street, Bristol BS2 8HW, UK

C. Van Broeckhoven Department of Molecular Genetics, Flanders Interuniversity Institute for Biotechnology (VIB) and Neurogenetics Laboratory, Born-Bunge Foundation (BBS), University of Antwerp (UIA), Department of Biochemistry, Universiteitsplein 1, B-2610 Antwerp, Belgium

W.H. Van den Hurk Department of Molecular Animal Physiology, University of Nijmegen, Toernooiveld 1, 6525 ED Nijmegen, The Netherlands

F.W. Van Leeuwen Graduate School for Neurosciences Amsterdam, Netherlands Institute for Brain Research, Meibergdreef 33, 1105 AZ Amsterdam, The Netherlands

D. Wahlsten Department of Psychology, University of Alberta, Edmonton, Alberta T6G 2E9, Canada

P.E. Wainwright Department of Health Studies and Gerontology, University of Waterloo, Waterloo, Ont. N2L 3G1, Canada

K. Wallace Program in Neuroscience and Department of Psychology, University of Delaware, Newark, DE 19716, USA

E.C. Warburton Department of Anatomy, University of Bristol, University Walk, Bristol BS3 1HW, UK

J.B. Watson 48-241 NPI, Department of Psychiatry and Behavioral Sciences, Mental Retardation Research Center, Brain Research Institute, UCLA School of Medicine, 760 Westwood Plaza, Los Angeles, CA 90024-1759, USA

N.M. Williams Neuropsychiatric Genetics Unit, Divisions of Psychological Medicine and Medical Genetics, Tenovus Building, University of Wales College of Medicine, Heath Park, Cardiff CF4 4XN, UK

P.M. Williams Department of Cell Biology and Technology, Genentech Inc., 1 DNA Way, South San Francisco, CA 94080-4990, USA

J. Winer Department of Cell Biology and Technology, Genentech Inc., 1 DNA Way, South San Francisco, CA 94080-4990, USA

J. Winslow Department of Neuroscience, Genentech Inc., 1 DNA Way, South San Francisco, CA 94080-4990, USA

A. Wynshaw-Boris Department of Pediatrics, UCSD School of Medicine, La Jolla, CA 92093, USA

W.G. Young Department of Neuropharmacology, The Scripps Research Institute, La Jolla, CA 92037, USA

Contents

III. Manipulating Known Genes

3.1. Transgenic Approaches to Rodents

3.2. Systemic Manipulation

3.3. Invertebrates

5. Human Neurobehavioral Disorders: from Molecular Genetics to Genetic Animal Modes

6. Ethical Considerations

Section 1

Neurobehavioral Informatics and Statistics

W.E. Crusio and R.T. Gerlai (Eds.)
Handbook of Molecular-Genetic Techniques for Brain and
Behavior Research (Techniques in the Behavioral and Neural Sciences, Vol. 13)

CHAPTER 1.1

The multi-dimensional database requirements of brain information in the era of rapid gene identification

Floyd E. Bloom and Warren G. Young

Department of Neuropharmacology, The Scripps Research Institute, La Jolla, CA 92037, USA

Even before the recent revolutions in methods to detect Expressed Sequence Tags (ESTs) of brain enriched or brain specific genes (see Mara et al., 1998), the neurosciences had been moving rapidly to accumulate data at rates clearly exceeding even a wise scientist's ability to organize and recall. Today, the trend in neuroscience informational overload continues (see Bloom and Young, 1994; Bloom, 1995). Now more than ever, it seems impossible to maintain an active ongoing comprehension of the scientific literature in any special corner of neuroscience research, neither the broad awareness of new discoveries, or the in-depth awareness of any but the most narrowly defined field. Yet the fact remains, that there have been few changes to the traditional habits of scientific information gathering, sharing and analyzing practiced by neuroscientists: namely reading research journals and traveling to scientific meetings.

The accelerated pace of molecular discovery has produced estimates that perhaps 97% of the ESTs associated with the major brain genes of mice are already in hand (see Mara et al., 1998). That progress means that there is an enormous backlog of preliminary information on gene expression patterns and potentially, some of the clues needed to link genes to neurons, neurons to circuits, and circuits to systems that underlie or regulate behavior. What is needed next are ways to link that sequence information with brain cellular and circuitry properties. That accomplishment would be a highly effective means to increase our knowledge about neuronal and glial phenotypes and about brain functions and diseases. However, there now exist no ways which are remotely as facile as the gene sequence comparison methods to determine how such genes map onto the independently growing base of information on the molecular, cellular, circuit, and systems of the brains of experimental animals and humans. For this article, we note some of our initial steps to harness the flow of neuroscientific data. Our long term aim is to develop within the community of neuroscientists some means of information handling that can rival the sophistication of the instruments and methods by which molecular geneticists can now acquire and compare data of utmost pertinence to the brain (see e.g. Chapter 1.2).

A database of the brain

For the past 10 years, we have devoted considerable effort towards the development of a complex hierarchical, relational, object oriented database of published neuroscience information, built around the orientation of a brain atlas template (see Bloom, 1990) for additional background on neuroscience databases). We have taken this path

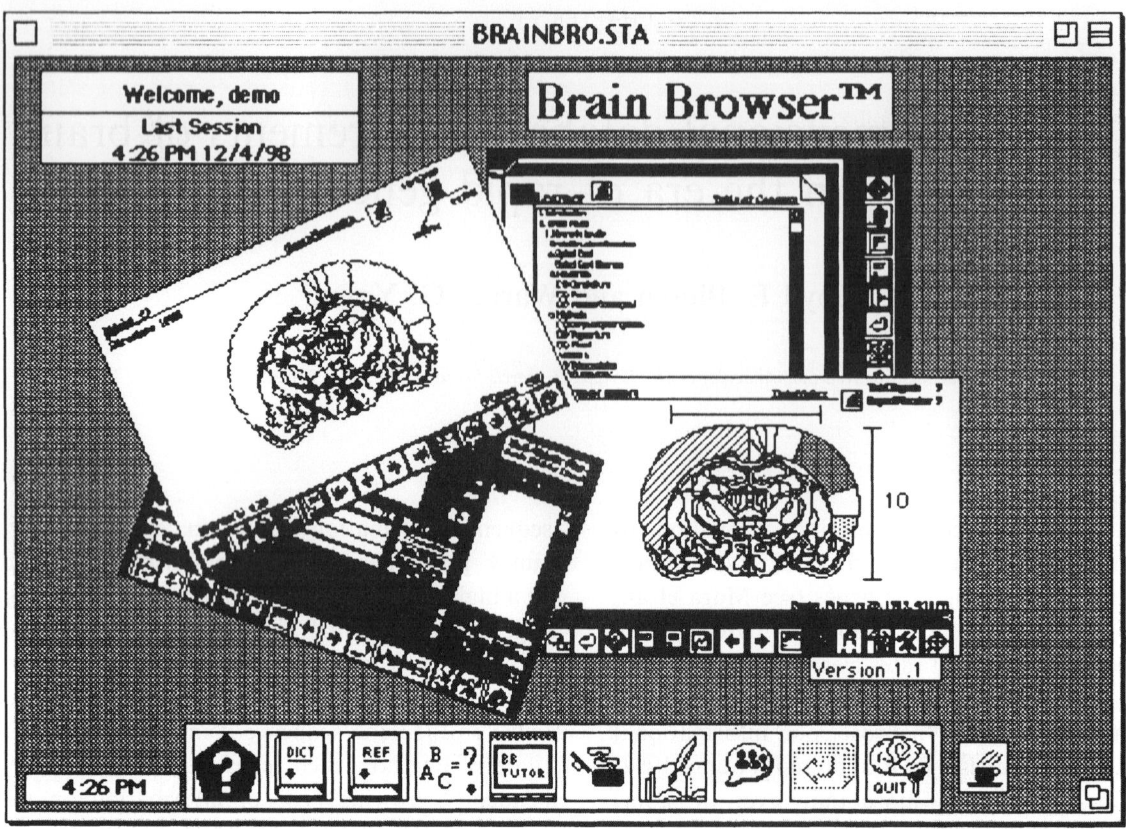

Fig. 1. Brain Browser is a Apple Hypercard® application that works on both Macintosh and Windows computers.

as our initial approach to an eventual super-highway of neuroscience information traffic control. Such an information handling system is, in our view, an absolute necessity to establish the normative parameters of molecular, cellular, and behavioral data (properties which in turn are composed of genetic, metabolic, physiologic and structural details of chemistry, circuitry, and physiology). The existence of such a normative database could then be used for the detection and definition of pathological variations.

Our initial publicly released product was termed "Brain Browser" (Bloom et al., 1989) a software application for the Macintosh, subsequently imported to both DOS and Windows 3.3 versions (Fig. 1). Brain Browser contained 4 modules. There was an introductory tutorial for those new to the terms and relationships of neuroscience. For more advanced users, the two most used modules were NeuroNavigator and the Linker. The NeuroNavigator (Fig. 2) contained re-drawn digital copies of every plate in the Paxinos and Watson rat brain atlas (Paxinos and Watson, 1987). A user could select quickly whether to display the plates in frontal, sagittal or horizontal planes. Each digital plate was employed as an active template to display the locations of every named location of neurons (nuclei, fields, areas, and layers) in the atlas, some ~900 locations. Passing the cursor over a location would display the name of the location or its stereotaxic location.

However, each location in the NeuroNavigator's atlas templates was directly linked

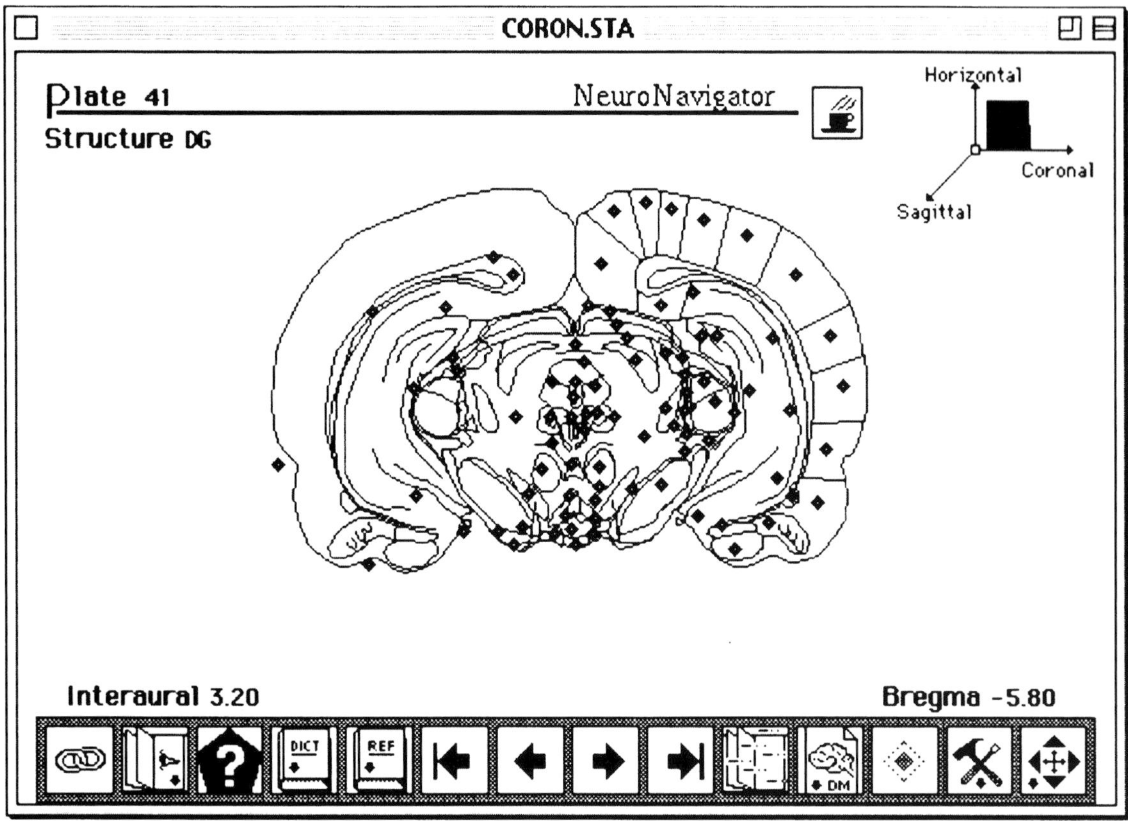

Fig. 2. The NeuroNavigator module of Brain Browser contains the three orthogonal planes of section of the rat brain (coronal, sagittal, and horizontal). The triaxial tool on the top right is used to control the navigation through the sections. Each diamond icon on the section represents a location that is known to Brain Browser. Clicking on the diamond opens the representative page in Linker where the attributes of the neuroanatomic place and its circuits can be examined further.

to a 'place card' in the Linker module (Fig. 3). By the conventions of hypertext, the display of a linked place card for a location in the atlas could then provide more detailed information. Each of the several hundred place cards would inform on the kinds of neurons in that location, their transmitters, receptors, afferents and efferents, as well as the behavioral functions with which that location had been associated. Selection of one of the afferent or efferent circuits would then take the user to a deeper level of information, displaying the details of the circuitry that linked to the original location, as well as the literature sources for the attributes described (Fig. 4). In principle, a user who made a potential new discovery in the identification of transmitter, receptor or circuitry could immediately spot whether such a finding existed in the database. A user in search of something important to search for could scan the database for gaps in small or large details, of which there were many.

Since the initial circuitry details were compiled by the authors from comprehensive recent reviews, it was hoped that users who were specialists in specific rat brain functions, circuits or transmitters would augment and share more detailed levels of analysis. For what ever reasons, that outcome was not realized, nor was the fourth module, Datamaker consistently used as intended. Datamaker (Fig. 5) was a means for users to

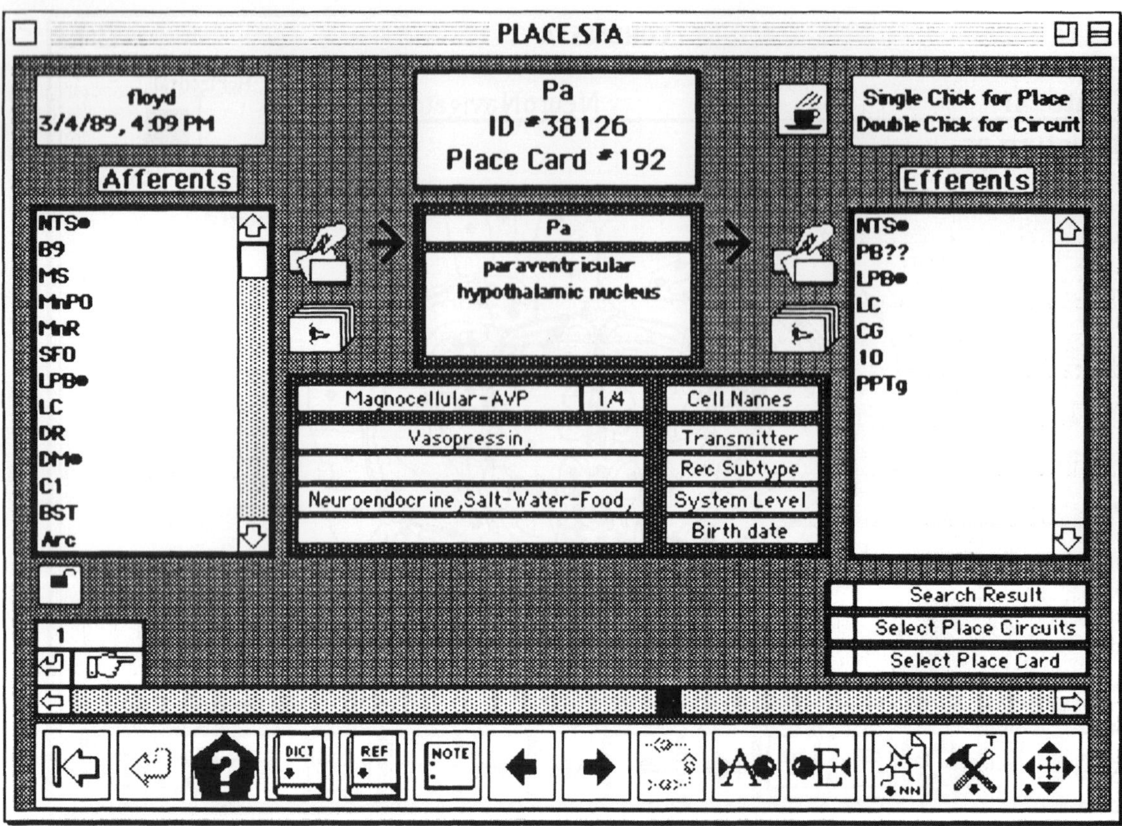

Fig. 3. The Linker module in Brain Browser is used to display known neuroanatomic places in the rat brain, afferent and efferent circuits of this place, attributes such as cell types located in this place, their transmitters, receptor subtypes, associated system levels, and their birthdates. Clicking on any displayed place will open Linker to that place. Double clicking on a afferent or efferent location will open up the circuit (see figure 4).

record their new research observations directly on one of the NeuroNavigators template pages, annotate the data, and even apply primitive morphometric analysis to their observations, and then share those data with other's of like interest. The fact that the entire Brain Browser package was wholly devoted to rat in an era when molecular neuroscience discoveries were also gaining significant momentum on murine and primate brains may have been a part of the reason why the numerous purchasers of the database system and its data failed to form a functional community.

A similar comprehensive neuronal circuitry database for each of the major vertebrate central nervous systems is also an essential tool for understanding the known molecular, cellular, and macroscopic features of tie brains and their inter-species relationships. We also take the pragmatic position that such a data management system is necessary to illuminate essential missing elements of information. As a consequence, our efforts have been dedicated to the development of a technology to combine informatics (the science of data collection, organization and interpretation) with neuroscience with the goal of improved management and distribution of neuroscience information. A program to encourage such research efforts has recently been promulgated by several of the NIH institutes with

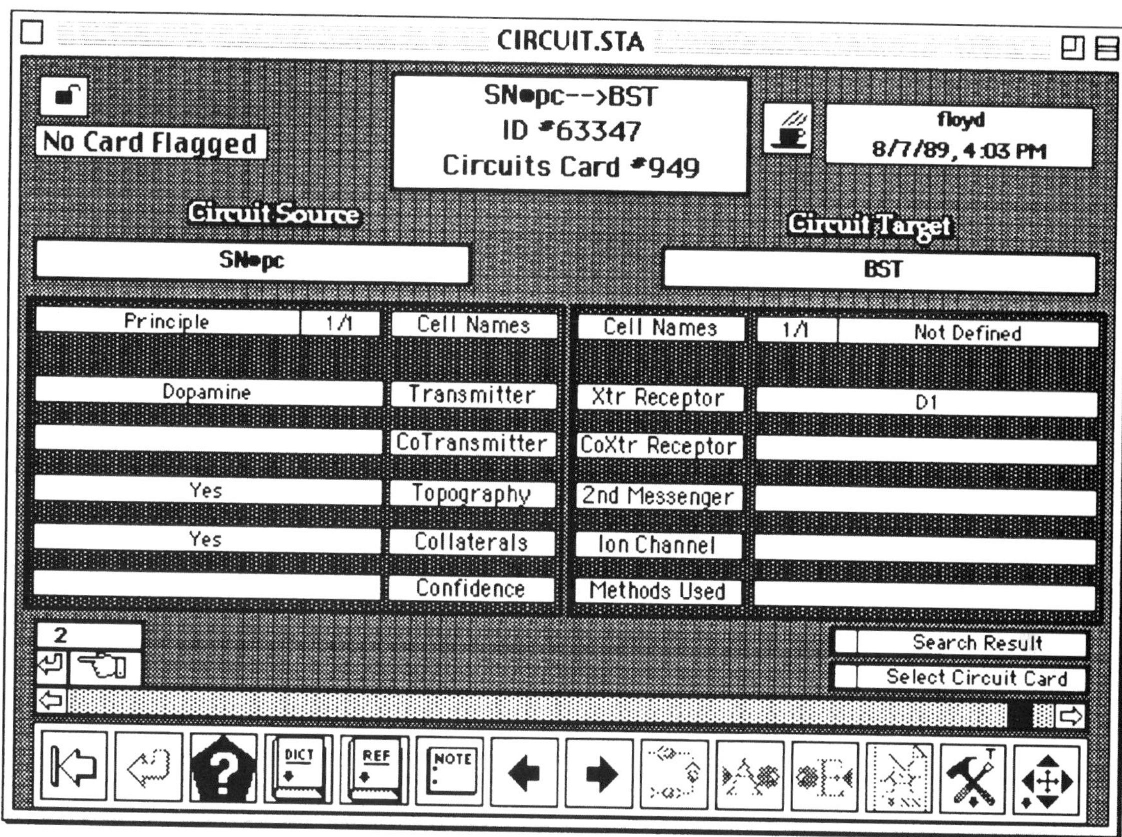

Fig. 4. The Linker module in Brain Browser can display the circuits between any two neuroanatomic locations. The cell types of both the source and target side are displayed, along with each cell type's attribute.

ultimate international applications in mind (see Huerta et al., 1993), no doubt based upon the experiences of those engaged in the Human Genome Project.

The growth of neuroscience information

Scientific interest in the neurosciences has grown enormously over the past several years. That growth is evidenced by the membership in national societies of neurosciences throughout the world, by the proliferation of scientific journals and magazines focused on the neurosciences, and by the programmatic interests of a wide governmental and non-governmental agencies. The sheer volume of accumulated published original reviewed articles in the neurosciences over the past 5 years probably rivals that over the entire previous history of neuroscience research. At present, one can expect that the existing molecular discovery momentum will reveal in rich detail far more reliable information on the detailed connections and mechanisms of interaction of neuronal circuitry at the cellular and molecular levels of understanding.

One might note, for example among recent discoveries, a wholly unexpected homolog of the now-classical hypothalamic neuropeptide, somatostatin, discovered by searching for genes selectively expressed within the hippocampus and cortex, and therefore termed cortistatin (De Lecea et al., 1997). Physiological analysis of synthetic cortistatin revealed several functional differences from somatostatin (De Lecea et al.,

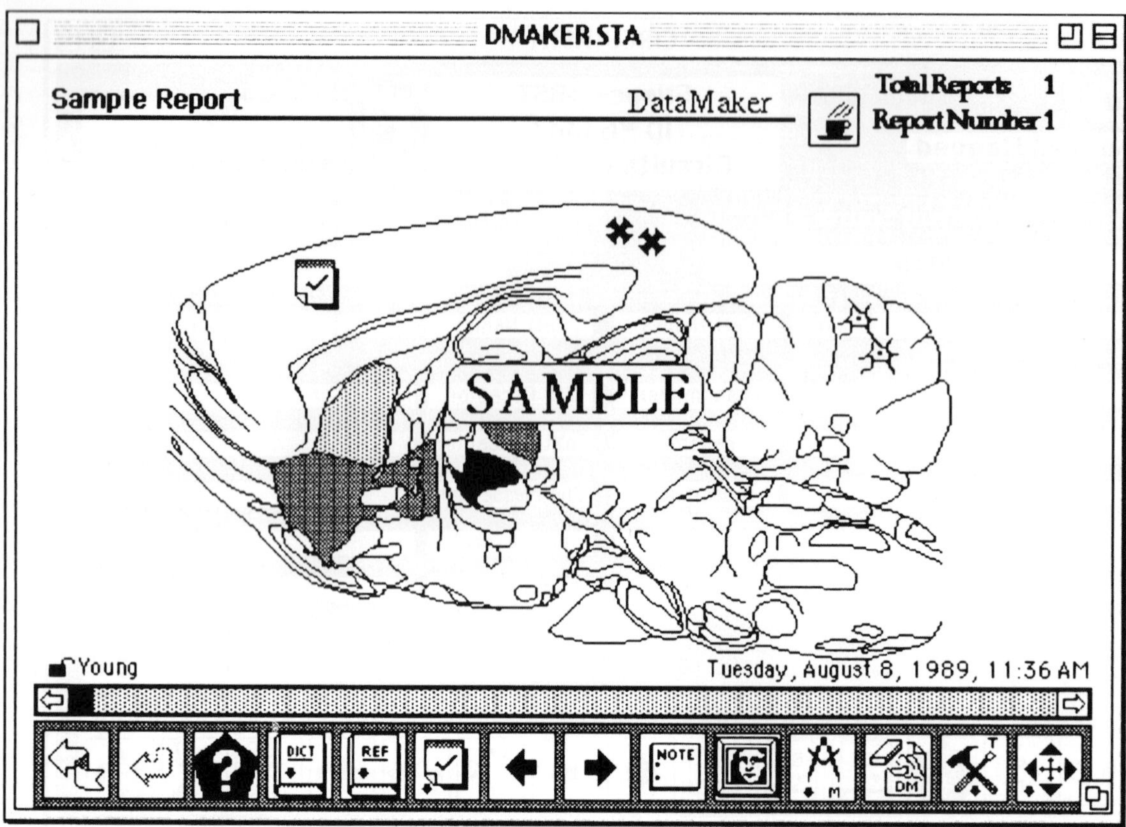

Fig. 5. The DataMaker module in Brain Browser is designed to make electronic copies of the NeuroNavigator plates of any orthogonal plane and allow for user annotations in the form of notes, locations, cells, and drawing graphics. A morphometics module also simple measurements such as stereotaxic location, lengths, and areas.

1996). In a similar, region-selective search for new hypothalamic gene products (see Gautvik et al., 1996) another neuropeptide was identified with a unique hypothalamic localization suggesting a role in visceral function (De Lecea et al., 1998), which was soon attributed to appetitive regulation and given the name orexin (see Sakurai et al., 1998). A third example of this nature are the discovery of yet another unexpected neuropeptide, named CART by virtue of its enhanced expressed following treatment with cocaine or amphetamine (see Couceyro et al., 1997; Koylu et al., 1997, 1998).

Thus, there can be little doubt that high quality data have expanded explosively in the neurosciences. However, there are several implicit barriers to the optimal utilization of this information. In particular, how can these myriad observations be converged into testable hypotheses of normal brain function or the molecular basis for any disease's neuropathogenesis. Because the transgenic mouse technology is providing tantalizing models for both Alzheimer's disease (see Games et al., 1995; Hsiao et al., 1996; Masliah et al., 1996; Sturchler-Pierrat et al., 1997) and amyotrophic lateral sclerosis (Ripps et al., 1995; Morrison et al., 1996, 1998), this may be one successful approach to that convergence.

Aside from these pathological considerations, there is a striking need for serious scholarly attempts to model human cognitive operations through incorporation of rigorous data from chemical neuroanatomy and neurophysiology into modeling algorithms. However, when one seeks to do so, there is an immediate awareness that we lack reliable quantitative information on most aspects of human and experimental neuroanatomy. As noted by Cherniak (1990), estimates published by highly regarded neuroscientists for such elemental factual considerations as the actual area or volume of the human cortex, the density of the neurons within this sheet, and the average number of synapses within the cortical neuropil differ by orders of magnitude. When one considers further the differences in brain shape between individuals, the difficulties in applying rigorously cytoarchitectonic and cortical connectivity criteria to define specific cortical regions, and the inability now to apply to human brain the connectivity tracing tools of experimental neuroanatomy (Crick and Jones, 1993), details of the structure and function of the human brain may appear unapproachable.

Quite apart from these pathological and quantitative correlates, there appear to be finite limits on the ability of any individual scientist to absorb, digest and interpret the existing studies and to monitor, evaluate and incorporate new data into one's appreciation for a given brain region, system or question. The characteristic motif of "neuroscience", namely the interdisciplinary merging of data acquired by anatomists, chemists and physiologists working at their preferred levels of resolution from the molecular to the organismic constitutes its own major barrier to substantive intellectual consolidation of the data.

Neuroscience tools for organized data gathering

Paralleling the data explosion in the neurosciences comes frustration. While it is generally possible to retrieve relevant reliable information on brain molecules, one may be quickly stymied for information to understand the cell systems which express these genes and then to relate those genes and cells to the pertinent behaviors governed by these cells and cell systems. One hungers for a means to perform such vertical integrations of information (from the molecular to the behavioral) in a manner that would meet rigorous scientific standards and yet permit individual scholars the intellectual opportunity to conduct investigation of the accumulated data for their own specific relationships and for hypothesis generation. In our view, several sorts of information management tools are most needed and therefore motivate our efforts in this activity.

High on our wish list is an integrated software system for the quantitative acquisition, display, and analysis of cellular and subcellular morphological information from the microscope. The initial tool set of software ensemble, NeuroZoom, have already produced abundant quantitative information on a variety of neuronal markers in both experimental (Huntley et al., 1994; Hof and Morrison, 1995; Hof et al., 1995b; Gazzaley et al., 1996a, 1996b, 1997; Nimchinsky et al., 1996, 1997) and clinical (Hof et al., 1994a, 1994b, 1995a, 1997; Hof and Morrison, 1996) studies.

NeuroZoom (Bloom et al., 1997; Young et al., 1997) software is a Macintosh-based program that permits mapping and quantification of structures from gross macroscopic measurements on whole brains or brain regions, through conventional optical or confocal microscopy of cells and cellular connections, down to the ultrastructural level of subcellular organelles, synapses, and dendritic spines (Fig. 6). NeuroZoom was designed to be used in conjunction with a microscope and a motorized stage, and can support any digital image capable of being scaled. Large-scale maps are easily executed, as are very high-magnification maps of individual neurons, with quantitative graphical representations of their morphologies (Fig. 7). Data are collected and stored with complete fidelity in three dimensions, permitting 3D rendering of detailed maps. Values such as neuron number, neuron volume, surface area and linear density (e.g. axonal arbor) are also obtainable using routines based on unbiased stereological principles, and systematic random sampling

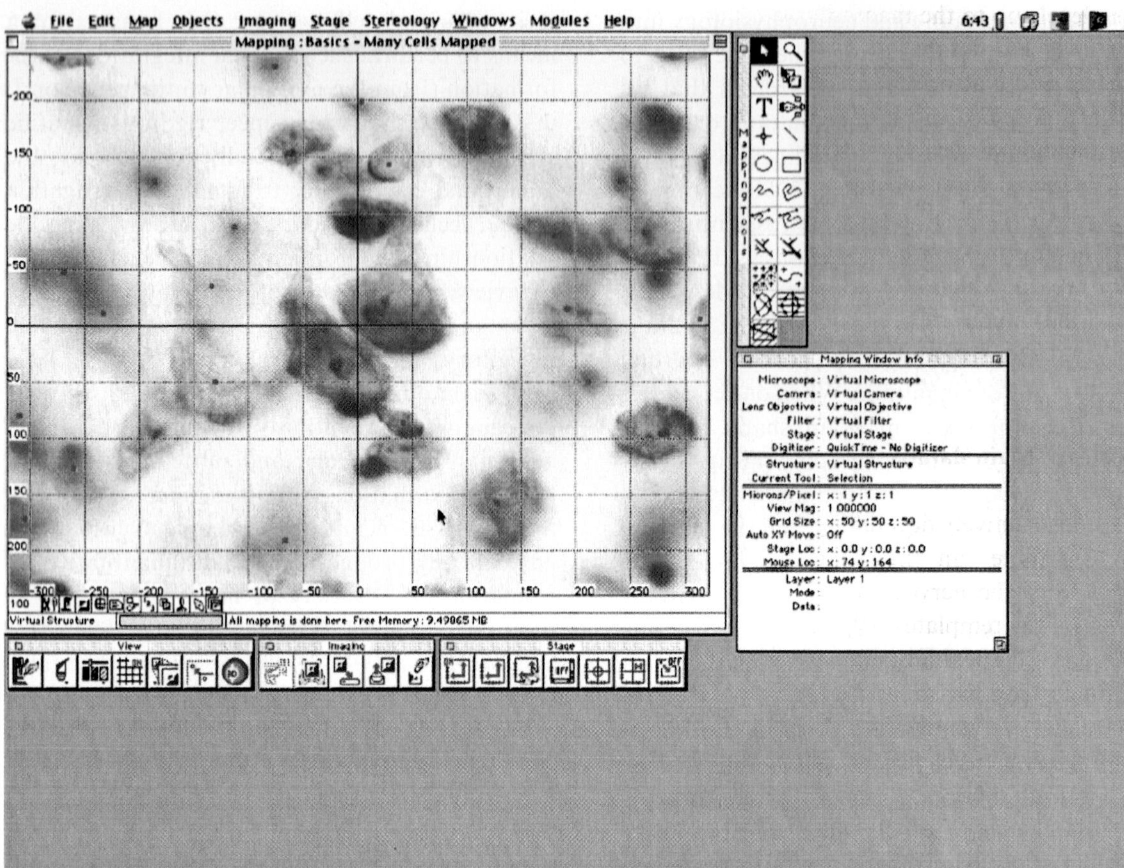

Fig. 6. NeuroZoom is a Macintosh based software application that is designed to support mapping and quantification of structures from gross macroscopic measurements on whole brains or brain regions, through conventional optical or confocal microscopy of cells and cellular connections, down to the ultrastructural level of subcellular organelles, synapses, and dendritic spines. This figure shows one of the stereological probes selected to measure mean linear length of cortical fibers.

schemes that help eliminate sampling bias (Fig. 8). The Macintosh interface is intuitive for users familiar with Macintosh or Windows-based programs, making accurate and rigorous quantification accessible even to small laboratories able to make only a limited investment in quantitative microscopy.

To realize its potential, such a data collection and analysis system should provide its data in a manner that can be integrated with a textual database management system. Only when these databases become an integral part of the literature awareness library of individual or working groups of scientists will improved scientific understanding begin. Comparable tools are also required for the data obtained by neurochemical and neurophysiological research strategies, similarly tied to the cells and regions in which those facts are acquired.

One eventual goal for such tools is to provide the neuroscience scholar, regardless of prior experience, access to a computer or microscope system of their selection and the capability to move from the synaptic level, through cellular, multi-cellular (like layers of specific cortical areas or nuclei of defined subcortical locations), and regional microscopic levels up to the macroscopic framework of our atlases and databases within a quantitatively accurate, and platform (i.e. computer type) independent graphic display environment. Since experimental verification of detailed structural and functional information on human brains is unlikely to be obtainable, we look to the non-human primate brain as a likely experimental route through which human brain scholars could access the much richer database of neuronal circuitry, chemistry and cellular function in other species.

A realistic brain database

We are also actively developing software to create and distribute an interactive Brain Object Database of the nervous system oriented within species on the templates representing the "pages" (electronically speaking) of a classical brain structural atlas. Comparisons of datasets across species are made with reference to the definable homologies between brain regional structures. Our working model is composed of many of the standard classes of objects that are encountered in neurosciences. Namely, we parse neuroanatomic structures at various levels of resolution (from the top down: areas, regions, groups, nuclei, cells, cellular organelles, and macromolecules). Similarly, we parse neurochemical objects (from gene and mRNA sequences "upwards" (in sizes) to proteins, organelles, and the regulatory molecular machinery for intracellular metabolic maintenance. In the same fashion, we parse intercellular transductive signaling) neurofunctional objects (cells, synapses, receptors, transductive mechanisms including ion channels, and their interactions on membrane properties) providing for cellcell interactions in the sense of defined circuits.

Realistic classes of generic neurons are initially encoded with the actual known details of their generalized features, and enhanced by their exceptional properties, which may be defined when determined. Such biologically based neurons may then be collected into defined assemblies of neurons that represent any of the several defined functional systems. This collection can be applied to both normal and pathological states and carried from the molecular specifications up to the behavioral levels. We also envision such a data representational system to encode other neurodata objects (classical data renditions EEG, and event-related potentials) as well as imaging modalities (MRI, PET, CAT, MEG). This Brain Object Database will be constantly extendible in the classes of objects (molecules, organelles, neurons, neuronal subtypes, etc) and derived classes (circuits, circuit operations) based upon the latest research information. The classes of objects and their relationships will be linked as life-like metaphors to their biological structures. Therefore, the system can be suitable not only for encoding and comparing data across levels of analysis and species, but should also be suitable for work at the theoretical level of cellular or systems simulations.

In this manner, we are working to establish a truly comprehensive database that can link graphic image sets as well as textually defined qualitative characteristics, based upon progressively accumulated and refined (and eventually quantitatively established). The data would range from the level of whole brains down to the level of DNA and protein sequence, along with archival reference lists to papers containing those and other data and would lend itself to other online commentary forums. Eventually one can envision an ongoing global neuroscience forum for informal and cooperative data analysis and concept formulation among those collecting the data and those hungry for data to interpret.

The software database model

Translating these metaphors into useful working software applicable to real life questions requires a brief software developer's digression. When we started on databases 10 years ago, our concerns were to incorporate user-friendliness, fast feed-

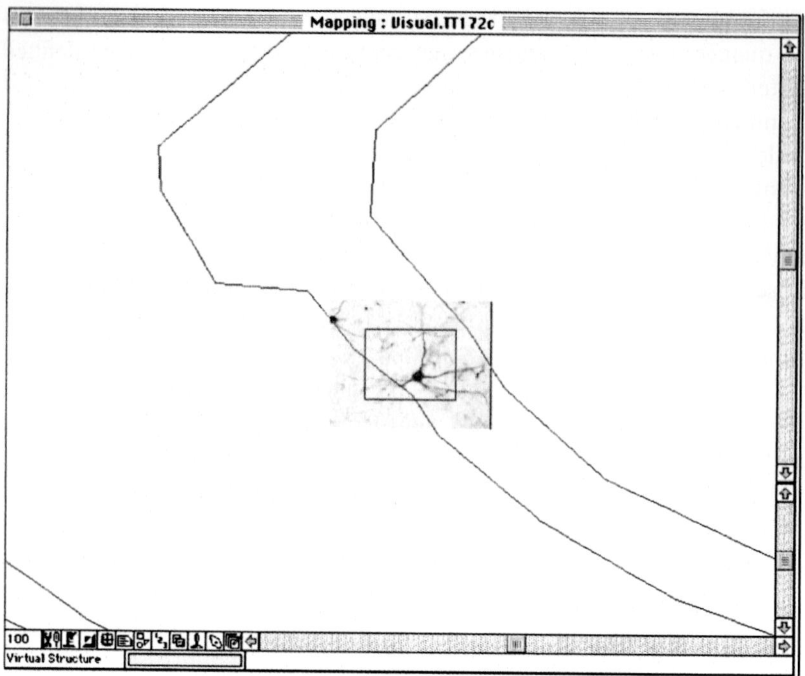

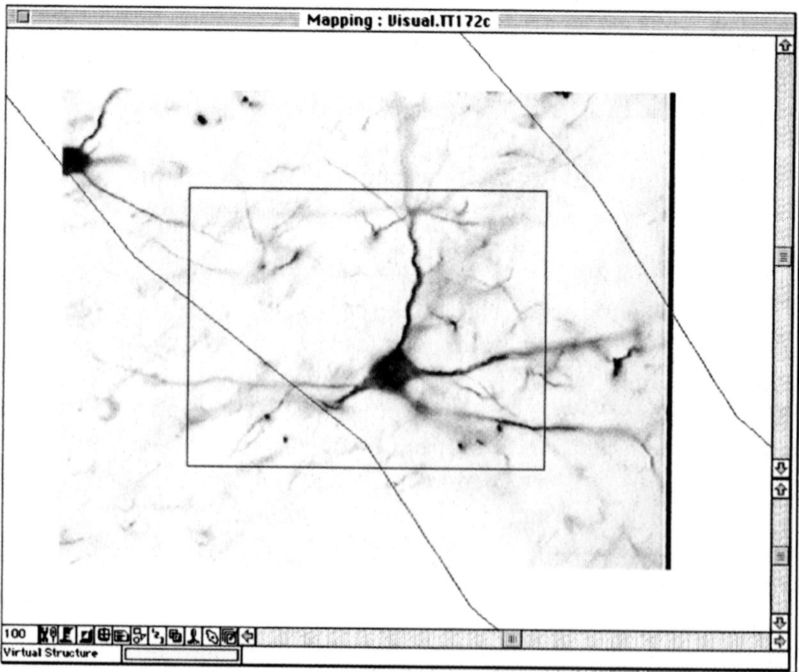

Fig. 7. The top image shows a fractionator study in NeuroZoom at low magnication where the outline of the cortical layers are captured. Each of the little red squares are fields of view from the microscope that are exactly aligned by using a motorized stage to the map. At the higher magnifications, the field of view reveals the actual image that is captured and stored in the NeuroZoom file for analysis of cell types.

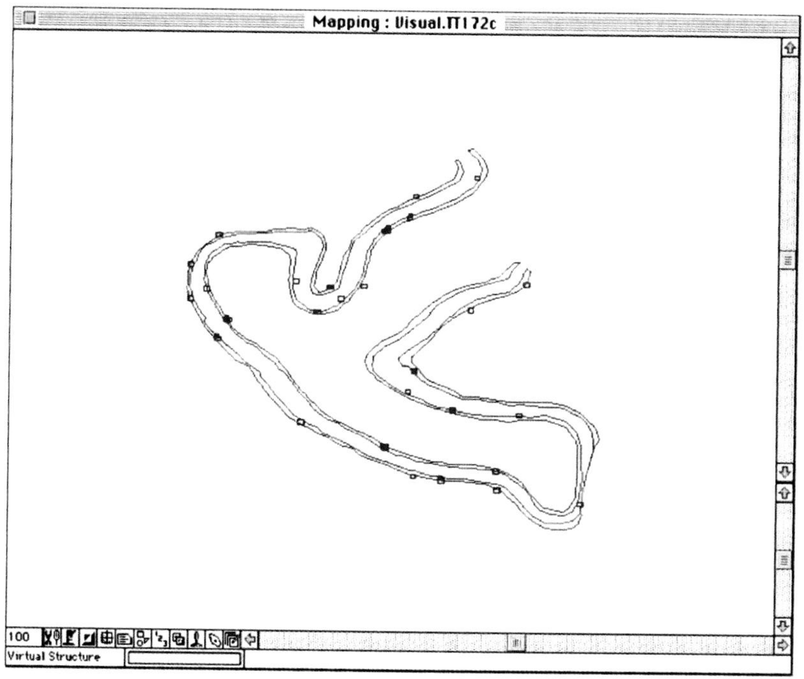
Mapping : Visual.TT172c
100
Virtual Structure

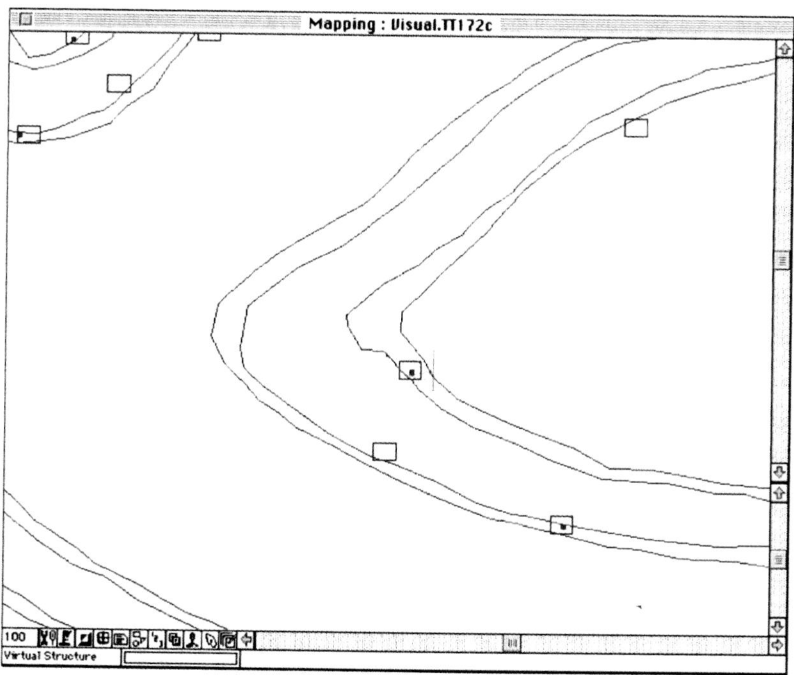
Mapping : Visual.TT172c
100
Virtual Structure

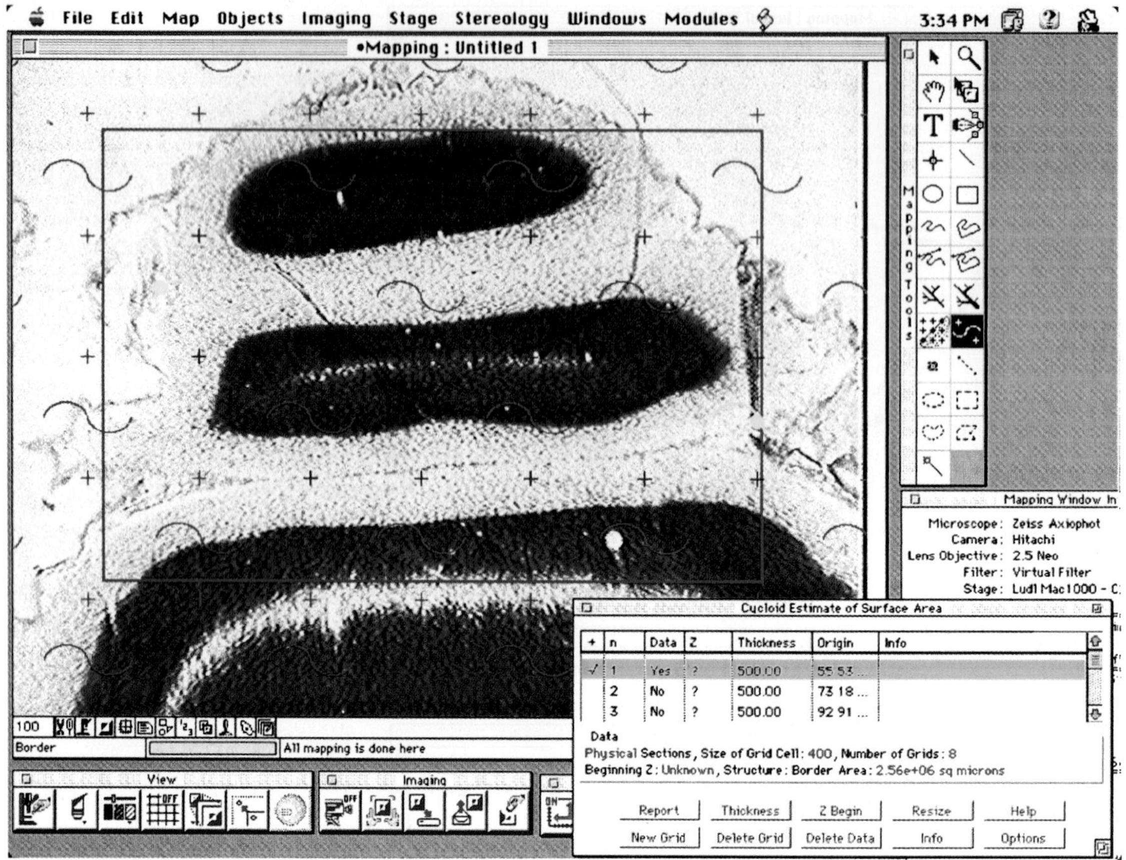

Fig. 8. The cycloid probe in NeuroZoom is used to estimate surface area of the cerebellum, and the Cavalieri point counting probe is simultaneously used to estimate volume. The microscope stage is active at all times and moves the stage automatically when data are entered outside of the red reference frame. Both display grids move with the tissue section as the stage moves, allowing the user to estimate surface area and volume simultaneously in optical sections. Furthermore, standard mapping tools can also be used at any time to gather information on distribution or other morphometrics.

back, and strong graphical capabilities into a brain circuitry database anyone could use and everyone would learn from. That goal was realized using the Apple Macintosh and the Hypercard software. Packaged with comprehensive circuitry from the literature and scaled to the Paxinos and Watson rat atlas, brain browser (Bloom et al., 1989) met with moderate success in 1980 standards for neuroscience software applications.

Buoyed by this success, a three-day workshop was convened in mid-1990 to gain feedback from the neuroscience community on how brain databases should evolve. Most scientists can agree on a common hierarchical arrangement at the anatomic level as a "top-down" approach: from levels of the neuroaxis, to cortical regions, to areas, subareas, layers, fields, groups, nuclei, and finally to cells. However, these divisions provide only the framework of the database. The content of a neuroscience database comes from the details that compose the data definitions. For example, pyramidal cells in the visual cortex are critical for describing visual function. The "data" content would be a definition lineage of "telencephalon–cortex–neocortex–primary visual cortex-layer VI".

However, cortical pyramidal cells as a neuronal class have many more attributes: apical dendrites that ascend to layer IV, an axon that descends to the thalamus, and recurrent axonal collaterals that ascend also to layer IV. The cortico-thalamic feedback loop is created from these circuitry connections between primary visual cortex pyramidal cells and the thalamus. The database model must also deal properly with these many dynamic forms comprised from the instances of the structural model.

A cell's definition, such as "cortical pyramidal cell", is also important when considering the organization of the brain across functional states, such as normal or diseased, young or old, male or female, etc. where the contrasting conditions are linked to a measurement of cell counts. These quantitative measurements provide a form of dynamic, living database model, whose statistical representations (mean, SD, variance, etc) represent in some way the instances of the structural model at any one time, much like a snapshot in time and space. Thus the software for managing the database must also allow for a model that is dynamic, in which the definitions of components vary over time, with flexibility in the implementation of the database.

Implementation of the database is also impacted by the circuitry data. The standard database model contains a framework in which specific data elements, as in spreadsheet cells, represent real objects (cells, nuclei, groups, area, etc). Linkages are created between these objects as circuits, in which each place can be any of the structural parts of the lineage as outlined above. Each place must then contain an additional attribute (database field) to act as a primary key to another place. However, several decades of research on neuronal circuitry make it clear that the inter-cellular structural and functional relationships already understood are far more complex than any existing database.

To search such a complex database for all known circuits falls into the classical highway algorithms in applications like telephone call routing. Brute force computation is required in this relational model to search all of the places, and to integrate this search until all end-nodes have been reached or some desired location is detected. Referential integrity of the database breaks down when data are redundant. Standard queries of circuits using relational technology are highly inefficient, because one rapidly reaches a point of diminishing returns. As the database content increases in depth with more and more samples, and in width with more attributes, the queries become much more time-consuming. Combine this with real-world access to the ideal database by multiple clients, and any desired database access will likely experience poor performance.

We envision object database techniques as a ray of hope, potentially. The database in the simplest sense might be made up of a core engine that understands how to store and retrieve data from computer storage devices. The data that are stored or retrieved are almost always language independent. Basic raw data types are exchanged — strings, numbers, dates, currency, long collections of binary data (BLOB — binary large object). The basic data types are morphed into and out of the object language instances. So, a pyramidal cell or the visual area might have a efferent connection from the thalamus, or conversely, the thalamus might have an afferent connection to the visual area. This circuit could be represented in the "database" as three numbers. The first number would be a specific instance of the thalamus. The second number would indicate the afferent connection to the visual area. The third number would be a specific instance of the visual area itself. The numbers are obviously language independent, but when put back into the context of the place objects, they have anatomical meaning. Indeed, this is the same as for relational technology, except for one important difference — in relational technology, the software application is completely decoupled from the database. These data (the three numbers) must be retrieved from tables and rows using a query language, then programmatically inserted into a skeletal object (e.g. thalamus and visual area) to create the connection. Imagine doing this while traversing every pathway between the known source and the desired target. The computational load on the database engine increases tremendously. Conversely, when it is time to store the object into the database, the three numbers must be removed from the object, a query created for the relational database, and the numbers inserted into the proper tables and rows. Because of this decoupled nature, the responsibility of the software developer increases as well, as validation of the object in memory and the object in the database are manually maintained. The chance for error due to synchronization problems becomes significant.

An object database has the mechanisms in place to build the objects automatically whenever the software application tries to reference or use the object. There is no need to "build" the object by making a query, creating a skeletal object, and inserting the real values. The object database does this automatically. Therefore, the software programmer needs only to concern himself with the needs of the biologist, creating the special functions that a scientist would want, and to spend much less time dealing with database mechanics.

We are currently investigating a specific object database, ObjectStore from Object Design, Inc that allows us to design the brain hierarchies completely using class definitions, represent the content with instances of each class, maintain cellular and circuitry specializations and variations with internal pointers in application memory, and have them stored persistently and transparently. To be optimally useful to society or a scientific community the database content may be distributed over a network in a transparent fashion. The database engines communicate with one another to determine where the data are located, and automatically forward references of the database objects to one another. Performance is very high. Storage capacities are in the pentabyte range. The circuit example outlined above can be at least 10100 times faster than relational technology. Furthermore, since object technology is used throughout the database design, there are no restrictions on the data types that can be stored. Any data object that can be represented by the object language can be stored in the database. The class structures will be written in Java, and initial access to the database will be Web enabled.

Attracting the users and data-producers

Ultimately the users of this or some subsequent iteration of a whole brain database may decide to create an online intellectual community. Such a User group would share their common interest in linking human brain data pertinent to neurological and psychiatric diseases with the collected wisdom deducible from experimental brain research. To build such a group of users, and more particularly to convince them that the overall database effort is sufficiently attractive to invest their own time in getting their data into a form that can be entered into the database will require a community-debated and harmoniously formulated database masterplan, and with community-accepted standard descriptors as well as standards of rigor for data inclusion. Our experience with prior efforts to develop even primitive databases of this type for the rat (Bloom et al., 1989) is that widespread user acceptance is essential to the effective participation of the community to get data analyzed in ways suitable for inclusion in a database, and that the only way to achieve effective and active user participation is through progressive iteration and modification with potential users.

Thus, a driving justification for establishing a Brain Object Database of realistic and modifiable objects representing real molecules in real linkages with other molecules, organelles, cells, circuits, and functions is that it will be realistic enough to attract data-producers and data-analysts. It should take on the significance of a "deep" knowledge system in which the pieces are not just mechanically linked, but rather contain realistic intellectual connections based on defined properties. Furthermore, the Brain Object Database can lead to a shared conceptual view of the brain, and of specific brains as a database suprastructure onto which new data, new linkages, and new concepts can be superimposed/incorporated.

The necessity for user acceptance, motivation, and participation also means to us that one cannot coerce users to any single form of computer platform. Thus, users should be free to use the device with which they are most comfortable, and in which they may already have substantial capital and intellectual investment. For that reason, we intend to use software that is portable across heterogeneous computer platforms such as the current favorite, Java. This developmental environment would make the entire database and its class structures executable on virtually any computer presently used by neuroscientists.

The database environment we envision should also provide means to integrate the efforts of many individual neuroscientists, regardless of where they work. There is no question that research institutions are already well connected via the Internet, and that having tasted this new communication capacity has generated a boundless demand for higher speeds on new data superhighways. We envision the databases as being distributed among the neuroscientists such that database actions (such as searching for a specific combination of facts or diseases) will automatically be empowered to reach out into the world-wide network to retrieve the desired data with no more delay than if those data were already present within the network of the local research laboratory.

A more informed future?

Perhaps in this fashion, one may begin to change the information gathering habits of the neuroscientific community and allow the science to rise to even more powerful means to capture the information residing within today's data and to ask more heuristic questions tomorrow. A user-friendly "deep knowledge" database is the ultimate goal. This would be a database of brain information that transcended levels of resolution (molecules to cells to systems to behaviors). It would also have to transcend grouped members of a species of brain. The Brain Object Database should also be operated in conjunction with an intelligent data filtering, indexing, and database entry system. In that way, interested users could in fact stay on top of current information that matches their user-defined profiles of interest and to which new interest elements can later be defined, dropped, or re-assigned. Both parts of this neuroscientific communications solution are necessary to solve the current information problem. While we have only spoken here about the database tool, the development of data gathering tools cannot lag far behind.

Acknowledgements

Supported by Grants MH52154 and MH57571.

References

Bloom, F.E. (1990) Databases of Brain Information. In: A. Toga (Ed.), *Three Dimensional Neuroimaging*. Raven Press, New York, pp. 273–306.

Bloom, F.E. (1995) Neuroscience-knowledge management: slow change so far. *Trends Neurosci.*, 18(2): 48–49.

Bloom, F.E. and Young, W.G. (1994) New solutions for neuroscience communications are still needed. *Prog. Brain Res.*, 100: 275–281.

Bloom, F.E., Young, W.G. and Kim, Y. (1989) *Brain Browser*. Academic Press, San Diego.

Bloom, F.E., Young, W.G., Nimchinsky, E.A., Hof, P.R. and Morrison, J.H. (1997) Neuronal vulnerability and informatics in human disease. In: S.H. Koslow and M.F. Huerta (Eds.), *Neuroinformatics — An Overview of the Human Brain Project*. Lawrence Erlbaum, Mahwah, pp. 83–123.

Cherniak, C. (1990) The bounded brain: toward quantitative neuroanatomy. *J. Cog. Neurosci.*, 2: 58–68.

Couceyro, P.R., Koylu, E.O. and Kuhar, M.J. (1997) Further studies on the anatomical distribution of CART by in situ hybridization. *J. Chem. Neuroanat.*, 12(4): 229–241.

Crick, F. and Jones, E.G. (1993) Backwardness of human neuroanatomy. *Nature*, 361: 109–110.

De Lecea, L., Criado, J.R., Prospero-Garcia, O., Gautvik, K.M., Schweitzer, P., Danielson, P.E., Dunlop, C.L., Siggins, G.R., Henriksen, S.J. and Sutcliffe, J.G. (1996) A cortical neuropeptide with neuronal depressant and sleep-modulating properties. *Nature*, 381(6579): 242–245.

De Lecea, L., del Rio, J.A., Criado, J.R., Alcantara,S., Morales, M., Danielson, P.E., Henriksen, S.J., Soriano, E. and Sutcliffe, J.G. (1997) Cortistatin is expressed in a distinct subset of cortical interneurons. *J. Neurosci.*, 17(15): 5868–5880.

De Lecea, L., Kilduff, T.S., Peyron, C., Gao, X., Foye, P.E., Danielson, P.E., Fukuhara, C., Battenberg, E.L., Gautvik, V.T., Bartlett, F.S. II, et al., (1998) The hypocretins: hypothalamus-specific peptides with neuroexcitatory activity. *Proc. Natl. Acad. Sci. USA*, 95(1): 322–327.

Games, D., Adams, D., Alessandrini , R., Barbour, R., Berthelette, P., Blackwell, C., Carr, T., Clemens, J., Donaldson, T., Gillespie, F. et al., (1995) Alzheimer-type neuropathology in transgenic mice overexpressing V717F beta-amyloid precursor protein (see comments). *Nature*, 373(6514): 523–527.

Gautvik, K.M., De Lecea, L., Gautvik, V.T., Danielson, P.E., Tranque, P., Dopazo, A., Bloom, F.E. and Sutcliffe, J.G. (1996) Overview of the most prevalent hypothalamus-specific mRNAs, as identified by directional tag PCR subtraction. *Proc. Natl. Acad. Sci. USA*, 93(16): 8733–8738.

Gazzaley, A.H., Siegel, S.J., Kordower, J.H., Mufson, E.J. and Morrison, J.H. (1996a) Circuit-specific alterations of *N*-methyl-D-aspartate receptor subunit 1 in the dentate gyrus of aged monkeys. *Proc. Natl. Acad. Sci. USA*, 93(7): 3121–3125.

Gazzaley, A.H., Weiland, N.G., McEwen, B.S., and Morrison, J.H. (1996b) Differential regulation of NMDAR1 mRNA and protein by estradiol in the rat hippocampus. *J Neurosci.*, 16(21): 6830–6838.

Gazzaley, A.H., Benson, D.L., Huntley, G.W. and Morrison, J.H. (1997) Differential subcellular regulation of NMDAR1 protein and mRNA in dendrites of dentate gyrus granule cells after perforant path transection. *J. Neurosci.*, 17(6): 2006–2017.

Hof, P.R. and Morrison, J.H. (1995) Neurofilament protein defines regional patterns of cortical organization in the macaque monkey visual system: a quantitative immunohistochemical analysis. *J. Comp. Neurol .*, 352(2): 161–186.

Hof, P.R. and Morrison, J.H. (1996) Hippocampal and neocortical involvement in normal brain aging and dementia: morphological and neurochemical profile of the vulnerable circuits. *J. Am. Geriatr. Soc.*, 44(7): 857–864.

Hof, P.R., Bouras, C., Perl, D.P. and Morrison, J.H. (1994a) Quantitative neuropathologic analysis of Pick's disease cases: cortical distribution of Pick bodies and coexistence with Alzheimer's disease (see comments). *Acta Neuropathol.*, 87(2): 115–124.

Hof, P.R., Perl, D.P., Loerzel, A.J., Steele, J.C. and Morrison, J.H. (1994b). Amyotrophic lateral sclerosis and parkinsonism-dementia from Guam: differences in neurofibrillary tangle distribution and density in the hippocampal formation and neocortex. *Brain Res.*, 650(1): 107–116.

Hof, P.R., Bouras, C., Perl, D.P., Sparks, D.L., Mehta, N. and Morrison, J.H. (1995a) Age-related distribution of neuropathologic changes in the cerebral cortex of patients with Down's syndrome. Quantitative regional analysis and comparison with Alzheimer's disease. *Arch. Neurol.*, 52(4): 379–391.

Hof, P.R., Nimchinsky, E.A. and Morrison, J.H. (1995b) Neurochemical phenotype of corticocortical connections in the macaque monkey: quantitative analysis of a subset of neurofilament protein-immunoreactive projection neurons in frontal, parietal, temporal, and cingulate cortices. *J. Comp. Neurol.*, 362(1): 109–133.

Hof, P.R., Vogt, B.A., Bouras, C. and Morrison, J.H. (1997) Atypical form of Alzheimer's disease with prominent posterior cortical atrophy: a review of lesion distribution and circuit disconnection in cortical visual pathways. *Vision Res.*, 37(24): 3609–3625.

Hsiao, K., Chapman, P., Nilsen, S., Eckman,C., Harigaya, Y., Younkin, S., Yang, F., Cole, G. (1996) Correlative memory deficits, A beta elevation, and amyloid plaques in transgenic mice (see comments). *Science*, 274(5284): 99–102.

Huerta, M.F., Koslow, S.H. and Leshner, A.I. (1993) The human brain project: an international resource. *Trends Neurosci.*, 16: 436–438.

Huntley, G.W., Vickers, J.C., Janssen, W., Brose, N., Heinemann, S.F. and Morrison, J.H. (1994) Distribution and synaptic localization of immunocytochemically identified NMDA receptor subunit proteins in sensory-motor and visual cortices of monkey and human. *J. Neurosci.*, 14(6): 3603–3619.

Koylu, E.O., Couceyro, P.R., Lambert , P.D., Ling, N.C., DeSouza, E.B. and Kuhar, M.J. (1997) Immunohistochemical localization of novel CART peptides in rat hypothalamus, pituitary and adrenal gland. *J. Neuroendocrinol.*, 9(11): 823–833.

Koylu, E.O., Couceyro, P.R., Lambert, P.D. and Kuhar, M.J. (1998) Cocaine- and amphetamine-regulated transcript peptide immunohistochemical localization in the rat brain. *J. Comp. Neurol.*, 391(1): 115–132.

Mara, M.A., Hillier, L. and Waterston, R.H. (1998) Expressed sequence tags- Establishing bridges between genomes. *Trends Genetics*, 14(1): 4–7.

Masliah, E., Sisk, A., Mallory, M., Mucke, L., Schenk, D. and Games, D. (1996) Comparison of neurodegenerative pathology in transgenic mice overexpressing V717F beta-amyloid precursor protein and Alzheimer's disease. *J. Neurosci.*, 16(18): 5795–5811.

Morrison, B.M., Gordon, J.W., Ripps, M.E. and Morrison, J.H. (1996) Quantitative immunocytochemical analysis of the spinal cord in G86R superoxide dismutase transgenic mice: neurochemical correlates of selective vulnerability. *J. Comp. Neurol.*, 373(4): 619–631.

Morrison, B.M., Janssen, W.G., Gordon, J.W. and Morrison, J.H. (1998) Time course of neuropathology in the spinal cord of G86R superoxide dismutase transgenic mice. *J. Comp. Neurol.*, 391(1): 64–77.

Nimchinsky, E.A., Hof, P.R., Young, W.G. and Morrison, J.H. (1996) Neurochemical, morphologic, and laminar characterization of cortical projection neurons in the cingulate motor areas of the macaque monkey. *J. Comp. Neurol.*, 374(1): 136–160.

Nimchinsky, E.A., Hof, P.R., Janssen, W.G.M., Morrison, J.H. and Schmauss, C. (1997) Expression of dopamine D3 receptor dimers and tetramers in brain and in transfected cells. *J. Biol. Chem.*, 272(46): 29229–29237.

Paxinos, G. and Watson, C. (1987). *The Rat Brain in Stereotaxic Coordinates*, 2nd Edn, Academic Press, San Diego.

Ripps, M.E., Huntley, G.W., Hof, P.R., Morrison, J.H. and Gordon, J.W. (1995) Transgenic mice expressing an altered murine superoxide dismutase gene provide an animal model of amyotrophic lateral sclerosis. *Proc. Natl. Acad. Sci. USA*, 92(3): 689–693.

Sakurai, T., Amemiya, A., Ishii, M., Matsuzaki, I., Chemelli, R.M., Tanaka, H., Williams, S.C., Richardson, J.A., Kozlowski, G.P., Wilson, S. et al., (1998) Orexins and orexin receptors: a family of hypothalamic neuropeptides and G protein-coupled receptors that regulate feeding behavior (see comments). *Cell*, 92(4): 573–585.

Sturchler-Pierrat, C., Abramowski, D., Duke, M., Wiederhold, K.H., Mistl, C., Rothacher, S., Ledermann, B., Burki, K., Frey, P., Paganetti, P.A., et al., (1997) Two amyloid precursor protein transgenic mouse models with Alzheimer disease-like pathology. *Proc. Natl. Acad. Sci. USA*, 94(24): 13287–13292.

Young, W.G., Nimchinsky, E.A., Hof, P.R., Morrison, J.H. and Bloom, F.E. (1997) *NeuroZoom Software User Guide and Reference Books*, YBM Inc., San Diego, 1038 pp. CD-ROM, Internet: http://www.neurozoom.com and http://neurozoom.scripps.edu

W.E. Crusio and R.T. Gerlai (Eds.)
Handbook of Molecular-Genetic Techniques for Brain and Behavior Research (Techniques in the Behavioral and Neural Sciences, Vol. 13)

CHAPTER 1.2

Bioinformatics and neuroscience in the post-genomic era

Bruno W.S. Sobral and Michael M. Harpold

1800-A Old Pecos Trail, National Center for Genome Resources, Santa Fe, NM 87505, USA

Introduction

The Human Genome Project and the associated projects focused on various microorganisms, including *Eschericia coli* and *Saccharomyces cerevisiae*, and model multicellular eucaryotic organisms, including *Caenorhabditis elegans*, *Arabidopsis thaliana*, *Drosophila melanogaster*, and mouse, all share the goal of sequencing the complete genomes of the respective organisms (Collins et al., 1998). These projects are among the primary initiatives of what has come to be called the "Genomics Era" of biological research. The main shift in approach that genomics has enabled is the capacity to analyze entire genomes at once, rather than only single (or few) genes at a time. During the past several years, significant technological advances in molecular biology and laboratory engineering (instrumentation, process and data management) have made possible the determination of the complete genome sequences of *E. coli* (Blattner et al., 1997), *S. cerevisiae* (Goffeau et al., 1996), and, most recently, essentially the complete sequence of the *C. elegans* genome (*C. elegans* Sequencing Consortium, 1998). Based upon these accomplishments, it now appears certain that the sequence of the complete *D. melanogaster*, mouse, *A. thaliana* and human genomes will be available within the next 5 years (Collins et al., 1998). Thus, it seems likely that the amount of publicly available sequence data will continue to grow exponentially for the foreseeable future. Clearly, the unprecedented amount of sequence data and associated information as well as its access and analysis is revolutionizing virtually all biological research, including research in neuroscience.

Bioinformatics has emerged from these projects as a discipline representing a coalescence of a variety of techniques using computers to address biological questions raised by the generation and availability of vast amounts of sequence data. Much of the initial focus of bioinformatics has involved the development of databases for the acquisition and storage of gene sequence data and algorithms for analysis of this sequence data to gain insights to gene function and phylogenetic relationships among organisms. In this article, bioinformatics will be used in a broad sense, to include computational and mathematical models used to develop predictive algorithms, as well as the more process-oriented areas of database development and data management in general. The former is also known as computational or theoretical biology.

To date, the genomic databases are of two primary types, generalized and specialized, and provide limited sets of analytical and computational tools. Importantly, most of these are now available to the research community via the Internet and the World Wide Web (WWW). The generalized genomic databases include the International Nucleotide Sequence Database Collaboration (IC) databases, GenBank (Benson et al., 1998), EMBL (Stoesser et al., 1998), and DDJB (Tateno et al., 1998). In addition to these databases, other primarily sequence-centric but organismally broad databases are GSDB (Harger et al., 1998) and the protein sequence databases such as Swiss-Prot (Bairoch and Apweiler, 1998) and PIR (Barker et al., 1998). Collectively, these databases are valuable archives of publicly available sequence data derived from any organism and submitted by researchers from around the world. Over the last decade, these generalized genomic databases have become essential resources for modern biological research. One of the primary reasons for this is that each of these databases provide general analysis methods (or 'tools'), typically BLAST (Altschul et al., 1990), for performing sequence similarity and sequence motif searches across sequence data from different species. This type of comparative sequence analysis has emerged as one of the most powerful approaches for identifying new genes, establishing evolutionary relationships between organisms, and predicting possible gene function.

A variety of specialized genomic-related databases have also been developed that, in general, are more restricted in scope but provide more depth concerning a specific topic. These databases frequently are organized around specific organisms, such as the model organism databases SGD (Cherry et al., 1998; *Saccharomyces*), EARLYDB (http://genome.wustl.edu/gsc/earlydb.shtml; *C. elegans*), FlyBase (FlyBase Consortium, 1998; *Drosophila*), AtDB (Flanders et al., 1998; *Arabidposis thaliana*) and MGD (Blake et al., 1998; mouse). Others are focused on specific data types such as physical and genetic mapping, e.g. GDB (Letovsky et al., 1998; human) and OMIM (http//www.ncbi.nlm.nih.gov/omim/; heritable human diseases), biological processes, e.g. EcoCyc (Karp et al., 1998; microorganism metabolic pathways), and gene/protein families, e.g. Nuclear Receptor Resource (Martinez et al., 1998) and Androgen Receptor Gene Mutations Database (Gottlieb et al., 1998).

At present there are a limited number of specialized genomic databases with a specific neuroscience focus. These include Flybrain (Armstrong et al., 1995; http://flybrain.uni-freiburg.de/), encompassing neuroanatomical information on the *Drosophila* central and peripheral nervous system that will ultimately provide data on the distribution of specific gene products such as RNAs and proteins, and the Interactive Fly (http://sdb.bio.purdue.edu/fly/aimorph/cns.htm), providing information concerning genes involved in *Drosophila* neurogenesis. Both of these specialized databases are also linked to FlyBase. Although, not strictly genomic-related there are Internet sites providing links to other useful neuroscience information sources such as Neurosciences on the Internet (http://www.neuroguide.com) and NeuroLinks (http://rsb.info.nih.gov/neurochem/bn5ht/neurolinks.html). These specialized databases are generally developed by experienced researchers in their respective topics and are designed to provide those specialists with a detail-oriented and valuable resource for further research. However, while data and information within these specialized databases may be of interest and useful to researchers outside the specific areas, the high level of detail and lack of data standardization and integration can limit their utility to such researchers.

With the realization that the sequences of entire genomes of many organisms can now be determined, increased attention is becoming focused on functional analysis and use of such data, through a combination of experimental and computer-assisted approaches. A major goal of this research is to describe and understand complex biological processes and organisms through predictive modeling of the relationship of genotype to phenotype. Many are referring to this quest for functional understanding of biological systems as the "post-genomic era" of biological research. Whatever the terms chosen, the impact of acquisition and analysis of large amounts of data from biological systems upon research in neuroscience has already been significant and promises to be even more dramatic in the coming years.

Selected applications of bioinformatics in neuroscience research

There are now numerous examples of the application of bioinformatics approaches to gain new insight to complex neurobiological problems. It should be noted that most represent only initial efforts given the relative novelty of bioinformatics tools and techniques. Rather than an exhaustive review, three examples will be discussed here with the goal of illustrating the current breadth of bioinformatics applications to neuroscience research and the potential power of using bioinformatics with multiple data types, and then we will discuss some potentially interesting areas of future exploration and opportunity. The following examples include the use of targeted gene sequencing and expression data together with physical and genetic mapping data to gain new insight to neurological disease, the analysis of large-scale gene expression mapping during development of the nervous system, and an initial application of whole genome analysis to neurobiology.

Sequence, mapping, and expression analysis-disease correlations

Over the past decade, efforts by a number of research groups have been targeted at the characterization of the genes encoding the subunits of voltage-gated calcium channels (VGCCs) expressed in the mammalian nervous system (Mori et al., 1991; Snutch et al., 1991), including human (Williams et al., 1992; Harpold et al., 1998). These and other studies have shown that the pore-forming subunits of high-voltage activated VGCCs expressed in the nervous system are encoded by at least five distinct genes, with α_{1A}, α_{1B}, α_{1C}, α_{1D}, and α_{1E} as the subunit designations, comprising a multigene family. Each of these genes also give rise to an extensive set of splice variants and each of the α_1 subunits confers unique biophysical and pharmacological properties to the heteromeric VGCC complexes containing a single distinct α_1 subunit. Using the sequences determined from cloned cDNAs encoding human α_1 subunits, Diriong et al. (1995) employed fluorescence in situ hybridization (FISH) techniques to map the gene encoding the α_{1A} subunit (designated CACNA1A) to human chromosome 19p13.1p13.2. This VGCC subunit is believed to be a component of the P/Q-type calcium channel that plays a significant role in the control of synaptic neurotransmitter release. Other independent studies using genetic mapping studies determined that two rare autosomal dominant neurological disorders in humans, familial hemiplegic migraine (FHM; Ophoff et al., 1994) and episodic ataxia type 2 (EA-2; Teh et al., 1995), also mapped to chromosome 19p13. Although the two disorders share some clinical similarities, other symptoms are distinct. The data derived from the cloned α_{1A} cDNA, showing that the gene was expressed in brain, encoded a subunit of a functional voltage-gated ion channel and its chromosomal localization correlated with the genetic mapping data, suggested that the gene was a strong candidate for involvement with either or both human disorders.

Ophoff et al. (1996) isolated and characterized by direct DNA sequencing the human chromosomal CACNA1A gene and identified four distinct missense mutations in sequences encoding conserved domains of the VGCC subunit in FHM and two mutations predicted to disrupt the reading frame of the same gene in EA-2. The identification of two distinct types of mutations in FHM and EA-2 predicts that different molecular mechanisms are involved in the two neurological disorders and provides a dramatic example of how different mutations within the same gene can lead to phenotypic heterogeneity. Importantly, not only have these studies led to the discovery of the involvement of the gene encoding the VGCC α_{1A} subunit in the pathogenesis of FHM and EA-2, they further suggest a novel target for investigation in the more common non-familial forms of migraine.

Interestingly, additional recent studies involving other heritable disorders in mice and humans have also converged on the gene encoding the VGCC α_{1A} subunit. Zhuchenko et al. (1997) recently discovered small CAG repeat expansions within this gene that are associated with another human autosomal dominant disorder SCA6, spinocerebellar ataxia type 6. It should also be noted that sequence analysis has revealed the existence of splice variants in which the CAG repeats are within translated sequences and encode polyglutamine expansions, whereas in other splice variants the CAG repeats are within the 3′-untranslated sequences. This may suggest a molecular mechanism for the observed selective neuronal loss.

In yet another study focused on the tottering and leaner mutant mice, models for human absence epilepsy, missense and premature translation termination mutations were also identified within the mouse gene homologous to the human gene encoding the VGCC α_{1A} subunit (Fletcher et al. 1996). The mouse mutations and gene map to a region of mouse chromosome 8 that has been shown to be syntenic with human chromosome 19p13. Again, there are quite distinct behavioral and histopathological differences between the two mutants, with more pronounced regional neuronal degeneration associated with the leaner mutation. It would be interesting to correlate differential expression mapping data for the transcripts carrying the mutations.

Large-scale transcriptional analysis

The increasing availability of sequence data for larger sets of genes from a wide variety of organisms and the sequence of complete genomes for selected model organisms together with new technologies for large-scale and genome-wide gene expression mapping, including Serial Analysis of Gene Expression (SAGE; Velculescu et al., 1995), oligonucleotide chips (Fodor et al., 1993), and DNA microarrays (Schena et al., 1995), is widely anticipated to provide deeper understanding of complex biological problems. To harness these large data sets and extract useful knowledge from them will require the development and application of computer-assisted methods of analysis to aid experimentation.

Although still early, several noteworthy large-scale gene expression studies have appeared in the scientific literature, including a genome-wide study of temporal gene expression changes involved in the metabolic shift from fermentation to respiration in *S. cerevisiae* (DeRisi et al., 1997) and an analysis of differential expression of almost 50,000 genes between human colorectal and pancreatic cancer cells (Zhang et al., 1997). The amount of data generated by such studies will likely exceed that deriving from the genome sequencing projects and underscores the requirements for new bioinformatics tools for the acquisition, storage, and analysis of large-scale gene expression data.

A large-scale gene expression study applied to the developing central nervous system in the rat was recently reported and illustrates the potential of this approach for neuroscience research (Wen et al., 1998). In this study, a temporal map of changes in mRNA expression of 112 genes over nine stages of development of the rat cervical spinal cord was produced. The 112 genes represented sets of genes involved in intra- and inter-cellular signaling, such as neurotransmitter receptors and peptide/growth factors, as well as those encoding various structural and metabolic proteins known to be expressed in the nervous system. To analyze the data, the gene expression time series was clustered to produce an Euclidean distance tree.

A number of interesting patterns and correlations emerged from this study. The results showed that the genes segregate into four "waves" of expression with the expression of a fifth group of genes remaining relatively constant at all stages.

The peptide/growth factor signaling genes primarily clustered to the constant expression group and to wave 1, i.e. high expression only during early developmental stages. In contrast, the genes encoding the neurotransmitter receptors generally mapped to waves 2 and 3. In wave 2 gene expression increases dramatically from E13 to E15 and remains constant into the adult, whereas in wave 3 gene expression also increases from E15 to E18 but declines from P0 into adult and in wave 4 gene expression continually increases from E11 through the adult stage. Most of the neurotransmitter receptors in waves 1 and 2 are ionotropic, whereas those in waves 3 and 4 are metabotropic. In general, the results indicated that functional gene families grouped together and that the Euclidean clusters also identified different phases of spinal cord development. For example, wave 1 included genes, such as cyclin B and the peptide/growth factors, involved in neural cell proliferation, while wave 2 included genes, such as neurotransmitter receptors and metabolizing enzymes, involved in neurogenesis. This study illustrates only an initial approach using a computational cluster analysis of large-scale temporal gene expression during development of a mammalian nervous system. It seems clear that similar studies, together with new computational methods, applied to the analysis of genetic signaling networks involved in other complex physiological and patho-physiological processes in the nervous system will provide new and unexpected insights.

Whole genome analysis

With the recent the determination of essentially the complete sequence of the *C. elegans* genome (*C. elegans* Sequencing Consortium, 1998), for the first time the research community has the ability to identify and analyze all the genes of a multicellular eucaryotic organism with a true nervous system. The availability of the complete genome sequence together with complete data concerning the cell lineage of each neuron as well as the synapses between all neurons in *C. elegans* provides an unprecedented opportunity for neuroscience research. Transforming this opportunity into reality will require, among other things, simple and intuitive ways for scientists to use existing data and to add their knowledge and interpretation to data in light of new experimentation. If this occurs, it will undoubtedly spur a more rapid and deeper understanding of significant aspects of the nervous system from the nematodes to humans.

As an initial approach to characterizing the complete set of *C. elegans* genes relating to the nervous system, Bargmann (1998) used sequence similarity searching techniques with sequences encoding ion channel, receptor, and signaling proteins against 18,452 predicted proteins, identified using gene-finding algorithms covering approximately 90% of the *C. elegans* genome. This study suggests that perhaps as much as one-third of the genome, or 6000 genes, may be involved in the function of the *C. elegans* nervous system. However, only a relatively small proportion of these genes is currently understood. The genes encoding many of the neurotransmitter receptors, neurotransmitter synthesis and release pathways, ion channels, and G protein-coupled receptors are highly conserved between *C. elegans* and higher vertebrates, including mammals. There are certain notable and interesting exceptions, including the apparent absence of genes encoding voltage-gated sodium channels.

The analysis of *C. elegans* genes potentially affecting nervous system function also reveals some unexpected findings. For example, approximately 5% of the nematode genes encode G protein-coupled receptors. About 100 of these genes are related to G protein-coupled receptors identified to date in other animals, whereas another 1000 appear to be nematode-specific and may encode chemoreceptors. Interestingly, while the *C. elegans* nervous system has over 500 neuronal gap junctions, the more than 20 genes encoding gap junction proteins have no identified relatives in vertebrate genomes. Thus, it appears that the genes encoding both gap junctions and a significant number of the chemosensory receptors evolved independently in nematodes and vertebrates. Finally, among the relatively highly conserved gene families, the *C. elegans* genome contains numerous divergent "outlier" genes. It will be of particular interest to determine whether these divergent members of the otherwise conserved gene families represent nematode-specific genes or can be used to identify new homologues in other organisms with novel functions.

The future of bioinformatics and applications in neuroscience

The examples described above, as well as other examples in this volume (e.g. see Chapter 1.1) and the recent scientific literature, demonstrate both the increases in the magnitude and diversity of data bearing on complex biological problems. It seems clear that future biological discovery will require significant advances in the organization, integration, and analysis of multiple data types. This will be particularly true in neuroscience where there are major efforts to correlate genes with behaviors. There are, however, a number of challenges and requirements that must be addressed to effectively achieve this goal.

The need for integrative approaches to bioinformatics

As physicists migrated into biology and contributed to the founding of molecular biology, a series of reductionist approaches and techniques to understand biological complexity were successfully devised and employed. Reductionist approaches to biological questions have provided important insights to genetic mechanisms and genome structures. The era of addressing all the genes in a single organism (genomics), or all the proteins in a single organism (proteomics), or all the metabolites in a single organism (metabolomics) was initiated by studies in individual laboratories on single genes, proteins, or metabolites. The necessary step to enable high-throughput approaches was the engineering of technologies that generate data, causing a shift from individual laboratories to large-scale research collaborations and industrial-scale biological data production. Thus, the rates of data production now far exceed our ability to comprehend their meaning. If biologists are going to address the fundamental questions concerning the complexity that underlies growth, development and phenotypic variability, however, there is an unmet and unprecedented need to acquire, understand, manipulate, and exploit high-value biological information that is coming out of these highly engineered biological data producing endeavors. It is the organization and analysis of these data sets that are the key to being able to query complex adaptive organisms. And, the interesting queries also encompass the organism's reaction to environmental variables. The key missing link then is to acquire and model the biological structures and responses observed in various sub-cellular, cellular, and multi-cellular (spatial) components through an information system that integrates this information and provides a systems approach to its querying and interpretation. To achieve such an information system will require application of computer science, mathematics, fuzzy logic, and software engineering to the problems of biological data and the biological processes themselves.

What we are alluding to is the need to identify and analyze the functional relationships between DNA, RNA, proteins, environment and phenotypes, including behavior, at various levels of granularity. In the case of the brain, this exquisite molecular and cellular-based computer, such a need also applies to acquiring and utilizing the information from the neural network of living organisms in real time. There are especially interesting possibilities when one considers that the complete genome of *C. elegans* has been elucidated recently (*C. elegans* Sequencing Consortium, 1998). Considering *C. elegans* possesses approximately 300 neurons, it is possible to envision acquiring data from each *C. elegans* neuron and dedicating a single processor of a massively parallel computer to deal with the outputs of each neuron. The possibilities start to become exciting, even though *C. elegans* produces relatively simple behavioral patterns. Furthermore, more complex animals are being studied and model systems, such as *Drosophila* with four orders of magnitude more neurons, are near-term targets for determination of their complete genomic sequence.

In addition to the integration of multiple data types, the development and application of new computational methods to modeling biological processes will lead to new insights and valuable and testable predictions. An exciting example is illustrated by a recent computational study of biochemical signaling networks thought to play a role in long-term potentiation (LTP; Bhalla and Iyengar, 1999). Rather than concentrating on individual, linear signaling pathways, models were developed, using existing experimentally derived data, for the analysis of interactions between up to four signaling pathways forming biochemical signaling networks. A number of unexpected properties emerged from the analysis of these networks, including self-sustaining feedback loops, comprised of a cascade of biochemical reactions, which can function as bistable switches. Such properties of these networks suggest novel mechanisms involved in the formation of LTP and memory. Further elaboration and extension of such computational methods to include additional signaling pathways and data types as well as to other pathways will likely provide further insight to molecular and cellular information storage.

Epistemologically, there are two main challenges in problem solving: failure to use known information and introduction of unnecessary constraints (Rubinstein, 1975). "The moment there were two different operating systems or two different languages or two different network interfaces, unnecessary constraints were introduced. These constraints led in addition to failure to use known information" (Ryan, 1996). Not all is lost, however. Fortunately, information systems (IS) are evolving to provide an integrative role in the business world, thus providing the biological research community with useful information technologies, products and experience. But the IS needed must be more than simply information archives, as have been built thus far. Ideally, such IS should enable scientific hypothesis generation and discovery in silico, thereby pushing the limits of discovery and innovation. In addition, such IS should have a cohesive interface permitting scientists to store, access, view and analyze diverse types of biological data and enable queries across diverse data resources in a manner that is transparent to the user. In this sense, such IS interfaces might look and feel like an Internet browser, through which a user connects and accesses diverse types of data distributed worldwide.

Challenges to integrating biological information

Most biological data repositories that support research now publish data through WWW servers. However, integration is not yet possible because WWW browsers do not process queries across multiple databases (DBs). Furthermore, the existing systems are composed of hierarchical, relational and object-oriented DB architectures. Because of the geographic and disciplinary distances and the heterogeneity of DB architectures among collaborators and DB resources, there is a 'tower of Babel' effect that limits the capacity of researchers to integrate and use the information to construct or challenge a hypothesis. Furthermore, there is a tendency toward bias in existing DB structures, typically reflecting the predominant views of the scientists responsible for data curation. Because science advances through competing hypotheses, this situation must be reversed. In principle, information systems must be open to accepting information (and knowledge) from diverse and distributed groups, and methods must exist to store competing constructs within them.

Future IS will not only need to be able to provide integration of different data types within an organism and support for data modeling components via visualization tools, but have the ability to compare biological information across organisms and data types. This would allow the development of true integrated biological information systems (IBIS).

The data integration model for such an IS requires a software architecture providing persistence for objects and object-oriented applications on top of distributed existing data resources. Considering the very short cycles of innovation observed in object-oriented technologies (Keller et al., 1998) and in the engineering of new high-throughput laboratory technologies modularization will be important to assure creation of an evolutionary system. Interfaces close to standards enable the exchange of custom-made parts with novel or commercially available software once better solutions appear. Clearly, a benefit from a component-based design approach and development and application of standards is that collaborators working on the same project (or related projects) may view the same information from multiple sources, at the same time and in the same way.

A strategy to consider for implementing an integrated biological information system (IBIS) is to develop it in parts, or components, *with an understanding of the whole system in mind.* It is logical to separate the development of such a system into component parts that, together, make up an IBIS, and make it a general requirement for the components to effectively function as a whole as the system evolves. This parallels the situation that we face with biological data generated on specific components of complex biological systems.

There are numerous critical technical challenges for implementation of any integrated IS that have not been thoroughly addressed herein because it was not our primary goal to do so. However, we do wish to outline several such challenges, recognizing that there are likely to be others we have not yet identified and that this list is not intended to be exhaustive.

First, there are many data types, such as structural sequences of DNA, RNA, and proteins, patterns of expression of genes and proteins, biochemical and signaling pathway information, concentrations of metabolites, patterns of neural behavior, and, ultimately, the traits, including behavior, we wish to study and many heterogeneous data sources. Because of this situation, there is a need to execute a parallel development of both the components and the whole of an IBIS. Additionally, such a system must be flexible and able to evolve in response to change, for example in the data producing technologies as well as in the methods of data analysis and modeling. These realities suggest an approach to system engineering that includes various interactive phases:

1. Conceptual definition of components and sub-components of the IBIS.
2. Definition and evolution of a data model for systems integration. This would include specification of how are the components and sub-components related and how should they be tied together to facilitate the complex queries.
3. Definition and iterative refinement of key use-cases of the IBIS. Use cases are intended to capture the thought processes and work and data flow of the scientists using the system to generate or test hypotheses.
4. Significant investment in heterogeneous database research. This is an essential requirement given that the current information resources are too valuable to simply re-write from scratch.
5. Definition of an architectural model for the IBIS. This underscores the requirements for professional software engineers.

Particular issues for the integrated system's engineering include the fact that there are two kinds of heterogeneity that relate to the system.

The first heterogeneity occurs within data types, i.e. multiple data resources for each data type, with distinct solutions for data modeling. The second type of heterogeneity is observed across data types, i.e. each data type, whether genomic sequences or phenotypic information, are gathered and modeled differently and without prior thought on integrative use of the data. Additionally, software engineers are faced with non-standard database management systems (DBMS) that have been employed to store existing data. Such DBMS typically suffer from limited query languages, unreliability and poor performance. Each of these limitations affects scalability of those systems in response to increased data production.

In addition, there are two levels of integration problems to be faced. A lower level problem relates to protocols for communication among generic components. For this there are standards emerging. The key technological platform on which integrated IS can be constructed is Distributed Object Technology (DOT). DOT "is the synergistic application of object-oriented principles to distributed, network-oriented application development and implementation" (Ryan 1996). The current standards and technologies that will permit DOT implementation of an IBIS include:

1. Common Object Request Broker Architecture (CORBA) as defined by the Object Management Group (OMG). This represents an attempt to foster cooperation between systems suppliers, software vendors and users. Activities related to genomics and bioinformatics that are in some way tied into OMG standards include:
   (a) The European Bioinformatics Institute (EBI) has developed several CORBA-based servers (http://corba.ebi.ac.uk/).
   (b) BBSRC: The Roslin Institute, has used CORBA in their server-side development relating to "ArkDB" that includes data relating to pig, chicken, sheep, cattle, horse, cat, and tilapia; (http://www.ri.bbsrc.ac.uk/bioinformatics/research.html).
   (c) The NetGenics SYNERGY system is a well-developed system based on CORBA. They have developed a 'white paper' on the use of CORBA in bioinformatics (http://www.netgenics.com/science.techno.html).
   (d) OMG's Life Sciences Research (http://www.omg.org/homepages/lsr) is a task force of the OMG specifically focused on applying CORBA in the life sciences. Within OMG there are working groups covering many specific areas, including sequence analysis, maps, and gene expression, involving approximately 100 members from industry, government, and academia.
2. Object Linking and Embedding/Component Object Model (OLE/COM): Invented to serve as a CORBA substitute and controlled by Microsoft.
3. World Wide Web (WWW): Provides access through Hypertext Transfer Protocol (HTTP), Hypertext Markup Language (HTML), Common Gateway Interface (CGI), and Java (a powerful new programming language developed by Sun Microsystems that allows developers to write a program once and deploy it on any platform).

Finally, at a higher level, though, there is a need for specification of the nature of particular, domain-specific objects, such as maps, chromosomes, biochemical and signaling pathways, etc. The latter is being addressed by OMG's LSR, for example.

Conclusion

At present, the application of bioinformatics to neuroscience research problems has been relatively limited. However, the power of bioinformatics techniques to compare, integrate, and analyze vast amounts of data as well as different data types will be an increasingly important element in all aspects of biological, including neuroscience, research in the future "post-genomic era". This will require continued development of new bioinformatics strategies and tools, including new databases and information management structures as well as analytical computational techniques, for integrating all the different data types necessary for understanding the relationship of genotype to phenotype. Further, we have argued here that for such integrated biological information systems to be successful and effective, it will be essential for participants, experimentalists included, to consider and require that the component parts function as a whole as the system is developed. With this in mind, our group has initiated the development of a component-based Integrated Biological Knowledge System encompassing multiple biological data types. Perhaps specific applications to neuroscience could be developed and incorporated within the Human Brain Project NeuroInformatics initiative (http://www.nimh.nih.gov/neuroinformatics/index.htm) with the goal of an integrated understanding of brain function, from genes to behaviors, facilitating hypothesis generation and distributed multidisciplinary research collaborations.

References

Altschul, S.F., Gish, W., Miller, W., Myers, E.W. and Lipman, D.J. (1990) Basic local alignment search tool. *J. Mol. Biol.*, 215: 403–410.

Armstrong, J.D., Kaiser, K., Muller, A., Fischbach, K.F., Merchant, N. and Strausfeld, N.J. (1995) Flybrain, an on-line atlas and database of the Drosophila nervous system. *Neuron.*, 15: 17–20.

Bairoch, A. and Apweiler, R. (1998) The SWISS-PROT protein sequence data bank and its supplement TrEMBL in 1998. *Nucleic Acids Res.*, 26: 38–42.

Bargmann, C.I. (1998) Neurobiology of the Caenorhabditis elegans genome. *Science*, 282: 2028–2033.

Barker, W.C., Garavelli, J.S., Haft, D.H., Hunt, L.T., Marzec, C.R., Orcutt, B.C., Srinivasarao, G.Y., Yeh, L-S.L., Ledley, R.S., Mewes, H-W., Pfeiffer, F. and Tsugita, A. (1998) The PIR-International protein sequence database. *Nucleic Acids Res.*, 26: 27–32.

Benson, D.A., Boguski, M.S., Lipman, D.J., Ostell, J. and Ouellette, B.F. (1998) GenBank. *Nucleic Acids Res.*, 26: 1–7.

Bhalla, U.S. and Iyengar, R. (1999) Emergent properties of networks of biochemical signaling pathways. *Science*, 283: 381–387.

Blake, J.A., Eppig, J.T., Richardson, J.E. and Davisson, M.T. and the Mouse Genome Informatics Group (1998) The Mouse Genome Database (MGD): a community resource. Status and enhancements. *Nucleic Acids Res.*, 26: 130–137.

Blattner, F.R., Plunkett, G., Bloch, C.A., Perna, N.T., Burland, V., Riley, M., Collado-Vides, J., Glasner, J.D., Rode, C.K., Mayhew, G.F., Gregor, J., Davis, N.W., Kirkpatrick, H.A., Goeden, M.A., Rose, D.J., Mau, B. and Shao, Y. (1997) The complete genome sequence of Escherichia coli K-12. *Science*, 277: 1453–1474.

C. elegans Sequencing Consortium (1998) Genome sequence of the nematode *C. elegans*: a platform for investigating biology. *Science*, 282: 2012–2018.

Cherry, J.M., Adler, C., Ball, C., Chervitz, S.A., Dwight, S.S., Hester, E.T., Jia, Y., Juvik, G., Roe, T.Y., Schroeder, M., Weng, S. and Botstein, D. (1998) SGD: Saccharomyces Genome Database. *Nucleic Acids Res.*, 26: 73–79.

Collins, F.S., Patrinos, A., Jordan, E., Chakravarti, A., Gesteland, R. and Walters, L. (1998) New goals for the U.S. Human Genome Project: 1998-2003. *Science*, 282: 682–689.

DeRisi, J.L., Iyer, V.R. and Brown, P.O. (1997) Exploring the metabolic and genetic control of gene expression on a genome scale. *Science*, 278: 680–686.

Diriong, S., Lory, P., Williams, M.E., Ellis, S.B., Harpold, M.M. and Taviaux, S. (1995) Chromosomal localization of the human genes for alpha 1A, alpha 1B, and alpha 1E voltage-dependent Ca^{2+} channel subunits. *Genomics*, 30: 605–609.

Flanders, D.J., Weng, S., Petel, F.X. and Cherry, J.M. (1998) AtDB, the Arabidopsis thaliana Database, and graphical-web-display of progress by the Arabidopsis Genome Initiative. *Nucleic Acids Res.*, 26: 80–84.

Fletcher, C.F., Lutz, C.M., O'Sullivan, T.N., Shaughnessy, J.D., Hawkes, R., Frankel, W.N., Copeland, N. and Jenkins, N.A. (1996) Absence epilepsy in tottering mutant mice is associated with calcium channel defects. *Cell*, 87: 607–617.

FlyBase Consortium (1998) FlyBase: a Drosophila database. *Nucleic Acids Res.*, 26: 85–88.

Fodor, S.P., Rava, R.P., Huang, X.C., Pease, A.C., Holmes, C.P. and Adams, C.L. (1993) Multiplexed biochemical assays with biological chips. *Nature*, 364: 555–556.

Goffeau, A., Barrell, B.G., Bussey, H., Davis, R.W., Dujon, B., Feldman, H., Hoheisel, J.D., Jacq, C., Johnston, M., Louis, E.J., Mewes, H.W., Murakami, Y., Philippsen, P., Tettelin, H. and Oliver, S.G. (1996) Life with 6000 genes. *Science*, 274: 546, 563–567.

Gottlieb, B., Lehvaslaiho, H., Beitel, L.K., Lumbroso, R., Pinsky, L. and Trifiro, M. (1998) The Androgen Receptor Gene Mutations Database. *Nucleic Acids Res.*, 26: 234–238.

Harger, C., Skupski, M., Bingham, J., Farmer, A., Hoisie, S., Hraber, P., Kiphart, D., Krakowski, L., McLeod, M., Schwertfeger, J., Seluja, G., Siepel, A., Singh, G., Stamper,

D., Steadman, P., Thayer, N., Thompson, R., Wargo, P., Waugh, M., Zhuang, J.J. and Schad, P.A. (1998) The Genome Sequence DataBase (GSDB): improving data quality and data access. *Nucleic Acids Res.*, 26: 21–26.

Harpold, M.M., Williams, M.E., Brust, P.F., Stauderman, K., Urrutia, A., Johnson, E.C. and Hans, M. (1998) Human neuronal voltage-gated calcium channels: splice variants, subunit interactions and subtypes. In: R.W. Tsien, J-P. Clozel and J. Nargeot (Eds.), *Low-voltage-activated T-type Calcium Channels*, Adis International, Chester, pp. 218–228.

Karp, P.D., Riley, M., Paley, S.M., Pellegrini-Toole, A. and Krummenacker, M. (1998) EcoCyc: Encyclopedia of Escherichia coli genes and metabolism. *Nucleic Acids Res.*, 26: 50–53.

Keller, W., Mitterbauer, C. and Wagner, K. (1998) Object-oriented data integration: Running several generations of database technology in parallel. In: A.B. Chaudhri and M. Loomis (Eds.), *Object Databases in Practice*, Prentice Hall, Englewood Cliffs, NJ.

Letovsky, S.I., Cottingham, R.W., Porter, C.J. and Li, P.W.D. (1998) GDB: the Human Genome Database. *Nucleic Acids Res.*, 26: 94–99.

Martinez, E., Moore, D.D., Keller, E., Pearce, D., Vanden Heuvel, J.P., Robinson, V., Gottlieb, B., MacDonald, P., Simons, S., Sanchez, E. and Danielsen, M. (1998) The Nuclear Receptor Resource: a growing family. *Nucleic Acids Res.*, 26: 239–241.

Mori, Y., Friedrich, T., Kim, M.S., Mikami, A., Ruth, P., Bosse, E., Hofmann, F., Flockerzi, V., Furuichi, T., Mikoshiba, K., Imoto, K., Tanabe, T. and Numa, S. (1991) Primary structure and functional expression from complementary DNA of a brain calcium channel. *Nature*, 350: 398–402.

Ophoff, R.A., Van Eijk, R., Sandkuijl, L.A., Terwindt, G.M., Grubben, C.P., Haan, J., Lindhout, D., Ferrari, M.D. and Frants, R.R. (1994) Genetic heterogeneity of familial hemiplegic migraine. *Genomics*, 22: 21–26.

Ophoff, R.A., Terwindt, G.M., Vergouwe, M.N., Van Eijk, R., Oefner, P.J., Hoffman, S.M., Lamerdin, J.E., Mohrenweiser, H.W., Bulman, D.E., Ferrari, M., Lindhout, D., Van Ommen, G.J., Hofker, M.H., Ferrari, M.D. and Frants, R.R. (1996) Familial hemiplegic migraine and episodic ataxia type-2 are caused by mutations in the Ca^{2+} channel gene CACNL1A4. *Cell*, 87: 543–552.

Rubinstein, M.F. (1975) *Patterns of Problem Solving*, Prentice Hall, Engelwood Cliffs, NJ.

Ryan, T.W. (1996) *Distributed Object Technology: Concepts and Applications*, Prentice Hall, Engelwood Cliffs, NJ.

Schena, M., Shalon, D., Davis, R.W. and Brown, P.O. (1995) Quantitative monitoring of gene expression patterns with a complementary DNA microarray. *Science*, 270: 467–470.

Snutch, T.P., Tomlinson, W.J., Leonard, J.P. and Gilbert, M.M. (1991) Distinct calcium channels are generated by alternative splicing and are differentially expressed in the mammalian CNS. *Neuron.*, 7: 45–57.

Stoesser, G., Moseley, M.A., Sleep, J., McGowran, M., Garcia-Pastor, M. and Sterk, P. (1998) The EMBL nucleotide sequence database. *Nucleic Acids Res.*, 26: 8–15.

Tateno, Y., Fukami-Kobayashi, K., Miyazaki, S., Sugawara, H. and Gojobori, T. (1998) DNA Data Bank of Japan at work on genome sequence data. *Nucleic Acids Res.*, 26: 16–20.

Teh, B.T., Silburn, P., Lindblad, K., Betz, R., Boyle, R., Schalling, M. and Larsson, C. (1995) Familial periodic cerebellar ataxia without myokymia maps to a 19-cM region on 19p13. *Am. J. Hum. Genet.*, 56: 1443–1449.

Velculescu, V.E., Zhang, L., Vogelstein, B. and Kinzler, K.W. (1995) Serial analysis of gene expression. *Science*, 270: 484–487.

Wen, X., Fuhrman, S., Michaels, G.S., Carr, D.B., Smith, S., Barker, J.L. and Somogyi, R. (1998) Large-scale temporal gene expression mapping of central nervous system development. *Proc. Natl. Acad. Sci. USA*, 95: 334–339.

Williams, M.E., Brust, P.F., Feldman, D.H., Patthi, S., Simerson, S., Maroufi, A., McCue, A.F., Velicelebi, G., Ellis, S.B. and Harpold, M.M. (1992) Structure and functional expression of an ω-conotoxin-sensitive human *N*-type calcium channel. *Science*, 257: 389–395.

Zhang, L., Zhou, W., Velculescu, V.E., Kern, S.E., Hruban, R.H., Hamilton, S.R., Vogelstein, B. and Kinzler, K.W. (1997) Gene expression profiles in normal and cancer cells. *Science*, 276: 1268–1272.

Zhuchenko, O., Bailey, J., Bonnen, P., Ashizawa, T., Stockton, D.W., Amos, C., Dobyns, W.B., Subramony, S.H., Zoghbi, H.Y. and Lee, C.C. (1997) Autosomal dominant cerebellar ataxia (SCA6) associated with small polyglutamine expansions in the alpha 1A-voltage-dependent calcium channel. *Nat. Genet.*, 15: 62–69.

W.E. Crusio and R.T. Gerlai (Eds.)
Handbook of Molecular-Genetic Techniques for Brain and Behavior Research (Techniques in the Behavioral and Neural Sciences, Vol. 13)

CHAPTER 1.3

PsycheMap – An online database of psychiatric genetics linkage, association, and genome mapping projects

Nigel M. Williams, Iain Fenton and Michael J. Owen

Neuropsychiatric Genetics Unit, Divisions of Psychological Medicine and Medical Genetics, Tenovus Building, University of Wales College of Medicine, Heath Park, Cardiff, CF4 4XN, UK

Introduction

One of the major goals of psychiatric genetics is to identify susceptibility genes for major psychiatric disorders such as schizophrenia and bipolar disorder (often known together as the functional psychoses). Most work to date has used either the candidate gene or positional cloning approach which have both been used successfully to identify genes for Mendelian disorders. The candidate gene approach (Collins, 1995) requires partial knowledge about the disease function from which a specific gene can be targeted because of its putative function. This approach therefore relies on the presence of previously identified genes whose features (e.g. pattern of expression, functional homology of the gene product) suggest that they may be suitable candidates. For the functional psychoses there are very few clues about the probable function of a disease gene and its product, so positional cloning strategies (Collins, 1995) have been chosen in many instances. Here, no information is required about the probable function of the defective gene product but instead the gene is located solely on its position in the genome. Most of these studies have involved linkage analysis on either large samples of nuclear families or large multiply affected families. Although major chromosomal abnormalities are uncommon in psychiatric disorders, the presence of cytogenetic rearrangements can aid positional cloning efforts by providing an initial clue to the location of a disease locus. In the functional psychoses the most significant attempts to identify regions that contain susceptibility genes have taken the form of complete genome screens (Moldin, 1997; Moldin and Gottesman, 1997). These have resulted in several reports of suggestive linkage from a number of chromosomal regions while reports of significant linkage have proved difficult to replicate in independent samples (Moldin, 1997; Kidd, 1997).

The Human Genome Project (http://www.ornl.gov/TechResources/Human_Genome/home.html) is an international project the main goal of which is to sequence the entire human genome and identify all human genes. Therefore, no matter what strategy is used, the Human Genome Project complements all studies that are aimed at identifying disease susceptibility genes. The project has already completed two of its initial aims, a whole genome physical map of 52,000 sequence tagged sites (short regions of genomic DNA for which a PCR assay is available) and a 1cM genetic map (Collins et al., 1998). The sequencing has been subjected to two different strategies, either large-scale genomic sequencing or just sequencing the coding DNA (cDNA) sequences. The latter strategy has been predominant as cDNA accounts for only 3% of the entire human genome and is the most interesting and medically relevant part, therefore the majority of genes can be sequenced and mapped far more quickly and cheaply than by sequencing the entire genome. Large scale sequencing of cDNA clones derived from different tissues has resulted in the generation of expressed sequence tags (ESTs) which have proved to be a valuable resource in gene cloning. ESTs are easily mapped and sequence homology searches can suggest the function of the gene product. At the time of writing over 1,000,000 human ESTs have already been generated (Borsani et al., 1998) with over 30,000 being mapped using radiation hybrids to give a map of human transcripts (Deloukas et al., 1998). One of the aims of the Human Genome Project is to produce a complete human sequence by 2003 while a complete genomic 'working draft' sequence is proposed for 2001 (Collins et al., 1998). While the project is far from completion, it has already resulted in a dramatic increase in the number of available cloned and mapped genes. This has given rise to the 'positional candidate' approach (Collins, 1995) where a disease locus is initially mapped to a chromosomal subregion (generally using linkage analysis) after which the genes previously mapped to the region are surveyed for their potential as disease candidates. The positional candidate approach will inevitably become the method of choice when identifying potential susceptibility genes for psychiatric disorders.

The success of the Human Genome Project has been responsible for an explosive growth in available genetic information which in turn has posed major challenges for the bioinformatics community. Genetic data are stored across a multitude of different databases, and prior knowledge of each is therefore required in order to retrieve the relevant genetic data, making the task non-trivial for a scientist unfamiliar with bioinformatics.

It is not possible for this chapter to cover in detail all features of bioinformatics required for genome analysis, as not only is this field vast but it is constantly evolving, with new data and software being added daily. We have decided to concentrate on a central database that we have developed which is specific to the functional psychoses and which can be used as a gateway to the major genetic databases that contain the relevant information for gene mapping. This should be an ideal start site for those new to the field of psychiatric genetics, and a useful resource for those more experienced.

PsycheMap
(http://www.uwcm.ac.uk/uwcm/mg/psychemap)

The number of studies reporting data in the field of psychiatric genetics has grown dramatically over recent years. With the absence of any clearly consistent and statistically significant regions of interest there is a need to collate and compare all studies in order to identify regions showing a high degree of replication. This has created a compelling need for tools allowing access to and comparison of the published data. We have developed 'PsycheMap' (Williams et al., 1998) which is an interactive World Wide Web based database providing easy access to the results of all of the linkage and association studies on schizophrenia, bipolar affective disorder and related phenotypes published since 1990.

The aim is to act as a central site where all studies can be accessed, offering a valuable tool for literature searching and for compiling studies for meta-analyses. PsycheMap presents all the studies in a graphical form providing a quick visual comparison of linkage, association and candidate gene studies (Fig. 1). Linkage studies are segregated into 3 groups, studies on schizophrenia, affective disorders and other studies such as those looking at a trait associated with a disorder, for example eye tracking dysfunction and schizophrenia. The length of each bar corresponds to the span of the markers looked at in the study with highlighted areas indicating the chromosomal locations of positive reports. The markers used in

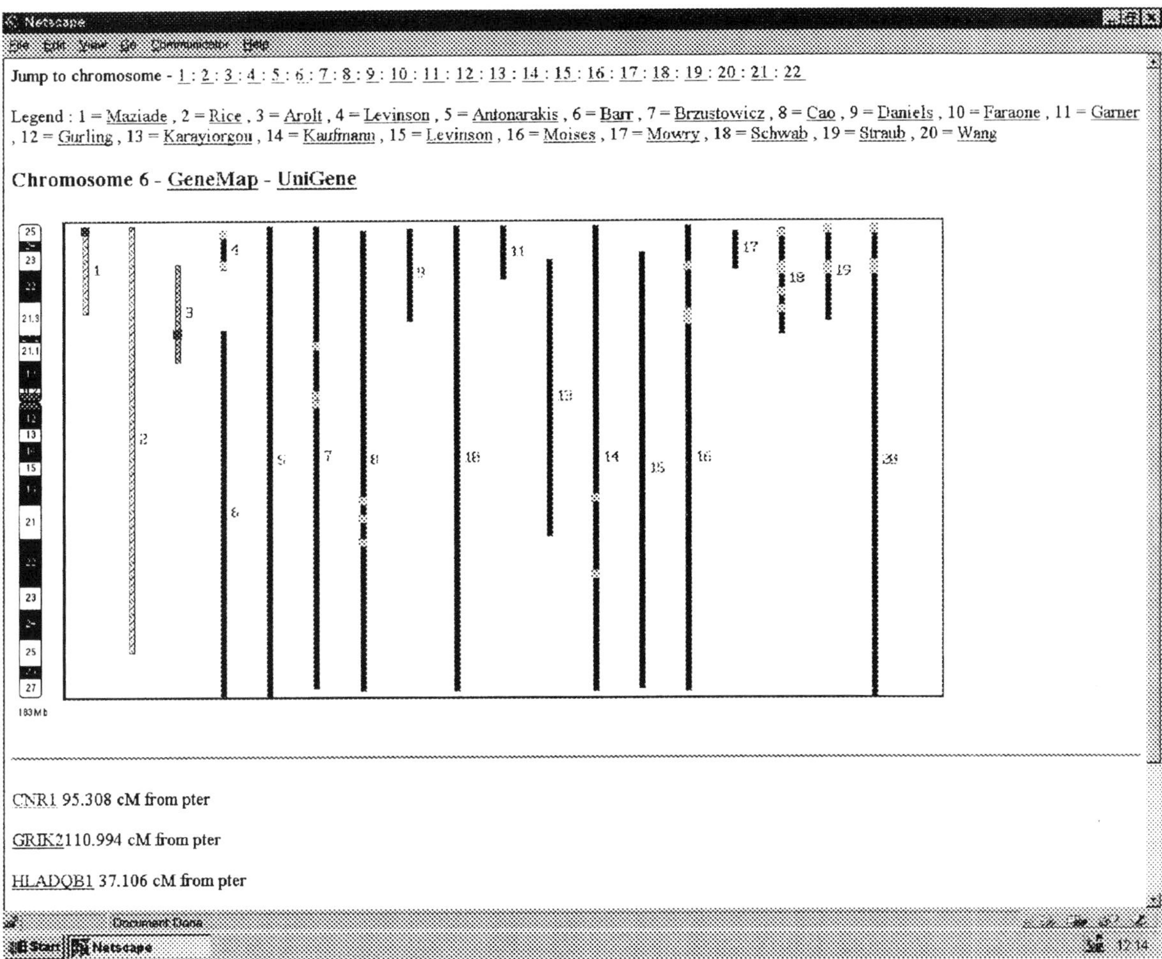

Fig. 1. Example of a page from PsycheMap. The span of markers for each published linkage study are represented by a bar, its colour being dependent on the phenotype used. Highlighted areas indicate the chromosomal locations of positive reports.

all studies are drawn according to the same reference marker map, the genetic location database (LDB) (Collins et al., 1996) (http://cedar.genetics.soton.ac.uk/public.html/). For each linkage study there are summary details, including the type of analysis that was used, a brief description of the markers run in the study, the sample and the diagnostic criteria used (Fig. 2). All studies stored in PsycheMap are linked to their respective location in MEDLINE. This allows the user to click on a study title and to be directed to the full reference including the abstract in PubMed. Similarly, all markers showing positive results are linked to their locations in GDB using the approved HUGO symbol, ensuring that the user avoids that database's occasionally unwieldy search facilities. All candidate genes with published genetic studies are linked to their locations in GDB, OMIM and GeneCards. PsycheMap sends the gene symbol for the current gene the respective search engine of each database, which then offers the correct gene location or a short list of possible options. Furthermore, PsycheMap displays the actual results for all linkage findings and large studies on candidate genes, and contains any reports of cytogenetic abnormalities in the

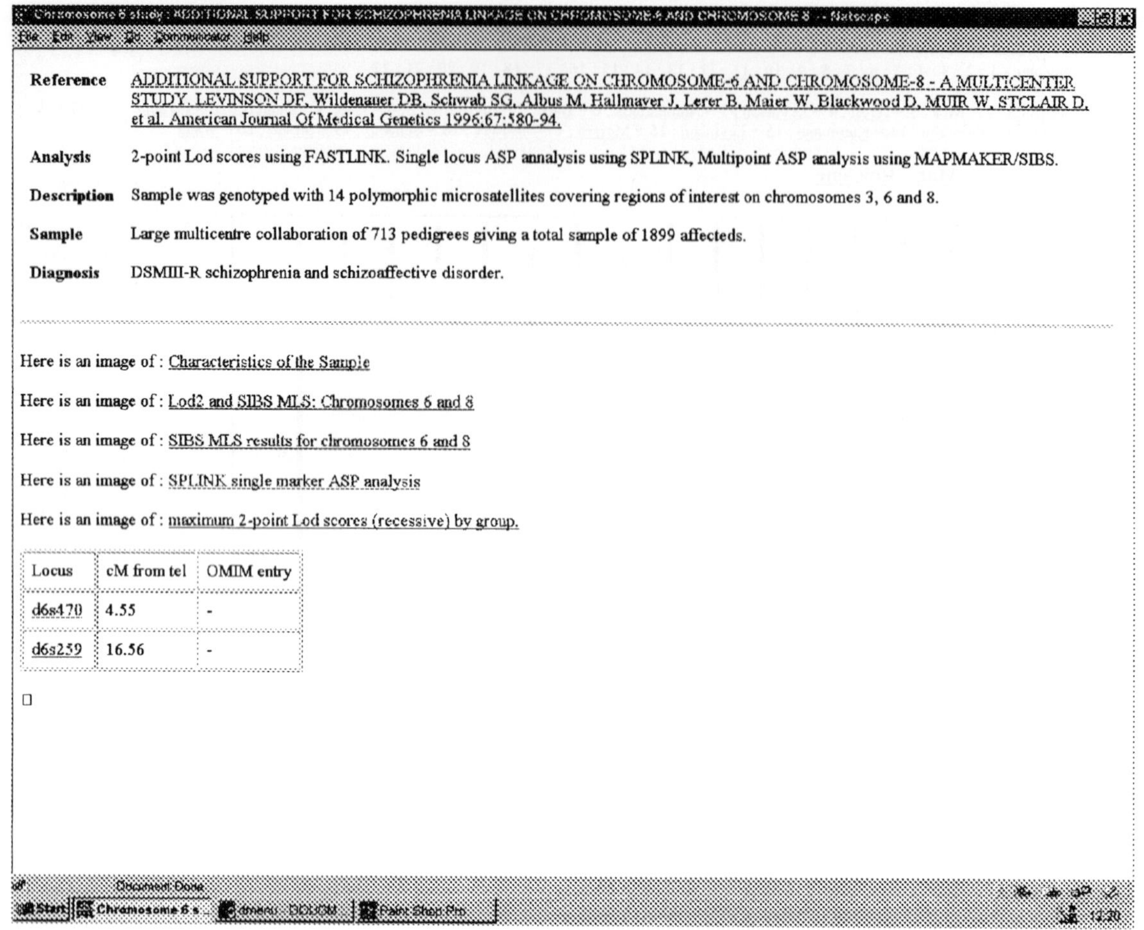

Fig. 2. Example of the summary information available for each linkage study in PsycheMap.

functional psychoses. A direct link is also offered to the GeneMap '98 database and UniGene allowing the rapid identification of any candidate genes that have been previously mapped to a region of interest.

Links provided by PsycheMap

The Genome Database (http://gdbwww.gdb.org/)

The Genome Database (GDB) serves as the central public repository for human gene mapping data. GDB contains extensive information regarding the location of human genes, clones, amplimers (PCR markers), breakpoints, cytogenetic markers, fragile sites, ESTs, syndromic regions, contigs and repeats. These features are included as part of graphically displayed maps which include cytogenetic, linkage (Généthon, CHLC, Marshfield), radiation hybrid (Stanford, Whitehead), and various integrated maps. GDB also contains relevant information about markers within the human genome including mutations and polymorphisms, and allele frequency data. As well as its extensive database GDB has excellent links to external databases such as the

SWISS-PROT protein database (http://www.expasy.ch/sprot/sprot-top.html), the Jackson Laboratory's MGD Mouse Database (http://www.informatics.jax.org/index.html), and the Human Gene Mutation Database (HGMD; http://www.uwcm.ac.uk/uwcm/mg/hgmd0.html). Whilst the mapping part of GDB is comprehensive, the method of viewing such data — using a large Java program — is cumbersome and slow, and also requires an up-to-date web browser (such as Netscape version 4 or Microsoft Internet Explorer 4) and additional software (namely a PostScript interpreter such as GNU's GhostScript) to view and print maps correctly. At the time of writing GDB is the most comprehensive human genome data repository, but because the focus of the Human Genome Project has changed from gene mapping to high throughput sequencing of cDNAs and genomic clones, GDB's primary sponsor (Department of Energy) has terminated the project. The future of GDB therefore remains unclear, although it will remain available to the scientific community without most data aquisition activities. There is though still a major requirement for a curatable database such as GDB whereby the mapping efforts of the scientific community can be merged and made accessible to all.

MEDLINE (http://www.nlm.nih.gov/databases/medline.html)

MEDLINE is the premier bibliographic database for medicine, nursing, dentistry, veterinary medicine and related subjects, and is based at the National Library of Medicine (NLM; http://www.nlm.nih.gov/) in the United States. One free online version is known as PubMed (http://www.ncbi.nlm.nih.gov/PubMed/), and all references stored in the PsycheMap database access this resource. The full paper can be ordered (for a charge) from the NLM if desired through a resource known as Loansome Doc (http://tendon.nlm.nih.gov/ld/loansome.html).

Online Mendelian Inheritance in Man (http://www.ncbi.nlm.nih.gov/Omim/)

OMIM is a catalog of human genes and genetic disorders written and edited by Dr Victor McKusick and colleagues at Johns Hopkins and elsewhere, and developed for the web by the National Center for Biotechnology Information (NCBI; http://www.ncbi.nlm.nih.gov/). OMIM is a huge database of clinical information with over 10,000 entries, produced by the relevant specialists for a particular disorder. Although the length of clinical data varies — the report of Wadia Swami eye tracking syndrome is a single paragraph with 3 references, whilst that for Huntington Disease is over 50 pages long and includes more than 230 references — there is usually an excellent clinical description of the disorder in question. There is a clinical synopsis section for most disorders and a gene map facility which while being simpler than any equivalent map in GDB does not require the sophisticated computer setup that is required by the GDB version. Note that OMIM is centred around clinical disorders, rather than disease-causing genes.

GeneCards (http://bioinformatics.weizmann.ac.il/cards/)

GeneCards is a database of human genes, their products and their involvement in diseases. It offers concise information about the functions of all human genes that have an approved symbol. It is especially useful for those who are searching for information about large sets of genes or proteins, such as scientists working in functional genomics and proteomics. Furthermore, GeneCards provides links to the Jackson Laboratory's Mouse Genome Database (MGD) for those genes that have a homologous murine locus. Whilst little data are stored at the GeneCards home at the Weizmann Institute of Bioinformatics in Israel, they collate data from many sources such as SWISS-PROT, GDB, MGD and OMIM. They uniquely also prepare search requests to the Excite online search engine. This can find obscure sources of data for some genes, but the user must be careful that their gene symbol is not also a command word or acronym elsewhere — little unique genetic information would be found searching for the Haptoglobin gene (whose approved gene symbol is HP), as one would have to wade through information about Hewlett Packard printers before any relevant entries were found.

GeneMap '98
(http://www.ncbi.nlm.nih.gov/genemap/)

The goal of the Human Genome Project is to generate the complete sequence for the entire human genome. This is an immense task which is proposed to be completed by 2003. It is though, unrealistic for the scientific community to wait until the completion of the final sequence before the release of a human gene map. Instead the consortium has chosen to release a series of increasingly comprehensive human gene maps at stages when there are significant improvements in the quality of available data. 1996 saw the release of the first human gene map (http://www.ncbi.nlm.nih.gov/SCIENCE96/) which described 16,000 genes mapped according to the existing framework of genetic and physical maps (Schuler et al., 1996). GeneMap '98 is the latest release from the consortium consisting of 41,644 human STSs representing 30,181 unique genes, which is a 100% increase in gene density and map accuracy from the initial 1996 map (Deloukas et al., 1998). All of these markers have been Radiation Hybrid mapped using the Genebridge 4 and Genebridge 3 panels. As the human genome has been estimated to contain 60,000 to 70,000 genes (Fields et al., 1994), then GeneMap '98 already contains transcript markers for approximately half of all human genes. The project is now improving the quality of the map by continuing to map all remaining cDNAs and also by redeveloping and retyping markers for the cDNAs that failed initial mapping attempts. GeneMap '98 is therefore an invaluable resource for the studies involved in identifying susceptibility genes for the functional psychoses. In particular it should prove essential for the positional candidate strategy by providing a very large validated set of mapped human gene sequences which we can use to design gene specific oligonucleotides or to select cDNA derived PCR products for population gene expression arrays, or both (Deloukas et al., 1998).

UniGene (http://www.ncbi.nlm.nih.gov/UniGene/Hs.Home.html)

The UniGene database was created by the Human Genome Project as a site where the products of large scale sequencing projects could be deposited and compared. At the time of writing the UniGene collection contains over 48,000 sequences, most of which are ESTs. All UniGene sequences have been screened against both the existing functionally cloned GenBank entries to eliminate any redundancies and also against each other in order to determine which ESTs are more likely to be derived from the same gene (Schuler et al., 1996). The UniGene sequence clusters have been radiation hybrid mapped to develop a comprehensive human transcript map. For each EST, data is provided on its mapping location, sequences characterisics, its expression pattern and the other ESTs that are part of the same cluster.

Positional cloning using PsycheMap

With such a huge volume of genetic data deposited in such a wide array of databases most stages of positional cloning can already be performed in silico. By using PsycheMap as a central site we can exploit the bioinformatics resources currently available to efficiently perform some early stages of positional cloning to identify potential susceptibility loci for the functional psychoses. For example, let us assume that we were interested in a schizophrenia susceptibility locus on chromosome 6p. Firstly, we could access the chromosome 6 page in PsycheMap and identify which published studies report data from chromosome 6, and in particular identify which studies had reported either positive or negative findings at the region of interest. By selecting a particular study we could access a summary page which would then allow us to see the citation and also the complete abstract via MEDLINE. We could then assess the results of the study by looking at the relevant summary information (such as sample size, study type, statistics used etc.) and by viewing images of the results. For studies reporting positive results a list of the markers with a positive signal are displayed at the foot of the page, and by selecting these we would be directed to GDB were we could access their relevent details (such as primer sequences, allele frequencies, heterozygosity). After reviewing the linkage studies, we could then look on the main page of PsycheMap to see if any candidate genes that map to the region had already been studied. This would supply us with a list of published studies for each gene, together with links to its location in GDB, OMIM and GeneCards. If we were still interested in the region, it would then be important to identify the number of ESTs already known to map within our critical region. PsycheMap offers links to the major resources, the Human Gene Map and UniGene. First, clicking on the GeneMap '98 link in the PsycheMap home page will direct us to the Human Gene Map home page. By entering key information about the critical region into the search option (such as flanking markers, cytogenetic location etc.) we would be returned with a list of the cDNA markers on the integrated map. Then, by clicking on an individual cDNA entry we will retrieve more detailed information such as sequence homology, the presence of overlapping ESTs and their tissue source, availability of full legnth cDNA sequences, and more refined mapping information. When searching for transcripts within a critical region any search in GeneMap should always be complemented by a similar search in the UniGene database, which is again accessible from PsycheMap. This is because the GeneMap'98 only contains 18,703 out of the 46,045 entries in UniGene and 4,684 of 6,000 human genes of known function (Deloukas et al., 1998). After such a first level analysis we should now have generated a list of transcripts mapped to our region of interest which would allow us to evaluate their sequence similarity and expression profiles so that we could select genes that represent suitable candidates for schizophrenia.

Bioinformatics resources

There are three main ways of locating bioinformatics resources on the web, (i) using a search engine, (ii) using a web directory, (iii) using a genetics collection of links. We shall discuss the pros and cons of each in turn.

Search engines

There are hundreds of different search engines on the web, although only a handful are truely useful. They all work in a broadly similar fashion, and each has slight quirks and problems, although these will not be discussed here. The biggest sites are currently:

AltaVista – www.altavista.com
Yahoo! – www.yahoo.com
Infoseek – www.infoseek.com
Lycos – www.lycos.com
Excite – www.excite.com

All of them allow the user to enter complex queries where whole phrases can be entered or banned from search results. For example, entering the word "Python" will inevitably return many webpages about the "Monty Python" television series rather than the slithering creature related to snakes. There are also problems finding information for common words, as many search pages automatically remove words such as "the", "and", "was", "is" and so on to prevent millions of pages being found unnecessarily. This is useful, unless you are looking for data on the gene causing Wiskott-Alrdich Syndrome, whose approved gene symbol is WAS. Also the user should realise that the web is dominated by computer businesses and enthusiasts, so entering a term such as "database" would return more computing related pages instead of the desired genetics pages. Also be careful that if you are using a word that is different in English and American (such as tumour/tumor, centre/center) then this can also affect search results.

Web directories

There are many thousands of web directories which attempt to categorise all possible websites, or at least a sizeable percentage of them, into a hierarchical structure to ease finding information. Many fail as the pace of change of the web ensures that only those directories that are updated often are worth browsing. That said, a handful stand out as being worthy of examination. Note that a number of these sites also have their own searching technology, and hence appear in the previous section.

Yahoo! – www.yahoo.com
Lycos – www.lycos.com
Netcenter – www.netscape.com
Excite – www.excite.com
NewHoo! – directory.mozilla.org

The first of these websites is known as possibly the first main web directory site, and is still one of the top 3 most popular of all websites today. They use automated programs to generate a number of sites which are then looked at by human beings (editors) who decide which sites to enter into their collection. They have a huge collection of links with many different sections for almost every area of biology and medical topics. One of the more interesting recent sites is NewHoo, which is entirely setup by editors who are specialists in their field. This may mean that if no-one has joined the NewHoo organisation then the list for a particular field may be poor, or it may be excellent.

Genetics collections

There are also a number of collections of website links that consist solely of bioinformatics or genetics or psychiatric pages. Looking at these collections has the advantage that you won't find lists of computer pages or links to Monty Python shows, and they are good places to start if you are looking for a popular genetics site. The list below is a small selection of the dozens of such sites around — they simply reflect the preferences of the authors.

GenomeWeb – www.hgmp.mrc.ac.uk/GenomeWeb/
GeneNet – www.genenet.com
GDB resources list – www.gdb.org/gdb/hgpResources.html
European Bioinformatics Institute – www.ebi.ac.uk

BioMedNet – www.biomednet.com

References

Borsani, G., Ballabio, A. and Banfi, S. (1998) A practical guide to orient yourself in the labyrinth of genome databases. *Human Molecular Genetics*, 7, 10 SISI, 1641–1648.

Collins, A., Frezal, J., Teague, J. and Morton, N.E. (1996) A metric map of humans: 23,500 loci in 850 bands. *Proc. Natl. Acad. Sci. USA*, 93: 14771–14775.

Collins, F.S. (1995) Positional cloning moves from perditional to traditional. *Nature Genetics*, 9, 4, 347–350.

Collins, F.S., Patrinos, A., Jordan, E., Chakravarti, A., Gesteland, R., Walters, L., Fearon, E., Hartwelt, L., Langley, C.H., Mathies, R.A., Olson, M., Pawson, A.J., Pollard, T., Williamson, A., Wold, B., Buetow, K., Branscomb, E., Capecchi, M., Church, G., Garner, H., Gibbs, R.A., Hawkins, T., Hodgson, K., Knotek, M., Meisler, M., Rubin, G.M., Smith, L.M., Smith, R.F., Westerfield, M., Clayton, E.W., Fisher, N.L., Lerman, C.E., McInerney, J.D., Nebo, W., Press, N. and Valle, D. (1998) New goals for the US Human Genome Project: 1998–2003. *Science*, 282, 5389, 682–689.

Deloukas, P., Schuler, G.D., Gyapay, G., Beasley, E.M., Soderlund, C., RodriguezTome, P., Hui, L., Matise, T.C., McKusick, K.B., Beckmann, J.S., Bentolila, S., Bihoreau, M.T., Birren, B.B., Browne, J., Butler, A., Castle, A.B., Chiannilkulchai, N., Clee, C., Day, P.J.R., Dehejia, A., Dibling, T., Drouot, N., Duprat, S., Fizames, C., Fox, S., Gelling, S., Green, L., Harrison, P., Hocking, R.,

Holloway, E., Hunt, S., Keil, S., Lijnzaad, P., LouisDitSully, C., Ma, J., Mendis, A., Miller, J., Morissette, J., Muselet, D., Nusbaum, H.C., Peck, A., Rozen, S., Simon, D., Slonim, D.K., Staples, R., Stein, L.D., Stewart, E.A., Suchard, M.A., Thangarajah, T., VegaCzarny, N., Webber, C., Wu, X., Hudson, J., Auffray, C., Nomura, N., Sikela, J.M., Polymeropoulos, M.H., James, M.R., Lander, E.S., Hudson, T.J., Myers, R.M., Cox, D.R., Weissenbach, J., Boguski, M.S. and Bentley, D.R. (1998) A physical map of 30,000 human genes. *Science*, 282, 5389, 744–746.

Fields, C., Adams, M.D., White, O. and Venter, J.C. (1994) How Many Genes In The Human Genome, *Nature Genetics*, 7, 3, 345–346.

Kidd, K.K. (1997) Can we find genes for schizophrenia? *Am. J. Med. Gen.*, 74:104–111.

Modin, S.O. (1997) The maddening hunt for madness genes. *Nature Genetics*, 17, 127–129.

Moldin, S.O. and Gottesman, I.I. (1997) At Issue: Genes, experience, and chance in schizophrenia-Positioning for the 21st century. *Schizophrenia Bulletin*, 23 (4) 547–561.

Schuler, G.D., Boguski, M.S., Stewart, E.A., Stein, L.D., Gyapay, G., Rice, K., White, R.E., RodriguezTome, P., Aggarwal, A., Bajorek, E., Bentolila, S., Birren, B.B., Butler, A., Castle, A.B., Chiannilkulchai, N., Chu, A., Clee, C., Cowles, S., Day, P.J.R., Dibling, T., Drouot, N., Dunham, I., Duprat, S., East, C., Edwards, C., Fan, J.B., Fang, N., Fizames, C., Garrett, C., Green, L., Hadley, D., Harris, M., Harrison, P., Brady, S., Hicks, A., Holloway, E., Hui, L., Hussain, S., LouisDitSully, C., Ma, J., MacGilvery, A., Mader, C., Maratukulam, A., Matise, T.C., McKusick, K.B., Morissette, J., Mungall, A., Muselet, D., Nusbaum, H.C., Page, D.C., Peck, A., Perkins, S., Piercy, M., Qin, F., Quackenbush, J., Ranby, S., Reif, T., Rozen, S., Sanders, C., She, X., Silva, J., Slonim, D.K., Soderlund, C., Sun, W.L., Tabar, P., Thangarajah, T., VegaCzamy, N., Vollrath, D., Voyticky, S., Wilmer, T., Wu, X., Adams, M.D., Auffray, C., Walter, N.A.R., Brandon, R., Dehejia, A., Goodfellow, P.N., Houlgatte, R., Hudson, J.R., Ide, S.E., Iorio, K.R., Lee, W.Y., Seki, N., Nagase, T., Ishikawa, K., Nomura, N., Phillips, C., Polymeropoulos, M.H., Sandusky, M., Schmitt, K., Berry, R., Swanson, K., Torres, R., Venter, J.C., Sikela, J.M., Beckmann, J.S., Weissenbach, J., Myers, R.M., Cox, D.R., James, M.R., Bentley, D., Deloukas, P., Lander, E.S., Hudson, T.J. (1996) A gene map of the human genome. *Science*, 274, 5287, 540–546.

Strachan, T and Read, A.P. (1996) *Human Molecular Genetics*, BIOS Scientific Publishers Ltd, pp. 335–365.

Williams, N.M., Fentonm I., Owen, M.J. (1998) PsycheMap — An interactive Internet-based database of psychiatric genetics linkage, association, and genome mapping projects. *Am. J. Med. Gen.*, 81, 6, 463–464.

W.E. Crusio and R.T. Gerlai (Eds.)
Handbook of Molecular-Genetic Techniques for Brain and
Behavior Research (Techniques in the Behavioral and Neural Sciences, Vol. 13)

CHAPTER 1.4

Experimental design and statistical inference

Douglas Wahlsten

Department of Psychology, University of Alberta, Edmonton, Alberta T6G 2E9, Canada

Introduction

After presenting a seminar at the University of Alberta in 1989 on the insensitivity of statistical methods to heredity-environment interaction (see Wahlsten, 1990), I was informed by the chairman of the Department of Genetics that probability no longer has a place in genetics because the field is going molecular. A similar opinion was expressed by a colleague doing in vitro work in neurobiology, who informed one of my graduate students that if statistics are needed to see an effect, it probably is not a real effect. Perusing recent issues of the Journal of Neuroscience and similar publications, it is apparent that many research projects do not employ statistical methods at all.

Is there a role for statistical inference in the era of molecular genetics, or can we do without probability theory and the angst it often inspires in our students and colleagues? Two things continue to make probability essential. First, the phenomena of meiotic cell division, crossing over of the chromatids, and union of gametes in fertilization are themselves inherently probabilistic; there is no way to predict the outcome of any particular case because the results are not specified by the antecedent conditions. Only the relative frequency of events in breeding populations can be anticipated. The laws of transmission of genes will forever be statistical. Second, genetics is not solely the study of the chemical genes spread neatly on a gel. For all but the narrowest of biochemists, genetics means the genetics of phenotypes above and beyond the molecular level. As a general principle, genotype does not map one to one onto phenotype, and organismic development is instead epigenetic and probabilistic (Gottlieb et al., 1998). Especially when behaviour is being examined and nongenetic variability is substantial, difficulties in replicating results of linkage studies or even inbred strain surveys are well known. Many of these difficulties originate with the probabilistic nature of heredity and development.

Planning experiments

The design and analysis of experiments are distinct processes. Design means planning a study, deciding the groups of subjects to be studied and the treatments to be administered, and choosing the numbers of individuals to be measured. However, a wise choice of sample size requires foresight about the kind of data analysis to be performed, and the processes of design and analysis are not mutually exclusive. Often the crucial issue in design of an experiment in neurobehavioural genetics is whether it effectively separates the influences of heredity (H) and environment (E), or whether one component of H or E is the sole difference between groups. Statistical analysis then estimates the magnitude of the influences and the possibility that any difference occurs merely by chance.

Research designs

Experiments often involve only two groups of subjects, and this simple design is sufficient to illustrate many issues. If the two groups differ with respect to only one thing, for example the genotype (G), and are equivalent in all other respects, we would like to infer that the difference in G is the cause of any difference in average value of the phenotype. Suppose two groups have sample means M_1 and M_2 and their difference is ΔM. In abbreviated form, we might say that the ΔG causes ΔM, or $\Delta G \rightarrow \Delta M$. This might be realized if a spontaneous mutation in an inbred strain creates coisogenic mice or if a targeted mutation (knockout) is produced in the inbred strain 129 and maintained on that strain background by mating heterozygotes (+/−). Mice of genotypes +/+ and −/− would then be littermates conceived and reared in the same maternal environment, and variation among animals of the same genotype could not reflect genetic variation. Environment would not and could never be homogeneous within a group, but there would be no consistent difference between genotypic groups in their environments. Although the logical interpretation of this study is $\Delta G \rightarrow \Delta M$, this does not imply that the precise magnitude of ΔM is specified solely by the genotypes (see Wahlsten, 1990). On the contrary, it is likely that the mean of each group also depends on the genetic background and the lab and maternal environments, and the value of ΔM might be substantially different on another strain background or with a different diet. To interpret results of this simple experiment, one need not presume the absence of epistatic or gene-environment interaction. At the same time, the group difference should not be attributed exclusively to ΔG when in fact it arises from ΔG existing in a specific genetic and environmental context. It is reasonable to conclude that ΔG is a participating but not a sufficient cause of ΔM.

If the mutation is produced in embryonic stem cells from the strain 129 and then crossed with another strain, two additional problems of interpretation arise for a study of two groups (+/+ versus −/−). There is a possibility that the difference may reflect nearby genes and not exclusively the one targeted by the experimenters (Gerlai, 1996), and there will be extensive genetic variation within a single group. The former instance imperils the validity of the conclusion $\Delta G \rightarrow \Delta M$, whereas the latter qualifies $\Delta G \rightarrow \Delta M$ but does not invalidate it. Genetic variation within a group leads to an average score M for a single genotype that is a mean of many kinds of epistatic and gene-environment interactions, but these complex developmental effects cannot be perceived when only two genotypic groups are studied.

A comparison of two inbred strains differing in many genes will entail nongenetic variation within a group. Although the difference between groups may be regarded as arising from different heredities, it will not necessarily be purely genetic in the Mendelian sense. Inbred strains may differ in their autosomes, sex chromosomes, cytoplasmic organelles (e.g. mtDNA), and maternal environments both pre- and postnatally. These components may be evaluated with a set of reciprocal crosses totaling 16 groups (Table 1). Comparisons between certain pairs of groups may reflect the difference in only one component. For example, the cytoplasmic organelles will differ consistently between groups 7 and 8, 11 and 12, 13 and 15, and 14 and 16. The origin of the Y chromosome of males will differ in groups 5 and 6, 9 and 10, 13 and 14, and 15 and 16. Groups 6 and 7 will differ substantially in the maternal environment, while groups 7 and 14 will differ only in the autosomes. The method can be further refined to study chromosomal imprinting effects. Maternal effects can also be detected using ovarian grafting, embryo transfer, and fostering (Carlier et al., 1992), and the reciprocal cross method can even be combined with ovarian grafting to test for complex interactions between components of

TABLE 1

Crosses to dissect the difference between two inbred strains into components[a]

Inbred strains	Reciprocal backcrosses to A	Reciprocal backcrosses to B	Reciprocal F_2 hybrids
1. A × A			
2. B × B	5. A × AB	9. B × AB	13. AB × AB
F_1 hybrids	6. A × BA	10. B × BA	14. AB × BA
3. A × B	7. AB × A	11. AB × B	15. BA × AB
4. B × A	8. BA × A	12. BA × B	16. BA × BA

[a] Strain of the mother is given first. The F_1 hybrid AB is A female crossed with B male. See Wahlsten (1979) and Sokolowski (1992) for further details.

heredity and the uterine environment (Carlier et al., 1991). These elaborate methods of analysis are designed to reveal the importance of a specific factor, but, as with the simple case of coisogenic mice, an apparent effect cannot be attributed exclusively to the factor that is manipulated.

Designs with four groups may either cross two factors (2 × 2 design) or form a one-way design where more than two factors exist (Table 2). For example, the adoption study of children's IQ by Capron and Duyme (1991) involved groups of adoptees closely comparable on many aspects of the family background but differing in socioeconomic status of the birth mother and the adopting parents. The post-adoption SES effect was clearly environmental, but the pre-adoption difference likely involved an unknown combination of genetic, uterine and perinatal environmental effects. Hence, the design could not separate the global effects of G and E, but it was able to assess one aspect of the environment. It also allowed a test of interaction between pre- and post-adoption SES by comparing the difference between groups 1 and 2 with the difference between groups 3 and 4. On the other hand, the one-way design with twins lends itself to three different kinds of questions. (1) Are twins generally inferior to singletons (group 1 vs 2, 3, and 4)? (2) Do MZ twins with separate chorions (dichorionic, DC) differ from those sharing one chorion (monochorionic, MC)? (3) Are DZ twins generally different from MZ twins? All three of these questions involve environmental rather than genetic sources of differences between group means, provided that the different types of twins are random events unrelated to genotype (see Machin, 1996). Because DZ twins are always dichorionic, possible genetic influences can be assessed by comparing twin correlations for groups 2 and 3, although the difference in correlation will not be purely genetic.

These examples of specialized research designs in neurobehavioural genetics show how difficult it is to identify the influence of a single factor

TABLE 2

Factorial and one-way designs with four groups[a]

		Post-adoption SES: Low	Post-adoption SES: High
Pre-adoption SES	Low	1	2
	High	3	4

Group	
Non-twins	1
DZ twins	2
MZ–DC twins	3
MZ–MC twins	4

[a] Abbreviations: SES, socioeconomic status; DZ, dizygotic; MZ, monozygotic; DC, dichorionic; MC, monochorionic.

on development, and they also show that a difference in phenotype caused by a difference in one factor cannot be attributed exclusively to that factor. Very complex designs are required to reveal interactions between the multifarious influences on organismic development, and it is wise to presume that a variety of biological and psychological interactions are also involved in simpler designs, even though interactions may not be visible in the statistical analysis.

Special control problems

When designing a study to reveal the influence of one or a few factors, we usually make special efforts to control the effects of many other things. This is done to avoid confounding the factor of interest with other variables and to increase the sensitivity of a test by reducing variation within a group. The best ways to achieve control depend on the species being studied and the design employed. In certain situations, it may not even be appropriate to control for a host of variables. For example, some psychologists want to find out what percentage of variance in IQ in the American population is associated with genetic variation, and they measure correlations between twins for this purpose. The Minnesota Twin Study (Bouchard et al., 1990) recruited MZ twins who had been reared apart for several years and then, several years after they were reunited, brought them to the same laboratory for extensive testing at the same time by examiners trained to use the same methods of test administration and scoring. The resulting data were statistically adjusted for age differences between twin pairs. Superficially, this appeared to be a well controlled study, but the controls in this instance exerted an upward bias on the estimated strength of genetic influence by expunging the effects of many environmental variables that contribute to phenotypic variance in IQ score among Americans at large. If the goal is to estimate a parameter for a population living outside the laboratory, one should not attempt to eliminate potentially important environmental influences. On the other hand, when seeking to demonstrate the importance of one among many factors, rigorous control of extraneous influences is helpful.

The realities of working with laboratory animals often pressure the researcher to take shortcuts at the expense of confounding the results with nuisance variables. Suppose one wants to compare the brains of two inbred strains of mice and their F_1 and F_2 hybrids. It is most efficient to purchase the inbreds from a commercial supplier, test some of them soon after arrival in the lab, and then mate them to produce the F_1 hybrids, which are later mated to obtain F_2 hybrids. This practice will minimize the scope and animal care costs of the study, but it will also confound group means with environmental effects on cohorts associated with season of testing and other experiences of their parents. The superior but more challenging method is to breed all mice in the lab so that every group can be tested at the same time (Fig. 1). This is extraordinarily difficult for a complex design with reciprocal crosses (Table 1) and is likely to be feasible only for a species like *Drosophila*, as achieved by Sokolowski (1992).

Animals bearing multiple offspring pose special challenges because of litter effects. An environmental treatment such as stress applied to the mother will impact the entire litter simultaneously and render the later measures of adult littermates

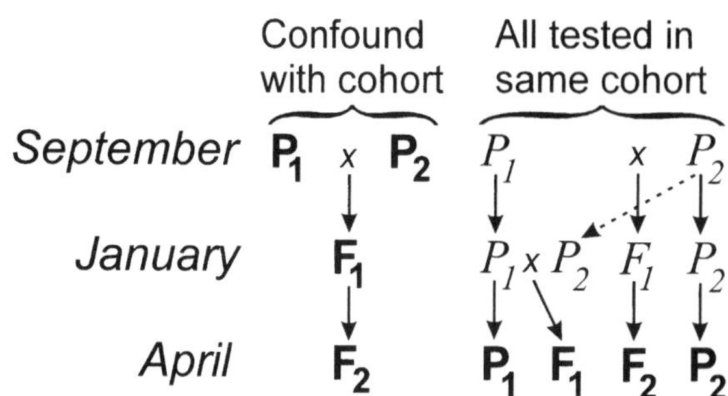

Fig. 1. Two methods to breed and test classical crosses. Bold type indicates the animals were tested. In the first method, they are tested and then bred, whereas in the second they are all tested at the same time and none of their parents were ever tested.

nonindependent. Litter sizes vary widely among mice and larger litters tend to develop more slowly and have smaller brains as adults. Culling the litter to a common size of perhaps 8 mice after birth is not very effective because so many litters are smaller than 8 from the outset, and culling controls only postnatal effects. Statistical control in a regression equation may be superior to experimental control by culling, provided the researchers are aware of possible group differences in the slopes of the litter size effects (see Bulman-Fleming and Wahlsten, 1988). After weaning into groups of same-sex sibs, many mice will be alone in the cage, and social isolation exerts a powerful influence on many adult behaviours.

Time of day when testing occurs is important for many behaviours, and different strains react differently to this factor (Peeler, 1995). For any but the simplest experiment, it will not be practical to restrict testing to a narrow window of time for all groups, and the researchers will instead be forced to use counter-balanced or randomized orders of testing to avoid confounding group with time effects. The variation in time of day will then inflate the variance within groups, unless time can be incorporated as a predictor in a multiple regression equation.

Numerous details of the test procedure will be important when studying behaviour or measuring the brains of live animals. If several tasks are administered to the same animals, the order in which tasks are given can have a major impact on results. The number of trials of a particular kind of test is also a crucial consideration. More trials generally yield a more sensitive test of group differences (Wahlsten, 1992), but different durations of testing may yield different patterns of results in a genetic linkage study (Poderycki et al., 1998).

The environment is certainly as complex as the genome and probably more so. Mice and humans possess about 70,000 different genes, and there may be at least that many features of the environment. Furthermore, the size of the genome is effectively constant over many generations, whereas the environment is always evolving without apparent limit. No comprehensive and standardized inventory of the environment exists at this time, but there are good reasons to believe such an enterprise would be a valuable contribution to future research in neurobehavioral genetics.

Basic statistical concepts

Most research in neurobehavioural genetics entails measurements made on a continuous scale, and these kinds of data are the focus of this chapter. The basic concepts apply equally to brain and behaviour. These will be presented in some detail for the simplest possible case, an experiment with only two groups, and then extended to more complex situations. The concepts involve *generic* statistics and avoid the unrealistic assumptions required by genetic statistics such as heritability and genetic correlation. Although genetic experiments employ a wide variety of specialized research designs unique to this field of study, they can sometimes be analysed with the same statistical methods that are applied in many areas of the biological and behavioural sciences. For certain specialized genetic experiments with animals, a quantitative genetic analysis as outlined by Crusio (1992) may also be useful.

In almost all studies, a sample from a population is observed and inferences are then made about the larger population. In the case of most laboratory research with animals, the population is conceptual. With mice and fruit flies, we usually breed only enough subjects to meet the needs of the current experiment, and the population is then the much larger group that *could* be bred from the same ancestors. It is important to distinguish between a parameter that is the true or hypothetical value for the population and its estimate based on the sample data. The population parameter is customarily represented by a Greek letter, whereas the sample statistic is conveniently written in the italicized Phonecian alphabet. For example, the population values of the mean, variance and standard deviation are μ, σ^2 and σ, and their sample estimates are M, S^2 and S. The sample estimate may be expressed as the Greek letter with a caret ($\hat{\mu}$) using the overstrike facility of a word processor.

Effect size

When comparing two groups, perhaps the most important number is the size of the group difference or the strength of the experimental treatment effect. This may be described conveniently as the difference between group means ($\Delta\mu$) if the scale of measurement is generally understood by consumers of the research. For example, a treatment may increase mouse brain weight by 25 mg. Knowing that mouse brains are usually in the range of 400–500 mg, one then has a rough idea of the impact of the treatment. Likewise, adoption of an infant from a poor into a wealthy family often increases the average IQ score by about 15 points, and one year of schooling increases IQ by about 4 or 5 points (Wahlsten, 1997). These effects can then be judged against the typical range from 70 to 130 points for most IQ scores in the population. However, not all intelligence tests are scaled the same way, and many other psychological tests vary widely in the range of scores. Few measures in neurobehavioural genetics involve scales of measurement that are readily understood by nonexperts.

A more useful index compares the difference between group means to the ubiquitous variation within a group. If the indicator of within-group differences is taken to be the standard deviation σ and these values are the same for the two groups, then the effect size δ may be expressed as the ratio of the difference between means to the standard deviation ($\delta = \Delta\mu/\sigma$). The sample estimate is commonly expressed as $d = \Delta M/S$, where S is the standard deviation based on the pooled variances of the two groups. The coefficient δ or d provides a clear impression of the size of the effect and a rough idea of the degree of overlap of the scores of the two groups. Standards may be devised for small, moderate, and large effects, and the single value of d for a study then provides a quick and effective idea of the results, even if the reader is not intimately familiar with the scale of the measuring instrument. The index also facilitates comparisons of studies of a similar topic that happened to employ somewhat different measures.

As shown in Fig. 2, when $\delta = 0.5$, there is a large degree of overlap of the scores and it would be difficult to perceive any difference between the groups simply by looking at frequency distributions of sample data. Only when δ reaches 2.0 or more will inspection of the sample frequency distributions reveal an obvious group difference where almost all scores of one group exceed the mean of the other group. Even then, there will be a large region of overlapping scores where group membership cannot be ascertained from the test score alone. A genuine dichotomy of scores will not exist unless the group means differ by 4 or more standard deviations. Although it is common practice among those studying sex differences in the brain to refer to a sexual "dimorphism" in their data, closer scrutiny of their findings usually reveals a rather small effect and a vast region of overlapping scores. Dimorphism clearly means two qualitatively different forms, and this will not occur for a continuous measurement unless $\delta = 4$ or more. By establishing widely accepted standards for effect size, we may prevent the

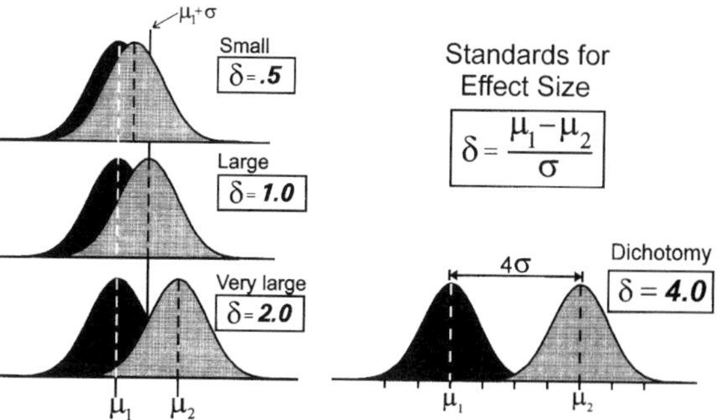

Fig. 2. Four instances where the true means of two independent groups differ by varying amounts.

TABLE 3

Standards for effect sizes in neurobehavioural genetics[a]

δ	ω^2	% of group 2 $> \mu_1$	% of group 2 $> \mu_1 + 2\sigma$	Descriptor
0.25	0.015	59.8	4.0	Very small
0.50	0.059	69.1	6.7	Small
0.75	0.123	77.3	10.6	Moderate
1.0	0.200	84.1	15.9	Large
2.0	0.500	97.7	50.9	Very large
4.0	0.800	99.99	97.7	Dichotomy

[a] Values for overlap of two groups apply when the data are normally distributed and groups have equal variances.

exaggeration of small differences that is sometimes inspired by ideology or politics.

Another good indicator of effect size partitions the total variance among scores of all subjects in the study into two components, the variance between the groups (σ_B^2) and the variance within the groups (σ_W^2). The ratio ω^2 indicates the fraction of total variance that is attributable to the difference between group means: $\omega^2 = \sigma_B^2/(\sigma_B^2 + \sigma_W{}^2)$. This index can also be extended to situations where there are more than two groups. For the case of two groups, it has a succinct relation to the δ index: $\omega^2 = \delta^2/(\delta^2 + 4)$. While ω^2 entails partitioning variance, it is not a proxy for a heritability ratio and requires no assumptions about the lack of interactions. The additivity of between- and within-group components of variation is an algebraic truth that applies to any experiment with independent groups. The proper interpretation of the estimated $\hat{\omega}^2$ depends on the specific experimental design and has no validity for any population beyond the confines of that experiment.

When the scores have a normal or bell-shaped distribution, the indicators δ and ω^2 also correspond to the degree of overlap of scores in the two groups, as represented in Table 3. Guttman's (1988) *disco* or discrimination coefficient is a superior indicator of the extent of overlapping scores; it expresses the probability of correctly predicting group membership from knowledge of the raw score, and it is valid even for groups differing markedly in variance and not normally distributed. *Disco* is implemented in the Hebrew University Data Analysis Package available from Dr. Reuven Amar of the Hebrew University of Jerusalem.

The verbal descriptors associated with each value of effect size are appropriate for research in behavioural neuroscience and neurogenetics where exquisite control of many variables is often achieved and potent treatments are administered. Cohen (1992) suggested standards for small, moderate, and large effect sizes should be 0.25, 0.5, and 0.8 based upon a review of studies in human social psychology, but in most research with laboratory animals these effects would be unacceptably small.

It happens that in small samples the value of d is slightly biased and tends to overestimate δ (Hedges 1982). When sample sizes are equal ($n = n_1 = n_2$), the unbiased estimate is given by

$$d_U = d\left(\frac{8n - 12}{8n - 9}\right) \tag{1}$$

For reasonably large samples where each group has at least 10 subjects, an approximation for the $100(1-\alpha)\%$ confidence interval for the true δ is provided by

$$d_U \pm z_{\alpha/2}\sqrt{\frac{2}{n}\left(1+\frac{d_U^2}{8}\right)} \qquad (2)$$

Suppose a control group and an experimental group each have 15 subjects and the sample $d=0.8$. The unbiased d_U is then 0.78 and the 95% confidence interval for δ is from 0.04 to 1.52. Because the confidence interval does not include $\delta=0$, there are grounds to believe that the treatment effect is real, but such a wide confidence interval indicates the true effect could range from very small to quite large.

Meta-analysis

Effect size is a useful index for combining data from a series of independent studies of the same subject matter using the methodology of meta-analysis (Hedges, 1982, Hedges and Olkin, 1985). For example, a recent survey of sex differences in the size of the human corpus callosum (Bishop and Wahlsten, 1997) found that individual studies seldom obtained evidence of a significant sex difference, but when the results were combined with meta-analysis there was clear indication that the average male corpus callosum is about 0.2 standard deviation larger than the female average. As shown in Fig. 3, studies with smaller sample sizes tend to yield more widely dispersed estimates of effect size, whereas larger studies tend to yield more similar values for this measure of the brain. Meta-analysis assigns more weight to results from larger studies, and it yields a confidence interval for the true value of δ that is much narrower than the interval for any one study because it is based on a much larger volume of data. Of course, this procedure is feasible only when there have been several attempts to replicate an interesting finding. The pooled estimate may not be dependable if the estimates in different studies are markedly heterogeneous, but heterogeneity may point to important variables differing between studies that influence the effect size of a treatment.

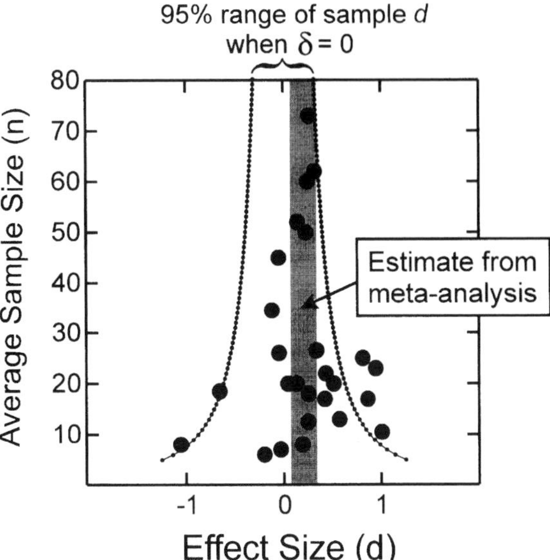

Fig. 3. Funnel plot of average sample size versus sample effect size from a meta-analysis of sex differences in the cross-sectional area of the human corpus callosum (Bishop and Wahlsten, 1997). Each large dot is the estimate from one published study. The small dots represent the theoretical 95% range of values expected if true effect size is zero (Hedges, 1982). The gray bar is the region of effect size indicated by the meta-analysis when data from all 27 studies are combined. Although individual studies usually did not reveal a significant sex difference, meta-analysis found that males had a slightly larger corpus callosum ($d=0.2$).

Meta-analysis is an excellent starting point for planning future studies on the same topic because it succinctly summarizes all previous findings and provides a plausible value for the effect size to be expected in the next study. The anticipated effect size is crucial for determining the number of subjects that should be tested in an experiment.

Errors of inference

Although effect size is the proper indicator of the magnitude of a group difference or the strength of a treatment, some researchers wrongly apply the concept of statistical "significance" in much the same way. Whereas effect size, like the mean

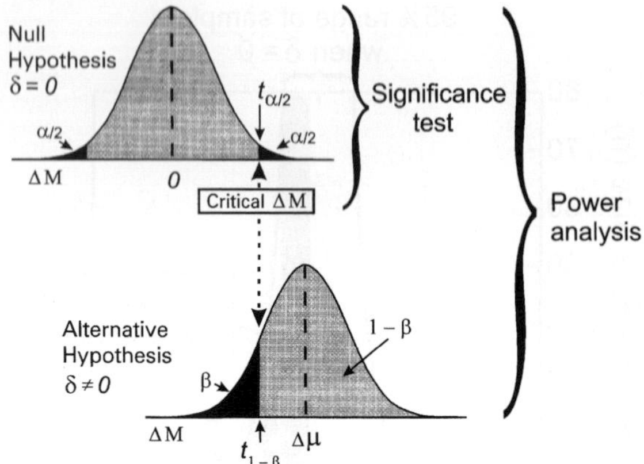

Fig. 4. Theoretical distributions of the difference between means (ΔM) of two groups when the null and alternative hypotheses are true. A sample value of ΔM greater than a critical value will cause the investigator to reject the null hypothesis. This example presumes that a two-tailed test of the null is used. The significance test of the null ignores the alternative hypothesis, whereas power analysis considers both hypotheses.

and standard deviation, is a descriptive statistic that helps the reader understand the importance of the findings, significance pertains to statistical inference about an hypothesis. There is always a risk that two groups will have different average scores merely by chance. Small differences could easily occur because of sampling error, whereas large differences are more likely to represent a real effect.

When comparing coisogenic mutants (−/−) and controls (+/+), we would like to know if the mutation is responsible for a change in brain or behaviour. One common approach is to posit a null hypothesis that the mutation actually has no effect. In symbols, this is H_0: $\mu_1 = \mu_2$, or $\Delta\mu = 0$, or $\delta = 0$. For continuous measures, the hypothesis is often evaluated with a t test having degrees of freedom $df = n_1 + n_2 - 2$. If the difference between sample means is larger than some critical value (see Fig. 4), then the investigator rejects the null hypothesis and concludes instead that there was a genuine effect. The critical value depends on the df and thus the sample sizes. It is set so that the probability of a Type I error of inference will have some fairly small and widely accepted value α. Type I error is a *false positive* result; this occurs when there really is no effect of a treatment, but the researcher wrongly rejects the null hypothesis because the group difference was fairly large as a consequence of sampling error.

The comparison of the sample value with the critical value may be made in four equivalent ways. (1) The sample t ratio may be compared with t_{critical} from a table. This is the most common method and is presumed to be familiar to most readers. If $t > t_{\text{critical}}$, then the null hypothesis is rejected and the group difference is said to be "significant" at the α level. (2) The area of the t distribution that occurs above the obtained t ratio is the *achieved significance level* or P value. The null is rejected if $P < \alpha$. If P is much less than α, the group difference is sometimes said to be "highly significant", but this wording is quite misleading because it implies that the effect size was large. It must be emphasized that d, not P, is the index of effect size. (3) The observed difference in means ΔM may be compared with $\Delta M_{\text{critical}}$, as shown in Fig. 4. If the standard error of the difference between means is $SE_{\Delta M}$, then $\Delta M_{\text{critical}} = t_{\text{critical}} SE_{\Delta M}$. (4) The sample d may be compared with the critical effect size d_{critical}. In many ways this is the best approach because it places greatest emphasis on effect size. When sample sizes are equal ($n_1 = n_2 = n$), the relation between t and d is also quite simple, as shown in Eqn. (3A). When n_1 and n_2 differ by a modest amount, their average value may be substituted for n.

$$\text{A. } t = d\sqrt{\frac{n}{2}} \qquad \text{B. } d_{\text{critical}} = t_{\text{critical}}\sqrt{\frac{2}{n}} \tag{3}$$

Knowing the number of subjects in a group, the critical value of d that would be required to warrant rejection of the null at the α level of significance can be readily determined with Eqn. (3B). As shown in Table 4, increasing the sample size has little impact on the critical t ratio but drastically lowers the magnitude of a treatment effect

TABLE 4

Critical values of t and d required to reject the null hypothesis when $\alpha = 0.05$

n per group	df	$t_{critical}$	$d_{critical}$
5	8	1.86	1.18
10	18	1.73	0.78
15	28	1.70	0.62
20	38	1.69	0.53
50	98	1.66	0.33
100	198	1.65	0.23

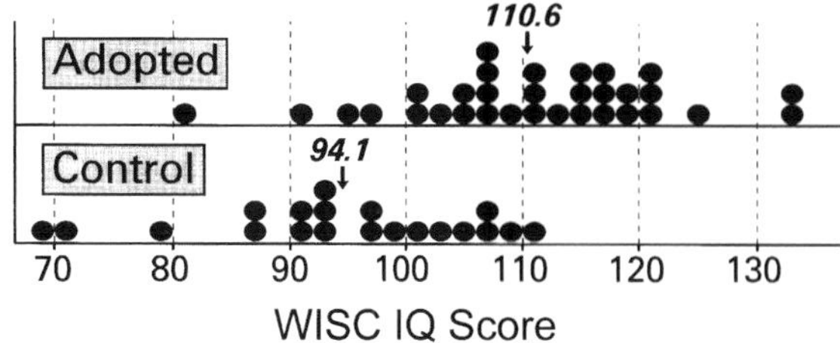

Fig. 5. IQ scores of children who had been adopted from poorly educated mothers into homes high in socioeconomic status, versus their control half-siblings who remained with the birth mother (Schiff et al., 1982). There is a large difference in the number of children scoring above 110.

that is necessary to celebrate a "significant" result. Likewise, if a study involves large samples, it may lead the investigator to proclaim significance for an effect that is really quite small and may even be trivial for that particular field of study.

Whereas the relation $\omega^2 = \delta^2/(\delta^2 + 4)$ holds for population values, the best way to estimate the percentage of variance attributable to the treatment effect for sample data is from the sample t ratio using Eqn. (4).

$$\hat{\omega} = \frac{t^2 - 1}{t^2 + df + 1} \tag{4}$$

The almost complete reliance on significance testing, especially in psychology and education research, has been severely criticized in recent years, and some authorities have even proposed that the null hypothesis be banned because it is so often misleading. It is unlikely that such a deeply entrenched methodology will ever be supplanted unless a better alternative is proposed. One contender is the humble confidence interval, which does not require the investigator to champion any hypothesis about the true difference between group means. We can make this method even more attractive and informative by presenting the confidence interval for effect size using Eqn. (2).

Consider the IQ scores from the study by Schiff et al. (1982) shown in Fig. 5. The customary null hypothesis test for these data would be presented to the reader as $t = 5.08$, $df = 50$, $P = 0.00001$, and this might be termed a highly significant effect of environment. However, the P value is a poor guide to effect size and it pertains to a null hypothesis than nobody believed in the first place. It was well known in 1982 that rearing in a privileged home rather than by a poverty-stricken single mother would have *some* impact on IQ score of the child. The question posed by Schiff et al. was not *whether* environment might influence IQ at all; rather, they asked *how much* IQ can be enhanced by better circumstances. The null hypothesis, that a large difference in environment would have no effect whatsoever on IQ, was a straw man in their study. On the other hand, effect size for their data was $d_U = 1.43$, and the 95% confidence interval for δ was from 0.89 to 2.01. Clearly, adoption into superior homes had a large effect on later IQ test score. There is no doubt about the "significance" of the effect because the lower limit of the confidence interval is far above zero. The confidence interval for δ, not the P value, answers the interesting question addressed by Schiff and colleagues. This interval reveals that the adoption effect was similar in magnitude to the difference between IQ scores of children born and reared in families of lower and upper social classes, a difference known for many years to be a little more than one standard deviation on the IQ scale (Anastasi, 1958).

Multiple hypothesis tests

An investigator rarely performs an experiment so modest in scope that the results can be analysed with only one significance test. Usually many measures are taken, and there may be many independent experiments published under the aegis of a single study. More than one research group around the globe may be examining the same phenomenon with similar measures. The consequence is that many significance tests are likely to be performed of essentially the same null hypothesis. Perhaps several research teams want to know if a knockout of the calmodulin kinase II α subunit gene (*Camk2a*) alters spatial memory, agonistic behaviour, or alcohol preference. If all tests of the effects of the mutation are conducted with the same $\alpha = 0.05$ criterion, the probability of Type I error will be elevated.

Consider an example where there are L laboratories studying the same mutation, and they do L independent tests of the same null hypothesis. For one test, P("significant" result | null true) $= \alpha$ and P(not significant | null true) $= 1 - \alpha$. For the L tests in aggregate, P(none "significant" | null true) $= (1-\alpha)^L$ and $P(\geq 1$ "significant" | null true) $= 1-(1-\alpha)^L$. As shown in Table 5, the probability of at least one false positive result increases rapidly as more research teams are recruited into the search for a gene effect. One remedy for this bloating of the Type I error rate is the Dunn-Bonferroni adjustment, whereby every team uses $\alpha' = \alpha/L$ for each test. This simple device maintains the overall Type I error rate at 5%.

TABLE 5

Type I error probabilities in relation to the number of independent laboratories (L) performing essentially the same significance test

L tests	α	$1-(1-\alpha)^L$	α/L	$1-(1-\alpha/L)^L$
1	0.05	0.05	0.05	0.05
2	0.05	0.10	0.025	0.049
5	0.05	0.23	0.01	0.049
10	0.05	0.40	0.005	0.049
20	0.05	0.64	0.0025	0.049

The same considerations apply within a study when many null hypothesis tests are performed. If a research team screens a knockout for a variety of behavioural abnormalities and employs independent groups of subjects for each of J tests, the Type I error rate may be controlled by using $\alpha' = \alpha/J$ for each test. This adjustment is not the best recourse when the multiple tests of different behaviours are administered to the same individuals. Then the measures are not independent, and a multivariate analysis of variance (MANOVA) is recommended.

Use of a more stringent α level is especially crucial in studies where a genome scan assesses many markers for linkage (Lander and Krugylak, 1995). The remedy in this situation is elaborated in chapters on linkage in this volume.

Adopting the confidence interval as an alternative to significance testing does not completely expunge the problem of multiple tests increasing the rate of false positive results. There is always a risk that the true value of a parameter will reside beyond the confidence limits, and the more intervals we compute, the more likely this will happen. Nevertheless, there is still a benefit from confidence intervals, because they may be applied in situations where the null hypothesis of no effect is not credible. The α level of significance can then be adjusted only for those tests where there really is doubt about the existence of any genuine effect.

Sensitivity of tests to real effects

An error of inference also occurs when the null hypothesis is false but we fail to reject it with the statistical test. The probability of this Type II error is represented by β, and the quantity $1-\beta$ is the power of the test or the probability of correctly rejecting a false null hypothesis. A synonym for power is sensitivity; a test with high power will be exquisitely sensitive to real group differences. Whereas the value of α is entirely arbitrary and is set by mutually agreed convention, the value of β depends on a specific alternative hypothesis that is thought to be reasonable in the circumstances. If the null is that $\delta=0$, the alternative must be some non-zero value of δ, as shown in Fig. 4. The probability of a Type I error does not depend on the sample size; even the largest study has a false positive rate of 1/20 if α is set at 0.05 and the null happens to be true. Type II error, however, can be greatly reduced by examining a large number of individuals, provided there is indeed a genuine effect awaiting detection.

Although many investigators focus on Type I errors when they analyse data, there are reasons to believe that Type II errors are much more common. Suppose the null is $\delta=0$ but the true value is actually $\delta=1.0$, a large effect size, and the researcher examines $n=10$ animals of each of two genotypes (+/+ versus −/−). If $\alpha=0.05$ and a one-tailed test is done, the Type II error probability is 0.36 and power is 64%. That is, in only a little more than half of the studies with $n=10$ will a correct conclusion about a large genetic effect be drawn. The problem is compounded when several tests are performed. Consider a scenario where the mutation is to be screened with two kinds of behavioural tests, one of which happens to be completely unaffected by the mutation ($\delta=0$) and the other involves a large effect ($\delta=1.0$). Suppose further that five labs independently assess the first behaviour and five others study the second behaviour, so that 10 significance tests are performed in total. For the five where the null is true, $P(\geq 1$ Type I error$)=0.23$; and for the five where the null is false, $P(\geq 1$ Type II error$)=0.89$. For all 10 tests combined, the probability of at least one error of inference is 0.92, and this will usually be a Type II error where the study fails to detect a real effect. It appears that one of the most common approaches to screening a mutation for behavioural abnormalities is almost certain to yield at least one mistaken inference, partly because of lax criteria for Type I error and partly because of inadequate sample size.

$$n=\frac{2(z_\alpha+z_\beta)^2}{\delta^2}+2 \tag{5}$$

The problem of sample size should be addressed in the planning phase of the study, not after the fact when it is too late to adopt an effective remedy for an insensitive test. When the null is $\delta=0$, the sample size n per group required to detect a specific effect size δ with power $1-\beta$ is specified in Eqn. (5) by a normal approximation to the non-central t distribution (Wahlsten, 1991). Table 6 stipulates the sample size required to detect various effect sizes when Type I and Type II error rates are both set at 0.05. It also shows the effect size that can be detected with a power of 95% when specific sample sizes are employed. It is common practice in psychology to set $\alpha=0.05$ and then power at 80%, but this practice perpetuates the undue emphasis on the null hypothesis and inflates the risk of what is really the most common kind of error. In my opinion, Type I and Type II errors are equally bad and should be treated as such when planning a study.

The reader may notice from Table 2 that, when $n=10$ per group, the null of $\delta=0$ will be rejected if the sample $d>0.78$, whereas in Table 6 we

TABLE 6

Sample size per group needed to detect various effect sizes when $\alpha=\beta=0.05$

Specify effect size, then find the required n		Specify the n, then find the detectable effect size	
δ	n	n	δ
0.10	2,167	5	2.69
0.25	348	10	1.64
0.50	89	15	1.29
0.75	41	25	0.97
1.00	24	50	0.67
2.00	8	100	0.47

are told that $n=10$ is adequate only to detect an effect of 1.64 standard deviations and 41 per group are needed to detect $\delta=0.75$ with power of 95%. The reason for this large discrepancy is that the power of the test of the null in Table 2, although unspecified in the example, is actually a pathetic 56% when $\delta=0.75$ and $n=10$.

Returning to the above example where five tests are done when $\delta=0$ and five are done with $\delta=1.0$, setting $\beta=0.05$ and using $\alpha=0.005$ adjusted for 10 independent tests, sample size should then be $n=38$ per group, and the probability of at least one error of inference will be reduced to 0.24 from 0.92 (when $n=10$ and $\alpha=0.05$). If power for each test is set at 99% and $\alpha=0.005$, sample size should be $n=51$ and the probability of at least one error of inference becomes a very respectable 0.07.

It should be apparent that the challenge of achieving reasonable sensitivity of a test of a mutation's effect can usually be met by using adequate sample sizes. The important question then becomes the smallest value of effect size that is meaningful to the researchers and readers. Studies that screen a mutation with numerous tests often proclaim that the knockout has significant effects on phenotypes 1 and 2 but is without impact on phenotypes 3 and 4. The significance test can demonstrate that a mutation has an effect, but it cannot prove there is no effect at all. A nonsignificant group difference does not prove $\delta=0$. This may simply indicate that the true effect is not very large and the sample is too small to reveal it. It is imperative that sample size in a study of a new mutation be decided not on the basis of what one expects its largest effect will be but what is the *smallest* effect size that the investigator would like to be able to detect, especially for those phenotypes where a large effect is not expected.

The confidence interval is clearly superior as a device for screening many aspects of brain and behaviour because it is less likely to mislead researchers by suggesting there is no effect. Suppose that a study with 10 subjects per group yields a sample effect size of $d=0.1$. This will not be even close to the criterion for rejecting the null hypothesis of no effect. However, the 95% confidence interval for the true value of δ found from Eqn. (2) will range from 0.78 to +0.97. This result should restrain the hubris of author and reader alike. When small samples are used, small group differences are uninformative. A large group difference may point to a real effect, but lack of a difference is consistent with quite substantial effects in either direction from the null.

Replication

The gold standard for proving the reality of an effect is generally believed to be replication in a second test. Suppose a study of two groups (+/+ versus −/−) with 10 per group yields $d=0.6$ and a tantalizing, "almost significant", $P=0.1$, and it is decided to do a second study in the same lab because the first results are "promising". Yet the 95% confidence interval for the true effect size ranges from 0.32 to +1.46 in the first study; hence, there is no good reason to expect the result to be similar on the next occasion. The probability of replicating such a finding depends on the true value of δ and the sample size in the second study. If the sample size was too small to yield adequate power in the first study, doing a replication with an equally small sample will not cure the malady. Suppose that the true effect size is $\delta=0.5$, a small effect close to what was observed in the first study. From Table 6 the sample size in the second study should be 89 per group if the researcher wants to have decent power to detect a significant group difference. The most important issue is therefore whether the researcher and the reader consider an effect size of $\delta=0.5$ to be worth detecting. If it is, then a proper sample size should be used to insure sufficient sensitivity of the test to a small effect.

Analysis of complex designs

All of the basic concepts of statistical inference apply equally to more complex designs as well as simply two groups. Only the mechanical details of calculations differ from a statistical perspective. The logic of the analysis, however, is considerably more involved.

The method of planned contrasts

Experiments with more than two groups are often examined with the analysis of variance (ANOVA) that separates the variance among groups into main effects and interactions. This may be quite informative for some designs, but neurobehavioral genetics often asks questions of the data that are not directly answered in the typical ANOVA table. A more flexible and informative approach involves planned contrasts among the various groups, where each contrast addresses one specific question. If there are J groups in the study, it is usually possible to formulate $J-1$ unique questions about the data.

Consider an experiment using ovarian grafting to evaluate the role of maternal environment in brain growth (see Carlier et al., 1992). The researchers suspect that mice of the inbred strain A will have larger brains if they are conceived and nurtured in a hybrid maternal environment, so they graft A ovarian follicle cells into the ovarian capsule of an AB female and later mate her with an A male. It happens that the grafting procedure typically results in smaller litter sizes, and this in itself might affect brain growth, so another group has A ovaries grafted into an inbred A female. Finally, an unoperated group of A mice is also observed to detect any effects of the grafting operation per se. This study with three groups can answer two questions. Q1: Is there an effect of the grafting operation itself? Q2: Does the hybrid maternal environment improve brain growth? The first question is addressed by comparing the two groups with A strain mothers. The second compares the F_1 hybrid maternal group with the groups having an inbred A mother, provided the answer to the first question is no. Otherwise, the second question is best answered by comparing the two grafted groups.

Table 7 shows how an experiment with four groups may be analysed in different ways with the method of contrasts. Inbred strains and their reciprocal F_1 hybrids are tested on the elevated

TABLE 7

Results for four groups of mice ($n=15$ per group) tested on an elevated "plus" maze, analysed with two sets of orthogonal contrasts[a]

Mother:	A	A	B	B			
Father:	A	B	A	B			
Mean:	20	40	46	40	$\hat{\psi}$	t	P(#tails)
Contrast analysis							
Inbreds:	−1	0	0	+1	20	2.58	0.006(1)
Reciprocals:	0	−1	+1	0	6	0.77	*0.445*(2)
Heterosis:	−0.5	+0.5	+0.5	−0.5	13	2.37	0.011(1)
Contrasts equivalent to ANOVA effects							
Female	−0.5	−0.5	+0.5	+0.5	13	2.38	0.011(1)
Male	−0.5	+0.5	−0.5	+0.5	7	1.28	*0.103*(1)
Interaction	+1	−1	−1	+1	26	2.37	0.011(1)

[a] These fictitious data indicate the percent of time in a 5 min test that the mice spend in the open arms rather than the arms enclosed with walls. The value of $MS_W=450.0$. The P values given in italics would not be considered statistically significant.

plus maze to assess murine anxiety or fear of open arms in the maze (Hogg, 1996). One set of contrasts involves three readily interpreted comparisons. Q1: Do the two inbred strains differ from each other? Q2: Do the reciprocal F_1 hybrids differ? Q3: Is there heterosis? That is, does the average of the F_1 hybrids exceed the average of the inbred parents? These three questions are independent or *orthogonal* in the sense that the answer to one does not in any way affect the possible answer to the other. For example, heterosis could exist, whether or not the inbred parents differ substantially. A fourth question could be asked, but the answer would depend on the outcome of one of the other comparisons. The investigator might like to know if there was *overdominance*, which compares the F_1 hybrids with the highest scoring inbred strain. This question overlaps with Q1 because overdominance is less likely to occur if there is a large difference between the inbreds, which will happen only if one of them has quite a high score. A second set of questions is equivalent to the usual ANOVA of a 2×2 factorial design. One factor is the strain of female (A or B), and the other is the strain of male (A or B). The two main effects are less readily interpreted than Q1 and Q2 in the first set of questions; the analysis tells us that the strain of female makes a difference but the strain of male does not. The interaction of the two factors in the ANOVA happens to be identical to the question of heterosis, however.

The reciprocal cross design in Table 1 is effectively analysed with a set of complex contrasts (see Wahlsten, 1979, Sokolowski, 1992), each addressing a unique issue such as the presence of a Y chromosome or cytoplasmic effect. A traditional ANOVA of this kind of experiment would be almost totally uninformative.

The method of contrasts assigns a coefficient c_j to each group with the stipulation that $\Sigma c_j = 0$. The contrast is then the sum of the mean of each group times c_j, using population values of μ to plan the study (Eqn. (6)) and sample values of M to analyse the data (Eqn. (7)).

$$\psi = c_1\mu_1 + c_2\mu_2 + \ldots + c_J\mu_J \qquad (6)$$

$$\psi = c_1 M_1 + c_2 M_2 + \ldots + c_J M_J \qquad (7)$$

The statistical significance of each contrast is assessed with Eqn. (8) using a t test having $df = \Sigma(n_j - 1)$, where the within group variance (MS_W) is obtained from a one-way ANOVA.

$$t = \frac{\hat{\psi} - 0}{SE_{\hat{\psi}}} = \frac{\hat{\psi}}{\sqrt{MS_W \Sigma \frac{c_j^2}{n_j}}} \qquad df = \Sigma(n_j - 1) \qquad (8)$$

The $100(1 - \alpha)\%$ confidence interval for the true value of ψ is given by Eqn. (9).

$$\hat{\psi} \pm t_{\alpha/2}\, SE_{\hat{\psi}} \qquad (9)$$

When there are more than two groups, the *partial* ω^2 for a specific contrast can be found from Eqn. (4) by substituting the value of t from Eqn. (8). This indicator of effect size compares the variance explained by the one contrast with the variance within the groups and does not include the variance attributable to the other contrasts in the denominator. The effect size d may not be meaningful for some contrasts, but for those such as a main effect that compares two sets of groups, one may take the ratio of the difference between the averages of the two sets to the pooled standard deviation.

The choice of the c_j coefficients is somewhat arbitrary in the sense that two groups could equally well be compared with (−1, +1) or (−0.5, +0.5). The advantage of (−1, +1) is that the contrast has an easily understood meaning, being the difference between group means (see Table 7). For the main effect of strain of female, we want to compare the average of two groups having a strain A mother with the average of groups having a strain B mother, and the (−0.5, +0.5) formulation yields a contrast that is the difference between these averages. The value of the t ratio will not be affected by the scale of the coefficients, however. Two contrasts are orthogonal when

$\Sigma c_{1j}c_{2j} = 0$, and this is a desirable criterion for a series of independent questions.

The sample size per group needed to detect a contrast with a desired degree of power is obtained from Eqn. (10) based on Wahlsten (1991). This gives the same result as Eqn. (5) when there are only two groups and contrast coefficients are (−1, +1). To use the equation, the researcher must have some idea of the likely value of variance within a group and then propose a set of mean values for the groups that will entail an effect of the kind tested by the contrast. The null hypothesis is always that the true contrast value is zero.

$$n = \frac{(z_\alpha + z_\beta)^2 \Sigma c_j^2}{\left(\frac{\psi}{\sigma}\right)^2} + 2 \tag{10}$$

Whether a one-tailed or two-tailed test of the null hypothesis ($\psi = 0$) is done depends on whether the investigator has good reason to expect a contrast will point in a specific direction. In Table 7 it is anticipated that strain B will exceed A and there will be hybrid vigour, but no hypothesis about which reciprocal F_1 hybrid will be superior is offered. If a two-tailed test is to be done, $z_{\alpha/2}$ should be substituted for z_α in Eqn. (10).

Interactions and sample size

Only when the experimental design entails four or more groups can the presence of interaction be fairly evaluated. Interaction in the statistical sense means that the influence of one variable depends on the specific level of the other variable(s) with which it is combined. If the effect is the same for all levels of the other variable, the effects are additive (Wahlsten, 1990; 1991). Interaction was tested in the adoption study of IQ by Capron and Duyme (1991), as summarized in Fig. 6. It was already well established in the study of Schiff et al. (1982) that adoption from low into high SES substantially enhances later IQ scores. The most interesting question posed by Capron and Duyme was whether this post-adoption environ-

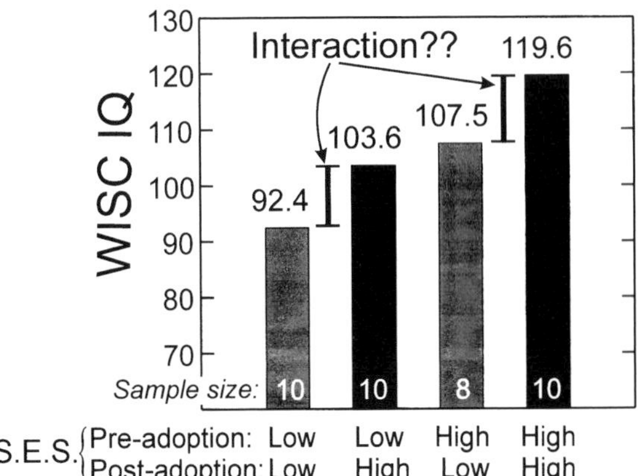

Fig. 6. Mean IQ scores of children in four groups differing greatly in parental socioeconomic status (S.E.S.) before and after adoption (Capron and Duyme, 1991). The question of interaction between the two factors asks whether the magnitude of the difference attributable to post-adoption S.E.S. depends on the pre-adoption S.E.S.

mental effect would also occur for children coming from high SES backgrounds. The difference between mean IQ scores of the two groups from low SES homes prior to adoption was 11.2 points, and the comparable difference between the two from high SES homes was 12.1 points. Interaction occurs if these two differences are themselves significantly different. The contrast analysis shows that they are not ($t = 0.1$, $df = 35$, $P > 0.5$), even though both main effects are significant. Considering that the standard deviation within groups is about 13.2 IQ points, the effect sizes (d_U) for the pre- and post-adoption effects are 1.16 and 0.86, respectively. Despite the minuscule interaction effect in the data, however, the 95% confidence interval for the interaction contrast ψ ranges from −16.6 to +18.4 IQ points! Owing to the low sample sizes, the study could not reject hypotheses of many kinds of very large interactions.

The sample size needed to detect the main effects can easily be determined from Eqn. (10) by knowing that the effect size is about $\delta = 1.0$, which means ψ will be about 13 IQ points and σ will also be about 13 points. The necessary n when $\alpha = \beta = 0.05$ is 13 subjects per group, which is not too far above what Capron and Duyme chose. As for the interaction, suppose we would like to be able to detect an interaction in which the effect size for the children coming from low SES homes is $\delta = 1.0$ but the effect size for those from high SES backgrounds is only $\delta = 0.5$. In this case the difference between the differences will be $\psi = 6.5$ IQ points, and the sample size needed for this purpose will be $n = 176$ subjects in each of the four groups, more than 10 times the number needed to detect the main effects!

Another situation of special interest to neurobehavioural genetics concerns interactions among genes. This field is now passing from the initial phase where genes were being discovered and characterized one at a time to a more advanced phase where combinations of genes are being studied. Experiments with double mutants are being used to assess serial and parallel pathways of gene functions. Suppose two recessive knockouts of genes A and B have been created. The minimal experiment to test for epistatic interaction involves four groups: +/+ +/+, *a/a* +/+, +/+ *b/b*, and *a/a b/b*. In Fig. 7 a model is proposed whereby each mutation on its own reduces performance on a behavoural test by 2 units. If effects of the two genes are strictly additive, the double mutant should score 4 units below +/+ +/+. Several other possible results shown in Fig. 7 would indicate interaction between the genes. Equation (10) can be used to find n_A, the sample size needed to detect the main effect of gene substitution at gene A, and $n_{A \times B}$, the sample size needed to detect the epistatic interaction. The approximate ratio of $n_{A \times B}$ to n_A is also shown in Fig. 7. If the double mutant involves a major sparing of effects, the sample size requirement is not onerous, but interactive effects in the other direction will require much larger samples.

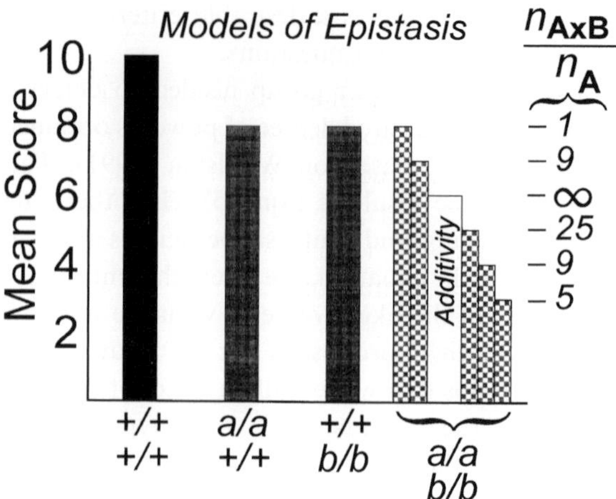

Fig. 7. Four groups to test for epistatic interaction using the double mutant *a/a b/b*. The model proposes that each mutation by itself reduces the score by 2 arbitrary units. If the gene effects are additive, the score of *a/a b/b* should be 6 units. Several possible deviations from additivity are indicated. For each, the approximate ratio is shown of sample size needed to detect the interaction versus sample size needed to detect the main effect of the mutation with the same degree of power.

Conclusion

Neurobehavioural genetics is a diverse field with elegant experimental designs available to study many components of the developing brain and behavioural domains. Most of these experiments can be analysed with generic statistical methods that require no assumptions about the absence of interactions between genes or between heredity and environment. There are real advantages to describing results in terms of effect size using confidence intervals rather than "significance", especially because knowledge of the likely effect size aids in choosing sample size for future studies. The sample sizes typically employed in a vast body of research in psychology as well as neurobehavioural genetics are too small to detect most kinds of interaction. These studies do not establish the absence of interactions in the data. Instead, they are wholly inconclusive on this issue. Sample size needed to detect many kinds of interactions tend to be much larger than those needed to detect average effects of a factor. Given the limited resources available in most laboratories, it therefore is wise to do experiments with relatively few groups but having enough subjects in a group to confer adequate sensitivity to tests of interaction. Grand experiments with dozens of groups to vary many factors simultaneously are possible in this field of study, but a given number of subjects spread too thinly over too many groups will tend to yield data that falsely appear to have a simple, additive structure. There is no valid reason to employ an experimental design capable of revealing interactions unless adequate sample sizes to detect the interactions statistically are also adopted.

References

Anastasi, A. (1958) *Differential Psychology*. Macmillan, NY.

Bishop, K. and Wahlsten, D. (1997) Sex differences in the human corpus callosum: myth or reality? *Neurosci. Biobehav. Rev.*, 21: 581–601.

Bouchard, T.J., Lykken, D.T., McGue, M., Segal, N.L. and Tellegen, A. (1990) Sources of human psychological differences: the Minnesota study of twins reared apart. *Science*, 250: 223–228.

Bulman-Fleming, B. and Wahlsten, D. (1988) Effects of a hybrid maternal environment on brain growth and corpus callosum defects of inbred BALB/c mice: A study using ovarian grafting. *Experimental Neurology*, 99: 636–646.

Capron, C. and Duyme, M. (1991) Children's IQs and SES of biological and adoptive parents in a balanced cross-fostering study. *Current Psychology of Cognition*, 11: 323–348.

Carlier, M., Nosten-Bertrand, M. and Michard-Vanhee, C. (1992) Separating genetic effects from maternal environment effects. In: D. Goldowitz, D. Wahlsten and R.E. Wimer (Eds.), *Techniques for the Genetic Analysis of Brain and Behavior. Focus on the Mouse*, Elsevier, Amsterdam, pp. 111–126.

Carlier, M., Roubertoux, P.L. and Pastoret, C. (1991) The Y chromosome effect on intermale aggression in mice depends on the maternal environment. *Genetics*, 129: 231–236.

Cohen, J. (1992) A power primer. *Psychol. Bull.*, 112: 155–159.

Crusio, W.E. (1992) Quantitative genetics. In: D. Goldowitz, D. Wahlsten, R. E. Wimer (Eds.), *Techniques for the Genetic Analysis of Brain and Behavior. Focus on the Mouse*. Elsevier, Amsterdam, pp. 231–250.

Gerlai, R. (1996) Gene-targeting studies of mammalian behavior: is it the mutation or the background genotype? *Trends Neurosci.*, 19: 177–181.

Gottlieb, G., Wahlsten, D. and Lickliter, R. (1998) The significance of biology for human development: a developmental psychobiological systems view. In: R.M. Lerner (Ed.), *Handbook of Child Psychology. Vol. 1: Theoretical Models of Human Development*, Wiley, NY, pp. 233–273.

Guttman, L. (1988) Eta, disco, odisco, and F. *Psychometrika*, 53: 393–405.

Hedges, L.V. (1982) Estimation of effect size from a series of independent experiments. Unknown, 490–499.

Hedges, L.V. and Olkin, I. (1985) *Statistical Methods for Meta-analysis*. Academic Press, Orlando.

Hogg, S. (1996) A review of the validity and variability of the elevated plus-maze as an animal model of anxiety. *Pharmacol. Biochem. Behav.*, 54: 21–30.

Lander, E. and Kruglyak, L. (1995) Genetic dissection of complex traits: Guidelines for interpreting and reporting linkage results. *Nature Genet.*, 11: 241–247.

Machin, G.A. (1996) Some causes of genotypic and phenotypic discordance in monozygotic twin pairs. *Am. J. Med. Gen.*, 61: 216–228.

Peeler, D.F. (1995) Shuttlebox performance in BALB/cByJ, C57BL/6ByJ, and CXB recombinant inbred mice: Environmental and genetic determinants and constraints. *Psychobiology*, 23: 161–170.

Poderycki, M.J., Simoes, J.M., Todorova, M.T., Neumann, P.E. and Seyfried, T.N. (1998) Environmental influences on epilepsy gene mapping in EL mice. *J. Neurogen.*, 12: 67–86.

Schiff, M., Duyme, M., Dumaret, A. and Tomkiewicz, S. (1982) How much could we boost scholastic achievement and IQ scores? A direct answer from a French adoption study. *Cognition*, 12: 165–196.

Sokolowski, M.B. (1992) Genetic analysis of behavior in the fruit fly, *Drosophila melanogaster*. In: D. Goldowitz, D. Wahlsten and R.E. Wimer (Eds.), *Techniques for the Genetic Analysis of Brain and Behavior. Focus on the Mouse*, Elsevier, Amsterdam, pp. 497–512.

Wahlsten, D. (1979) A critique of the concepts of heritability and heredity in behavior genetics. In: J.R. Royce and L. Mos (Eds.), *Theoretical Advances in Behavioral Genetics*, Sijthoff and Noorhoff, Alphen aan den Rijn, Netherlands, pp. 425–481.

Wahlsten, D. (1990) Insensitivity of the analysis of variance to heredity-environment interaction. *Behav. Brain Sci.*, 13: 109–161.

Wahlsten, D. (1991) Sample size to detect a planned contrast and a one degree-of-freedom interaction effect. *Psychol. Bull.*, 110: 587–595.

Wahlsten, D. (1992) The problem of test reliability in genetic studies of brain-behavior correlation. In: D. Goldowitz, D. Wahlsten and R.E. Wimer (Eds.), *Techniques for the Genetic Analysis of Brain and Behavior. Focus on the Mouse*, Elsevier, Amsterdam, pp. 407–422.

Wahlsten, D. (1997) The malleability of intelligence is not constrained by heritability. In: B. Devlin, S.E. Fienberg, D.P. Resnick and K. Roeder (Eds.), *Intelligence, Genes, and Success. Scientists Respond to The Bell Curve*, Copernicus, NY, pp. 71–87.

Section 2

Searching for New Genes

Section 2.1

Spontaneous and Induced Mutations with Effects on Neural and Behavioral Traits

W.E. Crusio and R.T. Gerlai (Eds.)
Handbook of Molecular-Genetic Techniques for Brain and Behavior Research (Techniques in the Behavioral and Neural Sciences, Vol. 13)

CHAPTER 2.1.1

Mapping single locus mutations in mice: towards gene identification of neurological traits

Wayne N. Frankel and Benjamin A. Taylor

The Jackson Laboratory, 600 Main St, Bar Harbor, ME 04609, USA

Rationale

Mouse mutants with inherited neurological deficits have been described for many years. They are discovered by observation of abnormal behaviors such as convulsions, ataxia, circling, shivering or shaking. The majority of these abnormalities are caused by spontaneous DNA mutations at the single nucleotide level, although many cases of spontaneous deletion (D'Arcangelo et al., 1995), insertion (Burgess et al., 1997) and retrotransposition (Letts et al., 1998) have also been observed as well as fortuitous insertion inactivation by transgenes (D'Arcangelo et al., 1995). Further analysis of these mutants may then reveal an anatomical defect, such as discrete cerebellar degeneration in ataxic mice (Herrup and Wilczynski, 1982), abnormal inner ears in circling mice (Steel, 1995) or abnormal myelin in shivering mice (Sidman et al., 1964). Dedicated secondary screening can reveal important but less obvious features, such as absence seizures (Noebels, 1986) or late-onset hearing loss (Erway et al., 1993). Alternatively, a careful phenotype screen when coupled with chemical mutagenesis to increase the frequency of mutation (and therefore the chance of detecting a specific type of mutant), can be a very powerful approach when the abnormality is not so obvious. A good example of the power of such an approach is the identification of the mouse *Clock* gene which is involved in the generation of the circadian rhythm (King et al., 1997).

Whether the mutation is spontaneous or induced, assigning the responsible locus to a specific chromosomal region gives it genetic definition, distinguishing it from similar loci that map elsewhere in the genome. That is, by mapping a new neurological mutation to a specific chromosomal region, allelism with phenotypically similar mutations may be excluded. For example, for many years the distinction between the phenotypically very similar ataxic and epileptic mouse mutants tottering and lethargic, now known to encode different subunits of an ion channel complex (Burgess et al., 1997; Fletcher et al., 1996), was that they mapped to different chromosomes (Dickie, 1964; Green and Sidman, 1962). Mapping trait loci can also facilitate genetic manipulation, for example, by accelerating the transfer of a recessive mutation onto a new genetic background to enable the identification of mutant mice prior to the appearance of phenotype.

Ultimately, however, genetic mapping of trait loci is the first step toward identification of the mutated genes. This is commonly known as *positional gene identification*, or *positional cloning* for short. In recent years, the availability of numerous mapped markers and large insert recombinant genomic DNA libraries, the ease of gene mapping and the existence of automated sequencing and computational analysis have enabled many mouse mutants to be positionally identified — success stories include mutants with "trade" names such as shaker 1, Snell's waltzer, weaver, staggerer, tottering, stargazer plus many not listed here and more to come. As the number of mapped and cloned genes increases in the mouse, positional identification is an increasingly powerful approach and can even work across species based on conserved synteny between chromosomes. For example, mapping the cerebellar mutation weaver (*wv*) to the distal part of Chromosome 16 and the subsequent mapping of expressed sequence tags (EST) to the corresponding region of human Chromosome 21 led to the discovery of a mutation in the gene encoding an inward rectifying potassium channel (Patil et al., 1995).

Genetic mapping in mice

When a trait locus is known only by a phenotype, mapping generally requires analysis of segregating crosses (i.e. meiotic mapping). Phenotypic trait loci vary from single, major mutations affecting simple qualitative trait differences, to multiple quantitative trait loci (QTLs) that exert small statistical effects on complex traits such as contextual fear conditioning (Wehner et al., 1997) or seizure frequency (Rise et al., 1991), to modifier loci which influence the penetrance of a major locus phenotype such as neurodegeneration (Cox et al., 1998; Hamilton et al., 1997). If the phenotypic trait variant is a new mutation, a new cross must be generated to detect linkage. On the other hand, if the new locus represents a variant that distinguishes inbred strains, special genetic systems, such as recombinant inbred strains, may provide an attractive alternative to a traditional linkage cross. In this chapter we will predominantly deal with new mutations.

Definitions

Phenotypic traits may be either under the control of a single segregating locus (*monogenic*) or two or more segregating loci (*polygenic* or *multigenic*). The number of loci contributing to variation in a trait is always specific to a particular genetic cross. Meiotic mapping depends on the independent reassortment of genes located on nonhomologous chromosomes as well as the genetic exchanges (*recombinations*) between different loci on the same chromosome due to *crossing over* between homologous chromosomes. When alleles at a pair of heterozygous loci are transmitted to offspring more frequently in *parental* combinations (alleles at both loci from the same grandparent) than in *recombinant* combinations (one allele from each grandparent), the loci are said to show *linkage.* The ratio of the number of recombinant progeny (A) to the total (N) is referred to as the *recombination frequency*, $r = A/N$. The genetic distance between two loci on the same chromosome is formally measured in *morgans*. A morgan (M) is defined as a length of chromosome in which the expected number of crossovers per meiosis is 1.0. A *centimorgan* (cM) is 1/100th of a morgan, and therefore, a segment within which the expected number of crossovers per gamete is 0.01. For small chromosomal segments (e.g. <10 cM), a given percentage recombination can be said to estimate the same number of cM, (e.g. 1% recombination $\approx$1 cM). For larger segments, this relationship may not hold because of the occurrence of multiple crossovers (chiefly doubles) in the same interval. Because of the phenomenon of genetic *interference*, the numbers of double and higher order crossovers in short segments are typically far fewer than would be predicted if different crossovers occurred independently. The trait locus to be mapped will be referred to here as the *target* locus, and any other previously mapped locus scored in a mapping experiment is referred to as a genetic *marker*. The distance on either side of a marker locus at which linkage with the target locus can be detected with a specified probability is the *swept radius*. The swept radius is determined by such factors as the type of cross and the number of progeny analyzed.

Mouse genome size

Laboratory mice have 19 autosomes pairs plus the XY pair. The estimated genetic lengths of individual chromosomes (based on 1998 Chromosome Committee reports; see URL: http://www.informatics.jax.org/) range from 55 cM for Chr 19 to 120 cM for Chr 2, with a total length of 1575 cM. However, genetic distances may vary between different crosses depending on the gender and genotype of the hybrid parent(s) (see below). Generally, it has been thought that recombination is somewhat more frequent in oogenesis than in spermatogenesis (Roderick et al., 1996). However, a recent mouse study suggests that recombination in spermatogenesis is equally frequent as in oogenesis, but more concentrated near chromosomal ends (Lawrie et al., 1995). Since there are $\sim 3 \times 10^9$ basepairs of DNA per haploid mouse genome, one cM is roughly equivalent to two megabases (Mb; 3000 Mb/1575 cM), although this varies considerably from one region to another because of nonrandomness of recombination.

Mapping crosses

Unlike human geneticists who must rely on available families for linkage analysis, mouse geneticists can plan and select matings to generate populations which are optimally informative. For a mating to be informative, at least one parent must be heterozygous for both the target locus and marker loci. By crossing mice of two inbred strains, F_1 hybrid parents are produced which are uniformly heterozygous at the same set of loci. Investigators will generally choose between one of two kinds of crosses: backcrosses (the N_2 generation) and intercrosses (F_2). In a backcross, F_1 hybrid mice are mated to mice of one of the parental strains to produce a segregating population. In an intercross, F_1 hybrid mice are mated inter se to produce the segregating population. Figure 1 illustrates the genetic consequences of backcrossing and intercrossing.

Identifying the phenotypic variant

The first step is to identify a phenotypic variant, such as a new mutation. Preliminary studies are usually done to establish that the variant is heritable, that is, genetically transmitted. These studies will usually indicate whether the variant is inherited as a recessive, dominant, or semidominant trait, and whether its inheritance conforms to predicted Mendelian segregation ratios (i.e. is monogenic). New mutations arising on inbred strains are generally expected to be monogenic, whereas those arising in non-inbred strains or transferred from one strain to another may be influenced by *modifier loci* (i.e. trait loci which show natural allelic variation between mouse strains), depending on the severity of the mutation and complexity of phenotype. Because the accuracy of phenotypic classification will be very important for mapping studies, it is also a good idea at this early stage to determine the phenotypic characteristics that best distinguishes mutant from normal mice. Modifier loci may be indicated when the phenotypic characteristics are unstable either between individuals or generations and the genetic background is not uniform.

Choosing a tester strain

A tester strain is one chosen to cross with the stock bearing the variant. The choice of a tester strain can be very important. First, it must carry a different target locus allele from the mutant strain, something that can usually be assumed. It also must carry different alleles from the variant-bearing stock at multiple, readily genotyped, marker loci. Initially, geneticists relied on phenotypic mutant genes as genetic markers. *Linkage testing stocks*, each carrying a handful

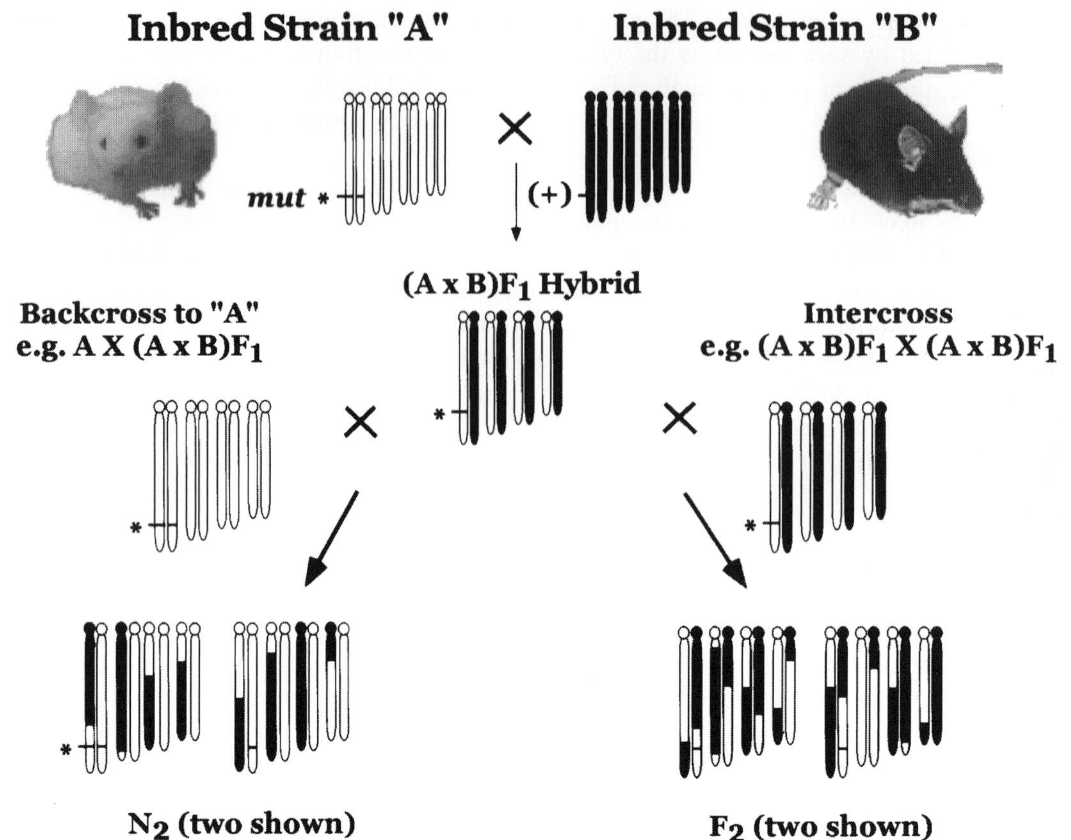

- **1/2 progeny affected for recessive mutation**
- **Each mouse informative "as is"**
- **Detect recombs in only F_1 parent**
- **Potential breeding problems for recessive mutations**
- **Calculate map distance directly from # recomb/# total**

Only 1/4 progeny affected for recessive mutation

Only affecteds informative "as is"

Can detect recombs in both parents, if "progeny-test" unaff. recombs

No breeding problems for recessive mutations, large litters

Cannot calculate map distance directly

Fig. 1. Backcross vs intercross design, with caveats for mapping recessive mutations. Hypothetical inbred strains "A" and "B" are shown, along with representations of their autosomal karyotype (only 4 of 19 autosomal pairs shown for convenience), and the location of a mutation, "*mut*," on the leftmost chromosome of the "A" strain. White chromosomes are derived from the "A" strain, black from the "B" strain. F_1 hybrids are genetically identical to each other, but inherit one of each chromosome from each parent. F_1 hybrids can be backcrossed to the recessive inbred strain (in this case "A") to generate the N_2 or backcross generation, and the genomes of two representative N_2 individuals are depicted. Note that one of each pair of chromosomes is of the parental type (rightmost) and the other may show a crossover; although crossovers occur in both F_1 hybrids and in parental inbred strains, they can only be followed when they occur in the F_1 hybrid. Alternatively, F_1 males and females may be crossed *inter se*, resulting in the F_2 where crossovers from both parents may be followed. Some pros and cons of each type of cross are listed.

Intraspecies	Intersubspecies	Interspecies
Mus musculus domesticus *Mus musculus musculus*	*M. m. castaneus* *M. m. molossinus*	*Mus spretus*
C57BL/6J-*mut* x DBA/2, A, BALB/c, C3H/He, FVB/N, etc.	C57BL/6J-*mut* x CAST/Ei MOLD/Rk	C57BL/6J-*mut* x SPRET/Ei
Easiest handling	Difficult to handle	Worst to handle
F_1 hybrids fertile	F_1 hybrids fertile	F_1 males sterile (no F_2)
Fixed for more 'modifier' loci	Modifier loci may vary	Modifier loci may vary
Limited marker polymorphism (identity in some regions)	Extensive marker polymorphism	Very extensive marker polymorphism

Fig. 2. Choices of tester strains for mapping. Listed are three different types of tester strain partners that are typically used by mouse geneticists, for a hypothetical mutation "*mut*" which arose on the inbred C_{57}BL/6 strain, along with some pros and cons regarding their use.

of useful marker genes (usually visible mutations), were constructed which allowed linkage testing of several chromosomes in a single cross (Carter and Falconer, 1951). These were replaced by polymorphic molecular markers, first protein electrophoretic variants, and then DNA variants. If both the target-bearing and tester stocks are inbred, one can be assured that the F_1 hybrid will be heterozygous for all loci at which the two strains carry distinct alleles. While any inbred strain, unrelated to the mutant- or variant-bearing stock, can be expected to exhibit multiple DNA polymorphisms on all chromosomes, the frequency and distribution of these variants in certain strain combinations may limit fine mapping (Dietrich et al., 1992). Consequently, geneticists usually select an evolutionarily divergent stock as the tester strain (see Fig. 2). The gene pool of most laboratory mice is derived principally from *Mus musculus domesticus*, the common house-mouse of western Europe, with a smaller contribution from related subspecies, probably *M. m. molossinus*, the Japanese house-mouse (Blank et al., 1986; Bonhomme and Guenet, 1996). Inbred strains have been derived from several *Mus* subspecies, including *M. m. castaneus* from Thailand (e.g. strain CAST), *M. m. molossinus* from Japan (e.g. strain MOLD), and *M. m. musculus* from eastern Europe (e.g. strain SKIVE) (Bonhomme and Guenet, 1996). These strains are excellent linkage testing strains when crossed with common laboratory stocks because of their marked genetic divergence, and also because both male and female F_1 hybrids are fertile. Crosses with *Mus spretus*, a distinct *Mus* species found in southern Spain and northern Africa, afford even greater genetic divergence (Robert et al., 1985), but are less useful because F_1 hybrid males between common mouse stocks and *Mus spretus* are sterile.

A major concern is that classification of genotypes at the target locus be unambiguous. Modifier genes contributed by the tester strain may either enhance or reduce expression of a mutant gene in segregants of a linkage cross. While modifier genes may be legitimate objects of study, their presence may obscure differences between target locus genotypes (DeJager et al., 1998; Friedman et al., 1991). Although there are some trends such as neurological phenotypes being more severe on the popular C57BL/6 background and less severe on the CAST background, the effect of a modifier on a specific locus in advance of actually making a cross is ultimately not predictable.

Choice of the tester stock also may affect the map distances obtained. Different crosses can show strikingly different distributions of crossovers in some regions (Reeves et al., 1991). Although inversion polymorphisms that distinguish common laboratory strains from wild-derived strains may be the source of variability in some cases (Himmelbauer and Silver, 1993), few such events have been documented. Other factors such as recombination hotspots may be involved (Shiroishi et al., 1993) although it is difficult to control for all the possible variables, such as parental age, parity, environmental conditions, in a single experiment.

Mating mutant and tester stock to produce F_1 progeny

In principle, intraspecies matings between the mutant stock and the tester stock may be made in either direction (i.e. mutant female × tester male or vice versa). In practice, however, whether to breed from tester males or tester females depends on availability and reproductive performance of males and females of both stocks. If the target mutation is recessive and if the mutation-bearing parent is heterozygous for the mutant allele, as is often the case due to lethality or infertility of homozygous neurological mutants, *progeny testing* (i.e. determining target locus genotype status by breeding an individual of unknown genotype to a known carrier) is necessary to identify F_1 carriers. Ovarian transplantation can be used to circumvent the problem posed by premature lethality or infertility of mutant mice in establishing a linkage cross (Davisson et al., 1991). In addition, by checking target locus phenotypes of F_1 males and females, it may be possible to test for X-linkage at this stage, if not already excluded.

Producing backcross progeny

The simplest type of mapping cross to follow is the *backcross* (illustrated in Fig. 3 are gene order and map distance determination for a recessive mutation causing albinism). To produce backcross progeny, selected F_1 progeny are mated to the recessive parent strain. (If the target locus is suspected to be X-linked, choose heterozygous F_1 females for mating.) When the target mutation is dominant, one can mate F_1 carriers to wild-type individuals of either parental stock. F_1 females tend to be very prolific, so it may be advantageous to use hybrid females for backcrossing to inbred males. If the goal is simply to map an autosomal mutation, the direction of the cross can be chosen arbitrarily, however, to be able to evaluate the effect of gender on recombination frequency, matings of both types are required. Record the gender of the F_1 parent for all backcross progeny, and report the resulting data accordingly. If F_1 parents of both genders are used, report the resulting linkage data separately by mating type, as well as combined. The size of the cross will depend on the goals of the study, as discussed further below.

Classifying progeny for the target locus

The next step is to determine the phenotype of backcross progeny and attempt to classify them as heterozygous vs homozygous at the target locus (i.e. infer target locus genotype). Accurate classification, including data recording, is one of

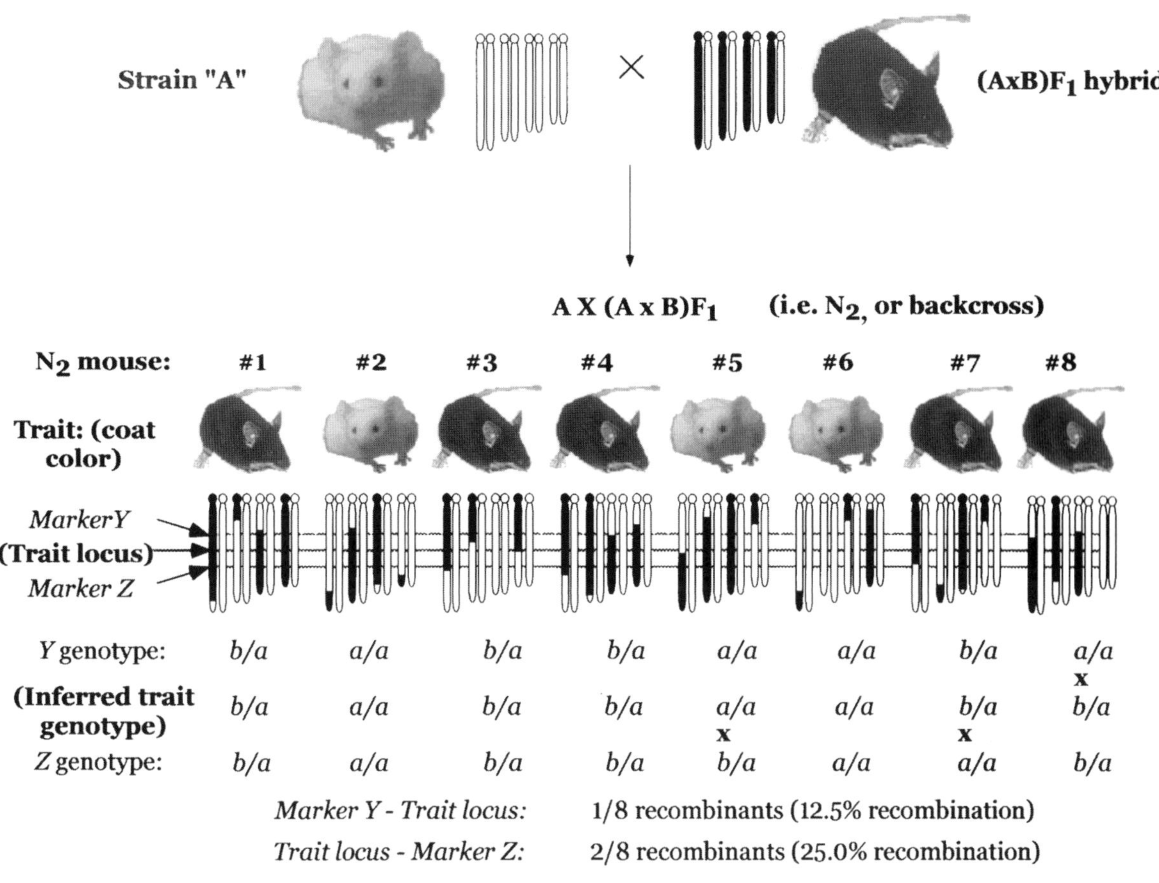

Fig. 3. Example of linkage analysis in a backcross. Illustrated is a hypothetical example of mapping a recessive trait for albinism (white coat color) in the backcross or N_2 generation, which maps to the leftmost chromosome similar to the example shown in Fig. 1. Here, eight individual N_2 mice are shown, approximately half of which are white (homozygous for the recessive allele) and the other half not white (heterozygous). Sites of recombination are shown in each, as are the locations of two flanking markers ("Y" and "Z"). Below the chromosomes the individual marker genotypes are written-out using generic allele code ("b" for the allele from the "B" strain parent and "a" for the allele from the "A" strain parent). Note that one recombinant exists between "Y" and the trait locus (mouse #8) and two between "Z" and the trait locus (mice #5 and #7). The respective recombination frequencies and inferred gene order are shown at the bottom. Additional sample calculations for backcross are shown on Table 1.

the most important steps in a mapping experiment. Examine the first few litters carefully to see whether any ambiguous or unexpected phenotypes are found. Deviations from standard phenotypes should be noted. If a range of phenotypes is encountered, it may be desirable to adopt a grading system for scoring progeny. Gender and any segregating visible marker genotypes (e.g. coat color) should also be recorded, as such information can be useful for detecting sample mix-ups. The target locus segregation data should be tested (using the χ^2 goodness-of-fit test) for conformity to a 1:1 ratio in the case of a backcross (or a 3:1 ratio in the case of a dominant/recessive gene segregating in an intercross). If significant deviations are found, one should suspect incomplete penetrance or reduced viability, phenocopies, transmission distortion, or some other disturbance such as allelic variation at modifier loci within the mapping cross.

Extracting a DNA sample

Extract a DNA sample from each backcross mouse, determine its concentration and purity, and store an aliquot at standard concentration, in a microtiter array. (Deep-well microtiter trays are available which will accommodate 0.8 ml samples [Marsh Biomedical Products, Rochester, NY]). Freeze-preserve a backup sample of tissue from each mouse in case a sample is lost or a mix-up is suspected, and it is a good precaution to extract a DNA sample from parents and grandparents as well. Tissues from several litters may be frozen, and processed as a batch. Kits for isolating DNA using DNA-binding resins in a 96-well format (Qiagen, Chatsworth, CA) can be used for mouse tails and other tissue (Schwarz, 1997); (http://www.elsevier.com/locate/tto) this format saves time and supplies, and reduces opportunities for sample mix-ups.

A linkage testing plan

It is important to develop a strategy for linkage detection, which requires an estimate of the *swept radius*, the distance between two loci such that linkage can be detected with a specified probability. The swept radius depends on the number of gametes tested (N), the type I error rate (α), the specified minimum probability of linkage detection (i.e. power), and the distribution of crossovers. Figure 4 shows the swept radius for different combinations of N, associated with four different probabilities of linkage detection and an α level of <0.025. Also shown is the critical number, A_m, the maximum number of recombinants consistent with the declaration of linkage at this significance level. It is almost equally important that the strategy has some practicality to it. For example, a simple plan would be to select 22 backcross progeny for genotyping (which fits in two rows of a microtiter tray, with

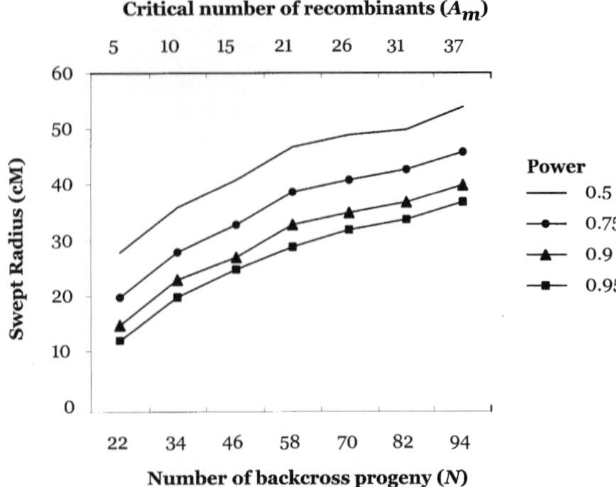

Fig. 4. Concept and usage of swept radius. The swept radius (Y axis) is the distance between a marker and the target locus at which the power to detect linkage attains a given level. Power levels, defined as the probability of rejecting the null hypothesis of no linkage when it is false, are specified as 0.5, 0.75, 0.9, and 0.95, respectively. The critical number (Am) is the maximum number of recombinants among N backcross progeny for which the null hypothesis of no linkage can be rejected ($\alpha \leq 0.025$). For example, if linkage is declared when 5 or fewer recombinants are obtained among 22 progeny, the probability of detecting linkage for a marker-target distance of 28 cM is 0.5. To improve the detection probability to 0.95, requires reducing the swept radius to 12 cM, and therefore the distance between adjacent markers would need to be reduced from 56 cM to 24 cM. Recombination frequency is assumed to relate to map distance according to the Kosambi mapping function. These different definitions of swept radius can be used to decide how many markers are needed to cover chromosomes of different lengths for a given number of backcross progeny.

space for two controls), and reject the null hypothesis if the number of recombinants with any marker (A_m) is five or less. This specific plan provides a type I error rate of 0.008, a minimal probability of linkage detection of 0.75, and an estimated swept radius of 20 cM. Thus, the autosomes may be screened with 44 markers, typed in 22 backcross progeny (plus controls), for a total of 1,056 tests. While a power level of 0.75 may not seem sufficiently high, it should be recognized that this is the level in the worst case, i.e. when the target locus happens to fall a full swept radius from the closest marker. For most sites, the target to marker distance will be less than the swept radius, thereby enhancing the probability of linkage detection. Also, if the target falls half-way between two markers, linkage may be detected with either. Thus, the effective swept radius is substantially underestimated in regions between two markers. Finally, some consensus chromosome map lengths are probably inflated.

Alternative screening plans can be developed using Fig. 4 as a guide; for example, given a higher N, complete autosomal coverage can be achieved with fewer markers. Individuals to be included in the screen may be chosen arbitrarily (e.g. the first progeny weaned) or randomly from a larger population. It should be evident that testing large numbers of progeny is unnecessary and counterproductive for detecting linkage of a monogenic, fully penetrant, trait locus. Note that the significance level recommended here (<0.025) is insufficient for establishing linkage beyond reasonable doubt. Firm proof requires a P-value of 0.0001 or less. However, the purpose of the initial linkage screen is to obtain primary evidence of linkage. Suspected linkages are easily confirmed (or excluded) by typing additional progeny and/or markers closer to the target locus. Sample probabilities at different rates of recombination for a sample size of 100 meiotic events are given in Table 1.

Genotyping a subset of backcross progeny for selected markers

Score the selected subset of backcross progeny sequentially for markers distributed on different chromosomes. (If there are obvious candidate genes for the mutation, one might as well test markers in their vicinity first). Since linkage may be detected with any given marker, it makes sense to carry out the genotyping in stages, a few markers at a time, looking for evidence of linkage before proceeding to test other markers — especially for single gene mutations. Test a single marker per chromosome until all chromosomes have been tested. But if a strong hint of linkage is found with some marker, test other markers

TABLE 1

Sample calculations for two-point linkage data

Recomb meioses	Non-recomb meioses	Total meioses[a]	Rec. freq. (r)	±SE	Lower 95% CI r	Upper 95% CI r	Kosambi (cM)	Probability (P)[b]
0	100	100	0.000	0.00	–	0.03	0.0	≪0.0001
5	95	100	0.050	0.022	0.016	0.113	5.0	≪0.0001
10	90	100	0.100	0.030	0.049	0.176	10.1	≪0.0001
15	85	100	0.150	0.036	0.087	0.235	15.5	≪0.0001
20	80	100	0.200	0.040	0.127	0.292	21.2	≪0.0001
25	75	100	0.250	0.043	0.169	0.347	27.5	≪0.0001
30	70	100	0.300	0.046	0.212	0.399	34.7	<0.0001
35	65	100	0.350	0.048	0.257	0.452	43.4	<0.0001
40	60	100	0.400	0.049	0.303	0.503	54.9	<0.01

[a] For a backcross, the number of meioses is equal to the number of progeny. For an intercross, the number of meioses is double the number of progeny. However, because in an intercross the number of recombinant meioses is a nominal estimate, (i.e. independent crossovers from each parent inherited in the same intercross mouse can go undetected), these estimates of recombination parameters work well in an intercross only for small intervals <5 cM).

[b] Probability of observing the data using 2×2 contingency χ^2 analysis. It is important to consider that although recombination frequencies up to 35% can be detected very confidently in 100 meioses, in fewer meioses the probability would drop (i.e. $P < 0.003$ for 17 recombinants in 50 backcross mice.)

on the same chromosome before proceeding. For example, if six recombinants were scored among 22 progeny in the first testing plan, it would make sense to test a second marker on the same chromosome. If no linkage is detected in this first pass (one marker per chromosome), test additional markers to fill in gaps. In a worst case scenario, one might screen all autosomes (44 markers) without detecting significant linkage. One would then review the data to see what regions are most consistent with linkage, and test additional markers in those regions. Often, linkage will be detected before one marker per chromosome has been typed.

Recording data

Data can be recorded in a spreadsheet or other computer file for storage and analysis by Map Manager (Manly, 1993), MAPMAKER (Lander et al., 1987), or similar computer program. Marker and target genotypes should be coded generically, e.g. heterozygous (H) or homozygous (A) to facilitate the recognition of linkage. Rescore the original data (e.g. gel photographs or autoradiograms) against a printout of the computer file to detect scoring and data entry errors. It is recommended that one individual do the original scoring and another the error checking.

Scanning data for linkage

The first step here is to scan the genotype data for clear linkage, i.e. obvious deficiency of recombinant progeny. Tabulate the number of recombinants between each marker and the target locus. If this is done manually, once the number of recombinants exceeds the critical number by two, the locus may be considered unlinked. A more general way to scan the data is to compute the estimated recombination frequency ($r = A/N$, as above) and its standard error $\{s_r = [r(1-r)/N]^{0.5}\}$ for each marker with the target locus. The statistic $[(0.5-r)/s_r]^2$ is distributed approximately as χ^2 with one degree of freedom, and tests for significant departures from free recombination. A simple 2×2 contingency χ^2 test provides a more exact test to judge whether there is a significant deviation from independent segregation (i.e. is linked). Probability (P) values of <0.025 should be pursued further. Again, linkage should not be considered firmly established until the probability of obtaining by chance the observed number, or fewer, recombinants is less than 0.0001.

Flank the target locus

Once clear evidence of linkage is obtained, type additional markers that are proximal and distal to the first linked marker. For example, if in the initial screen, a marker exhibits a recombination frequency of 0.1 with the target locus, try scoring markers 15 cM proximal and distal to the initial marker. If the suspected linkage is real, then one of the flanking markers should show much stronger evidence of linkage. Inspect or analyze the data to judge the likely three-point order. This involves trying the three possible locus orders and picking the one which requires the fewest double-crossovers. Computer programs such as MAPMAKER and Map Manager can compare multiple different orders and determine the order which is probabilitistically the favored one. However, with backcross data, involving correctly scored and well spaced loci, gene order is usually intuitive and unambiguous, even when there is only a single crossover between adjacent loci (see Fig. 3).

Getting even closer

Once closely linked (<10 cM) markers flanking the target locus are identified, try to identify other markers even closer. Then type the remainder of the backcross with these two flanking markers. Crossover progeny can then be typed with respect to other markers that fall between these flanking markers. (Noncrossover mice do not require further testing.) Based on the results of typing these markers in all recombinant progeny, additional markers can be selected. This process of genotyping selected recombinants with intervening markers is continued until (a) markers are identified that exhibit zero or single crossovers with the target locus in the entire cross, (b) the desired level of resolution is achieved, or (c) the supply of useful markers is exhausted. Such selective genotyping based on flanking markers is an efficient way to place the target locus relative to closely linked markers, but does not provide full information about the distances among partially typed markers. In addition, because markers that do not map to their originally assigned region are occasionally encountered, it is important to type each marker in enough progeny to verify linkage to the region. For the greatest precision from a given number of progeny, one should try to identify markers that exhibit no crossovers with the target locus as well as flanking markers that exhibit a single crossover, but this is certainly not obligatory, especially if the cross is large. If the cross involves an evolutionarily divergent strain, one can generally find MIT microsatellite markers that are closely linked to the target. Ultimately, how many markers one scores depends on the goal of mapping.

Analyzing complete data set to determine gene order and distance

Analyze the entire data set with a computer program such as MAPMAKER to check for best locus order and estimated distances among linked loci. Check any apparent double crossovers for possible errors, because in practice true double crossovers within smaller intervals are hardly ever observed. If one must postulate them, regardless of where the target locus is placed among the markers, *suspect phenotypic misclassification*. Recall that for markers typed only on the 'recombinant' subpanel, untyped progeny are assumed to be nonrecombinant, and these contribute to the denominator in computing recombination frequencies. It is legitimate to 'infer' genotypes based on the assumption that double crossovers do not occur within a short region — so long as this assumption is kept in mind and the data are reported accordingly. MAPMAKER uses maximum likelihood estimation to account for such missing data in an optimal way, making it unnecessary to provide inferred genotypes. Keep in mind that, due to interference, progeny selected for recombination in one interval will tend not to have crossovers in adjacent intervals. Thus, in such selected progeny, recombination between markers that define the selected interval and markers just outside the interval will be strongly reduced.

Reporting linkage

Traditionally, linkage data have been reported as the estimated recombination frequency with its standard error of estimate, $r \pm s_r$. The estimate of the standard error is based on a normal approximation of the binomial distribution, which is good when r is 0.2–0.5 and N is large. Unfortunately, this approximation is very bad when r is small — the most relevant linkages. A better way to report linkage results is to give the estimated recombination frequency along with the 95% confidence limits (CL). The latter can be obtained from the terms of the cumulative binomial distribution. To obtain the lower limit, find the value of r such that the probability of obtaining the observed (A) or greater number of recombinants among N progeny is 0.975. This is equivalent to finding the value of r such that one minus the probability of obtaining $A-1$ or fewer recombinants among N progeny equals 0.025. For the upper limit, find the value of r such that the probability of obtaining the observed (A) or fewer recombinants is 0.025. The terms of the cumulative binomial distribution have been published in tabular form (Diem et al., 1975), alternatively, computer spreadsheet programs such as Excel (Microsoft, Redmond, WA) have statistical functions for computing these terms.

For a pair of loci that fail to recombine, the 95% upper CL (i.e. a one-sided CL, since it is impossible to set a non-zero lower limit) of the recombination frequency may be computed more simply (based on the Poisson approximation of the binomial distribution) as $UCL_{95\%} = 1 - e^{(\ln\ 0.05)/N}$ or $\sim 3/N$. This approximation is good when N is large as in a high resolution mapping experiment towards positional cloning (e.g. for no recombinants between target locus and marker in 1000 N_2 mice, the 95% UCL is 0.003 cM, or about 650 kb).

Results may be reported as recombination frequencies directly or converted to cM (e.g. equating a recombination frequency of 0.04–4.0 cM). Because of the possibility of undetected double crossovers, recombination frequencies greater than 0.2 should not be converted into cM in this linear fashion, but may be converted using an appropriate *mapping function*. A good, commonly applied mapping function is Kosambi's (Kosambi, 1944), which, compared with the popular Haldane mapping function (Haldane, 1919) gives better estimates of map distance because it accounts for interference. Sample calculations for reporting two-point linkage data are given in Table 1.

Size of cross

The decision as to how many backcross progeny to analyze depends on how the data are to be used, and availability of resources. Mapping based on a 100-mouse backcross is generally considered adequate for reporting a new linkage. Such a cross permits one to look at the consensus map for potential candidate genes. Of course, to carry out a positional cloning effort would require a much larger cross (several thousand progeny).

The size of the cross determines the potential average mapping resolution. A backcross of N progeny can be used to generate a map in which crossovers occur on average every $100/N$ cM. Earlier we saw that the 95% upper confidence limit of the distance (in cM) beween nonrecombining loci is $\sim 300/N$. Thus, nonrecombining loci in a 100-mouse backcross are 95% certain to lie within three cM of each other. In this case, the target locus would be mapped within a six cM region. However, in most cases, the target locus will show recombinants with closely linked markers whose positions may be well established, thus defining specific proximal and distal bounds. This depends on prior knowledge about the position of the nearest markers in a consensus map. If such information is lacking or inadequate, consider mapping selected markers in one or more standard crosses, such as the BSS backcross panel available from The Jackson Laboratory (Rowe et al., 1994) or the EUCIB panel available from the Human Genome Mapping Panel Resource Center (Breen et al., 1994).

Mapping by intercross

Although the backcross design is straightforward, the intercross design has some key advantages, especially for fine-mapping towards positional cloning (see Fig. 1). An intercross results from mating males and females that are heterozygous for both the target locus and linkage markers. Generally, intercrosses are produced by mating F_1 hybrids between two inbred strains. The genotypes of the resulting F_2 progeny reflect recombination events in both parental gametes, but it is not possible to discriminate between crossovers in paternal and maternal gametes. Consequently, intercross data provides an estimate of the average recombination frequency in male and female gametes, but does not afford separate estimates of male and female recombination.

A major reason for choosing an intercross rather than a backcross is that an intercross is potentially twice as efficient as a backcross of the same size. This is simply the ability to detect crossovers in both parental gametes in intercross progeny. However, if the target mutation is dominant or recessive (as opposed to semidominant or codominant), this full efficiency is not necessarily realized, and the relative efficiency varies between equality with the backcross for values of r close to zero, and two-thirds as efficient, when the value of r is close to 0.5 (Green, 1981). Thus, for a dominant or recessive trait, the two-fold gain in efficiency applies only to the recessive class. However, for the purpose of linkage detection, one can selectively analyze only (or mostly) mutant homozygotes (in the case of a recessive), thus taking advantage of the two-fold gain in efficiency.

Nonetheless, the potential efficiency of the intercross can be recovered by progeny testing. First, the target locus is mapped approximately to identify closely linked flanking markers. The remainder of the cross is then typed with these flanking markers to identify crossovers near the target. Then recombinant mice, whose genotype at the target locus is uncertain, are progeny tested, e.g. by crossing back to an F_1 mouse. (If affected progeny are observed, the recombinant F_2 must have been heterozygous at the target locus). This requires keeping progeny with dominant phenotypes alive until flanking markers can be scored. Once the target gene has been mapped to a small region, the number of such progeny tests required is modest. This strategy is especially useful in positional cloning projects where high mapping resolution is desired, (Segre et al., 1995), but may not be practical when only moderate resolution is required.

Many neurological mutants are either juvenile lethals or infertile, making it necessary to choose an intercross rather than a backcross. Of course, in these cases it is necessary to progeny test the F_1 hybrids to first identify carriers of the mutant gene, or simply to retain and analyze only the expected 1 in 4 matings that produce affected progeny. Since most F_1 hybrids are excellent breeders, intercross matings tend to produce large and frequent litters (another advantage of an intercross design).

One problem with intercross matings is that recombination cannot be calculated from a simple algebraic equation. Instead, the method of maximum likelihood estimation is required. This involves a somewhat cumbersome, iterative estimation procedure, but one that can be carried out manually with the aid of published tables of scores which are applied to different genotypic classes (Green, 1963). These computational difficulties are circumvented entirely by the use of computer programs such as MAPMAKER that automate the estimation process. Another disadvantage of the intercross design is that mistyped progeny are not always apparent. Because interference is partially masked, misscored individuals are not as easily recognized as apparent 'double crossovers.' Since intercross progeny may by chance, inherit closely linked crossovers, one on each parental chromosome (i.e. *ABc*/*aBC*), identification of mistyped individuals is not simple. As crosses become larger, and markers become more dense, as for positional cloning, this limitation tends to disappear, but for small crosses and sparse markers, it is a distinct disadvantage.

Complexities of mapping: incomplete penetrance, phenocopies, and reduced viability

Some variants and mutations do not exhibit a consistent phenotype, a phenomenon referred to as incomplete or variable penetrance. A simple example would be a recessive mutation for which 90% of homozygous mutant (*mut*/*mut*) individuals exhibit a specific defect, but the remaining 10% are indistinguishable from normal (+/+ or +/*mut*) mice. Incomplete penetrance can seriously inflate estimates of recombination since nonrecombinant progeny are apt to be classified as recombinants. Often the degree of penetrance depends on the genetic background, in which case modifier loci presumably play a role as was the case for the mapping of the curly tail mutation which causes neural tube defects (Letts et al., 1995). A low level of incomplete penetrance should not seriously impede the detection of linkage, but it can result in inflated estimates of recombination frequencies and obscure gene order. Generally, if incomplete penetrance is suspected, it is very helpful to restrict analysis to mice that manifest the mutant phenotype for estimating genetic distances and determining gene order. However, it is important that some wild-type progeny be genotyped. Otherwise, one cannot discriminate between linkage and abnormal marker gene segregation due to causes unrelated to the mutation. Heterozygotes for the semidominant, loop tail mutation (*Lp*), another model for neural tube defects, show incomplete penetrance, with some carriers being indistinguishable from wild-type. Accurate mapping required restricting analysis to backcross progeny manifesting the mutation (Mullick et al., 1995). In human genetics, where incomplete penetrance is commonly encountered, special linkage analysis programs have been devised which provide maximum likelihood estimates of recombination frequency in the presence of incomplete penetrance. These programs provide estimates of the degree of penetrance from the data.

Phenocopies are the opposite of incomplete penetrance. Here, some mice which do not carry the mutant genotype nonetheless exhibit the phenotype associated with the mutant gene. The cause may be unknown environmental factors, or other genetic loci segregating in the cross. For example, a mouse runted because of environmental factors may be difficult to distinguish from a severe neurological mutation which causes difficulty to thrive, or one with an ear infection may show the same circling behavior of a mutant with an inherited inner ear disorder. If phenocopies are suspected in the data, one can try to restrict mapping analysis to wild-type mice. In general, it is useful to make note of any deviation from the standard phenotypes. However, if both incomplete penetrance and phenocopies are suspected, one cannot assume that either phenotype is exclusively associated with any specific genetic class. In this case, one needs to adopt a model that allows for the fact that phenotype is not strictly tied to genotype, and utilize one of the computer programs specifically designed for mapping complex traits. One should then think of loci as affecting the risk of a particular variant phenotype, rather than determining the phenotype directly.

In any case, if misclassification is suspected, the number of progeny screened for detecting linkage should be increased.

Reduced viability of mutant classes is not a major problem for linkage analysis unless recombinant and nonrecombinant classes have different viabilities. Inviability of certain classes was a greater problem when visible markers were utilized, and certain double mutant genotypes might be underrepresented.

Linkage markers and their use

Microsatellite markers are now used almost exclusively for general linkage analysis. Thousands of these sequences, consisting of mainly dinucleotide repeats (e.g. (CA)n) where n is between 10 and 60, are widely dispersed in mammalian genomes (Litt and Luty, 1989). The number of dinucleotides at any site is prone to mutation due to errors in DNA replication, resulting in abundant length variation in populations. Length variation is readily detected by amplifying microsatellite sequences by PCR (using site-specific primers) and separating the amplified DNA fragments by gel electrophoresis. Microsatellite loci are highly polymorphic, usually codominant, inexpensive and rapid to use. A total of 6331 microsatellite markers have been identified and mapped in the mouse genome by the Center for Genome Research at the Whitehead Institute (Dietrich et al., 1996) (URL: www.genome.wi.mit.edu/). These markers are denoted by symbols *D_Mit#*, where the *D* indicates an anonymous DNA marker, _ identifies the chromosome, *Mit* is the Laboratory Registration Code for the MIT genome center, and # is a serial number given to loci assigned to the particular chromosome (e.g. *D1Mit2*, is the second MIT microsatellite marker assigned to Chr 1). Their initial mapping, based on a 46-member set of (C_{57}BL/6J-*ob* × CAST)F_2 progeny, was not very exact (Dietrich et al., 1996). Some markers (1,245) also have now been mapped in 46 interspecific backcross progeny (Dietrich et al., 1996), and most also are assigned to YAC contigs. Also, many of these have been used by various investigators in diverse mapping projects. A panel of inbred strains have been characterized with respect to fragment sizes, and this information is on the MIT/Whitehead Web site. Primer pairs for amplifying these loci are available commercially (Research Genetics Inc., Huntsville, AL). It is a good idea to test the primers on the parental strain and F_1 hybrid DNAs prior to using them on the backcross DNAs.

One difficulty with microsatellite markers is that the DNA of some strains does not amplify well in competition with DNA from other strains (i.e. in heterozygotes), presumably due to sequence variation in the primer sites. Occasionally, primers will not amplify the DNA of a given strain at all. In some cases, reducing the annealing temperature of the amplification reaction may equalize amplification. Since the MIT microsatellite primers were designed to be complementary to strain C57BL/6J, amplification of the DNA of this strain is expected. The strong sequence divergence between the various *Mus* subspecies from laboratory means that microsatellite amplication from such strains is frequently a problem. This is especially true in the case of *Mus spretus*.

Microsatellite length polymorphisms are detected using either agarose or polyacrylamide gels. For agarose, ethidium bromide is used to visualize DNA fragments, while silver staining, radioisotopes, or fluorescent label is used to detect DNA fragments in acrylamide gels (Dietrich et al., 1996). Agarose gels are generally faster to set-up and analyze, but limit polymorphism detection to allelic fragment size differences of $\sim 8\%$ or greater. Several methods for improving the efficiency of microsatellite analysis using agarose gels has been described (Routman and Cheverud, 1995). Multiplexing, the amplification and/or analysis of multiple markers in a single assay, is possible. Access to an automated sequencing facility can greatly increase genotyping throughput. Fluorescently labeled primers increase the number of loci that can be scored in a single run (Reed et al., 1994).

Restriction fragment polymorphisms provide another means of mapping a marker locus relative to a phenotypic variant. Dispersed repetitive sequences, such as endogenous retrovirus genomes (Frankel et al., 1990; Lueders et al., 1993; Taylor et al., 1993), minisatellite sequences (Julier et al., 1990), and sequences flanked by B1 repeat elements (Hunter et al., 1996) can provide multiple markers involving single assays. However, visualization of these DNA variants requires blot hybridization, unless a PCR-based strategy is taken as was the case for certain classes of endogenous retroviruses (Kaushik and Stoye, 1994). Nevertheless, none of these approaches offer the same degree of choice regarding location as do microsatellite markers. However, with a little luck multilocus approaches can provide linkage very quickly.

In the future, more rapid means of discriminating genetic differences will undoubtedly become available. These will likely incorporate automated genotypic scoring and recording as well as efficient genotypic discrimination. A system is needed that would allow the scoring of 20–100 loci per mouse in a reliable, inexpensive, and automatable way. A system that does not require elaborate instrumentation would be desirable. Current research focuses on the development of oligonucleotide arrays synthesized on silicon chips as a means of detecting single nucleotide polymorphisms (SNPs) by hybridization (Wang et al., 1998).

DNA pooling

The time consuming and expensive task of genotyping linkage crosses can be drastically reduced by adopting the strategy of phenotypic DNA pooling. The rationale for pooling is that DNA pools made from phenotypically distinct backcross or intercross progeny will differ in the representation of marker alleles at loci linked to genes that determine phenotype. Linkage detection depends on recognition of this differential representation of allelic DNA fragments in pools. DNA pooling will generally shorten the search for linkage of major variants (Asada et al., 1994; Taylor et al., 1994), and can also be efficient in QTL detection (Collin et al., 1996; Darvasi and Soller, 1994; Mohlke et al., 1996; Taylor and Phillips, 1996).

For detecting linkage of monogenic traits by the DNA pooling strategy, two pools are made representing the two phenotypes for recessive/dominant traits segregating in an intercross or backcross, or the two homozygotes for semidominant or codominant trait loci segregating in an intercross. In general, the larger the number of individuals contributing to the two pools, the better the statistical efficiency of the test. Good statistical power is provided by representing at least 50 meioses in each DNA pool (e.g. 50 heterozygotes and 50 homozygotes in a backcross or 25 homozygous (recessive) mutants and 25 wild type progeny in an intercross). Relatively small improvement is achieved with larger numbers. However, one should not hesitate to use smaller numbers in pools (e.g. if the entire cross is small); one merely has to accept that there will be more regions showing a hint of linkage, and that a larger number of markers will be needed to cover the genome. Markers spaced at 40 cM (and within 20 cM of chromosome ends) are adequate for detecting linkage of monogenic loci. For an intercross, it is useful to include both parental strains and the F_1 hybrid as controls. For a backcross, appropriate DNA controls are an F_1 hybrid, the backcross parental strain, and a 1:1 mixture of these two samples. Codominant microsatellite markers whose allelic fragments are well resolved in heterozygotes work well for DNA pooling. Pooling is strongly recommended as a rapid, efficient strategy for mapping monogenic trait loci.

Positional gene identification

The ultimate goal of mapping a trait locus to a chromosomal region is positional gene identification. Positional gene identification is most easily accomplished by comparative mapping to find candidate genes that lie near the target locus, and subsequent mutation screening in affected individuals. Although candidate genes can be identified in the absence of a high resolution genetic map, the smaller the *critical interval* (i.e. the maximal genetically-defined region that contains the target mutation), the fewer candidate genes one needs to isolate and analyze for molecular variation amongst the alleles of interest (e.g. mutant vs wild-type). With modest effort, a good experimental design, such as a fully-tested intercross design as discussed earlier, and a little luck there is no reason why many mutations cannot be fine-mapped to the resolution of a single transcription unit, $\leq$100 kb. Thus, the aim of many positional cloning projects is to increase the resolution of the genetic map (by increasing the number of informative recombinants) such that only a small number of genes need be analyzed, typically to the order of 0.1 cM which corresponds to roughly 200 kb of genomic DNA.

However, in order to identify a trait gene by comparative mapping, even when most candidates have been excluded from a high resolution trait locus map, it is usually necessary to have at least an approximate map position for a candidate gene and an understanding of its function or expression pattern. Although in a few years a complete transcript map of human and mouse genomes is likely to be available, the biological function of most expressed transcripts is presently unknown. Therefore, while the positional candidate gene approach will become increasingly powerful, strict positional cloning is still very important. This requires the parallel construction of a physical map, typically by the tedious process of *chromosomal walking* across the critical interval, which entails the serial isolation of recombinant DNA clones (e.g. yeast or bacterial artificial chromosomes — YACs or BACs, respectively) containing large inserts of mouse genomic DNA. Two general approaches to the parallel construction of a physical genomic DNA map based

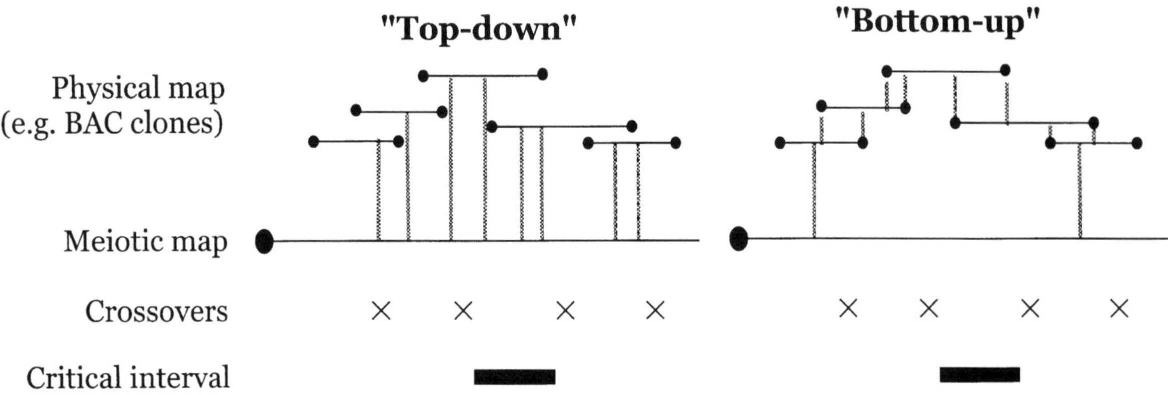

Fig. 5. Approaches to integrating meiotic maps with physical maps: "top-down" vs "bottom-up" strategies. In the top-down approach, a very high density of DNA markers is used to link large-insert recombinant DNA clones. The bottom-up approach begins with a collection of DNA markers too sparse to bring the physical map to "closure" and therefore one must "walk" from one clone to the next, using new markers generated from the clones themselves.

on a genetic map are illustrated in Fig. 5. Candidate genes can be readily isolated from, or identified by their presence on these DNA clones by techniques such as exon trapping (Buckler et al., 1991), direct cDNA selection (Morgan et al., 1992), or brute-force genomic DNA sequencing (Hamilton et al., 1996), and subsequently evaluated for defects in the mutant allele. None of the methods are foolproof, but when used together or in a complementary fashion they have proven to be quite effective as evidenced by the large number of genes positionally cloned from scratch to date. However, in the next five to ten years the sequence of the entire genome, together with coding genes, will be completely known and integrated with commonly used genetic markers. This accomplishment will radically accelerate the process of trait gene identification, by essentially eliminating the chromosomal walking and gene finding steps. Nevertheless, the initial step of high-resolution genetic mapping will always provide a cost benefit by reducing the number of candidate genes that need to be evaluated. Moreover, for new spontaneous or induced mutations, no improvement in technology can obviate the need for testing candidate genes for defects in the mutant allele.

Testing candidate genes and identifying mutations

When finding a candidate gene containing a putative causative alteration in the mutant allele of a target locus, it is important to gain as much evidence as possible that it is the true causative mutation and not merely a strain polymorphism or a secondary, inconsequential mutation (see Fig. 6). For this, it is important to consider that single gene mutations generally fall into three categories. The ideal mutation occured recently on a defined inbred mouse strain background, i.e. one that is *coisogenic* with normal individuals from the same colony (two strains that are genetically identical, i.e. isogenic, except for a difference at a single locus, are termed *coisogenic*). Thus, one would iso-

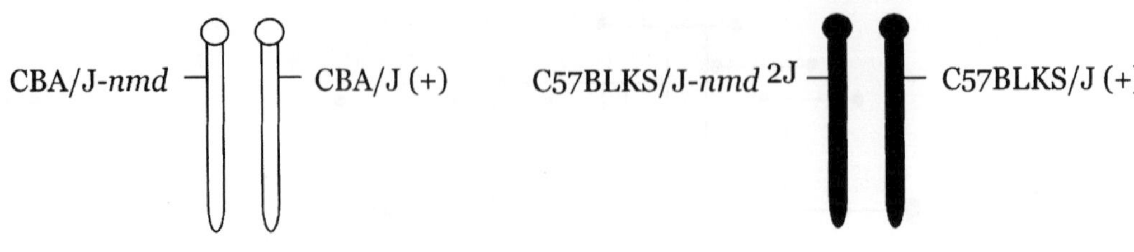

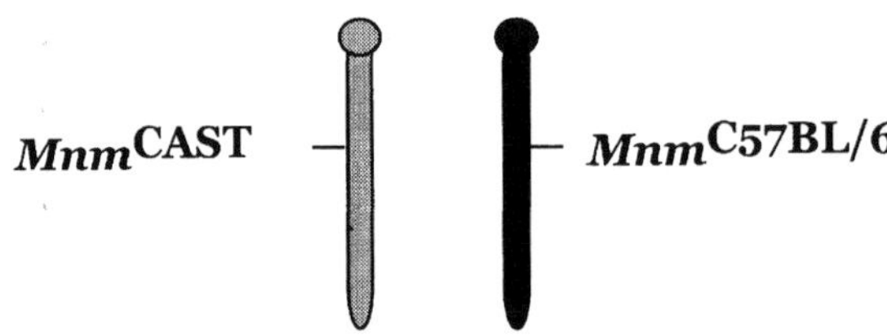

Fig. 6. Distinguishing causal mutations from strain polymorphisms. Spontaneous or induced mutants that arise on known strain backgrounds are considered to be 'co-isogenic' with that strain and therefore knowing the sequence from which the mutation arose is intuitive. Depicted here are the mutations *nmd* (neuromuscular degeneration) and nmd^{2J}, which arose recently on the strains CBA/J and C57BLKS/J, respectively (Cook et al., 1995). The presence of mutations in each mutant allele relative to its respective co-isogenic control (depicted by chromosomes of the same color) provided very strong circumstantial evidence that the mutations were causal (Cox et al., 1998). In contrast, the allelic forms of the semidominant modifier of *nmd* (*Mnm*), occur naturally in the strains C57BL/6J and CAST/EiJ (i.e. they occurred prior to the inception of each strain), and the sequence from which these mutations arose is unknown. Thus, when a candidate gene is identified for *Mnm* it will be more difficult to distinguish a putative causal mutation from a mere strain polymorphism distinguishing C57BL/6J from CAST/EiJ (depicted by chromosomes of different color). Consequenctly, functional tests may be necessary to obtain equally strong circumstantial evidence for the identity of this gene.

late the underlying gene using the strategies outlined above, find a DNA mutation which could potentially affect the structure or expression of the encoded protein, and show that the mutation was not present in the DNA from the strain of origin (Cox et al., 1997). Because the frequency of spontaneous mutation is very low, it is unlikely to be a chance occurrence unrelated to the phenotype. In this case, although such evidence should be considered circumstantial (because it cannot be formally disproven), it is ultimately very strong although it may be strengthened further by additional mutant alleles, when available (Cox et al., 1998).

A second, somewhat less satisfying case is when a mutation occurred on an unknown or uncertain strain background. Thus, one may find a candidate mutation but would not necessarily be sure that it was not merely a non-causative polymorphism (perhaps in a different gene) that was already present on the strain of origin. In such cases, researchers may be able to find additional mutant alleles, or even screen for them by chemical mutagenesis, which would greatly strengthen their evidence (Cordes and Barsh, 1994; Gibson et al., 1995). Alternatively, using linked DNA polymorphisms and a broad strain survey to gain some idea of the likely strain of origin (of course, together with a demonstration that the putative mutation itself cannot be found in any mouse strain other than the one which harbors the mutation), they may be able to gain additional circumstantial evidence for the strain of origin (Hamilton et al., 1996). Ultimately, however, in such cases it may be necessary to rely on transgenic gene rescue (Hamilton et al., 1997), or even gene targeting to establish formal proof. Finally, some single gene mutations exist only as natural strain variants (Bowes et al., 1990; Vidal et al., 1993). It is ultimately much more difficult to gain convincing circumstantial evidence that a given variant is causative for the phenotype. In some cases strain surveys for linked polymorphisms in common strains (Malo et al., 1994) can be convincing or functional evidence may be very compelling. However, it is much more satisfying to turn to transgenics or gene targeting for formal proof (Lem et al., 1992; Vidal et al., 1995).

Acknowledgements

We thank Muriel Davisson, Janan Eppig, Gregory Cox and Verity Letts for comments on earlier versions of this manuscript. This work was supported by NIH Research Grants GM18684, CA33093, DK50692, NS31348, NS32801 and DC03611.

References

Asada, Y., Varnum, D.S., Frankel, W.N. and Nadeau, J.H. (1994) A mutation in the *Ter* gene causing increased susceptibility to testicular teratomas maps to mouse Chromosome 18. *Nature Genetics*, 6: 363–368.

Blank, R.D., Campbell, G.R. and D'Eustachio, P. (1986) Possible derivation of the laboratory mouse genome from multiple wild Mus species. *Genetics*, 114: 1257–1269.

Bonhomme, F. and Guenet, J.-L. (1996) The laboratory mouse and its wild relatives. In: M.F. Lyon, M.F. Rastan and S.D.M. Brown (Eds.), Genetic Variants and Strains of the Laboratory Mouse, Oxford: Oxford University Press, pp. 1577–1596.

Bowes, C., Li, T., Danciger, M., Baxter, L.C., Applebury, M. L. and Farber, D.B. (1990) Retinal degeneration in the rd mouse is caused by a defect in the beta subunit of rod cGMP-phosphodiesterase. *Nature*, 347: 677–680.

Breen, M., Deakin, L., Macdonald, B., Miller, S., Sibson, R., Tarttelin, E., Avner, P., Bourgade, F., Guenet, J.-L., Montagutelli, X., Poirier, C., Simon, D., Tailor, D., Bishop, M., Kelly, M., Rysavy, F., Rastan, S., Norris, D., Shepard, D., Abbott, C., Pilz, P., Hodge, S., Jackson, S., Jackson, I., Boyd, Y., Blair, H., Maslen, G., Todd, J.A., Reed, P.W., Stoye, J., Ashworth, A., McCarthy, L., Cox, R., Schalkwyk, L., Lehrach, H., Klose, J., Gangadharan, U., and Brown, S. (1994) Towards high resolution maps of the mouse and human genomes — a facility for ordering markers to 0.1 cM resolution. *Hum. Mol. Genet.*, 3: 621–627.

Buckler, A.J., Chang, D.D., Graw, S.L., Brook, J.D., Haber, D.A., Sharp, P.A. and Housman, D.E. (1991) Exon amplification: A strategy to isolate mammalian genes based on RNA splicing. Proc. Natl. Acad. Sci. USA *88*, 4005–4009.

Burgess, D.L., Jones, J.M., Meisler, M.H. and Noebels, J.L. (1997) Mutation of the Ca^{2+} channel β subunit gene $Cchb_4$ is associated with ataxia and seizures in the lethargic (*lh*) mouse. *Cell*, 88: 385–392.

Carter, T.C. and Falconer, D.S. (1951) Stocks for detecting linkage in the mouse, and the theory of their design. J. Genet. *50*, 307–323.

Collin, G.B., Asada, Y., Varnum, D.S. and Nadeau, J.H. (1996) DNA pooling as a quick method for finding candidate linkages in multigenic trait analysis: an example involving susceptibility to germ cell tumors. *Mamm. Genome*, 7: 68–70.

Cook, S.A., Johnson, K.R., Bronson, R.T. and Davisson, M.T. (1995) Neuromuscular degeneration (*nmd*): a mutation on mouse chromosome 19 that causes motor neuron degeneration. *Mamm. Genome*, 6: 187–191.

Cordes, S.P. and Barsh, G.S. (1994) The mouse segmentation gene kr encodes a novel basic domain-leucine zipper transcription factor. *Cell*, 79: 1025–1034.

Cox, G.A., Lutz, C.M., Yang, C.-L., Biemesderfer, D., Bronson, R.T., Fu, A., Aronson, P.S., Noebels, J.L. and Frankel, W.N. (1997) Sodium/hydrogen exchanger gene defect in slow-wave epilepsy mutant mice. *Cell*, 91: 139–148.

Cox, G.A., Mahaffey, C.L. and Frankel, W.N. (1998) Identification of the mouse neuromuscular degeneration gene and mapping of a second site suppressor allele. Neuron 21: 1327–1337.

D'Arcangelo, G., Miao, G.G., Chen, S.C., Soares, H.D., Morgan, J.I. and Curran, T. (1995) A protein related to extracellular matrix proteins deleted in the mouse mutant reeler (see comments). *Nature*, 374: 719–723.

Darvasi, A. and Soller, M. (1994) Selective DNA pooling for determination of linkage between a molecular marker and a quantitative trait locus. *Genetics*, 138: 1365–1373.

Davisson, M.T., Guay-Woodford, L.M., Harris, H.W. and D'Eustachio, P. (1991) The mouse polycystic kidney disease mutation (*cpk*) is located on proximal chromosome 12. Genomics, 9: 778–781.

DeJager, P.L., Harvey, D., Polydorides, A.D., Zuo, J. and Heintz, N. (1998) A high-resolution genetic map of the nervous locus on mouse chromosome 8. *Genomics*, 48: 346–353.

Dickie, M.M. (1964) Lethargic (*lh*). *Mouse News Lett.*, 30: 31.

Diem, K., Lenter, C. and Eds. (1975) *Scientific Tables*, Ardsley, N.Y.: Geigy Pharmaceuticals.

Dietrich, W.F., Katz, H., Lincoln, S.E., Shin, H.-S., Friedman, J., Dracopoli, N. and Lander, E.S. (1992) A genetic map of the mouse suitable for typing intraspecific crosses. *Genetics*, 131: 423–447.

Dietrich, W.F., Miller, J., Steen, R., Merchant, M.A., Damron-Boles, D., Husain, Z., Dredge, R., Daly, M.J., Ingalls, K.A., O'Connor, T.J., Evans, C.A., DeAngelis, M.M., Levinson, D.M., Kruglyak, L., Goodman, N., Copeland, N.G., Jenkins, N.A., Hawkins, T.L., Stein, L., Page, D.C., and Lander, E.S. (1996) A comprehensive genetic map of the mouse genome. *Nature*, 380: 149–152.

Erway, L.C., Willott, J.F., Archer, J.R. and Harrison, D.E. (1993) Genetics of age-related hearing loss in mice: I. Inbred and F1 hybrid strains. *Hearing Res.*, 65: 125–132.

Fletcher, C.F., Lutz, C.M., O'Sullivan, T.N., Shaughnessy Jr, J.D., Hawkes, R., Frankel, W.N., Copeland, N.G. and Jenkins, N.A. (1996) Absence epilepsy in tottering mutant mice is associated with calcium channel defects. *Cell*, 87: 607–617.

Frankel, W.N., Stoye, J.P., Taylor, B.A. and Coffin, J.M. (1990) A linkage map of endogenous murine leukemia proviruses. *Genetics*, 124: 221–236.

Friedman, J.M., Leibel, R.L. and Bahary, N. (1991) Molecular mapping of obesity genes. *Mamm. Genome*, 1: 130–144.

Gibson, F., Walsh, J., Mburu, P., Varela, A., Brown, K.A., Antonio, M., Beisel, K.W., Steel, K.P. and Brown, S.D. (1995) A type VII myosin encoded by the mouse deafness gene shaker-1. *Nature*, 374: 62–64.

Green, E.L. (1981) *Genetics and probability in animal breeding experiments*. New York: Oxford University Press.

Green, M.C. (1963) Methods for testing linkage. In: W. J. Burdette (Ed.), *Methodology in Mammalian Genetics*, San Francisco: Holden-Day, pp. 56–82.

Green, M.C. and Sidman, R.L. (1962) Tottering — a neuromuscular mutation in the mouse. *J. Hered.*, 53: 79–94.

Haldane, J.B.S. (1919) The combination of linkage values, and the calculation of distances between the loci of linked factors. *J. Genet.*, 8: 299–309.

Hamilton, B.A., Frankel, W.N., Kerrebrock, A.W., Hawkins, T.L., FitzHugh, W., Kusumi, K., Russell, L.B., Mueller, K.L., Van Berkel, V., Birren, B.W., Kruglyak, L. and Lander, E.S. (1996) Disruption of the nuclear hormone receptor RORalpha in staggerer mice. *Nature*, 379: 736–739.

Hamilton, B.A., Smith, D.J., Mueller, K.L., Kerrebrock, A.W., Bronson, R.T., Van Berkel, V., Daly, M.J., Kruglyak, L., Reeve, M.P., Nemhauser, J.L., Hawkins, T.L., Rubin, E.M. and Lander, E.S. (1997) The vibrator mutation causes neurodegeneration via reduced expression of PITP alpha: positional complementation cloning and extragenic suppression. *Neuron*, 18: 711–722.

Herrup, K.X. and Wilczynski, S.L. (1982) Cerebellar cell degeneration in the leaner mutant mouse. *Neuroscience*, 7: 2185–2196.

Himmelbauer, H. and Silver, L.M. (1993) High-resolution comparitive mapping of mouse chromosome 17. *Genomics*, 17: 110–120.

Hunter, K.W., Riba, L., Schalkwyk, L., Clark, M., Resenchuk, S., Beeghly, A., Su, J., Tinkov, F., Lee, P., Ramu, E., Lehrach, H. and Housman, D. (1996) Toward the construction of integrated physical and genetic maps of the mouse genome using interspersed repetitive sequence PCR (IRS-PCR) genomics. Genome Res. *6*, 290–299.

Julier, C., De Gouyon, B., Georges, M., Guenet, J.L., Nakamura, Y., Avner, P. and Lathrop, G.M. (1990) Minisatellite linkage maps in the mouse by cross-hybridization with human probes containing tandem repeats. *Proc. Natl. Acad. Sci. USA*, 87: 4585–4589.

Kaushik, N. and Stoye, J.P. (1994) Intracisternal A-type particle elements as genetic markers: detection by repeat element viral element amplified locus-PCR. *Mamm. Genome*, 5: 688–695.

King, D.P., Zhao, Y., Sangoram, A.M., Wilsbacher, L.D., Tanaka, M., Antoch, M.P., Steeves, T.D., Vitaterna, M.H., Kornhauser, J.M., Lowrey, P.L., Turek, F.W., and Takahashi, J.S. (1997). Positional cloning of the mouse circadian clock gene. Cell, 89: 641–653.

Kosambi, D.D. (1944) The estimation of map distances from recombination values. *Ann. Eugen.*, 12: 172–175.

Lander, E.S., Green, P., Abrahamson, J., Barlow, A., Daly, M.J., Lincoln, S.E. and Newberg, L. (1987) MAPMAKER: an interactive computer package for constructing primary genetic linkage maps of experimental and natural populations. *Genomics*, 1: 174–181.

Lawrie, N.M., Tease, C. and Hulten, M.A. (1995) Chiasma frequency, distribution and interference maps of mouse autosomes. *Chromosoma*, 140: 308–314.

Lem, J., Flannery, J.G., Li, T., Applebury, M.L., Farber, D.B., and Simon, M.I. (1992) Retinal degeneration is rescued in transgenic *rd* mice by expression of the cGMP phosphodiesterase beta subunit. *Proc. Natl. Acad. Sci, USA*, 89: 4422–4426.

Letts, V.A., Felix, R., Biddlecome, G.H., Arikkath, J., Mahaffey, C.L., Valenzuela, A., Bartlett II, F.S., Mori, Y., Campbell, K.P. and Frankel, W.N. (1998) The mouse stargazer gene encodes a neuronal Ca2+ channel g subunit. *Nat. Genet.*, 19: 340–347.

Letts, V.A., Schork, N., Copp, A.J., Bernfield, M. and Frankel, W.N. (1995) A curly-tail modifier locus, *mct1*, on mouse Chromosome 17. Genomics *29*, 719–729.

Litt, M. and Luty, J.A. (1989) A hypervariable microsatellite revealed by in vitro amplification of a dinucleotide repeat within the cardiac muscle actin gene. *Am. J. Hum. Genet.*, 44: 397–401.

Lueders, K.K., Frankel, W.N., Mietz, J.A. and Kuff, E.L. (1993) Genomic mapping of intracisternal A-particle proviral elements. *Mamm. Genome*, 4: 69–77.

Malo, D., Vogan, K., Vidal, S., Hu, J., Cellier, M., Schurr, E., Fuks, A., Bumstead, N., Morgan, K. and Gros, P. (1994) Haplotype mapping and sequence analysis of the mouse *Nramp* gene predict susceptibility to infection with intracellular parasites. *Genomics*, 23: 51–61.

Manly, K.F. (1993) A Macintosh program for storage and analysis of experimental genetic mapping data. Mamm. Genome, 4: 303–313.

Mohlke, K.L., Nichols, W.C., Westrick, R.J., Novak, E.K., Cooney, K.A., Swank, R.T. and Ginsburg, D. (1996) A novel modifier gene for plasma von Willebrand factor level maps to distal mouse chromosome 11. *Proc. Natl. Acad. Sci. USA*, 93: 15352–15357.

Morgan, J.G., Dolganov, G.M., Robbins, S.E., Hinton, L.M. and Lovett, M. (1992) The selective isolation of novel cDNAs encoded by the regions surrounding the human interleukin 4 and 5 genes. *Nucleic Acids Res.*, 20: 5173–5179.

Mullick, A., Groulx, N., Trasler, D. and Gros, P. (1995) *Nhlh1*, a basic helix-loop-helix transcription factor, is very tightly linked to the mouse looptail (*Lp*) mutation. *Mamm. Genome*, 6: 700–704.

Noebels, J.L. (1986) Mutational analysis of inherited epilepsies. In: A.V. Delgado-Escueta, A.A. Ward, D.M. Woodbury and R.J. Porter (Eds.), *Basic Mechanisms of the Epilepsies: Molecular and Cellular Approaches*, Raven Press, New York, pp. 97–113.

Patil, N., Cox, D.R., Bhat, D., Faham, M., Myers, R.M., and Peterson, A.S. (1995) A potassium channel mutation in weaver mice implicates membrane excitability in granule cell differentiation. *Nature Genet.*, 11: 126–129.

Reed, P.W., Davies, J.L., Copeman, J.B., Bennett, S.T., Palmer, S.M., Pritchard, L.E., Gough, S.C.L., Kawaguchi, Y., Cordell, H.J., Balfour, K.M., Jenkins, S.C., Powell, E.E., Vignal, A. and Todd, J.A. (1994) Chromosome-specific microsatellite sets for fluorescence-based, semi-automated genome mapping. *Nat. Genet.*, 7: 390–395.

Reeves, R.H., Crowley, M.R., Moseley, W.S. and Seldin, M.F. (1991) Comparison of interspecific to intersubspecific backcrosses demonstrates species and sex differences in recombination frequency on mouse chromosome 16. *Mamm. Genome*, 1: 158–164.

Rise, M.L., Frankel, W.N., Coffin, J.M. and Seyfried, T.N. (1991) Genes for epilepsy mapped in the mouse. *Science*, 253: 669–673.

Robert, B., Barton, P., Minty, A., Daubas, P., Weydert, A., Bonhomme, F., Catalan, J., Chazottes, D., Guenet, J.-L. and Buckingham, M. (1985) Investigation of genetic linkage between myosin and actin genes using an interspecific mouse back-cross. *Nature*, 314: 181–183.

Roderick, T.H., Hillyard, A.L., Maltais, L.J. and Blake, C.S. (1996) Recombination percentages and chromosomal assignments. In: M.F. Lyon, M.F. Rastan and S.D.M. Brown (Eds.), *Genetic Variants and Strains of the Laboratory Mouse*, Oxford University Press, Oxford, pp. 929–1255.

Routman, E.J. and Cheverud, J.M. (1995) Polymorphism for PCR-analyzed microsatellites between the inbred mouse strains LG and SM. *Mamm. Genome*, 6: 401–404.

Rowe, L.B., Nadeau, J.H., Turner, R., Frankel, W.N., Letts, V.A., Eppig, J.T., Ko, M.S.H., Thurston, S.J. and Birkenmeier, E.H. (1994) Maps from two interspecific backcross DNA panels available as a community genetic mapping resource. *Mamm. Genome*, 5: 253–274.

Schwarz, H. (1997) Rapid high-throughput purification of genomic DNA from mouse and rat tails for use in transgenic testing. Technical Tips Online *(http://www.elsevier.com/locate/tto) T01146.*

Segre, J.A., Nemhauser, J.L., Taylor, B.A., Nadeau, J.H. and Lander, E.S. (1995) Positional cloning of the nude locus: genetic, physical, and transcription maps of the region and mutations in the mouse and rat. *Genomics*, 28: 549–559.

Shiroishi, T., Sagai, T. and Moriwaki, K. (1993) Hotspots of meiotic recombination in the mouse major histocompatibility complex. *Genetica*, 88: 187–96.

Sidman, R.L., Dickie, M.M. and Appel, S.H. (1964) Mutant mice (quaking and jimpy) with deficient myelination in the central nervous system. *Science*, 144: 309–311.

Steel, K.P. (1995) Inherited hearing defects in mice. *Ann. Rev. Genetics*, 29: 675–701.

Taylor, B.A., Navin, A. and Phillips, S.J. (1994) PCR-amplification of simple sequence repeat variants from pooled DNA samples for rapidly mapping new mutations of the mouse. *Genomics*, 21: 626–632.

Taylor, B.A. and Phillips, S.J. (1996). Detection of obesity QTLs on mouse Chromosomes 1 and 7 by selective DNA pooling. *Genomics*, 34: 389–398.

Taylor, B.A., Rowe, L. and Grieco, D.A. (1993) The MEV mouse linkage testing stock: mapping 30 novel proviral insertions and establishment of an improved stock. *Genomics*, 16: 380–394.

Vidal, S., Tremblay, M.L., Govoni, G., Gauthier, S., Sebastiani, G., Malo, D., Skamene, E., Olivier, M., Jothy, S. and Gros, P. (1995) The Ity/Lsh/Bcg locus: natural resistance to infection with intracellularparasites is abrogated by disruption of the *Nramp1* gene. *J. Exp. Med.*, 182: 655–66.

Vidal, S.M., Malo, D., Vogan, K., Skamene, E. and Gros, P. (1993) Natural resistance to infection with intracellular parasites: isolation of a candidate for *Bcg*. *Cell*, 73: 469–485.

Wang, D.G., Fan, J.B., Siao, C.J., Berno, A., Young, P., Sapolsky, R., Ghandour, G., Perkins, N., Winchester, E., Spencer, J., Kruglyak, L., Stein, L., Hsie, L., Topaloglou, T., Hubbell, E., Robinson, E., Mittmann, M., Morris, M.S., Shen, N., Kilburn, D., Rioux, J., Nusbaum, C., Rozen, S., Hudson, T.J., Lipshutz, R., Chee, M. and Lander, E.S. (1998) Large scale identification, mapping, and genotyping of single-nucleotide polymorphisms in the human genome. *Science*, 280: 1077–1082.

Wehner, J.M., Radcliffe, R.A., Rosmann, S.T., Christensen, S.C., Rasmussen, D.L., Fulker, D.W. and Wiles, M. (1997) Quantitative trait locus analysis of contextual fear conditioning in mice. *Nat. Genet.*, 17: 331–334.

W.E. Crusio and R.T. Gerlai (Eds.)
Handbook of Molecular-Genetic Techniques for Brain and
Behavior Research (Techniques in the Behavioral and Neural Sciences, Vol. 13)

CHAPTER 2.1.2

Experimental strategies for quantitative trait loci (QTL) analysis in laboratory animals

David A. Blizard[1] and Ariel Darvasi[2]

[1]*Center for Developmental and Health Genetics and Intercollege Graduate Program in Genetics, The Pennsylvania State University, University Park, PA 16802, USA*
[2]*Smith, Kline Beecham Pharmaceuticals, Harlow, Essex CM19 5AW, UK*

Introduction

Gene-mapping can be conducted in any species with a defined breeding structure and where there is a supply of genetic markers distributed throughout the genome. However, there are considerable advantages to using laboratory rats and mice. These include an abundant supply of inbred strains, extensive knowledge of phenotypic variation among them, as well as a large and commercially available supply of microsatellite markers distributed throughout the genome of both species. In addition, their biology has been found to be similar to other mammals including humans. Readers unfamiliar with either species are referred to widely consulted texts which provide invaluable, background information on their husbandry, biology, genetics and diseases (Baker et al., 1979, 1980; Green, 1966; Foster et al., 1981, 1982; Silver, 1995).

Our choice to organize our discussion around these two species reflects the versatility engendered by these genetic resources and also the fact that the bulk of neuroscientific research has used rats and mice as experimental subjects.

Evidence that genes contribute to variation in a phenotype

Before embarking on an enterprise which is both time-consuming and costly, a neuroscientist must first consider whether genes play an important role in the selected phenotype. A principal source of evidence are studies which illustrate significant differences between inbred strains because these provide *prima facie* evidence that genes contribute to variation in a trait (Broadhurst, 1978). Although neurobiological variation is for the most part covert, strain differences in selected behavioral phenotypes provide very useful indirect evidence about variation in neural processes.

Literature on strain differences in rats and mice

Aside from searching the literature for specific citations, useful references, particularly to research on laboratory mice, may be found in behavior genetic texts (Fuller and Thompson, 1960, 1978), monographs (Broadhurst, 1978), reviews (Crawley et al., 1997) and volumes focused on specific research domains (Sprott, 1975; Gershon et al., 1981; Lieblich, 1982; Crabbe and Harris, 1991; Driscoll, 1992; Goldowitz et al., 1992).

Bi-directional selection for differences in emotional behavior has been a feature of behavior genetic research on the laboratory rat. Compilations of the differences between some of these strain pairs may be found in Broadhurst (1975), Blizard (1981), and Fujita et al. (1994). Reviews of the findings on larger numbers of strains may be found in Broadhurst (1960, 1978) and Robinson (1965). Harrington (1981) also described phenotypic variation between 12 widely used rat strains on a large battery of behavioral tests. The value of the latter observations is enhanced because they were obtained from animals which were maintained in a well established colony and therefore experienced similar rearing and maintenance conditions. Festing's extensive compilations (e.g. Festing, 1979) of the phenotypic differences between inbred strains of rats and mice which have provided a vital record for many years are now being continually updated and may be consulted at http://www.informatics.jax.org.

Principles guiding strain surveys of new phenotypes

If evidence is lacking regarding the phenotype of interest then a comparison of several strains can be undertaken to provide that information. In planning such a screen, it is important to select strains which have different genetic origins to optimize the chances of finding genetic influences on the phenotype. Taylor identified the C57BL/6J (B6) strain as a genetic outlier compared with several other inbred strains of laboratory mice (Taylor, 1972). As a consequence, this strain has often been used in crosses to map genes and in the development of resources such as recombinant-inbred strains (Taylor, 1978, 1989, 1996). The derivation of existing strains of laboratory mice may also provide a useful guide to selecting appropriate strains for comparison and has been summarized in a number of places (Staats, 1981; Morse, 1981).

The origins of some of the principal rat strains has been described by Lindsey (1979) and Festing (1979). Analogous to the B6 mouse strain, the BN rat strain, which received a significant genetic contribution from wild rats (Billingham and Silvers, 1959), is a genetic outlier and is therefore an excellent candidate for inclusion in a phenotypic screen.

An empirical method to ensure genetic diversity is to use microsatellite polymorphism rate as a basis for selecting strains for screening. Dietrich et al. (1992) typed polymorphisms in more than 300 microsatellites across 10 inbred strains of laboratory mice and inbred strains of *Mus musculus castaneus* and *Mus spretus*. The inter-strain polymorphism rate among 9 of the strains is illustrated in Table 1 as well as the average polymorphism rate for each strain. Consistent with previous estimates of genetic relationships between strains, the B6 emerged as the most polymorphic, while strains known to share a common ancestry, e.g. DBA/2J (D2) and C3H/HeJ, differed to a lesser degree. Choosing strains for study on the basis of their polymorphism rate also has the practical advantage of providing access to ample numbers of polymorphic markers. This can be particularly important at the fine-mapping stage (see below) when the number of markers within a particular chromosomal region may be restricted. The polymorphism rates between *Mus mus castaneus*, *Mus spretus* and the other inbred strains are in excess of 90%. This degree of diversity can also be exploited during the fine-mapping stage.

In the rat, typing of more than 5,000 microsatellite polymorphisms in 48 inbred rat strains has been completed by Jacob and colleagues and polymorphism rates between pairs of strains can be assessed at http://waldo.wi.mit.edu/rat/public. A sample of polymorphism rates among 8 commonly used rat strains is provided in Table 2. As can be seen, two of the strains, BN/Ss/N and WKY/N, have exceptionally high polymorphism rates but the other strains also differ from each other by more than 50%. Although

TABLE 1

Microsatellite polymorphism rates (%) in nine inbred mouse strains (data from Dietrich et al. 1992)

	B6	DBA	A	C3H	BALB	AKR	NON	NOD	LP
B6	–								
D2	51.40	–							
A	52.70	45.80	–						
C3H	50.50	34.80	35.10	–					
BALB	49.30	45.20	31.60	38.00	–				
AKR	52.40	48.30	46.20	43.90	42.90	–			
NON	49.30	50.80	51.20	46.70	47.10	53.60	–		
NOD	54.50	53.60	51.20	48.10	51.00	51.00	43.90	–	
LP	57.40	53.40	54.50	53.80	49.30	56.70	55.40	55.90	–
MEAN	52.19	47.91	46.04	43.86	44.30	49.38	49.75	51.15	54.55

The data are based on comparisons of more than 300 microsatellites. The bottom row represents the mean polymorphism rate of the strain at the head of the column compared with the other 8 strains.
B6 = C57BL/6J, D2 = DBA/2J, A = A/J, C3H = C3H/HeJ, BALB = BalB/cJ, AKR = AKR/J, NON = NON/Lt, NOD = NOD/mrkTacBr, LP = LP/J.

TABLE 2

Microsatellite polymorphism rates (%) in 8 inbred rat strains. Data obtained from http://waldo.wi.mit.edu/rat/public

	ACI/N	BN/Ss/N	BUF/N	F344/N	M520/N	WN/N	MR/N	WKY/N
ACI/N	–							
BN/Ss/N	74	–						
BUF/N	58	71	–					
F344/N	57	71	49	–				
M520/N	57	71	50	43	–			
WN/N	62	71	56	54	55	–		
MR/N	61	71	55	55	56	55	–	
WKY/N	68	78	68	68	68	69	67	–
MEAN	62	72	58	57	57	60	60	69

The data are based on comparisons of more than 5,000 microsatellites typed by Jacob et al.
The bottom row represents the mean polymorphism rate of the strain at the head of the column compared with the other 7 strains

one cannot assume a one-to-one relationship between microsatellite polymorphism rates and variation in genes, judicious selection of a sub-set of strains of rats or mice using inter-strain polymorphism rate as a criterion would likely provide a sample with an excellent degree of heterogeneity.

Mapping with unfamiliar species

The amount of genetic variation in existing stocks of rats and mice appears to be more than adequate to confidently test for genetic influences when a new phenotype is studied. However, genetic diversity cannot be assumed for every laboratory species. Based on established records of capture and dissemination, the genetic origin of the Syrian hamster was believed to be narrow (Clark, 1987). Consistent with this, recent estimates show that inbred strains of laboratory hamsters exhibit polymorphism rates of between 5–6% (Okuizumi et al., 1997). Comparison among the existing inbred strains of hamsters will therefore provide only limited insight into the contribution of genes to phenotypic variation in this species. In light of this, before embarking on mapping studies with an unfamiliar species, an investigator should attempt to establish the origins of laboratory stocks, so that a judgment can be made about the degree of genetic diversity available for study.

Nature of the genetic architecture governing neurobiological processes

Simple genetic traits are those where a single gene determines the phenotype. In those instances a Mendelian segregation of 2 or 3 phenotypes is observed for the 3 possible genotypes at a specific locus. Complex traits are those where a 1 to 1 relationship between genotype and phenotype does not exist. They may be affected by a number (even hundreds) of genes, environmental factors and their possible interaction. Although the influence of single or major genes has been claimed for some behavioral characters (Collins and Fuller, 1968; Fuller, 1974), it is widely agreed that most neurobiological and behavioral phenotypes are influenced by a large number of biochemical, physiological and morphological processes and are most aptly considered under a complex or polygenic model (Frankel, 1995). In many instances, complex traits are quantitative in nature or can be measured in a quantitative manner. Therefore, theoretical aspects are often analyzed under a quantitative model and the genes studied are termed quantitative trait loci (QTL; Gelderman, 1975).

QTL mapping as a multi-stage procedure

The common paradigm used in such cases is a genome-wide scan utilizing anonymous markers. This needs to be tested in a segregating population so that a marker associated with the trait will indicate position on a specific chromosome. We will now describe the different strategies available for the identification of QTL. It is important to note that QTL analysis is a multi-stage procedure and that different approaches are required at the various stages of the enterprise. The first stage, QTL detection, tests the hypothesis that a marker or a set of markers is linked to the QTL. Sax (1923) was the earliest experiment to show association between a qualitative (seed-coat pigmentation) and a quantitative (bean size) trait. Since then, many examples have followed (Thoday, 1961; Thompson and Thoday, 1979).

The second stage, commonly done with the same data, estimates the position of the QTL on the chromosome where the QTL was detected. This procedure is conceptually different, involving parameter estimation (as opposed to hypothesis-testing for the first stage). Unless the number of animals in the cross is very large, or, unless special efforts have been made to increase recombination frequency in the study population, the second stage usually provides limited information as to the position of the gene (Darvasi et al., 1993. Therefore, a third stage of fine mapping will be necessary. In the following sections, we analyze the three stages separately.

QTL detection

Two experimental populations are classically used for QTL detection, an intercross (F_2) or backcross (BC). Figure 1 presents a schematic description of a BC population where only two genotypes are present for each locus. It is assumed that parents from two inbred lines are crossed to produce an F_1, parent 1 having MM and QQ genotypes at a marker and a linked QTL respectively; and parent 2 having mm and qq genotypes at the same sites. The F_1 offspring will all have Mm and Qq genotypes at the marker and QTL. The F_1 are crossed to P_2 to produce the segregating BC population. One chromosome of the BC will always have an m and q allele derived from P_2. The other chromosome will have one of 4 possible haplotypes: M-Q, M-q, m-Q and m-q derived from the F_1. The expected frequencies of the haplotypes are respectively $(1-r)/2$, $r/2$, $r/2$ and $(1-r)/2$, where r is the proportion of recombination between the marker and the QTL. The Q/q polymorphism cannot be directly tested since it is unknown. Therefore, the analysis is based on testing the M/m polymorphism. A test of significance can be done to test if the group of individuals with the M genotype differ significantly in their trait value from the group of individuals with the m genotype.

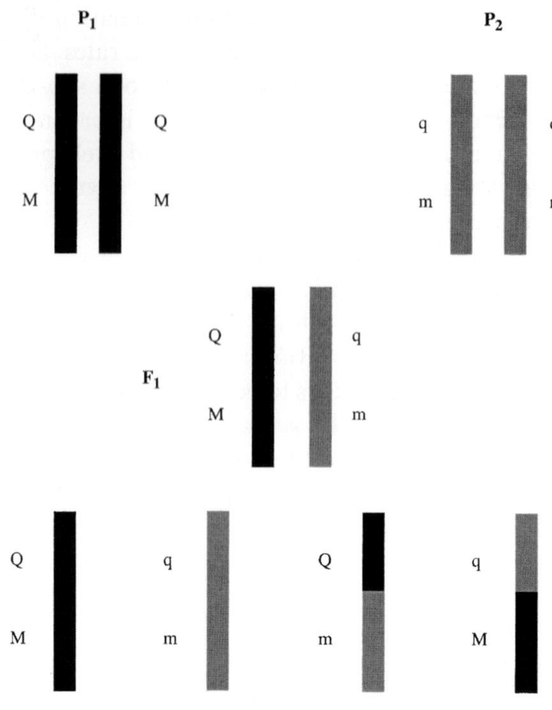

Fig. 1. Schematic to illustrate relationship between marker (M) and QTL (Q) in a standard backcross design. M and Q represent genotypes at a marker and linked QTL in parent 1 and m and q differential genotypes in parent 2. The F_1 which has Mm and Qq at marker and QTL sites respectively is crossed to P_2 to produce the segregating B_2 population. Thus one chromosome (originating from P_2) will always have m and q alleles. The other chromosome will have one of the 4 haplotypes shown with the frequency of each haplotype reflecting the proportion of recombination between marker and QTL (see text).

Soller et al. (1976) developed the theoretical aspects of this approach and provided the foundation of many subsequent improvements and refinements. Among the latter was the introduction of interval mapping (Lander et al., 1987; Lander and Botstein, 1989). Instead of looking at one marker at a time, one can look at pairs of markers and use maximum likelihood techniques to test for significance. A number of additional improvements and strategies for more cost-effective analysis have been suggested. When studying multiple correlated traits one can increase power by controlling residual variance (Korol et al., 1995; Changjian and Zeng, 1995). Selective genotyping is also frequently used to reduce the sample size that is genotyped by selecting the phenotypic extremes which carry most of the statistical power for QTL detection (Lebowitz et al., 1987; Lander and Botstein, 1989; Darvasi and Soller, 1992). Berretini et al. (1994) used this procedure to map 3 QTL which contributed to phenotypic variation in morphine preference in a backcross between B6 and D2 mice. Selective DNA pooling (Darvasi and Soller, 1994) can even further reduce genotyping by pooling the DNA of the phenotypic extremes and estimating allele frequencies from the DNA pools. Taylor and Phillips (1996) pooled DNA from the phenotypic extremes of a cross between EL/Suz and 129/Sv to map obesity QTL on chromosomes 1 and 7. Sequential sampling can increase efficiency by recurrently sampling and testing until significance is achieved (Motro and Soller, 1993). Following up the previous work of Melo et al. (1997), Peirce et al. (1998) used sequential sampling of backcross mice to increase their ability to detect alcohol preference QTL in the B6/D2 lineage. Finally, optimum marker spacing (Darvasi and Soller, 1994) can reduce overall costs of an experiment by optimally choosing markers. Generally 2-3 markers per chromosome will be all that is required for an initial screen.

When considering F_2 vs BC for QTL detection, it is important to define the precise purposes of the research. If it is desired to have a "general picture", that is, the number of QTL segregating and estimates of their additive and dominance effects, an F_2 is preferred. On the other hand, if one's objective is to detect at least some of the major QTL most efficiently, a BC should be used. For additive effects, an F_2 requires about 30% less progeny than BC. For dominance effects, however, a BC requires about half of an F_2 population. The source of the advantage for the BC is that the threshold statistic required for statistical significance is lower than for F_2 (Lander and Schork, 1994; Lander and Kruglyak, 1995) and the gene effect is relatively higher due to the reduction in residual genetic variance. Genetic variance caused by gene interactions is also expected to be reduced, making the BC design even more powerful (Silver, 1995). Two studies of 2-bottle alcohol preference illustrate the merits of the two approaches. Melo et al. (1996) and Peirce et al. (1998) used reciprocal backcrosses between B6/D2 F_1s and B6, and Tarantino et al. (1998) an F_2 intercross between B6 and D2 strains, to investigate the QTL architecture underlying 2-bottle alcohol preference. As discussed by Peirce et al. (ibid), many of the differences in the results obtained in the two studies reflect the fact that their crossing scheme could not detect B6-dominant alleles and had less power to detect B6 partially dominant alleles than the F_2 intercross employed by Tarantino et al. (ibid). The reader is referred to the original publications for detailed comparison of the two designs as well as discussion of the persuasive evidence for the gender-specificity underlying some of the QTL.

Recombinant inbred (RI) strains (Bailey, 1971; Taylor, 1978) have also been considered for QTL detection (Plomin et al., 1991). They can be conveniently viewed as a genetically fixed population of mice composed of homozygotes at all loci and which, compared to a standard F_2, are characterized by an enhanced frequency of recombination. They have the added advantage that RI sets have usually been typed on a large number of markers thus obviating the need for a researcher to carry out genotyping. Nevertheless, for most practical cases RI strains will be less powerful than BC or F_2 because of the limited number of strains in each set. If, however, a QTL explains a significant proportion of the genetic variation and the genetic variation itself is small compared to trait variation, RI strains can be considerably more efficient than BC or F_2 (Belknap et al., 1996). Although not widely recognized, RI strain means also reflect to an unknown degree the contribution of maternal factors, i.e. offspring are raised in the idiosyncratic environment provided by females in each RI strain. Thus, the possible contribution of maternal factors in RI analyses should be considered.

Although the genetic architecture (number of genes, range of effect sizes, nature of dominance relations at individual loci) is usually not available when planning mapping studies of new phenotypes, provisional assumptions must be made about them in order to calculate sample size and carry out power calculations. Plant geneticists have found that a relatively small number of genes (about 10) each account for a large proportion of the phenotypic variance (20–90%) in several different traits in rice and sorghum, with large numbers of genes (about 60) accounting for between 4–12% (Paterson, 1995). QTL studies of behavioral and physiological traits that have been carried out to date have also found relatively few QTL which account for large amounts of the phenotypic variance (Bachmanov et al., 1997; Blizard et al., 1999) but much larger numbers of QTL which account for between 5–15% of the phenotypic variation (Flint et al., 1995; Calderone et al., 1997; Wehner et al., 1997; Tarantino et al., 1998; Peirce et al., 1998). Although each neurobehavioral process will have its own genetic idiosyncrasies, it seems reasonable to expect that mapping experiments with adequate power to detect QTL within the same range of effect sizes will result in a productive QTL yield.

Simple tests of significance provide the most straightforward approach to QTL detection. Product-moment correlation coefficients calculated between number of alleles from one of the parental strains and the phenotypic score is a convenient way to assess QTL relationships with a large number of markers simultaneously. One-way analysis of variance using allele number as the different levels of the grouping factor is also appropriate. Whatever method is used, the level of significance should be adjusted by a standard procedure to provide protection against Type 1 errors (Lander and Kruglyak, 1995).

QTL mapping

Following QTL detection, the position of the QTL needs to be estimated. The popular Mapmaker/QTL software package has been most frequently used for this purpose and does an excellent job in estimating map position based on a maximum likelihood procedure (Lander et al., 1987). The procedure scans each chromosome looking at pairs of markers at a time. This procedure, termed interval mapping, assigns a probability value (LOD score) for the QTL occupying specific, regularly spaced points between the two markers. These LOD scores values are plotted and the obtained result is a LOD-score profile. The point with the highest LOD-score value is the point where the estimation procedure located the QTL. In parameter estimation theory, there are two important considerations. The first relates to whether the estimate is biased or unbiased. An unbiased estimate is deemed to have been made if the true value is obtained following the averaging of an infinite number of similar experiments. Interval mapping is a maximum likelihood procedure and is therefore asymptotically unbiased. The second issue concerns the distribution of estimates of QTL position, or, more specifically, the confidence interval (CI) of the QTL position. The CI has a defined probability to contain the gene (commonly, a 95% CI is used). Estimating the exact CI from the maximum likelihood procedure is a complex procedure. Therefore, Lander and Botstein (1989) suggested an approximation defined by the points on either side of the QTL peak where the LOD-score was 1 LOD score lower (usually called the 1-LOD support interval). Van Ooijen indicated that this method is not a consistent representation of a CI and in many cases at least a 2-LOD support interval is required. The limitations of the 1-LOD support interval as an estimate of the 95% C.I. were recently amply demonstrated by Bennett and Johnson (1998). Using markers flanking several F_2-derived 1-LOD support intervals to construct congenics targeted at QTL involved in the hypnotic influence of ethanol, 75% of the congenics failed to exhibit the predicted difference in phenotype, presumably because the QTL were outside the targeted range. Several other methods have been suggested to estimate the CI (Mangin et al., 1994; Visscher et al., 1996; Darvasi and Soller, 1997). For an a priori estimate of the accuracy expected, Darvasi and Soller (1997) provide a simple approximation:

$$\mathrm{CI} = k/Nd^2$$

Where k is a constant depending on experimental design and level of confidence required, N is sample size and d is the standardized QTL effect which also depends on experimental design.

As for QTL detection, BC provides greater accuracy for estimating the position of QTL with dominant effect whereas F_2 is preferred for estimation of QTL with additive effects. Larger sample size and larger QTL effect sizes increase power for both detection and mapping accuracy. The main parameter that differentially affects them is the amount of recombination present in the tested population. It is easier to detect QTL in situations where recombination frequency is low because markers will still be linked to the QTL even when they are at a considerable distance away. Mapping accuracy, however, is facilitated by increased recombination frequency: only markers which are closely linked to a QTL will detect an effect and therefore a smaller interval will be defined.

Advanced intercross (AI) lines were proposed as a means of systematically increasing the density of recombinations to facilitate mapping (Darvasi and Soller, 1995). AI lines are produced by intercrossing two parental strains to produce a standard F_2 generation. Starting with the F_2, animals are semi-randomly intercrossed to produce successive generations, the term semi-random being preferred because matings between relatives are systematically avoided. A parental population of 50 males and 50 females are maintained in each generation to minimize inbreeding. Each generation accumulates recombination at a rate of $r_t = [1-(1-r)^{t-2}(1-2r)]/2$, where r_t is the proportion of recombinants at the tth generation and r is the initial proportion of recombination. Using AI lines for QTL mapping will reduce the CI by a factor of approximately $t/2$. A recent study (Iraqi et al., 1998) successfully used the AI approach to fine map trypanosomiasis-resistance QTL. F_6 mice occupying the opposing extremes of response to trypanosomiasis infection were used to reduce the CIs associated with 3 QTL from 20–40 cMs (in F_2s) to 5–10 cMs which is approximately the degree of reduction predicted by Darvasi and Soller (1995).

AI lines may be particularly suitable in the presence of a genetic architecture characterized by many genes each with a small effect size because these can be mapped in a single cross. The presence of a QTL with a large effect size, however, may interfere with precise localization of QTL with small effect on other chromosomes unless the position of the former is localized with enough precision to permit appropriate statistical adjustment. Although the AI approach engenders great versatility, the methods described in the next section which approach fine-mapping via analysis of recombination defined intervals (RDI) may ultimately be required before embarking on positional cloning.

Fine mapping

Standard QTL mapping in F_2 and BC samples will usually locate genes to a large chromosomal interval. To identify the QTL, greater accuracy and greater confidence is necessary. Fine mapping, as promulgated in this section, consists of shifting from a genome-wide paradigm to a single-QTL paradigm. A QTL that has been previously detected and preferably confirmed will now be the focus of subsequent study. Darvasi (1998) described a series of strategies which will be presented below.

Selective phenotyping

This strategy may be especially attractive to neuroscientists who are involved in measurements of phenotypes e.g. electrophysiological recordings, which are arduous and time-consuming. A large F_2 or BC population is produced and only individuals recombinant at an interval previously defined to contain a QTL are selected for phenotyping. This strategy is based on the rationale that once a gene is mapped to a given interval only recombinant individuals within that interval contribute to further mapping accuracy. Considerable economies can be achieved by genotyping individual animals early in life and then retaining only those that qualify (as recombinants) for phenotyping as adults. Theoretically, selective phenotyping can be done in a sequential manner. That is, once an interval is defined to contain a QTL, recombinants within that interval are phenotyped. Any number of recombinants phenotyped will reduce the width of the QTL-containing interval and subsequently a smaller interval can be considered to look for new recombinants. For practical reasons, however, probably not more than two steps will be applied. At each stage of selective phenotyping, the total number of animals phenotyped is reduced by a factor of $1/2r(1-r)$ for an F_2 population and by $1/r$ for a BC population (r being the proportion of recombination between the markers bracketing the interval in question). With selective phenotyping, savings are in phenotyping only, the total number of animals produced being equal to that necessary with an F_2 or a BC.

Recombinant progeny testing

Individuals carrying a distinguishable recombinant chromosome at the region of interest are crossed to one of the parental strains to determine the location of the QTL relative to the recombination point. Reducing CI from y cM to x cM will require y/x recombinant individuals, each with a recombination at one of the y/x intervals covering the initial y-cM interval. The expected number of F_2 animals that will be screened, N_s, to detect these y/x recombinant individuals can be approximated as (Darvasi, 1997):

$$N_s = \frac{50}{x} \sum_{i=1}^{y/x} \frac{1}{i}$$

For example, if $y=25$ and $x=5$ then $N_s=23$, or if $y=5$ and $x=1$ then $N_s=114$. Moen et al. (1996) used the pattern of recombination present in a recombinant congenic strain to identify two QTL which influence susceptibility to 1,2-dimethyl hydrazine-induced colon tumors on mouse chromosome 2 and to map one of them (Sccl) by recombinant-progeny testing to a 2.4 cM region around D4Mit66.

Interval specific congenic strains (ISCS)

Similar to recombinant progeny testing, N_s individuals are screened to detect y/x recombinant individuals with recombinations equally distributed within the y-cM interval. These animals, however, are now backcrossed a number of times to the background parental strain to eliminate alleles from the donor parental strain at all other QTL affecting the trait. Then, animals are intercrossed and homozygotes for the recombinant haplotype are selected to establish one ISCS. Selection at the DNA level is done with the aid of markers, reducing significantly the number of generations required with little additional genotyping (Darvasi, 1997).

Recombinant inbred segregation test (RIST)

RIST is a novel experimental design aimed to take advantage of the theoretical high mapping resolution present in RI strains and applying it to QTL mapping (Darvasi, 1998). To reduce the QTL containing interval from y cM to x cM, y/x RI strains are selected with recombinations equally distributed within the y-cM interval. In some cases, fewer RI strains will be necessary since a single RI strain may have more than one recombination in the region of interest. It is expected that for $x=1$ cM such strains will be found for 50% of chromosomal regions when a set of 25 RI strains is available. The RIST population is constructed as outlined in Fig. 2. The F_2 or BC populations are phenotyped and genotyped with few markers. The $F_{2,1}$ or BC_1 population is genotyped with markers located in the region where P_2 alleles are present in the selected RI strain, and the $F_{2,2}$ or BC_2 is genotyped with markers located in the region where P_1 alleles are present in the selected RI strain. Since the QTL has been previously mapped to this region it will necessarily segregate in one of the F_2 or BC populations but not in the other. The analysis of the two populations will detect the one in which the QTL is segregating and accordingly locate the QTL above or below the recombination point. The overlapping results of all RI strains selected

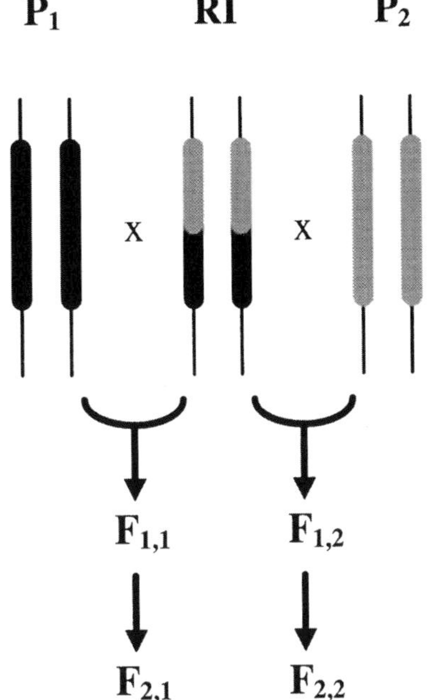

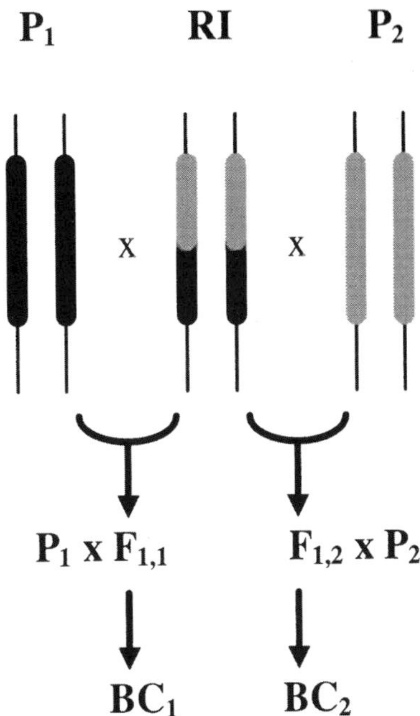

Fig. 2. Producing recombinant inbred segregation test (RIST) populations. P_1 and P_2 are two inbred strains that served as the parental strains of the recombinant inbred (RI) strains. A selected RI strain, with a recombinant haplotype in the region of interest is crossed to both parental strains to produce two separate F_1 populations, $F_{1,1}$ and $F_{1,2}$. Subsequently RIST-F_2 and RIST-BC populations are obtained through intercross and backross, respectively.

will locate the QTL to the desired interval. For additive effects RIST-F_2 is preferred. In that case, homozygous genotypes at the marker locus will contribute most of the information. Thus, only homozygous individuals will be selected for phenotyping. When a dominant effect is considered RIST-BC will be more efficient.

When considering which fine-mapping strategy to apply, one should look at a number of parameters. The effect size associated with the QTL, its dominant state, target accuracy required, availability of RI strains, time and money constraints, etc. will determine which strategy to use. Selective phenotyping is fast but not very cost efficient. Recombinant progeny testing is somewhat more efficient, in particular for dominant effects and it requires only one additional generation as compared to selective phenotyping. ISCS are cost effective but require a significant number of generations to produce. RIST is both cost effective and fast and limited only by the availability of appropriate RI strains.

Software for QTL analysis

As previously noted, the Mapmaker QTL program has been in existence for many years and has proven its reliability and versatility for QTL analysis. It has the capability of examining the distribution of sample data and, if these deviate from normality, providing a variety of data transformations before analysis. It can attempt to refine the statistical properties of a specific QTL by regressing on background variables known to be correlated with the phenotype (e.g. gender, weight). After the genetic architecture underlying a phenotype is established, the program also permits the experimenter to adjust for the influence of other QTL in specified positions elsewhere in the genome.

Several other programs e.g. Map QTL (Jensen, 1996) which carries out multiple QTL mapping have been developed to facilitate QTL analysis and provide important features such as the ability to perform permutation tests (Doerge and Churchill, 1996) so that significance levels can be based on the sample data and, in some cases, the practical convenience of a graphical user interface. It is beyond the scope of this chapter to review the various features of these programs in detail. However, this task has recently been carried out by Manly and Olson (1999) who, in providing a detailed introduction to Map Manager QT, compare and contrast the various features of nine other QTL analysis programs. To maintain familiarity with contemporary developments, the reader can also consult a Website (http://s27w007.pswfs.gov:80/qtl/software.html) which is dedicated to QTL- oriented software. Regardless of the software that is used Doerge et al. (1997) make the important point that the user must understand the underlying methodology, assumptions, as well as the specific implementation of these assumptions in the selected software. They also discuss QTL analyses of the same data set by Mapmaker and QTL Cartographer (Basten et al., 1997), another QTL analysis program in the public domain. Readers will also find the comprehensive treatment of QTL mapping by Liu (1998) a highly useful reference source."

Mapping with existing animal stocks

In the previous sections we described strategies that the experimentalist may adopt for a QTL mapping project. For the most part they consist of generating the entire population that one would need for the analysis. However, it would be wise to consider the use of existing stocks when developing a mapping strategy ab initio. Although they may not be the theoretically optimum population, they definitely have the advantage of being ready to use and may not require the sometimes lengthy generation procedure. Below we discuss some of the major types of resources available (or under development).

Consomics

Consomic strains are created by backcrossing individual chromosomes from a donor to a host strain. Until recently, they have been principally used to examine the role of the non-pairing region of the Y-chromosome (Y_{npr}) in biobehavioral processes. Hudgins et al. (1985) backcrossed the Y_{npr} of the BXSB strain to several different host strains to study autoimmunity and Maxson has pioneered the use of Y-chromosomal consomics to study the influence of the Y_{npr} on aggression (Maxson, 1996). More recently, in an attempt to develop a set of consomic strains which will encompass the entire genome, Nadeau (personal communication) is backcrossing individual A/J chromosomes to the B6 background by selection for appropriate microsatellite markers. An investigator may compare a consomic strain carrying a specific heterologous chromosome with the host strain if there is reason to suspect the presence of a gene on a specific chromosome, or conduct a whole genome scan by comparing the host strain with the entire set of 20 consomics (Nadeau et al., 1995).

As previously noted, the conceptually different stages of QTL detection and QTL mapping are often considered as one stage since in most cases the same population is used for both analyses. Consomics, however, can only locate the QTL to a specific chromosome. Therefore, they are excellent for QTL detection but without additional crosses have no use for QTL mapping.

Advanced intercross

AI lines were previously mentioned since their efficiency may justify their generation for a specific experiment. However, when the relevant stocks are already available, it obviously provides a significant additional advantage since the main drawback of the AI procedure, the lengthy period required to produce them, disappears and the advantage of fine mapping a number of QTL simultaneously can be realized. Colonies of advanced intercross mice derived from A/J and B6 parents have been developed to facilitate fine-mapping and were referred to previously (Iraqi et al., 1998).

Heterogeneous stocks (HS)

In addition to bilineally derived AI stocks, there exist stocks of mice and rats derived by systematic crossing of many inbred strains which have been maintained on a semi-random basis for many generations. McClearn et al. (1970) systematically intercrossed 8 inbred mouse strains to produce an HS stock which has now accumulated approximately 60 generations of intercrossing and a dense pattern of recombinations. It was recently used to map a QTL involved in emotional behavior on mouse chromosome 1 to an accuracy of less than 1 cM (Talbot et al., 1999).

A similar stock of rats designated the N/NIH heterogeneous stock derived from the 8 strains shown in Table 2 has been developed by Dr. Carl Hansen at the Small Animal Section at the National Institutes of Health (Li and Lumeng, 1984). So far it has not been used for QTL analysis but has been systematically intercrossed for approximately 20 generations (while maintaining a parental population of 60 pairs) and can be expected to greatly facilitate fine-mapping.

HS stocks of rats and mice derived from multiple inbred strains have the important additional advantage that they provide the opportunity to test polymorphisms from diverse ancestries thus ensuring a good measure of genetic heterogeneity. The observations of Talbot et al. (vide supra) were obtained by focussing on QTL whose location had been suggested by previous studies using standard crosses. However, it will be interesting to see if use of techniques such as DNA pooling of phenotypic extremes will permit HS stocks to be used for mapping of phenotypes whose QTL structure is undefined.

Recombinant-congenic strains (RCS)

RCS are derived by inbreeding strains after performing several independent backcrosses from a donor to a host strain (Demant and Hart, 1986). Those which have so far been developed each carry from 6.25–12.5% of donor genome alleles, depending on the degree of backcrossing which has been carried out before inbreeding, and sets of RCS strains (usually 20) are designed to sample 95% of the donor strain genome. Following identification of phenotypic differences in the relevant trait between the two progenitor strains, RCS studies usually continue with a comparison of an RCS set with the host strain. Discovery of a significant difference between host and individual RCS strains indicate the existence of one or more QTL residing on the donor chromosomal regions that have been fixed within that particular RCS strain. Correlation of the pattern of phenotypic differences across a set of RCS strains with the database of polymorphisms that have been typed in the RCS sets (Stassen et al., 1996; also available at http://www.informatics.jax.org) permits more specific hypotheses to be formulated about the regions containing the relevant polymorphisms. Usually, an additional intercross or backcross study will be necessary to assign a map location to the QTL (Moen et al., 1996). RCS strains have the added advantage that the greater homogeneity of their background will reduce genetic error variance and improve mapping accuracy as well as reducing the time needed to develop congenic strains.

Congenics

Congenic strains were originally developed by iterative backcrossing from a donor to a host strain accompanied by selection for major histocompatibility variants (Snell, 1948). Subsequent sets of congenics have been directed at both major and minor histocompatibility variants which are distributed on a number of different chromosomes (e.g. Klein, 1981). Flaherty (1981) estimated the length of the transferred segment in congenics after variable numbers of backcross generations. At 10 generations, the segment is expected to be 20 cM in length (on average). A beginning has been made via use of microsatellite polymorphisms to empirically define the regions transferred in some congenics (Jiang et al., 1995).

Congenic sets have been little used as a systematic method of mapping genes but Bailey calculated that the chromosomal inserts carried along with the selected variants in the C57BL/6By/Balbc/By bilineal congenic series would sweep approximately 1/3rd of the genome and suggested that they had general applicability for mapping of a variety of traits including behavior (Bailey, 1981a,b) and illustrated their use in a study of mandible morphology (Bailey, 1985). If an investigator only wishes to identify some of the genes affecting a phenotype, a set of congenics could be screened in the manner described for consomics (see above). Discovery of a significant difference between host and congenic strains at the appropriate level of significance (Bailey, 1981a) would indicate the presence of a gene in the region containing the variant.

Recombinant-inbred strains

RI strains originally proposed by Bailey (1971) and developed as a major mapping resource by Taylor (1978, 1989, 1996) were mentioned before since they are one of the oldest resources available. Nevertheless, one should maintain familiarity with the principal stocks because, in addition to their use for QTL detection, they enable use of the extremely efficient RIST protocol for fine mapping. The recently augmented, 35 strain BXD RI set (Taylor et al., 1998), derived from B6 and D2 strains, the recently revised, 27 strain AXB (Marshall et al., 1992; Sampson et al., 1998) derived from A/J and B6 are among the largest available and together with the CXB (Bailey, 1971; Panoutsakopoulou et al., 1997) derived from C57BL6/By and BALB/cBy, are the most well characterized genetically and phenotypically. Other sets have recently been described by Taylor (1996).

There are fewer RI sets among rat strains. However the SHR X BN set consisting of 36 strains (Pravenec et al., 1989; Pravenec et al., 1996) and the 23 strain set derived from F344 and LE/Stm (Hiai et al., 1997) have been well characterized genetically and are derived from parental strains which are highly polymorphic and play an important role in biomedical research on a variety of phenotypes.

As can be seen in Table 3, for certain strain combinations such as A/J X B6, there exist a wide range of genetic tools for fine-mapping including consomics, AI stocks and recombinant-inbred strains. In this lineage, detection of a QTL by the consomic approach could be followed up in a variety of ways. The specific path to be followed should be influenced by the knowledge of the genetic architecture which influences the phenotype under examination. If a gene with a large effect size is the object of the investigation, then the intra-chromosomal location of the QTL could be explored by phenotyping of the AXB recombinant-inbred set and correlation of the phenotype with the database of molecular markers that have been typed in this set (Gill et al., 1998). Following this determination, the same database of markers could be used to select suitable strains for use in the RIST protocol. Alternatively, if the trait is influenced by many genes, each with a small effect size, then fine-mapping using the AI approach might be attempted. Alternatively, one of the RDI approaches may be preferred.

TABLE 3

Genetic resources which facilitate QTL detection and mapping in specific lineages

Lineage	Consomics	Recombinant–Inbred	Recombinant–Congenic	Advanced Intercross	Congenics
A/J X C57BL/6J	Nadeau (personal communication)	27 strains, systematically typed[1]	Gill and Nadeau (1996)	Iraqi et al. (1998)	–
C57BL/6J X DBA/2J	–	35 strains, most densely typed of all RI sets[2]	–	Peirce et al. (1998)	Miscellaneous[4]
C57BL/6By X BALB/cBy	–	13 strains, systematically typed[3]	–	–	Numerous, covering 1/3rd of the genome[5]
O20 X B10.O20	–	–	20 strains[6]	–	–
BALBc/HeA X STS/A	–	–	20 strains[6]	–	–
C3H/DiSnA X C57BL/10ScSnA	–	–	20 strains[6]	–	–

[1] Marshall et al. (1992), Sampson et al. (1998).
[2] Taylor et al. (1998).
[3] Panoutsakopoulou et al. (1997).
[4] C57BL/6J is the standard host strain for mutations found in other strains or stocks, (see Lanelist at http//www.jax.org).
[5] Bailey (1981a and http//www.jax.org).
[6] (Demant and Hart, 1986).

In this case, the consomic strain containing the heterologous chromosome of interest would be crossed to the host strain and, following the production of F_2 or BC populations, the QTL mapped to a specific region of the chromosome using the standard mapping protocol. Efficiency of mapping of genes with small effect size would be facilitated because the homogeneity of the host strain background would ensure that the extraneous genetic variance would be effectively zero. Following completion of the mapping analyses, the relevant consomic strain could be used to derive congenic strains to implement one of the protocols described earlier. Again, the fact that the host genome has already been fixed in the consomic strain will greatly speed the stage of fine-mapping.

The BXD RI series has been the focus of many QTL-oriented studies of complex traits (Philips et al., 1994; Rodriguez et al., 1995) and the results obtained are a useful source to compare with the results of QTL studies obtained with other designs (Tarantino et al., 1998; Belknap et al., 1997). In addition, the characterization of the strains on multiple phenotypes permits genetic correlations to be calculated (Blizard and Bailey, 1979) and hypotheses concerning the pleiotropic effects of QTL to be entertained (Plomin et al., 1991).

As summarized in Table 3, aside from the design considerations that were discussed earlier, the availability of genetic resources in specific lineages can have a very large impact on progress toward QTL definition. Knowledge of these resources in the planning stage of mapping studies will therefore have a long-term impact on advancing the goals of the QTL research paradigm.

Overview

The genetic analysis of complex traits is definitely complex. In this chapter we attempted to describe the strategies for QTL mapping using two commonly used laboratory species. In pursuit of the understanding of complex genetic traits it is important to realise that a variety of tools can be applied and probably the integration of this variety will be the only means to disentangle the genetic complexity of these traits. In addition to the tools described in this chapter one can include transgenics, knock-out, expression analysis, proteomics, other model organisms (drosophila, yeast), human studies, etc. These tools combined with the enormously valuable bioinformatics resources which are available to manage the vast amount of information will undoubtedly be of significance in the upcoming genetic revolution.

Acknowledgement

David A. Blizard was supported by DC-02230 from the National Institute of Deafness and Communication Disorders and AG-14731 from the National Institute on Aging.

References

Bachmanov, A.A., Reed, D.R., Ninomiya, Y., Inoue, M., Tordoff, M.G., Price, R.A. and Beauchamp, G.K. (1997) Sucrose consumption in mice: major influence of two genetic loci affecting peripheral sensory responses. *Mamm. Genome*, 8: 545–548.

Baker, H.J., Lindsey, J.R. and Weisbroth, S.H. (Eds) (1979) *The Laboratory Rat. Vol. I: Biology and Diseases*, Academic Press, New York.

Baker, H.J., Lindsey, J.R. and Weisbroth, S.H. (Eds) (1980) *The Laboratory Rat. Vol. II: Research Applications*, Academic Press, New York.

Bailey, D.W. (1971) Recombinant-inbred strains: an aid to identify linkage and function of histocompatibility and other genes. *Transplantation*, 11: 325–327.

Bailey, D.W. (1981a) Recombinant inbred strains and bilineal congenic strains. In: H.L. Foster, J.D. Small, and J.G Fox (Eds), *The Mouse in Biomedical Research. Vol. 1, History, Genetics and Wild Mice*. Academic Press, New York, pp. 223–238.

Bailey, D.W. (1981b) Strategic uses of recombinant inbred and congenic strains in behavior genetics research. In: E.S. Gershon, S. Mattysse, X.O. Breakefield and R. D. Ciaranello (Eds), *Genetic Research Strategies for Psychobiology and Psychiatry*. The Boxwood Press, Pacific Grove, pp 189–198.

Bailey, D.W. (1985) Genes that affect the shape of the murine mandible. *J. Hered.*, 76: 107–114.

Basten, C. Weir, B.S. and Zeng, Z-B. (1997) Qtl Cartographer: a reference Manual and tutorial for QTL mapping. Department of Statistics, North Carolina State University, Raleigh, NC.

Belknap, J.K., Mitchell, S.R, OToole, L.A., Helms, M.L., and Crabbe, J.C. (1996) Type I and Type II error rates for quantitative trait loci (QTL) mapping studies using recombinant inbred strains. *Behav. Genet.*, 26: 149–160.

Belknap, J.K., Dubay, C., Crabbe, J.C. and Buck, K.J. (1997) Mapping quantitative trait loci for behavioral traits in the mouse. In: K. Blum and E.P. Noble (Eds), Handbook of Psychiatric Genetics. Chapter 26.

Bennett, B. and Johnson, T.E. (1998) Development of congenics for hypnotic sensitivity to ethanol by QTL-marker-assisted counter selection. *Mamm. Genome*, 9: 969–974.

Berrettini, W.H., Ferraro, T.N., Alexander, R.C., Buchberg, A.M. and Vogel, W.H. (1994) Quantitative Trait Loci Mapping of Three Loci Controlling Morphine Preference Using Inbred Mouse Strains. *Nat. Genet.* 7: 54–58.

Billingham R.E. and Silvers W.K. (1959) Inbred animals and tissue transplantation immunity. *Transplant. Bull.*, 6: 399–406.

Blizard, D.A. (1981) The Maudsley Reactive and NonReactive strains: a North American perspective. *Behav. Genet.*, 11: 469–489.

Blizard, D.A. and Bailey, D.W. (1979) Genetic Correlation between open-field activity and defecation. *Behav. Genet.*, 9: 349–357.

Blizard, D.A., Kotlus, B. and Frank, M. (1999) Quantitative trait loci (QTL) associated with short-term intake of sucrose, saccharin and quinine solutions in laboratory mice. Chem. Senses, in press.

Broadhurst, P.L. (1960) Experiments in Psychogenetics. In: H.J. Eysenck (Ed), *Experiments in Personality: Psychogenetics and Psychopharmacology*, Vol. 1, Routledge and Kegan Paul, London, pp. 3–102.

Broadhurst, P.L. (1975) The Maudsley Reactive and Non-Reactive strains: a survey. *Behav. Genet.*, 5: 299–319.

Broadhust, P.L. (1978) *Drugs and the inheritance of behavior*. Plenum Press, New York.

Calderone, B., Saavedra, C., Tartaglia, K., Wehner, J.M., Dudek, B.C. and Flaherty, L. (1997) Quantitative trait loci analysis affecting contextual fear conditioning in mice. *Nat. Genet.*, 17: 335–337.

Changjian, J. and Zeng, Z.B. (1995) Multiple trait analysis of genetic mapping for quantitative trait loci. *Genetics*, 140: 1111–1127.

Clark, J.D. (1987) Historical perspectives and taxonomy. In: G.L. Van Hoosier Jr., C.W. McPherson (Eds), *Laboratory hamster*. Academic Press, Orlando, FL: pp. 37.

Collins, R.L. and Fuller, J.L. (1968) Audiogenic seizure prone (asp): a gene affecting behavior in linkage group VIII of the mouse. *Science*, 162: 1137–1139.

Crabbe, J.C. and Harris, R.A. (Eds) (1991). *The Genetic Basis of Alcohol and Drug Actions*, Plenum, NY.

Crawley, J.N. et al. (1997) Behavioral phenotypes of inbred mouse strains: implications and recommendations for molecular studies, *Psychopharmacology*, 132: 107–124.

Darvasi, A. (1997) Interval-specific congenic strains (ISCS): an experimental design for mapping a QTL into a 1-centimorgan interval. *Mamm. Genome*, 8: 163–167.

Darvasi, A. (1998) Experimental strategies for the genetic dissection of complex traits in animal models. *Nat. Genetics*, 18; 19–23.

Darvasi, A., Soller, M. (1992) Selective genotyping for determination of linkage between a marker locus and a quantitative trait locus. *Theor. Appl. Genet.*, 85: 353–359.

Darvasi, A., Soller, M. (1994) Optimum spacing of genetic markers for determining linkage between marker loci and quantitative trait loci. *Theor. Appl. Genet.*, 89: 351–357.

Darvasi, A., Soller, M. (1994) Selective DNA pooling for determination of linkage between a molecular marker and a quantitative locus. *Genetics*, 138: 1365–1373.

Darvasi, A., Soller, M. (1995) Advanced intercross lines, an experimental population for fine genetic mapping. *Genetics*, 141: 1199–1207.

Darvasi, A. and Soller, M. (1997) A simple method to calculate resolving power and confidence interval of QTL map location. *Behav. Genet.*, 27: 125–132.

Darvasi, A., Weinreb, A., Minke, V., Weller, J.I. and Soller, M. (1993) Detecting marker-QTL linkage and estimating QTL gene effect and map location using a saturated genetic map. *Genetics*, 134: 943–951.

Demant, P. and Hart, A.A.M. (1986) Recombinant-congenic strains: a new tool for analyzing genetic traits determined by more than one gene. *Immunogenetics*, 24: 416–422.

Dietrich, W., Katz, H., Lincoln, S.E., Shin, H-S, Friedman J., Dracopoli, N.C. and Lander E.S. (1992) A Genetic map of the mouse suitable for typing intra-specific crosses. *Genetics*, 131: 423–447.

Doerge, R.W. and Churchill, G.A. (1996). Permutation tests for multiple loci affecting a quantitative character. *Genetics* 142: 285–294.

Doerge, R. W., Zeng, Z-B and Weir, B.S. (1997) Statistical issues in the search for genes affecting quantitative traits in experimental populations. *Stat. Sci.* 12: 195–219.

Driscoll, P. (1992) *Genetically defined Models of Neurobehavioral Dysfunction*, Birkhauser, Boston.

Festing, M. F.W. (1979) Inbred strains. In: H.J. Baker, J.R. Lindsey and S.H. Weisbroth (Eds), *The Laboratory Rat. Vol. 1: Biology and Diseases*. Academic Press. New York .

Flaherty, L. (1981) Congenic strains. In: H.L. Foster, J.D. Small. and J.G. Fox. (Eds), *The Mouse in Biomedical Research. Vol.1, History, Genetics and Wild Mice*. Academic Press, New York, pp. 215–222.

Flint, J., Corley, R., DeFries, J.C., Fulker, D.W., Gray, J.A., Miller, S. and Collins, A.C. (1995) A simple genetic basis for a complex psychological trait in laboratory mice. *Science*, 269: 1432–1435.

Foster, H.L., Small, J.D. and Fox, J.G. (1981) *The Mouse in Biomedical Research. Vol.1: History, Genetics, and Wild Mice*. Academic Press, New York.

Foster, H.L., Small, J.D. and Fox, J.G. (1982) *The Mouse in Biomedical Research. Vol. II: Diseases*. Academic Press, New York.

Frankel, W.N. (1995) Taking stock of complex traits in mice. Trends in Genetics. 11: 471–477.

Fujita, O., Annen, Y. and Kitaoka, A. (1994) Tsukuba High- and Low-Emotional strains of rats (Rattus norvegicus) an overview. *Behav. Genet.*, 24: 389–415.

Fuller, J.L. and Thompson, W.R. (1960) *Behavior Genetics*. John Wiley and Sons, Inc., New York.

Fuller, J.L. (1974) Single locus control of saccharin preference in mice. *J. Hered.*, 65: 33–36.

Fuller, J.L. and Thompson, W.R. (1978) *Foundations of Behavior Genetics*. C.V. Mosby Company, Saint Louis.

Gelderman, H. (1975) Investigations on inheritance of quantitative characters in animals by gene markers. I. Methods. *Theoret. Appl. Genet.*, 46: 319–330.

Gershon, E.S., Mattysse, S., Breakefield, X.O. and Ciaranello, R.D. (Eds) (1981) *Genetic Research Strategies for Psychobiology and Psychiatry*. The Boxwood Press.

Gill, K., Desaulniers, N., Desjardins, P. and Lake, K. (1998) Alcohol preference in AXB/BXA recombinant-inbred mice: gender differences and gender-specific quantitative trait loci. *Mamm. Genome*, 9: 929–935.

Gill, K. and Nadeau, J. (1996) Alcohol/drug-related traits in AXB/BXA recombinant-inbred and congenic mice. *Alc. Clin. Exp. Res. Suppl.* 20: 26a.

Goldowitz, D., Wimer, R.E. and Wahsten, D. (1992) *Techniques for the genetic analysis of brain and behavior*. Elsevier, Amsterdam.

Green, E.L. (1966) *Biology of the Laboratory Mouse*. McGraw-Hill, New York.

Harrington, G.M. (1981) The Har strains of rats: origins and characteristics. *Behav. Genet.*, 11: 445–468.

Hiai, H. (1997) The LEXF: a new set of recombinant inbred strains between LE/Stm and F344. *Mamm. Genome*, 8: 324–327.

Hudgins, C.C., Steinberg, R.T., Klinman, D.M., Patton Reeves, M.J. and Steinberg, A.D. (1985) Studies of consomic mice bearing the Y chromosome of the BXSB mouse. *J. Immunol.*, 134: 3849–3854.

Iraqi, F., Kemp, S. and Teale, A. (1998) Fine mapping of trypanosomiasis QTLs in mice using advanced intercross lines. International Society of Animal Genetics (ISAG), Auckland, New Zealand. August 915, 1998. *Animal Genetics*, in press.

Jansen, R.C. (1996) Complex plant traits: time for polygenic analysis. *Trends in Plant Science*, 1: 89–94.

Jiang, P.P., Hansen, T.H., Shreffler, D.C. and Miller, R.D. (1995) Mouse H2 congenic intervals: analysis and use for mapping. *Mammal. Genome*, 6: 586–591.

Klein, J. (1981) The histocompatibility-2 (H-2) complex. In: H.L. Foster, J.D. Small and J.G. Fox (Eds), *The Mouse in Biomedical Research*. Academic Press, New York.

Korol, A.B., Ronin, Y.I. and Kirzhner, V.M. (1995) Interval mapping of quantitative trait loci employing correlated trait complexes. *Genetics*, 140: 1137–1147.

Lander, E.S. and Botstein, D. (1989) Mapping Mendelian factors underlying quantitative traits using RFLP linkage maps. *Genetics*, 121: 185–199.

Lander, E.S., Green, P., Abrahamson, J., Barlow, A., Daly, M.J., Lincoln, S.E. and Newburg, L. (1987) MAPMAKER: An interactive computer package for constructing primary genetic linkage maps of experimental and natural populations. *Genomics*, 1: 174–181.

Lander, E. and Kruglyak, L. (1995) Genetic dissection of complex traits: guidelines for interpreting and reporting linkage results. *Nat. Genet.*, 11: 241–246.

Lander, E.S. and Schork, N.J. (1994) Genetic dissection of complex traits. *Science*, 265: 2037–2048.

Lebowitz, R.J., Soller, M. and Beckmann, J.S. (1987) Trait-based analyses for the detection of linkage between marker loci and quantitative trait loci in crosses between inbred lines. *Theor. Appl. Genet.*, 73: 556–562.

Li, T.-K. and Lumeng, L. (1984) Alcohol preference and voluntary alcohol intakes of inbred rat strains and the National Institutes of Health heterogeneous stocks of rats. *Alc. Clin. Exp. Res.*, 8: 485–486.

Lieblich, I. (1982) *Genetics of the Brain*. Elsevier, Amsterdam.

Lindsey, J.R. (1979) Historical Foundations. In: H.J. Baker, J.R. Lindsey and S.H. Weisbroth (Eds), *The Laboratory Rat. Vol. 1: Biology and Diseases*. Academic Press, New York.

Liu, B.H. (1996) Statistical Genomics: Linkage, Mapping, and QTL Analysis. CRC Press, Boca Raton, Florida.

Mangin, B., Goffinet, B. and Rebai, A. (1994) Constructing confidence intervals for QTL location. *Genetics*, 138: 1301–1308.

Manly, K.F. and Olson, J.M. (1999) Overview of QTL mapping software and introduction to Map Manger QT. *Mamm. Genome*, in press.

Marshall, J.D., Mu, J-L., Cheah, Y-C., Nesbitt, M.N., Frankel, W.N. and Paigen, B. (1992) The AXB and BXA set of recombinant-inbred mouse strains, *Mamm. Genome*, 3: 669–680.

Maxson, S.C. (1996) Searching for candidate genes with effects on an agonistic behavior, offense, in mice. *Behav. Genet.*, 26: 471–476.

McClearn, G.M., Wilson, J.R. and Meredith, W. (1970) The use of isogenic and heterogenic mouse stocks in behavioral research. In: G. Lindzey and D. Thiessen (Eds), *Contributions to behavior genetic analysis: the mouse as a prototype*, Appleton Century Crofts, New York, pp. 3–22.

Melo, J.A., Shendure, J., Pociask, K. and Silver, L.M. (1996) Identification of sex specific quantitative trait loci controlling alcohol preference in C57BL/6 mice. *Nat. Genet.*, 13: 147–153.

Melo, J., Shendure, J., Pociask, K. and Silver, L. (1996) Identification of sex-specific QTLs controlling alcohol preference by C57BL/6J mice. *Nat. Genet.*, 13: 147–153.

Moen, C.J.A., Groot, P.C., Hart, A.A.M., Snoek, M. and Demant, P. (1996) Fine mapping of colon tumor susceptibility (Scc) genes in the mouse, different from the genes known to be somatically mutated in colon cancer. *Proc. Natl. Acad. Sci. USA*, 93: 1082–1086.

Morse, H.C. III. (1981) The Laboratory Mouse: a historical perspective. In: H.L. Foster, J.D. Small and J.G. Fox (Eds). *The Mouse in Biomedical Research, Vol. 1. History, Genetics and Wild Mice*, Academic Press, New York.

Motro, U. and Soller, M. (1993) Sequential Sampling in Determining Linkage Between Marker Loci and Quantitative Trait Loci. *Theor. Appl. Genet.*, 85: 658–664.

Nadeau, J.H., Arbuckle, L.D. and Skamene, E. (1995) Genetic dissection of inflammatory diseases. *J. Inflamm.*, 45: 27–48.

Okuizumi, H. Ohsumi, T., Sasaki, N., Imoto, H., Mizuno, Y., Hanami, T., Yamashita, H., Kamiya, M., Takada, S., Kitamara, A., Muramatsu, M., Nishimura, M., Mori, M., Matsuda, Y., Tagaya, O., Okazaki, Y., Hayashizaki, Y. (1997) Linkage map of Syrian hamster with restriction landmark genomic scanning. *Mamm. Genome*, 8: 121–128.

Panoutsakopoulou, V., Spring, P., Cort, L., Sylvester, J.E., Blank, K.J. and Blankenhorn, E.P. (1997) Microsatellite typing of CXB recombinant inbred and parental mouse strains. *Mamm. Genome*, 8: 357–361.

Paterson, A.H. (1995) Molecular dissection of quantitative traits: progress and prospects. Genome Res., 5: 321–333.

Peirce, J.L., Derr, R., Shendure, J., Kolata, T. and Silver, L.M. (1998) A major influence of sex-specific loci on alcohol preference in C57BL/6 and DBA/2 inbred mice. *Mamm. Genome*, 9: 942–948.

Phillips, T.J., Buck, K. J., Wenger, C.D., Metten, P. and Belknap, J.K. (1994) Localisation of genes affecting alcohol drinking in mice. *Alc. Clin. Exp. R*es., 18: 931–941.

Plomin, R., McClearn, G.E., Gora-Maslak, G. and Neiderhiser, J.M. (1991) Use of recombinant inbred strains to detect quantitative trait loci associated with behavior. *Behav. Genet.*, 21: 99–116.

Pravenec, M., Klir, P., Kren, V., Zicha, J. and Kunes, J. (1989) An analysis of spontaneous hypertension in spontaneously hypertensive rats by means of new recombinant-inbred strains. *J. Hypertension*, 7: 217–222.

Pravenec, M., Gaugier, D., Schott, J-J, Buard, J., Kren, V., Bila, V., Szpirer, C., Szpirer, J., Wang, J-M., Huang, H., St. Lezin, E., Spence, M.A., Flodman, P., Printz, M., Lathrop, G.M., Vergnaud, G. and Kurtz, T.W. (1996) A genetic linkage map of the rat derived from recombinant-inbred strains. *Mamm. Genome*, 7: 117–127.

Robinson, R (1965) *Genetics of the Norway Rat*. Pergamon Press, London.

Rodriguez, L.A., Plomin, R. Blizard, D.A., Jones, B.C. and McClearn, G.E. (1995) Alcohol acceptance, preference and sensitivity in mice. II. Quantitative trait loci mapping analysis using BXD recombinant-inbred strains. *Alc. Clin. Exp. Res.*, 19: 367–373.

Sampson, S.B., Higgins D.C., Elliott, R.W., Taylor, B.A., Lueders, K.K., Koza, R.A. and Paigen, B. (1998) An edited linkage map for the AXB and BXA recombinant inbred mouse strains. *Mamm Genome*, 9: 688–694.

Sax, J. (1923) The association of size differences with seed-coat pattern and pigmentation in Phaseolus vulgaris. *Genetics*, 8: 552–560.

Silver, L.M. (1995) *Mouse Genetics Concepts and Applications*. Oxford University Press, New York and Oxford.

Snell, G.D. (1948) Methods for the study of histocompatibility genes. *J. Genet.*, 49: 87–103.

Soller, M., Genizi, A. and Brody, T. (1976) On the power of experimental designs for the detection of linkage between marker loci and quantitative loci in crosses between Inbred Lines. *Theor. Appl. Genet.*, 47: 35–39.

Sprott, R.L. (1975) Behavioral characteristics of B6, D2 and B6/D2 F_1 mice which are potentially useful for gerontological research, *Exp. Aging Research*, 1: 313–323.

Staats, J. (1981) Inbred and segregating inbred strains. In: H.L. Foster, J.D. Small. and J.G. Fox. (Eds), *The Mouse in Biomedical Research. Vol. 1, History, Genetics and Wild Mice*. Academic Press, New York, pp.177–210.

Stassen, A. P., Groot, P C., Eppig J. T. and Demant, P. (1996) Genetic composition of the recombinant congenic strains. *Mamm. Genome*, 7: 55–58.

Stuber, C.W. (1995) Mapping and manipulating quantitative traits in maize. *Trends Genet.*, 11: 477–481.

Talbot, C.J., Nicod, A., Cherny, S.S., Fulker, D.W., Church, A.C. and Flint, J. (1999) Fine structure mapping of quantitative trait loci in outbred mice. *Nat. Genetics*, in press.

Tarantino, L.M., McClearn, G.E., Rodriguez, L.A. and Plomin, R. (1998) Confirmation of quantitative trait loci for alcohol preference in mice. *Alc. Clin. Exp. Res.*, 22: 1099–1105.

Taylor, B.A. (1972). Genetic relationships between inbred strains of mice. *J. Hered.*, 63: 83–86.

Taylor, B.A. (1978) Recombinant inbred strains: use in gene mapping. In: H.C. Morse, III, (Ed). *Origins of Inbred Mice*, Academic Press, New York.

Taylor, B.A. (1989) Recombinant-inbred strains. In: M.F. Lyon and A.G. Searle (Eds), *Genetic variants and strains of the laboratory mouse*, 2nd Edn, Oxford University Press, Oxford, pp. 773–789,

Taylor B.A. (1996) Recombinant inbred strains. In: M.F. Lyon, S. Rastan and S. Brown, (Eds), *Genetic Variants and Strains of the Laboratory Mouse*, 3rd Edn, Vol. 2, New York, Oxford University Press, pp. 1597–1659.

Taylor, B.A. and Phillips, S.J. (1996) Detection of obesity QTLs on mouse chromosomes 1 and 7 by selective DNA pooling. *Genomics*, 34: 389–398.

Taylor, B.A., Wnek, C., Kotlus, B.S., Roemer, N., MacTaggart, T. and Phillips, S.J. (1998) Genotyping new BXD recombinant inbred mouse strains and comparison of BXD and consensus maps. *Mamm. Genome*, in press.

Thoday, J.M. (1961) Location of polygenes. *Nature*, 191: 368–370.

Thompson, J.N. Jr. and Thoday, J.M. (1979) *Quantitative Genetic Variation*, Academic Press, New York.

Van Ooijen, J.W. (1992) Accuracy of mapping a quantitative trait locus in autogamous species. *Theor. Appl. Genet.*, 84: 803–811.

Visscher, P.M., Thompson, R. and Haley, C.S. (1996) Confidence intervals in QTL mapping by bootstrapping. *Genetics*, 143: 1013–1020.

Wehner, J.M., Radcliffe, R.A., Rosann, S.T., Christensen, S.C., Rasmussen, D.L., Fulker, D.W. and Wiles, M. (1997) Quantitative trait locus analysis of contextual fear conditioning in mice. *Nat. Genet.*, 17: 331–334.

W.E. Crusio and R.T. Gerlai (Eds.)
Handbook of Molecular-Genetic Techniques for Brain and
Behavior Research (Techniques in the Behavioral and Neural Sciences, Vol. 13)

CHAPTER 2.1.3

Linkage strategies for mapping genes for complex traits in man

Laura Almasy and John Blangero

Department of Genetics, Southwest Foundation for Biomedical Research, San Antonio, TX, USA

Linkage analysis

The basic goal of linkage analysis is to localize one or more genes influencing a trait to specific chromosomal regions. This is accomplished by examining the co-segregation of the phenotype of interest with genetic markers. Relatives who are phenotypically alike will share in common alleles at markers surrounding the gene(s) influencing the phenotype whereas other relatives who are phenotypically dissimilar will not carry these alleles. The minimum data needed for linkage analyses is a set of families in which phenotyped individuals have known relationships to each other and the genotypes of these individuals at one or more genetic markers.

Complex traits

Unlike monogenic diseases whose genetic determinants have been successfully mapped in extended human pedigrees using classical penetrance-based linkage analysis methods, such as those described in Chapter 2.1, many behavioral and neurological phenotypes of interest are influenced by multiple genetic and environmental components and their interactions. The genetic analysis of such complex phenotypes requires different statistical approaches for the localization and evaluation of the relative importance of specific quantitative trait loci (QTLs). A QTL was originally defined as any locus affecting a quantitative trait, regardless of the magnitude of effect (Gelderman, 1975). Thus a QTL could be a major gene with a large effect on the quantitative phenotype, one of a few oligogenes each with moderate effects, or one of many polygenes each with very small effects. The QTL concept has been expanded to include loci influencing complex discrete traits, such as schizophrenia or diabetes, as the risk of developing these diseases is thought to be a function of a continuous quantitative underlying liability function. In some cases, there is a quantitative trait that directly indexes disease liability as with blood pressure and hypertension. In other cases, such as schizophrenia, there is no measurable indicator of liability other than diagnosis.

Classical penetrance-model based methods of linkage analysis require detailed knowledge of the underlying genetic model. For a disease trait, one must specify the prevalence of the trait, its mode of inheritance, the allele frequencies at the presumed disease gene, and the probability of being affected, or penetrance, for each genotype. For quantitative traits, one must specify allele frequencies and the mean trait value for each genotype. For complex phenotypes, which by definition are influenced by multiple genes with penetrance structures or genotypic means that may be contextually dependant on other genes or on unknown environmental factors, specification of the necessary parameters is problematic. Misspecification of these parameters reduces the power to detect linkage with the penetrance model-based methods (Greenberg and Hodge, 1989; Greenberg, 1990; MacLean et al., 1993) and can lead to spurious negative findings, including false exclusion of a region containing a QTL (Risch and Giuffra, 1992). One solution to this problem has been to perform classical penetrance-based linkage analyses with a variety of models of inheritance and a range of allele frequencies and penetrances in the hope that one of these sets of parameters will be close enough to the true model. However, this strategy has not met with much success in linkage studies of complex human traits (Barr et al., 1994; Bailey and Almasy, 1995; Risch and Botstein, 1996). It is in response to the need for methods specifically designed to address complex traits that penetrance model-free methods of analysis are being developed. This chapter gives an overview of the most widely used complex trait study designs and penetrance model-free linkage methods and their relative effectiveness for different classes of phenotypes. However, the best advice we can give investigators planning a linkage study of complex traits is to consult a statistical geneticist who can help you to select the most appropriate sample and method of analysis for your phenotype and can provide you with estimates of power to detect linkage given your proposed study design.

The phenotype determines the design

Penetrance model-free methods are often referred to as 'non-parametric' linkage methods, although this is something of a misnomer as these methods involve the estimation of various parameters other than those having to do with the mode of inheritance. Penetrance model-free methods for linkage analysis can be broken down into two basic categories both of which are based on identical by descent (IBD) allele sharing in relatives. Two

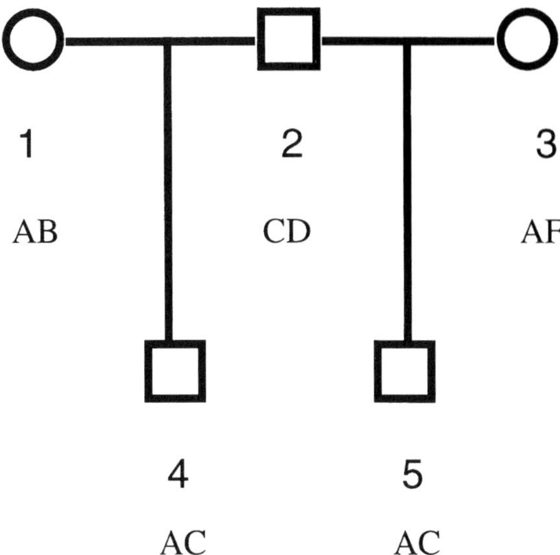

Fig. 1. An example of IBD and IBS

alleles are considered IBD if they originate from the same ancestral source. In Fig. 1, individuals 4 and 5 share the allele 'C' IBD because both received it from their common father. Although they both have an 'A' allele, these alleles are not IBD because they came from the unrelated mothers of the half-siblings and can not be traced back to a common ancestor. Allele sharing is also sometimes discussed in terms of identity by state (IBS) in which the ancestral source of the alleles is not considered. Two alleles are IBS if they are of the same form, regardless of their origin. In Fig. 1, individuals 4 and 5 share both of their alleles IBS. Additionally, while individuals 1 and 3 can not share any alleles IBD, as they are unrelated, they share the 'A' allele IBS. IBD allele sharing provides information about linkage while IBS allele sharing provides information about population level association between the marker and the trait of interest. However, in population isolates with a modest number of founders IBS allele sharing very closely approximates IBD allele sharing for highly polymorphic markers as each allele is likely to have entered the population through only one founder.

In the absence of inbreeding, a pair of individuals can share 0, 1, or 2 alleles IBD and the degree of IBD sharing is usually expressed as the proportion of alleles shared (π) which can take the values 0 1/2 and 1. If a pedigree is inbred (i.e. there are matings between related individuals), higher degrees of IBD sharing are possible when individuals are homozygous for an allele from a particular ancestor. In most situations commonly encountered in human studies, IBD proportions will not be higher than 1 and only siblings will have an IBD proportion greater than 1/2. The expected IBD sharing over the entire genome ($E(\pi)$) for a relative pair is equal to twice the pair's kinship coefficient (2ϕ). For siblings, $E(\pi)$ is 1/2, whereas it is 1/4 for half siblings and 1/8 for first cousins.

Calculation of IBD sharing for a genotyped marker locus is straightforward for nuclear families, but becomes more complicated for extended pedigrees in which the top few generations are often unavailable for genotyping. A number of methods have been proposed to calculate IBD probability matrices (Amos et al., 1990; Whittemore and Halpern, 1994; Curtis and Sham, 1994). One simple and effective approach described by Curtis and Sham (1994) uses maximum likelihood methods to sequentially estimate the IBD probability for each pair of individuals within a pedigree using the posterior probability of genotypes at a completely linked pseudo-marker. Essentially this approach works by creating an imaginary recessive disease carried by one member of the pair and using parametric linkage software to predict the risk to the other member of the pair. The risk of the relative being affected gives the probability for IBD sharing of 1 at the marker and the risk of the relative being a heterozygous carrier gives the probability of IBD sharing of 0.5. This method is relatively rapid for simple pedigrees without inbreeding or extensive marriage loops. For more complex extended pedigrees, a Monte Carlo method can be employed in which missing genotypes are imputed (Almasy and Blangero, 1998); IBD is calculated according to the algorithm of Davis et al. (1996); and a final IBD matrix is estimated based a weighted distribution produced over many Monte Carlo replicates.

The most basic division between the two most popular penetrance model-free linkage methods is whether IBD allele sharing is treated as an independant or dependant variable. Affected relative pair designs involve the comparison of observed versus expected IBD sharing, usually in sibling pairs. These pairs may be ascertained on shared affection status or concordance for high or low quantitative trait values or ascertained on discordance either for affection status or for a quantitative trait. Computer programs which assess deviations in IBD sharing of relatives pairs from that predicted by their kinship include MAPMAKER/SIBS (Kruglyak and Lander, 1995), GENEHUNTER (Kruglyak et al., 1996), and ASPEX (Hinds and Risch, 1998). The other major penetrance model-free method is the use of components of variance to model covariance among members of a family as a function of IBD sharing. This method can be used for either quantitative traits or affection status, assuming an underlying liability threshold, and is appropriate for the analysis of any type of pedigree whether randomly ascertained or ascertained through one or more individuals with a particular phenotype. Computer programs which utilize the variance component linkage method include SOLAR (Almasy and Blangero, 1998), ACT (Amos et al., 1998), and SEGPATH (Province et al., in press).

Regardless of the method used, penetrance model-free linkage studies require much larger sample sizes than do linkage studies of traditional Mendelian traits. Whereas 100 individuals is a large sample for a classical penetrance model-based linkage analysis, the minimum sample size for successful penetrance-model free linkage analysis of a complex trait is at least twice that and up to an order of magnitude larger depending on the sampling unit (pedigree, nuclear family, or relative pair), the number of genes involved, and disease prevalence or quantitative trait heritability.

The penetrance model-free linkage methods each have their advantages and disadvantages and the choice of which to use and in what type of sample depends heavily on the phenotype being analyzed. Recent methodological studies have suggested that quantitative traits are more informative for linkage analysis than dichotomozations of the same trait (Bailey and Almasy, 1995; Duggirala et al., 1997; Korczak and Goldstein, 1997; Wijsman and Amos, 1997). If the phenotype of interest can be measured quantitatively or has quantitative correlates, a linkage study of the quantitative trait directly will have more power than a study based on categorization of individuals into high and low or affected and unaffected groups. For traits that are necessarily discrete, the choice of sample and method is most dependent on the prevalence of the phenotype. As discussed in the sections on the respective discrete trait methods, the power of affected relative pair methods is maximized when population prevalence of disease is low whereas variance component methods are most powerful as disease prevalence approaches 50%.

Quantitative traits

Variance component methods

The variance of a quantitative trait describes the spread of the phenotype values around the population mean. The basic idea behind the variance component linkage method is to attribute that population variance to a variety of genetic and non-genetic causes. The variance component method seeks to explain the correlations in phenotype among members of a family by partitioning the phenotypic variance into components due to the effect of a specific QTL linked to a genotyped marker, to other QTLs unlinked to the region under consideration, to environmental factors shared among family members such as diet, and to individual-specific environmental sources such as measurement error (Hopper and Mathews, 1982; Amos, 1994). These methods have recently been extended to allow multipoint linkage analysis in pedigrees of arbitrary size and complexity (Almasy and Blangero, 1998) and to oligogenic analysis of multiple QTLs simultaneously (Blangero and Almasy, 1997; Almasy and Blangero, 1998). For a simple model with n QTLs, the covariance among pedigree members is modeled as:

$$\Omega = \sum_{i=1}^{n} \Pi_i \sigma_{qi}^2 + 2\Phi\sigma_a^2 + I\sigma_e^2$$

where Π_i is a matrix of IBD sharing among family members at marker i, σ_{qi}^2 is the additive genetic variance due to a QTL linked to marker i, Φ is a matrix of kinship values, σ_{a}^2 is the residual additive genetic variance, σ_{e}^2 is the individual-specific environmental variance, and I is an identity matrix. One advantage of this method is that it provides not only an estimate of QTL location, but also an estimate of the magnitude of effect of the QTL, through σ_{qi}^2. The maximum likelihood of a model in which one or more QTL components is estimated is compared to that of a model in which these components are constrained to zero, testing the hypothesis that σ_{qi}^2 is significantly greater than zero. The difference in $\log_e$ likelihood between these models is asymptotically distributed as a mixture of chi-square distributions, with the mixing proportions depending on the number of QTL components estimated (Self and Liang, 1987). For a single locus analysis, the difference in the $\log_{10}$ likelihood between the two models is equivalent to the LOD score obtained from classical parametric linkage analyses. These maximum likelihood analyses assume that the phenotype has a multivariate normal distribution, meaning that the plot of trait values follows a Gaussian or bell curve. While the method is robust to violations of normality, such as skewness in the distribution (Amos, 1994), it is sensitive to outliers whose phenotypic values are discontinuous from the rest of the distribution. However, the normality of the distribution and the presence or absence of outliers are easily assessed.

An advantage of the variance component method is that measured covariates, such as age or sex or environmental exposures (e.g. cigarette smoking, oral contraceptive use), are incorporated into the model as modifiers of the phenotypic mean and covariate effects are estimated simultaneously with the components of variance. Incorporation of covariate effects reduces the unexplained variance in the phenotype and increases the proportion of that variance due to the QTLs being sought, thus increasing power to detect linkage.

The variance component method has been extended to censored or truncated quantiative traits, such as age of onset (Duggirala et al., 1999). The age at onset in a presently unaffected person must be greater than their current age, resulting in a left truncated trait, whereas the age at onset of affected individuals is right truncated at or before the age at diagnosis. Covariates are incorporated into this analysis, as into the quantitative trait analyses described above, as effects on the mean age of onset.

It is also simple to incorporate more complex genetic models in variance component analyses. These include additional sources of variance such as dominance genetic effects, where the genotypic mean of heterozygotes does not fall halfway between that of the two homozygotes, or shared environmental components common to members of a household such as diet or exposure to pathogens (Almasy and Blangero, 1998). Interactive effects can be accommodated for both gene by gene (epistatic) interactions (Mitchell et al., 1997) and gene by environment interactions (Towne et al., 1997). A common example of gene by environment interaction is genotype by sex interaction, where a gene has different effects in males and females due to the different hormonal environments of the two sexes. The variance component method can also exploit genetic and environmental correlations between phenotypes through multivariate analyses to improve power to detect linkage and localization of QTLs (Almasy et al., 1997).

The power of the variance component method depends on the proportion of the phenotypic variance of the trait accounted for by the gene being sought (the QTL-specific heritability), the size of the sample, the distribution of those individuals across families, and whether the sample has been ascertained on individuals with a disease or an extreme quantitative phenotype. Heritability in the broad sense is the proportion of the phenotypic variance that is attributable to genetic effects. Typically, dominance effects are disregarded and additive genetic, or narrow sense, heritabilities are calculated. QTL-specific heritability is the proportion of the phenotypic variance attributable to the additive genetic effects of a specific QTL. Higher QTL-specific heritability and larger sample size yield greater power to detect linkage. Generally, the larger the sampling unit, the more powerful the variance component analysis (Williams et al., 1997). On a per person basis, nuclear families provide more linkage power than sib pairs and extended pedigrees provide more power than nuclear families. Whereas theoretical power calculations can be made for nuclear families of specified size, quasi-analytical power calculations on extended pedigrees require some empirical simulation (Williams and Blangero, (1999b)).

Old and new Haseman–Elston methods

Before the advent of variance component methods, the most widely used quantiative trait technique was the Haseman–Elston method (Haseman and Elston, 1972). This method is based on regressing the squared difference in sib pair trait values on the proportion of alleles shared IBD. The basic idea behind this test is that a significant negative regression coefficient should exist because sibs who are phenotypically more alike should share more alleles IBD at marker loci linked to a QTL than sibs who are phenotypically dissimilar. Numerous studies have demonstrated that for quantitative trait analysis the variance component method is more powerful than methods based on sib pair differences in traits (Amos et al., 1997; Pugh et al., 1997; Williams and Blangero, 1999a). Recently, the Haseman–Elston method has been revised (Elston et al., 1998) to use the product of the sibs' trait values rather than the difference. This new method addresses the phenotypic covariance between sibs and is functionally similar to variance component methods except in that it uses regression rather than maximum likelihood. Least squares regression is computationally more efficient than maximum likelihood and the new HasemanElston method will be more rapid than variance component methods. However, the variance component method explicitly deals with the non-independence of the elements of the phenotypic covariance matrix and since it is likelihood based yields a more powerful test than the Haseman–Elston method.

Common discrete traits — variance component methods

The variance component linkage method has recently been extended to accommodate analysis of dichotomous traits by use of a threshold model (Duggirala et al., 1997). The threshold model assumes that an individual belongs to a specific disease category if an underlying genetically determined risk or liability exceeds a certain threshold on a normally distributed liability curve. Unaffected individuals are assumed to have underlying liability values below the threshold and the threshold is set according to the disease prevalence so that a proportion of the liability curve equal to the disease prevalence lies above the threshold. The liability is assumed to have an underlying multivariate normal distribution, but as with quantitative traits the test is robust to non-normality. The correlation in liability between pairs of individuals is estimated using the affection status of unrelated individuals and various classes of relatives. For a simple model, the correlation in liability (ρ) between individuals i and j is given by:

$$\rho_{ij} = \pi_{mij} h_m^2 + 2\phi_{ij} h^2 + I_{ij} e^2$$

where π_{mij} is the proportion of alleles that individuals i and j share IBD at a marker (m) linked to a QTL; h_m^2 is the QTL-specific heritability attributed to a locus linked to marker m; h^2 is the residual additive genetic heritability; I_{ij} is the coefficient for the individual-specific environmental component; and e^2 is $1 - h^2 - h_m^2$ (i.e. the remaining variance not due to genetic effects). Measured covariates that may affect liability are incorporated into these analyses using a threshold for affection that varies with the vector of covariates. Thus males and females or smokers and non-smokers may have different prevalences of disease.

The power of the variance component method for discrete traits is dependent on the sample size and QTL heritability, as for quantitative traits, but also on the relative informativeness of affected and unaffected individuals with regard to the unobserved liability distribution. If the population prevalence of the disease is 1%, the risk threshold on the unobserved quantitative liability distribution is placed such that 99% of the curve is below the threshold. Thus unaffected individuals could have a wide range of true liability values and knowing that an individual is unaffected reveals very little about their unobserved liability, whereas knowing that an individual is affected severely restricts their potential liability values. In these cases, when disease prevalence is very low, unaffected individuals provide little information and analyses restricted to affected individuals, such as those described in the next section, will be most powerful. However, when the population prevalence of a disease is high, unaffected individuals can provide as much information as affected individuals. The power of the discrete trait variance component method is maximized as prevalence approaches 50% and theoretical calculations suggest that it will be more efficient than affected-only methods when the prevalence is over 10% (J.T. Williams and J. Blangero, unpublished data).

Rare discrete traits — relative pair methods

The relative pair method is based on detecting deviations in IBD allele sharing from that expected based on a relative pair's kinship (Suarez et al., 1978; Risch, 1990). The basic idea is similar to that of the Haseman–Elston method (Haseman and Elston, 1972) described above. That is, relatives who are phenotypically alike (e.g. concordant for disease) should share more alleles IBD at a marker locus linked to a QTL than expected by chance. This can be parameterized in terms of mean IBD allele sharing ($E(\pi)$) or in terms of the expected proportion of relative pairs sharing 0, 1, or 2 alleles IBD (z_0, z_1, and z_2). Sibling pairs are the most commonly studied relative pairs, with expectations of ½ for π and ¼, ½, and ¼ for z_0, z_1, and z_2 respectively. Phenotypically concordant sib pairs (whether affected or unaffected) are expected to share more than ½ of their alleles IBD whereas phenotypically discordant sib pairs are predicted to share less than ½ of their alleles IBD at markers linked to a QTL. Similarly, concordant sib-pairs are expected to share two alleles IBD more than ¼ of the time whereas discordant pairs are expected to share two alleles IBD less often. An important parameter for relative pair analyses is the relative risk, λ_R, which is K_R/K where K_R is the risk of disease in a relative of type R and K is the population prevalence. The relative risk is a measure of how much more likely a relative of an affected person is to develop the disease as compared to a randomly chosen individual from the population. Just as heritability can be discussed in QTL-specific terms as well as generally for unspecified QTLs, the relative risk can also be thought of in general for the trait and for specific QTLs. The proportion of alleles an affected relative pair is expected to share IBD at a QTL influencing disease status is a function of the relative risk above population prevalence attributable to that QTL.

TABLE 1

Nomial P-values for LOD scores

LOD	P-value
0.588	0.05
1.175	0.01
1.900	0.0015
3.000	0.0001
5.000	0.000001

A number of test statistics have been developed to evaluate the significance of relative pair deviations from IBD expectation. A one-sided test based on normal deviates may be used to assess deviations of z_2 from ¼ in sib-pairs (Day and Simmons, 1976; Suarez et al., 1978) or deviations of π from its expectation in any type of relative pair. Risch (1990) has developed a maximum likelihood-based test that compares the likelihood of the observed IBD distribution given predicted IBD sharing at a linked QTL vs the likelihood of the observed IBD data given expected IBD sharing under a recombination fraction of ½ (i.e. no linkage). While the tests of z_2 and π based on the normal distribution produce P-values, the likelihood based test produces the classical LOD score linkage statistic. In comparing linkage results using these different test statistics, it is important to note the correspondence between LOD scores and their nominal P-values (Table 1). The standard P-value significance cut-off of 0.05 does not have the correction for genome-wide testing that the accepted LOD score significance cut-off of 3.0 does. While a P-value of 0.01 sounds impressive, it corresponds to a LOD score of only 1.175. Regardless of the linkage method employed, a significance cutoff should be chosen that results in a type I error rate of one false positive QTL every 20 genome scans, or a genome-wide P-value of 0.05. Lander and Kruglayk (1995) have developed such standards for affected relative pair test statistics. However, their suggested criteria may be somewhat conservative as their theoretical calculations assume an infinitely dense map of genotyped markers whereas most human studies have at best a 5–10 cM map.

The extension of the relative pair framework to incorporate more complex genetic models, including covariates, oligogenic analysis, epistasis, or gene by environment interaction, is not well developed. In practice, attempts are usually made to address these issues through subsetting of the data. For example, in order to include sex or age-specific disease prevalences, male and female or older and younger relative pairs may be analyzed separately. This strategy has a number of problems. Subsetting the data set reduces sample size, decreasing power, and increases the number of parameters being estimated, decreasing statistical efficiency. Additionally, when a number of stratification schemes are explored, corrections must be made for multiple testing, potentially reducing the power further. Multilocus relative pair analysis is also usually addressed through subsetting the data into families who do and do not show linkage at a first locus and reassessing linkage at other loci in these subsets. Although they are not widely used, more sophisticated oligogenic methods which consider the likelihood of the joint IBD distribution of a relative pair at multiple markers have been developed (Cordell et al., 1995).

One obvious assumption of the relative pair methods is that the expected IBD sharing, $E(\pi)$, is known without error. In inbred populations where distant relatives inter-marry, such as the Amish, $E(\pi)$ is no longer 1/2 for sibling pairs and depends on the level of inbreeding in the population. When inbreeding exists, affected relative pair analyses which do not take this into account have been shown to have an inflated false positive rate while those which do consider it have increased power to detect linkage (Génin and Clerget-Darpoux, 1996). In practice, $E(\pi)$ can be easily confirmed for a class of relative pairs or for one particular relative pair by examining the average IBD sharing across all genotyped markers. Another common approach to this problem is to analyze both concordant and discordant pairs and demonstrate that the deviations from expectation in IBD sharing are in opposite directions.

A second important assumption of the relative pair method is that the pairs are independent. Although the unit of analysis is a relative pair, these pairs are commonly drawn from larger pedigrees, making them non-independent as knowledge of the IBD sharing of a subset of the pairs completely predicts IBD sharing in the remaining pairs. The sampling of pedigrees or nuclear families is an efficient relative pair study design because additional family members are needed to establish IBD sharing in a relative pair. When parents are not genotyped, estimates of IBD allele sharing in their offspring become dependant on estimates of population allele frequencies. Drawing multiple relative pairs from the same families requires that fewer total individuals be genotyped to establish IBD but the non-independence among the relative pairs must be taken into account in the analyses or the type I error rate will be greatly inflated. The correction for non-independence is accomplished in a number of ways. One scheme uses all $n(n-1)/2$ possible relative pairs from each family with n affected, but weights the pairs by $2/n$ giving each family an effective contribution of $n-1$. A second approach is to select as many independent pairs from the families as possible by pairing an index case with each available relative. However, it has recently been demonstrated that the particular index case selected can have dramatic effects on the outcome of the linkage analyses, suggesting that the use of all possible pairs with an appropriate weighting scheme is the more desirable method of correction (Van Eerdewegh et al., in press).

The power of affected relative pair methods depends on the sample size, the population prevalence of the disease, and the risk to relatives of an affected individual. Risch (1990) has demonstrated that the power of affected pair methods is a function of the relative risk, λ_R. In order to calculate the power to detect linkage for a given sample size and λ_R, assumptions must be made about the number of QTLs which exist and the interactions between them.

λ_R can be expressed as a function of heritability, depending on the kinship of the pair. For sibling pairs:

$$\lambda_{Sibs} = 1 + \frac{1-K}{K}(\tfrac{1}{2}h^2 + \tfrac{1}{4}d^2)$$

where h^2 is the additive genetic heritability of the trait on the binary scale and d^2 is the heritability due to dominance. Dominance occurs when the disease penetrance for heterozygotes does not fall halfway between that of the two relevant homozygotes. Although dominance is well documented for many Mendelian diseases, there is little evidence of dominance in complex trait systems and most complex trait linkage analyses assume only additive genetic effects. The most influential factor in λ is the prevalence. The term $(1-K)/K$ can take any positive value depending on the magnitude of K whereas the term $(1/2\ h^2 + 1/4\ d^2)$ has a theoretical maximum of 1/2. As K decreases, λ can become very large even when the contribution of the QTL (i.e. its heritability) is very small.

Alternative methods

While most relative pair methods make use of only one type of relative pair at a time (i.e. concordant or discordant), the classical Haseman–Elston method (Haseman and Elston, 1972), described above for quantitative traits, makes use of both types of relative pairs in the same analysis. The Haseman–Elston method treats the dichotomous trait as a quantitative trait with values of 0 and 1 and regresses the squared relative pair difference on IBD, testing for a significant negative regression. If reliable unaffecteds are genotyped and available, the Haseman–Elston method outperforms the affected relative-pair method (Goldin and Weeks, 1993). Thus if both concordant and discordant pairs are available or if unaffected individuals have been gentoyped to aid in reconstructing IBD sharing among the affecteds, the Haseman–Elston method may be more appropriate than methods which examine only concordant or only discordant relative pairs.

An alternative parameterization developed by Morton and colleagues (Collins et al., 1996), also uses both concordant and discordant pairs simultaneously. This method, called the Beta model, gives expectations for IBD sharing probabilities at a QTL as a function of a continuous underlying liability as opposed to the strictly binomial QTL probability expectations of affection status considered purely on the presence/absence scale. This may represent a more reasonable model for gene action.

Issues common to all methods

Choice of markers

In designing a linkage study, there are a number of issues that are common to all sampling designs and methodological approaches. One of these is the choice of markers to genotype. When candidate genes can be identified on the basis of known function and hypotheses about the biological pathways underlying the phenotype of interest, these are often genotyped and analyzed first. In these cases, it is difficult to interpret weak evidence of linkage as the prior hypothesis of functionality must be balanced against the necessity of correcting for testing of numerous candidate genes. A complete genome screen with highly polymorphic anonymous markers is obviously a more time consuming undertaking than a candidate gene study. However, a complete genome screen has the advantage that the likelihood of successfully localizing one or more genes of interest is dependent on the linkage power of the sample rather than the chance selection of the correct functional candidate. A hybrid approach, incorporating known functional candidates as part of a complete genome screen, provides the best of each method.

In linkage analyses, the ideal marker has many alleles of relatively equal frequency, leading to a high proportion of heterozygotes, or heterozygosity, and maximizing the likelihood of unambiguously resolving the segregation of alleles through a family. Assuming Hardy–Weinberg equilibrium, the expected heterozygosity for a marker with n alleles is:

$$1 - \sum_{A=1}^{n} (p_A)^2,$$

where p_A is the frequency of allele A. The short tandem repeat markers (STRs) that make up current commercially available screening sets are selected for high heterozygosity and even chromosomal distribution. The informativeness of this marker data is further improved in multipoint analyses in which all genotyped markers in a region or all genotyped markers on an entire chromosome are used simultaneously to estimate IBD allele sharing at a particular chromosomal location. A number of methods have been developed which use multiple marker genotypes directly and provide exact multipoint IBD estimates (Kruglyak and Lander, 1995; Sobel and Lange, 1996). However, while these methods work well in nuclear families and small pedigrees, they become prohibitively computationally expensive for extended pedigrees. In these cases, approximate multipoint IBD methods which use linear functions of the exact IBD estimates for each marker can accommodate pedigrees of any size and complexity (Almasy and Blangero, 1998).

Recently scientists have begun to develop another variety of genetic marker, called single nucleotide polymorphisms (SNPs). Although SNPs generally have only two alleles, and therefore a maximum heterozygosity of 50%, they have generated great interest. Due to their binary nature, it is possible to design extremely efficient high throughput assays for SNPs making them easier and cheaper to genotype than STRs. It is also anticipated that SNPs will be found more frequently in the human genome than STRs. The increased frequency of SNPs is expected to lead to a more dense marker map than can be achieved with STRs and thus the ability to more finely localize QTLs. However, because of the decreased heterozygosity, 2.5 times as many SNPs need to be genotyped to yield the same information content as a map of STRs with a heterozygosity of 0.75 (Kruglyak, 1997). For example, a 2 cM SNP map with biallelic markers of equal frequency has the same multipoint linkage information content as a 5 cM STR map with 75% heterozygosity (Kruglyak, 1997).

It has been suggested that SNPs may eventually be used to screen the genome through association methods based on linkage disequilibrium rather than linkage (Risch and Merikangas, 1996). Considerable reservations about this approach exist as the number of statistical tests required creates the potential of many false positive associations and more problematically, linkage disequilibrium need not exist even for markers within the gene being sought. If disequilibrium does not exist between a marker and a QTL, association analyses will fail to detect the QTL no matter how large the sample size or how strong the genetic effect. Recently, a SNP-based association analysis testing the likely utility of this method failed to identify the β-globin mutation known to cause sickle cell anemia, despite the availability of dozens of SNPs in and around the gene (Pennisi, 1998).

A strategy that combines STRs and SNPs and uses a two-stage design of both linkage and association may ultimately prove most effective. Linkage analyses with STRs can be used to identify candidate regions which may then be narrowed through association-based analyses with SNPs, should the necessary disequilibrium exist. The positional candidate regions identified through linkage analyses are often prohibitively large, on the order of 10 cM or more, and typing additional markers may not narrow the region to a manageable size for intensive molecular studies such as positional cloning. A known gene of functional relevance within the linked region may be chosen as a positional candidate gene. Alternatively, the region might be narrowed through linkage disequilibrium which, when present, exists over a much smaller chromosomal distance than does linkage. In the same sample used for linkage analyses, disequilibrium can be assessed through family-based association analyses such as the transmission disequilibrium test for discrete traits in nuclear families (Spielman et al., 1993) or the measured genotype test for quantitative traits in large pedigrees (Hopper and Matthews, 1982). If disequilibrium is detected, future work may focus on the immediate vicinity of the associated polymorphism(s).

Ascertainment

In any analysis of a quantitative trait, selected samples will always have greater power to detect linkage *for the trait on which they were ascertained*. Power decreases rapidly as the correlation between the focal ascertainment trait and other traits of interest declines. Similarly, concordant and discordant relative pair analyses increase in power as individuals are ascertained from higher and lower percentiles of the trait distribution. If a trait is rare, an ascertained sample is obviously necessary. However, if a quantitative trait or a common discrete trait are being studied, the increased power for localizing genes for a focal ascertained trait must be balanced against the efficiency of studying multiple phenotypes in a slightly larger, randomly selected sample. If ascertainment is undertaken, it should follow a clearly defined, systematic, set of rules. Until recently, it was commonly accepted that ascertainment must be carefully tracked for segregation analyses but not for linkage studies. While this is true for parametric analyses of Mendelian traits, it does not necessarily hold for non-parametric methods and complex traits. It has been demonstrated that correction for ascertainment, achieved by conditioning the likelihood of a pedigree on the likelihood of the proband(s), can dramatically affect power to detect linkage in the variance component method (Comuzzie and Williams, (1999)).

Summary

Linkage strategies for analyzing complex traits in man differ from the more classical Mendelian linkage methods in that no knowledge of the inheritance pattern of the trait or the characteristics of the responsible gene are required. The so-called non-parametric statistical methods developed to study complex traits derive their genetic information from IBD allele sharing among relatives. The choice of which method to use depends on the nature and prevalence of the phenotype.

If the trait of interest can be measured quantitatively that will provide more power to detect linkage than dichotomous versions of the same trait. For quantitative traits or common discrete traits, variance component methods in large pedigrees will provide the most power. For discrete traits with a prevalence of less than 10%, affected relative pair methods are most efficient. Ascertainment will maximize power for a single focal trait, but should be conducted in a systematic manner with identifiable probands. For studies of multiple related traits that are common or quantitative, random sampling is most efficient.

While candidate genes of known functional relevance should be tested, a complete genome screen with evenly-spaced, highly polymorphic markers presents the greatest chance of success. Positional candidate regions identified through such a screen may then be followed up with family-based disequilibrium analyses for finer QTL mapping.

Acknowledgements

The authors' current research on linkage methods for the analysis of complex traits is supported by NIH grants GM18897, MH59490, HL45522, GM31575, and HL28972.

List of abbreviations

IBD	Identical by descent
IBS	Identical by state
MCMC	Markov chain Monte Carlo
QTL	Quantitative trait locus
SNP	Single nucleotide polymorphism
STR	Short tandem repeat

References

Almasy, L., Dyer, T.D. and Blangero, J. (1997) Bivariate quantitative trait linkage analysis: Pleiotropy versus coincident linkage. *Genet. Epidemiol.* 14: 953–958.

Almasy, L. and Blangero J. (1998) Multipoint quantitative trait linkage analysis in general pedigrees. *Am. J. Hum. Genet.*, 62: 1198–1211.

Amos, C.I. (1994) Robust variance-components approach for assessing genetic linkage in pedigrees. *Am. J. Hum. Genet.*, 54: 535–543.

Amos, C.I., Dawson, D.V. and Elston, R.C. (1990) The probabilistic determination of identity-by-descent sharing for pairs of relatives from pedigrees. *Am. J. Hum. Genet.*, 47: 842–543.

Amos, C.I., Krushkal, K., Thiel, T.J., Young, A., Zhu, D.K., Boerwinkle, E. and De Andrade, M. (1997) Comparison of model-free linkage mapping strategies for the study of a complex trait. *Genet. Epidemiol.*, 14: 743–748.

Amos, C.I., De Andrade, M., Zhu, D. and Yu, L. (1998) ACT — Analysis for complex traits. ftp://request.mdacc.tmc.edu/pub/Programs/Act/

Bailey, J.N. and Almasy, L. (1995) A brute force dichotomization approach to quantitative trait linkage analysis. *Genet. Epidemiol.*, 12: 719–722.

Barr, C.L., Livingston, J., Williamson, R., Sandor, P., Kidd, K.K., Pauls, D. and Tsui, L.-C. (1994) Genome scan for linkage to Gilles De la Tourette syndrome. *Am. J. Hum. Genet.*, 55: A345.

Blangero, J. and Almasy, L. (1997). Multipoint oligogenic linkage analysis of quantitative traits. *Genet. Epidemiol.*, 14: 959–964.

Collins, A., MacLean, C.J. and Morton, N.E. (1996) Trials of the beta model for complex inheritance. *Proc. Natl. Acad. Sci.*, 93: 9177–9181.

Comuzzie, A.G. and Williams, J.T. (1999) Correcting for ascertainment bias in the COGA data set. *Genet. Epidemiol.*, in press.

Cordell, H.J., Todd, J.A., Bennett, S.T., Kawaguchi, Y. and Farrall, M. (1995) Two-locus maximum lod score analysis of a multifactorial trait: joint consideration of IDDM2 and IDDM4 with IDDM1 in type 1 diabetes. *Am. J. Hum. Genet.*, 57: 920934.

Curtis, D. and Sham, P.C. (1994) Using risk calculation to implement an extended relative pair analysis. *Ann. Hum. Genet.*, 58: 151–162.

Davis, S., Schroeder, M., Goldin, L.R. and Weeks, D.E. (1996) Nonparametric simulation-based statistics for detecting linkage in general pedigrees. *Am. J. Hum. Genet.*, 58: 867–880.

Day, N.E. and Simons, M.J. (1976) Disease susceptibility genes — their identification by multiple case family studies. *Tissue Antigens*, 8: 109–119.

Duggirala, R., Williams, J.T., Williams-Blangero, S. and Blangero, J. (1997) A variance component approach to dichotomous trait linkage analysis using a threshold model. *Genet. Epidemiol.*, 14: 987–992.

Duggirala, R., Blangero, J., Almasy, L., Dyer, T.D., Williams, K.L., Leach, R.J., O'Connell, P. and Stern, M.P. (1999) Linkage of type 2 diabetes mellitus and age of diabetes onset to a genetic location on chromosome 10q in Mexican Americans. *Am. J. Hum. Genet.* in press.

Elston, R.C., Buxbaum, S., Jacobs, K.B. and Olson, J.M. (1998) Haseman and Elston revisited. *Genet. Epidemiol.*, 15: 546.

Geldermann, H. (1975) Investigations on inheritance of quantitative characters in animals by gene markers. I. Methods. *Theoret. Appl. Genet.*, 46: 319–330.

Génin, E. and Clerget-Darpoux, F. (1996) Consanguinity and the sib-pair method: An approach using identity by descent between and within individuals. *Am. J. Hum. Genet.*, 59: 1149–1162.

Goldin, L.R. and Weeks, D.E. (1993) Two-locus models of disease: Comparison of likelihood and nonparametric linkage methods. *Am. J. Hum. Genet.*, 53: 908–915.

Greenberg, D.A. (1990) Linkage analysis assuming a single-locus mode of inheritance for traits determined by two loci: Inferring mode of inheritance and estimating penetrances. *Genet. Epidemiol.*, 7: 467–479.

Greenberg, D.A. and Hodge, S.E. (1989) Linkage analysis under "random" and "genetic" reduced penetrance. *Genet. Epidemiol.*, 6: 259–264.

Haseman, J.K. and Elston, R.C. (1972) The investigation of linkage between a quantitative trait and a marker locus. *Behav. Genet.*, 2: 319.

Hinds, D. and Risch, N. (1998) The ASPEX package: affected sib-pair mapping, v1.78. ftp://lahmed.stanford.edu/pub/aspex.

Hopper, J.L. and Mathews, J.D. (1982) Extensions to multivariate normal models for pedigree analysis. *Ann. Hum. Genet.*, 46: 373–383.

Korczak, J.F. and Goldstein, A.M. (1997) Sib-pair linkage analyses of nuclear family data: Quantitative versus dichotomous disease classification. *Genet. Epidemiol.*, 14: 827–832.

Kruglyak, L. (1997) The use of a genetic map of biallelic markers in linkage studies. *Nat. Genet.*, 17: 21–24.

Kruglyak, L. and Lander, E.S. (1995) Complete multipoint sib-pair analysis of qualitative and quantitative traits. *Am. J. Hum. Genet.*, 57: 439–454.

Kruglyak, L., Daly, M.J., Reeve-Daly, M.P. and Lander, E.S. (1996). Parametric and nonparametric linkage analysis: A unified multipoint approach. *Am. J. Hum. Genet.*, 58: 1347–1363.

Lander, E. and Kruglyak, L. (1995) Genetic dissection of complex traits: guidelines for interpreting and reporting linkage results. *Nat. Genet.*, 11: 241–247.

MacLean, C.J., Bishop, D.T., Sherman, S.L. and Diehl, S.R. (1993) Distribution of Lod scores under uncertain mode of inheritance. *Am. J. Hum. Genet.*, 52: 354–361.

Mitchell, B.D., Ghosh, S., Schneider, J.L., Birznieks, G. and Blangero, J. (1997) Power of variance component linkage analysis to detect epistasis. Genet. Epidemiol., 14: 1017–1022.

Pennisi, E. (1998) A closer look at SNPs suggests difficulties. *Science*, 281: 1787–1789.

Province, M.A., Rice, T., Borecki, I.B., Gu, C. and Rao, D.C. (1998) A multivariate and multilocus variance components approach using structural relationships to assess quantitative trait linkage via SEGPATH. *Genet. Epidemiol.*, in press.

Pugh, E.W., Jaquish, C.E., Sorant, A.J.M., Doetsch, J.P.,Bailey-Wilson, J.E. and Wilson, A.F. (1997) Comparison of sib-pair and variance-components methods for genomic screening. *Genet. Epidemiol.*, 14: 867–872.

Risch, N. (1990) Linkage strategies for genetically complex traits. II. The power of affected relative pairs. *Am. J. Hum. Genet.*, 46: 229–241.

Risch, N. and Botstein, B. (1996) A manic depressive history. *Nat. Genet.*, 12: 351–353.

Risch, N. and Giuffra, L. (1992) Model misspecification and multipoint linkage analysis. *Hum. Hered.*, 42: 77–92.

Risch, N. and Merikangas, K. (1996) The future of genetic studies of complex human diseases. *Science*, 273: 1516–1517.

Self, S.G. and Liang, K.-Y. (1987) Asymptotic properties of maximum likelihood estimators and likelihood ratio tests under nonstandard conditions. *J. Am. Stat. Assoc.*, 82: 605–610.

Sobel, E. and Lange, K. (1996) Descent graphs in pedigree analysis: applications to haplotyping, location scores, and marker sharing statistics. *Am. J. Hum. Genet.*, 58: 1323–1337.

Spielman, R.S., McGinnis, R.E. and Ewens, W.J. (1993) Transmission test for linkage disequilibrium: the insulin gene region and insulin-dependent diabetes mellitus. *Am. J. Hum. Genet.*, 52: 506–516.

Suarez, B.K., Rice, J. and Reich, T. (1978) The generalized sib pair IBD distribution: its use in the detection of linkage. *Ann. Hum. Genet.*, 42: 87–94.

Towne, B., Siervogel, R.M. and Blangero, J. (1997) Effects of genotype-by-sex interaction on quantitative trait linkage analysis. *Genet. Epidemiol.*, 14: 1053–1058.

Van Eerdewegh, P., Dupuis, J., Santangelo, S.L., Hayward, L.B. and Blacker, D.B. The importance of watching our weights: How the choice of weights for non-independant sibpairs can dramatically alter results. *Genet. Epidemiol.*, in press.

Whittemore, A.S. and Halpern, J. (1994) Probability of gene identity by descent: computation and applications. *Biometrics*, 50: 109–117.

Wijsman, E.M. and Amos, C.I. (1997) Genetic analysis of simulated oligogenic traits in nuclear families and extended pedigrees: summary of GAW10 contributions. *Genet. Epidemiol.*, 14: 719–735.

Williams, J.T., Duggirala, R. and Blangero, J. (1997) Statistical properties of a variance-components method for quantitative trait linkage analysis in nuclear families and extended pedigrees. *Genet. Epidemiol.*, 14: 1065–1070.

Williams, J.T. and Blangero, J. (1999a) Comparison of variance components and sibpair-based approaches to quantitative trait linkage analysis in unselected samples. *Genet. Epidemiol.*, 16: 113–124.

Williams, J.T. and Blangero, J. (1999b) Asymptotic power of likelihood ratio tests for detecting quantitative trait loci using the COGA data. *Genet. Epidemiol.*, in press.

W.E. Crusio and R.T. Gerlai (Eds.)
Handbook of Molecular-Genetic Techniques for Brain and Behavior Research (Techniques in the Behavioral and Neural Sciences, Vol. 13)

CHAPTER 2.1.4

Genetic association studies in behavioral neuroscience

Philip Gorwood[1,2]

[1] *INSERM 155 Unity, Genetics Epidemiology, Paris, France*
[2] *Hôpital Louis Mourier, Colombes, France*

Definition and mechanisms of association studies

Association describes the occurrence together in a population or kindred of two or more features with a frequency greater than would be anticipated on the basis of chance alone. Association studies are widely used in genetic epidemiology, searching for the involvement (and/or the localisation) of a gene in the risk for some disorders. In practice, association studies may consist in analysing the distributions of genotypes in different groups of patients, with or without a certain phenotype.

The measure of association can be qualitative (presence or absence of a significant association) or quantitative (strength of association evaluated by risk difference, relative risk, odds ratio and attributable risk). These parameters are explained in Table 1. The χ^2 test detects the existence of a significant association (usually with a fixed 5% risk of type I error). The risk difference evaluate the presence of the genetic marker in the sample having the phenotype, but taking into account the existence of subjects without the phenotype who also have this marker. The relative risk is a ratio of these probabilities (instead of a difference), and the Odds ratio is the ratio of odds for each population. It is possible to compute the significance of the association based on the Odds Ratio (is the OR significantly different from 1 ?) with the Mantel Haenszel test (1959). Lastly, attributable risk is frequently used in genetic epidemiology as not only giving the strength of the association, but also reflecting the impact of the allele in explaining the phenotype (as it depends on the odds ratio and the allele frequency in the general population).

The approach of association studies is based on phenotype/genotype correlation, and is mainly devoted to detecting the impact of candidate gene(s) in the risk for a feature or a disorder. Association studies are also used in different samples to test the allele frequencies of some genetic polymorphisms, which are potentially closed to, and in linkage disequilibrium with, the gene involved in the feature analysed. This second type of analysis is thus devoted to localisation studies, whether it be for a systematic screening of the genome or for a more precise localisation in certain candidate regions.

According to linkage disequilibrium, if a certain phenotype is influenced by a certain allele, then the polymorphisms of a closed (and available) genetic marker should be associated with the phenotype. Association studies in genetics are, in this view, based on the core concept of linkage disequilibrium, which requires explanation.

Mendel's third law stipulated that segregation of two different pairs of alleles is independent. Soon after Mendel's laws were rediscovered, Batelson (Batelson et al., 1908) found certain combinations in the Vetch, Lathyrus odoratus, more

TABLE 1

Measures of association

Phenotype	Genetic marker		
	Present	Absent	Total
Affected	a	b	a+b
Controls	c	d	c+d
Total	a+c	b+d	N

χ (qualitative approach) $= \Sigma(\text{obs-calc})^2/\text{calc}$. Where a,b,c,d are the observed (obs) numbers, and $(a+b)*(a+c)/N$ is the calculated (calc) value for a, and so on for b, c, d.

RD (Risk Difference) $= a/(a+b)] - [c/(c+d)]$
RR (Relative Risk) $= [a/(a+c)]/[b/(b+d)]$
OR (Odds Ratio) $= ad/bc$
AR (Attributable Risk) $= [a/(a+c)]*[(OR-1)/OR]$

often than expected. Batelson et al. coined the terms "coupling", and "repulsion", for excesses, or deficits, of certain haplotypes. Jennings (1917) and Robbins (1918) confirmed this observation in following the evolution of two loci in the general population. Morgan (Morgan, 1910) recognised that coupling and repulsion are two aspects of the same phenomenon which he coined "linkage", i.e. two genes on the same or homologous chromosomes are transmitted together more often than by chance only. If linkage is complete, only two types of progeny are observed, but frequently four types are observed, with two types in smaller number. Morgan explained this finding by exchange of chromosome pieces between homologous chromosomes during meiotic crossing over, the frequency of crossing-over depending on the distance between two gene loci in one chromosome.

In the absence of selection, population heterogeneity, and other confounding phenomena, linkage disequilibrium should theoretically be monotonely related to distance. If the number of meioses is large enough, and if we consider that crossover occurs independently of gender and the locus of the region analysed, then there is a correspondence between genetic distance (which is related to frequency of recombination) and physical distance (number of base pairs) between two loci. In humans, 30 crossing-over (Morgans) are observed during the first meiosis for the 3×10^9 base pairs. It is thus generally considered that 10^8 base pairs represent 1 Morgan, then 1 cM is evaluated at 10^6 base pairs. Different conditions need to be respected, and many methodological limitations have been raised in the utilisation of the linkage disequilibrium concept in association studies, such as the gap between physical distance and linkage disequilibrium. Looking for a strong and specific excess of an allele, genotype or haplotype in a group of subjects compared to unaffected matched controls according to the presence of linkage disequilibrium is exposed to numerous limitations.

Methodological limitations of association studies and their consequences

The theoretically regular relation between physical distance and linkage disequilibrium has major exceptions. Recombinational hot spots can be detected (Chakravarti et al., 1982) with frequent recombinations on small distance of chromatides. Telomeric regions are, furthermore, known to have a higher frequency of recombinations than the rest of the genome, and particularly to centromeric regions. Chromosome 12 and 13, lastly, are longer in women than in men. The different exceptions lead to the estimation of a variation of the equivalence between genetic and physical distance, 1 centimorgan being between 5.10^5 and 1.10^7 base pairs. Furthermore, the decay of linkage disequilibrium in succeeding generations is governed by the recombination frequency (i.e. the frequency of crossovers between the relevant loci during meiosis) but also the number of generations since the introduction of the mutation into the population (Jorde, 1995). For example, at a recombination fraction of 0.01, the "half life" of an association would be 69 generations or about 2,000 years (Owen and

McGuffin, 1993). Nevertheless, the shared segments extended over large genetic distances (over 1 cM) in a Finnish sample established over 2,000 years ago (Peterson et al., 1995). In conclusion, linkage disequilibrium may not be used as a constant and uniform concept, and linkage disequilibrium studies should be more productive among genetically isolated populations with a well-documented history (Nimgaonkar, 1997).

The second kind of limitation of association studies is based on the required comparison of groups (affected versus unaffected). First of all, association studies have to control the stratification effect, i.e. one of the sample, affected or controls may derive from an ethnic group sharing other DNA variants for reasons unrelated to illness. For example, the association between alleles of the HLA DQB1 gene has been observed widely in Caucasian populations, but is undetectable among the Japanese (Trucco, 1992). Control and affected groups have thus to be matched for ethnic origins.

The question of the best type of proband to analyse is still unresolved. For some authors, severely affected subjects with numerous familial cases, for others, much more representative probands are required. The question is also conflicting for controls (1) extracted form the general population, regardless of their phenotype (if the mutation analysed is involved, affected probands should be different from the whole population); (2) extracted from the general population, but excluding affected subjects, and (3) without any potentially linked disease, in the proband interview and also in his first (and sometimes second) degree family. This last group control (labelled "super-control") has been criticised (Kendler, 1990) as favouring the observation of coaggregation of different diseases in the affected group.

The control group based on volunteers in human association studies is also subject to bias as volunteer subjects have rates of psychopathology exceeding the population expectations (Gibbons et al., 1990; Risch et al., 1990; Thaker et al., 1990; Shtasel et al., 1991; Buckley et al., 1992). The use of unrelated husbands or spouses is also a potential artefact as they are more frequently affected, an artefact which is called "assortative mating" (Suarez and Hamp, 1994).

For multifactorial disorders and complex traits, the existence of incomplete penetrance (i.e. subjects with the vulnerability gene may not all express the phenotype), render the choice of the control group even more important. For example, for research on a common disorder such as alcoholism, it may be more relevant to recruit controls that drink alcohol regularly without developing alcohol abuse or dependence, instead of abstainers, who may quickly develop alcoholism if they were exposed to alcohol consumption.

Another set of problems is related to statistical power and specificity of the association detected or rejected. Small sample size and low heterogeneity of allele frequencies are important limitations in the opportunity to show linkage disequilibrium between two closely linked markers (Hill, 1974; Thompson et al., 1988; Gershon et al., 1989; Greenberg, 1993). The chances of detecting the association are further reduced if the allele conferring susceptibility is relatively common among unaffected individuals (Cox and Bell, 1989). Estimations of the sample size required to detect a significant association are given in the following Table 2. These estimations are relevant for testing a single hypothesis (i.e. is *one* allele significantly involved in *one* phenotype ?), without prior notion of the type of association (association or exclusion), and with reasonably low type I and type II errors ($\alpha = 5\%$ and $\beta = 10\%$. The number of patients that have to be included are given according to the frequencies of the studied allele in the control population (from 0.01 to 0.50), and the expected difference in allele frequency between patients and controls (usually based on previous publications).

On the other hand, Crowe (1993) raised the importance of the likelihood of false-positive results in the genetics of behaviour and psychiatric disorders, on the basis of the number of candidate genes (namely all genes expressed in the brain).

TABLE 2

Estimation of each sample size needed to show a significant association according to allele frequency in controls and the expected difference between patients and controls

N	Expected difference in allele frequency between patients and controls							
	1%	2%	5%	10%	20%	30%	40%	50%
1%	2578	822	207	80	32	19*	13*	10*
2%	4332	1270	284	100	38	22	15*	11*
5%	9312	2515	485	223	68	34	21	14*
10%	16862	4386	778	223	68	34	21	14*
20%	29254	7442	1249	334	92	44	26	17*
30%	39052	9598	1574	407	106	48	27	17*
40%	43259	10854	1754	443	111	48	26	14*
50%	44875	11214	1790	443	106	44	21	8*

* Sample size lower than 20 subjects require special test corrections.

If the 0.05 first risk error is divided by the number of genes to be tested (Bonferoni correction), a statistical significance of 0.00001 should be achieved to avoid a false-positive rate of 5%. On the other hand, if we were to use a significant level of 0.0001, 80% of positive findings would be false, and the traditional α level of 0.05 would yield 99.5% false-positive results! According to such a false-positive rate, and considering that half of the genes are expressed in the brain (thus entitled candidate genes), there are potentially 100 000 association studies which could be carried out for each neurobehavioral trait studied. Regarding such a nearly impossible-to-reach level of significance, alternative approaches have been proposed. Kidd (1993) suggested that consistent replication would be the best evidence for a true association (we may add, from different researches, on independent samples). Other techniques for association studies were proposed, in order to control one of the major problems of case/control studies, namely the stratification bias. There are two family based association studies, not requiring a control sample, which are the "haplotype relative risk" and the "transmission disequilibrium test".

The "haplotype relative risk" (HRR) strategy (Falk and Rubinstein, 1987; Ott, 1989) compares marker-allele frequencies in affected offspring with frequencies in control "individuals" consisting of parental alleles not transmitted to affected offspring (Fig. 1).

The "transmission disequilibrium test" (TDT) (Spielman et al., 1993) consider parents who are heterozygous for an allele associated with the disease, and evaluates the frequency with which this allele (or haplotype) is transmitted to affected offspring. The TDT is also protected from stratification bias, and is considered as being a test for linkage in presence of association (Spielman and Ewens, 1996).

Considering all the limitations previously described of association studies, the role and

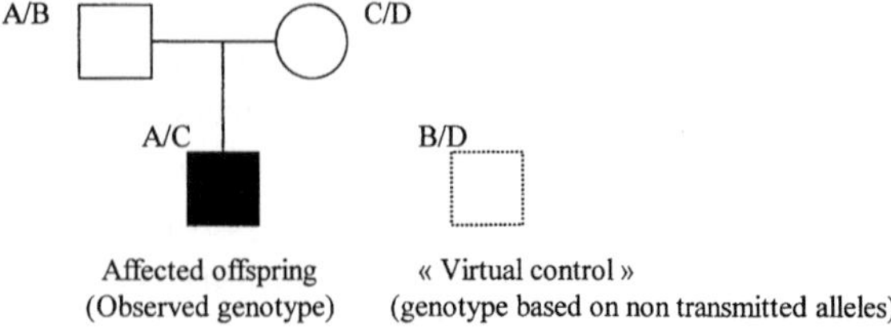

Fig. 1. The Haplotype Relative Risk (HRR) method.

the impact of association studies needs to be discussed compared to the other techniques. The frequent criticisms raised each time a positive association study is not further replicated (a phenomenon which appears quite frequently), raising the danger of "*throwing the baby out with the bathwater*". For example, some authors propose to refuse to accept any more case-control genetic association studies of complex traits (Paterson, 1997). Association studies are nevertheless extremely powerful in detecting small, partial or complex genetic effects, although their high sensibility is associated with low specificity. Often, association studies are in fact specifically required, and different methodological improvements may easily increase the validity of association studies.

Specificity of association studies compared to linkage studies

Linkage analysis is generally considered to be one of the most powerful methods for mapping genes responsible for hereditary diseases. Linkage analyses were successfully used for more than 200 diseases. Nevertheless, these diseases all have a Mendelian inheritance pattern, and the qualities of linkage analysis may not be applicable for common disorders that have a more complex pattern of inheritance. In fact, linkage analyses are exposed to methodological limitations (Risch, 1990) when the assumption of a single major gene is incorrect, when genetic heterogeneity and sporadic cases are present, when penetrance of susceptibility genotypes is reduced, and when the genetic aetiology is not well defined. These features all characterise the genetics of animal and human neurobehaviour, and association studies may thus be frequently more relevant.

For the genetics of animal and human behaviour, association studies are opposite to linkage analysis on many points. Firstly, association studies deal with discrete traits and alleles (found together with greater frequency than by chance alone), and linkage analyses are based on the observation of loci (transmitted together with greater frequency than 50%). Secondly, association studies look for causal relationship, as linkage analyses try to detect physical proximity on chromosome. Lastly, association studies require the studying of unrelated individuals within a population, while linkage analyses need the examining of multigenerational families. According to these characteristics, association studies should be (initially) preferentially chosen, rather than linkage studies, for polygenic (i.e. many additive genes, each of small effect, are likely), multiloci (i.e. two or more genes are necessary), multifactorial (i.e. epigenetic factors such as environment play a significant role in the individual vulnerability), heterogeneous (the phenotype analysed contains different entities which are dependent on different vulnerability factors) disorders or traits. From this viewpoint association studies are relevant for genetic susceptibility (or liability), not genetic determinism. A recent comparison of the power of linkage studies vs association studies to detect a significant role of a gene involved in complex human disease has been undertaken (Rish and Merikangas, 1996). When the genotypic risk ratio is low, 2.0 for example, then the number of families needed to detect a significant linkage is so large that it is not practically achievable (between 12,000 and 300,000 for allele frequencies between 0.01 and 0.50). On the contrary, the number of triplets to be investigated with the TDT method is much lower, from 340 to 5800, for allele frequencies between 0.01 and 0.50 (Rish and Merikangas, 1996).

On the other hand, association studies have a lower chance to throw light on the biological mechanism underlying the association. For instance, a major disease gene at another locus in linkage disequilibrium with the associated marker locus or a minor gene possibly located at the marker locus itself, cannot be distinguished.

Instead of preferring sensibility to specificity or vice-versa, a step-by-step strategy could be chosen, especially as it is in accordance with the recruitment of affected subjects. Beginning studies with a case-control analysis is strong, but the recruitment of parents may be done later. DNA from parents can be used to see if the allele found in excess in the affected sample is more frequently transmitted to the affected proband than not transmitted from the parents (HRR and TDT methods). Affected sibs can be later recruited (sibpair analysis), and once multiplex families are detected, linkage studies are easier to perform (lod score linkage analysis). A combined association-linkage analysis can also help to clarify this issue, for example, dividing families into two groups based on whether the index-affected case has the associated marker allele, and then comparing the linkage results between the two groups (Hodge, 1994; Greenberg, 1993; Baron, 1997).

Screening the genome with association studies?

The mutational analysis (denaturing gradient gel electrophoresis techniques, single stranded conformation polymorphism paradigms, and chemical cleavage of mismatches) can be applied for identifying new sequence variations in selected genes, in a series of affected individuals (Gejman and Gelernter, 1993). The alternative method to use association studies for detecting the genes involved in a complex disorder is to use anonymous markers and rely upon linkage disequilibrium. Linkage disequilibrium mapping by association studies might therefore prove useful in complex disorders assuming low mutation rates of both marker and susceptibility loci. It might be feasible to undertake a systematic search of the entire genome for association with susceptibility loci using regularly spaced markers, with highly informative polymorphisms, and roughly evenly spaced at about 2 cM intervals. This strategy has been successfully used in detecting the key role of the TGF alpha gene in cleft lip and palate (Holder et al., 1992), the glucokinase and glycogen synthase genes in non-insulin dependent diabetes (Chiu et al., 1992; Groop et al., 1993), the insulin gene and insulin dependent diabetes (Bell et al., 1984), and the myelin basic protein gene in multiple sclerosis (Tienari et al., 1992). The magnitude of OR at neighbouring loci may be more useful to map the disease gene, than "*P* values" that are influenced by sample size (Feder et al., 1996).

There are different ways to use random polymorphisms in association studies for detecting a region of interest. One of them may be particularly more relevant for complex disorders, namely DNA pooling in allelic association studies (Daniels et al., 1998). The advantage of this method is that the number of genotypings and gels is drastically reduced. This advantage may be particularly important in neurobehavior genetic studies in which the number of tests is going to be large before positive and real association is found.

Sobell et al. (1992, 1993) proposed another approach to mutation screening entitled VAPSE-based case-control association studies. Instead of anonymous random markers, the VAPSE is a Variation of the DNA sequences that Affects Protein Structure or Expression. These authors describe five phases of a VAPSE study. Firstly, select a candidate gene. Secondly, use a subset of patients to search for VAPSEs. Thirdly, if you find a VAPSE, use a large case-control association study to see whether the disease is associated with the VAPSE. Fourthly, replicate the observed association, and determine if the VAPSE and disease are linked using linkage analysis in families that exhibit the VAPSE. Lastly, estimate the population attributable risk of the VAPSE. However, as has been mentioned (Owen and McGuffin, 1993), the VAPSE method might not be so straightforward, as it is supposed to prove that any variation found in association with a disease is itself responsible for increasing susceptibility rather than another locus close by and in linkage disequilibium with it. Furthermore, attempts to confine studies to functional polymorphisms are confounded by the sensitivity of molecular pharmacological techniques and the problems of transferring findings from cultured cells to whole organisms.

TABLE 3

Recommendation for association studies in neurobehavioral genetics

Measuring phenotype

1 Analyse the phenotype with reliable instruments.

2 Adopt also a quantitative trait locus approach (severity, age at onset).

3 Potentially related phenotypes should also be analysed
(Drug abuse and antisocial personality disorder, schizophrenia and other psychotic disorders...).

Recruiting adequate sample

4 Look for ethnical homogeneity (for patients and controls, their parents and grandparents).

5 Screen familial history, mainly for disorder with high frequency, for first degree above a certain age. With direct interview, or with familial inventory (indirect interview) when relatives are not available.

6 Compute the size of the sample needed according to:
– the number of polymorphisms used (2 alleles, 3 genotypes, 5 markers).
– the number of tests to be realised (Bonferonni corrections).

Using relevant statistical methods

7 Power to detect the association is evaluated, and compatible with this particular study.

8 Analyse the "effect size" of the DNA variant under study, and the proportion of those affected who have disease due to that DNA variant.

9 Meta-analysis (between different recruitment centers, or with other studies already performed) estimate heterogeneity (which should be low), and the specific significancy of the association (which should be high).

10 Splitting the sample (patients and controls) into two halves (first and second part of the collected sample), should give non significant difference (except statistically because of smaller samples).

11 Evaluate the odds ratio (strength of the association) and the attributable risk (impact of the genetic factor on the risk for the analysed phenotype).

12 Use regression analysis (detecting the variables involved in explaining presence vs absence of the phenotype, or of the susceptibility allele).

13 Use censored data analysis (such as Kaplan–Meyer) when controls may develop the disease later on. This should not change significantly the results.

Choosing the most appropriate genetic marker

– When the genetic marker is supposed to be the involved mutation

14 It is expressed, or mainly expressed, in the brain.

15 The association confirms, or is in accordance with, some knowledge in neuroscience.

16 It is preferentially in an exon part of a gene, or in the promoter region. The polymorphism is affecting protein structure or expression (VAPSE).

– When the marker is used as being potentially in linkage disequilibrium with the involved mutation

17 The marker is highly informative.

18 The marker has a mutation rate low enough to permit the maintenance of linkage disequilibrium.

19 Other markers within the regions should be tested.

Performing alternative and complementary approaches

20 Test other markers within the region of interest, and use haplotypes (containing the region of interest).

21 Collect blood samples from both undiagnosed parents of part of the sample
Use non transmitted alleles from parents as "virtual controls" (HRR).
Use the transmission/disequilibrium test (TDT).

22 Collect blood samples from affected sibs
Use sibpair approach.

23 Collect blood samples from multiplex families
Use linkage studies, when the case/control association study is positive and confirmed by parents/affected children TDT.

Conclusions

Association studies are based on the simplest characteristic of genetic disorders, namely that if an allele increases the risk for a disorder, then the frequency of this allele must be more than that expected by chance only in the affected population. Furthermore, if an allele is found in excess in an affected population, then this allele must be close to (or part of) the gene which is involved in the risk for the disorder. This simple statement can nevertheless be easily copied, transformed, biased or disqualified. Association studies are thus one of the easiest but most dangerous ways to put to the fore genes which may be involved, sensibility being largely more important than specificity. Association studies are nevertheless frequently used, and sometimes inevitable, for example when major genes are only responsible for a minority of cases. Furthermore, many simple characteristics of the way association studies are performed could considerably increase their characteristics, and may help to evaluate the level of confidence for a positive (or negative) result, or to enrich the methodology of future association studies. Table 3 lists various topics that modify the validity of association studies, concerning the inclusion criteria, the way phenotype can be measured, the statistical methods used, the kind of polymorphisms tested, the required marker for the association with probes flanking the candidate gene, the utility of complementary methods. Analysing the potential validity of any association study, already performed or to be performed, may avoid "*throwing the baby out with the bathwater*".

References

Baron, M. (1997) Association studies in psychiatry: a season of discontent. *Molecular Psychiatry*, 2(4): 278–279.

Batelson, W., Saunders, R. and Punnett, R. (1908) Confirmations and extensions of mendel's principles in other animals and plants. Report to the evolution committee of the Royal Society, London.

Bell, G.I., Horita, S. and Karam, J.H. (1984) A polymorphic locus near the human insulin gene is associated with insulin-dependent diabetes mellitus. *Diabetes*, 33: 176–183.

Buckley, P., O'Callaghan, E., Larkin, C. and Waddington, J. (1992) Schizophrenia research: the problem of controls. *Biological Psychiatry*, 32: 215–217.

Chakravarti, A. and Nei, M. (1982) Utility and efficiency of linked marker genes for genetic counseloing. II. Identification of linkage phase by offspring phenotypes. *Am. J. Hum. Genet.*, 34: 531–551.

Chiu, K., Province, M. and Permutt, A. (1992) Glucokinase gene is genetic marker for NIDDM in american blacks. *Diabetes*, 141: 843–9.

Cox, N.J. and Bell, G.I. (1989) Disease associations. Chance, artifact, or suscepti- bility genes? *Diabetes,* 38: 947–950.

Crowe, R.R. (1993) Candidate genes in psychiatry: an epidemiological perspective. *Am. J. Med. Genet.* (*Neuropsych. Genet.*), 48: 74–77.

Daniels, J., Holmans, P., Williams, N., Turic, D., McGuffin, P., Plomin R. and Owen, M.J. (1998) A simple method for analyzing microsatellite allele image patterns generated from DNA pools and its application to allelic association studies. *Am. J. Hum. Genet.*, 62:5, 1189–97.

Falk, C.T. and Rubinstein P. (1987) Haplotype relative risk: an easy reliable way to construct a proper control sample for risk calculation. *Ann. Hum. Gen.* 51: 227–233

Feder, J.N., Gnirke, A., Thomas, W., Tsuchihashi, Z., Ruddy, D.A., Basava, A., Dormishian, F., Domingo, R. Jr, Ellis, M.C., Fullan, A., Hinton. M.L., Jones, N.L., Kimmel, B.E., Kronmal, G.S., Lauer, P., Lee, V.K., Loeb, D.B., Mapa, F.A., McClelland, E., Meyer, N.C., Mintier, G.A., Moeller, N., Moore, T., Morikang, E., Wolff, R.R., et al (1996) A novel MHC class I-like gene is mutated in patients with hereditary haemochromatosis. *Nat. Genet,* 13: 399–408.

Gejman, P. and Gelernter, J. (1993) Mutational analysis of candidate genes in psychiatric disorders. *Am. J. Med. Genet.*, 48: 184–191.

Gershon, E.S., Martinez, M., Goldin, L., Gelernter, J., Silver, J. (1989) Detection of marker associations with a dominant disease gene in genetically complex and heterogeneous diseases. *Am. J. Hum. Genet.*, 45: 578–585.

Gibbons, R., Davis, J., Hedeker, D. (1990) A comment on the selection of healthy controls for psychiatric experiments. *Arch. Gen. Psych.*, 47: 785–786.

Greenberg, D.A. (1993) Linkage analysis of 'necessary' disease loci versus 'susceptibility' loci. *Am. J. Hum. Genet.*, 52: 135–143.

Groop, L., Kankuri, M., Schalling, C. et al. (1993) Association between polymorphism of the glycogen synthase gene and non-insulin dependent diabetes mellitus. *N. Eng. J. Med.*, 328: 10–14.

Hill, W.G. (1974) Estimation of linkage disequilibrium in randomly mating populations. *Heredity*, 33(2): 229–39.

Hodge, S.E. (1994) What association analysis can and cannot tell us about the genetics of complex disease. *Am. J. Med. Genet.*, 54: 318–323.

Holder, S., Vintiner, G., Farren, B., Malcolm, S. and Winter R (1992) Confirmation of an association between RFLPs at the transforming growth factor — alpha locus and nonsyndromic cleft lip and palate. *J. Med. Genet.*, 29: 393–397.

Jennings, H.S. (1917) The numerical result of diverse systems of breedings with respect to two pairs of characters. *Genetics*, 2 : 97–106.

Jorde, L.B. (1995) Linkage disequilibrium as a gene-mapping tool. *Am. J. Hum. Genet.*, 56: 1114. Kendler, K. (1990) The super-normal control group in psychiatric genetics: possible evidence for coaggregation. *Psych. Genet.*, 1: 45–53.

Kidd, K.K. (1993) Association of disease with genetic markers: deja vu all over again. *Am. J. Med. Genet.*, 48: 71–72.

Mantel, N. and Haenzel, W. (1959) Statistical aspects of the analysis of data from retropective studies of disease. *J. Natl. Cancer Inst.*, 22: 719–748.

Morgan, T. (1910) Sex-limited inheritance in drosophila. *Science*, 32: 120–122.

Nimgaonkar, V. (1997) In defense of association studies. *Mol. Psych.*, 2(4): 275–276.

Ott, J. (1989) Statistical properties of the haplotype relative risk. *Genet. Epidemiol.*, 6: 127–130.

Owen, M., McGuffin, P. (1993) Association and linkage: complementary strategies for complex disorders. *J. Med. Genet.*, 30: 638–639.

Paterson, (1997) Case control association studies in complex traits-the end of an era? *Mole. Psych.*, 2(4): 277.

Peterson, A.C., Di Rienzo, A., Lehesjoki, A.E., de la Chapelle, A., Slatkin, M. and Freimer, N.B. (1995) The distribution of linkage disequilibrium over anonymous genome regions. *Hum. Mol. Genet.*, 4: 887–894.

Rish, N. and Merikangas, K. (1996) The future of genetic studies of complex human diseases. *Science*, 273: 1516–1517.

Risch, S., Lewine, R., Jewart, R., Eccard, M., McDaniel, J. and Risby, E. (1990) Ensuring the normality of "normal" volunteers. *Am. J. Psychiatry*, 147: 682–683.

Robbins, R. (1918) Some applications of mathematics to breeding problems III. *Genetics*, 3: 375–89.

Sobell, J.L., Heston, L.L. and Sommer, S.S. (1993) Novel association approach for determining the genetic predisposition to schizophrenia: case-control resource and testing of a candidate gene. *Am. J. Med. Genet.*, 48(1): 28–35.

Sobell, J.L., Heston, L.L. and Sommer, S.S. (1992) Delineation of genetic predisposition to multifactorial disease: a general approach on the threshold of feasibility. *Genomics*, 12(1): 1–6.

Spielman, R.S., McInnis, R.E. and Ewens, W.J. (1993) Transmission Test for Linkage Desequilibrium: The insulin gene region and insulin-dependent diabetus mellitus (IDDM). *Am. J. Hum. Genet.*, 52: 506–516.

Spielman, R. and Ewens, W. (1996) The TDT and other family-based tests for linkage disequilibrium and association. *Am. J. Hum. Genet.*, 59: 983–989.

Shtasel, D., Gur, R., Mozley, D., Richards, J., Taleff, M., Heimberg, C., Gallacher, F. and Gur, R. (1991) Volunteers for biomedical research. Recruitment and screening of normal controls. *Arch. Gen. Psychiatry*, 48: 1022–1025.

Suarez, B.K. and Hampe, C.L. (1994) Linkage and association. *Am. J. Hum. Genet.*, 54: 554–559.

Thaker, G., Moran, M., Lahti, A., Adami, H. and Tamminga, C. (1990) Psychiatric morbidity in research volunteers. *Arch. Gen. Psychiatry*, 47: 980.

Thompson, E.A., Deeb, S., Walker, D. and Motulsky, A.G. (1988) The detection of linkage disequilibrium between closely linked markers: RFLPs at the AI-CIII apolipoprotein genes. *Am. J. Hum. Genet.*, 42(1): 113–124.

Tienari, P., Wikstrom, J., Sajantila, A. et al. (1992) Genetic susceptibility to multiple sclerosis linked to myelin basic protein gene. *Lancet*, 340: 987–991.

Trucco, M. (1992) To be or not to be ASP 57, that is the question. *Diabetes Care*, 15: 705–715.

W.E. Crusio and R.T. Gerlai (Eds.)
Handbook of Molecular-Genetic Techniques for Brain and Behavior Research (Techniques in the Behavioral and Neural Sciences, Vol. 13)

CHAPTER 2.1.5

DNA pooling in allelic association studies

Nigel M. Williams and Michael J. Owen

Neuropsychiatric Genetics Unit, Divisions of Psychological Medicine and Medical Genetics, Tenovus Building, University of Wales College of Medicine, Heath Park, Cardiff, CF4 4XN

Introduction

Allelic association studies are a powerful way of locating genes for complex traits (Owen and McGuffin, 1993; Risch and Merikangas, 1996). They can be used both for fine mapping susceptibility loci as part of positional cloning strategies and for examining candidate genes. Large scale systematic searches for allelic association have not been carried out to date because of the belief that they are not technically feasible due to the fact that dense marker maps and large numbers of subjects are required. This could be overcome by adopting a DNA pooling strategy whereby the individual DNA samples from each of the two test groups are pooled and amplified together. This allows a large number of markers to be quickly screened and those that are most likely to show an allelic association to be identified. DNA pooling has been used successfully in quantitative trait loci (QTL) association studies in animals. Here it has been combined with "selective genotyping", whereby subjects are only genotyped if their scores on the trait under study lie at either extreme of the continuum. DNA from "high" and "low" groups respectively are then pooled, allowing a considerable saving in genotyping (Asada et al., 1994; Darvasi and Soller, 1994). In such a selective DNA pooling study Lipkin et al. (1998) were able to map QTL affecting milk protein percentage, demonstrating that the equivalent statistical power of 45,600 individual genotypings could be obtained from only 328 pool genotypings. DNA pooling has also been used to study recessive diseases in inbred human populations (Carmi et al., 1995). In both these cases it was a shift towards homozygosity of particular markers that was sought in pooled DNAs.

DNA pooling using outbred populations

So far DNA pooling has not been successfully applied to case-control association studies in outbred populations. This is mainly due to several problems which can occur when simple sequence repeats (SSRs) are applied to pooling. Two of these result from PCR artefacts commonly associated with SSRs, namely stutter banding and differential amplification which confound attempts to estimate allele frequencies directly from pooled samples.

Stutter bands are produced by the amplification of products one or two repeat units shorter than the correct sized amplimere due to slippage of Taq polymerase on the repeated sequence. When pooled samples are studied, these bands summate with the correct sized amplimeres of alleles one or two repeat units smaller. This confounds attempts to estimate the frequencies of different alleles directly from the amplified products of the pooled samples. Stutter is prevalent in dinucleotide repeats, the most common form of SSR marker, but occurs to a lesser extent in tri- and tetranucleotide repeats. Differential amplification is observed in heterozygotes and is due to the preferential amplification of the smaller allele. This is observed in all classes of microsatellite and is believed to be due to the larger alleles reannealing at a faster rate (as they contain more repeat units) which reduces the efficiency of PCR amplification (Demers et al., 1995).

Approaches to DNA pooling using outbred populations

Despite the problems described, several groups using different approaches have attempted to estimate allele frequencies of microsatellite markers from pooled samples. Khatib and colleagues (1994) selected dinucleotide microsatellites that did not produce stutter bands and predicted allele frequencies and standard errors from pools amplified with the selected markers. Pacek et al. (1994) also chose markers, a tetranucleotide and a 16bp repeat, that did not produce stutter bands, in order to demonstrate that allele frequencies can be accurately estimated from pools. However they observed differential amplification, which in the resulting analysis of pooled genotypes gave moderately inflated estimates of the smallest alleles and underestimates of the frequency of larger alleles.

LeDuc et al. (1995) developed a mathematical method to correct for stutter based on measuring the ratio between the amounts of true product and stutter bands for each marker. Reasonable estimates of allele frequencies were obtained for one marker but comparison between groups was not attempted.

Perlin et al. (1995) described a more accurate method for the mathematical correction of stutter. The relative amount of product for the true amplimere and stutter bands was calculated individually for each allele of a marker and then used to compose a matrix. This was used to transform the results from a pool where many alleles were represented. This method was more sophisticated than its predecessors because it took into account the observation that each allele of a particular marker has its own unique stutter pattern. Barcellos et al. (1997) have combined this method with mathematical correction for differential amplification and obtained promising data from four markers. However each individual marker required a lengthy work-up which will hinder large-scale studies where many markers will be needed.

The ΔAIP DNA pooling method

The ΔAIP (Allele Image Pattern Difference) DNA pooling method (Daniels et al. 1998) uses pooling as a screening step to identify markers for individual genotyping where the goal is not to estimate absolute allele frequencies for a single group, but simply to compare the uncorrected products of pooled SSR amplification. By ignoring the problems of stutter and differential amplification we are left with an efficient, accurate and versatile method which allows the application of DNA pooling to large-scale association studies, without the need for any lengthy analysis of each marker prior to genotyping.

The ΔAIP approach to DNA pooling is summarised in Fig. 1. Patient and control samples are pooled separately and markers are typed in the two pools. Using primers with fluorescent 5′ ends, PCR products can be analysed on an automated sequencing apparatus. The essence of the method is the observation that when SSRs are amplified in pooled DNA samples the images they produce when analysed by the GENESCAN™/GENOTYPER™ software on the ABI fluorescent system are highly stable and replicable (Fig. 2). Allele image patterns produced for the two groups are overlaid and differences in pattern area between pools computed. From this a ΔAIP statistic is calculated from the difference in areas between the two allele image patterns expressed as a function of the total shared and

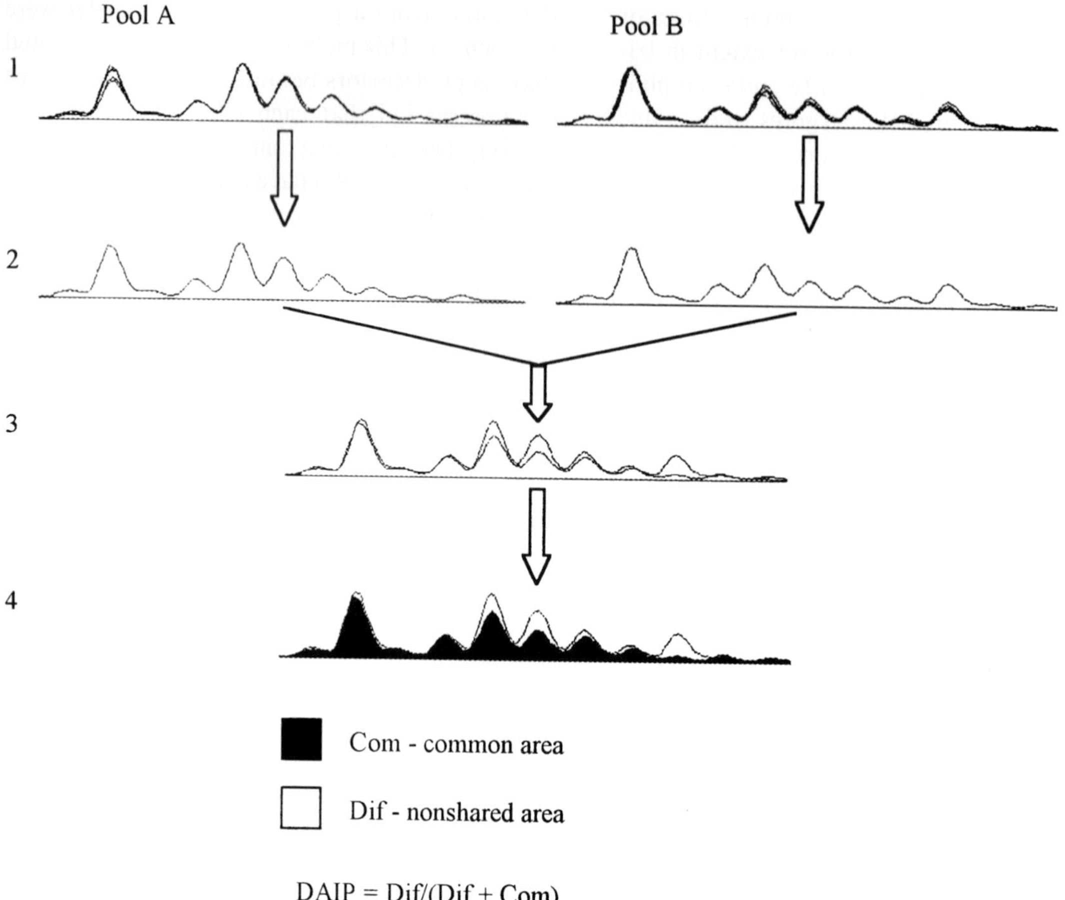

Fig. 1. Summary of the ΔAIP DNA pooling strategy. 1. Six AIPs for pools A and B are overlayed. 2. An AIP is selected as a consensus image representing each pool. 3. The consensus AIPs are overlayed and (4) then the nonshared (Dif) and shared (Com) areas between the two AIPs determined from which the ΔAIP statistic can be calculated.

non-shared area. This is an estimate of the relative difference in allele frequencies between the two groups and can therefore be used to select which markers are more likely to show a genuine significant allelic association and therefore those that need to be individually genotyped.

Technical aspects

Quantification of genomic DNA

Genomic DNA can be extracted using standard protocols from blood or mouth swabs. Accurate quantitation of each individual DNA sample is essential for DNA pooling. Trace residues such as phenol that can remain in the genomic DNA following its extraction can interfere with spectrophotometric readings, so we recommend a fluorometric assay using a reporter dye such as PicoGreen™ (Molecular Probes) as the preferred method to quantify DNA. We suggest the following procedure to ensure accurate quantitation of DNA samples.

1. Dilute each individual sample of genomic DNA to 80 ng/ul.
2. Allow to solublise by rotating for 72 h at 4°C.

D6S285

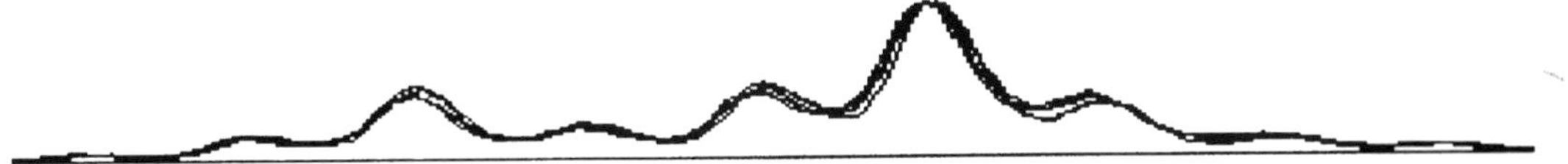

D6S1653

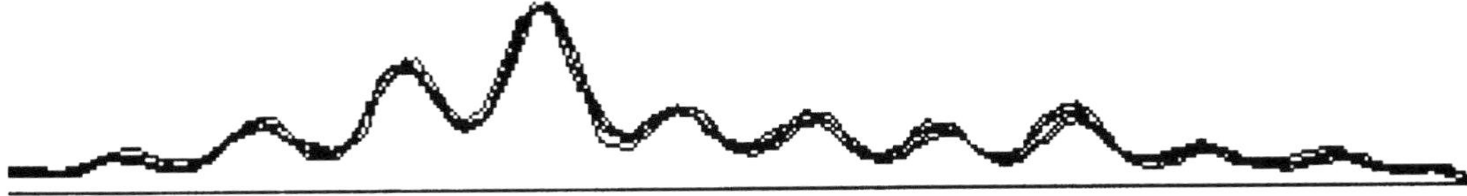

D6S1593

Fig. 2. DNA from 100 individuals was pooled in triplicate. Each of the three pools underwent PCR amplification in duplicate. Data from three microsatellite markers (D6S285, D6S1653 and D6S1539) are shown. This process generated 6 AIPs for each marker (3 pools × 2 PCRs). Figure shows the results of overlaying the 6 AIPs for each marker.

3. Quantify each sample using a spectrofluorimeter then dilute each sample to 8 ng/ul in TE buffer.
4. Allow each dilution to solublise by rotating for 72 h at 4°C.
5. Quantify each dilution using a spectrofluorimeter and adjust the concentration of any samples that are not at 8 ng/ul ± 2.
6. Amplify each samples using 2 microsatellite markers and run on an ABI373/377 automated sequencer.
7. Genotype all samples and reject any individuals that either failed to amplify or whose alleles appear unusually large or small.
8. Only individual samples that reach this stage are included in the pool.

Construction of DNA pools

For each test group, equal volumes of each individual sample are combined in the same tube. Pools should be constructed in triplicate to allow the accuracy and reproducibility of pool construction to be tested. Each pool should then be amplified in duplicate with the same 2 microsatellite markers that were used to test quality of each DNA sample. After running on an ABI373/377 automated sequencer the allele image patterns for each marker can be overlaid using GENOTYPER. Pools should only be accepted if all 6 traces (3 triplicate pools amplified in duplicate) overlay perfectly. Moreover it is essential that the amplified pools are approximately representative of the allele frequencies of both microsatellite markers as determined by individual genotyping.

Analysis of the pooled DNA

Amplification of microsatellite markers
Microsatellite markers can be amplified in the pooled DNA in exactly the same way that they would be for individual DNA. But, as the ΔAIP method compares the areas of each allele image pattern it is essential to eliminate any 'plus A' which can seriously affect the sensitivity of the method. This is caused by the terminal transferase activity of Taq Polymerase which promotes the 3′ nontemplated addition of adenine ('plus A'). This does not occur with all markers and is thought to be influenced by a number of factors, including the sequence at the 3′ end of the PCR product and the temperature and length of the final extension step of the PCR reaction. This problem can be removed in a number of ways. Firstly, the post-PCR products can be incubated at 37°C with either Klenow fragment or T4 polymerase which correct the problem by removing the extra A (Ginot et al. 1996). An alternative, is to optimise the PCR reaction for each marker to force either the complete addition of the extra base. This can be successfully achieved by modifying the 5′ end of the reverse, nonfluorescent primer by either adding a Guanine (G) (Magnuson et al., 1996) or a specific seven base tail ('PIG-tail') (Brownstein et al., 1996) which both promote adenylation of up to 100% of the PCR products.

Although it is not a requirement of the ΔAIP pooling method, a biological approach to reducing stutter and differential amplification would make the allele image patterns easier to analyse and the ΔAIP test statistic more accurate. Stutter can be reduced to some extent by higher fidelity Taq polymerase that is less prone to slippage on the repeat units. Preferential amplification of the smaller allele can be reduced by use PNA oligomers (peptide nucleic acids) that can sufficiently block DNA template inhibiting reassociation whilst allowing primer extension. PNA has been used with great efficiency in the amplification of a 16bp repeat microsatellite, D1S80, amplifying the larger allele with 4.2 times greater efficiency (Demers et al., 1995). This is not, however, a perfect solution to the problem of differential amplification. The concentration of PNA added to the PCR would have to be optimised for every marker, also the degree of enhancement of different alleles of the same marker is variable.

After amplification, the products of up to 9 markers (3 of each fluorescent label) can be combined and then electrophoresed in the same lane of a 6% polyacrilamide gel on an ABI 373/377 automated sequencer, using standard run conditions.

Comparison of the allele image patterns
After analysis of the gel using the GENESCAN™ software the allelic profile of each marker can be viewed in GENOTYPER™. As each pool is constructed in triplicate and each marker amplified in duplicate, the resulting 6 pooled images can then be overlaid in GENOTYPER™ which scales all traces so that the height of the largest peak is equal in each. The baseline is determined automatically in GENOTYPER™ and there is no need to modify the images in any way prior to analysis. If all 6 images overlay then a consensus image can be chosen to be representative of the allele frequencies in that pool. In our experience, in over 90% of marker/pool combinations all 6 patterns align completely. Figure 2 shows the 6 allele image patterns overlaid for 3 dinucleotide repeat markers. It is evident that there is almost 100% alignment between the 6 images produced for each marker and this clearly demonstrates the reproducibility of both the pooling process and of PCR amplification of pooled DNA.

In occasional cases where 1, or rarely 2, pools misalign the remaining 4 or 5 images that did align are taken as representative of the allele image pattern from which a consensus image can be chosen. Providing the same pool has amplified and overlaid perfectly for other markers we attribute this to PCR failure rather than to errors in pool construction. The consensus images representing each of the two pools to be compared can then be overlaid in GENOTYPER™ and stored using the Apple Macintosh screen capture facility (apple+shift+3).

Calculation of the ΔAIP

After isolating the pooled allele image in Simple Text™ (Apple Computer, Inc.) it can then be imported into the graphics manipulation software Debabelizer™ (Equilibrium) for analysis. To compare the results of pooled genotyping of the two consensus images the areas that are shared (COM) and not shared (DIF) by both image patterns are measured by shading the shared and non-shared areas in different colours and calculating the pixel count for each. A test statistic called ΔAIP (Allele Image Pattern Difference) can then be calculated from the expression DIF/(DIF+COM). Calculation of ΔAIP is illustrated in Fig. 1.

Estimation if the significance level

It is not necessarily the case that the largest ΔAIP values will correspond to the most statistically significant differences. This is because the value of ΔAIP will be influenced by factors such as the number of marker alleles and the number of cases and controls in the sample. In general, we have found that the test criteria for a ΔAIP depends on the informativity of the marker (measured here by the Polymorphism Information Content; PIC); the higher the PIC, the higher the criteria. In addition, markers with larger numbers of alleles require a higher criteria whereas the criteria is reduced for larger samples. In summary, simulations have suggested that for a marker with PIC of 0.5, a ΔAIP of 0.2 gives an approximate 5% test for samples containing 50 cases and 50 controls, a ΔAIP of 0.15 is required for 100 cases and controls, and a ΔAIP of 0.1 for 200 cases and controls. For a marker with PIC of 0.7, the corresponding ΔAIP values are 0.25, 0.2 and 0.15, and for a marker with PIC of 0.9 they are 0.35, 0.27 and 0.2 (Daniels et al., 1998). However, care must be taken with the interpretation of the test criteria because simulations were based on realistic approximations to stutter bands and differential amplification but it is possible that different values of these could lead to different results. Therefore the test criteria given above are intended as a rough indication only.

A more accurate estimate of the *P*-value may be obtained by simulating case and control samples from a population using allele frequencies estimated from the peak heights of the control sample. A computer program has been written for this process and will be made available for interested users (contact owenmj@cf.ac.uk). While such "simulated *P*-values" are not the true significance levels of the difference in allele frequencies in the two pools, they do indicate which markers have the more different allele image patterns when amplified in the pools. Since we recommend that the ΔAIP method is used simply as an initial screening procedure prior to individual genotyping, precise significance levels are not usually required. In our experience we have found that in general significant ΔAIP values occurred when chi-squares obtained from individual genotyping were significant. This was demonstrated in a series of experiments using 13 constructed pools from 100 individuals who had been typed with the marker D6S1279 (Daniels et al., 1998). The pools were made such that the smallest, middle and largest alleles of the pool were represented at different frequencies (10–50% at 10% intervals). The ΔAIP method successfully detected all the differences among pools, with four instances where differences were detected that were not present on individual genotyping (false positives). These results suggests that the most appropriate use of the ΔAIP DNA pooling method in association studies is as part of a rapid initial screen in which a liberal test criterion is employed

TABLE 1

40 microsatellite markers spanning approximately 41 cM of chromosome 6p24-21.1 were genotyped in pools constructed from 30 patients with haemachromatosis and 30 ethnically matched controls

Marker	cM*	Δ-AIP *P*-value	Δ-AIP	Chi-squared *P*-value
D6S1674	–	0.26	0.08	
D6S309	0.0	0.13	0.78	
D6S277	0.6	0.28	0.15	
D6S1034	4.1	0.25	0.20	
D6S1263	3.1	0.20	0.06	
D6S1955	1.0	0.14	0.42	
D6S1006	2.1	0.09	0.96	
D6S1279	1.0	0.19	0.27	
D6S1593	1.9	0.16	0.58	
D6S1653	0.2	0.15	0.29	
D6S259	1.2	0.20	0.52	
D6S1605	3.6	0.20	0.46	
D6S1584	1.7	0.18	0.51	
D6S1700	1.8	0.16	0.61	
D6S285	0.0	0.09	0.39	
D6S422	1.5	0.15	0.69	
D6S1665	0.7	0.29	0.09	
D6S1597	1.6	0.11	0.60	
D6S1588	0.1	0.35	0.08	
D6S1686	1.2	0.19	0.14	
D6S1660	0.9	0.23	0.13	
D6S1554	1.8	0.23	0.08	
D6S1691	0.7	0.39	0.01	0.19
D6S265	1.6	0.37	0.01	0.06**
D6S1558	0.0	0.28	0.02	0.02
D6S276	0.6	0.25	0.15	
D6S1666	0.5	0.24	0.25	
D6S1568	2.2	0.31	0.04	0.88
D6S439	0.6	0.29	0.07	
D6S1629	0.2	0.26	0.07	
D6S291	1.4	0.10	0.48	
D6S1602	2.6	0.07	0.95	
D6S1548	0.2	0.29	0.03	0.84
D6S1641	1.5	0.27	0.04	0.33
D6S1562	4.1	0.22	0.13	
D6S1616	1.4	0.15	0.79	
D6S426	1.1	0.44	0.01	0.65
D6S1549	1.6	0.17	0.28	
D6S1582	2.9	0.44	0.72	
D6S282	1.2	0.03	1.00	

(say α of < 0.05 or 0.10). "Positive" markers should then be subjected to individual genotyping. We have noted that rarely false negatives may occur when the associated allele is large due to the effects of differential amplification. However, we believe that that this problem can be avoided if the overlaid AIPs are inspected at the time of ΔAIP calculation. If even small differences are apparent in the portion of the AIPs deriving from larger alleles then individual genotyping should be undertaken as a precaution.

Fine mapping of the heamochromatosis gene using the ΔAIP method of DNA pooling

In order to demonstrate the effectiveness of the ΔAIP method in systematic searches for linkage disequilibrium we conducted a study to determine whether it would detect the known linkage disequilibrium between haemochromatosis and markers at 6p21.3 (Raha-Chowdhury et al., 1995). Forty closely-spaced microsatellite markers spanning approximately 41 cM of chromosome 6p 24-21.1 were genotyped in pools constructed from 30 patients with haemachromatosis and 30 ethnically matched controls. All markers were dinucleotide repeats apart from D6S1034, D6S1263, D6S1006 (trinucleotides), D6S1955 and D6S1279 (tetranucleotides). Pools were made and the ΔAIP method applied according to the methods described above. Individual genotyping of haemochromatosis cases and controls was undertaken with the 8 markers that yielded a ΔAIP simulated *P*-value of < 0.05. The results are shown in Table 1.

* Approximate distance between makers (ftp://ftp.Genethon.fr/pub/gmap/nature-1995 and http://cedar.genetics.soton.ac.uk/public.html/index.html). Pools were made and the ΔAIP method applied according to the methods described above. Individual genotyping of haemochromatosis cases and controls was undertaken with the 8 markers that yielded a ΔAIP simulated p value of <0.05. Chi-squared values for comparison of individual genotypes from patients and controls were calculated using CLUMP (Sham and Curtis, 1995).

The ΔAIP pooling method successfully detected the previously observed linkage disequilibrium between haemochromatosis and the 122bp allele of marker D6S265 (Raha-Chowdhury et al., 1995) as well as association with the closely adjacent marker D6S1558. Interestingly D6S1691 which lies only 1.6 cM the other side of D6S265 also gave a ΔAIP simulated P-value of < 0.05, though the results of individual genotyping failed to reach statistical significance. The remaining 5 markers for which ΔAIP analysis of pooled genotypes gave a $P < 0.05$, did not show significant differences between cases and controls on individual genotyping.

These results show that ΔAIP analysis is sufficiently sensitive to detect moderate associations, and that, while its false-positive rate may be slightly higher than its nominal value, it is not excessively so (here, 5 "false-positives" out of 40 tested loci).

Application to complex traits

The ΔAIP DNA pooling method should potentially be useful for the next rate limiting step in the dissection of complex traits, the fine mapping of putative loci in large critical regions. Putative linkage for complex traits often spans large genetic distances. In the situation where a region of interest spans 20 cM, at least 21 markers would have to be typed to cover the interval at a spacing of 1 cM, a reasonable marker density for identifying allelic association with microsatellites. Using the ΔAIP method once the pools have been constructed (approximately 2 weeks) only 12 PCRs (6 each for patient and control pools) are needed for each marker. Nine markers can be run per lane on an ABI 373/377 (due to the different sizes of the amplimers and the differential labelling of the forward primers with the 3 fluorophores available), hence all markers can be run on a typical 36 lane ABI gel. Therefore the screen for the whole 20 cM region can be run and analysed on one gel. In comparison standard genotyping of 100 patient and control individuals with 21 markers would require running at least 84 gels. Whilst the construction of the pools is time consuming the ΔAIP DNA pooling method offers a substantial time saving which will improve further with increasing sample size and marker number. Analysis is also far simpler for the ΔAIP method as allele calling (which in our hands has an error rate of approximately 1%) is not required and the ΔAIP test statistic is calculated from unmanipulated raw data.

Furthermore, DNA pooling makes it feasible to study a large number of markers in relatively large samples required to identify loci of small effect. The ΔAIP method has already been used to identify susceptibility loci for general cognitive ability (Fisher et al., 1999). Risch and Merikangas (1996) have calculated that a sample size of 1000 is generally sufficient to detect a locus with a genotype relative risk of only 1.5. This is well within the sample size limits of DNA pooling which are constrained by the total number of DNA molecules in the PCR reaction.

Discussion — the future

The ΔAIP DNA pooling method facilitates a rapid screen for allelic association over large genetic distances and has the capacity to handle large sample sizes whilst reducing costs in terms of time and consumables. Despite this, a limitation is that although this method can predict overall differences it is unable to estimate the actual allele frequencies in the pool and therefore does not indicate which alleles are associated with the difference. These have to be ascertained by following up with individual genotyping. Furthermore, not correcting for differential amplification results in the need for each allele image pattern to be subjectively inspected to allow some markers with non-significant ΔAIPs to be individually genotyped if the differences appear in the portion of the AIP representing larger alleles.

These limitations could all be overcome by improving the sensitivity of the ΔAIP test statistic by incorporating a more sophisticated mathematical correction which could result in accurate estimation of allele frequencies from analysis of pools by GENOTYPER^{TM} (Perlin et al., 1995; Barcellos et al., 1997). However, this is likely to require extensive data collection to allow determination of allele specific stutter patterns and the degree of differential amplification for each marker prior to the analysis of pooled samples. Therefore until mathematical correction methods are sophisticated enough to deal with the complexities of amplification of many alleles in pools, the ΔAIP test statistic will be the simplest and most powerful approach to screening for frequency differences between pools.

The development of this method means that we can now look forward to genome-wide allelic association studies which until now have not been considered feasible (Risch and Merikangas, 1996). Assuming a sex-averaged genome size of 3700 cM, approximately 2000 markers will be required for a genome scan at 1 cM intervals. Conventional genotyping of 400 patient and control samples would require 1,600,000 genotypes. This is reduced to just 4000 using the ΔAIP method. For larger sample sizes the relative efficiency of DNA pooling is even greater.

References

Asada, Y., Varnum, D.S., Frankel, W.N. and Nadeau, J.H. (1994) A mutation in the Ter gene causing increased susceptibility to testicular teratomas maps to mouse chromosome 18. *Nat. Genet.*, 6:363–368.

Barcellos, L.F., Klitz, W., Field, L., Tobias, R., Bowcock, A.M., Wilson, R., Nelson, M.P., Nagatomi, J. and Thompson, G. (1997) Association mapping of disease loci by use of a pooled DNA genomic screen. *Am. J. Hum. Genet.*, 61: 734–747.

Brownstein, M.J., Carpten, J.D. and Smith, J.R. (1996) Modulation of non-templated nucleotide addition by Taq DNA polymerase: Primer modifications that facilitate genotyping. *Biotechniques*, 20: 1004–1010.

Carmi, R., Rokhlina, T., Kwitek-Black, A.E., Elbedour, K., Nishimura, D., Stone, E.S. and Sheffield, V.C. (1995) Use of human DNA pooling strategy to identify a human obesity syndrome locus on chromosome 15. *Hum. Mol. Genet.*, 4: 9–13.

Darvasi, A. and Soller, M. (1994) Selective DNA pooling for determination of linkage between a molecular marker and a quantitative trait locus. *Genetics*, 138: 1365–1373.

Daniels, J., Holmans, P., Williams, N., Turic, D., McGuffin, P., Plomin, R. and Owen, M.J. (1998) A simple method for analyziing allele image patterns generated from DNA pools and its application to allelic association studies. *Am. J. Hum. Genet.*, 62: 1189–1197.

Demers, D.B., Curry, E.T., Egholm, M. and Sozer, A.C. (1995) Enhanced PCR amplification of VNTR locus D1S80 using peptide nucleic acid (PNA). *Nucelic Acids Research*, 23: 3050–3055.

Fisher, P.J., Turic, D., McGuffin, P., Asherson, P., Ball, D., Craig, I., Eley, T., Hill, L., Chorney, K., Chorney, M.J., Benbow, C.J., Lubinski, D., Plomin, R. and Owen, M.J. (1999) DNA Pooling Identifies QTLs for General Cognitive Ability in Children on Chromosome 4. *Hum. Mol. Genetics*, submitted.

Ginot, F., Bordelais, I., Nguyen, S. and Gyapay, G. (1996) Correction of some genotyping errors in automated fluorescent microsatellite analysis by enzymatic removal of one base overhangs. *Nucl. Acids Res.*, 24: 549–541.

Khatib, H., Darvasi, A., Plotski, Y. and Soller, M. (1994) Determining relative Microsatellite allele frequencies in pooled DNA samples. *PCR Methods and Applications*, 4: 13–18.

LeDuc, C., Miller, P., Lichter, J. and Parry, P. (1995) Batched analysis of genotypes. *PCR Methods and Applications*, 4: 331–336.

Lipkin, E., Mosig, M.O., Darvasi, A., Ezra, E., Shalom, A., Friedmann, A. and Soller, M. (1998) Quantitative trait locus mapping in dairy cattle by means of selective milk DNA pooling using dinucleotide microsatellite markers: Analysis of milk protein percentage. *Genetics*, 149: 1557–1567.

Magnuson, V.L., Ally, D.S., Nyland, S.J., Karanjawala, Z.E., Rayman, J.B., Knapp, J.I., Lowe, A.L., Ghosh, S., Collins, F.S. (1996) Substrate nucleotide-determined Non-templated addition of adenine by Taq DNA polymerase: Implications for PCR-based genotyping and cloning. *Biotechniques*, 21: 700–709.

Owen, M.J. and McGuffin, P. (1993) Association and linkage: Complementary strategies for complex disorders. *J. Med. Genet.*, 30: 638–639.

Pacek, P., Sajantila, A. and Syvanne, A-C. (1994) Determination of allele frequencies at loci with length polymorphism by quantitative analysis of DNA amplified from pooled samples. *PCR Method and Applications*, 2: 313–317.

Perlin, M.W., Lancia, G. and Ng, S-K. (1995) Toward fully automated genotyping: genotyping microsatellite markers. *Am. J. Hum. Genet.*, 57: 1–12.

Raha-Chowdhury, R., Bowen, D.J., Stone, C., Pointon, J.J., Terwilliger, J.D., Shearman, J.D., Robson, K.J.H., Bomford, A. and Worwood, M. (1995) New polymorphic microsatellite markers place the haemochromatosis gene telomeric to D6S105. *Hum. Mol. Genet.*, 4: 1869–1874.

Risch, N., Merikangas, K. (1996) The future of genetic studies of complex human diseases. *Science*, 273: 1516–1517.

Sham, P. and Curtis, D. (1995) Monte Carlo test for association between disease and alleles at highly polymorphic loci. *Ann. Hum. Genet.*, 59: 97–105.

W.E. Crusio and R.T. Gerlai (Eds.)
Handbook of Molecular-Genetic Techniques for Brain and Behavior Research (Techniques in the Behavioral and Neural Sciences, Vol. 13)

CHAPTER 2.1.6

Research strategies for the analysis of neurological mutants of the mouse

Constantino Sotelo[1] and Jean Mariani[2]

[1]*Neuromorphologie: Développement-Evolution, INSERM U. 106, Hôpital de la Salpêtrière, 75651 Paris Cedex 13, France*
[2]*Laboratoire de Neurobiologie du Développement, Institut des Neurosciences, Université Pierre et Marie Curie et CNRS UMR 7624, 75005 Paris, France*

Introduction

For years, the main link between genes and behavior has been the study of neurological mutant mice. With the technological advances introduced by targeted genetic mutations in mice (Mansour et al., 1988), most investigators interested in genes and behavior are using the new tools offered by molecular biology, with the aim of associating the inactivation of a precise gene with a defined behavior. Nevertheless, the behavioral alterations which characterize most spontaneous mutations can orient the search of new genes with known behavioral function, thus maintaining the interest of spontaneous mutations for behavioral genetic studies.

Although the interest in neurological mutants of the mouse has a long history (Grüneberg, 1952), the golden era — when the majority of the most interesting existing mutations were discovered or analyzed — was during the sixties and seventies. Thus, while in the first published "Catalog of the Neurological Mutants of the Mouse" by Sidman et al. (1965) only about 90 mutants were listed, in the "Genetic Variants and Strains of the Laboratory Mouse", edited by Lyon and Searle (1989), the number of mutants affecting the nervous system (including sensory organs and the neuromuscular system) is over 200.

Because detection of a new mutation requires the identification of the affected mice, even before determining the hereditary character of the disease, the vast majority of the known mutations give rise to either neurological symptoms or anatomical deformities which permit the recognition of the mutants by direct inspection. In fact, only a small proportion of the known mutant mice have been identified fortuitously, in the course of studies of brain morphology in adult or developing mice. This important requirement greatly biases the chances of detecting mutations specifically affecting cognitive functions or complex behaviors which will need the use of precise tests to detect their impairment. From the 210 listed mutants, practically half of them belong to mice with defects in sensory organs, particularly the eye and the inner ear, giving rise to anomalies in the size of the eye, deafness or cycling behavior. Of the other half, the majority (66%) are mice with more or less severe deterioration of locomotion, either because of muscular paralysis (neuromuscular mutants), disturbance of the gait, and/or tremor. A few of the mutant mice have been identified by their convulsive behavior, often as a consequence of audiogenic seizures.

The aim of this chapter is not to provide a catalog of techniques that can be used to analyze the phenotype of a neurological mutant mouse, but the research strategy that could be followed to identify newly appearing mutations. Because the subject of our own research has been the analysis of mutations affecting the cerebellum, most of the examples given for this research strategy will be taken from cerebellar mutants, which constitute about 40% of all mutations affecting locomotion (26 cerebellar mutations are known, that is to say 12% of all listed neurological mutant mice, see Table 1).

Search for behavioral changes

The presence of clinical symptoms is of course the first and simplest indication that the cerebellum and associated pathways might be affected in a given mouse: clinical inspection can reveal for instance abnormal position of the limbs, gait abnormalities, and/or tremor of variable importance in the several cerebellar mutants that have been identified. Of course these abnormalities are transmissible over generations, an essential feature to signalling the genetic nature of these problems. Besides cautious examination of the animals, simple measures can be used such as footprints made with waterproof black ink on white paper (ink is applied to the hind paws of individual mice and the mice induced to walk forward; stride lengths can even be measured). For all the spontaneous cerebellar mutant mice, the clinical abnormalities were easy to recognize and were in fact the basis of the screening procedure of these mutants. They are sometimes associated with other symptoms (such as seizures etc, see below) if additional damage exists in other brain regions. A variety of sensorimotor tests have been applied to analyze quantitatively the motor performance and motor learning deficits of several cerebellar mutant mice. These tests use different types of apparatus such as stationary beam, vertical or inclined grids, coat hanger, rotating rod or treadmill; they allow the analysis of several motor features such as muscular strength, motor activity, static or dynamic equilibrium, motor coordination and motor learning (for instance learning of synchronized walking). None of these tests allows the precise diagnosis of a given mutation among the others. On the other hand, characterizing similarities and differences in the behavior of various mutants, allows us to obtain information about the role of the cerebellum in sensorimotor performance and learning (for a recent review see Caston et al., 1998 and the Chapter 4.3).

Search for morphological targets

The morphological analysis of the CNS and its skeletal coverings is obviously the first step in the process of identification of a mutant phenotype. In the most favorable cases, gross anatomical inspection, together with weight of various large brain regions (i.e. cerebellum, forebrain) provide essential clues to identify the defect, orienting further research towards a distinct central region. This is the case for mutants with visible bone deformations, such as those with apparent hydrocephalus (cerebral degeneration, congenital hydrocephalus, hydrocephalus, hydrocephalus with hop gait, obstructive hydrocephalus), or with spina bifida (curly tail, loop tail, vacuolated lens). Distinct regions of the brain can also be affected macroscopically. For instance, for cerebellar mutants, in some of them (lurcher, reeler, scrambler, staggerer, weaver) the cerebellum is severely reduced in size and the foliation is abnormal. In a few others (rostral cerebellar malformation, swaying), the border between the cerebellum and the mesencephalon has disappeared, and the anterior part of the cerebellar cortex fuses with the colliculi. Nevertheless, in most of the mutants the gross anatomy of the CNS remains almost normal, and the localization of pathological alterations requires further morphological studies. In any case, for the precise localization and characterization of the pathology induced by the mutation, and for completion of the mutant's morphological phenotype, microscopical analysis of the CNS is the next step. First,

TABLE 1

Cerebellar mutant mice

Symbol	Name	Allele	Cloned gene	Knockout	References
ag	agitans				
ax	ataxia				
cdf	cerebellar deficient folia				
cod	cerebellar outflow degeneration				
dt	dystonia musculorum		dystonin (BPAG-1 gene)		Brown et al., 1995
du	ducky				
hpc	hyperspiny Purkinje cell				
lc	lurcher	lurcher J hot foot (ho)	GRID2 (glutamate receptor ionotropic delta 2 gene)		Zuo et al., 1997
mea	meander tail				
Mo^{br}	brindled				
nd	nodding				
nr	nervous				
pcd	Purkinje cell degeneration				
rcm	rostral cerebellar malformation		unc-5 H3 gene	unc-5 H3	Ackerman et al., 1997
rl	reeler	Orleans reeler	reelin gene		D'Arcangelo et al., 1995
scm	scrambler	yotari (yot)	mDabl1 (disabled gene)	mdabl1	Sheldon et al., 1997
sg	staggerer		RORα(orphan nuclear hormone receptor gene)	RORα	Hamilton et al., 1996
stg	stargazer		Cacng2 (g subunit-voltage gated Ca^{2+} channel gene)		Letts. et al., 1998
stu	stumbler				
sw	swaying		wnt-1 gene	wnt-1	Thomas et al., 1991
swe	slow-wave epilepsy		Nhe 1 (sodium/hydrogen exchanger gene)		Cox et al., 1996
tbl	tambaleante				
tg	tottering	rolling Nagoya (tg^{rol}) leaner (tg^{la})	Cacnl 1a4 (α subunit of a proposed P-type calcium channel gene)		Fletcher et al., 1996
vb	vibrator		PITPα (phosphatidylinositol transfer protein α gene)		Hamilton et al., 1997
wst	wasted				
wv	weaver		GIRK2 (G-protein-gated inward rectifying K+ channel gene)		Patil et al., 1995

with cytoarchitectonic methods (Nissl staining), thereafter with specific neuroanatomical (Golgi impregnation, axonal tracing methods) and chemoanatomical methods (immunohistochemistry and/or in situ hybridization with selective neural markers). Finally, with electron microscopy, particularly to study synaptic connections.

The characterization of the phenotype of the nodding (nd) mouse: as an example of the sequence in the morphological approach

The study of a mutant phenotype, either in a spontaneously appearing mutation or in the case of a knockout mouse, follows the sequential steps just mentionned above, and this will be further illustrated taking as an example our work on the nodding (*nd*) mutant mouse.

Neurological symptoms
The nodding recessive mutation arose spontaneously in the 129/Sv strain at the Pasteur Institute in Paris (Sotelo and Guénet, 1983). The first neurological signs appeared 8–10 d after birth, when the mouse pups started to develop a cerebellar syndrome, with ataxia and spasticity, the hind limbs being the most affected.

Light microscopic analysis
The simple inspection of the nodding CNS did not show visible alterations. Thus homozygous mutants, at different ages from one week to 6 months, were fixed by intracardiac perfusion with aldehydes. Serial sections were cut in the coronal and parasagittal planes, and used for:

Nissl stained preparations. With exception of the cerebellum, the study of the CNS failed to disclose noticeable pathological alterations. Concerning the cerebellum, it was somewhat reduced in size but showed normal foliation and lamination. In mutants aged over 3 months, some Purkinje cells were missing, and some granule cells exhibited a dark nucleus, with condensed chromatin like in apoptotic cell death.

Silver impregnation methods. The Golgi impregnation revealed that Purkinje cells (the pivotal elements of the cerebellar cortex, and its only output) had atrophic dendritic trees, with numerous ectopic spines emerging from the large dendritic branches (the proximal dendritic compartment). More interestingly, with the Cajal's reduced silver, it was possible to disclose a rather specific change which gave its uniqueness to the nodding cerebellum : basket cell axons were hypertrophic, particularly their pinceaux formations over the initial segment of the Purkinje cell axon (Figs 2 and 3). The main features of the nodding cerebellum were further studied by using:
Immunocytochemical methods. The immunolabeling with a Purkinje cell marker, especially the calcium binding protein calbindin allowed us not only to corroborate the loss of some of these neurons, mostly in the posterior vermis (Fig. 1), but also to reveal the occurrence of axonal torpedoes. Finally, the hypertrophy of the basket cell axons was corroborated by using either anti-neurofilaments antibodies (Fig. 4), or antibodies anti-glutamic acid decarboxylase (GAD), the rate-limiting synthetic enzyme for GABA, since basket cells are inhibitory GABAergic interneurons. All these sequential light microscopic studies showed the cerebellum as the target domain of the *nd* gene. However, the simultaneous implication of several categories of cerebellar neurons (Purkinje, granule and basket cells) did not facilitate the identification of the cellular target of the nodding mutation. With this aim the ultrastructural analysis was undertaken.

Electron microscopic analysis
In mice aged about one month, the numerous ectopic spines emerging from large dendritic segments of Purkinje cells remained free of innervation (Fig. 5), as if they were not mature enough to receive synaptic inputs. Whereas, spines emerging from spiny branchlets (distal dendritic compartment) were normally innervated by parallel fiber varicosities. Morevoer, owing to

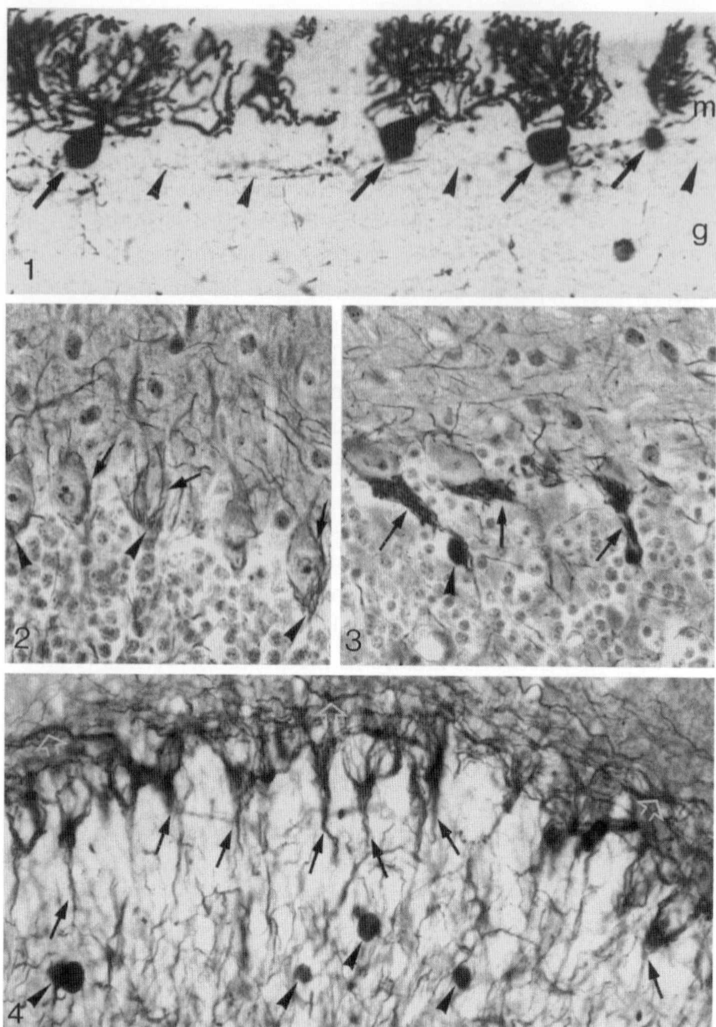

Plate 1. Light microscopy of the adult nodding cerebellum.

Fig. 1. Calbindin-immunostained Purkinje cells in lobule VIII. Parasagittal section. The arrows point to the remaining cell bodies of nodding Purkinje cells. Note the atrophy of their dendritic trees, and the loss (arrowheads) of Purkinje cells that do not form a continuous row of cell bodies at the interface of the molecular and granular layers, as normally occurs in control cerebellum. (m = molecular layer; g = granule cell layer).

Fig. 2. Cerebellar cortex of a control mouse stained with the reduced silver impregnation of Cajal. Parasagittal section. The arrowheads mark the basket cell axons around Purkinje cell bodies. The arrows point to the pinceaux formations around the initial segment of Purkinje cell axons. Lobule VI vermis.

Fig. 3. Similar lobule and staining to those in Fig. 2, but taken for an adult nodding cerebellum. Note that the baskets around Purkinje cell bodies and, particularly, the pinceaux formations are much denser and longer (arrows) than in control cerebellum. The arrowhead points to an axonal varicosity.

Fig. 4. Cerebellar cortex of an adult nodding mouse immunostained with a monoclonal antibody which recognizes phosphorylated neurofilaments. Note the varicosities of Purkinje cell axons (arrowheads), and the particularly long pinceaux formations (arrows) of the basket cell axons. The white arrows point to the transversally oriented basket axons.

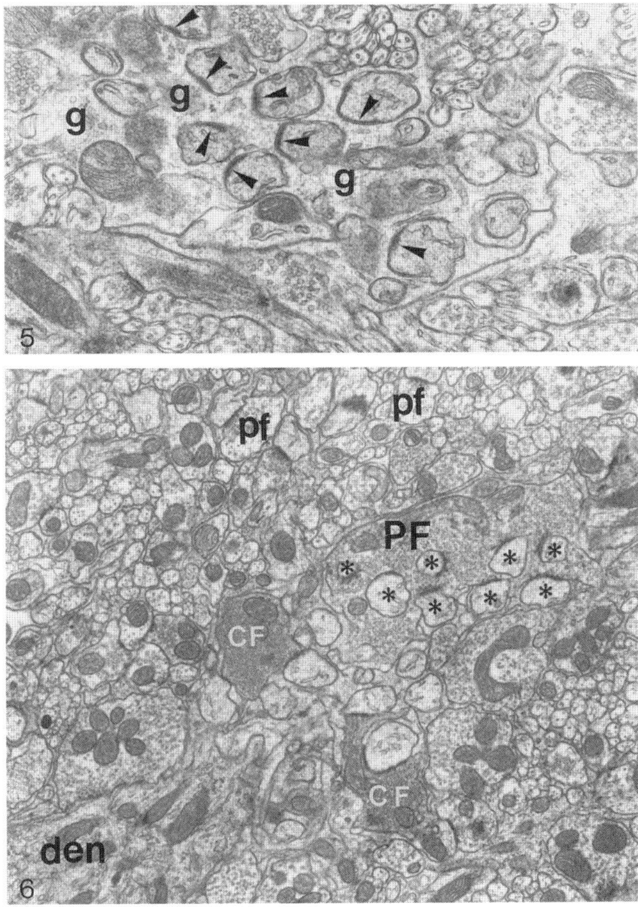

Plate 2. Electron micrographs of the nodding mouse cerebellar cortex.

Fig. 5. Deep third of the molecular layer of a 28-day-old nodding mouse. This micrograph illustrates a cluster of Purkinje cell dendritic spines cut in a tangential plane to a thick denditric branch (not seen in the micrograph). The ectopic spines bear postsynaptic differentiations (arrowheads) facing glial processes (g), since they are free of innervation.
Fig. 6. Similar field to that in Fig. 5 but taken from the cerebellum of a 230-day-old nodding mouse. A large branch of a Purkinje cell dendrite (den) is visible at the lower left corner. At this age the vast majority of the ectopic spines are innervated by a gigantic parallel fiber varicosity (PF), which established synaptic contacts with eight spines (asterisks).
CF = climbing fiber varicosities; pf = normal parallel fiber varicosities synapsing on distal spines emerging from Purkinje cell spiny branchlets.

the Purkinje cell dendritic atrophy, these neurons offered a reduced postsynaptic space to parallel fibers, explaining the death of some granule cells through a retrograde mechanism, secondary to the effect of the *nd* gene on Purkinje cells. Finally, in mutants aged between 6 and 14 months, the ectopic spines had reached maturity and were innervated by nearby axon terminals, particularly parallel fibers that progressively increased in size, and in the number of synaptic contacts with dendritic spines. These presynaptic elements become giant boutons, up to five times larger than those found in the molecular layer of control cerebellum (Fig. 6).

Such a kind of thorough morphological analysis (Sotelo, 1990), although not providing direct evidence that Purkinje cells were the cellular targets of the *nd* gene, strongly suggested that these neurons were the closest to the target cells: First, the dendritic atrophy and delayed maturation of the ectopic spines could explain the limited granule cell death. Secondly, the late maturation of the ectopic spines could induce a process of "axon terminal sprouting", and provide late innervation to the ectopic spines. Thus, the *nd* gene appeared to be acting on the process of Purkinje cell spinogenesis that, through the secondary mechanisms discussed above, could modulate the size and shape of the axon terminals of the cerebellar interneurons impinging on Purkinje cell dendritic spines.

Aggregation chimeras as a tool to determine which are the target cells of particular mutant genes

Not all detectable abnormalities — as discussed above for the nodding mutation — revealed by the morphological analysis are systematically the direct effect of a single-gene mutation, just as not all developmental events are exclusively genetically regulated. During the ontogenesis of the brain, cell-to-cell interactions are as important as genetic information. For instance, neurogenesis can be regulated by interactions between dividing progenitor cells and postmitotic, nearby located neurons, as is the case for progenitor cells in the cerebellar external granular layer, whose mitotic activity is regulated by Purkinje cells (Mariani et al., 1977; Smeyne et al., 1995). Similarly, postmitotic neurons need to migrate to reach their final locations, and in many regions of the brain premigratory neurons interact with radial glial cells for their migration (Rakic, 1971). Finally, differentiating neurons extend axons that, interacting with local cues and gradients of chemoattractant and/or chemorepulsive molecules, will reach their terminal domains and establish synaptic connections with their postsynaptic counterparts. In agreement with the neurotrophic hypothesis, which postulates that neuronal survival depends on the occurrence of available target cells (Levi-Montalcini, 1987), the loss — during development — of one class of neurons provokes a transsynaptic degeneration "en cascade" of their presynaptic neurons.

One main interest of the neurological mutant mice has been to provide direct evidence on the capital role of cell-to-cell interactions in the formation of neuronal networks. Cerebellar mutants, for instance, have elegantly illustrated the role of postsynaptic neurons in the survival of presynaptic ones. In those mutants where Purkinje cells were either affected early in development, like the staggerer (Herrup and Mullen, 1979a), or degenerated massively in the postnatal mice, like lurcher (Caddy and Biscoe, 1979; Dumesnil-Bousez and Sotelo, 1992) or Purkinje cell degeneration (Mullen et al., 1976), neurons presynaptic to Purkinje cells also degenerated. This was the case for granule cells (Sotelo and Changeux, 1974a; Caddy and Biscoe, 1979; Herrup, 1983; Herrup and Sunter, 1987), the origin of parallel fibers, and for inferior olivary neurons (Caddy and Biscoe, 1979; Shojaeian et al., 1985; Zanjani et al., 1990), origin of the climbing fibers, the two main inputs to Purkinje cells. The transsynaptic degeneration occurred somewhat later and at a slower pace than the Purkinje cell degeneration, thus differentiating neurons that were primary targets of the mutated loci from those secondarily affected. However, sometimes it is extremely difficult to distinguish between primary and secondary abnormalities: in cerebellar mutant mice, the distinction between primary and secondary targets can be inferred but not always proven. The weaver mutation is a good example of these difficulties, since it has been a matter of debate for almost 10 years. The major phenotype of the weaver cerebellum is its agranularity: granule cell progenitors proliferate normally but most of the postmitotic, premigratory granule cells die. Rakic and Sidman (1973) attributed the loss of these neurons to a primary defect of the *wv* gene action on Bergmann fibers, and concluded that since the Bergmann fibers (the radial glial axes for granule cell migration) were very much reduced in number and abnormal, premigratory granule cells could not migrate and died : the death of these neurons was, therefore, the consequence of migration impairment, and not of the direct action of the *wv* gene. A year later, Sotelo and Changeux (1974b) — using metallic impregnation to visualize astrocytes and electron microscopy — reported that Bergmann glial cells were numerous in the weaver cerebellum, and that their abnormalities were those of reactive astrocytes, as expected after massive granule cell death. More importantly, some granule cells left the inner zone of the external granular layer and died, despite their successful migration. For Sotelo and Changeux, the granule cells "are closer than the glial cells" to the primary cellular target of the weaver mutation. It was in 1982 that Goldowitz and Mullen (1982) applied the aggregation chimeras' approach to the weaver cerebellum, and reported some evidence that granule cells, and not Bergmann fibers, were the primary target of the mutation. Finally, the cloning of the weaver gene (Patil et al., 1995) has proven the primary action of the gene on granule cells (see below).

The best strategy to complement the morphological analysis of the phenotype, and to determine on which cells a mutant gene was acting, is to create a mosaic system in which wild-type and mutant cells could randomly aggregate, providing a normal environment and, particularly, normal cell-to-cell interactions to the genetically mutant cells. In mice, this type of mosaic systems can be generated by producing aggregation chimeras, also named tetraparental mice. Since the work of Tarkowski (1961), it is known that mouse aggregation chimeras can be obtained by mixing together, in vitro, two eight-cell-embryos: one from a normal strain and the other from a neurological mutant strain. After one night or one day in culture, the two embryos form a single blastocyst (doubled sized morula), which is surgically implanted into the uterus of a host pseudo-pregnant female, producing in this way a mouse whose cells derive from wild-type and mutant genotypes. The major point in the preparation of chimeric mice is the selection of one of the embryos (the wild-type or the mutant) from a strain of mice carrying an intrinsic cell marker, independent from the mutated gene. This cell marker will identify genotypically the cells arising from the embryo carrying the marker. In the chimeric mice, the comparison of the genotype of single cells (via the cell marker) with their phenotype (normal or mutated) will permit the determination of whether these cells are or are not affected by the mutant gene. If genotypically wild-type cells keep a normal phenotype, the mutant gene acts intrinsically, whereas if these cells exhibit a mutant phenotype, the gene acts extrinsically.

The selected intrinsic cell marker must fulfill several requirements. It must be: (i) cell autonomous and cell localized; (ii) selectively neutral, particularly without influencing the neurogenesis; (iii) present in all cells and easely detectable with neurocytological methods. At least four categories of cell markers have been used for the study of neurological chimeric mice:

1. *Markers based upon variants in the activity of enzymes*, that can be visualized either with histochemical or immunocytochemical methods. Variants in the activity of the β-glucuronidase, using strain of mice homozygous for the *Gus*^{b} allele (strong enzymatic reaction) combined with mice homozygous for the *Gus*^{h} allele (weak reaction), have been useful to study the fate of large neurons, such as Purkinje cells, in the chimeric mice (Mullen et al., 1997). Because mutations affecting the number (lurcher, staggerer, Purkinje cell degeneration) or the location (reeler) of Purkinje cell appeared in mice strains homozygous for the *Gus*^{h} allele, their chimeric analysis has been simplified. In these 4 types of cerebellar mutant chimeras, Purkinje cell mosaicisms were formed, with a very large variability in the respective percentages of cells from each genotype within each chimera. In the lurcher (Vogel et al., 1989) and

Purkinje cell degeneration chimeras (Mullen, 1978; Mullen et al., 1997), all β-glucuronidase positive Purkinje cells disappeared. The coincidence of the mutant phenotype with the cell marker showed that these two genetic *loci* act intrinsically on Purkinje cells. That is to say: Purkinje cells are the cellular target of the *Lc* and *pcd* genes. In staggerer chimeras (Herrup and Mullen, 1979b), Purkinje cells with mutant phenotype (small and ectopic) were mixed with normal Purkinje cells, but only those with mutant phenotype were stained with the β-glucuronidase enzyme reaction product. Here again, the coincidence of the mutant phenotype with the selected cell marker indicated that the *sg* gene acted intrinsically on Purkinje cells. On the contrary a normally proportioned granule cell layer was present. Thus, the severe granule cell loss, which also characterized the staggerer cerebellum, was secondary to the Purkinje cell deficiency. In contrast, in the reeler chimeras (Mullen, 1978; Mullen and Herrup, 1979), the majority of Purkinje cells were normally located independently of their genotype. More importantly, between those ectopically located Purkinje cells, β-glucuronidase negative cell were present. Thus, there were normal Purkinje cells in ectopic position and mutant ones in normal position. This mismatching between genotype and phenotype clearly indicated that Purkinje cells were not the cellular targets of the reeler mutation. Thus, the failure of neuronal migration characterizing the reeler cerebellum could be due to altered cell-to-cell interactions between migrating neurons and their radial glial axes (see below).

2. *Naturally occurring interspecies nuclear cell markers* have been widely used to study embryonic development, for instance in quail-chick chimeras (Le Douarin, 1982).

The occurrence of an abnormal nuclear morphology in the lymphocytes of the hair-deficient mutant mouse ichthyosis (*ic/ic*), prompted Goldowitz and Mullen (1982) to use this marker in heterozygous weaver ichthyosis chimeras. In this chimeras, genotypically non-weaver granule cells (normal from the point of view of the analyzed mutation) exhibited nuclei which did not resemble normal granule cell nuclei because their heterochromatin formed a central clump. In the cerebellum of the weaver ichthyosis chimeric mice Goldowitz and Mullen, 1982), granule cell mosaicisms were also formed. Based upon statistical analyses, it was found that ichthyosis granule cells had no migratory problems and formed an inner granule cell layer, whereas weaver granule cells did not migrate normally. These results suggest that the weaver gene acts intrinsically on granule cells.

3. *Interspecies-specific molecular markers* can be used as cell marking systems. The molecular specificity defining the genotype can be based upon:

(a) species-specific satellite DNA sequences which, with the in situ hybridization technique, will allow us the identification of the cellular genotype in histological sections of *M. musculus* ↔ *M. caroli* chimeras. With this approach, homozygous *wv/wv* (*M. musculus*) ↔ +/+ (*M. caroli*) chimeras have been produced (Goldowitz, 1989). Their analysis showed that the genotype of Purkinje cells and Bergmann glia was irrelevant for the death or survival of granule cells. Thus, extending the work with ichthyosis chimeras, it was possible to conclude that granule cells and not Bergmann fibers were the primary cellular target of the weaver *locus*.

(b) strain-specific antigens, as in the case of *rl/rl* ↔ +/+ chimeras, using mice of the C3H strain for +/+ donor embryos (Yoshiki and Kusakabe, 1998). In these chimeras, the wild type genotype can be recognized with immunocytochemical staining with an anti-C3H strain-specific antigen (CSA) antibody. The results obtained with these chimeras corroborated those reported above, because they also show that the position of cerebellar neurons and glial cells is controled by extrinsic signals, but not by their own genotype.

4. *Markers based on the use of transgenic mice* for the construction of chimeras. Such murine chimeras have been used, for instance, to identify

the primary cellular targets of the dystonia musculorum (*dt*) mutation in the spinal cord (Campbell and Peterson, 1992). In *dt/dt* mice, the most relevant histopathological lesion is the degeneration of primary sensory axons. Nevertheless, degeneration of second order sensory neurons also occurs, particularly of spinocerebellar neurons (Sotelo and Guenet, 1988), allowing us to consider this mutation as cerebellar. Campbell and Peterson (1992) used mice homozygous for a transgene encoding the human 68 kd neurofilament-light protein (hNF-L) as wild-type embryos for chimera construction. Thus, spinal neurons and axons with wild type genotype can be identified immunocytochemically by using the DP-5 antibody, which recognizes human but not mouse NF-L. In these chimeras, the axons of wild-type neurons were normal. Thus, the *dt* mutation acts via a mechanism intrinsic to the affected neurons. Another interesting example of transgenic lines used as +/+ donor embryos in chimeric analysis concerns the study of the meander tail mutation (Hamre and Goldowitz, 1997). In these chimeras, the +/+ embryos came from mice transgenic for a PBR-globin insert, that could be identified by nuclear labelling using in situ hybridization. With this approach it was concluded that the abnormal morphologies of Purkinje cells observed in the anterior lobe of the vermis in the *mea/mea* cerebellum were not due to the *mea* gene; whereas the loss of granule cells in the anterior lobe was the result of an intrinsic action of the *mea* gene.

Pleiotropic effects of some mutated genes

In the cerebellar mutants, fortuitous observations followed by more systematic studies have revealed in some cases more subtle phenotypes due to a pleiotropic effect of the mutated gene. Studies conducted by several groups during the last few years on the staggerer mutant provide a good example of these searches for "unexpected phenotypes".

Pleiotropic effects of the mutated staggerer gene

In 1986 Trenkner and Hoffmann were the first to describe that the homozygous *sg/sg* mutant presents immune abnormalities consisting in an atrophy of the thymus in the young adult mutant associated with a prolonged abnormal humoral response. This pioneer study has stimulated others to search for immune abnomalies in cerebellar mutants: the levels of several inflammatory cytokines (Interleukines 1 or 6, TNF α) are elevated not only in the cerebellum (ie where the neurodegeneration occurs) (Vernet-der-Garabedian et al., 1998) but, more surprisingly, outside the brain in the peripheral immune system (spleen, peritoneal macrophages); both the basal levels and those induced after inflammatory stimulation (for instance by peripheral injection of lipopolysaccharides, LPS) are abnormaly elevated compared to sex and strain matched controls, in several cerebellar mutants especially in staggerer and lurcher (Kopmels et al., 1992). These results have suggested that the corresponding genes might be involved not only in the survival of cerebellar neurons but also in the control of inflammatory responses in the organism and especially the production of inflammatory cytokines. This hypothesis, raised before the genes were identified, has led to the discovery of several new and unexpected phenotypes in these mutant mice: increased "sickness behavior" (Bluthé et al., 1997), enhanced glucocorticoid response to either inflammatory stimulus or novel environment stress (Frederic et al., 1997). Given the importance of inflammation in several degenerative pathologies, it was even found that the staggerer mutant develops, in response to a fat diet, dramatic atherosclerotic lesions in the aortic sinus and coronary arteries. This increased susceptibility is related to the inflammatory abnomalies but also to decreased circulating levels of HDL-cholesterol and its main proteic component, Apolipoprotein A-1, secondary to a decreased synthesis of this protein in the staggerer intestine (Mamontova et al., 1998). While this work was being performed, the staggerer mutation was identified as a deletion in the *RORα* transcription factor gene (Hamilton et al., 1996 and see below) and further analysis has identified this gene as a direct transactivator of the APO A-1 gene, providing at least part of the molecular explanation of the atherosclerotic phenotype (Vu-Dac et al., 1997). The expression of the *RORα* gene in various cells of the immune system will certainly allow the deciphering of the molecular basis of the many pathological defects present in the staggerer mutant. This illustrates how the recent identification of some of the mutated genes will be decisive in characterizing the molecular basis of the phenotype observed (see next paragraph).

Search for molecular targets

The expectations that the mouse would become the ultimate mammalian model for the study of neurogenetics and behavior, as well as for the analysis of the molecular mechanisms involved in brain development, were primarily based upon the good knowledge of and stability of the mouse genetic background and, particularly, upon the occurrence of spontaneous or provoked mutations. However, without the cloning of the genes responsible for those mutations, it was hopeless to try and understand how the gene products may work, and the initial enthusiasm for the neurological mutant mice slowly vanished. Recent technical advances of molecular biology have reinvigorated the field, and a number of genes implicated in the known neurological mutations have been cloned. These new techniques have put the mouse back at the center of research aimed at piecing together gene action and behavior or brain development. Here again, we will illustrate the search for molecular targets with some examples taken from mutations primarily or secondarily affecting the mouse cerebellum (see Table 1).

The cloning of genes responsible for spontaneous cerebellar mutant mice has been, in some instances, greatly helped by comparisons between the phenotype of mice with targeted gene inactivations and the known phenotypes of spontaneous mutant mice. Thus, after the inactivation of the *wnt-1* gene (Thomas and Capecchi, 1990), it was clear that the swaying mouse (*sw*) — characterized by atrophy of the anterior cerebellar lobe which was fused directly with the inferior colliculi — was phenotypically identical to the *wnt-1* null-mutant. Thomas et al. (1991) were the first to identify the gene mutated in a cerebellar mutant mouse, demonstrating that *sw* is a deletion of a single base pair from the proto-oncogene Wnt-1. These two studies have emphasized the major role of Wnt-1 in the cascade of molecular events leading to the ontogenesis of the cerebellum (see in Wassef and Joyner, 1997). Somewhat different has been the case of the staggerer mutation (*sg*) that, as reported above, affects Purkinje cells. The staggerer mutation was first identified by positional cloning as a deletion in the ROR-α gene, a retinoid-related orphan nuclear receptor (Hamilton et al., 1996). In a collaborative work with the group of Becker-André, who generated null-mutant mice for the ROR-α gene by targeted insertion of the lac Z reporter gene, we have shown (Steinmayr et al., 1998) that the morphological abnormalities in the cerebellum, and the behavioral phenotype of the homozygous ROR-α −/− mice were similar to those characterizing the staggerer mouse, thus proving the identity between the ROR- null-mutants and the staggerer mice. Moreover the insertion of the lac Z reporter gene in the mutated *sg* locus of the recombinant mouse has allowed the definition of the cell categories in which the ROR-α protein is present both inside and outside the brain.

The understanding of the molecular cascade leading to neuronal cell migration, one of the essential processes for the normal development of the CNS, has partially emerged from the cloning of the genes disrupted in 3 spontaneous mutant mice: reeler (*rl*), scrambler (*scm*) and its allele yotari (*yo*), and rostral cerebellar malformation (*rcm*). Since its early description (Caviness and Sidman, 1972), the reeler mouse has been considered as a model to study molecular mechanisms underlying neuronal cell migration. In this mutant, characterized by malposition of neurons in all laminated brain structures, neuronal birthdates are normal, but in the neocortex the "outside-in" positioning of neurons is somewhat inverted (Caviness and Sidman, 1972), and in the cerebellum about 90% of Purkinje cells do not reach the cortex, remaining ectopically in a large central cerebellar mass within the white matter (Mariani et al., 1977). The reeler gene has been recently decoded (D'Arcangelo et al., 1995) by chance discovery of a reeler allele when a c-fos supra-expressor transgene was inserted randomly into the reeler locus, providing not only a reeler knock-out but more importantly a marker for the reeler gene. The wild-type gene encodes a very large protein of 3461 amino acids, with a relative molecular mass of 388 kd. It is important to recall that Reelin does not have transmembrane domains and that, most probably, it is a released protein. The occurrence of a repeating sequence of 8 EGF motifs gives to Reelin some homology with tenascin, restrictin and β-integrin, and suggests that it is an extracellular matrix protein involved in cell adhesion. The mapping of mRNA expression for Reelin has been reported (see in Alcantara et al., 1998): in the neocortex, it is first expressed by pioneer neurons, the Cajal-Retzius cells that, owing to their location at the surface of the developing cortex, have been considered to coordinate cell positioning of the migrating cortical plate neurons.

The study of two alleles of a new spontaneous mutation, scrambler and yotari, has been critical to the identification of a gene probably involved in the Reelin signaling pathway (Sheldon et al., 1997). The CNS phenotypes of the scrambler and yotari mice were identical to the one of the reeler mouse. Nevertheless, in these mice expression of reelin mRNA and protein was normal, and Cajal-Retzius cells were immunostained with the CR-50 monoclonal antibody which is specific for Reelin, suggesting that this protein was synthesized and released normally. Sheldon et al. (1997) have provided evidence that scrambler and yotari arise from mutations of the *mdab1* gene, a mouse gene homologous to the Drosophila gene disabled (*dab*). The gene encodes a cytoplasmic phospho-protein which seems to function as an intracellular adaptor in protein kinase pathways. Moreover, the analysis of the pattern of expression of *mdab1* mRNA strongly suggests that the phosphoprotein is present in neurons exposed to Reelin. These results indicate that during the radial migration in the neocortex and cerebellum, Reelin may be the first step in a molecular cascade of events that will lead to the proper localization of pyramidal neurons and Purkinje cells among others. The finding of the Reelin receptor or receptors, is one of the urgent tasks needed to understand the molecular cascade. The fact that the tyrosine kinase signaling adaptor molecule Disabled is part of the cascade will be very helpful in understanding the Reelin signaling pathway.

The study of the *rcm* mice indicates that a totally different molecular pathway is followed by the precursors of the cerebellar granule cells during their early tangential migration. In this mutation granule and Purkinje cells migrate beyond their normal cerebellar boundaries, to ectopic locations in rostral areas of the mutant brainstem, invading the lateral regions of the pons and the inferior colliculus (Przyborski et al., 1998). The gene affected in this mutation has been decoded. It belongs to the unc-5H3 family, a mouse gene homologue of the *C. elegans unc-5*, which encodes a transmembrane receptor protein for netrin, which may provoke chemorepulsive signaling (Ackerman et al., 1997). The study of the expression patterns for netrin-1 and *unc-5H* mRNAs indicate that only the tangential migration of granule cell precursors is affected, the ectopia of Purkinje cells being a secondary effect of the mutation. Moreover, this mutation emphasizes the role of Unc-5H3 and Netrin-1 in determining the rostral cerebellar boundary.

Finally, a number of spontaneous mutations affecting the mouse cerebellum are produced by mutated genes normally encoding ionic and/or receptor channels. This is the case in the lurcher mouse which is induced by a gain of function mutation of the glutamate receptor ionotropic delta2 (*Grid2* gene) (Zuo et al., 1997). The same gene (*Grid2*) but with a different deletion has been proven to be the molecular target of the hotfoot (*ho*) mutant mouse (Lalouette et al., 1998), which exhibits a completely different phenotype than the lurcher mouse (Guastavino et al., 1990) even though they are alleles. Mutations in the 1A subunit of the P/Q-type voltage-sensitive calcium channel (Fletcher et al., 1996) are responsible for the tottering and leaner mutant mice, associated with epileptic seizures, and cerebellar degeneration. Very recently a novel gene has been described, *Cacng2* (Letts et al., 1998), whose expression is disrupted in the stargazer mouse, a mutant which has spike-waves seizures characteristic of absence epilepsy, associated with defects in the cerebellum and inner ear (Noebels et al., 1990). This gene encode a brain specific protein (Stargazin) stucturally similar to the g subunit of skeletal muscle voltage-gated calcium channels. The last of the mutations cited here will be weaver, whose mutated gene is that encoding the G protein-gated inwardly rectifying potassium channel protein (Girk2) (Patil et al., 1995). This protein is broadly expressed by many neuronal populations but, nevertheless, the weaver phenotype is characterized by loss of cerebellar granule cells and of dopaminergic nigral neurons. Thus, the molecular cloning of the weaver gene raises more questions than it provides answers: why does Girk2 have a wide distribution of expression, but only provokes the death of a few neuronal populations? What are the functional roles of the mutated Girk2 protein? Does it play a role in human Parkinson's disease?

References

Ackerman, S.L., Kozak, L.P., Przyborski, S.A., Rund, L.A., Boyer, B.B. and Knowless, B.B. (1977) The mouse rostral cerebellar malformation gene encodes and Unc-5-like protein. *Nature*, 386: 838–842.

Alcantara, S., Ruiz, M., D'Arcangelo, G., Ezan, F. De Lecea, L., Curran, T., Sotelo, C. and Soriano, E. (1998) Regional and cellular patterns of reelin mRNA expression in the forebrain of the developing and adult mouse. *J. Neurosci.*, 18: 7779–7799.

Bluthé, R. M., Michaud, B., Delhaye-Bouchaud, N., Mariani, J. and Dantzer, R. (1997) Hypersensitivity of the lurcher mutant mice to the depressing effects of lipopolysaccharide and interleukin-1 on behavior. *Neuroreport*, 8: 1119–1122.

Brown, A., Bernier, G., Mathieu, M., Rossant, J. and Kothary, R. (1995) The mouse dystonia musculorum gene is a neural isoform of bulbous pemphigoid antigen 1. *Nat. Genet.*, 10: 301–306.

Caddy, K.W.T. and Biscoe, T.J. (1979) Structural and quantitative studies on the normal C3H and Lurcher mutant mouse. *Phil. Trans. R. Soc. Lond.*, B287: 167–201.

Campbell, R.M. and Peterson, A.C. (1992) An intrinsic neuronal defect operates in dystonia musculorum: A study of dt/dt ↔ +/+ chimeras. *Neuron*, 9: 693–703.

Caston, J., Lalonde, R., Delhaye-Bouchaud, N. and Mariani J. (1988) The cerebellum and postural sensorimotor learning in mice and rats. *Behav. Brain Res.*, 95: 17–22

Caviness, V.S., Jr. and Sidman, R.L. (1972) Olfactory structures of the forebrain in the reeler mutant mouse. *J. Comp. Neurol.*, 145: 85–104.

Cox, G.A., Lutz, C.M., Yang, C.L., Biemesderfer, D., Bronson, R.T., Fu, A., Aronson, P.S., Nobels, J.L. and Frankel, W.N. (1997) Sodium/hydrogen exchanger gene defect in slow-wave epilepsy mutant mice. *Cell*, 91: 139–148.

D'Arcangelo, G., Miao, G.G., Chen, S.C., Soares, H.D., Morgan, J.I. and Curran T. (1995) A protein related to extracellular matrix proteins deleted in the mouse mutant reeler. *Nature*, 374: 719–723.

Dumesnil-Bousez, N. and Sotelo, C. (1992) Early development of the Lurcher cerebellum : Purkinje cell alterations and impairment of synaptogenesis. *J. Neurocytol.*, 21: 506–529.

Fletcher, C.F., Lutz, C.M., O'Sullivan, T.N., Shaughnessy, J.D., Jr., Hawkes, R., Frankel, W.N., Copeland, N.G. and Jenkin, N.A. (1996) Absence epilepsy in tottering mutant mice is associated with calcium channel defects. *Cell*, 87: 607–617.

Frederic, F., Chautard, T., Brochard, R., Chianale, C. , Wollman, E.E., Oliver, C., Delhaye-Bouchaud, N. and Mariani, J. (1997) Enhanced endocrine response to novel environment stress and endotoxin in Lurcher mutant mice. *Neuroendocrinology*, 66: 341–347.

Goldowitz, D. (1989) The weaver granuloprival phenotype is due to intrinsic action of the mutant locus in granule cells: Evidence from homozygous weaver chimeras. *Neuron*, 2: 1565–1575.

Goldowitz, D. and Mullen, R.J. (1982) Granule cell a a site of gene action in the weaver mouse cerebellum : evidence from heterozygous mutant chimeras. *J. Neurosci.*, 2: 1474–1485.

Grüneberg, H. (1992) *The Genetics of the Mouse*, 2nd Edn, Martinus Nijhoff, The Hague.

Guastavino, J.-M., Sotelo, C. and Damez-Kinselle, I. (1990) Hot-foot murine mutation : behavioral effects and neuroanatomical alterations. *Brain Res.*, 523: 199–210.

Hamilton, B.A., Frankel, W.N., Kerrebrock, A.W., Hawkins, T.L., FitzHugh, W., Kusumi, K., Russell, L.B., Mueller, K.L, Van Berkel, V., Birren, B. W., Kruglyak, L. and Lander, E.S. (1996) Disruption of the nuclear hormone receptor RORalpha in staggerer mice. *Nature*, 379: 736–739.

Hamilton, B.A., Smith, D.J., Müller, K.L., Kerrobrock, A.W., Bronson, R.T., Van Berkel, V., Daly, M.J., Kruglyak, L., Reeve, M.P., Nemhouser, J.L., Hawking, T.L., Rubin, E.M. and Lander, E.S. (1997) The vibrator mutation causes neurodegeneration via reduced expression of PITP alpha: positional complementation cloning and extragenic suppression. *Neuron*, 18: 711–722.

Hamre, K.M. and Goldowitz, D. (1997) meander tail acts intrinsic to granule cell precursors to disrupt cerebellar development: analysis of meander tail chimeric mice. *Development*, 124: 4201–4212.

Herrup, K. (1983) Role of staggerer gene in determining cell number in cerebellar cortex. I. Granule cell death is an indirect consequence of staggerer gene action. *Dev. Brain Res.*, 11: 267–274.

Herrup, K. and Mullen, R.J. (1979a) Regional variation and absence of large neurons in the cerebellum of the staggerer mouse. *Brain Res.*, 172: 1–12.

Herrup, K. and Mullen, R.J. (1979b) Staggerer chimeras: intrinsic nature of Purkinje cell defects and implications for normal cerebellar development. *Brain Res.*, 178: 443–457.

Herrup, K. and Sunter, K. (1987) Numerical matching during cerebellar development: quantitative analysis of granule cell death in staggerer mouse chimeras. *J. Neurosci.*, 7: 829–836.

Kopmels, B., Mariani, J., Delhaye-Bouchaud, N., Audibert, F., Fradelizi, D. and Wollman, E.E. (1993) Evidence for an hyperexcitability state of staggerer mutant mice macrophages. *J. Neurochem.*, 58: 192–199.

Lalonde, R. and Strazielle, C. (1999) Motor performance of spontaneous murine mutations with cerebellar atrophy. In: this book (Techniques in the Behavioral and Neural Sciences).

Lalouette, A., Guénet, J.L. and Vriz, S. (1998) Hot-foot mouse mutations affect the delta 2 glutamate receptor gene and are allelic to lurcher. *Genomics*, 15: 9-13.

Le Douarin, N. (1982) *The Neural Crest*, Cambridge University Press, Cambridge.

Letts V. A., Felix, R., Biddlecome, G.H., Arikkath, C.L., Mahaffey, C.L., Valenzuela, A., Bartlett II, F.S., Mori, Y., Campbell, K.P. and Frankel, W.N. (1998) The mouse stargazer gene encodes a neural Ca^{2+}-channel gamma subunit. *Nat. Genet.*, 19: 340–347.

Levi-Montalcini, R. (1987) The nerve growth factor 35 years later. Science, 237: 11541162.

Lyon, M.F. and Searle, A.G. (1989) *Genetic Variants and Strains of the Laboratory Mouse*, 2nd Edn, Oxford University Press, Oxford.

Mamontova, A., Duverger, N., Séguret-Macé, S., Esposito, B., Chaniale, C., Delhaye-Bouchaud, N., Staels, B., Mariani, J. and Tedgui, A. (1998) Severe atherosclerosis and hypoalphalipoproteinemia in the *staggerer* mouse, a mutant of RORα gene. *Circulation*, in press.

Mansour, S.L., Thomas, K.R. and Capecchi, M.R. (1988) Disruption of the proto-oncogene int-2 in mouse embryo-derived stem cells: a general strategy for targeting mutations to non-selectable genes. *Nature*, 336: 348–352.

Mariani, J., Crepel, F., Mikoshiba, K., Changeux, J.P. and Sotelo, C. (1977) Anatomical, physiological and biochemical studies of the cerebellum from *reeler* mutant mouse. *Phil. Trans. R. Soc. Lond. B*, 282: 1–28.

Mullen, R.J. (1978) Genetic dissection of the CNS with mutant-normal mouse and rat chimeras. In: W.M. Cowan and J.A. Ferrendelli (Eds.), *Society for Neuroscience*, Symposium 2, Society for Neuroscience, Bethesda, pp. 47–65.

Mullen, R.J. and Herrup, K. (1979) Chimeric analysis of mouse cerebellar mutants. In: X.O. Breakefield (Ed.), *Genetic Approaches to the Nervous System*, Elsevier, New York, pp. 173–196.

Mullen, R.J., Eicher, E.M. and Sidman, R.L. (1976) Purkinje cell degeneration a new neurological mutation in the mouse. *Proc. Natl. Acad. Sci. USA*, 263: 67–69.

Mullen, R.J., Hamre, K.M. and Goldowitz, D. (1997) Cerebellar mutant mice and chimeras revisited. *Perspect. Dev. Neurobiol.*, 5: 43–55.

Noebels, J.L., Quiao, X., Bronson, R.T., Spencer, C. and Davisson, M.T. (1990) Stargazer: a new neurological mutant on Chromosome 15 in the mouse with prolonged cortical seizures. *Epilepsy Res.*, 7: 129–135.

Patil, N., Cox, D.R., Bhat, D, Faham, M., Myers, R.M. and Peterson, A.S. (1995) A potassium channel mutation in weaver mice implicates membrane excitability in granule cell differentiation. *Nat. Genet.*, 11: 126–129.

Przyborski, S.A., Knowless, B.B. and Ackerman, S.L. (1998) Embryonic phenotype of Unc5h3 mutant mice suggests chemorepulsion during the formation of the rostral cerebellar boundary. *Development*, 125: 41–50.

Rakic, P. (1971) Neuron-glia relationship during granule cell migration in developing cerebellar cortex. A Golgi and electron microscopic study in macacus rhesus. *J. Comp. Neurol.*, 141: 283–312.

Rakic, P. and Sidman, R.L. (1973) Sequence of developmental abnormalities leading to granule cell deficit in cerebellar cortex of weaver mutant mice. *J. Comp. Neurol.*, 152: 103–132.

Sheldon, M., Rice, D.S., D'Arcangelo, G., Goldowitz, H., Nakajima, K., Mikoshiba, K., Howell, B.W., Cooper, J.A., Goldowitz, D. and Curran, T. (1997) Scrambler and yotari disrupt the disabled gene and produce a reeler-like phenotype in mice. *Nature*, 389: 730–733.

Shojaeian, H., Delhaye-Bouchaud, N. and Mariani, J. (1985) Decreased number of cells in the inferior olivary nucleus of the developing staggerer mouse. *Dev. Brain Res.*, 21: 141–146.

Sidman, R.L., Green, M.C. and Apple, S.H. (1965) *Catalog of the Neurological Mutants of the Mouse*, Harvard University Press, Cambridge, MA.

Smeyne, R.J., Lewin, A., Bian, F., Kunsch, C., Lira, S.A. and Oberdick, J. (1995) Local control of granule cell proliferation by cerebellar Purkinje cells. *Mol. Cell. Neurosci.*, 6: 230–251.

Sotelo, C. (1990) Cerebellar synaptogenesis : what we can learn from mutant mice. In: A. Aguayo and E. Howes (Eds.), Synapse Formation, *J. Exp. Biol.*, 153: 225–249.

Sotelo, C. and Changeux, J.P. (1974a) Trans-synaptic degeneration "en cascade" in the cerebellar cortex of staggerer mutant mice. *Brain Res.*, 67: 519–526.

Sotelo, C. and Changeux, J.P. (1974b) Bergmann fibers and granule cell migration in the cerebellum of the homozygous weaver mutant mice. *Brain Res.*, 77: 484–491.

Sotelo, C. and Guénet, J.L. (1983) Nodding, a new mutant of the mouse with cerebellar abnormalities. *Neurosci. Lett.* (suppl.), 14: S 353.

Sotelo, C. and Guénet, J.L. (1988) Pathologic changes in the CNS of dystonia musculorum mutant mouse: an animal model for human spinocerebellar ataxia. *Neuroscience*, 27: 403–424.

Steinmayr, M., André, E., Conquet, F., Rondi-Reig, L., Delhaye-Bouchaud, N., Auclair, N., Daniel, H., Crépel, F., Mariani, J., Sotelo, C. and Becker-André, M. (1998) *Staggerer* phenotype in retinoid-related orphan receptor α-deficient mice. *Proc. Natl. Acad. Sci. USA*, 95: 3960–3965.

Tarkowski, A.K. (1961) Mouse chimaeras developed from fused eggs. *Nature*, 190: 857–860.

Thomas, K.R. and Capecchi, M.R. (1990) Targeted disruption of the murine int-1 proto-oncogene resulting in severe abnormalities in midbrain and cerebellar development. *Nature*, 346: 847–850.

Thomas, K.R., Musci, T.S., Neumann, P.E. and Capecchi, M.R. (1991) Swaying is a mutant allele of the proto-oncogene Wnt-1. *Cell*, 67: 969–976.

Trenkner, E. and Hoffmann, M.K. (1986) Defective development of the thymus and immunological abnormalities in the neurological mouse mutation "staggerer". *J. Neurosci.*, 6: 1733–1737.

Vernet-der-Garabedian, B., Lemaigre-Dubreuil, Y., Delhaye-Bouchaud, N. and Mariani, J. (1998) Abnormal IL-1b cytokine expression in the cerebellum of the ataxic mutant mice staggerer and Lurcher. *Mol. Brain Res.*, 62: 224–227.

Vogel, M.W., Sunter, K. and Herrup, K. (1989) Numerical matching between granule and Purkinje cells in lurcher chimeric mice: a hypothesis for the trophic rescue of granule cells from target related cell death. *J. Neurosci.*, 9: 3454–3462.

Vu-Dac, N., Gervois, Ph., Grötzinger, T, De Vos, P., Schoonjans, K., Fruchart, J. C., Auwerx, J., Mariani, J., Tedgui, A. and Staels, B.1997) Transcriptional regulation of apolipoprotein A-I gene expression by the nuclear receptor RORα. *J. Biol. Chem.*, 272: 22401–22404.

Wassef, M. and Joyner, A. (1997) Early mesencephalon/metencephalon patterning and development of the cerebellum. *Perspect. Dev. Neurobiol.*, 5: 3–16.

Yoshiki, A. and Kusakabe, M. (1998) Cerebellar histogenesis in identified cells of normal-reeler mouse chimeras. *Int. J. Dev. Biol.*, 42: 695–700.

Zanjani, H.S., Mariani, J. and Herrup, K. (1990) Cell loss in the inferior olive of the staggerer mutant mouse is an indirect effect of the gene. *J. Neurogenet.*, 6: 229–241.

Zuo, J., De Jager, P.L., Takahashi, K.A., Jiang, W., Linden, D.J. and Heintz, N. (1997) Neurodegeneration in Lurcher mice caused by mutation in delta2 glutamate receptor. *Nature*, 338: 769–773.

W.E. Crusio and R.T. Gerlai (Eds.)
Handbook of Molecular-Genetic Techniques for Brain and
Behavior Research (Techniques in the Behavioral and Neural Sciences, Vol. 13)

CHAPTER 2.1.7

Genetic dissection of mouse behavior using induced mutagenesis

Lawrence H. Pinto[1] and Joseph S. Takahashi[2]

[1]Department of Neurobiology and Physiology, Northwestern University, 2153 North Campus Drive, Evanston, IL 60208-3520, USA
[2]Howard Hughes Medical Institute, Northwestern University, 2153 North Campus Drive, Evanston, IL 60208-3520, USA

Introduction

Complex biological processes often present an intractable problem for analysis because the identity of their molecular components is not known. The study of behaviors is still more complicated because each behavior may depend on several biological systems. Genetic dissection is perhaps the most feasible way to approach these analyses. The principle of genetic dissection is to identify the molecular components of a behavior by using a three-step process. First, random mutations are induced within a genome and animals demonstrating defects in the behavior of interest are identified by screening the mutagenized individuals. Second, the affected gene is genetically mapped or localized in the genome. Lastly, the gene is identified at the molecular level using candidate gene and positional cloning approaches.

Genetic dissection and the requirements for its application

Genetic dissection is the use of a series of mutations that affect a biological process in order to identify the functional elements of the process (Hotta and Benzer, 1969). This permits elucidation of their function and interactions (Konopka and Benzer, 1971). Genetic dissection has been used routinely to study biochemical pathways. A series of single gene mutations affecting enzymes within a particular pathway, can each perturb the product of the pathway, e.g. by reducing the function of the essential enzymes. Each affected enzyme can then be identified biochemically by taking advantage of knowledge of the substrates and products in the pathway.

Behaviors, however, are more difficult to study than biochemical pathways. They require communication between cell types within the nervous system and the appropriate interaction of multiple pathways and the molecular components of these pathways such as neurotransmitters, ion channels, and their transcription factors. Moreover, the identification of all of the genes which underlie a particular behavior usually cannot be assessed biochemically. This is because certain neuronal signals that control behaviors spread rapidly throughout the nervous system, making biochemical identification difficult. One powerful way to study behavioral genes is to clone them and study their gene products.

The use of mutant phenotypes to identify essential genes is also known as forward genetics. This approach has facilitated rapid progress in diverse areas (Takahashi et al., 1994). The demonstration that the *Shaker* gene encoded a potassium channel (Wu et al., 1983) and the subsequent cloning of this gene (Papazian et al., 1987; Tempel et al., 1987) facilitated the identification of several other families of ion channels and led to a detailed knowledge of their structure-function relationship (Doyle et al., 1998). Likewise, the field of circadian biology was advanced from descriptive to mechanistic analyses with the identification of the *per* gene (Konopka and Benzer, 1971). The forward genetic approach has also been of great utility in the elucidation of developmental processes. The identification of mutants with aberrant pattern (body plan) formation in the fly (Nusslein-Volhard and Wieschaus, 1980) and zebrafish (Brand et al., 1996) has permitted an exponential increase in our understanding of the processes involved. Forward genetics differs from reverse genetics, in which a gene is modified (e.g. with a "knockout"; see Keverne, 1997) in that knowledge of the gene sequence in not needed for the forward genetic approach.

Two principal requirements need to be satisfied in order to apply forward genetics to the study of a behavior. First, the population of animals used for the study should be genetically homogeneous, or isogenic, thereby permitting an induced mutation in the DNA sequence to be identified readily. Isogenicity is also helpful in reducing the variability of the phenotype under study, thus allowing animals with an abnormal phenotype to be identified reliably. Second, a reliable quantitative assay needs to be established for the behavior to be studied. Ideally, this assay should be neither labor intensive nor time consuming. This is desirable, though not always possible. It is still possible, but more difficult, to use this approach if the assay is lethal or renders the animal sterile.

The mouse satisfies both of the above requirements for many behaviors. First, inbred strains are available for the mouse. Inbred strains are produced by 20 or more generations of crosses either between siblings or between offspring and the older parent. Thus, each member of a strain is considered to be homozygous at each locus across the genome. This also reduces variation among mice under investigation. However, it should be kept in mind that the reproducibility from animal-to-animal within a given inbred strain varies among the strains. For example, in our experience the circadian period of animals of the C57BL/6J strain is much more consistent than that of either the 129 strain commonly used for gene targeting or the BALB/cJ strain. The second requirement, for a reliable assay, is more readily satisfied for some behaviors than for others. The mouse's size and consistent pattern of activity makes it ideal for studying circadian behavior. Learning and memory may be studied in mice using several tests, including context-dependent fear conditioning (Bourtchuladze et al., 1994; Owen et al., 1997). Mice are also amenable to the study of auditory-driven (Gibson et al., 1995) and olfactory-driven (Zhao et al., 1998) behavior, and they can be tested for addiction to alcohol and drugs of abuse (Grisel et al., 1997; Picciotto et al., 1998). However, the study of visually-driven behaviors needs to be approached with care. Several mouse strains (e.g. C3H/HeJ) carry the mutant allele at the *rd* locus, resulting in the progressive degeneration of photoreceptors starting about PND (postnatal day) 10. The eyes of mice track moving visual targets using the optokinetic nystagmus reflex (Mangini et al., 1985), but no rapid assays for visual perception have as yet been reported. In addition, several common mutations that cause ataxia actually result from degeneration of the inner ear as the primary phenotype (Steel and Brown, 1994). Thus, mice satisfy the requirements for many behaviors, but caution must be exercised in design of the screening assay and in choosing the appropriate inbred strain for each particular study.

Advantages and disadvantages of the forward genetic approach

There are several advantages to the forward genetic approach:

1. It requires no prior knowledge of the mechanism or components of the behavior to

be studied and it makes no assumptions about the genes that are involved.

2. It is often possible to isolate a number of mutant alleles of one gene that alter the function of the gene product in different ways. From these various alterations it is often possible to understand the function of the gene product better than by studying only the wild-type gene product.
3. Once a single gene that is essential for a behavior has been found, it is possible to use this gene to facilitate the search for other genes that are essential (see Modifier Screens below).
4. In some instances point mutations may be more informative than targeted null mutations.
5. Gain-of-function alleles can be identified.
6. Finally, it is possible to achieve certain economies in the process of screening for mutants. For example, one set of mutagenized animals can be screened for several behaviors.

There are three primary disadvantages to the forward genetic approach, two theoretical and one practical. Theoretically, should the behavior under study require the activity of a protein that is essential for life, then mutation of the gene encoding the essential protein could have a negative effect on the general state of health of the animal. Therefore, it may be difficult to tell if the reduced performance of the animal in the behavior under study is due to a specific effect in the behavioral mechanism or is due to general deterioration. An example of this is seen in the *Wheels* mutation, which was identified by screening of mutagenized animals. Heterozygous animals demonstrate a lengthened circadian period (Pickard et al., 1995). However, the *Wheels* mutation, which is homozygous lethal, also demonstrates abnormal development of the inner ear, leading to the conclusion that the circadian defects are secondary to the inner ear defects (Nolan et al., 1997). In brief, even for a behavior as basic as the circadian clock, a mutation that alters the behavior may not be fully informative about the mechanism for the behavior. However, the presence of morphological abnormalities in a mutant animal does not necessarily preclude the possibility that a disrupted gene product participates in a specific behavior. This has been demonstrated for the role of Fyn tyrosine kinase in hippocampal LTP. In fyn-deficient mice hippocampal LTP is impared, but the analysis of this finding is complicated by the presence of a number of neurological defects such as uncoordinated hippocampal architecture and reduced neurite outgrowth (Grant et al., 1992; Beggs et al., 1994). To address this complication fyn was expressed using the calcium/calmodulin-dependent protein kinase promoter to achieve high levels of Fyn expression in the forebrain but not the rest of the brain. The result was restoration of hippocampal LTP in the presence of persistent morphological abnormalities of the hippocampus (Kojima et al., 1997). One way to distinguish a primary from a secondary defect is to characterize several behaviors. If only one is altered, this finding strengthens the argument that the single defect is primary. A second theoretical shortcoming of the approach is that one function may be performed by two gene products, either of which is sufficient to sustain the function. Thus, removal of either gene product will not alter the function subserved by both. In other words, a gene must be vital in order for it to be detected by this approach. The number of vital genes in a region can be estimated by studying the frequency of lethal mutations linked to markers in the region (Shedlovsky et al., 1986). The principal practical shortcoming is that, having identified a mutant phenotype, the mutated gene still needs to be cloned and its gene product identified.

Induction of mutations

In order to serve as the basis for a forward genetic approach, a mutagen needs to satisfy several requirements. These include: high mutation rate

TABLE 1

Survey of mutagens in mice

Agent	Best dose	Best target	Mutation rate per locus ($\times 10^{-5}$)	Predominant mutation
X-rays	6 Gy	Spermatogonia	13	Small deletions
	5 Gy+% Gy	Spermatogonia	50	Small deletions
	3 Gy	Postmeiotic	33	Deletions, translocations
	4 Gy	Oocytes	19	Deletions, translocations
ENU	250 mg/kg	Spermatogonia	66	Intragenic point mutations
	4×100 mg/kg	Spermatogonia	150	Intragenic point mutations
Procarbazine	600 mg/kg	Spermatogonia	5	Small, intragenic
		Postmeiotic	22	Deletions, translocations
Chlorambucil	10 mg/kg	Postmeiotic	127	Deletions, translocations
		Transgene	~0.1	
Gene trap			~1	
None			0.5–1.0	

References for the various mutation rates are given in the text.

in premeiotic cells, allow survival and fertility of the mutagenized animal, and make only small perturbations in the genome (to facilitate detection of the altered nucleotide sequence). The first large-scale induced mutation experiments in mice were done using X-radiation with the intention of determining the susceptibility to ionizing radiation. Unfortunately, this method of mutagenesis not only produces mutants at a low rate but also induces large chromosomal perturbations such as deletions and inversions (Russell, 1951; Russell et al., 1959; Russell et al., 1989) (see Table 1). Other mutagens such as chlorambucil behave in a similar manner (Russell et al., 1989; Rinchik et al., 1990a). A wide range of mutagens is available (see Table 1). However, ENU (*N*-ethyl-*N*-nitrosourea) most closely satisfies the above requirements. It induces mutations at a high rate (Russell et al., 1979; Russell et al., 1982a; Russell et al., 1982b; Shedlovsky et al., 1993) and acts premeiotically. This results in the production of many gametes per mutagenized animal. One important advantage of ENU is that it induces point mutations, most often by transversion (Favor et al., 1991; Pearce et al., 1995; Favor et al., 1997) (see Fig. 1). As a result, both gain- and loss-of-function mutants can be produced. Lethal mutations, which are typically null or loss-of-function alleles, are less frequent with ENU because the mutations are more subtle (i.e. intragenic point mutations). The frequency of occurrence of lethal mutations has been studied for the T/t-H-2 region of chromosome 17 of the mouse (Shedlovsky et al., 1986) and found to be 11 lethal mutations in 280 gametes studied. The induction of point mutations offers the advantages that the mutation will most likely have an effect on the transcription unit in which it occurs and the base substitution can be identified by sequence comparison. This contrasts with deletion mutations which can exert an effect far from the deleted sequence.

ENU mutagenesis is successful in adult males but not females (Ehling and Neuhauser-Klaus, 1988), probably owing to the differing mechanisms for gametogenesis in male and female mammals.

Transversion of DNA Induced by ENU

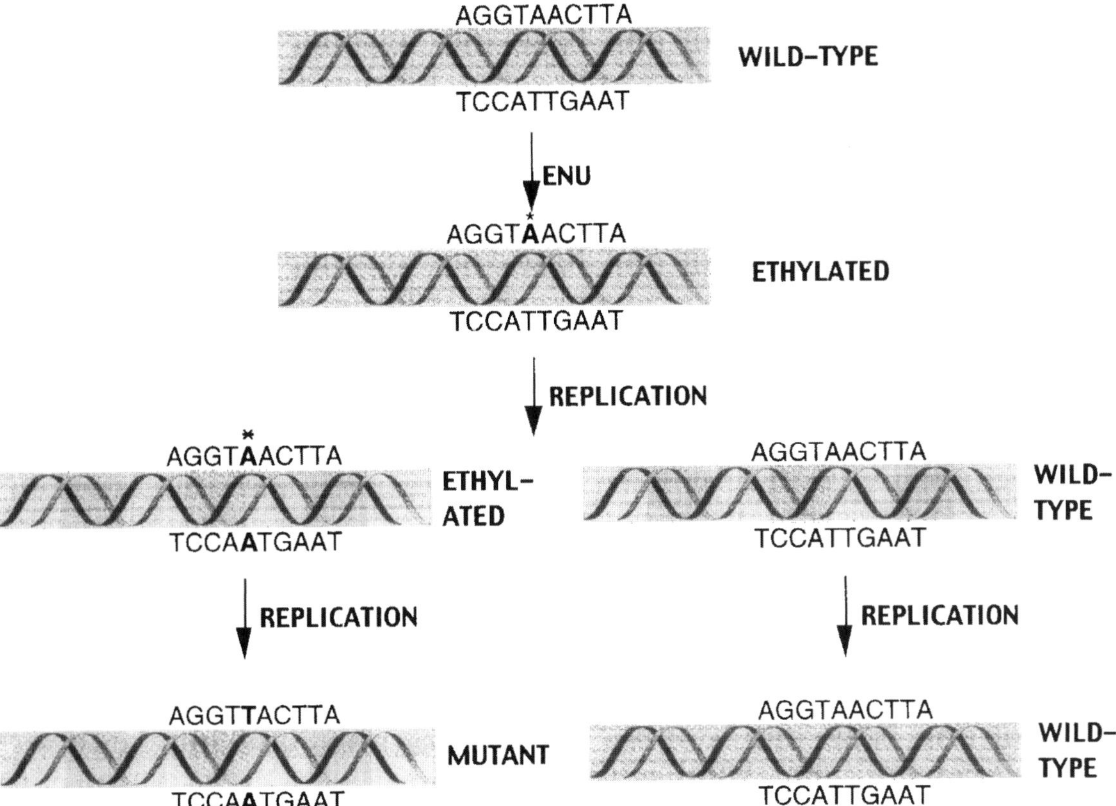

Fig. 1. Mechanism for transversion induced by ENU mutagenesis. The key to the mechanism is that ENU ethylates adenosine (second line) and causes it to hybridize with adenosine (third line) rather than thymine, resulting in a transversion from adenine to thymine. This is the most common mechanism found for *in vivo* mutagenesis.

Sperm are produced by division and differentiation of spermatagonial stem cells throughout the life of the animal, whereas ova are produced by maturation of oocytes which undergo meiosis I in the perinatal period. Thus, administration of ENU to adult males will affect spermatogonial stem cells, but administration to females will affect oocytes, in which DNA repair mechanisms are very active.

Sperm are not effectively mutagenized by ENU, and thus it is necessary to eliminate from the mutagenized animal those sperm which are mature or are in the process of maturation at the time ENU is administered. This can be achieved by allowing 1–2 month passage of time after administration of the mutagen and then testing for (temporary) sterility.

When administering ENU it is necessary to measure the actual concentration of the compound spectrophotometrically (Shedlovsky et al., 1986), as purity varies from lot-to-lot. It is also possible to take advantage of the increase in induced mutation rate that results from repetition of doses of ENU (Hitotsumachi et al., 1985). The investigator also needs to be observant of the strain that is being injected with ENU, as some strains of mice are more tolerant of the compound than others, and a dose established for one strain may result in permanent sterility when administered to another strain.

Detection of mutations

Breeding schemes

Mutations induced by ENU must be detected in the progeny of the mutagenized male. ENU will affect only the paternally-derived chromosomes. Consequently, only semidominant and dominant mutations can be found by screening first generation (G_1) progeny. This section will consider various breeding schemes, their utility and efficiency in identifying ENU-induced mutations.

Semidominant mutations are those for which the phenotype of the heterozygote is intermediate between wild-type and homozygous mutants. An example of semidominance can occur with oligomeric proteins. If one monomer of the oligomer is abnormal, it is easy to visualize how the entire oligomer might have reduced function. Two examples which illustrate this are the homotetrameric K channel (MacKinnon et al., 1993) and the heteromeric CLOCK-BMAL transcription factor complex (Gekakis et al., 1998). Semidominant mutations can be detected within G_1 progeny. However, for some phenotypes, the very nature of a semidominant mutation requires that the screening test be quantitative. It is noteworthy that animals hemizygous for a mutation within the Y-chromosome also can be identified within G_1 progeny.

Recessive mutations will not be detected in the heterozygous mutant progeny of a single generation cross. Instead, the third generation progeny (G_3) must be tested. These animals are most conveniently generated using the breeding scheme illustrated in Fig. 2(A). However, not every G_3 animal will be homozygous for the recessive mutation of the founder animal. The probability of finding one homozygote in this generation, termed the "Efficiency of scanning", "E" is related to the number of G_2 matings and the number of G_3 mice tested per G_2 mating (Shedlovsky et al., 1986).

$$E = 1 - [0.5 + (0.75^n)]^k, \qquad (1),$$

where k = the number of G_2 females, and n = number of progeny tested per G_2 female. For five progeny from each of four G_2 matings the efficiency is 85%.

Modifier screens

A modifier screen is based on the principle that the presence of one mutation in a pathway will enhance the sensitivity to a second mutation in that pathway. If a gene involved in a behavior, m, is identified, this gene can be used as a starting point to identify other genes involved in the same behavior. This can be done using a modifier screen that induces a new mutation, "*". Animals homozygous for the known mutation that affects the behavior under study (m/m) are crossed with mutagenized animals ($*/+$) to produce a fraction of offspring that are compound heterozygotes ($+/*$; $+/m$). If the newly mutated gene (*) and the known gene (m) interact, these compound heterozygotes will have a behavior which differs from that of both the homozygous mutant (m/m) and the wild type ($+/+$). This can be illustrated with a simple example. Assume that a behavior requires ca 50% throughput of a pathway that includes the product of gene m, and that heterozygotes demonstrate ca 50% activity of the product of gene m. Animals heterozygous for this recessive mutation ($m/+$) will have a phenotype that can not be readily distinguished from wild type. The newly induced mutation (*) reduces the throughput of the steps for which it is responsible by an additional 50%. The combination of the defects of genes m and * will reduce overall throughput of the pathway to 25% of normal, which is less than the amount required for normal behavior.

Consequently, the double heterozygotes (+/*; +/*m*) can be detected in the screening test. This approach facilitated the elucidation of the series of gene products needed to form the specialized seventh photoreceptor of the fly eye, starting with one mutation, *sevenless*, that served as the "sensitizing" mutation (Simon et al., 1991; Fortini et al., 1992; Simon et al., 1993) or starting point. One allele of the *tim* gene, a component of the circadian oscillator, was also identified as a result of its interaction with the another circadian gene *per* (Rutila et al., 1996). Modifier screens have the additional advantage of being able to reveal alleles of the original gene.

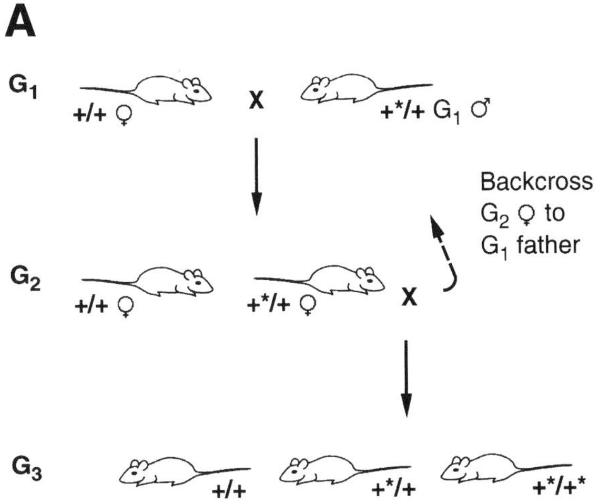

Breeding for modifier screens

To detect modifying mutations, the first generation progeny of the mutagenized male (+/+) and a female that is homozygous for the sensitizing mutation (*m*/*m*) are examined (Fig. 2(B)). In order to detect a modifying gene, sufficient progeny must be examined to ensure the generation of one compound heterozygote (+/*; +/*m*). For a 0.9 probability of detecting one such animal, it is

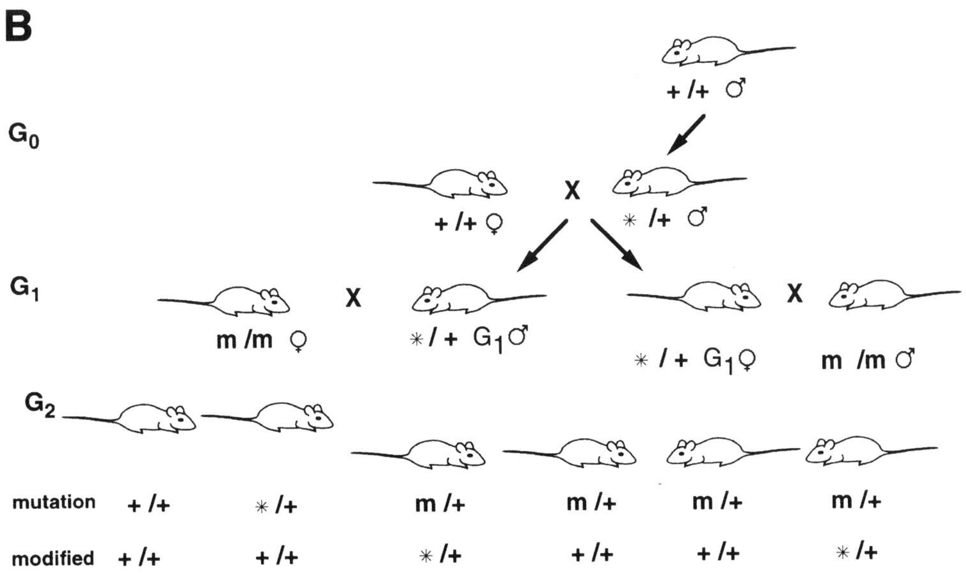

Fig. 2. (A) Breeding scheme to detect recessive mutations in third-generation (G_3) offspring. In this scheme, G_1 male progeny are mated with wild-type females to produce second-generation (G_2) progeny. The female G_2 progeny are then bred with their G_1 father to produce G_3 progeny. Compared to intercrossing G_2 females and males, this scheme doubles the likelihood of producing G_3 progeny that are homozygous for a mutation. Dominant and semidominant mutations are expressed in an average of 75% of the G_3 offspring. *Denotes altered allele or mutation. (B) Breeding scheme to detect mutations that modify the effect of a previously-discovered mutation, m. A G_1 animal that is heterozygous for a newly induced mutation (*) is bred to a mouse that bears the previously-discovered or "sensitizing" mutation m. One-half of the progeny of this mating will be compound heterozygotes.

necessary to screen 8 progeny. Although this appears to be tedious, an advantage results from this breeding scheme, based on the principle that a mutagenized parent carrying mutation *m* will give birth to more than one progeny bearing the mutation if enough progeny are examined. Thus, if one putative mutant is identified in the screen, then it ought to be possible to find additional affected progeny from the same breeding pair. This finding will provide evidence that the first individual was correctly identified as a mutant.

In most cases, modifiers alter quantitative variables, and false positives may occur if the experimental conditions are not constant. For phenotypes that are characterized quantitatively, it is imperative that the assay employed in a modifier screen be quantitative and that the conditions of rearing and testing be constant. However, other phenotypes, for example, the *vibrator* phenotype (Hamilton et al., 1997) allow modifiers to be scored with visual observation. Extra care must be exercised in the choice of the strain of mouse to be mutagenized in modifier screens. It is possible that the strain in which the sensitizing mutation occurs will differ from the strain chosen for mutagenesis. In this case the sensitizing mutation will be detected in a hybrid animal, and mapping the mutation on a hybrid background might prove difficult (see below). For this reason, it is advisable to perform a control experiment, without mutagen, to be sure that the choice of strains will permit mapping.

Selection of an appropriate mutagenesis scheme

The efficiency of the strategy to generate and detect mutations must be carefully considered. It is important to consider what fraction of the genes intrinsic to a process will be found by screening a given number of mutagenized mice. The goal of many genetic studies is to achieve saturation, the theoretical definition of which is the point at which each gene in the genome has been mutated at least once in a fashion that produces an identifiable phenotype. However, the induced mutation rate varies widely from locus-to-locus (Hitotsumachi et al., 1985), and thus it is possible that many mutations will occur at one locus before a single mutation appears at another locus.

The number of animals needed to be screened in a genome-wide scan

A simplified description of the process of mutagenesis can be given by the Poisson distribution in which there are many target genes (N), each of which has a small probability of being mutagenized (p). The number of mutagenized genes in the genomes of a population of animals, $M_a(n)$, will thus be given by the Poisson distribution.

$$M_a(n) = \frac{a^n e^{-a}}{n!} \qquad (2),$$

where $a = \mathrm{Np}$ and n is the number of mutations for which the distribution is being evaluated.

Let us estimate the number of animals required to be screened in order to find one mutant. The simplest case is that of a one generation screen for semidominant mutations. In this case the probability of finding one or more mutations in a single gene when a number of G_1 animals equal to (1/p) is screened equals (1 – the probability of finding no mutants), or ($1-M_1(0)$, or about 0.73, i.e., ($1-1/e$). Thus, if M genes are thought to be involved in a particular behavior, only (1/pM) animals would need to be screened for a 0.73 probability of finding the first mutant.

How many mutagenized animals need to be screened in order to identify every gene that is essential in a given process? Because the mutation rate varies from gene to gene, it is not possible to induce a single mutation in every gene without inducing many mutations in the most mutable genes. Thus, it is more useful to consider how many mice need to be screened in order to find mutations in 95% of the genes that have a probability *P* of being mutagenized. In order to achieve mutations with an identifiable phenotype in 95% of the genes $[(1-M_3(0))\sim 0.95]$, it is necessary to screen (3/p) animals (Eqn. (2)). Thus, to find 95% of the genes involved in a process which has a high induced mutation rate of $P\sim 2\times 10^{-3}$ (Shedlovsky et al., 1993), it is necessary to screen only about 600 gametes. However, by screening this number of gametes there is only a small chance of finding a gene with a low mutation rate of $P\sim 10^{-4}$.

Is it more efficient to screen for semidominant or for recessive mutations? For semidominant mutations, the induced mutation rate is lower than for recessive mutations (Favor, 1986; Favor et al., 1991). Assuming an induced mutation rate of $P\sim 1.5\times 10^{-4}$ per locus per gamete for semidominant mutations, it would be necessary to screen 15,000 mice to identify 95% of the genes involved. The basis for this calculation is as follows: Favor (1986) and Favor et al., 1991 found rates of about 1.5×10^{-3} for all dominant mutations that cause cataracts; the rate per locus will be much less, for example, if 10 genes are involved the rate will be 1.5×10^{-4}. A calculation of the number of animals required to detect a recessive mutation can yield a surprising result. The induced mutation rate for some recessive mutations at some loci is rather high, and for these mutations $P\sim 2\times 10^{-3}$ (Shedlovsky et al., 1993). It requires the screening of about 20 G-3 mice in each pedigree (see Fig. 2) to detect a recessive mutation (Eqn. (1)); thus, to identify a single mutation in such a highly mutable gene would require screening 20/p or about 10,000 mice. This gives rise to the ironic conclusion that detecting a recessive mutation in a readily mutable gene that is essential to a given process may require less effort than finding the dominant mutations that can affect the same process, even though additional breeding is required to render the recessive mutation homozygous. Moreover, finding a recessive mutation in one of several highly mutable genes might require fewer mice than finding a single gene with a dominant mutation. Two other complications that must be kept in mind when considering mutagenesis schemes are that there may exist genes for which it is not practically possible to generate dominant mutations, and that certain recessive mutations may be lethal.

What is the most efficient mutagenesis strategy?

The answer to this question depends on the goal of the study and the starting information available to the investigator. For example, if it is known that genes relevant to a behavior of interest are found in a particular region of the genome, then directing mutagenesis to that region may be sensible (Schimenti and Bucan, 1998). This can be done by breeding mutagenized males to mice that are hemizygous for a deletion in the appropriate region. This method has the advantage that the use of overlapping deletions can aid in the localization of genes, but the disadvantage that mice hemizygous for the deletion must be identified with molecular techniques before screening (Rinchik et al., 1990b; Brown and Peters, 1996; Justice et al., 1997). However, if nothing is known about the genes controlling the behavior under study, then genome-wide mutagenesis will reveal these genes with less effort than targeted mutagenesis to a series of regions (Table 2).

Gene trap strategies offer the advantage of leaving a unique DNA marker that can facilitate the cloning of the affected gene (Evans et al., 1997) and the detection of its expression pattern. However, this strategy requires manipulation of embryonic stem cells and a three generation breeding scheme to detect recessive mutations, and the presently available technology does not offer a high mutation rate in mice (Friedrich and Soriano, 1993; Hill and Wurst, 1993; Wurst et al., 1995). In spite of this, many gene trap mutants are available (Hill and Wurst, 1993; Evans et al., 1997; Hicks et al., 1997; Zambrowicz et al., 1998) because the number of genes involved in complex processes such as development is large, and the rate at which mutants are found is proportional to the number of genes involved

TABLE 2

Efficiency of various breeding schemes in mutagenesis

SCREEN	No. of gametes for 3-fold Poisson coverage[a]	No. of mice per gamete	Total no. of mice	Genes scanned per gamete	Genes scanned per mouse
Dominant, G_1	2000	1	2000	150	150
	20,000[c]	1	20,000	15	15
Specific Locus, G_1[a]	2000	1	2000	150	150
Genome-Wide Recessive, G_3[a]	2000	20	40,000	150	7.5
Recessive over Deletion, G_1[a] (10 cM deletions with 160 deletion lines)	$2000 \times 160 =$ 320,000	2[b]	640,000	0.63	0.31
Gene Trap, G_3					
assuming 10^{-4} forward mutation frequency	30,000	20	600,000	10	0.5
assuming 10^{-5} forward mutation frequency	300,000	20	6,000,000	1	0.05

[a] Assuming forward mutation frequency of 1.5×10^{-3} per locus per gamete with ENU; this is unrealistically low for dominant mutations but reasonable for recessive mutations at some loci (see text). Genes scanned per gamete calculated on the assumption of 100,000 genes per genome.

[b] Only half of the animals produced will be hemizygous for the deletion.

[c] Assuming a dominant mutation frequency of 1.5×10^{-4} per locus per gamete (see text).

(Eqn. (2)). However, this approach is less likely to be useful for studying processes for which the number of genes is small. Thus, although the efficiency of screening (the number of genes screened per animal tested) is low for the gene trap strategy (Table 2), this approach has great utility for the study of many processes.

Distinguishing a mutant from a false positive animal

False positives present a great practical problem for genetic screening projects. The first step, after identifying a putative mutant, is to establish whether its phenotype is heritable. A breeding program which will render the putative mutant allele in the heterozygous (semidominant screen) or homozygous (recessive screen) condition will achieve this. Testing of sufficient progeny will permit demonstration of the expected ratio of mutant: wild-type phenotypes in the population. This generally requires the testing of 20–30 animals. However, if the rate of false positives is high the number of mice needing to be retested may become excessive. Fortunately, two breeding schemes present an opportunity to confirm heritability of a phenotype by simply screening additional animals from the cross that produced the original putative mutant. In the first of these, the three-generation (recessive) screen, each G_3 animal is the progeny of a G_2 dam and the mutant G_1 sire. If one putative mutant arises from such a cross, then additional mutants should occur when its siblings are tested. In fact, each sibling will have a 25% chance to be homozygous for the same mutation. The second of these, a modifier screen, will also produce many offspring from the mutant G_1 sire. If the abnormal phenotype of a putative mutant results from a mutation and not from an environmental variable, then each sibling will have a 50% chance to inherit the same mutant allele. It is not possible to apply a similar approach to a semidominant G_1 screen, since each G_1 animal results from a different gamete in the mutagenized G_0 animal. Thus, two commonly-used breeding schemes provide a ready guard against the consequences of incorrectly identifying an animal as a mutant.

When should an animal with an abnormal phenotype be considered a putative mutant? The answer to this question is complex and depends on several considerations. First, if the abnormal phenotype is qualitatively different from wild-type (e.g. moves its eyes opposite to, rather than in the same direction as, a visual target) then the animal should be considered a putative mutant. If the animal simply does not perform the behavior at all, it is possible that the health of the animal has been compromised or that the animal has a problem with an essential physiological system that prevents it from performing the behavior. If a quantitative index of the animal's performance differs from that of wild-type, then the question can only be resolved by determining where the animal's performance lies with respect to the statistical distribution of that index for wild-type. Thus, it is imperative to possess thorough quantitative information about wild-type animals prior to beginning a screening process. If the putative mutant's performance does not differ by more than three standard deviation units from the mean for wild-type, it is likely that establishing heritability and mapping the mutant gene will prove difficult.

Genetic mapping

This procedure is necessary to demonstrate conclusively that the trait under study is indeed heritable and is essential to the next steps in identifying the gene by positional cloning. Simple sequence polymorphisms (SSLPs) provide a rapid, sensitive approach to genetic mapping in the mouse. Over 6500 SSLPs have been identified between mouse strains most frequently used in genetic mapping (Dietrich et al., 1992; Dietrich et al., 1994; Dietrich et al., 1996). These SSLPs permit localization of new markers to an interval of 0.1 cM, on the average (Dietrich et al., 1996). One breeding strategy that is used to take advantage of these polymorphisms is the cross-backcross system (Green, 1981). The initial mating is a cross of the mutant animal on the founder strain with an animal from another strain for which there exists a set of SSLPs that are polymorphic with the founder strain (the counterstrain). This cross will produce F_1 animals that are heterozygous for both the mutation and the polymorphisms. To map semidominant mutations the F_1 animals are then backcrossed to a wild-type animal of the founder strain to produce second generation offspring that will bear one chromosome of each pair that results from recombination in the F_1 parent. These progeny are then tested for the mutant phenotype and a set of polymorphic SSLPs located uniformly over the genome are characterized. This characterization is usually displayed as the haplotype of each animal, and these haplotypes will fall into groups. The mutant gene can be assigned to a region of the genome by noting the association of a haplotype carrying polymorphisms characteristic of the founder strain in a certain region with the mutant phenotype. This then allows assignment of the mutant gene to one certain region of the genome. For mapping recessive mutations, either a backcross to a homozygous mutant or an intercross is used.

There is a practical problem that occurs in the mapping of mutations. It is possible that the genetic background of the counterstrain may modify the phenotype that results from the mutant gene in such a way as to make difficult the establishment of the genotype of an individual animal from its phenotype. Each of the N_2 progeny of the F_1 animals in a mapping cross will possess a different combination of genes from the founder strain and counterstrain, and thus, if different alleles of genes capable of modifying the mutant gene exist in these strains, the resulting offspring will exhibit more variable phenotypes. However, this problem can usually be solved by the appropriate breeding of animals of questionable genotype (i.e. "test crosses"). For example, an animal in a mapping cross may be thought to be heterozygous for a semidominant mutation, but certain aspects of the phenotype may render this assignment ambiguous. If so, then a test cross of the questionable animal with a wild-type animal of the founder strain should produce progeny, half of which bear the mutant phenotype. It should be noted that there is a potential benefit that can arise from the ambiguity of mapping crosses. It is very possible that modifying alleles can be identified, and even mapped, from these crosses (Hamilton et al., 1997).

Molecular identification of mutants

Two approaches are currently used to identify mutant genes. The first is the candidate gene approach. If the mutant gene maps to a region of the genome in which a gene that is likely to participate in the process under study is found (e.g. a certain neurotransmitter receptor), then it is possible to test the hypothesis that the mutation affects this gene. The sequences of the genomic DNA of the wild-type gene and the mutant gene can be compared if the gene has already been cloned. If not, the sequences of their cDNAs can be compared. If these results demonstrate a difference in the encoding gene, the conclusion that the mutation exerts its effect through this gene can be strengthened by manipulation of the gene with gene targeting. However, in many cases the mutation will not map to a genomic region that contains candidate genes, and it will be necessary to employ positional cloning to identify the mutant gene.

Positional cloning

The gene of interest may map to a region of the genome containing genes which, when mutated, might be expected to yield the mutant phenotype. However, if no such candidate genes are identified, it will be necessary to employ the approach of positional cloning to find the mutant gene. The strategy for cloning genes positionally starting from information contained in their map positions has been reviewed (Stubbs, 1992; Copeland et al., 1993; Takahashi et al., 1994; Silver, 1995). The genomic resources available for the mouse are substantial. Among them are ample DNA polymorphisms, five YAC libraries covering the genome ca 20-fold, and BAC libraries covering the genome ca 40-fold, all of which are in the public domain (Table 3). The following is a brief review of one set of steps leading to positional cloning that we have found useful.

High resolution mapping

The first step in positional cloning is to map the gene with high enough resolution to allow the identification of cloned DNA fragments that span the region containing the gene, e.g. YAC and BAC clones. It is necessary to use about 1000 meioses in order to obtain sufficient map resolution to identify a YAC that spans the mutant gene (Guenet, 1996). However, the greater the number of meioses used, the smaller the genomic interval that will have to be examined at the molecular level (see below).

Functional rescue of the mutation

The advantage of functional rescue of a mutation lies in the demonstration that the unit of DNA used for rescue contains the genetic information that was altered by the mutation. Thus, it is best to achieve functional rescue using the smallest possible sequence of DNA. If the gene of interest

TABLE 3

The mouse as a genetic organism

Genome size:	3000 Mb, 1600 cM
Genetic maps:	>20,000 loci mapped, >6500 SSLP, >11,000 STS
Physical maps:	5 YAC libraries (~20X, 240–820 kb aver.) 2 P1 libraries (6.5X, 70–80 kb aver.) 5 BAC/PAC libraries (~40X, 120–195 kb aver.)
	YAC contig map Radiation hybrid map Wash U/HHMI Mouse EST Project
Informatics:	MGD, The Jackson Laboratory (www.informatics.jax.org) MIT Mouse Genome Center (www.genome.wi.mit.edu)
Synteny and rapid transfer to human genome	
YAC/BAC germ-line expression and mutant rescue in mice	

is very large, this functional unit must be a YAC (Huxley, 1998). However, for most genes, bacterial artificial chromosome (BAC) clones are large enough to contain the transcription unit. In fact, mutations in three important behavioral genes were rescued functionally as a prelude to their cloning: the per gene of *Drosophila* (Bargiello et al., 1984; Zehring et al., 1984) the *Clock* gene of the mouse (Antoch et al., 1997), and the vibrator gene of the mouse (Hamilton et al., 1997). Functional rescue in mice can be achieved by constructing transgenic animals with a mutant genotype into which the YAC or BAC (Antoch et al., 1997) has been incorporated. Because even high resolution mapping lacks the precision necessary to fix the position of the mutant gene within the region represented by a BAC clone, trial and error will be necessary to identify a BAC mapping near the mutated gene which is capable of functional rescue. An important control is to be sure that incorporation of other BACs that map nearby does not provide functional rescue. These experiments make possible the demonstration that the rescuing BAC does indeed encode the gene of interest.

Cloning the gene

Having confirmed that the BAC (or YAC) contains the gene of interest, all effort can be focused on sequences contained in the BAC in order to obtain the sequence of the gene and deduce the function of its gene product from the sequence. The details of this are beyond the scope of the present review, but it is useful to outline the steps that are often used (see King et al., 1997). First, shotgun sequencing of the BAC will identify sequences that have homology to known genes. These sequences can be identified by searching several databases, including that for expressed sequence tags. However, even if the sequence of the entire BAC were known, it is likely that several transcription units will occur within the BAC, and sequence information alone will not allow the gene of interest to be identified. Since ENU induces point mutations, it would not be reasonable to attempt to find a single point mutation among the 100-200 kb contained in the BAC. A more feasible approach is to analyze the mRNA that is expressed in the tissue which is likely to be affected by the mutation. This can be done is by sequencing the cDNAs corresponding to mRNAs that are both expressed in this tissue and encoded by sequences contained in the BAC. Two practical ways for doing this are to screen cDNA libraries using the BAC as a probe and to selectively enrich the cDNA clones by hybridization to the BAC. It must be demonstrated that cDNAs found in this way map to the physical interval of the genome that includes the BAC. These cDNAs can be tested by Northern or *in situ* hybridization to tissue known to be involved in the behavior under study. However, a wide pattern of expression does not preclude the possibility that the gene is involved in the behavior under study, for the same protein can participate in different processes in different tissues. Thus, the combined approach of genomic sequencing, hybridization to the rescuing BAC and enrichment of cDNAs will identify the genes expressed in the critical region of the genome and expressed in the tissue of interest.

To find the nucleotide substitution in the mutant gene, it may be sufficient to sequence the cDNA of the mutant animal and compare it with that of the wild-type animal or to sequence genomic DNA from the mutant animal and compare this sequence with that from the wild-type animal. However, the finding of a single base substitution is not sufficient to demonstrate that this substitution is the cause of the mutant phenotype. One such proof would be the identification of a second allele of the mutant gene and the demonstration of an alteration in the sequence of the same cDNA. Alternatively, the predicted gene products of wild-type and mutant must be compared to determine if the function of the altered gene product can be understood in terms of the known biology of the process under study. For example, if the base substitution that is found causes a truncation of the predicted gene product and that truncation would be expected to result in the observed phenotype, then the investigator is on solid ground to state that the gene has been found and that the mutation has been identified. However, this latter evidence would need to be supported with genomic rescue in order to constitute proof.

Case study: genetic dissection in the analysis of circadian behavior

Circadian behavior in mammals is driven by a primary oscillator located in the suprachiasmatic nucleus, the SCN (Moore, 1995; Hastings, 1997). This oscillator does not depend on the generation of action potentials (Schwartz et al., 1987), and the electrical activity of dissociated SCN neurons in culture has a period that is intrinsic to each cell, independent of its neighbors (Welsh et al., 1995; Liu et al., 1997; Herzog et al., 1998). These observations have led to the hypothesis that the primary circadian oscillator is a property of individual cells. The finding of circadian variations in mRNA of a circadian gene that is described below, *period* (*per*), (Hardin et al., 1990), and the transcriptional regulation of this gene's expression by its own gene product (Hardin et al., 1992; So and Rosbash, 1997) led to the conclusion that this intracellular oscillator results from an intracellular feedback loop. Support for the existence of an intracellular oscillator comes from the observation that application of protein synthesis inhibitors to mammals can alter the circadian period and phase, and that there is a critical period of protein synthesis early in the subjective day (Takahashi and Turek, 1987). This finding further suggests that the intracellular feedback loop involves protein synthesis, and subsequent mutagenesis studies have identified many of these proteins.

The first mutation known to affect the circadian clock in mammals, *Tau*, occurred spontaneously in the golden hamster (Ralph and Menaker, 1988). Unfortunately, this gene has still not been cloned because of inadequate genetic resources for the hamster such as genetic markers and inbred strains with well-characterized DNA polymorphisms. In contrast, a study using ENU mutagenesis in the mouse allowed a circadian gene, *Clock*, to be identified (Vitaterna et al., 1994). When mutated, this gene alters the circadian period, or in the homozygote, abolishes circadian rhythmicity after about 2 weeks in constant darkness. The gene responsible was cloned in a recent study (King et al., 1997). Transgenic technology in the mouse allowed functional rescue by a single BAC (Antoch et al., 1997). This latter study confirmed that the BAC contained the sequence of the wild-type allele of the gene and permitted identification of the *Clock* gene. The gene encodes a transcription factor with a basic helix–loop–helix DNA binding domain, a PAS domain and a Q-rich activation domain. The partner of the CLOCK protein, BMAL1, has been identified using the yeast two-hybrid system (Gekakis et al., 1998). It has been demonstrated that the CLOCK-BMAL1 heterodimer is capable of activating transcription of two other clock genes, *per* and *tim* in *Drosophila* (Darlington et al., 1998) and the *mPer1* gene in mammals (Gekakis et al., 1998; Sangoram et al., 1998). The PER and TIM proteins also form a heterodimer (Gekakis et al., 1995) and interactions between the two proteins permit translocation of the heterodimer to the nucleus (Saez and Young, 1996). Once in the nucleus, the PER-TIM heterodimer inhibits the activation of transcription of *per* and *tim* genes by the CLOCK-BMAL1 heterodimer (Darlington et al., 1998) in *Drosophila* and inhibits the transcription of the *mPer1* gene in mammals (Sangoram et al., 1998). Thus, these four proteins provide a framework for a transcriptional autoregulatory feedback loop oscillator. This system requires that the half-life of the PER-TIM protein complex to be short, so that transcriptional regulation can alter the levels the levels of the complex. In *Drosophila* the rapid turnover of cytoplasmic PER protein monomers results from the association of the monomers with a kinase (Kloss et al., 1998) encoded by the *double-time* (*dbt*) gene. The first circadian gene to be found, *per*, was identified by forward genetics in *Drosophila* (Konopka and Benzer, 1971). Subsequently, the *tim* (Sehgal et al., 1994; Myers et al., 1995) and *dbt* (Price et al., 1998) genes were identified in *Drosophila* by forward genetics. The human and mouse orthologs of *per* have been identified and cloned in recent studies (Albrecht et al., 1997; Shearman et al., 1997; Sun et al., 1997; Tei et al., 1997). Thus, almost all of the proteins that are currently thought to play a role in the primary intracellular circadian oscillator were first identified by forward genetics.

Synopsis of steps to identify a behavioral gene

The following steps outlines one approach to cloning a behavioral gene. The actual steps that will be followed will depend on the problem under study and the nature of the mutation that is isolated:

1. Choose a strain of mouse which demonstrates the behavior robustly.
2. Develop a rapid screening test for the behavior and establish the failure rate of the test in the strain of interest. Attempt to alter the behavior with non-genetic intevention to show that the test is sensitive.
3. Choose a counterstrain with SSLP polymorphisms and apply mapping crosses to determine whether the behavior is modified significantly by the genetic background of the counterstrain.
4. Inject ENU into mice of the target strain and perform breeding to produce progeny that are appropriate for the semidominant or recessive mutations being sought.
5. Screen progeny. Use test crosses to establish heritability.
6. Map the gene to within about 5 cM.
7. Decide if the phenotype is of great enough interest to merit cloning the gene and if the phenotype is robust enough to allow the next steps.
8. Perform high resolution mapping (1000-3000 informative meioses).
9. Search for candidate genes that map within the region.
10. Construct a YAC and BAC contig spanning the critical interval containing the mutation.

11. Perform functional rescue of the mutation.
12. Clone the gene with combination of sequencing, hybridizing to the rescuing BAC, cDNA selection, and Northern blotting of tissues expected to express the gene. The predicted gene product must have a predicted structure consistent with the function that was altered by the mutation.
13. Identify the mutation by comparing the sequence of genomic DNA from wild-type and mutant mice. The predicted mutant gene product must be consistent with the observed defect in the mutant animal.
14. Confirm with a second allele if available.

Acknowledgments

We thank Drs William Dove, Alexandra Shedlovsky and Martha Hotz Vitaterna for critical discussions and Dr Andrew McCallion for critically reading the manuscript. Supported by the NSF, NIH and an Unrestricted Grant in Neuroscience from the Bristol-Meyers Squibb Foundation. J.S.T. is an Investigator in the Howard Hughes Medical Institute.

References

Albrecht, U., Sun, Z.S., Eichele, G. and Lee, C.C. (1997) A differential response of two putative mammalian circadian regulators, mper1 and mper2, to light. *Cell*, 91: 1055–1064.

Antoch, M.P., Song, E.J., Chang, A.M., Vitaterna, M.H., Zhao, Y., Wilsbacher, L.D., Sangoram, A.M., King, D.P., Pinto, L.H. and Takahashi, J.S. (1997) Functional identification of the mouse circadian Clock gene by transgenic BAC rescue. *Cell*, 89: 655–667.

Bargiello, T.A., Jackson, F.R. and Young, M.W. (1984) Restoration of circadian behavioural rhythms by gene transfer in Drosophila. *Nature*, 312: 752–754.

Beggs, H.E., Soriano, P. and Maness, P.F. (1994) NCAM-dependent neurite outgrowth is inhibited in neurons from Fyn-minus mice. *J. Cell Biol.*, 127: 825–833.

Bourtchuladze, R., Frenguelli, B., Blendy, J., Cioffi, D., Schutz, G. and Silva, A.J. (1994) Deficient long-term memory in mice with a targeted mutation of the cAMP-responsive element-binding protein. *Cell*, 79: 59–68.

Brand, M., Heisenberg, C.P., Warga, R.M., Pelegri, F., Karlstrom, R.O., Beuchle, D., Picker, A., Jiang, Y.J., Furutani-Seiki, M., van Eeden, F.J., Granato, M., Haffter, P., Hammerschmidt, M., Kane, D.A., Kelsh, R.N., Mullins, M.C., Odenthal, J. and Nusslein-Volhard, C. (1996) Mutations affecting development of the midline and general body shape during zebrafish embryogenesis. *Development*, 123: 129–142.

Brown, S.D. and Peters, J. (1996) Combining mutagenesis and genomics in the mouse–closing the phenotype gap. *Trends Genet.*, 12: 433–435.

Copeland, N.G., Jenkins, N.A., Gilbert, D.J., Eppig, J.T., Maltais, L.J., Miller, J.C., Dietrich, W.F., Weaver, A., Lincoln, S.E., Steen, R.G. et al. (1993) A genetic linkage map of the mouse: current applications and future prospects. *Science*, 262: 57–66.

Darlington, T.K., Wager-Smith, K., Ceriani, M.F., Staknis, D., Gekakis, N., Steeves, T.D.L., Weitz, C.J., Takahashi, J.S. and Kay, S.A. (1998) Closing the circadian loop: CLOCK-induced transcription of its own inhibitors per and tim. *Science*, 280: 1599–1603.

Dietrich, W., Katz, H., Lincoln, S.E., Shin, H.S., Friedman, J., Dracopoli, N.C. and Lander, E.S. (1992) A Genetic Map of the Mouse Suitable for Typing Intraspecific Crosses. *Genetics*, 131: 423–447.

Dietrich, W.F., Miller, J., Steen, R., Merchant, M.A., Damron-Boles, D., Husain, Z., Dredge, R., Daly, M.J., Ingalls, K.A., O'Connor, T.J., Evans, C.A., DeAngelis, M.M., Levinson, D.M., Kruglyak, L., Goodman, N., Copeland, N.G., Jenkins, N.A., Hawkins, T.L., Stein, L., Page, D.C. and Lander, E.S. (1996) A comprehensive genetic map of the mouse genome. *Nature*, 380: 149–152.

Dietrich, W.F., Miller, J.C., Steen, R.G., Merchant, M., Damron, D., Nahf, R., Gross, A., Joyce, D.C., Wessel, M. and Dredge, R.D. (1994) A genetic map of the mouse with 4,006 simple sequence length polymorphisms. *Nat. Genet.*, 7: 220–245.

Doyle, D.A., Cabral, J.M., Pfuetzner, R.A., Kuo, A., Gulbis, J.M., Cohen, S.L., Chait, B.T. and MacKinnon, R. (1998) The structure of the potassium channel: molecular basis of K+ conduction and selectivity. *Science*, 280: 6977.

Ehling, U.H. and Neuhauser-Klaus, A. (1988) Induction of specific-locus mutations in female mice by 1-ethyl-1-nitrosourea and procarbazine. *Mutat. Res.*, 202: 139–146.

Evans, M.J., Carlton, M.B. and Russ, A.P. (1997) Gene trapping and functional genomics. *Trends Genet.*, 13: 370–374.

Favor, J. (1986) The frequency of dominant cataract and recessive specific-locus mutations in mice derived from 80 or 160 mg ethylnitrosourea per kg body weight treated spermatogonia. *Mutat. Res.*, 162: 69–80.

Favor, J., Neuhauser-Klaus, A. and Ehling, U.H. (1991) The induction of forward and reverse specific-locus mutations and dominant cataract mutations in spermatogonia of treated strain DBA/2 mice by ethylnitrosourea. *Mutat. Res.*, 249: 293–300.

Favor, J., Neuhauser-Klaus, A., Ehling, U.H., Wulff, A. and van Zeeland, A.A. (1997) The effect of the interval between dose applications on the observed specific-locus mutation rate in the mouse following fractionated treatments of spermatogonia with ethylnitrosourea. *Mutat. Res.*, 374: 193–199.

Fortini, M.E., Simon, M.A. and Rubin, G.M. (1992) Signalling by the sevenless protein tyrosine kinase is mimicked by Ras1 activation. *Nature*, 355: 559–561.

Friedrich, G. and Soriano, P. (1993) Insertional mutagenesis by retroviruses and promoter traps in embryonic stem cells. *Methods Enzymol.*, 225: 681–701.

Gekakis, N., Saez, L., Delahaye-Brown, A.M., Myers, M.P., Sehgal, A., Young, M.W. and Weitz, C.J. (1995) Isolation of Timeless by PER protein interaction: defective interaction between Timeless protein and long-period mutant PERL. *Science*, 270: 811–815.

Gekakis, N., Staknis, D., Nguyen, H.B., Davis, F.C., Wilsbacher, L.D., King, D.P., Takahashi, J.S. and Weitz, C.J. (1998) Role of the CLOCK protein in the mammalian circadian mechanism. *Science*, 280: 1564–1569.

Gibson, F., Walsh, J., Mburu, P., Varela, A., Brown, K.A., Antonio, M., Beisel, K.W., Steel, K.P. and Brown, S.D. (1995) A type VII myosin encoded by the mouse deafness gene shaker-1. *Nature*, 374: 62–64.

Grant, S.G., O'Dell, T.J., Karl, K.A., Stein, P.L., Soraino, P. and Kandel, E.R. (1992) Impaired long-term potentiation, spatial learning, and hippocampal development in fyn mutant mice. *Science*, 258: 760–763.

Green, E.L. (1981) *Genetics and Probability in Animal Breeding Experiments*, Oxford University Press Inc., New York.

Grisel, J.E., Belknap, J.K., O'Toole, L.A., Helms, M.L., Wenger, C.D. and Crabbe, J.C. (1997) Quantitative trait loci affecting methamphetamine responses in BXD recombinant inbred mouse strains. *J. Neurosci.*, 17: 745–754.

Guenet, J.-L. (1996) The Mouse Genome. In: M. J. Browne and P. L. Thurlby (Eds.), *Genomes, Molecular Biology and Drug Discovery*, Academic Press Inc., San Diego, pp. 27–51.

Hamilton, B.A., Smith, D.J., Mueller, K.L., Kerrebrock, A.W., Bronson, R.T., van Berkel, B., Daly, M.J., Kruglyak, L., Reeve, M.P., Nemhauser, J.L., Hawkins, T.L., Rubin, E.M. and Lander, E.S. (1997) The *vibrator* Mutation Causes Neurodegeneration via Reduced Expression of PITPα: Postitional Complementation Cloning and Extragenic Suppression. *Neuron*, 18: 711–722.

Hardin, P.E., Hall, J.C. and Rosbash, M. (1990) Feedback of the *Drosophila* period gene product on circadian cycling of its messenger RNA levels. *Nature*, 343: 536–540.

Hardin, P.E., Hall, J.C. and Rosbash, M. (1992) Circadian oscillations in period gene mRNA levels are transcriptionally regulated. *Proc. Natl. Acad. Sci. USA*, 89: 11711–11715.

Hastings, M.H. (1997) Central clocking. *Trends Neurosci.*, 20: 459–464.

Herzog, E.D., Takahashi, J.S. and Block, G.D. (1998) CLOCK controls circadian period in isolated suprachiasmatic nucleus neurons. *Nat. Neurosci.*, 1: 708–713.

Hicks, G.G., Shi, E.G., Li, X.M., Li, C.H., Pawlak, M. and Ruley, H.E. (1997) Functional genomics in mice by tagged sequence mutagenesis. *Nat. Genet.*, 16: 338–344.

Hill, D.P. and Wurst, W. (1993) Screening for novel pattern formation genes using gene trap approaches. *Methods Enzymol.*, 225: 664–681.

Hitotsumachi, S., Carpenter, D.A. and Russell, W.L. (1985) Dose-repetition increases the mutagenic effectiveness of *N*-ethyl-*N*-nitrosourea in mouse spermatogonia. *Proc. Natl. Acad. Sci. USA*, 82: 6619–6621.

Hotta, Y. and Benzer, S. (1969) Abnormal electroretinograms in visual mutants of *Drosophila*. *Nature*, 222: 354–356.

Huxley, C. (1998) Exploring gene function: use of yeast artificial chromosome transgenesis. *Methods*, 14: 199–210.

Justice, M.J., Zheng, B., Woychik, R.P. and Bradley, A. (1997) Using targeted large deletions and high-efficiency *N*-ethyl-*N*-nitrosourea mutagenesis for functional analyses of the mammalian genome. *Methods*, 13: 423–436.

Keverne, E.B. (1997) An evaluation of what the mouse knockout experiments are telling us about mammalian behaviour. *Bioessays* Vol. 19, pp. 1091–1098.

King, D.P., Zhao, Y., Sangoram, A.M., Wilsbacher, L.D., Tanaka, M., Antoch, M.P., Steeves, T.D., Vitaterna, M.H., Kornhauser, J.M., Lowrey, P.L., Turek, F.W. and Takahashi, J.S. (1997) Positional cloning of the mouse circadian Clock gene. *Cell*, 89: 641–653.

Kloss, B., Price, J.L., Saez, L., Blau, J., Rothenfluh, A., Wesley, C.S. and Young, M.W. (1998) The Drosophila Clock Gene Double-Time Encodes a Protein Closely Related to Human Casein Kinase I-Epsilon. *Cell*, 94: 97–107.

Kojima, N., Wang, J., Mansuy, I.M., Grant, S.G.N., Mayford, M. and Kandel, E.R. (1997) Rescuing impairment of long-term potentiation in fyn-deficient mice by introducing Fyn transgene. *Proc. Natl. Acad. Sci.*, 94: 4761–4765.

Konopka, R.J. and Benzer, S. (1971) Clock mutants of Drosophila melanogaster. *Proc. Natl. Acad. Sci. USA*, 68: 2112–2116.

Liu, C., Weaver, D.R., Strogatz, S.H. and Reppert, S.M. (1997) Cellular construction of a circadian clock: period determination in the suprachiasmatic nuclei. *Cell*, 91: 855–860.

MacKinnon, R., Aldrich, R.W. and Lee, A.W. (1993) Functional stoichiometry of Shaker potassium channel inactivation. *Science*, 262: 757–759.

Mangini, N.J., Vanable, J.W., Jr., Williams, M.A. and Pinto, L.H. (1985) The optokinetic nystagmus and ocular pigmentation of hypopigmented mouse mutants. *J. Comp. Neurol.*, 241: 191–209.

Moore, R.Y. (1995) Organization of the mammalian circadian system. In: Ciba Foundation Symposium, pp. 88–99.

Myers, M.P., Wager-Smith, K., Wesley, C.S., Young, M.W. and Sehgal, A. (1995) Positional cloning and sequence analysis of the Drosophila clock gene, timeless. *Science*, 270: 805–808.

Nolan, P.M., Kapfhamer, D. and Bucan, M. (1997) Random mutagenesis screen for dominant behavioral mutations in mice. *Methods*, 13: 379–395.

Nusslein-Volhard, C. and Wieschaus, E. (1980) Mutations affecting segment number and polarity in *Drosophila*. *Nature*, 287: 795–801.

Owen, E.H., Logue, S.F., Rasmussen, D.L. and Wehner, J.M. (1997) Assessment of learning by the Morris water task and fear conditioning in inbred mouse strains and F1 hybrids: implications of genetic background for single gene mutations and quantitative trait loci analyses. *Neurosci.*, 80: 1087–1099.

Papazian, D.M., Schwarz, T.L., Tempel, B.L., Jan, Y.N. and Jan, L.Y. (1987) Cloning of genomic and complementary DNA from Shaker, a putative potassium channel gene from *Drosophila*. *Science*, 237: 749–753.

Pearce, S.R., Peters, J., Ball, S., Morgan, M.J., Walker, J.I. and Faik, P. (1995) Sequence characterization of ENU-induced mutants of glucose phosphate isomerase in mouse. *Mamm. Genome*, 6: 858–861.

Picciotto, M.R., Zoli, M., Rimondini, R., Lena, C., Marubio, L.M., Pich, E.M., Fuxe, K. and Changeux, J.P. (1998) Acetylcholine receptors containing the beta2 subunit are involved in the reinforcing properties of nicotine. *Nature*, 391: 173–177.

Pickard, G.E., Sollars, P.J., Rinchik, E.M., Nolan, P.M. and Bucan, M. (1995) Mutagenesis and behavioral screening for altered circadian activity identifies the mouse mutant, Wheels. *Brain. Res.*, 705: 255–266.

Price, J.L., Blau, J., Rothenfluh, A., Abodeely, M., Kloss, B. and Young, M.W. (1998) Double-Time Is a Novel Drosophila Clock Gene That Regulates Period Protein Accumulation. *Cell*, 94: 83–95.

Ralph, M.R. and Menaker, M. (1988) A mutation of the circadian system in golden hamsters. *Science*, 241: 1225–1227.

Rinchik, E.M., Bangham, J.W., Hunsicker, P.R., Cacheiro, N.L., Kwon, B.S., Jackson, I.J. and Russell, L.B. (1990a) Genetic and molecular analysis of chlorambucil-induced germ-line mutations in the mouse. *Proc. Natl. Acad. Sci. USA*, 87: 1416–1420.

Rinchik, E.M., Carpenter, D.A. and Selby, P.B. (1990b) A strategy for fine-structure functional analysis of a 6- to 11-centimorgan region of mouse chromosome 7 by high-efficiency mutagenesis. *Proc. Natl. Acad. Sci. USA*, 87: 896–900.

Russell, L.B., Hunsicker, P.R., Cacheiro, N.L.A., Bangham, J.W., Russell, W.L. and Shelby, M.D. (1989) Chlorambucil effectively induces deletion mutations in mouse germ cells. *Proc. Natl. Acad. Sci. USA*, 86: 3704–3708.

Russell, W.L. (1951) *X-ray-induced mutations in mice*, Vol. 16 Cold Spring Harbor, New York.

Russell, W.L., Hunsicker, P.R., Carpenter, D.A., Cornett, C.V. and Guinn, G.M. (1982a) Effect of dose fractionation on the ethylnitrosourea induction of specific-locus mutations in mouse spermatogonia. *Proc. Natl. Acad. Sci. USA*, 79: 3592–3593.

Russell, W.L., Hunsicker, P.R., Raymer, G.D., Steele, M.H., Stelzner, K.F. and Thompson, H.M. (1982b) Dose-response curve for ethylnitrosourea-induced specific-locus mutations in mouse spermatogonia. *Proc. Natl. Acad. Sci. USA*, 79: 3589–3591.

Russell, W.L., Kelly, E.M., Hunsicker, P.R., Bangham, J.W., Maddux, S.C. and Phipps, E.L. (1979) Specific-locus test shows ethylnitrosourea to be the most potent mutagen in the mouse. *Proc. Natl. Acad. Sci. USA*, 76: 5818–5819.

Russell, W.L., Russell, L.B. and Cupp, M.B. (1959) Dependence of mutation frequency on radiation dose rate in female mice. *Proc. Natl. Acad. Sci. USA*, 45: 18–23.

Rutila, J.E., Zeng, H., Le, M., Curtin, K.D., Hall, J.C. and Rosbash, M. (1996) The timSL mutant of the *Drosophila* rhythm gene timeless manifests allele-specific interactions with period gene mutants. *Neuron*, 17: 921–929.

Saez, L. and Young, M.W. (1996) Regulation of nuclear entry of the Drosophila clock proteins Period and Timeless. *Neuron*, 17: 911–920.

Sangoram, A.M., Saez, L., Antoch, M.P., Gekakis, N., Stanknis, D., Whiteley, A., Fruechte, E.M., Vitaterna, M.H., Shimomura, K., King, D.P., Young, M.W., Weitz, C.J. and Takahashi, J.S. (1998) Mammalian circadian autoregulatory loop: a Timeless ortholog and mPER1 interact and negatively regulate CLOCK-BMAL1-induced transcription. *Neuron*, 21: 1101–1113.

Schimenti, J. and Bucan, M. (1998) Functional genomics in the mouse: phenotype-based mutagenesis screens. *Gen. Res.*, 8: 698–710.

Schwartz, W.J., Gross, R.A. and Morton, M.T. (1987) The suprachiasmatic nuclei contain a tetrodotoxin-resistant circadian pacemaker. *Proc. Natl. Acad. Sci. USA*, 84: 1694–1698.

Sehgal, A., Price, J.L., Man, B. and Young, M.W. (1994) Loss of circadian behavioral rhythms and per RNA oscillations in the Drosophila mutant timeless. *Science*, 263: 1603–1606.

Shearman, L.P., Zylka, M.J., Weaver, D.R., Kolakowski, L.F. Jr. and Reppert, S.M. (1997) Two period homologs: circadian expression and photic regulation in the suprachiasmatic nuclei. *Neuron*, 19: 1261–1269.

Shedlovsky, A., Guenet, J.L., Johnson, L.L. and Dove, W.F. (1986) Induction of recessive lethal mutations in the T/t-H-2 region of the mouse genome by a point mutagen. *Genet. Res.*, 47: 135–142.

Shedlovsky, A., McDonald, J.D., Symula, D. and Dove, W.F. (1993) Mouse models of human phenylketonuria. *Genetics*, 134: 1205–1210.

Silver, L.M. (1995) Mouse Genetics: concepts and applications. Oxford University Press, New York.

Simon, M.A., Bowtell, D.D., Dodson, G.S., Laverty, T.R. and Rubin, G.M. (1991) Ras1 and a putative guanine nucleotide exchange factor perform crucial steps in signaling by the sevenless protein tyrosine kinase. *Cell*, 67: 701–716.

Simon, M.A., Dodson, G.S. and Rubin, G.M. (1993) An SH3-SH2-SH3 protein is required for p21Ras1 activation and binds to sevenless and Sos proteins in vitro. *Cell*, 73: 169–177.

So, W.V. and Rosbash, M. (1997) Post-transcriptional regulation contributes to Drosophila clock gene mRNA. *EMBO J.*, 16: 7146–7155.

Steel, K.P. and Brown, S.D. (1994) Genes and deafness. *Trends Genet.*, 10: 428–435.

Stubbs, L. (1992) Long-range walking techniques in positional cloning strategies. *Mamm. Genome*, 3: 127–142.

Sun, Z.S., Albrecht, U., Zhuchenko, O., Bailey, J., Eichele, G. and Lee, C.C. (1997) RIGUI, a putative mammalian ortholog of the Drosophila period gene. *Cell*, 90: 1003–1011.

Takahashi, J.S., Pinto, L.H. and Vitaterna, M.H. (1994) Forward and reverse genetic approaches to behavior in the mouse. *Science*, 264: 1724–1733.

Takahashi, J.S. and Turek, F.W. (1987) Anisomycin, an inhibitor of protein synthesis, perturbs the phase of a mammalian circadian pacemaker. *Brain. Res.*, 405: 199–203.

Tei, H., Okamura, H., Shigeyoshi, Y., Fukuhara, C., Ozawa, R., Hirose, M. and Sakaki, Y. (1997) Circadian oscillation of a mammalian homologue of the Drosophila period gene. *Nature*, 389: 512–516.

Tempel, B.L., Papazian, D.M., Schwarz, T.L., Jan, Y.N. and Jan, L.Y. (1987) Sequence of a probable potassium channel component encoded at Shaker locus of *Drosophila*. *Science*, 237: 770–775.

Vitaterna, M.H., King, D.P., Chang, A.M., Kornhauser, J.M., Lowrey, P.L., McDonald, J.D., Dove, W.F., Pinto, L.H., Turek, F.W. and Takahashi, J.S. (1994) Mutagenesis and mapping of a mouse gene, Clock, essential for circadian behavior. *Science*, 264: 719–725.

Welsh, D.K., Logothetis, D.E., Meister, M. and Reppert, S.M. (1995) Individual neurons dissociated from rat suprachiasmatic nucleus express independently phased circadian firing rhythms. *Neuron*, 14: 697–706.

Wu, C.-F., Ganetzky, B., Haugland, F. and Liu, A.-X. (1983) Potassium currents in *drosophila:* different components affected by mutations of two genes. *Science*, 220: 1076–1078.

Wurst, W., Rossant, J., Prideaux, V., Kownacka, M., Joyner, A., Hill, D.P., Guillemot, F., Gasca, S., Cado, D., Auerbach, A. et al. (1995) A large-scale gene-trap screen for insertional mutations in developmentally regulated genes in mice. *Genetics*, 139: 889–899.

Zambrowicz, B.P., Friedrich, G.A., Buxton, E.C., Lilleberg, S.L., Person, C. and Sands, A.T. (1998) Disruption and sequence identification of 2,000 genes in mouse embryonic stem cells. *Nature*, 392: 608–611.

Zehring, W.A., Wheeler, D.A., Reddy, P., Konopka, R.J., Kyriacou, C.P., Rosbash, M. and Hall, J.C. (1984) P-element transformation with period locus DNA restores rhythmicity to mutant, arrhythmic Drosophila melanogaster. *Cell*, 39: 369–376.

Zhao, H., Ivic, L., Otaki, J.M., Hashimoto, M., Mikoshiba, K. and Firestein, S. (1998) Functional expression of a mammalian odorant receptor. *Science*, 279: 237–242.

W.E. Crusio and R.T. Gerlai (Eds.)
Handbook of Molecular-Genetic Techniques for Brain and
Behavior Research (Techniques in the Behavioral and Neural Sciences, Vol. 13)

CHAPTER 2.1.8

Mutagenesis in zebra fish: studying the brain dopamine systems

Su Guo[1], Wolfgang Driever[2] and Arnon Rosenthal[1]

[1]Department of Neuroscience, Genentech Inc., 1 DNA Way, South San Francisco, CA 94080, USA
[2]Biology 1, University of Freiburg, Hauptstraße 1, D-79110 Freiburg, Germany

Introduction

During development of the nervous system, distinct groups of neurons populate different brain areas, and subsequently may assume specific behavioral tasks. Genetic alterations in this process can lead to behavioral abnormalities. For instance, mutations of the *minibrain* gene in *Drosophila* causes reduction of the optic lobes and central brain hemispheres, and as a consequence, the mutant adult flies exhibit abnormalities in visual and olfactory behavior (Tejedor et al., 1995). Likewise, in the mouse mutation *weaver*, CNS precursors in the external germinal layer of the cerebellar cortex fail to differentiate, and this leads to motor abnormalities (Gao et al., 1992). Such mutations demonstrate that genes involved in the development of the nervous system can be tied to specific behaviors when their disruption does not cause early embryonic lethality, and allow important links to be made between genes, neurons and behavior. These genes can be subsequently identified by molecular tools. In this chapter, we will discuss the use of a vertebrate model system, the zebra fish, *Danio rerio* to identify genes that control neuronal development, survival and function in the nervous system. We first describe the advantages of using zebra fish for genetic studies, then we give an example of applying this approach to study the brain dopamine systems. The importance of the dopamine systems has been underscored by their involvement in human neurological disorders such as Parkinson's disease, schizophrenia and addictions. Thus, identification of genes involved in the dopamine systems will not only help understand brain development and function on the biochemical level, but could also provide basis for novel therapeutics.

Zebra fish: an emerging vertebrate genetic model system

Classical genetic model systems such as yeast (*Saccaromyces cerevisiae*), nematode (*Caenorhabditis elegans*), fruit fly (*Drosophila melanogaster*) and house mouse (*Mus musculus*) have been instrumental to our understanding of biological processes in general. Two genetic approaches, designated forward and reverse, have been commonly used in the genetic studies. The forward genetic approach is aimed at identifying novel genes by performing genome-wide mutagenesis to isolate mutants with desired phenotypes. This approach has been successfully applied to invertebrate model systems such as *C. elegans* (Brenner,

1974) and *Drosophila* (Nusslein-Volhard and Wieschaus, 1980). Many important genes including those involved in the development and function of the nervous system have thus been identified and characterized. The reverse genetic approach involves ablation of single known genes (gene targeting) to study their functions. Roles of vertebrate genes, many identified by their homology to *C. elegans* or *Drosophila* counterparts, have been studied in mice using this approach (Capecchi, 1989).

Due to the mode of intrauterine development and the large expenses required for carrying out a genome-wide mutagenesis screen in the mouse, it has proven difficult to identify novel genes in vertebrates by forward genetic means. Recently zebra fish, a freshwater teleost, has emerged as a vertebrate model system for forward genetic studies. Zebra fish was first used for this purpose by Streisinger and colleagues more than a decade ago (Streisinger et al., 1981), and many of its salient features have ever since attracted researchers around the world. Like *C. elegans* and *Drosophila*, zebra fish is amenable to genome wide mutagenesis screens. The adult fish is small, only about 5 cm long, thereby can be housed in large numbers with relatively little expense. The adult females can lay about one hundred eggs on a weekly basis, and this greatly facilitates genetic analysis. Furthermore, zebra fish develop externally and the embryo is transparent, allowing easy access to the embryo and direct visualization of its development. By 24 h post fertilization (24 hpf), the embryo has already developed a central nervous system typical of that of vertebrates, and simple behaviors such as twitching and wiggling. The 5-day-old fry possesses many patterns of behavior, for instance, feeding, swimming and complex escape reactions.

These characteristics, especially the easy access to a large quantity of transparent embryos have made zebra fish an excellent model system for studying vertebrate development. Two recent large-scale mutagenesis screens based on visible morphological defects have identified 1858 mutations affecting many aspects of embryonic development, from early patterning of the embryonic axes through organogenesis and control of behavior (Driever et al., 1996; Haffter et al., 1996). The zebra fish system has also been useful in recovering mutations affecting neural development, including patterning of the neural plate and CNS (Schier et al., 1996; Brand et al., 1996; Heisenberg et al., 1996), neural crest (Schilling et al., 1996; Piotrowski et al., 1996), control of neuronal specification (Jiang et al., 1996), stratification of neuroepithelia (Maliki et al., 1996), control of neuronal survival (Abdelilah et al, 1996; Furutani-Seiki et al., 1996; Rodruigez and Driever, 1997), axonal pathfinding (Trowe et al., 1996; Baier et al., 1996; Karlstrom et al., 1996), as well as locomotion behavior (Granato et al., 1996). With all these mutations in hand, promising progress in understanding vertebrate neural development shall be made in the foreseeable future.

Given the functional significance of dopaminergic neurons, we are interested in understanding the molecular mechanisms underlying their development, survival and function. Zebra fish have provided us with an opportunity to apply genetic tools to study this problem. Since mutations affecting a specific class of neurons such as the dopaminergic neurons may not lead to obvious morphological defects, they are likely to be missed in screens for visible defects. Therefore, we have carried out a mutagenesis screen in combination with immunohistochemistry to search for such mutations (Guo et al., 1999). In the following paragraphs, we first introduce what is known about the vertebrate dopamine systems, and the developing dopamine systems in the zerbrafish. We will then explain the mutagenesis approach that we have undertaken to isolate mutations that affect the dopaminergic neurons, and describe the preliminary phenotypes of one of the mutants which we have isolated. Finally, we discuss strategies that can be used to clone genes responsible for the mutant phenotypes.

Distribution and functional significance of dopamine systems in vertebrates

Dopaminergic neurons use dopamine as their neurotransmitter. Dopamine belongs to the family of biogenic monoamines, and is synthesized from the aromatic amino acid tyrosine. The synthesis of dopamine requires tyrosine hydroxylase (TH), a rate-limiting enzyme in converting tyrosine to dopamine. Dopamine is then converted to noradrenaline in the presence of the enzyme dopamine β-hydroxylase (DβH). Antibodies raised against these biosynthetic enzymes have been widely used to localize dopaminergic neuronal groups in the nervous system. For example, dopaminergic neurons are immunoreactive to antibodies against TH but immunonegative with antibodies against DβH.

Extensive immunohistochemical studies using antibodies raised against TH, DβH and dopamine have been carried out in adult vertebrates of various species (Smeets and Reiner, 1994). The major populations of dopamine-containing cell bodies reside in the mesencephalon and hypothalamus. In addition, dopamine-containing neurons can be found in the olfactory bulb and retina. While the retinal and olfactory DA cells are conserved among all vertebrates, the hypothalamic and midbrain DA cells show variations among vertebrate species. Dopaminergic amacrine cells in the retina link the inner and outer plexiform layers, and function in transducing information from photoreceptors to ganglion cells. Olfactory DA cells link mitral cell dendrites in adjacent glomeruli, and are involved in modulating the olfactory nerve input. The hypothalamic DA cells have been implicated in feeding and reproductive behaviors, but their exact roles remain elusive. The hypothalamus of amniotes (birds, reptiles and mammals) contains relatively few DA cells, while numerous hypothalamic DA neurons are detected in anamniotes (fish and amphibians). The midbrain tegmental DA neurons play a key role in controlling movement. They are present in large numbers in amniotes but are not detected in most anamniotes. These neurons send out input to the striatum, and their degeneration leads to Parkinson's disease in humans and movement deficits in mammals, birds and reptiles. Although anamniotes lack midbrain DA neurons, it has been proposed that pathways similar to the nigrostriatal system are present (Reiner and Northcutt, 1992). Interestingly, a Parkinsonian syndrome can be elicited in the teleost goldfish by treatment with MPTP, a neurotoxin that causes a Parkinson's disease-like syndrome in humans and lower primates (Poli et al., 1990; Pollard et al., 1992). This suggests that pathways similar to the nigrostriatal projections may also function in anamniotes to regulate movements.

Despite the wealth of information regarding the development and function of dopaminergic neurons on the cellular level, few genes controlling these processes have been identified. In order to elucidate the molecular mechanisms, we have first studied the DA systems in the zebra fish, *Danio rerio*.

The developing dopamine systems in zebra fish

In the developing zebra fish, dopaminergic neurons are detected in the forebrain with major populations present in the hypothalamus, telecephalon (olfactory bulb) and retina.

TH immunoreactive (ir) and DβH immunonegative neurons are detected in the ventral diencephalon very early during development (about 1 d post fertilization) Even at such early stages of development, long axons with many varicosities are seen extending from these neurons through the ventral midbrain and hindbrain (Fig. 1(A)).

By 3 d after fertilization, the number of hypothalamic DA neurons have increased, and they start to form discrete clusters. The rostrally located neurons are situated between ventral thalamus and hypothalamus, while the more caudal neurons are positioned along the border between the diencephalon and mesencephalon, a region called the posterior tuberculum. From this

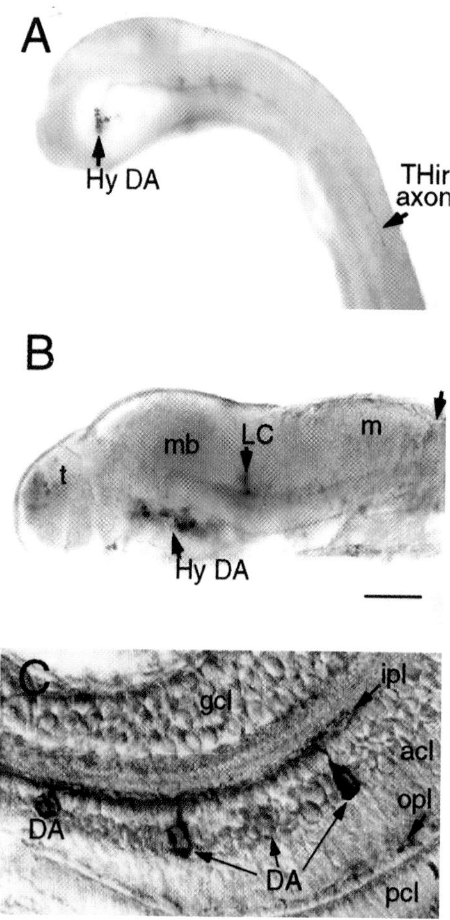

Fig. 1. Dopaminergic neurons in the developing zebra fish. (A) Lateral view of a prim-12 ($\sim$30 h post fertilization) zebra fish embryo showing hypothalamic dopaminergic neurons and their THir axons extending through midbrain, hindbrain to the rostral spinal cord. (B) A 4-day-old zebra fish fry showing hypothalamic and telencephalic DA neurons. The locus coeruleus and caudal hindbrain noradrenergic neurons are also indicated. (C) DA amacrine neurons in the retina of a 4-day-old zebra fish fry. Abbreviations: acl, amacrine cell layer; gcl, ganglion cell layer; Hy DA, hypothalamus DA neurons; ipl, inner plexiform layer; LC, locus coeruleus; m, medulla; mb, midbrain; opl, outer plexiform layer; t, telencephalon. Scale bar, 180 μm (A,B), 30 μm (C).

stage onward, THir axons are seen throughout the ventral lateral medulla terminating at the area postrema, where the medulla closes (Fig. 1(B)). In addition, there are THir processes in the midbrain tegmentum. Around 3 d after fertilization, weakly stained THir DA neurons also appear in the rostral/dorsal telencephalon, some but not all are within the olfactory regions of the telencephalon (Fig. 1(B)). THir DA neurons are also detected in the retina (Fig. 1(C)). These cells are located in the amacrine layer and have processes that extend into the inner and outer plexiform layer.

In summary, consistent with findings in other teleost fishes, such as the three-spined stickleback (Ekstrom et al., 1992) and the goldfish (Hornby and Piekut, 1990), major groups of dopaminergic neurons are detected in the hypothalamus, and minor groups of dopaminergic neurons are found in the telencephalon and retina.

The mutagenesis screen for zebra fish mutations affecting DA neurons

A general mutagenesis approach in a diploid organism such as zebra fish is outlined in Fig. 2. Mutations are induced in the germline of G_0 males. In order to be able to analyze phenotypes of recessive mutations, mutagenized chromosomes have to be driven to homozygosity by a two generation breeding scheme (Fig. 2). Mutations that cause lethality can be bred from F_2 carriers and chromosomes carrying the mutations are maintained for further study in a heterozygous state.

The chemical mutagen used for mutagenesis in zebra fish is ENU (ethyl nitrosourea), which has the highest mutagenic frequency among the reagents tested (Mullins et al, 1994; Solnica-Krezel et al., 1994). At the dosage used (3 mM), ENU generates predominantly single nucleotide changes, and induces mutations that have visible phenotypes at embryonic and early larval stages at a rate of approximately one morphologically visible mutation per haploid genome (Mullins et al., 1994; Driever et al., 1996).

The zebra fish breeding conditions and the mutagenesis procedures have been previously described (Mullins et al, 1994) (Solnica-Krezel et al., 1994). In brief, adult male fish were mutagenized with ENU (Sigma) by placing them

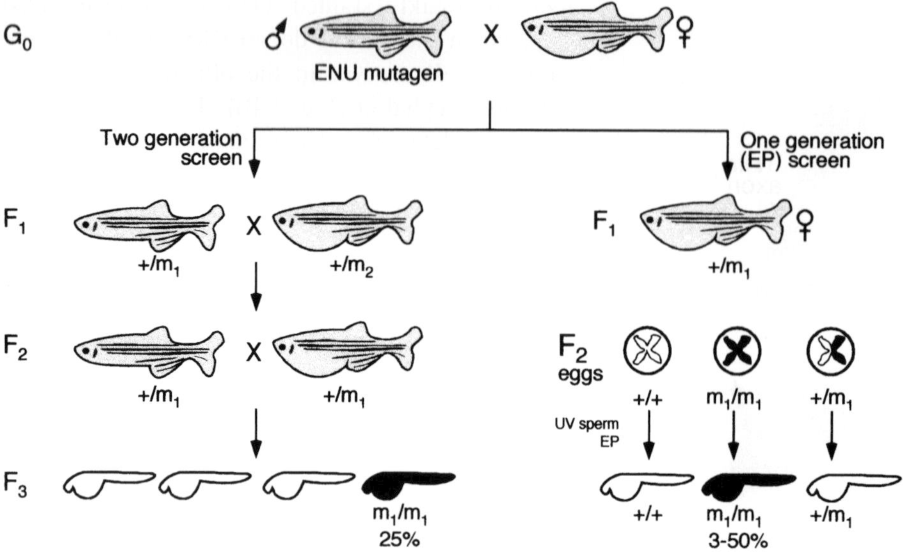

Fig. 2. Schematic diagrams illustrating two-generation and one-generation (Early Pressure) mutagenesis screens. G_0 males are mutagenized with chemical mutagen ENU, and bred with wildtype female fish to produce F_1 founder fish. The left is the scheme for a two-generation screen: F_1 males and females carrying different mutations are crossed to produce F_2 population. pair matings are performed among F_2 males and females to generate F_3 embryos. About 25% mutant embryos (for a recessive mendelian trait) can be identified in the F_3 generation. The right is the scheme for a one-generation EP screen: F_2 eggs are collected from individual F_1 fish and activated by UV-irradiated sperm in an in vitro fertilization process. These embryos are further subjected to hydrostatic pressure, which prevents second meiotic division of gametes (+/+, m1/m1, and meiotic recombinant +/m1). Homozygous mutant embryos (ranging from 3 to 50% due to meiotic recombination) can be identified in the F_2 generation.

into a 3 mM aqueous solution for three consecutive 1 h periods at weekly intervals. Starting three weeks after the last ENU treatment, males were mated with wildtype females on a weekly basis, and progeny from these crosses were raised. All the progeny derived from each mutagenized male were kept separate so the origin of individual mutations could be traced back to the founder male. These progeny are designated F_1 generation.

Screening for desired recessive mutations in most diploid vertebrates can only be carried out using two generation breeding schemes (Fig. 2). Such an approach requires breeding fish for two generations (one generation is about three months), so a longer time and a larger facility space are required to accommodate a two-generation screen. Therefore, an alternative genetic technique (Fig. 2) based on induction of gynogenetic development (development of an egg without genetic contribution from sperm) was applied, which produces homozygous mutant embryos from F_1 females (Streisinger et al., 1981). This technique is called "early pressure" treatment (EP, see Methods), since a pressure pulse disrupts the second meiotic division. The principal idea is to activate eggs with UV-irradiated sperms which stimulate the egg developmental program and cell division but do not contribute male genetic material. Such embryos, left alone, would develop as haploid individuals. However, eggs are further subjected to high hydrostatic pressure (EP) shortly after fertilization. The pressure dissociates microtubules and causes the suppression of second meiotic division and generates gynogenetic diploid embryos in which both homologous chromosomes are derived from the F_1 mother. Since the F_1 mother is heterozygous, and meiotic recombination products do not segregate due to EP, a portion of 3% (telomeric mutant locus) to 50% (centromeric mutant locus) of the embryos will be homozygous for mutations present in the F_1 mother (refer to Fig. 2). About 10–20% of EP-treated embryos may be damaged due to the pressure treatment and develop abnormally. Such epigenetic defects can be misleading in a visual screen, but it is less of a concern when a specific molecular marker such as TH is used. The only disadvantage of EP screen is that genes which are at the distal end of chromosomes will be harder to identify since they are underrepresented in embryo clutches due to meiotic recombination: for centromere-linked genes ~50% of EP embryos will be homozygous mutant, and percentage decreases as the gene locates further away from the centromere. Although slightly biased with respect to saturation of certain portions of the genome, such EP approach allows rapid identification of mutations, since breeding of an F_2 to sexual maturity generation is not required.

About 2-day-old embryos are processed for immunohistochemical staining with an antibody against tyrosine hydroxylase (see Methods). To identify and recover mutations after staining, individual embryos are examined under a light dissecting microscope with low magnification (about 25×), and those with alterations in DA neurons are further examined and photographed under DIC (Differential Interference Contrast) using Zeiss microscopes. If an identified mutation is of interest, the F_1 founder is mated with a wildtype male. Progeny from such a mating are raised to adulthood. About 50% of these progeny are heterozygous for the identified mutation. Pair matings can be performed within this generation to identify heterozygous founders and generate homozygous offsprings.

Given the efficiency of ENU mutagenesis in zebra fish (Solnica-Krezel et al., 1994; Driever et al., 1996), one can estimate that about 4000–5000 mutagenized genomes need to be analyzed in order to recover mutations in more than 80% of the genes affecting a certain aspect of development. The saturation achieved in an EP based screen is somewhat lower, depending on the average clutch size of EP diploid embryos analyzed. Our genetic screen was targeted at sampling a significant portion of the zebra fish genome for genes important for the development of dopamine systems, albeit it was not saturating for such mutations. After screening about 700 F_1 founders (each representing one mutagenized genome), at three genetic loci we have identified mutations that affect dopaminergic neurons. Two of the mutations disrupt other neuronal types in addition to affecting dopaminergic neurons, and are embryonic lethal. The third one which we named *too few* appears to affect the hypothalamic dopaminergic neurons quite specifically: the *too few* mutant embryos have fewer THir dopaminergic neurons in the hypothalamus when compared to wildtype siblings (Fig. 3). Two-day old wildtype embryos have about 10–14 DA neurons on each side of the midline in the hypothalamus, while *too few* mutant embryos have only one or two THir neurons on each side of the midline (Fig. 3(A), (B)). In 5-day-old fry, there are only ~10–30% DA neurons in the mutant comparing to the wildtype siblings (Fig. 3(C), (D)). Despite the defects in the hypothalamic DA neurons, the mutant embryo and fry are morphologically indistinguishable from their wildtype siblings, and can apparently develop to adulthood. This mutant gives us an opportunity not only to address developmental mechanisms that lead to the specification of these DA neurons, but also to assess the possible functions of these hypothalamic DA neurons in adults.

From mutations to gene identity

The ultimate goal of isolating mutations is to identify genes affected by the mutations. Three general methods have been successfully used in *C. elegans*, *Drosophila* or mice, and are now being applied to zebra fish (also see Chapters 2.1.1, 2.1.6). One is called candidate gene approach. This method works by testing linkage between mutations and genes that have already been cloned and may functionally relate to the mutant phenotypes. Quite a few genes have been cloned

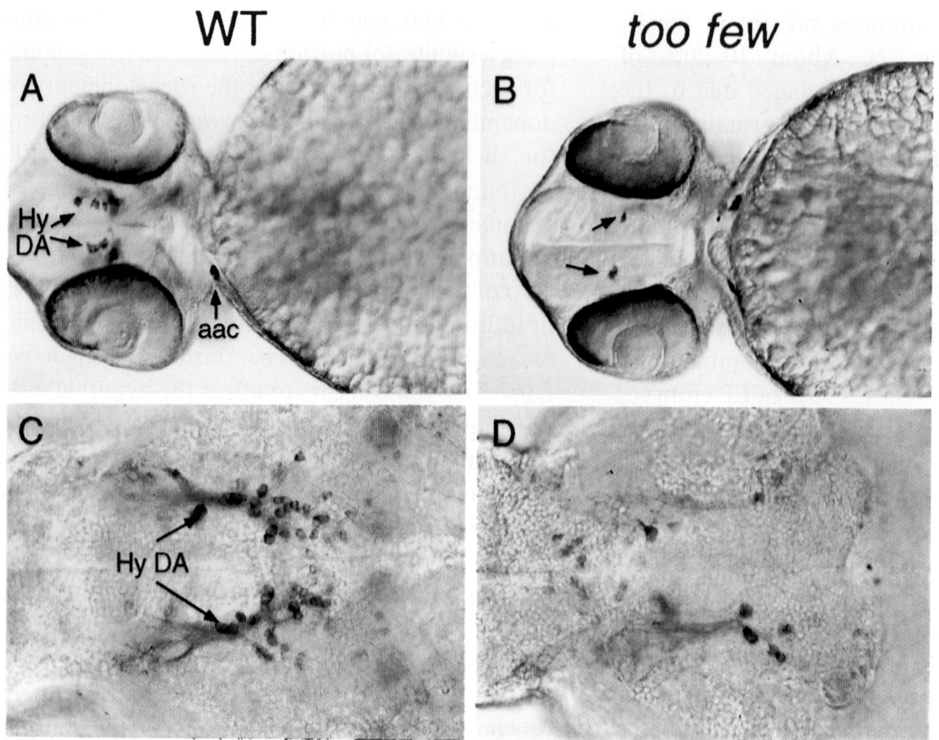

Fig. 3. *too few* mutant embryos have reduced number of hypothalamic neurons. (A–B) Ventral views of 2-day-old wildtype (left) and *too few* mutant embryos (right) labeled with anti-TH antibody, showing reduced number of dopaminergic neurons in the hypothalamus. (C–D) 4-dayold fry (left, widltype; right, *too few*) showing the reduction of hypothalamic DA neurons persist during development. Abbreviation: Hy DA, hypothalamic DA neurons. Scale bar, 150 μm (A–B), 60 μm (CD).

by this approach in zebra fish (for example, Schulte-Merker et al., 1994; Talbot et al., 1995), and many known zebra fish genes have been mapped (Postlethwait et al., 1998). However, currently this method has its limitations. For processes that are not well understood on the molecular level, it is not quite possible to match mutant phenotypes with known genes, due to the inavailability of such candidate genes. Such limitations can be overcome by the current zebra fish EST (Expressed Sequence Tag) project, which aims at identifying, mapping and examining expression patterns of a large number of transcripts.

Another approach starts with mutants generated by insertional mutagenesis. This approach allows rapid cloning of disrupted genes (Gaiano et al., 1996). However, the current insertional mutagenesis method in zebra fish is about 20-fold less efficient than chemical mutagenesis (Gaiano et al., 1996). As a result, not many insertional mutants have been isolated to date. In contrast, thousands of mutants generated by chemical mutagenesis are available and many have interesting phenotypic alterations.

Therefore, a third method called positional cloning is being employed to clone genes defined by chemically-induced mutations. The positional cloning approach is based on the existence of a large number of polymorphic DNA markers that span the entire genome: The gene of interest is first mapped using linkage analysis (see also Chapter 2.1.1). Then, large insert DNA clones within the mapped region are isolated, and the gene is identified by either functional complementation or DNA sequence analysis and identification of nucleotide alterations. Such a direct approach becomes feasible with the development of increasingly dense maps of thousands of molecular markers (Knapik et al., 1998). Given that no closely-linked markers are available on the map to the gene of interest, another technique called bulked segregant analysis can be used to specifically identify molecular markers that are tightly linked to the gene of interest. One such analysis is called AFLP (Amplified Fragment Length Polymorphism) (Vos et al., 1995). This method uses recombination analysis and PCR to identify DNA polymorphisms that are closely linked to the gene of interest (Fig. 4). The mutant of interest is crossed to fish from a different genetic background. Heterozygotes from this cross are established into F_1 founders. These founder fish are bred to generate embryos, among which DNA polymorphisms will segregate due to chromosomal segregation and recombination. These embryos are the source for AFLP analysis. Genomic DNA is extracted from a pool of wildtype and a pool of mutant embryos (the pool size is 10 to 40 individuals) and digested with appropriate enzymes (usually *Eco*RI and MseI). DNA adapters (double-stranded DNA of ~15 base pairs) are ligated to the digested genomic DNA. Primers with sequence of adapters plus restriction enzyme site plus NNN (random nucleotides) are used to amplify DNA fragments. Polymorphisms that are not linked to the gene of interest will segregate randomly, thus are present in both wildtype and mutant pools for example, marker b in Fig. 4). In contrast, DNA fragments that are physically linked to the wildtype copy of the mutated gene (for example, marker a in Fig. 4) will be present in the wildtype pool and absent in the mutant pool, allowing selective identification of polymorphic fragments that are linked to the gene of interest. Once identified, the genetic distance of such DNA fragments to the gene of interest can be defined by doing AFLP on individual mutant embryos.

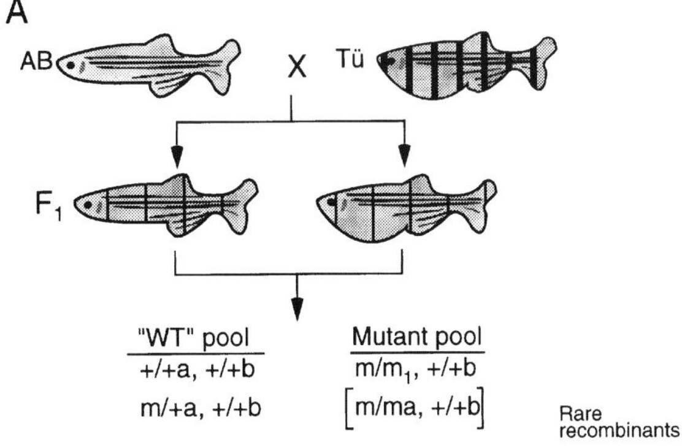

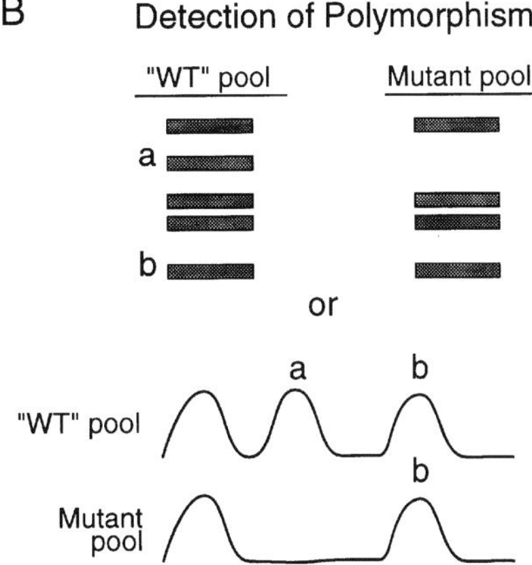

Fig. 4. Diagram illustrating strategies for positional cloning in zebra fish. (A) The strain (AB) carrying the mutation of interest (m) is crossed with a different background strain (Tu), and F_1 progeny are raised to adulthood. Heterozygous carriers are identified by pairmating among F_1 population, and mated to generate homozygous mutant embryos and their wildtype siblings. DNAs are made from pools of wildtype embryos and pools of mutant embryos, and subjected to PCR amplification using AFLP primers (see text) or primers from the microsatellite map (Knapik et al., 1998). (B) Polymorphisms are detected by either manual gel electrophoresis (top) or automated sequencing analysis (bottom).

In zebra fish, such method is particularly advantageous due to the availability of large number of embryos. Thus, DNA fragments as close as 0.1 cM. (corresponds to ~60 kb or more in zebra fish) to the gene of interest can be isolated. Mapping at this resolution greatly facilitates cloning. Once closely linked DNA fragments are identified, they can be used as probes to isolate genomic

DNA and cDNAs. Two genes have been identified by positional cloning (Zhang et al., 1998; Brownlie et al., 1998) in zebra fish.

Summary

We have discussed in this chapter the potential of applying the forward genetic approach to a newly established vertebrate model organism, the zebra fish, *Danio rerio*. We also described the use of zebra fish to isolate mutations that affect the brain dopamine systems, and strategies for identifying genes responsible for the mutant phenotypes. In the future, phenotypic characterization of these mutations, positional cloning of the genes identified by these mutations as well as more screens including behavioral screens in zebra fish will tell us how these important neurotransmitter systems contribute to our brain function and behavior.

Methods

Early pressure technique

Procedures for Early Pressure are the following (Streisinger et al., 1981; Westerfield, 1994). Two pairs of F_1 female fish are placed with one wildtype male fish to stimulate egg production. Before lights come on the next morning, females are separated from the male. For in vitro fertilization: Sperm is obtained by squeezing males gently, and is then irradiated using a Stratalinker (Stratagene). The length and strength of irradiation should be optimized for each machine to obtain genetically inactivated sperm. We irradiated the sperm in a watch glass for 1/2 min at 2800 uW/cm^2. Eggs are obtained by squeezing the females gently. The eggs are mixed with UV-irradiated sperm, and subjected to "early pressure" by application of a 4 1/2 min 8000 psi pressure pulse starting 84 s after fertilization. Post-EP embryos are allowed to develop. Around two days after fertilization, embryos are processed for immunostaining with antibody against TH (see bleow). F_1 founder female fish are kept individually until it is verified whether they carry mutations of interest.

Immunohistochemistry

Stages of development as well as general morphology of the brain have been previously described for zebra fish embryos (Kimmel et al., 1995; for review, Kimmel, 1993). Antibodies against TH and DβH (Chemicon) are used to identify dopaminergic neurons in the zebra fish embryo. Zebra fish embryos are collected by natural matings, and incubated at 28.5°C. At different developmental stages, embryos are fixed with 4% paraformaldehyde at room temperature for about 3 h. After blocking with 10% goat serum for one hour, embryos are incubated in TH or DβH antibodies (1:1000) at 4°C overnight. After washes with PBT (phosphate-buffered saline with 0.1% Tween) for about 2 h, the embryos are incubated with biotinylated secondary antibody (Vector Laboratories) at 4°C overnight. After incubation, embryos are washed with PBT for ~2 h, and incubated for 1 h at room temperature with Vecstain Elite ABC reagent (Vector laboratories). Embryos are rinsed with PBT for 1 hour, and color reaction is carried out using 3,3′-Diaminobenzidine tetrahydrochloride (DAB) and hydrogen peroxide (H_2O_2)(Sigma): embryos are incubated in PBT containing 0.5 mg/ml DAB for 30 min, and H_2O_2 was added to final concentration of 0.003%. Color development usually takes ~30 min. Subsequently, embryos are rinsed with PBT, cleared with 75% glycerol.

References

Abdelilah, S., Mountcastle-Shah, E., Harvey, M., Solnica-Krezel, L., Schier, A.F., Stemple, D.L., Malicki, J., Neuhauss, S.C.F., Zwartkruis, F., Stainier, D.Y.R., Rangini, Z. and Driever, W. (1996) Mutations affecting neural survival in the zebra fish, Danio rerio. *Development*, 123: 217–227.

Baier, H., Klostermann, S., Trowe, T., Karlstrom, R. O., Nusslein-Volhard, C., and Bonhoeffer, F. (1996) Genetic dissection of the retinotectal projection. *Development*, 123; 415–425.

Brand, M., Heisenberg, C.-P., Jiang, Y.-J., Beuchle, D., Lun, K., Furutani-Seiki, M., Granato, M., Haffter, P., Hammerschmidt, M., Kane, D., Kelsh, R., Mullins, M., Odenthal, J., Van Eeden, F.J.M. and Nusslein-Volhard, C. (1996) Mutations in zebra fish genes affecting the formation of the boundary between midbrain and hindbrain. *Development*, 123: 179–190.

Brenner, S. (1974) The genetics of *Caenorhabditis elegans*. *Genetics*, 77: 71–94.

Brownlie, A., Donovan, A., Pratt, S.J., paw, B.H., Oates, A.C., Brugnara, C., Witkowska, H.E., Sassa, S. and Zon, L.I. (1998) Positional cloning of the zebra fish *sauternes* gene: a model for congenital sideroblastic anaemia. *Nat. Genet.*, 20: 244–250.

Capecchi, M.R. (1989) Altering the genome by homologous recombination. *Science*, 244: 1288–1292.

Driever, W., Solnica-Krezel, L., Schier, A.F., Neuhauss, S.C., Malicki, J., Stemple, D.L., Stainier, D.Y., Zwartkruis, F., Abdelilah, S., and Rangini, Z. et al. (1996) A genetic screen for mutations affecting embryogenesis in zebra fish. *Development*, 123: 37–46.

Ekstrome, P., Honkanen, T., and Borg, B. (1992) Development of tyrosine hydroxylase-, dopamine-, and dopamine-B-hydroxylase-immunoreactive neurons in a teleost, the three-spined stickleback. *J. Chem. Neuroanat.*, 5: 481–501.

Furutani-Seiki, M., Jiang, Y.-J., Brand, M., Heisenberg, C.-P., Houart, C., Beuchle, D., Van Eeden, F. J. M., Granato, M., Haffter, P., Hammerschmidt, M., Kane, D. A., Kelsh, R. N., Mullins, M. C., Odenthal, J. and Nusslein-Volhard, C. (1996) Neural degeneration mutants in the zebra fish, *Danio rerio*. *Development*, 123: 229–239

Gao, W., Liu, X., and Hatten, M. (1992) The weaver gene encodes a nonautonomous signal for CNS neuronal differentiation. *Cell*, 68: 841–854.

Granato, M., Van Eeden, F. J. M., Schach, U., Trowe, T., Brand, M., Furutani-Seiki, M., Haffter, P., Hammerschmidt, M., Heisenberg, C.-P., Jiang, Y.-J., Kane, D. A., Kelsh, R. N., Mullins, M. C., Odenthal, J. and Nusslein-Volhard, C. (1996) Genes controlling and mediating locomotion behavior of the zebra fish embryo and larva. *Development*, 123: 399–413.

Guo, S., Wilson, S.W., Cooke, S., Chitnis, A.B., Driever, W., and Rosenthal, A. (1998) Mutations in the zebra fish unsmask shared regulatory pathways controlling the development of catecholaminergic neurons. *Dev. Biol.*, 208, 473–487.

Haffter, P., Granato, M., Brand, M., Mullins, M.C., Hammerschmidt, M., Kane, D.A., Odenthal, J., Van Eeden, F.J.M., Jiang, Y.J., and Heisenberg, C.P. et al. (1996) The identification of genes with unique and essential functions in the development of the zebra fish, *Danio rerio*. *Development*, 123: 1–36.

Heisenberg, C.-P., Brand, M., Jiang, Y.-J., Warga, R. M., Beuchle, D., Van Eeden, F.J. M., Furutani-Seiki, M., Granato, M., Haffter, P., Hammerschmidt, M., Kane, D.A., Kelsh, R.N., Mullins, M.C., Odenthal, J. and Nusslein-Volhard, C. (1996) Genes involved in forebrain development in the zebra fish, Danio rerio. *Development*, 123: 191–203.

Hornby, P.J. and Piekut, D.T. (1990) Distribution of catecholamine-synthesizing enzymes in goldfish brains: presumptive dopamine and norepinephrine neuronal organization. *Brain Behav. Evol.*, 35: 49–64.

Jiang, Y.-J., Brand, M., Heisenberg, C.-P., Beuchle, D., Furutani-Seiki, M., Kelsh, R.N., Warga, R.M., Granato, M., Haffter, P., Hammerschmidt, M., Kane, D.A., Mullins, M.C., Odenthal, J., Van Eeden, F.J.M. and Nusslein-Volhard, C. (1996) Mutations affecting neurogenesis and brain morphology in the zebra fish, Danio rerio. *Development*, 123: 205–216.

Karlstrom, R.O., Trowe, T., Klostermann, S., Baier, H., Brand, M., Crawford, A.D., Grunewald, B., Haffter, P., Hoffmann, H., Meyer, S.U., Muller, B.K., Richter, S., Van Eeden, F.J.M., Nusslein-Volhard, C. and Bonhoeffer, F. (1996) Zebra fish mutations affecting retinotectal axon pathfinding. *Development*, 123: 427–438

Kimmel, C.B., Ballard, W.W., Kimmel, S., Ullmann, B. and Schilling, T.F. (1995) Stages of the embryonic development of the zebra fish. *Developmental Dynamics*, 203: 253–310.

Kimmel, C.B. (1993) Patterning the brain of the zebra fish embryo. *Annu. Rev. neurosci.*, 16: 707–732.

Knapik, E.W., Goodman, A., Ekker, M., Chevrette, M., Delgado, J., Neuhauss, S., Shimoda, N., Driever, W., Fishman, M.C. and Jacob, H.J. (1998) A microsatellite genetic linkage map for zebra fish (*Danio rerio*). *Nature Genet.*, 18: 338–343.

Malicki, J., Neuhauss, S.C.F., Schier, A.F., Solnica-Krezel, L., Stemple, D.L., Stainier, D.Y.R., Abdelilah, S., Zwartkruis, F., Rangini, Z. and Driever, W. (1996) Mutations affecting development of the zebra fish retina. *Development*, 123: 263–273.

Mullins, M.C., Hammerschmidt, M., Haffter, P. and Nusslein-Volhard, C. (1994) Large-scale mutagenesis in the zebra fish: in search of genes controlling development in a vertebrate. *Curr. Biol.*, 4: 189–202.

Neuhauss, S.C.F., Solnica-Krezel, L., Schier, A.F., Zwartkruis, F., Stemple, D.L., Malicki, J., Abdelilah, S., Stainier, D.Y.R. and Driever, W. (1996) Mutations affecting craniofacial development in zebra fish. *Development*, 123: 357–367.

Nusslein-Volhard, C. and Wieschaus, E. (1980) Mutations affecting segment number and polarity in *Drosophila*. *Nature*, 287: 795–801.

Piotrowski, T., Schilling, T.F., Brand, M., Jiang, Y.-J., Heisenberg, C.-P., Beuchle, D., Grandel, H., Van Eeden, F.J.M., Furutani-Seiki, M., Granato, M., Haffter, P., Hammerschmidt, M., Kane, D.A., Kelsh, R.N., Mullins, M.C., Odenthal, J., Warga, R.M. and Nusslein-Volhard, C. (1996) Jaw and branchial arch mutants in zebra fish II: anterior arches and cartilage differentiation. *Development*, 123: 345–356.

Pollard, H.B., Dhariwal, K., Adeyemo, O.M., Markey, C.J., Caohuy, H., Levine, M., Markey, S. and Youdim, M.B.H. (1992) A parkinsonian syndrome induced in the goldfish by the neurotoxin MPTP. *The FASEB Journal*, 6: 3108–3116.

Poli, A., Guarnieri, T., Facchinetti, F. and Villani, L. (1990) Effect of 1-methyl-4-phenyl-1,2,3,6-tetrahydropyridine (MPTP) in goldfish brain. *Brain Research*, 534: 45–50.

Postlethwait, J.H., Yan, Y-L., Gates, M.A., Horne, S., Amores, A., Brownlie, A., Donovan, A., Egan, E.S., Force, A., Gong, Z., Goutel, C., Fritz, A., Kelsh, R., Knapik, E., Liao, E., Paw, B., Ransom, D., Singer, A., THomson, M., Abduljabbar, T.S., Yelick, P., Beier, D., Joly, J-S., Larhammar, D., Rosa, F., Westerfield, M., Zon, L.I., Johnson, S.L. and Talbot, W.S. (1998) Vertebrate genome evolution and the zebra fish gene map. *Nature Genet.*, 18: 345–349.

Reiner, A. and Northcutt, R.G. (1992) An immunohistochemical study of the telencephalon of the Senegal Bichir (polypterus senegalis). *J. Comp.* Neurol., 319: 359–386.

Rodriguez, M. and Driever, W. (1997) Mutations resulting in transient and localized degeneration in the developing zebra fish brain. *Biochemistry and Cell Biology*, 75: 7–28.

Schier, A.F., Neuhauss, S.C.F., Harvey, M., Malicki, J., Solnica-Krezel, L., Stainier, D.Y.R., Zwartkruis, F., Abdelilah, S., Stemple, D.L., Rangini, Z., Yang, H. and Driever, W. (1996) Mutations affecting the development of the embryonic zebra fish brain. *Development*, 123: 165–178.

Schilling, T.F., Piotrowski, T., Grandel, H., Brand, M., Heisenberg, C.-P., Jiang, Y.-J., Beuchle, D., Hammerschmidt, M., Kane, D.A., Mullins, M.C., Van Eeden, F.J.M., Kelsh, R.N., Furutani-Seiki, M., Granato, M., Haffter, P., Odenthal, J., Warga, R.M., Trowe, T. and Nusslein-Volhard, C. (1996) Jaw and branchial arch mutants in zebra fish I: branchial arches. *Development*, 123: 329–344.

Schulte-Merker, S., Van Eeden, F.J.M., Halpern, M.E., Kimmel, C.B. and Nusslein-Volhard, C. (1994) *No tail (Ntl)* is the zebra fish homologue of the mouse *T (Brachyury)* gene. *Development*, 120: 1009–1015.

Smeets, W.J.A.J. and Reiner, A. (1994) *Phylogeny and development of catecholamine systems in the CNS of vertebrates.* Cambridge University Press, 1994.

Solnica-Krezel, L., Schier, A.F. and Driever, W. (1994) Efficient recovery of ENU-induced mutations from the zebra fish germline. *Genetics*, 136: 1401–1420.

Streisinger, G., Walker, C., Dower, N., Knauber, D. and Singer, F. (1981) Production of clones of homozygous diploid zebra fish (*brachydanio rerio*). *Nature*, 291: 293–296.

Talbot, W.S., Trevarrow, B., Halpern, M.E., Melby, A.E., Farr, G., Postlethwait, J.H., Jowett, T., Kimmel, C.B. and Kimelman, D. (1995) A homeobox gene essential for zebra fish notochord development. *Nature*, 378: 150–157.

Tejedor, F., Zhu, X.R., Kaltenbach, E., Ackermann, A., Baumann, A., Canal, I., Heisenberg, M., Fischbach, K.F. and Pongs, O. (1995) *minibrain*: a new protein kinase family involved in postembryonic neurogenesis in *Drosophila. Neuron*, 14: 287–301.

Trowe, T., Klostermann, S., Baier, H., Granato, M., Crawford, A.D., Grunewald, B., Hoffmann, H., Karlstrom, R.O., Meyer, S.U., Muller, B., Richter, S., Nusslein-Volhard, C. and Bonhoeffer, F. (1996) Mutations disrupting the ordering and topographic mapping of axons in the retinotectal projection of the zebra fish, *Danio rerio. Development*, 123: 439–450.

Westerfield, W. (1994) *The Zebra fish Book*, University of Oregon Press, Eugene, OR USA.

Vos, P., Hogers, R., Bleeker, M., Reijans, M., Lee, T.V.D., Horner, M., Frijters, A., Pot, A., Peleman, J., Kuiper, M. and Zabeau, M. (1995) AFLP: a new technique for DNA finerprinting. *Nucl. Acids Res.*, 23: 4407–4414.

Zhang, J., Talbot, W.S. and Schier, A.F. (1998) Positional cloning identifies zebra fish one-eyed pinhead as a permissive EGF-related ligand required during gastrulation. *Cell*, 92: 241–251.

W.E. Crusio and R.T. Gerlai (Eds.)
Handbook of Molecular-Genetic Techniques for Brain and
Behavior Research (Techniques in the Behavioral and Neural Sciences, Vol. 13)

CHAPTER 2.1.9

Behavioral and electrophysiological screens for isolating zebra fish mutants with visual system defects

John E. Dowling

Department of Molecular and Cellular Biology, The Biological Laboratories, Harvard University, Cambridge, MA 02138, USA

Thirty years ago, Seymour Benzer (1967) showed that with simple behavioral tests it was possible to detect specific eye and brain mutations in the fruitfly, *Drosophila melanogaster*. A number of these behavioral mutants have proven to be exceptionally useful for unravelling neural and signal transduction mechanisms, (reviewed in Zuker, 1992 and Pak, 1995) as well as mechanisms underlying neural development in *Drosophila* (reviewed in Benzer, 1991). The sevenless mutant, for example, detected because the fly fails to respond to ultraviolet light, has provided definitive evidence on the role of cell-cell interaction in the development of the *Drosophila* eye. The fruitfly is an invertebrate, however, and we have not had a vertebrate with which to carry out similar mutagenesis experiments.

Zebra fish (*Danio rerio*) are highly visual animals, exhibiting light responses after just 3 days of development (Easter and Nicola, 1996), making them ideal for a genetic analysis of the visual system in a vertebrate. Zebra fish are small ($\sim$3 cm), freshwater teleosts that can be grown inexpensively in large numbers. They can be inbred successfully, and they have a relatively fast generation time ($\sim$3 months). Female zebra fish can lay as many as 100–200 eggs at weekly intervals, and efficient chemical mutagenesis techniques employing ethylnitrosurea (ENU) have recently been developed (Haffter et al., 1996; Driever et al., 1996; see also Chapter 2.1.8). Zebra fish have large eyes and are tetrachromatic, possessing ultraviolet-sensitive cones as well as red-, green- and blue-sensitive cones (Robinson et al., 1993). They also have abundant rods, and like other fish, their retinas continue to grow for the life of the animal (reviewed in Easter, 1983).

Our group has recently developed two behavioral tests that can be used to uncover visual system specific mutations in zebra fish. One test enables us to isolate recessive visual system mutations in zebra fish larvae that otherwise appear normal (Brockerhoff et al., 1995, 1997), whereas the other test reveals dominant visual system mutations in adults (Li and Dowling, 1997). We use electrophysiological recordings from the eye as a secondary screen to localize mutations to the retina, and in a number of cases, these recordings suggest the nature of the defect.

Recessive mutations — the optokinetic reflex

The challenge in uncovering recessive mutations is to be able to evaluate visual performance within the first week of life. Recessive mutations in homozygous form occur only in about one-quarter of the crosses of F_2 generation siblings, requiring the analysis of large numbers of F_3 generation fish. If F_3 larvae must be raised for several weeks before

Fig. 1. Apparatus for measuring optokinetic responses. See text for details. (From Brockerhoff et al., 1995.)

analysis, the cost and labor would be prohibitive. Larval zebra fish, for example, do not need to be fed for the first 5–7 days of development; the egg yolk provides sufficient nutrition for this period of time.

At 5–7 days of age, zebra fish are just 3–4 mm in length, but their eyes are 350–400 μm in diameter. Thus, the eyes are easily observable under the dissecting microscope and the animals begin to move their eyes in response to visual stimuli at 3–4 days. By 5 days of age, 98% of wild-type fish will respond to a moving stripe pattern with a smooth pursuit eye movement followed by a rapid saccade in the opposite direction. This is the well-known optokinetic reflex (OKR) and is the response we have chosen to measure visual performance in young zebra fish (Brockerhoff et al., 1995).

The apparatus for measuring optokinetic responses is shown in Fig. 1. Ten to twenty 5- to 7-day mutagenized F_3 larvae are placed in a 35 mm petri dish containing ~4% methylcellulose to immobilize the fish. (Larvae kept in methylcellulose for >1 h continue to develop normally when returned to fish water.) Larvae are aligned and arranged for optimal viewing with a dissecting needle. The dish is placed in the center of a microscope stage around which a circular drum is mounted. The drum has 18 black and white vertical stripes on the inside and is turned at 6 rpm by a belt attached to an adjacent motor. For each larva, the drum is rotated in two directions, and the eye movements are analyzed by watching the larva via a dissecting microscope. A response is considered positive if a single smooth pursuit and saccade eye movement in the proper direction is observed after starting drum rotation in each direction. A larvae is considered abnormal if it shows no eye movements at all or if the eye movements are unusual, such as too fast, too small, or in the wrong direction.

The drum can be illuminated either from above or below with white light (typically 3 μW/cm^2) from a fiber optic light source or with monochromatic light. Monochromatic illumination is generated using a fiber optic with a diffuser attached to a monochromator. When using monochromatic light or dim white illumination, the stage is usually illuminated from below with 750 nm light, which is detected by a near-infrared video camera. The eye movements are then observed on a video monitor. Larvae between 3 and 10 days of age do not move their eyes in response to rotating stripes illuminated with 750 nm light.

On average, 1 min is required to analyze each larva including time spent placing the fish in the petri dish, aligning and observing it, and returning it to its tank. Thus, an experimenter can screen up to ~60 larvae per hour or ~500 larvae per day. Optimally, at least 10 larvae are examined from each mating cross, and larvae from at least 6 crosses are examined from each F_2 family. We encounter mutants that have relatively specific eye defects in approximately 1 out of 10 genomes that have been chemically mutagenized (for our chemical mutagenesis methodology, see Brockerhoff et al., 1995). Most of these mutants show a morphological defect (small eyes, for example); however, in about 1 out of 40 genomes, we encounter a mutant fish without any detectable morphological alterations that shows a optokinetic response defect.

Dominant mutations — the escape response

Dominant mutations that cause slow retinal degeneration have been found in a variety of species including man (reviewed in Berson, 1993). Forty percent of the cases of human retinitis pigmentosa, for example, are dominantly inherited (Bunker et al., 1984). We have asked whether dominantly inherited degenerations that affect the retina might also be found in zebra fish. To search for such mutations requires the examination of adult fish. The optokinetic response assay is not effective for this purpose because adult zebra fish cannot be readily restrained non- invasively and adult fish must have water continually flowing over their gills. Thus, to evaluate visual sensitivity in adult zebra fish, we developed a new behavioral test based on the escape response exhibited by fish when they encounter a threatening object.

The test is conducted as follows: Individual fish are placed in a circular container (10 cm in diameter) that has clear sides and a 3 cm post in the middle (Fig. 2A). Surrounding the container is a rotating drum on which there is marked a black segment (5 × 5 cm) which serves as a threatening object. The drum is illuminated from above with a white light source (unattenuated intensity = 4.25×10^2 μW/cm^2) and turned at 10 rpm by a belt attached to an adjacent motor. The fish is viewed directly or on a television monitor attached to a video camera.

Fish normally swim slowly around the container, but when they "see" the black segment, they immediately flee to the other side of the container, by rapidly turning away from the approaching black segment or by speeding up their swimming if the segment is coming from behind. We have found that zebra fish will respond positively 85% of the time when the rotating drum is turned at a rate such that the fish encounter the threatening object 20–25 times a min. This means they are encountering the threatening object every 2–3 s. It is possible, therefore, to decide if a fish is seeing the black segment in 5–10 s (i.e. the fish responds at least two out of three times

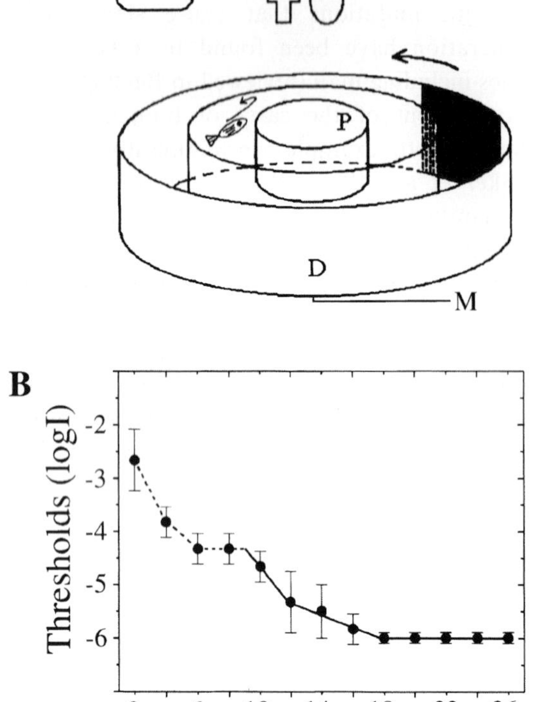

Fig. 2. (A) Apparatus for the behavioral analysis of visual sensitivity of adult zebra fish. TV, TV monitor; C, infrared video camera; L, light source; P, post; D, drum; M, motor. (B) Dark adaptation curve of adult wild-type zebra fish determined by behavioral testing. The initial portion (2–8 min) of the dark adaptation curve (dashed line) reflects cone adaptation; thereafter (continuous line), the curve reflects rod dark adaptation. Data represent the mean ± SD. (From Li and Dowling, 1997.)

when it encounters the black segment). Thus, by varying the intensity of light illuminating the drum with neutral density filters or a neutral density wedge, we can measure the time course of dark adaptation in zebra fish as well as determine the absolute thresholds of both the rod and cone systems.

Figure 2B shows an average dark adaptation curve obtained from 13 wild-type fish. The fish were first brightly light adapted for 20 min, and then placed in the dark. Thresholds were determined at 2-min intervals over the course of 26 min. Zebra fish do not habituate significantly to repeated encounters with the black segment, so that similar results are obtained whether the visual threshold of a single fish is monitored every 2 min for the 25–30 min it takes to complete dark adaptation or a different fish is used for each time point (which was the case for Fig. 2B).

Initially after light adaptation, the visual threshold was raised on average about 3.5 log units. The threshold rapidly fell to a first plateau after about 6 min of dark adaptation and then to a second and final plateau after about 18 min of dark adaptation. The initial, fast phase of dark adaptation is attributed to cone system adaptation; the second, slower phase to rod system adaptation. In accord with this notion we have found that zebra fish are more sensitive to red light during the first 6–8 min of dark adaptation than they are later, i.e. after 15 min. The absolute cone visual threshold is determined at 6–8 min after the start of dark adaptation whereas absolute rod threshold is measured after at least 20 min of dark adaptation.

While carrying out these control experiments we found that wild-type zebra fish are considerably more sensitive to light stimuli in the late afternoon hours (before dusk) than they are in the early morning hours (before dawn). Absolute rod thresholds, for example, may be as much as 2 log units lower when measured at 6 p.m. as compared to 4 a.m. These data suggest that both rod and cone visual sensitivity in zebra fish is under circadian control and this notion was confirmed by showing that the rhythmic alteration in visual sensitivity that occurs as a function of time of day continues when the animals are maintained in constant darkness or constant light (Li and Dowling, 1998). We also found that the rhythm can be altered by giving light pulses during the dark phase of the cycle, another feature of circadian rhythms. Thus, when measuring visual threshold responses in zebra fish, it is important to carry out the experiments at approximately the same time each day, optimally from mid- to late afternoon when the animals are most light sensitive. In our zebra fish facility, the lights are turned on in the morning at about 8 a.m. and the lights are turned off at 10 p.m. Since the visual sensitivity of zebra fish is significantly altered by a circadian mechanism, it is also important to test animals with raised visual thresholds at different times during the day and night to distinguish visual system mutations from possible circadian mutations.

To uncover dominant mutations, adult F_1 generation fish derived from chemically mutagenized males are tested for their rod system sensitivity. A threshold criterion approximately 1 log unit above absolute rod threshold is used. Fish who fail to respond at this light level are isolated and retested on at least three subsequent days. If a fish consistently fails to respond at the criterion light level, it is outcrossed to wild-type fish and the F_2 generation fish studied at length. We have so far isolated four dominant visual system mutations from a pool of 625 F_1 generation fish.

Figure 3 shows the dark adaptation curves of a wild-type fish and two individual *n*ight *b*lindness *a*, or *nba* mutant fish at 5.5 months of age. For the first 2–3 months of age, F_2 generation *nba* fish demonstrated normal visual thresholds. By 4 months of age, about 20% of the F_2 fish failed to respond at the criterion light level (log I = −5). By 13 months of age, 55% of the fish were night blind, i.e. failed to respond at the criterion light level, as expected if the mutation is dominant. As seen in Fig. 3, both rod and cone thresholds are elevated at 5.5 months of age in *nba* fish.

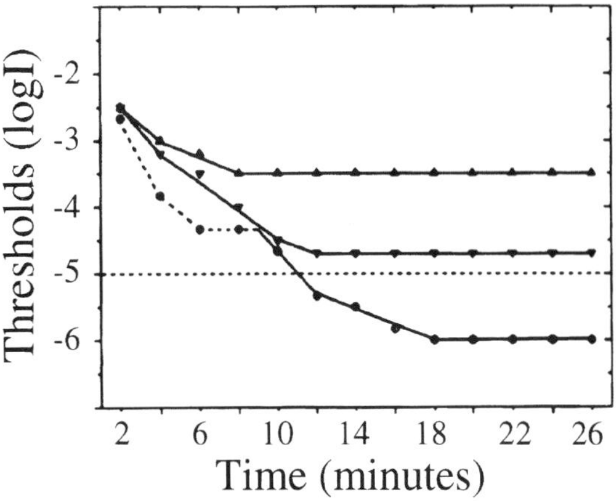

Fig. 3. Dark adaptation curves of a wild-type fish (circles) and two *nba* fish (triangles). The horizontal dashed line drawn at log I – 5.0 indicates the criterion light level used for mutant screening. (From Li and Dowling, 1997.)

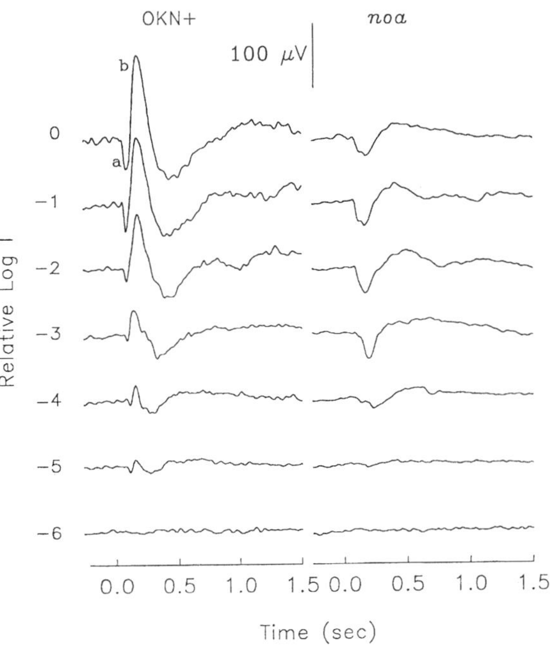

Fig. 4. Electroretinograms (ERGs) recorded from a 7-day-old *noa* larva (right) and its heterozygous sibling OKN+ (left). The responses were elicited with short, 0.01 s flashes of white light presented at time zero. The test flash intensity (Log I) decreased by factors of 1 log unit from top to bottom. The a- and b-wave of the ERG are indicated in the top recording on the left. (From Brockerhoff et al., 1995.)

Secondary screens — the electroretinogram

Once an animal with a visual behavioral defect is isolated, the next task is to determine where in the visual system the defect resides. The focus of our laboratory over the years has been on retinal studies and so for a secondary screen, we routinely record the electroretinogram (ERG), an easily obtained field potential that arises from the outer retina. The a-wave of the ERG derives from the photoreceptors, whereas the b-wave reflects second-order cell activity, particularly the responses of the ON-bipolar cells (see Dowling, 1987). To record ERGs from 5–7 day old larvae, dark-adapted (>1 h) animals are anesthetized in 3-aminobenzoic acid methyl ester (100 μg/ml) in fish water for 2 min at room temperature. In dim red light, the anesthetized fish are covered with ~3% methylcellulose containing anesthetic and positioned on top of a sponge so that one eye points toward a light source (100 W halogen light source; maximum illuminance on head = ~30,000 lux). A glass suction electrode (tip diameter, 10–20 μm) filled with fish Ringer's solution is applied to the cornea and a silver-silver chloride ground electrode placed in the bath. Flashes (0.01–0.5 s) produced with an electronically controlled shutter are attenuated by neutral density filters. The responses are amplified, averaged ($n = \sim 3$–10) and recorded. For bright flashes, only single responses are recorded unless it is clear that the first flash does not attenuate subsequent responses. When recording an intensity-response series, dim flashes are always presented first.

Figure 4 shows ERGs recorded from one recessive mutant, *noa* (*no optokinetic response a*), which failed to show any optokinetic response under any lighting condition used. On the left are ERGs recorded over a light range of 6 log units from a 7-day-old heterozygous animal; on the right are shown the ERGs of the homozygous *noa* mutant at the same age.

The ERGs recorded from the heterozygous animal are typical of wild-type fish and show both an a- and b-wave over 5 log units of intensity. In contrast, the ERGs recorded from the *noa* animal show only an a-wave over the same intensity range; no b-wave is evident at any stimulus intensity. The conclusion drawn from these findings is that the photoreceptors are functioning in the *noa* mutant, but that the visual signal is not being transmitted to the second-order cells or that the signal is not being received by the second-order cells properly. A failure of normal synaptic transmission between photoreceptors and second-order cells could account for the failure of the mutant animals to show an optokinetic response. Experiments by Victoria Connaughton at the National Institutes of Health have now shown that the bipolar cells in the *noa* mutant are abnormal; they do not respond to the photoreceptor neurotransmitter as do normal bipolar cells (personal communication).

ERGs are recorded from adult animals in basically the same way as from larval animals: Animals are dark adapted for ~1 h, anesthetized in 4% 3-amino benzoic acid methylester and immobilized with 10% gallamine triethiodide. Under dim red light illumination, they are placed on their side on a sponge with one eye facing toward the light source. A slow stream of fish water is directed into the mouth of the fish via plastic tubing to keep the animal well oxygenated during the experiment. A glass suction pipette (tip diameter of 25 μm) filled with a Ringer's solution is placed on the cornea of the fish, and a silver-silver chloride ground electrode placed in the bath. The light source is the same as used for the larval animals as are the stimulation protocols. The signals from adult animals are less noisy than those recorded from larval animals and typically require less averaging of the responses.

Figure 5A shows the ERGs of an adult wild-type fish and an *nba* mutant fish at one stimulus intensity (log I = −3) while Fig. 5B shows averaged voltage-intensity relationships (V-log I curves) for the b-wave of a number of animals (wild-type, $n = 10$; *nba*, $n = 11$). The ERG of the wild-type fish shows both an a- and b-wave whereas the ERG of the mutant fish at this intensity shows only a b-wave. The latency (implicit time) of the b-wave is also increased in the recording from the *nba* fish. As shown in Fig. 5B, the V-log I curves of *nba* fish are shifted to the right on the intensity axis, and the maximum b-wave amplitudes recorded are considerably less in the *nba* fish as compared with the wild-type fish. The increase in the ERG threshold to light was found to be comparable to the rise in the behavioral thresholds measured in *nba* animals.

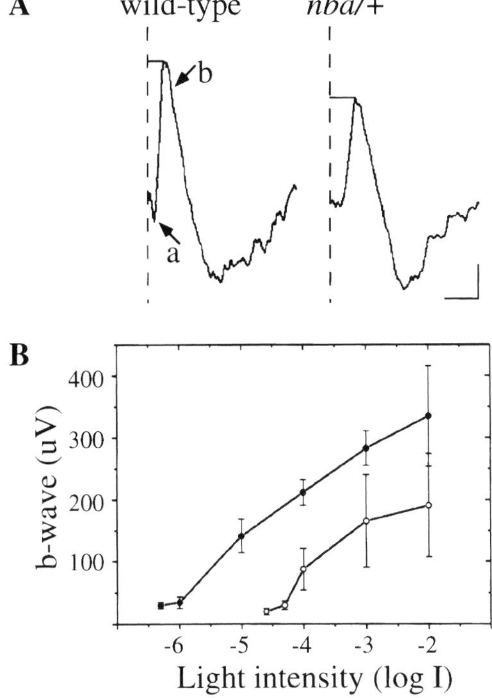

Fig. 5. (A) Electroretinograms (ERGs) of wild-type (left) and *nba* (right) fish to white light (log I = −3) stimuli: a, a-wave; b, b-wave. The vertical dashed lines indicate the onset of the 0.01 s light stimulus, and the horizontal lines indicate the implicit times (latency) of the b-waves. Calibration bars (right lower) signify 200 ms horizontally and 50 μV vertically. (B) Voltage-intensity (V-log I) curves of wild-type (solid circles) and *nba* (open circles) fish. The ERG b-wave amplitudes are plotted on a linear scale as a function of log light intensity. Data in B represent the mean ± SD. (From Li and Dowling, 1997.)

That the ERG is abnormal in *nba* fish indicates that the visual defect in these animals resides in the outer retina. Histology shows that the photoreceptors degenerate in *nba* fish — first the rods and then the cones — which can account for the behavioral and electrophysiological alterations seen in this animal.

Behavioral and ERG screens for color-blind mutants

As noted above, in addition to observing eye movements of larval zebra fish directly through the dissecting microscope, it is also possible to illuminate the fish with infrared light (which they do not see) and then to observe them on a television monitor via an infrared-sensitive TV camera. This allows the illumination of the striped drum with any intensity or wavelength of light. Using this procedure, it is possible to isolate mutants that have partial visual defects or defects involving a subtype of retinal cell. For example, by illuminating the drum with red light at an intensity of 5 μW/cm^2, it was possible to find a mutant that was red blind (Brockerhoff et al., 1997). This animal, *pob* (*p*artial *o*ptokinetic response *b*), responds to the moving stripes in white light, but not in red light. When ERGs were recorded from these fish, they were found to be considerably less red-sensitive than were the ERGs of a normal animal.

To record the spectral sensitivity of larval zebra fish via the ERG, interference filters are used to generate various wavelengths of light and neutral density filters used to attenuate the stimuli. A b-wave response of 20 μV is arbitrarily set as the criterion for a threshold response. Photon fluxes at each wavelength are determined using a radiometric detector, and the number of photons required to generate the criterion response at each wavelength is determined.

Figure 6 shows the ERG spectral sensitivity of *pob* fish compared to wild-type fish. At 6 days of age, the spectral sensitivity function of normal fish is relatively flat from 400 nm to 600 nm, reflecting mainly the summed spectral sensitivity functions of the four types of zebra fish cones. At 6 days of age, rods make little contribution to the zebra fish spectral sensitivity function — the rods mature more slowly across the retina than do the cones (Schmitt and Dowling, 1996). The ERG of the *pob* mutant is as sensitive to blue (400 nm) and green (500 nm) light as the ERG of wild-type fish, but is about 2 log units less sensitive to red light of 600 nm.

That the ERG of the *pob* mutant is red-insensitive indicates that the defect in this animal resides in the outer retina. Histological examination of the *pob* retina reveals a relatively

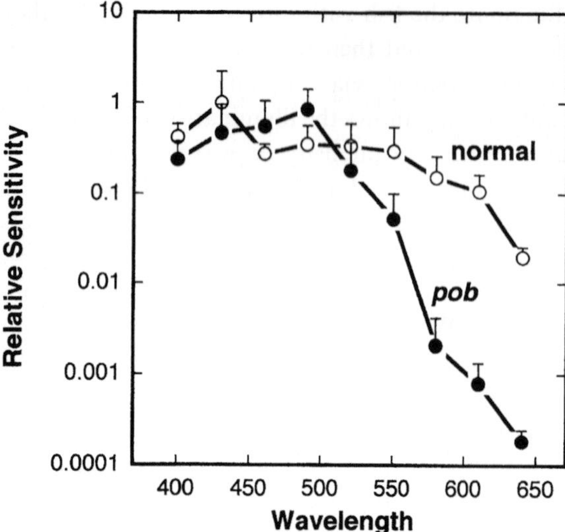

Fig. 6. Electroretinogram spectral sensitivity functions for 7-day-old *pob* and normal sibling larvae. The inverse of the number of photons required to generate a threshold (20 μv) b-wave was calculated for each wavelength. All data were normalized to the sensitivity of the normal larvae at 430 nm. (From Brockerhoff et al., 1997.)

normal appearing retina with, however, a somewhat reduced photoreceptor layer. For example, the number of nuclei in the outer nuclear layer in *pob* fish is about 20% lower than in normal animals, suggesting that red cones are missing in the *pob* retina.

To confirm this possibility, the expression of cone opsins in *pob* retinas was examined by in situ hybridization using probes against mRNA coding for the red-, blue-, and UV-sensitive opsin proteins. Whereas clear expression of mRNA for blue and UV opsins was observed in *pob* retinas at 5 days of age, virtually no red-opsin mRNA expression was seen except for some occasional staining in the far periphery of the retina. The conclusion drawn from these observations was that the red blindness of the *pob* mutant stems from a lack of red receptors.

Specific color vision defects in humans attributable to photoreceptor deficits have been linked to alterations in the visual pigment genes (Nathans et al., 1986). To test whether the same holds in zebra fish, James Hurley and his colleagues at the University of Washington identified a polymorphism in the red opsin gene in wild-type fish to use as a marker. They then looked for linkage between the polymorphism and the *pob* mutation. No linkage was found indicating that the mutation causing the loss of the red photoreceptors in *pob* fish is not in the red opsin gene. This finding indicates that the *pob* mutation is a new form of genetically inherited color-blindness (Brockerhoff et al., 1997).

Future directions — analyzing for inner retinal mutations

In addition to using our optokinetic reflex behavioral test to determine whether an animal is simply seeing or not in white or monochromatic light, it is possible to modify the testing apparatus to seek animals with more subtle visual defects. For example, we have recently modified our optokinetic reflex apparatus to isolate mutants that have deficits in movement sensitivity. Basically we test whether mutagenized fish can respond to slow and fast drum speeds as well as wild-type fish. Three such mutants have been tentatively identified so far (C. Chung and J. E. Dowling, unpublished observations). Two of these respond as well to slowly moving stripes as do wild-type fish, but then they fail to track the drum at higher speeds which normal fish can do readily. The third mutant fails to respond as well as normal fish to both slow and fast moving stripes, although it swims normally and can move its eyes. The ERGs of these fish are normal in terms of absolute sensitivity, voltage-intensity relations and waveform. In addition, we tested the flicker fusion frequency of the ERG in mutant animals to test whether these fish can respond to flickering light as well as wild-type animals. This is done by presenting flickering light to the eye of increasing frequency until the ERG no longer responds to the flickering light; i.e. the trace becomes flat. The flicker fusion frequency of the ERGs recorded from the mutants was found to be normal indicating that the defect in these animals is not in the outer retina, but deeper in the retina or elsewhere along the visual pathways.

To analyze mutations occurring in the inner retina requires recordings to be made from inner retinal cells. Ganglion cell responses can be recorded fairly readily extracellularly with metal microelectrodes from the surface of the retina in an eyecup preparation, from the optic nerve in an intact animal, or even from the surface of the tectum where the ganglion cell axons project. So far we have made recordings from the optic nerve in anaesthetized fish prepared as for ERG recording (L. Li and J. E. Dowling, unpublished observations). Connective tissue is removed from around the eye and the eye pushed out of the orbit slightly (1–2 mm) and held with glass rods. The optic nerve is then exposed and metal microelectrodes obtained commercially are inserted into the optic nerve with the aid of a micromanipulator. Both single and multi-unit recordings can be obtained, and it is also possible to make simultaneous ERG and ganglion cell recordings from such a preparation. So far we have looked at the response properties of the ganglion cells to full-field flashes of light and have determined the light thresholds for the units (which was the question we were interested in for one particular mutant). We found that there are ON-units, OFF-units and ON–OFF units that can be easily distinguished when dim lights are presented to the eye. With brighter lights, most units tended to give ON–OFF activity, suggestive of inhibitory surrounds coming into play for many of the units. We also have succeeded in making simultaneous ERG and single unit recordings which are useful for comparing response thresholds at different levels of the retina in normal and mutant fish. The next step is to map out the receptive field properties of the retinal ganglion cells in zebra fish and to use this information for the analysis of retinal mutants. As yet, little has been done in terms of characterizing the receptive fields of retinal ganglion cells in zebra fish.

Summary

Behavioral testing of both larval and adult zebra fish can uncover visual system mutations in animals that morphologically appear perfectly normal. We have succeeded in isolating mutants that appear completely blind as well as mutants that are partially sighted. Electrophysiological recordings from the eyes of the mutant fish localize the defect to the retina and can suggest the nature of the defect. More sophisticated behavioral testing and electrophysiological recordings will undoubtedly enable the isolation of more subtle retinal and visual system mutations and the eventual uncovering of retinal specific genes.

Acknowledgements

The research described in this was supported in part by grants from The National Institutes of Health (EY-00811, EY-00824 and NS-09611).

References

Benzer, S. (1967) Behavioral mutants of *Drosophila* isolated by countercurrent distribution. *Proc. Natl. Acad. Sci. USA*, 58: 1112–1119.

Benzer, S. (1991) The 1990 Helmerich Lecture: The Fly and Eye. In: D.M-K. Lam and C.J. Shatz (Eds.), *Development of the Visual System: Proceedings of the Retina Research Foundation Symposia*, Vol. 3, The MIT Press, Cambridge, MA.

Berson, E. (1993) Retinitis pigmentosa. The Friedenwald Lecture. *Invest. Ophthalmol. Vis. Sci*, 34: 1655–1676.

Brockerhoff, S.E., Hurley, J. B., Janssen-Bienhold, U., Neuhauss, S.C.F., Driever W. and Dowling J.E. (1995) A behavioral screen for isolating zebra fish mutants with visual system defects. *Proc. Natl. Acad. Sci.*, 92: 10544–10549.

Brockerhoff, S.E., Hurley, J.B., Niemi, G.A. and Dowling, J.E. (1997) A new form of inherited red-blindness identified in zebra fish. *J. Neurosci.*, 17: 4236–4242.

Bunker, C.H., Berson, E.L., Bromley, W.C., Hayes, R.P. and Roderick, T.H. (1984) Prevalence of retinitis pigmentosa in Maine. *Am. J. Ophthalmol.*, 97: 357–365.

Dowling, J.E. (1987) The Retina: An Approachable Part of the Brain. Harvard University Press, Cambridge, MA.

Driever, W., Solnica-Krezel, L., Schier, A.F., Neuhauss, S.C.F., Malicki, J., Stemple, D.L., Stainier, D.Y. R., Zwartkruis, F., Abdelilah, S., Rangini, Z., Belak, J. and Boggs, C. (1996) A genetic screen for mutations affecting embryogenesis in zebra fish. *Development*, 123: 37–46.

Easter, S.S. (1983) Postnatal neurogenesis and changing connections. *Trends Neurosci.*, 6: 53–56.

Easter, S.S., Nicola, G.N. (1996) The development of vision in the zebra fish (*Danio rerio*). *Develop. Biol.*, 180: 646–663.

Fadool, J.M., Brockerhoff, S.E., Hyatt, G.A. and Dowling, J.E. (1997) Mutations affecting eye morphology in the developing zebra fish (*Danio rerio*). *Develop. Genetics*, 20: 288–295.

Gaiano, N., Allende, M., Amsterdam, A., Kawakami, K. and Hopkins, N. (1996) Insertional mutagenesis and rapid cloning of essential genes in zebra fish. *Nature* (*London*), 383: 829–832.

Haffter, P., Granato, M., Brand, M., Mullins, M.C., Hammerschmidt, M., Kane, D.A., Odenthal, J., Van Eeden, F.J.M., Jiang, Y-J., Heisenberg, C-P., Kelsh, R.N., Furutani-Seiki, M., Vogeisang, E., Beuchle, D., Schach, U., Fabian, C. and Nüsslein-Volhard, C. (1996) The identification of genes with unique and essential functions in the development of the zebra fish (*Danio rerio*). *Development*, 123: 1–36.

Johns, P. R. and R. Fernald (1981) Genesis of rods in teleost fish retina. Nature (London) 293: 141–142.

Johns-Raymond P. (1977) Growth of the adult goldfish eye: III. Source of new retinal cells. *J. Comp. Neurol.*, 176: 343–357.

Li, L. and Dowling, J.E. (1997) A dominant form of inherited retinal degeneration caused by a non-photoreceptor cell-specific mutation. *Proc. Natl. Acad. Sci. USA*, 94: 11645–11650.

Li, L. and Dowling, J.E. (1998) Zebra fish visual sensitivity is regulated by a circadian clock. *Visual Neurosci.*, 15: 851–857.

Nathans, J., Thomas, D. and Hogness, D.S. (1986) Molecular genetics of inherited variation in human color vision. *Science*, 232: 203–210.

Pak, W.L. (1995) *Drosophila* in vision research. The Friedenwald Lecture. *Invest. Ophthalmol. Vis. Sci.*, 36: 2340–2357.

Raymond, P.A. and Rivlin, P.K. (1987) Germinal cells in the goldfish retina that produce rod photoreceptors. *Develop. Biol.*, 122: 120–138.

Robinson, J.R., Schmitt, E.A., Harosi, F.I., Reece, R.J. and Dowling, J.E. (1993) Zebra fish ultraviolet visual pigment: Absorption spectrum, sequence, and localization. *Proc. Natl. Acad. Sci., USA*, 90: 6009–6012.

Schmitt, E.A. and Dowling, J.E. (1996) Comparison of topographical patterns of ganglion and photoreceptor cell differentiation in the retina of the zebra fish, *Danio rerio*. *J. Comp. Neurol.*, 371: 222–234.

To, K.W., Adamian, M., Jakobiec, F.A. and Berson, E.L. (1993) Olivopontocerebellar atrophy with retinal degeneration: An electroretinographic and histopathologic investigation. *Ophthalmology*, 100: 15-23.

Zuker, C.S. (1992) Phototransduction in *Drosophila*: a paradigm for the genetic dissection of sensory transduction cascades. *Current Opinion in Neurobiology*, 2: 622–627.

Section 2.2

Finding Genes with Phenotypical Effects on Neural and Behavioral Phenotypes on the Basis of Gene-Expression

W.E. Crusio and R.T. Gerlai (Eds.)
Handbook of Molecular-Genetic Techniques for Brain and Behavior Research (Techniques in the Behavioral and Neural Sciences, Vol. 13)

CHAPTER 2.2.1

Subtractive cDNA hybridization and brain: then, now and tomorrow

Joseph B. Watson

48-241 NPI, Department of Psychiatry and Biobehavioral Sciences, Mental Retardation Research Center, Brain Research Institute, UCLA School of Medicine, 760 Westwood Plaza, Los Angeles, CA 90024-1759, USA

Subtraction and brain

In the early 1980s, just a few investigators began to identify cDNA clones of the approximately 30,000 unique mRNAs expressed in the mammalian brain (Milner and Sutcliffe, 1983; also see reviews by McKay, 1983; Sutcliffe, 1988). In the next decade or so, a collective effort involving numerous laboratories spanning the globe capitalized on emerging technologies in molecular biology and genetics to develop highly sensitive differential cDNA screening approaches to brain mRNA expression (see Chapter 2.2.2). Despite a continuing concerted effort in this regard, clones for a large portion of the potentially 20,000 brain-specific messages have remained elusive primarily due to their very low levels of expression (1–20 copies per cell, on average; $<0.02\%$ mRNA mass) (Milner and Sutcliffe,1983).

Because subtractive cDNA hybridization can reliably identify subsets of rare mRNAs in a given target tissue, a variety of subtraction-based approaches to differential screening have played a prominent role in the isolation of cDNA clones of novel proteins expressed in a specific brain region or neural cell type (reviewed in Watson and Margulies, 1993; also see Usui et al., 1994; Gautvik et al., 1996; Soares, 1997). In this Chapter, I will first provide a historical discussion of three generations of cDNA subtraction, each of which built on lessons from the previous applications and capitalized on emerging state-of-the-art technologies in molecular biology. I will then outline a general protocol for current uses of Directional Tag PCR (DTP)-Subtraction (Usui et al., 1994; Kilduff et al., 1998) and make strategic comments based on first-hand experience with the method. Lastly, I will suggest ways in which newer technologies for subtraction such as Representational Difference Analysis (RDA) (Hubank and Schatz, 1994) and Suppression Subtractive Hybridization (SSH) (Diatchenko et al., 1996) can converge with DTP to identify a comprehensive set of brain cDNAs on high density microarrays.

A historical perspective on subtraction

Classic subtraction

Historically the advent of subtractive cDNA hybridization in the early 1980s (e.g. Hedrick et al., 1984) initially grew out of the marriage of two prominent molecular biology reagents commonly used in conventional cDNA cloning. The first reagent is Reverse Transcriptase (RT), which copies mRNA into single-stranded complementary DNA (cDNA). The second reagent is oligo

$(dT)_n$ which is used both to prime cDNA synthesis by RT at the 3′ end of mRNAs as well as to enrich for ample chemical amounts of mRNA-containing poly $(A)^+$ RNA fractions. To perform what I will call Classic subtractive hybridization (see Fig. 1), radioactive antisense cDNA is synthesized from a target poly $(A)^+$ RNA fraction ($\sim$1–5 μg) using RT, oligo $(dT)_n$ and an α ^{32}P-labeled dNTP and this fraction is hybridized to an excess amount (50–100 fold) of cold, unlabeled (i.e. non-radioactive) driver poly $(A)^+$ RNA. The target is defined as the tissue of interest in which you wish to identify enriched expression of a subset of mRNAs relative to mRNAs present in a control tissue, called the driver. The radiolabeled target single-stranded cDNA is separated from complementary hybrids of target cDNA/driver mRNA by column chromatography (e.g. hydroxylapatite, HAP) and used to screen at high density 10^4–10^6 recombinants from a target cDNA library (either plasmid or λ phage) on nitrocellulose or nylon membrane filters. Candidate positives are picked and screened in low-density arrays differentially with either radiolabeled target or driver single-stranded cDNAs.

An example of a Classic subtraction screen is shown in Fig. 2A in which a representative set of candidate clones was screened differentially either with a radioactive cerebellar (Driver) cDNA probe, cortex (Target) cDNA probe or with a second cortex-minus-cerebellum (Subtraction) cDNA probe (Watson et al., 1990). Despite only screening 5000 clones, this screen identified the first cDNA clone for a novel forebrain-enriched protein which is called RC3/neurogranin (white circles, Fig. 2(A)). There were a number of important technical lessons learned in this screen: First, the implementation of a second subtracted probe (right panel, Fig. 2(A)) in the differential screen reduced the false positive rate when examined subsequently by Northern blot analyses and in situ hybridization. Second the identification of RC3/neurogranin ($\sim$0.02% mRNA abundance) provided an important positive control to assess quantitatively the sensitivity of detection of newer evolving subtractive techniques that sampled RC3-expressing brain regions (e.g. striatum-minus-cerebellum, see Fig. 2(B) and(C)).

Phenol-emulsion reassociation technique (PERT) subtraction

Although Classic Subtraction remained a viable approach, it was clear that much more sensitive methods were required to discover the rare class of mRNAs expressed in brain. The Phenol-Emulsion Reassociation Technique (PERT) of Subtraction (see Travis and Sutcliffe 1988, Travis et al., 1990) was a major advance in this regard and identified numerous novel neural proteins including the mouse retinal degeneration (*rd*) and retinal degeneration slow (*rds*) gene products, a G-protein γ7 subunit, a novel CaM kinase 1G5, cortexin and numerous others (Travis and Sutcliffe, 1988; Travis et al., 1989; Bernal et al., 1990; Bowes et al., 1990; Watson et al., 1992; Coulter et al., 1993; Godbout et al., 1994; Baba et al., 1994, 1995; see Travis et al., 1990 for detailed protocol). Under the PERT conditions, radiolabeled target cDNA is hybridized to an excess of denatured double-stranded cDNA as driver rather than RNA and importantly in the presence of phenol and high salt, which increased the kinetics of hybridization by orders of magnitude. With PERT, even rare mRNAs shared between target and driver are subtracted out.

An example of a PERT rat striatum-minus-cerebellum differential screen is shown in Fig. 2(B). Consistent with its enriched expression in striatum and absence from cerebellum, there were five isolates of RC3/neurogranin (white circles) detected in this screen. This screen also identified a single isolate of a clone initially called CPu1 (Watson et al., 1992; Watson et al., 1994) encoding a striatum-enriched G-protein γ7 subunit (squares, Fig. 2(B)), two isolates of IG5 (triangles, Fig. 2(B)) isolated in a previous PERT screen (Travis and Sutcliffe, 1988; Godbout et al., 1994) and numerous unknown clones including two novel cortex-enriched proteins, cortexin and a

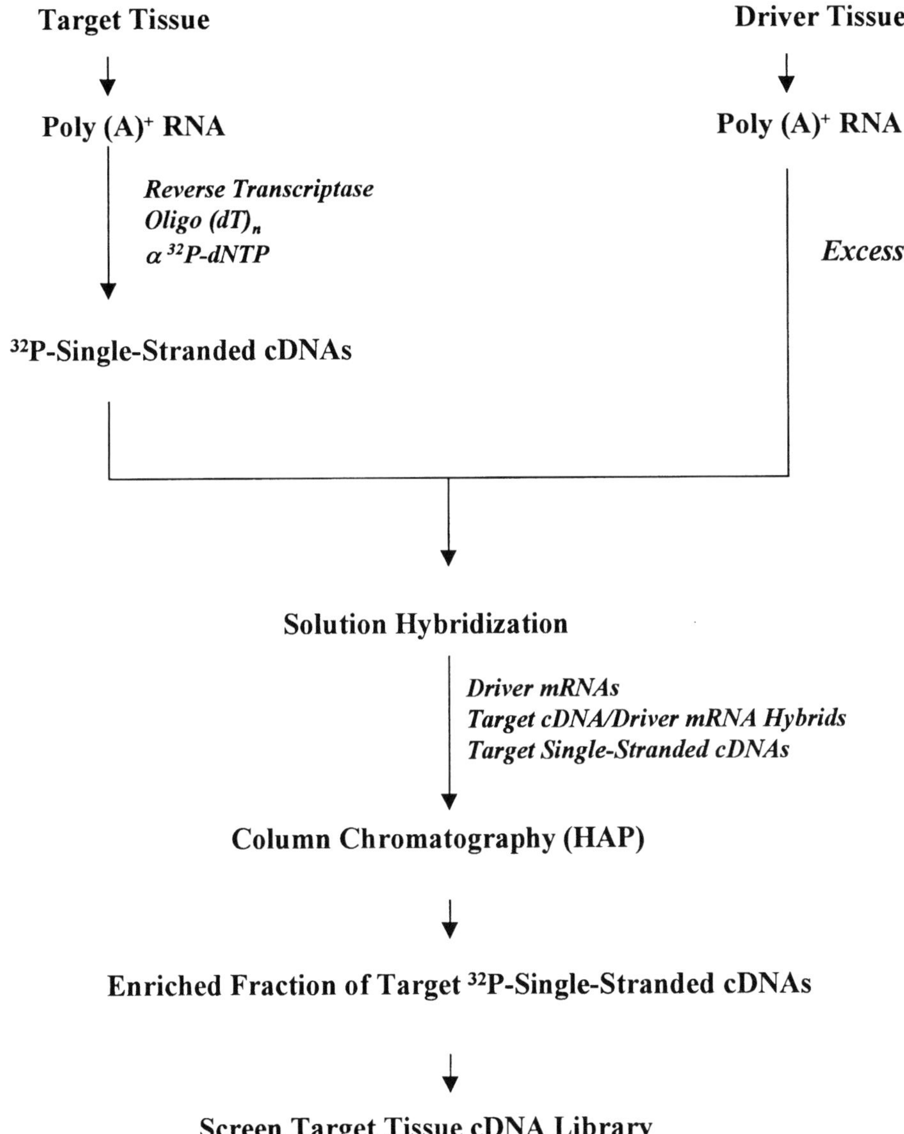

Fig. 1. Classic subtractive hybridization. In earlier subtractive approaches, a poly $(A)^+$ RNA enriched fraction is isolated from a target neural tissue and copied into single-stranded cDNAs by Reverse Transcriptase and oligo $d(T)_n$ primers in the presence of ^{32}P-dNTP (usually dCTP). Target cDNAs are subsequently hybridized in solution to an excess amount (50–100 fold) of cold non-radioactive poly $(A)^+$ RNA from a driver neural tissue. Target cDNAs that do not have a complementary template in the driver RNA fraction are enriched in a single-stranded fraction by column chromatography (usually HAP). Enriched radiolabeled target cDNAs are used as a complex probe to screen a target cDNA library of recombinant plasmids or λ phage at relatively high density (10^4–10^6 recombinants on multiple filters).

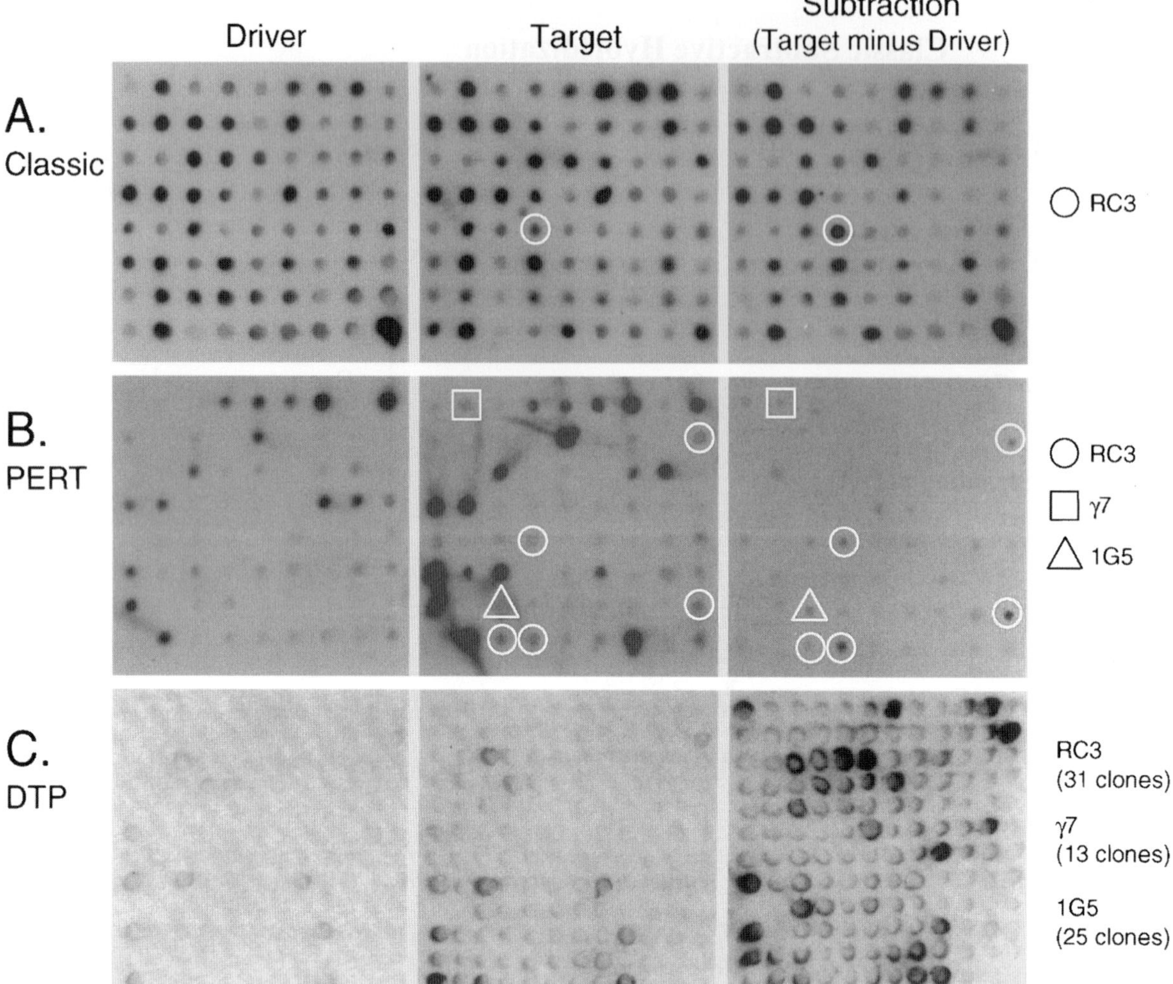

Fig. 2. Three generations of subtraction. Representative colony lifts of plasmid cDNA arrays on nylon filters were screened differentially either with a radiolabeled Driver, Target or Subtraction (Target minus Driver) complex rat cDNA probes. In Parts (A) and (B), candidate clones were picked in a previous high density screen of a Target library using subtracted probes and were screened at low density. Arrays, shown in Parts (A) and (B), have not appeared previously in print but composed part of the data set for earlier publications (Watson et al., 1990, 1992). Part (A). *Classic*: Driver = cerebellar cDNA; Target = cortex cDNA. Part (B). Phenol-Emulsion Reassociation Technique (*PERT*): Driver = cerebellar cDNA; Target = striatum cDNA. Part (C). Directional Tag PCR (*DTP*): Driver = cerebellar cDNA; Target = striatum cDNA. Identical target/driver combinations were used as in the PERT subtraction in Part (B), but a DTP-subtracted striatum-minus-cerebellum library was screened from the outset. This panel is modified from Usui et al. (1994) with permission from the corresponding author.

SH3 domain-containing protein (Coulter et al., 1993; Margulies et al., 1997).

Again, there were a number of important technical lessons learned in the PERT screens. First, the use of cDNA as Driver obtained from an amplified plasmid library reduced considerably the starting amount of brain tissue needed for poly $(A)^+$ RNA isolation. Second, PERT enhancement of the kinetics of hybridization reduced false positives. Third, it became readily apparent that it is critical to analyze more carefully clones with little or no signal above background, because they represent potential clones of rare mRNAs. For example, a handy trick which we used was to screen cDNA arrays of such clones (right filter, Fig. 2(B)) yet a second time with the subtracted probe but now in the absence of clones of obviously abundant mRNAs (also see Travis and Sutcliffe, 1988). In essence, this was a crude but effective way to normalize mRNA abundances which can now be addressed more systematically via the generation of normalized subtracted libraries (Bonaldo et al., 1996).

Directional Tag PCR subtraction

The discovery of the polymerase chain reaction (PCR) (Saiki et al., 1988) drastically changed the way recombinant molecular biology is done and not surprisingly spawned the development of multiple methods of PCR-based subtractive hybridization (e.g. Hara et al., 1991; Usui et al., 1994; Hubank and Schatz, 1994). We recently embarked on a set of subtraction-based differential screening studies to identify genes expressed selectively in the early segment of the oligodendrocyte lineage utilizing immortalized oligodendrocyte cell lines (Foster et al., 1995; Campagnoni et al., 1998, also see Fig. 4). We chose to use the Directional Tag PCR (DTP)-based subtraction method (Usui et al., 1994) rather than the PERT-based subtractive approach used in our previous studies, because it was much more "user-friendly". Moreover the DTP method offered increased sensitivity of detection by adding a critical step in which PCR-amplified subtracted cDNA libraries were screened differentially with PCR-amplified subtracted cDNA probes.

The DTP technique is outlined in Fig. 3 and examples of low density differential cDNA arrays from a DTP rat striatum-minus-cerebellum screen are shown in Fig. 2(C) (This is modified from Usui et al., 1994). Compared to an identical PERT-based screen (Fig. 2(B)), DTP subtraction dramatically increased the sensitivity of detection as exemplified by the presence of 31 clones of RC3, 13 clones of $\gamma 7$, and 25 clones of IG5 and numerous other novel clones detected in striatum-enriched libraries (Usui et al., 1994; also see Thomas et al., 1998).

There are two excellent protocols currently available (Wan and Erlander, 1997; Kilduff et al., 1998), which describe in detail how to perform DTP-subtraction, covering step by step construction of target and driver plasmid cDNA libraries, generation of target antisense cDNA and driver sense cRNA, subtractive hybridization, HAP-column purification, and differential screening of cDNA arrays. Here I will outline DTP-subtraction in more general terms (see Fig. 3) and point out some of the major tricks and pitfalls of the technique that we have encountered and minor modifications which we have incorporated.

RNA preparation

It cannot be stressed enough that the most important ingredient of a successful subtractive cDNA hybridization is the starting RNA. The mRNA complexity of initial poly $(A)^+$ RNA isolations will ultimately decide which mRNAs are identifiable through subtractive cDNA enrichments for a given target tissue relative to driver tissue. This is paramount for the large number of rare brain mRNAs which remain to be isolated. If at all possible, it is highly recommended that investigators isolate a cytoplasmic total RNA fraction either from fresh tissue (Shibler et al., 1980) or from cells in tissue culture (Penman et al., 1963; Watson and Sutcliffe, 1987), prior to enrichment for poly $(A)^+$ RNA. Use of cytoplasmic RNA fractions minimizes incompletely processed nuclear RNA and interestingly also decreases mitochondrial RNA contamination (unpublished observations). Unfortunately, similar cytoplasmic RNA fractions cannot be prepared from frozen tissue. When using the large variety of established methods for whole cell total RNA fractionation from frozen samples, caution must be exercised in the downstream analyses of candidate clones of differentially expressed abundant mRNAs

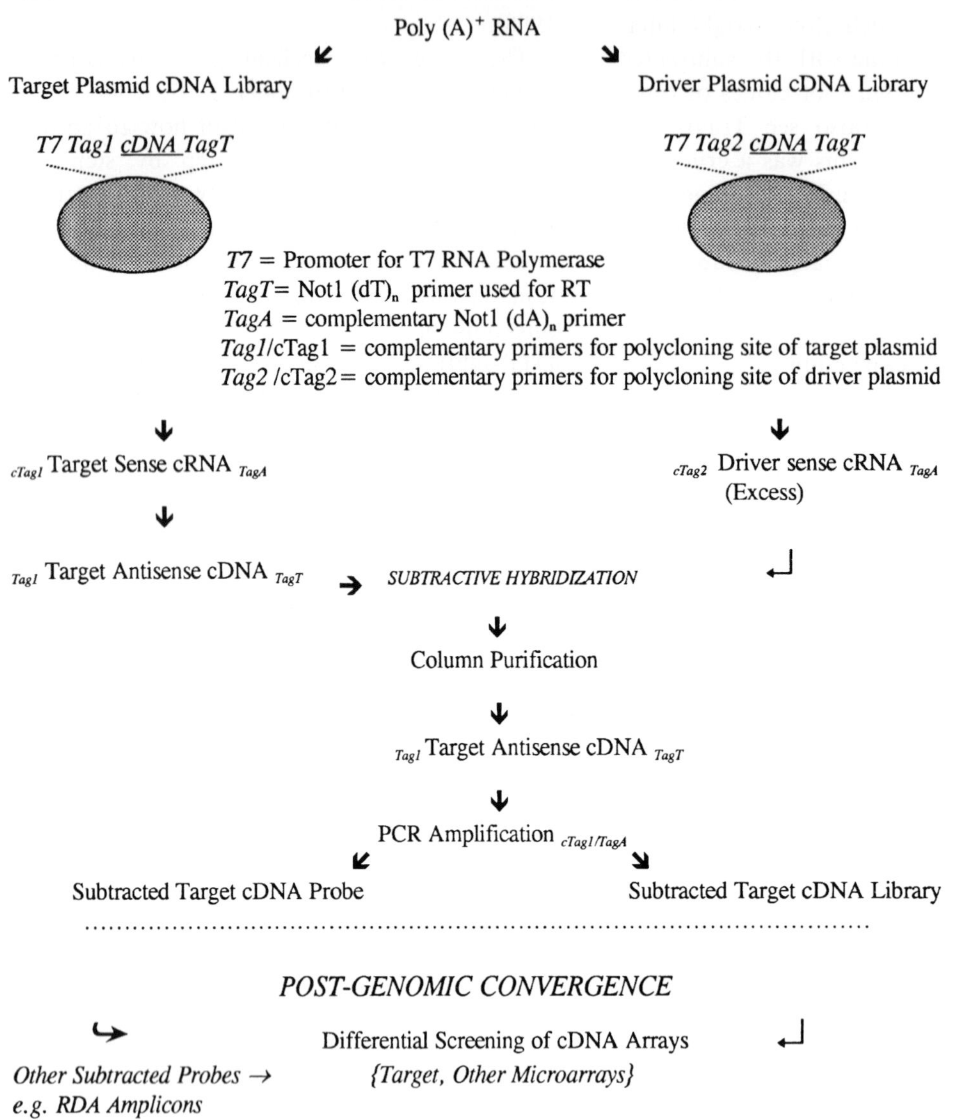

Fig. 3. Directional Tag PCR (DTP) subtraction. In the top half of the figure, a general outline is given for the DTP protocol. First plasmid libraries are constructed with the designated *Tag* primers and cDNA prepared from target or driver poly $(A)^+$ RNA (preferentially cytoplasmic). When using plasmids pT7T3D and pGEM11ZF(-), the exact sequences of primers can be obtained in the original DTP publication (Usui et al., 1994). Both plasmid libraries are amplified in liquid bacterial cultures, purified and used as complex templates for in vitro transcription to generate synthetic cRNAs. The Target cRNAs are reverse transcribed with RT into radioactive cDNAs and hybridized to an excess of cold driver cRNA, which in essence substitutes for the poly $(A)^+$ RNA fraction previously described in the Classic Subtraction protocol (see Fig. 1). Target cDNAs are isolated by HAP chromatography and amplified by PCR using the desginated *Tag* primers. The target cDNA fraction can be used either directly as a probe to screen a library deposited on arrays or made into an enriched subtracted target cDNA library for later analysis on arrays. In the bottom half of the figure, the potential post-genomic convergence of DTP subtractions with other PCR-based subtractions is denoted. For example, probes generated by RDA or SSH from other target tissue sources could be used to screen the same DTP-generated libraries for new gene discoveries.

which can often correspond to mitochondrial contaminants (see Watson et al., 1992; Campagnoni et al., 1998).

Plasmid library construction

The need to construct an initial set of cDNA libraries may be the biggest obstacle for many beginning investigators interested in using the DTP method or other subtraction-based, differential cDNA screening approaches for that matter. Indeed I would submit that "subtraction in general is not for everyone". However it is now feasible to prepare highly representative, plasmid cDNA libraries using highly efficient competent cells for electroporation-based transformations. If at all possible, it is also a good idea for investigators to make at this time highly representative target λ phage cDNA libraries, because ultimately they will need to pull out full-length clones of potentially interesting mRNAs. This becomes even more of an issue when one considers that most of these are likely rare mRNAs, which tend to be much longer (5–10 kb, see Sutcliffe, 1988). In this regard, the advent of enriched 5′ CAP-based full-length cDNA libraries is worth pursuing (Carninci et al., 1996).

To construct plasmid libraries, cDNA is prepared from cytoplasmic poly $(A)^+$ RNA isolated from either a target or driver neural tissue using Superscript RT (GibcoBRL) and a *TagT* primer [Not1 (dT)n, see Fig. 3], ligated with EcoR1 linkers and cloned into the Not1/EcoR1 sites of two different plasmid vectors (e.g. pT7T3D, Pharmacia; pGEM 11Zf(-), Promega; pBCSK$^+$, Stratagene). Cloning into two different vectors is critical, since it creates non-hybridizing sites for primer tags depicted simply as *Tag1, Tag2* (Fig. 3) within polycloning regions unique to each vector.

In our hands, the choice of competent *E.Coli* cells for transformation with a given plasmid cDNA library ligation mix was an important consideration. Using a BioRad Gene Pulser (0.1 cm cuvette, 25 μF capacitance, 1.80 KV, 200 S2 pulse controller), electroporation of BioRad MC1061 competent cells with pT7T3D as a target plasmid and pGEM 11ZF(-) as driver plasmid were optimal, consistent with the original recommendations of Usui et al. (1994). It is important, however, to note that the MC1061 cells contain a peculiar endogenous endonuclease activity that can rapidly degrade plasmid DNA so phenol extractions are critical. The relative mRNA complexity of each library should be determined based on the number of recombinant colonies titered on agar plates as well as the percentage of colonies containing plasmids with cDNA inserts. For our oligodendrocyte lineage studies, we generated cDNA libraries from two different immortalized mouse oligodendrocyte cell lines: the N19 cell line (early-like progenitors) and the N20.1 cell line (more mature forms). The libraries contained $11.6 \times 10^6 - 14.3 \times 10^6$ recombinants, while 12 of 16 randomly picked clones contained cDNA inserts of 1.1–1.4 KB on average. It is important to make note of the cDNA insert sizes, because minimally inserts >0.5 KB are required to ensure efficient subtractive hybridizations between target and driver nucleotide sequences as well as to maximize the resolution of enriched single-stranded target cDNA from fragmented target cDNA/driver cRNA hybrids during HAP-column purification (see below). Once libraries are generated, they are amplified by large-scale plasmid preparation in liquid bacterial cultures (see Travis et al., 1990), purified on Cesium Chloride gradients (recommended), and stored for subsequent use as templates for in vitro transcription to generate cRNAs for subtractive hybridization.

Subtractive hybridization

The next step is to generate radiolabeled target antisense cDNA and hybridize to an excess of cold driver sense cRNA. We followed basically to the letter DTP methods described previously (Wan and Erlander, 1997; Kilduff et al., 1998) and they worked quite well in our hands, probably due to the fact that most of the reagents were readily available from commercial sources. In general, both the target and driver plasmid libraries are linearized by Notl digestion, and transcribed in vitro with T7 RNA polymerase to generate sense cRNAs with complementary tags at the 5′ ends *(cTag1 or cTag2)* and 3′ ends (*TagA*). Approximately 1 μg of target sense cRNA is used as template for RT and the *TagT* primer in the presence of α^{32}P-dCTP to generate target antisense cDNA. This is hybridized in solution in an Eppendorf tube to an excess amount (~20 μg) of driver sense RNA followed by hydroxylapatite (HAP) column purification to elute single-stranded and double-stranded fractions with Phosphate buffers of increasing ionic strength.

For HAP column chromatography, we used a constant temperature water bath (Lauda, Brinkman) at 60°C linked to a BioRad jacketed Econo column and connected to a multistatic pump (Buchler) and micro-fraction collector (Gilson), as first described in Travis et al. (1990) and later refined in Usui et al. (1994) (also see Wan and Erlander, 1997). As an extra guard against RNAse contamination, the column and tubing was prewashed with a diluted solution of Absolve (New England Nuclear) and rinsed thoroughly with RNAse-free water taken directly from a Millipore Milli-Q water filtering system. The single-stranded fraction normally elutes in 120–150 mM Phosphate and contains about 20% of the radioactive counts based on scintillation counting, indicating subtraction of at least 80% of the nucleotide sequences in common between the target and driver cDNAs. For further enrichment, the single-stranded target cDNA fractions are collected, phenol extracted, concentrated with desalting filters (Centricon 100) and hybridized a second time to an excess amount of driver cRNA followed by a second HAP column. Because there can be a diminishing amount of radiolabeled target cDNA after the initial subtraction, we found that it is important to minimize losses in the subsequent extraction/concentration steps prior to the second subtraction. Enrichment can also be demonstrated by dot blot analysis (see Usui et al., 1994). Replicate dot blots containing cDNAs of both target-specific mRNAs and commonly expressed mRNAs are screened differentially with unsubtracted and subtracted probes to demonstrate relative fold enrichment and depletion. In the end approximately 85–90% of the cDNA sequences common to the target and driver neural tissue are depleted by the DTP subtraction protocol.

As a word of caution, never use a putative subtracted cDNA fraction for use either as a complex probe or to prepare a subtracted library unless you have first quantitatively demonstrated enrichment for your target cDNAs. It is important to resist the temptation to plunge ahead because too much precious time will be spent down the road to want to go back and start over again.

Subsequently, the enriched target single-stranded antisense cDNA fraction is PCR-amplified with compatible end-tagged primers (*cTag1, TagA,* Fig. 3) and cloned into the Notl/EcoR1 sites of target plasmid (e.g. pT7T3D) to generate a subtracted library with cDNA inserts of 0.5 KB size on average. The library is amplified in bacterial liquid cultures and plasmids are isolated, stored indefinitely at −20°C for use either as templates for successive sense cRNA/ antisense cDNA synthesis or to be plated out on 7296 well arrays of colony lifts on nylon membrane filters for differential screening (see DTP bottom panels, Fig. 2(C)). Eventually candidate clones must be confirmed as true positives by differential gene expression utilizing either rapid cDNA library Southern blotting (Usui et al., 1994), conventional Northern blot analysis (Watson et al., 1992, 1994), or ideally by multiple in situ hybridizations in brain sections or in cultures (Baba et al., 1994; Campagnoni et al., 1998). In our DTP-subtraction screen of the N19 early oligodendrocyte cell line, we identified three differentially expressed mRNAs, two of which encoded known proteins (T-cadherin, K-19 keratin) and a third which encoded a novel protein 3G5 of 80 amino acids (see Fig. 4; Campagnoni et al., 1998).

Fig. 4. Northern blot analysis of 3G5, an early oligodendrocyte mRNA. Nylon membranes containing poly $(A)^+$ RNA (4 μg) from either the mouse N19 immortalized early oligodendrocyte cell line, the N20.1 immortalized late oligodendrocyte cell line, mouse 2-day-old brain, 18-day old brain, liver or spleen were probed with a ^{32}P-labeled cDNA insert of clone 3G5. The 3G5 clone detects an early oligodendrocyte mRNA (0.64 kb) encoding a novel 80 amino acid protein, highly conserved between mouse and man (Campagnoni et al., 1998). Small amounts of 3G5 mRNA are detected in mouse brain (P2, P18) and spleen at longer film exposures (not shown).

A 21st Century view of subtraction

I have described technical experiences with the DTP-subtraction method but there are other attractive subtraction-based methodologies currently available such as Representational Difference Analysis (RDA) (Hubank and Schatz, 1994; Vician et al., 1997) and Suppression Subtractive Hybridization (SSH) (Diatchenko et al., 1996). I suspect that both the RDA, SSH and newly emerging subtraction methods will converge with DTP to identify cDNAs of novel brain mRNAs deposited on high density microarrays (filters, glass, chips) (see bottom, Fig. 3). Such arrays are now becoming increasingly available from commercial sources or are being generated by individual investigators (Schena et al., 1995). There are already examples of SSH-based and RDA-based robotic screening of non-brain microarrays (Von Stein et al., 1997; Welford et al., 1998) and this will likely expand to screening of brain microarrays in the Post-Genomic era. In the beginning of the 21st Century, the collective goal of subtraction together with other differential approaches (Differential Display, single-cell amplification) is to identify the entire set of cDNA clones of brain mRNAs including the elusive rare class. Ultimately the availability of these clones will complement the gene mapping field in the discovery of genes related to neurological, psychiatric, and neurodegenerative disease and also will aid the functional genomics field in the discovery of new diagnostic and therapeutic reagents for a variety of neural diseases.

Acknowledgements

I wish to thank Celia W. Campagnoni, J. Gregor Sutcliffe, Stanley F. Nelson, Julia. W. Chang, and Anthony T. Campagnoni for helpful comments on the manuscript. Thanks also to the International Order of the Rainbow for Girls for their support.

List of Abbreviations

Directional Tag PCR (DTP)
Representational Difference Analysis (RDA)
Suppression Subtractive Hybridization (SSH)

Reverse Transcriptase (RT)
Complementary DNA (cDNA)
Complementary RNA (cRNA)
Hydroxylapatite, HAP
Phenol-Emulsion Reassociation Technique (PERT)
Mouse retinal degeneration (*rd*)
Retinal degeneration slow (*rds*)
Polymerase Chain Reaction (PCR)

References

Baba, H., Fuss, B., Watson, J.B., Zane, L.T. and Macklin. W.B. (1994) Identification of novel mRNAs expressed in oligodendrocytes. *Neurochem. Res.*, 19: 1091–1099.

Baba, H., Fuss, B., Urano, J., Poullet, P., Watson, J.B., Tamanoi, F. and Macklin, W.B. (1995) GapIII, a new brain-enriched member of the GTPase-activating protein family. *J. Neurosci. Res.*, 41: 846–858.

Bernal, J., Godbout, M., Hasel, K.W., Travis, G.H. and Sutcliffe, J.G. (1990) Patterns of cerebral cortex mRNA expression. *J. Neurosci. Res.*, 27: 153–158.

Bonaldo, M.F., Lennon, G. and Soares, M.B. (1996) Normalization and subtraction: two approaches to facilitate gene discovery. *Genome Res.*, 6: 791–806.

Bowes, C., Li T., Danciger M., Baxter, L.C., Aplebury, M.L. and Farber, D.B. (1990) Retinal degeneration in the *rd* mouse is caused by a defect in the β-subunit of rod cGMP-phosphodiesterase. *Nature*, 347: 677–680.

Campagnoni, C.W., Landry, C.F., Pribyl, T.M., Schonmann, V., Kampf, K., Handley, V.W., Watson, J.B. and Campagnoni, A.T. (1999) Identification of stage-specific genes in the oligodendrocyte lineage through the analysis of conditionally immortalized cell lines, submitted.

Carninci, P., Kvam, C., Kitamura, A., Oshumi, T., Okazaki, Y., Itoh, M., Kamiya, M., Shibata, K., Sasaki, N., Izawa, M., Muramatsu, M., Hayashizaki, Y. and Schneider, C. (1996) High- efficiency full-length cDNA cloning by biotinylated CAP trapper. *Genomics*, 37: 327–336.

Coulter II, P.M., Bautista, E.A., Margulies, J.E. and Watson, J.B. (1993) Identification of cortexin: a novel, neuron-specific, 82-residue membrane protein enriched in rodent cerebral cortex. *J. Neurochem.*, 61: 756–759.

Diatchenko, L., Lay, Y.F., Campbell, A.P., Moqadam, F., Huang, B., Lukyanov, K., Gursakaya, N., Sverdlov, E.D. and Siebert, P.D. (1996) Suppression subtractive hybridization: a method for generating differentially regulated or tissue-specific cDNA probes and libraries. *Proc. Natl. Acad. Sci. USA*, 93: 6205–6030.

Foster, L.M., Landry, C., Phan, T. and Campagnoni, A.T. (1995) Conditionally immortalized oligodendrocyte cell lines migrate to different brain regions and elaborate 'myelin-like' membranes after transplantation into neonatal *Shiverer* mouse brains. *Dev. Neurosci.*, 17: 160–170.

Gautvik, K.M., De Lecea, L., Gautvik, V.T., Danielson, P.E., Tranque, P., Dopazo, A., Bloom, F.E. and Sutcliffe, J.G. (1996) Overview of the most prevalent hypothalamus-specific mRNAs, as identified by directional tag PCR subtraction. *Proc. Natl. Acad. Sci. USA*, 93: 8733–8738.

Godbout, M., Erlander, M.G., Hasel, K.W., Danielson, P.E., Wong, K.K., Battenberg, E.L.F., Bloom, F.E. and Sutcliffe, J.G. (1994) A calmodulin-binding, vessicle-associated, protein kinase-like protein enriched in forebrain neurites. *J. Neurosci.*, 14: 1–13.

Hara, E., Kato, T., Nakada, S., Sekiya, S. and Oda, K. (1991) Subtractive cDNA cloning using oligo $(dT)_{30}$-latex and PCR: Isolation of cDNA clones specific to undifferentiated human embryonal carcinoma cells. *Nucl. Acids Res.*, 19: 7097–7104.

Hedrick, S.H., Cohen, D.I., Nielsen, E.A. and Davis, M.M. (1984) Isolation of cDNA clones encoding T cell-specific membrane-associated protein. *Nature*, 308: 149–153.

Hubank, M. and Schatz, D.G. (1994) Identifying differences in mRNA expression by representational difference analysis of cDNA. *Nucl. Acids Res.*, 22: 5640–5648.

Kilduff, T.S., De Lecea, L., Usui, H. and Sutcliffe, J.G. (1998) Isolation and identification of specific transcripts by subtractive hybridization. In: R. Lydic (Ed.), *Molecular Regulation of Arousal States*, CRC Press, Bacon Raton, pp. 103–118.

Margulies, J.E., Spirina, O. and Watson, J.B. (1997) Cortex-enriched expression of a novel protein with an SH3 domain. *Soc. Neurosci. Abstr.*, 23: 985.

McKay, R.D.G. (1983) Molecular approaches to the nervous system. *Ann. Rev. Neurosci.*, 6: 527–546.

Milner, R.J. and Sutcliffe, J.G. (1983) Gene expression in rat brain. *Nucl. Acids Res.*, 11: 5497–5520.

Penman, S., Scherrer, K., Becker, Y. and Darnell Jr, J.E. (1963) Polyribosomes in normal and poliovirus-infected HeLa cells and their relationship to messenger-RNA. *Proc. Natl. Acad. Sci. USA*, 49: 654–662.

Saiki, R.K., Gelfrand, D.H., Stoffel, S., Scharf, S.J., Higuchi, R., Horn, G.T., Mullis, K.B. and Erlich, H.A. (1988) Primer-directed enzymatic amplification of DNA with thermostable DNA polymerase. *Science*, 239: 487–491.

Schena, M., Shalon, D., Davis, R.W. and Brown, P.O. (1995) Quantitative monitoring of gene expression patterns with a complementary DNA microarray. *Science*, 270: 467–470.

Shibler, K., Tosi, M., Pitter, A.C., Fabiani, L. and Wellauer, P.K. (1980) Tissue-specific expression of mouse α-amylase genes. *J. Mol. Biol.*, 142: 93–116.

Soares, M.B. (1997) Identification and cloning of differentially expressed genes. *Curr. Opin. Biotech.*, 8: 542–546.

Sutcliffe, J.G. (1988) mRNA in the mammalian central nervous system. *Annu. Rev. Neurosci.*, 11: 157–198.

Thomas, E.A., Danielson, P.E. and Sutcliffe, J.G. (1998) RGS9: A regulator of G-protein signaling with specific expression in rat and mouse striatum. *J. Neurosci. Res.*, 52: 118–124.

Travis, G.H. and Sutcliffe, J.G. (1988) Phenol emulsion enhanced DNA-driven subtractive cDNA cloning: Isolation of low-abundance monkey cortex-specific mRNAs. *Proc. Natl. Acad. Sci.*, 85: 1696–1700.

Travis, G.H., Brennan, M.B., Danielson, P.E., Kozak, C.A. and Sutcliffe, J.G. (1989) Identification of a photoreceptor specific mRNA encoded by the gene responsible for retinal degeneration slow (*rds*). *Nature*, 338: 70–73.

Travis, G.H., Milner, R.J. and Sutcliffe, J.G. (1990) Preparation and use of subtractive cDNA hybridization probes for cDNA cloning: In: A.A. Boulton, G.B. Baker and A.T. Campagnoni (Eds.), *Neuromethods: Molecular Neurobiological Techniques*, Vol. 16, The Humana Press, Clifton, pp. 49–78.

Usui, H., Falk, J.D., Dopazo, A., De Lecea, L., Erlander, M.G. and Sutcliffe, J.G. (1994) Isolation of clones of rat striatum-specific mRNAs by directional tag PCR subtraction. *J. Neurosci.*, 14: 4915–4926.

Vician, L., Basconcillo, R. and Herschman, H.R. (1997) Identification of genes preferentially induced by nerve growth factor vs epidermal growth factor in PC12 pheochromocytoma cells by means of representational difference analysis. *J. Neurosci. Res.*, 50: 32–43.

Von Stein, O.D., Thies, W.G. and Hofmann, M. (1997) A high throughput screening for rarely transcribed differentially expressed genes. *Nucl. Acids Res.*, 25: 2598–2602.

Wan, J.S. and Erlander, M.G. (1997) Cloning differentially expressed genes by using differential display and subtractive hybridization. *Meth. Mol. Biol.*, 85: 45–68.

Watson, J.B. and Sutcliffe., J.G. (1987) Primate brain-specific cytoplasmic transcript of the *Alu* repeat family. *Mol. Cell Biol.*, 7: 3324–3327.

Watson, J.B., Battenberg, E.F., Wong, K.K., Bloom, F.E. and Sutcliffe, J.G. (1990) Subtractive cDNA cloning of RC3, a rodent cortex-enriched mRNA encoding a novel 78 residue protein. *J. Neurosci. Res.*, 26: 397–408.

Watson, J.B., Coulter II, P.M. and Sutcliffe, J.G. (1992) Identification of genes enriched in expression in mammalian neostriatum and neocortex. *Brain Dysfunct.*, 5: 94–105.

Watson, J.B. and Margulies, J.E. (1993) Differential cDNA screening strategies to identify novel stage-specific proteins in the developing mammalian brain. *Devel. Neurosci.* 15: 77–86.

Watson, J.B., Coulter II, P.M., Margulies, J.E., De Lecea, L., Danielson, P.E., Erlander, M.G. and Sutcliffe, J.G. (1994) G-protein $\gamma 7$ subunit is selectively expressed in medium-sized neurons and dendrites of the rat neostriatum. *J. Neurosci. Res.*, 39: 108–116.

Welford, S.M., Gregg, J., Chen, E., Garrison, D., Sorensen, P.H., Denny, C.T. and Nelson, S.F. (1998) Detection of differentially expressed genes in primary tumor tissues using representational differences analysis coupled to microarray hybridization. *Nucl. Acids Res.*, 26: 3059–3065.

W.E. Crusio and R.T. Gerlai (Eds.)
Handbook of Molecular-Genetic Techniques for Brain and
Behavior Research (Techniques in the Behavioral and Neural Sciences, Vol. 13)

CHAPTER 2.2.2

Applying differential display to brain research

Claudio V. Mello[1] and Erich D. Jarvis[2]

[1]*Lab. Animal Behavior, The Rockefeller University, 1230 York Ave., New York, NY 10021, USA*
[2]*Department of Neurobiology, Duke University Medical Center, Durham, NC 27710, USA*

Introduction

The differential display method utilizes reverse transcription PCR (RT-PCR) amplification with specific oligo-dT primers plus a collection of arbitrary primers to identify differentially expressed genes in two or more distinct biological samples. The PCR fragments are analyzed on a sequencing gel for the presence of differential bands, which are then cloned and tested for confirmation of differential expression. In this chapter we present and discuss the strategy we have been following to search for differentially expressed genes in the brain of songbirds. We focus our comments on issues that are critical for the successful outcome of applying this methodology, and give emphasis to aspects that are of particular interest to neurobiologists.

Liang and Pardee (1992, 1997b) developed differential display (DD) as a method to identify differentially expressed genes in biological samples of practically any origin. The high appeal of the technique is based on the following:

1. One can simultaneously compare multiple samples. This is particularly helpful for screenings that involve several brain regions, or for analysis of drug dose-response and time-course curves;
2. The DD technique is based on PCR amplification and requires only small amounts of tissue during the initial screening phase. This is of considerable interest to neurobiologists, who often study molecular changes that occur in small brain areas or nuclei;
3. DD can detect genes expressed at both high and low levels;
4. DD permits a relatively quick assessment of whether a particular comparison is worth pursuing into the more labor-intensive stages of screening. The reason is that candidate differentially expressed PCR fragments can be detected within a few days after RNA isolation from the tissue samples, and comparisons that yield no identifiable candidates can be discarded at this stage. This is particularly advantageous for studies of behavior, as it is often difficult to determine which behaviorally meaningful paradigms are best suited for studying brain gene regulation.

We have been interested in identifying genes differentially expressed in the brains of songbirds, and have been conducting two classes of experiments: (a) Screening for genes expressed in specific brain areas or nuclei in the search for molecular markers for these areas (Denissenko et al., 1995); and (b) Investigating changes in gene expression under different stimulation or behavioral conditions, in the search for genes that

I. Reverse Transcription:

II. PCR amplification:

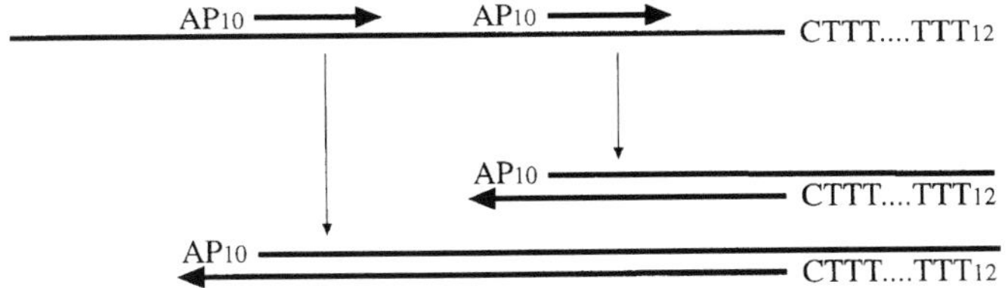

III. Selection of differential bands (sequencing gel):

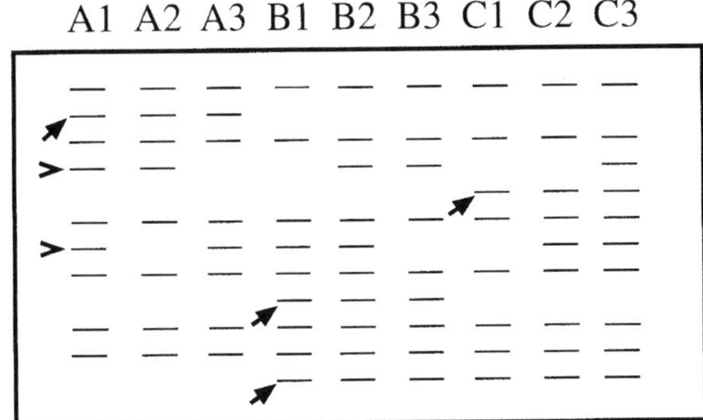

Fig. 1. Schematic diagram of the main steps involved in differential display. (**I**) A fraction of the messenger RNA population is reverse transcribed with a one-base anchored oligo-dT primer; represented are only the mRNAs ending in GAn, which are primed with $T_{12}C$. (**II**) The resulting cDNAs are PCR-amplified with the primer above plus a primer of arbitrary sequence (AP). The AP primer will hybridize to different cDNAs at varying distances from the polyA end, resulting in a collection of PCR fragments of different sizes (**III**) The PCR products are resolved on a sequencing gel for selection of candidate differential bands (arrows), while inconsistent bands (arrowheads) are discarded. In this hypothetical example, three experimental groups (A, B and C) are compared and the reactions are run in triplicates. See text for details.

are regulated by perceptual and motor aspects of singing behavior (Mello et al., 1992, 1997; Jarvis and Nottebohm, 1997).

Figure 1 illustrates the basic steps involved in detecting and isolating candidate differentially expressed genes by DD. First, a subpopulation of the total RNAs from two or more tissue samples are reverse transcribed using one- or two-base anchored oligo-dT primers (such as $T_{12}C$, or $T_{12}GC$). Second, the resulting cDNAs are PCR-amplified using the same primer of the reverse transcription step plus another primer of arbitrary sequence (AP), in the presence of radioactive nucleotides. The frequency at which an AP primer will anneal to different cDNAs, and thus the number of cDNA fragments amplified for each AP/oligo-dT primer combination, depends on the primer size and the stringency conditions of the PCR reaction (Liang and Pardee, 1992). Because an AP typically anneals at various locations upstream of the oligo-dT site of individual cDNAs, many fragments of different sizes are amplified in each PCR reaction. Third, the radioactive PCR products are resolved by electrophoresis on a sequencing gel, which is then exposed to an X-ray film. A couple of hundred fragments generated from a given primer combination can be usually visualized in the resulting autoradiogram. The patterns are then examined for the presence of differentially expressed candidate bands, i.e. bands that are present (or absent) in one of the conditions being compared. These bands are excised from the gel, eluted, and reamplified. The differential expression then needs to be confirmed by other methods, which may include Northern and reverse Northern hybridization, dot blots, and in situ hybridization. Candidates whose differential expression is confirmed are then sequenced for establishing their identity.

Below we first present a summarized version of the DD protocol we have been utilizing in our experiments in songbirds; detailed methodological descriptions have been presented elsewhere (Mello et al., 1997; several chapters in Liang and Pardee, 1997a). We then review and discuss the difficulties that are most often encountered, as well as critical issues that need to be addressed at each step of the procedure. Some considerations are generally applicable, while others are more related to studying the brain and its function. A flow chart depicting our general screening strategy is shown in Fig. 2 for reference.

Summary of protocol

RNA extraction

We follow essentially the protocol of Chomczynski and Sacchi (1987) to isolate total RNA from brain tissue, with few modifications (Mello et al., 1997). The tissue samples are homogenized in a guanidinium isothiocyanate solution and extracted with acid phenol:chloroform:isoamyl alcohol in the presence of sodium acetate. After phase separation, the RNAs are precipitated from the aqueous phase with isopropanol.

DNase treatment

The RNA preparations are incubated with RNase-free DNase I at 37°C for 30 min in the presence of RNasin. The reactions are phenol-extracted, the RNAs precipitated with ethanol and resuspended in DEPC-treated water.

cDNA synthesis

The DNase-treated RNAs (50–200 ng) are incubated with reverse transcriptase (RT) at 37°C for 10–30 min, in the appropriate buffer containing dNTPs and RNasin. A one- or two-base anchored oligo-dT is used as primer. Control reactions are run without adding RT.

PCR amplification

The resulting cDNAs are amplified with AmpliTaq DNA polymerase in the presence of dNTPs, using the same oligo-dT primer of the preceding step, one primer of arbitrary sequence (AP, usually a 10mer), radioactively labeled (^{35}S or ^{33}P) dATP, and appropriate buffer. Usual PCR conditions are 40 cycles of 94°C for 40 s, 40°C for 2 min, and 72°C for 30 s. Control reactions are run without adding cDNAs.

Selecting, eluting and reamplifying PCR fragments

The products of the PCR reaction are resolved on a 6% polyacrylamide gel. The gel is then dried without fixation and exposed to X-ray film. The resulting autoradiogram is analyzed for the presence of differentially expressed PCR fragments. After careful alignment of the gel with the X-ray film autoradiogram, the selected fragments are cut out of the gel, eluted,

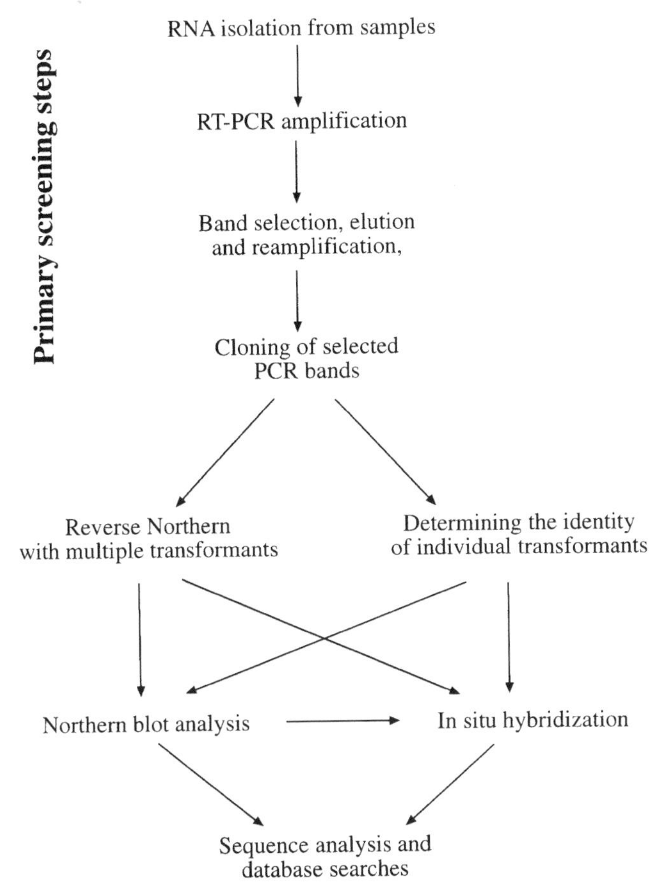

Fig. 2. Flow chart showing our overall screening strategy using DD.

precipitated, and reamplified with the same combination of primers and thermocycler programming used in the first PCR amplification, now omitting the radioactive isotope. The products of the reamplification are examined in an agarose gel for yield and size.

Cloning DD fragments

The reamplified PCR products of the correct size are cloned into a plasmid vector, using commercially available kits. Several transformant colonies (at least 4–6) are picked for each PCR fragment; the colonies are grown and plasmid DNAs prepared.

Confirming differential expression

The cloned PCR fragments from the previous step are analyzed for confirmation of differential expression, using one or all of the following methods:

Reverse Northern
The plasmid DNAs from the transformant colonies picked are denatured and dot-blotted onto replica nylon filters. PolyA+ RNAs isolated from the initial RNA samples being compared are used to generate radioactively-labeled cDNA probes, which are hybridized to the dot blots to identify the true differential colonies for each candidate fragment cloned. The differential colonies ident-

ified are then analyzed further by Northern blots and/or in situ hybridization.

Northern blots

Radioactively-labeled probes generated from individual differential colonies are hybridized to blots containing total or polyA+ RNA from the initial samples being compared. For increased sensitivity, we synthesize single-stranded DNA probes utilizing an oligo-dT primer, or riboprobes using the appropriate RNA polymerases.

In situ *hybridization*

Radioactively labeled riboprobes derived from individual colonies identified by one of the methods above are hybridized against frozen brain sections containing the brain region(s) of interest under the various conditions being compared. We utilize a highly efficient protocol (Mello et al., 1997) and, whenever possible, avoid the use of RNase for improved Nissl-staining of brain sections.

Sequence analysis

Clones whose differential expression is confirmed by the methods described above are then sequenced to establish their identities. Homology searches are run against a database such as GenBank. If necessary, the cloned DD fragments are used for screening cDNA libraries in search of full-length clones.

Discussion

Starting material and sample variability

Careful consideration needs to be given to the nature of the starting materials for DD experiments. While this probably applies to most biological investigations, it is a critical issue for the brain, due to the high complexity of the nervous tissue in terms of cellular and mRNA composition (Milner and Sutcliffe, 1983; Babity et al., 1997). Because of the diversity of factors that can influence brain gene expression, it is important to utilize paradigms that maximize possible effects on gene regulation, particularly for experiments involving behavioral manipulations. It is equally important to keep extraneous factors as constant as possible. Variability can be quite frustrating and is a factor frequently blamed for failed attempts at applying DD (Liang et al., 1993; Mello et al., 1997). Two basic sources of variability need to be addressed: (a) That which is intrinsic to the biological samples; and (b) That which is associated with the DD procedure.

Interanimal variability intrinsic to the biological samples is partly due to genetic differences, but can also be related to differences in hormonal, experiential, motivation, circadian and behavioral states. For instance, an animal that is generally more active may show increased gene expression relative to other animals, but not because of the specific behavior or stimulus being investigated (Jarvis et al., 1997). Similarly, repeated exposure to a particular experimental set-up will attenuate gene expression related to that exposure, while maximizing changes related to stimulus novelty (Campeau et al., 1991; Jarvis et al., 1995; Mello et al., 1995); therefore the animals' familiarity to the experimental paradigm needs to be well controlled for. Sample variability can also be introduced if not enough attention is given to how the tissues are collected and stored. In the case of nervous tissue, small differences in dissection can be critical, considering the high heterogeneity of cellular composition and the diversity encountered across different brain areas. Procedural variability associated with the DD method (Liang et al., 1993; Liang and Pardee, 1997b; Mello et al., 1997) is either due to pipetting and mixing errors, differences in preparation of solutions, contamination of tube or pipette tips etc., or intrinsic to the PCR amplification process. The latter can be rather pronounced, due to the low stringency of the PCR conditions that are normally used in DD (Liang and Pardee, 1992, 1997b).

In our experience, the best way to address interanimal variability in DD experiments is to use multiple animals per group, analyze them individually, and only consider the resulting PCR bands that occur consistently within a particular group (Mello et al., 1997). This approach allows one to assign a spurious PCR band to a specific animal within a group, and thus correlate its presence with a behavioral peculiarity or state presented by that animal, or simply to its genetic background. Since many groups and primer combinations are often compared simultaneously, the total number of reactions poses a restriction on how many individual samples can be compared within a given experiment. We find that triplicates (i.e. 3 animals per group) are a good compromise between diversity and economy. In some cases, to solve the problem of tissue availability from small brain nuclei (see below), tissue samples from several animals within each group compared have been pooled together (for instance, see Babity et al., 1997). While individual differences may be diluted in the pool if a high enough number of samples is combined, such differences may still be detected by the PCR amplification and interfere with the DD screening as "false positives".

To reduce procedural variability, it is critical that all the samples be processed in parallel, utilizing the same batches of reagents. Great care should be exercised in pipetting and mixing. One should first prepare fresh mixes containing all common reagents, and then divide the mixes equally into the individual reaction tubes, before adding specific components such as RNA samples and primers. To address the issue of intrinsic variability in the RT and PCR steps, one can directly compare duplicate RT-PCR reactions starting from the same RNA samples, but such an approach is also limited by the total number of reactions one can run in a single experiment. Moreover, the issue is already partly addressed by running single RT-PCR reactions for multiple independent samples (i.e. different animals) within each group compared, as described in the preceding paragraph. All the pipetting is done on ice, and a variant of a hot start protocol should be used for the PCR step (see below).

Tissue dissection and RNA isolation

Brain regions of interest to neurobiologists and the corresponding RNA yields are often small. However, DD protocols do not require the isolation of polyA+ RNA, and as little as 50–100 ng of total RNA are sufficient for the reverse transcription step. Therefore, large amounts of tissue are not necessary, at least for the initial screening stages. The material we normally work with consists of small tissue fragments punched out of brain sections with glass capillaries, or dissected out with a fine scalpel. Our typical RNA yield for an avian song or auditory nucleus ranges from 1 to 20 μg of total RNA. When possible, we evaluate the quality of the RNA by running 1–2 μg on an agarose/formaldehyde gel.

Brain nuclei of interest are often embedded within other regions and are difficult to access. A clean sample dissection is critical, however, if differential gene expression occurs in opposite directions in immediately adjacent areas. In this case, even a small amount of contaminating tissue can compromise an experiment. A possible way to address this problem is to carefully punch out specific nuclei from brain slices. If the procedure requires considerable time before the samples can be frozen, some RNA degradation may occur, but utilizing frozen sections (Babity et al., 1997) can minimize this problem. It is our impression, however, that the DD method is still compatible with a small degree of RNA degradation, as long as it occurs uniformly across the samples compared. The reason is that DD PCR fragments are typically short (mostly between 150 and 600 bp) and tend to represent the 3′ end of the respective mRNAs.

As an example, the song nucleus HVC (high vocal center) and adjacent shelf region can be identified in fresh unstained sections generated with a tissue chopper (200–300 μm thick), viewed under dark-field illumination with a dissecting microscope. The time required for preparing the slices, and then punching out and freezing HVC samples can be as long as 20 min after sacrifice. RNAs isolated from such samples have been used successfully to identify mRNAs with high expression in HVC (Denissenko et al., 1995), even though their integrity was not directly controlled for on agarose gels due to the little amount of material available.

The single step method of Chomczynski and Sacchi (1987) yields good preparations of total RNA with very low DNA and protein contamination, as long as the two following points are observed: (a) The recommended proportion of denaturing solution (solution D, containing guanidinium isothiocyanate) to tissue weight should be followed closely; and (b) Critical attention must be paid to the separation of the aqueous and organic phases. As there is only one phenol extraction step in the protocol, remaining contaminants from the interface may contain significant amounts of RNase, which would result in RNA degradation, or phenol, which would partially inhibit the RT reaction. The best approach to minimize such contaminants is to leave behind a portion of the aqueous phase during phase separation. The TRIzol method (GIBCO) offers a convenient version of the protocol, and is equally effective, as long as the guidlines discussed above are observed.

DNase treatment

In case contaminant DNA is present in the RNA preparations, genomic sequences recognized by the oligo-dT and/or AP primers will be amplified during the PCR reaction. DNA contamination can be minimized by using fresh and carefully prepared reagents and following closely the instructions for RNA isolation (see above). If persistent, the DNA contamination can be removed by DNase-treating the RNA samples. In our experience, however, a small degree of DNA contamination that is comparable across the samples usually does not interfere in a major way with the DD procedure, and usually appears as non-regulated PCR fragments in the sequencing gel. Most importantly, when RNA yields are small it may be desirable to avoid DNase treatment altogether, to avoid extra loss of RNA. To test for the effect caused by DNA contamination, one can run control reverse transcription reactions where the reverse transcriptase is omitted and only contaminant DNA is amplified in the subsequent PCR step. Running one such control per RNA sample is usually enough for a general assessment of DNA contamination.

Reverse transcription

In this step, cDNAs are generated from the RNA samples to be compared using an oligo-dT primer. One-base anchored primers (such as $T_{12}C$, $T_{12}A$ or $T_{12}G$) will each prime about 1/3 of the mRNA population (Liang and Pardee, 1992, 1997b). One can choose to decrease the portion of the mRNA population to be transcribed and the complexity of the patterns on sequencing gels by using two-base anchored primers like $T_{12}GC$ or $T_{12}CG$, each of which will prime about 1/12 of the mRNA population. There seems to be no reason for using two-base anchored primers with degeneracy at the penultimate base (for instance, $T_{12}MC$) as originally used (see discussion in Liang and Pardee, 1992, 1997b). It is also possible to use primers linked to restriction sites to facilitate subsequent cloning of the PCR products (Liang and Pardee, 1997b), although presently available kits provide very satisfactory solutions for PCR cloning (see below).

Using an excessive amount of RNA (>200 ng) for reverse transcription is undesirable, as inclusion of high amounts of cDNA in the subsequent PCR reaction often results in smearing of the final products (Mello et al., 1997; Denissenko, N., unpublished observations). Several reverse transcriptases are equally effective at this step; we prefer to use Superscript II (Gibco/BRL), as it maximizes transcript size due to the absence of exonuclease activity.

Inclusion of RNase inhibitor helps prevent RNA degradation from possible contaminant RNase activity. The RNAs should be denatured immediately before being added to the reaction, to help maximize the effectiveness of the RT step. Finally, we find that the PCR reactions tend to work best when fresh cDNA preparations are used; we therefore avoid prolonged storage of the cDNAs before performing the PCR amplification.

PCR amplification

The cDNAs obtained in the previous step are amplified using the same oligo-dT primer plus another primer of arbitrary sequence (AP). The AP primers are typically of small size (10mers), and the PCR reaction is performed at low stringency. Under the conditions that are normally utilized (annealing at 40°C) the primers hybridize approximately as 7mers (Liang and Pardee, 1992). The number of resulting PCR fragments is determined by the frequency at which target sequences for a particular AP primer occur along the length of the cDNAs being amplified (Liang and Pardee, 1992, 1997b). We currently favor the use of ^{33}P (dATP) as the labeling isotope, as it yields a higher signal and is less volatile than ^{35}S, reducing the risk of contamination (Trentmann et al., 1995). We recommend that all reaction tubes for a particular experiment be loaded in a thermocycler already equilibrated at the temperature for the denaturing step (hot start), to help reduce initial mispriming.

Selecting, eluting and reamplifying PCR fragments

In order to minimize the likelihood of false positives due to variability and other factors discussed above, one should use stringent criteria for selecting candidate differentially expressed bands. Preferably, these bands will be present in all or most reactions belonging to a particular group or sample. We recommend as a criterion the presence (or absence) of the band in at least 2 of triplicate reactions in one group, and in none of the reactions from the other group(s) (Mello et al., 1997). Before concluding that a specific band is absent from a particular lane, it is advisable to analyze closely that lane in comparison with other lanes for the presence of bands above and below the band in question. In case several or all PCR products from one reaction have significantly lower intensity than the products in other lanes, it may be safer to discard that reaction as non-informative. It is often rather difficult to detect differences on the sequencing gels, as DD bands do not always appear sharp. Besides the usual care involved in running such gels (as detailed in Sambrook et al., 1989), it is sometimes useful to optimize the parameters of the previous steps so as to minimize mispriming and band smearing.

The range of sizes of PCR fragments obtained in DD experiments is determined by the range of fragments typically resolved in sequencing gels (Sambrook et al., 1989). The upper limit is commonly about 700 bp, although some systems (such as Genomyx's sequencing apparatus) have a higher range of resolution and can separate bands above 1 kb in length. It is advisable to focus on fragments that are at least 150–200 bp long, as it is difficult to detect signal for smaller fragments during the subsequent confirmatory steps, even when these are optimized (Mello, C.V., Jarvis, E.D., Denissenko, N. and Rivas, M., unpublished observations).

Great care should be taken in aligning the sequencing gel and the respective X-ray film autoradiogram. After cutting out the selected PCR fragments, the gel should be re-exposed to confirm that the selected region was indeed cut out of the gel, with minimal contamination from adjacent bands. It is also important to store the gel clear of dust and dirt, and preferably extended within a cassette or another flat device to prevent its folding onto itself, if further selection of bands is to be performed later in the same gel.

One should pay close attention to the sizes of the fragments resulting from reamplification. A particular fragment will sometimes yield two bands upon reamplification. A second band of lower size than expected usually signifies the presence of an additional internal binding site for one of the primers. If a band is not detected upon reamplification, one can proceed through a second round of reamplification, whereupon the vast majority of PCR fragments are usually detected.

Cloning DD fragments

At this point, it is highly advisable to clone the bands obtained before proceeding with further examination of differential expression and sequence analysis. This is undoubtedly a most critical step in determining the successful outcome of DD experiments. The main reason is that the fragments resulting from PCR reamplification are often heterogeneous, due to one of the following factors: (a) Contamination with adjacent bands: this occurs while cutting out the selected fragment from the DD gel (see, for instance Fig. 2 in Mello et al., 1997); (b) Intrinsic heterogeneity of reamplified bands: this occurs if two or more distinct cDNAs generate a similar sized PCR fragment using a particular combination of primers. Also, cloning PCR fragments offers the practical advantage of providing a large supply of the product for expression and sequence analysis. In some cases, non-cloned PCR products have been successfully utilized for confirmatory steps with in situ hybridization (Berke et al., 1998). One should be aware, though, that only a fraction of differential PCR fragments are likely to be detected by this approach, while differential fragments that are contaminated with a non-regulated band are likely to be discarded as false positives.

It is desirable to use an efficient cloning strategy that guarantees a large number of independent transformants for each reamplified PCR fragment. In our hands, the TA- (InVitrogen) and pCR-Script (Stratagene) cloning kits, particularly the latter, yield very good results while allowing for the generation of riboprobes for Northern and in situ hybridization analysis (see below). Due to the heterogeneity issue mentioned above, at least 4–6 colonies representing independent transformants should be picked and analyzed individually for each cloned PCR fragment.

Confirming differential expression

The method of choice for confirmation of differential expression of the cloned bands will depend on factors such as the number of clones to be analyzed, the availability of RNA for Northern blots and/or for generating cDNA probes, and the abundance of the corresponding message in the tissue of interest. Because of the need to isolate several independent transformants after cloning the PCR fragments, it is typical for DD experiments to rapidly generate a large number of clones for testing. Whenever possible, our strategy has been to start with a method such as reverse Northern (Shoham et al., 1996; Zhang et al., 1997), which allows the simultaneous testing of a large number of colonies, and later conduct further analysis of selected colonies with more involved procedures.

Reverse Northern

A practical solution for testing a large number of transformant colonies for differential expression is to blot the respective plasmid DNAs in replicate filters, hybridize each replica with a radioactively-labeled cDNA probe derived from one of the initial RNA samples, and compare the individual colonies for differential signal intensity across the replicate filters. Some variants of this technique avoid the need for plasmid DNA minipreps. For instance, one can pick individual colonies, grow them in the appropriate medium, and then blot (in a dot blot apparatus) equal volumes of the bacterial suspension directly onto the filters, which are then denatured according to standard protocols for colony screening (Sambrook et al., 1989). This is a simple and rapid procedure, but its sensitivity is limited by the amount of plasmid DNA that can bind to the filter under such conditions. An interesting alternative is to PCR amplify the cloned inserts directly from colony lysates, using vector-specific primers that flank the cloned insert. This may be preferable, as there is no need to grow the colonies in liquid medium, and only insert sequences are PCR amplified and blotted onto the filters, increasing the chances of detection of low level signals.

When preparing the cDNA probes, it is desirable to use the same oligo-dT primer as in the reverse transcription step, in order to enrich the probes for the relevant sequences. We use standard protocols for DNA hybridization to nylon filters (Sambrook et al., 1989); as the optimal conditions may vary with different clones, it is useful to wash and expose the filters at increasingly higher stringency. Typically, only a fraction of the blotted colonies (those corresponding to messages of medium to high abundance) give detectable signal with this procedure. Clones that show differential expression are further analyzed using Northern blots or in situ hybridization, while those that do not are discarded (false positives). Clones that show no signal may represent low-abundance messages and need to be analyzed by a more sensitive method.

The approach above is only applicable if enough RNA is available from the starting brain samples. A possible alternative that circumvents this problem is to hybridize the replicate blots with the original radioactively labeled differential display reactions (Vögeli-Lange et al., 1997). Another approach is to proceed directly with a method such as in situ hybridization. In that case, however, it is important to first establish the identity of the transformants derived from each cloned fragment, so as to eliminate redundancy and minimize the number of colonies to be tested. This can be done by: (a) digesting the plasmid DNAs with a frequent cutter restriction enzyme and comparing the resulting fragments on an agarose gel; or (b) generating radioactively-labeled DNA probes from one of the transformant colonies obtained for a particular cloned fragment and hybridizing it to a Southern blot containing plasmid DNA from the other transformants. Based on either of the procedures above (for protocols see Sambrook et al., 1989), one can determine whether all the transformants obtained for each candidate PCR fragment contain the same insert, or whether they can be divided into subgroups, indicating that the cloned fragment was heterogeneous. In that case, one representative colony from each subgroup is picked for the subsequent confirmatory steps.

Northern blots

This is a more sensitive technique than reverse Northern for detecting messages of low to medium abundance. It also confirms probe specificity and allows the determination of size and relative abundance of specific mRNAs. However, the method is not practical as a confirmatory step for a very large number of clones, and it is not feasible when the RNA amount is limiting. To maximize the chances of detecting signal on Northern blots, we recommend the use of either riboprobes, or single-stranded antisense DNA probes generated with the specific one- or two-base anchored oligo-dT primer used for reverse transcription, rather than random primed probes. Such probes increase signal detection on Northern blots (and in situs; see below) dramatically, and minimize the number of DD clones that yield no signal (Mello et al., 1997). When generating riboprobes from the TA vector (InVitrogen), one must use a high concentration SP6 RNA polymerase (Stratagene), as that vector has a weak SP6 promoter site.

In situ hybridization

This is our method of choice for confirming differential expression of candidate differential DD clones. Although more labor intensive, in situ hybridization can detect expression within a small population of cells, or within single cells, particularly with the use of radioactively labeled riboprobes. As the conditions for optimal hybridization vary with the probe, information from Northern analysis, when available, provides a convenient guide for determining the stringency of hybridization and washes. Otherwise, one needs to conduct systematic tests for stringency, utilizing sense strand probes as controls (Mello et al., 1997).

Sequence analysis

We feel that it is only worthwhile to invest time and effort in sequencing and performing database searches for those clones whose differential expression has been confirmed. There is a bias in DD toward generating 3′ fragments, which often do not contain a segment of the Open Reading Frame (ORF). Database searches can therefore be non-informative, especially for work done on species that are not commonly represented in the databases, or for unknown splice variants. In such cases, it will be necessary to screen a cDNA library for isolation of larger clones and identification of ORFs.

Evaluating DD and our strategy

DD was originally devised and tested for comparing different homogeneous cell lines (Liang and Pardee, 1992). A growing body of evidence has demonstrated the applicability of the method to in vivo studies as well as heterogeneous tissues (several chapters in Liang and Pardee, 1997a). Regarding a complex organ such as the brain, DD has been particularly successful for identifying large or qualitative differences in gene expression. Some examples, among numerous others, include comparisons of different brain areas (Denissenko et al., 1995), genetic strains (Qu et al., 1996), ages and/or developmental stages (Wu et al., 1997), circadian times (Green and Besharse, 1996), or the investigation of pathological states (Wang et al., 1995; Babity et al., 1997; Tsuda et al., 1997). At present, due to the non-linearity of the PCR reaction, the method seems less well suited for detecting more subtle quantitative differences in mRNA levels. Indeed, we have found it rather difficult to detect quantitative changes in gene expression associated with perceptual states (Mello, C.V., and Jarvis, E.D., unpublished observations). While this is not necessarily an absolute limitation, further developments are still necessary to expand the range of applicability.

As discussed in the introduction, DD offers several potential advantages over other screening methods. However, the confirmatory steps for the candidate PCR fragments can be very labor intensive. This problem is acute when working with small-sized fragments, as signal detection is difficult and determination of identity often depends on screening a cDNA library for larger clones. In addition, neglecting the issues we discussed above, particularly concerning variability and the presence of "false positives", may result in failed attempts to fully implement the method. We believe that our overall strategy represents a useful approach to the problems that are typically encountered by neurobiologists, but alternative strategies may also be effective (Martin et al., 1997). Lastly, to our knowledge there has only been one direct comparison of efficacy between DD and subtractive hybridization (Wan et al., 1997). The main conclusion of that study is that both techniques are useful and may be complementary if used in conjunction. Given the points discussed in the introduction regarding tissue availability and flexibility of the method, DD will on occasion be the preferable approach.

References

Babity, J.M., Newton, R.A., Guido, M.E. and Robertson, H.A. (1997) The application of differential display to the brain. In: P. Liang and A.B. Pardee (Eds.), *Methods in Molecular Biology*, Vol. 85: Differential Display Method and Protocols, Humana Press Inc., Totoya, NJ, pp. 285–295.

Berke, J.D., Paletzki, R.F., Aronson, G.J., Hyman, S.E. and Gerfen, C.R. (1998) A complex program of striatal gene expression revealed by dopaminergic stimulation. *J. Neurosci.*, 18: 5301–5310.

Campeau, S., Hayward, M.D., Hope, B.T., Rosen, J.B., Nestler, E.J. and Davis, M. (1991) Induction of the c-fos proto-oncogene in rat amygdala during unconditioned and conditioned fear. *Brain Res.*, 565: 349–352.

Chomczynski, P. and Sacchi, N. (1987) Single-step method of RNA isolation by acid guanidinium thiocyanate:phenol:chloroform extraction. *Anal. Biochem.*, 162: 156–159.

Denissenko, N., Nottebohm, F. and Mello, C. (1995) PCR-based mRNA differential display reveals enrichment of aldehyde dehydrogenase in the high vocal center and in two other nuclei of the song system of songbirds. *Soc. Neurosci. Abstr.*, 21(2): 259.

Green, C.B. and Besharse, J.C. (1996) Use of a high stringency differential display screen for identification of retinal mRNAs that are regulated by a circadian clock. *Brain Res. Mol. Brain Res.*, 37: 157–165.

Jarvis, E. and Nottebohm, F. (1997) Motor-driven gene expression. *Proc. Natl. Acad. Sci. USA*, 94: 4097–4102.

Jarvis, E.D., Mello, C.V. and Nottebohm, F. (1995) Associative learning and stimulus novelty influence the song-induced expression of an immediate early gene in the canary forebrain. *Learning and Memory*, 2: 62–80.

Jarvis, E.D., Schwabl, H., Ribeiro, S. and Mello, C.V. (1997) Brain gene regulation by territorial singing behavior in freely ranging songbirds. *Neuroreport*, 8: 2073–2077.

Liang, P. and Pardee, A.B. (1992) Differential display of eukaryotic messenger RNA by means of the polymerase chain reaction. *Science*, 257: 967–971.

Liang, P and Pardee, A.B. (1997a) *Methods in Molecular Biology*, Vol. 85: Differential Display Method and Protocols, Humana Press Inc., Totoya, NJ.

Liang, P. and Pardee, A.B. (1997b) Differential display: a general protocol. In: P. Liang and A.B. Pardee (Eds.), *Methods in Molecular Biology*, Vol. 85: Differential Display Method and Protocols, Humana Press Inc., Totoya, NJ, pp. 311.

Liang, P. Averboukh, L. and Pardee, A.B. (1993) Distribution and cloning of eukaryotic mRNAs by means of differential display: refinements and optimization. *Nucleic Acid Res.*, 21: 3269–3275.

Martin, K.J., Kwan, C.-P. and Sager, R. (1997) A direct-sequencing-based strategy for identifying and cloning cDNAs from differential display gels. In: P. Liang and A.B. Pardee (Eds.), *Methods in Molecular Biology*, Vol. 85: Differential Display Method and Protocols, Humana Press Inc., Totoya, NJ, pp. 77–85.

Mello, C.V., Vicario, D.S. and Clayton, D.F. (1992) Song presentation induces gene expression in the songbird forebrain. *Proc. Natl. Acad. Sci USA*, 89: 6818–6822.

Mello, C.V., Nottebohm, F. and Clayton, D. (1995) Repeated exposure to one song leads to a rapid and persistent decline in an immediate early gene's response to that song in the zebra finch telencephalon. *J. Neurosci.*, 15: 6919–6925.

Mello, C.V., Jarvis, E.D., Denissenko, N. and Rivas, M. (1997) Isolation of song-regulated genes in the brain of songbirds. In: P. Liang and A.B. Pardee (Eds.), *Methods in Molecular Biology*, Vol. 85: Differential Display Method and Protocols, Humana Press Inc., Totoya, NJ, pp. 205–217.

Milner, R.J. and Sutcliffe, J.G. (1983) Gene expression in the brain. *Nucleic Acid Res.*, 11: 5497–5520.

Qu, D., Ludwig, D.S., Gammeltoft, S., Piper, M., Pelleymounter, M.A., Cullen, M.J., Mathes, W.F., Przypek, J., Kamarek, R. and Maratos-Flier, E. (1996) A role for melanin-concentrating hormone in the central regulation of feeding behavior. *Nature*, 380: 243–247.

Sambrook, J. Fritsch, E.F. and Maniatis, T. (1989) *Molecular Cloning: A Laboratory Manual*, 2nd Edn, Cold Spring Harbor Laboratory, Cold Spring Harbor, NY.

Shoham, N.G. Arad, T., Rosin-Abersfeld, R., Mashiah, P., Gazit, A. and Yaniv, A. (1996) Differential display assay and analysis. *BioTechniques*, 20: 182–184.

Trentmann, S.M., Knaap, P., Kende, H., Liang, P. and Pardee, A.B. (1995) Alternatives to ^{35}S as a label for the differential display of eukaryotic messenger RNA. *Science*, 267: 1186–1187.

Tsuda, M., Imaizumi, K., Katayama, T., Kitagawa, K., Wanaka, A., Tohyama, M. and Takagi, T. (1997) Expression of zinc transporter gene, ZnT-1, is induced after transient forebrain ischemia in the gerbil. *J. Neurosci.*, 17: 6678–6684.

Vögeli-Lange, R., Bürckert, N., Boller, T. and Wiemken, A. (1997) Screening for positive clones generated by differential display. In: P. Liang and A.B. Pardee (Eds.), *Methods in Molecular Biology*, Vol. 85: Differential Display Method and Protocols, Humana Press Inc., Totoya, NJ, pp. 95–103.

Wan, J.S., and Erlander, M.G. (1997) Cloning differentially expressed genes by using differential display and subtractive hybridization. In: P. Liang and A.B. Pardee (Eds.), *Methods in Molecular Biology*, Vol. 85: Differential Display Method and Protocols, Humana Press Inc., Totoya, NJ, pp. 45–68.

Wang, X., Yue, T.-L., Barone, F., White, R.F., Clark, R.K., Willeth, R.N., Sulpizio, A.C., Aiyar, N.V., Ruffolo Jr, R.R. and Feuerstein, G.Z. (1995) Discovery of adrenomedullin in rat ischemic cortex and evidence for its role in exacerbating focal brain ischemic damage. *Proc. Natl. Acad. Sci. USA*, 92: 11480–11484.

Wu, H.C. and Lee, E.H. (1997) Identification of a rat brain gene associated with aging by PCR differential display method. *J. Mol. Neurosci.*, 8: 13–18.

Zhang, H., Zhang, R. and Liang, P. (1997) Differential screening of differential display cDNA products by reverse Northern. In: P. Liang and A.B. Pardee (Eds.), *Methods in Molecular Biology*, Vol. 85: Differential Display Method and Protocols, Humana Press Inc., Totoya, NJ, pp. 67–93.

W.E. Crusio and R.T. Gerlai (Eds.)
Handbook of Molecular-Genetic Techniques for Brain and
Behavior Research (Techniques in the Behavioral and Neural Sciences, Vol. 13)

CHAPTER 2.2.3

Brain region-specific genes: the hippocampus

Benjamin S. Pickard, Ben J. Davies, Ken A. Rose, Genevieve Stapleton, Muriel Steel and Richard Lathe*

Centre for Genome Research and Centre for Neuroscience, University of Edinburgh, King's Buildings, West Mains Road, Edinburgh EH9 3JQ, UK

How does the brain store information? Such a question presents a formidable challenge, not only because of the undoubted complexity of the brain, but also because we are limited in the analytical techniques we can bring to bear. At one level our understanding of what individual brain regions might do derives from selective lesion studies. For instance, in patients, lesions to the hippocampus lead to a pronounced failure of new declarative or explicit memory, while other types of memory, and recall of events well prior to damage, are substantially unaffected (Scoville and Milner, 1957). In rodents, where the water maze affords a measure of spatial memory (Morris et al., 1986a), that shares common aspects with declarative memory (Cohen and Eichenbaum, 1993), ability to perform this navigation task is also abolished by hippocampal lesions (Morris et al., 1982).

At another level, identifying the role a brain region fulfils is only a first step towards understanding how it might fulfil that role. Neurons intercommunicate chemically and electrically, but what aspects of this intercommunication might underlie memory processes? The celebrated hypothesis put forward by Hebb, on purely theoretical grounds, held that if one neuron regularly communicates with another, then information could be stored by strengthening the synaptic connection between them. The discovery that hippocampal synapses display just such a form of synaptic plasticity, known as long-term potentiation (LTP; Bliss and Lømo, 1973), raises the important question of whether LTP might represent the molecular and cellular embodiment of certain types of memory. The answer will surely not be simple, but in favor of this contention Morris et al. (1986b) were able to show that blockade of NMDA receptors prevents LTP and also impairs performance of rats in the watermaze.

Such pharmacologic studies, as well as the lesion studies mentioned earlier, lack complete specificity. Surgical or neurotoxic lesions can damage brain regions adjacent to the target structure, while pharmacologic agents are seldom if ever fully specific in their activities. More recent investigations have brought the powerful technique of transgenesis to bear on the problem. By homologous recombination in cultured embryonal cell lines it is possible to disrupt or modify the activity of a single gene within the mouse genome. Using this highly selective procedure it was found that animals deficient in a subunit of the NMDA receptor show impairments in both LTP and spatial learning (Sakimura et al., 1995). Although mice lacking the core NMDA-R1 subunit of the

*Author for correspondence.

receptor (without which no NMDA receptor function is observed) fail to survive (Li et al., 1994), their specific deficits argue that this receptor is also necessary for the developmental formation of synaptic contacts.

Synaptic LTP in region CA1 is initiated by an ingress of Ca^{2+} through NMDA receptors (Bliss and Collingridge, 1993; Roberson et al., 1996); others have therefore focussed on generating mutations affecting calcium-dependent signalling molecules acting downstream of the NMDA receptor. Studies on a variety of mutant mice, including those lacking the α subunit of calcium/calmodulin protein kinase type II (CaMKII) (Silva et al., 1992a, 1992b) or a brain isoform of protein kinase C (PKC-γ) (Abeliovich et al., 1993a, 1993b), revealed abnormalities of LTP and/or learning (reviewed by Chen and Tonegawa, 1997).

Despite the elegance of these transgenic experiments, the anatomic and temporal non-specificity of the genetic lesions complicates interpretation. The gene disruption is present in all tissues and at all stages of development. Thus, neither the impairments of memory nor those of LTP can be ascribed uniquely to loss of gene function in the hippocampal formation in the adult animal at the time of testing (Lathe and Morris, 1994; Mayford et al., 1995; Lathe, 1996). Behavioral and electrophysiological deficits could also be due to alterations in developmental processes, or indeed to dysfunction of other body or brain systems.

To address some of these problems efforts have been made to exploit inducible transgene expression in brain (Mansuy et al., 1998) or a line of transgenic mice in which expression of the site-specific recombinase CRE was serendipitously restricted to the hippocampus (Tsien et al., 1996a).

This latter approach generated mice in which a target gene (NMDA-R1 subunit) that had been flanked by CRE recognition sequences was deleted only in hippocampus (Tsien et al., 1996b). Nonetheless, in such an approach the site of integration of the transgene expressing CRE is not predetermined, a large number of different transgenic lines will need to be screened for each construct; and with no assurance of reproducing the region-specificity. Importantly, the integration of multiple copies, as usually occurs, leads to unreliable transgene expression (Dobie et al., 1997).

To avoid these difficulties we have sought genes expressed selectively in hippocampus and subregions thereof. Not only would the identity of genes specifically expressed in this brain region provide new insights into hippocampal function but, more importantly, such genes would furnish essential tools for the specific analysis of the role of the hippocampus in learning and memory — they could be used to target reproducibly the expression of probe genes (exemplified by but not restricted to CRE) to hippocampal regions. That such region-specific genes might exist was was encouraged by the previous description of restricted gene expression in other brain regions, for instance of tyrosine hydroxylase in the locus coeruleus and substantia nigra (Berod et al., 1987).

Here we discuss four approaches to the identification of genes and transcripts whose expression is restricted to the hippocampus, and the advantages and drawbacks of the techniques employed. Unexpectedly, each approach highlighted a different and non-overlapping subset of genes, but the expression of at least some of these was largely if not exclusively restricted to the hippocampus.

Differential hybridization

The first and most basic technique we attempted comprised the screening a cDNA library from adult rat hippocampus with probes derived from hippocampus (HPC) RNA or from material prepared from brain from which the hippocampus had been surgically removed ('rest of brain' — ROB). A cDNA library from adult rat HPC (10^6 clones) was prepared in lambda-ZAPII, and plaque-lifts were screened with HPC or ROB cDNA probes (Stapleton et al., 1995). From 10^6 primary cDNA clones, 361 hybridized signifi-

TABLE 1

Differential hybridization

Clone	Transcript sizes	Identity	Expression HPC	ROB
3.22a	4.8	novel	++	(+/−)
3.28a	nd[a]	PDGF receptor[a]	nd[a]	
3.3a	2.4	novel	++	++
4.5a		mitochondrial gene		
4.5b	1.6	novel	++	++
6.19d		β-adaptin		
9.9a		calreticulin		
11.2a	2.9	mouse ECA39	++	+
12.10a	7[b], 3.9, 3.5, 2.7, 1.9	novel; homology to human fetal liver tropomodulin	++[b]	(+/−)
13.9c		ribosomal protein S13		
14.4a		calmodulin RCM3		
14.5a	5.0, 2.0, 1.8	novel, dubbed Cyp7b	++	(+/−)
14.15a		rat amyloidogenic glycoprotein (rAG)		
14.15b	2.25	novel	++	++
15.13a	4.4[c], 3.2, 2.5	novel	++[c]	(+)
16.26b		SNAP-25		
16.26c	1.8	novel	++	(+/−)

[a] For known genes the transcript sizes and expression patterns were not determined (nd);
[b] The 7.0 kb transcript of 12.10a was specific to the hippocampus;
[c] The 4.4 kb transcript corresponding to 15.13a was specific to the hippocampus.

cantly better to the HPC probe than the ROB probe. Of these, approximately 75% gave no significant signal with the ROB probe while 25% gave a weak signal on rescreening. A representative sample of 41 differentially hybridizing plaques was further analyzed by DNA sequencing. For reasons we do not understand, 11 contained repetitive elements. Why such sequences are particularly abundant in hippocampus is not known, but being inappropriate for transgenic experimentation were not studied further. Of the remaining, 9 were identified in the database while 21 were novel. Analysis of 17 selected clones is presented in Table 1.

Many identify transcripts expressed abundantly in rat hippocampus. Rat amyloidogenic glycoprotein (rAG) is the ortholog of the human amyloid precursor protein (APP) associated with the pathophysiology of Alzheimer's disease. The adult expression pattern of rAG mRNA is widespread, with particularly robust expression in hippocampus (Shivers et al., 1988). Transcripts for SNAP-25 (synaptosomal-associated protein) are particularly enriched in hippocampal mossy fibres and the molecular layer of the dentate gyrus as well as in several other brain regions (neocortex, piriform cortex, anterior thalamic nuclei, pontine nuclei, and cerebellar granule cells; Oyler et al., 1989). Other clones identified, however, did not show marked evidence of enriched expression in the hippocampus. These include β-adaptin, the receptor for platelet derived growth factor (PDGF), calmodulin (CaM), and RCM3. We have no explanation why these clones gave clear differential signals on primary and secondary screens, but are not reported to be enriched in hippocampus.

We focussed our attention instead on the novel cDNA species revealed by the differential screen. Northern analysis was performed using RNA from a number of brain regions. Of the clones analysed, several hybridized preferentially to hippocampal RNA, or identified transcript of sizes present only in hippocampus (Fig. 1). Expression of Clone 15.13a identified a number of transcripts of which one, at 4.4 kb, appeared exclusively in hippocampus (see Fig. 1) while clone 12.10a revealed a hippocampal- specific transcript at 7 kb (not shown). We have not established whether these transcripts are due to an alternative promoter that is only active in hippocampus, or is instead due to hippocampal-specific mRNA processing.

mRNA identified by the clone 14.5a was highly enriched in hippocampus; further analysis of this clone and further cDNAs corresponding to it revealed that 14.5a encodes a novel cytochrome P450 of the steroidogenic type, dubbed Cyp7b (Stapleton et al., 1995). Cyp7b expression in mouse and rat was confirmed by Northern blotting to be very highly enriched in the hippocampus, but was not exclusive to the formation (Fig. 1; Stapleton et al., 1995). In situ hybridization using full-length cDNA probles to both mouse and rat brain sections has now shown that the hippocampus is the predominant site of brain expression (unpublished data, see Fig. 3).

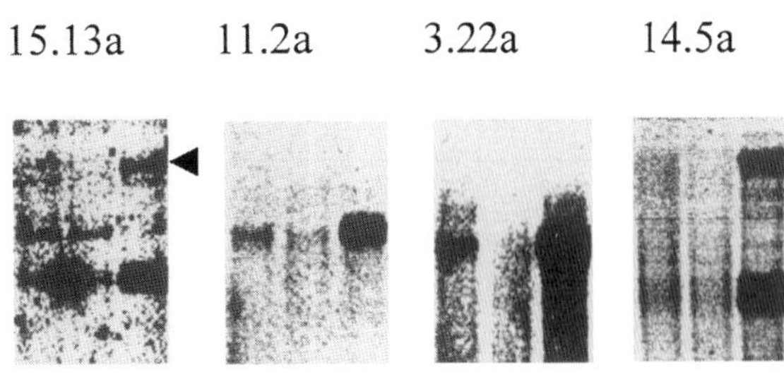

Fig. 1. Northern analysis of clones identified by differential hybridization screening. RNA samples were: Br, brain; Rb, rest of brain (ROB), Hi, hippocampus (HPC). 15.13a identifies several transcripts including one specific to hippocampus (4.4 kb, arrowed).

Subtractive hybridization

Hybridization can be used to remove sequences common to two starting preparations, so enriching for sequences only present in one or other pool. In the first technique we employed, chemical cross-linking subtraction (Hampson et al., 1992), difficulties were encountered: none of the 'enriched' clones demonstrated specificity for the hippocampus. The second technique tried, the multiple hybridization selection and reamplification technique of Wang and Brown (1991), was more successful. HPC and ROB mRNA preparations from adult rat brain were converted to double-stranded cDNA; each cDNA pool was divided in two, the aliquots were digested with frequent cutting restriction enzymes, and repooled. Oligonucleotide linkers were ligated to the cDNA fragments from each separate pool, permitting PCR amplification. The subtraction process was then carried out in tandem, one to identify HPC-specific cDNAs and the other to identify ROB-specific cDNAs. At each stage cDNA from one pool (the 'tracer') was hybridized with an excess of cDNA from the other pool (the 'driver') that had been biotinylated (using photobiotin in the presence of light); excess driver together with the tracer/driver hybrids was removed by streptavidin binding in conjunction with phenol-chloroform extractions, leaving unhybridized (subtracted) tracer cDNAs in the aqueous phase. Several rounds of subtraction were carried out (three sets of long and short hybridizations that favor the removal of cDNA species common to both pools and present at relatively high abundance, such species are usually refractory to subtraction because of their quantity; Wang and Brown, 1991). After each hybridization and subtraction the tracer pools were PCR amplified.

Initial experiments suffered from cross-contamination of driver and tracer pools, and a modification was made (see Balzer and Baumlein, 1994) in which the different cDNA pools are

TABLE 2

Subtractive hybridization

Clone	Identity	cDNA Southern		Northern	
		HPC	ROB	HPC	ROB
H4M6,22	novel	++	(+)	nd	
H4M10	hippocalcin	+++	−	+++	+/−
H4M19	novel	+	(+)	nd	
H4M21,25	novel	+++	(+)	nd	
H4M27	Na/H exchange protein 1	nd		nd	
H4M28,29	novel	+	(+)	++	++
H4M33	5-HT1a receptor	nd		nd	
H4M35	novel	++	−	++	++
H4M37	hEST T04934	++	(+)	nd	

nd, not determined.

made using different linker/primer sets, preventing coamplification of driver and tracer. This protocol was applied to four rounds of subtraction of hippocampal (HPC) vs 'rest of brain' (ROB) cDNA; the products were cloned into a single-stranded M13 bacteriophage vector taking advantage of restriction sites in the PCR primers.

Duplicate plaque lifts were screened for hybridization to the fourth round subtracted HPC cDNA (H+4) and the primary ROB cDNA pools. Eighteen individual clones appeared to hybridize the HPC-enriched material, but not to ROB cDNA; these were used as hybridization probes for Northern blots and for the original mixed cDNA pools resolved by agarose gel electrophoresis ('cDNA Southern' analysis).

This yielded contradictory results. According to cDNA Southern data, many clones represented transcripts which were very highly enriched in the hippocampus and, in at least two cases, appeared to be almost specific to this region. Northern analysis, however, for the most part contradicted these results (Table 2), with one exception. Clone H4M10, identified as hippocalcin, appeared relatively specific to the hippocampus, as reported (Kobayashi et al, 1992). cDNA Southern analysis concurred with the hippocampal specificity, but this was very much less pronounced when the clone was used to probe Northern blots of mRNA from different brain regions; expression was also seen in lesser amounts in the cortex (Fig. 2), confirming the results of Saitoh et al. (1993).

Our results demonstrate that the subtraction process is sufficiently thorough to enrich for sequences specific to or enriched in one starting pool. However, the protocols are not without their problems. First, specific cDNA species emerged during successive rounds of subtraction, and gave discrete bands on agarose gel electrophoresis. These appear to be due to cDNA species with extensive secondary structure such that they refold upon themselves, so avoiding removal by hybridization with the driver (Pickard, 1996). Second, in cases where the starting pools are not wholly representative of the cDNA populations, particular sequences will be selected for that, purely by chance, were present in one starting pool but not the other. Furthermore, any gene- or sequence-specific alterations that might skew the representation of a given sequence in the starting cDNA populations (due, for instance, to differential splicing, polyadenylation, hybridization to other endogenous RNAs or

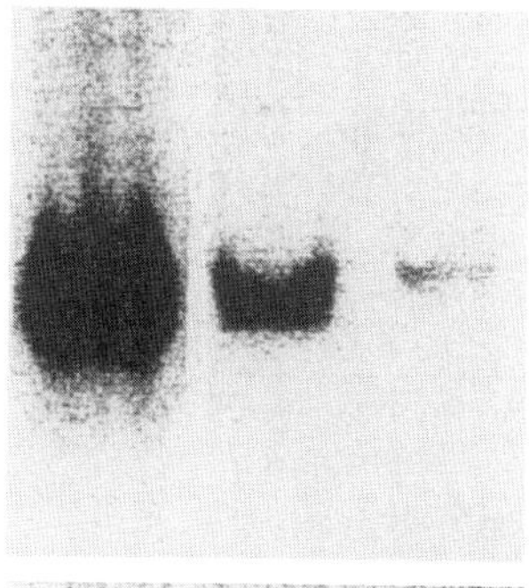

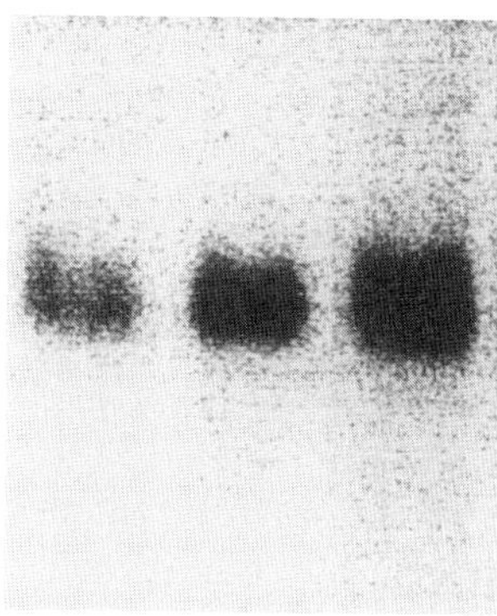

Fig. 2. Northern analysis of clone H4M10 identified by subtractive hybridization. RNA samples were Hi, hippocampus (HPC), Cx, cortex; Rb, rest of brain (ROB). The control probe corresponded to ribosomal protein S26, revealing overloading in two lanes that emphasises the hippocampal specificity of H4M10.

indeed covalent modification) will generate robust positives that are not confirmed by Northern hybridization (discussion).

Candidate gene approaches

Polymerase chain reaction (PCR) can address directly the spectrum of gene family members in a particular tissue, taking advantage of oligonucleotide primers corresponding to amino acid sequence motifs conserved within the family. Because amino acid sequence motifs may be encoded by a diversity of nucleotide sequences, a mixture of oligonucleotides (so-called 'degenerate' primers) representing all possible coding sequences is usually employed. Often, where any one of the bases A, C, T, or G could be used in a codon, the nucleotide inosine is specified to prevent excessive degeneracy (inosine can potentially hydrogen bond with all bases although it has a preference for G and A). Rat hippocampus cDNA was used as a template in PCR reactions designed to find members of five gene families which were selected for their apparent large sizes and potential roles in neuronal function (Table 3). PCR products were purified by agarose gel electrophoresis and cloned, by the 'TA overhang' method, into plasmid pCRII (Invitrogen).

G-protein-coupled receptors

This family comprises a diversity of cell-surface receptors characterized by the presence of 7 transmembrane domains and which, on ligand binding, alter cellular metabolism through the modulation of the function of heterotrimeric G (GTP binding) proteins. Using degenerate PCR based on conserved motifs we identified 9 previously described receptors (adenosine-a1 and -a3, adrenergic-α1b and -β2, dopamine-d1, cannabinoid, endothelin-1, rat testis GPCR, R334 receptor, and a candidate vasoactive intestinal peptide receptor). However, Northern blotting revealed that in no case was expression restricted to the hippocampus.

Metabotropic glutamate receptors

These represent a subclass of the G-protein coupled receptors. PCR amplification predominantly revealed mGluR3, though clones were obtained corresponding to mGluR4 amd mGluR5. Because mGluR3 was represented most highly in the products, ways of increasing the abundance of other members were sought. One method attempted relied on the fact that, of the clones obtained, only mGluR3 contained a TaqI restriction site within the PCR product sequence. The PCR products were cut with this enzyme

TABLE 3

Conserved motifs for candidate gene PCR amplification

	Sequence motif 1	Sequence motif 2
1. G protein-coupled receptors	LCVIALDRY	FILCWLPFF
	NLAVADL	FALCWLPYH
	S L EF	VI LQ
		FV N
		F
2. Metabotropic glutamate receptors	SGEELSY	TMYTTC
	S	
3. G protein a subunits	RLLLLG	DVGGQR
	K	
4. Protein tyrosine phosphatases	KC-QYWP	HCSAGAG
		I
	FWRMVIWE	
	I D	
	Q	
5. Serine proteases	VLTAAHC	GDSGGPLV

and then reamplified prior to cloning. Nevertheless, no new mGluRs were identified using this approach; none of the known mGluR subunits appears to be restricted to the hippocampus.

Heterotrimeric G-proteins: α subunits

These participate in downstream signalling from the receptor groups discussed above and cycle between active (GTP-bound) and inactive (GDP-bound) forms. PCR amplification of G α subunit sequences from rat HPC cDNA subunits α-q, -i1, -s, -olf, and -i3. The presence of the olfactory G protein α subunit was particularly surprising as this had been reported to be specifically expressed in the olfactory bulb (Zigman et al., 1993). However, even though a specifically expressed 3′ untranslated region exists, the coding region is very widely expressed in brain (not presented).

Protein tyrosine phosphatases (PTPs)

These enzymes are either membrane-associated or cytoplasmic polypeptides and modulate the activation of target proteins by tyrosine dephosphorylation. Nine different PTPs were identified — α, γ, δ, H1, zeta, P19, Tcell PTP, and two novel sequences, PTP-4 and PTP-24 (Pickard, 1996). Sequence analysis revealed substantial homology between PTP-4 and the large multidomain PTPs μ and κ (Gebbink et al., 1991; Jiang et al., 1993) while PTP-24 shows homology within the amplified catalytic domain to a number of PTPs, most convincingly with hPTP-β (Krueger et al., 1990). Northern blotting was performed for the two new species. PTP-4 demonstrates a major transcript size of approximately 5.5 kb, principally localized to brain but with peripheral expression in lung and testis. PTP-24 detects brain transcripts of approximately 6 kb, but either 4 kb or 5 kb in the other tissues examined. Overall,

14.5 a (Cyp7b) Kin (arg)

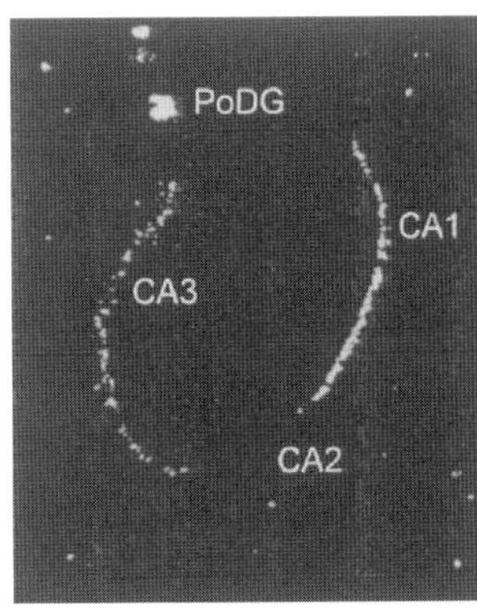

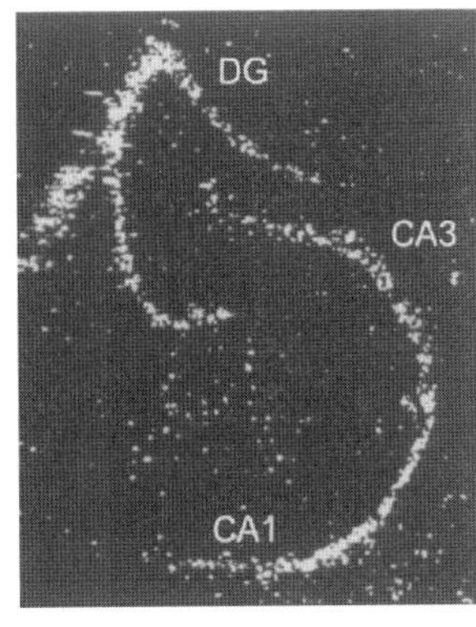

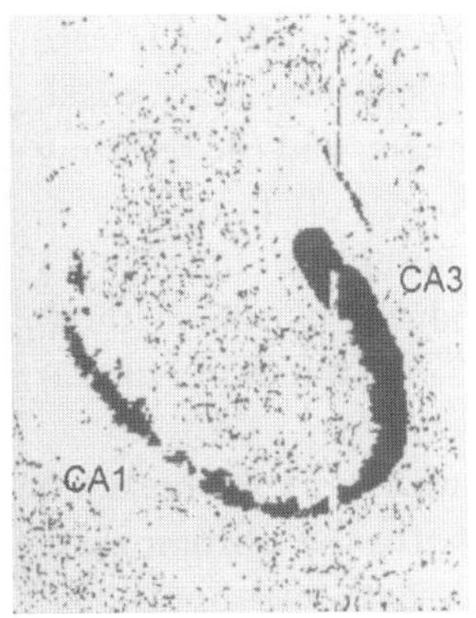

Fig. 3. Regional localization of the expression of three transcripts identified in this study. A, in situ hybridization (dark field) to rat brain of novel serine protease BSP1; CA1-3 and are hippocampal subregions, PoDG are polymorphic cells of the dentate gyrus. B, in situ hybridization of the novel steroid metabolizing enzyme Cyp7b to mouse brain showing expression throughout CA1-3 and DG. C, staining (light field) of the hippocampus of *kin* animals with the chromatogenic β-galactosidase (β-geo) substrate X-gal, showing highly specific coloration in CA1-3.

PTP-24 is less restricted in expression than PTP-4 but neither was specific in their expression to the hippocampus (Pickard, 1996).

Of the other PTPs, the candidate tumor suppressor PTP-γ was found to be expressed fairly selectively in hippocampus but, earlier in development, expression is reported to be widespread in septal/thalamic nuclei, cortex and hippocampus (Barnea et al., 1993)

Serine proteases

These enzymes are characterized by a conserved serine residue in the active site and are held to play a variety of roles in the brain, including substratum clearance during neuronal migration and synapse formation, and growth factor activation by proteolytic cleavage of precursor molecules. PCR primers were designed according to sequence motifs conserved within the major (chymotryptic) clan of enzymes. The predominant species obtained by amplification, and confirmed by Northern blotting, were tissue-type plasminogen activator (t-PA), and RNK-Met-1, a lymphocyte protease not previously reported in brain, and two new family members, BSP1 (brain serine protease 1) and BSP2.

In situ hybridization revealed BSP1 transcripts in the hippocampal CA fields, with expression in CA1 and CA3 but only low-level expression in CA2. Expression was also detected in the polymorphic layer of the dentate gyrus and within deep layers of the entorhinal cortex (Fig. 3). No significant expression was detected in other regions of the adult rat brain. Overall, expression of BSP1 was substantially restricted to the hippocampal formation while BSP2 expression was more widespread (Davies, 1998).

Gene-trapping

The final approach was gene-trapping. A promoterless reporter cassette, comprising a splice-acceptor site derived from *en*-2 preceding the '*β-geo*' (a β-galactosidase [*lacZ*]-neomycin resistance fusion) (Skarnes et al., 1992) was introduced into mouse ES cells. The *β-geo* open-reading frame lacks a translation initiation site: application of G418 selected for insertion events where the reporter has integrated into an endogenous gene locus, generating a fusion linking the N-terminus of an endogenous protein to the reporter polypeptide. G418-resistant ES cells were introduced into the cavity of mouse blastocysts, reimplanted into foster mothers, and the resulting chimeric males mated to produce offspring heterozygous for the gene-trap insertion. Initially 15 independent lines were produced by this method, as described (Skarnes, 1993, Wilson et al., in preparation; Steel et al., 1998).

To explore expression of the reporter gene, brain sections were incubated with the chromogenic β-galactosidase substrate X-gal. In 4 lines, *obn*, *kin*, *hpk* and *glnC*, distinct staining was present in the hippocampus. *glnC* expressed ubiquitously and was not analyzed further. The remaining 11 lines, with the exception of *glnA* (that expressed at barely detectable levels in cerebellum) failed to exhibit reporter gene expression in adult brain (Steel et al., 1998), but of which 6 were negative in all tissues. A frequency of 4/15 lines expressing in hippocampus suggests that the gene-trap procedure is a feasible method to identify genes expressed in this brain structure.

In *obn* and *hpk* reporter activity was in several brain regions, but in brain of *kin* mice staining was substantially, but not exclusively, restricted to the CA1-CA3 subfields of the hippocampus (Fig. 3). Weak staining was also detected in the molecular layer of the dentate gyrus and the lateral septum, with further weak reporter activity in amygdala, striatum and olfactory bulb (Steel et al., 1998).

Identification of the trapped genes employed a RACE protocol (rapid amplification of cDNA ends; Frohman et al., 1988) as modified (Skarnes et al., 1992). This involves priming reverse transcriptase with a *lacZ*-specific primer, tailing the product with poly-dA and subsequent PCR. We were initially unable to obtain cDNA clones corresponding to the hybrid transcripts and used a new modification that permits amplification of refractory sequences (Steel et al., 1998). This identified *kin* as an insertion into a gene encoding the Abelson-related non-receptor tyrosine kinase, *arg* (Kruh et al., 1990). In the *kin* line the *β-geo* reporter is inserted downstream of an alternative 5′ exon of the *arg* gene (Steel et al., 1998), and reports only expression from this promoter. Disappointingly, reporter mRNA was widespread in hippocampus and adjacent brain structures. The widespread pattern of mRNA expression, together with the restricted localization of the translation product, argues that the hybrid transcript is only translated efficiently within CA1–3.

Discussion

We have sought genes whose expression is restricted to the hippocampus, with a view to generating transgenic animals in which only the hippocampus is modified. We discuss below the different approaches employed, their merits and drawbacks, and possible means to improve them.

In our hands the most robust method was also the simplest – differential hybridization. This yielded a gene, Cyp7b, whose expression is largely but not exclusively restricted to the hippocampal formation (Stapleton et al., 1995) and encoding a novel enzyme metabolizing neurosteroids (Rose et al., 1997). Subtractive hybridization, though technically more demanding, highlighted clone H4M10 (Pickard, 1996) that turned out to be identical to hippocalcin, a gene whose expression was previously suggested to be specific to the hippocampus (Koboyashi et al., 1992). However, while enriched in the hippocampus, brain expression outwith the formation was also detected.

The candidate gene family approach yielded a new serine protease, BSP1 (Davies et al., 1998), with a highly restricted pattern of expression that will permit exploitation in a transgenic context, and protein tyrosine phosphatase (PTP) gamma, whose restricted pattern of expression warrants further study (Pickard, 1996). Finally, gene-trapping uncovered an insertion, kin, directing reporter gene expression almost exclusively to hippocampus (Steel et al., 1998). However, kin transcripts were expressed widely in brain, but without detectable reporter activity, suggesting that the fusion transcript is widely expressed but only translated in the hippocampus. A more recent gene-trap line, Ex-194, has identified a gene whose expression is highly restricted to the dentate gyrus (C. Jarvis and W. Skarnes, pers. comm.), underlining the power of the gene-trap approach.

The two latter techniques (candidate gene families and gene-trapping) owe much of their success to serendipity. There are emerging suggestions that the hippocampus is an unusually rich site of gene expression (Steel et al., 1998); further experiments will be required to address whether these approaches can be applied to other brain regions. The gene-trap approach does offer one important advantage. Because gene-trap insertions tend to be near the start of the trapped gene, the technique can identify alternative promoter sites whose activity is restricted to a given tissue. This is illustrated by the *kin* insertion that identifies an alternative 5′ promoter/exon of the targeted gene (Steel et al., 1998).

Despite the power of the two former techniques (differential and subtractive hybridization), as employed here, the differential complexity of the tissue samples may be a complicating factor. Here we compared gene expression of one small brain region, the hippocampus, with that of the entire brain devoid of hippocampus ('remainder of brain'). One improvement we suggest would be to perform first differential screening or subtraction using tissues of more similar complexity (for instance hippocampus vs cortex or cerebellum).

The most serious problem encountered is that many of the screens detect cDNAs from genes that are widely expressed. Although differential splicing, polyadenylation site selection, and promoter utilization, can all yield short cDNA segments whose abundance differs between different tissues, this was not a major complicating factor. In contrast, and for reasons we do not understand, cDNAs were routinely detected that are differentially represented in the starting cDNA populations but, on Northern analysis, did not differ significantly in their levels of expression between the different tissues examined (in this case hippocampus and other brain regions). This result is not understood, but we surmise that any post-transcriptional modification of an mRNA molecule that might differentially affect its representation in cDNA will reveal itself artefactually as a tissue-specific cDNA. We have considered several complex explanations, for instance, the possibility that tissue-specific complementary sequences could hybridize to an mRNA and prevent first strand cDNA synthesis, so producing biased representation of cDNA sequences.

A further caveat concerns the in situ hybridization technique as a means to confirm regional specificity. In the rodent hippocampus the cell bodies are very tightly aligned, while elsewhere in brain are often more evenly distributed. Thus, in situ hybridization, at short exposure times, tends to highlight the hippocampus. Caution is warranted in the interpretation of such experiments: even widely-expressed genes can appear to be 'specific' to the hippocampus. Indications of hippocampus-specificity from in situ hybridization must be considered unproven until confirmed by other techniques including Northern analysis.

Our analyses are also limited by their bias towards well-expressed genes. Partly this is an advantage, because genes expressed poorly are less attractive for transgenic exploitation, but at the same time region-specific genes that are expressed at relatively low levels, such as embryo brain kinase (*ebk*), may be missed. *ebk* expression is widespread during development becomes progressively more restricted to the hippocampus during maturation (Ellis et al., 1995). However, our preliminary experiments failed to confirm this specificity (not presented).

Other procedures that deserve consideration include differential display (Liang and Pardee, 1992). We attempted to use this method but found many artefacts: we estimate that a transcript whose abundance differs only by a factor of 5 between two different tissues will yield a candidate 'tissue-specific' band on display. However, differential display in combination with subtractive hybridization may offer a more powerful approach than provided by either technique alone.

For the future two further techniques may be envisaged. The first, 'virtual subtraction', reflects the increasing availability of databanks of randomly sequenced cDNA products from different tissues. This will permit 'subtraction' between two sequence banks, pinpointing sequences abundant in one tissue but sparsely represented in the other. The second involves the use of gridded cDNAs or expressed sequences. In principle, all the genes in the mammalian genome can be represented as 100,000 unique cDNA sequences. With emerging technologies these could be arrayed at very high density on filters for hybridization. Parallel hybridization with cDNA probes prepared from two (or more) different tissues will identify sequences expressed in one tissue but not another.

Finally, the question arises — do truly region-specific genes exist? On the one hand, it may be argued that an anatomically distinct region such as the hippocampus must require specific gene expression. The hippocampus could be, in terms of gene expression, as distinct from cortex or cerebellum as liver is different from lung. On the other hand, it may be that a pyramidal neuron in hippocampus and one in, for instance, auditory cortex, differ predominantly in their location and much less so in their gene expression. Our data are not consistent with this latter contention, and within the hippocampus we find different patterns of gene expression in the pyramidal cells of regions CA1, CA2, CA3. In addition, the dentate gyrus and the dentate hilus appear to represent distinct cell populations both in gene expression and morphology. Although the low resolution of our in situ hybridization data would not easily permit the detection of subpopulations of pyramidal neurones within a *single* hippocampal subregion, their existence is certainly not ruled out. Finally, in both CA1 and CA3 we observe intriguing gradients of gene expression (both transversely within a section and along the dorso-ventral axis) — indicating that location and gene expression do, at least to some extent, go hand in hand.

Even though we provide evidence for region-specific gene expression within the hippocampus, it is likely that all genes are expressed at some level in all tissues, and no gene may be considered to be 100% tissue-specific. However, even if not totally restricted to a given tissue, a gene expressed 100-fold more strongly in one tissue or brain region than another will still be of great utility. Further, we have uncovered examples where developmental expression is widespread, but later becomes restricted to the hippocampus; the exploitation of such genes will best be performed in combination with inducible systems such as conditional ablation or drug-activated transcriptional activators and recombination enzymes.

Despite these caveats, our analysis has pinpointed genes whose exploitation in a transgenic context may be envisaged (work in progress). It may be hoped that similar approaches may be employed to identify genes whose expression is restricted to other brain structures.

Acknowledgements

This project was part funded by the MRC (to RL and R.G.M. Morris), a BBSRC/CGR bursary to BP and by MRC studentships to GS and BD, by a Charitable Foundation, and by BBSRC funding to the Centre for Genome Research. We thank R.G.M. Morris for helpful suggestions on the manuscript.

References

Abeliovich, A., Chen, C., Goda, Y., Silva, A.J., Stevens, C.F. and Tonegawa, S. (1993a) Modified hippocampal long-term potentiation in PKC gamma-mutant mice. *Cell*, 75: 1253–1262.

Abeliovich, A., Paylor, R., Chen, C., Kim, J.J., Wehner, J.M. and Tonegawa, S. (1993b) PKC gamma mutant mice exhibit mild deficits in spatial and contextual learning. *Cell*, 75: 1263–1271.

Balzer, H.J. and Baumlein, H. (1994) An improved gene expression screen. *Nucleic Acids Res.*, 22: 2853–2854.

Barnea, G., Silvennoinen, O., Shaanan, B., Honegger, A.M., Canoll, P.D., D'Eustachio, P., Morse, B., Levy, J.B., Laforgia, S., Huebner, K. et al (1993) Identification of a carbonic anhydrase-like domain in the extracellular region of RPTP gamma defines a new subfamily of receptor tyrosine phosphatases. *Mol. Cell. Biol.*, 13: 1497–1506.

Berod, A., Biguet, N.D., Dumas, S., Bloch, B. and Mallet, J. (1987) Modulation of tyrosine hydroxylase gene expression in the central nervous system visualized by in situ hybridization. *Proc. Natl. Acad. Sci. USA*, 84: 1699–1703.

Bliss, T.V. and Collingridge, G.L. (1993) A synaptic model of memory: long-term potentiation in the hippocampus. *Nature*, 361: 31–39.

Bliss, T.V. and Lømo, T. (1973) Long-lasting potentiation of synaptic transmission in the dentate area of the anaesthetized rabbit following stimulation of the perforant path. *J. Physiol.*, 232: 331-356.

Cohen, N.J. and Eichenbaum, H. (1993) *Memory, Amnesia and the Hippocampal System.* MIT Press, Cambridge, MA.

Chen, C. and Tonegawa, S. (1997) Molecular genetic analysis of synaptic plasticity, activity-dependent neural development, learning, and memory in the mammalian brain. *Annu. Rev. Neurosci.*, 20: 157-184.

Davies, B. J. (1998) PhD thesis, The University of Edinburgh.

Davies, B.J., Pickard, B.S., Steel, M., Morris, R.G. and Lathe, R. (1998) Serine proteases in rodent hippocampus. *J. Biol. Chem.*, 273: 23004–23011.

Dobie, K.W., Mehtali, M., McClenaghan, M. and Lathe, R. (1997) Variegated gene expression in mice. *Trends Genet.*, 13: 127-130.

Ellis, J., Liu, Q., Breitman, M., Jenkins, N.A., Gilbert, D.J., Copeland, N. G., Tempest, H.V., Warren., S., Muir, E., Schilling, H., Fletcher, F.A., Ziegler, S.F. and Rogers, J.H. (1995) Embryo brain kinase: A novel gene of the eph/elk receptor tyrosine kinase family. *Mech. Dev.*, 52: 319–341

Frohman, M.A., Dush, M.K. and Martin, G.R. (1988) Rapid production of full-length cDNAs from rare transcripts: amplification using a single gene-specific oligonucleotide primer. *Proc. Natl. Acad. Sci. USA*, 85: 8998–9002.

Gebbink, M.F., Van Etten, I., Hateboer, G., Suijkerbuijk, R., Beijersbergen R.L., Geurts Van Kessel, A. and Moolenaar, W.H. (1991) Cloning, expression and chromosomal localization of a new putative receptor-like protein tyrosine phosphatase. *FEBS Lett.*, 290: 123–130.

Hampson, I.N., Pope, L., Cowling, G.J. and Dexter, T.M. (1992) Chemical cross linking subtraction (CCLS): a new method for the generation of subtractive hybridisation probes. *Nucleic Acids Res.*, 20: 2899.

Jiang, Y.P., Wang, H., D'Eustachio, P., Musacchio, J.M., Schlessinger J. and Sap J. (1993) Cloning and characterization of R-PTP-kappa, a new member of the receptor protein tyrosine phosphatase family with a proteolytically cleaved cellular adhesion molecule-like extracellular region. *Mol. Cell. Biol.*, 13: 2942–2951.

Kobayashi, M., Takamatsu, K., Saitoh, S., Miura, M. and Noguchi, T. (1992) Molecular cloning of hippocalcin, a novel calcium-binding protein of the recoverin family exclusively expressed in hippocampus. *Biochem. Biophys. Res. Commun.*, 189: 511–517.

Krueger, N.X., Streuli, M. and Saito, H. (1990) Structural diversity and evolution of human receptor-like protein tyrosine phosphatases. *EMBO J.*, 9: 3241–3252.

Kruh, G.D., Perego, R., Miki, T. and Aaronson, S.O. (1990) The complete coding sequence of arg defines the Abelson subfamily of cytoplasmic tyrosine kinases. *Proc. Natl. Acad. Sci. USA*, 87: 5802–5806.

Lathe, R. (1996) Mice, gene targeting and behaviour; more than just genetic background. *Trends Neurosci.*, 19: 183–186.

Lathe, R. and Morris, R.G.M. (1994) Analysing brain function and dysfunction in transgenic animals. *Neuropathol. Appl. Neurobiol.*, 20, 350–358.

Li, Y., Erzurumlu, R.S., Chen, C., Jhaveri, S. and Tonegawa, S. (1994) Whisker-related neuronal patterns fail to develop in the trigeminal brainstem nuclei of NMDAR1 knockout mice. *Cell*, 76: 427–437

Liang, P. and Pardee, A.B. (1992) Differential display of eukaryotic messenger RNA by means of the polymerase chain reaction. *Science*, 257: 967–971.

Mansuy, I.M., Winder, D.G., Moallem, T.M., Osman, M., Mayford, M., Hawkins, R.D. and Kandel, E.R. (1998) Inducible and reversible gene expression with the rtTA system for the study of memory. *Neuron*, 21: 257–265.

Mayford, M., Abel, T. and Kandel, E.R. (1995) Transgenic approaches to cognition. *Curr. Opin. Neurobiol.*, 5: 141–148.

Morris, R.G.M, Garrud, P., Rawlins, J.N.P. and O'Keefe, J. (1982) Place navigation impaired in rats with hippocampal lesions. *Nature*, 297: 681–683.

Morris, R.G.M., Hagan, J.J. and Rawlins, J.N.P. (1986a) Allocentric spatial learning by hippocampectomised rats: a further test of the "spatial mapping" and "working memory" theories of hippocampal function. *Q. J. Exp. Psychol.*, 38B: 365–395.

Morris, R.G., Anderson, E., Lynch, G.S. and Baudry, M. (1986b) Selective impairment of learning and blockade of long-term potentiation by an *N*-methyl-D-aspartate receptor antagonist, AP5. *Nature*, 319: 774–776.

Oyler, G.A., Higgins, G.A., Hart, R.A., Battenberg, E., Billingsley, M., Bloom, F.E. and Wilson, M.C. (1989) The identification of a novel synaptosomal-associated protein, SNAP-25, differentially expressed by neuronal subpopulations. *J. Cell Biol.*, 109: 3039–3052.

Pickard, B.S. (1996) PhD thesis, The University of Edinburgh.

Roberson, E.D, English, J.D. and Sweatt, J.D. (1996) A biochemist's view of long-term potentiation. *Learning Memory*, 3: 1–24.

Rose, K.A., Stapleton, G., Dott, K., Kieny, M.P., Best, R., Schwarz, M., Russell, D.W., Björkhem, I., Seckl, J. and Lathe, R. (1997) Cyp7b hydroxylates neurosteroids dehydroepiandrosterone (DHEA) and pregnenolone at the 7alpha Position. *Proc. Natl. Acad. Sci. USA*, 94: 4925-4930.

Saitoh, S., Takamatsu, K., Kobayashi, M. and Noguchi, T. (1993) Distribution of hippocalcin mRNA and immunoreactivity in rat brain. *Neurosci. Lett.*, 157: 107–110.

Sakimura, K., Kutsuwada, T., Ito, I., Manabe, T., Takayama, C., Kushiya, E., Yagi, T., Aizawa, S., Inoue, Y., Sugiyama, H. et al. (1995) Reduced hippocampal LTP and spatial learning in mice lacking NMDA receptor epsilon 1 subunit. *Nature*, 373: 151–155.

Scoville, W.B. and Milner, B. (1957) Loss of recent memory after bilateral hippocampal lesions. *J. Neurol. Neurosurg. Psychiat.*, 20: 11–21.

Shivers, B.D., Hilbich, C., Multhaup, G., Salbaum, M., Beyreuther, K. and Seeburg PH. (1988) Alzheimer's disease amyloidogenic glycoprotein: expression pattern in rat brain suggests a role in cell contact. *EMBO J.*, 7: 1365–1370.

Silva, A.J., Paylor, R., Wehner, J.M. and Tonegawa, S. (1992a) Impaired spatial learning in alpha-calcium-calmodulin kinase II mutant mice. *Science*, 257: 206–211.

Silva, A.J., Stevens, C.F., Tonegawa, S. and Wang, Y. (1992b) Deficient hippocampal long-term potentiation in alpha-calcium-calmodulin kinase II mutant mice. *Science*, 257: 201–206.

Skarnes, W.C., Auerbach, B.A. and Joyner, A.L. (1992) A gene trap approach in mouse embryonic stem cells: the *lac*Z reporter is activated by splicing, reflects endogenous gene expression, and is mutagenic in mice. *Genes. Dev.*, 6: 903–918.

Stapleton, G. (1994) PhD thesis, University of Edinburgh.

Stapleton, G., Steel, M., Richardson, M., Mason, J.O., Rose, K.A., Morris, R.G.M. and Lathe, R. (1995) A novel cytochrome P450 expressed primarily in brain. *J. Biol. Chem.*, 270: 29739–29745.

Steel, M., Moss, J., Clark., K.A., Kearns, I.R., Davies, C.H., Morris, R.G.M., Skarnes, W.C. and Lathe, R. (1998) Gene-trapping to identify and analyze genes expressed in the mouse hippocampus. *Hippocampus*, 8: 444–457.

Sung, L.A., Fowler, V.M., Lambert, K., Sussman, M.A., Karr, D. and Chien, S. (1992) Molecular cloning and characterization of human fetal liver tropomodulin. A tropomyosin-binding protein. *J. Biol. Chem.*, 267: 2616–2621

Takahashi, Y. (1992) Gene expression in cells of the central nervous system. *Prog. Neurobiol.*, 38: 523–569.

Tsien, J.Z., Chen, D.F., Gerber, D., Tom, C., Mercer, E.H., Anderson, D.J., Mayford, M., Kandel, E.R. and Tonegawa, S. (1996a) Subregion- and cell type-restricted gene knockout in mouse brain. *Cell*, 87: 1317–1326.

Tsien, J.Z., Huerta, P.T. and Tonegawa, S. (1996b) The essential role of hippocampal CA1 NMDA receptor-dependent synaptic plasticity in spatial memory. *Cell*, 87: 1327–1338.

Wang, Z. and Brown, D.D. (1991) A gene expression screen. *Proc. Natl. Acad. Sci. USA*, 88: 11505–11509.

Zigman, J.M., Westermark, G.T., LaMendola, J., Boel, E. and Steiner DF. (1993) Human G(olf) alpha: complementary deoxyribonucleic acid structure and expression in pancreatic islets and other tissues outside the olfactory neuroepithelium and central nervous system. *Endocrinology*, 133: 2508–2514.

W.E. Crusio and R.T. Gerlai (Eds.)
Handbook of Molecular-Genetic Techniques for Brain and Behavior Research (Techniques in the Behavioral and Neural Sciences, Vol. 13)

CHAPTER 2.2.4

Application of real-time RT-PCR for quantification of gene expression

Jane Winer,[1] Natasha Shinsky,[2] Robert Gerlai[2] and P. Mickey Williams[1]

[1]*Department of Cell Biology and Technology, Genentech, Inc., 1 DNA Way, South San Francisco, CA 94080, USA*
[2]*Department of Neuroscience, Genentech, Inc., 1 DNA Way, South San Francisco, CA 94080, USA*

Introduction

Scientists undertaking molecular biology research are finding a wealth of new sequence information to examine (see e.g. Chapter 1.2). Often a new expression sequence tag (i.e. est) or gene is described which is expressed in a tissue or cell of interest, but little is known as to the gene's function. Examination of gene expression as it correlates with biological response or disease is a step in the understanding of gene function and a point from which to develop further testable hypotheses. New technologies useful for gene expression analysis are emerging rapidly, driven by the need for tools to analyze the explosion of genome sequence information.

These technologies can be grouped into two application areas: (a) differential gene profiling; and (b) precise quantitative expression. Gene profiling methods are used to determine which genes are expressed differentially in two different biological states of a cell or tissue (e.g. disease vs normal, treatment vs non-treatment, etc). Profiling methods must be robust to produce qualitatively reproducible results, but do not require exquisite assay precision or accuracy. Gene profiling methods are generally designed to simultaneously analyze many genes (thousands) from any given sample. Methods used for gene profiling include: quantitative sequencing (Velculescu et al., 1995), differential display (Liang and Pardee, 1992), amplification fragment length polymorphism (i.e. AFLP (Bachem et al., 1996)) and nucleic acid arrays (Schena et al., 1995; Lockhart et al., 1996).

Quantitative gene expression technologies are applied when the goal is to monitor precise expression levels of a gene resulting from a biological state change. Quantitative methods require assay precision (i.e. the assay must yield identical results when identical replicate samples are analyzed). If absolute quantities are desired, assay accuracy is also essential (e.g. how many virus particles are present in 1 ml of plasma or how many mRNAs for a gene are expressed in a cell?). Technologies used for precise gene expression analysis include; quantitative competitive polymerase chain reaction, i.e. QC-PCR (Becker-Andre, 1989; Piatak et al., 1993; Williams et al., 1996), real-time or kinetic PCR (Heid et al., 1996; Ririe et al., 1997; Wittwer et al., 1997), branched DNA (Shen et al., 1998), scintillation proximity assays (Kenrick et al., 1997) and the traditional RNase protection assays. Quantitative methods are generally applied when moderate numbers of genes are of interest (hundreds or less) and precise measurements are required. These methods often permit moderate to high sample throughput (hundreds to thousands of samples). Quantitative applications require thorough knowledge of the method, its limitations and data interpretation. Considerable thought is necessary for successful assay design, especially when tissues rather than cultured cells comprise the biological sample. This chapter will focus on one of the newer quantitative methods, real-time PCR (a.k.a. kinetic PCR), and its application to precise quantitative gene expression analysis.

A previous stumbling block to quantitative gene expression experiments has been the necessity for large amounts of purified RNA (micrograms of total RNA) required for Northern blot or RNase protection analysis. Often experiments are performed in animal models in which very specialized regions of tissue or cells are the preferred sample source. Obviously, such samples do not usually supply sufficient amounts of RNA for traditional gene expression approaches. The desire to work with small starting samples, even possibly using microdissection, lends itself to PCR based amplification technologies. PCR has proven to be a powerful tool for analysis of very limited sample. A single target molecule is detectable in a well optimized PCR amplification. Although PCR permits extremely sensitive detection technology, it requires several hurdles be overcome for quantitative applications (Ferre, 1992; Becker-Andre, 1993; Piatak et al., 1993; Williams et al., 1996; Heid et al., 1996). Quantitative applications of PCR are designed to use the power of exponential target amplification for the purpose of increasing the starting target to a level of molecules easily detected. The difficulties reside in performing this exponential amplification in such a way that the amplified targets which are detected reflect starting target quantity. Quantitative competitive PCR (or reverse transcription-PCR; i.e. RT-PCR) has been developed as a means of overcoming the hurdles and utilizing PCR for quantitative purposes. This approach has proven to be a very robust method, but suffers from the enormous amount of effort required to develop and validate new assays (for review see Heid et al., 1996). Real-time PCR has recently been described and permits very rapid assay development and validation, leading to a methodology suited to moderately high through-put of samples and genes analyzed using minimal starting material.

Real-time PCR was originally described by (Heid et al., 1996) and (Gibson et al., 1996). The method utilizes novel instrumentation (Model 7700 Sequence Detector; ABIPerkin Elmer or LightCycler; Boehringer Mannheim) and TaqMan chemistry to achieve rapid and quantitative results. TaqMan chemistry (Holland and Abramson, 1991) utilizes a labeled hybridization probe (directed against the PCR product sequence) in addition to the forward and reverse primers. During primer extension the DNA polymerase reaches the site of probe hybridization and a conformation dependent nucleolytic activity in the polymerase cleaves the probe (Fig. 1). The probes are currently labeled with two fluorescent dyes (Lee et al., 1993; Livak et al., 1995). One is a reporter dye (often FAM; ABIPerkin Elmer) on the 5′ end of the probe. The other dye is a quenching dye (often TAMRA; ABIPerkin Elmer) on the 3′ end of the probe. The probes are phosphorylated on the 3′ end, such that polymerase extension does not occur from the probe. The instrumentation monitors reaction vessel fluorescence during the PCR amplification, hence real-time. As PCR proceeds, an accumulation of product with each successive cycle provides more target for the fluorescent probe to hybridize. This in turn results in an increase in probe cleavage with successive cycles of PCR. The uncleaved probe is designed such that the reporter dye and quenching dye are sufficiently close enough to decrease fluorescent emission of the reporter dye (via fluorescent resonant engery transfer to the quenching dye). As the probe is cleaved the reporter dye emission intensity increases, which is measured by the instrumentation. A threshold reporter fluorescent intensity is chosen (based on reaction and instrument background fluorescence). The cycle of PCR at which a particular sample has cleaved enough probe to reach the threshold is termed

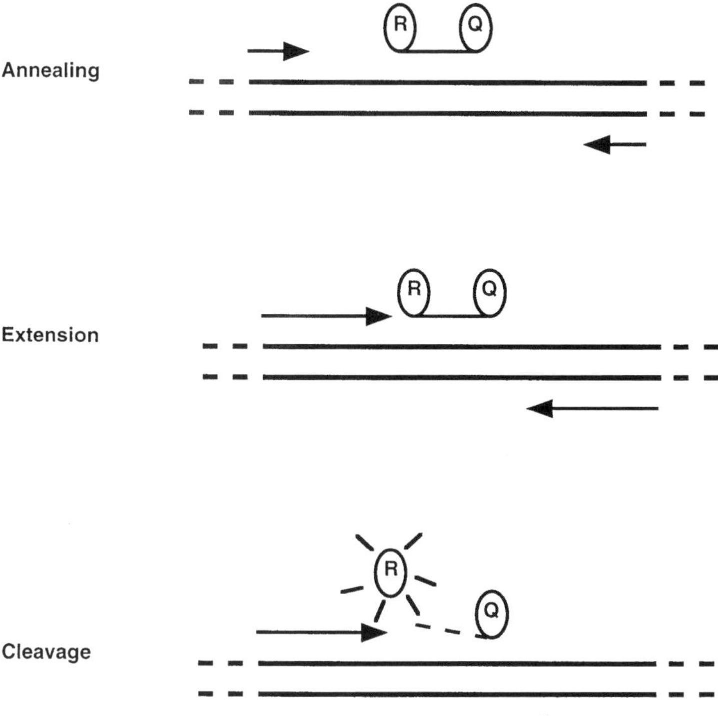

Fig. 1. TaqMan chemistry. Annealing step: The primers and probe anneal to target molecule. Extenison step: The primers are extended by the action of DNA polymerase. Cleavage step: The primers reach the site where the probe is hybridized. The DNA polymerase cleaves the hybridization probe. The reporter dye molecule is no longer in close enough proximity to the quenching dye. This results in an increase in reporter fluorescent emmission.

Ct (i.e. threshold cycle). The Ct value for a sample is proportional to the amount of starting target in that sample. Figure 2(A) demonstrates a two fold serial dilution of RNA which is analyzed for glyceraldehyde alkaline phosphate dehydrogenase (i.e. GAPDH) gene expression. It should be noted that as RNA concentration is lowered, thus decreasing the amount of target GAPDH mRNA, the Ct values increase (i.e. more cycles of PCR are required to degrade sufficient probe as target decreases). This dilution series results in a linear relationship between the Ct value calculated and the amount of target analyzed (Fig. 2(B)). From this data a standard curve can be generated from which a line equation can be derived. Such a standard curve can be used to determine the amount of target mRNA found in a unknown sample relative to the RNA used to generate the standard. We have previously demonstrated that this method is precise (coefficient of variance calculated from Cts of replicate samples generally result in CVs $< 1\%$) and permits very large assay dynamic range (linear for approximately 6 logs of input target molecules) (Heid et al., 1996).

Experimental design

Primer and probe design

The proper selection primers is critical to the success of any PCR application. Real-time PCR also requires another oligonucleotide, the probe, which must be designed following several important rules. Several software applications (Primer Express, Perkin Elmer–ABS or Oligo, National

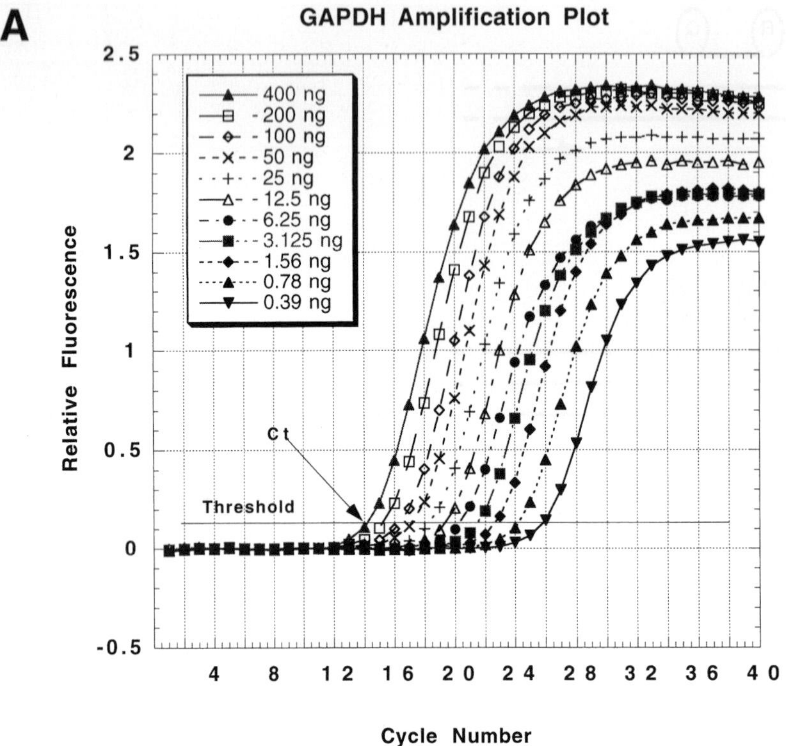

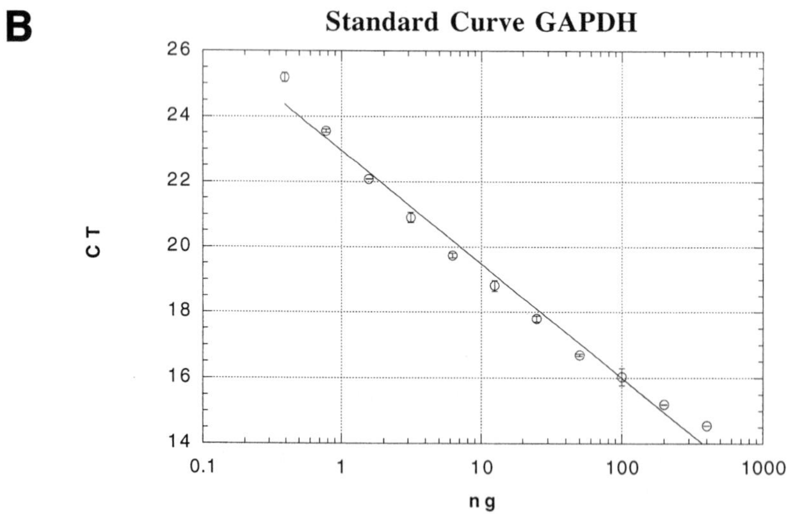

Fig. 2. Serial dilution of RNA sample. (A) A reference RNA sample was 1:2 serially diluted and analyzed by real-time PCR. Amplification plots are depicted of the dilution series. (B) A standard curve is generated by plotting the calculated threshold cycles (Ct) versus the RNA mass analyzed in each of the diluted reference samples. A line equation is calculated from the data; $y = mx + b$. Actual RNA mass is calculated from experimental samples by solving the line formula for RNA mass (x), knowing the mean Ct (y) of the actual sample (y is the known mean Ct value of the actual sample, m is the calculated slope of the standard curve, b is the y intercept calculated from the standard curve and x is the unknown RNA mass which needs to be solved for each actual samples).

Biosciences, Inc.) are recommended for the task of primer and probe design. Primers should be selected such that they maintain appropriate specificity and do not contain potential hairpin or primer-dimer generating sequences. The predicted primer melting temperature, Tm, should be matched within 2° for each primer. Smaller PCR products generally result in higher efficiency product accumulation which will yield better results. Product size is ideally in the range of 100–200 base pairs, although longer products can be used. The sequence of the probe should have a Tm approximately 10 degrees higher than the matching primer pair. The extension temperature used in PCR should always be below the Tm of the probe, otherwise the probe will melt off of the template and not be cleaved during the extension. The probe length should be no greater than 40 bases maximum. The probe should not contain a G base on the 5′ end, as this will act as an acceptor for the FAM reporter fluorescence energy. This will diminish the overall reporter fluorescent intensity achieved upon probe cleavage. The probe should also be blocked on the 3′ end such that it will not serve as a primer. This is usually accomplished by adding a 3′ phosphate group. Primers and probes are synthesized using common oligonucleotide chemisty or can be purchased from a variety of commercial sources.

Real-time RT-PCR components

There are many different approaches to performing reverse transcription (RT)-PCR. One approach is the single-tube format. There are many commercial sources available for combined RT-PCR enzymes and buffers. This approach uses two enzymes blended in a reaction buffer (e.g. MuLV reverse transcriptase and AmpliTaq Gold DNA polymerase, Perkin Elmer–ABS) or a single enzyme which has both activities (e.g. rTth, Perkin Elmer–ABS). Generally a portion of the RNA sample is added to the reaction tube containing all reagents necessary for real-time RT-PCR. The reverse PCR primer is used to initiate the reverse transcription. After reverse transcription is complete, the PCR amplification is performed without the need to open tubes. Another approach uses a two-tube format, where the entire sample is reverse transcribed (usually by using random hexamers or oligo dT primers). The cDNA generated from this reaction can than analyzed in a real-time PCR reaction which is done in a separate tube. Both single tube and two tube approaches are quite acceptable and yield similar results. The only critical requirment for real-time PCR is that the DNA polymerase must contain the native conformation dependent nucleolytic activity. This activity has been removed from some commercial DNA polymerase enzymes which have been specifically designed for other applications.

RNA used for real-time RT-PCR can be polyA selected or total RNA. The amount of RNA used in a reaction will depend on the purity of the sample, the total number of target mRNA molecules present in the sample and the volume of the reaction. The range of total RNA used per reaction is generally somewhere between 10 pg to 400 ng in a 50 μl reaction.

Listed below is the reaction mixture that was used for the data presented in this chapter:

50 μl
1× RT-PCR Buffer
 (Promega, Acesss RT-PCR System)
2.5 mM MgC12
0.1 U/μl T*fl* DNA polymerase
 (Promega, Acesss RT-PCR System)
0.1 U/μl AMV Reverse Transcriptase
 (Promega, Acesss RT-PCR System)
0.2 mM (each dNTP)
100 nM (fluorescent probe)
0.5 μM of each primer (forward and reverse)
10 μl of RNA sample
Water is added to bring the final volume up to 50 μl

Reaction conditions were:
reverse transcription performed for 45 min at 48°
40 thermal cycles of (30 s at 94°, 1 min at 60°, and 2 min at 69°)

This is an example, but as with any PCR reaction optimization is important. Parameters which should be optimized include the $MgCl_2$ concentration and reaction temperatures. Too little $MgCl_2$ results in no product generation and too much $MgCl_2$ decreases specificity of the amplification. The reaction temperatures should permit hybridization of the primers and probe during the annealing step and maintain probe hybridization during the elongation step.

Tissue versus cell culture

The source of biological sample determines the strategies used for quantitative mRNA analysis. Clonal populations of cultured cells maintained in strictly controlled environments are obviously the easiest source from which to begin experiments. The effort is compounded when animal tissue is the experimental sample source. The investigator must be aware that tissues require extra effort to recover intact RNA. All sample preparation methods used must rapidly and efficiently lyse the tissue and inactivate contaminating RNase activity. If tissues are harvested and not immediately used for sample preparation, they should be snap frozen using liquid nitrogen. These frozen samples can later be quickly crushed and thawed in the presence of the lysis reagent. Further, tissues are composed of different cell types and different cell types may regulate expression of the gene of interest in different ways or not at all. Therefore it is critical to consider this and analyze data in a manner that accounts for this. It is possible to use microdissection to obtain biological samples which are enriched for particular cell types. Also if the biological state changes occurring in the experiments do not radically effect the cell type distribution (e.g. change in number of inflammatory cells in the tissue etc.), RNA mass can be used for normalization of the gene of interest, although changes in gene expression will be diluted in the background of other cell types. Traditionally these inherent difficulties in gene expression analysis from tissue samples have been overlooked.

Sample preparation

The quality and purity of RNA samples are extremely important in the arena of quantitative PCR. Contaminating salt, organic solvents (i.e. phenol) or certain components of the biological matrix (i.e. heme) can and do inhibit PCR reactions, resulting in decreased amplification efficiency which will complicate successful quantitative PCR (see section of sample normalization below). Also the removal of genomic DNA is critical for reproducible results (contaminating DNA can provide a target for RT independent amplification confounding results). Finally, the purity of the RNA is crucial for RNA stability upon storage. Validation of the RNA preparation method is highly recommended, to assure that equivalent yields are obtained from identical samples, the DNA content is minimal (see below) and the resulting RNA sample contains minimal contaminating PCR inhibitors. Assessment of RNA is accomplished by many methods. Optical absorbance (OD_{260} and OD_{280}) yields information on the purity and concentration of RNA, but does not address potential DNA contamination or degradation of the sample. This method also generally requires a significant amount of a sample. Fluorescent dyes offer greater sensitivity but do not give any measurement of protein contamination. Electrophoretic analysis can give only information on RNA degradation and DNA contamination. This approach also requires significant expenditure of sample. In short there is no current method that provides sensitive RNA quantitation and measurements of all of the desired qualities. In our lab we have validated several methods of RNA preparation (e.g. Rneasy; Qiagen, for total or cytoplasmic RNA and PolyAtract; Promega, for polyA mRNA preparation; (Winer et al., submitted)) from a variety of biological sample sources. This was accomplished by monitoring quantity and quality of recovery from a biological sample prepared in replicate, utilizing optical absorbance (i.e. OD_{260} and OD_{280}) and real-time quantitative mRNA analysis of a gene. Ideal sample preparation methods should yield near identical $OD_{260,280}$ values from identical samples. If the preparation method yields pure RNA, real-time quantitative analysis of any given gene will demonstrate identical quantitative values from replicate sample preparations, thus demonstrating equal mass recovery and amplification efficiency. The sample RNA used in this manuscript was prepared as total RNA using the Qiagen, RNeasy protocol (as per the manufacturers recommendations).

Normalization of sample

Two potential variables must be normalized for the accurate application of PCR for quantitation. First, it is important to normalize each sample for the actual mass analyzed. Second, one must correct for any difference in amplification efficiency between samples. As discussed previously, robust validated sample preparation methods will yield samples minimally effected by PCR inhibiting contaminants. Although validated sample preparation methods are critical to success, it is advised that attempts be made to normalize each sample. Optical absorbance measurements are the first step toward assuring that equal RNA sample mass is analyzed from each sample. It is best to demonstrate equal mass was analyzed and demonstrate equal amplification efficiency of each sample. These goals are most easily met if a gene, invariant in expression, is analyzed from each sample. Housekeeping genes which do not undergo altered expression during the biological state changes of an experiment can be selected for RNA mass normalization. Glyceraldehyde phosphate dehydrogenase (i.e. GAPDH), beta-actin, cyclosporin, 18S rRNA have often been used for this purpose. However, there is no universal "housekeeping gene" whose expression levels remain unchanged in all biological samples and biological states. Hence it is imperative to demonstrate that any gene chosen for RNA mass normalization maintains invariant expression in the biological system of study. Once a method of RNA preparation has been selected and validated, real-time PCR permits rapid analysis for invariance of expression of a chosen "housekeeping gene". RNA samples prepared from the different biological states of interest can be used to compare quantitative expression of a "housekeeping gene". If these results demonstrate equal expression from the different biological states, this gene can be used to normalize RNA mass. However, sometimes the experimental model is so complex (e.g. many different cell types within a tissue and no known "housekeeping gene" demonstrates invariance) that the "housekeeping gene" approach is not valid. In such instances reliable optical absorbance measurements for normalization to RNA mass is the only recourse. If optical measurements for RNA mass are the sole source for sample mass normalization, the use of synthetic RNA doped into the sample is suggested for amplification efficiency normalization.

Absolute (i.e. accurate) quantitation

If absolute quantitation (e.g. how many virus particles in 1 ml of blood?) is desired synthetic RNA calibrators (i.e. spikes) are usually necessary. Synthetic RNA spikes can be designed for any gene of interest. The synthetic RNA spike is designed to utilize the same primer set used for the mRNA of interest, but contains a unique internal sequence from which to design a detection probe, thus differentiating the endogenous mRNA of quantitative interest from the synthetic RNA spike. It is imperative that such a synthetic RNA maintain identical amplification efficiency with the sequence of the gene of interest. This is usually accomplished by maintaining similarity of amplicon size and G+C content between the synthetic RNA and endogenous mRNA. Once a synthetic RNA has been produced, a known quantity of the synthetic RNA (e.g. 100 molecules) can

TABLE 1

Ephrin A5 expression in DBA/2 mice

Animal	Ct GAPDH	Mean Ct	SD	CT ephrin A5	Mean Ct	SD	Normalized fold expression
#2	16.04 16.14	16.09	0.07	22.09 22.30	22.20	0.15	1.08
#4	16.35 16.36	16.36	0.01	22.57 22.57	22.57	0.00	1.00

Ct values of replicate RNA samples of GAPDH and ephrin-A5. mRNA was harvested from hippocampal sections of two DBA/2 mice and analyzed in duplicate for GAPDH and ephrin-A5 expression. The GAPDH mean Ct value was used to calculate mean RNA mass using the GAPDH standard curve generated from Fig. 2(B). The mean Ct of ephrin A-5 analysis was used to calculate relative fold ephrin-A5 expression from the standard curve generated in Fig. 3. Next, ephrin-A5 expression for each was normalized to RNA mass load (GAPDH) and reported relative to the arbitrary 1-fold expression of animal #4

be spiked (doped) into all RNA samples prior to real-time quantitative analysis. The unknown quantity of the endogenous mRNA of interest can be determined by comparison to the synthetic RNA results (Gibson et al., 1996).

Examples of quantitative real-time RT-PCR

In order to demonstrate the application of quantitative real-time RT-PCR, we will describe several examples. In the first example we analyze ephrin-A5 gene expression in the hippocampus of mice. As mentioned above, quantitative gene expression methods require normalization for the mass amount of RNA analyzed in each sample and monitoring amplification efficiency from each RNA sample. For this exercise we have chosen to utilize the housekeeping gene approach. So the first task is the selection and validation of an appropriate housekeeping gene. We chose GAPDH for this exercise and set out to monitor its expression level from hippocampal tissue from two different animals. We prepared total RNA from our microdissected hippocampal tissue using Qiagen Rneasy protocols (which we had previously determined to yield reproducible RNA mass recoveries and amplification efficiencies). After total RNA from each hippocampus was prepared, the RNA was analyzed by optical absorbance to calculate purity and concentration. Based on the OD_{260}, 50 ng of each sample was analyzed in duplicate for GAPDH Ct values. As is seen in Table 1 the Ct values from each replicate for a given animal were very similar (standard deviation < 0.16). This indicates the precision of the real-time method. Second, the mean Ct values from each animal were also very similar (16.09 and 16.36). This demonstrates that GAPDH values do reflect the RNA mass analyzed in our samples.

Now that we have a means to monitor RNA mass load, we are prepared to perform a real-time analysis for ephrin-A5 expression in the hippocampus of two DBA/2 mice. To do this we obtain an RNA sample which will serve as a reference standard for the experimental samples. We routinely chose a tissue or cell source which is similar to our experimental samples. We have chosen to use mouse brain total RNA (Clontech, Palo Alto, CA), for the standard RNA sample. One can also generate a standard RNA from any experimental animal source which would normally express the genes of interest. This standard RNA is serially diluted and analyzed for housekeeping gene expression and also the gene of interest, ephrin-A5. The data are plotted as standard curves

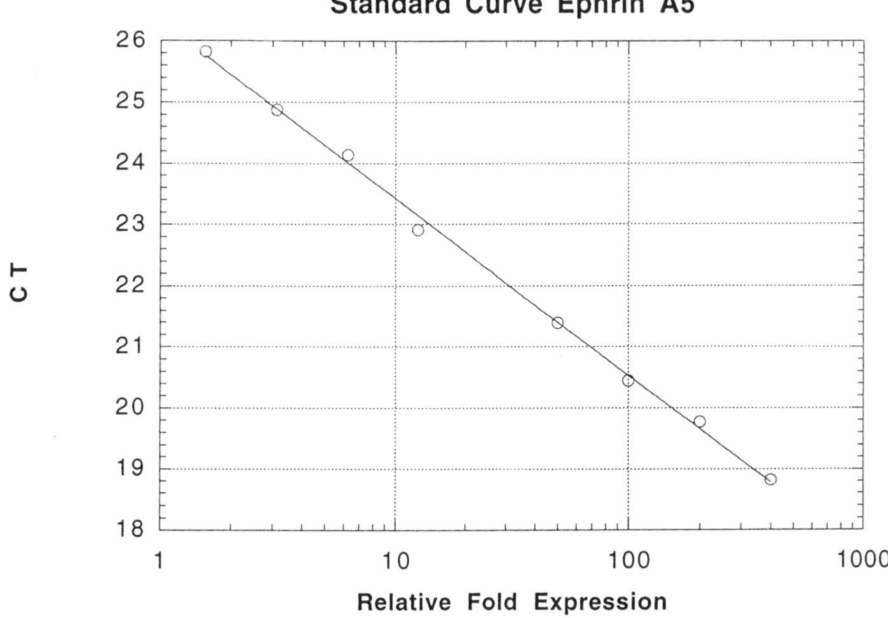

Fig. 3. Standard curve for the gene of interest (i.e. ephrin-A5). A reference RNA sample was 1:2 serially diluted and analyzed by real-time PCR for expression of ephrin-A5. A line equation is derived from this data, plotting Ct values (y) vs the relative fold expression of the gene of interest (x). Relative fold expression (x) is calculated for actual samples using this line equation and the known mean Ct value (y).

as shown in Fig. 2(B) (GAPDH) and Fig. 3 (ephrin-A5). The housekeeping gene reference sample data is plotted as Ct vs total mass of RNA loaded. The data for the gene of interest, ephrin-A5, are plotted as Ct vs relative fold expression. Line formulas are calculated for each gene's standard curve. The unknown experimental samples are analyzed simultaneously and the resultant Ct values for each unknown are used to solve for the relative level of expression. First the mean Ct for GAPDH expression from each experimental animal sample is solved for actual RNA mass using the GAPDH standard curve line formula. Second the mean Ct of ephrin-A5 for each experimental sample is used to calculate the relative fold expression, using the ephrin-A5 standard curve. Finally, the fold expression is normalized to the actual RNA mass value. In this experiment, the ephrin-A5 expression levels normalized to GAPDH values from each hippocampus sample are very similar, as expected (1.08 vs 1.00; Table 1). Using this type of approach one can easily see how more sophisticated experiments would be undertaken, such as measuring ephrin-A5 levels in response to a biological stimulus.

In a similar experiment, we wanted to compare the level of expression of several genes in two different strains of mice, DBA/2 and C57BL/6. RNA was prepared from hippocampal sections from mice of each strain. When 50 ng of total RNA is analyzed for expression of GAPDH mRNA, it is observed that they are very similar; mean Ct values of 16.09 and 16.36 for two DBA/2 animals (Table 2). These values are also very close to the two C57BL/ 6 animals' values of 16.12 and16.15. Using a GAPDH standard curve these Ct values are converted to RNA mass values (Table 2). Next we analyzed the mRNA expression levels of two genes; ephrin-A5 and N-CAM,

TABLE 2

Ephrin-A5 and N-CAM expression in DBA/2 and C57/BL6 strains of mice

	Animal	Ct GAPDH	RNA mass	Ct ephrin A5	Relative fold expression	Normalized fold expression ephrin-A5	Ct N-CAM	Relative fold expression	Normalized fold expression N-CAM
DBA/2	2	16.09	93.18	22.20	25.00	1.13	19.70	73.10	1.18
DBA/2	4	16.36	78.12	22.60	18.50	1.00	20.10	51.50	1.00
C57/BL6	17	16.12	91.30	22.50	20.40	0.94	19.80	63.70	1.06
C57/BL6	20	16.15	89.50	22.20	25.00	1.18	19.70	70.00	1.18

determining relative expression. As seen in Table 2, the normalized values for both ephrin-A5 and N-CAM are very similar between all animals, demonstrating no strain difference exists for expression of these genes in hippocampal tissues.

In a final demonstration of a quantitative application, gene expression is monitored in prostaglandin F2α (i.e. PGF2α) treated rat cardiac myocytes. Prostaglandin F2α has been demonstrated to cause hypertrophy of rat neonatal myocytes (Lai et al., 1996). For this application rat myocytes were cultured from neonatal hearts (King et al., 1998). The cultured myocytes were treated with 10-6* M PGF2α. Poly-A RNA was harvested at various time points after treatment (PolyAtract; Promega Corp. Madison, WI). Each sample represents approximately 5000 cells cultured in one well of a 96 well plate. In this instance extremely small amounts of RNA are recovered preventing measurements of RNA quantity. Invariant housekeeping genes, such as GAPDH, for sample normalization are essential (Winer et al., submitted). Figure 4 depicts the normalized expression of the ANF gene (atrial natiurectic factor) which is known to be induced in the cardiac hypertrophic response. ANF expression begins to increase above non-stimulated levels approximately 2 h after PGF2α treatment. Expression continues to increase throughout the time course, reaching approximately a 254-fold increase at 48 h. This data was generated using 4 replicate PCR reactions for each sample.

Conclusions

The use of real-time RT-PCR methodology permits rapid assay development from validation of RNA sample preparation methods, to validation of the invariance of housekeeping genes. Real-time PCR also provides a quantitative method for moderate to high throughput analysis of gene expression. The average instrument run time is 3 h, after which all of the data is ready for viewing. The ABI instrument permits the analysis of 96 tubes simultaneously, allowing for multiple samples and genes to studied in single PCR experiment. We have found this system to produce precise and reliable data which correlates with traditional methods (i.e. Northern blots). Additionally, PCR amplification enables the study of very small starting samples. For example, the average hippocampal weight used in this study was 50 mg. The RNA recovered from these tissues samples has been analyzed for expression of 20 different genes in replicate experiments (data not shown) and we still have sufficient RNA sample for future studies.

In conclusion, we believe that real-time RT-PCR is a technology that will play a major role

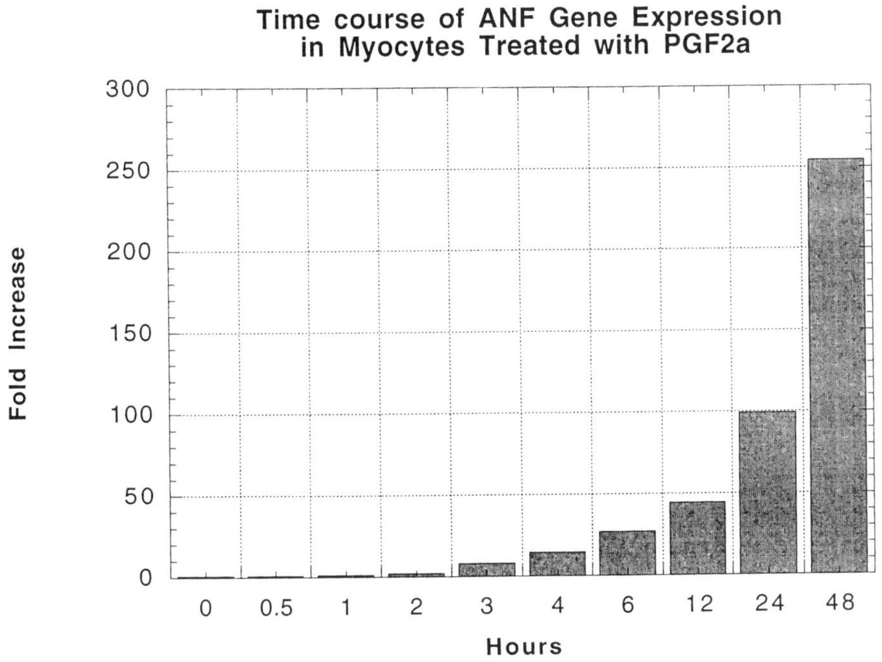

Fig. 4. Normalized fold expression of ANF mRNA. mRNA was harvested at various times after addition of PGF2α and analyzed for GAPDH and ANF mRNA levels. Standard curves using serial dilutions of reference RNA were used to calculate relative expression levels for GAPDH (RNA mass) and ANF (relative fold expression) as described above. ANF values for each time point were then normalized to the mass load of RNA (GAPDH). Finally fold expreesion is calculated by assigning time = 0 to be a 1-fold expression and reporting each time point relative to time = 0.

in the future of quantitative gene expression analysis. As new gene sequences are found, expression studies will follow. The use of a reliable precise method is critical to the success of understanding gene function.

References

Bachem, C. et al. (1996) Visualization of differential gene expression using a novel method of RNA fingerprinting based on AFLP: analysis of gene expression during potato tuber development. *Plant Journal*, 9: 745–753.

Becker-Andre, M. (1993) Absolute levels of mRNA by polymerase chain reaction-aided transcript titration assay. *Meth. Enzymol.*, 218: 420–445.

Becker-Andre, M. (1989) Absolute mRNA quantification using the polymerase chain reaction (PCR). A novel approach by a PCR aided transcript titration assay (PATTY). *Nucl. Acids Res.*, 17: 9437–9446.

Ferre, F. (1992) Quantitative or semi-quantitative PCR: reality vs myth. [Review]. *Pcr Methods and Applications*, 2: 1–9.

Gibson, U.E.M., Heid, C. and Williams, P.M. (1996) A Novel Method for Real Time Quantitative RT-PCR. *Genome Res.*, 6: 995–1001.

Heid, C., Stevens, J., Livak, K. and Williams, P.M. (1996) Real Time Quantitative PCR. *Genome Res.*, 6: 986–994.

Holland, P.M., Abramson, R.D., Watson, R. and Gelfand, D.H. (1991) Detection of specific polymerase chain reaction product by utilizing the 5′,3′ exonuclease activity of Thermus aquaticus DNA polymerase. *Proceedings of the Nat. Acad. of Sciences of the USA*, 88: 7276–7280.

Kenrick, M. et al. (1997) A homogeneous method to quantify mRNA levels: a hybridization of Rnase protection and scintillation proximity assay technologies. *Nucl. Acids Res.*, 25: 2947–2948.

King, K. et al. (1998) Phenylephrine, endothelin, prostaglandin F2alpha' and leukemia inhibitory factor induce different cardiac hypertrophy phenotypes in vitro. *Endocrine*, 9: 44–55.

Lai, J. et al. (1996) Prostaglandin $F_{2\alpha}$ induces cardiac myocyte hypertrophy *in vitro* and cardiac growth in vivo. *Am. J. Physiol. (Heart Circ. Physiol.)*, 2197–2208.

Lee, L.G., Connell, C.R. and Bloch, W. (1993) Allelic discrimination by nick-translation PCR with fluorogenic probes. *Nucl. Acids Res.*, 21: 3761–3766.

Liang, P. and Pardee, A. (1992) Differential display of eukaryotic messenger RNA by means of the polymerase chain reaction. *Science*, 257: 967–971.

Livak, K.J., Flood, S.J., Marmaro, J., Giusti, W. and Deetz, K. (1995) Oligonucleotides with fluorescent dyes at opposite ends provide a quenched probe system useful for detecting PCR product and nucleic acid hybridization. *PCR Methods and Applications*, 4: 357–362.

Lockhart, D.J. et al. (1996) Expression Monitoring by Hybridization to High-Density Oligonucleotide Arrays. *Nat. Biotech.*, 14: 1675–1680.

Piatak, M.J., Luk, K.C., Williams, B. and Lifson, J.D. (1993) Quantitative competitive polymerase chain reaction for accurate quantitation of HIV DNA and RNA species. *Biotechniques*, 14: 70–81.

Ririe, K., Rasmussen, R. and Wittwer, C. (1997) Product differentiation by analysis of DNA melting curves during the polymerase chain reaction. *Anal. Biochem.*, 245: 154–160.

Schena, M., Shalon, D., Davis, R.W. and Brown, P.O. (1995) Quantitative Monitoring of Gene Expression Patterns with a Complimentary DNA Microarray. *Science*, 270: 467–470.

Shen, L. et al. (1998) Quantification of cytokine mRNA in peripheral blood mononuclear cells using branched DNA (bDNA) technology. *J. Immunolog. Methods*, 215: 123–134.

Velculescu, V., Zhang, L., Vogelstein, B. and Kinzler, K. (1995) Serial analysis of gene expression. *Science*, 270: 484–487.

Williams, S. et al. (1996) Quantitative Competitive PCR: Analysis of Amplified Products of the HIV-1 gag Gene by Capillary Electrophoresis with Laser Induced Fluorescence Detection. *Anal. Biochem.*, 236: 146–152.

Winer, J., Jung, K., Shackel, I. and Williams, P. (submitted). Development and validation of real-time quantitative RT-PCR for monitoring gene expression in cardiac myocytes in vitro.

Wittwer, C. et al. (1997) The LightCycler: a microvolume multisample fluorimeter with rapid temperature control. *Biotechniques*, 22: 176–181.

W.E. Crusio and R.T. Gerlai (Eds.)
Handbook of Molecular-Genetic Techniques for Brain and
Behavior Research (Techniques in the Behavioral and Neural Sciences, Vol. 13)

CHAPTER 2.2.5

Analyzing genomic DNA discordance between monozygotic twins

J. Bouchard,[1] C. Foulon,[2] N. Storm,[1] G.H. Nguyen[1] and C.L. Smith[1]

[1]Center for Advanced Biotechnology and Departments of Biomedical Engineering, Biology and Pharmacology, Boston University, 36 Cummington Street, Boston, MA 02215, USA

[2]Institute des Sciences et Techniques de L'Ingénieur d'Angers, 62 Rue Notre Dame du Lac, 49100 Angers, France

Introduction

Monozygotic (MZ) or "identical" twins result from the fertilization of one ovum by one sperm. For unknown reasons, the embryo divides into two genetically "identical" embryos during the early stages of development following fertilization. Presently, sex, placentation, cord blood type, HLA antigens, and DNA fingerprinting are used for zygosity testing (Derom et al., 1987; Hill and Jeffreys 1985; Akane et al., 1991). Thus far, no method is considered the "gold standard" and all have the possibility for error.

Intra-uterine environmental differences in the allocation of cells in the placental vascular supply to each twin, as well as stochastic developmental events, may lead to major phenotypic discordance within a MZ twin pair (Hall, 1996). Application of new molecular techniques to MZ twins reveals that they do not possess identical genomes when individual loci are established (Jansen et al., 1994; Kruyer et al., 1994; Machin, 1996; Reyniers et al., 1993). Hence, it is important to reassess the role of genomic differences between MZ twins in producing phenotypic discordances. For example, the study of MZ twins is particularly valuable when only one twin is affected by disease, because a comparison of the almost identical genomes can lead to the identification of genetic abnormalities.

Various methods, including subtractive hybridization (Lisitzyn et al., 1993a), serial analysis of gene expression (SAGE; Velculescu et al., 1995), comparative genomic hybridization (CGH; Kallioniemi et al., 1992), and differential display (DD: Liang and Pardee, 1992) can be used for comparing genomes. However, these methods do not allow direct global comparisons between genomic DNA from complex genomes. Instead, they focus on comparing cDNAs. Working with cDNAs allows comparative studies to focus on important functional units (genes) and reduce sample complexity. For example, when cDNA is analyzed, the human genome complexity is reduced from 3×10^9 base pairs (bp) to an estimated 1×10^8 bp ($\sim$100,000 genes $\times$ $\sim$1000 bp/gene).

Targeted Genomic Differential Display (TGDD) was developed by us to compare complex genomes using genomic DNA rather then cDNA (Broude et al., 1997, 1999). In TGDD, it is essential that genomic complexity be reduced to an analyzable level. Targeting in TGDD is used to reduce genome complexity and to focus analysis on, and nearby, sequences of interest. The target sequence can be a simple repeating sequence (e.g. $(CAG)_n$ or $(TG)_n$), a sequence coding for a protein motif, a transcriptional regulator element, or any other important genomic region. In TGDD, as in DD, DNAs are fractionated (displayed) by size to produce a DNA fingerprint. Essentially, TGDD detects restriction fragment length polymorphisims (RFLPs), where each difference must be characterized to understand its origin. RFLPs may arise from different causes including DNA sequence variation, insertions, deletions recombination events or methylation changes.

Two TGDD protocols have been developed. Method I (Broude et al., 1997) uses a capture and PCR protocol whereas Method II (Broude et al., 1999) solely uses PCR (Figs 1 and 2, respectively). Here, the TGDD protocols are described and illustrated with results obtained from the genomic analysis of twin pairs.

Method

Isolation of genomic DNA (Method I and II)

TGDD can use genomic DNA isolated from human blood lymphocytes, buccal scrapes, or sperm. Blood samples were collected in 6 ml ACD (Becton-Dickenson) tubes to maximize the allowable time between collection and DNA isolation (i.e. 5 d maximum). The buccal scrape was performed by wiping each side of the oral cavity 10 times with a cytobrush (Medscand #1101). The brush was then shaken in 300 μl cell lysis solution and immediately removed. DNA from buccal scrape samples was collected and stored in the cell lysis solution included in the Puregene DNA Isolation Kit (Gentra Systems, Minneapolis). Freshly ejaculated sperm was mixed with 0.3X ESP (1X ESP = 1 mg/ml ethelene diamine tetraacetic acid (EDTA), 1% sodium lauroylsarcosine (Sigma #L5000), and 1 mg/ml proteinase K; Sigma), and 7% (V:V) β-mercaptoethanol as described by Smith et al., 1993. The high concentration of β-mercaptoethanol is necessary because of the large number of disulfide bonds in the sperm coat proteins. Samples stored in ESP may be shipped and stored at room temperature until further use.

The experimental data shown in this chapter were obtained from blood samples. The MZ twin samples were obtained from Terry Reed of the National Heart, Lung, and Blood Institute (NHLBI) and E. Fuller Torrey of the National Institute of Mental Health. The sibling samples were obtained from Clinton Baldwin of the Boston University Medical Center. The DNA was extracted using the protocol accompanying the Puregene DNA cell line isolation kit (Gentra System, Minneapolis). The red blood cells were lysed to facilitate their separation from white blood cells by adding 18 ml RBC lysis solution to 6 ml whole blood. After 10 min incubation at room temperature, the samples were centrifuged for 10 min at 3000 rpm and the supernatant was removed. To the remaining white pellets, 6 ml of Lysis Solution was added and pipeted up and down to lyse the white blood cells. Proteins were sedimented by centrifugation at 3000 rpm for 10 min after vigorous mixing of 2 ml protein precipitation solution. The supernatant was then transferred to a new tube and the genomic DNA was precipitated with 6 ml isopropanol by inverting the tube until the white threads of DNA formed a visible clump. The white DNA threads were scooped out with a heat-sealed Pasteur pipet and dissolved in a DNA hydration solution contained in a microcentrifuge tube. The tube was then incubated overnight in a 37°C oven. The sample was treated with RNase (1:1 (wt:wt) RNase:DNA;Boehringer Mannheim) at 37°C for 30 min, and then the DNA was stored at 4°C until used.

Genomic DNA concentration (Method I and II)

It is of great importance that the concentrations of DNAs being compared are well matched. The number of false differences between samples increases proportionally with differences in DNA template concentrations used in the PCR reaction. The DNA concentrations should be

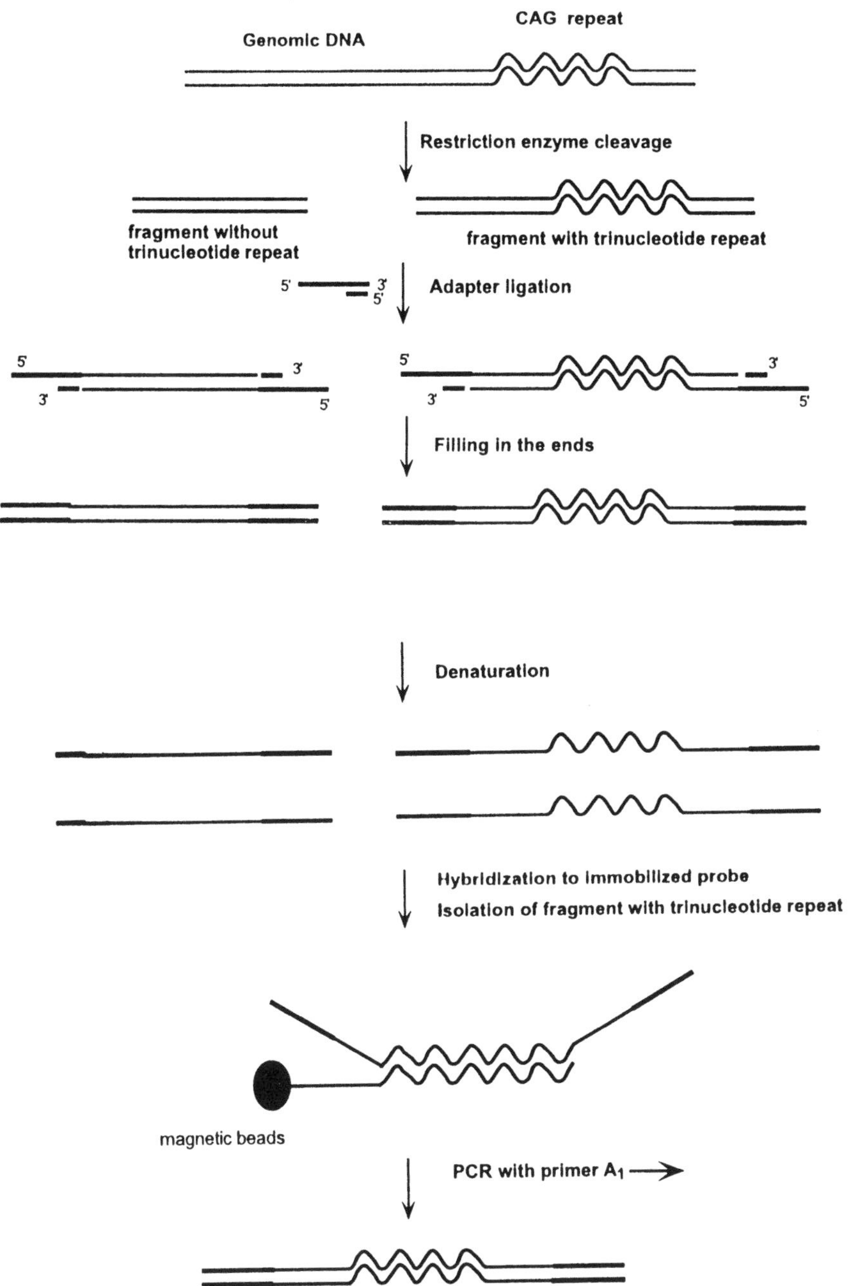

Fig. 1. General protocol for TGDD Method I (see text for description).

Method II

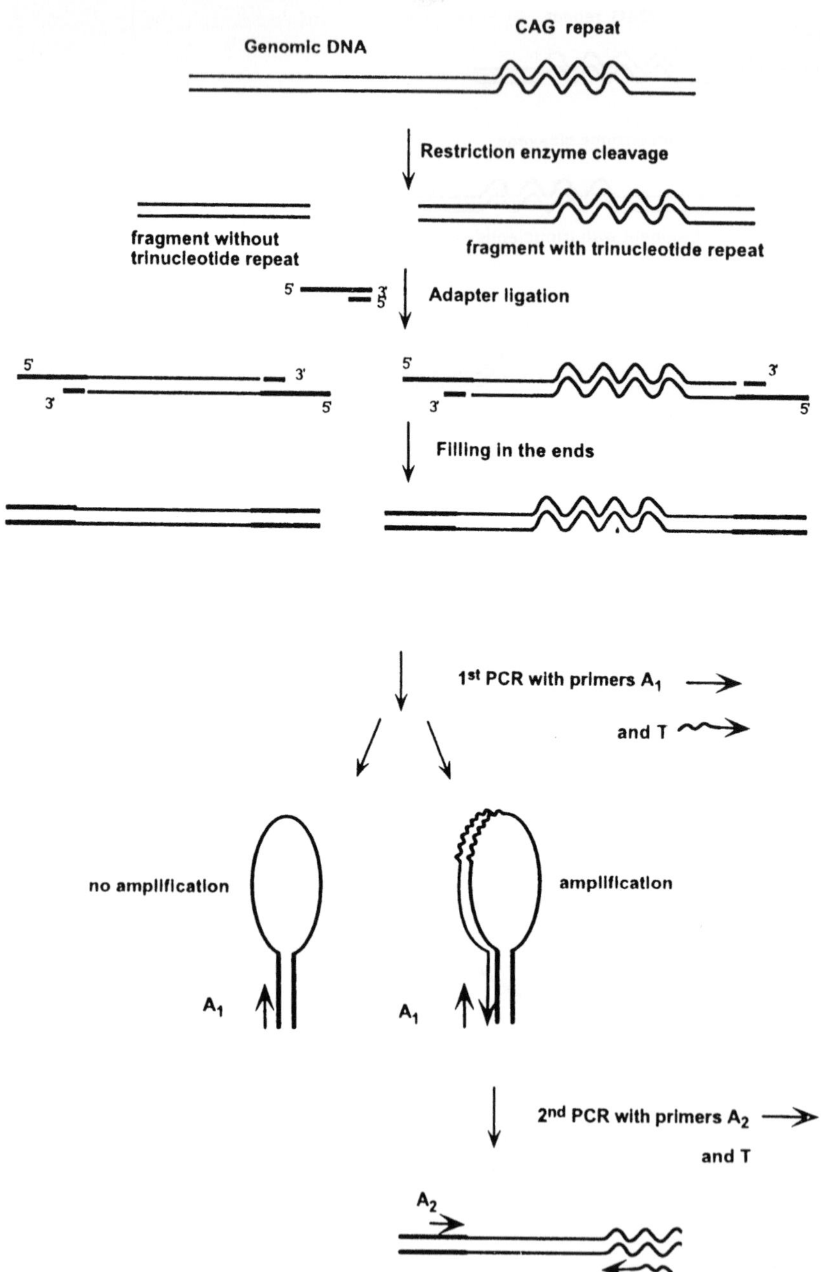

Fig. 2. General Protocol for TGDD Method II (see text for description).

matched throughout the protocol. In some cases, these differences in concentration may only be apparent after the products of TGDD are examined.

There are a number of ways to determine DNA concentrations. All methods have problems. Hence, two goals must be kept in mind. The DNA concentrations used must be well-matched and within an acceptable range. The acceptable range before the first PCR is ~5–20 ng, but some adjustment in the number of PCR cycles may be necessary to insure good signal-to-noise ratios are obtained.

Matching DNA concentrations is best determined on an agarose gel stained with ethidium bromide. The fluorescence of the samples are compared to each other and a standard DNA sample. The samples are briefly electrophoresed to ensure that the RNA component of the sample is separated from the DNA. DNA concentration has also been determined using a DipStick Kit (Invitrogen) as recommended by the manufacturer. Matching fluorescence on agarose gels, however, appears to be the most reliable method for DNA concentration determination and matching. It is important that an individual be very consistent to insure reproducibility.

DNA preparation (Method I and II)

Approximately 1 μg of MZ twin genomic DNA was digested in reactions containing 20 units of *Sau3AI* or *Hae III* (Method I and II respectively; New England Biolabs) and 1X New England Biolabs Buffer *Sau3AI* (100 mM NaCl; 10 mM Bis Tris Propane-HCl; 10 mM $MgCl_2$; 1 mM dithiothreitol, pH 7.0 at 25C) or New England Biolabs Buffer #2 (50 mM NaCl; 10 mM Tris-HCl; 10 mM $MgCl_2$; 1 mM dithiothreitol, pH 7.9 at 25°C) respectively supplemented with 100 μg/ml bovine serum albumin (BSA). Reactions were incubated at 37°C overnight, and the restriction fragments were purified using Phase Lock Gel Tubes (5 Prime-3 Prime, Boulder, CO) and ethanol precipitated.

Ligation (Method I and II)

The digested genomic DNA was then ligated to oligonucleotides of a known sequence (adapters) in a 20 μl reaction. Thus far, the adapter sequences used for Method I and Method II were different. Method I and II used oligonucleotides 1 and 2 or 3 and 4 respectively (Table 1). The reaction consisted of 10 μl of restricted genomic DNA (approximately 1 μg), 2 μl of 10X T4 ligation buffer from New England Biolabs (1X solution: 50 mM Tris-HCl, pH 7.5; 10 mM dithiothreitol; 1 mM ATP; 25 μg/ml BSA), 400 units (1 μl) T4 ligase (New England Biolabs), 2 μl of each 10 μM adapter mixture (equimolar amounts of oligonucleotides 1 and 2 or 3 and 4 in Methods I and II respectively; Table 1), and 3 μl of H_2O. Samples were incubated overnight at 16°C. DNA was phenol-extracted, precipitated with ethanol, washed with 70% ethanol, dried, and dissolved in TE buffer (10 mM Tris HCl, pH 8.0; 1 mM EDTA).

Capture/PCR targeting (Method I)

A biotinylated oligonucleotide (10 pmol) containing a target sequence (i.e. $(CTG)_{12}$; oligonucleotides 5, Table 1) was mixed with 50 ng of ligation products in 50 μl of TE buffer containing 2 μM of the corresponding adapter oligonucleotides to prevent annealing of the fragment ends to each other. After the addition of mineral oil, the sample was heated to 95°C, slowly cooled to room temperature, added to 100 μg of prewashed streptavidin coated magnetic beads M-280 [as directed by Dynal (Oslo)] using a 3-fold molar excess of biotin binding capacity over biotinylated oligonucleotides, and incubated at room temperature for 1 h with gentle rotation. The beads were collected with a magnet, washed twice at 55–60°C for 20 min with 3X standard saline citrate (SSC; 1X SSC = 0.15 M NaCl, 15 mM sodium citrate) and 0.5% SDS and, at room temperature, twice each, with TE containing 1 M NaCl and with

TABLE 1

Synthetic oligonucleotides used in this work

Number	Description	Sequence (5′ → 3′)[a]
1	*Sau3AI* adapter 24	CGGGAATTCTGGCTCTGCGACATG
2	*Sau3AI* adapter 10	GATCCATGTC
3	*Hae III* adapter 43	TGTAGCGTGAAGACGACAGAAAGGGCGTGGTGCGGAGGGCGGT
4	*Hae III* adapter 11	ACCGCCCTCCG
5	CTG-12	b-GATGATCCGACGCAT$(CAG)_{12}$
6	CTG-A T-primer	$(CTG)_6A$[b]
7	CTG-G T-primer	$(CTG)_6G$[b]
8	CTG-T T-primer	$(CTG)_6T$[b]
9	Na21 A-primer	TGTAGCGTGAAGACGACAGGA
10	ST19*HaeIII* A-primer[c]	AGGGCGTGGTGCGGAGGGCGGTCC
11	ST19*HaeIII*G A-primer[c]	AGGGCGTGGTGCGGAGGGCGGTCCG
12	ST19*HaeIII*GG A-primer[c]	AGGGCGTGGTGCGGAGGGCGGTCCGG
13	ST19*HaeIII*TG A-primer[c]	AGGGCGTGGTGCGGAGGGCGGTCCTG
14	ST19*HaeIII*AC A-primer[c]	AGGGCGTGGTGCGGAGGGCGGTCCAC

[a] b = biotin; [b] Cy5 labeled; [c] Note the presence of 3′ CC bases. These bases anneal to genomic GG sequences remaining from the *HaeIII* recognition site; 3′ anchored bases are terminal to the CC dinucleotide.

TE alone. Beads with captured DNA were stored in TE buffer at 4°C.

One-fifth of the captured DNA was amplified by PCR in a PTC-100 thermal cycler (MJ Research, Cambridge, MA). The 50 μl reaction contained 67 mM Tris HCl, (pH 8.8); 4 mM $MgCl_2$; 16 mM $(NH_4)_2SO_4$; 10 mM 2-mercaptoethanol; 300 μM of each dNTP; 2 units of AmpliTaq DNA polymerase; and 5 μM fluorescent labeled adapter primer (A-primer). After a hot start (see below), samples were subjected to 20–23 PCR cycles, each consisting of 1 min at 94°C and 3 min at 72°C, and a final incubation at 72°C for 5 min (see results for discussion of PCR primer choices).

PCR/targeting (Method II)

Two consecutive hot start PCRs (see below) were performed. The first and second PCRs targeted a specific sequence of interest, whereas the second uses a semi-nested PCR to further reduce the complexity of the amplified fragments. The products of the first PCR were diluted 1/100 and used as templates for a second PCR. Here, conditions for the second PCR were the same as in the first PCR, except that a different oligonucleotide (10–14; Table 1) is used in place of oligonucleotide 9 (Table 1) and 20–25 cycles were performed. Usually the 25 μl PCR contained 10–12 ng DNA, 5 mM $MgCl_2$, 1 mM of each dNTP, 5 pM of each primer and 1.25 Units of enzyme.

Hot start PCR (Method I and II)

Usually, TaqStart (Clontech, Palo Alto, CA) antibody was used to insure hot start PCR. Hot start PCR insures that primer annealing and elongation does not occur until the complementary sequence is found. The hot start mix consisted of 0.25 μl TaqStart Antibody (1.1 mg/μl; 7 μM), 0.25 μl AmpliTaq DNA polymerase (5 units/μl; 0.25 μM) and 1 μl dilution buffer, 50 mM KCl, 10 mM Tris-HCl (pH 7.0) was incubated at room temperature (20–22°C) for 5 min and then added to a 25 μl PCR reaction. The PCR was centrifuged and subjected to the PCR conditions listed above.

Hot start PCR was also performed by withholding one of the reaction components until the sample had gone through it first long denaturating step. It is best to withhold the enzyme to maximize its activity. For instance, the half-life of Amplitaq is 40 min at 95°C. For hot start PCR, the enzyme was diluted into 10 μl of 1X PCR buffer and the other components were made up in a total volume of 15 μl 1X PCR buffer. The 15 μl was overlaid with mineral oil, denatured and brought to 80°C. The diluted enzyme dropped on top of the mineral oil sedimented to the aqueous bottom layer to complete the reaction components. The PCR was then performed as described above.

TGDD fractionation (Method I and II)

Most times the amplification products were analyzed on a 2% agarose gel containing ethidium bromide. The approximate lengths of the amplified fragments were determined by comparison with a 50 bp ladder (Boehringer Mannheim). Then, the DNA fragments were analyzed on a high-resolution DNA sequencing gel. Usually, two and a half microliters of the second PCR were mixed in 3.5 μl of a stop solution (6 mg/μl of dextran blue and 0.1% Sodium dodecyl sulfate in deionized formamide; Pharmacia Biotech, Upsala, Sweden;) and denatured for 5 min at 94°C. After denaturation, the sample was immediately quenched on ice and loaded on a 6% denaturing polyacrylamide gel (PAGE) in 0.6X TBE (54 mM Tris-borate, 1.2 mM EDTA). The samples were fractionated using the ALFexpress Sequencer (Pharmacia Biotech). A Cy-5 labeled 50 bp ladder (ALFexpress sizer 50-500, 27-4539-01, Pharmacia Biotech) was used as a standard to determine the fragment length.

Data analysis

The results were visualized and compared using the Fragment Manager Software provided with the ALFexpress. This software displays band intensity as a function of elution time, which is approximately equal to bp length. The data can be viewed in two different formats: fullscale and autoscale. Full-scale data is unprocessed intensity measurements, where band intensity is shown relative to background intensity. Most of the data presented here is autoscaled with the highest intensity peak in each lane set to 100%. The lower intensity peaks are scaled relative to the largest peak. All data was examined in both formats. Full-scale analysis allows signal-to-noise ratios to be examined, while autoscaled analysis amplifies peak height and is especially useful with low signal-to-noise data. Differences found in twin pairs due to varying signal-to-noise ratios must be ignored in both cases. Figs. 4, 8 and 11 show autoscaled data, while Figs. 6, 9, 10 and 12 show fullscale data.

Isolation of polymorphic fragments (Method I and II)

Specific fragments were isolated from a 3% 3:1 (wt:wt) Nusieve: LE agarose (FMC Bioproducts) by fractionation of 20 μl of sample for 5 h at 130 V in a 1X TAE (40 mM Tris-acetate, 1.0 mM EDTA) buffer. Specific bands of interest were picked from the gel with a pipet tip and reamplified. These products were then analyzed on a 6% denaturing polyacrylamide gel as described above. The gel purification may be repeated for direct genomic sequencing or the fragments may be cloned and sequenced.

Results

Two methods for TGDD are described. These methods are very similar; however, there are important technical details that are described in the Materials and Methods that differentiate Method I from II. The goal of both methods is to create a library composed of restriction fragments that share a target sequence. This means that each member of the library has a common sequence and a unique sequence. The experiments described here focused on $(CAG)_n$ repeats containing restriction fragment libraries. These libraries can be used to create a complex DNA fingerprint as described in this work or may be used in other applications, such as creating clone libraries containing sequences useful for genetic mapping experiments (Oliveria et al., 1998).

The target libraries are created by a capture and PCR protocol (Method I) or a PCR protocol alone (Method II). Currently, most of our experiments use Method II because it is easier to perform. However, both methods are more than 90% effective in specific sequence targeting (Broude et al., 1997, 1999). It should be noted that the products of the methods are not the same. Specifically, the fragment products of Method I have a target sequence surrounded by unique sequence and end-tagged with the adapter sequence. The ends of the fragment products of Method II are the target and adapter sequences. Hence, additional experiments must be done to isolate the single copy sequence flanking the other side of the target sequence (see below and Figs. 1 and 2).

DNA preparation

Both methods can use the same preparation of DNA. The first decision to make in applying this procedure is the choice of restriction enzyme to fragment the genome. Our restriction enzyme choices were determined empirically. However, as more of the human genome sequence becomes available, it will be possible to choose the appropriate restriction enzyme a priori based not only on the frequency of occurrence of a specific recognition site, but also on the distance of restriction enzyme cleavage sites from specific target sequences. In fact, ongoing TGDD modeling experiments using the sequences of the *Saccharomyces cerevisiae* and *Caenorhabditis elegans* genomes are doing just this (Bouchard et al., manuscript in preparation).

Cleaved genomic restriction fragments are ligated to oligonucleotides (adapters) of a known sequence. A-primers complementary to the adapter sequences allow PCR amplification of the restriction fragments. Thus far, the adapters used in Methods I and II have been of different lengths. Method I uses adapters that are 20-mer or shorter in order to avoid the annealing of the complementary ends, since fragment end-annealing may interfere with A-primer annealing and extension. Method II uses adapter lengths of 40-mer to promote end annealing and inhibition of A-primer annealing and extension. In Method II, the first round of PCR amplification must occur from the target primer (T-primer) complementary to the target sequence (see below).

Method I

In Method I, the tagged restriction fragments are hybridized to an immobilized single-stranded DNA complementary to the target sequence (Fig. 1). The captured genomic fragments are eluted and used as template DNA for PCR. The single-stranded template DNA has the target sequence surrounded by unique sequences and is end-tagged with the known adapter sequence. PCR amplification may be done using different primer combinations (Fig. 3). The entire template DNA is amplified when a single A-primer is used. Alternatively, an A-primer may be used in combination with a T-primer. In the latter instance, the region between the target sequence and the adapter sequence on one strand is amplified along with the intervening unique sequence. Use of the target sequence complement and an A-primer amplifies the region between the target sequence and the adapter on the other strand. The use of a T-primer also increases the robustness of the targeting process. Most of our experiments used a labeled T-primer, but any primer in the experiment may be labeled. A TGDD comparison of an alleged MZ twin pair, generated by Method I, is shown in Fig. 4.

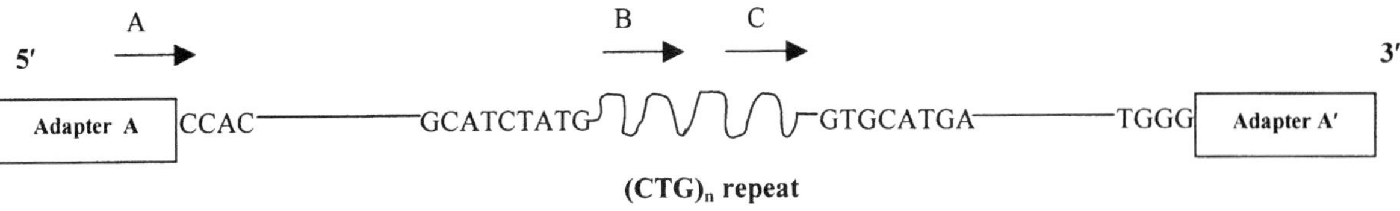

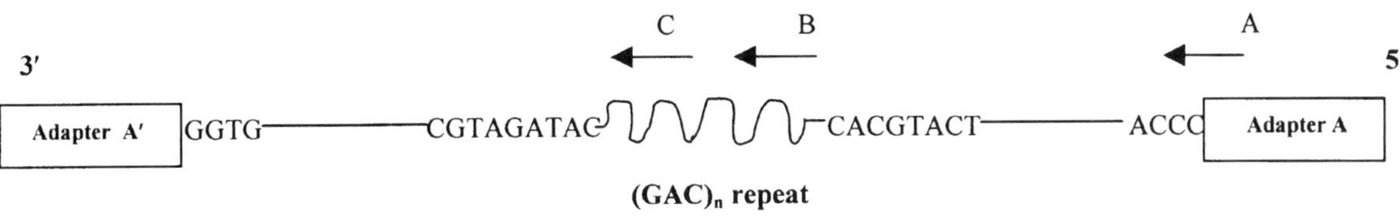

Fig. 3. Possible TGDD primer combinations. A-primers are indicated by A and T-primers are indicated by B or C. Note that the possible primer combinations are A alone, A and B, and A and C for Method I and A and B or A and C for Method II.

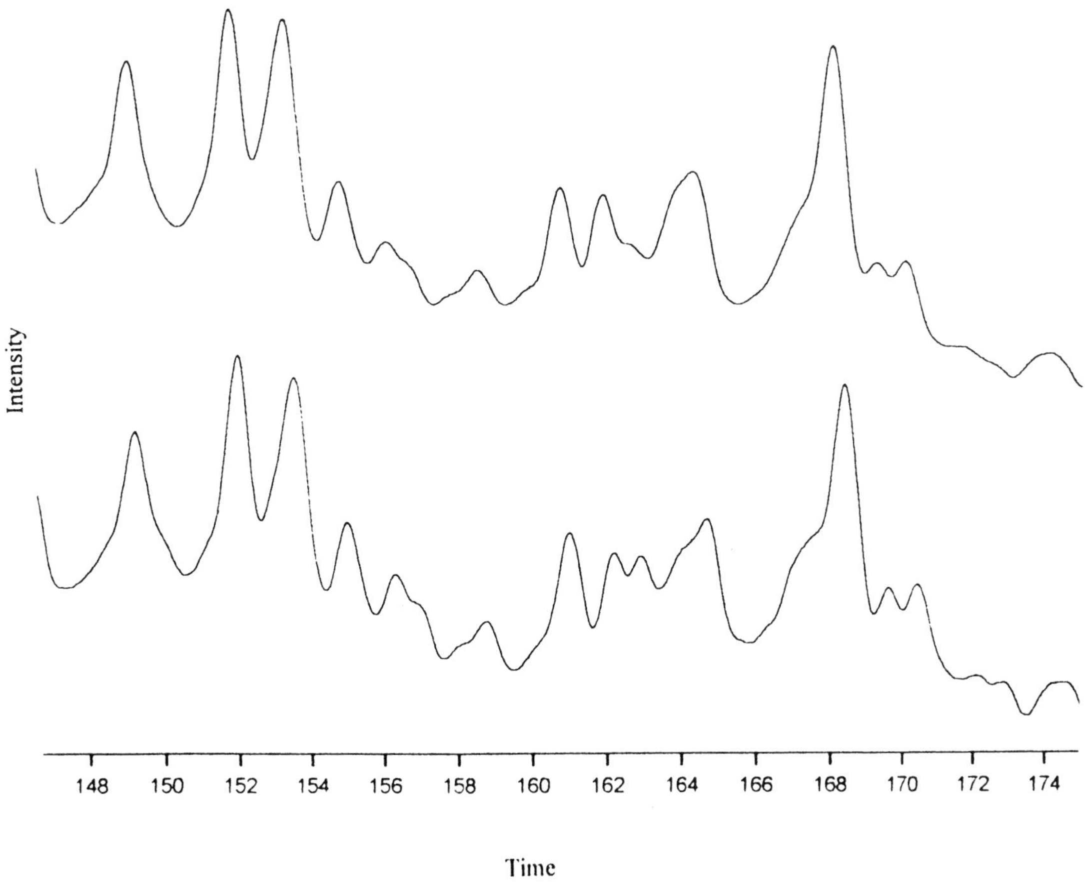

Intensity

148 150 152 154 156 158 160 162 164 166 168 170 172 174

Time

Fig. 4. A Method I comparison of MZ twins. The immobilized capture probes and T-primer were both a $(CTG)_n$ repeating sequences (oligonucleotide 5 and 6 respectively, Table 1).

Method II

Method II uses PCR alone for targeting (Fig. 2). This means that the biotin capture step is eliminated. Targeting is performed using a T-primer under PCR suppression conditions (PS: Siebert et al., 1995). PS takes advantage of the fact that the adapters ligated to the ends of restriction fragments are self-complementary, which enhances targeting. Annealing of the ends of a fragment to each other can interfere with the annealing and extension of a complementary A-primer. The interference is enhanced by using long (e.g. 40 bp) GC-rich adapters. Efficient amplification of such templates occurs only when a T-primer target is located on the internal single-stranded region of the template that is not annealed. Elongation of T-primer will produce a product with non-complementary ends (i.e. one end will contain the target sequence and the other end will contain the adapter sequence). This is the same structure that is obtained with Method I when a T-primer is used in the PCR amplification step. All of the remaining TGDD data presented in this paper used Method II (Figs 6, 8–12).

The T-primer

The T-primer can be composed of any sequence of interest in the genome. Thus far, our focus has centered on $(CAG)_n$ and $(CA)_n$ repeating sequences (Broude et al., 1997, 1999; Oliveria et al., 1998), long terminal repeats (LTR) of human endogenous retroviruses (Lavrentieva et al., 1999), and one Zn-finger gene family (Foulon et al., manuscript in preparation).

The design of the T-primer is a fundamental aspect of TGDD. A major question is whether or not a further reduction in sample complexity is necessary beyond that afforded by the selected T-primer. A sample that is too complex will be difficult to analyze. Sample complexity is reduced by adding unique bases (anchors) to the 3′ or 5′ end of the T-primer and/or A-primer (Fig. 5). It is important to remember that 3′ anchors added to A-primers must be terminal to bases needed to anneal to any remaining restriction enzyme recognition sequences.

The unique bases also anchor the T-primer to either the 3′ or 5′ end of a T-primer for a simple repeating sequence. Anchor bases anneal to the bases adjacent to the target sequence in genomic DNA. PCR amplification selects those target-containing restriction fragment pool that contain the bases complementary to anchors. Anchor T-primers must be used when the target is a simple repeating sequence to ensure amplification from the same location. A 3′ anchored T-primers amplifies genomic DNA adjacent to simple repeat sequence, and in general, do not provide information about repeat length. This is due to the fact that the PCR products all have the same 5′ end made from the T-primer sequence. A 5′ anchored T-primer should amplify the entire repeat sequence. The effectiveness of 5′ anchoring may be improved by increasing the number of bases (anchors) added to the 5′ end.

It should be noted that sequences such as $(CAG)_n$ repeats are capable of forming alternative structures which may interfere with T-primer annealing and/or extension (see Broude et al., 1997, 1999 for discussion). In fact, long $(CAG)_n$ repeat lengths form hairpin structures which appear to inhibit PCR with a $(CAG)_n$ containing T-primer. Hence, long versus short $(CAG)_n$ repeating sequences within the Huntington's disease locus were distinguished by TGDD (Broude et al., 1997).

In a random sequence, the addition of a single anchor would reduce the complexity of the sample by four-fold, whereas the addition of two bases would reduce the complexity by sixteen-fold. An example of complexity reduction using anchors is shown in Fig. 6.

Designing a T-primer to target a gene family is more complex. Conventionally, gene families have been identified by amino acid homologies. An example of amino acid alignments in different gene families is shown in Table 2. Some families (e.g.

Unanchored T-primer:

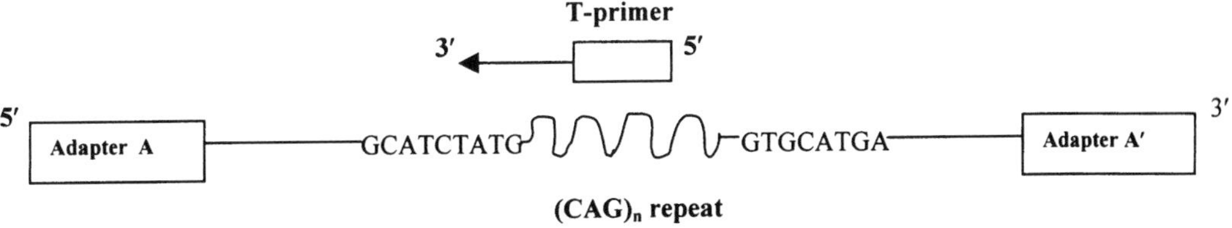

3′ C anchored T-primer:

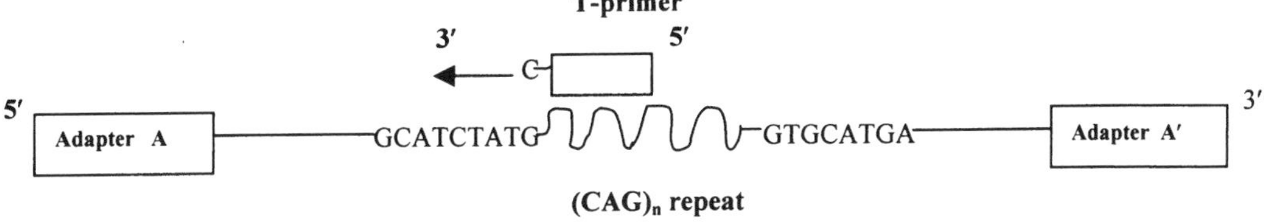

5′ C anchored T-primer:

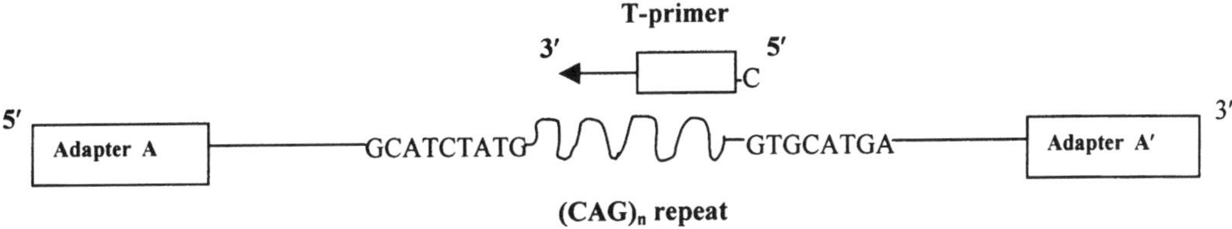

Fig. 5. Principles of anchoring (see text for description).

heat shock 70 proteins) have a well-defined amino acid consensus region while others do not (e.g. protein kinase C family terminal domain). Some gene families may also be subdivided into groups with different consensus sequences. Note that only 6 to 7 amino acids are needed for the PCR primer design.

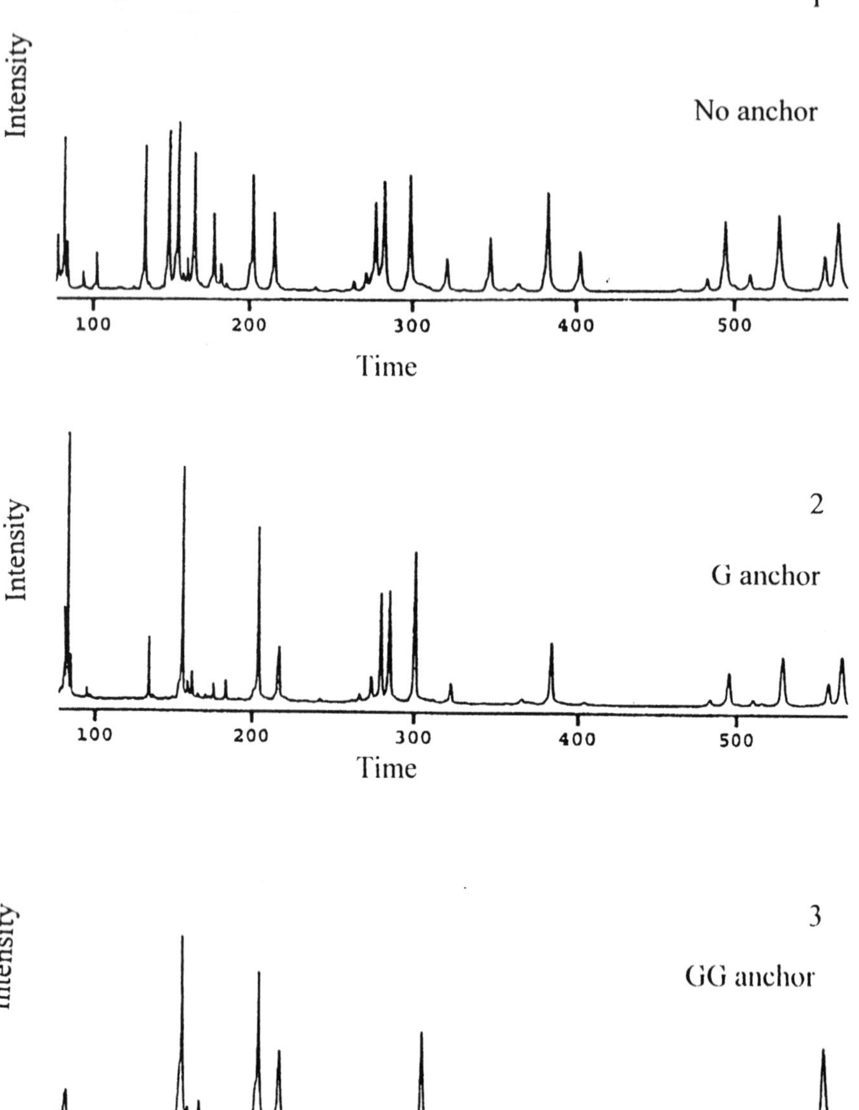

Fig. 6. The effects of 3′ anchoring on TGDD (Method II). The $(CAG)_n$ containing fragments were targeted using oligonucleotide 7 (Table 1). The unanchored A-primers were ST19*HaeIII* (lane 1) and the anchored A-primer was ST19*HaeIII*G (lane 2) or ST19*HaeIII*GG (lane 3); (oligonucleotides 10, 11, and 12 respectively; Table 1).

TABLE 2

Examples of consensus amino acid sequences of protein families. This table was adapted from information obtained from the Sanger Center web site: (http://www.sanger.ac.uk/Pfam/)

Family	Identification number	Selected alignments	Consensus % identity[a]	Total % identity[b]
Heat shock Hsp 70 proteins	DNAK_BACME/4-574	IIGIDLGTTNSCVAVLEGGEPKV	100	92
	DNAK_METMA/4-578	ILGIDLGTTNSCVAVMEFFEAVV	100	88
	DNAK_BORBU/4-599	IIGIDLGTTNSCVAIMEHGKPVV	100	87
	DNAK_CLOAB/4-576	VIGIDLGTTNSCVAVMEGGDPAV	100	85
	DNAK_CHLPN/10-603	IIGIDLGTTNSCVSVMEFFQAKV	100	82
	consensus:	II<u>GIDLGTT</u>NSCVAVMEGGEPVV		
Protein Kinase C terminal domain	KPC1_YEAST/1084-1149	RNINFDDILNLRVKPPYIPEIKSP	100	79
	PCK2_SCHPO/943-1008	SNINWDDIYHKRTQPPYIPSLNSP	83	77
	KPC1_CANAL/1030-1095	HDVNFDDVLNCRIPAPYIPEVQSE	83	54
	KPC1_HUMAN/601-667	RYIDWEKLERKEIQPPYKPKARDK	66	52
	PCK1_SCHPO/924-988	ASIVWDDLYNKLYEPSYKPLINDP	50	48
	consensus:	RNINWDDLLNRR8Q<u>PPYIPE</u>INSPIY		

[a] % identity to underlined consensus sequence
[b] % identity of total consensus

The consensus amino acid sequence is only a part of the puzzle. When deciding on which part of the consensus amino acid sequence to use for primer synthesis, it is best to look for amino acids with as few codons as possible (Table 3). For instance, tryptophan and Methiomne are coded for by only one codon. Most, but not all of the codon variability lies in the third position of the codon. Thus, variable bases used in the primer can compensate for these ambiguities and allow entire gene families or subgroups to be targeted. However, the introduction of ambiguity can have adverse effects on the PCR reactions. Each variable base included in the primer decreases the primer concentration by half. Hence, many times it is not effective to have a broad T-primer. An example is shown in Figure 7.

Instead, it may be useful to use a chimeric primer consisting of a variable and a constant region. This is called Tagged PCR (T-PCR; Grothues et al., 1993). In this approach, several initial PCR cycles are performed with the chimeric primer. The remaining PCRs are performed using a constant region primer, which usually has a higher T_m than the chimeric primer.

Another important consideration is codon usage. Not all codons are used equally within a particular genome. Table 3 shows general codon

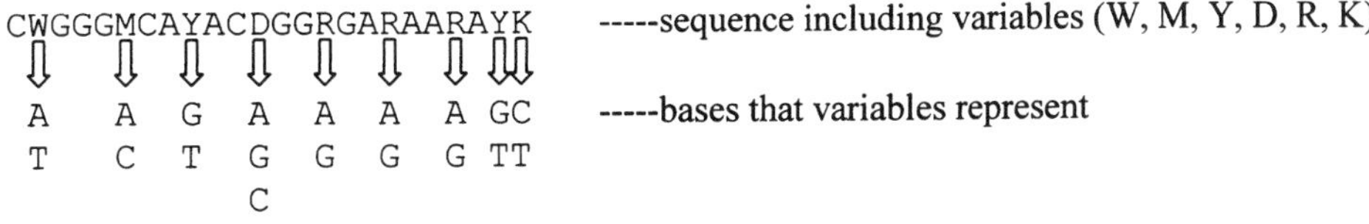

Fig. 7. Variable base options in a T-primer to a protein motif. Eight positions have 2 variables bases and 1 position has three variable bases. If all variability is built into the primer the concentration of any single sequence will be 768 fold ($2^8 \times 3^1$) less than the total concentration. For example, a 1.5 μM primer solution would only have the effectiveness of a 1 nM primer concentration. The best way to compensate for this problem is not obvious.

TABLE 3

The universal genetic code and human codon usage frequency. Information was adapted from www.nih.go.jp/ ~jun/research

Amino acid	Codons	Frequency	Amino acid	Codons	Frequency
Phe	TTT	0.43		GCC	0.40
	TTC	0.57		GCA	0.22
Leu	TTA	0.06		GCG	0.10
	TTG	0.12	Tyr	TAT	0.42
	CTT	0.12		TAC	0.58
	CTC	0.20	His	CAT	0.41
	CTA	0.07		CAC	0.59
	CTG	0.43	Gln	CAA	0.27
Ile	ATT	0.35		CAG	0.73
	ATC	0.52	Asn	AAT	0.44
	ATA	0.14		AAC	0.56
Met	ATG	1.00	Lys	AAA	0.40
Val	GTT	0.17		AAC	0.56
	GTC	0.25	Asp	GAT	0.44
	GTA	0.10		GAC	0.56
	GTG	0.48	Glu	GAA	0.41
Ser	TCT	0.18		GAG	0.59
	TCC	0.23	Cys	TGT	0.42
	TCA	0.15		TGC	0.58
	TCG	0.06	Trp	TGG	1.00
	AGT	0.14	Arg	CGT	0.09
	AGC	0.25		CGC	0.19
Pro	CCT	0.29		CGA	0.10
	CCC	0.33		CGG	0.19
	CCA	0.27		AGA	0.21
	CCG	0.11		AGG	0.22
Thr	ACT	0.23	Gly	GGT	0.18
	ACC	0.38		GGC	0.33
	ACA	0.27		GGA	0.26
	ACG	0.12		GGG	0.23
Ala	GCT	0.28			

usage in humans. However, different classes of genes use different codons. As the human genome project provides more sequence data, considerations such as codon usage will become clearer. A more complicating fact is that some gene families share structural motifs rather than protein sequence motifs. Presently, computational methods are being developed to address the identification of such families. It is not yet clear how the structural families relate to DNA sequence.

Decreasing the complexity of a targeted gene family fragment pool can be accomplished using anchored bases as described for targeting a repeated sequence. Alternatively, the T-primer can be designed to anneal only to a subset of the gene family sequence.

TGDD analyzes data rich Gaussian patterns

It should be noted that our TGDD products are viewed as data-rich Gaussian distributions, rather

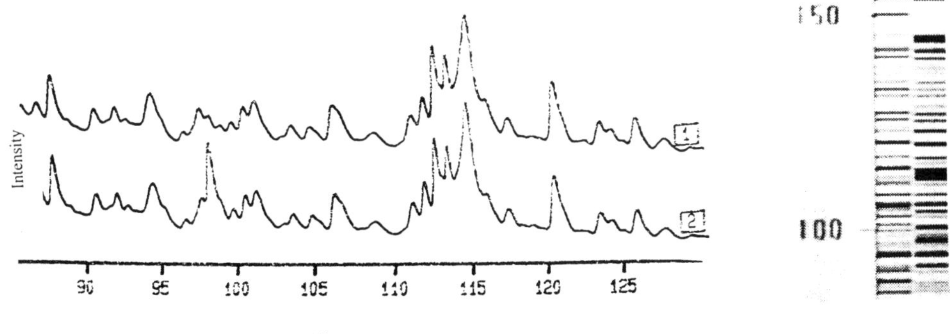

Fig. 8. Comparison of a high resolution TGDD fingerprint with low resolution TGDD fingerprint of an alleged MZ twin pair (Method II). Most implementations of DD by others use the low-resolution method of analysis. The $(CAG)_n$ containing fragments were targeted using oligonucleotide 8 (Table 1).

than low-resolution banding pattern. The high-resolution Gaussian analysis eliminates many false differences between samples. In Fig. 8, low and high-resolution analysis of the same data is shown. It is quite clear that many differences suggested in the low-resolution analysis do not survive high-resolution scrutiny.

Genomic DNA discordance within MZ twin pairs

The reliable de novo detection of genomic differences between MZ twin pairs depends strongly upon adherence to technical details described in materials and methods of this paper and elsewhere (Broude et al., 1997, 1999; Foulon et al., manuscript in preparation; Bouchard et al., manuscript in preparation). A variety of conditions were tested in order to optimize TGDD including: number of PCR cycles, primer concentrations, sample concentrations, and DNA extraction techniques. It was found that one of the most important factors is the matching of the DNA concentrations (see methods for details).

The existence of genomic differences between MZ twins is well established. Here, we describe a method that allows for the de novo detection of genomic differences in the absence of any knowledge of causation or location in the genome. TGDD also allows a quantitative assessment of genomic differences between individuals. An example of genomic discordances within a monozygotic twin pair is shown in Fig. 9. Here, the focus was directed on and nearby $(CAG)_n$ repeating sequences. Changing anchors on the T-primer and A-primers (oligonucleotides 6–8 and 10–14 respectively, Table 1) allowed a search through the pool of $(CAG)_n$ containing restriction fragments. This search was done until differences were identified. The fragments of interest were then picked from a high percentage agarose gel and reamplified. Figure 10 shows the isolated 250 and 390 bp fragments that will be sequenced.

Zygosity testing

As previously stated (see above), there is no "gold standard" zygosity test. All conventional methods are error prone. The zygosity of the twin samples studied here was determined to be MZ through conventional methods by other researchers. For example, the zygosity of the twin pair presented in Fig. 8 was determined by serology (22 antigens) in the following systems: ABO, Rh, P, Kell, Duffy, Kidd, and Lewis (Feinleib et al., 1977).

Ongoing studies are examining TGDD's application in zygosity testing. TGDD can effectively present a comparison of over 100 loci from a twin pair simultaneously. Our studies have shown that

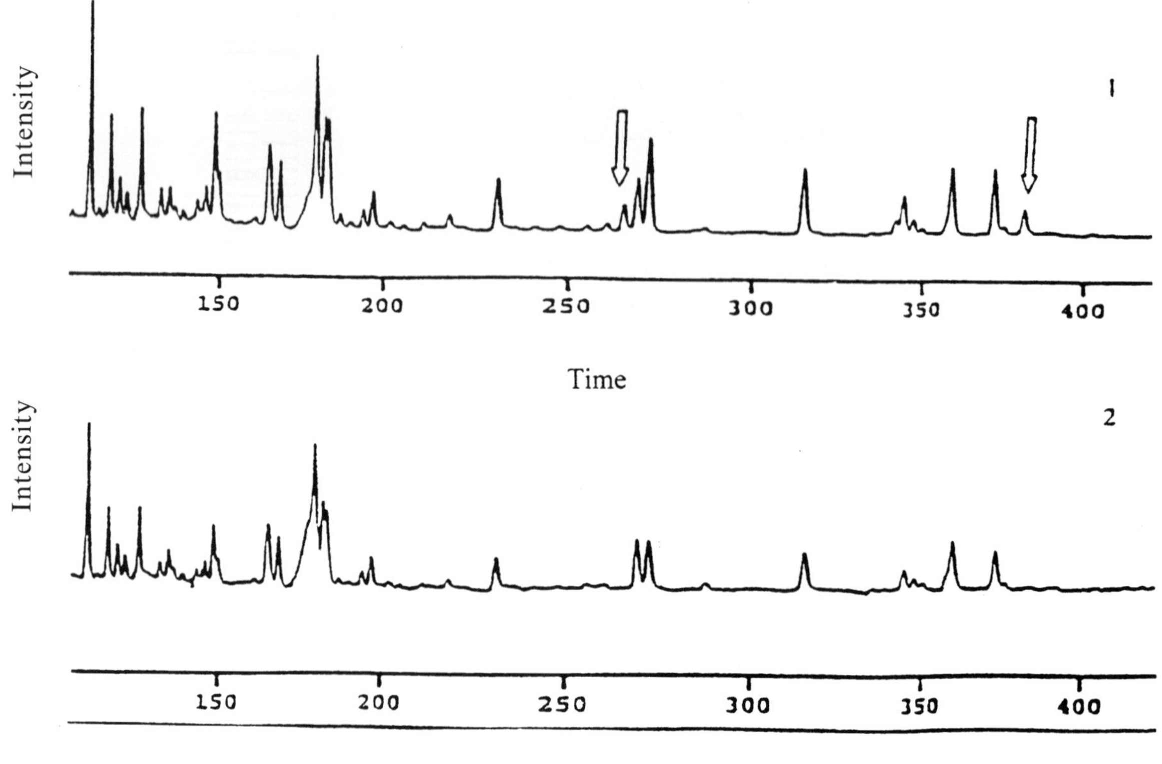

Fig. 9. TGDD (Method II) comparison between an alleged monozygotic twin pair. The A-primer was ST19*HaeIII*AC and the T-primer was $(CTG)_6A$ (oligonucleotides 14 and 6 respectively; Table 1). Arrows indicate discordant fragments.

at (CA6)n targets siblings and DZ twins have a higher level of variation than MZ twins. An example of TGDD targeting of an alleged MZ twin pair and a sibling pair is shown in Fig. 11(a) and (b) respectively. The exact variations in the level of genomic DNA discordance between MZ twin pairs and siblings is not yet fully understood. However, a preliminary assessment of our results indicates the presence of at least one incorrect zygosity determination.

Test/retest reliability of TGDD

Nine of the twelve MZ twin pairs analyzed by TGDD were found to have distinct and reproducible differences. The reproducible differences were seen when the samples were run in quadruplicate and also when different sets of experiments using the same samples were compared. The final conformation of the differences is done after obtaining the sequence of the restriction fragments. Once the sequence is obtained, a new primer that is specific for the difference is designed and used in a PCR. Such a conformation of a TGDD identified difference in a MZ twin pair is shown in Fig. 12. After the conformation of the difference, the two sequences are compared in order to understand how the difference arose and its impact on phenotype. For instance, the samples used in ongoing studies are discordant for schizophrenia or hypertension. Hence, a genomic discordance may impact an expression of these diseases.

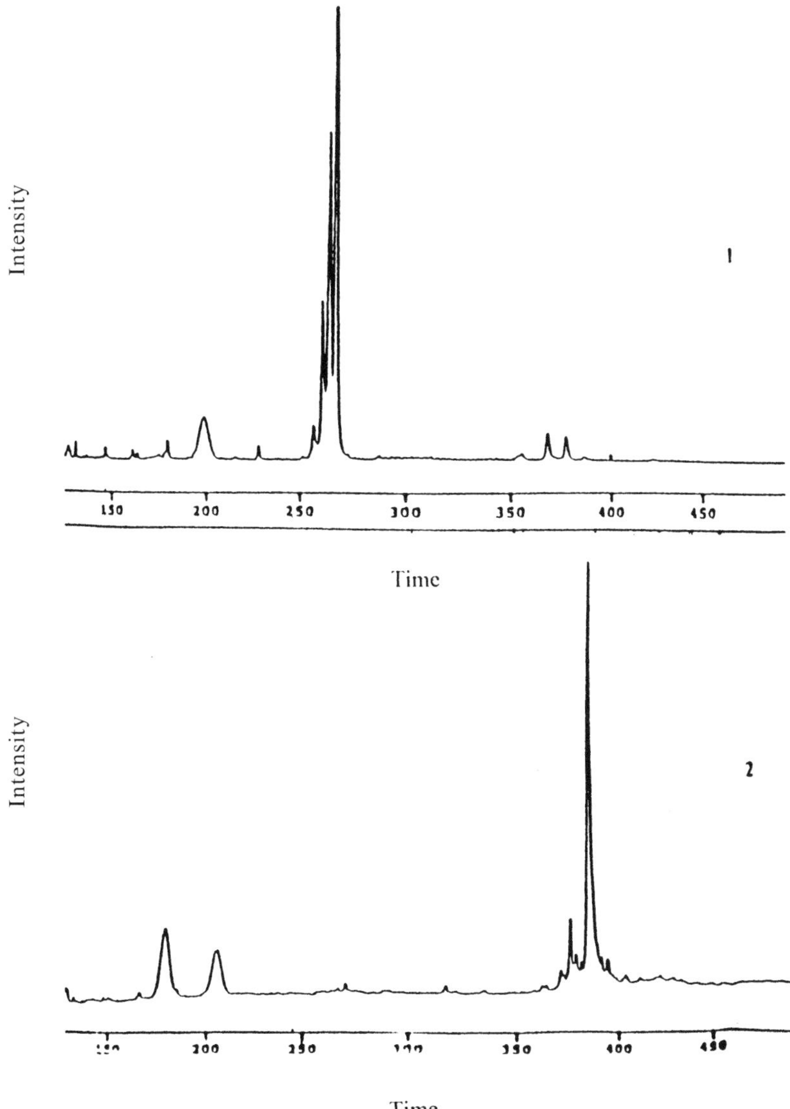

Fig. 10. The isolation of the discordant fragments shown in Fig. 9. The fragments were isolated and reamplified before fractionation on the Alfexpress as discussed in the Materials and Methods.

Prospectus

Currently, TGDD is being used to study monozygotic twins with schizophrenia or hypertension. Other studies are establishing the level of genomic identity of MZ twins and (re)assessing zygosity determination of others. The results shown here demonstrate that zygosity testing should be done with a large number of DNA sequences since the genomes of MZ twin pairs are very similar but not identical. Our preliminary work has detected an unexpected amount of genomic discordance in monozygotic twins (Nguyen et al., manuscript in preparation).

Understanding how and when the genomic DNA differences arise is the subject of ongoing

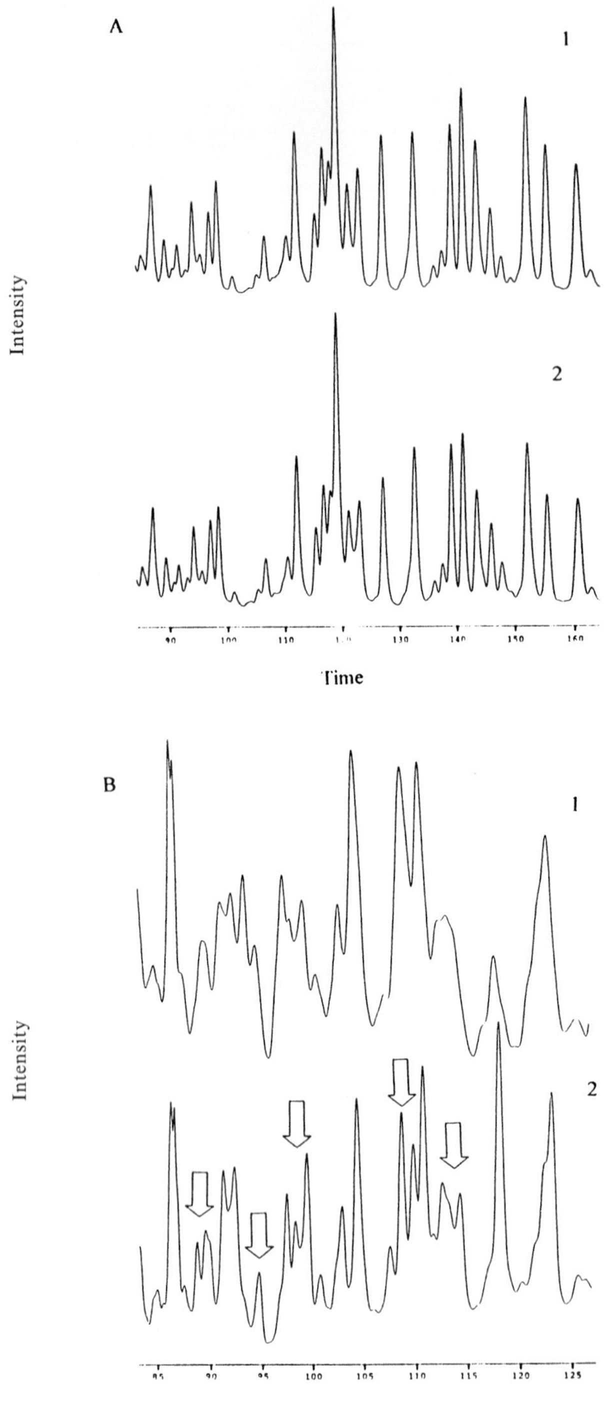

Fig. 11. A TGDD (Method II) comparison targeting $(CAG)_n$ repeat containing fragments using oligonucleotide 7 (Table 1). Samples being compared were (A) an alleged MZ twin pair and (B) a sibling pair.

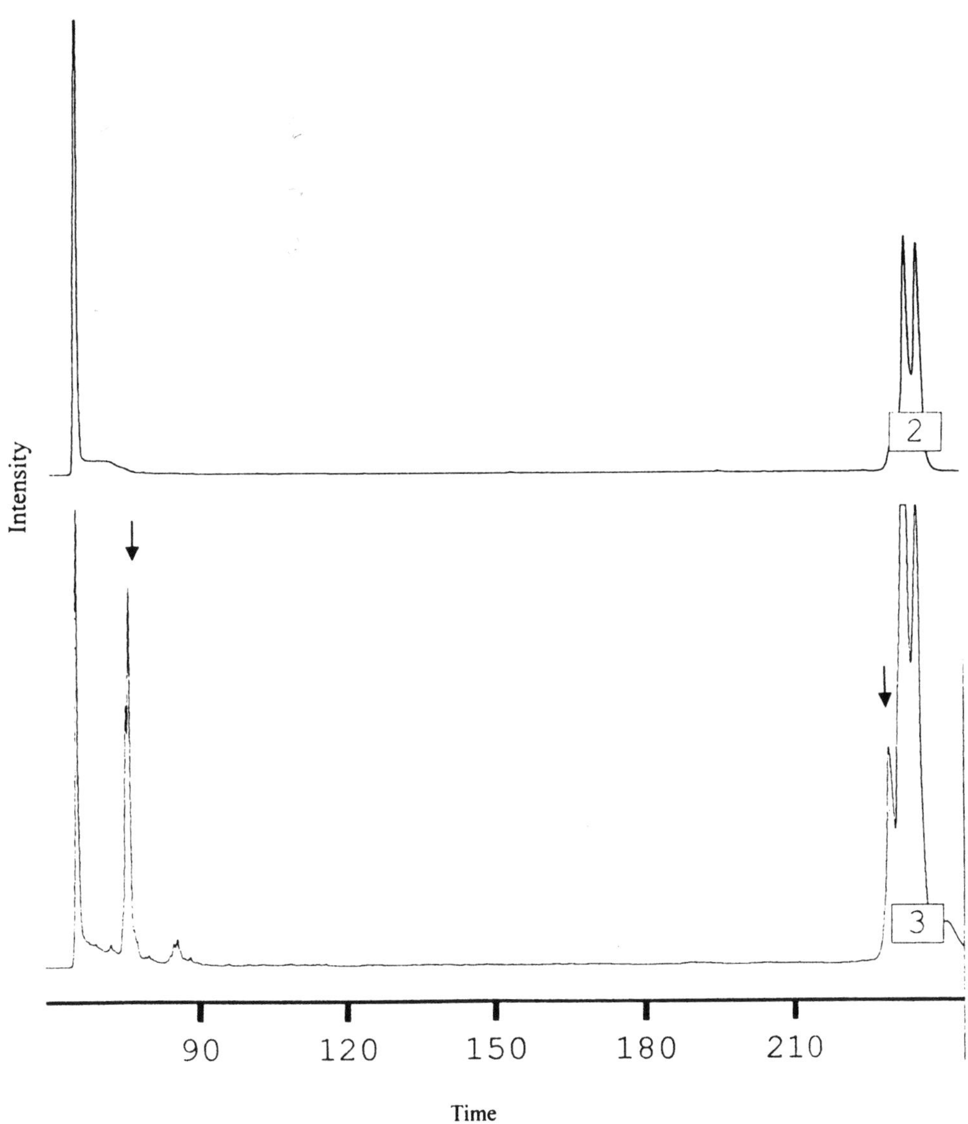

Fig. 12. Confirmation that TGDD identifies discordant genomic restriction fragments. A discordant fragment was isolated, cloned and sequenced (details in Nguyen et al., manuscript in preparation. Specific PCR primers were used to amplify homologous sequences from the genomic DNA of the MZ twin pair. Note comparison between the twin pair confirms differences, which are indicated by arrows.

experiments. It will be of special interest to determine the impact of these differences on phenotype, especially on disease discordance. The results also question the interpretation of disease discordance due to environmental factors. A great deal of the previous and ongoing research is focused on developing a quantitative method with a minimum of false positive and false negatives. This has led to the development of a model TGDD system using genomes whose entire sequence is known (Bouchard et al., manuscript in preparation). The model systems allowed experimental TGDD results to be compared to theoretical results. Thus, errors can be identified and eliminated.

Also underway is work focused on developing automated methods of analysis. This will allow comparisons between a highly complex mixture of fragments efficiently. These methods are also being tested in the model system.

Our long term goal is to analyze TGDD with DNA array microchip technology. Here, pools of fragments can be sorted by unique sequence adjacent to the target sequence by hybridization to DNA arrays. The use of targeting should greatly enhance this technology.

Acknowledgements

The authors would like to thank Haralambos Gavras, E. Fuller Torrey, Louis Keith, Terry Reed, David Shepro, and Karen Pimpis and the NHLBI twin study for their support. This work was supported by grants DOA (DAMD17-94-J-414) and NIH (1P50 HL55001) to CLS.

References

Akane, A., Matsubara, K., Shiono, H., Yamada, M. and Nakagome, Y. (1991) Diagnosis of twins zygosity by hypervariable RPLP markers. *Am. J. Med. Genet.*, 41, 96–98.

Broude, N.E., Chandra, A. and Smith, C. L. (1997) Differential display of genome subsets containing specific interspersed repeats. *Proc. Natl. Acad. Sci. USA*, 94: 4548–4553.

Broude, N.E., Storm, N., Malpel, S. and Smith, C.L. (1999) A PCR-based Targeted genomic and cDNA differential display method. *Genetic Analysis (Biomolecular Engineering)*, 15, 51–63.

Derom, C., Vlietinck, R., Derom, R., Van Den Berghe, H. and Thiery, M. (1987) Increased monozygotic twinning rate after ovulation induction. *Lancet*, 1: 1236–1238.

Feinleib, M., Garrison, R.J., Fabsitz, R., Christian, J.C., Hrubec, Z., Borhani, N.O., Kannel, W.B., Rosenman, R., Schwartz, J.T. and Wagner, J.O. (1977) The NHLBI twin study of cardiovascular disease risk factors: methodology and summary of results. *Am. J. Epidemiol.*, 106(4): 284–285.

Grothues, D., Canter C.R. and Smith, C. L. (1993) PCR amplification of megabase DNA tagged random primers (T-PCR). *Nucleic Acid Research*, 21: 1321–1322.

Hall, J.G. (1996) Twinning: mechanisms and genetic implications. *Curr. Opin. Genet. Dev.*, 6: 343–347.

Hill, A.V.S. and Jeffreys, A. (1985) Use of minisatellite DNA probes for determination of twin zygosity at birth. *Lancet*, 2: 1394–1395.

Jansen, G., Willems, P., Coerwinkel, M., Nillesen, W., Smeets, H., Vits, I., Howeler, C., Brunner, H. and Wieringa, B. (1994) Gonosomal mosaicism in myotonic dystrophy patients: involvement of mitotic event in $(CTG)_n$ repeat variations and selections against extreme expansion in sperm. *Am. J. Hum. Genet.*, 54: 575–585.

Kruyer, H., Mila, M., Glover, G., Carbonell P., Ballesta, F. and Estivill, X. (1994) Fragile X syndrome and the $(CGG)_n$ mutation: two families with discordant monozygotic twins. *Am. J. Hum. Genet.*, 54: 437–442.

Liang, P. and Pardee, A.B. (1992) Differential display of eukaryotic messenger RNA by means of the polymerase chain reaction. *Science*, 257: 967–971.

Lisitzyn, N. and Wigler, M. (1993) Cloning the differences between two complex genomes. *Science*, 259: 946–951.

Kallioniemi, A., Kallioniemi, O.P., Sudar, D., Rutovitz, D., Gray, J. W. and Waldman, F. (1992) Comparative genomic hybridization for molecular cytogenetic analysis of solid tumors. *Science*, 258: 818–821.

Lavrentieva, I., Broude, N.E., Lebedev, Y., Gottesman, I.I., Lukyanov, S.A., Smith, C.L. and Sverdlov, E.D. (1999) High polymorphism level of genomic sequences flanking insertion sites of human endogenous retroviral long terminal repeats. *FEBS Lett.*, 443: 341–347.

Machin G.A. (1996) Some causes of genotypic and phenotypic discordance in monozyotic twin pairs. *Am. J. Med. Genetics*, 61: 216–228.

Oliveira, R.P., Broude, N.E., Macedo, A.M., Cantor, C.R., Smith, C.L. and Pena, S.D. (1998) Probing the genetic population structure of *Trypanosoma cruzi* with polymorphic microsatellites. *Proc. Nat. Acad. Sci.*, 95: 3776–3780.

Reyniers, E., Vits, L., De Boulle, K.,Van Roy, B., Van Velzen, D., De Graaff, E., Verkerk. A.J. Jorens, H.Z., Darby, J.K., Oostra, B. and Willems., P.J. (1993) The full mutation in the FMR-1 gene of male fragile X patients is absent in their sperm. *Nat. Genetics*, 4: 143146

Siebert, P.D., Chenchik, A., Kellogg, D.E., Lukyanov, K.A. and Lukyanov, S.A. (1995) An improved PCR method for walking in uncloned genomic DNA. *Nucleic Acids Res.*, 23: 1087–1088.

Smith, C., Klco, S., Zhang, T., Fang, H., Rafael, O., Wang, D., Bremer, M. and Lawrance, S. (1993) Analysis of megabase DNA using pulsefield gel electrophoresis techniques. *Methods in Molecular Genetics*, 2: 155–175.

Velculescu, V.E., Zhang, L., Vogelstein, B. and Kinzler, K.W. (1995). Serial Analysis of gene expression. *Science*, 270: 484–487.

Section 3

Manipulating Known Genes

Section 3.1

Transgenic Approaches in Rodents

W.E. Crusio and R.T. Gerlai (Eds.)
Handbook of Molecular-Genetic Techniques for Brain and
Behavior Research (Techniques in the Behavioral and Neural Sciences, Vol. 13)

CHAPTER 3.1.1

Embryonic stem cells and gene targeting

Anthony Wynshaw-Boris[1], Lisa Garrett, Amy Chen and Carrolee Barlow[2]

Genetic Disease Research Branch, National Human Genome Research Institute, National Institutes of Health, Bethesda, MD 20892, USA
[1]Department of Pediatrics, UCSD School of Medicine, La Jolla, CA 92093, USA
[2]Laboratory of Genetics, The Salk Institute for Biological Studies, La Jolla, CA 92037, USA

Introduction

Techniques for manipulating the mammalian genome have been refined over the past decade, so that the modification of the mouse genome via transgenic technology is now fairly routine throughout the world. Genetic manipulation in the whole animal has been an extremely valuable approach to study gene function in the context of the whole organism. For example, valuable insights into the genetic control of neurologic function and behavior have been gained by either inactivating or inappropriately expressing genes with suspected roles in neurologic function or complex behaviors using gene targeting and transgenic technologies in mice. Many examples of the value of these approaches are included in this book (see following chapters in this section).

Neurologic and behavioral abnormalities are associated with known genetic defects in humans as well. The use of powerful linkage strategies for the mapping of genetic disease genes (see Chapters 2.1.1–2.1.6) has led to the positional cloning of a number of such genes associated with human genetic diseases. Mouse models for such diseases can be produced via transgenic technology, and the neurologic and behavioral testing of such mice can then be used to understand the function of the disease gene in normal animals, as well as its role in the pathophysiology of the disease. The genetic and biochemical pathways altered by mutation of such genes can also be dissected and investigated in animal models.

The mouse is a good experimental system for the creation of models of human genetic diseases (Wynshaw-Boris, 1996), and for producing animals with specific genetic alterations. The genome size and number of genes are similar between the human and the mouse, as are patterns of development. It is relatively inexpensive to house mice, in comparison to other mammals. Mice have a relatively short gestation period (19–21 d), brief time of maturation to sexual maturity (4–6 weeks), and large litter size (5–6 for inbred strains, 10–12 for outbred strains). Other mammalian species are amenable to pronuclear injection to create transgenic animals that over-express a particular protein (discussed in the next chapter), but the mouse is the only species where it is currently possible to perform germline knock-outs in embryonic stem cells (ES cells), thus making it the best mammalian species for genetic manipulation.

This chapter will provide a general discussion of how genes can be deleted in the mouse in order to study their impact on neurologic function and behavior in a mammalian organism. We will outline technical aspects of the generation of knock-out mice. For further information about these techniques, the reader is referred to several excellent laboratory manuals describing gene targeting (Hogan et al., 1994; Joyner, 1993; Wasserman and DePamphilis, 1993).

Origin of ES cells

ES cells are primary cells derived from the inner cell mass of mouse blastocysts. To generate an ES cell line, blastocysts are plated on feeder cell layers, which provide factors to maintain the undifferentiated character of these cells. Inner cell mass explants are dissociated, and occasionally clones of ES cells arise from these dissociated explants. ES cells are small, with very little cytoplasm, and divide rapidly (generation time of 14–16 h). They pack tightly into round colony nests of cells during growth, such that cell borders are difficult to distinguish. Female ES cell lines often lose one X chromosome, so male ES cells are almost always used for gene targeting. Karyotype analysis of newly made cells are performed to insure that the appropriate number of mouse chromosomes (40) are present, and the sex of the lines are determined by testing for the presence of Y chromosome-specific genes such as SRY by PCR. ES cells are pluripotent, which means that they can contribute to all cell lineages in mice, including the germline. ES cells are grown strictly under conditions that will maintain their pluripotency. Factors important for maintaining pluripotency are provided from feeder layers of primary embryonic fibroblasts, from fibroblast cell lines, or from growth factors added to the medium such as leukemia inhibitory factor (LIF). Once established, ES cells grow rapidly, and must be cared for meticulously. Contamination of ES cells with mycoplasma seriously compromises germline transmission potential. Fetal calf serum lots are tested to make sure that they will support ES cell growth, and not promote differentiation. Medium is changed daily, cells are kept in logarithmic growth phase by frequent passage, and cells are never allowed to morphologically differentiate (display distinct cell borders or flat colony morphology).

Fortunately, several ES cell lines have been developed and are currently in use in several laboratories, such as D3 (made by Tom Doetschman), AB1 (made by Allan Bradley), J1 (made by En Li in Rudolph Jaenisch's laboratory), R1 (made by Andras Nagy), and TC1 (made by A. W-B. and Chuxia Deng). These lines are available to the community, and appear to work well in numerous laboratories.

Feeder cell lines resistant to the antibiotic G418, used to select ES cells that express the *neo* gene after transfection to select for homologous recombinants (see below), have been made from immortalized fibroblasts that produce LIF, such as SNL (made by Allan Bradley). Conversely, primary embryonic fibroblast feeder lines can be made from mice that contain an active *neo* gene. Our protocol uses knock-out males for β_2 microglobulin, and are commercially available from the Jackson Laboratories.

Neo-resistant fibroblast feeder cell isolation

Primary fibroblast production

Feeder Media: 500 ml DMEM (high glucose, w/o L-glutamine, w/o sodium pyruvate, GibcoBRL #11960-010)
15% Heat-inactivated Fetal Calf Serum (FCS)
penicillin/streptomycin
2% L-glutamine

Freezing media: DMEM
25% FCS (heat inactivated)
10% DMSO

1. Set up crosses with β_2 microglobulin homozygous knockout males and two or three female mice without superovulation. Dissect the embryos at 14–15 d of gestation.

2. Using sterile technique, remove the uteri from pregnant mice, taking care to keep the uterus within the body cavity, not on the fur (to maintain sterility). Rinse the uteri in a Petri dish containing

PBS and penicillin/streptomycin (pen/strep) to remove blood and any hair or dander. Remove the uteri to a second dish with PBS and pen/strep. *The remainder of the procedure is performed in a cell culture hood.*

3. Remove the embryos carefully from the uterus. Remove the amniotic sac and placenta from each embryo, and place the embryos into another dish of PBS with pen/strep.

4. Remove the heart, liver and spleen, and discard. Place each embryo body in a separate empty Petri dish with DMEM with pen/strep. Mince the embryos as finely as possible. Place the minced embryo's into a 50 ml tube with 10 ml of diluted 0.04% trypsin in DMEM (prepared from 0.25% stock). Rinse the dish with another 10 ml of trypsin and place this aliquot into the 50 ml tube.

5. Place the tube in the 37°C incubator on a rocker for 20 min.

6. Triterate the solution gently with a 5 ml pipette. Allow the tissue pieces to settle to the bottom. Remove the supernatant to a fresh tube, leaving the pieces of tissue behind, and add twice the volume of media with 15% FCS, to the liquid. Centrifuge the tube at 1200 rpm for 5 min, remove the supernatant and resuspend the pellet up in feeder medium. Place on ice.

7. Add a second 20 ml of 0.04% trypsin in DMEM to the tissue in the tube and repeat steps 3, 4 and 5. Combine cells from same embryo in same 50 ml tube.

8. Repeat steps 3 and 4 (for the third trypsin incubation). There should be very little tissue remaining after this step. Collect the cells, avoiding the remaining tissue, and combine with cells from the same embryo.

9. (Passage 0) Plate the cells into the same number of 150 mm dishes as there were embryos. ($\sim 3 \times 10^6$ cells/dish) in the above media. Incubate until the cells are confluent. Some morphologically different cells may be growing along with the fibroblasts. These will differentiate and be lost during the passages that follow.

10. Passage the confluent cells by trypsinization, splitting the cells 1:3 onto three 150 mm plates. Grow these cells to confluence, trypsinize and count. There should be about 1×10^7 cells per starting embryo. Freeze these at 3×10^6 cells per vial in freezing medium. Label each vial "FEEDER SEED passage 1". The word "Seed" indicates that they have not been mitomycin C treated.

Expansion and mitomycin C treatment

(Each feeder seed vial will produce forty 150 mm plates for mitomycin treatment, and enough mitotically-inactivated feeder fibroblasts for about 160 10 cm plates.)

1. Thaw one vial into feeder media, and plate it onto one 150 mm plate (passage 2).

2. Grow to confluence (3–4 d), changing the media every 2 d. Trypsinize and replate onto seven 150 mm plates (passage 3).

3. Grow to confluence, changing the media every 2 days. Trypsinize and replate the seven plates of passage 3 to forty 150 mm plates (passage 4).

4. Grow to tight confluence. Fibroblasts will survive the mitomycin/freezing treatment better if they are tightly confluent and not in an exponential growth stage.

5. Dissolve 2 mg of mitomycin C in 1 ml of PBS, and add 1 mg (500 μl) to 100 ml of growth media (final concentration 10 mM).

6. Incubate confluent plates of feeder cells in mitomycin C media for 2 h.

7. Remove the mitomycin C media and wash the plates three times gently and rapidly with PBS. Handling time is critical, and take care that the cells do not wash off the plates.

8. Remove the cells from each plate with 3 ml of 0.25% trypsin/EDTA, neutralize with 6 ml of growth media, and collect the cells in a clean tube.

9. Count the cells, collect by centrifugation, count, and resuspend all cells in 40 ml cold freezing media on ice. There should be approximately 5×10^5 per vial.

10. Freeze at 0.25 ml per vial in a Styrofoam box at −80°C overnight. There should be 160 vials. Transfer to liquid nitrogen in the morning.

11. To use these feeders to support ES cells, thaw a vial, wash once with growth media, and plate into one 100 mm gelatin coated plate or one 6 well gelatin coated plate. There will be some (5–10%) cell death of the feeders as a result of the mitomycin C treatment and freezing. The dead feeders will not attach to the plates. ES cells plated onto feeder layers will carry feeder cells along with them. For this reason, it is optimal to be on the sparse side of confluence with your new feeder layer. If the feeders are too dense, they will peel off of the plate.

Homologous recombination

In contrast to the random integration of DNA, DNA can also integrate into the genome in a site-specific manner by homologous recombination. Foreign DNA introduced into mammalian cells generally integrates into the host genome via non-homologous recombination, with a frequency of homologous to non-homologous recombination of 1 in 1000. A major technical hurdle was overcome with the ability to grow cells in culture and to select for those cells that have undergone homologous recombination. The development of cell culture methods for growing, transfecting, and selecting for homologous recombination in ES cells while maintaining the ES cells in an undifferentiated, pluripotent state has allowed for the extensive manipulation of the mouse genome.

ES cells are grown in culture and manipulated DNA of interest is transferred by electroporation (Fig. 1). The manipulated DNA contains a marker gene that allows the selection of cells which have stabily incorporated the DNA, whether it be via homologous or non-homologous recombination. Following electroporation, cells are subcloned so that DNA can be extracted after selection. Genomic DNA is screened using standard molecular biology procedures, such as Southern blot hybridization, or PCR, in order to identify the clones that have undergone directed homologous recombination at one allele.

Several factors have been defined that enhance the frequency of homologous recombination and simplify screening procedures. Some of these factors are: the use of isogenic DNA (DNA from the identical strain of mice as the ES cells) in the targeting construct; the use of linearized DNA; and the use of positive-negative selection to improve the ratio of homologous to non-homologous clones surviving selection to the range of 1 in 15–300. In positive-negative selection, a positive selectable marker (usually *neomycin phosphotransferase*, or *neo*) is introduced within the region of homology, and imparts resistance to a drug toxic to mammalian cells (G418). The negative selectable marker (herpes simplex virus *thymidine kinase*) is placed outside of the region of homology in the targeting vector, and will remain in genomic DNA only if random integration has occurred. The negative selectable drug (gancyclovir or FIAU) can then be used to reduce the background caused by random integration. In addition, by ensuring that the total amount of homology used in the construct is at least 7 kb and that homologous sequence flanking the selectable marker is at least 1.5 kb, rates of obtaining ES clones that have undergone homologous recombination are in the range of 1 in 25–100 clones.

The protocol that we use for ES cell transfection is the following.

Transfection of ES cells

ES Medium: DMEM high glucose w/o L-glutamine and w/o sodium pyruvate
15% heat inactivated fetal calf serum (ES-cell tested)
1% non-essential amino acids
2% L-glutamine
1% of a 10 mM 2-mercaptoethanol stock (final 100 μM)
1% penicillin/streptomycin

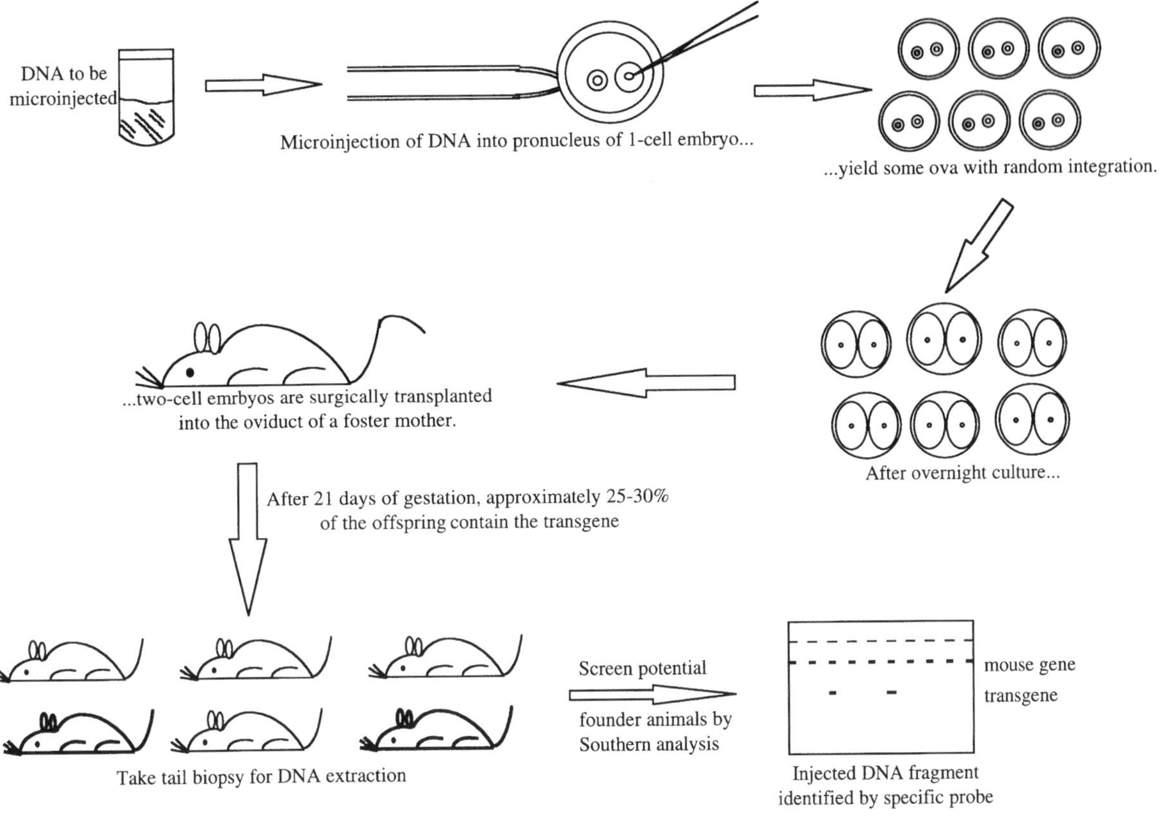

Fig. 1. Producing gene targeted mice in ES cells. See text for details.

50 μl per 500 ml of 10^7 units/ml of LIF stock (final 10^3 units/ml)

Freezing Medium: DMEM
25% heat inactivated fetal calf serum (ES cell tested)
10% DMSO

DNA: 25 μg of linear DNA for each transfection is optimal. Qiagen column kits can be used to prepare plasmid DNA, but cesium chloride banding protocols also work well for purification (see next chapter for details). The plasmid is linearized with an appropriate restriction enzyme, then phenol/chloroform extracted, chloroform extracted, and ethanol precipitated and washed with 70% ethanol. DNA is resuspended in 10 mM Tris, 1 mM EDTA, pH 8.0 (TE) at a concentration of 1 μg/μl.

1. Thaw frozen mitomycin-treated mouse embryonic fibroblasts (MEF) and plate five 10 cm gelatin coated plates with 5×10^5 MEF/plate one day prior to thawing ES cells. Prior to plating, rinse cells once in medium to remove DMSO, spin for 5 min at 1000 rpm, and plate in feeder medium.

2. Thaw one vial of TCI 129/SvEv embryonic stem cells (about 2–4 × 10^6 cells), at passage seven from the freezer (enough for up to three transfections). Rinse thawed cells with ES medium, collect by centrifugation, and plate the cells onto one of the 10 cm MEF feeder plates in 10 ml ES medium. Change the media daily.

3. After 3 or 4 d, the cells should be confluent (2.0×10^7 cells). Trypsinize the cells, using 1 ml of trypsin / EDTA per plate. Neutralize the trypsin with 9 ml of media, collect at 1000 rpm for 5 min.

4. Resuspend by triteration 20 times to disperse the cells. Replate them onto the remaining four gelatin coated plates MEF plates in 10 ml ES medium per plate, at a concentration of $\sim 4.0 \times 10^6$ cells / 100 mm dish.

5. Change the media the next day. Plate six 10 cm mitomycin treated MEF feeder plates per transfection, plus two extra plates for untransfected control cells. For example, if two transfections are being done, plate 2×6 plates, plus two extra plates, for a total of 14 plates of MEF cells.

6. The next day (2 d after plating the ES cells), several hours before electroporation, change the media on the ES cells.

7. Trypsinize the cells with 2 ml trypsin/EDTA per plate. Neutralize with 9 ml medium per plate. Collect at 1000 rpm for 5 min. Add 10 ml of phosphate-buffered saline (PBS) to cells. Triterate 20 times, and bring cells to 50 ml in PBS.

8. Count cells and collect by centrifugation. Resuspend at a concentration of 20×10^6 cells/ml in PBS.

9. Remove one or two aliquots of 2×10^6 ES cells to plate as controls for G418 killing, and plate each aliquot on one feeder plate in 10 ml ES medium.

10. Place 25 μg of linearized DNA (at 1 μg/μl in TE) in a 0.4 cm electroporation cuvette, add 1 ml of cells in PBS. Mix with a 1 ml pipette.

11. The cells are electroporated in a BioRad Genepulser. For 25 μg DNA in 1 ml PBS with 2×10^7 ES cells, use 600 V/25 μFd (0.6 kV). Hold down both "Pulse" buttons on the Genepulser simultaneously. Charge will accumulate in the capacitor until it reaches the programmed charge. At that point it will discharge with a beep. Check the display by pressing the buttons for volts, capacitance, and time constant (which will usually be 0.4 s), to make sure that the machine performed properly.

12. Keep cells in the cuvette for 3–10 min to allow the DNA to enter the cells.

13. Plate each transfection onto six 10 cm plates with feeder layers in 10 ml of media per plate. It is convenient to have one 50 ml conical tube labeled for each transfection. Add 30 ml of ES medium to each tube, then add the electroporated ES cells from the cuvette to the 50 ml conical tube, rinsing the cuvettes 2–3 times. Then, change each feeder plate to 10 ml ES medium, and plate the ES cells by adding 5 ml of the transfected cell mixture to each plate.

14. After 24 h, remove the media and add selection media. If *neo* is the selectable marker (which will almost always be the case), use 140 mg of G418/500 ml media (or 168 mg of G418 per 600 ml), for a final concentration of 280 μg/ml. If a TK gene is used for positive/negative selection then add 50 μl of 10,000× stock FIAU solution per 500 ml of medium (final concentration 0.2 μM). If other selectable markers are used, use the appropriate antibiotic. Change five of the plates per transfection to G418 + FIAU-containing medium, and change the remaining plate to G418-containing medium without FIAU, to determine FIAU enrichment. Change the untransfected control plates to G418-containing medium. (Different promoters provide different levels of resistance to selection, so this concentration may have to be varied empirically. Depending upon the experiment, the level of G418 may rarely have to be varied during the course of selection.)

15. Change the media each day with selection media. ES cell killing should be obvious by three days, and non-resistant cells will clear by seven days. One day prior to picking the colonies, plate five 24 well plates per transfection with mitomycin-treated MEF as above.

16. When resistant colonies are evident (7 d) rinse the plate with PBS to remove the serum. Place 10 ml of PBS on the plate. Select the resistant clones using a 200 μl pipette, set at 50 μl and using sterile, plugged tips. (The clones will need to be scraped off the plate with some of its feeder layer.) Place each clone into individual wells of a numbered, 96 well MEF-coated plate.

17. Add one drop of 1/2 strength trypsin/EDTA (0.125%), diluted in PBS, and warm for 3 min in the incubator. Add 100 μl of ES media without selection drugs, triterate twenty times with a 200 μl pipette set at 200 μl, remove the whole volume to one well of the MEF-coated 24 well plate, which has been changed to 1 ml/well of ES media. Allow the colonies to expand, changing the ES media daily.

18. After 3–4 d, the colonies will have expanded, and will be ready to freeze. Some colonies may be ready at 3 d, and some at 4 d. Change the cells on the morning of the day of freezing to ES medium. Later, remove the media, wash with PBS, add 2 drops of 0.25% trypsin/EDTA. Incubate 3 min at 37°C. Add 1 ml of ice cold freezing media to each well, and place the plate on ice. Triterate 20 times using a 1000 μl pipette set at 1000 μl, with a sterile plugged tip. Remove 200 μl to one well of a 6 well gelatin coated plate without feeder cells in ES medium, to be grown up to isolate DNA for genotyping. Freeze the remaining cells in cryo vials slowly, in Styrofoam containers at −80°C overnight. Place cells in liquid N_2 the next day.

19. Change the cells plated for DNA isolation the next day to ES medium, and let them expand until the medium begins to turn yellow. DNA is prepared by proteinase K/SDS digestion of the cells. DNA samples are analyzed for the presence of the targeted allele. Positive clones will be expanded for injection.

Types of genetic alterations

Using homologous recombination in ES cells, it is possible to make a variety of modifications to specific genomic loci. Most commonly, gene targeting has been used to completely inactivate a specific gene. A selectable marker such as *neo* is expressed from a promoter active in ES cells (such as the phosphoglycerate kinase promoter, or PGK), and can be placed into a convenient restriction enzyme site of an exon, introducing stop codons into all three reading frames of the gene. This will completely inactivate the gene as long as a stable truncated protein is not produced, or splicing does not occur around the exon containing the inserted *neo* gene. Alternatively, *neo* can be used to replace exons (deletion) containing critical protein domains. The goal of these types of disruptions is to create complete loss-of-function or null alleles.

It is also possible to make more subtle changes to genes using gene targeting. Various strategies have been used to introduce point mutations at specific residues of several genes. For example, a point mutation can be introduced in an exon of a gene, closely linked to the selectable marker. Clones that have undergone homologous recombination can be screened for the presence of the point mutation by restriction mapping or PCR. Conditional knock-outs are designed to inactivate a gene in specific tissues at specific times of development or adulthood. These methods are discussed in later chapters.

Finally, it is possible to use targeted mutagenesis to produce mice with large deletions (200 kb to >2 Mb) in specific chromosomal loci (Ramirez-Solis et al., 1995). Recombinase recognition sites can be introduced that surround defined regions of the genome by targeted mutagenesis, and the intervening region can be removed by applying recombinases to ES cells or in vivo. Such methods will be useful to modify regions of the genome where complex traits reside, and may facilitate the identification of genes responsible for complex traits such as behavioral differences between inbred strains of mice.

ES cell injection into blastocysts

Once correctly manipulated ES cell clones have been identified, they are then injected into blastocysts (host blastocysts) obtained by flushing the uterus of a three and one half day pregnant female (Fig. 1). The injected blastocysts are then transplanted into the uterus of a pseudopregnant female where the embryo continues to develop. Pseudopregnant females are produced by mating females with vasectomized males (see next chapter). The act of mating initiates the hormonal cascade required to prepare the uterus for implantation of embryos, but since the male is sterile, the only embryos that develop and implant are those that have been injected and transferred. Once implanted, the manipulated embryo is carried to term, and delivered. Offspring are screened for the presence of the altered allele. Such offspring are referred to as chimeras, as some of the resultant animal is derived from the host blastocyst and some from the injected ES cells which contain the altered allele or transgene. If the manipulated ES cells contribute to the germline, then the chimera can be bred and some of its offspring will be heterozygous for the altered allele. This is referred to as germline transmission. Germline transmission is essential for establishing a transgenic line.

Blastocysts can be isolated from superovulated or naturally mated C57BL/6J females by the following protocol.

Super-ovulation and blastocyte harvesting

Media: D-MEM powder (GibcoBRL cat. #12800-017) high glucose, with L-glutamine, with 110 mg/ml sodium pyruvate, w/o sodium bicarbonate

add: 25 ml 1M.HEPES
2.6 g NaCl
Penicillin/streptomycin

Bring to 1 liter with H_2O, then bring pH to 7.3–7.4 Refrigerate at 4°C. Before use add 2% L-glutamine and 10% FCS.

Day 6: Inject C57BL/6 3–4-week-old female mice with 2.5 units of PMS (Pregnant mares serum) (2.5 units will be 50 μl.) Calbiochem-Novabiochem Corp., PMS cat. #36722

Day 4: Inject the same mice with 2.5 units of HCG (Human Chorionic Gonadotropin). Organon Inc., "Pregnyl" cat. #5310122. Immediately after injecting, mate the mice with C57BL/6 male mice, one female to a male.

Day 3: The following morning, check the females for vaginal copulation plugs for evidence of mating. The male seminal fluid will form a white waxy plug. Conversely, natural matings can be set up as well at Day 4, for checking of plugs on Day 3.

Day 0: Sacrifice the mice no more than five at a time, using cervical dislocation. Flood the abdomen with ethanol. Make an incision laterally in the abdominal epidermis. Cut away the subcutaneous tissue to expose the abdominal cavity. Place the intestine anteriorly. Two uterine horns can be seen extending down from the ovaries. The oviducts emerge from two seperate fat pads in the dorsal pelvis. The ovaries are imbedded in the fat pads. Dissect the uterus from the surrounding tissue, removing as much of the associated fat as possible. Place the uterine horns connected by the cervix onto a 35 mm dish. Cut out the cervix (posterior portion of the uterine horns) and place the uterus into another 35 mm dish with injection media. Using a 25 gauge needle and 3 cc syringe, flush out the horns by inserting the needle in the anterior end of the uterus (this is a larger opening). Allow 3–4 drops to flush through and place the flushed uterus onto another dish with a small amount of media. Blastocysts are recovered by mouth pipette and placed into a clean drop of injection media covered with mineral oil. Let the blastocysts remain at 37°C until injection. Before blastocyst microinjection, ES cell clones are trypsinized from a 6-well dish, centrifuged and resuspended with injection media, as in the next protocol.

Thawing and plating ES cells for mircoinjection
ES Medium, Feeder Medium and Freezing Medium as above.

1. Thaw one frozen vial of mitomycin-treated mouse embryonic fibroblasts (MEF) and plate onto one 24-well gelatin coated plate one day prior to thawing ES cells. Prior to plating, rinse cells once in medium to remove DMSO, spin for 5 min at 1000 rpm, and plate in Feeder medium.

2. Thaw one vial of one or two of the targeted clones. Each clone was frozen from about a 75% confluent 24-cell well. Rinse thawed cells with media to remove DMSO, since DMSO is harmful to cells and will cause differentiation of ES cells. Collect each clone by centrifugation, and plate each clone onto one well of the 24-well MEF feeder plate in 1 ml ES cell medium. Prior to adding cells to the well, remove the old MEF media by aspiration and replace with fresh ES cell media.

3. Change the media daily. If the clone is dense, it may have to be trypsinized and replated the next day. If it is not dense, it needs to be passaged after 3–4 d. Trypsinize the cells, using two drops of trypsin/EDTA per plate. Neutralize the trypsin with 1 ml of media, collect at 1000 rpm for 5 min.

4. Resuspend in 1 ml ES medium by pipetting 20 times to disperse the cells. Plate half of cells to one 6-cell well with MEF feeders, prepared the day before.

5. Collect the remaining one half of cells by centrifugation at 1000 rpm for 5 min. Add 0.5 ml freezing medium, gently resuspend, and freeze in one cryovial.

6. Change the media daily. After 2 or 3 d, the cells will be confluent. Trypsinize the cells using 0.5 ml 0.25% trypsin/EDTA, neutralize with 10 ml media, and collect by centrifugation at 1000 rpm for 5 min. Resuspend in 2 ml of ES media. Freeze half of the cells (1 ml) as above. If cells are being used for injection, the rest of the cells can be plated on the remaining 5 wells of unused MEFs. Plate cells at a number of densities. Usually it is sufficient to plate 2, 4, 6, and 8 drops per well of the 2 ml ES cell suspension.

7. Change the cells daily. It is best to use cells 2 d after plating for injection, although you can also use cells one day after plating. *Do not use cells after 3 days in culture.* Instead, cells can be trypsinized and replated as needed to maintain wells of cells that are 1–2 d in culture for injection. The appropriate well can be selected for trypsinization on the day of injection, by choosing the well that has the best cell density, and shows the least differentiation.

8. Change the media on ES cells in the morning of the day of injection. Save at least one well of unused cells to collect DNA to test that the injected clone was in fact a targeted clone. Freeze the remainder of unused cells to have stocks of the targeted clones.

To inject ES cells into blastocysts, a Zeiss Axiovert 135 inverted microscope with Narishige hydraulic manipulators or Eppendorf motorized manipulators is placed on an air table or vibration resistant table. One manipulator is attached to a holding pipette, which is linked to a syringe that controls gentle vacuum pressure to stabilize the fertilized egg for injection. The other manipulator is attached to the injection pipette, which is attached to an independent syringe to control gentle pressure for picking up and injecting cells. Holding and injection pipettes are made from glass pipettes using a programmable pipette puller, to create glass needles of the appropriate diameters. To make the holding pipette, a microforge is used to break the pipette to the appropriate size, and to polish the end to make it smooth, so as not to damage the blastocyst. Injection needles are made cylindrical with an opening the size of an ES cell, which is approximately 20–25 μm in diameter (shown schematically in the Fig. 1). To allow for ease of penetration of the zona pellucida, a sharp barb can be placed on the end of the opening with the microforge. ES cells for injection are trypsinized and placed on the bottom of a small tissue cultuire plate, along with blastocysts. Round, small ES cells are systematically picked up into the barrel of the injection pipette, until 12–15 in number are in the pipette. The blastocyst is then held by the holding pipette near the inner cell mass. The injection pipette is inserted into the blastocyst, trying to enter between trophoblast cells, and the ES cells are expelled gently into the blastocyst cavity. Once the injection pipette is removed, the blastocyst collapses and the blastocyst cavity disappears. However, after 2–3 h, the blastocyst reexpands, and these blastocysts can then be introduced into the uterus of pseudopregnant females.

Transfer to recipients

Equipment/materials needed

Slide warmer for recipients and eggs (37°C)
#5 Dumont forceps (two)
Scalpel (#11 or #15)
Serrafin clip
Microdissecting forceps with teeth
Microdissecting forceps with serrated, blunt tip
4–0 suture
9 mm surgical wound clips and applier
Sterile drapes
Avertin anesthetic agent. Avertin stock is a solution of 1.0 g 2,2,2-tribromoethanol dissolved in 0.5 g tertiary amyl alcohol. Store the stock solution in the dark in a glass container. Dilute the Avertin to the final working concentration by dissolving 1.2 ml of the stock in 100 ml of boiling PBS. Shake to dissolve. Aliquot and store at −20°C.
Syringe (1/2 to 1 ml) and needle (25–28 1/2 ga)
Dissecting microscope with fiber optic light source
Glass micropipettes
Mouth pipetter

1. Following intraperitoneal injection of anesthetic agent into the pseudopregnant females and induction of deep anesthesia, the back of the mouse is washed with ethanol.

2. Place the mouse on a 4 × 4 gauze pad on the dissecting scope platform with the right side towards you. Approximately one-third of the way of the back, make an incision in the skin approximately 1 cm in length and approximately 1 cm to the right of the spine. To locate the region of the ovary, identify a vessel and a nerve running diagonally across the ovarian fat pad (the fat pad should appear easily as a white mass just below the surface of the peritoneum). The ovary is just caudal to the fat pad. Grasp the peritoneum with the micro-dissecting forceps with teeth and make an incision approximately 0.5 cm in length above the fat pad. With blunt forceps, pull out the fat pad with the ovary attached through the incision and place on a sterile drape. Anchor the fat pad and ovary in place with a Serrafin clip. Do not grasp the ovary or oviduct, but only the fat pad.

3. Adjust the magnification on the dissecting microscope so that the uterus near the ovary is in the field of view.

4. Make a small hole in the uterus near the ovary with a 26 g needle, without tearing any blood vessels.

5. Collect 5–6 blastocysts in the glass transfer pipette with an absolute minimum of liquid. Rate of loading and volume can be controlled by the introduction of air bubbles. Insert the pipette into the hole and gently blow the blastocysts into the uterus. Also blow a small bubble of air into the uterus to be sure that all the embryos have been expelled.

6. Gently push the ovary and uterus back into the peritoneal cavity with the blunt forceps.

7. Place one suture into the musculature and close the skin with 2 wound clips.

8. When transfer is complete place the recipient in a warm recovery cage.

9. When recovery is complete transfer the recipients to a clean cage.

Implanted pseudopregnant females give birth after 17–18 d, depending upon the strain used. Once the animals are old enough, coat color chimerism will be evident, demonstrating that ES cells have contributed to the offspring. The most commonly used mouse strain for derivation of ES cells is one of the 129 strains. Most 129 strains carry the agouti coat color locus which is dominant over the recessively inherited black coat color locus of the C57BL/6J mouse strain from which the host blastocysts are obtained. Offspring in which the 129 ES cells have contributed to the embryo will have agouti and black fur due to the difference in origin of the cells, agouti from the ES cells or black from the cells of the host blastocyst. This is referred to as coat color chimerism. If ES cells have also contributed to the germline, then mating of the chimera to a female with a recessive coat color results in offspring with the coat color of the ES cell-derived strain. ES cell-derived (agouti) offspring can then be screened for the presence of the mutated allele. Once animals heterozygous for the recombined allele have been identified, they can be intercrossed to generate animals homozygous for the mutated allele. In this way, null mutations or "knock outs", of a specific gene can be obtained.

Strain of origin of ES cells

The vast majority of knock-outs have been produced using ES cells derived from 129 substrains of mice. For unclear reasons, 129 mice are particularly useful for producing ES cells at reasonable frequency, and these cells maintain their pluripotency. Consequently, nearly all published mouse knock-outs have been done in 129-derived ES cells, and it is likely that most knock-outs in the future will be in such cells as well.

A complicating factor is that there is extensive genetic heterogeneity among substrains of 129, as well as ES cells derived from these substrains (Simpson et al., 1997). As a result, there is a remarkably wide range of characteristics that each substrain displays. Of particular relevance to the topic of this book, there is a wide range of behavioral phenotypes displayed by 129 substrains (Crawley et al., 1997).

Once produced, knock-outs are often outcrossed to C57BL/6 or other inbred strains. Thus, the genetic background of knock-outs may be quite complicated and vary within litters (also see Chapter 3.1.10 for further discussion of problems associated with genetic background). Therefore, it is of utmost importance when performing these types of experiments to keep strict records of the strain background during mating.

Although most ES cells are derived from 129 strains, behavioral analysis of these strains has been limited in the past. In contrast, there is an extensive literature describing in detail the behavioral characteristics of a number of other strains of mice, in particular C57BL/6 and DBA/2, and there has been a consensus that ES cells derived from these strains should be developed and tested (Crawley et al., 1997). A number of ES lines derived from other inbred strains or F_1 hybrids are being tested in several laboratories with successful germline transmission. These new ES cells can be used when producing mice with inactive genes that will likely result in behavioral phenotypes.

Gene targeting and creation of *Dvl*1 deficient mice

We will now give an example from our laboratory of the construction of a knock-out mouse, to illustrate the points made in this chapter. We have chosen to describe the production of *Dvl*1 deficient mice, since the analysis of the phenotype of these mice are the subject of another chapter by Paylor et al. (Chapter 3.1.9).

The murine *Dvl*1 gene contains 15 exons, and is located on mouse chromosome 4 (Fig. 2, top). To create a null allele for *Dvl*1, we replaced part of exon 2, and all of exons 3 and 4, with PGK*neo* in the opposite orientation relative to *Dvl*1 transcription. This disruption eliminates amino acids 131 to 225 (13.5% of the total coding sequence of 695 amino acids), and introduces multiple stop codons from PGK*neo* 3′ of exon 2 sequences. A

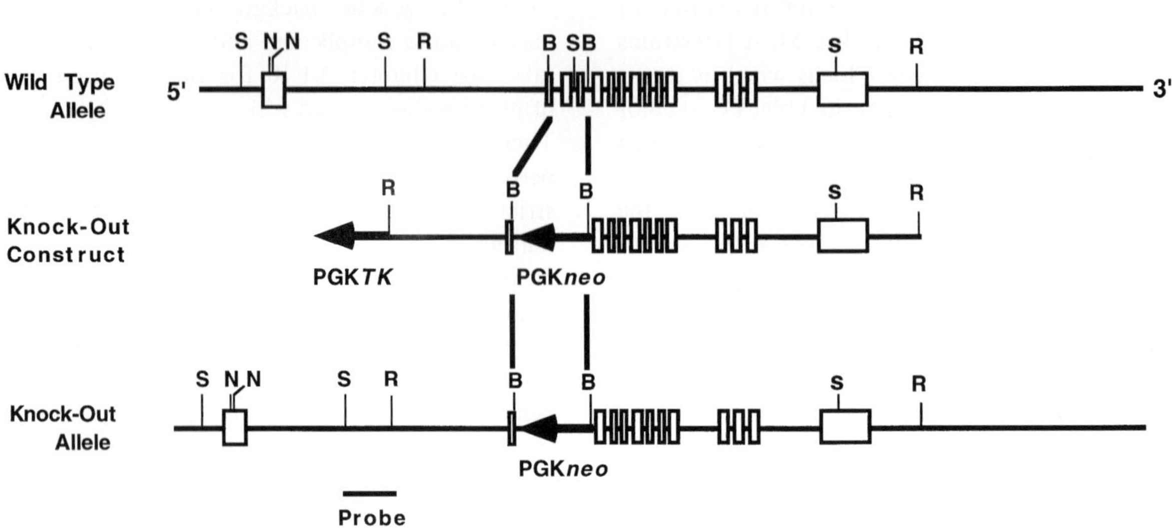

Fig. 2. Targeting vector used to produce *Dvl*1-deficient mice. The *Dvl*1 genomic locus (top), targeting vector (middle) and targeted allele (bottom). The location of exons, and PGK*neo* gene (with transcriptional orientation denoted by the arrow) are shown. B, *Bam*HI; Bg, *Bgl*II; R, *Eco*RI; S, *Sac*I.

129Sv *Dvl*1 genomic clone was used to make the *Dvl*1 targeting construct in the vector pPNT (Tybulewicz et al., 1991). A 2.5 kb *Eco*RI-*Bgl*II *Dvl*1 fragment, from within the first intron to the middle of exon 2, was inserted into the *Bam*HI and *Eco*RI sites of pPNT. A 6 kb *Bgl*II-*Eco*RI fragment extending from the middle of intron four to beyond exon 15 was inserted in the *Xho*I and *Not*I sites of pPNT vector to generate pPNT-*Dvl*1 (Fig. 2, middle). This resulted in the replacement of a 0.5 kb *Bgl*II fragment in *Dvl*1 genomic clone with a PGK*neo* gene in the opposite orientation relative to *Dvl*1 transcription, and the loss of most of exon 2 and all of exons 3 and 4 (Fig. 2, bottom). This deletion removed a *Sac*I site. In the construct, a PGK-TK gene was included outside of the region of homology to use as a negative selectable marker, reducing the background from random integration (Fig. 2, middle). pPNT-*Dvl*1 was linearized with *Not*I and transfected into TC1 embryonic stem cells as described previously (Deng et al., 1994; Deng et al., 1996). We screened 100 G418- and FIAU-resistant clones for targeted disruption of *Dvl*1 by Southern blot analysis. FIAU enrichment was 5–6 fold, determined by comparing the number of colonies per plate of G418 only versus G418 + FIAU-treated plates. Targeting was identified in three clones after *Sac*I digestion with the flanking probe A (wild-type allele 4.5 kb, and the targeted allele 9.5 kb). An internal probe B was used with several restriction enzymes to confirm correct targeting in these clones.

Two of the targeted ES cell clones were injected into C57BL/6J blastocysts. Germline transmission was obtained with both clones at high frequency, allowing us to establish the *Dvl*1 mutant allele ($Dvl1^{del131\text{-}225}$) in mixed (129SvEv x NIH Black Swiss) and inbred (129/SvEv) backgrounds. Genotyping was performed by Southern blotting, using probe B and *Bam*HI digestion (Fig. 2(C)).

We examined *Dvl*1 expression in cell lines derived from *Dvl*1 −/− and +/+ mice by immunoblot analysis, using a mouse monoclonal antibody (1-3F12) to a GST fusion protein containing the 45 carboxy-terminal amino acids of *Dvl*1. *Dvl*1 was identified in +/+ samples as a 75 kD protein, but no *Dvl*1 protein was detected in the mutant mice (Lijam et al., 1997), demonstrating that the *Dvl*1 mutant allele is a null allele.

Conclusions

In this chapter, we have described in general terms how ES cells are used to produce mice with specific alterations in the genome. The next chapter will describe methods used to introduce DNA randomly into the germline of mice through pronuclear injection. The use of these powerful transgenic techniques to modify specific gene function in vivo in mice has provided us with crucial insights into brain function and behavior. Since these techniques are becoming more routine in their use throughout the world, we will likely see many other important observations concerning brain function immediately and in the not-too-distant future.

References

Barlow, C., Hirotsune, S. Paylor, R., Liyanage, M. Eckhaus, M., Collins, F., Shiloh, Y., Crawley, J.N., Ried, T., Tagle, D. and Wynshaw-Boris, A. (1996) *Atm*-deficient mice: a paradigm of ataxia-telangiectasia. *Cell*, 86: 159–171.

Crawley, J.N., Belknap, J.K., Collins, A., Crabbe, J.C., Frankel, W., Henderson, N., Hitzemann, R.J., Maxson, S.C., Miner, L.L., Silva, A.J., Wehner, J.M., Wynshaw-Boris, A. and Paylor, R. (1997) Behavioral phenotypes of inbred mouse strains. *Psychopharmacology*, 132: 107–124.

Deng, C., Wynshaw-Boris, A., Zhou, F., Kuo, A. and Leder, P. (1996) Fibroblast growth factor receptor 3 is a negative regulator of bone growth. *Cell*, 84: 911–921.

Deng, C. X., Wynshaw-Boris, A., Shen, M. M., Daugherty, C., Ornitz, D. M. and Leder, P. (1994) Murine FGFR-1 is required for early postimplantation growth and axial organization. *Genes Dev*., 8: 3045–3057.

Hogan, B., Beddington, R., Costantini, F. and Lacy E. (1994) *Manipulating the mouse embryo: a laboratory manual*. Cold Spring Harbor Laboratory Press, Cold Spring Harbor.

Joyner, A.L. (1993) *Gene targeting: a practical approach*. Oxford University Press, Oxford.

Lijam, N., Paylor, R., McDonald, M.P., Crawley, J.N., Deng, C., Herrup, K., Stevens, K.E., Maccaferri, G., McBain, C.J., Sussman, D.J. and Wynshaw-Boris, A. (1997) Social interaction and sensorimotor gating abnormalities in mice lacking *Dvl*1. *Cell*, 90: 895–905.

Ramirez-Solis, R., Liu, P. and Bradley, A. (1995) Chromosome engineering in mice. *Nature*, 378: 720–724.

Simpson, E.M., Linder, C.C., Sargent, E.E., Davisson, M.T., Mobraten, L.E. and Sharp, J.J. (1997) Genetic variation among 129 substrains and its importance for targeted mutagensis in mice. *Nat. Genet*., 16: 19–27.

Tybulewicz, V.L., Crawford, C.E., Jackson, P.K., Bronson, R.T. and Mulligan, R.C. (1991) Neonatal lethality and lymphopenia in mice with a homozygous disruption of the *c-abl* proto-oncogene. *Cell*, 65: 1153–1163.

Wassarman, P.M. and DePamphilis, M.L. (1993) Guide to techniques in mouse development. *Methods in Enzymology*, Vol. 225, Academic Press, San Diego.

Wynshaw-Boris, A. (1996) Model mice and human disease. *Nat. Genet*., 13: 259–260.

W.E. Crusio and R.T. Gerlai (Eds.)
Handbook of Molecular-Genetic Techniques for Brain and Behavior Research (Techniques in the Behavioral and Neural Sciences, Vol. 13)

CHAPTER 3.1.2

Generation of transgenic mice by pronuclear DNA injection

Anthony Wynshaw-Boris,[1] Lisa Garrett, Amy Chen and Carrolee Barlow[2]

Genetic Disease Research Branch, National Human Genome Research Institute, National Institutes of Health, Bethesda, MD 20892, USA
[1]Department of Pediatrics, UCSD School of Medicine, La Jolla, CA 92093, USA
[2]Laboratory of Genetics, The Salk Institute for Biological Studies, La Jolla, CA 92037, USA

Introduction

Transgenic mice can be produced in one of two ways: by direct pronuclear injection of cloned DNA into the male pronucleus of a fertilized mouse egg; or by targeted introduction of transgenes by homologous recombination using ES cells. These techniques have allowed for the stable transfer of normal, altered, or chimeric genes into the mouse germ line, thus providing powerful tools with which to study mechanisms underlying gene expression within a physiologic context. The previous chapter (3.1.1) outlined methods for producing knock-out mice using ES cell technology. This chapter will describe methods for generating mice by pronuclear injection. In particular we will discuss technical aspects of creating transgenic mice with emphasis on issues that impact the study of behavior and the brain. For further information about these techniques in general, the reader is referred to several excellent laboratory manuals describing techniques in transgenic manipulation and gene targeting (Hogan et al., 1994; Wasserman and DePamphilis, 1993).

Construction of transgenes

Specific genetic alterations can be made in all cells or in specific populations of cells within an animal using transgenic techniques. There are several unique requirements for the use of constructs to produce transgenic mice in comparison to those used for expression in cell lines. In both situations, a promoter and polyadenylation signal are required, to provide signals for initiating and terminating transcription, respectively. However, in transgenic mice, it has been shown that the inclusion of an intron in the construct leads to much higher levels of expression of the transgene. Inclusion of an intron, along with splicing sequences, may stabilize the resulting transgene mRNA after splicing. It has also been shown that plasmid sequences in commonly used cloning vectors interfere with the expression of the gene of interest and that linearized DNA integrates more readily. Therefore, the transgene can be surrounded by unique restriction sites in the plasmid construct. When preparing the transgene DNA for injection, the construct can be digested with these restriction enzymes. The transgene can then be separated from the plasmid sequences as described below.

For this approach to be useful in vivo, the gene of interest must be expressed in the appropriate cell type within the mouse, and at an appropriate time in development to have a desired effect. In order to accomplish this, promoter and other regulatory elements must be included in the transgene construct. It is often difficult to determine a priori whether the chosen promoter will actually lead to expression of your gene of interest in the appropriate spatial and temporal pattern within the mouse. For example, not all promoters used in cell culture will result in predictable patterns of expression in analogous mouse tissues, and various promoters are sensitive to integration dependent misexpression. Several techniques have been employed in order to increase the chance that appropriate spatial and temporal transgenic expression will occur. As a first step, specific cell lines can be used to identify the smallest region of the promoter that gives tightly controlled and high levels of expression of the gene of interest. Fibroblast lines, such as NIH/3T3 cells, or differentiated neural cell lines such as PC12 cells, can be used to test for promoter activity. Neural cell lines should be the best ones to test neural-specific promoters. Once a promoter and other regulatory elements have been identified in cells, the defined region can be linked to a reporter gene, such as the β-galactosidase-encoding *LacZ* gene. The transgene can be prepared and used for production of transgenic mice, and these mice can then be quickly screened for appropriate patterns of expressions. Such "reporters" are useful because the expression can be monitored in vivo, in whole embryos or tissues, by simple histochemical stains. This type of experiment has also been used to identify and study promoter elements in vivo.

It often takes several modifications of endogenous promoters to identify sequences that will give appropriate levels of tissue specific transcription of transgenes in animals, particularly in nervous tissue. For example, the neurofilament nestin is expressed in the developing CNS. However, it is also expressed in the developing tooth and limb. Therefore, the endogenous promoter would not be appropriate for use in a CNS-specific experiment. By using the *LacZ* reporter gene assay in transgenic mice, researchers were able to identify regulatory regions (enhancer) in the second intron of the human nestin gene required for expression in embryonic CNS stem cells and in the adult spinal cord after injury. By using this region as well as a portion of the promoter region, CNS specific expression could be obtained in the developing mouse (Lothian and Lendahl, 1997).

The standard linear DNA fragments ranging from 5 kb to 50 kb for pronuclear microinjection are generally prepared by growing large cultures of bacteria containing the plasmid or cosmid and purifying the extrachromasomal DNA by double cesium chloride banding or commercial plasmid preparation columns such as Qiagen. However, we have had best success during microinjection with cesium chloride-prepared DNA. Nicked DNA, which reduces the efficiency of making transgenic animals, is removed by cesium banding. The plasmid DNA is then digested with restriction enzymes and the fragment for injection is purified away from any contaminating vector sequences either on a gel or on a sucrose gradient. If the DNA fragment is gel purified, the DNA can be either electroeluted or purified with glass milk beads. If the DNA is purified on a gradient, aliquots must be run on a gel to identify the sample without vector contamination. Vector-free aliquots are pooled and dialyzed against the microinjection buffer containing 10 mM Tris and 0.25 mM EDTA. Our laboratory protocol for purification of transgene DNA follows.

Plasmid DNA preparation

Day 0
1. Begin overnight culture of bacteria in 35 ml of 2xYT or LB broth.
2. Mix up and autoclave 1–2 l of LB broth per prep, cool overnight.

Day 1
1. Add 50 mg ampicillin per 1 liter LB broth for plasmids that contain the ampicillin resistance gene (amp is the most widely used antibiotic, but appropriate concentration of any other anti-

biotic can be used to grow plasmids with other antibiotic resistance genes).
2. Add 10 ml overnight culture to the LB broth with ampicillin.
3. Incubate at 37°C until the culture reaches an optical density at 600 nm (OD) of ~0.8 (3–6 h).
4. Pellet bacteria in a Sorvall centrifuge, 20 min at 5000 rpm.
5. Resuspend pellet in 12.5 ml 25% sucrose-0.05 M Tris (pH 8.0). Mix and place in Oak Ridge centrifuge tubes (35–50 ml).
6. Add 1 ml 10 mg/ml lysozyme in 0.25 M Tris pH 8.0, incubate for 5 min at room temperature to digest the bacterial cell wall.
7. Add 1.5 ml 250 mM EDTA pH 8.0, keep at room temperature for 5 min, to inhibit lysozyme and DNase.
8. Add slowly 6 ml Triton X-100 solution. (1% Triton X-100, 50 mM Tris pH 8.0, 62.5 mM EDTA), and shake vigorously to lyse bacteria.
9. Add 10 ml 10 mM Tris, 1 mM EDTA, pH 8.0 (TE) if necessary to bring up volume.
10. Centrifuge at 19 K rpm for 1 h at 4°C.
11. Remove supernatnant (decant) into 50 ml plastic Corning tubes. Bring to 30 ml with TE.
12. Add 27 g CsCl and dissolve (shake vigorously).
13. Add 2 ml 10 mg/ml ethidium bromide (EtBr), and bring to 42 ml final volume with 0.9 g/ml CsCl in TE.
14. Load into ultracentrifuge tube (16 gauge needle and 30 cc syringe) and seal with caps.
15. Spin at 47,000 rpm for 18–24 h using a vertical ultracentrifuge rotor.

Day 2 1. Visualize plasmid (lower band) and bacterial chromosomal DNA (upper band) with an ultraviolet light.
2. Use a needle and syringe to remove lower plasmid band into 50 ml centrifuge tube.
3. Add 2 ml EtBr, and bring to 42 ml with 0.9 g/ml CsCl in TE.
4. Load into ultracentrifuge tubes as above and spin at 47,000 rpm 18–24 h (second banding).

Day 3 1. Remove plasmid band into 50 ml centrifuge tube as above.
2. Extract at least three times with an equal volume of isopropanol saturated with 9 g/ml Cs/Cl in TE. Spin at 3000 rpm for 5 min. Remove all but 1 ml of the top pink phase each time and discard.
3. After the top phase is no longer pink in color, place the bottom phase in a dialysis bag (BRL# 15961-022 3/4 inch 12,000–40,000 Da exclusion weight) equilibrated with TE.
4. Dialyse against 4 l of TE, changing three times over 24 h at 4°C 24.

Day 4 1. Quanitate plasmid DNA for restriction enzyme digestion by absorbance at 260 nm.
2. Digest 50–100 μg of plasmid containing the insert for microinjection, to liberate the transgene from the plasmid backbone.
3. Separate the transgene band from the plasmid band by electrophoresis on a low melting point ultrapure agarose gel.
4. Excise band from the gel and purify fragment by Gene Clean (glass mild beads) using the protocol of the manufacturer and TE for elution. It is important to remove all glass beads.
5. Quanitate DNA with a known standard (preferably a series of dilutions of a reference plasmid between ~2 ng–50 ng) on an agarose gel.
6. Dilute DNA to store for microinjection to 100 μg/ml in 10 mM Tris, 0.25 mM EDTA (pH 7.5).

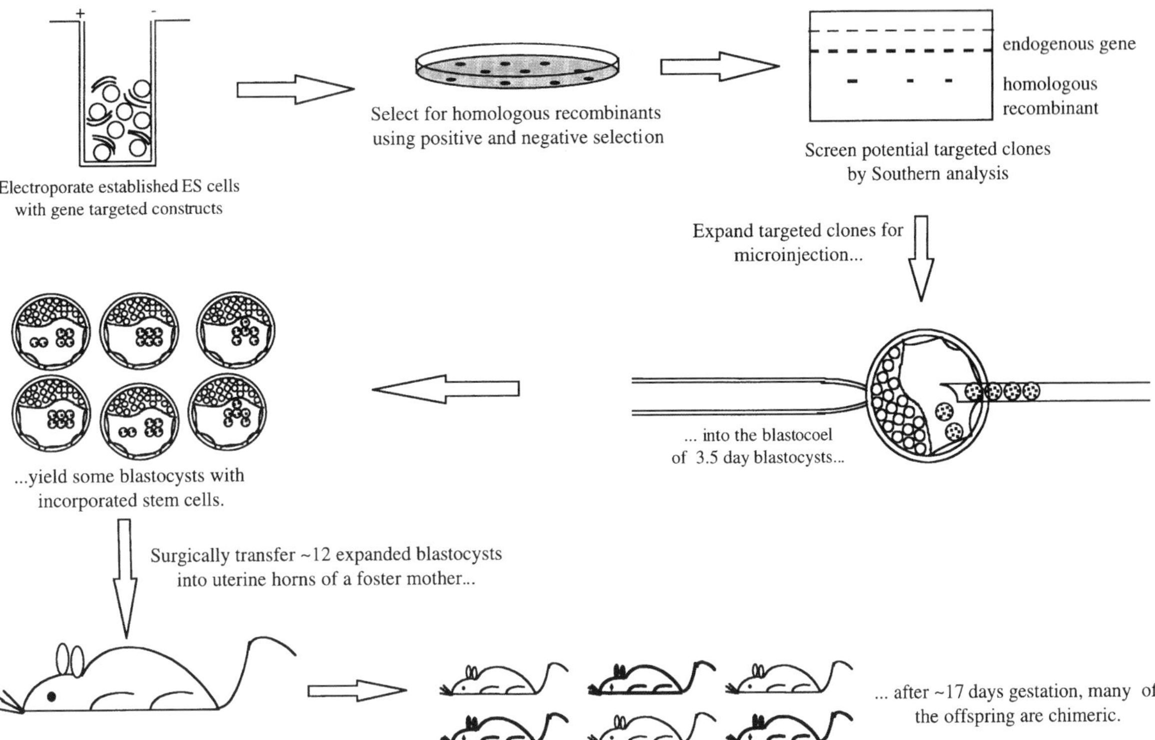

Fig. 1. Producing transgenic mice by pronuclear injection. See text for details.

Pronuclear injection

Transgenic mice are produced by microinjection of DNA directly into pronuclei of fertilized mouse eggs, resulting in the random integration of foreign DNA into the mouse genome (Fig. 1). Fertilized eggs are isolated the day after mating female mice (induced to superovulate by hormonal treatment) with fertile males. The male and female pronuclei do not fuse immediately after fertilization. The male pronucleus is large and, using micromanipulators to stabilize the egg, more than 100 copies of the purified DNA can be injected into the male pronucleus. The DNA generally integrates at the one cell stage and therefore foreign DNA will be present in every cell of the "transgenic" animal. The manipulated embryo can be implanted at this time, or maintained in culture until the two cell stage, and is then transferred into the oviduct of a pseudopregnant female where the embryo will continue to develop. Pseudopregnant females are produced by mating females with vasectomized males. The act of mating initiates the hormonal cascade required to prepare the uterus for implantation of embryos, but since the male is sterile, the only embryos that develop and implant are those that have been injected and transferred. Once implanted, the manipulated embryo is carried to term, and delivered. Offspring are analyzed by Southern blotting or PCR, and transgenic animals which have integrated the gene into their DNA are referred to as founders. Several founders can be created during each day of injection. These founders, in which DNA has integrated into the germ cells, are able to transmit the transgene as a heritable trait and hence a transgenic line can be established.

We produce fertilized eggs for microinjection by superovulating FVB/n females by the following protocol.

Superovulation of FVB/n mice and harversting fertilized eggs

Day 3: 1. Inject 15-20 FVB/n donor females (3–4 wk) with 5 units of PMS (pregnant mares serum, Calbiochem 1000 I.U./vial). The time of hormone injection can be from 11:30 a.m. until 1:00 p.m. It is best to inject earlier in this interval.

Day 1: 1. Inject 15–20 FVB/n females (previously injected with PMS) with 5 units of HCG (Organon, 10,000 units/vial) and mate with stud FVB/n males, one female per male.
2. On the same day, mate 20 CB6F1 female recipients (for the production of foster females for reimplantation of fertilized eggs) that are in estrus (moist, pink vagina, slightly gaping introitus) and mate with 20 vasectomized CB6F1 males. Vasectomized males can be purchased from Jacson laboratories, or made surgically in the lab of the investigator. Following induction of anesthesia (Avertin — see below) and sterilizing the surface of the skin with ethanol, a 1 cm midline incision is made through the skin, then through body wall. One testis is pushed up from scrotal sac into the abdomen and the testis fat pad is pulled through the incision. The vas deferens is identified and ligated in two places. A segment of vas between two ligatures is removed. This is repeated on the other side. The body wall is sutured and the skin is clipped or sutured.

Day 0: 1. Separate FVB/N females for egg harvest. Check for mating by the presence of a hard white copulation plug at the vaginal introitus. Separate "plugged" CB6F1 females from vasectomized males.
2. On the day of harvest, prepare six 35 mm dishes with 2 ml of M2 media (PGC-cat. #MR-015-D) each. Add 0.75 ml of 10 mg/ml hyaluronidase in DMEM to two of the dishes. Place on 37°C slide warmer under a 150 mm dish with an extra 35 mm dish filled with water (for humidity).
3. Sacrifice mice by anesthesia and cervical dislocation and expose abdominal cavity. Remove both ovaries by cutting above the ovary and below the fallopian tubes and avoid taking as much fat as possible. Transfer the ovaries to the two dishes containing M2/hyaluronidase (divide the number of ovaries between the 2 dishes, 20 per dish).
4. Using a dissecting scope, release eggs from the swollen ampulla area and transfer torn ovary out of dish. (It is easiest to use Dumont #5 forceps to tear open the ampulla).
5. Do not leave eggs in the media with the hyaluronidase for more than 10 min. Transfer the eggs to a new M2 dish without hyaluronidase. Transfer again to a second dish of M2 and get an approximate count. Eggs are now ready for injection.
6. To prepare DNA for microinjection, dilute construct DNA to approximately 2 ng/μl in a 200 ml total volume. Spin diluted DNA for 10 min at 14,000 rpm using Eppendorf microcentrifuge. Pull off all but 20 μl and aliquot centrifuged DNA into 4 × 1.5 ml tubes. Place pulled needles

into DNA solution and allow to backfill up to the tip of needle.

7. For microinjection, prepare a clean glass depression slide as follows: place a 25 μl drop of M2 media in the center of the depression and cover the entire depressed area with light weight mineral oil. Transfer ~20–50 eggs into center of the drop. It is best to start out with a fewer number of eggs for injection. You do not want the eggs to sit under the microscope for more than 20 min at a time. Transfer injected eggs into a new dish of M2 media. If you are not going to implant the embryos on the same day (1-cell), transfer the unlysed, injected eggs into BMOC-2 media (PGC- cat. #MR-013-D) which does not contain HEPES and incubate with CO_2 at 37°C for over night culture.

The apparatus used for pronuclear injection of DNA solutions is similar to that used for microinjection of ES cells, as described in the previous chapter. A Zeiss Axiovert 135 inverted microscope with Narishige hydraulic manipulators or Eppendorf motorized manipulators is placed on an air table or vibration resistant table. One manipulator is attached to a holding pipette, which is linked to a syringe that controls gentle vacuum pressure to stabilize the fertilized egg for injection. The other manipulator is attached to the injection pipette, which is attached to an air supply to push the DNA solution out of the injection needle, or to an automatic microinjector (produced by Eppendorf, for example). Holding and injection pipettes are made from glass pipettes using a programmable pipette puller, to create glass needles of the appropriate diameters. Injection needles are made to a fine point, with a narrow opening to allow the DNA solution to be injected (shown schematically in the figure). To make the holding pipette, a microforge is used to break off the pipette to the appropriate size, and to polish the end to make it smooth, so as not to damage the egg. Plasmid or cosmid DNA is microinjected at a concentration ranging from 1–5 ng/μl, with a needle approximately 1–2 μm in diameter. The fertilized egg is held by the holding pipette to expose the two haploid pronuclei. The female pronucleus is smaller, while the male pronucleus, derived from the sperm, is larger. Because of the size difference, the egg is positioned to allow the injection pipette to penetrate to zona pelucida, and the male pronucleus. Injection of the DNA solution into the pronucleus causes swelling, a visual confirmation that DNA has been injected into the nucleus.

Injected fertilized eggs are reimplanted into pseudopregnant foster females by the following protocol.

Transfer to recipients:

Equipment/materials needed

Slide warmer for recipients and eggs (37°C)
#5 Dumont forceps (two)
Scalpel (#11 or #15)
Serrafin clip
Microdissecting forceps with teeth
Microdissecting forceps with serrated, blunt tip
4–0 suture
9 mm surgical wound clips and applier
Sterile drapes
Avertin anesthetic agent. Avertin stock is a solution of 1.0 g 2,2,2-tribromoethanol dissolved in 0.5 g tertiary amyl alcohol. Store the stock solution in the dark in a glass container. Dilute the Avertin to the final working concentration by dissolving 1.2 ml of the stock in 100 ml of boiling PBS. Shake to dissolve. Aliquot and store at −20°C.
Syringe (1/2 to 1 ml) and needle (25–28 1/2 ga)
Dissecting microscope with fiber optic light source
Glass micropipettes
Mouth pipetter

1. Following intraperitoneal injection of anesthetic agent into the pseudopregnant females

and induction of deep anesthesia, the back of the mouse is washed with ethanol.

2. Place the mouse on a 4 × 4 gauze pad on the dissecting scope platform with the right side towards you. Approximately one-third of the way of the back, make an incision in the skin approximately 1 cm in length and approximately 1 cm to the right of the spine. To locate the region of the ovary, identify a vessel and a nerve running diagonally across the ovarian fat pad (the fat pad should appear easily as a white mass just below the surface of the peritoneum). The ovary is just caudal to the fat pad. Grasp the peritoneum with the micro-dissecting forceps with teeth and make an incision approximately 0.5 cm in length above the fat pad. With blunt forceps, pull out the fat pad with the ovary attached through the incision and place on a sterile drape. Anchor the fat pad and ovary in place with a Serrafin clip. Do not grasp the ovary or oviduct, but only the fat pad.

3. Adjust the magnification on the dissecting microscope so that the ovary and the oviduct are visible in the field of view.

4. Focus so that the interior of the infundibulum is visible. In the infundibulum, look for the oviduct opening at the end of the oviduct. With #5 Dumont forceps, carefully tear the mesovarium away from the ovary and oviduct, without tearing any blood vessels.

5. Collect 15–20 one cell embryos in the glass transfer pipette with an absolute minimum of M2 liquid. Rate of loading and volume can be controlled by the introduction of air bubbles. Insert the pipette into the infundibular opening and gently blow the embryos into the oviduct. Also blow a small bubble of air into the oviduct to be sure that all the embryos have been expelled.

6. Gently push the ovary and uterus back into the peritoneal cavity with the blunt forceps.

7. Place one suture into the musculature and close the skin with 2 wound clips.

8. When transfer is complete place the recipient in a warm recovery cage.

9. When recovery is complete transfer the recipients to a clean cage.

Implanted pseudopregnant females give birth after 19–21 d, depending upon the strain used. Once the animals are old enough, a tail biopsy is performed to isolate DNA, and founder lines containing the transgene are identified. Under ideal conditions, the efficiency for generating transgenic founder mice containing 5–50 kb fragments is between 20–30%. In general, the transgene integrates as 1–50 copies arranged in a head-to-tail orientation (tandem repeats). Founders must then be screened for expression of the transgene RNA or protein, using the various techniques described above, as sequences surrounding the integration site may result in undesired patterns of expression or may even prevent the gene from being expressed. In general, the expression pattern from integrated transgenes remains stable over many generations.

Analysis of transgene expression in founder lines and offspring

The expression of most transgenes is not as easily detectable as the *lac*Z marker gene. Consequently, it may be difficult to confirm that the transgene is expressed in the appropriate tissue. This can be particularly important when studying transgene expression in the brain, as it is likely necessary to be able to precisely identify subsets of neuronal populations that express a transgene. It may be possible to develop an antibody that is specific for the transgene product and does not cross react with any endogenous form of the protein in the mouse. Such an antibody could be used to perform immunohistochemistry on sections, but it is rare that such antibodies can be made with that degree of specificity to the transgene relative to homologous endogenous mouse genes. RNA in situ hybridization can also be used if there are unique sequences in the transgene that are not present in any endogenous gene, such as the junction between the polyA tail and the cDNA.

As an alternative approach, several types of modifications of transgenic constructs have been employed to allow expression of the transgene to be followed carefully in vivo. It is possible to link the expression of the gene of interest to a "reporter" gene to mark the cells where the transgene is expressed. For example, one can create a fusion protein by linking the gene of interest to a reporter fragment or gene. Unique protein domains, or "tags," such as hemagluttinin or c-myc, have been placed at the amino or carboxy terminus of transgenic proteins. These domains often do not alter the activity of the protein, and good antibodies have been developed to detect tagged proteins by Western blotting or immunohistochemistry. One disadvantage of this approach is that these mutant fusion proteins may not have the same biological properties as unmodified proteins. An alternative approach is to take advantage of various viral internal ribosome-entry site (IRES) which permit the effective internal initiation of translation in mammalian cells by directing efficient mRNA cap-independent entry of the translation apparatus. The promoter/regulatory region is linked to a transgene that includes the gene of interest, followed by the IRES sequence and finally a reporter gene. A single mRNA is produced, but two proteins are coded for, and the IRES allows both to be translationally initiated. Several sensitive histological markers, such as the aforementioned β-galactosidase encoding *lacZ* gene and the human placental alkaline phosphatase (hpAP) gene, have been used (for a recent review of the technique see Li et al., 1997). In addition, use of the IRES system eliminates the need to construct a transgene fusion that has the correct translational reading frame for efficient production of the reporter.

By using transgene tags, fusion proteins or dicistronic IRES-containing constructs, one can identify precisely which cells express the transgene by evaluation of the presence of the reporter using standard immunohistochemistry techniques which are especially important when studying the brain.

Transgenic mice produced with large DNA fragments

Recently, it has become possible to create transgenic mouse strains from the injection of large DNA vectors, such as yeast artificial chromosomes (YACs), bacterial artificial chromosomes (BACs) and P1 artificial chromosomes (PACs) (Peterson et al., 1993; Peterson et al., 1995; and Gnirke et. al., 1993; for additional examples see Chapters 5.8 and 5.9). Although technically more difficult than the injection of smaller transgenes, the introduction of large DNA fragments raises the possibility of doing functional genomics in the mouse (Smith and Rubin, 1997). For example, mice that are trisomic for a region of human chromosome 21 have been created by injecting YAC clones (Smith et al., 1997). These mice have been used to examine sequences which influence learning and memory when present in three copies.

YAC transgenesis has also been used to rescue a specific neurologic phenotype and verify that the transgenic fragment contains the mutant gene. For example, the mouse mutant gene *vibrator* (*vb*) results in an early-onset progressive action tremor, degeneration of brain stem and spinal cord neurons, and juvenile death. This gene was cloned using an in vivo positional complementation strategy. The *vb* locus was mapped to a region of the mouse genome, and the region containing the gene was narrowed to a 76 kb region by transgenic complementation of the phenotype by injecting YACs from the region. Once the region was narrowed, the *vb* mutation was found to be an intracisternal A particle insertion in intron 4 of the phosphatidylinositol transfer protein alpha gene (Hamilton et al., 1997).

We have found that there are several important points to consider when preparing and microinjecting DNA over 100 kb in size: mechanical shearing and denaturation of the DNA; DNA concentration; and purification of the DNA to avoid clogging the microinjection needle. YAC DNA is especially large and can easily be sheared in solution. Therefore, yeast containing a YAC to be microinjected are generally embedded in agarose plugs. Enzymatic digestion by yeast lytic enzymes within the agarose plug breaks open the yeast cell wall, releasing the YAC and yeast chromosomes, and plugs are treated with detergents for deproteinization. The YAC DNA is purified on pulse field gels in order to separate the YAC from contaminating yeast chromosomal DNA. It is important to have a high concentration of yeast per agarose plug to purify the YAC DNA. YAC DNA is carefully excised from the pulse field gel and treated gently with agarase, concentrated by ultrafiltration and quantified. High molecular weight DNA must be handled carefully to avoid shearing. It is usually manipulated with large bore pipette tips and diluted with a microinjection buffer that contains additional NaCl to stabilize the DNA. For microinjection, the YAC is injected at a concentration of 1 ng/μl. The microinjection needle is prepared with a much larger opening than a needle used for plasmid or cosmid injection, and injection of YAC DNA generally requires frequent needle changes. The efficiency for production of intact YAC transgenics ranges from 5–20%.

BAC or PAC DNA can be grown in large bacterial cultures and purified similarly to plasmid DNA (Yang et al., 1997). Chromosomal DNA is separated from extrachromasomal DNA by cesium chloride banding or sucrose gradients and then dialyzed against the microinjection buffer. BAC and PAC DNA is usually microinjected as circular DNA, but linear DNA can be injected as well, and careful handling is required as described above for YACs. Sepharose chromatography can be used to separate circular from linear DNA. It is not necessary to remove the plasmid backbone DNA from BAC or PAC DNA, since these sequences have not been found to interfere with transgene expression. Larger needles are used as well as large bore pipet tips for transferring the DNA. The efficiency of generating transgenic founder animals with BAC or PAC DNA is 10–20%.

Use of chicken retroviral receptors to target gene transfer and expression

It has recently been reported that by using a combination of transgenics and retroviral infection, high rates of delivery of target genes into adult nervous tissue has been observed (Holland et al., 1998). In these experiments a transgene encoding a specific receptor for an avian retrovirus (A avian leukosis virus ALV-A) under the control of an astrocyte-specific glial fibrillary acidic protein (GFAP) promoter was used to make a line of transgenic mice expressing the retrovirus receptor in glial tissue. These transgenic mice were then infected with a replication incompetent ALV-A virus that expressed a particular gene of interest, in this case a basic fibroblast growth factor by direct injection of virus into the ventricles of newborn mice. These viruses were made with a retroviral expression vector (RCAS), and virus was produced in chicken viral packaging cell lines. The only tissues infected were those that express the retroviral receptor, namely glial cells, sparing all other tissues from the consequences of infection. The authors were able to demonstrate glial specific infection and concomitant expression of the retrovirally delivered gene. Multiple viruses can infect the same cell, since transgenically expressed chicken receptor is not blocked by ALV-A infection. Consequently, it is possible to infect each cell with different viruses, each containing a different gene, to assay the effects of multiple genes simultaneously on a defined cell population in vivo.

Using this type of methodology, one could envision that several different types of promoters, such as nestin, or other neuron specific promoters, could be used to allow for infection of neurons as well as glia. This would allow for testing the effects of overexpression of a wide variety of products directly in the nervous system of adult animals, and for following their effects on neurologic function and behavior.

Strain issues

The use of specific strains for the production of transgenic animals has implications in their use for studies of behavior (e.g. see Chapter 3.1.10). The most commonly used mouse strains for isolating fertilized eggs are C57BL/6J, FVB/n, or F_1 hybrids such as C57BL/6J × C3H. These mice, especially FVB/n, have practical advantages for pronuclear injection, such as large egg yield, large and easily visualized pronuclei, and resistance to lysis after injection. However, there are no absolute technical limitations to using virtually any mouse strain for producing transgenic mice, since the pronuclei of all mouse fertilized eggs can be visualized microscopically. Of particular relevance to the topic of this book, there is a wide range of neurologic and behavioral phenotypes displayed by different strains of mice (Crawley et al., 1997). The choice of mouse strain for a particular transgenic experiment should be guided by the phenotypes that one wishes to address.

Conclusions

It is now feasible to manipulate the mouse genome to make precise genetic alterations, and to create gain-of-function and loss-of-function alleles of specific genes. In these two chapters, we have briefly reviewed the techniques required to create genetically manipulated mice using either pronuclear injection or targeted mutagenesis in ES cells Other chapters in this book will provide further details and examples of the use of these techniques to study neurologic and behavioral function in the whole animal.

References

Crawley, J.N., Belknap, J.K., Collins, A., Crabbe, J.C., Frankel, W., Henderson, N., Hitzemann, R.J., Maxson, S.C., Miner, L.L., Silva, A.J., Wehner, J.M., Wynshaw-Boris, A. and Paylor, R. (1997) Behavioral phenotypes of inbred mouse strains. *Psychopharmacology*, 132: 107–124.

Gnirke, A., Huxley, C., Peterson, K. and Olson, M.V. (1993) Microinjection of intact 200–500 kb fragments of YAC DNA into mammalian cells. *Genomics*, 15: 659–667.

Hamilton, B.A., Smith, D.J., Mueller, K.L., Kerrebrock, A.W., Bronson, R.T., Van Berkel, V., Daly, M.J., Kruglyak, L., Reeve, M.P., Nemhauser, J.L., Hawkins, T.L., Rubin, E.M. and Lander, E.S. (1997) The *vibrator* mutation causes neurodegeneration via reduced expression of PITP alpha: positional complementation cloning and extragenic suppression. *Neuron*, 18: 711–722.

Hogan, B., Beddington, R., Costantini, F. and Lacy E. (1994) *Manipulating the mouse embryo: a laboratory manual*. Cold Spring Harbor Laboratory Press, Cold Spring Harbor.

Holland E.C. and Varmus, H.E. (1998) Basic FGF induces cell migration and proliferation after glia-specific gene transfer in mice. *Proc. Natl. Acad. Sci. USA*, 95: 1218–1223.

Lothian C. and Lendahl U. (1997) An evolutionarily conserved region in the second intron of the human nestin gene directs gene expression to CNS progenitor cells and to early neural crest cells. *Eur. J. Neurosci.*, 9: 452–462.

Li, X., Wang, W. and Lufkin, T. (1997) Dicistronic LacZ and alkaline phosphatase reporter constructs permit simultaneous histological analysis of expression from multiple transgenes. *Biotechniques*, 23: 874–878, 880, 882.

Peterson, K.R., Clegg, C.H., Huxley, C., Josephson, B.M., Haugen, H.S., Furukawa, T. and Stamatoyannopoulos, G. (1993) Transgenic mice containing a 248-kb yeast artificial chromosome carrying the human β-globin locus display proper developmental control of human globin genes. *Proc. Natl. Acad. Sci. USA*, 90: 7593–7597.

Peterson, K.R., Li., Q.L., Clegg, C.H., Furukawa, T., Navas, P.A., Norton, E.J., Kimbrough, T.G. and Stamatoyannopoulos, G. (1995) Use of yeast artificial chromosomes (YACs) in studies of mammalian development: production of β-globin locus YAC mice carrying human globin developmental mutants. *Proc. Natl. Acad. Sci. USA*, 92: 5655–5695.

Smith, D.J., Stevens, M.E., Sudanagunta, S.P., Bronson, R.T., Makhinson, M., Watabe, A.M., O'Dell, T.J., Fung, J., Weier, H.U., Cheng, J.F. and Rubin, E.M. (1997) Functional screening of 2 Mb of human chromosome 21q22.2 in transgenic mice implicates minibrain in learning defects associated with Down syndrome. *Nat. Genet.*, 16: 28–36.

Smith, D.J. and Rubin, E.M. (1997) Functional screening and complex traits: human 21q22.2 sequences affecting learning in mice. *Hum. Mol. Genet.*, 6: 1729–1733.

Wasserman, P.M. and DePamphilis, M.L. (1993) Guide to techniques in mouse development. *Methods in Enzymology*, Vol. 225. Academic Press, San Diego.

Yang, X.W., Model, P. and Heintz, N. (1997) Homologous recombination based modification in *Esherichia coli* and germline transmission in transgenic mice of a bacterial artificial chromosome. *Nat. Biotech.*, 15: 859–865.

W.E. Crusio and R.T. Gerlai (Eds.)
Handbook of Molecular-Genetic Techniques for Brain and Behavior Research (Techniques in the Behavioral and Neural Sciences, Vol. 13)

CHAPTER 3.1.3

Brain region-specific and temporally restricted gene knockout using the Cre recombinase system

Joe Z. Tsien

Department of Molecular Biology and Neuroscience Program, Princeton University, Princeton, NJ 08544, USA

Introduction

Gene targeting in embryonic stem cells allows the production of mice to contain a deletion in a predefined gene of interest (Cappecchi, 1989). This is a powerful technique for the functional characterization of the gene product and has been widely used in biomedical sciences. In the field of cognitive neuroscience, gene knockout technology has been recently applied and begun to uncover molecular and cellular bases of cognitive behaviors (Chen and Tonegawa, 1995; Lipp and Wolfer, 1998). Compared to the traditionally used pharmacological approach, gene knockouts afford the highest degree of molecular specificity in that single subtypes of a gene family (e.g. receptors, kinases) can be deleted with certainty. Moreover, it offers a unique opportunity to study the function of targeted molecules (e.g. cytoskeleton proteins, transcription factors) for which highly specific pharmacological inhibitors are not yet available. In addition, it allows the production of a large number of animals all carrying the identical molecular lesion for correlative studies at the cellular, electrophysiological, and behavioral levels.

However, several shortcomings accompany the conventional gene knockout approach to behavioral neuroscience. Conventional gene knockout produces animals that inherit genetic deletions in all tissues throughout development. This regionally and temporally unrestricted genetic deletion may lead to severe developmental defects or premature death (Joyner, 1994), which can preclude analysis of post developmental gene functions. For example, null mutation of the NMDA receptor 1 gene resulted in neonatal death (Li et al., 1994; Forest et al., 1994), and therefore the role of NMDA receptor in adult mutant animals can not be studied. If mutant mice do complete development, interpretation of the observed physiological and behavioral phenotypes often encounters two kinds of uncertainties. First, the conventional gene knockout makes it difficult to attribute behavioral phenotypes to a particular type of neuron or brain region. Second, it is often difficult to exclude the possibility that the abnormal phenotype observed in adult animals has arisen indirectly from a developmental defect. For instance, in fyn tyrosine kinase knockout mice, it is not clear to what extent that the aberrant formation of the dentate gyrus and altered cytoarchitecture of the pyramidal cell layer contribute to the impairment in the hidden-platform water maze (Grant et al., 1993). It would therefore be highly desirable to have the ability to control the regional and temporal specificity of the genetic manipulations.

Cre recombinase and strategy for brain region-specific gene knockout

To augment the utility of gene knockout technology in cognitive neuroscience, it is necessary to develop conditional gene knockout techniques that impose regional and/or temporal restrictions in the brain (Tsien et al., 1996a). Toward that end, we have exploited the Cre/*lox*P recombination system derived from bacteriophage P1. Cre is a 38 kDa recombinase and belongs to a member of the Int family of recombinases (Argos et al., 1986). It can excise efficiently DNA fragment flanked by two *lox*P sites oriented in the same direction. The *lox*P site is a unique 34-bp sequence that includes two inverted 13-bp repeats (in shade) and an 8-bp core region.

ATAACTTCGTATAGCATACATTATACGAAGTTAT.

It is known that the 13-bp repeats serves as binding sites for Cre recombinase, and the recombination takes place in the 8-bp core region which is also responsible for the directionality of the *lox*P site (Abremski and Hoess, 1985). Previous studies have suggested that this site-specific recombination system allows the Cre recombinase to delete DNA fragment flanked by two *lox*P sites in vitro and in vivo mammalian cells (Sauer and Henderson, 1988; Orban et al., 1992; Lasko et al., 1992; Gu et al., 1993). Moreover, it appears that the size of the DNA fragment which can be efficiently deleted by Cre recombination can vary from several base pairs to 3-4 cM of chromosomal fragment (Ramirez et al., 1995).

The strategy to generate mutant mice in which a gene of interest is deleted in a brain region-specific manner is composed of two parts (Fig. 1): first, one needs to produce a mouse line in which the gene has been flanked by two *lox*P sites (termed a "floxed" gene) through ES cell targeting technique (Bradley, 1987; Hogan et al., 1994; Joyner, 1993). Second, one has to generate Cre transgenic mice that harbor the Cre transgene under the control of a brain region — and/or cell type-specific transcriptional promoter. Upon crossing homozygous floxed mice with Cre transgenic mice, in the progeny carrying both the floxed gene and Cre transgene, the floxed gene will be deleted by Cre/*lox*P recombination, but only in those cell types or regions in which the Cre transgene is expressed by the transciptional activity of the promoter (Tsien et al., 1996a, 1996b).

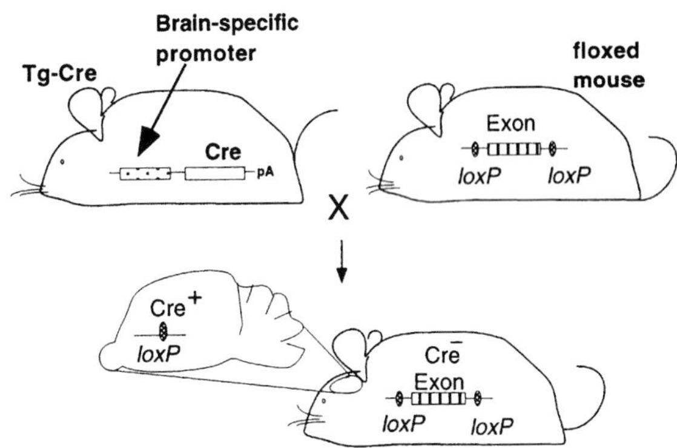

Fig. 1. Strategy for region-specific gene knockout using Cre/*lox*P recombination-mediated gene knockout system. The *lox*P sequences are inserted into the genome to flank one or more exons of any gene through ES gene targeting method. The homozygous mice (termed "floxed mice") are crossed to another transgenic mouse in which Cre recombinase is under control of a brain region-specific promoter such as αCaMKII promoter. In progeny, the floxed gene will be deleted only in these regions in which Cre is expressed.

There are two major considerations before performing the Cre/*lox*P-mediated knockout experiments. First, the *lox*P insertion locus should be carefully chosen during the design of targeting vectors for the production of floxed mice, so that the *lox*P insertions will not interfere with normal expression of the gene. For example, it would be wise to avoid critical sequences containing known promoter elements, splicing acceptor and donor sites, and transcription termination signals etc. Whenever feasible, one should always carry out diagnostic experiments in the targeted ES cells to check the integrity of the *lox*P sites by transfecting Cre-expressing construct. Second, promoters for the production of Cre transgenic animals should be selected to meet each experiment's need. As an alternative, one might also knock-in the Cre transgene into a promoter region of a known gene to achieve the desired expression pattern. In our work described below, the *α-CaM kinase II* promoter was used to drive expression of the Cre transgene in the forebrain, regions known to be involved in learning and memory.

Establishment of brain subregion-specific Cre transgenic lines

We have chosen the *α-CaMKII* promoter to direct the specific expression of Cre transgene to the mouse brain. This promoter has been used to restrict transgene expression to forebrain regions such as the cortex and the hippocampus (Mayford et al., 1995), and it has no activity in mid- and hind-brain regions. This is important consideration because gene knockout in the brainstem region and cerebellum could complicate the interpretation of behavioral experiments aimed at studying learning and memory. Moreover, the lack of transcriptional activity of the *α-CaMKII* promoter during prenatal and neonatal periods should eliminate the possibility of developmental defects caused by a gene knockout during embryonic development.

To generate Cre transgenic mouse lines, we made a Cre transgene expression vector under control of the CaM kinase promoter. This vector was constructed by inserting the 2.6 kb *Not*1 fragment of plasmid pZQCRE into the unique *Not*1 site of pMM279 containing a 8.5 kb *α-CaMKII* promoter sequence (Mayford et al., 1995). The 2.6 kb *Not*1 fragment of pZQCRE was composed of the CRE transgene with a nuclear localization signal (pBS317, a gift from Brian Sauer at NIH) and an exon-intron splicing and polyadenylation signals (pNN265, a gift from Nobuki Nakanishi at Harvard). The Cre expression vector was digested with Sal1 and the CaMKII promoter-Cre-Ploy(A) transgene fragment was removed isolated away from plasmid vector on 0.7% low-melting agarose gel electrophoresis. It was further purified using either 3 times of phenol extraction method or glass-milk method (GeneClean II, Bio101) followed by ethanol precipitation. The purified DNA construct (10 ng/μl) was used for pronuclear injection.

A total of fourteen Cre founders were produced, and eleven of them had successful germline transmission. Genotypes of all offspring were performed by clipping a short piece of mouse tails, followed by protease K digestion for extraction of tail DNA. The DNA was then subjected to southern blot and PCR analysis. For analysis by PCR, the 5′ and 3′ primers, 5-AGATGT-TCGCGATTATC-3′ and 5′-AGCTACACCAG-AGACGG-3′ respectively, were used to amplify a 490 bp band specific to Cre transgene. We used mouse tail DNAs (about 1 μg) as starting material and amplified for 30 cycles (1 min, 94°C; 45 s, 55°C; and 1 min, 72°C) on a thermal cycler.

To identify transgenic lines with correct expression pattern, we have conducted northern blot analysis for Cre mRNAs. We have identified five transgenic lines in which the expression of the Cre transgene mRNA is specific to the forebrain regions including the cortex and hippocampus. No Cre mRNA was detected from hindbrain regions such as the brain stem and cerebellum. Moreover, the level of Cre mRNA in the forebrain increased during the postnatal weeks (e.g. low at postnatal day 16 and higher at P23 and adult). This is in agreement with previous reports that *α-CaM kinase II* mRNA regulated by *α-CaMKII* promoter was barely detectable at P4 and became increasingly active over the next two weeks, reaching adult level by P30 (Burgin et al., 1990). The postnatal expression of Cre mRNA in these trangenic lines ensures that no Cre recombinase protein is available for the site-specific recombination during embryonic development. Thus, these Cre transgenic lines possess both regional and temporal specificity for the Cre/*lox*P-mediated gene deletion, suitable for the genetic dissection of brain function and behavior.

Region-specific knockout of the NR1 gene via Cre/*lox*P recombination

To test whether Cre/*lox*P recombination-mediated gene knockout actually works in the mouse brain, we decided to knockout NMDA receptor 1 (NR1) gene which encodes a key subunit responsible for the channel activity of the NMDA receptor (Moriyoshi et al., 1991). Previous genetic inactivation of the NR1 gene through conventional gene targeting method had resulted in developmental lethality in homozygous mice. The pups carrying null mutation died within 15–20 hr after birth, perhaps due to lack of suckling reflex as a result of disruption of the NMDA receptor function in the brainstem region (Li et al., 1994; Forrest et al., 1994). Therefore, it was not possible to analyze the function of NMDA receptors in adult mutants.

Our forebrain region-specific Cre lines has enabled us to genetically disrupt NMDA receptor functions in the cortex and hippocampal regions. The neonatal lethality of conventional NMDA knockouts thus allowed us to check phenotypically whether insertion of *lox*P sites into NR1 gene disrupts the NMDA receptor function.

To generate conditional NMDA receptor knockout mice, we have screened the 129sv genomic library with a cDNA probe specific to NMDA R1 gene and isolated a 40 kb genomic fragment spanning 21 out of 22 exons of the NMDA R1 gene. We constructed a targeting vector in which two *lox*P sequence were inserted into the NMDA receptor 1 (NR1) gene as follows: The first *lox*P sequence was placed in the middle of the largest intron (about 5 kb) that lies between exon 10 and 11, whereas the second *lox*P sequence was introduced in the downstream region of the 3′ end of the NR1 gene (Tsien et al., 1996b). In this manner, two *lox*P sequences flank a 12 kb region of the NR1 gene that encodes all four transmembrane domains as well as the entire C-terminal sequence of the polypeptide chain, thus ensuring the complete disruption of NMDA receptor functions. We also reasoned that the insertion of *lox*P sequences in the middle of the largest intron (5 kb) and 3 kb downstream of the last exon would likely minimize the risk of potential disruption of exon-intron splicing sites near the intron/exon boundary or transcription termination signals, respectively.

Mice homologous for the *lox*P-NR1-*lox*P gene (floxed NR1) were generated using ES cell targeting techniques (Joyner, 1993, Hogan et al., 1994). We first linearized targeting vector before electroporation into J1 ES cells derived from 129/terSv. Neomycin-resistant ES cells colonies maintained on subconfluent embryonic fibroblasts were picked and expanded. ES cells harbouring homologous recombination were determined by Southern blotting. The targeted ES cells were injected into C57BL/6 blastocytes. These blastocyts were transferred into pseudopregnant mothers. Chimeric mice with more than 90% agouti were bred against C57BL/6 mice. The F_1 mice with a germline transmission of the floxed NR1 gene were bred to homozygous.

We crossed these floxed NR1 mice with a Cre transgenic line (T29.1) showing CA1-specific Cre/*lox*P recombination via a Cre/lox-mediated *LacZ* reporter system (Tsien et al., 1996a). Mice carrying the Cre transgene and the homozygous floxed NR1 gene (CA1-specific knockout, or simply CA1-KO mice) were viable and grew normally (Tsien et al., 1996b). The overall general appearance is indistinguishable from wild-type and other control littermates. These characteristics of CA1-KO mice are in stark contrast to the neonatal lethality of the NR1 knockout mice produced by conventional gene knockout technique. Histochemical examinations showed that the brains from the CA1-KO mice did not exhibit any gross anatomical abnormalities at the macroscopic levels.

Our in situ hybridization of NR1 mRNA, with a probe whose sequence should be deleted at the DNA level by the Cre/*lox*P recombination, revealed that knockout of the NR1 gene is CA1 region-specific (Fig. 2(A)). Electrophysiological recordings further confirmed lack of NMDA

A

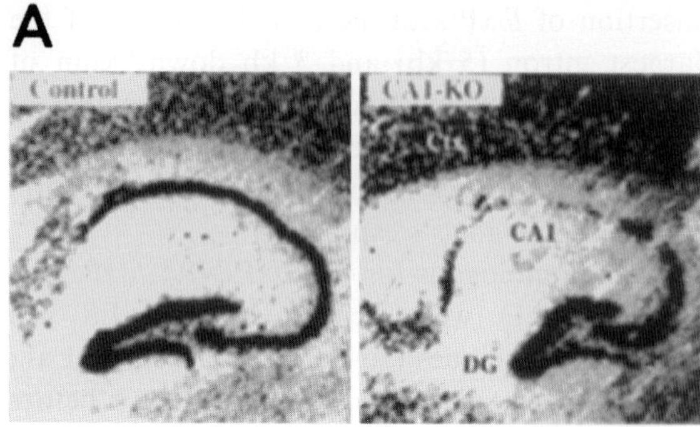

B

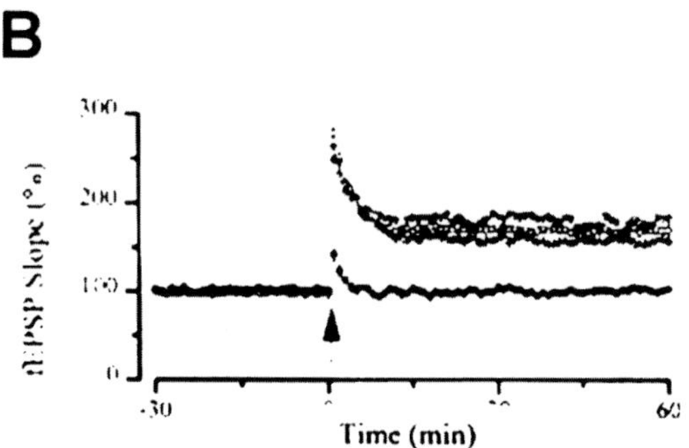

Fig. 2. (A) Lack of NMDAR1 mRNA in the CA1 region of CA1-KO mice. In situ hybridization of NR1 mRNA from wild type and CA-KO brains. Ctx, cortex; DG, dentate gyrus. (B) lack of NMDA-dependent LTP in the CA1 region from CA1-KO mice. The mean fields EPSPs in the four groups tested for LTP induction in CA1. The CA1-KO (closed circles, n = 21) did not show LTP, whereas others have clear LTP (T29-1, open circles, n = 12; floxed, upward open triangles, n = 4; wild type, downward open triangles, n = 5.

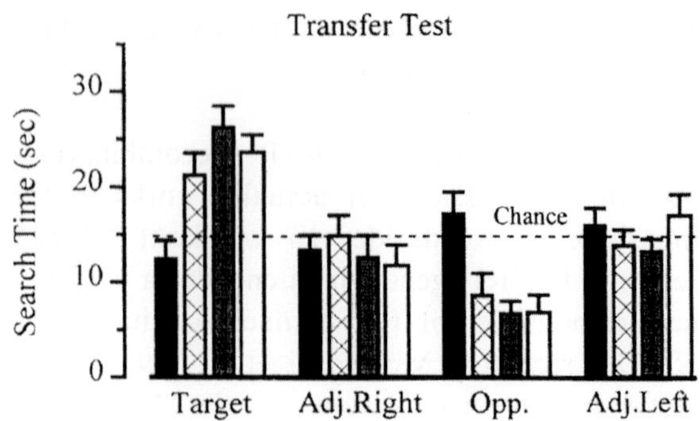

Fig. 3. CA1-KO mice show a deficient performance during the transfer test of the hidden platform water maze. During the transfer test, the CA1-KO mice spent equally amounts of time in every quadrant, whereas control groups spent sigificantly more time than chance in the target quadrant. Average time for the four groups (closed bars, CA1-KO; hatched bars, T29-1 transgenic; shade bars, floxed NMDA mice; open bar, wild-type).

receptors-mediated postsynaptic currents, STP, LTD and LTP specifically in the CA1 pyramidal cells (Fig. 2(B)).

Generation of CA1-specific NR1 knockout mice have allowed us to examine the correlation of NMDA receptor-mediated responses in the CA1 region, with CA1 place cell activity, and with spatial learning. We have tested these mutant mice for their performance in water mazes. We found that CA1-KO mice were deficient in learning the hidden-platform water maze task (Fig. 3), suggesting that NMDA receptor-mediated synaptic responses are essential for spatial learning (Tsien et al., 1996b). We have further analyzed these animals using a multi-electrode recording and has revealed that place fields represented by CA1 pyramidal cells in the conditional mutants well was diffused and less specific in comparison to control animals (McHugh et al., 1996). Moreover, these CA1 cells with overlapping place fields lack coordinated firing, thus suggesting that the ensemble coding of space in the CA1 region were severely impaired in the conditional mutants. Our results have demonstrated that the Cre/*lox*P-mediated, region-specific gene knockout can be achieved in the mouse brain, and that NMDA receptor-mediated synaptic plasticity in the hippocampal CA1 region plays a pivotal role in spatial learning and internal coding of space in the brain.

Functional in vivo assay for Cre/*lox*P-mediated gene knockout

Southern blot, in situ hybridization, and electrophysiological recordings are very effective and necessary for detailed characterization of the region- or tissue-specific gene knockout. However, these methods may not be the most convenient methods for the large-scale, simple quantitative evaluation of distribution and efficiency of Cre/*lox*P recombination in the various transgenic lines. We have employed a simple functional assay that can facilitate the characterization of various Cre transgenic lines (Tsien et al., 1996a). This in vivo assay uses a *LacZ* reporter gene whose activation requires Cre/*lox*P-mediated recombination. A reporter mouse was generated by incorporating a transgene in which the chicken β-actin promoter drives the expression of a *LacZ* gene, whose translation is prevented by a 1.5 kb "stop" sequence flanked by *lox*P sites (see Fig. 4(A)). Upon crossing of the reporter mouse to the Cre transgenic mouse lines, in mice carrying both Cre and *LacZ* transgenes, Cre-mediated recombination removes the stop sequence only in cells or regions where Cre is expressed at a sufficiently high level, thus allowing translation of β-galactosidase. In the remaining regions of the brain and in other tissues, expression of the Cre transgene should be absent due to the specificity of the CaM kinase II promoter, so that no Cre/*lox*P recombination and no expression of β-galactosidase occur. Therefore, the Cre/*lox*P-mediated recombination and its pattern in the double transgenic mouse brain can be functionally detected by simple X-gal staining.

Using this assay, we have identified several Cre transgenic lines with different forebrain region-specific recombination patterns (Tsien et al., 1996a). For example, in one line, T29-2, the Cre-recombination is forebrain-specific whereas in T17, Cre recombination is CA1/CA3-specific. In three lines (T29.1, T40 and T50), Cre/*lox*P-mediated gene deletion is restricted exclusively to the CA1 region of the hippocampus (Fig. 4(B)).

However, as a precautionary note, investigators should not always assume the pattern and efficiency of the Cre recombination detected by *LacZ* reporter are necessarily identical to the knockout pattern and efficiency of an actual gene of interest. This possibility can not completely rule out since Cre recombination could vary with each target gene due to different genome loci or different accessibility in the chromosome. It is thus prudent to examine the exact pattern of Cre/*lox*P-mediated gene knockout in every case. So far, based on our available experiments, we have found that the Cre-recombination pattern and efficiency demonstrated by *LacZ* reporter are similar with the knockout patterns of two genes, namely NMDA R1 receptor (Tsien, et al, 1996b) and NT3 (Prosser, Tsien, and Tonegawa, unpublished observation).

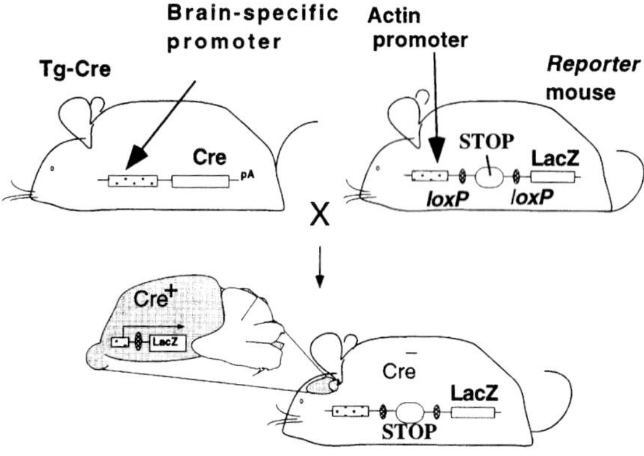

Fig. 4. (A) Functional detection of Cre/*lox*P-mediated gene deletion using *LacZ* reporter mice in which the β-galactosidase expression can occur only after the transcription and translational STOP signal is removed by the Cre/*lox*P-mediated deletional recombination. (B) Demonstration of the CA1 region-specific gene deletion by *LacZ*/X-gal blue staining in one of the CA1-specific line (line T29-1). The dark arc is the X-gal stained CA1 region on a sagittal brain section.

Parameters for Cre/*lox*P-mediated recombination

Since the application of the Cre/*lox*P system to study gene function in the mouse brain is new to the neuroscience field, we have decided to investigate further parameters that might affect the efficiency and patterns of Cre/*lox*P recombination in the brain. In particular, we were surprised by the CA1-specificity observed in some of Cre transgenic lines since the CaM kinase promoter is not only active in the CA1-region but also in other forebrain regions. Since we obtained three CA1-specific Cre lines (T29.1, T40 and T50 lines), it appears to us that the genomic locus effect may not play a major role in determining the CA1-specificity. Therefore, we have examined the relationship between Cre/*lox*P-mediated recombination efficiency and Cre expression levels in those transgenic lines. The amount of Cre mRNA assessed by Northern blot did not seem to always correlate with the distribution of Cre/*lox*P recombination. For example, in T29-1 mice, Cre mRNA appears to be as abundant in the cortex as in the hippocampus, whereas Cre/*lox*P recombination was highly restricted to the CA region of the hippocampus. However, staining the brain slices with an anti-Cre monoclonal antibody showed a greater amount of signal in the CA1 region in the T29-1 mouse line, demonstrating the frequency of Cre recombinase-positive neurons correlates fairly well with the frequency of Cre/*lox*P recombination-positive cells (Tsien et al., 1996a). These observations suggest that the Cre/*lox*P recombination-mediated gene knockout event appears to require a threshold level of Cre protein expression. In addition to promoter activities, the efficiency of the Cre/*lox*P-mediated recombination could be also affected by the steady state of Cre protein in the nucleus, accessibility to the DNA substrate, and the size of the cells etc. Therefore, theoretically, Cre recombination pattern may change if Cre protein is accumulated above the recombination threshold level when animals reach at a particular stage. We have looked into this possibility in T29.1 Cre line and found that CA1-specificity is maintained in 5-month-old animals but begin to spread to entire forebrain regions in aged animals. Thus different gene knockout patterns in two age groups could be advantageous for investigating the roles of those regions in various learning behaviors. Of course, this also points to the need for having promoters whose transcriptional activity is truly region specific and its expression levels are constant throughout the animal's life.

Since the previous Cre/*lox*P-mediated knockout/recombination were conducted in mitotic cells such as ES cells and immature T-cells (Gu et al., 1993; Gu et al., 1994), it was not clear whether Cre/*lox*P-mediated recombination requires active DNA replication and whether it can occur in postmitotic neurons in vivo. Therefore, we have investigated this issue by examining the time course of Cre/*lox*P-mediated gene deletion in all our Cre transgenic lines (both forebrain-specific lines and the CA1-specific lines). We have found that Cre/*lox*P-mediated genetic ablation occurred between the 3rd and 4th postnatal week. CA1 pyramidal cells undergo neurogenesis between E10 and E18 (Angevine, 1965) and enter the postmitotic state by P0. They are well differentiated by P7, with fully established synaptic connections (Stanfield and Cowan, 1979; Pokorny and Yamamoto, 1981a, 1981b). That time course of Cre/*lox*P recombination suggests that this type of recombination is independent of DNA replication and not inhibited by the highly differentiated, postmitotic state of cortical and hippocampal neurons.

Future directions

The above Cre/*lox*P-based conditional knockout is both brain region-specific and temporally restricted, and its value in studying the molecular and neural mechanisms of learning and memory have clearly been demonstrated (Tsien et al., 1996a, 1996b; McHugh et al., 1996). However, this system is not suitable for studying all stages of memory formation such as memory consolidation, transfer and retrievals. Therefore, it would be desirable to further improve the region-specific knockout system by incorporating inducible features (Gossen et al., 1992; No et al., 1996). For example, the recently reported ecdysone-inducible gene expression system could be used to regulate the expression of Cre recombinase. This inducible system is based on the ability of insect hormone 20-OH ecdysone to activate gene expression via activation of the modified ecdysone fusion receptor. This chimeric fusion receptor is formed by heterodimerization of RXR (retinoid X receptor which binds to a modified ecdysone response element) and VgEcR subunits (a fusion protein between V16 transactivation domain and Drosophila ecdysone receptor). In the presence of hormone, the heterodimerized receptor transactivates the ecdysone response element-containing promoter which can drive the expression of any transgene (No et al., 1996).

By using a CaM kinase II promoter to drive the bicistronic expression of RXR and VgEcR gene, one can achieve inducible expression of Cre recombinase in the forebrain regions (Fig. 5). Since the Cre/*lox*P-mediated gene deletion can occur efficiently in the postmitotic neurons in the brain, this ecdysone system should allow inducible gene knockout at any stages in the adulthood. Such inducible and region-specific knockout system will further enhance the power of genetic approaches for dissecting brain function and behaviors.

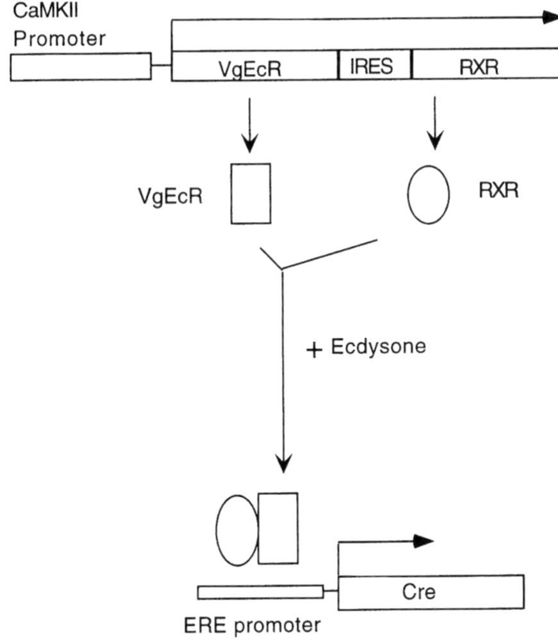

Fig. 5. Inducible and region-specific expression of Cre recombinase in the brain. The αCaMKII promoter is used to drive the expression of the modified ecdysone receptor heterodimers RXR and VcEcR in the forebrain regions. In the presence of ecdysone hormone, the heterodimerized receptor binds to the ecdysone response element and thus transactivates the expression of Cre recombinase.

References

Abremski, K. and Hoess, R. (1985) Phage P1 Cre-*lox*P site-specific recombination. Effects of DNA supercoiling on catenation and knotting of recombinant products. *J. Mol. Biol.*, 184: 211–220.

Angevine, J.B. (1965) Time of neuron origin in the hippocampal region. *Exp. Neurol.*, Suppl. 2: 1–70.

Argos, P., Landy, A.,Landy, Abremski, K., Egan, J.B., Ljungquist, E.H., Hoess, R.H., Kahn, M.L., Kalionis, B., Narayana, S.V.L., Pierson, L.S. Sternberg, N. and Leong, J.M. (1986) The integrase family of site-specific recombinases: regional similarities and global diversity. *EMBO J.* 5: 433–440.

Bradley, A. (1987) Production and analysis of chimeric mice. In: E.J. Robertson (Ed.), *Teratoarcinomas and embryonic stem cells: A practical approach*, Oxford: IRL Press, pp 113–151.

Burgin, K.E., Waxham, M.N., Rickling, S., Westgate, S.A., Mobley, W.C. and Kelly, P.T. (1990) In situ hybridization histochemistry of Ca^{2+}/Calmodulin-dependent protein kinase in developing rat brain. *J. Neurosci.*, 10: 1788–1798.

Capecchi M.R. (1989) Altering the genome by homologous recombination. *Science*, 244: 1288–1292.

Forrest, D., Yuzaki, M., Soares, H.D., Luk, D.C., Sheng, M., Stewart, C.L., Morgan, J. I., Connor, J.A. and Curran, T. (1994) Targeted disruption of NMDA receptor 1 gene abolishes NMDA response and results in neonatal death. *Neuron*, 13: 325–338.

Gossen, M. and Bujard, H. (1992) Tight control of gene expression in mammalian cells bytetracycline-responsive promoters. *Proc. Nat. Acad. Sci. USA*, 89: 5547–5551.

Grant, S.G.N., O'Dell, T.J., Karl, K.A., Soriano, P. and Kandel, E.R (1992) Impaired long-term potentiation, spatial learning and hippocampal development in fyn mutant mice. *Science*, 258: 1903–1910.

Gu, H., Marth, J.D., Orban, P.C. Mossman, H. and Rajewsky, K. (1994) Deletion of a DNA polymerase β gene segment in T cells using cell-type-specific gene targeting. *Science*, 265: 103–106.

Hogan, B., Beddington, R., Costantini, F. and Lacy, E. (1994) *Manipulating the mouse embryo: A laboratory manual*, 2nd Edn, Cold Spring Hourbor Press.

Joyner, A.L. (1994) Gene targeting and development of the nervous system. *Curr. Opin. Neurobio.*, 4: 37–42.

Li Y., Erzurumlu, R., Chen., C., Jhaveri, S. and Tonegawa, S. (1994) Whisker-related neuronal patterns fail to develop in the trigeminal brainstem nuclei of NMDAR1 knockout mice. *Cell*, 76: 427–437.

Mayford, M., Wang, J. Kandel, E.R. and O'Dell, J. (1995) CaMKII regulates the frequency response function of hippocampal synapses for the production of both LTP and LTD. *Cell*, 81: 891–904.

McHugh, T.J., Blum, K. I. Tsien, J.Z., Tonegawa, and Wilson, M. (1996) Impaired hippocampal representation of space in CA1-specific NMDAR1 knockout mice. *Cell*, 87: 1339–1349.

Moriyoshi, K.Masu, M., Ishii, T. Shigemoto, R., Mizuno, N. and Nakanish, S. (1991) Molecular cloning and characterization of the rat NMDA receptor, 354: 31–37.

No. D., Yao, T. and Evens, R. M. (1996) Ecdysone-inducible gene expression in mammalian cells and transgenic mice. *Proc. Natl. Acad. Sci. USA*, 93: 3346–335.

Orban, P.C., Chui, D. and Marth, J. (1992) Tissue- and site-specific DNA recombination in transgenic mice. *Proc. Natl. Acad. Sci. USA*, 89: 6861–6865.

Pokorny, J. and Yamamoto, T. (1981a) Postnatal ontogenesis of hippocampal CA1 area in rats. I. Development of dendritic arhorisation in pyramidal neurons. *Brain Res. Bull.*, 7: 113–120.

Pokorny, J. and Yamamoto, T. (1981b) Postnatal ontogenesis of hippocampal CA1 area in rats. I. Development of ultrastructural in stratum Iacunosum and moleculare. *Brain Res. Bull.*, 7: 121–130.

Sauer, B. and Henderson, N. (1988) Site-specific DNA recombination in mammalian cell by the Cre recombinase of bacteriophage P1. *Proc. Natl. Acad. Sci. USA*, 85: 5166–5170.

Stanfield, B.B. and Cowan, W. (1979) The development of the hippocampus and dentate gyrus in normal and reeler mice. *J. Comp. Neurol.*, 185: 423–459.

Tsien, J.Z.; Chen, D.F., Gerber, D., Tom, C; Mercer, E., Anderson, D.J., Mayford, M., Kandel, E.R. and Tonegawa, S. (1996a) Subregion- and cell type-restricted gene knockout in mouse brain. *Cell*, 87: 1317–1326.

Tsien, J.Z., Huerta, P.T. and Tonegawa, S. (1996b) The essential role of the hippocampal CA1 NMDA receptor-dependent synaptic plasticity in spatial learning. *Cell*, 87: 1327–1338.

W.E. Crusio and R.T. Gerlai (Eds.)
Handbook of Molecular-Genetic Techniques for Brain and Behavior Research (Techniques in the Behavioral and Neural Sciences, Vol. 13)

CHAPTER 3.1.4

Regulated temporal and spatial expression of mutants of CaMKII and calcineurin with the tetracycline-controlled transactivator (tTA) and reverse tTA (rtTA) systems

Isabelle M. Mansuy,[1] Mark Mayford[2] and Eric R. Kandel[3]

[1] *Institute of Cell Biology, Department of Biology, ETH Hönggerberg HPM D24, CH-8093 Zurich, Switzerland*
[2] *Department of Neurosciences 0691, University of California San Diego, School of Medicine, 9500 Gilman Drive, La Jolla, CA 92093, USA*
[3] *Howard Hughes Medical Institute, Center for Neurobiology and Behavior, College of Physicians and Surgeons of Columbia University and New York State Psychiatric Institute, 722 West 168th Street, New York, NY, 10032, USA*

A question in neuroscience that has been the focus of intense research in the last decade is how genes contribute to behavior. A number of studies have suggested that there is no direct link between genes and behavior. Rather, it seems that specific neural circuit(s) are involved in particular behavior(s) and that operation of this or these circuit(s) is governed by a combination of genes. Identification of these genes and detailed examination of their specific roles are essential steps towards the understanding of the molecular mechanisms of behavior.

Genetic approaches have facilitated this task in vivo by either loss of function of candidate gene(s) by homologous recombination in embryonic stem (ES) cells or expression of transgenes coding for dominant negative mutants or specific inhibitors of the molecule of interest. Conversely, gain of function experiments were performed by over-expressing the gene itself or an intrinsically active mutant form of the molecule.

These experiments have allowed the study of how particular genes are involved in synaptic transmission and synaptic plasticity in specific regions of the brain and how alterations in these synaptic functions affect behavior, in particular learning and memory.

In mice, transgene expression is usually restricted to specific tissues by promoter sequences that confer regional specificity. Most promoters, however, are active throughout development and generally in many more cell types than desired. This lack of temporal and cellular specificity has complicated the interpretation of defects observed in transgenic mice. Thus, it has been difficult to distinguish direct effects of a transgene from a developmental anomaly and to attribute an observed defect to a particular cell type in a tissue.

Recently, progress has been made towards the establishment of regulatable expression systems that allow both temporal and spatial control over transgene expression in mice.

The focus of this chapter is on two of these systems, the tetracycline-controlled transactivator (tTA) and the reverse tTA (rtTA) systems. The tTA and rtTA systems will be presented and the strategy employed to adapt them to the mouse brain will be described in detail.

Their application to the study of the molecular mechanisms of brain plasticity and behavior will be discussed and in particular, how both the tTA and rtTA systems have been used to examine specific forms of synaptic plasticity, learning and memory.

Regulation of transgene expression in the mouse brain with the tTA and rtTA systems

The tTA system

The tTA system, established in the laboratory of Herman Bujard, has been used to achieve regulated gene expression in the mouse brain (Gossen and Bujard, 1992).

Based on a hybrid transcription factor composed of the tetracycline repressor (tetR) fused to the transcriptional activation domain of the Herpes Simplex viral protein VP16, the tTA factor is able to activate transcription of a target gene placed downstream of the tetracycline operator tetO, a promoter element specific for tTA.

The tetO promoter is a fusion between a minimal promoter from human cytomegalovirus (hCMV) and sequences from the *E. Coli* transposon Tn10 tetracycline-resistance operon. Binding of tTA to the tetO sequences leads to transcriptional activation of the transgene. This transcriptional activation (transactivation) is blocked by tetracycline or its derivatives such as doxycycline.

Recently, the tTA system was used to achieve region specific transgene expression in the mouse brain by combining it with a brain-specific promoter for the α calcium/calmodulin-dependent kinase II (CaMKIIα) (Mayford et al., 1996b).

Two types of transgenic mice were generated (Fig. 1(A)).

> One mouse (Mouse 1) expresses tTA under the control of the CaMKIIα promoter that's restricts expression of the tTA factor to neurons in the postnatal forebrain.
> A second mouse (Mouse 2) carries the tetO promoter linked to the gene of interest.

By breeding mouse 1 with mouse 2, offspring were generated that carry both the tTA and tetO-driven transgenes (Fig. 1(A)). In these double transgenic mice, transcription of the gene of interest is activated by tTA and repressed by doxycycline.

When a *LacZ* reporter gene was used in place of the gene of interest, its expression was found to be activated by tTA and this activation occurred in restricted areas of the mouse brain including the neocortex, hippocampus, dentate gyrus, amygdala and striatum (Mayford et al., 1996a).

Interestingly, various patterns of *LacZ* gene expression were obtained when independent lines of *tet*O-*LacZ* mice were crossed with the same line of tTA-expressing mice. For example in hippocampus, *LacZ* gene expression was activated in areas CA1, CA2 and CA3 in one line of *tet*O-*LacZ* mice whereas it was found only in CA1 and CA2 areas but not in CA3 in another line of mice (Mayford et al., 1996a).

Regulated expression of mutant forms of CaMKII and calcineurin with the tTA system

The kinase CaMKII is a central component of the control of the activity-dependent synaptic modifications that underlie the formation and the storage of memory (Lisman, 1994). In neurons, active CaMKII was suggested to lead to an increase in synaptic efficacy. By contrast, decreased synaptic efficacy is thought to be controlled by a cascade of phosphatases composed essentially by the Ca^{2+}/calmodulin-dependent phosphatase calcineurin (PP2B) and the protein phosphatase 1 (PP1).

To address the role that these kinase and phosphatases play in the mechanisms of synaptic plasticity, learning and memory, both CaMKII and calcineurin were overexpressed in the mouse brain in a regulated manner using the tTA system.

Activation of transgene expression

Overexpression of a CaMKII mutant. To achieve regulated overexpression of CaMKII in the mouse

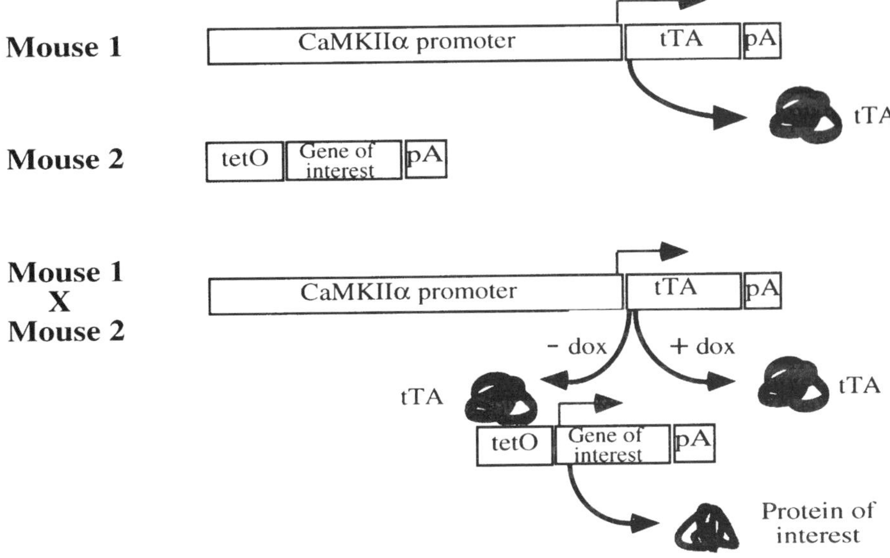

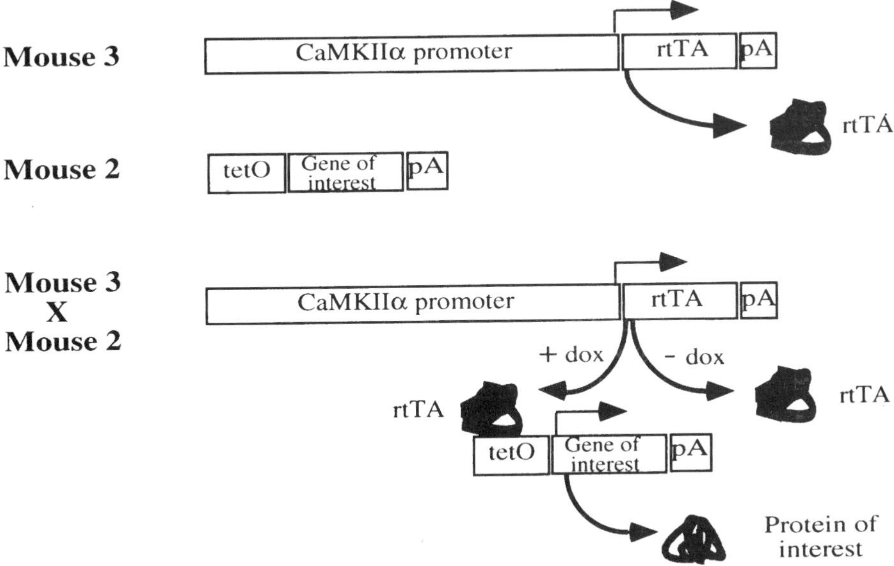

Fig. 1. The tTA and rtTA systems. Strategy to obtain doxycycline-regulated gene expression in the mouse brain. Mouse 1 and 3 express respectively tTA (A) or rtTA (B) under the control of the CaMKIIα promoter. Mouse 2 carries the tetO promoter linked to the gene of interest. Double transgenic mice are generated by breeding. In double transgenic mice (Mouse 1 × Mouse 2 or Mouse 3 × Mouse 2), expression of the gene of interest is either (A) activated by tTA and suppressed by administration of doxycycline or (B) induced by rtTA in the presence of doxycycline and not expressed in the absence of doxycycline.

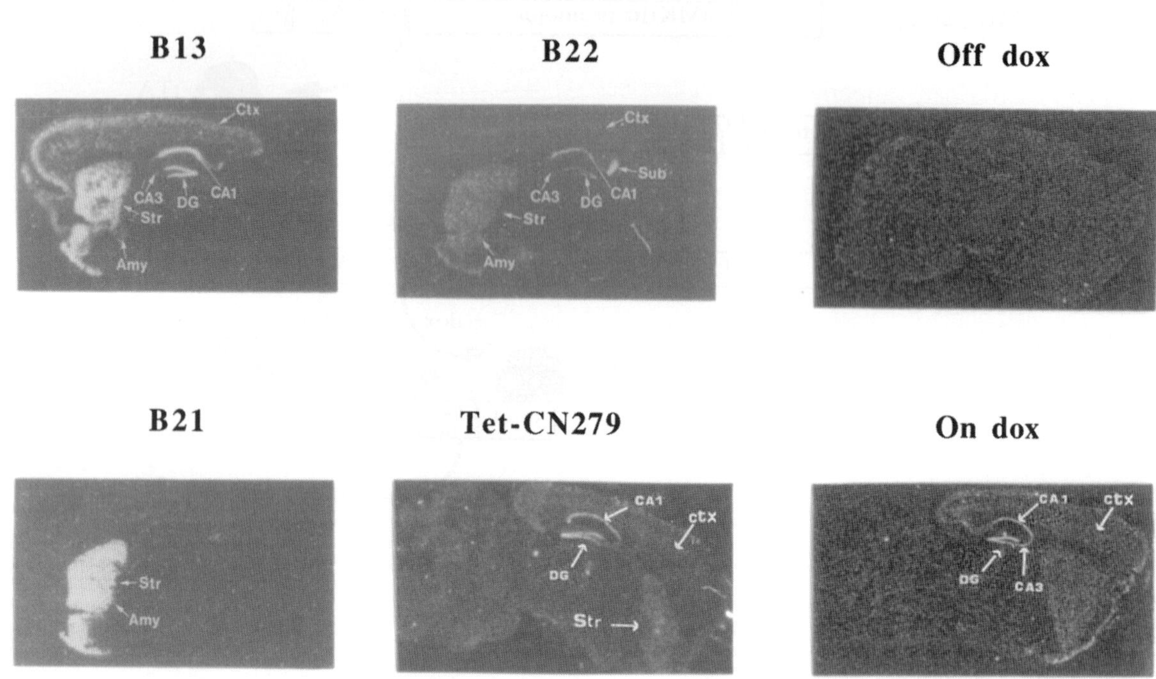

Fig. 2. Regulated expression of CaMKII-Asp286 and ΔCaM-AI mutants with the tTA and rtTA systems. Sagital sections of adult brain from double transgenic mice showing the pattern of expression of (A) CaMKII-Asp286 transgene in B13, B22, B21 mutant mice and ΔCaM-AI transgene in Tet-CN279 mutant mice; (B) ΔCaM-AI transgene in rTet-CN279 mutant mice either not treated (Off dox) or treated with doxycycline for 6 days with 6 mg/g food (On dox). Amy, amygdala; CA1 and CA3, CA1 and CA3 cell body layers in hippocampus; Ctx, cortex; DG, dentate gyrus; Str, striatum.

brain with the tTA system, a Ca^{2+}-independent mutant form of CaMKII, CaMKII-Asp286, was used. This mutant had previously been shown to lead to specific defects in synaptic plasticity and memory when expressed constitutively in the mouse brain (Bach et al., 1995; Mayford et al., 1995).

Several lines of mice carrying the tetO promoter linked to CaMKII-Asp286 were generated. These mice were then crossed with the mice expressing tTA under the control of the CaMKIIα promoter (line B, Mayford et al., 1996a).

In double transgenic animals, CaMKII-Asp286 expression was activated by tTA in forebrain neurons. Similarly to that observed with the *LacZ* reporter gene, various patterns of CaMKII-Asp286 expression were obtained with independent lines of tetO-CaMKII-Asp286 mice.

Thus with one line, CaMKII-Asp286 expression was found throughout the forebrain, in the neocortex, area CA1 of hippocampus with little or no expression in area CA3, dentate gyrus, amygdala and striatum (line B13, Fig. 2(A)). A similar pattern of expression was obtained with a second line of tetO-CaMKII-Asp286 mice but the levels of expression were lower (line B22, Fig. 2(A)).

Interestingly, a more restricted pattern of expression was observed in a third line of tetO-CaMKII-Asp286 mice (line B21, Fig. 2(A)). In this line, transgene expression was strong in the striatum and amygdala nuclei but was almost absent from the neocortex and hippocampus. In the striatum, transgene expression was accompanied by a 7-fold increase in Ca^{2+}-independent CaMKII activity when compared to control mice.

Overexpression of a calcineurin mutant. The tTA system was subsequently used to overexpress the Ca^{2+}-dependent phosphatase calcineurin. For this, a truncated mutant of the calcineurin catalytic subunit Aα, ΔCaM-AI, that has minimal Ca^{2+} requirement for activation was used (Mansuy et al., 1998a). Constitutive expression of this mutant in the mouse brain was shown to cause specific impairments in synaptic plasticity and memory (Mansuy et al., 1998a; Winder et al., 1998).

Similarly to the CaMKII mutant, mice carrying the tetO promoter linked to the calcineurin mutant ΔCaM-AI were generated. These mice were then crossed to the mice expressing tTA in brain to generate double transgenic mice (line B, Mouse 1, Fig. 1(A)).

With two independent lines of tetO-ΔCaM-AI mice, calcineurin transgene expression was activated in the striatum, area CA1 of the hippocampus with little or no expression in CA3 and dentate gyrus (Tet-CN279, Fig. 2(A)). In hippocampus, transgene expression resulted in a 2-fold increase in calcineurin activity.

Suppression of transgene expression with doxycycline

In addition to being spatially restricted, the expression of both CaMKII-Asp286 and calcineurin mutants was temporally regulated with the tTA system. Thus, transgene expression could be suppressed by administration of doxycycline to the animals in the drinking water.

In both B21 and Tet-CN279 mouse lines, several weeks of treatment with doxycycline at 1 mg/ml (in 5% sucrose) led to a reduction in Ca^{2+}-independent CaMKII and calcineurin activity, respectively, to basal levels.

Further, in the B21 mice, withdrawal of doxycycline after suppression of transgene expression resulted in re-expression of the transgene followed by an increase in Ca^{2+}-independent CaMKII activity similar to that observed before doxycycline treatment.

Variability and reversibility

The results presented here show that the CaMKIIα promoter in combination with the tTA system allows reversible activation of transgene expression in the adult mouse brain.

As illustrated above, various patterns of transgene expression are obtained when independent lines of mice carrying a tetO-linked transgene are used. The variability in pattern of expression is most probably due to the site of chromosomal integration of the tetO transgene, which is unique for each line of mice, and/or the number of integrated copies of the transgene (Palmiter and Brinster, 1986; Martin and Whitelaw, 1996).

However it has been reported that, when the neuron specific enolase (NSE) promoter is driving tTA expression in brain, no variability in the pattern of transgene expression activated by tTA is observed in double transgenic animals from different lines of tetO mice (Chen et al., 1998).

Methods to reduce the variability in expression pattern should be considered in future studies. For instance, use of cis-acting elements such as matrix attachment regions (MAR) or locus control regions (LCR) (McKnight et al., 1992; Talbot et al., 1994; Attal et al., 1995, Kalos and Fournier, 1995; Phi-Van and Stratling, 1996, Kioussis and Festenstein, 1997; Garrick et al., 1998) as well as use of strategies to control the number of integrated copies (see Jasin et al., 1996; Dobie et al., 1997 for review) may help eliminate position effects.

Variability of the onset of transgene expression may also be observed with the tTA system. For instance, whereas some lines of mice showed postnatal transgene expression with the CaMKIIα promoter and the tTA system, other lines were found to express the transgene during embryogenesis (Yamamoto, A., and Hen, R., unpublished results).

In the initial report on the use of the tTA system in the mouse brain, high doses of doxycycline were administered to suppress transgene expression (1 mg/ml in drinking water, Mayford et al., 1996a). More recently, we and others have found that lower doses (10–100 μg/ml) could repress transgene expression to the same extent (Mayford, M. unpublished results; Chen et al., 1998). Lower doses of doxycycline allow faster re-activation of transgene expression after suppression and reduce the likelihood of drug-induced side effects.

Transgene expression with the rtTA system

The rtTA system

The tTA system is a powerful expression system for brain-specific and suppressible transgene expression. However, to study the effect of a transgene in adult, its expression has to be repressed throughout development. Chronic administration of doxycycline, even at low doses, may have deleterious effects on brain development that would preclude further behavioral studies.

Moreover, since doxycycline accumulates in bone and soft tissue, it may take a long time to reactivate gene expression after long periods of gene suppression (Cohlan, 1977; Furth et al., 1994; Kistner et al., 1996; Mayford et al., 1996).

To circumvent these limitations, a novel tetracycline-controlled transactivator, the reverse tTA (rtTA) recently developed in the laboratory of Herman Bujard, was adapted to the mouse brain (Gossen et al., 1995; Mansuy et al., 1998b).

The factor rtTA was derived from tTA by random mutagenesis and four point mutations were introduced into the TetR gene. When compared to tTA, rtTA has reversed DNA binding properties and requires tetracycline derivatives for binding to tetO and for activation of gene expression.

In vitro studies have shown that gene expression can be activated by rtTA in various cell types (Gossen et al., 1995; Freundlieb et al., 1997). In transgenic mice, rtTA also induces transgene expression in various organs but only low levels of expression are obtained in the brain (Kistner et al., 1996; Ray et al., 1997).

Brain-specific and inducible transgene expression with the rtTA system

To achieve inducible transgene expression in the mouse brain, the rtTA system was used in combination with the CaMKIIα promoter. First, mice expressing the rtTA gene under the control of the CaMKIIα promoter were generated (Mouse 3, Fig. 1(B), Mansuy et al., 1998b).

Then, similarly to the tTA system, these mice were crossed with mice carrying the tetO promoter linked to the gene of interest (Mouse 2, Fig. 1(B)). Double transgenic animals were obtained by breeding (Mouse 3 × Mouse 2, Fig. 1(B)).

When the *LacZ* gene was used as a reporter, administration of doxycycline to double transgenic mice was found to induce its expression in the adult mouse brain. The levels and pattern of *LacZ* gene expression were found to depend on the dose of doxycycline administered and on the duration of the treatment.

Thus, 3 mg/g of doxycycline administered in the food for 6 days induced *LacZ* gene expression in striatum and septum. A dose of 6 mg/g however induced broader expression. Six days of treatment with 6 mg/g of doxycycline led to the induction of *LacZ* gene expression in hippocampus, dentate gyrus, neocortex and striatum in adult mouse brain. In contrast, 3 days of a treatment with 6 mg/g of doxycycline induced expression only in the striatum and septum. A dose of 12 mg/g led to a similar pattern of gene expression than with 6 mg/g of doxycycline given for 6 days.

Induced calcineurin overexpression with the rtTA system

When the rtTA system was used with the calcineurin mutant ΔCaM-AI, transgene expression was induced in hippocampus areas CA1, CA2 and CA3, dentate gyrus, neocortex and striatum after 6 days of treatment with doxycycline (line rTet-CN279, on dox, Fig. 2(B)). No expression was detected in mutant mice not treated with doxycycline (off dox, Fig. 2(B)). This expression was accompanied by a 1.5-fold increase in calcineurin activity in hippocampus.

Furthermore, transgene expression with the rtTA system was reversible. Two weeks after doxycycline removal, no transgene expression could be detected and the levels of calcineurin activity were reduced to basal levels.

About inducibility

The rtTA system applied to the mouse brain as described above allows rapid induction of transgene expression. The levels of expression are however moderate when compared to those obtained with the tTA system (see for instance Chen et al., 1998).

More rapid induction and higher levels of expression may be required in some experiments. The use of other analogs of tetracycline able to cross the blood brain barrier more easily and allow higher levels of drug in brain cells may help faster and better induction of transgene expression.

Application of the tTA and rtTA systems to the study of brain plasticity and behavior

Expression of CaMKII-Asp286 with the tTA system results in a reversible defect in synaptic plasticity

Constitutive expression of the CaMKII-Asp286 mutant in the hippocampus was shown to produce a shift in the stimulation frequency required for the production of long-term potentiation (LTP) and long-term depression (LTD) in the CA1 Schaffer collateral pathway. Whereas LTP induced by 100 Hz stimulation was not altered by CaMKII-Asp286 expression, the synaptic response to a low frequency stimulation was systematically shifted in size and direction in favor of LTD (Mayford et al., 1995).

When expressed in the CA1 area of the hippocampus using the tTA system (line B13), the CaMKII-Asp286 mutant led to a similar defect in LTP. Whereas a 10 Hz stimulation resulted in a long-lasting potentiation in control mice, it induced a slight depression in mutant mice (Fig. 3(A)).

Moreover, suppression of the transgene expression by administration of doxycycline in the drinking water for two to three weeks reversed this defect, suggesting that it was a direct effect of CaMKII-Asp286 mutant expression but not a developmental anomaly.

Regulated overexpression of calcineurin leads to a reversible impairment in I-LTP

The constitutive overexpression of calcineurin in the mouse hippocampus was shown to lead to a defect in a novel intermediate phase of LTP (I-LTP) that depends on the cAMP-dependent kinase (PKA) (Winder et al., 1998).

When expressed with either the tTA system in the absence of doxycycline or the rtTA system in the presence of doxycycline, the calcineurin transgene caused a similar impairment in I-LTP than that observed in the constitutive expresser line (Fig. 3(B)). This defect was reversed after suppression of the calcineurin transgene expression by either administration or withdrawal of doxycycline with the tTA and rtTA systems, respectively (Winder et al., 1998).

Specific forms of memory are affected depending on the pattern of transgene expression

Taking advantage of the variability in the pattern of CaMKII and calcineurin transgenes expression observed with the tTA system, it was possible to examine different forms of memory in the mutant mice.

Spatial memory

The constitutive expression of CaMKII-Asp286 and the calcineurin mutant in the mouse brain was shown to result in a specific impairment in hippocampal-dependent spatial memory on the Barnes circular maze (Bach et al., 1995; Mansuy et al., 1998a). Similar defects were observed in

A

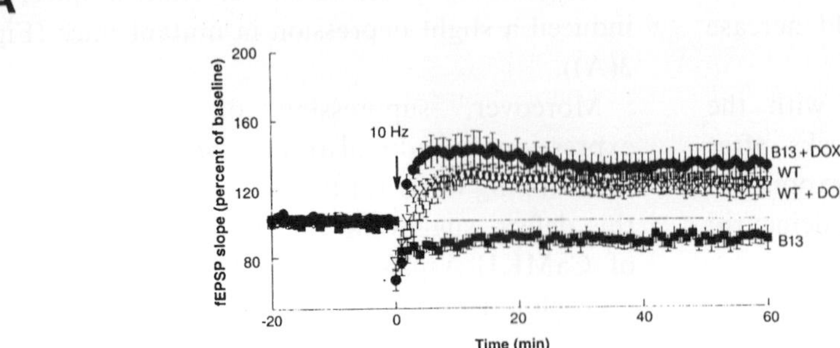

B

C

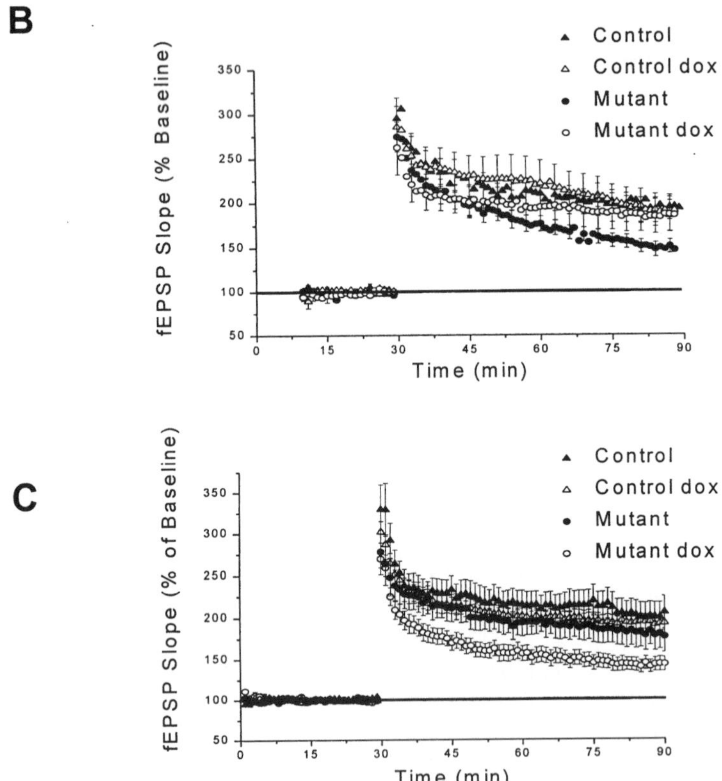

Fig. 3. Regulated overexpression of CaMKII and calcineurin results in specific defects in LTP. (A) LTP induced by 1 train of 10 Hz in B13 hippocampal slices. The stimulation induced a transient depression followed by potentiation in wild-type mice (WT) whereas in mutant mice (B13), a slight depression was observed. Doxycycline treatment reversed the defect in mutant mice (B13 + DOX) and had no effect on LTP in wild-type mice (WT + DOX). (B) I-LTP induced by 2 trains of 100 Hz in CA1 Schaffer collateral pathway from Tet-CN279 hippocampal slices. The defect in I-LTP observed in mutant mice (Mutant) is reversed by doxycycline (1 mg/ml in drinking water and 1 ng/ml in superfusate, Mutant dox). 2-train LTP is normal in control mice not treated (Control) or treated with doxycycline (Control dox). (C) I-LTP induced by 2 trains of 100 Hz in CA1 Schaffer collateral pathway from rTet-CN279 hippocampal slices. The I-LTP defect is induced by doxycycline (6 mg/g in the food and 6 ng/ml in superfusate) in mutant mice (Mutant dox) whereas I-LTP is normal in mutant mice not treated with doxycycline (Mutant) and in control mice not treated (Control) or treated with doxycycline (Control dox).

the mice expressing the same mutants with the tTA system.

The Barnes maze is a circular platform with 40 holes at the periphery and an escape tunnel placed under one of the holes. On a spatial version of the maze, mice must use the relationship between distal landmarks placed in the experimental room to efficiently locate the hole leading to the escape tunnel.

Typically, mice progress through three consecutive search strategies: random, serial and spatial (Fig. 4(A)). The first two search strategies consist in looking for the tunnel by checking holes either randomly or serially in a clockwise or counter-clockwise manner. The spatial search strategy, in contrast to the other two is hippocampal-dependent and defined as direct navigation to the tunnel with very few errors.

Both lines of mice expressing CaMKII-Asp286 or the calcineurin mutant (line B22 and Tet-CN279) did not learn the spatial version of the Barnes maze (Fig. 4(B) for line B22, similar results for line Tet-CN279). In these mice, the defect was due to impaired utilization of the spatial search strategy (Fig. 4(C) for line Tet-CN279, similar results for line B22) and was similar to the defect observed in mice constitutively expressing the transgenes (Mayford et al., 1996a; Mansuy et al., 1998b).

In both lines of mice, the defect in spatial memory was reversed when transgene expression was suppressed by doxycycline (Fig. 4(B) and (C)). These results suggest that the defects observed are a direct consequence of transgene expression.

Fear conditioning

A form of associative learning, contextual and cued fear conditioning, was also examined in the CaMKII mutant mice. Whereas contextual fear conditioning depends on both the hippocampus and the amygdala, cued fear conditioning requires only the amygdala (Kim and Fanselow, 1992; Phillips and LeDoux, 1992).

Experimentally, contextual and cued fear conditioning are elicited by pairing a footshock with a novel environment (context) or a tone (cue). Memory for the association footshock-context or footshock-tone is assessed by measuring the amount of freezing in response to fear elicited by the context alone (context conditioning) or the tone alone (cued conditioning).

On the fear conditioning task, mutant mice expressing high levels of the CaMKII-Asp286 transgene in the amygdala and striatum (line B21) exhibited a severe impairment in both contextual and cued conditioning. The defects were reversed when transgene expression was suppressed with doxycycline (Fig. 4(D)).

In contrast, mice from line B22 that expressed only low levels of the CaMKII transgene did not show any defects on this task (Fig. 4(D)). These results suggest that low levels of CaMKII-Asp286 expression are insufficient to disrupt fear conditioning and that the defect observed is dose-dependent.

These results combined with those obtained on the spatial memory task suggest a dissociation between spatial and contextual memory since both types of memory are impaired in the high expresser line and contextual memory is spared in the low expresser line.

Temporal regulation of gene expression with the tTA and rtTA systems allows to study memory retrieval

In addition to providing a way to assess different types of memory, the tTA and rtTA systems have allowed the examination of distinct phases of memory. In particular, the retrieval of memory was examined in both CaMKII and calcineurin mutant mice.

In CaMKII mutant mice from line B21, memory retrieval was assessed using the fear conditioning task (Mayford et al., 1996a). For this, expression of the CaMKII-Asp286 transgene was suppressed by doxycycline during training to allow normal learning of the task and was then re-activated by removal of doxycycline after learning. Retrieval

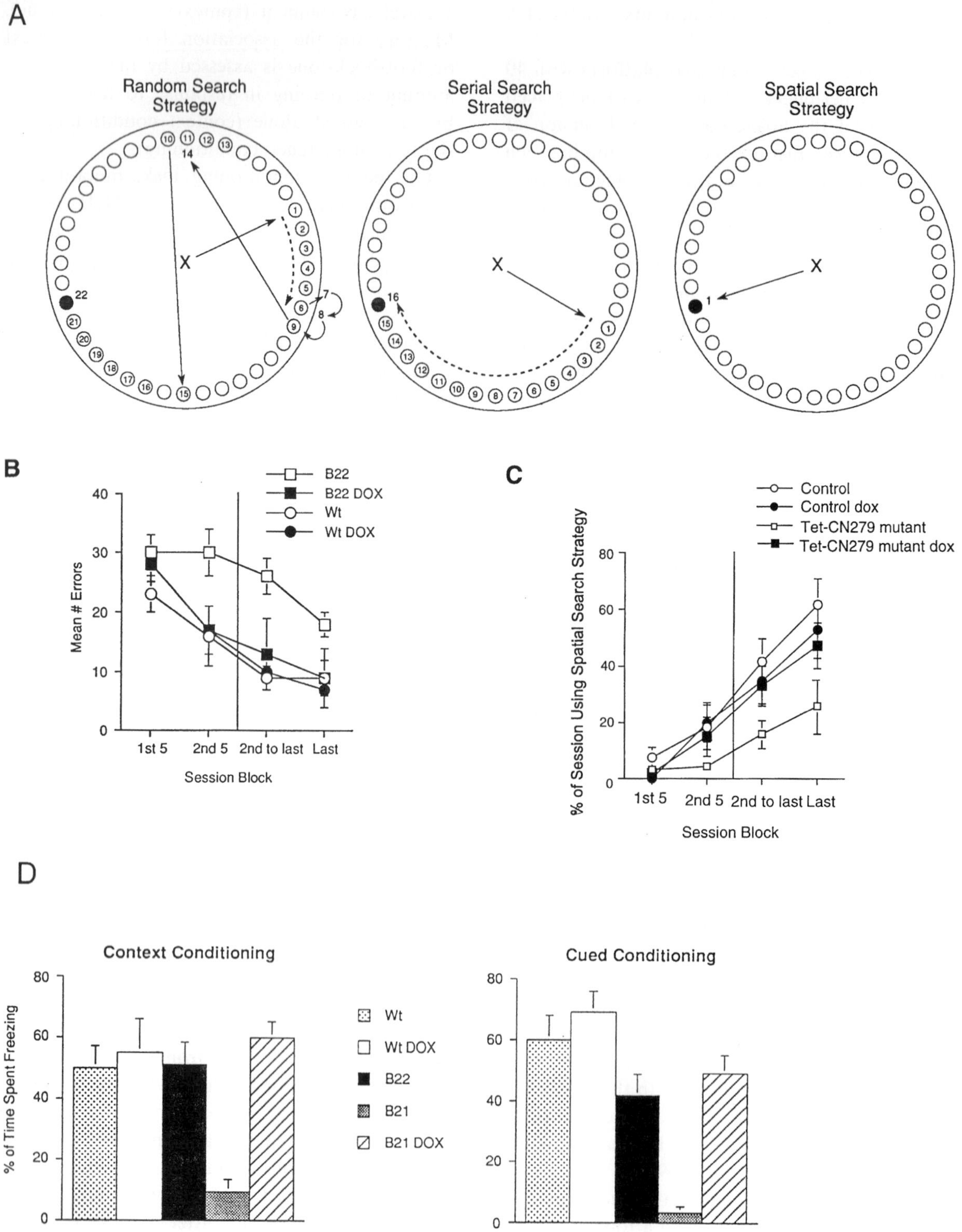

A

Random Search Strategy

Serial Search Strategy

Spatial Search Strategy

B

B22
B22 DOX
Wt
Wt DOX

Mean # Errors

40
30
20
10
0

1st 5
2nd 5
2nd to last
Last

Session Block

C

Control
Control dox
Tet-CN279 mutant
Tet-CN279 mutant dox

% of Session Using Spatial Search Strategy

80
60
40
20
0

1st 5
2nd 5
2nd to last
Last

Session Block

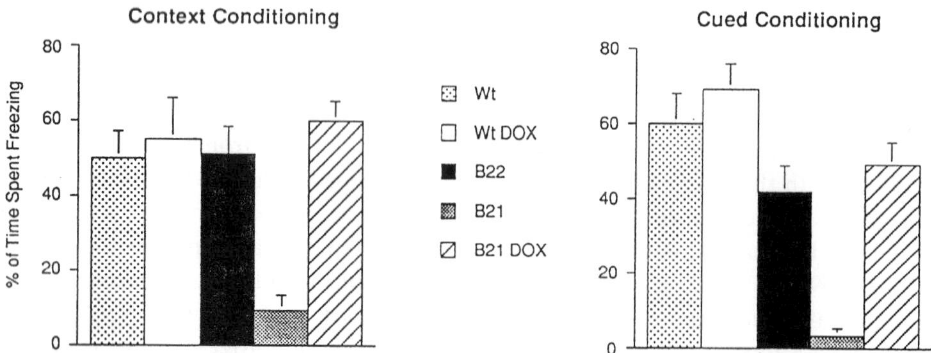

of memory on this task was tested several weeks after the end of training.

It was found that retrieval of memory for both contextual and cued conditioned fear was impaired when the CaMKII-Asp286 transgene was expressed after learning, suggesting that CaMKII is involved in the mechanisms of retrieval of fear conditioning.

The rtTA system was similarly used to examine the effect of the calcineurin transgene on the retrieval of spatial information using the Morris water maze task (Mansuy et al., 1998b).

The Morris water maze consists in a round tank filled with opaque water and containing in one of its four quadrant (training quadrant) an escape platform hidden under the surface of the water. Similarly to the Barnes maze, mice have to use the relationship between external and distal cues placed in the experimental room to locate the platform.

Mice are first trained to locate the position of the platform (5-day training with four trials per day) then a probe trial is performed during which the platform is removed from the pool. Memory for the platform position is measured by the time the mice spend searching for the platform in the training quadrant.

Mutant mice in which the calcineurin transgene expression was induced during training had impaired performance both during training and during a probe trial performed immediately after training (Fig. 5(A)). The mice did not spend more time searching for the platform in the training quadrant than in other quadrants.

When the expression of the calcineurin transgene was induced only at the end of training, after the information about the task was learned, mutant mice showed a similar deficit on a second probe trial performed two weeks after training. This result indicates that, on the second probe trial, mutant mice expressing the calcineurin transgene could not remember the position of the platform they had learned during training.

However, when the expression of the transgene was again suppressed by removal of doxycycline after the second probe trial, the mice spent significantly more time in the training quadrant on a third probe trial performed two weeks after the second one (Fig. 5(B)).

Thus, memory for the platform position was intact in mutant mice but retrieval of that memory was impaired by expression of the calcineurin transgene. These results suggest that, similarly to CaMKII, calcineurin is implicated in the mechanisms of memory retrieval.

Conclusions

The tTA and rtTA systems are very powerful tools for spatial and temporal restriction of transgene expression in the adult mouse brain. These systems allow reversible suppression or induction of the expression of a transgene by simple administration of doxycycline in the drinking water or in the diet.

The use of these systems has, however, revealed a number of limitations.

First, a certain variability in the pattern of transgene expression has been observed. This variability may provide, in some cases, unique and valuable patterns of transgene expression that may produce specific electrophysiological and

Fig. 4. Memory defects are reversed by doxycycline in B13 and Tet-CN279 mice. (A) Schematic representation of the Barnes circular maze. (B) Mean number of errors across 4 blocks of 5 trials on the Barnes circular maze in B22 control and mutant mice not administered doxycycline (Wt and B22) or administered 1 mg/ml doxycycline in the drinking water (Wt DOX and B22 DOX). Errors are defined as searching a hole with no tunnel beneath it. (C) Use of the spatial search strategy across 4 blocks of 5 trials on the Barnes circular maze by Tet-CN279 control and mutant mice not treated with doxycycline (Control and Tet-CN279 mutant) or treated with 1 mg/ml doxycycline (Control dox and Tet-CN279 mutant dox). (D) Context and cued conditioning. Time spent freezing to context or cue 24 hours after training in control mice not treated (Wt) or treated with doxycycline (Wt DOX), in B22 mutant mice not treated with doxycycline (B22) and in B21 mutant mice not treated (B21) or treated with doxycycline (B21 DOX).

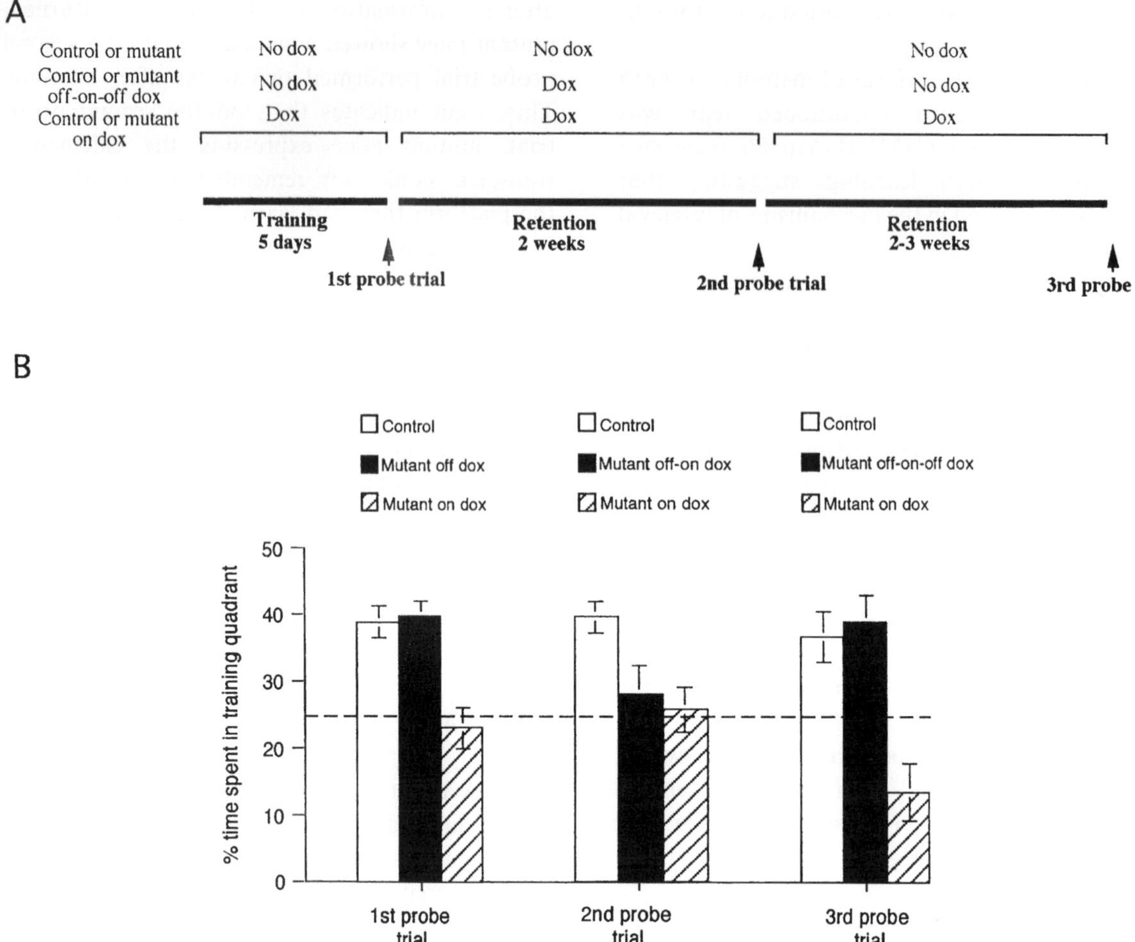

Fig. 5. Doxycycline-induced spatial memory defects in rTet-CN279 mice. (A) Diagram illustrating behavioral training, testing and retrieval on the hidden platform version of the Morris water maze task. (B) Summary of performance of rTet-CN279 mice in the training quadrant during the first, second and third probe trials. The time spent in the training quadrant was plotted across probe trials.

behavioral defects. However in most cases, a tight control over the spatial specificity of transgene expression is required.

Second with the rtTA system, the lack of spatial restriction observed in the line described limits the interpretation of the defects observed. Additional lines of rtTA-expressing mice need to be generated and screened to obtain more restricted patterns of transgene expression.

Alternatively, local injection of doxycycline using stereotaxic techniques or implanted mini-pumps to deliver the drug may help to gain restriction. Local infection of brain tissue with recombinant viruses carrying the transgenes may also provide fine spatial restriction of transgene expression (Ho et al., 1996, Hofmann et al., 1996, Watsuji et al., 1997; Harding et al., 1998).

Finally, the tTA and rtTA systems may be combined with other genetic means to manipulate gene expression. For instance with the Cre-loxP system, they should allow temporal and spatial control over gene inactivation (Kuhn et al., 1995; St Onge et al., 1996, Sauer, 1998).

Overall, the use of doxycycline-controlled regulatory systems will greatly contribute to the study of genes related to behavior.

References

Attal, J., Cajero-Juarez, M., Petitclerc, D., Theron, M.C., Stinnakre, M.G., Bearzotti, M., Kann, G. and Houdebine, L.M. (1995) The effect of matrix attached regions (MAR) and specialized chromatin structure (SCS) on the expression of gene constructs in cultured cells and in transgenic mice. *Mol. Biol. Rep.*, 22: 37–46.

Bach, M.E., Hawkins, R.D., Osman, M., Kandel, E.R. and Mayford, M. (1995) Impairment of spatial but not contextual memory in CaMKII mutant mice with a selective loss of hippocampal LTP in the range of the theta frequency. *Cell*, 81: 905–915.

Chen, J., Keltz, M.B., Zeng, G., Sakai, N., Steffen, C., Shockett, P.E., Picciotto, M.R., Duman, R.S. and Nestler, E.J. (1998) Transgenic animals with inducible, targeted gene expression in brain. *Mol. Pharmacol.*, 54: 495–503.

Cohlan, S.Q. (1977) Tetracycline staining of teeth. *Teratology*, 15: 127–130.

Dobie, K., Mehtali, M., McClenaghan, M. and Lathe, R. (1997) Variegated gene expression in mice. *Trends Genet.*, 13: 127–130.

Efrat, S., Fusco-Demane, D., Lemberg, H., Al Emran, O. and Wang, X. (1995) Conditional transformation of a pancreatic β-cell line derived from transgenic mice expressing a tetracycline-regulated oncogene. *Proc. Natl. Acad. Sci. USA*, 92: 3576–3580.

Freundlieb, S., Baron, U., Bonin, A.L., Gossen, M. and Bujard, H. (1997) Use of tetracycline-controlled gene expression systems to study mammalian cell cycle. *Methods Enzymol.*, 283: 150–173.

Furth, P.A., St Onge, L., Böger, H., Gruss, P., Gossen, M, Kistner, A., Bujard, H. and Hennighausen, L. (1994) Temporal control of gene expression in transgenic mice by a tetracycline-responsive promoter. *Proc. Natl. Acad. Sci. USA*, 91: 9302–9306.

Garrick, D., Fiering, S., Martin, D.I. and Whitelaw, E. (1998) Repeat-induced gene silencing in mammals. *Nature Genet.*, 18: 56–59.

Gossen M. and Bujard, H. (1992) Tight control of gene expression in mammalian cells by tetracycline-responsive promoters. *Proc. Natl. Acad. Sci. USA*, 89: 5547–5551.

Gossen, M., Freundlich, S., Bender, G., Müller, G., Hillen, W. and Bujard, H. (1995) Transcriptional activation by tetracyclines in mammalian cells. *Science*, 268: 1766–1768.

Harding, T., C., Geddes, B., J., Murphy, D., Knight, D. and Uney, J. B. (1998). Switching transgene expression in the brain using an adenoviral tetracycline-regulatable system. *Nat. Biotechnol.*, 16: 553–555.

Ho, D.Y., McLaughlin, J.R. and Sapolski, R.M. (1996) Inducible gene expression from defective herpes simplex virus vectors using the tetracycline-responsive promoter system. *Mol. Brain Res.*, 41: 200–209.

Hofmann, A., Nolan, G.P. and Blau, H.M. (1996) Rapid retroviral delivery of tetracycline-inducible genes in a single autoregulatory cassette. *Proc. Natl. Acad. Sci. USA*, 93: 5185–5190.

Kalos, M. and Fournier, R.E. (1995) Position-independent transgene expression mediated by boundary elements from the apolipoprotein B chromatin domain. *Mol. Cell. Biol.*, 15: 198–207.

Kim, J.J. and Fanselow, M.S. (1992) Modality-specific retrograde amnesia of fear. *Science*, 256: 675–677.

Kioussis, D. and Festenstein, R. (1997) Locus control regions: overcoming heterochromatin-induced gene activation in mammals. *Curr. Opin. Genet. Dev.*, 7: 614–619.

Kistner, A., Gossen, M., Zimmermann, F., Jerecic, J., Ullmer, C., Lübbert, H. and Bujard, H. (1996) Doxycycline-mediated quantitative and tissue-specific control of gene expression in transgenic mice. *Proc. Natl. Acad. Sci. USA*, 93: 10933–10938.

Lisman, J. (1994) The CaM kinase II hypothesis for the storage of synaptic memory. *Trends Neurosci.*, 17: 406–412.

Mansuy, I.M., Mayford, M., Jacob, B., Kandel, E.R. and Bach, M.E. (1998a) Restricted and regulated overexpression reveals calcineurin as a key component in the transition from short-term to long-term memory. *Cell*, 92: 39–49.

Mansuy, I.M., Winder, D.G., Moallem, T.M., Osman, M., Mayford, M., Hawkins, R.D. and Kandel, E.R. (1998b) Inducible and reversible gene expression with the rtTA system for the study of memory. *Neuron*, 21: 257–265.

Martin, D.I. and Whitelaw, E. (1996) The vagaries of variegated transgenes. *BioEssays*, 18: 919–923.

Mayford, M., Wang, J., Kandel, E.R. and O'Dell, T. (1995) CaMKII regulates the frequency-response function of hippocampal synapses for the production of both LTD and LTP. *Cell*, 81: 891–904.

Mayford, M., Bach, M.E., Huang, Y.-Y., Wang, L., Hawkins, R. and Kandel, E.R. (1996a) Control of memory formation through regulated expression of a CaMKII transgene. *Science*, 274: 1678–1683.

Mayford, M., Baranes, D., Podsypanina, K., and Kandel, E.R. (1996b). The 3'-untranslated region of CaMKIIα is a cis-acting signal for the localization and translation of mRNA in dendrites. *Proc. Natl. Acad. Sci. USA*, 93: 13250–13255.

McKnight, R.A., Shamay, A., Sankaran, L., Wall, R.J. and Hennighausen, L. (1992) Matrix-attachment regions can impart position-independent regulation of a tissue-specific gene in transgenic mice. *Proc. Natl. Acad. Sci. USA*, 92: 6991–6995.

Palmiter, R.D. and Brinster, R.L. (1986) Germ line transformation of mice. *Annu. Rev. Genet.*, 20: 465–499.

Phi-Van, L. and Stratling, W.H. (1996) Dissection of the ability of the chicken lysozyme gene 5' matrix attachment region to stimulate transgene expression and to dampen position effects. *Biochem.*, 35: 10735–10742.

Phillips, R.G. and LeDoux, J.E. (1992) Differential contribution of amygdala and hippocampus to cued and contextual fear conditioning. *Behav. Neurosci.*, 106: 274–285.

Ray, P., Tang, W., Wang, P., Homer, R., Kuhn, C. III, Flavell, R.A. and Elias, J.A. (1997) Regulated overexpresssion of interleukin 11 in the lung. Use to dissociate development-dependent and independent phenotypes. *J. Clin. Invest.*, 100: 2501–2511.

St. Onge, L., Furth, P. and Gruss, P. (1996) Temporal control of the Cre recombinase in transgenic mice by a tetracycline responsive promoter. *Nucleic Acids Res.*, 24: 3875–3877.

Talbot, D., Descombes, P. and Schibler, U. (1994) The 5' flanking region of the rat LAP (C/EBP beta) gene can direct high-level, position-independent, copy number-independent expression in multiple tissues in transgenic mice. *Nucleic Acids Res.*, 22: 756–766.

Watsuji, T., Okamoto, Y., Emi, N., Katsuoka, Y. and Hagiwara, M. (1997) Controlled gene expression with a reverse tetracycline-regulated retroviral vector (RTRV) system. *Biochem. Biophysi. Res. Commun.*, 234: 769–773.

Winder, D.G., Mansuy, I.M., Osman, M., Moallem, T.M. and Kandel, E.R. (1998) Genetic and pharmacological evidence for a novel, intermediate phase of long-term potentiation suppressed by calcineurin. *Cell*, 92: 25–37.

W.E. Crusio and R.T. Gerlai (Eds.)
Handbook of Molecular-Genetic Techniques for Brain and Behavior Research (Techniques in the Behavioral and Neural Sciences, Vol. 13)

CHAPTER 3.1.5

The use of targeted point mutants in the study of learning and memory

Karl Peter Giese

Department of Anatomy and Developmental Biology, University College London, London WC1E 6BT, UK

Introduction

In 1992 Silva and colleagues started a new era of studying learning/memory. They showed that the targeted disruption ("knock-out") of the mouse gene encoding the α-isoform of the Ca^{2+}/calmodulin kinase II (αCaMKII) results in impairments in long-term potentiation (LTP) in the hippocampal CA1 region and in spatial learning deficits (Silva et al., 1992a, 1992b; see also Grant et al., 1992). The finding that targeted gene disruptions can be used for studies of learning/memory motivated many scientists to utilize this technique (for a review, see Silva et al., 1997a). Recently, a targeted point mutation has been used for studying specific mechanisms underlying learning/memory (Cho et al., 1998; Giese et al., 1998). In this chapter I will illustrate the reasons and summarize the techniques required for generating such point mutations.

Why generate targeted point mutations?

As discussed below the generation of targeted point mutations is frequently the only possible approach to (i) accurately model inherited human diseases; and (ii) test specific molecular models. Furthermore, phenotypic comparisons between targeted disruptions (which lead to a loss of the target protein) and targeted point mutations (which can lead to a single amino acid substitution in the target protein) can deepen the insights into molecular and cellular mechanisms.

In many cases the generation of targeted point mutations is the only way for modeling accurately inherited human diseases, since a lot of these diseases result from missense mutations, i.e. mutations that change the amino acid sequence (e.g. Herrup and Noebels, 1997). These missense mutations may alter, but not disrupt the function of the protein. For example, episodic ataxia is a condition where patients develop intermittent incoordination associated with stress and fatigue. Episodic ataxia results from missense mutations in the gene encoding the voltage-gated K^+ channel subunit Kv1.1 (Browne et al., 1994; Adelman et al., 1995). These missense mutations alter, but do not block ion channel function (e.g. Zerr et al., 1998). Therefore, modeling episodic ataxia in mice requires the introduction of identified missense mutations into the Kv1.1 gene rather than generating a targeted disruption of the Kv1.1 gene.

The generation of targeted point mutants may sometimes be the only possibility for testing specific molecular models. For example, the model that CaMKII acts as a "memory molecule" (Miller and Kennedy, 1986; Lisman and Goldring, 1988; Lisman, 1994) was only able to be addressed by generating and analyzing a targeted point mutant (Giese et al., 1998). CaMKII is a

$Ca2^+$/calmodulin (CaM)-dependent kinase which can acquire CaM-independent activity upon autophosphorylation (for review, see Hanson and Schulman, 1992). In the case of αCaMKII, the autophosphorylation at threonine 286 was shown to be necessary and sufficient to switch the kinase into a CaM-independent activity state (Fong et al., 1989; Hanson et al., 1989; Waxham et al., 1990; Ohasako et al., 1991). CaMKII had been suggested to act as a "memory molecule" directly at the synapse, since it would "remember" previous Ca^{2+}-influx by being switched into the CaM-independent activity state (Miller and Kennedy, 1986; Lisman and Goldring, 1988; Lisman, 1994). To test the idea that CaMKII is a "memory molecule", it is hereby necessary to block specifically the switch-like property of CaMKII, leaving the CaM-dependent activity of the enzyme intact. Targeted disruptions as well as pharmacological blockade experiments could not address the memory molecule hypothesis, since they eliminate the CaM-dependent activity of enzyme in addition to its switch-like property. However, the introduction of a point mutation into the αCaMKII gene could test the memory molecule hypothesis (Giese et al., 1998). The design of the point mutation was based on the previous finding that an exchange of threonine 286 to alanine (T286A) of αCaMKII results in a kinase which has normal CaM-dependent activity, but cannot be switched into the CaM-independent state (Fong et al., 1989; Hanson et al., 1989; Waxham et al., 1990; Ohasako et al., 1991). The memory molecule hypothesis was tested by introducing the T286A mutation into the mouse αCaMKII gene (Giese et al., 1998). The $\alpha CaMKII^{T286A}$ mutants have no *N*-methyl-D-aspartate-receptor (NMDAR)-dependent LTP in the hippocampal CA1 region, exhibit unstable place cells and impaired spatial learning (Cho et al., 1998; Giese et al., 1998). Thus, the analysis of the $\alpha CaMKII^{T286A}$ mutants provided strong support of the memory molecule hypothesis.

As an alternative to the targeted T286A mutation, overexpression of a point mutated form of αCaMKII in a transgenic mouse was suggested to be suitable for testing the memory molecule hypothesis (Bach et al., 1995; Mayford et al., 1995, 1996; Rotenberg et al., 1996). Mayford and colleagues generated transgenic mice overexpressing αCaMKII with T286 exchanged to aspartate (T286D) under the control of the αCaMKII promoter (Mayford et al., 1995, 1996). Previous biochemical experiments showed that the T286D mutation mimics autophosphorylation at T286, resulting in a kinase with CaM-independent activity (Fong et al., 1989; Hanson et al., 1989; Waxham et al., 1990; Ohasako et al., 1991). The transgenic mice (Tg[αCaMKII-T286D]) show for low-frequency stimulations (5–10 Hz) long-term depression (LTD) instead of LTP, in addition to unstable place cells as well as impaired spatial learning (Bach et al., 1995; Mayford et al., 1995, 1996; Rotenberg et al., 1996). The electrophysiological phenotypes of the $\alpha CaMKII^{T286A}$ and the Tg[αCaMKII-T286D] mutants lead to different conclusions for the function of the autophosphorylation of αCaMKII at T286 in synaptic plasticity. The analysis of the $\alpha CaMKII^{T286A}$ mutants shows that the autophosphorylation of αCaMKII at T286 is required for NMDAR-dependent LTP (Giese et al., 1998). This conclusion is consistent with the results of acute overexpressions of CaM-independent forms of αCaMKII, leading to a potentiation of synaptic transmission which is occlusive to LTP (Pettit et al., 1994; Lledo et al., 1995). In contrast, the analysis of the Tg[αCaMKII-T286D] mutants suggests that the autophosphorylation of αCaMKII at T286 influences the balance between LTD and LTP (Mayford et al., 1995, 1996). It is possible that the Tg[αCaMKII-T286D] mutants may have uncovered a function of the autophosphorylation of αCaMKII at T286, which could not be revealed by other approaches. Alternatively, the electrophysiological phenotype of the Tg[αCaMKII-T286D] mutants may be artificial due to unphysiological phosphorylations. In conclusion, the phenotypic comparison between the $\alpha CaMKII^{T286A}$ and the Tg[αCaMKII-T286D] mutants shows that transgenic overexpression experiments cannot substitute for targeted point mutation approaches in testing specific molecular models.

Finally, the phenotypic comparison of targeted point mutants with the corresponding targeted disruptions can result in deeper mechanistic insights. For example, the targeted disruption of the αCaMKII gene leads to increased augmentation (Chapman et al., 1995), a short-lasting increase in neurotransmitter release after a tetanus (Zucker, 1989), whereas the αCaMKII^{T286A} mutants have normal augmentation (Giese et al., 1998). Thus, the CaM-dependent, but not the CaM-independent activity of the kinase is strongly suggested to participate in augmentation.

Strategies to introduce point mutations into target genes

The strategies for introducing point mutations into target genes are more complicated than the commonly used strategies for generating targeted disruptions. Targeted disruptions are usually generated by inserting a resistance gene into the target gene (the resistance gene is also used for selection of stably transfected cells). For the generation of targeted point mutations it is necessary to remove the resistance gene from the target gene after the initial homologous recombination (HR) event. The removal of the resistance gene is desirable, since its presence could interfere with the expression of the target gene and it may even influence the expression of neighboring genes (for example, see Olson et al., 1996). Three strategies for introducing point mutations into target genes have been described: the "hit and run procedure", the "tag and exchange strategy" and the "pointlox procedure". In comparison with strategies to generate targeted disruptions, the generation of targeted point mutations includes an additional manipulation step of embryonic stem (ES) cells. This additional step extends the time of culturing the ES cells which can compromise their pluripotency. Especially under sub-optimal ES cell culture conditions it can be difficult to obtain germline transmission with point-mutated ES cell clones.

The hit and run procedure (Hasty et al., 1991a) uses an insertion vector (Thomas and Capecchi, 1987) which integrates into the target gene when HR occurs (Fig. 1). The "hit and run vector" contains the desired point mutation and two gene expression cassettes for selection (Fig. 1). One of these gene expression cassettes serves for positive selection to obtain stably transfected ES cell clones. Most frequently a neomycin gene expression cassette (neo) is used for positive selection, having a strong promoter such as the phosphoglycerate kinase 1 (PGK) promoter (e.g. Soriano et al., 1991) to avoid position-dependent expression of the neo. Neo encodes a bacterial phosphotransferase which can transfer phosphate groups to aminoglycosides such as G418. G418 blocks protein synthesis, but becomes inactivated in the presence of the neo gene product by transfer of a phosphate group. In addition to the neo the "hit and run vector" contains a gene expression cassette for negative selection. A negative selection is survived by cells which have lost the gene expression cassette. Frequently, the Herpes simplex virus thymidine kinase cassette (HSV-tk) is used for negative selection (e.g. Mansour et al., 1988). For selection with HSV-tk a nucleoside analog such as FIAU (1-[2-deoxy, 2-fluoro-β-D-arabinofuranosyl]-5 iodouracil) is used. In the presence of HSV-tk the nucleoside analog becomes phosphorylated (but not by endogenous thymidine kinases) so it can be used for DNA synthesis during replication. Once the nucleoside analog is added to the synthesized DNA strand further DNA synthesis is blocked. In the "hit step" of the procedure ES cell clones with HR events are identified after transfection and positive selection. The targeted clones have a partial duplication of the target gene. The duplicated regions are separated by the plasmid and the gene expression cassettes for selection. In the "run step" the targeted clones are used for a negative selection. During the negative selection the clones expressing

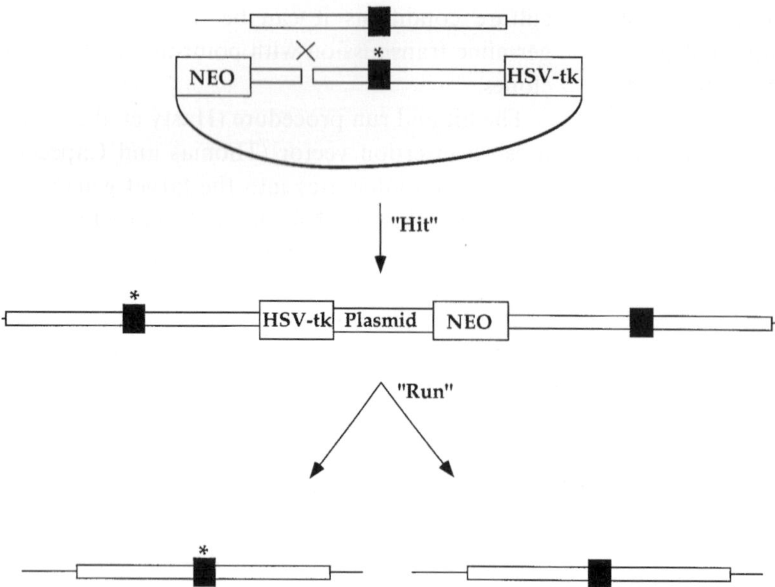

Fig. 1. The hit and run procedure is shown. The insertion vector contains the desired point mutation (asterisk) in an exon (black box), a neo gene expression cassette for positive selection and a HSV-tk gene expression cassette for negative selection. Homologous sequences between the insertion vector and the target gene are shown as white boxes. In the "hit step" the insertion vector inserts into the target gene, resulting in a partial duplication of the target gene. In the "run step" clones with intrachromosomal recombination events survive negative selection. Such clones have either the desired point mutation or the wild-type allele.

HSV-tk die. Most of the surviving ES cell clones lose HSV-tk by intrachromosomal recombination, resulting in either wild-type clones (in most of the cases) or in clones with the desired point mutation. ES cell clones with the desired point mutation are identified by Southern analyses. The hit and run procedure has been applied successfully by several groups (Hasty et al., 1991a; Ramirez-Solis et al., 1993; Wu et al., 1994; Van Doorninck et al., 1995; Lakhlani et al., 1997). However, for some genes there may be difficulties with intrachromosomal recombination so that only wild-type clones are obtained.

The tag and exchange strategy (Askew et al., 1993; Stacey et al., 1994; Wu et al., 1994) uses two replacement vectors (Thomas and Capecchi, 1987) (Fig. 2). A replacement vector replaces part of the target gene when HR occurs. The "tag vector" contains a neo and HSV-tk next to each other in the place where the point mutations are to be located. After ES cell clones which have homologously recombined with the "tag vector" are identified, they are transfected with the "exchange vector" containing the desired point mutation. For the "exchange step" a negative selection enriches for clones with the desired HR event. However, HSV-tk is frequently lost in a non-homologous way (by deletion), which may make the identification of HR events difficult (Vazquez et al., 1998). The tag and exchange strategy may be advantageous for generating several alleles with different point mutations.

The pointlox procedure (Reaume et al., 1996; Giese et al., 1998) uses a replacement vector containing the desired point mutation and a neo inserted in an intron (Fig. 3). The neo is flanked by loxP sites, so it can be removed by the bacteriophage Cre-loxP recombination system (for review, see Sauer, 1993). A loxP site is 34 bp long and has two 13 bp inverted repeats as well as a 8 bp asymmetric core. Cre recombinase excises DNA between two loxP sites of the same orientation

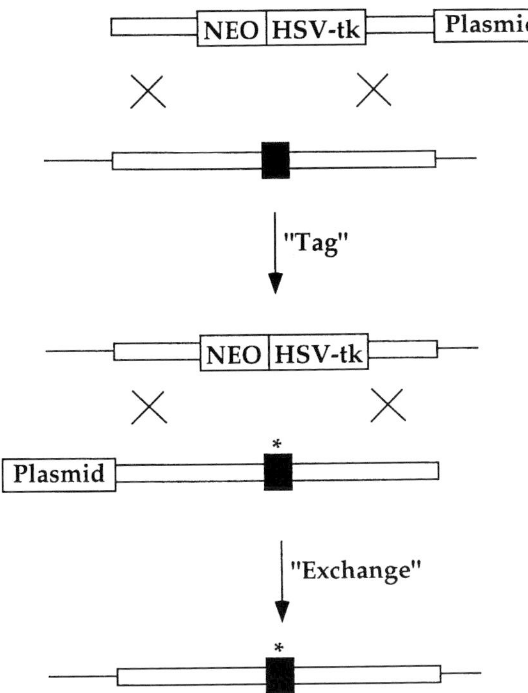

Fig. 2. The tag and exchange strategy is illustrated. The symbols are identical to Fig. 1. The first replacement vector contains a neo and a HSV-tk at the location where the point mutation is to be introduced. In the "tag step" the homologous region of the target gene is replaced by the vector sequences. In the "exchange step" targeted clones are transfected with a second replacement vector which contains the desired point mutation in an exon. Negative selection enriches for clones with the desired mutation.

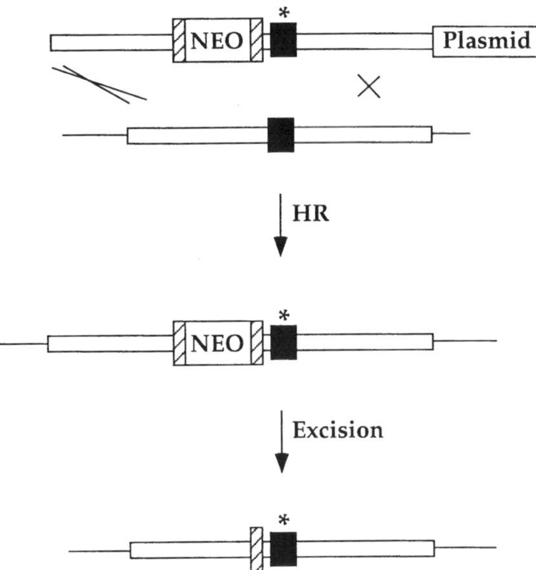

Fig. 3. The pointlox procedure is indicated. The symbols are identical to Fig. 1. The replacement vector contains the desired point mutation in an exon and a neo flanked by loxP sites (striped boxes) in an intron. Clones with HR events are transiently transfected with a plasmid expressing Cre recombinase to remove the neo, leaving behind one loxP site in the intron.

leaving one loxP site behind. Thus, after the identification of ES cell clones where HR events have occurred, these clones are transiently transfected with a plasmid expressing Cre recombinase. A transient transfection uses circular DNA, so it cannot integrate into the genome. It is possible to enrich for ES cell clones with excision events by negative selection, if the construct contains a neo/HSV-tk flanked by loxP sites (e.g. Brusa et al., 1995, Vazquez et al., 1998). However, the drugs used for negative selection can compromise ES cell pluripotency. Even without negative selection the efficiency of Cre recombination is usually high (4–20%), especially if Cre expression is under the control of an appropriate promoter (Araki et al., 1997) and if the recombinase contains a nuclear localization signal (Kellendonk et al., 1996). After the excision of the neo the residual loxP site in the intron should not interfere with splicing of the target gene. The pointlox procedure appears to be the most straightforward technique to introduce point mutations into target genes, and it is discussed in more detail below.

Designing a pointlox construct

Standard cloning techniques (Sambrook et al., 1989) are required for the generation of pointlox constructs. The point mutation can be generated via PCR (e.g. Hemsley et al., 1989) and an efficient cloning strategy needs to consider the screening strategy (see below). An example for a pointlox construct is shown in Fig. 4.

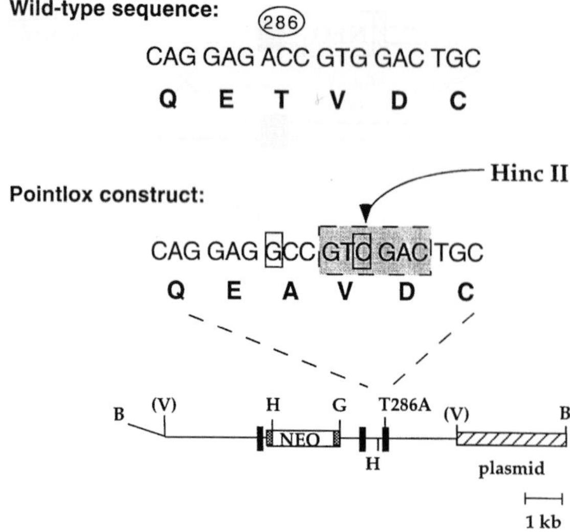

Fig. 4. The pointlox construct for introducing the T286A mutation into the αCaMKII gene (Giese et al., 1998) is shown. Part of the nucleic acid sequence with the corresponding amino acid sequence in the one letter code is presented. In comparison to the wild-type sequence, the pointlox construct contains two point mutations. The missense mutation (framed G) represents the T286A mutation and the marker mutation (framed C) gives rise to a diagnostic *Hinc*II restriction site without altering the amino acid sequence. The distance between the missense and marker mutation is only 4 bp, so that recombination events between these two mutations is very unlikely. The pointlox construct contains a neo gene expression cassette which is flanked by loxP sites. The neo is localized in an intron, 60 bp downstream of an exon (exons are black boxes) and after the excision of the neo 79 bp of foreign sequences remain in the intron. To obtain a high HR frequency, each arm of homology is about 3 kb long. A relatively small distance between the neo and the point mutations (about 1.2 kb) avoids the loss of the point mutations during most HR events.

For practical reasons it is desirable to obtain high HR frequencies with the pointlox construct. The factors influencing HR are not completely understood. Nevertheless, based on years of experience the following may be advised. As for every gene targeting construct, the homologous sequences in the pointlox construct determine the HR frequency. To obtain high HR frequencies, the homologous sequences should be isogenic (from the same mouse strain as the ES cells) (TeRiele et al., 1992) and they should be as long as possible (Hasty et al., 1991b). High HR frequencies (1–40%) can be expected if each of the two homology arms surrounding the neo is 2.5–5 kb long. It is important to know that the HR frequency can drop drastically if one homology arm is shorter than 2 kb (Hasty et al., 1991b).

It is advantageous if the pointlox construct has a marker mutation in addition to the missense mutation. A marker mutation gives rise to or destroys a restriction site, so the missense mutation can be identified easily. The marker mutation should be silent (not altering the amino acid sequence) and close (<10 bp) to the missense mutation (to make a recombination event between the two mutations unlikely). The distance between the point mutations and the neo should be relatively small (smaller than 1.5 kb is suggested), to keep the probability of recombination between the point mutations and the neo low (Deng et al., 1993), which would result in the loss of the point mutations during the HR event.

The neo should be inserted into an intron in such a way that the loxP site, which remains in the intron after Cre recombination, does not interfere with splicing of the target gene. Thus, the insertion of the neo should not destroy the splice junctions at the exon/intron borders nor the sequences required for lariat formation.

Pointlox screening strategies

The pointlox procedure consists of HR and excision steps. The screening strategies which distinguish the various possible integration events are schematically illustrated in Fig. 5. Southern analysis should be used to identify appropriate ES cell clones for both steps, because of the high reliability of the method.

In the initial screen ES cell clones with HR events are identified using an external probe, i.e. a probe that does not hybridize with the pointlox construct. For the identified ES cell clones additional random integrations of the pointlox construct should be excluded using an internal probe. Since replacement vectors can sometimes integrate homologously (Hasty et al., 1991c),

A

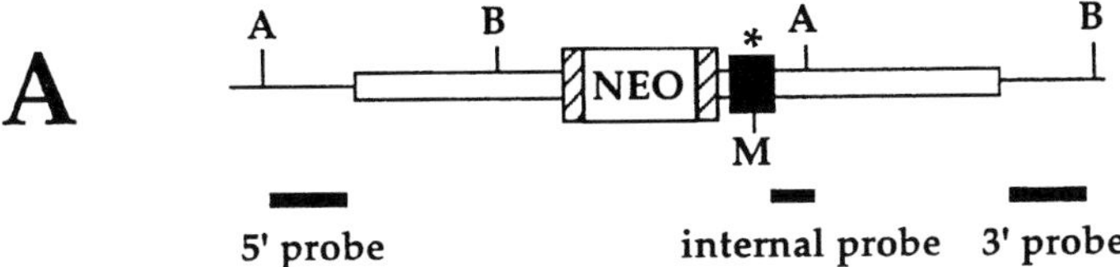

B

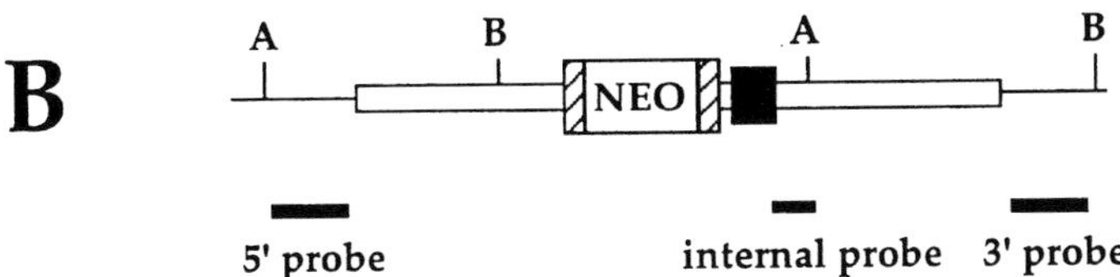

C

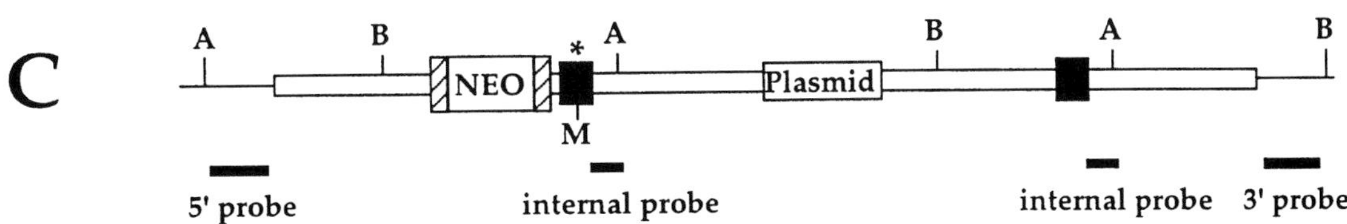

Fig. 5. Different HR events with a pointlox construct and the screening strategies are shown. (A) The desired HR event, (B) HR event without the desired point mutation, (C) homologous integration of the pointlox construct. These three HR events are identified using Southern analysis and a 5′ probe following DNA digestion with restriction enzyme "A". In comparison to the wild-type gene the length of the "A"-fragment is increased by the neo insertion. Because of the marker mutation (M) the presence of the point mutation can be identified with the internal probe and "A" + "M" digested DNA, leading to the exclusion of (B). Thereafter, screening with the 3′ probe (or the internal probe) with "B"-digested DNA distinguishes between (A) and (C).

the candidate ES cell clones should be screened with a second external probe, hybridizing on the other side of the targeting construct than the first external probe. Finally, the identified ES cell clones need to be screened for the presence of the point mutations taking advantage of the marker mutation.

ES cell clones with excision events should be identified with an external probe. It is important to test the identified clones for the presence of the point mutations to distinguish them from wild-type clones.

Culturing embryonic stem cells

There are many reviews about culturing ES cells (e.g. see Chapter 3.1.1) and the generation of chimeric mice (e.g. Joyner, 1993). Several ES cell lines are stable enough for two-step manipulations. For example, the R1 cell line (Nagy et al., 1993) has been used successfully for such manipulations (e.g. Brusa et al., 1995; Giese et al., 1998). The precise culturing conditions depend on the particular ES cell lines. In general, ES cells are cultured under conditions which avoid differentiation (culturing high ES cell densities on monolayers of division-inactivated fibroblasts; keeping the passage number as low as possible) and mycoplasma contamination (culturing without antibiotics). Since the conditions for a transient transfection are not yet reviewed, the procedure is:

1. Expand targeted ES cell clones until approx. 5×10^6 cells are obtained;
2. Electroporate the cells in 800 μl medium with 40 μg circular DNA in 10 μl H_2O in BioRad Gene Pulser at 240 V, 500 μF;
3. After 20 min on ice plate 500–1500 cells on a 9 cm dish containing a feeder layer;
4. After 8–9 d colonies can be picked.

It is recommended to transiently transfect several targeted ES cell clones independently (may be up to 4) for obtaining clones with excision events. This would spare the testing of the targeted ES cell clones for germline transmission, and it would be more likely to obtain clones with excision events which contribute to the germline.

Genetic background

The genetic background in which a mutation is studied can alter the phenotype dramatically, because, for example, the level of interacting proteins may be different in several backgrounds (for further discussion of this topic see Chapter 3.1.10). Problems concerning genetic background and possible solutions to these problems are reviewed by Silva and colleagues (Silva et al., 1997b). To have a well defined genetic background and to standardize the genetic background, it has been recommended to study mutations in the F_1 or F_2 generation between 129/J and C57BL/6J (Silva et al., 1997b). Nevertheless, a precise description of the genetic background should be given in any publication describing studies with mutant mice.

Acknowledgments

I thank Ofelia Carvalho, Susan M. Darling, and Paul W. Frankland for helpful suggestions.

References

Adelman, J.P., Bond, C.T., Pessia, M. and Maylie, J. (1995) Episodic ataxia results from voltage-dependent potassium channels with altered functions. *Neuron*, 15: 1449–1454.

Araki, K., Imaizumi, T., Okuyama, K., Oike, Y. and Yamamura, K. (1997) Efficiency of recombination by Cre transient expression in embryonic stem cells: Comparison of various promoters. *J. Biochem.*, 122: 977–982.

Askew, G.R., Doetschman, T. and Lingrel, J.B. (1993) Site-directed point mutations in embryonic stem cells: a gene targeting tag-and-exchange strategy. Mol. Cell. Biol., 13: 4115–4124.

Bach, M.E., Hawkins, R.D., Osman, M., Kandel, E.R. and Mayford, M. (1995) Impairment of spatial but not contextual memory in CaMKII mutant mice with a selective loss of hippocampal LTP in the range of the θ frequency. *Cell*, 81: 905–915.

Browne, D.L., Gancher, S.T., Nutt, J.G., Brunt, E.R.P., Smith, E.A., Kramer, P. and Litt, M. (1994) Episodic ataxia/myokymia syndrome is associated with point mutations in the human potassium channel gene, KCNA1. *Nat. Genet.*, 8: 136–140.

Brusa, R., Zimmermann, F., Koh, D.-S., Gass, P., Seeburg, P.H. and Sprengel, R. (1995) Early-onset epilepsy and postnatal lethality associated with an editing-deficient GluRB allele in mice. *Science*, 270: 1677–1680.

Chapman, P.F., Frenguelli, B., Smith, A., Chen, C.-M. and Silva, A.J. (1995) The α-calcium-calmodulin kinase II: a bidirectional modulator of pre-synaptic plasticity. *Neuron*, 14: 591–597.

Cho, Y.H., Giese, K.P., Tanila, H., Silva, A.J. and Eichenbaum, H. (1998) Abnormal hippocampal spatial representation in αCaMKII^{T286A} and CREB$^{\alpha\Delta-}$ mice. *Science*, 279: 867–869.

Deng, C., Thomas, K.R. and Capecchi, M.R. (1993) Location of crossovers during gene targeting with insertion and replacement vectors. *Mol. Cell. Biol.*, 13: 2134–2140.

Fong, Y.-L., Taylor, W.L., Means, A.R. and Soderling, T.R. (1989) Studies of the regulatory mechanism of Ca^{2+}/calmodulin-dependent protein kinase II. *J. Biol. Chem.*, 264: 16759–16763.

Giese, K.P., Fedorov, N.B., Filipkoski, R.K. and Silva, A.J. (1998) Autophosphorylation at Thr^{286} of the α calcium-calmodulin kinase II in LTP and learning. *Science*, 279: 870–873.

Grant, S.G.N., O'Dell, T.J., Karl, K.A., Stein, P.L., Soriano, P. and Kandel, E.R. (1992) Impaired long-term potentiation, spatial learning and hippocampal development in fyn mutant mice. *Science*, 258: 1903–1910.

Hanson, P.J., Kapiloff, M.S., Lou, L.L., Rosenfeld, M.G. and Schulman, H. (1989) Expression of a multifunctional Ca^{2+}/calmodulin-dependent protein kinase and mutational analysis of its autoregulation. *Neuron*, 3: 59–70.

Hanson, P.I. and Schulman, H. (1992) Neuronal Ca^{2+}/calmodulin-dependent protein kinases. *Annu. Rev. Biochem.*, 61: 559–601.

Hasty, P., Ramirez-Solis, R., Krumlauf, R. and Bradley, A. (1991a) Introduction of a subtle mutation into the Hox-2.6 locus in embryonic stem cells. *Nature*, 350: 243–246.

Hasty, P., Rivera-Perez, J. and Bradley, A. (1991b) The length of homology required for gene targeting in embryonic stem cells. *Mol. Cell. Biol.*, 11: 5586–5591.

Hasty, P., Rivera-Perez, J., Chang, C. and Bradley, A. (1991c) Target frequency and integration pattern for insertion and replacement vectors in embryonic stem cells. *Mol. Cell. Biol.*, 11: 4509–4517.

Hemsley, A., Arnheim, N., Toney, M.D., Cortopassi, G. and Galas, D.J. (1989) A simple method for site-directed mutagenesis using the polymerase chain reaction. *Nucl. Acids Res.*, 17: 6545–6551.

Herrup, K. and Noebels, J. (1997) Ion channel mutants and their impact on behavior and disease. Society for Neuroscience Short Course 2 Syllabus.

Joyner, A. L. (1993) *Gene targeting: A practical approach*. IRL Press (Oxford).

Kellendonk, C., Tronche, F., Monaghan, A.P., Angrand, P.O., Stewart, F. and Schütz, G. (1996) Regulation of Cre recombinase activity by the synthetic steroid RU 486. *Nucl. Acids. Res.*, 24: 1404–1411.

Lakhlani, P.P., MacMillan, L.B., Guo, T.Z., McCool, B.A., Lovinger, D.M., Maze, M. and Limbird, L.E. (1997) Substitution of a mutant alpha2a-adrenergic receptor via "hit and run" gene targeting reveals the role of this subtype in sedative, analgesic, and anesthetic-sparing responses in vivo. *Proc. Natl. Acad. Sci. USA*, 94: 9950–9955.

Lisman, J.E. and Goldring, M.A. (1988) Feasibility of long-term storage of graded information by the Ca^{2+}/calmodulin-dependent protein kinase molecules of the postsynaptic density. *Proc. Natl. Acad. Sci. USA*, 85: 5320–5324.

Lisman, J. (1994) The CaM kinase II hypothesis for the storage of synaptic memory. *Trends Neurosci.*, 17: 406–412.

Lledo, P.-M., Hjelmstad, G.O., Mukherji, S., Soderling, T.R., Malenka, R.C. and Nicoll, R.A. (1995) Calcium/calmodulin-dependent kinase II and long-term potentiation enhance synaptic transmission by the same mechanism. *Proc. Natl. Acad. Sci. USA*, 92: 11175–11179.

Mansour, S.L., Thomas, K.R. and Capecchi, M.R. (1988) Disruption of the proto-oncogene int-2 in mouse embryo-derived stem cells: a general strategy for targeting mutations to non-selectable genes. *Nature*, 336: 348–352.

Mayford, M., Wang, J., Kandel, E.R. and O'Dell, T.J. (1995) CaMKII regulates the frequency-response function of hippocampal synapses for the production of both LTD and LTP. *Cell*, 81: 891–904.

Mayford, M., Bach, M.E., Huang, Y.-Y., Wang, L., Hawkins, R.D. and Kandel, E.R. (1996) Control of memory formation through regulated expression of a CaMKII transgene. *Science*, 274: 1678–1683.

Miller, S.G. and Kennedy, M.B. (1986) Regulation of brain type II Ca^{2+}/calmodulin-dependent protein kinase by autophosphorylation: a Ca^{2+}-triggered molecular switch. *Cell*, 44: 861–870.

Nagy, A., Rossant, J., Nagy, R., Abramow-Newerly, W. and Roder, J.C. (1993) Derivation of completely cell culture-derived mice from early-passage embryonic stem cells. *Proc. Natl. Acad. Sci. USA*, 90: 8424–8428.

Ohasako, S., Nakazawa, H., Sekihara, S.-I., Ikai, A. and Yamauchi, T. (1991) Role of threonine-286 as autophosphorylation site for appearance of Ca^{2+}-independent activity of calmodulin-dependent protein kinase II α subunit. *J. Biochem.*, 109: 137–143.

Olson, E.N., Arnold, H.-H., Rigby, P.W.J. and Wold, B.J. (1996) Know your neighbors: three phenotypes in null mutants of the myogenic bHLH gene MRF4. *Cell*, 85: 1–4.

Pettit, D.L., Perlman, S. and Malinow, R. (1994) Potentiated transmission and prevention of further LTP by increased CaMKII activity in postsynaptic hippocampal slice neurons. *Science*, 266: 1881–1885.

Ramirez-Solis, R., Zheng, H., Whiting, J., Krumlauf, R. and Bradley, A. (1993) Hoxb-4 (Hox-2.6) mutant mice show homeotic transformation of a cervical vertebra and defects in the closure of the sternal rudiments. *Cell*, 73: 279–294.

Reaume, A.G., Howland, D.S., Trusko, S.P., Savage, M.J., Lang, D.M., Greenberg, B.D., Siman, R. and Scott, R.W. (1996). Enhanced amyloidgenic processing of the ?-amyloid precursor protein in gene-targeted mice bearing the Swedish familial alzheimer's disease mutations and a "humanized" a? sequence. *J. Biol. Chem.*, 271: 23380–23388.

Rotenberg, A., Mayford, M., Hawkins, R.D., Kandel, E.R. and Muller, R.U. (1996) Mice expressing activated CaMKII lack low frequency LTP and do not form stable place cells in the CA1 region of the hippocampus. *Cell*, 87: 1351–1361.

Sambrook, J., Fritsch, E.F. and Maniatis, T. (1989) *Molecular cloning: A laboratory manual*. CSHL Press.

Sauer, B. (1993) Manipulation of transgenes by site-specific recombination: use of Cre recombinase. *Methods Enzymol.*, 225: 890–900.

Silva, A.J., Stevens, C.F., Tonegawa, S. and Wang, Y. (1992a) Deficient hippocampal long-term potentiation in alpha-calcium calmodulin kinase II mutant mice. *Science*, 257: 201–206.

Silva, A.J., Paylor, R., Wehner, J.M. and Tonegawa, S. (1992b) Impaired spatial learning in alpha-calcium calmodulin kinase II mutant mice. *Science*, 257: 206–211.

Silva, A.J., Smith, A. and Giese, K.P. (1997a) Gene targeting and the biology of learning and memory. *Annu. Rev. Genet.*, 31: 527–546.

Silva, A.J., Simpson, E.M., Takahashi, J.S., Lipp, H.-P., Nakanishi, S., Wehner, J.M., Giese, K.P., Tully, T., Abel, T., Chapman, P.F., Fox, K., Grant, S., Itohara, S., Lathe, R., Mayford, M., McNamara, J.O., Morris, R.J., Picciotti,

M., Roder, J., Shin, H.-S., Schlesinger, P.A., Storm, D.R., Stryker, M.P., Wang, Y. and Wolfer, D.P. (1997b) Mutant mice and neuroscience: recommendations concerning genetic background. *Neuron*, 9: 755–759.

Soriano, P., Montgomery, C., Geske, R. and Bradley, A. (1991) Targeted disruption of the c-src proto-oncogene leads to osteopetrosis in mice. *Cell*, 64: 693–702.

Stacey, A., Schnieke, A., McWhir, J., Cooper, J., Colman, A. and Melton, D.W. (1994) Use of double-replacement gene targeting to replace the murine α-lactalbumin gene with its human counterpart in embryonic stem cells and mice. *Mol. Cell. Biol.*, 14: 1009–1016.

TeRiele, H., Maandag, E.R. and Berns, A. (1992) Highly efficient gene targeting in embryonic stem cells through homologous recombination with isogenic DNA constructs. *Proc. Natl. Acad. Sci. USA*, 89: 5128–5132.

Thomas, K.R. and Capecchi, M.R. (1987) Site-directed mutagenesis by gene targeting in mouse embryo-derived stem cells. *Cell*, 51: 503–512.

Van Doorninck, J.H., French, P.J., Verbeek, E., Peters, R.H.P.C., Morreau, H., Bijman, J. and Scholte, B.J. (1995) A mouse model for the cystic fibrosis ΔF508 mutation. *EMBO J.*, 14: 4403–4411.

Vazquez, J.C., Nogues, C., Rucker, E.B. and Piedrahita, J.A. (1998) Factors affecting the efficiency of introducing precise genetic changes in ES cells by homologous recombination: tag-and-exchange versus the Cre-loxP system. *Transgenic Res.*, 7: 181–193.

Waxham, M.N., Aronowski, J., Westgate, S.A. and Kelly, P.T. (1990) Mutagenesis of Thr-286 in monomeric Ca^{2+}/calmodulin-dependent protein kinase II eliminates Ca^{2+}/calmodulin-independent activity. *Proc. Natl. Acad. Sci. USA*, 87: 1273–1277.

Wu, H., Liu, X. and Jaenisch, R. (1994) Double replacement: strategy for efficient introduction of subtle mutations into the murine Colla-1 gene by homologous recombination in embryonic stem cells. *Proc. Natl. Acad. Sci. USA*, 91: 2819–2823.

Zerr, P., Adelman, J.P. and Maylie, J. (1998) Episodic ataxia mutations in Kv1.1 alter potassium channel function by dominant negative effects or haploinsufficiency. *J. Neurosci.*, 18: 2842–2848.

Zucker, R.S. (1989) Short-term synaptic plasticity. *Annu. Rev. Neurosci.*, 12: 13–31.

W.E. Crusio and R.T. Gerlai (Eds.)
Handbook of Molecular-Genetic Techniques for Brain and Behavior Research (Techniques in the Behavioral and Neural Sciences, Vol. 13)

CHAPTER 3.1.6

Genetic dissection of a postsynaptic multiprotein complex controlling synaptic plasticity and learning in the mouse

Seth G.N. Grant

Centre for Neuroscience and Centre for Genome Research, University of Edinburgh, King's Building, West Main Road, Edinburgh EH9 3JQ, UK

Introduction

We are now in the era when learning and memory, like so many other areas of biology that were tackled with the tools of genetics and molecular biology, will begin to reveal its most intimate secrets. By comparison with the molecular biology of cancer biology or immunology, learning and memory is still in a relatively early phase. For example, we now know that the mechanisms of gene transcription, RNA splicing and replication of DNA are conducted by highly organised and elaborate multi-protein complexes which themselves are regulated in the cell cycle by regulatory enzymes driven by signal transduction pathways. The discovery of these basic molecular mechanisms provides explanations for many of the broad biological features of cancer. Similarly, in immunology the clonal selection theory and the subsequent identification of its underlying mechanisms along with a molecular understanding of interactions between different immune cell types provides a mechanistic understanding of acquired immunity. In these and other examples, nature has provided us with elegant molecular machinery to perform these biological functions.

Psychological thinking has driven learning research since the broad description of short and long-term memory outlined in the 19th Century. These psychological descriptions promoted mechanistic explorations initially at the cellular and later at the molecular level, by attempting to define a simple parallel with psychological phenomena. For example, in 1949, Hebb postulated that the coincident activity between two neurons reinforced its connectivity (Hebb, 1949), and much later electrophysiologists studying synaptic mechanisms sort to extend this model of coincidence to the NMDA receptor (Bliss and Collingridge, 1993). Although it was unlikely that the behaviour of learning could be tied up in a single molecule in this way, these psychologically driven investigations set the stage for the discovery of fundamental mechanisms of learning and memory akin to those of replication, transcription or clonal selection.

The discovery of synaptic plasticity in vertebrates and invertebrates was rapidly recognised as a possible mechanism underlying learning. The observation that particular patterns of neuronal firing lead to long-term changes in synaptic efficiency combined with the idea that these changes in synaptic efficiency may represent a mechanism for storage of information in neural networks presents a very plausible mechanism for learning. Thus molecular mechanistic studies of learning have focussed on understanding synaptic plasticity and assays of synaptic plasticity have been used extensively by pharmacologists and molecular biologists to identify potential molecular mechanisms. For example, using the hippocampus slice preparation it is possible to test how various drugs or gene mutations interfere with the induction or maintenance of synaptic plasticity, and the same drugs and mutations may interfere with learning.

The availability of drugs for various classes of ion channels and second messenger systems led to the first basic molecular description of synaptic plasticity in the mammalian hippocampus. The cloning of genes encoding ion channels, kinases and second messenger pathways as well as genes induced by synaptic activity has led to a plethora of potential relevant molecular components. An important step toward testing the physiological significance of these genes was shown in 1992 when the first studies using gene targeting in the mouse tested the function of 5 different kinases in synaptic plasticity and discovered that two of them were required for the induction of long-term potentiation in the hippocampus (Grant et al., 1992; Silva et al., 1992a, 1992b). Moreover, in these mutant mice, there were impairments in hippocampus dependent forms of learning, that had been previously linked to LTP. This paradigm of mouse genetics in the study of synaptic plasticity and learning has set the stage for a large number of subsequent studies — many aimed at technical improvements including conditional knockouts — and others aimed at understanding the function of the mutant protein.

A particularly important role of genetics will be in the identification of molecular pathways or signalling cascades. The characterisation of a molecular cascade controlling a specific phenotype requires interruption of molecules at sequential steps in the cascade — in the simplest case these (null) mutations should produce the same phenotype and animals with mutations in more than one protein should also produce the same phenotype. Genetics alone is not sufficient to define the function of a particular protein or of a particular pathway (Brenner, 1973). Biochemistry of the individual proteins and the mutant proteins and the pathway in which those proteins act is necessary to understand the function of the gene. For example the physical interactions between these proteins will also help understand where in the pathway these two mutant proteins normally act. It should therefore be clear that a truly integrative biological approach is required combining genetics, biochemistry, synaptic physiology and psychology of learning and memory to identify the real mechanisms of learning and memory.

In this paper we will attempt to illustrate how genetics and biochemistry in a model organism that lends itself to an integrated approach — the mouse — has led to the identification of new mechanisms underlying synaptic plasticity and learning. This paper will not attempt to review gene targeting or new technical approaches to learning in mice but will focus attention on a genetic analysis of a multiprotein complex. This complex — the NMDA Receptor Signalling Complex (NRSC) appears to be an important integrator of synaptic second messenger pathways as well as a device for coupling the NMDA receptor to downstream pathways.

A primer on NMDA receptor in synaptic plasticity and learning

Within the intact brain the NMDA receptor plays an important role in modulation of many activity-dependent synaptic and neuronal properties. The functions of the NMDA receptor are first recruited for activity dependent developmental processes and later used in the adult to contribute to forms of behaviour including learning and memory (Bliss and Collingridge, 1993; Constantine-Paton and Cline, 1998). The most extensively studied model system of NMDA receptor dependent signalling is in the hippocampus CA3-CA1 synapses. Here, low frequencies of synaptic stimulation lead to a long lasting depression of synaptic transmission (Long-Term Depression, LTD) and high frequency stimulation results in long-term potentiation of transmission (LTP) (Bliss and Collingridge, 1993; Bear and Malenka, 1994). LTD is more readily observed in younger animals whereas LTP is the predominant form of plasticity in adults. Blockade of NMDA receptor by pharmacological (Morris et al., 1986) and genetic mutations (Tsien et al., 1996a, 1996b) interfere with spatial learning — a hippocampus dependent form of learning. This suggests that NMDA receptor activates synaptic plasticity (LTP or LTD) which is then required for learning.

The NMDA receptor signalling complex — NRSC

The NMDA receptor is an ion-channel formed by the assembly of a common NR1 subunit with one or more of four different NR2 subunits, NR2A-D (Nakanishi et al., 1994; Mori and Mishina, 1995; Seeburg et al., 1995). The original cloning and expression of recombinant NMDA receptor subunits in Xenopus oocytes and mammalian cell lines showed the NR1-NR2 complex comprises many of the basic features of NMDA receptors measured electrophysiologicaly in vivo. Biophysical studies combined with extensive pharmacological studies have clearly demonstrated that the NMDA receptor initiates signalling events at the synapse by allowing calcium influx through the pore of the NMDA receptor channel ion complex. Manipulations of the function of the NMDA receptor have indicated that the extent of calcium coming through the receptor determines whether the particular synaptic activity leads to a long-term potentiation (high calcium influx) or a long-term depression (low calcium influx) of synaptic efficacy (Bear and Malenka, 1994). This calcium influx has been shown to result in activation of downstream kinases and phosphatases which then act on a variety substrates that may participate in the expression of synaptic plasticity.

This view of the NMDA receptor ion channel as simply a complex of ion channel subunits that allows calcium into the cell which then acts on second messenger systems is now the subject of revision: the NMDA receptor ion channel subunits are normally bound to a set of cytoplasmic proteins in a large multiprotein complex and this complex controls signalling. We refer to this complex of ion channel and associated proteins as the NRSC: the NMDA Receptor Signalling Complex (Fig. 1).

The NR2 subunits have long cytoplasmic C-terminal domains that contain sites for phosphorylation (Smart, 1997) and interaction with cytoplasmic proteins. In vitro binding studies show the C-terminus of NR2 subunits bind to PSD-95/SAP90, Chapsyn-110/PSD-93 and other related members of the MAGUK family (membrane-associated guanylate kinases) (Kornau et al., 1995; Niethammer et al., 1995; Kim et al., 1996; Muller et al., 1996). PSD-95/SAP90 (Cho et al., 1992; Kistner et al., 1993; Massague, 1996) is an abundant postsynaptic density protein and is comprised of several domains mediating protein–protein interactions including 3 PDZ/DHR (PSD-95, Dlg, ZO-1/Dlg Homologous Region) domains, an SH3 (Src Homology 3) domain and a guanylate kinase homology (GK) domain. The second PDZ domain (PDZ2) is capable of directly binding the C-terminus of the NR2 subunits (Kornau et al., 1995; Niethammer et al., 1995). PSD-95 lacks any detectable enzymatic activity and acts as an adapter molecule through protein–protein interactions mediated by the discrete domains. Through these domains PSD-95 can bind a variety of proteins including neuronal Nitric Oxide Synthase (nNOS) (Brenman et al., 1996), SynGAP (Chen et al., 1998; Kim et al., 1998), Neuroligin (Irie et al., 1997), GKAP/SAPAP (Kim et al., 1997; Takeuchi et al., 1997) and other proteins.

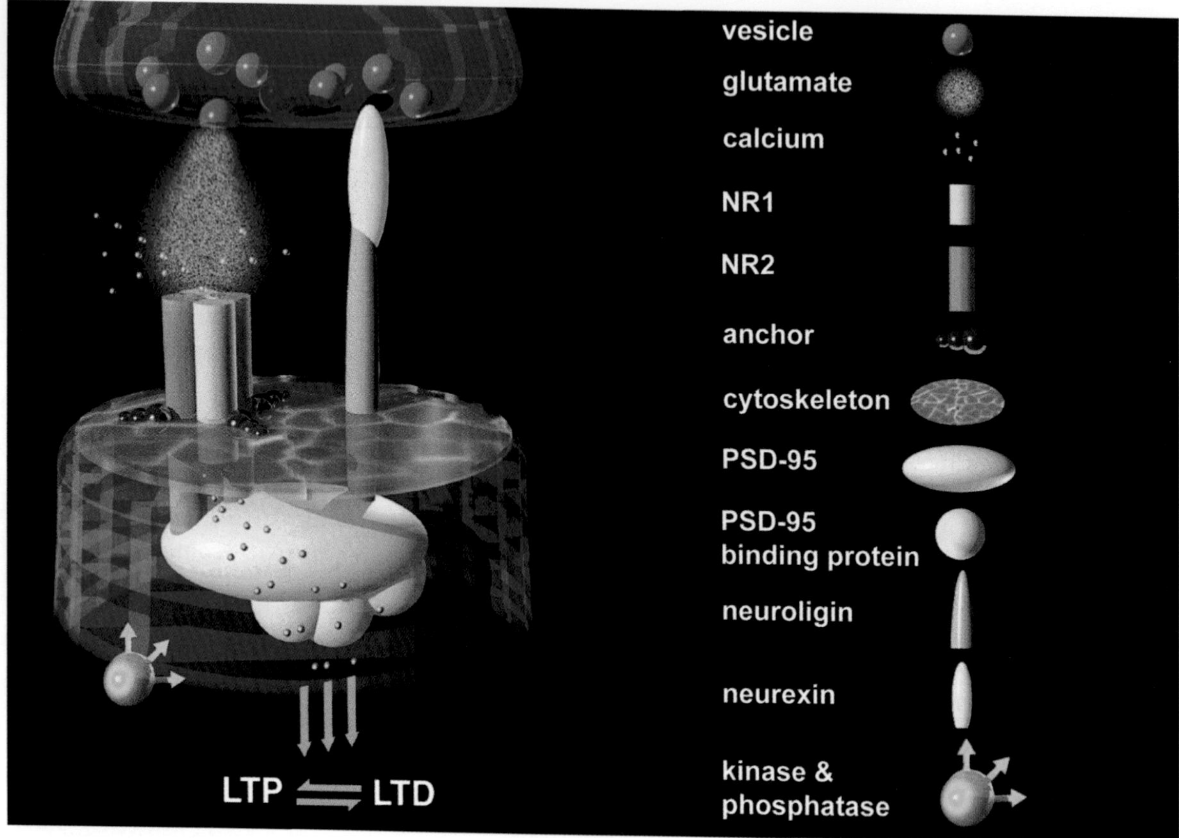

Fig. 1. Schematic diagram of the NRSC (NMDA Receptor Signalling Complex) at the synapse. The NMDA receptor/PSD-95 complex at the synapse with presynaptic terminal with synaptic vesicles (top) and dendritic spine (bottom). NR2 subunits bind PSD-95 which binds proteins including nNOS, SynGAP, GKAP/SAPAP and Neuroligin which binds the presynaptic protein Neurexin. Anchoring proteins tether the transduction complex via NMDA receptor subunits to the postsynaptic cytoskeleton. Kinases and phosphatases can phosphorylate both the channel subunits and the associated proteins. Glutamate release from the presynaptic terminal leads to Ca++ influx, which together with PSD-95 and associated proteins can regulate the balance between LTP and LTD, as indicated by arrows.

The NR1 subunit also interacts with several proteins including calmodulin (Ehlers et al., 1996), actinin (Wyszynski et al., 1997) and spectrin (Wechsler and Teichberg, 1998) which may contribute to signalling and localisation of the receptor. From a technical standpoint, it will be important to purify the NRSC and define the set of proteins associated with the NMDA receptor and PSD95. It is likely that there will be a large number of proteins, many of which were not identified by yeast 2-hybrid screens. As will be demonstrated and discussed later, the function of the complex can be addressed by creating mutations in the interacting proteins such as PSD-95.

There is extensive evidence indicating that kinases and phosphatases are required for the expression of synaptic plasticity and NMDA receptor function (Bliss and Collingridge, 1993; Bear and Malenka, 1994; Grant and O'Dell, 1994). For example, postsynaptic tyrosine (O'Dell et al., 1991) or serine/threonine kinase (Malenka et al., 1989; Malinow et al., 1989) inhibitors block the induction of LTP and mice carrying targeted mutations in Fyn tyrosine kinase (Grant et al., 1992; Kojima et al., 1997), Cam Kinase II (Silva et al., 1992; Silva et al., 1992) and PKC (Abeliovich et al., 1993) serine/threonine kinases also shift the threshold for LTP. Src family kinases also enhance NMDA receptor currents (Kohr and Seeburg, 1996; Lu et al., 1998) which may be due to tyrosine phosphorylation of NR2 subunits which are abundant tyrosine phosphoproteins (Moon et al., 1994; Lau and Huganir, 1995). The tyrosine phosphorylation of NR2 subunits is increased following the induction of LTP (Rosenblum et al., 1996; Rostas et al., 1996) which suggests that phosphorylation may contribute to the expression of LTP during the first 30 min following the activation of the NRSC. We have found that Fyn tyrosine kinase phosphorylates the NMDA receptor subunits as well as the associated NRSC proteins, and this phosphorylation may be important for NMDA receptor function (Webster et al., unpublished). The notion that phosphorylation affects the ion channel and associated proteins is likely to be important for NMDA receptor signalling.

Genetic evidence that the NRSC multiprotein complex contributes to NMDA receptor signalling in synaptic plasticity and learning has been obtained from mice carrying mutations that affect either the composition or phosphorylation of this complex. Mutant mice affecting channel subunits, associated proteins and kinases that phosphorylate the complex will now be discussed.

NMDA receptor channel subunit genes

The null (knockout) mutations generated in NR1 (Forrest et al., 1994; Li et al., 1994) resulted in homozygote mice that died shortly after birth. This unfortunately precluded analysis of NR1 function in the adult hippocampus although the analysis of these mice provided evidence that NR1 was involved in developmental plasticity of the barrell cortex (Li et al., 1994). The perinatal lethal phenotype was circumvented by delaying and restricting the knockout of NR1 to the CA1 neurones of the adult hippocampus using the Cre-loxP recombination system (Tsien et al., 1996b). These conditional knockout mice showed a phenotype very similar to that observed when NMDA receptor antagonists are perfused into the hippocampus of rats — blockade of LTP and LTD and impaired spatial learning.

Within the adult mouse forebrain, NR2A and NR2B are the predominant NR2 isoforms, yet show distinct developmental profiles of expression (Monyer et al., 1994; Nakanishi et al., 1994; Mori and Mishina, 1995; Seeburg et al., 1995). In contrast to NR2A which becomes expressed at 2 weeks of age, NR2B is coexpressed with NR1 at birth and all postnatal ages. Perhaps not surprisingly, the NR2B subunit knockout has a similar phenotype to the NR1 knockout and the NR2A knockout is viable to adulthood. Consistent with the view that NMDA receptor paricipates in normal plasticity and learning, the NR2A knockout showed reduced LTP and impaired spatial learning (Sakimura et al., 1995).

In the NR1, NR2A and NR2B knockouts, the function of the entire subunit is being examined, and therefore does not distinguish between the multiple and distinct functions of the subunits such as channel function and interaction with cytoplasmic proteins. Knockin mutations that delete the cytoplasmic carboxyl-terminal domains of NMDA receptor provide an insight into the function of these domains, which have been shown to participate in protein–protein interactions as well as provide phosphorylation sites for intracellular kinases and phosphatases. Expression of recombinant subunits lacking the cytoplasmic carboxyl-terminal domains of NR2 subunits indicate that these domains are not essential for the expression of the channel but participate in modulation of some channel functions. Mice lacking the C-terminus of NR2A or NR2B have a very similar phenotype to the respective null mutant mice (Mori et al., 1998; Sprengel et al., 1998). These results are consistent with a key signaling function for the C-terminus either in regulation of interacting proteins and kinases or

Long-term potentiation

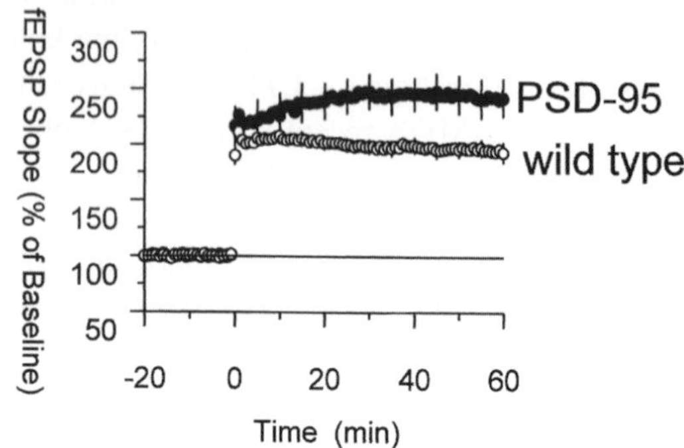

Spatial learning

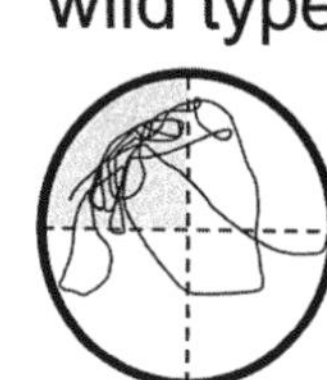

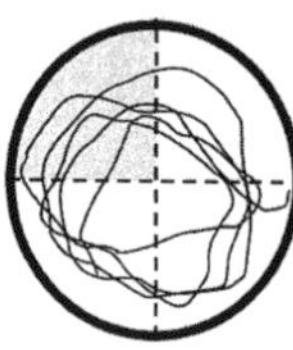

Fig. 2. Synaptic plasticity and learning in PSD-95 mutant mice. PSD-95 mutant mice showed enhanced LTP and impaired learning. Left had panel shows schematic of NRSC and indicates absence of PSD-95 protein in mutant mice. Top right panel shows LTP induced with two trains of 100 Hz stimulation delivered at time = 0. Note the larger potentiation seen 60 min after 100 Hz stimulation in slices from PSD-95 mutant animals (filled symbols) compared to slices from wild-type animals (open symbols). Lower right panel shows learning deficits in the spatial learning task in a water maze. Mice were trained to find a hidden platform in the shaded quadrant and after training were placed in the pool without the platform and respresentative swim paths for individual mice are shown. Wild-type mice search consistently in the correct location while PSD-95 mutants swim in many areas of the pool. Futher details available in Migaud et al., 1998.

in regulating synaptic localisation of the mutant subunits.

Postsynaptic density 95 mutant mice

PSD-95 binds to the terminal 4 residues of the cytoplasmic domain of NR2A and NR2B. Mice carrying a knockin mutation in PSD-95 provide strong evidence for a role in signal transduction from the NMDA receptor (Migaud et al., 1998) (Fig. 2). The frequency function of NMDA dependent LTP and LTD was shifted to produce strikingly enhanced LTP at all frequencies of synaptic stimulation tested. Moreover the time course of LTP indicated that PSD-95 was important in the first 30 min after tetanic stimulation which is consistent with it altering a downstream signalling pathway. The components of this downstream pathway are likely to include PSD-95 binding proteins such as neuronal Nitric Oxide Synthase (nNOS) (Brenman et al., 1996), SynGAP (Chen et al., 1998; Kim et al., 1998) , Neuroligin

(Irie et al., 1997), GKAP/SAPAP (Kim et al., 1997; Takeuchi et al., 1997) and other proteins.

Importantly the enhanced LTP in the PSD-95 mutant mice may arise because of a separation of distinct downstream signalling pathways from the NMDA receptor. The enhanced LTP suggests that PSD-95 couples 'negative effector' pathway to the NMDA receptor and that the 'positive effector' pathways are intact. This idea is supported by the enhancement in synaptic strength seen at low frequencies of synaptic stimulation which normally recruit 'negative effector' pathways to induce LTD.

In addition to signal transduction functions, the NMDA receptor appears to have specific mechanisms for synaptic localisation. Although earlier in vitro studies suggested PSD-95 may localise NMDA receptors to the synapse (Kim et al., 1996), the mutant mice show NMDA receptor was synaptically localised in the absence of synaptic PSD-95. This raises the possibility that PSD-95 is more important for signal transduction, and interactions independent of PSD-95 may be more important for synaptic localisation. Biochemical chararacterisation of the interactions of NMDA receptor — PSD95 and interacting proteins from the brains of the mutant mice should provide insights into the function of the NRSC in vivo.

Phosphorylation of the NRSC

The emerging picture that the NMDA receptor channel subunits form components of a much bigger multiprotein complex including PSD-95 and associated proteins raises several important issues about the function of phosphorylation. There is an extensive literature showing phosphorylation is involved with both synaptic plasticity and with modulation of NMDA receptor currents (Smart, 1997). Expression of recombinant channel subunits in non-neuronal cells with kinases and phosphatases indicate that direct phosphorylation of NMDA receptor subunits can modulate the function of these subunits. Indeed, the subunits are phosphorylated on serine, threonine and tyrosine residues invivo (Moon et al., 1994; Lau and Huganir, 1995; Omkumar et al., 1996). However, it is clear that PSD-95 associated proteins and other NMDA receptor associated proteins are also phosphorylated (Chen et al., 1998). This raises the possibility that modulation of phosphorylation invitro may lead to simultaneous changes in phosphorylation of several proteins within the NRSC.

Fyn tyrosine kinase can directly phosphorylate NR2A and NR2B as well NMDA receptor associated proteins (Webster et al., unpublished). Fyn mutant mice were reported to show impaired LTP and spatial learning (Grant et al., 1992; Kojima et al., 1997) (Fig. 3). Calcium calmodulin kinase II (CamKII) also phosphorylates NMDA receptor subunits (Omkumar et al., 1996) and the PSD-95 associated protein SynGAP (Chen et al., 1998), and CamKII mutant mice also show impairments in LTP and spatial learning (Silva et al., 1992a, 1992b). The NMDA receptor is also a substrate for serine-threonine kinases, protein kinase C, protein kinase A and serine-threonine phosphatases, which influence synaptic plasticity (Malinow et al., 1988; Malenka et al., 1989; Mulkey et al., 1993; Ehlers et al., 1995; Tong et al., 1995; Omkumar et al., 1996; Raman et al., 1996; Leonard and Hell, 1997). It seems likely that mutations in the kinases and phosphatases that act on the NRSC substrate may produce some of their effects on plasticity and learning through alterations in the properties of the NRSC. It will ultimately be necessary to show that the phosphorylation sites on the respective substrate proteins are involved by creating site specific mutations.

Phosphorylation of the NRSC may provide two interesting functions for the signalling properties of the NRSC. First, it may modify the NR channel itself and thereby regulate calcium influx. Second it may modify the associated signalling proteins so that they have altered function with respect to downstream signalling. In this way, the NRSC could act as a convergence point or integrative

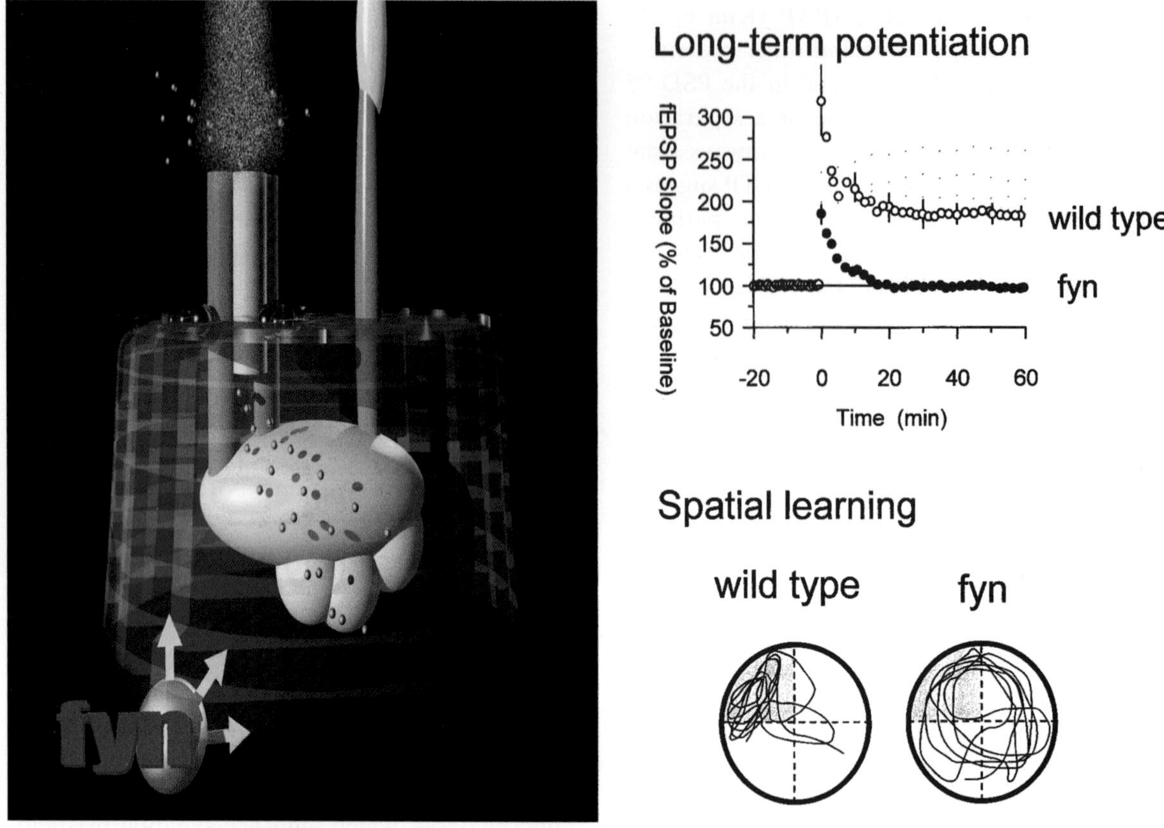

Fig. 3. Synaptic plasticity and learning in Fyn mutant mice. Fyn mutant mice show reduced LTP and impaired learning. Left panel shows schematic of NRSC and Fyn kinase which can phosphorylate NR2 subunits and some associated proteins. Top right panel shows 100 Hz induced LTP in control (open symbols) and Fyn mutant mice (closed symbols) which returns to baseline. The lower right panel shows learning deficits in the spatial version of the water maze. Mice were trained to find a hidden platform in the shaded quadrant and after training were placed in the pool without the platform and respresentative swim paths for individual mice are shown. Wild-type mice search consistently in the correct location while PSD-95 mutants swim in many areas of the pool. Figures modified from Grant et al., 1992.

structure responding to second messenger pathways controlling phosphorylation of the NRSC in response to other receptors. For example, other neurotransmitter receptors (Smart, 1997) or trophic factor (Kang and Schuman, 1995; Lessmann, 1998) receptors could activate kinases or phosphatases that act upon the NRSC. Thus the NRSC could act as a convergence point for signals from multiple enzymes and signalling pathways involved with synaptic plasticity, and PSD-95 may influence the balance of these signals and set thresholds for the NMDA receptor signalling pathways.

The NRSC as a postsynaptic transduction complex involved with learning

The demonstration that impairments in NMDA receptor mediated synaptic plasticity occur in mice carrying mutations in integral components of the NRSC (NR1, NR2A, NR2B, PSD-95) and in kinases that phosphorylate this complex (Fyn,

CaMKII, PKC, PKA) and the identification of other signalling proteins within the complex such as nNOS and SynGAP make a strong case that the complex is a key signalling device for the control of synaptic strength. Moreover, the specific mutations appear to alter the threshold between LTP and LTD as well as the level of these forms of plasticity, which suggests that the complex has a 'tuning' function. The importance of this tuning function may be in optimally regulating the strength of synapses for the purpose of information storage in neural networks.

Although it is clear that NMDA receptor blockade interferes with spatial learning and much attention has been levelled at the correlation of this blockade of LTP, it has been suggested that LTP alone is not sufficient, and that LTD may also confer advantages in storage of information in neural networks. In the case of the PSD-95 mutant mice, which showed a robust enhancement of LTP associated with a severe impairment in spatial learning, it may be that the failure to learn reflects a lack of LTD or an inappropriate amount of synaptic strengthening (Migaud et al., 1998). Thus mutations that shift the optimal threshold between LTP and LTD are associated with learning impairments (Willshaw et al., 1990; Hancock et al., 1991), which supports the view that the NRSC is involved with learning (Fig. 4).

Although the available data strongly supports an important function for the NRSC in both synaptic plasticity and learning, it may be premature to conclude that synaptic plasticity is the core mechanism of learning from these experiments. An insight obtained from studies of PSD-95 mutant mice is that there are likely to be multiple downstream pathways from the NRSC which could link to separate cellular mechanisms (Fig. 5). In the simplest case the NRSC may drive synaptic plasticity which itself is essential for learning. An alternative may be that the NRSC drives synaptic plasticity through one set of pathways which is separate to another core learning pathway. Both these models are consistent with data showing blockade of NMDA receptor interferes with learning and synaptic plasticity. The genetic dissection of these putative downstream pathways may lead to separation of the phenotypes of synaptic plasticity and learning. For example, a mutation in these downstream pathways could lead to a blockade of synaptic plasticity with no effect on learning, or vice versa. The genetic separation of phenotypes is a particularly powerful way to identify core mechanisms from associated phenomena and further understanding of various pathways and related mutations may separate plasticity and learning in important ways.

Concluding statements

Mouse mutations provide an enormously powerful tool for examining the physiological significance of specific proteins in the brain. There are now many examples where in vivo functions hitherto unknown from in vitro studies have emerged. This may be particulary important for the examination of synaptic proteins (other than channel subunits) which need to be studied in the context of mature synapses.

The concept of the NRSC acting as a signalling apparatus is analogous to the 'transducisome' as found in Drosophila photoreceptor G-protein coupled cascades where PDZ-domain proteins are important for the assembly of a signalling complex (Tsunoda et al., 1997; Huber et al., 1998; Scott and Zuker, 1998). In the fly photoreceptor the Trp calcium channel is coimmunoprecipitated with phospholipase C, protein kinase C, calmodulin, rhodopsin and a PDZ containing protein InaD (Huber et al., 1996). Channel signalling and regulation by multiprotein complexes is likely to be a general principle and apply to other channels that are associated with PDZ domain containing proteins, such as the Homer and GRIP proteins that interact with non-NMDA receptor glutamate channels (Brakeman et al., 1997; Dong et al., 1997). The PSD-95 mutant mice demonstrate that these signaling complexes play an

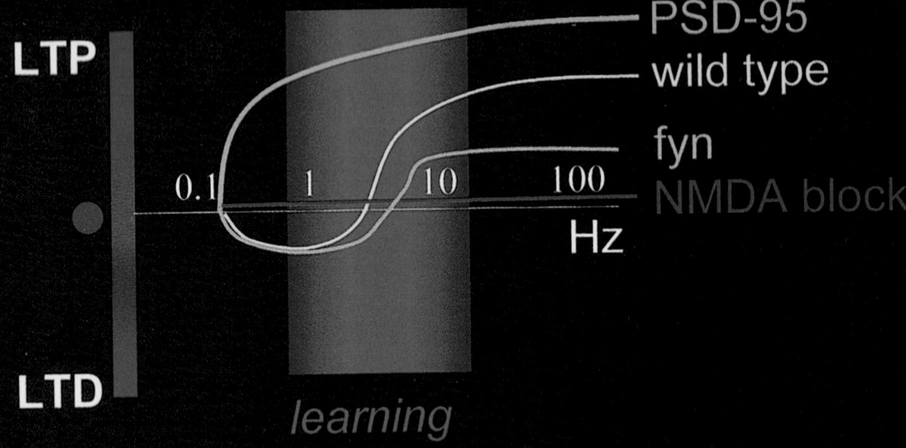
Frequency function of synaptic plasticity
LTP
LTD
0.1
1
10
100
Hz
PSD-95
wild type
fyn
NMDA block
learning

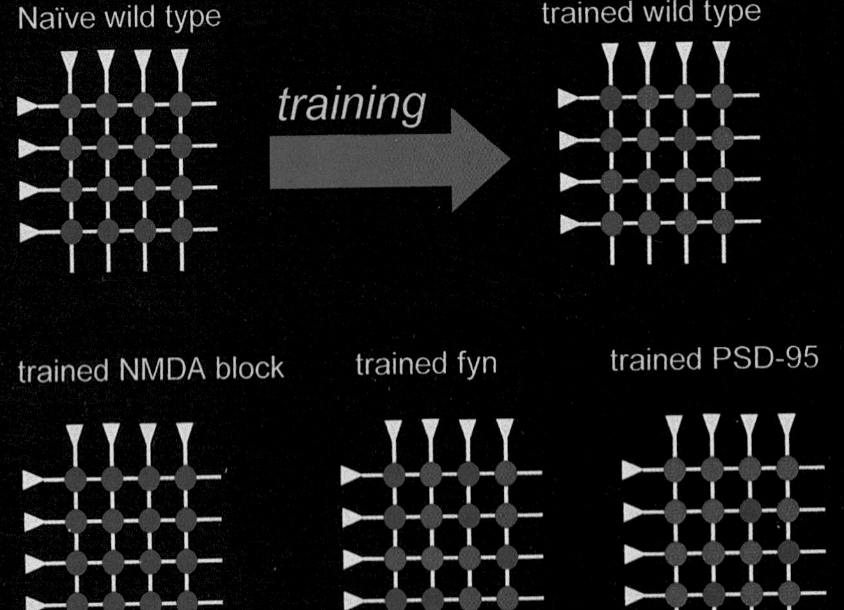
Neural network learning models
Naïve wild type
training
trained wild type
trained NMDA block
trained fyn
trained PSD-95

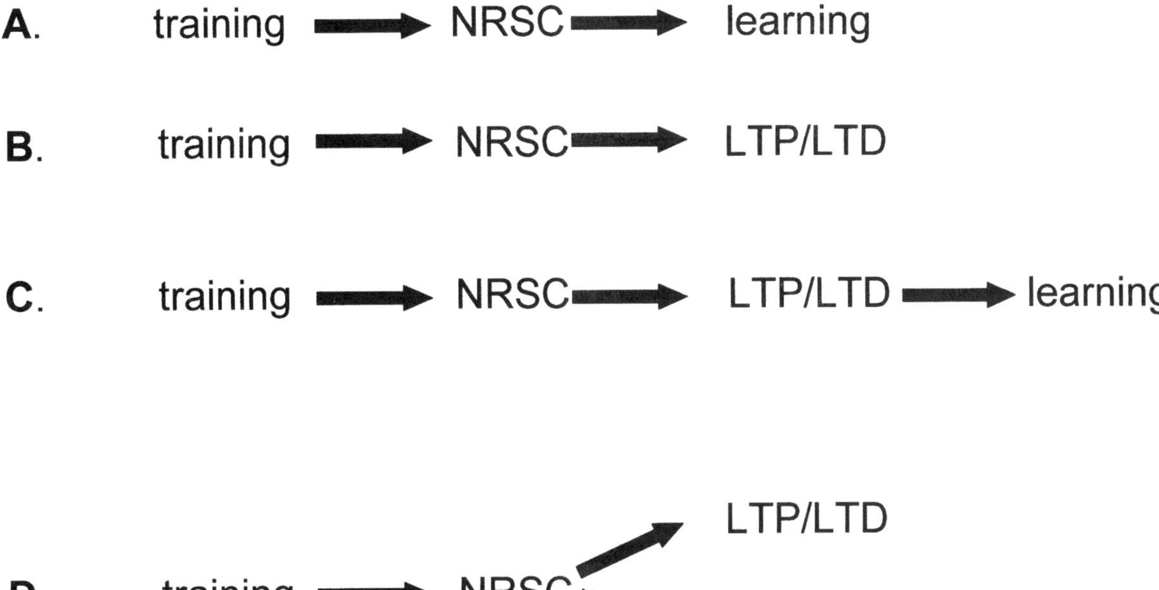

Fig. 5. Possible links between the NRSC and pathways controlling synaptic plasticity and learning. Panel (A): In the simplest case, training induces learning and this requires the function of the NRSC. Panel (B): Similarly, stimulation (training) of presynaptic pathways in the hippocampus activates the NRSC which is required for expression of either LTP or LTD. Panel (C) and (D) are possible combinations of (A) and (B), where (C) is the version most widely accepted in the field, namely: training activates the NRSC which induces LTP/LTD which modifies circuits and results in learning. In contrast, Panel (D) shows an alternative situation where training induces both LTP/LTD and learning but LTP/LTD is not the core mechanism of learning. In this case, it may be that mutations in pathways downstream from the NMDA receptor might only affect learning or might only affect LTP/LTD. Both (C) and (D) are consistent with the most tests which correlate plasticity and learning including NMDA receptor blockade and synaptic saturation experiments. Genetic experiments have the potential to show that either (C) or (D) is correct.

Fig. 4. Relationship between synaptic plasticity and network learning models in NRSC mutant mice. The frequency function for synaptic plasticity in hippocampus CA3-CA1 synapses is schematically represented where the level of postsynaptic activity (as a function of frequency of trains of synaptic stimulation, (Hz) on the X-axis is plotted against synaptic strength in the Y-axis (baseline in purple), where the corresponding degree of long-term potentiation LTP (shown in red) and long-term depression LTD (shown in blue). In wild type mice, low frequencies (around 1 Hz) can produce LTD and at higher frequencies LTP is produced (yellow plot); in PSD-95 mutant mice, there is large LTP at all frequencies of stimulation (green plot); in fyn mutants there is reduced LTP (mauve plot); and NMDA receptor blockade produces no change in synaptic strength (purple plot). During exploratory behaviour the firing frequency of hippocampal CA1 neurones is in the range indicated by the blue box and overlaps with the threshold between LTP and LTD. Therefore in wild-type mice, it would be expected that some synapses are strengthened and other weakened and this is schematically indicated in the neural network learning models. Grey triangles represent neuronal cell bodies in a 4×4 matrix and synaptic connections and are indicated by coloured circles. The colour corresponds to the Y-axis of the frequency function: purple is basal synaptic strength and red is LTP and blue is LTD. In a naïve network (labelled 'naïve wild type') the synapses are of equal basal strength (purple) and after training (blue arrow) the network shows some synapses with LTP, LTD or no change (network labelled 'trained wild type'). At the bottom are three networks that might result from training when the NRSC is disrupted by NMDA receptor blockade (no change in synaptic strengths), fyn mutants (low number of synapses becoming potentiated) and PSD-95 mutants (high number of synapses becoming potentiated). Acquisition, storage and recall of information requires an optimal number of potentiated synapses and shifts from that optimum, where there are too few or too many synapses potentiated may impair learning.

important role at the synapse and control synaptic strength.

The dissection of the postsynaptic mechanisms involved with synaptic plasticity and learning requires gene mutations, biochemistry, electrophysiology and psychological methods in an integrative setting. If the basic machinery can be identified and its function understood at the molecular level, transgenic techniques may allow this knowledge to be utilised for engineering of neural pathways. The study of these pathways may lead us to an understanding of how learning, memory and other cognitive functions work and how they came into being.

References

Abeliovich, A., Chen, C. et al. (1993) Modified hippocampal long-term potentiation in PKCgamma-mutant mice. *Cell*, 75: 1253–1262.

Bear, M.F. and Malenka, R.C. (1994) Synaptic plasticity: LTP and LTD. *Curr Opin Neurobiol*, 4(3): 389–99.

Bliss, T.V.P. and Collingridge, G.L. (1993) A synaptic model of memory; long-term potentiation in the hippocampus. *Nature*, 361: 31–39.

Brakeman, P.R., Lanahan, A.A. et al. (1997) Homer: a protein that selectively binds metabotropic glutamate receptors. *Nature*, 386: 284–288.

Brenman, J.E., Chao, D.S. et al. (1996) Interaction of nitric oxide synthase with the postsynaptic density protein PSD-95 and alpha1-syntrophin mediated by PDZ domains. *Cell*, 84: 757–767.

Brenner, S. (1973) The genetics of behaviour. *Br. Med. Bull.*, 29(3): 269–271.

Chen, H.J., Rojas-Soto, M. et al. (1998) A synaptic Ras-GTPase activating protein (p135 SynGAP) inhibited by CaM kinase II. *Neuron*, 20(5): 895–904.

Cho, K.-O., Hunt, C.A. et al. (1992) The rat brain postsynaptic density fraction contains a homolog of the drosophila discs-large tumor suppressor protein. *Neuron*, 9: 929–942.

Constantine-Paton, M. and Cline H.T. (1998) LTP and activity-dependent synaptogenesis: the more alike they are, the more different they become. *Curr. Opin. Neurobiol.*, 8(1): 139–148.

Dong, H., O'Brien, R.J. et al. (1997) GRIP: a synaptic PDZ domain-containing protein that interacts with AMPA receptors. *Nature*, 386: 279–284.

Ehlers, M.D., Tingley, W.G. et al. (1995) Regulated subcellular distribution of the NR1 subunit of the NMDA receptor. *Science*, 269(5231): 1734–1737.

Ehlers, M.D., Zhang, S. et al. (1996) Inactivation of NMDA receptors by direct interaction of calmodulin with the NR1 subunit. *Cell*, 84(5): 745–755.

Forrest, D., Yuzaki, M. et al. (1994) Targeted disruption of NMDA receptor 1 gene abolishes NMDA response and results in neonatal death. *Neuron*, 13(2): 325–238.

Grant, S.G.N. and O'Dell, T.J. (1994) Targeting tyrosine kinase genes and long-term potentiation. *Seminars in the neurosciences*, 6: 45–52.

Grant, S.G.N., O'Dell, T.J. et al. (1992) Impaired long-term potentiation, spatial learning, and hippocampal development in fyn mutant mice. *Science*, 258: 1903–1910.

Hancock, P.J.B., Smith, L.S. and Phillips, W.A. (1991) A biologically supported error-correcting learning rule. *Neural Comp.*, 3: 201–212.

Hebb, D.O. (1949) *The organization of behavior; a neuropsychological theory*. Wiley, New York.

Huber, A., Sander, P. et al. (1998) The TRP Ca2+ channel assembled in a signaling complex by the PDZ domain protein INAD is phosphorylated through the interaction with protein kinase C (ePKC). *FEBS Lett.*, 425(2): 317–322.

Huber, A., Sander, P. et al. (1996) The transient receptor potential protein (Trp), a putative store- operated Ca2+ channel essential for phosphoinositide-mediated photoreception, forms a signaling complex with NorpA, InaC and InaD. *EMBO J.*, 15(24): 7036–7045.

Irie, M., Hata, Y. et al. (1997) Binding of neuroligins to PSD-95. *Science*, 277: 1511–1515.

Kang, H.J. and Schuman, E.M. (1995) Neurotrophin-induced modulation of synaptic transmission in the adult hippocampus. *J. Physiol. Paris*, 89(1): 11–22.

Kim, E., Cho, K.O. et al. (1996) Heteromultimerization and NMDA receptor-clustering activity of chapsyn-110, a member of the PSD-95 family of proteins. *Neuron*, 17(1): 103–113.

Kim, E., Naisbitt, S. et al. (1997) GKAP, a novel synaptic protein that interacts with the guanylate kinase-like domain of the PSD-95/SAP90 family of channel clustering molecules. *J. Cell Biol.*, 136: 669–678.

Kim, J.H., Liao, D. et al. (1998) SynGAP: a synaptic RasGAP that associates with the PSD-95/SAP90 protein family. *Neuron*, 20: 683–691.

Kistner, U., Wenzel, B.M. et al. (1993) SAP90, a rat presynaptic protein related to the product of the Drosophila tumor suppressor gene dlg-A. *J. Biol. Chem.*, 268(7): 4580–4583.

Kohr, G. and Seeburg, P.H. (1996) Subtype-specific regulation of recombinant NMDA receptor-channels by protein tyrosine kinases of the src family." *J. Physiol.*, 492(2): 445–452.

Kojima, N., Wang, J. et al. (1997) Rescuing impairment of long-term potentiation in fyn-deficient mice by introducing Fyn transgene. *Proc. Natl. Acad. Sci. USA*, 94(9): 4761–4765.

Kornau, H.-C., Schenker, L.T. et al. (1995) Domain interaction between NMDA receptor subunits and the postsynaptic density protein PSD-95. *Science*, 269: 1737–1740.

Lau, L.F. and Huganir, R.L. (1995) Differential tyrosine phosphorylation of *N*-methyl-D-aspartate receptor subunits. *J. Biol. Chem.*, 270(34): 20036–20041.

Leonard, A.S. and Hell, J.W. (1997) Cyclic AMP-dependent protein kinase and protein kinase C phosphorylate *N*-methyl-D-aspartate receptors at different sites. *J. Biol. Chem.*, 272(18): 12107–12115.

Lessmann, V. (1998) Neurotrophin-dependent modulation of glutamatergic synaptic transmission in the mammalian CNS. *Gen. Pharmacol.*, 31(5): 667–674.

Li, Y., Erzurumlu, R.S. et al. (1994) Whisker-related neuronal patterns fail to develop in the trigeminal brainstem nuclei of NMDAR1 knockout mice. *Cell*, 76(3): 427–437.

Lu, Y.M., Roder, J.C. et al. (1998) Src activation in the induction of long-term potentiation in CA1 hippocampal neurons. *Science*, 279: 1363–1367.

Malenka, R.C., Kauer, J.A. et al. (1989) An essential role for postsynaptic calmodulin and protein kinase activity in long-term potentiation. *Nature*, 340(6234): 554–557.

Malinow, R., Madison, D.V. et al. (1988) Persistent protein kinase activity underlying long-term potentiation. *Nature*, 335(6193): 820–824.

Malinow, R., Schulman, H. et al. (1989) Inhibition of postsynaptic PKC or CaMKII blocks induction but not expression of LTP. *Science*, 245(4920): 862–866.

Massague, J. (1996) Crossing receptor boundaries. *Nature*, 382: 29–30.

Migaud, M., Charlesworth, P. et al. (1998) Enhanced long-term potentiation and impaired learning in mice with mutant postsynaptic density-95 protein (see comments). *Nature*, 396(6710): 433–439.

Monyer, H., Burnashev, N. et al. (1994) Developmental and regional expression in the rat brain and fuctional properties of four NMDA receptors. *Neuron*, 12: 529–540.

Moon, I.S., Apperson, M.L. et al. (1994) The major tyrosine-phosphorylated protein in the postsynaptic density fraction is *N*-methyl-D-aspartate receptor subunit 2B. *Proc Natl. Acad. Sci. USA*, 91(9): 3954–3958.

Mori, H., Manabe, T. et al. (1998) Role of the carboxy-terminal region of the GluR epsilon2 subunit in synaptic localization of the NMDA receptor channel. *Neuron*, 21(3): 571-580.

Mori, H. and Mishina, M. (1995) Structure and function of the NMDA receptor channel. *Neuropharmacology*, 34(10): 1219–1237.

Morris, R.G.M., Anderson, E. et al. (1986) Selective impairment of learning and blockade of long-term potentiation by an NMDA receptor antagonist, AP5. *Nature*, 319: 774.

Mulkey, R.M., Herron, C.E. et al. (1993) An essential role for protein phosphatases in hippocampal long-term depression. *Science*, 261(5124): 1051–1055.

Muller, B.M., Kistner, U. et al. (1996) SAP102, a novel postsynaptic protein that interacts with NMDA receptor complexes in vivo. *Neuron*, 17: 255–265.

Nakanishi, S., M. Masu, et al.,(1994). "Molecular diversity of glutamate receptors and their physiological functions." Exs **71**: 71-80.

Niethammer, M., Kim, E. et al. (1995) Interaction between the C terminus of NMDA receptor subunits and multiple members of the PSD-95 family of membrane-associated guanylate kinases. *J. Neurosci.*, 16(7): 2157–2163.

O'Dell, T.J., Kandell, E.R. et al. (1991) Long-term potentiation in the hippocampus is blocked by tyrosine kinase inhibitors. *Nature*, 353: 558–560.

Omkumar, R.V., Kiely, M.J. et al. (1996) Identification of a phosphorylation site for calcium/calmodulindependent protein kinase II in the NR2B subunit of the *N*-methyl-D-aspartate receptor. *J. Biol. Chem.*, 271(49): 31670–31678.

Raman, I.M., Tong, G. et al. (1996) Beta-adrenergic regulation of synaptic NMDA receptors by cAMP-dependent protein kinase. *Neuron*, 16(2): 415–421.

Rosenblum, K., Dudai, Y. et al. (1996) Long-term potentiation increases tyrosine phosphorylation fo the *N*-methyl-D-aspartate receptor subunit 2B in rat dentate gyrus in vivo. *Proceedings of the National Academy of Sciences of the USA*, 93: 10457–10460.

Rostas, J.A., Brent, V.A. et al. (1996) Enhanced tyrosine phosphorylation of the 2B subunit of the *N*-methyl-D-aspartate receptor in long-term potentiation. *Proc. Natl. Acad. Sci. USA*, 93(19): 10452–10456.

Sakimura, K., Kutsuwada, T. et al. (1995) Reduced hippocampal LTP and spatial learning in mice lacking NMDA receptor epsilon 1 subunit. *Nature*, 373(6510): 151–155.

Scott, K. and Zuker, C.S. (1998) Assembly of the Drosophila phototransduction cascade into a signalling complex shapes elementary responses. *Nature*, 395(6704): 805–808.

Seeburg, P.H., Burnashev, N. et al. (1995) The NMDA receptor channel: molecular design of a coincidence detector. *Recent Prog. Horm. Res.*, 50: 19–34.

Silva, A.J., Paylor, R. et al. (1992a) Impaired spatial learning in alpha-calcium-calmodulin kinase II mutant mice. *Science*, 257: 206–211.

Silva, A.J., Stevens, C.F. et al. (1992b) Deficient hippocampal long-term potentiation in alpha-calcium-calmodulin kinase II mutant mice. *Science*, 257: 201–206.

Smart, T.G. (1997) Regulation of excitatory and inhibitory neurotransmitter-gated ion channels by protein phosphorylation. *Current Opinion in Neurobiology*, 7: 358–367.

Sprengel, R., Suchanek, B. et al. (1998) Importance of the intracellular domain of NR2 subunits for NMDA receptor function in vivo. *Cell*, 92(2): 279–289.

Takeuchi, M., Hata, Y. et al. (1997) SAPAPs A family of PSD-95/SAP90-associated proteins localized at postsynaptic density. *J. Biol. Chem.*, 272(18): 11943–11951.

Tong, G., Shepher, D. et al. (1995) Synaptic desensitization of NMDA receptors by calcineurin. *Science*, 267: 1510–1512.

Tsien, J.Z., Chen, D.F. et al (1996a) Subregion and cell type-restricted gene knockout in mouse brain. *Cell*, 87: 1317–1326.

Tsien, J.Z., Huerta, P.T. et al. (1996) The essential role of hippocampal CA1 NMDA receptor-dependent synaptic plasticity in spatial memory. *Cell*, 87: 1327–1338.

Tsunoda, S., Sierralta, J. et al. (1997) A multivalent PDZ-domain protein assembles signalling complexes in a G-protein-coupled cascade. *Nature*, 388(6639): 243–249.

Wechsler, A. and Teichberg, V.I. (1998) Brain spectrin binding to the NMDA receptor is regulated by phosphorylation, calcium and calmodulin. *EMBO J.*, 17(14): 3931–3939.

Willshaw, D. and Dayan, P. (1990) Optimal plasticity from matrix memories: what goes up must come down. *Neural Comp.*, 2: 85–93.

Wyszynski, M., Lin, J. et al. (1997) Competitive binding of alpha-actinin and calmodulin to the NMDA receptor. *Nature*, 385: 439–442.

W.E. Crusio and R.T. Gerlai (Eds.)
Handbook of Molecular-Genetic Techniques for Brain and Behavior Research (Techniques in the Behavioral and Neural Sciences, Vol. 13)

CHAPTER 3.1.7

Molecular genetic analysis of glutamate receptor function in long-term potentiation in the mouse hippocampus

Zhengping Jia, You Ming Lu, Nadia Agopyan and John Roder

Samuel Lunenfeld Research Institute, Mount Sinai Hospital, 600 University Avenue, Toronto, Ont. M5G 1X5, Canada

Introduction

The ability to critically test the connection between glutamate receptors and downstream signalling pathways in LTP and learning and memory (Bliss and Collingridge, 1993) was greatly enhanced by the advent of gene targeting technology. For the last eight years various mutant mice with specific gene deletions and mutations have been made and analyzed. The following description is a summary of two null mutations in glutamate receptors mGluR5 and GluR2. The limitations of conventional genetic alterations in mice will also be discussed.

The role of mGluR5 in LTP

Class I metabotropic glutamate receptors have been postulated to play a role in synaptic plasticity (Bliss and Collingridge, 1993). In order to test the involvement of one member of this class, we have recently generated mutant mice which express no mGluR5, but normal levels of other glutamate receptors (Jia et al., 1998). The CNS revealed normal development of gross anatomical features, as shown by light microscopy of a complete set of serial sections through the entire CNS. More subtle, as yet undetected, defects may be present in theory. To examine synaptic functions we measured evoked postsynaptic field potentials (fEPSPs) in the hippocampal slice (Lu et al., 1997). Measures of presynaptic function, such as paired pulse facilitation showed normal responses in mutant CA1 neurons. The response of mutant CA1 neurons to low concentrations of ACPD (10 μM) agonist was missing, which suggests that mGluR5 may be the primary high affinity ACPD receptor in these neurons. LTP, following tetanic stimulation in mGluR5 mutants was significantly reduced in the NMDAR-dependent pathways such as the CA1 region ($177.6 \pm 10.2\%$, $n=19$ vs $150.5 \pm 5.7\%$, $n=19$) and dentate gyrus ($166.9 \pm 9.2\%$, $n=5$ vs $140.4 \pm 5.7\%$, $n=5$) of the hippocampus, whereas LTP remained intact in the mossy fibre synapses onto the CA3 region ($150 \pm 5.8\%$, $n=9$ vs 160.3 ± 14.79), an NMDAR-independent pathway. Some of the difference in CA1 LTP could lie at the level of expression of NMDA receptors since the reduction of LTP in the mutants was no longer observed 20 min after tetanus in the presence of AP5. We propose that mGluR5 plays a key regulatory role in NMDA-R dependent LTP (see below). These mutant mice were also impaired in the acquisition and utilization of spatial information in both the Morris water maze and contextual information in the fear conditioning test (see below). This is consistent with the hypothesis that

LTP in the CA1 region may underlie spatial learning and memory.

The role of mGluR5 in spatial learning and memory

If LTP is involved in learning and memory, then the mGluR5 mutants should show some impairment in learning and memory tasks, since hippocampal NMDAR-dependent LTP was significantly reduced in the mGluR5 mutant mice. Our results show that indeed mGluR5 mutant mice were significantly impaired in two different spatial learning tasks, which are known to depend on an intact hippocampus (Philips and LeDoux, 1992; Morris, 1990). In the water maze (Morris, 1990), subjects must actively escape from an aversive situation, while in contextual fear conditioning the subjects show passive fear response, (freezing) in the context in which they previously experienced shock (Philips and LeDoux, 1992). Each of the above paradigms includes nonspatial control tests in which nonspatial learning was evaluated. mGluR5 mutants performance in all nonspatial control tests was comparable to control animals. Since we did not find any changes in sensory, motor and exploratory activities in mGluR5 mutants, or in simple cue association, the impairment in their spatial learning seems to be selective. The learning acquisition impairment in the water maze also seen in the probe trial. While controls showed persistent search for the platform, spent significantly longer time in the target quadrant and more often crossed the platform side, the mutants did not stay significantly longer in the target quadrant, and never persistently searched the area of the platform site. The longer latencies of mutants were not caused by their floating in water, inactivity near the wall, or thigmotaxic swimming along the wall. Also, their motivation, visual acuity, swimming abilities and learning of the association between a single cue and the platform position was the same as in control animals when the platform was marked by a flag. Mutants also showed impairment in the fear conditioning test, which requires the association of background contextual information with the US, but showed normal learning of association between a tone (CS) and a shock (US). Animals from both groups explored a shock chamber during training at the similar rate and responded equally to US. Our results are compatible with the injection of the Class I mGluR5 antagonist, MCPG into rats, which reduced spatial learning; whereas a Class I agonist applied after learning facilitated memory recall (Riedel et al., 1996).

Hippocampal pathways involved in LTP and learning and memory

A variety of other studies have tried to test correlations between LTP and learning and memory (Martinez and Derrick, 1996). Pharmacological, physiological and surgical perturbations have been performed, and although the results are variable, in general they support a connection between hippocampal LTP and spatial learning. For example, all genetic disruptions (9/9) that impaired spatial learning and memory also impaired LTP in the Schaeffer collateral pathway to CA1. This includes the genes for fyn (Grant et al., 1992); CaMKII (Silva et al., 1992); CREB (Bourtchuladze et al., 1994); PKC (Abeliovitch et al., 1993); NMDAR2A (Sakimura et al., 1995); CaMKII-D286 (Bach et al., 1995 and Mayford et al., 1996); calbindin D-28 (Molinari et al., 1996); and NMDAR1 (Tsien et al., 1996). Gene deletion systems that excised the NMDAR1 gene only in the CA1 region, several weeks after birth, also yielded mice that showed impaired LTP and spatial learning and memory (Tsien et al., 1996). Conversely, a null mutant in the nociception gene had enhanced LTP in CA1 and enhanced learning and memory in the Morris water maze (Manabe et al., 1998). Gene disruptions that did not impair spatial learning and memory did not alter the Schaeffer collateral pathway in the spatial LTP on CA1 cells (Huang et al., 1995; Nosten-Bertrand et al., 1996). This emphasizes the importance of the Schaeffer collateral pathway in learning and memory. Two studies tended to dissociate LTP and learning and memory in the perforant path and mossy fibre pathways of the hippocampus. The disruption of the thy-1 gene disrupted granule cell LTP in the perforant pathway of the hippocampus (Nosten-Bertrand et al., 1996), whereas learning and memory were normal. Disruption of PKA selectively decreased CA3 LTP in the mossy fibre pathway, but learning and memory was also normal (Huang et al., 1995). The above evidence and our own data here in which mGluR5 mutants showed normal mossy fibre LTP, but decreased spatial learning, suggests that CA3 LTP may not be critical for spatial learning. In the absence of perforant paths, or mossy fibre LTP in the system, information relevant to spatial learning could reach CA1 directly from the entorhinal cortex (Huang, 1995). However, all these studies should be repeated in the freely-moving intact mice to confirm if LTP can be induced in various pathways.

These genetic correlations between LTP and some forms of learning and memory are also supported by a locally applied NMDA receptor (NMDA-R) antagonist (APV), which blocked induction of LTP in hippocampus, and appeared to block "spatial" learning but not a simple visual association task (Davis, 1992; Morris, 1990). However, more recent studies show that APV only impairs spatial learning in task-naive animals, while subjects pretrained in a spatial task resist this inhibition (Bannerman, 1995). Since both NMDA antagonists APV and NPC 17742, blocked LTP but failed to block spatial learning in pretrained rats (Saucier, 1995), the relationship between LTP and spatial learning is not direct. Whether or not NMDA-R is necessary for learning spatial strategies, or simply refining motor skills rather than spatial maps, requires more work. In addition, recent findings show that null mutations in PSD-95 show enhanced LTP but impaired "spatial" learning and memory in the water maze (Migaud et al., 1998).

Selective loss of the NMDA component of LTP in mGluR5 −/− mice and its rescue by PKC

The mechanisms underlying the differential expression of LTP by AMPA and NMDA receptors, are unknown, but could involve G-protein linked metabotropic glutamate receptors. To investigate this hypothesis we created mutant mice which expressed no mGluR5, but showed normal development. In the above studies of these mice we analyzed field excitatory post-synaptic potential (fEPSPs) in CA1 region of the hippocampus and found a small decrease; possibly arising from changes in the NMDAR-mediated component of synaptic transmission (Lu et al., 1997). In a more recent study we used whole-cell patch clamp recordings of evoked excitatory post-synaptic currents (EPSCs) in CA1 pyramidal neurons to identify the AMPAR- and NMDAR-mediated components of LTP (See Fig. 1 for details). Recordings from control mice following tetanus, or agonist application (IS, 3R-1-amino-cyclopentane 1,3-dicarboxylic acid) (ACPD), revealed equal (50:50) relative enhancement of the AMPA and NMDA receptor-mediated components. In contrast, CA1 neurons from mGluR5-deficient mice showed a complete loss of the NMDA receptor-mediated component of LTP (LTP_{NMDA}), but normal LTP of the AMPA receptor-mediated component (LTP_{AMPA}) (Fig. 2) (Jia et al., 1998). This selective loss of LTP_{NMDA} was seen in three different genotypic backgrounds and was apparent at all holding potentials (−70 mV to +20 mV). Furthermore, the LTP_{NMDA} deficit in mGluR5 mutant mice could be rescued by stimulating PKC with 4β-phorbol-12,13-dibutyrate (PDBu). These results suggest that PKC may couple the post-synaptic mGluR5 to the NMDA receptor potentiation during LTP, and that this signalling mechanism is distinct from LTP_{AMPA} (Fig. 1). Differential enhancement of AMPAR and NMDA receptors by mGluR5 also supports a post-synaptic locus for LTP.

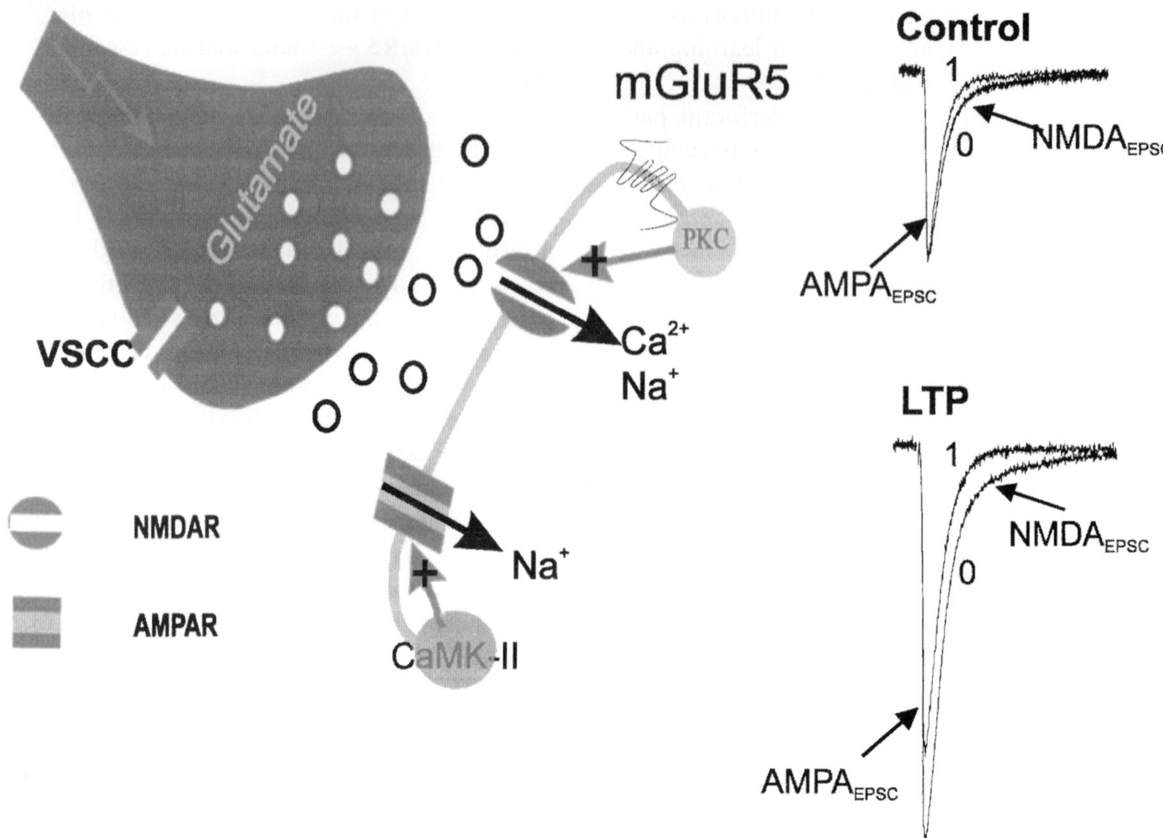

Fig. 1. *Left panel:* Postsynaptic receptor modifications in NMDAR-dependent LTP. Tetanus induces Ca^{2+} entry through NMDA channels. Ca^{2+} binds with CaM, activates CaMKII, which in turn phosphorylates and then up-regulates AMPARs, as shown by Soderling and others (Barria et al., 1997). mGluR5 is a G-coupled, glutamate receptor located at the edge of post-synaptic densities. G-protein activates PKC, which in turn generates IP3 and DAG. PKC is potentiated directly by DAG, and indirectly by IP3 inducted release of intracellular Ca^{2+} stores. PKC then phosphorylates NMDAR channels to increase single channel probability and current amplitude (J. MacDonald). This coupling could explain the loss of the NMDAR component of LTP that we observe in mGluR5 −/− mice. Note that mGluR5 is not coupled to the AMPAR. The loss in LTP_{NMDA} in the hippocampus may underlie the impairment we find in "spatial" learning in the Morris water maze and in context-dependent fear conditioning. *Right panel:* Traces show excitatory postsynaptic currents (EPSC) recorded in CA1 neurons under voltage clamp, following stimulation at 0.1 Hz of the Schaffer collateral inputs (control). The AMPAR current was taken as the downward slope (→) between 0.6–2 ms. This rapid current was blocked by the AMPAR antagonist CNQX (not shown). The late component at 100 ms (trace ○) was the NMDAR current since it was blocked by the NMDAR antagonist AP5 (trace 1). Following tetanic stimulation (LTP), both the $AMPAR_{EPSC}$ and $NMDAR_{EPSC}$ were increased.

Role of GluR2 in AMPA channels

In the mammalian central nervous system the differential distribution in mRNAs encoding different AMPAR subunits (GluR1, 2, 3, 4) account for the wide differences in Ca^{2+} permeability and gating which is seen between cells (Geiger, et al., 1995). Homomeric channels, assembled from GluR2 subunits alone, are not permeable to calcium and showed outwardly rectifying current–voltage (I–V) relations. In contrast, receptors assembled from GluR1,GluR3, and GluR4 subunits are highly permeable to calcium, and showed a doubly rectifying I–V relation (Hollmann et al., 1991; Verdoorn et al., 1991). Comparison of the amino acid sequence of the

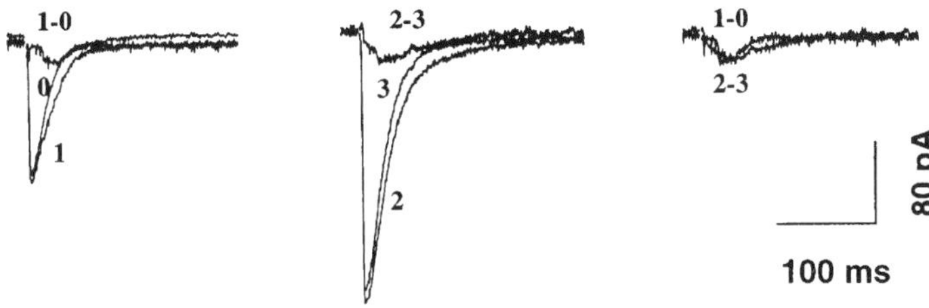

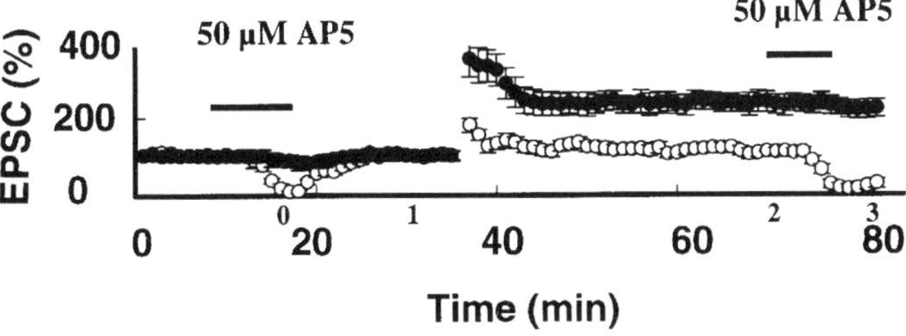

Fig. 2. Hippocampal slices were tetanized after a 35 min stable baseline was obtained. In wildtype mice we saw an equal relative enhancement of EPSC $_{AMPA}$ and $EPSC_{NMDAR}$ following tetanization (not shown). However, in homozygous mGluR5 −/− null mutants we saw a selective loss of $NMDAR_{EPSC}$ (○), whereas AMPAR EPSC (●) was normal. Bars show the addition of 50 TM AP5 which blocked the NMDAR currents. NBQX blocked the $AMPAR_{EPSC}$ (not shown). The superimposed traces were taken at the time shown on the axis. Some traces were subtracted as indicated. Holding potential −50 mV.

GluR2 subunit with that of the other subunits revealed a positively charged arginine (R) at position 586 of the transmembrane segment 2, instead of the neutral glutamine (Q) found at the homologous position in the other subunits. This position, termed the Q/R site, was subject to RNA editing (Sommer et al., 1991) and is the main site controlling divalent ion permeation of recombinant AMPA receptors (AMPARs) (Hume et al., 1991; Burnashev et al., 1992a; Dingledine et al., 1992). Co-expression of the GluR2 subunit with GluR1, 3, and 4 showed that AMPARs containing the Q/R edited GluR2 subunit exhibited lower Ca^{2+} permeability and outwardly rectifying I–V relationships. Conversely, overexpression of unedited GluR2 yielded higher Ca^{2+} permeability in the neurons of gene targeted mice (Brusa et al., 1995). The studies indicate that the channel properties of the GluR2 subunit dominate those of the heteromeric receptor complex. Therefore, one would predict that Ca^{2+} permeability in the CNS could be controlled by regulating expression of the GluR2 subunit. Indeed, the relative abundance of GluR2 mRNA correlates negatively with the Ca^{2+} permeability of native AMPARs in brain slices (Geiger et al., 1995). Thus, in principal neurons of the hippocampus (i.e. pyramidal and granule neurons), where GluR2 is highly expressed, AMPARs exhibit low Ca^{2+} permeability; whereas in cerebellar Bergman glial cells (Burnashev et al., 1992b), hippocampal basket cells, neocortical non-pyramidal cells (Koh et al., 1995) and spinal cord neurons (Gu et al., 1996), where expression of GluR2 is low or absent, AMPARs show higher Ca^{2+} permeability.

Targeted disruption of the GluR2 gene

In order to test whether GluR2 plays any role in the induction of LTP we generated mice which lacked this subunit. The procedure for generating mGluR5 and GluR2 knockouts is similar;

therefore, only that for GluR2 is described. To disrupt the GluR2 locus, an isogenic targeting vector was designed to delete transmembrane region 1 and the pore loop, which are essential for receptor (Hollmann and Heinemann, 1994). Use of isogenic DNA can enhance the frequency of homologous recombination by several fold in our hands. Our embryonic stem (ES) cell line, R1 (strain 129), was electroporated with this vector and selected in G418 and gancyclovir (Nagy, et al., 1993). Double resistant clones were screened for the desired homologous recombination by Southern blotting using a probe 5′ to exon 10. Four ES clones (from 1350 double resistant colonies) contained the targeting events, and were used to produce aggregation chimeras with CD1 morulae (Wood, 1993). Only one ES clone transmitted the GluR2 mutation through the germ line. Heterozygous mice from a CD1 x 129 cross were intercrossed to produce 477 F_2 offspring of which 22% were +/+, 54% +/− and 24 % −/−. This 1:2:1 Mendelian ratio suggests there was no embryonic lethality in the mutants. Mice from this cross were used in all experiments. Since a complex genetic background exists in 129 and CD1 strains, a large number of offspring were analyzed in order to minimize the effect of background (see below). Western blot analysis of brain protein with a GluR2 antibody (3A11), produced against the N-terminal domain (Puchalski et al., 1994), showed no detectable GluR2 protein, either full length or truncated even at a 100-fold excess loading and a 1 percent level of detection (data not shown). ECL detection showed that +/− mice had approximately 50% (10% ($n=3$) of GluR2 protein compared to +/+. In mutant mice, the level of the other glutamate receptors (GluR1, 4, 6, 7; NMDAR1, 2A, 2B) was not altered. Since there were no GluR3 antibodies available, we used one that recognizes the C terminus of both GluR2 and GluR3. GluR3 is known to be expressed at lower levels in brain than GluR2. The lane from GluR2 −/− mice showed a band indicating the presence of GluR3 in the GluR2 knockout mice. Based on the normal ratios of GluR2 and 3 it is unlikely that GluR3 is overexpressed, since the band is less intense than the GluR2 band. However, we cannot exclude the possibility that GluR3 is downregulated in the absence of GluR2. This experiment was repeated several times with similar results using enriched synaptic membranes as well as whole protein fractions from brain. Therefore, as best we can judge, the loss of GluR2 was selective and did not result in any developmental compensation or downregulation in other glutamate receptors. Whether or not the distribution of these receptors is altered is not known and is now under investigation. GluR2 −/− mice were also found to possess all major neuroanatomic loci and fiber pathways in grossly normal proportion (Jia et al., 1996). The hippocampus exhibited normal cellularity in the CA regions 1-3 and dentate gyrus. Golgi staining of a small number of individual pyramidal neurons also revealed somewhat smaller cells with less branched dendrites (higher n values are required). This was consistent with the increased input resistance measured in these neurons (Jia et al., 1996). Fibre pathways within the hippocampus, such as the perforant path also appeared normal, as did the structure of the cerebellum, the dorsal laminae of the spinal cord, and retina. Electron microscopy would provide more information on whether or not there are altered synaptic structures in these mutant mice.

Synaptic plasticity in the absence of GluR2

Long term potentiation (LTP) in the CA1 subfield of hippocampal slices was studied in 16–30 days old control and GluR2 mutant littermates. Tetanic stimulation (100 Hz for 200 ms delivered five times at 0.1 Hz) induced a long lasting (more than 2 h) increase in the synaptic strength in the control slices. The normalized EPSP slope for control mice at 60 min after tetanus was 167 15% of the average slope before stimulation ($n=20$). In slices obtained from GluR2 mutants, the extent of long lasting increase in the synaptic strength was enhanced by approximately 2-fold. The normalized EPSP slope for GluR2 mutant mice at 60 min after tetanus was $252 \pm 17.8\%$ ($P = 0.005$) of the average slope before stimulation ($n = 20$). When tetanic stimulation was delivered in the presence of bicuculline (i.e. in the absence of inhibitory GABAergic transmission), the magnitude of potentiation seen was indistinguishable from those obtained in the absence of bicuculline. At 60 min after tetanus, the normalized EPSP slope for the control and GluR2 mutant mice was $158 \pm 5.9\%$ ($n = 5$) and $295 \pm 17\%$ ($n = 12$; $P = 0.0003$), respectively. It was possible that neurons from GluR2 mutant mice were initially set at a more depressed level and as a consequence of this set value, the tetanus-induced enhancement in synaptic efficacy was greater. To investigate this possibility we first depotentiated the synapse and then induced LTP in both control and GluR2 mutant mice. Our data demonstrate that the magnitude of initial depression was not significantly different in control and GluR2 mutant mice, thereby suggesting that synapses in GluR2 mutant mice are not initially depressed. When averaged across slices, an initial transient depression following the standard low frequency stimulation (LFS) gave rise to a maintained LTD of $87.3 \pm 7.2\%$ ($n = 4$) in control slices and $82.6 \pm 9.9\%$ ($n = 8$; $P = 0.752$) in GluR2 mutant mice. Furthermore, our data demonstrated that the magnitude of enhancement in synaptic efficacy from de-potentiated level is significantly different between control and GluR2 mutant mice ($154.8 \pm 10.3\%$, $n = 4$ and $268 \pm 37.7\%$, $n = 8$; $P = 0.03$) respectively.

When the tetanic stimulation was delivered in the presence of 50–100 μM D-AP5 slices from control mice did not exhibit LTP, while those from GluR2 mice displayed a detectable residual LTP (normalized fEPSP slope at 60 min after tetanus was $154.8 \pm 10.3\%$, $n = 10$). This residual LTP in the presence of D-AP5 could be due to calcium influx via the high voltage activated calcium (HVAC) channels. However, when we induced LTP in the presence of D-AP5 and nifedipine (a HVAC channel antagonist), the normalized field EPSP slope at 60 min after tetanus was $132.7 \pm 7.6\%$ ($n = 8$), which was not significantly different from those recorded in D-AP5 alone ($P = 0.05$). Hence these results strongly suggest that GluR2 mutant mice have approximately 150% of enhancement in synaptic efficacy in the absence of NMDA and HVCA channel contribution. On the other hand, 5 μM NBQX or CNQX application, 20 min after tetanus, abolished all synaptic responses, implying that the residual LTP in GluR2 mutant mice is mediated by the AMPARs. Post-synaptic injection of the Ca^{2+} chelator BAPTA also blocked LTP induction in GluR2 −/− CA1 neurons, which suggests that Ca entry via AMPARs lacking GluR2 led to LTP (Mainen et al., 1998).

LTP is a saturable phenomenon (Bliss and Collingridge, 1993). If the LTP induction mechanism in GluR2 mutant mice is not altered, one would expect to see saturation of LTP after a plateau is reached. However, whereas LTP in slices from control mice saturates, those from GluR2 mutant mice can progressively produce more LTP until spreading depression develops. In 12 slices obtained from GluR2 mutant mice, the normalized fEPSP slope measured 5 min after the first tetanus was $214 \pm 18.2\%$, and that after the fifth tetanus was $397 \pm 21.7\%$, whereas in 5 slices obtained from control mice, the normalized fEPSP slopes were $150 \pm 8.6\%$ and $188 \pm 20.4\%$ respectively. These results, therefore, suggest that in GluR2 mutant mice, the mechanisms underlying LTP are modified.

Our data are compatible with the suggestion that the Ca^{2+}-permeable AMPARs can induce LTP. Hence, in the absence of GluR2, both LTP and Ca^{2+}-permeability (see below) are increased. Antagonists of AMPA receptors block all the enhanced LTP seen in GluR2 mutant mice. The data also show that Ca^{2+} influx via L-type calcium channels is not involved in LTP induction in GluR2 mutant mice. Therefore, the GluR2 subunit may play a crucial role in regulating both calcium influx and LTP.

Relative Ca^{2+} permeability of AMPA receptors

It has been shown in vitro that AMPARs containing the Q/R edited GluR2 subunit exhibit lower Ca^{2+} permeability, and distinct gating properties, compared to receptor channels assembled without this subunit (Hollmann and Heinemann, 1994; Burnashev et al., 1992a). As expected, the loss of this GluR2 subunit in individual CA1 pyramidal neurons from GluR2 mutant mice demonstrated a 9-fold increase in relative Ca^{2+} permeability following kainate application, when compared to that of control mice. Indeed, this shift was similar in magnitude to that observed for calcium-permeable NMDA receptors (Koh et al., 1995). These results, together with observations in hippocampal slices from GluR2 editing-deficient mice (Brusa et al., 1995), support a crucial role for the GluR2 subunit in inhibiting Ca^{2+} influx via AMPARs in vivo.

Therefore, our data raise the possibility that calcium influx via AMPARs devoid of GluR2 subunit is sufficient to induce LTP, in a NMDA-independent manner. A recent study on spinal cord neurons, where AMPA receptors are naturally devoid of GluR2 subunits, and therefore calcium permeable, showed that synaptic strength varied in an activity-dependent way (Gu et al., 1996). Furthermore, it is interesting to note that LTP in GluR2 mutant slices did not saturate, thus indicating that the mechanism(s) underlying LTP induction is altered. One reason for normal saturable LTP could be that Ca^{2+} influx through the NMDAR channel activates calmodulin, which is known to bind and inactivate the NMDAR channel (Ehlers et al., 1996). Therefore, normal LTP may be self-limiting. However, in GluR2 −/− mice the additional Ca^{2+} influx will not inactivate the AMPAR channel, thereby leading to an ever-increasing LTP at higher stimulation intensities. Alternatively, downstream signalling components may limit LTP in normal animals. In GluR2 −/− mice the additional Ca^{2+} could activate kinases that are normally rate-limiting, to generate nonsaturable LTP.

Our data raise the possibility that activation of Ca^{2+}-permeable AMPARs is sufficient to induce LTP; however, further experiments are needed to elucidate the mechanism(s) underlying this phenomenon. The generation of these GluR2 mutant mice provides one valuable model for determining the in vivo roles of the AMPARs and Ca^{2+} in synaptic physiology and behaviour. Our results indicate that GluR2 is not required for LTP but seems to play a negative role in its regulation. Therefore, GluR2 may play an important role in synaptic plasticity by regulating its expression in development.

Controls for genetic background

LTP in the CA1 region of the parental 129 and CD1 strains showed no significant differences (Jia et al., 1996). Therefore, the enhanced LTP seen in our genotyped GluR2 −/− mice from the F_2 cross could not be due solely to the inheritance of 129 genes conferring high LTP. To rule out the possibility that GluR2-linked 129 genes could contribute to the CA1-LTP phenotype in the context of a CD1 genetic background (see Chapter 3.1.10), we carried out experiments analyzing CA1 LTP distribution among an F_2 population ($n=36$) derived from wildtype F_1 intercrosses (i.e. CD1 x 129). The frequency histogram showed that the distribution of LTP was normal ($p=0.31$ Kolmogorov–Smirnov goodness-of-fit test). For the wildtype F_2 129sv X CD1 population, there is 0.99 probability that a randomly chosen subject from the population will yield LTP between 118 and 218%. If single recessive wildtype CD1 and 129 genes interact in a simple way as to mimic the high LTP seen in GluR2 −/−H, seen in GluR2 −/− mice from several litters compared to +/+ ($P<0.00001$). In this experiments there is no possibility that CD1 background genes are playing any role, since they were not used in this cross. These results show that enhanced LTP is seen in the absence of GluR2 on different genetic backgrounds. We feel these results together argue against the possibility that the genetic background is the only contributing factor in enhanced LTP in our GluR2 −/− mice.

Recruitment of silent synapses

Roberto Malinow has recently shown that in CA1 HPP-Spermine (Washburn et al., 1996) reduced (50%) fEPSP in our GluR2 −/− mice, but had no effect on wt neurons (Mainen et al., 1998). The block was use-dependent and reversible. The higher the stimulus frequency the faster the rate of HPP-SP blocking. Presynaptic enhancement of release probability (induced by PPF) increased the HPP-SP block by activating AMPARs more frequently. Postsynaptic enhancement (by aniracetam) of mean channel open time increased the block by HPP-SP. LTP on the other hand, did not show a greater block by HPP-SP. Therefore, LTP does not increase the rate at which synapses transmit and open postsynaptic AMPARs. This data supports the notion that LTP recruits new AMPARs. Our GluR2 −/− mice had the same quantal content as wt. HC slices transfected with GluR2 cells rescued the outward rectifying response to KA. (Malinow, personal comm.)

Summary

In GluR2 mutants, LTP in the CA1 region of hippocampal slices was markedly enhanced (2-fold) and non-saturating, whereas neuronal excitability and paired-pulse facilitation were normal. The 9-fold increase in Ca^{2+} permeability, in response to kainate application, suggests one possible mechanism for enhanced showed reduced exploration and impaired motor coordination. These results suggest an important role for GluR2 in regulating synaptic plasticity and behaviour.

Inducible gene expression

Gene function during mammalian development has been studied by making irreversible changes to the genome. This approach has a major drawback in that the function of the gene in question must be deduced from the phenotype of animals that have been deficient for the product of the disrupted gene throughout ontogeny. Compensation for the loss of the gene product could yield an apparently unaltered phenotype. Alternatively, the changes in the regulation of other genes could yield a misleading phenotype. If the genetic manipulation results in embryonic or neonatal lethality, gene function at later stages of development cannot be analyzed. It would thus be highly advantageous if the expression of a particular gene could be restricted (both temporally and spatially) through the use of an inducible genetic system (Chapter 3.1.4; Wilson and Tonegawa, 1997; Gingrich and Roder, 1998). The knockout of some glutamate receptors (GluR2, for example) resulted in generalized motor impairments that led to the inability of the mice to perform in behavioral tasks designed to measure learning and memory (Jia et al., 1996; Gerlai et al., 1998). Consequently, there has been considerable interest in designing an inducible, preferably neural-specific, genetic system for use in transgenic mice to circumvent the problem of neonatal lethality so that LTP and behaviour of adult animals lacking particular NMDA receptor subunits could be studied (see Chapters 3.1.3 and 3.1.4). These inducible and region-specific manipulations of a gene will now be applied to GluR2 to address more conclusively the role of this subunit after development and in specific regions of the brain.

Availability

Both the GluR2 and mGluR5 mutants have been deposited in the Jackson Lab's Induced Mutant Resource (www.jax.org/resources/documents/imr) and are available to users upon request. Genetic Resources Dept., Jackson Labs, 600 Main Street, Bar Harbor, ME 04609-1500, USA. Tel: (207) 288-6230; Fax: (207) 288-6149; e-mail: ccl@.aretha.jax.org

Acknowledgements

This work was supported by grants to John Roder from the Medical Research Council of Canada, the Ontario Mental Health Foundation and the Networks of Centres of Excellence in Neuroscience. The following people contributed key data: Chris Janus, Robert Gerlai, Franco Taverna, Jeff Henderson, Wanda-Abramow-Newerly and Martin Wojtowicz.

References

Abeliovitch, A., Paylor, R., Chen, C., Kim, J., Wehner, J.M. and Tonegawa, S. (1993) PKC -mutant mice exhibit mild deficits in spatial and contextual learning. *Cell*, 75: 1263–1271.

Ambros-Ingerson, J. and Lynch, G. (1993) Channel-gating kinetics and synaptic efficacy: a hypothesis for expression of long-term potentiation. *Proc. Natl. Acad. Sci. USA*, 90: 7903–7907.

Bach, M.E., Hawkins, R.D., Osuman, M., Kaendel, E.R. and Mayford, M. (1995) Impairment of spatial but not contextual memory in CaMKII mutant mice with a selective loss of hippocampal LTP in the range and theta frequency. *Cell*, 81: 905–910.

Bannerman, D.M., Good, M.A., Butcher, S.P., Ramsay, M. and Morris, R.G. (1995) Distinct components of spatial learning revealed by prior training and NMDA receptor bockade. *Nature*, 378: 182–186.

Barria, A., Muller, D., Derkach, V., Griffith, L.C. and Soderling, T.R. (1997) Regulatory phosphorylation of AMPA-type glutamate receptors by CaMKII during long-term potentiation. *Science*, 276: 2042–2045.

Bolshakov, V.Y. and Siegelbaum, S.A. (1995) Regulation of hippocampal transmitter release during development and long-term potentiation. *Science*, 269: 1730–1734.

Bourtchuladze, R., Frenguelli, B., Blendy, J., Cioffi, D., Schutz, G. and Silva, A.J. (1994) Deficient long term memory in mice with a targeted mutation of the cAMP-responsive element kindling protein. *Cell*, 79: 59–68.

Brusa, R., Zimmerman, F., Koh, D.S., Feldmeyer, D., Gass, P., Seeburg, P.H. and Sprengel, R. (1995) Early-onset epilepsy and postnatal lethality associated with an editing-deficient GluR-B allele in mice. *Science*, 270: 1677–1680

Burnashev, N., Monyer, H., Seeburg, P.H. and Sakmann, B. (1992a) Divalent ion permeability of AMPA receptor channels is dominated by the edited form of a single subunit. *Neuron*, 8: 189–198.

Burnashev, N., Khodorova, A., Jonas, P., Helm, P.J., Wisden, W., Monyer, H., Seeburg, P.H. and Sakmann, B. (1992b) Ca^{2+} permeable AMPA/KA receptors in fusiform cerebellar glial cells. *Science*, 256: 1566–1570.

Collingridge, G.L. and Singer, W. (1990) Excitatory amino acid receptors and synaptic plasticity. *Trends Pharmacol. Sci.*, 11: 290–296

Conquet, F., Bashir, Z.I., Davies, C.H., Daniel, H., Ferraguti, F., Bordi, F., Franz-Bacon, K., Reggiani, A., Matarese, V. and Conde, F. (1994) Motor deficit and impairment of synaptic plasticity in mice lacking mGluR1. *Nature*, 372: 237–243.

Davis, S., Butcher, S.P. and Morris, R.G. (1992) The NMDA receptor antagonist D-2-amino-phosphonopentanoate (D-AP5) impairs spatial learning and LTP in vivo at intracerebral concentrations comparable to those that block LTP in vitro. *J. Neurosci.*, 12(1): 21–34.

Dingledine, R., Hume, R.I. and Heinemann, S.F. (1992) Structural determinant of barium permeation and rectification in non-NMDA glutamate receptor channels. *J. Neurosci.*, 12: 4080–4087

Ehlers, M.D., Zhang, S., Bernhardt, J.P. and Huganir, R.L. (1996) Inactivation of NMDA receptors by direct interaction of calmodulin with the NR1 subunit. *Cell*, 84: 746–755.

Geiger, J.R.P., Melcher, T., Koh, D.S., Sakman, B., Seeburg, P.H., Jonas, P. and Monyer, H. (1995) Relative abundance of subunit mRNAs determines gating and Ca^{2+} permeability of AMPA receptors in principal neurons and interneurons in rat CNS. *Neuron*, 15: 193–204.

Gerlai, R., Henderson, J., Roder, J. and Jia, Z. (1998) Multiple behavioral abnormalities in GluR2 null mice exhibiting enhanced LTP. *Behav. Brain Res.*, 95(1): 37–45.

Gingrich, J. and Roder, J.C. (1998) Inducible gene expression in the nervous system of transgenic mice. *Ann. Rev. Neurosci.*, 21: 377–405.

Grant, S.G., O'Dell, T., Karl, K., Sleio, P.C., Soriano, P. and Kandel, E. (1992) Impaired long-term term potentiation, spatial learning and hippocampal development in fyn mutant mice. *Science*, 258: 1903–1910.

Greengard, P., Jen, J., Nairn, A.C. and Stevens, C.F. (1991) Enhancement of the glutamate response by cAMP-dependent protein kinase in hippocampal neurons. *Science*, 253: 1135–1138.

Gu, J.G., Albuquerque, C., Lee, C.J. and Macdermott, A.B. (1996) Synaptic strengthening through activation of Ca2+-permeable AMPA receptors. *Nature*, 381: 793–796.

Hollmann, M. and Heinemann, S. (1994) Cloned glutamate receptors. *Ann. Rev. Neurosci.*, 17: 31–108.

Hollmann, M., Hartley, M. and Heinemann, S. (1991) Ca^{2+} permeability of KA-AMPA glutamate receptor channels depends on subunit composition. *Science*, 52: 81–85.

Huang, Y.Y., Kendel, E.R., Varshavoksy, L., Brandon, E.P., Ming, Q., Idzerdo, E.P., McKnight, S.,Bourtchuladze, R. (1995) A genetic test of the effects of mutations in PKA on mossy fibre LTP and its relation to spatial and contextual learning. *Cell*, 83: 1211–1222.

Hume, I.R., Dingledine, R. and Heinemann, S.F. (1991) Identification of a site in glutamate receptor subunit that controls calcium permeability. *Science*, 253: 1028–1031.

Jia, Z.P., Agopyan N., Miu, P., Xiong, Z., Henderson, J., Gerlai, R., Taverna, F., Velumian, A., MacDonald, J., Carlen, P., Abramow-Newerly, W., Roder, J. (1996) Enhanced LTP in mice deficient in the AMPA receptor, GluR2. *Neuron*, 17: 945–956.

Jia, Z., Lu. Y.M., Henderson, J., Taverna, F., Romano, C., Abramow-Newerly, W., Wojtowicz, M. and Roder, J.C. (1998) Selective abolition of the NMDA component of long term potentiation in mice lacking mGluR5. Learning and Memory, 5: 331–343.

Koh, D.S., Geiger, J.R.P., Jonas, P. and Sakman, B. (1995) Ca^{2+} permeable AMPA and NMDA receptor channels in basket cells of rat hippocampal dendate gyrus. *J. Physiol.*, 485(2): 383–402.

Köhr, G. and Seeburg, P.H. (1996) Subtype-specific regulation of recombinant NMDA receptor-channels by protein tyrosine kinases of the src family. *J. Physiol. (Lond.)*, 492: 445(2): 445–452.

Kojima, N., Wang, J., Mansuy, I.M., Grant, S.G., Mayford, M. and Kandel, E. (1997) Rescuing impairment of long-term potentiation in fyn-deficient mice by introducing Fyn transgene. *Proc. Natl. Acad. Sci. USA*, 94(9): 4761–4765.

Kullman, D.M. and Nicoll, R.A. (1992) Long-term potentiation is associated with increases in quantal content and quantal amplitude. *Nature*, 357: 240–244.

Larkman, A., Hannay, T., Stratford, K. and Jack, J. (1992) Presynaptic release probability influences the locus of long-term potentiation. *Nature*, 360: 70–73.

Liao, D., Jones, A. and Malinow. R. (1992) Activation of postsynaptically silent synapses during pairing-induced LTP in CA1 region of the hippocampal slice. *Nature*, 375: 400–404.

Lisman, J. (1994) The CaM kinase II hypothesis for the storage of synaptic memory. *Trends Neurosci.*, 17(10): 406–412.

Lu, Y.M., Jia, Z., Janus, C., Henderson, J.T., Gerlai, R., Wojtowicz, J.M., Roder, J.C. (1997) Mice lacking metabotropic glutamate receptor 5 show impaired learning and reduced CA1 long-term potentiation (LTP) but normal CA3 LTP. *J. Neurosci.*, 17(13): 5196–205.

Mainen, Z.F., Jia, Z., Roder, J.C. and Malinow, R. (1998) Use-dependent AMPA receptor blockade indicates recruitment rather than increased use during LTP in CA1 hippocampus. *Nat. Neurosci.*, 1(7): 579–586.

Malenka, R.C., Kauer, J.A., Perkel, D.J., Mauk, M.D., Kelly, P.T., Nicoll, R.A. and Waxham, M.N. (1989) An essential role for postsynaptic calmodulin and protein kinase activity in long-term potentiation. *Nature*, 340: 554–557.

Malenka, RC., Kauer, J.A., Zucker, R.S. and Nicoll, R.A. (1988) Poststynaptic calcium is sufficient for potentiation of hippocampal synaptic transmission. *Science*, 242: 81–84.

Malenka, R.C., Lancaster, B. and Zucker, R.S. (1992) Temporal limits on the rise in postsynaptic calcium reuired for the induction of long-term potentiation. *Neuron*, 9: 121–128.

Malinow, R. and Tsien, R.W. (1990) Presynaptic enhancement shown by whole-cell recordings of long-term potentiation in hippocampal slices. *Nature*, 346: 177–180.

Malinow, R., Schulman, H. and Tsien, R.W. (1989) Inhibition of postsynaptic PKC or CaMKII blocks induction but not expression of LTP. *Science*, 245: 862–866.

Manabe, T., Renner, P. and Nicoll, R.A. (1992) Postsynaptic contribution to long-term Potentiation revealed by the analysis of miniature synaptic currents. *Nature*, 355: 50–55.

Manabe, T., Noda, Y., Mamiya, T., Katagiri, H., Houtani, T., Nishi, M., Noda, T., Takahashi, T., Sugimoto, T., Nabeshima, T. and Takeshima, H. (1998) Facilitation of long-term potentiation and memory in mice lacking nociceptin receptors. *Nature*, 394: 577–580.

Martinez, R. and Derrick, B. (1996) Long-term potentiation and learning. *Ann. Rev. Psychol.*, 47: 173–203.

Mayford, M., Bach, M.E., Huang, Y.Y., Wang, L., Hawkins, R. and Kandel, E. 1996) Control of memory formation through regulated expression of a CaMKII transgene. *Science*, 274: 1678–1683.

Migaud, M., Charlesworth, P., Dempster, M., Webster, L.C. et al. (1998) Enhanced long-term potentiation and impaired learning in mice with mutant postsynaptic density-95 protein. *Nature*, 396: 433–439.

Molinari, S., Battini, R., Ferrari, S., Pozzi, L., Koillcross, A., Robbins, T., Jouvenceau, A., Billard Dutor, P., Lamour, Y., Baker, W., Cox, H. and Emson, P. (1996) Deficits in memory and hippocampal LTP in mice with reduced calbindin D28K-expression. *Proc. Natl. Acad. Sci. USA*, 93: 8028–8033.

Morris, R.G.M. (1990) Toward a representative hypothesis of the role of hippocampal synaptic plasticity in spatial and other forms of learning. *Cold Spring Harbor Symposia on Quantitative Biology*, 55: 161–173.

Nagy, A., Rossant, J., Nagy, R., Abramow-Newerly, W. and Roder, J.C. (1993) Derivation of completely cell culture-derived mice from early-passage embryonic stem cells. *PNAS*, 90: 8424–8428.

Nicoll, R.A. and Malenka, R.C. (1995) Contrasting properties of two forms of long-term potentiation in the hippocampus. *Nature*, 377: 115–118.

Nosten-Bertrand, M., Errington, M.L., Murphy, K.P.S., Tokugawa, Y., Barloni, E., Kozlova, E., Michalovitch, D., Morris, R.G.M., Silver, J., Stewart, C.J., Bliss, T.V.P. and Morris, R.J. (1996) Normal spatial learning despite regional inhibition of LTP in mice lacking thy 1. *Nature*, 379: 826–829.

Philips, R.G. and LeDoux, J.E. (1992) Differential contribution of amygdala and hippocampus to cued and contextual fear conditioning. *Behav. Neurosci.*, 106: 274–285.

Riedel, G. (1996) Function of metabotropic glutamate receptors in learning and memory. *TINS*, 19: 219–224.

Sakimura, K., Kutsuwasle, T., Ho, I., Manabe, T., Takayama, C., Kushiya, E., Yago, T., Azawa, S., Inouye, Y., Suizyama, H. and Mishina, M. (1995) Reduced hippocampal LTP and spatial learning in mice lacking NMDA receptor /subunit. *Nature*, 373: 151–155.

Saucier, D. and Cain, D.P. (1995) Spatial learning without NMDA receptor-dependent long-term potentiation. *Nature*, 378: 186–189.

Silva, A.J., Stevens, C.F., Tonegawa, S. and Wang, Y. (1992) Deficient hippocampal long-term potentiation in -calcium-calmodulin kinase II mutant mice. *Science*, 257: 201–206.

Sommer, B., Kohler, M., Sprengel, R. and Seeburg, P.H. (1991) RNA editing in brain controls a determinant of ion flow in glutamate gated channel. *Cell*, 67: 11–19.

Stevens, C.F. and Wang, Y. (1994) Changes in reliability of synaptic function as a mechanism for plasticity. *Nature*, 371: 704–707.

Tsien, J.Z., Chen, D.F., Gerber, D., Tom, C., Mercer, E.H., Anderson, J.A., Mayford, M., Kandel, E.R. and Tonegawa, S. (1996a) Subregion-and cell type-restricted gene knockout in mouse brain. *Cell*, 87: 1317–1326.

Tsien, J.Z., Huerta, P.T. and Tonegawa, S. (1996b) The essential role of hippocampal CA1 NMDA receptor-dependent synaptic plasticity in spatial memory. *Cell*, 87: 1327–1338.

Verdoorn, T.A., Burnashev, N., Monyer, H., Seeburg, P.H. and Sakman, B. (1991) Structural determinants of ion flow through recombinant glutamate receptor channels. *Science*, 252: 1715–1718.

Washburn, M.S. and Dingledine, R. (1996) Block of alpha-amino-3-hydroxy-5-methyl-4-isoxazolepropionic acid (AMPA) receptors by polyamines and polyamine toxins. *J. Pharmacol. Exp. Therapeutics*, 278: 669–678.

Wilson, M. and Tonegawa, S. (1997) Synaptic plasticity, place cells and spatial memory: study with second generation knockouts. *Trends Neurosci.*, 20: 102–106.

Wood, S.A., Allen, N.D., Rossant, J., Auerbach, A. and Nagy, A. (1993) Non-injection methods for the production of embryonic stem cell embryo chimeras. *Nature*, 365: 87–89.

Wylie, D.J., Manabe, T. and Nicoll, R.A. (1994) A rise in postsynaptic Ca^{2+} potentiates miniature excitatory postsynaptic currents and AMPA responses in hippocampal neurons. *Neuron*, 12: 127–138.

W.E. Crusio and R.T. Gerlai (Eds.)
Handbook of Molecular-Genetic Techniques for Brain and Behavior Research (Techniques in the Behavioral and Neural Sciences, Vol. 13)

CHAPTER 3.1.8

Targeting aggression in mice

Randy J. Nelson and Lance J. Kriegsfeld

Departments of Psychology, Neuroscience, and Population Dynamics, The Johns Hopkins University, Baltimore, MD 21218-2686, USA

Introduction

Studies of mice with targeted disruption (knockout) of specific genes have been informative for behavioral neuroscience, generally, and aggressive behavior, specifically. Behavioral studies using knockouts have confirmed and extended our understanding of the mechanisms underlying aggression. For example, targeted deletion of genes that encode proteins essential to maintain brain serotonin concentrations or serotonin receptor numbers in the normal range affects aggression in ways often predicted by pharmacological manipulations. An unexpected benefit of the knockout technology is the serendipitous discovery of novel mechanisms involved in aggression. For example, targeted deletion of the gene encoding the neuronal isoform of nitric oxide synthase (nNOS) results in a behavioral phenotype exhibiting elevated aggression. Specific pharmacological manipulation of nNOS has confirmed the results from the nNOS gene knockout studies. Taken together, the two approaches of studying the behavioral phenotypes of knockout mice should continue to enhance our understanding of the mechanisms of aggression. The advent of so-called inducible knockouts should provide additional insights into the mechanisms of aggression in the absence of developmental confounds.

The use of mice with targeted disruption of specific genes has been a very important new tool in studies of behavioral neuroscience, generally, and aggressive behavior, specifically. Aggressive behavior (see Chapter 4.8 for behavioral tests) in knockout animals has been investigated for essentially one of two reasons. First, after specific genes have been deleted, animal caretakers or research personnel observe aggressive behavior in the animals' home cage, and researchers design studies to evaluate and quantify this enhanced aggressive behavior empirically. These types of studies often allow researchers to relate previously known mechanisms of aggression with the effects of the missing gene or gene product. In general, these types of knockout studies have confirmed and extended previous pharmacological and ablation studies. Additionally, pharmacological agents are often non-selective whereas knockout technology accomplishes a very precise removal of one particular molecule. Second, unexpected aggressive behavioral alterations are often revealed in knockout mice in which the missing gene product(s) were not previously associated with aggression, and aggressive behavior abnormalities are uncovered while evaluating animals on a number of sensorimotor and behavioral tasks. For example, targeted deletion of the genes for nitric oxide synthase (NOS), in addition to providing insights into the mechanisms involved in learning and memory, has suggested a role of nitric oxide (or a downstream mechanism) in aggressive behavior. Thus, these serendipitous findings have been valuable in revealing new mechanisms underlying aggression. When unexpected aggressiveness is observed as a behavioral phenotype of a knockout animal, careful behavioral studies must be conducted to rule out sensorimotor or other behavioral deficits that might contribute to the etiology of the aggression.

Although significant advances have been made in understanding the neuroendocrine and genetic factors regulating aggressive behavior, methodological problems still exist regarding a satisfactory definition of aggressive behavior patterns. Aggressive behavior is not a unitary phenomenon, but includes a range of behavior patterns. Typically, researchers record behaviors ranging from attacks (with and without biting), wrestling, boxing, pursuit, and tail rattling (Maxson, 1996). A number of experimental protocols have been designed to elicit aggressive behavior in rodents (see Chapter 4.8), and each of these experimental models is presumed to evaluate a specific biological function. Most behavioral tests are designed to assess aggression on the basis of how the behavior may be easily measured. However, caution is necessary when attributing an underlying biological function or adaptive significance to a particular type of aggression seen when using these models. For example, aggressive behavior is often elicited by isolating an animal (i.e. the resident) and introducing a conspecific animal (i.e. intruder) into the resident's home cage. The aggression elicited by this model presumably reflects territorial defense. Second, defensive postures are recorded after a pair of rodents (typically rats) are exposed to a painful stimulus (usually electric shock through a wire-grid floor). The aggression evoked in this paradigm is thought to represent aggression that would be seen in response toward a predator or large opponent. Third, predatory aggression is evaluated in rats by presenting the animals with mice and recording stalking, killing, seizing, and consuming behaviors (Miczek et al., 1984). Finally, aggression can be measured in a neutral arena to which both the experimental and stimulus animal has not been previously exposed. When evaluating aggression using this paradigm, researchers must be particularly careful to record the behavior of both the experimental and the stimulus animals. Often the experimental animal exhibits both offensive and defensive aggression dependent upon the behavior of the stimulus animal. To characterize only offensive aggression, stimulus animals rendered anosmic are used because these stimulus rodents elicit aggression from the experimental animal without exhibiting aggression towards the experimental animal. It is unlikely that a single neural substrate mediates the various forms of aggression, and the use of genetic manipulations provides an exciting opportunity to begin to unravel the mechanisms regulating different components of aggression.

Caveats and considerations

In contrast to recent suggestions (e.g. Good, 1996; Moran et al., 1996), increased aggressiveness is not a general behavioral phenotype of mice with a targeted disruption of any gene (see below). Some knockout mice display reduced aggression while other knockout animals display no alterations in aggressive behavior. Furthermore, some knockout animals exhibit dramatic, selective increases in aggressive behavior (e.g. $5\text{-HT}_{1B}-/-$ and nNOS−/− mice) in the absence of any other apparent behavioral abnormalities. In other cases, increases in aggressive behavior are noted, yet other behavioral abnormalities are observed that may be partially or entirely responsible for the alterations in aggression. As noted below, some knockout animals display increased fear responses associated with increased aggressive behavior. Because defensive aggression is thought to be motivated by fear (Miczek et al., 1984), increased fear responses likely account for increased defensive aggression in some knockout mice. The relationship between offensive aggression and fear is not as clear, but may account for some alterations in aggression among knockout mice. Taken together, these observations suggest the importance of evaluating knockout animals on a variety of behavioral, morphological, and neurological parameters when evaluating aggressive behavior in order to rule out the possibility that other abnormalities may be responsible for irregularities in aggression. Behavioral tests aimed at evaluating motor behavior, sensory abilities, and fear responses are essential. Likewise, morphological parameters such as body mass/size and endocrine measures such as plasma/serum androgen concentrations may provide some insight into alterations in dominance and aggression. The cases presented below are intended to provide the reader with some examples of alterations in aggression seen among knockout mice, and evaluate these alterations in the preceding context.

Targeting components of the mechanisms underlying aggression

Serotonin and aggression

Several lines of research have established that reduced brain serotonin concentrations result in increased aggression (reviewed in Albert and Walsh, 1984; Olivier et al., 1995; Olivier and Mos, 1992). For example, PCPA causes 5-HT depletion which leads to increased aggression in a variety of species in several different situations (reviewed in Vergnes et al., 1986). Depletion of 5-HT increases both offensive and defensive aggression (Vergnes et al., 1986). Shock-induced aggression in mice and rats results in defensive postures and aggression mediated by reduced serotonin concentrations in the brain; mouse-killing behavior in rats tends to be offensive (predatory) aggression, and both mice and rats show offensive and defensive aggression depending on their status as the intruder or resident, respectively, in a home cage (Barr et al., 1975; Nikulina, 1991; Olivier and Mos, 1992). It is noteworthy that brain 5-HT depletion increases behavioral responsiveness to sensory and painful stimuli (e.g., Lorens, 1978; Telner et al., 1979). Likewise, aggressive behavior in response to a wide variety of stimuli can be suppressed by 5-HT or experimental treatments that elevate 5-HT (Ieni and Thurmond, 1985).

The early literature on the inhibitory effects of 5-HT on aggressive behavior was based on nonselective 5-HT agonists and antagonists (reviewed in Bell and Hobson, 1994; Muehlenkamp et al., 1995). Radioligand binding studies have differentiated 5-HT receptors into 5-$HT_{1A,1B,\ 1D\alpha,\ 1D\beta,\ 1E,\ 1F}$, 5-$HT_{2A,\ 2B,\ 2C}$, 5-$HT_3$, 5-$HT_4$, and 5-$HT_5$ subtypes (Bell and Hobson, 1994). Recent studies have indicated that male-typical offensive aggression is reduced, in the absence of anxiolytic actions, by mixed 5-HT_{1A} and $_{1B}$ agonists (Cologer-Clifford et al., 1996).

The availability of knockout mice has allowed further elucidation of the role of 5-HT in aggressive behavior. Mice with targeted disruption of the 5-HT_{1B} receptor display elevated levels of offensive aggression when tested using the resident-intruder paradigm; mutant mice attack intruders faster and a greater number of times than WT mice (Saudou et al., 1994; Ramboz et al., 1996). This increase in aggressive behavior is independent of any other apparent behavioral abnormalities, including altered fear responses in an open field test. These data suggest that increased aggression due to depletion of serotonin may be mediated by the 5-HT_{1B} subtype of receptor. A class of 5-HT agonists called serenics, due to their anti-aggressive properties in numerous animals models of aggression, appear to act on the 5-HT_{1B} receptor (Olivier et al., 1986), thereby providing converging evidence for a role of this receptor subtype in mediating serotonin's effects on aggression.

Further evidence for a role of serotonin in aggression has come from the use of knockout mice either missing genes that directly or indirectly affect 5-HT concentrations or metabolism. Homozygous mutant mice missing the gene for α-calcium-calmodulin-dependent kinase II (α-CaMKII) display reduced offensive and defensive aggression (Chen et al., 1994). CaMKII mediates presynaptic transmitter release and activates trytophan hydroxylase, the rate-limiting enzyme in 5-HT metabolism (Silva et al., 1992; Ehret et al., 1989), and 5-HT release is reduced in the dorsal raphe of mice missing the gene for α-CaMKII (Chen et al., 1994). These data contrast sharply with pharmacological data reporting reduced 5-HT concentrations being associated with increased levels of aggression. As previously mentioned, because defensive aggressive behavior in rodents is thought to be motivated by fear, studies investigating knockout mice typically evaluate the mutants= response to fearful stimuli. Certainly, evaluation of mice with deficits in behaviors unrelated to fear need not evaluate fear responsiveness. α-CaMKII mutant mice display reduced fear when tested in a number of paradigms (Chen et al., 1994). Decreased fear likely explains the reductions in defensive aggression seen in α-CaMKII knockouts, and suggests that the offensive aggression data should be interpreted cautiously. These findings emphasize the importance of testing a variety of behaviors when evaluating phenotypes of mutant mice.

Whenever there is a danger that alternative behavioral deficits might account for the behavior of interest in a knockout animal, a battery of tests evaluating potentially confounding behaviors is necessary. A further cautionary lesson may be learned from the study of behavioral abnormalities in mutant mice with a random insertion of a transgene for monoamine oxidase A (MAOA) (Cases et al., 1995). MAOA is responsible for the degradation of norepinephrine and serotonin. Because 5-HT concentrations would be expected to be elevated in MAOA transgenic mice, one would also predict that aggression would be reduced in these animals. Contrary to expectation, MAOA transgenic mice exhibit increased levels of offensive aggression despite elevated 5-HT concentrations (Cases et al., 1995). However, these mice are not healthy, they tremble, have difficulty righting, and have an elevated fear response prior to adulthood. This finding suggests that there may often be pleiotropic phenotypical changes that should be considered in interpreting the results of behavioral tests. In adulthood, these knockout mice spend an increased amount of time in the center of an open field arena (typically indicative of reduced fear response) (Cases et al., 1995).

Animals missing the gene for pre-proenkephalin exhibit high levels of aggression relative to WT mice (Konig et al., 1996). These animals also exhibit a heightened supraspinal response to painful stimuli, and increased anxiety as measured by the open field test (Konig et al., 1996). Clearly these behavioral and physiological alterations can influence aggressive behavior in these mutants. Taken together, these studies suggest the importance of evaluating one behavioral phenotype only when considering a constellation of other relevant physiological and behavioral parameters, and have suggested that the relationship between brain levels of serotonin and aggression are not yet fully understood.

Hormones and aggression

As previously mentioned, aggressiveness is not a general behavioral phenotype of mice with a targeted disruption of any gene; mice with targeted disruption of the heme oxygenase-2 gene show no alterations in aggressive behavior (Kriegsfeld et al., unpublished data). Likewise, mice with targeted deletion of the oxytocin gene (OT−/−) display reduced aggression, as evidenced by a decreased duration spent engaging in agonistic encounters during a resident–intruder test and when tested in a neutral arena (DeVries et al., 1997). Male mice with targeted disruption of the gene for the estrogen receptor (ERKO) also display reduced aggression in a number of testing situations (Ogawa et al., 1997). Conversely, ERKO females exhibit increased levels of aggression towards other female mice relative to WT females, and ERKO females elicit aggression from WT males (Ogawa et al., 1996). Because estrogen is essential for the normal sexual differentiation of the CNS of male (and possibly female) mammals during development (Arnold, 1996), studies of adult behavior in ERKO mice are complicated by the inability to dissociate genetic from ontogenetic causes of behavior. Thus, when evaluating the results from studies with knockout mice, it is important to consider what effect the missing gene (or gene product) might have on normal development. Likewise, it is important to evaluate potential sex differences in behavior when systematically evaluating behavioral phenotypes of knockout animals.

Serendipitous targeting for aggression

Nitric oxide synthase

In a small proportion of cases, altered behaviors of knockout mice are often sufficiently obvious or unusual that they catch the attention of animal care personnel, who then notify the investigators. Currently, therefore, dramatic behaviors including increased aggression, altered maternal care, seizures, impaired motor coordination and sensory abilities are commonly reported for knockout mice (e.g. Barlow et al., 1996; Burnett et al., 1996; Chen et al., 1994; Nelson, 1997b; Saudou et al., 1994). Although it has been noted that, "... it is difficult ... to recognize minor neurological abnormalities in mice" (The Dutch–Belgium Fragile X Consortium, p. 25), presumably additional subtle behavioral changes of knockout mice await discovery by behavioral biologists. Thousands of genetic knockouts are being created, but few are examined for specific changes in aggression. Importantly, many knockouts have been examined in behavioral tests, but display no behavioral impairments (e.g. Lee, 1996; Pekny, 1995; Poss, 1995). Some examples of aggressive behavioral phenotypes of mice with targeted disruptions of specific genes tested are described in the following section.

Nitric oxide was initially identified as an endogenous regulator of blood vessel tone (Moncada and Higgs, 1993; Ignarro, 1990). NO also mediates the bactericidal and tumoricidal actions of macrophages (Nathan, 1992; Natham and Xie, 1994), and also serves as a putative neurotransmitter in the central and peripheral nervous system (Dawson and Snyder, 1994; Dawson and Dawson, 1996). NO has also been implicated as a retrograde messenger in long-term potentiation (O'Dell et al., 1994). Neurons containing the neuronal isoform of nitric oxide synthase (nNOS), the synthetic enzyme that produces NO and citrulline from arginine, are discretely localized throughout the brain. High densities of nNOS positive cells are localized in the cerebellum, as well as in areas of the limbic system that regulate emotional behavior, particularly the lateral septal nuclei, the posterior hypothalamus, the entorhinal cortex, and the amygdala (Vincent and Kimura, 1992).

As part of a multifactorial approach to determine the normal function of NO in the nervous system, mutant mice with targeted disruption of the nNOS gene were created (Huang et al., 1993). These mutant mice exhibit some morphological and physiological pathologies; enlarged stomachs, with hypertrophy of the pyloric sphincter, mimicking abnormalities in the human disease, infantile hypertrophic pyloric stenosis is a pronounced defect (Huang et al., 1993). Excessive NO release in the brain appears to mediate the glutamate neurotoxicity that may be responsible for vascular stroke damage (Dawson and Dawson, 1995, 1996); nNOS knockout animals (nNOS−/−) mice were previously evaluated for stroke damage after middle cerebral artery occlusion and have markedly reduced stroke damage (Huang et al., 1994). Two forms of long-term synaptic modulation, long-term depression (LTD) in the cerebellum and long-term potentiation (LTP) in the hippocampus, are sensitive to pharmacological inhibitors of NOS (Schuman and Madison, 1994). However, hippocampal LTP and cerebellar LTD are unaltered in nNOS−/− mice (O'Dell et al., 1994).

Two observations during the establishment of the breeding colony of nNOS−/− mice provoked additional behavioral assessments. Initially, five nNOS−/− mice were housed together in same-sexed groups and upon routine morning examinations, one or two dead males were often discovered (Nelson et al., 1995). Initial indirect observations suggested that the male nNOS−/− mice engaged in chronic aggressiveness, a behavioral phenotype not observed among the group-housed female nNOS−/− animals or the wild-type animals of either sex. Thereafter, all animals were housed individually, as soon as the cause of death was determined. Aggressive behaviors were compared in detail between WT and nNOS−/− animals; a battery of sensorimotor tasks was also conducted to discover the presence of any behavioral abnormalities that might affect aggressive and reproductive behaviors. Behavioral evaluations determined that the nNOS−/− males were much more aggressive, engaged in longer bouts of aggression, changed their bite target from the hindquarters to the nape of the neck, and reduced the incidence of submissive postures (Nelson et al., 1995).

Mice with targeted disruption of the gene encoding the endothelial isoform of nitric oxide synthase (eNOS−/−) have recently been created (Huang et al., 1995; Shesely et al., 1996). Because nitric oxide was originally identified as endothelium-derived relaxing factor (Furchgott and Vanhoutte, 1989; Ignarro, 1990), and eNOS is localized in the endothelial lining of vascular smooth muscle, blood pressure was the first phenotype investigated in these mutant mice (Huang et al., 1995; Shesely et al., 1996). eNOS−/− mice exhibit approximately a 35% increase in basal blood pressure relative to WT mice (110 mm Hg and 81 mm Hg, respectively; Huang et al., 1995). To our knowledge, no studies have been conducted on behavioral phenotypes in eNOS−/− mice or mice given specific pharmacological eNOS inhibitors.

Because nNOS−/− mice display elevated levels of aggressive behavior compared to WT mice, we recently investigated aggression in eNOS−/− mice generated by Paul Huang and colleagues (1995). Anecdotal observations indicated that these knockout animals were very docile. Animals were tested using two behavioral paradigms. First, animals were tested using the resident–intruder paradigm; eNOS−/− mice displayed fewer aggressive encounters and a decreased duration of agonistic encounters relative to WT mice when a WT intruder was placed into their homecage (Fig. 2). Likewise, when tested in a neutral arena with a WT stimulus male, eNOS−/− mice displayed fewer attacks and an increased latency to attack the stimulus male relative to WT mice (Demas et al., 1998) These data, in combination with the nNOS−/− data, suggest that the two isoforms of NOS may normally act to increase (eNOS−/−) and decrease (nNOS−/−) aggressive behavior in vivo. Thus, WT mice with normal concentrations of both isoforms of NOS display only moderate levels of aggression. Likewise, it may be the case that nNOS−/− mice exhibit a compensatory increase in eNOS resulting in elevated levels of aggression. Treatment with hydralazine, which normalized blood pressure in eNOS−/− mice, did not alter the low levels of aggression seen in eNOS−/− mice (Demas et al., 1998).

As previously mentioned, in studies of aggression, the nNOS−/− male mice exhibited a dramatic loss of behavioral inhibition reflected in persistent fighting and mounting behavior despite obvious signals of surrender or disinterest, respectively, by their test partners (Nelson et al., 1995). Female nNOS−/− mice displayed neither elevated aggressiveness, nor inappropriate mating behaviors (Fig. 1). Prolonged aggressiveness and mating behavior among males are often associated with elevated blood concentrations of testosterone; however, no differences in testosterone concentrations were detected between WT and nNOS−/− males. Numerous sensorimotor behaviors including open field and elevated plus maze measures of anxiety, gross motor ability,

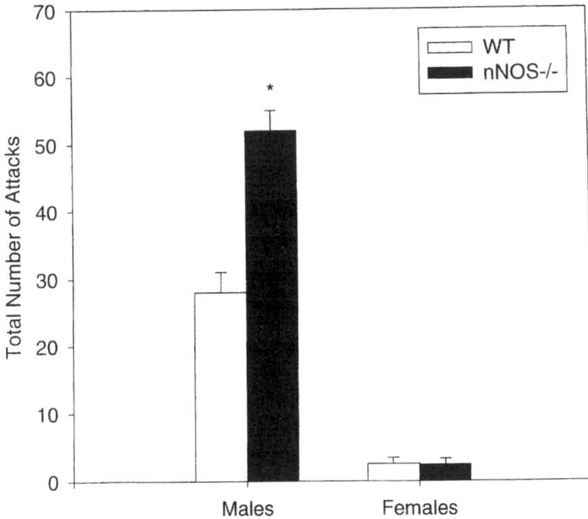

Fig. 1. The mean (±SEM) total number of attacks emitted by male and female WT (open bars) and nNOS−/− mice (filled bars). The nNOS−/− males engaged in significantly ($P < 0.01$) more attacks than WT males during behavioral testing. Males, regardless of phenotype, were much more aggressive than females.

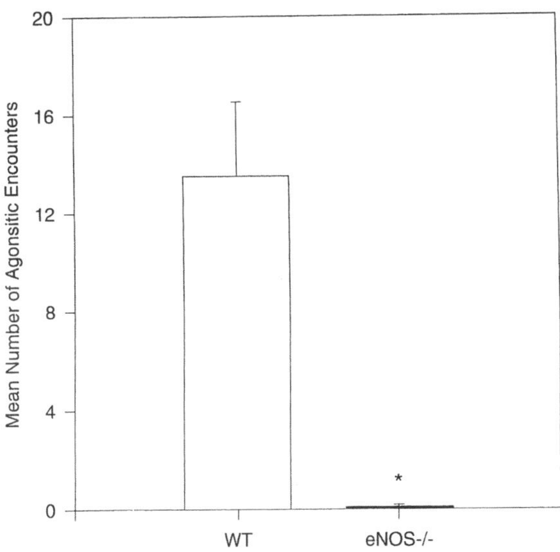

Fig. 2. Mean (±SEM) total number of agonistic encounters initiated by male WT (open bars) and eNOS−/− mice (filled bars) when tested using a resident-intruder paradigm. eNOS−/− mice exhibit far fewer attacks than WT controls during a 15 min test ($P < 0.05$).

olfactory and visual ability, balance and coordination, and forelimb strength were evaluated in nNOS−/− mice. No sensory or motor deficiencies were observed that could account for the elevated aggression among the nNOS−/− mice. Furthermore, no differences in open field behavior, a putative measure of anxiety, were detected among nNOS−/− and WT male and female mice.

Neural cell adhesion molecule

Both homozygous and heterozygous mice deficient in the neural cell adhesion molecule (NCAM−/−) display elevated aggression when tested in the resident-intruder paradigm (Stork et al., 1997). NCAM is thought to be important during development and in adult neural plasticity (Goridis and Brunet, 1992; Scholey et al., 1993). Testosterone concentrations are highly correlated with aggression in rodent species; castrated mice rarely display aggressive behavior, and testosterone replacement restores aggression to precastration values (reviewed in Gandelman, 1980). Studies of aggression in mice with targeted disruption of specific genes often fail to measure testosterone values, thereby complicating interpretation of the results from these studies. Pre- and post-test testosterone values in NCAM−/− mice are comparable to values seen in WT animals (Stork et al., 1997). Importantly, c-*fos* mRNA concentrations were elevated in limbic areas (i.e., septum, preoptic area, hypothalamus, amygdala, and dorsal raphe) in NCAM−/− mice relative to WT animals after the interaction with the intruder (Stork et al., 1997). This finding affords the possibility that NCAM−/− mice may experience a heightened "emotional" response to the presentation of a threatening stimulus, thereby leading to an increase in aggressive behavior. Converging evidence for this hypothesis was found when measuring pre- and post-test values of corticosterone in these mice. Although basal concentrations of corticosterone did not differ between genotypes, both homozygous and heterozygous NCAM mutants exhibit a dramatic post-test increase in corticosterone concentrations relative to WT mice (Stork et al., 1997). Studies of aggression in knockout mice should consider not only the transmitter systems most likely affected by the genetic mutation, but the endocrine profiles of these mice as well.

"Fierce" mutant

Putatively, the *Tlx* gene was deleted using homologous recombination (Moneghan et al., 1997; Simpson et al., 1998). The *Tlx* gene is the mouse homolog of the *Drosophila* organ nuclear receptor "tailless". Fierce is recessive and viable. Fierce mice appear normal at birth but are smaller than WT animals after 2–3 weeks and continue to remain small throughout adulthood. The Fierce mice display reduced brain size that is particularly obvious in the cerebrum and olfactory bulbs, while the cerebellum appears normal in size (Moneghan et al., 1997; Simpson et al., 1998). Both male and female Fierce mice display substantially increased aggressive behavior. This elevated aggressive behavior interrupts normal maternal care, and also disrupts mating behavior (Moneghan et al., 1997; Young et al., 1997). Presumably, the architectural development of the limbic system and rhinencephalic structures are impaired, but additional studies are required to assess the sensorimotor and endocrine status of these knockouts in order to design studies of the mechanisms underlying the enhanced aggression in both sexes of Fierce mice. *Tlx* knockout mice have normal plasma concentrations of testosterone and corticosterone. However, male and female knockouts exhibit reduced olfactory ability, abnormal postural reflexes, and poorer coordinated behavior relative to WT animals. Again, these observations suggest that caution should be applied when interpreting the aggression data in *Tlx* knockouts and alternative hypotheses for the increased aggression should be considered.

Conclusions

Although gene targeting has the potential to yield new insights into the study of aggression, there are some shortcomings of this technique that must be recognized when interpreting behavioral data of knockout animals (Nelson, 1997b). One conceptual problem with behavioral studies of mice with targeted genetic deletions, that is shared with all ablation studies, is that behavioral tests study the effects of the *missing* gene (and gene product), not the effects of the gene directly. This conceptual short-coming can be overcome in the same way that it is overcome in other types of ablation studies, by collecting converging evidence. For example, if similar behavioral deficits are obtained after pharmacological, lesion and genetic manipulations of the same factor, then it is reasonable to conclude that the missing factor is involved in the behavior, especially if the behavioral deficit is ameliorated when the missing factor is restored. Another disadvantage of using gene-targeting techniques in the study of aggression is that the products of many genes are essential to normal function, and inactivating the gene may prove lethal or induce gross morphological or physiological abnormalities that can complicate interpretation of discrete behavioral effects. Furthermore, unexpected compensatory or redundancy mechanisms might be activated when a gene is missing, and cloud interpretation of the normal contribution of the gene to behavior. Alternatively, mutant mothers may interact with their pups inappropriately (or merely differently) to cause elevated aggressiveness (Michel and Moore, 1995). This latter possibility has not been examined systematically, and this is probably an important issue because in virtually all behavioral studies of knockout animals, the mothers are also genetically-altered. Using pups with targeted deletion of specific genes that have been cross-fostered to WT mothers would be an excellent control procedure to parse out the effects of the gene from the effects of maternal care on subsequent behavioral alterations. Alternatively, the maternal effects could be controlled by having heterozygous parents for WT and mutant progeny. Finally, because mammalian genome mapping is currently focused on mice (*Mus musculus*), standardized behavioral testing of mice should be adopted, and new behavioral tests should be created as needed.

Against those disadvantages are several important advantages to using knockout mice in behavioral research:

1. Disabling a gene is often a very precise and "clean" ablation.
2. The effects of the gene product can be abolished without the side-effects of drugs
3. Genetic manipulations may be the only way to determine the precise role of many endogenous factors on behavior.

The use of new inducible knockouts, in which the timing and placement of the targeted gene disruption can be controlled (see Chapter 3.1.3), will be an extremely important tool in behavioral endocrinology research.

Taken together, the technique of targeted disruption, or knock-out, of genes provides an important new tool for behaviorists to ascertain the mechanisms underlying behavior. Importantly, the use of genetic tools to dissect behavioral mechanisms does not imply an acceptance of the notion that all behavior is "pre-wired" or completely determined by genes. Rather, behavior is an observable phenotype; as noted decades ago, phenotypes are always the result of the interaction of the genotype with the environment. An excellent approach to determine experiential effects on behavioral development is to start with a gene and understand the ontogenetic process that gives rise to a behavior. Although this approach has certainly been recognized intellectually for some time as one of the best ways to understand the environmental effects on behavior, it was technically impossible until recently. Behavioral neuroscientists should embrace this useful, new technology to continue their phenotypic studies of behavior, which should always be understood to represent the interaction of the genotype with the environment.

Acknowledgments

Preparation of this paper was supported by USPHS Grants MH22201, MH57535, and NSF Grant IBN97-23420. We thank Greg Demas for discussion of the data in this communication, access to unpublished data, and for bibliographic assistance. We also thank Drs Ted Dawson and Solomon H. Snyder for data discussion, and the nNOS knockout mice from their colony, and Drs Paul Huang and Arthur Burnett for access to eNOS knockout mice.

References

Albert, D.J. and Walsh, M.L. (1984) Neural systems and the inhibitory modulation of agonistic behavior: A comparison of mammal species. *Neurosci. Biobehav. Rev.*, 8: 5–24.

Arnold, A.P. (1996) Genetically triggered sexual differentiation of brain and behavior. *Horm. Behav.*, 30: 495–505.

Barlow, C., Hirotsune, S., Paylor, R., Liyanage, M., Eckhaus, M., Collins, F., Shiloh, Y., Crawley, J.N., Ried, T., Tagle, D. and Wynshaw,-Boris, A. (1996) A paradigm of ataxia telangiectasia. *Cell*, 86: 159–171.

Barr, G.A., Gibbons, J.L. and Moyer, K.E. (1975) The relationship between mouse killing and intraspecific fighting in the albino rat. *Behav. Biol.*, 14: 201–208.

Bell, R. and Hobson, H. 1994. 5-HT $_{1A}$ receptor influences on rodent social and agonistic behavior: A review and empirical study. *Neurosci. Biobeh. Rev.*, 18: 325–338.

Burnett, A.L., Nelson, R.J., Calvin, D.C., Liu, J-X., Demas, G.E., Klein, S.L., Kriegsfeld, L.J., Dawson, V.L., Dawson, T.M. and Snyder, S.H. (1996) Nitric oxide dependent penile erection in mice lacking neuronal nitric oxide synthase. *Mol. Med.*, 2: 288–296.

Cases, O., Seif, I., Grimsby, J., Gaspar, P., Chen, K., Pournin, S., Müller, U., Aguet, M., Babinet, C., Chen, J., and De Maeyer, E. (1995) Aggressive behavior and altered amounts of brain serotonin and norepinephrine in mice lacking MAOA. *Science*, 268: 1763–1766.

Chen, C., Rainnie, D.G., Green, R.W., Tonegawa, S. (1994) Abnormal fear response and aggressive behavior in mutant mice deficient for α-calcium-calmodulin-kinase II. *Science*, 266: 291–294.

Cologer-Clifford, A., Smoluk, S.A. and Simon, N.G. (1996) Effects of serotonergic 1A and 1B agonists and androgenic versus estrogenic systems for aggression. *Ann. NY Acad. Sci.*, 3339–3342.

Dawson, T.M. and Dawson, V.L. (1995) Nitric oxide neurotoxity. *The Neuroscientist*, 1: 7–20.

Dawson, T.M. and Dawson, V.L. (1996) Nitric oxide synthase: Role as a transmitter/mediator in the brain and endocrine system. *Ann. Rev. Med.*, 47: 219–245.

Dawson, T.M. and Snyder, S.H. (1994) Gases as biological messengers: Nitric oxide and carbon monoxide in the brain. *J. Neurosci.*, 14: 5147–5159.

Demas, G.E., Kriegsfeld, L.J., Drazen, D.L., Dawson, T.M., Dawson, V.L., Snyder, S.H. and Nelson, R.J. (1998) Reduced aggression in male mice lacking the gene for endothelial nitric oxide synthase. *Neurosci. Abst.*, 562(11): 1429.

DeVries, A.C., Young, W.S., and Nelson, R.J. (1997) Reduced aggressive behaviour in mice with targeted disruption of the oxytocin gene. *J. Neuroendocrinol.*, 9: 363–368.

Dutch–Belgium Fragile X Consortium. (1994) Fmrl knockout mice: A model to study Fragile X mental retardation. *Cell*, 78: 23–33.

Ehret, M., Cash, C.D., Hamon, M. and Maitre, M. (1989) Formal demonstration of the phosphorylation of rat brain tryptophan hydroxylase by Ca2+/calmodulin-dependent protein kinase. *J. Neurochem.*, 52: 1886–1891.

Furchott, R.F., and Vanhoutte, P.M. (1989) Endothelium-derived relaxing and factors. *FASEB J.*, 3: 2007–2018.

Gandelman, R. (1980) Gonadal hormones and the induction of intraspecific fighting in mice. *Neurosci. Biobehav. Rev.*, 4: 133–140.

Good, M. (1996) Targeted deletion of neuronal nitric oxide: a step closer to understanding its functional significance. *TINS*, 19: 83–84.

Goridis, C. and Brunet, J.F. (1992) NCAM: structural diversity, function, and regulation of expression. *Semin. Cell Biol.*, 3: 189–197.

Huang, P.L., Dawson, T.M., Bredt, D.S., Snyder, S.H. and Fishman, M.C. (1993) Targeted disruption of the neuronal nitric oxide synthase gene. *Cell*, 75: 1273–1286.

Huang, P.L., Huang, Z., Mashimo, H., Bloch, K.D., Moskowitz, M.A., Bevan, J.A., Fishman, M.C. (1995) Hypertension in mice lacking the gene for endothelial nitric oxide synthase. *Nature*, 377: 239–242.

Huang, Z., Huang, P.L., Panahian, N., Dalkara, T., Fishman, M.C. and Moskowitz, M.A. (1994) Effects of cerebral ischemia in mice deficient in neuronal nitric oxide synthase. *Science*, 265: 1883–1885.

Ieni, J.R. and Thurmond, J.B. (1985) Maternal aggression in mice: Effects of treatments with PCPA, 5-HTP, and 5-HT receptor antagonists. *Europ. J. Pharmacol.*, 111: 211–220.

Ignarro, L.J. (1990) Biosynthesis and metabolism of endothelium-derived nitric oxide. *Ann. Rev. Pharmacol. Toxicol.*, 30: 535–560.

König, M., Zimmer, A.M., Steiner, H., Holmes, P.V., Crawley, J.N., Brownstein, M.J. and Zimmer, A. (1996) Pain responses, anxiety, and aggression in mice deficient in pre-proenkephalin. *Nature*, 383: 535-538.

Lee, S.L., Sadovsky, Y., Swirnoff, A.H., Polish, J.A., Goda, P., Gavrilina, G. and Milbrandt, J. (1996) Luteinizing hormone deficiency and female infertility in mice lacking the transcription factor NGFI-A (Egr-1). *Science*, 273: 1219–1221.

Lorens, S. (1978) Some behavioral effects of serotonin depletion depend on method: A comparison of 5,7-dihydroxytryptomine, *p*-chlorophenylalanine, *p*-chloroamphetamine, and electrolytic raphe lesions. *Ann. NY Acad. Sci.*, 305: 532–555.

Maxson, S.C. (1996) Issues in the search for candidate genes in mice as potential animal models of human aggression. *Ciba Found. Symp.*, 194: 21–30.

Michel, G.F. and Moore, C. (1995) *Developmental Psychobiology*, Cambridge, MIT Press.

Miczek, K.A., DeBold, J.F. and Thompson, M.L. (1984) Pharmacological, hormonal, and behavioral manipulations in analysis of aggressive behavior. *Prog. Clin. Biol. Res.*, 167: 1–26.

Moncada, S. and Higgs, A. (1993) The L-arginine nitric oxide pathway. *N. Engl. J. Med.*, 329: 2002–2012.

Moneghan, A.P., Bock, D., Gass, P., Schwager, A., Wolfert, D.P., Lipp, H-P. and Schutz, G. (1997) Defective limbic system in mice lacking the tailless gene. *Nature*, 390: 515-517.

Moran, T.H., Reeves, R.H., Rogers, D. and Fisher, E. (1996) Ain't misbehavin' — it's genetic. *Nat. Genet.*, 12: 115–116.

Muehlenkamp, F., Lucion, A. and Vogel, W.H. (1995) Effects of selective serotonin agonists on aggressive behavior in rats. *Pharmacol. Biochem. Behav.*, 50: 671–674.

Nathan, C. (1992) Nitric oxide as a secretory product of mammalian cells. *FASEB J.*, 6: 3051–3064.

Nathan, C. and Xie, Q-W. (1994) Nitric oxide synthases: roles, tolls, and controls. *Cell*, 78: 915–918.

Nelson, R.J. (1997a) Sexual and aggressive behavior in male mice lacking neuronal nitric oxide synthase or heme-oxygenase-2. Excerpta Medica International Congress, in press.

Nelson, R.J. (1997b) The use of genetic "knock-out" mice in behavioral endocrinology research. *Horm. Behav.*, 31: 188–196.

Nelson, R.J., Demas, G.E., Huang, P.L., Fishman, M.C., Dawson, V.L., Dawson, T.M. and Snyder, S.H. (1995) Behavioral abnormalities in male mice lacking neuronal nitric oxide synthase. *Nature*, 378: 383–386.

Nikulina, E.M. (1991) Neural control of predatory aggression in wild and domesticated animals. *Neurosci. Biobehav. Rev.*, 15: 545–547.

O'Dell, T.J., Huang, P.L., Dawson, T.M., Dinerman J.L., Synder, S.H., Kandel, E.R. and Fishman, M.C. (1994) Endothelial NOS and the blockade of LTP by NOS inhibitors in mice lacking neuronal NOS. *Science*, 265: 542–546.

Ogawa, S., Lubahn, D.B., Korach, K.S. and Pfaff, D.W. (1997) Behavioral effects of estrogen receptor gene disruption in male mice. *Proc. Nat. Acad. Sci.*, 94: 1476–1481.

Ogawa, S., Taylor, J.A., Lubahn, D.B., Korach, K.S. and Pfaff, D.W. (1996) Reversal of sex roles in genetic female mice by disruption of estrogen receptor gene. *Neuroendocrinology*, 64: 467–470.

Olivier, B., and Mos, J. (1992) Rodent models of aggressive behavior and serotonergic drugs. *Prog. Neuropsychopharmacol. Biol. Psych.*, 16: 847–870.

Olivier, B., Mos, J., Van Oorschot, R. and Hen, R. (1995) Serotonin receptors and animal models of aggressive behavior. *Pharmacopsychiatry*, 28: 80–90.

Olivier, B., VanDalen, D. and Hartog, J. (1986) A new class of psychoactive drugs, serenics. *Drugs Future*, 11: 473–499.

Pekny, M., Leveen, P., Pekna, M., Eliasson, C., Berthold, C.H., Westermark, B. and Betsholtz, C. (1995) Mice lacking glial fibrillary acidic protein display astrocytes devoid of intermediate filaments but develop and reproduce normally. *EMBO J.*, 14: 1590–1598.

Poss, K.D., Thomas, M.J., Ebralidze, A.K., O'Dell, T.J., and Tonegawa, S. (1995) Hippocampal long-term potentiation is normal in heme-oxygenase-2 mutant mice. *Neuron*, 15: 867–873.

Ramboz, S., Saudou, F., Amara, D.A., Belzung, C., Segu, L., Misslin, R., Buhot, M-C., Hen, R. (1996) 5-HT_{1B} receptor knockout - behavioral consequences. *Behav. Brain Res.*, 73: 305–312.

Saudou, F, Amara, D.A, Dierich, A, LeMeurru, M., Ramboz, S., Segu, L., Buhot, M.C. and Hen, R.C.L. (1994) Enhanced aggressive behavior in mice lacking 5-HT-1β receptor. *Science*, 265: 1875–1878.

Schuman, E.M. and Madison, D.V. (1994) Nitric oxide and synaptic function. *Ann. Rev. Neurosci.*, 17: 153–183.

Scholey, A.B., Rose, S.P.R., Zamani, M.R., Bock, E. and Schachner, M. (1993) A role for the neural cell adhesion molecule in a late, consolidating phase of glycoprotein synthesis six hours following passive avoidance training of the young chick. *Neuroscience*, 55: 499–509.

Shesely, E.G., Maeda, N., Kim, H.S., Desai, K.M., Krege, J.H., Laubach, V.E., Sherman, P.A., Sessa, W.C. and Smithies, O. (1996) Elevated blood pressure in mice lacking endothelial nitric oxide synthase. *Proc. Nat. Acad. Sci.*, 93: 13176–13181.

Silva, A.J., Stevens, C.F., Tonegawa, S. and Wang, Y. (1992) Deficient hippocampal long-term potentiation in alpha-calcium-calmodulin kinase II mutant mice. *Science*, 257: 201–206.

Simpson, E.M., Mahaffey, C.L., Young, K.A., Hawes, A.L., Chang, B., Saionz, J.R., Smith, R.S., Rissman, E.F., Bronson, R.T., Nelson, R.J. and Berry, M.L. (1998) Fierce: a novel mouse mutant with developmental neurological and behavioral abnormalities is deleted for tailless. *Neurosci. Abst.*, 312(1): 780.

Stork, O., Welzl, H., Cremer, W.H. and Schachner, M. (1997) Increased intermale aggression and neuroendocrine response in mice deficient for the neural cell adhesion molecule (NCAM). *Eur. J. Neurosci.*, 9: 1117–1125.

Telner, J., Lepore, F. and Guillemot, J.P. (1979) Effects of serotonin content on pain sensitivity in the rat. *Pharmacol. Biochem. Behav.*, 10: 657–661.

Vergnes, M., Depaulis, A. and Boehrer, A. (1986) Parachlorophenylalanine-induced serotonin depletion increases offensive but not defensive aggression in male rats. *Physiol. Behav.*, 36: 653–658.

Vincent, S.R. and Kimura, H. J. (1992) Histochemical mapping of nitric oxide synthase in the rat brain. *Neuroscience*, 4: 755–784.

Young, K.A., Mahaffey, C.L., Johnson, K.A., Lerner, C.P., Perlick, H., Bronson, R.T., Nelson, R.J. and Simpson, E.M. (1997) Characterization of aggressive behavior in mice putatively lacking the autosomal zinc finger gene. Poster 99: Inaugural meeting of The Society for Behavioral Neuroendocrinology.

Zhang, J. and Snyder, S.H. (1995) Nitric oxide in the nervous system. *Ann. Rev. Pharmacol. Toxicol.*, 35: 213–233.

W.E. Crusio and R.T. Gerlai (Eds.)
Handbook of Molecular-Genetic Techniques for Brain and Behavior Research (Techniques in the Behavioral and Neural Sciences, Vol. 13)

CHAPTER 3.1.9

Behavioral analysis of *Dvl*1-deficient mice reveals a role for the *Dvl*1 gene in social behaviors and sensorimotor gating

Richard Paylor,[1] Nardos Lijam,[2] Michael P. McDonald,[1] Jacqueline N. Crawley,[1] Daniel J. Sussman[3] and Anthony Wynshaw-Boris[2]

[1]*Section on Behavioral Neuropharmacology, Experimental Therapeutics Branch, National Institute of Mental Health, National Institutes of Health, Bethesda, MD 20892, USA*
[2]*Laboratory of Genetic Disease Research, National Human Genome Research Institute, National Institutes of Health, Bethesda, MD 20892, USA*
[3]*Division of Human Genetics, University of Maryland School of Medicine, Baltimore, MD 21201, USA*

Introduction

Identifying behavioral phenotypes of gene-targeted mutant mice is a critical step in understanding the functional roles of genes. Using a test battery we have recently identified a unique set of behavioral phenotypes in mice deficient in the *Dishevelled*-1 (*Dvl*1) gene that suggests there is a role for this gene in complex central nervous system functions associated with certain aspects of neuropsychiatric disorders (Lijam et al., 1997).

The segment polarity gene, *Dishevelled* (*dsh*) was first cloned in *Drosophila* (Klingensmith et al., 1994; Theisen et al., 1994). *Dsh* genes are members of the wingless/Wnt pathway which is a highly conserved developmental pathway involved in cell fate determination (Moon et al., 1997). There are three closely related mouse *Dishevelled* (*Dvl*) genes (Klingensmith et al., 1996; Sussman et al., 1994; Tsang et al., 1996). There is 40–50% amino acid identity between the murine *Dvls* and the *Drosophila dsh*, and 60–70% identity among the murine *Dvls*. All of the *Dishevelled* proteins contain three highly conserved regions. The region of highest conservation contains the PDZ (or GLGF/DHR) motif, present in many proteins localized to tight junctions and cytoskeletal structures (Bryant et al., 1993; Ponting and Phillips, 1995). Other than PDZ and DEP domain (Ponting and Bork, 1996) homologies, *Dishevelled* proteins appear to be novel when compared with proteins in several sequence databases.

The murine *Dishevelled* genes (*Dvl*1, *Dvl*2 and *Dvl*3) are broadly expressed during embryonic development and in adult tissues (Klingensmith et al., 1996; Sussman et al., 1994; Tsang et al., 1996). The expression patterns overlap, suggesting redundancy of function among the *Dvl* genes. Although changes in subcellular localization of *Dvl* protein have been observed in *Wnt1*-overexpressing rat PC12 cells (Steitz et al., 1996), the function of mammalian *Dishevelled* is completely unknown.

Since there are no known naturally occurring murine *Dvl* mutants, we created mice completely deficient for *Dvl*1 to study the role of *Dvl* genes in the mouse. The strong pattern of expression of *Dvl*1 in the cerebellum, hippocampus and olfactory bulb (Sussman et al., 1994) suggested that *Dvl*1 might have a functional role in central nervous system functions. Findings from these studies (Lijam et al., 1997) suggest that *Dvl*1 plays a role in sensorimotor gating and social behaviors in mice.

Dvl-1 deficient mice

Complete details of the generation of the *Dvl*1-deficient mice can be found in Lijam et al. (1997), and described elsewhere in this volume (Chapter 3.1.1). The disruption of the gene was performed in TC1 embryonic stem cells (Deng et al., 1996) as described previously (Deng et al., 1994). TC1 ES cells were established from 129/SvEv mice (Taconic Farms), so completely inbred 129/SvEv mice were obtained by mating germline chimeras with 129/SvEv females. All experiments were carried out using the mice in the inbred background by experimenters that were blind to the genotype of the mice.

All three genotypes were detected in litters and the F_2 ratios for the wild-type, heterozygous, and mutant animals were not significantly different from the 1:2:1 Mendelian ratio. Immunoblot analysis (see Lijam et al., 1997) with a mouse monoclonal antibody (1–3F12) revealed a 75 kDa protein in +/+ mice, but no *Dvl*1 protein in −/− mice, demonstrating that the *Dvl*1 mutant allele was a null allele. Histological analysis of the *Dvl*1-deficient mice (Lijam et al., 1997) revealed no overt structural abnormalities in any peripheral tissue or in several brain regions. Sagittal and coronal sections through the hippocampus, olfactory bulbs, cerebellar vermis, and cerebellar hemispheres, were stained with either hematoxylin and eosin, or cresyl violet. These structures were chosen since they are areas of the brain where *Dvl*1 expression is high (Sussman et al., 1994). There were no histopathological differences in any of these brain regions between *Dvl*-deficient and wild-type mice. In addition, we looked at sections of other brain regions and observed no obvious differences between the mutant and wild-type mice. It is always important to note that this type of histopathology can detect severe morphological abnormalities, but does not determine if there are more subtle differences that could be detected with techniques designed to look at the fine morphology of individual neurons. Therefore, it is possible that there are indeed morphological differences between *Dvl*1-deficient and wild-type mice that are too subtle to be detected with the current type of histopathological analysis.

Locomotor activity, coordination and learning in Dvl*1-deficient mice*

Given the expression pattern in the CNS, in particular the cerebellum and hippocampus, it was important to evaluate locomotor activity, motor coordination, and learning and memory performance of *Dvl*1-deficient mice. To assess locomotor activity and motor coordination, open-field, wire-hang, and rotarod tests were employed. The spatial-learning version of the Morris (1981) water task was used to evaluate learning and memory performance of *Dvl*1-deficient mice.

Open-field activity

Exploratory locomotor activity of 11 wild-type and 11 mutant mice was measured in an open field (45 × 45 cm) by a Digiscan apparatus (model RXYZCM(8) Omnitech Electronics, Columbus, OH). Total horizontal activity for the 60 min period was used as a measure of open field activity.

Wire-hang

The same ten wild-type and mutant mice were tested for their ability to hang from wire bars (10 × 18 cm area, 1 mm in diameter spaced 1 cm apart). Mice were placed on the bars, turned upside down and latency to fall (maximum 60 s) was measured. Mice that fell in less than 10 s were given a second trial.

Rotarod

Mice were placed on a rotating (6.67 rpm) drum (10 cm in diameter) and latency to fall was measured up to 60 s. Mice that fell in less than 10 sec were given a second trial. Ten wild-type and 10 mutant mice were tested.

Spatial-learning in the Morris water task

Eight wild-type and 8 mutant mice were tested on the hidden-platform version of the Morris water maze task in a circular polypropylene (Nalgene) pool 105 cm in diameter. Each mouse was given 12 trials a day, in blocks of 4 trials for four consecutive days. The time taken to locate the escape platform (escape latency) was determined. After trial 36 and 48, each animal was given a 60 s probe trial. During the probe test, the platform was removed and quadrant search times and platform crossings were measured. The data for the two probe trials were averaged. To estimate long-term retention of this task, mice were given a probe test two weeks after training.

We believe that it is good practice to obtain some information about the behavioral responses of mice on tests used to assess basal activity, motor coordination and sensory abilities. This information is critical to understand behavioral phenotypes displayed by mutant mice on more complex tests because differences in activity, motor coordination, and/or sensory skills could interfere with an animal's ability to perform normally on tests that require more complex behavioral responses.

*Dvl*1-deficient mice were not impaired on the wire-hang test, rotarod test, or Morris water task. Locomotor activity was also similar between the mutant and wild-type mice. These findings demonstrate that *Dvl*1-deficient mice do not have any severe activity, motor coordination, or spatial learning impairments. More complete details of the results are presented in Lijam et al. (1997).

Sensorimotor gating impairments in Dvl*1-deficient mice*

Pizzuti et al. (1996) reported that a *Dvl*-like sequence mapped to human chromosome 22, within a region linked to schizophrenia. This finding suggested that *Dvl* genes may be candidates for schizophrenia loci, and that perhaps *Dvl*1-deficient mice may have a behavioral phenotype on tests thought to require normal neural functions that may be impaired in patients with neuropsychiatric disorders such as schizophrenia. To explore this possibility, levels of prepulse inhibition were determined in *Dvl*1-deficient mice.

Prepulse inhibition (PPI) is the phenomenon in which a weak prestimulus or prepulse suppresses the response to a startling stimulus (Ison et al., 1973; Graham, 1975). A number of studies have shown that patients with schizophrenia (Braff et al., 1978; Grillon, et al., 1992; McDowd et al., 1993) schizotypal personality disorder (Cadenhead et al., 1993), obsessive-compulsive disorder (Swerdlow et al., 1993), Tourette syndrome (Castellanos et al., 1996), and Huntington's disease (Swerdlow et al., 1995) have an impaired PPI response. The PPI impairment observed in these neuropsychiatric patients is thought to reflect an underlying problem with inhibitory mechanisms in neuronal systems used for sensorimotor gating (Braff and Geyer, 1990). PPI is one of the few paradigms in which humans and rodents are tested using similar methodologies. The PPI paradigm has quickly become the test of choice for scientists developing rodent models to study the mechanisms underlying the sensorimotor gating deficit observed in schizophrenia (Geyer and Braff, 1987; Swerdlow et al. 1994).

*Dvl*1-deficient and wild-type mice were tested for levels of PPI using SR-Lab Systems (San Diego Instruments, San Diego, CA). The background noise level in each chamber was 70 dB. Other details of the testing apparatus can be found in Paylor and Crawley (1997).

Each test session was started by placing a subject in the Plexiglas cylinder where it was left undisturbed for 5 min. After this acclimation period in TEST 1, each subject was presented with 56 trials. Each session consisted of seven trial types. Two startle trial types were 40 ms startle stimuli of either 100 or 115 dB (wide band width). There were four different acoustic prepulse plus acoustic startle stimulus trials presented with the onset of a prepulse stimulus 100 ms before the onset of the startle stimulus. Each 20 ms prepulse stimulus (either 74 or 90 dB) was presented before both acoustic startle stimuli. Finally, there were trials where no stimulus was presented to measure baseline movement in the cylinders. A diagram of the typical startle and prepulse trial types are presented in Fig. 1. The seven trial types were presented (15 s intertrial interval) 8 times in pseudorandom order such that each trial type was presented once within a block of seven trials. The startle response was recorded for 65 ms (measuring the response every 1 msec) starting with the onset of the startle stimulus. The maximum startle amplitude was used as the dependent variable. In TEST 1 there were 24 wild-type mice and 17 mutant mice.

The subjects (wild-type $n = 19$, mutant $n = 19$) in TEST 2, were presented 60 trials. Two startle trial types were 40 ms startle stimuli of either 100 or 120 dB. There were six different acoustic prepulse plus acoustic startle stimulus trial types. The 20 ms prepulse stimuli were sounds of 74, 82 or 90 dB. Each prepulse stimuli was presented before both acoustic startle stimuli. There were 3 prepulse-only trials. Finally, there were no stimulus trials. The 12 trial types were presented 5 times.

At least 3 d later, 41 of the wild-type subjects and 36 of the mutant subjects were tested for tactile startle response and PPI of the tactile startle response. Two male wild-type mice appeared ill and were not tested. Each session consisted of 7 trial types. One trial type was a 40 ms, 12 psi air puff. There were five acoustic prepulse plus tactile startle stimulus trials. The 20 ms prepulse stimuli were 74, 78, 82, 86, or 90 dB sounds. Finally, there were no stimulus trials. The 7 trial types were presented 6 times. The air puff was delivered directly to the back of the mouse through a piece of copper tubing. Assessing acoustic prepulse inhibition of the tactile startle response is important to determine if the auditory prepulse stimuli have the capability to inhibit a startle response elicited by a different sensory modality. These are the two most common modalities to elicit a startle response when studying prepulse inhibition, but visual prepulse stimuli have also been used.

The following formula was used to calculate % PPI of a startle response: 100 − [(startle response on acoustic prepulse and startle stimulus trials/startle response alone trials) × 100].

Results showed that in TEST 1, +/+ mice displayed significantly higher levels of prepulse inhibition than −/− mice. TEST 2 replicated these findings. In addition, +/+ mice displayed greater responses to acoustic startle stimuli than −/− mice in both TEST 1 and TEST 2. Thus, in two independent experiments wild-type mice displayed significantly higher overall levels of PPI and greater startle responses than *Dvl*1-deficient mice.

Similar to the results found for the acoustic startle response, *Dvl*1 mutant mice displayed overall significantly lower levels of PPI of the tactile startle response. The tactile startle response, however, was not significantly different between *Dvl*1-deficient and wild-type mice.

These findings clearly indicate that sensorimotor gating processes, as measured using the PPI paradigm, are attenuated in *Dvl*1-deficient mice. In addition, wild-type mice tend to have greater startle response to acoustic, but not tactile, startle stimuli. A PPI impairment can be associated with hearing deficits. However, it is unlikely that the observed differences in PPI between wild-type and *Dvl*1-deficient mice were the result of poor hearing in the mutant mice. *Dvl*1 mutant mice displayed PPI at prepulse sound intensities only 4 dB above background noise levels. In addition, auditory evoked responses in the hippocampus were similar in mutant and wild-type mice. Thus, *Dvl*1 mutant mice can detect a sound change of 4 dB, and neuronal processing of auditory stimuli — at least to the level of the hippocampus — is similar to wild-type mice. To confirm that the PPI impairment in *Dvl*1 mutant mice is not the result of hearing loss, auditory detection and discrimination studies will be employed in the future to directly examine the role of *Dvl*1 in audition.

As mentioned above these experiments were prompted by the findings that a *Dvl*-like sequence

Acoustic Prepulse Inhibition of the Acoustic Startle Response

Startle Stimulus Trial

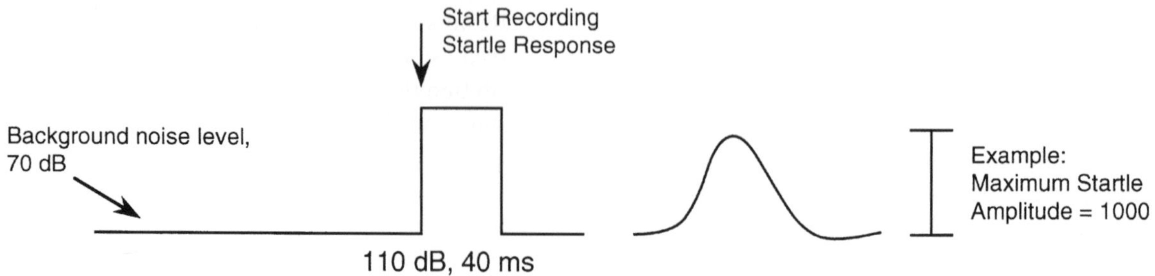

Prepulse Inhibition Trial

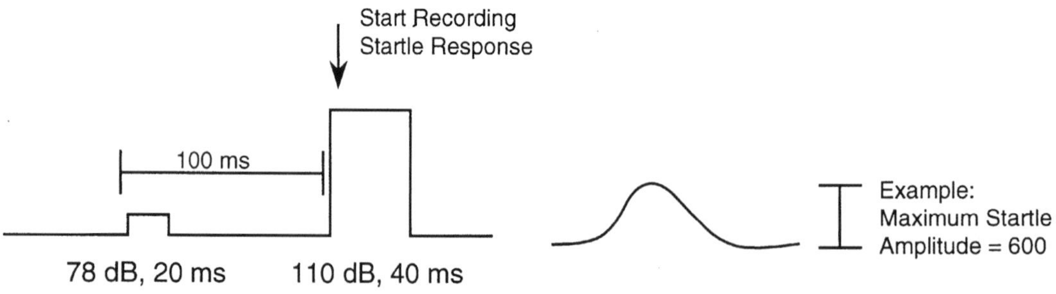

Example:
% PPI = 100 - [(600 / 1000) X 100] = 100 - 60
= 40 % PPI

Fig. 1. Representation of acoustic startle alone and prepulse inhibition trial types. In this example, on some trials animals would only receive a 40 ms, 110 dB acoustic startle stimuli. On other trials the animals would receive the same startle stimulus, but it would be preceded by a 20 ms, 78 dB acoustic prepulse. The onset of the prepulse is 100 ms prior to the onset of the startle stimulus. The maximum response measured in arbitrary units is recorded at the onset of the startle stimulus for both trial types. For all the trials the background noise level would be 70 dB. In this example, the maximum startle amplitude averaged across all startle stimulus alone trial types during a testing session was 1000. The maximum startle response during the prepulse inhibition trials with 78 dB prepulse stimuli averaged across all of these prepulse trial types was 600. Therefore, the percentage PPI to the 78 dB prepulse would be 40%.

mapped to human chromosome 22, within a region linked to schizophrenia (Pizzuti, et al., 1996). Further analysis of the sequence present in this region, however, is homologous only to the 3' untranslated region of the mouse *Dvl*1 gene. The *Dvl*1 mouse gene maps to chromosome 4 (Beier et al., 1992) which is syntenic to human 1p36. Though none of the mouse *Dvl* genes map to chromosome 22, we found that *Dvl*1-deficient mice have impaired behavioral responses in a test that is also impaired in a number of neuropsychiatric conditions including schizophrenia. Future studies will be necessary to determine if there are actual sequence differences in human *DVL* genes that are associated with impaired sensorimotor gating measured in-patients with neuropsychiatric disorders.

Practical considerations for measuring startle responses and PPI

When testing for startle response and PPI using more than one of the SR-Lab Systems it is essential to calibrate the sensitivity of the chambers using a vibrating standardization unit (San Diego Instruments) that emulates an animal's response to ensure that the sensitivity of the chambers are not different. In addition, it is important to report the actual readings using this standardization unit (see Paylor and Crawley, 1997) since differences in sensitivity can have a dramatic impact on the response output, and will impair the ability to compare data across laboratories. Also, it is important to consider that when a very low startle response is obtained using a system that is not sufficiently sensitive it becomes virtually impossible to detect subtle changes in that startle response produced by a prepulse. Thus, it is important to verify that any startle system is capable of detecting low level startle responses *and* attenuation of the startle response with prepulse stimuli. To verify that one's system is sufficiently sensitive to subtle changes in the startle response we would recommend testing a few inbred strains of mice that are known to have low levels of startle but that have substantial levels of PPI (Paylor and Crawley, 1997).

The amount of variability observed using the prepulse inhibition paradigm is similar to that for other behavioral tests commonly used to identify behavioral phenotypes in mutant mice. For that reason, we recommend that approximately 10 male and 10 female mice of each genotype be evaluated. This sample size should be large enough to detect reliable differences between mutant and wild-type mice.

A role for Dvl*1 in mouse social behaviors*

During the course of behavioral testing, we noticed that wild-type mice did not have whiskers. In contrast, every mutant mouse had a full compliment of whiskers. It is important to point out that *Dvl*1-deficient mice and wild-type mice derived from heterozygous matings were housed separately once they were genotyped. In order to determine if this peculiar difference in the presence of whiskers indicated another functional difference between wild-type and mutant mice we designed a set of experiments to answer the following questions based on the reports that whisker trimming is a social behavior (Strozik and Festing, 1981) in certain strains of mice (Hauschka, 1952; Long, 1972):

1. How prevalent was this difference in the presence of whiskers between wild-type and mutant mice?
2. Is the absence of whiskers in wild-type mice related to social interactions with other wild-type mice?
3. Since whisker-trimming can be related to social dominance, do wild-type mice display social dominance over mutant mice?

Whisker trimming

To determine the prevalence of the whisker difference between wild-type and mutant mice we recorded the presence or absence of whiskers in approximately 100 wild-type and mutant mice at various ages. At the age of weaning, all wild-type and mutant mice had full sets of whiskers, but with age, wild-type mice lost whiskers and facial hair. More specifically, at 2–3 months of age approximately 50% of the wild-type mice had no whiskers, and by 6 months of age that percentage was up to approximately 75%. In contrast, 100% of −/− mice had complete sets of whiskers and facial hair. Although some wild-type mice had whiskers, the overall difference in whisker pattern between wild-type and mutant mice was quite profound throughout the entire population of *Dvl*1-deficient and wild-type mice.

To determine if the whisker loss observed in the wild-type mice resulted from social interactions when housed with other wild-type mice, we housed one +/+ mouse that was devoid of whiskers and facial hair with one −/− mouse that had full whiskers and hair, in eleven cages, and observed the changes in whisker pattern after mixing. In all eleven pairs, 100% of the wild-type mice regrew full sets of whiskers and facial hair within 2–4 weeks. In addition, about 50% of the mutant mice lost all whiskers and facial hair. The wild-type mice were then returned to their homecage with their original wild-type littermates. Within two weeks, the whiskers were absent on every wild-type mouse. In addition, every whiskerless *Dvl*1-mutant mouse regrew whiskers within two weeks after being returned to its homecage. These findings clearly indicate that the absence of whiskers in wild-type mice results from being housed with other wild-type mice which suggests that wild-type mice interact differently in their homecage than mutant mice.

In addition, there were a few wild-type mice that had to be housed individually during the course of the study. Each wild-type mouse that was housed individually had a full set of whiskers which indicates that the whisker loss in wild-type mice does not result from excessive self grooming, and helps rule out the presence of bacterial or parasitic infections as the cause of the whisker loss in the wild-type mice.

Social dominance

Since whisker trimming is often associated with dominance (Long, 1972; Strozik and Festing, 1981), a social dominance tube test was performed (Lindzey et al., 1961; Messeri et al., 1975). Eleven wild-type and 11 mutant mice were tested using a white tube that was 30 cm long and 3.5 cm diameter (3.0 cm diameter for females). A wild-type and mutant mouse of the same gender, and similar weights were placed at opposite ends of the tube and released. A subject was declared a "winner" when its opponent had backed out of the tube, exposing the entire length of its tail.

Twenty-two trials were performed with eleven wild-type and eleven mutant mice. Each mouse was tested twice with different mice of the same gender. In 73% of the trials, the mutant mice backed out of the tube. Specifically, one *Dvl*1-deficient mouse won both of its trials, four *Dvl*1-mutants won one trial, and six mutants did not win either trial. In contrast, there were seven wild-types that won both their trials, two wild-types that won one trial, and two wild-types that did not win either trial. These results demonstrate that, when paired against the other genotype, wild-type mice are more dominant than the mutant mice. These findings are consistent with the hypothesis that there are differences in social behaviors between wildtype and *Dvl*1-deficient mice.

Practical considerations for using the tube-test

When using this test it is important that the tube is small enough that the two mice cannot walk over each other. The tube should be opaque so that mice are not easily distracted by stimuli or movement outside of the tube. Finally, prior to testing it is useful to establish that all the mice will readily walk through the tube when there is no other mouse present.

Home-cage observations

From anecdotal observations of mice in their homecages, it appeared that the *Dvl*1-deficient mice interacted less frequently than wild-type mice, suggesting that the mutant mice have attenuated social interactions. To assess homecage behaviors over extended periods of time, we videorecorded the behavior of two cages of wild-type and two cages of mutant mice for 30 total hours, and scored the mice for different social and non-social behaviors. During 2 h of behavioral scoring, wild-type mice displayed more social behaviors (social grooming, mounting, tail pulling and sniffing), than *Dvl*1-deficient mice. In contrast, the frequency of self-grooming behaviors was similar between wild-type and mutant mice. These preliminary observations suggested that the *Dvl*1 mutant mice displayed reduced social interactions, and that this decrease may be related to differences in whisker-trimming between wild-type and mutant mice.

Further examination of the videorecorded behaviors showed that wild-type mice generally slept huddled together more often than mutant mice. Based on these preliminary observations, sleeping patterns were analyzed in greater detail. We recorded the sleeping positions of mice in their homecages twice daily over five consecutive days to quantitate differences in sleeping patterns. Six cages of wild-type mice and five cages of mutant mice ($n = 4$ mice per cage) were observed in their homecage, and the position and behavior of each mouse was recorded. The percentage of subjects sleeping huddled in the same quadrant in each cage was determined. Nine observations were made over a 5-d test period. The results showed that mutant mice will sleep huddled, but that they do so significantly less than wild-type mice.

Practical considerations for evaluating sleeping patterns

It is important to have the cages situated in such a manner that it is easy to view the entire side without moving the cage. When a cage is moved during the observation interval the mice will often move around so that an accurate assessment of their position in the cage is difficult. Since cage changing increases the activity of mice it is essential to record when the cages are changed. When possible, cage observations should be made before the cages are changed.

In addition to differences in sleeping patterns between *Dvl*1-deficient and wild-type mice, pilot studies indicated that there were differences in nest building and sleeping in nests. To further characterize these apparent differences, a piece of cotton nesting material was placed in the bottom of six cages of wild-type mice and six cages of mutant mice. Forty-five minutes later, photographs were taken of each nest and the depth of the nest was measured. Wild-type mice built and huddled in fluffy, well-formed nests. In contrast, *Dvl*1-deficient mice did not build full nests and they tended to be huddled in scattered, random patterns on top of the nestlet material. Nest-building was quantified by measuring the depth of nests. Wild-type mice built nests that averaged 50 mm in depth, while *Dvl*1 mutant mice built significantly more shallow nests with depths that averaged less than 20 mm.

Practical considerations for immediate evaluation of nesting

Record the time it takes for mice to finish working with the nesting material prior to photographing the nest. Arrange the cages such that they can be photographed easily without disturbing the mice.

Taken together, the abnormal whisker-trimming behavior, the decrease in huddling while sleeping and the poor nest-building displayed by *Dvl*1-deficient mice demonstrate an important role for *Dvl*1 in normal homecage social behavior. Other experiments will be needed to establish the exact nature of these homecage behavioral differences, and to determine if social interaction abnormalities in *Dvl*1-deficient mice extend to other social behaviors such as aggression.

It is important to note that the differences in whisker trimming, social dominance, and nesting were observed in both male and female mice, so there appears to be no gender specificity to this phenotype.

Discussion

Table 1 presents a summary of the characterization of the *Dvl*1-deficient mice. Our analysis of *Dvl*1-deficient mice has revealed the surprising result that *Dvl*1 participates in complex behaviors in mammals. *Dvl*1 mutant mice displayed striking abnormalities in sensorimotor gating and social behaviors, but motor, sensory and spatial learning processes were grossly normal. In addition, no structural abnormalities were found in any tissue from the mutant mice. All experiments were performed in completely inbred mice, so strain background differences could not account for these defects. In addition, there were no observable phenotypic differences between mice generated from heterozygous or uniform-genotype crosses, which suggests that parental effects are unlikely to have had significant impact on the observed phenotype of the *Dvl*1-deficient mice. However, more systematic studies will be necessary to determine if differences in parental behavior between *Dvl*1-deficient and wild-type mice can account for the impaired prepulse inhibition and social interactions observed in the mutant mice.

There are two potential models to explain the participation of *Dvl*1 in complex behaviors such as social behavior and sensorimotor gating. *Dvl*1 could participate in developmental processes of central nervous system structures critical for these behaviors, or *Dvl*1 may have distinct post-developmental effects in the brain. These alternative hypotheses will be explored in future studies that include behavioral characterization of inducible and conditional *Dvl*1-deficient mice.

The genetics of social behavior have proved difficult to study, since interaction requires the participation of at least two individuals, and the behavioral output of a group cannot be easily attributed to the genetic complement of one individual (Hahn and Schanz, 1996). We were able to distinguish homecage social behaviors attributable to *Dvl*1 function for two reasons: first, the mice studied were of uniform 129/SvEv genetic background, except for the *Dvl*1 genotype; and second, mice were housed as uniform genotypes, so that long term observations could be made on individual genotypes. Social behavior was initially examined because of striking differences in the absence and presence of whiskers in wild-type and mutant mice housed as uniform genotypes. Video recorded observations suggested that mutant mice did not interact as frequently as wild-type mice, and we were able to design experiments to quantitate these differences.

TABLE 1

Phenotypic abnormalities in 129/SvEv *Dvl*1-deficient mice

Abnormal phenotypes	Normal phenotypes
Reduced acoustic startle response	Locomotor activity
Attenuated PPI of the acoustic startle response	Rotarod
Attenuated PPI of the tactile startle response	Wire hang
No whisker trimming	Spatial learning
Submissive response in tube test	Tactile startle response
Reduced huddled sleeping	Body weight
Poor nest building	Body temperature
	Brain morphology

Genotype mixing experiments, and the social dominance tube test, indicated that the differences in whisker-trimming behavior between wild-type and mutant mice likely resulted from differences in social interactions. In addition, differences in sleeping patterns and nest-building support the hypothesis that *Dvl*1-deficient mice have abnormal homecage behavior and social interaction, compared to wild-type controls.

The PPI deficits of *Dvl*1-deficient mice suggest that *Dvl*1 contributes to normal sensorimotor gating. The magnitude of the acoustic startle response was reduced in mutant mice, demonstrating that *Dvl*1 is important for the acoustic startle reflex. The depressed startle response in mutant mice is unlikely to have contributed to impairment of prepulse inhibition response for two reasons. First, the tactile startle response is similar between the two genotypes, yet acoustic prepulse inhibition of the tactile startle response was also depressed in the mutant mice. Second, results from pharmacological analysis (Bakshi et al., 1994; Johansson et al., 1995; Mansbach et al., 1988) and mouse inbred strains (Paylor and Crawley, 1997) have clearly dissociated the startle and prepulse inhibition responses. Thus, *Dvl*1 contributes to the normal auditory reflex as well as to the modulation of startle responses by higher brain structures.

*Dvl*1 is the first gene identified with effects on both sensorimotor gating and social behaviors. It is interesting to speculate that sensorimotor gating defects and abnormal social interactions may be related, based on the phenotype of *Dvl*1-deficient mice. Many patients with schizophrenia, Tourette's syndrome, and schizotypal personality disorder display both types of abnormalities (reviewed in Geyer and Braff, 1987; Pickar, 1995). Experimentally, social isolation of rats resulted in defective prepulse inhibition of startle (Geyer et al., 1993; Wilkinson et al., 1994), associated with enhancement of amphetamine-mediated dopamine release (Wilkinson et al., 1994). In addition, social interactions in mice can be altered by antipsychotics (see Dixon and Fisch, 1989). Thus, social interaction and sensorimotor gating may both be mediated by similar neurotransmitter pathways, and *Dvl*1 could conceivably participate in synaptic transmission through its PDZ domain (Bryant et al., 1993; Ponting and Phillips, 1995). Thus *Dvl*1-deficient mice should provide a genetic animal model to study factors which influence sensorimotor gating and social behaviors in mice. For example, *Dvl*1-deficient mice may be a useful model for screening drugs that modify abnormal social interaction and sensorimotor gating.

Genetic factors are important modifiers of a variety of simple and complex behaviors in virtually all organisms. Genetic effects have been inferred from inbred strain analysis in rodents, or from linkage analysis in rodents and humans, but these studies have not yet resulted in the isolation of specific genes involved in behavioral modification. More recently, genes influencing specific behaviors have been identified by analyzing behavioral abnormalities in mice with targeted gene disruption. In many cases, unexpected behavioral defects were found in mice (e.g. Chapter 3.1.8) that were generated by gene targeting to evaluate the general functions of genes expressed in the central nervous system, often providing novel insights into processes that influence behavioral variation. We have identified *Dvl*1 as a genetic factor influencing social behavior and sensorimotor gating. The availability of *Dvl*1 mutant mice will provide an opportunity to study the molecular basis for these defects, and to test whether other members of the *Wnt*-signalling pathway participate in similar complex behavioral traits. In addition, our data suggest that *Dvls,* and possibly other *Wnt* pathway members, may be candidate genes for neuropsychiatric disorders.

Comprehensive behavioral analysis of mutant mice is an essential step to better understand the functional role of genes. We believe that the findings with the *Dvl*1-mutant mice represents an excellent example of how unexpected behavioral phenotypical alterations can be discovered in mice with gene targeted mutations and how such discoveries can provide information about the functional relevance of certain genes.

References

Bakshi, V.P., Swerdlow, N.R. and Geyer, M.A. (1994) Clozapine antagonizes phencyclidine-induced deficits in sensorimotor gating of the startle response. *J. Pharmacol. Exp. Ther.*, 271: 787–794.

Beier, D.R., Dushkin, H. and Sussman, D.J. (1992) Mapping genes in the mouse using single-strand conformation polymorphism analysis of recombinant inbred strains and interspecific crosses. *Proc. Natl. Acad. Sci. USA*, 89: 9102–9106.

Braff, D., Stone, D., Callaway, E., Geyer, M.A., Glick, I., and Bali, L. (1978) Prestimulus effects on human startle reflex in normals and schizophrenics. *Psychophysiology*, 15: 339–343.

Braff, D.L., and Geyer, M.A. (1990). Sensorimotor gating and schizophrenia. *Archives of Gen. Psychiatry*, 47: 181–188.

Bryant, P.J., Watson, K.L., Justice, R.W. and Woods, D.F. (1993) Tumor suppressor genes encoding proteins for cell interactions and signal transduction in *Drosophila. Development*, (Suppl.) 118: 239–245.

Cadenhead, K.S., Geyer, M.A. and Braff, D.L. (1993) Impaired startle prepulse inhibition and habituation in patients with schizotypal personality disorder. *Am. J. Psychiatry*, 150: 1862–1867.

Castellanos, F.X., Fine, E.J., Kaysen, D., Marsh, W.L., Rapoport, J.L. and Hallett, M. (1996) Sensorimotor gating in boys with Tourette's syndrome and ADHD: preliminary results. *Biol. Psychiatry*, 39: 33–41.

Deng, C., Wynshaw-Boris, A., Zhou, F., Kuo, A. and Leder, P. (1996) Fibroblast growth factor receptor 3 is a negative regulator of bone growth. *Cell*, 84: 911–921.

Deng, C.X., Wynshaw-Boris, A., Shen, M.M., Daugherty, C., Ornitz, D.M. and Leder, P. (1994) Murine FGFR-1 is required for early postimplantation growth and axial organization. *Genes Dev.*, 8: 3045–3057.

Dixon, A.K. and Fisch, H.U. (1989) The ethopharmacological study of drug induced changes in behavior. In: R.J. Blanchard et al. (Eds), *Ethoexperimental Approaches to the Study of Behavior*. Kluwer Academic Publishers, The Netherlands, pp. 451–473.

Geyer, M.A. and Braff, D.L. (1987) Startle habituation and sensorimotor gating in schizophrenia and related animal models. *Schizophr. Bull.*, 13: 643–68.

Geyer, M.A., Wilkinsin, L.S., Humby, T. and Robbins, T.W. (1993) Isolation rearing of rats produces a deficit in prepulse inhibition of acoustic startle similar to that in schizophrenia. *Biol. Psychiatry*, 34: 361–372.

Graham, F.K. (1975) The more or less startling effects of weak prestimulation. *Psychophysiology*, 12: 238–248.

Grillon, C., Ameli, R., Charney, D.S., Krystal, J. and Braff, D. (1992) Startle gating deficits occur across prepulse intensities in schizophrenic patients. *Bio. Psychiatry*, 32: 939–943.

Hahn, M.D. and Schanz, N. (1996) Issues in the genetics of social behavior: revisited. *Behav. Genet.*, 26: 463–470.

Hauschka, T. (1952) Whisker-eating mice. *J. Hered.*, 43: 77–80.

Ison, J.R., McAdam, D.W. and Hammond, G.R. (1973) Latency and amplitude changes in the acoustic startle reflex of the rat produced by variation in auditory prestimulation. *Physiol. Behav.*, 10: 1035–1039.

Johansson, C., Jackson, D.M., Zhang, J. and Svensson, L. (1995) Prepulse inhibition of acoustic startle, a measure of sensorimotor gating: effects of antipsychotics and other agents in rats. *Pharmacol. Biochem. Behav.*, 52: 649–654.

Klingensmith, J., Nusse, R. and Perrimon, N. (1994) The *Drosophila* segment polarity gene dishevelled encodes a novel protein required for response to the wingless signal. *Genes Dev.*, 8: 118–130.

Klingensmith, J., Yang, Y., Axelrod, J.D., Beier, D.R., Perrimon, N. and Sussman, D.J. (1996) Conservation of *dishevelled* structure and function between flies and mice: isolation and characterization of *Dvl2*. *Mech. Devel.*, 58: 15–26.

Lijam, N., Paylor, R., McDonald, M.P., Crawley, J.N., Deng, C.-X., Herrup, K., Stevens, K.E., Maccaferri, G., McBain, C.J., Sussman, D.J. and Wynshaw-Boris, A. (1997) Social interaction and sensorimotor gating abnormalities in mice lacking *Dvl*1. *Cell*, 90: 895–905.

Lindzey, G., Winston, H. and Manosevitz, M. (1961) Social dominance in inbred mouse strains. *Nature*, 191: 474–476.

Long, S.Y. (1972) Hair-nibbling and whisker-trimming as indicators of social hierarchy in mice. *Anim. Behav.*, 20: 10–12.

Mansbach, R.S., Geyer, M.A. and Braff, D.L. (1988) Dopaminergic stimulation disrupts sensorimotor gating in the rat. *Psychopharmacology*, 94: 507–514.

McDowd, J.M., Fillon, D.L., Harris, M.J. and Braff, D.L. (1993) Sensory gating and inhibitory function in late-life schizophrenia. *Schizophr. Bull.*, 19: 733–746.

Messeri, P., Eleftheriou, B.E. and Oliverio, A. (1975) Dominance behavior: a phylogenetic analysis in the mouse. *Physiol. Behav.*, 14: 53–58.

Moon, R.T., Brown, J.D. and Torres, M. (1997) WNTs modulate cell fate and behavior during vertebrate development. *Trends Genet.*, 13: 157–162.

Morris, R.G.M. (1981) Spatial localization does not depend upon the presence of local cues. *Learning and Motivation*, 12: 239–260.

Paylor, R. and Crawley, J.N. (1997). Inbred strain differences in prepulse inhibition of the mouse startle response. *Psychopharmacology*, in press.

Pickar, D. (1995) Prospects for pharmacotherapy of schizophrenia. *Lancet*, 345: 557–562.

Pizzuti, A., Novelli, G., Mari, A., Ratti, A., Colosimo, A., Amati, F., Penso, D., Sangiuolo, F., Calabrese, G., Palka, G., Silani, V., Gennarelli, M., Mingarelli, R., Scarlato, G., Scambler, P. and Dallapiccola, B. (1996) Human homologue sequences to the *Drosophila* dishevelled segment-polarity gene are deleted in the DiGeorge syndrome. *Am. J. Hum. Genet.*, 58: 722–729.

Ponting, C.P. and Phillips, C. (1995) DHR domains in syntrophins, neuronal NO synthetase and other intracellular proteins. *Trends Biochem.*, 20: 102–103.

Ponting, C.R. and Bork, P. (1996) Pleckstrin's repeat performance: a novel domain in G-protein signaling? *Trends Biol. Sci.*, 21: 245–246.

Steitz, S.A., Tsang, M. and Sussman, D.J. (1996) Wnt-mediated relocalization of dishevelled proteins. *In Vitro Cell. Dev. Biol. Anim.*, 32: 441–445.

Strozik, E. and Festing, M.F. (1981) Whisker trimming in mice. *Lab. Anim.*, 15: 309–312.

Sussman, D.J., Klingensmith, J., Salinas, P., Adams, P.S., Nusse, R. and Perrimon, N. (1994) Isolation and characterization of a mouse homolog of the *Drosophila* segment polarity gene dishevelled. *Dev. Biol.*, 166: 73–86.

Swerdlow, N.R., Benbow, C.H., Zisook, S., Geyer, M.A. and Braff, D.L. (1993) A preliminary assessment of sensorimotor gating in patients with obsessive compulsive disorder. *Biol. Psychiatry*, 33: 298–301.

Swerdlow, N.R., Braff, D.L., Taaid, N. and Geyer, M.A. (1994) Assessing the validity of an animal model of deficient sensorimotor gating in schizophrenic patients. *Arch. Gen. Psychiatry*, 51: 139–154.

Swerdlow, N.R., Paulsen, J., Braff, D.L., Butters, N., Geyer, M.A. and Swenson, M.R. (1995) Impaired prepulse inhibition of acoustic and tactile startle response in patients with Huntington's disease. *J. Neurol. Neurosurg. Psychiatry*, 58: 192–200.

Theisen, H., Purcell, J., Bennett, M., Kansagara, D., Syed, A. and Marsh, J.L. (1994) Dishevelled is required during wingless signaling to establish both cell polarity and cell identity. *Development*, 120: 347–360.

Tsang, M., Lijam, N., Yang, Y., Beier, D.R., Wynshaw-Boris, A. and Sussman, D.J. (1996) Isolation and characterization of mouse *Dishevelled*-3. *Devel. Dyn.*, 207: 253–262.

Wilkinson, L.S., Killcross, S.S., Humby, T., Hall, F.S., Geyer, M.A. and Robbins, T.W. (1994) Social isolation in the rat produces developmentally specific deficits in prepulse inhibition of the acoustic startle response without disrupting latent inhibition. *Neuropsychopharmacology*, 10: 61–72.

W.E. Crusio and R.T. Gerlai (Eds.)
Handbook of Molecular-Genetic Techniques for Brain and Behavior Research (Techniques in the Behavioral and Neural Sciences, Vol. 13)

CHAPTER 3.1.10

Targeting genes associated with mammalian behavior: past mistakes and future solutions

Robert Gerlai

Genentech, Inc., Neuroscience Department, Mail stop # 72, 1 DNA Way, South San Francisco, CA 94080-4990, USA

Introduction

With the advancement of the human genome project and other projects aimed at the sequencing of the entire genome of other organisms one can expect a wealth of genetic information to be available for the scientific community. The challenge then will be to decipher the role the newly discovered genes may play. In mammals, approximately 30% of the genes are expressed in the brain. Therefore, I predict there will be a tremendous need for behavioral and neurobiological functional analysis. An elegant way to test gene function has been used in behavioral neuroscience. The approach is called gene targeting. Several chapters of Section 4 of this book discuss this technique and its utility in brain and behavior research. In this chapter I focus on critical issues important in gene targeting experiments that have been somewhat overlooked in the past. The problems I raise are not to imply that gene targeting is not useful. My intention is to elucidate the potential pitfalls of this technique in order to allow its better utilization and appropriate interpretation of data it generates.

Gene targeting allows one to create null mutations in mice and to analyze how the mutant organism responds to the lack of a single gene product. This has facilitated the molecular dissection of such complex characteristics as learning, memory, aggression, maternal behavior (for reviews see e.g. ; Grant and Silva, 1994; Mayford et al., 1995). However, the interpretation of the phenotypical changes which arise in null mutant mice has been shown to be subject to several caveats. The effect of other genes, the genetic background, has often been overlooked, and the importance of compensatory mechanisms has been underestimated leading to potential misinterpretation of results. It has been argued that first, the phenotypical abnormalities attributed to the null mutation in several molecular neurobiological studies could result from the effects of background genes and, second, that "compensatory" mechanisms may lead either to compensation (false negative) or to secondary phenotypical alterations (false positive) not directly related to the function of the gene of interest. The present review will discuss these points and review recent advances that both provide empirical proof for the above arguments and represent potential solutions to these caveats.

The gene targeting debate

Three years ago Gerlai (1996a) raised several points based on elementary classical genetic considerations that fundamentally questioned the interpretation of a large number of gene targeting studies. His debate article was followed by commentaries that confirmed and extended these arguments (Crawley, 1996; Crusio, 1996; Lathe, 1996). The problems pointed out in these papers did not prove that the cited gene targeting studies were wrong, but clearly indicated that there is room for alternative interpretation. Given the large amount of money and time invested in each of these gene targeting experiments and given the popularity and touted advantages of the technique, the 1996 debate papers attracted much attention and induced a considerable debate regarding the utility of gene targeting and the potential solutions to the problems raised (Lipp and Wolfer, 1998).

In this review I will focus on two issues both of which have general importance in gene targeting experiments concerned with brain and behavior function. First I will discuss the problem of compensatory mechanisms. This problem is rather difficult as there is no clear practical solution to avoid it. Furthermore, as it will become clear, "compensation" and "gene function" are terms that really depend on one's definition. Discussion on what these terms mean or what they may be is rather philosophical. Nevertheless, the reader may need to learn about this as it may help to understand the utility and limitations of the gene targeting technique. The second topic of this review is the genetic background and linkage (flanking region) problem and its relevance in gene targeting. Here I present several practical solutions as well. While the examples will be drawn mostly from the learning and memory field, the points they illustrate will be valid for any behavioral or neurobiological characteristics.

Compensatory mechanisms: the promise and the complexity

With gene targeting one can practically knock out a gene in vivo and create a mutant organism that completely lacks the gene product. Thus, the promise of gene targeting has been to reveal the in vivo function of the gene of interest (Grant et al., 1992; Silva et al., 1992; Abeliovich et al. 1993). For example, Grant et al. (1992) wrote: "...targeted disruption of genes provides a powerful tool for examining the role of specific proteins in the function of the brain". However, the functional relevance of gene targeting has been questioned (Morris and Kennedy, 1992; Rose, 1995; Routtenberg, 1995) because the mutation may lead to an avalanche of compensatory processes (up or down regulation of gene products) and resulting secondary phenotypical changes. Clearly, a null mutant organism may not only lack the product of a single gene but may also possess a number of developmental, physiological, or even behavioral processes that have been altered to compensate for the effect of the null mutation. Therefore, one may expect an array of complex phenotypical changes that may not be directly related to the function of the gene of interest.

The "helper" genes

Compensation may be due to genetic redundancy. Genetic redundancy in this context means that some putative "helper" genes may be able to take over the function of the targeted one and, for example, become upregulated and thus compensate for the absence of the targeted gene product. Crawley (1996) explains that compensatory changes can mask the functional outcome of the mutation, and indeed this is the most generally accepted interpretation prevalent in the literature. The phenomenon of compensatory mechanisms and the apparent absence of phenotypical effects of gene targeting in certain molecular neurobiological studies has led some to conclude that the targeted mutagenesis technique is "wholly inappropriate for resolving the issues for which it was intended" (Routtenberg, 1995). Others disagree with this view and argue that the inability to observe a phenotypical change may not mean that there is no phenotypical change to observe (Hochgeschwender and Brennan, 1995). This latter view is supported by the findings of Chen et al. (1995) who studied the phenotypical effects of a null mutation in protein kinase C γ

subtype (PKCγ) in mice. Interestingly, they found that both null mutant and control mice exhibited apparently normal long-term depression (LTD) in their cerebellum, a neurophysiological phenomenon thought to be dependent upon PKC activity. However, LTD could be blocked by a specific PKC inhibitor (PKC[19–36]) in control mice but not in the null mutants suggesting that LTD in the null mutant mice was mediated, at least partly, by non-PKC dependent processes. The authors suggest that some, yet unknown, alternative biochemical pathways might have been activated in the null mutant mice and these pathways were able to compensate for the lack of PKCγ and support apparently normal LTD. Gene targeting therefore may enable the investigator to reveal alternative biochemical pathways, functional interactions between the targeted gene and other genes.

Compensation for disrupted function 'A' leading to altered function in 'B'

It is also important to realize that a "compensatory" change (a form of epistatic gene interaction) may not always lead to an absence of phenotypical change (false negative findings) but, on the contrary, may be the *cause* of observed phenotypical abnormalities. For example, assume gene α serves hypothetical function '**A**'. Also assume that targeted gene α is compensated for by gene $\boldsymbol{\beta}$ which becomes upregulated in response to the absence of α gene product. The excess of gene $\boldsymbol{\beta}$ product is able to compensate for the lack of gene α product and no change is observed in function '**A**' at the phenotypical level. However, overexpression of gene $\boldsymbol{\beta}$ product may have some pleiotropic effects, i.e. may affect functions other than '**A**', similarly to the ways overexpression of genes alters brain function and behavior in transgenic mice (Chen et al., 1995; Gerlai et al., 1995). These functional alterations when observed at the phenotypical level by the investigator will be assigned to gene α. Although such phenotypical alterations may be due to the introduced mutation, they need not reveal *the* function of the gene of interest per se because they may be related to it *only indirectly*. A skeptic could say that it is really impossible to assign specific functions to specific genes: genes act in concert and a disruption, let it be as targeted as one would like it to be, will always lead to a complex systemic response. Crusio (1996) gives a simple, two component, example to show the interdependence of molecular components giving rise to complexity at the phenotypical level. He explains that pharmacogenetic analyses of inbred strains of mice (Van Abeelen et al., 1975; Van Abeelen, 1989), C57BL/6 and DBA/2, led to the conclusion that hippocampal dependent exploratory behaviors rely on a delicate balance of acetylcholinesterase and acetylcholine levels. Therefore, the functional effect of genetic disruption of either protein would depend upon the expression level of the other.

Compensation, however, may not necessarily originate at the molecular level. Behavioral compensation translating into secondary molecular changes is also possible. Assume you have a mouse with genetically disrupted olfaction. This alteration may force the mouse to prefer visual stimuli to olfactory, which in turn may lead to multiple changes in neural processes and brain areas involved in processing visual stimuli. An investigator then may conclude that the targeted gene plays a crucial role in vision. Gene targeting studies trying to understand gene function in the organism may face similar enigmas.

Teasing out the direct and indirect effects of the mutation is certainly not trivial and it is clear that dissection of the molecular/neurobiological mechanisms underlying complex behavioral characteristics will require meticulous studies in which all possible affecting factors need to be well controlled. Lathe (1996) discusses several confounding factors that must be taken into serious consideration in all molecular genetic studies of brain and behavior. Instead of reiterating his arguments point by point I will summarize what I think is the main issue, i.e. the necessity of using a "systemic approach" in gene targeting (Gerlai, 1996b).

Biological organization is the important issue

Although, the questions of genetic redundancy or compensatory mechanisms, or other factors confounding the effects of whatever artificial perturbance we introduce to the living system are important, the principal problem is really a systemic one that concerns biological organization and the functional units of this organization. From a geneticist's viewpoint the units of biological organization are clearly the genes and their function is to encode particular proteins. However, one may argue that when it comes to the question of phenotypical effects, genes may not be the "units" and the definition of their function may be more complicated. One may suggest that clusters of genes defined by higher organizational level phenomena, including developmental, physiological, or even behavioral ones, may represent the functionally relevant unit. Disruption of a single gene may alter biochemical processes within the functional gene cluster. Expression levels of the genes belonging to a functional cluster may change in concert. Investigation of such changes, for example by gene array technologies or other high throughput expression detectors, may allow the experimenter to reveal the biological organization of the brain. Perhaps the boundaries of these putative gene clusters are not sharp. Some genes may belong more, others less, to a specific functional gene group. This also implies that the gene group organization may not be orthogonal, i.e. some genes may belong to more than one functional group. It is also probable that there may be a hierarchical organization of such functional groups. A smaller number of genes may define subgroups that may make up groups that in turn may be organized into super-groups, etc. In any case, disrupting single genes will perturb the organism and will force it to respond in a way inherent to its biological organization. It is crucial to bear in mind, therefore, that the phenotypical changes we observe are the reflections of this organization. In addition to looking for the function of particular genes, I propose that investigators should also take this systemic organizational view into consideration when they search for phenotypical effects of null mutation or try to interpret the results of a gene targeting study. A conceptually similar approach, in which the effects of multiple system components are analyzed at the system properties level (see Metabolic Control Analysis) (Kacsér and Burns, 1973), has gained acceptance in biochemistry (Fell, 1992; Fell and Thomas, 1995).

Easier said than done

Indeed, it is easy to outline what we should do. But, can we, and how? For example, how can we study the complex system and analyze the functionally relevant unit. There is no simple answer. Several chapters in this book discuss techniques that can be applied to detect gene expression changes in multiple genes. These include differential display (Chapter 2.2.2), subtractive hybridization (Chapter 2.2.1), quantitative "TaqMan" RT-PCR (Chapter 2.2.4). These techniques or the new gene chip technology may enable one to see how genes respond in concert. One may also be able to manipulate, say knock out, the functional unit of several genes itself, instead of a single gene. An example for the latter is explained in Chapter 3.2.7 of this book and is called protein targeting because the manipulation is done at the protein level. This latter approach also enables the investigator to time the manipulation and, for example, limit its effects to post-developmental stages, or perhaps only to particular time windows of memory consolidation or recall.

Genetic background: an important confounding factor

The second topic of this chapter considers the effect of genetic background. It is important to realize that polymorphism in the genetic background may make the results of gene targeting studies difficult to interpret. Consider the following example. Assume that targeted disruption of gene α leads to a differential expression of alleles **b** and **B** of gene β, and a similar regulatory change (gene expression levels) of gene β leads to different phenotypical effects depending on which allele (**b** or **B**) is present in the α null mutant organism. Consequently, polymorphism in the genetic background, i.e. whether at the β locus **bb**, **bB** or **BB** alleles are found, will not allow one to conclude with certainty that a particular phenotypical change observed in a null mutant animal was indeed due to the null mutation or to the genetic background. And the above example involves the interaction of only two genes α and β. Imagine how multiple genes influence the situation. As the reader will see below, the genetic background of the null mutant animals is often different from that of their wild type, control, counterparts at a potentially large number of loci involving hundreds or thousands of genes. In order to appreciate the problem consider how gene targeting is carried out.

Null mutant mice of gene targeting studies are often hybrids of two mouse strains

Most gene targeting (Joyner, 1993) is currently carried out in cultured embryonic stem (ES) cells derived from the mouse strain '129' (e.g. see Gerlai, 1996a). The 129-type ES cells carrying the targeted mutation are introduced into a blastocyst stage embryo and the surviving chimeric embryos are allowed to develop to term, raised to adulthood, and mated to "wild type", i.e. non-mutated, mice. In the case of successful germline transmission, these matings produce an offspring generation in which heterozygous null mutant mice are found. Problems arise, however, if the genetic background of the ES cell and of the mice to which the chimeras are mated are not identical. In a number of studies (for examples see Gerlai, 1996a) the ES cells were derived from the mouse strain 129 but the chimeric mice were mated to, for instance, C57BL/6 (B6). The offspring of such matings (the F_1 generation) therefore are not only heterozygous for the null mutant allele but have one set of chromosomes from strain 129 and another from B6 (see Fig. 1). These heterozygous mice, when sibmated, will produce a segregating F_2 population in which, according to Mendel's Law, homozygous null mutant, heterozygous null mutant and wild type mice are found.

Hybrid null mutant mice are genetically different from their control littermates not only at the locus of the targeted gene but at other loci as well

Comparison of the homozygous mutant, heterozygous mutant and wild type littermates of an F_2 population appears to be an ideal way to reveal phenotypical changes brought about by the null mutation. However, it is important to remember that such a segregating population constitutes mice with recombinant genotypes derived from the two parental mouse strains (see Fig. 1). The difficulties arising from this are threefold. First, the recombination pattern, i.e. which locus contains strain 129 and which B6 alleles, and whether in a homozygous or heterozygous form, may be different between littermates. This implies that not even wild type littermates of their mutant counterparts represent a good control population since their alleles could be different from those of the mutants not only at the locus of the gene of interest but also at other loci. This may lead to false positive results. Second, due to the genetic variation resulting from the hybrid background, detecting significant effects of the mutant gene of interest may be difficult which may lead to false negative results. These two problems can be alleviated by measuring larger number of animals and thus increasing the power of statistical comparisons and decreasing the possibility of sampling error associated with recombination pattern differences between littermates. Increasing the sample size, however, will not solve the third problem which is associated with genetic linkage (see below).

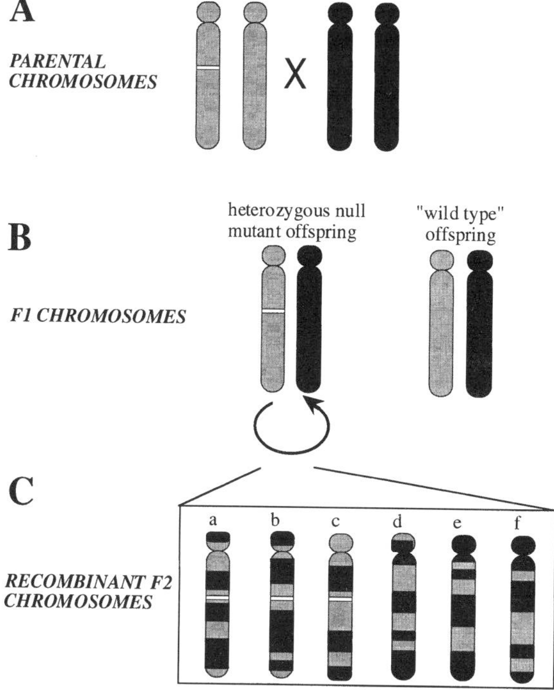

Fig. 1. Chromosomal constitution of mice generated by gene targeting. ES cells originating from mouse strain 129 carry one chromosome (grey) with the disrupted allele (white lesion) of the targeted gene. If these ES cells populate the germ-line in the chimeric mice, the mutation will be transmitted when the chimera is mated. A cross between a germline transmitting chimera and a C57BL/6 (B6) mouse (black chromosomes; panel A) will produce an F_1 population (panel B) in which 50% of the animals will have one copy of the mutant allele (heterozygous mutants) and 50% of them will have no mutant allele (wild type animals) at the targeted locus. Using Southern blotting or PCR (Polymerase Chain Reaction) one can detect the presence of the mutant allele and identify the heterozygous mutant animals. If these animals are mated with each other, according to Mendel's law, homozygous mutant (two mutant alleles), heterozygous mutant (one mutant and one wild type allele) and wild type (two wild type alleles) animals will be obtained. It is also important to remember, however, how genes at loci other than the targeted one will be inherited. Cross-over events during the meiotic process of gametogenesis will "shuffle" the alleles of these background genes and will create recombinant chromosomes (panel C) which will characterize the genotype of the sperm and the egg of the F_1 mice. The genotype of an F_2 individual, therefore, will be represented by a pair of such recombinant chromosomes. For example, a homozygous null mutant mouse may have chromosomes a and b, a and c, or b and c; a heterozygous mouse may have one of the recombinant chromosomes with the lesion (a, b, or c) and another without the lesion (d, e, or f); whereas a wild type control mouse may have chromosomes d and e, d and f, or e and f. Panel C shows that the null mutant allele of the targeted gene will be surrounded by 129-type genes, however, the wild type allele of the gene will be surrounded by B6 type genes. This linkage disequilibrium is simply due to the fact that the null mutant allele came from a strain 129 genetic background. In an F_2 animal of the above origin, the null mutant allele could be surrounded by B6 genes only if, during the meiotic processes of gametogenesis, cross-overs occured precisely flanking both sides of the targeted gene, events whose combined probability is infinitesimally small (modified from Gerlai, 1996a).

The alleles of genes that surround the targeted locus will be of 129-type in the null mutant mice and B6 in the wild type mice

If the targeted mutagenesis is made in ES cells from strain 129, the chromosome with the targeted locus will carry alleles of genes of 129-type. As the probability of genetic recombination is generally inversely related to the distance between the loci of the genes, the 129-type alleles of the genes whose loci are close to the locus of the mutated gene will remain together with the mutated allele of the gene of interest (see Fig. 1). That is, any time the mutation is detected in a mouse, e.g. by Southern blotting, that particular animal will also carry the linked 129-type genes with high probability. Conversely, a non-mutant, control, animal most probably will not carry these 129-type alleles and will have B6 alleles instead if the 129-ES cell chimera was crossed to B6. In effect, the mutation can be seen as a *marker* for the 129-type genes linked to the locus of the targeted gene. Consequently, any phenotypical differences observed between mutant and control littermates of the hybrid genetic origin may be due either to the introduced null mutation or to the background genes linked to the targeted locus or the interaction between these two genetic factors. Thus, one may find false positive results, a problem that has been ignored in a number of gene targeting studies (for examples see Gerlai, 1996a).

The behavioral alterations seen in null mutant mice may be due to the genetic background

Empirical evidence implies that one may not be able to dismiss the genetic background problem or to argue that the potential effect of linked genes, and the effect of the genetic variation, in relation to the effect of the mutation is negligible. For example, gene targeting studies investigating developmental consequences of the Epithelial Growth Factor (EGF) receptor disruption have demonstrated large background-genotype dependent null-mutation effects in mice (Sibilia and Wagner, 1995; Threadgill et al., 1995). The behavior genetics literature also provides ample evidence for large genetic differences between inbred mouse strains at the behavioral and neurobiology levels (Plomin et al., 1990). Furthermore, strain comparisons have revealed that, unluckily, strain 129, which has been the choice of ES cell–gene targeting studies, is one of the most unique strains in terms of behavior, neuroanatomy, or neurophysiology (Lipp et al., 1995; Crawley et al., 1997; Owen et al., 1997; Schauwecker and Steward, 1997; Wolfer et al., 1997). For example, these animals are severely impaired in spatial learning tasks, a behavioral paradigm frequently used in molecular neurobiology studies (Grant et al., 1992; Silva et al., 1992; Abeliovich et al., 1993; Cremer et al., 1994; Wu et al., 1995) and they are considered passive (Wolfer et al., 1997) a behavioral trait that can be a serious confounding problem in a number of other behavioral tests including the context dependent fear conditioning (Abeliovich et al., 1993), in which freezing response is measured, the open field exploration test or tests of motor function such as the rotorod (Aiba et al., 1994; Conquet et al., 1994; Kashiwabuchi et al., 1995). Moreover, strain 129 mice suffer from dysgenesis of the corpus callosum and possibly possess a number of other neuroanatomical peculiarities (Wahlsten et al., 1992; Lipp et al., 1995). 129 mice also exhibit a peculiar ability to resist kainic acid induced excitotoxic cell death (Schauwecker and Steward, 1997). In addition, some behavioral differences between 129 substrains have been reported (see Montkowski et al., 1997).

Interestingly, in several gene targeting studies (Silva et al., 1992; Abeliovich et al., 1993; Aiba et al., 1994; Conquet et al., 1994; Picciotto et al., 1995; Wu et al., 1995) in which the mice originated from a cross between a 129-ES cell chimera and another strain, the null mutant animals suffered from behavioral defects similar to those seen in strain 129 (for review see Gerlai 1996a). It is therefore possible that the differences observed between mutant and control mice were in fact due to the genetic differences (in the linked background genes) between the inbred strains used in the generation of null mutant animals and not to the null mutation. Direct empirical evidence supporting this argument comes form a recent analysis of D2 dopamine receptor-deficient (null mutant) mice (Kelly et al., 1998; for another example see Wolfer et al., 1997). These authors found that wild type 129 mice with functional D2 receptors were a virtual phenocopy of the predicted locomotor deficits caused by the loss of D2 receptors. When they compared the wild type 129 and B6 parental strains, their F_2 segregating generations, as well as a mutant (129 $\times$ B6) segregating F_2 generation, they found that genetic background contributed significantly to the phenotypical alteration seen in the null mutant mice. By studying mutant congenic strains (backcrossed to either B6 or 129 parental strains) they also showed a significant interaction between background genes and the targeted one, the former apparently exhibiting a greater effect on the behavioral phenotype (Kelly et al., 1998). On the other hand, Jia et al. (1996) found, in another gene targeting study, that genetic background did not contribute to the observed electrophysiological alterations (elevated long-long term potentiation, LTP) in GluR2 null mutant mice. These authors analyzed both parental strains (CD1 and 129), their F_2 hybrid population, and also a CD1 x 129 F_2 population in which the null mutation was also segregating. Although reassuring to some, interestingly, Jia et al. could not detect a difference in LTP between 129 and CD1 mice, a finding discordant with the results of another group (Abel et al., 1996). The cause of this discrepancy is unknown but it may be due to differences between substrains of 129 mice used or in the electrophysiological recording techniques applied.

Solutions for the genetic background problem

A classical solution, one might suggest, would be to decrease the probability of contribution of background genes by backcrossing the mutant hybrid animals several times to the strain of choice, e.g. to B6, and create a congenic strain that carries the mutation on the desired genetic background. Although, backcrossing will undoubtedly increase the representation of the desired background genotype in the mutant animals, eliminating 129-type genes that surround the locus of the gene of interest would be a considerable undertaking: with 12 backcrosses (approximately 2 years of breeding) to B6, the length of the 129-type chromosome segment introduced to the B6 genome would be, on average, about 16 centiMorgans (cM) (Festing, 1992). As the mouse genome covers about 1600 cM, this represents about 1% of the genome. Assuming the mouse genome contains 30,000 genes, then the introduced chromosome segment would be expected to have some 300 genes in it (Festing, 1992). Because a potentially very large number of genes are involved in neural processes, the chance that some of the introduced linked 129-type genes will have an influence on brain function and behavior is not negligible. Therefore, backcrossing is not the most optimal solution.

Zimmer (1996) suggested a simple breeding scheme to test the contribution of background genes linked to the targeted locus. His suggestion was to breed not only a mutant F_2 generation (by crossing two mutant heterozygous F_1 mice) but also a wild type F_2 generation (by crossing two wild type F_1 mice). F_1 in most cases means a hybrid between the 129 and B6 parental strains and both mutant heterozygous and wild type F_1 animals are obtainable after the founder chimeric (129 type sperm) mutant male is crossed to a B6 female. He then explains that if the behavioral alterations seen in the null mutants is really due to the mutation the mutant F_2 generation should show a segregation pattern whereas the wild type F_2 generation should not. Although correct in principle, Zimmer's solution will most probably not work in practice primarily because the behavioral traits are not qualitative and the segregation characteristics of the F_2 generations cannot be studied by simply counting the number of mice falling into three distinct categories (aa, ab, bb) but only by complicated quantitative genetic methods requiring very large sample sizes. Furthermore, Zimmer's example also assumes that there is no interaction (epistasis) between the background genes and the targeted one and also that the behavioral trait in question is monogenic in the wild type F_2 population, assumptions that are almost never fulfilled in behavioral genetics research.

Crusio et al. (1998) recently also presented some solutions to the genetic background and the flanking gene problem, all based on different breeding schemes. First, they suggested to use an approach known as "speed congenics", which is essentially a backcross system utilizing genetic markers to select those wild type recombinants that are most similar, in their recombination pattern, to the null mutant animals (see Fig 1 for recombination patterns). Although useful in speeding up the process of making mutant and wild type recombinants more similar in their genetic background, this approach requires a lot of breeding and genetic testing, and will never produce a control mouse that is identical to the mutant at all loci but the targeted one. The second breeding scheme, designated as "poor man's choice", suggested by Crusio et al. (1998) does eliminate the necessity of testing genetic markers. It basically entails crossing recombinant heterozygous mutants back to wild type 129 and also to themselves. These breedings will produce recombinant mice that are mutant for the targeted gene and the targeted locus is surrounded by 129 flanking genes, and mice that are wild type for the targeted locus and also have 129 alleles around it. The difficulty with the second scheme is that the two recombinants (mutant and wild type) are not identical, and, in fact, the wild type will have a larger 129 region flanking the targeted locus than the mutant and so this breeding scheme is also unable to produce 100% identical genetic background for wild type and mutant mice. Nevertheless, these suggestions are useful because they decrease the chance of false findings by significantly reducing genetic differences between null mutant and control in their genetic background.

An alternative approach to breeding schemes may be to do a rescue experiment which could rule out the potential effects of linked 129-type genes. In this experiment one could return the missing functional protein by introducing a transgene expressing the protein of the targeted gene or via a more direct systemic administration of the missing protein using osmotic minipumps for example. Then one could investigate whether such a manipulation restored the wild type phenotype in the null mutants. A positive answer would be a strong argument for the involvement of the targeted gene in the observed phenotypical alterations. Another potential solution would be to generate "knock in" mice in addition to the null mutant gene "knock out" animals. For instance, one could use homologous recombination (gene targeting) to insert a small DNA marker flanking the gene of interest, without disrupting it. The knock in mice generated this way would have a fully functional targeted gene and they would have, on average, the same linked genes as the knock out animals in which the gene of interest has actually been disrupted. Therefore, the knock in animals would represent the ideal control for the knock out null mutants.

Perhaps the most elegant solution to the background gene problem would be to use inducible knock out techniques. The first studies using an inducible tetracycline transactivator system (Mayford et al., 1996, see also Chapter 3.1.4) or the cell type restricted CRE recombinase KO system (Tsien et al., 1996; see also Chapter 3.1.3) to investigate brain function and behavior clearly show the future to come. These techniques will allow the investigator to switch off or on the gene of interest at particular times and particular places in the brain. Animals created with such inducible systems will represent internal controls for the phenotypical analysis of mutants since the experimenter will be able to compare the pre and post induction phenotypes.

Although the solutions suggested above are, or will soon be feasible, they are rather time consuming and expensive, and would at least double the efforts necessary to carry out a gene targeting study. A better, and more economical, solution would be to avoid using hybrid mice all together. Generating null mutant mice with a pure genetic background could be done if one could cross the chimeras carrying the null mutation to a mouse of genetic background identical to that of the ES cell line which was used for the targeted mutagenesis. Unfortunately, this apparently simple solution are usually not considered, mostly for practical reasons. Paradoxically, strain 129 from which ES cell lines are most easily derived is difficult to breed and is susceptible to a number of diseases. Thus, while the choice of ES cell studies is strain 129, fully grown strain 129 mice are often unwelcome in animal colonies. In addition, strain 129 mice are not preferred in animal behavioral studies either. It must be noted, however, that ES cell lines can be established from other strains of mice as well, including B6 (Joyner, 1993). Using a B6 cell line for gene targeting and crossing the created chimera to a B6 mouse would give rise to inbred B6 null mutant and wild type control mice in the F_2 generation with no genetic polymorphism or confounding background genotype effects. Since B6 mice are often preferred in animal behavioral studies and are fairly easy to keep and breed, establishing B6 ES cell lines represents the optimal solution. It is unclear at this moment whether generating and using B6 ES cell lines is more problematic than working with 129 ES cells. Most scientists have preferred 129 ES cells. However, this may simply be due to historical reasons: the cell culture and harvesting techniques were developed for 129 ES cells because 129 strain of mice were most used and useful in studying teratocarcinomas, a research line that eventually led to the development of ES cells (for examples see Joyner, 1993). Nevertheless, a recent study has demonstrated that B6 ES cells can be generated and successfully used (Lemckert et al., 1997). B6 ES cells has already enabled scientists to obtain a number of gene-knock out lines with a success rate similar to that reported for ES cells derived from the 129 mouse strains (Lemckert et al., 1997).

Finally, one could solve the problem associated with the usage of 129 ES cells by simply not having ES cells at all. The primary reason for using ES cells is that homologous recombination can be carried out in a petry dish and can be tested using the classical positive-negative selection scheme (for details see Joyner, 1993; and also Chapter 3.1.1). This selection scheme is regarded as an enrichment technique that enables the researcher to increase the detection ratio of true homologous recombinants (cells in which the integration of the transcript occured at the correct, targeted site) vs random integration recombinants (cells where the integration of the targeting construct was random, i.e. site independent). Procedures are being developed (e.g. Révet et al., 1993) that may enable one to achieve homologous recombination with high frequency without having to use the positive–negative selection (e.g. at PanGene Corporation). These procedures, currently not available to the public, hold the promise of doing gene targeting without ES cells. If this is achieved, one will not have to worry about the hybrid genetic background. Also, one could generate null mutant mice on any pure bred genetic background. Even more importantly, the procedure could be extended to other species for which currently gene targeting is not available as a result of lack of suitable ES cells.

Concluding remarks

In summary, I am convinced that gene targeting in combination with thorough molecular, neurobiological, and behavioral examination of mutant animals it generates will provide us with a high resolution and detailed understanding of the molecular mechanisms underlying complex behavioral phenomena including learning and memory. However, it must be remembered that behavioral and neurobiological traits are complex, often variable, and are most probably influenced by a large number of genes as well as environmental factors. In order to dissect such traits and to understand the complex web of interactions among the underlying biological mechanisms, it is crucial to control as many variables as possible. Understanding compensatory mechanisms and complex systemic responses to the absence of a gene product, and eliminating the confounding effects of background genes, are important steps forward that will facilitate our knowledge of how genes control brain and behavior.

References

Abel, T., Nguyen, P.V., Bourtchouladze, R., Bach, M.E., Gapridasvili, I., Jain, P. and Kandel, E.R. (1996) Strain-dependent differences in hippocampal LTP and spatial memory. *Society for Neuroscience Abstracts*, 19: 1510.

Abeliovich, A., Paylor, R., Chen, C., Krim, J., Wehner, J. and Tonegawa, S. (1993) PKCγ mutant mice exhibit mild deficits in spatial and contextual learning. *Cell*, 75: 1263-1271.

Aiba, A., Kano, M., Chen, C., Stanton, M.E., Fox, G.D., Herrup, K., Zwingman, T.A. and Tonegawa, S. (1994) Deficient cerebellar long-term depression and impaired motor learning in mGluR1 mutant mice. *Cell*, 79: 377–388.

Chamow S.M. and Ashkenazi, A. (1996) Immunoadhesins: principles and applications. *Trends Biotechn.*, 14: 52–60.

Chen C., Kano, M., Abeliovich, A., Chen, L., Bao, S., Kim, J.J., Hashimoto, K., Thompson, R.F. and Tonegawa, S. (1995) Impaired motor coordination correlates with persistent multiple climbing fiber innervation in PKC gamma mutant mice. *Cell*, 83: 1233–1242.

Conquet, F., Bashir, Z.I., Davies, C.H., Daniel, H., Ferraguti, F., Bordi, F., Franz-Bacon, K., Reggiani, A., Matarese, V., Condé, F., Collingridge, G.L. and Crépel, F. (1994) Motor deficit and impairment of synaptic plasticity in mice lacking mGluR1. *Nature*, 372: 237–243.

Crawley, J.N. (1996) Unusual behavioral phenotypes of inbred mouse strains. *Trends Neurosci.*, 19: 181–182.

Crawley J.N., Belknap J.K., Collins A., Crabbe J.C., Frankel W., Henderson N., Hitzeman R.J., Maxson S.C., Miner L.L. Silva A.J. et al. (1997) Behavioral phenotypes of inbred mouse strains — implications and recommendations for molecular studies. *Psychopharmacology*, 132: 107–124.

Cremer, H., Lange, R., Christoph, A., Plomann, M., Vopper, G., Roes, J., Brown, R., Baldwin, S., Kraemer, P., Scheff, S., Barthels, D., Rajewsky, K. and Wille, W. (1994) Inactivation of the N-CAM gene in mice results in size reduction of the olfactory bulb and deficits in spatial learning. *Nature*, 367: 455–459.

Crusio, W.E. (1996) Gene-targeting studies: new methods, old problems. *Trends Neurosci.*, 19: 186–187.

Crusio, W.E. and Gerlai, R (Eds.) (1999) *Handbook of Molecular Genetic Techniques in Brain and Behavior Research*. Elsevier, Amsterdam, in press.

Crusio W.E., Wolfer D.P. and Lipp H-P. (1998) Knock-out mice: Possible solutions to the genetic background and flanking gene problem. *Society for Neuroscience, Abstracts*, 28: 1203.

Fell, D.A. and Thomas, S. (1995) Physiological control of metabolic flux: the requirement for multisite modulation. *Biochem. J.*, 311: 35–39.

Fell, D.A. (1992) Metabolic control analysis: a survey of its theoretical and experimental development. *Biochem. J.*, 286: 313–330.

Festing, M.F.W. (1992) From character to gene: some strategies for identifying single genes controlling behavioral characters. In: D. Goldowiz, D. Wahlsten and R.E. Wimer (Eds.), *Techniques for the genetic analysis of brain and behavior: Focus on the mouse*. Elsevier, Amsterdam, pp. 17–38.

Friedman, G.C. and O'Leary, D.D.M. (1996) Eph receptor tyrosine kinases and their ligands in neural development. *Current Opinion in Neurobiology*, 6: 127–133.

Frisén, J., Yates, P., McLaughlin, T., Friedman, G.C., O'Leary, D.D.M. and Barbacid, M. (1998) Ephrin-A5 (AL-1/RAGS) is essential for proper retinal axon guidance and topographic mapping in the mammalian visual system. *Neuron*, 20: 235–243.

Gerlai, R., Cairns B., Van Bruggen N., Moran P., Shih A., Sauer, H., Phillips. H. S., Caras I. and Winslow J. (1998a) Protein targeting in the analysis of learning and memory: A potential alternative approach to gene targeting. *Exp.l Brain Res.*, in press.

Gerlai, R., Shinsky, N., Williams, P., Winer, J., Cairns, B., Mendoza, B., Gao, W-Q. and Phillips, H.S. (1998b) EphA5 and mammalian learning: Potential biological mechanisms. Soc. Neurosci. Ann. Mtng., Los Angeles, California. Abstracts, 172.15.

Gerlai, R., Shinsky, N. Caras, I., Gao, W.-Q., Phillips, H.S. and Winslow, J. (1997) The receptor tyrosine kinase, REK7, and its ligand, AL1, are involved in neural plasticity in vivo: A behavioral and electrophysiological study in mice. Soc. Neurosci., 27th Ann. Mtng, New Orleans, USA. Abstracts. 97.11.

Gerlai, R. (1996a) Gene targeting studies of mammalian behavior: Is it the mutation or the background genotype? *Trends Neurosci.*, 19: 177–181.

Gerlai, R. (1996b) Gene targeting in Neuroscience: The systemic approach. *Trends Neurosci.*, 19: 188–189.

Gerlai, R., Wojtowicz, J.M., Marks, A. and Roder, J. (1995) Over-expression of a calcium binding protein, S100β, in astrocytes alters synaptic plasticity and impairs spatial learning in transgenic mice. *Learning and Memory*, 2: 26–39.

Grant S.G.N., O'Dell T.J., Karl K.A., Stein P.L., Soriano P. and Kandel E.R. (1992) Impaired Long-Term Potentiation, Spatial Learning, and Hippocampal Development in *fyn* Mutant Mice. *Science*, 258: 1903-1910.

Grant, S.G.N. and Silva, A.J. (1994) Targeting learning. *Trends Neurosci.*, 17: 71–75.

Hochgeschwender, U. and Brennan, M.B. (1995) Knock out mice rule OK. *Nature*, 375: 543.

Jia, Z., Agopyan, N., Miu, P., Xiong, Z., Henderson, J., Gerlai, R., Taverna, F., MacDonald, J., Carlen, P., Abramow-Newerly, W. and Roder, J. (1996) Enhanced LTP in the absence of GluR2. *Neuron*, 17: 945–956.

Joyner, A.L. (1993) *Gene targeting. A practical approach*. Oxford: Oxford University Press. pp. **XXX**

Kacser, H. and Burns, J.A. (1973) The control of flux. *Symp. Soc. Exp. Biol.*, 27: 65–104.

Kashiwabuchi, N., Ikeda, K., Araki, K., Hirano, T., Shibuki, K., Takayama, C., Inoue, Y., Kutsuwada, T., Yagi, T., Kang, Y., Aizawa, S. and Mishina, M. (1995) Impairment of motor coordination, purkinje cell synapse formation, and cerebellar long-term depression in GluR?2 mutant mice. *Cell*, 81: 245–252.

Kelly, A.M., Rubinstein M., Phillips T.J., Lessov C.N., Burkhart-Kasch S., Zhang G., Bunzow J.R., Fang Y., Gerhardt G.A., Grandy D.K. and Low M.J. (1998) Locomotor activity in D2 dopamine receptor-deficient mice is determined by gene dosage, genetic background, and developmental adaptations. *J. Neurosci.*, 18: 3470–3479.

Lathe, R. (1996) Mice, gene targeting and behaviour: more than just genetic background. *Trends Neurosci.*, 19: 183–186.

Lemckert, F.A., Sedgwick, J.D. and Korner, H. (1997) Gene targeting in C57BL/6 ES cells. Successful germ line transmission using recipient BALB/c blastocysts developmentally matured in vitr. *Nucl. Acid Res.*, 25: 917–918.

Lipp, H-P. and Wolfer, D.P. (1998) Genetically modified mice and cognition. *Current Opinion in Neurobiology*, 8: 272–280.

Lipp, H-P., Stagliar-Bozizevic, M. and Wolfer, D.P. (1995) Behavior in large numbers of knockout mice: Results, caveats and perspectives. *Behavior Genetics*, 25: 275.

Mayford M., Bach M.E., Huang Y-Y., Wang L., Hawkins R. and Kandel E.R. (1996) Control of memory formation through regulated expression of a CaMKII transgene. *Science*, 274: 1678–1683.

Mayford, M., Abel, T., Kandel, E.R. (1995) Transgenic approaches to cognition. *Current Opinion in Neurobiology*, 5: 141–148.

Montkowski, A., Poettig, M., Mederer, A. and Holsboer, F. (1997) Behavioural performance in three substrains of mouse strain 129. *Brain Res.*, 762: 12–18.

Morris, R.G.M. and Kennedy, M.B. (1992) The pierian spring. *Current Biology: The biology of learning*. 2: 511–514.

Orioli, D. and Klein, R. (1997) The eph receptor family: Axonal guidance by contact repulsion. *Trends Genetics*, 13: 354–359.

Owen, E.H., Logue, S.F., Rasmussen D.L. and Wehner J.M. (1997) Assessment of learning by the Morris water task and fear conditioning in inbred mouse strains and F_1 hybrids: implications of genetic background for single gene mutation and quantitative trait loci. *Neuroscience*, 80: 1087–1099.

Picciotto, M.R., Zoli, M., Léna, C., Bessis, A., Lallemand, Y., LeNovère, N., Vincent, P., Pich, E.M., Brûlet, P. and Changeux, J-P. (1995) Abnormal avoidance learning in mice lacking functional high-affinity nicotine receptor in the brain. *Nature*, 374: 65–67.

Plomin, R., DeFries, J.C. and McClearn, G.E. (1990) *Behavioral Genetics. A primer*. 2nd Edn. Freeman and Company, New York, pp. 262–295.

Révet, B.M.J., Sena, E.P. and Zarling, D.A. (1993) Homologous DNA targeting with RecA protein-coated short DNA probes and electron microscope mapping on linear duplex molecules. *J. Mol. Biol.*, 232: 779–791.

Rose, S. (1995) The rise of neurogenetic determinism. *Nature*, 373: 380–383.

Routtenberg, A. (1995) Knockout mouse fault lines. *Nature*, 374: 314–315.

Schauwecker P.E. and Steward O. (1997) Genetic determinants of susceptibility to excitotoxic cell death: implications for gene targeting approaches. *Proceedings of the National Academy of Sciences of the United States of America*, 94: 4103–4108.

Sibilia, M. and Wagner, E. (1995) Strain-dependent epithelial defects in mice lacking the EGF receptor. *Science*, 269: 234–237.

Silva A. J., Paylor R., Wehner M. J. and Tonegawa S. (1992) Impaired spatial learning in α-Calcium-Calmodulin Kinase II mutant mice. *Science*, 257: 206–211.

Threadgill, D.W., Dlugosz, A.A., Hansen, L.A., Tennenbaum, T., Lichti, U., Yee, D., LaMantia, C., Mourton, T., Herrup, K., Harris, R.C., Barnard, J.A., Yuspa, S.H., Coffey, R.J. and Magnuson, T. (1995) Targeted disruption of mouse EGF receptor: Effect of genetic background on mutant phenotype. *Science*, 269: 230–234.

Tsien J.Z., Chen D.F., Gerber D., Tom C., Mercer E.H., Anderson D.J., Mayford M., Kandel E.R. and Tonegawa S. (1996) Subregion- and cell type-restricted gene knockout in mouse brain. *Cell*, 87: 1317–1326.

Van Abeelen, J.H.F., Ellenbroek, G.A. and Wigman, H.G.A.J. (1975) Exploratory behavior in two selectively-bred lines of mice after intrahippocampal injection of methylscopolamine. *Psychopharmacologia*, 41: 111–112.

Van Abeelen, J.H.F. (1989) Genetic control of hippocampal cholinergic and dynorphinergic mechanisms regulating novelty-induced exploratory behavior in house mice. *Experientia*, 45: 839–845.

Wahlsten, D., Ozaki, H.S. and Livy, D. (1992) Deficient corpus callosum in hybrids between ddN and three other abnormal mouse strains. *Neuroscience Letters*, 136: 99–101

Wolfer D.P., Muller U., Stagliar M. and Lipp H-P. (1997) Assessing the effects of the 129/Sv genetic background on swimming navigation learning in transgenic mutants: a study using mice with a modified beta-amyloid precursor protein gene. *Brain Res.*, 771: 1–13.

Wu, Z-L., Thomas, S.A., Villacres, E.C., Xia, Z., Simmons, M.L., Chavkin, C., Palmiter, R.D. and Storm, D.R. (1995) Altered behavior and long-term potentiation in type I adenylyl cyclase mutant mice. *Proc. Natl. Acd. Sci. USA*, 92: 220–224.

Zimmer, A. (1996) Gene targeting and behaviour: a genetic problem requires a genetic solution. *Trends Neurosci.*, 19: 470.

Section 3.2

Systemic manipulation

W.E. Crusio and R.T. Gerlai (Eds.)
Handbook of Molecular-Genetic Techniques for Brain and Behavior Research (Techniques in the Behavioral and Neural Sciences, Vol. 13)

CHAPTER 3.2.1

Gene transfer and therapy in the CNS

Marie-Claude Senut, Steven T. Suhr and Fred H. Gage

Laboratory of Genetics, The Salk Institute for Biological Studies, 10010 North Torrey Pines Road, La Jolla, CA 92037, USA

Introduction

In the last decade, progress has been made in identifying genes and factors which play a key role in neurodegenerative disorders. These advances have brought somatic gene therapy forward as an alternate approach to the treatment of a variety of genetic and multifactorial diseases affecting the mammalian central nervous system (CNS). Somatic gene therapy consists of introducing novel genetic material into non-germline cells to ameliorate damage or disease in the nervous system. The potential to transfer genes into CNS cells is not only important for therapeutic purposes, but is also a formidable tool for exploring the mechanisms underlying normal brain development, plasticity and function. Although application of gene transfer techniques to the adult mammalian CNS has been hampered by the anatomical and functional complexity of the brain and its disorders, many promising research reports have emerged over the last decade (see for review Suhr and Gage, 1993; Snyder and Senut, 1997). Two main strategies are widely used for delivering transgenes to the CNS: the direct injection of genetic material (in vivo approach), or the intracerebral transplantation of genetically modified cells (ex vivo approach) (Fig. 1). When the intracellular expression of a given gene is necessary, direct gene transfer with viral or chemical vectors is required. In contrast, when the primary goal is to provide an exogenous factor to the brain, the ex vivo method, where grafted cells act as a biological "pump" supplying the factor locally and chronically, may be the most reliable.

In this chapter, we will mainly focus on the ex vivo approach of CNS gene transfer, and will only briefly describe the current status of in vivo gene transfer technologies which are further analyzed in other chapters. In ex vivo applications, there are four primary considerations:

1. Identification and manipulation of a therapeutic transgene.
2. Identification of cells which can survive and retain gene expression following intracerebral transplantation.
3. Selection of suitable vectors for transferring novel transgenes into the selected cell type.
4. Intracerebral implantation of the engineered cells.

We will provide some protocols developed in our laboratory and examine the caveats and complications that we, and others, have encountered using ex vivo gene transfer. The potentials of ex vivo and in vivo gene transfer as a CNS gene therapy method through the analysis of an animal model of neurodegenerative disease will be discussed, and we will conclude with new techniques in development which hold some promises for successfull CNS gene transfer.

Manipulation of the transgene

In general, CNS gene therapy begins with the isolation and characterization of a specific cDNA or gene with potential beneficial properties. For gene transfer purposes, a transgene expression unit must have most of the same properties as genes do in situ, i.e. a regulatory or promoter region, and an expressed region with initiation and termination codons and a continuous open-reading-frame (ORF) encoding the protein. Each individual cDNA or gene has individual characteristics that may influence the selection of a gene transfer strategy; however, three general considerations should be briefly discussed. A first concern, which will be discussed further below, is the upper and lower limits of all viral vectors regarding transgene length. The second concern, of particular importance to transgenes destined for use in retroviral vectors, is the presence of internal polyadenylation (p(A)) signals that can result in premature p(A) of retroviral transcripts and low retroviral titers. For retroviral vectors, transgenes are generally truncated after termination codons to eliminate p(A) signals downstream. A third consideration is the relative strength of the initiation and termination codons of the protein ORF which can dramatically affect the level of functional protein produced from a given level of mRNA. In cases where maximal protein levels are desired, mutation of the initiation codon to Kozak (Kozak, 1984) consensus sequences can significantly boost translation levels, while proteins with potentially cytotoxic side effects at high levels can be significantly reduced by alteration of initiation codon sequences.

The promoter used to control transgene expression is either introduced as an individual element in the transgene expression unit by the researcher, or is donated by the viral vector in which it pulls double duty as a component of the viral life cycle. There are now four major categories of promoters used in the construction of gene transfer vectors: (a) strong-constitutive; (b) housekeeping; (c) tissue-specific; and (d) regulated. Strong promoters include the powerful viral promoters (i.e. CMV, RSV, SV40 — specific references on individual promoters may be found in Suhr and Gage, 1993) that are functional in a broad range of cell types and are used to provide the highest possible level of constant transgene expression. Housekeeping promoters (i.e. DHFR, TK, β-actin) function in a broad range of cell types and deliver a constant, but low level of expression. Tissue-specific promoters (i.e. GFAP, NF) and enhancers are designed to function optimally only in a narrow range of cell types and provide variable levels of expression which may be subject to changes in cell physiology. Regulated promoters (TetOs — Hoshimaru et al., 1996; Gal4RE — Wang et al., 1994; EcRE — Suhr et al., 1998; also see Chapters 3.1.4 and 3.2.3) are used in vectors in which ligand-gated transactivators are included, allowing regulation of transgene expression through exogenous drug administration. Selection of promoters for use in viral vectors depends largely on the specific application and there are few concerns in their selection and handling. The promoter, like the transgene itself, has distinct functional regions with which the researcher should have some familiarity. Knowledge of the localization of the core enhancers, basal promoter elements, and most importantly, transcription initiation site, will prevent errors in construction.

Since transgenes and their regulatory regions are as numerous as the number of genes in the living species of the world, it is not practical to devote lengthy discussion to techniques involving specific transgenes. There are, however, two "rules of thumb" that usually apply to the selection and use of a transgene: (a) *Know your transgene well.* An accurate sequence, restriction endonuclease map, and knowledge of functional regions (initiation codon, termination codon, p(A), etc.), of the complete DNA used is the single most valuable tool that the researcher can have; (b) *Smaller is better.* In most cases, the elimination of unnecessary sequences means that fewer things can go wrong. In general, this extends to the use of cDNAs over whole genes, which may splice inappropriately or contain unwanted activities that reside within the introns.

Which cells for transplantation?

The cells to be chosen should follow most, if not all, of the following criteria:

1. easily obtainable;
2. amenable to growth, maintenance and transfection in culture;
3. immunocompatible with the host organism;
4. prolonged survival after CNS implantation; and
5. non-cytotoxic or detrimental within the target area.

Two major types of cell are presently available for ex vivo gene transfer: cell lines and primary cells.

Cell lines

Initially, cell lines were used for intracerebral transplantation. They are readily available (often commercially), easy to maintain and to grow in culture, and amenable to gene transfer. A variety of cell lines of both neuronal and non-neuronal origin have been explored for ex vivo CNS gene transfer, such as immortalized fibroblasts (Rat-1, 3T3 and 208F lines), neuroblastoma lines, neuroendocrine cells, or glioma/schwannoma lines. The major pitfall inherent to their use is either the formation of brain tumors or destructuion of the transplant as a result of host immunological rejection. To avoid both of these eventualities, an alternative tactic is to encapsulate cell lines in a variety of substrates (Aebischer et al., 1991; Maysinger and Morinville, 1997) which allow the passage of small transgene products, while preventing the grafted cells from being eliminated by host inflammatory and immune cells. One advantage of this strategy is that the capsules can be retrieved easily if necessary, but one drawback to consider is the risk of leakage or rupture of the capsule.

In the past years, a new generation of cell lines has been established by immortalizing cells isolated from different brain areas using a variety of oncogenes (Snyder, 1994; Gage et al., 1995; Fisher, 1997; Martinez-Serrano and Björklund, 1997). Eventual tumorigenic properties of neural-derived cells can be controlled by infecting neural cells with retroviruses encoding a temperature-sensitive mutant of SV40 Large T antigen which is inactivated at the temperature of the mammalian CNS. Oncogene-immortalized CNS-derived cells are amenable to gene transfer, survive and incorporate well in the brain and, most importantly, are capable of forming neural connections with endogenous cells. Even though the potential for tumorigenicity is reduced with these specialized cell types, future research will determine if this type of cell poses significant long term risk. Furthermore, since the neural tissue used for preparing these cell lines may not be autologously-derived, immune responses from the host are still a consideration.

Primary cells

Until the problems associated with the use of cell lines are controlled, an alternative strategy for CNS gene transfer and therapy is to use autologous or isogenic cells. This strategy prevents or minimizes host immune rejection and allows better graft survival in the host brain. To this end, a number of primary cells can be used as effective transgene carriers into the CNS, such as fibroblasts, myoblasts, Schwann cells or astrocytes. There are several advantages of using primary fibroblasts: they can be easily obtained from skin biopsies, grown and maintained in culture, and they are amenable to stable gene transfer by viral infection. Primary fibroblasts integrate well into the host mammalian brain where they can survive up to 18 months in a contact-inhibited non-proliferative state (Senut et al., 1996). Primary muscle myoblasts have also been used as carriers of transgenes into the brain; however, recent studies suggest that intracerebral myoblast transplants might only be efficient for short-term gene delivery, since they survive and retain gene expression only for a few weeks (Lisovoski et al., 1997). Since fibroblasts and myoblasts do not migrate and cannot form connections in the brain, interest was brought in using cells of glial origin for ex vivo gene transfer. Astrocytes and Schwann cells can also be cultured, genetically modified and transplanted back into the CNS (Owens et al., 1991; Mucke and Rockenstein, 1993; Senut et al., 1996). More recently, several groups including ours have shown that primary neural progenitors isolated from embryonic or adult rodent CNS can be grown in culture with growth factors such as epithelial- or fibroblast- growth factors (Ray et al., 1993; Gage et al., 1995; Fisher, 1997). The potential of these cells to form neurons may make them particularly useful in repopulating a damaged brain.

Of all of the cell types used to date, among the least problematic and simplest to use is the primary skin fibroblast. The preparation of primary rat fibroblasts requires no specialized surgical or culture facilities to produce an essentially perpetual source of cells for gene transfer/transplantation studies.

Protocol for the generation of fibroblast cultures

Tissue preparation
A 1 cm^2 piece of skin is removed from the abdomen of an anesthetized rat, sterilized by repeated dipping in 70% ethanol, and placed in an Eppendorf tube containing DMEM, 10% fetal bovine serum, 2 mM L-glutamine, 2.5 μg/ml fungizone and 50 μg/ml gentamycin ($DMEM^+$).

Culture preparation
All culturing of cells is done in a tissue culture hood using sterile techniques. The skin fragment is washed twice in $DMEM^+$ and immersed into a petri dish with $DMEM^+$, where it is cut into 1–2 mm pieces using scalpels fitted with sterile blades. Four to six pieces are then placed in each well of a 6-well plate and covered with a sterile cover slip (22 mm^2) anchored in place by dipping the corners in silicone grease. $DMEM^+$ (3 ml) is added to each well, and the dish is incubated at 37°C/10% CO_2 atmosphere. Within 3–4 d, when the fibroblasts have started migrating, the medium is replaced with fresh $DMEM^+$ and the cultures are allowed to expand up to 80–90% confluency. When confluency is reached, cells can be passaged or frozen down for future use.

Passaging cultures
Aspirate the medium from the petri dish and rinse the cell monolayer twice with phosphate-buffered saline (PBS). Add enough trypsin-EDTA solution to barely cover the cells, swirl the plate and remove. Let the plate sit for 3–5 min (cells will round up) and give a sharp slap against the palm of the hand to dislodge the cells. Resuspend cells in $DMEM^+$ and passage to T75 flasks. Feed the cells twice a week depending on culture confluency.

Gene transfer vectors

Gene transfer vectors are generally complex structures which carry out two essentially independent functions: (a) delivery of the transgene to the target cell, and (b) proper expression of transgene after entry. Foreign genes can be introduced into cells or cell lines by non-viral and viral methods that all have been proven variably capable. Non-viral methods of gene transfer include chemical and physical techniques such as calcium phosphate, electroporation, or lipofection (Senut et al., 1995a). These methods have limited applications due to transient gene expression and will be described in this chapter only as they can be used in vitro to produce more efficient biological vectors. Over the last decade, several viral vectors with low toxicity, efficient infection rates, and persistent expression have emerged at the forefront of therapeutic gene transfer. They may be classified by whether they carry a DNA or RNA genome.

The DNA viruses most widely used in recent years are the adeno (AV — Le Gal La Salle et al., 1993) and adeno-associated (AAV — Xiao et al, 1997) viruses, both with their individual characteristics, complications, and benefits. The benefits of AV include the ability to carry large DNA inserts, to be produced at high titer with relative safety, and to infect a wide variety of cell types in vivo. The liabilities of AV, however, include a large, complex genome that must be modified by recombination, and the risk of significant immune response following introduction into the host. AAV may emerge as a scaled-down alternative to AV. AAV is a much simpler genetic unit encoding only two basic ORFs for a total of 4.5 kb. This simplicity, which is one of the assets of AAV biology, is also one of its primary drawbacks: transgene inserts into AAV including all gene regulatory sequences must be confined within 4.5 kb. Aside from this limitation, however, AAV, like AV, can be produced at very high titers and is capable of infecting a broad range of cell types. Further investigation of this vector will define its limitations and advantages. These vectors are described in greater detail in another chapter.

The RNA viruses are represented by retroviruses which are among the earliest and most thoroughly tested viral vectors used for gene transfer. Retroviral genomes are small enough to easily manipulate, yet are large enough to carry enough foreign genetic material for most gene transfer purposes (7–8 kb). Some special consideration (including removal of internal p(A) signals mentioned above) must be given to the construction of retroviral vectors since they go through a RNA intermediate before conversion to stably integrated double-stranded DNA in the host cell genome. In our laboratory, retroviral vectors derived from two retroviruses — the Moloney murine leukemia virus (MLV) (Rosenberg et al., 1988; Senut et al., 1995b), and human immunodeficiency virus (HIV; Naldini et al., 1996) — have been applied to questions of CNS gene transfer. MLV-based recombinant viruses have traditionally been produced through the use of stable packaging cell lines. Since retroviral infection depends on the expression of specific cellular receptors, packaging cell lines should be carefully chosen according to the cell type to be infected. Cell lines such as the Psi-2 (Mann et al., 1983) line produce virions with ecotropic envelope proteins and can infect only rodent cells whereas recombinant viruses produced from lines such as PA317 (Miller and Buttimore, 1986) are amphotropic and are capable of infecting cells from a wide range of species. The use of stable packaging cell lines can occasionally lead to the release of undesirable replication-competent helper virus through chance recombination between the integrated packaging construct and the retroviral vector. This necessitates the use of diagnostic assays of potential producer cell lines for both a good titer of recombinant retrovirus and the absence of contaminating helper viruses.

Recently, the process of producing replication-defective recombinant retroviruses has changed somewhat with the widespread acceptance and use of the 293 cell line and the VSV-G protein for "transient" production of pseudotyped retrovirus amenable to concentration to titers in excess of 1×10^8. Transiently produced retroviral particles can either be used directly for infection of cultured cells (ex vivo gene transfer), or concentrated for direct CNS injection (in vivo gene transfer) in the case of HIV-based vectors. Furthermore, transient production of retroviruses less frequently leads to the production of contaminating helper virus. One drawback of viral production by this method is that the transient transfection must often be performed each time virus is needed, whereas the use of stable retroviral producer cells results in an essentially limitless source of retrovirus once the original line is established. A detailed description of production of all virus types and methods of infection in vivo and in vitro would require a separate book, so we have selected to describe a common scenario of ex vivo application in which recombinant

MLV-based retroviruses are produced for in vitro infection of cultured cells (such as fibrobasts) that will subsequently be implanted into the CNS. Production of infectious recombinant retroviruses and their use to infect target cells in culture is described below.

Protocol for transfection of expression vectors into PA317 amphotropic cells by lipofection

PA317 cells are plated at 5×10^5 cells/60 mm petri dish and incubated at 37°C until the cultures reach 50–60% confluence. Since polypropylene tubes tend to bind the lipofectin DNA complex, polystyrene tubes should be used. Thirty μg of Lipofectin and 2–10 μg plasmid DNA are diluted separately in 1.5 ml Optimem medium and Lipofectin is then added dropwise to the DNA solution. The mixture is incubated at room temperature for 15–20 min. After washing the cells once with Optimem medium, the DNA-lipofectin mixture is added to the cells and incubated with the complex for 5–6 h in a 37°C incubator. The DNA-Lipofectin mixture is then replaced with fresh medium and 48 h later, cultures are put under G418 selection (400 μg/ml) for the selection of stable transformants. Prior to use of viral conditioned medium for infection, replace the medium with non selective medium so that target cells are not damaged by the antibiotic. Usually, we clone several helper virus-free stable colonies and expand them as established cell lines. Colonies that exhibit the highest viral titers are then selected for subsequent infection.

Protocol for "transient" transfection of expression vectors in 293 cells

Although any transfection technique can be used, we generally prefer calcium-phosphate ($CaPO_4$) co-precipitation since it is inexpensive and works quite well with 293 cells. Many different efficient protocols for $CaPO_4$ mediated-transfection have been described in the literature (i.e. Sambrook et al., 1989). For transient viral production, two or three supporting plasmids in addition to the plasmid encoding the retroviral vector are necessary. One plasmid must encode the retroviral *gag* and *pol* proteins (we use CMV-*gag-pol*, N. Somia and I. Verma, manuscript in preparation). A second plasmid encodes the *env* proteins (the VSV-G protein encoding plasmid pMDG is currently used in our laboratory, Naldini et al., 1996). Finally, a reporter construct may be included for assessing the efficiency of transfection (i.e. the β-galactosidase plasmid pCH110). For plasmids used in co-transfection, we generally follow this formula: 4-parts retroviral vector, 2-parts CMV-gag-pol plasmid, 1 part pMDG-VSV-G plasmid, 1-part reporter/internal control. If VSV-G protein is not used as the env, increase the *env* encoding plasmid to 2 parts. Good results with equamolar parts of all plasmids except the reporter have been described; however, we have found that higher levels of packaging plasmids frequently result in slightly lower titers and wastes plasmid DNA.

The 293 cells are plated in any size plate (ranging from one well of a 24-well Costar plate to a 10-cm petri dish) and allowed to grow until 25–40% confluency. To obtain 10^2–10^5 particles per ml, small scale production is more than sufficient, but when very higher titers are required, multiple large plates (and virus concentration) are necessary. The amount of retroviral vector used is 5–10 μg for a 10-cm plate, 1–3 μg for a 4–6 cm plate or 6-well plate, and 0.3–0.5 μg for a 24-well plate. Clean and RNA-free DNA should be used (the method of plasmid purification does not seem to matter). For transfection of one 10 cm plate, we use the following procedure (recipies for 2XHBS and 2M $CaCl_2$ are from Sambrook et al., 1989):

1. Mix the plasmids (10 μg vector, 5 μg CMV-gag-pol, 3 μg pMDG, 2 μg pCH110) with sterile water in a total volume of 1 ml.
2. Add 150 μl of 2M $CaCl_2$ slowly dropwise.
3. Add 1 ml of 2XHBS, and gently pipette up and down several times to mix.

4. Incubate 1.5–5 min at room temperature.
5. Add the transfection mixture dropwise to the plate of cultured 293 cells and return to the incubator for 36–48 h.

The culture medium from the transfected 293 cells is then collected from the well and filtered through a 0.45 μm filter to remove any contaminating cells. Transfected cells can be processed for expression of the internal control to determine whether transfection efficiency was sufficient to provide a useful titer. For high viral titers, >50% of the cells should be positive (>80% is not unusual). For low viral titers, >10% cell transfection should be usable; however, very low transfection efficiencies may only provide the researcher with few infected cells (or colonies).

Protocol for infection of target cells

Filtered, conditioned media from either stable producer cells or transient transfections can be frozen at −70°C or used immediately on target cells. Infection of cells is used for determination of viral titer and detection of helper virus contamination (see below), and production of cells for transplantation. For efficient infection, any target cell type should be actively dividing (with space for at least 2–3 d of active growth) and evenly distributed in the plate (as opposed to clumped up). For primary fibroblasts, we generally plate the cells at at 1×10^5 cells/10 cm plate and allow them to grow until 50–60% confluent.

The carrier compound polybrene increases infection efficacy of cells cultured in a monolayer, and it is usually added to the viral media prior to cell infection. We generally use polybrene at a final concentration of 8 μg/ml; however, since it can prove slightly toxic to some cell types, concentrations should be tested before final usage. Viral media produced by $CaPO_4$-transfection without media changes (as described above) do not require polybrene (as leftover precipitate has a similar function), although it may augment infection in some cases. Polybrene/viral medium is then added dropwise and in a variable quantity to the plate of target cells. Generally, the amount of viral medium is 10–30% of the final culture medium volume. If very low titers are known or predicted, the culture medium can be totally replaced with viral medium (because of possible cytotoxicity, this is only recommended for researchers very familiar with their viral stocks). Following an overnight infection, the viral medium is replaced with fresh culture medium to avoid toxicity from the polybrene or the residual $CaPO_4$. A day later, infected cells can be placed under selection (i.e. with G418 — generally 0.4–1.0 mg/ml).

Protocol for determination of titer and helper virus contamination

To determine the titer of produced virus and/or the presence of helper virus, fast growing fibroblastic cell lines such as 3T3 are usually used. The 3T3 cells are first plated at low density in a 6-well plate then 0.1–0.5 ml of viral medium supplemented with polybrene are added to the cultures (often multiple viral volumes in different wells are done simultaneously). Two days later, the infected 3T3 cells are put under selection. After 10–14 d, the culture medium is collected and the plate is lightly stained with a dilute solution of Comassie blue to count the resulting colonies. The number of colonies represents the functional viral titer which can be multiplied to provide colony forming units (CFUs)/ml. The collected medium is then used for a second round of infection on naive 3T3 cells which are then put under selection as described above. In the absence of helper virus, no resistant colonies should be obtained on the second plate. If resistant colonies are observed, the original clone tested should be discarded. Helper-free bulk populations or selected clonal colonies of infected cells may then be used to characterize transgene expression in vitro for future expansion and in vivo grafting.

Intracerebral transplantation

Prior to transfer in vivo, it is prudent to determine if the transgene and vector are functioning properly in vitro (Fig. 1). At the very least, PCR or RNA blot analysis of RNA isolated from transduced cells is recommended to insure that transcription is normal and at sufficient levels. Before in vivo applications, additional analysis of protein production and bioactivity by western blot, ELISA, or bioassays are almost always well worth the effort because they will reveal any unpredicted problems in transgene function as well as providing an estimate of protein expression levels on a per cell/time basis. Keep in mind, though, that in vitro studies are often poor predictors of efficacy after implantation.

The intracerebral area (ventricular, parenchymal or wound cavity) where the cells are to be implanted may be critical for graft survival and gene expression. For instance, myoblast grafts have been reported not to survive well following intraventricular injections (Lisovoski et al., 1997). When injected as a cell suspension into the ventricle, primary fibroblasts can survive well without major diffusion, although some cells can wash away. Access to the cerebrospinal fluid and space availability are important issues for cell implantation. Thus, cells that have been transfected and selected in serum conditions may require serum for correct transgene expression; consequently, an intralesional or intraventricular injection may help sustain stable in vivo gene expression. Genetically modified cells can be introduced into the host brain via three main strategies: cell suspensions, cells embedded in a collagen matrix and cells encapsulated in polymer fibers (Fig. 1). Cell suspensions are the preferred method into most cerebral sites, whereas collagen-embedded cells are chosen for grafting into wound cavities. Indeed, the collagen acts as a physical bridge favoring growing axons to reconnect to their target in the presence of appropriate tropic factors. Finally, as discussed previously, cell encapsulation should be used if potentially tumorigenic cell lines are used (Aebischer, et al., 1991; Hoffman et al., 1993). One should keep in mind that long-term survival of the implanted cells depends on a variety of factors associated with each experimental step, such as cell culture-related cellular damage (i.e. extensive culturing prior grafting or cell contamination), mechanics of cell implantation (too small or too large grafts), poor vascularization of the graft, and immune response of the host to foreign cells or foreign gene products. Our experience with primary fibroblasts is that using cells at lower passages and harvesting them at 80% confluence allow good survival of the in vivo grafts. In contrast, fibroblasts harvested in their quiescent state generally do not survive as well following intracerebral implantation.

Protocol for intracerebral injections of fibroblast suspensions

Preparation of cell suspensions
After removing the medium from the cell culture flasks, cells are rinsed with PBS, trypsinized and suspended in PBS. The cell concentration is determined by counting a sample of the cell suspension in a Coulter counter. Cells are then centrifuged to remove the PBS and resuspended at a final concentration of 5×10^4 to 1×10^5 cells/μl (optimal concentration) in a grafting solution containing PBS supplemented with 1 μg/ml $MgCl_2$, 1 μg/ml $CaCl_2$ and 0.1% glucose. We generally add to the grafting buffer a 2% concentration of serum from the same species to be grafted (rat serum, if the transplantation is to be performed on rats) to favor cell survival. For graft survival, cell suspensions should be used within 2–3 h and kept at room temperature during the grafting procedure. If a large number of animals are to be grafted, several batches of cells should be prepared during the course of the experiment.

Intracerebral implantation in the adult rat
Although recommended, sterile techniques are not necessary for successful grafting in rats. A 10 μl

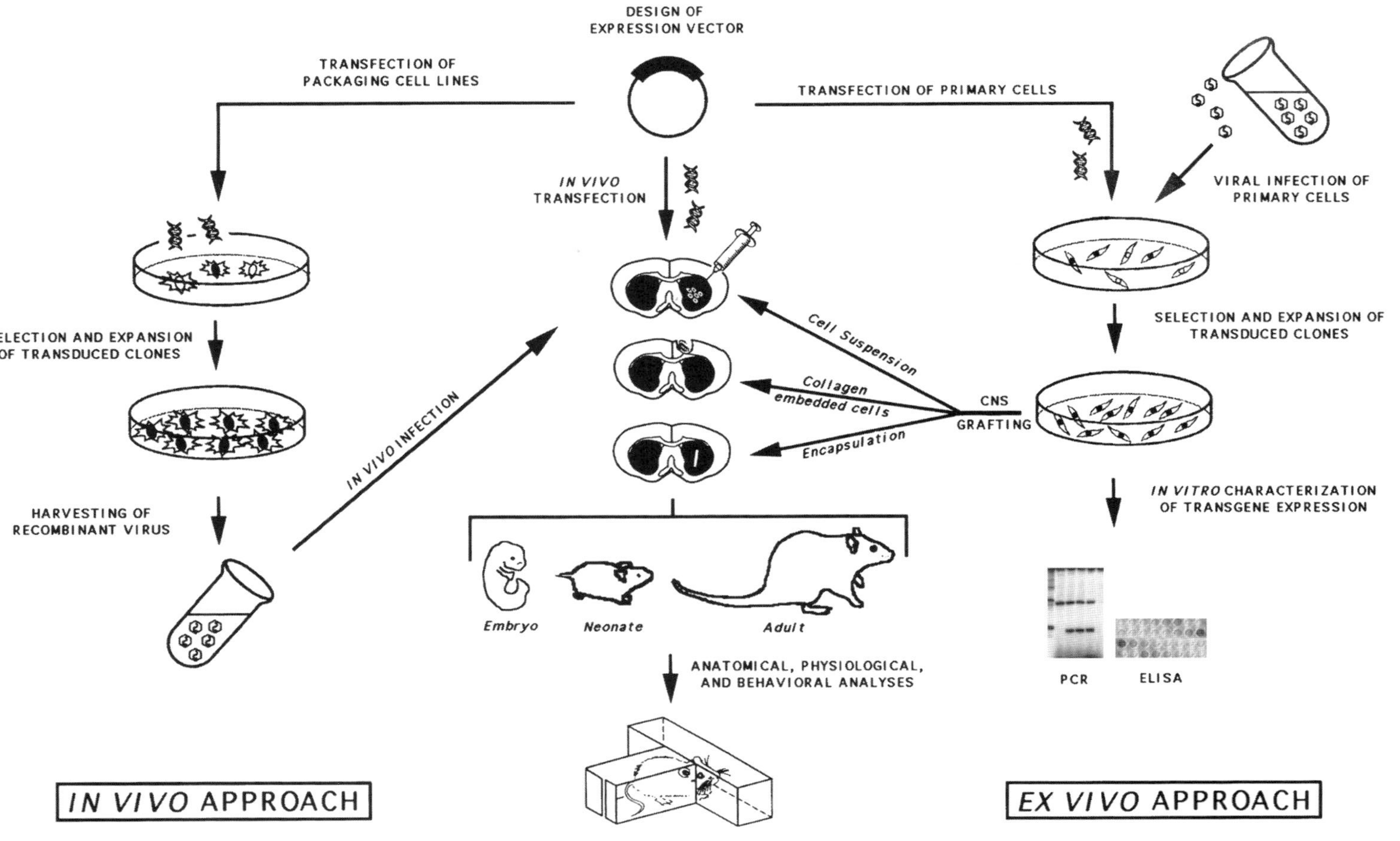

Fig. 1. Methods of CNS gene transfer.

Hamilton syringe is mounted on the stereotaxic frame; we prefer using beveled needles (26 or 30 gauge) which allow better penetration of the cerebral tissue. Rats are anesthetized with a general anesthetic and positioned in the stereotaxic frame after shaving their heads and applying antiseptic. A hole is drilled in the skull over the selected brain areas. Prior to loading the syringe with cell suspension, tap gently on the tube to resuspend the cells. The injection needle is then lowered to the predetermined stereotaxic coordinates and 1–3 μl is injected per site at a rate of 1 μl/min. The syringe is raised 1mm and left in place for 2 min and then slowly retrieved from the brain. This prevents the cell suspension from flowing back up along the needle track. If necessary, several deposits can be made along the needle tract. Syringes should be rinsed regularly with ethanol and then saline solution to prevent needle from clogging and allow efficient cell injections. The skin is then closed with metal wound clips and the rats are allowed to recover from surgery. At the end of the experiment we generally put back in culture the remaining cell suspension to determine cell viability. In the majority of our studies, the presence of the graft was detected at the anatomical level using either classical Nissl staining or immunohistochemical detection of fibronectin, a substance produced and secreted by fibroblasts and normally absent from the brain. The detection of a reporter gene such as ßgalactosidase also facilitates the detection of the grafted cells.

Protocol for collagen-embedded fibroblasts

When fibroblasts are grown in a dense rat tail collagen suspension, they form cell aggregates that are immobilized and encapsulated inside the collagen-derived gel. We usually prepare the collagen solution fresh the day before use by dissolving 10 mg of rat type I collagen (Sigma) into 3.3 ml of 0.1% glacial acetic acid (0.3% collagen solution). The resulting mixture is vortexed several times during the day and left shaking gently overnight. The next day, cells are washed, trypsinized and counted as described above. In a 1.5 ml conical Eppendorf tube, 1.2 $\times$ 10^5 cells are resuspend in 162 μl DMEM$^+$. To make one plug, add 6.9 μl of 0.1M NaOH to the cell mixture and pipette up and down to mix. Immediately add 81 μl of the collagen solution, pipette up and down again to mix, cap the tubes, and incubate overnight. The shape and the density of the plugs may be modified according to one's needs by manipulating the collagen concentration, the fibroblast density and/or the shape of the tubes used for preparing the plugs. The freshness of the collagen solution is important and it should not be kept too long before use. Anesthetized rats are placed in a Kopf stereotaxic frame after their heads have been shaved and antiseptic has been applied. A hole is drilled in the skull over the brain area of interest and a lesion cavity is carefully made by aspirating the tissue with a Pasteur pipette attached to a vacuum device. A piece of about 2–3 mm^3 in size of the collagen/fibroblast matrix is then implanted into the wound and maintained in place with hydrated gelfoam. The skull is treated with antibiotic powder, the skin sutured, and animals left to recover. The collagen-fibroblast matrix should be large enough to fill the lesion cavity and allow host/graft interactions.

Ex vivo and in vivo gene transfer for CNS gene therapy: the dopamine lesion model

Animal models of Parkinson's disease (PD) have been among the most utilized models for assessing the safety and efficacy of somatic gene transfer strategies. PD is characterized by a loss of the dopaminergic striatal innervation originating in the substantia nigra (SN) which results in bradykinesia in the affected individuals (see Fisher and Gage, 1995 for review). In animal models of PD, the nigrostriatal dopaminergic pathway is specifically depleted by utilizing neurotoxins such as 6-hydroxydopamine (6-OHDA) and 1-methyl-4-phenyl-1,2,3,6-tetrahydropyridine

(MPTP). Unilateral injections of 6-OHDA results in a hemispherical imbalance of forebrain dopamine which can be assessed behaviorally through the administration of dopamine-releasing drugs (i.e. amphetamine) or dopamine receptor agonists (i.e. apomorphine) and manifests in turning behavior (Fisher and Gage, 1995). MPTP is also commonly used to generate models of PD in rodents and non-human primates. Systemic injections of MPTP results in the rapid bilateral destruction of nigrostriatal dopaminergic neurons. This model, however, does not reproduce the progressive neurodegeneration typical of PD. Two general approaches have been developed for restoring normal function in lesioned animals. The neuroprotective approach consists in providing trophic factors known to promote survival of dopaminergic neurons (Table 1). The replacement approach aims at restoring normal intracerebral dopamine levels by introducing tyrosine hydroxylase (TH), the enzyme converting tyrosine to L-DOPA (precursor of dopamine) (Table 2). More complex strategies utilizing a combination of gene transfer and dopaminergic cell transplantation have also been tried (Senut et al., 1996).

As shown in Tables 1 and 2, most of the ex vivo gene transfer studies performed on animal models of PD have utilized genetically modified cell lines of both non-neuronal and neuronal origins. Following intrastriatal engraftment and with few exceptions, most of the cell lines used were capable of significantly increasing either neuronal survival and dopamine intracerebral levels, and/or of improving drug-induced rotational behavior. Unfortunately, the tumorigenic nature of some grafted cells led to tumor formation and subsequent graft rejection, preventing thereby the prospect of studying the long-term effects of the delivered transgenes (Tables 1 and 2). The work of Lindner et al. (1995) clearly demonstrates that the tactic of cell encapsulation may overcome the tumor formation problem and favor long-term cell survival and functional effects. However, the report of the rupture of the capsules in some animals raises questions about the safety of this technology. To date, the most promising results obtained with engineered cell lines in the dopaminergic lesion model seem to be those described with the genetic manipulation of neural cell lines. For example, Anton et al. (1994) used a temperature-sensitive immortalized neural cell line isolated from embryonic rat SN which was genetically modified to express TH. Following injection into the caudate of 6-OHDA-lesioned rats or MPTP-lesioned monkeys, TH-expressing immortal nigral cells succeeded in improving the apomorphine-induced rotational behavior. Cells survived up to 2 months and were reported not to show any sign of tumorigenesis or immune rejection. Despite their limited in situ neuronal differentiation, such cells appear as promising candidates for ex vivo gene transfer. Since extended survival of intracerebral grafts is crucial for establishing whether ex vivo gene transfer may be a potential therapeutical tool, genetically modified primary fibroblasts and astrocytes have also been grafted as an alternative to cell lines (Tables 1 and 2). TH- or growth factor-expressing fibroblasts and astrocytes survived intracerebral transplantation, did not show invasive growth, and promoted functional recovery. TH-expressing primary fibroblasts exhibited the longest survival times post-transplantation (10 weeks) (Fisher et al., 1991). Grafted TH-expressing astrocytes displayed some capability of migration in the host brain. Although encouraging, all studies using primary cells unfortunately reported a loss of transgene expression within weeks that resulted in the loss of beneficial effects. Although the exact cause of in vivo down-regulation of transgene expression is not totally clear, several factors may be involved, such as the progressive loss of grafted cells, the quiescent state of implanted cells, or perhaps more importantly, the host inflammatory response. Indeed, following CNS transplantation, the graft is infiltrated by inflammatory cells in response to the rupture of the blood brain barrier. These cells are known to secrete different cytokines

TABLE I

Parkinson's disease: neuroprotective strategy

Ex vivo cell type	In vivo vector	Transgene	Location	Effects	Survival time	Problems reported	References
BHK cells (Fibroblast cell line) (Encapsulated)	–	Rat GDNF	Striatum	Behavioral improvement	90 Days	Some capsules broke	Lindner et al., 1995
RAT1 (Fibroblast cell line)	–	Rat BDNF	Dorsal tegmentum (MPP Rat)	Increased survival of TH-positive SN cells	7 Days	Tumorigenic cell line	Frim et al., 1994
RAT1 (Fibroblast cell line)	–	Rat BDNF	Dorsal tegmentum (MPP Rat)	Increased dopamine levels Increased survival of TH-positive SN cells	7 Days	Tumorigenic cell line	Galpern et al., 1996
RN-22 (Schwannoma cell line)	–	FGF-4	Nigro-striatal path (Knife cut/Rat)	Axonal regeneration Behavioral improvement	10 Weeks	Tumorigenic cell line	Brecknell et al., 1996
Primary astrocytes	–	Human BDNF	Striatum (6-OHDA Rat)	Behavioral improvement No Axonal regeneration	32 Days	Loss of transgene expression	Yoshimoto et al., 1995
Primary fibroblasts	–	Human BDNF	Striatum (6-OHDA Rat)	Increased survival of dopaminergic terminals	5 Weeks	–	Levivier et al., 1995
–	AV	Human GDNF	Dorsal SN (6-OHDA Rat)	Increased survival of dopaminergic neurons	42 Days	Loss of transgene expression	Choi-Lundberg et al., 1997
–	AV	Rat GDNF	Striatum (6-OHDA Rat)	TH-positive striatal cells Increased survival of SN dopaminergic cells Behavioral improvement	3-4 Weeks	AV-related striatal inflammation and atrophy	Bilang-Bleuel et al., 1997

TABLE 2

Parkinson's disease: replacement strategy

Ex vivo cell type	In vivo vector	Transgene	Location (model)	Effects	Survival time	Problems reported	References
208F (Fibroblast cell line)	–	Rat TH	Caudate (6-OHDA Rat)	Behavioral improvement	2 Weeks	Tumorigenic cells	Wolff et al., 1989
NIH 3T3 (Fibroblast cell line)	–	Human TH	Dorsal tegmentum (MPP Rat)	Increased dopamine levels. No behavioral effect	9 Days	Tumorigenic cells	Horellou et al., 1990a
RIN (Endocrine cell line)	–	Human TH	Striatum (6-OHDA Rat)	Increased dopamine levels. No behavioral effect	9 Days	Tumorigenic cells	Horellou et al., 1990a
C6 glioma	–	Human TH	Striatum (6-OHDA Rat)	Behavioral improvement	10 Days	Tumorigenic cells	Uchida et al., 1992
AtT20 (Neuro-endocrine cell line)	–	Human TH	Striatum (6-OHDA Rat)	Behavioral improvement	2 Weeks	Tumor formation	Horellou et al., 1990b
NS20Y (Neuroblastoma cell line)	–	Rat TH	Striatum (6-OHDA Rat)	Behavioral improvement	9 Days	Tumor formation	Horellou et al., 1990b
SVG (Immortalized human fetal astrocyte cell line)	–	Human TH	Striatum (6-OHDA Rat)	Behavioral improvement	4 Weeks	Graft rejection	Tornatore et al., 1996
CSM 14.1.4 (Immortalized Nigro neural cell line)	–	Rat TH	Caudate (6-OHDA Rat) (MPTP non-human primate)	Behavioral improvement	2 Months	–	Anton et al., 1994
CSM 14.1.4 (Immortalized Nigro neural cell line)	–	Bovine TH	Caudate (MPTP non-human primate)	Behavioral improvement	1 Month	–	Anton et al., 1994
Primary astrocytes	–	Human TH	Caudate (6-OHDA Rat)	Behavioral improvement	2 Weeks	Loss of transgene expression	Lundberg et al., 1996

TABLE 2

Continued

Ex vivo cell type	In vivo vector	Transgene	Location (model)	Effects	Survival time	Problems reported	References
Primary fibroblasts	–	Human TH	Striatum (6-OHDA Rat)	Moderate behavioral improvement	2 Weeks	Loss of transgene expression	Bencsics et al., 1996
Primary fibroblasts	–	Rat TH	Striatum (6-OHDA Rat)	Behavioral improvement	10 Weeks	Loss of transgene expression	Fisher et al., 1991
–	Lipo-fectin	Rat TH	Striatum (6-OHDA Rat)	TH-positive cells Behavioral improvement	15 Days	Loss of transgene expression	Cao et al., 1995
–	HSV-1	Human TH	Striatum (6-OHDA Rat)	TH-positive cells Behavioral improvement	1 Year	Decrease in transgene expression Cytopathic effects	During et al., 1994
–	AAV	Human TH	Striatum (6-OHDA Rat)	TH-positive cells Behavioral improvement	4 Months	Decrease in the number of TH-positive cells	Kaplitt et al., 1994
–	AV	Human TH	Striatum (6-OHDA Rat)	TH-positive cells Behavioral improvement	2 Weeks	Tissue damage Gliosis inflammation	Horellou et al., 1994

which have been shown in vitro to down-regulate LTR-driven gene expression in primary fibroblasts (Schinstine et al., 1997).

In vivo gene transfer techniques have also been applied to the dopaminergic lesion model. Thus, intrastriatal transfer of TH or growth factors has been attempted either via replication-incompetent viral vectors (HSV, AV, AAV) or through non-viral vectors (lipofectin). All viral and non-viral vectors were reported to induce in vivo TH expression or increase neuronal survival, and to result in behavioral improvement (see refs in Tables 1 and 2). Regardless of the type of vector used, however, a more or less rapid time-dependent loss of transgene expression was observed in the infected brain cells. HSV- and AV-related cytopathic effects were also reported and included striatal inflammation and/or atrophy which with HSV resulted in animal death. In contrast, following intrastriatal AAV-TH injections, no tissue damage was reported up to 4 months (the last time-point examined), indicating that AAV vectors might be the safest in vivo.

Conclusion

The impediments to widespread sucessful usage of gene therapy today are most of the same problems hindering application of this technology over the past decade. Prolonged gene expression, high-efficiency and stable transduction of cells in vivo, immunological considerations, and safety, are as of much concern as they ever were; however, reports within recent years suggest that we might be on the verge of a new wave of useful vectors and methods that may be capable of surmounting some of the bigger problems. Expansion of the repertoire of retroviral vectors to include lentiviral vectors capable of infecting non-dividing cells with high efficiency is one area of vast potential (Naldini et al., 1996). The refinement of methods of producing recombinant adenoviruses and high-titer adenovirus-free AAV stocks is also of great potential utility and is described further in the chapter on adenoviral vectors. New regulatable vectors may be able to overcome questions of prolonged gene expression and simultaneously allow exogenous regulation of transgene expression using non-toxic drugs (Wang et al., 1994; Hoshimaru et al., 1996). The combined attributes of these new tools may bring gene therapy forward to the point where it is practical for application to the human CNS.

Acknowledgments

The authors thank Mary Lynn Gage for editorial comments on the manuscript. Supported by grants from the March of Dimes, the NIH and NINDS.

List of abbreviations

AAV: Adeno-associated virus
AV: Adenovirus
BDNF: Brain-derived neurotrophic factor
CMV: Cytomegalovirus promoter
CNS: Central nervous system
ELISA: Enzyme-linked immunoassay
GDNF: Glial-derived neurotrophic factor
HSV: Herpes virus
LTR: Long terminal repeat promoter
MPTP: 1-methyl-4-phenyl-1,2,3,6-tetrahydropyridine.
6-OHDA: 6-hydroxydopamine
ORF: open reading frame
p(A): polyadenylation
PD: Parkinson's disease
PCR: Polymerase chain reaction
SN: Substantia nigra
TH: Tyrosine hydroxylase

References

Aebischer, P., Winn S.R., Tresco, P.A., Jaeger, C.B. and Greene, L.A. (1991) Transplantation of polymer encapsulated neurotransmitter secreting cells: effect of the encapsulation technique, *J. Biomech. Eng.*, 113: 178–183.

Anton, R., Kordower, J.H., Maidment, N.T., Manaster, J.S., Kane, D.J., Rabizadeh, S., Schueller, S.B., Yang, J., Rabizadeh, S., Edwards, R.H., Markham, C.H. and Bredesen, D.E. (1994) Neural-targeted gene therapy for rodent and primate hemiparkinsonism. *Exp. Neurol.*, 127: 207–218.

Bencsics, C., Wachtel, S.R., Milstien, S., Htakeyama, K., Becker, J.B. and Kang, U.J. (1996) Double transduction with GTP cyclohydroxylase I and tyrosine hydroxylase is necessary for spontaneous synthesis of L-DOPA by primary fibroblasts. *J. Neurosci.*, 16: 4449–4456.

Bilang-Bleuel, A., Revah, F., Colin, P., Locquet, I., Robert, J-J., Mallet, J. and Horellou, P. (1997) Intrastriatal injection of an adenoviral vector expressing glial-cell-line derived neurotrophic factor prevents dopaminergic neuron degeneration and behavioral impairment in a rat model of Parkinson disease. *Proc. Natl. Acad. Sci. USA*, 94: 8818–8823.

Brecknell, J.E., Du, J.S., Muir, E., Fidler, P.S., Hlavin, M.L., Dunnett, S.B. and Fawcett, J.W. (1996) Bridge grafts of fibroblast growth factor-4 secreting schwannoma cells promote functional axonal regeneration in the nigrostriatal pathway of the adult rat. *Neuroscience*, 74: 775–784.

Cao, L., Zheng, Z-C., Zhao, Y-C., Jiang, Z-H., Liu, Z-G., Chen, S-D., Zhou, C-F. and Liu, X-Y. (1995) Gene therapy of Parkinson disease model rat by direct injection of plasmid DNA-lipofectin complex. *Hum. Gene Ther.*, 6: 1497–1501.

Choi-Lundberg, D.L., Lin, Q., Chang, Y-N., Chiang, Y.L., Hay, C.M., Mohajeri, H., Davidson, B.L. and Bohn, M.C. (1997) Dopaminergic neurons protected from degeneration by GDNF gene therapy. *Science*, 275: 838–841.

During, M.J., Naegele, J.R., O'Malley, K.L. and Geller, A.I. (1994) Long-term behavioral recovery in parkinsonian rats by an HSV vector expressing tyrosine hydroxylase. *Science*, 266: 1399–1403.

Fisher, L.J. (1997) Neural precursor cells: applications for the study and repair of the central nervous system. *Neurobiol. Dis.*, 4: 1–22.

Fisher. L.J. and Gage, F.H. (1995) Novel therapeutic directions for Parkinson's disease. *Molecular Medicine Today*, 1: 181–187.

Fisher, L.J., Jinnah, H.A., Kale, L.C., Higgins, G.A. and Gage, F.H. (1991) Survival and function of intrastriatally grafted primary fibroblasts genetically modified to produce L-dopa. *Neuron*, 6: 371–380.

Frim, D.M., Uhler, T.A., Galpern, W.R., Beal, M.F., Breakefield, X.O. and Isacson, O. (1994) Implanted fibroblasts genetically engineered to produce brain-derived neurotrophic factor prevent 1-methyl-4-phenylpyridinium toxicity to dopaminergic neurons in the rat. *Proc. Natl. Acad. Sci. USA*, 91: 5104–5108.

Gage, F.H., Ray, J. and Fisher, L.J. (1995) Isolation, characterization, and use of stem cells from the CNS. *Annu Rev Neurosci.*, 18: 159–192.

Galpern, W.R., Frim, D.M., Tatter, S.B., Altar, C.A., Beal, M.F. and Isacson, O. (1996) Cell-mediated delivery of brain-derived neurotrophic factor enhances dopamine levels in an MPP+ rat model of substantia nigra degeneration. *Cell Transplant.*, 5: 225–232.

Hoffman, D., Breakefield, X.O., Short, M.P. and Aebischer P. (1993) Transplantation of a polymer-encapsulated cell line genetically engineered to release NGF. *Exp. Neurol.*, 122: 100–106.

Horellou, P., Brundin P., Kalen P., Mallet J. and Björklund., A. (1990a) In vivo release of DOPA and dopamine from genetically engineered cells grafted to the denervated rat striatum. *Neuron*, 5: 393–402.

Horellou, P., Marlier, L. Privat, A. and Mallet, J. (1990b) Behavioral effect of engineered cells that synthesize L-dopa or dopamine after grafting into the rat neostriatum. *Eur. J. Neurosci.*, 2: 116–119.

Horellou, P., Vigne, E., Castel, M.N., Barneoud, P., Colin, P., Perricaudet, M., Delaere, P. and Mallet, J. (1994) Direct intracerebral gene transfer of an adenoviral vector expressing tyrosine hydroxylase in a rat model of Parkinson's disease. *Neuroreport*, 6: 49–53.

Hoshimaru, M., Ray, J., Sah, D.W.Y. and Gage, F.H. (1996) Differentiation of the immortalized adult neuronal progenitor cell line HC2S2 into neurons by regulatable suppression of the v-myc oncogene. *Proc. Natl. Acad. Sci. USA*, 93: 1518–1523.

Kaplitt, M.G., Leone, P., Samulski, R.J., Xiao, X., Pfaff, D., O'Malley, K.L. and During M.J. (1994) Long-term gene expression and phenotypic correction using adeno-associated virus vectors in the mammalian brain. *Nat. Gen.*, 8: 148–154.

Kozak, M. (1984) Point mutations close to the AUG initiator codon affect the efficiency of translation of rat preproinsulin in vivo. *Nature*, 308: 241–246.

Le Gal La Salle, G., Robert, J.J., Berrard, S., Ridoux, V., Stratford-Perricaudet, L.D., Perricaudet, M. and Mallet, J. (1993) An adenovirus vector for gene transfer into neurons and glia in the brain. *Science*, 259: 988–990.

Levivier, M., Przedborski, S., Bencsics, C. and Kang, U.J. (1995) Intrastriatal implantation of fibroblasts genetically engineered to produce brain-derived neurotrophic factor prevents degeneration of dopaminergic neurons in a rat model of Parkinson's disease. *J. Neurosci.*, 15: 7810–7820.

Lindner, M.D., Winn, S.R., Baetge, E.E., Hammang, J.P., Gentile, F.T., Doherty, E., McDermott, P.E., Frydel, B., Ullman, M.D., Schallert, T. and Emerich, D.F. (1995) Implantation of encapsulated catecholamine and GDNF-producing cells in rat with unilateral dopamine depletions and parkinsonian symptoms. *Exp. Neurol.*, 132: 62–76.

Lisovoski, F., Wahrmann, J.P., Pages, J.C., Cadusseau, J., Rieu, M., Weber, A., Kahn, A. and Peschanski, M. (1997) Long-term histological follow-up of genetically modified myoblasts grafted into the brain. *Molec.Brain Res.*, 44: 125–133.

Lundberg, C., Horellou P., Mallet, J. and Björklund, A. (1996) Generation of DOPA-producing astrocytes by retroviral transduction of the human tyrosine hydroxylase gene: in vitro characterization and in vivo effects in the rat Parkinson model. *Exp. Neurol.*, 139: 39–53.

Mann, R., Mulligan, R.C. and Baltimore, D. (1983) Construction of a retrovirus packaging mutant and its use to produce helper free defective retroviruses. *Cell*, 33: 153–159.

Martinez-Serrano, A. and Björklund, A. (1997) Immortalized neural progenitor cells for CNS gene transfer and repair. *TINS*, 20: 530–538.

Maysinger D. and Morinville A. (1997) Drug delivery to the nervous system. *TIBTECH*, 15: 410–418 .

Miller A.D. and Buttimore C. (1986) Redesign of retrovirus packaging cell lines to avoid recombination leading to helper virus production. *Mol. Cell. Biol.*, 6: 2895–2902.

Mucke, L. and Rockenstein, E. (1993) Prolonged delivery of transgene products to specific brain regions by migratory astrocyte grafts. *Transgene*, 1: 3–9.

Naldini, L., Blomer, U., Gallay, P., Ory, D., Mulligan, R., Gage, F.H., Verma, I.M. and Trono, D. (1996) In vivo gene delivery and stable transduction of nondividing cells by a lentiviral vector. *Science*, 272: 263–267.

Owens, G.C., Johnson, R., Bunge, R.P. and O'Malley, K.L. (1991) L-3-4-dihydroxyphenylalanine synthesis by genetically modified Schwann cells. *J. Neurochem.*, 56: 1030–1036.

Ray, J., Peterson, D.A., Schinstine, M. and Gage, F.H. (1993) Proliferation, differentiation, and long-term culture of primary hippocampal neurons. *Proc. Natl. Acad. Sci. USA*, 90: 3602–3606.

Rosenberg M.B., Friedmann T., Robertson R.C., Tuszynski M., Wolff J.A., Breakefield X.O. and Gage F.H. (1988) Grafting of genetically modified cells to the damaged brain: Restorative effects of NGF expression. *Science*, 242: 1575–1578.

Sambrook, J., Fritsch E.F. and Maniatis, T. (1989) *Molecular cloning. A laboratory manual*, 2nd Edn, Cold Spring Harbor Laboratory Press.

Schinstine, M., Ray, J. and Gage, F.H. (1997) Potential effect of cytokines on transgene expression in primary fibroblasts implanted into the rat brain. *Mol. Brain Res.*, 47: 195–201.

Senut M-C., Fisher L.J., Ray J. and Gage F.H. (1995a) Somatic gene therapy in the brain. In: R.G. Landers (Ed.), *Methods in Cell Transplantation*, CRC Press and the Mosby Year Book, pp.197–214.

Senut, M-C., Tuszynski, M.H., Raymon, H.K., Suhr, S.T., Liou, N.H., Jones, K.R., Reichardt, L.F. and Gage, F.H. (1995b) Regional differences in responsiveness of adult CNS axons to grafts of cells expressing human neurotrophin 3. *Exp. Neurol.*, 135: 36–55.

Senut, M-C., Suhr, S. T. and Gage, F. H. (1996) Transplantation of genetically modified non-neuronal cells in the central nervous system. In: D. Latchman (Ed.), *Genetic manipulation of the nervous system*, ISBN, Academic Press Limited, San Diego, pp. 181–202.

Snyder, E.Y. (1994) Grafting immortalized neurons to the CNS. *Curr. Opin. Neurobiol.*, 4: 742–751.

Snyder, E.Y. and Senut, M-C. (1997) The use of nonneuronal cells for gene delivery. *Neurobiol. Dis.*, 4: 69–102.

Suhr, S.T. and Gage, F.H. (1993) Gene therapy for neurologic disease. *Arch. Neurol.*, 50: 1252–1268.

Suhr, S.T., Gil, E.B., Senut, M-C. and Gage, F.H. (1998) High level transactivation by a modified *Bombyx* ecdysone receptor in mammalian cells without exogenous retinoid X receptor. *Proc. Natl. Acad. Sci. USA*, 95: 7999–8004.

Tornatore, C., Baker-Cairns, B., Yadid, G., Hamilton, R., Meyers, K., Atwood, W., Cummins, A., Tanner, V. and Major, E. (1996) Expression of tyrosine hydroxylase in an immortalized human fetal astrocyte cell line, in vitro characterization and engraftment into the rodent striatum. *Cell Transplantation*, 5: 145–163.

Uchida, K., Tsuzaki, N., Nagatsu, T. and Kohsaka, S. (1992) Tetrahydrobiopterin-dependent functional recovery in 6-hydroxydopamine-treated rats by intracerebral grafting of fibroblast transfected with tyrosine hydroxylase cDNA. *Dev. Neurosci.*, 14: 173–180.

Wang, Y., O'Malley Jr, B., Tsai, S.Y. and O'Malley, B.W. (1994) A regulatory system for use in gene transfer. *Proc. Natl. Acad. Sci. USA*, 91: 8180–8184.

Wolff, J.A., Fisher, L.J., Xu, L., Jinnah, H.A., Langlais, P.J., Iuvone, P.M., O'Malley, K.L., Rosenberg, M.B., Shimohama, S., Friedmann, T. and Gage, F.H. (1989) Grafting fibroblasts genetically modified to produce L-dopa in a rat model of Parkinson disease. *Proc. Natl. Acad. Sci. USA*, 86: 9011–9014.

Xiao, X., Li, J., McCown, T.J. and Samulski, R.J. (1997) Gene transfer by adeno-associated virus vectors into the central nervous system. *Exp. Neurol.*, 144: 113–124.

Yoshimoto, Y., Lin, Q., Collier, T.J., Frim, D.M., Breakefield, X.O. and Bohn, M.C. (1995) Astrocytes retrovirally transduced with BDNF elicit behavioral improvement in a rat model of Parkinson's disease. *Brain Res.*, 691: 25–36.

W.E. Crusio and R.T. Gerlai (Eds.)
Handbook of Molecular-Genetic Techniques for Brain and Behavior Research (Techniques in the Behavioral and Neural Sciences, Vol. 13)

CHAPTER 3.2.2

Adenovirus vectors for gene transfer into the central nervous system

Martine Barkats, Olga Corti and Jacques Mallet

Laboratoire de Génétique Moléculaire de la Neurotransmission et des Processus Neurodégénératifs, C.N.R.S., Hôpital de la Pitié Salpêtrière, 75013 Paris, France

Neuronal death is the common pathological event of neurodegenerative diseases. Neuropathological agents causing or involved in the cell death underlying some of these diseases have been identified. The β-amyloid protein in Alzheimer's disease (AD) (Hsiao et al., 1996; Geula et al., 1998; Soto et al., 1998), oxidative stress in amyotrophic lateral sclerosis (ALS) (Rosen et al., 1993), and CAG repeats in Huntington's disease (HD) (Snell et al., 1993) figure among these mechanisms. The discoveries of basic research are expected to lead to new therapeutic possibilities.

Vectors for CNS gene therapy

Macromolecules with therapeutic activity such as growth factors, cytokines, or enzymes cannot be easily delivered to the central nervous system (CNS) which is protected by the blood brain barrier (BBB). To reach the brain parenchyma, it is necessary to inject these molecules directly in situ and therapy may require repeated intracerebral injections or infusions using osmotic pumps. Because of the technical complexity of these procedures, and the risks of infection associated with repeated BBB disruption, the recent advances in knowledge of the pathogenesis of nervous diseases have not yet found clinical applications. Intracerebral grafting of embryonic brain cells is a palliative approach which could be an alternative to the pharmacological substitution of a deficient neurotransmitter. However, ethical considerations associated with the use of human fetal tissue and the large numbers of fetuses required to obtain sufficient material for the grafting of a single patient are major obstacles.

The recent development of gene delivery systems has opened the way to human gene therapy, a promising therapeutic method for neurological diseases for which classical treatments are unavailable. It may allow local production of the desired therapeutic protein in a given brain structure and make it available for prolonged periods after a single administration. This approach thereby avoids the problems associated with drug therapy.

To develop an efficient system for CNS gene therapy, several issues must be addressed:

The first is the identification of canditate factors for use in gene therapy and then the cloning of the corresponding genes. Neurotrophic factors including the glial cell line-derived neurotrophic factor (GDNF) and the nerve growth factor (NGF), are of particular interest because they considerably improve the survival of motor and dopaminergic (DA) neurons (Beck et al., 1995; Oppenheim et al., 1995), and of cholinergic neurons (Fischer et al., 1987), respectively. Oxidative metabolism enzymes such as superoxide dismutase (SOD), catalase, and glutathione peroxidase (GPx), are also potentially useful for pathological situations involving free radicals, e.g. ALS, AD, HD, Parkinson's disease (PD), Cerebral ischemia, and others (Olanow, 1993). Antioxidant-based gene therapy is particularly attractive as, in general, no toxicity is associated with the anti-oxidative enzymes which could protect neurons from neurodegeneration.

The second issue is the choice of an appropriate vector system in order to target the desired gene into the relevant brain area. Genes can be transferred either directly by intracerebral injection of a vector system (*in vivo* gene transfer), or indirectly by grafting genetically modified cells into the brain (*ex vivo* gene transfer). Because they naturally transfer their own genome into the cells that they infect, modified viruses are the most efficient biological vector systems for both these strategies.

Retroviruses, adenoviruses, adeno-associated viruses (AAV) and lentiviruses are the most promising types of virus for the transfer of therapeutic genes. The use of AAV is limited by several drawbacks (Anderson, 1998): the number of viral particles required for transduction of a single cell is very high; the viral genome is small, allowing the cloning of only 4.8 kb of added DNA; and these vectors (like retroviruses) integrate the DNA of infected cells in a non-specific manner (Russell et al., 1994) that may cause insertional mutagenesis. The herpes simplex virus, which can infect cells of the nervous system, is also used for gene transfer into neurons, but its use is limited by the toxicity of viral proteins. Recently, additional genes were deleted from the viral backbone to reduce the immunogenicity. These genetic manipulations allowed the genome to persist for longer periods of time, but also dramatically lowered the level of transgene expression (Samaniego et al., 1998). Adenoviruses seem to be appropriate vectors for CNS gene therapy because they transduce both quiescent and actively dividing cells with great efficacy and have little pathogenicity (Horwitz, 1996). Adenoviruses do not usually integrate into the host DNA, which limits the risks of mutagenesis by heterologous DNA insertion.

The ability of adenoviruses to efficiently transfer genes into neuronal cells was established a few years ago (Akli et al 1993; Bajocchi et al., 1993; Davidson et al., 1993; Le Gal La Salle et al., 1993), and their efficacy has recently been reported in several models of neurodegenerative diseases like PD (Barkats et al., 1997; Bilang-Bleuel et al., 1997; Choi-Lundberg et al., 1997, 1998) and motor neuron diseases (Baumgartner and Shine, 1997, 1998a, 1998b; Haase et al., 1997). These vectors have also proven their great interest in many other disorders of the CNS including lysosomal storage diseases (Li and Davidson, 1995; Ghodsi et al., 1998), Leisch-Nyhan disease (Plumb et al., 1996), and epilepsy (O'Connor et al., 1997, Robert et al., 1997).

The aim of this review is to document the potential of replication-deficient recombinant adenoviruses for gene transfer into nervous cells, and to discuss the recent advances of this technological approach in particular in models of neurodegenerative diseases.

Adenoviral vectors

Adenoviruses naturally infect humans inducing lower respiratory infections (Horwitz, 1996). The best biochemically and genetically characterized serotypes are serotypes 2 and 5, neither of which has been reported to be associated with oncogenicity. They are thus suitable vectors for the transfer of therapeutic genes to humans.

Production of recombinant adenoviruses

The genome of the human adenovirus is well characterized. It is large (36 kb of DNA) and contains many genes which are classified into early (from E1 to E4) and late (L1 to L5) genes, according to whether they are expressed prior to or after viral DNA replication. Both ends of viral DNA bear a short sequence called "Inverted Terminal Repeat" (ITR), which is essential for viral replication.

Wild-type adenovirus cause a lytic infection, so recombinant adenoviruses for gene transfer must be defective for some essential genes which are those of the E1 region (the E3 region which is not necessary for viral propagation in cell culture can also be deleted).

Recombinant adenoviruses are commonly obtained after co-transfection in the 293 cells (which are human kidney embryonic cells expressing the E1 gene) of a E1-deleted adenoviral DNA and a plasmid carrying the following essential sequences: the ITR and encapsidation sequences, a polylinker into which transgene sequences are introduced, and an adenoviral sequence allowing homologous recombination with the viral genome. To allow replication and amplification, E1 defective viruses are complemented in *trans* in 293 cells. Deletions in E1 and E3 can be combined in a single vector that should accomodate transgenes up to 8.3 kb in length and can be obtained at titers as high as 10^{11}–10^{12} particles per ml (Bett et al., 1994; Graham and Prevec, 1995). These replication-deficient adenoviruses have been referred to as "first-generation" recombinant adenoviruses (Fig. 1).

Delivering of recombinant adenoviruses

In vitro gene transfer into neural or non-neural cells
It is possible to use adenovirus to inoculate a variety of primary cell cultures including newborn rat superior cervical ganglia neurons, fetal rat cortical, striatal or mesencephalic neurons, newborn rat brain cerebellar granule neurons and astrocytes (Fig. 2). For each type of cultured cells, it is necessary to determine the optimal viral titer to infect a large percentage of cells with a minimal cytopathic effect. The duration of the exposure to the virus (one hour to several days) and the multiplicity of infection (MOI: plaque forming units per cell, or pfu/cell) must be optimized for each cell type. Generally, to infect cells *in vitro*, the culture medium is removed and saved. A thin layer of viral suspension without serum (1–1000 pfu) is added to the cells for 30–60 min at 37°C, then the medium is added back to the culture. The kinetics of transgene expression should also be optimized for each experiment and depends on the cell type, the transgene and the promoter: 48 h is usually necessary to obtain expression of β-galactosidase under the control of the Rous Sarcoma Virus Long Terminal Repeat (RSV promoter).

In vivo CNS gene transfer
Recombinant adenoviruses encoding for therapeutic agents can be delivered *in vivo* after direct intracerebral injection into particular brain areas. Several requirements must be fulfilled for successful *in vivo* gene transfer and expression: a substantial number of transgene-expressing cells, the precise and appropriate localization of the infected cells and the efficient targeting of the transgene to particular cells, no toxicity or side effects, and the long-term expression of the transgene.

A number of technical variables also need to be optimized for each experiment of *in vivo* gene transfer into the brain, mainly according to the brain region to be infected:

1. The injection rate of the adenoviral solution: slow and regular administration of the viral solution using a microinjection pump (0.2–0.5 μl/min) should cause less damage than more rapid manual injections. However, some authors have shown that varying the rate of infusion from 0.3 to 3 μl/min did not significantly change the volume of brain transduced, and that the lowest rate of infusion demonstrated the least amount of gene transfer (Betz et al., 1998). The rate of infusion is therefore an important factor that should be determined for each experimental situation.

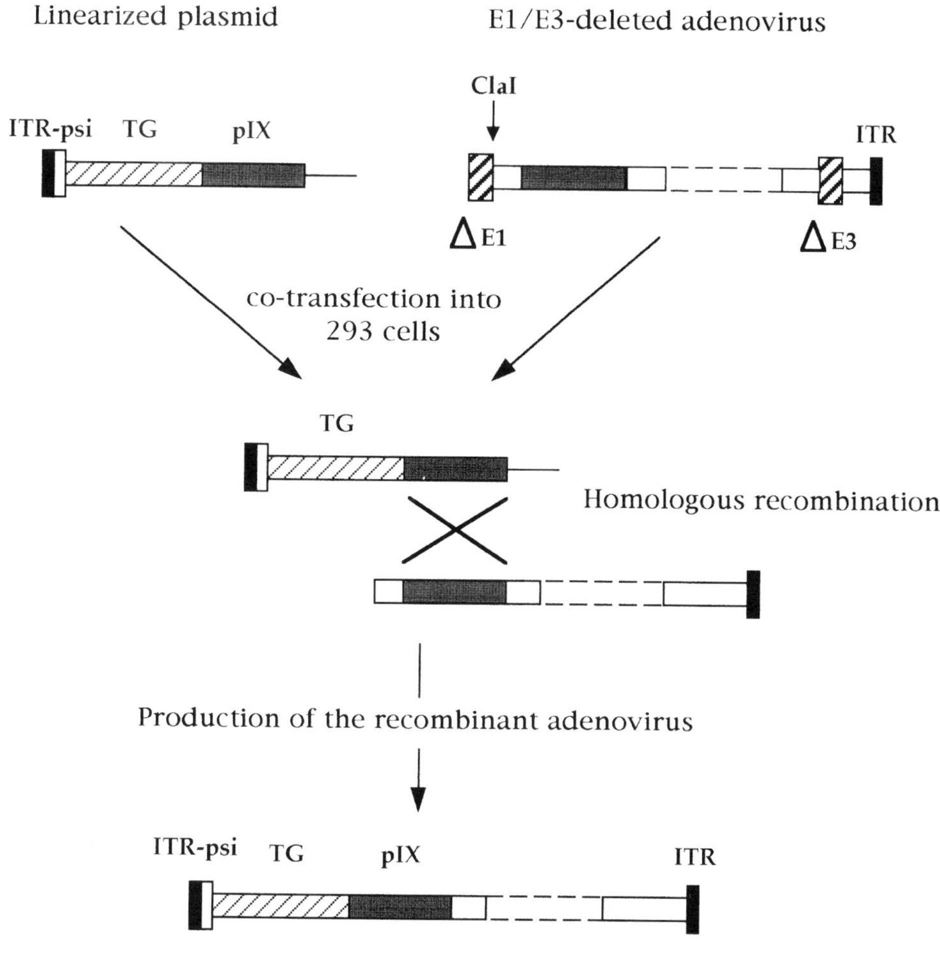

Fig. 1. Production of replication-defective recombinant adenoviruses. ΔE1: early region 1 deletion; ΔE3: early region 3 deletion; **ITR**: inverted terminal repeat; **pIX**: adenoviral sequence allowing the homologous recombination; **TG**: transgene; **psi**: encapsidation sequence.

2. The diameter of the cannula should be as small as possible to limit brain damage.

3. The number of injection sites should be sufficient for maximal dispersion of the virus: for example, nine injection sites were found to be required for diffusion of the virus through the whole striatum (Horellou et al., 1994).

4. The viral concentration needs to be adjusted to obtain an optimal level of expression with a minimal cytopathic effect (inflammation, gliosis, necrosis). About 10^7–10^8 pfu of virus are usually necessary for efficient *in vivo* infection of brain structures. It is recommended to keep the viral solution at 4°C during surgery to avoid inactivation of the virus.

5. The volume of injected virus has also to be evaluated: volumes of about one microliter seem appropriate to limit the damage to brain tissue.

Adenovirus can be transported in a retrograde manner from the injection site to the projection cell bodies offering promising applications for the specific targeting of selected neuronal populations not easily accessible by direct injection. This valuable characteristic of adenoviruses could be exploited in the design of gene therapy

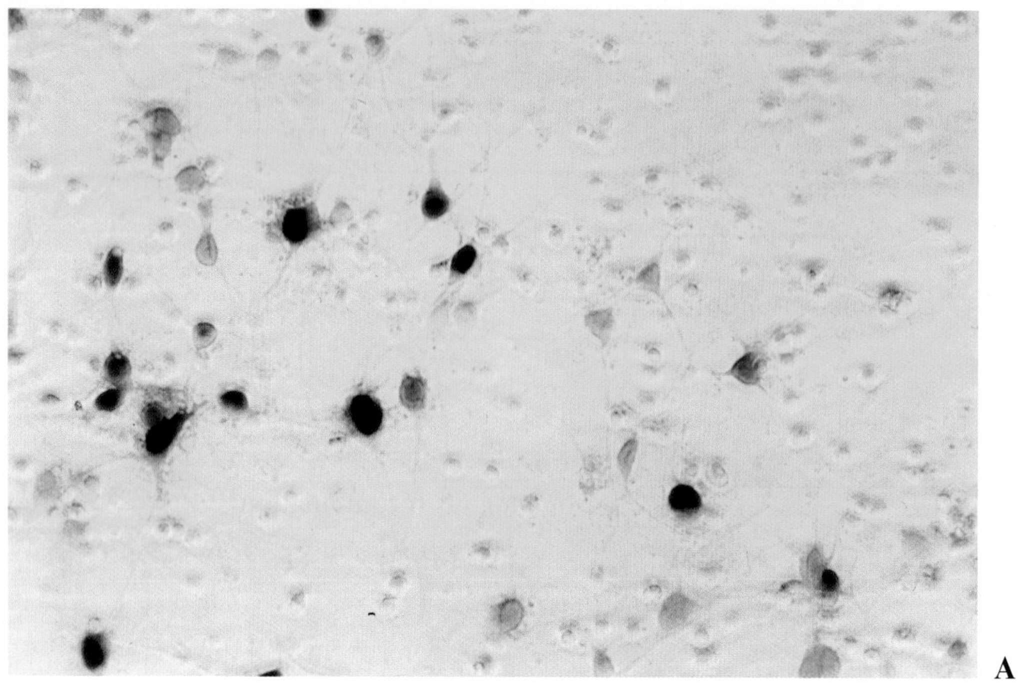

A

Fig. 2. *In vitro* infection of embryonic neurons from primary striatal cultures using replication-defective recombinant adenoviruses encoding the β-galactosidase (A, Xgal cytochemistry) and the human superoxide dismutase (B, hCuZnSOD immunostaining) under the control of the Rous sarcoma virus promoter.

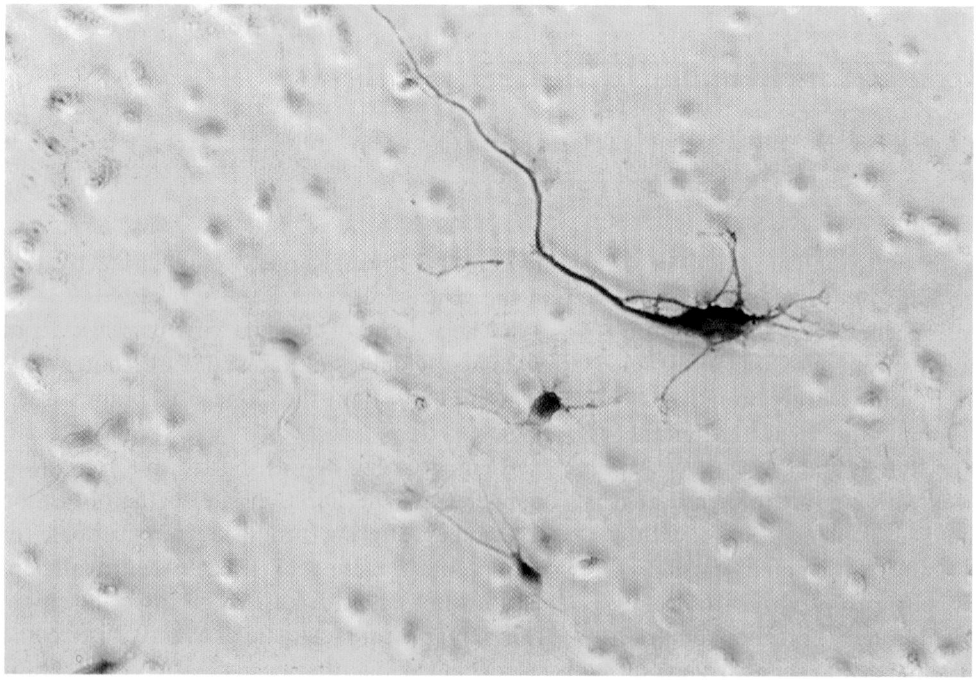

B

approaches to fatal neurodegenerative diseases of motor neurons, including ALS, Kennedy's disease, and spinal muscular atrophy. In particular, it has been shown that adenoviruses injected into muscles are transported along axons from the neuromuscular junctions of motor neurons to their cell bodies (Finiels et al., 1995; Ghadge et al., 1995). A simple intramuscular injection of 2.7×10^8 pfu of an adenovirus encoding β-galactosidase under the control of the RSV promoter (Ad-RSVβ-gal) led to specific transgene expression in 50–100% of the afferent motor neurons in the rat spinal cord (Finiels et al., 1995). Further examples of retrograde transport of adenovirus are described in following sections concerning neuroprotection in rat models of PD and axotomy-induced cell death.

Widespread delivery of a therapeutic factor may also be obtained by targeting the recombinant adenovirus to ependymal cells. In that case, injection is performed into the lateral ventricles which avoids the tissue damage inherent to direct intraparenchymal injections (Bajocchi, 1993).

Nasal instillation (Draghia et al., 1995) or injection into the carotid artery after BBB disruption with hyperosmotic mannitol (Doran et al., 1995; Muldoon et al., 1995; Nilaver et al., 1995) have been described as additional methods allowing the tranfer of adenoviruses to the brain

Ex vivo gene transfer

Adenoviral vectors are also powerful tools for *ex vivo* gene therapy. This approach involves the genetic modification of cells *in vitro* and their subsequent transplantation into the CNS. CNS-derived cells have obvious potential for intracerebral transplantation: being natural constituents of the brain, they are likely to integrate well in the long-term; the use of neuronal cells may allow delivery of neurotransmitters to synapses and therefore facilitate regulated release.

Astrocytes could be particularly well suited for brain repair because of their efficient secretory mechanisms and their role as neuronal support. Ridoux and co-workers reported the intracerebral grafting of rat primary astrocytes transduced with Ad-RSVβgal (Ridoux et al., 1994b). Grafts survived well and expressed the transgene for at least five months. Human adult astrocytes could be used for autologous transplantation, thus eliminating the problems of immunological rejection and the side effects of immunosuppressors. We have recently shown that these cells can be expanded and genetically modified *in vitro*. Astrocytes derived from human adult cerebral cortex were efficiently infected with an adenoviral vector encoding human tyrosine hydroxylase 1 (hTH-1) under the negative control of the tetracycline-(tet)-based regulatory system (tet-off; AdPGK-tet-hTH-1) (Fig. 3). The cells synthesized large amounts of active TH, responded to tet, and released L-DOPA (Ridet et al., 1999).

Similarly, we investigated the potential of human neural progenitors as vehicles for gene transfer. Amplification of these cells in culture may allow circumventing the ethical concerns raised by the clinical use of human fetal tissue, as well as the supply problems related to its extremely limited availability. It has been shown that neural cells explanted from germinative zones of the CNS from 6 to 10-week-old human fetuses proliferate in serum-free culture medium containing basic fibroblast growth factor (Buc-Caron et al., 1995). This allowed obtention of large numbers of progenitors in culture and their subsequent infection with Ad-RSVβgal adenovirus (at a MOI of 500). After transplantation to the rat brain a small percentage of neuroblasts were shown to survive and express β-galactosidase in 4 out of 13 grafted animals (Sabaté et al., 1995). In a recent extension of this study, these progenitor cells were engineered using AdPGK-tet-hTH-1 to synthesize TH in a conditional manner. After transplantation to the striatum of immunosuppressed hemiparkinsonian rats the infected neuroblasts expressed the TH trangene in a tet-dependent manner. Amelioration of the grafting conditions so as to transplant a large number of cells (8×10^5) at a high density (4×10^5/ml) allowed obtention of viable grafts con-

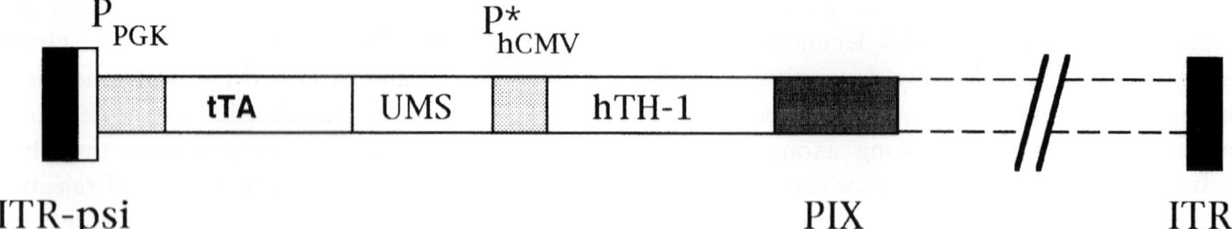

Fig. 3. Recombinant adenovirus expressing human tyrosine hydroxylase under the control of the tet-off regulatory system (AdPGKtethTH-1). ITR: Inverted Terminal Repeat; psi: Encapsidation sequence; P_{PGK}: murine phosphoglycerate kinase gene promoter; tTA: tetracycline-responsive transactivator; UMS: upstream mouse sequence; P^*_{hCMV}: minimal human cytomegalovirus tTA responsive promoter; hTH-1: human tyrosine hydroxylase 1. P_{IX}: sequence encoding the adenoviralprotein IX.

taining TH-immunolabeled cells in all animals for up to 3 months after transplantation (Corti et al., 1999). These encouraging results suggest that the use human neural progenitors genetically modified with recombinant adenoviruses is a promising model for the development of *ex vivo* restorative strategies for the treatment of neurodegenerative diseases.

Compared to *in vivo* gene transfer, additional issues are important for this type of gene transfer. They are the preparation of the cells and the conditions of infection before transplantation. For neuronal embryonic cells (which are less robust than glial cells), the *ex vivo* infection should be performed in a cell suspension rather than in culture, and the incubation time in the viral solution must be short (one hour instead of overnight) (for details, see Barkats et al., 1997).

Further examples of *ex vivo* gene transfer experiments using adenoviruses are illustrated below in the sections concerning adenoviral gene transfer for neuroprotection.

Adenovirus for neurodegenerative diseases

Ex vivo adenoviral gene transfer for neuroprotection

Neuroprotection of neural cells in patients suffering neurodegenerative diseases may be possible by *ex vivo* gene therapy. The idea is to protect degenerating cells by supplying neurotrophic factors into the brain. In the case of PD, which is characterized by the progressive loss of DA neurons from the substantia nigra (SN, a brain structure which innervates the striatum), an important aim is to protect the degenerating DA neurons. An attractive strategy would be to transfer a neuroprotective factor to dopaminergic fetal cells which could then be grafted into the brain. Adenovirus vectors may be particularly valuable for this strategy.

DA neurons are particularly exposed to oxidative stress (Olanow et al., 1992). The survival rate of ventral mesencephalon (VM) transplants from transgenic mice for the human copper-zinc SOD (hSOD-1), an antioxidative enzyme, was much higher than that of transplants from non-transgenic mice (Nakao et al., 1995). These observations led us to use an adenovirus vector encoding hSOD-1 to infect VM tissue and then graft it into the denervated striatum.

We recently constructed a replication-defective adenovirus containing the hSOD-1 cDNA (Ad-hSOD-1). This recombinant adenovirus efficiently directed the intracellular expression of hSOD-1, and protected neuronal cells in culture from glutamate-mediated oxidative stress (Barkats et al., 1996). Ad-hSOD-1 was used for transducing cell suspensions of rat VM which were then transplanted into the striatum of adult rats with a lesion of the mesostriatal pathway (this lesion causes depletion of DA in the denervated striatum) (Barkats et al., 1997). Beta-galactosidase (encoded by a control adenovirus, Ad-RSVßgal) and human SOD-1 were produced in the grafts

4 d after transplantation (Fig. 4), and the gene expression persisted for at least five weeks. Five weeks after grafting, functional recovery was more extensive in the SOD group than in the control group. The mean number of surviving DA cells in the SOD grafts was nearly double than that in controls although the difference was not statistically significant. The inflammatory consequences of the adenovirus gene transfer were minimal.

This study illustrates the potential of adenovirus for *ex vivo* gene transfer. Ad-hSOD-1 may thus be an efficient tool for enhancing survival and functional efficacy of grafted cells whether neuronal, or non-neuronal.

In vivo adenoviral gene transfer for restorative strategies

The conversion of part of the striatal cell population to TH-synthesizing cells obtained by direct infection with an adenovirus carrying a hTH-1 cDNA under the control of the RSV promoter (Ad-RSVhTH) can lead to a significant behavioral recovery in 6-hydroxydopamine (6-OHDA)-lesioned rats, a model of PD (Horellou et al., 1994). Three to four weeks after the 6-OHDA lesion, Ad-RSVhTH was used to inoculate the lesioned striatum. To promote the diffusion of the virus and the dispersion of transgene expression while minimizing tissue damage, an optimal dose of 15×10^7 particles was stereotactically injected into nine sites at a low delivery rate. The contralateral rotational behavior induced by apomorphine in this rat model was significantly decreased one and two weeks after injection of Ad-RSVhTH as compared to controls injected with Ad-RSVβgal. TH production was detected immunohistochemically in some neurons but most of the infected cells were reactive astrocytes. Thus, the observed behavioral improvement was likely to be due to synthesis of DOPA by the Ad-RSVhTH infected cells (Lundberg et al., 1996), and its subsequent decarboxylation to DA in the host striatum (Horellou et al., 1990). Recent advances in this field suggest that coinjection of Ad-RSVhTH with an adenovirus encoding GTP-cyclohydrolase catalyzing the rate-limiting step in the biosynthesis of the natural cofactor of TH (BH4), may further improve the extent of behavioral recovery in this animal model of PD (Mandel et al., 1998, Corti et al., unpublished results).

Behavioral improvement in 6-OHDA lesioned rats has also been obtained after direct intrastriatal injection of a herpes simplex virus 1 (HSV-1) and an adeno-associated vector (AAV) encoding TH (During et al., 1994; Kaplitt et al., 1994).

In vivo adenoviral gene transfer for neurotrophic strategies

Neuroprotective approach using adenovirus encoding GDNF

One way to protect DA nigral cells from the neurodegenerative process is to modify them genetically to produce a neurotrophic factor. The feasibility of this strategy has recently been demonstrated with a recombinant adenovirus vector (Bilang-Bleuel et al., 1997; Choi-Lundberg et al., 1997; Choi- Lundberg et al., 1998;). In these studies, an adenovirus encoding the rat GDNF (Ad-GDNF) protected DA neurons both in culture and in a rat model of progressive neurodegeneration.

The survival of embryonic DA neurons when co-cultivated with Ad-GDNF-infected astrocytes was better than that in control co-cultures. Moreover, the DA neurons appeared to be more differentiated, having larger cell bodies and more branched processes (Bilang-Bleuel et al., 1997).

Ad-GDNF was also tested *in vivo* using a model of progressive DA cell degeneration: the unilateral injection of 6-OHDA into the rat striatum (Sauer and Oertel, 1994). When Ad-GDNF was injected into the striatum 6 d before the 6-OHDA lesion, the production of GDNF was observed at both DA nerve terminals and nigral cell bodies following retrograde transport of the virus. Three weeks after the lesion, there were twice as many

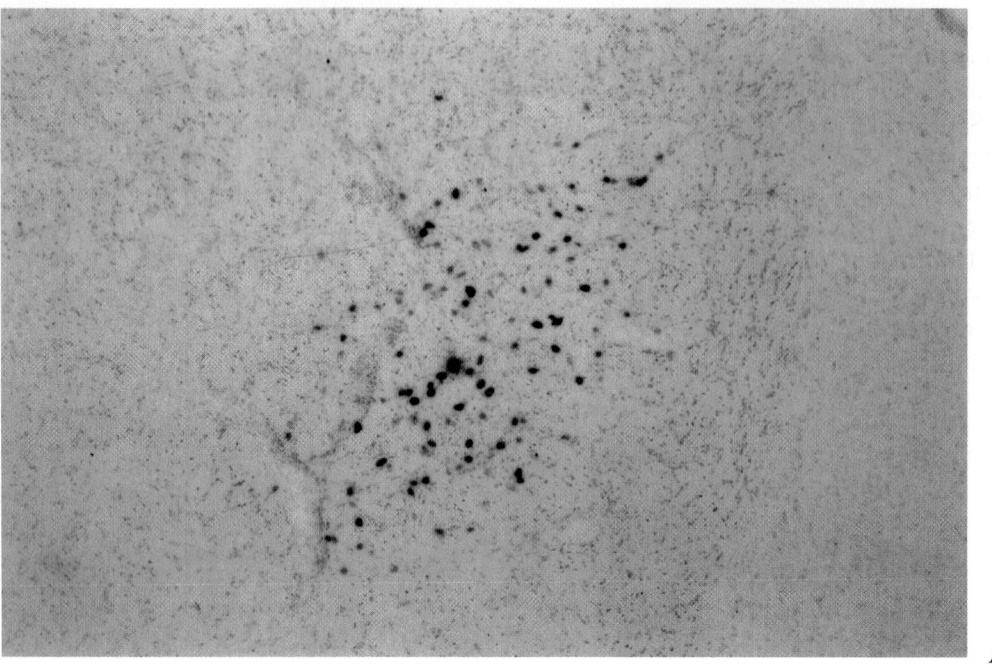

A

Fig. 4. *Ex vivo* infection of neural striatal grafts using replication-defective recombinant adenoviruses encoding the β-galactosidase (A, Xgal histochemistry) and the human superoxide dismutase (B, hCuZn immunostaining) under the control of the Rous sarcoma virus promoter.

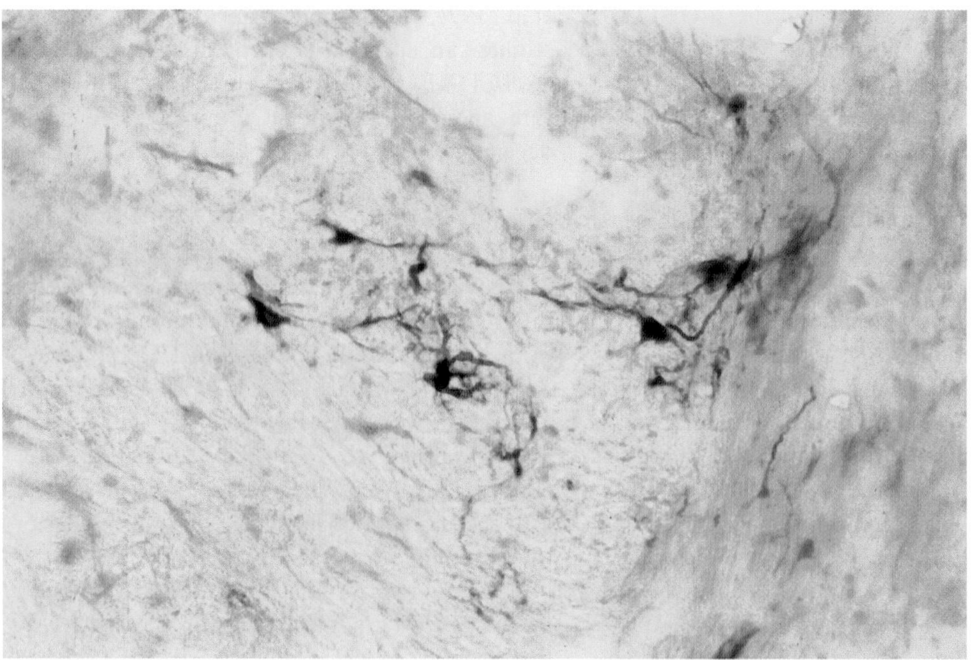

B

TH-immunoreactive cell bodies in the SN of the Ad-GDNF treated animals, than in that of control animals (Bilang-Bleuel et al., 1997). Furthermore, the amphetamine-induced behavioral asymmetry (behavioral marker for DA depletion) was markedly reduced in the Ad-GDNF treated rats.

A similar neuroprotective effect of an adenovirus encoding GDNF which was reported by Choi-Lundberg and collaborators in the 6-OHDA progressive lesion rat model of PD after direct injection of Ad-GDNF near the SN (Choi-Lundberg et al., 1997; Choi-Lundberg et al., 1998) also confirmed the observations of Bilang-Bleuel and coworkers after injection of Ad-GDNF into the striatum; however, in that case no striatal reinnervation was observed.

These studies clearly show that DA neurons are protected from degeneration caused by exposure of their terminals to 6-OHDA by an increased level of GDNF resulting from the expression of a recombinant adenovirus injected either into the striatum or near the SN. New treatments for PD based on adenovirus gene transfer technology are therefore plausible.

Therapeutic effect of an adenovirus encoding NT-3

There is virtually no therapy currently available for treating degenerative diseases of motor neurons. However, neurotrophic factors are potential candidates for the treatment of such diseases. The systematic administration of neurotrophic factors such as ciliary neurotrophic factor (CNTF), insulin-like growth factor (IGF-I), and brain-derived neurotrophic factor (BDNF), did not have a large beneficial effect in patients with ALS. These disappointing results may be because these compounds have only a poor access to motor neurons, or cause deleterious side-effects due to the ubiquitous distribution of binding proteins and receptors, like in the case of IGF-1. Alternatively, the binding sites may be down-regulated after repeated administration of the neurotrophic factor at high doses.

The strategy of continuous delivery of physiological amounts of neurotrophic factors by gene transfer technology has recently been successful in the mouse progressive motor neuronopathy mutant (*pmn*) using an adenovirus as the gene vector (Haase et al., 1997). An intramuscular injection of an adenovirus vector (1×10^9 pfu) carrying the neurotrophin-3 (NT-3, Ad-NT-3) gene had substantial therapeutic effects in this mutant. Expression of exogenous NT-3 correlated with the improvement of the mean life-span of mice which was of about 61 d for treated *pmn* and 40 d for untreated *pmn* control mice. Analysis of the motor unit potentials during spontaneous contraction of the diaphragm demonstrated an increase in the size of the motor units in Ad-NT-3 treated mice as compared to controls. It is therefore likely that expression of NT-3 had a stimulating effect on the sprouting of motor neurons that did not innervate injected muscles. This distance effect of Ad-NT-3 indicates that the neurotrophin was synthesized in the infected muscles and subsequently secreted into the circulation.

The effect on axon survival of a combined treatment with Ad-NT-3 and a recombinant adenovirus coding for CNTF (Ad-CNTF) was even more pronounced than that of each adenoviral vector alone. These studies suggest that peripheral administration of adenoviruses encoding NT-3 and CNTF could stimulate the maintenance and the sprouting of nerve terminals, and consequently improve motor function in patients suffering from neuromuscular diseases.

Prevention of axotomy-induced cell death using adenoviruses encoding GDNF, BDNF or CNTF

Adenoviruses are useful vectors for introducing neurotrophic factors into facial motor neurons of newborn rats, and thereby rescuing them from degeneration following axotomy (Baumgartner et al., 1997; Giménez y Ribotta et al., 1997; Gravel et al., 1997). Application of adenoviruses expressing neurotrophic factors to the cut stump of motor nerves of neonatal rats is one method to genetically modify motoneurons and test them for surviving to peripheral nerve damage. This strategy was used with success to confer neuroprotection to transduced and untransduced facial motoneurons using in particular an adenovirus containing the gene for GDNF (Baumgartner et al., 1997).

Giménez y Ribotta and collaborators studied the protective effect of injecting Ad-GDNF and a BDNF recombinant adenovirus (Ad-BDNF) into the nasolabial and lower lip muscles of newborn rats on the death of the axotomized facial neurons. Before axotomy, the adenoviruses were retrogradely transported to the motor neuron cell bodies to prevent the degenerative process. A significantly better survival of axotomized motor neurons was recorded one week after surgery in rats pretreated with Ad-BDNF (34.5%) or Ad-GDNF (41.2%), than in those pretreated with the control Ad-RSVßgal adenovirus. Similar results were reported by Gravel and collaborators using adenoviruses encoding Ad-CNTF or Ad-BDNF (Gravel et al., 1997). The same procedure of intramuscular injection of a recombinant adenovirus (Ad-GDNF) before axotomy was also used to confer neuroprotection to lumbar spinal motoneurons (Baumgartner et al., 1998).

The findings of these studies provide evidence of the potential of recombinant adenoviruses for the efficient long-term protection of motor neurons. They open the way for a possible treatment of severe degenerative diseases such as ALS, for which there is no classical treatment with proven efficacy.

CNS aging and Alzheimer's disease

The regulation and survival of the cholinergic neurons of the basal forebrain, the cells most affected by aging and AD, are dependent on the presence of neurotrophic factors. Therefore recombinant adenoviruses carrying growth factor genes may protect neurons against age-related degenerative processes. Castel-Barthe and coworkers investigated the trophic effect of NGF on the basal forebrain cholinergic neurons of aged rats after the unilateral injection of 10^6 pfu of an adenovirus encoding NGF (Ad-NGF) into the nucleus basalis magnocellularis (NBM) (Castel-Barthe et al., 1996). Three weeks after the adenoviral injection, the authors observed a significant ipsilateral increase of the cholinergic cell soma areas. This was not observed following injection of the control virus Ad-RSVβgal.

Although these results support the use of recombinant adenoviruses for the development of neuroprotective treatments, the adenovirus gene transfer technology has still not proven its efficacy for treating AD. Developing effective human gene therapy for AD is difficult because of the lack of a suitable animal model and the complexity of a disease characterized by a diffuse degeneration of many neuronal cell populations. Since oxidative stress has been widely implicated in amyloid-mediated toxicity (Behl et al., 1994; Bruce et al., 1996) which is believed to play a role in AD, adenoviruses encoding antioxidative enzymes may allow the prevention of the degenerative process associated with AD.

Longevity, regulation and targeting of transgene expression

Although adenovirus vectors are excellent candidates for gene transfer into nervous cells *in vivo*, several important issues must be resolved before gene therapy in brain can be used clinically. Among these issues are:

- the immune and inflammatory response to the virus
- the stability of transgene expression
- the appropriate regulation of gene expression
- the targeting of cells to avoid possible harmful effects of the delivered factors on non-targeted cells.

The immune reaction

First generation adenoviruses (e.g. adenoviruses that lack E1 and E3 sequences) induce an immune reaction (Yang et al., 1994a) when injected into immunocompetent humans or animals. Inflammation is undesired not only because it is detrimental to the patient's health, but also because it may cause a decline in transgene expression as a consequence of the destruction of the host cells. The low-level expression of early and late viral proteins and the foreign gene product may both trigger an immune reaction (Tripathy et al., 1996; Michou et al., 1997). This reaction against the virus involves both an early innate non-specific response, and a later adaptative immune response. During this second phase, the viral proteins are processed in the cytoplasm and presented on the cell surface associated with the major histocompatibility complex (MHC) class I antigens. This association induces the recognition process and the subsequent cell destruction by cytotoxic T cells (CTL). The later T-cell mediated immune response to adenovirus seems to be mainly directed against the viral proteins, since adenoviruses which do not express a transgene, or are irradiated with UV, trigger the early inflammatory reaction but not the late T-cell immune reaction (Byrnes et al., 1996).

The adenovirus vectors also elicit immune responses in the brain, both at the site of injection and at synaptically linked distant sites (Wood et al., 1996). However, the immune response following E1-deleted adenovirus injection into the brain is not sufficient to eliminate completely expression of genes encoded by the adenoviral construct (Byrnes et al., 1996; Kajiwara et al., 1997). Byrnes and collaborators suggest that the persistent expression in the brain may be a consequence of a less effective T cell response in this organ which enables the vector to persist; antigen-specific T cell-mediated immune response might therefore not account for the declining expression from adenoviral vectors (Byrnes et al., 1996). The mechanism causing transgene expression to decline over time in the brain is not fully understood, but the authors suggest that damage caused by the early T cell-independent inflammatory response to the virus would be responsible for the decline in transgene expression (Byrnes et al., 1996). In contrast, other authors have demonstrated the occurrence of a T cell-mediated immune response to adenoviral vectors after delivery to the eye, which reduces the level of intra-ocular transgene expression (Ali et al., 1998).

Because of the deleterious cytotoxic side-effects of the adenovirus vectors, the development of strategies to reduce the adenovirus-triggered immune response, through vector modification or immunomodulation, are required for the success of gene therapy in many organs, including the brain.

New generations of adenovirus vectors from which additional viral genes have been deleted are likely to be of lower immunogenicity. The early region 2 or early region 4 have been deleted from the genome to give new generations of vectors designed to prevent the production of viral proteins by the transduced host cells, and prolong transgene expression by elimination of cytopathic effects (Engelhardt et al., 1994; Yang et al., 1994b; Wang and Finer, 1996; Dedieu et al., 1997; Wang et al., 1997). An increased persistence of transgene expression is also expected from the use of "gutless" vectors from which essentially all of the viral genes have been eliminated, except the packaging sequence and the LTR elements (Parks et al., 1996; Chen et al., 1997; Hardy et al., 1997). Analysis of the properties of these improved adenoviral vectors could help to understand the nature of the immune reaction triggered by adenoviral vectors in the brain.

The immune reaction can be reduced by administration of immunosuppressive drugs such as cyclosporin simultaneously or before the virus infection. Cyclosporin treatment of mice infected with first generation adenoviruses increases stability of transgene expression (Engelhardt et al., 1994). To prolong transgene expression and to allow secondary vector administration, it is also possible to inhibit interactions between T cells and antigen presenting cells in order to prevent the costimulatory signals for T cell survival and proliferation (Kay et al., 1997). This has been recently achieved in the retina by the co-injection of an adenovirus encoding the secreted immunomodulatory protein CTL4-Ig with the adenovirus encoding β-gal, resulting in prolonged expression in retinal cells (Ali et al., 1998). In a previous study, the *in vivo* infection of mice muscles with a first-generation β-gal adenovirus was performed in combination with the administration of the recombinant protein CTL4-Ig (Guerette et al., 1996). This protein efficiently blocked the cellular immune response, and both cellular and humoral immune response were inhibited by co-administration of anti-CD4, anti-CD8 and anti-CD11a antibodies.

Recently, it was reported that injection of replication-deficient adenovirus vectors into the cerebral ventricles of rats caused a rapid increase in body temperature, occurring before any transgene expression, or with vectors encoding no transgene (Cartmell et al., 1999). The adenovirus-induced fever was linked to a significant increase of the cerebrospinal fluid levels of tumor necrosis factor and interleukine-1. This latter cytokine was demonstrated to be the main mediator of a very early inflammatory response to adenovirus in the brain. These findings should allow the elaboration of new strategies to reduce cytotoxic side-effects of adenovirus brain injections.

However, it is always preferable to manipulate the vector rather than the patients, already weakened by severe disease. Recently, the immunosuppressive viral interleukin-10 gene was introduced into an adenoviral vector (Qin et al., 1997). The transfer of this cytokine gene into murine cardiac allografts improved the persistence of the vector and prolonged gene expression of a co-injected *lac*Z adenovirus. These positive effects are probably due to the reduction of the adenovirus-specific cytotoxic T lymphocyte (CTL) response. This type of approach is also possible for adenovirus gene transfer into the CNS.

Because of the immunological problems linked to the use of first generation adenoviruses, these vectors could be particularly useful when the expression of a transgene is required only for short periods of time, such as for infection and killing of cancer cells. Recombinant adenovirus vectors expressing the herpes simplex virus thymidine kinase (HSV-tk) gene have been demonstrated to efficiently transduce rat C6 glioma cells, which rendered them sensitive to ganciclovir (GCV) (Chen et al., 1994). Recombinant adenoviruses that were further inoculated into tumors generated by stereotaxic injection of the glioma cells in nude mice, allowed the reduction of the tumor volume by >500-fold in mice treated with GCV (Chen et al., 1994). To determine the toxicity of this therapeutic strategy, adenovirus expressing HSV-TK was injected stereotactically into the normal brain of rats and rhesus monkeys in combination with ganciclovir. This treatment was reported to produce a dose-dependent inflammatory response in treated animals (Smith et al., 1997). The safety and efficacy of adenovirus-mediated transfer of the HSV-TK gene into primary human brain tumors is being evaluated in a phase one study (Eck et al., 1996).

The promoter

Targeting of gene expression

The promoter is an important element which can determine the duration of transgene expression. The gradual inactivation of the strong promoter elements from the RSV or the Cytomegalovirus (CMV) promoter that are generally used may explain the decline in expression over time following adenovirus infection (Palmer et al., 1991). The use of cellular house-keeping promoters or CNS-specific promoters which have the added benefit of allowing precise targeting of expression to particular cell types (Morelli et al., 1999) should prevent the fall off of transgene expression *in vivo*. The promoters specific for the CNS which could be valuable include the promoter of the neurofilament light chain (NFl) and that of the neuron-specific enolase (NSE), the glial acidic-fibrillary protein (GFAP) promoter specific for astrocytes, the myelin basic protein (MBP) promoter specific for oligodendrocytes, and the dopamine β-hydroxylase (DBH) and the TH promoters for the targeting of catecholaminergic and dopaminergic cells.

An adenoviral vector expressing the *E. coli* LacZ reporter gene under the control of the rat NSE promoter (Ad-NSEβgal) has been constructed (Navarro et al., in press). Ad-NSEβgal was injected into the rat hippocampus and the pattern of X-gal staining has been analyzed. Neurons were preferentially transduced and β-galactosidase expression persisted for six months (Navarro et al., in press). In an another study, we generated adenoviral constructs carrying neuron-restrictive silencer elements (NRSE) upstream from a ubiquitous promoter to target luciferase gene expression to neurons by repressing ectopic expression in non-neuronal cells. After intramuscular injection of adenoviral constructs containing 6 or 12 NRSE sequences, strong luciferase expression was obtained in neuronal cells after retrograde transport of the adenoviruses, whereas almost no expression was observed in non neuronal muscle cells (Millecamps et al., in press).

The recent report by Morelli et al., also discusses the use of neuronal and glial-specific promoters to target the expression of a transgene to predetermined brain cells (Morelli et al., 1999).

Control of gene expression

The design of safe gene delivery systems allowing tight control of the transgene product has become a major requirement before gene therapy can be applied to human diseases. The amount of protein produced from a transgene seems an important factor for the efficacy of gene therapy. Several regulatory circuits for the control of gene expression in mammalian cells have been developed (for a review see Saez et al., 1997). Of these, the tet-based systems are particularly versatile and therefore attractive for the development of regulatory techniques (Gossen and Bujard, 1992; Gossen et al., 1995). As other laboratories, we are interested in investigating the potential of these systems for gene therapy in the CNS. In an exploratory study, we have shown that the tet-off system allows a luciferase reporter gene to be switched off by tet in intracerebral grafts of a rat neural progenitor cell line (Corti et al 1996). This and other studies (Mayford et al., 1996; Yu et al., 1996) have provided the first demonstrations that the tet-based systems can function in the CNS, thus encouraging the development of regulatable vectors for brain repair. In particular, it has been recently shown that tet-regulated two-vector based adenoviral systems allow controlled expression of a EGFP reporter gene in the rat hippocampus until at least three months after direct viral injection (Harding et al., 1998). As was the case of several other studies, efficient regulation was only possible when the ratio of transactivator virus to reporter virus was adapted. We have simplified the original two-vector-based tet-off system for future clinical application, by designing one adenovirus for tet-regulated expression of therapeutic genes in the CNS (AdPGK-tet-hTH-1). This system allowed tet-controlled synthesis of human TH in cultured nerve cells. Moreover, efficient and reversible control of TH expression was obtained in a rat model of PD, both by an ex vivo gene transfer approach based on grafting of genetically modified human neural progenitors and by direct intracerebral injection of the regulatable adenovirus (Corti et al., 1999; Corti et al., submitted).

Recent advances in the development of new generation adenoviral vectors will allow the integration of regulatable systems into high capacity recombinant adenoviruses devoid of all viral coding sequences. A pioneer study in this field has demonstrated that a gutless adenovirus encoding the human growth hormone under the control of a mifepristone-regulatable system can mediate highly controlled liver-specific expression of the target gene in mice (Burcin et al., 1999). These results will encourage the design of more efficient and safer adenoviruses for controlled gene delivery to the brain.

Conclusion

In the past few years, gene transfer technologies based on adenovirus vectors have considerably progressed. However there are major stumbling blocks to be overcome before therapy can be widely used. The two main obstacles are targeting of transgene expression to specific cells and the maintenance of gene expression over time. Adenoviruses have considerable potential for both neuroprotection and restorative therapy. Further preclinical trials are still required to demonstrate the feasability, the efficacy, and the safety of adenoviral gene transfer.

Acknowledgements

The authors wish to thank Sue Orsoni for reading the manuscript. This work was supported by the Centre National de la Recherche Scientifique, the Institut National de la Santé et de la Recherche Médicale, Rhône-Poulenc Rorer, the Institut pour la Recherche sur la Moelle Epinière, the Association Française contre les Myopathies, the European Commission Biotechnology Program, the association française Retinis Pigmentosa and the Conseil Régional d'Ile de France.

Abbreviation list

AAV	adeno-associated virus
AD	Alzheimer's disease
Ad	adenovirus
ALS	amyotrophic lateral sclerosis
BBB	blood brain barrier
BDNF	brain-derived neurotrophic factor
βgal	β-galactosidase
BH4	tetrahydrobiopterin
CMV	cytomegalovirus
CNS	central nervous system
CNTF	ciliary neurotrophic factor
CTL	cytotoxic T lymphocytes
DA	dopamine, dopaminergic
GCV	ganciclovir
GDNF	glial cell line-derived neurotrophic factor
GPx	glutathione peroxidase
HD	Huntington's disease
HSV	herpes simplex virus
IGF-1	insulin-like growth factor 1
ITR	inverted terminal repeat
MHC	major histocompatibility complex
MOI	multiplicity of infection
NBM	nucleus basalis magnocellularis
NFl	neurofilament light chain
NGF	nerve growth factor
NRSE	neuron-restrictive silencer elements
NSE	neuron-specific enolase
NT-3	neurotrophin-3
OHDA	hydroxydopamine
PD	Parkinson's disease
pfu	plaque forming units
RSV	Rous sarcoma virus
SN	substantia nigra
SOD	superoxide dismutase
tet	tetracycline
TH	tyrosine hydroxylase
VM	ventral mesencephalon

References

Anderson, W.F. (1998) Human gene therapy. *Nature*, (Supp.) 392: 25–30.

Akli, S., Caillaud, C., Vigne, E., Stratford-Perricaudet, Poenaru, L., Perricaudet, M., Kahn, A. and Peschanski, M.R. (1993) Transfer of a foreign gene into the brain using adenovirus vectors. *Nat. Genet.*, 3: 224–228.

Ali, R.R., Reichel, M.B, Byrnes, A.P., Stephens, C.J., Thrasher, A.J., Baker, D., Hunt, D.M. and Bhattacharya, S.S. (1998) Co-injection of adenovirus expressing CTLA4-Ig prolongs adenovirally mediated *lacZ* reporter gene expression in the mouse retina. *Gene Ther.*, 5: 1561–1565.

Bajocchi, G., Feldman, S.H., Crystal, R.G. and Mastrangeli, A. (1993) Direct in vivo gene transfer to ependymal cells in the central nervous system using recombinant adenovirus vectors. *Nat. Genet.*, 3: 229–234.

Barkats, M., Bemelmans, A.-P., Geoffroy, M.-C., Robert, J.-J., Horellou, P., Revah, F. and Mallet, J. (1996) An adenovirus encoding CuZnSOD protects cultured striatal neurons against glutamate toxicity. *Neuroreport*, 7: 497–501.

Barkats, M., Nakao, N., Grasbon-Frodl, E.M., Bilang-Bleuel, A., Revah, F., Mallet, J. and Brundin, P. (1997) Intrastriatal grafts of embryonic mesencephalic rat neurons genetically modified using an adenovirus encoding human Cu/Zn superoxide dismutase. *Neuroscience*, 78: 703–713.

Baumgartner, B.J. and Shine, H.D. (1997) Targeted transduction of CNS neurons with adenoviral vectors carrying neurotrophic factor genes confers neuroprotection that exceeds the transduced population. *J. Neurosci.*, 17: 6504–6511.

Baumgartner, B.J. and Shine, H.D. (1998) Neuroprotection of spinal motoneurons following targeted transduction with an adenoviral vector carrying the gene for glial cell line-derived neurotrophic factor. *Exp. Neurol.*, 153: 102–112.

Beck, K.D., Valverde, J., Alexi, T., Poulsen, K., Moffat, B., Vandlen, R.A., Rosenthal, A. and Hefti, F. (1995) Mesencephalic dopaminergic neurons protected by GDNF from axotomy-induced degeneration in the adult brain. *Nature*, 373: 339–341.

Behl, C., Davis, J.B., Lesley, R. and Schubert, D. (1994) Hydrogen peroxide mediates amyloid ß protein toxicity. *Cell*, 77: 817–827.

Bett, A.J., Haddara, W., Prevec, L. and Graham, F.L. (1994) An efficient and flexible system for construction of adenovirus vectors with insertions or deletions in early regions 1 and 3. *Proc. Natl. Acad. Sci. USA*, 91: 8802–8806.

Betz, A.L., Shakui, P. and Davidson, B.L. (1998) Gene transfer to rodent brain with recombinant adenoviral vectors: effects of infusion parameters, infectious titer, and virus concentration on transduction volume. *Exp. Neurol.*, 150: 136–142.

Bilang-Bleuel, A., Revah, F., Colin, P., Locquet, I., Robert, J.-J., Mallet, J. and Horellou, P. (1997) Intrastriatal injection of an adenoviral vector expressing glial-cell-line-derived neurotrophic factor prevents dopaminergic neuron degeneration and behavioral impairment in a rat model of Parkinson disease. *Proc. Natl. Acad. Sci. USA*, 94: 8818–8823.

Bruce, A.J., Malfroy, B. and Baudry, M. (1996) ß-amyloid toxicity in organotypic hippocampal cultures: protection by EUK-8, a synthetic catalytic free radical scavenger. *Proc. Natl. Acad. Sci. USA*, 93: 2312–2316.

Buc-Caron, M.H. (1995) Neuroepithelial progenitor cells explanted from human fetal brain proliferate and differentiate in vitro. *Neurobiol. Dis.*, 2: 37–47.

Burcin, M.M, Schiedner, G., Kochanek, S., Tsai, S.Y. and O'Malley, B. W. (1999) Adenovirus-mediated regulable target gene expression in vivo. *Proc. Natl. Acad. Sci. USA*, 96: 355–360.

Byrnes, A.P., Wood, M.J.A. and Charlton, H.M. (1996) Role of T cells in inflammation caused by adenovirus vectors in the brain. *Gene Ther.*, 3: 644–651.

Cartmell, T., Southgate, T., Rees, G.S., Castro, M.G., Lowenstein, P.R. and Luheshi, G.N. (1999) Interleukin-1 mediates a rapid inflammatory response after injection of adenoviral vectors into the brain. *J. Neurosci.*, 19: 1517–1523.

Castel-Barthe, M.N., Jazat-Poindessous, F., Barneaoud, P., Vigne, E., Revah, F., Mallet, J. and Lamour, Y. (1996) Direct intracerebral nerve growth factor gene transfer using a recombinant adenovirus: effect on basal forebrain cholinergic neurons during aging. *Neurobiology of Disease*. 3: 76–86.

Cayouette, M. and Gravel, C. (1996) Adenovirus-mediated gene transfer to retinal ganglion cells. *Invest. Ophtalmol. Vis. Sci.*, 37: 2022–2028.

Chen, S.H., Shine, H.D., Goodman, J.C. and Grossman, R.G. (1994) Gene therapy for brain tumors: regression of experimental gliomas by adenovirus-mediated gene transfer in vivo. *Proc. Natl. Acad. Sci. USA*, 91: 3054–3057.

Chen, H.H., Mack, L.M., Kelly, R., Ontell, M., Kochanek, S. and Clemens, P.R. (1997) Persistence in muscle of an adenoviral vector that lacks all viral genes. *Proc. Natl. Acad. Sci. USA*, 94: 1645–1650.

Choi-Lundberg, D.L., Lin, Q., Chang, Y.-N., Chiang, Y.L., Hay, C.M., Mohajeri H., Davidson, B. and Bohn, M. (1997) Dopaminergic neurons protected from degeneration by GDNF gene therapy. *Science*, 275: 838–841.

Choi-Lundberg, D.L., Lin, Q., Schallert, T., Crippens, D., Davidson, B.L., Chang, Y.N., Chiang, Y.L., Qian, J., Bardwaj, L. and Bohn, M.C. (1998) Behavioral and cellular protection of rat dopaminergic neurons by an adenoviral vector encoding glial cell line-derived neurotrophic factor. *Exp. Neurol.*, 154: 261–275.

Corti, O., Horellou, P., Colin, P., Cattaneo, E. and Mallet, J. (1996) Intracerebral tetracycline-dependent regulation of gene expression in grafts of neural precursors. NeuroReport, 7: 1655–1659.

Corti, O., Sabaté, O., Horellou, P., Colin, P., Dumas, S., Buchet, D., Buc-Caron, M.-H. and Mallet, J. (1999) One adenovirus mediates doxycycline-controlled expression of tyrosine hydroxylase in brain grafts of human neural progenitors. *Nature Biotech*, in press.

Davidson, B.L., Allen, E.D., Kozarsky, K.F., Wilso, J.M. and Roessler, B.J. (1993) A model system for in vivo gene transfer into the central nervous system using an adenoviral vector. *Nat. Genet.*, 3: 219–223.

Dedieu, J.F., Vigne, E., Torrent, C., Jullien, C., Mahfouz, I., Caillaud, J.M., Aubailly, N., Orsini, C., Guillaume, J.M., Opolon, P., Delaere, P., Perricaudet, M. and Yeh, P. (1997) Long-term delivery into the livers of immunocompetent mice with E1/E4-defective adenoviruses. *J. Virol.*, 71: 4626–4637.

Doran, S.E., Ren, X.D., Betz, A.L., Pagel, M.A., Neuwelt, E.A., Roessler, B.J. and Davidson, B.L. (1995) Gene expression from recombinant viral vectors in the central nervous sytem after blood-brain barrier disruption. *Neurosurgery*, 36: 965–970.

Draghia, R., Caillaud, C., Manicom, R., Pavirani, A., Kahn, A. and Poenaru, L.(1995) Gene delivery into the nervous system by nasal instillation in rats. *Gene Ther.*, 2: 418–423.

During, M.J., Naegele, J.R., O'Malley, K.L. and Geller, A.I. (1994) Long-term behavioral recovery in parkinsonian rats by an HSV vector expressing tyrosine hydroxylase. *Science*, 266: 1399–1403.

Eck, S.L., Alavi, J.B., Alavi, A., Davis, A., Hackney, D., Judy, K., Mollman, J., Phillips, P.C., Wheeldon, E.B. and Wilson, J.M. (1996) *Hum. Gene Ther.*, 7: 1465–1482.

Engelhardt, J.F., Ye, X., Doranz, B. and Wilson, J.M. (1994) Ablation of *E2A* in recombinant adenoviruses improves transgene persistence and decreases inflammatory response in mouse liver. *Proc. Natl. Acad. Sci.*, 91: 6196–6200.

Finiels, F., Gimenez y Ribotta, M., Barkats, M., Samolyk, M.-L., Robert, J.-J., Privat, A., Revah, F. and Mallet, J. (1995) Specific and efficient gene transfer strategy offers new potentialities for the treatment of motor neuron diseases. *NeuroReport*, 6: 2473–2478.

Fischer, W., Wictorin, K., Björklund, A., Williams, L.R., Varon, S. and Gage, F.H. (1987) Amelioration of cholinergic neuron atrophy and spatial memory impairment in aged rats by nerve growth factor. *Nature*, 329: 65–68.

Geula, C., Wu, C.-K., Saroff, D., Lorenzo, A., Yuan, M. and Yankner, B.A. (1998) Aging renders the brain vulnerable to amyloid β-protein neurotoxicity. *Nat. Med.*, 4: 827–831.

Ghadge, G.D., Roos, R.P., Kang, U.J., Wollmann, R., Fishman, P.S., Kalynych, A.M., Barr, E. and Leiden, J.M. (1995) CNS gene delivery by retrograde transport of recombinant replication-defective adenoviruses. *Gene Ther.*, 2: 132–137.

Giménez y Ribotta, M., Revah, F., Pradier, L., Loquet, I., Mallet, J. and Privat, A. (1997) Prevention of motoneuron death by adenovirus-mediated neurotrophic factors. *J. Neurosci. Res.*, 48: 281–285.

Ghodsi, A., Stein, C., Derksen, T., Yang, G., Anderson, R.D. and Davidson, B.L. (1998) Extensive beta-glucuronidase activity in murine central nervous system after adenovirus-mediated gene transfer to brain. *Hum. Gene Ther.*, 9: 2331–2340.

Gossen, M. and Bujard, H. (1992) Tight control of gene expression in mammalian cells by tetracycline-responsive promoters. *Proc Natl Acad Sci USA*, 89: 5547–5551.

Gossen, M., Freudnlieb, S., Bender, G., Muller, G., Hillen, W. and Bujard, H. (1995) Transcriptional activation by tetracyclines in mammalian cells. *Science*, 268: 1766–1769.

Graham, F.L. and Prevec, L. (1995) Methods for construction of adenovirus vectors. *Mol. Biotechnol.*, 3: 207–220.

Gravel, C., Götz, R., Lorrain, A. and Sendtner, M. (1997) Adenoviral gene transfer of ciliary neurotrophic factor and brain-derived neurotrophic factor leads to long-term survival of axotomized motor neurons. *Nat. Med.*, 3: 765–770.

Guérette, B., Vilquin, J.T., Gingras, M., Gravel, C., Wood, K.J. and Tremblay, J.P. (1996) Prevention of immune reactions triggered by first-generation adenoviral vectors by monoclonal antibodies and CTLA4Ig. Hum. *Gene Ther.*, 7: 1455–1463.

Haase, G., Kennel, P., Pettmann, B., Vigne, E., Akli, S., Revah, F., Schmalbruch, H. and Kahn, A. (1997) Gene therapy of murine motor neuron disease using adenoviral vectors for neurotrophic factors. *Nat. Med.*, 3: 429–436.

Harding, T.C., Geddes, B.J., Murphy, D., Knight, D. and Uney, J.B. (1998) Switching transgene expression in the brain using an adenoviral tetracycline-regulatable system. *Nat. Biotechnol.*, 16: 563–555.

Hardy, S., Kitamura, M., Harris-Stansil, T., Dai, Y. and Phipps, M.L. (1997) Construction of adenovirus vectors through Cre-lox recombination. *J. Virol.*, 71: 1842–1849.

Horellou, P., Brundin, P., Kalén, P., Mallet, J. and Björklund, A. (1990) In vivo release of DOPA and dopamine from genetically engineered cells grafted to the denervated rat striatum. *Neuron*, 5: 393–402.

Horellou, P., Vigne, E., Castel, M.N., Barnéoud, P., Colin, P., Perricaudet, M., Delaère, P. and Mallet, J. (1994) Direct intracerebral gene transfer of an adenoviral vector expressing tyrosine hydroxylase in a rat model of Parkinson's disease. *NeuroReport*, 6: 49–53.

Horwitz, M.S. (1996) Adenoviruses. In: B.N. Fields, D.M. Knipe, P.M. Howley, et al. (Eds.), Fields Virology, 3rd Edn, Lipincott-Raven Publishers, Philadelphia, 2: 2149–2171.

Hsiao, K., Chapman, P., Nilsen, S., Eckman, C., Harigaya, Y., Younkin, S., Yang, F. and Cole, G. (1996) Correlative memory deficits, Aß elevation and amyloid plaques in transgenic mice. *Science*, 274: 99–102.

Kajiwara, K., Byrnes, A.P., Charlton, H.M., Wood, M.J. and Wood, K.J. (1997) Immune responses to adenoviral vectors during gene transfer in the brain. *Hum. Gene Ther.*, 10: 253–265.

Kaplitt, M.G., Leone, P., Samulski, R.J., Xiao, X., Pfaff, D.W., O'Malley, K.L. and During, M.J. (1994) Long-term gene expression and phenotypic cerrection using adeno-associated virus vectors in the mammalian brain. *Nat. Genet.*, 8: 148–154.

Kay, M.A., Meuse, L., Gown, A.M., Linsley, P., Hollenbaugh, D., Arufo, A., Ochs, H.D. and Wilson, C.B. (1997) Transient immunomodulation with anti-CD40 ligand antibody and CTLA4Ig enhances persistence and secondary adenovirus-mediated gene transfer into mouse liver. *Proc. Natl. Acad. Sci. USA*, 94: 4686–4691.

Le Gal La Salle, G., Robert, J.-J., Berrard, S., Ridoux, V., Stratford-Perricaudet, L.D., Perricaudet, M. and Mallet, J. (1993) An adenovirus vector for gene transfer into neurons and glia in the brain. *Science*, 259: 988–990.

Li, T. and Davidson, B.L. (1995) Phenotype correction in retinal pigment epithelium in murine mucopolysaccharidosis VII by adenovirus-mediated gene transfer. *Proc. Natl. Acad. Sci. USA*, 92: 7700–7704.

Lundberg, C., Horellou, P., Mallet, J. and Bjorklund, A. (1996) Generation of DOPA-Producing astrocytes by retroviral transduction of the human tyrosine hydroxylase gene: In vitro characterization and in vivo effects in the rat Parkinson model. *Exp. Neurol.*, 139: 39–53.

Mandel, R.J., Rendhal, K.G.., Spratt, S.K., Snyder, R.O., Cohen, L.K. and Leff, S.E. (1998) Characterization of intrastriatal recombinant adeno-associated virus-mediated gene tranfer of human tyrosine hydroxylase and human GTP-cyclohydrolase I in a rat model of Parkinson's disease. *J. Neurosci.*, 18: 4271–4284.

Mayford, M., Bach, M.E., Huang, Y.Y., Wang, L., Hawkins, R.D. and Kandel, E.R. (1996) Control of memory formation through regulated expression of a CamKII transgene. *Science*, 274: 1678–1683.

Michou, A.I., Santoro, L., Christ, M., Julliard, V., Pavirani, A. and Mehtali, M. (1997) Adenovirus-mediated gene transfer: influence of transgene, mouse strain and type of immune response on persistence of transgene expression. *Gene therapy*, 4: 473–482.

Millecamps, S., Kiefer, H., Navarro, V., Geoffroy, M.C., Robert, J.J., Finiels, F., Mallet, J. and Barkats, M. Neuron-restrictive silencer elements (NRSE) mediate neuron specificity of adenovirus gene expression in the rodent brain. *Nature Biotechnology*, in press.

Morelli, A., Larregina, A., Smith, G., Dewey, R., Southgate, T., Fontana, A., Castro, M.G. and Lowenstein, P.R. (1999) Neuronal and glial cell-type specific promoters within adenovirus recombinants restrict the expression of the apoptosis-inducing molecule Fas ligand to predetermined brain cell types, and abolish peripheral liver toxicity. *J. Gen. Virol.*, 80: 571–583.

Muldoon, L.L., Nilaver, G., Kroll, R.A., Pagel, M.A., Breakefield, X.O., Chiocca, E.A., Davidson, B.L., Weissleder, R. and Neuwelt, E.A. (1995) Comparison of intracerebral inoculation and osmotic blood-barrier disruption for delivery of adenovirus, herpesvirus, and iron oxide particles to normal rat brain. *Am. J. Pathol.*, 147: 1840–1851.

Nakao, N., Frodl, E.M., Widner, H., Carlson, E., Eggerding F.A., Epstein, C.J. and Brundin, P. (1995) Overexpressing Cu/Zn superoxide dismutase enhance survival of transplanted neurons in a rat model of Parkinson's disease. *Nat. Med.*, 1: 226–231.

Navarro, V., Millecamps, S., Geoffroy, M.C., Robert, J.J., Valin, A., Mallet, J. and Le Gal La Salle, G. Efficient gene transfer and long-term expression in neurons using a recombinant adenovirus with a neuron-specific promoter. *Gene Therapy*, in press.

Nilaver, G., Muldoon, L.L., Kroll, R.A., Pagel, M.A., Breakefield, X.O., Davidson, B.L. and Neuwelt, E.A. (1995) Delivery of herpesvirus and adenovirus to nude rat intracerebral tumors after osmotic blood-brain barrier disruption. *Proc. Natl. Acad. Sci. USA*, 92: 9829–9833.

O'Connor, W.M., Davidson, B.L., Kapplitt, M.G., Abbey, M.V., During, M.J., Leone, P., Langer, D., O'Connor, M.J. and Freese, A. (1997) Adenovirus vector-mediated gene transfer into human epileptogenic brain slices: propects for gene therapy in epilepsy. *Exp. Neurol.*, 148: 167–178.

Olanow, C.W. (1992) An introduction to the free radical hypothesis in Parkinson's disease. *Ann. Neurol.*, 32: S2–S9.

Olanow, C.W. (1993) A radical hypothesis for neurodegeneration. *TINS*, 16: 439–444.

Oppenheim, R.W., Houenou, L.J., Johnson, J.E., Lin, L.F., Li, L., Lo, A.C., Newsome, A.L., Prevette, D.M. and Wang, S. (1995) Developing motor neurons rescued from programmed and axotomy-induced cell death by GDNF. *Nature*, 373: 344–346.

Palmer, T.D., Rosman, G.J., William, R.A. and Miller, A.D. (1991) Genetically modified skin fibroblasts persist long after transplantation but gradually inactivate introduced genes. *Proc. Natl. Acad. Sci.*, 88: 1330–1334.

Parks, R.J., Chen, L., Anton, M., Sankar, U., Rudnicki, M.A. and Graham, F.L. (1996) A helper-dependent adenovirus vector system: removal of helper virus by Cre-mediated excision of the viral packaging signal. *Proc. Natl. Acad. Sci. USA*, 93: 13565–13570.

Plumb, T.J., Bosch, A., Roessler, B.J., Shewach, D.S. and Davidson, B.L. (1996) Hypoxanthine-guanine phosphoribosyltransferase (HPRT) expression in the central nervous system of HPRT-deficient mice following adenoviral-mediated gene transfer. *Neurosci. Lett.*, 214: 159–162.

Qin, L., Ding, Y., Pahud, D.R., Robson, N.D., Shaked, A. and Bromberg, J.S. (1997) Adenovirus-mediated gene transfer of viral interleukin-10 inhibits the immune response to both alloantigen and adenoviral antigen. *Hum. Gene Ther.*, 8: 1365–1374.

Ridet, J.L., Corti, O., Pencalet, P., Hanoun, N., Hamon, M., Philippon, J. and Mallet, J. (1999) Toward autologous ex vivo gene therapy for the central nervous system with human adult astrocytes. *Hum. Gene Ther.*, 10: 271–280.

Ridoux , V., Robert, J.J., Zhang, X., Perricaudet, M., Mallet, J. and Le Gal La Salle, G. (1994a) Adenoviral vectors as functional retrograde neuronal tracers. *Brain Res.*, 648: 171–175.

Ridoux, V., Robert, J.J., Zhang, X., Perricaudet, M., Mallet, J. and Le Gal La Salle, G. (1994b) The use of adenovirus vectors for intracerebral grafting of transfected nervous cells. *Neuroreport*, 5: 801–804.

Robert, J.J., Bouilleret, V., Ridoux, V., Valin, A., Geoffroy M.C., Mallet, J. and Le Gal La Salle, G. (1997) Adenovirus-mediated transfer of a functional GAD gene into nerve cells: potential for the treatment of neurological diseases. *Gene Ther.*, 4: 1237–1245.

Rosen, D.R., Siddique, T., Patterson, D. et al. (1993) Mutations in Cu/Zn superoxide dismutase gene are associated with familial amyotrophic lateral sclerosis. *Nature*, 362: 59–62.

Russell, D.W., Miller, A.D. and Alexander, I.E. (1994) Adeno-associated virus vectors preferentially transduce cells in S phase. *Proc. Natl. Acad. Sci. USA*, 91: 8915–8919.

Sabaté, O., Horellou, P., Vigne, E., Colin, P., Perricaudet, M., Buc-Caron, M.H. and Mallet, J. (1995) Transplantation to the rat brain of human neural progenitors that were genetically modified using adenoviruses. *Nat. Genet.*, 9: 256–260.

Sabel, B.A., Vick, A. and Hollt, V. (1995) Neurotoxic reactions of CNS cells following gene transfer with defective herpes simplex virus (HSV-1) vector. *Neuroreport*, 6: 2447–2449.

Saez, E., No, D., Wez, A. and Evans RM (1997) Inducible gene expression in mammalian cells and transgenic mice. *Curr. Opinion Biotechnol.*, 8: 608–616.

Samaniego, L.A., Neiderhiser, L. and DeLuca, N.A. (1998) Persistence and expression of the herpes simplex virus genome in the absence of immediate-early proteins. *J. Virol.*, 72: 3307–3320.

Sauer, H. and Oertel, W.H. (1994) Progressive degeneration of nigrostriatal dopamine neurons following intrastriatal terminal lesions with 6-hydroxydopamine: a combined retrograde tracing and immunocytochemical study in the rat. *Neuroscience*, 59: 401–415.

Smith, J.G., Raper, S.E., Wheeldon, E.B., Hackney, D., Judy, K., Wilson, J.M. and Eck, S.L. (1997) Intracranial administration of adenovirus expressing HSV-TK in combination with ganciclovir produces a dose-dependent, self-limiting inflammatory responses. *Hum. Gene Ther.*, 8: 943–954.

Snell, R.G., MacMillan, J.C., Cheadle, J.P., Fenton, I., Lazarou, L.P., Davies, P., MacDonald, M.E., Gusella, J.F., Harper, P.S. and Shaw, D.J. (1993) Relationship between trinucleotide repeat expansion and phenotypic variation in Huntington's disease. *Nat. Genet.*, 4: 393–397.

Soto, C., Sigurdsson, E.M., Morelli, L., Kumar, R.A., Castano, E.M. and Frangione, B. (1998) β-sheet breaker peptides inhibit fibrillogenesis in a rat brain model of amyloidosis: Implications for Alzheimer's therapy. *Nat. Med.*, 4: 822–826.

Tripathy, S.K., Black, H.B., Goldwasser, E. and Leiden, J.M. (1996) Immune responses to transgene-encoded proteins limit the stability of gene expression after injection of replication-defective adenovirus vectors. *Nat. Med.*, 2: 545–550.

Wang, Q. and Finer, M.H. (1996) Second-generation adenovirus vectors. *Nat. Med.*, 2: 714–716.

Wang, Q., Greenburg, G., Bunch, D., Farson, D. and Finer, M.H. (1997) Persistent transgene expression in mouse liver following in vivo gene transfer with a delta E1/delta E4 adenovirus vector. *Gene Ther.*, 4: 393–400.

Wood, M.J., Charlton, H.M., Wood, K.J., Kajiwara, K. and Byrnes, A.P. (1996) Immune responses to adenovirus vectors in the nervous system. *Trends Neurosci.*, 19: 497–501.

Yang, Y., Nunes, F.A., Berencsi, K., Furth, E.E., Gönczöl, E. and Wilson, J.M. (1994a) Cellular immunity to viral antigens limits E1-deleted adenoviruses for gene therapy. *Proc. Natl. Acad. Sci.*, 91: 4407–4411.

Yang, Y., Nunes, F.A., Berencsi, K., Gönczöl, E., Engelhardt, J.F. and Wilson, J.M. (1994b) Inactivation of E2a in recombinant adenoviruses improves the prospect for gene therapy in cystic fibrosis. *Nat. Genet.*, 7: 362–369.

Yu, J.S., Sena-Esteves, M., Paulus, W., Breakefield, X.O. and Reeves, S.A. (1996) Retroviral delivery and tetracycline-dependent expression of IL-1beta-converting enzyme (ICE) in a rat glioma model provides controlled induction of apoptotic death in tumor cells. *Cancer Res.*, 56: 5423–5427.

W.E. Crusio and R.T. Gerlai (Eds.)
Handbook of Molecular-Genetic Techniques for Brain and Behavior Research (Techniques in the Behavioral and Neural Sciences, Vol. 13)

CHAPTER 3.2.3

Regulatable adenoviral technology in behavioural neuroscience

E. Clea Warburton[2], Brad Geddes[1], Thomas Harding[1] and James B. Uney[1]

[1]*Department of Medicine, MRC Centre for Synoptic Plasticity, University of Bristol, Marlborough Street, Bristol BS2 8HW, UK*
[2] *Department of Anatomy, MRC Centre for Synoptic Plasticity, University of Bristol, University Walk, Bristol BS3 1HW, UK*

Until recently, behavioural research has to a large extent relied upon lesion techniques and pharmacological manipulations to identify the neural substrates of behaviour. These studies have provided valuable information concerning the role of specific brain regions and have identified key neurochemical systems underlying many behavioural processes. However, lesion studies can not provide any information regarding the intracellular mechanisms involved in behavioural processes and it is has been difficult to achieve appropriate specificity of drug action in pharmacological experiments. These problems can be overcome to some extent by overexpression and knockout transgenic techniques which allow the expression levels of specific intracellular proteins to be altered. Transgenic techiques, however, do not readily allow the temporal and spatial control of gene expression (but see Chapter 3.1.4). Therefore, a transfer system that enabled the efficient introduction of transgenes into neurons and the regulatable control of the expressed transgene would greatly facilitate behavioural and *in vivo* studies into neuronal gene function. In an attempt to develop such a system we incorporated the tetracycline-regulated transcriptional control system developed by Bujard and Gossen (Gossen and Bujard, 1992) into the replication-deficient type-5 adenoviral (Ad) transfection system (Harding et al., 1997, 1998). This regulatable adenoviral technology should be well suited to investigations into the cellular processes which underly forms of behaviour (e.g. learning and memory, and addictive behaviour) because: first, the viruses can be directed to specific populations of neurons (we have already shown that nuclei within the striatum, amygdala and hippocampus can be transfected); second, the genetic manipulation occurs in the adult, rather than at the embryonic stage, therefore compensatory genetic effects associated with transgenic animals are avoided; third, the transgene expression is highly regulatable, therefore once the virus is *in situ*, expression of the gene can be turned on or off, by the experimenter, without further surgical intervention. Fourth, the adenoviral manipulation is suited to use in rats, and not just in mice which, is of great benefit since both the behavioural tasks and neural regions which mediate certain types of behaviour are better characterised in rats, and the abundance of behavioural data obtained from various rat models provide the necessary comparative framework for such studies.

Components of the adenoviral-tetracycline (AdTet) system

The tetracycline regulatable gene activation system

The tetracycline On/Off gene control system (Gossen and Bujard, 1992; Gossen et al., 1995) is based on the Tn10 tetracycline resistance operon of *E. coli*. The tetracycline (Tet)-Off system is composed of a Tet controlled transactivator (termed the tTA) which, in the absence of Tet, binds a Tet regulatable element (TRE) and initiates transcription from a minimal promoter (Fig. 1(a)). However, in the presence of Tet, the tTA (or Tet-Off) transactivator undergoes a conformational change and is not able to initate transcription from the TRE. By modifying tTA by four amino acids, the reverse Tet-transactivator (rtTA or Tet-On transactivator) was generated (Gossen et al., 1995). The rtTA binds to the TRE in the presence of the drug and initiates transcription, however, in the absence of Tet, the rtTA cannot drive transcription (Fig. 1(b)). The rtTA transactivator was found to be 100 times more sensitive to the Tet analogue, doxycycline (Dox), and consequently Dox is used in to regulate gene expression in the Tet-On system.

This is the first mammalian gene expression system to: be highly inducible; exhibit no minimal basal leakiness; show no toxic or pleitropic effects of the inducer; maintain gene expression at very high levels in the on state (Yarranton et al., 1992; Gossen and Bujard, 1992; Harding et al., 1997). However, due to the difficulty in transfecting post-mitotic cells it has at present only been possible to apply this powerful new technology using conventional transgenic approaches in behavioural studies. (Kirstner et al., 1996; Mayford et al., 1996; Mantuy et al., 1998). Using the Tet-Off system, Mayford et al., 1996 was able to show that suppression of the CamKII gene by administration of Dox reversed a spatial memory deficit associated with expression of the transgene. To obtain rapidly inducible and reversible expression of transgenes in the forebrain of the mouse, Mansuy et al. (1998) used the Tet-On transactivator together with the CaMKII promoter. Using calcineurin as a test transgene, they showed that doxycycline-induced expression impaired both an intermediate form of LTP in the hippocampus and the storage of spatial memory. The reversibility of the rtTA system also allowed the effects of the transgene on memory retrieval after normal storage had occurred. However, there are a number of disadvantages to using knockout transgenics approaches. These include the technical difficulty and time involved in: propagating and selecting ES cells; continuously identifying and breeding knockout mice for experimentation.

Adenoviral vectors

The extensive characterisation of Type-5 adenoviruses (Ad) has allowed a replication-deficient Ad system to be developed which can accommodate relatively large pieces of DNA be grown to very high titres and infect non-replicating cells (Le Gal La Salle et al., 1993; Harding et al., 1997). Several studies have shown that these Ad vectors can be used to transfect a number of different neuronal cell types, and that they can mediate transgene expression for over three months in the CNS (reviewed by Neve et al., 1993; Le Gal La Salle 1993; Geddes et al., 1996, 1997; Harding et al., 1997). Studies are also being conducted with adenoviral expression systems which were developed to: further improve the safety (particularly important for peripheral delivery), increase insert capacity; simplify adenoviral production. He et al. (1998) developed a simplified production system which allows adenoviruses to be generated in *E. coli*, while Morsey et al. (1998) have developed a system which allows up to 37 kb of DNA to be incorporated into the vector. The deletion of some (or all) of the viral genome has led to a reduction in the antigenicity of the virus, which allows the virus to be expressed for greater periods of time. It is important to note that although Ad vectors are replication-deficient

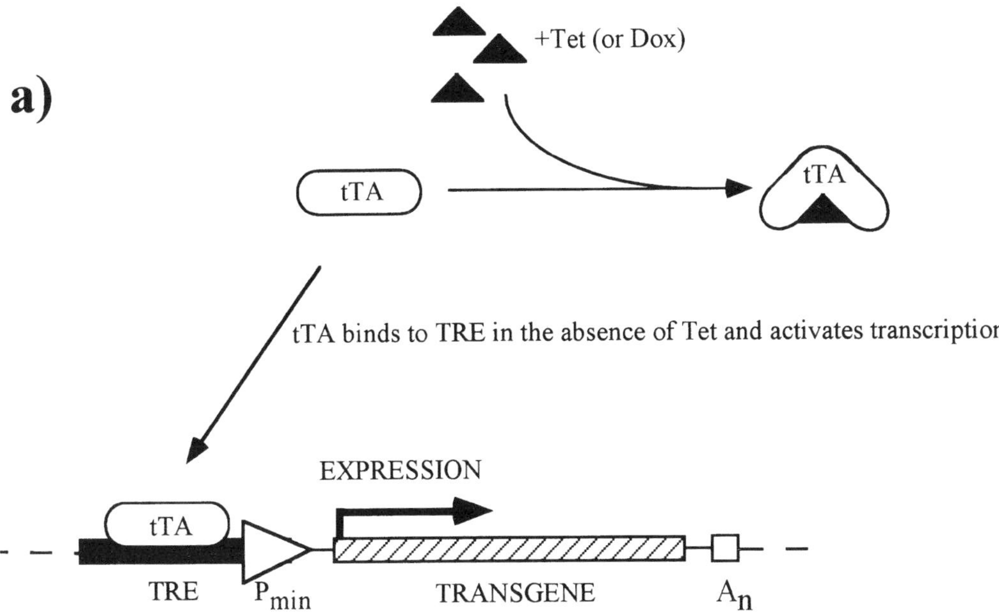

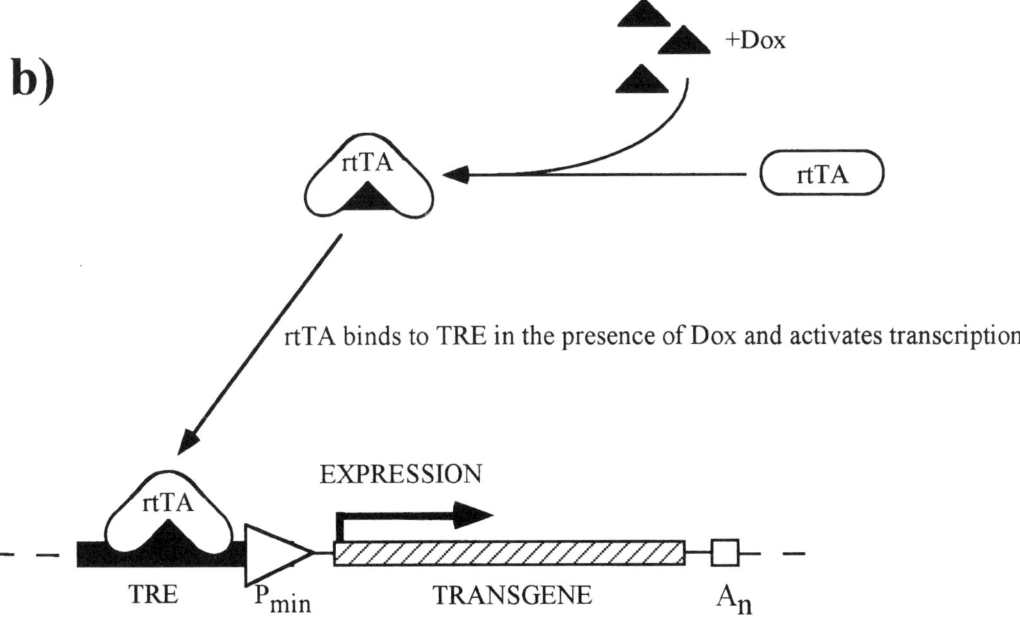

Fig. 1. Transgene regulation by the Tet-Off and Tet-On systems. (a) shows that the tTA (Tet-Off) transactivator can only induce transcription of the transgene from the TRE-P_{min} promoter in the absence of Tet. In the presence of Tet, a conformation change in the transactivator abolishes its ability to drive transcription. (b) shows that the rtTA (Tet-On) transactivator can only drive transcription from the TRE-P_{min} promoter in the presence of the Tet analogue, Dox. In the absence of Dox, the transactivator cannot bind to the TRE-P_{min} promoter and transcription is repressed.

experiments are normally carried out under genetic manipulation advisory group (GMAG) I or II conditions because the vectors will transfect most cell types. However, the ability to tightly regulate gene expression from Ad vectors would increase the safety of these vectors and would probably allow all experiments to be carried out under GMAG I conditions (most molecular biology laboratories are level I). In summary, the adenoviral system is able to transfect neurones with high efficiency, has been very well characterised, and has the potential for further rapid advances in technology. Therefore, to develop a highly regulatable transfection system which could be used in in vivo studies of neuronal gene function we combined the adenoviral system with the tetracycline regulatory system.

Construction and in vivo experiments with the adenoviral-tetracyline (AdTet) system

Generation of the adenoviral tetracycline (AdTet) vectors

Recombinant, E1 deleted, adenoviral constructs were produced according to standard techniques (McGrory et al., 1988; Graham and Prevec, 1991, 1995; Geddes et al., 1996). In these constructs the 5′ end of the E1 and a portion of the E3 regions are deleted (impairing the ability of the Ad-5 type virus to replicate and transform non-permissive cells), and a recombinant exogenous gene and promoter is inserted in place of the E1 deletion. The transgene, under the control of the CMV or Tet promoter is first cloned between flanking Ad sequences in a prokaryotic transfer vector (e.g. pXCX2). Recombinant virus is generated by homologous recombination between the transfer vector and pJM17 (a vector containing most of the AD-5 genome) following transfection of 293 cells (a trans-complementing cell line for E1 function, Graham et al. (1977)). After recombination the size of the viral genome is reduced below the packaging limit of the Ad particle and hence only recombinant virus is capable of generating plaques. This Ad suspension is then purified further by caesium chloride density centrifugation. The vectors described in this paper were made as follows. The tetracycline responsive element (TRE), minimal promoter (Pmin, a generous gift from H. Bujard and M. Gossen) and reporter gene, enhanced green fluorescent protein (EGFP) (Clontech laboratories Inc.) were inserted into the multiple cloning site of the plasmid pΔE1sp1A (Microbix Biosystems Inc.). The Tet-Off (tTA) and Tet-On (rtTA) transactivators (Clontech laboratories Inc.) were cloned into pXCXCMV under the control of the human cytomegalovirus (hCMV) promoter (Fig. 1). Recombinant virus was then generated by homologous recombination in HEK 293 cells (Graham and Prevec, 1995; Harding et al., 1996, 1998). This first entails the transfer vectors (pXCXCMV and pΔE1sp1A) and helper vector (PJM17; Microbix, Canada) plasmids being prepared and purified by CsCl density ultracentrifugation. For each 60 mm dish of HEK293 cells 6 μg of each plasmid was added to 0.5 ml of Hepes buffered saline (Hebs) in a sterile 5 ml bijou. The solution is then mixed by vortexing for 30 s and 30 μl of 2 M $CaCl_2$ added. The mixture was again briefly vortexed and left at room temperature for 25 min, at which point the calcium phosphate-DNA precipitate is visible. The suspension is then gently added drop-wise, with gentle swirling, to each dish. The dishes were incubated for 8–10 h at 37°C. After this period the medium was removed and the cells overlaid with medium +0.5% agarose. Plaques corresponding to recombinant adenovirus typically started to appear 7 d post-transfection at which point they are picked and analysed for the correct genetic structure. Recombinants are grown to high titre and purified by double CsCl density gradient centrifugation. Viral titres were then determined by plaque assay: Typical values obtained are 1×10^{10} plaque forming units/ml.

In our system the hybrid TRE-Pmin promoter and transactivator elements of the Tet-system were incorporated into separate viral vectors, so that the Pmin element was not in close proximity to the hCMV promoter which was used to constitutively drive expression of both transactivators. This strategy was chosen as there is evidence to suggest that the hCMV promoter when adjacent to the Pmin element may non-specifically induce expression, resulting in higher levels of background expression in the non-induced state (Yin et al., 1996). Furthermore, this strategy also allowed the transactivator to TRE-Pmin promoter ratio to be altered. A particularly important characteristic, since, in vivo it may only be possible to achieve the efficient regulation of transgene expression by altering the transactivator virus (AdrtTA and AdtTA) to reporter virus (AdTRE-EGFP) ratio. Indeed, this was subsequently found to be the case in our initial experiments which used a 1:1 ratio of AdrtTA/AdtTA to AdTRE-EGFP vectors and resulted in high basal expression levels. Subsequent experiments, however, carried out with a transactivator AdrtTA/AdtTA to AdTRE-EGFP virus ratio of 1:20, mediated undetectable EGFP background levels and highly regulatable transgene expression. This result is therefore probably due to virus being unable to diffuse substantially in the brain resulting in very high viral titres around the injection site. Consequently, this results in higher intracellular concentrations of the Tet-controlled transactivators which may non-specifically initiate transcription from the TRE-Pmin promoter (Bujard, 1996).

Surgery and stereotaxic injection of adenoviral vectors

Adult male Wistar-Kyoto rats were used in these experiments (although similar transfection results have been obtained with Lister-Hooded and Dark Agouti rats which are more commonly used in behavioural experiments). Following administration of anaesthetic, rats were placed in a Kopf stereotaxic frame. In these experiments animals received 1.5 μl of viral suspension (AdTRE-EGFP and the transactivating virus, Fig. 1) via a hand pulled glass needle using a Harvard infusion pump. In the experiments shown injections were directed into the CA1 or dentate gyrus region of the hippocampus (co-ordinates adapted from Paxinos and Watson, 1986). Following a number of preliminary experiments we have found that using glass needles drawn on a microelectrode puller greatly improves the precision of viral delivery and avoids tissue damage associated with wider bore metal needles. It is also essential to deliver the viral constructs slowly (a minimal twenty minute delivery period) to avoid any tissue damage and to avoid unwanted viral diffusion. To regulate expression animals were administered doxycycline hydrochloride (2 mg/ml, Sigma) in their drinking water (supplemented with 5% sucrose and changed every 3 d). In our studies animals were sacrificed 6 d–3 months post surgery (transgene expression remains strong for at least this period) and perfused intracardially with 400 mls of 4% paraformaldehyde in phosphate-buffered saline (PBS). Brains were post-fixed for 2 h and cryoprotected in 30% sucrose in PBS.

In vivo regulation of gene expression

Prior to using the AdTet system effectively in behavioural studies it is essential to demonstrate that transgene expression can be switched repeatedly (e.g. on, off then on) and to estimate over what time period switching can occur. In order to address these issues the AdTet-On and AdTet-Off systems were injected into the hippocampus and EGFP fluorescence was monitored over a period of three months, during which time Dox was administered and then withdrawn (following injection of the AdTet-Off system) or withdrawn and then readministered (following injection of the AdTet-On system). In vitro and in vivo experiments have been conducted with the AdTet-Off and AdTet-On systems expressing EGFP as a reporter gene (Harding et al., 1997, 1998). Using the AdTet-Off system, there was extensive EGFP fluorescence in hippocampal dentate gyrus neurons and hippocampal CA1

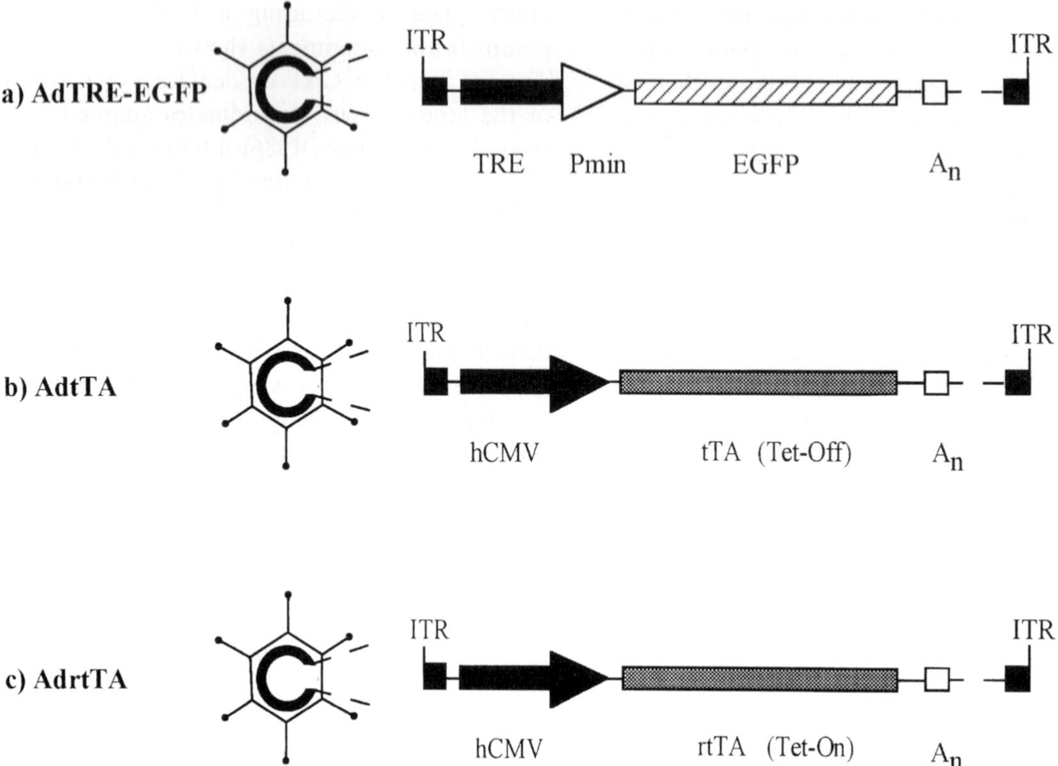

Fig. 2. The Adenoviral-Tetracycline (AdTet) system. Schematic representation of the replication-deficient type-5 adenoviral vectors which comprise the AdTet-On and AdTet-Off systems. (a) AdTRE-EGFP vector. This vector encodes the EGFP gene under the transcriptional control of the TRE/P_{min} promoter. The EGFP gene is followed by an SV40 polyA tail (An). Upto 7.8 kb of transgene DNA can be cloned into the multiple cloning site thus replacing the EGFP gene with a transgene. (b) The AdtTA and (c) AdrtTA vectors contain the tTA (tet-Off) and rtTA (Tet-On) transactivators respectively under the conrol of the constitutive hCMV promoter. These are also followed by an SV40 polyA tail. Mu denotes adenoviral map units and ITR indicates the adenoviral inverted terminal repeats. The functional AdTet-Off system is made by cotransfecting the AdTRE-EGFP/transgene construct with the AdtTA construct, while the AdTet-On system is made by combining the AdTRE-EGFP construct and the AdrtTA construct.

neurones 3 d after a single stereotaxic injection of virus into animals not treated with Dox (dentate gyrus neurones are shown in Fig. 3(A)). Following this 3-d period, if animals were administered Dox for 5 d, EGFP fluorescence was reduced in the dentate gyrus neurones (Fig. 3(B)) and hippocampal neurones (data not shown) and was completely abolished by ten days in both dentate gyrus neurons (Fig. 3(C)) and hippocampal CA1 neurones (data not shown). Three months after the removal of Dox, high levels of expression of EGFP could still be detected in both hippocampal CA1 neurones (Fig. 3(C)) and dentate gyrus neurones (data not shown).

Following transfection with the AdTet-Off systems transgene expression was tightly regulated *in vivo* simply by the oral administration of Dox. Similar experiments were conducted with the AdTet-On system and these also showed that dentate gyrus and CA1 neurones could be transfected and transgene expression turned on, off and on again weeks after the initial transfection event (Harding et al., 1998). Importantly, both the AdTet-On or AdTet-Off systems allowed transgene expression to be turned on months (three months was the longest time point measured) after the initial stereotaxic injection of the system simply by adding or removing

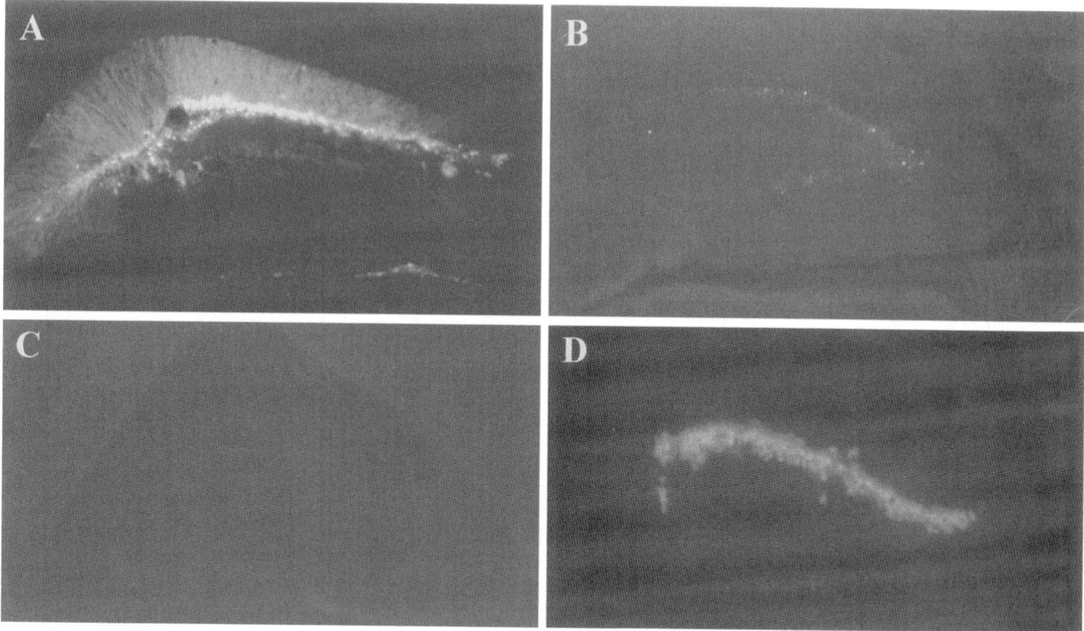

Fig. 3. Regulated control of EGFP in vivo following the stereotaxic delivery of the AdTet-Off system to neurones of the hippocampus. (a)–(d) are fluorescent photomicrographs of the hippocampus following the stereotaxic delivery of the AdTet-Off system to the dentate gyrus or hippocampal CA1 region of the hippocampus: (a) shows the high level of EGFP expression obtained in the dentate gyrus of a non-Dox treated rat following a single 1 μl injection of the AdTet-Off system; (b) demonstrates the almost complete dissappearance of EGFP expression from the hippocampus in an AdTet-Off injected animal which had been administered Dox for 5 d (commencing 3 d after the stereotaxic injection of the virus; (c) EGFP expression was completely abolished 10 d after the commencement of Dox adminstration; (d) shows the switching back on of EGFP fluorescence in the entire CA1 region of the hippocampus three months after the initial stereotaxic injection of the AdTet-Off system, as in (c) this group of animals was administered Dox for 10 d commencing 3 d after the initial injection.

Dox (Harding et al., 1998). We have now also shown that both the AdTet-Off and AdTet-On system can be used to mediate long term regulatable expression in the amygdala, striatum, hypothalamus and cerebellum.

Dox has been shown to cross the blood brain barrier and be effective at regulating gene expression in vivo at sub-toxic concentrations (Yin et al., 1996). However, we have also conducted preliminary experiments to assess the effects of chronic administration of Dox alone on several measures of behavioural performance, as any side effects on the animals locomotor activity or motivation for food reward would obviously preclude the use of the regulatable adenoviral system in many behavioural studies. The effects of Dox were examined in: (i) a T-maze alternation task, which taxes spatial working memory processes; (ii) in the Morris water maze task, a test of spatial navigation; (iii) on spontaneous locomotor activity. Male Lister Hooded rats (300 g) were food deprived to 85% of their body weight and Dox (2 mg/ml) and sucrose administered in the animals drinking water 3 d prior to the start of testing. Control animals received sucrose alone in the drinking water. The T-maze task is divided into two stages and was run as follows (for more detail see Aggleton et al., 1995; Warburton et al., 1997): During the ‘sample’ stage of the task, the rat is initially allowed to run into only one of the two arms of the maze, where it receives a food reward. In the ‘test’ stage, the rat is allowed into either arm of the maze, but the animal is only rewarded for chosing the arm not previously

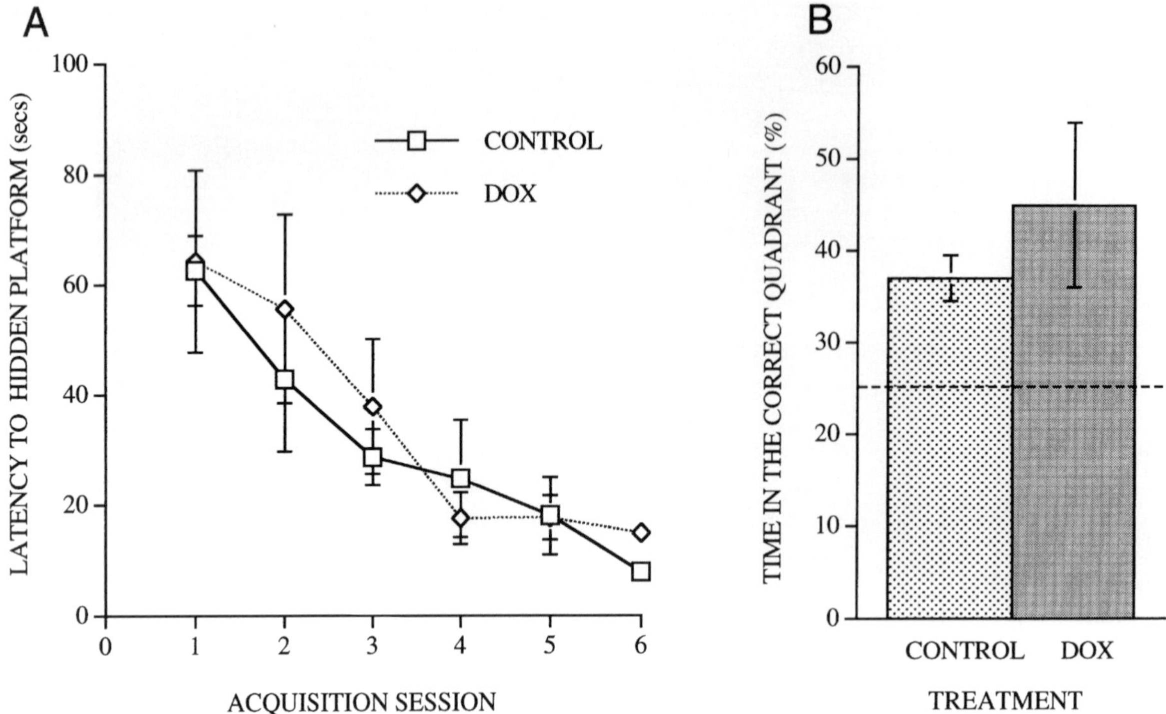

Fig. 4. (A) The mean (±S.E.M.) latency of doxycycline treated and control rats to find the hidden platform in the Morris water maze within each session (made up of four trials). On each trial the rat was placed in the water at the edge of the pool (2 m diameter) at a different location, and allowed a maximum of 2 min to find the platform. The animals were tested for six consecutive days, ANOVA confirmed that there was no significant difference between the groups ($F = < 1.0$, $P > 0.05$). On the seventh day the probe trial was conducted. (B) Mean (±S.E.M.) percentage of time spent swimming in the training quadrant during the probe trial. During the probe trial the platform is removed from the maze and the swim path in each of the four quadrants recorded over 1 min. The dotted line indicates chance performance and the graph clearly shows that both groups of rats spent significantly longer searching for the platform in the quadrant which had contained the platform during training, and that there was no difference between the two groups, confirmed by ANOVA ($F = < 1.0$, $P > 0.05$). S.E.M. values that are not visible of the figure are smaller than the symbol.

visited during the sample stage. If the rat choses the other arm no reward is obtained. The rats were tested for six trials a day, for a period of five consecutive days and Dox was found to have no effect on performance [mean number of correct responses: control = 5.9; Dox = 5.8]. In the Morris water maze task, the rats are required to learn the location of a hidden platform within a large water filled pool, using the distal cues around the room to navigate. This task has been shown to be highly sensitive to manipulations of the hippocampal formation and thus has been frequently used in genetic studies of learning and memory (Morris et al., 1982; Silva et al., 1992a, 1992b). Figure 4(a) illustrates that Dox administration had no effect on the rate of acquisition of the water maze task, nor on animals performance during the probe trial (see Fig. 4(b)). Importantly, Dox had no effect on other performance indices which may be measured in the water maze such as swim speed [mean swim speed (m/sec): control = 0.44; Dox = 0.44] or directionality (data not shown) and consistent with these findings Dox also had no effect on normal levels of locomotor activity as assessed in the photocell cages (data not shown).

These studies have therefore demonstrated that chronic administration of Dox alone has no effect on either normal locomotor activity levels nor on the baseline performance of two memory tasks which tap different spatial memory processes.

While these studies do not rule out the possibility that Dox may have some other side effects, the finding that several performance indices remain unaffected by this compound suggest that the use of the AdTet systems may be highly suitable for behavioural investigations. Indeed, a number of constructs have now been developed which will be tested in behavioural studies. In Glur2 knockout mice Gerlai et al. (1998) showed that long term potentiation was enhanced and that there were multiple behavioural anomalies. In collaboration with Dr. Jeremy Henley (University of Bristol) we have shown that the surface expression of the Glur2 AMPA receptor subunit can be regulated (Nishimune et al, 1998) following the transfection of neurones with Ad-Tet vectors expressing NSF binding proteins which alter receptor trafficking (Noel et al., 1999). We can therefore use the regulatable Ad system to study the mechanisms underlying Glur2 protein trafficking and its involvement in memory processing. Ad vectors expressing, hsp40, Jun-N-terminal kinase (JNK) — interacting protein (JIP, an inhibitor of JNK), protein kinase zeta dominant negatives, and modulators of the cyclic AMP response element pathway have been made and shown to be highly regulatable by Tet. It should also be noted that compared to the construction of transgenic knockouts, Ad vectors can be made relatively quickly (within six weeks) and cheaply. For example it is possible to buy all the vectors necessary to construct viruses for less than $600 and it will soon be possible to buy regulatable Ad plasmids and premade viruses.

Conclusion

The use of this regulatable adenoviral vector system which mediates relatively long-term expression should greatly facilitate studies of behaviour, since it offers a relatively simple and rapid method (compared to transgenesis) of manipulating neuronal gene expression in specific populations of neurones. It should be possible, for example, to use this system to manipulate neuronal gene expression following only a single stereotaxic injection of virus. The specific functional/behavioural outcome could then be correlated with the turning on or turning off of gene expression (by the administration or withdrawal of dox) weeks after the initial viral transfection. This allows the functional response to the genetic manipulation to be clearly dissected away from any effects mediated by viral tranfection. Furthermore, subtle intracellular genetic manipulations can be carried out in specific neuronal populations and the widespread destruction of neurones associated with conventional lesion studies and the ensuing complication in interpreting results is avoided. Importantly, behavioural testing of animals can be undertaken weeks after the initial stereotaxic surgery, so that, for example, the behavioural pretraining could be conducted 1–2 weeks after the injection of the AdTet system and the behavioural testing itself could start once Dox had been adminstered to turn transgene expression on (3–4 weeks after the initial surgery). Finally, the use of the AdTet system would permit the use of within-subjects experimental designs, in which every animal may serve as its own control as it may be tested both with transgene expression turned off and then on. Such experimental designs are advantageous as they require fewer subjects, and statistically are more powerful as the variability due to individual differences between animals is eliminated from the experimental error.

Thus, the use of adenoviral vectors, with the incorporation of the Tet On/Off system, should provide an extremely powerful tool for the investigation of cellular mechanisms underlying certain forms of behaviour.

References

Aggleton, J.P., Neave, N., Nagle, S. and Hunt, P.R. (1995) A comparison of the effects of anterior thalamic, mamillary body and fornix lesions on reinforced spatial alternation, *Behav. Brain Res.*, 68: 91–101.

Bujard, H. (1996) Controlling gene activities via the tet regulatory systems: a trouble shooting guide. www.zmbh.uniheidelburg.de/bujard/ homepage.html.

Geddes, B.J., Harding, T.C., Hughes, D.S., Byrnes, A.P., Lightman, S.L., Conde, G. and Uney J.B. (1996) Persistent transgene expression in the hypothalamus following stereotaxic delivery of a recombinant adenovirus: suppression of the immune response with cyclosporin. *Endocrinology*, 137(11): 5166–5169.

Geddes, B.J., Harding, T.C., Lightman, S.L. and Uney, J.B. (1997) Long-term gene therapy in the CNS: Reversal of hypothalamic diabetes insipidus in the brattleboro rat using an adenovirus expressing arginine vasopressin. *Nature Medicine* 3(12): 1402–1405.

Gerlai, R., Henderson, J.T., Roder, J.C. and Jia, Z. (1998) Multiple behavioral anomalies in GluR2 mutant mice exhibiting enhanced LTP. *Behav. Brain Res.*, 95(1): 37–45

Gossen, M. and Bujard, H. (1992) Tight control of gene expression in mammalian cells by tetracycline-responsive promoters. *Proc. Natl. Acad. Sci. USA*, 89: 5547–5551

Gossen, M., Freundlieb, S., Bender, G., Muller, G., Hillen, W. and Bujard, H. (1995) Transcriptional activation by tetracyclines in mammmalian cells. *Science*, 268: 1766–1769.

Graham, F.L. and Prevec, L. (1991) Manipulation of Adenovirus Vectors. In: Methods in Molecular Biology, Clifton, NJ: The Humana Press Inc. 7: 109–128.

Graham, F.L. and Prevec, L. (1995) Methods for construction of adenovirus vectors. *Mol. Biotechnol.*, 3: 207–220.

Graham, F.L., Smiley, J.S., Russell, W.C. and Nairn, R. (1977) Characteristics of a human cell line transformed by DNA from human adenovrius type 5. *J. Gen. Virol.*, 36: 59–72.

Harding, T.C., Geddes, B.J. Noel, J.D., Knight, D., Murphy. D. and Uney, J.B.(1997) Tetracycline-regulated transgene expression in hippocampal neurones following transfection with adenoviral vectors. *J. Neurochem.*, 69(6): 2620–2624.

Harding, T.C., Geddes, B.J., Murphy, D., Knight, D. and Uney, J.B. (1998) Switiching transgene expression in the brain following transfection by an adenoviral tetracycline-regulatable system. *Nature Biotechnology*, 16: 553–555.

He, T.C., Zhou, S., Da Costa, L.T., Yu, J., Kinzler, K.W. and Vogelstein, B. (1998) A simplified system for generating recombinant adenoviruses. *Proc. Natl. Acad. Sci. USA*, 95(5): 2509–2514.

Kandel, E.R. and Spencer, W.A. (1968) Cellular neurophysiological approaches in the study of learning. *Physiol. Rev.*, 48: 65–134.

Kistner, A., Gossen, M., Zimmermann, F., Jerecic, J., Ullmer, C., Lubbert, H. and Bujard, H. (1996) Dox-mediated quantitative and tissue-specific control of gene expression in transgenic mice. *Proc. Natl. Acad. Sci. USA*, 93: 10933–10938.

Le Gal La Salle, G., Robert. J.J., Berrard, S., Ridoux V, Stratford-Perricaudet, L.D., Perricaudet, M. and Mallet, J. (1993) An adenovirus vector for gene transfer into neurons and glia in the brain. *Science*, 259: 988–990

Mayford, M., Bach, M.E., Huang, Y., Wang, L., Hawkins, R.D. and Kandel, E.R. (1996) Control of Memory Formation Through Regulated Expression of a CaMKII Transgene. *Science*, 274: 1678–1683.

Mansuy, I.M., Winder, D.G., Moallem, T.M., Osman, M., Mayford, M., Hawkins, R.D. and Kandel, E.R. (1998) Inducible and reversible gene expression with the rtTA system for the study of memory. *Neuron*, 21(2): 257–265.

Morris, R.G.M., Garrud, P., Rawlins, J.N.P. and O'Keefe, J. (1982) Place navigation impaired in rats with hippocampal lesions. *Nature*, 297: 681–683.

Morsy, M.A., Gu, M., Motzel, S., Zhao, J., Lin, J., Su, Q., Allen, H., Franlin, L., Parks, R.J., Graham, F.L., Kochanek, S., Bett, A.J. and Caskey, C.T. (1998) An adenoviral vector deleted for all viral coding sequences results in enhanced safety and extended expression of a leptin transgene. *Proc. Natl. Acad. Sci. USA*, 95(14): 7866–7871.

McGrory, J., Baustista, D. and Graham, F.L. (1988) A simple technique for the rescue of early region I mutations into infectious human adenovirus type 5. *Virology*, 163: 614–617.

Neve, R.L. (1993) Adenovirus vectors enter the brain. *TINS*, 16(7): 251–253

Nishimune, A., Isaac, J.T., Molnar, E., Noel, J., Nash, S.R., Tagaya, M., Collingridge, G.L. and Nakanishi, S. (1998) NSF binding to GluR2 regulates synaptic transmission. *Neuron*, 21(1): 87–97

Noel, J., Ralph, G.S., Pickard, L., Williams, ?., Molnar, E., Uney, J.B., Collingridge, G.L., Henley, J.M. (1999) Surface expression of AMBA receptors in hippocampal neurons is regulated by NSF dependent mechanisms. *Neuron*, 23: 365–376.

Paxinos, G. and Watson, C. (1986) *The rat brain in stereotaxic coordinates*, 2nd Edn, Academic Press Inc., London.

Silva, A.J., Paylor, R., Wehner, J.M. and Tonegawa, S. (1992a) Impaired spatial learning in a a-calcium-calmodulin kinase II mutant mice. *Science*, 257: 206–211.

Silva, A.J., Stevens, C.F., Tonegawa, S. and Wang, Y. (1992b) Deficient hippocampal long-term potentiation in a a-calcium-calmodulin kinase II mutant mice. *Science*, 257: 201–206.

Warburton, E.C., Baird, A.L. and Aggleton, J.P. (1997) Assessing the magnitude of the allocentric spatial deficit associated with complete loss of the anterior thalamic nuclei in rats, *Behav. Brain Res.*, 87(2): 223–232..

Yarranton, G.T. (1992) Inducible vectors for expression in mammalian cells. *Curr. Opin. Biotechn.*, 3: 506–511.

Yin, D.X., Zhu, L. and Schimke, R.T. (1996) Tetracycline controlled gene expression system achieves high-level and quantitative control of gene expression. *Anal. Biochem.*, 235: 195–201.

Zhang, G.H., Gurtu, V. and Kain, S.R. (1996) An enhanced green fluorescent protein allows sensitive detection of gene-transfer in mammalian cells. *Biochem. Biophys. Res. Commun.*, 227(3): 707–711

W.E. Crusio and R.T. Gerlai (Eds.)
Handbook of Molecular-Genetic Techniques for Brain and
Behavior Research (Techniques in the Behavioral and Neural Sciences, Vol. 13)

CHAPTER 3.2.4

Antisense oligonucleotides to selectively suppress gene expression in the brain

Giuseppe Pollio and Adriana Maggi

Institute of Pharmacological Sciences, University of Milan, Via Balzaretti 9, 20133 Milan, Italy

Introduction

The use of specific drugs has greatly facilitated the progress in the understanding of the brain molecules involved in the manifestation of specific behaviours. These pharmacological tools, however, often suffer the disadvantage of inducing the onset of compensatory systems. For instance neuroleptics, antagonists of dopaminergic receptors, cause dopamine receptor up-regulation possibly determining the loss of effectiveness reported with their repeated use (e.g. dyskinesias known to occur in chronic treatment with neuroleptics) (Creese and Sibley, 1981).

Ideally, for the understanding of the functions of a given protein we should be able to control the expression of its respective gene. The availability of genetic engineering techniques and, consequently, the possibility to generate animal mutants carrying specific alterations in their genome represents a turning point in modern biology. By the use of these techniques, genes can be multiplied (in transgenic animals), or deleted (Koller and Smithies, 1992) (in gene knock out animals) at desire, facilitating the understanding of the physiological significance of their protein product. Thus, these techniques could be largely applied also to the study of the molecular basis of specific behaviours. The major limitation so far encountered in the generation of gene knock-out animals is linked to the fact that proteins exerting an important role in the control of behaviour may also be indispensable for the correct development of specific brain circuitries and their absence can be the cause of fatal errors in neural tissue development, thus impairing the future behavioural studies (Klein et al., 1993; Jones et al., 1994). On the other hand, their function can be substituted by other gene products, nullifying the effect of gene ablation on the phenotype. These shortcomings can be partially circumvented by the use of conditional mutants in which the transgene is expressed only under specific circumstances (pharmacological treatment, cell specific expression, etc.) or deleted only upon pharmacological intervention (e.g. LoxP) (Kilby et al., 1993; Golic, 1994). Yet, most of these systems have still limited application since each specific alteration in gene expression is permanent and cannot be abrogated.

Antisense oligonucleotides (AOs) seem to combine the advantage of both pharmacological and genetic engineering tools without suffering of their limitations. AOs may modulate gene expression and their action, similarly to what is occurring with drugs, is absolutely reversible (Rickman and Bowes Rickman, 1996).

AOs are short synthetic DNA molecules, generally composed of 15–24 bases, complementary to a specific messenger RNA sequence. Each

AOs hybridizes to a selected mRNA interfering with its stability or its translation and therefore affecting the production of a specific protein (Phillips et al., 1996). Antisense inhibition represents an extremely attractive pharmacological and investigative approach since it offers base to base specificity to the target protein and the versatility appropriate to the complexity of the genetic code.

So far, AOs have been successfully utilized for a number of studies related to brain functions (Pollio et al., 1993; Wahlestedt et al., 1993a, 1993b; Standifer et al., 1994; Zang et al., 1994; Weiss et al., 1997) since, apparently, their activity in the CNS is higher than in other tissues, probably because of the poor expression of DNases in neural cells. Yet, this pharmacological tool is to be considered in its infancy as investigative approach and many issues need to be clarified before they can be successfully utilized. The aim of this chapter is to analyse some of the issues related with the use of AOs in studies of brain functions.

Identification of an antisense oligodeoxynucleotide active both in vitro and in vivo

Selection of the target sequence

The paradigm of antisense inhibition implies that the antisense DNA molecules hybridize with the complementary mRNA coding for the target protein in the cytoplasm. The formation of this hybrid DNA/RNA molecule, termed heteroduplex, prevents either ribosomal assembly or mRNA read-through. Most oligonucleotides therefore are targeted to the initiation codon AUG, or part of the coding region just downstream of it. In particular, as indicated in the Fig. 1, the three regions can be indicated as best AOs target on the mRNA are the 5′ cap region, the AUG translation initiation codon and the 3′ untranslated region of the mRNA.

Besides translation inhibition, other mechanisms of AOs action have been proposed. Studies of cellular uptake of labelled oligonucleotides show that most AOs, upon intracellular penetration, rapidly migrate to the cell nucleus, suggesting an intranuclear site of action (Iversen et al., 1992). At this site, the action on gene expression might be more direct: AOs might inhibit pre-mRNA splicing, the transport of mRNA from the nucleus to the cytoplasm, or bind to the DNA, inhibiting its transcription. The observation that antisense oligonucleotides targeting exon-intron splicing sites (Colige et al., 1993) or the major groove of the DNA (Helene, 1993) are very effective would support this mechanistic hypothesis (Fig. 1).

Experimental evidence accumulated in the past few years of AOs research indicates that a large number of antisense oligonucleotides, targeted to different regions of mRNAs, were successful in down-regulating a large variety of proteins, although it is difficult to assess the number of failures.

When developing such an antisense oligo, one has to consider two important factors: the stability of the heteroduplex, which is dependent on the number and composition of complementary bases, and the availability of the target sequence, which is dependent on the folding of the mRNA molecule. Although there have been attempts to correlate these two factors with the efficiency of the AOs, there are no precise rules in selecting the oligo's sequence to be used. There is a general agreement that there is a positive correlation between affinity of AOs for the target mRNA and its effectiveness. With regard to the importance of the secondary structure of the target molecule, results are conflicting. It is well known that antisense oligo targeted to different regions of the mRNA have unequal efficiencies (Cowsert et al., 1993; Wakita and Wands, 1994). This could be due to the folding of the mRNA which influences the availability of the target sequence in spite of the potentially strong base pairing between the AO and its target sequence which

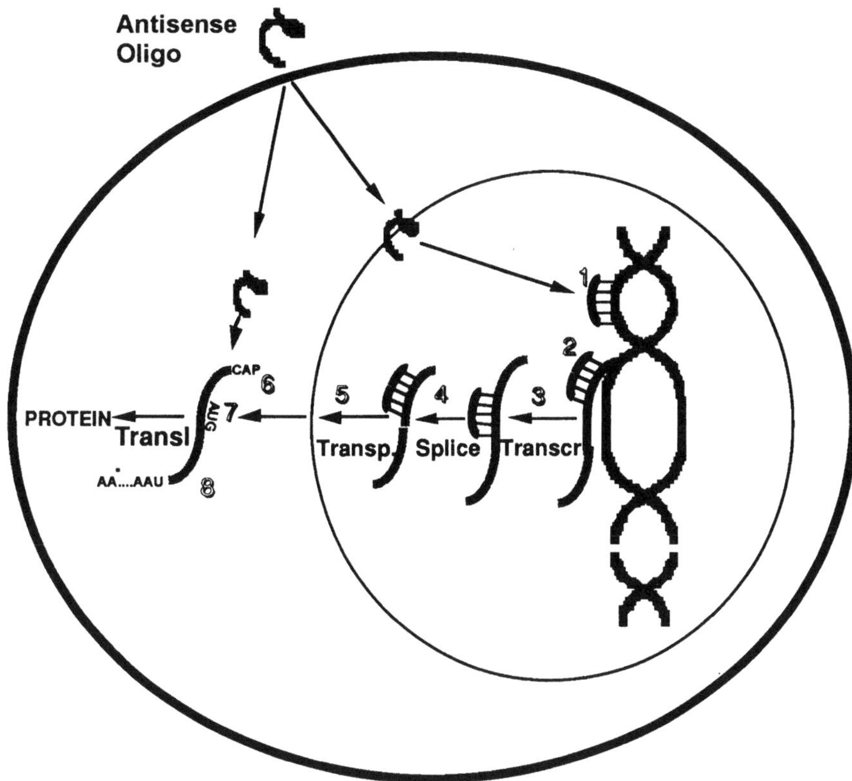

Fig. 1. Mechanisms of action of antisense oligos. (1) The antisense may form a triple helix with the DNA in the nucleus. (2) The antisense may bind the nascent RNA and prevent RNA elongation. (3) The antisense binds to exon and prevents maturation of the primary transcript. (4) The antisense binds to exon–intron junction and prevents splicing. (5) The antisense binding to the mRNA prevents transport to the cytoplasm. (6) The antisense binds to the 5′-CAP, or, (7) binds to the AUG initiation codon, or (8) binds to the 3′-untranslated region of the mRNA, all of which prevent translation. (Modified from Phillips et al., 1996.)

should be the most favourable structure from the thermodynamic point of view.

On the basis of the experimental evidences so far accumulated, it is hard to identify a general rule for the design of an antisense oligo effective on a particular target, possibly because different mechanisms of action may result effective in each specific experimental setting. On the other hand, a common denominator may exist in antisense action, but factors like differential cellular uptake and metabolism of AOs as well as intracellular feedback mechanisms on the synthesis of mRNA or protein incapacitate its identification.

There is no doubt that a systematic investigation is needed to better clarify the mechanisms of AOs activity in order to allow a rational approach to the design of AOs. To date, however we are bound to empirical approaches. Generally, the safest way for finding the optimal target sequence is to try a series of AOs aimed to hybridising different sites of the target mRNA, or better, to use an array of combinatorial oligonucleotides which allows to screen simultaneously for heteroduplex formation a large number of oligos and to correlate the high duplex yield with the activity both in vitro and in vivo (Milner et al., 1997).

In any case, the sequence of all AOs must be checked on GenBank to avoid any undesired activity due to sequence homologies with other mRNAs.

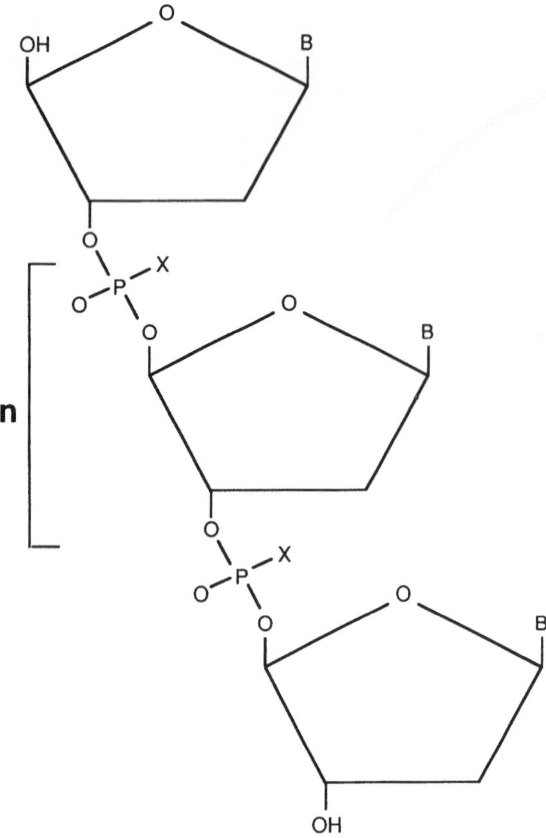

Fig. 2. Chemical structure of oligodeoxyribonucleotides analogues. B, purine or pyrimidine base; X = sulfur for Phosphotioate (S-AO); X = oxygen for Phosphodiester (O-AO); n = 15–20

Choice of the DNA analogue for in vivo experiments

The initial studies with AOs were carried out with natural oligos with phosphodiester bonds between the nucleotides to mime single strand DNA. However, these sequences are extremely susceptible to hydrolysis by nucleases making them unsuitable to in vivo studies. As a consequence, a large effort was made to develop various chemical analogues derived from phosphodiesters (Cook, 1993). Chemical modifications of naturally occurring nucleic acids can concern the base, the sugar ring, as well as internucleotide bonds. In brain research, together with the unmodified AOs, the phosphotioate analogues (S-AO) have been used almost exclusively. In S-AOs, the phosphodiester bonds are modified by introducing sulphur in place of non-bridging oxygen (Fig. 2). This modification results in improved resistance against nucleases (Stein et al., 1988). Unfortunately, highly active thiol groups of S-AOs have been reported to be responsible for the inhibition of cellular enzyme such as DNA polymerases, ribonuclease H and protein kinase C (Gao et al., 1992; Stein et al., 1993), as well as for general toxicity (Ehrlich et al., 1994). In an effort to preserve resistance against nucleases while minimising the toxicity, end capped oligos (EC-AOs) have been introduced (Ghosh et al., 1993). These oligos maintain the phosphodiester backbone and have a sulphur modification either at the extremity 3′ or at both 5′ and 3′. The next few years will presumably see further development of new generations of oligodeoxynucleotides with improved properties. New DNA analogues, already in use, include 2′-modified oligodeoxynucleotides (Monia et al., 1993), oligodeoxynucleotides with 5′-cholesteryl moieties (Krieg et al., 1993) and peptide nucleic acids (Nielsen et al., 1991). In addition to the modification of the oligo backbone a completely different approach involving RNA molecules with a sequence-specific DNase activity, called ribozyme, already offers interesting new possibilities for antisense research (Uhlenbeck et al., 1989).

Stability of oligodeoxynucleotides in in vivo experiments

Biodegradation has been shown to be a major obstacle in the in vivo application of AOs. Initial studies demonstrated that the half-life of unmodified oligo, infused in the rat circulatory system is less than 1 min, while, under the same conditions, S-AOs have been found to be much more resistant to the in vivo hydrolysis (Agrawal et al., 1991).

In the case of CNS, it has been shown that, ten hours after rat hippocampal infusion, the percentage of intact S-AOs was of 40% ± 23 and this was in striking contrast with values of about 6% observed with oligos unmodified or end capped (Szklarczyk and Kaczmarek, 1997). In fact, it has been described that major nucleases responsible for the degradation of AOs in the serum are 3′–5′ exonucleases, whereas in the cellular extracts AOs are degraded mostly by endonucleases (Hoke et al., 1991), therefore the 3′ end capped oligo should be well protected in the extracellular fluids but not inside the cells.

As previously mentioned, it is important to consider the possibility of toxic events resulting from AOs degradation inside the cell. In fact, it is conceivable that AOs metabolites are used as in the cellular mechanism of DNA repair and, in the case of phosphothioate or other analogues, the incorporation into the nuclear or mithocondrial genome of these synthetic substrates can lead to genotoxicity and other forms of cytotoxicity.

In vivo treatment with antisense analogues, the appropriate controls

Before utilising any AOs for in vivo studies it is advisable to analyse their activity in a model system: generally in cells expressing the mRNA/protein of interest at high concentration. Once proved that the AOs is active in this setting, studies in vivo can be initiated.

It is well known that oligonucleotides (or better contaminants retained after their purification) may cause a series of intracellular effects, it is therefore mandatory to prove, with the aid of appropriate controls, that the observed effects are linked to AOs activity. The specificity should be verified by testing in parallel a mismatched oligo with few bases changed. This will constitute a better control than scrambled sequences, represented by an oligo maintaining the same base composition as the antisense molecule but in a totally mismatched sequence, or sense oligos. The decrease in affinity associated with a mismatched base pair is dependent upon several factors and will be on average 500 times less than with the perfectly complementary oligonucleotide. The mismatched oligo or other controls need to be tested at the same concentrations range of the active one. Second, it is necessary to rule out possible oligo effects on proteins other than the target. For example, when targeting a specific receptor, one could assay for lack of activity on related receptor subtypes (e.g. by ligand-binding techniques). In addition, it may be informative to quantitate the concentration of the related gene transcript after administration of the AOs. As control of specificity of action, changes in the content of unrelated mRNAs can be evaluated. Third, to rule out general cytotoxicity effects of the AOs before starting the in vivo experiments, is a good habit to assess the effect of AOs treatment on the number of viable cells, as well as their protein content. The reversibility of the AOs effects will also constitute a reliable prove of lack of toxicity.

Antisense oligonucleotides in the CNS

Stability of the oligos in the CNS

In the brain, as above mentioned, nucleic acids biodegradation does not occur at high rate. In this organ RNA is stable for up to 48 h *post mortem* (Johnson et al., 1986), possibly because of the relatively low activities of both deoxyribonuclease I and endonucleases (Lacks, 1981; Giannakis et al., 1991). Yet, the data so far available do not provide us with a clear picture of AOs stability in brain tissue: in ex vivo experiments the half-life of S-AOs in the rat cerebro spinal fluid (CSF) is longer than 18 h (Campbell et al., 1990) while in vivo studies in which the unmodified AOs and S-AOs were administered into the rat ventricular system (i.c.v.), the half life was quite variable in relation to the type of AOs utilized (Withesell et al., 1993; Yee et al., 1994). Four hours after the infusion into the lateral ventricle of the rat brain the unmodified AOs can be recovered mostly in the form of free nucleotides or even their metabolites, while S-AOs are still present, in part, as full length oligos. The extent of degradation is not changed between 8 and 24 h after the infusion, suggesting that the most intensive cleavage occurs early after the administration, primarily by the action of blood nucleases, then, after the diffusion in the tissue milieu, by both the extracellular and intracellular enzymes.

Delivery of the antisense oligonucleotides into the CNS

The major obstacle to the use of systemic administration of AOs targeting CNS messages, is their inability to cross the blood-brain barrier. Agrawal et al., showed that after an intravenous or intraperitoneal injection of radiolabelled oligos, their concentration in the brain is significantly lower than in other tissues, e.g. up to 1000-fold lower than in the kidney or in the liver.

For topical delivery into the CNS, three approaches have been applied in animal models, all with considerable success:

1. Infusion into the cerebrospinal fluid (CSF), by implantation of *cannulae* in the lateral ventricle of rats or mice;
2. Direct intracerebral injections into different nuclei of the brain, also in this case a *cannula* is placed in situ according to stereotaxic co-ordinates; and,
3. Superinfusion of the spinal cord (Hou et al., 1997) after the intratecal introduction of a catheter via laminectomy at the L5 vertebra, locating the tip of the catheter near the lumbar enlargement of the spinal cord (Skilling et al., 1992).

To these methods we can add other systems of administration, as the exposed brain of Xenopus embryos (Retaux et al., 1996) in which the skin covering the brain of the frog embryos is removed and the tadpoles are treated with oligos simply by adding these to the buffered saline in which they are maintained; or intraocular injection of neonatal rats (Rickman and Bowes Rickman, 1996).

After a single injection of AOs into the brain lateral ventricle, oligos are rapidly cleared out, reflecting a fast entrance in the surrounding tissues and, possibly degradation. Continuous delivery via a miniosmotic pump has been successful used for a prolonged action. However, even this latter system allows the delivery of the antisense only to the proximity of the injection site and to the regions along the pathway of CSF drainage (Campbell et al., 1990; Withesell et al., 1993) Similarly, AOs infused intracerebrally infiltrate the tissue surrounding the injection site in various brain regions, such as the ventro medial nucleus of the hypothalamus (Pollio et al., 1993), the medial preoptic area (Ogawa et al., 1994), hippocampus (Tischmeyer et al., 1994), striatum (Sommer et al., 1993; Dragunow et al., 1994; Robertson et al., 1995) and basolateral amiygdala (Szklarczyk and Kaczmarek, 1995). According to Ogawa et al., the maximal diffusion of AOs throughout the medial preoptic area is 1.5 mm (Retaux et al., 1996). Finally, experiments with the unilateral superfusion of rat spinal cord with radiolabelled oligo have demonstrated that AOs penetrate into the neural tissue. The area reached by the radioactive oligo at 6 h was much larger if compared to 1 h after treatment (Gillardon et al., 1994).

In the case of site injections, metabolically stable oligodeoxynucleotides such as S-AOs should be used preferably because some tissue damage might activate the nucleases. O-AOs have also been successfully used in this context (Akabayashi et al., 1994; Pollio et al., 1993) even if, higher doses were required to obtain full response. Szklarczyk and Kaczmarek performed a series of studies with radiolabelled oligos injection into the centre of the dorsal part of the rat hippocampus. With autoradiography of the coronal sections obtained ten hours after AOs adminstration, they demonstrated the lack of difference in tissue penetration with all four oligonucleotides types (O-AOs, S-AOs, 3′EC-AOs and 3′5′EC-AO). Furthermore, in these experiments, the adjacent white matter, i.e. the cingulum, alveus hippocampi and external capsulae proved to represent a barrier for AOs propagation.

Uptake of the antisense analogues by neural cells

Cellular uptake of AOs has been intensely examined in cell culture (Akhtar et al., 1991; Neckers 1993). These studies led to propose that O-AOs and S-AOs bind to specific cell surface binding protein(s) before being internalized by endocytosis. The extent of internalized AOs was reported to range between 1 and 11% (Crook, 1991). However, because the experiments have been carried out without monitoring the stability of AOs, these numbers cannot be fully reliable. Furthermore, it is worth noting that once inside the cell, AOs can still be inactive because entrapped within endosome vesicles. AOs cellular uptake is relative to the cell type and the phase of cell cycle. It has been shown that cellular internalization of the oligo is more efficient in proliferating cells than in quiescent ones (Iwanaga and Ferriola, 1993; Krieg et al., 1991).

In the adult brain, neurones are arrested in the G0 phase of the cell cycle and therefore one can expect a poor capability to internalize oligos. Despite this fact most authors observed accumulation of the tracers linked to the AOs inside the cell bodies of neural cells, demonstrating that internalization occur, even if the integrity of labelled AOs has not been carefully evaluated. In one report the authors have found that the oligos could be recovered intact from the cytosol but not from the nuclear fraction (Whitesell et al., 1993). With another study (Agrawal et al., 1991), in which the radiolabelled oligos have been directly injected into the centre of the dorsal part of the rat hippocampus, has been possible to demonstrate that radiolabelled S-AOs could distribute uniformly within the hippocampal hilus, granule cells bodies as well as within their dendrites. Only a very small amount of these oligos could be recovered from the cells nuclei but, in the case of the S-AOs, in an undegraded form, demonstrating for the first time association of active oligo with the nuclei of cell in an in vivo experiment.

Not unexpectedly, experimental evidence suggests that various cell types differs greatly not only in their capability to uptake but also to degrade oligos. Neurones appear to have a relatively low capacity to degrade oligonucleotides and are therefore amenable to treatment with either O-AOs or nuclease-stable analogues. In contrast, other cell types inside and outside the CNS may have high degradative capacity, so the correct choice of the effective oligo analogue has to be carefully evaluated.

Antisense blockade of progesterone receptor: an example of antisense strategy for the study of the molecular basis of rat sexual behaviour

The use of antisense oligonucleotides allowed us to unravel the role played by progesterone in triggering a specific behaviour in female rats. The different steps involved in our study can be used as an example of application of antisense in behavioural studies and to underline the importance of what previously discussed with regard to the general use of antisense in neuroscience.

The importance of progesterone action in the display of lordosis in female rat was initially suggested by Moguilewsky and Raynaud (Moguilewsky and Raynaud, 1979). However, since that initial report, several authors had conflicting results on the real involvement of progesterone in the generation of lordotic behaviour. In order to prove that progesterone was indeed necessary for this behaviour, we planned to examine the effect of diminished number of progesterone receptors in the brain nucleus previously reported as essential for the lordosis to occur: the ventro medial hypothalamus (VMH). To affect PR synthesis we used specific AOs. However, before starting the in vivo experiments a number of studies, here presented, were done with the purpose of selecting and testing active AOs.

Selection of the target sequences

A set of oligonucleotides aimed to hybridize with various sequences of the progesterone receptor mRNA were synthesized. We chose oligos of eighteen base length targeting the AUG as well as splicing regions and poly-A signal region of the B-form of PR. Before doing cell culture experiments all the oligos were tested for:

1. Selectivity: by sequence homology searches in GenBank libraries.
2. Stability in serum for cell culture: Different batches of foetal bovine serum were tested for nucleases activity after heat inactivation at 65°C for 45 min to choose the medium with the lowest nucleases which was then utilized for the whole set of experiments in cell culture. In our conditions half of the amount of oligo incubated in medium containing 10% FBS was recovered intact after 24 hs at 37°C.

AO effects on PR content

To test the anti PR activity of our oligos we used the T47D mammalian cell line expressing PR at high concentration. A dose-response and a time-response series of experiments were performed in order to choose the conditions for the maximal AOs activity. The evaluation of the effect of AOs was done by immune-enzymatic assay, analising the receptor protein content after administration of the AOs. After this set of experiments only the oligonucleotide complementary to the AUG region of the PR-B mRNA showed the highest activity, decreasing the receptor content of more than 50% in 48 h of treatment. Sense and scrambled oligos did not cause any modification in progesterone receptor content demonstrating the sequence specificity of the antisense action.

Functional consequences of PR decrease

To test whether the reduced number of progesterone receptors lowered the cell response to progesterone, the effect of the AOs on the progesterone-dependent regulation of estrogen receptor (ER) synthesis was analysed. As reported in the literature (Alexander et al., 1990) the administration of progesterone to the medium of T47D cells determines a significant decrease in ER content. The pre-treatment of the cells with the active anti-PR AOs significantly decreased the effect of progesterone demonstrating a direct correlation between the PR content and hormonal activity.

Pharmacocynetic parameters

The rate and time of penetration of the oligonucleotides inside the cells was assessed with radioactive labelled oligos demonstrating a fast and saturable entrance of the AOs into the cells. The rate of degradation of the oligos inside the cells was assessed by electrophoresis, evaluating the percentage of full-length oligodeoxynucleotides on the total AOs extracted from cellular cytosol at different time following the treatment. General toxicity parameters were evaluated by morphological analysis and by counting of viable cells after antisense treatment. In addition, dose-dependent and time-dependent analyses were performed in order to choose the correct dose and time of treatment for the following in vivo studies.

Analysis of the antisense activity on progesterone induced sexual behaviour

Among the different systems of administration of the oligos in the CNS we chose the direct injection into the hypothalamic nuclei to minimize the degradation of the unmodified AOs we utilized, and to reduce the dispersion of the oligo in non-target tissues. The topical delivery of the antisense agent at the site of action in the CNS is generally performed by using microcannulae

TABLE 1

Behavioural effects of intracerebral administration of sense and antisense start AOs

Treatment	Number	Number of responders (LQ > 20)	LQ of responders (mean ± S.D.)
Controls	9	8	83 ± 4
Start Sense	4	3	90 ± 5
Start Antisense	10	3	46 ± 10*
Estrogens	3	–	–

* $P \leq 0.001$ vs control. (From: Pollio et al., 1993.)

stereotaxically implanted. To study the activity of our AOs in the CNS we implanted polyethylene heparinized cannulae (i.d. 0.28 mm; o.d. 0.61 mm) in the ventro medial nuclei of the hypothalamus using co-ordinates according to Pellegrino (Pellegrino et al., 1978).

The onset of the estrous response in female ovariectomized rats was induced by sub cutaneus sequential treatment with estradiol-benzoate and progesterone at doses mimicking the normal circulating hormones and effectiveness of AOs after intracerebral administration was evaluated. The sexual behaviour, progesterone-induced, was completely abolished by the antisense administration, while neither sense nor scrambled oligos were effective (Table 1). Animals were sacrificed by cervical dislocation, the brains were rapidly removed and frozen in liquid nitrogen and coronary sections were made. The assessment of the location of the cannulae tip and of the dye was accomplished using the stereotaxic atlas of the rat brain.

It is important to note that in our study we achieve the maximal activity with a dose of unmodified oligo of 0.08 nmol; following studied of other authors used antisense doses ranging from 0.08 to 5 nmol and in most of the cases these oligos were phosphothioate. As many authors pointed out, sequence-independent effects displayed by antisense agents should be carefully evaluated, expecially for the modified oligos (Landgraf et al., 1995; Chavany et al., 1995). In the case of in vivo administration of antisense oligonucleotides these non-specific effects can be ruled out by analysing sick behaviour parameters like body temperature, food intake, water consumption, circulating corticosterone concentrations or other inflammatory parameters (Schobitz et al., 1997).

Antisense oligodeoxynucleotides in the study of mammalian behaviour: problems and advantages over other techniques

Potential problems associated with AOs usage are summarized in the Table 2 (Wahlestedt, 1994). Oligos may evidence non specific or toxic action, especially with S-AOs, the rate of degradation of these agents is particularly fast and continuous or repeated administration are required, but the greatest drawback is the incomplete effect of AOs treatment every time a biological phenomenon is elicit by redundant mechanism, where, for example, only a fraction of receptors have to be occupied to obtain a full response, AOs treatment may become functionally silent. At the present time, to overcome the problem of this non complete knock down, different delivery systems are tested, improving the concentration of the antisense agent on its target and therefore the effectiveness of the treatment. As summarized in the Table 3, liposome carriers have been tested to increase the amount of oligonucleotide uptake in the cells and viral vectors like retrovirus or adenovirus are developed in order to obtain the synthesis of the AOs within the target cell. Among the viral vectors, the adeno-associated virus

TABLE 2

Comparison of antisense inhibition and knockout approaches in exp. animals

Advantages	Disadvantages
Antisense oligonucleotide	
Applicable to any stage of development	Incomplete treatment
A range of phenotypes can be created	Sequence-independent effect
Product of cloned gene from any species can be studied	Narrow experimental and therapeutic windows
The effect is reversible	Continuous or repeated administration necessary
Low cost to the laboratory	
Little specialised equipment is required	
Allows for "trial and errors"	
May have therapeutic potential	
Homologous recombination knockout	
Complete disappearance of gene product	Laborious
The effect is completely specific No variability between animals	Limited access to some manipulated animals
	Compensatory mechanism might be operative
	Only used in selected laboratories because of high costs
	Possibility of lethal phenotype

(From C. Wahlestedt, *TiPS*, February 1994, Vol. 15.

TABLE 3

Vector for oligonucleotides and genes transfer

Vector	Advantages	Disadvantages
Naked DNA	Administered directly into tissue by injection	Difficult to control site of uptake.
		Some tissues preferential, e.g. liver kidney, blood vessels
		DNA copy number unquantifiable
Liposome-encapsulated oligonucleotides	Easy to produce	DNA copy number unquantifiable
	Injectable	Preferential uptake in certain tissues
	May facilitate uptake	Toxic to some tissues, e.g. brain
Retrovirus	Efficient transfer	Only useful in dividing cells
	Single copy delivery	May damage nontarget dividing cells
	Efficient uptake	
Adenovirus	Produced without difficulty in high titers	Invokes inflammatory response to high titres
	infect broad range of cells	Copy number not controlled
		Does not integrate into the genome (episomal)
AAV	Integrates into the genome	Difficult to produce high titers
	Non pathogenic	Small loading capacity
	Effective	

(From M. Ian Phillips, *Hypertension*, 1997, 29.)

(AAV) (Phillips, 1997) is gaining interest for antisense treatment because, unlike other viruses, cannot be easily rescued and mobilized, thus is safe. In addition it is not pathogenous in man, does not evoke an inflammatory response and it integrates in a specific locus on chromosome nineteen, not disrupting endogenous genes.

In the future, viral vectors will undoubtedly prove to be useful to biologists seeking to disrupt gene function in vivo.

Antisense targeting is a potent pharmacological method to manipulate gene expression, which may represent a new direction in the analysis of neurobehavioural phenomena, manipulating neurotransmitters, receptors and second messenger systems and any specific protein in the activity of neural cells.

List of abbreviations

Aos	Antisense oligonucleotides
S-AOs	Phosphothioate antisense oligonucleotides
O-AOs	phosphodiester antisense oligonucleotides
EC-AOs	end capped antisense oligonucleotides
CNS	Central Nervous System
CSF	Cerebro Spinal Fluid
PR	Progesterone Receptor
VMH	Ventro medial hypothalamus
FBS	Foetal Bovine Serum
ER	Estrogen Receptor
AAV	Adeno-Associated Virus

References

Agrawal, S., Temsamani, J. and Taang, J.Y. (1991) Pharmacokinetics, biodistribution, and stability of oligodeoxynucleotide phosphorothioates in mice. *Proc. Natl. Acad. Sci. USA*, 88: 7595–7599.

Akabayashi, A., Wahlestedt, C., Alexander, J.T. and Leibowitz, S.F. (1994) Specific inhibition of endogenous neuropeptide Y synthesis in arcuate nucleus by antisense oligonucleotides suppresses feeding behavior and insulin secretion. *Mol. Brain Res.*, 21: 55–61.

Akhtar, S. Kole, R. and Juliano, RL. (1991) Stability of antisense DNA oligodeoxynucleotide analogs in cellular extracts and sera. *Life Sciences*, 49(24): 1793–801.

Alexander, I.E., Shine, J. and Sutherland, R.L. (1990) Progestin regulation of estrogen receptor messenger RNA in human breast cancer cells. *Mol. Endocrinol.*, 4: 828–828.

Campbell, J.M., Bacon, T.A. and Wickstrom, E.J. (1990) Oligodeoxynucleoside phosphorothioate stability in subcellular extracts, culture media, sera and cerebrospinal fluid. *Biochem. Biophys. Meth.*, 20: 259–267.

Chavany, C., Connell, Y. and Neckers, L. (1995) Contribution of sequence and phosphorothioate content to inhibition of cell growth and adhesion caused by c-myc antisense oligomers. *Mol. Pharmacol.*, 48: 738–746.

Colige, A., Sokolov, B.P., Nugent, P., Baserge, R. and Prockop, D.J. (1993) Use of an antisense oligonucleotide to inhibit expression of a mutated human procollagen gene (COL1A1) in transfected mouse 3T3 cells. *Biochemistry*, 32(1): 7–11.

Cook, P.D. (1993) Medicinal chemistry strategies for antisense research. In: *Antisense Research and Applications*, CRC Press, Ann Arbor, pp. 149–189.

Cowsert, L.M., Fox, M.C., Zon, G. and Mirabelli, C.K. (1993) In vitro evaluation of phosphorothioate oligonucleotides targeted to the E2 mRNA of papillomavirus: potential treatment for genital warts. *Antimicrob. Agents Chemother.*, 37(2): 171–177.

Creese, I., and Sibley, D.R. (1981) Receptor adaptations to centrally acting drugs. *Annu. Rev. Pharmacol. Toxicol.*, 21: 155–161.

Crooke, R.M. (1991) In vitro toxicology and pharmacokinetics of antisense oligonucleotides. *Anti-Cancer Drug Design*, 6: 609–646.

Dragunow, M., Tse, C., Glass, M. and Lawlor, P. (1994) c-fos antisense reduces expression of Krox 24 in rat caudate and neocortex. *Cell Mol. Neurobiol.*, 14: 395–405.

Ehrlich, G., Patinkin, D., Ginzberg, D., Zakut, H., Eckstein, F. and Soreq, H. (1994) Use of partially phosphorothioated "antisense" oligodeoxynucleotides for sequence-dependent modulation of hematopoiesis in culture. *Antisense Res. Dev.*, 4: 173–183.

Gao, W.Y., Han, F.S., Strorm, C., Egan, W. and Cheng, Y.C. (1992) Phosphorothioate oligonucleotides are inhibitors of human DNA polymerases and RNase H: implications for antisense technology. *Mol. Pharmacol.*, 41: 223–229.

Ghosh, M.K., Ghosh, K. and Cohen, J.S. (1993) Phosphorothioate-phosphodiester oligonucleotide co-polymers: assessment for antisense application. *Anti Cancer Drug Design*, 8: 15–39.

Giannakis, C., Forbes, I.J. and Zalewski, P.D. (1991) Ca2+/Mg(2+)-dependent nuclease: tissue distribution, relationship to inter-nucleosomal DNA fragmentation and inhibition by Zn2+. *Biochem. Biophys. Res. Commun.*, 181: 915–920.

Gillardon, F., Beck, H., Uhlmann, E., Herdegen, T., Sandkuhler, J Peyman, A. and Zimmerman, M. (1994) Inhibition of c-Fos protein expression in rat spinal cord by antisense oligodeoxynucleotide superfusion. *Eur. J. Neurosci.*, 6: 880–884.

Golic, K.G. (1994) Local transposition of P elements in *Drosophila melanogaster* and recombination between duplicated elements using a site-specific recombinase. *Genetics*, 137: 551–563.

Helene, C. (1991) The anti-gene strategy: control of gene expression by triplex-forming-oligonucleotides. *Anticancer Drug Res.*, 6(6): 569–584.

Hoke, G.D., Draper, K., Freier, S.M., Gonzales, C., Driver, V.B., Zounes, M.C. and Ecker, D.J. (1991) Effects of phosphorothioate capping on antisense oligonucleotide stability, hybridization and antiviral efficacy versus herpes simplex virus infection. *Nucl. Acids Res.*, 19: 5743–5748.

Hou, W.Y., Shyu, B.C., Chen, T.M., Lee, J.W., Shieh, J.Y. and Sun, W.Z. (1997) Intrathecally administered c-fos antisense oligodeoxynucleotide decreases formalin-induced nociceptive behavior in adult rats. *Euro. J. of Pharmacol.*, 329: 17–26.

Iversen, P.L., Zhu, S., Meyer, A. and Zon, G. (1992) Cellular uptake and subcellular distribution of phosphorothioate oligonucleotides into cultured cells. *Antisense Res. Dev.*, 2(3): 211–222.

Iwanaga, T. and Ferriola, P.C. (1993) Cellular uptake of phosphorothioate oligodeoxynucleotides is negatively affected by cell density in a transformed rat tracheal epithelial cell line: implication for antisense approaches. *Biochem. Biophys. Res. Commun.*, 191: 1152–1157.

Johnson, S.A., Morgan, D.G. and Finch, C.E. (1986) Extensive postmortem stability of RNA from rat and human brain. *J. Neurosci. Res.*, 16: 267–280.

Jones, K.R., Farinas, I.,Backus, C. and Reichardt, L.F. (1994) Targeted disruption of the BDNF gene perturbs brain and sensory neuron development but not motor neuron development. *Cell*, 76: 989–999.

Kilby, N.J., Snaith, M.R. and Murray, J.A. (1993) Site-specific recombinases: tools for genome engineering. *Trends Genet.*, 9: 413–421.

Klein, R., Smeyne, R.J., Wurst, W., Long, L.K., Auerbach, B.A., Joyner, A.L. and Barbacid, M. (1993) Targeted disruption of the trkB neurotrophin receptor gene results in nervous system lesions and neonatal death. *Cell*, 75: 113–122.

Koller, B.H. and Smithies, O. (1992) Altering genes in animals by gene targeting. *Annu. Rev. Immunol.*, 10: 705–30.

Krieg, A.M. Tonkinson, J. Matson, S. Zhao, Q. Saxon, M. Zhang, L.M. Bhanja, U. Yakubov, L. and Stein, C.A. (1993) Modification of antisense phosphodiester oligodeoxynucleotides by a 5′ cholesteryl moiety increases cellular association and improves efficacy. *Proc. Natl. Acad. Sci. USA*, 90: 1048–1052.

Krieg, A.M., Gmelig-Meyling, F., Gourley, M.F., Kish, W.J., Chrisley, L.A. and Steinberg, A.D. (1991) Uptake of oligodeoxyribonucleotides by lymphoid cells is heterogeneous and inducible. *Antisense Res. Dev.*, 1: 161–171.

Lacks, S.A. (1981) Deoxyribonuclease I in mammalian tissues. Specificity of inhibition by actin. *J. Biol. Chem.*, 256: 2644–2648.

Landgraf, R., Gerstberger, R., Montkowski, A., Probst, J.C., Wotjak, C.T., Holboer, F. and Engelmann, M. (1995). V1 vasopressin receptor antisense oligodeoxynucleotide into septum reduces vasopressin binding, social discrimination abilities, and anxiety-related behavior in rats. *J. Neurosci.*, 15: 4250–4258.

Milner, N., Mir, K.U. and Southern, M. (1997) Selecting effective antisense reagents on combinatorial oligonucleotide arrays. *Nat. Biotech.*, 15: 537–541.

Moguilewsky, M. and Raiynaud, J.P. (1979) The relevance of hypothalamic and hyphophyseal progestin receptor regulation in the induction and inhibition of sexual behavior in the female rat. *Endocrinology*, 105: 935–944.

Monia, B.P., Lesnik, E.A., Gonzalez, C., Lima, W.F., McGee, D., Guinosso, C.J., Kawasaki, A.M., Cook, P.D. and Freier, S.M. (1993) Evaluation of 2′-modified oligonucleotides containing 2′-deoxy gaps as antisense inhibitors of gene expression. *J. Biol. Chem.*, 268: 14514–14522.

Neckers, L.M. (1993) Medicinal chemistry strategies for antisense research. In: Crooke and Lebleu (Eds.), *Antisense Research and Applications*, CRC Press, Ann Arbor, pp. 452–459.

Nielsen, P.E., Egholm, M., Berg, R.H. and Buchardt, O. (1991) Sequence-selective recognition of DNA by strand displacement with a thymine-substituted polyamide. *Science*, 254: 1497–1500.

Ogawa, S., Olazabal, U.E., Parhar, I.S. and Pfaff, D.W. (1994) Effects of intrahypothalamic administration of antisense DNA for progesterone receptor mRNA on reproductive behavior and progesterone receptor immunoreactivity in female rat. *J. Neurosci.*, 14: 1766–1774.

Pellegrino, L.J., Pellegrino, A.S. and Cushman, A.J. (1978) *A Sterotaxic Atlas of the Rat Brain*, Plenum, New York.

Phillips, M.I. (1997) Antisense inhibition and adeno-associated viral vector delivery for reducing hypertension. *Hypertension*, 29: 177–187.

Phillips, M.I., Ambuhl, P. and Gyurko, R. (1996) Antisense oligonucleotides for in vivo studies of angiotensin receptors. In: M.K. Raizada et al. (Eds.), *Recent Advances in Cellular and Molecular Aspects of Angiotensin Receptors*, Plenum press, New York, pp. 79–92.

Pollio, G., Xue, P., Zanisi, M., Nicolin, A. and Maggi, A. (1993) Antisense oligonucleotide blocks progesterone-induced lordosis behavior in ovariectomized rats. *Molecular Brain Res.*, 19: 135–139.

Retaux, S., McNeill, L. and Harris, W.A. (1996) Engrailed, retinotectal targeting, and axonal patterning in the midbrain during Xenopus development: an antisense study. *Neuron*, 16: 63–75.

Rickman, D.W. and Bowes Rickman, C. (1996) Suppression of trkB expression by antisense oligonucleotides alters a neuronal phenotype in the rod pathway of the developing rat retina. *Proc. Natl. Acad. Sci. USA*, 93: 12564–12569.

Robertson, G.S., Tetzlaff, W., Bedard, A., St-Jean, M. and Wigle, N. (1995) C-fos mediates antipsychotic-induced neurotensin gene expression in the rodent striatum. *Neuroscience*, 67: 325–344.

Schobitz, B., Pezeshki, G., Probst, J.C., Reul, J., Skutella, T., Stohr, T., Holsboer, F. and Spanagel, R. (1997) Centrally administered oligodeoxynucleotides in rats: occurrence of non-specific effects. *Eur. J. Pharmacol.*, 331: 97–107.

Skilling, S.R., Sun, X., Kurtz, H.J. and Larson, A.A. (1992) Selective potentiation of NMDA-induced activity and release of excitatory amino acids by dynorphin: possible roles in paralysis and neurotoxicity. *Brain. Res.*, 575: 272–278.

Sommer, W., Bjelke, B., Ganten, D. and Fuxe, K. (1993) Antisense oligonucleotide to c-fos induces ipsilateral rotational behaviour to d-amphetamine. *Neuroreport*, 5: 277–280.

Standifer, K.M., Chien, C.C., Wahlestedt, C., Brown, G.P. and Pasternak, G.W. (1994) Selective loss of delta opioid analgesia and binding by antisense oligodeoxynucleotides to a delta opioid receptor. *Neuron*, 12: 805–812.

Stein, C.A. Subasinghe, C., Shinozuka, K. and Choen, J.S. (1988) Physicochemical properties of phosphorothioate oligodeoxynucleotides. *Nucl. Acids Res.*, 16: 3209–3219.

Stein, C.A., Tonkinson, J.L., Zhang, L.M. Yakubow, L., Gervasoni, J., Taub, R. and Rotenberg, S.A. (1993) Dynamics of the internalization of phosphodiester oligodeoxynucleotides in HL60 cells. *Biochemistry*, 32: 4855–4861.

Szklarczyk, A. and Kaczmarek, L. (1997) Pharmacokinetics of antisense analogues in the central nervous system. *Neurochem. Int.*, 31: 413–423.

Szklarczyk, A. and Kaczmarek, L. (1995) Antisense oligodeoxyribonucleotides: stability and distribution after intracerebral injection into rat brain. *J. Neurosci. Meth.*, 60: 181–187.

Tischmeyer, W., Grimm, R., Schicknick, H., Brysh, W. and Schlingensiepen, K.H. (1994) Sequence-specific impairment of learning by c-jun antisense oligonucleotides. *Neuroreport*, 5: 1501–1504.

Uhlenbeck, O.C., Dahm, S.C., Ruffner, D.E. and Fedor, M.J. (1989) Structure and mechanism of the hammerhead self-cleaving domain. *Nucleic Acids Symposium Series*, (21): 95–6.

Wahlestedt, C. (1994) Antisense oligonucleotide strategies in neuropharmacology. *Trends Pharmacol. Sci.*, 15: 42–46.

Wahlestedt. C., Pich. E.M., Koob. G.F., Yee. F. and Heilig. M. (1993a) Modulation of anxiety and neuropeptide Y-Y1 receptors by antisense oligodeoxynucleotides. *Science*, 259: 528–531.

Walhestedt, C., Golanov, E., Yamamoto, S., Yee, F., Ericson, H., Yoo, H., Inturrisi, C.E. and Reis, D.J. (1993b) Antisense oligodeoxynucleotides to NMDA-R1 receptor channel protect cortical neurons from excitotoxicity and reduce focal ischaemic infarctions. *Nature*, 363: 260–263.

Weiss, B., Zhang, S.P. and Zhou, L.W. (1997) Antisense strategies in dopamine receptor pharmacology. *Life Sciences*, 60: 433–455.

Wakita, T. and Wands, J.R. (1994) Specific inhibition of hepatitis C virus expression by antisense oligodeoxynucleotides. In vitro model for selection of target sequence. *J. Biol. Chem.*, 269(19): 14205–14210.

Whitesell, L., Geselowitz, D., Chavany, C., Fahmy, B., Walbridge, S., Alger, J.R. and Neckers, L.M. (1993) Stability, clearance, and disposition of intraventricularly administered oligodeoxynucleotides: implications for therapeutic application within the central nervous system. *Proc. Natl. Acad. Sci. USA*, 90: 4665–4669.

Yee, F., Ericson, H., Reis, D.J. and Wahlestedt, C. (1994) Cellular uptake of intracerebroventricularly administered biotin- or digoxigenin-labeled antisense oligodeoxynucleotides in the rat. *Cell. Mol. Neurobiol.*, 14: 475–487.

Zang, Z., Florijn, W. and Creese, I. (1994) Reduction in muscarinic receptors by antisense oligodeoxynucleotide. *Biochem. Pharmacol.*, 48: 225–228.

W.E. Crusio and R.T. Gerlai (Eds.)
Handbook of Molecular-Genetic Techniques for Brain and Behavior Research (Techniques in the Behavioral and Neural Sciences, Vol. 13)

CHAPTER 3.2.5

Application of recombinant proteins, peptides and antibodies in exploring the role of Src in regulating synaptic function

Michael W. Salter

Programmes in Brain and Behaviour and Cell Biology, Hospital for Sick Children, Department of Physiology, University of Toronto, Toronto, Ont., Canada M5G 1X8

Introduction

The individual neuron is the conduit through which the functions of molecules in the nervous system are expressed and translated into neural systems activity and ultimately into behaviour. Therefore, in order to gain a comprehensive understanding of how neuronal gene products function in the nervous system it is essential to determine their actions within individual neurons and in certain cases in subcellular compartments or specialized subcellular regions, such as synapses. Synapses are, of course, the fundamental points of information transfer between neurons where electrical signals, usually in the form of action potential discharge, in one neuron are transformed into an electrical signal in another neuron by means of diffusion of one or more chemical intermediaries. It is now appreciated that information transfer at synapses occurs not only through the electrical transmission of signals but also through the activation of intracellular biochemical pathways by these same chemical intermediaries, and in some cases, other molecules which are co-released. It is also beginning to be understood that rather than simply providing unidirectional transmission of information, signalling at synapses is bidirectional (Jessell and Kandel, 1993). The focus of this chapter is on using recombinant proteins, peptides and antibodies in individual neurons to probe the actions of known intracellularly localized gene products in regulating synaptic function. I will describe the general considerations which are important for the use of these types of reagent and then outline a specific example — elucidating the role of the protein tyrosine kinase, Src, in regulating synaptic transmission at glutamatergic synapses.

General considerations

Patch-clamp methods for intracellular delivery of macromolecules

The use of recombinant proteins, peptides and antibodies in cellular electrophysiological investigations has come about in large part because of the development over the past 20 years of patch-clamp techniques (Sakmann and Neher, 1984). These techniques have not only allowed for high fidelity recording of the activity of the ion channels that produce the electrical synaptic signals but also have established a means for delivering macromolecules, such as proteins, acutely into mammalian neurons. Thus, with appropriate reagents it is possible to explore the roles of particular gene products in electrical signalling at synapses. For technical reasons due to the small size of most presynaptic terminals in mammalian systems it is usually possible to administer macromolecules only into postsynaptic neurons. However, at several specialized mammalian synapses, for example the Calyx of Held, it is possible to use patch-clamp methods to record presynaptically (Takahashi et al., 1996) and thereby to deliver large molecules directly into the presynaptic terminal. The general considerations for the use of recombinant proteins, peptides and antibodies are the same whether one aims to investigate either the pre- or postsynaptic roles of a particular gene product.

Given that much of our current perspective on synaptic transmission relates to electrical signalling it is not surprising that the readout of synaptic function most often examined is the activity of one or more types of ion channel. Some types of channel — for example channels gated by fast neurotransmitters — are easier to study than other types but nevertheless it is generally possible by the clever use of pharmacological reagents and/or sophisticated voltage protocols, to study postsynaptic ion channel function in real time. The gating of an ion channel, which is mediated by changes in the conformation of the channel proteins, is one type of biochemical event that can be studied in real time. It is also possible to study other biochemical events important in synaptic functioning such as changes in the intracellular concentration of ions like Ca^{2+} (Murphy et al., 1995) or molecules like cyclic AMP (Adams et al., 1991). By contrast, the investigation of many other types of biochemical events — such as direct measurement of enzyme activation or of changes in protein-protein interactions — are thus far refractory to study in individual cells because the technologies have yet to be developed. However, given the progress that has been made with optical probes it seems likely that new methodologies for examining the functions and interactions of particular proteins in individual living neurons will be available soon. Thus, we will have the opportunity to develop a new level in our understanding of synaptic interactions beyond studying primarily electrical signalling.

The technical considerations for the use of patch-clamp techniques for studying ion channel function and regulation have been discussed extensively in many previous papers (for example Sakmann and Neher, 1984; Neher, 1988; Neher and Sakmann, 1992) and therefore, these techniques are only briefly considered here. Patch-clamp recordings can be made using one of a number of "configurations", each of which allow the activity of ion channels to be studied under various experimental conditions. The main configurations used in the work described below are illustrated diagrammatically in Fig. 1. In order that the ensemble activity of many ion channels of a particular type can be studied we typically use the whole-cell or perforated-patch configurations. These configuration can be used to investigate synaptically-evoked currents or currents stimulated by applying exogenous agonists, in the case of ligand-gated channels, or currents evoked by changing membrane potential, in the case of voltage-gated channels. The whole-cell configuration has the advantage of allowing the experimenter to introduce large molecules into the cell via the small hole created in the membrane enclosed by the pipette, and therefore peptides, proteins and antibodies can be introduced into the cell. Not all such reagents have equal access to the cell's interior and molecular weight is a major determinant of access — larger molecules diffuse into the cell more slowly and, in general, access rate varies in general with the cube root of the molecular weight (Neher, 1988). Molecular weight is not the only consideration and depending upon the particular reagent diffusion may be retarded by, for example, binding to intracellular sites which limit access to the channel of interest. One way to determine whether the reagent is accessing the cell, or region within the cell, is to couple the reagent to a fluorophore

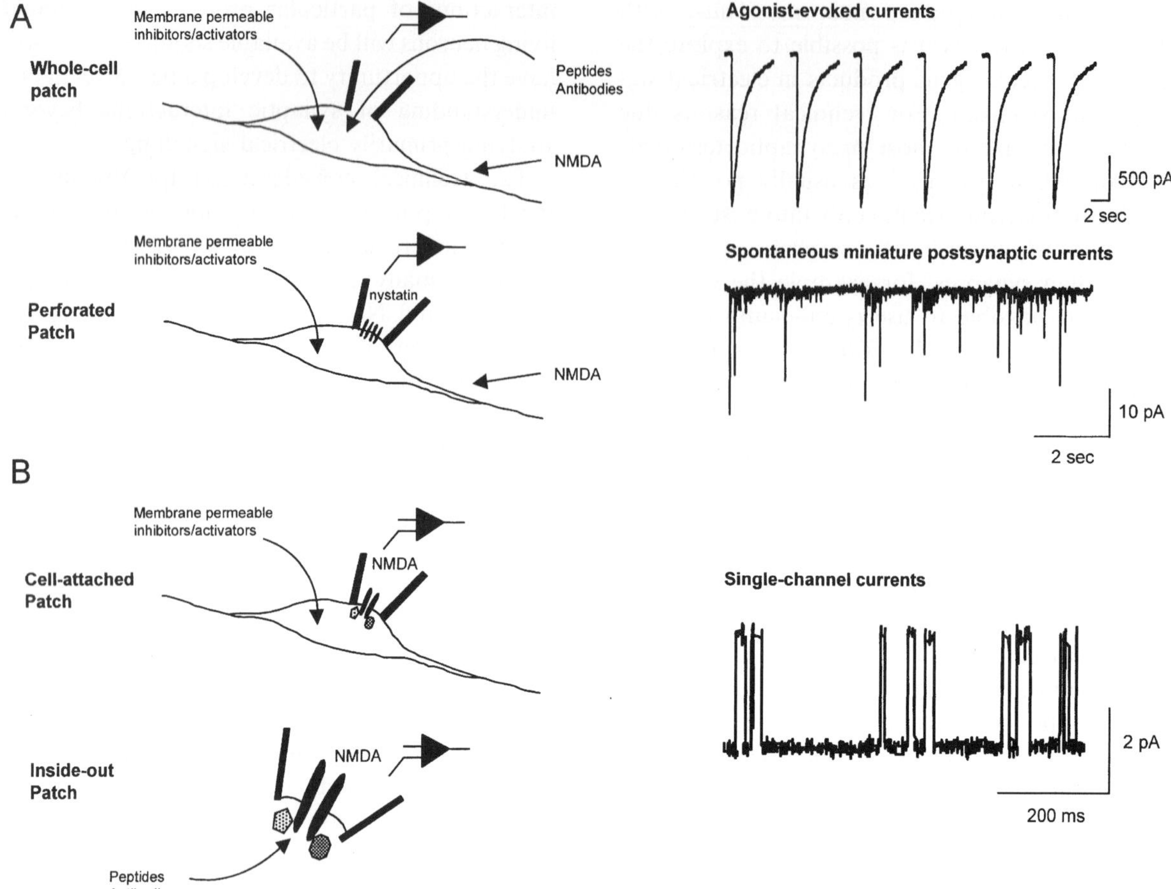

Fig. 1. Patch-clamp recording configurations and examples of ligand-gated ion channel activity that can be studied. (A) With either whole-cell or perforated patches one can record current responses evoked by exogenously applied agonists, in our case NMDA, (upper) or by synaptically released endogenous neurotransmitter. (B) Single channel activity can be recorded in cell-attached (upper), inside-out (lower) or outside-out (not illustrated) configurations. The small symbols near the inner side of the NMDA channel complex represent proteins associated with the channel. See also text.

and then to optically measure the accumulation and distribution of fluorescence. When using the whole-cell configuration one must also appreciate that important molecules, such as nucleotides and mobile Ca^{2+} buffers, may diffuse into the patch electrode and become depleted in the cell. Thus, one must supplement these essential components in order to avoid "rundown" of the currents. The necessity to supplement such components in intracellular solution may be a nuisance because reagents such as nucleotides are labile and must be added freshly on the day of recording. On the other hand, the rundown of the currents when such reagents is often the first important clue that function of a particular channel is regulated by intracellular processes like phosphorylation (eg. MacDonald et al., 1989).

With the perforated-patch method electrical access to the cell is gained by means of incorporating small pore-forming molecules, such as nystatin or amphotericin, into the membrane with an electrode attached to the cell. These pore-forming molecules are permeable to monovalent ions, which act as charge carriers for the currents measured, and are impermeable to divalent ions and to molecules with molecular weight larger than approximately 300 daltons. Thus, one can record the current responses of the cell with minimal disruption of the intracellular milieu and of second messenger systems. The disadvantage is that peptides and antibodies cannot be introduced while recording using this configuration. However, membrane-permeant reagents can be applied from the bathing solution. Thus, a particular signalling pathway can be studied if there are sufficiently selective membrane-permeant reagents and appropriate control compounds.

In order to study the activity of individual ion channels one of the single-channel patch configurations — cell-attached, inside-out or outside-out — can be used. The study of single-channel behaviour is important to characterize the changes that occur in the function of individual channels and thereby to determine whether, for example, the changes in the ensemble currents recorded with whole-cell or perforated-patch configurations are due to changing the single channel conductance or the gating properties of the channels. Studies of single-channels also allow analysis of the channel gating kinetics which can be used to attempt to determine the number of states of the channel and the rates of transitions to the various states. It is however important to understand that often patches contain more than one channel of the type of interest and there are analytical methods for determining whether the patch does indeed contain one or more than one channel.

The cell-attached configuration is similar to the perforated-patch in that the electrode remains attached to the cell but in this case a pore-forming compound is not included. Rather, in order to activate a ligand-gated channel one includes a selective agonist in the solution in the electrode. Voltage-gated channels can also be studied in cell-attached recordings and here one may need no special additives but in some cases it is necessary to include reagents to block the activity of channels with a voltage-dependence similar to that of the one of interest. The considerations for studying second-messenger systems and the regulation of channel function are the same as those described above for perforated patch recording. With cell-attached recordings one assumes that the intracellular environment is not altered except when explicitly done so by the investigator and diffusible factors within the cytoplasm are able to affect the function of the channels being studied. In contrast, such diffusible factors are lost in either of the configurations where the patches are excised from the cell, i.e. inside-out or outside-out patches. Thus, comparison of the modulation of channel activity in cell-attached vs excised patches can provide valuable insights into the regulation of channel function (e.g. Yu and Salter, 1998).

The inside-out configuration allows the investigator to apply reagents such as peptides or antibodies directly to the cytoplasmic face of the membrane containing the channel. Thus, one can determine whether the patch contains all of the components necessary to sustain a particular effect observed with the whole-cell or perforated-patch method. This may thereby narrow down the site of action to being close to, or the channel itself. However, because membrane patches contain components other than just the ion channel of interest, one must be cautious not to overinterpret effects observed with inside-out patches as indicating necessarily a direct action on the channel proteins themselves.

A principal advantage of using patch clamp methods to apply peptides or antibodies is that the channel activity can be studied before and during administration of a particular reagent. Thus, each cell or patch acts as its own control and this is important because there may be considerable cell-to-cell or patch-to-patch variability in the activity of the channels under study. Also, and especially with the inside-out patches, one may be able to remove the peptide or antibody by washing with solution free of the reagent. Thus, one can determine whether the effect is reversible. These two advantages are not found with other methods of delivering peptides and antibodies such as microinjection.

Probing function(s) of a gene product in wild-type neurons

Once a candidate gene product has been identified, administering appropriately selected recombinant proteins, peptides or antibodies directly into neurons or applying these reagents to membrane patches allows investigation of the role of that gene product in cells from genetically normal animals. That the function of a gene product can be studied in neurons from wild-type animals is an important point for physiological studies and contrasts with studies in which gene function is inferred from, for example, examining cells from animals with targeted deletion of the gene. There are potential issues of interpretation of results with using such gene deletions and these problems have been discussed in previous chapters. Some of these issues, particularly those related to developmental changes, can be overcome by inducible and/or cell-type specific gene targeting (Chapters 3.1.3 and 3.1.4 in this book). However, even these technologies require many hours or days to be effective, for example inducible knockouts require the turnover of existing proteins and the induction of inhibitory proteins may also be slow due to the time needed for protein synthesis, processing and transport. Thus, there is still an issue of interpretation especially of negative results because it is possible that the time required may be sufficient for the development of compensatory mechanisms.

With the acute administration of peptides or proteins effects are typically observed in seconds or minutes. As such, the possibility of compensatory changes in the levels of protein expression is typically not a concern especially when working with reduced preparations such as membrane patches. This is not to say that using peptides or proteins as probes is without potential problems of interpretation. As with all pharmacological reagents a crucial issue is that of proper controls, both positive and negative. For example, when administering a recombinant enzyme one needs to ensure that it is active under the conditions used and in the appropriate location in the cell. In terms of negative controls, it needs to be determined experimentally that the effect observed requires the activity of the enzyme used and that the effect is not due to a contaminant or to the vehicle, which can sometimes be a very non-physiological solution. In addition to concerns about liquid vehicles, there may be problems with salts, acids or bases retained in lyophilized proteins. We routinely add reagents to the intracellular solution immediately before use and then take a small aliquot to measure pH and osmolarity. If either of these changes we try to use the reagent at a lower concentration. Occasionally this is not possible and it is necessary to concentrate or to dialyze the reagent before use.

Questions that can be addressed

If care is taken to use proper controls then recombinant proteins, peptides and antibodies can be very valuable for addressing two key issues about the function of a given gene product: (i) is it sufficient for a particular process? and (ii) is it necessary? The question of necessity is often the more interesting and can be provocative. There can be controversies about the "necessity" of a particular gene or gene product which can arise because this issue can be seen from different perspectives. To get at the issue of necessity in the genetic sense one might eliminate the gene of interest or interfere with its expression whereas to examine physiological necessity one might block activity of the gene product. The consequence is that two different but related questions are being asked — in the genetic sense one is asking whether in the cell of interest a particular *gene* is necessary for a particular function whereas in the physiological sense one is asking given a particular repertoire of genes expressed, most often wild-type, is a particular *gene product* necessary for a particular function. These are not the same questions — a gene product may be necessary when it is expressed but when it is not expressed its place may be taken over by the product of another gene. Conversely, if the process examined still occurs despite blocking the gene product of interest, even in a carefully controlled experiment, this does not necessarily mean that the process does not normally proceed through that gene product. Thus, the problem of functional redundancy may occur in experiments on protein signaling networks as well. As such, acute physiological experiments with blocking reagents, even if well-controlled, are not exempt from giving false negative information. This information can however be useful in that it indicates where blockade of the function of a particular gene product will not be effective in preventing a particular process.

Types of protein, peptide and antibody tools that can be used

A great advantage of using patch-clamp methodologies is that one can administer membrane-impermeant reagents directly into the cell of interest. Thus, if the gene product to be studied can be made as a recombinant protein the question of sufficiency can be addressed by administering the protein itself. For some gene products there may be peptide activators which can be used to investigate the sufficiency of stimulating the endogenous protein. Such activators are especially useful for determining whether an endogenously expressed protein, as opposed to an exogenously applied one, is sufficient for causing the process under study. Antibodies typically either do not affect or block the function of intracellular signalling proteins. In contrast, it is not unusual for antibodies to activate cell surface receptors. It is possible that certain antibodies, or Fab fragments, might be designed so as to act as activators for particular proteins and could be used to address the issue of sufficiency.

The issue of necessity may be addressed in patch-clamp experiments by administering reagents in an attempt to block the function of the native protein. One type of protein that can be used for such experiments is a recombinant mutant which functions to suppress the activity of the native protein of interest, a so-called dominant negative protein. For some proteins there are peptide inhibitors that may be used and, as mentioned above, some antibodies can be function blocking. However, more often than not antibodies have no effect on protein function. This can be attributed to the fact that specificity is obtained by raising antibodies to the most non-conserved region of the protein. Binding of the antibody to such non-conserved regions often has no effect on the function of the protein but this does not mean that such antibodies, which can be highly specific, may not be useful in studying protein function in situ. One technique which has been developed and extensively used by the lab of Dan Jay for converting non-function blocking antibodies into function blocking reagents, involves coupling the chromophore Malachite Green to the antibody (Beermann and Jay, 1994). The labeled antibody is administered into the cell and the fluorophore is excited by brief pulses from a laser. The excited Malachite Green generates short-lived, highly-reactive hydroxyl radicals which destroy the protein to which the antibody is bound producing a rapidly inducible knockout. Moreover, the protein ablation can be restricted to targeted regions within the cell by using a narrow beam of laser light. Thus, for a protein with no known blockers but for which there is a very specific antibody, this technique offers the opportunity to investigate the functions of that protein in the cell.

Investigating the role of Src in synaptic transmission

Regulation of NMDA receptors by tyrosine phosphorylation

The work described below came about as a result of investigating the regulation of the NMDA subtype of glutamate receptor. NMDA receptors participate in rapid excitatory synaptic transmission throughout the CNS (McBain and Mayer, 1994) and have been implicated in a wide variety of physiological and pathophysiological processes ranging from development to learning and memory to epilepsy, chronic pain and neurodegeneration.

NMDA receptors are members of the superfamily of ligand-gated ion channels and a number of NMDA receptor subunit proteins (NR1, NR2A-D, NR3 [previously known as NR-L/χ1]) have been identified by molecular cloning (Gasic and Hollmann, 1992; Seeburg, 1993; Sucher et al., 1995; Ciabarra et al., 1995). The activated channels are permeable to monovalent cations, such as Na^+ and K^+, and also divalent cations the most important of which is Ca^{2+} (Mayer and Westbrook, 1987; Ascher and Nowak, 1988). NMDA receptors are hetero-ligomeric complexes (Sheng et al., 1994) with the most recent evidence indicating that the receptors are tetramers (Laube et al., 1998). The NR1 subunit is a common element that is necessary for forming functional channels whereas there may be various combinations of the other subunit proteins (Gasic and Hollmann, 1992).

Phosphorylation has been identified as a key intracellular process regulating NMDA receptor function (MacDonald et al., 1989). Subsequently the regulation by serine/threonine phosphorylation has been studied extensively (Chen and Huang, 1991, 1992; Durand et al., 1992; Kelso et al., 1992; Lieberman and Mody, 1994; Raman et al., 1996; Sigel et al., 1994; Tingley et al., 1993; Tong et al., 1995; Wang et al., 1994). Several years ago we began to explore whether the function of NMDA channels might be regulated by tyrosine phosphorylation (Tong et al., 1995).

To address this issue we studied the ensemble activity of NMDA channels on individual mammalian neurons in primary culture. In order to minimize the possibility of disturbing existing regulation by tyrosine phosphorylation, our first step was to record this ensemble activity under conditions where there would be little disruption of the intracellular milieu using the so-called perforated-patch method. With this approach we needed to administer reagents, in this case inhibitors of protein tyrosine kinases (PTKs), that would readily cross the plasma membrane when given extracellularly. A number of such membrane-permeant inhibitors were available and we found that the two inhibitors we tested, genistein and lavendustin A, depressed NMDA receptor-mediated currents.

Thus, we found similar effects with two structurally distinct PTK inhibitors, even though their corresponding inactive analogues had no effect. We therefore considered it crucial to demonstrate unequivocally that a PTK enzyme was sufficient to regulate the channels. For this purpose we intended to use a purified PTK catalytic domain that could be administered into the neurons using the whole-cell configuration of the patch clamp technique. The rationale for using only the catalytic domain was that other domains in PTKs can influence substrate binding and specificity. Since we did not have any a priori reason to suspect one kinase or another we sought to use a reagent that would have the broadest specificity possible. We were unable to acquire such a catalytic domain, or to convince anyone to make one for us, but we did find that a preparation of a recombinant purified PTK holoenzyme was available. This PTK was Src. We reasoned that if this one enzyme was found to upregulate NMDA channel function we would at least have proof of the principle that PTKs regulate the channels. When we found that administering the enzyme into the cells potentiated NMDA currents we had the first part of the evidence required.

The remaining part of our evidence was obtained to be sure that the potentiation was indeed due to the activity of Src and was not produced artefactually. It was possible that the vehicle had produced the increase, but this was eliminated since we found that applying the vehicle alone had no effect on NMDA currents. However, this was not sufficient for our conclusion because even though the preparation of Src we had obtained appeared to contain only a single main protein by HPLC analysis, it was possible that the increase in NMDA channel activity was caused by contaminants present in the sample. To control for this we immunoprecipitated Src from the solution just prior to use, using Western blotting to verify that this had been successful. We found that this Src-depleted solution did not affect the NMDA currents. This left us to determine whether Src needed to be active and we examined this by heat-inactivating Src just before use. Importantly, we found that boiling for more than 10 min was required to eliminate the enzymatic activity of Src. Also, to be a fair test the enzyme was boiled before adding it to the intracellular recording solution which contained the heat-labile component ATP, which was already known to be necessary for maintaining NMDA channel function (MacDonald et al., 1989). With these considerations, we found that the heat-inactivated Src did not affect NMDA currents. Thus, taking these results together we concluded that NMDA channel activity is indeed upregulated by a PTK.

It is important to point out that in these experiments NMDA receptors were activated by locally applying a selective agonist, L-aspartate, from a pressure pipette near the recorded cell. Although we took care to minimize synaptic transmission this type of application had the potential to cause effects indirectly: the applied agonist might activate receptors not only on the cell studied but also on terminals of other neurons synapsing with that cell or might act on other cells, like astrocytes, which are intimately associated with the neurons. It was possible that these could have been indirectly affected by applying Src into the cell studied. Also, it was conceivable that administering Src had unmasked a new population of receptors or altered the pharmacological properties of other existing receptors such that they became sensitive to L-aspartate. Thus, it was necessary to determine whether the Src-enhanced responses were fully mediated by NMDA receptors in the cell recorded. A simple approach was to show that the responses were abolished by the NMDA receptor antagonist, aminophosphonovalerate, but this did not eliminate the possibility of indirect effects of Src on other cells. However, we also found that the Src-enhanced responses were blocked by Mg^{2+} and that the degree of blockade depended upon the membrane potential of the cell recorded. This voltage-dependence would not be expected to have occurred if the effect of Src had been on cells other than the one into which the enzyme was administered. Thus, we concluded that the effect of applying Src was due to enhanced activity of NMDA channels in the cell studied.

In other work we followed a similar strategy to establish unequivocally that protein tyrosine phosphatase (PTP) function downregulates NMDA channel activity (Wang et al., 1996). This was determined using a broad spectrum PTP, a truncated form of T-cell PTP in which the regulatory domain has been deleted leaving only the catalytic domain. We found that applying this protein in the neurons depressed NMDA channels but there was no effect of either the vehicle or the heat-inactivated enzyme.

The experiments described above had been done using exogenously administered enzymes and indicated that the NMDA channel could be modulated by tyrosine phosphorylation. However, an important point was determining whether there are endogenous PTKs and PTPs which do in fact regulate NMDA channels. For this purpose we also needed to determine whether the endogenous enzymes are indeed localized in the neuron in which the NMDA currents were being recorded. That there were effects of the pharmacological blockers mentioned above points to regulation by endogenous PTKs and PTPs. However, because these blockers are membrane permeant it was conceivable that they were acting in cells other than the one in which the NMDA currents were being measured and that this generated a signal, such as a diffusible mediator, that was affecting the currents. Therefore, we took advantage of the whole-cell patch configuration to administer membrane impermeant inhibitors of PTKs or PTPs. We found that intracellularly applying the broad spectrum PTP inhibitor, orthovanadate, potentiated NMDA channel function. Orthovanadate may affect other enzymes but the enhancement of NMDA currents was observed at concentrations where its effects on other enzymes are known to be minimal and the enhancement was prevented by pretreating the neurons with the PTK inhibitor used previously, genistein. To test for endogenous PTKs we intracellularly applied a peptide, referred to as "peptide A", known to inhibit a range of PTKs including non-receptor PTKs and EGF receptor kinase (Sato et al., 1990; Fukami et al., 1993). We found that peptide A decreased NMDA currents (Wang et al., 1996). By comparing the effects of orthovanadate and peptide A, we determined that NMDA channel function is modulated over at least a 4-fold range by endogenous PTKs and PTPs.

The next step in our investigation was to determine where in the cell the endogenous PTKs and PTPs were localized. This was an important issue because in the previously described studies it was possible that the enzymes may have been located in the cytoplasm at a long distance from the NMDA channels or that the enzymes may have been located very close to the channels. Knowing the location of the enzymes was crucial in helping to narrow down the search for candidate enzymes. To this end we made use of recordings from membrane patches excised from the neurons using the so-called inside-out patch clamp configuration. With this configuration there is access to the cytoplasmic face of the membrane containing the channel and therefore one can administer a variety of peptides, proteins and antibodies, as well as other membrane impermeant reagents directly. Indeed this recording configuration has become the main preparation by which we examine the regulation of NMDA channels.

In our initial investigations with this recording configuration we found that applying recombinant Src enhanced the activity of NMDA channels whereas channel activity was decreased by applying truncated T-cell PTP (Wang et al., 1996). Our interpretation of these findings was that the membrane patches must contain the relevant tyrosine phosphorylated substrate protein, or proteins. Moreover, we found that applying orthovanadate enhanced NMDA channel activity and that this effect required MgATP and was prevented by PTK inhibitors. Taking these results together we concluded that not only must the patch contain the substrate protein(s) but it must also contain both the PTK and the PTP that regulate NMDA channel function.

Identifying Src as a tyrosine kinase regulating NMDA receptors

As our next line of investigation we undertook to identify the PTK that regulates NMDA channels. This may appear to be a daunting task given that the mammalian genome is predicted to contain more than 1000 PTKs (Hunter, 1994) many of which are likely to be expressed in the CNS. However, PTKs fall into distinctive families each with particular structural and regulatory characteristics. Our approach therefore was to take advantage of these features in order to first screen families of PTKs, so as to then implicate individual member(s). As it turned out we were successful with the first family we screened — Src kinases. We then proceeded to identify the particular family member as Src and along the way we gained insights into a novel regulatory mechanism for this kinase (Yu et al., 1997), a mechanism we are still investigating.

In order to understand the strategy we employed it is necessary to appreciate what was known at the time about the structural and regulatory features of the Src family of kinases (Superti-Furga and Courtneidge, 1995; Brown and Cooper, 1996). A total of nine members of the Src family have been discovered to date, five of which — Src, Fyn, Lyn, Lck and Yes — are known to be expressed in the CNS. Src is the prototype of this family with the catalytic, or Src-homology 1 (SH1), domain near the C-terminus and adjacent to a small region referred to as the tail (Brown and Cooper, 1996). On the N-terminal side of the catalytic domain are the SH2 and SH3 domains, which have received much attention as similar domains are found in numerous proteins. Protein–protein interactions which occur via SH2 or SH3 domains are widely involved in cell signalling (Pawson, 1995; Pawson and Scott, 1997). In Src family kinases the SH2 and SH3 domains not only mediate protein–protein interactions but control catalytic activity by intramolecular binding which results in folded conformations of the protein in which the enzymatic activity is suppressed (Brown and Cooper, 1996; Moarefi et al., 1997; Xu et al., 1997). The SH2 mediated suppression comes about upon phosphorylation of Tyr527, in the C-terminal tail, which then binds to the SH2 domain. Also, it has been found that intramolecular binding of the SH3 domain contributes to holding the kinase in a low activity conformation (Xu et al., 1997). Adjacent to the SH3 domain is the so-called unique domain, a region of approximately 70 amino acids with little sequence conservation among Src family members. At the N-terminus of Src is a small domain sometimes referred to as the SH4 domain which is a site for myristoylation, that may tether the kinase to the membrane.

The main isoform of Src expressed by cells is c-src but it was discovered many years ago that neurons express a high-activity isoform of Src called n-Src (Brugge et al., 1985; Martinez et al., 1987). N-Src was found to contain an alternatively spliced cassette of 6 amino acids inserted after amino acid 114, i.e. in the SH3 domain, of c-Src (Martinez et al., 1987) which is adjacent to the site for intramolecular SH3 mediated binding. More recently, it has been found that another distinct cassette may be inserted alone or in tandem generating a number of neuronal isoforms of src (Pyper and Bolen, 1990).

All members of the Src family of non-receptor PTKs have the general domain structure described above (Brown and Cooper, 1996). The catalytic, SH2, SH3, SH4 and C-terminal domains are highly homologous in the different Src family kinases. Where the various members differ most is in the unique domain, a region of low sequence conservation near the N-terminus. It has been suggested that the unique domain may simply be a spacer in the molecule (Superti-Furga and Courtneidge, 1995). However, the unique domain in some members of the Src family has been found to be crucial for directing the kinase and permitting the phosphorylation of particular substrates (Carrera et al., 1995; Vonakis et al., 1997). Thus, even though it is outside the catalytic region, the unique domain may regulate the function of Src kinases.

Src is expressed widely in the CNS with high levels of expression in cerebral cortex, hippocampus, pons, midbrain, cerebellum and spinal cord (Ross et al., 1988; Sugrue et al., 1990). Moreover, within CNS neurons Src is found both pre- and post-synaptically. The post-synaptic localization was particularly relevant with respect to our studies on the modulation of NMDA receptor function because Src was known to be found in the post-synaptic density (PSD) (Cudmore and Gurd, 1991; Atsumi et al., 1993). The PSD is the main structural component of excitatory synapses and is where NMDA and other glutamate receptors are concentrated. Thus, the presence of Src within the PSDs was consistent with our observations that the PTK regulating NMDA channels and the channels themselves were closely associated with one another.

Having established that an endogenous PTK is present in inside-out membrane patches we focused our efforts on using such patches as the most controllable preparation in which to study the regulation of native NMDA channels. This preparation also allowed us to characterize the actions of the reagents with the expectation of taking advantage of the most reliable reagents for use in subsequent studies with intact neurons. The first step in our strategy was to determine whether the PTK was a member of the Src family of non-receptor kinases. For this purpose we used a peptide, EPQ(pY)EEIPIA which has a high affinity for the SH2 domain of kinases in the Src family (Liu et al., 1993). As mentioned above, the SH2 domain is involved in negatively regulating Src kinase activity which occurs through intramolecular binding of the phosphorylated tyrosine in the tail of the kinase. The EPQ(pY)EEIPIA peptide competes for this intramolecular binding and thereby precludes the auto-inhibition of Src family kinases (Liu et al., 1993). We found that applying this "Src-activator" peptide to the cytoplasmic face of inside-out patches increased NMDA channel activity. As a control we tested the non-phosphorylated form of the peptide, EPQYEEIPIA, which does not affect kinase activity, and which had no effect on NMDA channels.

One of the most important considerations for using phosphopeptides is proper synthesis. Because the synthesis can be more difficult than for non-phosphorylated peptides we have phosphopeptides made by a facility with a long history of making such phosphopeptides *de novo*. Also, we have each peptide tested by mass spectrometry to ensure purity and correct synthesis. Non-phosphorylated control peptides are synthesized separately. An alternative approach would be to synthesize only the phosphorylated or dephosphorylated peptides and then either dephosphorylate or phosphorylate these, respectively. This approach requires a number of additional steps and rigorous purification to establish that the reaction is complete and, as importantly, that the final product is not contaminated with enzyme. For very long peptides this approach may be useful but for the peptides we have used we have determined that *de novo* synthesis of each peptide is more cost effective and efficient. We store peptides at −80°C either lyphophyilized or in single-use aliquots of stocks to be used at 100× or 1000× final dilution.

The binding of phosphotyrosine to particular SH2 domains is dependent upon the surrounding sequence of amino acids (Songyang and Cantley, 1995). Thus, an additional control peptide is one in which the phosphotyrosine is not in the appropriate location. This type of control has been used by Lancaster in studying effects of Src kinases in hippocampal slices (Lancaster and Rogers, 1998).

In order to complement the data from the experiments with the Src activator peptide we required a reagent to inhibit kinases in the Src family. For this purpose we used anti-cst1 (Roche et al., 1995b), an antibody directed against an epitope in the catalytic domain, which was shown to inhibit Src family kinases (Roche et al., 1995a). We found that anti-cst1, applied to the cytoplasmic face of inside-out patches, decreased NMDA channel activity. As a control in this experiment we used a non-specific IgG which had no effect. There are a number of important technical points relating to the use of antibodies in these types of experiments. The first is that the antibody should be as specific and work at as high a dilution as possible. Typically we prefer to use affinity purified antibodies; antisera can be used but should be controlled by pre-adsorption and, if possible, with pre-immune sera. We typically reject antibodies that recognize inappropriately sized bands on Western blots. However, this is not an absolute prohibition as it is well-known that some antibodies that do not work well for immunoblotting can work very specifically for other purposes, presumably due to differences in the epitope under different conditions. Pre-adsorption can help in this case but really only in the negative sense because loss of an effect by pre-adsorbing could be because the antibody is in fact cross-reacting in a specific way with an epitope present in protein(s) other than the intended target. Results using the antibody would be interpreted in such cases as providing supportive rather than definitive evidence. Conversely, even an antibody that recognizes a single band may have a broad spectrum of biological activity in situ. For some proteins (see below) in situ specificity can be examined in cells lacking the gene product of interest, for example obtained from animals with targeted gene deletions.

However, because of the potential for secondary changes in cells from gene-deleted animals even this approach on its own is not necessarily perfect.

After narrowing the search to the Src family of PTKs we next sought to identify which specific member or members of the family are necessary. For this purpose we made use of an antibody, anti--Src1, for which the biological activity was established and which was known to block selectively the function of Src but not other members of the Src family (Roche et al., 1995a). Anti-Src1 applied to the cytoplasmic face of inside-out patches caused a decrease in basal NMDA channel activity. Moreover, anti-src1 prevented the effect of EPQ(pY)EEIPIA, indicating that Src is necessary for the effect of the activating peptide. In other experiments, applying exogenous recombinant pp60$^{c\text{-}src}$ increased NMDA channel activity, but this was prevented by heat-inactivating the kinase just prior to use.

While anti-src1 has been used to inhibit Src function in cells (Roche et al., 1995a) this antibody does not bind to the catalytic domain of Src and has only a small effect on catalytic activity measured in vitro using small peptides as substrates (S. Courtneidge, personal communication). Thus, at first glance one might consider anti-src1 not to be a function-blocking antibody, and yet it affects function in cells and in membrane patches. We thus envisaged two potential mechanisms by which the antibody might act. The first is that the binding of the antibody, which is relatively large in size in comparison with Src, might sterically hinder enzymatic activity. Alternatively, it was possible that anti-Src1 acts by blocking an interaction between the region in Src which the antibody recognizes, amino acids 40–58, and a complementary region in a target protein. In order to differentiate between these possibilities we tested the peptide fragment, Src(40–58). We found that this peptide reduced NMDA channel activity when applied to the cytoplasmic face of the membrane. As a control for Src(40–58) we used a peptide with the same amino acid composition but in random order, "scrambled Src(40–58)". We found that scrambled Src(40–58) had no effect on NMDA channel activity. In other experiments we determined that Src(40–58) does not block in vitro phosphorylation of a small substrate peptide, EKIGEGTYGVKKK, by recombinant Src, indicating that Src(40–58) does not directly affect kinase catalytic activity. Thus, we concluded that the region 40–58 may interact with a component in the NMDA receptor complex and that this interaction is necessary for the effect of Src on NMDA channels. This interacting component might be an NMDA receptor subunit protein or one of the other proteins that associate with the channels.

Since EPQ(pY)EEIPIA may activate Src kinase family members in addition to Src itself we tested the involvement of Src in the upregulation of NMDA channel function by this peptide. We found that pre-administering Src(40-58) but not scrambled Src(40-58) prevented the effects of applying EPQ(pY)EEIPIA. We interpreted these findings as indicating that Src is necessary for the upregulation of NMDA channel activity by the activator peptide.

From the previous experiments it was not possible to determine whether Src is closely associated with NMDA channels because while membrane patches are small in comparison with cells, the patches are large in comparison with the size of the proteins involved. Thus, it is possible that within the patches Src and the NMDA receptor complex are not directly associated with one another. However, in other experiments we determined that Src and NMDA receptor subunit proteins co-precipitate with one another showing that Src can associate with the NMDA channel complex. Whether this association occurs via a direct interaction between Src and an NMDA receptor subunit protein or by means of an intervening adaptor protein is presently under investigation.

NMDA receptors are expressed on the cell surface both at synapses and extra-synaptically. The methods for recording single channel activity allow for direct measurements to be made only from extra-synaptically localized receptors, whereas for physiological phenomena it is the synaptic receptors that are more relevant. From analyzing NMDA channel activity in the patches it was determined that Src increases the channel gating during single activations of the receptors. Synaptic NMDA responses are known to be due to such single receptor activations (Edmonds et al., 1995), and therefore we predicted that if the synaptically stimulated receptors are affected similarly to the extrasynaptic receptors then synaptic NMDA responses should be increased by Src. This was investigated by studying miniature excitatory post-synaptic currents (mEPSCs) which are the responses evoked by individual quanta of glutamate. Using the peptides and antibodies that had been characterized in the studies of single channels we determined that the function of synaptic NMDA receptors is indeed upregulated by Src. Moreover, we found that Src differentially regulates synaptic NMDA but not the non-NMDA subclass of glutamate receptors which are colocalized at the same synapses (Yu et al., 1997).

What are the functional consequences of the regulation of NMDA receptors by Src?

NMDA receptors are widely present at central synapses and therefore these results raised the possibility that tyrosine phosphorylation could contribute to the plasticity of synaptic transmission throughout the CNS. The CA1 region of the hippocampus is one region where Src is highly expressed (Sugrue et al., 1990). Thus, we have recently examined the consequences of NMDA receptor upregulation by Src by studying CA1 neurons in an in vitro hippocampal slice preparation (Lu et al., 1998).

Tetanic stimulation of the Schaffer collateral inputs to CA1 pyramidal neurons causes a long-term potentiation (LTP) of synaptic transmission (Bliss and Collingridge, 1993; Malenka and Nicoll, 1993) that is known to depend upon both protein tyrosine kinase function (O'Dell et al., 1991) and NMDA receptor activation (Collingridge and Bliss, 1987). Also, it has been established that LTP in the CA1 region is induced by a cascade of biochemical steps occurring in the pyramidal neurons (Malenka et al., 1989; Nicoll and Malenka, 1995). Therefore, we explored whether Src participates in LTP by postsynaptically administering the reagents characterized previously.

We found that blocking Src, with Src(40-58) or anti-src1, but not the respective controls, prevented the induction of LTP indicating that Src is necessary. As mentioned above, the specificity of anti-src1 for blocking Src but not other Src family members had been established by Roche et al. (1995). By comparison Src(40-58) was relatively untested. This inhibitor peptide sequence is not found in any other protein in GenBank and this was one reason we considered it to be a useful tool. Using hippocampal slices provided us with the opportunity to similarly establish the specificity of Src(40-58): we tested this peptide in mice lacking Src, produced by targeted gene disruption (Soriano et al., 1991) and found that Src(40-58) does not affect LTP induction in CA1 neurons. Thus, in order for Src(40-58) to block LTP induction Src needs to be present, as it is in the wild-type animals.

Conversely, we found that administering recombinant Src or the Src activator peptide induced a long-lasting potentiation of synaptic responses which occluded the induction of LTP by tetanus. Moreover, we immunopurified Src from the portion of the slices around the stimulating electrode and determined that tetanic stimulation producing LTP enhanced the activity of Src, as measured in vitro using an immune-complex kinase assay. This enhancement of Src activity was observed at the earliest measurable time after the tetanic stimulation was delivered. Thus, Src is very rapidly upregulated by tetanic stimulation.

LTP in CA1 neurons is ultimately expressed as an enhancement of the component of synaptic responses mediated by non-NMDA receptors. Indeed our results showed that the non-NMDA synaptic currents, as well as those mediated by NMDA receptors, were potentiated by activating Src in the CA1 neurons. This apparent contradiction to the findings described in the preceding section was found to be due to the fact that in the prior studies intracellular Ca^{2+} was buffered highly whereas in the experiments in hippocampal slices it was necessary to use low intracellular Ca^{2+} buffering in order to allow LTP to be induced. When we buffered Ca^{2+} heavily in the CA1 neurons, activating Src no longer potentiated the non-NMDA responses, while synaptic NMDA responses were enhanced. We therefore concluded that the enhancement of non-NMDA synaptic responses produced by activating Src depended upon a rise in Ca^{2+}. However, it is known that Src is not Ca^{2+}-dependent (Erpel and Courtneidge, 1995) and therefore our findings indicated that Src upregulates the function of non-NMDA receptors indirectly by means of one or more Ca^{2+}-dependent steps.

In other experiments we determined that Src-induced potentiation of the non-NMDA responses is causally linked to NMDA responses because the potentiation was prevented by preblocking NMDA receptors. On the other hand, blocking NMDA receptors after the Src-induced potentiation had developed did not reverse it. Therefore, NMDA receptors are necessary to produce, but not to maintain, the potentiation of non-NMDA responses caused by activating Src as is the case with tetanus-induced LTP.

These findings have required us to propose a revised model for LTP induction in CA1 neurons as discussed in detail elsewhere (Salter, 1998). In brief, we hypothesize that during the induction of LTP Src is rapidly activated, leading to enhanced NMDA receptor function. The enhanced function of NMDA receptors promotes entry of Ca^{2+} which triggers the downstream signalling cascade that leads to potentiated activity of non-NMDA receptors (Barria et al., 1997; Lisman et al., 1997). Thus, activation of Src appears to be a biochemical mechanism for gating the induction of NMDA-receptor dependent LTP.

The possibility that tyrosine kinases are required for the induction of LTP was first suggested from work by O'Dell and colleagues who showed that LTP is blocked by bath-applying membrane permeable tyrosine kinase inhibitors (O'Dell et al., 1991). Subsequently, as mentioned above, they found tetanus-induced LTP in CA1 neurons from mutant mice lacking *src,* a genetic argument against the necessity of Src for LTP induction. Moreover, they found that LTP is blunted in mice lacking *fyn* (Grant et al., 1992) and this can be rescued by introducing the wild-type *fyn* transgene (Kojima et al., 1997). Interestingly, they also reported that the impairment in LTP is age-dependent in *fyn*−/− mice: in young *fyn*−/− animals LTP is comparable to that in the wild-type and the impairment of LTP in the *fyn*−/− mice correlates with a large decline in the level of expression of Src in the adult. It is known that Src and Fyn may substitute for each other in various processes (Thomas et al., 1995). Therefore, while our recent results indicate that in wild-type individuals Src is a required mediator of LTP induction, in animals that develop without Src it is possible another member of the Src family, likely Fyn, may substitute for Src. It is important to note that our results do not imply that Fyn might not also be necessary for LTP induction in the wild-type; this possibility has yet to be examined using Fyn-specific reagents. However, if Fyn is involved in LTP induction, our observations that the effects of the Src family activating peptide are prevented by the Src-specific blockers imply that, in a linear scheme, Fyn would be upstream of Src.

Conclusions

Recombinant proteins, peptides and antibodies provide useful tools for examining the role of intracellular signalling proteins in synaptic function in the CNS. With such tools we have identified Src as an endogenous PTK which enhances the function of NMDA channels in CNS neurons and have determined that activating Src is necessary for the induction of LTP in the CA1 region of the hippocampus. These observations highlight the importance of investigating physiological processes in wild-type animals when appropriately selective reagents are available. It has been found that mice developing with a targeted deletion of *src* show LTP and this, we suggest, is because when Src is absent its function is taken over by another PTK, possibly Fyn. Src is known to be expressed widely (Sugrue et al., 1990) and in certain brain regions, such as cerebellum (Boxall et al., 1996), tyrosine phosphorylation has been implicated in synaptic plasticity. Thus, it is conceivable that Src may participate in plasticity of excitatory synaptic transmission in many regions of the CNS. The present chapter has focused on postsynaptic roles for Src in regulating synaptic transmission but Src is also found presynaptically (Sugrue et al., 1990; Atsumi et al., 1993) and it is known that voltage-gated K^+ (Fadool et al., 1997; Holmes et al., 1996) and Ca^{2+} (Hu et al., 1998; Wijetunge and Hughes, 1996) channels are regulated by Src kinases. Using the reagents described above in recordings from terminals of mammalian central synapses it is now possible to determine whether there are presynaptic effects of Src and thus to gain a comprehensive understanding of the diverse roles of this kinase in synaptic transmission in the CNS.

Acknowledgments

Work of the author is supported by the Medical Research Council of Canada and the Nicole Fealdman Memorial Fund. Thanks to Xian-Min Yu, Yueqiao Huang and Jeff Gingrich for helpful discussions on this chapter. Many thanks also to the colleagues who have so generously provided reagents for this work — Joe Bolen, Sara Courtneidge, Edward Fisher, Teresa Perney, Robert Wenthold.

Abbreviations

CNS	central nervous system
LTP	long-term potentiation
MEPSC	miniature excitatory post-synaptic currents
NMDA	*N*-methyl-L-aspartate
NR	NMDA receptor
PSD	post-synaptic density
PTK	protein tyrosine kinase
PTP	protein tyrosine phosphatase
SH2	Src-homology 2
SH3	Src-homology 3

References

Adams, S.R., Harootunian, A.T., Buechler, Y.J., Taylor, S.S. and Tsien, R.Y. (1991) Fluorescence ratio imaging of cyclic AMP in single cells. *Nature*, 349: 694–697.

Ascher, P. and Nowak, L. (1988) The role of divalent cations in the NMDA responses of mouse central neurones in culture. *J. Physiol. (Lond.)*, 399: 247–266.

Atsumi, S., Wakabayashi, K., Titani, K., Fujii, Y. and Kawate, T. (1993) Neuronal pp60c-src(+) in the developing chick spinal cord as revealed with anti-hexapeptide antibody, *J. Neurocytol.*, 22: 244–258.

Barria, A., Muller, D., Derkach, V., Griffith, L.C. and Soderling, T.R. (1997) Regulatory phosphorylation of AMPA-type glutamate receptors by CaM-KII during long-term potentiation. *Science*, 276: 2042–2045.

Beermann, A.E. and Jay, D.G. (1994) Chromophore-assisted laser inactivation of cellular proteins. *Methods Cell Biol.*, 44: 715–732.

Bliss, T.V. and Collingridge, G.L. (1993) A synaptic model of memory: long-term potentiation in the hippocampus. *Nature*, 361: 31–39.

Boxall, A.R., Lancaster, B. and Garthwaite, J. (1996) Tyrosine kinase is required for long-term depression in the cerebellum. *Neuron*, 16: 805–813.

Brown, M.T. and Cooper, J.A. (1996) Regulation, substrates and functions of src. *Biochim. Biophys. Acta*, 1287: 121–149.

Brugge, J.S., Cotton, P.C., Queral, A.E., Barrett, J.N., Nonner, D. and Keane, R.W. (1985) Neurones express high levels of a structually modified, activated form of $pp60^{c\text{-}src}$. *Nature*, 316: 554–557.

Carrera, A.C., Paradis, H., Borlado, L.R., Roberts, T.M. and Martinez, C. (1995) Lck unique domain influences Lck specificity and biological function. *J. Biol. Chem.*, 270: 3385–3391.

Chen, L. and Huang, L.-Y.M. (1991) Sustained potentiation of NMDA receptor-mediated glutamate responses through activation of protein kinase C by a μ opioid. *Neuron*, 7: 319–326.

Chen, L. and Huang, L.-Y.M. (1992) Protein kinase C reduces Mg^{2+} block of NMDA-receptor channels as a mechanism of modulation. *Nature*, 356: 521–523.

Ciabarra, A.M., Sullivan, J.M., Gahn, L.G., Pecht, G., Heinemann, S. and Sevarino, K.A. (1995) Cloning and characterization of chi-1: A developmentally regulated member of a novel class of the ionotropic glutamate receptor family. *J. Neurosci.*, 15: 6498–6508.

Collingridge, G.L. and Bliss, T.V.P. (1987) NMDA receptors — their role in long-term potentiation. *Trends Neurosci.*, 10: 288–293.

Cudmore, S.B. and Gurd, J.W. (1991) Postnatal age and protein tyrosine phosphorylation at synapses in the developing rat brain. *J. Neurochem.*, 57: 1240–1248.

Durand, G.M., Gregor, P., Zheng, X., Bennett, M.V.L., Uhl, G.R. and Zukin, R.S. (1992) Cloning of an apparent splice variant of the rat *N*-methyl-D-aspartate receptor NMDAR1 with altered sensitivity to polyamines and activators of protein kinase C. *Proc. Natl. Acad. Sci. USA*, 89: 9359–9363.

Edmonds, B., Gibbs, R.A. and Colquhoun, D. (1995) Mechanisms of activation of glutamate receptors and the time course of excitatory synaptic currents. *Annu. Rev. Physiol.*, 57: 495–519.

Erpel, T. and Courtneidge, S.A. (1995) Src family protein tyrosine kinases and cellular signal transduction pathways. *Curr. Opin. Cell Biol.*, 7: 176–182.

Fadool, D.A., Holmes, T.C., Berman, K., Dagan, D. and Levitan, I.B. (1997) Tyrosine phosphorylation modulates current amplitude and kinetics of a neuronal voltage-gated potassium channel. *J. Neurophysiol.*, 78: 1563–1573.

Fukami, Y., Sato, K., Ikeda, K., Kamisango, K., Koizumi, K. and Matsuno, T. (1993) Evidence for autoinhibitory regulation of the c-src gene product. A possible interaction between the src homology 2 domain and autophosphorylation site. *J. Biol. Chem.*, 268: 1132–1140.

Gasic, G.P. and Hollmann, M. (1992) Molecular neurobiology of glutamate receptors. *Annu. Rev. Physiol.*, 54: 507–536.

Grant, S.G.N., O'Dell, T.J., Karl, K.A., Stein, P.L., Soriano, P. and Kandel, E.R. (1992) Impaired long-term potentiation, spatial learning, and hippocampal development in *fyn* mutant mice. *Science*, 258: 1903–1910.

Holmes, T.C., Fadool, D.A., Ren, R. and Levitan, I.B. (1996) Association of Src tyrosine kinase with a human potassium channel mediated by SH3 domain. *Science*, 274: 2089–2091.

Hu, X.Q., Singh, N., Mukhopadhyay, D. and Akbarali, H.I. (1998) Modulation of voltage-dependent Ca2+ channels in rabbit colonic smooth muscle cells by c-Src and focal adhesion kinase. *J. Biol. Chem.*, 273: 5337–5342.

Hunter, T. (1994) 1001 protein kinases redux — towards 2000. *Semin. Cell Biol.*, 5: 367–376.

Jessell, T.M. and Kandel, E.R. (1993) Synaptic transmission: A bidirectional and self-modifiable form of cell-cell communication. *Neuron*, 10(Suppl): 1–30.

Kelso, S.R., Nelson, T.E. and Leonard, J.P. (1992) Protein kinase C-mediated enhancement of NMDA currents by metabotropic glutamate receptors in *Xenopus* oocytes. *J. Physiol. (Lond.)*, 449: 705–718.

Kojima, N., Wang, J., Mansuy, I.M., Grant, S.G., Mayford, M. and Kandel, E.R. (1997) Rescuing impairment of long-term potentiation in fyn-deficient mice by introducing Fyn transgene. *Proc. Natl. Acad. Sci. USA*, 94: 4761–4765.

Lancaster, B. and Rogers, M.V. (1998) A peptide activator of endogenous tyrosine kinase enhances synaptic currents mediated by NMDA receptors. *Eur. J. Neurosci.*, 10: 2302–2308.

Laube, B., Kuhse, J. and Betz, H. (1998) Evidence for a tetrameric structure of recombinant NMDA receptors. *J. Neurosci.*, 18: 2954–2961.

Lieberman, D.N. and Mody, I. (1994) Regulation of NMDA channel function by endogenous Ca^{2+}- dependent phosphatase. *Nature*, 369: 235–239.

Lisman, J., Malenka, R.C., Nicoll, R.A. and Malinow, R. (1997) Learning mechanisms: the case for CaM-KII. *Science*, 276: 2001–2002.

Liu, X., Brodeur, S.R., Gish, G., Songyang, Z., Cantley, L.C., Laudano, A.P. and Pawson, T. (1993) Regulation of c-Src tyrosine kinase activity by the Src SH2 domain. *Oncogene*, 8: 1119–1126.

Lu, Y.M., Roder, J.C., Davidow, J. and Salter, M.W. (1998) Src activation in the induction of long-term potentiation in CA1 hippocampal neurons. *Science*, 279: 1363–1368.

MacDonald, J.F., Mody, I. and Salter, M.W. (1989) Regulation of N-methyl-D-aspartate receptors revealed by intracellular dialysis of murine neurones in culture. *J. Physiol. (Lond.)*, 414: 17–34.

Malenka, R.C., Kauer, J.A., Perkel, D.J., Mauk, M.D., Kelly, P.T., Nicoll, R.A. and Waxham, M.N. (1989) An essential role for postsynaptic calmodulin and protein kinase activity in long-term potentiation. *Nature*, 340: 554–557.

Malenka, R.C. and Nicoll, R.A. (1993) NMDA-receptor-dependent synaptic plasticity: multiple forms and mechanisms. *Trends. Neurosci.*, 16: 521–527.

Martinez, R., Mathey-Prevot, B., Bernards, A. and Baltimore, D. (1987) Neuronal $pp60^{c\text{-}src}$ contains a six-amino acid insertion relative to its non-neuronal counterpart. *Science*, 237: 411–415.

Mayer, M.L. and Westbrook, G.L. (1987) Permeation and block of *N*-methyl-D-aspartic acid receptor channels by divalent cations in mouse cultured central neurones. *J. Physiol. (Lond.)*, 394: 501–527.

McBain, C.J. and Mayer, M.L. (1994) *N*-Methyl-D-aspartic acid receptor structure and function. *Physiol. Rev.*, 74: 723–760.

Moarefi, I., LaFevre-Bernt, M., Sicheri, F., Huse, M., Lee, C.H., Kuriyan, J. and Miller, W.T. (1997) Activation of the Src-family tyrosine kinase Hck by SH3 domain displacement. *Nature*, 385: 650–653.

Murphy, T.H., Baraban, J.M. and Wier, W.G. (1995) Mapping miniature synaptic currents to single synapses using calcium imaging reveals heterogeneity in postsynaptic output. *Neuron*, 15: 159–168.

Neher, E. (1988) The use of the patch clamp technique to study second messenger- mediated cellular events. *Neuroscience*, 26: 727–734.

Neher, E. and Sakmann, B. (1992) The patch clamp technique. *Sci. Am.*, 266: 28–35.

Nicoll, R.A. and Malenka, R.C. (1995) Contrasting properties of two forms of long-term potentiation in the hippocampus. *Nature*, 377: 115–118.

O'Dell, T.J., Kandel, E.R. and Grant, S.G.N. (1991) Long-term potentiation in the hippocampus is blocked by tyrosine kinase inhibitors. *Nature*, 353: 558–560.

Pawson, T. (1995) Protein modules and signalling networks. *Nature*, 373: 573–580.

Pawson, T. and Scott, J.D. (1997) Signaling through scaffold, anchoring, and adaptor proteins. *Science*, 278: 2075–2080.

Pyper, J.M. and Bolen, J.B. (1990) Identification of a novel neuronal C-SRC exon expressed in human brain. *Mol. Cell Biol.*, 10: 2035–2040.

Raman, I.M., Tong, G. and Jahr, C.E. (1996) β-Adrenergic regulation of synaptic NMDA receptors by cAMP-dependent protein kinase. *Neuron*, 16: 415–421.

Roche, S., Fumagalli, S. and Courtneidge, S.A. (1995a) Requirement for Src family protein tyrosine kinases in G_2 for fibroblast cell division. *Science*, 269: 1567–1569.

Roche, S., Koegl, M., Barone, M.V., Roussel, M.F. and Courtneidge, S.A. (1995b) DNA synthesis induced by some but not all growth factors requires src family protein tyrosine kinases. *Mol. Cell Biol.*, 15: 1102–1109.

Ross, C.A., Wright, G.E., Resh, M.D., Pearson, R.C. and Snyder, S.H. (1988) Brain-specific src oncogene mRNA mapped in rat brain by in situ hybridization. *Proc. Natl. Acad. Sci. USA*, 85: 9831–9835.

Sakmann, B. and Neher, E. (1984) Patch clamp techniques for studying ionic channels in excitable membranes. *Annu. Rev. Physiol.*, 46: 455–472.

Salter, M.W. (1998) Src, *N*-methyl-D-aspartate (NMDA) receptors, and synaptic plasticity. *Biochem. Pharmacol.* 56: 789–798.

Sato, K., Miki, S., Tachibana, H., Hayashi, F., Akiyama, T. and Fukami, Y. (1990) A synthetic peptide corresponding to residues 137 to 157 of p60v- src inhibits tyrosine-specific protein kinases. *Biochem. Biophys. Res. Commun.*, 171: 1152–1159.

Seeburg, P.H. (1993) The molecular biology of glutamate receptor channels. *Trends Neurosci.*, 16: 359–365.

Sheng, M., Cummings, J., Roldan, L.A., Jan, Y.N. and Jan, L.Y. (1994) Changing subunit composition of heteromeric NMDA receptors during development of rat cortex. *Nature*, 368: 144–147.

Sigel, E., Baur, R. and Malherbe, P. (1994) Protein kinase C transiently activates heteromeric *N*-methyl-D-aspartate receptor channels independent of the phosphorylatable C-terminal splice domain and of consensus phosphorylation sites. *J. Biol. Chem.*, 269: 8204–8208.

Songyang, Z. and Cantley, L.C. (1995) Recognition and specificity in protein tyrosine kinase-mediated signalling. *TIBS*, 20: 470–475.

Soriano, P., Montgomery, C., Geske, R. and Bradley, A. (1991) Targeted disruption of the c-src proto-oncogene leads to osteopetrosis in mice. *Cell*, 64: 693–702.

Sucher, N.J., Akbarian, S., Chi, C.L., Leclerc, C.L., Awobuluyi, M., Deitcher, D.L., Wu, M.K., Yuan, J.P., Jones, E.G. and Lipton, S.A. (1995) Developmental and regional expression pattern of a novel NMDA receptor-like subunit (NMDAR-L) in the rodent brain. *J. Neurosci.*, 15: 6509–6520.

Sugrue, M.M., Brugge, J.S., Marshak, D.R., Greengard, P. and Gustafson, E.L. (1990) Immunocytochemical localization of the neuron-specific form of the c-src gene product, pp60c-src(+), in rat brain. *J. Neurosci.*, 10: 2513–2527.

Superti-Furga, G. and Courtneidge, S.A. (1995) Structure-function relationships in Src family and related protein tyrosine kinases. *Bioessays*, 17: 321–330.

Takahashi, T., Forsythe, I.D., Tsujimoto, T., Barnes-Davies, M. and Onodera, K. (1996) Presynaptic calcium current modulation by a metabotropic glutamate receptor. *Science*, 274: 594–597.

Thomas, S.M., Soriano, P. and Imamoto, A. (1995) Specific and redundant roles of Src and Fyn in organizing the cytoskeleton. *Nature*, 376: 267–271.

Tingley, W.G., Roche, K.W., Thompson, A.K. and Huganir, R.L. (1993) Regulation of NMDA receptor phosphorylation by alternative splicing of the C-terminal domain. *Nature*, 364: 70–73.

Tong, G., Shepherd, D. and Jahr, C.E. (1995) Synaptic desensitization of NMDA receptors by calcineurin. *Science*, 267: 1510–1512.

Vonakis, B.M., Chen, H., Haleem-Smith, H. and Metzger, H. (1997) The unique domain as the site on Lyn kinase for its constitutive association with the high affinity receptor for IgE. *J. Biol. Chem.*, 272: 24072–24080.

Wang, L.-Y., Orser, B.A., Brautigan, D.L. and MacDonald, J.F. (1994) Regulation of NMDA receptors in cultured hippocampal neurons by protein phosphatases 1 and 2A. *Nature*, 369: 230–232.

Wang, Y.T., Yu, X.-M. and Salter, M.W. (1996) Calcium-independent reduction of NMDA channel activity by protein tyrosine phosphatase. *Proc. Natl. Acad. Sci. USA*, 93: 1721–1725.

Wijetunge, S. and Hughes, A.D. (1996) Activation of endogenous c-Src or a related tyrosine kinase by intracellular (pY)EEI peptide increases voltage-operated calcium channel currents in rabbit ear artery cells. *FEBS Lett.*, 399: 63–66.

Xu, W., Harrison, S.C. and Eck, M.J. (1997) Three-dimensional structure of the tyrosine kinase c-Src. *Nature*, 385: 595–602.

Yu, X.M., Askalan, R., Keil, G.J. and Salter, M.W. (1997) NMDA channel regulation by channel-associated protein tyrosine kinase Src. *Science*, 275: 674–678.

Yu, X.-M. and Salter, M.W. (1998) Gain control of NMDA-receptor currents by intracellular sodium. *Nature*, 396: 469–474.

W.E. Crusio and R.T. Gerlai (Eds.)
Handbook of Molecular-Genetic Techniques for Brain and Behavior Research (Techniques in the Behavioral and Neural Sciences, Vol. 13)

CHAPTER 3.2.6

The use of immunoadhesins in neurobiology

David L. Shelton

Department of Neuroscience, Genentech, Inc., 1 DNA Way, South San Francisco, CA 94080-4990, USA

Structure and properties

The molecules that are the subject of this chapter now go by several names; immunoadhesins, receptorbodies, and ligand traps being the most common. Whatever the name, they are all recombinantly expressed fusion proteins consisting of at least a portion of a protein of interest, linked to the Fc portion of immunoglobulin heavy chain (see Fig. 1 from (Chamow and Ashkenazi, 1996). Most commonly, but not exclusively, the portion of the "protein of interest" is the extracellular domain of a single transmembrane receptor with the resulting chimera being used to bind the ligand of the receptor. Hence the name, ligand trap. These proteins are made by fusing, in frame, the cDNAs encoding the areas of interest of the receptor with one encoding the immunoglobulin Fc region and then transfecting that chimeric DNA into cells which make the fused protein. What this leads to is a convenient, soluble form of what is normally an integral membrane receptor. The general history and structures of these molecules has been recently reviewed (Chamow and Ashkenazi, 1996), and so this chapter will focus on selected examples from neurobiology which serve to illustrate the utility and limitations of these tools.

In addition to their solubility advantages, proteins modified and expressed in this manner have a number of other very useful features. The presence of an essentially normal immunoglobulin Fc domain conveys a number of useful properties. For many practical purposes, these proteins behave like antibodies. This is due to the fact that many of the commonly used manipulations of antibodies rely on the presence of the Fc. There are a wide variety of commercially available reagents (protein A, protein G, anti-Fc antibodies) which will bind this Fc domain thereby providing an extremely convenient "handle" for purification, immobilization, or localization purposes. Thus, these proteins are simple to purify with affinity techniques from cell culture supernatants. This makes production relatively straightforward. The producing cells are allowed to grow in tissue culture and the "conditioned media" from this growth is passed over an affinity column which has one of these Fc binding proteins covalently attached. The immunoadhesin is bound to the column via its Fc and all other components of the media are rinsed away. The immunoadhesin is then freed from the column by passing through a solution (most commonly a mild acid, ph 3.0 citrate buffer works well) which disrupts the Fc binding, and the immunoadhesin is collected in a pure form. Another useful application of this immobilization of the immunoadhesin via its Fc is ability to bind it to plastic for use in ELISA (enzyme-linked immunosorbent assay) studies or ligand binding studies. These assays can be easily performed in 96-well plates by first coating the wells with the Fc binding protein (protein A, protein G or anti-Fc antibody), and then allowing a solution of immunoadhesin to bind to that. The ability to bind specifically by the Fc region allows one to orient the molecule during these applications, keeping the specific moiety of the fusion protein relatively free of interference due to binding to the plate. With a plate coated in this fashion, one can then titer an antisera, check for ligand binding or a variety of other biochemical tests of protein interactions.

During in vivo experiments, the Fc domain commonly conveys a vastly increased plasma half-life on the protein. This can be an incredibly useful feature, allowing treatment of animals for days with a single injection of immunoadhesin. In addition, the presence of the Fc makes it relatively simple to localize and/or measure the fusion protein during or after treatment. This can be done with an ELISA as discussed above, which allows for quantitative determination of the amount of immunoadhesin, or with immunocytochemistry, which allows a fine structural location of the administered protein. Immunocytochemistry involves fixing the tissue from the treated animal, usually with formaldehyde and then allowing a labeled antibody (either with a fluorescent tag or an enzyme for later histochemical detection) to react with the tissue, either after sectioning or in a whole mount preparation. Since the Fc domain from the immunoadhesin is usually from a different species than the experimental animal, one can use commercially available anti-Fc antibodies which are species specific and detect the foreign Fc domain even in the presence of the host antibody Fc domains. Subsequent microscopic examination allows one to observe the precise cellular location of the immunoadhesin.

Since the Fc domain of the IgG molecule is responsible for dimerization, these fusion proteins of Fc and protein domain of interest also dimerize, yielding a bivalent molecule. This can be a convenient or not so convenient feature, depending on the application. Because the specific protein domain of the immunoadhesin is bivalent, whatever protein with which it interacts will tend to be crosslinked as well. If that protein is a cell surface receptor, this crosslinking can activate its signal transduction pathways.

The other portion of the chimeric protein conveys its specific utility. In the most common conformation, with a receptor extracellular domain, the immunoadhesins can often bind ligand, frequently in a manner indistinguishable from that of the native receptor in a cell's membrane. Thus, these molecules are capable of equal affinity and specificity to that of the native receptor. For those immunoadhesins which do bind ligand in a native manner, their soluble form can make them a very useful blocking agent for the biological activity of the ligand. Antibodies are commonly used for this purpose, but immunodhesins can have advantages. There are at least two properties which contribute to blocking ability, affinity for the ligand and binding site specificity.

Obviously, proteins with higher binding affinity for the ligand can block the ligand more effectively than those with lower affinity. Often the affinity of the native receptor (as expressed in an immunoadhesin) is quite high and would be difficult to match even with a very good antibody.

Binding site specificity is a bit more subtle. In order for a protein to block the activity of a ligand, it indeed must bind tightly to that ligand, but it must do so at a site on the ligand which is required for interaction with the ligand's receptor. The site on the ligand where an antibody binds is determined only by what the immune system recognizes as foreign during the production of the antibody, and may or may not be near the receptor binding site on the ligand. On the other hand, an immunoadhesin made with the cell surface receptor for a ligand binds that ligand exactly at the receptor binding site. This gives the most effective steric blocking action for a given affinity.

A further advantage of these recombinant proteins is that they are completely defined and not subject to the vagaries of animal to animal variability inherent in production of polyclonal antisera. When one injects an antigen into a rabbit or goat for production of antiserum, each sample of serum from each individual animal may be different as the host animal undergoes its immune response to the antigen. This can lead to changes in affinity or specificity in the resultant serum. One can avoid this by immortalizing a single antibody producing B cell from an animal (usually a mouse) to achieve a monoclonal antibody. This approach has the disadvantage that one must sometimes screen through many clones to find the ones with the desired properties.

Immunoadhesins have another interesting property which is critical for some in vivo applications. If one wants to systemically administer a foreign protein to an animal for more than a few days, that animal's immune response becomes a concern. For instance, administration of a heterologous antibody into a rodent elicits a dramatic immune response to that foreign antibody in 10–14 d. This can lead to dramatically decreased levels of the administered antibody, as the host rapidly clears it from the circulation. An immunoadhesin based on a heterologous species will generate this same response but, one has the option of changing the species of an immunoadhesin quite readily. By remaking the immunoadhesin with the equivalent receptor and Fc portion both cloned from the desired species, one can generate a protein which is essentially no longer foreign. Further, this immunoadhesin will be very similar if not identical to the original, heterologous, immunoadhesin. This does make the task of identifying the delivered protein (described above) more difficult. The immunoadhesin Fc domain is now the same as that of the experimental animal and therefore harder to distinguish.

Immunoadhesins are not without disadvantages. These molecules are only as fit for the purpose of binding a specific ligand as the native receptor might be. Evolution may have seen fit to design a receptor with multiple specificities, and this commonly will be reflected in the immunoadhesin generated with it. These molecules will not be good reagents to block a single one of the receptor's ligands, no matter if the affinity is good. Since the immunoadhesin will bind to and block all of the cognate ligands, specificity is compromised. Likewise, if a ligand is promiscuous, one cannot use the receptor immunoadhesin to define that the ligand is working through that receptor. One can only say that removing the ligand(s) from the picture causes a certain biological effect.

Another potential disadvantage of these molecules is that design and production requires some expertise in techniques of molecular biology. Production requires cell culture facilities and, if one needs to generate large amounts of the pure protein, this can require industrial scale cell culture fermentors. Although immunoadhesins can and have been made in high expression baculovirus based insect cell systems, the resultant molecules are often not very stable in vivo. This is probably due to differences in post-translational processing between the insect and mammalian cell expression systems. There are a few immunoadhesins beginning to be offered commercially (R&D Systems), but they are expensive.

Many of these advantages and disadvantages can be illustrated by examples of how these immunoadhesin proteins have been used. After the pioneering work of Gascoigne, Capon, Taunecker, Byrn, and Zettlmeissl, who developed this method using fusions of the T cell receptor or CD4 (Gascoigne et al., 1987; Capon et al., 1989; Traunecker et al., 1989; Byrn et al., 1990; Zettlmeissl et al., 1990), the ease and convenience of preparing these molecules was recognized very quickly. Soon a wide variety of receptors were expressed in this way as they were cloned. I will focus here on the uses of immunoadhesins made with two families of receptors of neurobiological interest, the trk family of neurotrophin receptors and the Eph and ephrins. I believe these families demonstrate the potential benefits and some caveats to the use of these molecules.

Immunoadhesins of trk family neurotrophin receptors

The trk family of neurotrophin receptors consists of three members, now commonly referred to as trkA, trkB, and trkC. They serve as signal transducing tyrosine kinase receptors for the members of the neurotrophin family, NGF, BDNF, NT3 and NT4/5. NGF binds primarily to trkA, BDNF and NT4/5 bind primarily to trkB and NT3 binds primarily to trkC, but with somewhat lower affinity is also capable of binding to trkB (Ip et al., 1993; Thoenen, 1991). The neurotrophins have primarily been studied in terms of their effects during neural development, where they are required for the normal survival and maturation of specific sets of neurons. Recently, more has been learned about the roles that neurotrophins play in the developing and adult animal, and trkIgG immunoadhesins have played a significant role in this work.

There has been a growing body of evidence that implicated NGF in the control of pain sensitivity. It had been known for many years that NGF controlled the developmental survival of those sensory neurons primarily responsible for pain sensation. It was also known that these peptidergic neurons continued to respond to NGF in the adult, at least in culture (Lindsay and Harmar, 1989). With the discovery of the trk receptors, it became possible to show that these neurons still expressed the NGF receptor, trkA in the normal adult (Averill et al., 1995; McMahon et al., 1994). Furthermore, NGF protein or mRNA was known to increase dramatically at sites of peripheral inflammation, which was accompanied by hyperalgesia (Weskamp and Otten, 1987). In a number of other experiments, most convincingly in human volunteers, it has been shown that pharmacological doses of exogenous NGF can increase pain sensitivity (human: Dyck et al., 1997), (rat: Lewin et al., 1993; Woolf et al., 1994; Andreev et al., 1995; Amann et al., 1996)). This raised the question of the physiological relevance of the observed upregulation of endogenous NGF during inflammation. After all, there are many compounds which increase during inflammation and are capable of giving rise to hyperalgesia when administered experimentally. However, the question "Does endogenous NGF play a necessary role in modulating pain sensitivity?" was still very much an open one. Using trkA-IgG to block endogenous NGF activity, we were able to show that endogenous NGF was indeed critical to the hyperalgesia which accompanied inflammation (McMahon et al., 1995). Furthermore, endogenous NGF was responsible for modulating pain sensitivity even in the absence of frank inflammation. The use of trkA-IgG resolved the ambiguities from experiments performed using a polyclonal antisera to NGF (Lewin et al., 1994; Woolf et al., 1994). The favorable pharmacokinetics and in vivo stability of trkA-IgG administered either systemically or locally (Bennett et al., 1998) were useful features of the trkA-IgG. However, principal advantage of trkA-IgG in this series of experiments was probably affinity. In studies we have done, the affinity as measured by the dissociation constant (K_d) of NGF for trkA-IgG is around 30 picomolar (pM) (Shelton et al., 1995) whereas we find most "good" antibodies to NGF have K_ds in the 500 pM range. Woolf et al. were forced to use doses of the antisera much larger than those which were effective with trkA-IgG, probably reflecting both the polyclonal nature of the serum and likely a lower affinity.

A further advantage of the trkA-IgG approach is demonstrated by work done to examine the effects of chronic administration of an NGF sequestering agent. This type of experiment has been performed before, by making animals autoimmune to their own NGF (Rich et al., 1984), although pain sensitivity was not determined. Animals were immunized with NGF, creating an anti-NGF antibody with the experimental animal's own immune system. With this approach, however, one loses a great deal of control over dosing and characterization of the antisera, since each animal is making its own treatment. Since each animal is being treated with a different antiserum, it is difficult to interpret the results. This study could not be done with a heterologous antisera produced in a separate animal due to the immune response of the treated animal. With trkA-IgG, it was straightforward to make a new molecule composed of trkA and Fc domains both of rat origin and administer the protein to rats for 10 weeks. While one animal did have an immune response to the trkA-IgG during this time, five others showed no signs of an immune response, thereby allowing an evaluation of the long term effects of this treatment.

trk fusion proteins have also been used in an elegant series of experiments examining the role of neurotrophins in cortical development. Using in vitro cortical slices, individual neurons were labeled by gene transfer of beta galactosidase. This allowed the evaluation of the development of dendritic morphology. Treatment with exogenous neurotrophins clearly had effects on the development of dendrites, and these effects were selective for different cortical neuronal subtypes (McAllister et al., 1995). Interestingly, not all of these effects were positive for growth of processes. For instance, addition of BDNF to cortical slices in vitro stimulated the growth of basal dendrites from layer 4 cortical neurons, whereas exogenous NT3 had no effect on its own but inhibited the BDNF effect. However, in contrast, NT-3 stimulated basal dendritic growth from layer 6 neurons in these same preparations, and addition of BDNF inhibited this effect. Again, it was critical to determine if these effects were indicative of underlying physiological ones, or were purely pharmacological artifacts of adding exogenous factors. By using trkB-IgG and trkC-IgG fusion proteins, it was demonstrated that endogenous neurotrophins were playing an important role in modulating cortical neuron development (McAllister et al., 1997). In layer 4, addition of trkB-IgG almost totally abolished basal dendrite outgrowth, while control application of trkA-IgG had no effect. Application of trkC-IgG actually stimulated outgrowth and complexity of basal dendrites from layer 4 cells, showing convincingly that endogenous NT-3 was acting in these cultures to inhibit development of dendritic growth. Equivalent experiments examining basal dendrites from layer 6 cells showed that the observed effects of exogenous neurotrophins also reflected underlying roles for endogenous neurotrophins.

Another study examining the possible role of neurotrophins in the control of early cortical architecture used the well-characterized development of the ocular dominance columns in visual cortex. This work was done with local infusion in vivo, and initially showed that treatment with BDNF and NT4/5, but not NT3 could disrupt the normal formation of ocular dominance columns (Cabelli et al., 1995). In order to examine the physiological role of endogenous neurotrophins, their biological activity was blocked by infusing trkA or trkB immunoadhesins. This work quite clearly demonstrated that while treatment with trkA-IgG was without effect on ocular dominance column formation, infusion of trkB-IgG disrupted the normal segregation of input from the eyes (Cabelli et al., 1997).

Ocular dominance column formation is clearly activity dependent, and it has long been postulated that selective strengthening of synapses is at least partially responsible for their formation. In mammalian brain, this modulation of synaptic efficacy has been most extensively examined in the hippocampal slice by studying the phenomenon of long term potentiation. Here too, neurotrophins seem to play a role, and trkB-IgG has been instrumental in proving that endogenous neurotrophins are important (Kang et al., 1997).

Effects on synaptic efficacy have also been shown on developing neuromuscular junctions in culture, showing that the neurotrophins are involved outside the cortex (Wang and Poo, 1997). In all of these experiments, it is critical that the trk-IgG molecules are reasonably stable either in vitro or in vivo. This has been directly demonstrated in some of these studies (Cabelli et al., 1997; Kang et al., 1997), and the dramatic effects of the treatment with the trk-Ig strongly argues for the presence of active molecule in the others.

One drawback to the use of trkB-IgG is evident form all of the above studies. That is, that because of the dual specificity shown by the trkB receptor, (i.e. binds both BDNF and NT4/5) one cannot distinguish whether the endogenous neurotrophin responsible for the observed effects is BDNF and/or NT4/5. This is a limitation of settling for the receptors and their native specificities. There certainly are ways around this difficulty, one might resort to making antibodies with high affinity and specificity for the individual neurotrophins. It may also be possible to use site directed mutagenesis of an immunoadhesin to either increase or decrease the specificity of the native receptor.

Eph receptors and ephrin ligands

Another family of proteins which serve to illustrate some advantages and limitations of the immunoadhesin approach is that of the Eph and ephrins. This is a large family of quite promiscuous tyrosine kinase receptors (Eph) and ligands (ephrin) that has been implicated in a wide variety of developmental processes both within and outside of the nervous system (Barinaga, 1995; Drescher et al., 1995; Friedman and O'Leary, 1996; Drescher, 1997; Orioli and Klein, 1997; Pasquale, 1997). The ligands are all cell surface proteins, some attached via a GPI linkage, and others having a transmembrane domain. To make matters more complicated, several of the "ligands" has been shown to be capable of acting as a receptor as well, with the "receptor" acting as a signaling molecule or ligand in this system. (Holland et al., 1996; Brückner et al., 1997). The nomenclature of this family has recently been simplified (Eph Nomenclature Committee, 1997), and this new system will be used here. Immunoadhesins have been commonly used in many ways to investigate this family, we will focus here on methodologies not already mentioned in work on neurotrophins and trks. An interesting twist with this family is the use of ligand immunoadhesins. Since the ligands are all transmembrane or GPI-linked, these are structurally analogous to receptor immunoadhesins in that the extracellular domain is fused to an immunoglobin Fc domain.

Unlike the neurotrophins, immunoadhesins were available during the very early days of work on the Eph and ephrins, when the biochemical characterization was still being carried out. Much of this early work highlights the utility of the immunoadhesins as a tool for biochemistry. Many of the Eph family of receptors were initially cloned as "orphan" tyrosine kinase receptors, with no known ligand. In several cases, the immunoadhesins of these receptors played crucial roles in the identification and isolation of the cognate ligands (Beckmann et al., 1994; Davis et al., 1994; Shao et al., 1994, 1995; Bennett et al., 1995; Cerretti et al., 1995; Kozlosky et al., 1995; Winslow et al., 1995). The immunoadhesins were used as staining reagents during fluorescence activated cell sorter analysis to identify cells expressing the ligands, thereby identifying a promising source for purification or expression cloning efforts. This relies on the ability of the immunoadhesin to bind to the cell surface molecule with its specific Eph moiety and display its Fc portion for further staining, similar to the immunocytochemical localization studies referred to earlier. This use as staining reagents has extended into expression cloning projects as the assay for expression of cloned membrane bound ligand. During protein purification projects, Eph-Igs bound to a solid support have been used as affinity reagents to assist in purifying the solubilized ligands. This is a very powerful method which has been instrumental in identifying several of the ephrin ligands.

The Eph-IgG and ephrin-IgG molecules have been used to clarify the complex issue of binding specificity. With so many receptor-ligand interactions possible, the ability to perform in vitro binding studies in a microtiter plate format has greatly eased the task of establishing which ligands and receptors interact. This exemplifies the power of the same plate coating techniques used in the ELISA method. It is clear from these studies that each of the Eph receptors can bind many different ligands and each of the ligands can bind many different receptors. However, there is a simplifying theme that the Ephs and ephrins fall into two main specificity subclasses (Gale et al., 1996). The large number of different receptors and ligands in this family, coupled with their extreme promiscuity, gives rise to a difficult case of the specificity of the immunoadhesins being limited by their inherent biology. This is a case where antibodies, which may be made specific for individual Eph or ephrin types, may display some significant advantages over immunoadhesins.

Nevertheless, immunoadhesins have some characteristics which have been used to great advantage in investigating the function of this family. Many, if not all of the Eph receptors require their ligand to be "clustered" or crosslinked for effective signaling. This requirement is presumably normally met by the ephrins' attachment to the membrane of an adjacent cell. Indeed, use of ephrin containing cells or membrane fragments leads to effective tyrosine phosphorylation on the appropriate Eph receptor. While monovalent, soluble ephrin extracellular domains have been ineffective in stimulating the receptor, the dimeric immunoadhesins of the ephrin ligands have given some success. In several cases, the presence of the dimer has been sufficient to give maximal activation of tyrosine phosphorylation, while in others, the dimer appears to have no better activity than the monomer. The addition of an anti-Fc antibody to cause further aggregation of the ligand causes inactive ephrin immunoadhesin dimers to become active, and usually increases the efficacy of active ephrin immunoadhesin dimers. The presence of the Fc "handle" on the ephrin extracellular domain made these studies much more straightforward.

The ability of some of the ephrin extracellular domains, even as dimeric immunoadhesins, to bind to the appropriate Eph receptor and yet not, or only minimally, activate them, opens up the possibility of using the ephrin immunoadhesins themselves as a competitive antagonist. This is a departure from the immunoadhesin blocking agents used in the neurotrophin field, where the receptor immunoadhesins have been used to block ligand. This has indeed been used to evaluate the functions of the Eph-B3/ephrin-B1 system in neural crest migration in vivo and in vitro (Krull et al., 1997). Interestingly, in this case, the ephrin-B1 ECD or the ephrinB1 immunoadhesin were effective while the corresponding Eph-B3 immunoadhesin was not.

The use of the Eph immunoadhesin as a staining reagent for expressing cells was useful during the cloning of several of the ephrins. An extension of this technique is the use of the Eph and ephrin immunoadhesins to indicate the location of the corresponding ephrin or Eph in situ. This is essentially similar to immunocytochemical labeling, only relying on the specificity of the receptor-ligand interaction instead of antibody-antigen interaction. This has been applied to staining whole embryos to examine the distribution of the Eph-ephrin family during development (Gale et al., 1996). Because of the large number of members of the Eph family that bind an individual ephrin and vice versa, these investigators relied on the observation that they can be split into two major subclasses, as described earlier. This gave a general picture of the distribution of each of these subclasses, and allows the formulation of more specific questions about the distribution and role of individual members of the family.

An interesting outcome of this work, and perhaps illustrating a general caveat about interpreting data obtained from this method, is that, as judged by staining, the Eph and ephrin distribution was mutually exclusive. This would imply that the only signaling that could occur would take place at the quite sharp boundary between the zones of Eph and ephrin expression. This might predict an important role for these molecules in laying out major aspects of body plan. However a caveat in interpreting these data, and perhaps applicable to all data obtained by staining with receptor or ligand chimeras, is that one can only presumably stain unoccupied receptor. For instance, if a given Eph and one of its cognate ephrins actually are expressed in an overlapping fashion, the ephrin binding site on the Eph is presumably occupied by the ephrin. This would effectively block binding of the ephrin immunoadhesin used for staining, showing only that portion of the zone of Eph expression where there was little or no ephrin expression. The same would hold true for staining the zone of ephrin expression using the Eph immunoadhesin, thereby giving an apparent sharp boundary and a non-overlapping pattern of expression. This is a potentially inconvenient aspect of the same underlying properties that give rise to the utility of the immunoadhesins as blocking agents. Immunoadhesins bind to their partners via the same molecular sites that are used in the natural interaction of the two proteins. It will require more work using alternate techniques, such as antibody staining, or in situ hybridization, to uncover the true underlying expression pattern of the Eph and ephrin family members.

Conclusion

As the above examples demonstrate, immunoadhesins are versatile and powerful tools. Both their strengths and limitations arise from the same properties, which need to be kept in mind while considering their use. They are convenient, soluble forms of cell surface or secreted proteins. They contain an essentially normal Fc domain, which gives them many of the same properties as antibodies. They can be "made to order" from any species, as long as one has the necessary facilities. In many ways, their properties complement those of traditional antibodies. It seems quite likely that they will continue to be important reagents in neurobiological investigations.

References

Amann, R., Schuligoi, R., Herzeg, G. and Donnerer, J. (1996) Intraplantar injection of nerve growth factor into the rat hind paw: local edema and effects on thermal nociceptive threshold. *Pain*, 64: 323–329.

Andreev, N., Dimitrieva, N., Koltzenburg, M. and McMahon, S.B. (1995) Peripheral administration of nerve growth factor in the adult rat produces a thermal hyperalgesia that requires the presence of sympathetic post-ganglionic neurones. *Pain*, 63: 109–115.

Averill, S., McMahon, S.B., Clary, D.O., Reichardt, L.F. and Priestley, J.V. (1995) Immunocytochemical localization of trkA receptors in chemically identified subgroups of adult rat sensory neurons. *Euro. J. Neurosci.*, 7: 1484–1494.

Barinaga, M. (1995) Receptors find work as guides. *Science*, 269: 1668–1670.

Beckmann, M.P., Cerretti, D.P., Baum, P., Vanden Bos, T., James, L., Farrah, T., Kozlosky, C., Hollingsworth, T., Shilling, H., Maraskovsky, E., Fletcher, F.A., Lhotak, V., Pawson, T., and Lyman, S.D. (1994) Molecular characterization of a family of ligands for eph-related tyrosine kinase receptors. *EMBO J.*, 13: 3757–3762.

Bennett, B.D., Zeigler, F.C., Gu, Q., Fendly, B., Goddard, A.D., Gillett, N. and Matthews, W. (1995) Molecular cloning of a ligand for the EPH-related receptor protein-tyrosine kinase Htk. *Proc. Natl. Acad. Sci. USA*, 92: 1866–1870.

Bennett, D.L.H., Koltzenburg, M., Priestley, J.V., Shelton, D.L., and McMahon, S.B. (1998) Endogenous nerver growth factor regulates the sensitivity of nociceptors in the adult rat. *Euro. J. Neurosci.*, in press.

Brückner, K., Pasquale, E.B. and Klein, R. (1997) Tyrosine phosphorylation of transmembrane ligands for Eph receptors. *Science*, 275: 1640–1643.

Byrn, R.A., Mordenti, J., Lucas, C., Smith, D., Marsters, S.A., Johnson, J.S., Cossum, P., Chamow, S.M., Wurm, F.M., Gregory, T. et al. (1990) Biological properties of a CD4 immunoadhesin. *Nature*, 344: 667–70.

Cabelli, R.J., Hohn, A. and Shatz, C.J. (1995) Inhibition of ocular dominance column formation by infusion of NT-4/5 or BDNF. *Science*, 267: 1662–1666.

Cabelli, R.J., Shelton, D.L., Segal, R.A. and Shatz, C.J. (1997) Blockade of endogenous ligands of trkB inhibits formation of ocular dominance columns. *Neuron*, 19: 63–76.

Capon, D.J., Chamow, S.M., Mordenti, J., Marsters, S.A., Gregory, T., Mitsuya, H., Byrn, R.A., Lucas, C., Wurm, F.M., Groopman, J.E. et al. (1989) Designing CD4 immunoadhesins for AIDS therapy. *Nature*, 337: 525–531.

Cerretti, D.P., Vanden Bos, T., Nelson, N., Kozlosky, C.J., Reddy, P., Maraskovsky, E., Park, L.S., Lyman, S.D., Copeland, N.G., Gilbert, D.J., Jenkins, N.A. and Fletcher, F.A. (1995) Isolation of LERK-5: A ligand of the eph-related receptor tyrosine kinases. *Mol. Immunol.*, 32: 1197–1205.

Chamow, S.M. and Ashkenazi, A. (1996) Immunoadhesins: principles and applications. *Trends Biotechn.*, 14: 52–60.

Davis, S., Gale, N.W., Aldrich, T.H., Maisonpierre, P.C., Lhotak, V., Pawson, T., Goldfarb, M. and Yancopoulos, G.D. (1994) Ligands for EPH-related receptor tyrosine kinases that require membrane attachment or clustering for activity. *Science*, 266: 816–819.

Drescher, U. (1997) The eph family in the patterning of neural development. *Curr. Biol.*, 7: R799–R807.

Drescher, U., Kremoser, C., Handwerker, C., Löschinger, J., Noda, M. and Bonhoeffer, F. (1995) In vitro guidance of retinal ganglion cell axons by RAGS, a 25 kDa tectal protein related to ligands for Eph receptor tyrosine kinases. *Cell*, 82: 359–370.

Dyck, P.J., Peroutka, S., Rask, C., Burton, E., Baker, M.K., Lehman, K.A., Gillen, D.A., Hokanson, J.L., and O'Brien, P.C. (1997) Intradermal recombinant human nerve growth factor induces pressure allodynia and lowered heat-pain threshold in humans. *Neurology*, 48: 501–505.

Eph Nomenclature Committee (1997) Unified nomenclature for Eph family receptors and their lignads, the ephrins. *Cell*, 90: 403–404.

Friedman, G.C. and O'Leary, D.D. (1996) Eph receptor tyrosine kinases and their ligands in neural development. *Curr. Opin. Neurobiol.*, 6: 127–133.

Gale, N.W., Holland, S.J., Valenzuela, D.M., Flenniken, A., Pan, L., Ryan, T.E., Henkemeyer, M., Strebhardt, K., Hirai, H., Wilkinson, D.G., Pawson, T., Davis, S. and Yancopoulos, G.D. (1996) Eph receptors and ligands comprise two major specificity subclasses and are reciprocally compartimentalized during embryogenesis. *Neuron*, 17: 9–19.

Gascoigne, N.R., Goodnow, C.C., Dudzik, K.I., Oi, V.T. and Davis, M.M. (1987) Secretion of a chimeric T-cell receptor-immunoglobulin protein. *Proceedings of the National Academy of Sciences of the United States of America*, 84: 2936–2940.

Holland, S.J., Gale, N.W., Mbamalu, G., Yancopoulos, G.D., Henkemeyer, M. and Pawson, T. (1996) Bidirectional signalling through the EPH-family receptor Nuk and its transmembrane ligand. *Nature*, 383: 722–725.

Ip, N.Y., Stitt, T.N., Tapley, P., Klein, R., Glass, D.J., Fandl, J., Greene, L.A., Barbacid, M. and Yancopoulos, G.D. (1993) Similarities and differences in the way neurotrophins interact with the Trk receptors in neuronal and nonneuronal cells. *Neuron*, 10: 137–149.

Kang, H., Welcher, A.A., Shelton, D. and Schuman, E.M. (1997) Neurotrophins and time: different roles for TrkB signaling in hippocampal long-term potentiation. *Neuron*, 19: 653–664.

Kozlosky, C.J., Marakovsky, E., McGrew, J.T., VandenBos, T., Teepe, M., Lyman, S.D., Srinivasan, S., Fletcher, F.A., Gayle III, R.B., Cerretti, D.P. and Beckmann, M.P. (1995) Ligands for the receptor tyrosine kinases hek and elk: isolation of cDNAs encoding a family of proteins. *Oncogene*, 10: 299–306.

Krull, C.E., Lansford, R., Gale, N.W., Collazo, A., Marcelle, C., Yancopoulos, G.D., Fraser, S.E. and Bronner-Fraser, M. (1997) Interactions of Eph-related receptors and ligands confer rostrocaudal pattern to trunk neural crest migration. *Curr.t Biol.*, 7: 571–580.

Lewin, G.R., Ritter, A.M. and Mendell, L.M. (1993) Nerve growth factor-induced hyperalgesia in the neonatal and adult rat. *J. Neurosci.*, 13: 2136–2148.

Lewin, G.R., Rueff, A. and Mendell, L.M. (1994) Peripheral and central mechanisms of NGF-induced hyperalgesia. *Euro. J. Neurosci.*, 6: 1903–1912.

Lindsay, R.M. and Harmar, A.J. (1989) Nerve growth factor regulates expression of neuropeptide genes in adult sensory neurons. *Nature*, 337: 362–364.

McAllister, A.K., Katz, L.C. and Lo, D.C. (1997) Opposing roles for endogenous BDNF and NT-3 in regulating cortical dendritic growth. *Neuron*, 18: 767–778.

McAllister, A.K., Lo, D.C. and Katz, L.C. (1995) Neurotrophins regulate dendritic growth in developing visual cortex. *Neuron*, 15: 791–803.

McMahon, S.B., Armanini, M.P., Ling, L.H. and Phillips, H.S. (1994) Expression and coexpression of Trk receptors in subpopulations of adult primary sensory neurons projecting to identified peripheral targets. *Neuron*, 12: 1161–1171.

McMahon, S.B., Bennett, D.L., Priestley, J.V. and Shelton, D.L. (1995) The biological effects of endogenous nerve growth factor on adult sensory neurons revealed by a trkA-IgG fusion molecule (see comments). *Nat. Med.*, 1: 774–780.

Orioli, D. and Klein, R. (1997) The Eph receptor family: axonal guidance by contact repulsion. *TIG*, 13: 354–359.

Pasquale, E.B. (1997) The Eph family of receptors. *Current Opinion in Cell Biology*, 9: 608–615.

Rich, K.M., Yip, H.K., Osborne, P.A., Schmidt, R.E. and Johnson Jr, E.M. (1984) Role of nerve growth factor in the adult dorsal root ganglia neuron and its response to injury. *J. Comparative Neurol.*, 230: 110–118.

Shao, H., Lou, L., Pandey, A., Pasquale, E.B. and Dixit, V.M. (1994) cDNA cloning and characterization of a ligand for the Cek5 receptor protein-tyrosine kinase. *J. Biol. Chem.*, 269: 26606–26609.

Shao, H., Lou, L., Pandey, A., Verderame, M.F., Siever, D.A. and Dixit, V.M. (1995) cDNA cloning and characterization of a Cek7 receptor protein-tyrosine kinase ligand that is identical to the ligand (ELF-1) for the Mek-4 and Sek receptor protein-tyrosine kinases. *J. Biol. Chem.*, 270: 3467–3470.

Shelton, D.L., Sutherland, J., Gripp, J., Camerato, T., Armanini, M.P., Phillips, H.S., Carroll, K., Spencer, S.D. and Levinson, A.D. (1995) Human trks: molecular cloning, tissue distribution, and expression of extracellular domain immunoadhesins. *J. Neurosci.*, 15: 477–91.

Thoenen, H. (1991) The changing scene of neurotrophic factors. *Trends Neurosci.*, 14: 165–170.

Traunecker, A., Schneider, J., Kiefer, H. and Karjalainen, K. (1989) Highly efficient neutralization of HIV with recombinant CD4-immunoglobulin molecules. *Nature*, 339: 68–70.

Wang, X.H. and Poo, M.M. (1997) Potentiation of developing synapses by postsynaptic release of neurotrophin-4. *Neuron*, 19: 825–835.

Weskamp, G. and Otten, U. (1987) An enzyme-linked immunoassay for nerve growth factor (NGF): a tool for studying regulatory mechanisms involved in NGF production in brain and in peripheral tissues. *J. Neurochem.*, 48: 1779–1786.

Winslow, J.W., Moran, P., Valverde, J., Shih, A., Yuan, J.Q., Wong, S.C., Tsai, S.P., Goddard, A., Henzel, W.J., Hefti, F., Beck, K.D. and Caras, I.W. (1995) Cloning of AL-1, a ligand for an Eph-related tyrosine kinase receptor involved in axon bundle formation. *Neuron*, 14: 973–981.

Woolf, C.J., Safieh-Garabedian, B., Ma, Q.P., Crilly, P. and Winter, J. (1994) Nerve growth factor contributes to the generation of inflammatory sensory hypersensitivity. *Neuroscience*, 62: 327–331.

Zettlmeissl, G., Gregersen, J.P., Duport, J.M., Mehdi, S., Reiner, G. and Seed, B. (1990) Expression and characterization of human CD4:immunoglobulin fusion proteins. *DNA and Cell Biology*, 9: 347–353.

W.E. Crusio and R.T. Gerlai (Eds.)
Handbook of Molecular-Genetic Techniques for Brain and Behavior Research (Techniques in the Behavioral and Neural Sciences, Vol. 13)

CHAPTER 3.2.7

Protein targeting in the functional analysis of EphA receptors: the use of immunoadhesins

R. Gerlai,[1] N. Shinsky,[1] A. Shih,[1] P. Williams,[2] J. Winer,[2] M. Armanini,[1] P. Moran,[1] B. Cairns,[3] J. Winslow,[1] W.-Q. Gao[1] and H.S. Phillips[1]

[1]*Department of Neuroscience, Genentech, Inc., 1 DNA Way, South San Francisco, California 94080-4990, USA*
[2]*Department of Research BioAssay, Genentech, Inc., 1 DNA Way, South San Francisco, California 94080-4990, USA*
[3]*Department of Pathology, Genentech, Inc., 1 DNA Way, South San Francisco, California 94080-4990, USA*

The previous chapter summarized the utility of immunoadhesins in brain research. This chapter will present yet another application of immunoadhesins, perhaps most relevant for the behavioral neuroscientist. It will focus on a particular protein family (EphA receptor tyrosine kinases) and the investigation of their potential role in certain aspects of behavior and brain function. The chapter is intended to be an example to delineate the advantages as well as disadvantages of using immunoadhesins, a method novel in behavioral neuroscience.

The protein family in question, the EphA receptors, represents the largest family of receptor tyrosine kinases (TK) (Friedman and O'Leary, 1996). The ephrin-A ligands interact with EphA receptors to mediate repulsive axonal guidance in patterning the developing nervous system (Drescher et al., 1995; Winslow et al., 1995; Gao et al., 1996; Zhang et al., 1996; Meima et al., 1997; Orioly and Klein 1997; Zhou, 1997; Flanagan and Vanderhaegen, 1998; Frisén et al., 1998; Gao et al., 1998). It is not known whether EphA receptors play any role in the adult brain in vivo. However, given the role of EphA receptors in determining patterns of synapse formation in development (Drescher et al., 1995; Gao et al., 1996; Zhang et al., 1996; Zhou 1997; Flanagan and Vanderhaegen 1998; Frisén et al., 1998; Gao et al., 1998), and initial evidence for in vitro effects on synaptic physiology (Gao et al., 1998) thought to underlie learning and memory (Bliss and Collingridge 1993), it is plausible that these receptors regulate neuronal function in vivo in the mature brain and, in turn, influence learning. The studies presented in this chapter are aimed at investigating this possibility.

Gene targeting vs the alternatives

Gene targeting (see e.g. Joyner, 1993) using homologous recombination in embryonic stem (ES) cells in the mouse is a powerful tool that is claimed to be more specific than traditional pharmacological approaches in which the pharmacological agents have multiple targets (Tonegawa et al., 1995). However, gene targeting may not always be as specific as it appears. Although well defined at the genetic level, the mutation generated may lead to compensatory processes that result in multiple secondary changes causing, for example, altered neurophysological or behavioral characteristics (for discussion of the topic see Gerlai, 1996b and also Chapter 3.1.10 in this book). This may be a serious limitation if one is interested in, for example, the role particular brain areas play in learning and memory and the targeted gene is expressed during brain development and/or in multiple brain regions. To control the temporal and spatial specificity of the introduced mutation molecular geneticists have designed inducible gene expression or cell type restricted gene targeting systems (for review see Wilson and Tonegawa, 1997; also see Chapters 3.1.3, 3.1.4 in this book). The former has been successfully achieved in a neuroscience application using the tetracycline transactivator system (Mayford et al., 1996), and the latter was employed using the CRE-recombinase system (Tsien et al., 1996). Although these studies signal the coming of a new generation of knock out systems in neuroscience, they may possess imprecise, or "leaky", temporal (precision of on and off time of the tetracycline transactivator system) and spatial control (rarity of restricted location specific promoters for CRE recombinase expression) and they also potentially suffer from other problems related to genetic background (discussed by Gerlai, 1996a) as well as from side effects, e.g. associated with the administration of tetracycline analoges. Moreover, development of mutant mice with the second generation gene targeting systems is extremely time consuming and labor intensive. Therefore, in addition to gene targeting, it may be useful to apply alternative research approaches that are faster to employ and offer comparable or better spatial and temporal control over the introduced perturbation of molecular systems under study. Such approaches could serve as an initial screen for a more elaborate gene targeting study.

Several such approaches are available. Classical pharmacological studies have been proven to yield useful initial information. However, as mentioned above, pharmacological agents often lack the specificity of single gene manipulation. Application of antisense oligonucleotides has some merits as it allows specific alteration of gene function at the transcription level (see Chapter 3.2.4), an approach useful for proteins with high turn over rate. Antibodies raised against particular proteins may be successfully applied to target particular proteins (see Chapters 3.2.5, 3.2.6). Antibodies may be systemically administered but may also be cloned and their genetic material used for transfection thus allowing manipulation of intracellular targets (Carlson, 1988). In the present paper we show yet another potentially useful approach we termed "protein targeting" with the use of immunoadhesins. We suggest it will be particularly useful in the analysis of molecular mechanisms involved in learning and memory and neural plasticity.

Protein targeting: the advantages

As explained in the previous chapter, immunoadhesins are fusion proteins generated by recombinant DNA techniques. The prototypic immunoadhesin is made by ligating two pieces of DNA one encoding the Fc portion of the heavy chain of the immunoglobulin IgG1 and another encoding for example the extracellular domain of a receptor of interest. This DNA construct can be transfected and expressed in a stable manner in cell lines, for example in chinese hamster ovary (CHO) cells (for other examples see review by Chamow and Ashkenazy, 1996). The fusion protein produced by the cell line can then be harvested and purified. Importantly, the fusion protein translated from the chimeric DNA will posses the proper ligand binding domain of the receptor but will have a signaling domain replaced by IgG. Since the Fc IgG portion is located far from the receptor binding domain and since the Fc domain contains the hinge region of the immunoglobulin, the receptor domain of the chimeric immunoadhesin maintains the natural conformation and binding specificity of the endogenous receptor of interest (Chamow and Ashkenazi, 1996) and will bind the endogenous ligand with high specificity and affinity. Winslow et al. (1995) utilized the strong and specific binding ability of the immunoadhesin and used

EphA5-IgG to discover and purify a ligand, ephrin-A5, for the "orphan" receptor tyrosine kinase, EphA5.

The ability of EphA5-IgG to specifically bind ephrin-A5 is very important in the present study. This immunoadhesin, by scavenging the endogenous ephrin-A5 ligand, can effectively inhibit the activation of the endogenous EphA5 receptor (Winslow et al., 1995, Meima et al., 1997). Interestingly, the opposite, i.e. receptor activation, can also be achieved by using immunoadhesins. Receptor activation requires dimerization of the EphA5 receptor (see e.g. Winslow et al., 1995, and Meima et al., 1997). The flexibility of the hinge region of the Fc domain and the fact that the prototypic immunoadhesin is a homodimer may allow appropriate conformation for a ligand-IgG fusion protein to activate the receptor. Indeed we show below that EphA5 can be activated by the immunoadhesin, ephrinA5-IgG.

An important feature of immunoadhesins is that they specifically bind to the endogenous ligand (or receptor) of interest via an unaltered natural binding site of the immunoadhesin molecule (Chamow and Ashkenazi, 1996). This is a crucial characteristic of the immunoadhesins because if the endogenous receptor of interest is promiscuous in vivo and can bind more than one ligand, the immunoadhesin will retain this feature and will also bind all functionally relevant ligands just as the endogenous receptor would. Thus, immunoadhesin application has an advantage over gene targeting in which a lack of a particular protein can be compensated for by upregulation of genes encoding functionally compatible sister proteins. Immunoadhesins may also be superior compared to antibodies. Antibodies raised against a particular protein may recognize domains of the target protein other than the one determining the specificity of binding interactions between ligand and receptor. Immunoadhesins bind to the relevant site. Immunoadhesins share several biological properties of monoclonal antibodies, including pharmacokinetic characteristics and target recognition capacity. When high affinity monoclonal antibodies are difficult to obtain but the gene of receptor, or ligand, of interest is known, generation of immunoadhesins is a viable alternative. Finally, since immunoadhesins are stable proteins that retain biological activity (Chamow and Ashkenazi, 1996) and can directly influence the biological activity of cell surface proteins they may be superior to antisense oligonucleotide application. Antisense oligonucleotides can alter translation of proteins by binding to corresponding mRNA species. However, antisense oligonucleotides may be difficult to deliver to the nucleus of target cells, or may degrade quickly. Furthermore, oligonucleotides may not be the best choice if the targeted molecule represents a protein with a low turnover rate: oligonucleotides do not influence the function of the protein of interest which is already translated.

Protein targeting vs gene targeting in the adult brain

Compared to gene targeting, protein targeting with immunoadhesins may be particularly useful when one is interested in adult brain function but the gene of interest is expressed both during development and in the adult brain. Protein targeting allows the manipulation of adult neural function without inducing developmental changes. Recent evidence suggests that a large number of genes, previously thought to play roles only in the development of the central nervous system may also be involved in neural plasticity and the formation of memory. Neurotrophins (for review see Thoenen, 1995; Schuman, 1997), cell adhesion molecules (for review see Fields and Itoh, 1996) and protein kinases (e.g. Grant et al., 1992; Martin et al., 1997) represent the best examples. Several members of these large protein families have been shown to play roles in a form of synaptic plasticity, long-term potentiation (LTP), which, in mammals, is most robustly observed, and perhaps plays a unique role, in the hippocampus. Targeting these proteins with immunoadhesins may be advantageous over the classical gene targeting approach in which developmental and adult functional effects are difficult to separate. It is important to realize, however, that immunoadhesins delivered to the brain as proteins can interact only with cell surface proteins, therefore their application, unlike gene targeting, is limited to this class of molecules. In principle, however, intracellular targets may be manipulated by transfection of expression vector constructs encoding the immunoadhesin but this approach has not been attempted in behavioral neuroscience applications and will not be discussed here.

Protein targeting has never been used in behavioral neuroscience research. Nevertheless, in principle, it is a typical pharmacological approach and it is also not substantially different from classical antibody application. Practically different but principally also similar are the studies in which not the protein itself, but cells or tissue grafts producing a particular protein are implanted (see e.g. Silver et al., 1996). Furthermore, it is also notable that immunoadhesins have recently been used in the analysis of biological function, in, for example, the study of effects of neurotrophins on long-term potentiation in hippocampal explants (Kang et al., 1997) and also in vivo in the development of ocular dominance columns in the cat (Cabelli et al., 1997). Thus, immunoadhesins appear to have gained some acceptance in the analysis of brain function. In the following pages we present our methods and findings demonstrating that direct infusion of immunoadhesins into the hippocampus is feasible in vivo and will enable researchers to analyze the role particular cell surface proteins may play in such complex behavioral phenomena as learning and memory.

Prior to protein targeting: the pilot experiments

Protein targeting entails the following steps: the design of an expression vector encoding an immunoadhesin, transfection of an appropriate cell line (e.g. CHO) with the vector, growth of the cells, purification of the immunoadhesin made by the cells, in vivo infusion of the immunoadhesin, and analysis of brain function. This chapter will focus on the last two steps, however, before one can embark on such a research project one has to conduct several pilot studies.

First, one has to know where the protein to be targeted is expressed. In the present study we wanted to investigate the function of EphA5 and wanted to know where in the brain it may be found. We used in situ hybridization to test mRNA expression. Brains of 4 month old DBA/2 and C57BL/6 inbred mice were fresh frozen with powdered dry ice. Sections were processed by a method described previously (Phillips et al., 1990; Melton et al., 1984). Probes were synthesized according to D.A. Melton et al. (1984) from a 407 DNA fragment that included nucleotides 3040–3446 (I.W. Caras et al., Patent: WO9613518-A1, Accession: P.T18893, 1996). We found strong expression of EphA5 mRNA in pyramidal neurons of all hippocampal CA fields and in dentate gyrus granule cells in the hippocampal formation in adult mice from inbred strains C57BL/6 (Fig. 1(A)) and DBA/2 (not shown). In addition, low levels of expression were detected in the cortex, amygdala, and thalamus.

Once we knew EphA5 mRNA was expressed strongly in the hippocampus, we focused our attention to this brain structure. We wanted to see if EphA5 protein can be found in the hippocampus and also whether it is in a phosphorylated form there. The presence of a phosphorylated protein would be an indication that EphA5 is translated from its gene and is, in fact, in an activated state in the brain. We applied a two step Western blot as described in detail by Winslow et al. (1995) and Gerlai et al. (1998). Briefly, freshly harvested hippocampal tissue was homogenized and EphA5 was immunoprecipitated using an anti-EphA5 antibody (Genentech, Inc.). Phosphorylation levels were tested using anti-phosphotyrosine kinase antibody (4G10, Upstate Biotechnology). The Western blot revealed that EphA5 protein is present in the mature hippocam-

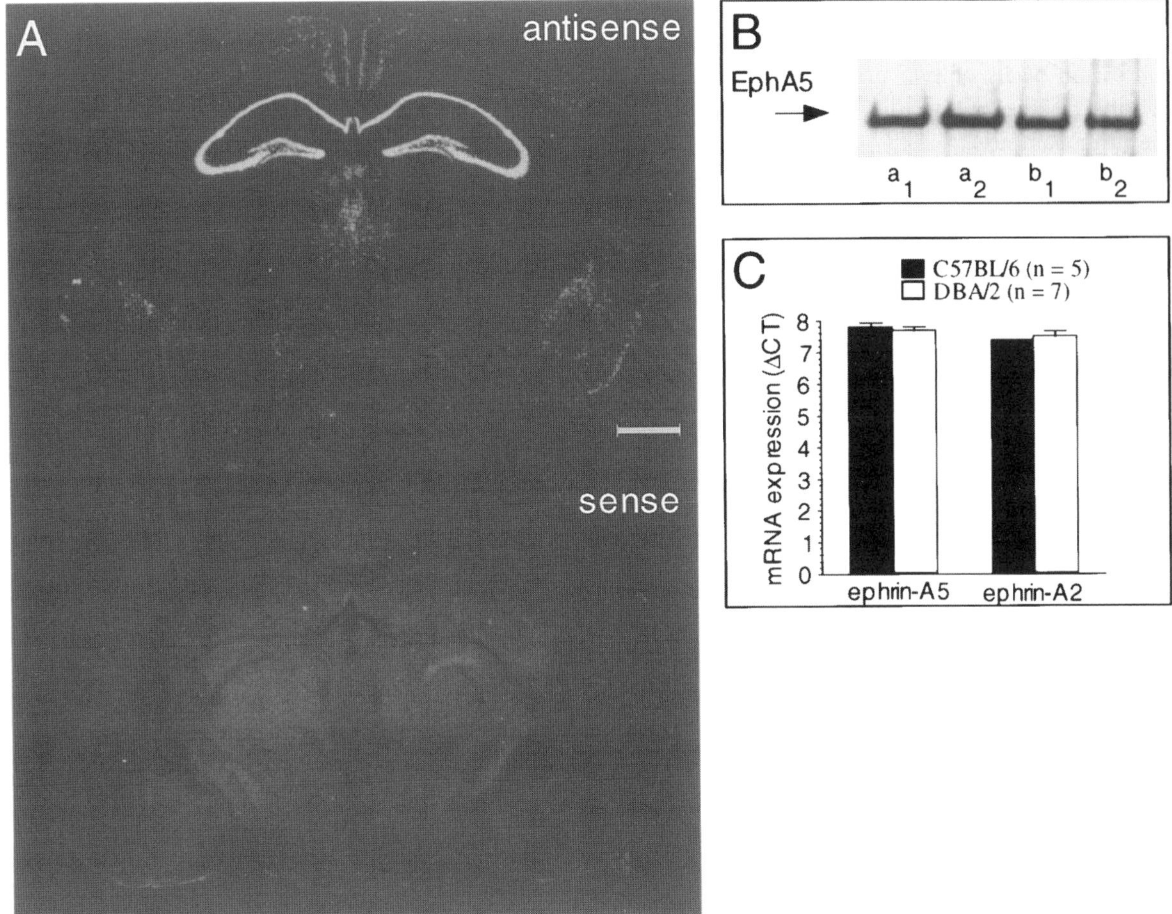

Fig. 1. The EphA5 receptor and its ligands are expressed in the adult mouse hippocampus. (A) In situ hybridization for EphA5. Sense and antisense probes (as indicated) were hybridized to sections of adult mouse hippocampus. Both C57BL6 (shown) and DBA/2 (not shown) mice display prominent expression of EphA5 mRNA in hippocampus and dentate gyrus. Scale bar is 1000 μm. (B) Western blot for phosphorylated EphA5. Each lane represents a hippocampal tissue sample from an individual mouse. Both DBA/2 (samples $a_{1,2}$) and C57BL/6 (samples $b_{1,2}$) strain of mice exhibit a prominent signal. (C) Ephrin-A5 and -A2 ligands are expressed in the hippocampus of both inbred strains of mice analyzed. Reverse-transcriptase polymerase chain reaction (RT-PCR) for mRNA of ligands ephrin-A5 and -A2 in the hippocampi of C57BL/6 (black bars) and DBA/2 (white bars). The data (means ± S.E.M.) are based upon the number of PCR amplification cycles required to reach a threshold level (cycles to threshold, CT) of cleavage of a fluorescent reporter probe (9) and are normalized to GAPDH house keeping gene transcript (ΔCT). Sample sizes (n) indicate the number of mice analyzed. Note that larger values mean smaller original mRNA amount in the hippocampal tissue sample. Also note that amplification characteristics are unique to each gene, therefore, comparison from one gene to another is not valid.

pus in both strains (C57BL/6 and DBA/2) analyzed and is in a phosphorylated form (Fig. 1(B)) implying that EphA5 is involved in some aspect of neural function in the adult mouse CNS. This result also implied that the ligand(s) for EphA5 must also be present in the hippocampus. This information was crucial as in situ hybridization could not detect a significant ephrin-A5 mRNA expression anywhere in the adult mouse brain. However, one must remember that in situ hybridization may not be the most sensitive method with which mRNA expression can be tested. Therefore, we used TaqMan quantitative RT-PCR (see Chapter 2.2.4) and analyzed the expression levels of two proteins ephrin-A5 and ephrin-A2, the preferred ligands of EphA5, in the hippocampus.

The methods of this highly sensitive and quantitative PCR are described in a separate Chapter (2.2.4).

Methods of TaqMan reverse transcriptase polymerase chain reaction (RT-PCR)

Total RNA was isolated using the Qiagen RNeasy Mini Kit from 50 mg mouse hippocampi freshly frozen in liquid nitrogen. RNA was aliquoted and stored at −70°C until used. Oligonucleotide probes and primers were designed to recognize the genes using the Primer Express software (ABI–Perkin Elmer). Real-time quantitative RT-PCR (TaqMan) was performed and analyzed using the Promega Access RT-PCR System in the ABI Model 7700 Sequence Detection system (Heid et al., 1996; Gibson et al., 1996). Samples were run in duplicate with standard curves of Clontech mouse brain total RNA on every plate for all genes and the linearity of amplification was confirmed (Heid et al., 1996; Gibson et al., 1996). The data shown are calculated from the standard curves and each sample is normalized to GAPDH, a house keeping gene, as described previously (Heid et al., 1996; Gibson et al., 1996).

Our results indicated significant ephrin-A5, and -A2 mRNA expression in the hippocampus of both strains (Fig. 1(C)) and the expression levels were not different between strains (ephrin-A5 $t = 0.447$, df $= 10$, $P > 0.60$; ephrin-A2 $t = 0.008$, df $= 10$, $P > 0.99$). These results convinced us that EphA5 plays some role in the hippocampus of adult mice. To investigate its function we designed antagonist (EphA5-IgG) and agonist (ephrinA5-IgG) immunoadhesins that could impair or improve EphA5 activity and we also used a neutral, control immunoadhesin (CD4-IgG).

Generation and purification of immunoadhesins

The first steps involved in the generation of immunoadhesins, i.e. the design and generation of DNA constructs, are not substantially different from the molecular genetic work involved in making transgenic or KO mice. First, a vector containing the appropriate DNA sequences of the immunoadhesin is designed and than this vector is transfected into a cell line, instead of an ES cell (gene targeting) or microinjection into the zygote (generation of transgenic mice). Perhaps the only major difference compared to these latter techniques is the protein purification step which is necessary to obtain a pure immunoadhesin solution. Purification of immunoadhesins, however, is rather easy and is not different from the typical procedures worked out for purification of any immunoglobulin molecule.

Methods of generation and purification of immunoadhesins

CD4-IgG immunoadhesin protein was previously generated by the fusion of residues 1–180 of the mature human CD4 receptor ectodomain and a human IgG1 Fc-domain from residue 216 (first residue of the IgG1 hinge region) to residue 441 (Capon et al., 1989). EphA5-IgG immunoadhesin construct was generated starting with the full length cDNA of EphA5 (Sadjii and Pasquale, 1993; Maisonpierre et al., 1993; Zhou et al, 1994). Owing to apparent sequences within EphA5 cDNA which promote rearrangements and deletions during DNA manipulation, a four step procedure was devised for the assembly of the EphA5-IgG expression vector described in detail elsewhere (Winslow and Caras, 1995). The final construct contained the EphA5 receptor ectodomain encoded by the initiation Met through to residue 462 (residue just prior to the transmembrane domain). EphrinA5-IgG immunoadhesin was constructed as described previously (Meima et al., 1997) and contained the ectodomain of ephrin-A5 up to residue 203, the beginning of the consensus GPI addition sequence (Winslow et al., 1995). All immunoadhesin constructs were cloned into a mammalian expression vector containing a CMV promoter and SV40 polyadenylation sequence (Eaton et al., 1986). Chinese Hamster Ovary (CHO) cell lines were transfected with the vector. Proteins were produced by collection of conditioned media of stably transfected CHO cell lines, and purified by Protein A affinity chromatography (Pharmacia). Protein concentration was verified by amino acid analysis, OD_{280}, and SDS-PAGE/Commassie staining.

Before administration of the immunoadhesins the proteins were concentrated and the Protein A elution buffer (0.1 M sodium citrate, pH 3.5, neutralized to pH 6.8 with K_3PO_4) was exchanged. Approximately 15–20 mg of CD4-IgG (5 mg/ml), EphA5-IgG (1.33 mg/ml), and AL-1-IgG (1.7 mg/ml) were added to Centriprep 30s (Amicon; 30,000 MW cutoff) which had been rinsed and centrifuged with sterile and endotoxin-removed artificial cerebral spinal fluid buffer (ACSF, see below). The samples were concentrated 8–10 fold by centrifugation at 4°C, and diluted with ACSF 2–3 times for $\geq$ 100-fold dilution of the Protein A elution buffer. Potential contaminating endotoxin was removed by passage of buffer-exchanged protein samples through Detoxi-gel (polymixin resin, Pierce) and sterile filtered through a 0.22 μm sterile micro-spin filter (Alltech). The final concentrations of CD4-IgG and EphA5-IgG were approximately 4.7–5.2 mg/ml and 1.5–3.0 mg/ml for ephrinA5-IgG by OD_{280} and SDS gels.

Characterization of binding ability of immunoadhesins: ELISA essay

Once the immunoadhesins were produced it was necessary to test whether they bind the appropriate receptor or ligand under conditions (body temperature) similar to those of the in vivo application we were planning. We used ELISA essay to test appropriate binding capacity of the immunoadhesins. This essay is based upon trapping the immunoadhesin by allowing it to bind to another protein attached to the surface of a plate and then testing how much immunoadhesin is bound after washing. The trap molecule contains the proper ligand or receptor binding sites and it binds the immunoadhesins only if they recognize and bind to these sites. We were planning to infuse the immunoadhesins from micro-osmotic pumps (ALZET).

Binding activity: ELISA essay

ALZET micro osmotic pumps (ALZA Corporation, Palo Alto, California) were filled with the immunoadhesin solution and the flow moderator of the pump was sealed. The pumps were then placed in a thermostat at 37°C. Samples were taken at the time of filling the pump (day 0), and 1, 3, and 9 d later. In order to measure binding activity of EphA5-IgG, plates were coated with 2 μg/ml goat anti-mouse IgG Fc (Jackson Laboratory cat. #115-005-071) in 50 mM carbonate buffer, pH 9.6 at 4°C over night. The plates were washed with a buffer {0.05% polysorbate 20, 0.01% thimerosol in 137 mM NaCl, 2.7 mM KCl, 8.0 mM Na2HPO4 and KH2PO4, pH 7.2 (PBS)} and were blocked with 0.5% BSA, 0.05% Tween 20, and 0.01% Thimerosal in PBS (Assay Diluent) for 1 h at 20°C. EphrinA5-IgG was diluted to 2 μg/ml in assay diluent, then added to the plates and incubated for 1 h at room temperature. Standards (4.0–0.125 ng/ml of EphA5-IgG) and EphA5-IgG samples, diluted in Assay Diluent, were incubated on the plates for 1 h at 20°C. Bound EphA5-IgG was detected using a goat anti-human Fc conjugated to horse radish peroxidase (Jackson laboratory cat. #109-036-098) at a 1:15,000 dilution in assay diluent, and incubated at 20°C for 1 h. Plates were washed between each step. TM-Blue (TSI-CDR cat. #TM102) was used as substrate and the reaction was stopped with an equal volumes of 1 N HCl. Absorbance was read at 450 nm on V_{max} plate reader (Molecular Devices). The standard curve was fitted using a four-parameter nonlinear regression curve-fitting program (Softmax program, Molecular Devices). Data points which fell in the range of the standard curve were used for calculating the EphA5-IgG concentration. EphA5-IgG concentrations reported in the present paper were the average of several dilutions performed in duplicate.

To measure binding activity of ephrinA5-IgG, the ELISA plate was coated with goat anti-human Fc antibody (Jackson Laboratory 109-005-098) in 50 mM carbonate buffer pH 9.6. After washing the plates were blocked with 0.5% BSA, 0.05% Tween 20 and 0.01% thimerosal in PBS (assay diluent) for 1 h at 20°C. Subsequently, the plate was incubated with EphA5-IgG at 2 μg/ml for 1 h at 20°C and washed. EphrinA5-IgG standards (4.0–0.125 ng/ml of ephrinA5-IgG) and samples, diluted in Assay Diluent, were incubated on the plates for 1 h at 20°C. Bound ephrinA5-IgG was detected using a goat anti-mouse Fc conjugated to horse radish peroxidase (Jackson Laboratory cat. #115-035-071) at a 1:15,000 dilution in assay diluent, and incubated at 20°C for 1 h. Termination of reaction, substrate used, reading of absorbance and calculations were done as described above.

The results are shown on Fig. 2. Both EphA5-IgG and ephrinA5-IgG exhibited a declining binding affinity with time (EphA5-IgG $F(3, 12) = 70.61$, P %lt; 0.001; ephrinA5-IgG $F(3, 6) = 26.14$, $P < 0.001$), which is most probably due to aggregation of the fusion proteins at 37°C or to adhesion to the internal surface of the minipump (Chamow pers comm). Nevertheless, both proteins showed a binding rate significantly above 0% even after 10 d of exposure to 37°C in the pump (EphA5-IgG t = 14.38, df = 4, $P < 0.001$; ephrinA5-IgG, $t = 35.00$, df = 2, $P < 0.001$) suggesting that these immunoadhesins retained their binding specificity. By the end of the 7th day (the time of behavioral testing) approximately 65% of EphA5-IgG and 18% of ephrinA5-IgG appears to be still actively binding

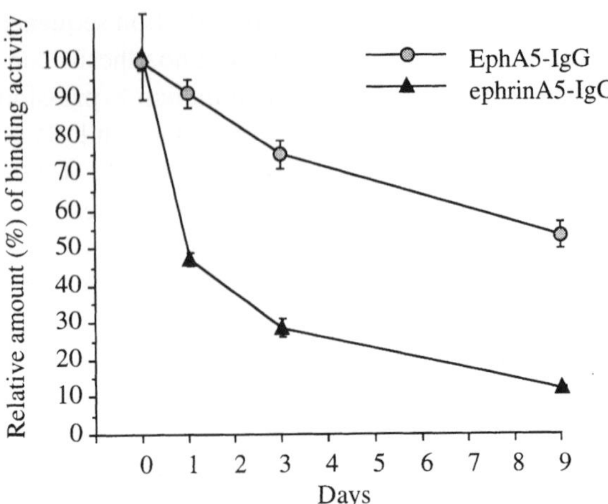

Fig. 2. Binding activity of immunoadhesins EphA5-IgG and ephrinA5-IgG as measured by ELISA assay. Note that although ephrinA5-IgG exhibited a less robust binding activity, both immunoadhesins show a significant binding activity even after 7 d (the time of behavioral experiments) of exposure to 37°C in the minipump.

the appropriate ligand or receptor. Assuming a linear pumping rate and decrease of binding capacity, we calculate that, on average, approximately 77% of the infused EphA5-IgG and 37% of the infused ephrinA5-IgG was active in the hippocampus of the mice during the 7-d period of infusion.

Infusion of immunoadhesins into the brain

The immunoadhesin (CD4-IgG, EphA5-IgG, or ephrinA5-IgG) was infused directly into the hippocampus bilaterally using the ALZET 1007D micro-osmotic pump with a standard ALZET brain infusion kit. These pumps allow water to diffuse through their surface, which contains a unique semipermeable polymer, into their internal cavity that houses a flexible sac filled with the solution to be pumped. As a result of the increasing osmotic pressure in the internal cavity, the content of the sac is emptied. In our study the internal volume of the osmotic pump was 99 μl and the pumping rate was 0.51 μl/h allowing the pump to deliver the immunoadhesin for an approximate period of 8 d.

Method of stereotaxic operation

Using a stereotaxic frame mice received bilateral intrahippocampal implantation of cannulae as described in detail elsewhere (Gerlai et al., 1998) at position AP — 1.5, ML ±1.8, as measured in mm from Bregma and DV 1.8 mm from flat skull surface. This cannula positioning places the tip of the cannula in the middle of the dorsal hippocampal formation just above the dorsal dentate gyrus granule cell layer. Canulae, cemented onto the skull by dental acrylic, were connected to a pair of ALZET 1007D micro-osmotic pumps placed subcutaneously on the back of the mouse. For C57BL/6 mice, CD4-IgG (control) or EphA5-IgG (antagonist) immunoadhesins were infused (4.8 mg/ml in artificial cerebrospinal fluid [ACSF] containing 124 mM NaCl, 3 mM KCl, 2.4 mM CaCl2, 2.4 mM MgSO4.7H2O, 1.25 mM KH2PO4, 26 mM NaHCO3, and 10 mM D-glucose). For DBA/2 mice, CD4-IgG or ephrinA5-IgG was infused (0.5 mg/ml in ACSF).

To make sure that the infusion described above can be carried out successfully and the immunoadhesins really infuse to the appropriate brain area we wanted to target, we conducted an immuno-histochemical analysis and studied where exactly the immunoadhesins diffused in the brain.

Method of immuno-histochemical analysis

The animals were anaesthetized with sodium pentobarbital, transcardially perfused with saline, followed by 4% paraformaldehyde solution (pH 7.4). The brains were removed, post-fixed in the paraformaldehyde solution overnight, and cryoprotected in 30% sucrose. The tissue was embedded in OCT compound and frozen in isopentane cooled by liquid nitrogen. Coronal sections (35 μm) were cut using a cryostat (Leica CM3000) throughout the hippocampus and were immunostained for the human IgG portion of the infused immunoadhesin. Sections were washed with PBS and quenched for endogenous peroxidase using Blocking Solution (Kirkegaard and Perry Laboratories) 1:10 in PBS for 30 min then rinsed in PBS containing 0.3% Triton X-100 (PBS/T), and incubated in 5% normal horse serum in PBS/T for 1 h, followed by an overnight incubation at 4°C in biotinylated donkey anti-human IgG antibody (Jackson ImmunoResearch, code 709-65-149) 1:200 in PBS/T. After rinses in PBS, the sections were incubated in avidin-biotin-peroxidase complex (Elite ABC kit, Vector) for 1 h at RT and rinsed again in PBS, then developed using a standard diaminbenzidine reaction (Pierce Enhanced Metal). All the sections were counter-stained in Mayer's hematoxylin, dehydrated and coverslipped.

The result of infusion showed homogeneous immunoadhesin distribution within the hippocampus (Fig. 3). Cortical areas within 1 mm surrounding the canula insertion site were also stained in

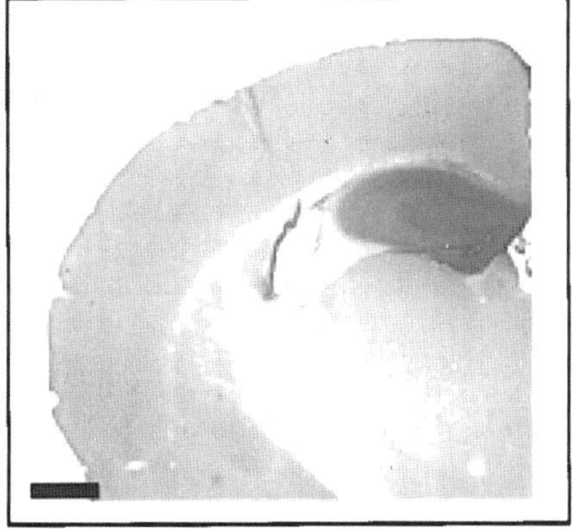

Fig. 3. Immunohistochemical staining for the IgG domain of the immunoadhesin reveals pronounced hippocampal diffusion of the protein after 7 d intrahippocampal infusion from micro-osmotic pump. The right hemisphere anterior to the canula insertion site (Bregma — 1.35 mm) with EphA5-IgG infusion is shown. Note the strong peroxidase (dark) staining observable in the hippocampus and the lack of staining in infrahippocampal areas. Note that staining of supra-hippocampal cortical areas within 1 mm from the canula insertion site was also observed in some specimens. Scale bar is 700 μm.

some samples but other cortical areas or infra-hippocampal structures remained uninfused.

This result suggested that the immunoadhesin was properly delivered and no, or very little part (only small ventral areas), of the hippocampus remained unaffected. This is an important finding because lesion experiments showed that even after substantial lesioning the hippocampus may function normally at the behavioral level unless the majority of its dorsal portion is disrupted (Moser et al., 1995). Thus one may need to affect the entire, or at least the majority of the dorsal portion of this structure in order to see behavioral changes.

In addition to studying where the immunoadhesins diffused in the brain we conducted a number of control experiments to investigate whether the infusion procedure itself disrupted brain function at the behavioral level. We also studied whether immunoadhesin infusion led to gross anatomical disruption, edema, or inflammation in the brain by using magnetic resonance imaging (MRI) and immunostaining for specific neuronal and glial markers. Although these studies are not described here, one has to emphasize the importance of conducting such control experiments. Fortunately, we found no indication of behavioral changes in response to the operating procedure and vehicle infusion, nor did we detect any other changes in response to immunoadhesin infusion by MRI and immunostaining (Gerlai et al., 1998). These results suggest that canulation and intrahippocampal infusion of immunoadhesins is safe, does not lead to non-specific changes in the brain, and thus may be used to test function of EphA receptors.

Finally, we wanted to study whether in vivo infusion of immunoadhesins really alters receptor activity. Phosphorylation of EphA5 receptor, similarly to other tyrosine kinase receptors, is thought to underlie receptor activation. Consistently with its antagonist action, EphA5-IgG was previously shown to impair phosphorylation of the EphA5 receptor (Winslow et al., 1995; Meima et al., 1997). But the evidence for the contrary, i.e. for an agonist action was lacking. Using

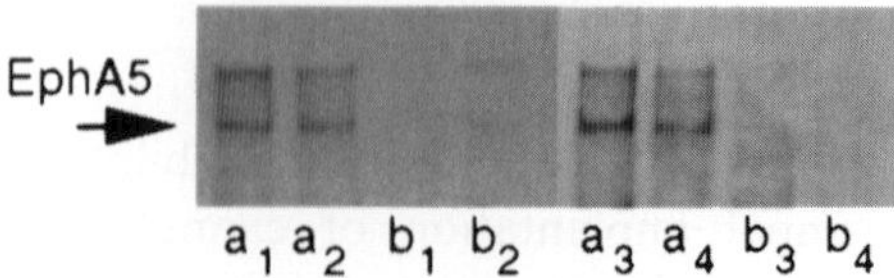

Fig. 4. EphA5 receptor phosphorylation is induced by in vivo infusion of the agonist ephrinA5-IgG in hippocampal tissue samples. After immunoprecipitation with anti-EphA5 antibody, phosphorylation levels were detected by anti-phosphotyrosine kinase antibody. Each lane represents hippocampal tissue from an individual mouse: (a) ephrinA5-IgG infusion; (b) CD4-IgG infusion.

Western blot analysis, we showed that EphA5 phosphorylation can be induced in vivo by ephrinA5-IgG infusion in the hippocampus (Fig. 4).

Immunoadhesin antagonist and agonist of EphA function elicits impairment and improvement in two behavioral tasks

The major aim of the present research was to investigate the role EphA receptors may play in the mouse adult brain. The hippocampal expression pattern of EphA5 and its ligands ephrin-A5 and ephrin-A2 suggested that this receptor may be involved in hippocampal function. To determine whether EphA receptors mediate neural function we focused our attention to the hippocampus because of this expression pattern and also because of the availability of behavioral paradigms and electrophysiological methods sensitive to detect dysfunction of this structure. We put considerable emphasis on behavioral analysis because it offers a sensitive way to detect neural functional changes (Gerlai, 1996c; Gerlai and Clayton, 1999).

We utilized the availability of two inbred strains of mice, C57BL/6 and DBA/2, that exhibit striking differences in hippocampal function at the behavioral level (Van Abeelen, 1989; Paylor et al., 1994; Gerlai, 1998a, 1998b). First we infused EphA5-IgG or a control immunoadhesin, CD4-IgG, bilaterally into the hippocampus of adult C57BL/6 mice for 8 d and then tested behavioral performance in T-maze continuous spontaneous alternation task (T-CAT), a newly developed paradigm that is sensitive to detect hippocampal dysfunction and abnormalities that affect spatial learning and working memory (Gerlai 1998a).

Method of T-CAT, the T-maze continuous alternation task

The T shaped maze (length of start and goal stems = 75 cm, width = 12 cm, height = 20 cm) was made of transparent acrylic glued to a matte black acrylic square bottom piece. The maze was equipped with three removable guillotine doors. One separated a 24 cm compartment at the beginning of the start arm. The other two were placed at the entrance of each goal arm and could be lowered to block entry. The apparatus was placed in a testing room rich in visual stimuli illuminated by fluorescent light tubes (12 W/m^2 area) from the ceiling. The extra-maze visual cues were 50–200 cm away from the maze. A fan provided a constant background noise to mask unwanted auditory stimuli. The T-maze was thoroughly cleaned with clydox spray and then dried after each individual mouse but not between trials of a test session (see below). T-maze alternation procedure was newly developed (Gerlai, 1998a) to exclude extensive handling of experimental animals usually required in spontaneous alternation tasks. This allowed measuring exploratory behavior better and in an unbiased way. The procedure does not use artificial motivators such as food reward or air puff punishment but relies on the animal's natural tendency to explore novelty (for the importance of ethologically relevant behavioral research see Chapter 4.1). Mice were allowed to alternate between the left and right goal arms of the T-maze throughout a 15 trial session. Once they have entered a particular goal arm, a guillotine door was lowered to block entry to the opposite arm. The door was removed only after the mice returned to the start arm, thus allowing a new alternation trial to be started. Alternation rate was calculated as a ratio between the alternating choices and total number of choices (50%, random choice; 100%, alternation at every trial; 0%, no alternation). Time to complete 15 choices was also measured.

C57BL/6 mice infused with the control CD4-IgG immunoadhesin exhibited an excellent alternation rate (Fig. 5(A)), a performance typical for intact mice of this strain (Gerlai, 1998a; Douglas, 1990). In contrast, mice infused with EphA5-IgG exhibited a significantly impaired alternation rate (Fig. 5(A)). The impairment is unlikely to be due to altered motor performance or motivation levels, because both groups of mice finished the task within the same amount of time (Fig. 5(B)). Furthermore, the impairment is also unlikely to be due to altered sensory capabilities as the task can be solved using multiple modalities of cues whose processing is distributed in several brain areas left unaffected by the immunoadhesin infusion.

The deficit in EphA5-IgG infused mice was further characterized utilizing context-dependent fear conditioning (CDFC), a configural learning task also found sensitive to hippocampal dysfunction (Kim and Fanselow, 1992; Phillips and LeDoux, 1992; Abeliovich et al., 1993; Aiba

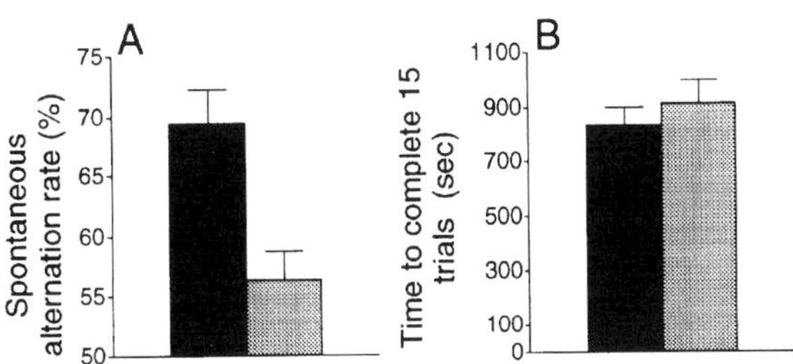

Fig. 5. EphA5-IgG infusion impairs T-maze continuous alternation in C57BL/6 mice in a 15 trial alternation session. EphA5-IgG infusion: hatched bar, $n = 23$; CD4-IgG infusion: black bar, $n = 22$. One choice was allowed at each trial. Alternation rate is a ratio between the alternating choices and total number of choices. (A) Significant difference was found in alternation rate ($t = 3.528$, df = 43, $P < 0.001$). (B) No significant difference was found between groups in time spent to complete 15 choices ($t = 0.915$, df = 43, $P > 0.36$). Error bars represent S.E.M.

et al., 1994; Bach et al., 1995; Gerlai, 1998b). The reason to conduct this additional behavioral test is that CDFC and T-CAT are associated with different motor (T-maze: activity; CDFC: passivity), perceptual (T-maze: extra maze visual cues; CDFC: intra- maze visual, olfactory, tactile, auditory cues) and motivational (T-maze: spontaneous curiosity; CDFC: negatively reinforced fear) requirements, yet both tests are sensitive to hippocampal dysfunction. Thus, comparison of T-maze and CDFC performance may allow one to dissociate simple performance deficits, or improvements, from alterations in hippocampal (cognitive) function. In the CDFC paradigm, hippocampal damage results in the loss of freezing (a natural response to pain or fear) to contextual cues while sparing response to a simple associative cue. The methods of this learning paradigm have been described in detail elsewhere (Gerlai, 1998b). Note that these methods were designed to maximize the context dependency, and hippocampal sensitivity, of the paradigm.

Method of CDFC, the context dependent fear conditioning

The CDFC paradigm had three phases: a training phase, a context dependent test, and a cue dependent test. For training, mice received 3 electric foot shocks (1 s, 0.7 mA) each preceded by an 80 dB, 2900 Hz, 20 s long tone cue in the Gemini avoidance apparatus (San Diego Instruments) as shown in Figs 5 and 7. Context and cue test apparati received olfactory cues different from those of training. The context test was performed in the training chamber but no shock or tone was delivered. The cue test was carried out in another chamber identical in size but different in visual, olfactory, and tactile cues from those of the training chamber. Tone signals identical to the one used in training but without a shock were given. Behavior was video-recorded and later quantified using an event recording computer program (Noldus, Observer). Behavior elements, as indicated on Figs 5 and 7 were analyzed.

After training, control (CD4-IgG-infused) C57BL/6 mice exhibited a substantial freezing response to both a simple associative cue (cued task) and complex contextual cues (context-dependent task) (Fig. 6(B) and (C)). As compared to controls, mice treated with EphA5-IgG showed a significant deficit in the context-dependent task, but not in the cued task (Fig. 6(B) and (C)). The performance deficit of the EphA5-IgG treated mice is unlikely to be due to sensory, motor, or motivational disturbances, as no differences were seen between responses of EphA5-IgG and CD4-IgG mice during either the training session or in the cued task (Fig. 6(A) and (C)). To further ascertain that changes in the freezing response was not simply due to altered motor performance, other behavioral elements associated with fear (Gerlai et al., 1993) including locomotion and grooming (negative correlation with fear) and long-body posture (positive correlation with fear, Fig. 5(A) insets) were also measured. Analysis of these elements confirmed the context-specific nature of the learning performance deficit (insets Fig. 6(B) and (C)).

It is unlikely that a context specific learning deficit induced by the EphA antagonist may result from non-specific disruption of brain function unrelated to learning. Nevertheless, to rule out this possibility we investigated whether improved learning performance in response to EphA activation could be elicited. Thus, ephrinA5-IgG, the agonist immunoadhesin, was bilaterally infused into the hippocampus of DBA/2 mice, a strain which displays deficits in hippocampal learning tasks and other measures of hippocampal function (Gerlai, 1998b; Crusio et al., 1990; Matsuyama et al., 1997). DBA/2 mice infused with ephrinA5-IgG exhibited an improvement in performance in both T-CAT and CDFC paradigms (Figs 7 and 8). While control DBA/2 mice infused with CD4-IgG did not alternate above chance level on the T-CAT, the alternation rate of mice treated with ephrinA5-IgG was increased significantly above chance and the rate of the control mice (Fig. 7(A)) but the time to complete

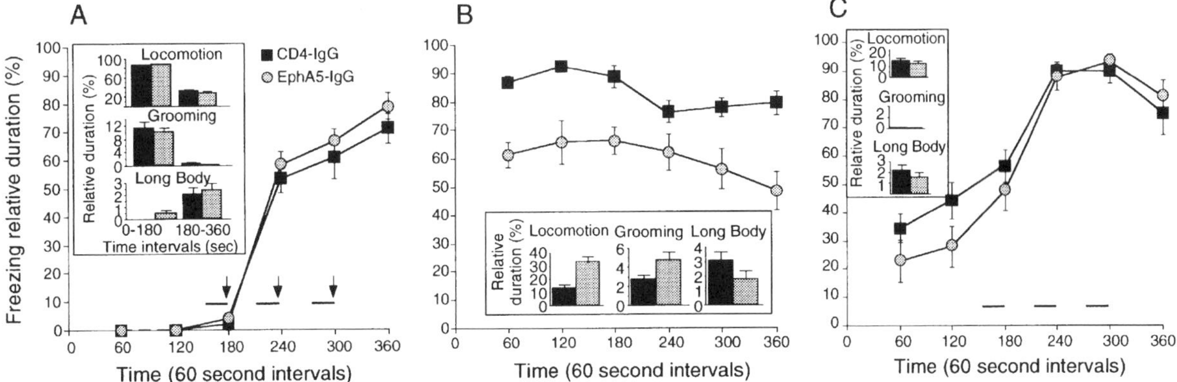

Fig. 6. EphA5-IgG infusion impairs learning performance in a context-specific manner in fear conditioning in C57BL/6 mice. In the CDFC paradigm, mice associate two substantially different types of cues with a negative reinforcer, an electric foot shock. The shock is paired with a simple associative cue, a tone, in a shock chamber characterized by multiple contextual cues. (A) During training, mice were given three 20 s tone signals (shown as solid horizontal bars) that coterminated with 1 s long electric shocks (0.5 mA, shown as arrows). Both CD4-IgG ($n = 22$) or EphA5-IgG infused mice ($n = 23$) responded to training with increased freezing, a natural response to painful stimuli, and no significant differences were seen between groups. Relative duration (%, time per 60 s) of freezing behavior is shown for 60 s intervals. In addition to freezing, 3 other behavioral elements (bar graphs in the inset) are also shown. Note that fear correlates negatively with locomotion and grooming and positively with long body. No significant differences were detected between the mice in any of the behavioral measures either before (interval 0–180 s), or after shock (180–360 s). (B) A randomly-assigned subset of trained mice ($n = 16$ for CD4-IgG infused; $n = 17$ for EphA5-IgG infused) was tested in the shock chamber for response to contextual stimuli. No tone cues or shock were given. The freezing behavior of EphA5-IgG infused mice was significantly impaired compared to that of the CD4-IgG infused animals ($F(1, 31) = 24.926$, $P < 0.0001$). In addition to freezing, relative duration of 3 other behavioral elements is also shown for the entire session. EphA5-IgG infused mice were found to exhibit an increased amount of locomotion ($t = 4.315$, df = 31, $P < 0.0001$) and grooming ($t = 2.133$, df = 31, $P < 0.05$) and exhibited decreased long body posture ($t = 2.100$, df = 31, $P < 0.05$), all suggesting decreased level of fear. (C) A randomly assigned subset of trained mice ($n = 12$ for CD4-IgG; $n = 13$ for EphA5-IgG) was tested in the cued test. The cued test was conducted in a chamber which lacked the olfactory, visual, and tactile cues (the contextual stimuli) of the shock chamber. Mice received three tone signals alone (solid horizontal bars), but no shock. Both groups of mice responded to the tone cue with a robust increase in freezing, and no significant differences were found between the two groups of mice on freezing ($F(1, 23) = 2.068$, $P > 0.15$) or any of the other behavioral measures analyzed ($t < 0.99$, df = 23, $P > 0.30$). Data obtained in fear conditioning are shown as means ± S.E. M.

the task remained unchanged (Fig. 7(B)). In the CDFC paradigm, ephrinA5-IgG infusion elicited a selective improvement on the context-dependent task (Fig 8(B)), a performance enhancement most probably due to cognitive rather than sensory, motor, or motivational effects (Fig. 8(A) and (C)). Again, the changes were observable not only in freezing but in the other behavioral measures of fear (Fig. 8 insets).

Given the pre- and post-synaptic expression of EphA receptors (Gao W-Q et al., 1998) and their role in synapse formation during development, to investigate potential mechanisms underlying the observed behavioral changes we tested several neuronal and synaptic markers at the immunostaining level including NMDA-R1,

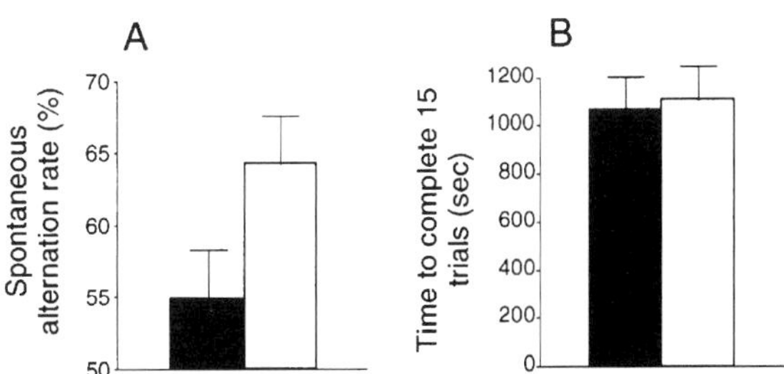

Fig. 7. Infusion of ephrinA5-IgG in DBA/2 mice improves spontaneous alternation rate in the T-maze. (A) EphrinA5-IgG treated mice (white bar, $n = 19$) exhibited higher levels of alternation on the TCAT test compared to CD4-IgG infused mice (black bar, $n = 17$; $t = 2.85$, df = 34, $P < 0.01$) (B) Time spent to complete 15 alternation trials did not differ between treatment goups ($t = 0.365$, df = 34, $P > 0.710$) suggesting that the improved alternation performance is not due to motoric or motivational influences. Error bars represent S.E.M. Methods as in Fig. 4.

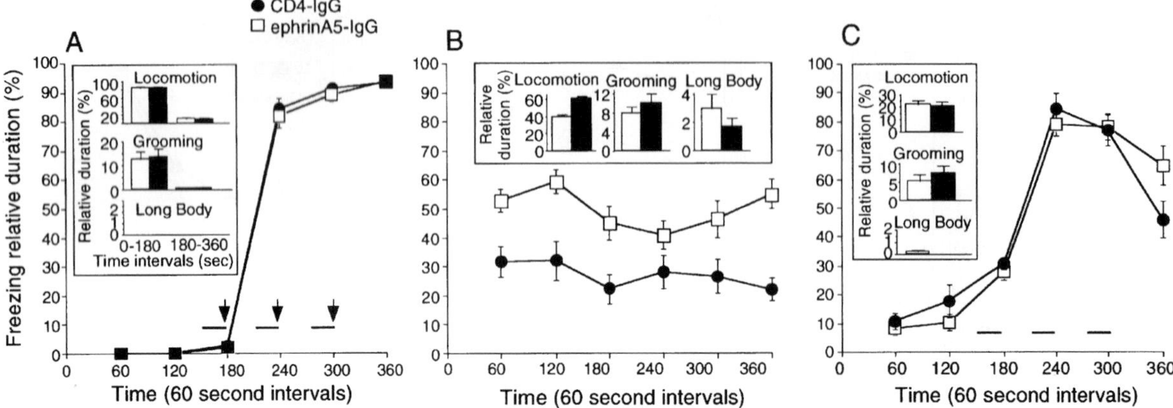

Fig. 8. EphrinA5-IgG infusion significantly improves learning performance in a context-specific manner in fear conditioning. Methods as described in Gerlai (1998b). (A). In the training session, no significant differences were found between the two groups of mice in any of the behaviors measured suggesting normal perceptual and motor performance. (B) Response to contextual stimuli is shown. The performance of ephrinA5-IgG infused mice was significantly improved compared to the CD4-IgG infused animals throughout the test session ($F(1, 34) = 33.434$, $P < 0.0001$). In addition to freezing, increased level of fear in EphrinA5-IgG infused mice is also indicated by changes in other behavioral elements (inset). (C) In the cued test, mice were placed in a chamber whose contextual stimuli were different from those of the shock chamber as explained in Fig. 6(C). The mice received three tone signals alone (solid horizontal bars), but no shock. All mice responded to the tone cue with a robust increase in freezing, and no significant difference was found between the two groups of mice in freezing ($F(1, 34) = 0.004$, $P > 0.95$) or in any of the other behavioral measures analyzed.

GluR1, PKC, GABA, synaptobrevin, synapsin and synaptotagmin. No alterations were found at the light microscope level (data not shown). Nissl stain also suggested normal cytoarchitecture and an absence of gross morphological difference between hippocampi infused with EphA5-IgG, ephrinA5-IgG, or CD4-IgG (not shown). RT-PCR for synaptophysine and NMDA-R1 performed on hippocampal tissue samples from mice infused with the immunoadhesins also showed no differences in expression levels. As EphA receptors are involved in neuronal cytostructural mechanisms during ontogenesis (Friedman and O'Leary, 1996; Orioli and Klein, 1997; Meima et al., 1997; Frisén et al., 1998; Gao et al., 1996; Zhang et al., 1996; Zhou, 1997; Gao et al., 1998; Flanagan and Vanderhaegen, 1998), we also investigated mRNA expression of the tubulin gene (M-α1), whose product is a cytosolic microtubular protein involved in dendritic and axonal cytoskeletal processes. Furthermore, we studied gene expression of another cytoskeletal protein the microtubule associated protein, MAP2, which is involved in neuronal activity dependent dendritic structural changes (Quinlan and Halpain, 1996). We detected significant immunoadhesin induced changes in tubulin mRNA expression by RT-PCR suggesting that EphA5-IgG increases and ephrinA5-IgG decreases transcription (Fig. 9(A)), a finding consistent with the growth arresting repulsive effects of the ephrin-A5 ligand. A similar trend, although non-significant, was seen in MAP2 expression (Fig. 9(B)). Note that mRNA expression is shown as the number of PCR cycles required to reach a threshold level of cleavage of a fluorescent reporter probe (Heid et al., 1996; Gibson et al., 1996; also see Chapter 2.2.4) and this measure is inversely related to the original amount of mRNA present in the sample.

Hippocampal long-term potentiation (LTP) is a cellular mechanism proposed to underlie spatial and configural learning in rodents (Bliss and Collingridge, 1993). Therefore, we also studied whether potential alterations in LTP may underlie the observed behavioral changes.

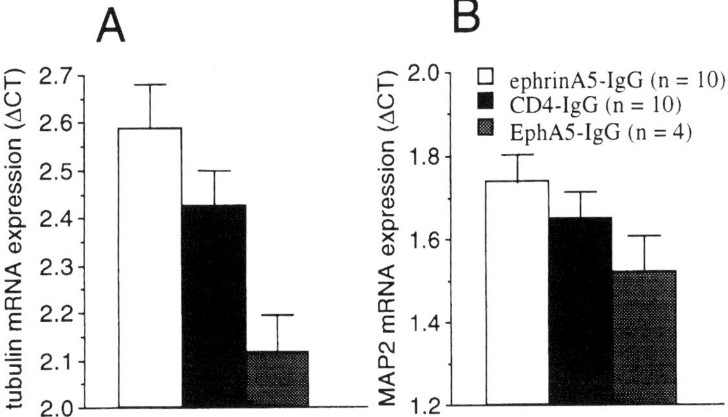

Fig. 9. Transcriptional changes induced by immunoadhesins in two cytoskeletal proteins, tubuline (A), and MAP2 (B). No strain differences were found. The data (means ± S.E.M.) are pooled for the strains and are based upon the number of PCR amplification cycles required to reach a threshold cleavage level of fluorescent reporter probe[18,19]. Results normalized to GAPDH house keeping gene transcript are shown (?CT). Sample sizes (n) are also indicated. Note that larger values represent smaller original mRNA amount in the hippocampal tissue sample analyzed. Note that expression of tubulin mRNA is increased by EphA5-IgG and decreased by ephrinA5-IgG infusion (ANOVA $F(2, 21) = 5.02$, $P < 0.02$). Expression changes in MAP2 show a similar, but non-significant, trend (ANOVA $F(2, 21) = 1.75$, $P > 0.19$).

Methods of electrophysiological analysis of long-term potentiation (LTP)

Transverse 300 μm hippocampal slices were submerged in a recording chamber (Fine Science Tools Inc., Foster City, CA), continuously perfused with 30°C, oxygenated (95% O2/5% CO2) ACSF for at least 1 h before recording. A bipolar glass electrode filled with ACSF stimulated Schaffer collaterals. Field excitatory postsynaptic potentials (fEPSP), elicited by single-pulse stimulation at 0.2 Hz, were recorded in CA1 stratum radiatum by a glass electrode filled with ACSF. Input/output (I/O) characteristic and ratio of the fEPSP slope to the presynaptic fiber volley (PSFV) amplitude was used to estimate basal synaptic transmission. I/O characteristics, recorded by applying gradually increased (in 20 steps increments) stimulus intensity to evoke fEPSP from minimum to maximum slope, were estimated by Michaelis-Menten Sigmoid curve fit. Km50 was taken as a 50% point between the threshold and maximal response (A/DVANCE software, McKellar Designs). Baseline was recorded for 1 h with the stimulus intensity set to evoke a fEPSP that was 30–50% of the maximal slope. Paired-pulse facilitation (PPF) was evoked by applying paired pulses of the same intensity with interpulse intervals of 50, 100, 150, 200 and 250 ms. Long term potentiation (LTP) was elicited by applying 4 trains of 100 Hz tetanus (1 s duration, 20 s apart), with the same stimulus intensity as for the baseline and PPF. Recordings were made with an Axoprobe-1A amplifier (Axon Instruments, Inc.) interfaced with a Power Macintosh 7100/66 computer. Data were acquired, digitized and analyzed using A/DVANCE software (McKellar Designs).

Hippocampal slices prepared from mice infused in vivo with EphA5-IgG or ephrinA5-IgG for 8 d were analyzed. In slices from C57BL/6 mice infused with EphA5-IgG the induction of LTP appeared normal, but the potentiated response decayed significantly faster compared to slices obtained from CD4-IgG control C57BL/6 mice (Fig. 10(A)). Other synaptic parameters such as paired-pulse facilitation, and basal synaptic trans-

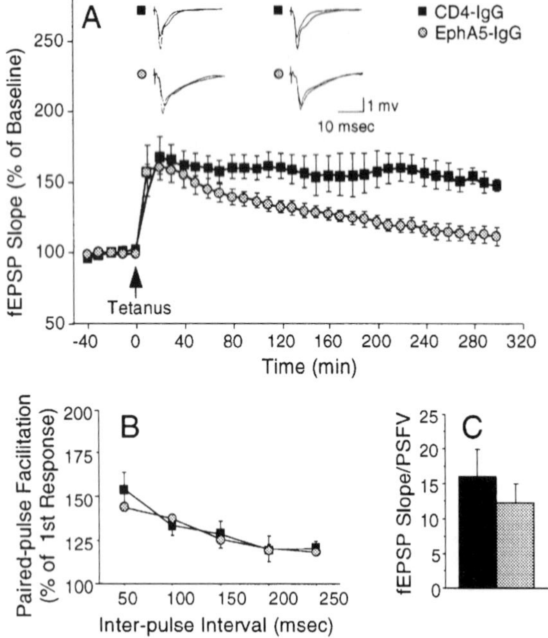

Fig. 10. EphA5-IgG impairs LTP maintenance in hippocampal slices prepared from C57BL/6 mice. (A) Field EPSP is expressed as percentage of baseline. Twenty sequential responses were averaged and plotted as one point. ANOVA revealed a significant immunoadhesin effect ($F(1, 8) = 8.09$, $P = 0.02$) and a significant immunoadhesin time interaction ($F(29, 232) = 4.98$, $P < 0.0001$). Tukey's HSD test showed that the groups became significantly different ($P < 0.05$) 90 min after tetanization. Representative traces before and after tetanization at corresponding time points, as indicated, are also shown. (B) No immunoadhesin effect was observed on paired pulse facilitation ($F(1, 11) = 0.164$, $P > 0.690$), which was assessed by applying paired pulses of equivalent intensity at interpulse intervals as indicated. Facilitation ratios are calculated by expressing the slope of the second fEPSP as a percentage of the slope of the first fEPSP. (C) Basal synaptic transmission, estimated by ratio of the fEPSP slope to the presynaptic fiber volley (PSFV) amplitude, was not altered by EphA5-IgG (hatched) compared to CD4-IgG infused mice (black) ($t = 1.119$, $df = 14$, $P > 0.28$). Estimation of basal synaptic transmission by input/output (I/O) characteristics using Michaelis–Menten Sigmoid curve fit revealed no significant differences (CD4-IgG mean $= 1.30 \pm 0.368$, EphA5-IgG mean $= 1.54 \pm 0.219$, $P > 0.80$).

mission (Fig. 10(B) and (C)) remained unaltered by EphA5-IgG infusion. Furthermore, the enhancement of cognitive function by ephrinA5-IgG infusion seen in DBA/2 mice is also accompanied by changes in hippocampal synaptic function. Slices from DBA/2 mice infused with ephrinA5-IgG exhibited a significant augmentation of LTP, beginning at induction and persisting up to 4.5 h (Fig. 11(A)). EphrinA5-IgG treatment also increased paired-pulse facilitation (PPF; Fig. 11(B)) suggesting the involvement of a presynaptic mechanism, while basal synaptic transmission was unchanged (Fig. 11(C)).

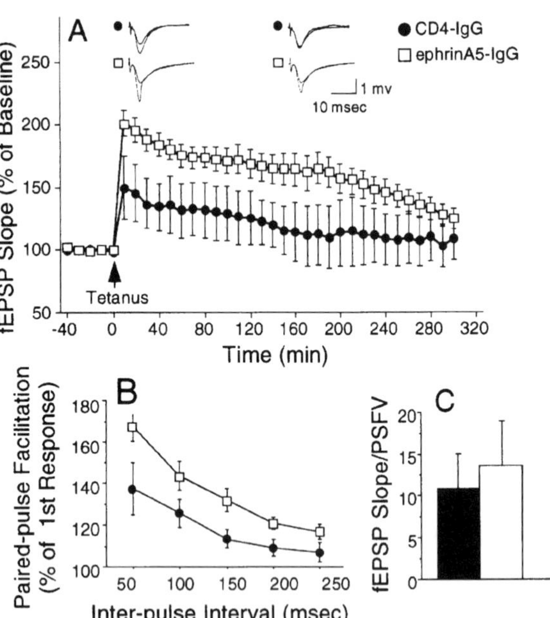

Fig. 11. EphrinA5-IgG improves LTP and increases magnitude of PPF in hippocampal slices prepared from DBA/2. (A) Field EPSP expressed as percentage of baseline. Twenty sequential responses were averaged and plotted as one point. ANOVA including data from all time points revealed an immunoadhesin effect that bordered significance ($F(1, 12) = 4.23$, $P = 0.06$), while Tukey's HSD test showed that the groups were significantly different ($P < 0.05$) up to 260 min. post-tetanization. Representative traces before and after tetanization at corresponding time points, as indicated, are also shown. (B) PPF varied as a function of immunoadhesin treatment ($F(1, 21) = 6.189$, $P < 0.03$), and no significant immunoadhesin interval interaction ($F(4, 84) = 1.654$, $P > 0.16$) was seen. (C). No significant differences were found in basal synaptic transmission between ephrinA5-IgG (white) and CD4-IgG (black) infused mice ($t = 0.416$, $df = 18$, $P > 0.68$). Estimation of basal synaptic transmission by input/output (I/O) characteristics using Michaelis–Menten Sigmoid curve fit revealed no significant differences (CD4-IgG mean $= 2.49 \pm 0.33$, ephrinA5-IgG mean $= 2.80 \pm 0.97$, $P > 0.70$).

Discussion

Our in situ hybridization findings demonstrating strong EphA5 expression in neurons of the adult mouse hippocampus, RT-PCR results showing hippocampal expression of two relevant ligands of this receptor, ephrin-A2 and -A3, and our Western blot analysis revealing the presence of EphA5 receptor in a phosphorylated, active, form in hippocampal tissue from adult mice strongly suggest that EphA receptor function is not restricted to the developing nervous system. Confirming this suggestion, intra-hippocampal infusion of an EphA antagonist led to significant impairment in behavioral function in adult mice whereas infusion of an agonist resulted in improved behavioral performance. Furthermore, infusion of these agents led to correlated changes both in the expression of a cytostructural protein, tubulin, and in electrophysiological measures of hippocampal synaptic plasticity.

These results were achieved employing protein targeting, a novel strategy in behavioral neuroscience. The applied genetically engineered immunoadhesins recognize the ligand, or the receptor, on the basis of high affinity natural ligand-receptor binding interaction thus they obviated the lack of EphA selective pharmacological agents. Furthermore, as a result of the unaltered binding sites, these immunoadhesins were capable of binding all relevant proteins the endogenous EphA5 or ephrin-A5 molecules would. As EphA receptors are promiscuous and may respond to several ephrin-A ligands, and conversely a ligand may activate several receptors, these immunoadhesins allowed us to manipulate all functionally relevant ligands, or receptors, without the confounding effects of compensation by closely related ligands, or receptors, as occur in gene targeting (Frisén et al., 1998; Gerlai, 1996a). Furthermore, stereotaxic delivery of the immunoadhesins into the hippocampus eliminated the possibility of inducing developmental defects during ontogenesis and also reduced the confounding effects of non-specific changes in tissues not associated with hippocampal function.

Our results indicate that inhibition and augmentation of EphA function by intrahippocampal infusion of the immunoadhesins in adult mice facilitates and impairs, respectively, behavioral responses in two tasks, T-CAT and CDFC. These tasks depend upon the hippocampus but have different motivational, perceptual, and motor requirements. Therefore, the consistent behavioral changes we observed imply an underlying mechanism associated with hippocampal function and not with idiosyncratic characteristics unique to one or the other task, a notion also supported by the lack of indication of altered performance factors detected in these tasks. We infer, therefore, that the observed behavioral changes were due to altered cognitive processes. Taken together, these results support a role for EphA in adult neural function associated with learning, one which can be augmented by ephrinA5-IgG and impaired by EphA5-IgG.

The potential neurobiological mechanism underlying the observed behavioral effects is speculative at this point. The recent observation showing that Eph receptors and their ephrin ligands contain PDZ recognition motifs and are bound and clustered by PDZ proteins at pre and postsynaptic sites of neuronal synapses in vitro suggests that Eph receptors may mediate synaptic plasticity (Torres et al., 1998; Hsueh and Sheng, 1998). This suggestion is now confirmed by the electrophysiological findings of the present study. Nevertheless, the question whether these findings explain the observed behavioral alterations will need to be addressed in the future.

Our results showing bidirectional gene expression changes in tubulin, a component of the cytoskeleton, suggest that EphA receptors may mediate cytostructural mechanisms in the adult brain. This notion is compatible with the known in vitro actions of EphA receptors on reorganization of actin (Meima et al., 1997; Drescher et al., 1995; Hayashi et al., 1996), a component of the cytoskeleton that interestingly has also been shown to play role in structural changes underlying synaptic plasticity (Hayashi et al., 1996; Fifkova and Delay, 1982; Westrum et al., 1980; Landis and Reese, 1983). Furthermore, the EphA5-IgG (antagonist) induced overexpression and ephrinA5-IgG (agonist) induced underexpression of tubulin is consistent with the known growth arresting effects of ephrin-A ligands on axons and dendrites (Winslow et al., 1995; Meima et al., 1997; Frisén et al., 1998). Taken together with our behavioral and electrophysiological results, these findings are consistent with the notion that tubulin, perhaps mediating axonal and dendritic transport or microstructural localization of gene products required for synaptic remodeling, plays a role in learning as suggested by Rose and associates (Mileusnic et al., 1980; Scholey et al., 1992; Rose, 1993). This raises the intriguing possibility that the effects of EphA receptors on cognitive function and/or LTP may be mediated by structural alterations in synaptic connectivity, e.g. development of perforated synapses, as observed in response to LTP induction (Edwards, 1995; Geinisman et al., 1993).

Regardless of the precise identity of the molecular mechanisms, the present findings now have revealed a role for EphA receptor tyrosine kinases in cognitive function in the adult mammalian brain. These findings open an unexpected avenue into the functional analysis of this large receptor protein family and may lead to novel targets for therapeutic intervention in human brain and behavioral disorders. They also demonstrate that protein targeting with the application of immunoadhesins may be a useful research strategy in the analysis of molecular components involved in such complex behavioral traits as learning and memory.

Acknowledgments

We thank Ingrid Caras, Alisha Eisert, Nerrisa Mendoza, Hans Sauer, Nick Van Bruggen, Simon P. Williams for technical help and discussions on the topic; Louis Tamayo, Allison Bruce for figures; Steven Chamow for CD4-IgG; Nicola S. Clayton and two anonimous internal reviewers for comments on an earlier version of this manuscript.

References

Abeliovich, A., Paylor, R., Chen, C., Krim, J., Wehner, J. and Tonegawa, S. (1993). PKCγ mutant mice exhibit mild deficits in spatial and contextual learning. *Cell*, 75: 1263–1271.

Aiba, A., Chen, C., Herrup, K., Rosenmund, C., Stevens, C. and Tonegawa, S. (1994) Reduced hippocampal long-term potentiation and context-specific deficit in associative learning in mGluR1 mutant mice. *Cell*, 79: 365–375.

Bach, M.E., Hawkins, R.D., Osman, M., Kandel, E.R. and Mayford, M. (1995) Impairment of spatial but not contextual memory in CaMKII mutant mice with a selective loss of hippocampal LTP in the range of the theta frequency. *Cell*, 81: 905–915.

Bliss, T.V.P. and Collingridge G.L. (1993) A synaptic model of memory: long-term potentiation in the hippocampus. *Nature*, 361: 34–39.

Cabelli, R.J., Shelton, D.L., Segal, R.A. and Shatz, C.J. (1997) Blockade of endogenous ligands of TrkB inhibits formation of ocular dominance columns. *Neuron*, 19: 63–76.

Capon, D.J., Chamow, S.M., Mordenti, J., Marsters, S.A., Gregory, T., Mitsuya, H., Byrn, R.A., Lucas, C., Wurm, F.M., Groopman, J.E., Broder, S. and Smith, D.H. (1989) Designing CD4 immunoadhesins for AIDS therapy. *Nature*, 337: 525–531.

Caras, I.W. et al. (1996) Patent: WO9613518-A1, Accession: P.T18893, 1996

Carlson, J.R. (1988) A new means of inducibly inactivating a cellular protein. *Molecular and Cellular Biology*, 8: 2638–2646.

Chamow S.M. and Ashkenazi, A. (1996) Immunoadhesins: principles and applications. *Trends Biotech.*, 14: 52–60.

Crusio, W.E., Bertholet, J.Y. and Schwegler, H.(1990) No correlations between spatial and non-spatial reference memory in a T-maze task and hippocampal mossy fibre distribution in the mouse. *Behav. Brain Res.*, 41: 251–259.

Douglas, R.J. (1990) Spontaneous alternation behavior and the brain. In: W.N. Denber, and L.L. Richman (Eds.), *Spontaneous alternation behavior*, Springer Verlag, New York, pp. 73–109.

Drescher, U., Kremoser, C., Handwerker, C., Loschinger, J., Noda, M. and Bonhoeffer, F. (1995) In vitro guidance of retinal ganglion cell axons by RAGS, a 25 kDa tectal protein related to ligands for Eph receptor tyrosine kinases. *Cell*, 82: 359–370.

Eaton, D.L., Wood, W.I., Eaton, D.L., Hass, P.E., Hollingshead, P., Wion, K., Mather, J., Lawn, R.M., Vehar, G.A. and Gorman, C. (1986) Construction and characterization of an active factor VIII variant lacking the cnetral one-third of the molecule. *Biochemistry*, 25: 8343–8347.

Edwards, F.A. (1995) LTP — a structural model to explain the inconsistencies. *Trends Neurosci.*, 18: 250–255.

Fifkova, E. and Delay, R.J. (1982) Cytoplasmic actin in neuronal processes as a possible mediator of synaptic plasticity. *J. Cell Biol.*, 95: 345–350.

Fields, R. D. and Itoh, K (1996) Neural cell adhesion molecules in activity-dependent development and synaptic plasticity. *Trends Neurosci.*, 19: 473–480.

Flanagan, J.G. and Vanderhaeghen, P. (1998) The ephrins and Eph receptors in neural development. *Ann. Rev. Neurosci.*, 21: 309–345.

Friedman, G.C. and O'Leary, D.D.M. (1996) Eph receptor tyrosine kinases and their ligands in neural development. *Curr. Opin. Neurobiol.*, 6: 127–133.

Frisén, J., Yates, P.A., McLaughlin, T., Friedman, G.C., O'Leary, D.D.M. and Barbacid, M. (1998) Ephrin-A5 (AL-1/RAGS) is essential for proper retinal axon guidance and topographic mapping in the mammalian visual system. *Neuron*, 20: 235–243.

Gao, P.P., Zhang, J.H., Yokoyama, M., Racey, B., Dreyfus, C.F., Black, I.B. and Zhou R. (1996) Regulation of topographic projection in the brain: Elf-1 in the hippocamposeptal system. *Proc. Natl. Acad. Sci. USA*, 93: 11161–11166.

Gao, P.P., Yue, Y., Zhang, J.H., Cerretti, D.P., Levitt, P. and Zhou, R. (1998) Regulation of thalamic neurite outgrowth by the Eph ligand ephrin-A5: implications in the development of thalamocortical projections. *Proc. Natl. Acad. Sci. USA*, 95: 5329–5334.

Gao, W-Q., Shinsky, N., Armanini, M.P., Moran, P., Zheng, J.L., Mendoza-Ramirez, J.L., Phillips, H.S., Winslow, J.W. and Caras, I.W. (1998) Regulation of hippocampal synaptic plasticity by the tyrosine kinase receptor, REK7/EphA5, and its ligand, AL-1/Ephrin-A5. *Mol. Cell Neurosci.*, 11: 247–259.

Geinisman, Y., deToledo-Morrell, L., Morrell, F., Heller, R.E., Rossi, M. and Parshall, R.F. (1993) Structural synaptic correlate of long-term potentiation: Formation of axospinous synapses with multiple, completely partitioned transmission zones. *Hippocampus*, 3: 435–446.

Gerlai, R. (1998a) A new continuous alternation task in T-maze detects hippocampal dysfunction in mice: A strain comparison and lesion study. *Behav. Brain Res.*, 95: 91–101.

Gerlai, R. (1998b) Contextual Learning and Cue Association in Fear Conditioning in Mice: A Strain Comparison and a Lesion Study. *Behav. Brain Res.*, 95: 191–203.

Gerlai, R. (1996a) Gene targeting studies of mammalian behavior: Is it the mutation or the background genotype? *Trends Neurosci.*, 19: 177–181.

Gerlai, R. (1996b) Gene targeting in Neuroscience: The systemic approach. *Trends Neurosci.*, 19: 188–189.

Gerlai, R. (1996c) Molecular genetic analysis of mammalian behavior and brain processes: Caveats and perspectives. *Sem. Neurosci.*, 8: 153–161.

Gerlai, R. and Clayton, N.S. (1999) Analysing hippocampal function in transgenic mice: An ethological perspective. *Trends Neurosci.*, 22: 47–51.

Gerlai, R., Cairns B., Van Bruggen N., Moran P., Shih A., Sauer, H., Phillips. H.S., Caras I. and Winslow J. (1998) Protein targeting in the analysis of learning and memory: A potential alternative approach to gene targeting. *Exp. Brain Res.*, 123: 24–35.

Gerlai, R. Friend, W., Becker, L., O'Hanlon, R., Marks, A. and Roder, J. (1993) Female transgenic mice carrying the human gene for S100β are hyperactive. *Behav. Brain Res.*, 55: 51–59.

Gibson, U.E.M., Heid, C.A. and Willimas, P.M. (1996) A novel method for real time quantitative RT-PCR. *Genome Methods*, 6: 995–1001.

Grant, S.G.N., O'Dell, T.J., Karl, K.A., Stein, P.L., Soriano, P. and Kandel, E.R. (1992) Impaired Long-Term Potentiation, Spatial Learning, and Hippocampal Development in *fyn* Mutant Mice. *Science*, 258:1903–1910.

Hayashi, K., Ishikawa, R., Ye, L.H., He, X.L., Takata, K., Kohama, K. and Shirao, T. (1996) Modulatory role of drebrin on the cytoskeleton within dendritic spines in the rat cerebral cortex. *J. Neurosci.*, 16: 7161–7170.

Heid, A.C., Stevens, J., Livak, K.J. and Williams, P.M. (1996) Real time quantitative PCR. *Genome Methods*, 6: 986–994.

Hsueh, Y-P. and Sheng, M. (1998) Eph receptors, ephrins, and PDZs gather in neuronal synapses. *Neuron*, 21: 1227–1229.

Joyner, A.L. (Ed.) (1993) *Gene Targeting. A Practical Approach*. Oxford University Press, Oxford, pp. 234.

Kang, H., Welcher, A.A., Shelton, D. and Schuman, E. (1997) Neurotrophins and time: Different roles for TrkB signaling in hippocampal long-term potentiation. *Neuron*, 19: 653–664.

Kim, J.J. and Fanselow, M.S. (1992) Modality-specific retrograde amnesia of fear. *Science*, 256: 675–677.

Landis, D.M. and Reese, T.S. (1983) Cytoplasmic organization in cerebellar dendritic spines. *J. Cell Biol.*, 97: 1169–1178.

Maisonpierre, P.C., Barrezuela, N.X. and Yancopoulos, G.D. (1993) Ehk-1 and Ehk-2: two novel members of the Eph receptor-like tyrosine kinase family with distinctive structures and neuronal expression. *Oncogene*, 8: 3277–3288.

Martin, K.C., Michael D., Rose J.C., Barad M., Casadio A., Zhu H. and Kandel E.R. (1997) MAP kinase translocates into the nucleus of the presynaptic cell and is required for long-term facilitation in *Aplysia*. *Neuron*, 18: 899–912.

Matsuyama, S., Namgung, U. and Routtenberg, A. (1997) Long-term potentiation persistence greater in C57BL/6 than DBA/2 mice: predicted on basis of protein kinase C levels and learning performance. *Brain Res.*, 763: 127–130.

Mayford, M., Bach, M.E., Huang, Y., Wang, L., Hawkins, R.D. and Kandel, E.R. (1996) Control of memory formation through a regulated expression of a CaMKII transgene. *Science*, 274: 1678–1683.

Meima, L., Kljavin, I.J., Shih, A., Winslow, J.W. and Caras, I.W. (1997) AL-1-induced growth cone collapse of rat cortical neurons is correlated with REK7 expression and rearrangement of the actin cytoskeleton. *Eur. J. Neurosci.*, 9: 177–188.

Melton, D.A., Krieg, P.A., Rebagliati, M.R., Maniatis, T., Zinn, K. and Green, M.R. (1984) Efficient in vitro synthesis of biologically active RNA and RNA hybridization probes from plasmids containing a bacteriophage SP6 promoter. *Nucleic Acids Res.*, 12: 7035–7056.

Mileusnic, R., Rose, S.P.R. and Tillson, P. (1980) Passive avoidance learning results in region-specific changes in concentration of and incorporation into colchicine-binding proteins in the chick forebrain. *J. Neurochem.*, 34: 1007–1015.

Moser, M-B., Moser, E.I., Forrest, E., Andersen, P. and Morris, R.G.M. (1995) Spatial learning with a minislab in the dorsal hippocampus. *Proc. Natl. Acad. Sci. USA*, 92: 9697–9701.

Orioli D. and Klein R. (1997) The Eph receptor family: axonal guidance by contact repulsion. *Trends. Genet.*, 13: 354–360.

Paylor, R., Tracy, R., Wehner, J. and Rudy, J.W. (1994) DBA/2 and C57BL/6 mice differ in contextual fear but not auditory fear conditioning. *Behav. Neurosci.*, 108: 810–817.

Phillips, R.G. and LeDoux, J.E. (1992) Differential contribution of amygdala and hippocampus to cued and contextual fear conditioning. *Behav. Neurosci.*, 106: 274–285.

Phillips, H.S., Hains, J.M., Laramee, G.R., Rosenthal, A. and Winslow, J.W. (1990) Widespread expression of BDNF but not NT-3 by target areas of basal forebrain cholinergic neurons. *Science*, 250: 290–294.

Quinlan, E.M. and Halpain, S. (1996) Postsynaptic mechanisms for bidirectional control of MAP2 phosphorylation by glutamate receptors. *Neuron*, 16: 357–368.

Rose, P.R.S. (1993) Synaptic Plasticity, Learning and Memory. In: M. Baudry, R.F. Thompson and J.L. Davis (Eds.), *Synaptic Plasticity*. MIT Press, Cambridge, USA, pp. 209–229.

Sadjii, F.G. and Pasquale, E.B. (1993) Five novel avian Eph-related tyrosine kinases are differentially expressed. *Oncogene*, 8: 1807–1813.

Scholey, A., Bullock, S. and Rose, S.P.R. (1992) Passive avoidance learning in the young chick results in time- and locus-specific elevations of alpha-tubulin immunoreactivity. *Neurochem. Int.*, 21: 343–350.

Schuman, E. (1997) Growth factors sculpt the synapse. *Science*, 275: 1277–1278.

Silver, R., LeSauter, J., Tresco, P. A. and Lehman, M.N. (1996) A diffusible coupling signal from the transplanted suprachiasmatic nucleus controlling circadian locomotor rhythms. *Nature*, 382: 810–813.

Thoenen, H. (1995) Neurotrophins and neuronal plasticity. *Science*, 270: 593–598.

Tonegawa, S., Li, Y., Erzurumlu, R.S., Jhaveri, S., Chen, C., Goda, Y., Paylor, R., Silva A., Kim, J.J., Wehner, J.M. Stevens, C.F. and Abeliovich, A. (1995) The gene knockout technology for the analysis of learning and memory, and neural development. *Progress Brain Res.*, 106: 3–14.

Torres, R., Firestein, B.L., Dong, H., Staudinger, J., Olson, E.N., Huganir, R.L., Bredt, D.S., Gale, N.W. and Yancopoulos, G.D. (1998) PDZ proteins bind, cluster, and synaptically colocalize with Eph receptors and their ephrin ligands. *Neuron*, 21: 1453–1463.

Tsien, J.Z., Chen, D.F., Gerber, D., Tom, C., Mercer, E.H., Anderson, D.J., Mayford, M., Kandel, E.R. and Tonegawa, S. (1996) Subregion and cell type-restricted gene knockout in the mouse brain. *Cell*, 87: 1317–1326.

Van Abeelen, J.H.F. (1989) Genetic control of hippocampal cholinergic and dynorphinergic mechanisms regulating novelty-induced exploratory behavior in house mice. *Experientia*, 45: 839–845.

Westrum, L.E., Jones, D.H., Gray, E.G. and Barron, J. (1980) Microtubules, dendritic spines and spine appratuses. *Cell Tissue Res.*, 208: 171–181.

Wilson, M.A. and Tonegawa, S. (1997) Synaptic plasticity, place cells and spatial memory: study with second generation knockouts. *Trends Neurosci.*, 20: 102–106.

Winslow, J.W., Moran, P., Valverde, J., Shih, A., Yuan, J.Q., Wong, S.C., Tsai, S.P., Goddard, A., Henzel, W.J., Hefti, F., Beck, K.D. and Caras, I.W. (1995) Cloning of AL-1, a ligand for an eph-related tyrosine kinase receptor involved in axon bundle formation. *Neuron*, 14: 973–981.

Winslow, J.W. and Caras, I.W. (1995) AL-1 neurotrophic factor. Patent filing WO 96/13518 WIPO patent publication, May 9, 1996.

Zhang, J.-H., Cerretti, D., Yu, T., Flanagan, J.G. and Zhou, R. (1996) Detection of ligands in regions anatomically connected to neurons expressing the eph receptor Bsk: potential roles in neuron-target interaction. *J. Neurosci.*, 16: 7182–7192.

Zhou, R. (1997) Regulation of topographic projection by the Eph family receptor Bsk (EphA5) and its ligands. *Cell Tissue Res.*, 290: 251–259.

Zhou, R., Copeland, T.D., Kromer, L.F. and Schultz, N.T. (1994) Isolation and characterization of Bsk, a growth factor receptor-like tyrosine kinase associated with the limbic system. *J. Neurosci. Res.*, 37: 129–143.

Section 3.3

Invertebrates

W.E. Crusio and R.T. Gerlai (Eds.)
Handbook of Molecular-Genetic Techniques for Brain and
Behavior Research (Techniques in the Behavioral and Neural Sciences, Vol. 13)

CHAPTER 3.3.1

A novel approach to *Drosophila* neurophysiology: the targeted expression of aequorin

Philippe Rosay[1,2], Kim Kaiser[1,2] and J. Douglas Armstrong[1]

[1]Division of Molecular Genetics, IBLS, University of Glasgow, 56 Dumbarton Road, Glasgow G11 6NU, UK
[2]Neuropa Ltd., 54 Dumbarton Rd., Robertson Building, Glasgow G11 6NU, UK

Drosophila neurobiology

Correlating brain structure and function, across a range of levels of description, is one of the most challenging scientific goals for the next century. Advances in neural imaging techniques and molecular genetic tools are providing exciting new methods that can contribute towards this goal.

Due to its small size, yet behavioural complexity, the brain of the fruit fly *Drosophila* is now largely accepted as an important model for brain function. Study of *Drosophila* neuroanatomy benefits from a relatively simple nervous system. Comprising less than one hundred thousand neurons in the adult, the intact brain is small enough to be examined by traditional optical or confocal microscopy. Yet despite its small brain, the organism displays a wide range of behaviours, including associative learning processes in response to tactile, visual and olfactory cues. Simple assays (Wolf et al., 1998) for these behaviours allow researchers to investigate structure/function relationships to a high degree of specificity. Whilst most techniques that can be applied to larger organisms can also be used in *Drosophila* (see for example Baines and Bate, 1998), some are technically very difficult due to small brain and neuron size. Physiological analysis of *Drosophila* neurons is particularly difficult. These difficulties are in part compensated for by the wide range of molecular genetic techniques available for *Drosophila*, which have been of particular significance for not only marking cells for anatomical study, but also provide means for manipulating neurons in the living organism.

P{GAL4} system

The P{GAL4} enhancer-trap system (see Fig. 1), has been used extensively for both visualisation and manipulation of *Drosophila* neurons in recent years (Brand and Dormand, 1995; Dormand and Brand, 1998).The P{GAL4} system effectively allows us to target transgene expression to particular neuron populations. These neuronal populations can be visualised using one set of transgenes and then their function in the living organism manipulated using others. Techniques addressing neuroanatomy, and for modifying neuron function, have been the subject of several recent book and literature reviews (Armstrong et al., 1999). The focus of this article is on the application of more recently developed P{GAL4} technology for physiological analysis of *Drosophila* neurons.

Aequorin

In the *Drosophila* brain, as in many tissues, small cell size precludes direct calcium (Ca^{2+}) measure-

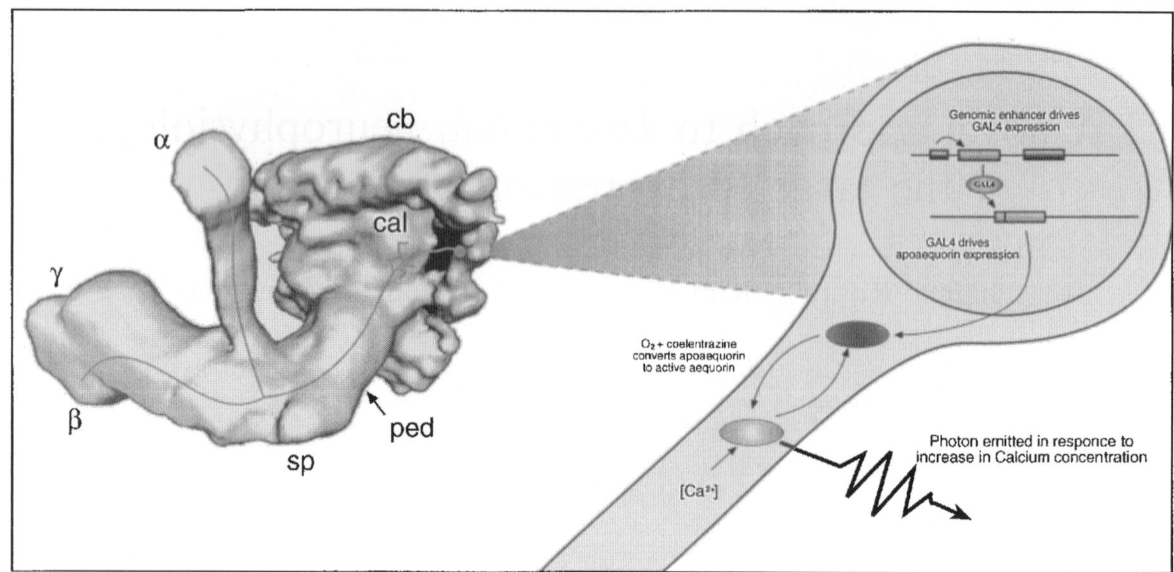

Fig. 1. The targeted expression of aequorin. 3D reconstruction of a mushroom body (left), a closely packed array of neurons that in the insect brain is required for olfactory associative learning and schematic drawing of a neuron (right) in which expression of the apoaequorin takes place under the control of the GAL4/UAS_G system. The reporter of P{GAL4} enhancer trap element is a yeast transcriptional activator functional in *Drosophila* that is used here to direct expression of a transgene coding for the apoaequorin. This transgene is under control of a GAL4-dependent promoter called UAS_G. Conversion of the apoaequorin into (active) aequorin is a slow (min range) process (reconstitution phase), and requires the presence of the co-factor coelenterazine and oxygen. The converse reaction used to monitor the changes in the concentration of calcium in the cytoplasm is extremely fast (msec). The luminescent emission is λ_{max} 470 nm. Abbreviations: α, α-lobe; β, β-lobe; γ, γ-lobe; sp, spur; ped, pedunculus; cal, calyx; cb, cell body layer.

ment with ion-selective microelectrodes. Ca^{2+}-sensitive fluorescent dyes have a number of drawbacks; they may interfere with Ca^{2+}-mediated functions and intracellular loading can be problematic in intact tissues. It is thus difficult to study cell-specific regulation of intracellular calcium concentration $[Ca^{2+}]_i$ by classical techniques. We have developed a P{GAL4} directed Ca^{2+} reporter system in *Drosophila* in order to measure cell-type-specific $[Ca^{2+}]_i$ changes in vivo, in the context of an intact organism or tissue.

Aequorin is a Ca^{2+}-sensitive luminescent protein isolated from the coelenterate *Aequorea victoria*. It is a complex of apoaequorin, a 21 kD polypeptide, and coelenterazine, a hydrophobic luminophore. Aequorin has previously been used in several vertebrate and invertebrate contexts for monitoring $[Ca^{2+}]_i$ changes, both by micro injection of purified protein and by transgene expression (Brini et al., 1995). In *Drosophila*, we have employed the binary GAL4/UAS_G system (Brand and Perrimon, 1993) to target an aequorin transgene in a neuron-type specific manner.

We have successfully targeted aequorin expression to the different cellular components of the brain (Rosay et al., unpublished obs.) and of the Malpighian tubule (O'Donnell et al., 1998; Rosay et al., 1997). The basic parameters of the aequorin expression system were established using the tubule model: functionality and sensitivity of aequorin in *Drosophila*, cytoplasmic and cell-specific expression of aequorin. Using this assay, it has been possible to correlate cell-specific intracellular signalling mechanisms with the tubule's physiological characteristics.

Vector construction

The apoaequorin coding region of cDNA clone pAEQ1 (Prasher et al., 1985) was extracted from the plasmid PAQ2 (Knight et al., 1991) by double digestion with SmaI and PstI, and was inserted into similarly cleaved pBKS$^+$. Re-cleavage with *Not*I and KpnI allowed insertion into the similarly cleaved pUAST transformation vector (Brand and Perrimon, 1993), downstream of the five UAS_G (GAL4-binding) sites, hsp70 TATA box and transcriptional start site (Fig. 2). Germ-line transformed *Drosophila* were then generated and two functionally equivalent lines, UAS_G-aeq30D and UAS_G-aeq49A, were obtained that have the transgene at the indicated cytogenetic locations. Insertions were also generated on the X, CyO and third chromosomes.

Direct visualisation of aequorin transgene expression at high resolution is not currently possible (see below). Antibodies are available and have been used in western blots. However, at least in our hands, the antibodies currently available do not appear to be good enough for immunohistochemistry. Suitable antibodies may become available as the use of aequorin as a calcium reporter becomes more widespread. We are currently in the process of developing epitope tagged aequorin transgenes to help us address this problem.

Protocol: brain preparation for luminometry

Equipment and reagents

• Dissection forceps, e.g. Dumont No. 5 Biologie.
• Ringer's saline medium (derived from Ashburner, 1989; Dow et al., 1994) consisting (in mmol^{-1}) Na^+ 132; K^+ 20; Ca^{2+} 2; Cl^- 158; Mg^{2+} 8.5; H_2PO^{4-} 4.3; HCO_3^- 10.2; Glucose 5.0 pH adjusted to 7.0 with NaOH.
• One tube luminometer, e.g. luminometer LB9507 (Berthold Wallac).
• One clean transparent tube for luminometer, e.g. No. 55.476, 5 ml × 75 12 mm (Sarstedt).
• One dark box for preincubation.
• Coelenterazine (Molecular Probes) is solubilised before use in ethanol (instead of methanol to minimise possible effects on intact living brains). Manipulation of coelenterazine: Protect from light at all times. Coelenterazine is swiftly solubilised at a mM concentration (0.25 g/l) in ethanol then split into 10 μl aliquots, dried (Speedvac) and then stored at −70°C. No loss of activity has been noticed even after 6 months storage. Aliquots are redissolved in 10 μl ethanol just before use,

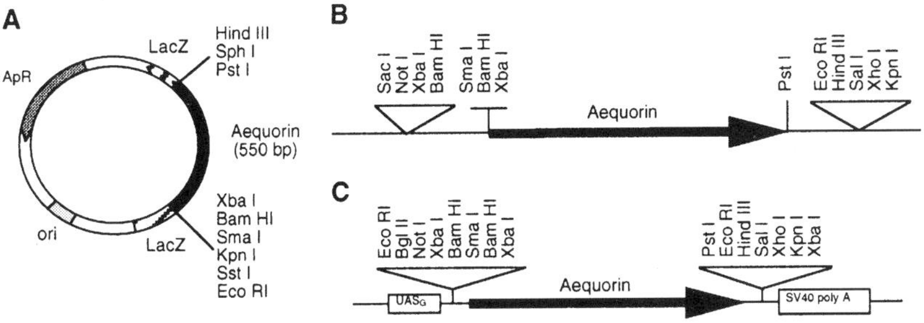

Fig. 2. Cloning of cDNA coding for the aequorin protein into the pUAST transformation vector. (A) pAQ2; (B) Bluescript pBKS$^+$; and (C) pUAST vectors containing the cDNA clone pAEQ1 coding for aequorin (black arrow). See text for details.

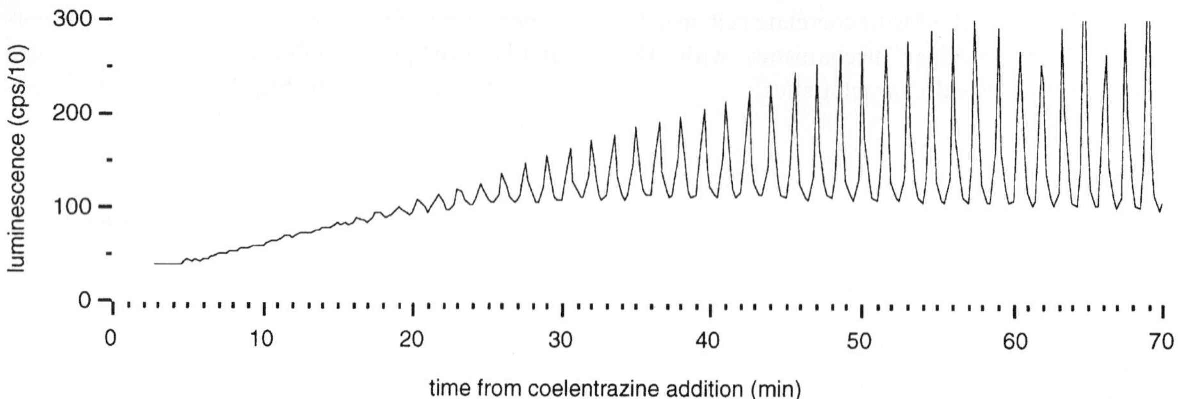

Fig. 3. A luminometer trace of P{GAL4} directed aequorin activity in intrinsic neurons of the mushroom bodies (see also Fig. 1). The brain was removed from the head capsule and bathed in insect saline containing coelenterazine. We observed robust oscillation of aequorin luminescence with a mean period in the minute range. Oscillation reaches full amplitude by 30–60 min, consistent with the known time-dependence of the formation of active aequorin from apoaequorin and coelenterazine, and can persist for 15 h or more. Though difficult to quantify, the amplitude of the effect clearly implies the superimposed luminescence of many neurons oscillating in synchrony.

and coelenterazine is added to the preparation at the final concentration of 2.5 μM (0.25% v/v).
• Lysis solution: 1% (v/v) Triton X-100, 100 mM $CaCl_2$

Method

• Dispense 100 μl Ringer's solution into a clean tube.
• Dissect brain in Ringer's solution.
• Transfer brain to tube and ensure the brain is floating (for proper oxygenation).
• Add coelenterazine (Ensure minimal exposure to light).
• Incubate for 20–40 min in a dark box to allow formation of active aequorin (or directly in luminometer, see trace in Fig. 3).
• Record data from the luminometer. In our case, data are collected every 0.5–5 s, however, most luminometers have sampling rates from 100 ms. Luminescence is expressed in arbitrary units (au) equivalent to cps/10 for the system we use.
• After recording luminescence, the brain is disrupted in 160 μl lysis solution causing discharge of the remaining aequorin and allowing estimation of the total amount of aequorin in the sample and $[Ca^{2+}]_i$ levels represented.

Variations on the protocol

1. Other tissues
This system can be applied to other *Drosophila* tissues although our experience is limited to the developing and adult brain, and the malpighian (renal) tubules (O'Donnell et al., 1998; Rosay et al., 1997). Previous studies, in which aequorin was injected into the embryo (see the book *Cellular calcium : a practical approach*, 1991), suggest that with appropriate enhancer-trap lines, this, or any other tissue would also be amenable.

2. Other buffers
Ringers is just one of several suitable buffers we have tested. Schneider medium also gives similar results, but complex media may interfere with the normal brain neurophysiology, since many of these buffers contain neurotransmitters (for example glycine, glutamate) at millimolar concentrations. For detailed overview of the charac-

teristics of different buffers used in *Drosophila* cell culture, see Echalier (1997).

3. Test compounds

The techniques described above allow visualisation of underlying neuronal activity via changes in $[Ca^{2+}]_i$. It can also be used to test the pharmacological effects of various compounds, and extracellular ionic conditions, in the context of both wild-type and mutant backgrounds. Simply by addition of test compounds to the buffer, we have been able to investigate the roles of various ion channels and neurotransmitter receptors on the underlying neuronal activity in the mushroom bodies (see Figs. 1, 3 and 4).

However, in all cases where results appear negative, accessibility of the test compound into the target tissues should be taken into consideration.

4. Solute Options

When testing compounds, aqueous solutions are used whenever possible. Ethanol or DMSO can be used as alternatives and do not affect brain physiology at relatively low (final) concentrations (Fig. 4). Actual working concentrations of ethanol or DMSO for the compounds we have investigated have been around 0.005% or less. This is a full order of magnitude below levels that were shown to have no measurable effects (see Fig. 4).

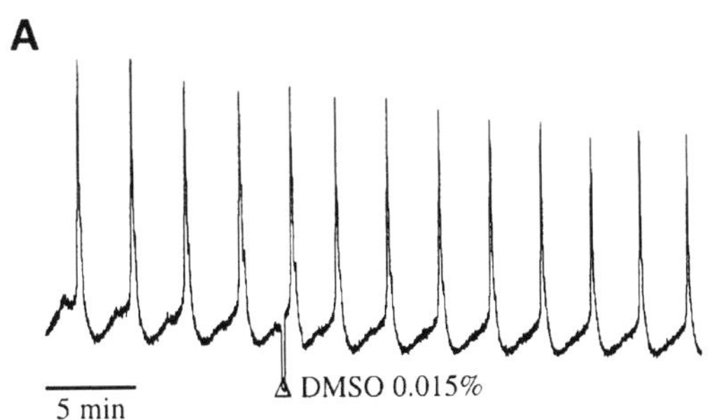

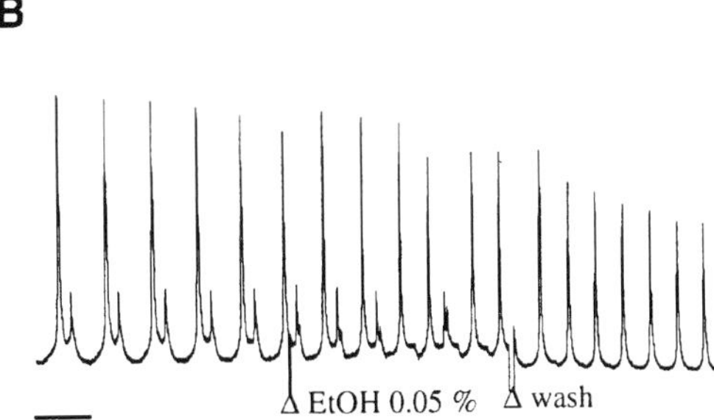

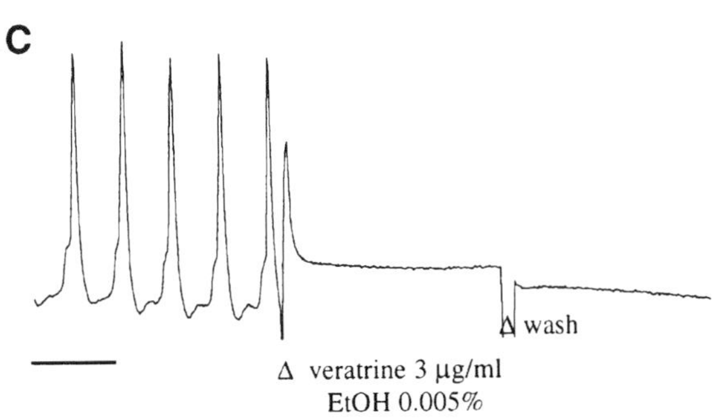

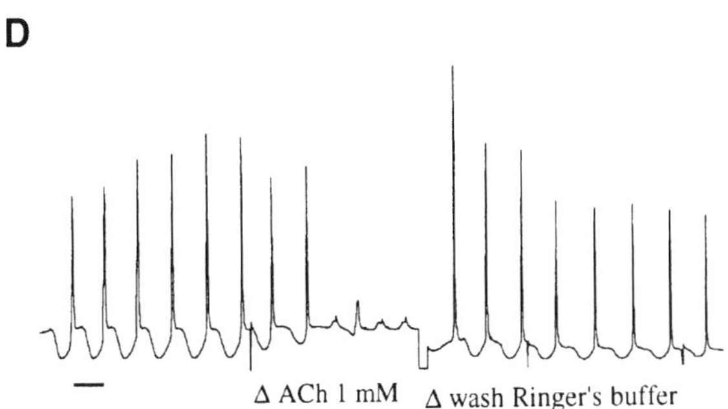

Fig. 4. Pharmacological stimulation of calcium oscillations in the *Drosophila* brain. Solvent controls. (A) Addition of DMSO 0.015% or (B) ethanol 0.05% has no visible effect on amplitude nor on frequency (the same applies for ethanol concentrations up to 0.5%, not shown). (C) Sodium channel pharmacology. GAL4-*aeq* adult brains treated as indicated with a sodium channel opener. Similar results were obtained with the mixture veratrine and with pure veratridine. After a transient peak, oscillation stops and the basal level of intracellular calcium is raised at a higher value than the previous base line. (D) Cholinergic pathway. The effect of 1 mM acetylcholine is shown. It reduces the oscillation amplitude in a reversible (washable) manner. All horizontal bars, 5 min.

Quantitative analysis

[Ca^{2+}]$_i$ levels

After recording [Ca^{2+}]$_i$ levels, the brain can be disrupted by addition of 160 μl lysis solution; 1% (v/v) Triton X-100, 100 mM $CaCl_2$, This causes discharge of the remaining aequorin into a calcium rich environment and allows estimation of the total amount of aequorin in the sample. The luminescence intensity can be calibrated using a suitable Ca^{2+} Calibration Buffer Kit with Magnesium II (Molecular Probes). The calibration curve (see Fig. 5) correlates free Ca^{2+} concentrations (pCa) to the rate (k) of aequorin consumption (Cobbold and Rink, 1987) and fits the equation $pCa = 0.415(-\log k) + 5.27$ for *hcp*-coelenterazine. The analogue *hcp*-coelenterazine was selected on the basis of increased sensitivity to Ca^{2+} in comparison with normal coelenterazine. For normal coelenterazine the calibration curve is $pCa = 0.475(-\log k) + 4.55$. Ca^{2+} concentrations are calculated as described by Button and Eidsath (1996).

Rapid kinetics

We have not yet applied this system to the investigation of rapid calcium dynamics. This is due to the 5–10 Hz sampling rate limit of most commercial tube luminometers. Given our present photon yield with this system it seems likely that our equipment would be sensitive enough to sample significant data at a much quicker rate. However, a range of PMT systems is available (e.g. Hamamatsu Photonics) that can sample data at rates in excess of 100 Hz. These could be adapted to study more rapid calcium dynamics than those described here.

Software

Most manufacturers of tube luminometers supply DOS or MS Windows software and cables that allow data capture to an IBM compatible PC. We have also captured data on Macintosh and Silicon Graphics (SGI) based systems. A Macintosh serial modem cable is usually all that is needed to connect the luminometer to either of these systems (in some SGI systems, IBM style PC cables are compatible and are usually supplied). Software is however somewhat lacking. On a Macintosh, we have managed to use the shareware terminal emulation package 'Black-Night' (Raine Storm Softworks) to capture data on most luminometer systems. Our experience with capturing on the SGI is more limited, however we have established communication with the **cu** UNIX command normally used to communicate with modems.

All of the multiwell plate luminometers we have examined used a dedicated PC and specialised software packages. The plate luminometers we have examined tended to be less sensitive than tube luminometers but allow a degree of sample automation. The mechanics of plate movement also upset our 'floating brain' preparations. This may not be relevant to sample preparation techniques where the tissue does not need to float. However, software for these is again somewhat limited. Most systems would appear to be ideally suited to recording a short base level, injecting one or more test compounds and recording subsequent effects on sequential samples.

Qualitative analysis

Direct visualisation of the photons emitted from tissue preparations is not a trivial matter. The photon yield from luminophores is many orders of magnitude lower than that of fluorescent calcium reporters. At all stages, the most efficient and highest quality optics must be used (these also tend to be the most expensive).

The use of wavelength specific filters to cut out background light is not advisable given the lowered efficiency. It is preferable to set the microscope up in a dark room. Spend at least 10 minutes in the room to allow your eyes to acclimatise, then survey the room for breaches

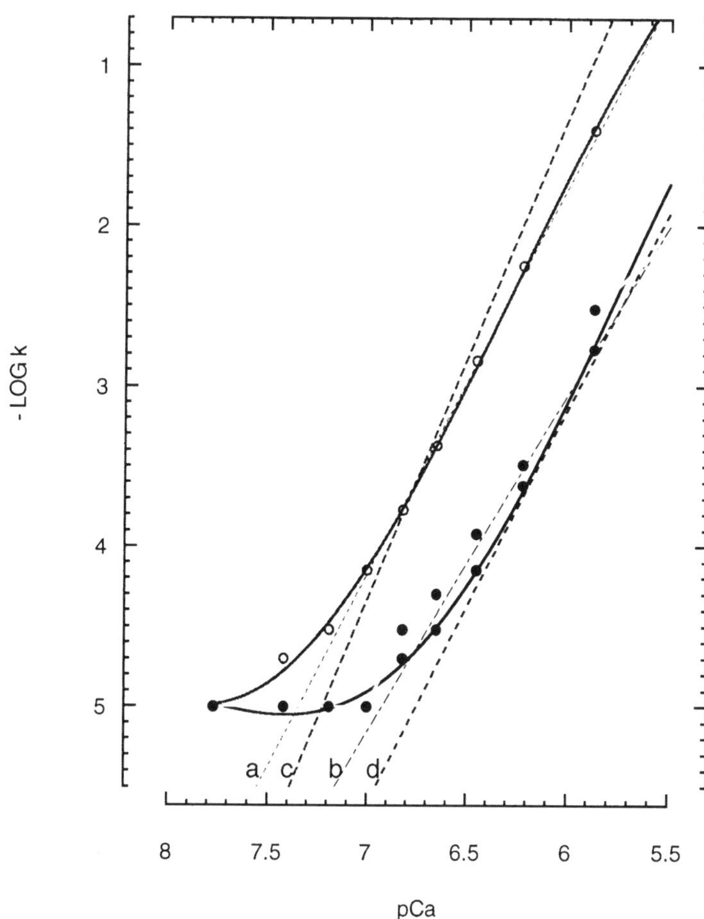

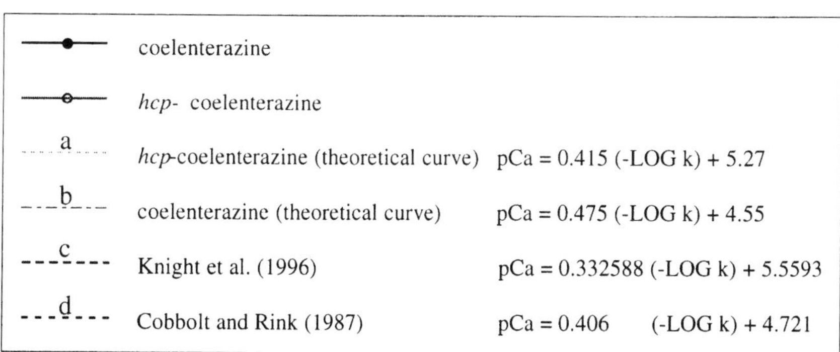

Fig. 5. Double logarithmic relationship between the calcium concentration (e.g. the free calcium in the cell cytoplasm) and the amount of total remaining aequorin discharged at any moment in time. To perform such a calibration, the amount of aequorin left at the end of an experiment is estimated by burning-out with a large excess of calcium. Then, at each time point of the experiment, the ratio between the luminescence at this point and the total remaining count is determined. This ratio, k, is the rate of aequorin consumption, and can be represented by the equation $pCa = -a \log k + b$, where pCa is the concentration of calcium. The parameters a and b are determined empirically using aequorin extract and known calcium solutions.

in the blackout coverings and the various LEDs that are likely to be around. Cover them all.

Photons (i.e. data) are lost at each optical interface. Therefore, the number and quality of these between the sample and image capture device must be optimised. We have obtained good results with a Zeiss Axiovert TV100 microscope using NeoFluar objectives. This particular model has a camera port directly beneath the objective and thus reflectors are not required to redirect the photons to the camera.

A very sensitive camera is required to capture the image. We have tried several CCD cameras from EG&G Systems and Hamamatsu Photonics. Best results were obtained using 'back lit' cooled CCD cameras. Comparable images were obtained using the most sensitive cameras available from both manufacturers. Image intensification systems, whilst theoretically more sensitive, produced higher levels of background noise. The cooling system required depends largely on the exposure time required. The EG&G 'Nightowl' camera was air-cooled and this proved sufficient for capture times we required (from seconds to several minutes). The Hamamatsu camera system was water-cooled and again this was sufficient for our purposes. Longer exposure times may be required and more extreme (liquid nitrogen) cooling systems are available, at a cost.

Alternative approaches

Whilst aequorin gives a direct means to visualise calcium activity in intact tissue, there are two main drawbacks: The photon yield is very low and, aequorin expression is difficult to visualise directly. One solution is to use 'traditional' fluorescent calcium probes. However, these would suffer from being non-specific and, in any case, loading them into intact brain preparations can be problematic.

A possible solution is to target expression of the new range of 'chameleon' calcium reporters using the P{GAL4} system. Basically these use two different modified green fluorescent proteins (GFP) linked together by a calcium sensitive linker, such as a calmodulin binding site. Each of the GFPs has different fluorescent characteristics such that the emission frequency of one is matched to the excitation frequency of the other; known as 'fluorescent resonance energy transfer' (FRET). Under specific calcium concentrations, conformation changes promote photon transfer between the two GFPs (Miyawaki et al., 1997; Romoser et al., 1997). Calcium concentration in the cell may be estimated by ratiometry.

We are in the process of investigating the use of chameleon systems in *Drosophila*. However, we do not yet know if such techniques will actually work.

Conclusions

Direct visualisation of neuronal activity using calcium sensitive probes is likely to prove a very useful tool in studies of *Drosophila* neurophysiology. At present this technology is not as sensitive as electrophysiological techniques, in particular patch clamping. However, it does allow us to investigate ensemble properties of neuron populations. We can assess the effect of both pharmaceutical agents and genetic factors.

Suppliers

BlackNight, Raine Storm Softworks, C. Swan,
http://www.kagi.com/raine

Carl Zeiss Ltd., PO Box 78, Woodfield Road,
Welwyn Garden City, Herts AL7 1LU.
http://www.zeiss.co.uk/home.e.html

EG&G Wallac, 20 Vincent Avenue, Crowhill
Buisness Centre, Crowhill, Milton Keynes,
MK8 OAB, UK.
http://www.berthold-online.de/NightOWL4.htm

Hamamatsu Photonics Ltd, Lough Point, 2
Gladbeck Way, Windmill Hill, Enfield,
Middlesex EN2 7JA, UK.
http://www.hamamatsu.com/

Molecular Probes, Inc., distributed by Cambridge Bioscience, 24-25 Signet Court, Newmarket Rd., Cambridge CB5 8LA, UK.
http://www.probes.com/

Thorn EMI Electron tubes, Bury Street Ruislip, Middlesex, HA4 7TA, UK.
http://www.electron-tubes.co.uk/

References

Cellular calcium: a practical approach (1991) Oxford, England; New York: IRL Press at Oxford University Press.

Armstrong, J.D., Yang, M.Y. and Kaiser, K. (1999) The P{GAL4} enhancer-trap technique in *Drosophila*. In: G. Coupland (Ed.), Insertional Mutagenesis: A practical Approach, Oxford University Press, in press.

Ashburner, M. (1989) Drosophila: *a laboratory manual*, Cold Spring Harbour Laboratory Press, Cold Spring Harbor, NY.

Baines, R.A. and Bate, M. (1998) Electrophysiological development of central neurons in the Drosophila embryo. *J. Neurosci.*, 18: 4673–4683.

Brand, A. and Perrimon, N. (1993) Targeted gene expression as a means of altering cell fates and generating dominant phenotypes. *Development*, 118: 401–415.

Brand, A.H. and Dormand, E.L. (1995) The GAL4 system as a tool for unravelling the mysteries of the Drosophila nervous system. *Curr. Opin. Neurobiol.*, 5: 572–578.

Brini, M., Marsault, R., Bastianutto, C., Alvarez, J., Pozzan, T. and Rizzuto, R. (1995) Transfected aequorin in the measurement of cytosolic Ca^{2+} concentration ($[Ca^{2+}]_c$). A critical evaluation. *J. Biol. Chem.*, 270: 9896–9903.

Button, D. and Eidsath, A. (1996) Aequorin targeted to the endoplasmic reticulum reveals heterogeneity in luminal Ca^{++} concentration and reports agonist- or IP_3-induced release of Ca^{++}. *Mol. Biol. Cell*, 7: 419–434.

Cobbold, P. and Rink, T. (1987) Fluorescence and bioluminescence measurement of cytoplasmic free calcium. *Biochem. J.*, 248: 313–328.

Dormand, E.L. and Brand, A.H. (1998) Runt determines cell fates in the Drosophila embryonic CNS. *Development*, 125: 1659–1667.

Dow, J.A., Maddrell, S.H., Gortz, A., Skaer, N.J., Brogan, S. and Kaiser, K. (1994) The malpighian tubules of Drosophila melanogaster: a novel phenotype for studies of fluid secretion and its control. *J. Exp. Biol.*, 197: 421–428.

Echalier, G. (1997) *Drosophila cells in culture*, Academic Press, San Diego.

Knight, M.R., Campbell, A.K., Smith, S.M. and Trewavas, A.J. (1991) Transgenic plant aequorin reports the effects of touch and cold-shock and elicitors on cytoplasmic calcium. *Nature*, 352: 524–526.

Miyawaki, A., Llopis, J., Heim, R., McCaffery, J.M., Adams, J.A., Ikura, M. and Tsien, R.Y. (1997) Fluorescent indicators for Ca^{2+} based on green fluorescent proteins and calmodulin. *Nature*, 388: 882–887.

O'Donnell, M.J., Rheault, M.R., Davies, S.A., Rosay, P., Harvey, B.J., Maddrell, S.H., Kaiser, K. and Dow, J.A. (1998) Hormonally controlled chloride movement across Drosophila tubules is via ion channels in stellate cells. *Am. J. Physiol.*, 274: R1039–1049.

Prasher, D., McCann, R.O. and Cormier, M.J. (1985) Cloning and expression of the cDNA coding for aequorin, a bioluminescent calcium-binding protein. *Biochem. Biophys. Res. Commun.*, 126: 1259–1268.

Romoser, V.A., Hinkle, P.M. and Persechini, A. (1997) Detection in living cells of Ca^{2+}-dependent changes in the fluorescence emission of an indicator composed of two green fluorescent protein variants linked by a calmodulin-binding sequence. A new class of fluorescent indicators. *J. Biol. Chem.*, 272: 13270–13274.

Rosay, P., Davies, S.A., Yu, Y., Sozen, A., Kaiser, K. and Dow, J.A. (1997) Cell-type specific calcium signalling in a Drosophila epithelium. *J. Cell Sci.*, 110: 1683–1692.

Wolf, R., Wittig, T., Liu, L., Wustmann, G., Eyding, D. and Heisenberg, M. (1998) *Drosophila* mushroom bodies are dispensable for visual, tactile, and motor learning. *Learning and Memory*, 5: 166–178.

W.E. Crusio and R.T. Gerlai (Eds.)
Handbook of Molecular-Genetic Techniques for Brain and
Behavior Research (Techniques in the Behavioral and Neural Sciences, Vol. 13)

CHAPTER 3.3.2

Behavior-genetic and molecular analysis of naturally occurring variation in *Drosophila* larval foraging behavior

Marla B. Sokolowski and Craig A. L. Riedl

Biology Department, University of Toronto, Erindale Campus, Missassauga, Ont. L5L 1C6, Canada

Introduction

In this chapter we present an overview of our genetic and molecular studies of natural variation in *Drosophila* larval behavior. Almost twenty years ago, when this work began, there was little known about larval behavior and even less known about the genes involved in naturally occurring behavioral variation. As a result, we have sometimes felt like pioneers in a new land. Our long term aim is to understand the mechanistic and evolutionary basis of natural variation in behavior. We use the foraging behavior of *Drosophila melanogaster* larvae as a model to accomplish this. In this chapter we describe the major hurdles that we encountered during the identification, localization and cloning of *foraging* (*for*), the major gene for rover and sitter differences in *Drosophila melanogaster* larval foraging behavior. The chapter begins with a review of the research which has been successful in identifying single genes involved in naturally occurring behavioral variation. We then ask why so little is known about these types of genes. We describe the research strategies and techniques used to identify, localize and clone the *foraging* gene. Finally, we discuss the cGMP/cGMP-dependent protein kinase (PKG) signal transduction pathway now known to be responsible for differences in larval foraging behavior.

Natural variants

Why has finding genes for normal behavioral variation been so difficult even in genetically well characterized and relatively simple model organisms such as the fruit fly, *Drosophila melanogaster*? There are several reasons, the first of which is historical. Geneticists interested in identifying individual genes for behavior induced genetic mutants in the laboratory to study the mechanistic basis of behavior. They assumed that there was one normal "wild-type" phenotype for behavior and mutants were generated which disrupted this "wild-type" function. However, it has recently become clear that natural variation exists for many behaviors and variation in behavior can exist within and between so called "wild-type" strains. Although genetic mutants are indispensable to the analysis of behavior they also have limitations. Many of them are so severely affected that they would not survive in nature and thus have not provided insight into the evolution of individual differences in behavior. Combined studies of natural behavior variants and mutations in genes that encode these variants have begun to provide insight into both the mechanistic and evolutionary significance of normal variation in behavior. The second reason that so few genes for natural behavioral variation have been identified lies in the inherent complexity of behavioral phenotypes. Normal behavioral variation cannot be readily placed into qualitatively determined discrete categories since it is continuously distributed and therefore must be quantified (characterized by means and variances). Natural behavioral phenotypes also show incomplete penetrance. Penetrance can be thought of as the proportion of individuals in a population who exhibit a phenotype in accordance with its genotype. One can think of it as the alleles at the "behavioral locus" giving the animal a probability of behaving a certain way in a given environment. In addition, the expression of behavioral phenotypes are often influenced by the internal and external environments of the animal.

The third reason arises from the underlying genetic etiology of natural behavioral variants. When a genetic component to behavioral differences can be demonstrated, it is usually polygenic (more than one gene). The type of polygenic inheritance ranges along a continuum from the case of classic polygenic inheritance with many genes (polygenes) all with small equal and additive effects on the phenotype to that of one, or several, major genes with minor modifiers (Falconer and Mackay, 1996). The latter type of genetic architecture more readily lends itself to gene localization and identification which facilitates cloning of the gene. The former type of genetic architecture has made it extremely difficult to identify and map the location of individual polygenes because their individual effects on behavior are small. Two approaches to analysing polygenic inheritance have been Biometrical Analysis (Mather and Jinks, 1982) and Quantitative Genetic Analysis (Falconer and Mackay, 1996). Both have been applied to naturally-occurring behavioral variation (Sokolowski and Bauer, 1989). The Biometrical approach allows one to speculate about the past history of selection that has acted on the trait (e.g. was it directional selection) but does not facilitate the identification or localization of individual genes that affect the behavioral phenotype. Biometricians tend to perform the same analyses on a variety of different phenotypes. The other approach, Quantitative Genetic Analysis has, in recent years given rise to Quantitative Trait Locus Mapping and the identification of quantitative trait loci (QTL). Methods for mapping polygenes in *Drosophila* have been developed that map polygenes to regions of the chromosomes using molecular markers such as transposable elements (roo elements) or restriction fragment length polymorphisms (Long et al., 1995). Once a polygene has been roughly localized to a region of the chromosome, previously cloned candidate genes are searched for in the gene banks. QTL mapping has not yet been used to map loci that segregate for naturally occurring behavioral traits, however, it has identified QTLs for other polygenic traits such as bristle number (Long et al., 1995) and longevity (Nuzhdin et al., 1997). An instructive case in point is that of bristle number in *Drosophila* which was touted as the "classic" unmappable polygenic character. The advent of molecular biology along with better quantitative genetic mapping techniques has made it possible to map and identify individual polygenes, even for bristle number! These approaches have been successful for mapping polygenes because they eliminated the confounding pleiotropic effects that morphological markers and new genetic backgrounds can have on the accuracy of gene localization. Genetic backgrounds are defined here as all the genes in the genome other than the one under study, in this case other than the gene being mapped. All of these factors have contributed to the past difficulty in localizing genes for naturally occurring behavioral variation.

Only very recently have a handful of studies emerged which successfully identify genes responsible for naturally occurring behavioral variants. They provide the stepping stones necessary to investigate the molecular and neuronal basis of naturally occurring behavioral variation. The studies can be classified into two groups from the point of view of how the research evolved. In the first case, natural behavior variants were found, and a single gene responsible for these behavioral differences was localized and cloned

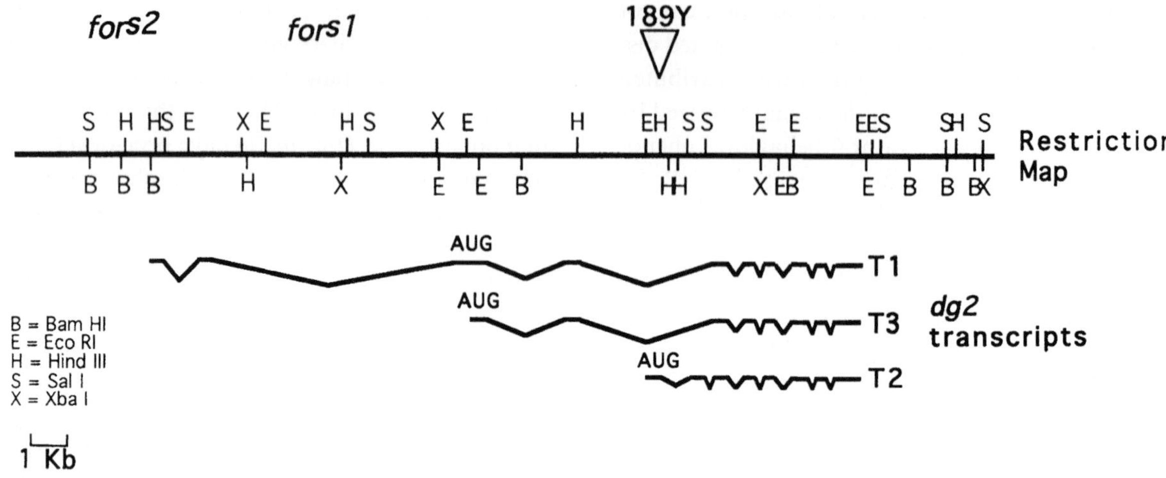

Fig. 1. The three major transcripts (T1, T2, and T3) of *for* (*dg2*) are shown against a restriction map of the region. Each encodes a common 3′ kinase domain. The sitter mutants (*for*^{s1} and *for*^{s2}) were generated in *for*^{R} flies and were mapped by RFLP analysis to the 5′ region of the gene. 189Y represents the P[GAL4] transposable element inserted into *for*. (From: Osborne et al., 1997.)

[e.g. rover and sitter larval behavior in *Drosophila* (Osborne et al., 1997) and social and solitary foraging behavior in *C. elegans* (De Bono and Bargmann, 1998)]. In the second case, variation in the protein of a previously cloned gene was identified, and later naturally occurring behavioral variation that resulted from these protein differences was found [the number of threonine/glycine repeats in the *period* gene of *Drosophila* (Sawyer et al., 1997)]. The first approach proceeds from "natural variants to gene to protein" and the second from "gene to protein to natural variant". These studies are briefly reviewed below.

Rover and sitter foraging in Drosophila

Sokolowski (1980) showed that, in the presence of food, *D. melanogaster* larvae behave either as rovers, exhibiting relatively long foraging paths, or sitters, having shorter foraging paths. In the absence of food, both variants move equally rapidly indicating that sitters are not simply sluggish animals (Pereira et al., 1995). Both variants are found in significant frequencies (30% sitters : 70% rovers) in orchards in the Toronto area (Sokolowski, 1980; Sokolowski et al., 1997). Rovers have higher fitness in crowded environments and sitters do better in uncrowded environments (Sokolowski et al., 1997). A single major gene called *foraging* (*for*) accounts for differences in these variants. In larvae the rover allele (*for*^{R}) shows complete genetic dominance to the sitter one (*for*^{s}) (De Belle and Sokolowski, 1987). The cloning of *for* demonstrated that it is identical to a gene called *dg2* which encodes one of two PKG genes in *D. melanogaster* (Osborne et al., 1997). Evidence for this was:

1. *for* mutants mapped in, or 5′ to, the *dg2* transcripts (Fig. 1),
2. a transposable element insertion in *dg2* changed behavior from rover to sitter and excision of the element reverted the behavior back to rover,
3. adult heads of rovers had higher PKG enzyme activity levels than did sitters and mutant sitters; similar results were found when the PKG activity of dissected larval central nervous systems were assayed,
4. rovers had higher *dg2* transcript abundance than sitters and sitter mutants, and finally,
5. *dg2* cDNA driven by a heat shock promoter in transgenic sitter larvae changed larval

behavior and PKG activity from sitter to rover demonstrating rescue of the phenotype in transgenic flies.

This was the final convincing evidence that *for* is synonymous with *dg2*, and thus the research identified a behavioral function for PKG in naturally occurring variation. We now have evidence that the larval precursor of the adult brain central complex is involved in foraging behavior (Varnam et al., 1996). Along with identifying other tissues involved, we would also like to learn more of the precise molecular differences in *for* that give rise to the rover and sitter differences in larval behavior.

Social and solitary foraging in C. elegans

De Bono and Bargman (1998) showed that *C. elegans* also exhibit a naturally occurring polymorphism in foraging behavior. Social worms forage in groups and exhibit higher rates of locomotion compared to solitary foragers. Remarkably, this difference in behavior results from a single amino acid substitution (from valine to phenylalanine) in the *npr-1* gene which encodes a predicted G protein-coupled receptor of the neuropeptide Y (NPY) receptor class. This amino acid substitution was responsible for differences in all of the twelve social and five solitary strains so far collected worldwide. Social worms exhibit rover behavior while foraging whereas solitary worms exhibit sitter behavior. This difference in movement rate (like the one found in flies) was conditional on the presence of food in the environment. It is not known how these two variants (social and solitary foragers) are maintained in nature. It will be interesting in the future to determine whether the *for* gene in flies (which encodes PKG) and the *npr-1* gene in *C. elegans* (encoding a neuropeptide Y-receptor homologue) are members of the same biochemical pathway important to foraging behavior (see Sokolowski, 1998, for discussion).

Circadian rhythms in Drosophila

The temperature compensation ability of the circadian clock has been shown to result from a molecular polymorphism within the *period* (*per*) gene (Sawyer et al., 1997). The *per* gene in *Drosophila* was originally identified in a mutagenesis screen for eclosion rhythm by Konopka (1987). Later it was shown to affect a number of rhythms in the fly including circadian locomotor activity rhythms and courtship song ultradian rhythms (see Konopka, 1987, for review). Molecular analysis of *per* and another clock gene in *Drosophila* called *timeless* has shown that both proteins are involved in an autoregulatory feedback loop critical for circadian rhythmicity (Hall, 1995). The *per* gene has a polymorphic threonine-glycine (Thr-Gly) encoding repeat which varies in length. Two of the major variants (Thr-Gly)17 and (Thr-Gly)20 show latitudinal clinal variation in Europe. These variable repeats are related to how well a fly can maintain its circadian rhythmicity as temperature varies; this is referred to as the ability of the circadian clock to temperature compensate (Hall, 1997).

Rover and sitter foraging behavior in *D. melanogaster*: a case study

At this point in the chapter we describe the research strategies and techniques we used to identify, localize and clone the *foraging* gene. The foraging behavior of *Drosophila* maggots (called larvae) is assayed in a circular arena covered in a thin yeast and water paste (for more detailed methods, see Sokolowski et al., 1997). Briefly, we collect third-instar larvae (96 ± 2 h post-hatching, 25°C), quickly rinse them in distilled water, and place them individually in shallow wells (0.5 mm deep × 4.25 cm diameter) in which a nutritive yeast paste (2 parts water, 1 part yeast, by weight) has been spread. After a larva has been placed in a well, the well is covered by a Petri dish lid to prevent drying of the paste. While larvae feed and move in the yeast paste, they leave a visible trail, the length of which is measured after a 5 min test interval. The wells are then rinsed and re-spread with clean yeast paste before new larvae are tested. As a control, larval locomotion is measured in a similar assay on a non-nutritive agar surface. This gives us an indication of the length of the larva's locomotory path in the absence of food. Over the years, the foraging behavior assay has evolved to enable us to quantify the behavior of 800 larvae in a single day. This has facilitated our genetic analysis and behavioral screens for mutations that affect foraging behavior. *D. melanogaster* larval foraging path length has a bimodal distribution. Some larvae move long distances while feeding whereas others move relatively shorter distances, feeding more locally. We call those with long foraging trails "rovers" and those with relatively shorter trails "sitters". Both rovers and sitters move similarly long distances on the control non-nutritive agar surface (Table 1). Thus, the differences in path length between rovers and sitters is conditional on the presence of food (yeast) in the environment and sitters are not simply "slow" moving animals.

In *Drosophila* it is possible to generate isogenic lines of rovers and sitters. These are lines that are essentially homozygous at all loci for a large percentage of the genome (96%). They are generated using balancer chromosomes (Greenspan, 1997) which limit recombination and facilitate the inheritance of intact chromosomes from one generation to another. Mendelian analyses were done once homozygous rover and sitter lines were generated. We performed a 16 reciprocal cross analysis of larval foraging behavior and showed that there was a major gene effect on the autosomes (De Belle and Sokolowski, 1987). Using this analysis we were able to rule out major effects of the X and Y chromosomes and maternal or cytoplasmic effects. We also found that the rover larval phenotype showed complete genetic dominance to the sitter one. Chromosome substitution analysis enabled us to isolate the rover/sitter difference in behavior to the second pair of autosomes (Sokolowski, 1980). This was done by generating new strains of flies with combinations of the major autosomes (second and third) from the rover and sitter strains (Sokolowski, 1992).

We are interested in the sources of variation that contribute to the differences in the distributions of larval foraging trail lengths (Fig. 2). We generated homozygous lines of rovers and sitters. The distributions and amount of overlap between the two lines are shown in Fig. 2. In simple terms, the rover and sitter natural variants could result from either genetic differences, environmental differences or a combination of both. In order to determine if there was a genetic component to the behavioral differences it was important to maximize the environmental conditions so that there was minimum overlap between the distributions of the two variants. We aimed for <10% overlap in the two distributions in order to classify animals of unknown phenotype without too much error. In some cases where the overlap between the two distributions was greater than 10% it was necessary to do repeated testing of a larva's behavior or progeny testing (Pereira et al., 1995). Progeny testing is when the fly of unknown genotype is mated to flies of the homozygous recessive type (in our case, sitter) and the larval behavior of their progeny is tested. If all of the

TABLE 1

The distances traveled by rover- and sitter-type larvae and adult *D. melanogaster* in both nutritive and non-nutritive environments

		Distances traveled: (mean ± SE) (cm) ($14 < n < 25$)	
		Nutritive environment (foraging)	Non-nutritive environment (locomotion)
Larvae:	Rover	10.8 ± 0.6	12.0 ± 0.7
	Sitter	4.5 ± 0.3	13.7 ± 1.
Adults:	Rover	37.6 ± 3.3	27.6 ± 5.3
	Sitter	15.3 ± 2.4	27.8 ± 5.3

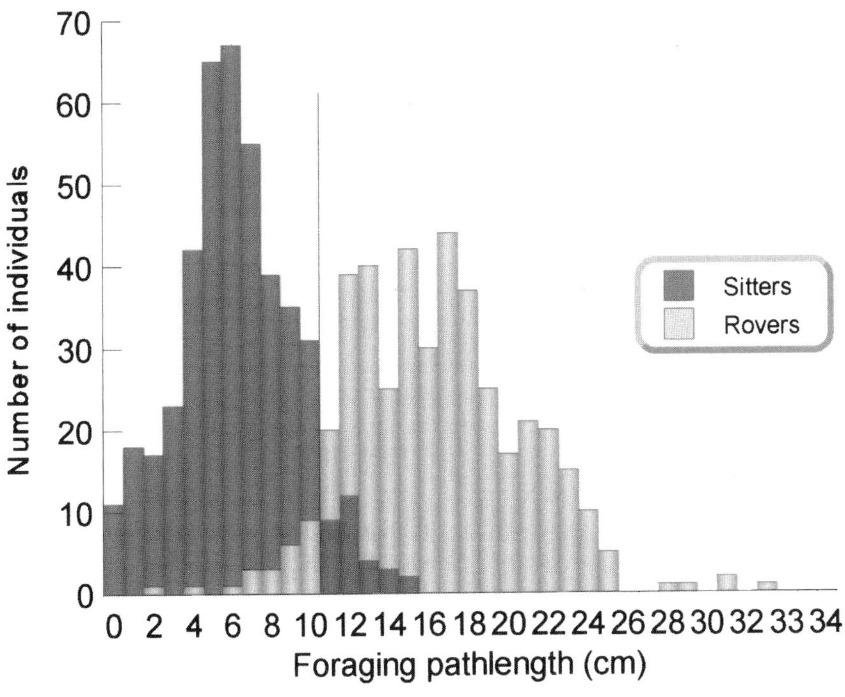

Fig. 2. Frequency distributions of sitter and rover larval foraging pathlengths. The dark bars indicate the pathlengths of homozygous sitter larvae, while the light bars represent homozygous rover pathlengths. Notice the area of overlap. (From: Sokolowski, 1992.)

progeny behave as sitter then the test animal was sitter; if the progeny are half rover (the genetically dominant phenotype) and half sitter then the test animal was heterozygous. We performed numerous genetic analyses to determine whether the differences in larval foraging path length between rovers and sitter had a genetic basis (for review see Partridge and Sgro, 1998).

One question of interest is whether the overlap between the two distributions resulted from genetic differences within the population of one variant or from incomplete penetrance. One way to address this was to allow individuals whose pathlengths lay in the area of overlap to mature and test cross them to sitter stock. The resultant progeny could then be assayed and the mean and distribution of their pathlengths compared to the values representing the original population. In our case we found that the overlapping distributions resulted from incomplete penetrance of the rover or sitter alleles. In simple terms, having one rover allele gave the larva a certain probability (>90%) of performing as a rover and similarly, homozygous sitter alleles gave the larva a >90% chance of behaving as a sitter.

When the results of the genetic analysis of behavior of the natural variants fit a simple genetic model [e.g. one or two genes with major effects along with many genes with minor effects (minor modifiers)] then one can proceed toward the identification, localization, and possible cloning of the major gene(s) involved. This was the case with the *for* gene. However, we should point out that genetic analyses often provide us with the "simplest" genetic model that the data will fit. Further gene localization may uncover a more complex genetic etiology; we found this to be the case when we tried to localize a putative major gene for natural variation in ovarian diapause in *Drosophila* (e.g. Williams and Sokolowski, 1993; Williams, unpublished).

One of the biggest hurdles involved the genetic localization of *for*. Chromosomal analysis and subsequent compound autosome analysis enabled us to localize *for* to the left arm of chromosome-*2* in *D. melanogaster* (Sokolowski, 1980; De Belle and Sokolowski, 1989). Further schemes which employed recombination mapping to localize the gene were not successful. Genetic markers, genetic background, and uncontrolled variation in the internal or external environments of the larva can act individually or in consert to cause us to misclassify the genotype of a larva [as rover (for^R) or sitter (for^s)] based on its measured phenotype. First, markers used in recombination mapping often have pleiotropic effects on behavior. For example, if a rover larva with a marker (a visible mutation at another genetic locus) had a reduced developmental rate, then it would have a shorter foraging trail due to its small size and could be incorrectly classified as a sitter. Second, the genetic background (all the other genes in the genome aside from the one being investigated) could have effects on the expression of penetrance of the rover and/or sitter *for* allele, thereby increasing the overlap between the distributions (Fig. 2). Third, the environment, which often plays a significant role in the expression of behavioral differences, could also affect the distributions of the phenotypes shown in Fig. 2. For example, hunger levels and food quality are known to affect larval and adult foraging behavior (Graf and Sokolowski, 1989; Tortorici and Bell, 1988; Pereira and Sokolowski, 1993). Table 2 shows how the environment (time since the last meal) can act to modify the expression of larval and adult foraging behavior. It is of interest to correlate changes in expression of larval and adult foraging behavior with changes in the level of expression of the *for* gene product. This would address the question of whether the *for* gene product affects both allelic variation between individuals and plasticity within individuals.

To map *for*, we developed the lethal tagging technique (De Belle et al., 1989). This technique has also been used to map the *Chaser* gene, a dominant suppressor of sitter found on chromosome-*3* (Pereira et al., 1995). The idea was to tag the behavioral gene which we were trying to localize with a discrete, easily quantified phenotype. We chose recessive lethality, which enabled us to categorize flies simply as being dead or alive. We designed a mutagenesis and screening procedure that would both lethal tag the *for* gene and generate mutant sitter alleles from a rover strain. We mutagenized our natural rover males (with gamma rays to induce chromosome rearrangements and later ethyl methanosulphonate (EMS) to induce single base substitutions) and crossed them to natural sitter females (De Belle et al., 1989, 1993). The F_1 larvae normally behave as rovers since for^R is completely dominant to for^s. We screened for larvae that behaved as sitters and carried a recessive lethal mutation on the second chromosome. We did a double screen for a change in behavior from rover to sitter and the induction of a recessive lethal on chromosome-*2*. In a series of crosses which used balancer chromosomes we were able to identify three independent sitter-behaving recessive pupal-lethal lines. These lines did not complement with each other for behavior or lethality and

TABLE 2

Foraging behaviour is influenced by the pre-test environmental conditions. Larvae (i) and adults (ii) of both sitters and rovers have shorter foraging pathlengths when starved for a period immediately preceding the behaviour test

Starvation time: (h)	Sitters	Rovers
	(i) Larval foraging pathlengths (mean ± SE [sample size]) (cm)	
0	2.0 ± 0.6 [25]	11.0 ± 1.3 [25]
4	1.0 ± 0.3 [25]	5.2 ± 1.0 [25]
	(ii) Adult foraging pathlengths (mean ± SE [sample size]) (cm)	
4	22.38 ± 3.7 [12]	36.17 ± 3.5 [12]
24	16.20 ± 3.4 [10]	25.6 ± 4.3 [10]

did not complement with the original natural sitter allele. This strongly suggested that we had caused alterations in each of these lines that affected the same gene. The chances of making three independent lethal mutations that did not complement for behavior and lethality, and did not tag *for*, were very small since there are approximately 2000 lethal complementation groups on chromosome-2 in *D. melangaster*. We concluded that either we had made recessive lethal alleles of *for* or we had caused a mutation in *for* and an adjacent vital gene. Nevertheless, we were able to easily map the lethal phenotype that tagged *for* using recombination and deletion mapping to region 24 of the *Drosophila* polytene chromosome map. Further mapping using chromosome rearrangements that we generated or received from other labs enabled us to localize *for* to 24A3-5 on the polytene chromosome map, a span of approximately 150–200 kb of DNA.

Our attempts to clone *for* began with a chromosome walk. In the late, 1980s little genetic material was available within this region of DNA. The material in hand was: seven lethal tagged sitter alleles, three viable sitter alleles, and DNA from a gene, *dg2*, found in this cytological region. *dg2*, which encodes a *Drosophila* PKG, was cloned by Kalderon and Rubin (1989); no mutants were available and the function of this kinase in *Drosophila* was unknown. A probe from the 5′ end of *dg2* was used to initiate our chromosome walk. Indeed, *dg2* was a candidate gene for *for* which we later showed to be identical to *for* (Osborne et al., 1997). Kate Osborne in our lab showed that several of the putative lethal alleles mapped immediately 5′ to *for*. They were likely in the *for* promoter region. At that time we incorrectly thought that the *for* gene was immediately 5′ to *dg2*. In fact, it was the lethal alleles that we had generated in lethal tagging that misguided us.

In Osborne et al. (1997) we showed that the viable *for* mutants mapped in, or 5′ to, the *dg2* transcripts (Fig. 1). We obtained a P[GAL4] element from K. Kaiser and showed that it was inserted in the 5′ end of the T2 transcript of *dg2*. This transposable element insertion in *dg2* caused a change in behavior from rover to sitter and excision of the element reverted the behavior back to rover. Our collaborators, Alain Robichon and Ralph Greenspan showed that adult heads of rovers had higher PKG enzyme activity levels than did natural or mutant sitters; similar results were found when the PKG activity of dissected larval central nervous systems were assayed. Natural rovers had higher *dg2* transcript abundance than natural sitters and sitter mutants; this suggested that the differences in the natural variants may arise from differences in the regulation of *for* rather than from structural differences in the protein. Future experiments will address this issue. *dg2* cDNA driven by the leaky expression of a heat shock promoter in transgenic sitter larvae changed larval behavior and PKG activity from sitter to rover, demonstrating rescue of the phenotype in transgenic flies. This suggested that a small increase in PKG (<30%) was sufficient to restore rover larval behavior. The transgenic rescue of rover behavior in sitter larvae using a *dg2* transgene was the final evidence that *for* is synonymous with *dg2*.

We will now present a brief discussion of PKG function and how it may pertain to foraging behavior.

The for *product: PKG*

for encodes one of two known *D. melanogaster* PKG isoforms (Kalderon and Rubin, 1989; Osborne et al., 1997). The genetic structure of *for* is quite complex, with the gene encoding up to six different polypeptides. To date, very little is known of the expression or function of the *for* product. In fact, the idea that it encodes a PKG was based almost entirely on sequence homology to known PKG genes, however, we have recently shown that PKG activity is reduced in *for* mutants (Osborne et al., 1997).

The *for* gene shares sequence homology and splice-site conservation with mammalian PKG genes (Ørstavik et al., 1997; Witczak et al., 1998), and the *for* product seems most closely related to the mammalian type I PKG (PKG-I) which functions as a soluble homodimer (Jarchau et al., 1994). Although the observed cellular localization of PKG-I is cytoplasmic, nuclear localization-like sequences have been identified on the protein (Gudi et al., 1996). The nuclear localization of PKG has not been widely observed, but PKG antibody immunofluorescence has been observed in macrophage nuclei (Pryzwansky et al., 1995). PKG-I is fully activated by the co-operative binding of two guanosine 3′,5′-cyclic monophosphate (cGMP) molecules and also binds Mg^{2+}-ATP (Døskeland et al., 1987). Each subunit has autoregulatory domains which may be autophosporylated by one or the other subunit to alter enzymatic activity (reviewed in: Hobbes and Ignarro, 1997; see also: Chu et al., 1998). PKG-I is expressed in many tissues including: smooth muscle, brain (especially cerebellum), heart, kidney, and blood platelets (reviewed in: Francis and Corbin, 1994). Originally, much of the neural expression was attributed to PKG in vascular tissues imbedded in the structures, however, recent work on sheep anterior pituitary glands shows that functional PKG is expressed and plays a regulatory role in the hormone-secreting cells, *per se* (Li et al., 1996).

The main physiological control of soluble PKG appears to be via manipulation of cellular cGMP levels. cGMP has other intracellular targets of action besides PKG, these include: cGMP-regulated cyclic nucleotide phosphodiesterases (PDE) and cGMP-gated ion channels. As a result, we must constantly remind ourselves not to assume that any cGMP-mediated signal will involve PKG activity.

PKG substrates are notoriously non-specific with many also being phosphorylated by adenosine 3′,5′-cyclic monophosphate dependent protein kinase (PKA). Nevertheless, a few substrates have been shown to be preferentially phosphorylated by PKG (some reviews of PKG structure and function are: Butt et al., 1993; Francis and Corbin, 1994; Hobbes and Ignarro, 1997; Lohmann et al., 1997; Wang and Robinson, 1997). A large number of other proteins are susceptible to PKG mediated phosphorylation, at least in vitro (Wang and Robinson, 1995), but the physiological significance of this is unclear.

The regulation of PKG

PKG activity is regulated by the binding of cGMP and hence linked to cellular cGMP levels. cGMP is synthesized by guanylyl cyclases (GC). *D. melanogaster*, like mammals, has at least two main types of GC: soluble GC (sGC) and particulate GC (pGC) (Yoshikawa et al., 1993; Liu et al., 1995; Shah and Hyde, 1995). pGCs are monomeric or homomultimeric membrane receptors whose ligands include natriuretic peptide hormones and related molecules (reviewed in: Goy, 1991). These hormones are produced by many tissues, effecting various results in the target cells (for review see Levin et al., 1998). Certain odorants may also bind directly as ligands to activate specific types of pGC in olfactory tissues (Juilfs et al., 1997; Yu et al., 1997). Many isoforms of pGC have been identified in various animals.

sGCs are heterodimers with α and β subunits (Foerster et al., 1996). There are probably many different α and β subunit types and different combinations may have altered regulation and activity in an individual (Shah and Hyde, 1995). sGCs are activated by gaseous neurotransmitters such as nitric oxide (NO) and carbon monoxide (CO), as well as by arachidonic acid. CO is produced by heme oxygenase, and little is known about its signalling processes in vivo (Verma et al., 1993; Friebe et al., 1996; Maines, 1996). NO is produced by nitric oxide synthetases (NOS) (reviewed in: Bredt and Snyder, 1992). Although there are three types of mammalian NOS (brain NOS, endothelial NOS, and macrophage NOS) only one form has so far been observed in *D. melanogaster* (dNOS) (Regulski and Tully, 1995). Activation of sGC by arachidonic acid is important for insect eclosion hormone signalling (Morton and Giunta, 1992).

The existence of a third type of GC has recently been reported. This is a soluble NO-insensitive form which appears to have structural similarity to pGCs but its mechanism of activation remains unknown. This form appears to be the predominate GC in the lobster central nervous system (Kojima et al., 1995; Prabhakar et al., 1997). A degree of expressional and functional overlap may exist for the various forms of GC.

Some actions of PKG

A variety of roles for PKG signalling have been identified in many tissues. The most well characterized of these is the role of PKG in mammalian smooth muscle contraction, or more specifically, relaxation (reviewed in Lincoln et al., 1996). PKG activity elicits a number of effects in smooth muscle cells (SMCs), most of these involve ion flux, especially Ca^{2+} (Pfeifer et al., 1998). In SMC's, PKG is believed to affect: both voltage- and ligand-gated Ca^{2+} channels, cell membrane and endoplasmic reticulum Ca^{2+} channels, the Na^+/Ca^{2+} exchanger, inositol 1,4,5-triphosphate (IP_3) signaling, phospholambin activity, and Ca^{2+}-dependent K^+ channels (reviewed in Hobbes and Ignarro, 1997).

PKG may also influence: G protein signalling (Pfeifer et al., 1995), vacuolar ATPase activity (Davies et al., 1997), ATP-sensitive K^+ channel function (Ferrer et al., 1998), cyclic-nucleotide phosphodiesterase activity (Wyatt et al., 1998), and protein transcription (Gudi et al., 1996). Besides smooth muscle, PKG is expressed in: blood platelets, kidney, intestine, heart, skeletal muscle and nervous tissue (see below). In these tissues it also generally affects ion flux. In the whole organism, PKG signalling has been implicated to affect hypertension, diuresis (reviewed in Lincoln et al., 1996), and learning and memory (Bernabeu et al., 1997) in mammals, as well as food search behavior (Osborne et al., 1997) and fluid secretion (Davies et al., 1997) in *D. melanogaster*.

Often, in vivo studies indicate a role for PKG by first observing a physiological reaction to NO donors or cGMP analogs and then inducing the loss of this response by administering specific GC or PKG inhibitors (for examples see: Bernabeu et al., 1997; or Komalavilas and Lincoln, 1996). In vitro studies may be more direct although their physiological implications can be difficult to assess.

Assaying PKG

PKG activity may be assayed using a synthetic peptide substrate (Kemp and Pearson, 1991) in the presence of [γ-^{32}P]ATP and inhibitors of PKA (Li et al., 1996). ^{32}P incorporation into the substrate can be assayed and comparisons to known standards will give an indication of PKG activity. Antibodies to PKG will assist in this assay by allowing affinity column purification of PKG to help reduce interference from PKA activity. More recently, a novel method for assaying PKG activation has been developed (Smolenski et al., 1998). This test also involves the use of a PKG-specific substrate, but in this assay, antibodies specific for the PKG-phosphorylated form of the substrate are used to help quantify PKG activity.

Presently, there are no antibodies to *D. melanogaster for*-encoded peptides, but we are working diligently to produce these. Once ready, they will be valuable tools for the elucidation of the various levels and foci of *for* expression and possible function.

PKG and foraging behavior

Despite the paucity of information regarding *for* expression and function in *Drosophila*, emerging information from other animal models can help us develop theories of a role for PKG in foraging. Foraging behavior can be functionally divided into four components:

1. monitoring the external environment;
2. monitoring the internal environment;
3. central integration of information; and
4. the response (Fig. 3).

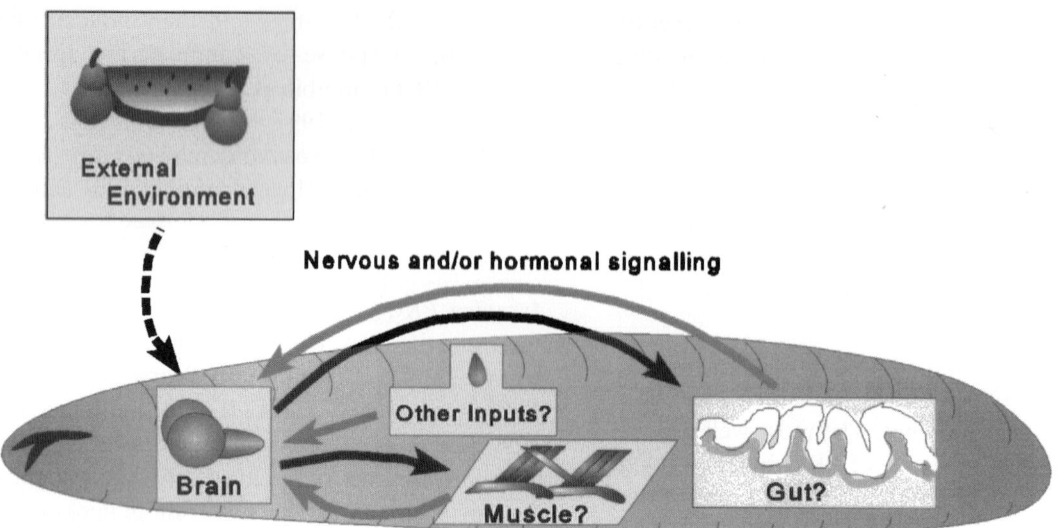

Fig. 3. The functional components of larval foraging behaviour: (1) The animal must monitor the external environment to detect the presence of food (dashed arrow). (2) Various internal inputs (grey arrows) may also influence the foraging strategy. (3) PKG activity in the brain and CNS may alter the integration and interpretation of the internal and external cues. (4) Differential PKG activity in the tissues that receive the efferent foraging signals (black arrows) may alter the final response to these cues. The influence of *for* on foraging behaviour may be the result of PKG function in any or all of these components as well as the nervous or hormonal signalling between them.

PKG signalling may be active in one or more of these components and any one or all may be the main point of PKG influence over foraging behavior.

1. PKG and the external environment. Since the behavioral differences between rover and sitter larvae are observable only in a nutritive environment, one way in which PKG may influence foraging is to affect the primary detection of external food. Many studies have demonstrated a role for cGMP and possible PKG signalling in chemoreception (Amakawa and Ozaki, 1989; Amakawa et al., 1990; Breer and Shepherd, 1993; Elphick et al., 1995; Kroner et al. 1996; Dittman et al., 1997; Moon et al., 1998). For example, activation of PKG in isolated rat olfactory cilia attenuates the response to odorant stimulation (Kroner et al., 1996). It is possible that differential PKG function in chemoreceptor cells results in alterations in the signals sent to the brains of rover and sitter flies, in turn leading to variation in perception of, and response to, similar external environments.

2. PKG and the internal environment. The feeding behavior of *D. melanogaster* is affected by the amount of food in the gut (Edgecomb et al., 1994). Research in the blowfly shows that neuron-mediated signals indicating the nutritive state of the fly are sent from the gut to the brain (Moss and Dethier, 1983). In mammals, gut signals travel to the brain via cranial nerves VII, IX and X (Haines and Mihailoff, 1997). NO-cGMP signalling has been implicated in the neurotransmission of cranial nerve X, the vagus nerve (Hatanaka et al., 1997). NO-cGMP-PKG signalling molecules are also expressed in the mammalian hypothalamus (Liu et al., 1997; Bhat et al., 1996), a tissue important in appetite regulation. At least some of the actions of the appetite affecting hormone neuropeptide Y (NPY) are mediated by PKG (Lemos et al., 1997). Recently, NPY signalling has also been implicated in the feeding behavior in *C. elegans* (De Bono and Bargmann, 1998). PKG function may be important for the internal feedback control of foraging behavior.

3. PKG and sensory information integration. In insects, the brain structures thought to be responsible for olfactory information integration are the antennal lobes (Grünbaum and Müller, 1998) and the mushroom bodies (Heisenberg et al., 1985). NO-cGMP signalling is observed in the both these tissues and may indicate a role for PKG in olfactory information processing (Bicker et al., 1996; Müller and Hildebrandt, 1995). PKG is also expressed in various other neural tissues (reviewed in: Lohmann et al., 1981, 1997; Wang and Robinson, 1997) and may act to increase neural excitation (Gammie and Truman, 1997). In mammals, roles for PKG in the regulation of some GABA-ergic (Leidenheimer, 1996; Wexler et al., 1998) and cholinergic (Rae et al., 1998) neurotransmission have recently been demonstrated and PKG is also involved in hippocampal long-term potentiation (Zhuo et al., 1994) as well as long-term depression in the dentate gyrus (Wu et al., 1998). Thus, differences in PKG activity in neural tissues may result in differences in brain activity and the interpretation and response to foraging cues.

4. PKG and the response. After the brain receives the cues and sends the signals to forage, the various targets for these signals may respond differently depending on their PKG expression. For example, PKG has been implicated in the respiration and energy usage of skeletal muscle (Young and Leighton, 1998). It also may play a role in: efferent nervous transmission (Ouedraogo et al., 1994; Sluka and Willis, 1998; also, for a review see Krukoff, 1998), muscle contraction (reviewed in: Lincoln et al., 1996), gastric emptying (Orihata and Sarna, 1996), digestive enzyme secretion (Fiorucci et al., 1995), and modulation of various hormonal responses (Pfeifer et al., 1995). Thus, differential PKG activity in the specific target tissues may alter the transmission of, or the reaction to, foraging signals descending from the brain.

It is therefore conceivable that PKG signalling may play a regulatory role at any or all of the four functional units of foraging behavior, as well as in the nervous and hormonal connections between them. Different *for* transcripts may be expressed in different domains and the natural variation may be limited to alterations of a single transcript. One way to investigate which transcript in which functional domain is important is to create a series of genetic mosaics using the GAL4-UAS system (Brand and Perrimon, 1993). In this system, the yeast transcriptional activator gene, GAL4, is inserted behind a promoter that shows a specific expression pattern of interest. The gene of interest, in our case a PKG transcript cDNA, is then placed behind a UAS (GAL4 target) sequence and transfected into the genome. We can then cross flies with the UAS-PKG transgene to those with the GAL4 gene and, this way, the GAL4 will drive transgene expression only in the specific tissues of interest. We can use this system to express the various *for* transcripts in different tissues (*e.g.* chemoreceptors, mushroom bodies, gut, etc.) to determine the behavioral significance of each transcript at each location. One limitation of this technique is the inability to temporally direct expression, that is, the transgene will be expressed constitutively.

One exciting new technique which may be very useful is the use of the tetracycline-dependent transactivator (tTA) system (Bello et al., 1998). In this system the tTA gene is inserted behind a promoter with a desired expression pattern and the transgene is put behind a tTA-responsive promoter. Tetracycline in the medium inhibits tTA activity and when transgene function is required one can simply transfer the animal to tetracycline-free medium allowing tTA expression and resultant transgene activation in both the tissue and time desired.

The effect of the *for* signalling cascade on the foraging behavior of *D. melanogaster* presents itself as a unique opportunity for the study of natural genetic influences on behavior. Questions that remain are: Where and when must *for* be expressed to produce rover behavior? Do the alternatively spliced *for* transcripts have different behavioral and/or developmental functions? Is *for* a vital gene? What is the precise molecular basis of rover and sitter differences in nature and how did they evolve? The fact that the original *for* alleles originated and were maintained in nature indicates the importance of this trait to the fitness of the individuals. By studying this trait we will learn not only about the genetics of feeding behavior in fruit flies, but also about the genetic origin and maintenance of natural behavioral variation in general.

List of abbreviations

cGMP	guanosine 3′,5′-cyclic monophosphate
GC	guanylyl cyclase
NO	nitric oxide
NOS	nitric oxide synthetase
NPY	neuropeptide Y
PDE	phosphodiesterase
pGC	particulate guanylyl cyclase
PKA	cyclic AMP-dependent protein kinase
PKG	cGMP-dependent protein kinase
QTL	quantitative trait loci
sGC	soluble guanylyl cyclase
SMC	smooth muscle cell
tTA	tetracycline transactivator

References

Amakawa, T. and Ozaki, M. (1989) Protein kinase C-promoted adaptation of the sugar receptor cell of the blowfly *Phormia regina*. *J. Insect Physiol.*, 35: 233–237.

Amakawa, T., Ozaki, M. and Kawata, K. (1990) Effects of cyclic GMP on the sugar taste receptor cell of the fly *Phormia regina*. *J. Insect Physiol.*, 36: 281–286.

Bello, B., Resendez-Perez, D. and Gehring W.J. (1998) Spacial and temporal targeting of gene expression in *Drosophila* by means of a tetracycline-dependent transactivator system. *Development*, 125: 2193–2202.

Bernabeu, R., Schroder, N., Quevedo, J., Cammarota, M., Izquierdo, I. and Medina, J.H. (1997) Further evidence for the involvement of a hippocampal cGMP/cGMP-dependent protein kinase cascade in memory consolidation. *NeuroReport*, 8: 2221–2224.

Bhat, G., Mahesh, V.B., Aguan, K. and Brann, D.W. (1996) Evidence that brain nitric oxide synthase is the major nitric oxide synthase isoform in the hypothalamus of the adult female rat and that nitric oxide potently regulates hypothalamic cGMP levels. *Neuroendocrinology*, 64: 93–102.

Bicker, G., Schmachtenberg, O. and DeVente, J. (1996) The nitric oxide/cGMP messenger system in olfactory pathwys of the locust brain. *Eur. J. Neurosci.*, 8: 2635–2643.

Brand, A.H. and Perrimon, N. (1993) Targeted gene expression as a means of altering cell fates and generating dominant phenotypes. *Development*, 118: 401–415.

Bredt, D.S. and Snyder, S.H. (1992) Nitric oxide, a novel neuronal messenger. *Neuron*, 8: 3–11.

Breer, H. and Shepherd, G.M. (1993) Implications of the NO/cGMP system for olfaction. *Trends Neurosci.*, 16: 5–9.

Butt, E., Geiger, J., Jarchau, T., Lohmann, S.M. and Walter, U. (1993) The cGMP-dependent protein kinase — gene, protein, and function. *Neurochem. Res.*, 18: 27–42.

Chu, D.M., Francis, S.H., Thomas, J.W., Maksymovitch, E.A., Fosler, M. and Corbin, J.D. (1998) Activation by autophosphorylation or cGMP binding produces a similar apparent conformational change in cGMP-dependent protein kinase. *J. Biol. Chem.*, 273: 14649–14656.

Davies, S.A., Stewart, E.J., Huesmann, G.R., Skaer, N.J.V., Maddrell, S.H.P., Tublitz, N.J. and Dow, J.A.T. (1997) Neuropeptide stimulation of the nitric oxide signaling pathway in *Drosophila melanogaster* malpighian tubules. *Am. J. Physiol.*, 273(*Regulatory Integrative Comp. Physiol.*, 42): R823–R827.

De Belle, J.S., Hilliker, A.J. and Sokolowski, M.B. (1989) Genetic localization of *foraging* (*for*): A major gene for larval behaviour in *Drosophila melanogaster*. *Genetics*, 123: 157–164.

De Belle, J.S. and Sokolowski, M.B. (1987) Heredity of rover/sitter: Alternative foraging strategies of *Drosophila melanogaster*. *Heredity*, 59: 73–83.

De Belle, J.S. and Sokolowski, M.B. (1989) Rover/sitter foraging behaviour in *Drosophila melanogaster*: Genetic localization to chromosome-2L using compound autosomes. *J. Insect Behav.*, 2: 291–299.

De Belle, J.S., Sokolowski, M.B. and Hilliker, A.J. (1993) Genetic analysis of the *foraging* microregion of *Drosophila melanogaster*. *Genome*, 36: 94–101.

De Bono, M. and Bargmann, C.I. (1998) Natural variation in a neuropeptide Y receptor homolog modifies social behavior and food response in *C. elegans*. *Cell*, 94: 679–89.

Dittman, A.H., Quinn, T.P., Nevitt, G.A., Hacker, B. and Storm, D.R. (1997) Sensitization of olfactory guanylyl cyclase to a specific imprinted odorant in coho salmon. *Neuron*, 19: 381–389.

Døskeland, S.O., Vintermeyr, O.K., Corbin, J.D. and Øgreid, D. (1987) Studies on the interactions between the cyclic nucleotide-binding sites of cGMP-dependent protein kinase. *J. Biol. Chem.*, 262: 3534–3540.

Edgecomb, R.S., Harth, C.E. and Schneiderman, A.M. (1994) Regulation of the feeding behavior in adult *Drosophila melanogaster* varies with feeding regime and nutritional state. *J. Exp. Biol.*, 197: 215–235.

Elphick, M.R., Kemenes, G., Staras, K. and O'Shea, M. (1995) Behavioral role for nitric oxide in chemosensory activation of feeding in a mollusc. *J. Neurosci.*, 15: 7653–7664.

Falconer, D.S. and Mackay, T.F.C. (1996) *Introduction to Quantitative Genetics*, 4th Edn, Longman, England.

Ferrer, M., Tejera, N., Marín and Balfagón, G. (1998) Effect of age on the vasorelaxation elicited by cromakalim. Role of K^+ channels and cyclic GMP. *Life Sciences*, 63: 2071–2078.

Fiorucci, S., Distrutti, E., Chiorean, M., Santucci, L., Belia, S., Fano, G., DeGiorgio, R., Stanghellini, V, Corinaldesi, R. and Morelli, A. (1995) Nitric oxide modulates pepsinogen secretion induced by calcium-mediated agonist in guinea pig gastric chief cells. *Gastroenterology*, 109: 1214–1223.

Foerster, J., Harteneck, C., Malkewitz, J., Schultz, G. and Koesling, D. (1996) A functional heme-binding site of soluble guanylyl cyclase requires intact N-termini of α_1 and β_1 subunits. *Eur. J. Biochem.*, 240: 380–386.

Francis, S.H. and Corbin, J. (1994) Progress in understanding the mechanism and function of cyclic GMP-dependent protein kinase. *Adv. Pharmacol.* 26: 115–170.

Friebe, A., Schultz, G. and Koesling, D. (1996) Sensitizing soluble guanylyl cyclase to become a highly CO-sensitive enzyme. *EMBO J.*, 15: 6863–6868.

Gammie, S.C. and Truman, J.W. (1997) An endogenous elevation of cGMP increases the excitability of identified insect neurosecretory cells. *J. Comp. Physiol. A*, 180: 329–337.

Goy, M.F. (1991) cGMP: the wayward child of the cyclic nucleotide family. *Trends Neurosci.*, 14: 293–299.

Graf, S.A. and Sokolowski, M.B. (1989) The rover/sitter *Drosophila* foraging polymorphism as a function of larval development, food patch quality and starvation. *J. Insect Behav.*, 2: 301–313.

Greenspan, R.J. (1997) *Fly Pushing: The Theory and Practice of Drosophila Genetics*, Cold Spring Harbor Laboratory Press, Cold Spring Harbor, NY.

Grünbaum, L. and Müller, U. (1998) Induction of a specific olfactory memory leads to a long-lasting activation of protein kinase C in the antennal lobe of the honeybee. *J. Neurosci.*, 18: 4384–4392.

Gudi, T., Huvar, I., Meinecke, M., Lohmann, S.M., Boss, G.R. and Pilz, R.B. (1996) Regulation of gene expression by cGMP-dependent protein kinase. *J. Biol. Chem.*, 271: 4597–4600.

Haines, D.E. and Mihailoff, G.A. (1997) An overview of the brainstem. In: D.E. Haines (Ed.), *Fundamental Neuroscience*, Churchill Livingstone, Inc., New York, pp. 143–148.

Hall, J.C. (1995) Tripping along the trail to the molecular mechanisms of biological clocks. *Trends Neurosci.*, 18: 230–240.

Hall, J.C. (1997) Circadian pacemakers blowing hot and cold-but they're clocks, not thermometers. *Cell*, 90: 9–12.

Hatanaka, S., Niijima, A. and Furuhama, K. (1997) Possible mechanisms underlying the suppression of gastric vagal afferents due to ecabapide (DQ-2511), a gastroprokinetic agent, in rats. *Jpn. J. Pharmacol.*, 74: 105–108.

Heisenberg, M., Borst, A., Wagner, S. and Byers, D. (1985) *Drosophila* mushroom body mutants are deficient in olfactory learning. *J. Neurogenet.*, 2: 1–30.

Hobbes, A.J. and Ignarro, L.J. (1997) The nitric oxide-cyclic GMP signal transduction system. In: W.M. Zapol and K.D. Bloch (Eds.), *Nitric oxide and the lung*, Marcel Dekker, Inc., New York, pp. 1–57.

Jarchau, T., Häusler, C., Markert, T., Pöhler, D., Vandekerckhove, J., De Jonge, H.R., Lohmann, S.M. and Walter, U. (1994) Cloning, expression, and in situ localization of rat intestinal cGMP-dependent protein kinase II. *Proc. Natl. Acad. Sci. USA*, 91: 9426–9430.

Juilfs, D.M., Fülle, H.-J., Zhao, A.Z., Houslay, M.D., Garbers, D.L. and Beavo, J.A. (1997) A subset of olfactory neurons that selectively express cGMP-stimulated phosphodiesterase and guanylyl cyclase-D define a unique olfactory signal transduction pathway. *Proc. Natl. Acad. Sci. USA*, 94: 3388–3395.

Kalderon, D. and Rubin, G.M. (1989) cGMP-dependent protein kinase genes in *Drosophila*. *J. Biol. Chem.*, 264: 10738–10748.

Kemp, B.E. and Pearson, R.B. (1991) Design and use of peptide substrates for protein kinases. *Methods Enzymol.*, 200: 121–134.

Kojima, M., Hisaki, K., Matsuo, H. and Kangawa, K. (1995) A new type soluble guanylyl cyclase, which contains a kinase-like domain: its structure and expression. *Biochem. Biophys. Res. Commun.*, 217: 993–1000.

Komalavilas, P. and Lincoln, T.M. (1996) Phosphorylation of the inositol 1,4,5-triphosphate receptor. Cyclic GMP-dependent protein kinase mediates cAMP and cGMP dependent phosporylation in the intact rat aorta. *J. Biol. Chem.*, 271: 21933–21938.

Konopka, R.J. (1987) The Genetics of biological rhythms in *Drosophila*. *Ann. Rev. Genet.*, 21: 227–236.

Kroner, C., Boekoff, I., Lohmann, S.M., Genieser, H.-G. and Breer, H. (1996) Regulation of olfactory signalling via cGMP-dependent protein kinase. *Eur. J. Biochem.*, 236: 632–637.

Krukoff, T.L (1998) Central regulation of autonomic function: no brakes? *Clin. Exp. Pharmacol. Physiol.*, 25: 474–478.

Leidenheimer, N.J. (1996) Effect of PKG activation on recombinant GABAA receptors. *Brain Res. Mol. Brain Res.*, 42: 131–134.

Lemos, V.S., Bucher, B., Côrtes, S.F. and Takeda, K. (1997) Inhibition of $[Ca^{2+}]_i$ transients in rat chromaffin cells by neuropeptide Y: role for a cGMP-dependent protein kinase-activated K^+ conductance. *Eur. J. Neurosci.*, 9: 1144–1152.

Levin, E.R., Gardner, D.G. and Samson, W.K. (1998) Natriuretic peptides. *New Engl. J. Med.*, 339: 321–328.

Li, H., Liu, J.P., Smith, R. and Robinson, P.J. (1996) Identification of cGMP-dependent protein kinase and its specific substrates in the anterior pituitary. *Mol. Cell. Endocrinol.*, 122: 159–171.

Lincoln, T.M., Cornwell, T.L., Padmini, K., MacMillan-Crow, L.A. and Boerth, N. (1996) The nitric oxide–cyclic GMP signalling system. In: M. Bárány (Ed.), *Biochemisty of smooth muscle contraction*, Academic Press, Inc., San Diego, pp. 257–268.

Liu, C., Ding, J.M., Faiman, L.E. and Gillette, M.U. (1997) Coupling of muscarinic cholinergic receptors and cGMP in nocturnal regulation of the suprachiasmatic circadian clock. *J. Neurosci.*, 17: 659–666.

Liu, W., Yoon, J., Burg, M., Chen, L. and Pak, W.L. (1995) Molecular characterization of two *Drosophila* guanylate cyclases expressed in the nervous system. *J. Biol. Chem.*, 270: 12418–12427.

Lohmann, S.M., Vaandrager, A.B., Smolenski, A., Walter, U. and DeJonge, H.R. (1997) Distinct and specific functions of cGMP-dependent protein kinases. *Trends Biochem. Sci.*, 22: 307–312.

Lohmann, S.M., Walter, U., Miller, P.E., Greengard, P. and De Camilli, P. (1981) Immunohistochemical localization of cyclic GMP-dependent protein kinase in mammalian brain. *Proc. Natl. Acad. Sci. USA*, 78: 653–657.

Long, A.D., Mullaney, S.L., Reid, L.A., Fry, J.D., Langley, C.H. and Mackay, T.F.C. (1995) High resolution mapping of genetic factors affecting abdominal bristle number in *Drosophila melanogaster*. *Genetics*, 139: 1273–1291.

Maines, M.D. (1996) Carbon monoxide and nitric oxide homology: differential modulation of heme oxygenases in brain and detection of protein and activity. *Methods Enzymol.*, 268: 473–488.

Mather, K. and Jinks, J.L. (1982) *Biometrical Genetics*, 3rd Edn, Chapman and Hall, N.Y.

Moon, C., Jaberi, P., Otto-Bruc, A., Baehr, W., Palczewski, K. and Ronnett, G.V. (1998) Calcium-sensitive particulate guanylyl cyclase as a modulator of cAMP in olfactory receptor neurons. *J. Neurosci.*, 18: 3195–3205.

Morton, D.B. and Giunta, M.A. (1992) Eclosion hormone stimulates cyclic GMP levels in *Manduca sexta* nervous tissue via arachidonic acid metabolism with little or no contribution from the production of nitric oxide. *J. Neurochem.*, 59: 1522–1530.

Moss, C.F. and Dethier, V.G. (1983) Central nervous system regulation of finicky feeding by the blowfly. *Behav. Neurosci.*, 97: 541–548.

Müller, U. and Hildebrandt, H. (1995) The nitric oxide\cGMP system in the antennal lobe of *Apis mellifera* is implicated in integrative processing of chemosensory stimuli. *Eur. J. Neurosci.*, 7: 2240–2248.

Nuzhdin, S.V., Pasyukova, E.G. Dilda, C.L. Zeng, Z.-B. and Mackay, T.F.C. (1997) Sex-specific quantitative trait loci affecting longevity in *Drosophila melanogaster*. *Proc. Natl. Acad. Sci. USA*, 94: 9734–9739.

Orihata, M. and Sarna, S.K. (1996) Nitric oxide mediates mechano- and chemoreceptor-activated intestinal feedback control of gastric emptying. *Dig. Dis. Sci.*, 41: 1303–1309.

Ørstavik, S., Natarajan, V., Taskén, K., Jahnsen, T. and Sandberg, M. (1997) Characterization of the human gene encoding the type Iα and type Iβ cGMP-dependent protein kinase (PRKG1) *Genomics*, 42: 311–318

Osborne, K., Robichon, A., Burgess E., Butland, S., Shaw, R.A., Coulthard, A. Pereira, H.S., Greenspan, R.J. and Sokolowski, M.B. (1997) Natural behavior polymorphism due to a cGMP-dependent protein kinase of *Drosophila. Science*, 277: 834–836.

Ouedraogo, S., Tschopl, M., Stoclet, J.C. and Bucher B. (1994) Effects of cyclic GMP and analogues on neurogenic transmission in the rat tail artery. *Br. J. Pharmacol.*, 112: 867–872.

Partridge, L. and Sgro, C.M. (1998) Behavioural genetics: molecular genetics meets feeding ecology. *Current Biol.*, 8: R23–R24.

Pereira, H.S., MacDonald, D.E., Hilliker, A.J. and Sokolowski, M.B. (1995) *Chaser* (*Csr*), a new gene affecting larval foraging behaviour in *Drosophila melanogaster*. *Genetics*, 140: 263 - 270.

Pereira, H.S. and Sokolowski, M.B. (1993) Mutations in the larval *foraging* gene affect adult locomotory behavior after feeding in *Drosophila melanogaster*. *Proc. Natl. Acad. Sci USA*, 90: 5044–5046.

Pfeifer, A., Klatt, P., Massberg, S., Ny, L., Sausbier, M., Hirneiß, C., Wang, G.-X., Korth, M., Aszódi, A., Andersson, K.-E., Krombach, F., Mayerhofer, A., Ruth, P., Fässler, R. and Hofmann, F. (1998) defective smooth muscle regulation in cGMP kinase I-deficient mice. *EMBO J.*, 11: 3045 -3051.

Pfeifer, A., Nürnberg, B., Kamm, S., Uhde, M., Schultz, G., Ruth, P. and Hofmann, F. (1995) Cyclic GMP-dependent protein kinase blocks pertussis toxin-sensitive hormone receptor signaling pathways in chinese hamster ovary cells. *J. Biol. Chem.*, 270: 9052–9059.

Prabhakar, S., Short, D.B., Scholz, N.L. and Goy, M.F. (1997) Identification of nitric oxide-sensitive and -insensitive forms of cytoplasmic guanylate cyclase. *J. Neurochem.*, 69: 1650–1660.

Pryzwansky, K.B., Kidao, S., Wyatt, T.A., Reed, W. and Lincoln, T.M. (1995) Localization of cyclic GMP-dependent protein kinase in human mononuclear phagocytes. *J. Leuko. Biol.*, 57: 670–678.

Rae, M.G., Khoyi, M.A. and Keef, K.D. (1998) Modulation of cholinergic neuromuscular transmission by nitric oxide in canine colonic circular smooth muscle. *Am. J. Physiol.*, 275: G1324–G1332.

Regulski, M. and Tully, T. (1995) Molecular and biochemical characterization of dNOS: a *Drosophila* Ca^{2+}/calmodulin-dependent nitric oxide synthase. *Proc. Natl. Acad. Sci. USA*, 92: 9072–9076.

Sawyer, L.A., Hennessy, J.M., Peixoto, A.A., Rosato, E., Parkinson, H., Costa, R. and Kyriacou, C.P. (1997) Natural variation in the *Drosophila* clock gene and temperature compensation. *Science*, 278: 2117–2120.

Shah, S. and Hyde, D.R. (1995) Two *Drosophila* genes that encode the á and â subunits of the brain soluble guanylyl cyclase. *J. Biol. Chem.*, 270: 15368–15376.

Sluka, K.A. and Willis, W.D. (1998) Increased spinal release of excitatory amino acids following intradermal injection of capsaicin is reduced by a protein kinase C inhibitor. *Brain Res.*, 798: 281–286.

Smolenski, A., Bachmann, C., Reinhard, K., Honig-Liedl, P., Jarchau, T., Hoschuetzky, H. and Walter, U. (1998) Analysis and regulation of vasodilator-stimulated phosphoprotein serine 239 in vitro and in intact cells using a phosphospecific monoclonal antibody. *J. Biol. Chem.*, 273: 20029–20035.

Sokolowski, M.B. (1980) Foraging strategies of *Drosophila melanogaster*: A chromosomal analysis. *Behav. Genet.*, 10: 291–302.

Sokolowski, M.B. (1992) Genetic analysis of behaviour in the fruit fly *Drosophila* melanogaster. In: D. Goldowitz, D. Winer and D. Wahlsten (Eds.), *Techniques for the Genetic Analysis of Brain and Behavior: Focus on the Mouse*, Elsevier Science Publishers B.V., Amsterdam, pp. 497-521.

Sokolowski, M.B. (1998) Genes for Normal Behavioral Variation: Recent clues from flies and worms. *Neuron*, 21: 1–4.

Sokolowski, M.B. and Bauer, S.J. (1989) Genetic analysis of pupation distance in *Drosophila melanogaster*. *Heredity*, 62: 177–183.

Sokolowski, M.B., Pereira, H.S. and Hughes, K. (1997) Evolution of foraging behavior in *Drosophila* by density dependent selection. *Proc. Natl. Acad. Sci. USA*, 94: 7373–7377.

Tortorici, C. and Bell, W.J. (1988) Search orientation in adult *Drosophila melanogaster*: responses of rovers and sitters to resource dispersion in a food patch. *J. Insect Behav.*, 1: 209–223.

Varnam, C.J., Strauss, R., De Belle, J.S., and Sokolowski, M.B. (1996) Larval behavior of *Drosophila* central complex mutants: interactions between *no bridge*, *foraging*, and *Chaser*. *J. Neurogenet.* 11: 99–115.

Verma, A., Hirsh, D.J., Glatt, C.E., Ronnett, G.V. and Snyder, S.H. (1993) Carbon monoxide: a putative neural messenger. *Science 259:* 381–384. Erratum: *Science*, 263: 15.

Wang, X. and Robinson, P.J. (1995) Cyclic GMP-dependent protein kinase substrates in rat brain. *J. Neurochem.*, 65: 595–604.

Wang, X. and Robinson, P.J. (1997) Cyclic GMP-dependent protein kinase and cellular signalling in the nervous system. *J. Neurochem.*, 68: 443–456.

Wexler, E.M., Stanton, P.K. and Nawy, S. (1998) Nitric oxide depresses GABAA receptor function via coactivation of cGMP-dependent kinase and phosphodiesterase. *J. Neurosci.*, 18: 2342–2349.

Williams, K. and Sokolowski, M.B. (1993) Diapause in *Drosophila melanogaster* females: a genetic analysis. *Heredity*, 71: 312–317.

Witczak, O., Ørstavik, S., Natarajan, V., Frengen, E., Jahnsen, T. and Sandberg, M. (1998) Characterization of the gene encoding the human type II cGMP-dependent protein kinase (PRKG2). *Biochem. Biophys. Res. Commun.*, 245: 113-119.

Wu, J., Wang, Y., Rowan, M.J. and Anwyl R. (1998) Evidence for involvement of the cGMP-protein kinase G signaling system in the induction of long-term depression, but not long-term potentiation, in the dentate gyrus in vitro. *J. Neurosci.*, 18: 3589–3596.

Wyatt, T.D., Naftilan, A.J., Francis, S.H. and Corbin, J.D. (1998) ANF elicits phosphorylation of the cGMP phosphodiesterase in vascular smooth muscle cells. *Am. J. Physiol.*, 274(*Heart Circ. Physiol.*, 43): H448–H455.

Young, M.E. and Leighton, B. (1998) Fuel oxidation in skeletal muscle is increased by nitric oxide/cGMP–evidence for involvement of cGMP-dependent protein kinase. *FEBS Lett.*, 424: 79–83.

Yoshikawa, S., Miyamoto, I., Aruga, J., Furuichi, T., Okano, H. and Mikoshiba, K. (1993) Isolation of a *Drosophila* gene encoding a head specific guanylyl cyclase. *J. Neurochem.*, 60: 1570–1573.

Yu, S., Avery, L., Baude, E. and Garbers, D.L. (1997) Guanylyl cyclase expression in specific sensory neurons: a new family of chemosensory receptors. *Proc. Natl. Acad. Sci. USA*, 94: 3384–3387.

Zhuo, M., Ylinghe, H., Schultz, C., Kandel, E.R. and Hawkins, R.D. (1994) Role of guanylyl cyclase and cGMP-dependent protein kinase in long-term potentiation. *Nature*, 368: 635–639.

W.E. Crusio and R.T. Gerlai (Eds.)
Handbook of Molecular-Genetic Techniques for Brain and
Behavior Research (Techniques in the Behavioral and Neural Sciences, Vol. 13)

CHAPTER 3.3.3

Structure–function analysis of the *Drosophila* optic lobes

Gert O. Pflugfelder

Lehrstuhl für Genetik, Theodor-Boveri-Institut, Biozentrum, Universität Würzburg, Am Hubland, D-97074 Würzburg, Germany

Structure-function analysis of the *Drosophila* optic lobes

By the time this book appears in print the *Drosophila* genome may be sequenced (Venter et al., 1998). This will lighten the burden of the molecularly oriented *Drosophila* neurogeneticist and allow him to raise his glance to the wonderful world of brain structure and behavior. This chapter, therefore, does not deal with the cloning of genes, but rather attempts to illustrate the potential and pitfalls of genetic and reverse genetic techniques in establishing structure-function relationships in the fly's optic lobes. The chapter, after reviewing optic lobe structure, its development, and visual behavior focuses on two behavioral domains: phototaxis and optomotor response.

Optic lobe structure

The optic lobes underneath the compound eyes are the main computation stations of visual information in the fly head. Their overall structure is conserved in the *Diptera* and other insect orders. Certainly within the *Diptera*, and possibly beyond, this conservation (but also departures therefrom) is also evident in the Golgi gestalt of individual neurons (Buschbeck and Strausfeld, 1996). This holds similarly for behavioral and neurophysiological phenomena. This chapter will, therefore, also contain glimpses across the species fence. For *Drosophila*, a catalogue of optic lobe Golgi gestalten was established by Fischbach and Dittrich (1989). The characterisitic features of neuronal structure, including cell body position, number, level, and width of dendritic and terminal arborization are difficult to render verbally. Fig. 1, showing predominantly small field neurons that connect lamina and medulla, may serve as a relatively simple example of neuronal complexity in the optic lobes. The figure contains most neurons that will be dealt with in this text. For further visual information, the reader is referred to the literature and to the Flybrain database (Armstrong et al., 1995, www.flybrain.org). Electronic media will be of great help (and probably essential) to perform the integration of disparate information eventually required for an in-depth structure-function analysis of the brain.

A description of the optic lobes is incomplete without, and most organically starts with the compound eye whose architecture is reflected in the optic lobes. Each of the approximately 800 facets (or ommatidia) of each eye contains 8 photoreceptor (retinula, R) cells. On structural and physiological grounds, three main retinula cell types can be recognized: the "outer" R cells R1–6, and the two "inner" R cell types R7 and R8. The photoreceptive specializations of these cells, the

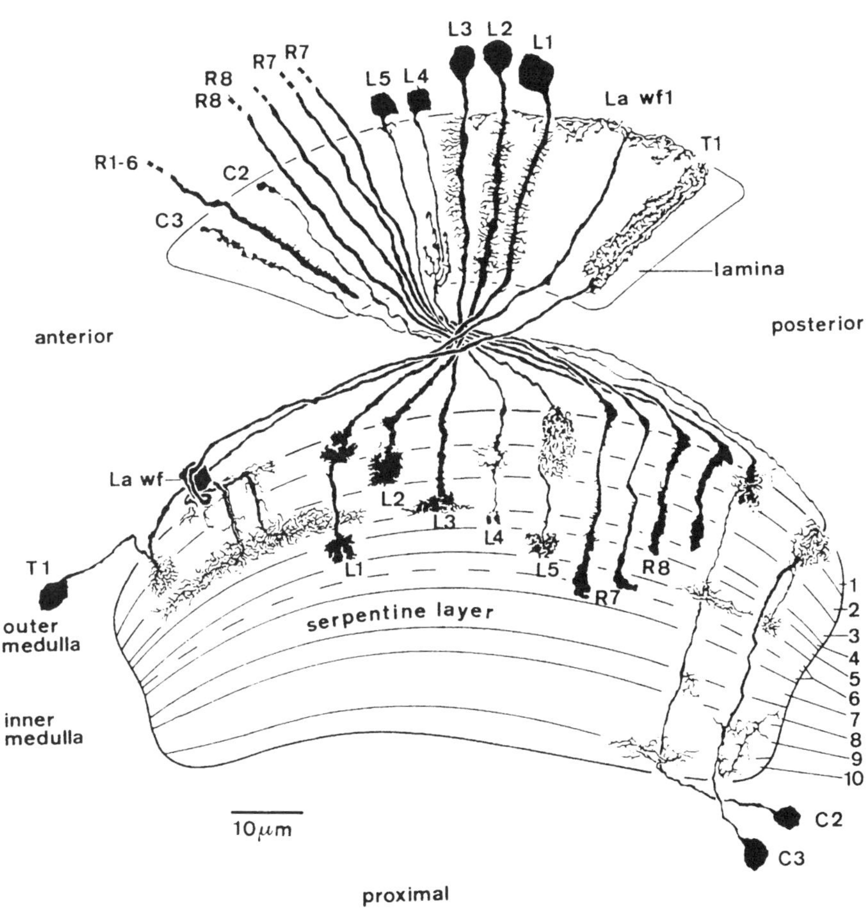

Fig. 1. Receptor and first order interneurons connecting lamina and medulla. Camera lucida drawings of Golgi-stained neurons are assembled as if seen in a horizontal section. R1–6 cell axons are collectively represented by one specimen, R7 and R8 by two axons each. With the exception of the lamina wide field neuron Lawf1, all neurons shown occur once per column. Medulla layers 1–10 are indicated on the right. (From Fischbach and Dittrich, 1989, by permission.)

rhabdomeres, are arranged in a stereotyped pattern within each ommatidium. The rhabdomeres of R8 and R7 are stacked upon each other in the center of the ommatidium, the rhabomeres of R1–6 are arranged in a trapezoidal pattern around this central element. The position of the distal tip of each rhabdomere in the focal plane of the ommatidial lense defines the optical axis of R1–7 (Kirschfeld, 1967). Within each ommatidium adjacent optical axes diverge by $\Delta\phi$ (about 5°; its precise value depends upon the ommatidial position within the eye, Heisenberg and Wolf, 1984). A trapezoidal group of 7 adjacent ommatidia contains a complete set of retinula cells with parallel optical axes. R1–6 cells from such a trapezoidal set are jointly presynaptic to a set of interneurons in one retinotopic unit ("cartridge") in the first optic lobe neuropil ("lamina"), whereas R1–6 cells from one ommatidium synapse onto 6 different cartridges ("neural superposition", Braitenberg, 1967; Kirschfeld, 1972). No mutation has yet been found that leads to a systematic alteration of this intricate arrangement; many mutations that affect retinula cell development lead to a disorganized crossing pattern, probably as a consequence of the inductive influence of eye on optic lobe development (see below) (Meinertzhagen and Hanson, 1993).

The lamina is the output station of the outer photoreceptors R1–6 but not of R7 and R8 which project to the second neuropil, the medulla, without engaging in synaptic contacts in the lamina. The lamina monopolar cells L1–L5 and the medulla cell T1 provide afferent connections between lamina and medulla (neuropil allocation of these cells is based on cell body localization). Centrifugal connections are provided by C2 and C3. Each of these cells as well as a lamina-intrinsic (amacrine) neuron are present in each of the about 800 lamina cartridges. Additional neurons and glia cells are probably present at a lower periodicity (Fischbach and Dittrich, 1989). The synaptic organization of the lamina is known in detail from serial electron microscopic reconstructions (Boschek, 1971; Meinertzhagen and O'Neil, 1991). The terminals of R1–6 are interconnected by gap junctions which leads to an averaging of the photoreceptor responses. The large monopolar cells (LMC) L1 and L2 reside in the center of the lamina cartridge and receive approximately symmetrical input from R1–R6 because all output from outer photoreceptor cells onto L1 and L2 occurs in tetrad synapses. These contain, in addition to L1 and L2, two further postsynaptic partners. Most elements in the lamina cartridge, including the R1–6 cells, are involved in local feedback circuits. This involvement is different for L1 and L2, thus breaking the symmetry between these two cells. While most synaptic connections are confined to the lamina cartridge (which is ensheathed by glia cells, Saint Marie and Carlson, 1983) there are also connections between cartridges (Strausfeld and Campos-Ortega, 1977; Meinertzhagen and O'Neil, 1991). The variety of different microcircuits in the lamina also indicates that the lamina is no simple feedforward relay station of retinal information. The six centripetal elements connecting lamina with medulla (L1–5, T1) could, in principle, provide 6 output channels from each lamina cartridge.

The medulla is considerably more complex, both in total neuron number (estimated at 40000 vs 6000 in the lamina, Hofbauer and Campos-Ortega, 1990) and number of cell types per cartridge. In *Musca*, 120 cell types were identified in the medulla, 34 of these with strict columnar periodicity, 74 did not occur in every column (Campos-Ortega and Strausfeld, 1972). There is no evidence for a simpler assembly in *Drosophila*, neither from Golgi analysis (Fischbach and Dittrich, 1989) nor from the average number of medulla neurons per column ($40000/800 = 50$). As yet there is little direct experimental data describing synaptic connectivity in the medulla (Strausfeld and Campos-Ortega, 1972). The medulla in silver-stained preparations has a clearly stratified structure, the strata lying orthogonally to the long axis of the columns (Campos-Ortega and Strausfeld, 1972). Tangential elements and the discrete arborization pattern of columnar elements contribute to the stratification (Fischbach and Dittrich, 1989). Both in *Musca* (Strausfeld and Lee, 1991) and *Drosophila* (Bausenwein et al., 1992) the depth relationship between medulla afferents and efferents to the lobula complex has been evaluated by quantifying the arborization densities of Golgi-stained neurons. While overlapping arborizations are no proof of synaptic contact, such contact can at least be considered in the light of other experimental evidence. In addition to the visual input from retina and lamina, innervation by a number of neurosecretory neurons which could exert a modulatory influence on medulla function has been recognized (e.g. Nässel, 1993; Helfrich-Förster and Homberg, 1993). The predominant output of the medulla is to the lobula complex but an as yet ill defined number of tangential elements leave the medulla through

Cucatti's bundle and project to the central brain or the contralateral medulla (Fischbach and Dittrich, 1989).

The lobula complex in *Dipteran* flies is split into two distinct neuropils, the lobula and the lobula plate. Its cell number lies between that of lamina and medulla (Hofbauer and Campos-Ortega, 1990). The lobula complex is intimately connected to the medulla by various types of relay neurons, more than 40 of which may exit each medulla column (Campos-Ortega and Strausfeld, 1972; Strausfeld and Lee, 1991). Some of the cell types connecting medulla and lobula or lobula plate, or the two parts of the lobula complex consist of several isomorphic neurons which only differ in the terminal arborization depth within the lobula plate neuropil (Fischbach and Dittrich, 1989; Strausfeld and Lee, 1991). Both parts of the lobula complex contain characteristic output elements. The lobula carries palisades of isomorphic neurons with a columnar periodicity lower than in the medulla (Braitenberg, 1972; Strausfeld, 1979). They project to the ventro-lateral deutocerebrum where they are presynaptic to neurons that descend through the cervical connective (Strausfeld and Lee, 1991). The lobula plate is characterized by an electrophysiologically well characterized set of giant tangential neurons many of which terminate more dorsally in the deutocerebrum (Hausen, 1984).

Optic lobe development

Optic lobe (OL) development was exquisitely documented by Meinertzhagen and Hanson (1993). Therefore, only a rough sketch of recent findings pertinent to the neurogenetic analysis of OL function will be given. The OL anlage (about 70 cells) invaginates from the posterior procephalic region of stage 12 embryos and fuses with the brain which forms from individually delaminated neuroblasts (Green et al., 1993). Whether the OL anlage has a specific segmental origin is controversial and was debated on the basis of the expression pattern of the segment polarity gene *engrailed* and of the phenotypes of head gap gene mutants (Schmidt-Ott et al., 1995; Younossi-Hartenstein et al., 1997). Null mutants of the *sine oculis* (*so*) gene, which is expressed in several invaginating embryonic tissues, fail to invaginate the OL anlage. Viable alleles show that *so* is required at several stages of visual system development (Cheyette et al., 1994; Serikaku and O'Tousa, 1994). The expression of an early marker of OL development, *optomotor-blind* (*omb*), is maintained in the non-invaginating OL primordium of *so* embryos (Cheyette et al., 1994). In wildtype development, OL precursor cells resume proliferation in the first larval instar and continue to proliferate till the early pupal stages. Reentry of the precursor cells into the cell cycle is under both negative and positive control (Ebens et al., 1993; Datta, 1995). The number of neuroblasts is initially greatly expanded by symmetrical division. Asymmetrical stem cell divisions set in at the beginning of the third larval instar.

Already in the middle of the first larval instar an inner and outer anlage can be discerned which will give rise to the various parts of the OL. With the resumption of proliferative activity, the inner and outer proliferation zones can be distinguished morphologically from the surrounding tissue (Hofbauer and Campos-Ortega, 1990). The approximately 10 cells of the outer anlage are also conspicuous by not yet expressing the horseradish peroxidase (HRP) antigen which marks the surrounding larval brain (Kaphingst and Kunes, 1994). Between mid and late first instar, cells at the ventral and dorsal poles of the outer anlage start to express *wingless* (*wg*) and continue to express this diffusible protein throughout larval development. The gene *decapentaplegic* (*dpp*), which also encodes a diffusible protein, is expressed in a neighbouring domain. Conditional or regulatory mutations in these genes lead to the absence or severe disorganization of the entire structure that derives from the outer optic anlage (Kaphingst and Kunes, 1994). *omb*, too, is expressed at the dorsal and ventral poles of the outer anlage (Balles, 1995). *omb* null mutations lead to similar structural defects as mutations in *wg* or *dpp* (Pflugfelder et al., 1992a). All three genes endow the expressing tissue with organizing function, since loss of gene expression from a small number of cells affects the development of the entire anlage. *wg*, *dpp*, and *omb* have been shown previously to encode components that aid in organizing wing development (Grimm and Pflugfelder, 1996).

At a later stage, optic lobe development is dependent on retinal innervation. This has long been suspected and was firmly established in mosaic animals (Meyerowitz and Kankel, 1978; Fischbach and Technau, 1984). There is, for instance, extensive cell degeneration in the optic lobes (but not the central brain) of eyeless so^1 pupae, a phenotype caused by the genotype of the eye and not by the genotype of the optic lobes. The developmental reasons for this dependency are beginning to become clear. In the late larval and early pupal eye anlage the arrival of the retinula cell axons triggers the final division cycle of the lamina precursor cells (LPCs), which otherwise remain arrested in the G1 phase of the cell cycle (Selleck and Steller, 1991; Selleck et al., 1992). The signal that promotes G1-S transition is the diffusible protein Hedgehog (Hh) (Huang and Kunes, 1996). Ectopic Hh expression can only partially substitute for axonal innervation; it does not suffice for terminal lamina cell differentiation, indicating the necessity of another retinular factor recently identified as the epidermal growth factor receptor ligand Spitz (Huang et al., 1998). The final G1-S transition of LPCs also requires *dally*. *dally* encodes a membrane attached proteoglycan (Nakato et al., 1995; Jackson et al., 1997). Proteoglycans can facilitate the action of several cytokines in a wide range of tissues (e.g. Schlessinger et al., 1995). It is conceivable that Dally plays such a role in Hh signaling. In *dally* mutant larvae, the final LPC division does not occur; nonetheless LPCs express the terminal differentiation marker gene *elav*. In eyeless flies, both LPC division and *elav* expression are suppressed (Nakato et al., 1995). This again indicates the necessity of additional retinula cell-derived signals for neuronal differentiation.

Degeneration in eyeless flies is not confined to lamina and medulla (which are directly innervated by retinula cell axons) but also occurs in the lobula plate cortex and in the inner optic chiasm (Fischbach and Technau, 1984). This suggests that the retinula cells, and possibly (by proxy) interneurons, exert a trophic influence on post-synaptic cells. In eyeless so^1 flies the residual neuropils are still retinotopically organized, however, indicating that this organizational feature develops independently of the eye (Fischbach, 1983).

Pupal cell divisions in the optic lobe anlage (except the lamina) of the moth *Manduca sexta* were shown to require a minimum concentration of the humoral signal 20-hydroxyecdysone (20-HE) (Champlin and Truman, 1998). It is not yet known whether this holds true for *Drosophila*. In this species, activity in the visual proliferation zones ceases after about 1 d of pupal development (Hofbauer and Campos-Ortega, 1990). That ecdysteroids play a role in Drosophila metamophic optic lobe development is obvious from the visual system phenotypes in mutants of the 20-HE inducible locus *Broad Complex* (Restifo and White, 1991; Liu and Restifo, 1998). Whether the structural disorganizations of the visual neuropils can be accounted for by a late proliferation defect remains to be shown.

There is still little information about lineage relationships during optic lobe development. The small number of neuroblasts at the beginning of the larval proliferation period and the large number of cell types in the adult brain in most cases do not allow monotypic lineages (for a possible exception see HS/VS cell development below). If cells of a given cell type with columnar periodicity were clonally related, visualization of such a clone by marking the descendants of a progenitor, should yield a stratified appearance. This has not been observed, so far, for neuronal clones. Rather, neuronal clones generated in the outer optic anlage early in larval life and inspected at the end of larval development, were found to have a radial appearance and encompass cells from both the medulla and lamina proliferation centers (Perez and Steller, 1996). While it was not demonstrated in these experiments that the clone encompassed all neuronal cell types of high columnar periodicity, the data clearly show that the marked precursor had a high developmental potency. The reverse is true for the precursor cells of the various identifiable glia cell types in the lamina. The progeny of glia precursors are confined to layers in the lamina, indicating their relation by lineage (Winberg et al., 1992; Perez and Steller, 1996). In the central brain of the fly, a lineage relationship was shown for the neurons within the four subunits of the mushroom body (Ito et al., 1997a). The determination of many neurons in the OL by cell-cell interaction or other local signals does not preclude the application of genetic tools in network dissection. This is clearly evident from the eye, in which cell fates are exclucively controlled by cell–cell interactions (Yamamoto, 1996).

The establishment of neuronal connectivity is a multi-step process and can be expected to be under the control of many genes. Screens for defects in embryonic axonogenesis have already identified a large number of complementation groups (e.g. Kolodziej et al., 1995; Winberg et al., 1998). Similarly, in the eye-optic lobe system, several mutants affecting axon outgrowth, target recognition, or retinotopy were isolated in histological screens (Kretzschmar et al., 1992; Schneider et al., 1995; Martin et al., 1995). Screens for mutants in visual system development are hampered by the fact that null mutations in many of the genes involved may be lethal at an earlier stage. Thaker and Kankel (1992) estimated that two thirds of all essential *Drosophila* genes are involved in eye development. It seems reasonable to assume that the percentage of essential genes acting in the development of the optic lobes is even higher. The fact that the eye-OL projection continues to develop in vitro offers the possibility to investigate processes that are not (yet) accessible genetically (Gibbs and Truman, 1998). Using a pupal brain preparation, it was shown that nitric oxide (NO) is required for the establishment of correct neuronal connectivity. NO is a locally acting, instructive signal produced during the first half of pupal development by cells of the optic lobe anlage and leads to cGMP synthesis in retinula cell axons. NO responsiveness is highest after arrival of the axons in their target tissue but before synaptogenesis. Pharmacological disruption of NO or cGMP production or action prevents the correct target (layer) selection of the retinula cell growth cones. Consequently, their axons project beyond their normal neuropil or neuropil layer (Gibbs and Truman, 1998). In vertebrates, there is ample evidence that in the last stages of address selection refinement of contact requires electrical activity in the participant neurons (Goodman and Shatz, 1993). Whether this holds for the lamina-OL projection in the fly is not yet known.

Visual behavior

The adult fly has 5 canonical eyes, the two compound eyes proper and the three dorsal ocelli. *Drosophila* mutants lacking these eyes still can be entrained a circadian rhythm by light–dark cycles (Helfrich, 1986). This suggests the existence of internal photosensitive organs. Evidence for such has been found with photoreceptor-specific antibodies (e.g. Hofbauer and Buchner, 1989; Cymborowsky and Korf, 1995). Photosensitivity may even be a ubiquitous cell-autonomous property which does not depend on nervous activity (Plautz et al., 1997; Yang et al., 1998). For orientation behavior, which requires a representation of the visual surround at high spatial resolution (Coppola et al., 1998), only input from the compound eye (and to a lesser extent from the ocelli, Goodman, 1981) can be relevant. Input from internal photosensitive organs may have a modulatory role.

A wealth of visually evoked behaviors can be observed in the laboratory in simplified settings, where the degrees of translational and rotational freedom of the fly can be reduced and where visual input can be controlled. The flexibility of the fly's CNS can even be challenged in evolutionarily absurd situations, for instance by an inverted coupling of the fly's torque with its visual surround in a flight simulator (Heisenberg and Wolf, 1984). Examples of behaviors that have been studied are: phototaxis, orientation to and following of stationary and moving targets, course and height control, and landing response. All of these cannot be treated in detail here. Only phototaxis and optomotor behavior are discussed as exemplary applications of the neurogenetic approach.

Phototaxis

Phototaxis screens

Seymour Benzer in, 1967 inaugurated the mutational dissection of *Drosophila* behavior and at the same time introduced a very simple but highly efficient device for the isolation of flies disturbed in phototaxis (Benzer, 1967). After an initial jolt, wildtype flies in a dark, horizontal tube run toward the light source within a few seconds ("fast phototaxis"). In "slow" paradigms, startling conditions are avoided and the fly is given more time to choose. In judging "fast" vs "slow" one should try and adopt a musco-centric attitude ("the random distribution of flies by gentle shaking" (Hu and Stark, 1977) to a fly may appear as an oxymoronic description of an event high up on its Richter scale). Phototaxis is a simple behavior only at first sight. Many parameters have an effect, not all of them visual. For instance, eliminating the wings genetically or surgically greatly reduces fast phototaxis, indicating a sensory feedback from the wings (Benzer, 1967). High light intensity and other factors can lead to photophobic behavior (Heisenberg and Buchner, 1977; Jacob et al., 1977; Markow and Merriam, 1977; Ballinger and Benzer, 1988). The older literature details more than a dozen parameters that can affect phototactic behavior, only one of them being the genetic status (Dürrwächter, 1957; Hadler, 1964; Rockwell and Seiger, 1973; Menzel, 1979). Results obtained in different laboratories or in different devices are, therefore, not necessarily directly comparable (Heisenberg and Götz, 1975). Since, in general, only the fly distribution at the end of the experiment was evaluated, it is not even clear whether the underlying behavior can be classified as phototaxis (and not, for instance, as photokinesis) (definitions e.g. in Siebeck, 1977). Irrespective of these qualifications, Benzeŕs fast photoaxis apparatus led to the isolation of a large number of mutants. Several of these have been molecularly characterized by now. Additional alleles to many mutants from Benzer's laboratory were isolated in unrelated screens, indicating in most cases a limited specificity of the behavioral defect or of the genes' functions. Examples of characterized mutants are:

no receptor potential A (*norpA*, phospholipase C, Bloomquist et al., 1988);
sevenless (*sev*, receptor tyrosine kinase, Banerjee et al., 1987; Hafen et al., 1987);
small optic lobes (*sol*, calpain protease and zinc finger domain protein, Delaney et al., 1991);
retinal degeneration B (*rdgB*, phosphatidylinositol transfer protein, Vihtelic et al., 1993);
no on transient A (*nonA*, nuclear RNA binding protein, Von Besser et al., 1990; Jones and Rubin, 1990);
sluggish A (*slgA*, mitochondrial proline oxidase, Hayward et al., 1993),
division abnormally delayed (*dally*, proteoglycan, Nakato et al., 1995),
orthodenticle (*otd*, homeodomain protein, Vandendries et al., 1996).

Mutants in many other genes are defective in fast photoaxis (e.g. *couch potato* (*cpo*, Bellen et al., 1992), *amyloid precursor protein-like gene* (*Appl*, Luo et al., 1992), and numerous others (Benzer,

1967; Pak et al., 1969; Heisenberg and Götz, 1975; Markow and Merriam, 1977)). The mutant designations of *couch potato* and *sluggish* (not to mention *comatose*) highlight a problem inherent in any behavioral screen. Behavioral performance always relies on some kind of motor activity. Sluggish phenotypes are no rare occurrence; it has been estimated that there are about 150 loci in the genome in which mutations can cause impaired phototaxis due to sluggishness (Hayward et al., 1993). The mutant focus (Hotta and Benzer, 1972) of *slgA* maps to the thoracic ganglion, remote from the visual processing centers (Hayward et al., 1993). Also the focus of a mutant in the (Na^+, K^+)ATPase α-subunit gene that was selected in a fast phototaxis screen because of its bang sensitivity (Schubiger et al., 1994) is unlikely to reside in the visual system. The central complex of the brain is another region that affects phototactic behavior (Strauss and Heisenberg, 1993). This indicates that many defects can lead to the loss of a phototactic response and that phototaxis is no simple reflex circuit beyond central control.

Mutants used in the analysis of phototaxis

At the network level, all cell types of the compound eye and the ocelli could be demonstrated by neurogenetic means to be involved in fast phototaxis. In the compound eyes these are R7 and R8 (the former not developing in *sev*) and R1–6 (degenerating in *rdgB*). Both mutants were isolated in a fast phototaxis mutant screen (Harris et al., 1976) (even though they are vigorously phototactic under other conditions, Ballinger and Benzer, 1988; Harris et al., 1976). The mutant *neither inactivation nor afterpotential E* (*ninaE*) was also used to eliminate R1-6 function. Ocelli were eliminated with the mutation *ocelliless* (*oc*). In the following, knowledge about these mutants will be shortly summarized to raise the awareness of potential pitfalls in the mutant approach.

sev mutants lack the R7 cell (Harris et al., 1976; Campos-Ortega et al., 1979). The elucidation of the molecular role of *sev* in R7 cell ontogeny is one of the major success stories of modern developmental biology and need not concern us here (Raabe, 1998). In addition, *sev* was instrumental in assessing the role of R7 in various visual behaviors. Wildtype flies in both slow and fast photochoice experiments prefer UV (350 nm) over green (550 nm) light over a wide range of intensity ratios. In *sev* this preference is reversed (Heisenberg and Buchner, 1977; Banerjee et al., 1987; Ballinger and Benzer, 1988). Coinciding alterations in R7 development and color choice are also observed in the mutant *bendless* (Oh et al., 1994). The lack of the R7 photoreceptor cell has secondary effects on the optics of the other retinula cells (Harris et al., 1976; Heisenberg and Wolf, 1984). R7 serves as a light guide for R8, so that in *sev* much less light is received by the R8 rhabdomer (Franceschini, 1975). Nonetheless, it is likely that the primary morphological eye defect is responsible for the phototaxis phenotype. Evidence for this comes from mutants of the gene for Ras GTPase activating protein (*gap1*). This protein antagonizes the action of the SEV receptor tyrosine kinase cascade. Loss of gene function in *gap1* leads to an excessive recruitment of R7 cells into the ommatidium, without restoring the original R7 cell lost in a *sev* background. The *sev gap1* double mutant in a fast photochoice test has an even higher UV preference than wildtype, indicating that the ectopic R7 cells not only rescue the deficit resulting from the loss of the original R7 cell but even contribute to a more pronounced response (Buckles et al., 1992; Ashley and Katz, 1994). *gap1* promotes the differentiation into R7 of a different cell type than does *sev*. It is, therefore, likely that the *sev* behavioral phenotype is caused by the conspicuous loss of the peripheral R7 cell and not by a more central and as yet undetected neuronatomical defect. This remote assumption would require that this hypothetical defect, too, could be rescued by *gap1*. In spite of this apparently unambiguous structure-function correlation, there is indirect evidence for a central function of *sev*. In the photochoice experiment described above, the wildtype UV preference is restored in *sev* by the presence of the dominant mutation *Photophobe* (*Ppb*), without restoring the original or providing ectopic R7 cells (Ballinger and Benzer, 1988). *sev*; *Ppb*/+ or *Ppb*/*Ppb* are negatively phototactic over a wide range of light intensities. The interaction of *sev* with *Ppb* is allele-specific which can be taken as a hint for a direct interaction between the two (mutant) gene products. *sev* is strongly expressed in the adult head (Banerjee et al, 1987) and appears to be weakly expressed during larval and pupal brain development (Banerjee et al, 1987; Basler and Hafen, 1989). Since the genetic interactions occur in a *sev* mutant background, i.e. in flies lacking the R7 cell, one has to assume a function of *sev* in phototaxis beyond the promotion of R7 cell development.

rdgB leads to a light-promoted posteclosional degeneration of the R1–6 photoreceptor cells. The mutant electroretinogram (ERG) deteriorates rapidly after the first light exposure and before any anatomical defect can be detected. In a *white* background, R1–6 degeneration was reported to be complete within a few days of light/dark cycle (Harris et al., 1976) whereas in red-eyed flies a few percent of surviving outer rhabdomeres were observed even in old flies (Heisenberg and Buchner, 1977). Anatomical and electrophysiological investigations of $rdgB^{9}$, the allele used in most behavioral experiments reported so far, found no hint of R7/R8 degeneration (Harris et al., 1976). *rdgB* is, however, also expressed at a low level in R7/8 cells and elsewhere in the brain (Vihtelic et al., 1993) and, at least in some alleles, degeneration of the central photoreceptor cells occurs (Lindsley and Zimm, 1992). This central R-cell degeneration is cell-autonomous and not induced by the surrounding decay (Milligan et al., 1997). *rdgB* alleles are thus best described as selective rather than specific mutants.

ninaE encodes opsin Rh1 of the outer photoreceptors. Loss-of-function mutations cause defects in pupal R1–6 rhabdomere development and lead to their posteclosional degeneration (Kumar et al., 1997). Distal remnants of the rhabdomeres remain, however (Harris et al., 1976) and appear to be functional (Coombe, 1984). In the older literature, *ninaE* is referred to under its original mutant designation *outer rhabdomeres absent* (ora^{JK84}) (Koenig and Merriam, 1977). ora^{JK8} turned out to be a double mutant of two closely linked genes, *ninaE* and *ort* (*ora transientless*), both of them, ironically, causing an ERG defect, by which method the mutant was isolated (O'Tousa et al., 1989; Lindsley and Zimm, 1992). Such co-isolation of functionally related genes in mutant screens is no astronomical rarity (see e.g. Salz et al., 1982 or Brunner et al., 1994). Extensive chromosome purification or the use of independent alleles is, therefore, recommended when structure-function analysis is extended to the gene. ora^{JK8} showed no phototaxis in a fast Y-maze photochoice test in which *rdgB* responded vigorously (Harris et al., 1976). In a slow phototaxis paradigm *ora* behaved similar to *rdgB* (Fischbach and Reichert, 1978). It is conceivable that the compound deficit of *ninaE* and *ort* is responsible for the fast phototaxis defect (Lindsley and Zimm, 1992).

oc has been used for the elimination of the dorsal ocelli (e.g. Fischbach and Reichert, 1978; Hu and Stark, 1980). *oc* is a regulatory allele of *orthodenticle* , hence otd^{oc} (Finkelstein et al., 1990). In otd^{oc} the ocelli and the ocellar nervous system are eliminated while the compound eyes appear normal (Stark et al., 1989). In addition, the allele causes a disruption of the protocerebral bridge of the adult central brain (Hirth et al., 1995), a structure not entirely innocuous with regard to phototaxis (Strauss and Heisenberg, 1993). *otd* is essential for the embryonic development of the first embryonic brain neuromere, the future protocerebrum (Hirth et al., 1995). A late developmental function of *otd* in the morphogenesis of the adult photoreceptor rhabdomeres has also been recognized by the isolation of a viable allele (otd^{uvi}) that is defective in fast phototaxis. otd^{oc} and otd^{uvi} show full complementation with regard to their individual phenotypes. The *otd*^{uvi} penotype is not enhanced over a deficiency (Vandendries et al., 1996). *otd*^{uvi} and *otd*^{oc} thus provide examples of viable regulatory amorphic alleles of an essential gene that have proved so useful in behavioral neurogenetics (Pflugfelder, 1998). Removal of the ocelli by *otd*^{oc} is not, however, equivalent to a loss of ocellar input caused for example by a coating with black paint (Fischbach and Reichert, 1978). In the dark, spontaneous activity could be monitored from the ocellar nerve in several insects, which was inhibited by illumination (Horridge, 1965). This spontaneous dark activity cannot, of course, be mimicked by removal of the ocellar nerve. One would rather expect occlusion and *oc* to have opposite effects on behavior. Indeed, *oc* seems to have a stimulatory effect on general behavioral activity (Leng and Strauss, 1998) as was proposed for the functional ocelli (e.g. Médioni et al., 1971; Goodman, 1981).

In several behavioral investigations (e.g. Hu and Stark, 1980) the *otd*^{oc} chromosome was marked with a *pentagon* (*ptg*) allele. This may have introduced an unnecessary complication, because several *ptg* alleles exhibit an olfactory phenotype in addition to the thoracic pigment patch (Helfand and Carlson, 1989). Because of the overlap of gene function in visual and olfactory pathways (Riesgo-Escovar et al., 1995) a certain wariness is recommended. A behavioral phenotype in pigmentation mutants may be the rule rather than an exception (see e.g. *ebony*, *yellow*, or *tan*, Lindsley and Zimm, 1992).

All the mutants discussed above result in a reduction of visual input. An increase in input is also feasible, however. The mutation *gigas* cell-autonomously produces larger cells. *gigas* clones in the eye result in larger photoreceptor cells which engage in an increased number of synaptic contacts with their genotypically wildtype postsynaptic partners (demonstrated in the lamina) (Canal et al., 1994). Flies which receive unequal visual input through their two compound eyes follow a helical path in phototaxis experiments, turning the "poor" eye towards the light source, apparently in an attempt to balance the input from both eyes (Benzer, 1973). In *gigas* mosaic animals with a mutant and a wildtype eye, the wildtype eye is turned toward the light source (Canal et al., 1994). Most mutants that were isolated by Markow and Merriam (1977) in Benzer's fast phototaxis pardigm as poorly phototactic turned out to be more photo-positive than the parent wildtype strain in a slow photochoice test. This indicates the action of negative control in the latter behavior.

Why no R8 cell mutants?

There are mutants selectively affecting R1–6 and R7. No mutant that in the mature eye selectively affects R8 has yet been isolated and the involvement of R8 in phototaxis has so far only been indirectly inferred. Because of the pioneering and inductive role R8 plays in ommatidial assembly (Yamamoto, 1996) no selective developmental mutant, which affects R8 but no other retinular cell, can be expected to exist. Selective mutations that affect R8 maintenance can, however, well be envisioned in analogy to *ninaE*. *ninaE* is the only opsin gene identified by classical genetic means. All other opsin genes were identified by homology screens, most recently *Rh5* and *Rh6* which are expressed in complementary sets of R8 cells (Chou et al., 1996; Huber et al., 1997). Another opsin gene, *Rh3*, which is predominantly expressed in R7 cells, is also expressed in a subset of R8 cells restricted to a single row of cells at the dorsal eye margin (Fortini and Rubin, 1991). This region in other flies was shown to be specialized for the detection of polarized light. Comparable to *Rh5* and *Rh6*, *Rh3* and *Rh4* are expressed in complementary but intermingled sets of R7 cells. This partial redundancy may have prevented the isolation of mutants defective in R7 or R8 cell opsins by behavioral genetic means (see also below for interaction between receptor subsystems). Another, more trivial reason may be the autosomal location of all opsin genes;

historically, it was predominantly the *X* chromosome that was screened for behavioral mutants. Targeted mutagenesis (Ballinger and Benzer, 1989; Kaiser and Goodwin, 1990) may lead to the isolation of opsin mutants irrespective of phenotype.

Analysis of photoreceptors feeding into the phototaxis response

In addition to the genetic tools outlined above, investigations of the input to the phototactic reponse made use of accessory techniques. The absorption spectra of *Dipteran* R1–6 and R7 are known from microspectrophotometry (Kirschfeld et al., 1978; Kirschfeld et al., 1983). Spectral sensitivities have been determined by ERG analysis (Harris et al., 1976), direct intracellular recording (Hardie, 1977), and by a variety of behavioral assays, phototaxis being only one of them. Behavioral experiments indicated that the response threshold to visible light of R1–6 is about 3 log units lower than that of R7/8 (Bicker and Reichert, 1978; Fischbach, 1979). R1–6 and R7 receptors can be reversibly inactivated by intense irradiation with blue and violet light, respectively. This leads to a large net conversion of the respective rhodopsin into a metastable metarhodopsin associated with a prolonged photoreceptor depolarisation that can last for hours (PDA, prolonged depolarizing afterpotential, reviewed in Hardie, 1983).

There are, thus, several ways of addressing a subset of the visual input channels: genetic elimination of R1–6 and/or R7 function, PDA inactivation of R1–6 or R7, stimulation of only R1–6 by low light intensity, occlusion or genetic elimination of the ocelli. It is the consistency of results obtained by the application of independent techniques which allowed the proposal of meaningful models of the organization of the visual input in a given phototaxis paradigm. Jacob et al. (1977) (in agreement with other reports) presented evidence for an antagonistic action of R7/8 on R1–6 in their slow photochoice procedure, which at medium intensities of visible light can lead to negative phototaxis (photophobic behavior). A similar antagonistic interaction between R1–6 and R7/8 in neighbouring ommatidia (lateral inhibition) was described in optomotor behavior (Kirschfeld and Lutz, 1974). Since R7 and R8 do not engage in synaptic contact in the lamina, the necessary interactions probably occur at a more proximal level. Interactions between R8 and LMC L2 have been described anatomically (Strausfeld and Campos-Ortega, 1972) but there is currently no direct information regarding the involvement of higher order neurons in phototaxis. Most phototactic mutants whose structural or physiological basis has been elucidated affect the development, maintenance or function of the sensory periphery (see above). This is why visual behavior mutants probably have been most useful in the elucidation of the cellular mechanism of the first step of visual information processing: signal transduction in photoreceptors (reviewed in Pak, 1975; Ranganathan et al., 1995).

Most mutants that were isolated in fast phototactic screens respond to light with a normal ERG (Pak et al., 1969; Heisenberg and Götz, 1975; Markow and Merriam, 1977), indicating normality of the retina–lamina connection. No mutant, so far, has pinpointed the requirement of a specific interneuron for a phototactic response. The mutant *extra lamina fiber* (*elf* aka *H37*) displays an altered ERG (Heisenberg and Götz, 1975) and has reduced UV sensitivity in a slow photochoice experiment. Most lamina cartridges of *elf* contain an additional fiber profile with a diameter characteristic of a large lamina monopolar cell (Heisenberg and Buchner, 1977). *elf*, thus, appears to have a gain-of-function structure which is difficult to interpret. *small optic lobes* (*sol*) is currently the only complementation group that holds promise to eventually identify neurons required for fast phototaxis. In sol^1 the volume of medulla and lobula complex is reduced to half, the lamina being unaffected. The volume reduction is due to a loss of cells during pupal development, suggesting that a defined class of cells is missing from each cartridge (Fischbach and Heisenberg, 1981; Fischbach and Technau, 1984). The lack of a particular cell is difficult to prove by Golgi analysis. This method can, however, demonstrate the presence of a cell type in a mutant. Thus, Fischbach and Heisenberg (1981) could identify several medulla cell types in *sol*[1]. Possibly as a consequence of the large structural alteration, they found, however, aberrant arborization of several identified neurons both within and between visual neuropils. Structural abnormalities in mosaic animals can appear in genotypically wildtype cells that are in contact with mutant tissue indicating nonautonomy of the structural phenotype (Garen and Kankel, 1983). Developmental plasticity is an inescapable complication in the classical neurogenetic approach. *sol*[1] nonetheless has a normal optomotor response, indicating that the perturbed visual system is still capable of performing part of its repertoire.

Optomotor response

Behavior elicited by moving visual stimuli is collectively called optomotor behavior. The term "optomotor response" more specifically refers to actions that can be evoked by large field moving visual stimuli. If not intentionally performing a turn, flies tend to follow the direction of motion of such stimuli. This reduces the image flow across the fly's retina and in effect serves to stabilize the course the fly had followed before the perturbation. Optomotor behavior has generally been investigated with flies walking or flying stationarily in a simulator surrounded by a panorama, the movement of which can be controlled by either the fly ("closed loop") or by external influence ("open loop") (Götz, 1964; Buchner, 1976; Heisenberg and Wolf, 1984). In traditional closed loop devices the force exerted by the immobilized fly during its intended turns are measured and fed back into the panorama movement (Reichardt, 1973; Götz and Wenking, 1973; Heisenberg and Wolf, 1979). Freely moving flies generally experience closed loop conditions: their actions result in a corresponding visual (and mechanosensory) feedback. A recently developed device makes "virtual" open loop experiments possible with flies walking freely within an arena. Here, the panorama image is formed by light emitting diodes. Fly position and orientation is continuously measured, evaluated, and used to control the apparent panorama movement in order to suppress the locomotion-induced visual feedback (Strauss et al., 1997).

Motor responses to rotating visual stimuli have been studied intensively in *Drosophila* and other arthropods. This overview will mainly be focused on the impact of genetic and molecular techniques on our understanding of structure-function relationships in these behaviors. The reader is referred to a large body of review articles for more general accounts (Hall and Greenspan, 1979; Götz, 1983; Buchner, 1984; Fischbach and Heisenberg, 1984; Heisenberg and Wolf, 1984; Strausfeld, 1984; Borst and Egelhaaf, 1989; Fischbach et al., 1989; Franceschini et al., 1989; Hausen and Egelhaaf, 1989; Strausfeld, 1989; Hengstenberg, 1991; Borst and Egelhaaf, 1993; Collett et al., 1993; Egelhaaf and Borst, 1993a, 1993b). Movement detection requires the sampling of visual information from at least two points in space. Phenomenological and more recently mechanistic models have been proposed that try to account for the measurable relationships between visual input parameters and motor output (see Egelhaaf and Borst (1993b) for references). The central element in the powerful "correlation model" is the elementary movement detector (EMD) that correlates temporally non-coincident visual input from photoreceptors that sample different points in space (Reichardt, 1957; Buchner, 1976). In experiments, comparable to the ones described for phototaxis, R1–6 were shown to be the input channels for the optomotor response (Heisenberg and Buchner, 1977; Bülthoff, 1986). Front-to-back (progressive) and back-to-front (regressive) pattern motion presented unilaterally elicit syndirectional turning responses of similar strength. The stimulation of different rostro-caudal parts of one eye with horizontal pattern movement elicits similar responses (Götz and Wenking, 1973; Götz, 1983; Heisenberg and Wolf, 1984; Bausenwein et al., 1986). This suggests that input elements to the EMDs are rather evenly distributed over the surface of the eye. This is supported by the optomotor response that could be measured in *so* flies that had less than 10 ommatidia per eye (Götz, 1970; Götz, 1983) and also by the fact that antidirectional horizontal stimulation in different parts of one eye will cancel when both stimuli cover the same number of ommatidia (Götz, 1975). There are, in fact, sets of EMDs that receive input from neighbouring photoreceptor cells within one facet. As outlined in the description of the structure of the fly's visual system, the 6 outer rhabdomeres of one facet view different points in space and are presynaptic onto different cartridges in the lamina. Kirschfeld (1972) elicited a turning response in *Musca* by sequentially stimulating two photoreceptors cells within one ommatidium.

Two screens have been described in the literature that were devised to select for mutant flies with aberrant optomotor behavior ("barber pole" and "optomotor maze", (Götz, 1970; Heisenberg and Götz, 1975). No mutants were isolated in the barber pole apparatus even though the device was able to fractionate wildtype from an incidentally chosen control strain which was later shown to have an aberrant translatory response to bilateral progressive or regressive stimulation (Götz and Wenking, 1973). About 200,000 mutant flies were screened in the optomotor maze to yield 49 mutants belonging to 6 complementation groups (Heisenberg and Götz, 1975; Pflugfelder and Heisenberg, 1995). All mutants had abnormal ERGs indicating defects in retina or lamina. This provoked the question of why no mutants with exclusive proximal CNS defects were found (in contrast, all 10 mutants isolated in a fast phototaxis screen had normal ERGs (Markow and Merriam, 1977)). Apart from blind mutants such as *no receptor potential A* (*norpA*) only two mutant alleles from this screen were characterized in depth at the behavioral and molecular level: *no on transient* A^{H2} (*nonA*^{H2}) and *optomotor-blind*^{H31} (*omb*^{H31}).

nonA^{H2} is an allele of the complex gene *nonA-dissonance,* alleles of which were isolated due to defects in optomotor behavior, fast phototaxis, or aberrant courtship song (Lindsley and Zimm, 1992; Rendahl et al., 1992). *nonA*^{H2} is peculiar because it has an allele-specific optomotor defect. In flight, its response to progressive stimuli of sufficient pattern wavelength is nearly completely eliminated with little effect on the regressive stimulus response (Heisenberg, 1972; Stanewsky et al., 1993). *nonA* encodes a putative RNA binding and chromatin associated protein (Von Besser et al., 1990) which is nearly ubiquitously expressed throughout development (Rendahl et al., 1992). The *nonA* and *dissonance* mutations are caused by amino acid changes in different parts of the protein (Rendahl et al., 1996). The various alleles of the *nonA-diss* complementation groups appear to be selective protein hypomorphs (Pflugfelder, 1998). The molecular analyis has not yielded any hint so far of where in the visual system the *nonA*^{H2} mutation exerts its effect.

Optomotor-blind

omb^{H31} presents a simpler case. Even though it is associated with an ERG defect, this is of a subtle nature. Unlike the situation in wildtype flies, unilateral stimulation of *omb*^{H31} with horizontally rotating patterns does not elicit equal response for progressive and regressive motion. In flight, regressive stimulation of the caudo-lateral eye leads to no or even anti-directional yaw torque, whereas progressive stimulation still yields about half of the wildtype response (Götz, 1983; Heisenberg and Wolf, 1984; Bausenwein et al., 1986; Brunner et al., 1992). The anti-directional yaw torque toward a regressive motion stimulus in the caudal visual field does not show a pattern wavelength dependence and thus is probably not motion-induced but rather caused by the position of the stimulus (Bausenwein et al., 1986). Different parts of the eye show different responses to progressive motion. Stimulation of the frontal part yields wildtype response, the lateral part appears unresponsive, and in the caudal part strong anti-directional yaw is elicited (Götz, 1983). This residual optomotor response in *omb*^{H31} is comparable to the reaction of wildtype to a single rotating stripe. The stripe elicits a strong response only during the progressive phase of the motion, most strongly in the frontal part of the visual field (Bausenwein et al., 1986). The wildtype optomotor response to a moving grating thus appears to be made up of two components. The horizontal "large field response" is most effectively elicited by stimulation of larger parts of the visual field, but independent of pattern position and stimulus direction (progressive or regressive). The "object response" is evoked by small rotating objects in progressive motion, and most effectively in the frontal visual field (Bausenwein et al., 1986). Large field response and object reponse to horizontal motion are channeled through different lobula plate interneurons, the HS (see below) and FD (figure detection) cells, respectively (Hausen and Wehrhahn, 1989; Hausen and Wehrhahn, 1990). *omb*^{H31} equally affects rotations about the other body axes (Blondeau and Heisenberg, 1982) and even more strongly compensatory head rotations (Hengstenberg, 1988; Pflugfelder and Heisenberg, 1995). *omb*^{H31} was investigated in many behavioral paradigms and found to be normal in pattern-induced behaviors such as the lift/thrust or the landing response, as well as in many other paradigms (reviewed in Pflugfelder and Heisenberg, 1995).

omb^{H31} has a conspicuous neuroanatomical phenotype. The mutant lacks a set of 10 large interneurons in the lobula plate, the horizontal and vertical system cells (HS and VS cells) (Heisenberg et al., 1978). This correlation between neuroanatomical and behavioral phenotypes in itself does not establish a causal relationship between them. There is, however, ample evidence from larger flies, obtained by a number of independent methods, that these highly conserved neurons are important integration and output elements of optomotor turning reactions (Pflugfelder and Heisenberg, 1995; Krapp and Hengstenberg, 1996). In addition to the HS/VS defect, *omb*^{H31} displays a number of more subtle neuroanatomical alterations affecting the anterior optic tract (AOT) and the inner optic chiasm (IOC) (Brunner et al., 1992). It is conceivable that these defects contribute to some of the other behavioral abnormalities which were observed in *omb*^{H31}, like the unusual orientation behavior in a forced choice paradigm (Bülthoff, 1982) or the fast (but not slow) phototaxis defect (Pflugfelder and Heisenberg, 1995). Nonetheless, the fact that in *omb*^{H31} the incorporation pattern of the activity label deoxyglucose in response to motion stimuli appears wildtype confirms the assumption that the behavioral phenotype is caused at the level of the HS/VS cells and not distal to it (Bülthoff and Buchner, 1985).

omb is an essential gene encoding a T domain transcription factor (Pflugfelder et al., 1992b; Papaioannou, 1997). Roles in *Drosophila* visual system, appendage, and abdominal development have been recognized (Grimm and Pflugfelder, 1996, 1998; Kopp and Duncan, 1997). Behavioral phenotypes are caused by the elimination of regulatory sequences downstream of the 3′ end of the transcription unit (Pflugfelder et al., 1992a; Brunner et al., 1992). The *omb*^{H31} phenotype is caused by an inversion breakpoint about 5 kb downstream of the last known *omb* exon. The regulatory sequences essential for HS/VS development are located a further 30–40 kb downstream, about 100 kb from the *omb* promoter. Elimination of this "optic lobe regulatory region 3" (OLR3) causes the loss of the 10 HS and VS cells. Surprisingly, optomotor defects in OLR3 mutants are less severe than in *omb*^{H31}. Responses to progressive large field motion are wildtype and there is considerable reaction to regressive motion (Brunner et al., 1992). Neither the AOT nor the

IOC defects can account for the behavioral differences between the two genotypes. omb^{H31} appears to cause additional structural defects which have not yet been recognized. In Musca a set of twin fibers to the HS and VS cells was described (Pierantoni, 1976), the loss of which might account for the increase in phenotypic severity from OLR3 mutants to omb^{H31}.

Comparison of the *omb* expression patterns in wildtype and omb^{H31} suggested that the loss of late embryonic expression from the optic lobe anlage in the mutant is responsible for the adult neuroanatomical phenotype (Poeck et al., 1993). Full expression of the HS/VS cell phenotype in OLR3 mutants indicated that the 10 kb OLR3 region should harbor an enhancer for embryonic *omb* OL expression. A systematic enhancer expression screen of OLR3 subfragments in transgenic flies confirmed this assumption (Hofmeyer and Pflugfelder, 1998). A 2.5 kb OLR3 subfragment directs transient reporter gene expression in about 7 cells/hemisphere of the OL anlage in late embryogenesis. Enhancer-driven reporter gene expression is thus approximately reciprocal to the loss of omb expression in omb^{H31}. Laser ablation experiments with *Musca* larvae suggested that the HS/VS cells derive from a single larval neuroblast (Geiger and Nässel, 1981, 1982), indicating a lineage relationship. It is, therefore, likely that one of the seven cells marked by OLR3 enhancer expression is the HS/VS precursor. Systematic genetic and molecular analysis of various *omb* mutant alleles thus have brought us close to an understanding of the developmental aberrations underlying the adult neuroanatomical and behavioral defects. Several questions remain to be answered, however. It is not clear, how loss of *omb* expression during embryogenesis affects the cell fate of larval neuroblasts. Nor is it known, whether in omb^{H31} the presumptive HS/VS precursor fails to resume proliferation during larval development or whether its cell fate is switched to a different lineage (or to cell death). Identification of the OLR3 enhancer provides the basis for the implementation of reverse genetic tools (see below) by which these and other questions could be solved.

Models for optomotor pathways

Optomotor behavior was also investigated in several mutants isolated by other means. The dominant mutation *Vacuolar medulla* (*Vam*) (Heisenberg and Böhl, 1979) lacks reponses to moving patterns (both gratings and single objects) but can still fixate broad bars. In *Vam* adults, the large monopolar cells L1 and L2 (and possibly other cells) vacuolate soon after emergence (Coombe and Heisenberg, 1986). The poor correlation between L1/2 degeneration and residual optomotor response makes it doubtful, however, that L1/2 are the sole first order interneurons in the optomotor pathway (Coombe et al., 1989). Several other mutants that appear to reduce neuronal complexity in the higher visual neuropils are remarkable for displaying normal optomotor behavior. Mutations in *sol* (which were discussed above in the context of phototaxis) reduce the medulla cell number by about half, most likely through the elimination of sets of columnar elements (Fischbach and Heisenberg, 1981; Fischbach and Technau, 1984). Similarly, the number of axons running in the AOT, which connects lobula and central brain, is reduced from about 1200 to 500 (Fischbach and Lyly-Hünerberg, 1983). Yaw torque and deoxyglucose incorporation into the distal optic neuropils in response to horizontal motion is wildtype in these mutant flies (Fischbach and Heisenberg, 1984; Bülthoff, 1986). *sol* flies, however, exhibit behavioral flaws in addition to the phototaxis defect described above (Heisenberg and Wolf, 1984; Götz, 1985). In the mutant *reduced optic lobes* (*rol*), medulla and lobula complex are reduced by about 30% while optomotor behavior is essentially normal (Lindsley and Zimm, 1992). *rol* and *sol* appear to affect different subsets of medulla columnar neurons and AOT fibers because the double mutant has additive defects (Fischbach et al.,

1989). Since neither set of missing cells appears to be essential for optomotor behavior one would expect that the double mutant is optomotor proficient. This is, however, not the case (Wolf and Heisenberg, 1986). These results are strong evidence for the existence of at least partially redundant optomotor pathways through the medulla (see below) and may be one of the reasons why no mutations affecting medulla elements were isolated in optomotor screens.

Neurogenetics has so far contributed little to close the knowledge gap in the optomotor pathways between photoreceptors and lobula plate integrators. The easy electrophysiological accessibilty of directionally sensitive lobula plate interneurons in large flies has allowed the application of a spectrum of ingenious techniques (e.g. microsurgery, in vivo calcium imaging, neuropharmacology) to probe the structural and functional organization of these output elements (Hausen and Wehrhahn, 1990; Borst and Egelhaaf, 1992; Brotz and Borst, 1996). This led to the identification of cholinergic and GABAergic receptors on HS/VS cells (Brotz and Borst, 1996) as was suspected from genetic mosaic studies in *Drosophila* (Greenspan et al., 1980). Distinct pathways between retina and lobula plate were proposed to exist based on a densitometric evaluation of Golgi profiles in *Musca* and *Drosophila* (Strausfeld and Lee, 1991; Bausenwein et al., 1992). Two of the proposed pathways appear to be coactivated by visual motion stimuli as judged from deoxyglucose accumulation: LMC L1–medulla intrinsinc cell 1–T4 cell–lobula plate, and LMC L2–transmedulla cell 1–T5 cell–lobula plate (Bausenwein and Fischbach, 1992a, 1992b). This is in agreement with the conclusion drawn from *rol* and *sol*. Douglass and Strausfeld (1995, 1996) were the first to record intracellularly (in Calliphorid flies) from several of the neurons postulated to lie on optomotor pathways. While these studies cannot establish the essential nature of a cell within a given functional pathway, they provide crucial building blocks for bottom-up models of visual motion processing.

Perspectives

The above discussions of two behavioral phenomena, phototaxis and optomotor reponse, have tried to make obvious the usefulness of behavioral mutants. However, they have also revealed how far we are from a thorough understanding of structure-function relationships in the optic lobes. Clearly, additional mutants with specific neuroanatomical and behavioral phenotypes are highly desirable. The evaluation of extensive collections of enhancer trap lines indicates that such mutants may not be easy to find by conventional means. Datta et al. (1993) concluded from an analysis of about 600 larval CNS expression patterns that few, if any, of these patterns were restricted to the optic lobes. Han et al. (1996), analyzing a 10-fold larger sample for adult head expression, arrived at a similar conclusion. In an even more comprehensive study (cited in Miklos and Rubin, 1996) only one of 20,000 enhancer trap lines was found to be specifically expressed in the visual system during development. Little cell type specificity is also found when only OL expression is considered. Interneurons are predominantly specified with regard to their structure, connectivity, and the appropriate cocktail of communication agents. This specification apparently (and necessarily) occurs by some combinatorial logic. Neuron numbers in *Caenorhabditis elegans* and *Drosophila melanogaster* (302 vs 200,000) differ by nearly 3 orders of magnitude but the two species were estimated to possess a comparable number of genes (12–19,000) (Miklos and Rubin, 1996, *C. elegans* Sequencing Consortium, 1998). Inactivation of one gene will, therefore, generally not result in the specific loss of one cell type. This expectation was borne out by an evaluation of a set of well-characterized behavioral mutants, in which the behavioral shortcomings are not obviously caused by defects in the sensory or motor periphery. The majority of them is caused by a selective inactivation of gene function (at the level of DNA, RNA, or protein). Since ethyl methanesulfonate (EMS) has been the predominant mutagen in *Drosophila* neurogenetics in the past, possibly the greatest source of obtaining selective mutations, the functional elimination of gene enhancers and silencers, has not been tapped effectively. Because of the internal redundancy of enhancers, they will generally not become inactivated by EMS-produced point mutations (Pflugfelder, 1998). *omb* may serve to illustrate the point. omb^{H31} is one of the most specific behavioral mutants currently known, despite the fact that *omb* is essential and functional in many tissues. The specificity is caused by the elimination of a set of regulatory elements from the gene. Even though omb^{H31} was isolated in an EMS mutagenesis screen, it probably arose spontaneously (Pflugfelder and Heisenberg, 1995). The transposable element *P* preferentially inserts around promoters (Spradling et al., 1995). *P* elements (generally as genetically modified derivatives, see Bellen et al., 1990) proved to be highly successful tools for the generation of behavioral mutants (e.g. Anholt et al., 1996; Pflugfelder, 1998). It is likely that the incorporation of enhancer blocking sequences into the transposon, for example the *gypsy* chromatin insulator (Gdula et al., 1996), will create an even superior tool (Roseman et al., 1995). Other mobile elements like *hobo* or *I*, that transpose with a different insertional specificity (Blackman et al., 1987; Smith et al., 1993; Dimitri et al., 1997; Busseau et al., 1998), may also eventually find their way into the neurogenetic tool box. Genetically engineered mutagens can have additional advantages such as carrying enhancer reporters (or other functions, see below) or allowing easy cloning of the tagged gene (Bellen et al., 1990).

Transposable elements have been modified such that genomic enhancers in the vicinity of the insertion drive the expression of the yeast transcription factor Gal4 gene. The transcription of any reporter or effector gene under control of the Gal4 response element ("upstream activating sequence", UAS) can be activated by a Gal4 driver of the desired spatio-temporal expression profile (Brand and Perrimon, 1993). The Gal4/UAS technology is versatile, allowing for instance the visualization of neuronal shape with UAS:tau (claimed to be superior to the Golgi method, Ito et al., 1997b), or the ablation or functional inactivation of Gal4-expressing cells (Zhou et al., 1997; Sweeney et al., 1995). The limited expression selectivity holds for β-galactosidase reporter insertions as well as for Gal4 enhancer trap lines. Combinatorial systems which introduce an additional layer of temporal control have been developed. In the tetracycline-dependent transactivator system, spatial expression is under enhancer control whereas the temporal expression profile can be influenced by the feeding of tetracycline (Bello et al., 1998). A combination of Gal4/UAS with flippase/FRT technology allows the induction of gene transcription in mosaic tissue spots by heat shock (Basler and Struhl, 1994). Other control systems are under development and will provide increasingly finer tools for the manipulation of gene activity and cellular function (Phelps and Brand, 1998). The complex system that is under study in behavioral neurogenetics will certainly become more accessible with these techniques.

The appropriate mutagen will reduce the problem of pleiotropy associated with most developmental genes (Hall, 1994; Pflugfelder, 1998) but cannot circumvent the difficulties caused by redundancy. Genetic redundancy within a given functional realm, but even complete redundancy can be evolutionarily stable under certain conditions (Nowak et al., 1997). While it is debatable whether true redundancy exists in established species in the wild outdoors, the problem is a real one for the narrow-focus screens in the laboratory (Miklos and Rubin, 1996). Functions that are genetically redundant are not directly accessible by standard loss-of-function mutageneses. Genes in such systems can be identified by their gain-of-function phenotype (e.g. Rorth et al., 1998) or by carrying out the mutagenesis in a susceptible genetic background. This was attempted for defective optomotor behavior in the background of *nightblind A* (Heisenberg and Götz, 1975) but might be repeated more successfully in the background of *rol* or *sol*.

Behavior can be analyzed, genes can be cloned, but the identification of the neuroanatomical or neurophyiological cause of mutant behavior is generally still the most difficult aspect of mutant characterization. Reverse genetics may provide the tools. By manipulating cells or cell types that at the same time are singled out by the expression of an appropriate marker, structure-function analysis may eventually become an easier exercise.

Acknowledgements

The author is indebted to E. Buchner, M. Heisenberg, A. Hofbauer, A. Keller, T. Zars, and two referees for improving the manuscript. Work on *optomotor-blind* in the author's laboratory was supported by Deutsche Forschungsgemeinschaft (DFG).

List of abbreviations

AOT	anterior optic tract
cGMP	guanosine 3′: 5′ cyclic monophosphate
EMS	ethyl methanesulfonate
ERG	electroretinogram
FD cell	figure detection cell
HS cell	horizontal system cell
IOC	inner optic chiasm
LMC	large monopolar cell
LPC	lamina precursor cell
NO	nitric oxide
OL	optic lobe(s)
OLR	optic lobe regulatory region
PDA	prolonged depolarizing afterpotential
R cell	retinula cell
VS cell	vertical system cell
20-HE	20-hydroxyecdysone

References

Anholt, R.R.H., Lyman, R.F. and Mackay, T.F.C. (1996) Effects of single P-element insertions on olfactory behavior in *Drosophila melanogaster*. *Genetics*, 143: 293–301.

Armstrong, J.D., Kaiser, K., Müller, A., Fischbach, K.-F., Merchant, N. and Strausfeld, N. J. (1995) Flybrain, an on-line atlas and database of the *Drosophila* nervous system. Neuron, 15: 17–20.

Ashley, J.A. and Katz, F.N. (1994) Competition and position-dependent targeting in the development of the *Drosophila* R7 visual projections. *Development*, 120: 1537–1547.

Balles, J. (1995) Das optomotor-blind Gen von *Drosophila*. Versuch zur entwicklungsbiologischen Funktion und zur molekularen Charakterisierung von Mutanten. Universität Würzburg.

Ballinger, D.G. and Benzer, S. (1988) Photophobe, (Ppb), a *Drosophila* mutant with a reversed sign of phototaxis; the mutation shows allele-specific interaction with sevenless. *Proc. Natl. Acad. Sci. USA*, 85: 3960–3964.

Ballinger, D.G. and Benzer, S. (1989) Targeted gene mutations in *Drosophila*. *Proc. Natl. Acad. Sci. USA*, 86: 9402–9406.

Banerjee, U., Renfranz, P.J., Pollock, J.A. and Benzer, S. (1987) Molecular characterization and expression of sevenless, a gene involved in neuronal pattern formation in the *Drosophila* eye. *Cell*, 49: 281–291.

Basler, K. and Hafen, E. (1989) Ubiquitous expression of sevenless: position-dependent specification of cell fate. *Science*, 243: 931–934.

Basler, K. and Struhl, G. (1994) Compartment boundaries and the control of *Drosophila* limb pattern by hedgehog protein. *Nature*, 368: 208–214.

Bausenwein, B., Dittrich, A.P.M. and Fischbach, K.-F. (1992) The optic lobe of *Drosophila melanogaster* II. Sorting of retinotopic pathways in the medulla. *Cell Tiss. Res.*, 267: 17–28.

Bausenwein, B. and Fischbach, K.-F. (1992a) Activity labeling patterns in the medulla of *Drosophila melanogaster* caused by motion stimuli. *Cell Tiss. Res.*, 270: 25–35.

Bausenwein, B. and Fischbach, K.-F. (1992b) Separation of functional pathways in the fly's medulla: combination of 2-deoxyglucose studies with anatomical fine analysis. In: R. N. Singh, (Ed.), *Nervous systems: principles of design and function*, Wiley Eastern, New Delhi, 223–239.

Bausenwein, B., Wolf, R. and Heisenberg, M. (1986) Genetic dissection of optomotor behavior in *Drosophila melanogaster*. Studies on wild-type and the mutant optomotor-blind. *J. Neurogenetics*, 3: 87–109.

Bellen, H.J., Vaessin, H., Bier, E., Kolodkin, A., D'Evelyn, D., Kooyer, S. and Jan, Y. N. (1992) The *Drosophila* couch potato gene: an essential gene for normal adult behaviour. *Genetics*, 131: 365–375.

Bellen, H.J., Wilson, C. and Gehring, W.J. (1990) Dissecting the complexity of the nervous system by enhancer detection. *BioEssays*, 12: 199–204.

Bello, B., Resendez-Perez, D. and Gehring, W.J. (1998) Spatial and temporal targeting of gene expression in *Drosophila* by means of a tetracycline-dependent transactivator system. *Development,* 125: 2193–2202.

Benzer, S. (1967) Behavioral mutants of *Drosophila* isolated by countercurrent distribution. *Proc. Natl. Acad. Sci. USA*, 58: 1112–1119.

Benzer, S. (1973) Genetic dissection of behavior. *Scientific American*, December: 24–37.

Bicker, G. and Reichert, H. (1978) Visual learning in a photoreceptor degeneration mutant of *Drosophila melanogaster. J. Comp. Physiol.*, 127: 29–38.

Blackman, R.K., Grimaila, R., Koehler, M.M.D. and Gelbart, W.M. (1987) Mobilization of hobo elements residing within the decapentaplegic complex: suggestion of a new hybrid dysgenesis system in *Drosophila melanogaster. Cell*, 49: 497–505.

Blondeau, J. and Heisenberg, M. (1982) The three-dimensional optomotor torque system of *Drosophila melanogaster*. Studies on wildtype and the mutant optomotor-blind[H31]. *J. Comp. Physiol.*, 145: 321–329.

Bloomquist, B.T., Shortridge, R.D., Schneuwly, S., Perdew, M., Montell, C., Steller, H., Rubin, G. and Pak, W.L. (1988) Isolation of a putative phospholipase C gene of *Drosophila*, norpA, and its role in phototransduction. *Cell*, 54: 723–733.

Borst, A. and Egelhaaf, M. (1989) Principles of visual motion detection. *TINS*, 12: 297–306.

Borst, A. and Egelhaaf, M. (1992) In vivo imaging of calcium accumulation in fly interneurons as elicited by visual motion stimulation. *Proc. Natl. Acad. Sci. USA*, 89: 4139–4143.

Borst, A. and Egelhaaf, M. (1993) Detecting visual motion: theory and models. In: F.A. Miles and J. Wallman, (Eds.), *Visual motion and its role in the stabilization of gaze*, Elsevier Science Publishers B.V., pp. 3–27.

Boschek, C.B. (1971) On the fine strucuture of the peripheral retina and lamina ganglionaris of the fly, *Musa domestica. Z. Zellforschung*, 118: 369–409.

Braitenberg, V. (1967) Patterns of projection in the visual system of the fly I. retina lamina projections. *Exp. Brain Res.*, 3: 271–298.

Braitenberg, V. (1972) Periodic structures and structural gradients in the visual ganglia of the fly. In: R. Wehner, (Ed.), *Information processing in the visual system of arthropods*, Springer, Berlin.

Brand, A.H. and Perrimon, N. (1993) Targeted gene expression as a means of altering cell fates and generating dominant phenotypes. *Development,* 118: 401–415.

Brotz, T. and Borst, A. (1996) Cholinergic and GABAergic receptors on fly tangential cells and their role in visual motion detection. *J. Neurophysiol.*, 76: 1786–1799.

Brunner, A. Twardzik, T. and Schneuwly, S. (1994) The *Drosophila* giant lens gene plays a dual role in eye and optic lobe development: inhibition of differentiation of ommatidial cells and interferference in photoreceptor axon guidance. *Mech. Dev.*, 48: 175–185.

Brunner, A., Wolf, R., Pflugfelder, G.O., Poeck, B. and Heisenberg, M. (1992) Mutations in the proximal region of the optomotor-blind locus of *Drosophila melanogaster* reveal a gradient of neuroanatomical and behavioural phenotypes. *J. Neurogenetics*, 8: 43–55.

Buchner, E. (1976) Elementary movement detectors in an insect visual system. *Biol. Cybernetics*, 24: 85–101.

Buchner, E. (1984) Behavioural analysis of spatial vision in insects. In: M. A. Ali (Ed.), *Photoreception and vision in invertebrates*, Plenum Press, New York and London.

Buckles, G.R., Smith, Z.D.J. and Katz, F.N. (1992) mip causes hyperinnervation of a retinotopic map in *Drosophila* by excessive recruitment of R7 photoreceptor cells. *Neuron*, 8: 1015–1029.

Bülthoff, H. (1982) *Drosophila* mutants disturbed in visual orientation. *Biol. Cybernetics*, 45: 63–70.

Bülthoff, I. (1986) Deoxyglucose mapping of nervous activity induced in *Drosophila* brain by visual movement III. outer rhabdomeres absent^{JK84}, small optic lobes ^{KS58} and no object fixation E^{B12} visual mutants. *J. Comp. Physiol. A*, 158: 195–202.

Bülthoff, I. and Buchner, E. (1985) Deoxyglucose mapping of nervous activity induced in *Drosophila* brain by visual movement. *J. Comp. Physiol. A*, 156: 25–34.

Buschbeck, E.K. and Strausfeld, N.J. (1996) Visual motion-detection circuits in flies: small-field retinotopic elements responding to motion are evolutionarily conserved across taxa. *J. Neurosci.*, 16: 4563–4578.

Busseau, I., Malinsky, S., Balakireva, M., Chaboissier, M.-C., Teninges, D. and Bucheton, A. (1998) A genetically marked I element in *Drosophila melanogaster* can be mobilized when ORF2 is provided in trans. *Genetics*, 148: 267–275.

Campos-Ortega, J.A., Jürgens, G. and Hofbauer, A. (1979) Cell clones, and pattern formation: studies on sevenless: a mutant of *Drosophila melanogaster. Roux's Arch.*, 186: 27–50.

Campos-Ortega, J.A. and Strausfeld, N.J. (1972) Columns and layers in the second synaptic region of the fly's visual system: the case for two superimposed neuronal architectures. In: R. Wehner (Ed.), *Information processing in the visual system of arthropods*, Springer.

Caenorhabditis elegans Sequencing Consortium (1998). Genome sequence of the nematode C. elegans: a platform for investigating biology. *Science*, 282: 2012–2018.

Canal, I., Farinas, I., Gho, M. and Ferrús, A. (1994) The presynaptic cell determines the number of synapses in the *Drosophila* optic ganglia. *Eur. J. Neurosci.*, 6: 1423–1431.

Champlin, D.T. and Truman, J.W. (1998) Ecdysteroid control of cell proliferation during optic lobe neurogenesis in the moth Manduca sexta. *Development,* 125: 269–277.

Cheyette, B.N.R., Green, P.J., Martin, K., Garren, H., Hartenstein, V. and Zipursky, S.L. (1994) The *Drosophila* sine oculis locus encodes a homeodomain-containing protein required for the development of the entire visual system. *Neuron*, 12: 977–996.

Chou, W.-H., Hall, K.J., Wilson, B., Wideman, C.L., Townson, S.M., Chadwell, L.V. and Britt, S.G. (1996) Identification of a novel *Drosophila* opsin reveals specific patterning of the R7 and R8 photoreceptor cells. *Neuron*, 17: 1101–1115.

Collett, T., Nalbach, H.-O. and Wagner, H. (1993) Visual stabilization in arthropods. In: F.A. Miles and J. Wallman (Eds.), *Visual motion and its role in the stabilization of gaze*, Elsevier, Amsterdam, pp. 239–263.

Coombe, P.E. (1984) The role of retinula cell types in fixation behaviour of walking *Drosophila melanogaster*. *J. Comp. Physiol. A.*, 155: 661–672.

Coombe, P.E., and Heisenberg, M. (1986) The structural brain mutant Vacuolar medulla of *Drosophila melanogaster* with specific behavioral defects and cell degeneration in the adult. *J. Neurogenet.*, 3: 135–158.

Coombe, P.E., Srinivasan, M.V. and Guy, R.G. (1989) Are the large monopolar cells of the insect lamina on the optomotor pathway? *J. Comp. Physiol. A.*, 166: 23–35.

Coppola, D.M., Purves, H.R., McCoy, A.N. and Purves, D. (1998) The distribution of oriented contours in the real world. *Proc. Natl. Acad. Sci. USA*, 95: 4002–4006.

Cymborowsky, B. and Korf, H.-W. (1995) Immunocytochemical demonstration of S-antigen (arrestin) in the brain of the blowfly *Calliphora vicina*. *Cell Tiss. Res.*, 279: 109–114.

Datta, S. (1995) Control of proliferation activation in quiescent neuroblasts of the *Drosophila* nervous system. *Development*, 121: 1173–1182.

Datta, S., Stark, K. and Kankel, D. R. (1993) Enhancer detector analysis of the extent of genomic involvement in nervous system development in *Drosophila melanogaster*. *J. Neurobiol.*, 24: 824–841.

Delaney, S.J., Hayward, D.C., Barleben, F., Fischbach, K.-F. and Miklos, G.L.G. (1991) Molecular cloning and analysis of small optic lobes, a structural gene of *Drosophila melanogaster*. *Proc. Natl. Acad. Sci. USA*, 88: 7214–7218.

Dimitri, P., Arcà, B., Berghella, L. and Mei, E. (1997) High genetic instability of heterochromatin after transposition of the LINE-like factor in *Drosophila melanogaster*. *Proc. Natl. Acad. Sci. USA*, 94: 8052–8057.

Douglass, J.K. and Strausfeld, N.J. (1995) Visual motion detection circuits in flies: peripheral motion computation by identified small-field retinotopic neurons. J. Neurosci., 15: 5596–5611.

Douglass, J.K. and Strausfeld, N.J. (1996) Visual motion-detection circuits in flies: parallel direction- and non-direction-sensitive pathways between the medulla and lobula plate. *J. Neurosci.*, 16: 4551–4562.

Dürrwächter, G. (1957) Untersuchungen über Phototaxis und Geotaxis einiger *Drosophila*-Mutanten nach Aufzucht in verschiedenen Lichtbedingungen. *Z. f. Tierpsychologie*, 14: 1–28.

Ebens, A.J., Garren, H., Cheyette, B.N.R. and S.L., Z. (1993) The *Drosophila* anachronism locus: a glycoprotein secreted by glia inhibits neuroblast proliferation. *Cell*, 74: 15–27.

Egelhaaf, M. and Borst, A. (1993a) Movement detection in arthropods. In: F.A. Miles and J. Wallman (Eds.), *Visual motion and its role in the stabilization of gaze*, Elsevier, Amsterdam, pp. 53–77.

Egelhaaf, M. and Borst, A. (1993b) A look into the cockpit of the fly: visual orientation, algorithms, and identified neurons. *J. Neurosci.*, 13: 4563–4574.

Finkelstein, R., Smouse, D., Capaci, T.M., Spradling, A.C. and Perrimon, N. (1990) The orthodenticle gene encodes a novel homeodomain protein involved in the development of the *Drosophila* nervous system and ocellar visual structures. *Genes Dev.*, 4: 1516–1527.

Fischbach, K.-F. (1979) Simultaneous and successive colour contrast expressed in "slow" phototactic behaviour of walking *Drosophila melanogaster*. *J. Comp. Physiol.*, 130: 161–171.

Fischbach, K.-F. (1983) Neural cell types surviving congenital sensory deprivation in the optic lobes of *Drosophila melanogaster*. *Dev. Biol.*, 95: 1–18.

Fischbach, K.-F. and Dittrich, A.P.M. (1989) The optic lobe of *Drosophila melanogaster* I. A Golgi analysis of wild-type structure. *Cell Tiss. Res.*, 258: 441–475.

Fischbach, K.-F. and Heisenberg, M. (1981) Structural brain mutant of *Drosophila melanogaster* with reduced cell number in the medulla cortex and with normal optomotor yaw response. *Proc. Natl. Acad. Sci. USA*, 78: 1105–1109.

Fischbach, K.-F. and Heisenberg, M. (1984) Neurogenetics and behaviour in insects. *J. Exp. Biol.*, 112: 65–93.

Fischbach, K.-F. and Lyly-Hünerberg, I. (1983) Genetic dissection of the anterior tract of *Drosophila melanogaster*. *Cell Tiss. Res.*, 231: 551–563.

Fischbach, K.-F. and Reichert, H. (1978) Interactions of visual subsystems in *Drosophila melanogaster*. *Biol. Behav.*, 3: 305–317.

Fischbach, K.-F. and Technau, G. (1984) Cell degeneration in the developing optic lobes of the sine oculis and small optic lobes mutants of *Drosophila melanogaster*. *Dev. Biol.*, 104: 219–239.

Fischbach, K.-F., Barleben, F., Boschert, U., Dittrich, A. P.M., Gschwander, B., Houbé, B., Jäger, R., Kaltenbach, E., Ramos, R.G.P. and Schlosser, G. (1989) Developmental studies on the optic lobe of *Drosophila melanogaster* using structural brain mutants. In: R.N. Singh and N.J. Strausfeld (Eds.), *Neurobiology of Sensory Systems*, Plenum Publishing Corporation, pp. 171–194.

Fortini, M.E. and Rubin, G.M. (1991) The optic lobe projection pattern of polarization-sensitive photoreceptor cells in *Drosophila melanogaster*. *Cell Tiss. Res.*, 265: 185–191.

Franceschini, N. (1975) Sampling of the visual environment by the compound eye of the fly: fundamentals and applications. In: A.W. Snyder and R. Menzel (Eds.), *Photoreceptor optics*, Springer, Berlin.

Franceschini, N., Riehle, A. and Le Nestour, A. (1989) Directionally sensitive motion detection by insect neurons. In: D.G. Stavenga and R.C. Hardie (Eds.), Facets of vision, Springer, Berlin.

Garen, S.H. and Kankel, D.R. (1983) Golgi and genetic mosaic analysis of visual system mutants in *Drosophila melanogaster*. *Dev. Biol.*, 96: 445–466.

Gdula, D.A., Gerasimova, T.I. and Corces, V.G. (1996) Genetic and molecular analysis of the gypsy chromatin insulator of *Drosophila*. *Proc. Natl. Acad. Sci. USA*, 93: 9378–9383.

Geiger, G. and Nässel, D.R. (1981) Visual orientation behaviour of flies after selective laser beam ablation of interneurons. *Nature*, 293: 398–399.

Geiger, G. and Nässel, D.R. (1982) Visual processing of moving single objects and wide-field patterns in flies: Behavioural analysis after laser-surgical removal of interneurons. *Biol. Cybern.*, 44: 141–149.

Gibbs, S.M. and Truman, J.W. (1998) Nitric oxide and cyclic GMP regulate retinal patterning in the optic lobe of *Drosophila*. *Neuron*, 20: 83–93.

Goodman, C.S. and Shatz, C.J. (1993) Developmental mechanisms that generate precise patterns of neural connectivity. *Cell*, 72: 77–98.

Goodman, L.J. (1981) Organisation and physiology of the insect dorsal ocellar system. In: (Ed H. Autrum.), *Handbook of sensory physiology*, Vol. VII/6C, Springer, Berlin, pp. 281–286.

Götz, K.G. (1964) Optomotorische Untersuchung des visuellens Systems einiger Augenmutanten der Fruchtfliege *Drosophila*. *Kybernetik*, 2: 77–92.

Götz, K. G. (1970). Fractionation of *Drosophila* populations according to optomotor traits. J. Exp. Biol., 52: 419–436.

Götz, K.G. (1975) Hirnforschung am Navigationssystem der Fliegen. *Naturwissenschaften*, 62: 468–475.

Götz, K.G. (1983) Genetic defects of visual orientation in *Drosophila*. In: *Verhandlungen der Deutschen Zoologischen Gesellschaft*, Gustav Fischer Verlag, Stuttgart, pp. 83–99.

Götz, K.G. (1985) Loss of flexibilty in an optomotor flight control system of the *Drosophila* mutant "small optic lobes". *Biol. Chem. Hoppe-Seyler*, 366: 116–117.

Götz, K.G. and Wenking, H. (1973) Visual control of locomotion in the walking fruitfly *Drosophila*. *J. Comp. Physiol.*, 85: 235–266.

Green, P., Hartenstein, A.Y. and Hartenstein, V. (1993) The embryonic development of the *Drosophila* visual system. *Cell Tiss. Res.*, 273: 583–598.

Greenspan, R. J., Finn, J.A. and Hall, J.C. (1980) Acetylcholinesterase mutants in *Drosophila* and their effects on the structure and function of the central nervous system. *J. Comp. Neurology*, 189: 741–774.

Grimm, S. and Pflugfelder, G.O. (1996) Control of the gene optomotor-blind in *Drosophila* wing development by decapentaplegic and wingless. *Science*, 271: 1601–1603.

Grimm, S. and Pflugfelder, G.O. (1998) optomotor-blind in *Drosophila* visual system development. In: N. Elsner and R. Wehner (Eds.), Göttingen Neurobiology report, Vol. II, Thieme, Stuttgart, 727.

Hadler, N.M. (1964) Genetic influence on phototaxis in *Drosophila melanogaster*. *Biol. Bull.*, 126: 264–273.

Hafen, E., Basler, K., Edstroem, J.-E. and Rubin, G.M. (1987) Sevenless, a cell-specific homeotic gene of *Drosophila*, encodes a putative transmembrane receptor with a tyrosine kinase domain. *Science*, 236: 55–63.

Hall, J.C. (1994) Pleiotropy of behavioural genes. In: Flexibility and Constraint in Behavioural Systems, R.J. Greenspan and C.P. Kyriacou (Eds.), John Wiley and Sons Ltd, pp. 15–27.

Hall, J.C. and Greenspan, R.J. (1979) Genetic analysis of *Drosophila* neurobiology. *Ann. Rev. Genet.*, 13: 127–195.

Han, P.-L., Meller, V. and Davis, R.L. (1996) The *Drosophila* brain revisited by enhancer detection. *J. Neurobiol.*, 31: 88–102.

Hardie, R.C. (1977) Electrophysiological properties of R7 and R8 in Dipteran Retina. *Z. Naturforsch.*, 32c: 887–889.

Hardie, R.C. (1983) Functional organization of the fly retina. In: D. Ottoson (Ed.), *Progress in sensory physiology*, Vol. 5, Springer, Berlin, pp. 1–79.

Harris, W.A., Stark, W.S. and Walker, J.A. (1976) Genetic dissection of the photoreceptor system in the compound eye of *Drosophila melanogaster*. *J. Physiol.*, 256: 415–439.

Hausen, K. (1984) The lobula complex of the fly: structure, function and significance in visual behaviour. In: M.A. Ali (Ed.), *Photoreception and vision in invertebrates*, Plenum Press, New York and London.

Hausen, K. and Egelhaaf, M. (1989) Neural mechanisms of visual course control in insects. In: D.G. Stavenga and R.C. Hardie (Eds.), *Facets of visions*, Springer Verlag, Berlin, pp. 391–424.

Hausen, K. and Wehrhahn, C. (1989) Neural circuits mediating visual flight control in flies. I. Quantitative comparison of neural and behavioral response characteristics. *J. Neurosci.*, 9: 3828–3836.

Hausen, K. and Wehrhahn, C. (1990) Neural circuits mediating visual flight control in flies. II. Separation of two control systems by microsurgical brain lesions. *J. Neurosci.*, 10: 351–360.

Hayward, D.C., Delaney, S.J., Campbell, H.D., Ghysen, A., Benzer, S., Kasprzak, A.B., Cotsell, J.N., Young, I.G. and Miklos, G.L.G. (1993) The sluggish-A gene of *Drosophila melanogaster* is expressed in the nervous system and encodes proline oxidase, a mitochondrial enzyme involved in glutamate synthesis. *Proc. Natl. Acad. Sci. USA*, 90: 2979–2983.

Heisenberg, M. (1972) Comparative behavioral studies on two visual mutants of *Drosophila*. *J. Comp. Physiol.*, 80: 119–136.

Heisenberg, M. and Böhl, K. (1979) Isolation of anatomical brain mutants of *Drosophila* by histological means. *Z. Naturforsch.*, 34c: 143–147.

Heisenberg, M. and Buchner, E. (1977) The role of retinula cell types in visual behavior of *Drosophila melanogaster*. *J. Comp. Physiol.*, 117: 127–162.

Heisenberg, M. and Götz, K.G. (1975) The use of mutations for the partial degradation of vision in *Drosophila melanogaster*. *J. Comp. Physiol.*, 98: 217–241.

Heisenberg, M. and Wolf, R. (1979) On the fine structure of yaw torque in visual flight orientation of *Drosophila melanogaster*. *J. Comp. Physiol.*, 130: 113–130.

Heisenberg, M. and Wolf, R. (1984) *Vision in* Drosophila. *Genetics of Microbehaviour*. Springer Verlag, Berlin.

Heisenberg, M., Wonneberger, R. and Wolf, R. (1978) optomotor-blind[H31] — a *Drosophila* mutant of the lobula plate giant neurons. J. *Comp. Physiol.*, 124: 287–296.

Helfand, S.L. and Carlson, J.R. (1989) Isolation and characterization of an olfactory mutant in *Drosophila* with a chemically specific defect. *Proc. Natl. Acad. Sci. USA*, 86: 2908–2912.

Helfrich, C. (1986) Role of the optic lobes in the regulation of the locomotor activity rhythm of *Drosophila melanogaster*: behavioral analysis of neural mutants. *J. Neurogenetics*, 3: 321–343.

Helfrich-Förster, C. and Homberg, U. (1993) Pigment-dispersing hormone immuno-reactive neurons in the nervous system of wild-type *Drosophila melanogaster* and of several mutants with altered circadian rhythmicity. *J. Comp. Neurol.*, 337: 177—190.

Hengstenberg, R. (1988) Mechanosensory control of compensatory head role during flight in the blowfly *Calliphora erythrocephala* Meig. *J. Comp. Physiol.*, 163: 151–165.

Hengstenberg, R. (1991) Gaze control in the blowfly *Calliphora*: a multisensory, two-stage integration process. *Seminars in the Neurosciences*, 3: 19–29.

Hirth, F., Therianos, S., Loop, T., Gehring, W., Reichert, H. and Furukubo-Tokunaga, K. (1995) Developmental defects in brain segmentation caused by mutations of homeobox genes orthodenticle and empty spiracles in *Drosophila. Neuron*, 15: 769–778.

Hofbauer, A. and Buchner, E. (1989) Does *Drosophila* have seven eyes? *Naturwissenschaften*, 76: 335–336.

Hofbauer, A. and Campos-Ortega, J.A. (1990) Proliferation pattern and early differentiation of the optic lobes in *Drosophila melanogaster. Roux's Arch. Dev. Biol.*, 198: 264–274.

Hofmeyer, K. and Pflugfelder, G.O. (1998) Reverse-genetic dissection of the optic lobe regulatory region of the *Drosophila* optomotor-blind gene. In: N. Elsner and R. Wehner (Eds.), *Göttingen Neurobiology report*, Vol. II, Thieme, Stuttgart, 726.

Horridge, G.A. (1965) Arthropoda: receptors for light, and optic lobe. In: T.H. Bullock and G.A. Horridge (Eds.), *Structure and function in the nervous systems of invertebrates*, Vol. II, Freeman, W.H., San Francisco and London.

Hotta, Y. and Benzer, S. (1972) Mapping of behavior in *Drosophila* mosaics. *Nature*, 240: 527–535.

Hu, K.G. and Stark, W.S. (1977) Specific receptor input into spectral preference in *Drosophila. J. Comp. Physiol.*, 121: 241–252.

Hu, K.G. and Stark, W.S. (1980) The roles of *Drosophila* ocelli and compound eyes in phototaxis. *J. Comp. Physiol.*, 135: 85–95.

Huang, Z. and Kunes, S. (1996) Hedgehog, transmitted along retinal axons triggers neurogenesis in the developing centers of the *Drosophila* brain. *Cell*, 86: 411–422.

Huang, Z., Shilo, B.-Z. and Kunes, S. (1998) A retinal axon fascicle uses Spitz, an EGF receptor ligand, to construct a synaptic cartridge in the brain of *Drosophila. Cell*, 95: 693–703.

Huber, A., Schulz, S., Bentrop, J., Groell, C., Wolfrum, U. and Paulsen, R. (1997) Molecular cloning of *Drosophila* Rh6 rhodopsin: the visual pigment of a subset of R8 photoreceptor cells. *FEBS Lett.*, 406: 6–10.

Ito, K., Awano, W., Suzuki, K., Hiromi, Y. and Yamamoto, D. (1997a) The *Drosophila* mushroom body is a quadruple structure of clonal units each of which contains a virtually identical set of neurones and glial cells. *Development*, 124: 761–771.

Ito, K., Sass, H., Urban, J., Hofbauer, A. and Schneuwly, S. (1997b) Gal4-responsive UAS-tau as a tool for studying the anatomy and development of the *Drosophila* central nervous system. *Cell Tiss. Res.*, 290: 1–10.

Jackson, S.M., Nakato, H., Sugiura, M., Jannuzi, A., Oakes, R., Kaluza, V., Golden, C. and Selleck, S.B. (1997) dally, a *Drosophila* glypican, controls cellular responses to the TGF-beta-related morphogen, DPP. *Development*, 124: 4113–4120.

Jacob, K.G., Willmund, R., Folkers, E., Fischbach, K.-F. and Spatz, H.C. (1977) T-maze phototaxis of *Drosophila melanogaster* and several mutants in the visual systems. *J. Comp. Physiol.*, 116: 209–225.

Jones, K.R. and Rubin, G.M. (1990) Molecular analysis of *no-on-transient A*, a gene required for normal vision in *Drosophila. Neuron*, 4: 711–723.

Kaiser, K. and Goodwin, S.F. (1990) "Site-selected" transposon mutagenesis in *Drosophila. Proc. Natl. Acad. Sci. USA*, 87: 1686–1690.

Kaphingst, K. and Kunes, S. (1994) Pattern formation in the visual centers of the *Drosophila* brain: wingless acts via decapentaplegic to specify the dorsoventral axis. *Cell*, 78: 437–448.

Kirschfeld, K. (1967) Die Projektion der optischen Umwelt auf das Raster der Rhabdomere im Komplexauge von *Musca. Exp. Brain Res.*, 3: 248–270.

Kirschfeld, K. (1972) Optics of the compound eye 1. The visual system of Musca: studies on optics, structure and function. In: R. Wehner (Ed.), *Information processing in the visual system of arthropods*, Springer, pp. 61–74.

Kirschfeld, K., Feiler, R. and Franceschini, N. (1978) A photostable pigment within the rhabdomere of fly photoreceptor no. 7. *J. Comp. Physiol.*, 125: 275–284.

Kirschfeld, K., Feiler, R., Hardie, R.K.V. and Franceschini, N. (1983) The sensitizing pigment in fly photoreceptors. *Biophys. Struct. Mech.*, 10: 81–92.

Kirschfeld, K. and Lutz, B. (1974) Lateral inhibition in the compound eye of the fly, Musca. *Z. Naturforschung*, 29c: 95–97.

Koenig, J. and Merriam, J.R. (1977) Autosomal ERG mutants. *Dros. Inf. Serv.*, 52: 50–51.

Kolodziej, P.A., Jan, L.Y. and Jan, Y.N. (1995) Mutations that affect the length, fasciculation and ventral orientation of specific sensory axons in the *Drosophila* embryo. *Neuron*, 15: 273–286.

Kopp, A. and Duncan, I. (1997) Control of cell fate and polarity in the abdominal segments of *Drosophila* by optomotor-blind. *Development*, 124: 3715–3726.

Krapp, H.G. and Hengstenberg, R. (1996) Estimation of self motion by optic flow processing in single visual interneurons. *Nature*, 384: 463–466.

Kretzschmar, D., Brunner, A., Wiersdorff, V., Pflugfelder, G.O., Heisenberg, M. and Schneuwly, S. (1992) giant lens, a gene involved in cell determination and axon guidance in the visual system of *Drosophila melanogaster*. *EMBO J.*, 11: 2531–2539.

Kumar, J.P., Bowman, J., O'Tousa, J.E. and Ready, D.F. (1997) Rhodopsin replacement rescues photoreceptor structure during a critical developmental window. *Dev. Biol.*, 188: 43.47.

Leng, S. and Strauss, R. (1998) Ethograms of three *Drosophila* mutant strains with structural defects in the protocerebral bridge. In: N. Elsner and R. Wehner (Eds.), *Göttingen Neurobiology Report*, Thieme, Stuttgart, New York, 259.

Lindsley, D.L. and Zimm, G.G. (1992) The genome of *Drosophila melanogaster*, Academic Press, Inc., San Diego.

Liu, E. and Restifo, L.L. (1998) Identification of a Broad Complex-regulated enhancer in the developing visual system of *Drosophila*. *J. Neurobiol.*, 34: 253–270.

Luo, L., Tully, T. and White, K. (1992) Human amyloid precursor protein ameliorates behavioural deficit of flies deleted for Appl gene. *Neuron*, 9: 595–605.

Markow, T.A. and Merriam, J. (1977) Phototactic and geotactic behavior of countercurrent defective mutants of *Drosophila melanogaster*. *Behav. Genet.*, 7: 447–455.

Martin, K.A., Poeck, B., Roth, H., Ebens, A.J., Ballard, L.C. and Zipursky, S.L. (1995) Mutations disrupting neuronal connectivity in the *Drosophila* visual system. *Neuron*, 14: 229–240.

Médioni, J., Campan, R. and Quéinnec, Y. (1971) Sur le role activateur des ocelles fronteaux chez les insects. Tachycardie induite par stimulation ocellaire chez *Calliphora vomitoria*. *Science*, 3: 485–503.

Meinertzhagen, I.A. and Hanson, T.. (1993). The development of the optic lobe. In: A. Martinez Arias and M. Bate (Eds.), *The development of* Drosophila melanogaster, Cold Spring Harbor Laboratory Press, Cold Spring Harbor, 1363–1491.

Meinertzhagen, I.A. and O'Neil, S.D. (1991) Synaptic organization of columnar elements in the lamina of the wild type in *Drosophila melanogaster*. *J. Comp. Neurol.*, 305: 232–263.

Menzel, R. (1979) Spectral sensitivity and color vision in invertebrates. In: H. Autrum (Ed.), Comparative physiology and evolution of vision in invertebrates. A. Invertebrate photoreceptors, Springer, Heidelberg, New York, pp. 504–580.

Meyerowitz, E.M. and Kankel, D.R. (1978) A genetic analysis of visual system development in *Drosophila melanogaster*. *Dev. Biol.*, 62: 112–142.

Miklos, G.L.G. and Rubin, G.M. (1996) The role of the genome project in determining gene function: insights from model organisms. *Cell*, 86: 521–529.

Milligan, S.C., Alb, J.G.J., Elagina, R.B., Bankaitis, V.A. and Hyde, D.R. (1997) The phosphatidylinositol transfer protein domain of *Drosophila* retinal degeneration B protein is essential for photoreceptor cell survival and recovery from light stimulation. *J. Cell Biol.*, 139: 351–363.

Nakato, H., Futch, T.A. and Selleck, S.B. (1995) The division abnormally delayed (dally) gene: a putative integral membrane proteoglycan required for cell division patterning during postembryonic development of the nervous system in *Drosophila*. *Development*, 121: 3687–3702.

Nässel, D.R. (1993) Neuropeptides in the insect brain: a review. *Cell Tiss. Res.*, 273: 1–29.

Nowak, M.A., Boerlijst, M.C., Cooke, J. and Maynard Smith, J. (1997) Evolution of genetic redundancy. *Nature*, 388: 167–171.

Oh, C.E., McMahon, R., Benzer, S. and Tanouye, M.A. (1994) Bendless, a *Drosophila* gene affecting neuronal connectivity, encodes a ubiquitin-conjugating enzyme homolog. *J. Neurosci.*, 14: 3166–3179.

O'Tousa, J.E., Leonard, D.S. and Pak, W.L. (1989) Morphological defects in oraJK84 photoreceptors caused by mutation in R1–6 opsin gene of *Drosophila*. *J. Neurogenet.*, 6: 41–52.

Pak, W.L. (1975) Mutations affecting the vision of *Drosophila melanogaster*. In: R.C. King (Ed.), *Handbook of Genetics*, Vol. 3, Plenum Press, New York, pp. 703–733.

Pak, W.L., Grossfield, J. and White, N.Y. (1969) Nonphototactic mutants in a study of vision of *Drosophila*. *Nature*, 222: 351–354.

Papaioannou, V.E. (1997) T-box family reunion. *TIG*, 13: 212–213.

Perez, S. E. and Steller, H. (1996) Migration of glial cells into retinal axon target field in *Drosophila melanogaster*. *J. Neurobiol.*, 30: 359–373.

Pflugfelder, G.O. (1998) Genetic lesions in *Drosophila* behavioural mutants. *Behav. Brain Res.*, 95: 3–15

Pflugfelder, G.O. and Heisenberg, M. (1995) *optomotor-blind* of *Drosophila melanogaster*: A neurogenetic approach to optic lobe development and optomotor behaviour. *Comp. Biochem. Physiol.*, 110A: 185–202.

Pflugfelder, G.O., Roth, H., Poeck, B., Kerscher, S., Schwarz, H., Jonschker, B. and Heisenberg, M. (1992a) The lethal(1)optomotor-blind gene of *Drosophila melanogaster* is a major organizer of optic lobe development: isolation and charcterization of the gene. *Proc. Natl. Acad. Sci. USA*, 89: 1199–1203.

Pflugfelder, G.O., Roth, H. and Poeck, B. (1992b) A homology domain shared between *Drosophila melanogaster* optomotor-blind and mouse Brachyury is involved in DNA binding. *Biochem. Biophys. Res. Comm.*, 186: 918–925.

Phelps, C.B. and Brand, A.H. (1998) Ectopic gene expression in *Drosophila* using Gal4 systems. *METHODS: A Companion to Methods in Enzymology*, 14: 367–379.

Pierantoni, R. (1976) A look into the cock-pit of the fly. The architecture of the lobular plate. *Cell Tiss. Res.*, 171: 101–122.

Plautz, J.D., Kaneko, M., Hall, J.C. and Kay, S.A. (1997) Independent photoreceptive circadian clocks throughout *Drosophila. Science*, 278: 1632–1635.

Poeck, B., Hofbauer, A. and Pflugfelder, G.O. (1993). Expression of the *Drosophila* optomotor-blind gene transcript in neuronal and glial cells of the developing nervous system. *Development,* 117: 1017–1029.

Raabe, T. (1998) Genetic analysis of sevenless tyrosine kinase signaling in *Drosophila.* In: A.J. Pawson (Ed.), *Protein modules in signal transduction*, Vol. 228, Springer, Berlin, pp. 343–361.

Ranganathan, R., Malicki, D.M. and Zuker, C.S. (1995) Signal transduction in *Drosophila* photoreceptors. *Annu. Rev. Neurosci.*, 18: 283–317.

Reichardt, W. (1957) Autokorrelations-Auswertung als Funktionsprinzip des Zentralnervensystems. *Z. Naturforschung*, 12b: 448–457.

Reichardt, W. (1973) Musterinduzierte Flugorientierung. Verhaltensversuche an der Fliege *Musca domestica. Naturwissenschaften*, 60: 122–138.

Rendahl, K.G., Gaukhshteyn, N., Wheeler, D.A., Fry, T.A. and Hall, J.C. (1996) Defects in courtship and vision caused by amino acid substitutions in a putative RNA-binding protein encoded by the no-on-transient A (nonA) gene of *Drosophila. J. Neurosci.*, 16: 1511–1522.

Rendahl, K.G., Jones, K.R., Kulkarni, S.J., Bagully, S.H. and Hall, J.C. (1992) The dissonance mutation at the no-on-transient A locus of *D. melanogaster*: a genetic control of courtship song and visual behaviors by a protein with putative RNA-binding motifs. *J. Neurosci.*, 12: 390–407.

Restifo, L.L. and White, K. (1991) Mutations in a steroid hormone-regulated gene dirupt the metamorphosis of the central nervous system. *Dev. Biol.*, 148: 174–194.

Riesgo-Escovar, J., Raha, D. and Carlson, J.R. (1995) Requirement for a phospholipase C in odor response: overlap between olfaction and vision in *Drosophila. Proc. Natl. Acad. Sci. USA*, 92: 2864–2868.

Rockwell, R.F. and Seiger, M.B. (1973) Phototaxis in *Drosophila*: a critical evaluation. *American Scientist*, 61: 339–345.

Rorth, P., Szabo, K., Bailey, A., Laverty, T., Rehm, J., Rubin, G.M., Weigmann, K., Milan, M., Benes, V., Ansorge, W. and Cohen, S.M. (1998) Systematic gain-of-function genetics in *Drosophila. Development,* 125: 1048–1057.

Roseman, R.R., Johnson, E.A., Rodesch, C.K., Bjerke, M., Nagoshi, R.N. and Geyer, P.G. (1995) A *P* element containing suppressor of Hairy-wing binding regions has novel properties for mutagenesis in *Drosophila melanogaster. Genetics,* 141: 1061–1074.

Saint Marie, S.L. and Carlson, S.D. (1983) The fine structure of neuroglia in the lamina ganglionaris of the housefly, *Musca domestica L. J. Neurocytol.*, 12: 213–241.

Salz, H.K., Davis, R.L. and Kiger Jr, J.A., (1982) Genetic analysis of chromomere 3D4 in *Drosophila melanogaster*: the dunce and sperm-amotile genes. *Genetics,* 100: 587–596.

Schlessinger, J., Lax, I. and Lemmon, M. (1995) Regulation of growth factor activation by proteoglycans: what is the role of the low affinity receptors? *Cell*, 83: 357–360.

Schmidt-Ott, U., González-Gaitán, M. and Technau, G. (1995) Analysis of neural elements in head-mutant *Drosophila* embryos suggests segmental origin of the optic lobes. *Roux's Arch. Dev. Biol.*, 205: 31–44.

Schneider, T., Reiter, C., Eule, E., Bader, B., Lichte, B., Nie, Z., Schimansky, T., Ramos, R.G.P. and Fischbach, K.-F. (1995) Restricted expression of the irreC-rst protein is required for normal axonal projections of columnar visual neurons. *Neuron*, 15: 259–271.

Schubiger, M., Feng, Y., Fambrough, D.M. and Palka, J. (1994) A mutation of the *Drosophila* sodium pump a subunit gene results in bang sensitive paralysis. *Neuron*, 12: 373–381.

Selleck, S.B. and Steller, H. (1991) The influence of retinal innervation on neurogenesis in the first optic ganglion of *Drosophila. Neuron*, 6: 83–99.

Selleck, S.E., Gonzalez, C., Glover, D.M. and White, K. (1992) Regulation of the G1-S transition in postembryonic neuronal precursors by axon ingrowth. *Nature*, 355: 253–255.

Serikaku, M.A. and O'Tousa, J.E. (1994) sine oculis is a homeobox gene required for *Drosophila* visual system development. *Genetics,* 138: 1137–1150.

Siebeck, O. (1977) Lichtorientierungen. In: F. Gessner (Ed.), *Das Tier und seine Umwelt*, Vol. V, Akademische Verlagsgesellschaft Athenaion, Wiesbaden, 173–178.

Smith, D., Wohlgemuth, J., Calvi, B.R., Franklin, I. and Gelbart, W.M. (1993) hobo enhancer trapping mutagenesis in *Drosophila* reveals an insertion sepcificity different from P elements. *Genetics,* 135: 1063–1076.

Spradling, A.C., Stern, D.M., Kiss, I., Roote, J., Laverty, T. and Rubin, G.M. (1995) Gene disruptions using P transposable elements: an integral component of the *Drosophila* genome project. *Proc. Natl. Acad. Sci. USA*, 92: 10824–10830.

Stanewsky, R., Rendahl, K.G., Dill, M. and Saumweber, H. (1993) Genetic and molecular analysis of the X-chromosomal region 14B17-14C4 in *Drosophila melanogaster* : loss of function in NONA, a nuclear protein common to many cell types, results in specific physiological and behavioral defects. *Genetics,* 135: 419–442.

Stark, W.S., Sapp, R. and Carlson, S.D. (1989) Ultrastructure of the ocellar visual system in normal and mutant *Drosophila melanogaster. J. Neurogenet.*, 5: 127–153.

Strausfeld, N. (1984) Functional neuroanatomy of the blowfly's visual system. In: M. A. Ali (Ed.), *Photoreception and vision in invertebrates*, Plemum Press, New York and London, pp. 483–522.

Strausfeld, N.J. (1979) The representation of a receptor map within a retinotopic neuropil of the fly. *Verh. Dtsch. Zool. Ges.*, 1979, 167–179.

Strausfeld, N.J. (1989) Beneath the compound eye: neuroanatomical analysis and physiological correlates in the study of insect vision. In: D.G. Stavenga and R.C. Hardie (Eds.), *Facets of vision*, Springer, Berlin.

Strausfeld, N.J. and Campos-Ortega, J.A. (1972) Some interrelationships between the first and second synaptic region of the fly's (*Musca domestic L.*) visual system. In: R. Wehner (Ed.), *Information processing in the visual system of arthropods*, Springer, Berlin, pp. 23–30.

Strausfeld, N.J. and Campos-Ortega, J.A. (1977) Vision in insects: pathways possibly underlying neural adaptation and lateral inhibition. *Science*, 195: 894–897.

Strausfeld, N. J. and Lee, J.-K. (1991) Neuronal basis for parallel visual processing in the fly. *Vis. Neurosci.*, 7: 13–33.

Strauss, R. and Heisenberg, M. (1993) A higher control center of locomotor behavior in the *Drosophila* brain. *J. Neurosci.*, 13: 1852–1861.

Strauss, R., Schuster, S. and Götz, K.G. (1997) Processing of artificial visual feedback in the walking fruit fly *Drosophila melanogaster*. *J. Exp. Biol.*, 200: 1281–1296.

Sweeney, S.T., Broadie, K., Keane, J., Niemann, H. and O'Kane, C.J. (1995) Targeted expression of tetanus toxin light chain in *Drosophila* specifically eliminates synaptic transmission and causes behavioural defects. *Neuron*, 14: 341–351.

Thaker, H.M. and Kankel, D.R. (1992) Mosaic analysis gives an estimate of genomic involvement in the development of the visual system in *Drosophila melanogaster*. *Genetics*, 131: 883–894.

Vandendries, E.R., Johnson, D. and Reinke, R. (1996) orthodenticle is required for photoreceptor cell development in the *Drosophila* eye. *Dev. Biol.*, 173: 243–255.

Venter, C., Adams, M.D., Sutton, G.G., A.R., K., Smith, H.O. and Hunkapiller, M. (1998) Shotgun sequencing of the human genome. *Science*, 280: 1540–1542.

Vihtelic, T.S., Goebl, M., Milligan, S., O'Tousa, J.E. and Hyde, D.R. (1993) Localization of *Drosophila* retinal degeneration B, a membrane-associated phosphatidylinositol transfer protein. *J. Cell Biol.*, 122: 1013–1022.

Von Besser, H., Schnabel, P., Wieland, C., Fritz, E., Stanewsky, R. and Saumweber, H. (1990) The puff-specific *Drosophila* protein Bj6, encoded by the gene no-on-transient A, shows homology to RNA-binding proteins. *Chromosoma*, 100: 37–47.

Winberg, M.L., Mitchell, K.J. and Goodman, C.S. (1998) Genetic analysis of the mechanisms controlling target selection: complementary and combinatorial functions of netrins, semaphorins, and IgCAMs. *Cell*, 93: 581–591.

Winberg, M.L., Perez, S. and Steller, H. (1992) Generation and early differentiation of glial cells in the first optic ganglion of *Drosophila melanogaster*. *Development*, 115: 903–911.

Wolf, R. and Heisenberg, M. (1986) Visual orientation in motion-blind flies is an operant behavior. *Nature*, 323: 154–156.

Yamamoto, D. (1996) *Molecular Dynamics in the Developing* Drosophila *Eye*. R.G. Landes Company/Springer, Austin, Texas, USA.

Yang, Z., Emerson, M., Su, H.S. and Sehgal, A. (1998) Response of the Timeless protein to light correlates with behavioral entrainment and suggests a nonvisual pathway for circadian photoreception. *Neuron*, 21: 215–223.

Younossi-Hartenstein, A., Green, P., Liaw, G.-J., Rudolph, K.J.L. and Hartenstein, V. (1997) Control of early neurogenesis of the *Drosophila* brain by the head gap genes tll, otd, ems, btd. *Dev. Biol.*, 182: 270–283.

Zhou, L., Schnitzler, A., Agapite, J., Schwartz, L.M., Steller, H. and Nambu, J.R. (1997) Cooperative functions of the reaper and head involution defective genes in the programmed cell death of *Drosophila* central nervous system midline cells. *Proc. Natl. Acad. Sci. USA*, 94: 5131–5136.

W.E. Crusio and R.T. Gerlai (Eds.)
Handbook of Molecular-Genetic Techniques for Brain and Behavior Research (Techniques in the Behavioral and Neural Sciences, Vol. 13)

CHAPTER 3.3.4

Testing associative learning in *Drosophila*

Thomas Préat

Institut Alfred Fessard, CNRS, 1 Avenue de la Terrasse, 91190 Gif-sur-Yvette, France

Introduction

Of the few organisms that can be readily studied by molecular genetics, *Drosophila melanogaster* represents a powerful choice for the study of learning and memory. The *Drosophila* brain is composed of only approximately 200,000 neurons. Furthermore, thanks to the enhancer-trap approach (O'Kane and Gehring, 1987; Brand and Perrimon, 1993), hundreds of genetic cellular markers are available to track neuronal pathways (Yang et al., 1995; Han et al., 1996; Ito et al., 1997; Simon et al., 1998). Secondly, *Drosophila* can learn under a wide range of conditions. Thus, although the *C. elegans* nervous system can be better studied at the level of individual cells, *Drosophila* can show learned behaviors that are highly sophisticated and have long lasting effects (Tully et al., 1994; Xia et al., 1997). Finally, the power of the molecular genetic tools available in *Drosophila*, combined with the fact that a large number of gene functions have been at least partially conserved through evolution, make *Drosophila* a strong model for the study of basic brain function. On the negative side, direct electrophysiological studies of the adult fly brain are difficult to perform, due to small size of cell bodies and to the organization of the brain: neuron cell bodies are located at the periphery of the brain, whereas synapses are found centrally in the neuropile.

The present chapter is neither an extensive description of the conditioning protocols that have been developed for *Drosophila melanogaster*, nor a global analysis of mutants with defective learning or memory. Readers seeking such information should refer for example to the review of Davis (1996). I will give examples of stimulating achievements in the field and then emphasize some important practical and theoretical factors which must be considered in studying associative learning in *Drosophila*. I will also analyze why, despite important findings, this field has not yet blossomed as have the molecular genetic studies of embryonic development, but how existing limitations could be overcome.

Testing learning and memory in "normal" *Drosophila*

Examples of associative conditioning protocols

In associative conditioning a conditioned stimulus (CS) is presented paired temporally with a positive or negative salient unconditioned stimulus (US). The reaction of trained animals to the presentation of the CS will differ from that of naive animals. If the association of the CS and the US during training occurs independently of animal's behavior, the conditioning is called classical (or Pavlovian).

If the animal's behavior triggers US presentation during training, the conditioning is called operant (or instrumental). In this latter case the stimulus preceding the US is called a cue rather than a CS.

The association of an odor with electric shocks (Tully and Quinn 1985) is a widely used classical conditioning protocol (Fig. 1(A)). In the regular procedure the CS^+ odor is presented for 1 min in association with twelve pulses of shocks, followed by presentation of an unshocked odor for 1 min. During the test the flies must choose between two tubes, each filled with one of the odors but in the absence of shock. This procedure yields an initial performance index (PI) of about 0.85. The PI is calculated following conditioning and testing and allows the internal status of animal to be evaluated. The PI normally ranges from 0 (no learning or memory displayed) to +1 (perfect learning or memory). In mass tests, where the behavior of a group of genotypically identical flies is analyzed (as in the olfactory protocol described above) the PI corresponds to the number of flies having made the correct choice at the end of the test period (this number is normalized so that a 50–50 distribution at the end of the test corresponds to a PI of 0). In protocols where single flies are conditioned, the PI represents the percentage of time spent making the right choice during the full sequence of the test.

The advantage of the Pavlovian olfactory protocol is that it is efficient and relatively easy to set up. Its disadvantage is that the behavior of individual animals cannot be followed in detail during either training or testing. This olfactory assay has been used to isolate learning and memory mutants (Boynton and Tully, 1992; Dura et al., 1993), together with the original olfactory operant protocol in which a light source was used to attract the freely moving flies into the grid tube (Quinn et al., 1974; Dudai et al., 1976; Quinn et al., 1979; Dudai et al., 1984).

In classical olfactory conditioning the PI is high immediately after training but declines to 0.20 after 24 h (Tully and Quinn, 1985), and after a few days no memory at all can be detected. If

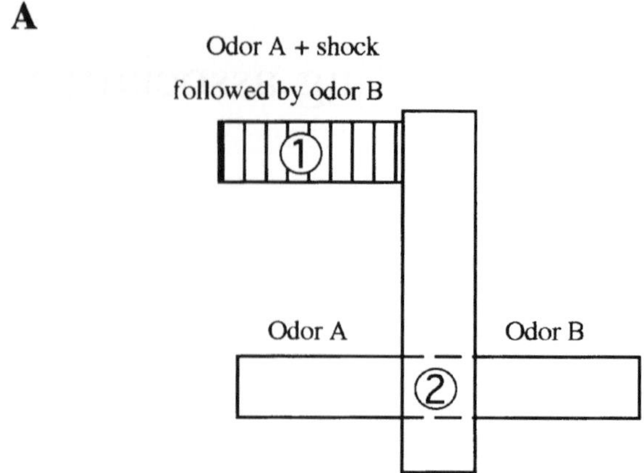

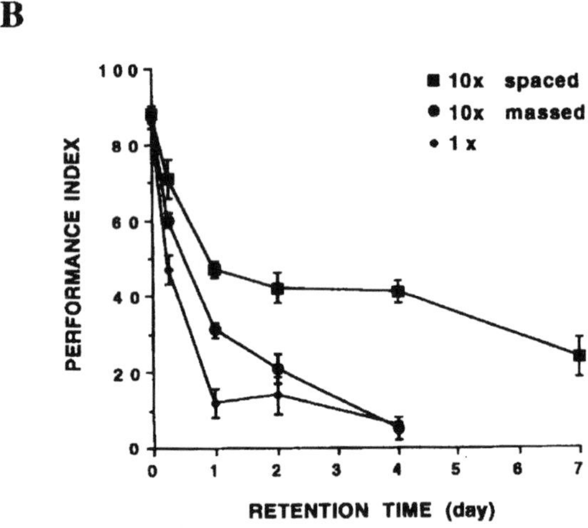

Fig. 1. (A) Schematic representation of the olfactory conditioning assay developed by Tully and Quinn (1985). During training (position 1), groups of 50–100 flies are first exposed for 60 s to a first odor (either undiluted 3-octanol or 4-methylcyclohexanol), during which time they receive electric shocks (1.5 s pulses of 60 V DC). After a 45 s rest period, flies are exposed for 60 s to the second odor, which is not paired with electric shock. For memory testing flies are kept in a vial with regular solid food, and then transported to the choice point of a T-maze (position 2), allowed to choose between the two odors for 120 s, and counted. (B) Retention curves of wild-type Canton-S flies following one cycle of training (small circles), 10 cycles of massed training (large circles) or 10 cycles of spaced training (squares), in which a 15 min rest interval occurred between each cycle (reprinted with permission from Tully et al., 1994).

the same group of flies is submitted to intensive training consisting of 10 regular sessions with 15 min rest intervals between each session, the 24 h PI increases to 0.45 and stable memory can be detected for at least a week (Tully et al., 1994) (Fig.1(B)).

The association of visual landmarks with heat (Wolf and Heisenberg, 1991) is a well studied operant conditioning protocol (Fig. 2(A)). In this assay a single fly is tethered to a torque meter. The torque meter measures the yaw torque imposed by the flying animal. A computer then feeds back movement to the arena surrounding the fly. If the arena carries landmarks, the animal can have the visual impression that it is turning (although it would not perceive the wind pressure changes normally associated with turning). In the normal conditioning procedure (Wolf and Heisenberg, 1991), the arena is divided in four quadrants, with identical geometrical pattern in opposite quadrants. The fly is punished with heat if it orientates toward one of the two patterns. Animals learn very quickly to avoid the heat. After conditioning, the heat is turned off and a PI of 0.35 can be measured. If repetitive training is performed, long term memory is induced (Xia et al., 1997) (Fig. 2(B)). The advantage of this protocol is that the behavior of a single fly can be analyzed in great detail during both training and testing. However, this system is probably too complex to allow the screening of mutants.

One of the few natural situations described so far in *Drosophila* which involves associative learning is courtship suppression. Females tend to mate only once. A male will actively court a virgin female, but this courtship is decreased by pre-exposure of the male to a mated female (Siegel and Hall, 1979). This is due to the aversive chemicals produced by the mated female which become associated in the male brain with the presence of a female. This situation is interesting but the parameters involved — intensity of the aversive the stimulus, etc. — cannot be controlled as thoroughly as with "artificial" conditioning protocols.

The level of learning and retention varies greatly among protocols

From a technical point of view, the degree of learning in a given protocol is an important parameter if one wants to isolate mutants: if normal flies already show poor performance it leaves less room to identify mutants with low PIs, given that variability does not necessarily decrease with performance. Also, stronger learning generally implies that memory traces will be detected for a longer time, thus allowing consolidated memory phases to be studied. However, it may be biologically important that the animal retains some behavioral diversity. Thus even after inhibitory conditioning with a strongly aversive stimulus, animals can make "errors". Another limiting factor to the level of learning is that species are likely to learn less efficiently when the modalities used for the test are not those that provide them with information from their natural environment. Thus while it is useful for scientists to work with protocols yielding high scores, it may be that in some situations learning remains low for fundamental reasons. A comparison of performance indices in discriminative learning assays is shown in Table 1. The following observations can be made:

1. *Drosophila* can learn many kinds of associations. In some cases the performance indices are quite low. This might reflect a fundamental difficulty for the animal to learn or it can reflect a limitation of the protocol itself. Thus the association of odor with electric shock yields PI of 0.34 in the operant protocol, possibly because flies must inhibit their attraction toward light during the test in order to avoid the odor previously associated with shock.
2. The inhibitory olfactory protocol induces the strongest learning. This protocol is one of the few which has been used in many laboratories (Tully and Quinn, 1985; Cowan and Siegel, 1986; Asztalos et al., 1993; Qiu and Davis, 1993; De Belle and Heisenberg, 1994; Préat 1998). Although

A

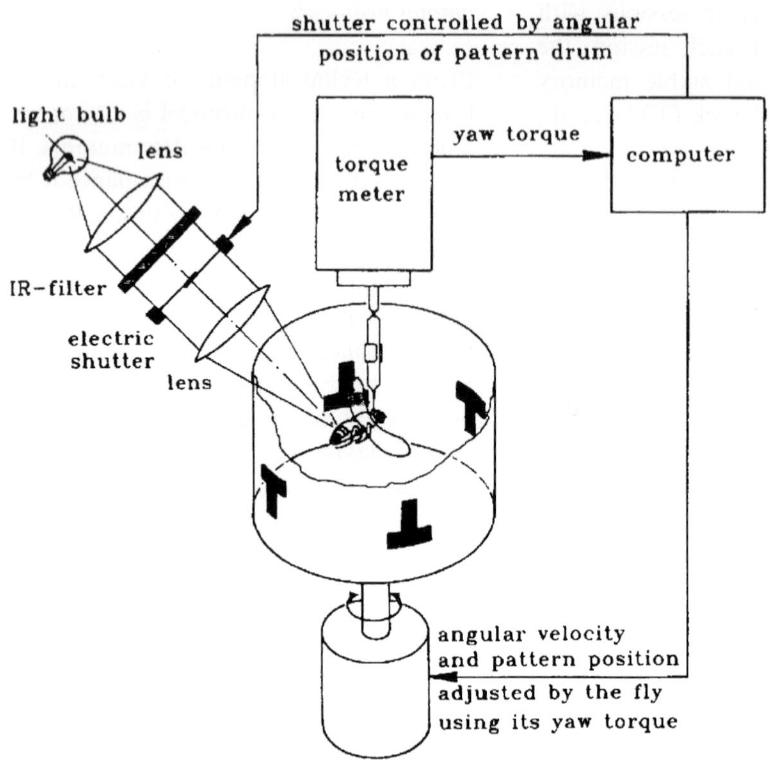

B

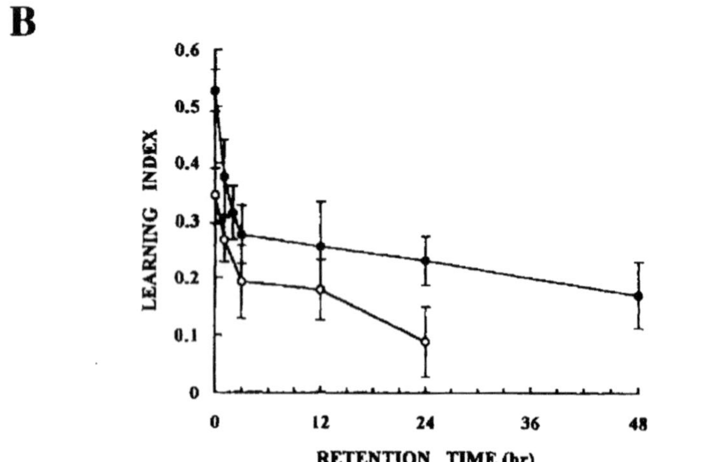

Fig. 2. (A) Diagram of flight stimulator (reprinted with permission from Wolf and Heisenberg, 1991). A single fly, attached to a torque meter, is flying stationary in the center of the arena. Yaw torque is transduced into DC voltage by the torque meter. The arena is rotated by the computer to simulate free flight. An infrared beam heats the fly when it flies toward one of the two patterns. Yaw torque and flight direction are recorded continuously. (B) Retention curves of wild-type Berlin flies after massed training (open circles) and spaced training (closed circles). (See Xia et al., 1997 for procedure details; reprinted with permission from Xia et al., 1997.)

TABLE 1

Discriminative associative learning in adult *Drosophila*

CS or Cue	US	Initial PI	Half life	Condition	Reference
odor	electric shock	0.34	1 hour	operant	Quinn et al., 1974
odor	electric shock	0.85	5 hours	classical	Tully and Quinn, 1985
odor	electric shock	0.85	24 hours	classical (spaced training)	Tully et al., 1994
odor	sucrose	0.36	about 24 hours	classical (starved animals)	Tempel et al., 1983
odor	sucrose	0.30	ND	classical (starved animals)	Heisenberg et al., 1985
colored light	quinine	0.09	ND	operant	Quinn et al., 1974
colored light	electric shock	0.10	ND	operant	Spatz et al., 1974
colored light	shaking	0.28	4 hours	classical	Folkers and Spatz, 1981
colored light	heat	0.55	few minutes	operant	Wolf and Heisenberg, 1997
visual landmark	heat	0.35	12 hours	operant	Wolf and Heisenberg, 1991
visual landmark	heat	0.53	24 hours	operant (spaced training)	Xia et al., 1997
tactile and idiothetic cues	heat	0.45	few minutes	operant	Wustmann and Heisenberg, 1997

(CS: conditioned stimulus; US: unconditioned stimulus; ND: not determined)

this protocol is relatively easy to set up, obtaining consistently high scores requires several parameters to be controlled (see next paragraph).
3. As previously mentioned, long term memory is induced after spaced training with both the olfactory and the visual assay. These forms of consolidated memory involve protein synthesis following conditioning (Tully et al., 1994; Xia et al., 1997).
4. The learning level induced by the excitatory conditioning protocol is low (0.36), but memory is very stable (Tempel et al., 1983). Thus the PI at 24 h is still 50% of the initial value.

Parameters influencing score variability

Another important aspect to be considered in studying learned behaviors in *Drosophila* is the stability of the protocol (or lack thereof). The screening and careful analysis of learning and memory mutants is time-consuming, and it is essential to work consistently. This situation is certainly not specific to *Drosophila*, but unfortunately many parameters — external as well as internal to the animals — can affect performance. This variability is one of the reasons why so few protocols have been used successfully by several laboratories. Exporting a *Drosophila* conditioning assay to another laboratory generally involves months of (human) training.

Some of the variability-inducing parameters have been identified, and they must be strictly controlled. For example, the following factors are known to affect learning:

1. Nature of the flies: the genetic background can have an enormous effect on performance. In the olfactory Pavlovian assay, the initial PI of the wild-type strains Berlin and Oregon-R are 40% lower compared to that of the wild-type stock Canton-S (Tully and Quinn, 1985). In a sucrose reward assay, Canton-S PI is 60% lower compared to that of the wild-type Berlin stock (Heisenberg et al., 1985). As a result, the search for single gene learning and memory mutants, as well as the behavioral characterization of these mutants, must be performed in a controlled background. Five generations of "outcrossing" (background replacement) are usually sufficient to place a mutation in a Canton-S background

(Dura et al., 1993). Note that different wild-type reference stocks are used for different protocols.

The age of flies tested can also be important. For example, 24–58-h-old flies show no learning in the operant visual protocol, whereas 72–108-h-old flies show PIs of around 0.3 (Guo et al., 1996). Such a strong age dependence in young flies is not observed with the olfactory conditioning protocol.

2. Preparation of flies: the general health of the flies used for conditioning is critical. Flies reared in severely infected bottles (bacteria, fungi) will perform poorly. Similarly flies raised on poor medium will not learn well. Interestingly, in the visual assay, the effect of "starvation" on poor medium can be detected for several generations after transfer to rich medium (Guo et al., 1996). This influence of food means that larval density must be strictly controlled: starvation will occur in overcrowded bottles. Also, larval density was shown to influence the development of mushroom bodies (MBs) (Heisenberg et al., 1995; Hitier et al., 1998), an Insect structure involved in learning and memory (see below).

I have observed that raising flies in a room controlled not only for temperature but also for humidity leads to more stability in the olfactory assay (we use a room with 25°C–60% humidity to rear flies).

3. Conditioning parameters: small variations in the experimental set up can have major effects, so conditions should be kept constant. Some parameters are relatively easy to measure. For example humidity influences olfactory inhibitory conditioning (possibly because it affects the way flies sense electric shock). In an uncontrolled environment PIs of 0.55 were obtained (Qiu and Davis, 1993) whereas PIs of 0.75–0.85 were observed in 70–80% humidity conditioning rooms (Dura et al., 1993; Skoulakis and Davis, 1996). Some other parameters are unfortunately more difficult to control. For example in the olfactory assay "blank" flies must be run in new conditioning tubes before they are fully operational (animals leave excretion spots that apparently have a positive effect). However at some point — which is best determined through experience — these tubes should be changed because they are too dirty.

In the case of the visual assay it has been shown that allowing the fly to manipulate the apparatus in advance of the training session will improve subsequent learning (Guo et al., 1996).

A consequence of these important sources of variability is that experiments should be prepared with great care (for practical informations see Connolly and Tully, 1998). They should be performed several times before their results are considered valid, even if the first round of experiments yielded statistically significant results.

Defining and using learning and memory mutants

Drosophila can learn in various conditions, showing that its behavioral repertoire is highly plastic. Furthermore, despite its rather short life span and high fecundity — which would tend to suggest that very stable information storage mechanisms play a negligible role in species survival — long term memory can be induced in *Drosophila* which can last for days. The reason for this plasticity in the wild remains unknown — although learning is clearly involved in sexual behaviour as previously mentioned (Siegel and Hall, 1979). Nevertheless, this set of laboratory protocols allow us to use *Drosophila* as a model system to tackle the relation between the genome and learning and memory.

It is sometimes argued that specific "learning genes" — i.e. genes encoding products whose sole role in the nervous system is learning — might not exist. Even in such a pessimistic scenario "clean" learning mutants can nevertheless be isolated. For example a particular protein isoform might be recruited for learning in a given structure so a mutation that only affects expression in this subdomain would be learning-specific. And even in cases where learning mutants show additional nervous system phenotypes, they can prove very useful for revealing biochemical or neuronal pathways activated during conditioning. In any case once a mutant is isolated which performs poorly in a conditioning assay it is essential to prove first that the learning defect does not derive from developmental or sensory anomalies.

A biochemical pathway, the cAMP pathway, and a brain structure, the mushroom bodies, play a central role in *Drosophila* learning and memory. The MBs are a symmetrical structure which in *Drosophila* are composed of about 5000 parallel neurons. Anatomical studies linked with behavioral studies provided the first evidence that *Drosophila* MBs are involved in olfactory learning and memory. Thus, mutants with abnormal MBs show poor olfactory learning (Heisenberg et al., 1985). More recently, chemical ablation of the MBs was shown to block olfactory learning completely (De Belle and Heisenberg, 1994).

At the biochemical level, the first two learning mutants isolated in *Drosophila*, *dunce* (*dnc*) and *rutabaga* (*rut*), are deficient in a phosphodiesterase (PDE) (Byers et al., 1981; Davis and Kiger 1981; Kauvar 1982; Chen et al., 1986) and an adenylyl cyclase (AC) (Livingstone et al., 1984; Levin et al., 1992) respectively. Both of these enzymes are involved in cAMP metabolism. Furthermore, these two enzymes are expressed preferentially (but not exclusively, see below) in the MBs (Nighorn et al., 1991; Han et al., 1992). A third protein involved in the response to cAMP, protein kinase A (PKA), is also strongly expressed in MBs (Skoulakis et al., 1993), and inhibition or partial lack of this enzyme reduces learning (Drain et al., 1991; Skoulakis et al., 1993). Current models suggest that olfactory conditioning result in a cAMP increase in MB neurons, leading to the activation of the cAMP dependent PKA. This kinase may regulate a K+ channel, leading to short term memory, and also be involved in gene regulation through the phosphorylation of transcription factors, thus leading to long term memory. Several other results have implicated MBs in olfactory learning:

1. Lack of the Leonardo 14-3-3 protein (Skoulakis and Davis, 1996) and of the Volado integrin (Grotewiel et al., 1998), two proteins preferentially accumulated in the MBs, result in a learning defect.

2. Activation of a constitutively active form of a Gs-protein in the MBs prevents learning (Connolly et al., 1995).

3. The size of MBs varies with experience in the adult, with richer living conditions resulting in bigger calices (Technau, 1984), and *dnc* and *rut* show an abnormal plasticity (Balling et al., 1987).

Characterization of learning and memory mutants: two-step sensory controls

In order to learn animals need to perceive the conditioning stimuli correctly. When a putative learning or memory mutant is isolated it is therefore essential to ensure that the mutation does not affect the ability to sense and react to stimuli. This sounds trivial, but these sensory controls can prove complex because on the one hand they must be performed in a situation as close as possible to the conditioning protocol, but on the other hand one has to make sure that no learning is measured in these controls.

As mentioned, the conditioning protocols used to isolate and characterize most *Drosophila* learning and memory mutants associate an odor with electric shock (Quinn et al., 1974; Tully and Quinn, 1985). It was originally shown that untrained mutants reacted normally to the stimuli used for conditioning, and in particular that their odor avoidance was normal (Dudai et al., 1976; Quinn et al., 1979; Dudai et al., 1984; Dura et al., 1993; Skoulakis and Davis, 1996; Grotewiel et al., 1998). Following conditioning, however, the flies' response to the odors is measured after they have received a strong shock. Therefore it is more appropriate to control mutants' odor avoidance after exposure to an electric shock. In the case of *latheo*, odor avoidance was tested after presentation of electric shock and was shown to be normal (Boynton and Tully, 1992). However in these experiments flies were not naive to the odor since they had been exposed to it after shock stim-

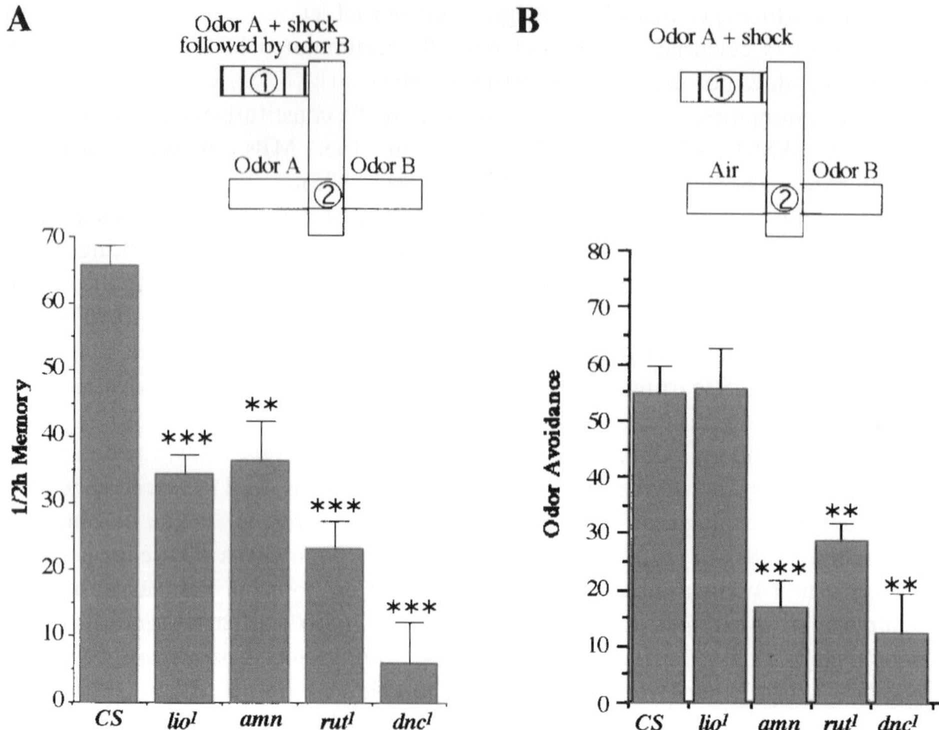

Fig. 3. (A) Half hour memory of normal (Canton-S) and mutant stocks after classical olfactory conditioning. (B) Odor avoidance of normal and mutant stocks after electric shock. A first repellent odor (octanol) is presented to the flies in association with electric shock. The flies are then brought to the choice point, where their reactivity to a second repellent odor (methylcyclohexanol) is measured (reprinted from Préat, 1998).

ulation. This situation is not ideal because pre-exposure to odor can lead to habituation and thus to a decreased response at the second presentation. If a mutant habituates slower than normal flies, its avoidance of the odor at the second presentation will tend to be higher than that of normal flies. In such a scenario an abnormal odor avoidance after shock might be masked in the mutant by a decreased habituation, thus leading to an overall normal score.

A more relevant control involves presentation of electric shocks with the first odor, while odor avoidance to a second odor is measured. In such test the mutants *amnesiac*, *dnc*, and *rut* displayed strongly reduced avoidance of the second odor in comparison with normal flies (Préat, 1998) (Fig. 3). This was in contrast to *lio* mutants which behaved in the same way as wild-type flies.

In the case of *dnc*, memory defects could be differentiated from unrelated deficits. The Dnc phosphodiesterase is preferentially expressed in the mushroom bodies, but this enzyme is also found in other parts of the brain (Nighorn et al., 1991). The odor avoidance after electric shock of two *dnc* deficiencies which remove different sets of *dnc* transcripts (Qiu and Davis, 1993) was studied (Préat 1998). The first deficiency, $Df(1)N^{79f}$, dramatically decreases Dnc expression in the mushroom bodies. Individuals carrying this deficiency display a normal odor avoidance after electric shock (Préat 1998), and a specific memory defect after Pavlovian conditioning. On the contrary, $Df(1)N^{64j15}$ affects the "general" Dnc expression as well as the mushroom body expression. These animals show a strong decrease in their odor avoidance after electric shocks, and an apparent learning defect after Pavlovian conditioning. This suggests that in *dnc* flies the learning deficit observed in the Pavlovian assay may be a secondary consequence of the abnormal odor response induced by electric shock. The Dnc phosphodiesterase accumulated in the mushroom bodies would thus be involved specifically in memory formation. These results confirm previous conclusions drawn from experiments where mutant flies were submitted to reduced stress: First, *dnc* shows near-normal learning in the operant olfactory protocol when tested immediately after training (Dudai, 1979; Dudai, 1983), and second, *dnc* learns normally when a positive stimulus (sugar) is used in association with odors (Tempel et al., 1983). These observations show the need for new non-associative controls in inhibitory conditioning experiments, in which the effect of the aversive unconditioned stimulus on the animals' ability to perceive the conditioned stimulus should be investigated.

Characterization of learning and memory mutants: brain function versus brain development

A major issue when a mutant deficient in a learning or memory assay has been isolated is to find whether the behavioral defect results from the defective functioning of neural circuitry in the adult, as opposed to a wiring defect caused by the lack of a developmental function. Although anatomical defects have proven useful in identifying learning and memory centers (Heisenberg et al., 1985), they are less likely to reveal direct mechanisms about how information is processed. This issue of brain development versus brain function can be answered by expressing in mutant individuals a normal copy of the gene only during the adult stage. If such individuals display normal learning or memory despite the fact that the gene of interest was inactive during development, it can then be assumed that the behavioral defect of the mutant arises from a dysfunction of the adult nervous system.

Such an approach was successfully applied to *Volado* (Grotewiel et al., 1998). Overexpressing the *Volado* gene three hours before training led to the rescue of the learning defect. In the case of *dnc*, partial rescue of behavior using either the *Drosophila dnc* gene or a rat homologue has been successfully carried out (Dauwalder and Davis, 1995). It should be noted, however, that only hypomorphic *dnc* mutants could be partially rescued (i.e. mutants showing a reduced but not absent Dnc enzymatic activity). The behavior of the null mutant dnc^{M14} was not rescued, suggesting that the Dnc PDE is involved both in MB development and in adult function. The fact that cultured *dnc* neurons show an abnormal branching pattern fits with this interpretation (Zhong et al., 1992).

An alternative approach used to prove that a gene product is involved in the physiology of learning consists in preventing activity solely in the adult. Thus ubiquitous expression in the adult of a peptide which inhibits the catalytic subunit of the PKA leads to a learning deficit (Drain et al., 1991).

The case of the *lio* mutant demonstrates that this brain development issue can be complex. lio^1 is a *P*-element induced mutant showing decreased associative learning and memory (Dura et al., 1993). Two genes lie in the vicinity of the transposable P-element in lio^1 (Dura et al., 1995). The first gene encodes a protein of unknown function and expression pattern (Bolwig et al., 1995). Ubiquitous expression of this gene in the adult has been reported to lead to full rescue of the behavior of lio^1 mutants (Bolwig et al., 1995). This result apparently contradicts our observation that the expression of this first gene appears normal in lio^1 mutants (Dura et al., 1995), suggesting that it is not misexpression of this first gene which underlies the learning and memory defect. However, expression of the second gene is severely affected in lio^1. This second gene encodes a putative tyrosine kinase (Callahan et al., 1995; Dura et al., 1995) involved in central brain development, and in particular in MB formation

(Moreau-Fauvarque et al., 1998; Simon et al., 1998). For example a strong MB defect can be observed in lio^1/Df animals. Since the MBs have been implicated in olfactory learning and memory, these results suggest that the learning defect of lio^1 individuals is a secondary consequence of the central brain anatomical defect, implying that the behavioral defect of lio^1 is not linked to the first gene. Replicating the behavioral rescue experiment with the first gene (Bolwig et al., 1995) would be a step in solving this enigma. Rescuing the lio^1 behavioral defect with the second gene would also be useful, as a dysfunction in both genes might be involved in generating the learning and memory defect observed in the lio^1 mutant.

Studying learning and memory mutants: how to link genetic with behavioral properties

The main advantage of testing learning in *Drosophila* lies in the powerful genetic tools and concepts which have been developed over the past decades. However, some important concepts which are relatively straightforward in studying developmental mutants, are more problematic when applied to learning and memory mutants.

Pleiotropy — the fact that mutating a single gene can lead to multiple phenotypes — is relatively easy to deal with in developmental studies. For example if a mutation affects both wing development and ovary formation, these phenotypes are not mutually dependent (even though similar biochemical pathways may underlie both defects). This in contrast to behavioral pleiotropy, where a first behavioral defect can produce the *illusion* that another behavioral phenotype is displayed by the mutant. This observation is especially important in the study of learning, because the phenotype can only be measured indirectly through behavioral modification. Thus a mutant which cannot smell properly is likely to be impaired for olfactory learning. Such an effect on learning is very indirect and studying the corresponding gene is unlikely to provide information about learning processes. These cases of "artifactual pleiotropy" are difficult to assess because, as seen previously in the case of *dnc*, the studied gene may indeed be involved in various neuronal functions.

There are two other important concepts in genetics, namely penetrance — the fact that all individuals bearing a mutation do not necessarily display a mutant phenotype — and expressivity — the fact that the strength of the defect varies between individual animals from the same stock. These factors are difficult to deal with in learning mutants because learning is by essence very sensitive both to the internal status of the brain — which can vary in a short period of time independently of the mutation — and to external parameters. More, the test itself can affect the way stored information is expressed. Thus the learning phenotype of a mutant individual is only *estimated* at a given time and under a particular condition. The same animal might have shown a quite different phenotype if it had been tested at a different time or under slightly different conditions.

The global performance of a learning or memory mutant stock is calculated by averaging the score of many heterogeneous individuals. This can have an important impact on conclusions drawn from memory retention curves, especially when these are interpreted in terms of memory phases. How penetrance and expressivity of mutant phenotypes interfere with the dynamics of memory decay remains to be taken into account. Monitoring the behavior of single flies rather than that of groups would help that goal.

Testing associative learning in *Drosophila*: perspectives

The classical genetic approach (the search for randomly generated mutants) is uniquely valuable as it can reveal a role for previously unsuspected proteins in generating properties of the neural network. The "reverse genetic" approach can also prove useful but it requires that the implication of a given gene in learning is suspected beforehand. Even though screening for learning and memory mutants in *Drosophila* requires extreme rigor and patience, this approach, which has had some success, still holds much promise. Several points are critical for the future of this field.

At the behavioral level, since screening procedures are difficult and may be subject to variability, machines should be developed which are simple enough for untrained researchers to use successfully. The protocols used so far to isolate *Drosophila* learning and memory mutants associate odors with electric shocks. Partially automated machines have been used which allow automatic conditioning but manual testing (Tully et al., 1994). We are currently developing fully automated machines.

To broaden our understanding of brain functioning, in particular with a view to structure/function mapping, it would be interesting to screen mutants using non-olfactory procedures. An apparatus has been recently designed in which individual flies learn to avoid the heated side of a small chamber (Wustmann et al., 1996; Wustmann and Heisenberg, 1997). The power of this protocol relies on its simplicity and the fact that many flies can be studied individually in parallel. However memory retention vanishes within a few minutes, so it might not be possible to isolate specific memory mutants using this method.

At the genetic level, genes involved in learning and memory need to be identified using more specific procedures. To circumvent the difficulty of developmental pleiotropy — the fact that a product involved in memory formation might be recruited for another function during development — one solution might be to use mutagenic agents that are more likely to disrupt specific regulatory elements (Pflugfelder, 1998). This should make it possible to disrupt an adult function while leaving developmental expression unaffected.

Deregulation mutagenesis can also be used (Rorth, 1996). This enables random genes to be over-expressed specifically in a given part of the adult brain. This method involves producing mutants with a transposable *P*-element carrying the upstream activating sequence (UAS) recognized by the yeast transcription activator Gal4. Once generated, these P(UAS) lines can be crossed to a line that expresses Gal4 specifically in substructures of the *Drosophila* brain (enhancer-trap line). In P(UAS)/+, Gal4/+ animals the gene located near the P(UAS) insertion site can be tissue-specifically activated by Gal4. The behavior of the corresponding animals can then be tested. The advantage of this approach, currently used in our laboratory, is that genes involved in learning and memory can be detected even if they are also normally expressed during development, as the over-expression is adult-specific.

At the anatomical level, the MBs have been widely implicated in learning and memory in *Drosophila* as well as in other insects. Based on a study of output circuitry (the MBs are close to the sensory input sites but further away from motor output centers) it has recently been suggested that MBs could serve merely as a relay for odor processing, rather than being the site of olfactory memory itself (Ito et al., 1998). However the lack of Dnc product in the MBs leads to a memory defect, not a learning defect (Dudai, 1979, 1983; Préat, 1998), which does not support that interpretation.

Enhancer-trap selection of genes expressed preferentially in the MB has proved to be a powerful way of carrying out a "preliminary screen" for genes involved in learning (Skoulakis and Davis, 1996; Grotewiel et al., 1998). However this strategy could be biased. Some of the proteins preferentially expressed in the MB might play no specific role in learning but instead be required in this structure for some other neuronal function. The lack of such protein would alter MB activity. If the MBs are indeed the main center for olfactory learning and memory in *Drosophila*, such a biochemical defect would result in a learning or memory phenotype, even though the protein is not particularly involved in learning. Analyzing the role of other brain centers in learning and memory remains essential.

At the molecular level, we need to find a direct correlate of conditioning, much as the modulation of the Period protein is a molecular correlate of the circadian rhythm of locomotor behavior (Zerr et al., 1990). In the case of associative learning this molecular correlate could be a coincidence detector which would provide a molecular link between the CS signal (the odor) and the US (the shock). It has been proposed that the Rut adenylyl cyclase might fulfill this function as these cyclases are stimulated by calcium/calmodulin and by Gs protein (Davis, 1993). MB depolarization following odor perception (resulting in an increase in intracellular $Ca2^{+}$ in MB neurons), combined with G protein activation following neurotransmitter release in the US pathway, would lead to an overactivation of the Rut AC. Such coincidence detection has been described for Aplysia AC (Abrams et al., 1991), but it remains to be observed in *Drosophila*.

A central role has been proposed for cAMP-responsive element-binding (CREB) proteins in *Drosophila* long term memory formation. Thus an activator form of the CREB protein has been shown to facilitate long term memory (Yin et al., 1995), whereas an inhibitor form blocks long term memory formation after intensive training (Yin et al., 1994). Unfortunately few results have come out since these original publications (i.e. what are the genes expressed downstream of CREB transcriptions factors during long term memory formation? In what region of the brain does the CREB activator form have to be expressed to generate long term memory after one training cycle?).

More generally, although a general biochemical model has been proposed to include the data from *Drosophila*, we still lack detailed molecular data about how these "learning" proteins function in the central brain during and following conditioning. Thus despite the exciting finding that the first two genes shown to be involved in learning and memory in *Drosophila* are preferentially expressed in the same structure and involved in the cAMP pathway, the way the Dnc and Rut proteins function in the MBs is still poorly understood. In particular we do not know why cAMP levels are 8 times higher in *dnc* mutants (Byers et al., 1981; Davis and Kiger, 1981) but only marginally reduced in *rut* (Livingstone et al., 1984). Moreover, *rut* partially suppresses the effect of *dnc* since double mutants show near-normal cAMP concentration, but the behavior of these animals is still strongly impaired (Livingstone et al., 1984). This result suggests the existence of some unknown dynamic interaction between Dnc and Rut.

In conclusion, careful tests must be performed before a mutant is proven to be a bona fide learning or memory mutant. Experience has shown me that these controls can take as long to perform as the initial screening of the mutants. Detailed molecular and cellular data must be combined with the behavioral data. Naturally, this observation is not specific to *Drosophila*. The advantage of the fly lies in the unrivaled genetic tools available, and in the fact that the insect brain must display less redundancy than that of mammals, thus increasing the possibility of dissecting behavior using mutations which affect the function of precise structures.

Acknowledgments

I thank Raphaël Hitier and Florian Petit for their fruitful comments on the manuscript.

References

Abrams, T.W., Larl, K.A. and Kandel, E.R. (1991) Biochemical studies of stimulus convergence during classical conditioning in *Aplysia*: dual regulation of adenylate cyclase by Ca^{2+}/calmodulin and transmitter. *J. Neurosci.*, 11: 2655–2665.

Asztalos, Z., Von Wegerer, J., Wustmann, G., Dombradi, V., Gausz, J., Spatz, H.C. and Friedrich, P. (1993) Protein phosphatase 1-deficient mutant *Drosophila* is affected in habituation and associative learning. *J. Neurosci.*, 13(3): 924–930.

Balling, A., Technau, G. M., and Heisenberg, M. (1987) Are the strutural changes in adult *Drosophila* mushroom bodies memory traces? Studies on biochemical learning mutants. *J. Neurogenet.*, 4: 65–73.

Bolwig, G.M., Del Vecchio, M., Hannon, G. and Tully, T. (1995) Molecular cloning of linotte in *Drosophila*: a novel gene that functions in adults during associative learning. *Neuron*, 15: 829–842.

Boynton, S. and Tully, T. (1992) Latheo, a new gene involved in associative learning and memory in *Drosophila melanogaster*, identified from P element mutagenesis. *Genetics*, 131: 655–672.

Brand, A.H. and Perrimon, N. (1993) Targeted gene expression as a means of altering cell fates and generating dominant phenotypes. *Development*, 118(2): 401–415.

Broadie, K., Rushton, E., Skoulakis, E.M. and Davis, R.L. (1997) Leonardo, a *Drosophila* 14-3-3 protein involved in learning, regulates presynaptic function. *Neuron*, 19(2): 391–402.

Byers, D., Davis, R.L. and Kiger, J.A. (1981) Defecting cyclic AMP phosphodiesterase due to the dunce mutation of learning in *Drosophila melanogaster*. *Nature*, 289: 79–81.

Callahan, C.A., Muralidhar, M.G., Lundgren, S.E., Scully, A. L. and Thomas, J.B. (1995) Control of neuronal pathway selection by a *Drosophila* receptor protein-tyrosine kinase family member. *Nature*, 376: 171–174.

Chen, C.N., Denome, S. and Davis, R.L. (1986) Molecular analysis of cDNA clones and the corresponding genomic coding sequences of the *Drosophila dunce*$^{+}$ gene, the structural gene for cAMP phosphodiesterase. *Proc. Natl. Acad. Sci. USA*, 83: 9313–9317.

Connolly, J.B., Roberts, I.J., Armstrong, J.D., Kaiser, K., Forte, M., Tully, T. and O'Kane, C.J. (1996) Associative learning disrupted by impaired Gs signaling in *Drosophila* mushroom bodies. *Science*, 274(5295): 2104–2107.

Connolly, J.B. and Tully, T. (1998) Behaviour, learning and memory. In: D.B. Roberts (Ed.), *Drosophila: A practical approach*, 2nd Edn,. IRL Press, Oxford, pp. 265–317.

Cowan, T.M. and Siegel, R.W. (1986) *Drosophila* mutations that alter ionic conduction disrupt acquisition and retention of a conditioned odor avoidance response. *J. Neurogenet.*, 3(4): 187–201.

Dauwalder, B. and Davis, R.L. (1995) Conditional rescue of the *dunce* learning/memory and female fertily defects with *Drosophila* or rat transgenes. *J. Neurosci.*, 15: 3490–3499.

Davis R.L. (1993) Mushroom bodies and *Drosophila* learning. *Neuron*, 11: 1–14.

Davis, R.L. and Kiger Jr, J.A. (1981) *dunce* mutants of *Drosophila melanogaster:* mutants defective in the cyclic AMP phosphodiesterase enzyme system. *J. Cell Biol.*, 90: 101–107.

Davis, R.L. (1996) Physiology and biochemistry of *Drosophila* learning mutants. *Physiological Reviews*, 76: 299–317.

De Belle J.S. and Heisenberg M. (1994) Associative odor learning in *Drosophila* abolished by chemical ablation of mushroom bodies. *Science*, 263: 692–695.

Drain, P., Folkers, E. and Quinn, W.G. (1991) cAMP-dependent protein kinase and the disruption of learning in transgenic flies. *Neuron*, 6: 71–82.

Dudai, Y., Jan, Y.N., Byers, D., Quinn, W.G. and Benzer, S. (1976) *Dunce*, a mutant of *Drosophila* deficient in learning. *Proc. Natl. Acad. Sci. USA*, 5: 1684–1688.

Dudai, Y. (1979) Behavioral plasticity in a *Drosophila* mutant, dunceDB276. *J. Comp. Physiol.*, 130: 271–275.

Dudai, Y. (1983) Mutations affect storage and use of memory differentially in *Drosophila melanogaster*. *Proc. Natl. Acad. Sci. USA*, 80(17): 5445–5448.

Dudai, Y., Zvi, S. and Segel, S. (1984) A defective conditioned behavior and a defective adenylate cyclase in the *Drosophila* mutant *rutabaga*. *J. Comp. Physiol. A*, 155: 569–576.

Dura, J.M., Préat, T. and Tully, T. (1993) Identification of *linotte*, a new gene affecting learning and memory in *Drosophila melanogaster*. *J. Neurogenet.*, 9: 1–14.

Dura, J.-M., Taillebourg, E. and Préat, T. (1995) The *Drosophila* learning and memory gene linotte encodes a putative receptor tyrosine kinase homologous to the human RYK gene product. *FEBS Lett.*, 370: 250–254.

Folkers, E. and Spatz, H.C. (1981) Visual learning behaviour in *Drosophila melanogaster* wildtype AS. *J. Insect Physiol.*, 27(9): 615–622.

Grotewiel, M.S., Beck, C.D.O., Wu, K.H., Zhu, X.R. and Davis, R.L. (1998) Integrin-mediated short-term memory in *Drosophila*. *Nature*, 391: 455–460

Guo, A., Liu, L., Xia, S., Feng, C., Wolf, R. and Heisenberg, M. (1996) Conditioned visual flight orientation in *Drosophila*: dependence on age, practice, and diet. *Learning and Memory*, 3: 49–59.

Han, P.L., Levin, L.R., Reed, R.R. and Davis, R.L. (1992) Preferential expression of the *Drosophila rutabaga* gene in mushroom bodies, neural centers for learning in insects. *Neuron*, 9: 619–627.

Han, P.L., Meller, V. and Davis, R.L. (1996) The *Drosophila* brain revisited by enhancer detection. *J. Neurobiol.*, 31(1): 88–102.

Heisenberg, M., Borst, A. and Byers, D. (1985) *Drosophila* mushroom body mutants are deficient in olfactory learning. *J. Neurogenet.*, 2: 1–30.

Heisenberg, M., Heusipp, M. and Wanke, C. (1995) Structural plasticity in the *Drosophila* brain. *J. Neurosci.*, 15(3): 1951–1960

Hitier, R., Heisenberg, M. and Préat, T. (1998) Abnormal mushroom body plasticity in the memory mutant *amnesiac*. *NeuroReport*, 9(12): 2717–2719.

Ito, K., Awano, W., Suzuki, K., Hiromi, Y. and Yamamoto, D. (1997) *Drosophila* mushroom body is a quadruple structure of clonal units each of which contains a virtually identical set of neurones and glial cells. *Development*, 124(4): 761–771.

Ito, K., Suzuki, K., Estes, P., Ramaswami, M., Yamamoto, D. and Strausfeld, N.J. (1998) The organization of extrinsic neurons and their implications in the functional roles of the mushroom bodies in *Drosophila melanogaster* Meigen. Learning and Memory, 5(1–2): 52–77.

Livingstone, M.S., Sziber, P.P. and Quinn, W.G. (1984) Loss of calcium/ calmodulin responsiveness in adelynate cyclase of rutabaga, a *Drosophila* learning mutant. Cell, 37: 205–215.

Kauvar, L.M. (1982) Defective cyclic adenosine 3':5'-monophosphate phosphodiesterase in the *Drosophila* memory mutant *dunce*. *J. Neurosci.*, 2: 1347–1358.

Levin, L.R., Han, P.L., Hwang, P.M., Feinstein, P.G., Davis, R.L. and Reed, R.R. (1992) The *Drosophila* learning and memory gene *rutabaga* encodes a Ca2+/calmodulin-responsive adenylyl cyclase. *Cell*, 68: 479–489.

Moreau-Fauvarque, C., Taillebourg, E., Boissoneau, E., Mesnard, J. and Dura, J.-M. (1998) The receptor tyrosine kinase gene *linotte* is required for neuronal pathway selection in *Drosophila* mushroom bodies. *Mec. Dev.*, 78: 47–61.

Nighorn, A.M., Healy, M.J. and Davis, R.L. (1991) The cyclic AMP phosphodiesterase encoded by the *Drosophila dunce* gene is concentrated in the mushroom body neuropil. *Neuron*, 6: 455–467.

O'Kane, C. and Gehring, W.J. (1987) Detection in situ of genomic regulatory elements in *Drosophila*. *Proc. Natl. Acad. Sci. USA*, 84: 9123–9127.

Pflugfelder, G.O. (1998) Genetic lesions in *Drosophila* behavioural mutants. *Behav. Brain Res.*, 95: 3–15.

Préat, T. (1998) Decreased odor avoidance after electric shock in *Drosophila* mutants biases learning and memory tests. *J. Neurosci.*, 18(20): 8534–8538.

Qiu Y. and Davis R.L. (1993) Genetic dissection of the learning/memory gene *dunce* of *Drosophila melanogaster*. *Genes Dev.*, 7: 1447–1458.

Quinn, W.G., Harris, W. and Benzer, S. (1974) Conditioned behavior in *Drosophila melanogaster*. *Proc. Natl. Acad. Sci. USA*, 71: 708–712.

Quinn, W.G., Sziber, P.P. and Booker, R. (1979) The *Drosophila* memory mutant *amnesiac*. *Nature*, 277: 212–214.

Rorth, P. (1996) A modular misexpression screen in *Drosophila* detecting tissue specific phenotypes. *Proc. Natl. Acad. Sci. USA*, 93: 12418–12422.

Siegel, R.W. and Hall, J.C. (1979) Conditioned responses in courtship behavior of normal and mutant *Drosophila*. *Proc. Natl. Acad. Sci. USA*, 76: 3430–3434.

Simon, A.F., Boquet, I., Synguélakis, M. and Préat, T. (1998) The *Drosophila* putative kinase Linotte (Derailed) prevents central brain axons from converging on a newly described interhemispheric ring. *Mech. Dev.*, 76: 45–55.

Skoulakis, M.C., Kalderon, D. and Davis, R.L. (1993) Preferential expression in mushroom bodies of the catalytic subunit of protein kinase A and its role in learning and memory. *Neuron*, 11: 97–208.

Skoulakis, E.M. and Davis, R.L. (1996) Olfactory learning deficits in mutants for leonardo, a *Drosophila* gene encoding a 14-3-3 protein. *Neuron*, 17(5): 931–944.

Spatz, H.C., Emanns, A. and Reichert, H. (1974) Associative learning of *Drosophila melanogaster*. *Nature*, 248: 359–361.

Technau, G.M. (1984) Fiber number in the mushroom bodies of adult *Drosophila melanogaster* depends on age, sex and experience. *J. Neurogenet.*, 1: 113–126.

Tempel, B.L., Bonini, N., Dawson, D.R. and Quinn, W.G. (1983) Reward learning in normal and mutant *Drosophila*. *Proc. Natl. Acad. Sci. USA*, 80: 1482–1486.

Tully, T. and Quinn, W.G. (1985) Classical conditioning and retention in normal and mutant *Drosophila melanogaster*. *J. Comp. Physiol. A*, 157: 263–277.

Tully, T., Préat, T., Boyton, S.C. and Del Vecchio, M. (1994) Genetic dissection of consolidated memory in *Drosophila*. *Cell*, 79: 35–47.

Wolf, R. and Heisenberg, M. (1991) Basic organization of operant behavior as revealed in *Drosophila* flight orientation. *J. Comp. Physiol. A*, 169: 699–705.

Wolf, R. and Heisenberg, M. (1997) Visual space from visual motion: turn integration in tethered flying *Drosophila*. *Learning and Memory*, 4: 318–327.

Wustmann, G., Rein, K., Wolf, R. and Heisenberg, M. (1996) A new paradigm for operant conditioning of *Drosophila melanogaster*. *J. Comp. Physiol. A*, 179: 429–436.

Wustmann, G. and Heisenberg, M. (1997) Behavioral manipulation of retrieval in a spatial memory task for *Drosophila melanogaster*. Learning and Memory, 4: 328–336.

Xia, S., Liu, L., Feng, C. and Guo, A. (1997) Memory consolidation in *Drosophila* operant visual learning. Learning and Memory, 4: 205–218.

Yang, M.G., Armstrong, J.D., Vilinsky, I., Strausfeld, N.J. and Kaiser, K. (1995) Subdivision of the *Drosophila* mushroom bodies by enhancer-trap expression patterns. *Neuron*, 15: 5–54.

Yin, J.C.P., Wallach, J.S., Del Vecchio, M., Wilder, E.L., Zhou, H., Quinn, W.G. and Tully, T. (1994) Induction of a dominant negative CREB transgene specifically blocks long-term memory in *Drosophila*. *Cell*, 79: 49–58.

Yin, J.C.P., Del Vecchio, M., Zhou, H. and Tully, T. (1995) CREB as memory modulator: induced expression of a dCREB2 activator isoform enhances long-term memory in *Drosophila*. *Cell*, 81: 107–115.

Zerr, D.M., Hall, J.C., Rosbash, M. and Siwicki, K.K. (1990) Circadian fluctuations of period protein immunoreactivity in the CNS and the visual system of *Drosophila*. *J. Neurosci.*, 10(8): 2749–2762.

Zhong, Y., Budnik, V. and Wu, C.-F. (1992) Synaptic plasticity in *Drosophila* memory and hyperexcitable mutants : role of cAMP cascade. *J. Neurosci.*, 12(2): 644–651.

W.E. Crusio and R.T. Gerlai (Eds.)
Handbook of Molecular-Genetic Techniques for Brain and Behavior Research (Techniques in the Behavioral and Neural Sciences, Vol. 13)

CHAPTER 3.3.5

Caenorhabditis elegans and the genetics of learning

Kevin R. Peters, Jennifer A. Galloway and Catharine H. Rankin

Department of Psychology, University of British Columbia, Vancouver, BC V6T 1Z4, Canada

Behavioral issues in the genetics of learning

What are the requirements of a learning mutant?

Since the first demonstration by Quinn et al. (1974) that fruit flies (*Drosophila melanogaster*) were capable of associative learning, the notion of genetically dissecting a behavior such as learning has become a viable and realistic one (Dudai, 1988). The wealth of knowledge on the genetics of *Drosophila* along with the information known about the cellular and molecular mechanisms of learning in other invertebrates made for an exciting combination.

Definition of a learning mutant
There are several levels at which a gene may exert its influence on learning: the acquisition, storage, or the retrieval of information. According to Dudai (1988), in order to be a learning mutant, there must be a single gene mutation that exerts its affect at one of these three levels and is not simply the result of other developmental, sensory, or motor deficits that may indirectly affect learning. It is important to note that this definition does not necessitate that the gene be involved only in learning — it may be involved in learning as well as some other cellular processes. The important criteria is that the learning effect can be reasonably dissociated from other defects through experimental manipulation.

Pleiotropy: do gene-specific behaviors exist?
The flexibility of this definition of a learning mutant stems in large part from the increasing awareness of the phenomenon of pleiotropy (Dudai, 1988). In relation to the genetic dissection of learning and memory, pleiotropy occurs when neural structures and/or physiological processes in addition to learning may be affected by mutations of a single gene. Research on the cellular basis of learning and memory in a variety of organisms has shown that these processes do not involve novel gene activity, rather they reflect modifications of on-going expression of genes involved in other processes such as development or structural maintenance of the cell. Furthermore, if structures or physiological processes are disrupted early in development, the animal will not develop normally, which may explain other behavioral deficits that may exist. Through the process of evolution, nervous systems have become highly conservative and efficient. Thus, it is not surprising that mutants that are defective in a behavior that is as complex as learning will also be defective in other processes that may affect learning directly or indirectly. However, as Tully (1994) points out, although a single-gene mutation may produce defects in several biological processes, it is incorrect to attribute any causal relationship to the presence of these defects. In other words, the co-occurrence of structural and functional abnormalities may represent indepen-

dent effects of mutations of the same gene. Furthermore, abnormalities in structure may lead to abnormalities in function indirectly rather than directly. The problem of pleiotropy is not insurmountable — methods exist which allow us to overcome the issues raised by pleiotropy. These methods focus on ways in which gene expression can be restricted both temporally and spatially rather than focusing simply on the presence or absence of the genes themselves.

***Caenorhabditis elegans* and why it is an ideal model to study the genetics of learning**

Brenner (1974) chose *C. elegans* as a model organism in which to study the way that genes interact to produce a nervous system and in turn, how this nervous system produces behavior. He used *C. elegans* because it possesses (a) a small genome size, (b) a relatively simple nervous system, and (c) a variety of behaviors.

Genome size

The genome of *C. elegans* is very small with only 100×10^6 nucleotide pairs in six haploid chromosomes (Coulson et al., 1991). This is approximately 20 times larger than the genome of *Escherichia coli* (*E. coli*) and approximately half the size of *Drosophila*'s genome. The size of the mouse's genome is approximately 30 times that of *C. elegans* (D.L. Baillie, personal communication, 1998). There is an on-going effort to provide a complete physical map and sequence of the animal's genome (Coulson et al., 1986; Edgely and Riddle, 1987), which is scheduled to be finished by the end of 1998. In addition, the identification and subsequent cloning of many of its genes is an on-going venture by a number of different laboratories, which will greatly facilitate our study of the role of genetics in learning. The Caenorhabditis Genetics Center, located at the University of Missouri, Columbia, contains a large central library of mutant lines frozen in liquid nitrogen that are available to researchers interested in studying a particular identified mutant.

Size of nervous system

C. elegans also has the advantage of a very small nervous system in comparison to *Drosophila* (302 neurons compared to approximately 10^5 neurons). The localization, cell lineage and complete anatomical connectivity of each of its 302 neurons have been mapped across development using serial electron microscopy (Sulston and Horvitz, 1977; Sulston et al., 1983; White et al., 1986). In addition, due to the transparency of its cuticle, methods to laser ablate single cells in living animals have been developed, which make the identification of neural circuits underlying various behaviors a plausible venture (Chalfie et al., 1985; Bargmann and Avery, 1995; Wicks and Rankin, 1995). Although electrophysiological recordings of neurons in *C. elegans* has been problematic in the past due to their small size, methods have been recently developed to make such recordings (Avery et al., 1995; Goodman et al., 1998).

Behavioral repertoire

The nematode *C. elegans* possesses a rich enough set of behaviors to allow for the study of the role of genes in behavior. It has been possible to isolate behavioral phenotypes that arise from mutations and manifest themselves in abnormal locomotion, feeding, defecation, and egg-laying behavior (Rankin and Beck, 1992; Bargmann, 1993). More germane to the present discussion, *C. elegans* has been shown to demonstrate various forms of learning including: nonassociative learning, i.e. habituation and sensitization (Rankin et al., 1990), and associative learning, i.e. classical conditioning (Wen et al., 1997), as well as both short- and long-term memory (reviewed in Gannon and Rankin, 1995). On-going work has been initiated to identify mutants that are defective in either of these forms of behavioral plasticity. A thorough behavioral assay for the characterization of such mutants will be presented later in this chapter.

Additional advantages in using *C. elegans* as a model system include that it is a self-reproducing hermaphrodite, although males do arise as a result of X-chromosome nondisjunction mutations that occur at a relatively low frequency, i.e. about one in 500 (Hodgekin et al., 1979; Rose and Baillie, 1979). Hermaphroditic reproduction makes the isolation of recessive mutants on each chromosome easier and facilitates the true breeding of homozygous mutants. Males allow for the heterozygous balancing of deleterious genes. Furthermore, the animal has a short life span and thus a short reproductive cycle (Brenner, 1974). The ability to freeze *C. elegans* for prolonged periods of time also gives it a practical advantage over *Drosophila*, in that once a mutant strain has been identified it can be frozen and the additional work to continually maintain colonies is no longer necessary (Riddle et al., 1997).

Learning in Caenorhabditis elegans

Learning refers to an organism's ability to modify its behavior across time as a result of previous experience (Gannon and Rankin, 1995). *C. elegans* has been shown to demonstrate a variety of forms of learning, including nonassociative and associative learning, and long-term memory.

Nonassociative learning

C. elegans has been shown to demonstrate several forms of nonassociative learning: habituation, dishabituation and sensitization. Habituation is the decrease in response that is observed following repeated presentations of a given stimulus (Thompson and Spencer, 1966; Groves and Thompson, 1970). Habituation is not synonymous with either sensory adaptation or response fatigue. Despite the reduction in response, the animal is still capable of sensing the stimulus and responding to it following the presentation of a noxious stimulus (dishabituation); whereas in both sensory adaptation and response fatigue, the animal is not capable of sensing or responding and is only able to do so after a given period of time. The earliest demonstrations of habituation to mechanosensory stimuli in *C. elegans* came from observations of response decrements in animals that occurred after repeatedly bumping into glass beads (Croll, 1975) and following repeated touches to the head or tail with a fine hair (Chalfie and Sulston, 1981). A detailed analysis of habituation in *C. elegans* came from the tap-withdrawal response protocol first described by Rankin et al. (1990). This protocol involves administering stimuli that are of a consistent nature (mechanical taps to the side of the petri-dish containing the animal rather than hand-delivered head or tail touches) and therefore allows for a quantitative assessment of habituation. In addition, dishabituation has been demonstrated in this protocol by delivery of a brief electric shock to the agar upon which the worm rests, followed by additional mechanical taps.

The other form of nonassociative learning is that of sensitization, which has been defined as an increase in responsiveness that is produced by the application of a noxious or strong stimulus (Groves and Thompson, 1970). Although sensitization has not been studied to the same extent as habituation, it has been demonstrated to exist in *C. elegans* (Rankin et al., 1990).

Associative learning

C. elegans is also capable of demonstrating associative learning by modifying its behavior on the basis of past experience through the pairing of events. For example, Hedgecock and Russell (1975) demonstrated that when presented with a thermogradient of several different temperatures, *C. elegans* will tend to move towards the temperature at which it was raised. However, if conditions of starvation or overcrowding are associated with its "native" temperature, the nematode will move away from this temperature, thus modifying its behavior on the basis of experience. In addition, *C. elegans* can learn to associate food with a particular chemoattractant. Wen et al. (1997) exposed animals to either sodium or chloride ions, one of which was associated with the presence of food (*E. coli*), the other with the absence of food. Later, when placed onto a plate with both ions available, the animals migrated towards the ion that was paired with the food. Worms also demonstrated conditioned aversion to an ion (sodium or chloride) that had been previously paired with garlic. Therefore, it appears as though *C. elegans* is able to be conditioned to either approach or avoid certain ions on the basis of past experience. Using this protocol, Van der Kooy and colleagues have isolated two mutant strains, *lrn-1* and *lrn-2*, that do not appear to display associative learning (Wen et al., 1997). The mutations involved in *lrn-1* and *lrn-2* have been mapped to chromosomes; however, the gene products have not yet been identified.

Long-term memory

The previous discussion demonstrated that *C. elegans* has been shown to be capable of exhibiting both nonassociative and associative learning. In addition to these tests of short-term memory, habituation of the tap-withdrawal response has been shown to persist for at least 24 h. Beck and Rankin (1997) have demonstrated long-term habituation following distributed training blocks of stimuli presented at long, i.e. 60 s, interstimulus intervals (ISIs) but not short, i.e. 10 s, ISIs. Long-term habituation was not present following massed training of stimuli, regardless of the ISI used. Enhancement of long-term memory following distributed training blocks has been observed in many species and is considered to represent a consistent factor affecting memory formation (DeZazzo and Tully, 1995). Furthermore, heat shock has been shown to disrupt long-term memory consolidation in *C. elegans*, presumably due to the production of heat shock proteins and the cessation of all other protein synthesis (Beck and Rankin, 1995).

The rest of this chapter will focus on habituation of the tap-withdrawal response and the ways that we use this protocol to investigate the genetics of learning. Of all forms of learning in *C. elegans*, habituation of the tap-withdrawal response has been studied in the most detail and the neural circuit underlying this behavior has been identified. In addition, the tap-withdrawal response has the advantage of allowing measurement of both acquisition and retention of learning.

Tap-withdrawal response protocol: measuring and assessing habituation

Equipment and protocol

The tap-withdrawal response is elicited when the side of the petri-dish is tapped. The petri-dish contains Nematode Growth Medium agar (see Epstein and Shakes, 1995) upon which a single worm is locomoting and the tap causes a vibration of this substrate. In most cases (90–95%), the adult worm will move backwards (termed a "reversal") in response to the tap, although they do move rapidly forward (termed an "acceleration") or pause on occasion. The taps are administered mechanically with a Grass stimulus generator and are in the range of 1–2 N of force. Habituation of this response is demonstrated by a reduction in both the magnitude and frequency of this reversal response. For example, the reversal to the first tap is usually at least the length of the animal's body; however, as the number of taps increases, the likelihood of reversing to the tap diminishes and the reversals that do occur are much smaller in magnitude.

Figure 1 shows the experimental apparatus that is used to examine the tap-withdrawal response. As can be seen, a 4-cm petri-dish (filled with 10 ml of Nematode Growth Medium agar) held in a plastic arm is attached to a micro-manipulator (Marzhauser Model MM33). Also attached to the micro-manipulator is a wire tapper that is connected to a Grass S88 stimulus generator via an electromagnetic relay that is used to deliver the taps to the side of the petri-dish. To facilitate recording of the responses, a dissecting microscope (Wild M3Z, Wild Leitz Canada) has a video camera (Panasonic Digital 5100) attached to it, which is also connected to a color monitor (NEC Model No. PM-1271A) and videocassette recorder (Panasonic AG1960). A time-date generator (Panasonic WJ-810) superimposes the date and time of the experiment.

The behavior of the animal and its responses to all stimuli are recorded on videotape. The responses are then analyzed using stop-frame video analysis in order to permit the precise tracing of each reversal response onto an acetate sheet.

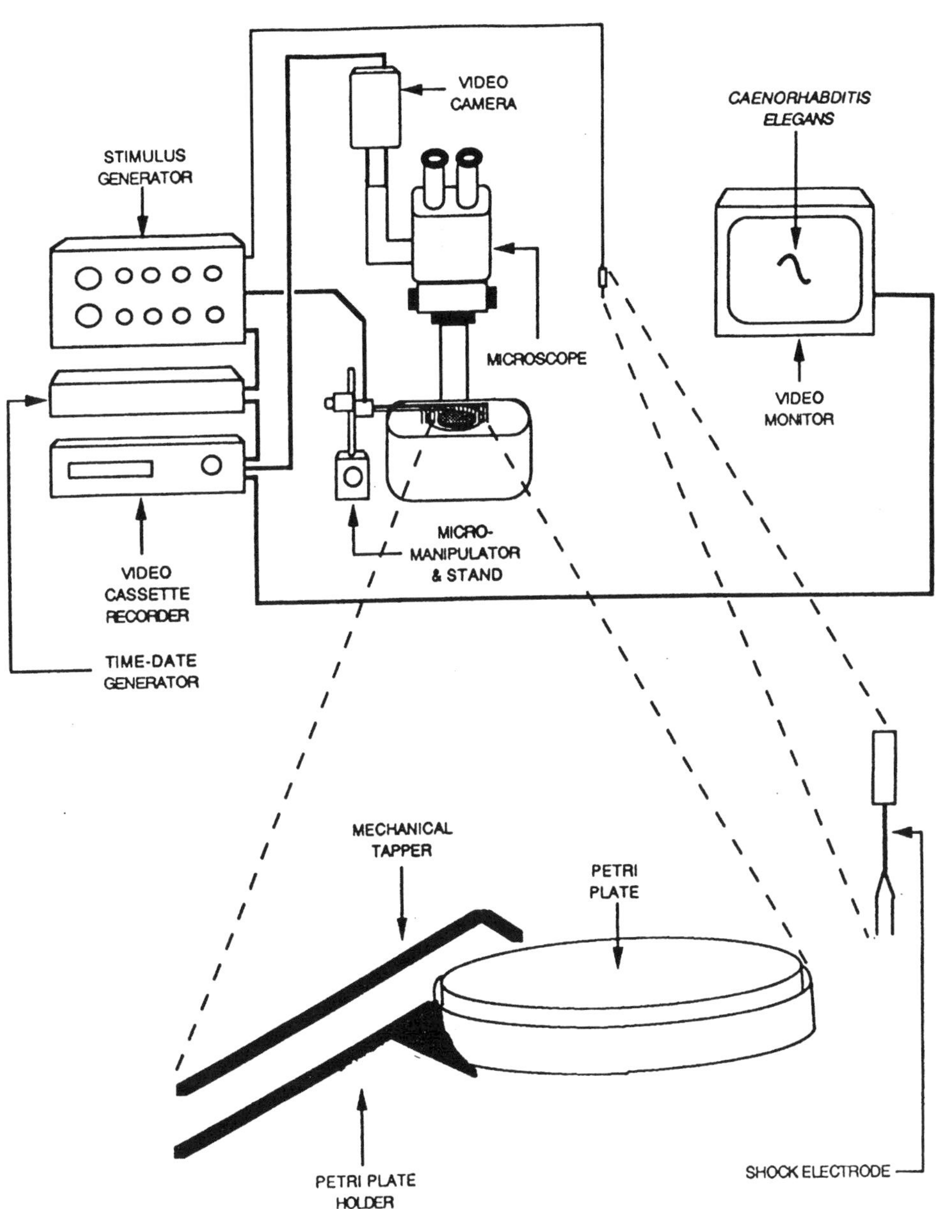

Fig. 1. Apparatus used to study habituation in *C. elegans*. The stimulus generator is used to deliver the taps to the petri-plate at a constant intensity. The worm's response to tap is displayed, in real-time, on a video monitor and is recorded using a videocassette recorder for future analysis. Below is a close-up of the mechanical tapper and petri-plate along with the electrode that is used to deliver shocks to assess for dishabituation. Adapted, with permission, from Mah (1991).

Each of the tracings from the acetate sheets are then scanned into a Macintosh 6200CD microcomputer using a Hewlett Packard ScanJet 3C scanner and the software package DeskScan II (Version 2.2) and the distance travelled in each of the responses is measured (in pixel lengths) using the NIH Image (Version 1.57) software package. The digitized responses are then imported directly into a statistical package for the purposes of data analysis.

Important factors to consider when assessing habituation

There are a number of important factors to control when assessing habituation in *C. elegans*. The age of the animals tested should be kept constant, as the response to tap has been shown to change as the animal progresses through its development (Chiba and Rankin, 1990; Beck and Rankin, 1993). Presumably, this change in response to tap is due to alterations in the underlying circuitry that mediates the tap-withdrawal response. In our studies of habituation we use 4-day-old adult animals from age-synchronized plates (see Gannon and Rankin, 1995). The decision to place animals onto plates with or without food should also be kept constant as should the length of time that animals are pre-plated onto individual testing plates. This decision will depend upon how long testing will last, as starved animals will behave differently than satiated animals. For example, to enhance classical conditioning of taste with food, Wen et al. (1997) used a paradigm in which worms were food-deprived for 5 h prior to training. For all behavioral studies it is important that each phase of the experiment be kept constant. This includes such things as the amount of time since removal from the incubator, the amount of time the plate is on the microscope stage, the method of transferring worms from plate to plate, etc.

It is also important to maintain as much control as possible over environmental factors such as temperature and relative humidity as both of these factors will alter the animal's behavior. We have observed that reversal responses to taps are much larger when the relative humidity is over 50%. Increased relative humidity may affect the properties of the agar, upon which the animals are kept, making it easier for them to move backwards. In addition, it is recommended that when bacterial or viral contamination is present on a worm culture, the animals from that plate should not be used. Techniques for cleaning contaminated plates have been described elsewhere (Brenner, 1974; Gannon and Rankin, 1995) and should be undertaken whenever contamination is present. Finally, it is critical to maintain a continuous supply of frozen worm cultures so that a fresh culture can be thawed each month. This is done to control for random genetic variability that results from the relatively high rate of spontaneous mutation in the N2 wild-type strain of worm (Gannon and Rankin, 1995; Johnsen and Baillie, 1997).

The tap-withdrawal circuit

To fully understand a behavior it is important to identify the neural circuit mediating the response and to determine the role of individual neurons in producing the response. Using the touch-withdrawal circuit identified through laser ablation techniques by Chalfie et al. (1985) as a starting point, Wicks and Rankin (1995) determined the neural circuit responsible for the tap-withdrawal response. A simplified version of this neural circuit is presented in Fig. 2. There are five primary sensory neurons: two ALM cells, one AVM cell, and two PLM cells and in addition, there are two putative stretch receptors, the PVD cells. There are nine interneurons: two AVA cells, two AVD cells, two AVB cells, two PVC cells, and one DVA cell. The interneurons AVA and AVB synapse onto two motorneuron pools and thus act as the driver cells responsible for initiating movement. As can be seen, there are two components to this circuit: an anterior subcircuit (which initiates backward movement) and a posterior subcircuit (which initiates forward movement). These two subcircuits, both activated by

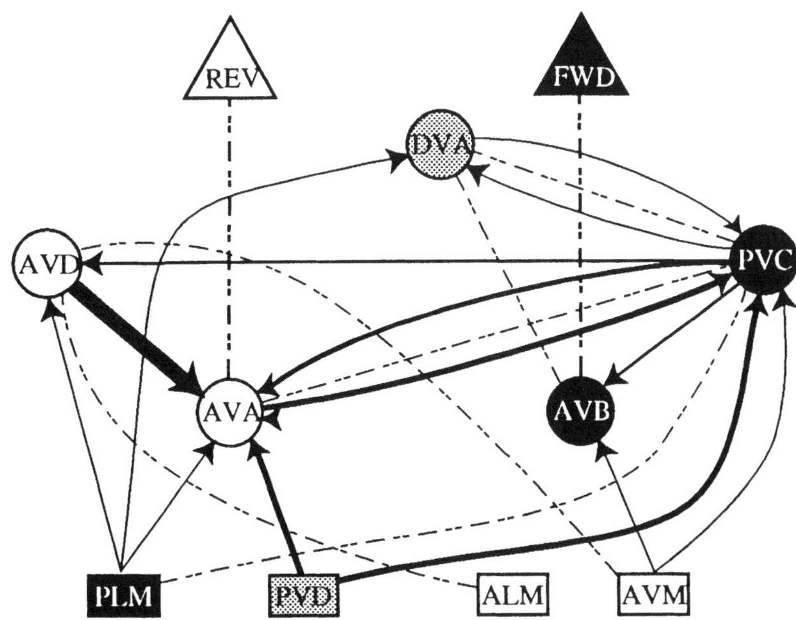

Fig. 2. A simplified version of the nematode tap-withdrawal neural circuit showing only unilateral connections. The circuit is comprised of seven sensory neurons (squares), nine interneurons (circles) and two pools of motoneuron pools (represented by triangles) which are responsible for producing forward and backward movement (Wicks and Rankin, 1995). With the exception of the midline cells AVM and PVD, all cells are bilaterally represented in the worm. Dotted lines indicate gap junctions. Arrows indicate chemical connections, the width of which are proportional to the number of synaptic contacts. Black cells with white lettering are involved in forward movement, white cells are involved in backward movement, and gray cells are involved in both. Modified from Wicks and Rankin (1995).

the tap, are integrated to produce the tap-withdrawal response. Thus, the tap-withdrawal response is more complex than the response to either head or tail touch alone. Interestingly, it has been shown that when isolated via single-cell laser ablation, these two subcircuits habituate at different rates (Wicks and Rankin, 1996a). Identification of the underlying circuitry is an important step in the attempt to identify the genetic basis of any behavior: there should be concordance between cells implicated in the behavior by laser methods and cells implicated by genetic methods.

A series of experiments by Wicks and Rankin (1997) demonstrated that the interneurons AVA, AVB, AVD and PVC make up a command system for forward and backward locomotion and are not affected by habituation to tap. This allows the localization of habituation of the tap-withdrawal response to the ALM, AVM, and PLM sensory cells and/or their synapses onto the interneurons (Wicks and Rankin, 1997). This finding can be applied to the genetic analysis of habituation to tap as there are cell-specific gene promoters, called the *mec* promoters that are expressed predominately in the touch cells. The use of these promoters will allow us to gain spatial control over gene expression, and to specifically target the cells in which we believe the learning is occurring.

The application of genetic techniques to examine learning in *Caenorhabditis elegans*

Forward genetics: how to characterize mutants that affect learning

Until recently, most of the genetic research using *C. elegans* has relied upon classic forward genetic techniques (Plasterk, 1992). This approach involves exposing groups of animals to chemical or X-ray irradiation mutagenesis and then identifying any interesting behavioral phenotypes that may result (see Epstein and Shakes, 1995 for more details regarding these procedures). The gene(s) involved in these mutations are then isolated and characterized. In order to characterize mutants that may be involved in learning we need a behavioral assay or screen that is sensitive to defects that may affect any of the different facets of learning. In addition, we need to rule out other deficits that may affect learning directly or indirectly. The rest of this section outlines assays that will characterize mutants that are defective in habituation. We have taken two approaches to discovering habituation mutants — one is to examine identified mutations in genes that have been implicated in plasticity in other organisms, i.e. PKC, PKA, potassium and calcium channels, CAM-kinase, et cetera; the other is to carry out a screen to discover new mutations that affect habituation.

Existing mutations

Once a mutant strain has been selected for analysis, there are a number of aspects of behavior that must be examined to determine the effect of the mutation on habituation. As with any genetic analysis of behavior it is important to first demonstrate that the behavioral background is unchanged and that the mutation only targets the behavior of interest. In the case of the tap-withdrawal response it is important to make certain that the central pattern generator (composed of the 4 pairs of interneurons) is functioning normally, or in a way that does not compromise the response to tap. Thus, the first step is to examine the spontaneous locomotory behavior of the mutant. This is done by recording the behavior of single worms (at least 20 worms per mutant strain and N2 wild-type for control) in the absence of taps and determining the overall activity level of the animal. In addition, the frequency and magnitude of spontaneous reversals (when the animal stops forward movement and moves backwards in the absence of any obvious stimulus) of the animal during this time are also recorded. When placed into a new environment, worms typically move around a great deal and exhibit a high number of spontaneous reversals; however, both of these measures of activity level tend to decrease as the worm adapts to its new environment. In addition, it is useful to examine how rapidly the mutant moves about on the plate. Speed of the animal may affect the assessment of habituation, particularly if the mutant is unusually slow or fast. Following our assessment of spontaneous behavior we measure the worm's response to a single tap stimulus. We look at the frequency and the magnitude of the response to tap and compare that to wild-type worms. If we find a mutant that shows an abnormality in one or more of the spontaneous measures, or in the response to tap we must take that into consideration when examining habituation. In some extreme cases, i.e. in uncoordinated strains, we may need to develop additional behavioral measures to assess the effects of repeated tap stimulation. For example, if a smooth backward moving response is not observed in a mutant strain, we may measure the number of sine waves produced in the response to tap, or the amount of time a worm is activated by the tap stimulus, etc.

After having determined whether there are any behaviors that are affected by the mutation that may influence habituation indirectly, habituation itself is examined, again using single worms with at least 20 animals for each strain tested and their results pooled. This is necessary because there is a high level of individual variation in habituation in all species, fewer worms would not give reliable results. The rate of stimulus presentation or interstimulus interval (ISI) is a very important parameter that affects both the rate and depth of habituation, as well as the rate of spontaneous recovery from habituation. As with other organisms, habituation in *C. elegans* is faster and more complete with shorter ISIs, and recovers more rapidly than with longer ISIs (Rankin and Broster, 1992). Rankin (1994) hypothesized that short and long ISIs may recruit some similar and some different cellular processes, thus we may find that some mutations affect only short ISIs and some only long ISIs, while a third class of mutation may affect both. In addition, habituation and spontaneous recovery may reflect different processes. Therefore it is necessary to assess habituation of a mutant with 30 stimuli at both a short (usually 10 s) and long (usually 60 s) ISI and then administer taps at 30 s, 5, 10, and at 15 min from the last tap to assess spontaneous recovery from habituation.

In analyzing habituation we typically measure the magnitude of reversal responses; however, to evaluate mutants, we also look at additional measures including a frequency distribution of response magnitudes as well as the frequency of reversal responses across stimuli. We have examined a number of mutants using this behavioral assay and have found some interesting results. For example, one of the more intensely-studied mutants in our laboratory has been the *eat-4 (ky5 III)* mutation. The *eat-4* mutant is defective in feeding, sensitivity to nose touch, chemotaxis, and thermotaxis. Using both *lac*Z and GFP translational fusions, Lee, Sawin, Chalfie, Horvitz and Avery (1999) demonstrated that *eat-4* is expressed in a number of pharyngeal neurons in addition to the ALM, AVM and PLM mechanosensory neurons and suggested that *eat-4* plays a role in glutamatergic transmission in the mechanosensory neurons. We have found that *eat-4* shows normal spontaneous behavior and the initial response to tap is normal in both frequency and magnitude (with a trend towards responses slightly larger than N2). In tests of habituation *eat-4* exhibits a number of features that may allow it to be considered a "learning" mutant. For example, Fig. 3 shows the habituation curves for both *eat-4* and N2 wild-type worms at 10 s and 60 s ISIs. *eat-4* worms demonstrate an accelerated rate of habituation, a depressed asymptotic level of habituation, and a rather retarded level of recovery from habituation when compared to N2 wild-type worms. Thus, *eat-4* worms may be characterized as "fast" learners and appear to retain such learning for longer periods of time than normal worms. It will be interesting to see whether this "increased retention" is also reflected in long-term habituation.

We have also found another class of mutants that show an alteration in the frequency distribution of response magnitudes to tap. An example of this is the *ced-3 (n717)* mutation. The *ced-3* mutants made significantly fewer spontaneous reversals than N2 wild-type animals; however, the magnitude of these spontaneous reversals was significantly larger than the spontaneous reversals of N2 wild-type animals. There were fewer initial responses to tap for the *ced-3* worms, and the initial responses were smaller than those of wild-type worms.

When tested for habituation, the response magnitude curves for the *ced-3* worms were not different from the curves for N2 wild-type worms (Fig. 4). However, an analysis of the frequency of response magnitudes showed many more "no responses" or "0" response magnitudes for the *ced-3* worms (mean number of response magnitudes of "0"" per worm out of 30 stimuli: *ced-3* $M = 25.3$; sd $= 2.78$ vs N2 $M = 6.67$; sd $= 3.42$). When the "0" responses were omitted and an analysis was done on the frequency of response magnitudes remaining there was an interesting difference in the frequency distributions for the *ced-3* worms. N2 animals showed a large number of small reversal responses (1–49 pixels) at a 10 s (Fig. 4); in contrast, *ced-3* animals demonstrated very few small reversal responses. The same findings were found for both groups of animals at a 60 s ISI. Therefore, when *ced-3* animals do reverse in response to tap, it is usually a moderately sized reversal rather than a small one. Thus, it appears that in *ced-3* worms both evoked and spontaneous reversals were affected by the mutation. We do not yet understand the reason for these changes in behavior; *ced-3* is a cell

A

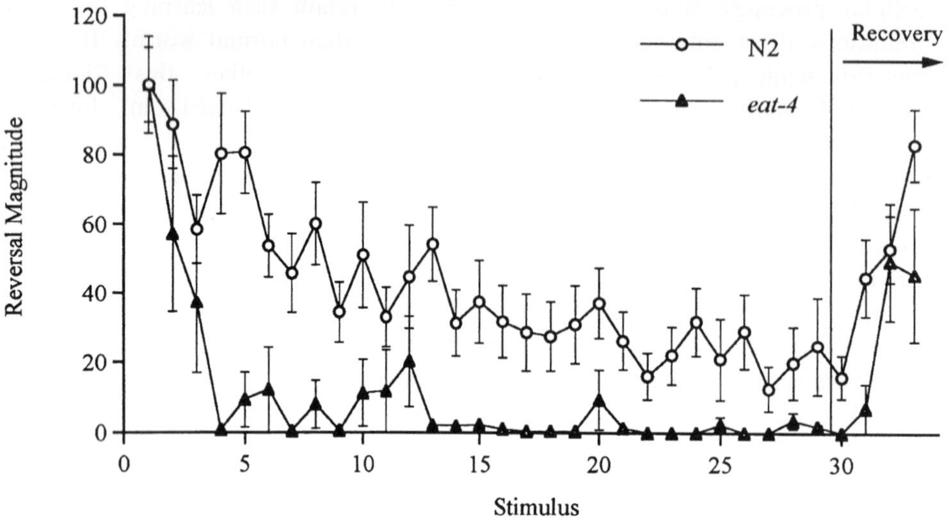

B

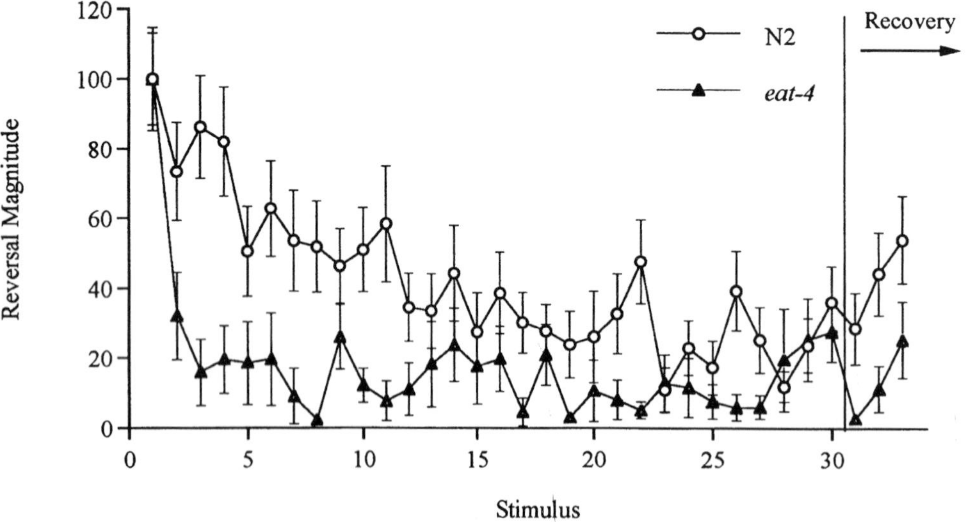

Fig. 3. Comparison of habituation and recovery from habituation of N2 wild-type and *eat-4 (ky5 III)* worms following repeated stimuli at 10 s ISI (A) and 60 s ISI (B). N2 animals are represented by open circles. *eat-4* animals are represented by darkened triangles. Reversal response magnitude is standardized to mean initial response for each group. Error bars indicate the standard error of the mean. The vertical line separates the responses to the 30 stimuli used to habituate the animals from responses to the three recovery taps, delivered at 30 s, 5 , and 10 min after habituation training. Habituation of the *eat-4* animals occurs at a much faster rate and is more pronounced than control animals at both the 10 s and 60 s ISIs.

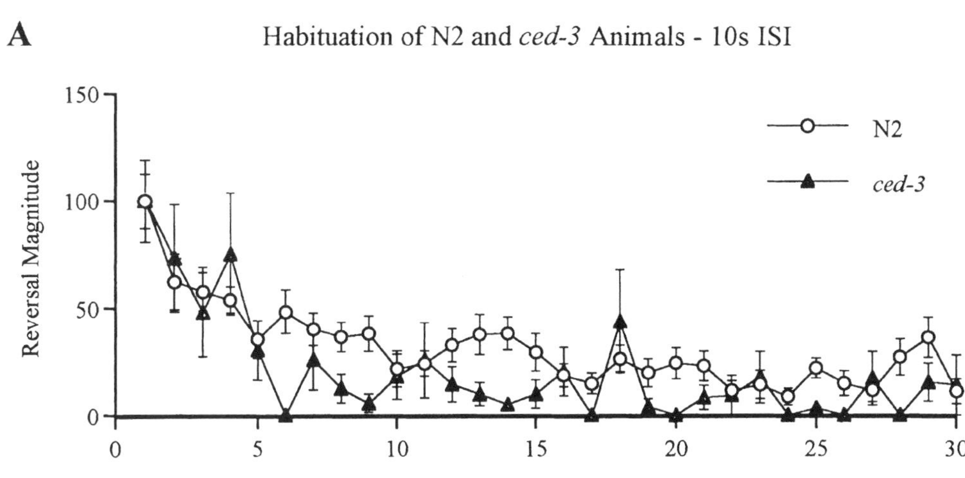

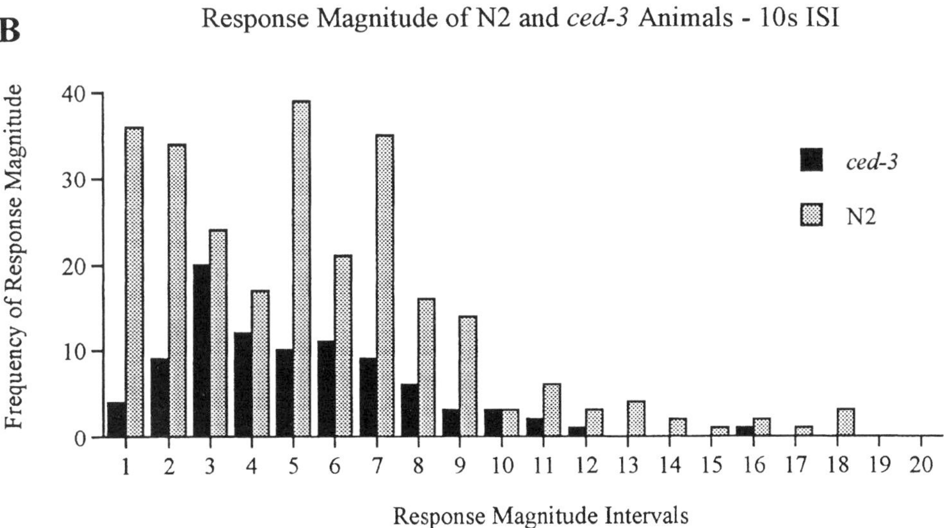

Response Magnitude Intervals (pixel lengths):

1= 1-49	7= 301-350	13= 601-650
2= 50-100	8= 351-400	14= 651-700
3= 101-150	9= 401-450	15= 701-800
4= 151-200	10= 451-500	16= 801-850
5= 201-250	11= 501-550	17= 851-900
6= 251-300	12= 551-600	18= 901-950

Fig. 4. (A) Habituation curves and (B) frequency distributions of response magnitudes for N2 wild-type and *ced-3 (n717)* animals at a 10s ISI. N2 animals are represented by open circles and gray hatched bars. *ced-3* animals are represented by darkened triangles and solid black bars. Reversal magnitude is standardized to mean initial response for each group. Error bars indicate the standard error of the mean. Below is the corresponding pixel length of each response magnitude interval derived from the NIH Image (Version 1.57) software package. The *ced-3* animals demonstrate very few small (1–49 pixels) reversal responses; in contrast, the control animals exhibit a large frequency of such reversal responses.

death mutant in which a number of cells that would normally die during development do not, including some cells that resemble touch cells (see Hengartner, 1997 for review of *ced* mutants). A detailed analysis of the neural circuit in these mutant worms might be necessary to understand the behavioral differences that we see. Given that both evoked and spontaneous reversals are affected similarly, our interpretation of the mutation would be that it does not affect habituation directly, but instead it alters the central pattern generator (the four interneurons) rather than the interface between the sensory cells and the interneurons, where habituation is presumably localized.

Thus, the results of the analyses of spontaneous behavior, of response to tap and of habituation to tap will allow us to categorize a variety of different alterations in behavior and to focus on those that appear to directly affect the processes of habituation rather than those that affect other components of the behavior.

Screen for new habituation mutants

The method of investigating mutant strains described above is very labor intensive and while useful for studying already identified mutants, it would be useless as a screen for new mutants affecting habituation. The time taken to carry out the assays using a single worm at a time would be prohibitive for screening the hundreds of thousands of strains necessary. Thus, a new approach had to be developed in order to screen for worms showing deficits in habituation. To do this we are developing a computer program that is able to track a number of worms on a plate simultaneously. Worms are mutagenized using ethymethylsulfonate (EMS), and adults allowed to lay eggs on a plate. As the eggs hatch, each of the progeny is then plated onto a seeded plate by itself to reproduce. When the offspring of these animals reach maturity they will form a homozygous group of worms with a specific mutation. It is these F_2 worms that will be tested in our screen. Worms ($\sim$50 at a time) will be taken from the F_2 plate and placed onto a blank Nematode Growth Medium agar plate on the microscope stage for testing. The plate containing the worms will then be given habituation training at either the 10 s or the 60 s ISI. The computer will track the movement of the worms on the screen, and calculate the number of animals reversing to each tap and the magnitude of the responses to tap. At the end of each habituation run, the computer will generate two curves, one showing the number of worms reversing to each tap and the other showing the magnitude of the reversal to the taps. These curves will then be compared to the wild-type curves to determine whether there are any differences in either the frequency or the response magnitudes between the mutant strains and the wild-type worms. Once strains suspected to show mutations in habituation have been identified, the defects in behavior will be more closely investigated as described above for existing mutations.

Learning vs memory: addressing the nature of the acquisition or rate of learning

The behavioral assay described above allows us to assess the processes of learning and memory independently by examining, in detail, the rate and features of the acquisition of habituation along with the retention of habituation. Typically, work with other models, i.e. *Drosophila*, involves training populations of animals during a single or multiple trials and then assaying the proportion of animals that learned and those that did not learn. However, by recording and analyzing behavior during training, we are able to examine the acquisition of learning in addition to memory for that learning. Through the application of this behavioral assay, we can potentially identify mutants that are "slow" or "fast" learners as well as those that are capable or incapable of retention of habituation training. Thus, we may detect genes that have a more subtle effect on the learning process than will assays that simply test retention.

Reverse genetics: methods of transforming DNA in Caenorhabditis elegans *and their application to the genetic dissection of learning*

In studies of learning and memory a number of compounds, i.e. PKC, PKA, CAM-kinase, CREB, potassium and calcium channels, etc, have been implicated in the cellular mechanisms underlying learning and memory. Reverse genetic techniques allow us to specifically target these compounds and either disrupt or enhance their function and to test the effect of that manipulation on habituation. A number of techniques have been developed to achieve transformation of DNA in *C. elegans*. For example, techniques used to interrupt specific genes via transposon (Tc1) insertion followed by isolation of mutants and maintaining frozen mutant libraries have been developed (Zwaal et al., 1993; Plasterk, 1995). Gene expression can also be affected by a technique known as RNA interference in which the wild-type hermaphrodite mother is treated with RNA from the desired gene (Tabara et al., 1998). Following this treatment, most of her offspring show the phenotype of the targeted gene. In addition, techniques for DNA microinjection and the subsequent assays required for determining DNA transformation and gene expression have been described in detail elsewhere (see Mello and Fire, 1995). The remainder of this chapter will highlight those aspects of transformation techniques that relate specifically to our discussion and make the genetic dissection of learning easier and more efficient. Methods used for exerting control over the spatial and temporal expression of genes will then be addressed.

Co-transforming mutants: choosing a mutant to rescue

When DNA is injected into *C. elegans* it is useful to have a fairly obvious phenotypic marker to identify individuals that have been transformed. By co-injecting the gene sequence of interest and the rescue sequence for the phenotypic marker, it is rather simple/easy to identify transformed worms. Mello and Fire (1995) discuss several genetic markers that have been used for the identification and subsequent selection of transformed lines of worms, these include: *rol-6 (su1006)* which produces a distinctive "rolling" phenotype (Kramer et al., 1990) and disruption of the *unc-22* gene via antisense RNA constructs which produce an easily recognizable "twitching" phenotype (Fire et al., 1991). Mutant rescue assays, in which the DNA sequence of interest is injected into the gonad of a known mutant strain along with the wild-type copy of the gene that particular mutant is missing, are ideal for this purpose. A mutant is "rescued" if the injected DNA is taken up by the oocytes and the defect present in the injected animal is not observed in the animal's progeny. Furthermore, in addition to the successful suppression of the mutation, it can be reasonably assumed that the progeny also possesses the DNA sequence of interest that was co-injected. However, as it will be pointed out later, this is not always the case. Due to the nature of the habituation behavioral assay, it is important to choose a genetic marker that will not produce a phenotype that will affect movement in a way that will interfere with the assessment of habituation.

When deciding which mutant and subsequently which gene to use as a marker phenotype, the mutant gene needs to have been cloned and a rescue sequence must be available. Mello and Fire (1995) suggest that recessive alleles with distinctive phenotypes are the most desirable. For example, the *unc-4* gene has been used to rescue a healthy but uncoordinated phenotype (Miller et al., 1992). However, as mentioned earlier, when assessing habituation it is important to use a mutation that will not affect the worm's movement should there be incomplete penetrance in expression of the injected DNA. A number of such mutant strains and rescue genes are available, including: *lin-12*, *rol-6*, *dpy-19*, and *dpy-20*. In our first experiment, we chose to use *dpy-20*, which is a temperature-sensitive mutant that exhibits an obvious "dumpy" phenotype. We have successfully rescued this mutant by co-injecting the *dpy-20* gene with a construct containing a hsp-16 promoter and CREB response element (CRE isolated from *Drosophila*). When rescued and raised at the appropriate temperature this transformed strain shows normal frequency and magnitude of reversals to tap, as well as normal habituation (to date we have no evidence of an effect of the *Drosophila* CRE). The choice of which mutant to use in an experiment will ultimately depend upon the behavioral assay that is required.

It is important to note that although there is often complete rescue of the mutant phenotype with the wild-type gene, there may be instances of partial rescue. Two ways in which partial rescue may manifest itself are with complete rescue that is transient, occurring in only a small proportion of the animals, or with incomplete rescue occurring in all of the transformed animals (Mello and Fire, 1995). This appears to be a general problem with virtually all attempts to transform DNA via microinjection: the DNA construct is often not taken up by a sufficient number of animals to allow transmission down through the germ line (Plasterk, 1992). In outlining the probable fates of injected DNA sequences into *C. elegans*, Mello and Fire (1995) suggest that the majority (10–100%) of the F_1 generation animals tend to be mosaic, with the DNA expressed in only a small proportion of cells (Fire and Waterston, 1989). An even smaller percentage of the DNA (1–10%) form long extrachromosomal tandem arrays, which then appear to be transmitted at a relatively stable rate to subsequent generations (Strinchcomb et al., 1985). Random integration of the DNA occurs even less frequently (0.1–0.5%), followed by homologous recombination (0.001–0.01%); however, the integrated DNA in such cases is then transmitted down along the germ line according to the principles of Mendelian inheritance (Fire, 1986; Broverman et al., 1993). The problem of infrequent homologous recombination of transgenic DNA in *C. elegans* has led to the development of other techniques such as target-selected gene inactivation using the Tc1 transposon (Plasterk, 1992, 1995). Although homologous recombination of genes and transgenic DNA has not proved to be as successful as one would like, an optimistic attitude should not be abandoned as developments in DNA transformation technology in the future should make this approach more viable (Mello and Fire, 1995).

Examining patterns of gene expression

Once a gene of interest has been identified, the cells in which the gene is expressed can be determined. The most common method of examining patterns of gene expression in *C. elegans* is the use of *lac*Z reporter vectors. These vectors are comprised of the promoter region of the gene of interest, the *E. coli* gene *lac*Z responsible for encoding *β*-galactosidase, and a gene sequence of interest. Following introduction of the vector, the transgene can be located by staining for the presence of *β*-galactosidase activity (McGhee and Krause, 1997). Although *lac*Z reporter vectors have been the most commonly used constructs, there are occasional problems of blocked expression occurring in the germ line (Mello and Fire, 1995). A further disadvantage of using constructs with the *lac*Z gene are that expression patterns cannot be identified in living animals — staining for *β*-galactosidase activity is performed on dead animals (Mello and Fire, 1995). A relatively recent method of tracking gene expression patterns in vivo is the use of aequorin or green fluorescent protein (GFP) reporters (Chalfie et al., 1994). Originally isolated from the jellyfish *Aequorea victoria*, GFP does not appear to interfere with cell growth and offers the advantage of requiring only irradiation by UV or blue light to be detected. When illuminated with light in the 450–490 nm range, expression of GFP in *C. elegans* is relatively stable, usually lasting for at least 10 min (Chalfie et al., 1994). Regardless of whether *lac*Z or GFP are used, one must always consider the possibility that introducing a reporter construct may somehow interfere with transcriptional and post-transcriptional factors to produce abnormal proteins. Therefore, the expression pattern must be confirmed independently through endogenous enzyme activity, antibody staining, or by in situ hybridization (McGhee and Krause, 1997).

Spatial control over gene expression and habituation in Caenorhabditis elegans

A fundamental question in the reductionistic approach to examining habituation is what level in the neural circuit is responsible for mediating the tap-withdrawal response? More specifically, is the integration of information provided by the two competing subcircuits carried out at the level of the sensory neurons or the interneurons? Through the use of reporter constructs, specific gene promoters that ensure gene expression occurs only in specified cells and targeted gene mutations, we can address this issue.

Fire et al. (1990) generated a number of modular *lac*Z fusion vectors that were useful in the study of gene expression in *C. elegans*. Since this time, new vectors have been created that allow for cell-specific or well-defined temporal patterns of expression through the use of a specific RNA or protein product. These ectopic expression vectors are described in Mello and Fire (1995) and a number of such vectors with specific *C. elegans* promoters are available from A. Fire. Of particular interest for the assessment of habituation of the tap-withdrawal response, are those ectopic vectors containing *mec* gene promoters. A number of mutants with abnormal mechanosensation have been identified and a model for the transduction of mechanostimuli that involves a number of MEC proteins has been developed (Driscoll and Kaplan, 1997). As a result of the analysis of the genes involved in mechanosensation a number of promoter sequences for the genes that are expressed only in the touch cells are available. The advantage of using these *mec* gene promoters is the spatial control of gene expression that can be obtained — DNA can be injected into an animal, with only the touch cells being affected. Since habituation of the tap-withdrawal response has been localized to the touch cells through behavioral and laser ablation techniques, the use of *mec* promoters will enable an analysis of the role of a gene of interest disrupted in only the cells involved in the behavior.

Temporal control of gene expression and habituation in Caenorhabditis elegans

The use of *mec* promoter genes can also allow some temporal control over gene expression. For example, the *mec-7* gene is expressed later in development than the *mec-3* gene, as the MEC-3 protein appears to be involved in the differentiation of the touch cells (Hamelin et al., 1992). In this regard, the *mec-7* promoter may be more valuable to the study of habituation in that it is expressed later and therefore reduces the developmental problems of pleiotropy discussed earlier in the chapter. A more conventional method of controlling the temporal expression of a gene is through the use of the promoter sequences of small heat shock (*hsp16*) genes. Using *hsp16::lac*Z fusions, Stringham et al. (1992) exposed worms to heat shock ($\sim 33°$ Celcius) and observed expression of *lac*Z in a variety of cell types. Thus, the use of heat shock promoters can allow for temporal control over gene expression in neurons and overcomes the problems of pleiotropy discussed earlier in the chapter.

In investigating the genes involved in habituation, it is important to use both tissue-specific promoters and heat shock promoters in order to control for the problems associated with one another. By using *mec* promoters we can limit gene expression primarily to the touch cells; however, this comes at the expense of the possible developmental problems of pleiotropy. On the other hand, using heat shock promoters we are able to exert temporal control, but this is at the expense of ectopic resolution. By using both types of promoters we get a more accurate picture of the role of the gene we are expressing in habituation.

Summary

The first paper to demonstrate that *C. elegans* could change its behavior as a result of experience and could learn and remember information for periods ranging from minutes to days, despite its short lifespan, tiny size, and small nervous system, was only published in, 1990 (Rankin et al., 1990). In the few short years since then, a number of learning assays have been described (see review in Gannon and Rankin, 1995), neural circuits underlying behaviors that show plasticity have been investigated (Wicks and Rankin, 1995, 1997), and mutant strains that show a variety of deficits in learning have been isolated (Wicks and Rankin, 1996b; Wen et al., 1997). The amenability of *C. elegans* to both forward and reverse genetic techniques, as well as the completion of a fully sequenced and mapped genome means that *C. elegans* offers great promise as a system in which a full understanding of the genetics of learning and memory is possible.

List of Abbreviations

CAM	calmodulin
ced	cell death (mutant gene)
C. elegans	*Caenorhabditis elegans*
cm	centimeters
CRE	cyclic AMP response element
CREB	cyclic AMP response-element binding protein
dpy	dumpy (mutant gene)
DNA	deoxyribronucleic acid
E. coli	*Escherichia coli*
eat	eating defect (mutant gene)
F1	first filial generation
GFP	green fluorescent protein
hsp	heat shock protein
ISI	interstimulus interval
lacZ	gene that encodes the enzyme β-galactosidase
lin	lineage (mutant gene)
lrn	learn (mutant gene)
M	mean
mec	mechanosensation (mutant strain)
min	minute
ml	millilitre
N	Newton
nm	nanometer
PKA	protein kinase A
PKC	protein kinase C
RNA	ribronuceic acid
rol	roller (mutant gene)
s	second
SD	standard deviation
unc	uncoordinated (mutant strain)
UV	ultraviolet

Note: Regarding cell names (ALM, AVA, AVD, AVM, DVA, PLM, PVC and PVD), some, but not all of these names did start out as abbreviations (e.g. ALM: anterior lateral microtubule); however, it is now common convention that their three letter designations are their names rather than an abbreviated version.

References

Avery, L., Raizen, D. and Lockery, S. (1995) Electrophysiological methods. In: H.F. Epstein and D.C. Shakes (Eds.), *Caenorhabditis elegans: Modern Biological Analysis of an Organism*, Academic Press, San Diego, pp. 251–269.

Bargmann, C.I. (1993) Genetic and cellular analysis of behavior in *C. elegans*. *Annu. Rev. Neurosci.*, 16: 47–71.

Bargmann, C.I. and Avery, L. (1995) Laser killing of cells in *Caenorhabditis elegans*. In: H.F. Epstein and D.C. Shakes, *Caenorhabditis elegans: Modern Biological Analysis of an Organism,* Academic Press, San Diego, pp. 225–250.

Beck, C.D.O. and Rankin, C.H. (1993) The effects of aging on behavioral plasticity in *C. elegans*. *Behav. Processes,* 28: 145–164.

Beck, C.D.O. and Rankin, C.H. (1995) Heat shock disrupts long-term memory consolidation in *Caenorhabditis elegans*. *Learnin and Memory*, 2: 161–177.

Beck, C.D.O. and Rankin, C.H. (1997) Long-term habituation is produced by distributed training at long ISIs and not by massed training or short ISIs in *Caenorhabditis elegans*. *Anim. Learn. Behav.*, 25: 446–457.

Brenner, S. (1974) The genetics of *Caenorhabditis elegans*. *Genetics*, 77: 71–94.

Broverman, S., MacMorris, M. and Blumenthal, T. (1993) Alteration of *Caenorhabditis elegans* gene expression by targeted transformation. *Proc. Nat. Acad. Sci. USA*, 90: 4359–4363.

Chalfie, M. and Sulston, J.E. (1981) Developmental genetics of the mechanosensory neurons of *C. elegans*. *Dev. Biol.*, 82: 358–370.

Chalfie, M., Sulston, J.E., White, J.G., Southgate, E., Thomson, J.N. and Brenner, S. (1985) The neural circuit for touch sensitivity in *Caenorhabditis elegans*. *J. Neurosci.*, 5: 956–964.

Chalfie, M., Tu, Y., Euskirchen, G., Ward, W. and Prasher, D. (1994) Green fluorescent protein as a marker for gene expression. *Science*, 263: 802–805.

Chiba, C.M. and Rankin, C.H. (1990) A developmental analysis of spontaneous and reflexive reversals in the nematode *C. elegans*. *J. Neurobiol.*, 21: 543–554.

Coulson, A.R., Kozono, Y., Lutterbach, B., Shownkeen, R., Sulston, J. and Waterson, R. (1991) YACs and the *C. elegans* genome. *BioEssays*, 13: 413–417.

Coulson, A.R., Sulston, J., Brenner, S. and Karn, J. (1986) Toward a physical map of the genome of the nematode *C. elegans*. Proc. Nat. Acad. Sci., 83: 7821–7825.

Croll, N.A. (1975) Components and patterns of behavior of the nematode *Caenorhabditis elegans*. *J. Exp. Zool.*, 176: 159–176.

DeZazzo, J. and Tully, T. (1995) Dissection of memory formation: From behavioral pharmacology to molecular genetics. *Trends Neurosci.*, 18: 212–218.

Driscoll, M. and Kaplan, J. (1997) Mechanotransduction. In: D.L. Riddle, T. Blumenthal, B.J. Meyer and J.R. Priess (Eds), *C. elegans* II, Cold Spring Harbor Laboratory Press, pp. 645–677.

Dudai, Y. (1988) Neurogenetic dissection of learning and short-term memory in *Drosophila*. *Annu. Rev. Neurosci.*, 11: 537–563.

Edgely, M.L. and Riddle, D.L. (1987) *Caenorhabditis elegans* genetic map. *Genetic Maps*, 4: 351–356.

Epstein, H.F. and Shakes D.C. (1995) *Caenorhabditis elegans: Modern biological analysis of an organism*, Academic Press, San Diego

Fire, A. (1986) Integrative transformation of *Caenorhabditis elegans*. *EMBO J.*, 5: 2673–2680.

Fire, A., Albertson, D., Harrison, S.W. and Moerman, D.G. (1991) Production of antisense RNA leads to effective and specific inhibition of gene expression in *C. elegans* muscle. *Development*, 113: 503–514.

Fire, A., Harrison, S. and Dixon, D. (1990) A modular set of *lac*Z fusion vectors for studying gene expression in *Caenorhabditis elegans*. *Gene*, 93: 189–198.

Fire, A. and Waterston, R. (1989) Proper expression of myosin genes in transgenic nematodes. *EMBO J.*, 8: 3419–3428.

Gannon, T. N. and Rankin, C.H. (1995) Methods of studying behavioral plasticity in *Caenorhabditis elegans*. In: H.F. Epstein and D.C. Shakes (Eds.), *Caenorhabditis elegans: Modern biological analysis of an organism*, Academic Press, San Diego, pp. 205–223.

Goodman, M.B., Hall, D.H., Avery, L. and Lockery, S.R. (1998) Active currents regulate sensitivity and dynamic range in *C. elegans* neurons. *Neuron*, 20: 763–772.

Groves, P.M. and Thompson, R.F. (1970) Habituation: A dual process theory. *Psychol. Rev.*, 77: 419–450.

Hamelin, M., Scott, I.M., Way J.C. and Culotti, J.G. (1992) The *mec*-7 beta-tubulin gene of *Caenorhabditis elegans* is expressed primarily in the touch receptor neurons. *EMBO J.*, 11: 2885–2893.

Hedgecock, E.M. and Russell, R.L. (1975) Normal and mutant thermotaxis in the nematode *Caenorhabditis elegans*. *Proc. Nat. Acad. Sci. USA*, 72: 4061–4065.

Hengartner, M.O. (1997) Cell death. In: D.L. Riddle, T. Blumenthal, B.J. Meyer and J.R. Priess (Eds.), *C. elegans* II, Cold Spring Harbor Laboratory Press, pp. 383–416.

Hodgekin, J., Horvitz, H.R. and Brenner, S. (1979) Nondisjunction mutants of the nematode *Caenorhabditis elegans*. *Genetics*, 91: 67–94.

Johnsen, R.C. and Baillie, D.L. (1997) Mutation. In: D.L. Riddle, T. Blumenthal, B.J. Meyer and J.R. Priess (Eds.), *C. elegans* II, Cold Spring Harbor Laboratory Press, pp. 79–95.

Kramer, J., French, R.P., Park, E. and Johnson, J.J. (1990) The *Caenorhabditis elegans rol-6* gene, which interacts with the *sqt-1* collagen gene to determine organismal morphology, encodes a collagen. *Mol. Cell Biol.*, 10: 2081–2090.

Lee, R.Y.N., Sawin, E.R., Chalfie, M., Horvitz, H.R. and Avery, L. (1999) EAT-4, a homolog of a mammalian sodium-dependent inorganic phosphate co-transporter, is necessary for glutamatergic neurotransmission in *Caenorhabditis elegans*. J. Neurosci., 19, 159–167.

Mah, K.B. (1991) An analysis of the Tap Withdrawal Response in Male *Caenorhabditis elegans*. Unpublished master's thesis, Department of Psychology, University of British Columbia, Vancouver, Canada.

McGhee, J.D. and Krause, M.W. (1997) Transcription factors and transcriptional regulation. In: D.L. Riddle, T. Blumenthal, B.J. Meyer and J.R. Priess (Eds.), *C. elegans* II, Cold Spring Harbor Laboratory Press, pp. 147–184.

Mello, C. and Fire, A. (1995) DNA transformation. In: H.F. Epstein and D.C. Shakes (Eds.), *Caenorhabditis elegans: Modern biological analysis of an organism*, Academic Press, San Diego, pp. 451–482.

Miller, D.M., Shen, M.M., Shamu, C.E., Burglin, T.R., Ruvkun, G., Dubois, M.L., Ghee, M. and Wilson, L. (1992) *C. elegans unc-4* gene encodes a homeodomain protein that determines the pattern of synaptic input to specific motor neurons. *Nature*, 355: 841–845.

Plasterk, R.H.A. (1992) Reverse genetics of *Caenorhabditis elegans*. *BioEssays*, 14: 629–633.

Plasterk, R.H.A. (1995) Reverse genetics. In: H.F. Epstein and D.C. Shakes (Eds.), *Caenorhabditis elegans: Modern biological analysis of an organism*, Academic Press, San Diego, pp. 59–79.

Quinn, W.G., Harris, W.A. and Benzer, S. (1974) Conditioned behavior in *Drosophila melanogaster*. *Proc. Nat. Acad. Sci. USA*, 71: 708–712.

Rankin, C.H. (1994) Mechanistic questions raised by a behavioral analysis of habituation in *Caenorhabditis elegans*. *Seminars in the Neurosciences*, 6: 3–9.

Rankin, C.H. and Beck, C.D.O. (1992) *Caenorhabditis elegans*: A simple systems approach to the genetics of behavior. In: D. Goldowitz, D. Wahlsten and R.E. Wimer (Eds.), *Techniques for the genetic analysis of brain and behavior. Focus on the mouse*, Elsevier Science Publishers, Amsterdam, pp. 445–463.

Rankin, C.H., Beck, C.D.O. and Chiba, C.M. (1990) *C. elegans*: A new model system for the study of learning and memory. *Behav. Brain Res.*, 37: 89-92.

Rankin, C.H. and Broster, B.S. (1992) Factors affecting habituation and recovery from habituation in the nematode *Caenorhabditis elegans. Behav. Neurosci.*, 106: 239–242.

Riddle, D.L., Blumenthal, T., Meyer, B.J. and Priess, J. R. (1997) Introduction to *C. elegans*. In: D.L. Riddle, T. Blumenthal, B.J. Meyer and J.R. Priess (Eds.), *C. elegans* II, Cold Spring Harbor Laboratory Press, pp. 1-22.

Rose, A.M. and Baillie, D.L. (1979) The effect of temperature and parental age on recombination and nondisjunction in *Caenorhabditis elegans. Genetics*, 92: 409–418.

Stinchcomb, D.T., Shaw, J.E., Carr, S.H. and Hirsch, D. (1985) Extrachromosomal DNA transformation of *Caenorhabditis elegans*. Mol. Cell Biol., 5: 3484–3496.

Stringham, E.G., Dixon, D.K., Jones, D. and Candido, E.P.M. (1992) Temporal and spatial expression patterns of the small heat-shock (HSP16) genes in transgenic *Caenorhabditis elegans. Mol. Biol. Cell*, 3: 221–233.

Sulston, J.E. and Horvitz, H.R. (1977) Post-embryonic cell lineages of the nematode *Caenorhabditis elegans. Dev. Biol.*, 56: 110–156.

Sulston, J.E., Schrierenberg, E., White, J.G. and Thomson, J.N. (1983) The embryonic cell lineage of the nematode *Caenorhabditis elegans*. Dev. Biol., 100: 64–119.

Tabara, H., Grishok, A. and Mello, C. (1998) RNAi in *C. elegans*: Soaking the genome sequence. *Science*, 282: 430–431.

Thompson, R.F. and Spencer, W.A. (1966) Habituation: A model phenomenon for the study of neuronal substrates of behavior. *Psychol. Rev.*, 173: 16–43.

Tully, T. (1994) Gene disruption of learning and memory: a structure-function conundrum? *Seminars in the Neurosciences*, 6: 59–66.

Wen, J.Y.M., Kumar, N., Morrison, G., Rambaldini, G., Runciman, S., Rousseau, J. and Van der Kooy, D. (1997) Mutations that prevent associative learning in *C. elegans. Behav. Neurosci.*, 3: 354–368.

White, J.G., Southgate, E., Thomson, J.N. and Brenner, S. (1986) The structure of the nervous system of *Caenorhabditis elegans. Philos. Trans. R. Soc. London B*, 314: 1–340.

Wicks, S.R. and Rankin, C.H. (1995) Integration of mechanosensory stimuli in *Caenorhabditis elegans. J. Neurosci.*, 15: 2434–2444.

Wicks, S.R. and Rankin, C.H. (1996a) The integration of antagonistic reflexes revealed by laser ablation of identified neurons determines habituation kinetics of the *Caenorhabditis elegans* tap withdrawal response. *J. Comp. Physiol. A*, 176: 675–685.

Wicks, S.R. and Rankin, C.H. (1996b) A study of habituation of eat-4, a prospective learning mutant in *C. elegans. Soc. Neurosci. Abstr.*, 22: 694.

Wicks, S.R. and Rankin, C.H. (1997) Effects of tap-withdrawal response to habituation on other withdrawal behaviors: The localization of habituation in the nematode *Caenorhabditis elegans. Behav. Neurosci.*, 111: 342–353.

Zwaal, R.R., Broeks, A., Van Meurs, J., Groenen, J.T.M. and Plasterk, R.H.A. (1993) Target-selected gene inactivation in *Caenorhabditis elegans* by using a frozen transposon insertion mutant bank. *Proc. Nat. Acad. Sci. USA*, 90: 7431–7435.

W.E. Crusio and R.T. Gerlai (Eds.)
Handbook of Molecular-Genetic Techniques for Brain and
Behavior Research (Techniques in the Behavioral and Neural Sciences, Vol. 13)

CHAPTER 3.3.6

Forward genetic approaches in the analysis of *Caenorhabditis elegans*

Anne C. Hart

Massachusetts General Hospital Cancer Center, Harvard Medical School, Department of Pathology, CNY149-7202 13th Street, Charlestown, MA 02129, USA

Introduction

Understanding how the nervous system develops and functions is an extraordinary challenge which will require research on multiple levels, diverse approaches and a wide variety of model systems. Genetic approaches to nervous system development and function in model organisms excel at the identification and analysis of the relevant basic biochemical and cellular mechanisms *in vivo*. One of the premiere genetic model organisms is the nematode *C. elegans*, a small free living animal amenable to molecular and genetic techniques. Due to a suprisingly large behavioral repertoire, ease of cultivation in the laboratory, a relatively compact genome, and techniques which have been developed to map, characterize and clone genes, the analysis of the *C. elegans* nervous system using genetic and molecular techniques has yielded insights into fundamental mechanisms of nervous system function (see *e.g.* Chapter 3.3.5). In the following pages, techniques used for classical (forward) genetic analysis in *C. elegans* will be surveyed. (Reverse genetic techniques and fundamentals of the *C. elegans* life cycle are reviewed by C. Li in Chapter 3.3.7.) Examples will be drawn from the analysis of mechanotransduction in both the body and the nose touch neurons — two distinct, but complementary systems for response to mechanical stimuli.

C. elegans: an introduction

The *C. elegans* nervous system, composed of only 302 neurons in the adult hermaphrodite and 381 neurons in the male, is strikingly similar to vertebrate nervous systems at both the cellular and molecular level. As in vertebrates, distinct sensory, interneuron and motorneuron populations exist. *C. elegans* hermaphrodites have 39 classes of sensory neurons, at least 21 classes of motorneurons and up to 58 classes of interneurons based on morphological analysis from an electron microscope level reconstruction of the nematode nervous system (White et al., 1986). Each individual *C. elegans* neuron can be unambiguously identified in an animal. The ability to consistently identify specific individual neurons has facilitated analyses of development and function, *i.e.* analyses of circuit function based on individual cellular ablation by laser microsurgery. The electron microscope reconstruction data (White *et al.*, 1986) was also used to establish the patterns of synaptic and gap junctions made between individual neurons — and established that these connectivity patterns are reproducible from animal to animal. The recent innovation of patch clamp recording from *C. elegans* neurons (Goodman et al., 1998) gives the nematode neurobiologist the opportunity to directly assess the function of neurons in both wild type and mutant animals.

The nervous system of *C. elegans* has been a subject of intense and productive genetic analysis. Both development and function have been analyzed in great detail. Several points of similarity with vertebrate nervous systems have emerged. First, fundamental concepts in nervous system development, structure, and function have been conserved between invertebrates and vertebrates. Illustrative examples include axon guidance (Hamelin *et al.*, 1993; Culotti, 1994; Wadsworth *et al.*, 1996), molecular and functional basis of synaptic release (Riddle *et al.*, 1997) and distinct classes of glia, support cells and neurons (White *et al.*, 1986). Second, the molecules involved in neuronal development and function are directly conserved — including classical (Riddle *et al.*, 1997) and peptidergic (De Bono and Bargmann, 1998; Nelson *et al.*, 1998) neurotransmitters and their receptors, cytoskeletal proteins (Savage *et al.*, 1989; Du *et al.*, 1996; Khan *et al.*, 1997), signal transduction molecules (Kornfeld, 1997), ligand or voltage gated ion channels (Kornfeld, 1997) and chemosensory G-protein coupled receptors (Troemel *et al.*, 1995). All these molecules are found in *C. elegans* performing homologous roles in appropriate neurons.

Genes which encode molecules which are critical for neuronal function or development in *C. elegans* are often identified using genetic techniques coupled with behavioral and/or morphological assays (see Table 1). The nematode responds to a wide variety of sensory cues which modulate intrinsic behaviors including locomotion, egg laying, feeding, defecation, and mating. *C. elegans* is capable of learning, chemotaxis, chemoavoidance, and mechanosensation. In many cases, research in *C. elegans* has played a crucial role in elucidating the function/interactions of molecules critical for nervous system function or development. Genes which are critically involved in these processes have been identified through genetic screens.

Genetic screens in *C. elegans* and other model organisms proceed through characteristic stages:

1. identification of a reproducible, rapid and robust behavioral or morphological assay
2. screening for mutations which are defective
3. mapping and characterizing identified mutations
4. cloning the corresponding gene.

Additional steps follow, including determining the pattern of gene expression and the function of the corresponding protein. These and other techniques will be discussed below, followed by a discussion of results of forward genetic analyses of mechanosensation

Identification of mutations in C. elegans

Most of the *C. elegans* nervous system is not essential for viability or fertility in the laboratory, allowing the identification of many of the genes directly involved in nervous system function or development as homozygous viable mutations. Even mutations in genes required for synaptic release can be propagated (Nonet et al., 1993). Mutations have also been identified which specifically perturb response to sensory stimuli using behavioral assays (Chalfie and Au, 1989) using genetic techniques first described by Brenner (Brenner, 1974). One example is the identification of genes required for response to nose touch. Wild type hermaphrodites are mutagenized with chemicals or radiation, *i.e.* 50 mM ethylmethanesulfonate (EMS) or 3400 rads of gamma radiation, then allowed to self fertilize. The second generation (F_2) progeny are screened several days later for recessive or dominant mutations which perturb response to touch (see Fig. 1). 800–1200 animals can be tested for response to touch in a single day of behavioral screening. At standard dosages of mutagen, screening roughly 4000 animals is statistically equivalent to testing each gene in

TABLE 1

Behavioral assays for *C. elegans*

Behavior	Assay	Reference
Locomotion	forward coordination	(Brenner, 1974)
	backward coordination	(Brenner, 1974)
	spontaneous forward initiation	(Brenner, 1974)
	spontaneous reverse initiation	(Brenner, 1974)
	dispersal relative to food	(Segalat et al., 1995)
	halting at food source	(Sawin, 1997)
	bordering/clumping	(De Bono and Bargmann, 1998)
Mating	searching	(Loer and Kenyon, 1993)
	turning	(Loer and Kenyon, 1993)
	copulation	(Loer and Kenyon, 1993)
Egg laying	rate of laying	(Waggoner et al., 1998)
	periodicity of laying	(Waggoner et al., 1998)
Defecation	anterior contraction	(Liu and Thomas, 1994)
	posterior contraction	(Liu and Thomas, 1994)
	expulsion	(Liu and Thomas, 1994)
Eating	rate of pumping	(Avery, 1993a)
	electrical recording (pharynx)	(Avery, 1993b)
Foraging	rate and range of nose motion	(Hart, 1995)
Chemosensation	soluble attractants	(Ward, 1973)
	soluble repellants	(Dusenbery et al., 1975)
	volatile attractants	(Bargmann and Horvitz, 1991)
	volatile repellents	(Bargmann and Horvitz, 1991)
	osmotic avoidance	(Culotti and Russell, 1978)
Mechanosensation	body touch	(Chalfie et al., 1985)
	nose touch	(Kaplan and Horvitz, 1993)
	head withdrawal	(Hart, 1995)
	tap reflex	(Rankin, 1991)
	harsh touch	(Way and Chalfie, 1989)
Dauer	dauer formation defective	(Riddle, 1977)
	dauer formation constitutive	(Cassada and Russell, 1975)
	pheromone induced dauer	(Golden and Riddle, 1982)
	dauer recovery	(Albert et al., 1981)
	dauer nictation	(Croll and Matthews, 1977)
	dauer lethargus	(Croll and Matthews, 1977)
Morphological	dye filling	(Hedgecock et al., 1985)
	visualization (GFP, *lac*Z or antisera)	(Chalfie et al., 1994; Epstein and Shakes, 1995)
Drug assays	serotonin (egg laying and locomotion)	(Horvitz et al., 1982; Segalat et al., 1995)
	octopamine (egg laying)	(Horvitz et al., 1982)
	aldicarb (locomotion, synaptic release)	(Rand and Russell, 1985)
	AMPA (foraging and pumping)	(Avery and Horvitz, 1990; Hart, 1995)
	dopamine (egg laying and locomotion)	(Schafer and Kenyon, 1995; Sawin, 1997)
	avermectins (pumping and survival)	(Avery and Horvitz, 1990; Novak and Vanek, 1992)
	muscimol (pumping, survival)	(Avery and Horvitz, 1990; McIntire et al., 1993)
Thermosensation	thermotaxis	(Hedgecock and Russell, 1975)
	cryotaxis	(Hedgecock and Russell, 1975)
	heat avoidance	*
Learning/adaptation	isopotential tracking- memory	(Hedgecock and Russell, 1975)
	conditioned learning, habituation/adaptation	(Rankin *et al.*, 1990; Wen *et al.*, 1997)

C. elegans behavioral assays. A representative, but not exhaustive, list of assays used in the analysis of the *C. elegans* nervous system. For drug assays, only one of many drugs perturbing each neurotransmitter function is listed. *Animals avoid a point source of intense heat by initiating backward locomotion. (GFP, green fluorescent protein)

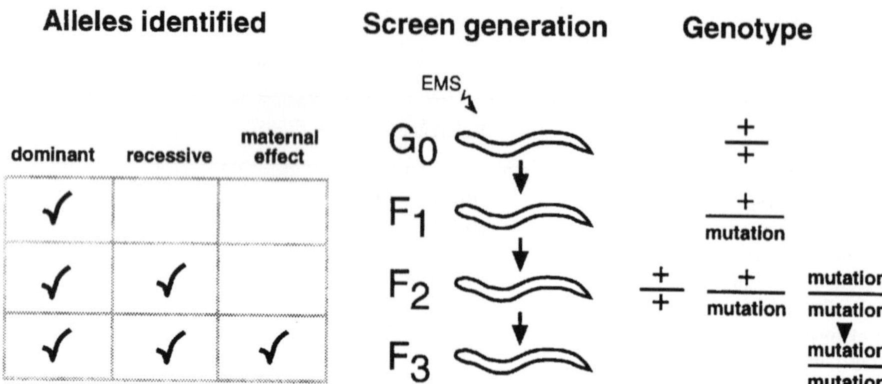

Fig. 1. Diagram of genetic screening in *C. elegans*. Identification of mutant alleles in *C. elegans*. Hermaphrodites are mutagenized with EMS or other mutagens in the first (G0) generation resulting in mutant germ cells. Self-progeny (F_1 generation) are heterozygous for mutations; dominant mutations can be identified in this F_1 generation. The F_1 animals self fertilize to yield the F_2 generation animals. One quarter of these F_2 animals are homozygous for a specific mutation. Recessive or dominant mutations can be identified in this F_2 generation. F_2 homozygous mutant animals yield all homozygous mutant progeny (arrowhead in right section). If the gene is only required maternally, then the mutant phenotype is only visible in the progeny of homozygous mutant animals and the F_3 generation must be examined. Screens of F_3 animals can also identify recessive or dominant mutations in nonmaternal effect genes as well.

the genome once for a role in response to touch (Brenner, 1974). Individual F_2 hermaphrodites which fail to respond to touch are allowed to self fertilize and their progeny are retested for a defect. Homozygous mutant animals yield all homozygous self progeny; animals carrying heterozygous dominant mutations yield one quarter homozygous mutant progeny and one half heterozygous mutant progeny. Protocol for mutagenesis are available at: http://www.dartmouth.edu/artsci/bio/ambros/protocols/MGH_protocols/koelle_prot/X0004_EMS_mutagenesis.html.

Of course, some genes required for behavior may also be essential for fertility or viability. However, these essential genes are frequently isolated either as partial loss of function alleles or in more sophisticated genetic enhancer or suppressor screens (Ferguson and Horvitz, 1989). Enhancer or suppressor screens differ from the screening strategy described above in that, (a) the animals mutagenized already carry a pre-existing mutation in a gene of interest and (b) the phenotype selected in the progeny is either less or more like wild type in enhancer or suppressor screens, respectively. Such synthetic phenotypes are more difficult to analyze, but can elegantly elucidate interactions in complex genetic pathways.

Various mutagens are used for inducing mutations; the choice of mutagen determines the physical nature of the change induced in the gene. Ideally, more than one mutagen is used to balance efficiency vs ease of DNA polymorphism detection. The most widely used chemical mutagen (Wood, 1988) is EMS due to its propensity to induce small deletions and point mutations (single base changes) with high efficiency (Anderson and Brenner, 1984; Dibb et al., 1985). However, most EMS alleles are not visible using standard Southern blots. But, the relatively small mutations induced by EMS are likely to perturb just one gene in contrast to the mutations induced by irradiation (gamma, X-ray or ultraviolet) which are more likely to perturb many genes and are less efficient than chemical mutagens. Radiation induced mutations are sometimes due to single base pair changes but as double stranded breaks in chromosomes are induced, deletions, translocations, duplications and other rearrangements often result (Greenwald and Horvitz, 1980). The third most common *C. elegans* mutagen is transposable elements — including the endogenous *C. elegans* Tc1 thru Tc7 transposable elements which can perturb gene function either via insertion or imprecise excision (Eide and Anderson, 1985; Plasterk, 1991). Usually individual genes are affected, although large deletions or rearrangements can result. Frequently, radiation and transposon induced alleles can be detected by Southern analysis facilitating gene identification during cloning. It should be remembered that at the high concentration/high level of mutagenesis in these protocols, multiple mutations are generated in genes on many chromosomes. Therefore, after mutagenesis, mutant strains are always backcrossed to unmutagenized animals to remove extraneous mutations in unrelated genes which may complicate phenotypic or genetic analyses. In addition, at a low but reproducible rate, mutant animals are identified in screens which carry synergistic (or antagonistic) mutations in more than one gene.

Cloning genes in C. elegans. Characterizing mutant alleles

First steps in the analysis of mutations identified in genetic screens include characterization of phenotypes and identification of genetic complementation groups. Phenotypic characterization includes quantification of the severity and penetrance of the defect induced and assessment of any pleiotropic defects. Severity refers to the magnitude of the defect seen in a specific animal whereas penetrance refers to the percentage of animals in a population which are affected. For example, not all mutations cause a complete loss in a behavioral response. These incompletely defective mutations can be grouped into two broad categories: mutations in genes in redundant pathways or incomplete loss of function for a specific gene. Although genetic arguments are often made based on the number of alleles of a specific phenotype identified, distinguishing definitively between these two possibilities is difficult without determining the complete loss of function phenotype for the genes involved.

Next, the development and function of relevant neurons/cells in mutant animals is examined for every mutant strain. Although not as daunting a task as that which faces a researcher in characterizing a mutant mouse strain, there are a large number of behavioral and morphological assays for *C. elegans* (see Table 1). A defect in the behavior of an animal can result from a defect in the generation or differentiation of the neurons involved. Genes involved in cell fate decisions are frequently identified using behavioral assays. The example of *odr-7* which encodes a steroid receptor homolog required for differentiation of chemosensory neurons is illustrative (Sengupta et al., 1994). In addition to confirming the presence and proper differentiation of the neurons involved, the animals are examined for defects in apparently unrelated behaviors. Frequently mutations identified using a given behavioral assay are defective for numerous other behaviors — which can shed light on the role of the encoded protein.

All recessive mutations generated in a screen can be tested for complementation. In the simplest scenario, males from one mutant strain are allowed to mate to hermaphrodites of another mutant strain. The resulting cross progeny carry one mutant copy of each gene. If the cross progeny are wild type for the phenotype in question, then the mutations are complementary and unlikely to be mutant alleles of the same gene. Alternatively, if the cross progeny are defective, then both mutations are likely to perturb just one gene (see Fig. 2). However, there are rare example of mutations in different genes which noncomplement (do not complement yielding a mutant phenotype in double heterozygotes) and examples of genes with exceptional alleles which complement (yielding a normal phenotype in double heterozygotes) (Huang et al., 1995). These exceptions are sorted out in the next phase of analysis: determining the genetic and physical location of the induced mutations.

If *mutant 1, mutant 2* and *mutant 3* have the same phenotype, but...

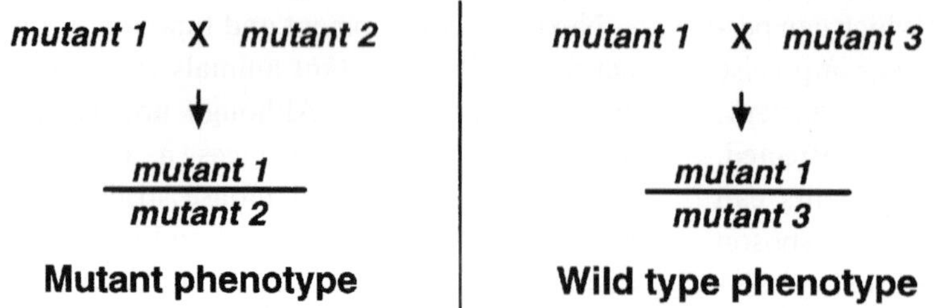

***mutant 1* and *mutant 2* do not complement and are likely alleles of the same gene, *mutant 3* complements *mutant 1* and is likely an allele of a different gene.**

Fig. 2. Complementation testing using recessive mutations in *C. elegans*. As many genes can mutate to the same phenotype, alleles of many genes are isolated in the same genetic screen. Complementation tests are done to learn which mutations correspond to same gene. Males for one mutant allele are crossed to hermaphrodites of another mutant allele. The phenotype of the progeny is assessed. If wild type, then the genes complement and are unlikely to be mutations in the same gene. If the cross progeny are mutant, then the two mutations may correspond to the same gene.

Cloning genes in C. elegans. Genetic mapping and rescue

The approximate physical location of most genes is determined by recombination and meiotic mapping relative to other genes (and their mutant alleles). Specifically, animals are generated which carry one copy of the mutation of interest and one copy of a mutation in a gene of known location. If the gene of interest is located near the known gene, then self-progeny are rarely identified which are homozygous for both mutations. Numerous mapping strategies have been developed, the first in the seminal paper by Sydney Brenner (Brenner, 1974). More recently, in addition to classical mapping with visible mutations, techniques have been developed to map using DNA polymorphisms (Williams et al., 1992). A large number of transposable elements are present in a few *C. elegans* isolates. These transposons are scattered throughout the genome, but many sites of insertion have been identified and used as sequenced tagged sites (STS) for mapping.

Once a general map position is known, recombination mapping is used to fine map the mutation in the gene of interest (Brenner, 1974) (see Fig. 3). Previously characterized deletions and duplications with defined end points can also be useful in mapping. Once mutation has been mapped to a relatively small interval, approximately 100–1000 kilobases, direct molecular detection techniques can be applied. As genomic DNA is generally available already cloned into cosmid, P1, YAC and other vectors (Sulston et al., 1992), usually no chromosome walking is required and Southern analysis of alleles can rapidly identify the gene of interest. Specific subclones of *C. elegans* genomic DNA listed in the GENBANK databases can be directly requested from Dr Alan Coulson at "alan@sanger.ac.uk", but cDNAs are available from Dr Yuji Kohara at "ykohara@lab.nig.ac.jp".

Alternatively, the gene of interest can be identified (or confirmed) by phenotypic rescue of the mutant phenotype in transgenic animals (Fire and Waterston, 1989). Transgenic *C. elegans* are readily generated by microinjection of DNA into the syncitial gonad of the adult hermaphrodite. Each gonad arm contains tens of uncellularized nuclei which are accessible to the injected DNA. Injected DNA, whether circular

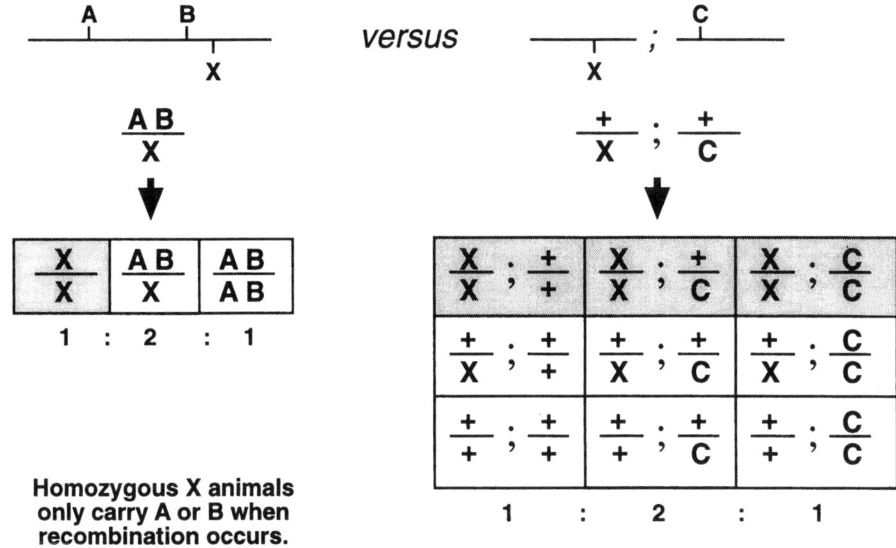

Fig. 3. Linkage and recombination mapping in *C. elegans*. The location of *C. elegans* mutant alleles on the chromosomes is determined by genetic mapping. Succinctly, the segregation of the mutant allele vs previously mapped mutations is determined. In the illustrated case, mutant allele of interest, X, is mapped vs previously mapped mutations A, B, and C. The genes corresponding to A, B and X are located on one chromosome, C is on another. Heterozygous hermaphrodites (AB/X on the left and C/+; +/X on the right) are generated and allowed to self fertilize. Homozygous X progeny (shaded cells) are scored for homozygous A, B or C. If A and B are located near X, then animals homozygous for X are rarely also homozygous for A or B as few recombinations between markers occur. But, one quarter of animals homozygous for X are homozygous for C as these genes are on different chromosomes and segregate independently. Many other genetic mapping strategies have been developed but all rely on these basic principles.

or linear, is converted into linear extrachromosomal arrays through a process which involves recombination (Mello et al., 1991). The extrachromosomal arrays are heritable, but are occasionally lost during cellular division. Conveniently, multiple DNA constructs can be coinjected and each extrachromosomal array generated usually carries functional copies of all the injected constructs. *dpy-20*, *lin-15*, *rol-6* and GFP (green fluorescent protein) reporter constructs are typical markers for array identification and propagation.

For example, *glr-1* mutant animals cannot respond to touch to the nose (described below). A double mutant strain of genotype *glr-1; dpy-20* can be constructed which is nose touch insensitive and dumpy (short in length) due to the *glr-1* and *dpy-20* mutations, respectively. Wild type genomic DNA for both genes (at 30 and 150 ng/μl, respectively) is injected into syncitial gonads of adult hermaphrodites. Injection of 15 *glr-1; dpy-20* animals generally yields between 50 and 100 self-progeny which no longer appear dumpy. The progeny of individual nondumpy F_1 animals are examined for heritable rescue of the dumpy marker and therefore the presence of a extrachromosomal array. Three to 20 independent arrays are usually generated and can be assayed for restoration of the ability to respond to nose touch due to the coinjected *glr-1* gene. A discussion of injection techniques can be found at http://www.dartmouth.edu/artsci/bio/ambros/protocols/MGH_protocols/koelle_prot/X0009_microinjection.1.html

It should be noted that each array generated is unique and although most arrays contain many copies of the DNA of interest, few copies may be actually expressed, presumably due to repetitive DNA silencing or mutation during array formation. Some genes are poorly rescued using standard techniques. Often only a percentage of arrays confer wild type gene activity. Although most somatically expressed genes can be rescued by arrays containing a wild type copy of the gene of interest, many genes expressed in the germline are difficult or impossible to rescue using standard techniques. It is often helpful in these cases to coinject the DNA of interest at low concentration (1–3 ng/μl) with high concentrations of wild type genomic DNA (100–300 ng/μl) — presumably preventing multicopy suppression of gene on the array (Kelly *et al.*, 1997). Extrachromosomal arrays can also be inserted into a chromosome yielding an integrated array (Mello *et al.*, 1991) — which usually contains multiple copies of all the original DNAs injected. Integration relies on repair of a mutagen-induced, double stranded break in the chromosome by insertion of the array. The protocols either require several days of work for each integrated array or high levels of mutagen with correspondingly high numbers of background mutations. The expression level of poorly expressing or relatively unstable arrays is sometimes improved after integration. For additional discussion of protocols and details see http://www.dartmouth.edu/artsci/bio/ambros/protocols/MGH_protocols/koelle_prot/X0007_integrating_arrays.html.

Cloning genes in C. elegans. Confirmation of identification

Although phenotypic rescue in transgenic animals is usually a reliable indicator of correct gene identification, additional evidence must be obtained for unequivocal proof. Additional evidence should include: sequencing mutant alleles, correlation of expression pattern, biochemical evidence and determination of the complete loss of function phenotype.

Critical information is obtained from determining allele-specific changes at the DNA sequence level. Genomic or cDNA is amplified by PCR from mutant alleles and directly sequenced. The phenotypic severity of the allele often correlates with the predicted severity based on sequence analysis, confirming correct gene identification. As determining the complete loss of function phenotype for a gene is often critical for assessment of protein function, determining the nature of the induced mutation is a requirement. A deletion which removes most or all of the DNA encoding the relevant protein (but not others) usually results in a complete loss of gene function. Alleles which contain a very early stop codon or frame shift frequently also result in complete loss of function. However, in the either case, the possibility of reinitiation at a novel methionine codon or alternative splicing must be considered. A variety of other changes, including transposon insertion, small deletions, misspliced exons, and single amino acid substitutions, can eliminate part or all of protein/gene function, but the severity of the defect varies with the change induced. Null (complete loss of function) alleles are often confirmed by both genetic and immunohistochemical analysis.

Determining the expression pattern of a gene in wild type and relevant mutant backgrounds is critical to both confirmation of gene identification and initial analyses of protein function *in vivo*. Usually at least two techniques are used to independently assess gene expression patterns. The most frequently and easily used technique is transcriptional and translational fusion to the green fluorescent protein (GFP) (Mello et al., 1991) or (historically) *lacZ*. Translational fusion of GFP to the gene of interest is often created by in frame insertion the GFP coding sequences into the full-length gene. As the average *C. elegans* gene is approximately 4 kb in length, including 5′ and 3′ regulatory sequences, introns and exons, a putative full length construct is usually easy to construct. Subcellular localization of wild type protein function is often correctly determined by determining the subcellular localization of a GFP fusion protein which retains wild type gene function. Transcriptional reporter constructs containing only putative 5′ regulatory sequences often recapitulate the cellular expression pattern for a given gene. A variety of useful vectors for the former experiments and for ectopic expression in *C. elegans* were generated by Dr. Andrew Fire and associates and are utilized by most *C. elegans* laboratories. (An FTP database of constructs and protocols is available at http://elegans.swmed.edu/Worm_labs/Fire/).

Of course, reporter construct data is subject to several caveats in interpretation — including high copy number effects, interruption of regulation, insufficient inclusion of regulatory sequences and lack of germline/embryonic expression from arrays. Therefore, the predicted expression pattern is usually confirmed independently using immunohistochemistry, *in situ* hybridization or, less frequently, mosaic analysis (Epstein and Shakes, 1995). Immunohistochemical data, although more laborious, expensive and technically difficult, usually closely approximates the endogenous gene expression pattern. RNA in situ hybridization is used in the embryo as reporter construct data is usually unavailable for confirmation and as RNA localization/transcriptional control has been shown to be critical for embryonic development in many species.

Occasionally, mosaic analysis is used to confirm the relevant cells/tissue expressing the gene of interest. Used and interpreted correctly, mosaic analysis directly addresses not just expression patterns, but also the actual cellular site of gene function (Herman, 1984). Mosaic animals are generated in *C. elegans* as homozygous mutant animals carrying a wild type gene copy on a free duplication (or extrachromosomal array) which is lost at a low rate during somatic mitosis during development resulting in mosaic individuals. Phenotype is correlated with the cellular pattern of duplication/array loss to draw conclusions regarding the cellular site of gene activity. Usually either GFP expression or *ncl-1* rescue is used to score array loss on a cell by cell basis. *ncl-1* is required for normal nucleolar morphology and gene function is cell autonomous (Hedgecock and Herman, 1995).

A relatively new technique for determining the loss of function phenotype for *C. elegans* genes is RNA mediated interference (RNAi) (Fire *et al.*, 1991; Tabara *et al.*, 1998). Double stranded RNA for the gene/s of interest is injected into adult hermaphrodites. The progeny are examined for phenotypic changes, which are occasionally heritable for several generations. Particularly effective for genes required during development, judicious analysis of affected animals can reveal both the phenotype resulting from both maternal rescue and the loss of function phenotype for a given gene. The cellular basis of RNAi mediated interference is under active investigation but annealing of sense and antisense strands is suspected to play a role. A protocol is available at (http://elegans.swmed.edu/Worm_labs/Fire/).

Mechanotransduction: an introduction

Research into mechanotransduction in *C. elegans* provides an excellent example of forward genetic analysis of nervous system function based on behavioral screens. There are two well-defined circuits which allow the animal to respond to gentle touch with an eyelash or hair, based both on cellular and on genetic analyses. Light touch to the body is primarily detected by the microtubule-containing mechanosensory neurons, ALM and PLM (Chalfie and Sulston, 1981); light touch to the nose is primarily detected by the ASH sensory neurons (Kaplan and Horvitz, 1993) (see Fig. 4). Each neuron class is bilaterally symmetric and characterized by specialized sensory processes for stimulus detection in the appropriate region of the body (White et al., 1986). Anterior touch to the body is detected by the ALML and ALMR neurons, on the left and right

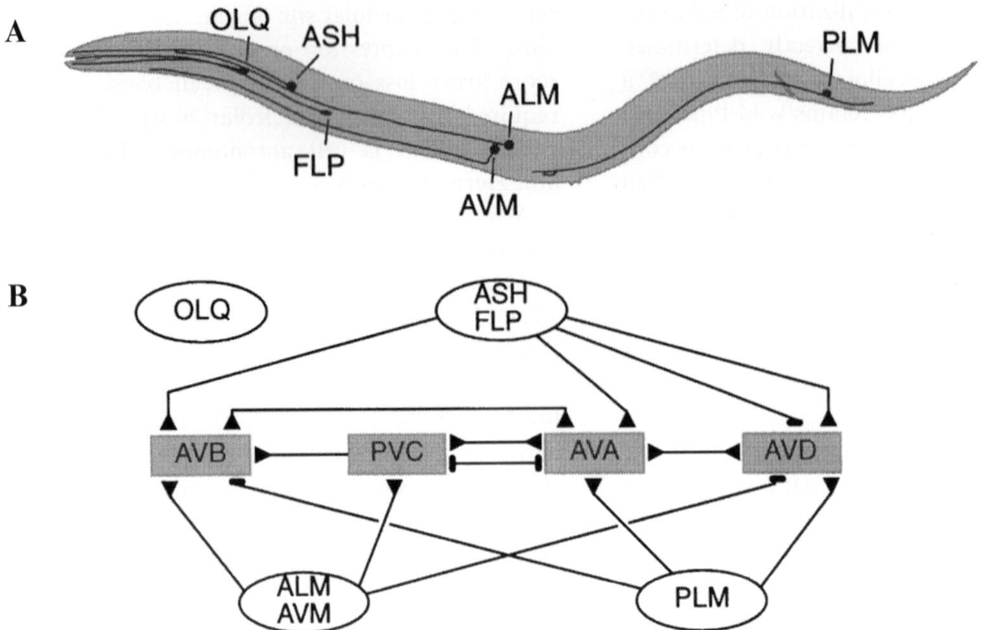

Fig. 4 (A) Diagram of *C. elegans* mechanosensory neurons. Position of cell bodies and sensory processes of mechanosensory neurons involved in response to body and nose touch are illustrated. ALM and PLM are required for response to body touch in the anterior and posterior regions of the body (Chalfie et al., 1985); ASH, OLQ and FLP for response to nose touch. ASH is primarily responsible for detection of nose touch (Kaplan and Horvitz, 1993) (about 63% of wild type response); FLP and OLQ play minor roles (about 30% and 7%, respectively). Only cells on the left side of the animal are illustrated. Each neuron is bilaterally symmetric (ASHL and ASHR) except the OLQ neurons which show four-fold symmetry (OLQDL, OLQDR, OLQVL, OLQVR). PVD neurons which mediate response to harsh touch are not shown (Way and Chalfie, 1989). Positions, morphology and connections (below) are from White et al. (1986). (B) Connections between mechanosensory neurons and interneurons. Interneurons are illustrated as boxes, sensory neurons as ovals. Synaptic connections are shown as lines with triangles; gap junctions as lines with bars. ASH and FLP make the same connections except only FLP is connected to AVD with a gap junction. ALM and AVM are connected by gap junction; only AVM synapses onto AVB. Synapses from both AVB and PVC onto AVD are not shown. OLQ has neither direct synaptic nor gap connections to the interneurons illustrated.

lateral side of the body respectively. Similarly, touch to the posterior body is detected by the PLM neurons whose activation results in cessation of backward locomotion and initiation of forward locomotion. The ALM and PLM neurons are also the sensory neurons which respond in the tap reflex, wherein a diffuse mechanical stimulus results in either backward locomotion or forward acceleration (Wicks et al., 1996). The tap reflex is mediated by the same interneurons which mediate response to body touch.

The ciliated sensory endings of the ASH neurons are located at the tip of the nose where they detect light touch to the nose, high osmolarity and volatile repellent chemicals (Kaplan and Horvitz, 1993; Troemel et al., 1995). Other ciliated sensory neurons play minor roles in the detection of nose touch (FLP and OLQ) and volatile repellents (ADL). Each of the stimuli detected by ASH evokes backward locomotion. Animals respond to body or nose touch in more than 90% of trials for each of these rapid and robust behavioral assays.

Both nose touch and high osmolarity are likely to be detected by ASH sensory neurons as a mechanical stimulus. Nose touch presumably causes deformation of the cell membrane vs the cytoskeleton of the ciliated sensory ending resulting in the activation of putative stretch-activated mechanosensitive receptors expressed by the ASH neurons. Changes in extracellular osmolarity are proposed to cause cell membrane stretch (or relaxation) resulting in the activation of stretch-activated receptors (French, 1992). Although both stimuli are classified as mechanosensory, the calculated magnitude of the mechanical force generated in the nose touch assay is significantly weaker than the force generated in osmotic avoidance assays (Hart *et al.*, 1999). It seems unlikely that the same receptor complex is detecting both stimuli. As described below, previously characterized *C. elegans* mechanosensory proteins are not required for ASH sensory function, suggesting that either homologs or a different class of proteins detect mechanical stimuli in the ASH neurons. One aim of genetic analysis in this circuit is to identify the proteins involved in mechanosensation and osmosensation in the ASH sensory neurons.

The direct synaptic targets of ALM, PLM and ASH sensory neurons include the interneurons which coordinate locomotion: AVA, AVB, AVD and PVC (White *et al.*, 1986). There are distinct patterns of synaptic and gap connections between the sensory and interneurons for each circuit (see Fig. 4(B)). ALM and PLM connect to the interneurons primarily through gap junctions; ASH primarily through chemical synapses. The interneurons are bilaterally symmetric and synapse onto the motorneurons which directly evoke locomotion. Laser ablation experiments demonstrated that AVA and AVD are primarily required for backward locomotion, AVB and PVC for forward locomotion (Chalfie *et al.*, 1985). Interestingly, there are multiple connections between these interneurons suggesting that their interactions may be complex. Another level of signalling complexity resides at the level of sensory synapses onto the interneurons. There is evidence that output from the ASH sensory neurons and activity in the interneurons varies with the stimulus detected by ASH (Hart, 1995). As the ASH neurons are invertebrate polymodal nociceptors, analogous to the human cutaneous nociceptive neurons, analysis of stimulus encoding in ASH may shed light on how painful stimuli are encoded by vertebrate nociceptors.

Genetic analysis of nose touch response

The analysis of this response (Kaplan and Horvitz, 1993) illustrates several basic principles in the genetic analysis of the nervous system. Mutations which perturb response to nose touch are isolated in behavioral genetic screens as animals which cannot respond. (Hart *et al.*, 1995). The corresponding genes may be required for ASH neuron development, morphology or function. Distinguishing between these possibilities prior to molecular analysis allows researchers to focus on genes which are specifically required for sensory signalling in ASH. For example, many of the nose touch defective mutations identified perturb the morphology of most or all neuronal ciliated sensory processes (Perkins *et al.*, 1986) (ASH is one of 25 classes of ciliated sensory neurons). *osm-3* mutant animals are defective for response to nose touch and high osmolarity due to a distal truncation in ciliated sensory processes of ASH and other sensory neurons. As all ciliated sensory endings are truncated in *osm-3* mutant animals, it is unlikely that *osm-3*, which encodes a kinesin motor related protein (Shakir et al., 1993), plays a specific role in mechanosensation.

Alternatively, mutations are identified which affect response to all stimuli detected by the ASH neurons although the neurons are morphologically normal. *eat-4* mutant animals are readily identified in behavioral screens based on their nose touch defective (Not) phenotype (Berger et al., 1998); but *eat-4* mutant animals are defective in their response to all other stimuli detected by ASH and to stimuli detected by many other neurons. *eat-4* animals are also eating defective (Avery, 1993a). The pleiotropic nature of the *eat-4* defect suggests that the corresponding protein is not directly involved in mechanosensation. The EAT-4 protein encodes a putative transport protein which is suggested to play a role in loading glutamate into vesicles for synaptic release (R. Lee and L. Avery, personal communication). This and other data described below suggests that ASH and other sensory neurons use glutamate as a neurotransmitter.

Given the large number of genes required for development, morphogenesis and general function of ciliated neurons, research has focused on the analysis of genes which specifically required for response to only one or two of the three stimuli detected by ASH. For example, *glr-1* (Hart *et al.*, 1995) animals are defective in their response to nose touch, but normal in their response to high osmolarity and volatile repellant chemicals which are also detected by ASH. *glr-1* encodes a AMPA class glutamate receptor which is specifically required for response to nose touch in the direct synaptic targets of ASH. This data strongly suggests that ASH releases glutamate in response to nose touch thereby activating the interneurons which are synaptic targets of ASH. Osmotic or chemical stimuli are likely to cause release of additional glutamate and as yet unidentified neuropeptide neurotransmitters. Analysis of *glr-1* suggests that sensory neuron output can vary with the strength or type of stimulus detected (Hart *et al.*, 1995).

The strengths of the ASH circuit analysis include relative specificity of mutations identified, apparent dissimilarity to body touch neuron sensory signalling pathways and potential molecular similarity to ciliated neurons which detect stimuli in vertebrates.

Genetic analysis of mechanosensory defective (mec) genes

Many laboratories have identified mutations which perturb response to body touch, most notably by M. Chalfie and coworkers (Chalfie and Sulston, 1981). The strengths of the *mec* analysis of mechanosensation include near saturation for genes which mutate to this mutant phenotype and insights into development and function of specific mechanosensory neurons. Roughly 50,000 animals have been screened for mec mutations using a variety of mutagens. More than 15 genes have been identified based on their mutation to a Mec phenotype (Chalfie and Sulston, 1981). The cellular and molecular processes affected by these mutations range from determination of

TABLE 2

Characterized genes required for *C. elegans* mechanosensation

Gene	Encoded protein
unc-86	POU-homeodomain transcription factor
mec-1	unknown
mec-2	stomatin homolog
mec-3	LIM-homeodomain transcription factor
mec-4	degenerin channel subunit
mec-5	collagen
mec-6	unknown
mec-7	β-tubulin
mec-8	novel with RNA binding domains
mec-9	novel with EGF and Kunitz domains
mec-10	degenerin channel subunit
mec-12	α-tubulin
mec-14	unknown
mec-15	unknown
mec-17	unknown
mec-18	unknown
glr-1	AMPA class neurotransmitter receptor
eat-4	possible synaptic vesicle transporter (glutamate)
osm-9	capsaicin receptor homolog
odr-3	G protein, alpha subunit
osm-10	novel

Characterized genes required for *C. elegans* mechanosensation, either in response to body touch or nose touch are listed. Genes listed as unknown have not yet been molecularly characterized. *mec* genes and *unc-86* are required for response to body touch as described in the text. *eat-4*, *osm-9* and *odr-3* (Roayaie *et al.*, 1998) mutant animals are defective in response to all stimuli detected by ASH; *osm-9* (Colbert *et al.*, 1997) mutations perturb structure of the ASH ciliated sensory ending. *osm-10* is specifically required for osmosensation mediated by ASH (Hart *et al.*, 1999). Many mutations which perturb response to multiple stimuli detected by ASH are not listed. *osm-10* perturbs only response to high osmolarity. The molecular basis of osmosensation in *C. elegans* is unclear, but mechanosensitive ion channels detect osmotic stimuli in *E. coli* (Sukharev *et al.*, 1994).

ALM/PLM cell fate to mechanotransduction (see Table 2). Prior to cloning, of course, the roles of a subset of these genes in ALM and PLM development, differentiation or survival was clear from phenotypic analysis of mutant animals as explained below.

unc-86 and *mec-3* encode POU- and LIM-homeodomain transcription factors (Finney *et al.*, 1988; Way and Chalfie, 1988) which are required for ALM and PLM neuron generation and differentiation, respectively. ALM and PLM neurons contain unique isoforms of tubulin encoded by the *mec-7* and *mec-12* genes (Savage *et al.*, 1989; Hamelin *et al.*, 1992). They combine to form 15 protofilament microtubules which fill the sensory processes of ALM and PLM and the microtubules serve as the structural scaffold for mechanotransduction. Loss of function mutations in any these genes alter the structure, morphology or differentiation of the ALM and PLM neurons.

In addition, genes have been identified which are part of a putative mechanosensory channel (see Table 2). Research by various researchers on *mec* genes, including M. Chalfie and M. Driscoll, suggests that a novel class of sodium channels, called degenerins (Chalfie and Wolinsky, 1990), may act as mechanoreceptor proteins in *C. elegans*. The *C.elegans* genes *mec-4* and *mec-10* encode degenerin proteins (Driscoll and Chalfie, 1991; Huang and Chalfie, 1994) and are required in the ALM and PLM neurons for response to body touch. Loss of function in either gene prevents response to light touch mediated by morphologically normal ALM or PLM neurons. Rare, gain of function mutations result in misregulated channels, mechanosensory cell swelling and necrotic cell death. Other degenerin proteins are encoded by the *C. elegans deg-1* (Chalfie and Wolinsky, 1990), *unc-105* (Liu *et al.*, 1996; Shreffler and Wolinsky, 1997), and *unc-8* (Tavernarakis *et al.*, 1997) genes but are expressed in other *C. elegans* cells. The degenerin proteins encoded by these genes have been proposed to function as mechanoreceptors in muscle cells (*unc-105*) and in neurons (*deg-1* and *unc-8*). But although *deg-1* and *unc-8* are expressed in ASH mechanosensory neurons, *deg-1* and *unc-8* function is not required for ASH sensory responses (Hart et al., 1999). The evidence that the degenerin proteins *mec-4* and *mec-10* function as mechanoreceptors in ALM and PLM is based on homology to vertebrate channels, genetic loss of function phenotype, mechanically-gated channel activity in heterologous cells and on genetic interactions with other *mec* genes.

Additional proteins have been characterized which may couple the *mec-2/mec-4* channel to both the cytoskeleton and the extracellular matrix. *mec-2* is proposed to couple the channel subunits to the cytoskeleton based on genetic analysis in *C. elegans* (interactions with *mec-10, mec-4, mec-12 and mec-7*) (Huang, 1994) and on MEC-2 homology with stomatin (Huang *et al.*, 1995), an integral membrane protein associated with human red blood cell cytoskeleton. *mec-5, mec-1* and *mec-9* gene products are proposed to form a complex in the extracellular matrix coupling the channel proteins activity to cuticle deformation caused by touch. Although *mec-1* is not molecularly characterized, *mec-5* encodes an extracellular collagen protein and *mec-9* encodes a secreted novel protein containing EGF and Kunitz domains (Du *et al.*, 1996). Mutations in *mec-6* and *mec-14* suppress the cell death induce by gain of function mutations of *mec-4* and *mec-10*. *mec-6* and *mec-14* may encode channel regulatory subunits or other molecules involved in channel function, processing or degradation

In summary, genetic analysis of mechanosensation in *C. elegans* has identified and characterized a putative mechanosensory channel complex which is required for mechanosensory neurons to respond to touch. Proceeding from a simple behavioral assay, genetic screening permitted the unbiased identification of the relevant proteins. The ability of genetic screens in simple organisms to identify *in vivo* the critical components for specific nervous system functions is one of the greatest strengths of genetic analyses of the nervous system.

Other resources

Additional information on *C. elegans* culture and protocols is available in either: *The nematode C. elegans*, W.B. Wood (Ed.), Cold Spring Harbor Laboratory Press, 1988. or in *Methods in Cell Biology, Caenorhabditis elegans: Modern Biological Analysis of an Organism*, H.F. Epstein and D. C. Shakes (Eds.), Academic Press, 1995.

Relatively comprehensive recent reviews of research in *C. elegans* chemosensation, mechanosensation, synaptic function and many other areas is found in: *C. elegans* II, D.L. Riddle, T. Blumenthal, B.J. Meyer and J.R. Priess (Eds.), Cold Spring Harbor Laboratory Press, 1997.

The *C. elegans* home page on the WWW can be found at http://elegans.swmed.edu/. Useful resources include online searching of abstracts from *C. elegans* journal articles, meeting abstracts and Worm Breeders Gazette abstracts. Note that abstracts from meetings and the Gazette articles are considered to be personal communications until published in journals and the abstracts cannot be cited in journal articles, reviews or grant applications without obtaining consent of the authors.

All information from genomic and cDNA sequence projects has been deposited in GENBANK and is available using BLAST and other programs at http://www.ncbi.nlm.nih.gov/.

References

Albert, P.S., Brown, S.J. and Riddle, D.L. (1981) Sensory control of dauer larva formation in *Caenorhabditis elegans*. *J. Comp. Neurol.*, 198: 435–451.

Anderson, P. and Brenner, S. (1984) A selection for myosin heavy chain mutants in the nematode *Caenorhabditis elegans*. *Proc. Natl. Acad. Sci. USA*, 81: 4470–4474.

Avery, L. (1993a) The genetics of feeding in *Caenorhabditis elegans*. *Genetics*, 133: 897–917.

Avery, L. (1993b) Motor neuron M3 controls pharyngeal muscle relaxation timing in *Caenorhabditis elegans*. *J. Exp. Biol.*, 175: 283–297.

Avery, L. and Horvitz, H.R. (1990) Effects of starvation and neuroactive drugs on feeding in *Caenorhabditis elegans*. *J. Exp. Zool.*, 253: 263–270.

Bargmann, C.I. and Horvitz, H.R. (1991) Chemosensory neurons with overlapping functions direct chemotaxis to multiple chemicals in *C. elegans*. *Neuron*, 7: 729–742.

Berger, A.J., Hart, A.C. and Kaplan, J.M. (1998) G alphas-induced neurodegeneration in *Caenorhabditis elegans*. *J. Neurosci.*, 18: 2871–2880.

Brenner, S. (1974) The genetics of *Caenorhabditis elegans*. *Genetics*, 77: 71–94.

Cassada, R.C. and Russell, R.L. (1975) The dauer larva, a post-embryonic developmental variant of the nematode *Caenorhabditis elegans*. *Dev. Biol.*, 46: 326–342.

Chalfie, M. and Au, M. (1989) Genetic control of differentiation of the *C. elegans* touch receptor neurons. *Science*, 243: 1027–1033.

Chalfie, M. and Sulston, J. (1981) Developmental genetics of the mechanosensory neurons of *Caenorhabditis elegans*. *Dev. Biol.*, 82: 358–370.

Chalfie, M., Sulston, J.E., White, J.G., Southgate, E., Thomson, J.N. and Brenner, S. (1985) The neural circuit for touch sensitivity in *C. elegans*. *J. Neurosci.*, 5: 956–964.

Chalfie, M., Tu, Y., Euskirchen, G., Ward, W. and Prasher, D. (1994) Green fluorescent protein as a marker for gene expression. *Science*, 263: 802–805.

Chalfie, M. and Wolinsky, E. (1990) The identification and suppression of inherited neurodegeneration in *Caenorhabditis elegans*. *Nature*, 345: 410–416.

Colbert, H.A., Smith, T.L. and Bargmann, C.I. (1997) OSM-9, a novel protein with structural similarity to channels, is required for olfaction, mechanosensation, and olfactory adaptation in *Caenorhabditis elegans*. *J. Neurosci.*, 17: 8259–8269.

Croll, N.A. and Matthews, B.E. (1977) *Biology of Nematodes*. Halsted Press, New York.

Culotti, J.G. (1994) Axon guidance mechanisms in *Caenorhabditis elegans*. *Curr. Opin. Genet. Dev.*, 4: 587–595.

Culotti, J.G. and Russell, R.L. (1978) Osmotic avoidance defective mutants of the nematode *Caenorhabditis elegans*. *Genetics*, 90: 243–256.

De Bono, M. and Bargmann, C.I. (1998) Natural variation in a neuropeptide Y receptor homolog modifies social behavior and food response in *C. elegans*. *Cell*, 94: 679–689.

Dibb, N.J., Brown, D.M., Karn, J., Moerman, D.G., Bolten, S.L. and Waterston, R.H. (1985) Sequence analysis of mutations that affect the synthesis, assembly and enzymatic activity of the unc-54 myosin heavy chain of *Caenorhabditis elegans*. *J. Mol. Biol.*, 183: 543–551.

Driscoll, M. and Chalfie, M. (1991) The *mec-4* gene is a member of a family of *C. elegans* genes that can mutate to induce neuronal degeneration. *Nature*, 349: 588–593.

Du, H., Gu, G., William, C.M. and Chalfie, M. (1996) Extracellular proteins needed for *C. elegans* mechanosensation. *Neuron*, 16: 183–194.

Dusenbery, D.B., Sheridan, R.E. and Russell, R.L. (1975) Chemotaxis-defective mutants of the nematode *Caenorhabditis elegans*. *Genetics*, 80: 297–309.

Eide, D. and Anderson, P. (1985) Transposition of Tc1 in the nematode *Caenorhabditis elegans*. *Proc. Natl. Acad. Sci. USA*, 82: 1756–1760.

Epstein, H.F. and Shakes, D.C. (Eds.) (1995) *Caenorhabditis elegans*: Modern Biological Analysis of an Organism. In: *Methods in Cell Biology*, Acadmic Press, New York.

Ferguson, E.L.Horvitz, H.R. (1989) The multivulva phenotype of certain *Caenorhabditis elegans* mutants results from defects in two functionally redundant pathways. *Genetics*, 123: 109–121.

Finney, M., Ruvkun, G. and Horvitz, H.R. (1988) The *C. elegans* cell lineage and differentiation gene unc-86 encodes a protein with a homeodomain and extended similarity to transcription factors. *Cell*, 55: 757–769.

Fire, A., Albertson, D., Harrison, S.W. and Moerman, D.G. (1991) Production of antisense RNA leads to effective and specific inhibition of gene expression in *C. elegans* muscle. *Development*, 113: 503–514.

Fire, A. and Waterston, R.H. (1989) Proper expression of myosin genes in transgenic nematodes. *EMBO J.*, 8: 3419–3428.

French, A.S. (1992) Mechanotransduction. *Ann. Rev. Physiol.*, 54: 135–152.

Golden, J.W. and Riddle, D.L. (1982). A pheromone influences larval development in the nematode *Caenorhabditis elegans*. Science, 218: 578-80.

Goodman, M.B., Hall, D.H., Avery, L. and Lockery, S.R. (1998) Active currents regulate sensitivity and dynamic range in *C. elegans* neurons. *Neuron*, 20: 763–772.

Greenwald, I.S. and Horvitz, H.R. (1980) *unc-93(e1500)*: A behavioral mutant of *Caenorhabditis elegans* that defines a gene with a wild-type null phenotype. *Genetics*, 96: 147–164.

Hamelin, M., Scott, I. and Culotti, J. (1992) The *mec-7* β-tubulin gene of *C. elegans* is expressed primarily in the touch receptor neurons. *EMBO J.*, 11: 2885–2893.

Hamelin, M., Zhou, Y., Su, M.W., Scott, I.M. and Culotti, J.G. (1993) Expression of the UNC-5 guidance receptor in the touch neurons of *C. elegans* steers their axons dorsally. *Nature*, 364: 327–330.

Hart, A., S. Simms and J.M. Kaplan (1995). Synaptic code for sensory modalities revealed by *C. elegans* GLR-1 glutamate receptor. Nature, 378: 82-85.

Hart, A.C., Kass, J., Shapiro, J. and Kaplan, J.M. (1999) Distinct signaling pathways mediate mechanosensation and osmosensation in a polymodal sensory neuron. *Journal of Neuroscience*, 16(6): 1952–1958.

Hedgecock, E.M., Culotti, J.G., Thomson, J.N. and Perkins, L.A. (1985) Axonal guidance mutants of *Caenorhabditis elegans* identified by filling sensory neurons with fluorescein dyes. *Dev. Biol.*, 111: 158–170.

Hedgecock, E.M. and Herman, R.K. (1995) The *ncl-1* gene and genetic mosaics of *Caenorhabditis elegans*. *Genetics*, 141: 989–1006.

Hedgecock, E.M. and Russell, R.L. (1975) Normal and mutant thermotaxis in the nematode *Caenorhabditis elegans*. *Proc. Natl. Acad. Sci. USA*, 72: 4061–4065.

Herman, R.K. (1984) Analysis of genetic mosaics of the nematode *C. elegans*. *Genetics*, 108: 165–180.

Horvitz, H.R., Chalfie, M., Trent, C., Sulston, J. and Evans, P. (1982) Serotonin and octopamine in the nematode *C. elegans*. *Science*, 216: 1012–1014.

Huang, M. and Chalfie, M. (1994) Gene interactions affecting mechanosensory transduction in *Caenorhabditis elegans*. *Nature*, 367: 467–470.

Huang, M., Gu, G., Ferguson, E.L. and Chalfie, M. (1995) A stomatin-like protein necessary for mechanosensation in *C. elegans*. *Nature*, 378: 292–295.

Huang, M. and Chalfie, M.C. (1994) Gene interactions affecting mechanosensory transduction in *C. elegans*. *Nature*, 367: 467–470.

Kaplan, J.M. and Horvitz, H.R. (1993) A dual mechanosensory and chemosensory neuron in *C. elegans*. *Proc. Natl. Acad. Sci. USA*, 90: 2227–2231.

Kelly, W.G., Xu, S., Montgomery, M.K. and Fire, A. (1997) Distinct requirements for somatic and germline expression of a generally expressed Caernorhabditis elegans gene. *Genetics*, 146: 227–238.

Khan, M.L., Gogonea, C.B., Siddiqui, Z.K., Ali, M.Y., Kikuno, R., Nishikawa, K. and Siddiqui, S.S. (1997) Molecular cloning and expression of the *Caenorhabditis elegans* klp-3, an ortholog of C terminus motor kinesins Kar3 and ncd. *J. Mol. Biol.*, 270: 627–639.

Kornfeld, K. (1997) Vulval development in *Caenorhabditis elegans*. *Trends Genet.*, 13: 55–61.

Liu, D.W. and Thomas, J.H. (1994) Regulation of a periodic motor program in *C. elegans*. *J. Neurosci.*, 14: 1953–1962.

Liu, J., Schrank, B. and Waterston, R.H. (1996) Interaction between a putative mechanosensory membrane channel and a collagen (see comments). *Science*, 273: 361–364.

Loer, C. and Kenyon, C. (1993) Serotonin-deficient mutants and male mating behavior in the nematode *C. elegans*. *J. Neurosci.*, 13: 5407–5417.

McIntire, S.L., Jorgensen, E. and Horvitz, H.R. (1993) Genes required for GABA function in *Caenorhabditis elegans* (see comments). *Nature*, 364: 334–337.

Mello, C., Kramer, J., Stinchcomb, D. and Ambros, V. (1991) Efficient gene transfer in *C. elegans*: extrachromosomal maintenance and integration of transforming sequences. *EMBO J.*, 10: 3959–3970.

Nelson, L.S., Kim, K., Memmott, J.E. and Li, C. (1998) FMRFamide-related gene family in the nematode, *Caenorhabditis elegans*. *Brain Res. Mol. Brain Res.*, 58: 103–111.

Nonet, M.L., Grundahl, K., Meyer, B.J. and Rand, J.B. (1993) Synaptic function is impaired but not eliminated in *C. elegans* mutants lacking synaptotagmin. *Cell*, 73: 1291–1305.

Novak, J. and Vanek, Z. (1992) Screening for a new generation of anthelminthic compounds. In vitro selection of the nematode *Caenorhabditis elegans* for ivermectin resistance. *Folia Microbiol.*, 37: 237–238.

Perkins, L.A., Hedgecock, E.M., Thomson, J.N. and Culotti, J.G. (1986) Mutant sensory cilia in the nematode *Caenorhabditis elegans*. *Dev. Biol.*, 117: 456–487.

Plasterk, R.H. (1991). The origin of footprints of the Tc1 transposon of *Caenorhabditis elegans*. *EMBO J.*, 10: 1919–1925.

Rand, J.B. and Russell, R.L. (1985) Molecular basis of drug-resistance mutations in *C. elegans*. *Psychopharmacol. Bull.*, 21: 623–630.

Rankin, C.H. (1991) Interactions between two antagonistic reflexes in the nematode *Caenorhabditis elegans*. *J. Comp. Physiol. A*, 169: 59–67.

Rankin, C.H., Beck, C.D. and Chiba, C.M. (1990) *Caenorhabditis elegans*: a new model system for the study of learning and memory. *Behav. Brain Res.*, 37: 89–92.

Riddle, D.L. (1977) A genetic pathway for dauer larva formation in *Caenorhabditis elegans*. *Stadler Genet. Symp.*, 9: 101–120.

Riddle, D.L., Blumenthal, T., Meyer, B. and Preiss, J.R. (Eds.) (1997). *C. elegans* II, Cold Spring Harbour Monograph Series, Cold Spring Harbour Laboratory Press.

Roayaie, K., Crump, J.G., Sagasti, A. and Bargmann, C.I. (1998) The G alpha protein ODR-3 mediates olfactory and nociceptive function and controls cilium morphogenesis in *C. elegans* olfactory neurons. *Neuron*, 20: 55–67.

Savage, C., Hamelin, M., Culotti, J.G., Coulson, A., Albertson, D.G. and Chalfie, M. (1989) *mec-7* is a beta-tubulin gene required for the production of 15-protofilament microtubules in *C. elegans*. *Genes Dev.*, 3: 870–881.

Sawin, B. (1997) Doctoral Thesis MIT, Department of Biology.

Schafer, W.R. and Kenyon, C.J. (1995) A calcium-channel homologue required for adaptation to dopamine and serotonin in *Caenorhabditis elegans*. *Nature*, 375: 73–78.

Segalat, L., Elkes, D.A. and Kaplan, J.M. (1995) Modulation of serotonin-controlled behaviors by Go in *Caenorhabditis elegans* (see comments). *Science*, 267: 1648-51.

Sengupta, P., Colbert, H.A. and Bargmann, C.I. (1994) The *C. elegans* gene *odr-7* encodes an olfactory-specific member of the nuclear receptor superfamily. *Cell*, 79: 971–980.

Shakir, M.A., Fukishige, T., H, H.Y., Miwa, J. and Siddiqui, S.S. (1993) *C. elegans* osm-3 gene mediating osmotic avoidance behavior encodes a kinesin-like protein. *NeuroReport*, 4: 891-894.

Shreffler, W. and Wolinsky, E. (1997) Genes controlling ion permeability in both motorneurons and muscle. *Behav. Genet.*, 27: 211–221.

Sukharev, S.I., Blount, P., Martinac, B., Blattner, F.R. and Kung, C. (1994) A large-conductance mechanosensitive channel in E. coli encoded by mscL alone. *Nature*, 368: 265–268.

Sulston, J., Du, Z., Thomas, K., Wilson, R., Hillier, L., Staden, R., Halloran, N., Green, P., Thierry-Mieg, J., Qui, L., Dear, S., Coulson, A., Craxton, M., Durbin, R., Berks, M., Metzstein, M., Hawkins, T., Ainscough, R. and Waterston, R. (1992) The *C. elegans* genome sequencing project: a beginning. *Nature*, 356: 37–41.

Tabara, H., Grishok, A. and Mello, C.C. (1998) RNAi in *C. elegans*: Soaking in the genome sequence. *Science*, 282: 430–431.

Tavernarakis, N., Shreffler, W., Wong, S. and Driscoll, M. (1997) *unc-8*, a DEG/ENaC family member, encodes a subunit of a candidate mechanically gated channel that modulates *C. elegans* locomotion. *Neuron*, 18: 107–119.

Troemel, E.R., Chou, J.H., Dwyer, N.D., Colbert, H.A. and Bargmann, C.I. (1995) Divergent seven transmembrane receptors are candidate chemosensory receptors in *C. elegans*. *Cell*, 83: 207–218.

Wadsworth, W.G., Bhatt, H. and Hedgecock, E.M. (1996) Neuroglia and pioneer neurons express UNC-6 to provide global and local netrin cues for guiding migrations in *C. elegans*. *Neuron*, 16: 35–46.

Waggoner, L.E., Zhou, G.T., Schafer, R.W. and Schafer, W.R. (1998) Control of alternative behavioral states by serotonin in *Caenorhabditis elegans*. *Neuron*, 21: 203–214.

Ward, S. (1973) Chemotaxis by the nematode *Caenorhabditis elegans*: identification of attractants and analysis of the response by use of mutants. *Proc. Natl. Acad. Sci. USA.*, 70: 817–821.

Way, J.C. and Chalfie, M. (1988) *mec-3*, a homeobox-containing gene that specifies differentiation of the touch receptor neurons in *C. elegans*. *Cell*, 54: 5–16.

Way, J.C. and Chalfie, M. (1989) The *mec-3* gene of *Caenorhabditis elegans* requires its own product for maintained expression and is expressed in three neuronal cell types. *Genes Dev.*, 3: 1823–1833.

Wen, J.Y., Kumar, N., Morrison, G., Rambaldini, G., Runciman, S., Rousseau, J. and Van der Kooy, D. (1997) Mutations that prevent associative learning in *C. elegans*. *Behav. Neurosci.*, 111: 354–368.

White, J.G., Southgate, E., Thomson, J.N. and Brenner, S. (1986) The structure of the nervous system of *Caenorhabditis elegans*. *Philos. Trans. R. Soc. Lond.*, 314: 1–340.

Wicks, S.R., Roehrig, C.J. and Rankin, C.H. (1996) A dynamic network simulation of the nematode tap withdrawal circuit: predictions concerning synaptic function using behavioral criteria. *J. Neurosci.*, 16: 4017–4031.

Williams, B.D., Schrank, B., Huynh, C., Shownkeen, R. and Waterston, R.H. (1992) A genetic mapping system in *Caenorhabditis elegans* based on polymorphic sequence-tagged sites. *Genetics*, 131: 609–624.

Wood, W.B. (Ed.) (1988) *The nematode* C. elegans, Cold Spring Harbor Laboratory, Cold Spring Harbor.

W.E. Crusio and R.T. Gerlai (Eds.)
Handbook of Molecular-Genetic Techniques for Brain and Behavior Research (Techniques in the Behavioral and Neural Sciences, Vol. 13)

CHAPTER 3.3.7

Analyzing neuropeptide function in *Caenorhabditis elegans* by reverse genetics

Chris Li

Department of Biology, Boston University, 5 Cummington Street, Boston, MA 02215, USA

Introduction

Neuropeptides are short sequences of amino acids that can function as chemical messengers between cells in the nervous system. A large variety of neuropeptides has been isolated from diverse organisms throughout the animal kingdom, where they exert a wide range of actions (Brown, 1990; Schwartz, 1991; Brownlee and Fairweather, 1999). The presence of neuropeptides in primitive organisms such as coelenterates (Grimmelikhuijzen et al., 1988) suggests that neuropeptides are among the first molecules used for chemical communication in the nervous system.

Like other proteins, neuropeptides that share a common motif are grouped into families. For instance, peptides in the FMRFamide (Phe-Met-Arg-Phe-NH_2)-related family range from four to 20 amino acids, but all have a common Arg-Phe-amide sequence at their C-termini (Price and Greenberg, 1977; Nambu et al., 1988; Yang et al., 1985; Brownlee and Fairweather, 1999). Neuropeptides are derived from larger precursor proteins by cleavage at mono-, di-, or tribasic amino acids, and may undergo other forms of post-translation modifications (such as amidation) before becoming neuroactive (Schwartz, 1991). One precursor molecule may give rise to multiple peptides that are identical or distinct in sequence; although distinct in sequence, the peptides may still share some sequence motifs (Schwartz, 1991).

Peptides within the same family can have similar or antagonistic effects at a synapse, suggesting that peptides within the same family either act via distinct receptors or act via the same receptor to elicit different responses (Wang et al., 1995; Brownlee and Fairweather, 1999). To date, only one neuropeptide-gated ion channel receptor has been identified; this receptor is activated by binding of a FMRFamide-related peptide (FaRP) (Lingueglia et al., 1995). Other neuropeptide receptors, including other FaRP receptors (Wang et al., 1995; Tensen et al., 1998), are G protein-coupled, seven transmembrane domain proteins (Strader et al., 1994; Blomqvist and Herzog, 1997; Gudermann et al., 1997).

Deciphering the functions of the diverse neuropeptides has been challenging, particularly as the full extent of neuropeptide gene families is still unknown. The nematode *Caenorhabditis elegans* is the first multicellular organism for which the entire genome has been determined (The *C. elegans* Sequencing Consortium, 1998). As such, it is the first organism for which the function of entire neuropeptide gene families can be elucidated.

The nervous system of *C. elegans*

As discussed in other chapters (3.3.5, 3.3.6), the nervous system of *C. elegans* is relatively small, consisting of 302 or 381 neurons in the adult hermaphrodite or male, respectively (Sulston and Horvitz, 1977; Sulston et al., 1983). This number corresponds to roughly one-third of the somatic cells in *C. elegans*. The entire connectivity of the hermaphrodite nervous system has been determined by serial section electron microscopic reconstructions (White et al., 1986). Each neuron, therefore, can be uniquely identified based on its morphology, synaptic connectivity, and position. Using these criteria, White et al. (1986) classified the neurons into 118 subgroups, which are composed of one to 13 cells.

C. elegans offers several experimental advantages for determining the function of neuropeptide gene families. As mentioned previously, the entire genome has been determined, so all members of a gene family can be found by database searches. Second, production of knockout and transgenic *C. elegans* animals is a routine and fairly rapid procedure (Rushforth et al., 1993; Zwaal et al., 1993; Jansen et al., 1997). Third, *C. elegans* is easy to maintain in the laboratory; they grow on agar plates streaked with their food source, bacteria, and the generation time (from egg through four larval stages to adult) is only 3 d (Brenner, 1974). Lastly, *C. elegans* wild-type and mutant strains are available from the *Caenorhabditis* Genetics Center in St. Paul, MN (stier@tc.umn.edu). The neuropeptide gene family encoding FMRFamide-related peptides (FaRPs) will be used as an illustrative example of how neuropeptide gene function can be analyzed in *C. elegans*.

Identifying genes that encode neuropeptides in *C. elegans*

Screening cDNA and genomic libraries

The traditional method to identify neuropeptide genes in *C. elegans* is to design oligonucleotides against a peptide sequence of interest and use labeled oligonucleotides to screen a cDNA library (see Sambrook et al., 1989). The oligonucleotides are usually degenerate (i.e. they have multiple nucleotide possibilities at certain positions) because of the degeneracy of the genetic code. In some cases, particularly if the peptide sequence is arginine-rich (arginine has 6 possible codons), several sets of degenerate oligonucleotides are made. If the peptide has not been previously isolated from *C. elegans*, the sequence of related peptides from other organisms should also be considered when designing the oligonucleotides.

After a positive clone has been isolated to homogeneity, the DNA sequence of the cDNA is determined. The sequence will reveal the number of encoded peptides and whether multiple copies of a single peptide or different peptide types are encoded. The cDNA can then be used as a probe to screen genomic libraries to isolate the genomic fragment corresponding to the cDNA. Genomic (S. Emmons, pers. comm.) and cDNA (Barstead and Waterston, 1989) libraries are available from many *C. elegans* researchers. Through this method, one gene encoding FaRPs was identified in *C. elegans* (Rosoff et al., 1992). According to *C. elegans* nomenclature convention, a gene name is three letters followed by a number. The FaRP-encoding gene was named *flp-1* for **F**MRFamide-**l**ike **p**eptide-1 gene.

Screening the genomic database

One of the major advantages in working in the *C. elegans* field is that the DNA sequence for the entire genome is known, and the genetic and physical maps have been cross-correlated (*C. elegans* Sequencing Consortium, 1998). The simplest way to identify neuropeptide genes in *C. elegans* is to search the genomic database, which can be accessed through the *C. elegans* server at University of Texas (http://elegans.swmed.edu) or through Genbank (http://www.ncbi.nlm.nih.gov/) on the World Wide Web. The *C. elegans* genome contains over 19,000 open reading frames

(ORFs) predicted by Genefinder programs. Cosmids containing the neuropeptide genomic regions can be requested from the *C. elegans* Sequencing Consortium (alan@sanger.ac.uk).

Because neuropeptides are relatively small and the Genefinder programs require a minimal number of amino acids in their search parameters, concatenated peptide sequences separated by putative cleavage sites must sometimes be used in the BLAST searches. Through such database searches, 22 genes encoding FaRPs (*flp-2* through *flp-22*) were identified in *C. elegans* (Nelson et al., 1998a; C. Li, unpubl. obs.). In total, 59 distinct FaRPs are encoded by the 22 *flp* genes (Nelson et al., 1998a; C. Li, unpubl. obs.).

Determining whether a neuropeptide gene is expressed

Since there is a possibility that some predicted genes are pseudogenes or that predicted ORFs are incorrect, verification of the Genefinder predictions must be performed. This verification can be done in several ways. Small segments [expressed sequence tags (ESTs)] of cDNAs are currently being sequenced by the *C. elegans* Sequencing Consortium (1998). If the gene of interest has a sequence in the *C. elegans* EST database, the gene is transcribed and presumably translated. The appropriate cDNA may be requested (the cm series from alan@sanger.ac.uk or rw@nematode.wustl.edu and the yk series from ykohara@lab.nig.ac.jp), and used to probe cDNA libraries to isolate a full-length cDNA.

If the EST database does not reveal a sequence corresponding to the gene of interest, the simplest method to verify expression of a gene is by the reverse transcription (RT)-polymerase chain reaction (PCR) method. RNA is isolated by standard procedures (many kits are now available commercially), and reverse-transcribed to make cDNA (kits are also available for this procedure). The cDNA is amplified using primers predicted by the sequences in the database. The amplification product is either directly sequenced or the product is inserted into a cloning vector and subsequently sequenced. In the example of the *flp* genes, in a few instances, the predicted ORF was different than the isolated cDNA; in addition, alternative transcripts were isolated for several genes (Nelson et al., 1998a).

To determine the developmental expression pattern of the gene, total RNA is isolated from animals from each of the different developmental stages and screened by RT-PCR (as described above) or by Northern blot analysis (see Sambrook et al., 1989).

Determining the cell-specific expression pattern of a neuropeptide gene

Localization of transcripts by in situ hybridization

To determine the cells in which a gene is transcribed, in situ hybridization is used to localize mRNA transcripts. In this method, DNA or RNA probes complementary to the target mRNA are generated; the probes are hybridized against target tissue and visualized. Probes are labeled with either a radioactive, fluorescent, or enzymatic marker. In general, riboprobes, RNA molecules transcribed from DNA templates by RNA polymerases, give the strongest signals for RNA localizations, and can be generated with commercially available kits.

The labeled probes are applied to whole mounts of fixed *C. elegans*, to fixed animals that have been cracked open after dry ice freezing, or less frequently, to sections of fixed animals (Albertson et al., 1995; Seydoux and Fire, 1995). For detection of radioactive probes, the slides are dipped into emulsion and allowed to expose for several weeks (time of exposure is determined empirically); radioactive signals, though sensitive, give limited cellular resolution. For non-radioactive probes, detection is generally with fluorescent or enzymatic markers, such as alkaline phosphatase. Although detection by alkaline phosphatase is highly sensitive, the cellular resolution of the signal is limited. By contrast, cellular resolution of a fluorescent signal is excellent, but the sensitivity of this method is lower than the enzymatic marker (Seydoux and Fire, 1995).

Thus far, the in situ hybridization technique has not worked optimally in *C. elegans*. The technique has been used most successfully in the localization of transcripts in embryos (Seydoux and Fire, 1995). Localization of transcripts in larval or adult animals, however, is still highly variable.

Cis-*acting elements directing expression of a reporter gene*

A more commonly used method for determining the cell type-specific expression pattern for a given gene makes use of reporter constructs in transgenic animals. The most widely used reporter genes in *C. elegans* are *lac*Z, whose product is β-galactosidase (β-gal), and green fluorescent protein (GFP; Chalfie et al., 1994); standard *C. elegans* vectors containing these reporter genes, generic 3′ untranslated regions, and polyadenlyation signals are available (Fire et al., 1990; Mello and Fire, 1995). To construct a transcriptional fusion, the promoter region of a gene of interest is placed upstream of the reporter gene. To construct a translational fusion, the coding region of a gene is positioned in-frame with the reporter gene. For the analysis of neuropeptide genes, a transcriptional fusion construct is usually made so that the transgene product is not secreted outside the cell.

Procedures for generating transgenic animals are now fairly routine in *C. elegans* (Fire, 1986; Mello et al., 1991). The reporter construct is generally co-injected with a marker plasmid to facilitate selection of transgenic animals. Co-injection markers are of two types: 1. markers that confer a visible phenotype to the animal. The most commonly used marker plasmid of this type, pRF4, carries a dominant mutation in the *rol-6* collagen gene (Kramer et al., 1990), which confers an obvious rolling phenotype to the animals; and 2. markers that rescue a phenotype. In this case, a mutant animal, such as a dumpy animal, is used for microinjection; animals that are rescued for the dumpy phenotype (i.e. now wild-type length) are selected.

DNA (50–100 ng/ml of reporter plasmid; 50 ng/ml marker plasmid) is injected into the syncytial gonad of a hermaphrodite animal with an electrode needle under an inverted microscope. F_1 progeny that show the marker phenotype are selected, and their progeny are in turn screened for the marker phenotype to determine whether heritable transmission of the transgenes has occurred. The time from the microinjection step to the isolation of a heritable, transgenic line takes about one week.

The injected plasmids undergo homologous recombination to form extrachromosomal arrays, which contains multiple, concatenated copies of the reporter and marker plasmids (Mello and Fire, 1995). Inheritance levels of the extrachromosomal arrays range from 10 to 90%, with an average of 50% inheritance in each generation. For this reason, arrays are often integrated into the genome by gamma- or ultraviolet-induced integration. Thirty to 50 fourth larval stage animals from a transgenic strain with low transmission rate (about 20%) are irradiated (3600–4800 Rads from a cobalt source or 300 Joules/m^2 in a Stratalinker). About 500 F_1 animals showing the marker phenotype are individually plated and allowed to self. F_2 animals from plates showing over 75% of the progeny with the marker phenotype are individually plated and allowed to self. Plates on which 100% of the F_3 progeny show the marker phenotype are likely to have integrated arrays.

Once a transgenic line has been established, the location of the reporter gene products is determined. β-gal expression can be monitored in two ways. The first method localizes β-gal using antibodies against the β-gal protein (available from Sigma) and an appropriate secondary antibody tagged with a fluorochrome (see Antibody Localization below for protocol). Alternatively, cells in which β-gal is active as an enzyme can be visualized; cleavage by β-gal of a substrate such as X-gal results in a blue precipitate that can be visualized in cells under brightfield or DIC optics. To visualize where β-gal is active, animals are fixed in a drop of acetone in a depression well slide. After evaporation of the acetone, the animals are incubated in a few drops of X-gal solution [100–125 mM phosphate buffer (pH 7.2–7.5), 10 mM $MgCl_2$, 0.004% SDS, 5 mM potassium ferrocyanate, 5 mM potassium ferricyanate, 0.024% X-gal, mixed fresh day of use] overnight at 4°C or several hours at 37°C (the higher temperature increases the rate of reaction); the reaction can be monitored under a stereomicroscope. The antibody localization method for β-gal has the advantage that it can be used in conjunction with other fluorochrome markers, such as a fluorescent nuclear marker or another antibody marker.

More recently, *GFP* has been the reporter gene of choice. GFP has intrinsic fluorescence properties when excited at a certain wavelength, and can, therefore, be visualized in living animals. Furthermore, axonal projections can be clearly distinguished over long distances with the GFP marker; this visualization greatly helps in making identifications of specific neurons, particularly if the animals are viewed with a confocal microscope. Live animals can be mounted on a 2% agar pad (with 1 mM sodium azide to anesthesize the animals) for viewing.

The patterns of gene expression seen with reporter constructs, however, do not necessarily correspond to the expression pattern of the protein or transcripts (Mello and Fire, 1995). For instance, if the reporter construct does not contain all *cis*-acting regulatory elements, the expression pattern may be incomplete. Reporter constructs are most often made with 3–4 kbp of upstream sequence; if transcriptional regulatory elements lie further upstream or lie within intronic or 3′ untranslated sequences, an incomplete or incorrect expression pattern will be seen. Second, the reporter construct may be poorly expressed. Certain transgenes are silenced and show either no or partial expression patterns. Third, expression of the reporter construct can have toxic effects on the cells. For instance, overproduction of GFP appears to cause ectopic projections of axons of certain neurons (C. Li, unpubl. obs.). Fourth, expression of many reporter constructs is blocked in the germline. Expression patterns seen in transgenic animals, therefore, should be verified by independent methods, such as in situ hybridization (see above) or antibody localization (see below).

Antibody localization of neuropeptides

To determine the precise location of neuropeptide gene products, antibodies specific for the neuropeptides are used. Both monoclonal and polyclonal antibodies have been used to detect nematode neuropeptides (Sithigorngul et al., 1989; Schinkmann and Li, 1992). The choice of the procedure depends on several factors. For monoclonal antibodies, a screen must be developed to identify the interesting antibodies; in the past, such antibodies were isolated based solely on their staining pattern in fixed animals. To generate polyclonal antibodies, sufficient quantities of purified protein must be produced. Because most proteins are not available in large quantities from extracts of *C. elegans*, recombinant proteins are produced in bacteria to isolate sufficient quantities. However, neuropeptides are relatively short, and can be synthesized inexpensively by commercial firms.

Many antibodies against neuropeptides are already available commercially or from individual researchers. Some of the antibodies against mammalian neuropeptides, such as Substance P and cholecystokinin, cross-react with peptides in *C. elegans*. Anti-FMRFamide antisera are available from several sources; however, since some of the antisera recognize only the C-terminus of FMRFamide (Marder et al., 1987), the antisera cross-react with FaRPs encoded by multiple genes (Nelson et al., 1998a). To differentiate between the different gene products, antibodies against the pre-propeptide precursor molecule or the specific neuropeptides need to be generated. Neuropeptides are generally conjugated to a carrier molecule, such as thyroglubulin, to generate a better immunologic response. Antibody production is now routinely performed by commercial firms.

For easier identification of neurons, whole animals are used for immunochemistry. Animals are fixed (generally 4% paraformaldehyde in standard phosphate buffer) overnight at 4°C in a microcentrifuge tube. To permeabilize the animal, animals are incubated with gentle agitation in 5% β-mercaptoethanol, 0.5% Triton X-100 in 0.1 M Tris buffer (pH 7.2 at room temperature) for 20–30 h at 37°C, rinsed in Tris buffer, and then incubated with gentle agitation with Type IV collagenase (Sigma) at 900 U/ml in 0.1 M Tris buffer (pH 7.7 room temperature), 1 mM $CaCl_2$ for 5–10 h (depending on the batch of collagenase) at 37°C. Animals are rinsed in phosphate buffer, and incubated overnight in primary antibody with a blocking buffer (10% goat serum, 0.5% Triton X-100 in 0.1 M phosphate buffer). After rinsing in phosphate buffer, animals are incubated in secondary antibody for 1–2 h, rinsed in phosphate buffer, and mounted in a drop of 70% glycerol, 30% phosphate buffer, 5% (by weight) *n*-propyl gallate. To visualize nuclei, diamidinophenolindole (DAPI; 1 μg/ml) can be added to the mounting solution.

Determining the function of a specific neuropeptide gene in *C. elegans*

To determine the function of a specific gene, three general strategies are used for *C. elegans*. The first approach is to isolate a mutant animal in which the neuropeptide gene is inactivated (i.e. a "knockout" animal) and to characterize the phenotypes seen in the knockout animals. A second approach is to inactivate a gene by RNA-mediated interference. The third approach is to overexpress the neuropeptide gene and determine the phenotypes in these animals. The expectation is that the phenotypes in animals overexpressing the neuropeptide gene are reciprocal to those seen in animals that express no neuropeptides.

Isolation of knockout animals by Tc1 transposon excision

To inactivate genes by transposon excision, a strain in which a transposon has inserted into the gene of interest must first be isolated. Several *C. elegans* strains are transposition-proficient due to genes, known as *mutator* or *mut* genes, that confer high rates of transposon transposition and excision. Strains carrying *mut-2* mutations have been frequently used to found populations of animals containing high copy numbers of Tc1 transposons. To isolate a transposon-insertion animal (Fig. 1), a population is started with 5–10 high copy number Tc1 animals, and their progeny are allowed to populate an agar plate. When the food source is depleted, the animals are washed off the plate. Part of the population is frozen for future use [*C. elegans* can be frozen in a freezing solution (0.58 g NaCl, 0.68 g KH_2PO_4, 30 g glycerol, 0.56 ml 1 M NaOH, water to 100 ml; after autoclaving, add 0.3 ml 0.1 M $MgSO_4$) and thawed to yield viable animals], part of the population is re-plated onto a fresh agar plate, and part of the population is lysed to isolate DNA. The population DNA is screened by PCR with a Tc1-specific and a gene-specific primer to determine whether a Tc1 transposon has inserted into the gene. Populations are screened until a transposon insertion in the gene of interest is isolated. When such a population is identified, the plate corresponding to this population is subdivided onto fresh agar plates, and the population from each of these plates is re-screened as above. This subdivision and re-testing of the population continues until a single animal containing the transposon in the gene of interest is isolated (see Fig. 1). Based on the insertional frequency of Tc1, at least 100 populations (representing 1000 genomes) must be screened to isolate a Tc1 insertion strain (Rushforth et al., 1993; Zwaal et al., 1993).

This method of screening populations, while fruitful, is somewhat laborious. For each gene of interest, new populations are made and screened. To decrease the amount of work, Ronald

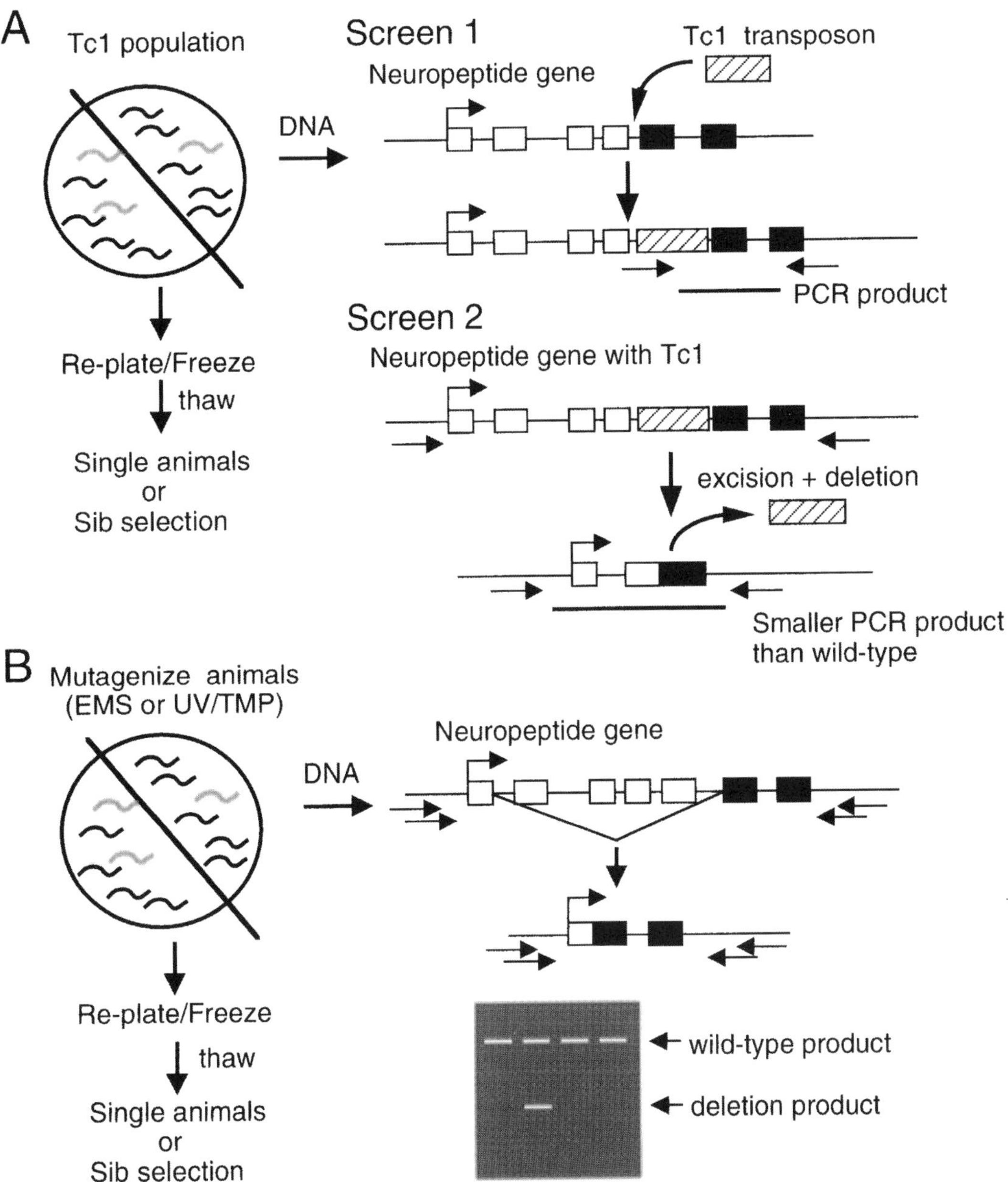

Fig. 1. Screening for knockout animals. (A) Screening by transposon excision method. Five to 10 animals are used to found a population of Tc1 animals. Animals are washed off the plate and divided for re-plating, freezing, or DNA lysis (animals carrying a Tc1 insertion are indicated in grey). DNA is screened by PCR with a Tc1- and gene-specific primer (horizontal arrows). A positive population will have a PCR product. Animals with a Tc1 insertion (indicated in the gene by hatched lines) are used to found new populations, which are divided as before. DNA is screened with primers flanking the gene to identify a population with a deletion (see (B)). (B) Screening by chemical (EMS or UV/TMP) mutagenesis. Five to 10 mutagenized animals are used to found a population. Animals are washed off the plate and divided for re-plating, freezing, or DNA lysis. DNA is screened by PCR with nested primers (horizontal arrows) flanking the gene. A positive population will have a smaller PCR product in addition to the wild-type product. Exons are shown as boxes, introns as lines.

Plasterk and coworkers refined the method by designing a transposon insertion bank of Tc1-insertion populations (Zwaal et al., 1993). They started with 5000 cultures, each founded by 1–10 animals. One-half of each population was frozen in triplicate, while the other half of the population was lysed to isolate DNA. DNA from each population was pooled into a three-dimensional matrix, with each population lysate contributing to three pools of DNA in the matrix. The matrix address of a positive Tc1 insertion revealed which population needed to be thawed. A single animal carrying the transposon insertion in the gene of interest is then isolated from the thawed population. The Plasterk transposon insertion bank has been screened repeatedly with different primer sets to isolate Tc1 insertions in numerous genes for researchers in the *C. elegans* field. Using this transposon insertion bank, animals carrying transposons in several *flp* genes (*flp-1*, *flp-6*, and *flp-9*) have been isolated (Nelson et al., 1998b; C. Li, unpubl. obs.). Occasionally, a Tc1 insertional allele could not be isolated from the transposon insertion bank.

To determine the site of insertion, the PCR product amplified with the Tc1- and gene-specific primers is directly sequenced or subcloned into a vector and subsequently sequenced. Because the integration site for Tc1 requires a TA sequence and because *C. elegans* introns are TA-rich, the transposon is likely to insert into an intron of the gene and be spliced from the mature RNA. Furthermore, in many cases, Tc1 insertions into exons are also spliced from the mature RNA (Rushforth et al., 1993). In these cases, the transposon-insertion strains are unlikely to represent knockout lines.

To isolate deletion derivatives from transposon-insertion strains, Plasterk and co-workers took advantage of the finding that Tc1 excision in homozygous Tc1-carrying animals generally results in a genomic "footprint" of several nucleotides, presumably due to imprecise double strand gap repair following Tc1 excision (Plasterk and Groenen, 1992); at a low frequency, the imprecise repair results in a deletion. Therefore, to isolate a deletion strain, populations of transposon-insertion animals in a transposition-proficient background (such as *mut-2*) are screened for a deletion by PCR with primers flanking the gene; the primers are generally 3–5 kbp apart. As before, each population is divided, such that part of the population is frozen, part is plated, and part is lysed for DNA isolation. Once a population containing a deletion is identified, it is subdivided and re-screened until a single deletion animal is isolated. During the population screening, several deletion derivatives may be isolated. Using this method, two deletion alleles of the *flp-1* neuropeptide gene were isolated (Nelson et al., 1998b). The two alleles of *flp-1* show varying degrees of phenotypic severity, presumably because a few of the *flp-1* neuropeptides are still active in one of the alleles.

Isolation of knockout animals by chemical mutagenesis

While the Tc1 insertion/excision method has yielded numerous knockout animals, the method is dependent on two somewhat laborious steps: 1. isolation of a Tc1 insertion animal; and 2. isolation of a deletion animal from the Tc1 insertion animal. Therefore, Plasterk, Barstead, and co-workers have used a more directed approach to isolate knockout animals (Fig. 1). Ethylmethane sulfonate (EMS; average mutational frequency 5×10^{-4}/*C. elegans* gene with 50 mM EMS) is a commonly used mutagen, which generally causes GC to AT transitions in DNA (Brenner, 1974). However, about 13% of the EMS-induced mutations are deletions in the DNA. Trimethylpsoralen (TMP) exposure (30 μg/ml for 15 min), followed by ultraviolet radiation (340 $\mu W/cm^2$ for 60 s), has an average mutational frequency of 3×10^{-5}/gene, of which about 50% of the induced mutations are deletions (Yandell et al., 1994). Plasterk, Barstead, and co-workers have successfully isolated animals containing deletions by screening chemically-mutagenized populations by PCR with primers flanking the gene of interest (Jansen et al., 1997; Dernburg et al., 1998).

As with the Tc1-insertional animals, Plasterk and co-workers created a deletion bank of the DNA isolated from the mutagenized populations and of the frozen animals (representing about 1 $\times$ 10^6 genomes); the deletion bank, therefore, can be screened repeatedly and after finding the correct address, a population of animals can be thawed and the deletion mutant isolated. By contrast, Barstead and co-workers create new mutagenized populations each time. Because chemical mutagenesis causes deletions at a low frequency, the number of genomes that must be screened is much larger than with the Tc1-insertion method. However, a deletion animal is obtained in one screening step with the chemical mutagensis procedure, while the Tc1 procedure requires two screening steps.

To determine the function of the different *flp* genes, the Plasterk deletion bank was screened for deletions in *flp-2* through *flp-12* using one primer pair per gene. Single mutants carrying deletions in *flp-3*, *flp-8*, and *flp-10* were isolated, and a population of animals that harbored a *flp-7* deletion animal was identified. The extent of the deletions was determined by DNA sequence analysis. The entire coding regions were deleted in the *flp-3* and *flp-8* mutants; the *flp-10* deletion removed the start site of transcription and part of the coding region. As an example of the speed in which knockout animals can be isolated, the time from the initial PCR screen to the isolation of the single animal containing the *flp-8* deletion was one week (C. Li, unpubl. obs.).

Gene inactivation by RNA-mediated interference

Recently, a simple method, RNA-mediated interference (RNAi), to disrupt endogenous gene function in *C. elegans* has been described (Fire et al., 1998). Sense and antisense RNA are synthesized from phagemid clones with either T3 or T7 RNA polymerase. Because double stranded RNA has been found empirically to be more efficient for gene interference than either sense or antisense RNA alone, the synthesized sense and antisense RNA populations are annealed (37°C for 10–30 min). The RNA populations (0.5 to 1 $\times$ 10^6 RNA molecules) are microinjected into the gonad, body cavity, or intestinal cells of adult hermaphrodites; alternatively, the animal can be soaked in RNA (C. Mello, pers. comm.). F_1 progeny are screened for defects (see below); occasionally, F_2 progeny will also show phenotypes. Thus far, this method has not been extensively used to examine neural genes; some data indicate that RNAi does not effectively inactivate neural gene expression (A. Fire, pers. comm.).

Production of animals overexpressing a neuropeptide gene

A simple prediction is that the phenotypes of animals overexpressing a gene are reciprocal to those seen in animals carrying deletions of a gene. To generate animals in which a gene is overexpressed, a DNA fragment containing the genomic region corresponding to the gene must be isolated. This fragment must contain the regulatory as well as coding regions of the gene. The fragment is used for germline transformation into wild-type or deletion animals to generate transgenic strains, similar to the procedure described above. Care must be taken in the choice of the marker that is used for selection of transgenic animals; many of the commonly used markers interfere with the behavior of the animal. Once a transgenic line is established, the transgene can be integrated into the genome as described above. Overexpression of the gene can be verified by Northern or Western blot analysis.

Occasionally, no transgenic lines can be isolated, suggesting that overexpression of the gene causes a lethal phenotype. Two strategies are used to overcome this problem. First, the amount of DNA injected can be decreased (to 1–10 ng/ml). Second, an inducible promoter, such as the heat-shock promoter (Stringham et al., 1992), can be used to drive expression of the gene. Use of an inducible promoter such as the heat-shock promoter, however, will drive expression in all cells, not only the cell in which the gene of interest is expressed.

Animals that overexpress the *flp-1* gene have many phenotypes that are opposite to those seen in the *flp-1* deletion animals (Nelson et al., 1998b). Interestingly, the phenotypes of the overexpression animals are more severe when the animals are in a *flp-1* deletion background than in a wild-type background. The increased severity of the phenotypes may be due to a compensatory up-regulation of the peptide receptors, which would result in an increased sensitivity to the overexpression of the FLP-1 neuropeptides.

Phenotypic characterization of deletion and overexpression animals

An isolated deletion animal is in a Tc1-rich or EMS-mutagenized background. Similarly, overexpression animals in which the transgenes have been integrated by irradiation may have unlinked irradiation-induced mutations. To remove unlinked Tc1 transposons or chemical- or radiation-induced mutations, deletion or overexpression animals are backcrossed several times into a wild-type background before phenotypic characterization is performed. Once the animal is backcrossed at least 5 times, the phenotypic analysis can begin. Because neuropeptides are generally expressed in the nervous system, behavioral phenotypes are examined first. In addition, the localization of the peptides to specific neurons may suggest functions to examine. Several of the most common behavioral phenotypes in *C. elegans* are described below.

Sensory behaviors

C. elegans responds to its environment through several sensory behaviors. These behaviors mediate responses to mechanical stimuli, such as body, nose, and head touch, chemical stimuli, and temperature stimuli. Distinct sensory neurons are responsible for transducing different types of external stimuli; a rare sensory neuron is responsive to multiple sensory modalities (Kaplan and Horvitz, 1993).

Mechanosensation

C. elegans has several touch responses that are mediated by distinct sensory neurons. *C. elegans* responds to light touch to the head region by moving backwards and touch to the tail region by moving forwards; this response is referred to as body touch (Chalfie, 1984). To assay body touch, an animal is gently stroked with an eyebrow hair (glued onto the end of a toothpick); an animal is tested three times in quick succession for the response. Body touch is mediated by six mechanosensory cells located in the anterior and posterior regions of the animal (Chalfie, 1984). The mechanosensory neurons make gap and chemical synapses with four types of interneurons, some of which have direct synaptic connections with the motoneurons (White et al., 1986).

Light touch to the nose of *C. elegans* causes an immediate recoil of the animal (Kaplan and Horvitz, 1993). Nose touch responses can be tested by placing an eyebrow hair in the path of an animal, and letting the animal bump its nose against the hair; each animal is tested at least three times. Nose-touch behavior is slightly more difficult to test than body touch, as one must predict where the animal will move next. However, after testing a wild-type animal for nose touch a few times, the response becomes fairly obvious. Nose touch behavior is mediated by the ASH, FLP, and OLQ neurons, which synapse onto some of the same interneurons as the mechanosensory neurons that mediate light body touch. *flp-1* deletion animals are nose-touch insensitive, but show wild-type responses to body touch (Nelson et al., 1998b).

The head of *C. elegans* moves constantly, presumably because the animal is continuously sampling its environment. While the arrangement of the body musculature allows waves only along the dorsal-ventral axis, the head musculature allows 360° movement (White, 1988). Light touch to the side of the nose activates an immediate head withdrawal reflex and interrupts foraging behavior. This response is mediated by two sets of sensory neurons, OLQ and IL1, which have direct synaptic connections to head motoneurons (Hart et al., 1995).

Chemosensation

Response to chemical stimuli can be subdivided into two categories: response to water-soluble molecules (taste) and response to airborne (volatile) molecules (smell). Taste and smell are mediated by different sensory neurons (Bargmann and Horvitz, 1991a; Bargmann et al., 1993), although both types of sensory neurons are similar in structure and belong to the same neuronal subclass (amphidial neurons) (Ward et al., 1975; Perkins et al., 1986). *C. elegans* shows positive and negative chemotaxis to different chemicals (Ward, 1973; Bargmann and Horvitz, 1991a).

To test responses to water-soluble molecules, a gradient of the molecule is established in a 10-cm petri dish. Petri dishes are poured with 10 ml of 2% agar, and a point source of the molecule is placed at one end of the dish. The molecule is allowed to diffuse overnight, thereby establishing the gradient. The following day, a control solution is put at the opposite pole to the point source of the test molecule; the point source is also replaced with fresh solution. At least 100 adult animals, which have been rinsed several times in water to rid the animals of salts and bacteria, are put equidistant between the two solutions. After 1 h, animals are counted at the control and test solution (only adult animals are counted). A chemotaxis index (CI) is calculated as follows: number of animals at the test spot minus number of animals at the control spot divided by the total number of animals (Bargmann and Horvitz, 1991a). A CI of zero indicates no preference for the test molecule; a CI of 1 indicates a positive preference, whereas a CI of −1 indicates a negative preference for the test molecule.

Alternatively, preferences for water-soluble molecules can be tested by using petri dishes that are divided into four quadrants (S. Wicks, pers. comm.). Control solutions are mixed into a 2% agar solution, and poured into two opposite quadrants; the test molecule is similarly mixed with agar, and poured into the remaining two quadrants. Animals are placed in the center. After 1 h, animals in the different quadrants are counted and a CI is determined.

To test responses to volatile substances, a modified procedure for testing water-soluble solutions is used. Instead of setting up a gradient overnight, the point source of the test molecule is applied at the same time as the control solution. Animals are immediately placed equidistant from the two solutions. As before, animals at the two solutions are counted after 1 h, and a CI is determined.

Using these protocols, Bargmann and co-workers (1991a, 1993) determined the wild-type preferences for many molecules. In addition, Bargmann and co-workers identified the specific amphidial neurons that were responsible for mediating responses to specific chemicals (1991a, 1993). A given mutant, therefore, can be tested for its chemotaxic responses to a variety of solutions, and variations from the wild-type response can indicate the pathway that is affected (such as a specific sensory neuron or the entire sensation of smell or taste).

Osmosensation

C. elegans avoids areas of high osmolarity, such as high concentrations of glucose or glycerol (Culotti and Russell, 1978). This response is mediated by the ASH neuron, which also mediates nose touch and response to volatile repellents (Kaplan and Horvitz, 1993; Hart et al., 1995). To test for this behavior, a ring of high osmolarity is made on a 6 cm-petri dish filled with agar containing standard *C. elegans* salts (Wood, 1988). A high osmolarity ring can be made by dipping the cap of a culture tube into a solution of high osmolarity (such as 60% glycerol) and then stamping the agar with the cap. If a dye has been added to the glycerol, the high osmolarity ring can be visualized. Eight to 10 animals (rinsed in water to remove salts and bacteria) are placed into the center of the ring. The time it takes for a given animal to cross the ring is measured. Wild-type animals take about 10–20 min, depending on the dryness of the plates, to cross a high osmolarity barrier such as 60% glycerol. Animals that are insensitive to high osmolarity will cross the barrier within a much shorter time. For instance, *flp-1* deletion animals will cross the 60% glycerol barrier within 5 min (Nelson et al., 1998b).

Thermotaxis

C. elegans is usually cultured at temperatures between 10°C and 25°C. When exposed to a radial (Mori and Ohshima, 1995) or linear (Hedgecock and Russell, 1975) thermal gradient, animals will move towards an optimal temperature, which generally corresponds to the temperature at which they were cultivated. In a radial temperature gradient, animals tend to circle around the plate at their optimal temperature.

A thermal gradient can be established by freezing a vial of glacial acetic acid, and placing the frozen vial against an agar plate. The cold will diffuse from the center, and a radial gradient from 17°C to 25°C can be made (Mori and Ohshima, 1995); tiny temperature sensors can be placed in the agar to measure the temperature at different times and places. Animals cultured at a certain temperature can be placed anywhere along the gradient, and their movement can be recorded.

Dauer formation

C. elegans undergoes an alternative life cycle, the dauer pathway, under adverse environmental conditions. Entry into the dauer pathway is determined by three factors: food supply, temperature, and a pheromone that is a population density marker (Riddle and Albert, 1997). These chemical signals are transduced by four sensory (amphidial) neurons, ADF, ASG, ASI and ASJ (Bargmann and Horvitz, 1991b). Animals in the dauer state have depressed pharyngeal pumping and locomotion, are relatively thin and dense due to radial shrinkage, and have an increased lifespan (Cassada and Russell, 1975). These phenotypes are readily visible with a stereomicroscope. When environmental conditions become favorable again, animals exit the dauer pathway and continue development into an adult animal. At least 30 genes are involved in the dauer pathway (Riddle and Albert, 1997). Mutations in these genes cause either constitutive dauer formation or no dauer formation.

Motor behaviors

Locomotion

In *C. elegans*, body wall musculature responsible for forward and backward movement is arranged in four longitudinal strips, two along the dorsal and two along the ventral surfaces (White, 1988). This muscle arrangement constrains body bends to the dorsal–ventral axis, resulting in sinusoidal waveforms. The waveform of animals from different larval stages is relatively uniform. Quantitative measurements of waveform can be determined by placing animals on a fresh bacterial lawn, and measuring the amplitude (usually measured from peak to peak) and wavelength of the waveforms left by the animals in the bacteria. Wild-type adult animals have an average amplitude of 0.20 ± 0.01 mm, while the amplitude of fourth larval stage animals is 0.15 ± 0.01 mm (Nelson et al., 1998b).

Several types of motoneurons innervate the body wall muscles. Because neuropeptides often have modulatory actions, disruption of neuropeptide expression may affect waveform. For example, inactivation of the *flp-1* neuropeptide gene, which is normally expressed in interneurons in the locomotion circuit, causes an exaggerated waveform in the animals (Nelson et al., 1998b). This phenotype is readily observable with a stereomicroscope.

Activity
Wild-type animals move at a certain rate, which is higher in the absence of food (bacteria) than in the presence of food. Activity of animals can be quantitated in several ways. First, the number of body bends per min can be counted either on unseeded plates (i.e. in the absence of bacteria) or on plates with a confluent bacterial lawn. Second, animals can be placed in a drop of physiological buffer in a microtiter well, and the number of body bends per min can be counted. Wild-type animals are more active in liquid than on a solid medium.

As with waveform, disruption of neuropeptide levels can result in animals that are hyperactive or hypoactive, behaviors that can be readily distinguished with a stereomicroscope. Altering the level of *flp-1* expression changes the activity levels of animals (Nelson et al., 1998b).

Pharyngeal pumping
The pharynx is a muscular, bi-lobed organ that is continuous with the hypodermis of the animal. The anterior end of the pharynx, which includes the first bulb of the pharynx, is called the corpus. The isthmus connects the anterior and posterior (or terminal) pharyngeal bulbs. Contraction of the corpus and anterior isthmus muscles causes liquid and bacterial particles to flow inwards; this initial contraction happens just prior to contraction of the terminal bulb muscles, which passes debris into the intestine. Relaxation of the muscles brings muscles back to their resting position and causes the lumen of the corpus to close. Relaxation is followed by isthmus peristalsis, which carries bacteria from the anterior to posterior isthmus. Every relaxation, however, is not followed by a peristalsis; on average, one peristalsis occurs for every four pumps (Avery and Thomas, 1997).

As its name implies, the grinder in the terminal bulb of the pharynx grinds food and passes it into the intestine. Under the stereomicroscope, the grinder is visible as a black dot, which moves back and forth during cycles of muscle contraction and relaxation. The number of pharyngeal pumps per 15 s, therefore, can be determined by counting the movement of the grinder. Wild-type animals average about 55 pumps per 15 sec. Many mutants in which the pumping rate is affected have been isolated (Avery and Thomas, 1997). Several pharyngeal neurons stain for FaRPs (Schinkmann and Li, 1992), suggesting that FaRPs may modulate pumping activity.

Defecation
As with feeding, defecation is the result of a series of coordinated muscle contractions (Croll and Smith, 1978). The cycle begins with contraction of all posterior body wall muscles, which causes the fluid contents of the intestine to move anteriorly. This posterior contraction is followed by a relaxation and contraction of anterior body wall muscles, both of which cause intestinal contents to move posteriorly. Lastly, muscles around the anus contract to release the contents of the intestine to the exterior. Defecation occurs every 45–50 s in wild-type animals (Liu and Thomas, 1994). The entire defecation cycle can be visualized with a stereomicroscope.

Animals that have defects in each of the defecation steps have been isolated (Avery and Thomas, 1997). Many of the mutants have muscle-related defects. Muscles responsible for the expulsion step are innervated by the DVB and AVL neurons, which presumably use GABA as their primary transmitter (McIntire et al., 1993). *unc-25* mutants, which do not synthesize GABA, are expulsion-defective, but have residual expulsions (Thomas, 1990); however, animals in which DVB and AVL are removed completely are expulsion-defective. Since the DVB neurons also stain for FaRPs (Schinkmann and Li, 1992), the residual expulsion activity may be due to FaRPs co-released with GABA. Overexpression of the *flp-1* gene, for instance, causes a longer defecation cycle (L. Nelson, pers. comm.).

Egg-laying
The egg-laying system in *C. elegans* consists of the gonad and vulva, which join to form a channel

through which fertilized eggs exit the animal (Sulston and Horvitz, 1977; Kimble and Hirsh, 1979), uterine and vulval muscles, which propel eggs out of the uterus and control the opening of the vulva, and two neuronal cell types, the HSN and VC cells, which provide the major innervation to the muscles (White et al., 1986). The HSN and VC neurons both express FaRPs (Schinkmann and Li, 1992) and acetylcholine (J. Duerr. pers. comm.); HSN cells also express serotonin (Desai et al., 1988). Wild-type hermaphrodites lay about 300 eggs over 4 d (Wood, 1988). Eggs are laid about 3 h after fertilization, at about the 30-cell stage (Hirsh et al., 1976).

The number of eggs an animal lays can change in response to environmental stimuli. For instance, animals lay more eggs in the presence than absence of food (Trent et al., 1983). In addition, animals placed in a liquid medium are transiently inhibited from egg laying for 1 h; this inhibition can be reversed by the addition of serotonin (5 mg/ml), imipramine (0.75 mg/ml), which blocks uptake of serotonin, levamisole (0.1 mg/ml), a cholinergic agonist, or phentolamine (10 mg/ml), an octopamine antagonist (Trent et al., 1983).

Over 50 genes that affect egg-laying behavior have been identified (Trent et al., 1983; Ferguson and Horvitz, 1985). Egg-laying mutants include: (a) lineage mutants that form no functional vulva (Ferguson and Horvitz, 1985). Eggs hatch internally and larval animals devour the parent; (b) muscle mutants (Waterston et al., 1980); (c) lineage or cell specification mutants that disrupt formation of the neurons (Trent et al., 1983).

Mutations that are more likely to be neural-related can be identified by pharmacological tests. Muscle mutants, for example, will not lay eggs even in the presence of serotonin, while HSN-defective mutants will. Some mutants respond to serotonin, but not to imipramine (Trent et al., 1983). To determine whether a mutant has egg-laying defects, the following characteristics can be scored: (a) number of eggs laid. The brood size is about 300 for wild-type animals; (b) stage of eggs released. Eggs are released at around the 30-cell stage. Some mutants release eggs immediately after fertilization (egg-laying constitutive). These animals have no or few fertilized eggs in their gonad. Other mutants delay egg release and become bloated with fertilized eggs (Trent et al., 1983); and (c) responses to serotonin, imipramine, levamisole, and phentolamine.

A pharmacological response to the neuropeptide can also be tried, but the cuticle surrounding *C. elegans* presents a permeability barrier. High concentrations (millimolar range) of the neuropeptide should be tried, and a cuticle defective mutant, such as *srf-3* (Politz et al., 1990), which allows greater peptide penetration, should also be used. Using the *srf-3* mutant, Schinkmann and Li (1992) found that addition of FLRFamide could potentiate the serotonin response. *flp-1* deletion animals have several egg-laying defects, including a smaller brood size (C. Li, unpubl. obs.) and an insensitivity to serotonin-induced egg-laying (Nelson et al., 1998b).

Rescue of the mutant phenotypes with the wild-type gene

To confirm that phenotypes seen in the deletion animal are due to inactivation of the gene, a DNA fragment containing the genomic region corresponding to the gene is used for germline transformation of the deletion animals, similar to the procedure described previously. Deletion animals rescued with the wild-type gene should show wild-type or overexpression phenotypes.

Summary

With the completion of the genome sequencing project and the wealth of information about its nervous system, *C. elegans* presents one of the most attractive systems in which neuropeptide gene families can be examined. The *flp* neuropeptide gene family is one of the largest neuropeptide families identified to date. Why are there so many *flp* genes and what are the functions of the individual FaRPs? Unraveling the functions of the different FaRPs and the mechanisms whereby these neuropeptides exert their actions in *C. elegans* is likely to give insights into the role of neuropeptide gene families in other organisms as well.

Acknowledgments

I am grateful to the worm community for sharing information, protocols, strains, and reagents, Ronald Plasterk and his laboratory for help with deletion screens, and Thomas Gilmore for comments on the manuscript.

References

Albertson, D.G., Fishpool, R.M. and Birchall, P.S. (1995) Fluorescence in situ hybridization for the detection of DNA and RNA. In: H.F. Epstein and D.C. Shakes (Eds.), *Caenorhabditis elegans: Modern biological analysis of an organism*, Academic Press, New York, pp. 339–364.

Avery, L. and Thomas, J.H. (1997) Feeding and Defecation. In: D.L. Riddle, T. Blumenthal, B.J. Meyer and J.R. Priess (Eds.), *C. elegans II*, Cold Spring Harbor Press, Plainview, pp. 679–716.

Bargmann, C.I. and Horvitz, H.R. (1991a) Chemosensory neurons with overlapping functions direct chemotaxis to multiple chemicals in C. elegans. *Neuron*, 7: 729–742.

Bargmann, C.I. and Horvitz, H.R. (1991b) Control of larval development by chemosensory neurons in *Caenorhabditis elegans*. *Science*, 251: 1243–1246.

Bargmann, C.I., Hartweig, E. and Horvitz, H.R. (1993) Odorant-selective genes and neurons mediate olfaction in C. elegans. *Cell*, 74: 515–527.

Barstead, R.J. and Waterston, R.H. (1989) The basal component of the nematode dense-body is vinculin. *J. Biol. Chem.*, 264: 110177–10185.

Blomqvist, A.G. and Herzog, H. (1997) Y-receptor subtypes-how many more? *Trends Neurosci.*, 20: 294–298.

Brenner, S. (1974) The genetics of *Caenorhabditis elegans*. *Genetics*, 77: 71–94.

Brown, M.R. (1990) Peptide biology: Past, present, and future. *Ann. NY Acad. Sci.*, 579: 8–16.

Brownlee, D.J.A. and Fairweather, I. (1999) Exploring the neurotransmitter labyrinth in nematodes. *Trends Neurosci.*, 22: 16–24.

Cassada, R.C. and Russell, R.L. (1975) The dauer larva, a post-embryonic developmental variant of the nematode *Caenorhabditis elegans*. *Dev. Biol.*, 46: 326–342.

The *C. elegans* Sequencing Consortium (1998) Genome sequence of the nematode *C. elegans*: A platform for investigating biology. *Science*, 282: 2012–2018.

Chalfie, M. (1984) Genetic analysis of nematode nerve-cell differentiation. *BioScience*, 34: 295–299.

Chalfie, M., Tu, Y., Euskirchen, G., Ward, W. and Prasher, D. (1994) Green fluorescent protein as a marker for gene expression. *Science*, 263: 802–805.

Croll, N.A. and Smith, J.M. (1978) Integrated behavior in the feeding phase of *Caenorhabditis elegans* (Nematoda). *J. Zool.*, 184: 507–517.

Culotti, J.G. and Russell, R.L. (1978) Osmotic avoidance defective mutants of the nematode *Caenorhabditis elegans*. Genetics, 90: 243–256.

Dernburg, A.F., McDonald, K., Moulder, G., Barstead, R., Dresser, M. and Villeneuve, A.M. (1998) Meiotic recombination in *C. elegans* initiates by a conserved mechanism and is dispensable for homologous chromosome synapsis. *Cell*, 94: 387–398.

Desai, C., Garriga, G., McIntire, S.L. and Horvitz, H.R. (1988) A genetic pathway for the development of the *Caenorhabditis elegans* HSN motor neurons. *Nature*, 336: 638–646.

Ferguson, E.L. and Horvitz, H.R. (1985) Identification and characterization of 22 genes that affect the vulval cell lineages of the nematode *Caenorhabditis elegans*. *Genetics*, 110: 17–72.

Fire, A. (1986) Integrative transformation of *Caenorhabditis elegans*. *EMBO J.*, 5: 2673–2680.

Fire, A., Harrison, S. and Dixon, D. (1990) A modular set of *lacZ* fusion vectors for studying gene expression in *Caenorhabditis elegans*. *Gene*, 93: 189–198.

Fire, A., Xu, S., Montgomery, M.K., Kostas, S.A., Driver, S.E. and Mello, C.C. (1998) Potent and specific genetic interference by double-stranded RNA in *Caenorhabditis elegans*. *Nature*, 391: 806–811.

Grimmelikhuijzen, C.J., Hahn, M., Rinehart, K.L. and Spencer, A.N. (1988) Isolation of pyroGlu-Leu-Leu-Gly-Gly-Arg-Phe-NH2 (Pol-RFamide), a novel neuropeptide from hydromedusae. *Brain Res.*, 475: 198–203.

Gudermann, T., Schöneberg, T. and Schultz, G. (1997) Functional and structural complexity of signal transduction via G-protein-coupled receptors. *Ann. Rev. Neurosci.*, 20: 339–427.

Hart, A.C., Sims, S. and Kaplan, J.M. (1995) Synaptic code for sensory modalities revealed by *C. elegans* GLR-1 glutamate receptor. *Nature*, 378: 82–85.

Hedgecock, E.M. and Russell, R.L. (1975) Normal and mutant thermotaxis in the nematode *Caenorhabditis elegans*. *Proc. Natl. Acad. Sci. USA*, 72: 4061–4065.

Hirsh, D., Oppenheim, D. and Klass, M. (1976) Development of the reproductive system of *Caenorhabditis elegans*. *Dev. Biol.*, 49: 200–219.

Jansen, G., Hazendonk, E., Thijssen, K. and Plasterk, R. (1997) Reverse genetics by chemical mutagenesis in *C. elegans*. *Nat. Genet.*, 17: 119–121.

Kaplan, J.M., and Horvitz, H.R. (1993) A dual mechansensory and chemosensory neuron in *Caenorhabditis elegans*. *Proc. Natl. Acad. Sci.*, 90: 2227–2231.

Kimble, J. and Hirsh, D. (1979) The postembryonic cell lineages of the hermaphrodite and male gonads in *Caenorhabditis elegans*. *Dev. Biol.*, 70: 396–417.

Kramer, J., French, R.P., Park, E. and Johnson, J.J. (1990) The *Caenorhabditis elegans rol-6* gene, which interacts with the *sqt-1* collagen gene to determine organismal morphology, encodes a collagen. *Mol. Cell. Biol.*, 10: 2081–2089.

Lingueglia, E., Champigny, G., Lazdunski, M. and Barbry, P. (1995) Cloning of the amiloride-sensitive FMRFamide peptide-gated sodium channel. *Nature*, 378: 730–733.

Liu, D.W.C. and Thomas, J.H. (1994) Regulation of a periodic motor program in *C. elegans*. *J. Neurosci.*, 14: 1953–1962.

Marder, E., Calabrese, R.L., Nusbaum, M.P. and Trimmer, B. (1987) Distribution and partial characterization of FMRFamide-like peptides in the stomatogastric nervous systems of the rock crab, *Cancer borealis*, and the spiny lobster, *Panulirus interrutpus*. *J. Comp. Neurol.*, 259: 150–163.

McIntire, S.L., Jorgensen, E., Kaplan, J. and Horvitz, H.R. (1993) The GABAergic nervous system of *Caenorhabditis elegans*. *Nature*, 364: 337–341.

Mello, C. and Fire, A. (1997) DNA Transformation. In: H.F. Epstein and D.C. Shakes (Eds.), *Caenorhabditis elegans: Modern biological analysis of an organism*, Academic Press, New York, pp. 451–482.

Mello, C.C., Kramer, J.M., Stinchcomb, D. and Ambros, V. (1991) Efficient gene transfer in *C. elegans*: Extrachromosomal maintenance and integration of transforming sequences. *EMBO J.*, 10: 3959–3970.

Mori, I. and Ohshima, Y. (1995) Neural regulation of thermotaxis in *Caenorhabditis elegans*. *Nature*, 376: 344–348.

Nambu, J.R., Murphy-Erdosh, C., Andrews, P.C., Feistner, G.J. and Scheller, R.H. (1988) Isolation and characterization of a *Drosophila* neuropeptide gene. *Neuron*, 1: 55–61.

Nelson, L.S., Kim, K., Memmott, J.E. and Li, C. (1998a) FMRFamide-related gene family in the nematode, *Caenorhabditis elegans*. *Mol. Br. Res.*, 58: 103–111.

Nelson, L.S., Rosoff, M.L. and Li, C. (1998b) Disruption of a neuropeptide gene, *flp-1*, causes multiple behavioral defects in *Caenorhabditis elegans*. *Science*, 281: 1686–1690.

Perkins, L.A., Hedgecock, E.M., Thomson, J.N. and Culotti, J.G. (1986) Mutant sensory cilia in the nematode *Caenorhabditis elegans*. *Dev. Biol.*, 117: 456–487.

Plasterk, R.H.A. and Groenen, J.T.M. (1992) Targeted alterations of the *Caenorhabditis elegans* genome by transgene instructed DNA double strand break repair following Tc1 excision. *EMBO J.*, 11: 287–290.

Politz, S.M., Philipp, M., Estevez, M., O'Brien, P.J. and Chin, K.J. (1990) Genes that can be mutated to unmask hidden antigenic determinants in the cuticle of the nematode *Caenorhabditis elegans*. *Proc. Natl. Acad. Sci. USA*, 87: 2901–2905.

Price, D.A. and Greenberg, M.J. (1977) The structure of a molluscan cardioexcitatory neuropeptide. *Science*, 197: 670–671.

Riddle, D.L. and Albert, P.S. (1997) Genetic and environmental regulation of dauer larva development. In: D.L. Riddle, T. Blumenthal, B.J. Meyer and J. R. Priess (Eds.), *C. elegans II*, Cold Spring Harbor Press, Plainview, pp. 739–768.

Rosoff, M. L., Bürglin, T.R. and Li, C. (1992) Alternatively spliced transcripts of the *flp-1* gene encode distinct FMRFamide-like peptides in *Caenorhabditis elegans*. *J. Neurosci.*, 12: 2356–2361.

Rushforth, A.M., Saari, B. and Anderson, P. (1993) Site-selected insertion of the transposon Tc1 into a *Caenorhabditis elegans* myosin light chain gene. *Mol. Cell. Biol.*, 13: 902–910.

Sambrook, J., Fritsch, E.F. and Maniatis, T. (1989) *Molecular cloning: A laboratory manual*, Cold Spring Harbor Laboratory, Cold Spring Harbor, New York.

Schinkmann, K. and Li, C. (1992) Localization of FMRFamide-like peptides in *Caenorhabditis elegans*. *J. Comp. Neurol.*, 316: 251–260.

Schwartz, J. H. (1991) Chemical messengers: Small molecules and peptides. In: E. R. Kandel, J. H. Schwartz and T. M. Jessell (Eds.), *Principles of Neural Science*, 3rd Edn, Elsevier, New York, pp. 213–234.

Seydoux, G. and Fire, A. (1997) Whole-mount in situ hybridization for the detection of RNA in *Caenorhabditis elegans* embryos. In: H.F. Epstein and D.C. Shakes (Eds.), *Caenorhabditis elegans: Modern biological analysis of an organism*, Academic Press, New York, pp. 323–337.

Sithigorngul, P., Cowden, C., Guastella, J. and Stretton, A.O.W. (1989) Generation of monoclonal antibodies against a nematode peptide extract: Another approach for identifying unknown neuropeptides. *J. Comp. Neurol.*, 284: 389–397.

Strader, C.D., Fong, T.M., Tota, M.R., Underwood, D. and Dixon, R.A. (1994) Structure and function of G protein-coupled receptors. *Annu. Rev. Biochem.*, 63: 101–132.

Stringham, E.G., Dixon, D.K., Jones, D. and Candido, E.P.M. (1992) Temporal and spatial expression patterns of the small heat shock (*hsp16*) genes in transgenic *Caenorhabditis elegans*. *Mol. Biol. Cell*, 3: 221–233.

Sulston, J.E. and Horvitz, H.R. (1977) Post-embryonic cell lineages of the nematode *Caenorhabditis elegans*. *Dev. Biol.*, 56: 110–156.

Sulston, J.E., Schierenberg, E., White, J.G. and Thomson, J.N. (1983) The embryonic cell lineage of the nematode *Caenorhabditis elegans*. *Dev. Biol.*, 100: 64–119.

Tensen, C.P., Cox, K.J., Smit, A.B., Van der Schors, R.C., Meyerhof, W., Richter, D., Planta, R.J., Hermann, P.M., Van Minnen, J., Geraerts, W.P., Knol, J.C., Burke, J.F., Vreugdenhil, E. and Van Heerikhuizen, H. (1998) The *Lymnaea* cardioexcitatory peptide (LyCEP) receptor: A G-protein-coupled receptor for a novel member of the RFamide neuropeptide family. *J. Neurosci.*, 18: 9812–9821.

Thomas, J.H. (1990) Genetic analysis of defecation in *Caenorhabditis elegans*. *Genetics*, 124: 855–872.

Trent, C., Tsung, N. and Horvitz, H.R. (1983) Egg-laying defective mutants of the nematode *Caenorhabditis elegans*. *Genetics*, 104: 619–647.

Wang, Z., Lange, A.B. and Orchard, I. (1995) Coupling of a single receptor to two different G proteins in the signal transduction of FMRFamide-related peptides. *Biochem. Biophys. Res. Commun.*, 212: 531–538.

Ward, S. (1973) Chemotaxis by the nematode *Caenorhaditis elegans*: Identification of attractants and analysis of the response by use of mutants. *Proc. Natl. Acad. Sci. USA*, 70: 817–821.

Ward, S., Thomson, N., White, J.G. and Brenner, S. (1975) Electron microscopical reconstruction of the anterior sensory anatomy of the nematode *Caenorhabditis elegans*. *J. Comp. Neurol.*, 160: 313–337.

Waterston, R.H., Thomson, J.N. and Brenner, S. (1980) Mutants with altered muscle structure of Caenorhabditis elegans. *Dev. Biol.*, 77: 271–302.

White, J.G. (1988) The Anatomy. In: W.B. Wood (Ed.), *The Nematode* Caenorhabditis elegans, Cold Spring Harbor Laboratory, Cold Spring Harbor, pp. 81–122.

White, J.G., Southgate, E., Thomson, J.M. and Brenner, S. (1986) The structure of the nervous system of the nematode *Caenorhabditis elegans*. *Philos. Trans. R. Soc. Lond. B. Biol. Sci.*, 314: 1–340.

Wood, W.B. (1988) *The Nematode* Caenorhabditis elegans, Cold Spring Harbor Laboratory, Cold Spring Harbor.

Yandell, M.D., Edgar, L.G. and Wood, W.B. (1994) Trimethylpsoralen induces small deletion mutations in *Caenorhabditis elegans*. *Proc. Natl. Acad. Sci. USA*, 91: 1381–1385.

Yang, H.-Y., Fratta, W., Majane, E.A. and Costa, E. (1985) Isolation, sequencing, synthesis, and pharmacological characterization of two brain neuropeptides that modulate the action of morphine. *Proc. Natl. Acad. Sci. USA*, 82: 7757–7761.

Zwaal, R.R., Broeks, A., Van Meurs, J., Groenen, J.T.M. and Plasterk, R.H.A. (1993) Target-selected gene inactivation in *Caenorhabditis elegans* by using a frozen transposon insertion mutant bank. *Proc. Natl. Acad. Sci. USA*, 90: 7431–7435.

Section 4

Evaluating Behavioral Phenotypes in Rodents

W.E. Crusio and R.T. Gerlai (Eds.)
Handbook of Molecular-Genetic Techniques for Brain and Behavior Research (Techniques in the Behavioral and Neural Sciences, Vol. 13)

CHAPTER 4.1

Ethological approaches in behavioral neurogenetic research

Robert Gerlai

Neuroscience Department, GENENTECH Inc., 1 DNA Way, South San Francisco, CA 94080-4990, USA

Introduction

Advances in recombinant DNA and transgenic technology have led to the dawn of a new era for neuroscience: manipulation of single genes and analysis of gene expression changes make it possible to dissect the complexities of neurobiological phenotypes and to understand many of the intricacies of the molecular and neurobiological mechanisms underlying brain and behavior, even in mammals. Thus the goal of bridging the gap from gene to behavior appears closer to reach. The dazzling technological advances in recombinant DNA methodology, however, should not diminish the fact that both pillars of the bridge, i.e. genetics *and* behavioral science, should stand on firm ground. In this chapter I focus on behavior and argue that appropriate behavioral experimentation is just as important as the genetic techniques. I critically review part of the behavioral literature and argue that principles of ethology may best guide the behavioral scientist in designing behavioral experiments and selecting behavioral phenotypes to measure in a neurobehavioral genetic study.

I will focus my attention on the phenotypical analysis of the effects of mutations, introduced by transgenesis or gene targeting in mice, that affect hippocampal function. These studies have yielded fascinating results and may also be viewed as examples that allow the reader to draw general conclusions. I will attempt to persuade the reader that behavioral approaches that take species-specific characteristics into account and use ethological methods may be most useful for analyzing the genetic and neurobiological mechanisms underlying behavior, and without ecological relevance interpreting the behavioral findings and understanding the biological mechanisms of brain function may be difficult.

The role of the hippocampus in learning and memory has fascinated neuroscientists ever since the discovery that HM and other patients with hippocampal damage suffer devastating learning and memory deficits (Scoville and Milner, 1957). Since then the hippocampus has been studied at almost every level of analysis from detailed behavioral studies of hippocampal dysfunction to the physiology and molecular mechanisms of neural plasticity. Several gene targeting and transgenic studies have also been carried out to study the behavioral effects of mutations on hippocampal function. However, behavioral analyses can be complicated, and since the experimenters of these gene targeting studies do not always have extensive training in animal behavior, it is a particularly apt time to reconsider some important principles about behavior, and in particular the ethological relevance of these experiments.

As several chapters of this book demonstrate, transgenic and gene targeting techniques are powerful tools for assessing how genes influence brain function in vivo. One can express a gene in elevated amount (Gerlai, 1995), decrease or eliminate gene expression using "knock outs" (Grant et al., 1992), and introduce mutations in a targeted manner (Bach et al., 1995). Much of this technology has been carried out on mice because they live and breed readily in captivity and have a short generation time. A large number of transgenic mice have been generated in the hope that analyzing their phenotype will shed light on functional aspects of the mutated gene and the molecular, biochemical pathways in which that gene may be involved. Unfortunately, it is not always possible to foresee what phenotypical changes the introduced mutation may cause because the disruption of a single gene may lead to a cascade of events, e.g. there may be compensatory effects of other genes, the presence or absence of which may vary depending on the mouse strain used (Gerlai, 1996).

Behavioral analysis may reveal abnormal brain function

Of the many approaches used to analyze phenotypical effects of mutations on brain function (Grant et al., 1992; Bach et al., 1995; Gerlai, 1995; 1996), behavioral analysis is perhaps the most tricky: experiments appear easy to execute and quantitative data can be collected quickly, yet there are several difficulties with analyzing the behavioral effects of a mutation on hippocampal function. An obvious problem is that certain basic performance factors may influence behavioral results: A mouse with a motor impairment may finish a test more slowly; and a short-sighted rat may not see the visual cue to be learnt. These problems have been extensively dealt with in the literature and the reader will also find several excellent answers to these questions in the following chapters of this book.

The present chapter is centered around another problem, the ethological relevance of the behavioral task. As Nadel (1995) states, many studies of hippocampal function have been "nature-blind" in the sense that they "bear no obvious relation to what the animal does to survive and flourish in the real world", an approach that "has led the field astray in the past, and will continue to do so in the future". Why is it important not to be nature-blind?

Clever experimenters and "dumb" rats: it all boils down to genetic predisposition

In the laboratory, animals may behave in ways that are difficult to interpret, in part because they are genetically predisposed to selectively attend to, process, and recall certain specific stimuli that may be altered under artificial conditions. Early examples demonstrating the importance of designing ethologically-relevant tasks came from studies of learning in rats. For example, rats learned classical conditioning avoidance tasks quickly when food was associated with nausea-inducing substances, and when light was associated with foot shock, but found it much harder to learn the reverse associations (Garcia and Koelling, 1966), i.e. when food was paired with shock and light with nausea, demonstrating the salience of associating appropriate stimuli. Why cannot all stimuli be associated equally well? Consider that animal behavior has been shaped by natural selection over a long evolutionary past of the species and as a result genetic predispositions may facilitate or inhibit making associations between certain stimuli. Returning to the above example, it is clear that encountering a predator can rarely cause nausea in rats and food substances with unusual taste can hardly perform a painful attack, whereas the reverse situation (sight of predator → pain; food → nausea) is much more plausible, i.e. ecologically.

A skeptic might argue that the ethological relevance is not important as long as all animals are exposed to the same task in a controlled way. This argument ignores the fact that artificial tasks may be less sensitive to differences between mutant and wild type animals. The problem is threefold. First, animals may respond to stimuli that seem irrelevant to the experimenter, or may not attend to those judged crucial for solving the task. Second, irrespective of genotype, experimental animals may find ethologically-inappropriate tasks difficult to solve. If performance is poor in all subjects then mutation-related impairments will go undetected. Third, a learning paradigm which ignores the behavioral ecology of the species may be swamped by environmental error variation (phenotypical behavioral variance arising from random variations in the environment during rearing or during the experiment) (Gerlai, 1996), and will be less sensitive to detecting behavioral differences that arise as a result of genetic manipulation (Van Abeelen, 1979; Gerlai, 1996). Finally, as the goal of any neurobehavioral genetic study is to understand the biological mechanisms underlying brain and behavior, it makes sense to focus on behaviors that are not artificially construed but represent ecologically relevant phenes. Why should the analysis of ecologically relevant behaviors facilitate understanding underlying genetic/biological mechanisms?

Consider again the effects of natural selection on brain and behavior. Natural selection operates on the genetic level, i.e. it changes allele frequencies in the population. In the case of brain function, the basis of favoring or disfavoring a genotype is the behavioral performance of the given organism. It is the adaptive nature of behavioral traits which makes an animal fit and not the underlying neurobiological or molecular machinery. Clearly, genes that affect brain function have been under selection pressure on the basis of how they influenced the behavioral performance of the animals harboring them. Thus in order to study such genes and their role in brain function, it may be advantageous to analyze appropriate behaviors that correspond to ecologically relevant traits favored by natural selection. But how can one define ecological relevance and design appropriate behavioral tests?

Know your subject: behavior is species-specific!

Species differences in learning are inevitable because each species faces different problems in nature. As Kamil and Maudlin point out "The effects of a learning procedure upon a species will depend on how the learning paradigm makes contact with the adaptations and response repertoire of the animal" (Kamil and Maudlin, 1988). Consider the performance of various rodent species which have been tested on the Morris watermaze, a spatial and relational learning task, that was originally designed for rats (Morris, 1984) which inhabit wetlands (Nowak and Paradiso, 1983). Transgenic studies of the molecular mechanisms of learning use the house mouse *(Mus musculus domesticus)* rather than the rat because a wealth of information is available on the genetics of mice and most transgenic and gene targeting procedures have been worked out on this species. Although the mouse is closely related to the rat, its natural history and habitat is different. For instance, mice burrow in dry fields such as savannas, grasslands and forests as opposed to wetlands (Sage, 1981; Nowak and Paradiso, 1983; Bonhomme, 1992).

In the laboratory, mice have been found to perform below the level of rats in the watermaze (Whishaw et al., 1995), an impairment that cannot be attributed to body size alone. Under similar training conditions and performance requirements mice needed approximately five times more trials (Grant et al., 1992; Gerlai, 1995) than rats (Eichenbaum et al., 1988) to reach a plateau in their learning curves. Given their differences in habitat, mice may not be so well adapted to water as rats, and so the difference in performance may reflect species differences in task suitability rather than memory ability. Support for this claim comes from comparative studies of how other rodents perform on the watermaze. For example, meadow voles *(Microtus pennsylvanicus)* which inhabit wetlands (Kavaliers and Galea, 1994) perform well, and island populations of deer mice *(Peromyscus maniculatus)* that frequent marshes in the wild perform better than those that originated from an arid mainland population (Galea et al., 1994).

Designing the right task for the right species

An understanding of the species' natural history allows the experimenter to design an ethologically appropriate task by predicting what sort of motor responses may occur, what kind of stimuli the experimental subjects are more sensitive to, and what kind of cognitive constraints characterize their learning abilities. There are numerous examples of species-specificity in motor responses, and the salience of associating the appropriate stimulus and response, e.g. rats learn to bar-press for food and to jump out of a box to avoid shock, but learn the opposite associations with extreme difficulty (Timberlake and Lucas, 1989).

Species also differ in the type of sensory stimuli they attend to. Many studies show human-bias. For example, much of the behavioral research on rodents has been concerned with the development of tasks which depend upon the ability to process visuospatial information during daylight (Lee et al., 1998) yet rodents are nocturnal, having poorer visual acuity than humans but possessing excellent olfactory capabilities. It is only relatively recently that researchers have begun to consider olfactory cues in their spatial learning tasks (Eichenbaum et al., 1988; Reid and Morris, 1992; 1993; Lavenex and Schenk, 1995). Such studies now show that nocturnal rodents do not necessarily rely on olfactory cues for solving spatial tasks (Reid and Morris, 1992; Reid and Morris, 1993), but the degree to which they rely on memory of olfactory versus visual cues depends on the test conditions (Lavenex and Schenk, 1995). It is therefore important to test which cues the subjects use for each particular paradigm.

Animals may also depend on different sensory modalities at different times of day or season, e.g. yellow pine chipmunks *(Tamia amoenus)* rely on spatial memory to locate hidden seed caches during the dry season but use olfactory cues instead of memory during rainy season (Vanderwall, 1991). The degree of hippocampal involvement in learning and memory can depend upon the types of cues used and when, as well as task difficulty and level of motivation, so performance on conventional learning tasks may reveal only part of the story unless they are designed to tap into the animal's natural learning abilities.

It is also important to realize that every time a new mutant is tested, novel behavioral responses may be expected. These responses may not necessarily be detected in "standard behavioral assays". Behavior is difficult to assay because it consists of a conglomeration of different but interrelated actions and also because it is affected by a potentially large number of environmental factors that may influence it in a genotype dependent manner. To determine how the mutation alters behavior, pre-existing tests may need to be modified or new ones designed. I suggest that an experimenter armored with knowledge of species-specific characteristics of the animals will be better able to design such tests.

Nevertheless, it is tempting to suggest a test that one may consider the best among those available. For hippocampal function, the Morris water maze has generally become the standard (Fig. 1), although some laboratories prefer to use other tasks (Bach et al., 1995), e.g. the Barnes maze, a dry equivalent of the water maze spatial learning task, or the context dependent fear conditioning paradigm (see below). Which task is better? There is no single answer to this question. Although both the Morris water maze and the Barnes dry maze are intended to tap into the animals' spatial abilities, these two tasks are associated with different motivational factors and motor requirements, and they probably evoke different adaptive behavioral responses in the mice being tested. For example, mouse strains that usually exhibit high levels of anxiety are almost impossible to train in the Barnes dry maze because they simply freeze in the large open arena (personal observation), a response probably adaptive in the wild in the avoidance of predators. By contrast, in the Morris water maze, the mouse will be more motivated to actively move and escape from the cold water. It is important to realize, however, that the latter may not always be better. For instance, the Barnes maze has been successfully used in the analysis of the CaMKII-Asp-286 point mutant transgenic mice and has also proven to be superior in investigating alternative strategies of exploratory behavior during spatial learning (Bach et al., 1995). The radial maze may also have some advantages over the above mazes and could be suggested as an appropriate task ecologically relevant under certain circumstances (see Chapter 4.4).

The above behavioral tasks are based on decades of experimentation and provide an excellent means with which to study behavior. It is important to remember, however, that there is no one way to run these behavioral tasks, and that behavior is influenced by numerous factors. The question becomes one of not what test to choose, but how to run it.

Exploring novelty, exploring space: conflicting motivational forces

The common feature of spatial learning tests applied in recent molecular neurobiology studies, apart from the fact that they were intended to reveal hippocampal dysfunction, is that they represent novel situations in which exploratory behavior is evoked. Exploratory behavior, however, is a complex response to novelty resulting from a compromise between the motivation to gather information about the surroundings, and avoiding predation (Suarez and Gallup, 1981; Crusio and Van Abeelen, 1986). So laboratory conditions which influence hunger, sexual drive, fear and chronic stress could all alter the behavioral performance of the subjects and should be controlled in a learning paradigm in which animals are exposed to novelty. For instance, increased sex drive may increase active exploration and lead to better learning performance. Dominant males typically exhibit elevated sexual drive compared to subordinates. If mutant males are subordinate to their wild type cage mates (Gerlai, 1996), a significant "spatial learning impairment" might be recorded not because the mutant animals are necessarily less able to learn but simply because of their suppressed sexual drive. Indirect support for this comes from studies which show that changes in circulating levels of sex hormones can have a dramatic effect on spatial learning (Sherry and Hampson, 1998). In meadow voles and deer mice, for example, changes in reproductive status have been shown to influence spatial learning in the Morris watermaze: male deer mice learned significantly faster during the breeding season than during the non-breeding season, whereas the reverse was true for females (Galea et al., 1996).

Changes in steroid hormones arise not only as a consequence of changes in reproductive status, but also as a result of changes in various environmental stressors such as competition for food or dominance status, and these changes in hormone levels can have a dramatic impact on hippocampal structure and function (Sapolsky, 1996; McEwan, 1997). Since the position in the dominance hierarchy may determine how well the animal responds to such stress and how much neurological damage is incurred as a result of steroid hormone levels (Sapolsky, 1996), these factors should not be ignored. Furthermore, stress does not only influence the organism's behavior while under test, but also has long term effects on how the animal performs in subsequent tests. Every attempt should be made to minimize the amount of stress to the animal, and to test for differences between mutant and control mice in the way they respond to stress. In the absence of careful control, differences in behavioral performance may reflect differences in the way that animals respond to stress and how these differences affect hippocampal structure and function rather than genetic differences in learning and memory. It is also useful to know what stimuli may elicit stress or what stimuli the experimental species are sensitive to under stressful conditions. For example, a pair of horizontal eye-like spots has been found to elicit aversive behaviors in a number of species from fish through birds to rodents (Topál and Csányi, 1994), responses that resemble predator avoidance. Dark bird like patterns above a maze of a rodent is also not a good idea, neither is high level of illumination, etc.

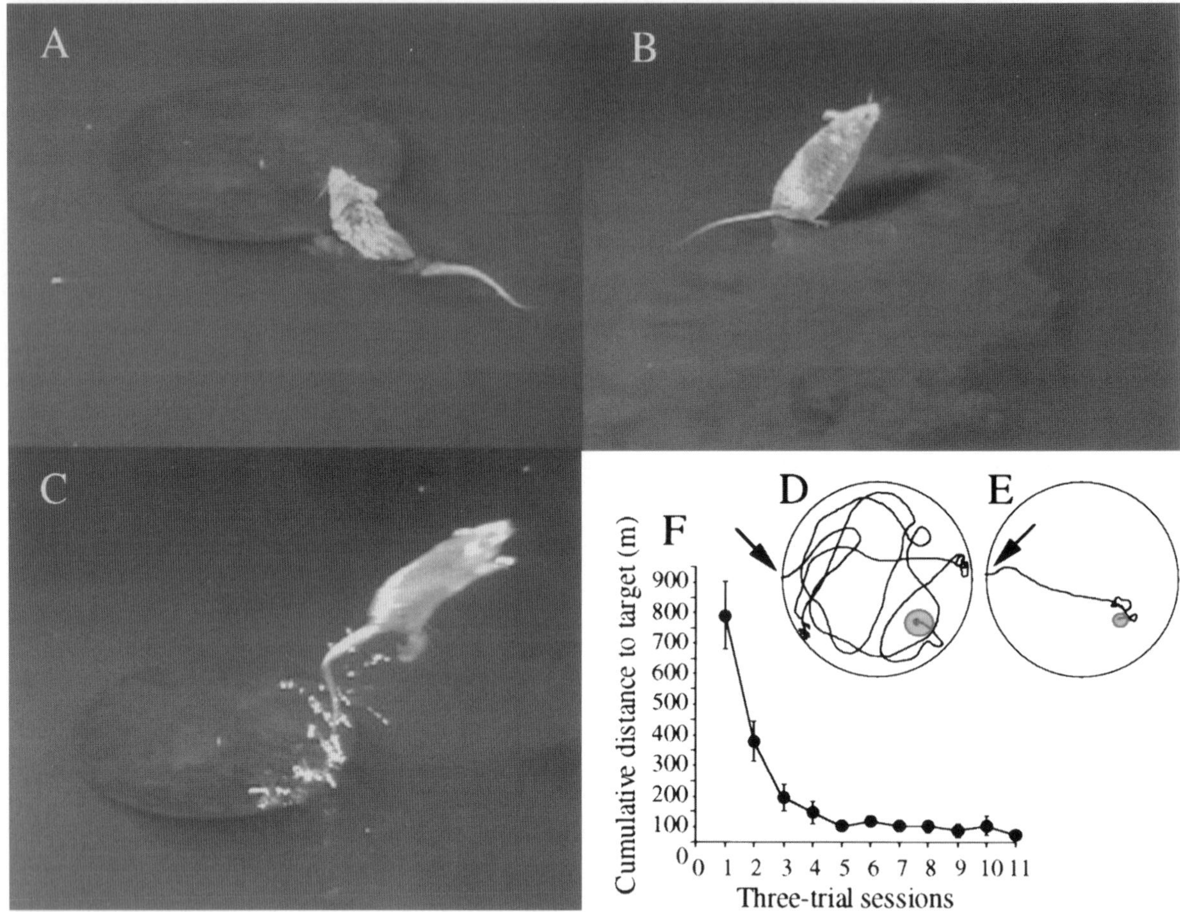

Fig. 1. In the Morris water maze rodents are trained to locate and climb onto a safe platform from water. In the spatial version of the task, the platform is hidden just beneath the surface of the water and the experimental subjects need to learn its position in relation to external visual cues around the maze. (A) a mouse has just found the hidden platform and is climbing onto it. Note that the water maze is black (Chen et al., 1997) and that a low illumination level is used in order to reduce stress. The water maze is 2 m in diameter, an unusually large maze for mice compared to those typically used, and this increases the spatial feature of the task. (B) An appropriately trained mouse often rears on the platform. (C) Mice that are not habituated to handling or are improperly handled during training (Gerlai et al., 1995), e.g. picked up from the platform by the tail, exhibit predator avoidance behaviour and actually escape from the platform. (D) An example of the swim path of a mouse at the beginning of training. The platform is indicated by the shaded circle, the starting location of the mouse by the arrow. (E) An example of the swim path of the same mouse 33 trials later. Note the change in the size of the target platform: in order to facilitate learning, at the beginning of training, larger platforms are used and as training proceeds their size is gradually decreased (Eichenbaum et al., 1990). (F) Learning performance in the hidden platform task (Gerlai unpublished data) of C57BL/6 × 129/SV F_1 hybrid mice ($n = 10$) as expressed in cumulative distance from target. Note that this measure was found to reflect spatial accuracy and spatial learning performance better (Gallagher et al., 1993) than the frequently used escape latency measure. (From Gerlai and Clayton, 1999, with permission.)

In summary, knowledge of the natural habitat and behavior of the species under study may be helpful.

Stress induced fear responses may complicate the interpretation of spatial learning test results. Mutant and control mice may differ in their fear response, or in ethological terms, their predator avoidance behavior (Topál and Csányi, 1994). Stimuli that are involved in predator recognition and avoidance, including the sight of a human observer, may evoke active or passive defense in novel situations, and if mutant and control mice differ in their responses then this may represent a confounding factor. For example, Silva et al. (1992) claimed that αCaMKII null mutant mice showed spatial learning deficits. However, the apparent "learning" deficit seen in these mutants was later attributed to differences in their fear response (Deutsch, 1993) because the mutants were more anxious when handled (Silva et al., 1992). Thus impaired "learning" performance of αCaMKII mutants might reflect an increase in sensitivity to human handling. Variation in response to human handling is rarely mentioned in the literature and yet it should not be overlooked: in the Morris watermaze, mice may learn to avoid the target platform from which the experimenter picked them up if they are not habituated to the experimenter or not handled appropriately during training (Fig. 1).

Depending on the level of fear or pain, the types of stimuli encountered during the experiment, and previous handling procedures, different forms of predator avoidance behavior may be evoked. These responses may range from species-specific passive avoidance (freezing) to vocalization, defensive attack posture, biting, or active avoidance (Blanchard et al., 1989). Given this variation in avoidance response, results could easily be misinterpreted especially if only one behavioral response is measured like in the context-dependent fear conditioning task (Bach et al., 1995), in which freezing is used as a measure of memory of a feared place. A decrease in freezing response is usually interpreted as a loss of memory of the place where the rodent received the shock, but it could also reflect the fact that the animal responds with behavioral reactions other than passive avoidance. Therefore, recording multiple behavioral responses, e.g. several elements of the natural behavioral repertoire of the species (ethogram) may be advisable.

The marriage of field and lab studies

Throughout this paper, I have emphasized the importance of considering the natural behavior of the animal when designing behavioral tests. In order to design ethologically appropriate tasks for the species in question, we need information on how these animals perform under naturalistic conditions. Surprisingly, it is only very recently that such an approach has been adopted to study hippocampal function in rodents in the field (Lavenex et al., 1998; Shiftlett and Jacobs, in press). Ethologists seem to have reserved the right to study more exotic species and neurobiologists appear uninterested in the natural behavior of their laboratory rodents. It would be illuminating to evaluate the spatial memory capabilities of free ranging mice, to determine which cues mice rely on naturally in the wild and how similar their responses are to those of laboratory-raised mice. Observing natural behavior in the wild would possibly also allow experimenters to design novel behavioral tasks in the laboratory that are ecologically relevant.

In naturalistic studies spontaneous behavior is observed. There is no artificially controlled punishment or reinforcement to shape the behavior of the animals under study, and no pretraining is enforced by human experimenters (Lavenex et al., 1998). In contrast, laboratory animals are usually confined to the apparatus during the test, and require numerous pretraining trials to teach them which cues to attend to, which motor responses to make, and which stimuli they should learn to associate. By allowing variables to be manipulated in a carefully constrained and controlled way, laboratory tests provide valuable information about what animals are capable of learning. However, they do not tell us how untrained animals would naturally solve these problems. As I argued above, naturalistic studies can inform us about the cognitive processes that are sensitive to natural selection, ones that might have been responsible for the evolution of spatial learning, and such studies may be used in designing behavioral experiments more appropriate for the analysis of neurobiological and genetic mechanisms underlying the behavior in question.

The food-caching paradigm serves as a striking example of how an understanding of the species and its natural history can be employed to develop novel approaches to the study of animal learning and memory in the laboratory (Clayton and Dickinson, 1998). Using this paradigm Clayton and Dickinson were able to test, and prove the existence of, episodic-like memory in birds, a type of memory that was previously attributed to our own species only. The advantage of naturalistic/semi-naturalistic tests is that they can be conducted in the laboratory but capitalize on species-specific behavioral capabilities whose ethological relevance is known. A closer synergism between laboratory and field studies holds great promise for the future: The goal is to apply the findings from naturalistic studies to laboratory tests of behavior, and thus achieve rigorous laboratory control while maintaining ethological validity.

Concluding remarks

Tempting though it may be to employ a general "behavioral assay" for investigating phenotypical changes that a single gene manipulation may have caused, the studies cited above demonstrate that we need to consider the ethological relevance of the tasks for the species in question and that we may need to modify existing tests or design new ones, depending on the question asked and the genetic manipulation we employ. Whilst the importance of using ethologically-relevant tasks does not imply that there are no general features and common molecular mechanisms underlying behavioral phenomena, a knowledge of the species-specific characteristics of behavior is essential for discovering these commonalities and generalizing the findings of animal research to human. The challenge is to design appropriate behavioral tests which tap into the natural memory capabilities of the species in question, and thereby make these tests sensitive enough to detect the genetic effects which we wish to study. In the following chapters of the book the reader will find numerous examples in which the ethological approach is combined with laboratory testing.

Acknowledgments

This chapter is based on two previously published papers (Gerlai, 1996; Gerlai and Clayton, 1999). I thank Pam Banta, Nicky Clayton, Wim Crusio, Tony Dickinson, Dan Griffiths, Martin Kavaliers, Sandra Kelly, Pierre Lavenex and Stephen Maxson for their useful comments on an earlier version of this paper and Kevin Ling for video processing.

References

Bach, M.E., Hawkins, R.D., Osman, M., Kandel, E.R. and Mayford, M. (1995) Impairment of spatial but not contextual memory in CaMKII mutant mice with a selective loss of hippocampal LTP in the range of the theta frequency. *Cell*, 81: 905–915.

Blanchard, R.J., Blanchard, D.C. and Hori, K. (1989) An ethoexperimental approach to the study of defense. In: R.J. Blanchard, P.F. Brain, D.C. Blanchard and S. Parmigiani (Eds.), *Ethoexperimental Approaches to the Study of Behavior*, NATO ASI Series. Dordrecht. Kluwer Academic Press, pp. 114–136.

Bonhomme, F. (1992) In: D. Goldowiz, D. Wahlsten and R.E. Wimer (Eds.), *Techniques for the Genetic Analysis of Brain and Behavior: Focus on the Mouse*, Amsterdam, Elsevier, pp. 41–56.

Chen, K.S., Nishimura, M.C., Armanini, M.P., Crowley, C., Spencer, S.D. and Phillips, H.S. (1997) Disruption of a single allele of the nerve growth factor gene results in atrophy of basal forebrain cholinergic neurons and memory deficits. *Journal of Neuroscience*, 17(19): 7288–7296.

Clayton, N.S. and Dickinson, A. (1998) Episodic-like memory during cache recovery by scrub jays. *Nature*, 395: 272–274.

Crusio, W.E. and Van Abeelen, J.H. (1986) The genetic architecture of behavioural responses to novelty in mice. *Heredity*, 56: 55–63.

Deutsch, J.A. (1993) Spatial learning in mutant mice. *Science*, 262: 760–763.

Eichenbaum, H., Fagan, A., Mathews, P. and Cohen, N.J. (1988) Hippocampal system dysfunction and odor discrimination learning in rats: impairment or facilitation depending on representational demands. *Behav. Neurosci.*, 102: 331–339.

Eichenbaum, H., Stewart, C. and Morris, R.G. (1990) Hippocampal representation in place learning. *Journal of Neuroscience*, 10(11): 3531–3542.

Galea, L.A., Kavaliers, M. and Ossenkopp, K.P. (1996) Sexually dimorphic spatial learning in meadow voles (*Microtus pennsylvanicus*) and deer mice (*Peromyscus maniculatus*). *J. Exper. Biol.*, 199: 195–200.

Galea, L.A., Kavaliers, M., Ossenkopp, K.P., Innes, D. and Hargreaves, E.L. (1994) Sexually dimorphic spatial learning varies seasonally in two populations of deer mice. *Brain Res.*, 635: 18–26.

Gallagher, M., Burwell, R. and Burchinal, M. (1993) Severity of spatial learning impairment in aging: development of a learning index for performance in the Morris water maze. *Behavioral Neuroscience*, 107(4): 618–626.

Garcia, J. and Koelling, R.A. (1966) Relation of cue to consequence in avoidance learning. *Psychonom. Sci.*, 4: 123–124.

Gerlai, R. (1995) Over-expression of a calcium binding protein, S100b, in astrocytes alters synaptic plasticity and impairs spatial learning in transgenic mice. *Learning and Memory*, 2: 26–39.

Gerlai, R. (1996) Molecular genetic analysis of mammalian behavior and brain processes: Caveats and perspectives. *Sem. Neurosci.*, 8: 153–161.

Gerlai, R. and Clayton, N.S. (1999) Analysing hippocampal function in transgenic mice: an ethological perspective. *Trends Neurosci.*, 22: 47–51.

Grant, S.G., O'Dell, T.J., Karl, K.A., Stein, P.L., Soriano, P. and Kandel, E.R. (1992) Impaired long-term potentiation, spatial learning, and hippocampal development in fyn mutant mice. *Science*, 258: 1903–1910.

Kamil, A.C. and Maudlin, J.E. (1988) In: R.C. Bolles and M.D. Beecher (Eds.), *Evolution and Learning*, Lawrence Erlbaum, pp. 117–134.

Kavaliers, M. and Galea, L.A. (1994) Spatial water maze learning using celestial cues by the meadow vole (*Microtus pennsylvanicus*). *Behav. Brain Res.*, 61: 97–100.

Lavenex, P. and Schenk, F. (1995) Influence of local environment olfactory cues on place learning in rats. *Physiol. Behav.*, 58: 1059–1066.

Lavenex, P., Shiftlett, M.W., Lee, R.K. and Jacobs, L.F. (1998) Spatial versus nonspatial relational learning in free-ranging fox squirrels. *J. Comp. Psychol.*, 112: 127–136.

Lee, D.W., Miyasato, L.E. and Clayton, N.S. (1998) Neurobiological bases of spatial learning in the natural environment: neurogenesis and growth in the avian and mammalian hippocampus. *Neuroreport*, 9: R15–27.

McEwan, B. (1997) Ovarian steroids and the brain: Implications for cognition and aging. *Neurology*, 48: 8–15.

Morris, R. (1984) Developments of a water-maze procedure for studying spatial learning in the rat. *J. Neurosci. Methods*, 11: 47–60.

Nadel, L. (1995) In: E. Alleva, A. Fasolo, H.-P. Lipp, L. Nadel and L. Ricceri, (Eds.) *Behavioral Brain Research in Naturalistic and Semi-Naturalistic Settings*, Kluwer Academic Press, pp. 323–342.

Nowak, R.M. and Paradiso, J.L. (1983) *Walker's Mammals of the World*, 4th Edn, The John Hopkins University Press, pp. 569–1362.

Reid, E.N. and Morris, R.G.M. (1992) Smells are no surer: rapid improvement in olfactory discrimination is not due to the acquisition of a learning set. *Proc. Roy. Soc. Lond. B*, 247: 137–143.

Reid, E.N. and Morris, R.G.M. (1993) The enigma of olfactory learning. *Trends Neurosci.*, 16: 17–20.

Sage, R.D. (1981) In: H.L. Foster, J.D. Small and J.G. Fox, (Eds.), *The Mouse in Biomedical Research*, Academic Press, pp. 34–57.

Sapolsky, R.M. (1996) Why stress is bad for your brain. *Science*, 273: 749–750.

Scoville, W.B. and Milner, B. (1957) Loss of recent memory after bilateral hippocampal lesions. *J. Neurol. Neurosurg. Psych.*, 20: 11.

Sherry, D.W. and Hampson, E. (1998) Evolution and the hormonal control of sexually-dimporphic spatial abilities in humans. *Trends Cogn. Neurosci.*, 1: 17–22

Shiftlett, M. and Jacobs, L.F. Spatial orientation on a vertical maze in free-ranging fox squirrels. *J. Comp. Psychol.*, in press.

Silva, A.J., Paylor, R., Wehner, J.M. and Tonegawa, S. (1992) Impaired spatial learning in alpha-calcium-calmodulin kinase II mutant mice. *Science*, 257: 206–211.

Suarez, S.D. and Gallup, G.G. (1981) An ethological analysis of open-field behaviour in rats and mice. *Learning and Motivation*, 12: 342–363.

Timberlake, W. and Lucas, G.A. (1989). In: S.B. Klein and R.R. Mowrer (Eds.), *Contemporary Learning Theories: Instrumental Conditioning Theory and the Impact of Biological Constraints on Learning*, Lawrence Erlbaum, pp. 237–275.

Topál, J. and Csányi, V. (1994) The effect of eye-like schema on shuttling activity of wild house mice (*Mus musculus domesticus*): Context-dependent threatening aspects of the eyespot patterns. *Anim. Learn. Behav.*, 22: 96–102.

Van Abeelen, J.H.F. (1979) In: J.R. Royce and L.P. Mos (Eds.), *Theoretical Advances in Behavior Genetics*, Sijthoff and Noordhoff, pp. 101–119.

Vanderwall, S.B. (1991) Mechanisms of cache recovery by yellow pine chipmunks. *Animal Behav.*, 41: 851–863.

Whishaw, I.Q., Cassel, J.C. and Jarrad, L.E. (1995) Rats with fimbria-fornix lesions display a place response in a swimming pool: a dissociation between getting there and knowing where. *J. Neurosci.*, 15: 5779–5788.

W.E. Crusio and R.T. Gerlai (Eds.)
Handbook of Molecular-Genetic Techniques for Brain and
Behavior Research (Techniques in the Behavioral and Neural Sciences, Vol. 13)

CHAPTER 4.2

What animals remember about past events: an ethological approach

Nicola S. Clayton

Section of Neurobiology, Physiology and Behavior, Briggs Hall, University of California Davis, 1 Shields Avenue, Davis, CA 95616, USA

Introduction

In the preceding chapter (4.1), Gerlai argues that appropriate behavioral methods are just as important as genetic techniques for ultimately understanding the genetic and neurobiological bases of behavior, and discusses the benefits of using ethological methods for studying hippocampal function. In that chapter and elsewhere (e.g. Gerlai and Clayton, 1999) he emphasizes the importance of designing tasks that are based upon a knowledge of the species-specific characteristics and exploit an animal's natural memory capabilities, ones it would rely on for its survival in the wild and hence be sensitive to natural selection.

Probing the cognitive functions of an organism by any means will provide some insight into the animal's behavior, and because that organism is a product of evolution such research will also reveal some information about the cognitive processes that are shaped by natural selection. Nonetheless, ethologically designed tasks that tap into the animal's natural talents may be especially good at detecting genetic variance relevant for fitness, and therefore facilitate the study of genetic mechanisms underlying the behavioral phenomena of interest (Gerlai and Clayton 1999b). In this chapter, I will continue with this theme and describe a series of experiments on the memory capabilities of food-caching birds to illustrate how an ethogical approach can help us understand hippocampal function, and more generally the learning and memory capabilities of animals.

In humans, the hippocampus is thought to play a crucial role in the recall of episodic information (Fletcher et al., 1995; Nyberg et al., 1996a, 1996b, 1996c; Vargha-Khadem et al., 1997), i.e. the ability to retrieve unique, personal experiences (Tulving, 1972). Studies of episodic memory in human patients suggest that it is the most fragile type of memory: it is the latest to be acquired during childhood (Pillemer and White, 1989; Perner and Ruffman, 1995); and it is the first to be lost in old age (Herlitz and Forsell, 1996; Nilsson et al., 1997a, 1997b, 1997c). It is also the most susceptible to neurodegenerative diseases such as Alzheimer's and Parkinson's disease (e.g. Sherwin, 1988; Henderson et al., 1994; Kimura, 1995) and so tests of episodic memory capabilities are therefore used as behavioral markers for the onset of these debilitating diseases of the mind. Given the fragility of episodic memory, such tests are likely to be highly sensitive to genetic dysfunction and therefore particularly appropriate when testing the effects of a particular mutation on brain and behavior.

To date virtually all studies of episodic memory have been conducted on human patients, and there is currently a huge gap between the human and animal models of memory. This is largely because tests of animal memory have not distinguished between episodic and other forms of declarative memory (events and facts). Indeed, episodic memory was thought to be unique to humans (e.g. Tulving, 1972, Tulving and Markowitsch, 1998). The inability to conduct interventional experiments on the human brain has imposed a major limitation on the analysis of the neural mechanisms that mediate these processes. Developing an appropriate animal model is essential for analyzing both the neural mechanisms and molecular genetic bases of episodic memory.

Until recently there was little or no evidence that animals could recall a specific past experience and respond appropriately, nor was there any reason to believe animals should need such an ability, given the types of laboratory tasks on which they were tested (see Clayton et al. 1999). Recent work on food-storing jays, however, suggests that several features of episodic memory may not be as exclusive to humans as previously thought (Clayton and Dickinson, 1998) and provides an exciting new model for unraveling the underlying neural and molecular mechanisms that mediate episodic recall, thereby bridging the crucial gap between human and animal studies of memory.

Animal analogues of memory

Although there are many laboratory procedures for investigating memory for discrete past episodes in animals, the tasks that have been used do not distinguish between episodic memories of trial-unique, personal experiences and semantic memories which are based on knowledge of impersonal declarative facts about the world (Tulving, 1983). Most of the animal studies have been conducted on rats or monkeys in a laboratory setting. Typically, the declarative memory tasks require the animal to retrieve information about only a single feature of the episode as opposed to testing its ability to recall and integrate information about several stimuli present during that event, or the animal is given multiple training experiences thereby removing the trial-uniqueness of the task. Moreover, it is becoming increasingly clear that performance in the standard animal memory tasks can be mediated by a simple discrimination of relative familiarity, a process which is dissociable both psychologically (Mandler, 1980) and neurobiologically (Aggleton and Brown in press) from episodic memory recall.

Griffiths et al. (1999) provide a review of possible examples of episodic memory in animals and their alternative accounts. In this chapter, I will describe just one example to illustrate the types of alternative non-episodic explanations. Consider the case of a monkey that has been trained to choose between two complex visual scenes on the basis of whether one of them contained a specific object at a particular location within the scene (e.g. Gaffan, 1992). This task appears to require retrieval of both spatial ('where') and object ('what') information. The discrimination was learned over many trials, however, and so the monkey need not have used specific episodic memories to solve the task. It seems likely that the monkey recalled more general declarative information that a particular what-where configuration was associated with a reward. By only requiring the monkeys to recognize a stimulus, but not recollect where and when it was previously seen, these matching tasks are most readily solved by familiarity rather than episodic recall (Griffiths et al., 1999). Indeed, the fact that monkeys learned more rapidly when novel objects are used supports the latter account (Mishkin and Delacour, 1975).

Most of the laboratory tasks of animal memory that have been used to date can be explained in terms other than episodic recall. A different strategy to testing whether or not animals are capable of episodic memory is to adopt an ethological approach, considering cases in nature in which an animal might benefit from the capacity to remember unique episodes that occurred in the past. As mentioned previously, according to Tulving's classical definition, episodic memory provides information about the 'what' and 'when' of a specific, unique event and about 'where' it happened (Tulving, 1983). The advantage of this definition is that animals can be tested for their ability to simultaneously retrieve and integrate information about these three features of a single, unique experience.

Ethological guidelines for studying episodic memory

What kinds of behavior may depend on the ability to encode and later recall information about the 'what, where and when' of an event? One potential example is the brood parasitic behavior seen some cowbirds, cuckoos and ducks. In some cowbird species, for example, females may keep track of the current status of hosts' nests in which they lay their eggs: based on just one previous visit to the host's nest they can later return to the correct place at the appropriate time. Interestingly females of species which engage in this behavior have larger hippocampal volumes relative to overall telencephalon size than those which do not, but this difference is only present during the breeding season when females are searching for host nests (Reboreda et al., 1996; Clayton et al., 1997).

Some aspects of mating behavior, such as seeking and guarding mates, may also depend on an episodic recall of 'where' an event happened, 'when' and to 'whom' (a variant of 'what'). For example, a polygynous male may need to remember not only the location of his various females' breeding sites, but also when each of his females come into estrous in order to maximize his chance of fertilizing their eggs and guarding against the threat of extra pair paternity by rival males. Presumably, species which live in complex social groups and those which defend territories whose boundaries are fluid may also be expected to be particularly good at remembering the 'what, where and when' of a particular past event. A comparative approach may be particularly powerful for testing hypotheses about extent to which species with different types of life history and behavioral ecology rely on various episodic memories to survive the trials and tribulations of life (e.g. Balda and Kamil, 1989; Jacobs et al., 1990; Sherry et al., 1992).

While the aforementioned candidates for episodic recall may be difficult to study in a laboratory setting, memory-based recovery of caches by food storing birds and mammals can be investigated in both the laboratory and field (e.g. Clayton, 1998; Van der Wall, 1990; Shettleworth, 1995). In the next section I will describe the results of recent experiments which demonstrate that food-storing scrub jays, *Aphelocoma coerulescens*, can remember what they cached where and when based on a single caching episode, and in a way that cannot be explained by relative familiarity (Clayton and Dickinson, 1998, 1999c). This memory capability by jays fulfills Tulving's classic behavioral criteria for episodic memory, and is thus referred to as 'episodic-like' (Clayton and Dickinson, 1998; Tulving pers. comm.). In fact the food-caching paradigm serves as a striking example of how an understanding of the species and its natural history can be employed to develop novel approaches to the study of episodic memory in animals. One strength of this ethological approach is that the tests can be designed to tap into the natural talents of the animal. Another advantage is that these studies can be conducted in the laboratory but capitalize on species-specific memory capabilities whose ethological relevance is known, thereby achieving rigorous laboratory control yet maintaining ethological validity (for discussion see Gerlai and Clayton, 1999a, 1999c).

Episodic-like memory for food caches in scrub jays

The experiments conducted in my laboratory on the episodic-like memories of food-storing birds were guided by two key features of their natural behavior. First, birds rely, at least in part, on memory to recover their caches. A number of field studies have shown that food-storing birds can recover a large proportion of their caches (e.g. Tomback, 1980) and in the laboratory, food-storing birds show accurate retention of many different spatial locations of cache sites for at least 285 days in Clark's nutcrackers, *Nucifraga columbiana*, (Balda and Kamil, 1992). Second, some food-storing species including scrub jays cache insects and other perishable items in the wild as well as seeds (Van der Wall, 1990). It may be adaptive, therefore, for them to encode and recall information about what has been cached when, as well as where. We therefore captialized on the jays' natural propensity to cache and recover perishable items in designing our experimental test of memory of "what, where and when".

To test if scrub jays are capable of episodic-like memory recall, birds were allowed to cache and recover perishable "wax worms" (wax moth larvae) and non-perishable peanuts. The logic is as follows. Jays show a very distinct preference to cache, recover and eat wax fresh worms when given both worms and peanuts. Worms degrade rapidly over time, however, so that if worms are left for a period of five days or so they become rotten and unpalatable. If birds can remember when they cached as well as what they cached and where, then they should recover worms when they were cached just a few hours ago. They should avoid the worms, however, if the worms were cached several days ago and have had time to rot.

Sixteen hand-raised scrub jays were given worms and peanuts which they could cache in and recover from sand-filled plastic ice-cube trays, containing a 2 × 7 array of 'cache sites' (ice cube molds). Each tray was attached to a wooden board and surrounded by a visuospatially-distinct structure of Lego® bricks that was placed next to one of the long sides of the tray (Clayton and Dickinson, 1998, 1999a, 1999b, 1999c). For each trial, the trays were trial-unique and placed in different places in the cage, to ensure that birds could not learn general rules about the contents of caching trays but instead must remember the 'what, where and when' of each individual caching episode. This is a crucial control given the importance of testing the animal's ability to remember a unique past experience.

The birds were divided into two groups, Degrade and Replenish, which differed in whether or not birds had the opportunity to learn that worms degrade and become unpalatable over time. Being hand-raised, these scrub jays had not had any previous opportunity to learn perishability. In order to learn that worms decay, in Experiment 1(a) jays in the Degrade group were given a series of pretraining trials in which they cached peanuts in one tray and worms in another tray, and then recovered their hidden caches from both trays either 4 h or 124 h later (Clayton and Dickinson 1999c). Birds in the Replenish group received the same treatments as those in the Degrade group except that the old wax worms were removed and replaced by fresh ones just before the start of the cache recovery phase so that these birds never had the opportunity to learn that worms decay over time. The birds in the Replenish group serve as an important control to test whether any switch in preference from worms to peanuts after the long retention interval cannot be explained in terms of either a genetic predisposition to prefer worms in some instances and peanuts in others, or simply that memories for perishable worms are forgotten more quickly than memories for non-perishable nuts.

During these pretraining trials, birds could rely on the smell of their caches as cues about where to search during cache recovery. To test whether or not jays could remember what, where and when the worms and peanuts had been cached in the absence of olfactory cues, each bird received a pair of probe test trials in which all food items were removed prior to the recovery phase of each test trial and fresh sand was placed in the tray. In terms of assessing memory-based recovery performance, both the total number of looks made to each site and where the birds search on their first look could be analyzed. As no caches are present during the recovery phase of a probe test trial, in terms of memory performance the critical data is where

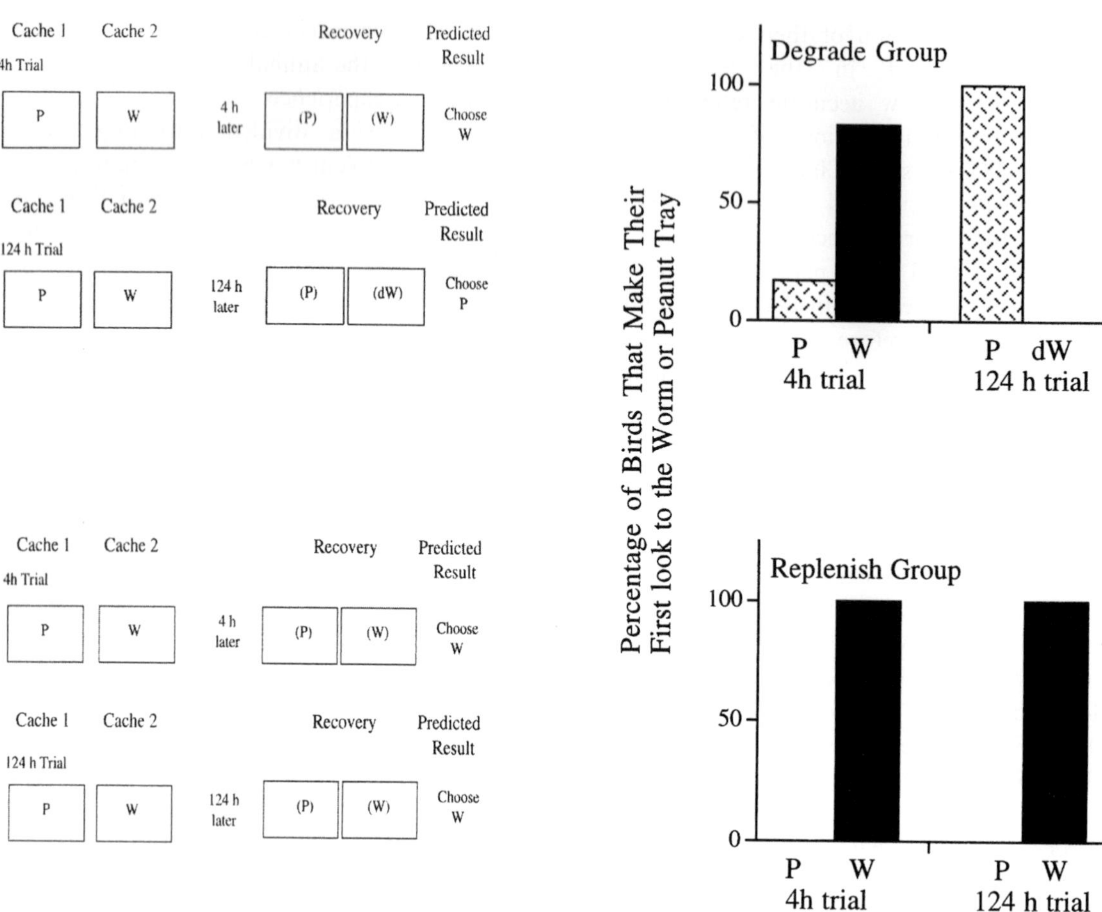

Fig. 1. The experimental design, predictions and observed results for birds in the Degrade (Fig. 1(a)) and Replenish (Fig. 1(b)) groups in Experiment 1a. Birds cached worms in one tray and peanuts in the other tray and then recovered these caches with in 4 h or 124 h later. Key: P = peanuts; W = fresh wax worms; dW = decayed wax worms; and ( ) = food items pilfered on test. The histograms show the percentage of birds that make their first look to the tray containing worms and peanuts during the recovery phase of 4 h and 124 h probe test trials.

the birds search on their first look, before they have had the opportunity to realize that their caches have been stolen by the experimenter.

The results of the probe test trials are summarized in Fig. 1. Eighty-three percent of birds in the Degrade group make their first visit to sites in which they had cached worms after the 4 h retention interval. After the 124 h retention interval, however, the same birds selectively made their first look to sites in which they had cached peanuts first and all birds in the Degrade group avoided the sites where they had cached worms (Clayton and Dickinson, in press). By contrast, 100% birds in the Replenish group preferred to search first in sites in which they had cached a worm, irrespective of when the worms had been cached.

In terms of demonstrating the ability to remember 'what, where and when', the critical result is the switch in preference for recovering worms after 4 h but peanuts after 124 h shown by birds in the Degrade group. The Replenish group, however, serves as a crucial control: As the switch in preference is shown only by the Degrade group, these results cannot be explained by a genetic predisposition to recover particular foods at particular times or in terms of worm caches being forgotten more quickly than peanut caches, but instead must have arisen as a result of learning during pretraining trials about the fate of worms (Clayton and Dickinson 1999c).

In order to recover peanuts preferentially during the 124-h test trials and worms during the 4-h test trials, birds in the degrade group must have been able to recognize the trays in which they had cached wax worms and peanuts at the time of recovery. Recognition can be achieved in two ways: recollection and familiarity. Familiarity is an automatic process based on the perceptual characteristics of the stimuli that allows the individual to distinguish previously experienced stimuli from novel ones but does not require any recollection of the original presentation of the stimuli. Thus the jays may have based their decision of when to search for worms on their relative familiarity with the two trays as opposed to actively recalling what they cached where and when. If this were the case then birds in the degrade group could have learned to search in the tray associated with the wax worms on the 4-h trials when the two trays are relatively familiar, whereas they learn to search in the tray associated with the peanuts when the trays have not been seen for 124 h and are therefore relatively unfamiliar.

To test if jays could remember what, where and when in the absence of familiarity as a cue, in Experiment 1b the same birds were then given a pair of training trials in which they were required to cache peanuts in one side of a single tray during one caching phase and wax worms in the opposite side of the same tray during the other caching phase 120 h later (Clayton and Dickinson, 1998). To ensure that the birds had access to only one side of the tray during each caching phase of a trial, the other side of the tray was covered with a Plexiglas strip that was secured with bulldog clips. The Plexiglas strip was removed during the recovery phase so that birds could recover their caches from both sides of the caching tray 4 h after the second caching phase. Thus at the time of recovery the birds had cached one food type 4 h ago and the other food type 124 h ago.

The important feature of this procedure is that it equates the familiarity of the trays across the two trials. On each trial, the birds experience the same tray 124 h and 4 h prior to recovery and therefore cannot base their choice on the relative familiarity of the tray. If the pattern of recovery observed in Experiment 1a was based on recollection as opposed to relative familiarity with the trays, then birds in the degrade group should show a preference for peanut side of the tray on the W/P trials but not on the P/W trials. To examine whether the pattern of searches with this procedure is based on encounters with the food items at recovery rather than memory for the caching periods, we also presented a pair of probe test trials in which the food items were removed prior to recovery, as in Experiment 1(a).

Birds thus received a pair of training trials followed by a pair of probe test trials. On the training and probe test P/W trial birds cached peanuts first, whereas on W/P trials birds cached worms first. Novel trays were used on each trial, and the trays were placed in different spatial locations within the cage, to ensure that the cache sites were trial-unique. Once the birds had received one P/W and W/P pretraining trial, each bird was given a pair of probe test trials in which their caches were removed prior to recovery. As in the first pair of probe tests, this allowed us to ascertain whether the birds were relying upon memory of the caching episode as opposed to using olfactory cues about the worm and peanut caches.

If jays can remember what and where they stored the food items, then during P/W trials birds in both groups should search for the worms that they had cached 4 h earlier because jays prefer fresh wax worms over peanuts. During W/P trials, however, birds in the Degrade group should search for peanuts rather than worms because they should have learned that worms would have decayed and

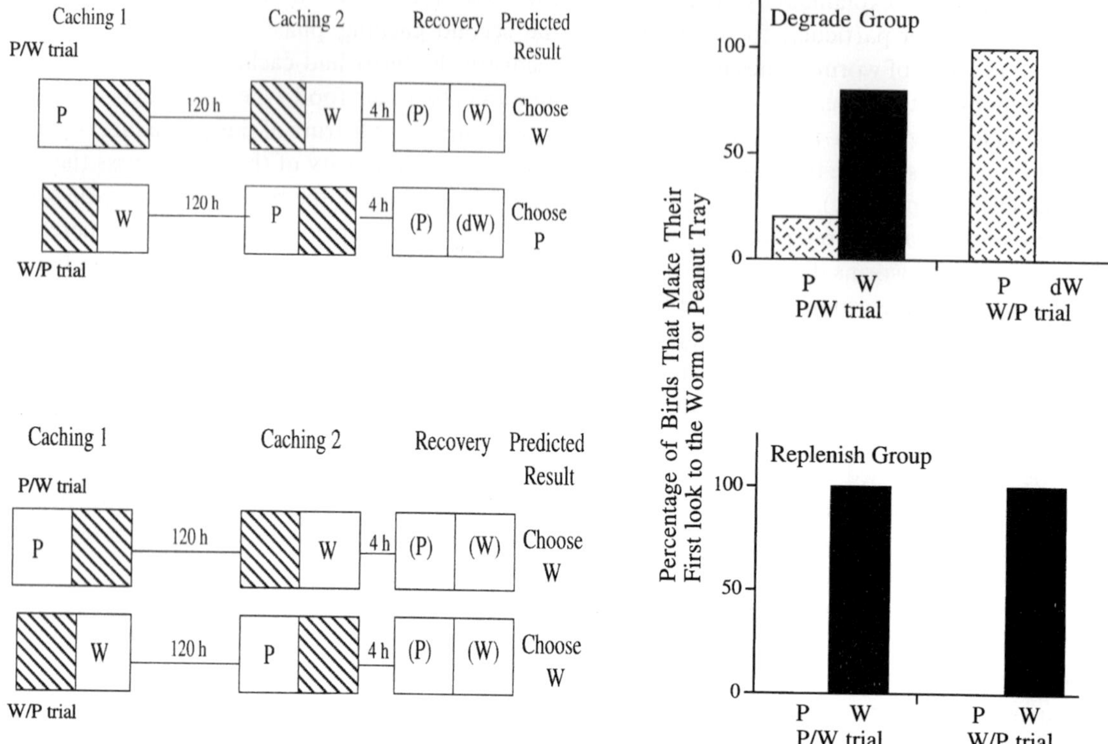

Fig. 2. The experimental design, predictions and observed results for birds in the Degrade (Fig. 2(a)) and Replenish (Fig. 2(b)) groups in Experiment 1b. During the caching phases birds were prevented from storing food items in the shaded halves of the tray by a cover (unavailable cache sites), while being free to cache in the open, non-shaded halves (available cache sites). Key: P = peanuts; W = fresh wax worms; dW = decayed wax worms; and () = food items pilfered on test. In P/W trials birds cached peanuts first, followed by worms 120 h later. The test phase followed 4 h after the end of the second caching phase. In W/P trials the order in which worms and peanuts were cached is reversed. The histograms show the percentage of birds that make their first look to the worm vs peanut sides of the tray during the recovery phase of P/W and W/P probe test trials.

become unpalatable by that time. As birds in the Replenish group only ever encountered fresh worms during pretraining and training trials, they had no opportunity to learn that worms Degrade and should continue to search for worms irrespective of whether the worms were cached first or last.

As predicted, all birds in the Replenish group searched first in the worm side of the caching tray on both W/P and P/W trials whereas birds in the Degrade group showed a switch in preference between P/W and W/P trials. The results are summarized in Fig. 2. Eighty percent of birds in the Degrade group searched the worm side of the caching tray first on the P/W probe test trial, whereas all birds inspected the peanut side first on the W/P probe test trial. Furthermore, birds made 79% of their searches on average to cache sites, demonstrating that the birds had an accurate memory of where their caches were located.

To summarize the results from both types of probe test trial (i.e. with two trays or with different sides of the same tray), a comparison of performance by birds in the Degrade and Replenish group shows that (a) the peanut-side preference shown by the Degrade group was not simply due to differential forgetting of worm caches; and (b) that the preference to search for worms 4 h after caching and for peanuts 124 h after caching does not reflect a genetic predisposition because this strategy was only adopted by birds in the Degrade group, which had the opportunity to learn that worms decay over time.

In terms of evidence for episodic-like memory, the critical result is the reversal of cache recovery preference shown by the Degrade group. The switch in preference shown by birds in the Degrade group requires the birds to recognize a particular cache site in terms of both its contents and the relative time that has elapsed between caching and recovery. This result can only be explained by recall of information about 'what' items (peanuts or worms) were cached; 'where' each type of item was stored (specific cache sites within the left or right sides of the tray); and 'when' (4 h or 124 h) the worms were cached. It is important to note that the information was formed on the basis of a single, trial-unique personal experience, and that to recover edible caches (worms when fresh, peanuts when worms are rotten) birds in the Degrade group must remember and integrate features about what happened, where the event occurred and when.

The results of this experiment therefore suggest that the cache recovery pattern of scrub jays relies on episodic-like memory (Clayton and Dickinson, 1998). It is important to note that this result cannot be explained by the simple rule 'search in the side of the tray in which food was stored most recently, regardless of food type' because during training and testing with two different trays (Clayton and Dickinson 1999c) the birds were also capable of remembering what had been cached, where and when. Since both food types were cached at the same time, i.e. at 4 h and 124 h ago, using a rule about recency would not allow the birds to solve this version of the task. Furthermore, their accuracy of searching for worm and peanut caches as opposed to simply on the worm versus peanut sides of the tray suggests that these birds have a highly accurate memory for the location of both peanuts and worms (see Clayton and Dickinson 1999c for further discussion).

The book keeping problem: remembering, forgetting and updating

In a second experiment, Clayton and Dickinson tested the jays' ability to recall specific past experiences during a cache recovery episode as opposed to during a caching episode (Clayton and Dickinson, 1999a). This experiment was also instigated by thinking about the behavioral ecology of the species: In the wild jays cache and recover many different food items during the autumn and winter months, and should therefore be able to remember which sites have been depleted by cache recovery and subsequently return only to those sites where their caches remain.

The aim of Experiment 2 was to test whether the jays can update their memories about cache sites, based on whether or not they have recovered the food from those sites. The experiment consisted of four caching phases and two recovery phases followed by a recovery test phase. Birds cached three peanuts in the left-hand sides, and three Iams® minichunk dog kibbles in the right hand sides, of two visuospatially distinct trays. Three hours later, the birds were presented with one side of one tray from which they recovered all the peanuts, and then one side of the other tray so that they could recover all the kibbles from the second tray. This meant that at the end of these recovery phases, one tray contained only kibbles and the other tray contained only peanuts. During each of these caching or recovery phases the birds had access to only half a tray at a time by using Plexiglas strips to cover the other side of each tray. During the recovery test, the jays were presented with both caching trays simultaneously and we recorded where the birds searched.

As in the previous experiment, caches were removed prior to the recovery probe test. This ensured that no extraneous cues were present and therefore the birds had to rely on memory about what they had cached and what they had recovered from the left and right hand sides of each tray. If birds can remember not only what type of food they cached in each of the sites, but also what they recovered from each site, and can integrate these two sources of information, then during the probe test birds should go to the tray that should still contain their preferred food.

Rather than relying on the intrinsic preference for one food over another as in Experiment 1(a) and 1(b), Clayton and Dickinson (1999a) manipulated the relative preference of the foods at recovery by prefeeding the birds with one of the two food types. The reason for this manipulation is that prefeeding with one food type, ether in whole seed or finely powdered form, selectively reduces the subsequent value of that food in terms of both eating and caching (Clayton and Dickinson, 1999b). Jays were given either kibbles or peanuts in powdered form during the prefeeding stage of this experiment to ensure that the birds would not cache during the prefeeding phase.

Birds that were prefed powdered kibbles should therefore search more in the tray that still contains peanuts whereas birds that had been prefed powdered peanuts should preferentially search in the sites where they had cached but not recovered the kibbles. While the control probe test and counterbalancing of the various treatments are essential, the most important feature of this design is that the pattern of probe test choices cannot be explained in terms of avoiding recently visited trays. The reason for this is that the cache sites in both trays are visited during the recovery phases and so both trays should be equally affected by any tendency to avoid recently visited locations.

The trays can be categorized in terms of whether they contained food items that were the *same* as or *different* from the prefed food at the end of the two recovery phases. If jays can also remember what they have cached where, then after prefeeding on one food type these birds should preferentially search in the different tray. The two sides of the tray can also be distinguished by whether those items had been *recovered* during the recovery phases or whether the caches remained *intact* (i.e. unrecovered). If birds can remember what happened during the two recovery phases, and integrate all of the information about caching and recovery, then they should preferentially search in the intact side of the different tray. These predictions were upheld, as summarized in Fig. 3. Nine of the 12 birds searched first in the intact side of the different tray. Furthermore, the birds were highly accurate in directing their searches to the specific sites within each tray where they had cached food items during the recovery test.

Four main conclusions can be drawn from this second experiment. First, scrub jays encode information about the type of food they store in cache sites. Without this knowledge they could not have directed their searches selectively to the tray that contained the non-prefed food. Second, the birds appear to know when to cache and when to recover in the sense that when food is presented ad libitum they begin to cache the peanuts or kibbles in the caching tray as well as eating the food, but when the trays are returned to them at a later date they immediately begin to recover their caches. Note that the birds are not trained to do this, but simply begin caching and recovering automatically. Third, they can update their memory of whether or not a caching location currently contains a food item following recovery, and in a way that cannot be explained by familiarity of the location given that they cached and recovered from both trays. Fourth, they can integrate information of the content of a cache at recovery with information about the specific location of the cache. Without the capacity for such integration, these birds could not have directed their searches selectively to the intact sides of the tray, let alone to the particular cache sites in which they had stored the food items (Clayton and Dickinson, 1999a).

Bridging the gap between animal models and human episodic memory

The results of Experiments 1(a) and 1(b) show that jays remember what, where, and when a particular caching event occurred in the past, and in a way

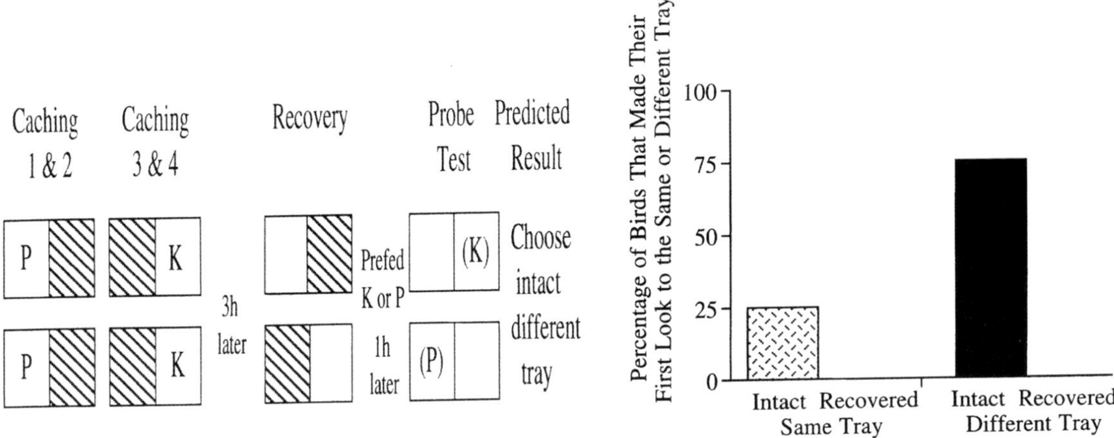

Fig. 3. The experimental design, predictions and observed results for Experiment 2. The shaded areas represent the covered halves of the trays (unavailable cache sites) and non-shaded areas the open halves of the trays (available cache sites). Key: P = peanuts; K = dog food kibbles; and ( ) = food items pilfered prior to the recovery probe test. The histogram shows the results of the recovery probe test trial in terms of the percentage of birds that search first in the intact (unrecovered) and recovered side of each caching tray. The trays are referred to as "same" or "different" based on whether the food items remaining in the tray after the two recovery phases were the same as or different from the type of food prefed before the recovery test.

that cannot be explained in terms of familiarity. The results of Experiment 2 demonstrate that these birds can remember not only what they cached where, but also what they recovered and from where. The data also suggest that the jays know when to cache and when to recover, even though the contextual cues other than time do not differ between caching and recovery phases of a trial. Not only can the birds differentiate between memories of caching and recovering, but they can also update the information about the current status of their caches, based on whether or not they have already recovered that particular cache.

These results fulfill the behavioral criteria for episodic memory (Tulving, 1983): the jay remembers a series of facts about an object (the food item), a place (where they stored it), a time (how long it was since they stored the item) and an action (caching vs cache recovery) that allow the bird to subsequently recall that information and execute the appropriate behavior. Each item could be considered as a semantic fact but when all the facts are integrated, the jay has sufficient information to isolate what was cached and what was recovered, where and how long ago: functionally, the animal has enough information to recall the episode of caching a specific item (Griffiths et al., 1999).

If animals can remember where-and-when and what-and-where associations then is the ability to remember what-where-and-when anything more than a complex declarative memory? In other words, what must an animal demonstrate that it has learned about a specific event to persuade us that it has a specific memory of a particular episode, as opposed a more general knowledge of a series of semantic facts? How much information an animal must encode and subsequently recall for the memory to be classed as episodic, and not just the sum of a series of semantic facts? Clearly the distinction between recollection and simple discrimination based on relative familiarity is crucial here, as is the ability to ensure that we test the animal's memory for a single, trial-unique past experience as opposed to its ability to acquire a rule based on associative learning over a series of trials or it's ability to remember a single feature of the episode.

Tulving and Markowitsch's contemporary definition of episodic memory invokes the involvement of "autonoetic" consciousness (Tulving and Markowitsch, 1998). As there are no agreed behavioral markers of conscious experience in animals this definition presents an insurmountable barrier to animal tests of episodic memory. The alternative approach is to characterize episodic memory in terms of the type of information encoded and recalled. According to the latter definition, the results of the scrub jay experiments suggest that there are naturally occurring situations where an animal's ability to recall specific past events about what happened, where it occurred and when, is of vital importance to its survival. An advantage of adopting this second approach is that we can test which animals show episodic-like memory and under what circumstances: In so doing we can ask questions about the evolution of episodic memories and the extent to which human memories resemble those of our ancestors as well as our present-day animal relatives.

Hopes for the future

To what extent can other animals also remember trial-unique personal experiences of "what, where and when", and in a manner that cannot be explained by a simple discrimination based on relative familiarity? How general a phenomenon is this? For example, within the food-storing domain some animals avoid the problems of perishability that jays have. Gray squirrels, for example, excise the cotyledon from acorns to prevent them from germinating and thus spoiling as a food source (Hadj-Chikh et al., 1996). Are these squirrels less able to perform a "what, where, and when" memory task, because they have taken a different route to solve the problem? (Griffiths et al., 1999). What of the memory capabilities of species that cache and recover in very harsh climates such as Alaska and Siberia, where the food is essentially cached in a freezer and has no chance to degrade over time? Such comparative studies may be extremely useful in developing a framework of the circumstances under which episodic memory may evolve.

Clearly there is a pressing need to develop other experimental tests of episodic-like memory in animals. As stated previously one approach is to consider other cases in nature in which animals might be expected to rely on episodic-like memory in nature and test whether animals engaged in these behaviors are capable of episodic-like memory. A second approach would be to try to adapt the current mammalian laboratory tasks that are used on rats and monkeys (e.g. delayed match to sample or place). Can such tasks be designed to unequivocally demonstrate an ability to recall the "what, where and when" of a specific event, or does the animal's behavior need to have an adaptive "requirement" for this type of recall, as in the food-storing example? (see Griffiths et al., 1999).

The other obvious approach is to adapt the food-storing scrub jay task so that other species can be tested in the laboratory. Given the excitement and potential power of transgenic and knock-out procedures for investigating the genetic and molecular bases of learning and memory in laboratory rodents, not to mention the wealth of knowledge about the neuroanatomy and neurophysiology of the rodent hippocampus, an obvious first step would be to develop a rat or mouse model of episodic-like memory based on the food-storing bird paradigm. The development of a rodent model system could make an enormous contribution to our understanding of the neural, molecular and behavioral mechanisms of mammalian episodic memory.

A word of warning, however: In designing new versions of the food-caching scrub jay task or developing adaptations of current laboratory tasks of declarative memory it is vital that the modifications are based upon a knowledge of the species-specific characteristics of the animal being tested and designed to exploit the animal's natural memory capabilities. The food-caching paradigm serves as a striking example of how an understanding of the species and its natural history can be employed to develop novel approaches to the study of animal learning and memory and to discover complex cognitive capabilities that had previously been undetected.

Acknowledgment

I thank the Whitehall Foundation, NIA, NIH and AFAR for financial support, and Robert Gerlai for comments on the manuscript.

References

Aggleton, J.P. and Brown, M.W. (1999) Episodic memory, amnesia and the hippocampal-anterior thalamic axis. *Behav. Brain Sci.*, in press.

Balda, R.P. and Kamil, A.C. (1989) A comparative study of cache recovery by three corvid species. *Anim. Behav.*, 38: 486–495.

Balda, R.P. and Kamil, A.C. (1992) Long-term spatial memory in Clark's nutcrackers, *Nucifraga columbiana*. *Anim. Behav.*, 44: 761–769.

Clayton, N.S. (1998) Memory and the hippocampus in food-storing birds: a comparative approach. *J. Neuropharmacol.*, 37: 441–452.

Clayton, N.S. and Dickinson, A.D. (1998) What, Where and When: Evidence for Episodic-Like Memory during cache recovery by scrub jays. *Nature*, 395: 272–274.

Clayton, N.S. and Dickinson, A.D. (1999a) Memory for the content of caches by Scrub Jays. *J. Exp. Psych.: Anim. Behav. Proc.*, 25: 82–91.

Clayton, N.S. and Dickinson, A.D. (1999b) Motivational control of food storing in the Scrub jay *Aphelocoma coerulescens*. *Anim. Behav.*, 57: 435–444.

Clayton, N.S. and Dickinson, A.D. (1999c) Scrub jays (*Aphelocoma coerulescens*) remember when as well as where and what food items they cached. *J. Comp. Psychol.*, in press.

Clayton, N.S., Griffiths, D. and Dickinson, A.D. (1999) Declarative and Episodic-Like Memory in Animals: Personal Musings of a Scrub Jay or When Did I Hide that Worm Over There? In: C.M. Heyes, L. Huber and A. Heschl (Eds.), *The Evolution of Cognition*, Altenberg Series in Theoretical Biology, MIT, in press.

Clayton, N.S., Reboreda, J.C. and Kacelnik, A. (1997) Seasonal changes of hippocampal volume in parasitic cowbirds. *Behav. Proc.*, 41: 237–243.

Eichenbaum, H. (1997) How does the brain organise memories? *Science*, 277: 330–332.

Fletcher, P.C., Dolan, R.J. and Frith, C.D. (1995) The functional anatomy of memory. *Experientia*, 51: 1197–1207.

Gaffan, D. (1992) Amnesia for complex naturalistic scenes and for objects following fornix transection in the rhesus monkey. *Eur. J. Neurosci.*, 4: 381–388.

Gerlai, R. and Clayton, N.S. (1999a) Analyzing hippocampal function in transgenic mice: an ethological perspective. *Trends Neurosci.*, 22: 47–51.

Gerlai, R. and Clayton, N.S. (1999b) Of mice and men, and models of memory: A reply to Steckler. *Trends Neurosci.*, in press.

Gerlai, R. and Clayton, N.S. (1999c) Tapping in to natural talents: A reply to Sarter and Berntson. *Trends Neurosci.*, in press.

Griffiths, D.P., Dickinson, A. and Clayton, N.S. (1999) Declarative and episodic memory: what can animals remember about their past? *Trends Cog. Sci.*, 3: 74–80.

Hadj-Chikh, L.Z., Steele, M.A. and Smallwood, P.D. (1996) Caching decisions by grey squirrels: a test of the handling time and perishability hypotheses. *Anim. Behav.*, 52: 941–948.

Henderson, V.W., Paganini-Hill, A., Emanuel, C.K., Dunn, M.E. and Buckwalter, J.G. (1994) Estrogen replacement therapy in older women: comparisons between Alzheimer's disease cases and nondemented control subjects. *Arch. Neurol.*, 51: 896–899.

Herlitz, A. and Forsell, Y. (1996) Episodic memory deficit in elderly adults with suspected delusional disorder. *Acta Psychiatry Scand.*, 93: 355–361.

Jacobs, L.F., Gaulin, S.J.C., Sherry, D.F. and Hoffman, G.E. (1990) Evolution of spatial cognition: Sex-specific patterns of spatial behavior predict hippocampal size. *Proc. Natl. Acad. Sci. USA*, 87: 6349–6352.

Kimura, D. (1995) Estrogen replacement therapy may protect against intellectual decline in postmenopausal women. *Hormones and Behav.*, 29: 312–321.

Mandler, G. (1980) Recognising: the judgement of previous experience. *Psych. Rev.*, 87: 252–271.

Mishkin, M. and Delacour, J. (1975) An analysis of short-term visual memory in the monkey. *J. Exp. Psychol.: Anim. Behav. Proc.*, 1: 326–334.

Nilsson, L.-G., Backman, L., Erngrund, K. and Nyberg, L. (1997a,b,c) The Betula prospective cohort study: Memory, health and aging. *Aging and Cognition*, 1: 1–36.

Nyberg, L., Cabeza, R. and Tulving, E. (1996a) PET studies of encoding and retrieval: The HERA model. *Psychonomic Bull. Rev.*, 3: 135–148.

Nyberg, L., McIntosh, A.R., Cabeza, R., Habib, R., Houle, S. and Tulving, E. (1996b) General and specific brain regions involved in encoding and retrieval of events: what, where and when. *Proc. Natl. Acad. Sci. USA*, 93: 11280–11285.

Nyberg, L., McIntosh, A.R., Houle, S., Nilsson, L.-G. and Tulving, E. (1996c) Activation of medial temporal structures during episodic memory retrieval. *Nature*, 380: 715–717.

Perner, J. and Ruffman, T. (1995) Episodic memory and autonoetic consciousness: Developmental evidence and a theory of childhood amnesia. *J. Exp. Child Psychol.*, 59: 516–548.

Pillemer, D.B. and White, S.H. (1989) Childhood events recalled by children and adults. *Adv. Child Dev. Behav.*, 21: 297–340.

Reboreda, J.C., Clayton, N.S. and Kacelnik, A. (1996) Species and sex differences in hippocampus size between parasitic and non-parasitic cowbirds. *NeuroReport*, 7: 505–508.

Sherry, D.F., Jacobs, L.F. and Gaulin, S.J.C. (1992) Spatial memory and adaptive specialization of the hippocampus. *Trends Neurosci.*, 17: 298–303.

Sherwin, B.B. (1988) Estrogen and/or androgen replacement therapy and cognitive functioning in surgically menopausal women. *Psychoneuroendocrinol.*, 13: 345–357.

Shettleworth, S.J. (1995) Memory in food-storing birds: From the field to the Skinner box. In: E. Alleva, H.-P. Fasolo and L. Nadel (Eds.), *Behavioural Brain Research in Naturalistic and Semi-naturalistic settings*, Klewer Academic Publishers, The Hague, pp. 158–179.

Squire, L.R. and Zola, S.M. (1996) Structure and function of declarative and non-declarative memory systems. *Proc. Natl. Acad. Sci. USA*, 93: 13515–13522.

Tomback, D.F. (1980) How nutcrackers find their stores. *Condor*, 82: 10–19.

Tulving, E. (1972) Episodic and semantic memory. In: E. Tulving and W. Donaldson (Eds.), *Organisation of Memory*, Academic Press, New York, pp. 381–403.

Tulving, E. (1983) *Elements of Episodic Memory*, Clarendon Press, Oxford.

Tulving, E. and Markowitsch, H.J. (1998) Episodic and declarative memory: Role of the hippocampus. *Hippocampus*, 8: 198–204.

Vander Wall, S.B. (1990) *Food Hoarding in Animals*, University of Chicago Press, Chicago.

Vargha-Khadem, F., Gadian, D.G, Watkins, K.E, Connelly, A., Van Paesschen and W., Mishkin, M. (1997) Differential effects of early hippocampal pathology on episodic and semantic memory. *Science*, 277: 376–380.

W.E. Crusio and R.T. Gerlai (Eds.)
Handbook of Molecular-Genetic Techniques for Brain and Behavior Research (Techniques in the Behavioral and Neural Sciences, Vol. 13)

CHAPTER 4.3

Motor performance of spontaneous murine mutations with cerebellar atrophy

R. Lalonde[1,2] and C. Strazielle[1,3]

[1]*Université de Rouen, Faculté des Sciences, UPRES PSY.CO-EA 1780, Laboratoire de Neurobiologie de l'Apprentissage, 76821 Mont-Saint-Aignan Cedex France*
[2]*Centre Hospitalier Universitaire de Montréal, Pavillon Hôtel-Dieu, Service de Neurologie, Montréal, Canada*
[3]*Université de Nancy 1, Nancy, France*

Introduction

Mutant mice with cerebellar degeneration differ in terms of whether damage is limited to the cerebellum or is accompanied by additional damage to other brain regions. The onset of degeneration occurs at different developmental stages, with involvement of selective cell populations. By characterizing similarities and differences in the behavior of various mutants, information may be obtained about the role of the cerebellum in sensorimotor performance and learning. Additional information may be gained by correlating motor performance with regional brain metabolism, as assessed by cytochrome oxidase (CO) activity.

Neuropathology in spontaneous cerebellar mutations

Table 1 illustrates the main neuropathological characteristics of those spontaneous murine mutations with cerebellar atrophy that have been evaluated for motor activity, motor coordination, and sensorimotor learning. All mutations are recessive except for the semidominant *Lurcher* mouse. *Lurcher* (Caddy and Biscoe, 1979) and *staggerer* mutant mice (Sax et al., 1968; Herrup and Mullen, 1979) have Purkinje cell-specific abnormalities (Soha and Herrup, 1995) causing secondary damage to granule cells during the developmental period. The Purkinje cell loss is accompanied by retrograde degeneration of the inferior olive (Blatt and Eisenman, 1985b; Shojaeian et al., 1985; Heckroth and Eisenman, 1991). By contrast, the deep nuclei are much less affected: a 30% decrease in cell number being described in *Lurcher* mutants (Heckroth, 1994) and no decrease at all in *staggerer* mutants (Roffler-Tarlov and Herrup, 1981). In *Lurcher*, there is a gain of function point mutation of the $\delta 2$ glutamate receptor, while in *staggerer* a deletion occurs in the ligand-binding domain of the RORα receptor (Sotelo and Mariani, 1999).

Two other mutants with damage to the cerebellum have been described: *nervous* (Sidman and Green, 1970) and *Purkinje cell degeneration* (Mullen et al., 1976). The main morphological phenotype in both mutants is degeneration of Purkinje cells with relative sparing of cerebellar granule cells. In *pcd* mutants, retrograde degeneration of inferior olive neurons was observed (Ghetti et al., 1987). The extent of cerebellar cell loss is much less extensive in *hot-foot* mutants than the four previously described mutants. These mutants are characterized by a deletion of the

TABLE 1

Cell types undergoing degeneration in cerebellar mutant mice

Mutant	Cerebellum	Cerebellar afferent regions
nervous	Purkinje	untested
Lurcher	Purkinje, granule	inferior olive
staggerer	Purkinje, granule	inferior olive
pcd	Purkinje	inferior olive
hot-foot	granule	untested
dt	untested	spinal cord
weaver	granule	substantia nigra

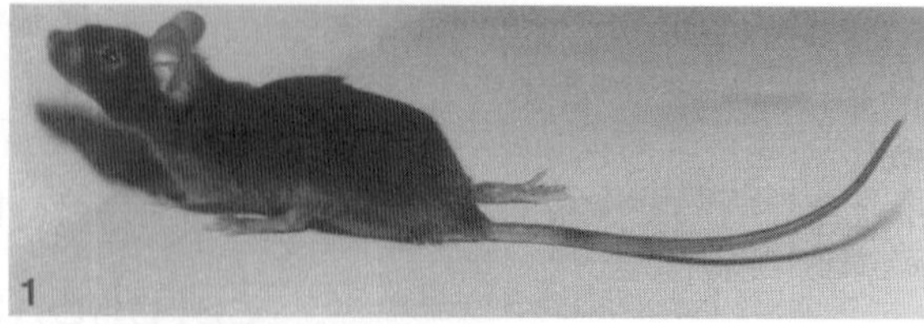

Fig. 1. Locomotion by crawling in a *dt* mutant.

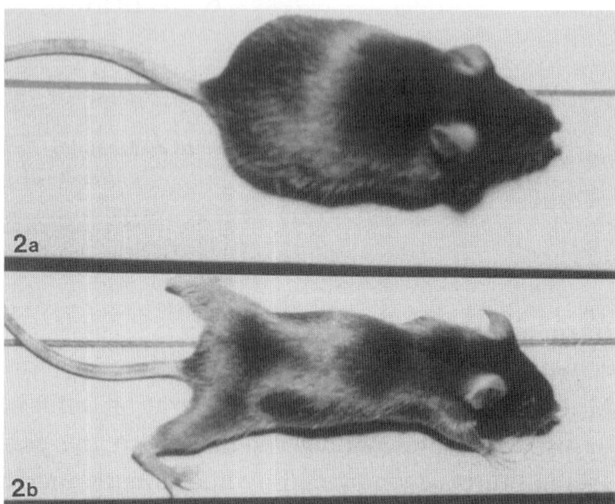

Fig. 2. (a) Normal posture in a control mouse; (b) Ataxic posture in a *dt* mutant.

same molecule affected in *Lurcher* (Sotelo and Mariani, 1999), resulting in defective innervation of Purkinje cells by parallel fibers and mild losses of cerebellar granule cells (Guastavino et al., 1990).

By contrast to the selective nature of olivocerebellar atrophy in the above-named mutants, *weaver* mutant mice have massive damage of cerebellar granule cells (Sotelo and Changeux, 1974), together with concomitant damage to substantia nigra pars compacta neurons (Triarhou et al., 1988). The mutant gene is GIRK2, a G-protein-gated inward rectifier K^+ channel (Sotelo and Mariani, 1999). The loss of cerebellar granule cells is due to their defective migration during the developmental period (Sotelo and Changeux, 1974). The loss of Purkinje cells is less severe than that of granule cells, probably the main reason why, contrary to other mutants, the inferior olive is preserved (Blatt and Eisenman, 1985a). Degeneration of substantia nigra pars compacta cells cause a reduction of dopamine concentrations and dopamine uptake sites in the dorsal striatum and to a more limited extent in the nucleus accumbens (Roffler-Tarlov and Graybiel, 1986; Strazielle et al., 1998).

A spinocerebellar mutant is available, *dystonia musculorum* (*dt*), characterized by degeneration of spinocerebellar tracts (Sotelo and Guenet, 1988) and by the presence of dystrophic peripheral sensory fibers (Janota, 1972). In addition to ataxia and loss of equilibrium, the limb positioning of *dt* mutants is dystonic to such an extent that in many instances the animals advance by crawling rather than by walking or running (Figs. 1 and 2).

Motor activity

These seven mutant mice, displaying signs of cerebellar ataxia, were compared to littermate controls of the same background strain on motor activity in a T-maze (Table 2). Motor activity was decreased in *staggerer*, *hot-foot*, *weaver*, and *dt* mutants (Lalonde et al., 1988a, 1988b, 1994; Filali et al., 1996), but not in *nervous*,

TABLE 2

Motor activity and motor coordination in cerebellar mutant mice

Mutant	Activity	Rectangular beam	Vertical grid
nervous	normal	untested	untested
Lurcher	normal	impaired	impaired
staggerer	reduced	impaired	impaired
pcd	normal	normal	normal
hot-foot	reduced	untested	untested
dt	reduced	impaired	impaired
weaver	reduced	impaired	impaired

Lurcher, and *pcd* mutants (Lalonde et al., 1986a, 1986b; Lalonde et al., 1989). Thus, massive depletion of Purkinje cells was not sufficient to cause hypoactivity. Although *Lurcher* mutants and *staggerer* mutants share many neuropathological features, such as a loss of Purkinje cells, granule cells, and inferior olive cells, only *staggerer* mutants were impaired in terms of motor activity. These results may be explained by differences in the neurochemical consequences of cerebellar-related pathways between the two mutants.

The *hot-foot* mutation is characterized by hypoactivity (Filali et al., 1996), despite a much less severe necrosis of cerebellar neurons (Guastavino et al., 1990) than that seen either in *Lurcher* (Caddy and Biscoe, 1979) or *staggerer* (Sax et al., 1968; Herrup and Mullen, 1979). The relative preservation of cerebellar cortical neurons in *hot-foot* by comparison to *Lurcher* appears to be detrimental to motor activity, indicating that a more severe behavioral phenotype may result from dysfunctional cerebellar cortical neurons than the absence of the same neurons. In support of this hypothesis, the motor performance of cerebellectomized *dystonic* rats was improved by comparison to non-cerebellectomized *dystonic* rats (LeDoux et al., 1993).

As in *staggerer* and *hot-foot* mutants, *weaver* mutants were found to be hypoactive by comparison to their control strain (Lalonde et al., 1988b). This hypoactivity may be ascribed at least in part to the loss of midbrain dopamine neurons (Triarhou et al., 1988). By contrast, no decrease of dopamine concentrations was found in *Lurcher* (Reader et al., 1998) and *staggerer* (Roffler-Tarlov and Graybiel, 1986). The reduced activity of *dt* mutants by comparison to controls (Lalonde et al., 1994) is ascribed to degeneration of spinocerebellar tracts (Sotelo and Guenet, 1988), causing dystonic posturing and crawling.

Motor coordination

Motor coordination of spontaneous murine mutations with cerebellar atrophy has been measured in several apparatus, including a stationary beam (Lalonde, 1987a, 1987b; Lalonde et al., 1993), a grid (Lalonde, 1994; Lalonde et al., 1994), a coat-hanger (Lalonde et al., 1992), and unsteady platform (Hilber et al., 1999), a rotorod (Caston et al., 1995; Lalonde et al., 1995; Gerlai et al., 1996), a rotating grid (Lalonde et al., 1996), and a treadmill (Le Marec and Lalonde, 1998, 1999). The principal measure in all these tests is the amount of time spent on the apparatus before falling.

In the stationary beam test (Fig. 3A), the mice must be placed on a round or square beam of a sufficient height from a cushioned table in order to avoid deliberate jumping. In addition to latencies before falling, the number of segment crossings on the beam may be measured. The latter measure may distinguish a mutant group from controls whenever the mutants adopt a freezing strategy in order to avoid a fall.

In the inclined grid test, the mice face downward on a surface with varying slopes. In addition to latencies before falling, a loss of equilibrium may be detected by measuring latencies before turning upward or latencies before reaching the top of the grid.

In the unsteady platform test (Fig. 3B), mice are placed on a narrow surface and must remain immobile in order to prevent a fall. This test

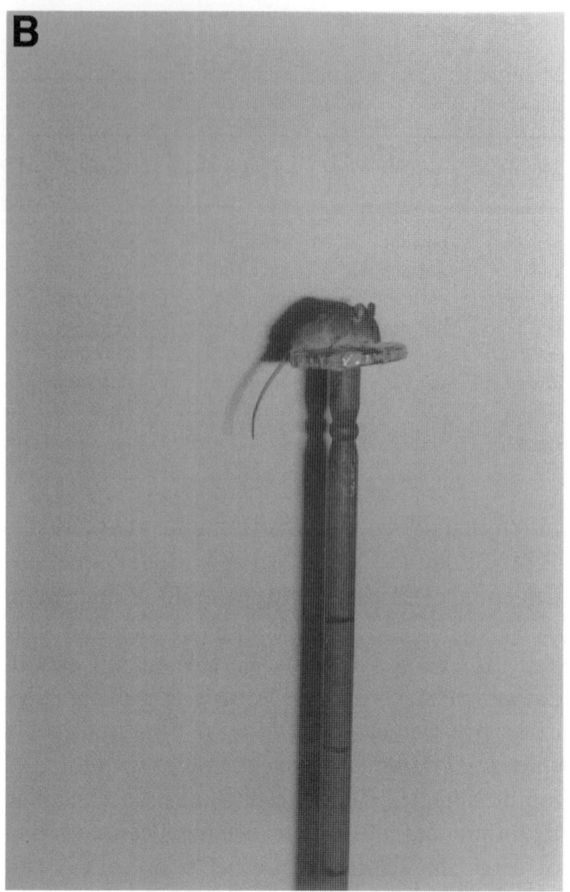

Fig. 3. Normal mice on: (A) the stationary beam; (C) the coat-hanger; (B) the unsteady platform; (D) the rotorod.

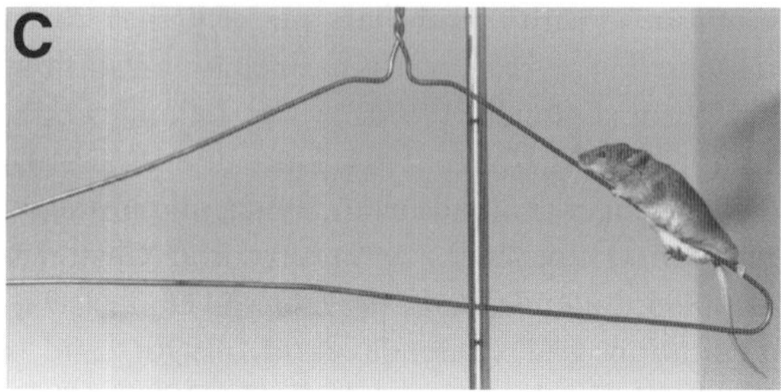

measures static as opposed to dynamic equilibrium, which prevails when the animal is free to move. In addition to latencies before falling, the number of slips is measured.

In the coat-hanger test (Fig. 3C), the mice are suspended upside down in the middle of a thin horizontal string and are motivated to move toward either of two diagonal side-bars in order to attain a right-side-up position. Care must be taken to release the mouse when the paws are in contact with the string in a stable body position in order to avoid a fall due to incorrect limb placement caused by struggling movements. In addition to latencies before falling, movement times before reaching the side-bar may be recorded with different paw criteria.

By contrast to the above-named tests, the rotorod, the rotating grid, and the treadmill are mobile apparatus, in which the mice must adjust their posture as a function of a moving surface. In the rotorod test (Fig. 3D), the mice are placed on a circular beam turning round its longitudinal axis at different speeds and must therefore continuously walk or run forward in synchrony with beam movement. Provided the traction is sufficient and no muscle weakness is evident, neurologically impaired mice can often cling to the rod without moving (passive rotation) or falling. Therefore, in addition to latencies before falling, it is necessary to measure the amount of time spent in motion. The rotating grid is a variation of the rotorod test, in which the mice are placed on a surface set in motion around its longitudinal axis at different speeds.

In the treadmill test, the mice are placed on a moving belt and, as on the rotorod or the rotating grid, must move forward in order to avoid a fall. The treadmill may be positioned at different slopes and set in motion at various speeds. In addition to latencies before falling, the time spent in motion and the onset before the first movement may be recorded.

The motor performance of five cerebellar mutant mice has been shown to differ relative to controls on stationary beam tests. Latencies before falling were lower in *staggerer* (Lalonde, 1987b), *weaver* (Lalonde, 1987a), and *dt* (Lalonde et al., 1994) than their respective controls on a rectangular-shaped beam. On the contrary, *Lurcher* (Lalonde et al., 1993) and *pcd* (Le Marec and Lalonde, 1997), despite unstable body positions, were not deficient in terms of latencies before falling from the same beam. This differential performance according the type of mutation is identical to that reported for T-maze motor activity. In the vertical grid test, while the performance of *staggerer* (Lalonde, 1987b) *weaver* (Lalonde, 1987a), and *dt* (Lalonde et al., 1994) was once again deficient, the performance of *pcd* was not (Le Marec and Lalonde, 1997). However, contrary to their normal performance on motor activity and stationary beam tests, *Lurcher* mutants were impaired on the vertical grid, a result that may be ascribed to their more severe cerebellar cortical atrophy than *pcd* (Lalonde et al., 1992; Le Marec and Lalonde, 1997). By contrast, all five mutants had lower latencies before falling from the coat-hanger than controls (Lalonde et al., 1992, 1994, 1996; Le Marec and Lalonde, 1997).

Postural sensorimotor learning

In view of evidence that synaptic modifications occur in the cerebellar cortex during the course of postural sensorimotor learning in normal mice (Anderson et al., 1996; Black et al., 1990; Kleim et al., 1997), it is of interest to examine the learning abilities of cerebellar mutant mice with different types of cerebellar cortical atrophy. *Lurcher*, *staggerer*, and *hot-foot* mutants were compared to controls on a round stationary beam, the coat-hanger, the rotating grid, and the rotorod (Lalonde et al., 1996). Although impaired relative to controls, the performance of *Lurcher* mutants improved with repeated trials on all four tasks, while the performance of *staggerer* mutants did not. This pattern of results is similar to that reported in motor activity, grid, and rectangular beam tests mentioned above, with *staggerer* mutants being relatively more impaired than *Lurcher* mutants.

By contrast, *hot-foot* mutants showed an intermediary level of performance in this series of tests. Their performance improved on the coat-hanger and on the rotating grid but not on the round beam or on the rotorod. These results indicate that although the cerebellar cortical atrophy is more severe in *Lurcher* than in *hot-foot* (Caddy and Bisco, 1979; Guastavino et al., 1990), the learning capabilities of the former mutant are superior. Thus, defective innervation of Purkinje cells caused poorer sensorimotor learning than the absence of Purkinje cells.

The learning performance of *Lurcher* mutants could be distinguished from that of *pcd* mutants on the treadmill. While acquisition of the task was impaired in *Lurcher* mutants (Le Marec and Lalonde, 1999), it was not in *pcd* mutants (Le Marec and Lalonde, 1998). As in the vertical grid test, the differential performance of these two mutants may be ascribed to the more severe atrophy of the cerebellar cortex in *Lurcher*.

Regional brain metabolism

The study of brain-behavior relationships has been enhanced by the development of metabolic mapping, based on the fundamental concept that energy utilization in brain tissue is tightly coupled to neuronal functional activity. The activity of cytochrome oxidase (CO), a mitochondrial energy-generating enzyme, has been validated as a marker of neuronal activity with high spatial resolution (Wong-Riley, 1979, 1989; Horton and Hubel, 1981). By contrast to 2-deoxyglucose autoradiography which records physiological changes for a period as short as 5 min, CO alterations develop on a scale of hours and days, in relation to plasticity and development (Mjaatvedt and Wong-Riley, 1988; Tuor et al., 1994), or aging and disease (Nobrega et al., 1993; Chagnon et al., 1995; Vila et al., 1996; De La Torre et al., 1997; Gonzalez-Lima et al., 1997). Because of the chronic and selective nature of atrophy in cerebellar mutant mice, CO histochemistry was the method of choice for determining long-term regional brain metabolic activity throughout the brain (Krémarik et al., 1998; Strazielle et al., 1998). Moreover, quantitative measurements of CO activity permit the possibility of determining linear correlations between enzyme activity and behavior.

Quantitative measurements of cytochrome oxidase (CO) activity

Although fixed tissue has been used for CO histochemistry, freshly frozen tissue is preferable. CO activity in the spinal cord after perfusion-fixation with a 2.5% paraformaldehyde and 1.5% glutaraldehyde solution was decreased by approximately 10% in comparison to freshly frozen tissue (Chalmers and Edgerton, 1989), perhaps because the fixatives caused protein cross-linking (Liu et al., 1993). Moreover, fixation caused an unequal distribution of changes, as some neurons retained their activity while others did not.

In our laboratory, we have used a slightly modified version of the protocol described by Wong-Riley (1979). Slides are incubated in a solution of 0.1 M phosphate buffer (pH 7.4) containing 50 mg of DAB (3,3′-diaminobenzidine), 20 mg of horse-heart cytochrome *c* (Type III purchased from Sigma-Aldricht), 4 g of sucrose, and 18 mg of catalase per 90 ml, stirred continously in the dark for a period varying between 60 and 90 min depending of the type of tissue and the desired staining intensity. Catalase is used to eliminate the production of hydrogen peroxyde that causes non-specific staining when reacting with DAB. Darkness is necessary in order to avoid the production of spontaneous DAB oxidation, leading to non-specific staining. Constant agitation is also necessary in order to prevent the formation of a staining gradient. The incubation is stopped at the moment when a clear contrast between highly reactive and non-reactive portions can be discerned. If weak or absent staining occurs after 1 h of incubation, the incubation time and the concentration of cytochrome *c* can be increased, depending on the type of tissue and on the species of animals used. The slides are washed in cold buffer for 5 min and immersed in a 10% buffered formalin solution with 10% sucrose for 30 min. The slides are then washed a second time in buffer (2×5 min), dehydrated in successive ethanol and xylene baths, and coverslipped with eukitt. Additional sections can be incubated for the same duration, in the absence of DAB or with 0.01 M potassium cyanide added to the incubation medium in order to verify that no visible CO reaction product is seen in these sections.

Several complete sets of CO activity standards, cut at a similar thickness during the same period, are stained simultaneously with each set of brain sections. As CO histochemical stains have variable optical densities in different incubation reactions, it is imperative to incubate the standards and the slides together, at the same time and under the same conditions, thereby providing the assurance that no uncontrolled factor affects the optical density readings.

Sections 10, 20 and 40 μm in thickness are used to cover the range of activity measured in the different brain regions by histochemistry, a standardization method previously validated by Gonzalez-Lima and Jones (1994). Under our experimental conditions, the intensity of the staining is proportional to the thickness of the standard sections. The linear function (r correlation coefficient) must be verified for each experiment.

The analysis of CO labelled images is performed by means of a video-based computer-assisted optical densitometer. Standards are used to convert optical density readings into enzymatic activity in μmol/min/g of tissue. The computer converts directly the densitometric data into enzymatic activity, thereby providing a quantitative determination of enzymatic potential. CO staining permits high resolution brain mapping. Nissl (cresyl violet 1%) or methylene blue (1%) stainings are generally performed on adjacent slides in order to identify neuronal types and to determine the possible cell density variations in the different structures examined.

Under normal conditions, CO staining is heterogeneously distributed at the regional (Fig. 4) and at the single cell level. For example, CO activity is generally higher in distal dendrites and in the cell body, depending on the spontaneous activity and the synaptic inputs of the neuron (Wong-Riley, 1989). An excitatory input requires a higher level of oxidative metabolism than an inhibitory input (Mjaatvedt and Wong-Riley, 1988). Because oxidative metabolism is sensitive to altered neuronal activity and to transsynaptic changes (Wong-Riley, 1989), measurements of CO activity are of great interest in neurodegenerative diseases. By such means, the effects of lesions in selective brain areas on oxidative metabolism throughout the brain can be estimated.

The consequences of chronic cerebellar lesions on regional brain metabolic activity were determined in *Lurcher* (Strazielle et al., 1998) and in *hot-foot* (Krémarik et al., 1998) mutant mice. CO activity (reported as a function of tissue weight) in the shrunken but still identifiable molecular and granule layers of the cerebellum in *Lurcher* mutants was unchanged by comparison to control mice of the same background strain. CO activity in atrophied tissue reflects the metabolic activity of the remaining neurons, as the contribution of glial cells to oxidative metabolism is minimal (Wong-Riley, 1989). By contrast, in comparison to controls, CO activity in the molecular layer of the cerebellum was increased in *hot-foot* mutants (Krémarik et al., 1998). However, the CO activity of this layer was unchanged at the intracellular level. Therefore, the higher metabolic activity detected at lower magnification is probably due to an upregulation in the activity of cerebellar afferent fibers in response to the defective dendritic organization of Purkinje cells.

An increase of CO activity was found in all three cerebellar deep nuclei (Fig. 5) of *Lurcher* (Strazielle et al., 1998) and *staggerer* mutants (unpublished results). This result is explained by the loss of the GABAergic inhibitory input from the missing Purkinje cells to these nuclei. By contrast, the CO activity of the deep nuclei was unchanged in *hot-foot*, explainable by the absence of a massive loss of Purkinje cells (Guastavino et al., 1990). An increase of CO activity was also observed in cerebellar efferent regions of *Lurcher* mutants, such as the lateral vestibular nucleus, the magnocellular red nucleus, the ventrolateral thalamus, and the dorsal raphe. On the contrary, CO activity was decreased in the inferior olive, probably as a secondary consequence to the Purkinje cell loss.

A positive linear correlation was observed between CO activity in the magnocellular red nucleus and latencies before falling from the rotating grid and from the rotorod in *Lurcher* mutants. In view of the fact that the rubrospinal tract discharges in phase with the locomotor cycle (Orlovsky, 1972), it is suggested that the red nucleus takes over from impaired cerebellar function, possibly as a result of the increased cerebellar nuclear activity and/or of the increased sero-

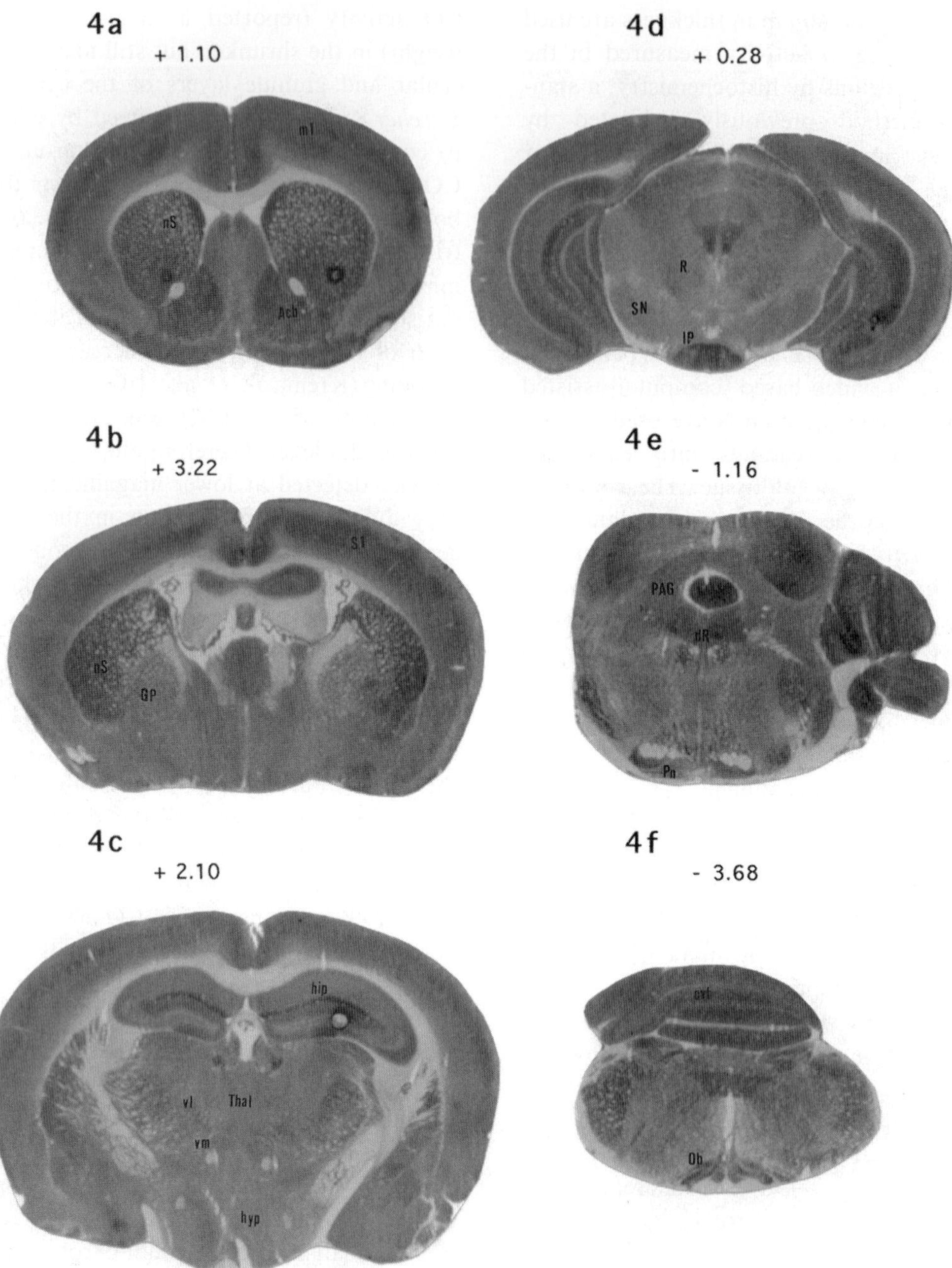

Fig. 4. Cytochrome oxidase labelling in serial coronal sections of normal mouse brain extending from anterior (+1.1 mm) to posterior (−3.68 mm); the plane of section passes through the neostriatum (a), the caudal striatum and anterior thalamus (b), the thalamus, hypothalamus, and dorsal hippocampus (c), the red nucleus and substantia nigra (d), the inferior colliculus and the pontine nuclei (e), and the inferior olive (f), magnification × 9. Abbreviations: m1 = primary motor cortex; ns = neostriatum; Acb = nucleus accumbens; Gp = globus pallidus; hip = hippocampus; vl = ventrolateral and vm = ventromedial thalamus; hyp = hypothalamus; R = red nucleus, SN = substantia nigra; IP = interpeduncular nucleus; PAG = periaqueductal grey region; dR = dorsal raphe; Pn = pontine nuclei; cvt = ???; Ob = occipital cortex.

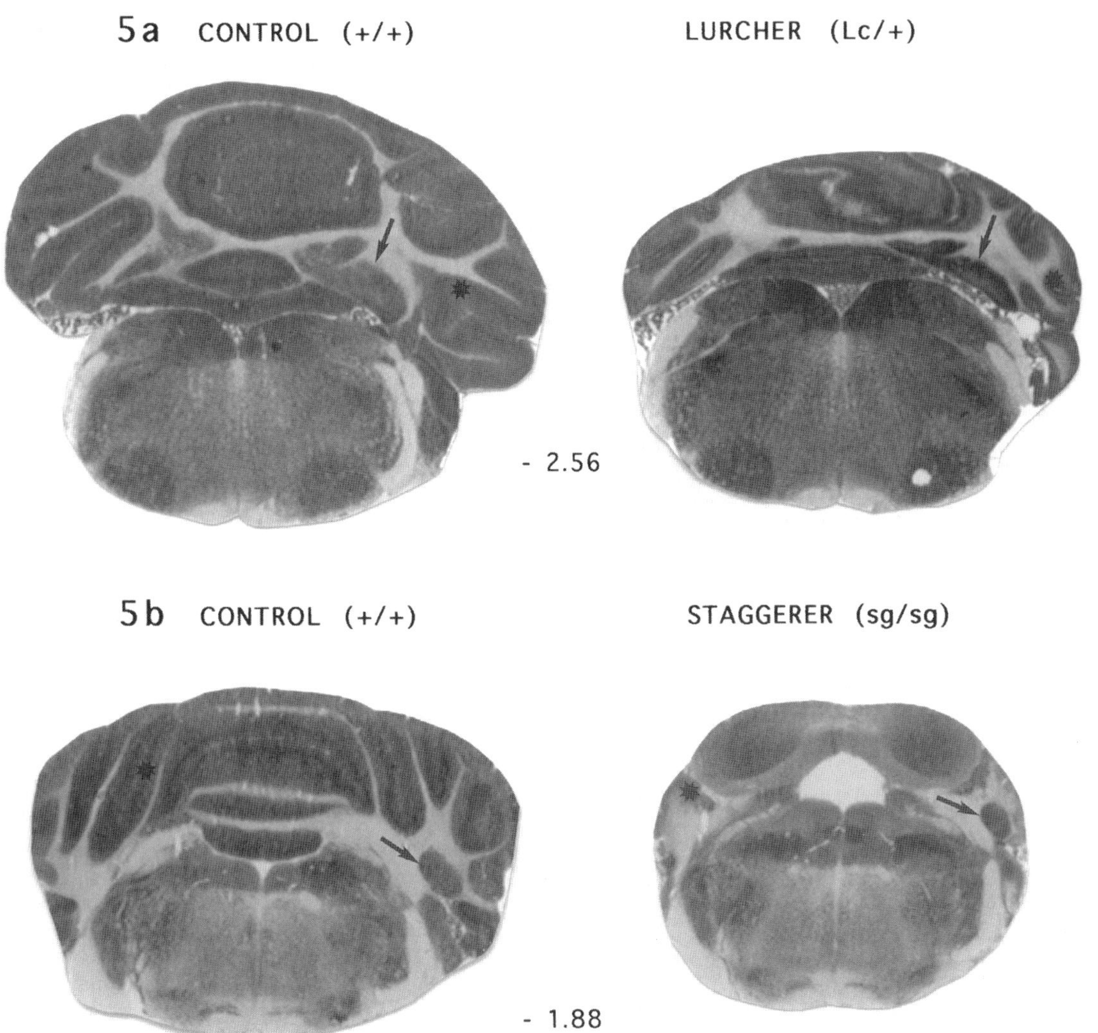

Fig. 5. CO labelling of the cerebellum in control and *Lurcher* (a) or *staggerer* (b) mutants. The deep cerebellar nuclei are indicated by arrows; note the increase of CO labelling in both mutants.

toninergic input in this nucleus (Strazielle et al., 1996). Lesion studies in monkeys had previously demonstrated that an intact red nucleus facilitates functional recovery of paralysis seen after section of the corticospinal tract (Lawrence and Kuypers, 1968). Future studies with the CO technique will permit understanding of the relation between regional brain metabolism and motor performance in cerebellar and other neurogenerative diseases and the delineation of those brain regions involved in motor recovery.

References

Anderson, B.J., Alcantara, A.A. and Greenough, W.T. (1996) Motor-skill learning: changes in synaptic organization of the rat cerebellar cortex. *Neurobiol. Learn. Mem.*, 66: 221–229.

Black, J.E., Isaacs, K.R., Anderson, B.J., Alcantara, A.A. and Greenough, W.T. (1990) Learning causes synaptogenesis, whereas motor activity causes angiogenesis, in cerebellar cortex of adult rats. *Proc. Nat. Acad. Sci. USA*, 87: 5568–5572.

Blatt, G.J. and Eisenman, L.M. (1985a) A qualitative and quantitative light microscopic study of the inferior olivary complex of normal, *reeler*, and *weaver* mutant mice. *J. Comp. Neurol.*, 232: 117–128.

Blatt, G.J. and Eisenman, L.M. (1985b) A qualitative and quantitative light microscopic study of the inferior olivary complex in the adult *staggerer* mutant mouse. *J. Neurogenet.*, 2: 51–66.

Caddy, K.W.T. and Biscoe, T.J. (1979) Structural and quantitative studies on the normal C3H and *Lurcher* mutant mouse. *Philos. Trans. Roy. Soc. Lond. (Biol.)*, 287: 167–201.

Caston, J., Jones, N. and Stelz, T. (1995) Role of preoperative and postoperative sensorimotor training on restoration of the equilibrium behavior in adult mice following cerebellectomy. *Neurobiol. Learn. Mem.*, 64:, 195–202.

Chagnon, P., Bétard, C., Robitaille, Y., Cholette, A. and Gauvreau, D. (1995) Distribution of brain cytochrome oxidase activity in various neurodegenerative diseases. *NeuroReport*, 6: 711–715.

Chalmers, G.R. and Edgerton, V.R. (1989) Marked and variable inhibition by chemical fixation of cytochrome oxidase and succinate dehydrogenase in single motoneurons. *J. Histochem. Cytochem.*, 37: 899–901.

De la Torre, J.C., Cada, A., Nelson, N., Davis, G., Sutherland, R.J. and Gonzalez-Lima, F. (1997) Reduced cytochrome oxidase and memory dysfunction after chronic brain ischemia in aged rats. *Neurosci. Lett.*, 223: 165–168.

Filali, M., Lalonde, R., Bensoula, A.N., Guastavino, J.-M. and Lestienne, F. (1996) Spontaneous alternation, motor activity, and spatial learning in *hot-foot* mutant mice. *J. Comp. Physiol. A*, 178: 101–104.

Gerlai, R., Millen, K.J., Herrup, K., Fabien, K., Joyner, A.L. and Roder, J. (1996) Impaired motor learning performance in cerebellar En-2 mutant mice. *Behav. Neurosci.*, 110: 126–133.

Ghetti, B., Norton, J. and Triarhou, L.C. (1987) Nerve cell atrophy and loss in the inferior olivary complex of "*Purkinje cell degeneration*" mutant mice. *J. Comp. Neurol.*, 260: 409–422.

Gonzalez-Lima, F. and Jones, D. (1994) Quantitative mapping of cytochrome oxidase activity in the central auditory system of the gerbil: a study with calibrated activity standards and metal-intensified histochemistry. *Brain Res.*, 660: 34–49.

Gonzalez-Lima, F., Valla, J. and Matos-Collazo, S. (1997) Quantitative cytochemistry of cytochrome oxidase and cellular morphometry of the human inferior colliculus in control and Alzheimer's patients. *Brain Res.*, 752: 117–126.

Guastavino, J.-M., Sotelo, C. and Damez-Kinselle, I. (1990) *Hot-foot* murine mutation: behavioral effects and neuroanatomical alterations. *Brain Res.*, 523: 199–210.

Heckroth, J.A. (1994) Quantitative morphological analysis of the cerebellar nuclei in normal and *Lurcher* mutant mice. I. Morphology and cell number. *J. Comp. Neurol.*, 343: 173–182.

Heckroth, J.A. and Eisenman, L.M. (1991) Olivary morphology and olivocerebellar atrophy in adult *Lurcher* mutant mice. *J. Comp. Neurol.*, 312: 641–651.

Herrup, K. and Mullen, R.J. (1979) Regional variation and absence of large neurons in the cerebellum of the *staggerer* mouse. *Brain Res.*, 172: 1–12.

Hilber, P., Lalonde, R. and Caston, J. (1999) An unsteady platform test for measuring static equilibrium in mice. *J. Neurosci. Meth.*, in press.

Horton, J.C. and Hubel, D.H. (1981) Regular patchy distribution of cytochrome oxidase staining in primary visual cortex of macaque monkey. *Nature*, 292: 762–764.

Janota, I. (1972) Ultrastructural studies of an hereditary sensory neuropathy in mice (*dystonia musculorum*). *Brain*, 95: 529–536.

Kleim, J.A., Vij, K., Ballard, D.H. and Greenough, W.T. (1997) Learning-dependent synaptic modifications in the cerebellar cortex of the adult rat persist for at least four weeks. *Neurobiol. Learn. Mem.*, 17: 717–721.

Krémarik, P., Strazielle, C. and Lalonde, R. (1998) Regional brain variations of cytochrome oxidase activity and motor coordination in *hot-foot* mutant mice. *Eur. J. Neurosci.*, 10: 2802–2809.

Lalonde, R. (1987a) Motor abnormalities in *weaver* mutant mice. *Exp. Brain Res.*, 65: 479–481.

Lalonde, R. (1987b) Motor abnormalities in *staggerer* mutant mice. *Exp. Brain Res.*, 68: 417–420.

Lalonde, R. (1994) Motor learning in *Lurcher* mutant mice. *Brain Res.*, 639: 351–353.

Lalonde, R., Bensoula, A.N. and Filali, M. (1995) Rotorod sensorimotor learning in cerebellar mutant mice. *Neurosci. Res.*, 22: 423–426.

Lalonde, R., Botez, M.I. and Boivin, D. (1986a) Spontaneous alternation and habituation in a T-maze in *nervous* mutant mice. *Behav. Neurosci.*, 100: 350–352.

Lalonde, R., Joyal, C.C. and Botez, M.I. (1994) Exploration and motor coordination in *dystonia musculorum* mutant mice. *Physiol. Behav.*, 56: 277–280.

Lalonde, R., Manseau, M. and Botez, M.I. (1988a) Spontaneous alternation and exploration in *staggerer* mutant mice. *Behav. Brain Res.*, 27: 273–276.

Lalonde, R., Manseau, M. and Botez, M.I. (1988b) Spontaneous alternation and exploration in *weaver* mutant mice. *Behav. Brain Res.*, 31: 111–114.

Lalonde, R., Manseau, M. and Botez, M.I. (1989) Spontaneous alternation and exploration in *Purkinje cell degeneration* mutant mice. *Brain Res.*, 479: 201–203.

Lalonde, R., Botez, M.I., Joyal, C.C. and Caumartin, M. (1992) Motor abnormalities in *Lurcher* mutant mice. *Physiol. Behav.*, 51: 523–525.

Lalonde, R., Filali, M., Bensoula, A.N. and Lestienne, F. (1996) Sensorimotor learning in three cerebellar mutant mice. *Neurobiol. Learn. Mem.*, 65: 113–120.

Lalonde, R., Joyal, C.C., Guastavino, J.-M. and Botez, M.I. (1993) Hole poking and motor coordination in *Lurcher* mutant mice. *Physiol. Behav.*, 54: 41–44.

Lalonde, R., Lamarre, Y., Smith, A.M. and Botez, M.I. (1986b) Spontaneous alternation and habituation in *Lurcher* mutant mice. *Brain Res.*, 362: 161–164.

Lawrence, D.G. and Kuypers, H.G.J.M. (1968) The functional organization of the motor system in the monkey: II. The effects of lesions of the descending brain-stem pathways. *Brain*, 91: 15–38.

LeDoux, M.S., Lorden, J.F. and Ervin, J.M. (1993) Cerebellectomy eliminates the motor syndrome of the genetically *dystonic* rat. *Exp. Neurol.*, 120: 302–310.

Le Marec, N. and Lalonde, R. (1997) Sensorimotor learning and retention during equilibrium tests in *Purkinje cell degeneration* mutant mice. *Brain Res.*, 768: 310–316.

Le Marec, N. and Lalonde, R. (1998) Treadmill performance of mice with cerebellar lesions. 1. *Purkinje cell degeneration* mutant mice. *Behav. Neurosci.*, 112: 225–232.

Le Marec, N. and Lalonde, R. (1999) Treadmill performance of mice with cerebellar lesions. 2. *Lurcher* mutant mice. *Neurobiol. Learn. Mem.*, in press.

Liu, Y., Gu, Q. and Cynader, M.S. (1993) An improved staining technique for cytochrome C oxidase. *J. Neurosci. Meth.*, 49: 181–184.

Mjaatvedt, A.E. and Wong-Riley, M.T.T. (1988) Relationship between synaptogenesis and cytochrome oxidase activity in Purkinje cells of the developing rat cerebellum. *J. Comp. Neurol.*, 277: 155–182.

Mullen, R.J., Eicher, E.M. and Sidman, R.L. (1976) *Purkinje cell degeneration*: a new neurological mutation in the mouse. *Proc. Natl. Acad. Sci. USA*, 73: 208–212.

Nobrega, J.N., Raymond, R., DiStefano, L. and Burnham, W.M. (1993) Long-term changes in regional brain cytochrome oxidase activity induced by electroconvulsive treatment in rats. *Brain Res.*, 605: 1–8.

Orlovsky, G.N. (1972) Activity of rubrospinal neurons during locomotion. *Brain Res.*, 46: 99–112.

Reader, T.A., Strazielle, C., Botez, M.I. and Lalonde, R. (1998) Brain dopamine and amino acid concentrations in *Lurcher* mutant mice. *Brain Res. Bull.*, 45: 489–493.

Roffler-Tarlov, S. and Graybiel, A.M. (1986) Expression of the *weaver* gene in dopamine-containing neuronal systems is dose-dependent and affects both striatal and nonstriatal regions. *J. Neurosci.*, 6: 3319–3330.

Roffler-Tarlov, S. and Herrup, K. (1981) Quantitative examination of the deep cerebellar nuclei in the *staggerer* mutant mouse. *Brain Res.*, 215: 49–59.

Sax, D.S., Hirano, A. and Shofer, R.J. (1968) *Staggerer*, a neurological murine mutant. *Neurology*, 18: 1093–1100.

Shojaeian, H., Delhaye-Bouchaud, N. and Mariani, J. (1985) Decreased number of cells in the inferior olivary nucleus of the developing *staggerer* mouse. *Dev. Brain Res.*, 21: 141–146.

Sidman, R.L. and Green, M.C. (1970) "Nervous", a new mutant mouse with cerebellar disease. In: M. Sabourdy (Ed.), *Les mutants pathologiques chez l'animal*, CNRS, Paris, pp. 69–79.

Soha, J.M. and Herrup K. (1995) Stunted morphologies of cerebellar Purkinje cells in *Lurcher* and *staggerer* mice are cell-intrinsic effects of the mutant genes. *J. Comp. Neurol.*, 357: 65–75.

Sotelo, C. and Changeux, J.P. (1974) Bergmann fibers and granular cell migration in the cerebellum of homozygous *weaver* mutant mouse. *Brain Res.*, 77: 484–491.

Sotelo, C. and Guenet, J.L. (1988) Pathologic changes in the CNS of *dystonia musculorum* mutant mouse: an animal model for human spinocerebellar ataxia. *Neuroscience*, 27: 403–424.

Sotelo, C. and Mariani, J. (1999) Research strategies for the analysis of neurological mutants of the mouse. In: W. Crusio and R. Gerlai (Eds.) *Molecular and Genetic Techniques in Behavioral Neuroscience*, Elsevier, Amsterdam. pp. ???.

Strazielle, C., Krémarik, P., Ghersi-Egea, J.-F. and Lalonde, R. (1998) Regional brain variations of cytochrome oxidase activity and motor coordination in *Lurcher* mutant mice. *Exp. Brain Res.*, 121: 35–45.

Strazielle, C., Lalonde, R., Riopel, L., Botez, M.I. and Reader, T.A. (1996) Regional distribution of the 5-HT innervation in the brain of normal and *Lurcher* mice as revealed by [^{3}H]citalopram quantitative autoradiography. *J. Chem. Neuroanat.*, 10: 157–171.

Strazielle, C., Lalonde, R., Amdiss, F., Botez, M.I., Hébert, C. and Reader, T.A. (1998) Distribution of dopamine transporters in basal ganglia of cerebellar ataxic mice by [^{125}I]RTI-121 quantitative autoradiography. *Neurochem. Int.*, 32: 61–68.

Triarhou, L.C., Norton, J. and Ghetti, B. (1988) Mesencephalic dopamine cell deficit involves areas A8, A9 and A10 in *weaver* mutant mice. *Exp. Brain Res.*, 70: 256–265.

Tuor, U.I., Kurpita, G. and Simone C. (1994) Correlation of local changes in cerebral blood flow, capillary density, and cytochrome oxidase during development. *J. Comp. Neurol.*, 342: 439–448.

Vila, M., Levy, R., Herrero, M.T., Faucheux, B., Obeso, J.A., Agid, Y. and Hirsch, C. (1996) Metabolic activity of the basal ganglia in Parkinsonian syndromes in human and non-human primates: a cytochrome oxidase histochemistry study. *Neuroscience*, 71: 903–912.

Wong-Riley, M.T.T. (1979) Changes in the visual system of monocularly sutured or enucleated cats demonstrable with cytochrome oxidase histochemistry. *Brain Res.*, 171: 11–28.

Wong-Riley, M.T.T. (1989) Cytochrome oxidase : an endogenous metabolic marker for neuronal activity. *Trends Neurosci.*, 12: 94–101.

W.E. Crusio and R.T. Gerlai (Eds.)
Handbook of Molecular-Genetic Techniques for Brain and Behavior Research (Techniques in the Behavioral and Neural Sciences, Vol. 13)

CHAPTER 4.4

Methodological considerations for testing learning in mice

Wim E. Crusio

Génétique, Neurogénétique et Comportement, CNRS UPR 9074, 3B Rue de la Férollerie, 45071 Orléans Cedex 02, France

Introduction: General considerations

Comparative psychologists have been studying rodent learning since the early decades of this century. Almost from the very beginning, genetics played some role: Tryon performed his now classical selection studies, leading to the establishment of the Tryon maze-bright and maze-dull rat lines, already in the late twenties (Tryon, 1929). Nevertheless, despite regular publications on the topic, the genetics of rodent learning did not receive much attention until relatively recently, with the advent of molecular-genetic techniques that enable direct intervention in the genome of animals and the production of transgenic and knock-out animals. Most other research on learning was not carried out with a genetics perspective and most of the time rats were the experimental subjects of choice. In consequence, many different learning tests have been developed for rats, whereas only very few have been conceived specifically for mice. However, recent years have seen a veritable spate of publications in which molecular neurobiologists have started to study the effects on learning performance of adding or deleting genes from the mouse genome. In these studies, researchers have mostly used learning tasks that have been developed for rats, sometimes (but certainly not always) in an adapted form. However, as we will see below, learning and memory cannot be measured directly and, consequently, the phenotypical trait will often depend largely on the testing method used. In the present chapter I will therefore concentrate on some methodological considerations for testing learning and memory in mice. For the purpose of this chapter, we will define learning as the process by which subjects collect and treat information enabling them to adjust their behavior in certain situations.

When testing learning behavior in mice, we must of course adhere to all those points that are important for testing behavior of any kind. For instance, we should realize that learning and memory can only be inferred and cannot be measured directly. All we can measure are the presence or absence of motor responses, from which we then have to draw conclusions about the underlying cognitive processes. Accordingly, Hölscher and O'Mara operationally defined learning as "an enduring change in behavior as a result of experience, that is not attributable to extraneous factors such as aging or drugs, etc. but is attributable to the interaction between the organism and environmental events that it experiences" (1997, p. 81). In practice, this often means that changes in behavioral responses over repeated trials are attributed to learning, but it is important to keep in mind that this need not necessarily be true.

Many factors, such as arousal, sensitization, fatigue, illness, etc., may influence performance and distinguishing changes resulting from the acquisition of information from those due to other influences is not always simple (McGaugh, 1989). For instance, experiments should be carried out at the same time of day as much as possible, as circadian influences on mouse learning have been described (Hoffmann and Balschun, 1992; Peeler, 1995). Of course, in general it will be impossible to test all animals at exactly the same time of day. To avoid finding spurious differences between groups, due solely to differences in the time-of-day that tests were carried out, we should randomize testing times over groups. One result of this will be that the larger the testing window is that we use, the more variation will be induced in our experimental groups. The time range during which animals are tested should be kept as short as possible to counter this. However, it may well be possible that some groups are subject to stronger variations over the day than others, differences in circadian rhythms between strains having been known for a long time. In extreme cases, group rankings may therefore depend on the part of the day-night cycle at which tests took place. For the moment, not much is known about the probability that such confounding effects may occur as not much research has been carried out on circadian influences on mouse learning behavior.

Another important general consideration is that handling-induced stress should be avoided as much as possible (see also Chapter 4.1). The amount of handling-induced stress in mice should not be underestimated and is generally much larger than in rats. Laboratory rats have been selected for docility ever since domestication began, albeit not on purpose, something which is much less the case for mice. Especially repeated handling, as is often necessary when the learning task used requires multiple trials, can induce severe stress in mice. High levels of stress might prevent animals from performing correctly and, of course, as strain differences in susceptibility to stress are well-known to occur (see Chapter 4.6), interactions between stress and learning capability might be different across genotypes. If experimental groups differ in stress susceptibility, then we might conclude on a difference in learning performance between the groups that might, in fact, be completely due to differences in susceptibility to stress. Keeping levels of stress as low as feasible will therefore be crucial in any experiment in which learning is being studied. Although this potential problem was signalled already over 25 years ago (Duncan et al., 1971), not much systematic research has been done on this topic. Handling-induced stress is of course proportional to the number of times an animal has to be handled. Most maze procedures require the animal to be handled every trial. It has been shown in rats that fewer trials per day increase learning efficiency in some tasks (Means and Dent, 1991). If this is true also for mice, then it might be expedient to give lower numbers of trials over a longer period.

A further general point concerns tests in which food is used as a reward. As some strains may differ in their taste sensitivity (Lush, 1981), it may be prudent not to rely on the supposed appetitiveness of certain food items. Instead, food deprivation should be used where possible (in studies of old mice, this may not always be advisable, however). As mouse strains, and even individuals within strains, may have very different body weights and metabolic rates, the amount of food given each day to each animal in order to maintain a certain level of the free-feeding body weight should be determined individually. Rats are generally deprived to 80% of their free-feeding body weight in most learning studies. But mice are much smaller than rats, so that if one would make an erroneous estimate of just 0.5 g of the amount of food needed on a given day for a certain individual, the animal's body weight may be much lower than the target weight on the next day. In our experience, mice whose weight falls below 75% show symptoms of illness (piloerection, "hunched" way of walking) and if their body weight falls even more, they may refuse food altogether and starve to death. In our own experiments, we have therefore chosen to food-deprive mice to 85% of their free-feeding body weight, which in our experience motivates them enough to search for food rewards. Again, mice being only small animals, these rewards should be much smaller than those used generally with rats, to avoid excessive feeding times during the experiment and to avoid satiation to occur before the end of the trial. In our experience, small crumbs (about 10–15 mg) of standard mouse pellets will suffice.

A final problem is how to deal with animals that do not learn a given task. Many behavioral neuroscientists habitually remove subjects from their study if they do not perform satisfactorily. This may be admissible if one wants to determine, for instance, whether it may be possible for some animals to remember a certain task when under the influence of some drug or when some brain structure has been damaged. However, in genetic studies the question most of the time is not whether animals can master some task or not, but whether different groups of animals have different learning capabilities. Evidently, it will be inadmissible to exclude animals from some experiment based on their performance. This includes situations in which certain individuals refuse to perform the task at hand at all, for instance, if animals refuse to run a maze. The important point here is that it cannot be known without further experimentation whether they refuse to perform for some reason unrelated to learning or not: perhaps some individuals get so frustrated with not being able to learn a task that they stop trying. Or, the task has been designed in such a way that (some) animals develop a predator-avoidance response during testing (see Chapter 4.1), which interferes with correct performance. In summary, excluding non-performing animals from the analysis may bias the learning scores of experimental groups differentially, i.e. some groups will and others will not be affected, and such a procedure may thus render the comparison of these groups invalid.

In what follows, I will discuss some of the learning tasks that are currently used most often for mice. It should be noted that strain differences have been reported for all these tasks.

Response conditioning

In associative conditioning procedures a *conditioned stimulus* (CS) is presented to the experimental subjects. The CS can be any neutral stimulus, such as a light or tone signal, that does not normally evoke a response from the animal. Over the course of training, subjects have to learn that the CS is associated with an *unconditioned stimulus* (US), which can be a food reward or the opportunity to avoid or to escape from an unpleasant situation (for instance, an electrical shock). Animals have to learn to exhibit a certain response upon presentation of the CS in order to obtain the reward or escape the punishment, or, alternatively, subjects must learn to suppress a response to the CS. Examples of the former are active avoidance conditioning and lever-press tasks in a Skinner box. An example of the latter is passive avoidance conditioning. Finally, in some tasks, such as contextual fear paradigms, the normal species-specific behavior of the animals is studied to provide an indication of their memory capacity. With the exception of the Skinner box, which has rarely been used with mice, these tasks will be briefly described below.

Avoidance conditioning

The associative conditioning task that has been applied most frequently with genetically defined mice is probably two-way active avoidance learning in a shuttle-box. A shuttle box (Fig. 1) consists of two compartments separated by a barrier or a hole in the separating wall, enabling subjects to move from one compartment to the other. The floor (usually a grid) is constructed in such a way, that animals can be given a foot-shock. The shock intensity ideally should be high enough to be unpleasant, but below the pain threshold. In other words, the shock intensity should be the lowest possible that can be perceived by the subjects and still can provide a punishment, but not so high as to cause physical damage. The latter is necessary not only to avoid the test being

Fig. 1. Shuttle box used for testing two-way active avoidance behavior.

unnecessary cruel, but also to avoid inducing high levels of stress in the experimental subjects. Basically, animals have to learn that presentation of some CS (either a sound or a light signal) will be followed by the US, a footshock, which they can avoid by changing compartments. When a trial begins, the CS signal is turned on in the compartment in which the animal is at that moment. After some time (varying from a few to up to 10 s), a foot shock is given and the animal will experience the punishment, unless it has changed compartments in the meantime. The foot shock (US) is stopped after a fixed amount of time or when the animal changes compartments, whichever comes first. After an appropriate intertrial interval, the procedure is repeated. Sometimes animals are tested until they reach a set criterion, and learning is quantified as the number of trials needed to reach that criterion. Alternatively, one can also count the number or percentage of avoidance responses per session. Other variables that can be measured are number of escapes (compartment changes after the footshock has started), number of intertrial crossings, and response latencies.

Many different protocols have been described in the literature. In most procedures, compartment changes between trials are allowed, sometimes they are punished. The number of trials administered per session, the number of sessions per day, and the number of days on which animals are tested also vary greatly between different studies. In addition, much variation is found in the length of the CS, length and intensity of the US, size and exact design of the apparatus used, etc. These variations are not without consequences. Peeler showed that depending on time of day and the type of partitioning between compartments, completely

Box I: Sample protocol two-way active avoidance

Apparatus: See Fig. 1. Dimensions: 27.5 × 12.2 × 14.0 cm. The floor consists of closely spaced metal bars, through which a footshock can be (35 V, 60 μA, pulse duration 2 ms, 50 Hz). The apparatus is placed in a sound-attenuating hood and testing is done automatically.

Procedure: Five daily test blocks of 80 trials each. A trial starts when the light goes on in the compartment where the mouse is at that moment. This is followed by a footshock after 10 s, unless the animal has changed compartments (avoidance response). The duration of the footshock is 10 s, which is interrupted by a compartment change (escape reaction). After an intertrial interval of 10 s, the procedure is repeated. Compartment changes during the intertrial interval are counted as a measure of intertrial activity.

Source: Crusio (unpublished, modified after Schwegler et al., 1981).

different rankings were obtained in a battery of inbred mouse strains (Peeler, 1995).

In a one-way active avoidance task, the animal is punished in one compartment only, the other compartment being always "safe". Weinberger et al. (1992) used an apparatus with a dark and a light compartment. Normal mice prefer the dark compartment over the light one (this species-specific characteristic of mice is also used in some anxiety tests, see Chapter 4.6). At the beginning of a trial, the animal was placed by the experimenter in the dark compartment, with the door leading to the other compartment closed. After some time the door separating the two compartments was opened and if animals did not spontaneously move into the other compartment, a foot shock was applied. A disadvantage of this task design is that before another trial can be done, the subjects have to be picked up by the experimenter and placed back in the start box. This repeated handling is not only time--consuming, but may also induce unwanted stress (in addition to the stress already elicited by foot-shock). The variables to be analyzed may be the number of avoidance and escape responses and the response latencies.

In active avoidance tasks, the animal has to do something in order to avoid the US. In contrast, in passive-avoidance tasks animals have to remain in the place where they are in order to avoid punishment. For instance, an animal is placed on an elevated platform within a cage (Wimer et al., 1968; Sprott, 1972). Mice have the natural tendency to avoid high, exposed places (again a tendency that is exploited to test anxiety, see Chapter 4.6). However, if the animal steps down, the platform is pulled-up against the wall to prevent the subject's return and the US is applied. Variables to be analyzed may be the difference in step-down latencies between the subsequent exposures to the apparatus or the number of trials to reach a certain criterion. One potential problem is that some individuals spontaneously remain seated on the elevated platform for long periods, without ever experiencing footshock (Sprott, 1972).

Fear conditioning

Conditioned freezing tests have been applied to mice since only recently (for instance, Paylor et al., 1994; Caldarone et al., 1997). Different versions of this test are being used. Contextual, but not auditory, fear conditioning has been shown to be hippocampus dependent in rats (Phillips and LeDoux, 1992). The procedure has the advantage of being brief, because in principle just two trials suffice (Paylor et al., 1994). The basic procedure is simple: animals are exposed to a novel environment, generally a conditioning chamber, and are presented with an auditory and/or visual CS, followed by a footshock (US). In the auditory and visual versions of the task, subjects are exposed to the CS, but in a different environment. In the contextual version of the task, subjects are placed in the same conditioning chamber as before, but the CS is now not presented. The duration of freezing is taken as a measure of conditioned fear. It has been shown that the inbred strains DBA/2 and C57BL/6 differ on the contextual version of the task, but not on the auditory one (Paylor et al., 1994), with DBA/2 mice showing much lower levels of freezing than C57BL/6 animals. Interestingly, it has been known for a long time (Van Abeelen, 1974) that the hippocampus of DBA/2 mice functions less well than that of C57BL/6 animals, supporting the above mentioned hippocampus dependency of the contextual task only. A word of caution is at his place here, however. Recently, Gerlai (1998a) showed that whether or not DBA/2 animals were impaired on the contextual version of this task depended on whether or not they were given a tone cue during training. In C57BL/6 animals, the presence or absence of a tone cue during training had no effect on performance in the contextual task. This indicates that the contextual and the cued learning components of this task cannot always be separated as clearly as is usually assumed and, even more importantly, that training procedure may interact with genotype (whether mutant or inbred).

Box II: Sample protocol contextual learning and cue association

---

Apparatus: Shock chamber (23 × 28 × 16.5 cm) with three opaque and one transparent wall and illuminated by a 10 W light bulb. Floor made of metal bars delivering a scrambled footshock (1 s, 70 μA). Tone cue duration 20 s (80 dB, 2900 Hz). Cue test chamber: similar in dimensions to the shock chamber, but otherwise very different: all walls transparent, no metal bars on the bottom, placed in a different part of the experimental room with different extramaze visual cues. Tone generator similar to previous. Apparatus cleaned with clydox between trials to reduce potential similarity in olfactory cues.

Procedure: Training trial: after 160 s tone on; 179 s shock on; 180 s tone and shock off; 220 s tone on; 239 s shock on; 240 s tone and shock off; 280 s tone on; 299 s shock on; 300 s tone and shock off; 360 s recording end. Tone is delivered to both the "cue" and "context" groups. Test trial: 24 h after training. "Cue group": placed in the test chamber, "context group" placed in the shock chamber. "Cue" group is presented with the tone cue, according to the previous schedule, "context" group receives contextual stimuli but no tone. Neither group receives any shock. In addition to freezing duration, other behavioral variables can also be measured.

Source: modified after Gerlai (1998a).

As said above, this task has not yet been applied very often, although it is becoming increasingly popular in research with transgenic and KO animals. It remains unknown for the moment, whether inbred strains perhaps differ in their baseline levels of freezing in this type of situations. In the above-cited examples, baseline levels were very similar (indeed, animals showed hardly any freezing at all in a control situation without footshock; Paylor et al., 1994) so that the conditioned association made by the animals between the CS or the context could be inferred directly from the duration of induced freezing. However, should baseline levels of freezing differ between groups, then the interpretation of results might be less straightforward, as in such cases absolute or relative changes of freezing might conceivably lead to different conclusions.

Maze learning

Mazes are used to test many different kinds of learning, the most common ones being discrimination learning and spatial learning. In what follows, we will discuss the different types of tasks according to the configuration of the maze. Although many different mazes have been used in the past, we limit ourselves here to those that are currently used most extensively: T mazes, radial mazes, and the Morris navigation task, respectively.

T mazes

A T-maze consists of three runways that are connected to each other at 90° angles, sometimes the angles are larger (Y-maze). T mazes have mostly been employed in discrimination learning tasks, which have lost some of their initial popularity in recent years after having been widely employed in the late sixties and the seventies. Most such procedures are based on some form of black/white discrimination. For instance, mice were required to choose between two arms of a T-shaped water maze, in which the walls of one arm were white and those of the other arm black (Van Abeelen et al., 1973). Only in the white arm was there a possibility to escape. The white arm was being reinforced, because mice have a natural tendency to avoid strongly illuminated places and will therefore generally display a spontaneous preference for the black arm. An automated version of a T-maze discrimination task, in which animals can be trained to distinguish between two visual or vibrissotactile stimuli has been developed by Lipp and Van der Loos (1991), but has not yet been applied very often. In this case, the reinforcement is negative: an unpleasant air-blow that is used as a punishment. It should perhaps be noted at this point that some authors distinguish between reward and reinforcement. The former is then taken as an environmental stimulus that has the opportunity to elicit approach responses, whereas the latter refers to the tendency of certain stimuli to strengthen learned stimulus-response tendencies (White, 1989).

T-mazes have occasionally also been used to test spatial learning. Generally, subjects are being trained to either go left or right to obtain a food reward or, in water mazes, to escape from the water. However, animals may solve such a task using either of two possible strategies: by always making the same turn or by always going to the same place. Only the last strategy would be a purely spatial one in the sense of O'Keefe and Nadel (1978), whereas the former appears to be a taxon strategy. The problem can be solved by turning the maze 180° on certain trials. To avoid a bias by possible spontaneous side preferences, this should be done following a semi-random schedule, meaning a non-systematic pattern in which exactly 50% of trials are in either of the two possible positions. Animals can now be trained to go always to the same place, so that depending on the position of the maze they have to make either a left- or a right-hand turn. Alternatively, animals can be trained to make the same turn at every trial, regardless of the position of the maze, so that the food reward is obtained in different places, depending on the start position of the maze (Crusio et al., 1990; see Fig. 2).

In addition to being used for testing learning, T-mazes are also used to test a special kind of

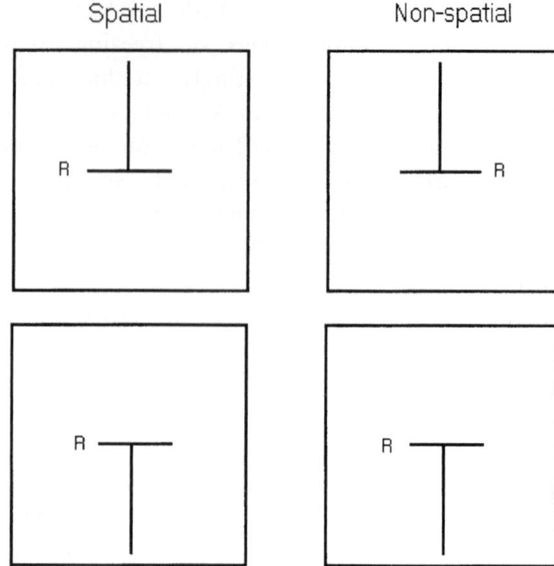

Fig. 2. Schematic diagram of spatial and non-spatial T-maze tasks. The quadrangles represent the walls of the experimental room, which should be rich in visual cues. The upper and lower panels of the figure represent the two possible positions of the maze. R is the location of the food reward: in the spatial task this was always located at the same place within the room, whereas in the non-spatial task the reward could be found by always taking the same turn. (Adapted from Crusio et al., 1990.)

Box III: Sample protocol black-white discrimination in a T-maze

Apparatus: Metal T-shaped maze with curved arms, one black, the other white. Stem of the "T" painted grey. The maze has no bottom or top and is partly immersed in a tank of water (15 cm, about 25°C). Wire mesh escape ramp present in the white arm.

Procedure: Training session: massed trials, one trial per 5 min. Session continued until criterion of 9 correct out of 10 consecutive trials was reached (maximum 40 trials). Error scores and escape latencies are registered. The procedure is repeated 24 h later (retention test). A second retention test can be added 14 d after training.

Source: Van Abeelen et al. (1973).

exploratory behavior: spontaneous alternation (see Chapter 4.6 and Gerlai, 1998b).

The radial maze

A radial maze consists of a more-or-less circular central platform from which a number of arms radiate outwards (Fig. 2). The number of arms may vary from 3 to 16, the most popular configuration being an 8-arm maze. Animals are food deprived and generally receive one or two habituation trials in which they can roam freely through the maze without any food rewards being present before training starts. Many different types of task can be devised for this maze. The simplest one is where animals can find a small food reward at the end of each arm. If mice can move around the maze without restriction, they will very often adopt a kinesthetic strategy, visiting adjacent arms in a clockwise or counterclockwise fashion (Crusio et al., 1987). This may be undesired if spatial learning is to be tested, because such chaining responses do not necessarily reflect a spatial strategy. By confining animals to the central platform for at least 5 s between subsequent arm choices, this kinesthetic strategy is interrupted (Bolhuis et al., 1986; Schwegler et al., 1990). However, it appears that similar underlying neuronal mechanisms are involved regardless of whether animals use a kinesthetic strategy, or not (Crusio and Schwegler, 1997). In more complex tasks only some arms are rewarded and others are not. The latter procedure allows to distinguish between two types of errors: repeated entries to previously visited arms, called working memory errors, and entries into arms that are never baited, called reference memory errors (Olton et al., 1979). However, there is evidence that in mice these two types of errors do not reflect the action of different underlying memory processes (Crusio et al., 1993).

Both spatial and nonspatial tasks can be devised in the radial arm maze. For spatial tasks it is necessary to prevent subjects as much as possible from using within-maze cues, such as olfactory traces. This can be done by turning the maze between trials, but keeping the rewarded place constant with regard to extra-maze cues (Crusio et al., 1993). To enable mice to see the environment, the walls of the maze (if present at all) should be made of clear plexiglas. Some of the most-often used inbred strains are visually impaired, either because they are albino (e.g. BALB/c; several substrains of strain 129 are homozygous for the chinchilla allele, c^{ch}; Silvers, 1979; Staats, 1985) or carry the allele for *retinal degeneration, rd* (e.g. strains C3H and FVB; Griep et al., 1998), or are homozygous for other mutations influencing vision, such as *pink-eyed dilution, p* (Silvers, 1979). Although *rd/rd* animals will eventually become blind, they still can see until at least 100 d of age (Nagy and Misanin, 1970) and can still perceive light/dark differences at the age of 1 yr (Mrosovsky and Hampton, 1997). Nevertheless, their visual acuity is impaired (especially at lower light intensities, see Dräger and Hubel, 1978) and we should take this reduced visual acuity of many inbred strains into account in our experimental set-up by providing clear visual cues in close proximity to the maze. For nonspatial tasks, arms may be provided with individual cues, such as different black/white floor patterns (Ammassari-Teule et al., 1993; Crusio et al., 1993). Sometimes, arms are provided with opaque walls, even further reducing the amount of spatial information available to the animals (Crusio et al., 1993). It appears that spatial learning as measured in various radial maze tasks is a unitary process, whereas different independent forms of non-spatial learning exist (Crusio and Schwegler, 1997).

Most radial mazes have open (that is, non-enclosed) arms and are elevated above the floor. However, this is a very stressful situation for rodents: the validity of the plus maze as a test of anxiety is based upon this principle (see Chapter 4.6). There are indications that repeated exposures to an elevated maze, such as will be the case in a learning experiment, may lead to increasing levels of anxiety in mice (Espejo, 1997). In our own

Box IV: Sample protocol radial maze

Apparatus: see Fig. 3. Dimensions: arms 25 × 6 × 6 cm, central platform 20 cm in diameter. For spatial tasks clear plexiglas arms are used; opaque PVC arms for nonspatial tasks. Entries to the arms can be closed by Plexiglas doors. Several objects are placed close to the maze, to provide clear extramaze cues even to visually-impaired animals. Food pellets are deposited behind perforated walls at the end of each arm to prevent subjects from smelling the presence or absence of a food reward, which may be hidden behind a low barrier. Maze always orientated in space in the same way.

Procedure: 48 h prior to training subjects are deprived of food and receive two 10 min habituation trials, 24 h apart (all arms freely accessible, no food rewards). Animals are trained on 5 consecutive days, 1 trial per day. Food rewards (~10 mg) are placed in each arm. The first two trials are terminated after the subject has eaten all rewards with a maximum of 15 min. No time limit on later trials. Errors are noted when an animal enters an arm in which the reward has already been eaten (on the first 2 d animals occasionally enter arms without eating the reward). To assess working and reference memory, 10 trials are administered and only 4 arms out of 8 are rewarded (selected randomly for each animal). The maze is rotated 45° between trials to avoid the use of within-maze cues to identify the rewarded arms. Entries into arms that are never rewarded are counted as reference memory errors. Repeated entries into any arm are counted as working memory errors.

Source: Crusio et al. (1993; 1987).

experiments, we used an enclosed, "mouse-sized" maze (Fig. 2), that is placed on the floor. Mice from all inbred strains that have been tested in this device up till now (over a dozen) readily run through it without showing any signs of discomfort. In contrast, in an experiment with a large, elevated T-maze, many individuals from the same strains often refused to run the maze, even though refusals to run where punished (Crusio et al., 1990). Some of these motivational problems may be solved by using a water version of the radial maze (Hyde et al., 1998). However, this may introduce other problems as in my experience some mouse strains (especially BALB/c; see also Klapdor and Van der Staay, 1996) respond to immersion in water by immobility. A potential other alternative, in which the possibility to return to the home cage is presented as reinforcer, has been proposed, but up till now only one inbred strain (*ddY*) has been tested in this set-up (Masuda et al., 1994).

In the experimental set-up used in our own studies (e.g. Crusio et al., 1993) mice learn this task rather rapidly. For example, in the spatial task in which all 8 arms are rewarded and in which ani-

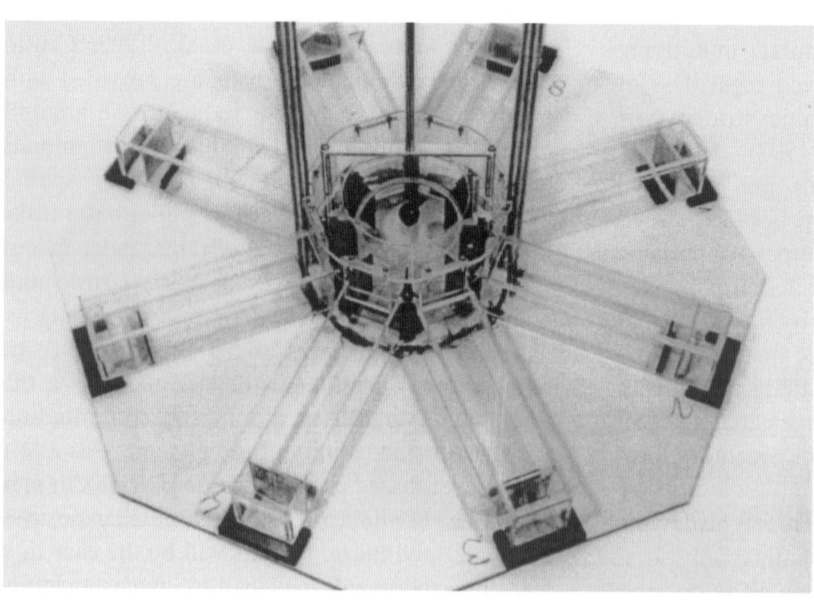

Fig. 3. Radial maze for use with mice. Plexiglass doors can be lowered to limit access to all arms simultaneously.

mals are confined to the central platform for 5 s between subsequent arm choices, mice showed satisfactory performance after only 5 d (1 trial per day). In fact, mice compare very favorably with rats on this type of task (Whishaw and Tomie, 1996).

Navigation tasks

In navigation tasks animals are free to roam unimpeded by maze alleys. Both water and dry-land versions exist, the former generally being known as the "Morris maze", although, stricly speaking, it is not a maze (Schenk, 1998).

The Morris water-navigation task

The Morris water-navigation task is a deceptively simple looking test: subjects are put in a circular basin filled with water from which the sole possibility to escape is a platform that is just submerged under water (Morris, 1981). The platform will be invisible to the subjects because of some white colorant (for instance, powdered milk) that renders the water opaque. Two general versions exist: a spatial one, in which the platform can only be found by using distal spatial cues, and a visual cue version, in which the place of the platform is indicated by some visual cue (the platform may be slightly above the water surface, or its place is indicated by an object attached to it or hanging just above it). The latter task is often used as a control for the former: if animals are not impaired on the visual task, then it is concluded that possible deficits on the spatial task are not due to non-specific motor, sensorial, or motivational impairments. However, this may not be as straightforward as it seems at first sight. For instance, it was recently shown that on the cued platform task blind rats cannot be clearly discriminated from rats that were rendered cognitively impaired by means of atropine injections (Lindner et al., 1997), an indication that results obtained with this task should be interpreted with caution.

A number of different measures have been used to assess learning in this task. The most often employed variable, escape latency, will not suffice as a measure of learning because some groups might exhibit different swimming speeds (Klapdor and Van der Staay, 1996) or because of other unspecific motor differences. A more reliable one will be the path length, which is relatively independent of the swimming capacity of the animals. Another popular way of assessing memory in this task is the so-called probe trial, given after a pre-determined number of acquisition trials. In such probe trials, the platform will be taken away and the time spent by the animal in the quadrant where the platform was located in previous trials can be compared with the time spent in other quadrants. In addition, one may count the number of annulus crossing, that is, the number of times a subject swims over the previous location of the platform. Sophisticated computer programs are available to analyze the performance of animals in this task (Wolfer and Lipp, 1992). One problem sometimes obtained with such computer programs is that the tracking systems used to obtain position coordinates have sometimes difficulties correctly identifying albino animals on a white background. A solution to this problem may be to paint the mice to increase contrast or the use of a black tank with a black platform (Klapdor and Van der Staay, 1996). In this set-up, there is no need to add a colorant to the water. Of course, the system used to track the mice may now have difficulties finding black mice. However, Klapdor and Van der Staay (1996) showed that whether a black or a white set-up was used had no effect on learning performance as measured by path length, at least in the CFW1, BALB/c, NMRI, and C57BL/6 strains used by them. Nevertheless, swimming speed per se was affected.

Many different testing protocols have been used in this task. In a typical procedure, subjects will undergo 3–6 trials a day, starting each time from a different point. Most mice will readily learn this task after about 15–20 trials (Schöpke et al., 1991). Gerlai (Chapter 4.1) has already discussed the

Box V: Sample protocol swimming navigation task

Apparatus: White Plexiglas circular pool of up to 150 cm diameter, 50 cm height, filled with 15 cm of water (24–26°C), made opaque by the addition of milk. Distant visual cues provided by posters on the walls. Wire mesh platform (14 × 14 cm) placed just below the water surface, its center 35 cm from the wall of the pool. A video camera fixed above the middle of the pool records swimming paths, which are analyzed with WinTrack (Wolfer and Lipp, 1992).

Procedure: Subjects are transferred from their cages to the pool in a white plastic cup from which they glide into the water (facing the wall) at one of eight symmetrically placed start positions on the pool perimeter. Subjects are allowed to swim for a maximum of 120 s until they find the platform, which is defined as staying on the grid at least 3 s. After staying on the plaform for about 10 s, mice are offered the opportunity to climb on a wire mesh attached to a stick, which is also used to pick-up animals that haven't found the platform. Mice are placed under infrared lamps between trials to dry and warm up. Intertrial interval 30–60 min, 6 trials per day, 5 d in total. Position of the platform constant during first 3 d (acquisition phase), placed in the opposite quadrant for the last 2 d (reversal phase). The first 30 s of the first trial serves as probe trial. Several variables are measured, for instance: swim path length, swim duration, escape latency, etc. Some other variables measuring, e.g. thigmotaxis or average path tortuosity may also be evaluated.

Source: Wolfer et al. (1998).

importance of avoiding handling-induced stress. This may be done by using a cup attached to a stick to place the mice in the pool and use a wire mesh attached to another stick to take the animals out of it (Schöpke et al., 1991). This will not resolve the above mentioned problems with strains like BALB/c that exhibit excessive freezing behavior in swimming tasks and actually display diminishing swimming speeds with increased training (Klapdor and Van der Staay, 1996). Unsurprisingly, it has recently been shown that stress impairs performance in this task (Hölscher, 1999).

In most versions of this procedure, reference memory is being tested: the place where the platform can be found is constant from trial to trial. Working memory versions of this task have also been devised. For instance, Whishaw (1995) tested mice in pairs of trials in a "matching to place" task. In each pair of trials a different platform location was selected, so that in the second trial of a pair animals had to remember the location of the platform in the first trial.

Several dry-land versions of the Morris maze have been developed for use with rats, such as the cone board (Van der Staay et al., 1990) and the circular platform or Barnes maze (Schenk, 1989; Barnes, 1991; Schenk et al., 1997). However, these devices have never, respectively only seldomly, been used with mice up till now.

Recommendations

The foregoing leads to the following 4 general recommendations for designing experiments to test learning behavior in rodents:

- One of the most important considerations will be the extent to which the testing situation is expected to induce stress in the subjects to be tested. To evaluate this question, the genetic background of our subjects will have to be taken into account. For instance, BALB/c animals have a strong tendency to freezing when put into the water, but other strains such as C57BL/6 do not appear to be overly stressed in this situation. In fact, it has been argued (Whishaw, 1995; Whishaw and Tomie, 1996) that water-maze tasks may be less suitable to test the learning and memory capacities of mice since they have evolved in dryer habitats than have rats.
- Another important point is that results obtained in one test cannot always be extrapolated to other test situations. The reason for this is twofold. On the one hand, different tasks may tap into different learning processes. On the other hand, it may be that the two tasks test the same underlying memory process, but differ in their requirements to non-cognitive factors related to perception, motor abilities, motivation, or stress. It appears to be advisable, therefore, to use at least two different tasks that

test similar learning abilities but differ in their non-cognitive requirements if one wishes to evaluate the learning capabilities of some inbred strain or some knockout or transgenic mutant mouse.

- If visually impaired animals have to be used in a spatial learning task, orientation by means of visual cues should be facilitated as much as possible. This is generally accomplished easier in, e.g. the radial maze than in set-ups for swimming navigation tasks.
- To improve comparability between experiments carried out in different laboratories, some commonly available inbred strains should always be included as controls. Two widely used strains that are often at opposite ends of the possible spectrum of performances are DBA/2 and C57BL/6J. An enormous body of data is available on these two strains. Any deviations from normal performance exhibited by animals from these strains may alert researchers to possible shortcomings in their experimental design. If possible, both positive and negative controls should also be included, that is a group in which an expected behavioral performance change is known in advance and another in which the lack of this behavioral change is known.

Acknowledgment

I would like to thank Jacques Guillot (Clichy, France) for preparing Fig. 1 and Robert T. Gerlai (South San Francisco) for critically reading the manuscript. WEC was supported by the Centre National de la Recherche Scientifique (UPR 9074), Ministry for Research and Technology, Région Centre, and Préfecture de la Région Centre. UPR 9074 is affiliated with INSERM and the University of Orléans.

References

Ammassari-Teule, M., Hoffmann, H.J. and Rossi-Arnaud, C. (1993) Learning in inbred mice: strain-specific abilities across three radial maze problems. *Behav. Genet.*, 23: 405–412.

Barnes, C.A. (1991) Memory changes with age: Neurobiological correlates. In: J.L. Martinez and R.P. Kesner (Eds.), *Learning and Memory*, 2nd Edn, Academic Press, San Diego, pp. 259–296.

Bolhuis, J.J., Bijlsma, S. and Ansmink, P. (1986) Exponential decay of spatial memory of rats in a radial maze. *Behav. Neural Biol.*, 46: 115–122.

Caldarone, B., Saavedra, C., Tartaglia, K., Wehner, J.M., Dudek, B.C. and Flaherty, L. (1997) Quantitative trait loci analysis affecting contextual conditioning in mice. *Nature Genet.*, 17: 335–337.

Crusio, W.E., Bertholet, J.-Y. and Schwegler, H. (1990) No correlations between spatial and non-spatial reference memory in a T-maze task and hippocampal mossy fibre distribution in the mouse. *Behav. Brain Res.*, 41: 251–259.

Crusio, W.E. and Schwegler, H. (1997) Hippocampal involvement in spatial and non-spatial radial maze learning in inbred mice. *Exp. Brain Res.*, 117: S3.

Crusio, W.E., Schwegler, H. and Brust, I. (1993) Covariations between hippocampal mossy fibres and working and reference memory in spatial and non-spatial radial maze tasks in mice. *Eur. J. Neurosci.*, 5: 1413–1420.

Crusio, W.E., Schwegler, H. and Lipp, H.-P. (1987) Radial-maze performance and structural variation of the hippocampus in mice: a correlation with mossy fibre distribution. *Brain. Res.*, 425: 182–185.

Dräger, U.C. and Hubel, D.H. (1978) Studies of visual function and its decay in mice with hereditary retinal degeneration. *J. Comp. Neurol.*, 180: 85–114.

Duncan, N.C., Grossen, N.E. and Hunt, E.B. (1971) Apparent memory differences in inbred mice produced by differential reaction to stress. *J. Comp. Physiol. Psychol.*, 74: 383–389.

Espejo, E.F. (1997) Effects of weekly or daily exposure to the elevated plus-maze in male mice. *Behav. Brain Res.*, 87: 233–238.

Gerlai, R. (1998a) Contextual learning and cue association in fear conditioning in mice: A strain comparison and a lesion study. *Behav. Brain Res.*, 95: 191–203.

Gerlai, R. (1998b) A new continuous alternation task in T-maze detects hippocampal dysfunction in mice. A strain comparison and lesion study. *Behav. Brain Res.*, 95: 91–101.

Griep, A.E., Krawcek, J., Lee, D., Liem, A., Albert, D.M., Carabeo, R., Drinkwater, N., McCall, M., Sattler, C., Lasudry, J.G. and Lambert, P.F. (1998) Multiple genetic loci modify risk for retinoblastoma in transgenic mice. *Invest. Ophthalmol. Vis. Sci.*, 39: 2723–2732.

Hoffmann, H.J. and Balschun, D. (1992) Circadian differences in maze performance of C57Bl/6 Ola mice. *Behav. Proc.*, 27: 77–84.

Hölscher, C. (1999) Stress impairs performance in spatial water maze learning tasks. *Behav. Brain Res.*, 100: 225–235.

Hölscher, C. and O'Mara, S.M. (1997) Model learning and memory systems in neurobiological research: Conditioning and associative learning procedures and spatial learning paradigms. In: M.A. Lynch and S.M. O'Mara (Eds.), *Neuroscience Labfax*, Academic Press, London, pp. 81–93.

Hyde, L.A., Hoplight, B.J. and Denenberg, V.H. (1998) Water version of the radial-arm maze: learning in three inbred strains of mice. *Brain. Res.*, 785: 236–244.

Klapdor, K. and Van der Staay, F.J. (1996) The Morris water-escape task in mice: strain differences and effects of intra-maze contrast and brightness. *Physiol. Behav.*, 60: 1247–1254.

Lindner, M.D., Plone, M.A., Schallert, T. and Emerich, D.F. (1997) Blind rats are not profoundly impaired in the reference memory Morris water maze and cannot be clearly discriminated from rats with cognitive deficits in the cued platform task. *Cogn. Brain Res*, 5: 329–333.

Lipp, H.-P. and Van der Loos, H. (1991) A computer-controlled Y-maze for the analysis of vibrissotactile discrimination learning in mice. *Behav. Brain Res.*, 45: 135–145.

Lush, I.E. (1981) The genetics of tasting in mice. I. Sucrose octaacetate. *Genet. Res.*, 38: 93–95.

Masuda, Y., Odashima, J., Murai, S., Saito, H., Itoh, M. and Itoh, T. (1994) Radial arm maze behavior in mice when a return to the home cage serves as the reinforcer. *Physiol. Behav.*, 56: 785–788.

McGaugh, J.L. (1989) Dissociating learning and performance: drug and hormone enhancement of memory storage. *Brain Res. Bull.*, 23: 339–345.

Means, L.W. and Dent, M.F. (1991) The effects of number of trials per day, retention interval, gender and time of day on acquisition of a two-choice, win-stay water-escape working memory task in the rat. *J. Neurosci. Methods*, 39: 77–87.

Morris, R.G.M. (1981) Development of a water-maze procedure for studying spatial learning in the rat. *J. Neurosci. Methods*, 11: 47–60.

Mrosovsky, N. and Hampton, R.R. (1997) Spatial responses to light in mice with severe retinal degeneration. *Neurosci. Lett.*, 222: 204–206.

Nagy, Z.M. and Misanin, J.R. (1970) Visual perception in the retinal degenerate C3H mouse. *J. Comp. Physiol. Psychol.*, 72: 306–310.

O'Keefe, J. and Nadel, L. (1978) The Hippocampus as a Cognitive Map. Clarendon Press, Oxford.

Olton, D.S., Becker, J.T. and Handelmann, G.E. (1979) Hippocampus, space, and memory. *Behav. Brain Sci.*, 2: 313–365.

Paylor, R., Tracy, R., Wehner, J. and Rudy, J.W. (1994) DBA/2 and C57BL/6 mice differ in contextual fear but not auditory fear conditioning. *Behav. Neurosci.*, 108: 810–817.

Peeler, D.F. (1995) Shuttlebox performance in BALB/cByJ, C57BL/6ByJ, and CXB recombinant inbred mice: Environmental and genetic determinants and constraints. *Psychobiol.*, 23: 161–170.

Phillips, R.G. and LeDoux, J.E. (1992) Differential contribution of amygdala and hippocampus to cued and contextual fear conditioning. *Behav. Neurosci.*, 106: 274–285.

Schenk, F. (1989) A homing procedure for studying spatial memory in immature and adult rodents. *J. Neurosci. Methods*, 26: 249–258.

Schenk, F. (1998) The Morris water maze (is not a maze). In: N. Foreman and R. Gillet (Eds.), *A Handbook of Spatial Research Paradigms and Methodologies*. Vol. 2: Clinical and Comparative Studies, Psychology Press, Hove, East Sussex, UK, pp. 145–188.

Schenk, F., Grobéty, M.C. and Gafner, M. (1997) Spatial learning by rats across visually disconnected environments. *Q. J. Exp. Psychol. B, Comp. Physiol. Psychol.*, 50: 54–78.

Schöpke, R., Wolfer, D.P., Lipp, H.-P. and Leisinger-Trigona, M.C. (1991) Swimming navigation and structural variations of the infrapyramidal mossy fibers in the hippocampus of the mouse. *Hippocampus*, 1: 315–328.

Schwegler, H., Crusio, W.E. and Brust, I. (1990) Hippocampal mossy fibers and radial-maze learning in the mouse: a correlation with spatial working memory but not with non-spatial reference memory. *Neurosci.*, 34: 293–298.

Schwegler, H., Lipp, H.-P., Van der Loos, H. and Buselmaier, W. (1981) Individual hippocampal mossy fiber distribution in mice correlates with two-way avoidance performance. *Science*, 214: 817–819.

Silvers, W.K. (1979) The Coat Colors of Mice. A Model for Mammalian Gene Action and Interaction. New York: Springer Verlag.

Sprott, R.L. (1972) Passive-avoidance conditioning in inbred mice: effects of shock intensity, age, and genotype. *J. Comp. Physiol. Psychol.*, 80: 327–334.

Staats, J. (1985) Standardized nomenclature for inbred strains of mice: Eighth listing. *Cancer Res.*, 45: 945–977.

Tryon, R.C. (1929) The genetics of learning ability in rats. *Preliminary report. Univ. Calif. Publ. Psychol.*, 4: 71–89.

Van Abeelen, J.H.F. (1974) Genotype and the cholinergic control of exploratory behaviour in mice. In: J.H.F. Van Abeelen (Ed.), *The Genetics of Behaviour*, North-Holland, Amsterdam, pp. 347–374.

Van Abeelen, J.H.F., Daems, J. and Douma, G. (1973) Memory storage in three inbred mouse strains after injection of cycloheximide. *Physiol. Behav.*, 10: 751–753.

Van der Staay, F.J., Krechting, B., Blokland, A. and Raaijmakers, W. (1990) The cone field: a spatial discrimination task for the automatic and simultaneous assessment of working and reference memory in rats. *J. Neurosci. Methods*, 31: 13–22.

Weinberger, S.B., Koob, G.F. and Martinez Jr, J. (1992) Differences in one-way active avoidance learning in mice of three inbred strains. *Behav. Genet.*, 22: 177–188.

Whishaw, I.Q. (1995) A comparison of rats and mice in a swimming pool place task and matching to place task: some surprising differences. *Physiol. Behav.*, 58: 687–693.

Whishaw, I.Q. and Tomie, J.A. (1996) Of mice and mazes: similarities between mice and rats on dry land but not water mazes. *Physiol. Behav.*, 60: 1191–1197.

White, N.M. (1989) Reward or reinforcement: what's the difference? *Neurosci. Biobehav. Rev.*, 13: 181–186.

Wimer, R.E., Symington, L., Farmer, H. and Schwartzkroin, P. (1968) Differences in memory processes between inbred mouse strains C57BL-6J and DBA-2J. *J. Comp. Physiol. Psychol.*, 65: 126–131.

Wolfer, D.P. and Lipp, H.-P. (1992) A new computer program for detailed off-line analysis of swimming navigation in the Morris water maze. *J. Neurosci. Methods*, 41: 65–74.

Wolfer, D.P., Mohajeri, H.M., Lipp, H.-P. and Schachner, M. (1998) Increased flexibility and selectivity in spatial learning of transgenic mice ectopically expressing the neural cell adhesion molecule L1 in astrocytes. *Eur. J. Neurosci.*, 10: 708–717.

W.E. Crusio and R.T. Gerlai (Eds.)
Handbook of Molecular-Genetic Techniques for Brain and
Behavior Research (Techniques in the Behavioral and Neural Sciences, Vol. 13)

CHAPTER 4.5

Drug and alcohol dependence-related behaviors

John C. Crabbe and Christopher L. Cunningham

Portland Alcohol Research Center, Portland, OR 97201, USA

Introduction

Drug or alcohol dependence, as defined in the current Diagnostic and Statistical Manual in use, is diagnosed from signs of drug withdrawal, tolerance, or maladaptive patterns of use (APA, 1994). The signs are various and, to a degree, specific to the particular drug at issue. Nonetheless, animal models have been developed with the goal of recapitulating either specific signs of drug dependence, or groups of signs. Specific reviews of genetic animal models of drug dependence have frequently appeared (Metten and Crabbe, 1995; Uhl et al., 1995).

The great utility of mice for genetic studies has led to the preferential use of this species for genetic animal models of drug dependence. In consequence, much of the material we discuss here is derived from experiments with mice. Because of their historical importance in neuroscience research, rats are also (and will continue to be) used. In some cases, the same behavioral test "behaves" differently in the two species, and such examples will also be discussed.

Genetic determinants of the effects of drugs

When a drug-naive organism is exposed to a drug for the first time, it may prove to be very responsive, or unresponsive, and many experiential and environmental factors contribute to individual differences in initial response. One important factor is genetic, and the others are usually termed "environmental," including such things as relative food and water deprivation state, handling immediately before the test, ambient temperature and lighting, and many others. The source of genetic contributions to individual differences can be of two sorts — allelic differences, and differences in which genes are expressed at the time of the test (and to what degree). Other chapters in the volume deal with the methods available for sorting out these sources of genetic variability. Most reviews of initial sensitivity to drugs report that about half of individual variation is probably due to allelic variability, and the rest to some combination of gene expression and environmental factors (Crabbe and Harris, 1991; Crabbe et al., 1994). This is a very broad generalization, as it spans many different drugs and responses.

Concepts of drug dependence

When drugs are repeatedly administered, and a behavioral response measured, one of two things generally occurs: the response becomes less pronounced, or it becomes more pronounced. The former case is called tolerance and is one of the defining features of human dependence. The latter case, sensitization, has also been postulated to play a role in drug dependence. Sensitization to withdrawal hyperexcitability following ethanol has been cited as a potentially important contributor to dependence (Ballenger and Post, 1978; Becker et al., 1997). More generally, it has been postulated that sensitization to the rewarding effects of drugs may be crucial (Koob and Bloom, 1988; Tabakoff and Hoffman, 1988; Robinson and Berridge, 1993; Phillips et al., 1997).

Humans can report a feeling of being "dependent" on a drug, but animals cannot. Dependence in animal models must be inferred from the occurrence of signs of withdrawal when the drug is eliminated. Often (and characteristically for drugs that depress the central nervous system), the withdrawal signs are opposite to the initial effect of the drug. Thus, alcohol acts acutely as an anticonvulsant, but alcohol withdrawal is characterized by overt seizures or, in milder cases, a reduction in the threshold required to elicit seizures by other means (McQuarrie and Fingl, 1959).

Theories of drug dependence frequently invoke alterations in the motivational effects of drugs as one contributor to dependence. These effects can be studied through assays of the drug's reinforcing or stimulus properties, and/or through allowing the animal to self-administer the drug (Cunningham, 1998).

We will consider in turn each of these aspects of the spectrum of behaviors related to drug dependence. Methods discussed are selected for their applicability to determination of genetically-based differences during drug withdrawal. Because they are covered in other chapters in this section, we will not discuss in detail methods for studying withdrawal-related alterations in several behaviors. Learning (Criswell and Breese, 1993), activity (Segal and Kuczenski, 1997), anxiety (Prather and Lal, 1992), and sensorimotor gating (Rassnick et al., 1992) all are known to be affected by drug dependence, and changes in these domains can be used as indices of drug withdrawal severity. Also, drugs can be administered by a variety of routes (e.g. inhalation of alcohol vapor and self-administration of cocaine via chronically indwelling intravenous catheters by performance of operant responses). Methods for inducing dependence are not treated, but are discussed in many of the reviews cited herein. Finally, withdrawal symptoms are obviously dependent to some degree upon the class of drug studied. For example, opioids have an idiosyncratic and characteristic set of withdrawal signs, and are treated in a separate section. Withdrawal from psychostimulant drugs is rather difficult to quantify, as it comprises mostly symptoms of depression (see Schindler et al., 1994). Studies of alcohol have always been preponderant in pharmacogenetics, and many examples are drawn from the literature on alcohol withdrawal (see Metten and Crabbe, 1995 for review). Behaviors used to index alcohol withdrawal can generally be used to study barbiturates and benzodiazepines as well.

Withdrawal reactivity

The handling-induced convulsion

In mouse, perhaps the most common sign of depressant drug withdrawal was proposed by Dr Dora B. Goldstein in 1971. She demonstrated that mice, when picked up by the tail, showed a characteristic "handling-induced convulsion," (or HIC), which could easily be scored on a rating scale (Goldstein and Pal, 1971). She showed that the severity of withdrawal was a joint function of dose and duration of alcohol vapor exposure (Goldstein, 1972), and that the severity of withdrawal was partly heritable (Goldstein, 1973). The HIC is taken as an index of central nervous system excitability: even in non-withdrawing animals, proconvulsant doses of convulsant drugs exacerbate the HIC (Crabbe et al., 1991b), and anticonvulsant drugs suppress it (Grant et al., 1992).

Most genetic work on drug withdrawal has studied alcohol, and the HIC is a sensitive index of withdrawal severity whether alcohol is given by vapor, by chronic liquid diet (Snell et al., 1996), or by injections (Crabbe et al., 1991a). In fact, some strains of mice show HIC scores that are non-zero in the absence of drugs (Crabbe et al., 1980). The HIC index is so sensitive that a single high-dose administration of alcohol (Crabbe et al., 1991a); barbiturates or benzodiazepines (Metten and Crabbe, 1994); inhaled nitrous oxide (Belknap et al., 1987); or acetaldehyde or synthetic barbiturates (Crabbe et al., 1991a) can reveal a rebound increase in HIC scores as the drug is eliminated from the brain. This acute drug withdrawal response is genetically determined (see above references), and QTLs underlying acute alcohol and pentobarbital withdrawal have recently been mapped (Buck et al., 1997; Buck et al. 1999). The current version of the HIC rating scale is given in Table 1.

The advantages of the HIC are its simplicity, reliability (test-retest and inter-rater reliability estimates typically exceed .85), and sensitivity. Its disadvantages are three-fold. Some genotypes of mouse (e.g. withdrawal seizure-resistant, or WSR mice, selectively bred to be resistant to alcohol withdrawal), display no measurable HIC before or after chronic drug treatment (e.g. Terdal and Crabbe, 1994). A more serious difficulty is that the sign does not appear in rats. Finally, despite its systematic use for more than 25 years, the underlying physiology of the HIC is not known.

TABLE 1

Handling-induced convulsion rating scale

Score	Symptom
7	Severe, tonic-clonic convulsion, with quick onset and long duration: spontaneous, or elicited by mild environmental stimulus such as lifting cage top
6	Severe, tonic-clonic convulsion when lifted by the tail, with quick onset and long duration, often continuing for several seconds after the mouse is released
5	Tonic-clonic convulsion when lifted by the tail, often with onset delayed by as much as 1–2 s
4	Tonic convulsion when lifted by the tail
3	Tonic-clonic convulsion after gentle 180 degree spin
2	No convulsion when lifted by the tail, but tonic convulsion elicited by gentle 180° spin
1	Only facial grimace after gentle 180° spin
0	No grimace or convulsion

Changes in seizure threshold

The HIC is very likely a seizure, but evidence of paroxysmal brain electrical activity has not been gathered. Many studies have shown that chronic treatment with drugs that suppress CNS function, such as alcohol, benzodiazepines, and barbiturates, induce dependence, which is inferred from withdrawal at a later point (dependent upon rate of drug metabolism). Withdrawal from depressants manifests as a decrease in the threshold dose of a convulsant required to produce overt seizures (Gonzalez, 1993; Watson and Little, 1995; Finn and Crabbe, 1999). If possible, such studies should (but routinely do not) control for changes in the brain level of the convulsant administered. Alternative versions of this test have used audiogenic seizures as the seizure-inducing treatment (Majchrowicz, 1975). An effective stimulus is shaking a set of keys, which will induce seizures in sensitive animals. This should be avoided, however, because the stimulus is highly variable and inherently unquantifiable. Alternatively, a tone of approximately 100 decibels can be administered in a sound-isolation chamber (Faingold and Riaz, 1994).

For mice, the most reliable method for administering convulsants is by timed tail-vein infusion. Methodological issues surrounding the choice of drug concentration and infusion rate are discussed elsewhere (Kosobud et al., 1992).

In many studies, different groups of animals are made dependent, then withdrawn and given a dose of a convulsant. In such an experiment, it is necessary to equate the withdrawing groups given different convulsant doses for the level of dependence. This can be difficult, as the latter variable itself must be inferred from the dose and duration of drug administered. If a liquid diet has been used, animals will typically self-administer different amounts of drug, and do so according to different temporal patterns. This method also requires many animals. Finally, in mice, isolating a lateral tail vein in mice is quite simple as they are close to the surface, but the procedure is considerably more difficult in rats, and it is not often used for this species.

A hybrid method is sometimes used with the drug pentylenetetrazol (PTZ), which elicits a particularly clear myoclonic twitch as its initial seizure sign. Animals in the drug-withdrawn and control group are given a sub-threshold dose of PTZ (e.g. 15 mg/kg, ip) and monitored for seizures. If none are seen after, e.g. 15 min, a second dose is given, and the procedure is repeated until all animals have eventually accumulated a dose of PTZ above threshold. This method is inherently more variable than either of the above: this is probably due in part to the inevitable increased emphasis on any differences in drug absorption and distribution, cumulated over multiple injections and a relatively long time.

Electrophysiological measures

Alcohol withdrawal has been reported to be accompanied by enhanced sensitivity to chemical or electrical stimuli in several electroencephalographic preparations (Chapin and Woodward, 1989; Faingold and Riaz, 1994; Ripley et al., 1996). The changes reported are specific to particular neurotransmitter systems, and thus do not appear to reflect simply an increase in "neural excitability." However, seizureform activity during alcohol withdrawal has also been recorded (Walker and Zornetzer, 1974). The rather specialized nature of electrophysiology has precluded wide use of such measures to index drug withdrawal severity.

Measures specific to opioid withdrawal

Opioid withdrawal is typically not robust in rodents unless it is elicited by injection with an antagonist, such as naloxone or naltrexone. When antagonist injections are given, signs of withdrawal are dramatic and sudden. For mice, withdrawal manifests as a jumping response of uncertain origin. Diarrhea characterizes both rats and mice, and can be quantified by volume or weight. Straub tail is a characteristic withdrawal sign for both rats and mice. A review of rodent withdrawal signs can be found in a paper describing dose-effect relationships among the various signs (Schulteis et al., 1994).

Index approaches

Alcohol withdrawal, particularly in rats, is difficult to quantify. This has led several investigators to propose "indexing" approaches. The best-known such system was proposed by Majchrowicz (1975). Rats were maintained in a dependent state for 4 d by 3–5 daily, spaced intragastric intubations, and followed throughout withdrawal. Waller et al. (1982) used a scale derived from the former experiment to index withdrawal in P rats allowed free access to alcohol. Animals from this line were selectively bred for their high preference for ethanol solutions over water. P rats displayed significant withdrawal scores. However, one cautionary note regarding such approaches is that it is often not made clear what individual rats are displaying; rather, average index scores are simply reported.

In an elegant series of papers, Boisse and Okamoto (1978) developed a "chronically equivalent" dosing procedure for studying tolerance and dependence in cats. The scale for indexing withdrawal from barbital and pentobarbital was also used to study alcohol and diazepam. Gallaher et al. (1986) have described the withdrawal syndrome from chronic benzodiazepines in mice using a battery of behavioral indices.

McClearn and colleagues attempted a novel set of experiments with the express goal of identifying an alcohol withdrawal "syndrome." Reasoning that alcohol dependence was complex, and therefore not likely to be captured in a single behavioral endpoint, they gave a large number of mice a liquid diet containing alcohol for several days. They then compared their behavioral responses on a number of behavioral tests with those of a control-diet group. Of the multiple tests studied, 7 were identified that discriminated withdrawing from control animals (Hutchins et al., 1981; McClearn et al., 1982). They then constructed a "factor" from the common variance in the withdrawal scores, and proceeded to breed selectively for the response. Details of the genetic selection have been reviewed (Metten and Crabbe, 1995). This multivariate approach to defining withdrawal severity for genetic studies remains unique in the animal model literature, and is deserving of further consideration.

Homeostatic dysregulation

Systems approaches to the study of drug withdrawal have also been attempted. The thermoregulatory system is the most frequently explored. Withdrawal from chronic ethanol has been reported to exert a confusing array of changes on thermoregulatory systems (for review, see Crawshaw et al., 1994). Opioid withdrawal is accompanied by thermoregulatory dysfunction (Pinelli and Trivulzio, 1997). Chronic drug administration can dysregulate circadian rhythms (Kakihana and Moore, 1976). Thermoregulation, locomotor activity, and ingestive behaviors are all under circadian control, so the use of measures such as these as withdrawal indices is complicated by the changing baseline associated with different points of the light-dark cycle (Dugovic et al., 1992; Stinus et al., 1998).

Motivational effects

Initial sensitivity to the motivational (rewarding or aversive) effects of a drug is believed to play an important role in determining whether an individual seeks that drug again (e.g. Tabakoff and Hoffman, 1988). As noted earlier, repeated administration of a drug can cause decreases (tolerance) or increases (sensitization) in drug responses, both of which are hypothesized to play a role in drug dependence. Thus, behavioral tasks that assess the motivational effects of drugs are of interest because of their relevance to understanding the development of the compulsive pattern of drug-taking that produces dependence. These tasks can also be used to study any changes in motivational drug effects that occur as a result of repeated drug administration. Such changes may be important for understanding long-term maintenance of drug-taking behavior and relapse to drug-taking after withdrawal and extended periods of abstinence (Cunningham, 1998).

The rewarding and aversive effects of drugs are not measured directly. Rather, they must be inferred by measuring an individual's tendency to approach and contact or escape and avoid the drug itself or stimuli that have previously been associated with the drug. The tasks most commonly used for this purpose are described in the following sections.

Drinking/preference tasks

Three general kinds of self-administration tasks have been used in rodents: (a) home-cage drinking/preference, (b) operant self-administration via the oral route, and (c) operant self-administration via implanted cannulae (e.g. intravenous, intragastric). By far, the most frequently used of these tasks is home-cage drinking and preference, especially in the case of drugs that are water-soluble or that are commonly consumed orally by humans (e.g. ethanol). This task has been used extensively to examine genetic differences in both rats and mice (e.g. McClearn and Rodgers,

1959; Belknap et al., 1993a, 1993b; Li et al., 1993) and to identify QTLs for drug intake and preference (e.g. Berrettini et al., 1994; Rodriguez et al., 1995; Belknap et al., 1997; Carr et al., 1998; Phillips et al., 1994, 1998; Tarantino et al., 1998). A typical procedure involves giving subjects access to two drinking tubes, one of which contains a drug solution while the other contains vehicle. In conducting such studies, it is important to minimize drug spillage or evaporation and to control for possible confounding effects of position bias (e.g. by counterbalancing the position of the drug tube and periodically reversing the positions of the drug and vehicle tube). Because much drinking is prandial, the proximity of the food source to each drinking tube should be equated. These and many more of the methodological and conceptual issues related to drinking/preference tasks have been addressed in the case of ethanol by Myers and Veale (1972).

Drinking studies can provide continuous 24-h access to the drug solution or access can be restricted to much shorter periods of time each day (e.g. 30 min). A primary consideration in the choice of continuous vs limited-access procedures is the pattern of drug intake over time. With 24 h access, it is possible for subjects to distribute their drug consumption in small, widely-spaced bouts that do not produce appreciable brain drug levels. Thus, in order to provide evidence of control by the drug's pharmacological effects, one must implement techniques that assess the "microstructure" of feeding and drinking of both the vehicle and drug solution and take blood or tissue samples at appropriate times. Information about the frequency, duration and size of drug bouts (e.g. using lickometers or devices that measure changes in the volume of the drug reservoir) and their relationship to eating and to drinking of the vehicle solution will facilitate interpretation of drug intake and preference in 24-h access tasks.

In contrast, limited-access procedures require the subject to ingest all of its daily drug dose in a single bout or small number of bouts over a relatively short period of time. This approach improves the experimenter's ability to gauge the pharmacological effects of each drinking episode and may increase the likelihood that a subject's drinking or preference is governed by the drug's pharmacological effects. However, in contrast to 24-h procedures, considerable training over a period of several weeks or longer may be required to establish stable levels of limited-access drinking (Lê et al., 1994). It is also possible that behavior observed in limited-access procedures may differ quantitatively or qualitatively from that obtained in 24-h paradigms, although data bearing on this issue are quite sparse.

Because oral intake of a drug may be limited by its bad taste or by other noxious orosensory effects (e.g. bad odor, burning sensation in mouth or throat), several strategies have been devised to enhance intake. Often, these strategies are introduced temporarily to allow subjects to ingest sufficient drug to experience its presumed postabsorptive rewarding effects, which are then expected to play a bigger role in determining subsequent drug taking. For example, drug may be introduced at a relatively low concentration and gradually increased over time to allow habituation to noxious effects or tolerance to drug effects that interfere with drinking. Another approach is to add a highly preferred flavor or sweetener (e.g. sucrose or saccharin) in order to overcome aversive gustatory effects of the drug solution. In such cases, the concentration of the added flavor may be gradually reduced ("faded out") over time (e.g. Gauvin and Holloway, 1992). Food or fluid deprivation are also sometimes used to increase drug intake. For example, in the alcohol "acceptance" test, alcohol intake after 24-h of fluid deprivation is compared to water intake measured before fluid deprivation (McClearn, 1968; Crabbe et al., 1983). Giving access to the drug solution during the daily feeding of food-deprived animals has also been used to facilitate drug intake (e.g. Cunningham and Niehus, 1997). However, the advantages offered by strategies involving added flavors or deprivation must be weighed against potential complications in interpretation and the need for additional control comparisons.

Historically, home cage drinking and preference tasks have been widely used because of their relative simplicity and convenience. However, problems can arise in interpreting differences in intake related to subject variables or experimental manipulations that are tested at only one drug concentration. Because the function relating drug intake to concentration is typically an inverted-U, one cannot determine whether higher intakes reflect a stronger or weaker motivational effect in tests involving a single concentration. For example, a mouse strain that is more sensitive to the positive motivational effects of a drug may drink a greater volume at low concentrations but a smaller volume at higher concentrations compared to a mouse strain that is less sensitive to those effects. Thus, conclusions about the nature of the motivational effects underlying drinking differences are strengthened when one examines a wide range of drug concentrations.

Operant self-administration via the oral route

In an operant task, delivery of drug is dependent upon completion of a specific behavior such as pressing a lever, pulling a chain, or poking a nosehole. Thus, operant tasks provide an opportunity to separate "drug-seeking behavior" from actual drug ingestion. This separation can be useful, especially in situations where continuous drug ingestion produces "side effects" (e.g. sedation, hyperactivity, motor impairment) that interfere with ongoing behavior. Much of the literature on operant procedures involving oral self-administration has been conducted in rats using alcohol as the reinforcer (Meisch, 1977; Samson, 1987; Samson and Hodge, 1996). However, there are many examples of operant responding for other orally-administered drugs by rats (see review by Meisch and Carroll, 1987), and several recent operant studies in mice, including studies of genetic differences in oral self-administration (e.g. Elmer et al., 1986, 1987, 1988; Risinger et al., 1998).

Because the drug is given orally, most of the methodological and conceptual issues raised earlier in the discussion of home-cage drinking studies apply to the design of operant studies involving oral administration. Although operant studies typically limit daily access to the drug, several recent studies have examined operant self-administration under conditions that allow nearly continuous access to the drug, food and water in the operant chamber (e.g. Samson et al., 1992; Risinger et al., 1998). As in the case of drinking studies, special "initiation" procedures are often used to establish orally-administered drugs as reinforcers. These procedures include such things as: (a) pretraining the operant with a known reinforcer (e.g. food or sucrose) and then switching to the drug reinforcer; (b) paired presentation of the drug reinforcer with another reinforcer (secondary reinforcement); (c) making the drug available during access to the daily food ration under food deprivation (prandial drinking); (d) induction of high fluid intakes by intermittent delivery of food pellets during food deprivation (schedule-induced polydipsia), (e) acclimatization to gradually increasing concentrations of drug, and (f) adulteration of the drug's taste by the addition of preferred flavors or sweeteners (e.g. sucrose or saccharin fading). More extensive discussions of procedures used to establish operant oral self-administration in rodents have been provided (Meisch and Carroll, 1987; Samson, 1987).

In addition to providing an opportunity to separate drug-seeking from drug ingestion, operant paradigms allow relatively precise characterization of within-session patterns of drug intake. Moreover, by manipulating the schedule of drug reinforcement, one can exercise experimental control over the pattern and total amount of drug intake. It is also possible that operant paradigms are more likely to provide the kinds of environmental stimulus conditions necessary to induce the compulsive, excessive drug intake observed in human drug addicts and alcoholics (Samson,

1987). However, when compared to home-cage drinking tasks, operant tasks may present disadvantages in terms of the need for specialized equipment, sophisticated behavioral expertise, and relatively long periods of training.

Operant self-administration via implanted cannulae

Intake-limiting gustatory effects can complicate interpretation of drug intake differences during oral self-administration, especially when making comparisons across genotypes that differ in sensitivity to gustatory stimuli. Such problems can be avoided by surgically implanting cannulae to allow direct intravenous, intragastric or intracranial delivery of drug contingent upon an operant response. Of these routes, the intravenous route has been used most often. Techniques for intravenous drug delivery to unrestrained rats were pioneered by Weeks and colleagues in the early, 1960s (e.g. Weeks, 1962) and summaries of current techniques and findings can be found in several recent reviews (e.g. Weeks and Collins, 1987; Yokel, 1987; Roberts and Goeders, 1989; Meisch and Lemaire, 1993). Of special interest in the present context is the increasing number of studies in which intravenous self-administration tasks have been used successfully to study genetic differences in mice (e.g. Carney et al., 1991; Grahame and Cunningham, 1995, 1997, 1998; Deroche et al., 1997; Rocha et al., 1998a, 1998b).

Development of intravenous self-administration in rodents generally requires less training than is needed to develop oral self-administration, although pretraining with a known reinforcer is sometimes used when the reinforcing efficacy of the target drug is unknown or weak. Other advantages of intravenous self-administration include the ability to specify the precise drug dose received by the animal and the ability to deliver drug independently of the animal's behavior. However, it is important to note that effects of non-contingent drug delivery may differ substantially from those of response-contingent drug delivery (e.g. Mutschler and Miczek, 1998).

Although technically demanding, intravenous studies can shed new light on findings obtained from drinking and oral operant studies. For example, a recent study has shown that DBA/2J mice, long characterized as alcohol avoiders in drinking paradigms, will learn to self-administer alcohol intravenously to intoxicating levels (Grahame and Cunningham, 1997). In fact, DBA mice acquired selective responding for the alcohol-reinforced response at the same rate as C57BL/6J mice, a strain well-known for its high intake and preference for alcohol in drinking tasks. In this case, the findings of the intravenous study support earlier suggestions that low drinking in DBA mice may be caused by their greater sensitivity to ethanol's aversive orosensory effects rather than insensitivity to postabsorptive pharmacological effects (Belknap et al., 1977).

Self-administration via implanted cannulae poses several unique challenges. It is clearly the most invasive and most demanding of the self-administration procedures. Highly specialized equipment and substantial surgical and behavioral expertise are required. Moreover, the overall duration of time that each subject can be studied is generally much shorter than in oral self-administration studies because of difficulties in maintaining cannula patency and avoiding infection. Nevertheless, as noted above, this approach may have advantages over oral self-administration in certain situations and can provide important new insights on determinants of drug intake.

Place conditioning

Self-administration tasks assess motivational effects of drugs by measuring drug intake or behaviors that allow drug intake. These tasks can be contrasted with place and taste conditioning tasks in which motivational effects of drugs are indexed by approach and contact or escape and avoidance from distinctive stimuli that have previously been paired with the drug's effects. Behavior measured in these tasks is often interpreted as a form of Pavlovian conditioning and it may have important implications for understanding the phenomena of drug craving and relapse to drug-taking after withdrawal (Cunningham, 1998). Place conditioning studies usually involve at least two phases. During the first phase, subjects receive one or more pairings of the drug (e.g. IP injection) with brief exposure (5–60 min) to a distinctive set of environmental stimuli (e.g. a box with salient visual or tactile cues). Each drug trial is matched by exposure to an alternative set of stimuli paired with exposure to the vehicle. During the second phase, which typically occurs without drug, subjects are given a choice between a spatial location containing the drug-paired stimuli and one or more other locations containing stimuli that have not been paired with drug. The amount of time spent in contact with drug-paired stimuli during this choice test is used to assess conditioned motivational effects of the drug. Longer contact with drug-paired stimuli is generally interpreted as evidence of a drug's rewarding effects whereas shorter contact with drug-paired stimuli is thought to be indicative of aversive drug effects.

The place conditioning literature reveals substantial variation in the apparatus, experimental design, and detailed procedures used to study this phenomenon. Detailed summaries of critical variables, control procedures and critiques of various approaches to place conditioning can be found elsewhere (Bozarth, 1987; Van der Kooy, 1987; Carr et al., 1989; Hoffman, 1989; Swerdlow et al., 1989; Cunningham, 1993; Schechter and Calcagnetti, 1993; Bardo et al., 1995). Recently, the place conditioning paradigm has been increasingly used as a tool to characterize genetic differences in the motivational effects of various drugs in mice, including knockout mice (e.g. Cunningham et al., 1991, 1992; Seale and Carney, 1991; Miner et al., 1995; Matthes et al., 1996; Risinger et al., 1996; Hiroi et al., 1997; Maldonado et al., 1997; Miner, 1997; Takahashi et al., 1997; Chester et al., 1998; Sora et al., 1998). Place conditioning has also been used to identify provisional QTLs for ethanol-induced reward in BXD recombinant inbred mice (Cunningham, 1995).

Because place conditioning most often involves giving drug by injection (IP or SC), this task has the same general advantage as intravenous self-administration in being able to avoid complications related to gustatory variables. Like intravenous self-administration, place conditioning can be established rapidly and allows precise control over drug dose. However, place conditioning has the additional advantage of not requiring an invasive surgical procedure or maintenance of implanted cannulae. Moreover, because tests for place conditioning typically occur without drug, one can avoid interpretive problems caused by drug-induced sensory or motor deficits. Also, the place conditioning task can be used to study both positive and negative drug effects, including instances where the same drug appears to produce both effects (e.g. Fudala and Iwamoto, 1990; Cunningham et al., 1997). Such effects are not easily studied in self-administration tasks. As in the case of intravenous self-administration, place conditioning has offered new insights on old findings from the drinking literature. For example, alcohol-nonpreferring DBA mice have been shown to develop stronger alcohol-induced conditioned place preference than alcohol-preferring C57 mice (Cunningham et al., 1992). Consistent with the previously mentioned IV self-administration study in these strains (Grahame and Cunningham, 1997), this finding argues against interpretations of low drinking in DBA mice based on insensitivity to postabsorptive rewarding effects of ethanol intoxication.

Although place conditioning has many advantages, it also has certain disadvantages. For example, in contrast to operant tasks, place conditioning is not well suited for studying multiple within-subject manipulations. Thus, place-conditioning studies often require relatively large between-subjects experimental designs. Another complication in place conditioning studies is the possibility that high levels of locomotor activity during testing may interfere with the expression of conditioned preference or aversion (see Cunningham, 1995). This issue must be given special consideration in genetic studies where strain differences in place conditioning may be jointly determined by genetic differences in sensitivity to conditioned drug effects and by genetic differences in basal activity levels. Another criticism often leveled at place conditioning is that it does not provide a direct measure of the primary reinforcing (or punishing) effects of a drug, i.e. place conditioning assesses secondary (conditioned) drug effects. However, this feature does not detract from use of place conditioning to study the motivational effects of drugs and their role in drug seeking behavior. In fact, the use of separate training and test phases enables one to address the possibility that the neurobiological bases for conditioned drug effects differ from those underlying primary drug effects (see Cunningham et al., 1995). Despite its limitations, place conditioning is a valuable tool for complementing and extending the information derived from drinking and operant self-administration studies.

Taste conditioning

The taste conditioning procedure is methodologically similar to place conditioning in that a drug's motivational effects are assessed by measuring behavioral changes to a stimulus that has previously been paired with the drug. The primary differences are that taste conditioning usually involves a novel-tasting fluid instead of an exteroceptive stimulus and the effect of conditioning is determined by measuring subsequent ingestion of the taste solution rather than a spatial choice between external stimuli. In the typical procedure, subjects are initially adapted for several days to a restricted schedule of fluid access. The duration of daily fluid access is selected to avoid severe dehydration, and is usually longer in mice (1–2 h) than in rats (10–30 min). After adaptation to this schedule, subjects receive one or more conditioning trials in which access to a novel taste solution (e.g. saccharin) is followed immediately by injection of drug. Subsequent taste-drug pairings are usually separated by 2 or more days to allow recovery from effects of the previous drug exposure, with access to water only on intervening days. The effects of taste-drug pairings are most often measured simply by recording the volume of taste solution consumed on consecutive conditioning trials. In some cases, however, taste conditioning is assessed in a two-bottle choice test (e.g. paired taste vs water) or during an extinction procedure in which subjects have repeated access to the taste solution but no longer receive drug injections. More detailed discussions of methodological and conceptual issues in the study of taste conditioning can be found elsewhere (Braveman and Bronstein, 1985; Goudie, 1987; Cunningham, 1993).

The most common outcome of taste-drug pairings is reduced intake of the paired taste in experimental subjects compared to control subjects (i.e. subjects who receive taste-vehicle pairings or unpaired taste and drug exposures). This phenomenon, called "conditioned taste aversion", has often been attributed to an association between the taste and aversive motivational effects of drug exposure. While this analysis seems reasonable in the case of emetic and other illness-inducing drugs, it does not always seem appropriate in the case of abused drugs that have positive motivational effects as indexed by self-administration or place conditioning tasks (Goudie, 1987). For example, studies of conditioned taste reactivity responses produced by abused drugs do not show conditioned aversive somatic and orofacial responses like those elicited by tastes paired with illness-inducing drugs (Parker, 1995). Such findings have been interpreted as suggesting that the reduced intake of tastes paired with abused drugs is not the result of an association with aversive motivational effects, but is caused by some other mechanism. One hypothesis is that "drug shyness" or novelty of the rewarding effects of abused drugs is responsible for taste avoidance induced by abused drugs (Hunt and Amit, 1987). Another recent theory, based on the notion of reward comparison, suggests that abused drugs produce taste avoidance because the initial value of the taste is reduced by anticipation of the impending availability of a highly rewarding drug (Grigson, 1997).

Studies of genetic influences on taste conditioning may eventually shed light on theoretical approaches to the paradox posed by the conditioning of taste avoidance with abused drugs. For example, a recent study of ethanol-induced taste conditioning in 20 of the BXD recombinant inbred strains found no significant genetic correlations between magnitude of taste avoidance (Risinger and Cunningham, 1998) and either ethanol intake or preference in a drinking task (Phillips et al., 1994) or ethanol-induced conditioned place preference (Cunningham, 1995). Thus, in the case of ethanol, there is little support for the hypothesis that a common motivational effect mediates taste avoidance and behavior in either of the other tasks. Rather, the data seem more consistent with the suggestion that alcohol has aversive motivational effects that are independent of its rewarding motivational effects. In order to explain taste avoidance, one must argue that taste stimuli are selectively associated with ethanol's aversive effects, whereas external cues are more readily associated with the drug's rewarding effects (Reicher and Holman, 1977). Future studies must determine whether similar conclusions will apply to other abused drugs.

Summary

Methods for investigating drug dependence and withdrawal are in every respect as complex as those directed at any other class of behaviors. One area that seems underrepresented in the literature reviewed is the simultaneous consideration of multiple aspects of dependence and withdrawal. We have generally considered two broad classes of behavior, those related directly to withdrawal and those reflecting changes in drug-related reinforcement. In several examples, we have shown that even commonly-studied inbred mouse strains do not necessarily show patterns of behavior that seem "consistent" upon casual observation. The full array of neurobehavioral genetic as tools can be used to increase our understanding of how these various types of drug dependence-related behaviors derive from the underlying neurobiological changes during chronic drug administration. Increased knowledge of these relationships will in turn then allow us to understand the effects of targeted gene manipulations more directly.

Acknowledgments

Preparation of this chapter was supported by a grant from the Department of Veterans Affairs, and by NIH grants AA10760, DA05228, and AA07702.

References

American Psychiatric Association (1994) *Dignostic and Statistical Manual of Mental Disorders*, 4th Edn, (DSM-IV), American Psychiatric Association, Washington, DC.

Ballenger, J.C. and Post, R.M. (1978) Kindling as a model for alcohol withdrawal syndromes. *Br. J. Psychiat.*, 133: 1–14.

Bardo, M.T., Rowlett, J.K. and Harris, M.J. (1995) Conditioned place preference using opiate and stimulant drugs: A meta-analysis. *Neurosci. Biobehav. Rev.*, 19: 39–51.

Becker, H.C., Diaz-Granados, J.L. and Weathersby, R.T. (1997) Repeated ethanol withdrawal experience increases the severity and duration of subsequent withdrawal seizures in mice. *Alcohol*, 14: 319–326.

Belknap, J.K., Laursen, S.E. and Crabbe, J.C. (1987) Ethanol and nitrous oxide produce withdrawal-induced convulsions by similar mechanisms in mice. *Life Sci.*, 41: 2033–2040.

Belknap, J.K., Belknap, N.D., Berg, J.H. and Coleman, R. (1977) Preabsorptive vs postabsorptive control of ethanol intake in C57BL/6J and DBA/2J mice. *Behav. Genet.*, 7: 413–425.

Belknap, J.K., Crabbe, J.C., Riggan, J. and O'Toole, L.A. (1993a) Voluntary consumption of morphine in 15 inbred mouse strains. *Psychopharmacol.*, 112: 352–358.

Belknap, J.K., Crabbe, J.C. and Young, E.R. (1993b) Voluntary consumption of ethanol in 15 inbred mouse strains. *Psychopharmacol.*, 112: 503–510.

Belknap, J.K., Richards, S.P., O'Toole, L.A., Helms, M.L. and Phillips, T.J. (1997) Short-term selective breeding as a tool for QTL mapping: Ethanol preference drinking in mice. *Behav. Genet.*, 27: 55–66.

Berrettini, W.H., Ferraro, T.N., Alexander, R.C., Buchberg, A.M. and Vogel, W.H. (1994) Quantitative trait loci mapping of three loci controlling morphine preference using inbred mouse strains. *Nat. Genet.*, 7: 54–58.

Boisse, N.R. and Okamoto, M. (1978) Physical dependence to barbital compared to pentobarbital. III. Withdrawal characteristics. *J. Pharmacol. Exp. Ther.*, 204: 514–525.

Bozarth, M.A. (1987) Conditioned place preference: A parametric analysis using systemic heroin injections. In: M.A. Bozarth (Ed.), *Methods of Assessing the Reinforcing Properties of Abused Drugs*, Springer-Verlag, New York, pp. 241–273.

Braveman, N.S. and Bronstein, P. (1985) Experimental assessment and clinical applications of conditioned food aversions. *Ann. NY Acad. Sci.*, Vol. 443.

Buck, K.J., Metten, P., Belknap, J.K. and Crabbe, J.C. (1997) Quantitative trait loci involved in genetic predisposition to acute alcohol withdrawal in mice. *J. Neurosci.*, 17: 3946–3955.

Buck, K.J., Metten, P., Belknap, J.K. and Crabbe, J.C. (1999) Quantitative trait loci affecting risk for pentobarbital withdrawal map near alcohol withdrawal loci on mouse chromosomes 1, 4, and 11. *Mammal. Genome*, 10: 431–437..

Carney, J.M., Landrum, R.W., Cheng, M.S. and Seale, T.W. (1991) Establishment of chronic intravenous drug self-administration in the C57BL/6J mouse. *NeuroReport*, 2: 477–480.

Carr, G.D., Fibiger, H.C. and Phillips, A.G. (1989) Conditioned place preference as a measure of drug reward. In: J.M. Liebman and S.J. Cooper (Eds.), *Neuropharmacological Basis of Reward*, Oxford, New York, pp. 264–319.

Carr, L.G., Foroud, T., Bice, P., Gobbett, T., Ivashina, J., Edenberg, H., Lumeng, L. and Li, T.-K. (1998) A quantitative trait locus for alcohol consumption in selectively bred rat lines. *Alc.: Clin. Exper. Res.*, 22: 884–887.

Chapin, J.K. and Woodward, D.J. (1989) Ethanol withdrawal increases sensory responsiveness of single somatosensory cortical neurons in the awake, behaving rat. *Alc.: Clin. Exper. Res.*, 13: 8–14.

Chester, J.A., Risinger, F.O. and Cunningham, C.L. (1998) Ethanol reward and aversion in mice bred for sensitivity to ethanol withdrawal. *Alc.: Clin. Exper. Res.*, 22: 468–473.

Crabbe, J.C. and Harris, R.A. (1991) *The Genetic Basis of Alcohol and Drug Actions*, Plenum Press, New York.

Crabbe, J.C., Belknap, J.K. and Buck, K.J. (1994) Genetic animal models of alcohol and drug abuse. *Science*, 264: 1715–1723.

Crabbe, J.C., Janowsky, J.S., Young, E.R. and Rigter, H. (1980) Handling induced convulsions in twenty inbred strains of mice. *Subst. Alcohol Actions Misuse*, 1: 159–163.

Crabbe, J.C., Kosobud, A., Young, E.R. and Janowsky, J.S. (1983) Polygenic and single-gene determination of responses to ethanol in BXD/Ty recombinant inbred strains. *Neurobehav. Toxicol. Teratol.*, 5: 181–187.

Crabbe, J.C., Merrill, C.D. and Belknap, J.K. (1991a) Acute dependence on depressant drugs is determined by common genes in mice. *J. Pharmacol. Exp. Ther.*, 257: 663–667.

Crabbe, J.C., Merrill, C.D. and Belknap, J.K. (1991b) Effects of convulsants on handling-induced convulsions in mice selected for ethanol withdrawal severity. *Brain Res.*, 550: 1–6

Crawshaw, L.I., O'Connor, C.S., Crabbe, J.C. and Hayteas, D.L. (1994) Temperature regulation in mice during withdrawal from ethanol dependence. *Am. J. Physiol.*, 267: R929–R934.

Criswell, H.E. and Breese, G.R. (1993) Similar effects of ethanol and flumazenil on acquisition of a shuttle-box avoidance response during withdrawal from chronic ethanol treatment. *Br. J. Pharmacol.*, 110: 753–760.

Cunningham, C.L. (1993) Pavlovian drug conditioning. In: F. Van Haaren (Ed.), *Methods in Behavioral Pharmacology*, Elsevier, Amsterdam, pp. 349–381.

Cunningham, C.L. (1995) Localization of genes influencing ethanol-induced conditioned place preference and locomotor activity in BXD recombinant inbred mice. *Psychopharmacol.*, 120: 28–41.

Cunningham, C.L. (1998) Drug conditioning and drug-seeking behavior. In: W. O'Donohue (Ed.), *Learning and Behavior Therapy*, Allyn and Bacon, Boston, pp. 518–544.

Cunningham, C.L., Dickinson, S.D. and Okorn, D.M. (1995) Naloxone facilitates extinction but does not affect acquisition or expression of ethanol-induced conditioned place preference. *Exper. Clin. Psychopharmacol.*, 3: 330–343.

Cunningham, C.L., Hallett, C.L., Niehus, D.R., Hunter, J.S., Nouth, L. and Risinger, F.O. (1991) Assessment of ethanol's hedonic effects in mice selectively bred for sensitivity to ethanol-induced hypothermia. *Psychopharmacol.*, 105: 84–92.

Cunningham, C.L. and Niehus, J.S. (1997) Flavor preference conditioning by oral self-administration of ethanol. *Psychopharmacol.*, 134: 293–302.

Cunningham, C.L., Niehus, D.R., Malott, D.H. and Prather, L.K. (1992) Genetic differences in the rewarding and activating effects of morphine and ethanol. *Psychopharmacol.*, 107: 385–393.

Cunningham, C.L., Okorn, D.M. and Howard, C.E. (1997) Interstimulus interval determines whether ethanol produces conditioned place preference or aversion in mice. *Anim. Learn. Behav.*, 25: 31–42.

Deroche, V., Caine, S.B., Heyser, C.J., Polis, I., Koob, G.F. and Gold, L.H. (1997) Differences in the liability to self-administer intravenous cocaine between C57BL/6 x SJL and BALB/cByJ mice. *Pharmacol. Biochem. Behav.*, 57: 429–440.

Dugovic, C., Meert, T.F., Ashton, D. and Clincke, G.H. (1992) Effects of ritanserin and chlordiazepoxide on sleep-wakefulness alterations in rats following chronic cocaine treatment. *Psychopharmacol.*, 108: 263–270.

Elmer, G.I., Meisch, R.A. and George, F.R. (1986) Oral ethanol reinforced behavior in inbred mice. *Pharmacol. Biochem. Behav.*, 24: 1417–1421.

Elmer, G.I., Meisch, R.A. and George, F.R. (1987) Mouse strain differences in operant self-administration of ethanol. *Behav. Genet.*, 17: 439–451.

Elmer, G.I., Meisch, R.A., Goldberg, S.R. and George, F.R. (1988) Fixed-ratio schedules of oral ethanol self-administration in inbred mouse strains. *Psychopharmacol.*, 96: 431–436.

Faingold, C.L. and Riaz, A. (1994) Increased responsiveness of pontine reticular formation neurons associated with audiogenic seizure susceptibility during ethanol withdrawal. *Brain Res.*, 663: 69–76.

Finn, D.A. and Crabbe, J.C. (1999) Chronic ethanol differentially alters susceptibility to chemically-induced convulsions in Withdrawal Seizure-Prone and -Resistant mice. *J. Pharmacol. Exp. Ther.*, 288: 782–790.

Friedman, H.J. (1980) Assessment of physical dependence on and withdrawal from ethanol in animals. In: H. Rigter and J.C. Crabbe (Eds.), *Alcohol Tolerance and Dependence*, Elsevier/North Holland Biomedical Press, Amsterdam, pp. 93–121.

Fudala, P.J. and Iwamoto, E.T. (1990) Conditioned aversion after delay place conditioning with amphetamine. *Pharmacol. Biochem. Behav.*, 35: 89–92.

Gallaher, E.J., Henauer, S.A., Jacques, C.J. and Hollister, L.E. (1986) Benzodiazepine dependence in mice after ingestion of drug-containing food pellets. *J. Pharmacol. Exp. Ther.*, 237: 462–467.

Gauvin, D.V. and Holloway, F.A. (1992) Historical factors in the development of EtOH-conditioned place preference. *Alcohol*, 9: 1–7.

Goldstein, D.B. (1972) Relationship of alcohol dose to intensity of withdrawal signs in mice. *J. Pharmacol. Exp. Ther.*, 180: 203–215.

Goldstein, D.B. (1973) Inherited differences in intensity of alcohol withdrawal reactions in mice. *Nature*, 245: 154–156.

Goldstein, D.B. and Pal, N. (1971) Alcohol dependence produced in mice by inhalation of ethanol: Grading the withdrawal reaction. *Science*, 172: 288–290.

Gonzalez, L.P. (1993) Sensitivity to strychnine seizures is unaltered during ethanol withdrawal. *Alcohol. Clin. Exp. Res.*, 17: 1029–1034.

Goudie, A.J. (1987) Aversive stimulus properties of drugs: The conditioned taste aversion paradigm. In: A.J. Greenshaw and C.T. Dourish (Eds.), *Experimental Psychopharmacology: Concepts and Methods*, Humana Press, Clifton, NJ, pp. 341–391.

Grahame, N.J. and Cunningham, C.L. (1997) Intravenous ethanol self-administration in C57BL/6J and DBA/2J mice. *Alc: Clin. Exper. Res.*, 21: 56–62.

Grahame, N.J., Low, M.J. and Cunningham, C.L. (1998) Intravenous self-administration of ethanol in beta-endorphin-deficient mice. *Alc.: Clin. Exper. Res.*, 22: 1093–1098.

Grahame, N.J., Phillips, T.J., Burkhart-Kasch, S. and Cunningham, C.L. (1995) Intravenous cocaine self-administration in the C57BL/6J mouse. *Pharmacol. Biochem. Behav.*, 51: 827–834.

Grant, K.A., Snell, L.D., Rogawski, M.A., Thurkauf, A. and Tabakoff, B. (1992) Comparison of the effects of the uncompetitive *N*-methyl-D-aspartate antagonist (+−+)-5-aminocarbonyl-10,11-dihydro-5H-dibenzo[a,d]cyclohepten-5,10-imine (ADCI) with its structural analogs dizocilpine (MK-801) and carbamazepine on ethanol withdrawal seizures. *J. Pharmacol. Exp. Ther.*, 260: 1017–1022.

Grigson, P.S. (1997) Conditioned taste aversions and drugs of abuse: A reinterpretation. *Behav. Neurosci.*, 111: 129–136.

Hiroi, N., Brown, J.R., Haile, C.N., Ye, H., Greenberg, M.E. and Nestler, E.J. (1997) *FosB* mutant mice: Loss of chronic cocaine induction of *Fos*-related proteins and heightened sensitivity to cocaine's psychomotor and rewarding effects. *Proc. Nat. Acad. Sci.*, 94: 10397–10402.

Hoffman, D.C. (1989) The use of place conditioning in studying the neuropharmacology of drug reinforcement. *Brain Res. Bull.*, 23: 373–387.

Hunt, T. and Amit, Z. (1987) Conditioned taste aversion induced by self-administered drugs: Paradox revisited. *Neurosci. Biobehav. Rev.*, 11: 107–130.

Hutchins, J.B., Allen, D.L., Cole-Harding, L.S. and Wilson, J.R. (1981) Behavioral and physiological measures for studying ethanol dependence in mice. *Pharmacol. Biochem. Behav.*, 15: 55–59.

Kakihana, R. and Moore, J.A. (1976) Circadian rhythm of corticosterone in mice: the effect of chronic consumption of alcohol. *Psychopharmacol.*, 46: 301–305.

Koob, G.F. and Bloom, F.E. (1988) Cellular and molecular mechanisms of drug dependence. *Science*, 242: 715–723.

Kosobud, A.E., Cross, S.J. and Crabbe, J.C. (1992) Neural sensitivity to pentylenetetrazol convulsions in inbred and selectively bred mice. *Brain Res.*, 592: 122–128.

Lê, A.D., Ko, J., Chow, S. and Quan, B. (1994) Alcohol consumption by C57BL/6, BALB/c and DBA/2 mice in a limited access paradigm. *Pharmacol. Biochem. Behav.*, 47: 375–378.

Li, T.-K., Lumeng, L. and Doolittle, D.P. (1993) Selective breeding for alcohol preference and associated responses. *Behav. Genet.*, 23: 163–170.

Majchrowicz, E.(1975) Induction of physical dependence upon ethanol and the associated behavioral changes in rats. *Psychopharmacol.*, 43: 245–254.

Maldonado, R., Saiardi, A., Valverde, O., Samad, T.A., Roques, B.P. and Borrelli, E. (1997) Absence of opiate rewarding effects in mice lacking dopamine D2 receptors. *Nature*, 388: 586–589.

Matthes, H.W.D., Maldonado, R., Simonin, F., Valverde, O., Slowe, S., Kitchen, I., Befort, K., Dierich, A., LeMeurs, M., Dolle, P., Tzavara, E., Hanoune, J., Roques, B.P. and Kieffer, B.L. (1996) Loss of morphine-induced analgesia, reward effect and withdrawal symptoms in mice lacking the m-opioid receptor gene. *Nature*, 383: 819–823.

McClearn, G.E. (1968) The use of strain rank orders in assessing equivalence of techniques. *Behav. Res. Meth. Instr. Comput.*, 1: 49–51.

McClearn, G.E. and Rodgers, D.A. (1959) Differences in alcohol preference among inbred strains of mice. *Quart. J. Stud. Alc.*, 20: 691–695.

McClearn, G.E., Wilson, J.R., Petersen, D.R. and Allen, D.L. (1982) Selective breeding in mice for severity of the ethanol withdrawal syndrome. *Subst. Alcohol Actions Misuse*, 3: 135–143.

McQuarrie, D.G. and Fingl, E. (1959) Effects of single doses and chronic administration of ethanol on experimental seizures in mice. *J. Pharmacol. Exp. Ther.*, 124: 264–271.

Meisch, R.A. (1977) Ethanol self-administration: Infrahuman studies. In: T. Thompson and P. Dews (Eds.), *Advances in Behavioral Pharmacology*, Academic Press, New York, pp. 35–84.

Meisch, R.A. and Carroll, M.E. (1987). Oral drug self-administration: Drugs as reinforcers. In: M.A. Bozarth (Ed.), *Methods of Assessing the Reinforcing Properties of Abused Drugs*, Springer-Verlag, New York, pp. 143–160.

Meisch, R.A. and Lemaire, G.A. (1993) Drug self-administration. In: F. Van Haaren (Ed.), *Methods in Behavioral Pharmacology*, Elsevier Science, Amsterdam, pp. 257–300.

Metten, P. and Crabbe, J.C. (1994) Common genetic determinants of severity of acute withdrawal from ethanol, pentobarbital and diazepam in inbred mice. *Behav. Pharmacol.*, 5: 533–547.

Metten, P. and Crabbe, J.C. (1995) Dependence and withdrawal. In: R.A. Deitrich and V.G. Erwin (Eds.), *Pharmacological Effects of Ethanol on the Nervous System*, CRC Press, Boca Raton, FL, pp. 269–290.

Miner, L.L. (1997) Cocaine reward and locomotor activity in C57BL/6J and 129/SvJ inbred mice and their F_1 cross. *Pharmacol. Biochem. Behav.*, 58: 25–30.

Miner, L.L., Drago, J., Chamberlain, P.M., Donovan, D. and Uhl, G.R. (1995) Retained cocaine conditioned place preference in D1 receptor deficient mice. *NeuroReport*, 6: 2314–2316.

Mutschler, N.H. and Miczek, K.A. (1998) Withdrawal from a self-administered or non-contingent cocaine binge: differences in ultrasonic distress vocalizations in rats. *Psychopharmacol.*, 136: 402–408.

Myers, R.D. and Veale, W.L. (1972) The determinants of alcohol preference in animals. In: B. Kissin and H. Begleiter (Eds.), *Biology of Alcoholism*, Plenum Press, New York, pp. 131–167.

Parker, L.A. (1995) Rewarding drugs produce taste avoidance, but not taste aversion. *Neurosci. Biobehav. Rev.*, 19: 143–151.

Phillips, T.J., Belknap, J.K., Buck, K.J. and Cunningham, C.L. (1998) Genes on mouse chromosomes 2 and 9 determine variation in ethanol consumption. *Mammal. Genome*, 9: 936–941.

Phillips, T.J., Crabbe, J.C., Metten, P. and Belknap, J.K. (1994) Localization of genes affecting alcohol drinking in mice. *Alc.: Clin. Exper. Res.*, 18: 931–941.

Phillips, T.J., Roberts, A.J. and Lessov, C.N. (1997) Behavioral sensitization to ethanol: genetics and the effects of stress. *Pharmacol. Biochem. Behav.*, 57: 487–493.

Pinelli, A. and Trivulzio, S. (1997) Quantitative evaluation of opioid withdrawal signs in rats repeatedly treated with morphine and injected with naloxone, in the absence or presence of the antiabstinence agent clonidine. *J. Pharmacol. Toxicol. Meth.*, 38: 117–131.

Prather, P.L. and Lal, H. (1992) Protracted withdrawal: sensitization of the anxiogenic response to cocaine in rats concurrently treated with ethanol. *Neuropsychopharmacol.*, 6: 23–29.

Rassnick, S., Koob, G.F. and Geyer, M.A. (1992) Responding to acoustic startle during chronic ethanol intoxication and withdrawal. *Psychopharmacol.*, 106: 351–358.

Reicher, M.A. and Holman, E.W. (1977) Location preference and flavor aversion reinforced by amphetamine in rats. *Anim. Learn. Behav.*, 5: 343–346.

Ripley, T.L., Whittington, M.A., Butterworth, A.R. and Little, H.J. (1996) Ethanol withdrawal hyperexcitability in vivo and in isolated mouse hippocampal slices. *Alcohol Alcohol.*, 31: 347–357.

Risinger, F.O., Bormann, N.M. and Oakes, R.A. (1996) Reduced sensitivity to ethanol reward, but not ethanol aversion in mice lacking 5-HT1b receptors. *Alc.: Clin. Exper. Res.*, 20: 1401–1405.

Risinger, F.O., Brown, M.M., Doan, A.M., Oakes, R.A. (1998) Mouse strain differences in oral operant ethanol reinforcement under continuous access conditions. *Alc.: Clin. Exper. Res.*, 22: 677–684.

Risinger, F.O. and Cunningham, C.L. (1998) Ethanol-induced conditioned taste aversion in BXD recombinant inbred mice. *Alc.: Clin. Exper. Res.*, 22: 1234–1244.

Roberts, D.C.S. and Goeders, N. (1989) Drug self-administration: Experimental methods and determinants. In: A.A. Boulton, G.B. Baker and A.J. Greenshaw (Eds.), *Neuromethods 13: Psychopharmacology*, Humana Press, Clifton, NJ, pp. 349–398.

Robinson, T.E. and Berridge, K.C. (1993) The neural basis of drug craving: an incentive-sensitization theory of addiction. *Brain Res. Brain Res. Rev.*, 18: 247–291.

Rocha, B.A., Odom, L.A., Barron, B.A., Ator, R., Wild, S.A. and Forster, M.J. (1998a) Differential responsiveness to cocaine in C57BL/6J and DBA/2J mice. *Psychopharmacol.*, 138: 82–88.

Rocha, B.A., Scearce-Levie, K., Lucas, J.J., Hiroi, N., Castanon, N., Crabbe, J.C., Nestler, E.J. and Hen, R. (1998b) Increased vulnerability to cocaine in mice lacking the serotonin-1B receptor. *Nature*, 393: 175–178.

Rodriguez, L.A., Plomin, R., Blizard, D.A., Jones, B.C. and McClearn, G.E. (1995) Alcohol acceptance, preference, and sensitivity in mice. II. Quantitative trait loci mapping analysis using BXD recombinant inbred strains. *Alc.: Clin. Exper. Res.*, 19: 367–373.

Samson, H.H. (1987) Initiation of ethanol-maintained behavior: A comparison of animal models and their implication to human drinking. In: T. Thompson, P.B. Dews and J.E. Barrett (Eds.), *Neurobehavioral Pharmacology*, Vol. 6, Lawrence Erlbaum Associates, Hillsdale, NJ, pp. 221–248.

Samson, H.H. and Hodge, C.W. (1996) Neurobehavioral regulation of ethanol intake. In: R.A. Deitrich and V.G. Erwin, *Pharmacological Effects of Ethanol on the Nervous System*, CRC Press, Boca Raton, FL, pp. 203–226.

Samson, H.H., Schwarz-Stevens, K., Tolliver, G.A., Andrews, C.M. and Files, F.J. (1992) Ethanol drinking patterns in a continuous-access operant situation: effects of ethanol concentration and response requirements. *Alcohol*, 9: 409–414.

Schechter, M.D. and Calcagnetti, D.J. (1993) Trends in place preference conditioning with a cross-indexed bibliography: 1957–1991. *Neurosci. Biobehav. Rev.*, 17: 21–41.

Schindler, C.W., Persico, A.M., Uhl, G.R. and Goldberg, S.R. (1994) Behavioral assessment of high-dose amphetamine withdrawal: importance of training and testing conditions. *Pharmacol. Biochem. Behav.*, 49: 41–46.

Schulteis, G., Markou, A., Gold, L.H., Stinus, L. and Koob, G.F.(1994) Relative sensitivity to naloxone of multiple indices of opiate withdrawal: a quantitative dose-response analysis. *J. Pharmacol. Exp. Ther.*, 271: 1391–1398.

Seale, T.W. and Carney, J.M. (1991) Genetic determinants of susceptibility to the rewarding and other behavioral actions of cocaine. *J. Addict. Dis.*, 10: 141–162.

Segal, D.S. and Kuczenski, R. (1997) Repeated binge exposures to amphetamine and methamphetamine: behavioral and neurochemical characterization. *J. Pharmacol. Exp. Ther.* 282: 561–573.

Snell, L.D., Szabo, G., Tabakoff, B. and Hoffman, P.L. (1996) Gangliosides reduce the development of ethanol dependence without affecting ethanol tolerance. *J. Pharmacol. Exp. Ther.*, 279: 128–136.

Sora, I., Wichems, C., Takahashi, N., Li, X.-F., Zeng, Z., Revay, R., Lesch, K.-P., Murphy, D.L. and Uhl, G.R. (1998) Cocaine reward models: Conditioned place preference can be established in dopamine- and in serotonin-transporter knockout mice. *Proc. Nat. Acad. Sci.*, 95: 7699–7704.

Stinus, L., Robert, C., Karasinski, P. and Limoge, A. (1998) Continuous quantitative monitoring of spontaneous opiate withdrawal: locomotor activity and sleep disorders. *Pharmacol. Biochem. Behav.*, 59: 83–89.

Swerdlow, N.R., Gilbert, D. and Koob, G.F. (1989) Conditioned drug effects on spatial preference: Critical evaluation. In: A.A. Boulton, G.B. Baker and A.J. Greenshaw (Eds.), *Neuromethods 13: Psychopharmacology*, Humana Press, Clifton, NJ, pp. 399–446.

Tabakoff, B. and Hoffman, P.L. (1988) A neurobiological theory of alcoholism. In: C.D. Chaudron and D.A. Wilkinson (Eds.), *Theories on Alcoholism*, Addiction Research Foundation, Toronto, pp. 29–72.

Takahashi, N., Miner, L.L., Sora, I., Ujike, H., Revay, R.S., Kostic, V., Jackson-Lewis, V., Przedborski, S. and Uhl, G.R. (1997) VMAT2 knockout mice: Heterozygotes display reduced amphetamine-conditioned reward, enhanced amphetamine locomotion, and enhanced MPTP toxicity. *Proc. Nat. Acad. Sci.*, 94: 9938–9943.

Tarantino, L.M., McClearn, G.E., Rodriguez, L.A. and Plomin, R. (1998) Confirmation of quantitative trait loci for alcohol preference in mice. *Alc.: Clin. Exper. Res.*, 22: 1099–1105.

Terdal, E.S. and Crabbe, J.C. (1994) Indexing withdrawal in mice: Matching genotypes for exposure in studies using ethanol vapor inhalation. *Alcohol. Clin. Exp. Res.*, 18: 542–547.

Uhl, G.R., Elmer, G.I., LaBuda, M.C. and Pickens, R.W. (1995) Genetic influences in drug abuse. In F.E. Bloom and D.J. Kupfer (Eds.), *Psychopharmacology: The Fourth Generation of Progress*, Raven Press Ltd., NY, pp. 1793–1806.

Van der Kooy, D. (1987) Place conditioning: A simple and effective method for assessing the motivational properties of drugs. In: M.A. Bozarth (Ed.), *Methods of Assessing the Reinforcing Properties of Abused Drugs*, Springer-Verlag, New York, pp. 229–240.

Walker, D.W., Zornetzer, S.F. (1974) Alcohol withdrawal in mice: electroencephalographic and behavioral correlates. *Electroencephalogr. Clin. Neurophysiol.*, 36: 233–243.

Waller, M.B., McBride, W.J., Lumeng, L. and Li, T.-K. (1982) Induction of dependence on ethanol by free-choice drinking in alcohol-preferring rats. *Pharmacol. Biochem. Behav.*, 16: 501–507.

Waller, M.B., McBride, W.J., Gatto, G.J., Lumeng, L. and Li, T.-K. (1984) Intragastric self-infusion of ethanol by ethanol-preferring and -nonpreferring lines of rats. *Science*, 225: 78–80.

Watson, W.P. and Little, H.J. (1995) Identification of distinct components, with different time courses, of the changes in response to convulsive stimuli during ethanol withdrawal. *J. Pharmacol. Exp. Ther.*, 272: 876-884.

Weeks, J.R. (1962) Experimental morphine addiction: Method for automatic intravenous injections in unrestrained rats. *Science*, 138: 143–144.

Weeks, J.R. and Collins, R.J. (1987) Screening for drug reinforcement using intravenous self-administration in the rat. In: M.A. Bozarth (Ed.), *Methods of Assessing the Reinforcing Properties of Abused Drugs*, Springer-Verlag, New York, pp. 35–43.

Yokel, R.A. (1987) Intravenous self-administration: Response rates, the effects of pharmacological challenges, and drug preference. In: M.A. Bozarth (Ed.), *Methods of Assessing the Reinforcing Properties of Abused Drugs*, Springer-Verlag, New York, pp. 1–33.

W.E. Crusio and R.T. Gerlai (Eds.)
Handbook of Molecular-Genetic Techniques for Brain and Behavior Research (Techniques in the Behavioral and Neural Sciences, Vol. 13)

CHAPTER 4.6

Evaluating anxiety in rodents

Jacqueline N. Crawley

Section on Behavioral Neuropharmacology, Experimental Therapeutics Branch, Intramural Research Program, National Institute of Mental Health, Building 10, Room 4D11, Bethesda, MD 20892-1375, USA

Introduction

Excellent rodent models of human generalized anxiety symptoms are available for analyzing the behavioral phenotype of mutant mice. While behavioral neuroscientists refrain from anthropomorphizing, since the human investigator cannot know whether a rodent experiences anxiety in the same way as humans, the natural behavioral repertoire of rodents does include a variety of behaviors which have been interpreted as anxiety-like. Approach-avoidance behaviors in rodents provide good measures of responses to conflicts in the natural environment and in the laboratory. Ethologists and operant psychologists have designed quantitative measures for scoring conflict behaviors in rats and mice. This chapter describes several of the best available methods for evaluating anxiety in rodents. Paradigms presented in this chapter are those which are widely used in basic research and in drug development, and which are well-validated in terms of specificity for anxiety-like behaviors, selectivity for compounds that show anxiolytic activity in humans, and reproducibility across many laboratories. Applicability to transgenic and knockout mice is discussed for each paradigm.

Elevated plus maze

The elevated plus maze is a conflict test based on the tendency of rodents to explore a novel environment vs the tendency of rodents to avoid the aversive properties of a brightly-lit, open, elevated space. This paradigm was developed by File and coworkers (Pellow et al., 1985), based on previous observations by Montgomery (1958). The apparatus consists of four long alleys of the same size and shape, at right angles from each other, forming the shape of a plus (+), raised to a height of 40–60 cm from the floor. Two of the alleys are simple open runways, which allow the animal to see the edge, i.e. providing a visual cliff. Two of the alleys are closed runways, with high side walls, which provide an enclosed environment, allow the animal to huddle against the walls and corners, and prevent the animal from seeing the edge of the cliff. In addition, the closed arms may be painted black or made of black plexiglas, while the open arms are brightly lit by overhead illumination and/or painted white or made of clear plexiglas. The animal is placed in the center, at the junction of the four arms, and allowed to explore the maze during a 5 min test session. A video camera mounted from the ceiling records the animal's movements. The videotape is later scored by investigators uninformed of the treatment condition.

The critical independent variables are:

1. number of entries into the open arms;
2. number of entries in the closed arms;
3. time spent in the open arms;
4. time spent in the closed arms; and
5. time spent in the center.

The most commonly used index of anxiety in the elevated plus maze is percentage entry into the open arms:

$$\frac{\text{number of entries into the open arms}}{\text{number of total entries}} \times 100\%$$

This index represents the tendency of the animal to explore the aversive open, brightly lit, elevated environment, rather than remaining in the preferred enclosed, dark environment. Anxiolytic drugs increase the percentage of open arm entries (Handley and Mithani, 1984; Pellow et al., 1986; Rodgers et al., 1992; Handley and McBlane, 1993; Dawson and Tricklebank, 1995), while anxiogenic drugs decrease the percentage of open arm entries (Treit et al., 1993; Dawson et al., 1995). Percentage of time spent in the open arms is used as a corroborating index.

Since this paradigm requires the animal to locomote over a considerable distance, any treatments which affect motor functions will confound the interpretation of the percentage of open arm entries and percentage of time on the open arms. The elevated plus maze paradigm includes internal controls for locomotor activity: number of entries into the closed arms, and total number of entries. If a drug treatment or genetic mutation changes the total number of entries into both the open and closed arms of the plus maze, then the data cannot be interpreted in terms of an anxiolytic or anxiogenic effect of the treatment. Independent paradigms which specifically analyze motor behaviors are necessary in these cases (Dawson et al., 1995).

Mice perform similarly to rats on the elevated plus maze. Strain distribution analyses in the elevated plus maze indicate a genetic substrate (Rodgers and Cole, 1993; Trullas and Skolnick, 1993).

Elevated zero maze

The zero maze, or O-maze, a variant on the elevated plus maze, was designed to eliminate the time spent in the center area, a confounding problem inherent in the plus maze configuration (Shepherd et al., 1992). The O-maze contains the open and closed arms, but configures them alternately around the perimeter of a circular runway, elevated 65 cm above the floor. This paradigm has been validated for anxiolytic drugs (Shepherd et al, 1994), although it has not yet gained the popularity of the elevated plus maze for genetic studies in transgenic and knockout mice.

Light↔dark exploration

The light↔dark exploration task is a conflict test which is conceptually similar to the elevated plus maze and the elevated O-maze, based on the conflict between the tendency of rodents to explore a novel environment versus the aversive properties of a brightly-lit open field (Crawley and Goodwin, 1980; Crawley, 1981). This paradigm has also been called the black–white box (Hughes et al., 1990). The apparatus is a standard plastic rat cage, divided by a photocell-equipped border panel into a large, open area which is illuminated by an overhanging lamp, and a smaller, enclosed area, painted black. The photocell array quantitates the number of transitions between the light and dark chambers, and the time spent in the light and dark chambers, over a 10 min session. Number of transitions is the most sensitive independent variable. Anxiolytic drugs increase the number of transitions, conceptualized as shifting the propensity of the animal to exit the small, dark, enclosed chamber and initiate exploration of the brightly lit, open chamber (Crawley, 1981; Crawley et al., 1981; 1986; Hughes et al., 1990; Pierrefiche et al., 1993; Chaouloff et al., 1997).

This paradigm is less useful for anxiogenic drugs, since low levels of general locomotor activity, due to sedation, motor dysfunctions, or seizures will produce false positives (Crawley et al., 1984). New classes of drugs require independent testing for motor functions, in locomotor paradigms such as open field activity, to detect stimulant or sedative properties that could confound interpretation of the light↔dark transition measure.

The light↔dark exploration paradigm has two advantages which are particularly useful for behavioral phenotyping of transgenic and knockout mice: (a) It is completely automated, such that videotaping and observer scoring is not required; (b) it is highly sensitive to genotype. Inbred mouse strain distribution studies found that the C57BL/6J inbred strain of mice shows high numbers of transitions, and increased numbers of transitions in response to anxiolytics such as diazepam, whereas low numbers of transitions characterize the BALB/c and A/J inbred strains of mice (Crawley and Davis, 1982; Mathis et al., 1994). Further, quantitative trait loci analyses have linked light↔dark exploration levels to several chromosomal loci (Mathis et al., 1995). Light ↔dark exploration has been successfully used to reveal an anxiogenic phenotype of transgenic mice overexpressing corticotropin releasing factor (Heinrichs et al., 1997).

Social interaction

The social interaction test is similarly based on the unconditioned aversion which rodents show to brightly-lit open areas (File and Hyde, 1979; File, 1980). In a highly illuminated open field environment, rats will spend less time engaged in social behaviors than rats placed in a dimly illuminated environment. An additional factor in this task is familiarity vs unfamiliarity of the environment.

Pairs of previously isolated rats engage in sniffing, following, and grooming each other, over a five or ten minute test session. Behaviors are videotaped and scored by the investigator. Low levels of illumination, familiar environment, and anxiolytic drugs increase the number of social interactions (File and Hyde, 1979; File, 1980; Dunn and File, 1987; Higgins et al., 1988).

The social interaction paradigm has many of the advantages of the elevated plus maze, O-maze, and light↔dark exploration, including ethological relevance to the natural conflicts experienced by rodents. Issues of motor function and non-specific behavioral sedation and activation can confound the interpretation of social interaction results, as described above for the related conflict paradigms. Social interaction has the advantage of adding a new set of behavioral substrates, interactions with a conspecific, which may prove relevant to specific hypotheses about the functions of genes relevant to social behaviors. However, the social interaction paradigm has been less useful for mice (Sandra File, personal communication; Paylor and Crawley, unpublished data), possibly due to the higher levels of exploration directed to the novel environment in mice as compared to rats (DeAngelis and File, 1979; Lister and Hilakivi, 1988).

Defensive burying

Conditioned defensive burying measures the time spent burying an aversive object, e.g. a small, wire-wrapped metal prod which had previously delivered a single electric shock (Pinel and Treit, 1978). Rodents will cover the shock source with cage litter in a stereotyped manner. Anxiolytic drugs suppress conditioned defensive burying, reducing the mean duration of burying (Treit et al., 1981; Treit 1985). This paradigm has the advantage of requiring much less ambulatory locomotor activity than the elevated plus maze, elevated O-maze, light↔dark exploration, and social interaction paradigms. Again, treatments that produce behavioral sedation can produce false positives on this task, such that independent measurement of general activity in a separate open field apparatus may be required.

Open field

Open field activity is the oldest and simplest measure of rodent emotional behavior (Hall, 1936; Henderson, 1967). Exploratory locomotion, proximity to the walls, and number of fecal boli deposited, are quantitated in a brightly-lit open field. Early versions employed human observers and scoring of real-time behaviors or videotapes. Fully automated open field systems are now widely used. A photocell-equipped automated apparatus, available from several commercial sources, measures horizontal ambulatory activity and vertical rearing activity.

Spontaneous exploratory locomotion is measured over a five or ten minute session in a novel open field arena. Habituation to novelty is measured by the decline in activity over a session length of at least 1 h. Thigmotaxis is measured by comparing activity in the center vs activity at the perimeter of the open field. High perimeter activity with low center activity indicates high levels of anxiety. Open field activity provides a useful first approximation of normal vs abnormal exploratory activity. However, open field activity is not a specific measure of anxiety, and not as specific a measure of anxiolytic drug response, compared to the more specialized paradigms described in this chapter.

Inbred mouse strain distributions have been extensively described for open field behavors (De Fries et al., 1978; Crabbe 1986). Further, quantitative trait loci analyses and targeted gene mutations have linked open field activity levels to chromosomal loci (Flint et al., 1995; Mathis et al., 1995). Knockout mice deficient in the dopamine transporter show increased open field activity as a major behavioral phenotype (Giros et al., 1996).

Geller-Seifter and Vogel conflict tests

The Geller–Seifter and the Vogel paradigms are conflict tests using operant responding (Geller and Seifter, 1960; Vogel et al., 1985). The approach component is a food reward in the Geller–Seifter paradigm, and a water reward in the Vogel paradigm, with electric shock as the aversive component in both. The independent variable is the number of lever presses for the food or water reward, in conjunction with a mild shock delivered on a fixed ratio schedule. Anxiolytic drugs increase the number of lever presses, i.e. the number of shocks accepted, for the appetitive reward. Unpunished responding represents the internal control for non-specific drug effects. These two paradigms have been widely used for anxiolytic drug discovery (Corda et al., 1983; Bernard et al., 1985; Crawley et al., 1986; Koob et al., 1986; Bing et al., 1993). The Geller–Seifter paradigm has the advantage of stable performance baselines and repeated use of the same animal as its own control, although several weeks of shaping and training on the lever press task are required. The Vogel paradigm requires no prior training, as the water reward is delivered through the standard drinking tube. Ethical considerations have limited the use of these shock paradigms, such that the less aversive, more naturalistic conflict paradigms described above are more commonly used today.

The Geller–Seifter and Vogel paradigms have been primarily used in rats. The Maudsley Reactive and non-reactive genetic strains of rats show differential conditioned suppression of drinking in the Vogel test, indicating a genetic determinant in this paradigm (Commissaris et al., 1986).

Applications to molecular genetics

Behavioral phenotyping of transgenic and knockout mice requires the application of established, well-validated behavioral paradigms (Crawley and Paylor, 1997). The anxiety-related paradigms described above are good choices for testing hypotheses about genes mediating anxiety. Since these tests can be influenced to varying degrees by factors unrelated to anxiety, the best approach is to use two or three different anxiety tests, which focus on different sensory and motor modalities, to characterize a new knockout or transgenic mouse. Corroboration of an anxiogenic profile across several different paradigms will greatly strengthen the interpretation of a behavioral phenotype for a mutant mouse as low or high on anxiety traits.

Equipment for most of the tests described in this chapter can be used for either mice or rats. The elevated plus maze and zero maze can be built with identical properties but smaller runways for mice, approximately half the length and width used for rats. The light↔dark exploration apparatus was developed for mice. Automated photocell-equipped open field equipment requires no modifications for use with mice. Operant tasks require a smaller food dispenser, along with smaller food pellets, or a smaller liquid dispenser. In addition, mice require levers which are smaller and more sensitive to the lighter touch of mice, or a photocell-equipped nose-poke unit which replaces the levers. All of these mouse adaptations to standard rat operant chambers are commercially available. As mentioned, the social interaction test is not useful for mice. Defensive burying has not yet been characterized for mice.

The tests described above generally require at least ten animals per treatment group. For knockout mouse studies, treatment groups should include $-/-$ knockouts, $+/-$ heterozygote littermates, and $+/+$ wild-type littermates. If gender differences are detected, at least ten of each gender of each genotype are required. For transgenics, treatment groups include the genotype expressing the transgene, heterozygotes or individuals with partial expression representing a gene-dose range, and wild-type littermate controls. When mutations are bred onto an outbred line of mice, much larger Ns may be required for each treatment group, based on the larger variation in genetic backgrounds among outbred mice. Wild-type littermates are the best controls for the potentially confounding variables related to genetic background. In addition, it is useful to conduct pilot experiments with commercially available mice from the parental strains used for the embryonic stem cells and for the breeding, to refine the conditions of the task as needed, in cases where the background genes of the parental strains contribute an unusual behavioral phenotype on the behavioral domain of interest (Crawley et al., 1997).

The availability of many good mouse models of anxiety offers a strong set of tools to molecular geneticists and behavioral neuroscientists interested in the genetic substrates of anxiety-related behaviors (Crawley and Paylor, 1997). The descriptions presented in this chapter represent a small subset, distilled from the long history of animal models of anxiety. The publications cited in this chapter provide a recommended sampling of the voluminous literature on anxiolytic drug treatments and on anxiety-related behaviors in genetic strains of animals (Crawley, 1989; Crawley et al., 1997; Crawley and Paylor, 1997; File, 1997). Collaboration with an established behavioral laboratory specializing in anxiety models will ensure the optimal application of these paradigms to the growing field of molecular genetics.

References

Bernard, P.S., Bennett, D.A., Pastor, G., Yokoyama, N. and Liebman, J.M. (1985) CGS 9896: Agonist-antagonist benzodiazepine receptor activity revealed by anxiolytic, anticonvulsant and muscle relaxation assessment in rodents. J. Pharmacol. Exper. Ther., 235(1): 98–105.

Bing, O., Möller, C., Engel, J.A., Söderpalm, B. and Heilig, M. (1993) Anxiolytic-like action of centrally administered galanin. *Neurosci. Lett.*, 164: 17–20.

Chaouloff, F., Durand, M. and Mormede, P. (1997) Anxiety- and activity-related effects of diazepam and chlordiazepoxide in the rat light/dark and dark/light tests. *Behav. Brain Res.*, 85: 27–35.

Commissaris, R.L., Harrington, G.M., Ortiz, A.M. and Altman, H.J. (1986) Maudsley reactive and non-reactive rat strains: Differential performance in a conflict test. *Physiol. Behav.*, 38: 291–294.

Corda, M.G., Blaker, W.D., Mendelson, W.B., Guiddotti, A. and Costa, E. (1983) β-carbolines enhance shock induced suppression of drinking in rats. *Proc. Natl. Acad. Sci. USA*, 80: 2072–2076.

Crabbe, J.C. (1986) Genetic differences in locomotor activity in mice. *Pharmacol. Biochem. Behav.*, 25: 289–292.

Crawley, J.N. (1989) Animal models of anxiety, *Curr. Opin. Psych.*, 2: 773–776.

Crawley, J.N. (????) Neuropharmacological specificity of a simple animal model for the behavioral actions of benzodiazepines. *Pharmacol. Biochem. Behav.*, 15: 695–699.

Crawley, J.N. and Davis, L.G. (1982) Baseline exploratory activity predicts anxiolytic responsiveness to diazepam in five mouse strains. *Brain Res. Bull.*, 8: 609–612.

Crawley, J.N. and Goodwin, F.K. (1980) Preliminary report of a simple animal behavior model for the anxiolytic effects of benzodiazepines. *Pharmacol. Biochem. Behav.*, 13: 167–170.

Crawley, J.N. and Paylor, R. (1997) A proposed test battery and constellations of specific behavioral paradigms to investigate the behavioral phenotypes of transgenic and knockout mice. *Hormones Behav.*, 31: 197–211.

Crawley, J.N., Glowa, J.R., Majewska, M.D. and Paul, S.M. (1986) Anxiolytic activity of an endogenous adrenal steroid. *Brain Res.*, 398: 382–385.

Crawley, J.N., Marangos, P.J., Paul, S.M., Skolnick, P. and Goodwin, F.K. (1981) Purine diazepam interaction: inosine reverses diazepam-induced stimulation of mouse exploratory behavior. *Science*, 22: 725–727.

Crawley, J.N., Skolnick, P. and Paul, S.M. (1984) Absence of intrinsic actions of benzodiazepines antagonists on a mouse exploratory model of anxiety. *Neuropharmacol.*, 25: 531–537.

Crawley, J.N., Belknap, J.K., Collins, A., Crabbe, J.C., Frankel, W., Henderson, N., Hitzemann, R.J., Maxson, S.C., Miner, L.L., Silva, A.J., Wehner, J.M., Wynshaw-Boris, A. and Paylor, R. (1997) Behavioral phenotypes of inbred mouse strains: Implications and recommendations for molecular studies. *Psychopharmacol.*, 132: 107–124.

Dawson, G.R., Crawford, S.P., Collinson, N., Iversen, S.D. and Tricklebank, M.D. (1995) Evidence that the anxiolytic-like effects of chlordiazepoxide on the elevated plus maze are confounded by increases in locomotor activity. *Psychopharmacol.*, 118: 316–323.

Dawson, G.R. and Tricklebank, M.D. (1995) Use of the elevated plus maze in the search for novel anxiolytic agents. *Trends Pharmacol. Sci.*, 16: 33–36.

DeAngelis, L. and File, S.E. (1979) Acute and chronic effects of three benzodiazepines in the social interaction anxiety test in mice. *Psychopharmacol.*, 64: 127–130.

De Fries, J.C., Gervais, M.C. and Thomas, E.A. (1978) Response to 30 generations of selection for open-field activity in laboratory mice. *Behav. Genet.*, 8: 3–13.

Dunn, A.J. and File, S.E. (1987) Corticotropin-releasing factor displays an anxiogenic action in the social interaction test. *Hormones Behav.*, 21: 193–202.

File, S.E. (1980) The use of social interaction as a method for detecting anxiolytic activity of chlordiazepoxide-like drugs. *J. Neurosci. Methods*, 2: 219–238.

File, S.E. (1997) Animal tests of anxiety. In: J.N. Crawley, C.R. Gerfen, R. McKay, M.A. Rogawski, D.R. Sibley and P. Skolnick (Eds.), *Current Protocols in Neuroscience*, John Wiley and Sons, New York, pp. 8.3.1–8.3.15.

File, S.E. and Hyde, J.R.G. (1979) A test of anxiety that distinguishes between the actions of benzodiazepines and those of other minor tranquillisers and of stimulants. *Pharmacol. Biochem. Behav.*, 11: 65–69.

Flint, J., Corley, R., DeFries, J.C., Fulker, D.W., Gray, J.A., Miller, S. and Collins, A.C. (1995) A simple genetic basis for a complex psychological trait in laboratory mice. *Science*, 269: 1432–1435.

Geller, I. and Seifter, J. (1960) The effects of meprobamate, barbiturates, d-amphetamine and promazine on experimentally induced conflict in the rat. *Psychopharmacol.*, 1: 482–492.

Giros, B., Jaber, M., Jones, S.R., Wightman, R.M. and Caron, M.G. (1996) Hyperlocomotion and indifference to cocaine and amphetamine in mice lacking the dopamine transporter. *Nature*, 379: 606–612.

Hall, C.S. (1936) Emotional behavior in the rat. III. The relationship between emotionality and ambulatory activity. *J. Comp. Physiol. Psychol.*, 22: 345–352.

Handley, S.L. and McBlane, J.W. (1993) An assessment of the elevated X-maze for studying anxiety and anxiety-modulating drugs. *J. Pharmacol. Toxicol. Methods*, 29: 129–138.

Handley, S.L. and Mithani, S. (1984) Effects of alpha-adrenoceptor agonists and antagonists in a maze exploration model of "fear-motivated" behaviour. *Naunyn Schcmiedeberg's Archives of Pharmacology*, 327: 1–5.

Heinrichs, S.C., Min, H., Tamraz, S., Carmouche, M. and Boehme, S.A. (1997) Vale WA, Anti-sexual and anxiogenic behavioral consequences of corticotropin-releasing factor overexpression are centrally mediated. *Psychoneuroendocrinol.*, 22(4): 215–224.

Henderson, N.D. (1967) Prior treatment effects on open field behavior of mice: a genetic analysis. *Animal Behav.*, 15: 364–376.

Higgins, G.A., Bradbury, A.J., Jones, B.J. and Oakley, N.R. (1988) Behavioural and biochemical consequences following activation of 5-HT1-like and GABA receptors in the dorsal raphe nucleus of the rat. *Neuropharmacol.*, 27: 993–1001.

Hughes, J., Boden, P., Costall, A.B., Domeney, A., Kelly, E., Horwell, D.C., Hunter, J.C., Pinnock, R.D. and Woodruff, G.N. (1990) Development of a class of selective cholecystokinin type B receptor antagonists having potent anxiolytic activity. *Proc. Natl. Acad. Sci. USA*, 87: 6728–6737.

Koob, G.F., Braestrup, C. and Britton, K.T. (1986) The effects of FG 7142 and RO 15-1788 on the release of punished responding produced by chlordiazepoxide and ethanol in the rat. *Psychopharmacol.*, 90: 173–178.

Lister, R.G. and Hilakivi, L.A. (1988) The effects of novelty, isolation, light and ethanol on the social behavior of mice. *Psychopharmacol.*, 96: 181–187.

Mathis, C., Paul, S.M. and Crawley, J.N. (1994) Characterization of benzodiazepine-sensitive behaviors in the A/J and C57BL/6J inbred strains of mice. *Behav. Genet.*, 24(2): 171–180.

Mathis, C., Neumann, P.E., Gershenfeld, H., Paul, S.M. and Crawley, J.N. (1995) Genetic analysis of anxiety-related behaviors and responses to benzodiazepine-related drugs in AXB and BXA recombinant inbred mouse strains. *Behav. Genet.*, 25: 557–568.

Montgomery, K.C. (1958) The relation between fear induced by novel stimulation and exploratory behavior. *J. Comp. Physiol. Psychol.*, 48: 254–260.

Pellow, S., Chopin, P., File, S.E. and Briley, M. (1985) Validation of open:closed arm entries in an elevated plus-maze as a measure of anxiety in the rat. *J. Neurosci. Methods*, 14: 149–167.

Pierrefiche, G., Zerbib, R. and Laborit, H. (1993) Anxiolytic activity of melatonin in mice: Involvement of benzodiazepine receptors. *Res. Commun. Chem. Pathol. Pharmacol.*, 82:2: 131–142.

Pinel, J.P.J. and Treit, D. (1978) Burying as a defensive response in rats. *J. Comp. Physiol. Psychol.*, 92: 708–712.

Rodgers, R.J. and Cole, J.C. (1993) Influence of social isolation, gender, strain, and prior novelty on plus-maze behavior in mice. *Physiol. Behav.*, 54: 729–736.

Rodgers, R.J., Cole, J.C., Cobain, M.R., Daly, P., Doran, P.J., Eells, J.R. and Wallis, P. (1992) Anxiogenic-like effects of fluprazine and eltoprazine in the mouse elevated plus maze: profile comparisons with 8-OH-DPAT, TFMPP and mCPP. *Behav. Pharmacol.*, 3: 621–634.

Shepherd, J.K. (1992) Preliminary evaluation of an elevated "zero-maze" as a model of anxiety in laboratory rats. J. Psychopharmacol., 6: 223.

Shepherd, J.K., Grewal, S.S., Fletcher, A., Bill, D.J. and Dourish, C.T. (1994) Behavioural and pharmacological characterisation of the elevated "zero-maze" as an animal model of anxiety. *Psychopharmacol.*, 116, 56–64.

Treit, D. (1985) The inhibitory effect of diazepam on defensive burying: Anxiolytic vs. analgesic effects. Pharmacol. Biochem. Behav., 22: 47–52.

Treit, D., Pinel, J.P.J. and Fibiger, H.C. (1981) Conditioned defensive burying: A new paradigm for the study of anxiolytic agents. *Pharmacol. Biochem. Behav.*, 15: 619–626.

Treit, D., Menard, J. and Royan, C. (1993) Anxiogenic stimuli in the elevated plus maze. *Pharmacol. Biochem Behav.*, 44: 463–469.

Trullas, R. and Skolnick, P. (1993) Differences in fear motivated behaviors among inbred mouse strains. *Psychopharmacol.*, 111: 323–331.

Vogel, J.R., Beer, B. and Clody, D.E. (1971) A simple and reliable conflict procedure for testing antianxiety agents. *Psychopharmacol.*, 21: 1–7.

W.E. Crusio and R.T. Gerlai (Eds.)
Handbook of Molecular-Genetic Techniques for Brain and
Behavior Research (Techniques in the Behavioral and Neural Sciences, Vol. 13)

CHAPTER 4.7

A neurobehavioral system approach in rats to study the molecular biology of fear

Jeffrey B. Rosen, Seema Malkani, Karin J. Wallace and Barbara L. Thompson

Program in Neuroscience and Department of Psychology, University of Delaware, Newark, DE, 19716, USA

Introduction

Fear is one of the most basic emotions and has tremendous consequences on behavior and psychopathology. Fear is a normal emotional state and behavioral response to danger or threat that motivates one to relieve this negative emotional state. When an animal is faced with danger, threat, or an aversive situation, an adaptive fear response is elicited which prepares the animal for defensive action. If danger increases and becomes inevitable, the heightened emotional fear response will motivate the animal to engage in evasive action, such as flight, or aggressive action, like fighting. Once the threat diminishes, the fear subsides. Thus, fear is functional and has evolved as a protective mechanism. However, fear can also become dysfunctional, exaggerated and inappropriate to the point where an individual is so fearful and anxious that it disrupts normal daily function. The exaggeration of fear can play a role in a number of anxiety and affective disorders. For example, social phobia is an anxiety disorder characterized by extreme fear of social interaction leading to a pathological withdrawal from social intercourse. High levels of co-morbidity of anxiety disorders and unipolar depression indicate that exaggerated fear may also be an underlying pathology of depression.

In order to understand and eventually better treat anxiety and affective disorders, the neurobiology of normal fear and pathological anxiety need to be elucidated at the molecular level. This can only be accomplished through a neurobehavioral systems approach which analyzes the molecular substrates of fear within the constraints of the neuroanatomy underlying fear behavior. Molecular changes correlated with fear may occur in many brain regions, some critical for fear and others not. Those changes found uniquely in neural fear circuits are probably most central to the learning, memory and expression of fear, and should be the focus of investigation. Therefore, elucidation of the behavioral phenotypes of fear and then delineation of the neuroanatomical circuits that mediate these phenotypes are necessary before a molecular analysis can yield relevant results. Fortunately, research over the last two decades has produced behavioral methods that have allowed substantial progress in defining neuroanatomical substrates of fear. The molecular mechanisms of fear behavior within these fear circuits are just beginning to be studied.

This chapter will present a framework for investigating the molecular biology of fear. The three levels of a neurobehavioral systems approach will begin with a discussion of phenotypic fear behavior in rats. A neuroanatomical circuit of fear that mediates these phenotypic fear responses will be presented. Finally, several paradigms will be offered as examples for studying gene expression of learned fear, innate fear, and exaggerated fear (model of pathological fear).

Fear behavior

Ethological and functional analyses of fear behavior has determined that animals have a set of genetically-determined species-specific defense responses (SSDR) that are employed during danger (Bolles, 1970). Humans and animals display similar evasive behaviors in the face of danger (Marks, 1987; Blanchard and Blanchard, 1989a; Fanselow, 1994). In many rodents including rats and mice, these behaviors include freezing (crouching with no movement except for breathing), potentiated startle, defensive fighting, threat displays, and jump attacks (Blanchard and Blanchard, 1989a; Fanselow, 1994). The level of perceived fear appears to be one of the more important determinants of the type of behavior that is displayed (Bindra, 1978; Bolles and Fanselow, 1980; Fanselow, 1994). The animal's perception of the probability of harm is the determinant of the particular response (Bolles and Fanselow, 1980). Freezing behavior is typically the initial response in many species to detection or anticipation of a potential danger (Blanchard and Blanchard, 1989a; Fanselow, 1994). Freezing or crouching prepares the animal for fight or flight if attacked and reduces detection by a predator or threatening conspecific (Blanchard and Blanchard, 1969). Concomitant with freezing are other autonomic, behavioral, and perceptual components of fear responses including heart rate, blood pressure and respiratory changes, potentiated startle responses and increased vigilance (Davis, 1992; Kapp et al., 1992; LeDoux, 1992; Gallagher and Holland, 1994). As danger becomes inevitable or if an actual attack occurs, the defensive response selection shifts (Blanchard and Blanchard, 1989a). Instead of freezing, rodents engage in fight or flight behaviors (escape, jump attack) to diminish harm. In severe, injurious and life-threatening attack, tonic immobility may predominate (Gallup, 1974). These species-specific fear and defensive reactions are innately and automatically available to the animal and are expressed on a continuum with different levels of threat; anticipated harm automatically elicits crouching and freezing, and inevitable or actual harm automatically elicits flight and fight.

Fear can either be innate or learned. Innate fear is the perception of a stimulus as fearful without having prior experience with the stimulus. Infants have an innate fear of strangers during a critical period of development. Many animals also have an innate fear of species-specific predators. As will be discussed in more detail later, rats are innately fearful of the odors of predators' urine and feces. Learned fear is learning that a previously neutral stimulus now signals the potential occurrence of a harmful event. Fear learning primarily develops through Pavlovian conditioning mechanisms. In the laboratory, this is accomplished by pairing a conditioned stimulus (e.g. a light or tone) with an unconditioned stimulus (usually a footshock) (LeDoux, 1987). In addition to a specific conditioned stimulus, animals can also learn to be fearful of the environment or context where a footshock was previously experienced. Subsequent presentation of the conditioned stimulus alone induces fear responses in anticipation of the footshock even though the shock is not delivered. Both cue-specific and contextual fear learning elicit increased freezing following the conditioned stimulus-footshock pairings. Fear-potentiated startle can also be used as a measure of conditioned fear. During presentation of the conditioned stimulus, rats will display an increased startle response to a loud noise (Davis et al., 1987). Fear-conditioned autonomic changes, such as heart-rate, can also be measured by presentation of conditioned stimuli (Kapp et al., 1992).

Neural circuit of fear

Research in the earlier part of the 20th century established that the amygdala is a critical brain region for the expression of fear (Kluver and Bucy, 1939). Humans, monkeys, rodents and other species become tame and do not display fearful behavior when the amygdala is damaged (Kling et al., 1970; Davis, 1992; Kapp et al., 1992; LeDoux, 1992; Adolphs et al., 1995). In the last 20 years, our understanding of the amygdala and fear has increased to the point of dividing the amygdala into various nuclei (Pitkanen et al., 1997; Swanson and Petrovich, 1998) that contribute differentially to fear and delineating the amygdala projections to other brain areas to control the expression of the multifaceted aspects of cognitive and behavioral fear responses (Davis, 1992; LeDoux, 1996; Rosen et al., 1998).

Primarily through the use of Pavlovian conditioning paradigms and the analysis of fear-related behaviors, it has been determined that lesions of the lateral, basolateral and central nuclei of the amygdala disrupt the expression of fear-induced bradycardia, respiration, freezing, startle and analgesia (Davis, 1992; Kapp et al., 1992; LeDoux, 1992). The lateral and basolateral nuclei of the amygdala receive highly processed sensory information from polymodal association cortices (e.g. perirhinal cortex, auditory cortex) (MacDonald and Jackson, 1987; Romanski and LeDoux, 1993; Burwell et al., 1995; Suzuki, 1996). Sensory (e.g. auditory and nociceptive) information directly from subcortical thalamic nuclei also project to the lateral and basolateral nuclei of the amygdala, suggesting that the lateral and basolateral nuclei of the amygdala are involved in the formation of associations between neutral and aversive stimuli. The lateral and basolateral nuclei have many large pyramidal cells, similar to cortical pyramidal cells, the major integrative neurons of the cortex (MacDonald, 1982). *N*-methyl-d-aspartate antagonists, which block various forms of learning, also block fear conditioning when injected into the lateral and basolateral nuclei of the amygdala at the time of conditioning (Miserendino et al., 1990; Campeau et al., 1992; Maren et al., 1996). In addition, neurons of the lateral nucleus respond to conditioned fear stimuli and display long-term potentiated responses following fear conditioning (Quirk et al., 1995; Rogan et al., 1997). These data indicate that the lateral and basolateral nuclei of the amygdala are critical sites for the formation of associations during fear conditioning (Fig. 1, top).

The central nucleus of the amygdala also plays a role in fear. It receives heavy projections from the lateral and basolateral nuclei of the amygdala (Pitkanen et al., 1997). In turn the central nucleus of the amygdala projects via the ventral amygdalofugal pathway directly to many diencephalic, midbrain and brain stem structures, such as the periaqueductal gray (Hostege, 1995), lateral hypothalamus (Krettek and Price, 1978; Price and Amaral, 1981), nucleus reticularis pontis caudalis (Rosen et al., 1991), parabrachial nucleus (Hopkins and Holstege, 1978; Veening et al., 1984), dorsal motor nucleus of the vagus and nucleus ambiguous (Hopkins and Holstege, 1978; Veening et al., 1984) (Fig. 1, bottom). The central gray or periaqueductal gray plays a central role as an output pathway in the generation of freezing and escape behavior (Carrive, 1993; Bandler and Shipley, 1994; Fanselow, 1994), while the hypothalamus (paraventricular nucleus and lateral hypothalamus) may activate endocrine and autonomic functions (Beaulieu et al., 1987; LeDoux et al., 1988). The nucleus reticularis pontis caudalis is part of the startle circuit (Davis et al., 1982; Lee et al., 1996) where input from the central nucleus of the amygdala impinges to enhance startle (Rosen and Davis, 1988; Rosen and Davis, 1990; Rosen et al., 1991). The parabrachial nucleus controls respiration (Von Euler et al., 1976), whereas the dorsal motor nucleus of the vagus and nucleus ambiguous control the parasympathetic responses such as bradycardia and respiratory-sinus arrhythmia (Porges, 1995).

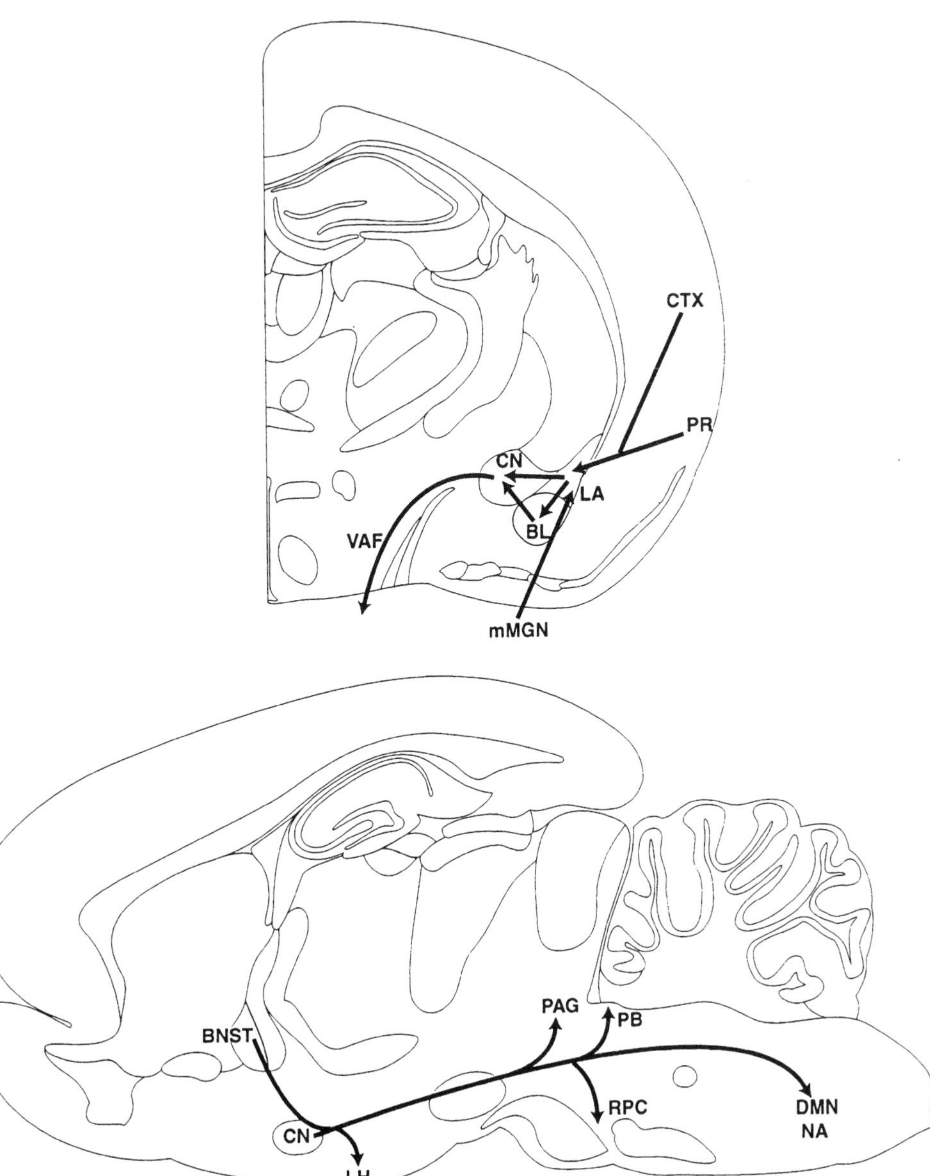

Fig. 1. Top: Schematic drawing of the neuronal fear circuit. Information flow through the amygdala. Information from conditioned fear stimuli enters the lateral nucleus of the amygdala (LN) from the medial portion of the medial geniculate nucleus (mMGN) and/or the perirhinal cortex (PR). (Other areas of cortex (CTX) also project to the amygdala, however they are not considered an inrinsic part of the fear circuit.) Information then travels to the basolateral (BL) and central nucleus (CN) of the amygdala. Efferents of the central nucleus then travel to lower brain structures via the ventral amygdalofugal pathway (VAF). Bottom: Efferents from the central nucleus of the amygdala and lateral bed nucelus of the stria terminalis (BNST) project to the periaqueductal gray (PAG) to induce freezing, to the reticularis pontis caudalis (RPC) to potentiate startle, to the parabrachial nucleus (PB) to alter respiration rate, and to the lateral hypothalamus (LH), dorsal motor nucleus of vagus (DMN) and nucleus ambiguous (NA) to influence heart rate and blood pressure (Rosen and Schulkin, 1998).

An important target region of the central nucleus of the amygdala is the periaqueductal gray in the midbrain which controls fear-induced freezing and defensive behaviors (Bandler and Shipley, 1994; Fanselow, 1994; Hostege, 1995). Amygdala projections innervate the ventral, but not the dorsal, periaqueductal gray (PAG) (Hostege, 1995). Lesions of the ventral PAG block fear-induced freezing (Fanselow, 1994). In contrast, lesions of the dorsolateral PAG have no effect on footshock-induced freezing but decrease the footshock-induced motor activity burst (Fanselow, 1994). In corroboration, stimulation of lateral regions of the PAG produces active defensive aggression, whereas ventral stimulation elicits hypoactivity and bradycardia (Bandler and Shipley, 1994). During low-levels of threat when fear-related behaviors like freezing predominate, the dorsal and dorsolateral PAG may be inhibited, but are released with more threatening stimuli (Graeff, 1994). Similarly, the dorsal PAG may inhibit fear-related responses (i.e. attenuation of fear-potentiated startle) under highly threatening situations since lesions of this area releases fear-potentiated startle from inhibition (Walker and Davis, 1997). Thus, although there are neuroanatomical distinctions in the PAG between the substrates of fear-related and defensive behaviors, the systems interact with each other so that behavior can switch from one to the other when necessary (Fanselow, 1991).

Projections from the basolateral nucleus of the amygdala to regions other than the central nucleus of the amygdala also mediate aspects of fear and fear responses. The basolateral nucleus projects to the ventral striatum and may be involved in learned operant and avoidance responses (Killcross et al., 1997). These are a class of responses that are learned after many trials and allow an animal to perform more intricate responses than SSDRs in order to avoid harmful stimuli or situations.

Other brain regions also play a role in fear and the expression of fear-related behaviors. The hippocampus plays a role in context-specific processing of fearful information (Kim and Fanselow, 1992; Phillips and LeDoux, 1992). The septohippocampal region may also be important for behavioral inhibition and vacillation of approach-avoidance behavior during times of risk-assessment (Gray and NcNaughton, 1996) The prefrontal cortex may play a role in extinction of behavior to fearful stimuli (Morgan and LeDoux, 1995); however, see (Gewirtz et al., 1997). The amygdala also projects to widespread regions of sensory cortex to influence sensory information processing (Amaral et al., 1992). Obviously the amygdala and its associated fear circuits do not act independently of other brain regions. Theories that integrate the amygdala with many brain regions and systems are necessary for a full understanding of emotions (e.g. Gray and NcNaughton, 1996; LeDoux, 1996; Panksepp et al., 1991).

The molecular biology of fear

We have laid out a behavioral and neuroanatomical framework to begin an investigation of the molecular biology of fear and anxiety. Behaviorally, fear is studied using ethologically-relevant species-specific fear behaviors such as freezing. Anatomically, changes in expression of genes and proteins relevant to fear should be sought in the amygdala (particularly in the lateral, basolateral and central nuclei) and other nuclei within fear circuits discussed above.

The number of neuron-specific genes that may be associated with fear and the learning of fear are potentially astronomical. However, some analyses have found that relatively few genes increase or decrease expression with learning (Abel et al., 1998), suggesting that the search for molecular substrates of learned fear may not be overwhelming. The products of the genes may be located in various compartments of neurons and play roles in numerous cellular functions. Receptors, enzymes, kinases, transcriptions factors, structural proteins are all targets of inquiry. What genes to study? Initially, we are focusing our research on a few immediate-early genes. These genes (e.g. *c-fos*, *egr-1*) are used extensively in neuroscience research as anatomical markers for transcriptional activation in neurons because their expression is very rapidly increased (within minutes) following neuronal excitation. Extracellular stimulation by glutamate NMDA agonists (Worley et al., 1990; Bading et al., 1995), psychostimulants (Wang and McGinty, 1994; Wang et al., 1994), electrical stimulation (Worley et al., 1993), and various behavioral manipulations induce *c-fos* and *egr-1*. In addition, these and related immediate-early genes are the first set of genes activated by external signals and do not require de novo synthesis of proteins for their induction suggesting they act as third messengers. Second messengers (e.g. calcium, cAMP) and signal transduction molecules (e.g. mitogenic-activating protein kinase, cAMP-responsive element binding protein) regulate the expression of the immediate-early genes (Changelian et al., 1989; Christy and Nathans, 1989; DeFranco et al., 1993). These immediate-early genes function as transcription factors that couple extracellular signals to changes in gene expression. This suggests that some immediate-early genes be an early step in a cascade of events in learning and memory. Indeed, expression of *egr-1* (also called Ngfi-A, zif268, krox-24, TIS-8) seems to correlate well with long-term potentiation in the hippocampus (Wisden et al., 1990; Abraham et al., 1991; Richardson et al., 1992; Worley et al., 1993) and visual paired associative learning in the anterior temporal lobe (Okuno and Miyashita, 1996).

The remainder of this monograph will focus on three examples of a neurobehavioral systems approach to the study of the molecular biology of fear, with an emphasis on expression of the immediate-early genes *egr-1* and *c-fos*. Each of these examples employ ethologically relevant fearful situations that have been modified for study in the laboratory. A major emphasis in these studies is that the search for changes in the molecular substrates of fear is pursued within the constraints of a neurobehavioral system. In other words, because the learning and display of fear is dependent on the amygdala, expression of immediate-early genes within the amygdala is the focus of our molecular analysis. Changes in expression of the immediate-early genes also occur in other brain regions during fear and are certainly performing important functions, but as shown by lesion studies, many of those brain areas are not crucial for expression of emotional behaviors. Molecular changes in the amygdala and fear circuits are more likely to be central to the emotion of fear.

Three examples of behavioral paradigms that our lab uses to study the molecular biology of the whole continuum of fear (i.e. learned, innate, and pathological fear) are discussed below.

Example 1: Immediate-early gene expression with contextual fear conditioning

Pavlovian conditioning is typically used to study the learning of fear in the laboratory. We are currently investigating contextual fear conditioning where the conditioned stimulus is the training context or environment and the unconditioned stimulus is a footshock. Contextual fear conditioning has been shown to be a very reliable and robust type of learning that takes place in one trial (i.e. with a single footshock) and is dependent on an intact amygdala. Because learning occurs in one trial, the temporal pattern of a change in gene expression can be determined and linked to the conditioning event. However, to conclude that a change in expression of a gene in the amygdala is specific for contextual fear conditioning, one must control for the non-specific effects of the footshock. One would need a procedure that delivers footshock but does not produce fear. A paradigm that incorporates a control group that receives footshock but does not display fear is the immediate-shock-deficit paradigm (Blanchard et al., 1976; Fanselow, 1986; Landeira-Fernandez et al., 1995). Typically in contextual fear conditioning experiments, rats are placed in a new environment, are given a few minutes to acclimatize, and then receive a footshock. Following the shock the animals freeze, defecate and are hypoalgesic — all indices of fear (Fanselow, 1994). If returned to the chamber at a later time, for example, 24 h later, the rats again display these fear-related behaviors indicating contextual fear learning. In contrast, the immediate-shock deficit is a phenomenon where rats are not given time to acclimate, but instead receive the footshock immediately upon entering the new environment. These rats do not display these species-specific fear responses (i.e. freezing, hypoalgesia, increased defecation). This deficit in fear responses to the footshock still persists even though the rats remain in this environment for several more minutes. Also, they do not demonstrate conditioned fear when returned to the same environment at a later time. Theoretically, the immediate-shock-deficit phenomenon suggests experiencing an aversive event is not sufficient for fear to develop. For successful fear conditioning, time must be given for an animal to form a representation of the to-be-conditioned stimuli (contextual stimuli in this case) so associations can be made with the unconditioned stimulus (Faneslow, 1990; Fanselow, 1986). Remaining in the environment after experiencing the shock is not sufficient for associations between contextual cues and the unconditioned stimulus to form. This is similar to the difficulties in obtaining backward conditioning in more traditional classical conditioning paradigms. Thus, contextual fear conditioning is a true form of classical conditioning.

In the experiments presented here, the immediate-shock-deficit paradigm (see Table 1) had four groups of rats:

1. a handled group that received pre-experiment handling like the other groups, but did not experience any other part of the procedure,
2. a context-no shock group that was handled and placed in the conditioning chamber, but did not receive footshock,
3. an immediate-shock group that was handled and received a 1 s footshock of either 1.0 or 1.5 mA immediately upon being placed in the chamber, and
4. a delayed-shock group that was handled, placed in the conditioning chamber and allowed to acclimate for 3 min before being given the 1 s, 1.0 or 1.5 mA footshock.

The percent of time spent freezing in a 4 min period after the delivery of the footshock was used as the measure of fear. In the context-no shock group, freezing was measured for 4 min after a 3 min acclimatizion period. Following the post-shock observation period the rats were returned to their home cages. Some rats were returned to the conditioning chamber 24 h after conditioning to test for retention of fear conditioning.

Fear, as measured by freezing behavior (crouch position with no movement except for breathing (Blanchard and Blanchard, 1969)), was sampled during the post-shock period and retention test. Every 5 s each rat was observed and scored to be freezing or not. The percentage of samples scored as freezing during the observation period was calculated for each rat. Group differences in the time spent freezing were compared by Kruskal-Wallis test followed by a Mann-Whitney U test for individual comparisons or ANOVA followed by a Student–Newman–Keuls test.

In an intial experiment 1.0 mA footshock was used as an unconditioned stimulus (Rosen et al., 1998). Only rats in the delayed shock group displayed freezing in the post-shock period or in a retention test 24 h after the context conditioning (mean of 22% of the test time spent freezing). The rats that received the immediate shock or no shock did not freeze. In order to increase the strength of conditioning as measured by an increase in the time spent freezing, the experiment was replicated with 1.5 mA footshock. Figure 2 demonstrates that the delayed group froze about 75% of the time and displayed significantly more freezing than the context-no shock and immediate-shock groups both in the post-shock

TABLE 1

Behavioral protocols

1. The immediate-shock deficit paradigm for contextual fear conditioning:

Rats are brought into the laboratory from the animal quarters on the 4 d before the conditioning and handled for 10–20 min.

On the contextual fear conditioning day, rats are divided into 4 groups:

1. Handled, no shock – Decapitated.
2. Context, no shock – A rat is placed in the test chamber for 7 min. We use startle chambers and sound attenuating boxes from San Diego Instruments, San Diego CA. A 70 dB white noise acts to mask noise and 25 W bulb illuminates the box. The front door of the box is left open to allow for observers to view and score freezing behavior (see below). Freezing is scored during the last 4 min the rat is in the chamber. The rat is then removed and placed in a holding cage (standard Nalgene Plexiglas rat cage) until decapitation.
3. Immediate-shock – A rat is placed in the test chamber and given a 1 s, 1.0 or 1.5 mA footshock. Freezing is scored for 4 min beginning immediately after the shock. The rat is removed and placed in a holding cage until decapitation.
4. Delayed-shock – A rat is placed in test chamber and allowed 3 min to acclimatize. A 1 s, 1.0 or 1.5 mA footshock then delivered. Freezing is scored for 4 min beginning immediately after the shock. The rat is removed and placed in a holding cage until decapitation.

The observation of freezing behavior (crouch position with no movement except for respiration) is scored yes or no every 5 s for the 4 min period indicated above. The observers sit in the same darkened room that houses the test chambers and use a yes/no check list to record the freezing behavior. The percent of time spent freezing is equal to:
number of observations of freezing divided by total observations (48) times 100.

2. Predator odor-induced freezing

1. Rats are placed in the test chambers for 15 min a day for 4 or 5 d, until the animals spend less that 20% of the time freezing. (The same startle chamber as described above are used, except they are placed in a fume hood without the sound attenuating enclosure.)
2. On testing day, a rat is placed in the chamber for 3 min and freezing behavior is observed. If freezing occurs less than 20% of the time, then an odor is presented by switching the doors with doors that have the odor. Freezing behavior is scored for 15 min in the presence of the odor.
3. The rat is immediately decapitated and the brain processed for in situ hybridization (see below).
4. The fox odor is trimethyl-thiazoline purchased from Phero Tech, Inc. (Delta, B.C., Canada). Butyric acid is purchase from ICN (Cleveland, OH).

3. Facilitation of contextually-conditioned fear by chronic corticosterone treatment

1. Rats are handled as described above in the immediate-shock deficit paradigm.
2. Rats are given 5 ds of 2 injections daily (9 am and 4 pm) of 2.5 mg/kg s.c. corticosterone (Sigma) suspended in vegetable oil or vegetable oil.
3. On the sixth day, rats are given their usual injection. However, some of the rats given oil, this time receive an injection of 2.5 mg/kg s.c. corticosterone. This is the acute CORT group in Fig. 8.
4. Two hours after this last injection, rats are placed in the text chambers as described in the immediate-shock deficit paradigm. After 3 min of acclimatization, a 1, s, 0.6 mA footshock is given. After another 4 min, the rats are returned to their home cage.
5. Five days after training, the rats are returned to the test chambers and scored for retention of contextual fear. Freezing is observed and recorded for 4 min as described above.

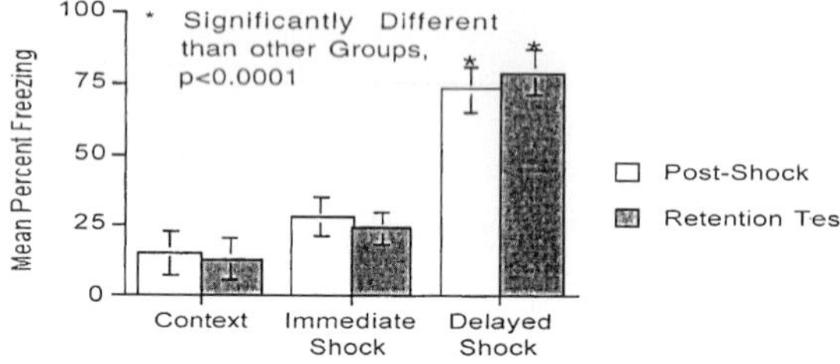

Fig. 2. Increase in fear (freezing) with contextual fear conditioning. Rats in the delayed-shock group spent significantly more time freezing immediately following conditioning (post-shock period) and in a retention test for contextually conditioned fear 24 h after conditioning.

period and the retention test. The amount of time spent freezing in the context-no shock and immediate-shock groups did not differ.

Contextual fear conditioning with the immediate-shock deficit paradigm offers several advantages for studying the expression of immediate-early genes or other genes in fear learning. Contextual fear conditioning occurs with one trial so the initiation of learning is known and can be controlled. Temporal changes in gene expression after the single trial can be determined. Importantly, the effects of footshock without fear conditioning (immediate-shock group) can be compared to those of footshock with fear conditioning (delayed-shock group). This comparison allows for the determination of gene expression that is specifically related to fear learning and not merely a result of the painful footshock.

For detection of changes in the levels of immediate early genes in the amygdala associated with fear conditioning, rats in the immediate- and delayed-shock groups were decapitated either 15 or 30 min after the delivery of the footshock and the brains were rapidly removed and frozen. Rats in a context-no shock group were also decapitated either 15 or 30 min after being placed in the test chamber without receiving footshock. Brains from a handled-no shock group of rats were also processed. The brains were sliced coronally on a cryostat, mounted on slides and prepared for in situ hybridization of *c-fos* and *egr-1* mRNA using 35^S-labeled cRNA probes and standard in situ hybridization procedures (see Table 2 for details of in situ hybridization protocol). Following in situ hybridization the brain sections were apposed to X-ray film. The autoradiograms of the in situ hybridization were digitized and subjected to image analysis. Group differences in the density of labeling (nCi/g tissue) in the lateral nucleus of the amygdala were analyzed by ANOVA followed by a Student–Newman–Keuls.

Significant increases in *egr-1* mRNA expression were found in the dorsal portion of the lateral nucleus of the amygdala in the delayed-shock group 15 and 30 min after fear conditioning (Fig. 3). At 15 min after the 1.0 mA footshock, *egr-1* mRNA in the delayed-shock group was significantly greater than *egr-1* levels of the handled- and context-no shock groups. At the same time, levels of *egr-1* mRNA expression in the immediate-shock group were not different from any other group. Stronger contextual fear conditioning with 1.5 mA compared to 1.0 mA footshock produced better separation of *egr-1* expression between the experimental groups. *Egr-1* mRNA expression of the delayed-shock group was statstically greater than all of the groups, including the immediate-shock group. The increase in expression in the delayed-shock group was about 50% greater than the handled- and context-no shock groups, and about 20% greater than the immediate-shock group. Additionally, the immediate-shock group had significantly greater levels of *egr-1* mRNA expression than the handled- and context-no shock groups, indicating that the footshock alone without fear conditioning induced *egr-1* mRNA expression 15 min after the footshock. Analysis of *egr-1* mRNA expression 30 min after the footshock found that the density of *egr-1* mRNA in the delayed-shock group remained 50% greater than the other groups including the immediate-shock group, while expression in the immediate-shock group returned to the handled- and context-no

TABLE 2

In situ hybridization protocol (a more detailed protocol can be found in Rosen et al. (1998).

Tissue preparation

1. Rats are decapitated, the brains are removed quickly and submerged for 15–20 seconds in −45°C isopentane. The brains are stored at −70° C until sectioned.
2. Fifteen μm whole brain coronal sections containing dorsal amygdala, hippocampus, and pyriform and parietal cortices (plate 28 in the atlas of Paxinos and Watson) are cut on a cryostat and thawed onto Plus-on (VWR) microscope slides. Two adjacent brain sections are placed on each slide. The slides are stored at −70°C until processed for in situ hybridization.

cRNA probe radioisotope labeling

1. cDNA plasmids containing a sense and antisense 900 bp sequence of the *c-fos* mRNA (gift form T. Curran) or 230 bp sequence of *egr-1* (gift from J. Milbrandt) are used to make cRNA probes (riboprobes). The riboprobes are transcribed and labeled in vitro with ^{35}S-UTP (approximately 106 DPM/μg) using Sp6 or T7 RNA polymerase Maxiscript kit according to manufacturer's instructions (Ambion, Cambridge, MA).

Pre-hybridization preparation (all solutions are made in DEPC-treated and autoclaved deionized water)

1. The sections on microscope slide are air thawed and fixed in 4% formaldehyde in 1× phosphate-buffered saline (PBS) for 10 min and then rinsed in PBS.
2. The sections are then treated with 0.25% acetic anhydride in 0.1 M triethanolamine for 10 min at room temperature to reduce nonspecific hybridization of the probes.
3. They are then dehydrated with increasing concentrations of ethanol, defatted in chloroform followed by another ethanol rinse and air dried.

Hybridization (all solutions are made in DEPC-treated and autoclaved deionized water)

1. The ^{35}S-labeled riboprobe (2×10^6 cpm) is mixed into 100 μl of hybridization buffer. The hybridization buffer contains 20 mM Tris-HCl (pH 7.4), 50% formamide, 300 mM NaCl, 1 mM EDTA (pH 8), 1× Denhardt's, 250 μg/ml yeast tRNA, 250 μg/ml yeast total RNA, 100 μg/ml salmon sperm DNA, 10% dextran sulfate, 100 mM dithiothreitol, 0.1% SDS, and 0.1% sodium thiosulfate.
2. The hybridization buffer (100 μl) and applied by pipette to each slide containing two brain sections.
3. The slides are covered with a glass coverslips and incubated in a humidified box overnight at 56°C for 18–24 h.

Post-hybridization (all solutions are make in deionized water)

1. The next day, the coverslips are removed by dipping them in 4× SSC. The slides are rinsed another 3 times for 5 min each in 4×SSC.
2. The sections are then treated with 20 μg/ml RNAse A (Boerhinger-Mannheim, Indianapolis, IN) in a solution containing 0.5 M NaCl, 1 mM EDTA, 10 mM Tris, pH 8 for 30 min at room temperature.
3. After rinsing for 5 min each in decreasing concentrations of 1×, 0.5×, 0.1× SSC at room temperature, the sections are washed in 0.1× SSC 2 times for 30 min at 65°C.
4. Finally the slides are rinsed in increasing concentrations of ethanol containing 300 mM ammonium acetate. There is a final rinse in 100% ethanol.
5. The slides are then exposed to Kodak Biomax MR film for 2–7 d (depending on the strength of the desired signal). 14C standards (Amersham) are also apposed to the film along with the experimental sections.

Image analysis

1. The films are developed for 4 minutes in GBX developer (Kodak), rinsed in water for 30 s, fixed with Kodak Fixer for 5 min, rinsed in flowing water for 10 min and then air-dried.
2. The autoradiograms of in situ hybridization of *c-fos* and *egr-1* mRNAs are digitized and converted to gray values using a Sierra Scientific CCD video camera with Image program v.1.59 (NIH, Bethesda, MD) on an Apple PowerMac 7600 and then analyzed with the same program.
3. The Image program was used to subtract background (2D-rolling ball option) and measure the mean density of pixels within a circumscribed area (amygdala in our case).
4. For each region analyzed, the sizes of the areas measured from individual rats are all within 25% of each other, so that any differences reflected differences in density of labeling in comparable regions.

TABLE 2 (*continued overleaf*)

TABLE 2 (*continued*)

5. The ^{14}C standards are used for two purposes. One is to ensure that the radiographic signals were in the linear portion of the photochemical response of the film. The second is to convert the measured mean gray values to nCi/g values.
6. One-way ANOVAs are performed to test for a statistically significant overall difference between the four experimental groups. This is followed by a Scheffe's S test. The data of the experiment groups can also be normalized to the handled group by dividing the nCi/g of the area measured in each subjects of the experimental groups by the mean of the handled group nCi/g levels. The density of mRNA labeling can then be expressed as a percent of the handled control group as shown in Fig. 3.

shock levels. The results demonstrate that the expression of *egr-1* mRNA in the lateral nucleus of the amygdala can discriminate between contextual fear conditioning, exposure to a new context and the unconditioned effects of an aversive, painful stimulus. Moreover, the data interestingly suggest that fear conditioning alters the processing of *egr-1* mRNA and not just the amount of transcription. It is possible that either *egr-1* mRNA continues to be transcribed or its catabolism is reduced for at least 30 min after footshock in the delayed-shock group, whereas in the immediate-shock group *egr-1* mRNA is reduced to baseline within 30 min.

This discriminative ability of *egr-1* mRNA expression is specific for the lateral nucleus of the amygdala. *egr-1* was also found increased in the neocortex and pyriform cortex of both the immediate- and delayed-shock groups (Fig. 3 images) compared to the handled and context-no shock groups. However, unlike *egr-1* in the lateral nucleus of the amygdala, there was no difference in the level of expression in these cortical regions between the immediate- and delayed-shock groups. This indicates that the effects of footshock can not be discriminated from those of fear conditioning by *egr-1* expression in the cortex. In addition, analysis of *egr-1* expression in the hippocampus did not reveal differences between any of the four groups. This reinforces the notion that expression of *egr-1* is a specific marker in the lateral nucleus of the amygdala for fear conditioning and, thus, may play a role in fear conditioning.

Additional evidence further suggests that *egr-1* in the lateral nucleus of the amygdala may play a role in the acquisition of learned fear, but not in its retrieval or expression (Rosen et al., 1998). When *egr-1* mRNA expression was analyzed 15 min after a retention test 24 h after fear conditioning, there was no increase in *egr-1* in the lateral nucleus of the amygdala in any of the groups (Fig. 4). Thus, the data reveal that *egr-1* mRNA is increased following the acquisition of fear conditioning, but not following a memory test of the conditioning. This suggests that *egr-1* in the lateral nucleus of the amygdala may play a role in the acquisition but not the expression of learned fear.

c-fos mRNA expression was also analyzed after acquisition and a retention test of contextual fear (Rosen et al., 1998). Following acquisition, there was widespread expression of *c-fos* mRNA throughout the cortex which was greater in the three groups that were placed in the experimental chamber (context-no shock, immediate-shock and delayed-shock) compared to the handled group. However, no differences were revealed between these three groups. In the amygdala, the only discernible nucleus with consistent *c-fos* mRNA expression was the medial nucleus. Post-hoc analysis revealed medial nucleus *c-fos* levels of the context-no shock, and immediate and delayed shock groups were each different from the handled control group *c-fos* levels. In addition, the immediate shock and delayed shock groups had different levels than the context-no shock group. However, the immediate shock and delayed shock *c-fos* levels were not significantly different (Fig. 5) indicating that *c-fos* is not a marker for fear conditioning like *egr-1* is.

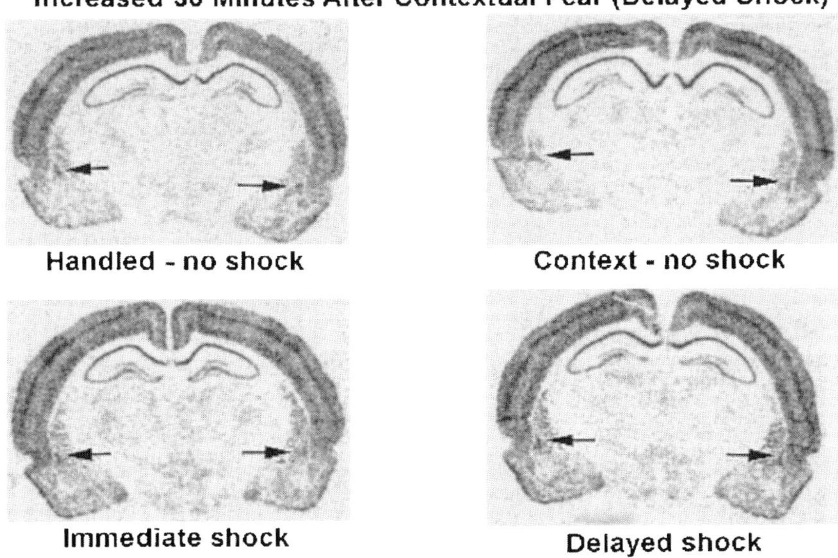

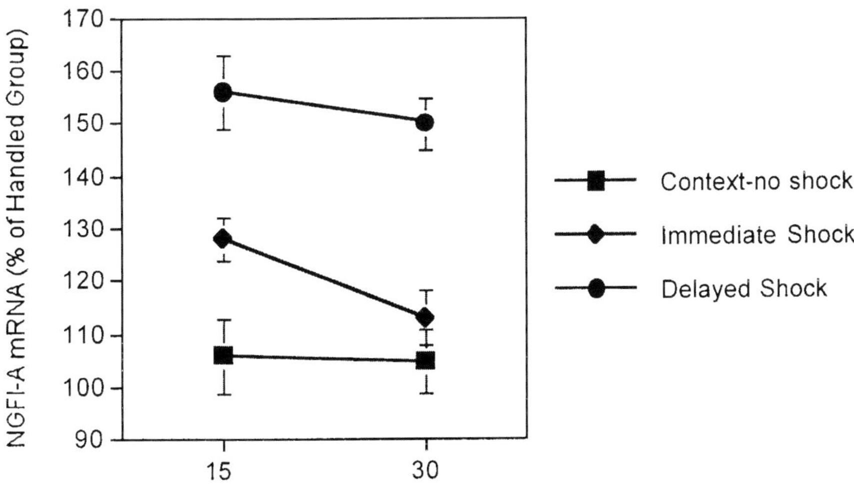

Fig. 3. (A) Digitized images of *egr-1* mRNA expression 30 min after footshock or sham footshock in the four groups of the immediate-shock deficit paradigm. Arrows point to lateral nucleus of the amygdala. (B) Graph of *egr-1* mRNA expression in the context, immediate-shock and delayed-shock groups in the lateral nucleus of the amygdala 15 and 30 min after footshock or sham footshock is expressed as a percent of expression of the handled-no shock group. At 15 min, expression in the immediate- and delayed-shock groups is significantly greater than the context- and handled-no shock groups. Expression of the delayed-shock group is also significantly greater than the immediate-shock expression. At 30 min, *egr-1* expression in the delayed-shock group is still significantly different than the other groups, while expression in the immediate-shock group returned to baseline (handled) levels.

A

NGFI-A mRNA Expression Following Retention Test

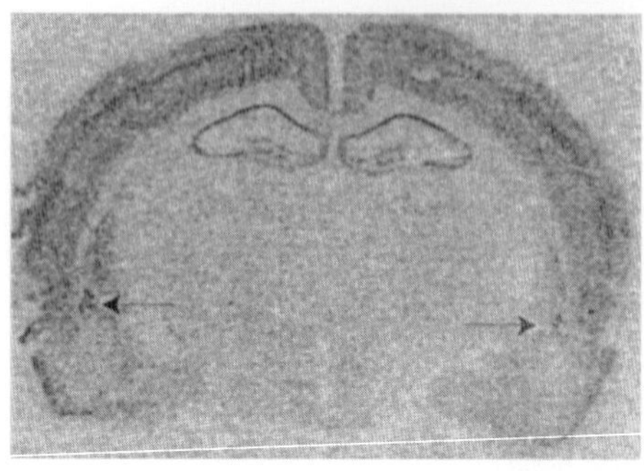

Handled-No Shock

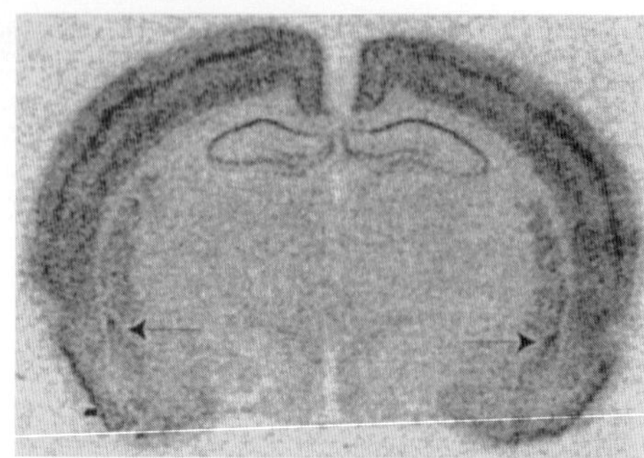

Context-No Shock

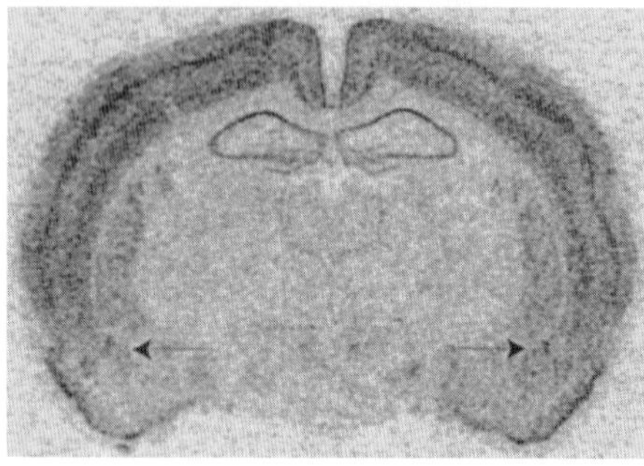

Immediate Shock

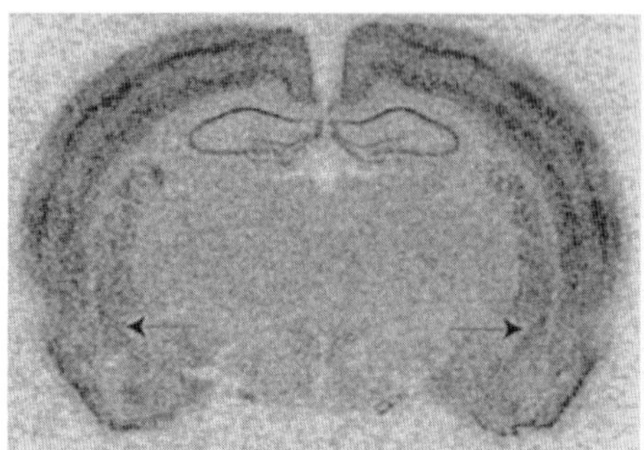

Delayed Shock

B

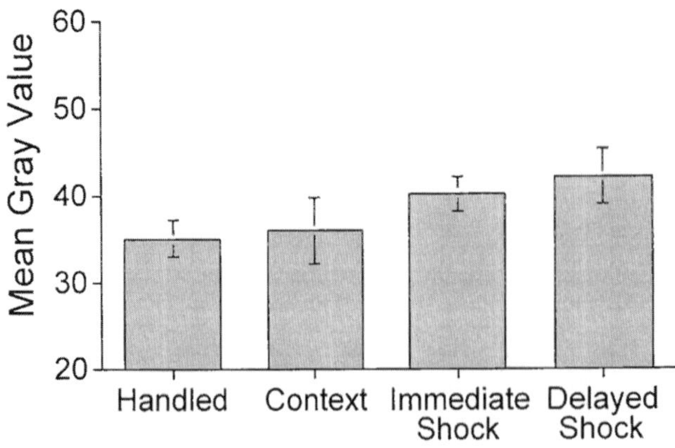

Fig. 4. Graphic representation of *egr-1* mRNA expression in the lateral nucleus of the amygdala 15 min following a retention test for contextual fear conditioning performed 24 h after conditioning. *egr-1* mRNA expression was not increased at this time (Rosen et al., 1998).

A

c-fos mRNA Expression 15 minutes After Footshock

Cg
Pr
Py

Handled-No Shock

Context-No Shock

Immediate Shock

Delayed Shock

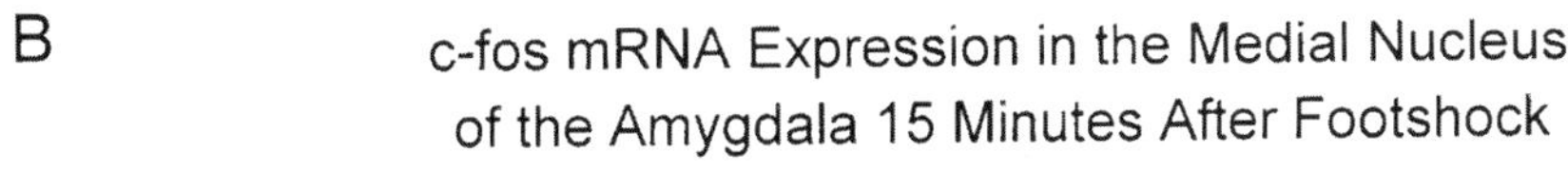

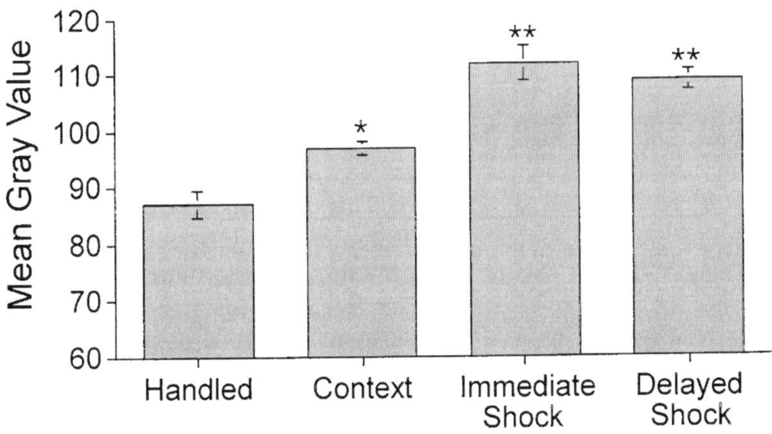

Fig. 5. (A) Digitized images of *c-fos* mRNA expression 15 min after footshock or sham footshock in the four groups of the immediate-shock deficit paradigm. Arrows point to medial nucleus of the amygdala (Rosen et al., 1998). (B) *c-fos* mRNA expression is greater in the immediate- and delayed footshock groups compared to the handled- and context-no shock groups (Rosen et al., 1998).

Example 2: Immediate-early gene expression induced by a predator odor

In addition to learned fear, fear can also be innate. Evolutionarily, it would be advantageous to be fearful of dangers in one's environmental niche without having to experience the danger. Most mammals have specific predators that inhabit their environment and prey on them. Being innately fearful of these predators, and thus avoiding them, would increase survival. Moreover, it would be even more valuable if one could detect the predator when it is still far away. Waiting until it is close enough to be seen or heard may be too late. Sensing odors from predators is a mechanism that would alert an organism that there is danger in the vicinity before it comes too close for comfort.

Most mammals are equipped with two olfactory systems that detect meaningful odors, the main olfactory and accessory olfactory systems (Dulac, 1997; Eisthen, 1997). Receptors in the main olfactory epithelium of the nasal cavity respond to airborne odorants. The neurons synapse with mitral cells in the main olfactory bulb which project to the olfactory (pyriform) cortex, several amygdala cortices, and the central nucleus of the amygdala (Swanson and Petrovich, 1998). The accessory olfactory system responds through the vomeronasal organ to pheromones that become dissolved in fluid of the vomeronasal organ. Nerve fibers from the vomeronasal organ innrevate cells of the accessory olfactory bulb which then project the medial nucleus, cortical nucleus, and central nucleus of the amygdala (Swanson and Petrovich, 1998). Pheromones are known to elicit species-specific sexual, defensive and aggressive behavior in many mammals (Eisthen, 1997), and possibly, at least, sexual behavior in humans (Weller and Weller, 1993; Cutler et al., 1998). Specific odors of predators may signal danger and innately elicit fear and defensive reactions (Blanchard and Blanchard, 1989b; Weldon, 1990; Heale et al., 1994).

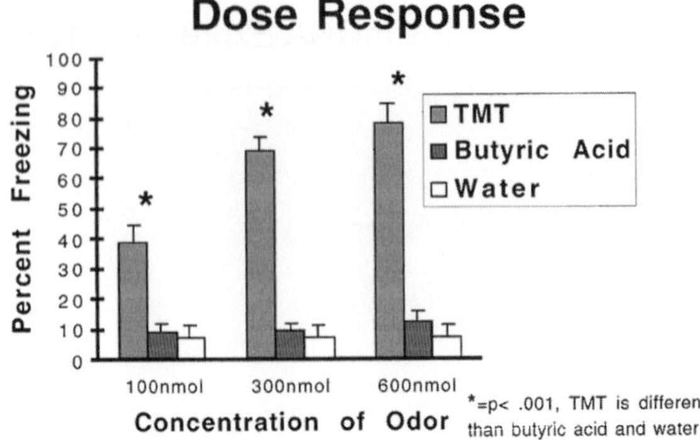

Fig. 6. Concentration-response curve to the fox predator odor, TMT, the noxious odor of butyric acid, and water. TMT increased freezing in a dose-related manner.

Fox and weasel are natural predators of rodents. Odors from waste products of fox (urine and feces) and anal secretions of weasel elicit unconditioned fear and escape responses from naive, laboratory bred rats (Weldon, 1990). We are using these odors to test whether they elicit freezing and the same increases of immediate-early gene expression in the amygdala found with conditioned fear.

To examine whether predator odors induce a fear response in naive rats, the predator odor, trimethylthiazoline (TMT), a component of fox feces, was presented to rats (see Table 1 for protocol). Rats were first acclimatized to the test chamber for 3 to 4 d. They were then exposed to various concentrations of TMT, butyric acid (the putrid smell from rancid butter) or water for 15 min in the test chamber. Exposure of 100–600 nmol TMT increased freezing in a concentration-related manner, whereas exposure to water or the noxious butyric acid did not (Fig. 6). This suggests that the increase in freezing elicited by TMT is due to TMT's fear-eliciting properties and not simply to its noxious qualities. The levels of freezing in response to 300 and 600 nmol TMT were comparable to freezing induced by the 1.5 mA footshock in our contextual fear conditioning paradigm. Furthermore, the level of freezing remained constant for the entire 15 min test period and did not habituate after three exposures to the TMT.

Personal communication with several researchers has revealed to us that it is very difficult to elicit freezing to predator odors in adult rats. In our experience, rats do not freeze to TMT if the odor is presented in a large cage that allows the rat to avoid the source of the odor. We have presented TMT on a cotton ball or in a small Eppendorf tube to rats in a Plexiglas 35 × 24 × 21 cm cage covered with wood bedding chips. The rat's initial response is to startle or jump away from the odor and then move to the opposite side of the cage with no discernible fear response. The rat does not freeze. After a few minutes the rat will move closer to the odor source by stretching and then sniffing. After a few sniffing episodes, the rat may bury the odor source with bedding chips. This sequence of behavior is very similar to that described by Blanchard and Blanchard (1989b) in their visual burrow living environment that has small burrows surrounding a central open area. When a cat is introduced to central open area of the environment for only a few minutes, the rats will scurry in the burrows and remain there for hours. Finally, one rat will investigate the central area by stretching its body into the area and sniffing. After several bouts of this it will venture further and further into the open area. The other rats will then begin to return to their normal actvity both in the burrows and in the central open area.

Our ability to elicit a freezing response to a predator odor in rats is likely due to method of odor presentation and the small test cage that limits the movement of the rat. The cage is a Plexiglas cylinder (8.6 cm diameter, 20 cm long) that is closed on both ends with doors. The rat is not restrained, it can turn around, but because of the cylindrical shape of the cage, the rat always faces one of the doors. The odors are pipetted onto a piece of tissue that is taped to the inside of each door. The rat's nose is almost continuously within a centimeter or so of the odor source (except when turning around), and as discussed above, TMT generates freezing for 40–80% of the entire 15 min test period, depending on the concentration presented.

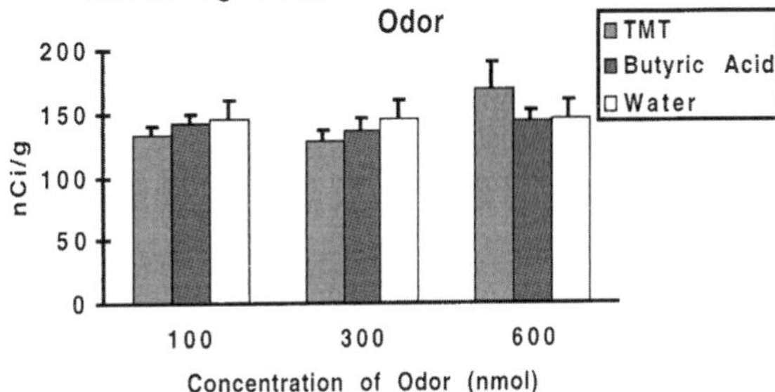

Fig. 7. Graphic representation of a lack of increased expression *egr-1* mRNA in the lateral nucleus of the amygdala and *c-fos* mRNA in the medial nucleus of the amygdala induced by TMT compared to butyric acid and water.

Immediately following a 15 min exposure to the odors, we examined the expression of *c-fos* and *egr-1* mRNA in the amygdala following the elicitation of innate fear to predator odor. The rats were decapitated, the brains removed and processed for in situ hybridization as described earlier. In contrast to fear conditioning and footshock, expression of *c-fos* or *egr-1* mRNA was not increased in any nuclei of the amygdala following this 15 min exposure to TMT (Fig. 7). On the other hand, *c-fos* expression was increased in the mitral layer of the olfactory bulb with exposure to TMT, indicating that the odor was detected. The data indicate that innate fear induced by a natural predator odor does not induce the same cascade of transcriptional processes in the amygdala as fear conditioning does. The difference in expression of *egr-1* with fear conditioning and innate fear may have important implications for a specific role of *egr-1* in the learning of fear.

Example 3: Facilitated retention of contextually-conditioned fear by chronic corticosterone treatment

The previous two examples demonstrate the use of ethologically and evolutionarily relevant stimuli and behavior to investigate molecular expression in the amygdala that may be important for normal, adaptive fear. The pathology of fear, that is, anxiety, can also be studied with a neurobehavioral system approach. It has been argued that exaggerated fear and hyperexcitability in the amygdala underlies much of the pathology of anxiety and affective disorders (Rosen et al., 1996; Rosen and Schulkin, 1998). Long-lasting hyperexcitability or hyperresponsivity in the amygdala may stem from genetic factors and/or from stressful experiences during development or as an adult (Rosen et al., 1998). Physiologically, the stress response is characterized by a rapid rise of circulating stress hormones (i.e. epinephrine, norepinephrine and the glucocorticoids (cortisol in humans, corticosterone in rodents)). Corticosterone is necessary for contextual-fear conditioning in rats (Pugh et al., 1997) and administration of high levels of exogenous corticosterone can facilitate conditioned-fear-induced freezing (Corodimas et al., 1994) and CRH-enhanced acoustic startle (Lee et al., 1994). These effects of corticosterone seem to be mediated through the amygdala and high, chronic levels of corticosterone induce expression of CRH mRNA exclusively in the central nucleus of the amygdala and bed nucleus of the stria terminalis (Swanson and Simmons, 1989; Makino et al., 1994). CRH in the amygdala and bed nucleus of the stria terminalis is thought to be an important modulator of fear (Lee and Davis, 1997; Merali et al., 1998; Wiersma et al., 1998). The data suggest that glucocorticoids are important regulators of excitability in the amygdala and may possibly facilitate lasting increases in amygdala function during fear and act as a precipitating factor for the development of anxiety disorders (Rosen and Schulkin, 1998).

We, in collaboration with Dr Jay Shulkin, Georgetown University, are just beginning to develop a model of lasting exaggerated fear induced by exposure to high, chronic levels of corticosterone (see Table 1 for details of protocol). Rats were given two 2.5 mg/kg, s.c. injections of corticosterone or oil daily for 5 d. This regimen of corticosterone administration is known to induce CRH mRNA in the central nucleus of the amygdala, while completely inhibiting CRH mRNA expression in the paraventricular nucleus of the hypothalamus (Makino et al., 1994). On the sixth day, the rats were given one injection of their usual treatment (corticosterone or oil). A group of rats that received 5 d of oil were given an injection of 2.5 mg/kg corticosterone. Two hours after the injection, the rats were contextually fear conditioned with a moderate level footshock (1 s, 0.6 mA) in the test chamber. The rats were then allowed 5 d without further injections before being tested for retention of conditioned fear. The results of our initial study is shown in Fig. 8. The 0.6 mA footshock induced moderate levels of freezing in all groups. The level of freezing was the same in rats that received oil, a single injection of corticosterone, or 5 d of corticosterone indicating that corticosterone had no effect on the perception of pain or the amount of fear induced by the footshock. However, when the rats were brought back to the chamber 5 d later to test for retention of contextually conditioned fear, the rats that earlier received 5 d of corticosterone displayed increased freezing (Fig. 8). The data indicate that memory of an emotional event is

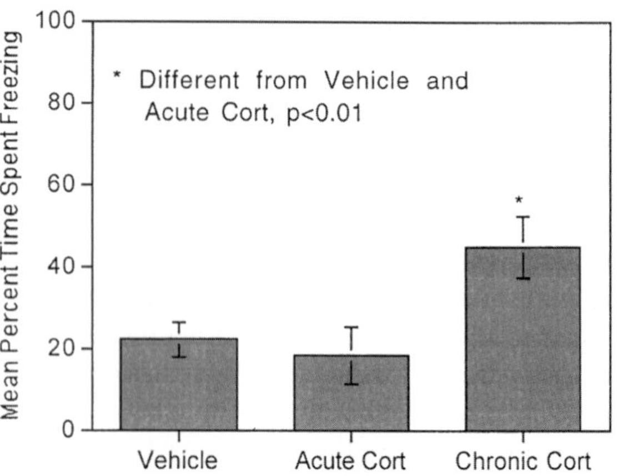

Fig. 8. Chronic, high levels of corticosterone (Cort) given for 5 d before contextual fear conditioning enhanced the memory of the learning tested 5 d after conditioning and cessation of corticosterone treatment.

facilitated when chronic, high levels of glucocorticoids are present at the time of learning. When the emotional event is encountered again, even though high levels of glucocorticoids may not be present, the fear response is enhanced or exaggerated.

Although these results are preliminary without confirmation of glucocorticoid levels, increased expression of CRH or immediate-early genes in the amygdala, we believe this method may have potential for understanding molecular responses in the amygdala and fear circuits during times of high stress, and therefore have applicability to the development of anxiety disorders such as post-traumatic stress disorder.

Conclusion

We have presented three examples of a neurobehavioral system approach to the study of gene expression in contextual fear conditioning, innate fear of a predator odor, and exaggerated contextually conditioned fear. All of the examples use a species-specific, ethologically-relevant fear response: freezing. Because fear-related freezing is dependent on the lateral, basolateral and central nucleus of the amygdala, the neurobehavioral systems approach compels us to investigate molecular changes in these nuclei to find the molecular basis of fear. We believe this approach has already yielded important differences in gene expression between fear conditioning and innate fear.

There are five caveats for this research we want to make in closing. First, although we rely on freezing as a measure of fear, other behavioral responses can and should be analyzed. Because freezing is an increase in the cessation of movement that is correlated with an increase in fear, it is particularly important to measure an increase in fear with an active phenotypic behavioral or autonomic fear response. Fear-potentiated startle is one response that satisfies this requirement since an increase in startle is a measure of an increase in fear (Davis et al., 1987). Second, in addition to contextual fear conditioning, gene expression with other types of fear conditioning needs to be investigated. Cue-specific fear conditioning, using simple conditioned stimuli such as lights or tones, or more complex stimuli such as spatial cues, may elicit their own unique patterns of gene expression. This would not be surprising because different extra-amygdaloid regions would participate in conditioning with different cues. Third, although for obvious reasons we have focussed our attention in the amygdala, gene expression in other parts of fear circuits need to be investigated. For example, *c-fos* expression has been shown to increase with fear conditioning in the ventral periaqueductal gray (PAG), a major recipient of central nucleus of the amygdala projections that controls fear-induced freezing (Carrive et al., 1997). Are the changes in *c-fos* expression in the PAG important for fear conditioning? Could changes in other areas other than the amygdala account for the behavior seen in innate fear? Fourth, other genes than the few immediate-early genes that we have studied, both those that increase and decrease in expression with fear, need to be investigated. We have found that *egr-1* is increased in the lateral nucleus of the amygdala with fear conditioning. What other genes within this nucleus are also expressed and with what temporal pattern? Also, the basolateral and central nuclei of the amygdala are critical for fear conditioning. What genes are expressed in these nuclei with fear? Finally, changes in the expression of genes that we are exploring in the neurobehavioral system of fear are only correlative. Studies which manipulate the expression of these genes are critical for determining functional significance of these gene. For instance, behavioral studies of fear conditioning in mutant mice lacking a particular gene, or over-expression of a gene by transfection into the amygdala can help address whether a gene identified within a neurobehavioral fear system is necessary for fear.

Further implementation and augmentation of neurobehavioral strategies like ones we have presented should produce more insights into the basic neurobiology of conditioned and innate fear, and have important implications for the neurobiology of anxiety disorders.

References

Abel, T., Martin, K.C., Bartsch, D. and Kandel, E.R. (1998) Memory suppressor genes: Inhibitory constraints on the storage of long-term memory. *Science*, 279: 338–341.

Abraham, W.C., Dragunow, M. and Tate, W.P. (1991) The role of immediate early genes in the stabilization of long-term potentiation. *Mol. Neurobiol.*, 5: 297–314.

Adolphs, R., Tranel, D., Damasio, H. and Damasio, A.R. (1995) Fear and the human amygdala. *J. Neurosci.*, 15: 5879–5891.

Amaral, D.G., Price, J.L., Pitkanen, A. and Carmichael, S.T. (1992) Anatomical organization of the primate amygdaloid complex. In: Aggleton, J.P. (Ed.) *The amygdala: Neurobiological aspects of emotion, memory, and mental dysfunction*, Wiley, New York, pp. 1–66.

Bading, H., Segal, M.M., Sucher, N.J., Dudek, H., Lipton, S.A. and Greenberg, M.E. (1995) *N*-methyl-d-aspartate receptors are critical for mediating the effects of glutamate on intracellular calcium concentration and immediate gene expression in cultured hippocampal neurons. *Neurosci.*, 64: 653–664.

Bandler, R. and Shipley, M.T. (1994) Columnal organization in the midbrain peraqueductal gray: Modules for emotional expression? *Trend. Neurosci.*, 17: 379–389.

Beaulieu, S., DiPaolo, T., Cote, J. and Barden, N. (1987) Participation of the central amygdaloid nucleus in the response of adrenocorticotropin secretion to immobilization stress: Opposing roles of the noradrenergic and dopaminergic systems. *Neuroendocrinol.*, 45: 37–46.

Bindra, D. (1978) How adaptive behavior is produced: A perceptual-motivational alternative to response-reinforcement. *Behav. Brain Sci.*, 1: 41–91.

Blanchard, D.C. and Blanchard, R.J. (1969) Crouching as an index of fear. *J. Comp. Physiol. Psychol.*, 67: 370–375.

Blanchard, D.C. and Blanchard, R.J. (1989a) Experimental animal models of aggression: What do they say about human behavior? In: J. Archer and K. Browne (Eds.), *Human aggression: Naturalistic approaches*, Routledge, New York, pp. 43–68.

Blanchard, R.J. and Blanchard, D.C. (1989b) Antipredator defensive behaviors in a visible burrow system. *J. Comp. Psychol.*, 103: 70–82.

Blanchard, R.J., Fukanaga, K.K. and Blanchard, D.C. (1976) Environmental control of defensive reactions to footshock. *Bull. Psychon. Soc.*, 8: 129–130.

Bolles, R.C. (1970) Species-specific defensive reactions and avoidance learning. *Psychol. Rev.*, 71: 32–48.

Bolles, R.C. and Fanselow, M.S. (1980) A percpetual-defensive-recuperative model of fear and pain. *Behav. Brain Sci.*, 3: 291–323.

Burwell, R.D., Witter, M.P. and Amaral, D.G. (1995) Perirhinal and postrhinal cortices of the rat: A review of the neuroanatomical literature and comparison with findings from the monkey brain. *Hippocampus*, 5: 390–408.

Campeau, S., Miserendino, M.J. and Davis, M. (1992) Intra-amygdala infusion of the *N*-methyl-d-Aspartate receptor anatagonist AP5 blocks acquisition but not expression of fear-potentiated startle to an auditory conditioned stimulus. *Behav. Neurosci.*, 106: 469–574.

Carrive, P. (1993) The periaqueductal gray and defensive behavior: Functional representation and neuronl organization. *Behav. Brain Res.*, 58: 27–47.

Carrive, P., Leung, P., Harris, J. and Paxinos, G. (1997) Conditioned fear to context is associated with increased Fos expression in the caudal ventrolateral region of the midbrain periaqueductal gray. *Neurosci.*, 78: 165–177.

Changelian, P.S., Feng, P., King, T.C. and Milbrandt, J. (1989) Structure of the NGFI-A gene and detection of upstream sequences responsible for its transcriptional induction by nerve growth factor. *Proc. Nat. Acad. Sci.*, 86: 377–381.

Christy, B. and Nathans, D. (1989) Functional serum response elements upstream of the growth factor-inducible gene zif268. *Mol. Cell. Biol.*, 9: 4889–4895.

Corodimas, K.P., LeDoux, J.E. and Schulkin, J. (1994) Corticosterone potentiation of learned fear. In: E.R. De Kloet, E.C. Azmatia and P.W. Longfield (Eds.), *Annals of the New York Academy of Sciences*, New York Academy of Sciences, New York, pp. 392–393

Cutler, W.B., Friedmann, E. and McCoy, N.L. (1998) Pheromonal influences on sociosexual behavior in men. *Arch. Sex. Behav.*, 27: 1–13.

Davis, M. (1992) The role of the amygdala in conditioned fear. In: J.P. Aggleton (Ed.), *The amygdala: neurobiological aspects of emotion, memory and mental dysfunction*, Wiley-Liss, New York, pp. 255–306.

Davis, M., Gendelman, D.S., Tischler, M.D. and Gendelman, P.M. (1982) A primary acoustic startle circuit: Lesion and stimulation studies. *J. Neurosci.*, 2: 791–805.

Davis, M., Hitchcock, J.M. and Rosen, J.B. (1987) Anxiety and the amygdala: Pharmacological and anatomical analysis of the fear-potentiated startle paradigm. In: G.H. Bower (Ed.), *The Psychology of Learning and Motivation*, Academic Press, San Diego, CA, pp. 264–306.

DeFranco, C., Damon, D.H., Endoh, M. and Wagner, J.A. (1993) Nerve growth factor induces transcription of NGFIA through complex regulatory elements that are also sensitive to serum and phorbol 12-myristate 13-acetate. *Mol. Endocrinol.*, 7: 365–379.

Dulac, C. (1997) How does the brain smell? *Neuron*, 19: 477–480.

Eisthen, H.L. (1997) Evolution of vertebrate olfactory systems. *Brain Behav. Evol.*, 50: 222–233.

Fanselow, M.S. (1986) Associative vs topographical accounts of the immediate shock freezing deficit in rats: Implication for the response selection rules governing species-specific defensive reactions. *Learn. Motiv.*, 17: 16–39.

Faneslow, M.S. (1990) Factors governing one trial contextual conditioning. *Anim. Learn. Behav.*, 18: 264–270.

Fanselow, M.S. (1991) The midbrain periaqueductal gray as a coordinator of action in response to fear and anxiety. In: A. Depaulis and R. Bandler (Eds.), *The midbrain periaqueductal gray: Functional, anatomical and immunohistochemical organization*, (Nato ASI Series A), Plenum, New York, pp. 151–173.

Fanselow, M.S. (1994) Neural organization of the defensive behavior system responsible for fear. *Psychon. Bull. Rev.*, 1: 429–438.

Gallagher, M. and Holland, P.C. (1994) The amygdala complex: Multiple roles in associative learning and attention. *Proc. Nat. Acad. Sci.*, 91: 11771–11776

Gallup, G.G. (1974) Animal hypnosis: Factual status of a fictional concept. *Psychol. Bull.*, 81: 836–853

Gewirtz, J.C., Falls, W.A. and Davis, M. (1997) Normal conditioned inhibition and extinction of freezing and fear-potentiated startle following electrolytic lesions of medial prefrontal cortex in rats. *Behav. Neurosci.*, 111: 712–726

Graeff, F.G. (1994) Neuroanatomy and neurotransmitter regulation of defensive behaviors and related emotions in mammals. *Braz. J. Med. Biol. Res.*, 27: 811–829

Gray, J.A. and NcNaughton, N. (1996) The neuropsychology of anxiety: Reprise. In: D.A. Hope (Ed.), *Perspectives on anxiety, panic and fear* (Nebraska symposium of motivation), Nebraska University press, Lincoln, pp. 61–134.

Heale, V.R., Vanderwolf, C.H. and Kavaliers, M. (1994) Components fo weasel and fox odors elicit fast wave bursts in the dentate gyrus of rats. Behav. Brain Res., 63: 159–165.

Hopkins, D.A. and Holstege, G. (1978) Amygdaloid projections to the mesencephelon, pons, and medulla oblongata in the cat. *Exp. Brain Res.*, 32: 529–547.

Hostege, G. (1995) The basic, somatic, and emotional components of the motor system in mammals. In: G. Paxinos (Ed.), *The rat nervous system*, Academic Press, San Diego, pp. 137–154.

Kapp, B.S., Whalen, P.J., Supple Jr, W.F. and Pascoe, J.P. (1992) Amygdaloid contributions to conditioned arousal and sensory information processing. In: J.P. Aggleton (Ed.), *The amygdala: Neurobiological aspects of emotion, memory, and mental dysfunction*, Wiley, New York, pp. 229–254.

Killcross, S., Robbins, T. and Everitt, B. (1997) Different types of fear-conditioned behaviour mediated by separate nuclei within amygdala. *Nature*, 24: 377–380.

Kim, J.J. and Fanselow, M.S. (1992) Modality-specific retrograde amnesia of fear. *Science*, 256: 675–677.

Kling, A., Lancaster, J. and Benitone, J. (1970) Amygdalectomy in the free-ranging vervet (Cercopithecus aethiops). *J. Psychiat. Res.*, 7: 191–198.

Kluver, H. and Bucy, P.C. (1939) Preliminary analysis of functions of the temporal lobes in monkeys. *Arch. Neurol. Psychiat.*, 42: 979–1000.

Krettek, J.E. and Price, J.L. (1978) Amygdaloid projections to subcortical structures within the basal forebrain and brainstem in the rat and cat. *J. Comp. Neurol.*, 78: 225–254.

Landeira-Fernandez, J., Fanselow, M.S., DeCola, J.P. and Kim, J.J. (1995) Effects of handling and context preexposure on the immediate shock deficit. *Anim. Learn. Behav.*, 23: 335–339.

LeDoux, J.E. (1987) *Emotion*. In Handbook of Physiology. Section 1: The Nervous System. Vol. 5: Higher Functions of the Brain, F. Plum, ed. Bethesda, MD: American Physiological Society, pp. 419– 460.

LeDoux, J.E. (1992) Emotion and the amygdala. In: J.P. Aggleton (Ed.), *The amygdala: neurobiological aspects of emotion, memory and mental dysfunction*, Wiley-Liss, New York, pp. 339–352.

LeDoux, J.E. (1996) *The emotional brain*, Simon and Schuster.

LeDoux, J.E., Iwata, J., Cicchetti, P. and Reis, D.J. (1988) Different projections of the central amygdaloid nucleus mediate emotional responses conditioned to acoustic stimuli. *J. Neurosci.*, 8: 17–29.

Lee, Y. and Davis, M. (1997) Role of the hippocampus, the bed nucleus of the stria terminalis, and the amygdala in the excitatory effect of corticotropin-releasing hormone on the acoustic startle reflex. *J. Neurosci.*, 17: 6434–6446.

Lee, Y., Schulkin, J. and Davis, M. (1994) Effect of corticosterone on the enhancement of the acoustic startle reflex by corticotropin releasing factor (CRF). *Brain Res.*, 666: 93–99.

Lee, Y., Lopez, D.E., Meloni, E.G. and Davis, M. (1996) A primary acoustic startle circuit: Obligatory role of cochlear root neurons and the nucleus reticularis pontis caudalis. *J. Neurosci.*, 16: 3775–3789.

MacDonald, A.J. (1982) Neurons of the lateral and basolateral amygdaloid nuclei: A golgi study in the rat. *J. Comp. Neurol.*, 212: 293–312.

MacDonald, A.J. and Jackson, T.R. (1987) Amygdaloid connections with posterior insular and temporal cortical areas in the rat. *J. Comp. Neurol.*, 262: 59–77.

Makino, S., Gold, P.W. and Schulkin, J. (1994) Effects of corticosterone on CRH mRNA and content in the bed nucleus of the stria terminalis: Comparison with the effects in the central nucleus of the amygdala and the paraventricular nucleus of the hypothalamus. *Brain Res.*, 657: 141–149.

Maren, S., Aharonov, G., Stote, D.L. and Fanselow, M.S. (1996) *N*-methyl-d-aspartate receptors in the basolateral amygdala are required for both acquisition and expression of conditional fear in rats. *Behav. Neurosci.*, 110: 1365–1374.

Marks, I.M. (1987) *Fears, phobias and rituals: Panic, anxiety and their disorders*, Oxford University Press, Oxford University Press.

Merali, Z., McIntosh, J., Kent, P., Michaud, D. and Anisman, H. (1998) Aversive and appetitive events evoke the release of corticotropin-releasing hormone and bombesin-like peptides at the central nucleus of the amygdala. *J. Neurosci.*, 15: 4758–4766.

Miserendino, M.J.D., Sananes, C.B., Melia K.R. and Davis, M. (1990) Blocking of acquisition but not expression of conditioned fear-potentiated startle by NMDA antagonists in the amygdala. *Nature*, 345: 716–718.

Morgan, M.A. and LeDoux, J.E. (1995) Differential contribution of dorsal and ventral medial prefrontal cortex to the acquisition and extinction of conditioned fear in rats. *Behav. Neurosci.*, 109: 681–688.

Okuno, H. and Miyashita, Y. (1996) Expression of the transcription factor Zif268 in the temporal cortex of monkeys during visual paired associate learning. *Eur. J. Neurosci.*, 8: 2118–2128.

Panksepp, J., Sacks, D.S., Crepeau, L.J. and Abbott, B.B. (1991) The pscyho- and neurobiology of fear systems in the brain. In: M.R. Denney (Ed.), *Fear, avoidance, and phobias*, Erlbaum, Hillsdale, NJ, pp. 7–59.

Phillips, R.G. and LeDoux, J.E. (1992) Differential contribution of amygdala and hippocampus to cued and contextual fear conditioning. *Behav. Neurosci.*, 106: 274–285.

Pitkanen, A., Savander, V. and LeDoux, J.E. (1997) Organization of intra-amygdaloid circuitries in the rat: an emerging framework for understanding functions of the amygdala. *Trends Neurosci.*, 20: 517–523.

Porges, S.W. (1995) Orienting in a defensive world: Mammalian modifications of our evolutionary heritage: A polyvagal theory. *Psychophysiol.*, 32: 301–318.

Price, J.L. and Amaral, D.G. (1981) An autoradiographic study of the projections of the central nucleus of the monkey amygdala. *J. Neurosci.*, 1: 1242–1259.

Pugh, C.R., Tremblay, D., Fleshner, M. and Rudy, J.W. (1997) A selective role for corticosterone in contextual-fear conditioning. *Behav. Neurosci.*, 111: 503–511.

Quirk, G.J., Repa, C. and LeDoux, J.E. (1995) Fear conditioning enhances short latency auditory responses of lateral amygdala neurons: Parallel recordings in the freely moving rat. *Neuron*, 15: 1029–1039.

Richardson, C.L., Tate, W.P., Mason, S.E., Lawlor, P.A., Dragunow, M. and Abraham, W.C. (1992) Correlation between the induction of an immediate early gene, zif/268 and long-term potentiation in the dentate gyrus. *Brain Res.*, 580: 147–154.

Rogan, M.T., Staubli, U.V. and LeDoux, J.E. (1997) Fear conditioning induces associative long-term potentiation in the amygdala. *Nature*, 390: 604–607.

Romanski, L.M. and LeDoux, J.E. (1993) Information cascade from primary auditory cortex to the amygdala: Corticocortical and corticoamygdaloid projections of the temporal cortex in the rat. *Cereb. Cor.*, 3: 515–532.

Rosen, J.B. and Davis, M. (1988) Temporal characteristics of enhancement of startle by stimulation of the amygdala. *Physiol. Behav.*, 44: 117–123.

Rosen, J.B. and Davis, M. (1990) Enhancement of electrically elicited startle by amygdaloid stimulation. *Physiol. Behav.*, 48: 343–349.

Rosen, J.B. and Schulkin, J. (1998) From normal fear to pathological anxiety. *Psychol. Rev.*, 105: 325–350.

Rosen, J.B., Hitchcock, J.M., Sananes, C.B., Miserendino, M.J.D. and Davis, M. (1991) A direct projection from the central nucleus of the amygdala to the acoustic startle pathway: Anterograde and retrograde tracing studies. *Behav. Neurosci.*, 105: 817–825.

Rosen, J.B., Fanselow, M.S., Young, S.L., Sitcoske, M. and Maren, S. (1998) Immediate-early gene expression in the amygdala following footshock stress and contextual fear conditioning. *Brain Res.*, 796: 132–142.

Rosen, J.B., Hamerman, E., Sitcoske, M., Glowa, J. and Schulkin, J. (1996) Hyperexcitability: exaggerated fear-potentiated startle produced by partial amygdala kindling. *Behav. Neurosci.*, 110: 43–50.

Suzuki, W.A. (1996) The anatomy, physiology, and functions of the perirhinal cortex. *Curr. Opin. Neurobiol.*, 6: 179–186.

Swanson, L.W. and Petrovich, G.D. (1998) What is the amygdala? *Trend. Neurosci.*, 21: 323–331.

Swanson, L.W. and Simmons, D.M. (1989) Differential steroid hormone and neural influences on peptide mRNA levels in CRH cells of the paraventricular nucleus: A hybridization histochemical study in the rat. *J. Comp. Neurol.*, 285: 413–435.

Veening, J.G., Swanson, L.W. and Sawchenko, P.E. (1984) The organization of projections from the central nucleus of the amygdala to brain stem sites involved in central autonomic regulation: A combined retrograde transport-immunohistochemical study. *Brain Res.*, 303: 337–357.

Von Euler, C., Martila, I., Remmers, J.E. and Trippenbach, J. (1976) Effects of lesions in the parabrachial nucleus on the mechanisms for central and reflex termination of inspiration in the cat. *Acta Physiol. Scan.*, 96: 324–337.

Walker, D.L. and Davis, M. (1997) Involvement of the dorsal periaqueductal gray in the loss of fear-potentiated startle accompnaying high footshock training. *Behav. Neurosci.*, 111: 692–702.

Wang, J.Q. and McGinty, J.F. (1994) NMDA receptors mediate amphetamine-induced upregulation of zif/268 and prepordynorphin mRNA expression in rat striatum. *Synapse*, 18: 343–353.

Wang, J.Q., Daunais, J.B. and McGinty, J.F. (1994) Role of kainate/AMPA receptors in induction of striatal zif/268 and preprodynorphin mRNA by a single injection of amphetamine. *Mol. Brain Res.*, 27: 118–126.

Weldon, P. (1990) Responses by vertebrates to chemicals from predators. In: D.W. Macdonald, D. Muller-Schwarze and S.E. Natynczuk (Eds.), *Chemical signals in vertebrates*, Oxford University Press, New York, pp. 500–521.

Weller, L. and Weller, A. (1993) Human menstrual synchrony: a critical assessment. *Neurosci. Biobehav. Rev.*, 17: 427–439.

Wiersma, A., Konsman, J.P., Knollema, S., Bohus, B. and Koolhaas, J.M. (1998) Differential effects of CRH infusion into the central nucleus of the amygdala in the Roman high-avoidance and low-avoidance rats. *Psychoneuroendocrinol.*, 23: 261–274.

Wisden, W., Errington, M.L., Dunnett, S.B., Waters, C., Hitchcock, D., Evan, G., Bliss, T.V. and Hunt, S.P. (1990) Differential expression of immediate early genes in the hippocampus and spinal cord. *Neuron*, 4: 603–614.

Worley, P.F., Cole, A.J., Saffen, D.W. and Baraban, J.M. (1990) Regulation of immediate early genes in brain: Role of NMDA receptor activation. *Prog. Brain Res.*, 86: 277–285.

Worley, P.F., Bhat, R.V., Baraban, J.M., Erickson, C.A., McNaughton, B.L. and Barnes, C.A. (1993) Thresholds for synaptic activation of transcription factors in hippocampus: Correlation with long-term enhancement. *J. Neurosci.*, 13: 4776–4786.

W.E. Crusio and R.T. Gerlai (Eds.)
Handbook of Molecular-Genetic Techniques for Brain and
Behavior Research (Techniques in the Behavioral and Neural Sciences, Vol. 13)

CHAPTER 4.8

Measuring aggression in the mouse

Pierre L. Roubertoux, Isabelle Le Roy, Stephane Mortaud, Fernando Perez-Diaz and Sylvie Tordjman

UPR CNRS 9074, Génétique, Neurogénétique, Comportement, Institut de Transgénose, 3 B rue de la Férollerie, and Université d'Orléans, 45071 Orléans Cedex 02, France

Introduction

Classical ethology has long considered aggression as the prime mover in the evolution of the species (Scott, 1992). The explanation not of why aggression occurs but how impulsive aggressive behavior happens is crucial. However, infering evolutionary mechanisms from inbred strains of laboratory mice remains a hazardous enterprise (Bruell, 1970). An inbred strain corresponds to one genotype and inference to the wide variety of genotypes of the species from 10 or less genotypes should be left to academics. The undisputed usefulness of mice experimental genetics in explaining behaviors and more specifically aggression, lies elsewhere: in the opportunity of identifying the functions of genes and the processes at work in the functioning of the brain.

Fifty-six years after the publication of two seminal papers on genetics and aggression (Ginsburg and Allee, 1942; Scott, 1942) several difficulties remain in measuring aggression using genetic analysis.

Spontaneous fights happen between adult mice housed in the same cage. Wounds or even death occur, the most severe injuries being when more than one male resides in the same cage. The number and the seriousness of wounds reflect the individuals' rank in the hierarchy indicating the social structure of the group. Change or stability of the social structure may be of interest within an ethological framework but genetic analysis of such change remains difficult. Defining aggression as a social behavior implies that it corresponds to several behavioral sequences occurring between two or more individuals following a complex network well described by ethogramms analyzable with Markow's chains. The same could be said of pup care, courtship or copulation. In the most simple case, involving two individuals, the signals produced by the first will modify the signals produced by the second, again modifying the signals from the first emitter. A network of behavioral sequences results from these social interactions. Neither this network, nor the resulting social structure have genes and cannot be used for genetic analysis. On the contrary, the ability to produce signals or to react by attacks, to specific cues of a partner, can be subjected to genetic analysis. In order to measure this, it is necessary to keep the effect of the partner constant or to compute the average reaction of one individual in several dyadic situations with a representative set of partners. Subsequent questions arise.

Several types of behavior occur in encounters that are not specific to these situations: grooming, digging, pushing. It is necessary to select relevant measures. Should these measures be analyzed separately or pooled to compute a global index?

A major difficulty with social behavior–genetic analysis resides in infering an individual ability from a situation that necessarily interacts with the individual. This difficulty culminates with aggression, several types of which have been described (offense, defense, predation, infanticide, female reactions against males attempting to mount or against lactating females). Are these behaviors correlated? Do they have common physiological mechanisms and functions?

The behavioral level of functioning integrates different levels of the nervous system. Testing rodents for aggression does not only encompass motivation or emotionality. Cognitive factors play a crucial role in initiating attack behavior. When exposed to an opponent for the first time, the mouse or the rat has a social history based on social interaction with parents, littermates or possibly other social partners previously met. From these preliminary experiences, the animal can bet on the tide of the encounter, deciding to fight or flee or to modify its strategy based on a partner's behavior. Strict control of the cognitive state of the tested animal via its social history is necessary in order to measure aggression.

Several genes implicated in murine aggression have been identified by overexpression, invalidation, insertional mutation or linkage (Maxson, 1998; Roubertoux et al., 1998;). Most of these studies used male mice and developed a dyadic encounter design. The present chapter focuses on this type of design, subsequently examining the means for controling the prerequisites presented above. Alternative protocols for testing female aggression will also be examined.

A tridimensional structure describes dyadic encounter protocols

Most of the designs used for measuring aggression in genetics have a tridimensional structure (Fig. 1).

The first dimension corresponds to the social status of the tested male, isolated vs non isolated and naive vs non naive (when the animal has been

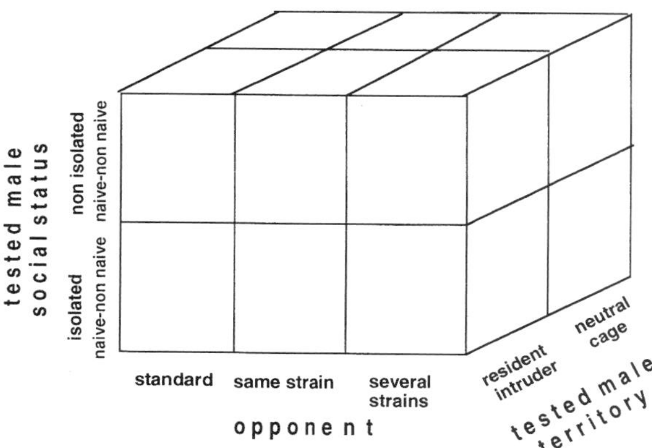

Fig. 1. Tridimensional representation of the most widely used tests for measuring attack behavior against conspecific males in experimental mouse genetics.

used for a dyadic encounter with other males). Duration of isolation varies over experiments from one week to several months. The second dimension refers to the status of the opponent with three possible levels: a standard opponent (one male is selected at random among a pool belonging to one strain, most usually different from the tested males), homogeneous set pair (a male coming from the strain of the tested male) or the round robin test (the same panel of strains for all tested males, implying several tests for one tested male). The third dimension is territoriality defined with two levels: neutral and resident cage. The neutral area with clean sawdust differs from the resident's cage.

Most authors use of the twelve possible combinations and usually do not justify the design selected. They assume that the other conditions would provide similar results. To test this hypothesis, we selected four of the designs combining the most frequent levels presented in the published literature to investigate for linkage or gene modifications. Independent male mice from eleven strains (listed in Table 1) were examined according to the procedures described below. All the strains were reared in our animal facilities for several generations prior to the beginning of the experiment and were produced under standardized conditions (see Roubertoux et al., 1994, for details). Each pregnant female was isolated from the mating cages. Litters having fewer than four pups were discarded and the others culled to six pups. At weaning, each male was housed with one female from its strain (generally a littermate) until testing (66 ± 4 d after birth). Several variables were measured but only the frequency of attacking males will be considered here.

Condition I — Standard opponent test, non isolated test males, and neutral cage

The conditions of observation and measurement (described by Carlier and Roubertoux, 1986) were identical for the 11 strains. Briefly, the test was performed in a *neutral area*: a transparent cage (42 × 26 × 18 cm) with a transparent lid. The floor of the cage was covered with clean dust free sawdust. Each test was a dyadic encounter with an A/JOrl (A/J) male as the *standard opponent* coming from a group male cage. This strain was chosen for its low scores of aggression. Indeed, A males rarely initiated attack. The males tested in such a situation were discarded. The behavioral records began when the tested male sniffed the A opponent for the first time. The test lasted 15 min in cases where no attack occurred. Tested males were *not isolated* since they were housed with a female from weaning to the day of the test. This procedure was applied 10 months later using an independent sample of the 11 strains to estimate the reliability of the measures (Condition I').

Condition II — Standard opponent test, non naive test males and neutral cage

Rearing and testing conditions were identical except for a slight modification. Tested males were exposed to the standard opponent procedure presented in the section below prior to being tested. The results of the second testing are shown in Table 1.

Condition III — Standard opponent test, isolated test males and resident intruder

Rearing conditions presented above were kept constant. At 54 ± 2 d the males were removed and housed in a clean cage. Thirteen days later a standard opponent (A/J male) was introduced in the resident cage.

Condition IV —Homogeneous set pair test, isolated test males and resident intruder

We used the same procedure with one exception, the opponent and the tested male belonged to the same strain within a pair. This non-isolated opponent came from a cage housing four males from the same strain.

Table 1 presents descriptive statistics corresponding to conditions and strains in the upper part and correlations between these measures in the lower part. These statistics will allow for a discussion of possible effects of the levels corresponding to the three dimensions previously described: social status, opponent effect and territory of tested males. The main effects and interaction were tested according to the SAS CATMOD procedure. Conditions, strains and interactions were included in this model and tested according to an analysis of variance design as indicated in the SAS package (1989). These three factors reached significance ($\chi^2_{(1)} = 215.07$, $P < 0.00001$; $\chi^2_{(10)} = 170.07$, $P < 0.00001$; $\chi^2_{(10)} = 137.78$, $P < 0.00001$ respectively).

Dimension I: Status of the tested male

Isolation vs *non isolation*

It has long been the belief that aggression needs to be induced because male mice do not spontaneously display aggression during the limited test period. Empirical observations made by fanciers and the simple idea that altrical species react by frustration to social deprivation led to the use of isolation as an inducer when investigations for the genetic correlates of aggression

TABLE 1

Upper half: percentage of attacking males (number of attacking/number of non attacking males) in 11 inbred strains of laboratory mice under different conditions. Lower half: Spearman rank coefficient

	Condition I • non isolated • standard opponent A/J • neutral area • tested once	Condition I′ • non isolated • standard opponent A/J • neutral area • tested once	Condition II • isolated (1 d) • standard opponent A/J • neutral area • tested twice	Condition III • isolated (13 d) • standard opponent A/J • resident intruder • tested once	Condition IV • isolated (13 d) • homogeneous set test • resident intruder • tested once
C57BL/10Bg	20 (4/16)	27.3 (3/8)	10 (2/18)	1.6 (3/15)	10 (2/18)
BA	47.3 (9/10)	42.1 (8/11)	50 (10/10)	20 (4/16)	19 (4/17)
C57BL/6By	15 (3/17)	16.6 (2/10)	52.3 (11/10)	10 (2/18)	14.3 (3/18)
CAST/Ei	66.6 (20/10)	71.1 (15/6)	100 (21/0)	94.1 (16/1)	88.2 (15/2)
BALB/cJ	28.1 (9/23)	3.1 (5/11)	65 (13/7)	45.4 (5/6)	10.5 (2/17)
CBA/H	0 (0/22)	0 0/15)	9(1/10)	12.5 (2/14)	69.2 (18(8)
CPB-K	29 (9/22)	26.3 (5/14)	42.1 (8/11)	47.8 (11/12)	60 (12/8)
DBA/2J	0 (0/22)	0 (0/14)	40 (8/12)	50 (10/10)	80 (16/4)
NZB/BINJ	80 (36/9)	78.9 (15/4)	0 (23/0)	0 (21/0)	100 (15(0)
XLII	122.7 (5/17)	17.4 (4/19)	13.6 (3/19)	25 (6/18)	0 (0/13)
C3H/He	4.7 (1/20)	11.1 (2/16)	14.2 (2/12)	29.1 (7/17)	75 (15/5)
Condition I • non isolated • standard opponent A/J • neutral area • tested once sample 1		0.954 $P < 0.0001$	0.741 $P < 0.009$	0.524 $P < 0.10$	0.173 $P < 0.611$
Condition I′ • non isolated • standard opponent A/J • neutral area • tested once sample 2			0.718 $P < 0.012$	0.451 $P < 0.164$	4.118 $P < 0.729$
Condition II • isolated (1 day) • standard opponent A/J • neutral area • tested twice				0.575 $P < 0.064$	0.384 $P < 0.244$
Condition III • isolated (13 days) • standard opponent A/J • resident intruder • tested once					0.655 $P < 0.030$

began. A rationale for the isolation procedure appeared later when the role of testosterone in aggression was suspected. Batty (1978a, 1978b) observed that social status induced variations in plasma testosterone concentration. Copulatory behavior reduced testosterone leading to the avoidance of housing a male with a female when planning observation of agonistic behavior in males. A stable social structure appears in groups of housed males: one is dominant exhibiting high plasma testosterone concentration while the others, who have low testosterone, are submissive. Since there is only one dominant male per group, the average plasma testosterone concentration and the correlated scores of aggression for the housed individuals are low. Isolation induces a high plasma testosterone concentration favoring initiation of attack behavior. The use of non conventional strains led us to discover that NZB/BlNJ and CAST/Ei males attack spontaneously when non isolated (reared from weaning to the aggression test with a female) (Carlier and Roubertoux, 1986; Le Roy et al., 1998). In addition the link between testosterone and aggression is controversial (Carlier et al., 1990) and non linear at least (Mortaud and Degrelle, 1996).

Interaction between strain and isolation further complicates any inferences from the measures of ability to initiate attack behavior. What is it we are measuring when using isolated males? Aggression, susceptibility to isolation, susceptibility to modification of neurotransmitters functioning?

Naive vs *non naive males*

Isolated or non isolated males can be naive (without any preliminary experience of dyadic encounter) or non naive (having at least one experience of dyadic encounter). This occurs when isolated males, or males reared with a female, are tested several times or when several males are reared together prior to aggression measuring. A single exposure to a dyadic encounter the day before the test (see Condition II) does not significantly increase the median percentage of attacking males. The results in Table 1 however show that previous experience in dyadic encounters increased aggression scores in C57BL/6J, CAST/Ei and BALB/CJ and decreased in C57BL/10Bg additional proof that the individual (or its genotype) interacts with the situation. Several general conclusions have been published about the effect of prior experience on aggression. A mouse who wins during a first test has a higher probability of winning during a retest, even if the second opponent belongs to a dominant strain compared to the strain of the tested male (Ginsburg and Allee, 1942). The reverse is true with a dominant male having lost once. When repeated, the first procedure generates long term winners and the second irreversible losers. Such effects occur when the encounter is long and when victory or defeat is blatant. Under usual conditions these effects depend on the strain. Several hypotheses could explain either this behavioral difficulty to be trained or sensorial impairments preventing perception of cues produced by the opponent.

Early events

Subtle events that cannot be controlled modify the social status of the male. In utero position of the fetuses (between two males or between two females) modifies the ability to attack (vom Saal, 1983). The large size of the litter reduces the frequency of the attacks (Carlier and Roubertoux, unpublished). Unexplained uterine factors (Sluyter et al., 1996a) interact with the strain. Carlier et al. (1991) demonstrated that the maternal post natal environment modifies the effect of the pairing region of the Y and X-chromosomes on initiation of attack behavior in crosses derived from NZB/BlNJ and CBA/H.

Recommendations

As a general rule, we suggest the factors that could produce differential reactions according to the genotypes of the mice be kept constant so as to avoid unexpected modifications of the physiological states.

Attention must be paid to early experience. Uterine position is impossible to keep constant in usual measures of aggression but carefully conducted experiments provide effective control for most other undesirable factors. Early control of the size of the litter reduces inequalities in food access during the preweaning period. The use of either embryo transfers on isogenic pseudo-gestant mothers (Carlier et al., 1991) or offspring from in situ grafted ovaries on isogenic mothers (Roubertoux et al., 1994) maintains uterine and post natal early environments constant.

Previous agonistic experiences (attacks from the father against pups, housing with several males or multiple testings) induce undesirable consequences. Isolating pregnant females from males before parturition keeps pups or juveniles from being attacked by sires. This procedure has another advantage. The need to accelerate production of rare and coveted genotypes very often leads to post partum fecondation. The females become pregnant which is not without possible impact on pups development. It has been shown that the physiological state of the mother modifies the development of embryos. Wahlsten (1982) demonstrated that the expression of genes implicated in the development of corpus callosum was magnified in embryos from lactating females. Use of post partum fecondation has irreversible effects on the developing brain and may also affect aggression.

Dimension II: The opponent

The opponent effect

This effect was reported in the early studies on aggression using inbred strains of mice (Ginsburg and Allee, 1942). An opponent effect is present when males from the same strain display different patterns or scores of aggression with opponents belonging to different strains (François et al., 1990). Table 1 illustrates this point when comparing the frequency of attacks in Conditions III and IV, as indicated by the low correlations shown in the lower part of Table 1. Several factors explain this effect. The opponent produces visual (movements, postures), auditory (ultrasounds and clicks) (Cohen-Salmon, 1988) or olfactory cues. Cues differ among strains in number, intensity and qualities (Roubertoux et al., 1996) and the tested male can perceive or ignore these cues. Several strains are deaf (DBA/2, A/J, see Zippelius and Schleidt, 1956; Cohen-Salmon et al., 1985) and some are blind (C3H/He, FVB that was not used in this study). Given the difficulty of investigating olfactory abilities very little is known about partial anosmy in inbred strains. Several lines of evidence indicate that this sensorial channel could be crucial as shown by Yamazaki's group that demonstrated the ability to discriminate partners depending on the strain of the emitter (Yamazaki et al., 1989; Yamazaki et al., 1982) or Monahan and Maxson (1998) showing that the number of attacks displayed by one strain depends on the origin of urine with which the opponent was painted. Rosenblatt, et al. (1979) concluded that experimental impairment of ability to smell (induced anosmia) was not linearly related to pup care. We could suspect the same for aggression. This difficulty can be overcome easily by selecting from the following three procedures.

Round robin test

It was suggested by Ginsburg and Allee (1942) and it still appears to be the best way to measure the ability to attack compared to the standard opponent or homogeneous set test. The test consists in averaging the scores of one tested male that is opposed to a large panel of opponents, in successive dyadic encounters. This procedure cannot be used with non reproductible genotypes (backcrosses or intercrosses) since multiple encounters may blur the measure of ability to initiate the attack. On the contrary, it should be used when the genotype survives the individual

(Recombinant Inbred Strains, congenic, transgenic mice). In this case, a global score and not an individual score is available.

Homogeneous set test

The opponent belongs to the strain of the tested male (Maxson et al., 1979). Each strain is tested with a different opponent and is tested under different conditions. With this procedure, it must be noted that two strains may attack or not for different reasons. This is all the more true when there is no co-selection between ability to produce and to perceive signals for a sensorial channel in inbred strains of mice. That the highest number of ultrasounds are produced by newborn mice in the strains where deaf adults are unable to hear them, illustrates this point very well (Cohen-Salmon et al., 1985). Olfactory cues considered as the chemical signature of an individual are more or less detected by the tested male according to its olfactory capacities (Ropartz, 1968, Novikov, 1993). Smith (1969), Scott (1960), Elwood and McCawley (1983) hypothezised that ultrasounds that are emitted during agonistic encounters act as dissuasive signals.

Standard opponent

One strain is selected for all the comparisons. Under this condition the individual abilities of the tested males are measured because the "stimulus" is kept constant across the population. A non passive standard opponent can first attack the tested male this being the sole information. With a passive opponent, on the contrary, several sequences are available (see below). We selected A/J mice for their passive behavior during dyadic encounters on the basis of previous reports (Southwick and Clark, 1968) and we confirmed that very few of the several thousands males from this strain displayed attack behavior during our tests. Most attempted to flee when the attacks started or adopted submissive postures. Induction of passivity remains possible from an initially non-dominated strain.

Recommendations

Differences in initiation of attack behavior may result from perception of sensorial cues by tested males. It would be advisable to screen for sensorial abilities in these males prior to measuring aggression.

The round robin test is preferable for measuring aggression with replicable genotypes. It is expensive however. The minimal design with transgenic mice is: two lines for one transgene and one control group with 20 mice per strain and a panel of 6 strains for the round robin test. Three hundred and sixty naïve tested males are necessary plus a sizeable number of opponents. Standard opponent design is preferable for financial reasons.

Dimension III: Territories

The following example sheds light on the impact of territoriality regarding aggressive behavior. Dominance across *Mus musculus domesticus* and wild mice (*M. spretus* or *M. spicilegus*) differs according to the size of the area that is used for testing (Thaler, 1981). Similar results were found with mice differing for t-haplotypes (Lenington et al., 1996). Size of the area where the encounter takes place, number of markings and familiarity of the odors in different situations, lead to different conditions and different results are expected. Two main areas are used generally: neutral area or rearing cage.

Neutral area

It consists of a cage with clean sawdust having two variants. The tested male is introduced to the cage 2 min before the opponent. Maxson's group uses another device. The cage is divided into two parts separated by a grid, each of the two males being placed in each part and the grid removed several minutes later. The male marks the sawdust during the time between his introduction into the cage and access to the opponent. Covering the floor of the cage with sawdust from several cages housing males of the tested groups accelerates the onset of the first attack but does not modify the other measures, including the percentage of attacking males (Carlier and Roubertoux, 1986).

Resident cage

The opponent is introduced into the cage occupied by the tested male, hence the label "resident intruder". Attack behavior might be considered as defensive behavior (protection of the territory) in the resident intruder procedure whereas the neutral area implies offensive behavior in the narrow sense. It is difficult to use this procedure with non isolated males. When the male is maintained with a female, the sawdust contains pheromones from male and female. The latter increase the frequency of attacks from the tested male.

Recommendations

The neutral area provides an indisputable measure of initiation of attack behavior but the latency of the attack may not be a good indicator of this aptitude since latency is probably correlated with the frequency of marking process. The two part apparatus limits this undesirable effect.

Correlations between conditions

Different conditions provide different measures of aggression

Comparison of the correlations between frequencies of attack behavior under different conditions indicates that measures of aggression are condition dependant. From the comparison of the correlations, a distinction appears between aggression measured with isolated vs non isolated tested males (Table 1, lower part). The correlations decrease much more when differences in isolation are paired with a standard opponent vs homogeneous set test (Condition IV). A factor analysis (principal component) after rotation (varimax) (SAS Institute Inc., 1989). shows two factors (Table 2). Loadings are higher than 0.40 for Conditions I (or I′) and II on the first factor whereas Condition III belongs to factor 1 and 2 with equivalent loadings. Condition IV has loadings on factor 2 only. Each factor is composite. Factor 1 can be defined by the tested male status (non isolated) and the status of the opponent (standard opponent). Factor 2 corresponds to the other side of the tested male (isolated) and the genotype of the opponent (same strain). Thus different conditions do not provide equal information about measuring ability to attack.

Reliability of aggression measurement

The detection of an experimental effect depends on the size of true variance as deduced from reliability. Correlations between measures performed twice on the same individuals indicate reliability. Produced moment correlations under strict assumptions provide the best estimates of true variance since they correspond to the correlations themselves and not to the squared value (Anastasi, 1988)

Attack behavior precludes the use of a test–retest design because of the possible consequences of previous exposure to dyadic situations. Hence we looked at correlations between independent samples of the same strains for Condition I. The nonparametric correlation indicates high reliability (0.954). Another strategy for estimating reliability was used for the other conditions because we did not wish to multiply the groups. The former considers the highest correlation observed for a measure as the lowest estimate of its reliability (Anastasi, 1988, for statistical arguments). Proportions of attacking males thus had reliability scores higher than 0.718 (Condition II) and 0.655 (Conditions II and III, see Table 1).

TABLE 2

Loadings for five testing conditions (principal component analysis, varimax rotated factor pattern)

Conditions	Factor 1	Factor 2
• non isolated, • standard opponent A/J, • neutral area, tested once, sample 1	0.97	0.12
Condition I′ • non isolated, • standard opponent A/J, • neutral area, • tested once, sample 2	6.97	0.15
Condition II • isolated (1 d), • standard opponent A/J, • neutral area, • tested twice	0.87	0.32
Condition III • isolated (13 d), • standard opponent A/J, • resident intruder, • tested once	0.72	0.64
Condition IV • isolated (13 d), • homogeneous set test, • resident intruder, • tested once	0.13	0.98

Recommendations

The reliability of the observations must be known before performing a genetic or an environmental analysis with new genotypes and conditions. A high reliability implies well-trained observers. Training observers requires the definition of common criteria and identical application. The agreement of their rankings is given by the τ coefficient due to Kendall (Hays and Winkler, 1971) that indicate to what degree criteria are common and application identical.

The condition "standard opponent test, non-isolated tested males and neutral cage" is probably the least ambiguous measure of inter-male aggression. The round robin condition provides valuable information but it is not suitable for studies dealing with non-replicable genotypes since it assumes that tested males are naive. It cannot be used for linkage detection with intercross or backcross, for which standard opponent is the only available procedure.

Interaction between strains and conditions

The statistical analysis shown in Table 1 (upper part) indicated that both strain and conditions were involved in aggression with a strain and condition interaction. This indicates that the strains did not react to conditions in the same way. It was possible to define several patterns of mice according to their behaviors under the conditions previously described. Standardized scoring coefficients for both conditions and strains were derived from the principal component analysis (SAS, 1989) presented above. The results are shown on Fig. 2.

A first pattern emerges with mice that rarely attack (CPB-K). The second group comprises NZB/BlNJ and CAST/Ei, the third C3H/He,

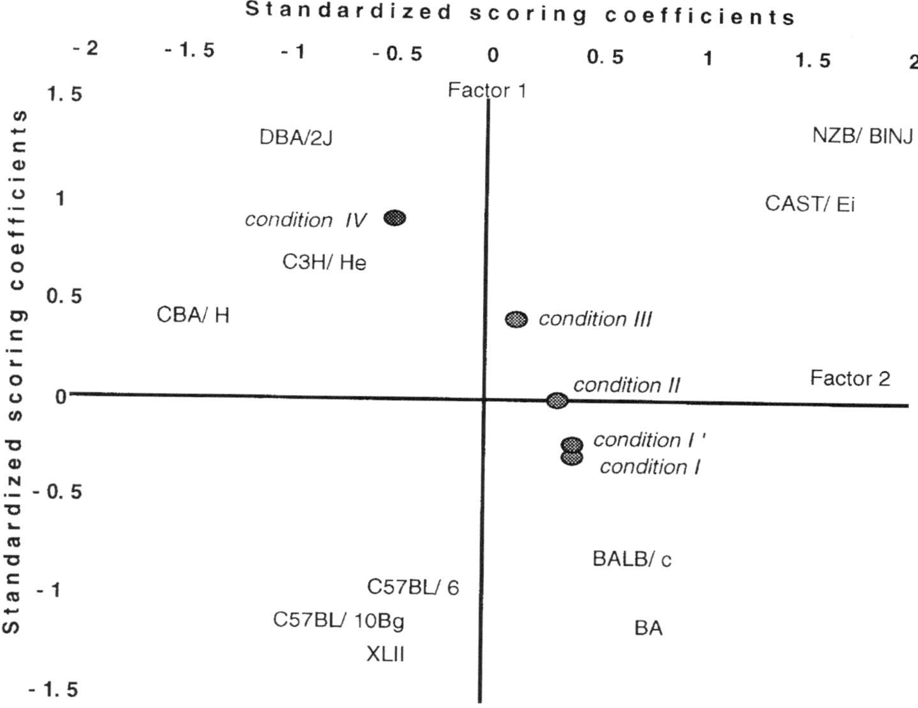

Fig. 2. Relative positioning of conditions and strains from factor analysis after deriving standardized scoring coefficients.

DBA/2J and CBA/H attacking when previously isolated. The fourth includes XLII, BA and C57BL/10Bg that did not attack when previously isolated.

Multiple measures of aggression within the same condition

We analyzed above the proportion of attacking males. Usually, several events are recorded during dyadic encounters. The question arises as to whether these measures were related together or related to aggression.

The difficulty with considering several measures during one dyadic encounter is that several of the measures are not independent by construct. The males attacking later show a reduced number of attacks when the duration of the test is limited. To overcome this difficulty we developed a simple procedure (Roubertoux and Carlier, 1988). Behavioral recording started when the tested male sniffed the opponent for the first time, and the test lasted 2 mins after the first attack or 6 min (or 15 in the studies reported in the Tables 1 and 2) after initial sniffing when no attack occurred. When no attack occured a latency corresponding to the maximal value was recorded for the individual. Under these conditions the frequency of different behaviors was measured for 2 min hence the risk of an artifactual correlation between the latency of the behavior and its frequency was reduced. This limited duration is scientifically acceptable and takes into consideration animal welfare since with some strains such as NZB/BlNJ, more than 2 min of fighting often results in mortal wounds.

We used this procedure using the 11 strains listed in Table 1, under "standard opponent test, non-isolated tested males and neutral cage" conditions. Each male was observed for 2 min after

TABLE 3

Phenotypic (line 1) and genetic correlations (line 2) obtained from 11 strains of laboratory mice (measures recorded during Condition I: Standard opponent test, non isolated test males, and neutral cage)

	Number of chases	Number of attacks	Latency first attack	Latency first rattling
Number of rattlings	0.62* 0.61	0.45 0.16	–0.38 0.17	–0.50 0.31
Number of chases		0.51 0.07	–0.46 0.21	–0.41 0.34
Number of attacks			–0.49 0.29	–0.25 0.05
Latency first attack				0.37 0.26

the first attack. The following variables were measured:

1. Sniffing latency;
2. First tail rattling latency;
3. First attack latency;
4. Number of tail rattlings;
5. Number of attacks.

Phenotypic and genetic correlations were estimated using the procedure developed by Hegmann and Possidente (1981). The results (Table 3) show that phenotypic and genetic correlations were low, indicating that the measures were geneticaly independent. These results are in accordance with a reanalysis from previously published data for 14 strains of mice (Southwick and Clark, 1968) by Michard and Carlier (1983), indicating that the measures performed during dyadic encounters were not correlated. This implies independent physiological bases and different functions for these behavioral measures.

Recommendations

These results undisputably indicate that there is no general factor of aggression within a condition. Hence avoid composite indexes should be avoided.

Measuring aggression in female mice

Usually, hormonal treatments are required to initiate attack in female mice. Early injection of testosterone is known to induce aggression in the adult female. The procedure was standardized as follows by Michard-Vanhee and Roubertoux (1990). Testosterone propionate dissolved in peanut oil was injected subcutaneously in 4 d old pups. At weaning the females were isolated then tested at 80 ± 10 d of age. Injections of estradiol benzoate 44–66 h before the test and progesterone 3–4 h before the test were used to increase female attractiveness. A male attempting to mount was presented to the female in a neutral area. Females from C57BL/6By, BALB/cBy and their Recombinant Inbred Strains differed for the measures of aggression presented in the previous section. Control mice who had received the vehicle and no testosterone propionate did not differ from control mice that displayed no attack thus indicating that isolation was not a sufficient condition in order to induce aggression in females. However, spontaneous aggression may appear in female mice. Mann et al., (1983) observed strain differences in infanticide. Females from the C57BL/6J between 25 and 45 d of age killed newborn pups placed in a neutral area whereas females from DBA/2 strain ignored the pups under identical conditions. The proportions were the same for the two strains when mice were 45 d old but they were modified after hormonal injections.

Recommendations

These investigations were carried out with a limited number of strains and more information is needed on a wide sample of inbred strains before performing more genuine genetic investigations. However female mice should be screened for sex-limited aggressive behavior if they carry either invalidated or overexpressed genes previously shown to be implicated in attack behavior in males.

Conclusion

This chapter was devoted to methodological difficulties encountered in measuring aggression in experimental genetics. It was not intended to present a general review on aggression nor the state-of-the-art in the field of genetic studies. The reader should refer to Scott (1992) or Karli (1991) for evolutionary and neuronal impacts respectively of genetic research in aggression. A recent paper and a book are of particular interest. Sluyter et al., (1996) presented an approach integrating genetics and neuroendocrinology of behavior and a volume published by Ciba Foundation (1996) includes several chapters on criminality and related behaviors in human species.

An evaluation of the genetic approach for the pre-molecular age will be found in Wimer and Fuller (1966), Maxson (1992 a, 1992b) and Carlier et al., (1990, with focus on the Y-chromosome). Perspectives offered by experimental molecular genetics (Hen, 1996) were disputed by Balaban et al., (1996). Two recent reviews attempted to synthetize the overexpression-invalidation gene approach and linkage analysis Maxson (1998), Roubertoux et al. (1998). They argued in favor of the use of identical procedures in measuring aggression in mice so as to arrive at an integrated perspective from overexpression-invalidation techniques and linkage analysis.

Acknowledgements

This research was supported by CNRS (UPR 9074), Ministry for Research and Technology (Université d'Orléans), Région Centre and Préfecture de la Région Centre. UPR CNRS 9074 is affiliated with INSERM and Université d'Orléans.

References

Anastasi, A. (1988) *Psychological testing*, Macmillan, New York.

Balaban, E., Alper, J. and Kasamon, Y.L. (1996) Mean genes and the biology of aggression: a critical review of recent animal and human research. *J. Neurogenet.*, 11: 1–43.

Batty, J. (1978a) Acute changes in plasma testosterone levels and their relation with measures of sexual behaviour in the male house mouse (*Mus musculus*). *Anim. Behav.*, 26: 339–348.

Batty, J. (1978b) Plasma levels of testosterone and male sexual behaviour in strains of the male house mouse (*Mus musculus*). *Anim. Behav.*, 26: 349–357.

Bruell, J. (1970) Behavioral population genetics and wild Mus musculus. In: G. Lindzey and D. Thiessen (Eds.), *Contribution to behavior-genetic analysis. The mouse as a prototype*, Appleton Century Crofts, New York.

Carlier, M. and Roubertoux, P.L. (1986) Differences between CBA/H and NZB mice on intermale aggression. I – Comparison between parental strains and reciprocal F1s. In: J. Medioni and G. Vaysse (Eds.), Genetic Approaches to Behavioral Phenotypes, Toulouse, Privat, pp. 47–57.

Carlier, M., Nosten-Bertrand, M. and Michard-Vanhée, Ch. (1991) Separating genetic effects from maternal environmental effects. In: D. Goldowitz, D. Wahlsten and R. Wimer (Eds.), *Techniques for the Genetic Analysis of Brain and Behavior: Focus on the Mouse*, Elsevier, Amsterdam, pp. 111–126.

Carlier, M., Roubertoux, P.L., Kottler, M.-L. and Degrelle H. (1990) Y chormosome and aggression in strains of laboratory mice. *Behav. Genet.*, 20: 137– 156.

Carlier, M., Roubertoux, P.L. and Pastoret, C. (1991) The Y chromosome effect on intermale aggression in mice depends on the maternal environment. *Genet.*, 129: 231–236.

Ciba Foundation (1996) *Genetics of criminal and antisocial behaviour*, Ciba Foundation Symposium, Chichester, J. Wiley.

Cohen-Salmon, Ch. (1988) What role does sensory perception play in the onset and maintenance of pup care behavior in laboratory rodents? *Eur. Bull. Cognit. Psychol.*, 8: 53–94.

Cohen-Salmon, Ch., Carlier, M., Roubertoux, P.L., Jouhaneau, J., Semal, C. and Paillette, M. (1985) Differences in patterns of pup care in mice. V – Pup ultrasonic emissions and pup care behavior. *Physiol. Behav.*, 35: 167–174.

Elwood, R.W. and McCawley, P. J. (1983) Communication in rodents: Infants to adults. In: R.W. Elwood (Ed.), *Parental Behavior of Rodents*, Wiley, New York, pp. 127–149.

François, M.H., Nosten-Bertrand, M., Roubertoux, P.L., Kottler, M.L. and Degrelle, H. (1990) Opponent strain effect on eliciting attacks in NZB mice: physiological correlates, *Physiol. Behav.*, 47: 1181–1185.

Ginsburg, B. and Allee, W.C. (1942) Somme effects of conditioning on social dominance and subordination in inbred strains of mice. *Physiol. Zool.*, 15: 485–506.

Hays, W.L and Winkler, R.L. (1971) Statistics, Holt, Rinehart, New York.

Hegmann, J.P. and Possidente, B. (1981) Estimating genetic correlations from inbred strains. *Behav. Genet.*, 11: 103–113.

Hen, R. (1996) "Mean genes". *Neuron*, 16: 17–21.

Karli, P. (1991) *L'homme agressif*, Odile Jacob, Paris.

Lenington, S., Drickamer, L. Robinson, A.S. and Erhart, E. (1996) Genetic basis for male aggression and survivorship in wild house mice (Mus domesticus). *Aggressive Behavior*, 27: 135–145.

Le Roy, I., Roubertoux, P.L., Jamot, L., Maarouf, F., Tordjman, S., Mortaud, S., Blanchard, C., Martin, B. and Duquenne, V. (1998) Neuronal and Behavioral Differences between *Mus musculus domesticus* (C57L6/JBy) and *Mus musculus castaneus* (CAST/Ei). *Behav. Brain Res.*, 95: 135–142.

Mann, M., Kingsley, C., Broida, J. and Svare, B. (1983) Infanticide exhibited by female mice: genetic, developmental and hormonal influences. *Physiol. Behav.*, 30: 697–702.

Maxson, S.C. (1998) Homologous genes, aggression and animal models. *Devel. Neuropsychol.*, 14: 143–156.

Maxson, S.C. (1992a) Methodological issues in genetic analyses of an agonistic behavior (offense) in male mice. In: D. Goldowitz, D. Wahlsten and R. Wimer (Eds.), *Techniques for the Genetic Analysis of the Brain and Behavior: Focus on the Mouse*, Elsevier, Amsterdam, pp. 349–374.

Maxson, S.C. (1992b) Potential genetic models of aggression and violences in males. In: P. Driscoll (Ed.), *Genetically Defined Animals Models of Neurobehavioral Dysfunctions*, Birkhäuser, Boston, pp. 174–188.

Maxson, S.C., Ginsburg, B. and Trattner A. (1979) Interaction of Y-chromosomal gene(s) in the development of intermale aggression in mice. *Behav. Genet.*, 9: 219–226.

Michard-Vanhee, C.H. and Roubertoux, P.L. (1990) Genetic analysis of differences in behavioral reactivity to neonatal injection of testosterone in female mice, *Behav. Genet.*, 19: 63–71.

Michard, Ch. and Carlier, M. (1983) Les conduites d'agression intaspécifiques chez la souris domestique, différences individuelles et analyses génétiques. *Biol. Behav.*, 10: 123–146.

Monahan, E.J. and Maxson, S.C. (1998) Y chromosome, urinary chemosignals and an agonistic behavior (offense) of mice. *Physiol. Behav.*, 64: 123–132.

Mortaud, S. and Degrelle, H. (1996) Steroid control of higher brain function and behavior. *Behav. Genet.*, 26: 367–372.

Novikov, S.N. (1993) The genetics of pheremonally mediated intermale aggression in mice: current status and prospect of the model. *Behav. Genet.*, 25: 505–508.

Ropartz, P. (1968) Olfaction et comportement social chez les rongeurs. *Mammalia*, 32: 550–569.

Rosenblatt, J.S., Siegel, H.I and Mayer, A.D. (1979) Progress in the study of maternal behavior in the rat: hormonal, nonhormonal sensoria and developmental aspects. *Advances in the Study of Behavior*, 10: 225–311.

Roubertoux, P.L., Mortaud, S., Tordjman, S., Le Roy, I. and Degrelle, H. (1998) Behavior-genetic analysis and aggression: the mouse as a prototype In: M. Sabourin, F. Craik and M. Robert (Eds.), *Advances in Psychological Science: Biological and Cognitive Aspects*, Psychology Press, London, pp. 3–29.

Roubertoux, P.L, Carlier, M., Degrelle, H., Haas-Dupertuis, M.C., Phillips, J. and Moutier, R. (1994) Co-segregation of the pseudoautosomal region of the Y chromosome with aggression in mice, *Genetics*, 135: 254–263.

Roubertoux, P.L. and Carlier, M. (1988) Differences between CBA/H and NZB on intermale aggression II – Maternal effects, *Behav. Genet.*, 18: 175–184.

Roubertoux, P.L., Nosten-Bertrand, M. and Carlier, M. (1990) Additive and interactive effects of genotype and maternal environment. *Advances in the Study of Behavior*, 19: 205–247.

Roubertoux, P.L., Martin, B., Le Roy, I., Beau, J., Marchaland, C., Perez-Diaz, F., Cohen-Salmon, Ch. and Carlier, M. (1996) Vocalizations in Newborn Mice: Genetic Analysis. *Behav. Genet.*, 26: 427–437.

SAS institute Inc. (1989) *SAS/STAT User's Guide*, Version 6,. Cary, NC, SAS institute Inc.

Scott, J.P. (1942) genetic differences in the social behavior in inbred strains of mice. *J. Hered.*, 33: 11–15.

Scott, J.P. (1960) *Aggression*, The University Press, Chicago.

Scott, J.P. (1992) Aggression: functions and control in social systems. *Aggress. Behav.*, 18: 1–20.

Sluyter, F., Van Oortmerssen, G.A., Ruiter, A.J.H. and Koolhaas, J.M. (1996) Aggression in wild house mice: Current state of affairs. *Behav. Genet.*, 26: 489–495.

Sluyter, F., Van der Vlugt, J., Van Oortmerssen, G.A., Koolhaas, J.M. Van der Hoeven, F. and De Boer P. (1996a) Studies in wild house mice: VII. Prenatal maternal environment and aggression, *Behav. Genet.*, 26: 513–518.

Smith, J.W. (1969) Messages in vertabrate communication. *Science*, 165: 145–150.

Southwick, C.H. and Clark, L. (1968) Interstrain differences in aggressive behavior and exploratory activity of inbred mice. *Comm. Behav. Biol. A*, 1: 49–59.

Thaler, L. (1981) Processes of speciation in the mouse house. *Symp. Zool. Soc. London*, 47: 27–41.

vom Saal, F.S (1983) Models of hormonal effects on intrasex aggression in mice. In: B. Svare (Ed.), *Hormones and Aggressive Behavior*, New York, Plenum Press.

Wahlsten, D. (1982) Mice in utero while their mother is lactating suffer higher frequency of deficient corpus callosum. *Develop. Br. Res.*, 5: 354–357.

Wimer, R.E. and and Fuller, J.L. (1966) Patterns of behavior. In: M. Green (Ed.), *Biology of the Laboratory Mouse*, New York, Pergamon.

Yamazaki, K., Beauchamp, G.K., Bard, J. and Boyse, E.A. (1989) Sex-chromosomal odor types influence the maintenance of early pregnancy in mice. *Proc. Natl. Acad. Sci. USA*, 86: 9399–9401.

Yamazaki, K., Beauchamp, G.K., Bard, J., Thomas, L. and Boyse, E.A. (1982) Chemosensory recognition of phenotypes determined by the Tla and H2K regions of chromosome 17 of the mouse. *Proc. Natl. Acad. Sci.*, 79: 7828–7831.

Zippelius, H.M. and Schleidt, W.M. (1956) Ultraschall-laute bei Jungen Mäusen. *Naturwissenchaften*, 43: 502–503.

W.E. Crusio and R.T. Gerlai (Eds.)
Handbook of Molecular-Genetic Techniques for Brain and Behavior Research (Techniques in the Behavioral and Neural Sciences, Vol. 13)

CHAPTER 4.9

Methodological issues in the assessment of behavioral development in laboratory mice

P.E. Wainwright

Department of Health Studies and Gerontology, University of Waterloo, Waterloo, Ont. N2L 3G1, Canada

Introduction

A behavioral phenotype, like any other, is the outcome of the complex interaction of the genotype with the environment during the developmental process. Thus developmental studies are intrinsic to our understanding of the multiple factors that contribute to the etiology of behavioral differences. When such studies derive from an ethological perspective, they tend to focus on factors that influence the development of species-typical behaviors (and, by implication, differences between species), whereas the emphasis of the psychological approach is usually on understanding the sources of individual differences in behavior within a species.

The laboratory mouse has proven to be a valuable tool in the latter endeavor, particularly because of the availability of well-characterized inbred strains. These strains differ among themselves genotypically, but, as a result of generations of brother × sister mating, all animals within a strain are expected to be homogeneous and genetically homozygous. They have been used most commonly in the study of spontaneously occurring neurological mutants, as well as in behavioral genetic studies that use quantitative methods to assess the extent to which the differences between individuals in a population reflect differences in genotype. The latter has historically been expressed most often in terms of estimates of heritability, but such estimates have been criticised strongly on various grounds, including that they are descriptive only, and may be limited to both the population and environment studied. As it is not the intent to deal specifically with these issues here, the reader interested in a more comprehensive discussion is referred to other sources in the literature (Crusio, 1992, Wahlsten, 1994). These concerns notwithstanding, the laboratory mouse continues to offer further exciting possibilities in terms of testing specific mechanistic hypotheses, particularly in the light of more recent advances, such as the development of congenic and recombinant inbred strains, as well as molecular genetic techniques which allow the production of transgenic and knockout mice (Goldowitz et al., 1992). As exemplified by the extensive work on the agenesis of the corpus callosum in the BALB/c mouse strain (Wahlsten, 1989), developmental genetic studies can serve not only to identify processes that are associated with anomalous brain development but, in so doing, also enhance our understanding of the normal developmental process. Another important reason for including developmental studies in our search for brain-behavior relationships is that of accounting for developmental plasticity. For example, there is evidence that substances which function as neurotransmitters in mature animals also act as trophic agents in the developing brain (Lauder, 1993). Thus a brain with a congenital disruption of one of these systems as a result of a targeted genetic manipulation may develop quite differently in terms of organization of function compared with that of a normal animal (Levitt et al., 1997). It would not therefore be surprising to see quite different behavioral outcomes in genetic studies of this type compared with those where neurotransmitter activity is manipulated directly in the brains of normal adult animals by pharmacological means. The implication of this is that the behavioral phenotype of the adult knockout animal cannot be related unequivocally to the concurrent neurochemical anomaly, but reflects rather the result of a developmental history of such a deficit. It therefore behooves us to consider the study of the behavioral development as intrinsic to our overall pursuit of understanding in these matters.

Whether one is interested in the effects of either genotype or environment on behavioral development, it is readily apparent that similar methodological considerations will apply. Thus, in comparing differences in behavioral development between the offspring of two inbred mouse strains, or between the offspring of two groups of pregnant animals exposed to an experimental treatment, such as a drug or dietary regimen, all aspects of the studies, other than the nature of the independent variable (genotype or treatment), would be the same. In fact, the most informative study designs for some questions are often those in which both factors are varied. Inbred strains can be very useful in studying differences in response to the neurotoxicological effects of developmental exposure to a drug such as ethanol and, if present, the unravelling of these genotype by environment interactions can prove informative in elucidating the mechanisms involved (e.g. Cassells et al., 1987). The importance of appropriate methodology and design for detecting the functional effects of environmental chemicals has led regulatory bodies such as the Environmental Protection Agency (EPA) in North America to develop stringent protocols for neurotoxicity testing in animals. Although there are inherent differences between the processes of basic research and regulatory monitoring that might in turn lead to differences in emphasis in terms of the actual protocols used, the basic experimental principles of developmental research do not differ. With this in mind, the objective of this chapter is to provide an overview of the methodological issues that pertain to the measurement of neural and behavioral development in mice. For more extensive discussion of issues of particular interest, the reader is encouraged to consult specific references (Roubertoux et al., 1992; Bates et al., 1996) as well more comprehensive texts in this area (Goldowitz et al., 1992; Kimmel and Buelke-Sam, 1994; Hood, 1996). The material in this chapter reflects our own familiarity with the mouse, as well as the fact that many of the genetic models have been developed in this species, but similar considerations will apply in many other rodent models, including the rat.

General experimental procedures in developmental studies

Behavior does not emerge de novo at birth, but is rather the functional manifestation of an ongoing developmental process that begins at conception. The time of conception is relatively easy to ascertain in mice because of the presence of copulatory plugs. A male is introduced into a cage of females at a specified time of day for a predetermined period, the latter depending on the accuracy of the time interval one wishes to study, and the females are then checked for vaginal plugs. The time of conception is judged to be the midpoint of the mating interval (Time 0, Day 0). Because mice are nocturnal animals, the dark-cycle is the optimum time for mating, as well as for behavioral testing. Thus it often becomes convenient to maintain the animals on a reversed dark-light cycle (e.g. lights off at 0900 h to

21:00 h). It is particularly important in developmental studies that all animals be tested at approximately the same time of day, which means that the number of mated animals that can be assigned to a study on any one day will depend on the resources available for the behavioral testing procedures. It is also important to ensure that groups are randomized to testing time within the daily testing interval, so that any circadian variation that is present will be included in the error term. Pregnant dams are usually housed separately in a cage with ample nesting material, and weighed at regular intervals, starting at conception, and usually every 2–3 d thereafter; in some cases, measures of food intake might also prove informative (see discussion of maternal effects below). Although monitoring of maternal weight is important in dietary and toxicological studies, this has the disadvantage of introducing additional handling effects, which may be stressful. Thus there may be some instances where the experimenter decides that it is better that this procedure be omitted. The gestation period in mice is normally between 18 and 20 d (counting conception as Day 0). Thus starting on Day 17, the dams are checked regularly for births. This should be done at least every 12 h, but this period can be shorter if more accurate estimates are needed. Litters are counted (live and dead animals), and weighed and sexed as soon as possible after birth. This initial measure of litter size is important, because of the possibility of selective mortality. This is illustrated by a study that we conducted to investigate differences between BALB/c and C57/BL6 mice in response to prenatal ethanol exposure (Cassells et al., 1987). These results indicated a significantly greater effect of 17.5% ethanol-derived calories on body weight at birth in C57s compared with BALBs. However, the prenatal ethanol treatment also reduced litter size in BALBs, which suggests that only the more robust BALBs survived to be compared with the less rigorously selected C57s, who in turn represented the entire range of the effects. So, by combining effects on litter size with those on birth weight, we were led to a different interpretation of the outcome.

Because the timing of birth can vary due to factors that are unrelated to the timing of events related to brain development, it is best to continue to use postconceptual age during the postnatal period, rather than counting age from birth. In our laboratory litters are usually culled on Day 21 postconception to a maximum of 6–8 pups (depending on the strain) with equal representation of males and females if possible. Normally animals to be culled are selected randomly, but special considerations might apply in studies of genetic mutants and their littermate controls (discussed below). If necessary, individual pups can be identified by marking with an indelible dye, such as Bouin's solution. Pups are weighed at regular intervals, at least bi-weekly, and more often if necessary. They can also be assessed on other developmental landmarks, such as the time of eye-opening or the attainment of sexual maturity. We customarily assess general sensorimotor capabilities on Day 32 postconception, using a standardized developmental scale developed for mice (discussed below). Activity measures are often informative, with different methods of assessment providing different information. Activity in the open field represents initially the response to a novel situation, with the decline seen over time indicative of habituation. The spatial open field, with objects positioned at various points in the field, also allows one to assess spatial learning capabilities by measuring the animals' behavioral response to the relocation of these objects (Roullet and Lasalle, 1990). Home-cage activity, on the other hand, optimally measured over a 24 h period to show circadian variation, is more indicative of spontaneous baseline activity. But an obvious obstacle in terms of the latter is the need for young pups to nurse regularly. More specific tests of particular functions also constitute important aspects of behavioral test batteries, and these can include the assessment of cognitive development using both classical and instrumental conditioning procedures, as well as social competence (Stanton and Spear, 1990). Given the relatively immature state of motoric capabilities in young animals, as well age-related differences in motivation or in response to reinforcement, it is particularly important to be cognizant of the possible role of performance factors in any behavioral testing situation (Spear and Rudy, 1991). Standardized behavioral test batteries, such as that described below, may be useful in identifying such shortcomings. In some instances it might be advantageous to scale down a testing apparatus to accommodate their needs, for example, using a smaller water maze so that the procedure is physically less taxing. It is also important in the case of testing very young animals to be aware that because their thermoregulatory capacity is not mature, they should be tested under thermoneutral conditions (Satinoff, 1991).

Litter effects

Developmental studies using multiparous species, i.e. a single birth resulting in more than one offspring, pose unique problems, not only in terms of experimental design, but also in terms of the experimental protocol followed (Rice and Ray, 1996). From a statistical viewpoint, the randomly assigned unit is the litter and not the individual pup, and it is important that litter effects be taken into account, both in the design of the study and the subsequent statistical analysis of the data (Wainwright, 1998). Because outcome measures on pups from the same litter are likely to be correlated, treating each pup as an independent unit will artificially inflate the power of the study, such that significance will be claimed for effects that are in truth non-significant. The most conservative approach is to measure all animals in a litter, and to then average these values to give a litter mean score. Alternatively, only one animal (or one male and one female) randomly selected from each litter, can be tested. Because the variability within litters might be quite large, this latter approach could require a large number of litters in order to have sufficient power to detect effects of a meaningful size (Zorrilla, 1997). Both these approaches assume a high level of correlation however, and the recent development of statistical techniques that allow one to assess the actual degree of correlation within litters, and then to adjust the degrees of freedom accordingly, might prove to be a viable alternative approach (Muthén and Curran, 1997). Nonetheless, to design studies with the appropriate sample size would still require some prior knowledge of the degree of correlation expected in the measures of interest. A major disadvantage of breeding and testing large numbers of litters over time is the demand placed on limited resources such as space and personnel. This may necessitate completing the study as a series of cohorts, in which case it is important to ensure that all experimental groups are represented in each of the cohorts.

There is one way in which the litter "problem" can be turned to an advantage in longitudinal studies. This is the opportunity it affords to test different animals from a litter at different time points, thereby allowing one to make multiple measures on an experimental unit, while at the same time avoiding the problem of handling/practice effects from one testing session to the next. However, there is one caveat to be observed should the experimental protocol require that animals be killed after behavioral testing for further measures of brain morphology or chemistry. This relates to the importance of keeping litter size constant throughout the study. This is because developmental indices of growth, such as body weight, are related to litter size (unless the number of pups falls below the minimum necessary to maintain adequate lactational performance in the dam) (Wainwright et al., 1989). By having available extra litters at the same stage of development during the postweaning period, one can then replace the pups lost to testing, as long as the replacement pups can be identified clearly so as not to be confused with those in the litter remaining to be tested (Wainwright et al., 1994). It is important to realize that testing of different animals in a litter in longitudinal studies is appropriate only in the case wherein all the animals have the same history with respect to the independent variable, i.e. genotype in the case of inbred strains, or when it is the dam that is subjected to an experimental treatment. Different considerations apply when one is working with mutants, when, in the case of a recessive character, some animals in a litter might be homozygous normal, others heterozygous carriers, and still others will be homozygous mutants. On the face of it, this may appear to be the ideal situation in that one can use these littermate control animals for comparison. But one issue to consider is whether animals of the same age are the correct comparison for the question of interest, or whether animals of the same developmental stage might be more appropriate. This will be discussed further below.

Maternal effects

Although behavioral measurements are made on individual pups, maternal factors become an important consideration in interpreting any effects that may be found (Hood and Miller, 1996). This is because an experimental manipulation (or condition, such as genotype) can either affect pups' behavior directly, or such effects might be mediated indirectly through effects on maternal physiology and behavior. Parity is an interesting example in this respect. It has been shown in mice, for example, that maternal behavior differs between primiparous and biparous dams (Cohen-Salmon, 1987), and that the development of behavioral responses in pups is delayed in those from third, compared with second, litters (Crusio and Schmitt, 1996). Similarly, in our own laboratory, we have shown that some of the effects of prenatal stress on development in mice can be attributed to stress-induced reductions in maternal food intake (Ward and Wainwright, 1988). This latter effect represents the direct outcome of the experimental manipulation on the dam, but such effects could also be mediated indirectly, as when the effect of the pup's genotype is to elicit particular behaviors from the mother, which in turn become instrumental in altering behavioral development. A classic example of this is the observation that rat mothers spend more time in anogenital licking of male, compared with female, pups (Moore, 1984). This serves to illustrate that, although observations of maternal behavior might not be included initially in an experimental protocol of behavioral development, they might prove crucially informative at a later stage of identifying the proximal mechanisms involved.

When one is working with inbred strains one is faced with an invariant correlation between pup and the maternal genotype, and hence the possibility of both pre- and postnatal maternal environmental effects confounding the interpretation of differences in developmental outcomes among the pups (Wainwright, 1980, Carlier et al., 1982). There are inventive ways not only of controlling such effects, but also of turning them into an advantage (Carlier et al., 1992). Postnatally these include fostering techniques, with the approach chosen dictated by the exact purpose of the study, i.e. in terms of whether one wanted to actually study maternal effects or merely to control for them. Cross-fostering, where litters of each strain are fostered at birth, either to dams of their own and the opposite strain, allows one to assess maternal effects. If one is also interested in the effects of fostering per se, this requires the inclusion of groups that are not fostered, but remain to be reared by their natural dam. Surrogate fostering involves fostering all litters in the study to dams of a common genotype, and thereby controlling for postnatal maternal effects by providing a common maternal environment (but see below). In the latter case the foster dams should be chosen on the basis of their maternal competence, and often the best choice for this is the F_1 hybrid animal. Fostering pups at birth is usually a fairly robust procedure, as long as it is done as soon as possible after birth, and one takes care to cause minimum disruption of the nest, and to rub some of the foster bedding material on the pups to impart the appropriate odor (these two latter points are important whenever pups are handled). Although surrogate fostering has been touted as a control for maternal effects, it is wise not to be too complacent about this, because of the possibility of interactions between pup characteristics and maternal behavior discussed above. Thus, if one does use surrogate fostering for this purpose, there are several potential aspects of maternal behavior one might choose to include as measures in the study for reassurance. Pup-retrieval is one, as described in Wainwright et al. (1994), but this may be too robust a phenomenon to reflect other more subtle, but nonetheless important, behavioral differences. There is a more detailed discussion of maternal behavior in this volume (Fleming, 1999) which may be useful to the investigator in deciding which measures are likely to be informative.

There is another way of controlling maternal behavioral effects during the postnatal period, and this is to rear animals artificially. This technique, which was first described by Messer (1969) and Hall (1975) in rats, involves the use of mechanical infusion pumps to deliver artificial rat's milk to the pups through gastrostomy tubes. It is often termed "pups-in-cups", because the pups are housed individually in Styrofoam cups that float in a warm water bath. Although this procedure avoids postnatal maternal influences, there is the concern that the artificial rearing condition may itself present a problem for normal behavioral development. That is why it is important that such studies also include a normally suckled control for comparison. We have recently used this method to study the effects of dietary long-chain essential fatty acids on brain and behavioral development in rats (Ward et al., 1998; Wainwright et al., 1998, 1999) but, to our knowledge, this technique has not been used in mice, no doubt because the small size of the pups precludes it. Because of this work, we have also started to develop some tests specific to the development of locomotor behavior in the rat, which are based on earlier work by Altman and Sudarshan (1975).

The importance of the prenatal period to subsequent behavioral organization in mammals is becoming increasingly appreciated as a result of work conducted in fetal rats (Smotherman and Robinson, 1990). Prenatal maternal effects can be studied in ways that are conceptually similar, although the procedures are technically more complicated than those implemented during the postnatal period. In addition to the transfer of ova or fertilized embryos between dams, these approaches include ovarian transplants to achieve ends that are analogous to those of fostering, i.e. either from one strain to another or to F_1 hybrid dams (Bulman-Fleming and Wahlsten, 1988). It is highly probable that even more sophistication in this area will become possible with the current rapid advances in reproductive technology.

Developmental behavioral test battery

This section begins with a discussion of the behavioral test battery that we have used in our developmental work in mice. The items that constitute this battery were derived from work done by Fox (1965), and have been further developed by Wahlsten (1974, 1975). The tests are always administered in the same sequence, an animal's performance is assigned a score of between 0 and 1.0 on each test, and these scores are then averaged over all the test items. One disadvantage of a composite average score such as this is that different animals (or groups) could attain the same overall score, but for different reasons. This is why it is important that preliminary analysis of the data also includes comparing animals on each of the individual tests in order to ascertain whether or not the patterns are similar (see further discussion of this issue in Roubertoux et al., 1992, p. 436). If it in fact becomes apparent that the use of the composite score is obscuring some interesting differences in the overall pattern, or that the differences seen on the composite score are being driven by large differences in only a few of the component scores, then it may be more informative to present the outcomes for the individual tests (with suitable statistical controls for multiple outcomes on the same subject).

Of necessity, behavioral assessments such as these involve some degree of subjectivity. Thus, in order to avoid an inadvertent testing bias, these tests should be conducted with the experimenter unaware of the experimental condition of the animal. Included in the description below is an indication of the judgement criteria that we have developed over the years of using this test battery. As these tests are very labor-intensive, it is often the case that more than one observer will be involved in the collection of data, and we have developed a videotape that we use to train observers on this test. In computing inter-observer reliability, it is important to be aware that a measure such as the Pearson product-moment correlation will be high as long as the relative ranking of the animals is consistent, but that this measure does not address possible discrepancies in terms of the actual numbers assigned by different observers. This can be avoided if one uses instead the intra-class correlation, which does take the differences between observers into account (Howell, 1997, p.490). In the event of a low score on this measure, it important to make every attempt to minimize such differences through further training. Moreover, in order to avoid introducing bias between the experimental groups, it is important to use assignment strategies such as stratified random assignment to ensure that each observer tests an approximately equal number of animals from each experimental condition. In this way any consistent difference between observers becomes part of the error term, rather than a bias that is introduced between the experimental groups. However, if this is large, it can also be a problem in increasing the probability of Type 2 error, i.e. of failing to identify a true effect.

The behavioral test battery consists of the following items:

1. *Righting reflex* — Does subject (S) return rapidly to its feet when placed on its back? Place S on its back, head over tail, three times.
   - 1.0 Rapid, consistent righting, always turns over
   - 0.9 Not quite rapid, inconsistent speed, always turns over
   - 0.8 Rapid first time, but then slower, always turns over
   - 0.7 Slow but consistent speed, always turns over
   - 0.6 Slow righting from beginning but does turn over on all trials, may struggle
   - 0.5 Slow and fatigues, gets slower with consecutive trials
   - 0.4 Turns over on two out of three trials
   - 0.3 Only turns over once
   - 0.2 Makes an attempt but never turns over
   - 0.0 Cannot right

2. *Cliff aversion* — Does S withdraw from the edge of a flat surface? Place S snout and forepaws just past the joints over the edge of plastic jar-lid (roughened to allow grip).
   - 1.0 Withdraws promptly
   - 0.9 Slow withdrawal, no struggle
   - 0.8 Withdrawals with little struggle
   - 0.5 Withdrawals with considerable struggle
   - 0.3 Struggles, but falls over the edge
   - 0.0 Falls over the edge or random movements

3. *Fore- and hind-limb grasp reflex* — How strongly does S grasp? Leave the forepaws on the lid of the jar and lift the hind limbs. Touch pad of both hind paws with 18-gauge needle barrel. Next, suspend the animal by its tail and, again, touch pad of both forepaws with the 18-gauge needle. Rate the fore- and hind limbs separately and average the scores for a final score. How strongly does the S grasp?
   - 1.0 Rapid, strong grasp by both paws
   - 0.8 Rapid, firm but not strong; can remove needle easily
   - 0.5 Slow grasp which does surround barrel; weak

0.4 Slow grasp which does surround barrel, but sometimes losses grip
0.3 Slow, weak, never completely closes paws
0.2 Weaker
0.0 No sign of closure at all

4. *Vibrissae placing reflex* — Does S place forepaw on a cotton swab that is stroked across the vibrissae? Touch vibrissae with Q-tip. Bring Q-tip from behind and brush past vibrissa once on each side. Repeat this three times, placing the animal on the ground for 5 s between each trial.
1.0 Forepaw is raised promptly and placed on Q-tip
0.7 Rapid placing only on some trials
0.5 Placing is slower and inconsistent
0.4 Some directed movement of head and arm
0.3 Head and occasionally arm movement
0.2 Head movement only
0.0 No directed movement of forepaw and head

5. *Screen level test* — Can S maintain grip on a level screen? Place on a level 288-mesh screen, pull tail firmly back and slightly up (on a 45° angle) for 5 s.
1.0 Strong grasp on screen with all paws; does not fatigue
0.9 Exploratory movements with forepaws but no slipping
0.8 May lose grip of hind paws; occasional slip back
0.5 Inconsistent grip; tug will dislodge
0.3 Never strong grip; may be dragged steadily across screen
0.0 No grip or resistance at all

6. *Screen vertical test* — Can S maintain grip on a vertical screen? As above, but orient screen vertically (80°, animal at base, head pointing up).
1.0 Strong grip; does not fall or slip at all
0.8 Good grip, but slips a few notches
0.5 Good grip at first, but fatigues and slips down
0.3 Very brief grip, falls regularly
0.0 No grip, falls immediately

7. *Screen test climb* — Does S climb a vertical screen using both fore- and hind limbs? Leave screen in vertical position; see if subject climbs.
1.0 Climbs promptly up screen in adult pattern, fore and hind
0.9 Climbs with some hesitation (more slowly)
0.8 Climbs well, but hind limbs are used for support, not thrust
0.6 Same as 0.8, but only moves a few steps
0.5 Climbs only a short distance, tends to turn downwards, little aid from hind limbs
0.4 Same as 0.5, but falls
0.3 Tries to climb, but does not move one step up
0.0 No attempt to climb or pull self up

8. *Pole grasp* — Can S grasp a 2.5 mm wooden shaft? Place forepaws on stick of Q-tip release hold on tail, allow animal to hold on for 10 s and examine use of fore- and hind limbs for grasp.
1.0 Strong grasp with fore- and hind-paws
0.8 Fore- strong; hind- present, but weak
0.7 Fore- strong; hind of minimal assistance
0.6 Fore- adequate; hind touch stick but not grasp
0.5 Fore- adequate; hind not touch
0.4 Fore- momentary, but will lose grip quickly
0.3 Fore- closes, but never gets good grip
0.0 No grasp at all

9. *Forelimb and hindlimb stick grasp* — Can S grip firmly a 9.5 mm-wide stick? Place fore- and hind-paws along axis of flat stick, rate fore- and hindlimbs separately and average score.
1.0 Strong, immediate grasp

0.8 Immediate grasp, but stick can be pulled away readily
0.7 Struggles, but grasp is strong
0.5 Must struggle to get grasp, grip is weak
0.3 Tries to grasp, never gets a good grip
0.1 Weak attempt
0.0 No attempt

10. *Eyes open* — Inspect eyes.
1.0 Both eyes wide open
0.9 Extremely slight squint in both
0.8 One wide open, other open but slight squint
0.7 Both slightly squinted
0.6 Both very squinted
0.5 Both slightly open
0.4 Between 0.5 and 0.3
0.3 Two slightly open (glimmer)
0.2 Between 0.3 and 0.1
0.1 One slightly open (glimmer) and other closed
0.0 Both eyes closed

11. *Visual placing* — Does S extend forelimbs when lowered towards a flat surface? Suspend by tail 5 cm above substrate of black and white checkers for 3 s and then bring animal back to the surface. Repeat for 4 trials, allows for a couple of seconds rest between trials.
1.0 Strong consistent outreach with forelimbs (4/4)
0.8 Outreach on 3/4 trials
0.5 Outreach on 2/4 trials
0.2 Outreach on 1/4 trials
0.0 No outreach on any trial

12. *Auditory startle* — Does S show whole body startle response to auditory stimulus? Snap a 2? paper clip within about 10 cm from a stationary animal.
1.0 Intense jump or hop, leaves substrate by >1 cm
0.5 Small jumpor hop, leaves substrate by <1 cm
0.4 Quick step backwards or forwards with one or more paws
0.3 More than ear twitch, but no step
0.1 Small twitch of ears only
0.0 No observable reaction

Developmental time scales: the question of age vs stage

There are two possible interpretations of the differences in particular aspects of brain development or behavior found in experimental studies. The first is that the experimental manipulation has merely retarded overall development, and the outcome is the same as that expected of a normal, but younger animal. Alternatively, the effects seen represent an aberration of normal development, such that the relationships among various developmental indices differ between the experimental groups. Thus, rather than compare animals of the same chronological age, it may be more informative to compare animals at the same developmental stage. This is where standardized developmental time scales, such as that developed by Wahlsten (1974, 1975) can be very useful. This scale was developed on the F_2 offspring of $B6D2F_1$ hybrid mice, which are characterized by robustness and reproductive vigor. Separate litters of the $B6D2F_1/B6D2F_1$ cross were measured on each of postconceptual days 27 through 36 on various developmental indices, including body weight, brain weight, thickness of the external granular layer (EGL) of the cerebellum, as well as the behavioral test battery described above. Then a quadratic regression equation of chronological age on each score was derived from the litter mean score. By substituting the actual score (x) of a developing animal into this equation, one is able to obtain its developmental age (y), which is the age at which the standard $B6D2F_2$ mice are expected to attain that value. These equations then allow one to compare, for example, developmental age as derived from body weight with that derived from behavioral development, in pups of the same chronological age. The advantage of this procedure over those based on percentage differences is that differences on all measures are indicated on a common scale, i.e. days of normal growth, thereby allowing empirical comparison of the magnitude of the retardation seen in each. Using this approach to investigate the effects of postnatal litter size on development, we found that, for each additional pup in the litter, body growth was retarded by the equivalent of 1.28 d, brain weight by 0.44 d, and behavioral development only by 0.07 d (Wainwright et al., 1989). Moreover, as long as an untreated $B6D2F_2$ group is included as a control in each experiment to support the reliability of the scale, it is also possible to make quantitative comparisons between experiments conducted at different times and in different laboratories. For example, we have used it to show effects of both pup and maternal genotype (Wainwright, 1980), prenatal ethanol exposure (Wainwright et al., 1990), and dietary n-3 fatty acids (Wainwright et al., 1991) on behavioral development. In these studies we have chosen to compare groups on day 32 postconception because consideration of the developmental curve shows that this is when rate of change is greatest, and when most animals score in the middle range; this then avoids the problem of insensitivity of measurement due to restriction in range.

Although the focus of this chapter is behavioral, it is worth mentioning that a similar time-scale has been developed based on criteria of morphological development during the prenatal period, and this scale has proven useful in studying hereditary differences in prenatal morphological development (Wahlsten and Wainwright, 1977). It has also been used to show that in the BALB/c strain, which sometimes shows anomalous development of the corpus callosum, this structure develops later relative to other developmental indices in comparison with the C57/BL6 strain (Wahlsten, 1984; Wainwright and Deeks, 1984).

Conclusion

Our intent in this chapter was to acquaint the reader with some of the methodological and interpretative pitfalls to be avoided when studying behavioral development in the mouse. The broad scope of such an undertaking has precluded extensive discussion of any one topic. It is therefore important that anyone considering working in this further consult material provided in the reference list, particularly the general reference texts edited by Goldowitz et al. (1992), Kimmel and Buelke-Sam (1994), and Hood (1996). In closing, we emphasize what we feel is the most important general principle in work of this nature. In order to reach a meaningful interpretation of variation in the development of some particular aspect of the nervous system, including behavior, this must be done in a context that includes the factors that contribute to normal development in the animal's species-typical ecological niche.

Acknowledgements

The author thanks Dr M. B. Bulman-Fleming and Dr G.R. Ward and for their helpful comments on an earlier draft of this chapter.

References

Altman, J. and Sudarshan, K. (1975) Postnatal development of locomotion in the laboratory rat. *Anim. Behav.*, 23: 896–920.

Bates, H.K., Cunny, H.C. and Kebede, G.A. (1996) Developmental neurotoxicity testing methodology. In: R.D. Hood (Ed.), *Handbook of Developmental Toxicology*, CRC Press, Boca Raton, pp. 291–324.

Bulman-Fleming, M.B. and Wahlsten, D. (1988) Effects of a hybrid maternal environment on brain growth and corpus callosum defects of inbred BALB/c mice: A study using ovarian grafting. *Exp. Neurol.*, 99: 636–646.

Carlier, M., Roubertoux, P. and Cohen-Salmon, C. (1982) Differences in patterns of pup care in *Mus Musculus Domesticus*. 1. Comparisons between eleven strains. *Behav. Neural Biol.*, 35: 209–220.

Carlier, M., Nosten-Bertrand, M. and Michard-Vanhée, C. (1992) Separating genetic effects from maternal environmental effects. In: D. Goldowitz, D. Wahlsten and R.E. Wimer (Eds.), *Techniques for the Genetic Analysis of Brain and Behavior: Focus on the Mouse*, Elsevier Science Publishers, Amsterdam, pp. 111–126.

Cassells, B., Wainwright, P. and Blom, K. (1987) Heredity and alcohol-induced brain anomalies: Effects of alcohol on anomalous prenatal development of the corpus callosum and anterior commissure in BALB/c and C57BL/6 mice. *Exp. Nerol.*, 95: 587–604.

Cohen-Salmon, C. (1987) Differences in patterns of pup care in *Mus musculus domesticus*. II. Effects of parity on eight inbred strains. *Biol. Behav.*, 7: 337–346.

Crusio, W.E. (1992) Quantitative Genetics. In: D. Goldowitz, D. Wahlsten and R.E. Wimer (Eds.), *Techniques for the Genetic Analysis of Brain and Behavior: Focus on the Mouse*, Elsevier Science Publishers, Amsterdam, pp. 231–250.

Crusio, W.E. and Schmitt, A. (1996) Prenatal effects of parity on behavioral ontogeny in mice. *Physiol. Behav.*, 59: 1171–1174.

Fleming, A. (1999) Chapter on maternal behavior in this volume.

Fox, M.W. (1965) Reflex ontogeny and behavioral development of the mouse. *Anim. Behav.*, 13: 234–241.

D. Goldowitz, D. Wahlsten and R.E. Wimer. (1992). *Techniques for the Genetic Analysis of Brain and Behavior: Focus on the Mouse*, Elsevier Science Publishers, Amsterdam.

Hall, W.G. (1975) Weaning and growth of artificially-reared rats. *Science*, 190: 1313–1315.

Hood, R.D. (1996) *Handbook of Developmental Toxicology*, CRC Press, Boca Raton.

Hood, R.D. and Miller, D.B. (1996) Maternally mediated effects on development. In: R.D. Hood (Ed.), *Handbook of Developmental Toxicology*, CRC Press, Boca Raton, pp.61–90.

Kimmel, C.A. and Buelke-Sam, J. (1994) *Developmental Toxicology*, 2nd Edn, Raven Press, New York.

Howell, D.C. (1997) *Statistical Methods for Psychology*, Duxbury Press, Toronto.

Lauder, J. (1993) Neurotransmitters as growth regulatory signals: role of receptors and second messengers. *Trends Neurosci.*, 16: 233–239.

Levitt, P., Harvey, J.A., Friedman, E., Simansky, K. and Murphy, E.H. (1997) New evidence for neurotransmitter influences on brain development. *Trends Neurosci.*, 20: 269–274.

Messer, M., Thoman, E.B. Terrassa, A.G. and Dallman, P.R. (1969) Artificial feeding of infant rats by continuous gastric infusion. *J. Nutr.*, 98: 404–410.

Moore, C.L. (1984) Maternal contributions to the development of masculine sexual behavior in laboratory rats. *Dev. Psychobiol.*, 17: 346–356.

Muthén, B. and Curran, P.J. (1997) General longitudinal modelling of individual differences in experimental designs: A latent variable framework for analysis and power estimation. *Psych. Methods*, 2: 371-402.

Rice, S.A. and Ray, R.M. (1996) Developmental neurotoxicity testing — data evaluation. In: R.D. Hood (Ed.), *Handbook of Developmental Toxicology*, CRC Press, Boca Raton, pp. 325–355.

Roubertoux, P.L., Nosten-Bertrand, M., Cohen-Salmon, C. and l'Hotellier, L. (1992) Behavioral development: a tool for genetic analysis in mice. In: D. Goldowitz, D. Wahlsten and R.E. Wimer (Eds.), *Techniques for the Genetic Analysis of Brain and Behavior: Focus on the Mouse*, Elsevier Science Publishers, Amsterdam, pp. 423–441.

Roullet, P. and Lasalle, J .M. (1990) Genetic variation, hippocampal mossy fibre distribution, novelty reactions and spatial representation in mice. *Behav. Brain Res.*, 41: 61–69.

Satinoff, E. (1991) Developmental aspects of behavioral and reflexive thermoregulation. In: H.N. Shair, G.A. Barr and M.A. Hofer (Eds.), *Developmental Psychobiology: New Methods and Changing Concepts*, Oxford University Press, New York, pp. 169–188.

Smotherman, W.P. and Robinson, S.R. (1990) The prenatal origins of behavioral organization. *Psycholog. Sci.*, 1: 97–106.

Spear, N. and Rudy, J. (1991) Tests of the ontogeny of learning and memory: Issues, methods and results. In: H.N. Shair, G.A. Barr and M.A. Hofer (Eds.), *Developmental Psychobiology: New Methods and Changing Concepts*, Oxford University Press, New York, pp. 84–113.

Stanton, M.E. and Spear, L.P. (1990) Workshop on the qualitative and quantitative comparability of human and animal developmental neurotoxicity, Work Group 1 report: Comparability of measures of developmental neurotoxicity in humans and laboratory animals. *Neurotoxicol. Teratol.*, 12: 261–267.

Wahlsten, D. (1974) A developmental time scale for postnatal changes in brain and behavior of B6D2F$_2$ mice. *Brain Res.*, 72: 251–264.

Wahlsten, D. (1975) Genetic variation in the development of mouse brain and behavior: Evidence from the middle postnatal period. *Dev. Psychobiol.*, 8: 371–380.

Wahlsten, D. (1984) Growth of the mouse corpus callosum. *Dev. Brain Res.*, 15: 59–67.

Wahlsten, D. (1989) Genetic and developmental defects of the mouse corpus callosum. *Experientia*, 45: 828–838.

Wahlsten, D. (1994) The intelligence of heritability. *Can. Psychol.*, 35: 244–259.

Wahlsten, D. and Wainwright, P. (1977) Application of a morphological time scale to hereditary differences in prenatal mouse development. *J. Embryol. Exp. Morph.*, 42: 79–92.

Wainwright, P. (1980) Relative effects of maternal and pup heredity on postnatal mouse development. *Dev. Psychobiol.*, 13: 493–498.

Wainwright, P.E. (1998) Issues of design and analysis relating to the use of multiparous species in developmental nutritional studies. *J. Nutr.*, 128: 661–663.

Wainwright, P. and Deeks, S. (1984) A comparison of corpus callosum development in the BALB/cCF and C57BL/6J inbred mouse strains. *Growth*, 48: 192–197.

Wainwright, P., Pelkman, C. and Wahlsten, D. (1989) The quantitative relationship between nutritional effects on preweaning growth and behavioral development in mice. *Dev. Psychobiol.*, 22: 183–195.

Wainwright, P.E., Ward, G.R., Winfield, D., Huang, Y-S., Mills, D.E., Ward, R.P. and McCutcheon, D. (1990) Effects of prenatal ethanol and long-chain n-3 fatty acid supplementation on development in mice. 1. Body and brain growth, sensorimotor development, and water T-maze reversal learning. *Alcohol. Clin. Exp. Res.*, 14: 405–412.

Wainwright, P.E., Huang, Y-S., Bulman-Fleming, M.B., Mills, D.E., Redden, P. and McCutcheon, D. (1991) The role of n-3 essential fatty acids in brain and behavioral development: A cross-fostering study in the mouse. *Lipids*, 26: 37–45.

Wainwright, P.E., Huang, Y-S., Coscina, D., Lévesque, S. and McCutcheon, D. (1994) Brain and behavioral effects of dietary n-3 deficiency in mice: A three generational study. *Dev. Psychobiol.*, 27: 467–487.

Wainwright, P.E., Xing, H-C., Girard, T., Parker, L. and Ward, G.R. (1998) Effects of dietary n-3 fatty acid deficiency on Morris water-maze performance and amphetamine-induced conditioned place preference in rats. *Nutr. Neurosci.*, 1: 281–293.

Wainwright, P.E., Xing, H-C., Ward, G.R., Huang, Y-S., Bobik, E., Auestad, N. and Montalto, M. (1999) Water maze performance is unaffected in artificially reared rats fed diets supplemented with arachidonic acid and docosahexaenoic acid. *J. Nutr.*, 129: 1079–1089.

Ward., G.R. and Wainwright, P.E. (1988) Reductions in maternal food and water intake account for prenatal stress effects on neurobehavioral development in B6D2F_2 mice. *Physiol. Behav.*, 44: 781–786.

Ward, G. R., Huang, Y-S., Bobik, E., Xing, H-CX., Mutsaers, L., Auestad, N., Montalto, M., and Wainwright, P. (1998) Long-chain polyunsaturated fatty acid levels in formulae influence deposition of docosahexaenoic acid and arachidonic acid in brain and red blood cells of artificially reared neonatal rats. *J. Nutr.*, 128: 2473–2487.

Zorrilla, E.P. (1997) Multiparous species present problems (and possibilities) to developmentalists. *Dev. Psychobiol.*, 30: 141–150.

W.E. Crusio and R.T. Gerlai (Eds.)
Handbook of Molecular-Genetic Techniques for Brain and Behavior Research (Techniques in the Behavioral and Neural Sciences, Vol. 13)

CHAPTER 4.10

Understanding maternal behavior: analyses of behavior, c-Fos expression and calmodulin binding proteins in the medial preoptic area and other areas of the rat brain

Alison S. Fleming[1] and Danton H. O'Day[2]

[1]*Department of Psychology and* [2]*Department of Zoology, University of Toronto at Mississauga, Mississauga, Ont. L5L 1C6, Canada*

Introduction

This chapter will discuss the multiple behavioral and neuroanatomical systems that are activated when a female rat becomes maternal at the time of parturition and that must be activated for the female to be maternal. Our approach has been to first analyze the natural ongoing behavior by observing animals interacting with pups in their home cages under a variety of conditions. We then consider 'motivational' influences on the behavior, by using tests and paradigms outside the maternal context that reveal these behavioral propensities. Finally, we attempt to delineate those brain mechanisms that must be recruited in service of those influences that interrelate with neural mechanisms underlying the behavior itself. These analyses involve looking for behavioral changes in animals that either have received brain lesions, hormones, or electrical stimulation and examining associated neurochemical and neurophysiological changes in brains of animals that have been subjected to various behavioral conditions.

In this chapter, we begin by describing what maternal behavior is and how it is measured within its functional context. We then describe procedures we have employed to show that with the birth of the litter, mothers become less fearful and more attracted to pups. Through the use of operant procedures, it is clear that with experience interacting with young, pups acquire reinforcing properties that facilitate the considerable learning that the mother does when she is interacting with pups. This learning is demonstrated in a series of acquisition, consolidation and retrieval paradigms that utilize both the natural behaviors and conditioned place preference techniques. These show us that mothers learn a great deal about pups and mothering along the way and that experience serves them well at a later point when hormones are no longer motivating or after subsequent births. Finally, we describe a series of approaches we are using to show how brain anatomy, chemistry, and physiology are involved in all of these behavioral changes and that they interface in interesting ways with the maternal circuit, for the functional expression of maternal behavior at the biologically relevant time.

Behavioral strategies in the analysis of maternal

Maternal behavior as a species-characteristic behavior

Rat maternal behavior involves a relatively stereotyped complex of behaviors that occur in general after the birth of the young and persists until the young are weaned 21 d later (e.g. Fig. 1). At parturition, the new mother rat eats the placenta, cleans off the amniotic membrane from around the pups, breaks the umbilicus and intensively licks the pups. She then retrieves young back into the nest, continues to lick them and eventually adopts a nursing posture over them for suckling (Wiesner and Sheard, 1933; Rosenblatt and Lehrman, 1963). This repertoire of behaviors can easily be observed, especially in the early postpartum days if the nest is disrupted. Tests of maternal behavior are usually done by separating the mother from the pups for a ½ h period. Then 6 pups are reintroduced into the cage, scattering them out of the nest so the mother has to retrieve them into the nest in order to engage in subsequent maternal pup-directed behaviors. In our laboratory behavioral observations take place once daily for 15 min of continuous observations, using either an event-recorder or a one-zero time-sampling procedure, with 5 s recording intervals (Fleming and Rosenblatt, 1974a). The event-recorder is computer based and permits the simultaneous and continuous recording of up to 35 behaviors in real time. There are 35 keys on the event-recorder. Hence we can get information on behavior frequencies, duration, latencies, intervals, and sequences of behavior (Fleming et al., 1988). The 0/1 time sampling procedure refers to the occurrence (or not) of each behavior that is recorded for each successive 5 s interval over an observation period of 10,15 or 60 min. Variations on this procedure are found in the work of others (e.g. Numan, 1994; Bridges, 1996; Stern, 1996).

Depending on the study, these maternal tests may be given to new mothers or to virgin females, juvenile animals, or males. While new mothers are immediately maternal to pups and retrieve, crouch over and lick them, males and virgin females are initially very unresponsive to pups and do not express these behaviors. However, both males and virgin females can be induced to become maternal through daily exposure to foster pups.

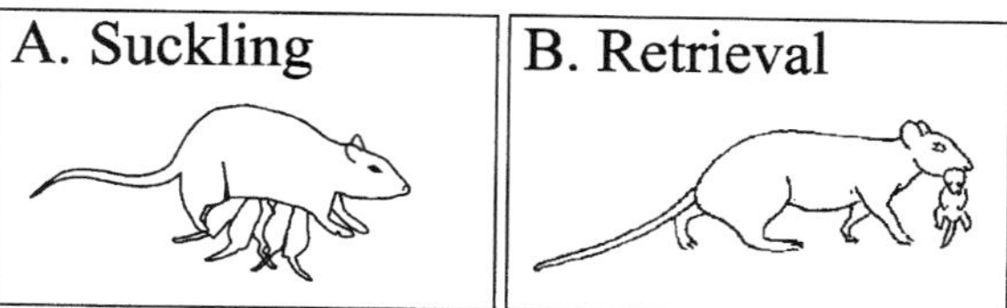

Fig. 1. Schematic of maternal rat showing suckling (A) and retrieving (B) of pups. Females are housed in plastic cages (22 × 4 × 15 cm), with Purina Rat Chow and water available ad lib. The temperature of the room is thermostatically regulated at 22°C with a 12:12 h illumination cycle, with lights on at 0800 h. Three days prior to maternal behavior testing, females are moved into large plastic cages (37 × 47 × 21 cm) and provided with wood shavings for bedding. *Maternal induction tests.* Maternal induction tests are given to all animals once they have given birth or as nulliparous animals. Females are tested for maternal behavior each morning between 0900 h and 1200 h. On the first day, four 1–5 d old foster pups are taken from a lactating "donor" mother and placed opposite the female in her cage. Using a 0–1 time sampling paper and pencil procedure, the occurrence of several behaviors is recorded for 180 5-s intervals (or 15 min). After the observation period 2 spot-checks are made at approximately 3-h intervals. The foster pups are left with the female overnight and removed the following morning. Pups are returned to their "donor" mothers and a new recently fed litter of four foster pups is placed in the female's cage to start the next 8-min observation period. During an observation the following behaviors are recorded: 1. "Pup retrieval," when the female picks a pup in her mouth and carries it to her nest site; 2. "Pup licking" is divided into "general body" and "genital," which occur when the female licks the pups' bodies and genital region; 3. "Pup sniffing" involves actual active sniffing of the pups by the female; 4. "Lying in contact with pups or crouching" occurs when the female is lying down in the nest on top of the pups, but not in a good lactating posture; 5. "Lactating posture" is recorded when the female positions herself over the pups and arches her back to allow the pups to "suckle"; 6. "Nest building," when the female brings paper toweling to her nest quadrant and shapes it into a nest; 7. "Cannibalizing," when a female attacks and/or kills one or more pups; cannibals are not tested again. A female is said to fulfill the criterion for maternal behavior if she retrieves all pups into the nest during the observation period and is observed to be with them in the nest either during the test or during spot-checks.

When non-postpartum animals are tested for their maternal responsiveness, they are tested with pups taken from a lactating mother. In these cases, recently fed foster pups are presented to the mother and are observed for 15 min and then during spot checks throughout the day. Pups are left with the animal throughout the day and returned to their own mothers on the next day, to be re-nourished, and replaced by a new recently fed litter (see Fleming and Rosenblatt, 1974a). In this way animals can be tested for their maternal responsiveness using 1–3 d old newborn pups that are healthy and nourished. With this regime, known as maternal induction testing, non postpartum adult animals in general respond maternally after a latency period of from 2–10 d, depending on the strain of rat being tested (for reviews, see Stern, 1989; Numan, 1994; Fleming and Corter, 1995). Interestingly, juveniles show behaviors similar to maternal behaviors, almost immediately. Whether these behaviors are truly maternal or are simply a part of the animals' play pattern is an interesting question. We have attempted to address it by observing juvenile behavior in response to foster pups, to other juveniles, and to small toys prior to and after brain manipulations that activate or inhibit brain regions known to affect adult maternal behavior (mpoa) or adult social behavior (basolateral amygdala) (Brunelli et al., 1986; Natterer and Fleming, 1999).

Maternal behavior involves a reduction in pup-avoidance and 'fear': the open-field test

In contrast to the juvenile, the adult virgin animal is very unresponsive to pups on initial exposure, but with continuous exposure to pups comes to be responsive. What is happening in the interim? We have good evidence now that the virgin animal initially withdraws from pups and is somewhat neophobic, showing fearfulness of novelty. To test this hypothesis, we placed pups into the nest of the virgin animal (rather than as usual at a distance) on each day and observed the frequency with which she changed her nest site. Mother rats remain with pups in the nest; virgins change their nest sites, leaving the pups behind. By daily test of this kind, we were able to show a gradual reduction in nest site changes and withdrawal responses, until one day pups were seen to be in the nest with the female. Within one day other maternal responses of licking, retrieving, etc. were observed (Fleming and Luebke, 1981). Fearfulness can also be demonstrated by testing animals in open-field apparatus (see Fig. 2). In this test animals are placed into a large novel arena, with the floor divided into a grid pattern, so that it is possible to trace the animal's path in the arena by recording where on the grid the animal is at any point in time. Animals are observed over a 5–10 minute period and their path is traced on an observation sheet continuously during the 10 min period. Animals that spend proportionally more time in the central squares are considered to be less fearful (Fleming and Luebke, 1981; Fleming et al., 1989; Morgan et al., 1997). Using this open-field procedure, it is possible to assess the animal's response to novelty, by placing novel objects, scents, and pups at strategic positions within the open-field, and observing the animals' investigative sniffing, licking, mouthing, and digging behaviors and withdrawal responses.

Postpartum animals spend proportionately more of their grid crossings in the central squares and investigate more than do virgin animals. Males have not been tested. These data once again suggest that new mothers are less timid or neophobic than are virgin animals. When comparing the behavior of new mothers and virgins we always use 1–3 d old foster pups in both cases. Mother rats respond similarly to their own offspring and to same aged foster offspring. Depending on the experiment in some cases we test animals with their own litters. Moreover, this fearfulness must be overcome in order for animals to respond maternally to pups. One job that the parturitional hormones have is to reduce the animals neophobia or fearfulness. Virgin animals injected with parturitional hormones show a pattern of grid crossing that is more similar to postpartum animals. If female rats are made less neophobic or fearful, by extensive daily handling, or brain manipulations that remove perception of the fear stimulus, (olfactory bulb removal) or fear itself (basolateral amygdala) animals express rapid maternal behavior towards foster pups

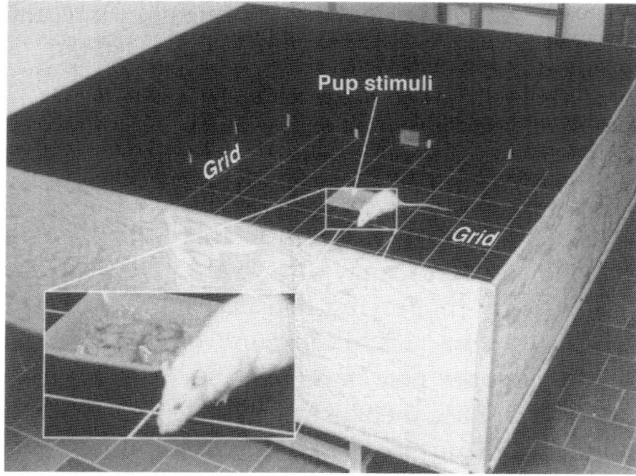

Fig. 2. The open-field test. The open-field test is given to all animals to determine 'emotionality'. *Apparatus*: This consists of a start box, identical to the home cage but with a 11 × 13 cm gate leading directly into a 150 × 150 cm open-field arena with a black Plexiglas floor and black wooden walls. The floor area is marked off into 100 15-cm squares. Illumination is provided by a single 40 W light bulb centered over the field and suspended 200 cm from the floor of the field. An AM radio, set at low volume is used to mask background noise during testing. Maternal nest odors are obtained by providing lactating donor animals with clean nesting material on the day of parturition and then removing 100 g of the wood shavings from the nest 2–3 d later at test time. *Emergence and open-field testing*. Open-field testing consists of testing animals over 4 consecutive days, such that test days 1 and 3 are 'no odor' conditions and 2 and 4 are 'odor' conditions. Prior to each test, the start box, sliding door, and arena floor and walls is thoroughly wiped with 70% alcohol. The glass jar containing either 100 grams of lactating nest material (odor condition) or 100 gm of clean nest material (no odor condition) is then placed into the arena, 3 squares in front of the door opening. *Emergence test*. The subject is placed into the start-box for a 5-min period, with the door adjoining the start box and arena. The *emergence test* is started when the door is raised and the female is allowed to enter the field. Once the female enters the field with all 4 feet, the door is lowered and the open-field test is initiated. If the female does not enter the field within 15 min, the animal is placed into the field directly in front of the closed door. The time from the start of the test to when the animal places all 4 feet into the field is called the *emergence latency*. A maximum score of 901 s occurred when a female fails to enter the field. *Open-field test*. During the 5-min open-field test, the movements of the animal are monitored using a data sheet that is divided into 100 squares. In this way it is possible to establish the frequency of square crossings and entries into any part of the field. An animal is said to have entered a square when all four feet have crossed over from an adjoining square. We also monitor entries into the ring of squares surrounding the odor source, (designated the 'critical region'). In addition, we record the number of times the animal groomed and reared up, and the number of urine drops and fecal pellets deposited. It is assumed that the higher the ambulation score obtained, the less 'timid' the animal; and the higher the critical/total crossing ratio the higher the animal's attraction to the odor.

(Fleming and Rosenblatt, 1974b, 1974c; Fleming et al., 1980).

Through experience pups become reinforcing to the mother: modified operant or 'Skinner' box

Another behavioral system that is activated or recruited when an animal becomes maternal is the hedonic system or reinforcement system. With parturition, mothers develop an attraction to pup-related odors and other cues and will work hard to gain access to pups. If females are given a 5 min test in a Y maze, with nest material from a lactating female's nest and her pups at the end of one arm and the nest of a nonlactating female at the other, new mothers prefer the lactating female's nest; virgins do not (Fleming, 1986). Another way to test this olfactory attraction is to look at responses to different odors presented in the context of an open field (Fleming et al., 1989); in this case, hormonally primed females spend more time in the critical odor squares than do the control animals. We have also found that animals that have been trained to bar-press in an operant chamber will bar-press at a greater rate for delivery of pups, after parturition than before and after maternal experience than in experienced animals (Lee et al., 1999; see Fig. 3). Bar pressing for food remains unchanged. The operant box contains a carousel divided into 12 compartment, each one containing a rat pup. Whenever the female presses the bar, it causes the carousel to advance one position and the floor of the carousel drops out and the pup is released into a shoot with a gentle incline that delivers the pup into the hopper. Among the factors that affect the bar-press rate are the number of hours of pup-deprivation, the accessibility of the pup after its delivery into the hopper,

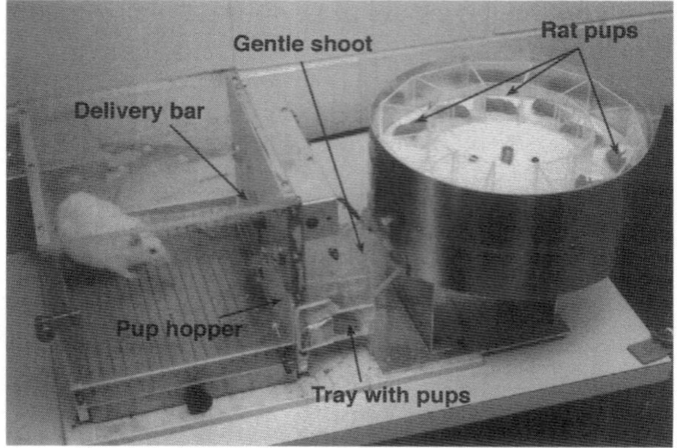

Fig. 3. Operant responding. *Apparatus*: Testing is carried out in a modified operant chamber (Grason Corporation, United States), measuring 26 × 26 × 30 cm and having one retractable lever that operates with a force of 10 g. The lever is located 7.5 cm above the Plexiglas transparent floor. The chamber is made entirely of transparent Plexiglass (6 mm thick) with a perforated roof. The feeder or pup dispenser opening, a hole 5.0 cm in diameter, is located 10.0 cm to the right of the lever, and provides a means for the subject to retrieve a 'reward' (i.e. a Froot Loop or 1–3 d old pup). A 45 angle shoot transports the pup or food reward from the 12 compartment carousel-type magazine (30.5 cm diameter) to the dispenser. The dispenser contains a removable 6.0 × 3.0 × 10 cm drawer; this arrangement permits removal of the pups from the dispenser in the event that the female does not herself retrieve the pups from the dispenser. An opening (8 cm in diameter) in the rear of the box allows the experimenter to remove the pups from the housing chamber in the event that the female removes the pups from the dispenser. A microcomputer (IBM PC-XT, Best Computers Inc. Canada) fitted with the control language RATINT (programmed by Erindale College Microelectronics Department), is used to control the operant chambers and record the data. *Procedures*: All studies consist of training the rats to bar press for food reinforcement (Kelloggs Froot Loops), conducted during the first half of the light-dark cycle (between 0900 and 1400 h). *Operant training habituation and magazine training*: Animals are placed on a controlled feeding regime that maintains their body weights at 90% of baseline levels. This regime is maintained throughout the training period prior to pregnancy. Habituation consists of placing rats in the inactive operant chamber once daily for a 1-h period. Ten pieces of Froot Loops are also placed into the chamber during the habituation phase. This procedure is designed to familiarize the rats with the chamber and the food. *Shaping*. Animals are trained to press the lever for portions of Froot Loops via successive approximations. This consists of simultaneously presenting the reinforcement for behaviors that resemble the bar press response. Initially, subjects are rewarded for rearing anywhere in the chamber; then for sniffing the lever or dispenser opening; then for rearing near the lever, until they acquire the bar press response. Shaping sessions are 30 min in length, and, consumption of the number of Froot Loops is recorded in order to adjust the amount of food pellets (Purina Rat Chow) given in the home cage necessary to maintain animals at 90% of their baseline body weight. *Conditioning*: Following bar press acquisition, a FR 1 schedule of reinforcement (one bar press delivered one reinforcement) is implemented to condition the lever-press response. The subject acquires the bar press response if it performs 50 lever presses within 30 min for 2 consecutive days. Any subject who does not meet criteria within 8 d is eliminated from the experiment. Immediately following acquisition criteria animals are placed in their home cage and are given food ad libitum. *Test phase*: Pup-test phase is conducted between 1400 and 2100 h. It consists of placing the female within the operant chamber for 30 min in which a bar press delivered a 1–3 d old pup as a reward. Twelve pups are placed in the rotating magazine compartments, which are thoroughly cleaned with 70% alcohol. The carousel is refilled with pups following the removal of 6 retrieved pups from the chamber by a rear opening, or the removal of 6 non-retrieved pups from the dispenser via the removable drawer. Behavioral observations are recorded at the commencement of each pup-test phase for 8 min at 10 s intervals. *Behavioral observations*. The following behavioral observations are analyzed. *Pup lick/anogenital lick*: Female licks pup's body or anogenital region; *Hover over/crouch*: Female adopts a crouch position over the pups with her ventrum directly above them. *Bar press*: Female rears up at the lever, with her forepaws usually pressed down on the lever to advance the carousel; *Pup-retrieval*: Female picks the pup up in her mouth from the dispenser and transports it to an area within the chamber.

the maternal status of the female, and the hormonal status of the female.

Mothers become conditioned to pup cues and associated stimuli: conditioned place preference

Pups are initially attractive to the mother. With experience interacting with pup, pups become more highly reinforcing and can sustain learned behavior. For instance pups are not only reinforcing to new mothers, but also having interacted with them, she becomes conditioned to other non pup-based cues with which pups have become associated. Mothers will come to prefer an initially novel and distinctive environment (box) that has been associated with pups over another equally distinctive but different box that has not been associated with pups (Fleming et al., 1994; Magnusson and Fleming, 1995; Morgan et al.,

1997). In this conditioned place paradigm mothers are given exposure to 2 environments over a 1 h period on each of 4 consecutive days (see Fig. 4). In one, pups are also present; in the other they are not. On the 5th d, mothers are placed into the apparatus and allowed to move freely between the two boxes. Preference is determined by the proportion of the test time mothers spend in the pup-associated box. Such conditioning can be shown in a number of ways and usually occurs when mothers have had an opportunity to physically interact with pups. Simple exposure to pups, without an opportunity to crouch over them and to retrieve them does not sustain conditioning. Manipulations that reduce the animals touch sensitivity, especially around the mothers mouth and nipple region (by application of a local anaesthetic), prior to conditioning, prevent it from occurring. Conversely, the presence of the hormones associated with parturition enhances this effect further. Processes of learning and reinforcement underlie the observation that second time mothers are more maternally responsive than first time mothers and that in the absence of hormones of parturition, animals will continue to respond maternally towards pups until weaning.

We see, then, that using a variety of standard psychological techniques we can establish something about what is occurring within the newly-parturient female rat that contributes to the timing and quality of her species-characteristic behavior.

Neuroanatomy of the maternal system

Lesions and stimulation of the brain disrupt, or facilitate, behavior

Once we gained an understanding of the behavioral systems implicated in maternal responding, we explored the neuroanatomy mediating these behaviors that intersect with the maternal circuit. A number of strategies have been adopted to establish the relevant sites and systems. Bilateral lesions of the brain result in a variety of behavioral changes in the new mother. Bilateral lesions are performed in some animals who are compared in their behavior to other animals who have sham lesions or lesions to other brain sites. For instance, lesions of the medial preoptic area, a diencephalic region anterior to the hypothalamus, eliminate retrieving and nest-building in all animals that have been studied including the new mother, the virgin animal and the juvenile animal (see Numan, 1994). Through lesions of other brain sites that project to the medial preoptic area, it has also been possible to demonstrate which regions of the brain mediate neophobia and emotional responsiveness to pups, as well as regions that mediate responses to pup odors, pup reinforcement and maternal memories. Thus, it has been possible to demonstrate that the medial nucleus of the amygdala mediates in the new mothers an aversion to the novel olfactory characteristics of pups, whereas the nucleus accumbens and basolateral amygdala mediate the reinforcement-learning systems, within this maternal context (Fleming and Corter1995). Not surprisingly, if the brain is electrically stimulated, as opposed to lesioned, the opposite behavioral effects are produced. Unilateral kindling-like stimulation of a brain region known to produce decrements in behavior after lesioning, are facilitated after electrical stimulation (Morgan et al., 1997). Thus electrical stimulation of the mpoa facilitates maternal behavior whereas stimulation of the medial amygdala, inhibits it.

c-fos *expression in mpoa is enhanced with maternal behavior*

Another very powerful tool in the analysis of brain areas implicated in maternal responding is through c-*fos* immunocytochemistry and the analysis of c-*fos* expression during the exhibition of maternal behavior at the time of parturition and after the formation of maternal memories. The proto-oncogene c-*fos* is one of a class of immediate-early genes that are expressed in response to a variety of stimulus conditions. Its protein product, Fos, can, in appropriate combination with other proteins, act as a transcriptional

Exposure: Pups in box A **Exposure: No pups in box B** **Test**

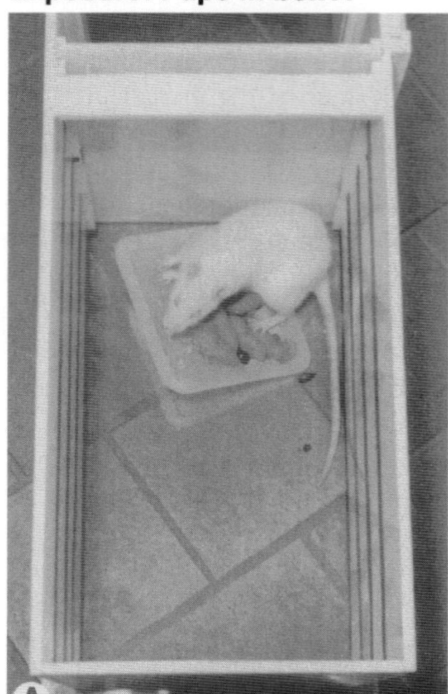

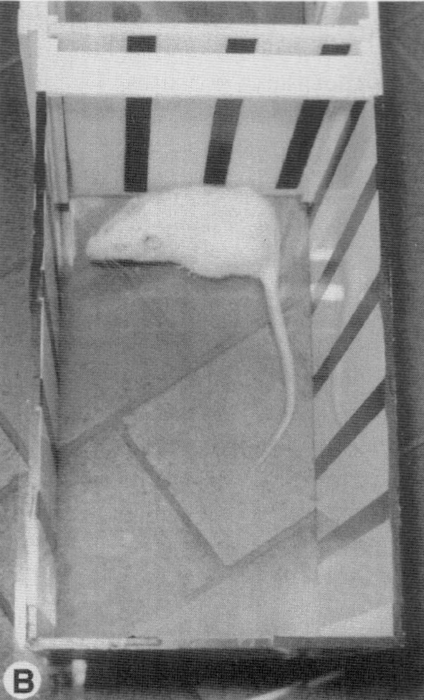

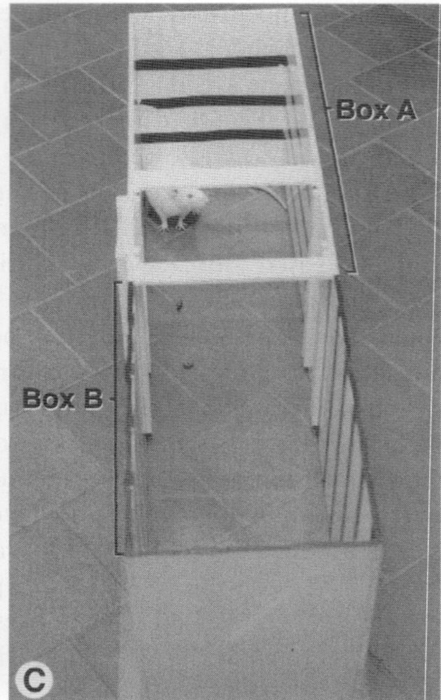

Fig. 4. The conditioned place preference task. The conditioned place preference task is given to all animals in adulthood to determine ability to be conditioned, using pup-reinforcement. *Apparatus*: The place preference boxes consists of two 21.6 × 40.6 × 30.5 cm White Plexiglas boxes, with either horizontal (Box A) or vertical (Box B) stripes made out of 1.9 cm electrical tape which are spaced approximately 1.9 cm apart. The tops of the boxes are open but covered with 0.64 cm wire mesh screen during exposure and testing. All inner surfaces of the boxes are cleaned with 70% alcohol between exposures and between tests. For testing, the vertical and horizontal striped boxes are connected to one another by means of a clear Plexiglas central connector (21.6 × 10.2 × 30.5 cm; see "Test phase" below). During *exposure*, all walls of the boxes are in place; during *test*, the end walls abutting the middle connector section are removed from each box so that the animal has free access to both boxes. The position in the room of the boxes is kept constant from exposure to exposure so that any room or light cues used by the animal to make the discrimination remains constant. *Procedure*: The rats are housed individually in standard polypropylene shoebox cages (48.3 × 26.7 × 16.5 cm). The conditioning procedure involves an exposure phase and a test phase. The 4-d conditioning procedure is described below. For the 2- and 8-d conditioning procedures, the number of exposures to each of the boxes are either reduced by one (2-d procedure) or increased by two (8-d procedure) and the day of test occurs on Day 4 or Day 9, respectively. *Exposure phase*: For all deprived groups, on Day 1 postpartum pups are removed from dams for a 23-h period. For non-deprived groups, pups are removed for only a 10-min period immediately prior to the exposures. On Days 2 and 4 postpartum, females are placed into either the vertical or the horizontal box with box 1–3 d old pups for a 1-h period (Box A). The pup-paired box is the same for the 2 pup exposure days but counterbalanced across animals within each group. On Days 3 and 5, again after 23 h (or 10 min, for nondeprived groups) of pup deprivation, the animals are placed into the non-pup-associated alternate box (either horizontal or vertical) for a 1-h period (Box B). On non-pup days, deprived females are given pups in their home cages for a 1-h period starting 1–2 h after the 1-h conditioning. This procedure is adopted to equate the duration of pup deprivation on the pup and non-pup days, so that all animals are equally "motivated" on the two types of days. Six spot checks, at 10 min intervals, are made over the 1-h exposure period. For each spot check, the occurrence of a number of different maternal behaviors is recorded, including hovering over pups, exhibiting a crouch over pups, licking pups, sniffing pups, and sleeping. If animals are seen either hovering or crouching over pups, they are described as being maternal on that spot check. Thus, over six spot-checks, a total of six indicates that animals are maternal throughout the exposure day. In addition, an overall maternal status rating is derived for each animal. This consists of a scale ranging from 4 to 1: 4 (*no hovering or crouching seen on any spot check*; see Stern and Johnson, 1990), 3 (*hovering seen on at least two spot checks*), 2 (*hovering seen on three consecutive spot checks on 1 exposure day*) and 1 (*hovering seen on three consecutive spot checks on each of 2 exposure days*). *Test phase*. Five minute tests are given. Females are placed into the apparatus containing Boxes A, B with a mid-connector region between them. All barriers are removed. The time spent in each box and in the central connector area is recorded.

regulator. It can, therefore, modulate cellular physiology and has been suggested to act as a "third messenger" in intracellular communication, especially in relation to long-term plastic changes in cells (Sagar et al., 1988). Fos has been shown to be expressed selectively in a variety of neurons responding to sensory, hormonal, neurochemical, and behavioral events (Bullitt, 1990; Robertson et al., 1991; Baum and Everitt, 1992). These observations suggest that nuclear Fos immunoreactivity can serve in some circumstances as a marker permitting mapping of functional pathways in the brain (Sagar et al., 1988). By using the distribution of cells that show c-*fos* expression and Fos staining it is possible to trace the neural structures that are activated during the expression of a particular behavior. We have previously detailed the Fos staining method as used in the studies described here (Fleming, et al., 1994b). In these experiments new mothers are separated from pups at birth and then are permitted a one hour interaction with pups. At the end of one, or at most, 2 h, animals are sacrificed and their brains are prepared for Fos immunocytochemistry. Subsequently, the number of cells within different brain regions that show the dark nuclear staining characteristic of the Fos protein can be quantified by direct visual examination or through use on an image analysis program. To facilitate identification of the brain regions of interest, we have developed an atlas that identifies the relevant landmarks for positioning the reticule and indicates the number and pattern grids within which counts are made. To insure that the activation pattern is in response to a specific stimulus or reflects the expression of a specific behavioral system, a number of additional groups are usually included to control for specificity of the stimulus, or general arousal. In our experiments we often include a group that is given a novel food to eat, or a novel odor to smell, or who are given the opportunity to interact with a conspecific.

These studies show that when new mother rats interact with pups there is an increase in the intensity of Fos staining. In keeping with these there is a concomitant increase in the number of stained cells in the mpoa, the amygdala, the nucleus accumbens, the olfactory, piriform, and parietal cortices and a number of other limbic sites (see Fig. 5). The only site that was specific to pups was the medial preoptic area. In contrast, if mothers are maternally inexperienced and are presented with pups in a wire-mesh cage (and hence are prevented from interacting with pups) elevated c-*fos* expression does not occur. In contrast to the mpoa, within amygdaloid and sensory structures (e.g. medial or cortical amygdala, parietal cortex) both pups and a social stimulus produce elevated

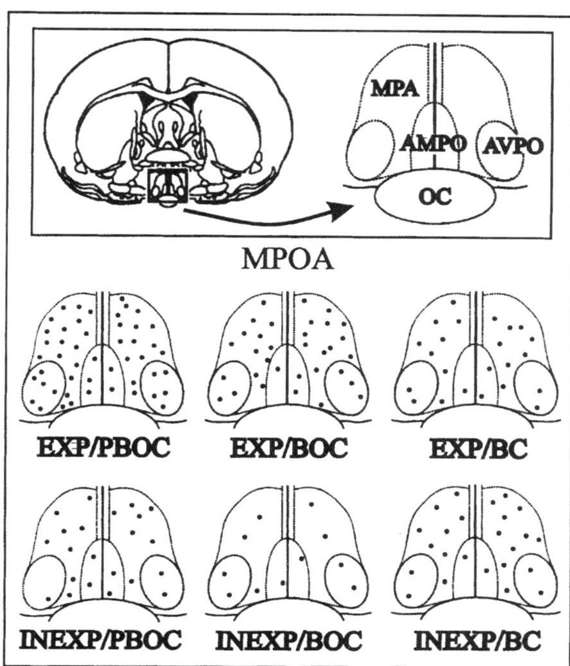

Fig. 5. Diagrammatic representation of Fos-stained cells in the medial preoptic area (MPA/MPOA) showing typical responses. The upper panel shows a cross-section of the rat brain in the region of the MPOA (boxed) with an arrow showing the enlarged MPOA with its corresponding regions. Below this the MPOA is diagrammed with dots to represent the amount of Fos-immunostaining. Each dot represents five labeled cells. Groups include rats experienced (EXP) or inexperienced (INEXP) with pups; box (B); odor (O); or cage (C). AMPO = anteromedial preoptic area; AVPO = anteroventral preoptic area (Fleming and Korsmit, 1996).

activation, indicating the nonspecific nature of the activation (Fleming et al., 1994b). Interestingly, to show that the mpoa is not involved in sensory processing and that the amygdala is, desensitization of either the parietal or the olfactory systems, does not prevent pup-induced elevations in c-*fos* expression in mpoa. It does, however, eliminate in an additive fashion activation of nuclear groups within the amygdala that receive both olfactory and somatosensory input (e.g. basolateral nucleus) (Walsh et al., 1996). Remaining undisturbed or receiving a food stimulus had little effect on c-*fos* expression.

Finally, to determine which brain systems are involved in the memory for pups or pup-associated cues, mothers were presented with pups for a 2 or 4 h period after parturition. To provide the opportunity to develop conditioned responses to pup-associated cues, pups were either scented with a novel odorant or not and presented in a novel or familiar chamber. After the exposure period, mothers were presented with either pups, odors, and/or novel cage conditions. Brain regions showing higher c-*fos* expression in experienced than in inexperienced animals and that responded to pup-associated cues included the medial preoptic area, the basolateral amygdala, and the parietal cortex. These sites seem to be more permanently changed by a prior maternal experience (Fleming and Korsmit, 1996).

Analysis of signal transduction components

Signaling elements that mediate Fos expression: calmodulin binding proteins

To this point we have discussed the role of parturitional hormones, the central importance of the MPOA and the expression of c-*fos* as a marker during maternal behavior in the rat. The question remains, "How are these components linked together?" To answer this question, we have been taking a multi-faceted approach involving behavioral, physiological, biochemical and pharmacological methodologies (Fig. 6). Once

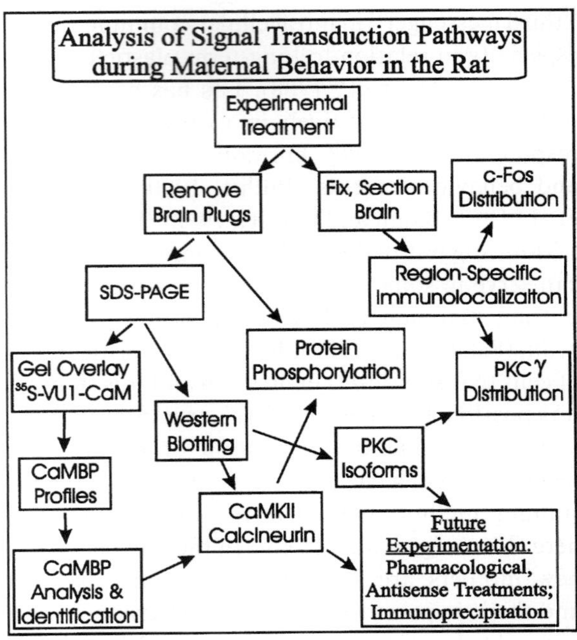

Fig. 6. A flow chart revealing the inter-relationship between the various experimental methods described in this review. Following specific experimental procedures the brain is either sectioned for the region-specific immunolocalization of the molecule under study (e.g. Fos, PKC isoform) or brain plugs are removed for biochemical analyses (e.g. SDS-PAGE or in vitro phosphorylation). After SDS-PAGE, the total CaMBP profile can be assessed via the CaM-overlay method or the separated proteins can be transferred to a membrane for western analysis. CaMBPs can be tentatively identified and quantified from the gel overlay method and these results can be compared to the results obtained with western blotting (e.g. as done for CNA in this study, CaMKII, PKCγ). Comparing the data obtained from these different methods will reveal if specific CaMBPs or PKC isoforms undergo changes implicating them in aspects of maternal behavior and can identify potential substrates (i.e. from phosphorylation studies). The roles of any signal transduction elements identified by these steps that appear to be intimately linked to events of maternal behavior will then be elucidated by future experimentation using pharmacological or antisense treatments (e.g. local injection of specific inhibitors or antisense RNA). Other molecules that associate with the component under analysis (e.g. CNA) would be isolated by immunoprecipitation with an antibody against that molecule (e.g. anti-CNA). In the long term, this approach will be extended to other signal transduction proteins to identify the critical signaling pathways that mediate the onset and maintenance of maternal behavior in the rat.

the linking elements are defined it may be possible to understand the critical steps that oversee maternal behavior in the rat and, possibly, other mammals. To date, the upstream signal transduction components that lead to *fos* gene expression and/or that mediate the onset and maintenance of maternal behavior have not been studied. Numerous neurohormones, neuropeptides and neurotransmitters that are operational during and are linked to maternal behavior lead to increases in neuronal, cytosolic free calcium levels. There are also many calcium-mediated signaling routes that lead to the expression c-*fos* (e.g. Ghosh and Greenberg, 1995; Rosen et al., 1995). As the primary intracellular receptor for calcium in all cells, calmodulin most often acts as the major mediator of increased calcium levels. Calmodulin is a small, essential calcium binding that exists in all eukaryotic cells. When it binds to Ca^{2+} it changes in conformation allowing it to bind to specific proteins. Thus, it exerts its downstream effects via a wide range of calmodulin binding proteins (CaMBPs) such as CaM-dependent kinase II (CaMKII) and the sole CaM-dependent protein phosphatase, calcineurin (CN). In this regard, CaMKII is an upstream regulator of c-*fos* expression in rat tissues (Miralem and Templeton, 1998). CaMKII and CN are both central players in models of long term potentiation (LTP) and long term depression (LTD) models for neuronal plasticity and learning (e.g. Malenka et al., 1989;Suzuki, 1994; Wang and Kelly, 1997). As one example that ties these ideas together, glutamate (L-glutamic acid), the most prevalent neurotransmitter in the vertebrate nervous system, binds to its receptor leading to an increase in cytoplasmic calcium. This increase in calcium, in turn, activates calmodulin leading to the stimulation of CaMKII activity and the subsequent induction of c-*fos* expression (Rosen et al., 1995). Glutamate also alters neuronal excitability during in LTP and in long term depression (LTD) (Lerea et al., 1995). The central role of calcium and calmodulin in nerve cell function coupled with the proposed function of such CaMBPs as CaMKII and CN in specific behaviors points to the potential importance of these and possibly other CaMBPs during maternal behavior and *fos* expression in the rat. Examination of the complete profile of CaMBPs present in selected brain regions associated with the acquisition of specific behaviors should reveal the presence and amounts of such proteins as CaMKII and CN that have been implicated in specific behavioral process. It could also reveal changes in other critical, novel CaMBPs that remain to be identified as well as pointing out known CaMBPs that have not been behaviorally linked. While there are over 3 dozen known CaMBPs, probing of mammalian tissue extracts for the total CaMBP population typically reveals less than a dozen (Antrobus et al., 1995; Colomer et al., 1994). On the other hand, a newer method using recombinant CaM unveils several dozen CaMBPs in a wide variety of cell types from microbes to mammals (Lydan and O'Day, 1993a,b; Nykamp et al., 1994; O'Day et al. unpublished).

CaMBP characterization with ^{35}S-VU1-CaM in SDS-PAGE gel overlays

Our first approach was to use this technique to examine the complete profile of CaMBPs in selected regions of the rat brain during the onset of maternal behavior in an attempt to identify likely candidates for further analysis (Fleming et al., 1996). To do this, brain tissue plugs were removed from the hippocampus, basolateral amygdala, parietal cortex and medial preoptic area from female rats exposed to pups and various controls (cage, box). After exposure, the rats were immediately sacrificed by decapitation and the brain was extracted followed by a 30 min incubation at 30°C. After sections were made at AP levels corresponding to the appropriate brain sites, samples (1.0 mm) were removed from specific areas using a neuropunch sampling tool (Fine Science Tools). Samples were individually frozen by placing vials in liquid nitrogen followed by storage at −70°C.

The punches were solubilized and subjected to 10% SDS-PAGE (Laemmli, 1970; Lydan and O'Day, 1993a). After electrophoresis, the gels were washed and overlaid with radiolabelled recombinant mammalian CaM which binds to calmodulin binding proteins separated in the gel (Lydan and O'Day, 1993b). The gel was then exposed to autoradiographic film which subsequently was developed revealing the CaMBPs as dark bands (Fig. 7).

Over 30 CaMBPs are evident in specific regions of the rat brain

Each of the brain regions examined revealed at least 30 CaMBPs by the gel overlay technique (Figs 7 and 8). Of these only a few were Ca^{2+}-independent CaMBPs demonstrating that the CaMBPs of rat brain are primarily Ca^{2+}-dependent (O'Day et al., unpublished results). This represents the most CaMBPs revealed to date by a single method and indicates that the mammalian brain is rich in a diversity of Ca^{2+}-dependent CaMBPs many of which remain to be identified. These are true CaMBPs because in the presence of trifluoperazine (TFP; Stelazine^{TM}), a specific inhibitor of CaM-binding to CaMBPs, no binding of ^{35}S-VU1-CaM occurs (Lydan and O'Day unpublished results). Furthermore, extensive studies with other CaM inhibitors and with ^{35}S-VU1-CaM binding to known purified CaMBPs verifies the sensitivity and accuracy of the method in revealing only true CaMBPs down to nanogram amounts (Lydan and O'Day, 1994; O'Day and Lydan, 1993). While more than one CaMBP can migrate to the same position in the same gel (e.g. CaMKII, CN have similar molecular weights), most often the bands visualized will represent a single CaMBP. Since the number of CaMBPs revealed by the method approximates the number of total CaMBPs that are known to exist in eukaryotic cells, this argument appears to be valid. Regardless, changes in the presence or amount of a specific "band" would indicate that a change in one or two

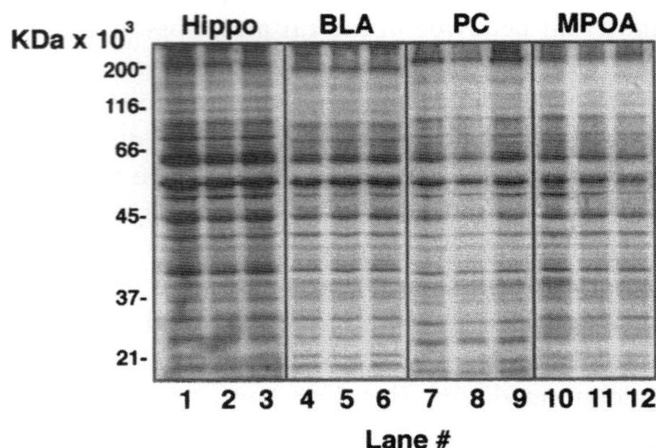

Fig. 7. Profiles of calmodulin binding proteins (CaMBPs) from selected regions of the rat brain. Brain punches were each solubilized in sodium-dodecyl sulfate-polyacrylamide gel electrophoresis (SDS-PAGE) sample buffer, prior to loading (10 μg) total protein/sample) the gels. Samples were subjected to SDS-PAGE in 10% mini-gels at 175 V until the bromophenol blue dye front reached the end of the gel (Laemmli, 1970; Lydan and O'Day, 1993a). Pre-stained protein standards were also run to determine the approximate molecular weight of all bands detected following the procedure (Gibco BRL, Canada). The gels were fixed for 30 min in 40% methanol/7%acetic acid followed by rinsing with dH_2O before being washed overnight in 10% ethanol. Gels were then rinsed in 10% ethanol, followed by 100 mM imidazole (pH 7.0) for 30 min, 20 mM imidazole (pH 7.0) for 15 min, and a final solution of probe buffer (200 mM KCl, 0.1% BSA plus either 1.0 mM CaCl2 (Ca^{2+}-dependent CaMBPs) or 1 mM EGTA (Ca^{2+}-independent CaMBPs) for 15 min. CaMBP probing occurred overnight at 4°C in the appropriate probe buffer plus 1 μCi/ml ^{35}S-VU1-CaM. The production of ^{35}S-VU1-CaM is detailed elsewhere (Lydan and O'Day, 1993b). After probing, the gels were washed 3–4 times for 2 h each in probe buffer then fixed overnight (40% methanol/7% acetic acid). Gels were then stained with Coomassie Blue, destained, dried onto filter paper and then exposed to Beta-Max autoradiographic film (Amersham) for 2 weeks prior to developing the film. After autoradiography, the gels can be scanned using a densitometer and each band (putative CaMBP) can be quantified as to its relative abundance compared to the total number of CaMBPs detected. Comparing the relative molecular weights of the bands to those of previously identified CaMBPs also allows the researcher to tentatively identify many of the CaMBPs revealed in the autoradiograph. Each band represents at least one CaMBP and each lane represents a different sample from each brain region revealing the specificity and reproducibility of the banding patterns produced by this method. Hippo = hippocampus; BLA = basolateral amygdala; PC = parietal cortex; MPOA = medial preoptic area.

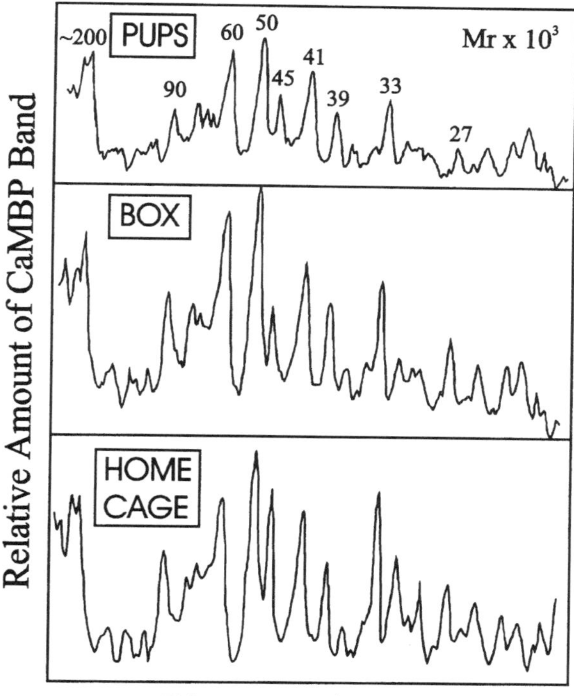

Fig. 8. Densitometric scans of CaMBP profiles reveal changes in the amounts of CaMBPs in the MPOA during the onset of maternal behavior. After exposing mothers to pups in a box (Pups) or control conditions (Box; Home Cage), proteins from plugs of tissue from the MPOA were separated by SDS-PAGE (Fig. 7) and probed for CaMBPs. After autoradiography the films were scanned densitometrically to reveal the amount of protein in each band. The numbers at the top of the Pups scan represent the Mr of the CaMBPs (e.g. 60 = CaMBP60, a putative CaMBP with a tentative molecular weight of 60 kDa). An overall decrease if seen in the CaMBPs in the MPOA of mothers exposed to pups.

CaMBPs was occurring. The extension of this work to other behavioral situations where Ca^{2+}/CaM signaling is important should be potentially rewarding.

Interestingly, while no significant changes in specific bands were evident in the CaMBP population in the hippocampus, parietal cortex and basolateral amygdala during the onset of maternal behavior, there was a major decrease in several of the CaMBPs in the MPOA of maternal females (Fig. 8; Fleming et al., 1996; O'Day et al., 1999). Preliminary quantification of the bands has verified that many bands that did not undergo any significant changes in either the hippocampus, parietal cortex or basolateral amygdala all decreased significantly in the MPOA in females exposed to pups. These CaMBPs did not decrease in amount in the MPOA when mothers were subjected to novel cages alone. Several CaMBPs have been identified previously in rats and based upon their known molecular weights they were tentatively identified in the CaMBP profiles. Of these, CaMBP60 migrates to the same general Mr as both CaMKII (MW 55–60 kDa) and Calcineurin (MW 61 kDa). The question remained, "Does CaMBP60 actually represent the levels of CaMKII and/or CN present in those brain regions?" If so, then we should be able to use the CaMBP profiles generated by our technique to identify and quantify other known CaMBPs and relate them to specific aspects of maternal behavior.

Western blotting identifies specific CaMBPs: calcineurin

To begin to answer the question, we carried out western blotting on control and maternal rat brain samples to determine the presence and amount of calcineurin A (CNA) the CaM-binding, catalytic subunit of calcineurin in specific areas. CN makes up approximately 1% of total brain tissue and, for this reason, CNA (MW 61 kDa) should be represented as a major band in our CaMBP gels (i.e. contribute significantly to band CaMBP60). Brain plugs and standard molecular markers were separated by 10% SDS-PAGE as detailed in the legend to Fig. 7, blotted to a PVDF membrane, blocked and probed with anti-CN followed by a secondary antibody and chemiluminscence substrates prior to exposing imaging film to reveal the CN protein as bands. This western blotting showed that calcineurin, considered one of the major CaMBPs of all brain tissue, was not present in all brain regions. This protein migrated to the same vicinity as CaMBP60 and it was present in the hippocampus, basolateral amygdala and the parietal cortex. In contrast, it was not detectable in the medial preoptic area under any condition (Fig. 9; O'Day et al., manuscript in preparation). This suggests that CN makes up different amounts of band CaMBP60 in different brain regions indicating that other CaMBPs with similar molecular weights are contributing to this band in different regions. Previous work has shown that calcineurin is present in different amounts and display different patterns of localization in different regions of the brain but it was always present in significant amounts to be detected histochemically (e.g. Morioka et al., 1997). One interpretation of these results is this lack of CNA indicates that calcineurin is not a critical signaling element in the MPOA and thus may not function during maternal behavior in the rat. Experiments involving Ca^{2+}/CaM dependent phosphorylation and dephosphorylation patterns so far have not provided any further more insight into this unexpected finding (O'Day et al., 1999).

The general decrease in Ca^{2+}-dependent CaMBP amounts in the MPOA seen in female rats exposed to pups may indicate that the maintenance of maternal behavior is associated with a loss in the dependence upon Ca^{2+}/CaM-dependent processes. While the initial events during the onset of maternal behavior may be Ca^{2+}/CaM-dependent, once the behavior is set in motion this dependence appears to be lost. Similar processes have been observed in other rat tissues for specific CaMBPs. For example, the elevation of intracellular calcium levels in rat tissues can lead to autophosphorylation of CaMKII resulting in the conversion of CaMKII into a Ca^{2+}/calmodulin-independent enzyme (Eason, et al., 1998). The phosphorlyated CaMKII then remains functional even in the absence of Ca^{2+}/CaM signaling. If this occurs during maternal behavior, the subsequent decrease in total CaMBPs might represent a final step for the removal of CN activity, which might dephosphosphorylate CaMKII or its substrates, for example. The decrease in CaMBPs could also reflect the loss of other Ca^{2+}/calmodulin-dependent kinases which might negatively affect maternal behavior. While these ideas are very speculative they provide further direction to research into the general decline in CaMBPs that specifically occurs in the MPOA, but not other brain regions, during the onset of maternal behavior in the rat.

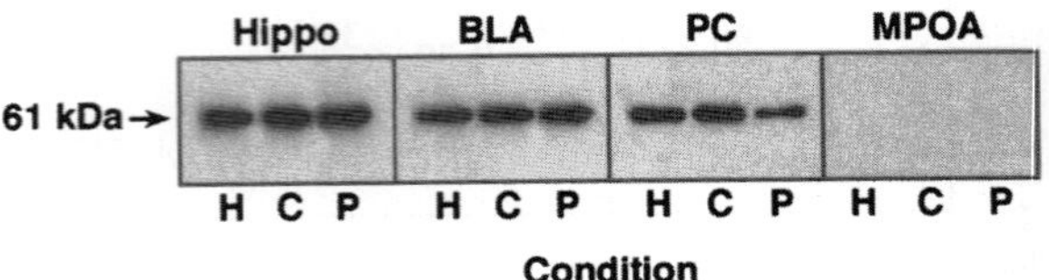

Fig. 9. Identification of calcineurin in specific regions of the rat brain by western blotting. Following electrophoresis as detailed in Fig. 7, gels were soaked in blotting buffer (25 mM Tris, 192 mM glycine, 20% methanol and 10% SDS) for 15 min. A mini-trans-blot cell (Bio-Rad) was used to transfer the separated proteins from the SDS-PAGE gel to a poly-vinylidene difluoride (PVDF) membrane at a constant voltage of 100 V for 1 h. Following the transfer, the PVDF membranes were blocked overnight at 4°C in 5% non-fat milk in TTBS (20 mM Tris, 150 mM NaCl, 0.05% Tween 20). We have found that blocking with skim milk is critical because it greatly diminishes the amount of non-specific staining as compared to other blocking agents (e.g. BSA). The blocked PVDF membranes were incubated with primary antibody (1/300 anti-CN, IgG2a Monoclonal, Cat. #C26920; Transduction Laboratories), diluted in blocking solution, for 1 h at 37°C. Blots were then washed with agitation for 90 min. The wash buffer (TBS plus 0.05% Tween) was changed every 10 min. Next, the blots were incubated with the secondary antibody (1/19,000 goat anti-mouse IgG HRP polyclonal, Transduction Laboratories), diluted in blocking buffer, for 30 min at 37°C. The membranes were then washed for 90 min with agitation, changing the wash buffer every 10 min. Excess buffer was drained from the PVDF membranes and premixed chemiluminescent detection substrates were added to the membranes (as per instructions detailed by Boehringer Mannheim, Canada). Scientific imaging X-OMAT AR film (Kodak, US) was placed over the membranes which were subsequently placed in an X-ray exposure cassette (Kodak, US) for the appropriate time. Hippo = hippocampus; BLA = basolateral amygdala; PC = parietal cortex; MPOA = medial preoptic area; P = Pups; B = Box; H = Home Cage.

On the other hand, the general decline in CaMBPs might reflect that other signaling pathways are more critical at this time. Another upstream signaling element that leads to an increase in c-*fos* expression is protein kinase C (PKC) a protein kinase that has also been linked to other behavioral processes involving plasticity, such as LTP (e.g. Malenka et al., 1989; Suzuki, 1994). For this reason, we have examined the profiles of PKC expression in various brain regions during the onset of maternal behavior (Drmic et al., unpublished results). Using tissue-specific immunolocalization and western blotting as detailed above for other proteins, we have found that there is a tight link between the expression of PKCγ and the expression of c-*fos* during the onset of maternal behavior (Fleming et al., 1996; manuscript in preparation). Furthermore, in retention experiments where c-*fos* expression remains low unless pups are presented, the levels of PKCγ remain elevated (especially in experienced animals) and thus might serve to elicit the more rapid re-expression of c-*fos* when pups are re-introduced. We are currently using pharmacological approaches to understand the importance of PKC in maternal behavior.

Maternal behavior: future research

Our long term goal is to show the critical importance to the onset and maintenance of maternal behavior of any putative upstream regulators of c-*fos* expression that we identify by the methods described above (Fig. 6). In addition to using pharmacological intervention (e.g. inhibitors, activators injected into the MPOA), we are in the process of adapting anti-sense technology for use in the rat for further critical studies. Another approach we have just initiated is to identify the G proteins that are localized to the MPOA during the onset of maternal behavior (Payne et al., unpublished results). While Gs is expressed at relatively low levels in all brain regions including the MPOA, Gq is present at very high levels in the MPOA. We intend to generate antibodies against Gq and use them to immunoprecipitate upstream receptors and downstream effectors in the MPOA that are present specifically during the onset and maintenance of maternal behavior. In total, the multifaceted cellular and molecular approaches coupled with in-depth behavioral analyses that have been described here should ultimately identify critical elements that mediate maternal behavior.

Acknowledgements

We would like to thank Jason Watchus, Ronda Smyth, Marc Korsmit, Mike Lydan, Irene Drmic and Lori Anne Payne for their technical assistance in portions of the research reviewed here. We also thank Alison Dias for producing the photographic figures. This work was supported by grants from the Medical Research Council of Canada (ASF) and Natural Sciences and Research Council of Canada (ASF; DHO'D).

References

Antrobus, S.D., Siaw, E.K.O. and Walters, M.R. (1995) Protracted 1,25-dihydroxyvitamin D treatment stimulates multiple calmodulin binding proteins in rat kidney. *Endocrinology*, 136: 1516–1522.

Baum, M.J. and Everitt, B.J. (1992) Increased expression of c-*fos* in the medial preoptic area after mating in male rats: Role of afferent inputs from the medial amygdala and midbrain central tegmental field. *Neurosci.*, 50: 627–646.

Bridges, R.S. (1996) Biochemical basis of parental behavior in the rat. In: J.S. Rosenblatt and C.T. Snowdon (Eds.), *Advances in the Study of Behavior*, Vol. 25, Academic Press, San Diego, pp. 215–242.

Brunelli, S.A., Shindledecker, R.D. and Hofer, M.A. (1986) Development of maternal behaviors in prepubertal rats at three ages: age-characteristic patterns and responses. *Dev. Psychol.*, 98: 91–99.

Bullitt, E. (1990) Expression of c-*fos* like protein as a marker for neuronal activity following noxious stimulation in the rat. *J. Comp. Neurol.*, 296: 517–530.

Colomer, J., Agell, N., Engel, P. and Bachs, O. (1994) Expression of calmodulin and calmodulin binding proteins in lymphoblastoid cells. *J. Cell Physiol.*, 159: 542–550.

Eason, R.A., Tarpley, J.L., Fuller, N.R. and Bhatt, H. (1998) Dephosphorylation and deactivation of Ca^{2+}/calmodulin-dependent kinase II in B-TC3 cells is mediated by Mg^{2+}- and okadaic-acid-sensitive protein phosphatases. *Biochem. J.*, 329: 283–288.

Fleming, A.S. (1986) Psychobiology of rat maternal behavior. In: B.R. Komsaruk, H.I. Siegel, M-F Cheng and H.H. Feder (Eds.), *Reproduction: A Behavioral and Neuroendocrine Perspective, Ann. NY Acad. Sci.*, 474: 234–251.

Fleming, A.S. and Corter, C. (1995) Psychobiology of Maternal Behavior in Non-Human Mammals: Role of Sensory, Experiential and Neural Factors. In: M. Bornstein (Ed.), *Handbook of Parenting*. Lawrence Erlbaum Associates, New Jersey, pp. 59–86.

Fleming, A.S. and Korsmit, M. (1996) Plasticity in the maternal circuit: Effects of maternal experience on Fos-lir in hypothalamic, limbic and cortical structures in the postpartum rat. *Behav. Neurosci.*, 110: 567–582.

Fleming, A.S. and Luebke, C. (1981) Timidity prevents the nulliparous female from being a good *mother. Physiol. Behav.*, 27: 863–868.

Fleming, A.S. and Rosenblatt, J. (1974a) Maternal behavior in the virgin and lactating rat. *J. Comp. Physiol. Psychol.*, 86: 957–972.

Fleming, A.S. and Rosenblatt, J. (1974b) Olfactory regulation of maternal behavior in rats: I. Effects of olfactory bulb removal in experienced and inexperienced lactating and cycling females. *J. Comp. Physiol. Psychol.*, 86: 221–232.

Fleming, A.S., and Rosenblatt, J.S. (1974c) II. Effects of peripherally-induced anosmia and lesions of the lateral olfactory tract in pup-induced virgins. *J. Comp. Physiol. Psychol.*, 86: 233–246.

Fleming, A.S., Vaccarino, F. and Leubke, C. (1980) Amygdaloid inhibition of maternal behavior in the nulliparous female rat. *Physiol. Behav.*, 25: 731–743.

Fleming, A.S., Flett, G.L., Ruble, D.N. and Shaul, D.A. (1988) Postpartum adjustment in first-time mothers: Relations between mood, maternal attitudes, and mother-infant interactions. Dev. Psychol., 24: 71–81.

Fleming, A.S., Cheung, U., Myhal, N. and Kessler, Z. (1989) Effects of maternal hormones on 'timidity' and attraction to pup-related odors in female rats. *Physiol. Behav.*, 46: 449–453.

Fleming, A.S., Korsmit, M. and Deller, M. (1994a) Rat pups are potent reinforcers to the maternal animal: Effects of experience, parity, hormones and dopamine function. *Psychobiology*, 22: 44–53.

Fleming, A.S., Suh, E.J., Korsmit, M. and Rusak, B. (1994b) Activation of Fos-like immunoreactivity in MPOA and limbic structures by maternal and social interactions in rats. *Behav. Neurosci.*, 108: 724–734.

Fleming, A.S., Watchus, J., Smyth, R., Lydan, M.A. O'Day, D.H. and Ivy, G. (1996) Neuroplasticity of the maternal brain: Changes in signal transduction and glial proteins associated with post-partum experiencce in rats. *Soc. Neurosci., Washington, D.C.*

Ghosh, **XXX** and Greenberg, M.E. (1995) Calcium signalling in neurons: molecular mechanisms and cellular consequences. *Science*, 268: 239–253.

Laemmli, U.K. (1970) Cleavage of structural proteins during the assembly of the head of bacteriophage T4. *Nature*, 227: 680–685.

Lee, A., Clancy, S. and Fleming, A.S. (1999) Mother Rats Bar-press for Pups: Rate influenced by reproductive status, maternal status, and integrity of the medial preoptic area. *Behav. Brain Res.*, **XXXXX**.

Lerea, L.S., Carlson, N.G. and McNamara, J.O. (1995) *N*-methyl-D-aspartate receptors activate transcription of c-*fos* and NGFI- A by distinct phospholipase A2-requiring intracellular signaling pathways. *Mol. Pharmacol.*, 47: 1119–1125.

Lydan, M.A. and O'Day, D.H. (1993a) Calmodulin and calmodulin-binding proteins in *Dictyostelium discoideum*: Developmental regulation by calcium ions. *Exp. Cell Res.*, 205: 134–141.

Lydan, M.A. and O'Day, D.H. (1993b) Production of ^{35}S-labelled proteins in E. coli and their use as molecular probes. In: A.J. Harwood (Ed.), *Methods in Molecular Biology Methods for Gene Analysis*, Vol. 31, Chap. 36, Humana Press Inc. Totawa, NJ, pp. 389–396.

Magnusson, J. and Fleming, A. (1995) Rat Pups are Reinforcing to the Maternal Rat: Role of Sensory Cues. *Psychobiology*, 23: 69–75.

Malenka, R.C., Kauer, J.A., Perkell, D.J., Mauk, M.D., Kelly, P.T., Nicoll, R.A. and Waxham, M.N. (1989) An essential role for postsynaptic calmodulin and protein kinase activity in long-term potentiation. *Nature*, 340: 554–557.

Miralem, T. and Templeton, D.M. (1998) Heparin inhibits Ca^{2+}/calmodulin-dependent kinase II activation and c-*fos* induction in mesangial cells. *Biochem. J.*, 330: 651–657.

Morgan, H.D., Watchus, J.A. and Fleming, A.S. (1997) The Effects of Electrical Stimulation of the Medial Preoptic Area and the Medial Amygdala on Maternal Responsiveness in Female Rats. The Integrative Neurobiology of Affiliation. *Ann. NY Acad. Sci.*, 807: 602–605.

Morioka, M., Nagahiro, S., Fukunaga, K., Miyamoto, E. and Ushio, Y., (1997) Calcineurin in the adult rat hippocampus: Different distribution in CA1 and CA3 subfields. *Neuroscience*, 78: 673–684.

Natterer, G. and Fleming, A.S. (1999) Effects of medial preoptic and amygdala lesions on maternal and social behaviors in the juvenile rat, manuscript submitted.

Numan, M. (1994) Maternal behavior. In: E. Knobil and J. Neill (Eds.), *Physiology of Reproduction*, 2nd Edn, Raven Press, New York, pp. 221–302.

Nykamp, D.A., Lydan, M.A., O'Day, D.H. and Lange, A. (1994) Calmodulin mediates contraction of the oviduct of *Locusta migratoria. Insect Biochem. Mol. Biol.*, 24: 507–516.

O'Day, D.H., Watchus, J., Drmic, I., Payne, L.-A. and Fleming, A.S. (1999) Analysis of calmodulin binding proteins and CaM-dependent phosphorylation patterns in the medial preoptic area and other areas of the rat brain during the onset of maternal behavior. Submitted.

Robertson, G.S., Pfaus, J., Atkinson, L., Matsumura, H. Phillips, A. and Fibiger, H. (1991) Sexual behavior increases c-*fos* expression in the forebrain of the male. *Brain Res.*, 54: 352–357.

Rosen, L.B., Ginty, D.D. and Greenberg, M.E. (1995) Calcium regulation of gene expression. *Adv. Second Messenger Phosphoprotein Res.*, 30: 225–253.

Rosenblatt, J.S. and Lehrman, D.S. (1963) Maternal Behavior of the Laboratory Rat. In: *Maternal Behavior in Mammals*, pp. 8–57.

Sagar, S.M., Sharp, F.R. and Curran, T. (1988) Expression of c-*fos* protein in the brain: Metabolic mapping at the cellular level. *Science*, 240: 132–133.

Stern, J.M. (1989) Maternal behavior: Sensory, hormonal, and neural determinants. In: F.R. Brush and S. Levine (Eds.), *Psychoendocrinology*, Academic Press, New York, pp. 105–226.

Stern, J. S. (1996) Somatosensation and maternal care in Norway rats. In: J.S. Rosenblatt and C.T. Snowdon (Eds.), *Advances in the Study of Behavior*, Vol. 25, Academic Press, San Diego, pp. 243–294.

Suzuki, T. (1994) Protein kinases involved in the expression of long-term potentiation. *Int. J. Biochem.*, 26: 735–744.

Walsh, C., Fleming, A.S., Lee, A. and Magnusson, J. (1996) The effects of olfactory and somatosensory desensitization on Fos-like immunoreactivity in the brains of pup-exposed postpartum rats. *Behav. Neurosci.*, 110: 1–20.

Wang, J.-H. and Kelly, P.T. (1997) Postsynaptic calcineurin activity downregulates synaptic transmission by weakening intracellular Ca2+ signaling mechanisms in hippocampal CA1 neurons. *J. Neurosci.*, 17: 4600–4611.

Weisner, B.P. and Sheard, N.M. (1933) *Maternal behavior in the rat*, Oliver and Boyd, Edinburgh.

W.E. Crusio and R.T. Gerlai (Eds.)
Handbook of Molecular-Genetic Techniques for Brain and Behavior Research (Techniques in the Behavioral and Neural Sciences, Vol. 13)

CHAPTER 4.11

Measuring rodent exploratory behavior

Catherine Belzung

Laboratoire d'Ethologie et de Pharmacologie du Comportement (LEPCO), Faculté des Sciences, Parc Grandmont, F-37200 Tours, France

Introduction

Exploration generally refers to behaviors that are triggered off by novelty. It has been extensively studied using many test paradigms such as confrontation with novel objects (Misslin and Ropartz, 1981a, 1981b, 1981c; Misslin, 1982; Poucet et al., 1986), unknown conspecifics (File, 1980; Figler and Einhorn, 1983) or novel places (Katzir, 1982; Miller et al., 1986; Misslin and Cigrang, 1986; Griebel et al., 1993a; Blois-Heulin and Belzung, 1995), especially in small rodents such as mice or rats (Berlyne, 1950; Hughes, 1968; Barnett and Smart, 1975; Barnett and Cowan, 1976; Misslin and Cigrang, 1986). When these animals have free access to a novel place, they first exhibit an orienting response (Berlyne, 1960; Sokolov, 1960; Vinogradova, 1970) followed by approach responses toward the novel stimulus, called novelty-seeking behavior, and then by marked preference for the novel place (Hughes, 1968; Misslin and Ropartz, 1981c). Many explanations of this behavior have been proposed. For example, exploration permits the animal to gain information about a novel environment, which may have an adaptative value because it allows an optimization of foraging and escape from predators (Birke and Archer, 1983; Renner, 1990). However, as exploration exists outside of the direct context of foraging or of threat, another possible motivation may be related to the reinforcing properties of this behavior. Indeed, the opportunity to discover complex novel places may act as an effective reward in learning tasks (Montgomery, 1954).

Exploration-related responses include a large number of behavioral items such as scanning, sniffing, walking, rearing, leaning, jumping, digging, dragging objects (Blois-Heulin and Belzung, 1995; Renner and Seltzer, 1991). However, the expression of such a rich behavioral repertoire depends upon the experimental paradigm employed. For example, approach toward novelty cannot be exhibited when a subject is forced into novelty. Digging or eating is not possible when sawdust or food are not available. Most laboratory studies have been undertaken using experimental paradigms allowing the expression of very few behaviors and in which only a limited number of items such as locomotion or rearing have been registered, thus permitting the automatic recording of behaviors using photoelectric cells.

Exploration depends upon different determinants, such as size of the apparatus (rodents are agoraphobic and therefore avoid large spaces), lighting conditions (exploration is inhibited under brightly-lit conditions) (File, 1980), presence of attractive or aversive stimuli within the environment (for example, exploration is inhibited by presence of a predator's olfactory marks) (Berton et al., 1998), free or forced access novelty (forced access is stressful, when compared to free access) (Misslin and Cigrang, 1986), complexity of the environment (Taylor, 1974), level of discrepancy between unknown and familiar space (Birke and Archer, 1983), degree of satiety (exploration is largely increased in food deprived rodents), etc.

This chapter describes several experimental paradigms useful for the study of exploration in rodents. These tests are widely used in behavioral neurobiology and differences between inbred strains have been repeatedly reported using these devices. Therefore, they may be useful to detect behavioral differences between transgenic, knockout or knockdown mice and their respective wildtype genotypes.

Exploration of novel places

Open fields (Fig. 1A)

The open-field test consists of the measurement of behaviors elicited by placing a rodent in a novel arena from which escape is prevented by surrounding walls (Walsh and Cummins, 1976). This paradigm has been originally described by Hall (1934) for the study of emotionality. Hall's device consisted of a brightly-illuminated circular arena of about 1.2 m diameter bound by a wall 0.45 m high, with linoleum floor marked out by radial lines and concentric circles. Hall placed rats in the outer ring of the arena and observed its movements for 2 min, during daily repeated trials. Sometimes, rats were tested after 24 or 48 h food deprivation. Hall observed that rats walked further when food deprived, but not all of them did eat. Rats that did not eat were called emotional. When compared to non-emotional ones, they exhibited lower number of entries in the central part of the arena and higher levels of defecation.

The open-field test is now one of the most popular devices used in animal psychology. Different versions are available, differing in their shape (circular, square or rectangular), lighting (zenithal lighting by placing a bulb above the open-field or underneath lighting using a bulb placed under a transparent floor), presence of objects within the arena such as platforms, columns, tunnels (see for example Takahashi and Kalin, 1989), etc. Procedure generally consists in forced confrontation of a rodent with the situation. The animal is placed in the center or the periphery of the apparatus and the following behavioral items are recorded, for a period ranging from 2 to 20 min (usually 5 min): horizontal locomotion (number of crossings of the lines marked on the floor), frequency of rearing or leaning (sometimes termed as vertical activity), grooming (long duration washing of the fur). In such a situation, rodents spontaneously exhibit a preference for the periphery of the apparatus when compared with activity in the central parts of the open-field. Indeed, mice and rats walk close to the walls, a behavior called thigmotaxis. Increase of time spent in the central part as well as of the ratio central/total locomotion or a decrease of the latency to enter the central part are an indication of anxiolysis. Some authors used a procedure in which subjects were allowed free access to the open-field, from a familiar cage (see for example Kopp et al., 1997). In this case, number of risk assessment postures toward the open-field may provide a good measure of the approach response toward novelty, that is, of exploration. All the other measures cited above are not really measures of exploration but rather of activity or anxiety.

Differences in inbred mouse strain behavior have been rather well described using the open-field (De Fries et al., 1978; Crabbe, 1986; Crusio et al., 1989; Mathis et al., 1994). It is to be noticed that generally albino inbred strains such as the BALB/c or the A/J exhibit low activity and high avoidance of the central part when compared to non albinos such as C57BL/6 (Defries et al., 1966; Defries, 1969). Furthermore, genotype × lighting interactions have been shown, the BALB/c and the C57BL/6 strain not differing any more when tested under red light conditions. Therefore, in the case of a mutation affecting pigmentation, a modification of open-field activity under brightly-lighted conditions should be interpreted with caution. Using recombinant inbred

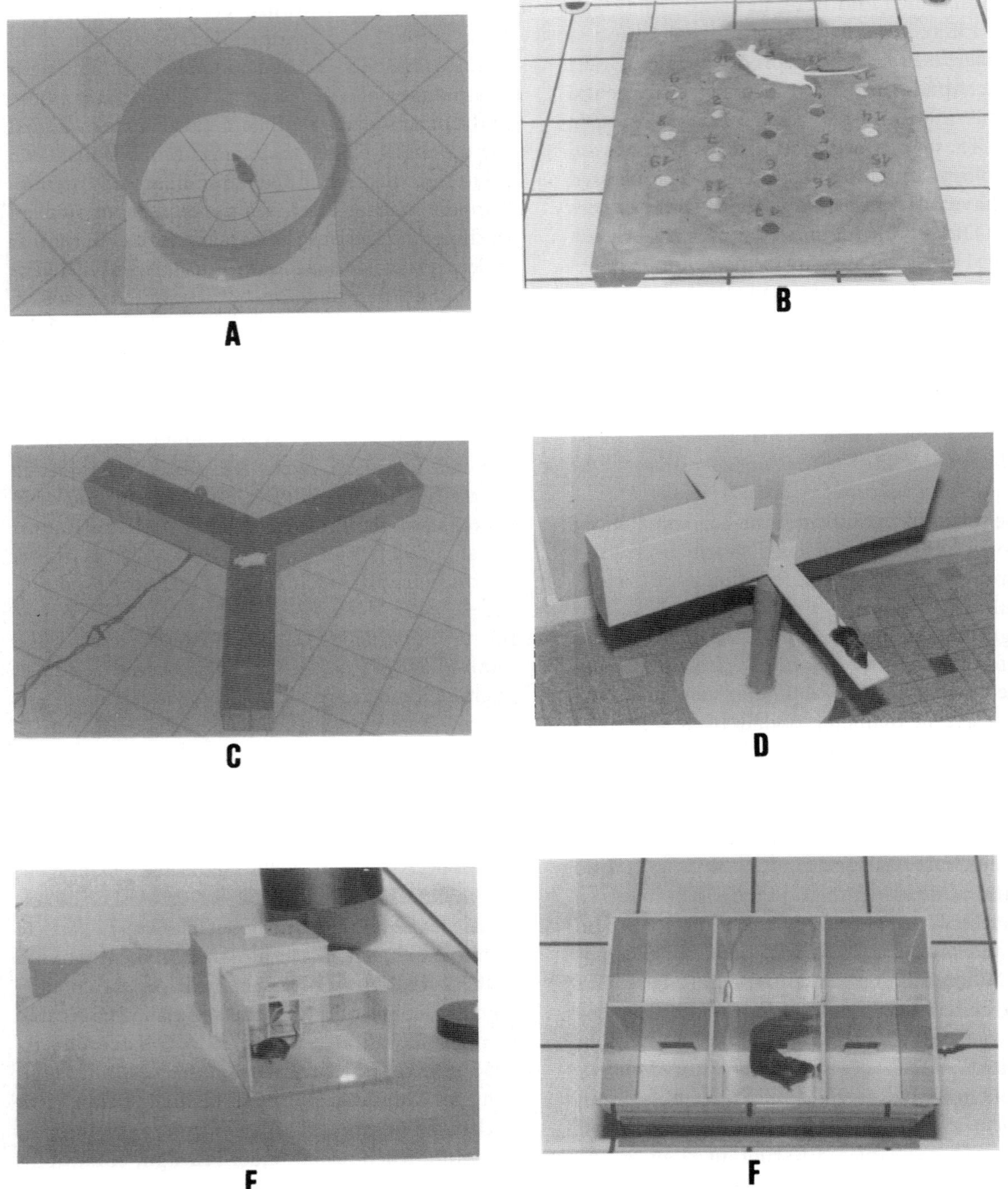

Fig. 1. Illustration of some testing situations used to study exploratory behavior in rodents. (A) Circular open-field. (B) Holeboard. (C) Y-maze used to study spontaneous exploration. (D) Elevated plus maze. (E) Light/dark choice test. (F) Free exploratory test. (Photographs from Serge Barreau.)

strains (Plomin et al., 1991; Mathis et al., 1995) or linkage-testing strains (Flint et al., 1995; Clément et al., 1995; 1997), different authors found some genetic markers associated with open-field activity. In some cases, behavioral consequences of the pharmacological inhibition of the function of a given protein parallels the behaviors induced by mutations of the gene coding for the same protein. For example, open-field activity is increased by specific dopamine transporter inhibitors (Garreau et al., 1997) as well as in knockout mice deficient for the dopamine transporter (Giros et al., 1996). However, this is not always the case. No difference in open-field behavior appears between mice lacking the serotonin 5-HT_{1B} receptor and their wildtype controls, but after challenge with a 5-HT_{1B} receptor agonist, activity is increased in controls and not in mutant mice (Saudou et al., 1994).

Holeboards (Fig. 1B)

The holeboard test has been first described by Boissier and Simon (1962). It consists of a wooden board, elevated above the floor of the testing room, in which sixteen holes (3 cm in diameter) have been drilled. During 3 or 5 min sessions, experimenters record the frequency of head dipping in the holes and the number of iterative head dipping in the same hole. It is to be noticed that all these variables are inter-correlated (Belzung and Le Pape, 1994) and therefore the concomitant recording of more than one of them may lead to redundancies.

Differences in holeboard behavior have been found between normal and Lurcher mutant mice (Lalonde et al., 1993), between different strains of rats (Rex et al., 1996), and between haloperidol sensitive and haloperidol resistant gerbils (Upchurch and Schallert, 1983), indicating a genotype influence. Huntington's disease transgenic mice exhibit a decrease in exploratory head-dips (File et al., 1998). Recently, this test has been used to show exploration alterations in transgenic mice expressing high levels of brain tumor necrosis factor-α (Fiore et al., 1998).

Spontaneous alternation procedures (Fig. 1C)

The natural tendency that rodents exhibit, when moving from one place to another, to enter the least visited area or an area which has changed in some way, has been demonstrated by Tolman (1925) and is known as "spontaneous alternation" (see Dember and Fowler, 1958; Berlyne, 1960; Barnett, 1975 for further details). Spontaneous alternation has been often studied using non-reinforced T-mazes: rats or mice are placed in the start arm and allowed to enter one of the goal arms. The subject is then replaced in the start arm and, if a second opportunity is given to enter one of the goal arms, the animal may enter the arm that had not been chosen during the first trial. If other trials are given the animal, an alternation between right and left arm choices will appear. More recently, Y-mazes are used to study spontaneous alternation because they eliminate the necessity of handling the rodents between two consecutive trials, the goal compartment becoming the start box for the next trial (Roullet and Lasalle, 1993).

The Y maze apparatus consists of three arms, with a 120° angle between two adjacent arms. Each arm is ended by a start / goal compartment that can be separated from the rest of the maze by a removable door. At the beginning of the session, animals are confined in the start compartment during 1 min. The door is then opened, providing free access to the maze. The rodents can then freely move until a given number of choices (ranging from 6 to 30) have been completed. An alternative procedure consists in a forced first choice followed by a second free choice. By imposing a delay between two consecutive choices or rotating the apparatus between two choices, it is also possible to use this paradigm to study memory or spatial processes (Bertholet and Crusio, 1991). An experimenter records the choice (right or left arm) made by the animal which will permit to calculate the percent of spontaneous alternation, the number of choices of the preferred side and the longest series of same side choices. Spontaneous alternation is usually observed in 60–70% of the choices.

Spontaneous alternation differences have been found between several inbred mouse strains (Bertholet and Crusio, 1991) and between the Roman high and Roman low avoidance strains of rats (Willig et al., 1992), suggesting that performance in this procedure is genotype-dependent. Furthermore, spontaneous alternation deficits have been observed in several cerebellar mutant mice such as nervous mutants (Lalonde and al., 1986a), lurcher mutants (Lalonde et al., 1986b; Lalonde et al., 1988; Caston et al., 1997), staggerer mutants (Lalonde, 1987), hot foot mutants (Filali et al., 1996) and Purkinje cell degeneration mutants (Lalonde et al., 1987). Note that all these cerebellar mutants exhibit motor deficits, which may explain spontaneous alternation deficits because animals cannot control their movements. Deficits are also observed in some animal models of human neurodegenerative disorders associated with mental retardation or dementia, such as transgenic mice carrying multiple copies of the human gene for S100β, which is a model of Down syndrome (Gerlai et al., 1994) or in models of Alzheimer's disease such as transgenic mice expressing the 751-amino acid isoform of the human beta-amyloid precursor protein (Moran et al., 1995) or mice carrying both amyloid precursor protein and presenilin 1 transgenes (Holcomb et al., 1998). However, in this procedure exploration deficits are difficult to distinguish from working memory deficits. This is emphasized by the fact that most of the behavioral genetic studies that showed differences between strains in spontaneous alternation have used models of mental retardation, dementia or of cerebellar dysfunction which exhibit amnesic deficits. In order to study exploration rather then memory, it is necessary not to impose a delay between two consecutive choices.

Exploration tests used to measure anxiety (Fig. 1D and E)

Some anxiety models are based upon the fear-induced inhibition of exploration. They involve forced confrontation with an unfamiliar environment characterized by an important contrast between two areas, the one of which is attractive and the other one aversive. In such a situation, the rodent is confronted with two opposite spontaneous motivations: on the one hand, to escape from the unknown paradigm (forced confrontation is stressful) and thus entering all parts of the apparatus to find an exit, and on the other hand avoiding the aversive part of the apparatus. Two such models have been designed for the measurement of exploration or anxiety: the elevated plus maze (Handley and Mithani, 1984; Pellow et al., 1985; Lister, 1987), based on the contrast between two arms closed by lateral walls and two perpendicular open arms, and the light/dark choice test (Crawley and Goodwin, 1980; Crawley, 1981), based on the contrast between a brightly lit (2/3 of the surface) and a dark part (1/3 of the surface) of an area (for further details on the methods, see the in Chapter by Crawley). The percent of entries in the open arms and the number of transitions from the dark to the lit arena have been described as good measures of anxiety: the higher the value of these variables, the lower the anxiety level of a given subject. Measurement of other variables permits to obtain further information. For example, in a modified version of the light/dark choice test which consists of two equal size boxes linked by a tunnel, the time spent in the lit area is a variable that is sensitive to the anxiolytic as well as to the anxiogenic effects of drugs (Belzung et al., 1987) and that does not depend upon activity. Therefore, if it is predicted that a given genotype may express an anxious phenotype, it is necessary to measure the time spent in the lit area and not only the transitions, because these last ones do not allow to measure anxiogenesis. In the elevated plus maze, animals initially exhibit a high level of risk assessment from the central platform toward the open arms (Rodgers et al., 1992). Interestingly, anxiolytic compounds acting via a GABAergic mechanism decrease the percentage of entries in the open arms as well as risk assessment while anxiolytic agents acting via other pharmacological targets such as serotoninergic function do not consistently modify the percentage of open arm entries at non sedative doses while they decrease risk assessment (Rodgers, 1997). Therefore, when testing the effects of a genetic manipulation interfering with serotoninergic function, it will be necessary to study effects on risk assessments as well. In the elevated plus maze or the light/dark choice tests, different factors can modify the baseline of controls such as illumination level, isolation or daytime of testing (Misslin et al., 1989; Griebel et al., 1993b). These factors must be controlled by the experimenter.

Strain distributions have been described for the elevated plus maze (Trullas and Skolnick, 1993) and the light/dark apparatus (Crawley and Davis, 1982; Mathis et al., 1994; Crawley et al., 1997). Furthermore, high level of anxiety has been reported in Fyn tyrosine kinase deficient mice (Miyakawa et al., 1994) and in mice lacking the adenosine A_{2a} receptor (Ledent et al., 1997) using the light/dark choice test. Using the elevated plus maze, high level of anxiety has been reported in transgenic mice overexpressing the corticotropin- releasing factor (Stenzel-Poore et al., 1994), in adenosine A_{2a} receptor knockout mice (Ledent et al., 1997), in mice lacking the serotonin 1A receptor (Parks et al., 1998; Ramboz et al., 1998) while a low level of anxiety has been found in young transgenic mice overexpressing the bovine growth hormone gene (Meliska et al., 1997), in dopamine D3 receptor deficient mice (Steiner et al., 1997), in transgenic mice with impaired glucocorticoid (type II) receptor function (Rochford et al., 1997), in Huntington's disease transgenic mice (File et al., 1998) or using antisense targeting of CRH (Skutella et al., 1996).

Free access to novelty (Fig. 1F)

These situations consist in permitting an animal free access to an unknown place from their home cage or another place to which they have been familiarized for at least 24 h. Various paradigms have been designed, including free access to a runway (Blois Heulin and Belzung, 1995; Berton et al., 1998) or to an open field (Kopp et al., 1997) from the home cage in which mice have previously been isolated for 24 h. However, the sole situation that has been extensively validated is the free exploratory test. This test has first been described by Hughes (1968) for rats and adapted to mice by Misslin and Ropartz (1981c). The paradigm used in the mouse is a rectangular box (20 × 30 × 20 cm) that can be divided in two equal parts (10 × 30 × 20 cm each) by removable doors. Each of these two parts is divided in three "exploratory boxes" (10 × 10 × 20 cm). Mice are introduced in one part of the apparatus 24 h before testing with the doors closed, in order to get familiarized with this side. The floor of this part of the apparatus is covered with sawdust, and food and water are available ad libitum. The other part remains empty. Twenty-four hours later, the animals are given the opportunity to choose between this familiar part and the other part by removal of the temporary partitions. During 10 min, an experimenter records the number of attempts (risk assessment posture the rodent exhibits toward the novel place, while remaining in the familiar part with at least one paw), the time spent on the novel side, the number of boxes entered (locomotion) and the number of rearings.

In such a situations, most strains of mice exhibit a preference for the novel side (Misslin and Cigrang, 1986; Beuzen and Belzung, 1995), probably because the procedure implies the entries in very small (10 × 10 cm) spaces and not in large areas. However, some strains such as BALB/c (Griebel et al., 1993a; Belzung et al., 1994; Belzung and Berton, 1997) show neophobic behavior characterized by a higher number of attempts and a preference for the familiar side. This neophobic behavior can be reversed by benzodiazepines such as diazepam or chlordiazepoxide (Griebel et al., 1993a; Belzung and Berton, 1997). Knock out for the $\gamma 2$ subunit of the $GABA_A$ receptor, which display a reduction of 25% of the number of benzodiazepine binding sites, exhibit increased neophobia in this situation (Crestani et al., 1996).

Exploration of novel objects

Behavioral reactions of mice toward novel objects has been studied in several experimental paradigms such as open-fields. However, exploration of novel objects depends upon the discrepancy between novelty of the object and novelty of the background. Indeed, a novel object placed in a familiar environment induces avoidance and burying behaviors (neophobic behavior) whereas the same object placed in a novel environment may induce approach responses (neophylic behavior) (Misslin, 1982; Misslin and Ropartz, 1981a).

Reactions of mice toward novel objects have been extensively studied using a small environment (10 × 30 cm), divided in three exploratory units (10 × 10 cm each). The floor of the apparatus is covered with sawdust and unlimited access to food and water is given the animals during a 24-h period. After this familiarization period, a novel object is introduced in this familiar environment and the behavior of the mice recorded under red light during 10 min. Strong avoidance responses such as burying of the novel object are observed (Misslin and Ropartz, 1981a, 1981b; Misslin, 1982). Snell Dwarf mice and staggerer mutants do not display this avoidance response (Misslin et al., 1986; Bouchon et al., 1987).

In another device called spatial open-field, object exploration is used to provide an index of long-term memory of spatial representation (Wiltz and Bolton, 1971; Thinus-Blanc et al., 1992). In such studies, a set of novel objects is introduced in an open-field: the habituation of object exploration over daily sessions as well as the reaction of mice to the spatial rearrangement of the objects on the last session are tested. Variation of response has been reported among inbred strains of mice (Lassalle et al., 1991; Roullet and Lassalle, 1990). Moreover, in such a situation, Janus et al. (1995) showed that transgenic mice over-expressing a human calcium-binding protein (the S100 β protein) exhibit a decrease in object exploration.

Exploration of novel conspecifics

Rodents are a social species which are able to identify individuals within a given group via their capacity to recognize the olfactory properties of a subject (Rawleigh et al., 1993). This implies that a rat or a mouse is able to distinguish a familiar from a non familiar animal. When confronted to an unknown conspecific, rodents can exhibit a large pattern of responses, ranging from investigatory to agonistic behaviors, depending on various factors such as social status (dominant or not), type of breeding (grouped vs isolated subjects), place of confrontation (familiar place vs new place, territory of a given animal), experimental design (high or low lighting), gender (male vs female) etc.

Some abnormalities of conspecific exploration have been found in S100β transgenic mice (Roder et al., 1996) using a T-maze. Moreover, female mice exhibiting corticotropin-releasing factor over-expression display active rejection of sexually experienced males, which has been interpreted as a reduction in social interactions (Heinrichs and al., 1997). However, a modification of social exploration needs some caution in interpretation. For example, some genetic modifications affect social behaviors in rodents, but cannot be interpreted in terms of modification of exploration of novel conspecifics (Lijam et al., 1997).

Methodological and interpretational cautions

Methodological cautions

First, it is very important to remember that exploration refers to behaviors triggered off by confrontation with novelty. So, every novelty-related experiment can interact with the rodent's exploratory behavior. Therefore, bedding, cage or room changing should be avoided during the 24 h that precede the test session. It is also necessary not to allow animals to move freely outside of that breeding cage before the experiments. Animals can only be tested once in a given apparatus. Furthermore, rodents are very sensitive to the rhythm of activity of persons working in the laboratory, which generally decreases during weekends. After experiencing two relative quiet days, animals generally behave differently on Mondays than on subsequent weekdays (Lassalle and Wahlsten, 1992). Other factors can directly modify exploration and therefore it is important to control them in the experimental designs. Indeed, open-field behavior is modified by isolation, handling, illumination, ceiling height, cross-fostering, floor texture, age of testing and litter size (Nagy and Glaser, 1970; Nagy and Holm, 1970; Whitford and Zipf, 1975; Deitchman et al., 1976; Dixon and Mayeda, 1976; Misslin et al., 1976; Le Pape and Lassalle, 1984). For example, a mutation could modify fertility: such litter size-induced modification of open-field behavior may not be interpreted in terms of modified exploration! As maternal effects have been demonstrated by cross-fostering for open-field behavior (Reading, 1966; Poley and Royce, 1970), it may be important that the experimental design used allows the distinction between genetic and epigenetic factors. Therefore, one may recommend when comparing two genotypes, to use mice issued from a +/− × +/− cross, rather then animals issued respectively from +/+ × +/+ and −/− × −/− pairings, which differ not only for genotype but also for maternal effects.

It should be noticed that important intra-group variability can appear in experiments investigating exploratory behavior. Therefore, it is required to use a parallel group design in which different genotypes are tested simultaneously. If it is not possible to run all animals at the same time, a small number of subjects of each group should be tested at each session. This should then be repeated until all animals have been observed. Furthermore, as circadian activity changes occur in these species, it is necessary to randomize time of test over genotypes. Animals should be kept under a 12/12 h reversed light/dark cycle, in order to observe them during their high activity period, that is when lights are off.

Cautions in interpretation of results

First of all, the word "exploration" can only be employed referring to behaviors measured when a subject is freely confronted to a novel environment. This excludes behaviors exhibited in the breeding cage and in situations in which rodents are forced into novel situations. Second, modifications of behavioral responses toward novel environments can be elicited by other factors than changes of exploration per se. This can include modification in sensorial processing or in motor execution. For example, a mutation that induces peripheral or central modifications of visual processing may alter behavior in a light/dark choice test. One may emphasize that several strains of mice used as genetic background in knock out experiments possess some sensorial deficiency. For example, C57BL/6 mice show a specific anosmia (Wysocki et al., 1977) while 129/Ev mice exhibit barbering behavior, which results in lack of facial whiskers and hairs (Lijam et al., 1997). Behavior in the free exploratory paradigm largely depends upon olfaction (Giebel et al., 1993a) while mice without vibrissae do not avoid open arms in the elevated plus maze (unpublished results). This may mask effects of a given mutation. Consequently, it is very important to study the effect of a given gene using several experimental paradigms based upon different sensory modalities and using at least two different genetic backgrounds.

References

Barnett, S.A. (1975) *The Rat: a study in Behavior*. Chicago: University of Chicago Press.

Barnett, S.A. and Cowan, P.E. (1976) Activity, exploration, curiosity and fear: an ethological study. *Interdisciplin. Sci. Rev.*, 1: 43–62.

Barnett, S.A. and Smart, J.L. (1975) The movements of wild and domestic house mice in an artificial environment. *Behav. Biol.*, 15: 85–94.

Belzung, C. and Berton, F. (1997) Further pharmacological validation of the BALB/c neophobia in the free exploratory paradigm as an animal model of anxiety. *Behav. Pharmacol.*, 8: 541–548.

Belzung, C. and Le Pape, G. (1994) Comparison of different behavioral test situations used in psychopharmacology for measurement of anxiety. *Physiol. Behav.*, 56: 623–628.

Belzung, C., Misslin, R., Vogel, E., Dodd, R.H. and Chapouthier, G. (1987) Anxiogenic effects of methyl-β-carboline-3-carboxylate in a light/dark choice situation. *Pharmacol. Biochem. Behav.*, 28: 29–33.

Belzung, C., Pineau, N., Beuzen, A. and Misslin, R. (1994) The effectiveness of PD 135158, a selective CCK-B receptor antagonist, in reducing "state" but not "trait" anxiety in BALB/c mice. *Pharmacol. Biochem. Behav.*, 49: 433–436.

Berlyne, D.E. (1950) Novelty and curiosity as determinants of exploratory behaviour. *Br. J. Psychol.*, 41: 68–70.

Berlyne, S.E. (1960) *Conflict, arousal and curiosity*, London, McGraw-Hill.

Bertholet, J.Y. and Crusio, W. (1991) Spatial and non-spatial spontaneous alternation and hippocampal mossy fibre distribution in nine inbred mouse strains. *Behav. Brain Res.*, 43: 197–202.

Berton, F., Vogel, E. and Belzung, C. (1998) Modulation of mice anxiety in response to cat odour as a consequence of predators diet. *Physiol. Behav.*, 65: 247–254.

Beuzen, A. and Belzung, C. (1995) Link between emotional memory and anxiety states: A study by principal component analysis. *Physiol. Behav.*, 58: 111–118.

Birke, L.I.A. and Archer, J. (1983) Some issues and problems in the study of animal exploration. In: J. Archer and L. Birke (Eds.), *Exploration in animals and humans*, Van Nostrand Reinhold, Berkshire, UK.

Blois-Heulin, C. and Belzung, C. (1995) Effects of previous familiarization on novelty reactions in mice (*Mus musculus*). *Behav. Proc.*, 34: 197–212.

Boissier, J.R. and Simon, P. (1962) La réaction d'exploration chez la souris. *Thérapie*, 17: 1225–1232.

Bouchon, R., Misslin, R., Vogel, E. and Ropartz, P. (1987) Total recovery of responses to novelty after ectopic pituitary isografts in Snell Dwarf mice (dw/dw). *Behav. Proc.*, 15: 155–165.

Caston, J., Vasseur, F., Delhaye-Bouchaud, N. and Mariani, J. (1997) Delayed spontaneous alternation in intact and cerebellectomized control and lurcher mutant mice: differential role of cerebellar cortex and deep cerebellar nuclei. *Behav. Neurosci.*, 111: 214–218.

Clément, Y., Martin B., Venault, P. and Chapouthier, G. (1995) Involvement of regions of the 4th and 7th chromosomes in the open-field activity in mice. *Behav. Brain Res.*, 70: 51–57.

Clément, Y., Proeschel, M.F., Bondoux, D., Girard, F., Launay, J.M. and Chapouthier, G. (1997) Genetic factors regulate processes related to anxiety in mice. *Brain Res.*, 752: 127–135.

Crabbe, J.C. (1986) Genetic differences in locomotor activity in mice. *Pharmacol. Biochem. Behav.* 25: 289–292.

Crawley, J.N. (1981) Neuropharmacological specificity of a simple animal model for the behavioral action of benzodiazepines. *Pharmacol. Biochem. Behav.*, 15: 695–699.

Crawley, J.N., Belknap, J.K., Collins, A., Crabbe, J.C., Frankel, W., Henderson, N., Hitzemann, R.J., Maxson, S.C., Miner, L.L., Silva, A.J., Wehner, J.M., Winshaw-Boris, A. and Paylor, R. (1997) Behavioral phenotypes of inbred mouse strains: implications and recommendations for molecular studies. *Psychopharmacology*, 132: 107–124.

Crawley, J.N. and Davis, L.G. (1982) Baseline exploratory activity predicts anxiolytic responsiveness to diazepam in five mouse strains. *Brain Res. Bull.*, 8: 609–612.

Crawley, J.N. and Goodwin, F.K. (1980) Preliminary report of a simple animal behaviour model for the anxiolytic effects of benzodiazepines. *Pharmacol. Biochem. Behav.*, 13: 167–170.

Crestani, F., Benke, D., Reyes, G., Fritschy, J.M., Lüscher, B. and Mohler, H. (1996) Impairment of $GABA_A$ receptor function by subunit gene targeting in mice results in neophobia. *Soc. Neurosci. Abstr., Washington.*

Crusio, W.E., Schwegler, H. and Van Abeelen, J.H. (1989) Behavioral responses to novelty and structural variation of the hippocampus in mice. I. Quantitative-genetic analysis of behavior in the open-field. *Behav. Brain Res.*, 32: 75–80.

De Fries, J.C. (1969) Pleitropic effects of albinism on open field behaviour in mice. *Nature*, 221: 65–66.

De Fries, J.C., Gervais, M.C. and Thomas, E.A. (1978) Response to 30 generations of selection for open-field activity in laboratory mice. *Behav. Gen.* 8: 3–13.

De Fries, J.C., Hegmann, J.P. and Weir, M.W. (1966) Open field behavior in mice: Evidence for a major gene effect mediated by the visual system. *Science*, 154: 1577–1579.

Deitchman, R., Lavine, A. and Burkholder, J. (1976) Litter size, emotionality, learning, and social interaction in C57BL/6J mice. *Psychol. Rep.*, 39: 27–31.

Dember, W.N. and Fowler, H (1958) Spontaneous alternation behavior. *Psychol. Bull.*, 55: 412–428.

Dixon, L. and Mayeda, D.V. (1976) Effects of floor textures on open field behavior in selected lines of mice. *Behav. Genet.*, 6: 87–92.

Figler, M.H. and Einhorn, D.M. (1983) The territorial prior residence effect in convit cichlids (*Cichlasoma Nigrofasciatum Günther*): temporal aspects of establishment and retention, and proximate mechanisms. *Behaviour*, 85: 157–183.

Filali, M., Lalonde, R., Bensoula, A.N., Guastavino, J.M. and Lestienne, F. (1996) Spontaneous alternation, motor activity, and spatial learning in hot-foot mutant mice. *J. Comp. Physiol. (A)*, 178: 101–104.

File, S.E. (1980) The use of social interaction as a method for detecting anxiolytic activity in chlordiazepoxide-like drugs. *J. Neurosci. Meth.*, 2: 219–238.

File, S.E., Mahal, A., Mangiarini, L. and Bates, G.P. (1998) Striking changes in anxiety in Huntington's disease transgenic mice. *Brain Res.*, 805: 234–240.

Fiore, M., Alleva, E., Probert, L., Kollias, G., Angeluci, F. and Aloe, L. (1998) Exploratory and displacement behavior in trnasgenic mice expressing high levels of brain TNF-α. *Physiol. Behav.*, 63: 571–576.

Flint, J., Corley, R., De Fries, J.C., Fulker, D.W., Gray, J.A., Miller, S. and Collins, A.C. (1995) A simple genetic basis for a complex psychological trait in laboratory mice. *Science*, 269: 1432–1435.

Garreau, L., Emond, P., Belzung, C., Guilloteau, D., Frangin, Y., Besnard, J.C. and Chalon, S. (1997) *N*-(3-iodoprop-2E-enyl)-2β-carbomethoxy-3β-(3',4'-dichlorophenyl) nortro-pane or β-CDIT, a tropane derivative: pharmacological characterization as a specific ligand for the dopamine transporter in rodent brain. *JPET*, 282: 467–474.

Gerlai, R., Marks, A. and Roder, J. (1994) T-maze spontaneous alternation rate is decreased in S100 beta transgenic mice. *Behav. Neurosci.*, 108: 100–106.

Giros B., Jaber, M., Jones, S.R., Wightman, R.M. and Caron, M.G. (1996) Hyperlocomotion and indifference to cocaine and amphetamine in mice lacking the dopamine transporter. *Nature,* 379: 606–612

Griebel, G., Belzung, C., Misslin, R. and Vogel, E. (1993a) The free exploratory paradigm: an effective method for measuring neophobic behaviour in mice and testing potential neophobia reducing drugs. *Behav. Pharmacol.*, 4: 637–644.

Griebel, G., Moreau, J.L., Jenck, F., Martin, J.R. and Misslin, R. (1993b) Some critical determinants of the behaviour of rats in the elevated plus-maze. *Behav. Proc.*, 29: 37–48.

Hall, C.S. (1934) Emotional behavior in the rat. I. Defecation and urination as measures of individual differences in emotionality. *J. Comp. Psychol.*, 18: 385–403.

Handley, S.L. and Mithani, S. (1984) Effects of alpha-adrenoceptor agonists and antagonists in a maze exploration model of 'fear'-motivated behaviour. *Naunyn-Schmiedeberg's Archives of Pharmacology*, 327: 1–5.

Heinrichs, S.C., Min, H., Tamraz, S., Carmouche, M., Boehme, S.A. and Vale, W.W. (1997) Anti-sexual and anxiogenic behavioral consequences of corticotropin-releasing factor over-expression are centrally mediated. *Psychoneuroendocrinol.*, 22: 215–224.

Holcomb, L., Gordon, M.N., McGowan, E., Yu, X., Benkovic, S., Jantzen, P., Wright, K., Saad, I., Mueller, R., Morgan, D., Sanders, S., Zehr, C., O'Campo, K., Hardy, J., Prada, C.M., Eckman, C., Younkin, S., Hsiao, K. and Duff, K. (1998) Accelerated Alzheimer-type phenotype in transgenic mice carrying both mutant amyloid precursor protein and presenilin 1 transgenes. *Nat. Med.,* 4: 97–100.

Hughes, R.N. (1968) Behaviour of male and female rats with free choice of two environments differing in novelty. *Anim. Behav.*, 16: 92–96.

Janus, C., Janus, M. and Roder, J. (1995) Spatial exploration in transgenic mice expressing human beta-S100. *Neurobiol. Learn. Mem.*, 64: 58–67.

Katzir, G. (1982) Relationships between social structure and response to novelty in captive jackdaws, Corvus monedula1.response to novel space. *Behaviour*, 81: 231–239.

Kopp, C., Misslin, R., Vogel, E., Rettori, M.C., Delagrange, P. and Guardiola-Lemaître, B. (1997) Effects of day-length variations on emotional responses towards unfamiliarity in Swiss mice. *Behav. Proc.*, 41: 151–157.

Lalonde, R. (1987) Exploration and spatial learning in staggerer mutant mice. *J. Neurogenet.*, 4: 285–291.

Lalonde, R., Botez, M.I. and Boivin, D. (1986a) Spontaneous alternation and habituation in a T-maze in nervous mutant mice. *Behav. Neurosci.*, 100: 350–352.

Lalonde, R., Joyal, C.C., Guastavino, J.M. and Botez, M.I. (1993) Hole poking and motor coordination in lurcher mutant mice. *Physiol. Behav.*, 54: 41–44.

Lalonde, R., Lamarre, Y., Smith, A.M. and Botez, M.I. (1986b) Spontaneous alternation and habituation in lurcher mutant mice. *Brain Res.*, 362: 161–164.

Lalonde, R., Lamarre, Y. and Smith, A.M. (1988) Does the mutant mouse lurcher have deficits in spatially oriented behaviours? *Brain Res.*, 455: 24–30.

Lalonde, R., Manseau, M. and Botez, M.I. (1987) Spontaneous alternation and habituation in Purkinje cell degeneration mutant mice. *Brain Res.*, 411: 187–189.

Lassalle, J.M.: Wahlsten, D. (1992) Behavioral paradigms: general procedures and spatial memory. In: D. Goldowitz, D. Wahlsten and R.E. Wimer (Eds.), *Techniques for the Genetic Analysis of Brain and Behavior*, Elsevier Science Publishers.

Lassalle, J.M., Bulman-Fleming, B. and Wahlsten, D. (1991) Hybrid vigour and maternal environment in mice. II. Water escape learning, open-field activity and spatial memory. *Behav. Proc.,* 23: 35–45.

Le Pape, G. and Lassalle, J.M. (1984) A developmental genetic analysis of locomotor activity in mice: Maternal effects in the BALB/c and C57BL/6 strains and heredity in F1 hybrids. *Behav. Genet.*, 14: 21–29.

Ledent, C., Vaugeois, J.M., Schiffmann, S.N., Pedrazzini, T., El Yacoubi, M., Vanderhaeghen, J.J., Costentin, J., Heath, J.K., Vassart, G. and Parmentier, M. (1997) Aggressiveness, hypoalgesia and high blood pressure in mice lacking the adenosine A_{2a} receptor. *Nature*, 388: 674–678.

Lijam, N., Paylor, R., McDonald, M.P., Crawley, J.N., Deng, C.X., Herrup, K., Stevens, K.E., Maccaferri, G., McBain, C.J., Sussman, D.J. and Wynshaw-Boris, A. (1997) Social interaction and sensorimotor gating abnormalities in mice lacking *Dvl1*. *Cell*, 90: 895–905.

Lister, R.G. (1987) The use of a plus-maze to measure anxiety in the mouse. *Psychopharmacology*, 92: 180–185.

Mathis, C., Neumann, P.E., Gershenfeld, H., Paul, S.M. and Crawley, J.N. (1995) Genetic analysis of anxiety-related behaviors and response to benzodiazepine-related drugs in AXB and BXA recombinant inbred mouse strains. *Behav. Genet.*, 25: 557–568.

Mathis, C., Paul, S.M. and Crawley, J.N. (1994) Characterization of benzodiazepine-sensitive behaviors in the A/J and C57BL/6J inbred strains of mice. *Behav. Genet.* 24: 171–180.

Meliska, C.J., Burke, P.A., Bartke, A. and Jensen, R.A. (1997) Inhibitory avoidance and appetitive learning in aged normal mice: comparison with transgenic mice having elevated plasma growth hormone levels. *Neurobiol. Learn. Mem.*, 68: 1–12.

Miller, C.L., Bard, K.A., Juno, C.J. and Nadler, R.D. (1986) Behavioural responsiveness of young chimpanzees (Pan troglodytes) to a novel environment. *Folia Primatol.*, 47: 128–142.

Misslin, R. (1982) Aspects du déterminisme des réactions de la Souris à un objet nouveau. *Biol. Comport.*, 3: 209–214.

Misslin, R., Belzung, C. and Vogel, E. (1989) Behavioural validation of a light/dark choice procedure for testing anti-anxiety agents. *Behav. Proc.*, 18: 119–132.

Misslin, R., Bouchon, R. and Ropartz, P. (1976) Signification de certains paramètres comportementaux chez la souris placée dans un open-field. *Physiol. Behav.*, 17: 767–770.

Misslin, R., Cigrang, M. and Guastavino, J.M. (1986) Responses to novelty in staggerer mutant mice. *Behav. Proc.*, 12: 51–56.

Misslin, R. and Cigrang, M (1986) Does neophobia necessarily imply fear or anxiety? *Behav. Proc.*, 12: 45–50.

Misslin, R. and Ropartz, P. (1981a) Responses in mice to a novel object. *Behaviour*, 78: 169–177.

Misslin, R. and Ropartz, P. (1981b) Olfactory regulation of responsiveness to novelty in mice. *Behav. Neur. Biol.*, 33: 230–236.

Misslin, R. and Ropartz, P. (1981c) Effects of lateral amygdala lesions on response to novelty in mice. *Behav. Proc.*, 6: 329–336.

Miyakawa, T., Yagi, T., Watanabe, S. and Niki, H. (1994) Increased fearfulness of Fyn tyrosine kinase deficient mice. *Brain Res. Mol. Brain Res.*, 27: 179–182.

Montgomery, K.C. (1954) The role of the exploratory drive in learning. *J. Comp. Physiol. Psychol.*, 47: 60–64.

Moran, P.M., Higgins, L.S., Cordell, B. and Moser, P.C. (1995) Age-related learning deficits in transgenic mice expressing the 751-amino acid isoform of human beta-amyloid precursor protein. *Proc. Natl. Acad. Sci. USA*, 92: 5341–5345.

Nagy, Z.M. and Glaser, H.D. (1970) Open field behavior of C57BL/6J mice: Effect of illumination, age and number of test days. *Psychon. Sci.*, 19: 143–145.

Nagy, Z.M. and Holm, M. (1970) Open field behavior in C3H mice: Effect of early handling, field illumination, and age at testing. *Psychon. Sci.*, 19: 273–275.

Parks, C.L., Robinson, P.S., Sibille, E., Shenk, T. and Toth, M. (1998) Increased anxiety of mice lacking the serotonin$_{1A}$ receptor. *Proc. Natl. Acad. Sci. USA*, 95: 10734–10739.

Pellow, S. Chopin, P., File, S.E. and Briley, M. (1985) Validation of open:closed arm entries in an elevated plus maze as a measure of anxiety in the rat. *J. Neurosci. Meth.*, 14: 149–167.

Plomin, R., MacClearn, G.E., Gora-Maslak, G. and Neiderhiser, J.M. (1991) Use of recombinant inbred strains to detect quantitative trait loci associated with behavior. *Behav. Genet.*, 21: 99–116.

Poley, W. and Royce, J.R. (1970) Genotype, maternal stimulation and factors of mouse emotionality. *J. Comp. Physiol. Psychol.*, 71: 246–250.

Poucet, B., Chapuis, N., Durup, M. and Thinus-Blanc, C. (1986) A study of exploratory behaviour as an index of spatial knowledge in hamsters. *Anim. Learn. Behav.*, 14: 93–100.

Ramboz, S., Oosting, R., Aït Amara, D., Kung, H.F., Blier, P., Mendelsohn, M., Mann, J.J., Brunner, D. and Hen, R. (1998) Serotonin receptor 1A knockout: An animal model of anxiety-related disorder. *Proc. Natl. Acad. Sci. USA*, 95: 14476–14481.

Rawleigh, J.M., Kemble, E.D. and Ostrem, J. (1993) Differential effects of prior dominance or subordination experience on conspecific odor preference in mice. *Physiol. Behav.*, 54: 35–39.

Reading, A.J. (1966) Effect of maternal environment on the behavior of inbred mice. *J. Comp. Physiol. Psychol.*, 62: 437–440.

Renner, M.J. (1990) Neglected aspects of exploratory and investigatory behavior. *Psychobiol.*, 18: 16–22.

Renner, M.J. and Seltzer, C.P. (1991) Molar characteristics of exploratory and investigatory behavior in the rat (*Rattus norvegicus*). *J. Comp. Psychol.* 105: 326–339.

Rex, A., Sondern, U., Voigt, J.P., Franck, S. and Fink, H. (1996) Strain differences in fear-motivated behavior of rats. *Pharmacol. Biochem. Behav.*, 54: 107–111.

Rochford, J., Beaulieu, S., Rousse, I., Glowa, J.R. and Barden, N. (1997) Behavioral reactivity to aversive stimuli in a transgenic mouse model of impaired glucocorticoid (type II) receptor function: effects of diazepam and FG-7142. *Psychopharmacol.*, 132: 145–152.

Roder, J.K., Roder, J.C. and Gerlai, R. (1996) Conspecific exploration in the T-maze: abnormalities in S100β transgenic mice. *Physiol. Behav.*, 60,: 31–36.

Rodgers, R.J. (1997) Animal models of 'anxiety': where next? *Behav. Pharmacol.*, 8: 477–496.

Rodgers, R.J., Cole, J.C., Cobain, M.R., Daly, P., Doran, P.J., Eells, J.R. and Wallis, P. (1992) Anxiogenic-like effects of fluprazine and eltoprazine in the mouse elevated plus maze: profile comparisons with 8-OH-DPAT, CGS12066B, TFMPP and mCPP. *Behav. Pharmacol.*, 3: 621–634.

Roullet, P. and Lassalle, J.M. (1990) Genetic variation, hippocampal fibres distribution, novelty reactions and spatial representation in mice. *Behav. Brain Res*, 41: 61–69.

Roullet, P. and Lassalle J.M., (1993) Spontaneous exploration, response plus position learning and hippocampal mossy fibre distribution: a correlational study. *Behav. Proc.*, 29: 217–228.

Saudou, F., Aït Amara, D., Dierich, A., Lemeur, M., Ramboz, S., Segu, L., Buhot, M.C. and Hen, R. (1994) Enhanced aggressive behavior in mice lacking 5-HT_{1B} receptor. *Science,* 265: 1875–1878.

Skutella, T., Probst, J.C., Behl, C. and Holsboer, F. (1996) Antisense targeting of corticotropin-releasing hormone and corticotropin-releasing hormone receptor type I. In: R.B. Raffa and F. Porreca (Eds.), *Antisense strategies for the study of receptor mechanisms*, Springer, R.G. Landes Company, Austin.

Sokolov, E.K. (1960) Neuronal models and the orienting reflex. In: M.A.B. Brazier (Ed.), *The Central Nervous System and Behaviour*, New York: Josiah Macy Jr. Foundation.

Steiner, H., Fuchs, S. AND Accili, D. (1997) D3 dopamine receptor-deficient mouse: evidence for reduced anxiety. *Physiol. Behav.*, 63: 137–141.

Stenzel-Poore, M.P., Heinrichs, S.C., Rivest, S., Koob, G.F. and Vale, W.W. (1994) Overproduction of corticotropin-releasing factor in transgenic mice: a genetic model of anxiogenic behavior. *J. Neurosci.*, 14: 2579–2584.

Takahashi, L.K. and Kalin, N.H. (1989) Role of corticotropin-releasing factor in mediating the expression of defensive behavior. In: R.J. Blanchard, P.F. Brain, D.C. Blanchard and S. Parmigiani (Eds.), *Ethoexperimental approaches to the study of behavior*, NATO ASI Series, Kluwer Academic Publishers, pp. 580–594.

Taylor, G.T. (1974) Stimulus change and complexity in exploratory behavior. *Anim. Learn. Behav.*, 2: 115–118.

Thinus-Blanc, C., Durup, M. and Poucet, B. (1992) The spatial parameters encoded by hamsters during exploration: a further study. *Behav. Proc.*, 26: 43–57.

Tolman, E.C. (1925) Purpose and cognition: the determiners of animal learning. *Psychol. Rev.*, 32: 285–297.

Trullas, R. and Skolnick, P. (1993) Differences in fear motivated behaviors among inbred mouse strains. *Psychopharmacol.*, 111: 323–331.

Upchurch, M. and Schallert, T. (1983) A behavior analysis of the offspring of "haloperidol-sensitive" and "haloperidol-resistant" gerbils. *Behav. Neural Biol.*, 39: 221–228.

Vinogradova, O.S. (1970) Registration of information and the limbic system. In: G. Horn and R.A. Hinde (Eds.), *Short term changes in Neural Activity and Behaviour*, Cambridge: Cambridge University Press.

Walsh, R.N. and Cummins, R.A. (1976) The open field test: A critical review. *Psychol. Bull.*, 83: 481–504.

Whitford, F.W. and Zipf, S.G. (1975) Open-field activity in mice as a function of ceiling height: A genotype-environment interaction. *Behav. Genet.*, 5: 275–280.

Willig, F., Van De Velde, D., Laurent, J., M'Harzi, M. and Delacour, J. (1992) The Roman strains of rats as a psychogenetic tool for pharmacological investigation of working memory: example with RU 41656. *Psychopharmacol.*, 107: 415–424.

Wilz, K.J. and Bolton, R.L. (1971) Exploratory behaviour in response to the spatial rearrangement of familiar stimuli. *J. Comp. Physiol. Psychol.*, 92: 312–321.

Wysocki, C.J., Whitney, C. and Tucker, D. (1977) Specific anosmia in the Laboratory Mouse. *Behav. Genet.*, 7: 171–188.

Section 5

Human Neurobehavioral Disorders: from Molecular Genetics to Genetic Animal Models

W.E. Crusio and R.T. Gerlai (Eds.)
Handbook of Molecular-Genetic Techniques for Brain and
Behavior Research (Techniques in the Behavioral and Neural Sciences, Vol. 13)

CHAPTER 5.1

Psychiatric genetics — a current perspective

Douglas F. Levinson

Department of Psychiatry, MCP Hahnemann University, MCPH-EPPI, Room 206, 3200 Henry Ave., Philadelphia, PA 19129, USA

Introduction

Over the past ten years, psychiatric disorders have been a major focus of genetic investigation. They are among the most common disorders for which predominantly genetic etiologies have been suggested by family and twin studies, they are causes of severe morbidity and mortality, and there are essentially no other specific pathophysiological clues, so that genetic linkage studies have seemed justified despite the many obstacles to the detection of loci underlying genetically complex disorders. The most intensively-studied of these disorders are schizophrenia and bipolar (manic-depressive) disorder, with panic disorder, autism and alcoholism also receiving substantial attention. Published estimates of lifetime risk in the general population, risk to siblings of probands, MZ concordance and heritability for some of these disorders are shown in Table 1. The discussion here will focus primarily on schizophrenia and bipolar disorder to illustrate the issues currently facing this field of research.

Methodological issues

Studies of psychiatric disorders have played a central role in promoting discussion and understanding of methodological issues which are important for all complex (non-mendelian) disorders:

Defining the phenotype

Over the past 30 years, substantial progress has been made in validating clinical criteria for syndromes (disorders) such as schizophrenia or bipolar disorder that predict familial aggregation of similar and related disorders. However, it has proven difficult to identify more specific familial subtypes based on symptom clusters, age of onset or other clinical variables, as has been possible in disorders such as breast cancer, Alzheimer's disease or diabetes, nor are there physiological tests for this purpose. One exception may be major depressive disorder, where increased recurrence risk in relatives is predicted by two proband characteristics, recurrent episodes and earlier age of onset (Bland et al., 1986; Weissman et al., 1993; Moldin et al., 1991). It is typically possible to define a narrow diagnostic category and a broader "spectrum" of disorders that co-aggregate in the same families, as summarized in Table 2. In general the broader categories permit the identification of a larger number of affected relatives per pedigree, but the relative risk to first-degree relatives vs that to the general population is typically higher for narrower than for broader phenotypes, suggesting that the former

TABLE 1

Characteristics of several major psychiatric disorders

Disorder	Population risk	Risk to sibs	MZ concordance	Heritability
Schizophrenia	0.003–0.008	0.06–0.10	0.50	0.65–0.85
Bipolar	0.002–0.01	0.05–0.10	0.80	0.80–0.90
Panic disorder	0.012–0.038	0.08–0.20	0.30–0.70	0.40–0.70
Autism	0.0001–0.003	0.03	0.60	0.90–1.00

References:

Schizophrenia: Gottesman and Shields, 1982; Tsuang et al, 1991, Kendler and Diehl, 1993a; Kendler et al, 1996a; Kendler et al, 1993b, 1993c, 1993d.

Bipolar disorder: Gershon et al, 1982, 1988; Goodwin and Jamison, 1990; Tsuang et al, 1994.

Panic disorder: Weissman, 1993; Maier et al, 1993b; Katerndahl and Realini, 1993; Dick et al, 1994; Eaton et al, 1994; Perna et al, 1997.

Autism: Smalley, 1991; Fombonne et al, 1992; Bolton et al, 1994; Bailey et al, 1995; Le Couteur, 1996; Piven et al, 1997.

For further review, see Nurnberger and Berrettini (1998).

may be more useful for linkage studies. Investigators variously choose to study a single narrow diagnostic category, or to perform several (and sometimes a larger number) of linkage analyses based on alternative diagnostic models. It is also possible to design linkage analyses that assign different weights to cases according to diagnosis (Terwilliger and Ott, 1994; Levinson, 1997a), although this approach has not been commonly used.

Psychiatric genetics investigators must work within the many limitations of available diagnostic methods. In other branches of medicine, clinical diagnostic methods can be supplemented by more objective tests, either in every case (i.e. glucose tolerance tests in diabetes) or in a proportion of cases in which clinical methods can be tested against such a gold standard (i.e. histopathological examination at autopsy in Alzheimer's disease). Because there is such reliance on clinical diagnosis, intensive efforts have been made over the past 30 years to develop valid and reliable methods for research (Spitzer et al., 1978; American Psychiatric Association, 1987, 1994; World Health Organization, 1990). Semi-structured interview schedules have been developed for interviewing probands and relatives (Endicott and Spitzer, 1978; Wing et al., 1990; Williams et al., 1992;

TABLE 2

Spectrum of disorders that co-aggregate in the same families

Disorder	Spectrum of co-aggregating disorders
Schizophrenia	schizoaffective, other non-affective psychoses (schizophreniform, atypical, delusional), schizotypal and paranoid personality disorders
Bipolar I	bipolar II, schizoaffective, recurrent major depressive (unipolar), single-episode major depressive disorders
Panic disorder	separation anxiety disorder, agoraphobia, atypical panic disorder, possibly major depression
Autism	pervasive development disorder, social and language deficits

See Table 1 for references.

Nurnberger et al., 1994) so that features of all relevant disorders are screened in all subjects. The consistency (reliability) of information obtained by different raters has been tested for observers of the same interview and for separate interviews over time, and the reliability of diagnoses assigned by different experts assessing all available information about each case has similarly been studied (Leckman et al., 1982; Williams et al., 1992; Numberger et al., 1994). In an iterative process, family studies have been conducted using reliable interview schedules and diagnostic criteria, and the familial aggregation and co-aggregation of various diagnostic categories has been used to modify diagnostic criteria and update diagnostic interviews, and the new criteria then used in subsequent family studies.

For example, the Research Diagnostic Criteria developed in the late 1960s included criteria for several types of schizoaffective disorder (concurrent schizophrenic and manic or depressive syndromes). Family studies using these criteria demonstrated that the "mainly schizophrenic" subtype, in which schizophrenia-like symptoms persist after remission of mania or depression, co-aggregates in families with schizophrenia, while the "mainly affective" subtype (without persistence of psychosis) generally co-aggregates with other mood disorders (Gershon et al., 1982, 1988; Baron et al., 1982; Levinson and Mowry, 1991). The DSM-IIIR and DSM-IV criteria, subsequently developed by the American Psychiatric Association (1987, 1994) and now used in most genetics research, have therefore applied definitions of schizoaffective disorder which are similar to the RDC mainly schizophrenic subtype, while including the mainly affective cases within the mood disorder categories. Selection of schizoaffective probands with DSM-IIIR criteria predicts increased familial risk of schizophrenia, as well as mood disorders (Kendler et al., 1993c, 1995), and indeed clinically, it can be difficult to differentiate reliably between schizophrenic and schizoaffective disorders (Faraone et al., 1996).

Although these efforts have resulted in substantial diagnostic consistency across research groups, it is likely that there is still an unknown degree of variation for some diagnoses. The extraordinarily labor intensive nature of the ascertainment and diagnostic procedures makes it impossible to determine cross study reliability in most cases. Further, it is not known how precise an approach to diagnosis is optimal for each disorder. For schizophrenia, for example, the co-aggregation of a range of psychotic diagnoses in families suggests that a very restrictive approach would be arbitrary and perhaps counter-productive — and also suggests that if different studies exclude a slightly different proportion of cases, there would be little effect on results. "Broader" diagnoses are typically less reliably diagnosed.

As discussed in more detail by Nosten-Bertrand et al. Chapter 5.2 of this volume, there are many possible alternative approaches. Symptom dimensions can be measured and used for quantitative linkage analysis, as was done by Brzustowicz et al. (1997) in a study supportive of linkage of chromosome 6p to psychotic symptoms in schizophrenia families. There is some evidence to suggest that relatives are likely to share the symptom clusters of probands. For example, a schizophrenic or a manic syndrome in the proband (whether observed alone or comorbid with the other syndrome) predicts increased risk to first-degree relatives for the same syndrome (Maier et al., 1993a); and bipolar probands with hypomania (bipolar I disorder) are more likely to have relatives with the same syndrome (Andreasen et al., 1987). However, dimensional ratings have only begun to receive careful study in family or linkage studies (Kendler et al., 1997; Cardno et al., 1997).

Another strategy is to study linkage of abnormal physiological responses that are more common in schizophrenic probands and their relatives. The most successful example of this strategy to date is the detection of strongly significant evidence for linkage of deficiency of pre-pulse inhibition of the P50 event-related brain potential response to markers on chromosome 15q (Freedman et al., 1997). However, significant linkage to schizophrenia has yet to be reported in this region, and the locus responsible for the electrophysiological defect has not been identified, although animal studies strongly suggest that the alpha-7 nicotinic receptor or its regulatory sites may be involved. Abnormal smooth-pursuit eye tracking has also been reported in both schizophrenic patients and unaffected relatives, but the relevant loci have not been detected (Levy et al., 1994).

Sampling strategies

Most of the early psychiatric linkage studies adopted sampling strategies that had proven successful in the study of mendelian disorders. For example, Egeland and colleagues collected members of a large, multigenerational Old Order Amish pedigree affected with bipolar and unipolar mood disorders (1987), and Kidd and colleagues studied a similar very large pedigree from northern Sweden with multiple cases of schizophrenia (Barr et al., 1994). Other investigators have studied samples of "moderate-size" pedigrees (typically 5–8 affected cases per pedigree), such as the two Icelandic schizophrenia samples collected respectively by Sherrington et al. (1988) and Moises et al. (1995), and Berrettini et al.'s 22 bipolar pedigrees (Berrettini et al., 1997), while most studies included 10–80 small pedigrees (two- or three-generation families with 2–5 affected cases) as discussed below. Any of these samples would have been sufficiently powerful to detect a single major locus causing disease in most or all families, and many samples could have detected a single locus with a major effect on disease susceptibility in a substantial minority of families. The one larger sample included 265 small schizophrenia pedigrees from Ireland, collected by Kendler et al. (1993b). Although segregation analyses of psychiatric disorders have never provided strong support for single major locus effects, it has frequently been argued that only linkage studies could determine whether such effects might be detected in samples of multiply-affected pedigrees.

As multiple genome scans have failed to detect replicable major locus effects (see below), the case for polygenic or oligogenic models has been strengthened. For example, in an influential series of papers, Risch (1990a) extended the hypothesis of James (1971) that the power to detect linkage depends primarily on the locus-specific contribution to the relative risk to sibs vs the general population (lambda-sibs). Based on relative risk estimates, Risch (1990a) demonstrated that schizophrenia family study data most closely support an oligogenic model in which at least 3 loci (and perhaps many minor loci) contribute to disease susceptibility, and no one locus produces greater than a three-fold increase in risk. Also, the failure to detect linkage in single very large pedigrees has suggested that heterogeneity and/or polygenic mechanisms may be present in these cases as well.

Consideration of oligogenic models has led to increased interest in larger samples of nuclear families or affected sibling pairs. Simulation studies demonstrate that a locus responsible for a two- to three-fold increase in risk to siblings of probands in a population could reliably be detected in samples of 100–200 affected sibling pairs, but when the locus-specific relative risk drops to 1.3, 500–800 affected sibling pairs would be needed (Hauser et al, 1996). It should be noted, however, that a small locus-specific effect on risk in a population can result from a relatively uniform effect in most or all families, or from a larger effect in a proportion of families. The effect of locus heterogeneity on detection of linkage has been well-studied for mendelian models of transmission (Goldin, 1992; Levinson, 1993), but not for explicitly oligogenic models or for loci with very small effects on risk.

The assumed model of disease transmission also has implications for the selection of individual pedigrees for study. Many early psychiatric linkage studies excluded pedigrees with various definitions of bilineal inheritance, on the assumption that dominant genetic effects are more likely than recessive effects (given the common occurrence of these disorders in successive generations). Spence et al. (1993) reported simulation studies suggesting no disadvantage to the inclusion of bilineal families, but only a narrow range of disease models was considered. McCarthy et al. (1998) simulated samples for common diseases with two-locus models of transmission and studied the effect of various types of rules for selecting affected sib pair, such as exclusion of bilineal pedigrees, inclusion only of sibships with a specified number of affected cases, etc. They reported that every strategy was advantageous for some models of transmission but disadvantageous for others, and recommended sampling all pairs.

An alternative to the linkage study is the association study of candidate gene regions. Most such studies have used the case-control design, but this strategy carries the possibility of spurious findings when case and control samples are drawn from different proportions of subpopulations which differ both in allele frequencies at the tested locus and in disease risk. Some studies have used family-based control designs, which consider the proportion of alleles transmitted or not transmitted from parents to probands (Falk and Rubinstein, 1987; Spielman et al., 1993; Terwilliger, 1995; Spielman and Ewens, 1996, 1998). This eliminates population stratification effects, but reduces the number of probands available for study. Candidate genes have also been identified on the basis of cytogenetic findings. For example, Shprintzen's syndrome (velo-cardio-facial syndrome or VCFS) is associated with an approximately 3 mB deletion on chromosome 22q, and can produce symptoms of both schizophrenia and mood disorders. The possible relevance of this region to psychiatric disorders has been reviewed by Levinson and Coon (1998). There is also a translocation breakpoint on chromosome 11q associated with schizophrenia in one large family (Muir et al, 1995), although linkage studies have generally been negative in this region. More recently, the effects of evolution and population genetics have received greater consideration in the design of psychiatric linkage studies, as discussed below.

Linkage analysis strategies

When psychiatric linkage studies were first initiated, most analyses were conducted using likelihood-based parametric methods for computing lod scores between a disease transmission model and a single marker (two-point analysis). Two-point methods were also available for affected sibling pair analysis, but appropriately large ASP samples were not available. Multipoint likelihood-based methods were also available to test linkage to several markers simultaneously, but were difficult to apply to many markers at a time. The failure of several early positive findings to replicate raised questions about the appropriateness of these methods. For example, Sherrington et al. (1988) reported highly significant evidence for linkage between chromosome 5 markers and schizophrenia, which could not be replicated (McGuffin et al, 1990). The lod score depended heavily on the marker status of unaffected individuals in the pedigrees, because the disease transmission model assumed a high penetrance for the disease allele, and Ott (1990) demonstrated that linkage would not have been supported if only affected individuals had been considered in the linkage analysis. More recently, rapid multipoint "non-parametric" (model-free) statistical methods have been developed for affected sibling pairs and for small pedigrees generally (for review and comparison of these methods, see Chapter 2.1.3; and Davis and Weeks, 1997). It has also been demonstrated that it can be more efficient to screen the genome with a more sparse map (10-20 cM) and perhaps with only the affected individuals in each pedigree, to select regions for further study which meet a very liberal statistical threshold, and then to study these regions with a finer map and a stringent statistical (e.g. 5 cM) (Elston et al, 1996; Hauser et al, 1996; Holmans and Craddock, 1997).

Interpretation of results

The determination of the statistical significance of linkage and association findings has been a particular problem in psychiatric genetics. There are multiple problems which compound each other. The mode of transmission of each disorder is unknown, so that the parameters selected for likelihood-based tests and the associated nominal significance levels must be incorrect. The traditional significance threshold — a cumulative lod score of 3.0 — cannot be accepted because it assumes that there is an a priori probability of 1 that linkage will be detected somewhere in the genome, and no such certainty exists for complex disorders because the sample and the statistical strategy of a given study may not be appropriate. Thus every independent linkage region represents an additional opportunity for false positive results. The use of multiple separate diagnostic models, with a separate statistical analysis for each model, further increases the type I error rate. There is currently a degree of polarization in the field concerning these issues. On the one hand, Lander and Kruglyak (1995) have computed the probability of detecting various maximum lod scores somewhere in the genome, assuming the use of a single statistical analysis per study and a very dense (essentially fully informative) map of markers. They proposed standards for significant linkage (a result expected to occur by chance in 5% of complete genome scans) and suggestive linkage (a result expected once per genome scan). This approach attempts to minimize the number of false positive results. Others have criticized this approach. For example, Elston (1997) has questioned the value of establishing any arbitrary threshold, and has advocated the use of multiple analyses in a studies of complex disorders to identify initial clues to location of susceptibility genes, i.e. permitting more false positives in order to minimize false negative results.

The issue has become more contentious than this author believes to be justified (Levinson, 1997b). There is obviously a value in computing the probability of observing a particular result by chance, to guide the field toward the most significant and replicable results. This requires taking into account all planned statistical tests and the effect of studying the entire genome. But it is also likely to be worthwhile to examine the pattern of positive results within and across studies. Given the difficulty of detecting susceptibility genes for complex disorders, regions which produce unusually positive results in multiple studies may be the most likely to contain such genes, even if arbitrary thresholds of significance are not reached in any study. It may be particularly difficult to achieve highly "significant" results in psychiatric studies because of the purely clinical, and probably inexact, nature of all psychiatric available diagnostic methods, as discussed above.

Suarez et al. (1994) presented some relevant simulation studies demonstrating that it is difficult to detect and to replicate the component loci of a disease-producing polygenic system. This is because nearby marker alleles segregate inconsistently with disease due to the variable combinations of susceptibility alleles causing disease in different families and even in different cases within families, so that stochastic variation in segregation produces stronger evidence for linkage in some samples than in others unless extremely large samples are available. Thus, when an inconsistent pattern of results is observed across studies, it is difficult to determine whether the partial convergence of findings in a given region represents a true or false positive. One approach is to carry out multicenter studies of the same markers in the most positive regions, as has been attempted for schizophrenia (Gill et al, 1996; Schizophrenia Linkage Collaborative Group, 1996). Another is to develop meta-analysis techniques to examine rank-ordered results across complete genome scans, as suggested by Lewis et al. (1998).

Similar issues complicate interpretation of association studies. Because in most samples long regions of linkage disequilibrium. cannot be expected, every polymorphism must be considered an entirely independent test, increasing type I error. A simulation problem studied in the ninth Genetic Analysis Workshop confirmed that, even with a very dense map, the number of false positive results at any given statistical threshold was perfectly predicted by the number of tests performed (Levinson, 1995). It must further be assumed that any research group reporting association data will study multiple regions, so that no single analysis truly stands alone. In this context, the interpretation of P values less than 0.05 or 0.01 as "significant" must be challenged. Because such results have frequently found their way into print, there is also the danger of publication bias, which complicates the interpretation of meta-analyses.

Progress in linkage studies to date: is the glass half-empty or half-full?

Numerous completed and nearly-completed genome scans of psychiatric disorders have been published over the past few years. Many of these studies were initiated 8–10 years ago, and most have collected samples of affected sibling pairs or multiply-affected small or moderate-sized pedigrees that would have been sufficiently powerful to detect loci with a moderate effect (two-to-three-fold increase in risk) in a large proportion of a sample of pedigrees, or a large effect (five-to-ten-fold increase in risk) in a smaller proportion of pedigrees. To date there is no instance of a study producing an unequivocally significant result in two separate samples, for any psychiatric disorder.

There were early reports of significant linkage for both bipolar disorder and schizophrenia, but in both cases the findings could not be replicated in other samples and further analyses of the original samples also proved negative (Egeland et al, 1987; Kelsoe et al, 1989; Pauls et al, 1991; Ginns et al, 1992; Baron et al, 1993; Sherrington et al, 1988; McGuffin et al, 1990). The reactions in the scientific community have been instructive: the initial findings were met with a degree of enthusiasm which facilitated the funding of many projects, but the ultimate failure of these findings seems to have produced a backlash against psychiatric linkage studies which (at least in the United States) required some years to overcome. Meanwhile, numerous genome scan projects proceeded to completion despite many obstacles including the slow pace of clinical recruitment and assessment, inconsistent funding, and multiple advances in genotyping and statistical methods which required re-thinking and re-tooling in every study. As the initial results of genome scans have started to become available, there again has been something of a backlash: neither definitive proof of linkage nor consistent replication has been observed, and this, combined with the statistical arguments suggesting that the relevant susceptibility loci might not be detectable in currently available samples, has again produced considerable skepticism about the prospects for success.

Yet, if one assumes that, as suggested by Suarez et al (1994), consistent replication is unlikely for studies of complex disorders, a more optimistic view may be justified. Given what is known about the complex transmission of these disorders, it was assumed by many investigators that at least a decade would be required to make substantial progress. During this first decade, the best available diagnostic methods have been applied, multiple samples have been collected — including publicly available samples for both bipolar disorder and schizophrenia (Cloninger et al., 1998; Anonymous, 1997) — using a variety of well justified sampling strategies, and several remarkable advances in statistical and genotyping methods have been successfully incorporated into study designs, so that the resulting genome scans have used more informative marker maps than could have been imagined a decade ago, and data have typically been analyzed using new and powerful methods. Positive findings appear to be converging in a small number of chromosomal regions for each of the most widely studied disorders, and susceptibility loci may yet be found in some of these regions.

TABLE 3

Convergence of results in schizophrenia genome scan projects

Reference	Whole scan reported	No. of peds	2p	5q	6p	6q	8p	9q	10p	10q	13q	22q
Coon et al., 1994	Y	9	X									x
Moises et al., 1995	Y	5 + 65[a]			x			x				
Gill et al., 1996 (multicenter)	–											x
Schizophrenia Linkage Collaborative Group, 1996 (multicenter)	–				x		x					
Williams et al., 1997	N	100[c]								x		
Straub et al., 1997a, 1997b, 1998; Kendler et al., 1996b	N	265[d]		X	XX		x		x			
Wildenauer et al., 1997; Schwab et al., 1995, 1997, 1998	N	72		x	x				x			
Cao et al., 1997; Martinez et al., 1998	N	53[e]				x						
Levinson et al., 1998; Martinez et al., in press	Y	43				x		x		x		
NIMH/Millennium:												
Faraone et al., 1998	Y	43							X	x		
Kaufmann et al., 1998	Y	30				x	x	x				
Blouin et al., 1998	Y	54 + 51[b]					x				XX	x
Coon et al., 1998	Y	(isolate)										
Shaw et al., in press	Y	70	x							x	x	x

Shown are the chromosomal regions in which the most positive linkage results were observed in each dataset.
XX – result with nominal $P < 0.00005$ (without correction for multiple tests)
X – result with nominal $P < 0.001$
x – most positive results in genome scan (above a threshold identified by authors, or selected from a longer list of "positives")

[a] Five moderate-size Icelandic pedigrees studied with genome map; 65 additional pedigrees from multiple datasets studied with 10 most positive markers.
[b] Genome scan completed in 54 pedigrees; most positive results followed up in a separate sample of 51 pedigrees.
[c] Genome scan (20 cM) completed in 100 affected sib pairs (all possible pairs), and follow up of regions with nominal $P < 5$ is proceeding in an additional 100 pairs.
[d] Genome scan completed by typing each marker using one-third of pedigrees, so that each one-third was scanned every 30 cM; largest positives studied in all pedigrees.
[e] Genome scan progressing in 53 independent affected sib pairs, and positives (such as 6q) studied in an additional 87 sib pairs from the NIMH Genetics Initiative dataset, and in 54 sib pairs (43 pedigrees) from the US/Australia dataset.

As examples, Tables 3 and 4 summarize genome scan and other major results for schizophrenia and bipolar disorder respectively. Neither table presents all "positive" results, and individual reports of negative results are omitted. Each table summarizes findings in the chromosomal regions in which at least one study reported a nominal P value associated with a 5% genome-wide threshold of significance (symbolized by "XX" in the tables) ignoring multiple testing issues, or at least one study reported a nominal P value less than 0.001 ("X") and at least one other study reported that the region produced one of its most positive scores ("x") by whatever threshold was used in that study, or at least three studies reported that the region exceeded this latter type of

TABLE 4

Convergence of results in bipolar disorder genome scan projects

Reference	Whole scan reported	No. of peds	4p	11p	13q	18p	18q	21q
Straub et al., 1994; Aita et al., 1999	N	47[a]						X
Stine et al., 1995; McMahon et al., 1997	N	28				X	X	
Coon et al., 1996	Y	8						
McInnes et al., 1996	Y	2		x		x	XX	
Blackwood et al., 1996	Y	12	XX					
Old Order Amish pedigrees:								
Polymeropoulos & Shaffer, 1996	Y	1	x					
Ginns et al., 1996; Philibert et al., 1997	Y	5			X			
Malafosse et al., 1997	N	11		XX				
Smyth et al., 1997	N	23		x				x
Detera-Wadleigh, 1997; Edenberg et al., 1997; Rice et al., 1997; Stine et al., 1997	Y	97	x		x			x
Berrettini et al., 1997; Detera-Wadleigh et al., 1996	Y	22		x		XX		X

Symbols as in Table 3.
[a] The positive result on chromosome 21q was in 1 pedigree only (lod = 3.41), n.s in entire 47.

threshold. Published complete genome scans have been included. In some cases the present author has exercised judgment in categorizing findings, given the variability in the statistical approaches across studies.

For each disorder, there are chromosomal regions which have produced unusually positive results in several studies. For schizophrenia, chromosomes 6p and 13q have produced significant results in one study each (ignoring the multiple testing issue within each study), and 6p has been supported by two other scans and by a large multicenter study (with a sample drawn largely from the studies listed here). The initial report of modestly positive results in chromosome 13q was by Lin et al. (1995), but is not shown in Table 3 which emphasizes genome scan results. Chromosome 22q has received modest support from three separate studies as well as a large multicenter study (again with a sample largely overlapping this list), and 8p from three separate studies and a multicenter study. Chromosomes 2p, 5q, 6q, 9q, 10p and 10q have also received support across studies. A multicenter study of 5q, 6q, 10p and 13q is in progress.

For bipolar disorder, multiple studies have produced positive results on chromosome 18, although in several different locations, with strong evidence for both 18p (centromeric) and 18q loci. The strongest single finding has been a 4p locus in a single Scottish pedigree, while multiple studies have reported modestly positive results in the same region of 21q, and also 11p near tyrosine hydroxylase — which is near the original "discredited" finding of Egeland et al. (1987), demonstrating the need for patience and caution in evaluating results in this field. Also, Barden has reported orally on genome-wide significant evidence for linkage to bipolar disorder in a region of chromosome 12q in the isolated Saguenay-Lac St. Jean population of French Canada (N. Barden, Park City Molecular Psychiatry Conference, February, 1998) but these data have not yet bee published; additional positive data on 12 been reported by Dawson et al. (1995). also been reports that evidence f implicated regions of chromosome 18 could be related to the sex of the transmitting parent, but the data have been inconsistent (Stine et al., 1995; Gershon et al., 1996; McMahon et al., 1997).

There are two rather different ways of looking at these patterns of results. One possibility is that they are all false positives reflecting the many opportunities for "positive" findings when many laboratories studying all or most of the genome. For schizophrenia, for example, with 12 scans reported, there would be 0.60 probability that one region would achieve a 5% genome-wide level of significance, without accounting for the additional opportunities due to multiple testing. Given that only two regions have achieved this level of statistical significance, and none in multiple studies, it is possible that the modest observed convergence of results could occur by chance alone.

On the other hand, once a strongly positive result has been reported for each region, the probability that two additional studies will produce one of its most positive findings within 30–40 cM is reasonably low, although difficult to compute precisely. The probability of observing this degree of convergence in multiple regions is still lower. The convergence of positive results is even more striking for bipolar disorder. One could hypothesize that there are fewer susceptibility loci for the latter disorder, that there are loci of larger effect, or that fewer complete genome scans have been reported such that there are fewer moderately positive findings. Given that a disease locus is the most likely to produce at least weakly positive results in any given study, it must be considered possible that for each disorder, the loci with the greatest effect on susceptibility may be among those discussed here.

This discussion has not focussed on a number of other approaches that have been applied to psychiatric disorders. Although association studies of polymorphisms in candidate gene regions have generally failed to produce replicable positive results, there have been some promising findings for schizophrenia as discussed by Jacobsen et al. in Chapter 5.4 of this volume. There has also been some support for anticipation (reduced age of onset and/or increased severity in successive generations), so that improved methods for detecting trinucleotide repeat expansions have been developed by groups engaged in studies of psychiatric disorders as discussed in Chapter 5.3 of this volume.

Implications of evolutionary and population genetics

Recent discussions of strategies in this field have focussed on two major issues: the need for larger samples, and the possible advantages of studying "isolated" or "homogeneous" populations. Two related but different rationales have been offered for recruiting samples from specific populations: one is that "founder" effects can be exploited by detecting linkage disequilibrium (LD) around a disease allele inherited from a common ancestor, with the expected length of the region of LD predictable from the age of the population in terms of the number of generations it has been isolated from other populations. A variation on this theme is that, in an isolated population of stable size, genetic drift can produce linkage disequilibrium at markers tightly linked to disease loci despite there being multiple founders for disease-related mutations in the same population. The second rationale is that, because gene duplication and subsequent alteration of function are major mechanisms of evolution, and because most disease-causing mutations occur only once, there may be many sets of loci that can produce the same disorder, and the frequency of disease-related alleles at any one such locus could vary substantially across different human populations, making detection of linkage and LD more likely in samples drawn from a single historical population group.

LD mapping was proposed by Lander and Botstein (1986) as a method for determining the location of disease-related genes. Finland has been suggested as an ideal population for this method

(De la Chapelle, 1993) because the current population descends primarily from an emigration of several thousand individuals about 100 generations ago. The average length of shared DNA around an allele inherited from a common ancestor is predicted to be 1 divided by the number of generations since the common ancestor (in Morgans) (A. Chakravarti, personal communication; Sanda and Ford, 1986; and see Terwilliger et al., 1997), so that in a population about 100 generations old, about 1 cM of DNA would be shared on average by individuals carrying the same disease ancestral mutation. Linked haplotypes, presumably due to founder effects, have been demonstrated for numerous mendelian disorders in Finland (e.g. Hastbacka et al., 1992; Haataja et al., 1993; Nikali et al., 1995). As genome-wide maps become more dense, LD mapping could be used as a primary method, i.e.1 regions of shared DNA could be detected in samples of "unrelated" (actually distantly related) probands from such populations, without the need for preliminary linkage study of multiply-affected pedigrees.

Other investigators have proposed that populations which have been isolated for shorter periods of time are ideal for the initial localization of disease genes using LD mapping (Houwen et al, 1994). For example, after ten generations, individuals with a common ancestral allele will share about 10 cM of DNA in the region on average, which can be detected by screening the genome with currently-available maps of microsatellite markers. Chakravarti and colleagues used this method to detect a locus for the complex disorder Hirschsprung's disease in an isolated Mennonite community (Puffenberger et al., 1994), and Houwen et al. (1994) mapped a gene for a recessive disorder, benign recurrent intermittent cholestasis, with just three subjects from a Dutch village that has been isolated for over 200 years. Freimer and colleagues have attempted to apply the same strategy to the study of bipolar disorder in the Central Valley region of Costa Rica, where a population of about two million is descended largely from under 200 Spanish and Amerindian couples about 300 years ago (Freimer et al., 1996a; Escamilla et al., 1996)). Their initial approach was to use linkage methods to study two extended multiply-affected pedigrees (McInnes et al., 1996), followed by a search for shared haplotypes in the most positive regions both in members of these pedigrees and in "unrelated" singleton probands. Evidence for linkage and LD was detected in one region of chromosome 18q (Freimer et al., 1996b).

Although LD mapping has been remarkably successful in the study of rare mendelian disorders, it is less clear how useful it will be with complex disorders. Hirschsprung's disease is not directly comparable to the psychiatric disorders because it is relatively rare and there is a high relative risk to siblings of probands. For a disorder to which susceptibility requires interaction of alleles at several separate loci, pathogenic variants at each locus would have to be quite common to produce the prevalences observed for most major psychiatric disorders. Thus, among thousands or even hundreds of population founders, there may be many pathogenic mutations alleles at each susceptibility locus, with each mutation on a different founder haplotype, so that each one could be quite difficult to detect. It might therefore be useful to study relatively recent populations with very small numbers of common founders (i.e. 10–30). In such a population, the phenomenon of genetic drift the random process by which some ancestral alleles are transmitted multiple times and become common in a population, while others die out completely — would be expected to have a particularly dramatic effect, and a small number of disease susceptibility alleles could become sufficiently common to be detected by linkage and/or LD mapping methods (Te Meerman and Van der Meulen, 1997). Terwilliger et al. (1997) have pointed out that in isolated populations with stable population size, linkage disequilibrium may be detectable despite the presence of multiple distinct founder mutations at a locus. An apparent dilemma here is that sample size is likely to be limited in the relatively recent, small populations that have descended from a small number of founders, while larger samples will typically come from populations with a sufficiently large number of founders that multiple ancestral alleles are present at each susceptibility locus. In any event, LD mapping of complex disorders will require the development and validation of statistical methods, which is now starting to occur.

A second perspective on complex disease genetics has grown out of the fields of population and evolutionary genetics. The study of genetic polymorphisms has, of course, contributed greatly to the understanding of evolution of species as well as the movement of human populations throughout the world. Weiss (1993) has summarized some possible implications of these fields of research for complex disease genetics. The traditional assumption of a wild-type and a single disease allele at each disease locus has given way to an understanding that most mutations occur only once, so that there is a theoretically infinite number of alleles possible at any locus, with variable phenotypic effects. Further, evolution rests heavily on the phenomenon of gene duplication events, such that a second copy of a gene (or a slightly altered version of a gene) is created on the same or another chromosome and then evolves a different function than its predecessor. Mutations in different members of these gene families can create different or quite similar phenotypic effects. From this perspective, for every locus that can contribute to a complex phenotype, there are likely to be other loci — some of them closely related sequences with slightly different functions, and others relevant to the same biochemical or neurophysiological pathway — which can have similar effects. The prediction of the theory is that even in very large populations, if they have bred independently for long periods of time, the frequencies of mutations in the various loci that contribute to the disease phenotype will become quite different because of the population-specific history of mutation and drift. Thus, any one locus would be most easily identified by studying a sample drawn entirely from a population in which mutations in that gene were more common, while it might be difficult to detect any disease loci in mixed populations. In any event, the short regions of linkage disequilibrium upon which fine mapping of disease genes depends will certainly be most easily detected in samples drawn from historically homogeneous populations.

Another lesson from population genetics is that even in large populations, allelic variants exist in the context of short tightly-linked haplotypes inherited from the ancient ancestor in whom the mutation occurred. Kidd and colleagues (e.g. Cubells et al., 1997) have demonstrated that association studies can be informed by knowledge of the population-specific haplotypes in the vicinity of each candidate gene. For example, if two alleles at tightly-linked markers are almost always found in the same haplotypes in a population under study, the results of a study must be viewed with caution if a disorder is found to be associated with the allele from that haplotype at the first marker but not at the second marker.

A full discussion of this perspective is beyond the scope of the present chapter. It would appear to hold great promise for the field of psychiatric genetics. However, it would not be justified to assume without empirical evidence that the locations and frequencies of susceptibility alleles for all psychiatric disorders will show substantial variation across world populations. For example, Risch (1990b) has pointed out that there are no observable familial or population-specific subtypes of schizophrenia, as would be expected if many different sets of genes could interact to produce similar phenotypes — i.e. although certain subtypes might be indistinguishable clinically, it would seem unlikely that all genetic subtypes would produce such a similar range of phenotypes. Thus the hypothesis of a single polygenic system underlying many cases of schizophrenia in most populations is as likely as the alternative hypothesis of population-specific systems, pending further empirical data. We do not know, in fact, whether the inconsistency of linkage results across studies is due to population-related variability, or to stochastic variation in segregation patterns or a random pattern of false positive findings.

Future prospects

Progress in psychiatric genetics is likely to occur as a result of the simultaneous exploitation of multiple avenues of research, including the following:

1. Efficient utilization of current samples and methods. A large number of pedigree, case-control and proband/parent trio samples have already been collected for several of the major psychiatric disorders. As discussed above for schizophrenia and bipolar disorder, there is a degree of convergence of findings that suggests a number of candidate regions for each disorder. The present author favors the "half-full" rather than the "half-empty" position, i.e. further study of these samples could yield important results. Luck may be required — definitive proof of linkage is unlikely even with large multicenter analyses, and definitive narrowing of a putatively linked region could require many more affected pedigrees than are available. But it would not be surprising for serendipity to permit the identification of disease-related loci as a result of further linkage and linkage disequilibrium studies of existing samples and subsequent mutation-detection and sequencing experiments.

2. Collection of new, very large samples. It has taken most of the past decade to achieve some clarity about the sample sizes that could be required to detect loci with very small genetic effects (or at least, small effects averaged across a sample). A major implication of this understanding is that much larger samples are likely to be useful — either with current or, more likely, with future molecular methods. Samples will be needed both for linkage and for genetic association studies (proband/parent trios and well-matched case-control samples). It is not clear whether affected sibships or more diverse constellations of affected individuals will be most useful for linkage, and investigators will continue to select different strategies.

3. Collection of samples from a variety of isolated populations, of varying historical "ages" and rates of expansion, with substantial founder effects. As genetic maps become more dense and improved methods are developed for detecting sequence similaries and differences among individuals and groups of individuals, power to detect founder effects (including haplotype sharing due to multiple founder effects) is likely to increase.

4. Collection of samples from larger but reasonably "homogeneous" populations. It is a reasonable hypothesis that, when a sample of similar size can be collected from a more or a less homogeneous population, the more homogeneous population might be preferred because the degree of heterogeneity in disease etiology might be reduced. However, we know too little to assume that homogeneity is always better than larger sample size, and the largest samples can typically be collected from the more hetereogeneous populations which have been formed from multiple subpopulations (ethnic, religious and/or language groups which developed independently for long periods of time prior to the formation of modern empires and states).

5. Application of new methods for measuring phenotypes. This is likely to include characterization of clinical features of disorders using dimensional rather than categorical systems, measurement of "personality" dimensions which may contribute to susceptibility to overt disorders, and measurement of physiological, chemical or anatomical variation which could be under more direct genetic control than the clinical syndromes.

6. Development of animal models. For example, animal models have been developed for the study of the genetics of addiction (Berrettini et al., 1994; Melo et al., 1996) and of deficiency of p50 pre-pulse inhibition (Freedman et al., 1996).

7. Application of new molecular methods including dense chip-based maps, rapid methods for detecting mutations, adapation of methods

related to genome mismatch scanning and representational difference analysis, improved methods for detecting regions with expanded tandem repeat motifs, and others.

In conclusion, substantial progress has been made in the study of the genetics of psychiatric disorders: although specific disease-related loci have not been identified, many conceptual and methodological issues have been successfully addressed, samples have been collected using appropriate clinical methods, and for several major disorders the available linkage evidence appears to be converging on a reasonably small number of chromosomal regions. There are many strategies available to facilitate further progress. In the present author's view, the main obstacle to progress is the tendency of the field to succumb to spurts of unjustified optimism and skepticism about various methods, so that funding is directed first toward one approach and then another. While it is an unattainable ideal to achieve an entirely logical strategy to exploit every plausible avenue, we can at least strive to remain open to conflicting evidence and multiple strategies.

References

Aita, V.M., Liu, J., Knowles, J.A., Terwilliger, J.D., Baltazar, R., Grunn, A., Loth, J.E., Kanyas, K., Lerer, B., Endicott, J., Wang, Z., Penchaszadeh, G., Gilliam, T.C. and Baron, M. (1999) A comprehensive linkage analysis of chromosome 21q22 supports prior evidence for a putative biopolar affective disorder locus. *Am. J. Hum. Genet.*, 64: 210–217.

Alda, M. (1997) Bipolar disorder: from families to genes. *Canadian Journal of Psychiatry — Revue Canadienne de Psychiatrie*, 42: 378–387.

American Psychiatric Association, (1987) *Diagnostic and Statistical Manual*, 3rd Edn — Revised, American Psychiatric Association, Washington, DC.

American Psychiatric Association, (1994) *Diagnostic and Statistical Manual*, 4th Edn, American Psychiatric Association, Washington, DC.

Andreasen, N.C., Rice, J., Endicott, J., Coryell, W., Grove, W.M. and Reich, T. (1987) Familial rates of affective disorder. A report from the National Institute of Mental Health Collaborative Study. *Arch. Gen. Psychiatry*, 44: 461–469.

Anonymous (1997) Genomic survey of bipolar illness in the NIMH genetics initiative pedigrees: a preliminary report. *Am. J. Med. Genet.*, 74: 227–237.

Bailey, A., Le Couteur, A., Gottesman, I., Bolton, P., Simonoff, E., Yuzda, E. and Rutter, M. (1995) Autism as a strongly genetic disorder: evidence from a British twin study. *Psychol. Med.*, 25: 63–77.

Baron, M. (1997) Genetic linkage and bipolar affective disorder: progress and pitfalls. *Mol. Psychiat.*, 2: 200–210.

Baron, M., Gruen, R., Asnis, L. and Kane, J. (1982) Schizoaffective illness, schizophrenia and affective disorders: morbidity risk and genetic transmission. *Acta. Psychiatri. Scand.*, 65: 253-262.

Barr, C.L., Kennedy, J.L., Pakstis, A.J., Wetterberg, L., Sjogren, B., Bierut, L., Wadelius, C., Wahlstrom, J., Martinsson, T., Giuffra, L., et al. (1994) Progress in a genome scan for linkage in schizophrenia in a large Swedish kindred. *Am. J. Med. Genet.*, 54: 51–58.

Berrettini, W.H., Ferraro, T.N., Alexander, R.C., Buchberg, A.M. and Vogel, W.H. (1994) Quantitative trait loci mapping of three loci controlling morphine preference using inbred mouse strains. *Nat. Genet.*, 7: 54–58.

Berrettini, W.H., Ferraro, T.N., Goldin, L.R., Detera-Wadleigh, SD., Choi, H., Muniec, D., Guroff, JJ., Kazuba, D.M., Nurnberger, J.I., Hsieh, W.T., Hoehe, M.R. and Gershon, E.S. (1997) A linkage study of bipolar illness. *Arch. Gen. Psych.*, 54: 27–35.

Blackwood, D.H., He, L., Morris, S.W., McLean, A., Whitton, C., Thomson, M., Walker, M.T., Woodburn, K., Sharp, C.M., Wright, A.F., Shibasaki, Y., St. Clair, D.M., Porteous, D.J. and Muir, W.J. (1996) A locus for bipolar affective disorder on chromosome 4p. *Nat. Genet.*, 12: 427–430.

Bland, R.C., Newman, S.C. and Orn, H. (1986) Recurrent and nonrecurrent depression. *Arch. Gen. Psych.*, 43: 1085–1089.

Blouin, J.L., Dombroski, B.A., Nath, S.K., Lasseter, V.K., Wolyniec, P.S., Nestadt, G., Thornquist, M., Ullrich, G., McGrath, J., Kasch, L., Lamacz, M., Thomas, M.G., Gehrig, C., Radhakrishna, U., Snyder, S.E., Balk, K.G., Neufeld, K., Swartz, K.L., DeMarchi, N., Papadimitriou, G.N., Dikeos, D.G., Stefanis, C.N., Chakravarti, A., Childs, B. and Pulver, A.E. (1998) Schizophrenia susceptibility loci on chromosomes 13q32 and 8p21. *Nat. Genet.*, 20: 70–73.

Bolton, P., Macdonald, H., Pickles, A., Rios, P., Goode, S., Crowson, M., Bailey, A. and Rutter, M. (1994) A case-control family history study of autism. *J. Child Psychol. Psychiat. Allied Discip*. 35: 877–900.

Cao, Q., Martinez, M., Zhang, J., Sanders, A.R., Badner, J.A., Cravchik, A., Markey, C.J., Beshah, E., Guroff, J.J., Maxwell, M.E., Kazuba, D.M., Whiten, R., Goldin, L.R., Gershon, E.S. and Gejman, P.V. (1997) Suggestive evidence for a schizophrenia susceptibility locus of chromosome 6q and a confirmation in an independent series of pedigrees. *Genomics*, 43: 1–8.

Cardno, A.G., Holmans, P.A., Harvey, I., Williams, M.B., Owen, M.J. and McGuffin, P. (1997) Factor-derived subsyndromes of schizophrenia and familial morbid risks. *Schizophrenia Res.*, 23: 231–238.

Cloninger, C.R., Kaufmann, C.A., Faraone, S.V., Malaspina, D., Svrakic, D.M., Harkavy-Friedman, J., Suarez, B.K., Matise, T.C., Shore, D., Lee, H., Hampe, C.L., Wynne, D., Drain, C., Markel, P.D., Zambuto, C.T., Schmitt, K. and Tsuang, M.T. (1998) A genome-wide search for schizophrenia susceptibility loci: the NIMH genetics initiative & Millennium consortium. *Am. J. Med. Genet.*, 81: 275–281.

Coon, H., Jensen, S., Hoff, M., Holik, J., Plaetke, R., Reimherr, F., Wender, P., Leppert, M. and Byerley, W. (1993) A genome-wide search for genes predisposing to manic-depression, assuming autosomal dominant inheritance. *Am. J. Hum. Genet.*, 52: 1234–1249.

Coon, H., Jensen, S., Holik, J., Hoff, M., Myles-Worsley, M., Reimherr, F., Wender, P., Waldo, M., Freedman, R., Leppert, M. and Byerley, W. (1994). Genomic scan for genes predisposing to schizophrenia. *Am. J. Med. Genet.*, 54: 59–71.

Coon, H., Hoff, M., Holik, J., Hadley, D., Fang, N., Reimherr, F., Wender, P. and Byerley, W. (1996) Analysis of chromosome 18 DNA markers in multiplex pedigrees with manic depression. *Biol. Psychiat.*, 39: 689–96.

Coon, H., Myles-Worsley, M., Tiobech, J., Hoff, M., Rosenthal, J., Bennett, P., Reimherr, F., Wender, P., Dale, P., Polloi, A. and Byerley, W. Evidence for a chromosome 2p 13-14 schizophrenia susceptibility locus in families from Palau, Micronesia. *Mol. Psychiatry*, in press.

Cubells, J.F., Kobayashi, K., Nagatsu, T., Kidd, K.K., Kidd, J.R., Calafell, F., Kranzler, H.R., Ichinose, H. and Gelernter, J. (1997) Population genetics of a functional variant of the dopamine beta-hydroxylase gene (DBH). *Am. J. Med. Genet.*, 74: 374–379.

Daly, M., Kirby, A. and Kruglyak, L. (1997) Do multiple data sets provide support for a bipolar illness susceptibility locus on chromosome 18? *Genetic Epidemiology*, 14: 599–604.

Davis, S. and Weeks, D. (1997) Comparison of nonparametric statistics for detection of linkage in nuclear families: single-marker evaluation. *Am. J. Hum. Genet.*, 61: 1431–1444.

Dawson, E., Parfitt, E., Roberts, Q., Daniels, J., Lim, L., Sham, P., Nothen, M., Propping, P., Lanczik, M. and Maier, W. (1995) Linkage studies of bipolar disorder in the region of the Darier's disease gene on chromosome 12q23-24.1. *Am. J. Med. Genet.*, 60: 94–102.

De Bruyn, A., Souery, D., Mendelbaum, K., Mendlewicz, J. and Van Broeckhoven, C. (1996) Linkage analysis of families with bipolar illness and chromosome 18 markers. *Biol. Psychiat.*, 39: 679–688.

De la Chapelle, A. (1993) Disease gene mapping in isolated human populations: the example of Finland. *J. Med. Genet.*, 30: 857–865.

Detera-Wadleigh, S.D., Badner, J.A., Goldin, L.R., Berrettini, W.H., Sanders, A.R., Rollins, D.Y., Turner, G., Moses, T., Haerian, H., Muniec, D., Nurnberger Jr, J.I. and Gershon, E.S. (1996) Affected-sib-pair analyses reveal support of prior evidence for a susceptibility locus for bopolar disorder, on 21q. *Am. J. Hum. Genet.*, 58: 1279–1285.

Detera-Wadleigh, S.D., Badner, J.A., Yoshikawa, T., Sanders, A.R., Goldin, L.R., Turner, G., Rollins, D.Y., Moses, T., Guroff, J.J., Kazuba, D., Maxwell, M.E., Edenberg, H.J., Foroud, T., Lahiri, D., Nurnberger, J.I., Stine, O.C., McMahon, F., Meyers, D.A., MacKinnon, D., Simpson, S., McInnis, M., DePaulo, J.R., Rice, J., Goate, AGS and Gershon, E.S. (1997) Initial genome scan of the NIMH genetics initiative bipolar pedigrees: chromosomes 4, 7, 9, 18, 19, 20, and 21q. *Am. J. Med. Genet.*, 74: 254–262.

Dick, C.L., Bland, R.C. and Newman, S.C. (1994) Epidemiology of psychiatric disorders in Edmonton. Panic disorder. *Acta Psychiatrica Scandinavica Supplementum.*, 376: 45–53.

Dorr, D.A., Rice, J.P., Armstrong, C., Reich, T. and Blehar, M. (1997) A meta-analysis of chromosome 18 linkage data for bipolar illness. *Genet. Epidemiol.*, 14: 617–622.

Eaton, W.W., Kessler, R.C., Wittchen, H.U. and Magee, W.J. (1994) Panic and panic disorder in the United States. *Am. J. Psychiat.*, 151: 413–420.

Edenberg, H.J., Foroud, T., Conneally, P.M., Sorbel, J.J., Carr, K., Crose, C., Willig, C., Zhao, J., Miller, M., Bowman, E., Mayeda, A., Rau, N.L., Smiley, C., Rice, J.P., Goate, A., Reich, T., Stine, O.C., McMahon, F., DePaulo, J.R., Meyers, D., Detera-Wadleigh, S.D., Goldin, L.R., Gershon, E.S., Blehar, M.C. and Nurnberger Jr, J.I. (1997) Initial genomic scan of the NIMH genetics initiative bipolar pedigrees: chromosomes 3, 5, 15, 16, 17, and 22. *Am. J. Med. Genet.*, 74: 238–246.

Egeland, J.A., Gerhard, D.S., Pauls, D.L., Sussex, J.N., Kidd, K.K., Allen, C.R., Hostetter, A.M. and Housman, D.E. (1987) Bipolar affective disorders linked to DNA markers on chromosome 11. *Nature*, 325: 783–787.

Ehlers, S. and Gillberg, C. (1993) The epidemiology of Asperger syndrome. A total population study. *J. Child Psychol. Psychiat. Allied Discipl.*, 34: 1327–1350.

Elston, R.C. (1997) Algorithms and inferences: the challenge of multifactorial diseases. *Am. J. Hum. Genet.*, 60: 255–262.

Elston, R.C., Guo Xiuqinq and Williams, L.V. (1996) Two-stage global search designs for linkage analysis using pairs of affected relatives. *Genet. Epidemiol.*, 13: 535–558.

Endicott, J. and Spitzer, R.L. (1978) A diagnostic interview: The schedule for affective disorders and schizophrenia. *Arch. Gen. Psychiatry*, 35: 837–844.

Escamilla, M.A., Spesny, M., Reus, V.I., Gallegos, A., Meza, L., Molina, J., Sandkuijl, L.A., Fournier, E., Leon, P.E., Smith, L.B. and Freimer, N.B. (1996) Use of linkage disequilibrium approaches to map genes for bipolar disorder in the Costa Rican population. *Am. J. Med. Genet.*, 67: 244–253.

Falk, C.T. and Rubinstein, P. (1987) Haplotype relative risks: an easy reliable way to construct a proper control sample for risk calculations. *Ann. Hum. Genet.*, 51: 227–233.

Faraone, S.V., Blehar, M., Pepple, J., Moldin, S.O., Norton, J., Nurnberger, J.I., Malaspina, D., Kaufmann, C.A., Reich, T., Cloninger, C.R., DePaulo, J.R., Berg, K., Gershon, E.S., Kirch, D.G. and Tsuang, M.T. (1996) Diagnostic accuracy and confusability analyses: an application to the Diagnostic Interview for Genetic Studies. *Psycholog. Med.*, 26: 401–410.

Faraone, S.V., Matise, T., Svrakic, D., Pepple, J., Malaspina, D., Suarez, B., Hampe, C., Zambuto, C.T., Schmitt, K., Meyer, J., Markel, P., Lee, H., Harkevy-Friedman, J., Kaufmann, C.A., Cloninger, C.R. and Tsuang, M.T. (1998) A genome scan of the European-American schizophrenia pedigrees of the NIMH Genetics Initiative. *Am. J. Med. Genet.*, 81: 290–295.

Fombonne, E. and du Mazaubrun, C. (1992) Prevalence of infantile autism in four French regions. *Social Psychiatry and Psychiatric Epidemiology*, 27: 203–210.

Freedman, R., Adler, L.E., Myles-Worsley, M., Nagamoto, H.T., Miller, C., Kisley, M., McRae, K., Cawthra, E. and Waldo, M. (1996) Inhibitory gating of an evoked response to repeated auditory stimuli in schizophrenic and normal subjects. Human recordings, computer simulation, and an animal model. *Arch. Gen. Psychiat.*, 53: 1114–1121.

Freedman, R., Coon, H., Myles-Worsley, M., Orr-Urtreger, A., Olincy, A., Davis, A., Polymeropoulos, M., Holik, J., Hopkins, J., Hoff, M., Rosenthal, J., Waldo, M.C., Reimherr, F., Wender, P., Yaw, J., Young, D.A., Breese, C.R., Adams, C., Patterson, D., Adler, L.E., Kruglyak, L.L.S. and Byerley, W. (1997) Linkage of a neurophysiological deficit in schizophrenia to a chromosome 15 locus. *Proc. Natl. Acad. Sci. USA*, 94: 587–592.

Freimer, N.B., Reus, V.I., Escamilla, M., Spesny, M., Smith, L., Service, S., Gallegos, A., Meza, L., Batki, S., Vinogradov, S., Leon, P. and Sandkuijl, L.A. (1996a) An approach to investigating linkage for bipolar disorder using large Costa Rican pedigrees. *Am. J. Med. Genet.*, 67: 254–263.

Freimer, N.B., Reus, V.I., Escamilla, M.A., McInnes, L.A., Spesny, M., Leon, P., Service, S.K., Smith, L.B., Silva, S., Rojas, E., Gallegos, A., Meza, L., Fournier, E., Baharloo, S., Blankenship, K., Tyler, D.J., Batki, S., Vinogradov, S., Weissenbach, J.B.H., Barondes, S. and Sandkuijl, L.A. (1996b) Genetic mapping using haplotype, association and linkage methods suggests a locus for severe bipolar disorder (BPI) at 18q22-q23. *Nat. Genet.*, 12: 436-441.

Gershon, E.S., Badner, J.A., Detera-Wadleigh, S.D., Ferraro, T.N. and Berrettini, W.H. (1996) Maternal inheritance and chromosome 18 allele sharing in unilineal bipolar illness pedigrees. *Am. J. Med. Genet.*, 67: 202–7.

Gershon, E.S., Badner, J.A., Goldin, L.R., Sanders, A.R., Cravchik, A. and Detera-Wadleigh, S.D. (1998) Closing in on genes for manic-depressive illness and shizophrenia. *Neuropsychopharmacology*, 18: 233–242.

Gershon, E.S., DeLisi, L.E., Hamovit, J., Nurnberger, J.I. Jr., Maxwell, M.E., Schreiber, J., Dauphinais, D., Dingman, C.W. II and Guroff, J.J. (1988) A controlled family study of chronic psychoses. *Arch. Gen. Psychiat.*, 45: 328–337.

Gershon, E.S., Hamovit, J., Guroff, J.J., Dibble, E., Leckman, J.F., Sceery, W., Targum, S.D., Nurnberger Jr., J.I., Goldin, L.R. and Bunney Jr, W.E. (1982) A family study of schizoaffective, bipolar I, bipolar II, unipolar, and normal control probands. *Arch. Gen. Psychiat.*, 39: 1157–1167.

Gill, M., Vallada, H., Collier, D., Sham, P., Holmans, P., Murray, R., McGuffin, P., Nanko, S., Owen, M., Antonarakis, S., Housman, D., Kazazian, H. Nestadt, G., Pulver, A.E., Straub, R.E., MacLean, C.J., Walsh, D., Kendler, K.S., DeLisi, L., Polymeropoulos, M., Coon, H., Byerley, W., Lofthouse, R., Gershon, E.S. et al. (1996) A combined analysis of D22S278 market alleles in affected sib-pairs: support for a susceptibility locus for schizophrenia at chromosome 22q12. Schizophrenia Collaborative Linkage Group (Chromosome 22). *Am. J. Med. Genet.*, 67: 40–45.

Ginns, E.I., Egeland, J.A., Allen, C.R., Pauls, D.L., Falls, K., Keith, T.P. and Paul, S.M. (1992) Update on the search for DNA markers linked to manic-depressive illness in the Older Order Amish. *J. Psychiatr. Res.*, 26: 305–308.

Ginns, E.I., Ott, J., Egeland, J.A., Allen, C.R., Fann, C.S., Pauls, D.L., Weissenbachoff, J., Carulli, J.P., Falls, K.M., Keith T.P. and Paul, S.M. (1996) A genome-wide search for chromosomal loci linked to bipolar affective disorder in the Old Order Amish. *Nat. Genet.*, 12: 431–435.

Goldin, L.R. (1992) Detection of linkage under heterogeneity: comparison of the two-locus vs. admixture models. *Genet. Epidemiol.*, 9: 61–66.

Goodwin, F.W. and Jamison, K.R. (1990) *Manic-depressive illness*. Oxford University Press, New York.

Gottesman, I.I. and Shields, J. (1982) *Schizophrenia: the epigenetic puzzle*, Cambridge University Press, Cambridge, England.

Haataja, R., Vaisanen, M.L., Li, M., Ryynanen, M. and Leisti, J. (1994) The fragile X syndrome in Finland: demonstration of a founder effect by analysis of microsatellite haplotypes. *Hum. Genet.*, 94: 479–483.

Hastbacka, J., De la Chapelle, A., Kaitkila, I., Sistonen, P., Weaver, A. and Lander, E. (1992) Linkage disequilibrium mapping in isolated founder populations: diastrophic dysplasia in Finland. *Nat. Genet.*, 2: 204–211.

Hauser, E.R., Boehnke, M., Guo, S.W. and Risch, N. (1996) Affected sib-pair interval mapping and exclusion for complex genetic traits: sampling considerations. *Genet. Epidemiol.*, 13: 117–138.

Holmans, P. and Craddock, N. (1997) Efficient strategies for genome scanning using maximum likelihood affected-sib-pair analysis. *Am. J. Hum. Genet.*, 60: 657–666.

Houwen, R.H., Baharloo, S., Blankenship, K., Raeymaekers, P., Juyn, J., Sandkuijl, L.A. and Freimer, N.B. (1994) Genome screening by searching for shared segments: mapping a gene for benign recurrent intrahepatic cholestasis. *Nat. Genet.*, 8: 380–386.

James, J.W. (1971) Frequency in relatives for an all-or-none trait. *Ann. Hum. Genet.*, 35: 47–49.

Katerndahl, D.A. and Realini, J.P. (1993) Lifetime prevalence of panic states. *Am. J. Psychiat.*, 150: 246–249.

Kaufmann, C.A., Suarez, B., Malaspina, D., Pepple, J., Svrakic, D., Markel, P.D., Meyer, J., Zambuto, C.T., Schmitt, K., Matise, T.C., Harkavy-Friedman, J.M., Hampe, C., Lee, H., Shore, D., Wynne, D., Faraone, S.V., Tsuang, M.T. and Cloninger, C.R. (1998) The NIMH genetics initiative Millennium schizophrenia consortium: linkage analysis of African-American pedigrees. *Am J. Med Genet.*, 282–289.

Kawada, Y., Hattori, M., Fukuda, R., Arai, H., Inoue, R. and Nanko, S. (1995) No evidence of linkage or association between tyrosine hydroxylase gene and affective disorder. *J. Affective Disorders*, 34: 89–94.

Kelsoe, J.R., Ginns, E.I., Egeland, J.A., Gerhard, D.S., Goldstein, A.M., Bale, S.J., Pauls, D.L., Long, R.T., Kidd, K.K., Conte, G., et al (1989) Re-evaluation of the linkage relationship between chromosome 11p loci and the gene for bipolar affective disorder in the Old Order Amish. *Nature*, 342: 238–243.

Kendler, K.S. and Diehl, SR. (1993) The genetics of schizophrenia: a current, genetic-epiderniologic perspective, *Schizophr. Bull.*, 19: 261–285.

Kendler, K.S., Gallagher, T.J., Abelson, J.M. and Kessler, R.C. (1996a) Lifetime prevalence, demographic risk factors, and diagnostic validity of nonaffective psychosis as assessed in a US community sample. The National Comorbidity Survey. *Arch. Gen. Psychiatry*, 53: 1022–1031.

Kendler, K.S., Karkowski-Shuman, L., O'Neill, F.A., Straub, R.E., MacLean, CJ. and Walsh, D. (1997) Resemblance of psychotic symptoms and syndromes in affected sibling pairs from the Irish Study of High-Density Schizophrenia Families: evidence for possible etiologic heterogeneity. *Am. J. Psychiatry*, 154: 191–198.

Kendler, K.S., MacLean, C.J., O'Neill, F.A., Burke, J., Murphy, B., Duke, F., Shinkwin, R., Easter, S.M., Webb, B.T., Zhang, J., Walsh, D. and Straub, R.E. (1996b) Evidence for a schizophrenia vulnerability locus on chromosome 8p in the Irish Study of High-Density Schizophrenia Families. *Am J Psychiatry*, 153: 1534–1540.

Kendler, K.S., McGuire, M., Gruenberg, A.M., O'Hare, A., Spellman, M. and Walsh, D. (1993a) The Roscommon Family Study. I. Methods, diagnosis of probands, and risk of schizophrenia in relatives. *Arch. Gen. Psychiatry*, 50: 527–540.

Kendler, K.S., McGuire, M., Gruenberg, A.M., Spellman, M., O'Hare, A. and Walsh, D. (1993b) The Roscommon Family Study. II. The risk of nonschizophrenic nonaffective psychoses in relatives. *Arch. Gen. Psychiatry*, 50: 645–652.

Kendler, K.S., McGuire, M., Gruenberg, A.M., O'Hare, A., Spellman, M. and Walsh, D. (1993c) The Roscommon Family Study. IV. Affective illness, anxiety disorders, and alcoholism in relatives. *Arch. Gen. Psychiatry*, 50: 952–960.

Kendler, K.S., McGuire, M., Gruenberg, A.M. and Walsh, D. (1995) Examining the validity of DSM-III-R schizoaffective disorder and its putative subtypes in the Roscommon Family Study. *Am. J. Psychiatry*, 152: 755–764.

Lander, E. and Kruglyak, L. (1995) Genetic dissection of complex traits: guidelines for interpreting and reporting linkage findings. *Nat. Genet.*, 11: 241–247.

Lander, E.S. and Botstein, D. (1986) Strategies for studying heterogenous genetic traits in humans by using a linkage map of restriction fragment length polymorphisms. *Proc. Nat. Acad. Sci. USA*, 83: 7353–7357.

Le Couteur, A., Bailey, A., Goode, S., Pickles, A., Robertson, S., Gottesman, I. and Rutter, M. (1996). A broader phenotype of autism: the clinical spectrum in twins. *J. Child Psychol. Psychiatry Allied Disciplines*, ???.

Leckman, J.F., Sholomskas, D., Thompson, W.D., Belanger, A. and Weissman, M.M. (1982) Best estimate of lifetime psychiatric diagnosis: a methodological study. *Arch. Gen. Psychiatry*, 39: 879–883.

Levinson, D.F. (1993) Power to detect linkage with heterogeneity in samples of small nuclear families. *Am. J. Med. Genet. (Neuropsychiatric Genetics)*, 48: 94–102.

Levinson, D.F. (1995) Detection of vulnerability loci by association and sib-pair methods. *Genet. Epidemiol.*, 12: 631–635.

Levinson, D.F. (1997a) Linkage analysis of complex disorders with multiple phenotypic categories: Simulation studies and application to bipolar disorder data. *Genet. Epidemiol.*, 14: 653–658.

Levinson, D.F. (1997b) Pragmatics and statistics in psychiatric genetics. *Am. J. Med. Genet.*, 74: 220–222.

Levinson, D.F. and Coon, H. (1998) Chromosome 22 workshop. *Psychiatr. Genet.*, 8: 115–120.

Levinson, D.F., Mahtani, M.M., Nancarrow, D.J., Brown, D.M., Kruglyak, L., Kirby, A., Hayward, N.K., Crowe, R.R., Andreasen, N.C., Black, D.W., Silverman, J.M., Endicott, J., Sharpe, L., Mohs, R.C., Siever, L.J., Walters, M.K., Lennon, D.P., Jones, H.L., Nertney, D.A., Daly, M.J., Gladis, M. and Mowry, B.J. (1998) Genome scan of schizophrenia. *Am. J. Psychiatry*, 155: 741–750.

Levinson, D.F. and Mowry, B.M. (1991) Defining the schizophrenia spectrum: Issues for genetic linkage studies. *Schizophr. Bull.*, 17: 491–514.

Levinson, D.F. (1995) Detection of vulnerability loci by association and sib pair methods. *Genet. Epidemiol.*, 12: 631-635.

Levy, D.L., Holzman, P.S., Matthysse, S. and Mendell, N.R (1994) Eye tracking and schizophrenia: a selective review. *Schizophr. Bull.*, 20: 47–62.

Lewis, C.M., Wise, L.H. and Lanchbury, J.S. (1998) Meta-analysis of genome search results. *Am. J. Hum. Genet.*, 63: A235 (abstract).

Lin, M.W., Curtis, D., Williams, N., Arranz, M., Nanko, S., Collier, D., McGuffin, P., Murray, R., Owen, M., Gill, M., et al. (1995) Suggestive evidence for linkage of schizophrenia to markers on chromosome 13q14.1-q32. *Psychiatr. Genet.*, 5: 117–126.

Maier, W., Lichtermann, D., Minges, J., Hallmayer, J., Heun, R., Benkert, O. and Levinson, D.F. (1993a) Continuity and discontinuity of affective disorders and schizophrenia. Results of a controlled family study. *Arch. Gen. Psychiatry*, 50: 871–883.

Maier, W., Lichtermann, D., Minges, J., Oehrlein, A. and Franke, P. (1993b) A controlled family study in panic disorder. *J. Psychiat. Res.*, 27 Suppl 1: 79–87.

Malafosse, A., Leboyer, M., d'Amato, T., Amadeo, S., Abbar, M., Campion, D., Canseil, O., Castelnau, D., Gheysen, F., Granger, B., Henrikson, B., Poirier, M.F., Sabate, O., Samolyk, D., Feingold, J. and Mallet, J. (1997) Manic depressive illness and tyrosine hydroxylase gene: linkage heterogeneity and association. *Neurobiology of Disease*, 4: 337–349.

Martinez, M., Goldin, L.R., Cao, Q., Zhang, J., Sanders, A.R., Nancarrow, D.J., Taylor, J.M., Levinson, D.F., Kirby, A., Crowe, R.R., Andreasen, N.C., Black, D.W., Silverman, J.M., Lennon, D.P., Nertney, D.A., Brown, D.M., Mowry, BJ, Gershon, E.S., Gejman, P.V. Follow-up study on a susceptibility locus for schizophrenia on chromosome 6q. *Am. J. Med. Genet.* (*Neuropsychiatric Genetics*) in press.

McCarthy, M.I., Kruglyak, L. and Lander, E.S. (1998) Sib-pair collection strategies for complex diseases. *Genet. Epidemiol.*, 15: 317–340.

McGuffin, P., Sargeant, M., Hetti, G., Tidmarsh, S., Whatley, S. and Marchbanks, R.M. (1990) Exclusion of a schizophrenia susceptibility gene from the chromosome 5q11.3 region: new data and a reanalysis of previous reports. *Am. J. Hum. Genet.*, 47: 524–535.

McInnes, L.A., Escamilla, M.A., Service, S.K., Reus, V.I., Leon, P., SIlva, S., Rojas, E., Spesny, M., Baharloo, S., Blankenship, K., Peterson, A., Tyler, D., Shimayoshi, N., Tobey, C., Batki, S., Vinogradov, S., Meza, L., Gallegos, A., Fournier, E., Smith, L.B., Barondes, S.H., Sandkuijl, L.A. and Freimer, N.B. (1996) A complete genome screen for genes predisposing to severe bipolar disorder in two Costa Rican pedigrees. *Proc. Nat. Acad. Sci. USA*, 93: 13060–13065.

McMahon, F.J., Hopkins, P.J., Xu, J., McInnis, M.G., Shaw, S., Cardon, L., Simpson, S.G., MacKinnon, D.F., Stine, O.C., Sherrington, R., Meyers, D.A. and DePaulo, J.R. (1997) Linkage of bipolar affective disorder to chromosome 18 markers in a new pedigree series. *Am J. Hum. Genet.*, 61: 1397–1404.

Melo, J.A., Shendure, J., Pociask, K. and Silver, L.M. (1996) Identification of sex-specific quantitative trait loci controlling alcohol preference in C57BL/6 mice. *Nat. Genet.*, 13: 147–153.

Mendelbaum, K., Sevy, S., Souery, D., Papadimitriou, G.N., De bruyn, Raeymaekers, P., Van Broeckhoven, C. and Mendlewicz, J. (1995). Manic-depressive illness and linkage analysis in the Xq27-Xq28 region of chromosome X. *Neuropsychobiology*, 31: 58–63.

Moises, H.W., Yang, L., Kristbjarnarson, H., Wiese, C., Byerley, W., Macciardi, F., Arolt, V., Blackwood, D., Liu, X., Sjogren, B., Aschauer, H.N., Hwu, H.-G., Jang, K., Livesley, W.J., Kennedy, J.L., Zoega, T., Ivarsson, O., Bui, M.-T., Yu, M.-H., Havsteen, B., Commenges, D., Weissenbach, J., Schwinger, E., Gottesman, I.I., Pakstis, A.J., Wetterberg, L., Kidd, K.K. and Helgason, T. (1995) An international two-stage genome-wide search for schizophrenia susceptibility genes. *Nat. Genet.*, 11: 321–324.

Moldin, S.O., Reich, T. and Rice, J.P. (1991) Current perspectives on the genetics of unipolar depression. *Behav. Genet.*, 21: 211–242.

Muir, W.J., Gosden, C.M., Brookes, A.J., Fantes, J., Evans, K.L., Maguire, S.M., Stevenson, B., Boyle, S., Blackwood, D.H., St. Clair, D.M., Porteous, D.J. and Weith, A. (1995) Direct microdissection and microcloning of a translocation breakpoint region, t(1;11)(q42.2;q21), associated with schizophrenia. *Cytogenet. Cell Genet.*, 70: 35–40.

Nikali, K., Suomalainen, A., Terwilliger, J., Koskinen, T., Weissenbach, J. and Peltonen, L. (1995) Random search for shared chromosomal regions in four affected individuals: the assignment of a new hereditary ataxia locus. *Am. J. Hum. Genet.*, 56: 1088–1095.

Nurnberger Jr, J.I., Blehar, M.C., Kaufmann, C.A., York-Cooler, C., Simpson, S.G., Harkavy-Friedman, J., Severe, J.B., Malaspina, D. and Reich, T. (1994) Diagnostic interview for genetic studies. Rationale, unique features, and training. *Arch. Gen. Psychiatry*, 51: 849–859.

Nurnberger Jr, J.I. and Berrettini, W. (1998) *Psychiatric Genetics*. Chapman and Hall Medical, London.

Ott, J. (1990) Genetic linkage and complex diseases: a comment. *Genet. Epidemiol.*, 7: 35–36.

Pauls, D.L., Gerhard, D.S., Lacy, L.G., Hostetter, A.M., Allen, C.R., Bland, S.D., LaBuda, M.C. and Egeland, J.A. (1991) Linkage of bipolar affective disorders to markers on chromosome 11p is excluded in a second lateral extension of Amish pedigree 110. *Genomics*, 11: 730–736.

Perna, G., Caldirola, D., Arancio, C. and Bellodi, L. (1997). Panic attacks: a twin study. *Psychiat. Res.*, 66: 69–71.

Philibert, R.A., St. Jean, P.L., Schork, N.J., Anderson, M.C., Dalwadi, H., Damschroder-Williams, P.J., Doherty, A.E., Dymarskaia, I., Ehlert, M.A., Galdzicka, M., Lau, E., Long, R.T., Patel, A., Paul, S.P., Remortel, B.G., Stubblefield, B.K., Martin, B.M., Allen, C.R., Pauls, D.L., Elston, R.C., Egeland, J.A., Paul, S.M. and Ginns, E.I. (1997) A genome-wide search for chromosome regions linked to bipolar affective disorder in the Old Ordre Amish identifies several candidate loci. *Am. J. Hum. Genet.* (*Neuropsychiatric Genetics*) 74: 585–586 (abstract).

Piven, J., Palmer, P., Landa, R., Santangelo, S., Jacobi, D. and Childress, D. (1997) Personality and language characteristics in parents from multiple-incidence autism families. *Am. J. Med. Genet.*, 74: 398–411.

Polymeropoulos, M.H. and Schaffer, A.A. (1996) Scanning the genome with 1772 microsatellite markets in search of a bipolar disorder susceptibility gene. *Mol. Psychiatry*, 1: 404–407.

Puffenberger, E.G., Hosoda, K., Washington, S.S., Nakao, K., deWit, D., Yanagisawa, M. and Chakravarti, A. (1994) A missense mutation of the endothelin-b receptor gene in multigenic Hirschsprung's disease. *Cell*, 79: 1257–1266.

Rice, J.P., Goate, A., Williams, J.T., Bierut, L., Dorr, D., Wu, W., Shears, S., Gopalakrishnan, G., Edenberg, H.J., Foroud, T., Nurnberger, J., Gershon, E.S., Detera-Wadleigh, S.D., Goldin, L.R., Guroff, J.J., McMahon, F.J., Simpson, S., MacKinnon, D., McInnis, M., Stine, O.C., DePaulo, J.R., Blehar, M.C. and Reich, T. (1997) Initial genome scan of the NIMH genetics initiative bipolar pedigrees: chromosomes 1, 6, 8, 10, and 12. *Am. J. Med. Genet.*, 74: 247–253.

Risch, N. (1990a) Linkage strategies for genetically complex traits I. Multilocus models. *Am. J. Hum. Genet.*, 46: 222–228.

Risch, N. (1990b) Genetic linkage and complex diseases, with special reference to psychiatric disorders, *Genet. Epidemiol.*, 7: 3–16.

Sanda, A.I. and Ford, J.P. (1986) Genomic analysis I: inheritance units and genetic selection in the rapid discovery of locus linked DNA markers. *Nucl. Acids Res.*, 14: 7265–7283.

Schizophrenia Linkage Collaborative Group for Chromosomes 3, 6 and 8 (1996) Additional support for schizophrenia linkage on chromosomes 6 and 8: a multicenter study. *Am. J. Med. Genet.*, 67: 580–59.

Schwab, S.G., Albus, M., Hallmayer, J., Honig, S., Borrmann, M., Lichtermann, D., Ebstein, R.P., Ackenheil, M., Lerer, B., Risch, N., et al. (1995) Evaluation of a susceptibility gene for schizophrenia on chromosome 6p by multipoint affected sib-pair linkage analysis. *Nat. Genet.*, 11: 325–327.

Schwab, S.G., Eckstein, G.N., Hallmayer, J., Lerer, B., Albus, M., Borrmann, M., Lichtermann, D., Ertl, M.A., Maier, W. and Wildenauer, D.B. (1997) Evidence suggestive of a locus on chromosome 5q31 contributing to susceptibility for schizophrenia in German and Israeli families by multipoint affected sib-pair linkage analysis. *Mol. Psychiatry*, 2: 156–160.

Schwab, S.G., Hallmayer, J., Albus, M., Lerer, B., Hanses, C., Kanyas, K., Segman, R., Borrman, M., Dreikon, B., Lichtermann, D., Rietschel, M., Trixler, M., Maier, W. and Wildenauer, D.B. (1998) Further evidence for a susceptibility locus on chromosome 10p14-p11 in 72 families with schizophrenia by nonparametric linkage analysis. *Am. J. Med. Genet.*, 81: 302–307.

Shaw, S.H., Kelly, M., Smith, A.B., Shields, G., Hopkins, P.J., Loftus, J., Laval, S.H., Vita, A., De Hert, M., Cardon, L., Crow, T.J., Sherrington, R. and DeLisi, L.E. A genome-side search for schizophrenia susceptibility genes. *Am. J. Med. Genet.*, in press.

Sherrington, R., Brynjolfsson, J., Petursson, H., Potter, M., Dudleston, K., Barraclough, B., Wasmuth, J., Dobbs, M. and Gurling, H. (1988) Localization of a susceptibility locus for schizophrenia on chromosome 5. *Nature*, 336: 164–167.

Smalley, S.L. (1991) *Genetic influences in autism*, Psychiatric Clinics of North America, 14: 125–139.

Smyth, C., Kalsi, G., Curtis, D., Brynjolfsson, J., O'Neill, J., Rifkin, L., Moloney, E., Murphy, P., Petursson, H. and Gurling, H. (1997) Two-locus admixture linkage analysis of bipolar and unipolar affective disorder supports the presence of susceptibility loci on chromosomes 11p15 and 21q22. *Genomics*, 39: 271–278.

Spence, M.A., Bishop, D.T., Boehnke, M., Elston, R.C., Falk, C., Hodge, S.E., Ott, J., Rice, J., Merikangas, K. and Kupfer, D. (1993) Methodological issues in linkage analyses for psychiatric disorders: secular trends, assortative mating, bilineal pedigrees. Report on the MacArthur Foundation Network I Task Force on Methodological Issues. *Hum. Heredity*, 43: 166–172.

Spielman, R.S., McGinnis, R.E. and Ewens, W.J. (1993) Transmission test for linkage disequilibrium: the insulin gene region and insulin-dependent diabetes mellitus (IDDM). *Am. J. Hum. Genet.*, 52: 506–516.

Spielman, R.S. and Ewens, W.J. (1996) The TDT and other family-based tests for linkage disequilibrium and association. *Am. J. Hum. Genet.*, 59: 983–989.

Spielman, R.S. and Ewens, W.J. (1998) A sibship test for linkage in the presence of association: the sib transmission/disequilibrium test. *Am. J. Hum. Genet.*, 62: 450–458.

Spitzer, R.L., Endicott, J. and Robins, E. (1978) Research Diagnostic Criteria: rationale and reliability. *Arch. Gen. Psychiatry*, 35: 773–782.

Stine, O.C., Xu, J., Koskela, R., McMahon, F.J., Gschwend, M., Friddle, C., Clark, C.D., McInnis, M.G., Simpson, S.G., Breschel, T.S. et al (1995) Evidence for linkage of bipolar disorder to chromosome 18 with a parent-of-origin effect. *Am. J. Hum. Genet.*, 57: 1384–1394.

Stine, O.C., McMahon, F.J., Chen, L., Xu, J., Meyers, D.A., MacKinnon, D.F., Simpson, S., McInnis, M.G., Rice, J.P., Goate, A., Reich, T., Edenberg, H.J., Foroud, T., Nurnberger, J.I., Detera-Wadleigh, S.D., Goldin, L.R., Guroff, J., Gershon, E.S., Blehar, M.C. and DePaulo, J.R. (1997) Initial genome screen for bipolar disorder in the NIMH genetics initiative pedigrees: chromosomes 2, 11, 13, 14, and X. *Am. J. Med. Genet.*, 74: 263–269.

Straub, R.E., Lehner, T., Luo, Y., Loth, J.E., Shao, W., Sharp, L., Alexander, J.R., Das, K., Simon, R., Fieve, R.R. et al (1994) A possible vulnerability locus for bipolar affective disorder on chromosome 21q22.3. *Nat. Genet.*, 8: 291–296.

Straub, R.E., MacLean, C.J., Martin, R.B., Myakishev, M.V., Harris, rr.C., O'Neill, F.A., Walsh, D. and Kendler, K.S. (1998) A schizophrenia locus may be located in region 10p15.1. *Am. J. Med. Genet.*, (abstract), 74: 559.

Straub, R.E., MacLean, C.J., O'Neill, F.A., Walsh, D. and Kendler, K.S. (1997b) Support for a possible schizophrenia vulnerability locus in region 5q22-31 in Irish families. *Mol. Psychiatry*, 2: 148–155.

Suarez, B.K., Hampe, C.L. and Van Eerdewegh, P.V. (1994) Problems of replicating linkage claims in psychiatry. In: E.S. Gershon and C.R. Cloninger (Eds.), *Genetic Approaches to Mental Disorders*, Washington, DC: American Psychiatric Press, pp. 23–46.

te Meerman, G.J. and Van der Meulen, M.A. (1997) Genomic sharing surrounding alleles identical by descent: effects of genetic drift and population growth. *Genet. Epidemiol.*, 14: 1125–1130.

Terwilliger, J.D. (1995) A powerful likelihood method for the analysis of linkage disequilibrium between trait loci and one or more polymorphic marker loci. *Am. J. Hum. Genet.*, 56: 777–787.

Terwilliger, J.D. and Ott, J. (1994) *Handbook of Human Genetic Linkage*, Johns Hopkins, Baltimore.

Terwilliger, J.D., Shannon, W.D., Lathrop, G.M., Nolan, J.P., Goldin, L.R., Chase, G.A. and Weeks, D.E. (1997) True and false positive peaks in genomewide scans: applications of length-biased sampling to linkage mapping. *Am. J. Hum. Genet.*, 61: 430–438.

Terwilliger, J.D., Zöllner, S., Laan, M. and Pääbo, S. (1998) Mapping genes through the use of linkage disequilibrium generated by genetic drift: "draft mapping" in small populations with no demographic expansion. *Hum. Heredity*, 48: 138–154.

Tsuang, M.T., Gilbertson, M.W. and Faraone, S.V. (1991) The genetics of schizophrenia: current knowledge and future directions. *Schizophr. Res.*, 4: 157–171.

Tsuang, M.T., Faraone, S.V. and Green, R.R. (1994) Genetic epidemiology of mood disorders. In D.F. Papolos and H.M. Lachman (Eds), *Genetic Studies in Affective Disorders*, New York, Wiley.

Weiss, K.M. (1993) *Genetic Variation and Human Disease*, Cambridge, England, Cambridge University Press.

Weissman, M.M. (1993) Family genetic studies of panic disorder. *J. Psychiat. Res.*, 27 Suppl 1: 69–78.

Weissman, M.M., Wickramaratne, P., Adams, P.B., Lish, J.D., Howrath, E., Charney, D., Woods, S.W., Leeman, E. and Frosch, E. (1993) The relationship between panic disorder and major depression. A new family study. *Arch. Gen. Psychiatry*, 50: 767–780.

Wildenauer, D.B., Albus, M., Schwab, S.G., Hallmayer, J., Hanses, C., Eckstein, G.N., Zill, P., Hönig, S., Lerer, B., Ebstein, R., Lichtermann, D., Trixler, M., Borrmann, M. and Maier, W. (1997) Searching for susceptibility genes in schizophrenia by affected sib-pair analysis (Germany). *Am. J. Med. Genet.* (abstract), 74: 558.

Williams, J.B.W., Gibbon, M., First, M.B., Spitzer, R.L., Davies, M., Borus, J., Howes, M.J., Kane, J., Pope, H.G. Jr, Rounsaville, B. and Wittchen, H.-U. (1992) The Structured Clinical Interview for DSM-III-R (SCID): II. Multisite test-retest reliability. *Arch. Gen. Psychiatry*, 78: 630–636.

Williams, N.M., Rees, M.I., Holmans, P., Daniels, J., Fenton, I., Cardno, A.G., Murphy, K.C., Jones, L.A., Asherson, P., McGuffin, P. and Owen, M.J. (1997) Genome search for schizophrenia susceptibility genes using a two-stage sib-pair approach. *Am. J. Med. Genet.* (abstract), 74: 559.

Wing, J.K., Babor, T., Brugha, T., Burke, J., Cooper, J.E., Giel, R., Jablenski, A., Regier, D. and Sartorius, N. (1990). SCAN. Schedules for Clinical Assessment in Neuropsychiatry. *Arch. Gen. Psychiatry*, 47: 589–93.

World Health Organisation (1992) *The ICD-10 Classification of Mental and Behavioural Disorders: Diagnostic criteria for research*, World Health Organisation, Geneva.

Wyszynski, D.F., Doetsch, J.P., Pugh, E.W. and Bailey-Wilson, J.E. (1997) Comparison of selected methods used to analyze bipolar disorder. *Genet. Epidemiol.*, 14: 705–710.

W.E. Crusio and R.T. Gerlai (Eds.)
Handbook of Molecular-Genetic Techniques for Brain and Behavior Research (Techniques in the Behavioral and Neural Sciences, Vol. 13)

CHAPTER 5.2

Defining phenotypes for psychiatric genetics

Marika Nosten-Bertrand,[1] Frank Bellivier[2] and Marion Leboyer[2–3]

[1]*Chargée de Recherche au CNRS, INSERM U 513, IM3, Faculté de Médecine, 8 Rue du Général Sarrail, 94000 Créteil Cedex, France*
[2]*Henri Mondor Hospital, Service de Psychiatrie Adulte, Avenue du Maréchal de Lattre de Tassigny, 94010 Créteil Cedex, France*
[3]*Albert Chenevier Hospital, Service de Psychiatrie Adulte, 40 Rue de Mesly, 94010 Créteil Cedex, France*

Introduction

Despite the astonishing progress made in molecular and epidemiologic genetics, and the long history of the ever growing field of psychiatric genetics, we still have very little, if any, insight into the nature of the genetic and environmental contributions to psychiatric disorders and how they may interact to trigger a predisposition to a particular illness. The early enthusiasm about the role that molecular genetics would play in our understanding of mental disorders was tempered in the late-1980s by failure to replicate previously observed linkage for bipolar affective disorder and schizophrenia (Risch and Botstein, 1996), and with the subsequent withdrawal of the original findings (Kelsoe et al., 1989).

Since the earliest stages of psychiatric genetics, epidemiological genetics has pointed out that phenotypic uncertainties in classical psychiatric nosology could hamper progress in psychiatric genetics (Gottesman and Shields, 1972; Kidd and Matthysee, 1978; Kendell, 1989). Despite their warning, to date classification of psychiatric illnesses in genetic studies still relies entirely on classical categorical diagnosis, based on clinical symptoms.

As already demonstrated for other complex disorders, narrowing the definition of a disease can help identification of traits that are appropriate for genetic analysis. Failure to understand the genetic basis of psychiatric disorders may thus result in part to the inadequate definition of the phenotypes used in genetic studies.

In this chapter, we discuss how phenotypic uncertainties do contribute to the current failure of genetic psychiatric studies and describe how a combination of strategies from clinical and epidemiological genetics may contribute to identify phenotypes more relevant for genetic studies of psychiatric disorders.

The different levels of phenotypic uncertainties

Without biological markers for psychiatric disorders, case definition rely on clinically defined categories. Classification criteria in psychiatry allow reliable standard diagnostic to be used in linkage studies on a world wide (Merikangas et al., 1989). This approach was based on the demonstrated inter-rater reliability of the assessment procedures (Nurnberger et al., 1994). However, it is demonstrated that giving priority to the reliability of a diagnosis might decrease its validity (Carey and Gottesman, 1978) and, to date, the genetic validity of the currently used diagnostic entities in psychiatry remains a moot point (Kendler, 1990).

Moreover, within each diagnostic category, the remaining substantial clinical and familial heterogeneity has also cast doubt on the biological validity of the currently used diagnostic criteria. One observes heterogeneity in symptom manifestations, age and pattern of onset, comorbid features, treatment response, course and outcome. In addition, this categorical approach accounts neither for the continuum between normal and pathological status nor for the existence of a spectrum of diseases observed in first degree relatives of patients.

These uncertainties lead to various types of phenotypic misclassification known to produce spurious linkage and association results (Tsuang et al., 1993). Changes in diagnosis of only one or two individuals can shift the position of the maximum lod score by several centiMorgans or can drop it dramatically. Labeling family members as " unaffected ", when affected (false-negative), can reduce the apparent penetrance. Labeling family members as " affected ", when unaffected (false-positive), can mask the presence of linkage because a recombination appears when in fact there is none. The currently used criteria do not allow the distinction of unaffected subjects who are carriers of the predisposing loci, from unaffected subjects who are non-carriers. For example, in monozygotic twins discordant for schizophrenia, the risk of developing schizophrenia is the same for the offspring of affected and non-affected co-twins (Gottesman and Bertelsen, 1989). This renders the detection of linkage spurious and specific localization of susceptible genes almost impossible.

Thus the following question remains unanswered: Do the contemporary definitions of clinical syndromes (considered as phenotypes in psychiatric genetics) accurately reflect some underlying genetic substrates?

Epidemiological geneticists have advocated the use of a specific "psychiatric genetic nosology" that seeks to classify patients corresponding to distinct genetic entities (Kendler, 1990; Merikangas, 1995). They have put forth strategies aimed at improving diagnosis "accuracy" that evaluates the degree to which a phenotype correctly classifies people with and without a putative genetic illness (Tsuang et al., 1993). In the literature, various nomenclature were used for such refined phenotypes that could help to bridge the gap between genetic and psychiatry. In the following sections, we present these strategies, considering successively those that have focused on the description of affected individuals and those describing non-affected relatives of affected individuals. We have arbitrarily adopted the nomenclature of "candidate symptom" approach for the former and "endophenotype" approach for the latter (Leboyer et al., 1998b).

Candidate symptom approach

The candidate symptom approach aims at identifying homogenous subgroups among affected individuals showing phenotypic characteristics more likely to be associated with a genotype and showing a simpler pattern of inheritance. In that sense, this strategy assumes that current diagnostic criteria properly discriminates affected vs non-affected individuals and it follows the currently accepted view that heterogeneity at the clinical level could be related to an etiological heterogeneity.

To improve genetic analysis, a candidate symptom (Gottesman and Shields, 1972), also defined as a diagnostic phenotypic indicator (Tsuang, 1993), should show good concordance rate among affected monozygotic twins and should be correlated in pairs of affected siblings. The relevance for genetic studies of such candidate symptoms can be tested through various genetic epidemiologic techniques (Tsuang, 1993). As demonstrated by Risch (Risch, 1990), the extent to which restriction of the phenotype definition simplifies the task of genetic mapping can be measured by the resulting increase of the relative risk, λ_R, (defined by the risk for a relative of a patient divided by the risk in the general population).

Narrowing the definition of a phenotype by using characteristics such as severity, age at onset, associated symptoms or family history have already been fruitful in genetic studies of complex somatic diseases such as colon cancer (Kinzler et al., 1991), hypertension (Williams et al., 1990) or coronary heart disease (Cambien et al., 1992; Stengard et al., 1996) and have significantly contributed to the identification of relevant genes. For example, subdivision according to age of onset and mode of inheritance has been particularly efficient to clarifying genetic heterogeneity in Alzheimer's diseases (Sandbrink et al., 1996).

In psychiatry, age of onset, severity and family history may also be helpful in the identification of more homogeneous subtypes. Early onset is associated with increased familial risk in schizophrenia (Sham et al., 1994), bipolar affective (Strober et al., 1988), major depressive disorder (Weissman et al., 1984) and obsessive compulsive disorder (Pauls et al., 1995). Age of onset in schizophrenia appears to be substantially influenced by familial factors, since correlation in age of onset in affected pairs of siblings ranges from 0.2 to 0.4 (Kendler et al., 1987; Leboyer et al., 1992) and those of monozygotic twins range from 0.5 to 0.8 (Kendler et al., 1987). Similarly, in bipolar disorder, a significant correlation for age of onset in affected siblings have been demonstrated (Leboyer et al., 1998a). Furthermore, the decrease of age of onset of schizophrenia in successive generations seems to be consistent with the phenomenon of genetic anticipation (Petronis and Kennedy, 1995). Of interest is that schizophrenic females with late onset and no family history of schizophrenia form a subgroup of patients that may share specific etiological factors (Gorwood et al., 1995; Seeman and Lang, 1990).

The occurrence of specific symptom constellation may also help in identifying a subgroup that may be etiologically homogeneous. Promising data have been obtained in psychopathology studies when investigations were shifted to each individual core compartment found to account for schizophrenia. For example, based on clinical, neuropsychological and neuro-imaging arguments, Carpenter has demonstrated that schizophrenic patients with primary deficit syndrome (Carpenter et al., 1988) constitute an homogeneous subgroup that may share common vulnerability factors (Carpenter et al., 1995). Similarly, schizophrenic patients with periodic catatonia was demonstrated to be a clinically homogenous subgroup with a homogeneous psychopathological picture in familial transmission: a high family morbidity risk with homogeneous psychosis in the first degree relatives (26.9%) and a pronounced vertical transmission with a significant pattern of anticipation (Stober et al., 1995). Such defined core compartment of schizophrenia is a promising candidate symptom for genetic studies (Stober et al., 1998).

In a family study, Eapen (Eapen et al., 1997) demonstrated that, among individuals with obsessive compulsive disorder (OCD), those who have aggressive and sexual obsessions, checking, compulsions of symmetry and exactness were more likely to have a positive family history of OCD than individuals with OCD characterized by contamination and compulsions. Thus, a clinical constellation of symptoms may help the identification of familial forms of the disease that could be selected for linkage studies.

To our knowledge, no linkage study using such candidate symptoms have been performed. By contrast, exploratory studies have reported association between candidate genes and subgroups of patients defined according to candidate symptoms. In bipolar depressive illness, polymorphism at the Apolipoprotein E gene was found to be associated with the early onset sub-group (Bellivier et al., 1997a) and the HUMTH01 polymorphism with the late onset (Bellivier et al., 1997b). Similarly, it has been suggested that the associations reported between polymorphism at the serotonin transporter gene with both unipolar and bipolar affective disorders was due to a subgroup of patients with suicidal behavior (Bellivier et al., in preparation). Interestingly, this association between the serotonin transporter gene and suicidal behavior was also observed in a sample of alcoholic patients (Gorwood et al., 1998). These preliminary data require caution in the interpretation in the context of strong methodological limitations (*a posteriori* studies and by chance findings) (Crowe, 1993). But prospective studies using such potential candidates should certainly be worth considering.

Endophenotype approach

Substantial evidence demonstrates that non-affected relatives of psychiatric patients manifest subtle biological and clinical differences when compared to control individuals. These subclinical traits, or endophenotypes, are neurophysiological, neuroanatomical, or neuropsychological variables that may reflect the genetic liability to a disorder in non-affected subjects. An endophenotype may be a variant expression of a disease genotype having pleiotropic effects. Or, it may be one of the several entities contributing to the disease, but is neither necessary nor sufficient for the disease outcome. In that case, an endophenotype might be valuable for identifying common alleles with non-specific and moderate effects on disease risk.

If such an endophenotype is a vulnerability trait for the illness, then identifying the genes that are important for the expression of this endophenotype will serve the purpose of identifying genes that also increase the susceptibility for the illness. One should not eliminate the possibility that an endophenotype may be associated with the illness because the responsible gene is in linkage disequilibrium with the disease gene. In that case, an endophenotype could help identify a candidate region for the location of the illness susceptibility loci.

Schizophrenia is the most widely studied psychiatric disorders for identification of potential endophenotypes. Group differences between relatives of schizophrenic patients and controls have been observed for many variables such as eye tracking dysfunction (Holzman et al., 1977; Iacono et al., 1981; Levy et al., 1994), attentional impairment (Cornblatt and Erlenmeyer-Kimling, 1985, Cornblatt, 1994), working memory (Park et al., 1995), neuropsychological impairments (Kremen et al., 1994), and auditory evoked potentials (Friedman and Squires-Wheeler, 1994). However, as demonstrated by epidemiologic genetic studies (Faraone et al., 1996), variables that significantly differ between relatives of patients and normal subjects do not necessarily constitute good potential endophenotypes: as for candidate symptoms, the extent to which an endophenotype will improve the power of the linkage study depends on its relative risk (Lander and Schork, 1994).

In an exploratory experiment, 10 neuropsychological functions, previously observed to significantly differ between relatives of schizophrenic patients and controls, were tested (Faraone et al., 1995). Three functions — abstraction, verbal memory and auditory attention — were found to meet criteria for optimal diagnostic accuracy and could thus constituted good endophenotypes for linkage analyses. In an expanded sample (Toomey et al., 1998), they duplicated these preliminary findings and illustrated how epidemiological genetic tools could be used to combine multiple endophenotypes for better identification of non-affected carriers.

It has long been observed that neuronal mechanisms responsible for filtering of sensory input were impaired in schizophrenic patients and several electrophysiological paradigms have also demonstrated brain function deficits in relatives of schizophrenic patients. Using the obligate carrier strategy, Waldo (Waldo et al., 1995) studied sensory gating deficits in schizophrenic pedigrees using the P50 auditory evoked response to repeated stimuli. All parents with a positive family history of schizophrenia had a deficit of the P50, whereas such a deficit was only observed in 7% of parents with no family history. Using the P50 auditory-evoked response as an endophenotype, a genome-wide linkage analysis was carried-out in nine families with multiple cases of schizophrenia (Freedman et al., 1997). The P50 deficit was found to be strongly linked to the α7-nicotinic receptor locus at chromosome

15q13-14. This study illustrates the value of the endophenotype strategy to provide possible insight, not only into the genetic risk but also into the underlying neuronal mechanisms of complex disorders.

Similarly, there were several lines of evidence that eye movement dysfunctions (EMDs) constitute a potential good endophenotype for schizophrenia (Levy et al., 1994). Whereas 51–85% of schizophrenic patients exhibit EMDs, a higher prevalence is observed in non-affected relatives of schizophrenic (45%) compared to the prevalence in the general population (8%) (Blackwood et al., 1991). Moreover, in twin studies discordant for schizophrenia, monozygotic twins were highly and significantly more concordant for EMDs ($r = 0.77$) than dizygotic twins ($r = 0.39$) (Holzman et al., 1980). In 1988, Holzman (Holzman et al., 1988) suggested that, in offsprings of monozygotic and dizygotic twins discordant for schizophrenia, a single autosomal dominant gene could account for EMDs and schizophrenia. Despite these strong arguments, it was only in 1996 that eye tracking dysfunction was used as an endophenotype in a linkage analysis (Arolt et al., 1996). EMDs were found to map to a chromosome 6 locus, closed to regions previously highlighted in several linkage studies of schizophrenia. Even if such preliminary data require replication on larger sample size, this study provides encouraging support for shifting genetic analysis of psychiatric disorders to refined phenotypes having a better diagnostic accuracy.

In order to overcome the absence of clear cut boundaries within the spectrum of a categorical disorder as well as between affected and non-affected individuals, the use of dimensional traits have been advocated to increase power of linkage studies (Risch, 1990; Tsuang et al., 1993). Following quantitative genetics (Falconer, 1965), several epidemiologic genetic strategies have been developed to test for the efficiency and informativeness of dimensional traits. One example is given by the concept of caseness developed to account for a dimensional measure of the degree to which a subject is truly ill. In other words, subjects are assigned weights (or index of caseness) indicating the probability that they are affected (Ott, 1991). The main advantage of such an index for psychiatric genetics is that it can be derived from multiple sources of diagnostic and neurobiological data thus taking into account several aspects of the disorder. Using these methods, phenotypes have been derived for depression (Blacker et al., 1993) and attention deficit disorder (Faraone et al., 1993) that were demonstrated to be more familial and more informative for genetic analysis. The extent to which a quantitative phenotype may increased the power of linkage analysis over categorical phenotype is illustrated in a recently published study (Brzustowicz et al., 1997). Schizophrenic families and sib-pairs were assessed using both diagnostic categories and quantitative-trait measures of positive and negative symptoms. While both parametric and sib-pair analysis failed to produce significant evidence of linkage with categorical disease definitions, they found significant evidence of linkage of positive symptom score to chromosome 6p. Similarly, a quantitative linkage analysis was carried out using the P3 event-related brain potential (ERP) as an endophenotype to identify individuals at genetic risk for alcoholism (Begleiter et al., 1998). The analysis yielded significant linkage on chromosomes 2 and 6 that point to possible candidate loci underlying a functional neuro-electrophysiological activity. These preliminary data was obtained in the context of a large scale and multidisciplinary research program initiated to investigate genetic susceptibility to alcohol abuse and dependence (Collaborative study of the Genetics of Alcoholism, COGA). COGA is unique in the effort that has been put forth to assess probands and their relatives for clinical and neurophysiological phenotypes (Begleiter, 1995). COGA's potential will certainly be a major contribution to better identify phenotypes in genetic studies of complex disorders.

In childhood psychiatry, the concordance rate for certain cognitive and language disorders was found to be higher among monozygotic autistic twins (82%) than among dizygotic twins (10%), thus suggesting that investigation of familial aggregation of a broader range of cognitive disorders was needed to better define the phenotype in autism (Folstein and Rutter, 1977). Since then, observation of the relatives of autistic children has pointed to cognitive, social, language and biochemical phenotypes that might also represent the variable expression of susceptibility genes to autism (Bailey et al., 1998).

As a conclusion, the joint use of candidate symptom and endophenotype approaches has been successful in the genetic analysis of epilepsy, for which a positive linkage to chromosome 6 was recently observed: a highly specific group of patients with the juvenile myoclonic form of epilepsy (candidate symptom) was first identified, and all pedigree members were then investigated for abnormal EEG used as a sub-clinical marker (endophenotype) (Greenberg et al., 1988). Epilepsy is a common disease for which the difficulties of nosology and phenotypic definition are similar to those found in psychiatric diseases. This result thus provides an encouraging illustration of the extent to which reshaped phenotypes can simplify the task of genetic analysis.

Perspectives

To view the clinical heterogeneity as an indicator of etiologic heterogeneity is the topic of a long lasting debate — the arguments of which are beyond the topic of this chapter (DeLisi, 1995; Gottesman, 1994). However, as it has been demonstrated for Alzheimer's diseases (Sandbrink et al., 1996), the genetic heterogeneity of a given clinical entity, even with a well described biological marker, can go from the most Mendelian form, up to polygenic forms involving multiple susceptibility genes. Reversibly, as demonstrated for the Von Recklinghausen's disease (Hinrichs et al., 1987), the phenotypic expression of a punctual mutation at a single locus can be unexpectedly diverse (pleiotropy). The strategies advocated here do not preclude the nature and the number of susceptibility factors, neither do they prejudge for or against heterogeneity in the etiology of psychiatric disorders.

Reshaping phenotypic definition will stimulate the emergence of a different model for psychiatric disorders. In particular, transnosographical studies may also lead to the identification of relevant phenotypes for genetic studies. For example, accumulating data have shown that serotonin neurotransmission is strongly associated with suicidal behavior, regardless of the associated psychiatric disorder (Mann, 1998). The identification of genetic susceptibility factors for phenotypes that are common of several mental disorders (transnosographical) would not only improve our understanding of clinical spectrum, but may also provide a clue to underlying neural mechanisms (Andreasen, 1997).

The debate on phenotypic definition for psychiatric genetics relaunches the confusing question of clinical validity (Carey and Gottesman, 1978). One might argue that the use of a specific nosology advocated here for psychiatric genetics might lead to the identification of relevant phenotypes to simplify the task of the geneticist but not that of the therapist. For example, age of onset might have good genetic validity but not a good treatment validity. It is because the elucidation of underlying mechanisms is often the best way of predicting, treating and preventing a disorder that, in modern medicine, the most valid diagnostics are those whose etiology is known. Validity, however, is not an absolute quality (Kendell, 1989). And if elucidation of etiology contributes to bridging the gap between fundamental and applied sciences, the strategies and means applied to attain this goal may well differ.

Disentangling classical nosographical entities and focusing on vulnerability traits should stimulate the development of useful animal models, in order to examine etiological factors of these associated phenotypes. For example,

Swerdlow (Swerdlow et al., 1994) demonstrated that prepulse inhibition in rat can be a useful animal model — with good face, predictive, and construct validity — for studying the neurobiology of impaired sensorimotor gating in schizophrenia. Goldman-Rakic (Goldman-Rakic, 1994) has proposed a model suggesting that working memory is the fundamental process impaired in schizophrenia. The large set of developmental, anatomical, physiological and behavioral experiments developed in nonhuman primates has already yielded fruitful information about the function of the prefrontal cortex, thus providing new insights into potential mechanisms involved in mental disorders. More recently, mapping quantitative trait loci (QTLs) has been at the forefront of significant progress in mouse models of polygenic traits such as substance abuse (Berrettini et al., 1994; Melo et al., 1996) and emotionality (Flint et al., 1995) that could be relevant for human psychiatric genetics (Bedell et al., 1997). Such an opportunity is worth considering since mouse models still constitute the gold-standard for genetic studies to pinpoint epistasis as well as genetic and environmental interacting factors.

Conclusion

It has long been claimed that defining phenotypes is the rate-limiting step in psychiatric genetic studies. Developing a specific psychiatric nosology requires specific experimental designs — including multidisciplinary and prospective follow-up studies — to be implemented in research projects. In our point of view, despite the difficulty and expense involved, this approach is a *sine qua non* condition for progress in psychiatric genetics.

Acknowledgements

We wish to thank Dr Sabrina Davis for helpful discussion. Preparation of this chapter was supported by the Assistance Publique des Hôpitaux de Paris (Délégation à la Recherche Clinique, CRC), the CNRS, INSERM, AFM, Rhône-Poulenc Rorer and la Fondation pour la Recherche Médicale.

References

Andreasen, N.C. (1997) Linking mind and brain in the study of mental illnesses: a project for a scientific psychopathology. Science, 275: 1586–1593.

Arolt, V., Lencer, R., Nolte, A., Muller-Myhsok, B., Purmann, S., Schurmann, M., Leutelt, J., Pinnow, M. and Schwinger, E. (1996) Eye tracking dysfunction is a putative phenotypic susceptibility marker of schizophrenia and maps to a locus on chromosome 6p in families with multiple occurrence of the disease. Am. J. Med. Genet., 67: 564–579.

Bailey, A., Palferman, S., Heavey, L. and Le Couteur, A. (1998) Autism: the phenotype in relatives. J. Autism Dev. Disord., 28: 369–292.

Bedell, M.A., Largaespada, D.A., Jenkins, N.A. and Copeland, N.G. (1997) Mouse models of human disease. Part II: recent progress and future directions. Genes Dev., 11: 11–43.

Begleiter, H. (1995) The collaborative study on the genetics of alcoholism. *Alcohol Health and Research World*, 19–3: 228–236.

Begleiter, H., Porjesz, B., Reich, T., Edenberg, H.J., Goate, A., Blangero, J., Almasy, L., Foroud, T., Van Eerdewegh, P., Polich, J., Rohrbaugh, J., Kuperman, S., Bauer, L. O., O'Connor, S.J., Chorlian, D.B., Li, T.K., Conneally, P.M., Hesselbrock, V., Rice, J.P., Schuckit, M.A., Cloninger, R., Nurnberger Jr, J., Crowe, R. and Bloom, F.E. (1998) Quantitative trait loci analysis of human event-related brain potentials: P3 voltage. *Electroencephalogr. Clin. Neurophysiol.*, 108: 244–250.

Bellivier, F., Laplanche, J.L., Szöke, A., Cordebar, F., Henry, C., Nosten-Bertrand, M., Féline, A. and Leboyer, M. Association between serotonin transporter gene polymorphism and suicidal behavior in major affective disorders. *Mol. Psychiat.*, in preparation.

Bellivier, F., Laplanche, J.L., Schurhoff, F., Feingold, J., Feline, A., Jouvent, R., Launay, J.M. and Leboyer, M. (1997a) Apolipoprotein E gene polymorphism in early and late onset bipolar patients. *Neurosci. Lett.*, 233: 45–48.

Bellivier, F., Nosten-Bertrand, M., Leboyer, M., Schürhoff, F., Feingold, J., Meloni, R., Allilaire, J.F. and Mallet, J. (1997b) Association between late-onset bipolar affective disorder and tyrosine hydroxylase gene polymorphism. *Am. J. Med. Genet.*, 74: 614.

Berrettini, W.H., Ferraro, T.N., Alexander, R.C., Buchberg, A.M. and Vogel, W.H. (1994) Quantitative trait loci mapping of three loci controlling morphine preference using inbred mouse strains (see comments). *Nat. Genet.*, 7: 54–58.

Blacker, D., Lavori, P.W., Faraone, S.V. and Tsuang, M.T. (1993). Unipolar relatives in bipolar pedigrees: a search for indicators of underlying bipolarity. *Am. J. Med. Genet.*, 48: 192–199.

Blackwood, D., St, C.D., Muir, W. and Duffy, J. (1991) Auditory P300 and eye tracking dysfunction in schizophrenic pedigrees. *Arch. Gen. Psychiat.*, 48: 899–909.

Brzustowicz, L.M., Honer, W.G., Chow, E.W., Hogan, J., Hodgkinson, K. and Bassett, A.S. (1997) Use of a quantitative trait to map a locus associated with severity of positive symptoms in familial schizophrenia to chromosome 6p. *Am. J. Hum. Genet.*, 61: 1388–1396.

Cambien, F., Poirier, O., Lecerf, L., Evans, A., Cambou, J.P., Arveiler, D., Luc, G., Bard, J.M., Bara, L., Ricard, S. et al. (1992) Deletion polymorphism in the gene for angiotensin-converting enzyme is a potent risk factor for myocardial infarction (see comments). *Nature*, 359: 641–644.

Carey, G. and Gottesman, I.I. (1978) Reliability and validity in binary ratings: areas of common misunderstanding in diagnosis and symptom ratings. *Arch. Gen. Psychiat.*, 35: 1454–1459.

Carpenter Jr, W.T., Buchanan, R.W. and Kirkpatrick, B. (1995) New diagnostic issues in schizophrenic disorders. *Clin. Neurosci.*, 3: 57–63.

Carpenter Jr, W.T., Heinrichs, D.W. and Wagman, A.M. (1988). Deficit and nondeficit forms of schizophrenia: the concept. *Am. J. Psychiat.*, 145: 578–583.

Cornblatt, B.A. and Erlenmeyer-Kimling, L. (1985) Global attentional deviance as a marker of risk for schizophrenia: specificity and predictive validity. *J. Abnorm. Psychol.*, 94: 470–486.

Cornblatt, B.A. and Keilp, J.G. (1994) Impaired attention, genetics, and the pathophysiology of schizophrenia. *Schizophr Bull.*, 20, 31–46.

Crowe, R.R. (1993) Candidate genes in psychiatry: an epidemiological perspective (see comments). *Am. J. Med. Genet.*, 48: 74–77.

DeLisi, L.E. (1995) Current controversies in schizophrenia research. I. Is schizophrenia a heterogeneous disorder? *Schizophren. Res.*, 17: 133.

Eapen, V., Robertson, M.M., Alsobrook 2nd, J.P. and Pauls, D.L. (1997) Obsessive compulsive symptoms in Gilles de la Tourette syndrome and obsessive compulsive disorder: differences by diagnosis and family history. *Am. J. Med. Genet.*, 74: 432–438.

Falconer, D.S. (1965). The inheritance of liability to certain diseases, estimated from the incidence among relatives. Ann. Hum. Genet., 29: 51–76.

Faraone, S., Blehar, M., Pepple, J., Moldin, S., Norton, J., Nurnberger, J., Malaspina, D., Kaufmann, C., Reich, T., Cloninger, C., DePaulo, J., Berg, K., Gershon, E., Kirch, D. and Tsuang, M. (1996) Diagnostic accuracy and confusability analyses: an application to the Diagnostic Interview for Genetic Studies. *Psychol. Med.*, 26: 401–10.

Faraone, S.V., Biederman, J., Sprich-Buckminster, S., Chen, W. and Tsuang, M.T. (1993) Efficiency of diagnostic criteria for attention deficit disorder: toward an empirical approach to designing and validating diagnostic algorithms. *J. Am. Acad. Child Adolesc. Psychiat.*, 32: 166–174.

Faraone, S.V., Seidman, L.J., Kremen, W.S., Pepple, J.R., Lyons, M.J. and Tsuang, M.T. (1995) Neuropsychological functioning among the nonpsychotic relatives of schizophrenic patients: a diagnostic efficiency analysis. *J. Abnorm. Psychol.*, 104: 286–304.

Flint, J., Corley, R., DeFries, J.C., Fulker, D.W., Gray, J.A., Miller, S. and Collins, A.C. (1995) A simple genetic basis for a complex psychological trait in laboratory mice. *Science*, 269: 1432–1435.

Folstein, S. and Rutter, M. (1977) Infantile autism: a genetic study of 21 twin pairs. *J. Child Psychol. Psychiat.*, 18: 297–321.

Freedman, R., Coon, H., Myles-Worsley, M., Orr-Urtreger, A., Olincy, A., Davis, A., Polymeropoulos, M., Holik, J., Hopkins, J., Hoff, M., Rosenthal, J., Waldo, M.C., Reimherr, F., Wender, P., Yaw, J., Young, D.A., Breese, C.R., Adams, C., Patterson, D., Adler, L.E., Kruglyak, L., Leonard, S. and Byerley, W. (1997) Linkage of a neurophysiological deficit in schizophrenia to a chromosome 15 locus. *Proc. Natl. Acad, Sci, USA*, 94: 587–592.

Friedman, D. and Squires-Wheeler, E. (1994) Event-related potentials (ERPs) as indicators of risk for schizophrenia. *Schizophr. Bull.*, 20: 63–74.

Goldman-Rakic, P.S. (1994) Working memory dysfunction in schizophrenia. *J. Neuropsychiat. Clin. Neurosci.*, 6: 348–357.

Gorwood, P., Boni, C., Batel, P. and Adès, J. (1998) 5HTT genetic polymorphisms, alcoholism, and suicidal behavior. *Am. J. Med. Genet.*, 81: 507.

Gorwood, P., Leboyer, M., Jay, M., Payan, C. and Feingold, J. (1995) Gender and age at onset in schizophrenia: impact of family history. *Am. J. Psychiat.*, 152: 208–212.

Gottesman, I.I. and Bertelsen, A. (1989) Confirming unexpressed genotypes for schizophrenia. Risks in the offspring of Fischer's Danish identical and fraternal discordant twins. *Arch. Gen. Psychiat.*, 46: 867–872.

Gottesman, I. (1994) Complications to the complex inheritance of schizophrenia. *Clin. Genet.*, 46: 116–23.

Gottesman, I.I. and Shields, J. (1972) *Schizophrenia and genetics: a twin study vantage point*, New York: Academic Press.

Greenberg, D.A., Delgado-Escueta, A.V., Widelitz, H., Sparkes, R.S., Treiman, L., Maldonado, H.M., Park, M.S. and Terasaki, P.I. (1988) Juvenile myoclonic epilepsy (JME) may be linked to the BF and HLA loci on human chromosome 6. *Am. J. Med. Genet.*, 31: 185–192.

Hinrichs, S.H., Nerenberg, M., Reynolds, R.K., Khoury, G. and Jay, G. (1987) A transgenic mouse model for human neurofibromatosis. *Science*, 237: 1340–1343.

Holzman, P.S., Kringlen, E., Levy, D.L. and Haberman, S.J. (1980) Deviant eye tracking in twins discordant for psychosis. A replication. *Arch. Gen. Psychiat.*, 37: 627–631.

Holzman, P.S., Kringlen, E., Levy, D.L., Proctor, L.R., Haberman, S.J. and Yasillo, N.J. (1977) Abnormal-pursuit eye movements in schizophrenia. Evidence for a genetic indicator. *Arch. Gen. Psychiat.*, 34: 802–805.

Holzman, P.S., Kringlen, E., Matthysse, S., Flanagan, S.D., Lipton, R.B., Cramer, G., Levin, S., Lange, K. and Levy, D.L. (1988) A single dominant gene can account for eye tracking dysfunctions and schizophrenia in offspring of discordant twins (see comments). *Arch. Gen. Psychiat.*, 45: 641–647.

Iacono, W.G., Tuason, V.B. and Johnson, R.A. (1981) Dissociation of smooth-pursuit and saccadic eye tracking in remitted schizophrenics. An ocular reaction time task that schizophrenic perform well. *Arch. Gen. Psychiat.*, 38: 991–996.

Kelsoe, J., Ginns, E., Egeland, J., Gerhard, D., Goldstein, A., Bale, S., Pauls, D., Long, R., Kidd, K., Conte, G., Housman, D. and Paul, S. (1989) Re-evaluation of the linkage relationship between chromosome 11p loci and the gene for bipolar affective disorder in the Old Order Amish (see comments). *Nature*, 342: 238–243.

Kendell, R.E. (1989) Clinical validity. *Psychol. Med.*, 19: 45–55.

Kendler, K.S. (1990) Toward a scientific psychiatric nosology. Strengths and limitations. Arch. Gen. Psychiat., 47: 969–973.

Kendler, K.S., Tsuang, M.T. and Hays, P. (1987) Age at onset in schizophrenia. A familial perspective. Arch. Gen. Psychiat., 44: 881–890.

Kidd, K.K. and Matthysee, S. (1978) Research designs for the study of gene-environment interactions in psychiatric disorders. Report of a Foundations Fund for Research in Psychiatry Panel. *Arch. Gen. Psychiat.*, 35: 925–932.

Kinzler, K.W., Nilbert, M.C., Su, L.K., Vogelstein, B., Bryan, T.M., Levy, D.B., Smith, K.J., Preisinger, A.C., Hedge, P., McKechnie, D. et al. (1991) Identification of FAP locus genes from chromosome 5q21. *Science*, 253: 661–665.

Kremen, W.S., Seidman, L.J., Pepple, J.R., Lyons, M.J., Tsuang, M.T. and Faraone, S.V. (1994) Neuropsychological risk indicators for schizophrenia: a review of family studies. *Schizophr. Bull.*, 20: 103–119.

Lander, E.S. and Schork, N.J. (1994) Genetic dissection of complex traits (published erratum appears in *Science*, 21 Oct. 1994; 266(5184): 353). *Science*, 265: 2037–2048.

Leboyer, M., Bellivier, F., McKeon, P., Albus, M., Borrman, M., Perez-Diaz, F., Mynett-Johnson, L., Feingold, J. and Maier, W. (1998a) Age at onset and gender resemblance in bipolar siblings. *Psychiat. Res.*, 81: 125–131.

Leboyer, M., Bellivier, F., Nosten-Bertrand, M., Jouvent, R., Pauls, D. and Mallet, J. (1998b) Psychiatric genetics: search for phenotypes. *Trends Neurosci.*, 21: 102–105.

Leboyer, M., Filteau, M.J., Jay, M., Campion, D., Rochet, T., D'Amato, T., Feingold, J., Des Lauriers, A. and Widlocher, D. (1992) No gender effect on age at onset in familial schizophrenia? (letter). *Am. J. Psychiat.*, 149: 1409.

Levy, D.L., Holzman, P.S., Matthysse, S. and Mendell, N.R. (1994) Eye tracking and schizophrenia: a selective review. *Schizophr. Bull.*, 20 : 47–62.

Mann, J.J. (1998) The neurobiology of suicide. *Nat. Med.*, 4: 25–30.

Melo, J.A., Shendure, J., Pociask, K. and Silver, L.M. (1996) Identification of sex-specific quantitative trait loci controlling alcohol preference in C57BL/ 6 mice. *Nat. Genet.*, 13: 147–153.

Merikangas, K.R. (1995) Contribution of genetic epidemiologic research to psychiatry. *Psychopathology*, 28(Suppl 1): 41–50.

Merikangas, K.R., Spence, M.A. and Kupfer, D.J. (1989) Linkage studies of bipolar disorder: methodologic and analytic issues. Report of MacArthur Foundation Workshop on Linkage and Clinical Features in Affective Disorders. *Arch. Gen. Psychiat.*, 46: 1137–1141.

Nurnberger, J.J., Blehar, M., Kaufmann, C., York-Cooler, C., Simpson, S., Harkavy-Friedman, J., Severe, J., Malaspina, D. and Reich, T. (1994) Diagnostic interview for genetic studies. Rationale, unique features, and training. NIMH Genetics Initiative. *Arch. Gen. Psychiat.*, 51: 849–859; discussion 863–864.

Ott, R.W. (1991) Genetic linkage analysis under uncertain disease definition. In: C.R. Cloniger and H. Begleiter (Eds.), *Genetics and Biology of Alcoholism*, Cold Spring Harbor, New York, pp. 327–331.

Park, S., Holzman, P.S. and Goldman-Rakic, P.S. (1995) Spatial working memory deficits in the relatives of schizophrenic patients. *Arch. Gen. Psychiat.*, 52: 821–828.

Pauls, D.L., Alsobrook 2nd, J.P., Goodman, W., Rasmussen, S. and Leckman, J.F. (1995) A family study of obsessive-compulsive disorder. *Am. J. Psychiat.*, 152: 76–84.

Petronis, A. and Kennedy, J.L. (1995) Unstable genes–unstable mind? *Am. J. Psychiat.*, 152: 164–172.

Risch, N. (1990) Linkage strategies for genetically complex traits. II. The power of affected relative pairs (see comments). *Am. J. Hum. Genet.*, 46: 229–241.

Risch, N. and Botstein, D. (1996) A manic depressive history (news). *Nat. Genet.*, 12 : 351–353.

Sandbrink, R., Hartmann, T., Masters, C.L. and Beyreuther, K. (1996) Genes contributing to Alzheimer's disease (see comments). *Mol. Psychiat.*, 1: 27–40.

Seeman, M.V. and Lang, M. (1990) The role of estrogens in schizophrenia gender differences. *Schizophr. Bull.*, 16: 185–194.

Sham, P.C., Jones, P., Russell, A., Gilvarry, K., Bebbington, P., Lewis, S., Toone, B. and Murray, R. (1994) Age at onset, sex, and familial psychiatric morbidity in schizophrenia. Camberwell Collaborative Psychosis Study. *Br. J. Psychiat.*, 165: 466–473.

Stengard, J.H., Pekkanen, J., Ehnholm, C., Nissinen, A. and Sing, C.F. (1996) Genotypes with the apolipoprotein epsilon4 allele are predictors of coronary heart disease mortality in a longitudinal study of elderly Finnish men. *Hum. Genet.*, 97: 677–684.

Stober, G., Franzek, E., Haubitz, I., Pfuhlmann, B. and Beckmann, H. (1998) Gender differences and age of onset in the catatonic subtypes of schizophrenia (In Process Citation). *Psychopathology*, 31: 307–312.

Stober, G., Franzek, E., Lesch, K.P. and Beckmann, H. (1995) Periodic catatonia: a schizophrenic subtype with major gene effect and anticipation. *Eur. Arch. Psychiat. Clin. Neurosci.*, 245: 135–141.

Strober, M., Morrell, W., Burroughs, J., Lampert, C., Danforth, H. and Freeman, R. (1988) A family study of bipolar I disorder in adolescence. Early onset of symptoms linked to increased familial loading and lithium resistance. *J. Affect. Disord.*, 15: 255–268.

Swerdlow, N.R., Braff, D.L., Taaid, N. and Geyer, M.A. (1994) Assessing the validity of an animal model of deficient sensorimotor gating in schizophrenic patients. *Arch. Gen. Psychiat.*, 51: 139–154.

Toomey, R., Faraone, S.V., Seidman, L.J., Kremen, W.S., Pepple, J.R. and Tsuang, M.T. (1998) Association of neuropsychological vulnerability markers in relatives of schizophrenic patients. *Schizophr. Res.*, 31: 89–98.

Tsuang, M.T. (1993) Genotypes, phenotypes, and the brain. A search for connections in schizophrenia. *Br. J. Psychiat.*, 163: 299–307.

Tsuang, M.T., Faraone, S.V. and Lyons, M.J. (1993) Identification of the phenotype in psychiatric genetics. *Eur. Arch. Psychiat. Clin. Neurosci.*, 243: 131–142.

Waldo, M., Myles-Worsley, M., Madison, A., Byerley, W. and Freedman, R. (1995) Sensory gating deficits in parents of schizophrenics. *Am. J. Med. Genet.*, 60: 506–511.

Weissman, M.M., Gershon, E.S., Kidd, K.K., Prusoff, B.A., Leckman, J.F., Dibble, E., Hamovit, J., Thompson, W.D., Pauls, D.L. and Guroff, J.J. (1984) Psychiatric disorders in the relatives of probands with affective disorders. The Yale University — National Institute of Mental Health Collaborative Study. *Arch. Gen. Psychiat.*, 41: 13–21.

Williams, R.R., Hopkins, P.N., Hunt, S.C., Wu, L.L., Hasstedt, S.J., Lalouel, J.M., Ash, K.O., Stults, B.M. and Kuida, H. (1990) Population-based frequency of dyslipidemia syndromes in coronary-prone families in Utah. *Arch. Intern. Med.*, 150: 582–588.

W.E. Crusio and R.T. Gerlai (Eds.)
Handbook of Molecular-Genetic Techniques for Brain and
Behavior Research (Techniques in the Behavioral and Neural Sciences, Vol. 13)

CHAPTER 5.3

Trinucleotide repeat disorders

Gunnar Sandberg,[1] Kerstin Lindblad,[1] Ben A. Oostra and Martin Schalling[1]

[1]Department of Molecular Medicine, Neurogenetic Unit, CMM L8:00, Karolinska Hospital, 171 76 Stockholm, Sweden
[2]Dept of Clinical Genetics, Erasmus University Rotterdam, PO Box 1738, Dr Molewaterplein 50, 3000 DR Rotterdam, The Netherlands

Introduction

In 1991 Fragile X mental retardation (Hagerman, 1996) was found to be caused by a novel mutational mechanism. A short repeat sequence expanded into a long repeat sequence in a gene termed the fragile X mental retardation gene (FMR1) was present only in affected individuals. To date, more than fifteen disorders have been shown to be caused by expanded repeat sequences (Wells et al., 1998). Most of these disorders affect the brain and several have behavioral consequences. Here, we will outline the knowledge obtained from the study of these fifteen disorders during the last seven years. These discoveries have stimulated the development of new techniques for isolation of expanded repeats as well as the study of disease mechanisms. In this chapter we will describe the methods developed to ascertain whether the expansion mutation and associated methylation impairs transcription and translation of the FMR1 gene. A large section is spent on detailing methods currently used to screen for, identify and clone disease genes containing expanded repeats. The methods include genome and cDNA screening, RED technology, DIRECT, RAPID and selective recognition of protein with polyglutamine tracts by monoclonal antibodies.

Fragile X mental retardation

Fragile X mental retardation (FRAXA) is the most common inherited form of mental retardation with a frequency of 1 in 4000 males (Murray et al., 1997). The syndrome is strongly correlated with a fragile site at Xq27.3 and results from loss of the *FMR1* gene activity. In most cases fragile X syndrome is due to an expansion of a CGG repeat located in the 5′-untranslated region (5′UTR) of the *FMR1* gene (Verkerk et al., 1991). In the normal population, the CGG repeat of the FMR1 gene is 2–54 units long. However, individuals with the fragile X syndrome show a repeat containing over 200 CGG units.

Gene structure

The *FMR1* gene spans over 40 kb and contains 17 exons (Eichler et at., 1994). Its mRNA is 3.9 kb, with a 1.8 kb long 3′-untranslated region (Verkerk et al., 1991). The gene is expressed at high levels in most cells during development, whereas the adult expression pattern is non-uniform with high levels in organs affected in fragile X syndrome such as brain and testis (Hinds et al., 1993). The promoter elements have recently been characterized by in vivo-footprinting analysis (Schwemmle et al., 1997). Structural and functional characterization of the FMR1 promoter has revealed similarities with the promoter region of the gene for heterogeneous nuclear ribonucleoprotein (hnRNP-A2) (Drouin et al., 1997).

FMRP the protein product of FMR1

Several of the 17 exons in the FMR1 gene are alternatively spliced resulting in different mRNAs and protein isoforms (Ashley et al., 1993, Verkerk et al., 1993). Western blot analysis have demonstrated FMR1 protein (FMRP) products ranging from 70 to 80 kDa (Devys et al., 1993, Verheij et al., 1993). Very little is known about the function of FMRP. However, the presence of two hnRNP K Homology (KH) domains and RGG boxes indicates that the FMR1 protein has RNA binding properties, which has also been demonstrated in vitro (Gibson et al., 1993; Siomi et al., 1993). RNA binding studies have shown that the FMR1 protein is able to bind to its own messenger RNA as well as 4% of human fetal brain mRNA (Ashley et al., 1993). The RNA binding can be disrupted by deletions of the C-terminus ablating the RGG boxes. A point mutation has been found in a fragile X patient altering an Ile to Asn at position 304 in the second KH domain (De Boulle et al., 1993). This mutant protein also show reduced in vitro RNA binding activity (Siomi et al., 1994; Tamanini et al., 1996). However, it is uncertain whether the selective RNA binding with in vitro-translated FMRP is an intrinsic property or require associated proteins (Brown et al., 1998). Recent reports have shown that FMRP is associated with translating ribosomes (Eberhart et al., 1996). The FMRP is a ribonucleoprotein containing both nuclear localization signal (NLS) and nuclear export signal (NES). Isoforms of FMRP lacking exon 14 have been reported to be localized to the nucleus (Sittler et al., 1996; Willemsen et al., 1996). Deletion study have localized the NLS to the amino terminal end of the FMRP and NES to exon 14 (Eberhart et al., 1996; Bardoni et al., 1997). Interestingly, the NES motifs in the HIV-1 Rev protein resemble the FMRP NES (Fischer et al., 1995). The current hypothesis is that FMRP shuttles between the nucleus and cytoplasm bringing specific RNAs to somatodendritic polyribosomes for translation (Feng et al., 1997). Furthermore, it has recently been demonstrated that FMRP interacts with the homologues FXR1 and FXR2. These proteins can associate with each other or form homomers (Zhang et al., 1995). In humans, FXR1 has been localized to 3q28 and FXR2 to 17p13.1. It has been hypothesized that these proteins may complement each other.

DNA methylation and CGG triplet expansion; a regulatory function of FMR1 gene expression

In mammalian cells DNA methylation is mainly considered to be part of the gene regulation system. The only methylated base detected so far is the 2′-deoxy-5-methylcytosine (5m-dCTP) in the dinucleotide sequence CpG. It is still not clear how the mechanism by which DNA methylation prevents gene expression is working in the eukaryotic cell. The general hypothesis is that hypermethylation of the promoter region can inactivate gene expression (Kruczek et al., 1982; Busslinger et al., 1983) for instance it has been found that methylation of the structural coding region of Herpes Simplex Virus thymidine kinase gene (HSV *tk*) can inactivate transcription (Keshet et al., 1985). Additionally, a recent study (Rountree and Selker., 1997) using nuclear run-on assays demonstrate that methylation significantly inhibits transcript elongation in Neurospora crassa. There are several reports that CpG are clustered in the promoter region of a eukaryotic gene, the so called CpG-islands (Bird, 1986). This has especially been observed in the promoters of housekeeping genes where hypomethylation frequently occurrs. However, decreased methylation of CG rich areas in the 5′-region of genes is not sufficient to allow gene expression to occur (Oliva and Dixon., 1989). Furthermore, several studies suggest that DNA methylation per se does not inhibit the RNA polymerase II (Boyes and Bird.,

1992). It is assumed that DNA methylation alters the specific DNA binding of either regulatory proteins or chromatin. The presence of 5 methylcytosine (5m-C) in the recognition motifs for the HeLa cell transcription factor can interfere with the binding to DNA (Watt and Molloy., 1988). In contrast, the methylation of the CpG dinucleotide within the binding site of Sp-1 transcription factor has no or little effect of its DNA binding efficiency (Höller et al., 1988). Since some transcription factors are not inhibited by methylated CpGs in their recognition sequence, it has been proposed that a mediator capable of binding methylated sites might be involved preventing the formation of a transcription complex (Cedar, 1988). In human placenta a protein has been identified that can bind methylated CpGs. Furthermore, it has been shown that this protein (MDBP-1) has a limited sequence specificity (Wang et al., 1986; Zhang et al., 1986; Khan et al., 1988; Antequera et al., 1989). Other proteins (MeCP-1 and MeCP-2) lack sequence specificity, but require clusters of methylated CpGs for DNA binding (Meehan et al., 1989; Buschausen et al., 1987). Studies also indicate that methylated DNA may have the potential to regulate transcription via changes in chromatin structure (Keshet et al., 1986; Yisraeli et al., 1986) It is known that MeCP-2 is involved in the deacetylation complex of histones and in that way is controlling the level of transcription (Nan et al., 1998).

Fragile X syndrome and myotonic dystrophy are both associated with altered chromatin, aberrant methylation and suppressed expression of the adjacent *FMR1* and *DMAHP* genes (Tapscott et al., 1998). In the FMR1 promoter region, one CpG island is positioned 250 base pairs upstream from the CGG repeat. This CpG island is frequently found to be methylated in fragile X patients, demonstrating an inverse correlation between DNA methylation and gene expression (Oberlé et al., 1991; Bell et al., 1991; Hansen et al., 1992; Smeets et al., 1995). In addition, it has recently been found that a fully mutated *FMR1* gene containing 300-800 repeats treated with the demethylating agent 5-azadC led to reactivation of specific mRNA and protein production (Chiurazzi et al., 1998). The role of the CGG expansions in establishing the methylation pattern in the CpG island is not known. However, some results indicate that expanded CGG or CCG repeats could display conformational changes leading to a hairpin structure, which probably induce de novo methylation (Smith et al., 1994). In this case, DNA methylation will serve to stabilize the structure and possibly label the repeat for repair (Fry et al., 1992). Several results suggest that the hairpin formation induces slippage and hypermethylation of the fragile X DNA repeats (Chen et al., 1995). It has recently been found that the deletion of all CGG repeats plus flanking sequences in *FMR1* does not abolish gene expression (Gronskov et al., 1997).

Methods for analyzing the FMR1 promoter correlated to gene expression and transcription

Loss of *FMR1* expression is widely accepted to be the cause of fragile X syndrome. However, it is controversial as to whether this loss is the result of a CGG expansion per se, or a commonly occurring hypermethylation of the promoter region (Feng et al., 1995; Smeets et al., 1995). Furthermore the role of CpG methylation of the FMR1 promoter in regulating the *FMR1* expression is still not clear. It was demonstrated that methylation of all CpG dinucleotides by *Sss I* methylase blocked the *FMR1* promoter activity 48 h after transfection in the absence of a repeat expansion (Hwu et al., 1993). On the other hand a fragile X patient with a lung tumor had *FMR1* expressing tumor cells despite methylation of both the EagI and the BssHII site in the CpG island of the FMR1 promoter (De Graaf et al., 1995). It is possible that a special CpG site, not recognized by a known restriction site, has to be methylated to reduce gene-activity. Such CpG sites have pre-

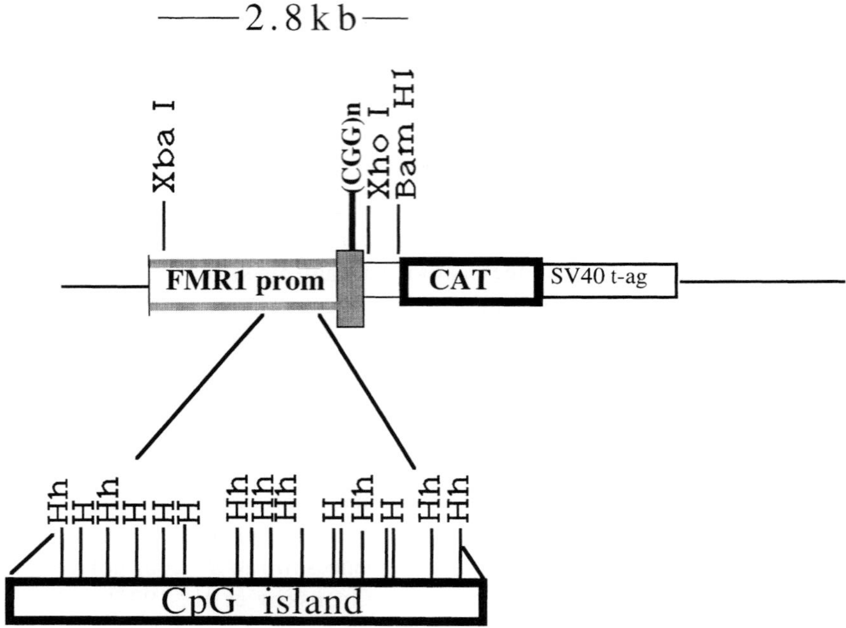

Fig. 1. Schematic map of pFXCAT. The positions of the HhaI (Hh), and HpaII (H) sites are shown in the CpG island of the FMR1 promoter. 1μg of pFXCAT was methylated with 15 U of M HpaII in (50 mM Tris-Cl pH 7.5, 10 mM EDTA, 5 mM 2-mercaptoethanol) and 0.2 mM S-adenosyl methionine. The methylation efficiency after M HpaII methylation was always controlled with HpaII and MspI digestion.

viously been identified (Wölfl et al., 1991) and may apply also to the regulation of the *FMR1* gene. Novel methylation sites may in fact be produced by the expansion itself and it has been demonstrated that DNA methyltransferase can recognize and selectively methylate unusual DNA structures such as trinucleotide CGG repeats in vitro.

The purpose of experiments described in this section is to determine the effect of in vitro methylation of the FMR1 promoter region and the effect of insertion of a CGG repeat expansion in the 5′UTR of *FMR1* on gene expression in cell culture. The first step is to construct an expression vector with the FMR1 promoter in front of a reporter gene. An ideal reporter gene for in vitro methylation studies is the HpaII methylation insensitive CAT gene (Doerfler et al., 1983) (Fig. 1). We determined that the FMR1 promoter was a relatively weak inducer of CAT activity in COS-1 cells despite the fact that these cells have been shown to be among the best cell-lines for the *FMR1* expression (Devys et al., 1993). Methylation per se did not inhibit transient protein-expression of pFXCAT-CH_3 plasmid in COS-1 cells during the first 48 h after transfection (Table 1). However, we did observe a reduced mRNA transcription in cells transfected with pFXCAT-CH_3 48 h after DNA transfer indicating that methylation may down regulate the *FMR1* gene. This was further supported by the increase in CAT activity seen using unmethylated pFXCAT in contrast to methylated DNA 72 h posttransfection (Fig. 2). The delayed difference in expression may also be due to chromatin assembly of the trans-DNA in the nucleus. It is well known that microinjected supercoiled plasmid DNA is assembled into minichromosomes 48–72 h after gene-transfer (Buschausen et al., 1987). This may lead to more efficient inhibition of the meth-

TABLE 1

Construct	% Acetylation		
	24 h	48 h	72 h
pFXCAT	2 ± 0.1	3 ± 0.1	10 ± 0.6
pFXCAT-CH3	2 ± 0.1	3 ± 0.1	3 ± 0.1
pSV2CAT	42 ± 2.1	61 ± 2.9	94 ± 2.4
pBCAT	0.5 ± 0.1	0.5 ± 0.1	1 ± 0.1

CAT enzyme activity was measured 24 h, 48 h and 72 h after transfection of 2 μg DNA/10^6 COS-1 cells, using the electroporation method (Biorad, 1050 μF, 220 V). Percentage acetylation was measured as acetylated amount ^{14}C-chloramphenicol divided by the total amount added. pFXCAT methylated with HpaII methyltransferase prior to transfection is indicated as pFXCAT-CH_3. pSV2CAT was used as positive control. pBCAT was used as control for basal transcription level. Data are means for three separate transfection experiments.

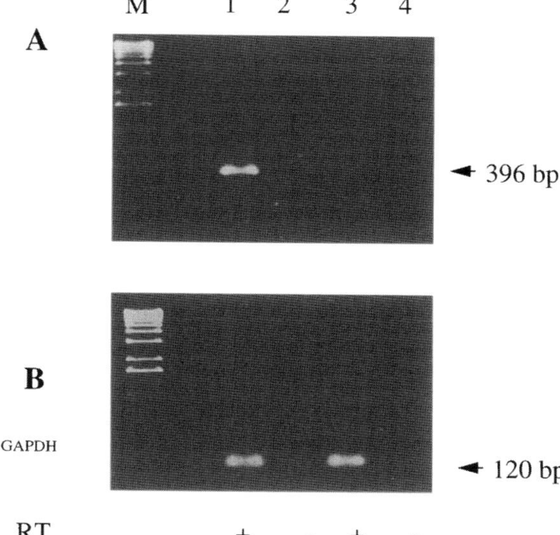

Fig. 2. RT-PCR from total RNA prepared from pFXCAT transfected COS-1 cells 48 h after DNA transfer. For first-strand cDNA synthesis 1 μg total RNA (24) dissolved in DEPC treated water from transfected cells was used with "Ready To Go" KIT (Pharmacia, Biotech, Uppsala, Sweden) containing M-MuLV reverse transcriptase and an oligo(dt) primer to generate first-strand cDNA. The RNA samples were DNase treated and denatured and placed at 37°C for 60 min. The complete first-strand reaction was heat inactivated at 90°C for 5 min. 5 μl of the first-strand reaction was used for PCR amplification. Amplifications were initially carried out using the primers (CAT-1) GAGGGCATTTCAGTCAGTTGC and (CAT-2) TGAAACTCACCCAGGGATTG corresponding to nucleotides' 4985–5007 and 5359–5378 of the CAT gene. 10 μl 10× Dynazyme buffer was added and 1 μl 20 mM dNTP mix and 30 pmol upstream primer (CAT-1) and 30 pmol downstream primer (CAT-2) and water to 100 μl and 2.5 U of Dynazyme enzyme. The PCR amplification profile was 94°C for 4 min, 4 cycles at 94°C for 1 min, 57°C for 2 min, 72°C for 3 min and 31 cycles at 94°C for 35 s, 57°C for 40 s, 72°C for 2.30 min. The PCR products were monitored on a 1% agarose gel. (A) Lane 1. pFXCAT with RT+, lane 2. pFXCAT without RT−, lane 3. pFXCAT methylated with Hpa II with RT+, lane 4. pFXCAT methylated without RT−. M. The 1 kb ladder from Promega was used as marker. (B) Primers for the housekeeping genes GAPDH was used as a control. This experiment was repeated with three separate transfection experiments.

ylated construct due to interaction between methyl group and histone octamers, histone 1 or other methyl CpG binding protein. The stability of the CAT enzyme in the cell might maintain a base level of gene-activity seen in the cells transfected with methylated pFXCAT even in the absence of de novo transcription (Sandberg and Schalling., 1997).

CGG repeat expansion in the first exon of the FMR1 gene is assumed to have an important role in development of the fragile X phenotype. To characterize the effect of a CGG repeat expansion on gene expression in general, we PCR amplified a DNA fragment from a male fragile X patient with an expanded repeat. PCR amplification of this CG rich sequence was difficult and different conditions were evaluated. The PCR amplification resulted in a product of 1000 bp containing approximately 200 CGG repeats inserted in the plasmids pFXCAT and pSV2CAT (Fig. 3).

The biological activity of pSV2$(CGG)_{200}$CAT was reduced as early as 24 h after gene-transfer and a complete loss of CAT activity was observed after transfection of pFX$(CGG)_{200}$CAT (Sandberg and Schalling., 1997). These findings taken together support the notion that a CGG repeat expansion alone can inhibit gene expression (Feng et al., 1995). The presence of similar

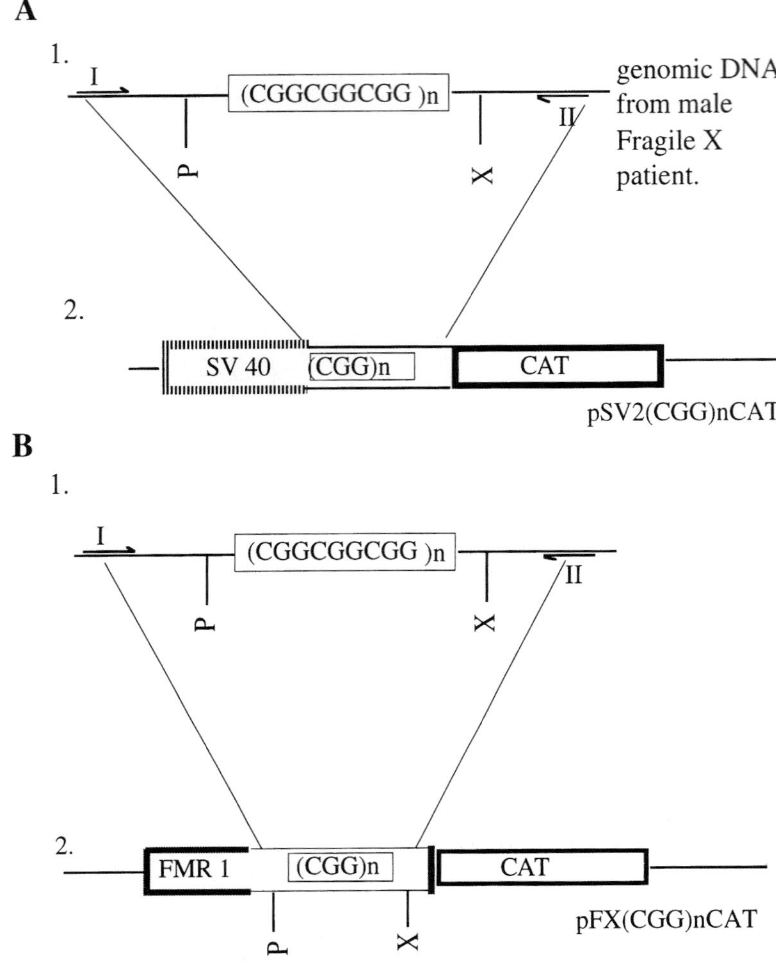

Fig. 3. Schematic description of the insert of a PCR amplified -CGG repeat fragment from a fragile X patient in pSV2CAT (A) and pFXCAT (B) *Hind*III (H). *Pst*I(P) and *Xho*I (X).

transcription levels in cells expressing pFX-$(CGG)_{200}$CAT and pFXCAT implies that translation rather than transcription is influenced by the change in the number of CGG repeat units at least in the short size range of pathological $(CGG)_n$ expansions. However, in vivo experiments using fragile X patients' lymphoblastoid cells have demonstrated reactivation of transcription and FMR1 protein production following treatment with 5-azadC (Chiurazzi et al., 1998).

In conclusion, with our methods we have demonstrated that a CGG expansion as well as in vitro methylation of the FMR1 promoter can regulate gene expression. The CGG repeat expansion affects in vitro cell system translation but not transcription and acts independent of methylation. Future studies using cell-lines transfected with pFX$(CGG)_n$CAT containing various repeat unit size may reveal in what way the CGG repeat expansion itself can influence the pattern of methylation of flanking CpG sequences. We will also determine the correlation of methylation and histone deacetylation in the biological activity of the FMR1 promoter.

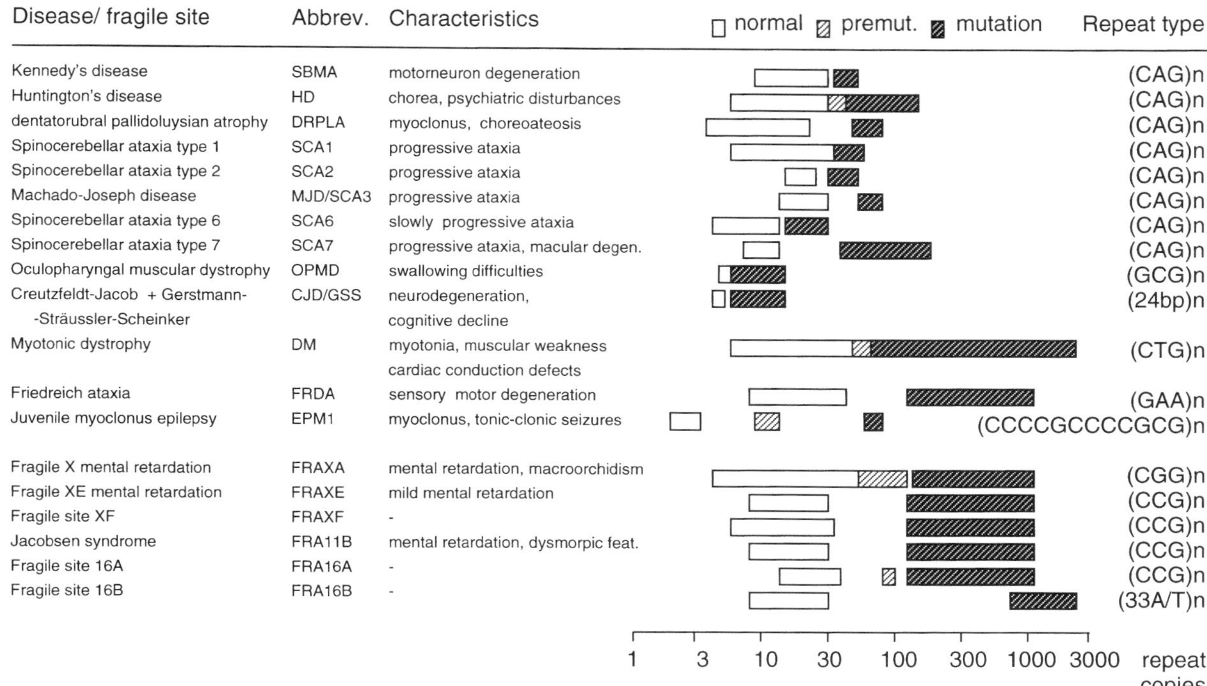

Fig. 4. Expanded repeats causing disease or fragile sites. Several of these mutations affect multiple organ systems. Note that only main characteristics of each disorder has been included due to space limitations. (From Lindblad K., 1998, *Genomic studies of expanded repeat sequences: focus on neuropsychiatric disorders*. Reproprint AB, Stockholm.)

Disorders caused by expanded repeats

Expanded repeat sequences that increase in size upon transmission, have been demonstrated in more than 15 disorders including FRAXA (Fig. 4). The majority of these disorders display anticipation seen as an earlier age at onset and/or an increased disease severity from one generation to the next. The repeat sequence motif (base composition), size and localization of the expansion within the respective gene varies between disorders and influence the way the repeat causes disease (Fig. 5).

Polyglutamine disorders

The most commonly found pathogenic expanded repeats are coupled to the triplet CAG. This repeat encoding polyglutamine tracts has been found in several disorders including Huntington's disease (The Huntington's Disease CRG, 1993), Kennedy's disease (SBMA) (La Spada et al., 1991), dentatorubral pallidoluysian atrophy (DRPLA) (Koide et al., 1994; Nagafuchi et al., 1994) spinocerebellar ataxia (SCA) type 1 (Orr et al., 1993), SCA2 (Imbert et al., 1996; Sanpei et al., 1996; Pulst et al., 1996), SCA6 (Ikeuchi et al., 1997), SCA7 (David et al., 1997) and Machado-Joseph disease/SCA3 (Kawaguchi et al., 1994). These diseases are all progressive disorders with degeneration of specific neuronal pathways. The onset usually occurs in adulthood but juvenile cases are also seen. Repeat sizes in affected individuals are between 36 and 200 repeat copies. There is a clear correlation between increased repeat size and decreased onset age in all polyglutamine disorders except, for SCA6

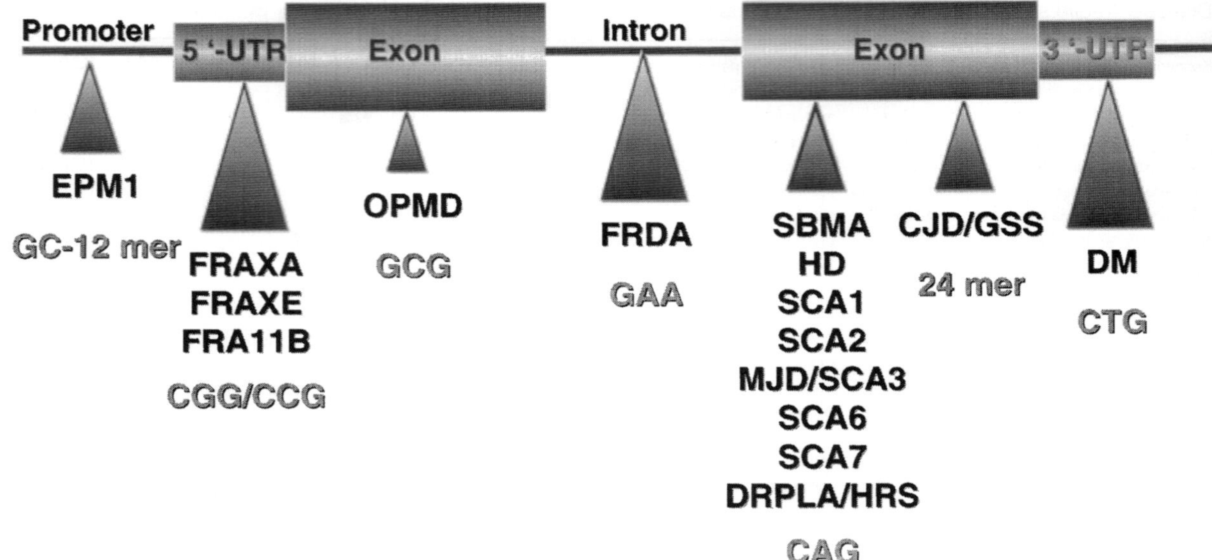

Fig. 5. Location within genes of pathogenic expansions. (From Lindblad K., 1998, *Genomic studies of expanded repeat sequences: focus on neuropsychiatric disorders.* Reproprint AB, Stockholm.)

where the expanded alleles are only between 21 and 27 copies (Ikeuchi et al., 1997). The function differs between each disease gene product. For example the androgen receptor found mutated in Kennedy's disease is involved in regulation of transcription (Simental et al., 1991) whereas the repeat causing SCA6 is localized in a gene coding for an ion channel (Ikeuchi et al., 1997). The localization of the proteins in the cell also vary, many are cytoplasmic (Gutekunst et al., 1995; Trottier et al., 1995a; DiFiglia et al., 1995; Yazawa et al., 1995; Trottier et al., 1995b; Paulson et al., 1997) but ataxin 1 (Servadio et al., 1995) and ataxin7 (Trottier et al., 1995b; Stevanin et al., 1996) are both localized to the nucleus. The androgen receptor shuttles between the cytoplasm and nucleus (Zhou et al., 1994). The pathogenic mechanism appears to be a gain of function mechanism related either to aberrant protein interactions dependent on polyglutamine size (Koshy et al., 1996; Burke et al., 1996; Matilla et al., 1997), or to the formation of insoluble aggregates of elongated polyglutamine chains (DiFiglia et al., 1997; Cummings et al., 1998; Paulson et al., 1997; Holmberg et al., 1998). Recent results indicate that both mechanisms may be important in pathogenesis. Protein aggregates have so far been seen as neuronal intranuclear inclusions (NII) in HD, SCA1, MJD/SCA3 and SCA7 (DiFiglia et al., 1997; Cummings et al., 1998; Paulson et al., 1997; Holmberg et al., 1998) and in DRPLA brain samples (Yazawa et al., 1995; Davies et al., 1997a). In addition, NIIs have been seen in transgenic mice and *Drosophila* containing expanded polyglutamine repeats in the known disease genes as well as in the *HPRT* gene (Davies et al., 1997a, 1997b; Ordway et al., 1997; Warrick et al., 1998). Since the *HPRT* gene does not normally contain a polyglutamine repeat, this shows that the formation of NII is dependent on the polyglutamine itself. The polyglutamines comprise a main component of the NIIs as detected by an antibody. Proteins involved in protein degradation and refolding have also been detected in NIIs, suggesting that the elongated polyglutamine chains disrupt or overload the proteolytic pathways in the cell (DiFiglia et al., 1997; Cummings et al., 1998).

Despite the general conclusion that polyglutamine-containing NIIs are a hallmark in these disorders, recent evidence shows that they may not be necessary for disease. Transgenic mice containing ataxin-1 (the SCA1-protein) with an expanded polyglutamine repeat but lacking the self association region of the protein (Burright et al., 1997) have been raised. These mice show the typical neurological and pathological alterations, but lack NIIs at 16 weeks of age (Klement et al., 1998). Thus, aberrant protein interactions appear to suffice to cause at least a large part of the phenotype. Still, SCA1 transgenic mice do not show as much neurodegeneration as humans leaving the possibility that NIIs contribute more to the neurodegenerative part of the phenotype.

Another mystery is why only certain regions of the brain are affected when many of the polyglutamine-containing proteins are widely expressed both in the brain and other tissues. The key is probably the expression pattern of the interacting proteins. As an example, the leucine-rich acidic nuclear protein (LANP) is expressed primarily in Purkinje cells in the cerebellum, the cells primarily affected in SCA1. The affinity of LANP for ataxin-1 increases dramatically with increasing polyglutamine size (Matilla et al., 1997). In addition, LANP is found in the inclusions in SCA1 transgenic mice suggesting that the increased binding may cause the LANP to associate with the inclusions thereby disrupting LANP's normal function. Similarly, several other proteins have been shown to interact with the disease proteins in a size dependent manner (Koshy et al., 1996; Burke et al., 1996). The expression pattern of these interacting proteins could determine the tissue specific pathology for the respective disorders.

Other gain of function disorders

Autosomal dominant oculopharyngal muscular dystrophy (OPMD) presents with swallowing difficulties, eyelid drooping and proximal limb weakness late in life (Tome et al., 1997). Unique nuclear filamentous inclusions as well as mitochondrial structural abnormalities have been reported (Tome et al., 1997; Blumen et al., 1996). Despite reports of anticipation (Blumen et al., 1997) in addition to the observed NII, the disease was not associated with expanded CAG/CTG repeats larger than 40 copies (Teh et al., 1997). Recently, the gene involved in OPMD, PABP2, was isolated and the mutation was identified as a short expanded GCG-repeat translated into a polyalanine chain (Brais et al., 1998). The repeat is 6 or 7 copies in normal individuals and 8–13 copies in affected individuals from many different regions of the world. Similar to this, enlarged polyalanine tracts causing disease are present in dominant cleidocranialdysplacia (Mundlos et al., 1997) and dominant synpolydactyly (Muragaki et al., 1996; Akarsu et al., 1996). As both genes are transcription factors the expanded polyalanine repeat may alter gene regulation. The expanded repeats in dominant cleidocranial dysplacia and dominant synpolydactyly are interrupted repeats on the DNA level and appear to have increased in size by gene-conversion (Muragaki et al., 1997). They are therefore stably inherited and not true dynamic mutations. However, the presence of enlarged polyalanines in all three disorders shows that homopolymers other than polyglutamines can cause disease. The mechanisms may be slightly different from that of polyglutamines, but the presence of NIIs in OPMD opens the possibility that also polyalanine chains form aggregates.

Myotonic dystrophy — a gain and loss of function disorder

Myotonic dystrophy is a dominant disorder caused by a CTG expansion found in the 3′-untranslated region of the *DMPK* gene (Brook et al., 1992). The expansions vary between 50 and 2700 repeat copies. The pathological function of the repeat is not yet clear, however effects both on DNA, RNA and protein levels have been reported (Lavedan et al., 1993). It has been demonstrated that CTG repeats are able to bind nucleosomes strongly altering the DNA conformation around the expanded repeat (Wang et al., 1994; Wang et al., 1996). The *DMAHP* gene is located downstream of the CTG expansion with its promoter spanning the repeat and is down-regulated by the expanded repeat (Boucher et al., 1995). This could either be due to DNA conformation effects or to disruption of the promoter by the repeat. Both up and down-regulation of the *DMPK* gene itself has been reported (Fu et al., 1993; Sabouri et al., 1993), but increasing evidence now indicates that DMPK transcripts from the expanded allele are trapped in discreet foci in the nucleus (Taneja et al., 1995; Davies et al., 1997b). The enlarged CTG expansion, transcribed into a CUG expansion in the RNA, may cause binding and accumulation of CUG-binding proteins (CUG-BP) in the nucleus (Timchenko et al., 1996a). The CUG-BP may then bind and retain expanded as well as normal DMPK transcripts in the nucleus (Timchenko et al., 1996b; Morrone et al., 1997). In addition, the accumulation of CUG-BP in the nucleus could interfere with normal splicing of other transcripts. Recent reports show that CUG-BP are involved in the alternative splicing of exon 5 in the chicken cTNT gene through binding to intronic regulatory elements containing short CUG sequences downstream of exon 5 (Philips et al., 1998). As cTNT is expressed in both skeletal and cardiac muscle this may be related to the cardiac conduction defects seen in myotonic dystrophy.

Loss of function disorders

Friedreich ataxia (FRDA) is the most common hereditary ataxia with a prevalence of 1/50,000 in the European population (Romeo et al., 1983). It is an autosomal recessive degenerative disease involving the sensory and cerebellar pathways as well as the heart (Harding et al., 1993). The gene, X25, encoding the frataxin protein has a GAA repeat within the first intron (Campuzano et al., 1996). The majority of FRDA patients (andap; 96%) are homozygotes for expansions of 200–1000 copies of this repeat. The expanded repeat is proposed to inhibit transcription of the X25 gene through formation of a stable intramolecular triple helix. This inhibited transcription leads to a partial or complete loss of protein and leads to disease (Campuzano et al., 1996; Cossée et al., 1997). The remaining disease alleles have point mutations. Loss of frataxin leads to mitochondrial iron accumulation rendering the mitochondria sensitive to oxidative stress (Campuzano et al., 1997). No clear anticipation had been reported prior to the identification of the disease gene, probably due to the recessive nature of the disorder. However, instability of the expanded repeats has been demonstrated in parent-offspring transmission (Monros et al., 1997). There is a correlation between the GAA repeat size of the *smallest* allele and age of onset, disease progression and cardiomyopathy (Filla et al., 1996; Dürr et al., 1996; Lamont et al., 1997; Montermini et al., 1997). This is logical as the allele within the smallest expanded repeat may retain some residual protein production. The identification of FRDA as an triplet repeat disorder was particularly interesting, as it showed that expanded repeats can cause disease in a large variety of ways and that also recessive disorders should be considered as candidate disorders.

In addition, an expanded dodecamer repeat, $(CCCCGCCCCGCG)_n$, located in the promoter of the cystatin B gene, has been shown to constitute the majority of mutations in progressive myoclonus epilepsy (Lafreniere et al., 1997; Lalioti et al., 1997; Virtaneva et al., 1997). The repeat down regulates cystatin B expression, but the mechanism is not known. Cystatin B is ubiquitously expressed and functions as a cysteine protease inhibitor (Turk and Bode, 1991). Lack of this protein may well lead to lack of proteinase inhibition and accumulation of degradation products causing degeneration of brain cells in EPM1 patients.

Animal models for trinucleotide diseases

No naturally occurring animal models for the trinucleotide diseases have been described. To gain more insight in the pathology and physiology of trinucleotide diseases, several groups have tried to generate animal models for the different diseases. The development of an animal model has major advantages. First, the unlimited supply of tissues gives the opportunity to study the effects on the morphological and molecular level. Second, the phenotype and behaviour of mice can be studied in order to understand the development of the disease phenotype found in human patients. Different animal models can be made. First, a knockout model in which the gene is modulated in such a way that the gene of interest has become inactive. Second, a transgenic model can be made in which a mutation is introduced in the mouse homologue of the human gene. Third, a copy of (part of) the human gene or even a complete YAC containing the human gene is introduced in which the human mutation is present. For this purpose the human gene is placed under the control of either its own promoter or another (tissue) specific promoter. This will give cell type specific expression and in most cases also the level of expression can be influenced. Expression can be manipulated by using promoters that can be switched on or off in a tissue or time-dependent way. This will allow studies on the influence of the contribution of different cell types and during different time points during embryology. Especially mutations of polyglutamine tracts of different length can be studied in such a way. All three types of models have been generated for trinucleotide diseases and an overview of the different mouse models is given in Table 2. Mouse models with a phenotype showing characteristics of the human disease have been generated for fragile X syndrome, HD, SCA1, and SCA3 (Bakker et al., 1994; Burright et al., 1995; Mangiarini et al., 1996; Ikeda et al., 1996). Surprisingly, one mouse model for HD only con-

TABLE 2

Summary of mouse models for trinucleotide repeat diseases.

Disease	Construct	Repeat length	Phenotype	Instability	Reference
Fragile X syndrome	KO	NR	+	NR	Bakker et al. 1994
Fragile X syndrome	Transgene	88	NR	–	Lavedan et al. 1997
Fragile X syndrome	Transgene	81	NR	–	Bontekoe et al. 1997
HD	Transgene cDNA	44	–	–	Goldberg et al. 1996
HD	Transgene exon 1	115–155	+	+	Mangiarini et al. 1996 Mangiarini et al. 1997
HD	KO	NR	Lethal	NR	White et al. 1997
AR	Transgene cDNA	45	–	–	Bingham et al. 1995
AR	transgene	45	–	+	LaSpada et al. 1998
SCA1	Transgene cDNA	82	+	–	Burright et al. 1995
SCA3/MJD	Transgene cDNA	79	+	–	Ikeda et al. 1996
MD	KO	NR	Minor	NR	Jansen et al. 1996
MD	transgene	Normal	Minor	NR	Jansen et al. 1996
MD	KO	NR	+	NR	Reddy et al. 1996
MD	Transgene genomic	??	-	+	Monckton et al. 1997
MD	Transgene genomic	55	-	+	Gourdon et al. 1997

NR: not relevant; KO: knockout

sists of a small part of the HD gene (Mangiarini et al., 1996), although there is some doubt now whether phenotype of the mouse is caused by an effect other than the introduction of the truncated HD gene. These mouse models are, and will be, used to study the pathology and physiology of the diseases in more detail and might help to shed light on the factors that determine the specificity of the neurodegeneration. The mouse models for MD showed only minor characteristics (Jansen et al., 1996; Reddy et al., 1996). Instability of the trinucleotide, a characteristic of this class of human diseases, has been found only for some disease genes (Mangiarini et al., 1996; LaSpada et al., 1998; Monckton et al., 1997; Gourdon et al., 1997). However, if instability of the repeat is seen, the instability is very subtle. These first mouse models of these diseases promise to shed light on the molecular mechanisms underlying repeat instability. Transgene analysis has revealed germline hypermutability, including expansions, deletions and parent-of-origin effects, somatic and early embryonic instability and segregation distortion. Mutational differences between lines and sexes demonstrate that stability, as in humans, is modulated by as yet unidentified cis and trans acting genetic elements.

Methods for detection and isolation of disease relate expanded repeats

Anticipation has been reported in a large number of disorders in addition to those already known to be related to expanded repeat sequences (for review see Paterson et al., 1998). In addition, the genome contains a large number of short repeat sequences suggesting that these could also expand and cause disease. To facilitate identification of novel expanded repeats, a number of new gene isolation strategies have been investigated. Additional repeat containing loci were isolated from genomic DNA (Gastier et al., 1995; Grierson et al., 1998) as well as cDNA libraries (Riggins et al., 1992; Margolis et al., 1996) and tested for expansions in affected individuals from different candidate disorders. To make the search more efficient, several methods for direct detection of expanded repeats in genomic DNA have now been developed. The first methods developed for this purpose was the repeat expansion detection (RED) technique (Schalling et al., 1993) (Fig. 6). This method can detect expanded repeats without any knowledge of the flanking sequence or their location in the genome. The DNA to be tested is used as a template for a repeat oligonucleotide in a two-step cycling procedure. Multimers of the original oligonucleotide are generated using a thermostable ligase. The products are size separated on a polyacrylamide gel and transferred onto a membrane followed by hybridization to a complementary radioactively labeled repeat probe. The longest product detected reflects the length of the longest repeat in the tested DNA. The RED method has been used for identification of expanded CAG/CTG repeat sequences in myotonic dystrophy (Schalling et al., 1993), MJD/SCA3 (Lindblad et al., 1996a) and SCA7 (Lindblad et al., 1996b) as well as to associate expanded repeats with bipolar affective disorder (Lindblad et al., 1995; O'Donovan et al., 1995, 1996; Oruc et al., 1997; Lindblad et al., 1998). RED has also been optimized for eight out of ten trinucleotide repeat sequence motifs (Lindblad et al., 1995). Practical information for the RED method can be found at the end of this section.

The disadvantage with RED is that it does not provide information about on what chromosome or in what gene the detected expanded repeat is located. However, the RED method can be used as a tool to isolate the expanded repeat as well as specific sequences flanking the repeat. In several strategies (Nakamoto et al., 1997) including RAPID (repeat analysis pooled isolation and detection) (Koob et al., 1998) RED is used for identification of fractions or pools containing the expanded repeat in each of a series of physical separation steps.

Nakamoto and coworkers used a sib selection strategy when they cloned the ERDA1 locus

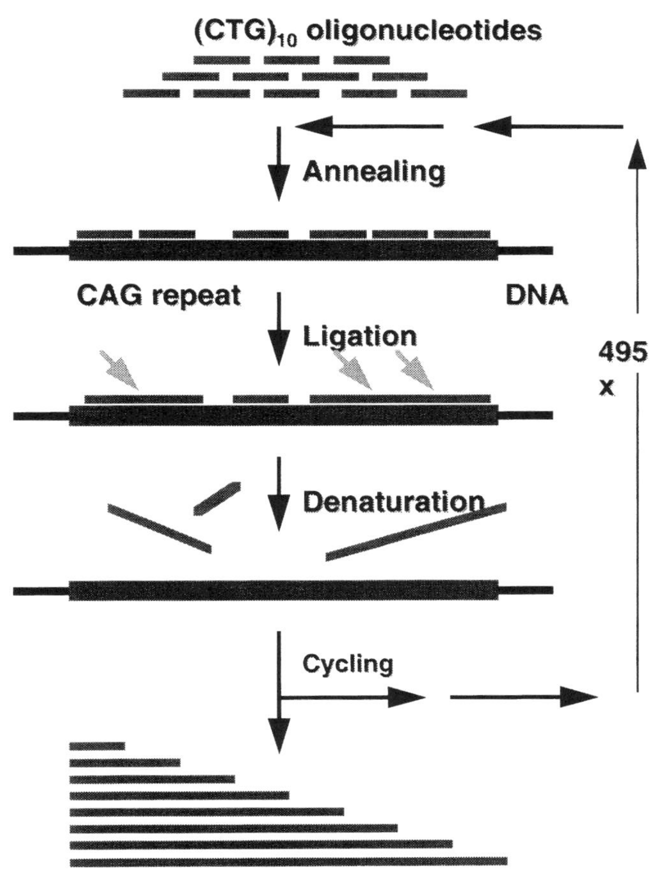

Fig. 6. The principle of RED. DNA is heat denatured and a repeat oligonucleotide of the motif of interest is allowed to anneal to the template. Ligation occurs when oligonucleotides anneal at adjacent positions. These reaction step are repeated several hundred times on a thermocycler yielding a pool of multimers,where the longest product corresponds to the longest repeat sequence in the tested DNA. (From Lindblad K., 1998, *Genomic studies of expanded repeat sequences: focus on neuropsychiatric disorders*. Reproprint AB, Stockholm.)

(Nakamoto et al., 1997). In principle, genomic DNA containing an expanded repeat was digested and a genomic library was made in λ-phage. The phages were eluted from each individual plate and screened for expanded repeats using RED. The eluates containing the expanded repeat were subdivided, replated and screened again using RED. If this procedure is repeated several times, a single plaque containing the expanded repeat could be obtained. As details have not been published, it is hard to comment on how many plaques were screened to identify the positive phage. However, as repeat sequences are known to be unstable in prokaryotic hosts (Kang et al., 1995; Ohshima et al., 1996b; Samadashwily et al., 1997) a very large number of phage usually have to be screened. A more elaborate protocol where the fragment containing the expanded repeat is physically enriched prior to cloning is the RAPID (repeat analysis pooled isolation and detection) method. In RAPID, size separation is performed on an agarose gel, followed by agarase treatment and RED detection of fractions containing long repeats (Koob et al., 1998). Positive fractions are cloned into λ phage and converted into a single stranded DNA library containing uracil. A $(CAG)_{10}$ oligonucleotide is used for primer extension from the ssDNA. Uracil-DNA glycosylase removes uracils from the original DNA strand. Upon transformation in *E. coli* CAG-containing double stranded DNA is repaired and replicated, whereas the single-stranded background is removed. Pools of clones are screened using RED and subdivided into smaller DNA pools until a single repeat contain clone exists. RAPID has been successfully used to isolate the expanded repeats in SCA7 and CTG18.1 loci (Koob et al., 1998). One advantage with RAPID may be the removal of non-repeat containing fragments prior to cloning. This may considerably improve the competitiveness of long repeat sequences for growth in *E coli.* However, Koob and coworkers reported that the RED product sizes of positive pools did not correspond well with the RED product size originally seen in the genomic sample. Thus, repeat instability may also represent a real problem in this strategy.

Another strategy, DIRECT (direct identification of repeat expansion and cloning technique) (Sanpei et al., 1996; Pujana et al., 1998) is based on stringent hybridization conditions allowing specific detection of long repeats on a Southern blot. Genomic DNA from both affected and unaffected family members is digested with a battery of 6-base cutter restriction enzymes and submitted to agarose gel electrophoresis and blotted onto a membrane. A radioactively labelled PCR probe with very high specific activity is generated. One normal and one biotinylated PCR primer is used to amplify a 55 copy CAG/CTG repeat from a plasmid. The probe is made single stranded using streptavidin-magnetic beads. After a very stringent hybridization and wash only the long repeats will be detected on the Southern blot. Theoretically, the restriction fragment sizes will vary between affected and unaffected individuals indicating the position of the expanded repeat on the gel. The corresponding gel slice is excised and cloned in λ phage. This method was successfully used to isolate SCA2 (Sanpei et al., 1996). However, it is relatively work intensive as many restriction enzymes may have to be tried to find one with a good resolution of the fragment of interest. In addition, generation of the probe is rather tricky as single stranded probes with high specific activity are unstable. A new protocol with two major improvements has been reported (Pujana et al., 1998). Firstly, 4-cutter enzymes are used for the restriction digest yielding smaller fragments and better resolution of size differences between affected and unaffected individuals. Thus, only a few enzymes need to be used in each experiment. Secondly, the probe is generated by radioactive PCR using $(CAG)_8$ and $(CTG)_8$ primers and no template. A double stranded probe containing only large CAG and CTG repeats is generated. Products of a certain size range, typically 200–500 bp, can be excised from a gel and used as a probe after careful denaturation. This probe is much more stable and easy to handle. A disadvantage with DIRECT is that multiple bands are often detected on blots, due to the many short repeats present in the genome. Still a stringent hybridization can minimize these problems. The main advantage with DIRECT is that smaller expansions can be detected. If the optimal restriction enzyme and gel resolution is used, relatively small size differences could be detected.

Detection of expanded polyglutamine repeats on the protein level has been achieved using the 1C2 antibody (Trottier et al., 1995a). This antibody has been used to detect the presence of large polyglutamine repeats in Huntington's disease, expansions in SCA2 and SCA7 as well as familial spastic paraplegia by Western blot (Trottier et al., 1995b; Stevanin et al., 1996; Nielsen et al., 1997). A problem with the antibody is that it detects a number of glutamine containing proteins present in most cells, making it difficult to determine if a protein is associated with disease. Furthermore, the main drawback of this technique is the need to analyze the tissue expressing of the patogenic protein. Comparison between several affected and unaffected individuals is therefore necessary. The 1C2 antibody has also been used to isolate the SCA2 gene from an expression library (Imbert et al., 1996). Here, also a disadvantage is that repeats are unstable during the cloning step.

In summary, a number of complementary strategies exist for the identification and isolation of expanded repeat sequences. However, care must be taken to keep the expanded repeats as stable as possible by choosing hosts and procedures carefully.

Acknowledgements

This research was funded by NARSAD, Magnus Bergvalls Foundation, Fredrik and Ingrid Thurings Foundation, BIOMED PL95-1663 and funds from the Karolinska Institute and Hospital.

References

Akarsu, A.N., Stoilov, I., Yilmaz, E., Sayli, B.S. and Sarfarazi, M. (1996) Genomic structure of HOXD13 gene: a nine polyalanine duplication causes synpolydactyly in two unrelated families. *Hum. Mol. Genet.*, 5(7): 945–952.

Antequera, F., Macleod, D. and Bird, A.P. (1989) Specific protection of methylated CpGs in mammalian nuclei. *Cell*, 58: 509–517.

Ashley, C.T., Sutcliffe, J.S., Kunst, C.B., Leiner, H.A., Eichler, E.E. Nelson, D.L. and Warren, S.T. (1993) Human and murine FMR-1 : alternative splicing and translational initiation downstream of the CGG -repeat. *Nat. Genet.*, 4: 244–251.

Bakker, C.E., Willemsen, R., van der Helm, R., Oerlemans, F., Vermey, M., Bygrave, A., Hoogeveen, A.T., Oostra, B.A., Reyniers, E., De Boulle, K., D'Hooge, R., Cras, P., van Velzen, D., Nagels, G., Martin, J-J., De Deyn, P.P., Darby, J.K. and Willems, P.J. (1994) Fmr1 knockout mice: a model to study fragile X mental retardation. *Cell*, 78: 23–33.

Bardoni, B.S.A., Shen, Y. and Mandel, J.-L. (1997) Analysis of domains affecting intracellular localization of the FMRP protein. *Neurobiol. Disease*, 4: 329–336.

Bell, M.V., Hirst, M.C., Nakahori, Y., MacKinnon, R.N., Roche, A., Flint, T.J., Jacobs, P.A., Tommerup, N., Tranebjaerg, L., Froster, I.U., Kerr, B., Turner, G., Lindenbaum, R.H., Winter, R., Pembrey, M., Thibodeau, S. and Davies, K.E. (1991) Physical mapping across the fragile X hypermethylation and clinical expression of the fragile X syndrome. *Cell*, 64: 861–866.

Bingham, P.M., Scott, M.O., Wang, S., McPhaul, M.J., Wilson, E.M., Garbern, J.Y., Merry, D.E. and Fischbeck, K.E. (1995) Stability of an androgen receptorgene in transgenic mice. *Nat. Genet.*, 9: 191–196.

Bird, A.P. (1986) CpG rich islands and the function of DNA methylation. *Nature*, 321: 209–213.

Blumen, S.C., Nisipeanu, P., Sadeh, M., Asherov, A., Blumen, N., Wirguin, Y., Khilkevich, O., Carasso, R.L. and Korczyn, A.D. (1997) Epidemiology and inheritance of oculopharyngeal muscular dystrophy in Israel. *Neuromuscular Disorders*.

Blumen, S.C., Sadeh, M., Korczyn, A.D., Rouche, A., Nisipeanu, P., Asherov, A. and Tome, F.M. (1996) Intranuclear inclusions in oculopharyngeal muscular dystrophy among Bukhara Jews. *Neurology*, 46(5): 1324–1328.

Bontekoe, C.J., de Graaff, E., Nieuwenkuizen, M., Willemsen, R. and Oostra, B.A. (1997) FMR1 premutation allele (CGG)81 is stable in mice. *Eur. J. Hum. Genet.*, 5: 293–298.

Boucher, C.A., King, S.K., Carey, N., Krahe, R., Winchester, C.L., Rahman, S., Creavin, T., Meghji, P., Bailey, M.E., Chartier, F.L. et al. (1995) A novel homeodomain-encoding gene is associated with a large CpG island interrupted by the myotonic dystrophy unstable (CTG)n repeat. *Hum. Mol. Genet.*, 4(10): 1919–1925.

Boyes, J. and Bird, A. (1992) Repression of genes by DNA methylation depends on CpG density and promoter strength:evidence for involvement of a methyl-CpG binding protein. *EMBO J.*, 11: 327–333.

Brais, B., Bouchard, J.P., Xie, Y.G., Rochefort, D.L., Chretien, N., Tome, F.M., Lafreniere, R.G., Rommens, J.M., Uyama, E., Nohira, O., Blumen, S., Korcyn, A.D., Heutink, P., Mathieu, J., Duranceau, A., Codere, F., Fardeau, M. and Rouleau, G.A. (1998) Short GCG expansions in the PABP2 gene cause oculopharyngeal muscular dystrophy. *Nat. Genet.*, 18(2): 164–167.

Breschel, T.S., McInnis, M.G., Margolis, R.L., Sirugo, G., Corneliussen, B., Simpson, S. G., McMahon, F.J., MacKinnon, D.F., Xu, J.F., Pleasant, N., Huo, Y., Ashworth, R.G., Grundstrom, T., Kidd, K.K., DePaulo, J.R. and Ross, C.A. (1997) A novel, heritable, expanding CTG repeat in an intron of the SEF2-1 gene on chromosome 18.21.1. *Hum. Mol. Genet.*, 6: 1855–1863.

Brook, J.D., McCurrach, M.E., Harley, H.G., Buckler, A.J., Church, D., Aburatani, H., Hunter, K., Stanton, V.P., Thirion, J.P., Hudson, T. et al. (1992) Molecular basis of myotonic dystrophy: expansion of a trinucleotide (CTG) repeat at the 3′ end of a transcript encoding a protein kinase family member. *Cell*, 68: 799–808.

Brown, V.S.K., Lakkis, L, Feng, Y, Gunter, C, Wilkinson, K.D. and Warren, S.T. (1998) Purified recombinant fmrp exhibits selective RNA binding as an intrinsic property of the fragile X mental retardation protein. *J. Biol. Chem.*, 273: 15521–15527.

Burke, J.R.E.J., Martin, M.E., Jou, Y.S., Myers, R.M. and Roses, A.D. et al. (1996) Huntingtin and DRPLA proteins selectively interact with the enzyme GAPDH. *Nat. Med.*, 2: 347–350.

Burright, E.N., Davidson, J.D., Duvick, L.A., Koshy, B., Zoghbi, H.Y. and Orr, H.T. (1997) Identification of a self-association region within the SCA1 gene product, ataxin-1. *Hum. Mol. Genet.*, 6(4): 513–518.

Burright, E.N.C.H., Servadio, A., Matilla, T., Feddersen, R.M., Yunis, W.S., Duvick, L.A. and Zoghbi, H.Y.O.H. (1995) SCA1 transgenic mice: a model for neurodegeneration caused by an expanded CAG trinucleotide repeat. *Cell*, 82: 937–948.

Buschausen, G., Wittig, B., Gräßmann, M. and Gräßmann, A. (1987) Chromatin structure is required to block transcription of the methylated herpes simplex virus thymidine kinase gene. *Proc. Natl. Acad. Sci. USA*, 84: 1177–1181.

Busslinger, M., Hurst, J. and Flavell, R.A. (1983) DNA methylation and the regulation of globin gene expression. *Cell*, 34: 197–206.

Campuzano, V., Montermini, L., Lutz, Y., Cova, L., Hindelang, C., Jiralerspong, S., Trottier, Y., Kish, S.J., Faucheux, B., Trouillas, P., Authier, F.J., Durr, A., Mandel, J.L., Vescovi, A., Pandolfo, M. and Koenig, M. (1997) Frataxin is reduced in Friedreich ataxia patients and is associated with mitochondrial membranes. Hum. Mol. Genet., 6(11): 1771–1780.

Campuzano, V., Montermini, L., Molto, M.D., Pianese, L., Cossee, M., Cavalcanti, F., Monros, E., Rodius, F., Duclos, F., Monticelli, A. et al. (1996) Friedreich's ataxia: autosomal recessive disease caused by an intronic GAA triplet repeat expansion. *Science*, 271(5254): 1423–1427.

Cedar, H. (1988) DNA methylation and gene activity. *Cell*, 53: 3–4.

Chen, X., Mariappan, S.V., Moyzis, R.K., Bradbury, E.M., Gupta, G. Chen, X., Mariappan, S.V., Moyzis, R.K., Bradbury, E.M. and Gupta, G. (1995) Hairpins are formed by the single DNA strands of the fragile X triplet repeats: structure and biological implications. *Proc. Natl. Acad. Sci USA*, 92: 5199–5203.

Chiurazzi, P., Pomponi, M.G., Willemsen, R., Oostra, B.A. and Neri, G. (1998) In vitro reactivation of the FMR1 gene involved in fragile X syndrome. *Hum. Mol. Genet.*, 7: 109–113.

Clark, S.J.H.J., Paul, C.L. and Frommer, M. (1994) High sensitivity mapping of methylated cytosines. *Nucl. Acids Res.*, 22: 2990–2997.

Cossée, M., Campuzano, V., Koutnikova, H., Fischbeck, K., Mandel, J.L., Koenig, M., Bidichandani, S.I., Patel, P.I., Molte, M.D., Canizares, J., De Frutos, R., Pianese, L., Cavalcanti, F., Monticelli, A., Cocozza, S., Montermini, L. and Pandolfo, M. (1997) Frataxin fracas. *Nat. Genet.*, 15(4): 337–338.

Cummings, C.J., Mancini, M.A., Antalffy, B., Defranco, D.B., Orr, H.T. and Zoghbi, H.Y. (1998) Chaperone suppression of aggregation and altered subcellular proteosome localization imply protein misfolding in SCA1. *Nat. Genet.*, 19: 148–154.

David, G., Abbas, N., Stevanin, G., Durr, A., Yvert, G., Cancel, G., Weber, C., Imbert, G., Saudou, F., Antoniou, E., Drabkin, H., Gemmill, R., Giunti, P., Benomar, A., Wood, N., Ruberg, M., Agid, Y., Mandel, J.L. and Brice, A. (1997) Cloning of the SCA7 gene reveals a highly unstable CAG repeat expansion. Nat. Genet., 17(1): 65–70.

Davies, M.B., McCurrach, M.E., Taneja, K.L., Singer, R.H. and Housman, D.E. (1997) Expansion of a CUG trinucleotide repeat in the 3′ untranslated region of the myotonic dystrophy protein kinase transcript results in nuclear retension of transcripts. Proc. Natl. Acad. Sci., 94: 7388–7393.

Davies, S.W., Turmaine, M., Cozens, B.A., DiFiglia, M., Sharp, A.H., Ross, C.A., Scherzinger, E., Wanker, E.E., Mangiarini, L. and Bates, G.P. (1997) Formation of neuronal intranuclear inclusions underlies the neurological dysfunction in mice transgenic for the HD mutation. *Cell*, 90(3): 537–548.

De Boulle, K., V.A.J.M.H., Reyniers, E., Vits, L., Hendrikx, J., Van Roy, B., Van Den Bos, F., de Graff, E., Oostra, B.A., Willems, P.J. (1993) A point mutation in the FMR-1 gene associated with fragile X mental retardation. *Nat. Genet.*, 3: 31–35.

de Graaf, E.W.R., Zhong, N., de Die-Smulders, C.E., Brown, W.T., Freling, G. and Oostra, B.A. (1995) Instability of the CGG repeat and expression of the FMR1 protein in a male fragile X patient with a lung tumor. *Am. J. Hum. Genet.*, 57: 609–618.

Devys, D., Lutz, Y., Rouyer, N., Belocq, J.P. and Mandel, J.L. (1993) The FMR-1 protein is cytoplasmic, most abundant in neurons and appears normal in carriers of fragile X premutation. *Nat. Genet.*, 4: 335–340.

DiFiglia, M., Sapp, E., Chase, K., Schwarz, C., Meloni, A., Young, C., Martin, E., Vonsattel, J.-P., Carraway, R., Reeves, S.A., Boyce, F.M. and Aronin, N. (1995) Huntingtin is a cytoplasmatic protein associated with vesicles in human and rat neurons. *Neuron*, 14: 1075–1081.

DiFiglia, M., Sapp, E., Chase, K.O., Davies, S.W., Bates, G.P., Vonsattel, J.P. and Aronin, N. (1997) Aggregation of huntingtin in neuronal intranuclear inclusions and dystrophic neurites in brain. *Science*, 277(5334): 1990–1993.

Doerfler, W. (1983) "DNA methylation and gene activity". *Annu. Rev. Biochem.*, 52: 93–124.

Drouin, R., Angers, M., Dallaire, N., Rose, T.M., Khandjian, W. and Rouseau F. (1997) Structural and funtional characterization of the human FMR1 promoter reveals similarities with the hnRNP-A2 promoter region. *Hum. Mol. Genet.*, 6: 2051–2060.

Dürr, A., Cossee, M., Agid, Y., Campuzano, V., Mignard, C., Penet, C., Mandel, J.L., Brice, A. and Koenig, M. (1996) Clinical and genetic abnormalities in patients with Friedreich's ataxia (see comments). *New Engl. J. Med.*, 335(16): 1169–1175.

Eberhart, D.E.M.H., Feng, Y., Warren, S. (1996) The fragile X mental retardation protein is a ribonucleoprotein containing both nuclear localization and nuclear export signals. *Hum. Mol. Genet.*, 5: 1083–1091.

Eichler, E.E., Holden, J.J., Popovich, B.W., Reiss, A.L., Snow, K., Thibodeau, S.N., Richards, C.S., Ward, P.A. and Nelson, D.L. (1994) Length of uniterrupted CGG repeats determines instability in the FMR1 gene. *Nat. Genet.*, 8: 88–94.

Feng, Y., Gutekunst, C.A., Eberhart, D.E., Yi, H., Warren, S.T. and Hersch, S.M. (1997) Fragile X mental retardation protein: nucleocytoplasmic shuttling and association with somatodendritic ribosomes. *J. Neurosci.*, 17(5): 1539–1547.

Feng, Y.Z.F., Lokey, L.K., Chastain, J.L., Lakkis, L., Eberhart, D. and Warren, S.T. (1995) Translational suppression by trinucleotide repeat expansion at FMR1. *Science*, 268: 731–734.

Filla, A., De, M.G., Cavalcanti, F., Pianese, L., Monticelli, A., Campanella, G. and Cocozza, S. (1996) The relationship between trinucleotide (GAA) repeat length and clinical features in Friedreich ataxia. *Am. J. Hum. Genet.*, 59(3): 554–560.

Fischer, U.H.J., Boelens, W.C., Mattaj, I.W. and Luhrmann, R. (1995) The HIV-1 Rev activation domain is a nuclear export signal that accesses an export pathway used by specific cellular RNAs. *Cell*, 82: 475–483.

Fry, M.a.L., L.A. (1992) The fragile X syndrome dCGGnnucleotide repeats form a stable tetrahelical structure. *Proc. Natl. Acad. Sci. USA*, 89: 763–767.

Fu, Y.H., Friedman, D.L., Richards, S., Pearlman, J.A., Gibbs, R.A., Pizzuti, A. et al. (1993) Decreased expression of myotonin-protein kinase messenger RNA and protein in adult form of Myotonic Dystrophy. *Science*, 260: 235–238.

Gastier, J.M., Pulido, J.C., Sunden, S., Brody, T., Buetow, K.H., Murray, J.C., Weber, J.L., Hudson, T.J., Sheffield, V.C. and Duyk, G.M. (1995) Survey of trinucleotide repeats in the human genome: assessment of their utility as genetic markers. *Hum. Mol. Genet.*, 4(10): 1829–1836.

Gibson, T.J., Thompson, J.D. and Heringa, J. (1993) KH domains within the FMR1 sequence suggest that fragile X syndrome stems from a defect in RNA metabolism. *FEBS Lett.*, 324: 361–366.

Goldberg, Y.P., Kalchman, M.A., Metzler, M., Nasir, J., Zeisler, J., Graham, R., Koide, H.P., O'Kusky, J., Sharp, A.H., Ross, G.A., Jirik, F. and Hayden, M.R. (1996) Absence of disease phenotype and intergenerational stability of the CAG repeat in transgenic mice expressing the human Huntington disease transcript. *Hum. Mol. Genet.*, 5: 177–185.

Gourdon, G.R.F., Lia, A.S., Duros, C., Blanche, M., Abitbol, M., Junien, C. and H.H.-R. (1997) Moderate intergenerational and somatic instability of a 55-CTG repeat in transgenic mice. *Nat. Genet.*, 15: 190–192.

Grierson, A.J., Van Groeningen, M., Groot, P.B.N., Lindblad, K., Schalling, M., Hoovers, J.M.H. and Baas, F. (1998) An integreted map of chromosome 18 CAG trinucleotide repeat loci. *Eur. J. Mol. Genet.*, in press.

Gronskov, K., Hjalgrim, H., Bjerager, M.O. and Brondum-Nielsen, K. (1997) Deletion of all CGG repeats plus flanking sequences in FMR1 does not abolish gene expression. *Am. J. Hum. Genet.*, 61: 961–967.

Gutekunst, C.A., Levey, A.I., Heilman, C.J., Whaley, W.L., Yi, H., Nash, N.R., Rees, H.D., Madden, J.J. and Hersch, S.M. (1995) Identification and localization of huntingtin in brain and human lymphoblastoid cell lines with anti-fusion protein antibodies. *Proc. Natl. Acad. Sci. USA*, 92(19): 8710–8714.

Hagerman, R.J. (1996) Physical and behavioral phenotype. In: R.J. Hagerman and A. Cronister (Eds.), *Fragile X syndrome: diagnosis treatment and research*, Johns Hopkins University Press, Baltimore, pp. 3–87.

Hansen, R.S., Gartler, S.M., Scott, C.R., Chen, S.H. and Laird, C.D. (1992) Methylation analysis of CGG sites in the CpG island of the human FMR1 gene. *Hum. Mol. Genet.*, 1(8): 571–578.

Harding, A.E. (1993) Clinical features and classification of the late onset autosomal dominant cerebellar ataxias. *Adv. Neurol.*, 61: 1–14.

Hinds, H.L., Ashley, C.T., Sutcliffe, J.S., Nelson, D.L., Warren, S.T., Housman, D.E. and Schalling, M. (1993) Tissue specific expression of FMR-1 provides evidence for a functional role in fragile X syndrome. *Nat. Genet.*, 3: 36–43.

Holmberg, M., Duyckaerts, C., Dürr, A., Cancel, G., Gourfinkel-An, I., Damier, P., Facheux, B., Trottier, Y., Hirsch, E.C., Agid, Y. and Brice, A. (1998) spinocerebellar ataxia type 7 (SCA7): a neurodegenerative disorder with neuronal intranuclear inclusions. *Hum. Mol. Genet.*, 7: 913–918.

Hwu, W.L.L.Y., Lee, S.C. and Wang, T.R. (1993) In vitro DNA methylation inhibits FMR-1 promoter. *Biochem. Biophys. Res. Commun.*, 193: 324–329.

Höller, M., Westin, G., Jiriciny, J. and Schaffner, W. (1988) Sp1 transcription factor binds DNA and activates transcription even when the binding site is CpG methylated. *Genes and Develop.*, 2: 1127–1135.

Ikeda, H., Yamaguchi, M., Sugai, S., Aze, Y., Narumiya, S. and Kakizuka, A. (1996) Expanded polyglutamine in the Machado-Joseph disease. Nat. Genet., 13: 196–202.

Ikeuchi, T., Takano, H., Koide, R., Horikawa, Y., Honma, Y., Onishi, Y., Igarashi, S., Tanaka, H., Nakao, N., Sahashi, K., Tsukagoshi, H., Inoue, K., Takahashi, H. and Tsuji, S. (1997) Spinocerebellar ataxia type 6: CAG repeat expansion in alpha1A voltage-dependent calcium channel gene and clinical variations in Japanese population. Ann. Neurol., 42(6): 879–884.

Imbert, G., Saudou, F., Yvert, G., Devys, D., Trottier, Y., Garnier, J. M., Weber, C., Mandel, J.L., Cancel, G., Abbas, N., Durr, A., Didierjean, O., Stevanin, G., Agid, Y. and Brice, A. (1996) Cloning of the gene for spinocerebellar ataxia 2 reveals a locus with high sensitivity to expanded CAG/glutamine repeats. Nat. Genet., 14(3): 285–291.

Jansen, G.G.P., Bachner, D., Jap, P.H., Coerwinkel, M., Oerlemans, F., van den Broek, W., Gohlsch, B.P.D., Plomp, J.J., Molenaar, P.C., Nederhoff, M.G., van Echteld, C.J., Dekker, M., Berns, A.H.H. and Wieringa, B. (1996) Abnormal myotonic dystrophy protein kinase levels produce only mild myopathy in mice. *Nat. Genet.*, 13: 316–324.

Kang, S., Jaworski, A., Ohshima, K. and Wells, R.D. (1995) Expansion and deletion of CTG triplet repeats from human disease genes are determined by the direction of replication. *Nat. Genet.*, 10: 213–218.

Kawaguchi, Y., Okamoto, T., Taniwaki, M., Aizawa, M., Inoue, M., Katayama, S., Kawakama, S., Kawakami, H., Nakamura, S., Nishimura, M., Akiguchi, I., Kimura, J., Narumiya, S. and Kakizuka, A. (1994) CAG expansions in a novel gene for Machado-Joseph disease at chromosome 14q32.1. *Nat. Genet.*, 8: 221–228.

Keshet, I., Yisralei, J. and Cedar, H. (1985) Effect of regional DNA methylation on gene expression. *Proc. Natl. Acad. Sci USA*, 82: 2560–2564.

Keshet, I., Lieman, J. and Cedar, H. (1986) DNA methylation affects the formation of active chromatin. *Cell*, 44: 535–543.

Khan, R., Zhang, X.-Y., Supakar, P.C., Ehrlich, K.C. and Ehrlich, M. (1988) Human methylated DNA-binding protein: determinants of a pBR322 recognition site. *J. Biol. Chem.*, 264: 14374–14383.

Klement, I.A., S.P.J., Kaytor, M.D., Yi, H., Hersch, S.M., Clark, H.B. and Zoghbi, H. Y., and Orr, H.T. (1998). Ataxin-1 Nuclear Localization and Aggregation: Role in Polyglutamine-Induced Disease in SCA1 Transgenic Mice. *Cell,* **XXXX**

Koide, R., Ikeuchi, T., Onodera, O., Tanaka, H., Igarashi, S., Endo, K., Takahashi, H., Kondo, R., Ishikawa, A., Hayashi, T. et al. (1994) Unstable expansion of CAG repeat in hereditary dentatorubral-pallidoluysian atrophy (DRPLA). Nat. Genet., 6(1): 9–13.

Koob, M.D., Benzow, K.A., Bird, T.D., Day, J.W., Moseley, M.L. and Ranum, L.P. (1998) Rapid cloning of expanded trinucleotide repeat sequences from genomic DNA. Nat. Genet., 18(1): 72–75.

Koshy, B.M.T., Burright, E.N., Merry, D.E., Fischbeck, K.H., Orr, H.T. et al. (1996) Spinocerebellar ataxia type-1 and spinobulbar muscular atrophy gene products interact with glyceraldehyde-3-phosphate dehydrogenase. *Hum. Mol. Genet.*, 5: 1311–1318.

Kruczek, I.a. D.W. (1982) The unmethylated state of the promoter/leader and 5′-regions of integrated adenovirus genes correlates with gene expression. *EMBO J.*, 1: 409–414.

La Spada, A.R., Wilson, E.M., Lubahn, D.B., Harding, A.E. and Fischbeck, K.H. (1991) Androgen receptor gene mutations in X-linked spinal and bulbar muscular atrophy. *Nature*, 352: 77–79.

La Spada, A.R., P.K., Meadows, S.A., McClain, M.E., Jeng, G., Chmelar, R.S., Haugen, H.A., C.K., Singer, M.J., Moore, D., Trask, B.J., Fischbeck, K.H., Clegg, C.H. and McKnight, G.S. (1998) Androgen receptor YAC transgenic mice carrying CAG 45 alleles show trinucleotide repeat instability. *Hum. Mol. Genet.*, 7: 959–967.

Lafreniere, R.G., Rochefort, D.L., Chretien, N., Rommens, J.M., Cochius, J.I., Kalviainen, R., Nousiainen, U., Patry, G., Farrell, K., Soderfeldt, B., Federico, A., Hale, B.R., Cossio, O.H., Sorensen, T., Pouliot, M.A., Kmiec, T., Uldall, P., Janszky, J., Pranzatelli, M.R., Andermann, F., Andermann, E. and Rouleau, G.A. (1997) Unstable insertion in the 5′ flanking region of the cystatin B gene is the most common mutation in progressive myoclonus epilepsy type 1, EPM1. *Nat. Genet.*, 15(3): 298-302.

Lalioti, M.D., Scott, H.S., Buresi, C., Rossier, C., Bottani, A., Morris, M.A., Malafosse, A. and Antonarakis, S.E. (1997) Dodecamer repeat expansion in cystatin B gene in progressive myoclonus epilepsy (see comments). *Nature*, 386(6627): 847–851.

Lamont, P.J., Davis, M.B. and Wood, N.W. (1997) Identification and sizing of the GAA trinucleotide repeat expansion of Friedreich's ataxia in 56 patients. Clinical and genetic correlates. *Brain*, **XXXX**.

Lavedan, C., Hofmann-Radvanyi, H., Shelbourne, P., Rabes, J.P., Duros, C., Savoy, D. et al. (1993) Myotonic-Dystrophy: Size and sex-dependent dynamics of CTG meiotic instability, and somatic mosaicism. *Am. J. Hum. Genet.*, 52: 875–883.

Lavedan, C.N., Garett, L. and Nussbaum, R.L. (1997) Trinucleotide repeats (CGG)22TGG(CGG)43TGG(CGG)-21 from the fragile X gene remain stable in transgenic mice. *Hum. Genet.*, 100: 407–414.

Lindblad, K.L.A., Maciel, P., Stevanin, G., Zander, C., Klockgether, T., Ratzlaff, T., Brice, A.R.G., Hudson, T., Auburger, G. and Schalling, M. (1996) Mutation detection in Machado-Joseph disease using repeat expansion detection. *Mol. Med.*, 2: 77–85.

Lindblad, K., N.P., De Bruyn, A., Sourey, D., Zander, C., Engstrom, C., Holmgren, G., Hudson, T., C.J., Mendlewicz, J. et al. (1995) Detection of expanded CAG repeats in bipolar affective disorder using the repeat expansion detection (RED) method. *Neurobiol. Dis.*, 2: 55–62.

Lindblad, K., N.P.-O., Zander, C., Yuan, Q.-P., Ståhle, L., Engström, C., Balciuniene, J., Pettersson, U., Breschel, T., McInnis, M., Ross, C.A., Adolfsson, R. and Schalling, M. (1998) Two commonly expanded CAG/CTG repeat loci: involvement in affective disorders? *Mol. Psychiat.*, 2.

Lindblad, K., S.M., Stevanin, G., Holmberg, M., Digre, K., Zander, C., Ehrsson, H., David, G., Benomar, A., Nikoskelainen, E., Trottier, Y., Holmgren, G., Ptacek, L.J., Anttinen, A., Brice, A. and Schalling, M. (1996) An expanded CAG repeat sequence in spinocerebellar ataxia type 7. *Genome Res.*, 6: 965–971.

Mangiarini, L., S.K., Seller, M., Cozens, B., Harper, A., Hetherington, C., Lawton, M., Trottier, Y., Lehrach, H., Davies, S.W. and Bates, G.P. (1996) Exon 1 of the HD gene with an expanded CAG repeat is sufficient to cause a progressive neurological phenotype in transgenic mice. *Cell*, 87: 493–506.

Mangiarini, L., S.K., Mahal, A., Mott, R., Seller, M. and Bates, G.P. (1997) Instability of highly expanded CAG repeats in mice transgenic for the Huntington's disease mutation. *Nat. Genet.*, 15: 197–200.

Margolis, R.L., Stine, O.C., McInnis, M.G., Ranen, N.G., Rubinsztein, D.C., Leggo, J., Brando, L.V., Kidwai, A.S., Loev, S.J., Breschel, T.S., Callahan, C., Simpson, S.G., DePaulo, J.R., McMahon, F.J., Jain, S., Paykel, E.S., Walsh, C., DeLisi, L.E., Crow, T.J., Torrey, E.F., Ashworth, R.G., Macke, J.P., Nathans, J. and Ross, C.A. (1996) cDNA cloning of a human homologue of the Caenorhabditis elegans cell fate-determining gene mab-21: expression, chromosomal localization and analysis of a highly polymorphic (CAG)n trinucleotide repeat. *Hum. Mol. Genet.*, 5(5): 607–616.

Matilla, T., Koshy, B., Cummings, C.J., Isobe, T., Orr, H.T. and Zoghbi, H.Y. (1997) The cerebellar leucine rich acidic nuclear protein interacts with ataxin-1. *Nature*, 389: 974–978.

Meehan, R.R., L.J., McKay, S., Kleiner, E.L. and Bird, A.P. (1989) Identification of a mammalian protein that binds specifically to DNA containing methylated CpGs. *Cell*, 58: 499–507.

Monckton, D.G., C.M., Ashizawa, K.T., Siciliano, M.J. and Caskey, C.T. (1997) Hypermutable myotonic dystrophy CTG repeats in transgenic mice. *Nat. Genet.*, 15: 193–196.

Monros, E., Molto, M.D., Martinez, F., Canizares, J., Blanca, J., Vilchez, J.J., Prieto, F., de Frutos, R. and Palau, F. (1997) Phenotype correlation and intergenerational dynamics of the Friedreich ataxia GAA trinucleotide repeat. Am. J. Hum. Genet., 61(1): 101–110.

Montermini, L., A.E., Labuda, M., Richter, A., Pandolfo, M., Cavalcanti, F. et al. (1997) The Friedrich ataxia GAA triplet repeat: premutation and normal alleles. *Hum. Mol. Genet.*, 6: 1261–1266.

Morrone, A., Pegoraro, E., Angelini, C., Zammarchi, E., Marconi, G. and Hoffman, E.P. (1997) RNA metabolism in myotonic dystrophy: patient muscle shows decreased insulin receptor RNA and protein consistent with abnormal insulin resistance. *J. Clin. Investigation*, 99(7): 1691–1698.

Mundlos, S., Otto, F., Mundlos, C., Mulliken, J.B., Aylsworth, A.S., Albright, S., Lindhout, D., Cole, W.G., Henn, W., Knoll, J.H., Owen, M.J., Mertelsmann, R., Zabel, B.U. and Olsen, B.R. (1997) Mutations involving the transcription factor CBFA1 cause cleidocranial dysplasia. *Cell*, 89(5): 773–779.

Muragaki, Y., Mundlos, S., Upton, J. and Olsen, B.R. (1996) Altered growth and branching patterns in synpolydactyly caused by mutations in HOXD13. Science, 272(5261): 548–551.

Muragaki, Y., Mundlos, S., Upton, J. and Olsen, B.R. (1997) Polyalanine expansion in synpolydactyly might result from unequal crossing-over of HOX13. *Science*, 275: 408–409.

Murray, J., C.H., Taylor, G. and Hewison, J. (1997) Screening for fragile X syndrome. *Health Technology Assessment*, 1: 1–71.

Nagafuchi, S., Yanagisawa, H., Sato, K., Shirayama, T., Ohsaki, E., Bundo, M., Takeda, T., Tadokoro, K., Kondo, I., Murayama, N., Tnaka, Y., Kikushima, H. et al. (1994) Dentatorubral pallidoluysian atrophy expansion of an unstable CAG trinucleotide on chromosome 12p. *Nat. Genet.*, 1: 14–18.

Nakamoto, M., Takebayashi, H., Kawaguchi, Y., Narumiya, S., Taniwaki, M., Nakamura, Y., Ishikawa, Y., Akiguchi, I., Kimura, J. and Kakizuka, A. (1997) A CAG/CTG expansion in the normal population. Nat. Genet., 17: 385–386.

Nan, X., N.H., Johnson, C.A., Laherty, C.D., Turner, B.M., Eisenman, R.N. and Bird, A. (1998) Transcriptional repression by the methyl-CpG-binding protein MeCP2 involves a histone deacetylase complex. *Nature*, 393: 386–389.

Nielsen, J.E., K.P., Abell, K., Hasholt, L., Eiberg, H., Fenger, K. et al. (1997) CAG repeat expansion in autosomal dominant pure spastic paraplegia linked to chromosome 2p21-p24. *Hum. Mol. Genet.*, 6: 1811–1816.

O'Donovan, M.C., Guy, C., Craddock, N., Bowen, T., McKeon, P., Macedo, A., Maier, W., Wildenauer, D., Aschauer, H.N., Sorbi, S., Feldman, E., Mynett, J.L., Claffey, E., Nacmias, B., Valente, J., Dourado, A., Grassi, E., Lenzinger, E., Heiden, A.M., Moorhead, S., Harrison, D., Williams, J., McGuffin, P. and Owen, M.J. (1996) Confirmation of association between expanded CAG/CTG repeats and both schizophrenia and bipolar disorder. *Psychol. Med.*, 26(6),: 1145–1153.

O'Donovan, M.C., Guy, C., Craddock, N., Murphy, K.C., Cardno, A.G., Jones, L.A., Owen, M.J. and McGuffin, P. (1995) Expanded CAG repeats in schizophrenia and bipolar disorder. *Nat. Genet.*, 10: 380–381.

Oberlé, I., Rousseau, F., Heitz, D., Kretz, K., Devys, D., Hanauer, A., Boué, J., Bertheas, M.F. and Mandel, J.L. (1991) Instability of a 550- base pair DNA segment and abnormal methylation of in fragile X syndrome. *Science*, 252: 1097–1102.

Ohshima, K., Kang, S., Larson, J.E. and Wells, R.D. (1996) Cloning, characterisation, and properties of seven triplet repeat DNA sequences. *J. Biol. Chem.*, 271: 16773–16783.

Ohshima, K., Kang, S. and Wells, R.D. (1996) CTG triplet repeats from human hereditary disease are dominant genetic expansion products in *E. coli*. *J. Biol. Chem.*, 271: 1853–1856.

Oliva, R. and Dixon, G.H. (1989) Chicken protamine genes are intronless. The complete genomic sequence and organization of the two loci. *J. Biol. Chem.*, 264: 12472–12481.

Ordway, J.M., Tallaksen, G.S., Gutekunst, C.A., Bernstein, E.M., Cearley, J.A., Wiener, H.W., Dure, L.T., Lindsey, R., Hersch, S.M., Jope, R.S., Albin, R.L. and Detloff, P.J.

(1997) Ectopically expressed CAG repeats cause intranuclear inclusions and a progressive late onset neurological phenotype in the mouse. *Cell*, 91(6): 753–763.

Orr, H.T., Chung, M.Y., Banfi, S., Kwiatkowski, T.J., Servadio, A., Beaudet, A.L., McCall, A.E., Duvick, L.A., Ranum, L.P. and Zoghbi, H.Y. (1993) Expansion of an unstable trinucleotide CAG repeat in spinocerebellar ataxia type 1. *Nat. Genet.*, 4(3): 221–226.

Oruc, L., Lindblad, K., Verheyen, G.R., Ahlberg, S., Jakovljevic, M., Ivezic, S., Raeymaekers, P., Van Broeckhoven, C. and Schalling, M. (1997) CAG repeat expansion in bipolar and unipolar disorders. *Am. J. Hum. Genet.*, 60: 730–732.

Paterson, A.D., Naimark, D.M.J., Vincent, J.B., Kennedy, J.L. and Petronis, A. (1998) Genetic anticipation in neurological and other disorders. In: R.D. Wells and S.T. Warren (Eds.), *Genetic instability and hereditary neurological diseases*, Academic Press, San Diego, pp. 413–428.

Paulson, H.L., Perez, M.K., Trottier, Y., Trojanowski, J.Q., Subramony, S.H., Das, S.S., Vig, P., Mandel, J.L., Fischbeck, K.H. and Pittman, R.N. (1997) Intranuclear inclusions of expanded polyglutamine protein in spinocerebellar ataxia type 3. *Neuron*, 19(2): 333–344.

Philips, A.V., Timchenko, L.T. and Cooper, T.A. (1998) Disruption of splicing regulated by a CUG-binding protein in myotonic dystrophy. *Science*, 280: 737–741.

Pujana, M.A., Volpini, V. and Estivill, X. (1998) Large CAG/CTG repeat templates produced by PCR, usefulness for the DIRECT method of cloning genes with CAG/CTG expansions. *Nucl. Acid Res.*, 5: 1352–1353.

Pulst, S.M., Nechiporuk, A., Nechiporuk, T., Gispert, S., Chen, X.N., Lopes Cendes, I., Pearlman, S., Starkman, S., Orozco, D.G., Lunkes, A., DeJong, P., Rouleau, G.A., Auburger, G., Korenberg, J.R., Figueroa, C. and Sahba, S. (1996) Moderate expansion of a normally biallelic trinucleotide repeat in spinocerebellar ataxia type 2. *Nat. Genet.*, 14(3): 269–276.

Reddy, S., S.D., Rich, M.M., Leferovich, J.M., Reilly, P., Davis, B.M., Tran, K., Rayburn, H., Bronson, R., C.D., Balice-Gordon, R.J. and Housman, D. (1996) Mice lacking the myotonic dystrophy protein kinase develop a late onset progressive myopathy. *Nat. Genet.*, 13: 325–335.

Riggins, G.J., Lokey, L.K., Chastain, J.L., Leiner, H.A., Sherman, S.L., Wilkinson, K.D. and Warren, S.T. (1992) Human genes containing polymorphic trinucleotide repeats. Nat. Genet., 2(3): 186–191.

Romeo, G., Menozzi, P., Ferlini, A., Fadda, S., Di Donato, S., Uziel, G. and Lucci, B. (1983) Incidence of Friedreich ataxia in Italy estimated from consanguinous marriages. Am. J. Hum. Genet., 35: 523–529.

Rountree, M.R. and Selker, E. (1997) DNA methylation inhibits elongation but not initiation of transcription in *Neurospora crassa*. *Genes Dev.*, 11: 2383–2395.

Sabouri, L.A., Mehedevan, M.S., Narang, M., Lee, D.S., Surh, L.C. and Korneluk, R.G. (1993) Effect of the myotonic dystrophy (DM) mutation on mRNA levels of the DM gene. *Nat. Genet.*, 4: 233–238.

Samadashwily, G.M., Raca, G. and Mirkin, S.M. (1997) Trinucleotide repeats affect DNA replication in vivo. *Nat. Genet.*, 17: 298–304.

Sandberg, G. and Schalling, M. (1997) Effect of in vitro promoter methylation and CGG repeat expansion on FMR-1 expression. *Nucl. Acids Res.*, 25(14): 2883–2887.

Sanpei, K., Takano, H., Igarashi, S., Sato, T., Oyake, M., Sasaki, H., Wakisaka, A., Tashiro, K., Ishida, Y., Ikeuchi, T., Koide, R., Saito, M., Sato, A., Tanaka, T., Hanyu, S., Takiyama, Y., Nishizawa, M., Shimizu, N., Nomura, Y., Segawa, M., Iwabuchi, K., Eguchi, I., Tanaka, H., Takahashi, H. and Tsuji, S. (1996) Identification of the spinocerebellar ataxia type 2 gene using a direct identification of repeat expansion and cloning technique, DIRECT. *Nat. Genet.*, 14(3): 277–284.

Schalling, M., Hudson, T.J., Buetow, K.H. and Housman, D.E. (1993) Direct detection of novel expanded trinucleotide repeats in the human genome. *Nat. Genet.*, 4(2): 135–139.

Schwemmle, S., De Graff, E., Deissler, H., Gläser, D., Wöhrle, D., Kennerknecht, I., Just, W., Oostra, B.A., Doerfler, W., Vogel, W. and Steinbach P. (1997) Characterization of FMR1 promoter elements by in vivo -footprinting analysis. *Am. J. Hum.*, 60: 1354–1362.

Servadio, A., Koshy, B., Armstrong, D., Antalffy, B., Orr, H.T. and Zoghbi, H.Y. (1995) Expression analysis of the ataxin-1 protein in tissues from normal and spinocerebellar ataxia type 1 individuals. *Nat. Genet.*, 10: 94–98.

Simental, J.A., Sar, M., Lane, M.V., French, F.S. and Wilson, E.M. (1991) Transcriptional activation and nuclear targeting signals of the human androgen receptor. *J. Biol. Chem.*, 266: 510–518.

Siomi, H., C.M., Siomi, M., Nussbaum, R.L. and Dreyfuss, G. (1994) Essential role for KH domains in RNA binding: Impaired RNA binding by a mutation ion the KH domainsof FMR1 that causes Fragile X Syndrome. *Cell*, 77: 33–39.

Siomi, H., S.M.C., Nussbaum, R.L. and Dreyfuss, G. (1993) The protein product of the fragile X gene, FMR1, has characteristics of an RNA-binding protein. *Cell*, 74: 291–298.

Sittler, A., D.D., Weber, C. and Mandel, J.-L. (1996) Alternative splicing of exon 14 determines nuclear or cytoplasmic localization of fmr1 protein isoforms. *Hum. Mol. Genet.*, 5: 95–102.

Smeets, H.J., S.A., Verheij, C.E., Theelen, J.P., Willemsen, R., van de Burgt, I., Hoogeveen, A.T., Oosterwijk, J.C. and Oostra, B.A. (1995). Normal phenotype in two brothers with a full FMR1 mutation. *Hum. Mol. Genet.*, 4: 2103–2108.

Smith, S.S., L.A., Lingeman, R.G., Baker, D.J. and Riley, J. (1994) Hypermethylation of telomere-like foldbacks at codon 12 of the human c-Ha-ras gene and the trinucleotide repeat of the FMR-1 gene of fragile X. *J. Mol. Biol.*, 243(2): 143–151.

Stevanin, G., Trottier, Y., Cancel, G., Durr, A., David, G., Didierjean, O., Burk, K., Imbert, G., Saudou, F., Abada, B.M., Gourfinkel, A.N.I., Benomar, A., Abbas, N., Klockgether, T., Grid, D., Agid, Y., Mandel, J.L. and Brice,

A. (1996) Screening for proteins with polyglutamine expansions in autosomal dominant cerebellar ataxias. *Hum. Mol. Genet.*, 5(12): 1887–1892.

Tamanini, F., M.N., Verheij, C., Willems, P.J., Galjaard, H., Oostra, B.A. and Hoogeveen, A.T. (1996): FMRP is associated to the ribosomes via RNA. *Hum. Mol. Genet.*, 5: 809–813.

Taneja, K.L., McCurrach, M., Schalling, M., Housman, D. and Singer, R.H. (1995) Foci of trinucleotide repeat transcripts in nuclei of myotonic dystrophy cells and tissues. *J. Cell Biol.*, 128: 9955–1002.

Tapscott, S.J., K.T., Widrow, R.J., Stoger, R. and Laird, C.D. (1998) Fragile-X syndrome and myotonic dystrophy: parallels and paradoxes. *Curr. Opin. Genet. Dev.*, 8: 245–253.

Teh, B.T., Sullivan, A.A., Farnebo, F., Zander, C., Li, F.Y., Strachan, N., Schalling, M., Larsson, C. and Sandstrom, P. (1997) Oculopharyngeal muscular dystrophy (OPMD) - report and genetic studies of an Australian kindred. *Clin. Genet.*, 51(1): 52–55.

The Huntington's Disease Collaborative Research Group (1993) A novel gene containing a trinucleotide repeat that is expanded and unstable on Huntington's disease chromosomes. *Cell*, 72: 971–983.

Timchenko, L.T., Miller, J.W., Timchenko, N.A., DeVore, D.R., Datar, K.V., Lin, L., Roberts, R., Caskey, C.T. and Swanson, M.S. (1996) Identification of a (CUG)n triplet repeat RNA-binding protein and its expression in myotonic dystrophy. *Nucl. Acids Res.*, 24(22): 4407–4414.

Timchenko, L.T., Timchenko, N.A., Caskey, C.T. and Roberts, R. (1996) Novel proteins with binding specificity for DNA CTG repeats and RNA CUG repeats: implications for myotonic dystrophy. Hum. Mol. Genet., 5(1): 115–121.

Tome, F.M., Chateau, D., Helbling, L.A. and Fardeau, M. (1997) Morphological changes in muscle fibers in oculopharyngeal muscular dystrophy. *Neuromuscular Disorders*.

Trottier, Y., Devys, D., Imbert, G., Saudou, F., An, I., Lutz, L., Weber, C., Agid, Y., Hirsch, E.C. and Madnel, J.-L. (1995) Cellular localization of the Huntington's disease protein and discrimination of the normal and mutated form. Nat. Genet., 10: 104–110.

Trottier, Y., Lutz, Y., Stevanin, G., Imbert, G., Devys, D., Cancel, G., Saudou, F., Weber, C., David, G., Tora, L., Agid, Y., Brice, A. and Mandel, J.-L. (1995) Polyglutamine expansion as a pathological epitope in Huntington's disease and four dominant cerebellar ataxias. *Nature*, 378(6555): 403–406.

Turk, V. and Bode, W. (1991) The cystatins: portein inhibitors of cystein proteinases. *FEBS Lett.*, 285: 213–219.

Verheij, C., B.C.E., de Graff, E., Keulemanns, J., Willemsen, R., Verkerk, A.J.M.H., Galjaard, H., Reuser, A.J.J., Hoogeveen, A.T. and Oostra, B.A. (1993) Characteruzation and localization of the FMR-1 gene product associated with the fragile X syndrome. *Nature*, 363: 722–724.

Verkerk, A.J.M.H., d.G.E., De Boulle, K., Eichler, E., Konecki, D.S., Reyniers, E., Manca, A. Poustka, A. Willems, P.J. Nelson, D.L. and Oostra, B.A. (1993) Alternative splicing in the fragile X gene FMR1. *Hum. Mol. Genet.*, 64: 399–404.

Verkerk, A.J.M.H., Pieretti, M., Sutcliffe, J.S., Fu, Y.H., Kuhl, D.P. and Pizzuti, A. et al. (1991) Identification of a gene (FMR-1) containing a CGG repeat coincident with a breakpoint cluster region exhibiting length variation in fragile X syndrome. *Cell*, 65: 905–914.

Virtaneva, K., D'Amato, E., Miao, J., Koskiniemi, M., Norio, R., Avanzini, G., Franceschetti, S., Michelucci, R., Tassinari, C.A., Omer, S., Pennacchio, L.A., Myers, R.M., Dieguez, L.J., Krahe, R., de la Chapelle, A. and Lehesjoki, A.E. (1997) Unstable minisatellite expansion causing recessively inherited myoclonus epilepsy, EPM1. *Nat. Genet.*, 15(4): 393–396.

Wang, R., Zhang, X.-Y. and Ehrlich, ?? (1986) A human DNA-binding protein is methylation -specific and sequence-specific. *Nucl. Acids. Res.*, 14: 1599–1614.

Wang, Y.-H., Gellibolian, R., Shimizu, M. and Wells, R.D. (1996) Long repeating CCG triplet repeat blocks exclude nucleosomes: a possible mechansim for the nature of fragile sites in chromosomes. *J. Mol. Biol.*, 263: 511–516.

Wang, Y.H., Amirhaeri, S., Kang, S., Wells, R.D. and Griffith, J.D. (1994) Preferential nucleosome assembly at DNA triplet repeats from the myotonic dystrophy gene. *Science*, 265(5172): 669–671.

Warrick, J.M., Paulson, H.L., Gray-Board, G.L., Bui, Q.T., Fischbeck, K.H., Pittman, R.N. and Bonini, N.M. (1998) Expanded polyglutamine protein forms nuclear inclusions and causes neural degeneration in *Drosophila*. *Cell*, 93: 939–949.

Watt, F. and Molloy, P.L. (1988) Cytosine methylation prevents binding to DNA of a Hela cell transcription factor required for optimal expression of the adenovirus major late promoter. *Genes Dev.*, 2: 1136–1143.

Wells, R.D. and Warren, S.T. (Eds.) (1998) *Genetic instabilites and hereditary neurological disease*, Academic Press, San Diego.

Willemsen, R., B.C., Tamanini, F., Galjaard, H., Hoogeveen, A. and Oostra, B. (1996) Association of FMRP with ribosomal precursor particles in the nucleolus. *Biochem. Biophys. Res. Commun.*, 5: 27–33.

Wolfl, S., S.M. and Wittig, B. (1991) Lack of correlation between DNA methylation and

transcriptional inactivation: the chicken lysozyme gene. *Proc. Natl. Acad. Sci. USA*, 88: 271–275.

Yazawa, I., Nukina, N., Hashida, H., Goto, J., Yamada, M. and Kanazawa, I. (1995) Abnormal gene product identified in hereditary dentatorubral-pallidoluysian atrophy (DRPLA) brain. *Nat. Genet.*, 10: 99–103.

Yazawa, I., Nukina, N., Hashida, H., Goto, J., Yamada, M. and Kanazawa, I. (1995) Abnormal gene product identified in hereditary dentatorubral-pallidoluysian atrophy (DRPLA) brain. *Nat. Genet.*, 10: 99–103.

Yisraeli, J., Adelstein, R.S., Melloul, D., Nudel, U., Yaffe, D., and Cedar, H. (1986) Muscle-specific activation of a methylated chimeric actin gene. *Cell*, 46: 409–416.

Zhang, X.Y., Ehrlich, K.C., Wang, R.Y.-H. and Ehrlich, M. (1986) Effect of site-specific DNA methylation and mutagenesis on recognition by methylated DNA binding protein from human placenta. *Nucl. Acids Res.*, 14: 8387–8397.

Zhang, Y., O.C.P., Siomi, M., Srinivasan, S., Dutra, A., Nussbaum, R.L., Dreyfuss, G. (1995) The fragile X mental retardation syndrome protein interacts with novel homologs FXR1 and FXR2. EMBO J. 14, 5358-5366.

Zhou, Z.-X., Wong, C.-I., Sar, M. and Wilson, E.M. (1994) The androgen receptor: an overview. *Recent Progress in Hormone Research*, 49: 249–274.

W.E. Crusio and R.T. Gerlai (Eds.)
Handbook of Molecular-Genetic Techniques for Brain and
Behavior Research (Techniques in the Behavioral and Neural Sciences, Vol. 13)

CHAPTER 5.4

Finding liability genes for schizophrenia

Nick J.O. Jacobsen, Nigel M. Williams and Michael J. Owen

Neuropsychiatric Genetics Unit, Divisions of Psychological Medicine and Medical Genetics, Tenovus Building, University of Wales College of Medicine, Heath Park, Cardiff CF4 4XN, UK

Introduction

It has been recognised for many years that schizophrenia tends to aggregate in families. From the results of some 40 family studies over seven decades, the lifetime risk of schizophrenia in first degree relatives of affected probands has been estimated at about 10%, which is approximately 10 times higher than the risk in the general population world-wide (Gottesman, 1991). The fact that an important role for genetics is now accepted owes much to a series of now classical twin and adoption studies that have demonstrated unequivocally that familial clustering reflects mainly genetic mechanisms. Thus, concordance rates for schizophrenia are consistently higher in monozygotic (MZ) than in dizygotic (DZ) twins, with probandwise MZ concordance rates averaging 46% vs DZ rates of about 14% (Gottesman and Shields, 1982). Also, adoption studies of a variety of designs have shown that the biological relatives of affected individuals have significantly greater rates of schizophrenia and related disorders than adoptive relatives (McGuffin et al., 1994).

While these findings imply that genes are major etiological factors, it is also apparent from the high rates of discordance in MZ twins that what is inherited is a predisposition or liability to schizophrenia rather than the certainty of developing the disorder. The exact nature of the genetic 'trait' remains elusive but it is likely to embrace a much more subtle spectrum of illness than simply 'full-blown' schizophrenia (Rosenthal et al., 1968; Farmer et al., 1987).

As well as difficulties in defining exactly what is transmitted, it is uncertain how 'it' is transmitted. In spite of evidence for the involvement of genes, inheritance of schizophrenia, like other common disorders does not follow simple Mendelian patterns. This has traditionally been attributed to a combination of multiple additive genetic effects (oligogenes or polygenes) and environmental factors (multifactorial inheritance) which together determine an individual's liability to disease, with illness occurring when a threshold of liability is exceeded (so-called multifactorial liability threshold models) (Lander and Schork, 1994; Plomin et al., 1994).

The location and identification of genes for complex traits using molecular genetic techniques is, in general, difficult. Schizophrenia, as well as other psychiatric disorders, is even more challenging because of problems defining the phenotype; no laboratory or pathological diagnostic tests exist and virtually nothing is known about the pathophysiology of the disease. However, schizophrenia can be diagnosed reliably using modern operational diagnostic systems which define syndromes with high heritability (McGuffin et al., 1994).

Animal models have been used in genetic studies of some complex disorders, for example, diabetes

(Todd et al., 1991). Animals such as rodents offer many advantages over humans for genetic studies because they have large families and short intervals between generations, and because experimental crosses can be set up. Mouse models of disease have been used to help in identification of human genes by mapping the relevant mouse gene and then determining its human homologue. This is possible because homologues of closely linked human genes are often syntenic (on the same chromosome) in mouse. There are obvious problems in obtaining animal models for the study of schizophrenia although some advances have been made. For example, rodents have been bred that display strain differences in susceptibility to developing 'learned helplessness' (Henn and Edwards, 1994) which may be of use in identifying genes involved in susceptibility to depression. Three loci that influence emotionality in mice, a trait that may be related to anxiety in humans, have recently been mapped (Flint et al., 1995). Rodents have also been used successfully to map genes of potential importance in substance-abuse (Berrettini, 1993). It seems likely that animal models will play an increasing role in future genetic studies (Collins, 1995).

Family studies — searching for genes of major effect

Traditional linkage studies ideally require large families with multiply affected members. Linkage analysis involves proposing a transmission model that will account for the observed inheritance pattern of genotypes and phenotypes in a pedigree, with the level of statistical significance being defined as a lod (logarithm of the odds) score. This approach therefore relies of the prior definition of specific genetic parameters, such as gene frequencies and disease gene penetrances, hence this form of analysis is known as parametric linkage analysis. As simple Mendelian traits have relatively few possible genetic models then these methods are ideally suited to them. For complex diseases such as schizophrenia where mode of inheritance is unknown it is difficult to define a precise inheritance pattern, so linkage analysis must be undertaken for several genetic models. This inevitably increases the potential of type I (false positive) errors due to multiple testing, and also type II (false negative) errors if the true mode of inheritance is not included within the parameter space examined. The approach is based upon the assumption that the large multiply affected families studied reflect the presence of disease genes of large effect. Linkage analysis requires that all individuals in the family are classified as affected or unaffected which is not easy for schizophrenia, where diagnosis can be strictly defined to include only patients meeting the DSM criteria for schizophrenia (American Psychiatric Association, 1994) or more loosely to include patients with other diagnoses including schizoid personality disorders. Moreover, the lod scores are very sensitive to any change in diagnoses of individuals in the family, increasing the chance of type II error. The realistic possibility of genetic heterogeneity, whereby different pedigrees will be linked to the different loci although they are phenotypically identical, means that any linkage model used for schizophrenia should allow for genetic heterogeneity in the sample. For these reasons it is clear that schizophrenia is a less than ideal substrate for this form of analysis, and it is perhaps not surprising that attempts to replicate positive linkage findings have been fraught with difficulty (O'Donovan and Owen, 1992). This method has, however, proved successful when applied to Mendelian forms of a number of common disorders, such as familial Alzheimer's disease (Goate et al., 1991; Sherrington et al., 1995), non-insulin dependent diabetes (Vionnet et al., 1992) and familial breast cancer (Hall et al., 1990; Miki et al., 1994).

Technical aspects

Molecular genetic analysis of families involves the use of microsatellite markers which are small regions of repetitive sequence The most abundant is the $(CA)_n$ dinucleotide repeat microsatellite of which there are an estimated 35,000–100,000 copies distributed throughout the human genome, giving an average density of one per 100 kilobases (Koreth et al., 1996). The polymorphic nature of many microsatellites makes them ideal molecular phenotypes for linkage analysis. Typically a 'grid' of evenly-spaced microsatellites (e.g. one marker per 20 cM) is selected and each marker analysed within the family or families being analysed. Genotyping is achieved using the polymerase chain reaction (PCR) which allows the specific and efficient amplification of a DNA sequence from a complex background (i.e. the human genome). Two oligonucleotides (approximately 20 bp each) are designed specific to the flanking regions of microsatellite sequences and amplification is achieved by repeated cycles of denaturation, annealing, and polymerisation using a thermostable DNA polymerase (e.g. *Taq* DNA polymerase)

Analysis of PCR products containing microsatellite sequences is achieved by polyacrylamide gel electrophoresis and detection by prior radioactive, or fluorescent dye labelling of oligonucleotides. Fluorescent dye labelling is particularly favourable, since modern technology allows the high-throughput analysis of many samples. Microsatellite PCR assays have been developed and genetically mapped on a large scale by organisations such as the Généthon laboratory in France (Weissenbach and colleagues). Microsatellite markers are now being placed on integrated maps to provide physical, as well as genetic, information concerning their localisations (Strachan and Read, 1996).

Application of family based linkage studies to schizophrenia

The first result that has been of sustained interest was reported after a complete genome scan using conventional linkage methods in 39 multiply affected families (Pulver et al., 1994b). Significant evidence for linkage was not obtained, but a post hoc analysis which maximised the lod score by varying the genetic model was suggestive of linkage to chromosome 22q12-q13. Later, the same group co-ordinated a four-centre replication study which was negative (Pulver et al., 1994a). Despite this, a number of groups have also reported weakly positive data from this region (Coon et al., 1994; Vallada et al., 1995). The second encouraging finding emerged from a study of small Irish pedigrees co-ordinated by a US team. In this large sample of 256 families, moderate evidence of linkage to markers from chromosome 6p24-p22 was provided in overlapping datasets (Straub et al., 1995; Wang et al., 1995). Estimates of genetic heterogeneity suggested that the disease allele was segregating in about 15–30% of the families, although the results were also consistent with oligogenic inheritance. Suggestive, but not significant, linkages to markers in the same region were subsequently reported by three independent groups (Antonarakis et al., 1995; Moises et al., 1995; Schwab et al., 1995) but not by others (Gurling et al., 1995; Mowry et al., 1995). The markers yielding strongest evidence for linkage varied between the four positive studies but there was overlap in the regions demonstrating linkage. Therefore, it seems likely that if there is a susceptibility gene for schizophrenia within 6p24-p22 it is probably of small effect ($\lambda_s = 1.25$) (Daniels et al., 1997). Rather weaker evidence has emerged that the short arm of chromosome 8 (Pulver et al., 1995; Moises et al., 1995) may also contain a susceptibility gene for schizophrenia, and preliminary evidence for linkage to chromosome 13q14-q32 (Lin et al., 1995) is gaining support from other datasets (Antonarakis et al., 1998). Currently, attempts to replicate these findings are ongoing using a multicentre collaborative approach based on that employed for chromosome 22.

Studies of large multiply affected families are not necessarily suitable for schizophrenia because they may not represent a Mendelian sub-form of the disease and could instead be a result of polygenic transmission. Furthermore, it has been shown that in heterogeneous diseases such as schizophrenia, large, densely affected pedigrees are actually more likely to have multiple disease genes than are smaller less densely affected ones (Durner et al., 1992). Moreover, if polygenic or oligogenic inheritance is operating, as seems likely, parents of several affected children have an increased probability of being homozygous at the disease locus thus making them uninformative for linkage. In contrast, collection of nuclear families with two affected siblings and preferably with both parents unaffected maximises the probability that the parents will be heterozygous at the disease locus and thus informative.

Sibling-pair studies

When the mode of inheritance of a disease such as schizophrenia is unknown or uncertain the type and size of the sample chosen for linkage analysis must provide optimal power to detect susceptibility genes over a range of genetic models. As previously described, interacting oligogenes or polygenes represent plausible models for schizophrenia (Risch, 1990c). This presents considerable problems for parametric methods of linkage analysis and because of this non-parametric methods of have become the focus of many current linkage studies in schizophrenia.

Allele sharing methods of linkage analysis are not concerned with predicting genetic models, but rather on rejecting a null hypothesis. The aim is to identify chromosomal regions that have transmission patterns which are not consistent with Mendelian segregation and are therefore inherited by affected relatives more often than expected by chance. Affected sib pair analysis is the simplest form of this method, having a number of advantages over traditional lod score methods. First, there is no requirement to specify any inheritance models and is therefore robust to these uncertainties. Second, the cases gathered are more representative of typical disease and it is possible to focus effort on collecting affected subjects who satisfy narrow and stable definitions of illness. Third, large samples are readily obtained and linkage results are less sensitive to laboratory or clinical errors. Finally, when compared to multiply affected families, affected sib pairs are more likely to be informative for linkage under oligogenic epistatic models which are plausible for complex traits such as schizophrenia (Risch, 1990a).

The use of affected sib pairs is more powerful when looking at complex traits as siblings would still be expected to show excess allele sharing even in the presence of incomplete penetrance, phenocopy, genetic heterogeneity, and if the disease allele exists at a high frequency in the general population. The degree of allele sharing that occurs at a particular locus depends upon the distance of the marker to the gene and also the recurrence risk to the population prevalence (λ_s) attributable to the gene in question. Although it is difficult to estimate the λ_s attributable to each of the putative genes involved in schizophrenia, plausible oligogenic models and those where two unlinked loci act multiplicatively (Holmans and Craddock, 1997) suggest that it is unlikely that any of the genes involved have a λ_s of more than 3. The power of an affected sib pair study to demonstrate linkage falls as the λ_s decreases (Risch, 1990b), and samples of 600–800 sib-pairs will be required to have sufficient power to detect genetic effects of the magnitude likely to be operating in schizophrenia (Hauser et al., 1996). The feasibility of accomplishing a complete genome scan using sibling pairs was first demonstrated in a study of type I diabetes where it is possible that five different susceptibility genes are involved (Davies et al., 1994).

Technical aspects of high-throughput genotyping

The aim of a genome screen is to genotype markers that are evenly spaced across the genome so that the distance from any gene will be no more than half the average marker separation. Increasing the marker density improves the power of the study, although the relationship is not linear and the marker grid used has to provide an optimal balance between power and quantity of work. For example, increasing the marker density from 20 cM to 2 cM doubles the power of the study but increases the amount of genotyping by a factor of 10 (Hauser et al., 1996). Two-stage strategies have therefore been proposed to allow an efficient design of a genome scan whereby the amount of laboratory work is minimised without adversely affecting power. In general, the stage one sample should have enough power to detect linkage to genes with a low significance threshold which reflects the nominal significance level of around 5%. Although most of the lod scores that exceed this threshold will be false positives, due to random excess sharing in the sample, it is hoped that at least a proportion will reflect genuine linkage and will remain significant when typed in the stage 2. Thus, the second stage sample should have enough power to detect linkage at high levels of significance (lod score >3). For example, one might genotype 100 sibling pairs using a wide 'grid' of markers (20 cM) in the first stage. The second stage may then involve the analysis of regions showing potential linkage in a full sample of 200 sibling pairs with a marker density of 5 cM. This two stage design thus reduces the amount of genotyping by concentrating on the interesting subchromosomal regions.

Even if a two stage approach is employed whereby 215 microsatellite markers at an average spacing of 20 cM are genotyped in the first 100 affected sib pairs, then approximately 43,000 PCR reactions are required for stage one alone. In order to complete a genome scan using a sibling pair approach in a realistic time frame high-throughput technology has been developed.

PCR reactions are routinely prepared by hand, or using a robotic sample processor. PCR-products containing microsatellites are labelled with different coloured fluorescent dyes and analysed using automated DNA sequencers, for example the ABI system. In this approach markers are batched according to the sizes of their PCR products. Typically, a batch will contain five markers with each PCR product having a different average size from each of the others (the size range being between 80 and 350 basepairs). By labelling each batch with one of 3 different colour fluorescent dyes (blue, green, or yellow) then 3 batches can be combined into a panel and analysed simultaneously. This means that up to 15 markers can be resolved according to their size and colour in a single lane of a polyacrylamide gel on an automated ABI sequencer. Data is automatically collected and analysed by computer software which sizes every fragment in relation to an internal size standard. Data is automatically genotyped using the 'Genotyper' software package which is able to discriminate between stutter bands and alleles. Data is then checked by the operator. The genotype data can then be electronically transferred to a phenotype database, thus reducing any manual intervention and therefore potential error. If each gel contains 48 lanes then potentially this system allows up to 720 genotypes to be generated from a single gel (Reed et al., 1994).

It is important to note that even low frequencies of genotyping errors (1.5%) can have profound influence on mapping outcomes (Buetow, 1991). Errors in genotyping can reduce power to detect genuine linkage, alter genetic marker order, and map length, and provide significant support for false data. In the case of family studies errors may result in misclassification of recombination events.

Application of the sibling pair linkage approach

With regard to schizophrenia, the sibling pair analysis approach was used to investigate the linkage finding on chromosome 22q12-q13 (Coon et al., 1994; Pulver et al., 1994a; Pulver et al., 1994b; Vallada et al., 1995). A UK based group co-ordinated a meta-analysis of the world data from 11 centres. Overall, the data suggested ($P = 0.001$) that a susceptibility locus is present in this subchromosomal region, although this still does not constitute definitive proof (Gill et al., 1996). We are currently completing analysis of this region in a sample of sibling pairs.

Association studies

Once linked regions containing susceptibility genes have been identified the next stage is to move from an approximate position on a genetic map to isolating the pathogenic mutation. This is likely to prove formidable, given the complex mode of transmission and the large number of sibling pairs required to map a disease locus to a relatively small region.

At present the best approach depends upon studies aimed at detecting linkage disequilibrium. This effect occurs when a particular genetic marker is situated in close proximity to a disease gene such that they are only very infrequently separated by recombination events during meiosis. This limited genetic shuffling gives rise to a small chromosomal segment that contains the disease gene in question and a non-random distribution of marker alleles. This can be detected in genetic studies as a difference in the frequency of marker alleles observed in a population of unrelated patients as compared to a population of control samples. Thus, unlike the linkage approach, association studies do not look at inheritance patterns within families, but they simply compare the frequency of alleles in samples of unrelated affected and unaffected individuals.

Linkage disequilibrium is an inherently short range phenomenon working upon the fact that individuals possessing the disease gene inherited it from a common ancestor and is usually only detectable over a distance of 1 cM (approximately 1 Megabase) or less. Therefore, the closer the two loci are the greater the chance that disequilibrium will be maintained. Consequently, this approach allows a much more precise localisation than linkage.

An important factor that confounds linkage disequilibrium studies is that the control group and patient group must be very well matched for demographic variables, including ethnic status, because allele frequencies may vary dramatically between populations (O'Donovan and Owen, 1992). This phenomenon is known as population stratification and can lead to false claims of linkage disequilibrium simply because there is a systematic difference in the genetic background of the patients and controls. It is possible to circumvent this problem by using internal controls, using the transmission disequilibrium test (TDT) developed by Spielman and Ewens (1998) and the haplotype relative risk method (HRR) developed by Falk and Rubinstein (1987) that study allele frequencies in a group of unrelated patients and their parents. Haplotype relative risk studies compare the allele frequencies in the patient group with those of a notional control group consisting of the parental alleles that were not passed on to the patients (Yang et al., 1993; Inayama et al., 1994). This ensures that both the experimental and control alleles are drawn from the same gene pool. Collection of so-called trios can be combined with collection of sibling pairs to produce a sample of use in both linkage studies using allele sharing approaches and association studies (Williams et al., 1996). The main disadvantage is that both parents need to be alive and willing to participate. The transmission disequilibrium test (TDT) introduced by Spielman and Ewens (1998) compares the frequency with which an allele is transmitted to affected individuals to the frequency with which it is not transmitted. A significant difference implicates linkage disequilibrium between the disease and the marker. An advantage of the TDT over the HRR is that it can be used to study multiple affected sibs within a sibship.

Association studies offer a number of practical advantages over linkage studies. They do not require large multiply affected families and the analysis requires no prior definition of the disease model. Furthermore, they have sufficient power to detect genes contributing only a small genetic effect using much smaller samples than those required by linkage analysis. For example, it has been calculated that an association study design using 700 trios should have sufficient power to detect an allele contributing to only 10% of the disease susceptibility in a population, whereas a sample of over 5000 ASP would be required to detect linkage at lod >3 (Risch and Merikangas, 1996).

The potential power and utility of linkage disequilibrium depends upon the population under investigation (De la Chapelle and Wright, 1998). Linkage disequilibrium can only be expected if a small number of single-origin founder mutations are common to many living affected individuals occurs. Also, expansion of a population should have occurred predominantly by growth rather than immigration. Many human populations fulfil these criteria, although the population has to be large enough to provide enough affected individuals. Also, if a high-resolution mapping study is being performed enough generations must have passed since the founding mutation in order to give the number of crossovers required. Conversely, for low-resolution mapping studies a relatively small number of generations must have passed. In isolated populations, especially those of relatively recent origin, linkage disequilibrium can be detectable over large distances. For instance, significant linkage disequilibrium was observed between familial hyperinsulinism and markers on chromosome 11 over a 10 cM region in an Ashkenazi Jewish population (Glaser et al., 1995).

Populations of individuals that are culturally or physically isolated are often genetically isolated (Sheffield et al., 1998). Use of isolated populations, therefore, make it probable that individuals affected with the disease in question will share common causative genes. Thus, for complex disorders, isolated populations can be ideal for detection of oligogenes. Examples of isolated populations in which schizophrenia is more prevalent have been characterised in Finland (Hovatta et al., 1997). A disadvantage of isolated populations is that markers which are highly informative in outbred populations can be much less so in isolated populations due to increases in homozygosity (Sheffield et al., 1998).

Technical aspects

Association studies are performed using similar technology to that employed for linkage analysis. Microsatellite markers are utilised and analysed by PCR and gel electrophoresis. An association-based, whole-genome screen to identify genes for schizophrenia would require a 1 cM grid of markers sufficient density to detect disequilibrium, and a prohibitively large sample of patient and controls. The logistical problems of typing over 3500 markers together with the inherent statistical problems caused by multiple testing mean that association studies tend to be confined to functionally significant variations within candidate genes (Sobell et al., 1993) or subchromosomal regions already implicated in standard linkage studies. An association based genome wide screen is a realistic possibility if DNA pooling strategies are adopted. Such an approach reduces the amount of genotyping by pooling the patient and control samples separately, and typing the markers only in the two pools as a single PCR reaction. In such an approach developed at Cardiff, it is possible to compare allele frequency profiles between two pooled reactions using fluorescent dye labelling of PCR products and detection using polyacrylamide gel electrophoresis. If a difference in allele frequency profiles is observed and found to have statistical significance standard single sample genotyping then follows in order to verify the results obtained from the pooling experiments (Daniels et al., 1995).

Although microsatellite markers are highly polymorphic and are therefore ideally suited to linkage analysis, they are not found in the genome at suitably high frequency to be useful for association studies. Single nucleotide polymorphisms (SNP) on the other hand are more common in the genome, occurring on average 1 every 1000 bp. Each SNP is only a biallelic polymorphism, therefore a number of SNP loci are required to provide an equivalent informativity as a multiallelic microsatellite marker (Kruglyak,

1997). Biallelic polymorphisms have been traditionally typed using relatively labour intensive techniques (allele specific PCR, RFLP analysis) although high throughput methods are being developed which include single nucleotide primer extension (SNuPE) (Greenwood and Burke, 1996) and the use of DNA microchip technology. In a study by Wang and colleagues (1998) prototype genotyping chips were developed that allowed the simultaneous genotyping of 500 SNPs, demonstrating the huge potential of this technology.

Application of the association approach

The association approach has been applied to the dopamine D3 receptor gene (DRD3), a candidate gene for schizophrenia. This gene maps to chromosome 3q13.3 (Giros et al., 1990). There has been considerable interest in a polymorphism in exon 1 of the DRD3 gene. This creates a *Bal*1 restriction endonuclease site and brings about an amino acid change (glycine-serine) in the N-terminal extracellular domain of the receptor. This is a single nucleotide polymorphism which brings about a change in amino acid sequence of the protein. Such polymorphisms are becoming an increasingly attractive method of identifying susceptibility genes (Sobell et al., 1993).

We and a number of groups have observed that homozygosity at the D3 receptor is more common in patients vs controls (Crocq et al., 1992; Nimgaonkar et al., 1993; Mant et al., 1994). While there has also been a number of negative findings (Yang et al., 1993; Nothen et al., 1993; Jonsson et al., 1993), a meta-analysis supports overall association with both the 1-1 genotype and combined homozygosity in 23 unpublished datasets, and confirms these effects are present in males but not females (Williams et al., 1998).

Recent interest has centred around association between schizophrenia and a T to C polymorphism at position 102 in the serotonin 2a (5HT2a) receptor gene, mapping to chromosome 13q14-21 (Sparkes et al., 1991). This was observed first in a small Japanese study (Inayama et al., 1994) and has been replicated in a large multicentre European collaborative study (Williams et al., 1996). Although the odds ratio is small (1.7 for possession of one or more copies of allele 2), the attributable fraction is relatively high (0.35) because allele 2 is relatively common in the population.

It should be stressed that 5HT2a and DRD3 only make very small contributions to the overall liability of developing schizophrenia. Also, the functional significance of the amino acid change in the DRD3 receptor is not known. In the case of the 5HT2a polymorphism, the significance of it is less clear since it does not occur within the coding sequence of this gene, although it may act post-transcriptionally through effects on mRNA structure or stability. It is possible that the polymorphisms studied in both genes are merely in linkage disequilibrium with other variants within these, or even adjacent genes, and that these polymorphisms are functionally significant. However, it does not appear that the coding sequences of either gene possess such variants associated with schizophrenia (Asherson et al., 1996; Erdmann et al., 1996) although the regulatory regions of these genes still require analysis.

Dynamic mutation

Recent research suggests that a novel mutational mechanism, the expanded trinucleotide repeat, may be involved in the pathogenesis of schizophrenia. The phenotypic signature of this type of mutation is 'anticipation', which is the phenomenon of decreasing age at onset or increasing severity of phenotype as a disease passes through successive generations of a family (Gorwood et al., 1996). Anticipation is largely due to the propensity of expanded trinucleotide repeats to increase in size between generations, a characteristic which has led to their description as 'dynamic mutations'. Instability of trinucleotide repeats is seen in myotonic dystrophy, caused by a

CTG expansion in the 3′ untranslated region of the myotonin gene (Brook et al., 1992), but both phenomena also occur between generations in Huntington's disease and the other expanded CAG repeat disorders (Willems, 1994).

A number of studies have all found that in families where schizophrenia occurs in multiple generations, age of onset decreases in successive generations (Asherson et al., 1994; Bassett and Honer, 1994; Thibaut et al., 1995; Yaw et al., 1996), thus awakening interest in dynamic mutation as a mechanism in psychiatric disorders. Unfortunately, while these data are consistent with genetic anticipation, and transmission through a trinucleotide repeat mechanism, they can be attributed, just as convincingly, to systematic sampling biases which are difficult to discount (Penrose, 1991; Asherson et al., 1994; Yaw et al., 1996).

Technical aspects

In order to scan the genome for trinucleotide repeat expansion the RED (repeat expansion detection) method was developed (Schalling et al., 1993), which determines the maximum size of trinucleotide repeats located anywhere in the genome. The CAG/CTG repeat is an example of a trinucleotide repeat that has received attention by virtue of the role of such repeats, coding for polyglutamine, in other neurological and psychiatric disorders such as Huntington's disease (Willems, 1994). The technique involves the use of the ligase chain reaction (LCR). In this situation genomic DNA samples are analysed by addition of a radioactively labelled oligonucleotide of repeat sequence under analysis (e.g. $(CAG)_{10}$), and a thermostable DNA ligase which joins adjacent ends of DNA molecules. The reaction is incubated through cycles of DNA denaturation, oligonucleotide annealing, and ligation. The products of the RED reaction are resolved by polyacrylamide gel electrophoresis, and detected by autoradiography. The longer the repeat, the more $(CAG)_{10}$ oligonucleotides will be able to join together at that repeat, and hence, the larger observed product.

Applications to schizophrenia

We and another group (Morris et al., 1995; O'Donovan et al., 1995) have used the RED method and found that the maximum length of the CAG/CTG repeat is greater is schizophrenics than in unaffected individuals. These findings have been replicated in a European multicentre study (O'Donovan et al., 1996), and suggests that dynamic mutation may play a role in the pathogenesis of schizophrenia.

Paradoxically, although the RED studies were undertaken because of apparent 'anticipation', the data from the largest studies suggest that there is no correlation between repeat size and age at onset (O'Donovan et al., 1996). This suggests that CAG or CTG repeat size is not a significant factor in determining the age at onset of schizophrenia, although it remains possible that it may yet explain a relatively small part of the total variance. Secondly, although it has been shown that there is an association between disease status and large repeat size, stratification has not been entirely excluded as a possible cause for this finding. Furthermore, it is not possible to exclude the possibility that the expansions are secondary to a different mutation that also causes susceptibility to schizophrenia or to an environmental variable associated with schizophrenia, such as drug treatment. Recent work has identified large CAG/CTG repeats on chromosome 17 and 18 (Breschel et al., 1997; Ikeuchi et al., 1998), and it is possible that this repeat may account for the expansions observed in schizophrenics. We are currently investigating this possibility. Therefore, until RED data are corroborated by alternative methods, the trinucleotide repeat hypothesis of schizophrenia is likely to remain interesting but controversial.

Vincent and colleagues (1998) hypothesised that unstable repeat sequences could explain discordance of phenotype for schizophrenia in pairs of monozygotic twins, with the affected twin possessing a greater number of repeats than the unaffected twin. This was investigated in 27 pairs of monozygotic twins using locus-specific PCR to search for expanded CAG/CTG repeats, and using RED and Southern hybridisation techniques to look for expanded GAA/TTC repeats. No evidence for repeat expansion was found, although the techniques employed had limits to their sensitivity, and discordance for other trinucleotide repeat sequences cannot be ruled out.

Chromosome 22 deletions and schizophrenia

Identification of a cytogenetic abnormality, such as a deletion, translocation, or inversion, associated with disease can provide a rapid route to isolating a disease gene, especially in the case of complex disorders where standard genetic mapping approaches may not yield conclusive evidence in favour of linkage. The site of the abnormality provides positional information to guide the choice of genetic marker. The abnormality itself may also be exploited using a range of techniques that allow the arduous experimental work surrounding the positional cloning of a disease gene to be circumvented.

Chromosome 22q12-q13 is of interest with regard to schizophrenia from the results of various linkage studies. Additional interest in long arm of chromosome 22 is provided by velocardiofacial syndrome (VCFS) — a disorder associated with deletion of chromosome 22q11. Pulver et al. (1994c) showed that there is a 29% incidence of schizophrenia and schizoaffective disorders in patients with VCFS. Karayiorgou et al (1995) estimated the size of deletions to be between 1.5 and 2 Megabases in a sample of schizophrenic patients. The implication is that a susceptibility gene lies within the region deleted in VCFS patients suffering from schizophrenia, and therefore allows for a more focused search for positional candidate genes.

Technical aspects — detection of chromosomal abnormalities

Several methods can be employed to screen for chromosomal abnormalities. Primarily, karyotypes are analysed. In this process metaphase chromosomes are stained with agents that produce characteristic banding patterns. A karyogram is then constructed displaying all chromosomes in a pictorial fashion and each chromosome can be analysed for abnormal banding patterns. This technique is suitable for detecting relatively large abnormalities of several megabases in size (Strachan and Read, 1996).

At higher levels of resolution, metaphase chromosome spreads can be hybridised against fluorescently labelled genomic clones from a clone contig that maps to the region of interest. This process, called fluorescence in situ hybridisation (FISH), can aid the localisation of chromosomal abnormalities that require a 50–500 kb resolution level. At higher resolution levels, FISH may be performed upon extended chromatin fibres (Senger et al., 1994) or DNA fibres producing levels of resolution between 5 kb and 700 kb (Heiskanen et al., 1995).

A microdeletion may cause the removal of a particular polymorphic marker or markers. Having localised the disease gene to a subchromosomal region using genetic mapping approaches subsequent analysis of families with all markers within the region may reveal a loss of heterozygosity (LOH) for specific markers segregating with the disease gene. This technique has the advantage of being a technically straight forward way of screening individuals for microdeletions (Strachan and Read, 1996).

Alternatively, genomic DNA can be screened at high resolution using long-range restriction endonuclease mapping. The human genome is GC poor (approximately 41%) and GC rich sequences tend to be confined to regions known as CpG islands. These regions of DNA sequence, typically 1–2 kb, surround a proportion of genes either at their 5′ or 3′ ends (Bird, 1986). A large proportion of CpG islands are methylated. By the use of methylation sensitive restriction endonucleases that have GC rich recognition sequences it is possible to digest human DNA to yield relatively large fragments up to several megabases in size. Large-fragment digests can be resolved using pulsed-field gel electrophoresis, Southern blotted, and hybridised to radioactive probes that are specific to loci within the region of interest. With a satisfactory coverage of probes a long-range restriction map of the region can be prepared and the information applied to the screening of patient cases for changes in hybridisation patterns that may be indicative of a chromosomal abnormality (Strachan and Read, 1996).

The future

Research into the molecular genetics of schizophrenia is entering a new phase of optimism. Currently, the main challenge for the field is how to convert tentative and suggestive findings of linkage into confident assignments to well-circumscribed regions of the genome. This will almost certainly involve the analysis of large samples of affected sibling-pairs and of large unrelated patient samples for linkage disequilibrium. A second challenge is the optimisation of techniques that allow large samples to be analysed in a realistic time frame. Finally, methods will also be required that allow the high-throughput mutational analysis of many genes within candidate regions. However, these difficulties should not prevail and it seems likely that we can now expect steady advances in the molecular genetics of schizophrenia.

References

American Psychiatric Association. (1994) *Diagnostic and Statistical Manual of Mental Disorders*, 4th Edn, American Psychiatric Association, Washington D.C.

Antonarakis, S.E., Blouin, J.L., Pulver, A.E., Wolyniec, P., Lasseter, V.K., Nestadt, G., Kasch, L., Babb, R., Kazazian, H.H., Dombroski, B., Kimberland, M., Ott, J., Housman, D., Karayiorgou, M. and Maclean, C.J. (1995) Schizophrenia susceptibility and chromosome 6p24-p22. *Nat. Gen.*, 11: 235–236.

Antonarakis, S.E., Blouin, J.L. and Curran, M. (1998) Linkage and sib-pair analysis reveal a potential schizophrenia susceptibility gene on chromosome 13q32. Presented at Cold Spring Harbour Symposium Function and Dysfunction in the Central Nervous System.

Asherson, P., Walsh, C., Williams, J., Sargeant, M., Taylor, C., Clements, A., Gill, M., Owen, M. and McGuffin, P. (1994) Imprinting and anticipation — are they relevant to genetic studies of schizophrenia? *B. J. Psych.*, 164: 619–624.

Asherson, P., Mant, R., Holmans, P., Williams, J., Cardno, A., Murphy, K., Jones, L., Collier, D., McGuffin, P. and Owen, M.J. (1996) Linkage, association and mutational analysis of the dopamine D3 receptor gene in schizophrenia. *Mol. Psych.*, 1: 125–132.

Bassett, A.S. and Honer, W.G. (1994) Evidence for anticipation in schizophrenia. *Am. J. Hum. Gen.*, 54: 864–870.

Berrettini, W. (1993) Quantitative trait loci mapping: a novel approach to candidate genes for psychiatric disease. *Psych. Gen.*, 3: 203–205.

Bird, A.P. (1986) CpG rich islands and the function of DNA methylation. *Nature*, 321: 209–213.

Breschel, T.S., McInnis, M.G., Margolis, R.L., Sirugo, G., Corneliussen, B., Simpson, S.G., McMahon, F., MacKinnon, D.F., Xu, J.F., Pleasant, N., Huo, Y., Ashworth, R.G., Grundstrom, C., Grundstrom, T., Kidd, K.K., DePaulo, J.R. and Ross, C.A. (1997) A novel, heritable, expanding CTG repeat in an intron of the SEF2-1 gene on chromosome 18q21.1. *Hum. Mol. Gen.*, 6: 1855–1863.

Brook, J.D., Mccurrach, M.E., Harley, H.G., Buckler, A.J., Church, D., Aburatani, H., Hunter, K., Stanton, V.P., Thirion, J.P., Hudson, T., Sohn, R., Zemelman, B., Snell, R.G., Rundle, S.A., Crow, S., Davies, J., Shelbourne, P., Buxton, J., Jones, C., Juvonen, V., Johnson, K., Harper, P.S., Shaw, D.J. and Housman, D.E. (1992) Molecular basis of myotonic dystrophy — expansion of a trinucleotide (CTG) repeat at the 3′ end of a transcript encoding a protein kinase family member. *Cell*, 68: 799–808.

Buetow, K.H. (1991) Influence of aberrant observations on high-resolution linkage analysis outcomes. *Am. J. Hum. Gen.*, 49(5): 985–994.

Collins, F.S. (1995) Positional cloning moves from perditional to traditional. *Nat. Gen.*, 9: 347–350.

Coon, H., Holik, J., Hoff, M., Reimherr, F., Wender, P., Myleswarsley, M., Waldo, M., Freedman, R. and Byerley, W. (1994) Analysis of chromosome 22 markers in 9 schizophrenia pedigrees. *Am. J. Med. Gen.*, 54: 72–79.

Crocq, M.A., Mant, R., Asherson, P., Williams, J., Hode, Y., Mayerova, A., Collier, D., Lannfelt, L., Sokoloff, P., Schwartz, J.C., Gill, M., Macher, J.P., McGuffin, P. and Owen, M.J. (1992) Association between schizophrenia and homozygosity at the dopamine D3 receptor gene. *J. Med. Gen.*, 29: 858–860.

Daniels, J.K., Williams, N.M., Plomin, R., McGuffin, P. and Owen, M.J. (1995) Selective DNA pooling using automated technology: a rapid technique for allelic association studies in quantitative traits and complex diseases. *Psych. Gen.*, 5: S39

Daniels, J.K., Spurlock, G., Williams, N.M., Cardno, A.G., Jones, L.A., Murphy, K.C., Asherson, P., Holmans, P., Fenton, I., McGuffin, P. and Owen, M.J. (1997) Linkage study of chromosome 6p in sib-pairs with schizophrenia. *Am. J. Med. Gen.*, 74: 319–323.

Davies, J.L., Kawaguchi, Y., Bennett, S.T., Copeman, J.B., Cordell, H.J., Pritchard, L.E., Reed, P.W., Gough, S.C.L., Jenkins, S.C., Palmer, S.M., Balfour, K.M., Rowe, B.R., Farrall, M., Barnett, A.H., Bain, S.C. and Todd, J.A. (1994) A genome-wide search for human type-1 diabetes susceptibility genes. *Nature*, 371: 130–136.

Durner, M., Greenberg, D.A. and Hodge, S.E. (1992) Inter and intrafamilial heterogeneity: effective sampling strategies and comparison analysis methods. *Am. J. Hum. Gen.*, 51: 859–870.

De la Chapelle, A. and Wright, F.A. (1998) Linkage disequilibrium mapping in isolated populations: The example of Finland revisted. *Proc. Natl. Acad. Sci. USA*, 95: 12416–12423.

Erdmann, J., Shimronabarbanell, D., Rietschel, M., Albus, M., Maier, W., Korner, J., Bondy, B., Chen, K., Shih, J.C., Knapp, M., Propping, P. and Nothen, M.M. (1996) Systematic screening for mutations in the human serotonin-2a (5-HT2a) receptor gene — identification of 2 naturally occurring receptor variants and association analysis in schizophrenia. *Hum. Gen.*, 97: 614–619.

Falk, C.T. and Rubinstein, P. (1987) Haplotype relative risks: An easy reliable way to construct a proper control sample for risk calculations. *Ann. Hum. Gen.*, 51: 227–233.

Farmer, A.E., McGuffin, P. and Gottesman, I.I. (1987) Twin concordance for DSM-III schizophrenia - scrutinizing the validity of the definition. *Arch. Gen. Psych.*, 44: 634–641.

Flint, J., Corley, R., DeFries, J.C., Fulker, D.W., Gray, J.A., Miller, S. and Collins, A.C. (1995) A simple genetic basis for a complex psychological trait in laboratory mice. *Science*, 269: 1432–1435.

Gill, M., Vallada, H., Collier, D., Sham, P., Holmans, P., Murray, R., McGuffin, P., Nanko, S., Owen, M., Antonarakis, S., Housman, D., Kazazian, H., Nestadt, G., Pulver, A.E., Straub, R.E., Maclean, C.J., Walsh, D., Kendler, K.S., Delisi, L., Polymeropoulos, M., Coon, H., Byerley, W., Lofthouse, R., Gershon, E., Golden, L., Crow, T., Freedman, R., Laurent, C., Bodeaupean, S., Damato, T., Jay, M., Campion, D., Mallet, J., Wildenauer, D.B., Lerer, B., Albus, M., Ackenheil, M., Ebstein, R.P., Hallmayer, J., Maier, W., Gurling, H., Curtis, D., Kalsi, G., Brynjolfsson, J., Sigmundson, T., Petursson, H., Blackwood, D., Muir, W., Stclair, D. et al. (1996) A combined analysis of D22S278 marker alleles in affected sib-pairs - support for a susceptibility locus for schizophrenia at chromosome 22q12. *Am. J. Med. Gen.*, 67: 40–45.

Giros, B., Martres, M.P., Sokoloff, P. and Schwartz, J.C. (1990) cDNA cloning of the human dopaminergic D3 receptor and chromosome identification. *C. Roy. Acad Sci.*, 311: 501–508.

Glaser, B., Chiu, K.C., Liu, L., Anker, R., Nestorowicz, A., Cox, N.J., Landau, H., Kaiser, N., Thornton, P.S., Stanley, C.A., Cerasi, E., Baker, L., Doniskeller, H. and Permutt, M.A. (1995) Recombinant mapping of the familial hyperinsulinism gene to an 0.8 cM region on chromosome 11p15.1 and demonstration of a founder effect in Ashkenazi Jews. *Hum. Mol. Gen.*, 4: 879–886.

Goate, A., Chartierharlin, M.C., Mullan, M., Brown, J., Crawford, F., Fidani, L., Giuffra, L., Haynes, A., Irving, N., James, L., Mant, R., Newton, P., Rooke, K., Roques, P., Talbot, C., Pericakvance, M., Roses, A., Williamson, R., Rossor, M., Owen, M. and Hardy, J. (1991) Segregation of a missense mutation in the amyloid precursor protein gene with familial Alzheimer's disease. *Nature*, 349: 704–706.

Gorwood, P., Leboyer, M., Falissard, B., Jay, M., Rouillon, F. and Feingold, J. (1996) Anticipation in schizophrenia: New light on a controversial problem. *Am. J. Psych.*, 153(9): 1173–1177.

Gottesman, I.I. (1991) *Schizophrenia Genesis. The Origins of Madness*, W H Freeman, New York.

Gottesman, I.I. and Shields, J. (1982) *Schizophrenia, the Epigenetic Puzzle*, Cambridge University Press, Cambridge.

Greenwood, A.D. and Burke, D.T. (1996) Single nucleotide primer extension — quantitative range, variability, and multiplex analysis. *Gen. Res.*, 6: 336–348.

Gurling, H., Kalsi, G., Chen, A.H.S., Green, M., Butler, R., Read, T., Murphy, P., Curtis, D., Sharma, T., Petursson, H. and Brynjolfsson, T. (1995) Schizophrenia susceptibility and chromosome 6p24-p22. *Nat. Gen.*, 11: 234–235.

Hall, J.M., Lee, M.K., Newman, B., Morrow, J.E., Anderson, L.A., Huey, B. and King, M.C. (1990) Linkage of early onset familial breast cancer to chromosome 17q21. *Science*, 250: 1684–1689.

Hauser, E.R., Boehnke, M., Guo, S.W. and Risch, N. (1996) Affected sib-pair interval mapping and exclusion for complex genetic traits — sampling considerations. *Genet. Epidemiol.*, 13: 117–137.

Heiskanen, M., Hellsten, E., Kallioniemi, O.P., Makela, T.P., Alitalo, K., Peltonen, L. and Palotie, A. (1995). Visual mapping by fiber-FISH. Genomics 30:31-36.

Henn, F.A. and Edwards, E. (1994) Animal models in the study of genetic factors in human psychopathology. In: *Genetic Studies of Affective Disorders. Overview of Basic Methods, Current Directions, and Critical Research Issues*, John Wiley, New York, pp. 177–192.

Holmans, P. and Craddock, N. (1997) Efficient strategies for genome scanning using maximum likelihood affected sib-pair analysis. *Am. J. Hum. Gen.*, 60: 657–666.

Hovatta, I., Terwilliger, J.D., Lichtermann, D., Makikyro, T., Suvisaari, J., Peltonen, L. and Lonnqvist, J. (1997) Schizophrenia in the genetic isolate of Finland. *Am. J. Med. Gen.*, 74(4): 353–360.

Ikeuchi, T., Sanpei, K., Takano, H., Sasaki, H., Tashiro, K., Cancel, G., Brice, A., Bird, T.D., Schellenberg, G.D., Pericakvance, M.A., WelshBohmer, K.A., Clark, L.N., Wilhelmsen, K. and Tsuji, S. (1998) A novel long and unstable CAG/CTG trinucleotide repeat on chromosome 17q. *Genomics*, 49: 321–326.

Inayama, Y., Yoneda, H. and Ishida, T. (1994) An association between schizophrenia and a serotonin DNA marker (5HTR2). *Neuropsychopharmacology*, 10: 56.

Jonsson, E., Lannfelt, L., Sokoloff, P., Schwartz, J.C. and Sedvall, G. (1993) Lack of association between schizophrenia and alleles in the dopamine D3 receptor gene. *Acta Psych. Scand.*, 87: 345–349.

Karayiorgou, M., Morris, M.A., Nestadt, G., Antonarakis, S.E., Pulver, A.E. and Housman, D.E. (1995) Schizophrenia susceptibility associated with interstitial deletions of chromosome 22q11. *Am. J. Hum. Gen.*, 57: 1245.

Koreth, J., O'Leary, J.J. and O'D.McGee, J. (1996) Microsatellites and PCR genomic analysis. *J. Path.*, 178: 239–248.

Kruglyak, L. (1997) The use of a genetic map of biallelic markers in linkage studies. *Nat. Gen.*, 17: 21–24.

Lander, E.S. and Schork, N.J. (1994) Genetic dissection of complex traits. *Science*, 265: 2037–2048.

Lin, M.W., Curtis, D. and Williams, N. (1995) Suggestive evidence for linkage of schizophrenia to markers on chromosome 13q14.1-q32. *Psych. Gen.*, 5: 117–126.

Mant, R., Williams, J., Asherson, P., Parfitt, E., McGuffin, P. and Owen, M.J. (1994) Relationship between homozygosity at the dopamine D3 receptor gene and schizophrenia. *Am. J. Med. Gen.*, 54: 21–26.

McGuffin, P., Owen, M.J., O'Donovan, M.C., Thapar, A. and Gottesman, I.I. (1994) *Seminars in Psychiatric Genetics*, Gaskell Press, London.

Miki, Y., Swensen, J., Shattuckeidens, D., Futreal, P.A., Harshman, K., Tavtigian, S., Liu, Q.Y., Cochran, C., Bennett, L.M., Ding, W., Bell, R., Rosenthal, J., Hussey, C., Tran, T., Mcclure, M., Frye, C., Hattier, T., Phelps, R., Haugenstrano, A., Katcher, H., Yakumo, K., Gholami, Z., Shaffer, D., Stone, S., Bayer, S., Wray, C., Bogden, R., Dayananth, P., Ward, J., Tonin, P., Narod, S., Bristow, P.K., Norris, F.H., Helvering, L., Morrison, P., Rosteck, P., Lai, M., Barrett, J.C., Lewis, C., Neuhausen, S., Cannonalbright, L., Goldgar, D., Wiseman, R., Kamb, A. and Skolnick, M.H. (1994) A strong candidate for the breast and ovarian cancer susceptibility gene BRCA1. *Science*, 266: 66–71.

Moises, H.W., Yang, L., Kristbjarnarson, H., Wiese, C., Byerley, W., Macciardi, F., Arolt, V., Blackwood, D., Liu, X., Sjogren, S., Aschauer, H.N., Hwu, H.G., Jang, K., Livesley, W.J., Kennedy, J.L., Zoega, T., Ivarsson, O., Bui, M.T., Yu, M.H., Havsteen, B., Commenges, D., Weissenbach, J., Schwinger, E., Gottesman, I.I., Pakstis, A.J., Wetterberg, L., Kidd, K.K. and Helgason, T. (1995) An international 2-stage genome wide search for schizophrenia susceptibility genes. *Nat. Gen.*, 11: 321–324.

Morris, A.G., Gaitonde, E., Mckenna, P.J., Mollon, J.D. and Hunt, D.M. (1995) CAG repeat expansions and schizophrenia — association with disease in females and with early age at onset. *Hum. Mol. Gen.*, 4: 1957–1961.

Mowry, B.J., Nancarrow, D.J., Lennon, D.P., Sandkuijl, L.A., Crowe, R.R., Silverman, J.M., Mohs, R.C., Siever, L.J., Endicott, J., Sharpe, L., Walters, M.K., Hayward, N.K. and Levinson, D.F. (1995) Schizophrenia susceptibility and chromosome 6p24-p22. *Nat. Gen.*, 11: 233–234.

Nimgaonkar, V.L., Zhang, X.R., Caldwell, J.G., Ganguli, R. and Chakravarti, A. (1993) Association study of schizophrenia with dopamine D3 receptor gene polymorphisms — probable effects of family history of schizophrenia. *Am. J. Med. Gen.*, 48: 214–217.

Nothen, M.M., Cichon, S., Propping, P., Fimmers, R., Schwab, S.G. and Wildenauer, D.B. (1993) Excess of homozygosity at the dopamine D3 receptor gene in schizophrenia not confirmed. *J. Med. Gen.*, 30: 708.

O'Donovan, M.C., Guy, C., Craddock, N., Murphy, K.C., Cardno, A.G., Jones, L.A., Owen, M.J. and McGuffin, P. (1995) Expanded CAG repeats in schizophrenia and bipolar disorder. *Nat. Gen.*, 10: 380–381.

O'Donovan, M.C., Guy, C., Craddock, N., Bowen, T., Mckeon, P., Macedo, A., Maier, W., Wildenauer, D., Aschauer, H.N., Sorbi, S., Feldman, E., Mynettjohnson, L., Claffey, E., Nacmias, B., Valente, J., Dourado, A., Grassi, E., Lenzinger, E., Heiden, A.M., Moorhead, S., Harrison, D., Williams, J., McGuffin, P. and Owen, M.J. (1996) Confirmation of association between expanded CAG/CTG repeats and both schizophrenia and bipolar disorder. *Psych. Med.*, 26: 1145–1153.

O'Donovan, M.C. and Owen, M.J. (1992) Advances and retreats in the molecular genetics of major mental illness. *Ann. Med.*, 24: 171–177.

Penrose, L.S. (1991) Survey of cases of familial mental illness. *Euro. Arch. Psych. Clin. Neuro.*, 240: 315–324.

Plomin, R., Owen, M.J. and McGuffin, P. (1994) The genetic basis of complex human behaviours. *Science*, 264: 1733–1739.

Pulver, A.E., Karayiorgou, M., Lasseter, V.K., Wolyniec, P., Kasch, L., Antonarakis, S., Housman, D., Kazazian, H.H., Meyers, D., Nestadt, G., Ott, J., Liang, K.Y., Lamacz, M., Thomas, M., Childs, B., Diehl, S.R., Wang, S.B., Murphy, B., Sun, C.E., Oneill, F.A., Nie, L., Sham, P., Burke, J., Duke, B.W., Duke, F., Kipps, B.R., Bray, J., Hunt, W., Shinkwin, R., Nuallain, M.N., Su, Y., Maclean, C.J., Walsh, D., Kendler, K.S., Gill, M., Vallada, H., Mant, R., Asherson, P., Collier, D., Parfitt, E., Roberts, E., Nanko, S., Walsh, C., Daniels, J., Murray, R., McGuffin, P., Owen, M., Laurent, C., Dumas, J.B. et al. (1994a) Follow up of a report of a potential linkage for schizophrenia on chromosome 22q12-q13.1 .2. *Am. J. Med. Gen.*, 54: 44–50.

Pulver, A.E., Karayiorgou, M., Wolyniec, P.S., Lasseter, V.K., Kasch, L., Nestadt, G., Antonarakis, S., Housman, D., Kazazian, H.H., Meyers, D., Ott, J., Lamacz, M., Liang, K.Y., Hanfelt, J., Ullrich, G., Demarchi, N., Ramu, E., Mchugh, P.R., Adler, L., Thomas, M., Carpenter, W.T.,

Manschreck, T., Gordon, C.T., Kimberland, M., Babb, R., Puck, J. and Childs, B. (1994b) Sequential strategy to identify a susceptibility gene for schizophrenia — report of potential linkage on chromosome 22q12-q13.1 .1. *Am. J. Med. Gen.*, 54: 36–43.

Pulver, A.E., Nestadt, G., Goldberg, R., Shprintzen, R.J., Lamacz, M., Wolyniec, P.S., Morrow, B., Karayiorgou, M., Antonarakis, S.E., Housman, D. and Kucherlapati, R. (1994c) Psychotic illness in patients diagnosed with velo-cardio-facial syndrome and their relatives. *J. Nerv. Men. Dis.*, 182: 476–478.

Pulver, A.E., Lasseter, V.K., Kasch, L., Wolyniec, P., Nestadt, G., Blouin, J.L., Kimberland, M., Babb, R., Vourlis, S., Chen, H.M., Lalioti, M., Morris, M.A., Karayiorgou, M., Ott, J., Meyers, D., Antonarakis, S.E., Housman, D. and Kazazian, H.H. (1995) Schizophrenia — a genome scan targets chromosomes 3p and 8p as potential sites of susceptibility genes. *Am. J. Med. Gen.*, 60: 252–260.

Reed, P.W., Davies, J.L., Copeman, J.B., Bennett, S.T., Palmer, S.M., Pritchard, L.E., Gough, S.C.L., Kawaguchi, Y., Cordell, H.J., Balfour, K.M., Jenkins, S.C., Powell, E.E., Vignal, A. and Todd, J.A. (1994) Chromosome specific microsatellite sets for fluorescence based, semiautomated genome mapping. *Nat. Gen.*, 7: 390–395.

Risch, N. (1990a) Linkage strategies for genetically complex traits .1. multilocus models. *Am. J. Hum. Gen.*, 46: 222–228.

Risch, N. (1990b) Genetic linkage and complex diseases, with special reference to psychiatric disorders. *Gen. Epid.*, 7: 3–16.

Risch, N. (1990c) Linkage strategies for genetically complex traits .2. the power of affected relative pairs. *Am. J. Hum. Gen.*, 46: 229–241.

Risch, N. and Merikangas, K. (1996) The future of genetic studies of complex human diseases. *Science*, 273: 1516–1517.

Rosenthal, D., Wender, P.H. and Kety, S.S. (1968) Schizophrenics offspring reared in adoptive homes. In: *The Transmission of Schizophrenia*, Pergamon Press, Oxford.

Schalling, M., Hudson, T.J., Buetow, K.H. and Housman, D.E. (1993) Direct detection of novel expanded trinucleotide repeats in the human genome. *Nat. Gen.*, 4: 135–139.

Schwab, S.G., Albus, M., Hallmayer, J., Honig, S., Borrmann, M., Lichtermann, D., Ebstein, R.P., Ackenheil, M., Lerer, B., Risch, N., Maier, W. and Wildenauer, D.B. (1995) Evaluation of a susceptibility gene for schizophrenia on chromosome 6p by multipoint affected sib-pair linkage analysis. *Nat. Gen.*, 11: 325–327.

Senger, G., Jones, T.A., Fidlerova, H., Sanseau, P., Trowsdale, J., Duff, M. and Sheer, D. (1994) Released chromatin — linearized DNA for high resolution fluorescence *in situ* hybridization. *Hum. Mol. Gen.*, 3: 1275—1280.

Sheffield, V.C., Stone, E.M. and Carmi, R. (1998) Use of isolated inbred human populations for identification of disease genes. *TIG*, 14(10): 391–396.

Sherrington, R., Rogaev, E.I., Liang, Y., Rogaeva, E.A., Levesque, G., Ikeda, M., Chi, H., Lin, C., Li, G., Holman, K., Tsuda, T., Mar, L., Foncin, J.F., Bruni, A.C., Montesi, M.P., Sorbi, S., Rainero, I., Pinessi, L., Nee, L., Chumakov, I., Pollen, D., Brookes, A., Sanseau, P., Polinsky, R.J., Wasco, W., Dasilva, H.A.R., Haines, J.L., Pericakvance, M.A., Tanzi, R.E., Roses, A.D., Fraser, P.E., Rommens, J.M. and Stgeorgehyslop, P.H. (1995) Cloning of a gene bearing missense mutations in early onset familial Alzheimer's disease. *Nature*, 375: 754–760.

Sobell, J.L., Heston, L.L. and Sommer, S.S. (1993) Novel association approach for determining the genetic predisposition to schizophrenia — case-control resource and testing of a candidate gene. *Am. J. Med. Gen.*, 48: 28–35.

Sparkes, R.S., Lan, N., Klisak, I., Mohandas, T., Diep, A., Kojis, T., Heinzmann, C. and Shih, J.C. (1991) Assignment of a serotonin 5HT-2 receptor gene (HTR2) to human chromosome 13q14-21 and mouse chromosome 14. *Genomics*, 9: 461–465.

Spielman, R.S. and Ewens, W.J. (1998) A sibship test for linkage in the presence of association: the sib transmission/disequilibrium test. *Am. J. Hum. Gen.*, 62: 450–458.

Strachan, T. and Read, A.P. (1996) *Human Molecular Genetics*, 1st Edn, Bios Scientific Publishers, Oxford.

Straub, R.E., Maclean, C.J., Oneill, F.A., Burke, J., Murphy, B., Duke, F., Shinkwin, R., Webb, B.T., Zhang, J., Walsh, D. and Kendler, K.S. (1995) A potential vulnerability locus for schizophrenia on chromosome 6p24-p22 — evidence for genetic heterogeneity. *Nat. Gen.*, 11: 287–293.

Thibaut, F., Martinez, M., Petit, M., Jay, M. and Campion, D. (1995) Further evidence for anticipation in schizophrenia. *Psych. Res.*, 59: 25–33.

Todd, J.A., Aitman, T.J., Cornall, R.J., Ghosh, S., Hall, J.R.S., Hearne, C.M., Knight, A.M., Love, J.M., Mcaleer, M.A., Prins, J.B., Rodrigues, N., Lathrop, M., Pressey, A., Delarato, N.H., Peterson, L.B. and Wicker, L.S. (1991) Genetic analysis of autoimmune type-1 diabetes mellitus in mice. *Nature*, 351: 542–547.

Vallada, H.P., Gill, M., Sham, P., Lim, L.C.C., Nanko, S., Asherson, P., Murray, R.M., McGuffin, P., Owen, M. and Collier, D. (1995) Linkage studies on chromosome 22 in familial schizophrenia. *Am. J. Med. Gen.*, 60: 139–146.

Vincent, J.B., Kalsi, G., Klempan, T., Tatuch. Y., Sherrington, R.P., Breschel. T., McInnis, M.G., Brynjolfsson, J., Petursson, H., Gurling, H.M.D., Gottesman, I.I., Torrey, E.F., Petronis. A. and Kennedy, J.L. (1998) No evidence of expansion of CAG or GAA repeats in schizophrenia families and monozygotic twins. *Hum. Gen.*, 103(1): 41–47.

Vionnet, N., Stoffel, M., Takeda, J., Yasuda, K., Bell, G.I., Zouali, H., Lesage, S., Velho, G., Iris, F., Passa, P., Froguel, P. and Cohen, D. (1992) Nonsense mutation in the glucokinase gene causes early onset non insulin dependent diabetes mellitus. *Nature*, 356: 721–722.

Wang, D.G., Fan, J.B., Siao, C.J., Berno, A., Young, P., Sapolsky, R., Ghandour, G., Perkins, N., Winchester, E., Spencer, J., Kruglyak, L., Stein, L., Hsie, L., Topaloglou, T., Hubbell, E., Robinson, E., Mittmann, M., Morris, M.S.,

Shen, N.P., Kilburn, D., Rioux, J., Nusbaum, C., Rozen, S., Hudson, T.J., Lipshultz, R., Chee, M. and Lander, E.S. (1998) Large-scale identification, mapping, and genotyping of single-nucleotide polymorphisms in the human genome. *Science*, 280(5366): 1077–1082.

Wang, S.B., Sun, C.E., Walczak, C.A., Ziegle, J.S., Kipps, B.R., Goldin, L.R. and Diehl, S.R. (1995) Evidence for a susceptibility locus for schizophrenia on chromosome 6pter-p22. *Nat. Gen.*, 10: 41–46.

Willems, P.J. (1994) Dynamic mutations hit double figures. *Nat. Gen.*, 8: 213–215.

Williams, J., Spurlock, G., McGuffin, P., Mallet, J., Nothen, M.M., Gill, M., Aschauer, H., Nylander, P.O., Macciardi, F. and Owen, M.J. (1996) Association between schizophrenia and T102C polymorphism of the 5-hydroxytryptamine type 2a-receptor gene. *Lancet*, 347: 1294–1296.

Williams, J., Spurlock, G., Holmans, P., Mant, R., Murphy, K., Jones, L., Cardno, A., Asherson, P., Blackwood, D., Muir, W., Meszaros, K., Aschauer, H., Mallet, J., Laurent, C., Pekkarinen, P., Seppala, J., Stefanis, C.N., Papadimitriou, G.N., Macciardi, F., Verga, M., Pato, C., Azevedo, H., Crocq, M.A., Gurling, H., Kalsi, G., Curtis, D., McGuffin, P. and Owen, M.J. (1998) A meta-analysis and transmission disequilibrium study of association between the dopamine D3 receptor gene and schizophrenia. *Mol. Psych.*, 3: 141–149.

Yang, L., Li, T., Wiese, C., Lannfelt, L., Sokoloff, P., Xu, C.T., Zeng, Z., Schwartz, J.C., Liu, X.H. and Moises, H.W. (1993) No association between schizophrenia and homozygosity at the D3 dopamine receptor gene. *Am. J. Med. Gen.*, 48: 83–86.

Yaw, J., Mylesworsley, M., Hoff, M., Holik, J., Freedman, R., Byerley, W. and Goon, H. (1996) Anticipation in multiplex schizophrenia pedigrees. *Psych. Gen.*, 6: 7–11.

W.E. Crusio and R.T. Gerlai (Eds.)
Handbook of Molecular-Genetic Techniques for Brain and Behavior Research (Techniques in the Behavioral and Neural Sciences, Vol. 13)

CHAPTER 5.5

Genetics of idiopathic epilepsy

Louise Bate and Mark Gardiner

University College London, Department of Paediatrics, The Rayne Institute, University Street, London WC1E 6JJ, UK

Introduction

The term epilepsy encompasses a heterogeneous group of disorders, which altogether have a lifetime cumulative incidence of 3% (Hauser et al., 1996). It has been estimated that genetic factors contribute to the etiology of epilepsy in up to 40% (Bird, 1992).

A clear understanding of the current terminology is an essential basis for understanding the genetics of this group of disorders.

Epileptic seizure

An epileptic seizure is a transient episode of abnormal cortical neuronal activity apparent to the patient or an observer. The abnormal cortical activity may be manifest as a motor, sensory, cognitive or psychic disturbance. A seizure is a clinical diagnosis, although electrophysiological data may be used in determining the precise seizure type.

Epilepsy

Epilepsy is a disorder of the brain characterized by recurrent, unprovoked epileptic seizures. Epilepsy may occur as a feature of a disease, in which there is a common etiology, or a syndrome in which a particular pattern of seizures, physical and investigative findings occur together.

Single epileptic seizures, occasional provoked seizures (such as febrile convulsions, eclamptic seizures and hypoglycemic seizures) or seizures that occur during an acute illness (e.g. encephalitis) are not, therefore classified as epilepsy.

Epilepsy can be broadly divided into two groups; partial epilepsy (localization related or focal epilepsy), and generalized epilepsy.

Partial epileptic seizures
Partial epileptic seizures arise from localized cortical foci. Consciousness may be retained (simple partial seizures), or lost (complex partial seizures), and any motor disturbance may be localized to a certain group of muscles. The ictal EEG shows discharges arising from a specific area. Simple partial seizures may evolve into complex partial seizures, in which consciousness is impaired.

Generalized seizures
Generalized seizures are those in which the initial clinical changes and ictal EEG patterns indicate involvement of both cerebral hemispheres. Any motor manifestations are bilateral. Consciousness may or may not be disturbed.

Epilepsy can be further classified depending on whether a cause is apparent, symptomatic epilepsy, or not, idiopathic epilepsy.

Symptomatic epilepsy
Symptomatic epilepsies are those caused by a known or suspected disorder of the central nervous

system. Epilepsy may be acquired as a result of environmental insults, such as head trauma, CNS infection, cerebrovascular disease or brain tumors, and is seen as part of the phenotype in over 120 Mendelian disorders.

Idiopathic epilepsy
Idiopathic epilepsies are not associated with an identifiable preceding disorder or underlying cause, other than a possible hereditary predisposition. Over 60% of all epilepsies may be classified as idiopathic (Annegers et al., 1996). The majority of idiopathic epilepsies show a complex pattern of inheritance. Mendelian idiopathic epilepsies, though rare, have proved an important resource in the genetic analysis of epilepsy.

Cryptogenic epilepsy
Cryptogenic epilepsies are presumed to be non-genetic, but with insufficient evidence to assign a specific etiology.

Mendelian idiopathic epilepsies

The molecular genetic basis of the idiopathic epilepsies is gradually being elucidated (Table 1). The genes mutated in three Mendelian idiopathic epilepsies have been identified. Mutations in the gene *CHRNA4*, encoding the $\alpha4$ subunit of the neuronal nicotinic acetylcholine receptor (nAChR) have been found in families with autosomal dominant nocturnal frontal lobe epilepsy (ADNFLE) (Steinlein et al., 1995; Steinlein et al., 1997). Mutations in the genes *KCNQ2* and *KCNQ3*, encoding two voltage-gated potassium channels, were discovered in families with benign familial neonatal convulsions (BFNC) (Biervert et al., 1998; Charlier et al., 1998; Singh et al., 1998). More recently, a mutation in *SCN1B*, which encodes the voltage gated sodium channel $\beta1$ subunit, was found in a large pedigree with generalized epilepsy with febrile seizures plus (GEFS$^+$) (Wallace et al., 1998).

Mapping projects in Mendelian idiopathic epilepsies have led to the localization of a gene for partial epilepsy with auditory symptoms to chromosome 10q (Ottman et al., 1995), benign familial infantile convulsions to chromosome, 19q (Guipponi et al., 1997), and a second GEFS$^+$ locus to chromosome 2 (Lopes-Cendes et al., 1996).

Autosomal dominant nocturnal frontal lobe epilepsy (ADNFLE)

ADNFLE was first described in six families in Australia, Canada and the UK (Scheffer et al., 1994). The predominant seizure pattern is that of brief (30–40 s) nocturnal motor seizures, characterized by tonic and/or hyperkinetic motor activity. The attacks are frequently preceded by an aura, and occur in clusters of up to twenty attacks. Individuals may retain consciousness throughout the seizure. Ictal EEG recordings show bifrontal spikes, rhythmic theta or delta activity, or may be normal. The inter-ictal EEG is usually normal. Clinical onset is usually in childhood, and the disease often persists through adult life. The disease gene penetrance is ~75%. ADNFLE is often misdiagnosed as nightmares or other sleep disorders, or a psychiatric disturbance.

Linkage analysis

Linkage analysis is based on the fact that the closer two loci are, the less likely it is that crossing over, observed as recombination, will occur during meiosis. One centiMorgan (cM) is the genetic distance between two loci, which recombine in 1% of meioses.

A genetic marker which co-segregates with the gene of interest, with significantly fewer recombinations that would be expected by chance, is 'linked' to the gene.

The locus for ADNFLE on chromosome 20q
The ADNFLE locus was mapped in a large Australian pedigree with 27 affected individuals spanning six generations by exclusion mapping. Regions with possible homology to the major

TABLE 1

Human idiopathic epilepsy genes

Idiopathic epilepsy	Epilepsy type	Mode of Inheritance	Gene location	Gene symbol (gene product)	(O)MIM	Key refs.
Benign familial neonatal convulsions (*EBN1*)	Generalized	AD	20q	*KCNQ2*	121200	(Biervert et al., 1998; Singh et al., 1998)
Benign familial neonatal convulsions (*EBN2*)	Generalized	AD	8q	*KCNQ3*	121201	(Charlier et al., 1998)
Generalized epilepsy with febrile seizures plus (*GEFS1*)	Generalized	AD	19q	*SCN1B*		(Wallace et al., 1998)
Generalized epilepsy with febrile seizures plus	Generalized	AD	2	unknown		(Lopes-Cendes et al., 1996)
Juvenile myoclonic epilepsy	Generalized	Complex	15q	*?CHRNA7*	254770	(Elmslie et al., 1997)
Juvenile myoclonic epilepsy +/− idiopathic generalized epilepsy (*EJM1*)	Generalized	Complex	6p	unknown	254770	(Durner et al., 1991); (Weissbecker et al., 1991); (Whitehouse et al., 1993); (Liu et al., 1995); (Elmslie et al., 1996)
Autosomal dominant nocturnal frontal lobe epilepsy	Partial	AD	20q	*CHRNA4*	600516	(Steinlein et al., 1995); (Steinlein et al., 1997)
Partial epilepsy with auditory symptoms	Partial	AD	10q	unknown	600512	(Ottman et al., 1995)
Benign familial infantile convulsions	Partial	AD	19q	unknown	601764	(Guipponi et al., 1997)

partial epilepsy gene in the mouse model (Rise et al., 1991), regions containing other epilepsy genes, and finally the rest of the genome were tested.

DNA samples from family members were genotyped using standard methodology. A maximum two point lod score of 9.29 at a recombination fraction, $\theta = 0.0$ was obtained at D20S19, which maps to chromosome 20q13.2-q13.3 (Phillips et al., 1995).

Analysis of linkage data

Various algorithms exist to analyze the data produced from linkage studies.

Parametric analysis requires assumptions to be made about mode of inheritance and disease gene penetrance, and is therefore most useful for the analysis of Mendelian traits.

Non-parametric analysis evaluates the inheritance pattern in pedigrees and calculates a scoring function to measure whether affected individuals share alleles identical by descent (or in some models identical by state) more often than would be expected under random segregation.

These models have been designed to overcome the problems of multiple modeling in the analysis of complex traits using parametric approaches.

A lod score is the $\log_{10}$ of the likelihood of two markers being linked, as opposed to their not being linked.

Two-point analysis calculates the lod score of two loci separated by a specified genetic distance.

Multipoint analysis calculates location lod scores at positions along a genetic map.

The conventional **threshold for significance** is taken as a lod score of 3 or greater. This corresponds to a 1 in 1000 probability of the observed inheritance pattern occurring by chance. Linkage is excluded when a negative lod score of −2 is obtained.

Identification of the gene for ADNFLE, CHRNA4

This region was then examined for candidate genes. The α4 subunit of the neuronal nicotinic acetylcholine receptor (*CHRNA4*) maps to the same region of chromosome 20. Neuronal nicotinic acetylcholine receptors are pentameric ligand-gated ion channels. They are comprised of various combinations of α and β subunits. Each subunit consists of an N-terminal extracellular domain involved in ligand binding, four hydrophobic transmembrane domains and a large cytoplasmic loop between transmembrane domains 3 and 4 (Fig. 1). The commonest configuration is of two α and three β subunits (Fig. 2).

Eight genes for human nAChR subunits have been mapped (Table 2) (Anand and Lindstrom, 1992; Chini et al., 1994). The neuronal nicotinic acetylcholine receptors are thought to be predominantly presynaptic, and may have a role in modulating neurotransmitter release. Nicotine has been shown to stimulate the release of glutamate in an α-BGT-blockable manner in rat hippocampal slice preparations (Gray et al., 1996).

α4 subunits most commonly combine with β2 subunits to form nAChRs with the subunit composition $(\alpha 4)_2$ $(\beta 2)_3$. The α4 subunit gene is expressed in all layers of the frontal cortex, making it a strong candidate gene for ADNFLE (Wevers et al., 1994).

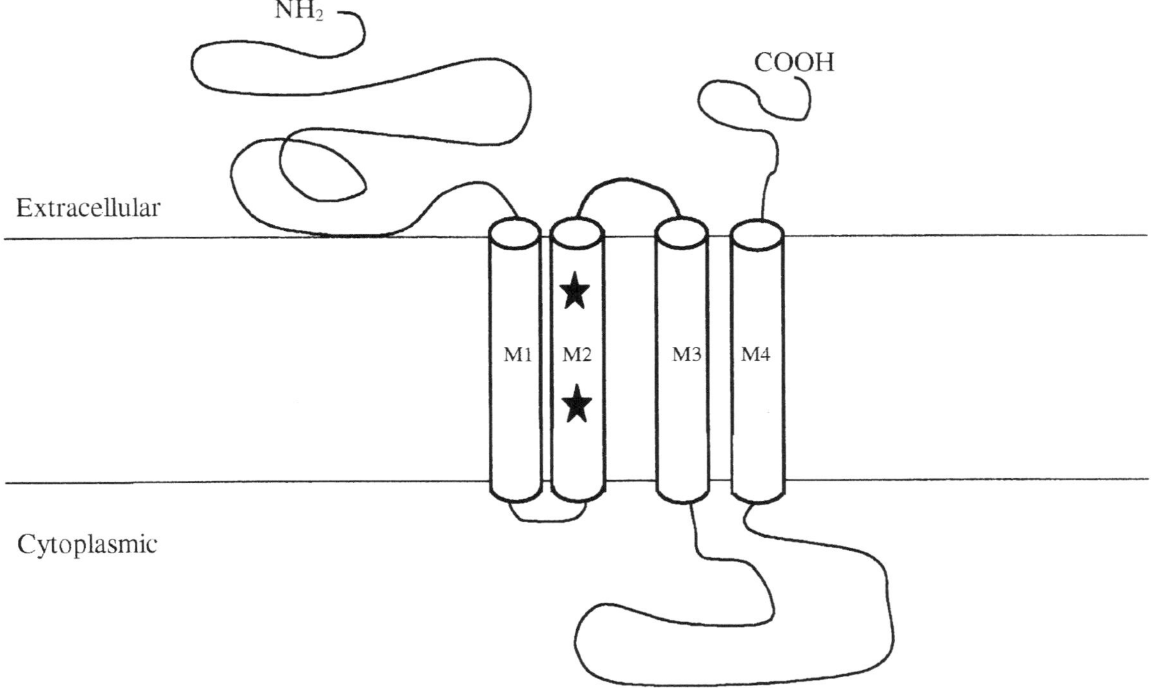

Fig. 1. Schematic representation of a single a4 nicotinic acetylcholine receptor subunit. M1–4 = transmembrane domains; M2 = pore forming region; ⋆ = sites of mutations in ADNFLE families.

TABLE 2

Neuronal nicotinic acetylcholine receptor subunit genes and their localization

nAChR subunit	Chromosomal localization
CHRNA2	8p21
CHRNA3	15q24
CHRNA4	20q13.3
CHRNA5	15q24
CHRNA7	15q14
CHRNB2	1p21.1-q21
CHRNAB3	8p11.22
CHRNB4	15q24

Mutational Analysis of CHRNA4

Mutational analysis of the candidate gene was then undertaken. Exon 5, which encodes the main coding region of *CHRNA4,* was amplified from genomic DNA of ADNFLE patients belonging to the chromosome 20-linked family. Single-strand conformational polymorphism analysis (SSCP or SSCA) detected altered mobility in two overlapping fragments.

Single-strand conformational polymorphism analysis (SSCA or SSCP)

Single stranded DNA (ssDNA) has a tendency to fold up and form complex structures stabilized by weak intramolecular bonds. The resultant conformation is determined by the DNA sequence. The electrophoretic properties of such structures on non-denaturing gels depend on the chain length and conformation. Differences in sequence between mutant and wild-type strands can be detected as band shifts on a gel.

Each PCR product consists of two strands of double stranded DNA (dsDNA), one from each parental chromosome. Each strand of dsDNA yields a *sense* and an *antisense* single strand.

In individuals homozygous for a wild-type allele, each *sense* strand will be identical, giving rise to a single band on SSCP. Identical *antisense* bands will be seen as a second band. Thus PCR products from an unaffected homozygous individual will give rise to 2 bands on SSCP analysis.

PCR products from individuals heterozygous for a disease allele mutation will yield four bands on SSCP: two bands from the normal chromosome and 2 bands from the *sense* and *antisense* strands derived from the chromosome carrying the mutation.

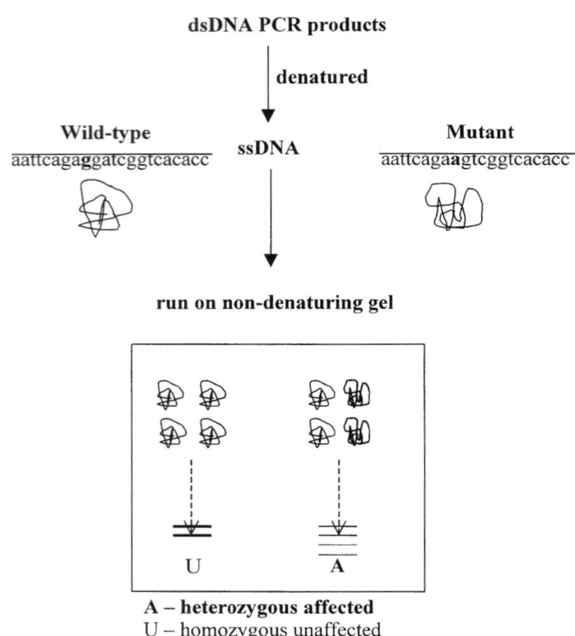

Semi-Automated Sequencing

In a sequencing reaction, a DNA template is polymerized using *Taq* polymerase, a sequencing primer and nucleotide precursors. A mixture of normal nucleotide precursors (dNTPs) and a small proportion of fluorescently labeled base-specific dideoxynucleotides (ddNTPs) are used. Each of the four ddNTPs is labeled with a different colored fluorophore.

ddNTPs differ from dNTPs in that they lack a 3′ hydroxyl group. This prevents phosphodiesterase bond formation at their 3′ end when they are incorporated into a growing DNA chain. This results in abrupt termination of chain synthesis.

Chain termination occurs randomly at only a small proportion of bases. The reaction generates a collection of DNA fragments of differing sizes. The fragments share a common 5′ end (defined by the sequencing primer), but vary at their 3′ ends, each of which terminates with a specific ddNTP.

The fragments are then separated on a denaturing polyacrylamide gel. A laser detects the fluorescence signal as the different sized fragments pass a fixed point on the gel. The output is generated by computer as dye-specific, and hence base-specific, intensity profiles.

The PCR fragments were either subcloned and sequenced using vector-specific primers, or sequenced directly. The SSCP band shift was

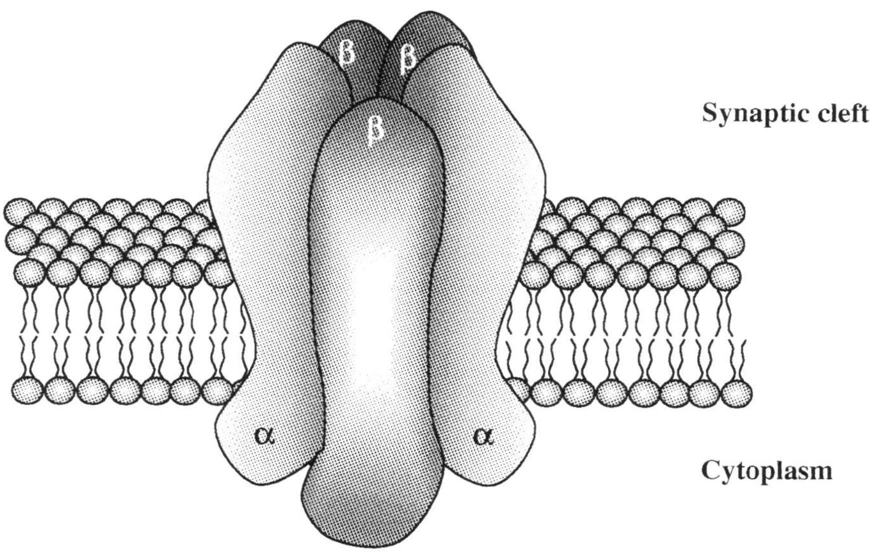

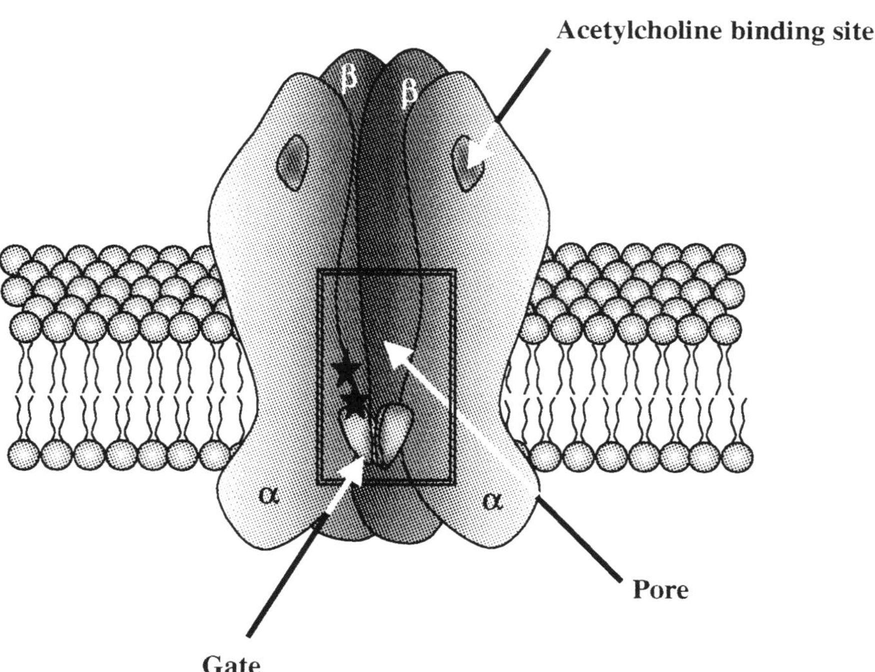

Fig. 2. Schematic representation of a neuronal nicotinic acetylcholine receptor. The acetylcholine binding sites are located in the α subunits. The M2 transmembrane domain of each subunit lines the pore. $\star$ = site of mutations in ADNFLE families. (Reproduced from Schmitz and Sander, 1999, with permission.)

shown to correspond to a C to T transversion, causing a serine to be replaced with phenylalanine at position 248, Ser248Phe. This replaces a highly conserved small hydrophilic amino acid with a large hydrophobic amino acid, at a region of the M2 transmembrane domain thought to be involved in cation selectivity of the channel (Figs 1 and 2). A PCR assay was developed to allow rapid mutation screening. A primer was designed with a one base pair mismatch which created a new *HpaII* restriction site in the wild type but not the mutant allele. The mutant allele co-segregates with ADNFLE in the pedigree, and was not found in over 600 control chromosomes.

Expression studies of the wild-type and mutant α4 β2 channels

Functional studies of the α4 β2 wild-type and mutant Ser248Phe channel, by expression in *Xenopus* oocytes, show that the functional capacity of the Ser248Phe α4 is reduced by four mechanisms: loss of Ca^{2+} permeability, increased desensitization, reduced channel opening time and reduced channel conductance. All of these mechanisms lead to a reduction in the net flow of ions through the activated channel (Weiland et al., 1996; Kuryatov et al., 1997).

Xenopus oocyte expression system

Xenopus oocytes are large unfertilized eggs from the clawed frog. When injected with messenger RNA (mRNA), they are able to perform all the biochemical steps required — from translation to insertion into the plasma membrane — to produce a functional protein.

cDNA encoding either a wild type or mutant channel is subcloned into a suitable vector and transcribed in vitro using RNA polymerase. The resultant cRNA (mRNA derived from cDNA) is injected into the oocyte.

The consequent expressed protein can be detected and characterized using a variety of techniques.

If the protein is able to generate an electrical signal, the **two microelectrode voltage clamp** which measures whole cell current, or **patch clamping**, which can detect single ion channel kinetics, may be used.

Patch clamping can also be used to investigate the transmitter of interest in the investigation of ligand gated channels.

Identification of a second mutation in CHRNA4

A second mutation in the M2 transmembrane domain has been found in a Norwegian pedigree with ADNFLE (Steinlein et al., 1997). The entire coding region of the *CHRNA4* gene was amplified from the DNA of the index patient. PCR fragments were then analyzed by SSCA and heteroduplex analysis.

Heteroduplex analysis

Heteroduplex bands are formed through cross-annealing between complementary wild-type and mutant DNA strands. The electrophoretic mobility of heteroduplexes in polyacrylamide gels is reduced compared to that of homoduplexes. This difference in band pattern allows heteroduplex detection.

Both methods revealed abnormal migration in some PCR fragments. These products were then subcloned and sequenced, or sequenced directly. On analysis, a three nucleotide insertion (GCT) was found following bp776/bp788 (using *torpedo* and human α4 subunit numbering respectively).

The insertion does not alter the open reading frame, but results in the insertion of a leucine residue near the C-terminal end of the M2 domain (Figs 1 and 2). The 776ins3 mutation segregates with ADNFLE in the pedigree, with affected individuals being heterozygous for the wild type/mutant alleles. The insertion mutation was not found in a panel of 254 controls.

Expression studies of the wild-type and mutant α4 β2 subunits

Expression studies of the wild-type and mutant α4 β2 subunits revealed that the extra leucine residue does not disrupt receptor function. No detectable abnormalities in current amplification or desensitization were found between wild-type and mutant receptors. Mutant receptors did show a 10 fold increase in ACh apparent affinity, and a significant reduction in calcium permeability. It is hypothesized that this mutation, indirectly, causes a loss of function of the α4 β2 channel.

Thus two mutations in the M2 transmembrane domain of the $\alpha 4$ subunit of the neuronal nicotinic acetylcholine receptor have been found in families with ADNFLE. Both mutations are thought to reduce the efficacy of the channel via a reduction in Ca^{2+} permeability, although the exact effect on channel gating is unknown. Not all families with ADNFLE are linked to *CHRNA4*, and mutations in other neuronal nicotinic acetylcholine receptors, or other ion channels may be responsible for the disease in these families.

Benign familial neonatal convulsions

Benign familial neonatal convulsions (BFNC) is an autosomal dominant idiopathic generalized epilepsy of the newborn. Seizures begin at the age of 2–4 days, and spontaneously remit between 2–15 weeks of age. Seizures typically start with a tonic posture, ocular symptoms and other autonomic features, which may progress to clonic movements and motor automatisms. Affected neonates are normal between seizures, and their subsequent neurological and intellectual development is normal. Seizures recur later in life in approximately 16% of patients (Ronen et al., 1993). Two susceptibility loci for BFNC have been mapped to chromosomes 8 and 20 (Leppert et al., 1989; Lewis et al., 1993).

The locus for BFNC on chromosome 20q

The locus on chromosome 20q, *EBN1* was originally mapped in, 1989, in a large four generation pedigree with, 19 affected individuals. Two polymorphic marker loci, *D20S19* and *D20S20* were found to be tightly linked to the disease locus, giving maximum lod scores of 2.87 ($\theta = 0.039$) and 3.12 ($\theta = 0.003$) respectively. Linkage to these markers was subsequently confirmed in additional families of European origin. A more distal marker, *D20S24* showed complete co-segregation with the BFNC phenotype in chromosome 20-linked families.

Identification of the gene for EBN1, KCNQ2

A positional cloning strategy was undertaken by two separate groups, in order to identify the gene for *EBN1*.

One group, using cosmid DNA from a previously described contig in the chromosomal region 20q13.3 (Steinlein et al., 1994), isolated a 3.4 kb partial cDNA clone from a human fetal brain cDNA library.

Positional cloning

This is a process by which disease genes are isolated on the basis of their chromosomal location, without prior knowledge of the defective protein.

The gene locus is initially mapped by linkage analysis, ideally to a region of interest of less than 2 cM. A large number of families are needed to achieve this. Chromosomal variants, such as deletions or translocations can also be used to define a small critical interval.

A series of markers from within the candidate interval are then used to identify overlapping clones containing genomic DNA from this region. A variety of genomic clones with different insert sizes are available:

- Yeast artificial chromosomes (YACs) ~0.2–2 Mb
- Bacterial artificial chromosomes (BACs) <300 kb
- P1 artificial chromosomes (PACs) 30–150 kb
- Cosmid vectors 30–44 kb

Polymorphic markers identified during this process of physical mapping may further define the candidate region.

Transcripts (coding sequences) are identified from genomic clones by a wide variety of techniques:

- cDNA selection
- Exon trapping
- Direct sequencing and exon prediction programs
- cDNA library screening with genomic clones

The identified transcripts are then sequenced, assembled into full length cDNAs and characterized (expression patterns, size, homology etc.).

The clone was then extended by 5′ RACE (Rapid Amplification of cDNA Ends). Adult human cDNA was amplified with gene-specific primer to produce a clone encoding an 844 amino acid protein, *KCNQ2*. A homology search using the GenBank database revealed that the clone showed ~50% identity to *KCNQ1*, a voltage gated delayed rectifying K^+ channel gene, and significant homology to *HNSPC* (a putative K^+ channel isolated from a human neuroblastoma cell line) (Yokoyama et al., 1996; Biervert et al., 1998).

RACE (Rapid Amplification of cDNA Ends)

The gene identification protocols previously discussed may result in the identification of incomplete cDNA fragments. RACE is a PCR-based protocol that allows the identification of previously unknown 5′ or 3′ sequence. The known cDNA sequence is extended in a 5′ and/or 3′ direction using 'anchored' cDNA, a gene-specific primer, and an anchor-specific primer:

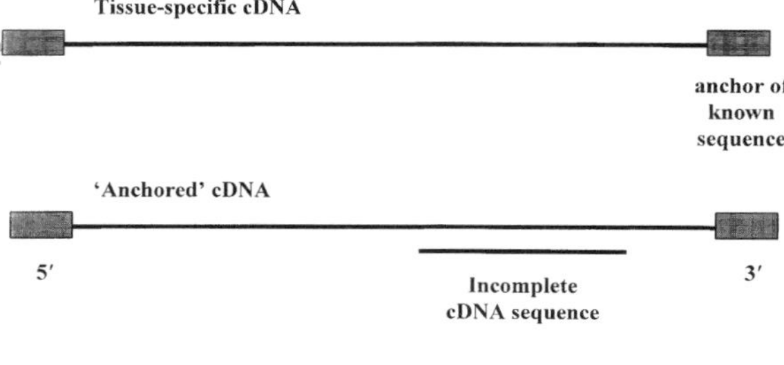

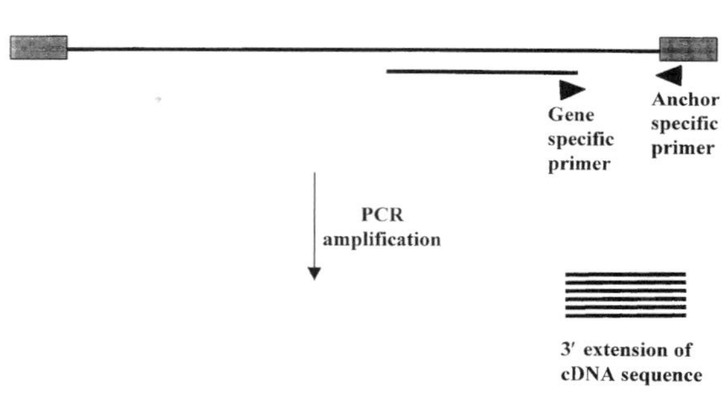

In the second study, a sub-microscopic deletion was identified in a single *EBN1*-linked family, where it was observed that a null allele for the 3 allele marker *D20S24* occurred exclusively in individuals with BFNC, and two non-penetrant family members.

The existence of a deletion was confirmed by fluorescence *in situ* hybridization (FISH) in family members. Using *D20S24* probes, hybridization to only one chromosome 20 homologue was seen in the affected individual tested, and to both chromosome 20 homologues in 93% of the normal metaphase spreads examined. The extent of the deletion was determined by comparing the FISH results of *D20S24* and overlapping P1 clones.

Fluorescence *in situ* hybridization (FISH)

Chromosomal *in situ* hybridization is a process in which denatured metaphase chromosomes are incubated with labeled nucleic acid probes. The probe hybridizes to the region of the chromosome containing complementary DNA.

The chromosome preparation is then washed, to remove excess probe, and chromosomal banding performed.

The chromosome-specific banding patterns can be detected using a microscope, and correlated with the signal obtained from the hybridized probe. This allows cytogenetic localization of the probe DNA.

In **FISH**, the nucleic acid probe is labeled by incorporation of a fluorescently labeled molecule. The signal produced can then be detected using a fluorescence microscope. If expressed in the tissue of interest, the cDNAs can then be screened for mutations in linked families.

The same probes were then used to screen a fetal brain cDNA library. Three cDNAs were isolated, which showed significant homology to *KCNQ1*, *HNSPC* and *nKQT1* (a *C. elegans* gene) (Wei et al., 1996; Altschul et al., 1990).

Additional sequence of the new gene, *KCNQ2*, was obtained using RACE, by amplifying adult brain, fetal brain and temporal lobe cDNAs with *KCNQ2*-specific primers. This led to the isolation of composite clones encoding 872 amino acids of *KCNQ2*. The complete 3′ untranslated region was not identified.

KCNQ2 encodes a protein of 844, 854 (Biervert et al., 1998), or 872 amino acids (Singh et al., 1998) depending on the splice variant considered. The gene displays the hallmarks of a *Shaker*-type potassium channel: six transmembrane domains, a pore region, and common charged amino acids in the S2, S3, and S4 transmembrane domains (Fig. 3).

KCNQ2 shows 60–70% identity to the transmembrane domains of *KCNQ1*, and a homologous region was also identified in the C terminal region.

To determine the tissue distribution of *KCNQ2*, northern blotting analysis was performed. *KCNQ2* is expressed in both fetal and adult brain, especially the cerebellum (Biervert et al., 1998), temporal lobe and putamen (Singh et al., 1998).

Northern blotting analysis

This is used to detect the tissue distribution pattern of a specific gene. RNA fragments from different tissues are size-fractionated by agarose gel electrophoresis, and transferred to a nitro-cellulose membrane. Each lane of the membrane contains mRNA from a specific tissue. cDNA encoding the gene under investigation is radioactively labeled, and hybridized to the prepared northern blot.

The probe will hybridize to mRNA from tissues in which the gene is expressed. X-ray radiography is used to detect the resultant hybridization patterns, from which the expression pattern of the gene can be determined.

Voltage gated K^+ channels repolarize neuronal membranes which have been depolarized by Na^+ and Ca^{2+} voltage gated channels. K^+ channels are thought to repolarise neuronal membranes after activation of excitatory neurotransmitter ion channels, including acetylcholine and glutamate.

Mutations in *KCNQ1* are responsible for the long QT syndrome, and Jervell Lange Nielsen cardioauditory syndrome (Wang et al., 1996; Neyroud et al., 1997). Both these diseases are paroxysmal disorders of cardiac rhythm.

KCNQ2 is therefore an excellent candidate gene for BFNC.

Mutational analysis of KCNQ2

In the first study, a large Australian family with BFNC, previously linked to chromosome 20 was analyzed. A five base pair insertion at the triplet encoding amino acid 534 was identified. This lies at the C terminal end, in a region highly conserved between *KCNQ1* and *KCNQ2* (Fig. 3). The resulting frameshift would result in a premature stop, truncating over 300 amino acids. This insertion mutation co-segregated with the BFNC phenotype in the family. The mutation was not found in a panel of 462 control chromosomes.

In the second study, mutation analysis of *KCNQ2* was performed on one affected member of each family in the resource (12 families). Using SSCP and direct sequencing, mutations were found in five BFNC families, in addition to the deletion found initially. Two missense mutations, two frameshift mutations and one splice site mutation were identified. These all occurred in the S6 domain, channel pore, and in or just before the highly conserved C terminal region (Fig. 3). None of the mutations were seen in 140 control chromosomes. In the families in which mutations were identified, the mutations were shown to segregate completely with affection status. The breakpoints of the deletion have not been determined. However, the plasmid which includes the marker locus *D20S24*, deleted in one family, contains at least 80 codons of sequence from the highly conserved 3′ region of the *KCNQ2* gene, indicating that this portion of the gene is deleted in affected individuals from this family.

Expression studies of the wild type and mutant KCNQ2 *channels*

Expression of wild-type *KCNQ2* protein in *Xenopus* oocytes was detected by two microelectrode voltage clamping. The currents produced are potassium-selective currents that activate slowly with depolarization. These currents resemble those of the delayed rectifier *KCNQ1* channel in their selective permeability, voltage dependence and kinetics (Barhanin et al., 1996; Sanguinetti et al., 1996; Wollnik et al., 1997).

Expression of the truncated *KCNQ2* protein resulting from the 5 bp insertion produced currents identical to those of negative controls. Assuming that synthesis of the mutant *KCNQ2* protein in the expression system was successful, this indicates that the mutation abolishes channel function.

Co-injection of mutant and wild-type cRNA at a 1:1 ratio, to mimic a heterozygous individual, was then performed. The currents produced were smaller than those recorded from oocytes injected with the same total amount of wild-type cRNA alone.

No obvious dominant negative effect of the mutant channel was observed. However, the dominant inheritance of the disorder may be explained by haploinsufficiency, differences in expression in oocytes and in vivo, or the fact that an auxiliary subunit is required for normal channel function in vivo.

KCNQ1 α subunits co assemble with minK β subunits to form heteromultimeric channels in the heart (Sanguinetti et al., 1996). Co-expression of wild-type *KCNQ2* with minK produced currents that did not differ significantly from those predicted, based on *KCNQ1*/minK expression studies (Wollnik et al., 1997). It may be that *KCNQ2* subunits form heteromultimers with min-K-like β subunits in the brain.

Expression studies of the six *KCNQ2* mutations found in the second study have not been performed. The predicted altered protein products resulting from these mutations may alter channel function by loss of function, gain of function or dominant-negative effect.

Insertion mutations in the pore region of K^+ channels have been shown to alter the length of the pore (crucial for channel function), and modify the signature sequence producing a truncated protein (Heginbotham et al., 1994; Tytgat, 1994; Hidalgo and MacKinnon, 1995; Nakamura et al., 1997). The S6 domain in the *Shaker* K^+ channel subtype, and the C-terminal in inwardly rectifying K^+ channels are involved in K^+ ion conduction (Lopez et al., 1994; Kukuljan et al., 1995), and may have the same function in KCNQ2 channels.

The locus for BFNC on chromosome 8q

A second locus, *EBN2*, was identified in, 1993. A large 3 generation Mexican–American family with 14 affected members was studied, in which linkage

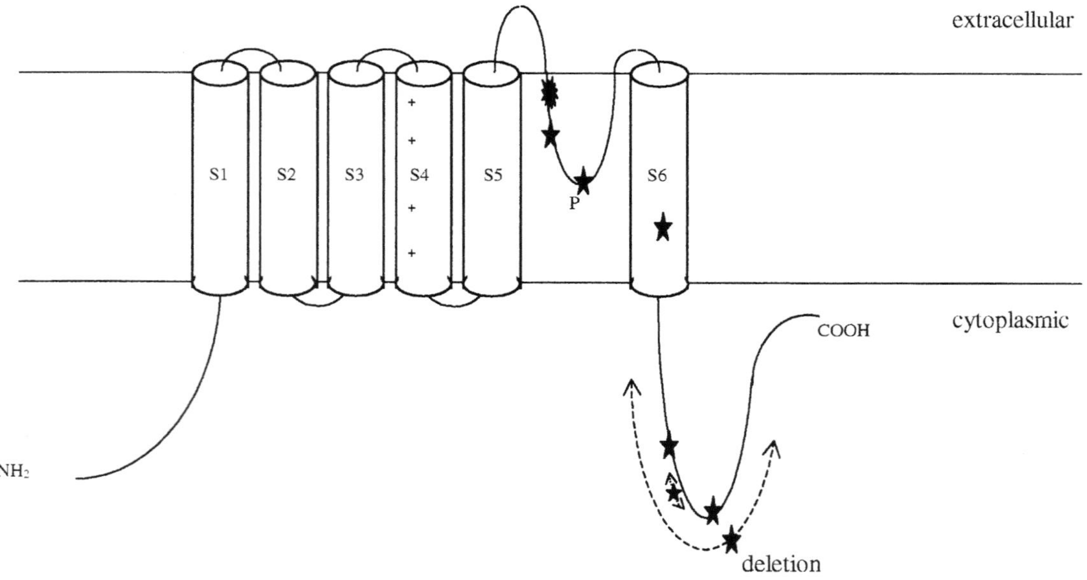

Fig. 3. Schematic representation of a KCNQ-type voltage gated potassium channel. P = pore forming region; S1–S6 = transmembrane domains; + = positively charged S4; ⋆ = sites of mutation in *KCNQ2*; ⟨⋆⟩ = 5 bp insertion; * = site of mutation in *KCNQ3*.

to the locus on chromosome 20 had previously been excluded.

A genome wide search was undertaken, with preliminary results showing weak evidence for linkage to chromosome 8q. This region was then analyzed with 12 additional markers. Significant pairwise lod scores (>3) were found with *D8S284, D8S256* and *MYC*, locating the gene to chromosome 8q24.

Identification of the gene for EBN2, KCNQ3
Following the identification of mutations in *KCNQ2* associated with *EBN1*, further members of the *KCNQ* family were sought. A tBLASTx search was performed with the full length *KCNQ2* cDNA against the expressed-sequence-tag (EST) database. Five human EST clones that presented homology with *KCNQ2* were identified. The clones came from a retina cDNA library and an infant brain cDNA library, and can be organized in two non-overlapping contigs, which belong to the same gene, named *KCNQ3* (Charlier et al., 1998).

The genomic localization of the clones was mapped to chromosome 8 using a somatic cell hybrid panel (Drwinga et al., 1993). The position was refined using a panel of 97 radiation hybrids previously constructed for determining the linear order and inter-marker distance of chromosome 8 loci (Lewis et al., 1995). Tight linkage of *KCNQ3* was observed with markers previously mapped to chromosome band 8q24. This localizes *KCNQ3* to the *EBN2* critical region, making it a strong positional candidate for the chromosome 8 BFNC locus.

Radiation Hybrid Mapping

A radiation hybrid is produced when human chromosome fragments, generated by lethal irradiation of somatic cell hybrids containing a single chromosome, are fused with recipient rodent cells. The resultant hybrid cells can be identified by screening for the *Alu* repeat sequence (which occurs ~ every 4 kb in the human genome). Continued culture of the hybrid cells selects for those cells in which the human fragments have become stably integrated into the rodent chromosomes. Such hybrid cells constitute a panel of **radiation hybrids**.

As in linkage analysis, 2 markers that are close together are likely to remain on the same fragment, and therefore be contained in the same hybrid.

To map a locus using a radiation hybrid panel *either*

- a locus-specific probe is hybridized to a southern blot of each hybrid *or*
- each hybrid is amplified by PCR with locus-specific primers.

Each hybrid is scored as either positive or negative for the locus. The data is analyzed using a computer program, and the locus is assigned a linear map position.

The EST sequence was extended by 5′ and 3′ RACE, to produce an incomplete *KCNQ3* cDNA sequence. This is predicted to encode an 825 amino acid protein with six putative transmembrane domains, a pore region, a stop codon and the 3′ untranslated region containing the poly-A^+ tail. The protein exhibits 69% similarity and 62% identity with *KCNQ2* for the entire coding region of the genes. Homology is also conserved with *KCNQ1* and *nKQT1*.

Mutation analysis of *KCNQ3* in the original *EBN2* family was performed. Using SSCP, a variant was identified in a PCR fragment containing the S5 transmembrane domain, and half of the pore-forming region. This SSCP variant co-segregated with the BFNC phenotype, and was also present in a non-penetrant individual with the disease-marker haplotype. The variant was not seen in a panel of 264 chromosomes from unaffected unrelated European and Mexican–American individuals.

The mutation was characterized by cloning and sequencing the PCR product. The mutation is a missense mutation (G→T), causing a glycine to be substituted by a valine in position 263 of the highly conserved pore-forming region (Fig. 3). By analogy with mutations in other K^+ channels, this mutation may affect channel gating properties (Dreyer et al., 1998; Guo and Kubo, 1998; Molina et al., 1998). Electrophysiological studies of *KCNQ3* are awaited.

Generalized epilepsy with febrile seizures plus

This autosomal dominant idiopathic generalized epilepsy was first described in a large Australian pedigree in, 1997 (Scheffer and Berkovic, 1997). The syndrome GEFS$^+$ involves a variety of phenotypic features. The commonest feature is termed 'febrile seizures plus'. This comprises childhood onset of multiple febrile seizures, with febrile or afebrile tonic-clonic seizures continuing beyond six years of age. Seizures usually cease by mid-childhood. Other phenotypic features include febrile seizures associated with absences, myoclonic seizures or atonic seizures, and myoclonic-astatic epilepsy. Within the pedigree 25 individuals over four generations were affected.

The locus for GEFS$^+$ on chromosome 2

The locus for in this pedigree was then mapped (Lopes-Cendes et al., 1996). A genome wide search using 210 polymorphic microsatellite markers spaced at 20 cM intervals on the human genome was undertaken. Forty-four family members including 23 affected individuals were typed. Positive lod scores were obtained with markers on chromosome 2, with a maximum two-point lod score of 4.7.

The locus for GEFS$^+$ on chromosome, 19q

A second GEFS$^+$ locus was mapped in another Australian pedigree, comprising 378 individuals from 6 generations (Wallace et al., 1998). Twenty-six family members were diagnosed as having GEFS$^+$, 3 had other forms of epilepsy, 4 had unclassified seizures and 9 had unconfirmed seizures and were classed as unaffected. One hundred and thirty-eight individuals including 20 with GEFS$^+$ and seven obligate carriers were analyzed by exclusion mapping. A penetrance of 64% and a phenocopy rate of 3% were assumed. The loci for febrile convulsions (Wallace et al., 1996; Johnson et al., 1998), regions containing genes for idiopathic epilepsy, and regions containing acetylcholine receptor subunits were examined and excluded. A genome wide scan was then undertaken by manual genotyping of microsatellite markers. A multipoint lod score of 3.85 was obtained between the markers *D19S882* and *D19S414*, providing strong evidence for a second locus for GEFS$^+$ on chromosome, 19q13, *GEFS1*. A gene for benign familial infantile convulsions has also been mapped to this region (Guipponi et al., 1997), which may be allelic to *GEFS1*.

Identification of the gene for GEFS1, SCN1B

Several possible candidate genes have been mapped to chromosome 19q13.1, including the voltage gated sodium channel $\beta 1$ subunit, *SCN1B* (Makita et al., 1994; Mohrenweiser et al., 1998).

Voltage gated sodium channels are responsible for the rapid rise in membrane Na^+ permeability during the initial phase of the action potential in most excitable tissues. They consist of a large α subunit of 4 homologous domains (I–IV), each with six membrane-spanning units, S1–S6. Two β subunits are associated with the α subunit (Fig. 4). The $\beta 1$ subunit is covalently linked to the α helices, and the $\beta 2$ subunit (not present in skeletal muscle voltage gated sodium channels) is linked by a disulfide bond. They act to modulate the gating properties of the channel (Isom et al., 1992; Makita et al., 1994). Expression studies show that although the α subunit alone can exhibit all the functional attributes of a voltage gated Na^+ channel, the β subunit is required for normal inactivation kinetics (Isom et al., 1992). *SCN1B* is expressed in the brain, heart, and skeletal muscle (Isom et al., 1992).

Mutations in the α subunit gene *SCN4A* (chromosome 17q23.1-25.3) are associated with hyperkalemic periodic paralysis, paramyotonia congenita, and atypical paramyotonia congenita. Mutations in the α subunit gene *SCN5A* (chromosome 3p21) cause long QT syndrome in some families (Bulman, 1997). Na^+ channel subunit genes are also attractive candidate genes for the epilepsies as anti-epileptic drugs, such as carbamazepine and phenytoin, act by enhancing

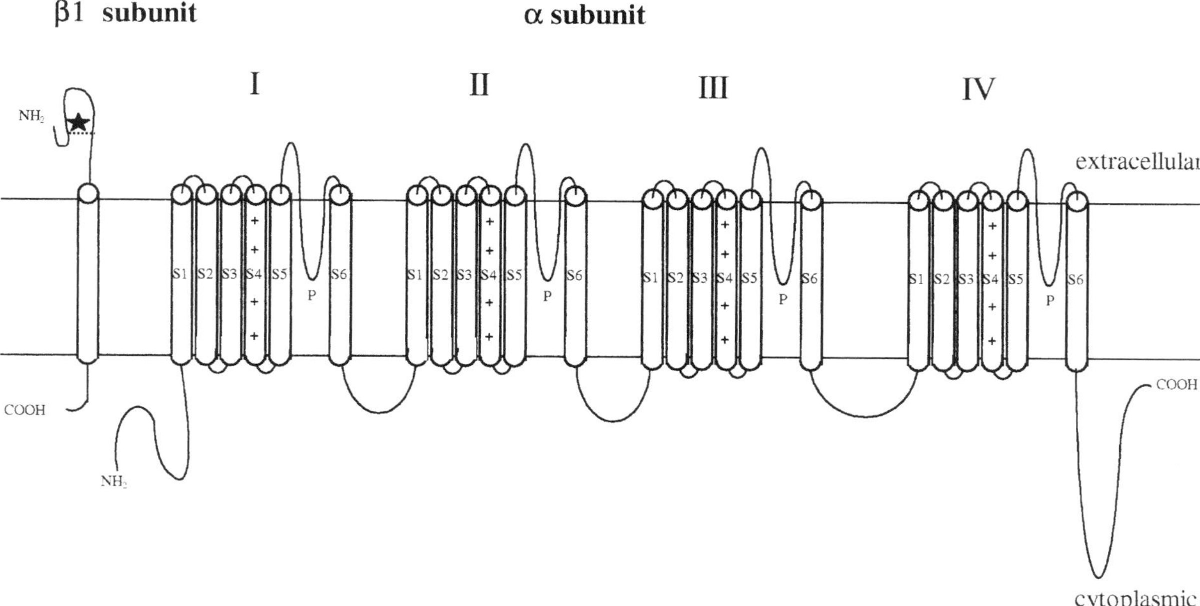

Fig. 4. Schematic representation of a voltage gated sodium channel. P = pore forming regions; S1–S6 = transmembrane domains of α subunit; + = positively charged S4 domain; ... = putative disulphide bond; ⋆ = site of mutation in GEFS$^+$ patients.

Na^+ channel inactivation (Macdonald and Kelly, 1994).

Mutation analysis of SCN1B

SCN1B consists of 5 translated exons contained within 9 kb of genomic DNA. Mutational analysis of the exons of the *SCN1B* gene was carried out using SSCP. An exon 3 PCR product revealed a bandshift which segregated with the GEFS$^+$ phenotype in all except two individuals, who were uninformative for most of the markers used to construct the haplotype. The same band shift was not seen in a panel of 192 control chromosomes. These data strongly suggest that the SSCP variant contains a causative mutation.

A cloned *SCN1B* exon 3 PCR product was sequenced. This revealed a C→G transversion within the coding sequence of the exon at nucleotide 387 (387 C→G). This results in the substitution of a highly conserved cysteine with a tryptophan (Cys121Trp). A restriction enzyme digest was performed on all family members with the SSCP variant. This demonstrated that each individual with the SSCP change had a 387 C→G mutation in the *SCN1B* gene.

This mutation disrupts the putative disulfide bond that maintains the extra-cellular fold motif of the β subunit, and may alter the secondary structure of the extracellular domain (Fig. 4).

Expression studies of the wild type and mutant β1 RBII channels

Functional studies of the wild type and mutant $\beta1$ subunit were then performed. The $\beta1$ subunits were co-expressed with a rat brain Na^+ channel α subunit (RBII) in *Xenopus* oocytes. Expression of the α subunit alone produced a functional channel. Co-expression with the wild type $\beta1$ subunit shortened inactivation time of the channel and accelerated recovery from inactivation.

Co-expression of the α and mutant $\beta1$ subunits produce Na^+ channels with slowed inactivation compared to those produced by expression of the α subunit alone. The mutant $\beta1$ subunit had no measurable effect on recovery from reactivation. Assuming that synthesis of the mutant $\beta 1$ subunit in the expression system was successful, this demonstrates that the mutation causes loss of function by affecting the ability of the $\beta 1$ subunit to modulate Na^+ channel function.

Disruption of the extracellular domain in *SCN1B* is predicted to cause reduced sodium channel expression, slower inactivation, and slower recovery from inactivation (Makita et al., 1996; McCormick et al., 1998).

These observed and predicted functional changes may cause persistent Na^+ influx in neuronal cells, resulting in a more depolarized membrane, and hyperexcitablity. The apparent temperature dependence of the $GEFS^+$ phenotype may reflect the effects of temperature on both conductance and gating of Na^+ channels.

Loci for two other monogenic idiopathic epilepsies have been mapped.

Benign familial infantile convulsions (BFIC)

Benign familial infantile convulsions (BFIC) is an autosomal dominant idiopathic epilepsy. The disease is characterized by onset of afebrile seizures between the ages of 3.5 and 12 months. Seizures are partial with secondary generalization, and clinically manifest as head and eye deviation (not always the same side in each attack), diffuse hypertonia followed by bilateral limb jerks. The interictal EEG is normal. Ictal EEG in the original Italian family described showed a diffuse discharge with onset in the central-occipital region. In common with other idiopathic epilepsies, laboratory, radiological, and neurological findings are normal (Vigevano et al., 1992).

The locus for benign familial infantile convulsions (BFIC) on chromosome, 19

Linkage analysis in five Italian BFIC pedigrees, by exclusion mapping of known candidate regions, followed by a genome wide search using polymorphic microsatellite markers was performed. Two-point, multipoint and haplotype analyses were carried out. A lod score >8 was obtained for the interval *D19S250-D19S245,* suggesting a locus for BFIC lies in this region (Guipponi et al., 1997).

Autosomal dominant partial epilepsy with auditory symptoms

Autosomal dominant partial epilepsy with auditory symptoms was described in a three generation American family, with 11 members classed as having idiopathic/cryptogenic epilepsy (Ottman et al., 1995). The epilepsy was clearly localization-related in ten of these cases. In these affected individuals, seizures began between the ages of 8 and 19 years, and occurred infrequently. Six affected individuals, and one family member with symptomatic epilepsy described non-specific auditory disturbances as a simple partial component of their seizures. All family members were of normal intelligence, except for one individual who had epilepsy associated with cerebral palsy.

The locus for autosomal dominant partial epilepsy with auditory symptoms on chromosome 10q

Linkage analysis was performed using this family, with affected status defined as idiopathic/ cryptogenic epilepsy. Individuals with symptomatic epilepsies were classified as unknown, as were two individuals with suspected epilepsy but insufficient clinical data.

An initial genome wide search using markers spaced at 10–20 cM intervals on the genome gave preliminary evidence for linkage with markers on chromosome 10q. Dense typing of the region with a further 12 markers yielded a maximum 2 point lod score of 3.99 for *D10S192.*

All ten living affected family members share a single haplotype for seven contiguous markers spanning 10 cM from *D10S200–D10S205.* The 2 point lod score for the seven locus haplotype was 4.83 (which approximates to the maximum multipoint lod score for these loci). Inspection of the haplotypes for recombination events placed the susceptibility locus for partial epilepsy with auditory symptoms within a 10 cM interval between markers *D10S185*and *D10S566*.

Idiopathic epilepsies with a complex mode of inheritance

The majority of idiopathic generalized epilepsies display a complex pattern of inheritance. This common group of epilepsies includes juvenile myoclonic epilepsy (JME), juvenile absence epilepsy (JAE), childhood absence epilepsy (CAE) and epilepsy with generalized tonic-clonic seizures on awakening. Active idiopathic generalized epilepsy (IGE) accounts for 90% of all generalized epilepsies (Hauser et al., 1991).

Relatives of probands with IGE have a 5–8% risk of developing epilepsy (Janz et al., 1989). Frequently, two or more different IGE phenotypes are found within a single pedigree, which suggests that there may be susceptibility loci that are common to all idiopathic generalized epilepsies. A common locus, or group of loci, may act to determine the seizure threshold, with additional loci contributing further phenotypic specificity.

Juvenile myoclonic epilepsy

Juvenile myoclonic epilepsy is an IGE with an age of onset between 7 and 26 years. The phenotype is characterized by generalized myoclonic seizures, and may include generalized tonic-clonic seizures and absence seizures. Inter-ictal EEG patterns may be normal, or show generalized spike and wave or polyspike and wave discharges. Neurological and neuroradiological examination is normal.

The locus for JME and other IGE phenotypes on chromosome 6p
Mapping of loci involved in idiopathic epilepsies that display a complex mode of inheritance has produced conflicting results. Studies have provided evidence both for and against the existence of a locus, designated *EJM1*, which predisposes to juvenile myoclonic epilepsy or related idiopathic generalized epilepsies on chromosome 6p (Greenberg et al., 1988; Durner et al., 1991; Weissbecker et al., 1991; Liu et al., 1995; Whitehouse et al., 1993; Elmslie et al., 1996). However it has not been possible to refine the localization of *EJM1,* or to clarify the exact phenotype to which it confers susceptibility.

The locus for JME on chromosome 15q
Following the identification of mutations in *CHRNA4* in families with ADNFLE, the genes encoding subunits of the neuronal nicotinic acetylcholine receptor have emerged as candidate genes for the inherited idiopathic epilepsies, including JME.

A major locus for juvenile myoclonic epilepsy has been mapped to chromosome 15q using the candidate gene approach (Elmslie et al., 1997).

Thirty-four European families, including two or more individuals with clinical JME were studied. These contained 165 individuals of which 73 were classified as affected, and five (with other epilepsy phenotypes) as unknown.

An initial screen with 22 polymorphic microsatellite markers spanning the chromosomal regions to which nAChR subunits map yielded negative lod scores in all regions except the *CHRNA7* region on chromosome 15q14. Further investigation was undertaken with 8 markers in this region, and the results analyzed using parametric and non-parametric linkage analysis (implemented using the GENEHUNTER program). Multipoint parametric analysis was carried out assuming either autosomal dominant or autosomal recessive inheritance with 50% penetrance, and allowing for locus heterogeneity. Three liability classes were specified to allow for age dependant penetrance.

Significant evidence in favor of linkage with heterogeneity was obtained. Multipoint parametric linkage analysis gave a maximum lod score of 4.42 at a point 1.7 cM telomeric to *D15S144* at $\alpha = 0.65$ (i.e. 65% of families linked), assuming autosomal recessive inheritance.

Non-parametric analysis of the data gave a maximum total score of $Z_{all} = 2.94$, $P = 0.00048$ at *ACTC* .

CHRNA7 encodes the α7 subunit of the neuronal nicotinic acetylcholine receptor, which forms homo-oligomeric channels when expressed in *Xenopus* oocytes (Couturier et al., 1990). α7 is expressed throughout the brain (Breese et al., 1997), and based on its position and putative modulatory role, is an excellent candidate gene. Mutational analysis of *CHRNA7* is currently being undertaken.

Idiopathic epilepsies — a family of ion channel disorders?

Seizures occur when there is an excess of excitatory processes in the brain over inhibitory processes. Changes in afferent excitation, disinhibition, alterations in extracellular ion concentrations, opening of voltage gated ion channels and enhanced neuronal synchrony all play a role in the initiation and propagation of seizure activity (Fisher, 1995). Neuronal activity is regulated by concentration of ions in the extracellular and intracellular spaces, and their selective flux across the neuronal membrane. Mutations in voltage gated or ligand gated ion channel genes which result in channel dysfunction, may alter ion concentrations across the cell membrane. This could result in reduced or increased neuronal excitability.

Mutations in four genes that encode voltage gated or ligand gated ion channels have been found to be associated with human idiopathic epilepsies. All are predicted to directly or indirectly increase neuronal excitability.

Spontaneously arising mouse models of epilepsies also provide evidence of epilepsy as an ion channel disease. Mutations in the pore forming α and regulatory β and γ subunits of voltage gated calcium channels have been identified in tottering, lethargic and stargazer mice (Fletcher et al., 1996; Burgess et al., 1997; Letts et al., 1998), neurological mouse mutants whose phenotypes all include generalized absence seizures.

Problems associated with the isolation of epilepsy genes

Difficulties may arise at every stage in the process of the isolation of epilepsy genes. Reliable and accurate clinical information is essential in a disease such as epilepsy where no definitive diagnostic test exists.

Incorrect diagnosis may occur for several reasons. Patients are often unaware of seizures occurring early in childhood and the prevalence of certain acquired epilepsy phenotypes in the population may lead to a high phenocopy rate.

Linkage analysis of non-Mendelian idiopathic epilepsies is itself very problematic. Parametric analysis requires assumptions to be made about mode of inheritance and disease penetrance. Conventional thresholds of significance may not apply to linkage data derived from testing multiple models of inheritance, phenotype or genome wide sets of markers (Lander and Kruglyak, 1995).

Once a disease-causing gene has been isolated, the inferred function of the wild-type and mutant protein products from expression studies in vitro may not accurately reflect their true properties in vivo.

Future techniques for the genetic analysis of the idiopathic epilepsies

The elucidation of the genetic basis of rare Mendelian idiopathic epilepsies and mouse models of epilepsy will provide a range of candidate genes for the analysis of idiopathic epilepsies which display a complex mode of inheritance.

The development of oligonucleotide arrays (DNA chips) will allow rapid genotyping, gene expression analysis, and candidate gene mutation screening in affected families (Chee, 1996).

The development of the biallelic single nucleotide polymorphism (SNP) map (Wang et al., 1998), and appropriate DNA chip technology, will allow the analysis of large numbers of small nuclear families by intra-familial association studies.

The progress made in recent years in epilepsy genetics, the development of the techniques discussed above, and continued advances in molecular genetic techniques (such as sequencing and SSCP) should all facilitate the isolation of the genes responsible for the idiopathic epilepsies.

Acknowledgements

Louise Bate is a Wellcome Trust Clinical Research Training Fellow. We thank Dr Jane Barclay and Dr Richard Thompson for advice and comments on the manuscript.

List of Abbreviations

ACh	acetylcholine
AD	autosomal dominant
ADNFLE	autosomal dominant nocturnal frontal lobe epilepsy
nAChR	neuronal nicotinic acetylcholine receptor
AR	autosomal dominant
BAC	bacterial artificial chromosome
BFIC	benign familial infantile convulsions
BFNC	benign familial neonatal convulsions
bp	base pair
Ca^{2+}	calcium ion
CAE	childhood absence epilepsy
cM	centiMorgan
CNS	central nervous system
Cys	cysteine
EEG	electroencephalogram
FISH	fluorescence *in situ* hybridization
DNA	deoxyribose nucleic acid
G-banding	giesma banding
cDNA	complementary deoxyribose nucleic acid
dsDNA	double stranded deoxyribose nucleic acid
ssDNA	single stranded deoxyribose nucleic acid
dNTP	deoxynucleotide triphosphate
ddNTP	dideoxynucleotide triphosphate
$GEFS^+$	generalized epilepsy with febrile convulsions plus
JAE	juvenile absence epilepsy
JME	juvenile myoclonic epilepsy
K^+	potassium ion
kb	kilobase
Na^+	sodium ion
PAC	P1 artificial chromosome
PCR	polymerase chain reaction
Phe	phenylalanine
(O)MIM	(online) Mendelian inheritance in man
RACE	rapid amplification of cDNA ends
RNA	ribose nucleic acid
cRNA	complementary ribose nucleic acid
mRNA	messenger ribose nucleic acid
SSCA } SSCP }	single-strand conformational polymorphism analysis
Ser	serine
SNP	single nucleotide polymorphism
STRs	short tandem repeats
Trp	tryptophan
YAC	Yeast artificial chromosome
θ	recombination fraction
α	proportion of linked families

References

Altschul, S.F, Gish, W., Miller, W., Myers, E.W. and Lipman, D.J. (1990) Basic local alignment search tool. *J. Mol. Biol.*, 215(3): 403–410.

Anand, R. and Lindstrom, J. (1992) Chromosomal localization of seven neuronal nicotinic acetylcholine receptor subunit genes in humans. *Genomics*, 13(4): 962–967.

Annegers, J.F, Rocca, W.A. and Hauser, W.A. (1996) Causes of epilepsy: contributions of the Rochester epidemiology project. *Mayo Clin. Proc.*, 71(6): 570–575.

Barhanin, J., Lesage, F., Guillemare, E., Fink, M., Lazdunski, M. and Romey, G. (1996) K(V)LQT1 and lsK (minK) proteins associate to form the I(Ks) cardiac potassium current. *Nature*, 384(6604): 78–80.

Biervert, C., Schroeder, B.C., Kubisch, C., Berkovic, S.F., Propping, P., Jentsch, T.J. and Steinlein, O.K. (1998) A potassium channel mutation in neonatal human epilepsy. *Science*, 279(5349): 403–406.

Bird, T.D. (1992) In: R.A. King, J.I. Rotter and A.G. Motulsky (Eds.), *Epilepsy. The Genetic Basis of Common Diseases*, Vol. 20, Ch 36, Oxford University Press, Oxford, pp. 732–752.

Breese, C., Adams, C., Logel, J., Drebing, C., Rollins, Y., Barnhart, M., Sullivan, B., Demasters, B., Freedman, R. and Leonard, S. (1997) Comparison of the regional expression of nicotinic acetylcholine receptor alpha7 mRNA and [125I]-alpha-bungarotoxin binding in human postmortem brain. *J. Comp. Neurol.*, 387(3): 385–398.

Bulman, D.E. (1997) Phenotype variation and newcomers in ion channel disorders. *Hum. Mol. Genet.*, 6(10): 1679–1685.

Burgess, D.L., Jones, J.M., Meisler, M.H. and Noebels, J.L. (1997) Mutation of the Ca^{2+} channel βsubunit gene *Cchb4* is associated with ataxia and seizures in the lethargic (*lh*) mouse. *Cell*, 88(3): 385–392.

Charlier, C., Singh, N., Ryan, S., Lewis, T., Reus, B., Leach, R. and Leppert, M. (1998) A pore mutation in a novel KQT-like potassium channel gene in an idiopathic epilepsy family. *Nat. Genet.*, 18(1): 53–55.

Chee, M. (1996) Accessing genetic information with high-density DNA arrays. *Science*, 274: 610–614.

Chini, B., Raimond, E., Elgoyhen, A.B., Moralli, D., Balzaretti, M. and Heinemann, S. (1994) Molecular cloning and chromosomal localization of the human α7-nicotinic receptor subunit gene (CHRNA7). *Genomics*, 19: 379—381.

Couturier, S., Bertrand, D., Matter, J., Hernandez, M., Bertrand, S., Millar, N., Valera, S., Barkas, T. and Ballivet, M. (1990) A neuronal nicotinic acetylcholine receptor subunit (alpha 7) is developmentally regulated and forms a homo-oligomeric channel blocked by alpha-BTX. *Neuron*, 5(6): 847–856.

Dreyer, I, Becker, D., Bregante, M., Gambale, F, Lehnen, M., Palme, K. and Hedrich, R. (1998) Single mutations strongly alter the K+-selective pore of the K(in) channel KAT1. *FEBS Lett.*, 430(3): 370–376.

Drwinga, H.L., Toji, L.H., Kim, C.H., Greene, A.E. and Mulivor, R.A. (1993) NIGMS human/rodent somatic cell hybrid mapping panels 1 and 2. *Genomics*, 16(2): 311–314.

Durner, M., Sander, T., Greenberg, D.A., Johnson, K., Beck-Mannagetta, G. and Janz, D. (1991) Localization of idiopathic generalized epilepsy on chromosome 6p in families of juvenile myoclonic epilepsy patients. *Neurology*, 41(10): 1651–1655.

Elmslie, F.V., Rees, M., Williamson, M.P., Kerr, M., Juel Kjeldsen, M., Pang, K.-A., Sundqvist, A., Friis, M.L., Chadwick, D., Curtis, D., Whitehouse, W.P. and Gardiner, R.M. (1997) Genetic mapping of a major susceptibility locus for juvenile myoclonic epilepsy on chromosome 15q. *Hum. Mol. Genet.*, 6(8): 1329–1334.

Elmslie, F.V., Williamson, M.P., Rees, M., Kerr, M., Juel Kjeldsen, M., Pang, K.A., Sundqvist, A., Friis, M., Richens, A., Chadwick, D., Whitehouse, W.P. and Gardiner, R.M. (1996) Linkage analysis of juvenile myoclonic epilepsy and microsatellite loci spanning 61 cM of human chromosome 6p in, 19 nuclear pedigrees provides no evidence for a susceptibility locus in this region. *Am. J. Hum. Genet.*, 59: 653–663.

Fisher, R.S. (1995) Cellular mechanisms of the epilepsies. In: A. Hopkins, S.D. Shorvon, and G. Cascino, (Eds.), *Epilepsy*, Chapman & Hall Medical, London; New York, pp. 35–58.

Fletcher, C.F., Lutz, C.M., O'Sullivan, T.N., Shaughnessy Jnr, J.D., Hawkes, R., Frankel, W.N., Copeland, N.G., and Jenkins, N.A., (1996) Absence epilepsy in *tottering* mutant mice is associated with calcium channel defects. *Cell*, 87: 607–617.

Gray, R., Rajan, A., Radcliffe, K., Yakehiro, M. and Dani, J. (1996) Hippocampal synaptic transmission enhanced by low concentrations of nicotine. *Nature*, 383(6602): 713–716.

Greenberg, D.A., Delgado-Escueta, A.V., Widelitz, H., Sparkes, R.S., Treiman, L., Maldonado, H.M., Park, M.S. and Terasaki, P.I. (1988) Juvenile myoclonic epilepsy may be linked to the BF and HLA loci on human chromosome 6. *Am. J. Med. Genet.*, 31(1): 185–192.

Guipponi, M., Rivier, F., Vigevano, F., Beck, C., Crespel, A., Echenne, B., Lucchini, P., Sebastianelli, R., Baldy-Moulinier, M. and Malafosse, A. (1997) Linkage mapping of benign familial infantile convulsions (BFIC) to chromosome, 19q. *Hum. Mol. Genet.*, 6(3): 473–477.

Guo, L. and Kubo, Y. (1998). Comparison of the open-close kinetics of the cloned inward rectifier K+ channel IRK1 and its point mutant (Q140E) in the pore region. *Receptors Channels*, 5(5): 273–289.

Hauser, W., Annegers, J. and Kurland, L. (1991) Prevalence of epilepsy in Rochester, Minnesota: 1940–1980. *Epilepsia*, 32(4): 429–445.

Hauser, W.A, Annegers, J.F. and Rocca, W.A. (1996) Descriptive epidemiology of epilepsy: contributions of population-based studies from Rochester, Minnesota. *Mayo Clin. Proc.*, 71(6): 576–586.

Heginbotham, L., Lu, Z., Abramson, T. and MacKinnon, R. (1994) Mutations in the K+ channel signature sequence. *Biophys. J.*, 66(4): 1061–1067.

Hidalgo, P. and MacKinnon, R. (1995) Revealing the architecture of a K+ channel pore through mutant cycles with a peptide inhibitor. *Science*, 268(5208): 307–310.

Isom, L.L., De Jongh, K.S., Patton, D.E., Reber, B.F., Offord, J., Charbonneau, H., Walsh, K., Goldin, A.L. and Catterall, W.A. (1992) Primary structure and functional expression of the beta 1 subunit of the rat brain sodium channel. *Science*, 256(5058): 839–842.

Janz, D., Durner, M. and Beck-Mannagetta, G. (1989) Family studies on the genetics of juvenile myoclonic epilepsy (epilepsy with petit mal). In: G. Beck-Mannagetta, V. Anderson, H. Doose and D. Janz (Eds.), *Genetics of the Epilepsies*, Springer-Verlag, Berlin, pp. 43–52.

Johnson, E.W., Dubovsky, J., Rich, S.S., O'Donovan, C.A., Orr, H.T., Anderson, V.E., Gil-Nagel, A., Ahmann, P., Dokken, C.G., Schneider, D.T. and Weber, J.L. (1998) Evidence for a novel gene for familial febrile convulsions, FEB2, linked to chromosome, 19p in an extended family from the Midwest. *Hum. Mol. Genet.*, 7(1): 63–67.

Kukuljan, M., Labarca P. and Latorre R. (1995) Molecular determinants of ion conduction and inactivation in K+ channels. *Am. J. Physiol.*, 268(3 Pt 1): C535–56.

Kuryatov, A, Gerzanich, V., Nelson, M., Olale F. and Lindstrom J. (1997) Mutation causing autosomal dominant nocturnal frontal lobe epilepsy alters $Ca2^+$ permeability, conductance, and gating of human alpha4beta2 nicotinic acetylcholine receptors. *J. Neurosci.*, 17(23): 9035–9047.

Lander, E. and Kruglyak, L. (1995) Genetic dissection of complex traits: guidelines for interpreting and reporting linkage results. *Nat. Genet.*, 11: 241–247.

Leppert, M., Anderson, V.E., Quattlebaum, T., Stauffer, D., O'Connell, P., Nakamura, Y., Lalouel, J.M. and White, R. (1989) Benign familial neonatal convulsions linked to genetic markers on chromosome 20. *Nature*, 337(6208): 647–648.

Letts, V.A., Felix, R., Biddlecome, G.H., Arikkath, J., Mahaffey, C.L., Valenzuela, A., Bartlett, F.S., Mori, Y., Campbell, K.P. and Frankel, W.N. (1998) The mouse stargazer gene encodes a neuronal $Ca2^+$-channel gamma subunit. *Nat. Genet.*, 19, 340–347.

Lewis, T.B., Leach, R.J., Ward, K., O'Connell, P. and Ryan, S.G. (1993) Genetic heterogeneity in benign familial neonatal convulsions: identification of a new locus on chromosome 8q. *Am. J. Hum. Genet.*, 53: 670–675.

Lewis, T.B., Nelson, L., Ward, K. and Leach, R.J. (1995) A radiation hybrid map of 40 loci for the distal long arm of human chromosome 8. *Genome Res.*, 5(4): 334–341.

Liu, A.W., Delgado-Escueta, A.V., Serratosa, J.M., Alonso, M.E., Medina, M.T., Gee, M.N., Cordova, S., Zhao, H.Z., Spellman, J.M., Ramos Peek, J.R., Rubio Donnadieu, F. and Sparkes, R.S. (1995) Juvenile Myoclonic Epilepsy locus in chromosome 6p21.2-p11:linkage to convulsions and electroencephalography trait. Am. J. Hum. Genet., 57: 368–381.

Lopes-Cendes, I., Scheffer, I.E., Berkovic, S.F., Rousseau, M., Andermann, E. and Rouleau, G. (1996) Mapping a locus for idiopathic epilepsy in a large multiplex family. *Epilepsia*, 37(S5): 127.

Lopez, G.A., Jan, Y.N. and Jan, L.Y. (1994) Evidence that the S6 segment of the Shaker voltage-gated K^+ channel comprises part of the pore. *Nature*, 367(6459): 179–182.

Macdonald, R.L. and Kelly, K.M. (1994) Mechanisms of action of currently prescribed and newly developed antiepileptic drugs. *Epilepsia*, 35(Suppl 4): S41–S50.

Makita, N., Bennett Jr, P.B. and George Jr, A.L. (1994) Voltage-gated Na+ channel beta 1 subunit mRNA expressed in adult human skeletal muscle, heart, and brain is encoded by a single gene. *J. Biol. Chem.*, 269(10): 7571–7578.

Makita, N., Bennett, P.B. and George Jr, A.L. (1996) Molecular determinants of beta 1 subunit-induced gating modulation in voltage-dependent Na^+ channels. *J. Neurosci.*, 16(22): 7117–7127.

Makita, N., Sloan-Brown, K., Weghuis, D.O., Ropers, H.H. and George A.L. Jr, (1994) Genomic organization and chromosomal assignment of the human voltage- gated Na^+ channel beta 1 subunit gene (SCN1B). *Genomics*, 23(3): 628–634.

McCormick, K.A., Isom, L.L., Ragsdale, D., Smith, D., Scheuer, T. and Catterall, W.A. (1998) Molecular determinants of Na+ channel function in the extracellular domain of the beta1 subunit. *J. Biol. Chem.*, 273(7): 3954–3962.

Mohrenweiser, H.W., Tsujimoto, S., Gordon, L. and Olsen, A.S. (1998) Regions of sex-specific hypo- and hyper-recombination identified through integration of 180 genetic markers into the metric physical map of human chromosome, 19. *Genomics*, 47(2): 153–162.

Molina, A., Ortega-Saenz, P. and Lopez-Barneo, J. (1998) Pore mutations alter closing and opening kinetics in Shaker K+ channels. *J. Physiol. (Lond.)*, 509(Pt 2): 327–337.

Nakamura, R.L., Anderson, J.A. and Gaber, R.F. (1997) Determination of key structural requirements of a K+ channel pore. *J. Biol. Chem.*, 272(2): 1011–1018.

Neyroud, N., Tesson, F., Denjoy, I., Leibovici, M., Donger, C., Barhanin, J., Faure, S., Gary, F., Coumel, P., Petit, C., Schwartz, K. and Guicheney, P. (1997) A novel mutation in the potassium channel gene KVLQT1 causes the Jervell and Lange-Nielsen cardioauditory syndrome (see comments). *Nat. Genet.*, 15(2): 186–189.

Ottman, R., Risch, N., Hauser, W.A., Pedley, T.A., Lee, J.H., Barker-Cummings, C., Lustenberger, A., Nagle, K.J., Lee, S.S., Scheuer, M.L., Neystat, M., Susser, M. and Wilhelmsen, K.C. (1995) Localization of a gene for partial epilepsy to chromosome 10q. *Nat. Genet.*, 10: 56–60.

Phillips, H.A., Scheffer, I.E., Berkovic, S.F., Hollway, G.E., Sutherland, G.R. and Mulley, J.C. (1995) Localization of a gene for autosomal dominant nocturnal frontal lobe epilepsy to chromosome 20q13.2. *Nat. Genet.*, 10: 117–118.

Rise, M.L., Frankel, W.N., Coffin, J.M. and Seyfried, T.N. (1991) Genes for Epilepsy mapped in the mouse. *Science*, 253: 669–673.

Ronen, G.M., Rosales, T.O., Connolly, M., Anderson, V.E. and Leppert, M. (1993) Seizure characteristics in chromosome 20 benign familial neonatal convulsions. *Neurology*, 43(7): 1355–1360.

Sanguinetti, M.C., Curran, M.E., Zou, A., Shen, J., Spector, P.S., Atkinson, D.L. and Keating, M.T. (1996) Coassembly of K(V)LQT1 and minK (IsK) proteins to form cardiac I(Ks) potassium channel. *Nature*, 384(6604): 80–83.

Scheffer, I.E. and Berkovic, S.F. (1997) Generalized epilepsy with febrile seizures plus. A genetic disorder with heterogeneous clinical phenotypes. *Brain*, 120(Pt 3): 479–490.

Scheffer, I.E., Bhatia, K.P., Lopes-Cendes, I., Fish, D.R., Marsden, C.D., Andermann, F., Andermann, E., Desbiens, R., Cendes, F., Manson, J.I. and Berkovic, S.F. (1994) Autosomal dominant frontal epilepsy misdiagnosed as sleep disorder. *Lancet*, 343: 515–517.

Schmitz, B. and Sander, T. (1999) The major susceptibility locus for juvenile myoclonic epilepsy on chromosome 15q. In: B. Schmitz and T. Sander (Eds.), *Juvenile Myoclonic Epilepsy*, Wrightson Biomed. Publ., Petersfield.

Singh, N.A., Charlier, C., Stauffer, D., DuPont, B.R., Leach, R.J., Melis, R., Ronen, G.M., Bjerre, I., Quattlebaum, T., Murphy, J., McHarg, M.L., Gagnon, D., Rosales, T.O.,

Peiffer, A., Anderson, V.E. et al. (1998) A novel potassium channel gene, *KCNQ2*, is mutated in an inherited epilepsy of newborns. *Nat. Genet.*, 18: 25–29.

Steinlein, O., Magnusson, A., Stoodt, J., Bertrand, S., Weiland, S., Berkovic, S., Nakken, K.O., Propping, P. and Bertrand, D. (1997) An insertion mutation of the CHRNA4 gene in a family with autosomal dominant nocturnal frontal lobe epilepsy. *Hum. Mol. Genet.*, 6(6): 943–947.

Steinlein, O., Smigrodzki, R., Lindstrom, J., Anand, R., Kohler, M., Tocharoentanaphol, C. and Vogel, F. (1994) Refinement of the localization of the gene for neuronal nicotinic acetylcholine receptor alpha 4 subunit (CHRNA4) to human chromosome 20q13.2-q13.3. *Genomics*, 22(2): 493–495.

Steinlein, O.K, Mulley, J.C., Propping, P., Wallace, R.H., Phillips, H.A., Sutherland, G.R., Scheffer, I.E. and Berkovic, S.F. (1995) A missense mutation in the neuronal nicotinic receptor α4 subunit is associated with autosomal dominant nocturnal frontal lobe epilepsy. *Nat. Genet.*, 11: 201–203.

Tytgat, J. (1994) Mutations in the P-region of a mammalian potassium channel (RCK1): a comparison with the Shaker potassium channel. *Biochem. Biophys. Res. Commun.*, 203(1): 513–518.

Vigevano, F., Fusco, L., Di Capua, M., Ricci, S., Sebastianelli, R. and Lucchini, P. (1992) Benign infantile familial convulsions. *Eur. J. Pediatr.*, 151(8): 608–612.

Wallace, R., Wang, D., Singh, R., Scheffer, I., George, A.J., Phillips, H., Saar, K., Reis, A., Johnson, E., Sutherland, G., Berkovic, S. and Mulley, J. (1998) Febrile seizures and generalized epilepsy associated with a mutation in the Na^+-channel beta1 subunit gene SCN1B. *Nat. Genet.*, 19(4): 366–370.

Wallace, R.H., Berkovic, S.F., Howell, R.A., Sutherland, G.R. and Mulley, J.C. (1996) Suggestion of a major gene for familial febrile convulsions mapping to 8q13-21. *J. Med. Genet.*, 33(4): 308–312.

Wang, D.G., Fan, J.B., Siao, C.J., Berno, A., Young, P., Sapolsky, R., Ghandour, G., Perkins, N., Winchester, E., Spencer, J., Kruglyak, L., Stein, L., Hsie, L., Topaloglou, T., Hubbell, E. et al. (1998) Large-scale identification, mapping, and genotyping of single- nucleotide polymorphisms in the human genome. *Science*, 280(5366): 1077–1082.

Wang, Q., Curran, M.E., Splawski, I., Burn, T.C., Millholland, J.M., VanRaay, T.J., Shen, J., Timothy, K.W., Vincent, G.M., De Jager, T., Schwartz, P.J., Toubin, J.A., Moss, A.J., Atkinson, D.L., Landes, G.M. et al. (1996) Positional cloning of a novel potassium channel gene: KVLQT1 mutations cause cardiac arrhythmias. *Nat. Genet.*, 12(1): 17–23.

Wei, A., Jegla, T. and Salkoff, L. (1996) Eight potassium channel families revealed by the *C. elegans* genome project. *Neuropharmacology*, 35(7): 805–829.

Weiland, S., Witzemann, V., Villarroel, A., Propping, P. and Steinlein, O. (1996) An amino acid exchange in the second transmembrane segment of a neuronal nicotinic receptor causes partial epilepsy by altering its desensitization kinetics. *FEBS Lett.*, 398: 91–96.

Weissbecker, K.A., Durner, M., Janz, D., Scaramelli, A., Sparkes, R.S. and Spence, M.A. (1991) Confirmation of linkage between juvenile myoclonic epilepsy locus and the HLA region on chromosome 6. *Am. J. Med. Genet.*, 38(1): 32–36.

Wevers, A., Jeske, A., Lobron, C., Birtsch, C., Heinemann, S., Maelicke, A., Schroder, R. and Schroder, H. (1994) Cellular distribution of nicotinic acetylcholine receptor subunit mRNAs in the human cerebral cortex as revealed by non-isotopic in situ hybridization. *Brain Res. Mol. Brain Res.*, 25(1–2): 122–128.

Whitehouse, W.P., Rees, M., Curtis, D., Sundqvist, A., Parker, K., Chung, E., Baralle, D. and Gardiner, R.M. (1993) Linkage analysis of idiopathic generalised epilepsy (IGE) and marker loci on chromosome 6p in families of patients with juvenile myoclonic epilepsy: no evidence for an epilepsy locus in the HLA region. *Am. J. Hum. Genet.*, 53(3): 652–662.

Wollnik, B., Schroeder, B.C., Kubisch, C., Esperer, H.D., Wieacker, P. and Jentsch, T.J. (1997) Pathophysiological mechanisms of dominant and recessive KVLQT1 K+ channel mutations found in inherited cardiac arrhythmias. *Hum. Mol. Genet.*, 6(11): 1943–1949.

Yokoyama, M., Nishi, Y., Yoshii, J., Okubo, K. and Matsubara, K. (1996) Identification and cloning of neuroblastoma-specific and nerve tissue- specific genes through compiled expression profiles. *DNA Res.*, 3(5): 311–320.

W.E. Crusio and R.T. Gerlai (Eds.)
Handbook of Molecular-Genetic Techniques for Brain and
Behavior Research (Techniques in the Behavioral and Neural Sciences, Vol. 13)

CHAPTER 5.6

Identification and functional analysis of genes and genetic risk factors in Alzheimer's disease

Chris De Jonghe and Christine Van Broeckhoven

Department of Molecular Genetics, Flanders Interuniversity Institute for Biotechnology (VIB) and Neurogenetics Laboratory, Born-Bunge Foundation (BBS), University of Antwerp (UIA), Department of Biochemistry, Universiteitsplein 1, B-2610 Antwerp, Belgium

Alzheimer's disease: clinicopathology

Alzheimer's disease (AD), by far the most common form of dementia in the elderly (Ott et al., 1995), is clinically characterized by a gradual, progressive loss in cognitive functioning and changes in personality, ultimately leading to death. The disease can be diagnosed with 90% reliability by excluding other causes of dementia, although a definite diagnosis requires post-mortem brain examination. As described originally by Aloïs Alzheimer in 1907, senile plaques and neurofibrillary tangles are the neuropathological hallmarks of AD that are found in the cerebral cortex, hippocampus and amygdala. Senile plaques (SP) represent compact extracellular fibrillar deposits, mainly consisting of amyloid β (Aβ), a 4 kDa peptide of 40–43 amino acids, surrounded by dystrophic neurites. Amorphous deposition of the same Aβ peptide also occurs in parenchyma of AD brains in the form of diffuse plaques. Moreover, Aβ is present in the blood vessel walls of leptomeninges and cerebral cortex where it is denoted as congophilic amyloid angiopathy (CAA). Neurofibrillary tangles (NFT) are intraneuronal inclusions of paired helical filaments (PHF) mainly composed of hyperphosphorylated tau, a microtubule binding protein. Although there is evidence that amyloid deposition is the earliest sign of AD pathogenesis, the severity of dementia correlates better with the presence of NFT.

Genetics of AD

Aging is the major risk factor for AD. This is reflected in the increasing prevalence of AD with advancing age: less than 1% of 60–64 year olds are affected, but up to 40% of those older than 85 years are diseased (Breteler et al., 1992). Although AD occurs mostly in the elderly population, in 10–20% of the AD cases, the first symptoms become apparent before the age of 60–65 years. By convention, 60 or 65 years is used as a cut-off to separate senile AD or late-onset AD (LOAD) from presenile AD or early-onset AD (EOAD). LOAD and EOAD cases are clinically and pathologically indistinguishable, although the disease progression is more rapid in EOAD cases.

The second most important risk factor for AD is a positive family history for dementia. Although AD aggregates within LOAD families as well as in EOAD families, EOAD cases tend to have more often a positive family history (van Duijn et al., 1994). In 10% of the familial cases AD is inherited as an autosomal dominant trait. Since the underlying biochemical defects in AD are unknown, the positional cloning approach was used to identify genes involved in AD pathogenesis (Chapter 2.1.5, this volume). This strategy has led to the identification of 3 genes in which mutations can cause AD: the amyloid precursor protein (APP) gene on chromosome 21, the presenilin 1 (PSEN1) gene on chromosome 14 and the presenilin 2 (PSEN2) gene on chromosome 1. A fourth gene, the apolipoprotein E (APOE) gene on chromosome 19 was identified as a risk factor in both familial and sporadic, presenile and senile AD. In addition to APOE a number of other potential susceptibility genes have been implicated as genetic risk factors for AD by using genetic association methods (Chapter 2.1.4, this volume).

Once the AD genes were identified, functional analysis of the (mutant) genes was performed both in vitro (in cell culture) and in vivo (transgenic animals (Chapter 3.1.10, this volume) to elucidate the biochemical role of the mutant geneproducts in the pathogenesis of AD.

The APP gene

The purification of Aβ as the major constituent of senile plaques (Glenner and Wong, 1984) led to the identification of its larger precursor protein, the amyloid β precursor protein (APP), which maps on chromosome 21 (Goldgaber et al., 1987; Kang et al., 1987; Tanzi et al., 1987; Robakis et al., 1987). The APP gene is a housekeeping gene comprising 18 exons and encoding a *N*- and *O*-glycosylated single membrane spanning protein with a large ectodomain and a small cytoplasmic tail (Ponte et al., 1988; Tanzi et al., 1988; Konig et al., 1992; Sandbrink et al., 1994) (Fig. 1). Due to alternative splicing, 10 different APP isoforms can be distinguished (Kitaguchi et al., 1988; Ponte et al., 1988; Tanzi et al., 1988; Konig et al., 1992; Sandbrink et al., 1994). APP695, lacking exon 7 encoding a Kunitz-type protease inhibitor domain, is the most prominent isoform in brain. Exons 16 and 17 of APP encode in part for Aβ, which is released from the full-length precursor protein by proteolytic processing.

The observation that middle-aged Down's syndrome patients, who carry an extra copy of chromosome 21, invariably develop AD-like pathology, led to the hypothesis that overexpression of a gene on chromosome 21 might lead to AD. This idea was further supported by the report of genetic linkage between DNA markers on chromosome 21 and AD in several kindreds (St-George Hyslop et al., 1987; Pericak-Vance et al., 1988; Goate et al., 1989). However, the first mutation in the APP gene was not found in AD patients, but in patients suffering from recurrent cerebral hemorrhages and progressive dementia of the multi-infarct type i.e. hereditary cerebral hemorrhages with amyloidosis — Dutch type (HCHWA-D) (Haan et al., 1991). HCHWA-D has been referred to as the vascular form of AD since the vascular deposits contain the same Aβ peptide as SP, however, no or very few dystrophic neurites or neurofibrillary tangles have been observed (van Duinen et al., 1987). The HCHWA-D mutation is a missense mutation resulting in an E to Q amino acid substitution at codon 693 (numbering according to APP770 isoform) within the Aβ sequence.

Subsequently, 6 different APP mutations were identified in less than 20 EOAD families worldwide: a double mutation K670N/M671L was detected in a Swedish family (Mullan et al., 1992), 3 different mutations were observed at codon 717: V717I (Goate et al., 1991), V717G (Chartier-Harlin et al., 1991) and V717F (Murrell et al., 1991) and recently an I716V mutation (Eckman et al., 1997) and a L723P mutation (Kwok et al., 1998) were found in the APP gene. Another APP mutation (A692G) was identified in patients with cerebral hemorrhages due to CAA and in EOAD patients belonging to the same Dutch family (Hendriks et al., 1992; Cras et al., 1998). Mutations in the APP gene are estimated to account for about 5% of all familial EOAD cases and therefore are a rare cause of AD (Van Broeckhoven, 1995).

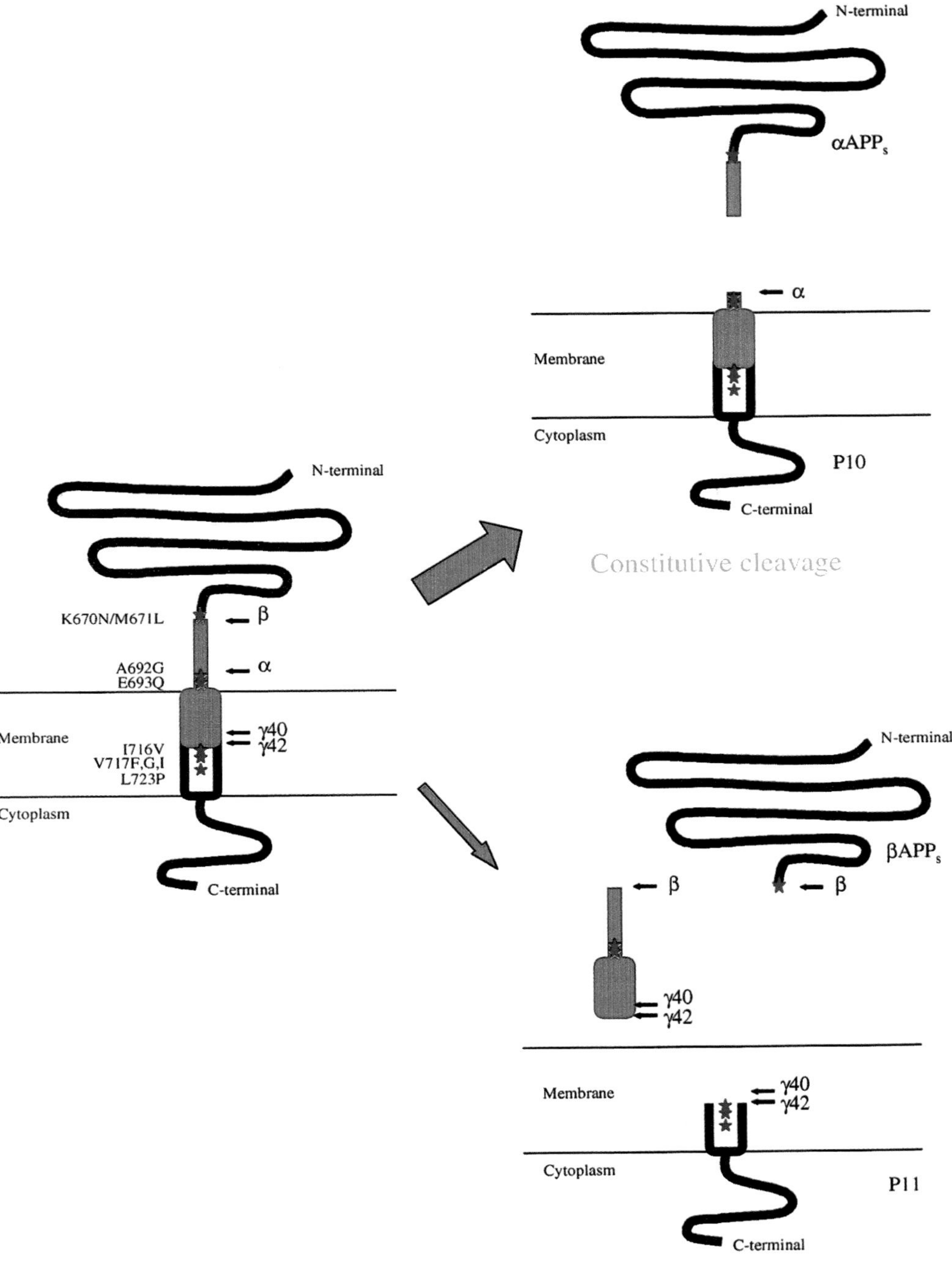

Fig. 1. The APP protein with the Aβ peptide indicated in blue. The mutations shown as red stars are numbered according to the largest APP transcript APP770. The proteases active on APP are named with their Greek symbols and their cleavage site is indicated with arrows. The result of APP cleavage by α-, respectively β- and γ-secretases is depicted on the right site of the figure.

The PSEN1 gene

Since mutations in the APP gene explain the development of AD only in a small number of cases, a genome-wide search in families segregating presenile AD was started in many labs. A second locus for AD was subsequently identified on chromosome 14 (Schellenberg et al., 1992), at 14q24.3 (Van Broeckhoven et al., 1992). Using genetic and physical mapping experiments, the S182 gene, later renamed as presenilin 1 (PSEN1), was isolated (Sherrington et al., 1995). Since then, over 40 different mutations have been detected in the PSEN1 gene in a large number of families of different ethnic origin (Cruts and Van Broeckhoven, 1998). Although the contribution of PSEN1 mutations to AD was initially estimated as high as 70% of all familial EOAD cases (Van Broeckhoven, 1995), in more recent studies this number was brought down to 18–50% (Cruts et al., 1998).

The PSEN1 gene encodes a 467 amino acids protein (Fig. 2) and comprizes 10 coding exons (Sherrington et al., 1995; Cruts et al., 1996), numbered 3–13 (Alzheimer's Disease Collaborative Group, 1995). Three additional exons were found in the 5′ untranslated region (5′UTR), the first two being alternatively transcribed (Sherrington et al., 1995). PSEN1 transcription is driven by 2 alternative promotors producing transcripts with different 5′ UTRs (Rogaev et al., 1997). The use of different polyadenylation signals leads to transcripts of respectively 2.7 and 7.5 kb (Rogaev et al., 1997). Several alternatively spliced transcripts have been reported for PSEN1. A major one uses a different splice donor site in exon 3, resulting in the presence or absence of a 12 bp sequence coding for the amino acids VRSQ (Cruts et al., 1995). Transcripts lacking exon 8 have been observed only in leukocytes (Sherrington et al., 1995).

All PSEN1 mutations reported so far are missense mutations with two exceptions. An in-frame splice site mutation results in the deletion of exon 9 (Δ9) and an amino acid substitution (S290L) at the splice junction of exon 8 and 10 (Perez-Tur et al., 1995). Moreover, a PSEN1 truncating mutation, resulting from a one basepair deletion in the splice donor site of intron 4 (Δ4), causes a frameshift and consecutively a premature termination codon (Tysoe et al., 1998).

Most PSEN1 mutations cause AD with onset ages between 35 and 55 years (Van Broeckhoven, 1995), although also PSEN1 mutations have been identified causing AD in the sixth decade (Cruts et al., 1998). Ages at onset are usually very tight within a family and unrelated patients, carrying the same PSEN1 mutation have similar ages at onset (Van Broeckhoven, 1995). The age at onset of AD is generally slightly lower in individuals carrying PSEN1 mutations than in APP mutation carriers.

The PSEN2 gene

Soon after the identification of PSEN1, screening of sequence databases revealed the presence of a homologous gene on chromosome 1, STM2, later renamed presenilin 2 (PSEN2) (Rogaev et al., 1995). Simultaneously, genetic linkage was found between DNA markers on chromosome 1q31-q42 and AD in a group of families, called the Volga-Germans (Levy-Lahad et al., 1995b). Linkage to chromosome 14 and 21 had previously been excluded in these families, that have a German origin and emigrated first to Russia and later to the USA. Sequencing of the PSEN2 gene in the Volga-Germans revealed the presence of a mutation at codon 141 substituting a N for an I (Levy-Lahad et al., 1995a). PSEN1 and PSEN2 share 67% overall homology. Like PSEN1, PSEN2 contains 10 coding exons, but only 2 exons were identified in the 5′UTR (Levy-Lahad et al., 1996). PSEN2 is transcribed into 2 messages of respectively 2.4 and 2.8 kb. In addition to a transcript lacking exon 8, an alternative transcript, lacking exons 3 and 4 was identified for PSEN2 (Prihar et al., 1996). Since exon 3 contains the start codon, the latter transcript is predicted to be translated from an alternative translation initiation codon.

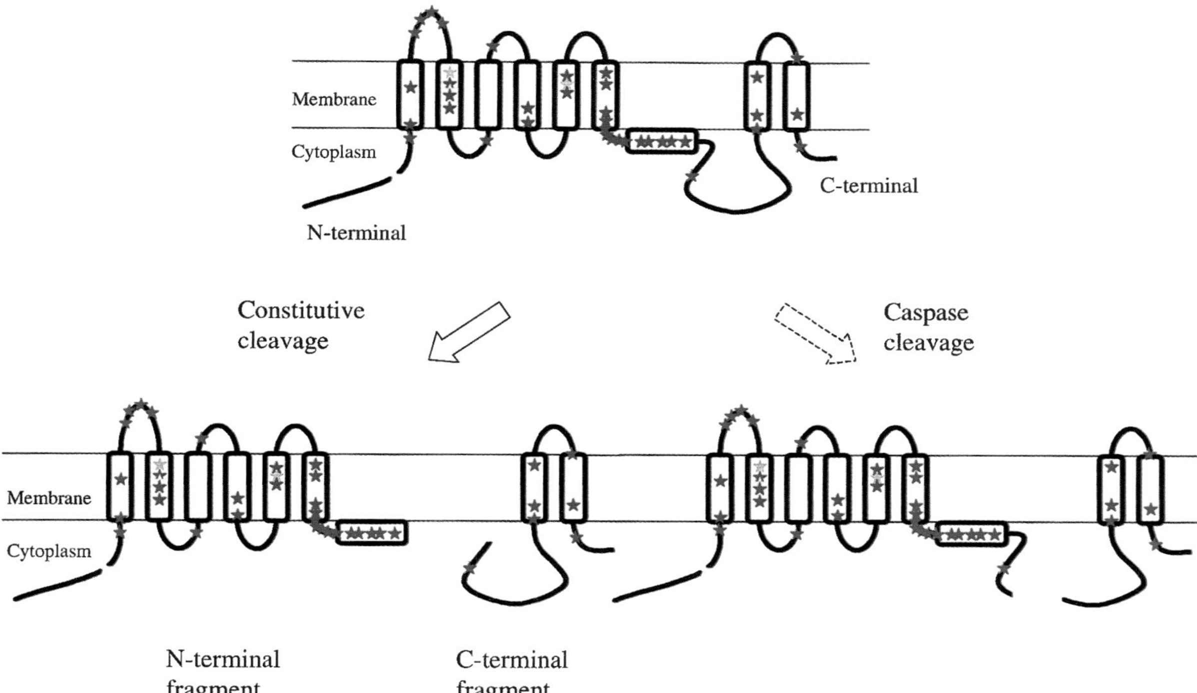

Fig. 2. Schematic representation of the PSEN protein structure. PSEN1 mutations are represented by red stars, PSEN2 mutations by yellow stars. In case PSEN1 and PSEN2 mutations coincide, we used orange stars. For the PSEN2 mutations, the mutated codons were converted to their corresponding conserved amino acids in PSEN1. The result of the constitutive cleavage as well as the caspase cleavage are depicted on the lower part of the figure.

Since the identification of the N141I mutation in PSEN2, only 3 other mutations have been detected (Cruts et al., 1998; Beyer et al., 1998) (Fig. 2), suggesting that PSEN2 mutations are a rare cause of AD. Unlike mutations in PSEN1, mutations in PSEN2 give rise to a broad range of onset ages and incomplete penetrance (Levy-Lahad et al., 1995a; Cruts and Van Broeckhoven, 1998). All AD causing mutations are located in regions that are conserved between PSEN1 and PSEN2.

The APOE gene

The identification of the PSEN genes in addition to the APP gene have been major breakthroughs in our understanding of AD pathogenesis in early-onset mendelian forms of the disease. However, these cases represent less than 1% of the total number of AD cases, the majority of which are sporadic cases with no family history of disease and have later onset ages (> 65 years). Linkage analysis in families with LOAD, revealed a locus on chromosome 19 (Pericak-Vance et al., 1991). The APOE gene was one of the genes located in the chromosomal region that co-segregated with the disease and became immediately a candidate gene since APOE immunoreactivity was found in SP, NFT and CAA. APOE has been studied extensively because of its role in cholesterol metabolism (Mahley, 1988). Three common alleles of the APOE gene ($\varepsilon 2$, $\varepsilon 3$ and $\varepsilon 4$) encode three APOE isoforms that differ from each other in one or two amino acids at positions 112 (C in ε2 and ε3, R in ε4) and 158 (C in ε2, R in ε3 and ε4). Since the APOE ε4 allele was observed with a higher frequency in LOAD patients than in a control population, it was identified as a genetic risk factor for LOAD (Strittmatter et al., 1993a; Roses and Saunders, 1994). The APOE ε4 risk effect has been extended to presenile familial and sporadic AD cases (van Duijn et al., 1994) and confirmed by many other groups (Roses and Saunders, 1994). There is evidence that the risk associated with the APOE ε4 is dose-dependent, reflecting in the age at onset of disease: individuals with 2 APOE ε4 alleles have lower onset ages than individuals with only one APOE ε4 allele, who have an earlier onset than those without APOE ε4 allele (Corder et al., 1993). The APOE genotype further modulates the age at onset of AD in APP717 and APP670/671 families (Alzheimer's Disease Collaborative Group, 1993; Sorbi et al., 1995). However, for the AD/CAA mutation APP692 and the HCHWA-D mutation APP693, no effect on age at onset, occurrence of dementia or number of strokes was observed (Haan et al., 1994). Also, no modification of onset age was observed for PSEN1 or PSEN2 mutation carriers (Cruts and Van Broeckhoven, 1998). A protective effect has been suggested for the APOE ε2 allele (Corder et al., 1994), but other studies contradict these results (van Duijn et al., 1995).

Although the increased risk to AD for APOE ε4 has been fully recognized, not everyone carrying an APOE ε4 allele will develop AD, while many individuals lacking an APOE ε4 allele will.

A candidate locus for familial LOAD was recently assigned to chromosome 12 (Pericak-Vance, 1997). Genetic association studies had already previously suggested the lipoprotein receptor like protein (LRP) gene at chromosome 12 as a susceptibility gene for LOAD. Recently, also the α2-macroglobulin gene at chromosome 12p was identified as a susceptibility gene for LOAD (Blacker et al., 1998). However, the genetic association of AD with α2-macroglobulin and LRP could not be confirmed in all studies.

Other susceptibility genes

The α1-antichymotrypsin gene (AATC), coding for a protease inhibitor which is a minor constituent of SPs, has been reported to modify the risk associated with APOE ε4 (Kamboh et al., 1995). Other genes that were genetically associated with AD include the very low-density lipoprotein receptor gene (VLDLR) (Okuizumi et al., 1995), CYP2D6, a gene coding for a protein involved in detoxifying environmental toxins (Saitoh et al., 1995), the apolipoprotein J gene (APOJ) (Tycko et al., 1996), the non-amyloid component (NACP) gene (Xia et al., 1996), the serotonin transporter gene 5HTT (Li et al., 1997b), HLA-DR antigens of the class II major histocompatibility complex (Payami et al., 1997; Curran et al., 1997), the gene for butyrylcholinesterase K variant (BCHE-K) (Lehmann et al., 1997) and the bleomycine hydroxylase (BH) gene (Montoya et al., 1998). However, their contribution to AD pathogenesis still needs to be further confirmed. Interestingly, also a polymorphism in intron 8 of the PSEN1 gene was suggested to be associated with an increased risk for senile AD (Wragg et al., 1996). We recently demonstrated association between EOAD and a polymorphism in the PSEN1 promotor (van Duijn et al., in press).

The mechanism of the disease

APP function and effect of the pathogenic mutations

Two major pathways have been described for APP proteolysis: in the constitutive pathway APP is cleaved between residue K687 and L688 (numbering according to the APP770 isoform) by a protease called α-secretase which has not yet been characterized (Esch et al., 1990) (Fig. 1). α-Secretase cleavage either takes place in the trans-Golgi network or post-Golgi vesicles (Sambamurti et al., 1992; De Strooper et al., 1993; Kuentzel et al., 1993; Tomita et al., 1998) or near the cell surface (Sisodia, 1992; Haass et al., 1992) within caveolae (Ikezu et al., 1998). Since the α-secretase cleavage site is located within the Aβ region, this pathway cannot lead to Aβ formation, but instead produces a 10 kDa membrane-bound APP fragment (P10) as well as a secreted APP fragment (αAPP_s) (Fig. 1). In the amyloidogenic pathway, cleavage by β- and subsequently by γ-secretase(s) releases Aβ from APP (Fig. 1). β-Secretase cleaves highly specifically between amino acids M671 and D672, secreting a truncated APP fragment (βAPP_s). The membrane-bound, Aβ containing remnant (P11) is subsequently cleaved by γ-secretase(s), which is less specific and cuts either before residue I712 (γ40) to yield a 40 amino acids long Aβ (Aβ40) or before residue T714 (γ42) to produce a 42 amino acids Aβ isoform (Aβ42) (Fig. 1). There is evidence that Aβ40 and Aβ42 are generated by 2 different proteolytic γ-activities, γ40 and γ42 (Citron et al., 1996; Klafki et al., 1996). It has been recognized that Aβ generation is not associated with pathological conditions, but is a normal cellular event (Haass et al., 1992; Seubert et al., 1993; Seubert et al., 1992; Shoji et al., 1992; Busciglio et al., 1993). In contrast to non-neuronal cells, neurons seem to process APP preferentially through amyloidogenic pathways (De Strooper et al., 1995; Simons et al., 1996). At least two different amyloidogenic pathways exist: an endosomal/lysosomal pathway was proposed requiring APP reinternalization from the cell surface in clathrin-coated vesicles (Cole et al., 1992; Estus et al., 1992; Golde et al., 1992; Haass et al., 1992a; Haass et al., 1992b; Ono et al., 1997). Besides this endocytosis-dependent pathway, it is also possible that APP or Aβ containing APP fragments are further cleaved in the Golgi or trans-Golgi network (Xu et al., 1997). Aβ produced in these pathways may be either secreted or retained intracellularly, as was shown particularly in neurons (Wertkin et al., 1993; Turner et al., 1996; Forman et al., 1997; Wild-Bode et al., 1997; Tienari et al., 1997; Hartmann et al., 1997; Lee et al., 1998), where it can form an insoluble pool (Skovronsky et al., 1998). Recent studies have provided evidence that intracellular and secreted Aβ are generated in different pathways (Tienari et al., 1997; Wild-Bode et al., 1997): Aβ42 produced in ER/intermediate compartment (Cook et al., 1997; Chyung et al., 1997; Wild-Bode et al., 1997; Hartmann et al., 1997; Skovronsky et al., 1998) does not contribute to the secreted pool of Aβ, while Aβ40 is produced in TGN and is more prevalent in the secreted Aβ pool (Xu et al., 1997; Wild-Bode et al., 1997; Hartmann et al., 1997). Pathogenic APP mutations leading to AD have been demonstrated to interfere with APP processing, which is not surprising since they are clustered in the vicinity of the secretase cleavage sites: the APP K670N/M671L double mutation resides near the β-secretase cleavage site and the APP V717I/G/F mutations as well as the APP I716V are located near the γ-secretase site. In contrast to the above mentioned 'pure AD' mutations, the APP A692G and E693Q mutations are located within the Aβ sequence, next to the α-secretase cleavage site.

The APP K670N/M671L Swedish (APP 670/671) mutation has been studied most extensively. Both in vitro (Citron et al., 1992; Cai et al., 1993; Felsenstein and Lewis-Higgins, 1993; Hendriks et al., 1995) and in vivo (Citron et al., 1994; Johnston et al., 1994; Scheuner et al., 1996) studies indicated that the APP 670/671 mutation increased secreted Aβ 3–8 fold. Most likely, this is due to an increased β-secretase activity, which has been suggested to result from sorting of APP into different intracellular compartments (Haass et al., 1995; De Strooper et al., 1995). Unlike the APP 670/671 mutation, the mutations located C-terminally from Aβ (APP 716 and 717), do not increase total Aβ (Cai et al., 1993; Felsenstein and Lewis-Higgins, 1993), but instead secrete relatively more Aβ42 compared to Aβ40, resulting in an elevated Aβ42/Aβ40 ratio (Suzuki et al., 1994; Tamaoka et al., 1994; Kosaka et al., 1997). The Flemish (APP 692) mutation generates more Aβ42 as well as Aβ40 (De Jonghe et al., 1998), presumably because the presence of the mutation hampers α-secretase cleavage (Haass et al., 1994). Furthermore, the mutation within Aβ might alter the biochemical properties of the peptide (Clements et al., 1993).

All APP mutations leading to AD have in common that they lead to an increased secretion of Aβ42, which has been shown, at least in vitro, to aggregate more rapidly into fibrils (Jarrett and Lansbury, 1993). Moreover, Aβ42 has been claimed to be the earliest and most predominant Aβ isoform in SPs (Iwatsubo et al., 1994; Gravina et al., 1995; Mann and Iwatsubo, 1996). As a result Aβ, in particular Aβ42, has been assigned a pivotal role in AD pathogenesis. However, transgenic mice, overexpressing mutant (V717F or K670N/M671L) APP, develop amyloid deposits in their brains, but fail to reveal a neuronal response and therefore do not mimic AD pathology entirely (Games et al., 1995; Hsiao et al., 1996).

Interestingly, APP secretion seems to be modulated by cholesterol in the presence of APOE (Racchi et al., 1997; Howland et al., 1998; Simons et al., 1998). APOE promotes Aβ fibril formation in vitro (Ma et al., 1994). Also, different APOE isoforms have been reported to have different binding affinities for Aβ (Strittmatter et al., 1993b; Zhou et al., 1996), however these in vitro data are controversial. Nevertheless, this might explain why the APOE genotype influences age at onset in APP mutation families. The only exception to this rule is the APP A692G mutation, which is located within that part of Aβ having the strongest avidity for APOE (Strittmatter et al., 1993a).

PSEN function and effect of the pathogenic mutations

PSEN localization and endoproteolysis

Like APP, the presenilins (PSENs) are ubiquitously expressed (Sherrington et al., 1995; Levy-Lahad et al., 1995a; Rogaev et al., 1997; Alzheimer's Disease Collaborative Group, 1995; Hutton et al., 1996; Levy-Lahad et al., 1995a; Rogaev et al., 1995), encoding proteins of 467 and 448 amino acids for PSEN1 and PSEN2 respectively. Based on hydrophobicity plots, a 6–9 transmembrane (TM) topology was predicted (Sherrington et al., 1995; Levy-Lahad et al., 1995a). More recent data suggest that there are 8 TM domains, with the N-terminus, a large hydrophilic loop between TM VI and TM VII and the C-terminus facing towards the cytoplasm (Li and Greenwald, 1996; Doan et al., 1996; De Strooper et al., 1997; Li and Greenwald, 1998). PSENs are localized in the endoplasmic reticulum (ER) and the Golgi complex (Kovacs et al., 1996; Moussaoui et al., 1996; Elder et al., 1996; Lee et al., 1996; Uchihara et al., 1996; Boissiere et al., 1996; Deng et al., 1996b; Cook et al., 1996; Walter et al., 1996; Culvenor et al., 1997; De Strooper et al., 1997; Weber et al., 1997; Levey et al., 1997; Benkovic et al., 1997; Kim, K. et al., 1997; Busciglio et al., 1997; Huynh et al., 1997; Hendriks et al., 1998; Xia et al., 1998; Zhang et al., 1998). However, also the cell surface (Takashima et al., 1996; Dewji and Singer, 1997), the nuclear envelop as wells as centrosomes and kinetochores (Li et al., 1997) were suggested to contain PSENs.

Immunohistochemical analysis of AD and normal brains using PSEN-specific antibodies, reveals a granular staining pattern localized in the perinuclear region of neurons (Kovacs et al., 1996; Moussaoui et al., 1996; Elder et al., 1996; Lee et al., 1996; Uchihara et al., 1996; Weber et al., 1997; Levey et al., 1997; Benkovic et al., 1997; Kim et al., 1997; Busciglio et al., 1997; Huynh et al., 1997; Hendriks et al., 1998; Xia et al., 1998). It is still a matter of debate whether PSEN immunostaining is associated or not with SPs (Wisniewski et al., 1995; Wisniewski et al., 1997; Uchihara et al., 1996; Busciglio et al., 1997; Weber et al., 1997; Xia et al., 1998) or with NFTs (Murphy et al., 1996; Uchihara et al., 1996; Busciglio et al., 1997; Giannakopoulos et al., 1997; Hendriks et al., 1998; Takashima et al., 1998; Chui et al., 1998; Xia et al., 1998).

Consistent with their localization in ER and Golgi, PSENs are not post-translationally modified (Cook et al., 1996). However, both PSENs can be phosphorylated, although very limited for PSEN1 (Walter et al., 1996; De Strooper et al., 1997). Full length PSENs are hardly detected, because the majority of the proteins is constitutively endoproteolytically processed both in vitro and in vivo (Thinakaran et al., 1996; Ward et al., 1996; Mercken et al., 1996; Podlisny et al., 1997; Hendriks et al., 1997; Kim et al., 1997a; Okochi et al., 1998). Full length PSEN are quickly turned over (within 60–90 min), while the proteolytic fragments have much longer half-lives (24 h) (Thinakaran et al., 1996; Ratovitski et al., 1997). PSEN1 endoproteolysis results in an ~28 kDa N-terminal fragment (NTF) and an ~18 kDa C-terminal fragment (CTF) (Thinakaran et al., 1996; Mercken et al., 1996; Podlisny et al., 1997). A protease which has not yet been characterized cuts PSEN1 between T291 and A299 (Podlisny et al., 1997). PSEN1 endoproteolysis seems to be a highly regulated process in which endoproteolytic fragments accumulate to saturable levels in a 1:1 stoechiometry (Thinakaran et al., 1996). Recent data suggest that in order to be cleaved, PSEN1 needs to be stabilized in a higher molecular weight protein complex (Thinakaran et al., 1997; Ratovitski et al., 1997; Capell et al., 1998), containing β-catenin (Yu et al., 1998). PSEN1 full-length protein that is not stabilized, is rapidly degraded by the 26S proteasome in a ubiquitin-dependent manner (Marambaud et al., 1998; Fraser et al., 1998), authors' data). PSEN2 proteolytic cleavage occurs between residues K306 and L307 (Shirotani et al., 1997), resulting in a ~35 kDa NTF and a ~20 kDa CTF (Kim et al., 1997a; Kim et al., 1997b; Loetscher et al., 1997; Grunberg et al., 1998). Ubiquitination and subsequent degradation of the full-length protein by the 26S proteasome was also described for PSEN2 (Kim et al., 1997a). The PSEN1 mutant Δ9 is of particular interest regarding proteolytic processing since the cleavage site is located within the deleted region. Therefore PSEN1 Δ9 cannot be cleaved and accumulates both in vitro and in vivo (Thinakaran et al., 1996; Borchelt et al., 1996). Remarkably, the expression of exogenous PSEN1 Δ9 downregulates the amount of endogenous PSEN1 fragments (Thinakaran et al., 1996; Lee et al., 1997), indicating that PSEN1 Δ9 competes with WT full length PSEN1 for entry in the stabilization/cleavage pathway (Ratovitski et al., 1997).

AD-linked missense PSEN1 mutants were shown to be normally processed (Thinakaran et al., 1996; Borchelt et al., 1996; Hendriks et al., 1997), although it has also been reported that some of the missense mutations prevent PSEN1 cleavage, regardless of their location in the molecule (Mercken et al., 1996; Murayama et al., 1997). Hyperaccumulation of mutant PSEN NTF and CTF, due to increased proteolytic cleavage, has been observed in brains of transgenic mice (Lee et al., 1997), while mutant full length PSEN1 was found to accumulate in human brain (Levey et al., 1997). However, others did not see such a phenomenon (Hendriks et al., 1997; Okochi et al., 1998).

PSEN and apoptosis

Apart from the conventional endoproteolytical cleavage, PSENs have also been shown to be death substrates undergoing cleavage by caspases during apoptosis (programmed cell death). Caspase-3 cleavage occurs in PSEN1 between D345 and D346, while in PSEN2 it cuts between D329 and S330 (Kim et al., 1997a) (Fig. 2) (Loetscher et al., 1997; Grunberg et al., 1998). Both full-length protein and conventional CTF are cleaved by caspases. Interestingly, AD-linked mutations can shift conventional PSEN cleavage towards caspase cleavage (Kim et al., 1997a).

Involvement of PSEN in apoptosis was first suggested for PSEN2 since PC12 cells, overexpressing PSEN2, showed increased sensitivity to apoptosis (Deng et al., 1996a; Wolozin et al., 1996). On the other hand apoptosis was antagonized in the presence of antisense PSEN2 or a C-terminal PSEN2 fragment (Vito et al., 1996), homologous to the murine ALG-3, an inhibitor of apoptosis. Mutant (I141N) PSEN2 showed enhanced capacity to induce apoptosis (Wolozin et al., 1996; Janicki and Monteiro, 1997), accompanied by increased caspase mediated PSEN2 cleavage (Kim et al., 1997a). The C-terminal PSEN2 caspase fragment has been suggested to execute an anti-apoptotic function as a negative feedback signal (Vito et al., 1997).

Mutations in PSEN1 on the other hand were claimed to render neuronal cells vulnerable to apoptosis by a mechanism involving destabilization of cellular calcium homeostasis, which leads to oxidative stress and mitochondrial dysfunction (Guo et al., 1996; Guo et al., 1997; Guo et al., 1998; Keller et al., 1998). In contrast to PSEN2, reduction of PSEN1 expression results in apoptosis (Roperch et al., 1998). Recent data also suggest that PSEN1 mediated apoptosis is closely connected with regulation of cancer-related pathways (Roperch et al., 1998; Guo et al., 1998). However, the role of apoptosis in neuronal death in AD remains to be proven.

PSEN and notch signaling

SEL-12, a *C. elegans* protein that facilitates LIN-12 and GLP-1-mediated signaling involved in the specification of cell fates during development, shows about 50% homology with PSEN1 (Levitan and Greenwald, 1995). PSENs and SEL-12 were shown to be mutually interchangeable, since the introduction of a PSEN transgene rescued the severe egg-laying phenotype in mutant SEL-12 worms (Levitan et al., 1996; Baumeister et al., 1997). The failure of different mutant PSEN to completely revert the egg-laying deficit in SEL-12 mutant worms (Levitan et al., 1996; Baumeister et al., 1997) led to the hypothesis that mutations in PSEN result in loss-of-function or haploinsufficiency of the WT allele. Proteolytic cleavage of PSEN1, both constitutive (Baumeister et al., 1997) and caspase-mediated (Brockhaus et al., 1998) is not required for rescuing the mutant SEL-12 phenotype and is therefore not believed to be a prerequisite for normal PSEN function.

Further evidence for the possible role of PSEN in notch signaling was provided by the study of PSEN1-deficient mice. PSEN1 knockout mice die during embryogenesis with abnormal somite segmentation, a phenotype that is shared with NOTCH1-deficient mice (Shen et al., 1997; Wong et al., 1997). Also, the expression pattern of PSEN during murine development almost completely matches this of NOTCH1 (Lee et al., 1996; Berezovska et al., 1997).

Although it might be difficult to see the relevance for a role of PSEN in notch signaling leading to an adult-onset disorder of the CNS, mutations in the NOTCH3 gene were identified in a dementia-associated illness with onset ages in adulthood, CADASIL (cerebral autosomal dominant arteriopathy with subcortical infarcts and leukoencephalopathy) (Joutel et al., 1996), indicating that impaired notch signaling may indeed play a role in this type of CNS diseases.

PSEN and amyloid

The first indication that PSEN might be involved in APP processing came from the observation that Aβ42 was selectively increased in plasma and conditioned medium of fibroblasts from individuals with PSEN1 mutations (Scheuner et al., 1996). This finding was confirmed in medium from cells, transfected with mutant PSEN (Borchelt et al., 1996; Citron et al., 1997; Xia et al., 1997a; Mehta et al., 1998). The amount of Aβ42 did not correlate, however, with the severity of the phenotype (Citron et al., 1997; Mehta et al., 1998). Increased Aβ42 levels were also found in brains of transgenic mutant PSEN overexpressing mice (Duff et al., 1996; Borchelt et al., 1996; Citron et al., 1997; Oyama et al., 1998). Although significantly elevated levels of Aβ42 are present in the brains of PSEN transgenic mice compared to their non-transgenic littermates, no amyloid deposits are formed. Only double transgenic animals, carrying both a mutant PSEN1 and a mutant APP transgene produce large numbers of fibrillar Aβ deposits (Borchelt et al., 1997; Holcomb et al., 1998). In brains from AD patients carrying PSEN mutations, the number of Aβ42 immunoreactive plaques is increased compared to sporadic cases

(Lemere et al., 1996; Mann et al., 1996; Mann et al., 1997; Gomez-Isla et al., 1997; Ishii et al., 1997).

The observation that APP cleavage by γ-secretase is inhibited in primary neurons from PSEN1 deficient mice provided further evidence for a role of PSEN in APP processing (De Strooper et al., 1998). Moreover, introduction of a mutant PSEN1 transgene rescues the embryonic lethal phenotype of PSEN1 knockout mice and simultaneously elevates Aβ42 levels (Davis et al., 1998; Qian et al., 1998). Taken together, these findings led to the hypothesis of a gain-of-misfunction for mutant PSEN by increasing Aβ42 and they also provided further evidence for the central role of Aβ, in particular Aβ42, in AD pathogenesis.

Apparently only full length PSEN is implicated in APP processing, since PSEN2 fragments, carrying AD-linked mutations were not able to increase secretion of Aβ42 (Tomita et al., 1998). Also, caspase-cleavage of PSEN was ascertained not to be a prerequisite for the effect of mutant PSEN on Aβ42 secretion (Brockhaus et al., 1998). The exact cellular mechanism by which APP and PSEN are associated is not yet fully understood. A direct interaction between APP and PSEN was suggested by yeast two hybrid and immunoprecipitation experiments (Weidemann et al., 1997; Waragai et al., 1997; Xia et al., 1997b), while this was contradicted by others (Kim, S. et al., 1997; Thinakaran et al., 1998). In the brain, APP as well as Aβ were detected together with endoproteolytic PSEN1 fragments in detergent-insoluble glycolipid-enriched membrane domains (DIGs), suggesting a direct function of PSEN1 in APP cleavage (Lee et al., 1998). Because of its homology to SPE-4, a *C. elegans* protein involved in spermatogenesis mediating protein trafficking in the Golgi (L'Hernault and Arduengo, 1992), PSEN was thought to play a role in intracellular APP routing. Since it was shown that the γ42 and γ40-protease activities reside in different subcellular compartments, PSEN mutants were believed to direct APP preferentially towards γ42-containing compartments. However, recent data suggest that PSEN1 mutations do not disrupt protein transport from the ER to the Golgi (Tan et al., 1998).

Conclusions

Although it is widely recognized that AD is a complex disorder of the CNS in which many genetic and environmental factors influence the pathogenesis, molecular genetic analysis of AD has provided us with major breakthroughs in the unraveling of the molecular causes for this dementia. With the identification of genes and genetic risk factors, the knowledge about the biochemical processes underlying this disease gradually continues to grow as well. Knowledge about the biological function and dysfunction of genes involved in AD, although still fragmentarily, may lead to understanding the primary disease mechanisms. This in turn can help to develop a more efficient therapy that can either decelerate or cure this devastating disorder of human mind.

List of abbreviations

AD	Alzheimer's disease
SP	senile plaques
Aβ	amyloid β
CAA	congophilic amyloid angiopathy
PHF	paired helical filaments
LOAD	late-onset Alzheimer's disease
EOAD	early-onset Alzheimer's disease
APP	amyloid precursor protein
PSEN1	presenilin-1
PSEN2	presenilin-2
APOE	apolipoprotein E
HCHWA-D	hereditary cerebral hemorrhages with amyloidosis – Dutch type
UTR	untranslated region
AATC	α1-antichymotrypsin
VLDLR	very low-density lipoprotein receptor
APOJ	apolipoprotein J

HLA-DR	human leukocyte antigen-DR
BCHE-K	butyrylcholinesterase K
BH	bleomycin hydroxylase
ER	endoplasmic reticulum
WT	wild type
CADASIL	cerebral autosomal dominant arteriopathy with subcortical infarcts and leukoencephalopathy
CNS	central nervous system
DIG	detergent-insoluble glycolipid-enriched membrane domains
K	lysine
N	asparagine
M	methionine
L	leucine
A	alanine
G	glycine
E	glutamic acid
Q	glutamine
I	isoleucine
V	valine
F	phenylalanine
P	proline
S	serine
D	aspartic acid

References

Alzheimer's Disease Collaborative Group (1993) Apolipoprotein E genotype and Alzheimer's disease. *Lancet*, 342: 737–738.

Alzheimer's Disease Collaborative Group (1995) The structure of the presenilin 1 (S182) gene and identification of six novel mutations in early onset AD families. *Nat. Genet.*, 11: 219–222.

Baumeister, R., Leimer, U., Zweckbronner, I., Jakubek, C., Grünberg, J. and Haass, C. (1997) Human presenilin-1, but not familial Alzheimer's disease (FAD) mutants, facilitate Caenorhabditis elegans Notch signalling independently of proteolytic processing. *Genes and Function*, 1: 149–159.

Benkovic, S.A., McGowan, E.M., Rothwell, N.J., Hutton, M., Morgan, D.G. and Gordon, M.N. (1997) Regional and cellular localization of presenilin-2 RNA in rat and human brain. *Exp. Neurol.*, 145: 555–564.

Berezovska, O., Xia, M.Q., Page, K., Wasco, W., Tanzi, R.E. and Hyman, B.T. (1997) Developmental regulation of presenilin mRNA expression parallels notch expression. *J. Neuropathol. Exp. Neurol.*, 56: 40–44.

Beyer, K., Lao, J.I., Fernandez-Novoa, L. and Cacabelos, R. (1998) Identification of a novel mutation (V148I) in the TM2 domain of the presenilin 2 gene in a patient with late-onset Alzheimer's disease. *Neurobiol. Aging*, 19: S87–S87(Abstract).

Blacker, D., Wilcox, M.A., Laird, N.M., Rodes, L., Horvath, S.M., Go, R.C., Perry, R., Watson, B.J., Bassett, S.S., McInnis, M.G., Albert, M.S., Hyman, B.T. and Tanzi, R.E. (1998) Alpha-2 macroglobulin is genetically associated with Alzheimer disease. *Nat. Genet.*, 19: 357–360.

Boissiere, F., Pradier, L., Delaere, P., Faucheux, B., Revah, F., Brice, A., Agid, Y. and Hirsch, E.C. (1996) Regional and cellular presenilin 2 (STM2) gene expression in the human brain. *NeuroReport.*, 7: 2021–2025.

Borchelt, D.R., Ratovitski, T., van Lare, R., Lee, M., Gonzales, V., Jenkins, N.A., Copeland, N.G., Price, D. and Sisodia, S.S. (1997) Accelerated amyloid deposition in the brains of transgenic mice coexpressing mutant presenilin 1 and amyloid precursor proteins. *Neuron*, 19: 939–945.

Borchelt, D.R., Thinakaran, G., Eckman, C.B., Lee, M.K., Davenport, F., Ratovitsky, T., Prada, C.M., Kim, G., Seekins, S., Yager, D., Slunt, H.H., Wang, R., Seeger, M., Levey, A.I., Gandy, S.E., Copeland, N.G., Jenkins, N.A., Price, D.L., Younkin, S.G. and Sisodia, S.S. (1996) Familial Alzheimer's disease-linked presenilin 1 variants elevate Abeta1-42/1–40 ratio in vitro and in vivo. *Neuron*, 17: 1005–1013.

Breteler, M.M., Claus, J.J., van Duijn, C.M., Launer, L.J. and Hofman, A. (1992) Epidemiology of Alzheimer's disease. *Epidemiol. Rev.*, 14: 59–82.

Brockhaus, M., Grunberg, J., Rohrig, S., Loetscher, H., Wittenburg, N., Baumeister, R., Jacobsen, H. and Haass, C. (1998) Caspase-mediated cleavage is not required for the activity of presenilins in amyloidogenesis and NOTCH signaling. *NeuroReport.*, 9: 1481–1486.

Busciglio, J., Gabuzda, D.H., Matsudaira, P. and Yankner, B.A. (1993) Generation of beta-amyloid in the secretory pathway in neuronal and nonneuronal cells. *Proc. Natl. Acad. Sci. USA*, 90: 2092–2096.

Busciglio, J., Hartmann, H., Lorenzo, A., Wong, C., Baumann, K., Sommer, B., Staufenbiel, M. and Yankner, B.A. (1997) Neuronal localization of presenilin-1 and association with amyloid plaques and neurofibrillary tangles in Alzheimer's disease. *J. Neurosci.*, 17: 5101–5107.

Cai, X.D., Golde, T.E. and Younkin, S.G. (1993) Release of excess amyloid beta protein from a mutant amyloid beta protein precursor. *Science*, 259: 514–516.

Capell, A., Grunberg, J., Pesold, B., Diehlmann, A., Citron, M., Nixon, R., Beyreuther, K., Selkoe, D.J. and Haass, C. (1998) The proteolytic fragments of the Alzheimer's disease-associated presenilin-1 form heterodimers and occur as a 100–150 kDa molecular mass complex. *J. Biol. Chem.*, 273: 3205–3211.

Chartier-Harlin, M.C., Crawford, F., Houlden, H., Warren, A., Hughes, D., Fidani, L., Goate, A., Rossor, M., Roques, P., Hardy, J. et al. (1991) Early-onset Alzheimer's disease caused by mutations at codon 717 of the beta-amyloid precursor protein gene. *Nature*, 353: 844–846.

Chui, D.H., Shirotani, K., Tanahashi, H., Akiyama, H., Ozawa, K., Kunishita, T., Takahashi, K., Makifuchi, T. and Tabira, T. (1998) Both N-terminal and C-terminal fragments of presenilin 1 colocalize with neurofibrillary tangles in neurons and dystrophic neurites of senile plaques in Alzheimer's disease. *J. Neurosci.* Res., 53: 99–106.

Chyung, A.C., Greenberg, B.D., Cook, D.G., Doms, R.W. and Lee, V.M. (1997) Novel beta-secretase cleavage of beta-amyloid precursor protein in the endoplasmic reticulum/intermediate compartment of NT2N cells. *J. Cell Biol.*, 138: 671–680.

Citron, M., Diehl, T.S., Gordon, G., Biere, A.L., Seubert, P. and Selkoe, D.J. (1996) Evidence that the 42- and 40-amino acid forms of amyloid beta protein are generated from the beta-amyloid precursor protein by different protease activities. *Proc. Natl. Acad. Sci. USA*, 93: 13170–13175.

Citron, M., Oltersdorf, T., Haass, C., McConlogue, L., Hung, A.Y., Seubert, P., Vigo, P.C., Lieberburg, I. and Selkoe, D.J. (1992) Mutation of the beta-amyloid precursor protein in familial Alzheimer's disease increases beta-protein production. *Nature*, 360: 672–674.

Citron, M., Vigo-Pelfrey, C., Teplow, D.B., Miller, C., Schenk, D., Johnston, J., Winblad, B., Venizelos, N., Lannfelt, L. and Selkoe, D.J. (1994) Excessive production of amyloid beta-protein by peripheral cells of symptomatic and presymptomatic patients carrying the Swedish familial Alzheimer disease mutation. *Proc. Natl. Acad. Sci. USA*, 91: 11993–11997.

Citron, M., Westaway, D., Xia, W., Carlson, G., Diehl, T., Levesque, G., Johnson, W.K., Lee, M., Seubert, P., Davis, A., Kholodenko, D., Motter, R., Sherrington, R., Perry, B., Yao, H., Strome, R., Lieberburg, I., Rommens, J., Kim, S., Schenk, D., Fraser, P., St-George, H.P. and Selkoe, D.J. (1997) Mutant presenilins of Alzheimer's disease increase production of 42-residue amyloid beta-protein in both transfected cells and transgenic mice. *Nat. Med.*, 3: 67–72.

Clements, A., Walsh, D.M., Williams, C.H. and Allsop, D. (1993) Effects of the mutations Glu22 to Gln and Ala21 to Gly on the aggregation of a synthetic fragment of the Alzheimer's amyloid beta/A4 peptide. *Neurosci. Lett.*, 161: 17–20.

Cole, G.M., Bell, L., Truong, Q.B. and Saitoh, T. (1992) An endosomal-lysosomal pathway for degradation of amyloid precursor protein. *Ann. N. Y. Acad. Sci.*, 674: 103–17: 103–117.

Cook, D.G., Forman, M.S., Sung, J.C., Leight, S., Kolson, D.L., Iwatsubo, T., Lee, V.M. and Doms, R.W. (1997) Alzheimer's A beta(1–42) is generated in the endoplasmic reticulum/intermediate compartment of NT2N cells. *Nat. Med.*, 3: 1021–1023.

Cook, D.G., Sung, J.C., Golde, T.E., Felsenstein, K.M., Wojczyk, B.S., Tanzi, R.E., Trojanowski, J.Q., Lee, V.M. and Doms, R.W. (1996) Expression and analysis of presenilin 1 in a human neuronal system: localization in cell bodies and dendrites. *Proc. Natl. Acad. Sci. USA*, 93: 9223–9228.

Corder, E.H., Saunders, A.M., Risch, N.J., Strittmatter, W.J., Schmechel, D.E., Gaskell, P.C., Rimmler, J.B., Locke, P.A., Conneally, P.M., Schmader, K.E. et a (1994) Protective effect of apolipoprotein E type 2 allele for late onset Alzheimer disease. *Nat. Genet.*, 7: 180–184.

Corder, E.H., Saunders, A.M., Strittmatter, W.J., Schmechel, D.E., Gaskell, P.C., Small, G.W., Roses, A.D., Haines, J.L. and Pericak, V.M. (1993) Gene dose of apolipoprotein E type 4 allele and the risk of Alzheimer's disease in late onset families. *Science*, 261: 921–923.

Cras, P., Van Harskamp, F., van Duijn, C., Hendriks, L., Stefanko, S.Z., Hofman, A., Kros, J.M., Ceuterick, C., Van Broeckhoven, C. and Martin, J.J. (1998) Presenile Alzheimer dementia characterized by amyloid angiopathy and a large amyloid core type of senile plaques in the APP 692Ala-Gly mutation. *Acta Neuropathol.*, 96: 253–260.

Cruts, M., Backhovens, H., Wang, S.Y., Van Gassen, G., Theuns, J., De Jonghe, C., Wehnert, A., De Voecht, J., De Winter, G., Cras, P. et al. (1995) Molecular genetic analysis of familial early-onset Alzheimer's disease linked to chromosome 14q24. 3. *Hum. Mol. Genet.*, 4: 2363–2371.

Cruts, M., Hendriks, L. and Van Broeckhoven, C. (1996) The presenilin genes: a new gene family involved in Alzheimer disease pathology. *Hum. Mol. Genet.*, 5 Spec No: 1449–1455.

Cruts, M. and Van Broeckhoven, C. (1998) Presenilin mutations in Alzheimer's disease. *Hum. Mut.*, 11: 183–190.

Cruts, M., van Duijn, C.M., Backhovens, H., van-den Broeck, M., Wehnert, A., Serneels, S., Sherrington, R., Hutton, M., Hardy, J., St-George Hyslop, P.H., Hofman, A. and Van Broeckhoven, C. (1998) Estimation of the genetic contribution of presenilin-1 and -2 mutations in a population-based study of presenile Alzheimer disease. *Hum. Mol. Genet.*, 7: 43–51.

Culvenor, J.G., Maher, F., Evin, G., Malchiodi-Albedi, F., Cappai, R., Underwood, J.R., Davis, J.B., Karran, E.H., Roberts, G.W., Beyreuther, K. and Masters, C.L. (1997) Alzheimer's disease-associated presenilin 1 in neuronal cells: evidence for localization to the endoplasmic reticulum-Golgi intermediate compartment. *J. Neurosci.* Res., 49: 719–731.

Curran, M., Middleton, D., Edwardson, J., Perry, R., McKeith, I., Morris, C. and Neill, D. (1997) HLA-DR antigens associated with major genetic risk for late-onset Alzheimer's disease. *NeuroReport.*, 8: 1467–1469.

Davis, J.A., Naruse, S., Chen, H., Eckman, C., Younkin, S., Price, D.L., Borchelt, D.R., Sisodia, S.S. and Wong, P.C. (1998) An Alzheimer's disease-linked PS1 variant rescues the developmental abnormalities of PS1-deficient embryos. *Neuron*, 20: 603–609.

De Jonghe, C., Zehr, C., Yager, D., Prada, C.M., Younkin, S., Hendriks, L., Van Broeckhoven, C. and Eckman, C. (1998) Flemish and Dutch mutations in amyloid β precursor protein have different effects on amyloid β secretion. *Neurobiol. Dis.*, 5,281–286:

De Strooper, B., Beullens, M., Contreras, B., Levesque, L., Craessaerts, K., Cordell, B., Moechars, D., Bollen, M., Fraser, P., St George-Hyslop, P.H. and Van Leuven, F. (1997) Phosphorylation, subcellular localization, and membrane orientation of the Alzheimer's disease-associated presenilins. *J. Biol. Chem.*, 272: 3590–3598.

De Strooper, B., Craessaerts, K., Dewachter, I., Moechars, D., Greenberg, B., Van Leuven, F. and Van Den Berghe, H. (1995) Basolateral secretion of amyloid precursor protein in Madin-Darby Canine Kidney cells is disturbed by alterations of intracellular pH and by introducing a mutation associated with familial Alzheimer's disease. *J. Biol. Chem.*, 270: 4058–4065.

De Strooper, B., Saftig, P., Craessaerts, K., Vanderstichele, H., Guhde, G., Annaert, W., Von Figura, K. and Van Leuven, F. (1998) Deficiency of presenilin-1 inhibits the normal cleavage of amyloid precursor protein. *Nature*, 391: 387–390.

De Strooper, B., Simons, M., Multhaup, G., Van Leuven, F., Beyreuther, K. and Dotti, C.G. (1995) Production of intracellular amyloid-containing fragments in hippocampal neurons expressing human amyloid precursor protein and protection against amyloidogenesis by subtle amino acid substitutions in the rodent sequence. *EMBO J.*, 14: 4932–4938.

De Strooper, B., Umans, L., Van Leuven, F. and Van Den Berghe, H. (1993) Study of the synthesis and secretion of normal and artificial mutants of murine amyloid precursor protein (APP): cleavage of APP occurs in a late compartment of the default secretion pathway. *J. Cell Biol.*, 121: 295–304.

Deng, G., Pike, C.J. and Cotman, C.W. (1996a) Alzheimer-associated presenilin-2 confers increased sensitivity to apoptosis in PC12 cells. *FEBS Lett.*, 397: 50–54.

Deng, G., Su, J.H. and Cotman, C.W. (1996b) Gene expression of Alzheimer-associated presenilin-2 in the frontal cortex of Alzheimer and aged control brain. *FEBS Lett.*, 394: 17–20.

Dewji, N.N. and Singer, S.J. (1997) Cell surface expression of the Alzheimer disease-related presenilin proteins. *Proc. Natl. Acad. Sci. USA*, 94: 9926–9931.

Doan, A., Thinakaran, G., Borchelt, D.R., Slunt, H.H., Ratovitsky, T., Podlisny, M., Selkoe, D.J., Seeger, M., Gandy, S.E., Price, D.L. and Sisodia, S.S. (1996) Protein topology of presenilin 1. *Neuron*, 17: 1023–1030.

Duff, K., Eckman, C., Zehr, C., Yu, X., Prada, C.M., Perez, T.J., Hutton, M., Buee, L., Harigaya, Y., Yager, D., Morgan, D., Gordon, M.N., Holcomb, L., Refolo, L., Zenk, B., Hardy, J. and Younkin, S. (1996) Increased amyloid-beta42(43) in brains of mice expressing mutant presenilin 1. *Nature*, 383: 710–713.

Eckman, C., Mehta, N., Crook, R., Perez-Tur, J., Prihar, G., Pfeiffer, E., Graff-Radford, N., Hinder, P., Yager, D., Zenk, B., Refolo, L., Prada, C.M., Younkin, S.G., Hutton, M. and Hardy, J. (1997) A new pathogenic mutation in the APP gene (I716V) increases the relative proportion of Aβ42(43) *Hum. Mol. Genet.*, 6: 2087–2089.

Elder, G.A., Tezapsidis, N., Carter, J., Shioi, J., Bouras, C., Li, H.C., Johnston, J.M., Efthimiopoulos, S., Friedrich-VL, J. and Robakis, N.K. (1996) Identification and neuron specific expression of the S182/presenilin I protein in human and rodent brains. *J. Neurosci.* Res., 45: 308–320.

Esch, F.S., Keim, P.S., Beattie, E.C., Blacher, R.W., Culwell, A.R., Oltersdorf, T., McClure, D. and Ward, P.J. (1990) Cleavage of amyloid beta peptide during constitutive processing of its precursor. *Science*, 248: 1122–1124.

Estus, S., Golde, T.E., Kunishita, T., Blades, D., Lowery, D., Eisen, M., Usiak, M., Qu, X.M., Tabira, T. and Greenberg, B.D. (1992) Potentially amyloidogenic, carboxyl-terminal derivatives of the amyloid protein precursor. *Science*, 255: 726–728.

Felsenstein, K.M. and Lewis-Higgins, L. (1993) Processing of the beta-amyloid precursor protein carrying the familial, Dutch-type, and a novel recombinant C-terminal mutation. *Neurosci. Lett.*, 152: 185–189.

Forman, M.S., Cook, D.G., Leight, S., Doms, R.W. and Lee, V.M. (1997) Differential effects of the swedish mutant amyloid precursor protein on β-amyloid accumulation and secretion in neurons and nonneuronal cells. *J. Biol. Chem.*, 272: 32247–32253.

Fraser, P.E., Levesque, G., Yu, G., Mills, L.R., Thirlwell, J., Frantseva, M., Gandy, S.E., Seeger, M., Carlen, P.L. and St-George, H.P. (1998) Presenilin 1 is actively degraded by the 26S proteasome. *Neurobiol. Aging*, 19: S19–S21.

Games, D., Adams, D., Alessandrini, R., Barbour, R., Berthelette, P., Blackwell, C., Carr, T., Clemens, J., Donaldson, T., Gillespie, F. et al. (1995) Alzheimer-type neuropathology in transgenic mice overexpressing V717F beta-amyloid precursor protein. *Nature*, 373: 523–527.

Giannakopoulos, P., Bouras, C., Kovari, E., Shioi, J., Tezapsidis, N., Hof, P.R. and Robakis, N.K. (1997) Presenilin-1-immunoreactive neurons are preserved in late-onset Alzheimer's disease. *Am. J. Pathol.*, 150: 429–436.

Glenner, G.G. and Wong, C.W. (1984) Alzheimer's disease: initial report of the purification and characterization of a novel cerebrovascular amyloid protein. *Biochem. Biophys. Res. Commun.*, 120: 885–890.

Goate, A., Chartier-Harlin, M.C., Mullan, M., Brown, J., Crawford, F., Fidani, L., Giuffra, L., Haynes, A., Irving, N., James, L. et al. (1991) Segregation of a missense mutation in the amyloid precursor protein gene with familial Alzheimer's disease. *Nature*, 349: 704–706.

Goate, A.M., Haynes, A.R., Owen, M.J., Farrall, M., James, L.A., Lai, L.Y., Mullan, M.J., Roques, P., Rossor, M.N., Williamson, R. et al. (1989) Predisposing locus for Alzheimer's disease on chromosome 21. *Lancet*, 1: 352–355.

Golde, T.E., Estus, S., Younkin, L.H., Selkoe, D.J. and Younkin, S.G. (1992) Processing of the amyloid protein precursor to potentially amyloidogenic derivatives. *Science*, 255: 728–730.

Goldgaber, D., Lerman, M.I., McBride, O.W., Saffiotti, U. and Gajdusek, D.C. (1987) Characterization and chromosomal localization of a cDNA encoding brain amyloid of Alzheimer's disease. *Science*, 235: 877–880.

Gomez-Isla, T., Wasco, W., Pettingell, W.P., Gurubhagavatula, S., Schmidt, S.D., Jondro, P.D., McNamara, M., Rodes, L.A., DiBlasi, T., Growdon, W.B., Seubert, P., Schenk, D., Growdon, J.H., Hyman, B.T. and Tanzi, R.E. (1997) A novel presenilin-1 mutation: increased beta-amyloid and neurofibrillary changes. *Ann. Neurol.*, 41: 809–813.

Gravina, S.A., Ho, L., Eckman, C.B., Long, K.E., Otvos, L., Younkin, L.H., Suzuki, N. and Younkin, S.G. (1995) Amyloid beta protein (A beta) in Alzheimer's disease brain. Biochemical and immunocytochemical analysis with antibodies specific for forms ending at A beta 40 or A beta 42(43) *J. Biol. Chem.*, 270: 7013–7016.

Grunberg, J., Walter, J., Loetscher, H., Deuschle, U., Jacobsen, H. and Haass, C. (1998) Alzheimer's disease associated presenilin-1 holoprotein and its 18–20 kDa C-terminal fragment are death substrates for proteases of the caspase family. *Biochemistry*, 37: 2263–2270.

Guo, Q., Fu, W., Xie, J., Luo, H., Sells, S.F., Geddes, J.W., Bondada, V., Rangnekar, V.M. and Mattson, M.P. (1998a) Par-4 is a mediator of neuronal degeneration associated with the pathogenesis of Alzheimer disease. *Nat. Med.*, 4: 957–962.

Guo, Q., Furukawa, K., Sopher, B.L., Pham, D.G., Xie, J., Robinson, N., Martin, G.M. and Mattson, M.P. (1996) Alzheimer's PS-1 mutation perturbs calcium homeostasis and sensitizes PC12 cells to death induced by amyloid beta-peptide. *NeuroReport.*, 8: 379–383.

Guo, Q., Robinson, N. and Mattson, M.P. (1998b) Secreted beta-amyloid precursor protein counteracts the proapoptotic action of mutant presenilin-1 by activation of NF-kappaB and stabilization of calcium homeostasis. *J. Biol. Chem.*, 273: 12341–12351.

Guo, Q., Sopher, B.L., Furukawa, K., Pham, D.G., Robinson, N., Martin, G.M. and Mattson, M.P. (1997) Alzheimer's presenilin mutation sensitizes neural cells to apoptosis induced by trophic factor withdrawal and amyloid beta-peptide: involvement of calcium and oxyradicals. *J. Neurosci.*, 17: 4212–4222.

Haan, J., Hardy, J.A. and Roos, R.A. (1991) Hereditary cerebral hemorrhage with amyloidosis — Dutch type: its importance for Alzheimer research. *Trends Neurosci.*, 14: 231–234.

Haan, J., Van Broeckhoven, C., van Duijn, C.M., Voorhoeve, E., van Harskamp, F., van Swieten, J.C., Maat-Schieman, M.L., Roos, R.A. and Bakker, E. (1994) The apolipoprotein E epsilon 4 allele does not influence the clinical expression of the amyloid precursor protein gene codon 693 or 692 mutations. *Ann. Neurol.*, 36: 434–437.

Haass, C., Hung, A.Y., Selkoe, D.J. and Teplow, D.B. (1994) Mutations associated with a locus for familial Alzheimer's disease result in alternative processing of amyloid beta-protein precursor. *J. Biol. Chem.*, 269: 17741–17748.

Haass, C., Koo, E.H., Mellon, A., Hung, A.Y. and Selkoe, D.J. (1992a) Targeting of cell-surface beta-amyloid precursor protein to lysosomes: alternative processing into amyloid-bearing fragments. *Nature*, 357: 500–503.

Haass, C., Lemere, C.A., Capell, A., Citron, M., Seubert, P., Schenk, D., Lannfelt, L. and Selkoe, D.J. (1995) The Swedish mutation causes early-onset Alzheimer's disease by beta-secretase cleavage within the secretory pathway. *Nat. Med.*, 1: 1291–1296.

Haass, C., Schlossmacher, M.G., Hung, A.Y., Vigo-Pelfrey, C., Mellon, A., Ostaszewski, B.L., Lieberburg, I., Koo, E.H., Schenk, D. and Teplow, D.B. (1992b) Amyloid beta-peptide is produced by cultured cells during normal metabolism. *Nature*, 359: 322–325.

Hartmann, H., Busciglio, J., Baumann, K.H., Staufenbiel, M. and Yankner, B.A. (1997a) Developmental regulation of presenilin-1 processing in the brain suggests a role in neuronal differentiation. *J. Biol. Chem.*, 272: 14505–14508.

Hartmann, T., Bieger, S.C., Bruhl, B., Tienari, P.J., Ida, N., Allsop, D., Roberts, G.W., Masters, C.L., Dotti, C.G., Unsicker, K. and Beyreuther, K. (1997b) Distinct sites of intracellular production for Alzheimer's disease A beta40/42 amyloid peptides. *Nat. Med.*, 3: 1016–1020.

Hendriks, L., Cras, P., Martin, J.J. et al. (1995) Alzheimer's disease and hemorrhagic stroke: their relationship to βA4 amyloid deposition. In: K.S. Kosik et al. (Eds.), *Alzheimer's disease: Lessons from Cell Biology*, Springer-Verlag, Berlin, pp. 37–48.

Hendriks, L., De Jonghe, C., Lubke, U., Woodrow, S., Vanderhoeven, I., Boons, J., Cras, P., Martin, J.J. and Van Broeckhoven, C. (1998) Immunoreactivity of presenilin-1 and tau in Alzheimer's disease brain. *Exp. Neurol.*, 149: 341–348.

Hendriks, L., Thinakaran, G., Harris, C.L., De Jonghe, C., Martin, J.J., Sisodia, S.S. and Van Broeckhoven, C. (1997) Processing of presenilin 1 in brains of patients with Alzheimer's disease and controls. *NeuroReport.*, 8: 1717–1721.

Hendriks, L., van Duijn, C.M., Cras, P., Cruts, M., Van Hul, W., Van Harskamp, F., Warren, A., McInnis, M.G., Antonarakis, S.E., Martin, J.J. et al. (1992) Presenile dementia and cerebral haemorrhage linked to a mutation at codon 692 of the beta-amyloid precursor protein gene. *Nat. Genet.*, 1: 218–221.

Holcomb, L., Gordon, M.N., McGowan, E., Yu, X., Benkovic, S., Jantzen, P., Wright, K., Saad, I., Mueller, R., Morgan, D., Sanders, S., Zehr, C., O'Campo, K., Hardy, J., Prada,

C.M., Eckman, C., Younkin, S., Hsiao, K. and Duff, K. (1998) Accelerated Alzheimer-type phenotype in transgenic mice carrying both mutant amyloid precursor protein and presenilin 1 transgenes. *Nat. Med.*, 4: 97–100.

Howland, D.S., Trusko, S.P., Savage, M.J., Reaume, A.G., Lang, D.M., Hirsch, J.D., Maeda, N., Siman, R., Greenberg, B.D., Scott, R.W. and Flood, D.G. (1998) Modulation of secreted beta-amyloid precursor protein and amyloid beta- peptide in brain by cholesterol. *J. Biol. Chem.*, 273: 16576–16582.

Hsiao, K., Chapman, P., Nilsen, S., Eckman, C., Harigaya, Y., Younkin, S., Yang, F. and Cole, G. (1996) Correlative memory deficits, Abeta elevation, and amyloid plaques in transgenic mice. *Science*, 274: 99–102.

Hutton, M., Busfield, F., Wragg, M., Crook, R., Perez, T.J., Clark, R.F., Prihar, G., Talbot, C., Phillips, H., Wright, K., Baker, M., Lendon, C., Duff, K., Martinez, A., Houlden, H., Nichols, A., Karran, E., Roberts, G., Roques, P., Rossor, M., Venter, J.C., Adams, M.D., Cline, R.T., Phillips, C.A., Goate, A. et al. (1996) Complete analysis of the presenilin 1 gene in early onset Alzheimer's disease. *NeuroReport.*, 7: 801–805.

Huynh, D.P., Vinters, H.V., Ho, D.H., Ho, V.V. and Pulst, S.M. (1997) Neuronal expression and intracellular localization of presenilins in normal and Alzheimer disease brains. *J. Neuropathol. Exp. Neurol.*, 56: 1009–1017.

Ikezu, T., Trapp, B.D., Song, K.S., Schlegel, A., Lisanti, M.P. and Okamoto, T. (1998) Caveolae, plasma membrane microdomains for alpha-secretase-mediated processing of the amyloid precursor protein. *J. Biol. Chem.*, 273: 10485–10495.

Ishii, K., Ii, K., Hasegawa, T., Shoji, S., Doi, A. and Mori, H. (1997) Increased A beta 42(43)-plaque deposition in early-onset familial Alzheimer's disease brains with the deletion of exon 9 and the missense point mutation (H163R) in the PS-1 gene. *Neurosci. Lett.*, 228: 17–20.

Iwatsubo, T., Odaka, A., Suzuki, N., Mizusawa, H., Nukina, N. and Ihara, Y. (1994) Visualization of A beta 42(43) and A beta 40 in senile plaques with end-specific A beta monoclonals: evidence that an initially deposited species is A beta 42(43) *Neuron*, 13: 45–53.

Janicki, S. and Monteiro, M.J. (1997) Increased apoptosis arising from increased expression of the Alzheimer's disease-associated presenilin-2 mutation (N141I) *J. Cell Biol.*, 139: 485–495.

Jarrett, J.T. and Lansbury, P.T.J. (1993) Seeding "one-dimensional crystallization" of amyloid: a pathogenic mechanism in Alzheimer's disease and scrapie? *Cell*, 73: 1055–1058.

Johnston, J.A., Cowburn, R.F., Norgren, S., Wiehager, B., Venizelos, N., Winblad, B., Vigo, P.C., Schenk, D., Lannfelt, L. and O'Neill, C. (1994) Increased beta-amyloid release and levels of amyloid precursor protein (APP) in fibroblast cell lines from family members with the Swedish Alzheimer's disease APP670/671 mutation. *FEBS Lett.*, 354: 274–278.

Joutel, A., Corpechot, C., Ducros, A., Vahedi, K., Chabriat, H., Mouton, P., Alamowitch, S., Domenga, V., Cecillion, M., Marechal, E., Maciazek, J., Vayssiere, C., Cruaud, C., Cabanis, E.A., Ruchoux, M.M., Weissenbach, J., Bach, J.F., Bousser, M.G. and Tournier-Lasserve, E. (1996) Notch3 mutations in CADASIL, a hereditary adult-onset condition causing stroke and dementia. *Nature*, 383: 707–710.

Kamboh, M.I., Sanghera, D.K., Ferrell, R.E. and DeKosky, S.T. (1995) APOE*4-associated Alzheimer's disease risk is modified by alpha 1- antichymotrypsin polymorphism *Nat. Genet.*, 10: 486–488.

Kang, J., Lemaire, H., Unterbeck, A., Salbaum, J.M., Masters, C.L., Grzeschik, K., Multhaup, G., Beyreuther, K. and Müller-Hill, B. (1987) The precursor of Alzheimer's disease amyloid A4 protein resembles a cell-surface receptor. *Nature*, 325: 733–736.

Keller, J.N., Guo, Q., Holtsberg, F.W., Bruce-Keller, A.J. and Mattson, M.P. (1998) Increased sensitivity to mitochondrial toxin-induced apoptosis in neural cells expressing mutant presenilin-1 is linked to perturbed calcium homeostasis and enhanced oxyradical production. *J. Neurosci.*, 18: 4439–4450.

Kim, K.S., Wegiel, J., Sapienza, V., Chen, J., Hong, H., Wisniewski, H.M. (1997) Immunoreactivity of presenilin-1 in human, rat and mouse brain. *Brain Res.*, 757: 159–163.

Kim, S.-S., Choi, Y.-M. and Suh, Y.-H. (1997) Lack of interactions between amyloid precursor protein and hydrophilic domains of presenilin 1 and 2 using the yeast two hybrid system. *J. Mol. Neuroscience*, 9: 49–54.

Kim, T.W., Pettingell, W.H., Hallmark, O.G., Moir, R.D., Wasco, W. and Tanzi, R.E. (1997a) Endoproteolytic cleavage and proteasomal degradation of presenilin 2 in transfected cells. *J. Biol. Chem.*, 272: 11006–11010.

Kim, T.W., Pettingell, W.H., Jung, Y.K., Kovacs, D.M. and Tanzi, R.E. (1997b) Alternative Cleavage of Alzheimer-Associated Presenilins During Apoptosis by a Caspase-3 Family Protease. *Science*, 277: 373–376.

Kitaguchi, N., Takahashi, Y., Tokushima, Y., Shiojiri, S., Ito, H. (1988) Novel precursor of Alzheimer's disease amyloid protein shows protease inhibitory activity. *Nature*, 331: 530–532.

Klafki, H.W., Abramowski, D., Swoboda, R., Paganetti, P.A. and Staufenbiel, M. (1996) The carboxyl termini of β-amyloid peptides 1–40 and 1–42 are generated by distinct β-secretase activities. *J. Biol. Chem.*, 271: 28655–28659.

Konig, G., Monning, U., Czech, C., Prior, R., Banati, R., Schreiter-Gasser, U., Bauer, J., Masters, C.L. and Beyreuther, K. (1992) Identification and differential expression of a novel alternative splice isoform of the beta A4 amyloid precursor protein (APP) mRNA in leukocytes and brain microglial cells. *J. Biol. Chem.*, 267: 10804–10809.

Kosaka, T., Imagawa, M., Seki, K., Arai, H., Sasaki, H., Tsuji, S., Asami, O.A., Fukushima, T., Imai, K. and Iwatsubo, T. (1997) The beta APP717 Alzheimer mutation increases the percentage of plasma amyloid-beta protein ending at A beta42(43) *Neurology*, 48: 741–745.

Kovacs, D.M., Fausett, H.J., Page, K.J., Kim, T.W., Moir, R.D., Merriam, D.E., Hollister, R.D., Hallmark, O.G., Mancini, R., Felsenstein, K.M., Hyman, B.T., Tanzi, R.E. and Wasco, W. (1996) Alzheimer-associated presenilins 1 and 2: neuronal expression in brain and localization to intracellular membranes in mammalian cells. *Nat. Med.*, 2: 224–229.

Kuentzel, S.L., Ali, S.M., Altman, R.A., Greenberg, B.D. and Raub, T.J. (1993) The Alzheimer beta-amyloid protein precursor/protease nexin-II is cleaved by secretase in a trans-Golgi secretory compartment in human neuroglioma cells. *Biochem. J.*, 295: 367–378.

Kwok, J.B., Hallupp, M., Milward, L., Whyte, S. and Schofield, P.R. (1998) Novel familial early-onset Alzheimer's disease mutation (Leu723Pro) in Amyloid Precursor Protein (APP) gene increases production of 42(43) amino-acid isoform of amyloid-beta peptide. *Neurobiol. Aging*, 19: S91–S91(Abstract).

L'Hernault, S.W. and Arduengo, P.M. (1992) Mutation of a putative sperm membrane protein in Caenorhabditis elegans prevents sperm differentiation but not its associated meiotic divisions. *J. Cell Biol.*, 119: 55–68.

Lee, M.K., Borchelt, D.R., Kim, G., Thinakaran, G., Slunt, H.H., Ratovitski, T., Martin, L.J., Kittur, A., Gandy, S., Levey, A.I., Jenkins, N., Copeland, N., Price, D.L. and Sisodia, S.S. (1997) Hyperaccumulation of FAD-linked presenilin 1 variants in vivo. *Nat. Med.*, 3: 756–760.

Lee, M.K., Slunt, H.H., Martin, L.J., Thinakaran, G., Kim, G., Gandy, S.E., Seeger, M., Koo, E., Price, D.L. and Sisodia, S.S. (1996) Expression of presenilin 1 and 2 (PS1 and PS2) in human and murine tissues. *J. Neurosci.*, 16: 7513–7525.

Lee, S.J., Liyanage, U., Bickel, P.E., Xia, W., Lansbury, P.T.J. and Kosik, K.S. (1998) A detergent-insoluble membrane compartment contains A beta in vivo. *Nat. Med.*, 4: 730–734.

Lehmann, D.J., Johnston, C. and Smith, A.D. (1997) Synergy between the genes for butyrylcholinesterase K variant and apolipoprotein E4 in late-onset confirmed Alzheimer's disease. *Hum. Mol. Genet.*, 6: 1933–1936.

Lemere, C.A., Lopera, F., Kosik, K.S., Lendon, C.L., Ossa, J., Saido, T.C., Yamaguchi, H., Ruiz, A., Martinez, A., Madrigal, L., Hincapie, L., Arango, J.C., Anthony, D.C., Koo, E.H., Goate, A.M. and Selkoe, D.J. (1996) The E280A presenilin 1 Alzheimer mutation produces increased A beta 42 deposition and severe cerebellar pathology. *Nat. Med.*, 2: 1146–1150.

Levey, A.I., Heilman, C.J., Lah, J.J., Nash, N.R., Rees, H.D., Wakai, M., Mirra, S.S., Rye, D.B., Nochlin, D., Bird, T.D. and Mufson, E.J. (1997) Presenilin-1 protein expression in familial and sporadic Alzheimer's disease. *Ann. Neurol.*, 41: 742–753.

Levitan, D., Doyle, T.G., Brousseau, D., Lee, M.K., Thinakaran, G., Slunt, H.H., Sisodia, S.S. and Greenwald, I. (1996) Assessment of normal and mutant human presenilin function in *Caenorhabditis elegans*. *Proc. Natl. Acad. Sci. USA*, 93: 14940–14944.

Levitan, D. and Greenwald, I. (1995) Facilitation of lin-12-mediated signalling by sel-12, a Caenorhabditis elegans S182 Alzheimer's disease gene. *Nature*, 377: 351–354.

Levy-Lahad, E., Wasco, W., Poorkaj, P., Romano, D.M., Oshima, J., Pettingell, W.H., Yu, C.E., Jondro, P.D., Schmidt, S.D., Wang, K. et al. (1995a) Candidate gene for the chromosome 1 familial Alzheimer's disease locus. *Science*, 269: 973–977.

Levy-Lahad, E., Wijsman, E.M., Nemens, E., Anderson, L., Goddard, K.A., Weber, J.L., Bird, T.D. and Schellenberg, G.D. (1995b) A familial Alzheimer's disease locus on chromosome 1. *Science*, 269: 970–973.

Levy-Lahad, E., Poorkaj, P., Wang, K., Fu, Y.H., Oshima, J., Mulligan, J. and Schellenberg, G.D. (1996) Genomic structure and expression of STM2, the chromosome 1 familial Alzheimer disease gene. *Genomics*, 34: 198–204.

Li, J., Xu, M., Zhou, H., Ma, J. and Potter, H. (1997a) Alzheimer presenilins in the nuclear membrane, interphase kinetochores, and centrosomes suggest a role in chromosome segregation. *Cell*, 90: 917–927.

Li, T., Holmes, C., Sham, P.C., Vallada, H., Birkett, J., Kirov, G., Lesch, K.P., Powell, J., Lovestone, S. and Collier, D. (1997b) Allelic functional variation of serotonin transporter expression is a susceptibility factor for late onset Alzheimer's disease. *NeuroReport.*, 8: 683–686.

Li, X. and Greenwald, I. (1996) Membrane topology of the *C. elegans* SEL-12 presenilin. *Neuron*, 17: 1015–1021.

Li, X. and Greenwald, I. (1998) Additional evidence for an eight-transmembrane-domain topology for *Caenorhabditis elegans* and human presenilins. *Proc. Natl. Acad. Sci. USA*, 95: 7109–7114.

Loetscher, H., Deuschle, U., Brockhaus, M., Reinhardt, D., Nelboeck, P., Mous, J., Grunberg, J., Haass, C. and Jacobsen, H. (1997) Presenilins are processed by caspase-type proteases. *J. Biol. Chem.*, 272: 20655–20659.

Ma, J., Yee, A., Brewer, H.B.J., Das, S. and Potter, H. (1994) Amyloid-associated proteins alpha 1-antichymotrypsin and apolipoprotein E promote assembly of Alzheimer beta-protein into filaments. *Nature*, 372: 92–94.

Mahley, R.W. (1988) Apolipoprotein E: cholesterol transport protein with expanding role in cell biology. *Science*, 240: 622–630.

Mann, D.M. and Iwatsubo, T. (1996) Diffuse plaques in the cerebellum and corpus striatum in Down's syndrome contain amyloid beta protein (A beta) only in the form of A beta 42(43) *Neurodegeneration.*, 5: 115–120.

Mann, D.M., Iwatsubo, T., Cairns, N.J., Lantos, P.L., Nochlin, D., Sumi, S.M., Bird, T.D., Poorkaj, P., Hardy, J., Hutton, M., Prihar, G., Crook, R., Rossor, M.N. and Haltia, M. (1996) Amyloid beta protein (Abeta) deposition in chromosome 14-linked Alzheimer's disease: predominance of Abeta42(43) *Ann. Neurol.*, 40: 149–156.

Mann, D.M., Iwatsubo, T., Nochlin, D., Sumi, S.M., Levy, L.E. and Bird, T.D. (1997) Amyloid (Abeta) deposition in chromosome 1-linked Alzheimer's disease: the Volga German families. *Ann. Neurol.*, 41: 52–57.

Marambaud, P., Ancolio, K., Lopez, P.E. and Checler, F. (1998) Proteasome inhibitors prevent the degradation of familial Alzheimer's disease-linked presenilin 1 and potentiate A beta 42 recovery from human cells. Mol. Med., 4: 147–157.

Mehta, N.D., Refolo, L.M., Eckman, C., Sanders, S., Yager, D., Perez, T.J., Younkin, S., Duff, K., Hardy, J. and Hutton, M. (1998) Increased Abeta42(43) from cell lines expressing presenilin 1 mutations. *Ann. Neurol.*, 43: 256–258.

Mercken, M., Takahashi, H., Honda, T., Sato, K., Murayama, M., Nakazato, Y., Noguchi, K., Imahori, K. and Takashima, A. (1996) Characterization of human presenilin 1 using N-terminal specific monoclonal antibodies: Evidence that Alzheimer mutations affect proteolytic processing. *FEBS Lett.*, 389: 297–303.

Montoya, S.E., Aston, C.E., DeKosky, S.T., Kamboh, M.I., Lazo, J.S. and Ferrell, R.E. (1998) Bleomycin hydrolase is associated with risk of sporadic Alzheimer's disease (letter). *Nat. Genet.*, 18: 211–212.

Moussaoui, S., Czech, C., Pradier, L., Blanchard, V., Bonici, B., Gohin, M., Imperato, A. and Revah, F. (1996) Immunohistochemical analysis of presenilin-1 expression in the mouse brain. *FEBS Lett.*, 383: 219–222.

Mullan, M., Crawford, F., Axelman, K., Houlden, H., Lilius, L., Winblad, B. and Lannfelt, L. (1992) A pathogenic mutation for probable Alzheimer's disease in the APP gene at the N-terminus of beta-amyloid. *Nat. Genet.*, 1: 345–347.

Murayama, O., Honda, T., Mercken, M., Murayama, M., Yasutake, K., Nihonmatsu, N., Nakazato, Y., Michel, G., Song, S., Sato, K., Takahashi, H. and Takashima, A. (1997) Different effects of Alzheimer-associated mutations of presenilin 1 on its processing. *Neurosci. Lett.*, 229: 61–64.

Murphy, G.M.J., Forno, L.S., Ellis, W.G., Nochlin, D., Levy, L.E., Poorkaj, P., Bird, T.D., Jiang, Z. and Cordell, B. (1996) Antibodies to presenilin proteins detect neurofibrillary tangles in Alzheimer's disease. *Am. J. Pathol.*, 149: 1839–1846.

Murrell, J., Farlow, M., Ghetti, B. and Benson, M.D. (1991) A mutation in the amyloid precursor protein associated with hereditary Alzheimer's disease. *Science*, 254: 97–99.

Okochi, M., Sahara, N., Kametani, F., Usami, M., Arai, T., Tanaka, K., Ishii, K., Yamamoto, A. and Mori, H. (1998) Presenilin 1 cleavage is a universal event in human organs. *Neurobiol. Aging*, 19: S3–10.

Okuizumi, K., Onodera, O., Namba, Y., Ikeda, K., Yamamoto, T., Seki, K., Ueki, A., Nanko, S., Tanaka, H. and Takahashi, H. (1995) Genetic association of the very low-density lipoprotein (VLDL) receptor gene with sporadic Alzheimer's disease. *Nat. Genet.*, 11: 207–209.

Ono, Y., Kinouchi, T., Sorimachi, H., Ishiura, S. and Suzuki, K. (1997) Deletion of an endosomal/lysosomal targeting signal promotes the secretion of Alzheimer's disease amyloid precursor protein (APP). *J. Biochem. (Tokyo)*, 121: 585–590.

Ott, A., Breteler, M.M., van Harskamp, F., Claus, J.J., van-der Cammen, T., Grobbee, D.E. and Hofman, A. (1995) Prevalence of Alzheimer's disease and vascular dementia: association with education. The Rotterdam study. *BMJ.*, 310: 970–973.

Oyama, F., Sawamura, N., Kobayashi, K., Morishima-Kawashima, M., Kuramochi, T., Ito, M., Tomita, T., Maruyama, K., Saido, T.C., Iwatsubo, T., Capell, A., Walter, J., Grunberg, J., Ueyama, Y., Haass, C. and Ihara, Y. (1998) Mutant presenilin 2 transgenic mouse: effect on an age-dependent increase of amyloid beta-protein 42 in the brain. *J. Neurochem.*, 71: 313–322.

Payami, H., Schellenberg, G.D., Zareparsi, S., Kaye, J., Sexton, G.J., Head, M.A., Matsuyama, S.S., Jarvik, L.F., Miller, B., McManus, D.Q., Bird, T.D., Katzman, R., Heston, L., Norman, D. and Small, G.W. (1997) Evidence for association of HLA-A2 allele with onset age of Alzheimer's disease. *Neurology*, 49: 512–518.

Perez-Tur, J., Froelich, S., Prihar, G., Crook, R., Baker, M., Duff, K., Wragg, M., Busfield, F., Lendon, C., Clark, R.F. et al. (1995) A mutation in Alzheimer's disease destroying a splice acceptor site in the presenilin-1 gene. *NeuroReport.*, 7: 297–301.

Pericak-Vance, M.A., Bass, M.P., Yamaoka, L.H., Gaskell, P.C., Scott, W.K., Terwedow, H.A., Menold, M.M., Conneally, P.M., Small, G.W., Vance, J.M., Saunders, A.M., Roses, A.D., Haines, J.L. (1997) Complete genomic screen in late-onset familial Alzheimer disease. Evidence for a new locus on chromosome 12. *JAMA*, 278: 1237–1241.

Pericak-Vance, M.A., Bebout, J.L., Gaskell-PC, J., Yamaoka, L.H., Hung, W.Y., Alberts, M.J., Walker, A.P., Bartlett, R.J., Haynes, C.A., Welsh, K.A., et al. (1991) Linkage studies in familial Alzheimer disease: evidence for chromosome 19 linkage. *Am. J. Hum. Genet.*, 48: 1034–1050.

Pericak-Vance, M.A., Yamaoka, L.H., Haynes, C.S., Speer, M.C., Haines, J.L., Gaskell, P.C., Hung, W.Y., Clark, C.M., Heyman, A.L., Trofatter, J.A. et al. (1988) Genetic linkage studies in Alzheimer's disease families. *Exp. Neurol.*, 102: 271–279.

Podlisny, M.B., Citron, M., Amarante, P., Sherrington, R., Xia, W., Zhang, J., Diehl, T., Levesque, G., Fraser, P., Haass, C., Koo, E.H., Seubert, P., St. Teplow, D.B. and Selkoe, D.J. (1997) Presenilin proteins undergo heterogeneous endoproteolysis between Thr291 and Ala299 and occur as stable N- and C-terminal fragments in normal and Alzheimer brain tissue. *Neurobiol. Dis.*, 3: 325–337.

Ponte, P., Gonzalez-DeWhitt, P., Schilling, J., Miller, J., Hsu, D., Greenberg, B., Davis, K., Wallace, W., Lieberburg, I. and Fuller, F. (1988) A new A4 amyloid mRNA contains a domain homologous to serine proteinase inhibitors. *Nature*, 331: 525–527.

Prihar, G., Fuldner, R.A., Perez, T.J., Lincoln, S., Duff, K., Crook, R., Hardy, J., Philips, C.A., Venter, C., Talbot, C., Clark, R.F., Goate, A., Li, J., Potter, H., Karran, E., Roberts, G.W., Hutton, M. and Adams, M.D. (1996) Structure and alternative splicing of the presenilin-2 gene. *NeuroReport.*, 7: 1680–1684.

Qian, S., Jiang, P., Guan, X.M., Singh, G., Trumbauer, M.E., Yu, H., Chen, H.Y., Van-de, P.L. and Zheng, H. (1998) Mutant human presenilin 1 protects presenilin 1 null mouse against embryonic lethality and elevates Abeta1-42/43 expression. *Neuron*, 20: 611–617.

Racchi, M., Baetta, R., Salvietti, N., Ianna, P., Franceschini, G., Paoletti, R., Fumagalli, R., Govoni, S., Trabucchi, M. and Soma, M. (1997) Secretory processing of amyloid precursor protein is inhibited by increase in cellular cholesterol content. *Biochem. J.*, 322: 893–898.

Ratovitski, T., Slunt, H.H., Thinakaran, G., Price, D.L., Sisodia, S.S. and Borchelt, D.R. (1997) Endoproteolytic processing and stabilization of wild-type and mutant presenilin. *J. Biol. Chem.*, 272: 24536–24541.

Robakis, N.K., Wisniewski, H.M., Jenkins, E.C., Devine, G.E., Houck, G.E., Yao, X.L., Ramakrishna, N., Wolfe, G., Silverman, W.P. and Brown, W.T. (1987) Chromosome 21q21 sublocalisation of gene encoding beta-amyloid peptide in cerebral vessels and neuritic (senile) plaques of people with Alzheimer disease and Down syndrome. *Lancet*, 1: 384–385.

Rogaev, E.I., Sherrington, R., Rogaeva, E.A., Levesque, G., Ikeda, M., Liang, Y., Chi, H., Lin, C., Holman, K., Tsuda, T. et al. (1995) Familial Alzheimer's disease in kindreds with missense mutations in a gene on chromosome 1 related to the Alzheimer's disease type 3 gene. *Nature*, 376: 775–778.

Rogaev, E.I., Sherrington, R., Wu, C., Levesque, G., Liang, Y., Rogaeva, E.A., Ikeda, M., Holman, K., Lin, C., Lukiw, W.J. de J.P., Fraser, P.E., Rommens, J.M. and St George-Hyslop, P.H. (1997) Analysis of the 5′ sequence, genomic structure, and alternative splicing of the presenilin-1 gene (PSEN1) associated with early onset Alzheimer disease. *Genomics*, 40: 415–424.

Roperch, J.P., Alvaro, V., Prieur, S., Tuynder, M., Nemani, M., Lethrosne, F., Piouffre, L., Gendron, M.C., Israeli, D., Dausset, J., Oren, M., Amson, R. and Telerman, A. (1998) Inhibition of presenilin 1 expression is promoted by p53 and p21WAF-1 and results in apoptosis and tumor suppression. *Nat. Med.*, 4: 835–838.

Roses, A.D. and Saunders, A.M. (1994) APOE is a major susceptibility gene for Alzheimer's disease. *Curr. Opin. Biotechnol.*, 5: 663–667.

Saitoh, T., Xia, Y., Chen, X., Masliah, E., Galasko, D., Shults, C., Thal, L.J., Hansen, L.A. and Katzman, R. (1995) The CYP2D6B mutant allele is overrepresented in the Lewy body variant of Alzheimer's disease. *Ann. Neurol.*, 37: 110–112.

Sambamurti, K., Refolo, L.M., Shioi, J., Pappolla, M.A. and Robakis, N.K. (1992) The Alzheimer's amyloid precursor is cleaved intracellularly in the trans-Golgi network or in a post-Golgi compartment. *Ann. N. Y. Acad. Sci.*, 674: 118–128.

Sandbrink, R., Masters, C.L. and Beyreuther, K. (1994) Beta A4-amyloid protein precursor mRNA isoforms without exon 15 are ubiquitously expressed in rat tissues including brain, but not in neurons. *J. Biol. Chem.*, 269: 1510–1517.

Schellenberg, G.D., Bird, T.D., Wijsman, E.M., Orr, H.T., Anderson, L., Nemens, E., White, J.A., Bonnycastle, L., Weber, J.L., Alonso, M.E. et al. (1992) Genetic linkage evidence for a familial Alzheimer's disease locus on chromosome 14. *Science*, 258: 668–671.

Scheuner, D., Eckman, C., Jensen, M., Song, X., Citron, M., Suzuki, N., Bird, T.D., Hardy, J., Hutton, M., Kukull, W., Larson, E., Levy, L.E., Viitanen, M., Peskind, E., Poorkaj, P., Schellenberg, G., Tanzi, R., Wasco, W., Lannfelt, L., Selkoe, D. and Younkin, S. (1996) Secreted amyloid beta-protein similar to that in the senile plaques of Alzheimer's disease is increased in vivo by the presenilin 1 and 2 and APP mutations linked to familial Alzheimer's disease. *Nat. Med.*, 2: 864–870.

Seubert, P., Oltersdorf, T., Lee, M.G., Barbour, R., Blomquist, C., Davis, D.L., Bryant, K., Fritz, L.C., Galasko, D. and Thal, L.J. (1993) Secretion of beta-amyloid precursor protein cleaved at the amino terminus of the beta-amyloid peptide. *Nature*, 361: 260–263.

Seubert, P., Vigo-Pelfrey, C., Esch, F., Lee, M., Dovey, H., Davis, D., Sinha, S., Schlossmacher, M., Whaley, J. and Swindlehurst, C. (1992) Isolation and quantification of soluble Alzheimer's beta-peptide from biological fluids. *Nature*, 359: 325–327.

Shen, J., Bronson, R.T., Chen, D.F., Xia, W., Selkoe, D.J. and Tonegawa, S. (1997) Skeletal and CNS defects in Presenilin-1-deficient mice. Cell, 89: 629–639.

Sherrington, R., Rogaev, E.I., Liang, Y., Rogaeva, E.A., Levesque, G., Ikeda, M., Chi, H., Lin, C., Li, G., Holman, K. et al. (1995) Cloning of a gene bearing missense mutations in early-onset familial Alzheimer's disease. *Nature*, 375: 754–760.

Shirotani, K., Takahashi, K., Ozawa, K., Kunishita, T. and Tabira, T. (1997) Determination of a cleavage site of presenilin 2 protein in stably transfected SH-SY5Y human neuroblastoma cell lines. *Biochem. Biophys. Res. Commun.*, 240: 728–731.

Shoji, M., Golde, T.E., Ghiso, J., Cheung, T.T., Estus, S., Shaffer, L.M., Cai, X.D., McKay, D.M., Tintner, R., Frangione, B. et al. (1992) Production of the Alzheimer amyloid beta protein by normal proteolytic processing. *Science*, 258: 126–129.

Simons, M., De Strooper, B., Multhaup, G., Tienari, P.J., Dotti, C.G. and Beyreuther, K. (1996) Amyloidogenic processing of the human amyloid precursor protein in primary cultures of rat hippocampal neurons. *J. Neurosci.*, 16: 899–908.

Simons, M., Keller, P., De Strooper, B., Beyreuther, K., Dotti, C.G. and Simons, K. (1998) Cholesterol depletion inhibits the generation of beta-amyloid in hippocampal neurons. *Proc. Natl. Acad. Sci. USA*, 95: 6460–6464.

Sisodia, S.S. (1992) Beta-amyloid precursor protein cleavage by a membrane-bound protease. *Proc. Natl. Acad. Sci. USA*, 89: 6075–6079.

Skovronsky, D.M., Doms, R.W. and Lee, V.M. (1998) Detection of a novel intraneuronal pool of insoluble amyloid beta protein that accumulates with time in culture. *J. Cell Biol.*, 141: 1031–1039.

Sorbi, S., Nacmias, B., Forleo, P., Piacentini, S., Latorraca, S. and Amaducci, L. (1995) Epistatic effect of APP717 mutation and apolipoprotein E genotype in familial Alzheimer's disease. *Ann. Neurol.*, 38: 124–127.

St-George Hyslop, P., Tanzi, R.E., Polinsky, R.J., Haines, J.L., Nee, L., Watkins, P.C., Myers, R.H., Feldman, R.G., Pollen, D., Drachman, D. et al. (1987) The genetic defect causing familial Alzheimer's disease maps on chromosome 21. *Science*, 235: 885–890.

Strittmatter, W.J., Saunders, A.M., Schmechel, D., Pericak, V.M., Enghild, J., Salvesen, G.S. and Roses, A.D. (1993a) Apolipoprotein E: high-avidity binding to beta-amyloid and increased frequency of type 4 allele in late-onset familial Alzheimer disease. *Proc. Natl. Acad. Sci. USA*, 90: 1977–1981.

Strittmatter, W.J., Weisgraber, K.H., Huang, D.Y., Dong, L.M., Salvesen, G.S., Pericak, V.M., Schmechel, D., Saunders, A.M., Goldgaber, D. and Roses, A.D. (1993b) Binding of human apolipoprotein E to synthetic amyloid beta peptide: isoform-specific effects and implications for late-onset Alzheimer disease. *Proc. Natl. Acad. Sci. USA*, 90: 8098–8102.

Suzuki, N., Cheung, T.T., Cai, X.D., Odaka, A., Otvos, L., Eckman, C., Golde, T.E. and Younkin, S.G. (1994) An increased percentage of long amyloid beta protein secreted by familial amyloid beta protein precursor (beta APP717) mutants. *Science*, 264: 1336–1340.

Takashima, A., Murayama, M., Murayama, O., Kohno, T., Honda, T., Yasutake, K., Nihonmatsu, N., Mercken, M., Yamaguchi, H., Sugihara, S. and Wolozin, B. (1998) Presenilin 1 associates with glycogen synthase kinase-3beta and its substrate tau. *Proc. Natl. Acad. Sci. USA*, 95: 9637–9641.

Takashima, A., Sato, M., Mercken, M., Tanaka, S., Kondo, S., Honda, T., Sato, K., Murayama, M., Noguchi, K., Nakazato, Y. and Takahashi, H. (1996) Localization of Alzheimer-associated presenilin 1 in transfected COS-7 cells. *Biochem. Biophys. Res. Commun.*, 227: 423–426.

Tamaoka, A., Odaka, A., Ishibashi, Y., Usami, M., Sahara, N., Suzuki, N., Nukina, N., Mizusawa, H., Shoji, S., Kanazawa, I. et al. (1994) APP717 missense mutation affects the ratio of amyloid beta protein species (A beta 1-42/43 and a beta 1-40) in familial Alzheimer's disease brain. *J. Biol. Chem.*, 269: 32721–32724.

Tan, Y., Hong, J., Doan, T., McConlogue, L. and Maltese, W.A. (1998) Presenilin-1 mutations associated with familial Alzheimer's disease do not disrupt protein transport from the endoplasmic reticulum to the Golgi apparatus. *Biochim. Biophys. Acta*, 1407: 69–78.

Tanzi, R.E., Gusella, J.F., Watkins, P.C., Bruns, G.A., St-George Hyslop, P.H., Van, K.M., Patterson, D., Pagan, S., Kurnit, D.M. and Neve, R.L. (1987) Amyloid beta protein gene: cDNA, mRNA distribution, and genetic linkage near the Alzheimer locus. *Science*, 235: 880–884.

Tanzi, R.E., McClatchey, A.I., Lamperti, E.D., Villa-Komaroff, L., Gusella, J.F. and Neve, R.L. (1988) Protease inhibitor domain encoded by an amyloid protein precursor mRNA associated with Alzheimer's disease. *Nature*, 331: 528–530.

Thinakaran, G., Borchelt, D.R., Lee, M.K., Slunt, H.H., Spitzer, L., Kim, G., Ratovitsky, T., Davenport, F., Nordstedt, C., Seeger, M., Hardy, J., Levey, A.I., Gandy, S.E., Jenkins, N.A., Copeland, N.G., Price, D.L. and Sisodia, S.S. (1996) Endoproteolysis of presenilin 1 and accumulation of processed derivatives in vivo. *Neuron*, 17: 181–190.

Thinakaran, G., Harris, C.L., Ratovitski, T., Davenport, F., Slunt, H.H., Price, D., Borchelt, D.R. and Sisodia, S.S. (1997) Evidence that levels of presenilins (PS1 and PS2) are coordinately regulated by competition for limiting cellular factors. *J. Biol. Chem.*, 272: 28415–28422.

Thinakaran, G., Regard, J.B., Bouton, C.M., Harris, C.L., Price, D.L., Borchelt, D.R. and Sisodia, S.S. (1998) Stable association of presenilin derivatives and absence of presenilin interactions with APP. *Neurobiol. Dis.*, 4: 438–453.

Tienari, P.J., Ida, N., Ikonen, E., Simons, M., Weidemann, A., Multhaup, G., Masters, C.L., Dotti, C.G. and Beyreuther, K. (1997) Intracellular and secreted Alzheimer beta-amyloid species are generated by distinct mechanisms in cultured hippocampal neurons. *Proc. Natl. Acad. Sci. USA*, 94: 4125–4130.

Tomita, S., Kirino, Y. and Suzuki T (1998a) Cleavage of Alzheimer's amyloid precursor protein (APP) by secretases occurs after O-glycosylation of APP in the protein secretory pathway. Identification of intracellular compartments in which APP cleavage occurs without using toxic agents that interfere with protein metabolism. *J. Biol. Chem.*, 273: 6277–6284.

Tomita, T., Tokuhiro, S., Hashimoto, T., Aiba, K., Saido, T.C., Maruyama, K. and Iwatsubo, T. (1998b) Molecular dissection of domains in mutant presenilin 2 that mediate overproduction of amyloidogenic forms of amyloid beta peptides. Inability Of truncated forms of ps2 with familial alzheimer's disease mutation to increase secretion of abeta42. *J. Biol. Chem.*, 273: 21153–21160.

Turner, R.S., Suzuki, N., Chyung, A.S., Younkin, S.G. and Lee, V.M. (1996) Amyloids beta40 and beta42 are generated intracellularly in cultured human neurons and their secretion increases with maturation. *J. Biol. Chem.*, 271: 8966–8970.

Tycko, B., Feng, L., Nguyen, L., Francis, A., Hays, A., Chung, W.Y., Tang, M.X., Stern, Y., Sahota, A., Hendrie, H. and Mayeux, R. (1996) Polymorphisms in the human apolipoprotein-J/clusterin gene: ethnic variation and distribution in Alzheimer's disease. *Hum. Genet.*, 98: 430–436.

Tysoe, C., Whittaker, J., Xuereb, J., Cairns, N., Cruts, M., Van Broeckhoven, C., Wilcock, G. and Rubinsztein, D.C. (1998) A presenilin-1 truncating mutation is present in two cases with autopsy-confirmed early-onset Alzheimer's disease. *Am J Hum. Genetics*, 62: 70–76.

Uchihara, T., el Hachimi, H.K., Duyckaerts, C., Foncin, J.F., Fraser, P.E., Levesque, L. and Hauw, J.J. (1996) Widespread immunoreactivity of presenilin in neurons of normal and Alzheimer's disease brains: double-labeling immunohistochemical study. *Acta Neuropathol. Berl.*, 92: 325–330.

Van Broeckhoven, C. (1995) Presenilins and Alzheimer disease. *Nat. Genet.*, 11: 230–232.

Van Broeckhoven, C., Backhovens, H., Cruts, M., De Winter, G., Bruyland, M., Cras, P. and Martin, J.J. (1992) Mapping of a gene predisposing to early-onset Alzheimer's disease to chromosome 14q24. 3. *Nat. Genet.*, 2: 335–339.

Van Broeckhoven, C.L. (1995) Molecular genetics of Alzheimer disease: identification of genes and gene mutations. *Eur. Neurol.*, 35: 8–19.

van Duijn, C.M., Clayton, D.G., Chandra, V., Fratiglioni, L., Graves, A.B., Heyman, A., Jorm, A.F., Kokmen, E., Kondo, K., Mortimer, J.A. et al. (1994) Interaction between genetic and environmental risk factors for Alzheimer's disease: a reanalysis of case-control studies. EURODEM Risk Factors Research Group. *Genet. Epidemiol.*, 11: 539–551.

van Duijn, C.M., Cruts, M., Theuns, J., Van Gassen, G., Backhovens, H., van den Broeck, M., Wehnert, A., Serneels, S., Hofman, A., Van Broeckhoven, C. Genetic association of the presenilin-I regulatory region with early-onset Alzheimer's disease in a population-based sample. *Eur. J. Hum. Genet.*, in press.

van Duijn, C., de Knijff, P., Wehnert, A., De Voecht, J., Bronzova, J.B., Havekes, L.M., Hofman, A. and Van Broeckhoven, C. (1995) The apolipoprotein E epsilon 2 allele is associated with an increased risk of early-onset Alzheimer's disease and a reduced survival. *Ann. Neurol.*, 37: 605–610.

van Duijn, C., de Knijff P., Cruts, M., Wehnert, A., Havekes, L.M., Hofman, A. and Van Broeckhoven, C. (1994) Apolipoprotein E4 allele in a population-based study of early-onset Alzheimer's disease. *Nat. Genet.*, 7: 74–78.

van Duinen, S.G., Castano, E.M., Prelli, F., Bots, G.T., Luyendijk, W. and Frangione, B. (1987) Hereditary cerebral hemorrhage with amyloidosis in patients of Dutch origin is related to Alzheimer disease. *Proc. Natl. Acad. Sci. USA*, 84: 5991–5994.

Vito, P., Ghayur, T. and D'Adamio, L. (1997) Generation of anti-apoptotic presenilin-2 polypeptides by alternative transcription, proteolysis, and caspase-3 cleavage. *J. Biol. Chem.*, 272: 28315–28320.

Vito, P., Wolozin, B., Ganjei, J.K., Iwasaki, K., Lacana, E. and D'Adamio, L. (1996) Requirement of the familial Alzheimer's disease gene PS2 for apoptosis. Opposing effect of ALG-3. *J. Biol. Chem.*, 271: 31025–31028.

Walter, J., Capell, A., Grunberg, J., Pesold, B., Schindzielorz, A., Prior, R., Podlisny, M.B., Fraser, P., St George-Hyslop, P.H., Selkoe, D.J. and Haass, C. (1996) The Alzheimer's disease-associated presenilins are differentially phosphorylated proteins located predominantly within the endoplasmic reticulum. *Mol. Med.*, 2: 673–691.

Waragai, M., Imafuku, I., Takeuchi, S., Kanazawa, I., Oyama, F., Udagawa, Y., Kawabata, M. and Okazawa, H. (1997) Presenilin 1 binds to amyloid precursor protein directly. *Biochem. Biophys. Res. Commun.*, 239: 480–482.

Ward, R.V., Davis, J.B., Gray, C.W., Barton, A.J., Bresciani, L.G., Caivano, M., Murphy, V.F., Duff, K., Hutton, M., Hardy, J., Roberts, G.W. and Karran, E.H. (1996) Presenilin-1 is processed into two major cleavage products in neuronal cell lines. *Neurodegeneration.*, 5: 293–298.

Weber, L.L., Leissring, M.A., Yang, A.J., Glabe, C.G., Cribbs, D.H. and LaFerla, F.M. (1997) Presenilin-1 immunoreactivity is localized intracellularly in Alzheimer's disease brain, but not detected in amyloid plaques. *Exp. Neurol.*, 143: 37–44.

Weidemann, A., Paliga, K., Durrwang, U., Czech, C., Evin, G., Masters, C.L. and Beyreuther, K. (1997) Formation of stable complexes between two Alzheimer's disease gene products: presenilin-2 and beta-amyloid precursor protein. *Nat. Med.*, 3: 328–332.

Wertkin, A.M., Turner, R.S., Pleasure, S.J., Golde, T.E., Younkin, S.G., Trojanowski, J.Q. and Lee, V.M. (1993) Human neurons derived from a teratocarcinoma cell line express solely the 695-amino acid amyloid precursor protein and produce intracellular beta-amyloid or A4 peptides. *Proc. Natl. Acad. Sci. USA*, 90: 9513–9517.

Wild-Bode, C., Yamazaki, T., Capell, A., Leimer, U., Steiner, H., Ihara, Y. and Haass, C. (1997) Intracellular generation and accumulation of amyloid beta-peptide terminating at amino acid 42. *J. Biol. Chem.*, 272: 16085–16088.

Wisniewski, T., Dowjat, W.K., Permanne, B., Palha, J., Kumar, A., Gallo, G. and Frangione, B. (1997) Presenilin-1 is associated with Alzheimer's disease amyloid. *Am. J. Pathol.*, 151: 601–610.

Wisniewski, T., Palha, J.A., Ghiso, J. and Frangione, B. (1995) S182 protein in Alzheimer's disease neuritic plaques. *Lancet*, 346: 1366.

Wolozin, B., Iwasaki, K., Vito, P., Ganjei, J.K., Lacana, E., Sunderland, T., Zhao, B., Kusiak, J.W., Wasco, W. and D'Adamio, L. (1996) Participation of presenilin 2 in apoptosis: enhanced basal activity conferred by an Alzheimer mutation. *Science*, 274: 1710–1713.

Wong, P.C., Zheng, H., Chen, H., Becher, M.W., Sirinathsinghji, D.J., Trumbauer, M.E., Chen, H.Y., Price, D.L., Van-der, P.L. and Sisodia, S.S. (1997) Presenilin 1 is required for Notch1 and DII1 expression in the paraxial mesoderm. *Nature*, 387: 288–292.

Wragg, M., Hutton, M. and Talbot, C. (1996) Genetic association between intronic polymorphism in presenilin-1 gene and late-onset Alzheimer's disease. Alzheimer's Disease Collaborative Group. *Lancet*, 347: 509–512.

Xia, M.Q., Berezovska, O., Kim, T.W., Xia, W.M., Liao, A., Tanzi, R.E., Selkoe, D. and Hyman, B.T. (1998) Lack of specific association of presenilin 1 (PS-1) protein with plaques and tangles in Alzheimer's disease. *J. Neurol. Sci.*, 158: 15–23.

Xia, W., Zhang, J., Kholodenko, D., Citron, M., Podlisny, M.B., Teplow, D.B., Haass, C., Seubert, P., Koo, E.H. and Selkoe, D.J. (1997a) Enhanced production and oligomerization of the 42-residue amyloid beta-protein by Chinese hamster ovary cells stably expressing mutant presenilins. *J. Biol. Chem.*, 272: 7977–7982.

Xia, W., Zhang, J., Perez, R., Koo, E.H. and Selkoe, D.J. (1997b) Interaction between amyloid precursor protein and presenilins in mammalian cells: implications for the pathogenesis of Alzheimer disease. *Proc. Natl. Acad. Sci. USA*, 94: 8208–8213.

Xia, Y., Rohan, dS.H., Rosi, B.L., Yamaoka, L.H., Rimmler, J.B., Pericak-Vance, M.A., Roses, A.D., Chen, X., Masliah, E., DeTeresa, R., Iwai, A., Sundsmo, M., Thomas, R.G., Hofstetter, C.R., Gregory, E., Hansen, L.A., Katzman, R., Thal, L.J. and Saitoh, T. (1996) Genetic studies in Alzheimer's disease with an NACP/alpha-synuclein polymorphism. *Ann. Neurol.*, 40: 207–215.

Xu, H., Sweeney, D., Wang, R., Thinakaran, G., Lo, A.C., Sisodia, S.S., Greengard, P. and Gandy, S. (1997) Generation of Alzheimer beta-amyloid protein in the trans-Golgi network in the apparent absence of vesicle formation. *Proc. Natl. Acad. Sci. USA*, 94: 3748–3752.

Yu, G., Chen, F., Levesque, G., Nishimura, M., Zhang, D.M., Levesque, L., Rogaeva, E., Xu, D., Liang, Y., Duthie, M., St George-Hyslop, P.H. and Fraser, P.E. (1998) The presenilin 1 protein is a component of a high molecular weight intracellular complex that contains beta-catenin. *J. Biol. Chem.*, 273: 16470–16475.

Zhang, J., Kang, D.E., Xia, W., Okochi, M., Mori, H., Selkoe, D.J. and Koo, E.H. (1998) Subcellular distribution and turnover of presenilins in transfected cells. *J. Biol. Chem.*, 273: 12436–12442.

Zhou, Z., Smith, J.D., Greengard, P. and Gandy, S. (1996) Alzheimer amyloid-beta peptide forms denaturant-resistant complex with type epsilon 3 but not type epsilon 4 isoform of native apolipoprotein E. *Mol. Med.*, 2: 175–180.

W.E. Crusio and R.T. Gerlai (Eds.)
Handbook of Molecular-Genetic Techniques for Brain and
Behavior Research (Techniques in the Behavioral and Neural Sciences, Vol. 13)

CHAPTER 5.7

Aging, Alzheimer's disease and frameshift mutations

Wilhelmina H. Van den Hurk,[1] Fred W. Van Leeuwen[2] and Gerard J.M. Martens[1]

[1]Department of Molecular Animal Physiology, University of Nijmegen, Toernooiveld 1, 6525 ED Nijmegen, The Netherlands
[2]Graduate School for Neurosciences Amsterdam, Netherlands Institute for Brain Research, Meibergdreef 33, 1105 AZ Amsterdam, The Netherlands

Introduction

To protect their cells, humans have developed cellular mechanisms that provide defensive capacity, e.g. against environmental stress. As these mechanisms and their backups gradually give up, the cells lose this capacity, leading to aging and eventually death. An example of an age-related disease is Alzheimer's disease (AD), a neurodegenerative disorder of the brain and the most frequent form of dementia at higher age. Two different types of AD can be discriminated: the early-onset (familial) and late-onset (sporadic) forms. The sporadic form accounts for more than 60% of all AD cases, but the cause of the development of these cases is unknown. A recent study provided a new view on the molecular mechanisms behind the process of aging, in particular the development of the sporadic forms of AD (Van Leeuwen et al., 1998a). The new view involves a novel type of dinucleotide deletion in the transcripts of two neuronal genes, leading to frameshift mutations and proteins with an aberrant carboxy-terminus. The same type of frameshift mutation may underlie other neuropathologies and age-related diseases. This opens new avenues for experimental analysis at the molecular and cell physiological level. In this chapter, aging models will be related to AD and the new findings will be put in a broader perspective.

The process of aging

Aging has been defined as 'the accumulation of changes responsible for the sequential alterations that accompany advancing age and the associated progressive increases in the chance of disease and death' (Harman, 1991). Major causal factors of aging have been postulated to be programmed cell death (apoptosis), telomere shortening, activation of senescence genes and the amount of oxidative damage (accumulated stress). Examples of physiological systems that change during aging are the hormonal system and the stress response to metabolic insults. A decline in the activities of a number of hormonal systems occurs during normal aging: (a) a decrease in growth hormone release by the pituitary gland causes a decrease in the production of insulin-like growth factor I by the liver and other organs (somatopause); (b) a decrease in the release of gonadotropin luteinizing hormone and follicle-stimulating hormone, together with a decreased secretion at the gonadal level (from the ovaries decreased estrogen; from the testicle decreased testosterone), causes menopause and andropause, respectively;

(c) a decrease in the production of dihydro-epiandosterone by the adrenocortical cells (adrenopause). The aging of endocrine systems may be regulated by changes in peripheral organs (testicles, ovaries, and adrenal cortex), together with a central pacemaker, hypothesized to be present in the hypothalamus or in higher brain areas (or both; review Lamberts et al., 1997). During aging, cells lose their ability to activate the stress response (Liu et al., 1989; Fargnoli, 1990). The cellular stress response helps to protect organisms against a variety of metabolic insults and involves the induction of heat-shock protein (Hsp) expression. These proteins act as molecular chaperones, i.e. they transiently interact with proteins that are being synthesized, folded and translocated into subcellular organelles, as well as with mature proteins that tend to unfold because of environmental insults such as heat shock. Recently, the hypothesis was put forward that molecular chaperones might also play a role in converting soluble proteins into aggregates (Welch and Gambetti, 1998). Schirmer and Lindquist (1997) showed highly specific interactions between the chaperone Hsp104 and the yeast prion protein Sup35, a mammalian prion protein (PrP) and amyloid-β (Aβ). These interactions stabilize a shift to β-sheet rich, multimeric conformations leading to aggregates of proteins. Age-associated changes can also be the result of changes at the level of DNA. Somatically acquired mutations and oxidative damage of the genome leads to fragmentation of mitochondrial DNA, cellular energy crisis, naturally occurring cell death (apoptosis), tissue degeneration and atrophy. In mitochondrial DNA, several age-related genetic and functional changes have been described (review Ozawa, 1997). Furthermore, investigators have tried to identify age-related genes that may speed up or slow down the rate of aging. In humans, one such gene (WRN) has recently been cloned and codes for a DNA helicase-like protein (Yu et al., 1996). The WRN gene causes Werner's syndrome, a disease that has several characteristics of premature aging (Salk et al., 1985).

Models of aging

The use of a number of model systems, such as *Drosophila*, *Saccharomyces cerevisiae* and *Caenorhabditis elegans*, has enabled studies on the genetics of aging (review Finch and Tanzi, 1997). In *Drosophila*, strains bearing extra copies of genes encoding either superoxide dismutase or catalase display increased life spans (Orr and Sohal, 1994). These enzymes detoxify superoxide anions by converting them into water. Oxidative damage may thus normally limit life span. The budding yeast *S. cerevisiae* has several starvation-resistant strains that turn out to have a prolonged life span due to mutations in genes encoding silent information regulator (Sir) proteins (Kennedy et al., 1995). Recently, several Sir proteins, which are distributed in the nucleolus, have been investigated thoroughly in mother cells at various ages. In nuclei of young cells, the Sir complex is at the perimeter of the nucleus and is bound to bundles of telomeres. After cells have completed one-half to two-thirds of their life spans, the Sir complex has redistributed to the nucleolus. In very old cells, the nucleolus enlarges and fragments into several nucleolar bodies around the perimeter of the nucleolus. The Sir complex is found in all of these bodies. This shows a relationship between the distribution and fragmentation of the nucleolus, and the limited number of cell divisions (Gotta et al., 1997; Kennedy et al., 1997; review Guarente, 1997). Another yeast gene involved in aging is SGS1, which encodes a DNA helicase with homology to the human Werner's syndrome WRN gene. A mutation in SGS1 causes premature aging in yeast mother cells (Sinclair et al., 1997). Since mutations in genes encoding similar proteins (DNA helicases) lead to premature aging of both yeast and humans, a conserved mechanism of cellular aging can be proposed. Since aging in yeast is linked to fragmentation of the nucleolus, in humans a relationship between aging and the nucleolar structure might exist as well. *C. elegans* is of special interest, because all cells of the adult worm are post-mitotic, thus providing an excellent model for aging in mammalian organs consisting of nondividing cells, and thus for neurodegenerative diseases. In these worms, several gene loci have been identified that, when mutated, can increase life span (review Kenyon, 1996; Hekimi et al., 1998). For instance, mutations in *daf-2* or *daf-23/age-1* (Friedman and Johnson, 1988; Kenyon et al., 1993; Malone et al., 1996; Morris et al., 1996) can turn on the 'Dauer' pathway that, when activated in adults, can cause *C. elegans* to live twice as long as wild type worms. *Daf-2* and *age-1* encode a receptor for an insulin-like ligand (Kimura et al. 1997) and a phosphatidylinositol-3-OH (PI3) kinase (Morris et al., 1996), respectively. The 'Dauer' pathway may thus be determined by an intercellular signalling pathway that slows metabolism in 'Dauer' animals. Another class of *C. elegans* mutants are the clock-mutants (*clk*) that have an extended life span (Wong et al., 1995; Lakowski and Hekimi, 1996). The *Clk-1* gene (Ewbank et al., 1997) is similar to the yeast gene CAT5 (Proft et al., 1995) that activates oxidative metabolism, resulting in less oxidative damage. This again shows a relationship between a decrease in oxidative damage and a longer life span. Also, a restriction in caloric intake lowers steady-state levels of oxidative stress and damage, retards the age-associated changes, and extends the maximum life span of mammals (review Sohal and Weindruch, 1996).

Most models of aging fall into one of two broad categories, which are not mutually exclusive. The first category involves the 'failure to repair damage' resulting from oxygen radicals or other agents. The second category concerns the 'genetic program', which includes apoptosis. Animal studies have led to four possible models of aging at the genomic level: telomere shortening (Bodnar et al., 1998; De Lange, 1998), loss of silencing (Sir proteins), accumulated DNA damage (due to oxidation) and defect in ribosomal DNA transcription (review Guarente, 1996). Which molecular events cause aging remains, however, still unclear.

Alzheimer's disease

An example of an age-related neurodegenerative disorder of the brain is AD. This disease was first described by Alois Alzheimer (1907). The prevalence of AD increases from 0.5% at age 65, and 10–20% at age 85 to 40% or more at higher ages (Evans et al., 1989; Breteler et al., 1992). Patients with AD can be characterized by the presence of neurofibrillary tangles, neuropil threads and neuritic plaques in the brain. An important constituent of these plaques is Aβ a peptide of 40–42 amino acids that is proteolytically derived from the Amyloid Precursor Protein (APP) (Glenner and Wong, 1984a; Selkoe, 1994). APP is a type I transmembrane protein encoded by a gene localized on chromosome 21, and exists in three major forms of 770, 751 and 695 amino acids, resulting from alternative splicing (Kang et al., 1987; Tanzi et al., 1988; Kitaguchi et al., 1988). Several possible functions for APP and its derivatives have been proposed, but their exact role(s) remain(s) elusive (review Mattson, 1997). APP is a major protein in neurons and is transported in a strictly polarized way, first to the axons and then by transcytosis to the dendrites (Simons et al., 1995; Yamazaki et al., 1995). Under non-pathological conditions APP is transported through the secretory pathway from the *trans*-Golgi network to the plasma membrane, where about 30% of the APP is cleaved proteolytically (Weidemann et al., 1989). Cleavage of APP by α-secretase results in the production of a form of APP that can be secreted ($APP_{\alpha s}$) and effectively prevents formation of Aβ (Palmert et al., 1989; Weidemann et al., 1989; Esch et al., 1990; Kametani et al., 1990; Wang et al., 1991). The remaining part of APP is internalized from the plasma membrane towards the endosomes, a process in which clathrin-coated pits are involved. From the endosomes, APP is returned to the plasma membrane or redirected to the lysosomes. Under pathological conditions APP routing is thought to be aberrant (Esch et al., 1990; Sisodia et al., 1990; Haass et al. 1992; Knops et al., 1992;

Shoji et al., 1992; Kuentzel et al., 1993; Haass et al., 1995; Munger et al., 1995). A link between a deviated APP routing/processing and Aβ formation has been proposed (Selkoe, 1994). β-Secretase cleavage of APP results in the release of secreted forms of APP (APP$_{\beta s}$) and leaves intact Aβ behind in a membrane-associated 100 amino acids carboxy-terminal fragment of APP. This fragment can be further processed by γ-secretase, presumably in an endosomal-lysosomal compartment, to release the toxic Aβ (Estus et al., 1992; Selkoe, 1993; Higaki et al., 1995; Knops et al. 1995). However, the secretases that cleave APP are not well characterized and have not been isolated (review Selkoe, 1993). The neuropathological hallmarks of AD (tangles, neuropil threads and plaques) can also be found in brains of patients suffering from Down syndrome (DS, trisomy 21), further implicating the involvement of APP in AD (Glenner and Wong, 1984b; Wisniewski et al., 1985).

Genetic and risk factors for AD

Two different types of AD can be discriminated: the early-onset (familial) and the late-onset (sporadic) forms. The familial types of AD have been investigated thoroughly and three genes account for all known cases of early-onset AD (see also Chapter 5.6). A small percentage of early-onset AD is due to mutations in the APP gene itself (Levy et al., 1990; Goate et al., 1991; Chartier-Harlin et al., 1991; Murrell et al., 1991; Hendriks et al., 1992; Mullan et al., 1992). The majority of the early-onset form of AD is, however, caused by mutations in the genes encoding presenilin 1 (PS1) and PS2, located on chromosome 14 and 1, respectively. The two presenilins are transmembrane proteins sharing 67% amino acid sequence identity (Levy-Lahad et al., 1995; Sherrington et al., 1995). The occurrence of a mutation in any of these three genes (APP, PS1, PS2) might cause an acceleration of the formation and accumulation of Aβ which is generally accepted to be involved in AD. Transgenic mice expressing mutant APP, PS1 or PS2 have been created to study the formation of plaques and tangles (review Hardy, 1997). Thus far, these mouse models for AD only show the neuritic plaques, and never the neurofibrillary tangles, suggesting that mutations in one of the three genes can not fully explain the characteristics of AD in human. Moreover, at least 60% of AD patients do not have a familial history of the disease. A protein involved in both types of AD is apolipoprotein E (ApoE), the gene of which is located on chromosome 19. ApoE is a risk factor in that individuals carrying allele ApoE4 have a higher risk to develop AD and the onset can be advanced (Strittmatter et al., 1993). The mechanisms explaining aging and neurodegeneration may, however, be more general than suggested by models based on the genetic and risk factors.

Relationship between AD and the process of aging

As AD is age-related, researchers have attempted to unravel the relationship between processes of normal aging and the induction of neurodegenerative disorders. In this section, the above-mentioned aging factors, such as the involvement of hormones, heat shock/stress response, reactive oxygen species (ROS) and apoptosis, will be discussed in relation to AD. For instance, hormones have been implicated to be also involved in AD: estrogen has a protective effect with respect to the onset of this disorder (Tang et al., 1996). Furthermore, multiple defects in energy metabolism have been identified in AD patients, e.g. in the temporoparietal region of the affected brain, a reduced glucose metabolism has been demonstrated (review Bowling and Beal, 1995) and skin fibroblasts from AD patients exhibit decreased glucose utilization (Peterson and Goldman, 1986).

The influence of a number of environmental factors (stress) and second messengers on the regulation of APP production has been investigated. APP production and processing was found to be

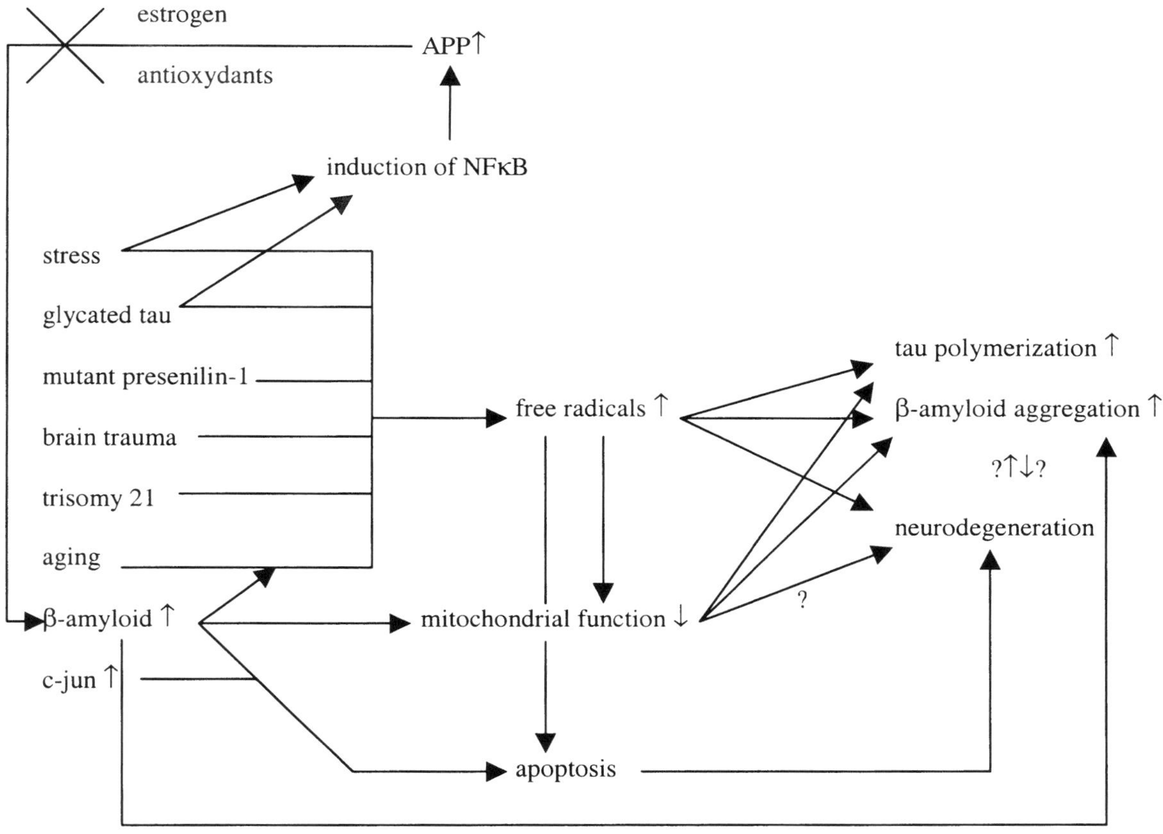

Fig. 1. Factors affecting neurodegenerative processes. APP = amyloid precursor protein; × = prevents; ↑ = increase; ↓ = decrease.

highly responsive to injury of the nervous system (review Mattson, 1997). In cultured human endothelial cells, a heat shock induced an increase in the level of APP-mRNA and the release of APP_s, either $APP_{\alpha s}$ or $APP_{\beta s}$ (Ciallella et al., 1995). In cells expressing recombinant APP proteins tagged with alkaline phosphatase, a heat shock caused altered compartmentalization of APP with abnormal accumulations in an unidentified subcellular compartment (Pappolla et al., 1995). The stress-induced expression of APP may be mediated at least in part by the stress-responsive transcription factor NFκB. Glycated tau, which generates oxyradicals, induces activation of NFκB which, in turn, increases APP levels and the release of Aβ from cultured neuroblastoma cells (Yan et al., 1995). The APP gene promoter contains two NFκB binding sequences (Salbaum et al., 1988; Grilli et al., 1995) and other stress-responsive elements, indeed suggesting that the APP gene can be induced by stress (Vostrov et al., 1995). During aging more free radicals may be produced, resulting in the induction of APP expression. A stressor may thus trigger higher expression of APP, leading to altered compartmentalization of APP, possibly followed by Aβ formation (Fig. 1).

The formation of free radicals and the resulting toxicity might be responsible for brain degeneration in AD. This hypothesis is based on the observation that some of the processes which lead to free radicals (e.g. brain trauma and aging) are risk factors for AD, and on the relationship between AD and Down's syndrome, the latter being characterized by an increase in free radical formation (Volicer and Crino, 1990). In studies on the oxidative phosphorylation system, AD has been associated with reduced complex IV activity in the mitochondria, presumably due to abnormal catalytic activity rather than to decreased enzyme levels. Mitochondrial dysfunction and oxidative damage facilitate amyloid aggregation and tau polymerization, which are thought to be toxic for surrounding neurons and induce apoptosis. The toxicity of APP fragments in PC12 cells is inhibited by the antioxidants vitamin E and propyl gallate. Aβ itself has the ability to generate free radicals, thereby inhibiting mitochondrial function, and facilitating amyloid aggregation and tau polymerization. Oxidation of tau proteins also appears to facilitate dimerization and polymerization into filaments (review Bowling and Beal, 1995).

Recent findings have suggested links between the deposition of Aβ oxidative stress, the disruption of ion homeostasis, and an apoptotic form of neuronal death (Fig. 1). Furthermore, Aβ can induce apoptosis in cultured neurons (Forloni et al., 1993; Loo et al., 1993). Expression of a human PS1-mutant in PC12 cells increases their vulnerability to apoptosis induced by trophic factor withdrawal and Aβ. The mechanism whereby mutant PS1 promotes apoptosis seems to involve the disruption of calcium homeostasis and increased oxidative stress (Guo et al., 1997). The process of apoptosis involves two phases: an activation phase, consisting of the intracellular signalling pathways that activate apoptosis and an execution phase, consisting of the machinery that is necessary for apoptosis to occur. ROS can activate apoptosis, but they are not required for the execution of the death programme (review Jacobson, 1996). Also, the expression of the immediate early gene c-jun has been associated with apoptosis. In AD, c-jun expression is elevated and cultured neurons undergoing Aβ-induced apoptosis exhibit a selective and prolonged induction of c-jun (Anderson et al., 1995). Recent studies have indicated that c-jun may participate in the regulation of early events associated with DNA damage and/or apoptosis (Anderson et al., 1996).

Despite all evidence that has been marshalled in favor of ROS involvement in AD pathogenesis, direct evidence for increased ROS activity in AD is lacking, as is evidence that such activity precedes the development of AD neuropathology (Choi, 1995). Whether ROS are a primary cause of the development of AD is thus still unclear.

Age-related frameshift mutations

Recently, an exciting new view on the cause of the development of age-related diseases has been proposed. This view was based on an observation by Richards et al. (1985) in adult di/di Brattleboro rats suffering from diabetes insipidus which was due to a single-base deletion in the vasopressin gene. Surprisingly, in these rats solitary neurons immunoreactive for wild-type vasopressin and an age-related reversion to a hemizygous wild-type vasopressin peptide phenotype were found (Richards et al., 1985; Van Leeuwen et al., 1989). Analysis of polymerase chain reaction (PCR) products generated from hypothalamic cDNA led to an explanation of the revertant peptide phenotypes, namely an age-related dinucleotide deletion in the vasopressin transcript (Evans et al., 1994; 1995). This mutation restored the normal reading frame, resulting in the production of wild type vasopressin in solitary neurons. The deletion (ΔGA) occurred at the mRNA level at two specific hotspots within GAGAG motifs that were downstream of the initial one-base deletion in the gene. The mutation was also found in a small percentage of the wild type vasopressin transcripts, generating a truncated protein with an altered carboxy-terminal region. On the basis of these findings the hypothesis was put forward that this kind of mutation might also occur in other transcripts, e.g. in those involved in human aging and age-related diseases. To test this hypothesis, two proteins linked to AD and DS were investigated (Van Leeuwen et al., 1998a), namely APP and the brain-specific garbage protein ubiquitin-B (ubi-B) that associates with misfolded proteins and targets the faulty proteins to the proteasomal pathway for degradation. Both proteins are common in AD tangles and plaques. The APP and ubi-B transcripts contain a number of GAGAG motifs. In APP, seven GAGAG motifs are present in regions corresponding to exons 4, 6, 9, 10 and 14. Since three motifs are clustered in exons 9 and 10, this region of the transcript was selected for the possible occurrence of a +1 frameshift mutation. A +1 frameshift in exon 9 or 10 results in a truncated APP (APP^{+1}) with a novel carboxy-terminal region. In two of the three repeats in ubi-B mRNA, a GAGAG motif is present. The predicted +1 frameshift results in an aberrant carboxy-terminus of the first or second repeat ($ubi\text{-}B^{+1}$). To examine the presence of the predicted +1 proteins, antibodies directed towards the novel carboxy-terminal sequences of APP^{+1} and $ubi\text{-}B^{+1}$ were generated. These antibodies were used for immunocytochemistry and Western blot analysis of cerebral cortex from AD, DS and control individuals. Immunoreactivity for APP^{+1} and $ubi\text{-}B^{+1}$ was prominent in neuritic plaques, tangles and neuropil threads in the brains of early- and late-onset AD and even more prominent in DS patients compared to non-demented controls matched for age, sex, postmortem delay and duration of fixation. Immunoreactivity was also found in aged non-demented controls with initial neuropathology, indicating that this phenomenon is an early event in AD. The APP^{+1} and $ubi\text{-}B^{+1}$ proteins displayed cellular colocalization, suggesting a common origin. The predicted molecular sizes of the immunoreactive products were confirmed by SDS-PAGE and Western blotting. To establish that the frameshift mutations had occurred at the RNA-level, a PCR-screening method was developed (Fig. 2). Total RNA was isolated from control, AD and DS brain material, random primed with hexamers and reverse transcribed. The PCR was performed with primers flanking the GAGAG motifs and containing AβmHI or HindIII sites. PCR-products were digested with *Bam*HI and HindIII, and subcloned into the prokaryotic expression vector pQE31. Electro-competent bacteria were used to accomplish an efficient transformation. Colonies were grown on a filter and screened for the expression of frameshift proteins by using the APP^{+1} and $ubi\text{-}B^{+1}$ antibodies. As a negative control, preimmune antiserum was used. Due to the low frequency of the frameshift mutations, many clones had to be generated and screened. The immunopositive clones were sequenced to establish the mutations at the nucleotide level. In the APP-transcripts, GA deletions were found in either exon 9 or exon 10, but with a higher frequency in exon 9. The motif in exon 9 is GAGAGAGA which suggests a possible relationship between the length of GA repeats and the occurrence of the dinucleotide deletion (see below). In ubi-B transcripts, not a GA but a GU deletion had occurred in the first repeat. This mutation was located directly adjacent to the GAGAG motif. In the third repeat of ubi-B, a CU-deletion was detected in a CUCU-region. A striking observation was that the frequency of $ubi\text{-}B^{+1}$ immunopositive clones was 6 to 600 times higher than that of APP^{+1} clones, also reflected by the more widespread distribution of $ubi\text{-}B^{+1}$ immunoreactivity relative to APP^{+1} reactivity in AD brains. The observed dinucleotide deletions were not the result of PCR or cloning artifacts, since these mutations did not occur in the brains of age-matched non-demented persons. Moreover, the presence of mutated ubi-B mRNA was confirmed by stringent in situ hybridization, using an oligoprobe that distinguished between mutant and wild type mRNA. The in situ RNA signal indeed colocalized with the $ubi\text{-}B^{+1}$ immunoreactivity. At the DNA level these mutations were not detected.

The fact that the two +1 transcripts and proteins displayed cellular colocalization suggests a common mechanism of origin. Since the chance that

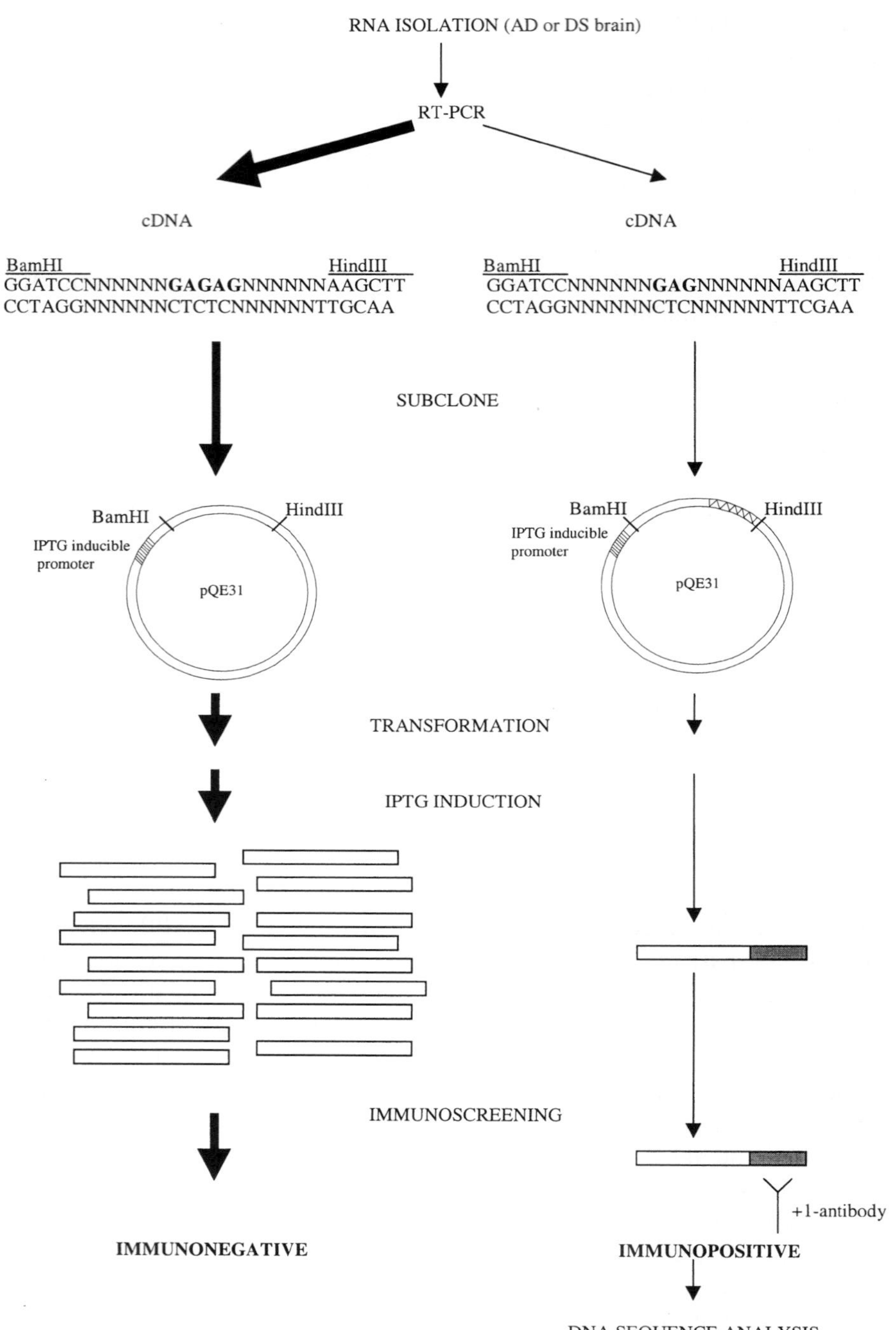

Fig. 2. PCR/immunoscreening strategy to illuminate frameshift mutations at the level of RNA. PCR-products were subcloned into the prokaryotic expression vector pQE31, transformed to bacteria and immunoscreened for the presence of +1 frameshift proteins. AD = Alzheimer's disease; DS = Down syndrome; RT-PCR = reverse transcription-polymerase chain reaction, IPTG = isopropylthiogalactosidase.

mutations are generated in two genes in one cell is rather remote, it is more likely that the mechanism responsible for the mutations operates at the transcriptional level or by posttranscriptional editing of RNA. The occurrence of this type of transcript mutation, (so called 'molecular misreading', Van Leeuwen et al., 1998b) represents a new view on AD, opens new avenues for future research and may well be an important mechanism to explain the widely occurring nonfamilial early- and late-onset forms of AD, and other neurodegenerative and age-related diseases.

Frameshift mutations related to other diseases

During the last decade many disease-linked mutations (point mutations, insertions, deletions and expansion of repeats) have been described at the level of DNA. For example, familial AD is caused by point mutations in the APP, PS1 or PS2 genes as described above (review Hardy, 1997). In McArdle's disease, a G is mutated to TT, resulting in a frameshift and causing the absence of functional muscle glycogen phosphorylase (Bartram et al., 1994). Tay-Sachs disease, a catastrophic neurodegenerative disorder, is caused by the absence of an isoenzyme of lysosomal β–hexosaminidase due to a deletion of a C in the gene encoding the α-subunit (Lau and Neufeld, 1989). In Chediak-Higashi syndrome, caused by a defective vesicular transport from the lysosomes and late endosomes, an extra G is introduced in a GA-rich region (GGGAGGAAGAAGAGGAGGAG-ACGCA-CATGGGCA) of the *Lyst* gene (Barbosa et al., 1996). Another example is a dinucleotide deletion (ΔGA) in the α-globin gene in hemoglobin H disease (Safaya and Rieder, 1988). Remarkable is the occurrence of this deletion in a GAGAG motif. These examples are not directly age related and are caused by mutations at the DNA level (Fig. 3).

Mono-, di-, and trinucleotide repeats in the human genome have been found to be unstable. Expansion of trinucleotide repeats (from a limited number to hundreds or even thousands of copies) does in itself not cause a frameshift, but is still associated with a number of heritable genetic diseases. These include myotonic dystrophy and Huntington's disease in which an expansion of a CTG-repeat and of a CAG-repeat occurs in the myotonin and huntingtin gene, respectively. The number of repeats is related to the severity or age of onset of the disease (review Sutherland and Richards, 1995). The molecular mechanism underlying simple-repeat sequence instability is complex and may differ between various repeats. Hereditary non-polyposis colon cancer (HNPCC) tumors have been used for the search of genes linked to such instabilities. In tumor tissue from patients with HNPCC, many simple tandem repeats are unstable and novel alleles are present resulting from increases or decreases in the copy number of repeats (Ionov et al., 1993; Aaltonen et al., 1993; Thibodeau et al., 1993). HNPCC loci were mapped genetically to chromosomes 2 and 3. The responsible genes turned out to be the human homologues of the bacterial *mutS* and *mutL* genes (yeast *MLH2* and *MLH1*, respectively). The bacterial genes are involved in mismatch repair. The HNPCC mutations allow normal flaws in the replication of simple tandem repeat sequences to manifest as changes in repeat copy number through a failure to repair slippage-induced mismatch (review Sutherland and Richards, 1995). The mismatch repair deficiency in HNPCC, bacteria and yeast is related to cell division. The frequency of tract alterations in yeast is higher during meiosis compared to mitosis. DNA polymerase showed a high rate of slippage on templates containing simple repeats. Most of these errors are not corrected through its proofreading activity, but by cellular mismatch repair systems. The instability of repeats may be a consequence of either an increased rate of DNA polymerase slippage or a decreased efficiency of mismatch repair (Strand et al., 1993). Although the instability of trinucleotide repeats at the DNA level in itself does not cause a frameshift, a relationship with the frameshift mutations found at the RNA level may still exist

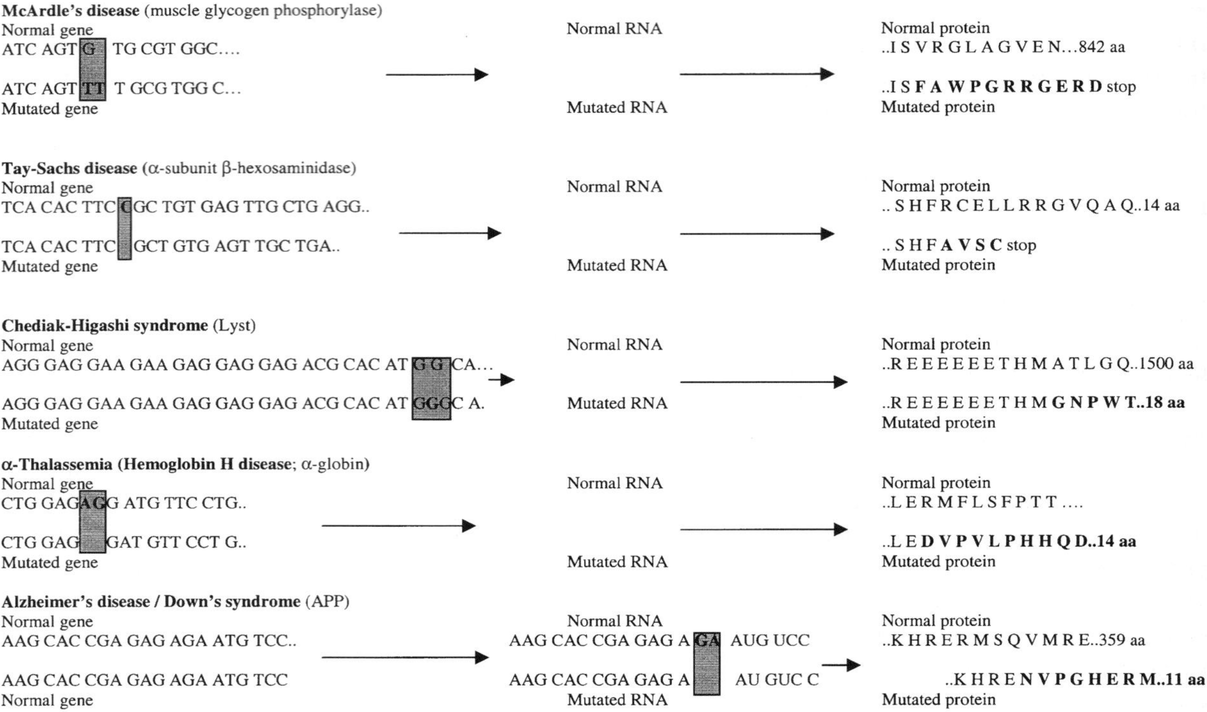

Fig. 3. Frameshift mutations at the level of DNA in a number of heritable diseases, and mutations at the level of RNA in Alzheimer's disease and Down syndrome; mutations are boxed.

because the enzymes operating at the transcriptional level might function in a similar way as the enzymes involved in the replication of DNA. Thus, at the transcriptional level errors due to simple repeats might also occur, leading to frameshift mutations. To obtain more insights into the introduction and repair of mutations at either the DNA or RNA level, bacteria (dividing cells) have been used as a model system with respect to their 'adaptive' mutations. The following plausible mechanisms have been proposed (review Finch and Goodman, 1997).

1. SOS error-prone repair. The SOS regulon contains approximately 25 genes induced by DNA damage and is under the control of the Lex A repressor and Rec A. A number of these SOS-inducible genes are involved in repairing the DNA damage. Among the SOS-induced proteins are Umu C and Umu D' with the surprising property that they significantly increase the level of mutations. One molecule of Umu C combines with two Umu D' molecules to form a tight Umu D'2C mutagenic complex. When encountering a DNA lesion, this complex is believed to allow the stalled replication proteins to copy past the damaged template sites. This may serve as an emergency act of 'last resort', which facilitates cell survival in the presence of extensive DNA damage.
2. The down-regulation of enzymes required for post-replication repair.
3. Transcription-coupled DNA repair. This reflects a recent conceptual breakthrough in molecular biology, namely that basic cellular reactions, like cell cycle, replication,

transcription, recombination and repair, previously thought to function more or less independently, have intimate interconnections.

Not only in bacteria, but also in humans a link between the transcriptional machinery and DNA nucleotide excision repair has been found. In three rare hereditary diseases (Xeroderma pigmentosum, Cockayne's syndrome, and trichothiodystrophy), gene products implicated in transcription-coupled repair are defective. Some factors necessary for transcription initiation, like TFIIH, are also required for nucleotide excision repair. When, during transcription, the transcriptional machinery recognizes a lesion in the DNA, the arrested RNA polymerase II undergoes a conformational change to initiate transcript shortening and reannealing of DNA strands at the lesion site. The coupling factor(s) recruits TFIIH and the additional elements required for incision, excision and repair synthesis. Following repair, transcription resumes, thereby completing the RNA transcript in progress and facilitating the passage of an advancing replication fork (review Hanawalt, 1994). In non-dividing cells, the transcription-coupled repair is also present. In some cases the mechanism may cause mutations. In replicating DNA, a signal might be present that designates which of the two strands has to be repaired. If this signal is absent from non-replicating DNA, the loss of strand specificity during error correction will lead to a significant increase in mutagenesis. In non-dividing cells, unrepaired DNA lesions might be copied leading to increased mutagenesis, perhaps by a process analogous to SOS error-prone repair in bacteria (review Finch and Goodman, 1997). Whether the mismatch-repair enzymes are indeed active in non-dividing cells is unclear.

The replication machinery has been investigated in more depth than the transcriptional machinery. Furthermore, the enzymatic activities and mechanisms that enhance fidelity during DNA and protein synthesis have been extensively studied (Kirkwood et al., 1986). The fidelity of RNA synthesis has received less attention, even though transcriptional errors have the potential to be even more deleterious than translational errors. Recently, Thomas et al. (1998) showed the involvement of a general elongation factor (SII), a protein that stimulates nuclease activity, in the transcriptional fidelity and proofreading of RNA polymerase II (pol II). A mechanism by which pol II can discriminate between the correct and incorrect base was described. Studies at the DNA level have shown that a misincorporation inhibits the rate of addition of the next nucleotide by DNA polymerase, allowing time for the exonucleolytic removal of the misincorporated base (Donlin et al., 1991; Goodman et al., 1993). The SII elongation factor, together with TFIIF, ELL and elongin, is thought to act directly on the pol II ternary elongation complex in order to prevent arrest and suppress transient pausing. In this way they function together to sustain elongation at physiological rates. SII could efficiently mediate error correction during rapid transcription. TFIIF, which increases the rate of elongation in vitro to approach estimated in vivo rates (Izban and Luse, 1992), did not interfere with SII-mediated error correction. Thus, SII stimulates proofreading of the nascent RNA during normal progressive elongation in vitro. Whether proofreading occurs in vivo remains an unanswered question. Recent investigations provide evidence that the activity of another transcription elongation factor, elongin, is regulated in mammalian cells, suggesting that the rate and fidelity of transcription by pol II may be carefully controlled in eukaryotic cells (Shilatifard et al., 1997; Thomas et al., 1998).

The above-mentioned (partially) heritable diseases are the result of mutations at the DNA level. No causal relationship between the onset of a disease and a frameshift mutation at the RNA level has been reported, although the same factors or enzymes might be involved in transcription as well as in replication. An example of an error in RNA has recently been reported at the level of splicing.

Alternative splicing is used to generate different transcripts with the splicing machinery recognizing AG-GU sequences. Under some circumstances transcripts are generated by cleavage and ligation using cryptic sites (still obeying the AG-GU rule), resulting in aberrant mRNAs. In EAAT2 transcripts, encoding a protein involved in the sporadic form of 'amyotrophic lateral sclerosis', another form of aberrant splicing has been observed. In this case exons are skipped and introns are retended. Furthermore, a completely new aberrant splicing variant of EAAT2 transcripts was found, namely resulting from the fact that the pre-mRNAs were not cleaved and ligated according to the AG-GU rule but a variety of dinucleotides was used as a signal for the formation of the EAAT2 mRNAs (Lin et al., 1998; review Bai and Lipton, 1998). This is an example of the generation of aberrant products, with the DNA still being intact. Aberrant RNA processing, either as a primary defect as in spinal muscle atrophy or secondary to other cellular defects, could constitute an important mechanism leading to neurodegenerative diseases. Whether the cause of this phenomenon lies in the damage of DNA encoding RNA processing enzymes or the enzymes themselves is at the present unclear. In addition, proteins involved in repair may not function optimally, resulting in mutations at the DNA level or at the transcriptional level.

Future prospects

The frameshift mutations observed in AD and DS patients have been found to occur at a simple repeat at the transcriptional level. Thus far simple repeats are known to be hotspots for mutations at the DNA level. However, since enzymes involved in DNA replication and repair may also be part of the transcriptional machinery, mutations may occur at simple repeats during transcription as well, with the DNA remaining unaffected. Alternatively, cellular stress may cause a distorded local microenvironment, leading to inaccurate DNA transcription.

One goal of future research will concern the elucidation of the machinery responsible for the frameshift mutations. It will be of interest to isolate the transcriptional machinery from affected cells and see via in vitro transcription of a template DNA with multiple copies of a hotspot, if frameshifts are introduced, at which positions/hotspots and the frequency of mutation. However, the molecular mechanisms underlying the transcriptional process itself are still unclear and first more information about this machinery is necessary. Furthermore, the effects of the expression of products resulting from the frameshift mutations should be studied. The mutated proteins can be transfected into neuronal cell lines to examine the influence of these proteins on cell physiology, e.g. Aβ-production, cell survival and senescence. Another approach concerns the generation of transgenic mice expressing the frameshift proteins. When using the mouse as a model system for an age-related disease like AD, one should realize that in an aged wild type mouse AD neuropathology (plaques and tangles) is absent and the mechanism leading to the frameshift mutations has not been shown to be present in this species. In the transgenic mice, the aberrant proteins are introduced to investigate their influence on normal cell functions. Finding a(n) (animal) model system in which the mechanism responsible for the occurrence of frameshift mutations is present, would be a step forward. Since Aβ and other proteins deposited in the plaques circulate in the cerebrospinal fluid (CSF), the +1 proteins may also be found (FWvL, unpublished results). Because the generation of +1 transcripts may well be an early event in AD, the use of specific antibodies directed against the frameshift proteins in an RIA to detect the +1 proteins (e.g. in the CSF) may lead to a diagnosis of AD at an early stage. At the molecular level other GAGAG-containing transcripts with a frameshift mutation may be identified and, furthermore, it is likely that the GAGAG-motif is not the only hotspot for such mutations. Reporter genes, such as luciferase, *lac*Z or green fluorescent protein, can be useful for this approach. To screen for frameshift mutations, one could clone PCR fragments produced from a randomly chosen transcript in front of a reporter-gene in such a way that the reporter gene is generated only when a mutation has occurred in the transcript.

In conclusion, the identification of a frameshift mutation at a typical hotspot in an RNA has led to an interesting new view and has offered exciting new possibilities to study the mechanism underlying neurodegenerative and age-related disorders.

List of abbreviations

AD	Alzheimer's disease
APP	amyloid precursor protein
ApoE	apolipoprotein E
Aβ	amyloid-β
CSF	cerebrospinal fluid
Hsp	heat-shock protein
HNPCC	hereditary non-polyposis colon cancer
PCR	polymerase chain reaction
PS1	presenilin 1
PS2	presenilin 2
ROS	reactive oxygen species
Sir	silent information regulator
ubi-B	ubiquitin-B

References

Aaltonen, L.A., Peltomaki, P., Leach, F.S., Sistonen, P., Pylkkanen, L., Mecklin, J.-P., Jarvinen, H., Powell, S.M., Jen, J., Hamilton, S.R., Peterson, G.M., Kinzler, K.W., Vogelstein, B. and De la Chapelle, A. (1993) Clues to the pathogenesis of familial colorectal cancer. *Science*, 260: 812–816.

Alzheimer, A. (1907) Über eine eigenartige Erkrankung der Hirnrinde. *Algemeine Zeitschrift für Psychiatrie und Psychisch-Gerichtliche Medizin*, 64: 146–148.

Anderson, A.J., Pike, C.J. and Cotman, C.W. (1995) Differential induction of IEGs in response to Aβ peptides in cultured neurons: association between c-Jun immunoreactivity and Aβ-mediated apoptosis. *J. Neurochem.*, 65: 1487–1498.

Anderson, A.J., Su, J.H. and Cotman, C.W. (1996) DNA damage and apoptosis in Alzheimer's disease: colocalization with c-jun immunoreactivity, relationship to brain area, and effect of postmortem delay. *J. Neurosci.*, 16: 1710–1719.

Bai, G. and Lipton, S.A. (1998) Aberrant RNA splicing in sporadic amyotrophic lateral sclerosis. Neuron, 20: 363–366.

Barbosa, M.D.F.S., Nguyen, Q.A., Tchernev, V.T., Ashley, J.A., Detter, J.C., Blaydes, S.M., Brandt, S.J., Chotai, D., Hodgman, C., Solari, R.C.E., Lovett, M. and Kingsmore, S.F. (1996) Identification of the homologous beige and Chediak-Higashi syndrome gene. *Nature*, 382: 262–265.

Bartram, C., Edwards, R.H.T., Clague, J. and Beynon, R.J. (1996) McArdle's disease: a rare frameshift mutation in exon 1of the muscle glycogen phosphyorylase gene. *Biochem. Biophys. Acta*, 1226: 341–343.

Bodnar, A.G., Ouelette, M., Frolkis, M., Holt, S.E., Chiu, C-P., Morin, G.B., Harley, C.B., Shay, J.W., Lichtsteiner, S. and Wright, W.E. (1998) Extension of life-span by introduction of telomerase into human cells. *Science*, 279: 349–352.

Bowling, A.C. and Beal, M.F. (1995) Bioenergetic and oxidative stress in neurodegenerative diseases. *Life Sciences*, 56: 1151–1171.

Breteler, M.M.B., Claus, J.J., Van Duijn, C.M., Launer, L.J. and Hofman, A. (1992) Epidemiology of Alzheimer's disease. *Epidemiol. Rev.*, 14: 59–82.

Chartier-Harlin, M., Crawford, M., Houldon, H., Warren, A., Hughes, D., Fidani, L., Goate, A., Rossor, M., Roques, P., Hardy, J. and Mullan, M. (1991) Early-onset Alzheimer's disease caused by mutations at codon 717 of the β-amyloid precursor protein gene. *Nature*, 353: 844–846

Choi, B.H. (1995) Oxidative stress and Alzheimer's disease. *Neurobiol. Aging*, 16: 675–678.

Ciallella, J., Rangnekar, V. and McGillis, J. (1995) Heat shock alters Alzheimer's β-amyloid precursor protein expression in human endothelial cells. *J. Neurosci. Res.*, 37: 769–776.

Donlin, M.J., Patel, S.S. and Johnson, K.A. (1991) Kinetic partitioning between the exonuclease and polymerase sites in DNA error correction. *Biochemistry*, 30: 538–546.

Esch, F., Keim, P., Beattie, E., Blacher, R., Culwell, A., Olterdorf, T., McClure, D. and Ward, P. (1990) Cleavage of amyloid β peptide during constitutive processing of its precursor. *Science*, 248: 1122–1124.

Estus, S., Golde, T., Kunishita, T., Blades, D., Lowery, D., Eisen, M., Usiak, M., Qu, X., Tabira, T., Greenberg, B.D. and Younkin, S.G. (1992) Potentially amyloidogenic carboxy-terminal derivatives of the amyloid protein precursor. *Science*, 255: 726–728.

Evans, D.A.P., Burbach, J.P.H. and Van Leeuwen, F.W. (1995) Somatic mutations in the brain. *Mutat. Res.*, 338: 173–182.

Evans, D.A., Funkenstein, H.H., Albert, M.S., Scherr, P.A., Cook, N.R., Chown, M.J., Hebert, L.E., Hennekens, C.H. and Taylor, J.O. (1989) Prevalence of Alzheimer's disease in a community population of older persons: higher than previously reported. *JAMA*, 262: 2551–2556.

Evans, D.A.P., Van der Kleij, A.A.M., Sonnemans, M.A.F., Burbach, J.P.H. and Van Leeuwen, F.W. (1994) Frameshift mutations at two hotspots in vasopressin transcripts in post-mitotic neurons. *Proc. Natl. Acad. Sci. USA*, 91: 6059–6063.

Ewbank, J.J., Barnes, T.M., Lakowski, B., Lussier, M., Bussey, H. and Hekimi, S. (1997) Structural and functional conservation of the *Caenorhabditis elegans* timing gene *clk-1*. *Science*, 275: 980–983.

Fargnoli, J., Kunisada, T., Fornace Jr, A.J. Schneider, E.L. and Holbrook, N.J. (1990) Decreased expression of heat shock protein 70 mRNA and protein after heat treatment in cells of aged rats. *Proc. Natl. Acad. Sci. USA*, 87: 846–850.

Finch, C.E. and Goodman, M.F. (1997) Relevance of 'adaptive' mutations arising in non-dividing cells of microorganisms to age-related changes in mutant phenotypes of neurons. *Trends Neurosci.*, 20: 501–507.

Finch, C.E. and Tanzi, R.E. (1997) Genetics of aging. *Science*, 278: 407–411.

Forloni, G., Chiesa, R., Smiroldo, S. and Verga, L. (1993) Apoptosis-mediated neurotoxicityinduced by chronic application of β-amyloid fragment 25–35. *NeuroReport*, 4: 523–526.

Friedman, D.B. and Johnson, T.E. (1988) A mutation in the age-1 gene in Caenorhabditis elegans lenghtens life and reduces hermaphrodite fertility. *Genetics*, 118: 75–86.

Glenner, G.G. and Wong, C.W. (1984a) Alzheimer's disease: Initial report of the purification and characterization of a novel cerebrovascular amyloid protein. *Biochem. Biophys. Res. Commun.*, 120: 885–890.

Glenner, G.G. and Wong, C.W. (1984b) Alzheimer's disease and Down's syndrome sharing of a unique cerebrovascular amyloid fibril protein. *Biochem. Biophys. Res. Commun.*, 122: 1131–1135.

Goate, A., Chartier-Harlin, M., Mullan, M., Brown, J., Crawford, F., Fidani, L., Giuffra, L., Haynes, A., Irving, N., James, L., Mant, R., Newton, P., Rooke, K., Roques, P., Talbot, C., Pericak-Vance, M., Roses, A., Williamson, R., Rossor, M., Owen, M. and Hardy, J. (1991) Segregation of a missense mutation in the amyloid precursor protein gene with familail Alzheimer's disease. *Nature*, 349: 704–706.

Goodman, M.F., Creighton, S., Bloom, L.B. and Petruska, J. (1993) Biochemical basis of DNA replication fidelity. *Crit. Rev. Biochem. Mol. Biol.*, 28: 83–126.

Gotta, M., Strahl-Bolsinger, S., Renauld, H., Laroche, T., Kennedy, B.K., Grunstein, M. and Gasser, S.M. (1997) Localization of Sir2p: the nucleolus as a compartment for silent information regulators. *EMBO J.*, 16: 3243–3255.

Grilli, M., Ribola, M., Alberici, A., Valerio, A., Memo, M. and Spano, P. (1995) Identification and characterization of a kappa B/Rel binding site in the regulatory region of the amyloid precursor protein gene. *J. Biol. Chem.*, 270: 26774–26777.

Guarente, L. (1996) Do changes in chromosomes cause aging? *Cell*, 86: 9–12.

Guarente, L. (1997) Link between aging and the nucleolus. *Genes Dev.*, 11: 2449–2455.

Guo, Q., Sopher, B.L., Furukawa, K., Pham, D.G., Robinson, N., Martin, G.M. and Mattson, M.P. (1997) Alzheimer's presenilin mutation sensitizes neural cells to apoptosis induced by trophic factor withdrawal and amyloid-β-peptide: involvement of calcium and oxyradicals. *J. Neurosci.*, 17: 4212–4222.

Haass, C., Capell, A., Citron, M., Teplow, D.B. and Selkoe, D.J. (1995) The vacuolar H^+ ATPase inhibitor bafilomycin A1 differentially affects proteolytic processing of mutant and wild type β-amyloid precursor protein. *J. Biol. Chem.*, 270: 6186–6192.

Haass, C., Koo, E.H., Mellon, A., Hung, A.Y. and Selkoe, D.J. (1992) Targeting of cell-surface β-amyloid precursor protein to lysosomes: alternative processing into amyloid-bearing fragments. *Nature*, 357: 500–503.

Hanawalt, P.C. (1994) Transcription-coupled repair and human disease. *Science*, 266: 1957–1958.

Hardy, J. (1997) Amyloid, the presenilins and Alzheimer's disease. *Trends Neurosci.*, 20: 154–159.

Harman, D. (1991) The aging process: Major risk factor for disease and death. *Proc. Natl. Acad. Sci. USA*, 88: 5360–5363.

Hekimi, S., Lakowski, B., Barnes, T.M., Ewbank, J.J. (1998) Molecular genetics of life span in *C. elegans*: how much does it teach us? *TIG*, 14: 14–20.

Hendriks, L., Van Duijn, C.M., Cras, P., Cruts, M., Van Hul, W., Van Harskamp, F., Warren, A., McInnis, M.G., Atonarakis, S.E., Martin, J-J., Hofman, A. and Van Broeckhoven, C. (1992) Presenile dementia and cerebral haemorrhage linked to a mutation at codon 692 of the β-amyloid precursor protein gene. *Nat. Genet.*, 1: 218–221.

Higaki, J., Quon, D., Zhong, Z. and Cordell, B. (1995) Inhibition of β-amyloid formation identifies proteolytic precursors and subcellular site of catabolism. *Neuron*, 14: 651–659.

Ionov, Y., Peinado, M.A., Malkhosyan, S., Shibata, D. and Perucho, M. (1993) Ubiquitous somatic mutations in simple repeated sequences reveal a new mechanism for colon carcinogenesis. *Nature*, 363: 558–561.

Izban, M.G. and Luse, D.S. (1992) Factor-stimulated RNA polymerase II transcribes at physiological elongation rates on naked DNA but very poorly on chromatin templates. *J. Biol. Chem.*, 267: 13647–13655.

Jacobsen, M.D. (1996) Reactive oxygen species and programmed cell death. *TIBS*, 21: 83–86.

Johnson, T. (1996), In: N. Holbrook, G. Martin and R. Lockshin (Eds.), *Cellular aging and death* Wiley-Liss, NewYork, pp. 1–17.

Kametani, F., Haga, S., Tanaka, K. and Ishii, T. (1990) Amyloid β-protein precursor (APP) of cultured cells; secretory and non-secretory forms of APP. *J. Neurol. Sci.*, 97: 43–52.

Kang, J., Lemaire, H., Unterbeck, A., Salbaum, J.M., Masters, C.L., Grzeschik, K., Multhaup, G., Beyreuther, K. and Müller-Hill, B. (1987) The precursor of Alzheimer's disease amyloid A4 protein resembles a cell-surface receptor. *Nature*, 325: 733–736.

Kennedy, B.K., Austriaco Jr, N.R., Zhang, J. and Guarente, L. (1995) Mutation in the silencing gene SIR4 can delay aging in *S.cerevisiae*. *Cell*, 80: 485–496.

Kennedy, B.K., Gotta, M., Sinclair, D.A., Mills, K., McNabb, D.S., Murthy, M., Pak, S.M., Laroche, T., Gasser, S.M. and Guarente, L. (1997) Redistribution of silencing proteins from telomeres to the nucleolus is associated with extension of life span in *S. cerevisiae*. *Cell*, 89: 381–391.

Kenyon, C. (1996) Ponce d'elegans: genetic quest for the fountain of youth. *Cell*, 84: 501–504.

Kenyon, C., Chang, J., Gensch, E., Rudner, A. and Tabtiang, R. (1993) A *C. elegans* mutant that lives twice as long as wild type. *Nature*, 366: 461–464.

Kimura, K.D., Tissenbaum, H.A., Liu, Y. and Ruvkun, G. (1997) Daf-2, and insulin receptor-like gene that regulates longevity and diapause in *Caenorhabditis elegans*. *Science*, 277: 942–946.

Kirkwood, T.B.L., Rosenberger, R.F. and Galas, D.J. (1986) *Accuracy in molecular processes*, Chapman and Hall, London.

Kitaguchi, N., Takahashi, Y., Tokushima, Y., Shiojiri, S. and Ito, H. (1988) Novel precursor of Alzheimer's disease amyloid protein shows protease inhibitory activity. *Nature*, 331: 530–532.

Knops, J., Lieberburg, I. and Sinha, S. (1992) Evidence for a nonsecretory acidic degradation pathway for amyloid precursor protein in 293 cells. *J. Biol. Chem.*, 267: 16022–16024.

Knops, J., Suomensaari, S., Lee, K.J.M., McConlogue, L., Seubert, P. and Sinha, S. (1995) Cell-type and amyloid precursor-type specific inhibition of $A\beta$ release by bafilomycin A_1, a selective inhibitor of vacuolar ATPases. *J. Biol. Chem.*, 270: 2419–2422.

Kuentzel, S.L., Ali, J.S.M., Ahman, R.D., Greenberg, B.D. and Raub, J.T.J. (1993) The Alzheimer β-amyloid protein precursor/protein nexin-II is cleaved by secretase in a *trans*-Golgi secretory compartment in human neuroglioma cells. *Biochem. J.*, 295: 367–378.

Lakowski, B. and Hekimi, S. (1996) Determination of life span in *Caenorhabditis elegans* by four clock genes. *Science*, 272: 1010–1013.

Lamberts S.W.J., Van den Beld, A.W. and Van der Lely, A-J. (1997) The endocrinology of aging. *Science*, 278: 419–424.

De Lange, T. (1998) Telomeres and senescence: ending the debate. *Science*, 279: 334–335.

Lau, M.M.H. and Neufeld E.F. (1989) A frameshift mutation in a patient with Tay-Sachs disease causes premature termination and defective intracellular transport of the α-subunit of β-hexosaminidase. *J. Biol. Chem.*, 264: 21376–21380.

Levy, E., Carman, M.D., Fernandez-Madrid, I.J., Power, M.D., Lieserburg,I., Van Duinen, S.G., Bots, G.T.A.M., Luyendijk, W. and Frangione, B. (1990) Mutation of the Alzheimer's disease amyloid gene in hereditary cerebral hemorrhage, Dutch type. *Science*, 248: 1124–1126.

Levy-Lahad, E., Wasco, W., Poorkaj, P., Romano, D.M., Oshima, J., Pettingell, W.H., Yu, C., Jondro, P.D., Schmidt, S.D., Wang, K., Crowley, A.C., Fu, Y-H., Guenette, S.Y., Galas, D., Nemens, E., Wijsman, E.M., Bird, T.D., Schellenberg, G.D. and Tanzi, R.E. (1995) Candidate gene for the chromosome 1 familial Alzheimer's disease locus. *Science*, 269: 973–977.

Lin, C-L.G., Bristol, L.A., Jin, L., Dykes-Hoberg, M., Crawford, T., Clawson, L. and Rothstein, J.D. (1998) Aberrant RNA processing in a neurodegenerative disease: the cause for absent EAAT2, a glutamate transporter, in amyotrophic lateral sclerosis. *Neuron*, 20: 589–602.

Liu, A.Y., Lin, Z., Choi, H.S., Sorhage, F. and Li, B. (1989) Attenuated induction of heat shock gene expression in aging diploid fibroblasts. *J. Biol. Chem.*, 264: 12037–12043.

Loo, D.T., Copani, A., Pike, C.J., Whittemore, E.R., Walencewicz, A.J. and Cotman, C.W. (1993) Apoptosis is induced by β-amyloid in cultured central nervous system neurons. *Proc. Natl. Acad. Sci. USA*, 90: 7951–7955.

Malone, E.A., Inoue, T. and Thomas J.H. (1996) Genetic analysis of the roles of daf-28 and age-1 in regulating Caenorhabditis elegans dauer formation. *Genetics*, 143: 1193–1205.

Mattson,M.P. (1997) Cellular actions of β-amyloid precursor protein and its soluble and fibrillogenic derivatives. *Physiol. Rev.*, 77: 1081–1132.

Morris, J.Z., Tissenbaum, H.A. and Ruvkin, G. (1996) A phophatidylinositol-3-OH kinase family member regulating longevity and diapause in *Caenorhabditis elegans*. *Nature*, 382: 536–539.

Mullan, M., Crawford, F., Axelman, K., Houldon, H., Lilius, L., Winblad, B. and Lannfelt, L. (1992) A pathogenic mutation for probable Alzheimer's disease in the APP gene a the N-terminus of β-amyloid. *Nat. Genet.*, 1: 345–347.

Munger, J., Haass, C., Lemere, C., Shi, G., Wong, W., Teplow, D., Selkoe, D.J. and Chapman, H. (1995) Lysosomal processing of amyloid precursor protein to $A\beta$ peptides: a distinct role for cathepsin S. *Biochem. J.*, 311: 299–305.

Murrell, J., Farlow, M., Ghetti, B. and Benson, M.D. (1991) A mutation in the amyloid precursor protein associated with hereditary Alzheimer's disease. *Science*, 254: 97–99.

Orr, W.C. and Sohal, R.S. (1994) Extension of life-span by overexpression of superoxide dismutase and catalase in *Drosophila melanogaster*. *Science*, 263: 1128–1130.

Ozawa, T. (1997) Genetic and functional changes in mitochondria assosiated with aging. *Physiol. Rev.*, 77: 425–464.

Palmert, M., Siedlak, S., Podlisny, M., Greenberg, B., Shelton, E., Chan, H., Usiak, M., Selkoe, D.J., Perry, G. and Younkin, S. (1989) Soluble derivatives of the β amyloid protein precursor of Alzheimer's disease are labeled by antisera to the β amyloid protein. *Biochem. Biophys. Res. Commun.*, 165: 182–188.

Pappolla, M., Sambamurti, K., Efthimiopoulos, S., Refolo, L., Omar, R. and Robakis, N. (1995) Heat shock induces abnormalities in the cellular distribution of amyloid precursor protein (APP) and APP fusion proteins. *Neurosci. Lett.*, 195: 105–108.

Peterson, C. and Goldman, J.E. (1986) Alterations in calcium content and biochemical processes in cultured skin fibroblasts from aged and Alzheimer donors. *Proc. Natl. Acad. Sci. USA*, 83: 2758–2762.

Proft, M., Kotter, P., Hedges, D., Bojunga, N. and Entian, K.-D. (1995) CAT5, a new gene necessary for derepression of glucoconeogenic enzymes in *Saccharomyces cerevisiae*. *EMBO J.*, 14: 6116–6126.

Richards, S.J., Morris, J.F. and Raisman, G. (1985) Solitary magnocellular neurons in the homozygous Brattleboro rat have vasopressin and glycopeptide immunoreactivity. *Neuroscience*, 16: 617–623.

Safaya, S. and Rieder, R.F. (1988) Dysfunctional α-globin gene in hemoglobin H disease in blacks. *J. Biol. Chem.*, 263: 4328–4332.

Salbaum, J.M., Weidemann, A., Lemaire, H.-G., Masters, C.L. and Beyreuther, K. (1994) The promotor of Alzheimer's disease amyloid A4 precursor gene. *EMBO J.*, 7: 2807–2813.

Salk, D., Fujiware, Y. and Martin, G.M. (1985) *Werner's syndrome and human aging*, Plenum Press, New York.

Schirmer, E.C. and Lindquist S. (1997) Interactions of the chaperone Hsp104 with yeast Sup35 and mammalian PrP. *Proc. Natl. Acad. Sci.*, 94: 13932–13937.

Selkoe, D.J. (1993) Physiological production of the β-amyloid protein and the mechanism of Alzheimer's disease. *Trends Neurosci.*, 16: 403–409.

Selkoe, D.J. (1994) Cell biology of the amyloid β-protein precursor and the mechanism of Alzheimer's disease. *Annu. Rev. Cell. Biol.*, 10: 373–403.

Sherrington, R., Rogaev, E.I., Liang, Y., Rogaeva, E.A., Levesque, G., Ikeda, M., Chi, H., Lin, C., Li, G., Holman, K., Tsuda, T., Mar, L., Foncin, J-F., Bruni, A.C., Montesi, M.P., Sorbi, S., Rainero, I., Pinessi, L., Nee, L., Chumakov, I., Pollen, D., Brookes, A., Sanseau, P., Polinsky, R.J., Wasco, W., Da Silva, H.A.R., Haines, J.L., Pericak-Vance, M.A., Tanzi, R.E., Roses, A.D. Fraser, P.E., Rommens, J.M. and St.George-Hyslop, P.H. (1995) Cloning of a gene bearing missense mutations in early-onset familial Alzheimer's disease. *Nature*, 375: 754–760.

Shilatifard, A., Conaway, J.W. and Conaway, R.C. (1997) Mechanism and regulation of transcriptional elongation and termination by RNA polymerase II. *Curr. Opin. Genet. Dev.*, 7: 199–204.

Shoji, M., Golde, T.E., Ghiso, J., Cheung, T.T., Estus, S., Shaffer, L.M., Cai, X.-D., McKay,D.M., Tintner, R., Frangione, B. and Younkin, S.G. (1992) Production of the Alzheimer amyloid β protein by normal proteolytic processing. *Science*, 258: 126–129.

Simons, M., Ikonen, E., Tienari, P.J., Cid-Arregui, A., Mönning, U., Beyreuther, K. and Dotti, C.G. (1995) Intracellular routing of human Amyloid Protein Precursor: axonal delivery followed by transport to the dendrites. *J. Neurosci. Res.*, 41: 121–128.

Sinclair, D.A., Mills, K. and Guarente, L. (1997) Accelerated aging and nucleolar fragmentation in yeast *sgs*1 mutants. *Science*, 277: 1313–1316.

Sisodia, S.S., Koo, E.H., Beyreuther, K. and Unterbeck, A. (1990) Evidence that β-amyloid protein in Alzheimer's disease is not derived by normal processing. *Science Wash. DC*, 248: 492–495.

Sohal, R.S. and Weindruch, R. (1996) Oxidative stress, caloric restriction, and aging. *Science*, 273: 59–63.

Strand, M., Prolla, T.A., Liskay, R.M. and Petes, T.D. (1993) Destabilization of tracts of simple repetitive DNA in yeast by mutations affecting DNA mismatch repair. *Nature*, 365: 274–276.

Strittmatter, W.J., Saunders, A.M., Schmechel, D., Pericak-Vance, M., Enghild, J., Salvesen, G.S. and Roses, A.D. (1993) Apolipoprotein E: high-activity binding to β-amyloid and increased frequency of type 4 allele in late-onset familial Alzheimer's disease. *Proc. Natl. Acad. Sci. USA*, 90: 1977–1981.

Sutherland, G.R. and Richards, R.I. (1995) Simple tandem DNA repeats and human genetic disease. *Proc. Natl. Acad. Sci. USA*, 92: 3636–3641.

Tang, M.X., Jacobs, D., Stern, Y., Mander, K., Schofield, P., Gurland, B., Andrews, S.H. and Mayeux, R. (1996) Effect of oestrogen during menopause on risk and age at onset of Alzheimer's disease. *Lancet*, 348: 429–432.

Tanzi, R.E., McClatchey, A.I., Lamperti E.D., Villa-Komaroff, L., Gusella, J.F. and Neve, R.L. (1988) Protease inhibitor domain encoded by an amyloid protein precursor mRNA associated with Alzheimer's disease. *Nature*, 331: 528–530.

Thibodeau, S.N., Bren, G. and Schaid, D. (1993) Microsatellite instability in cancer ofthe proximal colon. *Science*, 260: 816–819.

Thomas, M.J., Platas, A.A. and Hawley, D.K. (1998) Transcriptional fidelity and proofreading by RNA polymerase II. *Cell*, 93: 627–637.

Van Broeckhoven, C. (1995) Presenilins and Alzheimer's disease. *Nat. Genet.*, 11: 230–231.

Van Leeuwen, F.W., Van der Beek, E., Seger, M., Burbach, J.P.H. and Ivell, R. (1989) Age-related development of a heterozygous phenotype in solitary neurons of the homozygous Brattleboro rat. *Proc. Natl. Acad. Sci. USA*, 86: 6417–6420.

Van Leeuwen, F.W., Burbach, J.P.H. and Hol, E.M. (1998b) Mutations in RNA: a first example of molecular misreading in Alzheimer's disease. *Trends Neurosci.*, 21: 331–335.

Van Leeuwen, F.W., De Kleijn, D.P.V., Van den Hurk, H.H., Neubauer, A., Sonnemans, M.A.F., Sluijs, J.A., Köycü, S., Ramdjielal, R.D.V., Salehi, A., Martens, G.J.M., Grosveld, F.G., Burbach, J.P.H. and Hol, E.M. (1998a)

Frameshift mutants of β amyloid precursor protein and ubiquitin-B in Alzheimer's and Down patients. *Science*, 279: 242–247.

Volicer, L. and Crino, P.B. (1990) Involvement of free radicals in dementia of the Alzheimer's type. *Neurobiol. Aging*, 11: 567–571.

Vostrov, A., Quitschke, W., Vidal, F., Schwarzman, A. and Goldgaber, D. (1995) USF binds to the APB alpha sequence in the promotor of the amyloid β-protein precursor gene. *Nucl. Acid Res.*, 23: 2734–2741.

Wang, R., Meschia, J.F., Cotter, R.J. and Sisodia, S.S. (1991) Secretion of the b/A4 amyloid precursor protein. Identification of a cleavage site in cultured mammalian cells. *J. Biol. Chem.*, 266: 16960–16964.

Weidemann, A., Konig, G., Bunke, D., Fischer, P., Salbaum, J.M., Masters, C.L. and Beyreuther, K. (1989) Identification, biogenesis, and localization of precursors of Alzheimer's disease A4 amyloid peptide. *Cell*, 57: 115–126.

Welch, W.J. and Gambetti, P. (1998) Chaperoning brain diseases. *Nature*, 392: 23–24.

Wisniewski, K.E., Wisniewski, H.M. and Wen, G.Y. (1985) Occurrence of neuropathological changes and dementia of Alzheimer's disease in Down's syndrome. *Ann. Neurol.*, 17: 278–282.

Wong, A., Boutis, P. and Hekimi, S. (1995) Mutations in the *clk-1* gene of *Caenorhabditis elegans* affect developmental and behavioral timing. *Genetics*, 139: 1247–1259.

Yamazaki, T., Selkoe, D.J. and Koo, E.H. (1995) Trafficking of cell surface β-amyloid precursor protein: retrograde and transcytotic transport in cultured neurons. *J. Cell. Biol.*, 129: 431–442.

Yan, S.D., Yan, S.F., Chen, X., Fu, J., Chen, M., Kuppusamy, P., Smith, M.A., Perry, G., Godman, G.C., Nawroth, P., Zweier, J.L. and Stern D. (1995) Non-enzymatically glycated tau in Alzheimer's disease induces neuronal oxidant stress resulting in cytokine gene expression and release of amyloid β-peptide. Nat. Med., 1: 693–699.

Yu, C-E., Oshima, J., Fu, Y-H., Wijsman, E.M., Hisama, F., Alisch, R., Matthews, S., Nakura, J., Miki, T., Ouais, S., Martin, G.M., Mulligan, J. and Schellenberg, G.D. (1996) Positional cloning of the Werner's syndrome gene. *Science*, 272: 258–262.

W.E. Crusio and R.T. Gerlai (Eds.)
Handbook of Molecular-Genetic Techniques for Brain and Behavior Research (Techniques in the Behavioral and Neural Sciences, Vol. 13)

CHAPTER 5.8

Transgenic mouse models of Alzheimer's disease

Karen Duff

Nathan Kline Institute, 140 Old Orangeburg Road, Orangeburg, NY 10962, USA

Introduction

Alzheimer's disease (AD) is a major cause of dementia and death in the elderly, which, due to its incipient, mentally debilitating progression, causes great distress to patient and caregiver alike. The identification of AD causing genetic lesions, as well as the clearly defined pathological and behavioral phenotype, made the disease potentially very amenable to modeling in transgenic animals. Unfortunately, many of the potential pitfalls of reconstructing a complex human disease in mice were experienced by genetic engineers working in this field. The following article aims to demonstrate the technical considerations involved in making a transgenic mouse for a disease like AD, and to discuss how far we have come in generating a usable model for the disease.

AD pathology

Alzheimer's disease is characterized by the presence in the brain of extra-cellular plaques composed mainly of a deposited protein fragment called beta amyloid (Aβ), and intracellular neurofibrillary tangles (NFTs) composed of an abnormal form of the tau protein. There are four well-recognized risk factors in AD: heredity, Down's syndrome, environmental insult (such as head injury) and old age. The familial forms of AD were the first to be studied and have given us great insight into some of the causes of AD. Although we now know that the causes of AD are heterogeneous, the phenotype is remarkably consistent, with amyloid plaques and NFTs being the major pathognomic features. The relative contributions of amyloid plaques and NFTs to AD dementia have been contested since the pathology of the disease was first described in, 1904. Some patients with AD seem to be devoid of plaques (also known as deposits) (Bancher and Jellinger, 1994) which argues for NFTs being the causative lesion. No AD-causing genetic lesions have been published in the tau gene, but several genetic lesions in different genes have been identified that directly influence amyloid accumulation. This simple fact continues to convince most of the field that amyloid is at the root of disease etiology, with tau possibly having an indirect, secondary role.

Genetic analysis reveals a unifying mechanism

In, 1991, John Hardy and his colleagues used linkage analysis to demonstrate that a mutation at codon 717 of the amyloid precursor protein (APP) gene caused classic, early onset familial AD in a small number of families (Goate et al., 1991). Since then, several other mutations have been identified in APP; two in the same codon (Murrell et al., 1991; Chartier-Harlin et al., 1991), one in the adjacent codon (Eckman et al., 1997), and a double mutation at codons 670 and 671 (Mullan et al., 1992). Although the APP mutations only account for a total of 5% of early onset cases which in tern account for less than 1% of the total number of AD cases, they were the first indication that APP may be the culprit in AD. Several groups have subsequently identified AD causing mutations in genes other than APP. Mutations in the presenilin 1 (PS1) gene on chromosome 14 (Sherrington et al., 1995) account for the majority of the early onset cases of familial AD, whereas mutations in the closely related presenilin 2 (PS2) gene on chromosome 1 are associated with late onset AD (Levy-Lahad et al., 1995) (see Cruts et al., 1996 for review). The majority of the familial cases of late onset AD however are associated with a polymorphism in the apolipoprotein E (ApoE) gene on chromosome, 19 where it seems that certain allelic combinations of the ApoE protein are particular risky for AD (Corder et al., 1993, 1994).

Researchers first focused their attention on the APP protein for three simple reasons; mutations in the APP gene are known to cause AD, an extra copy of the APP gene due to trisomy 21 is linked to AD in Down's syndrome and the Aβ region of APP makes up the core of the AD amyloid plaque. The APP gene is located on chromosome 21 and is expressed as several alternatively spliced transcripts, the most abundant of which encode proteins of 770, 751 and 695 amino acids (Yoshikai et al., 1991). The precursor protein undergoes a complex series of post-translational processing events that release the bulk of the protein into the extra-cellular space. Events at the C terminal can generate a range of fragments including the potentially amyloidogenic beta amyloid fragment. This peptide exists in several forms but the most abundant is 40 amino acids long and is known as Aβ40. A less abundant species is 42 (or 43) amino acids long and has been designated Aβ42(43).

Under certain conditions both types of Aβ have the potential to form insoluble, fibrillar, beta pleated strands of amyloid that are deposited in specific regions of the brain. In vitro studies on synthetic Aβ have shown that this tendency to fibrillarize increases when hydrophobic residues are included at the carboxy terminal (i.e. extending the molecule from Aβ40 to Aβ42(43)) (Jarrett et al., 1993). Cells transfected with an APP cDNA containing the double mutation at codons 670 and 671 have been shown to produce an approximate six-fold increase in Aβ fragments compared to cells transfected with a non mutant gene (Citron et al., 1992; Cai et al., 1993). The 716 and 717 mutations on the other hand do not lead to an increase in the amount of Aβ but lead to an increase in the amount of the Aβ42(43) form relative to Aβ40 (Eckman et al., 1997; Suzuki et al., 1994). Mutations in the APP gene are not the only influencing factor on APP processing however. Researchers have found that the level of Aβ42(43) is elevated relative to Aβ40 in fibroblasts isolated from Alzheimer patients known to carry mutations in the PS-1 gene (Scheuner et al., 1996). It is therefore possible that in individuals with APP or PS mutations, AD is caused by perturbations in APP metabolism and that the elevation of the Aβ42(43) form of Aβ is the critical pathogenic event.

APP transgenic mice as models of AD

The race to generate an animal model for Alzheimer's disease began in, 1991 when the first APP mutation was discovered. Since then, several different approaches have been undertaken to produce a mouse model with the major hallmarks of AD; plaques, tangles, neuro-degeneration and cognitive deficits. These approaches fall into three main categories: cDNA over-expression models, whole gene over-expression models and gene targeted models. Although several models have now been created that model the amyloid accumulation and deposition process, a mouse with all the features of the phenotype, including tangle formation, cell loss and cognitive impairment has not yet been generated.

TABLE 1

Transgenic animals overexpressing AD related genes

	Construct	Elevated $A\beta$	Pathology	Other phenotypes
Mutant APP overexpressing mice			*Aβ deposits*	
Games et al., 1995 (PDAPP)		$A\beta42$ then $A\beta42 + 40$	yes	Hyperphosphorylated-tau.
Hsaio et al., 1996 (Tg2576)	huPrP:APP695:swe	$A\beta42 + 40$	yes	Oxidative damage, cognitive impairment
Sturchler-Pierrat et al., 1996 (line 23)	Thy1:APP751:swe	$A\beta42 + 40$	yes	Hyperphosphorylated-tau, cell loss
Nalbantoglu et al., 1997	NF-L:APP C-term	ND	yes	Cognitive impairment, cell loss
Mutant PS1 overexpressing mice				
Duff et al., 1996 (PDPS1)	PDGF:PS1M146L and M146V	$A\beta42$	No	Calcium homeostasis disruption
Borchelt et al., 1996	moPrP:PS1A246E	$A\beta42$	No	Calcium homeostasis disruption
Citron et al., 1997	hmPrP:PS1M146L & L286V	$A\beta42$	No	
Tau overexpressing mice			*Tangles*	
Gotz et al., 1995 (Alz7)	Thy 1:Tau40	N/D	No	Hyperphosphorylated-tau

cDNA transgenics

The basic approach to modeling AD aims to over-express $A\beta$ in the cortex and hippocampus using neuron directing promoters and mutant or normal sequence APP cDNA (see Quon et al., 1991; Kammesheidt et al., 1992; and Neurobiology of Aging, Vol. 17(2), 1996 for several articles). Most of the early models failed to achieve significant elevation of $A\beta$ levels and none described robust pathology changes although one did describe the formation of diffuse amyloid deposits (Quon et al., 1991; Higgins et al., 1994). The apparent failure of these models led investigators to believe that marginal over-expression of a single splice variant of APP in regions and cell types of the brain that were not necessarily involved in AD pathogenesis may have contributed to their failure to develop robust, abnormal pathology. The focus of several labs therefore moved to whole-gene transgenics.

In, 1995, the first transgenic mouse with amyloid deposits was reported (Games et al., 1995). This mouse had been created using a mutant (V717F) APP mini-gene under the control of a neuron preferential PDGF promoter. The APP cDNA contained intronic sequences between exons that were alternately spliced and was capable of generating all three of the major APP isoforms from the same cDNA. This mouse had very high levels of APP and $A\beta$, including elevated $A\beta42(43)$. It developed numerous extra-cellular thioflavin S-positive $A\beta$ deposits and neuritic plaques in the hippocampus, corpus collosum and cerebral cortex at 6–9 months of age and displayed other features of AD, such as astrocytosis and microgliosis. Classic NFTs and overt cell loss were absent however. In October, 1996, a second mouse with amyloid deposits was reported (Hsiao et al., 1996). This mouse had been created using a different promoter (the prion gene promoter and regulatory elements) and expressed a mutant

(K670N/M671L) isoform of APP (the 695 form only). The resulting mouse (known as tg2576) also had very high levels of APP and Aβ40/Aβ42(43) that increased during its lifetime. Pathology testing was limited but mature amyloid plaques had been confirmed in a mouse at 12 months of age. Again, NFTs were not seen. In, 1997, a third mouse with deposits was reported which also shows other features of AD pathology such as hyper-phosphorylated tau (but not NFTs) (Sturchler-Pierrat et al., 1997).

It therefore appears that mice are capable of forming amyloid deposits when challenged with sufficient levels of beta amyloid although the level is far greater than that required by humans. In this respect, both the PDAPP and tg2576 mice are excellent models of the deposition process. Additional plaque components such as murine ApoE have also been identified in deposits suggesting that the deposition process is very similar in mice and humans (C. Lemere pers. comm).

Whole-gene transgenics

One of the major criticisms of the early, non--depositing, APP cDNA mice was that the level of transgene expression was low relative to endogenous levels and one possible explanation for this was that they lacked vital regulatory regions within the construct. cDNA is derived from mRNA which consists mainly of the protein coding sequence. mRNA is produced from the genomic (nuclear) DNA during transcription. The genomic DNA contains all of the regulatory regions necessary to direct the efficient and correct transcription of the mRNA, in the correct spatial and temporal pattern. In mammals, the genomic DNA segment that represents a gene is usually far larger than the mRNA derived from it. It is possible however to identify the gene region, including the promoter and polyadenylation signal site associated with it, and clone it into one of several special vectors known as a yeast artificial chromosome (YAC), bacterial artificial chromosome (BAC) or P1 derived artificial chromosome (PAC). The choice of vector determines the size of genomic fragment included and for large genes such as APP, the gene is only available as a YAC. Numerous human genes have now been cloned into YACs, BACs and PACs, and previously uncharacterized genes can be isolated from gridded libraries available through companies such as Genome Systems or Research Genetics.

Although YACs can offer distinct advantages over cDNAs, YAC directed genetic engineering of mice was not possible until recently. Several articles published in, 1993 explored the various ways in which these large genomic fragments could be introduced into mouse eggs (Gnirke et al., 1993; Jakobovits et al., 1993; Schedl et al., 1993; Strauss et al., 1993; Choi et al., 1993). It was found that smaller YACs (up to 500 kb) could be injected directly into mouse pronuclei in the same way as cDNA transgenes (Gnirke et al.1993). At 650 kb, the human APP YAC was thought to be too large to pass through a standard microinjection needle intact. Consequently, researchers examined the possibility of introducing the APP YAC into mice via ES cells. To do this, the YAC was introduced into the ES cell nucleus by lipofection where a lipid complex binds to the YAC DNA and fuses with the lipid component of the nuclear membrane. In a few cells, the YAC DNA passes into the nucleus and integrates into the mouse's own genomic DNA. Because the YAC transgene carries all its regulatory sequences, the human APP gene is expressed at a similar level to the mouse's own gene. Having identified recombinant ES cells in culture, the cells are injected into the blastocyst from a 3.5 d donor embryo. The injected ES cells then differentiate and contribute to cell lines within the developing embryo including the germline. Unlike the mice derived by pro-nuclear injection where (in theory) every cell carries the transgene, the founder mice derived from ES cell manipulations are genetically chimeric. If the transgene is inherited, the offspring of the founder mice can be bred to genetic homogeneity.

Two research groups have introduced an intact human APP YAC into mice in the hope that integrating one or two copies of the human gene would result in a mouse over-expressing APP in the correct regions of the brain and with the correct ratios of the different splice variants. Both labs have so far reported that although elevated levels of Aβ were generated, AD pathology was not seen at the ages tested. (Lamb et al., 1993; Pearson and Choi et al., 1993; Buxbaum et al., 1993). Mice have subsequently been created with mutant YACs to boost the levels of Aβ40 and 42(43) (Lamb et al., 1997), and, although Aβ levels are elevated, there have been no reports of pathology yet.

Recently, an attempt has been made to generate a transgenic rat model for AD (Folkesson et al. unpublished data). Although the process of generating a transgenic rat is less efficient than for mice, rats have the added utility in that they are a better-characterized model system in terms of their behavioral and pharmacological properties. The rat model has been generated by the injection of a mutant APP YAC into rat embryos. Technically, the micro-injection procedure is easier for rats as the diameter of the micro-injection needle can be made slightly larger, thereby reducing the risk of DNA shearing. Although the YAC appears to be intact, data is not yet available on the level of Aβ generated, or whether the rats develop AD pathology.

Gene targeting

One potential drawback of both the cDNA and YAC models outlined above is that the transgene integrates into the mouse genome at a random position and often as several copies. There is therefore a risk that the transgene could have integrated either in the vicinity of, or within, another gene. If the transgene has integrated within a gene, it will almost certainly knock out the function of that gene which could be disastrous for the mouse. If the transgene integrates within the range of another gene's regulatory elements, there is a possibility that it will either not be expressed efficiently or correctly. One way to overcome this random integration liability is to target a construct to the mouse's own APP gene and replace the mouse portion with, for example, the human AD mutations. Another advantage of targeting the mouse's own gene is that it overcomes the possibly deleterious effect of expressing a human transgene in a background where the mouse's own gene is also being expressed, which may prevent the disease phenotype from being manifested.

Some researchers therefore decided that the genetically pure, targeted model that only contained a single mutant copy of the APP gene would mimic the human disease better than either the YAC or cDNA models. The human DNA sequence of the beta amyloid region of APP differs from the mouse sequence at three amino acids. Investigators "humanized' and mutated the murine APP gene by introducing the targeting transgene into the mouse genome in embryonic stem (ES) cells, using Cre-Lox technology to remove unwanted sequences from the gene after manipulation (Reaume et al., 1996). The resulting mice were examined for Aβ levels and pathological changes but despite elevated levels, no abnormal pathology was noted.

PS1 transgenic mice as models of AD

The amyloid based hypothesis of AD was re-inforced in, 1996 when data from transfected cells and transgenic mice showed that over--expression of mutant PS1 (but not wild type) led to a significant elevation in the levels of Aβ42(43) but not Aβ40 (Duff et al., 1996; Borchelt et al., 1996; Citron et al., 1997). This suggests that the PS mutations cause AD by elevating Aβ42(43) levels in the same way as the APP 716 and 717 mutations (Suzuki et al., 1994; Eckman et al., 1998). Furthermore, crossing mutant PS1 mice with over-expressing APP mice (such as tg2576) leads to a rapid acceleration of tg2576 driven pathology, with amyloid deposits becoming visible as young as 15 weeks, compared to 9–12 months for the parental, singly transgenic tg2576 mouse (Holcomb et al., 1998). This data, as well as data from another PS/APP cross, (Borchelt et al., 1997) suggests that the ability to deposit amyloid is dependent on the amount of Aβ in the cell. It appears that the PS1 directed elevation in Aβ42(43) brought the cell to the critical concentration faster thereby accelerating the deposition process.

Behavioral characteristics of AD transgenic mice

Alzheimer's disease is essentially characterized by age-linked memory loss and one of the features of the disease that should be apparent in a transgenic mouse model is hippocampal dysfunction. A number of tests have been applied to transgenic mice to assess for behavioral differences (reviewed extensively in Crawley and Paylor, 1997), but the most commonly applied behavioral paradigms applied to AD mice are changes in spontaneous alternation in T, Y or radial mazes and performance in the Morris water maze (spatial version) (Hsiao et al., 1995, 1996; Yamaguchi et al., 1996; Nalbantoglu et al., 1997; Moran et al., 1995; D'Hooge et al., 1996; Moechars et al., 1996). Applying rat behavioral methods to mice can be problematic — some tests (such as the water maze) need to be scaled down for mice, which may change the memory process being tested. Mouse sub-strains are also very variable in their response to behavioral tests and some are physically less able to perform than others (reviewed in Crawley and Paylor, 1997; Crawley, 1996). This is particularly relevant for the most commonly micro-injected inbred strain, FVB, which is essentially blind, or the 129 strain used for knockouts which often suffers agenesis of the corpus collosum (reviewed in Gerlai, 1996 also see Chapter 3.1.10). Assessing transgenic mice and their non-transgenic littermates for motor and sensory problems before embarking on a behavioral test is essential to remove poor performers that might skew results.

Strain-specific responses can also confound behavioral observations. For example, the FVB strain seems to be particularly sensitive to over--expressed APP and Aβ. One group found that FVB mice over-expressing an APP transgene show a strain dependent decrease in longevity and increased incidence of neophobia (Hsiao et al., 1995). Another group showed disturbed behavior, neuronal degeneration and decreased longevity when the C terminal (Aβ fragment) was over--expressed in FVB mice (Moechars et al., 1996). Neither mouse line forms deposits, most likely due to insufficient levels of Aβ. Although the neophobic behavior and decrease in longevity are not immediately reminiscent of human AD, it is possible that FVB mice will exhibit features of the disease that other mouse strains are more resistant to.

Many of the transgenic AD lines have now been assessed for behavioral changes. PS mutant transgenic mice do not show behavioral deficits (Holcomb et al., 1998). Although some of the earlier APP (or C-terminal fragment) over-expressing models did not develop robust plaque pathology, the transgenics performed worse in behavioral tests (Hsiao et al., 1995; Moechars et al., 1996; Moran et al., 1995; D'Hooge et al., 1996; Yamaguchi et al., 1991). Later mice that developed robust pathology were assessed in the same way, and one (tg2576) was reported to show age dependent decline in performance in the water maze (Hsiao et al., 1996). A second mouse also has reported decline in water maze performance and this has been related to amyloid deposition and cell loss (Nalbantoglu et al., 1997). Preliminary behavioral characterization (Dodart et al., 1997; Justice and Motter, 1997) has also been performed on the PDAPP mouse from Athena neuroscience (Games et al., 1995), which shows very extensive amyloid deposition. Several tests (Y and radial arm, water maze etc) were performed and the mice showed impairment in all tests at a young age, before amyloid deposition had occurred. Spontaneous alternation was already reduced to chance levels in the young transgenic mice verses their controls so changes on aging were not measured. In the water maze, young transgenic mice were impaired relative to controls but this did not increase with aging. Although these data have not yet been published in a peer reviewed journal, the analysis looks convincing.

Our own data also suggests that mice that over-express APP show deficits in spontaneous alternation and this deficit is obvious in young mice, in the absence of Aβ deposits (Holcomb et al., 1998). As the transgenic mice are already at chance levels, it was not possible to examine the effects of aging on this phenotype. Our mice are derived from a cross between tg2576 and PS1, and the deficit in spontaneous alternation seen in young tg2576 parental mice (Hsiao et al., 1996) is reminiscent of our results, although the tg2576 data never reached significance. Taking the data from all the behavioral tests together, it appears that some feature of APP over-expression, or the elevated levels of Aβ that result, leads to mild cognitive impairment and that deposition of Aβ is not required, at least for deficits in spontaneous alternation to become apparent. Other tests may be more sensitive to Aβ accumulation or more informative in the different mouse models created due to mouse strain influences.

The issue of cell loss in AD transgenic mice is controversial. Mice with reported behavioral problems (e.g. tg2576 and PDAPP) do not show overt cell loss (Irizarry et al., 1997a, 1997b) and most mice with reported cell loss (or neurodegeneration) have not been tested behaviorally (Oster-Granite et al., 1996, Sturchler-Pierrat et al., 1997). Cell loss and hippocampal dysfunction do not necessarily have to go together though (Rapp and Gallagher, 1996) as cell dysfunction rather than loss might explain behavioral impairment. Ideally, we would like to see some cell loss associated with cognitive decline, but, perhaps more importantly considering the late age of onset of human AD, we need to see an age dependent decrease in whatever parameter we chose to examine, be it cell number, synapse density, cell dysfunction or cognitive ability. The central question then is can this be correlated with Aβ accumulation and/or deposition or is this just an epi-phenomenon? It is crucial therefore that the one mouse reported to show deposition, cell loss and behavioral deficits be confirmed (Nalbantoglu et al., 1997).

Technical considerations in the generation of cDNA transgenic mice

The most straightforward approach to modeling a genetic disease is to over-express a human cDNA from the gene of interest and include pathogenic mutations. The AD phenotype has been modeled most successfully by over-expressing human APP cDNA containing mutations that flank the Aβ peptide to generate increased levels of Aβ40 and/or 42 (Games et al., 1995; Hsiao et al., 1996; Sturchler-Pierrat et al., 1997). Mice that over-express other genetic components, such as mutant presenilin genes or ApoE have also been examined for AD phenotype and have been crossed to APP transgenics to examine the effects of gene interaction on AD phenotype (Borchelt et al., 1997; Bales et al., 1997; Holcomb et al., 1998).

Promoters

The most important consideration is where the transgene is to be expressed: for APP and other AD related genes, neuronal expression is most relevant. A variety of promoters such as platelet derived growth factor (PDGF), thy1, prion (PrP) and neuron specific enolase (NSE) have been used to direct high level expression in brain neurons (Games et al., 1995; Sturchler-Pierrat et al., 1997; Hsiao et al., 1996; Higgins et al., 1994). Viral promoters do not give high level expression in neurons and should not be used. Promoter fragments from either rodent or human often work equally well. It is wise to use a well characterized promoter to drive transgene expression in the mouse as promoter elements such as enhancers or repressors can affect the expression of cDNAs in vivo and their location within a promoter fragment is often unknown. Promoters that work well in transfected cells do not necessarily work well in mice and vice versa, so testing a transgene in transfected cells before micro-injection is often not informative (Petitclerc et al., 1995).

A potential promoter fragment should be checked for robust and regionally correct expression in mice by using it to drive a reporter gene (luciferase or β-galactosidase for example) and examining transgenic mice for reporter gene product. Many tissue/cell specific promoters have already been identified and described in the literature, and they are often supplied upon request. A larger promoter is usually less influenced by position effects although negative regulatory elements may be included in the sequence leading to lower levels of expression.

cDNA

Choice of transcript may be a consideration if the message undergoes alternate splicing. For APP, there are three major transcripts in the brain that encode proteins of 695, 751 and 770 amino acids. The two longer proteins contain a Kurnitz protease inhibitor region of unknown functional significance. There has been some suggestion that the ratio between KPI containing (751 and 770) and KPI deficient (695) isoforms is important in AD pathogenesis as the KPI-deficient APP695 isoform has been reported to decrease in AD brain (Johnson et al., 1990). APP 695, however, is most abundant in neurons (Tanzi et al., 1988). The earliest mouse model that deposited amyloid used a mini gene approach to accommodate the alternate splice problems (Games et al., 1995). In APP, splicing occurs around exons 7, 8 and 9 to generate the three proteins. The mini-gene included intronic splice elements inserted into the cDNA between the relevant exons to allow the production of the three different protein isoforms. Because some of the introns were too big to manipulate, partial introns (essentially containing the splice donor and acceptor and the branch region) were inserted into the cDNA by PCR. Mice over-expressing just APP695 (Hsiao et al., 1996) or APP751 (Sturchler-Pierrat et al., 1997) later demonstrated that over-expressing single APP isoforms could also evoke amyloid deposition and that the level of amyloid was important, not the type. An important point has been demonstrated however, and different diseases may have different requirements.

We usually isolate the cDNA with gene specific primers from a human brain cDNA using a gene trapping kit. cDNA, complete with mutations can be isolated from RNA from affected patients if available. Alternatively, mutations, can be introduced into the cDNA by standard mutagenesis methods. Restriction sites can be added to the ends of the cDNA by incorporating them into primers used to amplify the cDNA. It is advisable to remove unwanted UTR sequence at the 5′ and 3′ ends of the cDNA and express only the coding region, from the ATG start to the TAG stop signal. The translational start site should be compared to an ideal Kozak sequence (A/G)-NNATGG) to ensure that the transgene will be translated with maximum efficiency and changes can be incorporated by PCR (when adding restriction sites) if necessary. PCR should be performed with high fidelity Taq (e.g. PFU from Stratagene or Expand High Fidelity from Boehringer) to prevent errors and the coding sequence (as well as critical regions of the promoter) should be sequenced to ensure that the translated protein will be correct.

Introns

It is general transgenic folklore that inclusion of an intron will increase the level of expression (Choi et al., 1991). Although the exact reasons are not clear, introns can have a great influence on attaining reliable, high level expression of the transgene. Many promoter fragments or cDNAs with 5′-UTRs already contain an intron, although it will not always be apparent unless the promoter/gene has been characterized. Inserting an 'artificial' intron either between the cDNA and promoter or between early coding exons is advisable if the intron status of the transgene is in doubt. Introns inserted at the 3′ end have been reported to be less efficient than 5′ introns (Choi et al., 1991). Occasionally, several exons and introns can be PCR amplified from genomic DNA if cloning sites are suitably located in cDNA exons, and the PCR product is reasonably small. The advantage of placing an intron between the promoter and the cDNA are two-fold; the intron may not be spliced out correctly if it is inserted between coding exons, leading to the generation of an aberrant protein and secondly, the promoter/intron cassette can be used to drive expression of different cDNAs inserted downstream.

Polyadenylation signal

The most commonly used artificial signal is from SV40. The signal site and some flanking sequence (approximately 100 bp on either side) can easily be removed from commercial vectors by PCRing across the site, using primers made to the sequence data usually supplied with the vector, and incorporating restriction sites for subsequent cloning. The poly-A site can be incorporated into the promoter/intron/cDNA construct. Occasionally, translation of a certain cDNA can be influenced by the length (or composition) of the 3′-UTR, and Poly-A signal choice (De Sauvage et al., 1991). Endogenous 3′-UTR and polyadenylation signal sites can be included in the transgene instead of an artificial site as long as the site is robust enough to polyadenylate the transgene mRNA when the transgene is inserted into the mouse genome.

Overall design

To improve integration efficiency, transgene DNA should be linearized prior to micro-injection. Vector sequence should also be removed as vector DNA has been shown to reduce transgene expression levels in mice (Kjer-Nielsen et al., 1992). The Stratagene pBluescript range of vectors are particularly useful in this respect as two unique BSSHII sites are located flanking the cloning region which can be used to isolate the transgene if BSSHII does not occur within the construct.

Example: generation of mutant PS1 cDNA mice (see also Duff et al., 1996 for primer sequences and PCR conditions)

1. Promoter cloning. The promoter from the PDGF gene (Sasahara et al., 1991) was chosen as it directs neuronal expression to a high level and it had been used successfully in the PDAPP line of mice (Games et al., 1995). A 1.4 kb fragment of the promoter was PCR amplified using High Fidelity Taq (Boerhringer), from a template provided by T. Collins. Xba 1 sites were included in the PCR primers to allow cloning into the corresponding site in pBluecriptII SK-(Stratagene).
2. Poly-A tail cloning. The poly-A signal site (AATAAA) and 100 bp of flanking sequence was PCR amplified using the expression vector, pβNASS, (Clontech) as template. Sal 1 sites were included in the primer sequence.
3. cDNA cloning. The PS1 cDNA was isolated from a human brain library (Gibco/BRL human brain library in pCMV.Sport) using a primer to the 5′-UTR for trapping (Gibco/BRL gene trapper kit). The coding region (from the ATG start to the TAG stop sites) was amplified from the cDNA using primers which incorporated a Not 1 site at each end of the product. In addition, the 5′ forward primer included a perfect Kozak sequence.
4. Mutagenesis: Three different constructs were prepared, one bearing the wild-type PS1 sequence and two bearing mutations that cause AD with onset in the early forties. The wild type over-expressing transgenic mouse is an important control to distinguish between the effect of mutations and

transgene over-expression on phenotype. Site directed mutagenesis was performed on the complete transgene using the transformer mutagenesis kit (Clontech) with primers incorporating the different mutations coupled to a selection primer provided in the kit. Mutagenesis was performed with good efficiency on the 9 kb pBluescript plasmid, although higher efficiencies have been attained with smaller constructs that just contained the cDNA.

5. Intron cloning: We were fortunate in finding two Bcl 1 sites in adjoining early coding exons of PS1. PCR primers were made to exon 4, preceding the Bcl 1 site, and to the sequence following the Bcl 1 site in exon 5. These primers were used to PCR amplify across the 2.8 kb PS1 intron using human genomic DNA (in the form of a PS1 PAC) as a template. The PCR product was digested with Bcl 1 and cloned into the corresponding site in the PS1 cDNA. This intron fragment (which includes the 3′ end of exon 4, the splice donor site, intron, splice acceptor and 5′ end of exon 5) has since been inserted, blunt ended, between the PDGF promoter and other cDNAs such as PS2 and α-synuclein, where it functions well as an artificial intron.
6. In general, primer restriction sites were preceded by three or four random bases (e.g. CAT) to facilitate cleavage by the restriction enzyme. Following PCR, the unpurified product was digested directly in 1x restriction enzyme buffer with 1–2 U of enzyme for 3 h. The digested DNA was purified using a Wizard clean up column (Promega). The pBluescript plasmid was digested with the relevant enzyme for 3 h and was phosphatased to prevent vector self re-ligation by the addition of 1 μl Calf Alkaline Phosphatase (Promega) in 1 buffer, with incubation at 37 for 30 min. The DNA was purified using a Wizard clean up kit (promega). The concentration of insert and vector DNA was assessed on an ethidium bromide stained agarose gel and a 3:1 (insert : vector) ratio of DNA was ligated using the Rapid DNA Ligation Kit from Boerhinger. One tenth of the ligation mix was used to transform sub-cloning efficiency competent DH5a cells (Gibco/BRL). Colonies that grew on ampicillin containing agar were picked, grown up in L broth and screened by diagnostic restriction digestion to check the presence and orientation of the cloned insert.
7. DNA preparation: Once all the transgene components had been assembled and sequenced to check for integrity, the plasmid was digested with BSSHII. The vector (2.8 kb) and construct (6.4 kb) were resolved by gel electrophoresis on 1% low melting point agarose. The larger band was excised and purified using a Wizard PCR clean up kit. DNA was eluted from the column in TE buffer heated to 60° (Tris 10 mM, pH 7.4, EDTA 0.1 mM, pH 8 made up in endotoxin-free water, e.g. Gibco/BRL). DNA concentration was assessed by comparison to concentration assessment markers (Gibco/BRL) on an agarose gel and the DNA was diluted in the same buffer to a concentration of 3–5 ng/ul. The diluted DNA (in a volume of approximately 100 μl) was centrifuged at 14 k RPM in a micro-fuge and the top 80 μl was carefully removed to a fresh tube. DNA was stored at 4° until required for micro-injection, but was centrifuged again after periods of storage.
8. Micro-injection: transgenic mice were generated by micro-injection of DNA into the pro-nucleus of a single celled mouse embryo. A triple hybrid strain (B6/D2F1 x Swiss Webster) was used for the micro-injections as this strain is known to support brain amyloidosis. Swiss Webster females were used as egg donors as they are an inexpensive out-bred strain that super-ovulate well and give easily injected eggs.

9. Founder pups were assessed for transgene status by Southern blot analysis to ensure that the transgene was intact and that the cDNA probe could distinguish between endogenous and transgene DNA. Genomic DNA was isolated using a genomic DNA kit (Promega) from 1 cm tail snips taken from 10 d old pups. Assessment of subsequent litters requires less scrupulous determination of transgene status and can be performed by routine PCR or dot-blot analysis. The genotype of each mouse used in experiments should be checked after sacrifice to confirm the transgene status.

Comparison of cDNA and whole gene over-expression models

Cloning and mutagenesis

As cDNAs are usually small enough to clone into standard plasmids, cDNA transgene cloning procedures are usually straightforward. Mutagenesis can be performed by one of several methods such as site directed mutagenesis, extension overlap PCR, or cassette swapping.

Whole gene cloning can be more difficult due to the lack of unique restriction sites in large genomic regions: YACs can be mutagenized by two step homologous recombination in yeast, which, although time-consuming, is usually very efficient (Duff and Huxley, 1996). PACS and BACs are more difficult to mutate if cloning sites are not available, but can be mutated through a process known as RARE (Boren et al., 1996). PACs and BACs are simple to grow up as they behave like plasmids and DNA yields are good, especially when using the newer genomic DNA extraction kits such as KB-100 (Genome systems). YACs require preparation in yeast and the DNA yield is usually relatively low. An excellent practical guide to the manipulation of YACs, PACs and BACs can be found in Current Protocols in Human Genetics, Chapter 5.

Linearization, vector removal and DNA purification

cDNA transgene linearization usually involves standard restriction digests and agarose gel purification. The large size of YACs, PACs and BACs means that DNA must usually be manipulated by pulsed field gel electrophoresis (PFGE) and extra care must be taken to avoid shearing. Difficulty increases with increasing size (see Current Protocols in Human Genetics, Chapter 5 for details of PFGE).

YACs do not need to be digested before micro-injection as they are already linear. Ideally, the YAC should be removed from the yeast chromosomes, which usually requires PFGE. If the YAC and the surrounding yeast chromosomes co-migrate, they may not resolve by PFGE. In this case, it may be necessary to co-inject them into the mouse, although this may result in yeast chromosome integration/expression problems.

Ideally, PACs and BACs should be linearised and have as much vector removed as possible. Linearization can be performed by restriction digestion (if sites are present in the vector but not the large insert) or by Cre-mediated recombination (Mullins et al., 1997. Most manipulations with YACs, PACs and BACs still require PFGE to isolate DNA for injection and the purification of intact DNA that is clean enough to pass through an injection needle can be difficult.

Injections

The micro-injection of cDNA transgenes is usually straightforward. Highly purified water (e.g. endotoxin-free cell culture water from Gibco/BRL) should be used to avoid embryo lethality. We have injected DNA at concentrations from 2 to 20 ng/μl with no obvious differences in integration efficiency or copy-number. Whole-gene sized DNA can be sheared during injection so larger diameter needles are advisable and the DNA integrity should be checked in founder mice by PFGE mapping. PCR is not suitable as fragments of the DNA may be scattered in the genome giving a positive PCR signal, even though the DNA is not contiguous.

Transgene stability

Transgene stability through the generations is difficult to predict, although it is usually better for smaller transgenes. Larger transgenes can undergo recombination if several copies are present. Chimerism can influence the transmission frequency of the transgene from the founder to the F_1s, but this is usually due to the timing of injections rather than the size or composition of the transgene. Low transmission frequency can usually be rectified by isolating eggs for micro-injection later, or earlier in the day, or by changing the light/dark cycle that can be considerably different for each strain.

Transgene expression level and regulation

For cDNA transgenes with small (<5 kb) promoter regions, integration position in relation to surrounding murine genes can be extremely influential. Both the level, and pattern, of expression can be affected and in situ/immunohistochemistry should be performed if regional expression patterns are important. Expression levels from small transgenes are usually not copy number dependent. Due to the buffering effect of sequence flanking the promoter, large promoters in cDNA transgenes or genes contained in PACs, BACs or YACs are usually not influenced by integration position and expression levels are copy number dependent (Lamb et al., 1993). If all the regulatory elements are included (enhancers, introns, polyadenylation signals etc) the transgene acts more like the endogenous gene and 1:1 levels of expression can be achieved. The main advantages of whole-gene transgenes are two-fold: firstly, the transgene will be expressed with the correct spatial and temporal distribution and secondly, the naturally occurring alternate splice forms will be generated in the correct relative ratio and distribution.

It should be remembered that both cDNA and whole-gene transgenes are expressed in addition to the endogenous gene. This can be problematic if the resulting gene over-expression cause physiological artifacts or lethality, if the endogenous protein interferes with the transgene derived protein or if the endogenous (normal) gene masks the effects of the transgene derived mutant protein. In this case, gene targeting may be a better option.

Overview

The history of the transgenic modeling of Alzheimer's disease illustrates how difficult it can be to model a seemingly straightforward disease. It was expected that over-expression of either normal APP (to model Down's associated AD) or mutant APP (to model the familial AD cases) would lead to the Aβ elevation and deposition into plaques, the formation of pathogenic tau and tangles, cell damage, neuro-degeneration and concomitant cognitive impairment. After many years, we have reliably achieved the first part of the process. Although there are reports of associated cell damage and/or cognitive impairment in these mice, further work is required to fully illustrate the extent and significance of these changes. Despite the fact that Aβ depositing mice such as tg2576 and PDAPP do not show all the features of the human disease, the link between Aβ levels and AD causing mutations in APP and the presenilins is highly suggestive of Aβ's central role in disease etiology. Fortunately, we have already generated a mouse with extensive amyloid perturbations. Drugs that are effective at breaking down Aβ deposits, or lowering intracellular Aβ in the mice, may be effective in humans, at least in the crucial first step of slowing the disease's progression. We may not need to achieve the full behavioral and histopathological phenotype in mice to successfully help humans and the mysteries of why mice fail to develop NFTs or undergo extensive neuro-degeneration like their human counter parts may become academic.

References

Bales, K.R., Verina, T., Dodel, R.C., Du, Y., Altstiel, L., Bender, M., Hyslop, P., Johnstone, E.M., Little, S.P., Cummins, D.J., Piccardo, P., Ghetti, B. and Paul, S.M. (1997) Lack of apolipoprotein E dramatically reduces amyloid beta-peptide deposition. *Nat. Genet.*, 3: 263–264

Bancher, C. and Jellinger, K.A. (1994) Neurofibrillary tangle predominant form of senile dementia of Alzheimer type: a rare subtype in very old subjects. *Acta Neuropathologica*, 88: 565–570,

Boren, J., Lee, I., Callow, M.J., Rubin, E.M. and Innerarity, T.L. (1996) A simple and efficient method for making site-directed mutants, deletions, and fusions of large DNA such as P1 and BAC clones. *Genome Res.*, 11: 1123–1130

Borchelt, D.R., Thinakaran, G., Eckman, C.B., Lee, M.K., Davenport, F., Ratovitsky, T., Prada, C.M., Kim, G., Seekins, S., Yager, D., Slunt, H.H., Wang, R., Seeger, M., Levey, A.I., Gandy, S.E., Copeland, N.G., Jenkins, N.A., Price, D.L., Younkin, S.G. and Sisodia, S.S. (1996) Familial Alzheimer's disease-linked presenilin 1 variants elevate Abeta1-42/1-40 ratio in vitro and in vivo. *Neuron*, 5: 1005–1013

Borchelt, D.R., Ratovitski, T., Van Lare, J., Lee, M.K., Gonzales, V., Jenkins, N.A., Copeland, N.G., Price, D.L. and Sisodia, S.S. (1997) Accelerated amyloid deposition in the brains of transgenic mice co-expressing mutant presenilin 1 and amyloid precursor proteins. *Neuron* 4: 939–945

Buxbaum, J.D., Christensen, J.L., Ruefli, A.A., Greengard, P. and Loring, J.F. (1993) Expression of APP in brains of transgenic mice containing the entire human APP gene. *Biochem. Biophys. Res. Commun.*, 2: 639–645

Cai, X.-D. et al. (1993). Excess amyloid β production is released from a mutant amyloid β protein precursor linked to familiar Alzheimer's disease. *Science*, 259, 514–516

Chartier-Harlin, M.-C. et al. (1991) Early onset Alzheimer's disease caused by a mutation at codon 717 of the β-amyloid precursor protein gene. *Nature*, 353: 844–846

Choi, T.K., Hollenbach, P.W., Pearson, B.E. et al. (1993) Transgenic mice containing a human heavy chain immunoglobulin gene fragment cloned in a yeast artificial chromosome. *Nat. Genet.*, 4: 117.

Choi, T., Huang, M., Gorman, C. and Jaenisch, R. (1991) A generic intron increases gene expression in transgenic mice. *Mol. Cell Biol.*, 6: 3070–3074

Cruts, M. et al (1996) The presenilin genes: a new gene family involved in Alzheimer's disease pathology. *Hum. Mol. Genet.*, 5: 1449–1455.

Citron, M. et al. (1992) Mutation of the β-amyloid precursor protein in familial Alzheimer's disease increases b-protein production. *Nature*, 360: 672–674

Citron, M., Westaway, D., Xia, W., Carlson, G., Diehl, T., Levesque, G., Johnson-Wood, K., Lee, M., Seubert, P., Davis, A., Kholodenko, D., Motter, R., Sherrington, R., Perry, B., Yao, H., Strome, R., Lieberburg, I., Rommens, J., Kim, S., Schenk, D., Fraser, P., St George Hyslop, P. and Selkoe, D.J. (1997) Mutant presenilins of Alzheimer's disease increase production of 42-residue amyloid beta-protein in both transfected cells and transgenic mice. *Nat. Med.*, 1: 67–72

Corder, E., Saunders, A., Strittmatter, W., Schmechel, D., Gaskell, P., Small, G., Roses, A., Haines, J. and Pericak-Vance, M.A. (1993) Gene dose of apolipoprotein E type 4 allele and the risk of Alzheimer's disease in late onset families. *Science*, 261: 921–923.

Corder, E.H. et al. (1994) Protective effect of apolipoprotein E type 2 allele for late onset Alzheimer disease. *Nat. Genet.,* (2): 180–184.

Crawley, J.N. and Paylor, R. (1997) A proposed test battery and constellations of specific behavioral paradigms to investigate the behavioral phenotypes of transgenic and knockout mice. *Horm. Behav.*, 3: 197–211.

Crawley, J.N. (1996) Unusual behavioral phenotypes of inbred mouse strains. *Trends Neurosci.*, 5: 181–182.

De Sauvage, F., Kruys, V., Marinx, O., Huez, G. and Octave, J.N. (1992). Alternative polyadenylation of the amyloid protein precursor mRNA regulates translation. *EMBO J.*, 8: 3099–3103.

D'Hooge, R., Nagels, G., Westland, C.E., Mucke, L. and De Deyn, P.P. (1996) Spatial learning deficit in mice expressing human 751-amino acid beta-amyloid precursor protein. *NeuroReport*, 7(15–17): 2807–2811.

Dodart, J., Meziane, H., Mathis, C., Bales, K., Paul, S. and Ungerer, A. (1997) *Memory and learning impairment precede amyloid deposition in the V717F PDAPP transgenic mouse*. Abs. 636.5, Society for Neurosciences, New Orleans.

Duff, K. and Huxley, C. (1996) Targeting mutations to YACs by homologous recombination. *Methods Mol. Biol.*, 54: 187–198

Duff, K., Eckman, C., Zehr, C., Yu, X., Prada, C.M.., Perez-tur, J., Hutton, M., Buee, L., Harigaya, Y., Yager, D., Morgan, D., Gordon, M.N., Holcomb, L., Refolo, L., Zenk, B., Hardy, J. and Younkin, S. (1996) Increased amyloid-β42(43) in brains of mice expressing mutant presenilin 1. *Nature*, 383: 710–713.

Eckman, C.B., Mehta, N.D., Crook, R., Perez-tur, J., Prihar, G., Pfeiffer, E., Graff-Radford, N., Hinder, P., Yager, D., Zenk, B., Refolo, L.M., Prada, C.M., Younkin, S.G., Hutton, M. and Hardy, J. (1997) A new pathogenic mutation in the APP gene (I716V) increases the relative proportion of A-beta 42(43). *Hum. Mol. Genet.*, 12: 2087–2089

Games, D. et al. (1995) Alzheimer-type neuropathology in transgenic mice over-expressing the V717F β-amyloid precursor protein. *Nature*, 373: 523–527.

Gerlai, R. (1996) Gene-targeting studies of mammalian behavior: is it the mutation or the background genotype? *Trends Neurosci.*, 5: 177–181.

Gnirke, A., Huxley, C., Peterson, K. and Olson, M. (1993) Microinjection of intact 200–500 kb fragments of YAC DNA into mammalian cells. *Genomics*, 15: 659–667.

Goate, A. et al. (1991) Segregation of a missense mutation in the amyloid precursor protein gene with familial Alzheimer's disease. *Nature*, 349: 704–706.

Higgins, L. et al. (1994) Transgenic mice brain histopathology resembles early Alzheimer's disease. *Annals of Neurology*, 35(5): 598–607.

Holcomb, L., Gordon, M.N., McGowan, E., Yu, X., Benkovic, S., Jantzen, P., Wright, K., Saad, I., Mueller, R., Morgan, D., Sanders, S., Zehr, C., O'Campo, K., Hardy, J., Prada, C.M., Eckman, C., Younkin, S., Hsiao, K. and Duff, K. (1998) Accelerated Alzheimer-type phenotype in transgenic mice carrying both mutant amyloid precursor protein and presenilin 1 transgenes. *Nat. Med.*, 1: 97–100.

Hsiao, K.K., Borchelt, D.R., Olson, K., Johannsdottir, R., Kitt, C., Yunis, W., Xu, S., Eckman, C., Younkin, S., Price, D. et al. (1995) Age-related CNS disorder and early death in transgenic FVB/N mice overexpressing Alzheimer amyloid precursor proteins. *Neuron*, 5: 1203–1218

Hsiao, K. et al. (1996) Correlative memory deficits, Aβ elevation, and amyloid plaques in transgenic mice. *Science*, 274: 99–102.

Irizarry, M.C., McNamara, M., Fedorchak, K., Hsiao, K. and Hyman, B.T. (1997a) APPSw transgenic mice develop age-related A-beta deposits and neuropil abnormalities, but no neuronal loss in CA1. *J. Neuropathol. Exp. Neurol.*, 9: 965–973.

Irizarry, M.C., Soriano, F., McNamara, M., Page, K.J., Schenk, D., Games, D. and Hyman, B.T. (1997b) A-beta deposition is associated with neuropil changes, but not with overt neuronal loss in thehuman amyloid precursor protein V717F (PDAPP) transgenic mouse. *J. Neurosci.*, 17(18): 7053–7059.

Jakobovits, A., Moore, A.L., Green, L., Vergara, G., Maynard-currie, C., Austin, H. and Klapholz, S. (1993) Germline transmission and expression of a human-derived yeast artificial chromosome. *Nature*, 18: 255–258.

Jarrett, J., Berger, E. and Lansbury, P. (1993) The carboxy terminus of β-amyloid protein is critical for the seeding of amyloid formation: implications for the pathogenesis of Alzheimer's disease. *Biochemistry*, 32: 4693.

Johnson, S.A., McNeill, T., Cordell, B. and Finch, C.E. (1990) Relation of neuronal APP-751/APP-695 mRNA ratio and neuritic plaque density in Alzheimer's disease. *Science*, 18;248(4957): 854–857.

Justice, A. and Motter, R. (1997) *Behavioral characterisation of PDAPP transgenic Alzheimer mice.* Abs 636.6, Society for Neurosciences, New Orleans.

Kammesheidt, A., Boyce, F.M., Spanoyannis, A.F., Cummings, B., Ortegon, M., Cottman, C., Vaught, J. and Neve, R.L. (1992) Deposition of beta/A4 immunoreactivity and neuronal pathology in transgenic mice expressing the carboxyl-terminal fragment of Alzheimer amyloid precursor in the brain. *Proc. Natl. Acad. Sci. USA*, 89: 10857–10861.

Kjer-Nielsen, L., Holmberg, K., Perera, J.D. and McCluskey, J. (1992) Impaired expression of chimaeric major histocompatibility complex transgenes associated with plasmid sequences. *Transgenic Res.*, 4: 182–187.

Lamb, B.T. et al. (1993) The introduction and expression of the 400 kb human genomic sequence of the amyloid precursor protein gene in embryonic stem cells and transgenic mice. *Nat. Genet.*, 5(1): 22–29.

Lamb, B.T., Call, L.M., Slunt, H.H., Bardel, K.A., Lawler, A.M., Eckman, C.B., Younkin, S.G., Holtz, G., Wagner, S.L., Price, D.L., Sisodia, S.S. and Gearhart, J.D. (1997) Altered metabolism of familial Alzheimer's disease-linked amyloid precursor protein variants in yeast artificial chromosome transgenic mice. *Hum. Mol. Genet.*, 9: 1535–1541.

Levy-Lahad, E. et al. (1995) Candidate gene for the chromosome 1 familial Alzheimer's disease locus. *Science*, 269: 973–977.

Moechars, D., Lorent, K., De Strooper, B., Dewachter, I. and Van Leuven, F. (1996) Expression in brain of amyloid precursor protein mutated in the alpha-secretase site causes disturbed behavior, neuronal degeneration and premature death in transgenic mice. *EMBO J.*, 5(6): 1265–1274.

Moran, P.M., Higgins, L.S., Cordell, B. and Moser, P.C. (1995) Age-related learning deficits in transgenic mice expressing the 751-amino acid isoform of human beta-amyloid precursor protein. *Proc. Natl. Acad. Sci. USA*, 12: 5341–5345.

Mullan, M. et al. (1992) A new mutation in APP demonstrates that pathogenic mutations for probable Alzheimer's disease frame the beta-amyloid sequence. *Nat. Genet.*, 1: 345–347.

Mullins, L.J., Kotelevtseva, N., Boyd, A.C. and Mullins, J.J. (1997) Efficient Cre-lox linearisation of BACs: applications to physical mapping and generation of transgenic animals. *Nucl. Acids Res.*, 12: 2539–2540.

Murrell, J. et al. (1991) A mutation in the amyloid precursor protein associated with hereditary Alzheimer's disease. *Science*, 254: 97–99.

Nalbantoglu, J., Tirado-Santiago, G., Lahsaini, A., Poirier, J., Goncalves, O., Verge, G., Momoli, F., Welner, S.A., Massicotte, G., Julien, J.P. and Shapiro, M.L. (1997) Impaired learning and LTP in mice expressing the carboxy terminus of the Alzheimer amyloid precursor protein. *Nature*, 387(6632): 500–505.

Oster-Granite, M.L., McPhie, D.L., Greenan, J. and Neve, R.L. (1996) Age-dependent neuronal and synaptic degeneration in mice transgenic for the C terminus of the amyloid precursor protein. *J. Neurosci.*, 21: 6732–6741.

Pearson, B.E. and Choi, T.K. (1993) Introduction of a yeast artificial chromosome containing the human beta-amyloid precursor protein gene into embryonic stem cells and transgenic mice. *Proc. Natl. Acad. Sci. USA*, 90: 10578–10582.

Petitclerc, D., Attal, J., Theron, M.C., Bearzotti, M., Bolifraud, P., Kann, G., Stinnakre, M.G., Pointu, H., Puissant, C. and Houdebine, L.M. (1995) The effect of various introns and transcription terminators on the efficiency of expression vectors in various cultured cell lines and in the mammary gland of transgenic mice. *J. Biotechnol.*, 3: 169–178.

Quon, D., Wang, Y., Catalano, R., Marian Scardina, J., Murakami, K. and Cordell, B. (1991) Formation of β-amyloid protein deposits in the brains of transgenic mice. *Nature*, 352: 239–241.

Rapp, P.R. and Gallagher, M. (1996) Preserved neuron number in the hippocampus of aged rats with spatial learning deficits. *Proc. Natl. Acad. Sci. USA*, 93(18): 9926–9930.

Reaume, et al. (1996) Enhanced amyloidogenic processing of the b-amyloid precursor protein in gene-targeted mice bearing the Swedish familial Alzheimer's disease mutations and a humanized Aβ sequence. *JBC*, 271: 23380–23388.

Sasahara, M., Fries, J.W., Raines, E.W., Gown, A.M., Westrum, L.E., Frosch, M.P., Bonthron, D.T., Ross, R. and Collins, T. (1991) PDGF B-chain in neurons of the central nervous system, posterior pituitary, and in a transgenic model. *Cell*, 64(1): 217–227.

Schedl, A., Montoliu, L., Kelsey, G. and Schulz, G. (1993) A yeast artificial chromosome covering the tyrosinase gene confers copy number dependent expression in transgenic mice. *Nature*, 362: 258–261.

Scheuner, D., Eckman, C., Jensen, M., Song, X., Citron, M., Suzuki, N., Bird, T.D., Hardy, J., Hutton, M., Kukull, W., Larson, E., Levy-Lahad, E., Viitanen, M., Peskind, E., Poorkaj, P., Schellenberg, G., Tanzi, R., Wasco, W., Lannfelt, L., Selkoe, D. and Younkin, S. (1996) The amyloid β protein deposited in the senile plaques of Alzheimer's disease is increased in vivo by the presenilin 1 and 2 and APP mutations linked to familial Alzheimer's disease. *Nat. Med.*, 2: 864–870.

Sherrington, R., Rogaev, E.I., Liang, Y., Rogaeva, E.A., Levesque, G., Ikeda, M., Chi, H., Lin, C., Li, G., Holman, K., et al. (1995) Cloning of a gene bearing missense mutations in early-onset familial Alzheimer's disease. *Nature*, 375: 754–760.

Strauss, W., Dausman, J., Beard, C., Johnson, C., Lawrence, J. and Jaenisch, R. (1993) Germ line transmission of a yeast artificial chromosome spanning the murine alpha (19) collagen locus. *Science*, 259: 1904–1908.

Sturchler-Pierrat, C., Abramowski, D., Duke, M., Wiederhold, K.H., Mistl, C., Rothacher, S., Ledermann, B., Burki, K., Frey, P., Paganetti, P.A., Waridel, C., Calhoun, M.E., Jucker, M., Probst, A., Staufenbiel, M. and Sommer, B. (1997) Two amyloid precursor protein transgenic mouse models with Alzheimer disease-like pathology. *Proc. Natl. Acad. Sci. USA*, 94(24): 13287–13292.

Suzuki, N., Cheung, T., Cai, X.-D., Odaka, A., Otvos, L., Eckman, C., Golde, T. and Younkin, S.G. (1994) An increased percentage of long amyloid β protein secreted by familial amyloid β protein precursor (βAPP717) mutants. *Science*, 264: 1336–1340.

Tanzi, R.E., McClatchey, A.I., Lamperti, E.D., Villa-Komaroff, L., Gusella, J.F. and Neve, R.L. (1988) Protease inhibitor domain encoded by an amyloid protein precursor mRNA associated with Alzheimer's disease. *Nature*, 331(6156): 528–530.

Yamaguchi, F., Richards, S.J., Beyreuther, K., Salbaum, M., Carlson, G.A. and Dunnett, S.B. (1991) Transgenic mice for the amyloid precursor protein 695 isoform have impaired spatial memory. *NeuroReport*, 12: 781–784.

Yoshikai, S., Sasaki, H., Doh-ura, K., Furuya, H. and Sakaki, Y. (1991) Genomic organization of the human-amyloid beta-protein precursor gene. *Gene*, 102(2): 291–292.

W.E. Crusio and R.T. Gerlai (Eds.)
Handbook of Molecular-Genetic Techniques for Brain and Behavior Research (Techniques in the Behavioral and Neural Sciences, Vol. 13)

CHAPTER 5.9

Modelling Down syndrome in mice

Mara Dierssen, Melanie Pritchard,[1] Cristina Fillat, Mariona Arbonés, Josep M. Aran, Jesús Flórez[2] and Xavier Estivill

Down Syndrome Research Group, Medical and Molecular Genetics Center-IRO, Hospital Duran i Reynals, Autov. Castelldefels km 2.7, 08907 L'Hospitalet de Llobregat, Barcelona, Spain
[1]IRD, Monash University, Clayton, Victoria, Australia
[2]Department of Physiology and Pharmacology, Faculty of Medicine, University of Cantabria, 39011 Santander, Spain

Introduction

Trisomy of HSA21 produces a variety of developmental anomalies recognized as Down syndrome (DS). The disorder affects one in 700 newborns and is the most significant genetic cause of mild to moderate mental retardation, due to alterations in neural development. Individuals with trisomy 21 may show several abnormalities to various extents, including craniofacial dismorphology, short stature, congenital defects of the heart and gastrointestinal tract, infertility (mostly in men), immune system alterations, thyroid dysfunction or hematopoietic disorders that include a higher incidence of childhood leukemias. At a gross morphological level, DS brains are smaller than normal, brain weight is reduced disproportionately, mostly affecting the cerebellum and brainstem, and the depth and number of sulci are reduced. Neuronal number is reduced in distinct regions, including the cochlear nuclei, cerebellum, hippocampus, the cholinergic neurons of the basal forebrain, the granular layers of the cerebral cortex, and areas of the brain stem, and abnormal neuronal morphology is observed, especially in the cerebral cortex. In addition, over the age of 35, DS individuals exhibit neuropathology typical of Alzheimer's disease (AD) and frequently show premature aging (reviewed in Pueschel, 1992; Flórez, 1992). There are other disorders involving HSA21 genes, such as familial amyotrophic lateral sclerosis, familial AD, Ewing's sarcoma and other primitive neuroectodermal tumors and acute myeloid leukemia.

Defining how an extra copy of all or part of HSA21 results in the phenotype of DS is a specific case of the more general problem of explaining how chromosomal imbalance produces abnormalities in function and morphology. No single mechanism can explain the deleterious consequences of aneuploidy, and therefore, there is no simple solution to counteract its phenotypic impact. While some loci may have a greater phenotypic effect, it is the cumulative effect of imbalance of many genes that determines the overall phenotype. Substantial progress towards regional characterization of these genes has emerged through the analysis of human individuals with segmental trisomy 21, arising from translocations or duplications, but the resolution of this approach is limited by the phenotypic variation among DS individuals. It has been proposed that genetic loci situated in the DS critical region (DSCR) could harbor genes with major effects. The study of the DSCR has been intensive and has led to the identification of numerous transcriptional units that are potentially involved in the phenotypic alterations observed in DS. The genetic and physical maps of HSA21 are well-developed (Antonarakis et al., 1993) and transcriptional mapping is rapidly progressing (Yaspo et al., 1995; Chen et al., 1996a; Guimerà et al., 1997). However, even though 112 genes of the estimated 300–500 (EMBO workshop 1999) genes on HSA21 have currently been identified (http://www-eri.uchsc. edu/chr21/eridna.html), the functions of most of these genes remain largely unknown, as does their contribution, if any, to the DS phenotype (Epstein, 1986, 1995; Kola and Hertzog, 1997). Even knowing the molecular defect, it is difficult to decipher the complex pathophysiology of the disease, the developmental consequences of the trisomy and the impact on behavior and cognitive function. The difficulty in defining the specificity of phenotypic expression and in elucidating the role of individual genes in DS has led to the development of many experimental strategies, including the generation of animal models that provide ready access to cells and tissues from different stages of the disease. The animal models are expected to help us understand some of the clinical alterations observed in DS and may facilitate the development of clinical interventions. Three major approaches have been used to study the effects of overexpression of HSA21 genes in mice:

1. transgenic animals overexpressing single or combinations of genes
2. mouse models that carry all or part of MMU16 which has regions of conserved homology with HSA21
3. transgenic mice with large foreign DNA pieces introduced on yeast artificial chromosomes (YACs) or bacterial artificial chromosomes (BACs) into mice (Fig. 1).

Modelling DS in mice: role of individual genes

An animal model should possess genetic and physiological systems that resemble the human condition. It is recognized that mice provide an experimental system for modelling human disorders and for testing therapies. Mice share anatomical and metabolic characteristics with humans but have the advantages of a short life span and a short generation time. They are relatively easy to maintain in large numbers and knowledge about their genetics is advanced (Mouse Genome Database: http://www. informatics.jax.org). Orthologous genes are frequently linked in similarly conserved chromosomal segments in the mouse and human genomes. For this reason aneuploidy for regions of the mouse genome that are conserved on human chromosomes can serve to study the concerted effect of over (or under) expression of multiple genes. We can alter the genetic constitution in mice by adding new genes by transgenesis, including human genes. We can modify the genes before introducing them, and we can also remove or alter existing genes through gene replacement techniques.

An important issue is the characterization of the phenotype of the mouse models generated. Attempts are being made to standardize the phenotypic characterization (Crawley and Paylor, 1997). Recently, a comprehensive phenotypic testing protocol was proposed (Rogers et al., 1997). The SHIRPA protocol consists of different sets of tests covering mouse neurological, behavioral, biochemical and histopathological aspects (http://www.mgu.har.ac.uk/). The phenotypic profile obtained can then be compared to other existing models on the basis that the same criteria were used. However, though necessary, this general characterization is only a first step, and a customized set of tests must be designed depending on the particular genetic manipulation produced.

Various approaches have been used to study the consequences of increased gene dosage in DS and to investigate phenotype/genotype relationships of genes (Table 1). Although over-expressing a single gene will not be a model of DS, the value of single gene transgenesis is to identify genes

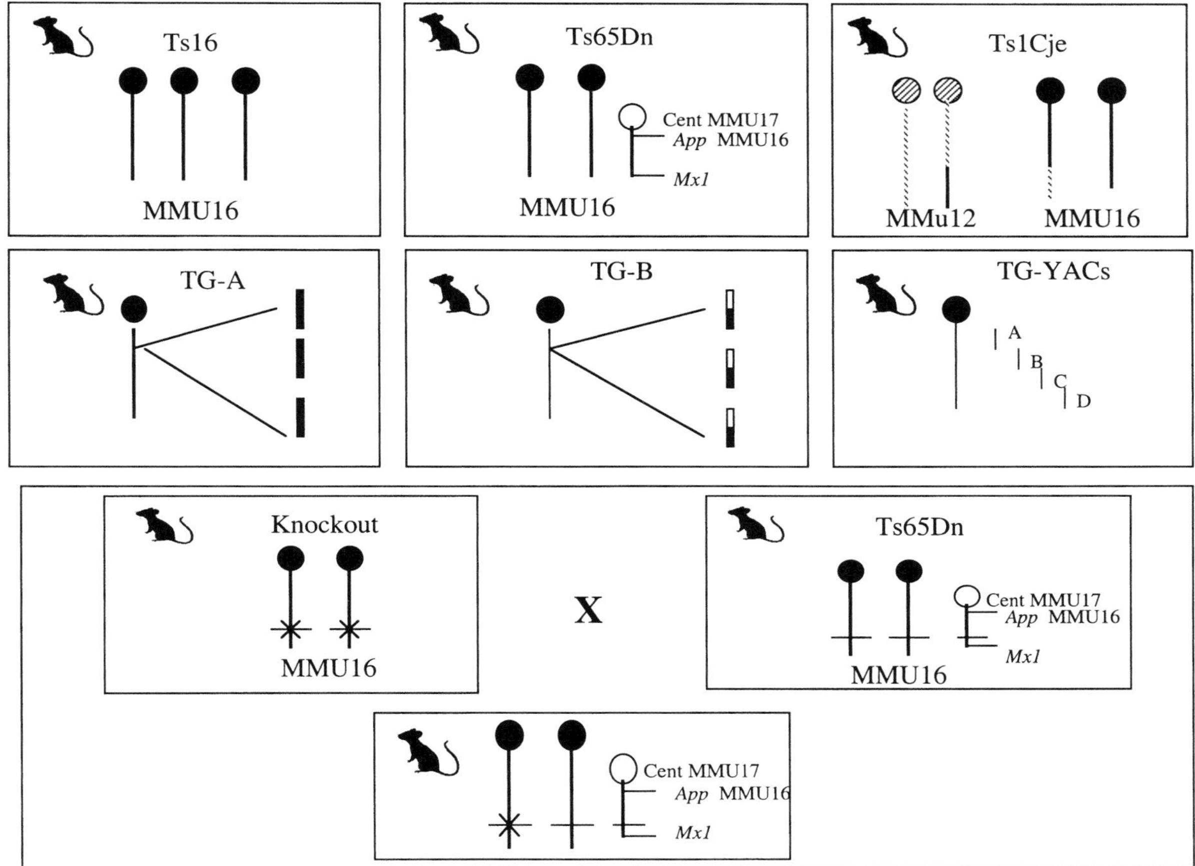

Fig. 1. Murine models for Down syndrome. Upper panel shows the murine trisomies; middle panel shows transgenic mice with homologous (TG-A) and heterologous (TG-B) promoter and YAC transgenics; lower panel represents the crossing of knockout mice with Ts65Dn for reduction of a particular gene dosage.

related to specific pathophysiological features. The development and analysis of transgenic mice that overexpress single genes, either with their own genomic regions for gene regulation and expression, or with heterologous promoter sequences that drive the expression of the genes in specific tissues (Fig. 1), permits, to some extent, the study of the contribution of each gene to the phenotype. The selection of candidate genes is done on the basis of their location in the chromosomal region that has been considered critical for the development of the syndrome, or due to previous knowledge of the functions of the proteins that they encode. To date only a few HSA21 encoded genes have been used to make transgenic mice. However, the identification in the last years of many novel genes through systematic exon trapping, cDNA selection and mass sequencing of the entire chromosome (Antonarakis, 1998) will inevitably lead to the generation of large numbers of transgenic mouse lines.

There are some limitations inherent in the approaches directed at dissecting individual gene function (see Chapter 3.1.10). Two properties of transgene integration, the randomness of the site of integration and the variability of copy number affect transgene expression so that it is not necessarily correlated with copy number. Integration is usually in a single chromosomal locus, but occasionally occurs into two chromosomal loci, requiring breeding to the next generation to segregate the transgenes. Other concerns are the

TABLE 1

Murine models for Down syndrome

Murine Model	Pathology	References
Single gene transgenics		
Tg*SOD1*	abnormal neuromuscular junctions abnormal LTP	Bar Peled et al., 1996; Epstein et al., 1987; Gahtan et al., 1998; Schwartz et al., 1998
Tg*APP*	AD-like neuropathology in mice with mutated APP in combination with TGFB	Games et al., 1995; Hsiao et al., 1996; Sturchler-Pierrat et al., 1997
Tg*S100beta*	astrocytosis and abnormal dendritic development	Gerlai and Roder,1993; Reeves et al., 1994; Whitaker-Azmitia et al., 1997
Tg*MNB*	ND	Bartsch et al., 1998
Tg*ETS2*	DS-like skeletal abnormalities	Sumarsono et al., 1996
Tg*PFKL*	Increased phosphofructokinase activity	Elson et al., 1994
P1 transgenic		
Tg 1017 (1109S-21ES0302)	ND	Smith et al., 1995
Tg 1121 (D21S395-21ES0305)	ND	Smith et al., 1995
Tg1048 (21ES0302-21ES0303)	ND	Smith et al., 1995
Tg0073 (21ES0305-D21S167)	ND	Smith et al., 1995
YAC transgenic		
Tg 152F7(570Kb) (D21S267-21ES0227)*MNB*	spatial learning deficits	Smith et al., 1997
Tg 152F7 tel (180Kb) (D21S394-21ES0227)*MNB*	spatial learning deficits	Smith et al., 1997
Tg 230E8(670Kb) (CBR-D21S334)	abnormal cortical neuronal density weak learning deficits	Smith et al., 1997
Tg 141G6(475Kb) (D21S167-21ES0203)	ND	Smith et al., 1995
Tg 285E6(430Kb) (21ES0291-21ES0123)	ND	Smith et al., 1995
Trisomies		
Ts 16	Congenital heart defects, abnormal brain development, immunologic and hematologic alterations, neurotransmitter alterations	Buselmaier et al., 1991; Coyle et al., 1991; Grausz et al., 1991; Miyabara et al., 1984; Oster-Granite and Lacey-Casem, 1995
Ts65Dn	spatial and non-spatial learning and memory deficits, BFCN atrophy, altered LTP, immunologic alterations, neurotransmitter alterations	Davisson et al., 1990; Dierssen et al., 1997; Dierssen et al. 1996; Escorihuela et al., 1995; Escorihuela et al., 1998; Holtzman et al., 1996; Insausti and Flórez, 1998
Ts1Cje	learning deficits	Sago et al., 1998

ND: non described; BFCN: basal forebrain cholinergic neurons.

possible disruption of an endogenous gene by a transgenic insertion event resulting in a mutant phenotype as a consequence of a disruption, deletion or translocation, rather than as a consequence of transgenesis. Finally, the use of heterologous promoters that allow the generation of inducible transgenic models may lead to non-physiological effects. Therefore, the best way to produce a transgenic is to introduce the gene under the transcriptional control of its own promoter. The transgene integrates randomly but is then expressed along with its endogenous counterpart in the appropriate tissues. However, in this approach, the gene must be cloned and characterized to determine that it contains all the regulatory sequences necessary for the normal tissue-specific and temporal expression. Before drawing conclusions about the relevance of the phenotype obtained to DS, the limitations outlined above should be taken into account.

Several genes mapped to HSA21 have been identified, whose overexpression may be responsible for some of the phenotypic characteristics observed in DS. These include the homologues of *Drosophila single minded* (*SIM2*) (Dahmane et al., 1998) and *minibrain* (*MNB*) (Guimerà et al., 1996), and the inwardly rectifying K+ channel gene, *GIRK2* (Patil et al., 1995). However, other interesting genes mapping outside the DSCR have also been identified, including a novel cell adhesion molecule, *DS-CAM* (Yamakawa et al., 1998), a gene highly expressed in fetal brain and adult heart (*DSCR1*) (Fuentes et al., 1995), a gene encoding a 61 amino acid peptide (*PEP19*) expressed during development in the Purkinje cells of the cerebellum, a cerebral region profoundly altered in DS (Chen et al., 1996b), and *GRIK1,* whose product, the GluR5 kainate receptor regulates inhibitory synaptic transmission (Clarke et al., 1997). Other previously known genes include *AML1* (a gene that is truncated in several leukemia related translocations) (Nucifora et al., 1994), the CuZn superoxide dismutase gene (*SOD1*) (Siddique and Deng, 1996) and *ETS2* (Sumarsono et al., 1996, see below).

The first transgenic mice harboring a candidate gene for DS were generated using the human *SOD1* gene (Epstein et al., 1987). SOD1 is a key enzyme in the metabolism of oxygen free radicals, so that its overexpression due to gene dosage, may disturb the steady-state equilibrium of reactive oxygen species within the cell, resulting in damage to biologically important molecules. Tg*SOD1* animals showed a 1.6–6-fold increase in activity of the enzyme in the brain and other tissues, as occurs in DS individuals. Gross anatomy was normal but Tg*SOD1* mice exhibited some abnormalities characteristic of DS patients, including abnormal neuromuscular junctions, decreased plasma levels of serotonin and increased biogenic amine function (reviewed in Groner, 1995). The recent discoveries that missense mutations in the human *SOD1* gene are involved in familial amyotrophic lateral sclerosis has focused interest on the role of free radicals in neurodegeneration. It has been observed that cultured neurons from transgenic mice were more susceptible to kainic acid-induced apoptotic cell death (Bar-Peled et al., 1996; Schwartz et al., 1998). Moreover, a reversible impairment in spatial learning and in hippocampal long term potentiation (LTP) due to increased formation of H_2O_2 evoked by tetanic stimulation has been reported (Gahtan et al., 1998).

Although for some loci, such as *SOD1*, gene dosage is directly related to expression levels, this may not be the case for others. For example, overexpression of the liver-type phosphofructokinase gene (*PFKL*) in DS patients leads to an increase in enzymatic activity higher than that expected on the basis of gene dosage, and Tg*PFKL* mice showed an increase in phosphofructokinase activity in blood and brain, but the rate of glucose metabolism was differentially affected, being slower in blood but faster in brain (Elson et al., 1994; Levanon et al., 1986).

In some cases, overexpression of a single gene can give rise to a marked phenotype. This is the case for the *Ets2* transcription factor (Sumarsono et al., 1996). Tg*Ets2* mice show DS like skeletal abnormalities, particularly affecting the craniofacial region, with altered head-shape and brachycephaly. These alterations are also observed in mice trisomic for MMU16, the Ts16 mouse (see Section 4) and could be due to the expression of *Ets2* in cartilaginous precursor and vertebral primordial cells.

DS individuals develop Alzheimer's-like amyloid plaques and neurofibrillary tangles by approximately the third decade of life (reviewed in Pueschel, 1992). This has been ascribed to the fact that the human *β-amyloid precursor protein* gene (*βAPP*), a gene involved in AD neuropathology, is located on HSA21. However, of the numerous attempts to generate valid experimental models for the expression of human *APP*, very few transgenic mouse lines have been obtained that develop Aβ deposits characteristic of AD (Games, 1995; Hsiao et al., 1996; Sturchler-Pierrat et al., 1997; see Section 3). In addition to the neuropathological hallmarks of AD, transgenic mice overexpressing *APP* exhibited aggressive behavior, which could be modified using serotonergic drugs (Moechars et al., 1998). Recently, a relationship was established between changes in acetylcholinesterase levels and glycosylation patterns and the alteration of APP metabolism, after the observation that acetylcholinesterase is increased in the brains of transgenic mice expressing the C-terminal fragment (CT100) of βAPP (Sberna et al., 1998).

One of the most important concerns in DS is the alteration in learning and memory that gives rise to mental retardation. No single neuropathological process can be identified as causing mental retardation, and therefore, it is probably the synergistic effect of various genes that induces the phenotype. It has been suggested that the alteration of the shape and number of dendritic spines could be responsible for the disruption of the learning process. S100β, a neurotrophic factor released by astroglial cells, is of particular interest because of its effect in vitro on neuritic outgrowth, its influence on the establishment of hippocampal LTP and its increased expression in some brain regions affected in DS. Mice bearing multiple copies of *S100β* tend to be less efficient in some hippocampal-dependent learning tasks (Gerlai and Roder, 1996) and female present some behavioral alterations (Gerlai and Roder, 1993). Overexpression of *S100β* leads to astrocytosis (Reeves et al., 1994) and abnormal dendritic development with an excess of hippocampal dendrites in younger animals followed by premature loss of dendrites (Whitaker-Azmitia et al., 1997).

Our research group is currently working on HSA21 encoded genes. We have isolated several genes, and are currently focusing on two candidate genes for CNS neuropathology: *MNB* and *DSCR1*. The *Drosophila mnb* gene encodes a serine-threonine protein kinase, which is expressed in distinct neuroblast proliferation centers during postembryonic neurogenesis. RNA in situ hybridization of *Mnb* in mice detected expression in several regions of the mouse brain, including the olfactory bulb, the cerebellum, the cerebral cortex, the pyramidal cell layer of the hippocampus and several hypothalamic nuclei (Guimerà et al., 1996). This expression pattern corresponds to the regions of the brain that are abnormal in individuals with DS and suggests that altered expression of *MNB* may be involved in the altered neuronal development observed in DS. In fact, transgenic mice containing YACs encompassing a 2 Mb region of HSA21 displayed behavioral and learning defects (Smith et al., 1997). The only gene subsequently found to map to this genomic fragment was *MNB* (see Section 3). Moreover *Mnb* is in triplicate in a mouse with segmental trisomy of MMU16, (Ts65Dn), that is considered a DS model (see Section 4, Fig. 2), thus enabling comparison between both models. Recently, a transgenic mouse was generated with doxycycline regulated *Mnb* transgene expression in the forebrain, and the results indicated that Mnb may play an important role in both developmental and adult synaptic plasticity (Bartsch et al., 1998).

DSCR1 encodes a protein which was proposed to be involved in signal transduction and/or transcriptional regulation. The pattern of expres-

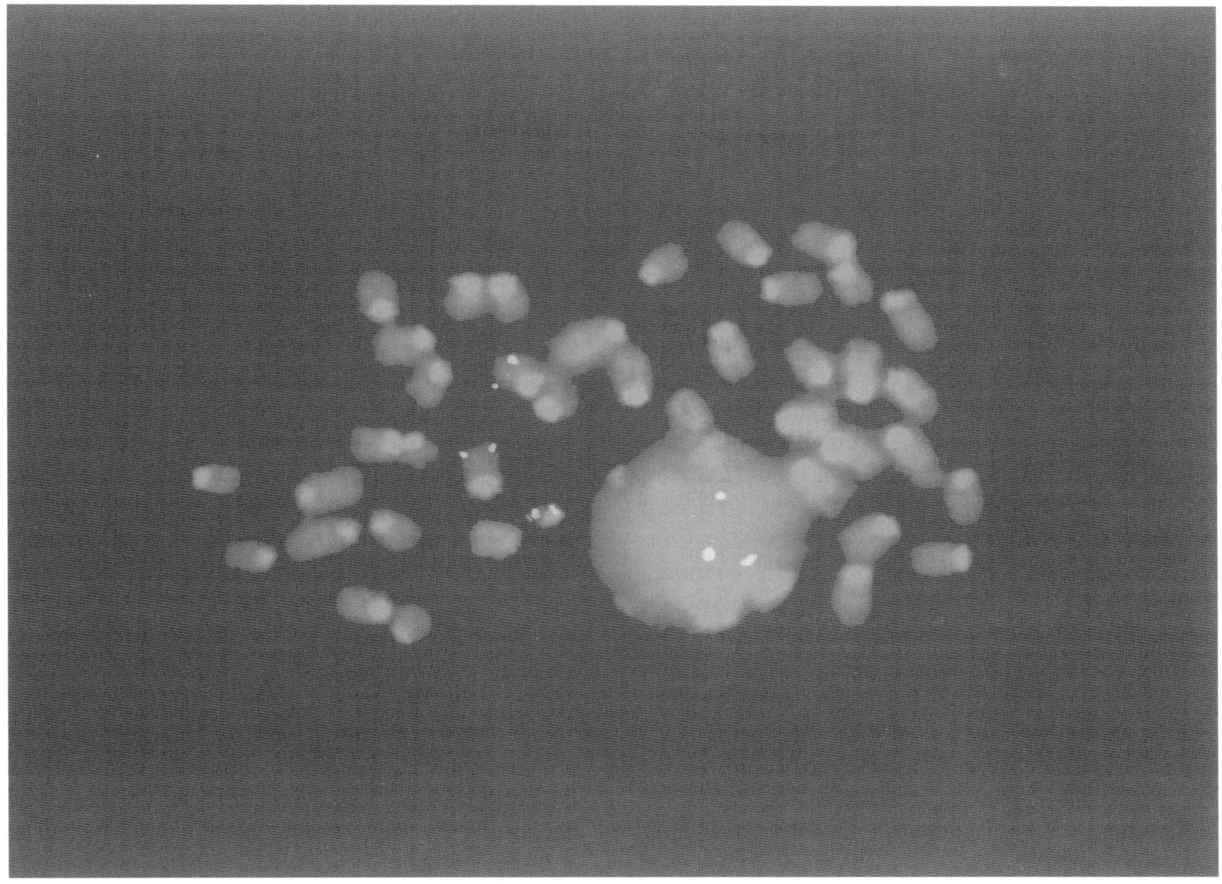

Fig. 2. FISH analysis of a Ts65Dn mouse showing trisomy for *Mnb*. Metaphase spread showing the hybridization of P1(pAd10-SacBII) clone containing *Mnb* sequences to both MMU16 and the marker chromosome (Marga Nadal, personal communication).

sion and tissue distribution of *DSCR1* suggests that its overexpression may result in some of the phenotypic consequences of DS (Fuentes et al., 1995). Expression is detected in the neocortex cell layers II, III and V, in the pyriform cortex, the dentate granule cell layer, the pyramidal cell layer of the hippocampus, the striatum and in the Purkinje cell layer of the cerebellar cortex (C. Casas, unpublished results). A study of the structure of *DSCR1* revealed four alternative first exons (Fuentes et al., 1997). The existence of alternative splicing determined the generation of several transgenic lines with chimaeric genes containing *DSCR1* cDNAs with the alternative first exons predominant in brain, under the control of the brain-specific *PDGF–B* promoter. Four independent transgenic lines were obtained containing different numbers of copies of the integrated *DSCR1* transgene. Preliminary behavioral data suggested that overexpression of *DSCR1* leads to hyperactivity and some alterations in motor coordination with respect to non-transgenic controls (Martínez-Cué et al., 1997). *DSCR1* is also in triplicate in the Ts65Dn mouse (see Section 4) thus providing two different models of overexpression. The comparison with Ts65Dn indicates a quite distinct behavioral profile in Tg*DSCR1*, but some aspects regarding learning and memory may overlap in both models. However, experiments were performed in only one transgenic line, and thus, until we have analyzed at least another independent line, we are unable to discount the possibility that the effect observed is due to the position of integration of the transgene (unpublished data).

Transgenic models with a multiple, yet well defined set of HSA21 genes (double and triple transgenics) would offer exciting opportunities to explore how several genes interact in the formation of certain features. These models represent a significant step closer to the goal of reconstructing the full scenario of gene-dosage effects for a group of HSA21 genes.

Although the DS phenotype is due to overdosage of a cluster of proteins, and thus, models of overexpression should be considered as the best strategy to elucidate the contribution of single genes to the phenotype, other genetic models may also be of relevant value. The availability of mouse embryonic stem (ES) cells and gene targeting systems allow the creation of loss of function (knock-out) mice (Crusio, 1996) (see Chapter 4.4). The physiological roles of a number of genes on HSA21 such as *βAPP* and *SOD1* have been assessed in mice using this technology. Gene targeted ES cells can also be used as a vehicle for the generation of mice carrying an additional copy of a particular gene. For example, mice bearing three functional copies of the angiotensin gene have been produced by embryo injections that have a complete tandem duplication of the gene (Smithies and Kim, 1994). In contrast to standard transgenesis, this method allows a precise control of both copy number and chromosome location of the additional gene.

The breeding of partial trisomy 16 mice (Ts65Dn or Ts1Cje, see Section 4) with mice that carry a disrupted gene gives rise to trisomic progeny with the gene of interest in disomy (Fig. 1). These progeny can be evaluated for the reversal of parts of the phenotype seen in the partial trisomic mice. It should be born in mind that some features are strongly influenced by the genetic background of the animals (Gerlai, 1996; see also Chapter 3.1.10 in this book), therefore, for phenotype comparison, it is important to produce a genetic background close to that of the partial trisomic mice. To overcome this problem, the use of antisense techniques to reduce the expression of one of several genes potentially involved in DS may prove an excellent genetic tool (see Chapter 3.2.4).

Antisense oligonucleotides can be delivered to adult Ts65Dn mice and also to pregnant trisomic females to assess the effects of reducing the expression of developmentally regulated genes during different embryonic stages. Several in vivo studies have shown that antisense oligonucleotides directly delivered to the brains of animals produced a modification in behavior (Zapata et al., 1997; Persico et al., 1998). This strongly supports the proposal that antisense methodology has the potential to be used as a powerful tool for the study of the molecular basis of behavior and for the treatment of associated disorders (Ogawa and Pfaff, 1996). Moreover, in addition to being able to dissect the contribution of a particular gene or gene subset to the overall DS phenotype, antisense technology holds potential for therapy.

YAC technology

The development of YAC technology for the generation of transgenic mice holds enormous potential for studying aneuploidy syndromes such as DS (also see Chapter 5.8 for application in modelling Alzheimer's disease). It is possible to transfer large pieces of DNA that may encompass several genes, thus allowing the study of the effects of multiple genes as opposed to single gene effects. YACs are cloning vectors that can carry large pieces of DNA up to 2 Mb in length and allow stable propagation in yeast (Burke et al., 1987). This means that multiple genes or large genes along with their accompanying regulatory sequences may be cloned in one YAC. Thus, YACs serve as vehicles with which whole genes can be transferred into mice where the gene may exhibit its normal expression pattern. Manipulating large genes is not possible using the single-gene transgene approach and it has been found that in some instances the results with smaller transgene constructs do not always mirror the results obtained using YAC transgenes. For instance, genomic context was critical for the correct regulation of expression of the imprinted gene *H19*. Genomic imprinting is an epigenetic process involving methylation of DNA which results in monoallelic expression of a gene, depending on its parental origin. Normally, the *H19* gene is only expressed from the maternal allele. When present on a 130 kb YAC clone, the *H19* gene was correctly expressed only from the maternal allele and appropriately repressed after paternal inheritance. When mini-*H19* transgenes were used correct expression was not observed (Ainscough et al., 1997).

Largely due to the Genome Project and the world wide efforts in genome mapping, there are several YAC libraries available. For many chromosomes the YACs have been physically mapped and ordered along the chromosome arms (Chumakov et al., 1995; http://www.cephb.fr). There is a YAC contig for the entire short arm of HSA21 containing clones with known insert sizes and genomic content. As more HSA21 genes are discovered, they along with their regulatory sequences are mapped back to the YACs. For chromosome 21, it is now as simple as choosing a YAC containing the desired gene from a database (http://eri.uchsc.edu).

YACs also have the enormous advantage that they can be readily manipulated in yeast. It is relatively easy to introduce mutations such as deletions, insertions or nucleotide substitutions into precise locations in the YAC by homologous recombination (Duff et al., 1994; Peterson et al., 1995; Lamb et al., 1997). Transgenic mice can be generated using segments of genomic DNA containing single genes or clusters of genes and the sequences responsible for producing a certain phenotype can be further localized by fragmentation of the YAC or by the deletion of specific sequences through homologous recombination, followed by reintroduction into mice.

One further advantage to using YACs for modelling the DS phenotype is their low copy number upon integration into the host genome. Constructs containing cDNAs are often integrated in huge tandem arrays, resulting in high levels of overexpression of the gene product. DS results from one extra copy of each gene, such that the level of expression of each gene is expected to be elevated by around 50%. YAC transgenics in general show this low level of overexpression.

Three different techniques have been used to successfully transfer YACs into the mouse germline (Fig. 3): lipofection of YAC clones into ES cells (Lamb et al., 1993; Strauss et al., 1993; Lee and Jaenisch, 1996), yeast spheroplast fusion with mouse ES cells (Davies et al., 1993; Jakobovits et al., 1993; Green et al., 1994) and microinjection of gel purified YAC DNA into the pronuclei of mouse zygotes (Peterson et al., 1993; Schedl et al., 1993; Maas et al., 1997). For all three methods, the expression of the YAC transgenes was in nearly all cases comparable to that of the endogenous counterpart (reviewed in Montoliu et al., 1994).

For lipofection, YAC DNA is purified away from the other yeast chromosomes by pulsed-field electrophoresis, complexed with a lipid reagent and transfected into ES cells. A selectable marker (neo) can be added to one of the YAC arms by homologous recombination in yeast to allow selection of ES cell clones with G418. Lipofection with YACs containing selectable cassettes results in about 10% of the cells incorporating an intact YAC.

YACs can be introduced into the male pronucleus of the single cell mouse embryo by microinjection. Embryos at the one cell-stage are collected from the oviducts of female mice and a solution containing the YAC is injected into the pronucleus of the fertilized egg. Eggs which survive microinjection are then transferred into

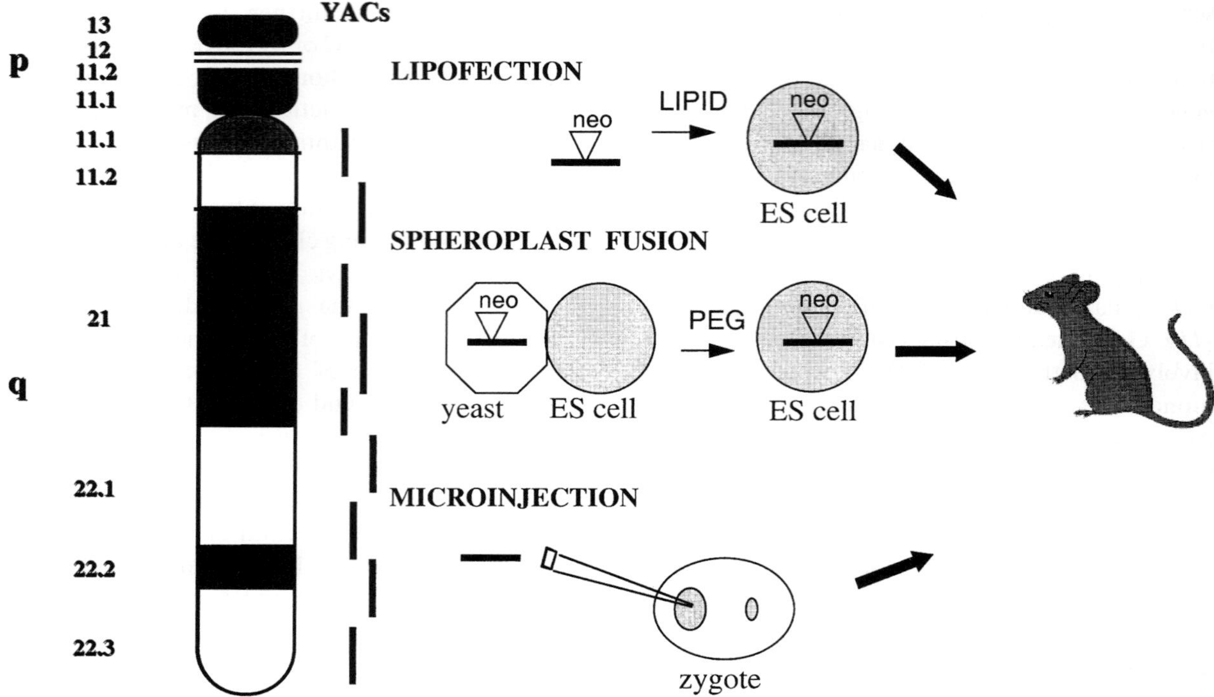

Fig. 3. Generation of transgenics overexpressing HSA21 encoded genes using YAC technology. Introduction of YAC DNA by transfection, lipofection and microinjection to create a transgenic mouse.

pseudo-pregnant females and the progeny tested for the presence of the YAC by the hybridization of a YAC-specific probe to a Southern blot of the DNA extracted from the tails of the progeny. As for lipofection, for microinjection the YAC DNA must be gel purified and concentrated. Because manipulation of the YAC causes DNA breakage and denaturation, various protective agents (e.g. polyamines) are required to stabilize the DNA. Solutions containing YAC DNA are quite viscous, so that another physical limitation is the low number of YAC molecules contained in a solution that may be injected through a microinjection needle. Due to the damaging forces placed upon the YAC during preparation and microinjection, the efficiency of successful transfer of an intact YAC into a mouse is around 1%, and this diminishes with the increasing size of the YAC.

Spheroplast fusion is the method of choice for the transfer of very large fragments of DNA (1–2 Mb) as the YAC DNA is not subjected to the same deleterious shearing forces as in the other two methods. A neo cassette is introduced into the YAC and then yeast spheroplasts (whole yeast cells with the rigid cell wall removed by enzymatic digestion with zymolyase) are fused with ES cells using polyethylene glycol. Selection on G418 results in about 40% of the cells retaining the YAC in either single or multiple copies. The major advantage with this method is that the arduous task of isolating intact and purified YAC DNA is avoided. However, yeast DNA is transferred along with the YAC and may have unknown effects which should be taken into account when interpreting the effect of the YAC in transgenic mice. However, Cabin and colleagues (Cabin et al., 1995) have studied the integration of yeast

DNA into mouse chromosomes after spheroplast fusion and detected a single site of integration in every mouse ES cell examined. This site was always at the same position as the YAC, meaning that a single insertion event occurs using this method resulting in the insertion of all the yeast genetic material into one spot.

For all of the transfer techniques, the integrity of the integration event and the copy number of the YAC must be determined. This can be achieved either by in situ hybridization or Southern blot using YAC vector probes in addition to gene probes, or by PCR STS-content mapping. For a discussion on these methods and their application in assessing copy number and the presence of intact and fragmented YACs see Smith et al., 1995 and, 1997. Each technique has inherent advantages and disadvantages and the technique of choice comes down to the expertise, materials and facilities available in the laboratory. Microinjection is probably the most straightforward option. The methods involving ES cells are slower, but the major advantage with the ES cell techniques is that the presence, copy number, and integrity of the YAC can be tested in cells before the generation of chimaeric mice, instead of having to test all the progeny resulting from a microinjection experiment. A final consideration is the need to develop more than one transgenic mouse line. Integration events can cause insertional mutations, therefore, the same phenotype must be observed in more than one line to ensure that the effect results from an extra copy of the genomic material provided by the YAC.

The ability to introduce large, well defined genomic regions into transgenic mice holds enormous potential for studying the effects of gene duplication in DS. Two complementary approaches are presently ongoing. One involves the transfer of a random panel of HSA21 derived YACs. This type of analysis will greatly refine the localization of genes responsible for a particular DS characteristic, and individual genes can then be isolated from the YACs. Two groups have reported the construction of such panels (Cabin et al., 1995; Smith et al., 1995), although to date very little data has been published. Smith and colleagues created a panel of transgenic mice containing YACs encompassing a 2 Mb region of HSA21. Since the presence of learning deficits and mental retardation is a characteristic feature of DS, these mice were subjected to behavioral and learning tests. One of the YACs had fragmented upon microinjection, and the mouse carrying this fragmented YAC, YAC152F7tel, showed the same phenotype as the mouse with the original intact YAC152F7. In the Morris water maze, YAC152F7tel transgenic mice were midly impaired in probe test crossings and severely impaired in the reverse versions of the tasks, that assess learning flexibility. The only gene subsequently found to map to this genomic fragment was *MNB* (Guimerà et al., 1996), suggesting that the small increase in minibrain dose leads to learning defects (Smith et al., 1997). Mice transgenic for YAC 230E8, that encompasses the carbonyl reductase region, were normal in spatial and non-spatial learning, but presented impaired learning flexibility. The phenotypes derived from the two YACs were different, suggesting that dosage imbalance for genes from at least two different regions on HSA21 gave rise to learning/behavioral defects.

Complementary to the transfer of largely uncharacterized pieces of genomic DNA is the introduction of YACs containing well characterized genes. Only one example of a YAC transgenic mouse containing a HSA21 encoded gene has been reported (Lamb et al., 1993). When a mouse carrying a 650 kb YAC containing the *APP* gene was generated, the gene was observed to be correctly expressed and processed. The extra copy of the *APP* gene has been implicated in the etiology of AD, and there is mounting evidence implicating oxidative stress in neurodegenerative diseases. The relationship between increased susceptibility to oxidative stress and altered APP metabolism (both of which occur in DS and in AD) is being tested in these mice (Calingasan et al., 1996; Murai et al., 1998).

Finally, phenotypic characteristics arising from complex gene interactions can be assessed by crossing mice carrying different YACs. In this way it is feasible to build up a library of mice, each carrying an increasing number of HSA21 genomic fragments in an attempt to reconstruct the entire DS phenotype. YAC transgenic technology is without doubt a promising tool for dissecting the complex phenotypic traits observed in DS.

Mouse trisomies

Comparative mapping between mice and humans has revealed that HSA21 shares a large region of genetic homology with MMU16 (Fig. 4). Based on this rationale, mice trisomic for MMU16 have been extensively used as a model for DS (Fig. 1). The trisomy models have the advantage that they overexpress the orthologues of multiple genes located on HSA21 which permits the study of how arrays of genes interact to control each other at the level of transcription. Hence their power lies in modelling and characterizing the syndrome (or a part of it) and in testing therapeutic strategies.

The first mice used had the complete MMU16 in three copies, but more recently two partial trisomic mice have been generated. Using double heterozygosity for homobrachial Robertsonian chromosomes murine trisomies can be produced at reasonably high frequencies. When male mice with balanced bilateral Robertsonian translocations of MMU16 (e.g. the fusion of MMU16 with another chromosome) are mated with normal females, approximately one-third of the progeny surviving to 15 d gestation have trisomy 16, whereas two-thirds are euploid (Gearhart et al., 1986). Ts16 mice exhibit some characteristics of DS, but their value is limited by several factors. First, Ts16 mice die in utero. This inability to survive is an insurmountable obstacle to the study of many of the major features that occur as part of the DS phenotype postnatal (e.g. mental retardation, thyroid disturbances) or even much later in life (e.g. Alzheimer's disease neuropathology). In addition, MMU16 contains a large number of genes whose human counterparts are located on chromosomes other than HSA21 (3, 16, 22) and several genes located on the distal end of the long arm of HSA21 which are thought to be essential for the DS phenotype are not located on MMU16. Thus, Ts16 mice are not trisomic for some HSA21 genes and are at a dosage imbalance for other genes that are not implicated in the pathogenesis of DS.

In spite of all these shortcomings, the Ts16 model has the advantage over other MMU16-related models (see below) in that the mice show characteristic cardiac as well as craniofacial and ocular malformations, much more like those observed in individuals with DS (Oster-Granite, 1986). With regard to the heart, the trisomy produces endocardial cushion defects (Miyabara et al., 1984) and vascular malformations related to the embryonic aortic-arch system (Buselmaier et al., 1991). Delays were also found in the development of the sphenoid, mandible and maxilla, as well as distortions of the cervical vertebrae, and alterations in the development of the inner ear. Significant and complex alterations are observed in the morphogenesis and neurochemistry of the brain, as well as in the properties of the cultured neurons (see Coyle, 1991; Oster-Granite and Lacey-Casem, 1995). Common alterations have been also described in the development of the cerebellum, hippocampus, cortical plate and ocular structures including a reduction of periorbital connective tissue and microophtalmia (Grausz et al., 1991). Many of these features are qualitatively similar to those appearing in individuals with DS.

A major advance in mouse models was the generation by Davisson et al. (1990) of a partial trisomy 16 mouse, the Ts(17^{16})65Dn or Ts65Dn model, that includes most of the MMU16 region homologous to HSA21 (from *App* to *Mx1*) but lacks the remaining approximately 40% of HSA21 genes. This model is viable and high percentage of the mice reach old age. Production of adult mice trisomic for specific segments was based on two

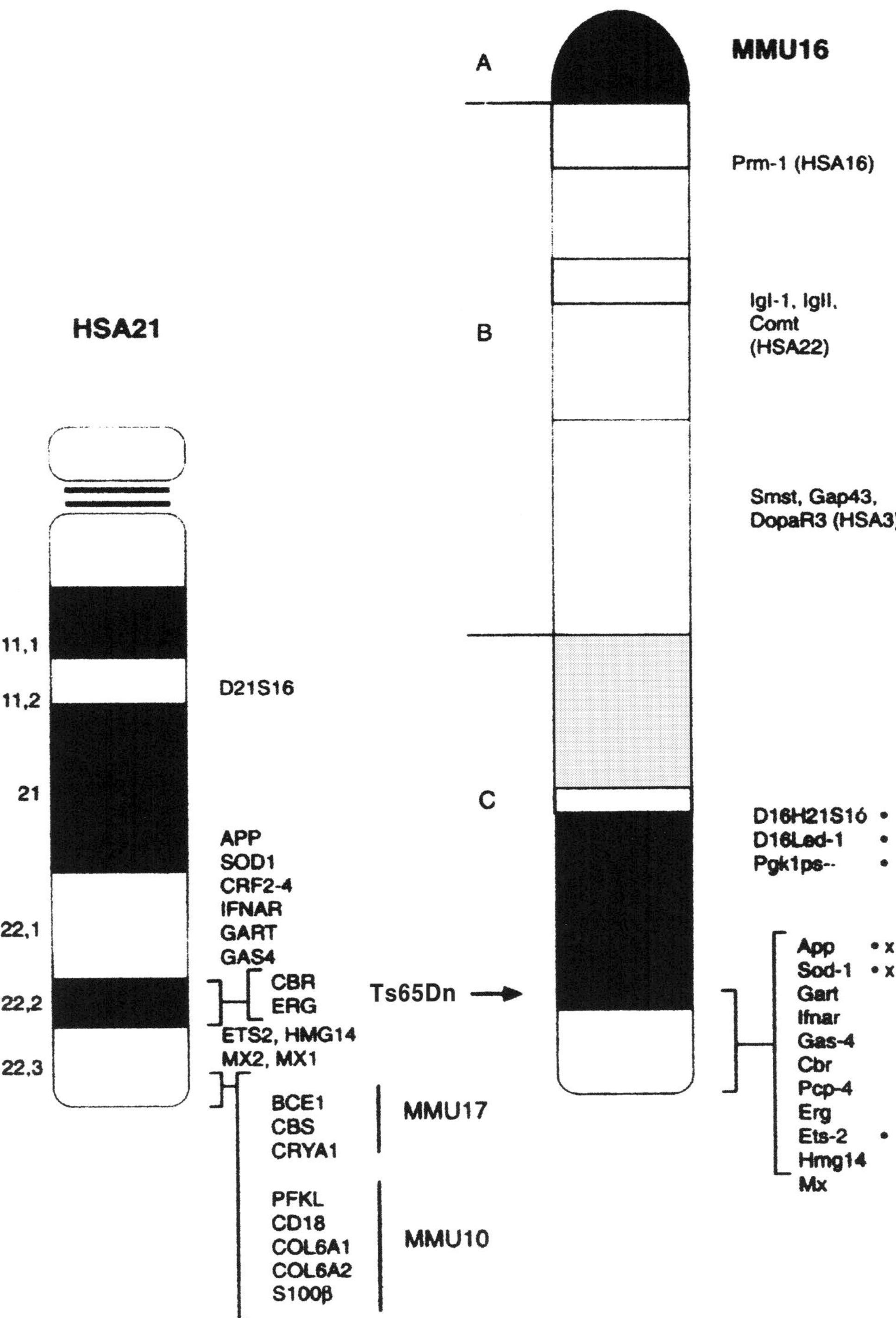

Fig. 4. Genetic comparative maps of HSA21 and MMU16.

facts: (a) mice heterozygous for reciprocal translocation chromosomes, where the translocation chromosomes are small, can produce progeny trisomic for the small translocation product; and (b) mice trisomic for small chromosomal segments usually survive to adulthood. Reciprocal translocations were induced by irradiating the testes of DBA/2J mice. The males were mated immediately to C57BL/6J females to recover radiation damage induced in spermatids and spermatozoa. F_1 progeny from these matings were screened for chromosomal aberrations in females (peripheral lymphocytes) and males (testis).

Several translocations involving MMU16 can be produced with translocation breakpoints spaced along the entire length of the chromosome. Ts65Dn has the distal tip (16C4→ter) of MMU16 with centromeric material from MMU17. Mice carrying chromosomal aberrations were mated to B6 mice to determine whether they were fertile and to recover the translocation products. Female mice heterozygous for reciprocal translocations in which one translocation product is very small often produce offspring trisomic for the genetic segments contained in the small marker chromosome. Stocks for the translocations are maintained by mating the carriers to B6 or a B6 × C3H/HeJ (B6C3H) F_1 hybrid.

Although many overexpressed genes in Ts65Dn are homologous to those included in the DSCR, these mice do not show many of the gross abnormalities frequently observed in DS, such as craniofacial deformities and cardiac valvular congenital abnormalities (Montero et al., 1996). They do exhibit, however, other features comparable to those found in DS. Cognitive deficits have been consistently demonstrated in the Ts65Dn model, including spatial learning that requires the integration of visual and spatial information (Escorihuela et al., 1995; Reeves et al., 1995; Holtzman et al., 1996), as well as working memory and long-term memory (Escorihuela et al., 1998). However, extensive training may facilitate the achievement of levels of performance as high as in control mice. Other behavioral anomalies are also present in Ts65Dn mice. They show consistent hyperactivity and reduced attention levels (Escorihuela et al., 1995; Reeves et al., 1995; Coussons-Read and Crnic, 1996; Holtzman et al., 1996). Recently (Martínez-Cué et al., 1999) a reduced responsiveness to nociceptive stimuli has been observed in these mice. The basis for these anomalies is not clear. Neurochemical studies have revealed functional abnormalities in the central noradrenergic system in the cerebral cortex and hippocampus but not in the cerebellum. In these regions, basal as well as stimulated production of cyclic AMP was reduced (Dierssen et al., 1996, 1997), and this reduction in adenyl cyclase efficacy may account for the disturbed cognitive achievement seen in Ts65Dn mice. In addition, alterations in LTP have been reported in young and old Ts65Dn mice (Siarey et al., 1997). Stereological morphometric studies have also demonstrated minor irregularities in some regions of the hippocampus of Ts65Dn mice, including reductions in the volume of CA2 and in the mean neuron number in the dentate gyrus (Insausti et al., 1998). Holtzman et al. (1996) have found an age-related degeneration of septohippocampal cholinergic neurons and astrocytic hypertrophy, markers of the Alzheimer's disease pathology present in DS individuals.

A second partial trisomy 16 model has been recently developed, termed Ts1Cje mouse (Sago et al., 1998). This model is trisomic for a smaller region of MMU16 from *Sod1* to *Mx1*. The Ts1Cje mouse was derived during *Sod1* gene targetting and involves an MMU12:16 reciprocal translocation and thus has a diploid chromosome number. Heterozygous *Sod1* mutant mice were found to segregate aberrantly when crossed to wild type animals. FISH studies of the heterozygous *Sod1* mutants revealed a translocation between MMU16 proximal to *Sod1* and the very distal region of MMU12. The balanced carriers for the translocation designated T(12;16)1Cje have one each of the normal chromosomes MMU12 and MMU16 in addition to the 12^{16} and 16^{12} translocation chromosomes. Partial trisomy of the distal region of MMU16, the Ts1Cje mouse, is generated when a gamete carrying $MMU12^{16}$ and MMU16 combines with a normal gamete carrying MMU12 and MMU16. In addition, few if any of the telomeric genes on MMU12 were missing from the 12^{16} chromosome. Therefore Ts1Cje appears to be trisomic only for the segment of the distal region of MMU16, but because the *Sod1* allele on MMU12 is disrupted by the neomycine resistance sequence, the Ts1Cje animals are not functionally trisomic for *Sod1*.

Ts1Cje mice survive to adulthood and present no apparent limb or facial malformations or other dysmorphic features. Contrary to Ts65Dn, Ts1Cje did not show hyperactivity and their learning deficits were also less severe. They performed efficiently in the Morris water maze when the platform was visible, but they showed impairment in the hidden platform and probe tests, and in the reverse platform test (Sago et al., 1998). In fact the learning impairment was stronger in Ts1Cje mice than in the YAC152Ftel transgenic mice, but less severe than those found in Ts65Dn animals. However, to draw meaningful conclusions about the comparisons, the mice should be on the same genetic background and the same characterization protocols should be applied. Ts1Cje mice do not demonstrate the neuronal atrophy found in Ts65Dn, suggesting that the region from *App* to *Sod1* is required for this neuropathological trait. The only known genes in this regions are *App*, associated with AD and *Grik1*, whose product, the GluR5 kainate receptor regulates synaptic transmission. By crossing Ts65Dn females to genetically balanced Ts(12 : 16) 1Cje males a new segmental trisomy, designated Ms1Ts65 has been generated that shows less severe abnormalities in spatial learning (Epstein, C.J.; EMBO Workshop 1999).

Finally, a new strategy has been reported that involves the transference of a human chromosome into mice, through the mouse germ line (Tomizuca et al., 1997). Efforts are being made to create mice that carry all or part of HSA21 as a freely segregating extra chromosome. To produce strains of these transgenomic animals, the whole or parts of HSA21 are transferred from a human somatic cell line into mouse embryonic stem cells using irradiation microcell-mediated chromosome transfer (Hernández et al., 1997). These transchromosomal mice could be particularly useful for modelling human aneuploidy syndromes and provide a qualitatively different approach for the study of the pathology of trisomy 21.

DS is a paradigmatic complex disease and the mechanism by which it produces phenotypic abnormalities is not understood. Despite intense investigation of the pathology, biochemistry and physiology of DS, it is still not known how the individual genes on HSA21, either singly or in concert produce the alterations associated with the trisomy. Several useful mouse models of DS have been generated, and new strategies are being initiated. The identification of a growing number of HSA21 genes will undoubtedly lead to the generation of many new single-gene transgenic models. In the YAC transgenics, the approach is rather different and involves the utilization of several overlapping or contiguous YACs that can cover a significant part of the genome. In a strategy termed 'in vivo libraries', large fragments of the genome can be propagated in vivo. These latter two approaches overcome some of the limitations of single-gene transgenics since the models generated harbor the introduced sequences at a low-copy number, similar to DS, and the regulatory elements that control the normal gene expression are present. Finally, the mouse trisomies constitute animal models for DS with different degrees of phenotypic impact, but they do not permit the role of each gene in the pathophysiology of the disease to be deciphered. Most likely, the strategies of crossing the trisomic models with knockout models, or the antisense technology will shed light on the role of single genes in a trisomic environment. The models described herein, are the best available at the present time for evaluating the efficacy of different pharmacological, genetic or environmental therapeutic strategies. To understand the molecular pathogenesis of DS, an overall analysis and comparative study of the existing models should be considered, but for this, mice should be on the same genetic background and experimental protocols for the characterization and description of the phenotype of the models generated must be standardized and cover every aspect of mouse neurological, behavioral, biochemical and histopathological features.

Acknowledgments

The authors are indebted to M. Nadal and C. Casas for FISH and in situ hybridization results. We thank C. Pucharcós for help with the figures. This work was supported in part by grants of CEC/BIOMED2 (BMH4-CT98-3039), Fundació Catalana Síndrome de Down/Marató TV3-1993, DGES (PM95-0106-CO2-01) and Fundación Marcelino Botín.

References

Ainscough, J.F., Koide, T., Tada, M., Barton, S. and Surani, M.A. (1997) Imprinting of Igf2 and H19 from a 130 kb YAC transgene. *Development*, 124: 3621–3632.

Antonarakis, S.E., Avramopoulos, D., Blouin, J-L., Talbot, C.C. and Schinzel, A.A. (1993) Mitotic errors in somatic cells cause trisomy 21 in about 4.5% of cases and are not associated with advanced maternal age. *Nat. Genet.*, 3: 146–150.

Antonarakis, S.E. (1998) 10 years of genomics, chromosome 21, and Down syndrome. *Genomics*, 51: 1–16.

Bar Peled, O., Korkotian, E., Segal, M. and Groner, Y. (1996) Constitutive overexpression of Cu/Zn superoxide dismutase exacerbates kainic acid-induced apoptosis of transgenic-Cu/Zn superoxide dismutase neurons. *Proc. Natl. Acad. Sci. USA*, 93: 8530–8535.

Bartsch, D., Patterson, S., Karl, K.A., Yang, X., Chen, A., Chen, M. and Kandel, E.R. (1998) Transgenic model of memory deficit in Down syndrome: role of the minibrain kinase. *Proc. Soc. Neurosci.* (*Los Angeles*), p. 441.

Burke, D.T., Carle, G.F. and Olson, M.V. (1987) Cloning of large segments of exogenous DNA into yeast by means of artificial chromosome vectors. *Science*, 236: 806–12.

Buselmaier, W., Bachus, C. and Sterz, H. (1991) Genesis and systematization of cardiovascular anomalies in murine trisomy 16. In: C.J. Epstein (Ed.), *The Morphogenesis of Down Syndrome*, Wiley-Liss, New York, pp. 203–214.

Cabin, D.E., Hawkins, A, Griffin, C. and Reeves, R.H. (1995) YAC transgenic mice in the study of the genetic basis of Down syndrome. *Prog. Clin. Biol. Res.*, 393: 213–226.

Calingasan, N.Y., Gandy, S.E. et al. (1996) Novel neuritic clusters with accumulations of amyloid precursor protein and amyloid precursor-like protein 2 immunoreactivity in brain regions damaged by thiamine deficiency. *Am. J. Pathol.*, 149: 1063–1071.

Chen, H., Chrast, R., Rossier, C., Morris, M.A., Lalioti, M.D. and Antonarakis, S.E. (1996a) Cloning of 559 potential exons of genes of HSA21 by exon trapping. *Genome Res.*, 6: 747–760.

Chen, H., Bouras, C. and Antonarakis, S.E. (1996c) Cloning the cDNA for a human homolog of the rat PEP-19 gene and mapping to chromosome 21q22.2–q22.3. *Human Genet.*, 98: 672–677.

Chumakov, I., Rigault, P., Le Gall et al and Cohen, D. (1995) A YAC contig map of the human genome. *Nature*, 377(Supp): 174–297.

Clarke, V.R., Ballyk, B.A., Hoo, K.H., Mandelzys, A., Pellizzari, A., Bath, C.P. et al. (1997) A hippocampal GluR5 Rainate receptor regulating inhibitory synaptic transmission. *Nature*, 389: 599–603.

Coussons-Read, M.E. and Crnic, L.S. (1996) Behavioral assessment of the Ts65Dn mouse, a model for Down syndrome: altered behavior in the elevated plus maze and open field. *Behav. Genet.*, 26: 7–13.

Coyle, J.T., Oster-Granite, M.L., Reeves, R., Hohmann, C., Corsi, P. and Gearhart, J. (1991) Down syndrome and the trisomy 16 mouse: impact of gene imbalance on brain development and aging. In: P.R. McHugh and V.A. McKusick (Eds.), *Genes, Brain and Behavior*, Raven Press, New York, pp. 85–99.

Crawley, J. and Paylor, R. (1997) A proposed test battery and constellations of specific behavioral paradigms to investigate the behavioral phenotypes of transgenic and knockout mice. *Horm. Behav.*, 31: 197–211.

Crusio, W.E. (1996) Gene-targetting studies: new methods, old problems. *Trends Neurol. Sci.*, 19: 186–187.

Dahmane. N., Ghezala, G.A. et al. (1998) Transcriptional map of the 2.5-Mb CBR-ERG region of HSA21 involved in Down syndrome. *Genomics*, 48: 12–23.

Davies, N.P., Rosewell, I.R., Richardson, J.C., Cook, G.P., Neuberger, M.S., Brownstein, B.H., Norris, M.L. and Bruggemann, M. (1993) Creation of mice expressing human antibody light chains by introduction of a yeast artificial chromosome containing the core region of the human immunoglobulin kappa locus. *Biotechnology*, 11: 911–914.

Davisson, M.T., Schmidt, C. and Akeson, E.C (1990) Segmental trisomy for murine chromosome 16: a new system for studying Down syndrome. In: D. Patterson and C.J. Epstein (Eds.), *Molecular Genetics of Chromosome 21 and Down syndrome*, Wiley-Liss, New York, pp. 263–280.

Dierssen, M., Vallina, I.F., Baamonde, C., García-Calatayud, S., Lumbreras, M.A. and Flórez, J. (1997) Alterations of central noradrenergic transmission in Ts65Dn mouse, a model for Down syndrome. *Brain Res.*, 749: 238–243.

Dierssen, M., Vallina, I.F., Baamonde, C., Lumbreras, M.A., Martínez-Cué, C., Calatayud, S.G. and Flórez, J. (1996) Impaired cyclic AMP production in the hippocampus of a Down syndrome murine model. *Dev. Brain Res.*, 95: 122–124.

Duff, K., McGuigan, A., Huxley, C., Schulz, F. and Hardy, J. (1994) Insertion of a pathogenic mutation into a yeast artificial chromosome containing the human amyloid precursor protein gene. *Gene Therapy*, 1: 70–75.

Elson, A., Levanon, D., Weiss, Y. and Groner, Y. (1994) Overexpression of liver-type phosphofructokinase (PKLF) in transgenic-PKLF mice: implications for gene dosage in trisomy 21. *Biochem. J.*, 299: 409–415.

Epstein C.J. (1986) *Consequences of Chromosome Imbalance: Principles, Mechanisms, and Models*, Cambridge Univ. Press, New York.

Epstein, C.J., Avraham, K.B., Lovett, M., Smith, S., Elroy-Stein, O., Rotman, G., Bry, C. and Groner, Y. (1987) Transgenic mice with increased Cu/Zn-superoxide dismutase activity: animal model of dosage effects in Down syndrome. *Proc. Natl. Acad. Sci. USA*, 84: 8044–8048.

Epstein, C.J. (1995) The metabolic and molecular basis of inherited disease. In: C.R. Scriver, A.L. Beaudet, W.S. Sly and D. Valle (Eds.), *Down syndrome*, Vol. 1, 7th Edn, McGraw-Hill, New York.

Escorihuela, R.M., Fernández-Teruel, A., Vallina, I.F., Baamonde, C., Lumbreras, M.A., Dierssen, M., Tobeña, A. and Flórez, J. (1995) Behavioral assessment of Ts65Dn mice: a putative DS model. *Neurosci. Lett.*, 199: 143–146.

Escorihuela, R.M., Vallina, I.F., Martínez-Cué, C., Baamonde, C., Dierssen, M., Tobeña, A. and Flórez, J., Fernández-Teruel, A. (1998) Impaired short- and long-term memory in Ts65Dn mice, a model for DS. *Neurosci. Lett.*, 247: 171–174.

Flórez, J. (1992) Neurologic abnormalities. In: S.M. Pueschel and J.K. Pueschel (Eds.), *Biomedical Concerns in Persons with Down Syndrome*, Paul H. Brookes Pub. Co., Baltimore, pp. 159–173.

Fuentes, J-J., Pritchard, M.A., Planas, A.M., Bosch, A., Ferrer, I. and Estivill, X. (1995) A new human gene from the Down syndrome critical region encodes a proline-rich protein highly expressed in fetal brain and heart. *Human Mol. Genet.*, 4: 1935–1944.

Fuentes, J-J., Pritchard, M.A. and Estivill, X. (1997) Genomic organisation, alternative splicing and expression patterns of the DSCR1 (DS candidate region 1) gene. *Genomics*, 44: 358–361.

Games, D., Adams, D. Alessandrini, R. et al. (1995) Alzheimer-type neuropathology in transgenic mice overexpressing V717F β-amyloid precursor protein. *Nature*, 373: 523–527.

Gahtan, E., Auerbach, J.M., Groner, Y. and Segal, M. (1998) Reversible impairment of long-term potentiation in transgenic Cu/Zn-SOD mice. *Eur. J. Neurosci.*, 10: 538–44

Gearhart, J.D., Davisson, M.T. and Oster-Granite, M.L. (1986) Autosomal aneuploidy in mice: generation and developmental consequences. *Brain Res. Bull.*, 16: 789–801.

Gerlai, R (1996) Gene targetting studies of mammalian behavior: is it the mutation or the background phenotype? *Trends Neurosci.*, 19: 177–181.

Gerlai, R. and Roder, J. (1993) Female specific hyperactivity in S100 beta transgenic mice does not habituate in the open-field. *Behav. Brain.* Res. 59: 119–124.

Gerlai, R. and Roder, J. (1996) Spatial and nonspatial learning in mice: effects of S100 beta overexpression and age. *Neurobiol. Learn. Mem.* 66: 143–154.

Grausz, H., Richtsmeier, J.T. and Oster-Granite, M.L. (1991) Morphogenesis of the brain and craniofacial complex in trisomy 16 mice. In: C.J. Epstein (Ed.), *The Morphogenesis of Down Syndrome*, Wiley-Liss, New York, pp. 169–188.

Green, L.L., Hardy, M.C. et al. (1994) Antigen-specific human monoclonal antibodies from mice engineered with human Ig heavy and light chain YACs. *Nat. Genet.*, 7: 13–21.

Groner, Y. (1995) Transgenic models for HSA21 gene dosage. *Prog. Clin. Biol Res.*, 393: 213–226.

Guimerà, J., Casas, C., Pucharcos, C., Solans, A., Domenech, A., Planas, A.M., Ashley, J., Lovett, M., Estivill, X. and Pritchard, M.A. (1996) A human homologue of Drosophila minibrain (MNB) is expressed in the neuronal regions affected in Down syndrome and maps to the critical region. *Hum. Mol. Genet.*, 5: 1305–1310.

Guimerá, J., Pucharcòs, C., Domènech, A., Casas, C., Solans, A., Gallardo, T., Ashley, J., Lovett, M., Estivill, X. and Pritchard, M.A. (1997) Cosmid contig and transcriptional map of three regions of human chromosome. *Genomics*, 45: 56–67.

Hernández, D., Tybulewicz, V., Fisher, E.M.C. (1997) Creating a panel of 'transgenomic' mouse embryonic stem cell lines that contain freely segregating portions of human chromosome 21. *Proc. Int. Chromosome 21 Workshop (Berlin)*, p. 40.

Holtzman, D.M., Santucci, D. et al. (1996) Developmental abnormalities and age-related neurodegeneration in a mouse model of Down syndrome. *Proc. Natl. Acad. Sci. USA*, 93: 13333–13338.

Hsiao, K., Chapman, P., Nilsen, S., Eckman, C., Harigaya, Y., Younkin, S., Yang, P. and Cole, G. (1996) Correlative memory deficits, Abeta elevation, and amyloid plaques in transgenic mice. *Science*, 274: 99–102.

Insausti, A.M., Megías, M., Crespo, D., Cruz-Orive, L.M., Dierssen, M., Vallina, I.F., Insausti, R. and Flórez, J. (1998) Hippocampal volume and neuronal number in Ts65Dn mice: a murine model of Down syndrome. *Neurosci. Lett.*, 253: 1–4.

Jakobovits, A., Moore, A.L., Green, L.L., Vergara, G.J., Maynard-Currie, C.E., Austin, H.A. and Klapholz, S. (1993) Germ-line transmission and expression of a human-derived yeast artificial chromosome. *Nature*, 362: 255–258.

Kola, I. and Hertzog, P. (1997) Animal models in the study of the biological function of genes on HSA21 and their role in the pathophysiology of Down syndrome. *Hum. Mol. Genet.*, 6: 1713–1727.

Lamb, B.T., Sisodia, S.S., Lawler, A.M., Call, L.M., Slunt, H.H., Kitt, C.A., Kearns, W.G., Pearson, P.L., Price, D.L. and Gearhart, J.D. (1993) Introduction and expression of the 400 kilobase amyloid precursor protein gene in transgenic mice. *Nat. Genet.*, 5: 22–30.

Lamb, B.T., Call, L.M., Slunt, H.H., Bardel, K.A., Lawler, A.M., Eckman, C.B., Younkin, S.G., Holtz, G., Wagner, S.L., Price, D.L., Sisodia, S.S. and Gearhart, J.D. (1997) Altered metabolism of familial Alzheimer's disease-linked amyloid precursor protein variants in yeast artificial chromosome transgenic mice. *Hum. Mol. Genet.*, 6: 1535–1541.

Lee, J.T. and Jaenisch, R. (1996) A method for high efficiency YAC lipofection into murine embryonic stem cells. *Nucl. Acids Res.*, 24: 5054–5055.

Levanon, D., Danciger, E., Dafni, N. and Groner, Y. (1986) Genomic clones of the human liver-type phosphofructokinase. *Biochem. Biophys. Res. Commun.*, 141: 374–380.

Maas, A., Dingjan, G.M., Savelkoul, H.F., Kinnon, C., Grosveld, F. and Hendriks, R.W. (1997) The X-linked immunodeficiency defect in the mouse is corrected by expression of human Bruton's tyrosine kinase from a yeast artificial chromosome transgene. *Eur. J. Immunol.*, 27: 2180–2187.

Martínez-Cué, C., Vallina, I.F., Baamonde, C., Dierssen, M., Fillat, C., Fuentes, J.J., Pritchard, M., Estivill, X. and Flórez, J. (1997) Transgenetic mice overexpressing the human *DSCR1* gene: behavioral characterization of two lines. *Proc. Int. Chromosome 21 Workshop (Berlin)*, p. 45.

Martínez-Cué, C., Baamonde, C., Vallina, I.F., Lumbreras, M.A., Dierssen, M. and Flórez, J. (1999) A murine model for Down syndrome shows reduced responsiveness to pain. *NeuroReport*, 10: 1119–1122.

Miyabara, S., Sugihara, H., Yonemitsu, N. and Yun, K. (1984) Comparative study of phenotypic expression of mice trisomy 16 by different female strains: Attempt at an animal model for human trisomy 21. *Cong. Anomal.*, 24: 283–292.

Moechars, D., Gilis, M., Kuiperi, C., Laenen, I. and Van Leuven, F. (1998) Aggressive behavior in transgenic mice expressing APP is alleviated by serotonergic drugs. *NeuroReport*, 9: 3561–3564.

Montero, J.J., Flórez, J., Baamonde, C., Vallina, I.F., García-Calatayud, S. and Dierssen, M. (1996) Valoración ecográfica mediante doppler pulsado de la función cardíaca en el ratón Ts65Dn, un modelo murino de síndrome de Down. *Rev. Síndrome Down*, 13: 31–32.

Montoliu, L., Schedl, A., Kelsey, G., Zentgraf, H., Lichter, P. and Schutz, G. (1994) Germ line transmission of yeast artificial chromosomes in transgenic mice. *Reprod, Fertil Dev.*, 6: 577–584.

Murai, H., Pierce, J.E. et al. (1998) Twofold overexpression of human beta-amyloid precursor proteins in transgenic mice does not affect the neuromotor, cognitive, or neurodegenerative sequelae following experimental brain injury. *Comp. Neurol.*, 392: 428–438.

Nucifora, G., Begy, C.R., Kobayashi, H., Roulston, D., Claxton, D., Pedersen-Bjergaard, J., Parganas, E., Ihle, J.N. and Rowley, J.D. (1994) Consistent intergenetic splicing and production of multiple transcripts between AML1 at 21q22 and unrelated genes at 3q26 in (3;21)(q26;q22) translocations. *Proc. Natl. Acad. Sci. USA*, 91: 4004–4008.

Ogawa, S. and Pfaff, D.W. (1996) Application of antisense DNA method for the study of molecular bases of brain function and behavior. *Behav. Genet.* 26: 279–292.

Oster-Granite, M.L. (1986) The neurobiologic consequences of autosomal aneuploidy in mice and men. *Brain Res. Bull.*, 16: 767–771.

Oster-Granite, M.L. and Lacey-Casem, M.L. (1995) Neurotransmitter alterations in the trisomy 16 mouse: a genetic model system for studies of Down syndrome. *Ment. Retard. Develop. Dis. Res. Rev.*, 1: 227–236.

Patil, N., Cos, D.R., Bhat, D., Faham, M., Myers, R.M. and Peterson, A.S. (1995) A potassium channel mutation in weaver mice implicates membrane excitability in granule cell differentiation. *Nat. Genet.*, 11: 126–129.

Persico, A.M., Schindler, C.W., Davis, S., Ambrosio, E. and Uhl, G. (1998) Medial prefrontal cortical injections of c-fos antisense oligonucleotides transiently lower c-Fos protein and mimic amphetamine withdrawal behaviors. *Neuroscience*. 82: 1115–1129.

Peterson, K.R., Clegg, C.H., Huxley, C., Josephson, B.M., Haugen, H.S., Furukawa, T. and Stamatoyannopoulos, G. (1993) Transgenic mice containing a 248 kb yeast artificial chromosome carrying the human beta-globin locus display proper developmental control of human globin genes. *Proc Natl Acad Sci USA*, 90: 7593–7597.

Peterson, K.R., Li, Q.L., Clegg, C.H., Furukawa, T., Navas, P.A., Norton, E.J., Kimbrough, T.G. and Stamatoyannopolous, G. (1995) Use of yeast artificial chromosomes (YACs) in studies of mammalian development: Production of b-globin locus YAC mice carrying huiman globin devopmental mutants. *Proc. Natl. Acad. Sci. USA*, 92: 5655–5659.

Pueschel, S.M. (1992) Phenotypic characteristics. In: S.M. Pueschel and J.K. Pueschel (Eds.), *Biomedical Concerns in Persons with Down Syndrome*, Paul H. Brookes Pub. Co., Baltimore, pp. 1–12.

Reeves, R.H., Yao, J., Crowley, M.R., Buck, S., Zhang, X., Yarowsky, P., Gearhart, J.D. and Hilt, D.C. (1994) Astrocytosis and axonal proliferation in the hippocampus of S100β transgenic mice. *Proc. Natl. Acad. Sci. USA* , 91: 5359–5363.

Reeves, R.H., Irving, N.G., Moran, T.H., Wohn, A., Kitt, C., Sisodia, S.S., Schmid, C., Bronson, R.T. and Davisson, M. (1995) A mouse model for Down syndrome exhibits learning and behaviour deficits. *Nat. Genet.*, 11: 177–184.

Rogers, D.C., Fisher, E.M., Brown, S.D., Peters, J., Hunter, A.J. and Martin, J.E. (1997) Behavioral and functional analysis of mouse phenotype: SHIRPA, a proposed protocol for comprehensive phenotype assessment. *Mamm. Genome.*, 8: 711–713.

Sago, H., Carlson, E.J., Smith, D., Kilbridge, J., Rubin, E.M., Mobley, W., Epstein, C.J. and Huang, T (1998) Ts1Cje, a partial trisomy 16 mouse model for Down syndrome, exhibits learning and behavioral abnormalities. *Proc. Natl. Acad. Sci. USA*, 95: 6256–6261.

Sberna, G., Saez-Valero, J., Li, Q.X., Czech, C., Beyreuther, K., Masters, C.L., McLean, C.A. and Small, D.H. (1998) Acetylcholinesterase is increased in the brains of transgenic mice expressing the C-terminal fragment (CT100) of the beta-amyloid protein precursor of Alzheimer's disease. *J. Neurochem.*, 71: 723–731.

Schedl, A., Larin, Z., Montoliu, L., Thies, E., Kelsey, G., Lehrach, H. and Schutz, G. (1993) A method for the generation of YAC transgenic mice by pronuclear microinjection. *Nucl.c Acids Res.*, 21: 4783–4787.

Schwartz, P.J., Reaume, A., Scott, R. and Coyle, J.T. (1998) Effects of over- and under-expression of Cu,Zn-superoxide dismutase on the toxicity of glutamate analogs in transgenic mouse striatum. *Brain Res*, 789: 32–39.

Siarey, R.J., Stoll, J., Rapoport, S.I. and Galdzicki, Z. (1997) Altered long-term potentiation in the young and old Ts65Dn mouse, a model for Down syndrome. *Neuropharmacol.*, 36: 1549–1554.

Siddique, T. and Deng, H.X. (1996) Genetics of amyotrophic lateral sclerosis. *Hum. Mol. Genet.*, 5: 1465–1470.

Smith, D.J., Zhu, Y., Zhang, J., Cheng, J.F. and Rubin, E.M. (1995) Construction of a panel of transgenic mice containing a contiguous 2-Mb set of YAC/P1 clones from HSA21q22.2. *Genomics*, 27: 425–434.

Smith, D.J., Stevens, M.E. et al. (1997) Functional screening of 2 Mb of HSA21q22.2 in transgenic mice implicates minibrain in learning defects associated with Down syndrome. *Nat. Genet.*, 16: 28–36.

Smithies, O. and Kim, H.-S. (1994) Targeted gene duplication and disruption for analyzing quantitative genetic traits in mice. *Proc. Natl. Acad. Sci. USA*, 91: 3612–3615.

Strauss, W.M., Dausman, J., Beard, C., Johnson, C., Lawrence, J.B. and Jaenisch, R. (1993) Germ-line transmission of a yeast artificial chromosome spanning the murine a1(I)Collagen locus. *Science*, 259: 1904–1907.

Sturchler-Pierrat, C., Abramowski, D., Duke, M. et al. (1997) Two amyloid precursor protein transgenic mouse models with Alzheimer disease-like pathology. *Proc. Natl. Acad. Sci. USA*, 94: 13287–13292.

Sumarsono, S.H., Wilson, T.J., Tymms, M.J., Venter, D.J., Corrick, C.M., Kola, R., Lahoud, M.H., Papas, T.S., Seth, A. and Kola. I. (1996) Down's syndrome-like skeletal abnormalities in Ets2 transgenic mice. *Nature*, 379: 534–537.

Tomizuca, K., Yoshida, H., Uejima, H., Kugoh, H., Sato, K., Ohguma, A., Hayasaka, M., Hanaoka, K., Oshimura, M. and Ishida, I. (1997) Functional expression and germline transmission of a human chromosome fragment in chimaeric mice. *Nat. Genet.*, 16: 133–143.

Whitaker-Azmitia, P.M., Wingate, M., Borella, A., Gerlai, R., Roder, J. and Azmitia, E.C. (1997) Transgenic mice overexpressing the neurotrophic factor S-100 beta show neuronal cytoskeletal and behavioral signs of altered aging proccesses: implications for Alzheimer's disease and Down's syndrome. *Brain Res.*, 776: 51–60.

Yamakawa, K., Huo, Y-K., Haendel, M.A., Hubert, R., Chen, X-N., Lyons, G.E. and Korenberg, J.R. (1998) DSCAM: a novel member of the immunoglobulin superfamily maps in a Down syndrome region and is involved in the development of the nervous system. *Hum. Mol. Genet.*, 7: 227–237.

Yaspo. M-L., Gellen, L., Mott, R., Dorn, B., Nizetic, D., Poustka, A.M. and Lehrach, H. (1995) Model for a transcript map of HSA21: isolation of new coding sequences from exon and enriched cDNA libraries. *Hum. Mol. Genet.*, 8: 1291–1304.

Zapata, A., Capdevila, J.L., Tarrason, G., Adan, J., Martínez, J.M., Piulat, J. and Trullás, R. (1997) Effects of NMDA-R1 antisense oligodeoxynucleotide administration: behavioral and radioligand binding studies. *Brain Res.*, 745: 114–120.

Section 6

Ethical Considerations

W.E. Crusio and R.T. Gerlai (Eds.)
Handbook of Molecular-Genetic Techniques for Brain and
Behavior Research (Techniques in the Behavioral and Neural Sciences, Vol. 13)

CHAPTER 6.1

Genes and human behavior: scientific and ethical implications of the human genome project

Jon Beckwith

Department of Microbiology and Molecular Genetics, Harvard Medical School, 200 Longwood Avenue, Boston, MA 02115, USA

Introduction

The spectacular advances in the science of genetics in the last 20 years have given great public prominence to its achievements. From the applications of gene cloning techniques so successfully pursued by the biotechnology industry to the mapping and characterization of genes for many human diseases, these accomplishments have emphasized the importance of genes in human life and health. One consequence of this widespread interest in genetics has been an increasing focus on genes as the explanatory factor for much of human behavior. Whether it is psychiatric conditions, the range of human aptitudes, or the normal variations in temperamental traits, the publicity given human behavioral genetics has suggested a substantial increase in our understanding.

I would like to focus on three aspects of this increasing belief in the power of genes in determining human behavior. The first is to emphasize the significant social impact that reports of human behavior genetic studies can have. That is, the appearance of such studies in the scientific literature and the subsequent media attention they receive can have consequences that most geneticists are not aware of. It is important that scientists do become aware of these consequences because it (a) may encourage behavior geneticists themselves to be more cautious in their presentation of results and (b) may induce others in the field of genetics to play a more active role in correcting scientific and public misrepresentations of such studies.

Second, I will cover some of the problematic aspects of the current research into genetics of human behavior. In addition, I will point out how our views on the relationship between genes and human traits are changing as a result of research being carried out on the human genome. With the availability of new gene mapping technologies, researchers are discovering unanticipated complexity in the expression of genes related to various *physical* diseases.

Finally, I will discuss what I consider to be the implications of the first two sections. In particular, I will discuss how the discordance between the public impressions of this science and the actual state of the science of human behavior genetics can be reconciled.

The social impact of human behavior genetic studies

When I was trained in genetics approximately 40 years ago, I learned essentially nothing about the history of the social uses of genetics. However, I believe that today there is much more awareness among scientists of such historical realities as the existence and impact of eugenics movements in Western European countries and North America earlier in this century. There is also much greater recognition of the contribution of geneticists to Nazi theories of genetic superiority and inferiority (Müller-Hill, 1988; Watson, 1997). However, many of us were still surprised when we learned of the existence of extensive sterilization programs still carried out relatively recently in some northern European countries. These programs were based on eugenic arguments (Butler, 1997).

A striking aspect of the public impact of human behavior genetics in both historical and contemporary examples is that it has often resulted from generalizations made from highly questionable data. Either scientists or the media or both have overstated the significance of such research and used it to explain major social problems. A second important component of the public face of genetics is the impression given that if something is said to be genetic or to have a strong genetic component, it is fated. This deterministic or fatalistic view is often used to inform debates over social policy (Alper and Beckwith, 1993). Ideas of biological determinism long precede the founding of modern genetics at the turn of the century. One such example comes from the works of the French novelist Emile Zola. A strong thread of fatalism pervades his novels, and, in one case, his account has striking parallels with an example from recent newspaper headlines. In "*La Bête Humaine*", Zola states that the main character, Jacques Lantier, suffers from a "fêlure héreditaire" — a hereditary flaw (Zola, 1977). This flaw causes him to fall into sudden murderous rages that lead him to unaccountably strangle or stab women. Zola speaks of him as "a man pushed to acts where his own *free will* stood for nothing".

A century after the publication of "*La Bête Humaine*", events occurred reminiscent of Zola's fictional case, but now strengthened by the apparent scientific backing that contemporary genetics provides. "Man's genes made him kill, his lawyers claim" proclaimed the newspaper headline from a 1994 issue of the *Wall St. Journal* (Felsenthal, 1994). In this case, a man convicted of murder in the state of Georgia was appealing his conviction and death sentence on the basis of a "fêlure héréditaire". He argued that the males in his family had a history of aggression and that he *might* have inherited his violent tendencies. According to the newspaper article, legal experts now "worry that genetic research could tear at the fabric of the criminal-justice system by allowing people to argue that they were born without control over their actions". Despite the appeal to modern genetics, the legal request was rejected because of the lack of direct evidence for a specific genetic defect.

More recently, a story appeared in the *New York Times* describing an unusual conference (Blakeslee, 1996). In July of 1996, at Woods Hole on Cape Cod, Massachusetts, 35 judges from both Federal and State courts met with 20 scientists to discuss the implications of the revolution in genetics for the legal system. Judges learned themselves how to sequence DNA in the lab and listened to lectures on the latest developments in DNA fingerprinting which has obvious relevance to the law. But they also participated in round-table discussions of the ethical issues raised by the new genetics.

This report in the *New York Times* cited one of these issues:

> "....judges asked what would happen if science demonstrated that genes controlled behavior or that bad early environments conspired with genes to turn some people inevitably into criminals- showing that free will did not exist in these situations."

Why the sudden interest in genes, criminal behavior and free will? In fact, the most recent impetus for this interest derives largely from two reports published in the *American Journal of Human Genetics* and *Science Magazine* (Brunner et al., 1993a, 1993b). These reports describe studies of a family from the Netherlands conducted by Han Brunner and his coworkers.

A number of the male members of this family were reported to exhibit both mental retardation and extremely aggressive behavior. These traits were found to be associated with a specific mutation in the gene for monoamine oxidase A (MAOA). While the study is somewhat difficult to evaluate because the researchers do not present direct evidence for the characterization of the behavior of these males, the indirect descriptions of behavior and the precision of the gene identification appear to present a reasonable case for the conclusions involving this particular family.

What are the broader implications of this study for research into aggressive behavior? I would suggest that they are likely to be quite limited. First, this family is thought to represent a case of the inheritance of an extremely rare mutation. Others have tried to find similar families without success. Consequently, this type of mutation itself is likely responsible for only a minuscule fraction of violent behavior in society. We might compare it to the case of phenylketonuria (PKU), a rare recessive genetic condition, where the mutations have a severe effect on cognitive ability. The cognitive problems arise because of the accumulation of a toxic compound that affects brain function. But, no one has argued that variation in this gene is a major contributor to the variation in cognitive ability in the population. That is, the finding of unusual families where a clear-cut correlation can be made between a specific genetic defect and a behavior usually cannot be extended to explain the variation in that behavior in the general population.

How did we move from this limited study to a court case and judges worrying about free will and responsibility? I would suggest that, initially, the authors of the research article themselves played a role in stimulating public interest in the social implications of the study. In the Discussion sections of their papers they include the following suggestions:

> "...should a mutation in the MAOA structural gene be identified in our family. this will have implications for the study of the biological mechanisms underlying disturbed aggression regulation in general..." (Brunner et al., 1993b)

> "given the wide range of variation of MAOA activity in the normal population, one could ask whether aggressive behavior is confined to complete MAOA deficiency." (Brunner et al., 1993a)

Thus, although carefully worded, the authors hint at a possible generalization of this finding with one family to the larger problem of aggression in society. This hint was not missed. A *Science Magazine* reporter suggested "it might be possible to identify people who are prone to violent acts by screening for MAOA gene mutations...." (Morell, 1993) The result of the provocative sentences in the papers and the Science news article was a tremendous media interest. This included a Newsweek article entitled "The genetics of bad behavior," in which the accompanying graphic showed a violent confrontation between Palestinians and Israelis, implying a genetic basis for world strife, in general (Cowlet and Hall, 1993); and a TV news report that used films of US street gang violence as a backdrop for a report on the MAOA study (X. Breakefield, personal communication)! This media attention was followed by the legal case in Georgia in which Dr Xandra Breakefield, one of the authors of the *Science* paper, was called upon to testify on the genetic basis of aggressive and violent behavior. Then the discussions at the conferences for judges. Even though the legal appeal to genetics was not accepted, this case and the publicity for the Dutch study illustrate the ways in which supposed genetic knowledge can rapidly be transmitted to the public, be misused and can be incorporated into discussions of social policy. Thus, on the most modest of genetic findings, major questions of public interest may be influenced.

To their credit, the scientists involved in the study have shown some concern about the public representations of their reports. Dr Breakefield was dismayed enough by the publicity to announce that she would no longer work on links between violence and genes (Breakefield, 1994). Dr Brunner's statement in a recent article that "...the notion of an 'aggression gene' does not make sense..." (Brunner, 1995) clearly reflected concern for the ways in which the study was interpreted by the media.

Another area of behavior genetics research with obvious social impact is that of genetics, race and IQ. Arguments from genetics have been repeatedly used to support calls for changes in social policy with regard to such issues as equal rights and immigration issues. In France, within the last few years, members of the National Front have used genetics to support their positions. The Mayor of Vitrolles, Catherine Mégret was fined by a French court for stating in an interview that "there are differences between the races.....in the genes". (Associated Press, 1997) Similar statements about genetic inequality between the races were made by Jean-Marie Le Pen, the leader of The National Front (Anonymous, 1996a).

In the United States in the last thirty years, there have been two major occasions when this issue has become prominent. The first was in 1969, when Arthur Jensen, a psychology professor from the University of California at Berkeley attracted a good deal of attention with his article stating that the differing intellectual performance of blacks and whites was due to genetically based differences in intelligence (Jensen, 1969). Relying largely on the results of identical twin studies, Jensen argued that a strong genetic component for intelligence indicated that these differences could not be overcome by improving educational programs. This deterministic rendering of the meaning of genetics received widespread media attention. Jensen's publication was read into the United States Congressional Record several times during debates over ending various remedial education programs. It was also cited in court cases during the battles over school desegregation.

In 1994, Harvard University psychologist Richard Herrnstein and political scientist Charles Murray published "*The Bell Curve*". (Herrnstein and Murray, 1994) This book amplified the claims of Jensen. It not only argued that genetically determined IQ led to social failure of certain groups, including blacks, but it also claimed that low IQ is correlated with a host of social problems, including crime, welfare and single parenting. Again, twin and adoption studies were the basis for these claims. Herrnstein and Murray worried that high birth rate among the poor was leading to an increasingly inferior and dysfunctional society. Thus, they proposed that women from the upper classes should be encouraged by new social programs to bear more children and that welfare and remedial education programs should be ended.

The claims of "*The Bell Curve*" received widespread attention in the United States with cover stories on the book in several popular magazines (Browne, 1994; Morganthau, 1994; Allman, 1994) and a feature article on Murray presented in the *New York Times Sunday Magazine* (De Parle, 1994). Charles Murray became an influential advisor to the Republican party on such issues as welfare policy. After the publication of this book in 1994 he was invited to be a featured speaker at a major session introducing freshman Republican congressman to the issues they might deal with in the next United States Congressional session.

Beyond the potential direct effect on policies, these genetic explanations for differences between groups can't help but support discrimination against certain groups and strengthen racist divisions. Perhaps more tragically, these kinds of arguments affect feelings of self-worth among people within those groups that are labelled as inferior. An example of this impact on people's self-confidence comes from a recent paper by Claude Steele of Stanford University and Joshua Aaronson of University of Texas at Austin. They have studied the behavior of different ethnic groups who were given tests of cognitive ability

(Steele and Aronson, 1995). They took a group of Stanford University undergraduate students that included both black and white students. They broke them up into smaller groups. But they always retained in each group the same mix of black and white students. They administered a test composed of questions from the examinations usually given to evaluate prospective graduate students. The *same* test was given to all groups. However, the researchers described the purpose of the test *differently* to different groups. To some groups they described the test as "diagnostic of intellectual ability" and to other groups it was described as a "laboratory problem-solving task that was nondiagnostic of ability". The results were as follows: on the test that students considered a test of their ability, the black students scored considerably lower than the white students. However, on the very same test that was only described as a "task", the black and white students did equally well. The authors presented evidence that the black students were influenced by stereotyped social attitudes about their innate ability. Thus, the black students experienced apprehension when they thought that this ability was being tested and this anxiety interfered with their performance. If this phenomenon can be extended beyond this one example, it suggests that the claims of genetic inferiority emanating from the scientific community and other sources generate feedback and a self-fulfilling prophecy. The more it is stated that blacks or some other group are inferior, the worse they will do on tests, thus reinforcing the arguments of Herrnstein and others.

The consequences of questionable reports that genes influence human aptitudes is also illustrated by an earlier example. In 1980, a study was published in the journal *Science* implying that girls were biologically inferior in math ability to boys (Benbow and Stanley, 1980). The journal featured this study in a news article, in which one of the authors, Camilla Benbow, stated that "girls would be better off accepting the differences". Benbow's statement and the media coverage implied a deterministic reading of the implications of biology. For example, a Newsweek article was headlined "Do males have a math gene?" Subsequent studies by sociologists showed that the publicity this message received had a direct influence on public attitudes on the issue, on parental expectations of their daughters' abilities, and on girls' own feelings about their potential to do well in math (Beckwith, 1983).

The issue here is not whether the arguments that genetic differences may play some role in the expression of different aptitudes are correct. While, up until now, the studies that make these claims are quite weak, it is certainly possible that in the future some genetic influences may be pinpointed. But, the deterministic position that some scientists and their interpreters have taken and extended into the realm of social policy is not warranted by what we know about the relationship between genes and phenotypes.

A final example of the extension of genetic studies into the social arena is the recent interest in the biological basis of homosexuality. Some researchers have pursued this issue with the following rationale (Marshall, 1992). They believe that if it can be shown that "homosexuality is genetic", then homosexual behavior is not a matter of choice. As a result, it is argued, the public would be more accepting of homosexuality since it would be considered as something natural. One of these researchers is Dean Hamer at the United States National Institutes of Health. In 1993, he and his colleagues presented evidence for an X-linked locus for male homosexuality from a sib-pair study in families where it appeared that the behavior might be inherited through the mother (Hamer et al., 1993). Recognizing the likely public interest in this finding, Dr Hamer called a press conference to announce the publication of his finding. The presentation at this conference was handled very cautiously- Hamer pointed out many qualifications to the conclusions of his study including the need for replication, that the X-linkage may reflect only a small proportion of all male homosexuality and that environment was very important (Hamer, 1994). Yet, within weeks, Dr Hamer was testifying in a court in Colorado in a gay rights case that his evidence showed that homosexuality was biological and that homosexual behavior was not a choice; therefore, homosexuals should not be discriminated against (Herbert, 1997). A *US News and World Report* reporter referred to the "strange twist" in which a liberal found himself "arguing the deterministic position".

Earlier genetic research that attracted much attention studied possible genetic factors in male homosexuality by surveying monozygotic and dizygotic twins and ordinary siblings. Despite quite serious questions about this study, the scientific coauthors, Michael Bailey and Richard Pillard, went on to argue in a *New York Times* opinion piece the following:

> "Homophobes sometimes justify their prejudice against homosexuals by alleging that homosexuality is contagious — that young homosexuals become that way because of older homosexuals... ...[But} if homosexuality is largely innate, this would prove that these claims are groundless." (Bailey and Pillard, 1991)

I agree with the aims of those scientists who are trying to reduce discrimination against homosexuals. But, to (a) use preliminary, unreplicated research to influence social policy or attitudes and (b) state or imply that a behavior that has a genetic component is fixed and unchangeable are both unwarranted and potentially dangerous.

Human genome research and behavior genetics

The last 20 years have witnessed the rapid development and improvement of techniques for readily mapping genes in any organism, including humans. Most recently, the pace of this research, facilitated by the Human Genome Project, has increased to the point where hardly a week goes by without a newspaper headline announcing the finding of a new gene correlated with a human health problem. The ease with which these advances have been made has encouraged the entry of molecular geneticists into the field of human behavioral genetics. The result of this influx of new researchers has also been a series of headlines announcing the mapping of genes for such behavioral traits as schizophrenia, bipolar manic depressive illness, alcoholism, homosexuality, risk-taking, happiness, anxiety, criminal behavior and others. What is less apparent to the public are the retractions of certain studies and, more generally, the contradictory and non-confirmatory studies that have appeared in the scientific literature (Kelsoe et al., 1989; Owen, 1992; Gelernter et al., 1993; Baron, 1998). The questionable nature of nearly all these studies is well-recognized within the field. Numerous review articles point out where these studies have gone wrong and suggest more stringent criteria for evaluating the data in this field. In particular, it is argued that the LOD scores used to "establish" significant correlations between a behavior and a genetic locus were too low and that the bar must be raised. Furthermore, these reviews suggest that the genetics of behaviors is turning out to be much more complex than assumed and requires more complex modes of analysis. In one of many recent reviews anguishing over these problems, Stanford geneticists Neil Risch and David Botstein stated:

> "The distress engendered by the numerous reversals and non-replications has led many to rethink the paradigm being employed." (Risch and Botstein, 1996)

In other words, faced with a puzzling inability to get direct evidence for genes contributing to behavior, scientists are questioning their assumptions and their approaches. You would hardly know this from reading the newspaper headlines.

Thus, for most thoughtful people in the field there is no convincing evidence yet for any of the genes for the list of behaviors I have recited.

In my opinion, the twin studies that have provided the basis for much of behavior genetics research until recently are so fraught with problems that they cannot be taken as definitive (Billings et al., 1992; Spitz and Carlier, 1996; Devlin et al., 1997; Beckwith and Alper, 1998). One of the approaches that has been subjected to serious criticism is the attempts to find associations between known "candidate" loci affecting brain function and behavioral characteristics or illnesses. Recent examples include studies claiming associations of alcoholism and "novelty seeking" behavior with alleles of dopamine receptor genes. Kidd (Kidd, 1993) and Baron (Baron, 1998) have argued that the "prior probability" of these studies being successful is extremely low. They point out that there are probably tens of thousands of genes involved in brain development and function and the likelihood that a meaningful association could be found with one of the very few genes already known is highly unlikely. Kidd also criticizes what he terms as "definition drift", which is quite common in this field. This "drift" refers to the attempts to find correlations by changing the definition of the trait until significant statistics appear. Calling them the "chief nemeses" of true positive results, Baron, in addition to the low prior probability issue, points to other problematic characteristics of such candidate gene studies as population stratification and multiple testing. Overall, the critics argue that there is no reason not to use total genome scans to approach gene mapping for complex traits.

Despite the failures of and questions about the various approaches to behavior genetics, I do want to emphasize that the ability to locate genes and do total human genome scans provides a much more precise and readily replicable way of approaching issues of genes and human behaviors than have been employed before. The new molecular approaches can put human behavior genetics on a more solid footing. Thus, I think it likely that we will be hearing of some well-established examples of genetic loci contributing to specific behavioral issues in the near future. For instance, although not yet definitive, there are recent studies on schizophrenia and manic depressive illness that have provided stronger evidence for such loci than have been obtained heretofore (Lander and Kruglyak, 1995).

The approaches to mapping genes for human behaviors and aptitudes are also being tempered by recent progress in the genetics of human disease, much of it facilitated by Human Genome programs. Contemporary research in human genetics is revealing unanticipated complexity in the relationship between genes and human traits. What is important about understanding this complexity is not just that simplistic genetic explanations for human disease don't pertain, but that deterministic concepts about genetics are facing even greater challenges.

We always knew that there was significant variation in the expression of genes for various human disease. But now, with all the new DNA technologies allowing the detection of anybody who carries a mutation known to be associated with a host of health conditions, new, unexpected findings are emerging. Consider Gaucher's Disease, a genetic disease found most commonly in Jewish populations. Until recently, it had been assumed that everyone who inherited two mutant copies of the gene for this condition would suffer from the disease. But, with the ability to test for the mutant gene in families where the disease is clearly inherited, it appears that perhaps as many as 1/2 of people with particular mutant alleles do not exhibit symptoms of the disease (Sidransky and Ginns, 1993). Although not as striking, surprising findings of a similar nature have been made with cystic fibrosis and Huntington's Disease. That is, people who would have been predicted on the basis of finding an altered gene to be suffering from the disease were asymptomatic. In the case of cystic fibrosis, this was found in an interesting way. Males with cystic fibrosis, in addition to suffering from the life-threatening symptoms of the disease, are sterile. The same type of sterility was found in a number of males who did not suffer from the disease (Meschede et al., 1993). When the DNA of these males was examined, they were found to carry exactly the same mutations as other males who represented typical cystic fibrosis cases.

In the case of Huntington's Disease, variation in onset and severity of the disease can, in some instances, be explained by the finding that the gene responsible amplifies in length as it is passed on from parent to child. The greater the amplification, the greater the chance of the disease being manifested and the age at which it is contracted may be affected. However, recent work has revealed that there are many individuals who do have the highly amplified versions of the gene, but are now over 90 years old and are showing no signs of the condition (Rubinsztein et al., 1996). Up until this study, these people would have been predicted to have been dead.

Most recently, similar findings have been reported with genes that cause breast cancer to run in families (Couch et al., 1997). It has been reported that, depending on the family, the same mutation can lead to most women in a family eventually developing breast cancer or to families in which the incidence of breast cancer is much less frequent. This study, if confirmed, adds yet another example to the contemporary puzzling findings of complexity in human genetics.

This substantial variability in the correlation between genetic mutations and manifestation of a disease remains unexplained. In some cases, it could be due to the effects of other genes within each individual on the expression of the trait; in others, it could be due to environmental factors, including prenatal conditions, diet, and even the actions of individuals themselves in directing their own lives. Clearly, these findings reinforce even more strongly the understanding that there is no simple deterministic relationship between mutations in a gene and human traits.

Conclusion

So, I have described a picture of emerging complexity of expression of disease genes and I have described the difficulties being encountered in attempts to find genes associated with various human behaviors. The last 10 or so years of hunts for genes for manic depressive illness and schizophrenia have been disappointing, in that studies based on simple assumptions have not worked. This complexity suggests a rich arena of research for scientists in this field. As Risch and Botstein have put it, these developments may necessitate a new paradigm for examining these questions. While these authors focus on more sophisticated genetic theories, others argue for a paradigm that integrates genetics, environment and gene-environment interactions.

What we already knew and what we are learning from genetics does not tell us that genes are the puppet strings that limit our movement and our achievements. Yet some researchers are still using simplistic assumptions and generalizing from their findings beyond what is warranted. I want to emphasize that while the media certainly shares the blame, much of the misrepresentation derives from the way scientists have presented this work, not from complete media distortion. By misstating the implications of unreplicated studies, by generalizing from limited data, and by making genetic determinist statements, they do a disservice to public understanding of science. To repeat, the social consequences of misrepresentation of genetic information and inferences of genetic determinism can be quite real and quite harmful. Yet, it is my hope that the new genetic knowledge that is developing will lead to some degree of closure to the controversies over genetics and human behavior — that the progress in this field, in fact, will be a force liberating us from the sterile arguments of the past. For there is nothing inherent in genetic knowledge to be feared. It is only the misuse and misrepresentation that I have described that should concern us.

However, there is one caution — whether this liberatory consequence of genetic progress is the case depends on whether scientists will participate in educating the public to the true meanings of genetics. This requires that scientists doing the research be much more cautious in wording their papers and presenting their results to the public. It also requires that others of us in genetics point out publicly where genetic information is being misrepresented. This has happened to some extent in France, for instance, after Le Pen made his statement a year ago. More than 600 scientists and historians signed a statement criticizing Le Pen from a scientific point of view (Anonymous, 1996b; Butler, 1996). Unfortunately, this is an exception; prior experience is not terribly promising. During the early part of the twentieth century era in the United States, leading geneticists were at first active participtants in the eugenics movement (Ludmerer, 1972). But, when they became disenchanted with the science and politics of this movement, they did little to blunt its social and political impact. Rather, it was mainly social scientists who spoke out against the eugenicists.

More recently, in the United States, in response to *The Bell Curve*, the American Psychological Association (Azar, 1994) and the American Sociological Association published critical statements on the arguments of the book; but, except for the National Society of Genetic Counselors, the various genetic societies refused to (Beckwith, 1997). These failures lead me to suggest that for scientists to be in a position to respond to the social impact of science they must receive an education that integrates sociological, historical, philosophical and other relevant studies of science with the study of scientific knowledge itself.

References

Allman, W. (1994) Why IQ isn't destiny. *US News and World Report*, 24 October, 73–80.

Alper, J.S. and Beckwith, J. (1993) Genetic fatalism and social policy: The implications of behavior genetics research. *Yale J. Biol. Med.*, 66: 511–524.

Anonymous (1996a) M. Le Pen demande à ses militants de se préparer à une "révolution". *Le Monde*, 17 Sept., 1.

Anonymous, (1996b) Inégalité des races: 600 scientifiques répondent a Le Pen. *L'Humanité*, 27 Sept.

Associated Press (1997) French mayor fined for racism. *The Boston Globe*, 9 Sept., A10.

Azar, B. (1994) Psychology weighs in on *Bell Curve* debate. *APA Monitor*, 1 Dec., 22–23.

Bailey, M. and Pillard, R. (1991) Are some people born gay? *NY Times*, 17 Dec., A21.

Baron, M. (1998) Mapping genes for personality: is the saga sagging? *Mol. Psychiatry*, 3: 106-108.

Beckwith, J. (1983) Gender and math performance: does biology have implications for educational policy? *J. Educ.* (*Boston Univ.*) 165: 158-174.

Beckwith, J. (1997) The responsibilities of scientists in the genetics and race controversies. In: E. Smith and W. Sapp (Eds.), *Plain Talk about the Human Genome Project*, Tuskegee University Press, Tuskegee, pp. 83–94.

Beckwith, J. and Alper, J.S. (1998) L'apport réel des études sur les jumeaux. *La Recherche*, July/Aug., 311: 72–76.

Benbow, C. and Stanley, J. (1980) Sex differences in mathematical ability: Fact or artifact? *Science*, 210: 1262–1264.

Billings, P.R., Beckwith, J. and Alper, J.S. (1992) The genetic analysis of human behavior: a new era? *Soc. Sci. Med.*, 35: 227–238.

Blakeslee, S. (1996) Genetic questions are sending judges back to classroom. *The New York Times*, 9 July, C1–C9.

Breakefield, X. (1994) Presentation at Harvard Medical School Conference entitled "*Genes that Make News: News that Makes Genes*", 3 Dec.

Browne, M.W. (1994) What is intelligence, and who has it? *NY Times Book Rev.*, 16 Oct., 3: 41–45.

Brunner, H.G., Nelen, M., Breakefield, X.O., Ropers, H.H. and Van Oost, B.A. (1993a) Abnormal behavior associated with a point mutation in the structural gene for monoamine oxidase A. *Science*, 262: 578–583.

Brunner, H.G., Nelen, M.R., vanZandvoort, P., Abeling, N.G.G.M., Van Gennip, A.H., Wolters, E.C., Kuiper, M.A., Ropers, H.H. and Van Oost, B.A. (1993b) X-linked borderline mental retardation with prominent behavioral disturbance: Phenotype, genetic localization, and evidence for disturbed monoamine metabolism. *Am. J. Human Gen.*, 52: 1032–1039.

Brunner, H.G. (1995) MAOA deficiency and abnormal behavior: perspectives on an association. In: *Genetics of criminal and antisocial behavior* (*Ciba Foundation Symposium 194*), Wiley, Chichester, pp. 155–164.

Butler, D. (1996) French scientists rally against racist claim. *Nature*, 383: 369.

Butler, D. (1997) Eugenics scandal reveals silence of Swedish geneticists. *Nature*, 389: 9.

Couch, F.J., DeShano, M.L., Blackwood, M.A., Calzone, K., Stopfer, J., Campeau, L., Ganguly, A., Rebbeck, T. and Weber, B.L. (1997) BRCA1 mutations in women attending clinics that evaluate the risk of breast cancer. *N. Engl. J. Med.*, 336: 1409–1415.

Cowlet, G. and Hall, C. (1993) The genetics of bad behavior. *Newsweek*, 1 Nov.

De Parle, J. (1994) Daring Science or 'Social Science Pornography'. *NY Times Mag.*, 9 Oct., 48–53, 62, 70–71, 78, 80.

Devlin, B., Daniels, M. and Roeder, K. (1997) The heritability of IQ. *Nature*, 388: 468–471.

Felsenthal, E. (1994) Man's genes made him kill, his lawyers claim. *The Wall Street Journal*, 1 Nov., B1–B5.

Gelernter, J., Goldman, D. and Risch, N. (1993) The A1 allele at the D_2 dopamine receptor gene and alcoholism: A reappraisal. *JAMA*, 269: 1673–1677.

Hamer, D. (1994) Presentation at Harvard Medical School Conference entitled "Genes that Make News: News that Makes Genes", 3 Dec.

Hamer, D.H., Hu, S., Magnuson, V.L., Hu, N. and Pattatucci, A.M.L. (1993) A linkage between DNA markers on the X chromosome and male sexual orientation. *Science*, 261: 321–327.

Herbert, W. (1997) How the nature vs. nurture debate shapes public policy-and our view of ourselves. *US News and World Report*, 21 April, 72–80.

Herrnstein, R.J. and Murray, C. (1994) *The Bell Curve*, Free Press, New York.

Jensen, A.R. (1969) How much can we boost IQ in scholastic achievement? *Harv. Ed. Rev.*, 33: 1–123.

Kelsoe, J.R., Ginns, E.I., Egeland, J.A., Gerhard, D.S., Goldstein, A.M., Bale, S.H., Pauls, D.L., Long, R.T., Kidd, K.K., Conte, G., Housman, D.E. and Paul, S.M. (1989) Re-evaluation of the linkage relationship between chromosome llp loci and the gene for bipolar affective disorder in the Old Order Amish. *Nature*, 342: 238–243.

Kidd, K.K. (1993) Associations of disease with genetic markers: déja vu all over again. *Am. J. Med. Genet.*, 48: 71–73.

Lander, E. and Kruglyak, L. (1995) Genetic dissection of complex traits: guidelines for interpreting and reporting linkage results. *Nat. Genet.*, 11: 241–247.

Ludmerer, K. (1972) *Genetics and American Society*, Johns Hopkins University Press, Baltimore.

Marshall, E. (1992) When does intellectual passion become conflict of interest? *Science*, 257: 620–621.

Meschede, D., Eigel, A., Horst, J. and Nieschlag, E. (1993) Compound heterozygosity for the deltaF508 and F508C cystic fibrosis transmembrane regulator (CFTR) mutations in a patient with congenital bilateral aplase of the vas deferens. *Am. J. Human Genet.*, 53: 292–293.

Morell, V. (1993) Evidence found for a possible 'aggression gene'. *Science*, 260: 1722–1723.

Morganthau, T. (1994) IQ:is it destiny? *Newsweek*, 24 Oct., 53–55.

Müller-Hill, B. (1988) *Murderous Science: Elimination by Scientific Selection of Jews, Gypsies and Others, Germany 1933–1945*, Oxford University Press, Oxford.

Owen, M.J. (1992) Will schizophrenia become a graveyard for molecular geneticists. *Psychol. Med.*, 22: 289–293.

Risch, N. and Botstein, D. (1996) A manic depressive history. *Nat. Genet.*, 12: 351–353.

Rubinsztein, D.C., Leggo, J., Coles, R. et al. (1996) Phenotypic characterization of individuals with 30–40 CAG repeats in the Huntington Disease (HD) gene reveals HD cases with 36 repeats and apparently normal elderly individuals with 36–39 repeats. *Am. J. Human Gen.*, 59: 16–22.

Sidransky, E. and Ginns, E.I. (1993) Clinical heterogeneity among patients with Gaucher's Disease. *J. Am. Med. Assoc.*, 269: 1154–1157.

Spitz, E. and Carlier, M. (1996) La Méthode des jumeaux de 1875 à nos jours. *Psychiatrie de l'enfant*, 39: 137–159.

Steele, C.M. and Aronson, J. (1995) Stereotype threat and the intellectual test performance of African Americans. *J. Personality Soc. Psychol.*, 69: 797–811.

Watson, J.D. (1997) *President's Essay: Genes and Politics*, Cold Spring Harbor Lab. Ann. Report 1–20.

Zola, E. (1977) *La Bête Humaine*, Gallimard Folio, Paris.

W.E. Crusio and R.T. Gerlai (Eds.)
Handbook of Molecular-Genetic Techniques for Brain and Behavior Research (Techniques in the Behavioral and Neural Sciences, Vol. 13)

CHAPTER 6.2

Ethical issues and psychiatric genetics

A.E. Farmer and P. McGuffin

Institute of Psychiatry, Social, Genetic and Developmental, Psychiatry Research Centre, Denmark Hill, London SE5 8AF, UK

Introduction

The other chapters in this book convey much of the excitement induced by the novel discoveries in the field of genetic research in the behavioural neurosciences. In this as for many areas in medicine, there is a real potential for enhanced understanding of biological mechanisms and therapeutic developments (Rutter and Plomin, 1997). Indeed the recency of the reference listings in the former chapters, also attests to the rapid pace of these advances. The media, both lay and professional has almost weekly accounts of new technologies and discoveries. Most dramatic from the viewpoint of the public have been mammalian cloning (Wilmut et al., 1997) but whole new scientific disciplines such as bioinformatics and pharmacogenomics have now emerged (Tanne, 1998).

However the past abuses undertaken in the name of "eugenics", cast a long shadow over these exciting developments and much concern has been expressed about the ethical aspects of genetic research. Much of the focus of this ethical concern has related to genetic research into mental disorders and the behavioural characteristics of human subjects, and such research can be considered one of the main catalysts in the recent growth of bioethics as an academic discipline (Meslin, 1997).

In this chapter, we will firstly consider the history of psychiatric genetics throughout the 20st Century. Secondly, we will trace the rise of the "eugenics movement" and the rapid acceptance of its principles throughout the world. Thirdly, we will consider the development of "bioethics" as an academic discipline and discuss why this has also led to some disquiet. Lastly we will consider the ethical implications of molecular genetic research into psychiatric disorder and aspects of human behaviour with particular reference to the Nuffield Council Report on Bioethics. (Nuffield Council on Bioethics, 1998)

A very short history of psychiatric genetics

Psychiatric genetics has had a difficult and often controversial history. Kraepelin (1896) was the first to develop a workable psychiatric nosology, with his recognition of two major psychotic disorders, dementia praecox (later termed "schizophrenia" by Bleuler in, 1911) and manic depressive disorder (now usually called bipolar disorder). This coincided with the rediscovery of Mendel's laws which had been ignored when first proposed in 1866, but which subsequently led to the birth of a new branch of science, "genetics" as it was termed by Bateson in 1905. Classic genetic methods of twin, family and adoption studies were applied in the major psychiatric disorders in the 1930s through to the 1970s (Slater and Cowie, 1971; McGuffin et al., 1994). Most of the important evidence supporting a substantial genetic contribution to schizophrenia and manic depression comes from studies published throughout this period. Although such studies demonstrated that genetic factors exist, disorders presenting with psychiatric symptoms did not show a straightforward Mendelian pattern of inheritance. For example, such studies demonstrated that genetic factors were involved in the aetiology of schizophrenia and manic depression, but the modes of inheritance were clearly complex and at that time seemed to have little bearing on clinical practice.

Thus in the third quarter of the twentieth century psychiatric genetic research became decidedly unfashionable and experts who maintained activities in the field were few. The recent renaissance of interest in psychiatric genetics has coincided with the rapid expansion of human molecular genetic research and its widespread influence on medicine as a whole (Owen and McGuffin, 1997). Consequently, as we approach the Millennium, genetics has become one of the fastest growing areas of psychiatric research with its own international society and no fewer than 3 specialist journals mainly or wholly devoted to the field.

The eugenics movement

The origins of eugenics and behavioural genetics can be traced back to the work of Francis Galton who contributed much to the development of modern statistics, population biology and genetics. Galton's monograph on "Hereditary Genius" in 1869 is arguably the first attempt to explore the inheritance of intellectual ability. Seemingly unaware of the work of his contemporary Gregor Mendel, Galton first introduced the term "eugenics" and considered it to be a branch of biology devoted to improvements in the human race by attention to optimal reproductive practices. The term eugenics is derived from the Greek for "well-born". The guiding principles of eugenics according to Galton are that the reproduction of individuals suffering from inherited defects should be curtailed, and he believed that this would lead to certain diseases being abolished (negative eugenics). On the other hand, positive eugenics was the improvement in the human stock due to the encouragement of fecundity among those with desirable attributes.

These essentially rather simplistic ideas became prevalent and popular in intellectual circles and rapidly spread to other strata in society in the first half of the twentieth century. Although usually associated with right wing politics, eugenic ideas were espoused by members of the socialist Fabian movement as well as by other socialist thinkers who found it quite reasonable that the "feeble minded" should be prevented from having children because otherwise they would "cost other people a lot of money". (Mitchison N, cited in *The Sunday Times*, 1997)

However, despite his dubious advocacy of eugenics, Galton also introduced the notion that studies of the familial clustering of traits could shed light on their causes and first suggested the idea that studies of twins could provide a method of teasing apart the effects of genes and environment. Geneticists of the Munich School, such as Seimens and Luxemberger developed and applied the now classic twin method. The first systematic family studies of schizophrenia (dementia praecox) were carried out by another member of the Munich School namely Ernst Rüdin (Kendler and Rüdin, 1996).

Rüdin's important contribution to the elucidation of the genetic influence in schizophrenia has, however, been overshadowed by his subsequent involvement with the Nazi party and his possible contributions to their views on how eugenic principles might be implemented (Gejman, 1997). Although Rüdin was arrested and tried by the Allies following the end of World II, he was judged to have been a "fellow traveller" and not a major contributor to Nazi war crimes (Gottesman and Bertelson, 1996). Whether this view of Rüdin is correct or over lenient still remains a contentious issue. A recent attempt to reassess Rüdin's work and present some of it in English translation resulted in a heated debate in the letters to the editor of one of the leading psychiatric genetics journals. (Lerer and Segman, 1997; Kendler, 1997; Gershon, 1997).

However what is certain is that the Nazi regime employed eugenic ideas as a justification for initially the forced sterilisation and ultimately the "euthanasia" of allegedly inferior ethnic groups such as Jews and gypsies as well as psychiatric patients. Thousands of individuals in psychiatric hospitals were put to death, including those suffering from Huntington's Chorea, alcoholism, and "feeble mindedness" as well as disorders such as schizophrenia and manic depression. Many leading scientists as well as doctors not only accepted these policies but encouraged their adoption, despite the fact that they are both morally repugnant as well as scientifically incorrect (Müller-Hill, 1988).

Some leading German psychiatric geneticists appalled by this grotesque misuse of their science refused to join the Nazi party and one, Luxemberger, at risk of his career, and possibly his personal safety, ridiculed some of the more extraordinary supposed scientific facts proposed by the Nazis.

It is erroneous to believe that only the Nazis were guilty of using eugenics to justify discriminatory laws. Harsh policies have been applied in several other countries as reflected in policies on sterilisation, immigration and the assimilation of ethnic minorities. In the United States, Charles B. Davenport, Director of the Cold Spring Harbor Laboratories made the simple assumption that much of insanity, feeble-mindedness and other undesirable behaviours were genetically determined, without seeking hard scientific evidence to support this. His views were highly persuasive in shaping United States immigration polices that favoured certain nationalities (e.g. immigrants from the British Isles, Scandinavia and Germany) compared to others (from Southern and Eastern Europe) (Watson, 1997).

Psychometric testing and in particular the development of the intelligence quotient (IQ tests) by Terman were originally assumed to provide a measure of innate ability. Large scale IQ testing commenced with World War I army conscripts in the United States. The poorer average scores of non-English speaking immigrants and of black army recruits compared with their white counterparts were interpreted as genetically based. These were therefore used as further justification for discrimination. Even to the present day investigation of the genetics of intelligence and other human behaviours is seen as controversial and provocative (Kamin, 1974; Harper, 1995).

Eugenic philosophies have continued to influence politics in various countries until relatively recently. For example until the 1960s, eugenics influenced Swedish policies on the sterilisation of the mentally handicapped, and also the Australian government introduced adoption policies in an attempt to eliminate Aboriginal people as a distinct ethnic group (Scott-Clark and Levy, 1997).

Particularly in the aftermath of the Second World War, psychiatric genetics became guilty by association with eugenic ideas. The Munich School ceased to be a centre for psychiatric genetic research. A diaspora of foreigners who had visited Munich or studied there as research fellows in the pre-war years kept the subject alive in their own countries. Franz Kallmann became a leading proponent of psychiatric genetics in the US, while Slater in the UK and Stromgren and Essen-Møller in Scandinavia continued to undertake psychiatric genetic research, although it was very much a minority interest (Slater, 1971; Gottesman and McGuffin, 1996). Despite this in the third quarter of the twentieth century, important refining and implementing of the methods of family, twin and adoption studies occurred. The advent of molecular genetics and the knowledge that genes could be localised and identified, even in diseases with a complex and multifactorial aetiology led to its renaissance in the past two decades.

Classic genetic studies have confirmed that with one or two exceptions, psychiatric disorders are polygenic and multifactoral in aetiology (McGuffin et al., 1994). Although genetic factors contribute substantially to the liability to develop major psychotic disorders (heritability estimates for schizophrenia and bipolar disorders are around 80%), what genes confer is a susceptibility to develop the disorder. Thus what is inherited is not the certainty of the disorder developing, but rather a vulnerability to become ill when other factors (e.g. environmental stressors) are also present. In this respect psychiatric disorders are no different from other common, complex diseases such as coronary artery disease, diabetes and hypertension.

The development of bioethics

The Director of the United States National Institutes of Health, Nobel Laureate James D. Watson announced in, 1988 that 3% of the budget for the Human Genome Project would be devoted to supporting a programme of research and discussion on the *ethical, legal and social implications* of the new genetic knowledge. This bioethics programme (ELSI), had a budget of US$1.5m in its first year and this increased by 5% annually. Thus by, 1996 the annual ELSI budget was approximately US$7m. Watson's motivation for this was once again the guilt by association of genetic research with the horrors unleashed by the misinterpretation of genetic principles. Watson stated that he wished to oppose the notion that he was in anyway a "closet eugenicist, having as my real long-term goal the unambiguous identification of genes that lead to social and occupational stratification as well as genes justifying racial discrimination" (Watson, 1997).

The ELSI programme, the largest single source of funding for ethical research in the world, covers all aspects of genetic research and not just those relating to psychiatric disorder and behavioural traits. Elsewhere, particularly in Europe, more modestly funded initiatives have taken place. France has had a National Bioethics Committee since, 1983 (Butler, 1997). Germany, which is particularly sensitive in this regard given its twentieth century history, is in the process of setting up a National Bioethics Advisory Group (Abbott, 1997).

In the UK, both major medical research funding bodies, the Wellcome Trust and the Medical Research Council (MRC) have held conferences and public consultation on genetic research, with the MRC focussing particularly on behavioural genetics. In 1998, the Nuffield Council on Bioethics; an independent organisation funded by the Wellcome Trust, the Medical Research Council and the Nuffield Foundation published the recommendations of its working party into the ethical aspects of the genetics of psychiatric disorders. These recommendations will be discussed in more detail later in this chapter (Nuffield Council on Bioethics, 1998).

However amidst all this discussion about the ethical aspects of genetic research have come some criticisms. An editorial in the *Lancet* coined the term "the bioethics industry" (*Lancet*, 1997). This was picked up in an editorial published by Nature which went on to point out that perceptions about new moral concepts resulting from medical developments in biology are often misconceived (Wadman et al., 1997). Clearly independent consideration of the ethical implications of research is beneficial in the long run. However, too many ethical controls and constraints on a difficult area of research and could lead to researchers becoming disenchanted with the whole area of endeavour. At the Fifth World Congress on Psychiatric Genetics held in Santa Fe in, 1997, a plenary session was devoted to the topic of bioethics. One of the speakers Eric Meslin, Director of the ELSI Research Program, emphasised the importance of not allowing genetic research in the sensitive areas of psychiatric disorder and behavioural traits to go "unbridled" (Meslin, 1997). Also a plenary speaker, one of us (AF) extended this metaphor to point out that a bridle allows the rider to steer the horse in a particular direction. However if the bridle is too strong, the horse may refuse to go anywhere and may alternatively bolt off. Clearly if genetic research into mental disorders is considered somehow different and is "singled out" for more constraints than other areas of medical genetics, it is possible that such research will simply not take place. Consequently any possible benefits genetic research could bring to those suffering from mental disorders would not occur. We would therefore suggest that what is required is not an elaborate structure but rather an attempt at "good enough" ethics, i.e. constraints on research which the intelligent lay person would consider reasonable given our current state of knowledge.

Current ethical issues relating to modern psychiatric genetic research

For the reasons alluded to above it is not surprising that there are those who are fundamentally opposed to research into the genetics of human behaviour (Goodey, 1996; Harper and Clarke, 1997). The fear is that such research promotes biological determinism and the negation of free will. Also that if a trait or disorder is inherited it is fixed, immutable and consequently untreatable. Thus the suspicion arises that what genetic research into psychiatric disorder is about is labelling and discrimination against the mentally ill.

However it has been recently pointed out by Rutter and Plomin (1997) that these fears are based on fallacies. Although such research can be regarded as rooted in reductionism, in that its ultimate aim is the understanding of molecular mechanisms, this does not equate with determinism, i.e. the research recognises that there is not a direct correspondence between genotype and phenotype. It is clear that even in identical twins, the concordance for severe disorders such as schizophrenia and bipolar disorder are considerably less than 100% and therefore environmental risk factors are clearly extremely important in the aetiologies of these disorders. It has been pointed out that many of the new insights into the role of environmental factors in mental disorders and behavioural traits have been provided by genetic research (Brown, 1996). Indeed the interaction and co-action between gene and environmental factors is an important aspects of current genetic research into behaviour (Plomin, 1994; Rutter and Sandberg, 1992).

The pace of technical development and subsequent new discoveries are constantly providing new ethical challenges. Some of the main topics of concern considered by the ELSI programme, as well as by the recent Nuffield Council report, have been those of informed consent, the role of institutional review boards (IRBs) (the equivalent of ethics committees found elsewhere), the role of individuals and families as participants in genetic research, privacy and confidentiality, and commercial interests.

Informed consent

As with any other research activity, participants in genetic research need to understand likely benefits, foreseeable risks and alternatives to their participation in the research. The physical risk of harm to the subject is generally minimal and relates to the need to obtain a sample of venous blood for the extraction of DNA. Even venepuncture can be avoided, since for some purposes, a sufficient amount of DNA can be obtained from a sample of buccal mucosa cells obtained by mouthwash or swabbing the inside of the cheek. Arguably what is less easy to evaluate is the degree of "non-physical" harm, particularly if participants fail to understand the difference between procedures relating to their clinical care and those which are undertaken merely to provide scientific data. Consequently individuals may consent to participate in genetic research because they believe they will learn more about their own risks for a genetic disorder. However, at present the genetic research is unable to provide such predictive clinical information. It is therefore important that it is made explicit to research participants that the results of DNA tests will not be revealed to them, and that the benefits of their participation are solely those of advancing scientific understanding.

The Nuffield Council report recommends that the arrangements regarding informed consent for the majority of those with a mental disorder should not be any different from those required for other illnesses. The report goes on to state that for those who are only intermittently competent, consent should only be sought when they are competent. Interestingly the report also suggests that even those who are permanently incompetent can participate in non-therapeutic research. However, strict safeguards must be in place for such research to be considered ethical. If genetic registers are to be set up, the Nuffield Council Working Group also recommended that informed consent guidelines are required (Nuffield Council on Bioethics, 1998).

Ethics committees

Most industrialised countries require the approval of a local ethics committee or IRB for any research project involving human subjects. The Nuremberg code provides the guiding principles, which include freely given voluntary consent, not to be subjected to suffering and that the experiment will provide "fruitful results for the good of society, unprocurable by other methods or means of study" (Faraone et al., 1997).

The major problem for such committees with such a rapidly advancing field as genetic research relates to the committee's competence to judge the scientific merit of a research proposal. In such situations, the committee can either take the investigator's proposal "on trust" or obtain an external referee's comments. It is not the place of the ethics committee to duplicate the work of a grants committee in this respect. However, if the research is highly novel or there has not been a prior peer review from a funding body, then clearly, an external opinion is warranted. Ethics committees therefore have to exercise good judgement as well as practical common sense in the context of their local knowledge.

As genetic research requires larger and larger sample sizes drawn from ever increasing geographical areas, the researcher may be faced with having to apply to a number of locally autonomous ethical committees. This has led to situations where a project has been considered ethical in one area but not in that served by an adjacent ethical committee. Also, the need to apply to a dozen or so different ethical committees is a time consuming and expensive exercise, which has to be undertaken before the research can begin. It would clearly make much sense for such large-scale research to receive ethical consideration from organisations with a wider geographical base. An attempt has been made in the UK by the Medical Research Council to set up such mechanisms in the context of multicentre clinical trials.

The impact of psychiatric genetic research on individuals and their families

Much genetic research includes collecting information on those affected by mental disorder as well as their unaffected relatives. There is clearly an ethical issue regarding recruitment. Most mental disorders remain stigmatising. Those suffering from the disorder may not wish their relatives to know about their illness. Even when family members are aware of mental disorder in one or more of its members, the family's perception of the illness may be altered by the family becoming involved in a genetic study. The main ethical consideration however relates to the method of recruiting relatives. Most would consider that it is not appropriate to approach relatives when the index subject either withholds their permission or is not made aware of the approach being made to their relatives. A more debatable point is that once permission has been given to approach relatives, whether this approach can only be made via the index case and not directly to the relatives themselves.

Privacy and confidentiality

Because of the stigma that is still associated with psychiatric disorder, issues relating to privacy and confidentiality are of great sensitivity in relation to psychiatric genetic research. For example, family members may be prepared to reveal symptoms to a researcher that they have concealed from their relatives and in such circumstances it is imperative that their confidentiality is strictly respected. The Nuffield Council report (1998) also comments on the confidential nature of genetic information but also recognises that in exceptional circumstances, disclosure to close family members might be justified, for example, if there are serious implications for them. The Working Party's recommendations is that such decisions should be judged on a case by case basis, and that doctors have "discretion" but not a duty to inform others exposed to risk. In these circumstances any subsequent disclosure would not be regarded as a culpable breach of confidentiality.

Commercial interests in psychiatric genetic research

Clearly genetic research can lead to important new commercial products. A number of genes have already been patented, for example the cystic fibrosis genes, the BRCA1 and a mutation on the amyloid precursor protein gene which leads to the early onset, Mendelian subform of Alzheimer's disease. The ethical debate concerns two issues. Firstly, disease mutation genes are spontaneously occurring biological anomalies not inventions, and consequently it may be difficult to justify their patenting. However, it can be argued that isolating a gene and finding a particular mutation is an endeavour with the same originality as an invention even if the discovery has occurred in someone else's DNA.

The second ethical consideration relates to who should profit from the research. The view on patenting adopted in North America is that the discoverer of the mutation is its sole owner, rather than the subject from whom the DNA came.

Another related issue is that of the impact of genetic research on drug discovery. Pharmaceutical companies are increasingly interested in the genetic bases of common disorders, since they see them as a route to identifying novel targets which will hopefully facilitate the development of therapeutic compounds. Several major pharmaceutical companies have invested in psychiatric genetic research projects. One such company, deCODE Genetics, have approached the Icelandic Ministry of Health to pass legislation to allow the company to compile a database containing the medical records of all Icelandic nationals together with genetic and genealogical information. The proposed bill would allow the licensee exclusive rights to the commercial exploitation of the data for 12 years. DeCODE Genetics' president, Kari Stefansson, has written enthusiastically of a "benign and fruitful collaboration between a genomics company and the population of Iceland" (Gulcher and Stefansson, 1998). However, it is clear that there is considerable opposition both within Iceland and outside the country who are highly critical of the legislative proposals. Admittedly, pharmaceutical companies would need to invest considerable amounts of money in genetic research, but the rewards are also potentially immense. The development of novel drugs as a result of research on a population's DNA, is potentially highly profitable, but the profits will go to the pharmaceutical company and not to those who have provided the DNA. What information or inducements should be given to human subjects involved in such research? At the very minimum would seem important to inform subjects that their clinical details and information derived from their DNA or even the DNA itself would be supplied to a pharmaceutical company, albeit anonymously. Unless the issue relating to profit can be satisfactorily determined between the pharmaceutical company and those whose DNA they wish to collect, it is highly likely that the latter will refuse to participate, as seems likely in Iceland.

Further issues relating to psychiatric genetic research

Because most psychiatric disorders and behavioural traits are contributed to by many genes as well as environmental risk factors, it is very unlikely that diagnostic genetic testing will be of any use. Even if a number of susceptibility genes were identified for a particular disorder, the Nuffield Council Report concluded that without an understanding their interactions, their identification would still be inadequate for predicting individual clinical risk. Similarly, prenatal testing or population screening are unlikely to be of any benefit in the case of psychiatric disorders. The Nuffield Council Report goes on to recommend that genetic testing for susceptibility genes which offer low predictive or diagnostic certainty should be discouraged unless there is clear medical benefit. The report also comments that the testing of children before they are able to give informed consent should require special safeguards, and that predictive testing or testing for carrier status for mental or other disorders in children should be strongly discouraged.

"Good enough" ethics

Thus far, psychiatric genetics has had little direct impact on clinical practice but it has been predicted that this will change (Farmer and Owen, 1996). Real gains in the understanding of the neurobiological basis of mental disorders are imminent. The methods and technologies are already available to facilitate the discovery of susceptibility genes contributing to the aetiology of mental disorders. In time, all will be located and their protein products identified. It is highly likely that this will lead to the development of better targeted drug treatments, thereby providing considerable benefits for those suffering from mental disorders. Also because the majority of disorders are polygenic, pre-natal and presymptomatic testing and population screening are likely to be of limited value. Consequently the complex ethical issues posed by such testing are likely to be avoided at least in the medium term.

Despite fears expressed by some patients and their relatives that exploring the genetic basis of psychiatric disorders will lead to labelling and discrimination, recent history suggests the opposite. When the public acknowledges that a mental disorder has a "physical" cause, there is a tendency to view it as a "real" disease and one worthy of care and attention, rather than of shame and stigma. This is already apparent in Alzheimer's disease, which was previously associated with considerable stigma. Since the discoveries of some genetic mutations associated with Alzheimer's disease, it is now acceptable for his spokespersons to publicly announce that a former US President is suffering from the disorder. Similarly, the substantial genetic contribution to conditions such as schizophrenia and autism has also meant that clinicians have moved away from blaming other family members for their causation (the terms "schizophrenogenic" mother, and "refrigerator" parents have previously been applied as causing these disorders). The main hope must be that with the elucidation of the genetic underpinnings to mental disorders that stigma and consequent discrimination will be reduced.

Despite expressing optimism, we suggest that ethical issues must continue to be debated especially given the current pace of technological developments. However, it is also imperative that the ethical constraints placed upon psychiatric genetic research are not more restricting than on any other branch of human genetics. There is a danger that disorders that involve abnormal behaviour rather than abnormal bodily function might become the focus of a specialised debate, and seen as an especially fruitful growth area for the fledging bioethics discipline. This could be encouraged by an old fashioned Cartesian mind-body dualism, as much as by the particular current and historical issues that relate to the genetics of mental disorders. We would suggest that the ethical issues relating to psychiatric genetics are largely the same as those that are central to mainstream human genetics with its increasing emphasis on common, complex, rather than rare Mendelian disorders. Furthermore the debate needs to be firmly grounded in the public's concerns and what is required are "good enough" ethics based on pragmatism and common sense and which deal with life issues as they arise.

References

Abbott, A. (1997) Germany's past still casts a long shadow. *Nature*, 389: 647.

Bleuler, E. (1950) *Dementia praecox or the group of schizophrenias*, (1911). Translated by J. Zinkin. International Universities Press, New York.

Brown, G.W. (1996) Genetics of depression: a social science perspective. *Int. Rev. Psychiat.*,8: 387–401.

Butler, D. (1997) France reaps benefits and costs of going by the book. *Nature*, 389: 661–662.

Editorial. (1997) The ethics industry. *Lancet*, 350: 897.

Faraone, S.V., Gottesman, I.I. and Tsuang, M.T. (1997) Fifty years of the Nuremberg Case: a time for retrospection and introspection (editorial). *Am. J. Med. Genet.*, (*Neuropsychiat. Genet.*) 74: 345–347.

Farmer, A.E. and Owen, M.J. (1996) Genomics: the next psychiatric revolution? *Brit. J. Psychiat.*, 169: 135–138.

Gejman, P.V. (1997) Ernst R. Rüdin and Nazi euthanasia: another stain on his career. *Amer. J. Med. Genet.* (*Neuropsychiat. Genet.*), 74: 455–456.

Goodey, C. (1996) Genetic markers for intelligence. *Bulletin of Medical Ethics*, August, pp. 13–16.

Gershon, E.S. (1997) Ernst Rüdin, a Nazi psychiatric and geneticist. *Amer. J. Med. Genet.* (*Neuropsychiat. Genet.*) 74: 457–458.

Gottesman, I.I. and Bertelson, A. (1996) Legacy of German psychiatric genetics: hindsight is always 20/20. *Amer. J. Med. Genet.* (*Neuropsychiat. Genet.*) 67: 343–346.

Gottesman, I.I. and McGuffin, P. (1996) Eliot Slater and the birth of psychiatric genetics in Great Britain. In: H. Freeman and G.E. Berrios (Eds.), *150 years of British psychiatry II*, pp. 537–548.

Gulcher, J. and Stefansson, K. (1998) Population genomics. Laying the groundwork for genetic disease modelling and targeting. *Clin. Chem. Lab. Med.*, 36(8): 523–527.

Harper, P.S. and Clarke, A.J. (1997) *Genetics, society and clinical practice*, Oxford, Bios Scientific Publishers.

Harper, P.S. (1995) DNA markers associated with high versus low IQ: Ethical considerations. *Behav. Genet.*, 25(2): 197–198.

Kamin, L.J. (1974) *The science and politics of IQ*, Chichester, Wiley.

Kendler, K. and Zerbin-Rüdin, E. (1996) Abstract and review of 'Zur Erbapthologie der Schizophrenic' (Contribution to the genetics of schizophrenia), 1916. *Amer. J. Med. Genet.* (*Neuropsychiat. Genet.*), 67: 338–342.

Kendler, K. (1997) Reply to Gejman, Gershon and Lerer and Segman *Amer. J. Med. Genet.* (*Neuropsychiat. Genet.*), 74: 461–463.

Kraepelin, E. (1896) Der psychologische versuch in der psychiatrie. *Psycholog Arbeit*, Vol. 1.

Lerer, B. and Segman, R.H. (1997) Correspondence regarding German psychiatric genetics and Ernst Rüdin. *Amer. J. Med. Genet.* (*Neuropsychiat. Genet.*), 74: 459–460.

McGuffin, P., Owen, M., O'Donovan, M., Thapar, A. and Gottesman, I.I. (1994) *Seminars in psychiatric genetics*, London, Gaskell Press.

Meslin, E. (1997) *Plenary address.* 5th World Congress of Psychiatric Genetics.

Mitchison, N. (1997) An outline for boys and girls. Cited in 'When Britain was for good breeding'. *The Sunday Times*, 31 Aug.

Muller-Hill, B. (1988) *Murderous science*, Oxford, Oxford University Press.

Nuffield Council on Bioethics (1998) Mental disorders and genetics: the ethical context. *Nuffield Council on Bioethics*, London.

Owen, M.J. and McGuffin, P. (1997) Genetics and psychiatry. *Brit. J. Psychiat.*, 171: 201-202.

Plomin, R. (1994) *Genetics and Experience: The interplay between Nature and Nurture*, Sage Publications, London.

Rutter, M. and Plomin, R. (1997) Opportunities for psychiatry from genetic findings. *Brit. J. Psychiat.*, 17: 209-219.

Rutter, M. and Sandberg, S. (1992) Psychosocial stressors: Concepts, causes and effects. *Euro. Child Adolesc. Psychiat.*, 1: 3–13.

Scott-Clark, C. and Levy, A. (1997) Little white lies. Article published in *The Sunday Times Magazine*, 31 Aug., 14–22.

Slater, E. (1971) Autobiographical sketch: the road to psychiatry, In: J. Shields and I.I. Gottesman (Eds.), *Man, mind and heredity*, Baltimore, Johns Hopkins Press, pp. 1–23.

Slater, E. and Cowie, V. (1971) *The genetics of mental disorders*, Oxford University Press, Oxford.

Tanne, J.H. (1998) The new word in designer drugs. *Brit. Med. J.*, 316: 1930.

Wadman, M., Levitin, C., Abbott, A. et al. (1997) Business booms for guides to biology's moral maze. *Nature*, 389: 658–659.

Watson, J.D. (1997) Genes and politics. *J. Mol. Med.*, 75: 624–636.

Wilmut, T., Schnieke, A.K., McWhir, J., Kind, A.J. and Campbell, K.H.S. (1997) Viable offspring derived from fetal and adult mammalian cells. *Nature*, 385: 810–813.

W.E. Crusio and R.T. Gerlai (Eds.)
Handbook of Molecular-Genetic Techniques for Brain and Behavior Research (Techniques in the Behavioral and Neural Sciences, Vol. 13)

CHAPTER 6.3

Ethical implications of knock-out and transgenesis techniques for animal research

Louis Marie Houdebine

Unité de Différenciation Cellulaire, Institut National de la Recherche Agronomique, 78352 Jouy-en-Josas Cedex, France

Introduction

Human communities have always had complex relationships with animals. The history of Noah's Ark, the paintings in prehistoric sites are two of the numerous examples of this reality. Human beings have always been impressed by animals, by their strength, their beauty or their cruelty. In most civilizations, various chimeric creatures being a mixture of human being and animals have been imagined and represented. These animals are supposed to cumulate the qualities of both species and they often have a sacred nature.

Human communities have hunted animals for thousand of years to get food (meat, milk, eggs) and a certain number of secondary products (wool, skin, ivory, gut, etc.). More recently, they were able to domesticate a certain number of animal species to use them more easily.

First of all, breeding animals appeared a more comfortable and more reliable way to get regularly meat, milk and eggs. An important contribution of animals was to provide humans with their muscular strength to pull carriages or to transport people. One may imagine that some animals lived in the vicinity of human communities obtaining from them some food or protection, and finally became pets. During the last centuries, some animal species started being used as experimental material.

Exploiting animals is in fact not an exclusivity of human beings. Other species, such as ants, breed some insects to get regular lymph from them. From a moral point of view, the behavior of humans and ants although essentially similar are profoundly different. Human beings have conscientiousness that ants do not have. Hence, human beings started breeding animal with the clear aim to kill them some day and to eat them. If it is a crime to use animals in this manner, the fault was initially committed at this time. The contemporary behavior of the modern human communities is morally similar to that of their ancestors. Just the techniques and the intensity of the animal exploitation changed.

In developed countries, food is more than sufficient and many people feel that the exploitation of animals is no longer indispensable for their survival. Moreover, most of the people are now living in cities where they have limited and quite particular contacts with animals. They just ignore many aspects of the reality concerning animal life and breeding. They consider, more and more, that human beings should live in harmony with nature rather than trying at any cost to dominate it. Animal welfare is thus a concern for an increasing number of people.

Transgenesis is a quite logical approach for biologists. For many people, transgenesis is gene manipulation which is associated with a certain number of mythical stories in which a mad and naughty scientist is creating monsters which can not be controlled. In this context, it cannot always be very comfortable for biologist to modify the genome of experimental mice when one of these animals is the very famous and attractive hero of popular cartoons. In democratic countries, a certain number of groups have asked to control or even to stop the use of animals for experimentation. This public debate is logical. It may even help scientists to reconsider their protocols. Whatever happens, many wrong arguments against animal transgenesis are commonly used. Obviously, numerous discussions are not based on a sufficient technical and scientific knowledge of what animals transgenesis really is. Emotion also often prevents people from considering all the aspects of the question: the benefit, the challenge and the real welfare problems. This chapter aims at evaluating how transgenesis could be used in a more controlled and efficient manner compatible with the respect of animals.

The methods to generate transgenic animals

The first transgenic multicellular organism was obtained in 1980 (Gordon et al., 1980). The full demonstration that transgenesis was a really utilizable tool came two years later when the transgene, the rat growth hormone gene, was not only stably integrated in mouse genome but was also expressed and responsible for an increase of the size of the mouse body (Palmiter et al., 1982). This experiment showed that transgenesis is a wonderful tool to put back an isolated gene in the complex environment in which it normally works. Using transgenesis is thus one of the best ways to escape from an exceeding reductionism which may result from the gene study carried out with cell-free or even cell systems.

Although gene addition to the genome of an animal rapidly appeared an indispensable tool and although it is more and more commonly used, gene replacement, especially to inactivate specifically a gene, was highly desired. This technique became available in 1989 (Capecchi, 1989a, 1989b). It is now widely used in the mouse in which almost one thousand genes have been knocked out.

The addition of a foreign gene by microinjection

The classical technique to generate transgenic mice harboring a foreign additional gene includes the following steps. The embryos are collected from the uterus of superovulated females the day after the mating with a normal male. The isolated gene, native or recombined, is microinjected into one or both pronuclei of embryos at the one cell stage. After microinjection, the embryos are transferred into recipient females which were pseudopregnant after the mating with vasectonized males.

About 1–3 transgenic mice are routinely obtained from 100 manipulated embryos. Usually 20–30% of the embryo survive to the microinjection and to the transfer into recipient females. The injected DNA interacts with the host genome generating various mutations which are often lethal (Bishop, 1997). Moreover, for various reasons a certain percentage of the recipient females do not become pregnant.

All these techniques are used in routine by the experimenters working on embryos and their impact on animals is generally considered as inevitable. A certain number of people not familiar with animal experimentation have a quite different perception of the protocol. They consider several of the steps of the method as extremely invasive for the animals. They also claim that many animals are sacrificed to generate only a few transgenics.

In birds, microinjection can be performed in the cytoplasm of single-cell embryos but not in the pronuclei which cannot be visualized. The embryo can pursue their development after having been transferred in eggs, but not into recipient females (Naito, 1997; Ono, 1997). The transfer of foreign genes with retroviral vectors is a valuable alternative in some cases (Ronfort et al., 1997).

In lower vertebrates and especially in fish, the microinjection into cytoplasm may be quite efficient (Collas et al., 1997; Devlin, 1997). It is very low in some laboratory fish and the use of retrotransposon seems to improve the situation greatly (Hackett et al., 1997).

In *Xenopus*, a stable integration after a microinjection occurs with a very low frequency (Gong and Etkin, 1997). A recent work has shown that a gene transfer through spermatozoa can be achieved with high efficiency (Kroll and Amaya, 1996). This technical advance now allows this important experimental model to be used in new experimental conditions.

In invertebrates microinjection into the cytoplasm is also the rule. However, conditions to generate transgenis are different according to species. In *Drosophila melanogaster*, the use of the transposon P has become the method of choice several years ago (Kaiser, 1997). In *Caenorhabaditis elegans*, the microinjection of linear DNA into the syncytium of the gonad proved to be quite efficient (Thierry-Mieg et al., 1997). In marine nonvertebrates the microinjection into the cytoplasm of embryos may be used with some success (Miahle et al., 1997).

The improvement of the microinjection technique

Since the technique was proposed in 1980 (Gordon et al., 1980), it has not fundamentally been improved. The technique was extended in 1985 to rabbit, sheep and pig (Hammer et al., 1985). It then appeared that the yield was still lower for rabbit, pig and mainly for ruminants.

It depends on the skillfulness of the experimenters. Ideally, they should not regularly prepare transgenic animals if they have not been sufficiently trained or it appeared they could not meet a reasonable success. Whatever happens, 5 transgenic mice per 100 manipulated embryos seem a mean maximum yield (and a lower in most of the other species). Interestingly however, a recent publication suggested that microinjecting DNA into both pronuclei of mouse embryo, generates twice as many transgenics (Kupriyanov et al., 1998). It should be mentioned that this work was published recently and not yet confirmed. On the other hand, the double microinjection did not significantly improve the yield of transgenic goats (Y. Echelard personal communication).

A reduction in the cost came from the manipulation of embryos. In cows, embryos can be generated after in vitro maturation of oocytes collected from ovaries which are by-products of slaughterhouses and after in vitro fertilization. One-cell embryos can then be microinjected and cultured until the blastocyst stage. This protocol has several advantages. The damaged embryos do not survive during the culture and less recipient females are needed. Embryos at the blastocyst stage can develop when transferred into the uterus whereas a transfer in the uterine horn is needed when the embryos are at the earliest stages of their development. In cows, embryo transfer in uterus can be performed by the vaginal way not involving surgery (Krimpenfort et al., 1991). In the best cases, the embryos at the blastocyst stage which carry the foreign gene can be identified using PCR with the DNA from one cell of the blastocyst or in a more reliable manner by measuring the expression of a reporter gene coinjected with the gene of interest (Thompson et al., 1995).

Theoretically, this approach could be extended to sheep, goat, pig, rabbit, rat, mouse etc. In practice, the quality of the embryos prepared in vitro is lower than in those obtained after in vivo fertilization. It remains justified for cows in which natural embryos are particularly expensive. In pig, rabbit, rat and mouse, the in vitro approach to generate embryos seems unattractive for experimenters. Indeed, these species have relatively short reproductive cycle and they are naturally prolific. The in vitro preparation of embryos would considerably complicate the task of experimenters. A reduction in the number of mice, rats, rabbits and pigs used to generate transgenics has little chance to become a reality in this way in a predictable future.

The addition of foreign genes through cultured cells

Embryonic totipotent cell which are named embryonic stem cell (ES) when they can be cultured as established cell lines can participate to the generation of chimeric animals after having been reintroduced into developing embryos at the morula or the blastocyst stage (Nagy, 1997). Primordial germ cells have essentially the same property and they can be used to establish lines (named EG cells) and to generate chimeric animals (Donovan et al., 1997). For unknown reasons, functional ES or EG cells have been obtained only in mouse. This may reflect the fact that the real nature of ES or EG cells has been only partially established (Fléchon, 1997). Recently however, it was shown that ES cells can be obtained in cow (Cibelli et al., 1998a).

DNA can be added by transfection in the cultured totipotent cells. The selected cells which harbor the foreign genes can then be used to generate transgenic animals which are chimeric and thus mosaic for the transgene at the first generation, but homogeneous at the following generations. This approach was successfully followed in cow, on condition however to transfer the foreign DNA into somatic cells which were used to generate embryos by the cloning technique (Cibelli et al., 1998a).

This protocol can be theoretically extended to other species. It implies a cloning step which requires enucleated oocytes. In cow, the mature oocytes can be obtained in vitro and no donor animals are strictly needed. This approach would be easier in mouse in which well-characterized and quite functional ES cells lines are available. Less animals might be necessary in this way than with the conventional microinjection technique. If this happened to be the case, the work to be done would be so much heavier that the ES approach has little chance to be adopted by experimenters just to add genes into animals.

Recent studies have shown that the classical cloning technique of embryo using freshly prepared embryonic cells can be extended to cultured fetal and even adult cells (Wilmut et al., 1997). Gene transfer can be achieved in the cultured cells which are used to generate cloned transgenic sheeps (Schnieke et al., 1997) and cows (Cibelli et al., 1998b). Undoubtedly, this protocol requires less experimental animals than the classical microinjection even if the efficiency of the cloning technique is still very low. This method has been extended to goat (Baguisi et al., 1999). It seems attractive for pig. However, the cloning of pig embryo by nuclear transfer has not been performed successfully so far. The cloning of rabbits is possible although not routinely achieved. It is impossible to postulate that microinjection in pig will be abandoned some day in favor of the cloning technique. The cloning of rat should be possible in the future and potentially substituted also to microinjection.

For years, cloning of mouse embryo was considered as extremely difficult, if at all possible. The invoked explanation was that the embryo genome is activated very early after fertilization in this species. This property is generally considered as not compatible with the reprogramming of the donor cell nucleus by the cytoplasm of the recipient oocytes. A recent work has shown that the cloning of mouse was possible on condition to use preferably quiescent cells from the cumulus and to inject isolated nuclei into the enucleated oocytes rather than to fuse the enucleated oocytes with the donor cells (Wakayama and Yanagimachi, 1998; Wakayama et al., 1998). This approach may be adopted in the future to replace microinjection. It is not certain that significantly less animals would be required. On the other hand, the cloning technique should be improved and popularized before it becomes a reality.

Gene replacement in animal genome

Gene replacement can suppress genetic information from a genome. This is a potent and subtle way to understand the role of a gene. A given gene may also be replaced by the same gene mutated in a chosen site or by a quite different gene. Gene replacement may also be targeted to a neutral region on the genome containing no genetic information. This may be done to choose the site of a gene in which a foreign DNA is integrated. Multiples vectors are used to induce the homologous recombination (Viville, 1997).

The technique used for cloning of embryos should be efficient not only for gene addition (Cibelli et al., 1998b; Schnieke et al., 1997) but also for gene replacement. If so, the possibility of generating clones from cultured fetal cells will have rendered possible gene replacement in all species in which cloning is possible. This means that animal models generated by gene replacement have good chance to become a reality not only in mouse but also in rat, rabbit, pig, ruminants and probably in some other species.

In a recent study, it was shown that homologous recombination can be induced directly in a one cell embryo after microinjection of a mutagenic construct containing the corresponding single stand DNA pre-associated with the bacteria recombinase Rec A (Pati, 1998). This new and elegant approach will facilitate the task of experimenters. It is by no means certain that it will reduce the number of experimental animals.

The effect of transgenes

The side-effects of gene addition

The unspecific recombination which occurs during gene additions is mutagenic (Bishop, 1997). This is a major cause of early embryonic death. In some cases, a small region of the chromosome is deleted at the site of integration. In other cases, the transgene is integrated within a transcribed region in the genome and this leads to the inactivation of a host gene. It seems difficult to reduce the impact of this phenomenon.

To prevent this random integration, animals containing a lox P sequence from P_1 phage in a given site might be prepared in a first step. The introduction of the lox P sequence might be performed after a homologous recombination. Alternatively, the lox P sequence could be integrated by the classical random heterologous recombination. Only the lines of animal having the lox P site in an appropriate site of their genomes not interfering with gene expression might be kept. The linear transgene flanked by two lox P sequences and introduced in embryos by microinjection will be integrated very preferentially in the lox P site of the host if the P_1 phage Cre recombinase is added in the embryo by a direct microinjection of the enzyme or of a circular plasmid containing a functional gene coding for the enzyme (Viville, 1997). Following this protocol, one may imagine that a great majority of the transgenic embryos will be protected against the mutagenic effect of the foreign DNA.

Introducing a foreign gene at a known site of the host genome would have other advantages. The expression of a transgene is frequently influenced by the regulatory sequences of host genes present in its vicinity. The integration sites might be chosen to keep the transgene out of the influence of the host genes.

Among the transgenic animals obtained after microinjection, a large proportion of them do not express the transgene in an appropriate manner and they cannot be included in the experimental protocol. This represents a significant wastage of time for experimenters but also the useless sacrifice of animals. A certain number of transgenes just do not work. This gene extinction is sometimes due to a position effect. Gene silencer in chromatin may definitively inhibit the transgenes. The extinction is in some cases an active process, which may be a cellular defense mechanism. This phenomenon seems more frequent when multiple copies of the transgene are cointegrated. The addition of a lox P sequence in the transgene and a treatment of the embryos by the Cre recombinase might induce an efficient and specific homologous recombination at the lox P sites leading to a reduction of the transgene number to one copy (Garrick et al., 1998). This approach proved to enhance quite significantly the proportion of the transgenes being transcriptionally active.

Alternatively, insulating sequences might be added to the gene construct before microinjection to reduce or even completely eliminate the position affects which extinguish the transgenes. Some of these insulators are gene and cell specific (Sippel et al., 1997) while others seem to work with most genes and in all types (Ortiz et al., 1997; Wang et al., 1997; Taboit-Dameron et al., 1999).

The problems of transgene expression as a function of the position effect have received relatively little attention for years. The majority of experimenters work with transgenic mice which can be generated with limited difficulty. Despite the fact that transgene expression is often unpredictable, most of the experimenters have accepted the situation without trying to improve it. Indeed, the establishment of multiple lines of mice with a given transgene often leads to obtain several of them working in a satisfactory manner. For large animals, the establishment of transgenic lines is much heavier. Vectors able to express transgenes efficiently are therefore still more necessary for these animals. Independently of these practical problems, the elements controlling transgene expression is under study in an increasing number of laboratories.

Controlling the expression of a transgene by an exogenous inducer which do not act with the endogenous genes is highly desirable. Several systems of that kind are available. The most popular includes two genes. One expresses a transcription factor activated or inactivated by tetracycline or doxycycline and the other is under the dependency of this transcription factor (Blau and Rossi, 1999). Such systems theoretically allow the activation or the silencing of the transgene at any time of the life of the animals. This tool is providing better models and it contributes to reduce the number of transgenic animals for the study a given problem.

The side-effects of gene replacement

Homologous recombination on animals is essentially used to knock genes out. The genome modification is then perfectly well known. The effect of the gene knock out is then reduced to the lack of a genetic information in the organism. Undoubtedly, in a certain number of cases, gene knock out alters the health of the animals.

In the original technique of gene knock out, the targeted gene is inactivated by homologous recombination performed in ES cells (Capecchi et al., 1989a). The effect of the knock out thus starts from the embryo stage. This may be lethal or heavily detrimental for the animals and prevent the use of the model. Conditional knock out has been carried out successfully in a certain number of cases. The homologous recombination takes place between lox P sites which were introduced in the targeted region of the genome. The recombination at the lox P sites which induces the knock out of the gene is targeted by the presence of the Cre recombinase. The expression of the Cre recombinase gene may be under the control of tissue specific expression, of hormonal inducers and even of the tetracycline system (Kühn et al., 1995).

Knock out is an irreversible manner to inactivate the expression of a gene. Inducing the knock out is a more subtle approach. Several other methods may conditionally block the expression of a gene. RNA forming a stable triple helix with the transcribed part of a gene or its regulatory region may totally prevent transcription. The expression of antisense RNA or ribozymes may inactivate the mRNA or a given gene at the nuclear or the cytoplasmic level (Vasquez and Wilson, 1998). Transdominant negative proteins, such as inactive receptors may prevent the action of a given protein. A recent study suggests that double strand RNA might mutate the homologous region of a given genome and inactivate the gene (Wagner and Sun, 1998).

All these RNA may be synthesized under the control of specific promoters including this which is tetracycline-dependent. These tools offer potent and subtle ways to modulate specifically the expression of a cellular gene.

The ethical problems of animal transgenesis

The use of transgenesis for research and especially for the study of human diseases is generally considered as inevitable. The level of acceptability is different according the social groups. The patients are strongly demanding everything which can help them, including to use transgenic animals.

A few groups of people refuse completely any utilization of transgenic organisms, considering that the living animals and plants were created as such by God and should not be modified by humans. These groups even ask that a distinction between pharmaceutical drugs obtained thanks to conventional methods and to transgenesis should be made. This demand includes not only the drugs directly obtained by transgenesis but also those resulting from knowledge brought by transgenic models. Such an opinion is as respectable as any other although not easily understandable. In practice, this distinction is impossible. The scientific knowledge results from multiple and complex interactions between experimenters and it cannot be compartmentalized.

A certain number of people evoke animal rights in a manner similar to human rights. Some philosophers strongly refute this attitude, considering that animals are not conscious of their existence and of justice for the individuals. In these conditions, animals cannot have specific rights. Human beings have consciousness and morality. They have therefore duties towards animals and they must respect them. This means that not all kind of animal exploitation is acceptable to the same degree (Schroten, 1997).

The integrity of living organisms, and mainly of animals is indispensable for some people. This seems strange for biologists who know that all the living organisms are in permanent evolution. Hence, what seems the most essential is the welfare of animals but not so much their integrity.

The transfer of a gene from one species to another is hardly acceptable for some people, especially that of the transfer of human genes to animals. Again, a biologist cannot agree easily with this view. The term "humanized" has often been used to characterize an animal, a cell or an organ harboring a human gene. It is true that the human gene is sometimes expected to give a human property to an animal. Whatever happens, the animals will remain an animal, unless the great majority of human genes replaced the animal genes. This is by no means the case and not ready to be so.

For obscure reasons, a certain number of people accept animal models obtained after natural mutations and selection or even after mutations experimentally induced by chemical substances but they refuse the transgenic approach. Obviously, the point to consider is the status of the animals and not the method implemented to obtain the models. This attitude reveals that gene manipulation still remains considered in some way as the action of the devil or men who try to play God. The reality is simpler and more trivial.

A certain number of people imagine that gene transfer, whatever it is, denatures the animals so much that they inevitably suffer. It is clear that these people have never visited facilities where transgenic animals are bred. Obviously, only the minority of the transgenic animals have altered health.

Whatever happens, it seems reasonable to limit the use of transgenic animals. The three Rs concept has been proposed to give a rational frame to this activity. All the involved people should reduce as much as possible the number of transgenic animals use, refine their experimental protocols and replace transgenesis as much as possible by other methods, namely by in vitro experiments. These rules are not at all fundamentally in opposition with the practice of experimenters. Some people imagine that scientists are fascinated by transgenesis and just want to develop this technique at all cost. They obviously ignore that working with experimental animals is uneasy, uncomfortable and costly and that most experimenters from public institutes or from private companies would certainly prefer not to use them.

Other expressions have been proposed to tentatively define the attitude we should have towards transgenesis. For some people, the rule should be "no unless". In this case, transgenesis is fundamentally forbidden but authorized exceptionally, for example for medical research. For other people, the rule should be "yes but". They consider that transgenesis is fundamentally accepted but strictly limited. None of these two tools seems able to take into account all the reality. Experimental animals are used essentially to get a new knowledge. Their number is limited and they are generally not the direct source of a profit. On the other hand, these animals are a new experimental material. All the effects of the transgenes cannot therefore be predicted. The "yes but" rule seems appropriate for this kind of animals.

The transgenic farm animals which will be produced for human consumption in the next century should be subjected to the "no unless" rule. Indeed, in this case the animals are physiologically known, and particularly the possible side effects of the transgenes. They are prepared in large number to generate a potential important profit. Transgenesis may improve the quality of animal products and reduce the impact of breeding on the environment. In their majority transgenic animals are not expected to be crucial for the survival of human communities. Hence, only the transgenic farm animals bringing a real improvement of animal production should be accepted.

Scientists usually claim that the use of transgenesis will reduce the number of experimental animals as soon as transgenics are more relevant models. Animal protectionists reply that, on the contrary, the number of transgenic animals is increasing with no parallel reduction of the non-transgenic animals. This is true but the point that should be considered is the number of animals used per studied human disease or per collected information. The increase of the number of transgenic animals results from a rapid general development of research in biology and not from a specific amplification of animal use.

The human communities, which are reluctant to gene manipulation in animals, may profoundly change their view of the problem as soon as things are clearly explained. The result of the referendum that took place recently in Switzerland is a clear illustration of this fact. Transgenesis appears far less unacceptable to most people when all the arguments are carefully examined.

A certain number of people claim that cellular tests should be used rather than transgenic animals. In the vast majority of cases, the experimenters have chosen to use transgenic animals because the relevant cellular tests are available. This is and will remain the case for the complex biological situations involving interactions between different tissues in the body. It remains that sometimes experimenters adopt an available model without considering the possibility of searching a substitute involving cells rather than animals.

It is more and more admitted that experimental animals are not a material just as any other. Considering an animal as a thing has never been and is less and less the common position of experimenters. The animals are clearly not equivalent. It is admitted that the same information cannot be obtained with primates or with *Caenorhabditis elegans*. Experimenters should therefore choose their experimental animals because they are expected to give an answer to a scientific question. They also increasingly take into account the fact that the same problem might be solved as well with an animal species having a less developed nervous system and thus less subjected to pain, stress etc.

The patenting of transgenic animal is the matter of multiple and complex debates, especially in the EU (Warcoin, 1997). In the EU, for years, the patenting of transgenic animals was not clearly authorized. The animals are not objects and for this reason they were not considered as patentable. Patenting is now possible in the EU but only if the transgenic animals bring enough relevant information for the health of human beings. Transgenic mice prepared to study baldness were thus not accepted as patentable, baldness being not considered as a disease. It may seem strange that the patenting is subjected to ethical consideration. It would probably have been simpler to accept or not a patent for transgenic animals on the strict basis of their scientific and technical novelty. Soon after or before patenting, ethical committees, independently, may or may not give their agreement for the use of these animals.

In practice, patenting of genetically modified animals is justified only in a certain number of cases. The lines of animals obtained after spontaneous mutations are traditionally not patented even if they can be interesting as experimental models or for breeding. Methods having global and reproducible effects such as an increase of ploidy to sterilize lower vertebrates or invertebrates are not considered as a matter for patenting.

Theoretically, patenting of transgenic animals is strictly required only when its biological properties are interesting but not reproducibly obtained. This is the case when the reasons why the animals have acquired new biological properties cannot be easily determined. This occurs for example when the foreign gene inactivated or modified the expression of host genes. This may also happen when subtle interferences between the transgene and the genetic background of the host generate animals having heritable new biological properties.

When the effects of the transgenic is reproducible, the gene construct, with its injected effect in animals, can be patented rather than animals themselves.

The use of transgenesis for human therapy is periodically evoked. This potential activity is subjected to a moratorium in most of the concerned countries. The vast majority of people would probably accept human transgenesis but only to prevent diseases and not to alter in some way human integrity. For this reason, random gene addition by microinjection is banished. Gene replacement leading to the controlled addition and expression of a foreign gene is not easily acceptable. Indeed, all the interference of the foreign gene with the host cannot be predicted. The only situation in with transgenesis would have very little chance to be harmful is the case when a mutated gene is replaced by the normal allele. To reach this goal, homologous recombination including ES cells or the cloning technique would be required. This is so complicated that, for long, it would appear more appropriate to eliminate the embryos harboring the mutated gene or to use a specific therapy in adults. Experimental transgenesis in human just does not make sense in the present time.

The optimization of transgenic animal use

As mentioned above, the current practice in animal transgenesis has not deeply searched to optimize the use of transgenic animals. Improvement to avoid wastage of animals results from several independent actions. The obtention of transgenic animals should be only in the hands of skillful and well-trained people. The transgenic animals should be obtained as much as possible after in vitro embryo manipulation which reduces the number of donors and recipients. Only the most relevant protocols should be retained. As much as possible, only gene constructs providing the highest chance to be properly expressed should be used to generate transgenic animals. As much as possible, only the DNA sequences useful for expression of transgenes should be kept in gene constructs to avoid putative unwanted side-effects.

Most of the animals that were used to evaluate a given hypothesis are not kept after the end of the experiments. They are eliminated although they might be used in other laboratories for other studies. Efforts should be done to keep the most relevant transgenic lines as living animals, frozen embryos, frozen sperm or even lyophilized sperm (Wakayama and Yanagimachi, 1998). In the

US, the Jackson Laboratory keeps commercially available relevant transgenic animals (Sharp and Morbraaten, 1997). In the EU a laboratory, which will soon be created in Rome, will have a similar function and will also be a repository for the transgenic lines.

The transgenic animals should be under the control of veterinaries who might determine when animals are suffering or in danger. Behaviour observations can reveal the suffering of animals. Only people familiar with the animals can reasonably give a correct interpretation of behaviour. Studies are being carried out to define biochemical parameters reflecting pain or stress. The level of glucocorticoids and of some interleukins is being measured for this purpose. These methods may be applied to transgenic animals.

The optimized use of transgenic animals implies that the data obtained is known by the scientific communities. This is presently only partially the case. Indeed, many data not relevant for a group might be very precious for another. This is specially the case for the transgenic animals which have unexpected biological properties due to the unique effects of a foreign gene. Only the data in relation with the hypothesis of the laboratory are usually published. Negative results, namely the experiments in which transgenic animals could not be obtained due to the toxicity of a gene, are generally not reported. This implies that other laboratories repeat the same experiments with the same negative results. A database (Tbase) for all transgenic animals is available. It contains all the information concerning the transgenes including the structure of the gene construct, the yield of transgenesis, the unexpected effects, the health status, the availability of the animal etc. (Anagnostopoulos, 1997). The Tbase contact information is accessible by dialing 1-410-955-9705 or by the e-mail: help@gbd.org.

The control of transgenic animal use

A good practice in transgenic animal use includes several complementary aspects. The biosafety rules must be respected. The most relevant protocols must be chosen. The specific ethical aspects must be taken into consideration.

Biosafety includes various points. In all cases, the dissemination of the transgenes must be strictly controlled. This implies that specific physical barriers are used to prevent the escape of animals or their uncontrolled mating with other animals in facilities. Other confinement rules taking into account the pathogenecity of transgenic animals must be followed. A classification into four categories has been proposed by the French committee of biorisk evaluation (Commission de Génie Génétique) (Houdebine, 1997). In class A_1, are animals harbouring no foreign gene having any pathogenic property. The confinement of these animals is then reduced to the specific physical barriers. When the transgenic animals harbor pathogenic genes such as those coding for receptors for viruses, prions, viral genomes etc., a classification from A_2 to A_4 has been proposed. The corresponding confinements are more and more stringent implying the use of biosafety cabinet, negative pressure in facilities, HEPA filters, double-entry auto-clave etc, and good practice of experimenters. The classification of transgenic animals is always done on a case by case basis taking into account the complexity and the diversity of the biological situations.

A control of the experimental activity in laboratories can be performed essentially at local level. Centralized committees would be overburdened by the dossiers of the experimenters. A protocol involving the use of transgenic animals may be decided rapidly after knowing the results of a previous experiment (involving or not transgenesis). It is essential for the scientists to be able to decide quickly how their experiments will be pursued. Local committees are therefore more appropriate to meet the demand of the experimenters.

Local committees exist in some countries but not in others. The rules to be followed by the different communities should be harmonized to allow a similar evaluation of the problems in all places.

A standard checklist might help the different committees in their appreciation of the ethical problems and the experimenters to take into account all the essential points to be considered. EVCAM (European Centre for the Validation of Alternative Methods) has proposed precise recommendations on the matter (Mepham et al., 1998).

The EU supports research in the form of grants, namely in the field of biotechnology. The scientific quality of the project is estimated by a panel of scientists. The accepted projects including potential ethical problems are later examined by a group of people formed by a few scientists but also by various persons not involved in research but concerned by animal welfare and ethics. This procedure is a response to a demand of European citizens. The transparency and democratic principles are thus respected. This procedure is however relatively heavy and costly. It may also in some cases lead to a mediocre evaluation of the real impact of particular projects. A superficial survey of the scientific part of a project by people not involved in research may tend to eliminate the proposal of excellent groups. To avoid this kind of misunderstanding scientists are more and more invited to justify their project also from an ethical point of view.

It is by no means certain that the ethical committees will have a so strong impact to improve the fate of transgenic animals. A comparison of the practice in countries where committees examine the projects of scientists with this in countries currently not having such committees would probably reveal only minor differences. The technical and scientific constraints but also a similar appreciation of animal welfare have already strongly uniformed the practice.

The evaluation of committees may significantly improve the fate of transgenic animals. It remains that one of the major impacts in this field will come from the experimenters themselves. The ethical as the biosafety constraints must henceforth be an integral part of the experimental protocols.

References

Anagnostopoulos, A.V. (1997) TBASE: the relationalized database of transgenic animals and targeted mutations. In: L.M. Houdebine (Ed.), *Transgenic Animal Generation and Use*, Harwood Academic Publishers, Amsterdam, pp. 533–553.

Baguisi, A., Behboodi, E., Melican, D.T., Pollock, J.S., Destrempes, M.M., Cammuso, C., Williams, J.L., Nims, S.D., Porter, C.A., Midura, P., Palacios, M.J., Ayres, S.L., Denniston, R.S., Hayes, M.L., Ziomek, C.A., Meade, H.M., Godke, R.A., Gavin, W.G., Overström, E.W. and Echelard, Y. (1999) Production of goats by somatic cell nuclear transfer. *Nat. Biotechnol.*, 17: 456–461.

Blau, H.M. and Rossi, M.V. (1999) Tet B or not tet B: Advances in tetracycline-inducible gene expression. *Proc. Natl. Acad. Sci. USA*, 96: 797–799.

Bishop, J.O. (1997) Chromosomal insertion of foreign DNA. In: L.M. Houdebine (Ed.), *Transgenic Animal Generation and Use*, Harwood Academic Publishers, Amsterdam, pp. 219–224.

Capecchi, M.R. (1989a) Altering the genome by homologous recombination. *Science*, 244: 1288–1292.

Capecchi, M.R. (1989b) The new mouse genetics: altering the genome by gene targeting. *Trends Genet.*, 5: 70–76.

Chung, J.H., Bell, A.C. and Felsenfeld, G. (1997) Characterization of the chicken β-globin insulator. *Proc. Natl. Acad. Sci. USA*, 94: 575–580.

Cibelli, J.B., Stice, S.L., Golueke, P.J., Kane, J.J., Jerry, J., Blackwell, C., Ponce De Leon, F.A. and Robl, J.M. (1998a) Transgenic bovine chimeric offspring produced from somatic cell-derived stem-like cells. *Nat. Biotechnol.*, 16: 642–646.

Cibelli, J.B., Stice, S.L., Golueke, P.J., Kane, J.J., Jerry, J., Blackwell, C., Ponce De Leon, F.A. and Robl, J.M. (1998b) Cloned transgenic calves produced from non quiescent fetal fibroblasts. *Science*, 280: 1256–1258.

Collas, P., Husebye, H. and Alestrom, P. (1997) Transferring foreign genes into Zebrafish eggs by microinjection. In: L.M. Houdebine (Ed.), *Transgenic Animal Generation and Use*, Harwood Academic Publishers, Amsterdam, pp. 119–123.

Devlin, R.H. (1997) Transgenic Salmonids. In: L.M. Houdebine (Ed.), Transgenic Animal Generation and Use, Harwood Academic Publishers, Amsterdam, pp. 105–119.

Donovan, P.J., Resnick, J.L., Cheng, L. and Lock, L.F. (1997) Towards the use of primordial germ cells for germline modification. In: L.M. Houdebine (Ed.), *Transgenic Animal Generation and Use*, Harwood Academic Publishers, Amsterdam, pp. 179–189.

Fléchon, J.-E. (1997) What are ES cells? In: L.M. Houdebine (Ed.), *Transgenic Animal Generation and Use*, Harwood Academic Publishers, Amsterdam, pp. 157–167.

Froll, K.L. and Amaya, E. (1996) Transgenic Xenopus embryos from sperm nuclear transplantations reveal FGF signaling requirements during gastrulation. *Development*, 122: 3173–3183.

Garrick, D., Fiering, S., Martin, D.I.K. and Whitelaw, E. (1998) Repeat-induced gene silencing in mammals. *Nat. Genet.*, 18: 56–59.

Gong, S.-G. and Etkin, L.D. (1997) Gene injection into *xenopus* embryos. In: L.M. Houdebine (Ed.), *Transgenic Animal Generation and Use*, Harwood Academic Publishers, Amsterdam, pp. 123–129.

Gordon, J.W., Scangos, G.A., Plotkin, D.J., Barbosa, J.A. and Ruddle, F.H. (1980) Genetic transformation of mouse embryos by microinjction of purified DNA. *Proc. Natl. Acad. Sci. USA*, 77: 7380–7384.

Hackett, P.B., Izsvak, Z., Ivics, Z. and Caldovic, L. (1997) *Development of genetic tolls for transgenic fish*, Transgenic animals in Agriculture Conf. Tahoe City California.

Hammer, R.E., Pursel, V.G., Rexroad Jr, C.E., Wall, R.J., Bolt, D.J., Ebert, K.M., Palmiter, R.D. and Brinster, R.L. (1985) Production of transgenic rabbits, sheep and pigs by microinjection. *Nature*, 315: 680–683.

Houdebine, L.M. (1997) The biosafety problems of transgenic animals. In: L.M. Houdebine (Ed.), *Transgenic Animal Generation and Use*, Harwood Academic Publishers, Amsterdam, pp. 559–562.

Kaiser, K. (1997) Gene transfer in Drosophila melanogaster. In: L.M. Houdebine (Ed.), *Transgenic Animal Generation and Use*, Harwood Academic Publishers, Amsterdam, pp. 133–137.

Krimpenfort, P., Rademakers, A., Eyestone, W., Van der Schans, A., Van den Broek, S., Kooiman, P., Kootwijk, E., Platenburg, G., Pieper, F and Strijker, R. (1991) Generation of transgenic dairy cattle using 'in vitro' embryo production. *Bio/Technology*, 9: 844–847.

Kühn, R., Schwenk, F., Aguet, M. and Rajewski, K. (1995) Inducible gene targeting in mice. *Science*, 269: 1427–1429

Kupriyanov, S., Zeh, K. and Baribault, H. (1998) Double pronuclei injection of DNA into zygotes increases yields of transgenic mouse lines. *Transgenic Res.*, 7: 223–226.

Mepham, T.B., Combes. R.D., Balls, M., Barbieri, O., Blokhuis, H.J., Costa, P., Crilly, R.E., De Cock Buning, T., Delpire, V.C., O'Hare, M.J., Houdebine, L.M., Van Kreijl, C.F., Van der meer, M., Reinhardt, C.A., Wolk, E. and Van Zeller, A.-M. (1998) The use of transgenic animals in the European Union. *Atla*, 26: 21–43.

Miahle, E., Boulo, V., Cadoret, J.-P., Cedeno, V., Rousseau, C., Motte, E., Gendreau, S. and Bachère, E. (1997) Gene transfer technology in marine invertebrates. In: L.M. Houdebine (Ed.), *Transgenic Animal Generation and Use*, Harwood Academic Publishers, Amsterdam, pp. 151–157.

Nagy, A. (1997) Formation of mouse chimeric embryos from ES cells. In: L.M. Houdebine (Ed.), *Transgenic Animal Generation and Use*, Harwood Academic Publishers, Amsterdam, pp. 167–173.

Naito, M. (1997) The microinjection of DNA into early chicken embryo. In: L.M. Houdebine (Ed.), *Transgenic Animal Generation and Use*, Harwood Academic Publishers, Amsterdam, pp. 69–75.

Ono, T. (1997) The complete in vitro development of quail embryo. In: L.M. Houdebine (Ed.), *Transgenic Animal Generation and Use*, Harwood Academic Publishers, Amsterdam, pp. 61–69.

Ortiz, B.D., Cado, D., Chen, V., Diaz, P.W. and Winoto, A. (1997) Adjacent DNA element dominantly restrict the ubiquitous activity of a novel chromatin-opening region to specific tissues. *EMBO J.*, 16: 5037–5045.

Palmiter, R.D., Brinster, R.L., Hammer, R.E., Trumbauer, M.E., Rosenfeld, M.G., Birnberg, N.C. and Evans, R.M. (1982) Dramatic growth of mice that develop from eggs microinjected with metallothionein-growth hormone fusion genes. *Nature*, 300: 611–615.

Pati, S. (1998) *Genetically engineering and cloning animals*, Park City Deer Valley, USA, 21–23 June.

Ronfort, C.M., Legras, C. and Verdier, G. (1997) The use of retroviral vectors for gene transfer into bird embryo. In: L.M. Houdebine (Ed.), *Transgenic Animal Generation and Use*, Harwood Academic Publishers, Amsterdam, pp. 83–95.

Schnieke, A., Kind, A.J., Ritchie, W.A., Mycock, K., Scott, A.R., Ritchie, M., Wilmut, I., Colman, A. and Campbell, K.H.S. (1997) Human factor IX transgenic sheep produced by transfer of nuclei from transfected fetal fibroblasts. *Science*, 278: 2130–2133

Schroten, E. (1997) From a moral point of view: ethical problems of animal transgenesis. In: L.M. Houdebine (Ed.), *Transgenic Animal Generation and Use*, Harwood Academic Publishers, Amsterdam, pp. 569–575.

Sharp, J.J. and Mobraaten, L.E. (1997) To save or not to save: the role of repositories in a period of rapidly expanding development of genetically engineered strains of mice. In: L.M. Houdebine (Ed.), *Transgenic Animal Generation and Use*, Harwood Academic Publishers, Amsterdam, pp. 525–533.

Sippel, A.E., Saueressig, H., Hubler, M.C., Faust, N. and Bonifer, C. (1997) Insulation of transgenes from chromosomal position effects. . In: L.M. Houdebine (Ed.), *Transgenic Animal Generation and Use*, Harwood Academic Publishers, Amsterdam, pp. 257–267.

Taboit-Dameron, F., Malassagne, B., Viglietta, C., Puissant, C., Leroux-Coyau, M., Chéreau, C., Attal, J., Weill, B. and Houdebine, L.-M. (1999) Association of the 5′HS4 sequence of the chicken β-globin locus control region with human EF1α gene promoter induces ubiquitous and high expression of human CD55 and CD59 cDNAs in transgenic rabbits. *Transgenic Research*, Kluwer Academic Publishers, Amsterdam, 2169: pp. 1–13.

Thierry-Mieg, D., Naert, K. and Bonnerot, C. (1997) Genetic transformation on Caenorhabditis elegans. In: L.M. Houdebine (Ed.), *Transgenic Animal Generation and Use*, Harwood Academic Publishers, Amsterdam, pp. 137–151.

Thompson, E.M., Adenot, P., Tsuji, F.I. and Renard, J.P. (1995) Real time imaging of transcriptional activity in live mouse preimplantation embryos using a secreted luciferase. *Proc, Natl. acd. Sci. USA*, 92: 1317–1321.

Vasquez, K.M. and Wilson, J.H. (1998) Triplex-directed modification of gene and gene activity. *TIBS*, 23: 4–9.

Viville, S. (1997) Mouse genetic manipulation via homologous recombination. In: L.M. Houdebine (Ed.), *Transgenic Animal Generation and Use*, Harwood Academic Publishers, Amsterdam, pp. 307–323.

Wagner, R.W. and Sun, L. (1998) Double stranded RNA pose puzzle. *Nature*, 391: 744–745.

Wakayama, T., Perry, A.C.F., Zucotti, M., Johnson, K.R. and Yanagimachi, R. (1998) Full-term development of mice from enucleated oocytes injected with cumulus cell nuclei. *Nature*, 394: 369–374.

Wakayama, T. and Yanagimachi, R. (1998) Development of normal mice from oocytes injected with freeze-dried spermatozoa. *Nat. Biotechnol.*, 16: 639–641.

Wang, Y., DeMayo, F.J., Tsai, S.Y. and O'Malley, B.W. (1997) Ligand-inducible and liver-specific target gene expression in transgenic mice. *Nat. Biotechnol.*, 15: 239–243.

Warcoin, J. (1997) The patenting of transgenic animals. In: L.M. Houdebine (Ed.), *Transgenic Animal Generation and Use*, Harwood Academic Publishers, Amsterdam, pp. 553–559.

Wilmut, I., Schnieke, A.E., McWhir, J., Kind, A.J., Campbell, K.H.S. (1997) Viable offspring derived from fetal and adult mammalian cells. *Nature* 385, 810–813.

Subject Index